PATH INTEGRALS

in Quantum Mechanics,
Statistics, Polymer Physics,
and Financial Markets

5th Edition

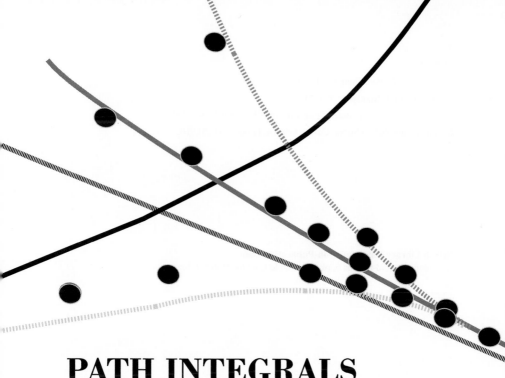

PATH INTEGRALS
in Quantum Mechanics, Statistics, Polymer Physics, and Financial Markets

5th Edition

Hagen KLEINERT

Professor of Physics
Freie Universität Berlin, Germany
ICRANet Pescara, Italy, and Nice, France

World Scientific

NEW JERSEY · LONDON · SINGAPORE · BEIJING · SHANGHAI · HONG KONG · TAIPEI · CHENNAI

Published by

World Scientific Publishing Co. Pte. Ltd.

5 Toh Tuck Link, Singapore 596224

USA office: 27 Warren Street, Suite 401-402, Hackensack, NJ 07601

UK office: 57 Shelton Street, Covent Garden, London WC2H 9HE

British Library Cataloguing-in-Publication Data
A catalogue record for this book is available from the British Library.

PATH INTEGRALS IN QUANTUM MECHANICS, STATISTICS, POLYMER PHYSICS, AND FINANCIAL MARKETS
(5th Edition)

ISBN-13 978-981-4273-55-8
ISBN-13 978-981-4273-56-5 (pbk)

Printed in Singapore by Mainland Press Pte Ltd.

To Annemarie and Hagen II

Nature alone knows what she wants.
GOETHE

Preface

The paperback version of the fourth edition of this book was sold out in Fall 2008. This gave me a chance to revise it at many places. In particular, I improved considerably Chapter 20 on financial markets and removed some technical sections of Chapter 5.

Among the many people who spotted printing errors and suggested changes of text passages are Dr. A. Pelster, Dr. A. Redondo, and especially Dr. Annemarie Kleinert.

H. Kleinert
Berlin, January 2009

Preface to Fourth Edition

The third edition of this book appeared in 2004 and was reprinted in the same year without improvements. The present fourth edition contains several extensions. Chapter 4 includes now semiclassical expansions of higher order. Chapter 8 offers an additional path integral formulation of spinning particles whose action contains a vector field and a Wess-Zumino term. From this, the Landau-Lifshitz equation for spin precession is derived which governs the behavior of quantum spin liquids. The path integral demonstrates that fermions can be described by Bose fields—the basis of Skyrmion theories. A further new section introduces the Berry phase, a useful tool to explain many interesting physical phenomena. Chapter 10 gives more details on magnetic monopoles and multivalued fields. Another feature is new in this edition: sections of a more technical nature are printed in smaller font size. They can well be omitted in a first reading of the book.

Among the many people who spotted printing errors and helped me improve various text passages are Dr. A. Chervyakov, Dr. A. Pelster, Dr. F. Nogueira, Dr. M. Weyrauch, Dr. H. Baur, Dr. T. Iguchi, V. Bezerra, D. Jahn, S. Overesch, and especially Dr. Annemarie Kleinert.

H. Kleinert
Berlin, June 2006

Preface to Third Edition

This third edition of the book improves and extends considerably the second edition
of 1995:

- Chapter 2 now contains a path integral representation of the scattering amplitude and new methods of calculating functional determinants for time-dependent second-order differential operators. Most importantly, it introduces the quantum field-theoretic definition of path integrals, based on perturbation expansions around the trivial harmonic theory.

- Chapter 3 presents more exactly solvable path integrals than in the previous editions. It also extends the Bender-Wu recursion relations for calculating perturbation expansions to more general types of potentials.

- Chapter 4 discusses now in detail the quasiclassical approximation to the scattering amplitude and Thomas-Fermi approximation to atoms.

- Chapter 5 proves the convergence of variational perturbation theory. It also discusses atoms in strong magnetic fields and the polaron problem.

- Chapter 6 shows how to obtain the spectrum of systems with infinitely high walls from perturbation expansions.

- Chapter 7 offers a many-path treatment of Bose-Einstein condensation and degenerate Fermi gases.

- Chapter 10 develops the quantum theory of a particle in curved space, treated before only in the time-sliced formalism, to perturbatively defined path integrals. Their reparametrization invariance imposes severe constraints upon integrals over products of distributions. We derive unique rules for evaluating these integrals, thus extending the linear space of distributions to a semigroup.

- Chapter 15 offers a closed expression for the end-to-end distribution of stiff polymers valid for all persistence lengths.

- Chapter 18 derives the operator Langevin equation and the Fokker-Planck equation from the forward–backward path integral. The derivation in the literature was incomplete, and the gap was closed only recently by an elegant calculation of the Jacobian functional determinant of a second-order differential operator with dissipation.

- Chapter 20 is completely new. It introduces the reader into the applications of path integrals to the fascinating new field of econophysics.

For a few years, the third edition has been freely available on the internet, and several readers have sent useful comments, for instance E. Babaev, H. Baur, B. Budnyj, Chen Li-ming, A.A. Drăgulescu, K. Glaum, I. Grigorenko, T.S. Hatamian, P. Hollister, P. Jizba, B. Kastening, M. Krämer, W.-F. Lu, S. Mukhin, A. Pelster, C. Öалır, M.B. Pinto, C. Schubert, S. Schmidt, R. Scalettar, C. Tangui, and M. van Vugt. Reported errors are corrected in the internet edition.

When writing the new part of Chapter 2 on the path integral representation of the scattering amplitude I profited from discussions with R. Rosenfelder. In the new parts of Chapter 5 on polarons, many useful comments came from J.T. Devreese, F.M. Peeters, and F. Brosens. In the new Chapter 20, I profited from discussions with F. Nogueira, A.A. Drăgulescu, E. Eberlein, J. Kallsen, M. Schweizer, P. Bank, M. Tenney, and E.C. Chang.

As in all my books, many printing errors were detected by my secretary S. Endrias and many improvements are due to my wife Annemarie without whose permanent encouragement this book would never have been finished.

H. Kleinert
Berlin, August 2003

Preface to Second Edition

Since this book first appeared three years ago, a number of important developments have taken place calling for various extensions to the text.

Chapter 4 now contains a discussion of the features of the semiclassical quantization which are relevant for multidimensional chaotic systems.

Chapter 3 derives perturbation expansions in terms of Feynman graphs, whose use is customary in quantum field theory. Correspondence is established with Rayleigh-Schrödinger perturbation theory. Graphical expansions are used in Chapter 5 to extend the Feynman-Kleinert variational approach into a systematic *variational perturbation theory*. Analytically inaccessible path integrals can now be evaluated with arbitrary accuracy. In contrast to ordinary perturbation expansions which always diverge, the new expansions are convergent for all coupling strengths, including the strong-coupling limit.

Chapter 10 contains now a new action principle which is necessary to derive the correct classical equations of motion in spaces with curvature and a certain class of torsion (gradient torsion).

Chapter 19 is new. It deals with relativistic path integrals, which were previously discussed only briefly in two sections at the end of Chapter 15. As an application, the path integral of the relativistic hydrogen atom is solved.

Chapter 16 is extended by a theory of particles with fractional statistics (*anyons*), from which I develop a theory of polymer entanglement. For this I introduce non-abelian Chern-Simons fields and show their relationship with various knot polynomials (Jones, HOMFLY). The successful explanation of the fractional quantum Hall effect by anyon theory is discussed — also the failure to explain high-temperature superconductivity via a Chern-Simons interaction.

Chapter 17 offers a novel variational approach to tunneling amplitudes. It extends the semiclassical range of validity from high to low barriers. As an application, I increase the range of validity of the currently used large-order perturbation theory far into the regime of low orders. This suggests a possibility of greatly improving existing resummation procedures for divergent perturbation series of quantum field theories.

The Index now also contains the names of authors cited in the text. This may help the reader searching for topics associated with these names. Due to their great number, it was impossible to cite all the authors who have made important contributions. I apologize to all those who vainly search for their names.

In writing the new sections in Chapters 4 and 16, discussions with Dr. D. Wintgen and, in particular, Dr. A. Schakel have been extremely useful. I also thank Professors G. Gerlich, P. Hänggi, H. Grabert, M. Roncadelli, as well as Dr. A. Pelster, and Mr. R. Karrlein for many relevant comments. Printing errors were corrected by my secretary Ms. S. Endrias and by my editor Ms. Lim Feng Nee of World Scientific. Many improvements are due to my wife Annemarie.

H. Kleinert
Berlin, December 1994

Preface to First Edition

These are extended lecture notes of a course on path integrals which I delivered at the Freie Universität Berlin during winter 1989/1990. My interest in this subject dates back to 1972 when the late R. P. Feynman drew my attention to the unsolved path integral of the hydrogen atom. I was then spending my sabbatical year at Caltech, where Feynman told me during a discussion how embarrassed he was, not being able to solve the path integral of this most fundamental quantum system. In fact, this had made him quit teaching this subject in his course on quantum mechanics as he had initially done.[1] Feynman challenged me: "Kleinert, you figured out all that group-theoretic stuff of the hydrogen atom, why don't you solve the path integral!" He was referring to my 1967 Ph.D. thesis[2] where I had demonstrated that all dynamical questions on the hydrogen atom could be answered using only operations within a *dynamical group* $O(4, 2)$. Indeed, in that work, the four-dimensional oscillator played a crucial role and the missing steps to the solution of the path integral were later found to be very few. After returning to Berlin, I forgot about the problem since I was busy applying path integrals in another context, developing a field-theoretic passage from quark theories to a collective field theory of hadrons.[3] Later, I carried these techniques over into condensed matter (superconductors, superfluid ^{3}He) and nuclear physics. Path integrals have made it possible to build a unified field theory of collective phenomena in quite different physical systems.[4]

The hydrogen problem came up again in 1978 as I was teaching a course on quantum mechanics. To explain the concept of quantum fluctuations, I gave an introduction to path integrals. At the same time, a postdoc from Turkey, I. H. Duru, joined my group as a Humboldt fellow. Since he was familiar with quantum mechanics, I suggested that we should try solving the path integral of the hydrogen atom. He quickly acquired the basic techniques, and soon we found the most important ingredient to the solution: The transformation of time in the path integral to a new path-dependent pseudotime, combined with a transformation of the coordinates to

---

[1]Quoting from the preface of the textbook by R.P. Feynman and A.R. Hibbs, *Quantum Mechanics and Path Integrals*, McGraw-Hill, New York, 1965: "Over the succeeding years, ... Dr. Feynman's approach to teaching the subject of quantum mechanics evolved somewhat away from the initial path integral approach."

[2]H. Kleinert, Fortschr. Phys. *6*, 1, (1968), and *Group Dynamics of the Hydrogen Atom*, Lectures presented at the 1967 Boulder Summer School, published in *Lectures in Theoretical Physics*, Vol. X B, pp. 427–482, ed. by A.O. Barut and W.E. Brittin, Gordon and Breach, New York, 1968.

[3]See my 1976 Erice lectures, *Hadronization of Quark Theories*, published in *Understanding the Fundamental Constituents of Matter*, Plenum press, New York, 1978, p. 289, ed. by A. Zichichi.

[4]H. Kleinert, Phys. Lett. B *69*, 9 (1977); Fortschr. Phys. *26*, 565 (1978); *30*, 187, 351 (1982).

"square root coordinates" (to be explained in Chapters 13 and 14).[5] These transformations led to the correct result, however, only due to good fortune. In fact, our procedure was immediately criticized for its sloppy treatment of the time slicing.[6] A proper treatment could, in principle, have rendered unwanted extra terms which our treatment would have missed. Other authors went through the detailed time-slicing procedure,[7] but the correct result emerged only by transforming the measure of path integration inconsistently. When I calculated the extra terms according to the standard rules I found them to be zero only in two space dimensions.[8] The same treatment in three dimensions gave nonzero "corrections" which spoiled the beautiful result, leaving me puzzled.

Only recently I happened to locate the place where the three-dimensional treatment went wrong. I had just finished a book on the use of gauge fields in condensed matter physics.[9] The second volume deals with ensembles of defects which are defined and classified by means of operational cutting and pasting procedures on an ideal crystal. Mathematically, these procedures correspond to nonholonomic mappings. Geometrically, they lead from a flat space to a space with curvature and torsion. While proofreading that book, I realized that the transformation by which the path integral of the hydrogen atom is solved also produces a certain type of torsion (gradient torsion). Moreover, this happens only in three dimensions. In two dimensions, where the time-sliced path integral had been solved without problems, torsion is absent. Thus I realized that the transformation of the time-sliced measure had a hitherto unknown sensitivity to torsion.

It was therefore essential to find a correct path integral for a particle in a space with curvature and gradient torsion. This was a nontrivial task since the literature was ambiguous already for a purely curved space, offering several prescriptions to choose from. The corresponding equivalent Schrödinger equations differ by multiples of the curvature scalar.[10] The ambiguities are path integral analogs of the so-called *operator-ordering problem* in quantum mechanics. When trying to apply the existing prescriptions to spaces with torsion, I always ran into a disaster, some even yielding noncovariant answers. So, something had to be wrong with all of them. Guided by the idea that in spaces with constant curvature the path integral should produce the same result as an operator quantum mechanics based on a quantization of angular momenta, I was eventually able to find a consistent *quantum equivalence principle*

---

[5]I.H. Duru and H. Kleinert, Phys. Lett. B *84*, 30 (1979), Fortschr. Phys. *30*, 401 (1982).

[6]G.A. Ringwood and J.T. Devreese, J. Math. Phys. *21*, 1390 (1980).

[7]R. Ho and A. Inomata, Phys. Rev. Lett. *48*, 231 (1982); A. Inomata, Phys. Lett. A *87*, 387 (1981).

[8]H. Kleinert, Phys. Lett. B *189*, 187 (1987); contains also a criticism of Ref. 7.

[9]H. Kleinert, *Gauge Fields in Condensed Matter*, World Scientific, Singapore, 1989, Vol. I, pp. 1–744, *Superflow and Vortex Lines*, and Vol. II, pp. 745–1456, *Stresses and Defects*.

[10]B.S. DeWitt, Rev. Mod. Phys. *29*, 377 (1957); K.S. Cheng, J. Math. Phys. *13*, 1723 (1972), H. Kamo and T. Kawai, Prog. Theor. Phys. *50*, 680, (1973); T. Kawai, Found. Phys. *5*, 143 (1975), H. Dekker, Physica A *103*, 586 (1980), G.M. Gavazzi, Nuovo Cimento *101*A, 241 (1981); M.S. Marinov, Physics Reports *60*, 1 (1980).

for path integrals in spaces with curvature and gradient torsion,[11] thus offering also a unique solution to the operator-ordering problem. This was the key to the leftover problem in the Coulomb path integral in three dimensions — the proof of the absence of the extra time slicing contributions presented in Chapter 13.

Chapter 14 solves a variety of one-dimensional systems by the new techniques.

Special emphasis is given in Chapter 8 to instability (*path collapse*) problems in the Euclidean version of Feynman's time-sliced path integral. These arise for actions containing bottomless potentials. A general stabilization procedure is developed in Chapter 12. It must be applied whenever centrifugal barriers, angular barriers, or Coulomb potentials are present.[12]

Another project suggested to me by Feynman, the improvement of a variational approach to path integrals explained in his book on statistical mechanics[13], found a faster solution. We started work during my sabbatical stay at the University of California at Santa Barbara in 1982. After a few meetings and discussions, the problem was solved and the preprint drafted. Unfortunately, Feynman's illness prevented him from reading the final proof of the paper. He was able to do this only three years later when I came to the University of California at San Diego for another sabbatical leave. Only then could the paper be submitted.[14]

Due to recent interest in lattice theories, I have found it useful to exhibit the solution of several path integrals for a finite number of time slices, without going immediately to the continuum limit. This should help identify typical lattice effects seen in the Monte Carlo simulation data of various systems.

The path integral description of polymers is introduced in Chapter 15 where stiffness as well as the famous excluded-volume problem are discussed. Parallels are drawn to path integrals of relativistic particle orbits. This chapter is a preparation for ongoing research in the theory of fluctuating surfaces with extrinsic curvature stiffness, and their application to world sheets of strings in particle physics.[15] I have also introduced the field-theoretic description of a polymer to account for its increasing relevance to the understanding of various phase transitions driven by fluctuating line-like excitations (vortex lines in superfluids and superconductors, defect lines in crystals and liquid crystals).[16] Special attention has been devoted in Chapter 16 to simple topological questions of polymers and particle orbits, the latter arising by the presence of magnetic flux tubes (Aharonov-Bohm effect). Their relationship to Bose and Fermi statistics of particles is pointed out and the recently popular topic of fractional statistics is introduced. A survey of entanglement phenomena of single orbits and pairs of them (ribbons) is given and their application to biophysics is indicated.

---

[11]H. Kleinert, Mod. Phys. Lett. A *4*, 2329 (1989); Phys. Lett. B *236*, 315 (1990).

[12]H. Kleinert, Phys. Lett. B *224*, 313 (1989).

[13]R.P. Feynman, *Statistical Mechanics*, Benjamin, Reading, 1972, Section 3.5.

[14]R.P. Feynman and H. Kleinert, Phys. Rev. A *34*, 5080, (1986).

[15]A.M. Polyakov, Nucl. Phys. B *268*, 406 (1986), H. Kleinert, Phys. Lett. B *174*, 335 (1986).

[16]See Ref. 9.

Finally, Chapter 18 contains a brief introduction to the path integral approach of nonequilibrium quantum-statistical mechanics, deriving from it the standard Langevin and Fokker-Planck equations.

I want to thank several students in my class, my graduate students, and my post-docs for many useful discussions. In particular, T. Eris, F. Langhammer, B. Meller, I. Mustapic, T. Sauer, L. Semig, J. Zaun, and Drs. G. Germán, C. Holm, D. Johnston, and P. Kornilovitch have all contributed with constructive criticism. Dr. U. Eckern from Karlsruhe University clarified some points in the path integral derivation of the Fokker-Planck equation in Chapter 18. Useful comments are due to Dr. P.A. Horvathy, Dr. J. Whitenton, and to my colleague Prof. W. Theis. Their careful reading uncovered many shortcomings in the first draft of the manuscript. Special thanks go to Dr. W. Janke with whom I had a fertile collaboration over the years and many discussions on various aspects of path integration.

Thanks go also to my secretary S. Endrias for her help in preparing the manuscript in LaTeX, thus making it readable at an early stage, and to U. Grimm for drawing the figures.

Finally, and most importantly, I am grateful to my wife Dr. Annemarie Kleinert for her inexhaustible patience and constant encouragement.

H. Kleinert
Berlin, January 1990

Contents

List of Figures

List of Tables

Ay, call it holy ground,
The soil where first they trod!
F. D. HEMANS (1793-1835), Landing of the Pilgrim Fathers

1

Fundamentals

Path integrals deal with fluctuating line-like structures. These appear in nature in a variety of ways, for instance, as particle orbits in spacetime continua, as polymers in solutions, as vortex lines in superfluids, as defect lines in crystals and liquid crystals. Their fluctuations can be of quantum-mechanical, thermodynamic, or statistical origin. Path integrals are an ideal tool to describe these fluctuating line-like structures, thereby leading to a unified understanding of many quite different physical phenomena. In developing the formalism we shall repeatedly invoke well-known concepts of classical mechanics, quantum mechanics, and statistical mechanics, to be summarized in this chapter. In Section 1.13, we emphasize some important problems of operator quantum mechanics in spaces with curvature and torsion. These problems will be solved in Chapters 10 and 8 by means of path integrals.[1]

1.1 Classical Mechanics

The orbits of a classical-mechanical system are described by a set of time-dependent generalized coordinates $q_1(t), \ldots, q_N(t)$. A Lagrangian

$$L(q_i, \dot{q}_i, t) \tag{1.1}$$

depending on $q_1, \ldots, q_N$ and the associated velocities $\dot{q}_1, \ldots, \dot{q}_N$ governs the dynamics of the system. The dots denote the time derivative d/dt. The Lagrangian is at most a quadratic function of $\dot{q}_i$. The time integral

$$\mathcal{A}[q_i] = \int_{t_a}^{t_b} dt\, L(q_i(t), \dot{q}_i(t), t) \tag{1.2}$$

of the Lagrangian along an arbitrary path $q_i(t)$ is called the *action* of this path. The path being actually chosen by the system as a function of time is called the *classical path* or the *classical orbit* $q_i^{\mathrm{cl}}(t)$. It has the property of extremizing the action in comparison with all neighboring paths

$$q_i(t) = q_i^{\mathrm{cl}}(t) + \delta q_i(t) \tag{1.3}$$

---

[1]Readers familiar with the foundations may start directly with Section 1.13.

having the same endpoints $q(t_b)$, $q(t_a)$. To express this property formally, one introduces the *variation* of the action as the linear term in the Taylor expansion of $\mathcal{A}[q_i]$ in powers of $\delta q_i(t)$:

$$\delta\mathcal{A}[q_i] \equiv \{\mathcal{A}[q_i + \delta q_i] - \mathcal{A}[q_i]\}_{\text{lin}} \ . \tag{1.4}$$

The extremal principle for the classical path is then

$$\delta\mathcal{A}[q_i]\Big|_{q_i(t) = q_i^{\text{cl}}(t)} = 0 \tag{1.5}$$

for all variations of the path around the classical path, $\delta q_i(t) \equiv q_i(t) - q_i^{\text{cl}}(t)$, which vanish at the endpoints, i.e., which satisfy

$$\delta q_i(t_a) = \delta q_i(t_b) = 0. \tag{1.6}$$

Since the action is a time integral of a Lagrangian, the extremality property can be phrased in terms of differential equations. Let us calculate the variation of $\mathcal{A}[q_i]$ explicitly:

$$\begin{aligned}
\delta\mathcal{A}[q_i] &= \{\mathcal{A}[q_i + \delta q_i] - \mathcal{A}[q_i]\}_{\text{lin}} \\
&= \int_{t_a}^{t_b} dt \ \{L\left(q_i(t) + \delta q_i(t), \dot{q}_i(t) + \delta\dot{q}_i(t), t\right) - L\left(q_i(t), \dot{q}_i(t), t\right)\}_{\text{lin}} \\
&= \int_{t_a}^{t_b} dt \ \left\{\frac{\partial L}{\partial q_i}\delta q_i(t) + \frac{\partial L}{\partial \dot{q}_i}\delta\dot{q}_i(t)\right\} \\
&= \int_{t_a}^{t_b} dt \ \left\{\frac{\partial L}{\partial q_i} - \frac{d}{dt}\frac{\partial L}{\partial \dot{q}_i}\right\}\delta q_i(t) + \frac{\partial L}{\partial \dot{q}_i}\delta q_i(t)\Big|_{t_a}^{t_b}.
\end{aligned} \tag{1.7}$$

The last expression arises from a partial integration of the $\delta\dot{q}_i$ term. Here, as in the entire text, repeated indices are understood to be summed (*Einstein's summation convention*). The endpoint terms (*surface* or *boundary terms*) with the time t equal to t_a and t_b may be dropped, due to (1.6). Thus we find for the classical orbit $q_i^{\text{cl}}(t)$ the *Euler-Lagrange equations*:

$$\frac{d}{dt}\frac{\partial L}{\partial \dot{q}_i} = \frac{\partial L}{\partial q_i}. \tag{1.8}$$

There is an alternative formulation of classical dynamics which is based on a Legendre-transformed function of the Lagrangian called the *Hamiltonian*

$$H \equiv \frac{\partial L}{\partial \dot{q}_i}\dot{q}_i - L(q_i, \dot{q}_i, t). \tag{1.9}$$

Its value at any time is equal to the energy of the system. According to the general theory of Legendre transformations [1], the natural variables which H depends on are no longer q_i and $\dot{q}_i$, but q_i and the generalized momenta p_i, the latter being defined by the N equations

$$p_i \equiv \frac{\partial}{\partial \dot{q}_i}L(q_i, \dot{q}_i, t). \tag{1.10}$$

In order to express the Hamiltonian $H(p_i, q_i, t)$ in terms of its proper variables p_i, q_i, the equations (1.10) have to be solved for $\dot{q}_i$,

$$\dot{q}_i = v_i(p_i, q_i, t). \tag{1.11}$$

This is possible provided the *Hessian metric*

$$H_{ij}(q_i, \dot{q}_i, t) \equiv \frac{\partial^2}{\partial \dot{q}_i \partial \dot{q}_j} L(q_i, \dot{q}_i, t) \tag{1.12}$$

is nonsingular. The result is inserted into (1.9), leading to the Hamiltonian as a function of p_i and q_i:

$$H(p_i, q_i, t) = p_i v_i(p_i, q_i, t) - L(q_i, v_i(p_i, q_i, t), t). \tag{1.13}$$

In terms of this Hamiltonian, the action is the following functional of $p_i(t)$ and $q_i(t)$:

$$\mathcal{A}[p_i, q_i] = \int_{t_a}^{t_b} dt \left[p_i(t)\dot{q}_i(t) - H(p_i(t), q_i(t), t) \right]. \tag{1.14}$$

This is the so-called *canonical form* of the action. The classical orbits are now specified by $p_i^{\mathrm{cl}}(t)$, $q_i^{\mathrm{cl}}(t)$. They extremize the action in comparison with all neighboring orbits in which the coordinates $q_i(t)$ are varied at fixed endpoints [see (1.3), (1.6)] whereas the momenta $p_i(t)$ are varied without restriction:

$$
\begin{aligned}
q_i(t) &= q_i^{\mathrm{cl}}(t) + \delta q_i(t), \qquad \delta q_i(t_a) = \delta q_i(t_b) = 0, \\
p_i(t) &= p_i^{\mathrm{cl}}(t) + \delta p_i(t).
\end{aligned}
\tag{1.15}
$$

In general, the variation is

$$
\begin{aligned}
\delta \mathcal{A}[p_i, q_i] &= \int_{t_a}^{t_b} dt \left[\delta p_i(t)\dot{q}_i(t) + p_i(t)\delta \dot{q}_i(t) - \frac{\partial H}{\partial p_i}\delta p_i - \frac{\partial H}{\partial q_i}\delta q_i \right] \\
&= \int_{t_a}^{t_b} dt \left\{ \left[\dot{q}_i(t) - \frac{\partial H}{\partial p_i} \right] \delta p_i - \left[\dot{p}_i(t) + \frac{\partial H}{\partial q_i} \right] \delta q_i \right\} \\
&\quad + p_i(t)\delta q_i(t) \Big|_{t_b}^{t_b}.
\end{aligned}
\tag{1.16}
$$

Since this variation has to vanish for the classical orbits, we find that $p_i^{\mathrm{cl}}(t)$, $q_i^{\mathrm{cl}}(t)$ must be solutions of the *Hamilton equations* of motion

$$
\begin{aligned}
\dot{p}_i &= -\frac{\partial H}{\partial q_i}, \\
\dot{q}_i &= \frac{\partial H}{\partial p_i}.
\end{aligned}
\tag{1.17}
$$

These agree with the Euler-Lagrange equations (1.8) via (1.9) and (1.10), as can easily be verified. The $2N$-dimensional space of all p_i and q_i is called the *phase space*.

As a particle moves along a classical trajectory, the action changes as a function of the end positions (1.16) by

$$\delta \mathcal{A}[p_i, q_i] = p_i(t_b)\delta q_i(t_b) - p_i(t_a)\delta q_i(t_a). \tag{1.18}$$

An arbitrary function $O(p_i(t), q_i(t), t)$ changes along an arbitrary path as follows:

$$\frac{d}{dt}O\left(p_i(t), q_i(t), t\right) = \frac{\partial O}{\partial p_i}\dot{p}_i + \frac{\partial O}{\partial q_i}\dot{q}_i + \frac{\partial O}{\partial t}. \tag{1.19}$$

If the path coincides with a classical orbit, we may insert (1.17) and find

$$
\begin{aligned}
\frac{dO}{dt} &= \frac{\partial H}{\partial p_i}\frac{\partial O}{\partial q_i} - \frac{\partial O}{\partial p_i}\frac{\partial H}{\partial q_i} + \frac{\partial O}{\partial t} \\
&\equiv \{H, O\} + \frac{\partial O}{\partial t}.
\end{aligned}
\tag{1.20}
$$

Here we have introduced the symbol $\{\ldots, \ldots\}$ called *Poisson brackets*:

$$\{A, B\} \equiv \frac{\partial A}{\partial p_i}\frac{\partial B}{\partial q_i} - \frac{\partial B}{\partial p_i}\frac{\partial A}{\partial q_i}, \tag{1.21}$$

again with the Einstein summation convention for the repeated index i. The Poisson brackets have the obvious properties

$$\{A, B\} = -\{B, A\} \qquad \text{antisymmetry,} \tag{1.22}$$

$$\{A, \{B, C\}\} + \{B, \{C, A\}\} + \{C, \{A, B\}\} = 0 \qquad \text{Jacobi identity.} \tag{1.23}$$

If two quantities have vanishing Poisson brackets, they are said to *commute*.

The original Hamilton equations are a special case of (1.20):

$$
\begin{aligned}
\frac{d}{dt}p_i &= \{H, p_i\} = \frac{\partial H}{\partial p_j}\frac{\partial p_i}{\partial q_j} - \frac{\partial p_i}{\partial p_j}\frac{\partial H}{\partial q_j} = -\frac{\partial H}{\partial q_i}, \\
\frac{d}{dt}q_i &= \{H, q_i\} = \frac{\partial H}{\partial p_j}\frac{\partial q_i}{\partial q_j} - \frac{\partial q_i}{\partial p_j}\frac{\partial H}{\partial q_j} = \frac{\partial H}{\partial p_i}.
\end{aligned}
\tag{1.24}
$$

By definition, the phase space variables p_i, q_i satisfy the Poisson brackets

$$
\begin{aligned}
\{p_i, q_j\} &= \delta_{ij}, \\
\{p_i, p_j\} &= 0, \\
\{q_i, q_j\} &= 0.
\end{aligned}
\tag{1.25}
$$

A function $O(p_i, q_i)$ which has no *explicit* dependence on time and which, moreover, commutes with H (i.e., $\{O, H\} = 0$), is a *constant of motion* along the classical path, due to (1.20). In particular, H itself is often time-independent, i.e., of the form

$$H = H(p_i, q_i). \tag{1.26}$$

Then, since H commutes with itself, the energy is a constant of motion.

The Lagrangian formalism has the virtue of being independent of the particular choice of the coordinates q_i. Let Q_i be any other set of coordinates describing the system which is connected with q_i by what is called a *local*[2] or *point transformation*

$$q_i = f_i(Q_j, t). \tag{1.27}$$

Certainly, to be of use, this relation must be invertible, at least in some neighborhood of the classical path,

$$Q_i = f^{-1}{}_i(q_j, t). \tag{1.28}$$

Otherwise Q_i and q_i could not both parametrize the same system. Therefore, f_i must have a nonvanishing Jacobi determinant:

$$\det\left(\frac{\partial f_i}{\partial Q_j}\right) \neq 0. \tag{1.29}$$

In terms of Q_i, the initial Lagrangian takes the form

$$L'\left(Q_j, \dot{Q}_j, t\right) \equiv L\left(f_i\left(Q_j, t\right), \dot{f}_i\left(Q_j, t\right), t\right) \tag{1.30}$$

and the action reads

$$\begin{aligned}
\mathcal{A} &= \int_{t_a}^{t_b} dt\, L'\left(Q_j(t), \dot{Q}_j(t), t\right) \\
&= \int_{t_a}^{t_b} dt\, L\left(f_i\left(Q_j(t), t\right), \dot{f}_i\left(Q_j(t), t\right), t\right).
\end{aligned} \tag{1.31}$$

By varying the upper expression with respect to $\delta Q_j(t)$, $\delta \dot{Q}_j(t)$ while keeping $\delta Q_j(t_a) = \delta Q_j(t_b) = 0$, we find the equations of motion

$$\frac{d}{dt}\frac{\partial L'}{\partial \dot{Q}_j} - \frac{\partial L'}{\partial Q_j} = 0. \tag{1.32}$$

The variation of the lower expression, on the other hand, gives

$$\begin{aligned}
\delta \mathcal{A} &= \int_{t_a}^{t_b} dt\, \left(\frac{\partial L}{\partial q_i}\delta f_i + \frac{\partial L}{\partial \dot{q}_i}\delta \dot{f}_i\right) \\
&= \int_{t_a}^{t_b} dt\, \left(\frac{\partial L}{\partial q_i} - \frac{d}{dt}\frac{\partial L}{\partial \dot{q}_i}\right)\delta f_i + \frac{\partial L}{\partial \dot{q}_i}\delta f_i\Big|_{t_a}^{t_b}.
\end{aligned} \tag{1.33}$$

If δq_i is arbitrary, then so is δf_i. Moreover, with $\delta q_i(t_a) = \delta q_i(t_b) = 0$, also δf_i vanishes at the endpoints. Hence the extremum of the action is determined equally well by the Euler-Lagrange equations for $Q_j(t)$ [as it was by those for $q_i(t)$].

---

[2]The word *local* means here *at a specific time*. This terminology is of common use in field theory where *local* means, more generally, *at a specific spacetime point*.

Note that the locality property is quite restrictive for the transformation of the generalized velocities $\dot{q}_i(t)$. They will necessarily be linear in $\dot{Q}_j$:

$$\dot{q}_i = \dot{f}_i(Q_j, t) = \frac{\partial f_i}{\partial Q_j}\dot{Q}_j + \frac{\partial f_i}{\partial t}. \tag{1.34}$$

In phase space, there exists also the possibility of performing local changes of the canonical coordinates p_i, q_i to new ones P_j, Q_j. Let them be related by

$$\begin{aligned} p_i &= p_i(P_j, Q_j, t), \\ q_i &= q_i(P_j, Q_j, t), \end{aligned} \tag{1.35}$$

with the inverse relations

$$\begin{aligned} P_j &= P_j(p_i, q_i, t), \\ Q_j &= Q_j(p_i, q_i, t). \end{aligned} \tag{1.36}$$

However, while the Euler-Lagrange equations maintain their form under *any* local change of coordinates, the Hamilton equations do not hold, in general, for any transformed coordinates $P_j(t)$, $Q_j(t)$. The local transformations $p_i(t), q_i(t) \rightarrow P_j(t), Q_j(t)$ for which they hold, are referred to as *canonical*. They are characterized by the form invariance of the action, up to an arbitrary surface term,

$$\begin{aligned} \int_{t_a}^{t_b} dt\,[p_i\dot{q}_i - H(p_i, q_i, t)] &= \int_{t_a}^{t_b} dt\,\left[P_j\dot{Q}_j - H'(P_j, Q_j, t)\right] \\ &\quad + F(P_j, Q_j, t)\Big|_{t_a}^{t_b}, \end{aligned} \tag{1.37}$$

where $H'(P_j, Q_j, t)$ is some new Hamiltonian. Its relation with $H(p_i, q_i, t)$ must be chosen in such a way that the equality of the action holds for *any* path $p_i(t), q_i(t)$ connecting the same endpoints (at least any in some neighborhood of the classical orbits). If such an invariance exists then a variation of this action yields for $P_j(t)$ and $Q_j(t)$ the Hamilton equations of motion governed by H':

$$\begin{aligned} \dot{P}_i &= -\frac{\partial H'}{\partial Q_i}, \\ \dot{Q}_i &= \frac{\partial H'}{\partial P_i}. \end{aligned} \tag{1.38}$$

The invariance (1.37) can be expressed differently by rewriting the integral on the left-hand side in terms of the new variables $P_j(t), Q_j(t)$,

$$\int_{t_a}^{t_b} dt\,\left\{p_i\left(\frac{\partial q_i}{\partial P_j}\dot{P}_j + \frac{\partial q_i}{\partial Q_j}\dot{Q}_j + \frac{\partial q_i}{\partial t}\right) - H(p_i(P_j, Q_j, t), q_i(P_j, Q_j, t), t)\right\}, \tag{1.39}$$

and subtracting it from the right-hand side, leading to

$$\begin{aligned} \int_{t_a}^{t_b} \left\{\left(P_j - p_i\frac{\partial q_i}{\partial Q_j}\right) dQ_j - p_i\frac{\partial q_i}{\partial P_j}dP_j \right. \\ \left. - \left(H' + p_i\frac{\partial q_i}{\partial t} - H\right) dt\right\} &= -F(P_j, Q_j, t)\Big|_{t_a}^{t_b}. \end{aligned} \tag{1.40}$$

The integral is now a line integral along a curve in the $(2N + 1)$-dimensional space, consisting of the $2N$-dimensional phase space variables p_i, q_i and the time t. The right-hand side depends only on the endpoints. Thus we conclude that the integrand on the left-hand side must be a total differential. As such it has to satisfy the standard Schwarz integrability conditions [2], according to which all second derivatives have to be independent of the sequence of differentiation. Explicitly, these conditions are

$$\frac{\partial p_i}{\partial P_k}\frac{\partial q_i}{\partial Q_l} - \frac{\partial q_i}{\partial P_k}\frac{\partial p_i}{\partial Q_l} = \delta_{kl},$$

$$\frac{\partial p_i}{\partial P_k}\frac{\partial q_i}{\partial P_l} - \frac{\partial q_i}{\partial P_k}\frac{\partial p_i}{\partial P_l} = 0, \tag{1.41}$$

$$\frac{\partial p_i}{\partial Q_k}\frac{\partial q_i}{\partial Q_l} - \frac{\partial q_i}{\partial Q_k}\frac{\partial p_i}{\partial Q_l} = 0,$$

and

$$\frac{\partial p_i}{\partial t}\frac{\partial q_i}{\partial P_l} - \frac{\partial q_i}{\partial t}\frac{\partial p_i}{\partial P_l} = \frac{\partial(H' - H)}{\partial P_l},$$

$$\frac{\partial p_i}{\partial t}\frac{\partial q_i}{\partial Q_l} - \frac{\partial q_i}{\partial t}\frac{\partial p_i}{\partial Q_l} = \frac{\partial(H' - H)}{\partial Q_l}. \tag{1.42}$$

The first three equations define the so-called *Lagrange brackets* in terms of which they are written as

$$(P_k, Q_l) = \delta_{kl},$$
$$(P_k, P_l) = 0,$$
$$(Q_k, Q_l) = 0. \tag{1.43}$$

Time-dependent coordinate transformations satisfying these equations are called *symplectic*. After a little algebra involving the matrix of derivatives

$$J = \begin{pmatrix} \partial P_i/\partial p_j & \partial P_i/\partial q_j \\ \partial Q_i/\partial p_j & \partial Q_i/\partial q_j \end{pmatrix}, \tag{1.44}$$

its inverse

$$J^{-1} = \begin{pmatrix} \partial p_i/\partial P_j & \partial p_i/\partial Q_j \\ \partial q_i/\partial P_j & \partial q_i/\partial Q_j \end{pmatrix}, \tag{1.45}$$

and the symplectic unit matrix

$$E = \begin{pmatrix} 0 & \delta_{ij} \\ -\delta_{ij} & 0 \end{pmatrix}, \tag{1.46}$$

we find that the Lagrange brackets (1.43) are equivalent to the Poisson brackets

$$\{P_k, Q_l\} = \delta_{kl},$$
$$\{P_k, P_l\} = 0,$$
$$\{Q_k, Q_l\} = 0. \tag{1.47}$$

This follows from the fact that the $2N \times 2N$ matrix formed from the Lagrange brackets

$$\mathcal{L} \equiv \begin{pmatrix} -(Q_i, P_j) & -(Q_i, Q_j) \\ (P_i, P_j) & (P_i, Q_j) \end{pmatrix} \qquad (1.48)$$

can be written as $(E^{-1}J^{-1}E)^T J^{-1}$, while an analogous matrix formed from the Poisson brackets

$$\mathcal{P} \equiv \begin{pmatrix} \{P_i, Q_j\} & -\{P_i, P_j\} \\ \{Q_i, Q_j\} & -\{Q_i, P_j\} \end{pmatrix} \qquad (1.49)$$

is equal to $J(E^{-1}JE)^T$. Hence $\mathcal{L} = \mathcal{P}^{-1}$, so that (1.43) and (1.47) are equivalent to each other. Note that the Lagrange brackets (1.43) [and thus the Poisson brackets (1.47)] ensure $p_i \dot{q}_i - P_j \dot{Q}_j$ to be a total differential of some function of P_j and Q_j in the $2N$-dimensional phase space:

$$p_i \dot{q}_i - P_j \dot{Q}_j = \frac{d}{dt} G(P_j, Q_j, t). \qquad (1.50)$$

The Poisson brackets (1.47) for P_i, Q_i have the same form as those in Eqs. (1.25) for the original phase space variables p_i, q_i.

The other two equations (1.42) relate the new Hamiltonian to the old one. They can always be used to construct $H'(P_j, Q_j, t)$ from $H(p_i, q_i, t)$. The Lagrange brackets (1.43) or Poisson brackets (1.47) are therefore both necessary and sufficient for the transformation $p_i, q_i \to P_j, Q_j$ to be canonical.

A canonical transformation preserves the volume in phase space. This follows from the fact that the matrix product $J(E^{-1}JE)^T$ is equal to the $2N \times 2N$ unit matrix (1.49). Hence $\det(J) = \pm 1$ and

$$\prod_i \int [dp_i \, dq_i] = \prod_j \int [dP_j \, dQ_j]. \qquad (1.51)$$

It is obvious that the process of canonical transformations is reflexive. It may be viewed just as well from the opposite side, with the roles of p_i, q_i and P_j, Q_j exchanged [we could just as well have considered the integrand (1.40) as a complete differential in P_j, Q_j, t space].

Once a system is described in terms of new canonical coordinates P_j, Q_j, we introduce the new Poisson brackets

$$\{A, B\}' \equiv \frac{\partial A}{\partial P_j} \frac{\partial B}{\partial Q_j} - \frac{\partial B}{\partial P_j} \frac{\partial A}{\partial Q_j}, \qquad (1.52)$$

and the equation of motion for an arbitrary observable quantity $O(P_j(t), Q_j(t), t)$ becomes with (1.38)

$$\frac{dO}{dt} = \{H', O\}' + \frac{\partial O}{\partial t}, \qquad (1.53)$$

by complete analogy with (1.20). The new Poisson brackets automatically guarantee the canonical commutation rules

$$\begin{aligned}
\{P_i, Q_j\}' &= \delta_{ij}, \\
\{P_i, P_j\}' &= 0, \\
\{Q_i, Q_j\}' &= 0.
\end{aligned} \tag{1.54}$$

A standard class of canonical transformations can be constructed by introducing a *generating function* F satisfying a relation of the type (1.37), but depending explicitly on half an old and half a new set of canonical coordinates, for instance

$$F = F(q_i, Q_j, t). \tag{1.55}$$

One now considers the equation

$$\int_{t_a}^{t_b} dt \, [p_i \dot{q}_i - H(p_i, q_i, t)] = \int_{t_a}^{t_b} dt \left[P_j \dot{Q}_j - H'(P_j, Q_j, t) + \frac{d}{dt} F(q_i, Q_j, t) \right], \tag{1.56}$$

replaces $P_j \dot{Q}_j$ by $-\dot{P}_j Q_j + \frac{d}{dt} P_j Q_j$, defines

$$F(q_i, P_j, t) \equiv F(q_i, Q_j, t) + P_j Q_j,$$

and works out the derivatives. This yields

$$\begin{aligned}
\int_{t_a}^{t_b} dt & \left\{ p_i \dot{q}_i + \dot{P}_j Q_j - [H(p_i, q_i, t) - H'(P_j, Q_j, t)] \right\} \\
&= \int_{t_a}^{t_b} dt \left\{ \frac{\partial F}{\partial q_i}(q_i, P_j, t) \dot{q}_i + \frac{\partial F}{\partial P_j}(q_i, P_j, t) \dot{P}_j + \frac{\partial F}{\partial t}(q_i, P_j, t) \right\}.
\end{aligned} \tag{1.57}$$

A comparison between the two sides renders for the canonical transformation the equations

$$\begin{aligned}
p_i &= \frac{\partial}{\partial q_i} F(q_i, P_j, t), \\
Q_j &= \frac{\partial}{\partial P_j} F(q_i, P_j, t).
\end{aligned} \tag{1.58}$$

The second equation shows that the above relation between $F(q_i, P_j, t)$ and $F(q_i, Q_j, t)$ amounts to a Legendre transformation.

The new Hamiltonian is

$$H'(P_j, Q_j, t) = H(p_i, q_i, t) + \frac{\partial}{\partial t} F(q_i, P_j, t). \tag{1.59}$$

Instead of (1.55) we could, of course, also have chosen functions with other mixtures of arguments such as $F(q_i, P_j, t), F(p_i, Q_j, t), F(p_i, P_j, t)$ to generate simple canonical transformations.

A particularly important canonical transformation arises by choosing a generating function $F(q_i, P_j)$ in such a way that it leads to time-independent momenta $P_j \equiv \alpha_j$. Coordinates Q_j with this property are called *cyclic*. To find cyclic coordinates we must search for a generating function $F(q_j, P_j, t)$ which makes the transformed H' in (1.59) vanish identically. Then all derivatives with respect to the coordinates vanish and the new momenta P_j are trivially constant. Thus we seek for a solution of the equation

$$\frac{\partial}{\partial t} F(q_i, P_j, t) = -H(p_i, q_i, t), \tag{1.60}$$

where the momentum variables in the Hamiltonian obey the first equation of (1.58). This leads to the following partial differential equation for $F(q_i, P_j, t)$:

$$\partial_t F(q_i, P_j, t) = -H(\partial_{q_i} F(q_i, P_j, t), q_i, t), \tag{1.61}$$

called the *Hamilton-Jacobi equation*.

A generating function which achieves this goal is supplied by the action functional (1.14). When following the solutions starting from a fixed initial point and running to all possible final points q_i at a time t, the associated actions of these solutions form a function $A(q_i, t)$. Due to (1.18), this satisfies precisely the first of the equations (1.58):

$$p_i = \frac{\partial}{\partial q_i} A(q_i, t). \tag{1.62}$$

Moreover, the function $A(q_i, t)$ has the time derivative

$$\frac{d}{dt} A(q_i(t), t) = p_i(t)\dot{q}_i(t) - H(p_i(t), q_i(t), t). \tag{1.63}$$

Together with (1.62) this implies

$$\partial_t A(q_i, t) = -H(p_i, q_i, t). \tag{1.64}$$

If the momenta p_i on the right-hand side are replaced according to (1.62), $A(q_i, t)$ is indeed seen to be a solution of the Hamilton-Jacobi differential equation:

$$\partial_t A(q_i, t) = -H(\partial_{q_i} A(q_i, t), q_i, t). \tag{1.65}$$

1.2   Relativistic Mechanics in Curved Spacetime

The classical action of a relativistic spinless point particle in a curved four-dimensional spacetime is usually written as an integral

$$\mathcal{A} = -Mc^2 \int d\tau L(q, \dot{q}) = -Mc^2 \int d\tau \sqrt{g_{\mu\nu}\dot{q}^\mu(\tau)\dot{q}^\nu(\tau)}, \tag{1.66}$$

where τ is an arbitrary parameter of the trajectory. It can be chosen in the final trajectory to make $L(q, \dot{q}) \equiv 1$, in which case it coincides with the *proper time* of the particle. For arbitrary τ, the Euler-Lagrange equation (1.8) reads

$$\frac{d}{dt}\left[\frac{1}{L(q, \dot{q})} g_{\mu\nu} \ddot{q}^{\nu}\right] = \frac{1}{2L(q, \dot{q})}\left(\partial_{\mu} g_{\kappa\lambda}\right) \dot{q}^{\kappa} \dot{q}^{\lambda}. \tag{1.67}$$

If τ is the proper time where $L(q, \dot{q}) \equiv 1$, this simplifies to

$$\frac{d}{dt}\left(g_{\mu\nu} \dot{q}^{\nu}\right) = \frac{1}{2}\left(\partial_{\mu} g_{\kappa\lambda}\right) \dot{q}^{\kappa} \dot{q}^{\lambda}, \tag{1.68}$$

or

$$g_{\mu\nu} \ddot{q}^{\nu} = \left(\frac{1}{2}\partial_{\mu} g_{\kappa\lambda} - \partial_{\lambda} g_{\mu\kappa}\right) \dot{q}^{\kappa} \dot{q}^{\lambda}. \tag{1.69}$$

At this point one introduces the *Christoffel symbol*

$$\bar{\Gamma}_{\lambda\nu\mu} \equiv \frac{1}{2}(\partial_{\lambda} g_{\nu\mu} + \partial_{\nu} g_{\lambda\mu} - \partial_{\mu} g_{\lambda\nu}), \tag{1.70}$$

and the Christoffel symbol of the second kind[3]

$$\bar{\Gamma}_{\kappa\nu}{}^{\mu} \equiv g^{\mu\sigma} \bar{\Gamma}_{\kappa\nu\sigma}. \tag{1.71}$$

Then (1.69) can be written as

$$\ddot{q}^{\mu} + \bar{\Gamma}_{\kappa\lambda}{}^{\mu} \dot{q}^{\kappa} \dot{q}^{\lambda} = 0. \tag{1.72}$$

Since the solutions of this equation minimize the length of a curve in spacetime, they are called *geodesics*.

1.3 Quantum Mechanics

Historically, the extension of classical mechanics to quantum mechanics became necessary in order to understand the stability of atomic orbits and the discrete nature of atomic spectra. It soon became clear that these phenomena reflect the fact that at a sufficiently short length scale, small material particles such as electrons behave like waves, called *material waves*. The fact that waves cannot be squeezed into an arbitrarily small volume without increasing indefinitely their frequency and thus their energy, prevents the collapse of the electrons into the nucleus, which would take place in classical mechanics. The discreteness of the atomic states of an electron are a manifestation of standing material waves in the atomic potential well, by analogy with the standing waves of electromagnetism in a cavity.

---

[3]In many textbooks, for instance S. Weinberg, *Gravitation and Cosmology*, Wiley, New York, 1972, the upper index and the third index in (1.70) stand at the first position. Our notation follows J.A. Schouten, *Ricci Calculus*, Springer, Berlin, 1954. It will allow for a closer analogy with gauge fields in the construction of the Riemann tensor as a covariant curl of the Christoffel symbol in Chapter 10. See H. Kleinert, *Gauge Fields in Condensed Matter*, Vol. II *Stresses and Defects*, World Scientific Publishing Co., Singapore 1989, pp. 744-1443 (http://www.physik.fu-berlin.de/~kleinert/b2).

1.3.1   Bragg Reflections and Interference

The most direct manifestation of the wave nature of small particles is seen in diffraction experiments on periodic structures, for example of electrons diffracted by a crystal. If an electron beam of fixed momentum $\mathbf{p}$ passes through a crystal, it emerges along sharply peaked angles. These are the well-known *Bragg reflections*. They look very similar to the interference patterns of electromagnetic waves. In fact, it is possible to use the same mathematical framework to explain these patterns as in electromagnetism. A free particle moving with momentum

$$\mathbf{p} = (p^1, p^2, \dots, p^D). \tag{1.73}$$

through a D-dimensional Euclidean space spanned by the Cartesian coordinate vectors

$$\mathbf{x} = (x^1, x^2, \dots, x^D) \tag{1.74}$$

is associated with a *plane wave*, whose field strength or *wave function* has the form

$$\Psi_{\mathbf{p}}(\mathbf{x}, t) = e^{i\mathbf{k}\mathbf{x} - i\omega t}, \tag{1.75}$$

where $\mathbf{k}$ is the *wave vector* pointing into the direction of $\mathbf{p}$ and ω is the *wave frequency*. Each scattering center, say at $\mathbf{x}'$, becomes a source of a spherical wave with the spatial behavior e^{ikR}/R (with $R \equiv |\mathbf{x} - \mathbf{x}'|$ and $k \equiv |\mathbf{k}|$) and the wavelength $\lambda = 2\pi/k$. At the detector, all field strengths have to be added to the total field strength $\Psi(\mathbf{x}, t)$. The absolute square of the total field strength, $|\Psi(\mathbf{x}, t)|^2$, is proportional to the number of electrons arriving at the detector.

The standard experiment where these rules can most simply be applied consists of an electron beam impinging vertically upon a flat screen with two parallel slits a with spacing d. At a large distance R behind these, one observes the number of particles arriving per unit time (see Fig. 1.1)

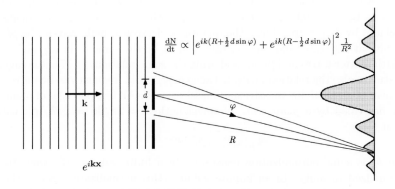

$$\frac{dN}{dt} \propto \left| e^{ik(R + \frac{1}{2}d\sin\varphi)} + e^{ik(R - \frac{1}{2}d\sin\varphi)} \right|^2 \frac{1}{R^2}$$

Figure 1.1 Probability distribution of particle behind double slit, being proportional to the absolute square of the sum of the two complex field strengths.

$$\frac{dN}{dt} \propto |\Psi_1 + \Psi_2|^2 \approx \left| e^{ik(R+\frac{1}{2}d\sin\varphi)} + e^{ik(R-\frac{1}{2}d\sin\varphi)} \right|^2 \frac{1}{R^2}, \tag{1.76}$$

where φ is the angle of deflection from the normal.

Conventionally, the wave function $\Psi(\mathbf{x}, t)$ is normalized to describe a single particle. Its absolute square gives directly the probability density of the particle at the place $\mathbf{x}$ in space, i.e., $d^3x \, |\Psi(\mathbf{x}, t)|^2$ is the probability of finding the particle in the volume element d^3x around $\mathbf{x}$.

1.3.2 Matter Waves

From the experimentally observed relation between the momentum and the size of the angular deflection φ of the diffracted beam of the particles, one deduces the relation between momentum and wave vector

$$\mathbf{p} = \hbar\mathbf{k}, \tag{1.77}$$

where $\hbar$ is the universal *Planck constant* whose dimension is equal to that of an action,

$$\hbar \equiv \frac{h}{2\pi} = 1.0545919(80) \times 10^{-27} \text{erg sec} \tag{1.78}$$

(the number in parentheses indicating the experimental uncertainty of the last two digits before it). A similar relation holds between the energy and the frequency of the wave $\Psi(\mathbf{x}, t)$. It may be determined by an absorption process in which a light wave hits an electron (for example, by kicking it out of the surface of a metal, the well-known *photoelectric effect*). From the threshold property of the photoeffect one learns that an electromagnetic wave oscillating in time as $e^{-i\omega t}$ can transfer to the electron the energy

$$E = \hbar\omega, \tag{1.79}$$

where the proportionality constant $\hbar$ is the same as in (1.77). The reason for this lies in the properties of electromagnetic waves. On the one hand, their frequency ω and the wave vector $\mathbf{k}$ satisfy the relation $\omega/c = |\mathbf{k}|$, where c is the light velocity defined to be $c \equiv 299\,792.458 km/s$. On the other hand, energy and momentum are related by $E/c = |\mathbf{p}|$. Thus, the quanta of electromagnetic waves, the *photons*, certainly satisfy (1.77) and the constant $\hbar$ must be the same as in Eq. (1.79).

With matter waves and photons sharing the same relations (1.77), it is suggestive to postulate also the relation (1.79) between energy and frequency to be universal for the waves of all particles, massive and massless ones. All free particles of momentum $\mathbf{p}$ are described by a *plane wave* of wavelength $\lambda = 2\pi/|\mathbf{k}| = 2\pi\hbar/|\mathbf{p}|$, with the explicit form

$$\Psi_{\mathbf{p}}(\mathbf{x}, t) = \mathcal{N}e^{i(\mathbf{p}\mathbf{x} - E_{\mathbf{p}}t)/\hbar}, \tag{1.80}$$

where $\mathcal{N}$ is some normalization constant. In a finite volume, the wave function is normalized to unity. In an infinite volume, this normalization makes the wave function vanish. To avoid this, the *current density* of the particle probability

$$\mathbf{j}(\mathbf{x}, t) \equiv -i\frac{\hbar}{2m}\psi^*(\mathbf{x}, t) \overleftrightarrow{\boldsymbol{\nabla}} \psi(\mathbf{x}, t) \tag{1.81}$$

is normalized in some convenient way, where $\overset{\leftrightarrow}{\nabla}$ is a short notation for the difference between right- and left-derivatives

$$
\begin{aligned}
\psi^*(\mathbf{x}, t) \overset{\leftrightarrow}{\nabla} \psi(\mathbf{x}, t) &\equiv \psi^*(\mathbf{x}, t) \overset{\rightarrow}{\nabla} \psi(\mathbf{x}, t) - \psi^*(\mathbf{x}, t) \overset{\leftarrow}{\nabla} \psi(\mathbf{x}, t) \\
&\equiv \psi^*(\mathbf{x}, t) \nabla \psi(\mathbf{x}, t) - [\nabla \psi^*(\mathbf{x}, t)] \psi(\mathbf{x}, t).
\end{aligned} \tag{1.82}
$$

The energy $E_{\mathbf{p}}$ depends on the momentum of the particle in the classical way, i.e., for nonrelativistic material particles of mass M it is $E_{\mathbf{p}} = \mathbf{p}^2/2M$, for relativistic ones $E_{\mathbf{p}} = c\sqrt{\mathbf{p}^2 + M^2 c^2}$, and for massless particles such as photons $E_{\mathbf{p}} = c|\mathbf{p}|$. The common relation $E_{\mathbf{p}} = \hbar\omega$ for photons and matter waves is necessary to guarantee conservation of energy in quantum mechanics.

In general, both momentum and energy of a particle are not sharply defined as in the plane-wave function (1.80). Usually, a particle wave is some superposition of plane waves (1.80)

$$
\Psi(\mathbf{x}, t) = \int \frac{d^3 p}{(2\pi\hbar)^3} f(\mathbf{p}) e^{i(\mathbf{p}\mathbf{x} - E_{\mathbf{p}} t)/\hbar}. \tag{1.83}
$$

By the Fourier inversion theorem, $f(\mathbf{p})$ can be calculated via the integral

$$
f(\mathbf{p}) = \int d^3 x \, e^{-i\mathbf{p}\mathbf{x}/\hbar} \Psi(\mathbf{x}, 0). \tag{1.84}
$$

With an appropriate choice of $f(\mathbf{p})$ it is possible to prepare $\Psi(\mathbf{x}, t)$ in any desired form at some initial time, say at $t = 0$. For example, $\Psi(\mathbf{x}, 0)$ may be a function sharply centered around a space point $\bar{\mathbf{x}}$. Then $f(\mathbf{p})$ is approximately a pure phase $f(\mathbf{p}) \sim e^{-i\mathbf{p}\bar{\mathbf{x}}/\hbar}$, and the wave contains all momenta with equal probability. Conversely, if the particle amplitude is spread out in space, its momentum distribution is confined to a small region. The limiting $f(\mathbf{p})$ is concentrated at a specific momentum $\bar{\mathbf{p}}$. The particle is found at each point in space with equal probability, with the amplitude oscillating like $\Psi(\mathbf{x}, t) \sim e^{i(\bar{\mathbf{p}}\mathbf{x} - E_{\bar{\mathbf{p}}} t)/\hbar}$.

In general, the width of $\Psi(\mathbf{x}, 0)$ in space and of $f(\mathbf{p})$ in momentum space are inversely proportional to each other:

$$
\Delta\mathbf{x}\,\Delta\mathbf{p} \sim \hbar. \tag{1.85}
$$

This is the content of *Heisenberg's principle of uncertainty*. If the wave is localized in a finite region of space while having at the same time a fairly well-defined average momentum $\bar{\mathbf{p}}$, it is called a *wave packet*. The maximum in the associated probability density can be shown from (1.83) to move with a velocity

$$
\bar{\mathbf{v}} = \partial E_{\bar{\mathbf{p}}}/\partial \bar{\mathbf{p}}. \tag{1.86}
$$

This coincides with the velocity of a classical particle of momentum $\bar{\mathbf{p}}$.

1.3.3 Schrödinger Equation

Suppose now that the particle is nonrelativistic and has a mass M. The classical Hamiltonian, and thus the energy $E_\mathbf{p}$, are given by

$$H(\mathbf{p}) = E_\mathbf{p} = \frac{\mathbf{p}^2}{2M}. \tag{1.87}$$

We may therefore derive the following identity for the wave field $\Psi_\mathbf{p}(\mathbf{x}, t)$:

$$\int \frac{d^3p}{(2\pi\hbar)^3} \, f(\mathbf{p}) \left[H(\mathbf{p}) - E_\mathbf{p} \right] e^{i(\mathbf{px} - E_\mathbf{p}t)/\hbar} = 0. \tag{1.88}$$

The arguments *inside* the brackets can be removed from the integral by observing that $\mathbf{p}$ and $E_\mathbf{p}$ inside the integral are equivalent to the differential operators

$$\begin{aligned} \hat{\mathbf{p}} &= -i\hbar\boldsymbol{\nabla}, \\ \hat{E} &= i\hbar\partial_t \end{aligned} \tag{1.89}$$

outside. Then, Eq. (1.88) may be written as the differential equation

$$[H(-i\hbar\boldsymbol{\nabla}) - i\hbar\partial_t)]\Psi(\mathbf{x}, t) = 0. \tag{1.90}$$

This is the *Schrödinger equation* for the wave function of a free particle. The equation suggests that the motion of a particle with an arbitrary Hamiltonian $H(\mathbf{p}, \mathbf{x}, t)$ follows the straightforward generalization of (1.90)

$$\left(\hat{H} - i\hbar\partial_t \right) \Psi(\mathbf{x}, t) = 0, \tag{1.91}$$

where $\hat{H}$ is the differential operator

$$\hat{H} \equiv H(-i\hbar\boldsymbol{\nabla}, \mathbf{x}, t). \tag{1.92}$$

The rule of obtaining $\hat{H}$ from the classical Hamiltonian $H(\mathbf{p}, \mathbf{x}, t)$ by the substitution $\mathbf{p} \to \hat{\mathbf{p}} = -i\hbar\boldsymbol{\nabla}$ will be referred to as the *correspondence principle*.[4] We shall see in Sections 1.13–1.15 that this simple correspondence principle holds only in Cartesian coordinates.

The Schrödinger operators (1.89) of momentum and energy satisfy with $\mathbf{x}$ and t the so-called canonical commutation relations

$$[\hat{p}_i, x_j] = -i\hbar, \qquad [\hat{E}, t] = 0 = i\hbar. \tag{1.93}$$

If the Hamiltonian does not depend explicitly on time, the Hilbert space can be spanned by the energy eigenstates $\Psi_{E_n}(\mathbf{x}, t) = e^{-iE_nt/\hbar}\Psi_{E_n}(\mathbf{x})$, where $\Psi_{E_n}(\mathbf{x})$ are

---

[4]Our formulation of this principle is slightly stronger than the historical one used in the initial phase of quantum mechanics, which gave certain translation rules between classical and quantum-mechanical relations. The substitution rule for the momentum runs also under the name *Jordan rule*.

time-independent *stationary states*, which solve the time-independent Schrödinger
equation

$$\hat{H}(\hat{\mathbf{p}}, \mathbf{x})\Psi_{E_n}(\mathbf{x}) = E_n\Psi_{E_n}(\mathbf{x}). \tag{1.94}$$

The validity of the Schrödinger theory (1.91) is confirmed by experiment, most
notably for the Coulomb Hamiltonian

$$H(\mathbf{p}, \mathbf{x}) = \frac{\mathbf{p}^2}{2M} - \frac{e^2}{r}, \tag{1.95}$$

which governs the quantum mechanics of the hydrogen atom in the center-of-mass
coordinate system of electron and proton, where M is the reduced mass of the two
particles.

Since the square of the wave function, $|\Psi(\mathbf{x}, t)|^2$, is interpreted as the probability
density of a single particle in a finite volume, the integral over the entire volume
must be normalized to unity:

$$\int d^3x \, |\Psi(\mathbf{x}, t)|^2 = 1. \tag{1.96}$$

For a stable particle, this normalization must remain the same at all times. If
$\Psi(\mathbf{x}, t)$ is to follow the Schrödinger equation (1.91), this is assured if and only if the
Hamiltonian operator is Hermitian,[5] i.e., if it satisfies for arbitrary wave functions
Ψ_1, Ψ_2 the equality

$$\int d^3x \, [\hat{H}\Psi_2(\mathbf{x}, t)]^* \Psi_1(\mathbf{x}, t) = \int d^3x \, \Psi_2^*(\mathbf{x}, t)\hat{H}\Psi_1(\mathbf{x}, t). \tag{1.97}$$

The left-hand side defines the *Hermitian-adjoint* $\hat{H}^\dagger$ of the operator $\hat{H}$, which satisfies the identity

$$\int d^3x \, \Psi_2^*(\mathbf{x}, t)\hat{H}^\dagger\Psi_1(\mathbf{x}, t) \equiv \int d^3x \, [\hat{H}\Psi_2(\mathbf{x}, t)]^* \Psi_1(\mathbf{x}, t) \tag{1.98}$$

for all wave functions $\Psi_1(\mathbf{x}, t), \Psi_2(\mathbf{x}, t)$. An operator $\hat{H}$ is *Hermitian*, if it coincides
with its Hermitian-adjoint $\hat{H}^\dagger$:

$$\hat{H} = \hat{H}^\dagger. \tag{1.99}$$

Let us calculate the time change of the integral over two arbitrary wave functions,
$\int d^3x \, \Psi_2^*(\mathbf{x}, t)\Psi_1(\mathbf{x}, t)$. With the Schrödinger equation (1.91), this time change vanishes indeed as long as $\hat{H}$ is Hermitian:

$$
\begin{aligned}
&i\hbar\frac{d}{dt} \int d^3x \, \Psi_2^*(\mathbf{x}, t)\Psi_1(\mathbf{x}, t) \\
&= \int d^3x \, \Psi_2^*(\mathbf{x}, t)\hat{H}\Psi_1(\mathbf{x}, t) - \int d^3x \, [\hat{H}\Psi_2(\mathbf{x}, t)]^* \Psi_1(\mathbf{x}, t) = 0.
\end{aligned} \tag{1.100}
$$

---

[5]Problems arising from unboundedness or discontinuities of the Hamiltonian and other
quantum-mechanical operators, such as restrictions of the domains of definition, are ignored here
since they are well understood. Correspondingly we do not distinguish between Hermitian and self-
adjoint operators (see J. v. Neumann, *Mathematische Grundlagen der Quantenmechanik*, Springer,
Berlin, 1932). Some quantum-mechanical operator subtleties will manifest themselves in this book
as problems of path integration to be solved in Chapter 12. The precise relationship between the
two calls for further detailed investigations.

This also implies the time independence of the normalization integral $\int d^3x \, |\Psi(\mathbf{x}, t)|^2 = 1$.

Conversely, if $\hat{H}$ is not Hermitian, one can always find an eigenstate of $\hat{H}$ whose norm changes with time: any eigenstate of $(H - H^\dagger)/i$ has this property.

Since $\hat{\mathbf{p}} = -i\hbar\boldsymbol{\nabla}$ and $\mathbf{x}$ are themselves Hermitian operators, $\hat{H}$ will automatically be a Hermitian operator if it is a sum of a kinetic and a potential energy:

$$H(\mathbf{p}, \mathbf{x}, t) = T(\mathbf{p}, t) + V(\mathbf{x}, t). \tag{1.101}$$

This is always the case for nonrelativistic particles in Cartesian coordinates $\mathbf{x}$. If $\mathbf{p}$ and $\mathbf{x}$ appear in one and the same term of H, for instance as $\mathbf{p}^2\mathbf{x}^2$, the correspondence principle does not lead to a unique quantum-mechanical operator $\hat{H}$. Then there seem to be, in principle, several Hermitian operators which, in the above example, can be constructed from the product of two $\hat{\mathbf{p}}$ and two $\hat{\mathbf{x}}$ operators [for instance $\alpha\hat{\mathbf{p}}^2\hat{\mathbf{x}}^2 + \beta\hat{\mathbf{x}}^2\hat{\mathbf{p}}^2 + \gamma\hat{\mathbf{p}}\hat{\mathbf{x}}^2\hat{\mathbf{p}}$ with $\alpha + \beta + \gamma = 1$]. They all correspond to the same classical $\mathbf{p}^2\mathbf{x}^2$. At first sight it appears as though only a comparison with experiment could select the correct operator ordering. This is referred to as the operator-ordering problem of *quantum mechanics* which has plagued many researchers in the past. If the ordering problem is caused by the geometry of the space in which the particle moves, there exists a surprisingly simple geometric principle which specifies the ordering in the physically correct way. Before presenting this in Chapter 10 we shall avoid ambiguities by assuming $H(\mathbf{p}, \mathbf{x}, t)$ to have the standard form (1.101), unless otherwise stated.

1.3.4 Particle Current Conservation

The conservation of the total probability (1.96) is a consequence of a more general *local conservation law* linking the *current density* of the particle probability

$$\mathbf{j}(\mathbf{x}, t) \equiv -i\frac{\hbar}{2m}\psi(\mathbf{x}, t) \overleftrightarrow{\boldsymbol{\nabla}} \psi(\mathbf{x}, t) \tag{1.102}$$

with the *probability density*

$$\rho(\mathbf{x}, t) = \psi^*(\mathbf{x}, t)\psi(\mathbf{x}, t) \tag{1.103}$$

via the relation

$$\partial_t\rho(\mathbf{x}, t) = -\boldsymbol{\nabla} \cdot \mathbf{j}(\mathbf{x}, t). \tag{1.104}$$

By integrating this *current conservation law* over a volume V enclosed by a surface S, and using Green's theorem, one finds

$$\int_V d^3x \, \partial_t\rho(\mathbf{x}, t) = -\int_V d^3x \, \boldsymbol{\nabla} \cdot \mathbf{j}(\mathbf{x}, t) = -\int_S d\mathbf{S} \cdot \mathbf{j}(\mathbf{x}, t), \tag{1.105}$$

where $d\mathbf{S}$ are the directed infinitesimal surface elements. This equation states that the probability in a volume decreases by the same amount by which probability leaves the surface via the current $\mathbf{j}(\mathbf{x}, t)$.

By extending the integral (1.105) over the entire space and assuming the currents to vanish at spatial infinity, we recover the conservation of the total probability (1.96).

More general dynamical systems with N particles in Euclidean space are parametrized in terms of $3N$ Cartesian coordinates $\mathbf{x}_\nu$ ($\nu = 1, \ldots, N$). The Hamiltonian has the form

$$H(\mathbf{p}_\nu, \mathbf{x}_\nu, t) = \sum_{\nu=1}^{N} \frac{\mathbf{p}_\nu^2}{2M_\nu} + V(\mathbf{x}_\nu, t), \tag{1.106}$$

where the arguments $\mathbf{p}_\nu, \mathbf{x}_\nu$ in H and V stand for all $\mathbf{p}_\nu$'s, $\mathbf{x}_\nu$ with $\nu = 1, 2, 3, \ldots, N$. The wave function $\Psi(\mathbf{x}_\nu, t)$ satisfies the N-particle Schrödinger equation

$$\left\{ -\sum_{\nu=1}^{N} \left[\frac{\hbar^2}{2M_\nu} \partial_{\mathbf{x}_\nu}{}^2 + V(\mathbf{x}_\nu, t) \right] \right\} \Psi(\mathbf{x}_\nu, t) = i\hbar \partial_t \Psi(\mathbf{x}_\nu, t). \tag{1.107}$$

1.4 Dirac's Bra-Ket Formalism

Mathematically speaking, the wave function $\Psi(\mathbf{x}, t)$ may be considered as a vector in an infinite-dimensional complex vector space called *Hilbert space*. The configuration space variable $\mathbf{x}$ plays the role of a continuous "index" of these vectors. An obvious contact with the usual vector notation may be established, in which a D-dimensional vector $\mathbf{v}$ is given in terms of its components v_i with a subscript $i = 1, \ldots D$, by writing the argument $\mathbf{x}$ of $\Psi(\mathbf{x}, t)$ as a subscript:

$$\Psi(\mathbf{x}, t) \equiv \Psi_{\mathbf{x}}(t). \tag{1.108}$$

The usual norm of a complex vector is defined by

$$|\mathbf{v}|^2 = \sum_i v_i^* v_i. \tag{1.109}$$

The continuous version of this is

$$|\Psi|^2 = \int d^3x \, \Psi_{\mathbf{x}}^*(t) \Psi_{\mathbf{x}}(t) = \int d^3x \, \Psi^*(\mathbf{x}, t) \Psi(\mathbf{x}, t). \tag{1.110}$$

The normalization condition (1.96) requires that the wave functions have the norm $|\Psi| = 1$, i.e., that they are unit vectors in the Hilbert space.

1.4.1 Basis Transformations

In a vector space, there are many possible choices of orthonormal basis vectors $b_i{}^a$ labeled by $a = 1, \ldots, D$, in terms of which[6]

$$v_i = \sum_a b_i{}^a v_a, \tag{1.111}$$

---

[6]Mathematicians would expand more precisely $v_i = \sum_a b_i{}^a v_a^{(b)}$, but physicists prefer to shorten the notation by distinguishing the different components via different types of subscripts, using for the initial components $i, j, k, \ldots$ and for the b-transformed components $a, b, c, \ldots$.

with the components v_a given by the scalar products

$$v_a \equiv \sum_i b_i^{a*} v_i. \tag{1.112}$$

The latter equation is a consequence of the *orthogonality relation*[7]

$$\sum_i b_i^{a*} b_i^{a'} = \delta^{aa'}, \tag{1.113}$$

which in a finite-dimensional vector space implies the *completeness relation*

$$\sum_a b_i^{a*} b_j^{a} = \delta^{ij}. \tag{1.114}$$

In the space of wave functions (1.108) there exists a special set of basis functions called *local basis functions* is of particular importance. It may be constructed in the following fashion: Imagine the continuum of space points to be coarse-grained into a cubic lattice of mesh size ϵ, at positions

$$\mathbf{x_n} = (n_1, n_2, n_3)\epsilon, \qquad n_{1,2,3} = 0, \pm 1, \pm 2, \dots . \tag{1.115}$$

Let $h^{\mathbf{n}}(\mathbf{x})$ be a function that vanishes everywhere in space, except in a cube of size ϵ^3 centered around $\mathbf{x_n}$, i.e., for each component x_i of $\mathbf{x}$,

$$h^{\mathbf{n}}(\mathbf{x}) = \begin{cases} 1/\sqrt{\epsilon^3} & |x_i - x_{\mathbf{n}i}| \leq \epsilon/2, \quad i = 1, 2, 3. \\ 0 & \text{otherwise.} \end{cases} \tag{1.116}$$

These functions are certainly orthonormal:

$$\int d^3x \, h^{\mathbf{n}}(\mathbf{x})^* h^{\mathbf{n}'}(\mathbf{x}) = \delta^{\mathbf{nn}'}. \tag{1.117}$$

Consider now the expansion

$$\Psi(\mathbf{x}, t) = \sum_{\mathbf{n}} h^{\mathbf{n}}(\mathbf{x})\Psi_{\mathbf{n}}(t) \tag{1.118}$$

with the coefficients

$$\Psi_{\mathbf{n}}(t) = \int d^3x \, h^{\mathbf{n}}(\mathbf{x})^* \Psi(\mathbf{x}, t) \approx \sqrt{\epsilon^3}\Psi(\mathbf{x_n}, t). \tag{1.119}$$

It provides an excellent approximation to the true wave function $\Psi(\mathbf{x}, t)$, as long as the mesh size ϵ is much smaller than the scale over which $\Psi(\mathbf{x}, t)$ varies. In fact, if $\Psi(\mathbf{x}, t)$ is integrable, the integral over the sum (1.118) will always converge to $\Psi(\mathbf{x}, t)$. The same convergence of discrete approximations is found in any scalar product, and thus in any observable probability amplitudes. They can all be calculated with arbitrary accuracy knowing the discrete components of the type (1.119) in the limit

---

[7]An orthogonality relation implies usually a unit norm and is thus really an *orthonormality relation* but this name is rarely used.

$\epsilon \to 0$. The functions $h^n(\mathbf{x})$ may therefore be used as an approximate basis in the same way as the previous basis functions $f^a(\mathbf{x}), g^b(\mathbf{x})$, with any desired accuracy depending on the choice of ϵ.

In general, there are many possible orthonormal basis functions $f^a(\mathbf{x})$ in the Hilbert space which satisfy the orthonormality relation

$$\int d^3x \, f^a(\mathbf{x})^* f^{a'}(\mathbf{x}) = \delta^{aa'}, \tag{1.120}$$

in terms of which we can expand

$$\Psi(\mathbf{x}, t) = \sum_a f^a(\mathbf{x}) \Psi_a(t), \tag{1.121}$$

with the coefficients

$$\Psi_a(t) = \int d^3x \, f^a(\mathbf{x})^* \, \Psi(\mathbf{x}, t). \tag{1.122}$$

Suppose we use other orthonormal basis $\tilde{f}^b(\mathbf{x})$ with the orthonormality relation

$$\int d^3x \, \tilde{f}^b(\mathbf{x})^* \tilde{f}^{b'}(\mathbf{x}) = \delta^{bb'}, \qquad \sum_b \tilde{f}^b(\mathbf{x}) \tilde{f}^b(\mathbf{x})^* = \delta^{(3)}(\mathbf{x} - \mathbf{x}'), \tag{1.123}$$

to re-expand

$$\Psi(\mathbf{x}, t) = \sum_b \tilde{f}^b(\mathbf{x}) \tilde{\Psi}_b(t), \tag{1.124}$$

with the components

$$\tilde{\Psi}_b(t) = \int d^3x \, \tilde{f}^b(\mathbf{x})^* \, \Psi(\mathbf{x}, t). \tag{1.125}$$

Inserting (1.121) shows that the components are related to each other by

$$\tilde{\Psi}_b(t) = \sum_a \left[\int d^3x \, \tilde{f}^b(\mathbf{x})^* f^a(\mathbf{x}) \right] \Psi_a(t). \tag{1.126}$$

1.4.2   Bracket Notation

It is useful to write the scalar products between two wave functions occurring in the above basis transformations in the so-called *bracket notation* as

$$\langle \tilde{b} | a \rangle \equiv \int d^3x \, \tilde{f}^b(\mathbf{x})^* f^a(\mathbf{x}). \tag{1.127}$$

In this notation, the components of the state vector $\Psi(\mathbf{x}, t)$ in (1.122), (1.125) are

$$\begin{aligned} \Psi_a(t) &= \langle a | \Psi(t) \rangle, \\ \tilde{\Psi}_b(t) &= \langle \tilde{b} | \Psi(t) \rangle. \end{aligned} \tag{1.128}$$

The transformation formula (1.126) takes the form

$$\langle \tilde{b} | \Psi(t) \rangle = \sum_a \langle \tilde{b} | a \rangle \langle a | \Psi(t) \rangle. \tag{1.129}$$

The right-hand side of this equation may be formally viewed as a result of inserting the abstract relation

$$\sum_a |a\rangle\langle a| = 1 \qquad (1.130)$$

between $\langle \tilde{b}|$ and $|\Psi(t)\rangle$ on the left-hand side:

$$\langle \tilde{b}|\Psi(t)\rangle = \langle \tilde{b}|1|\Psi(t)\rangle = \sum_a \langle \tilde{b}|a\rangle\langle a|\Psi(t)\rangle. \qquad (1.131)$$

Since this expansion is only possible if the functions $f^b(\mathbf{x})$ form a complete basis, the relation (1.130) is alternative, abstract way of stating the completeness of the basis functions. It may be referred to as completeness relation à la Dirac.

Since the scalar products are written in the form of brackets $\langle a|a'\rangle$, Dirac called the formal objects $\langle a|$ and $|a'\rangle$, from which the brackets are composed, *bra* and *ket*, respectively. In the bracket notation, the orthonormality of the basis $f^a(\mathbf{x})$ and $g^b(\mathbf{x})$ may be expressed as follows:

$$\begin{aligned} \langle a|a'\rangle &= \int d^3x \, f^a(\mathbf{x})^* f^{a'}(\mathbf{x}) = \delta^{aa'}, \\ \langle \tilde{b}|\tilde{b}'\rangle &= \int d^3x \, \tilde{f}^b(\mathbf{x})^* \tilde{f}^{b'}(\mathbf{x}) = \delta^{bb'}. \end{aligned} \qquad (1.132)$$

In the same spirit we introduce abstract bra and ket vectors associated with the basis functions $h^{\mathbf{n}}(\mathbf{x})$ of Eq. (1.116), denoting them by $\langle \mathbf{x_n}|$ and $|\mathbf{x_n}\rangle$, respectively, and writing the orthogonality relation (1.117) in bracket notation as

$$\langle \mathbf{x_n}|\mathbf{x_{n'}}\rangle \equiv \int d^3x \, h^{\mathbf{n}}(\mathbf{x})^* h^{\mathbf{n'}}(\mathbf{x}) = \delta_{\mathbf{nn'}}. \qquad (1.133)$$

The components $\Psi_{\mathbf{n}}(t)$ may be considered as the scalar products

$$\Psi_{\mathbf{n}}(t) \equiv \langle \mathbf{x_n}|\Psi(t)\rangle \approx \sqrt{\epsilon^3}\Psi(\mathbf{x_n}, t). \qquad (1.134)$$

Changes of basis vectors, for instance from $|\mathbf{x_n}\rangle$ to the states $|a\rangle$, can be performed according to the rules developed above by inserting a completeness relation à la Dirac of the type (1.130). Thus we may expand

$$\Psi_{\mathbf{n}}(t) = \langle \mathbf{x_n}|\Psi(t)\rangle = \sum_a \langle \mathbf{x_n}|a\rangle\langle a|\Psi(t)\rangle. \qquad (1.135)$$

Also the inverse relation is true:

$$\langle a|\Psi(t)\rangle = \sum_{\mathbf{n}} \langle a|\mathbf{x_n}\rangle\langle \mathbf{x_n}|\Psi(t)\rangle. \qquad (1.136)$$

This is, of course, just an approximation to the integral

$$\int d^3x \, h^{\mathbf{n}}(\mathbf{x})^* \langle \mathbf{x}|\Psi(t)\rangle. \qquad (1.137)$$

The completeness of the basis $h^{\mathbf{n}}(\mathbf{x})$ may therefore be expressed via the abstract relation

$$\sum_{\mathbf{n}} |\mathbf{x_n}\rangle\langle \mathbf{x_n}| \approx 1. \qquad (1.138)$$

The approximate sign turns into an equality sign in the limit of zero mesh size, $\epsilon \to 0$.

1.4.3   Continuum Limit

In ordinary calculus, finer and finer sums are eventually replaced by integrals. The same thing is done here. We define new continuous scalar products

$$\langle \mathbf{x}|\Psi(t)\rangle \approx \frac{1}{\sqrt{\epsilon^3}}\langle \mathbf{x_n}|\Psi(t)\rangle, \tag{1.139}$$

where $\mathbf{x_n}$ are the lattice points closest to $\mathbf{x}$. With (1.134), the right-hand side is equal to $\Psi(\mathbf{x_n}, t)$. In the limit $\epsilon \to 0$, $\mathbf{x}$ and $\mathbf{x_n}$ coincide and we have

$$\langle \mathbf{x}|\Psi(t)\rangle \equiv \Psi(\mathbf{x}, t). \tag{1.140}$$

The completeness relation can be used to write

$$\begin{aligned} \langle a|\Psi(t)\rangle &\approx \sum_{\mathbf{n}}\langle a|\mathbf{x_n}\rangle\langle \mathbf{x_n}|\Psi(t)\rangle \\ &\approx \sum_{\mathbf{n}} \epsilon^3 \langle a|\mathbf{x}\rangle\langle \mathbf{x}|\Psi(t)\rangle\big|_{\mathbf{x}=\mathbf{x^n}} , \end{aligned} \tag{1.141}$$

which in the limit $\epsilon \to 0$ becomes

$$\langle a|\Psi(t)\rangle = \int d^3x \, \langle a|\mathbf{x}\rangle\langle \mathbf{x}|\Psi(t)\rangle. \tag{1.142}$$

This may be viewed as the result of inserting the formal completeness relation of the limiting local bra and ket basis vectors $\langle \mathbf{x}|$ and $|\mathbf{x}\rangle$,

$$\int d^3x \, |\mathbf{x}\rangle\langle \mathbf{x}| = 1, \tag{1.143}$$

evaluated between the vectors $\langle a|$ and $|\Psi(t)\rangle$.

With the limiting local basis, the wave functions can be treated as components of the state vectors $|\Psi(t)\rangle$ with respect to the local basis $|\mathbf{x}\rangle$ in the same way as any other set of components in an arbitrary basis $|a\rangle$. In fact, the expansion

$$\langle a|\Psi(t)\rangle = \int d^3x \, \langle a|\mathbf{x}\rangle\langle \mathbf{x}|\Psi(t)\rangle \tag{1.144}$$

may be viewed as a re-expansion of a component of $|\Psi(t)\rangle$ in one basis, $|a\rangle$, into those of another basis, $|\mathbf{x}\rangle$, just as in (1.129).

In order to express all these transformation properties in a most compact notation, it has become customary to deal with an arbitrary physical state vector in a *basis-independent* way and denote it by a ket vector $|\Psi(t)\rangle$. This vector may be specified in any convenient basis by multiplying it with the corresponding completeness relation

$$\sum_a |a\rangle\langle a| = 1, \tag{1.145}$$

resulting in the expansion

$$|\Psi(t)\rangle = \sum_a |a\rangle\langle a|\Psi(t)\rangle. \tag{1.146}$$

This can be multiplied with any bra vector, say $\langle b|$, from the left to obtain the expansion formula (1.131):

$$\langle b|\Psi(t)\rangle = \sum_a \langle b|a\rangle\langle a|\Psi(t)\rangle. \tag{1.147}$$

The continuum version of the completeness relation (1.138) reads

$$\int d^3x\,|\mathbf{x}\rangle\langle\mathbf{x}| = 1, \tag{1.148}$$

and leads to the expansion

$$|\Psi(t)\rangle = \int d^3x\,|\mathbf{x}\rangle\langle\mathbf{x}|\Psi(t)\rangle, \tag{1.149}$$

in which the wave function $\Psi(\mathbf{x},t) = \langle\mathbf{x}|\Psi(t)\rangle$ plays the role of an $\mathbf{x}$th component of the state vector $|\Psi(t)\rangle$ in the local basis $|\mathbf{x}\rangle$. This, in turn, is the limit of the discrete basis vectors $|\mathbf{x_n}\rangle$,

$$|\mathbf{x}\rangle \approx \frac{1}{\sqrt{\epsilon^3}}\,|\mathbf{x_n}\rangle\,, \tag{1.150}$$

with $\mathbf{x_n}$ being the lattice points closest to $\mathbf{x}$.

A vector can be described equally well in bra or in ket form. To apply the above formalism consistently, we observe that the scalar products

$$\begin{aligned}
\langle a|\tilde{b}\rangle &= \int d^3x\,f^a(\mathbf{x})^*\tilde{f}^b(\mathbf{x}),\\
\langle\tilde{b}|a\rangle &= \int d^3x\,\tilde{f}^b(\mathbf{x})^*f^a(\mathbf{x})
\end{aligned} \tag{1.151}$$

satisfy the identity

$$\langle\tilde{b}|a\rangle \equiv \langle a|\tilde{b}\rangle^*. \tag{1.152}$$

Therefore, when expanding a ket vector as

$$|\Psi(t)\rangle = \sum_a |a\rangle\langle a|\Psi(t)\rangle, \tag{1.153}$$

or a bra vector as

$$\langle\Psi(t)| = \sum_a \langle\Psi(t)|a\rangle\langle a|, \tag{1.154}$$

a multiplication of the first equation with the bra $\langle\mathbf{x}|$ and of the second with the ket $|\mathbf{x}\rangle$ produces equations which are complex-conjugate to each other.

1.4.4 Generalized Functions

Dirac's bra-ket formalism is elegant and easy to handle. As far as the vectors $|\mathbf{x}\rangle$ are concerned there is, however, one inconsistency with some fundamental postulates of quantum mechanics: When introducing state vectors, the norm was required to be unity in order to permit a proper probability interpretation of single-particle states.

The limiting states $|\mathbf{x}\rangle$ introduced above do not satisfy this requirement. In fact, the scalar product between two different states $\langle\mathbf{x}|$ and $|\mathbf{x}'\rangle$ is

$$\langle\mathbf{x}|\mathbf{x}'\rangle \approx \frac{1}{\epsilon^3}\langle\mathbf{x_n}|\mathbf{x_{n'}}\rangle = \frac{1}{\epsilon^3}\delta_{nn'}, \tag{1.155}$$

where $\mathbf{x_n}$ and $\mathbf{x_{n'}}$ are the lattice points closest to $\mathbf{x}$ and $\mathbf{x}'$. For $\mathbf{x} \neq \mathbf{x}'$, the states are orthogonal. For $\mathbf{x} = \mathbf{x}'$, on the other hand, the limit $\epsilon \to 0$ is infinite, approached in such a way that

$$\epsilon^3\sum_{\mathbf{n'}}\frac{1}{\epsilon^3}\delta_{nn'} = 1. \tag{1.156}$$

Therefore, the limiting state $|\mathbf{x}\rangle$ is not a properly normalizable vector in the Hilbert space. For the sake of elegance, it is useful to weaken the requirement of normalizability (1.96) by admitting the limiting states $|\mathbf{x}\rangle$ to the physical Hilbert space. In fact, one admits all states which can be obtained by a limiting sequence from properly normalized state vectors.

The scalar product between states $\langle\mathbf{x}|\mathbf{x}'\rangle$ is not a proper function. It is denoted by the symbol $\delta^{(3)}(\mathbf{x} - \mathbf{x}')$ and called *Dirac δ-function*:

$$\langle\mathbf{x}|\mathbf{x}'\rangle \equiv \delta^{(3)}(\mathbf{x} - \mathbf{x}'). \tag{1.157}$$

The right-hand side vanishes everywhere, except in the infinitely small box of width ϵ around $\mathbf{x} \approx \mathbf{x}'$. Thus the δ-function satisfies

$$\delta^{(3)}(\mathbf{x} - \mathbf{x}') = 0 \quad \text{for} \quad \mathbf{x} \neq \mathbf{x}'. \tag{1.158}$$

At $\mathbf{x} = \mathbf{x}'$, it is so large that its volume integral is unity:

$$\int d^3x'\, \delta^{(3)}(\mathbf{x} - \mathbf{x}') = 1. \tag{1.159}$$

Obviously, there exists no proper function that can satisfy both requirements, (1.158) and (1.159). Only the finite-ϵ approximation in (1.155) to the δ-function are proper functions. In this respect, the scalar product $\langle\mathbf{x}|\mathbf{x}'\rangle$ behaves just like the states $|\mathbf{x}\rangle$ themselves: Both are $\epsilon \to 0$ -limits of properly defined mathematical objects.

Note that the integral Eq. (1.159) implies the following property of the δ - function:

$$\delta^{(3)}(a(\mathbf{x} - \mathbf{x}')) = \frac{1}{|a|}\delta^{(3)}(\mathbf{x} - \mathbf{x}'). \tag{1.160}$$

In one dimension, this leads to the more general relation

$$\delta(f(x)) = \sum_i \frac{1}{|f'(x_i)|}\delta(x - x_i), \tag{1.161}$$

where x_i are the simple zeros of $f(x)$.

In mathematics, one calls the δ-function a *generalized function* or a *distribution*. It defines a linear functional of arbitrary smooth *test functions* $f(\mathbf{x})$ which yields its value at any desired place $\mathbf{x}$:

$$\delta[f;\mathbf{x}] \equiv \int d^3x\, \delta^{(3)}(\mathbf{x} - \mathbf{x}') f(\mathbf{x}') = f(\mathbf{x}). \tag{1.162}$$

Test functions are arbitrarily often differentiable functions with a sufficiently fast falloff at spatial infinity.

There exist a rich body of mathematical literature on distributions [3]. They form a linear space. This space is restricted in an essential way in comparison with ordinary functions: products of δ-functions or any other distributions remain undefined. In Section 10.8.1 we shall find, however, that physics forces us to go beyond these rules. An important requirement of quantum mechanics is coordinate invariance. If we want to achieve this for the path integral formulation of quantum mechanics, we must set up a definite extension of the existing theory of distributions, which specifies uniquely integrals over products of distributions.

In quantum mechanics, the role of the test functions is played by the wave packets $\Psi(\mathbf{x}, t)$. By admitting the generalized states $|\mathbf{x}\rangle$ to the Hilbert space, we also admit the scalar products $\langle \mathbf{x}|\mathbf{x}'\rangle$ to the space of wave functions, and thus all distributions, although they are not normalizable.

1.4.5 Schrödinger Equation in Dirac Notation

In terms of the bra-ket notation, the Schrödinger equation can be expressed in a basis-independent way as an operator equation

$$\hat{H}|\Psi(t)\rangle \equiv H(\hat{\mathbf{p}}, \hat{\mathbf{x}}, t)|\Psi(t)\rangle = i\hbar\partial_t|\Psi(t)\rangle, \tag{1.163}$$

to be supplemented by the following specifications of the canonical operators:

$$\langle \mathbf{x}|\hat{\mathbf{p}} \equiv -i\hbar\boldsymbol{\nabla}\langle \mathbf{x}|, \tag{1.164}$$

$$\langle \mathbf{x}|\hat{\mathbf{x}} \equiv \mathbf{x}\langle \mathbf{x}|. \tag{1.165}$$

Any matrix element can be obtained from these equations by multiplication from the right with an arbitrary ket vector; for instance with the local basis vector $|\mathbf{x}'\rangle$:

$$\langle \mathbf{x}|\hat{\mathbf{p}}|\mathbf{x}'\rangle = -i\hbar\boldsymbol{\nabla}\langle \mathbf{x}|\mathbf{x}'\rangle = -i\hbar\boldsymbol{\nabla}\delta^{(3)}(\mathbf{x} - \mathbf{x}'), \tag{1.166}$$

$$\langle \mathbf{x}|\hat{\mathbf{x}}|\mathbf{x}'\rangle = \mathbf{x}\langle \mathbf{x}|\mathbf{x}'\rangle = \mathbf{x}\delta^{(3)}(\mathbf{x} - \mathbf{x}'). \tag{1.167}$$

The original differential form of the Schrödinger equation (1.91) follows by multiplying the basis-independent Schrödinger equation (1.163) with the bra vector $\langle \mathbf{x}|$ from the left:

$$\begin{aligned}\langle \mathbf{x}|H(\hat{\mathbf{p}}, \hat{\mathbf{x}}, t)|\Psi(t)\rangle &= H(-i\hbar\boldsymbol{\nabla}, \mathbf{x}, t)\langle \mathbf{x}|\Psi(t)\rangle \\ &= i\hbar\partial_t\langle \mathbf{x}|\Psi(t)\rangle.\end{aligned} \tag{1.168}$$

Obviously, $\hat{\mathbf{p}}$ and $\hat{\mathbf{x}}$ are Hermitian matrices in any basis,

$$\langle a|\hat{\mathbf{p}}|a'\rangle = \langle a'|\hat{\mathbf{p}}|a\rangle^*, \tag{1.169}$$

$$\langle a|\hat{\mathbf{x}}|a'\rangle = \langle a'|\hat{\mathbf{x}}|a\rangle^*, \tag{1.170}$$

and so is the Hamiltonian

$$\langle a|\hat{H}|a'\rangle = \langle a'|\hat{H}|a\rangle^*, \tag{1.171}$$

as long as it has the form (1.101).

The most general basis-independent operator that can be constructed in the generalized Hilbert space spanned by the states $|\mathbf{x}\rangle$ is some function of $\hat{\mathbf{p}}, \hat{\mathbf{x}}, t$,

$$\hat{O}(t) \equiv O(\hat{\mathbf{p}}, \hat{\mathbf{x}}, t). \tag{1.172}$$

In general, such an operator is called Hermitian if all its matrix elements have this property. In the basis-independent Dirac notation, the definition (1.97) of a Hermitian-adjoint operator $\hat{O}^\dagger(t)$ implies the equality of the matrix elements

$$\langle a|\hat{O}^\dagger(t)|a'\rangle \equiv \langle a'|\hat{O}(t)|a\rangle^*. \tag{1.173}$$

Thus we can rephrase Eqs. (1.169)–(1.171) in the basis-independent form

$$\hat{\mathbf{p}} = \hat{\mathbf{p}}^\dagger,$$
$$\hat{\mathbf{x}} = \hat{\mathbf{x}}^\dagger, \tag{1.174}$$
$$\hat{H} = \hat{H}^\dagger.$$

The stationary states in Eq. (1.94) have a Dirac ket representation $|E_n\rangle$, and satisfy the time-independent operator equation

$$\hat{H}|E_n\rangle = E_n|E_n\rangle. \tag{1.175}$$

1.4.6   Momentum States

Let us now look at the momentum $\hat{\mathbf{p}}$. Its eigenstates are given by the eigenvalue equation

$$\hat{\mathbf{p}}|\mathbf{p}\rangle = \mathbf{p}|\mathbf{p}\rangle. \tag{1.176}$$

By multiplying this with $\langle\mathbf{x}|$ from the left and using (1.164), we find the differential equation

$$\langle\mathbf{x}|\hat{\mathbf{p}}|\mathbf{p}\rangle = -i\hbar\partial_{\mathbf{x}}\langle\mathbf{x}|\mathbf{p}\rangle = \mathbf{p}\langle\mathbf{x}|\mathbf{p}\rangle. \tag{1.177}$$

The solution is

$$\langle\mathbf{x}|\mathbf{p}\rangle \propto e^{i\mathbf{p}\mathbf{x}/\hbar}. \tag{1.178}$$

Up to a normalization factor, this is just a plane wave introduced before in Eq. (1.75) to describe free particles of momentum $\mathbf{p}$.

In order for the states $|\mathbf{p}\rangle$ to have a finite norm, the system must be confined to a finite volume, say a cubic box of length L and volume L^3. Assuming periodic boundary conditions, the momenta are discrete with values

$$\mathbf{p^m} = \frac{2\pi\hbar}{L}(m_1, m_2, m_3), \quad m_i = 0, \pm 1, \pm 2, \ldots \quad . \tag{1.179}$$

Then we adjust the factor in front of $\exp(i\mathbf{p^m x}/\hbar)$ to achieve unit normalization

$$\langle \mathbf{x}|\mathbf{p^m}\rangle = \frac{1}{\sqrt{L^3}}\exp(i\mathbf{p^m x}/\hbar), \tag{1.180}$$

and the discrete states $|\mathbf{p^m}\rangle$ satisfy

$$\int d^3x \, |\langle \mathbf{x}|\mathbf{p^m}\rangle|^2 = 1. \tag{1.181}$$

The states $|\mathbf{p^m}\rangle$ are complete:

$$\sum_{\mathbf{m}} |\mathbf{p^m}\rangle\langle \mathbf{p^m}| = 1. \tag{1.182}$$

We may use this relation and the matrix elements $\langle \mathbf{x}|\mathbf{p^m}\rangle$ to expand any wave function within the box as

$$\Psi(\mathbf{x}, t) = \langle \mathbf{x}|\Psi(t)\rangle = \sum_{\mathbf{m}}\langle \mathbf{x}|\mathbf{p^m}\rangle\langle \mathbf{p^m}|\Psi(t)\rangle. \tag{1.183}$$

If the box is very large, the sum over the discrete momenta $\mathbf{p^m}$ can be approximated by an integral over the momentum space [4].

$$\sum_{\mathbf{m}} \approx \int \frac{d^3p\, L^3}{(2\pi\hbar)^3}. \tag{1.184}$$

In this limit, the states $|\mathbf{p^m}\rangle$ may be used to define a continuum of basis vectors with an improper normalization

$$|\mathbf{p}\rangle \approx \sqrt{L^3}|\mathbf{p^m}\rangle, \tag{1.185}$$

in the same way as $|\mathbf{x_n}\rangle$ was used in (1.150) to define $|\mathbf{x}\rangle \sim (1/\sqrt{\epsilon^3})|\mathbf{x^n}\rangle$. The momentum states $|\mathbf{p}\rangle$ satisfy the orthogonality relation

$$\langle \mathbf{p}|\mathbf{p'}\rangle = (2\pi\hbar)^3\delta^{(3)}(\mathbf{p} - \mathbf{p'}), \tag{1.186}$$

with $\delta^{(3)}(\mathbf{p} - \mathbf{p'})$ being again the Dirac δ-function. Their completeness relation reads

$$\int \frac{d^3p}{(2\pi\hbar)^3}|\mathbf{p}\rangle\langle \mathbf{p}| = 1, \tag{1.187}$$

such that the expansion (1.183) becomes

$$\Psi(\mathbf{x}, t) = \int \frac{d^3p}{(2\pi\hbar)^3} \langle \mathbf{x}|\mathbf{p}\rangle \langle \mathbf{p}|\Psi(t)\rangle, \tag{1.188}$$

with the momentum eigenfunctions

$$\langle \mathbf{x}|\mathbf{p}\rangle = e^{i\mathbf{p}\mathbf{x}/\hbar}. \tag{1.189}$$

This coincides precisely with the Fourier decomposition introduced above in the description of a general particle wave $\Psi(\mathbf{x}, t)$ in (1.83), (1.84), with the identification

$$\langle \mathbf{p}|\Psi(t)\rangle = f(\mathbf{p})e^{-iE_{\mathbf{p}}t/\hbar}. \tag{1.190}$$

The bra-ket formalism accommodates naturally the technique of Fourier transforms. The Fourier inversion formula is found by simply inserting into $\langle \mathbf{p}|\Psi(t)\rangle$ a completeness relation $\int d^3x |\mathbf{x}\rangle\langle \mathbf{x}| = 1$ which yields

$$\begin{aligned} \langle \mathbf{p}|\Psi(t)\rangle &= \int d^3x \, \langle \mathbf{p}|\mathbf{x}\rangle \langle \mathbf{x}|\Psi(t)\rangle \\ &= \int d^3x \, e^{-i\mathbf{p}\mathbf{x}/\hbar}\Psi(\mathbf{x}, t). \end{aligned} \tag{1.191}$$

The amplitudes $\langle \mathbf{p}|\Psi(t)\rangle$ are referred to as *momentum space wave functions*.

By inserting the completeness relation

$$\int d^3x |\mathbf{x}\rangle\langle \mathbf{x}| = 1 \tag{1.192}$$

between the momentum states on the left-hand side of the orthogonality relation (1.186), we obtain the Fourier representation of the δ-function

$$\begin{aligned} \langle \mathbf{p}|\mathbf{p}'\rangle &= \int d^3x \, \langle \mathbf{p}|\mathbf{x}\rangle\langle \mathbf{x}|\mathbf{p}'\rangle \\ &= \int d^3x \, e^{-i(\mathbf{p}-\mathbf{p}')\mathbf{x}/\hbar}. \end{aligned} \tag{1.193}$$

1.4.7   Incompleteness and Poisson's Summation Formula

For many physical applications it is important to find out what happens to the completeness relation (1.148) if one restrict the integral so a subset of positions. Most relevant will be the one-dimensional integral,

$$\int dx \, |x\rangle\langle x| = 1, \tag{1.194}$$

restricted to a sum over equally spaced points $x_n = na$:

$$\sum_{n=-N}^{N} |x_n\rangle\langle x_n|. \tag{1.195}$$

Taking this sum between momentum eigenstates $|p\rangle$, we obtain

$$\sum_{n=-N}^{N} \langle p|x_n\rangle\langle x_n|p'\rangle = \sum_{n=-N}^{N} \langle p|x_n\rangle\langle x_n|p'\rangle = \sum_{n=-N}^{N} e^{i(p-p')na/\hbar} \qquad (1.196)$$

For $N \to \infty$ we can perform the sum with the help of *Poisson's summation formula*

$$\sum_{n=-\infty}^{\infty} e^{2\pi i\mu n} = \sum_{m=-\infty}^{\infty} \delta(\mu - m). \qquad (1.197)$$

Identifying μ with $(p-p')a/2\pi\hbar$, we find using Eq. (1.160):

$$\sum_{n=-\infty}^{\infty} \langle p|x_n\rangle\langle x_n|p'\rangle = \sum_{m=-\infty}^{\infty} \delta\left(\frac{(p-p')a}{2\pi\hbar} - m\right) = \sum_{m=-\infty}^{\infty} \frac{2\pi\hbar}{a}\delta\left(p-p' - \frac{2\pi\hbar m}{a}\right). (1.198)$$

In order to prove the Poisson formula (1.197), we observe that the sum $s(\mu) \equiv \sum_m \delta(\mu - m)$ on the right-hand side is periodic in μ with a unit period and has the Fourier series $s(\mu) = \sum_{n=-\infty}^{\infty} s_n e^{2\pi i\mu n}$. The Fourier coefficients are given by $s_n = \int_{-1/2}^{1/2} d\mu\, s(\mu) e^{-2\pi i\mu n} \equiv 1$. These are precisely the Fourier coefficients on the left-hand side.

For a finite N, the sum over n on the left-hand side of (1.197) yields

$$\sum_{n=-N}^{N} e^{2\pi i\mu n} = 1 + \left(e^{2\pi i\mu} + e^{2\cdot 2\pi i\mu} + \ldots + e^{N\cdot 2\pi i\mu} + \text{c.c.}\right)$$

$$= -1 + \left(\frac{1 - e^{2\pi i\mu(N+1)}}{1 - e^{2\pi i\mu}} + \text{c.c.}\right) \qquad (1.199)$$

$$= 1 + \frac{e^{2\pi i\mu} - e^{2\pi i\mu(N+1)}}{1 - e^{2\pi i\mu}} + \text{c.c.} = \frac{\sin\pi\mu(2N+1)}{\sin\pi\mu}.$$

This function is well known in wave optics (see Fig. 2.4). It determines the diffraction pattern of light behind a grating with $2N + 1$ slits. It has large peaks at $\mu = 0, \pm 1, \pm 2, \pm 3, \ldots$ and $N - 1$ small maxima between each pair of neighboring peaks, at $\nu = (1 + 4k)/2(2N + 1)$ for $k = 1, \ldots, N - 1$. There are zeros at $\nu = (1 + 2k)/(2N + 1)$ for $k = 1, \ldots, N - 1$.

Inserting $\mu = (p - p')a/2\pi\hbar$ into (1.199), we obtain

$$\sum_{n=-N}^{N} \langle p|x_n\rangle\langle x_n|p'\rangle = \frac{\sin(p - p')a(2N + 1)/2\hbar}{\sin(p - p')a/2\hbar}. \qquad (1.200)$$

Let us see how the right-hand side of (1.199) turns into the right-hand side of (1.197) in the limit $N \to \infty$. In this limit, the area under each large peak can be calculated by an integral over the central large peak plus a number n of small maxima next to it:

$$\int_{-n/2N}^{n/2N} d\mu \frac{\sin\pi\mu(2N+1)}{\sin\pi\mu} = \int_{-n/2N}^{n/2N} d\mu \frac{\sin 2\pi\mu N \cos\pi\mu + \cos 2\pi\mu N \sin\pi\mu}{\sin\pi\mu}.$$

$$(1.201)$$

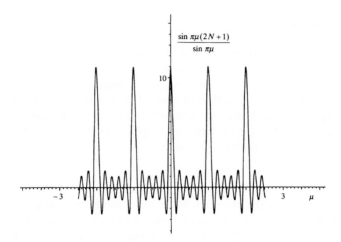

Figure 1.2   Relevant function $\sum_{n=-N}^{N} e^{2\pi i \mu n}$ in Poisson's summation formula. In the limit $N \to \infty$, μ is squeezed to the integer values.

Keeping a fixed ratio $n/N \ll 1$, we may replace in the integrand $\sin \pi \mu$ by $\pi \mu$ and $\cos \pi \mu$ by 1. Then the integral becomes, for $N \to \infty$ at fixed n/N,

$$\int_{-n/2N}^{n/2N} d\mu \frac{\sin \pi \mu (2N+1)}{\sin \pi \mu} \xrightarrow{N \to \infty} \int_{-n/2N}^{n/2N} d\mu \frac{\sin 2\pi \mu N}{\pi \mu} + \int_{-n/2N}^{n/2N} d\mu \cos 2\pi \mu N$$

$$\xrightarrow{N \to \infty} \frac{1}{\pi} \int_{-\pi n}^{\pi n} dx \frac{\sin x}{x} + \frac{1}{2\pi N} \int_{-\pi n}^{\pi n} dx \cos x \xrightarrow{N \to \infty} 1, \qquad (1.202)$$

where we have used the integral formula

$$\int_{-\infty}^{\infty} dx \frac{\sin x}{x} = \pi. \qquad (1.203)$$

In the limit $N \to \infty$, we find indeed (1.197) and thus (1.205), as well as the expression (2.464) for the free energy.

There exists another useful way of expressing Poisson's formula. Consider an arbitrary smooth function $f(\mu)$ which possesses a convergent sum

$$\sum_{m=-\infty}^{\infty} f(m). \qquad (1.204)$$

Then Poisson's formula (1.197) implies that the sum can be rewritten as an integral and an auxiliary sum:

$$\sum_{m=-\infty}^{\infty} f(m) = \int_{-\infty}^{\infty} d\mu \sum_{n=-\infty}^{\infty} e^{2\pi i \mu n} f(\mu). \qquad (1.205)$$

The auxiliary sum over n squeezes μ to the integer numbers.

1.5 Observables

Changes of basis vectors are an important tool in analyzing the physically observable content of a wave vector. Let $A = A(\mathbf{p}, \mathbf{x})$ be an arbitrary time-independent real function of the phase space variables $\mathbf{p}$ and $\mathbf{x}$. Important examples for such an A are $\mathbf{p}$ and $\mathbf{x}$ themselves, the Hamiltonian $H(\mathbf{p}, \mathbf{x})$, and the angular momentum $\mathbf{L} = \mathbf{x} \times \mathbf{p}$. Quantum-mechanically, there will be an observable operator associated with each such quantity. It is obtained by simply replacing the variables $\mathbf{p}$ and $\mathbf{x}$ in A by the corresponding operators $\hat{\mathbf{p}}$ and $\hat{\mathbf{x}}$:

$$\hat{A} \equiv A(\hat{\mathbf{p}}, \hat{\mathbf{x}}). \tag{1.206}$$

This replacement rule is the extension of the correspondence principle for the Hamiltonian operator (1.92) to more general functions in phase space, converting them into observable operators. It must be assumed that the replacement leads to a unique Hermitian operator, i.e., that there is no ordering problem of the type discussed in context with the Hamiltonian (1.101).[8] If there are ambiguities, the naive correspondence principle is insufficient to determine the observable operator. Then the correct ordering must be decided by comparison with experiment, unless it can be specified by means of simple geometric principles. This will be done for the Hamiltonian operator in Chapter 8.

Once an observable operator $\hat{A}$ is Hermitian, it has the useful property that the set of all eigenvectors $|a\rangle$ obtained by solving the equation

$$\hat{A}|a\rangle = a|a\rangle \tag{1.207}$$

can be used as a basis to span the Hilbert space. Among the eigenvectors, there is always a choice of orthonormal vectors $|a\rangle$ fulfilling the completeness relation

$$\sum_a |a\rangle\langle a| = 1. \tag{1.208}$$

The vectors $|a\rangle$ can be used to extract physical information concerning the observable A from arbitrary state vector $|\Psi(t)\rangle$. For this we expand this vector in the basis $|a\rangle$:

$$|\Psi(t)\rangle = \sum_a |a\rangle\langle a|\Psi(t)\rangle. \tag{1.209}$$

The components

$$\langle a|\Psi(t)\rangle \tag{1.210}$$

yield the probability amplitude for measuring the eigenvalue a for the observable quantity A.

The wave function $\Psi(\mathbf{x}, t)$ itself is an example of this interpretation. If we write it as

$$\Psi(\mathbf{x}, t) = \langle \mathbf{x}|\Psi(t)\rangle, \tag{1.211}$$

---

[8]Note that this is true for the angular momentum $\mathbf{L} = \mathbf{x} \times \mathbf{p}$.

it gives the probability amplitude for measuring the eigenvalues $\mathbf{x}$ of the position operator $\hat{\mathbf{x}}$, i.e., $|\Psi(\mathbf{x}, t)|^2$ is the probability density in $\mathbf{x}$-space.

The expectation value of the observable operator (1.206) in the state $|\Psi(t)\rangle$ is defined as the matrix element

$$\langle\Psi(t)|\hat{A}|\Psi(t)\rangle \equiv \int d^3x \langle\Psi(t)|\mathbf{x}\rangle A(-i\hbar\boldsymbol{\nabla}, \mathbf{x})\langle\mathbf{x}|\Psi(t)\rangle. \tag{1.212}$$

1.5.1 Uncertainty Relation

We have seen before [see the discussion after (1.83), (1.84)] that the amplitudes in real space and those in momentum space have widths inversely proportional to each other, due to the properties of Fourier analysis. If a wave packet is localized in real space with a width $\Delta\mathbf{x}$, its momentum space wave function has a width $\Delta\mathbf{p}$ given by

$$\Delta\mathbf{x}\,\Delta\mathbf{p} \sim \hbar. \tag{1.213}$$

From the Hilbert space point of view this uncertainty relation can be shown to be a consequence of the fact that the operators $\hat{\mathbf{x}}$ and $\hat{\mathbf{p}}$ do not commute with each other, but the components satisfy the canonical commutation rules

$$\begin{aligned}
[\hat{p}_i, \hat{x}_j] &= -i\hbar\delta_{ij}, \\
[\hat{x}_i, \hat{x}_j] &= 0, \\
[\hat{p}_i, \hat{p}_j] &= 0.
\end{aligned} \tag{1.214}$$

In general, if an observable operator $\hat{A}$ is measured sharply to have the value a in one state, this state must be an eigenstate of $\hat{A}$ with an eigenvalue a:

$$\hat{A}|a\rangle = a|a\rangle. \tag{1.215}$$

This follows from the expansion

$$|\Psi(t)\rangle = \sum_a |a\rangle\langle a|\Psi(t)\rangle, \tag{1.216}$$

in which $|\langle a|\Psi(t)\rangle|^2$ is the probability to measure an arbitrary eigenvalue a. If this probability is sharply focused at a specific value of a, the state necessarily coincides with $|a\rangle$.

Given the set of all eigenstates $|a\rangle$ of $\hat{A}$, we may ask under what circumstances another observable, say $\hat{B}$, can be measured sharply in each of these states. The requirement implies that the states $|a\rangle$ are also eigenstates of $\hat{B}$,

$$\hat{B}|a\rangle = b_a|a\rangle, \tag{1.217}$$

with some a-dependent eigenvalue b_a. If this is true for all $|a\rangle$,

$$\hat{B}\hat{A}|a\rangle = b_a a|a\rangle = ab_a|a\rangle = \hat{A}\hat{B}|a\rangle, \tag{1.218}$$

the operators $\hat{A}$ and $\hat{B}$ necessarily commute:

$$[\hat{A}, \hat{B}] = 0. \tag{1.219}$$

Conversely, it can be shown that a vanishing commutator is also sufficient for two observable operators to be simultaneously diagonalizable and thus to allow for simultaneous sharp measurements.

1.5.2 Density Matrix and Wigner Function

An important object for calculating observable properties of a quantum-mechanical system is the quantum mechanical density operator associated with a pure state

$$\hat{\rho}(t) \equiv |\Psi(t)\rangle\langle\Psi(t)|, \tag{1.220}$$

and the associated density matrix associated with a pure state

$$\rho(\mathbf{x}_1, \mathbf{x}_2; t) = \langle\mathbf{x}_1|\Psi(t)\rangle\langle\Psi(t)|\mathbf{x}_2\rangle. \tag{1.221}$$

The expectation value of any function $f(\mathbf{x}, \hat{\mathbf{p}})$ can be calculated from the trace

$$\langle\Psi(t)|f(\mathbf{x}, \hat{\mathbf{p}})|\Psi(t)\rangle = \mathrm{tr}\left[f(\mathbf{x}, \hat{\mathbf{p}})\hat{\rho}(t)\right] = \int d^3x \langle\Psi(t)|\mathbf{x}\rangle f(\mathbf{x}, -i\hbar\boldsymbol{\nabla})\langle\mathbf{x}|\Psi(t)\rangle. \tag{1.222}$$

If we decompose the states $|\Psi(t)\rangle$ into stationary eigenstates $|E_n\rangle$ of the Hamiltonian operator $\hat{H}$ [recall (1.175)], $|\Psi(t)\rangle = \sum_n |E_n\rangle\langle E_n|\Psi(t)\rangle$, then the density matrix has the expansion

$$\hat{\rho}(t) \equiv \sum_{n,m} |E_n\rangle\rho_{nm}(t)\langle E_m| = \sum_{n,m} |E_n\rangle\langle E_n|\Psi(t)\rangle\langle\Psi(t)|E_m\rangle\langle E_m|. \tag{1.223}$$

Wigner showed that the Fourier transform of the density matrix, the *Wigner function*

$$W(\mathbf{X}, \mathbf{p}; t) \equiv \int \frac{d^3\Delta x}{(2\pi\hbar)^3} e^{i\mathbf{p}\Delta\mathbf{x}/\hbar} \rho(\mathbf{X} + \Delta\mathbf{x}/2, \mathbf{X} - \Delta\mathbf{x}/2; t) \tag{1.224}$$

satisfies, for a single particle of mass M in a potential $V(\mathbf{x})$, the Wigner-Liouville equation

$$\left(\partial_t + \mathbf{v}\cdot\boldsymbol{\nabla}_\mathbf{X}\right)W(\mathbf{X}, \mathbf{p}; t) = W_t(\mathbf{X}, \mathbf{p}; t), \qquad \mathbf{v} \equiv \frac{\mathbf{p}}{M}, \tag{1.225}$$

where

$$W_t(\mathbf{X}, \mathbf{p}; t) \equiv \frac{2}{\hbar} \int \frac{d^3q}{(2\pi\hbar)^3} W(\mathbf{X}, \mathbf{p} - \mathbf{q}; t) \int d^3\Delta x \, V(\mathbf{X} - \Delta\mathbf{x}/2) e^{i\mathbf{q}\Delta\mathbf{x}/\hbar}. \tag{1.226}$$

In the limit $\hbar \to 0$, we may expand $W(\mathbf{X}, \mathbf{p} - \mathbf{q}; t)$ in powers of $\mathbf{q}$, and $V(\mathbf{X} - \Delta\mathbf{x}/2)$ in powers of $\Delta\mathbf{x}$, which we rewrite in front of the exponential $e^{i\mathbf{q}\Delta\mathbf{x}/\hbar}$ as powers of $-i\hbar\boldsymbol{\nabla}_\mathbf{q}$. Then we perform the integral over $\Delta\mathbf{x}$ to obtain $(2\pi\hbar)^3\delta^{(3)}(\mathbf{q})$, and perform the integral over $\mathbf{q}$ to obtain the classical *Liouville equation* for the probability density of the particle in phase space

$$\left(\partial_t + \mathbf{v}\cdot\boldsymbol{\nabla}_\mathbf{X}\right)W(\mathbf{X}, \mathbf{p}; t) = -F(\mathbf{X})\boldsymbol{\nabla}_\mathbf{p}W(\mathbf{X}, \mathbf{p}; t), \qquad \mathbf{v} \equiv \frac{\mathbf{p}}{M}, \tag{1.227}$$

where $F(\mathbf{X}) \equiv -\boldsymbol{\nabla}_\mathbf{X}V(\mathbf{X})$ is the force associated with the potential $V(\mathbf{X})$.

1.5.3   Generalization to Many Particles

All this development can be extended to systems of N distinguishable mass points with Cartesian coordinates $\mathbf{x}_1, \ldots, \mathbf{x}_N$. If $H(\mathbf{p}_\nu, \mathbf{x}_\nu, t)$ is the Hamiltonian, the Schrödinger equation becomes

$$H(\hat{\mathbf{p}}_\nu, \hat{\mathbf{x}}_\nu, t)|\Psi(t)\rangle = i\hbar\partial_t|\Psi(t)\rangle. \tag{1.228}$$

We may introduce a complete local basis $|\mathbf{x}_1, \ldots, \mathbf{x}_N\rangle$ with the properties

$$\langle \mathbf{x}_1, \ldots, \mathbf{x}_N | \mathbf{x}_1', \ldots, \mathbf{x}_N' \rangle = \delta^{(3)}(\mathbf{x}_1 - \mathbf{x}_1') \cdots \delta^{(3)}(\mathbf{x}_N - \mathbf{x}_N'),$$
$$\int d^3x_1 \cdots d^3x_N |\mathbf{x}_1, \ldots, \mathbf{x}_N\rangle\langle \mathbf{x}_1, \ldots, \mathbf{x}_N| = 1, \tag{1.229}$$

and define

$$\begin{aligned} \langle \mathbf{x}_1, \ldots, \mathbf{x}_N | \hat{\mathbf{p}}_\nu &= -i\hbar\partial_{\mathbf{x}_\nu}\langle \mathbf{x}_1, \ldots, \mathbf{x}_N|, \\ \langle \mathbf{x}_1, \ldots, \mathbf{x}_N | \hat{\mathbf{x}}_\nu &= \mathbf{x}_\nu\langle \mathbf{x}_1, \ldots, \mathbf{x}_N|. \end{aligned} \tag{1.230}$$

The Schrödinger equation for N particles (1.107) follows from (1.228) by multiplying it from the left with the bra vectors $\langle \mathbf{x}_1, \ldots, \mathbf{x}_N|$. In the same way, all other formulas given above can be generalized to N-body state vectors.

1.6   Time Evolution Operator

If the Hamiltonian operator possesses no explicit time dependence, the basis-independent Schrödinger equation (1.163) can be integrated to find the wave function $|\Psi(t)\rangle$ at any time t_b from the state at any other time t_a

$$|\Psi(t_b)\rangle = e^{-i(t_b-t_a)\hat{H}/\hbar}|\Psi(t_a)\rangle. \tag{1.231}$$

The operator

$$\hat{U}(t_b, t_a) = e^{-i(t_b-t_a)\hat{H}/\hbar} \tag{1.232}$$

is called the *time evolution operator*. It satisfies the differential equation

$$i\hbar\partial_{t_b}\hat{U}(t_b, t_a) = \hat{H}\,\hat{U}(t_b, t_a). \tag{1.233}$$

Its inverse is obtained by interchanging the order of t_b and t_a:

$$\hat{U}^{-1}(t_b, t_a) \equiv e^{i(t_b-t_a)\hat{H}/\hbar} = \hat{U}(t_a, t_b). \tag{1.234}$$

As an exponential of i times a Hermitian operator, $\hat{U}$ is a *unitary operator* satisfying

$$\hat{U}^\dagger = \hat{U}^{-1}. \tag{1.235}$$

Indeed,

$$\begin{aligned} \hat{U}^\dagger(t_b, t_a) &= e^{i(t_b-t_a)\hat{H}^\dagger/\hbar} \\ &= e^{i(t_b-t_a)\hat{H}/\hbar} = \hat{U}^{-1}(t_b, t_a). \end{aligned} \tag{1.236}$$

If $H(\hat{\mathbf{p}}, \hat{\mathbf{x}}, t)$ depends explicitly on time, the integration of the Schrödinger equation (1.163) is somewhat more involved. The solution may be found iteratively: For $t_b > t_a$, the time interval is sliced into a large number $N + 1$ of small pieces of thickness ϵ with $\epsilon \equiv (t_b - t_a)/(N + 1)$, slicing once at each time $t_n = t_a + n\epsilon$ for $n = 0, \ldots, N + 1$. We then use the Schrödinger equation (1.163) to relate the wave function in each slice approximately to the previous one:

$$|\Psi(t_a + \epsilon)\rangle \approx \left(1 - \frac{i}{\hbar} \int_{t_a}^{t_a + \epsilon} dt\, \hat{H}(t)\right) |\Psi(t_a)\rangle,$$

$$|\Psi(t_a + 2\epsilon)\rangle \approx \left(1 - \frac{i}{\hbar} \int_{t_a + \epsilon}^{t_a + 2\epsilon} dt\, \hat{H}(t)\right) |\Psi(t_a + \epsilon)\rangle, \tag{1.237}$$

$$\vdots$$

$$|\Psi(t_a + (N+1)\epsilon)\rangle \approx \left(1 - \frac{i}{\hbar} \int_{t_a + N\epsilon}^{t_a + (N+1)\epsilon} dt\, \hat{H}(t)\right) |\Psi(t_a + N\epsilon)\rangle.$$

From the combination of these equations we extract the evolution operator as a product

$$\hat{U}(t_b, t_a) \approx \left(1 - \frac{i}{\hbar} \int_{t_N}^{t_b} dt'_{N+1}\, \hat{H}(t'_{N+1})\right) \times \cdots \times \left(1 - \frac{i}{\hbar} \int_{t_a}^{t_1} dt'_1\, \hat{H}(t'_1)\right). \tag{1.238}$$

By multiplying out the product and going to the limit $N \to \infty$ we find the series

$$\hat{U}(t_b, t_a) = 1 - \frac{i}{\hbar} \int_{t_a}^{t_b} dt'_1\, \hat{H}(t'_1) + \left(\frac{-i}{\hbar}\right)^2 \int_{t_a}^{t_b} dt'_2 \int_{t_a}^{t_2} dt'_1\, \hat{H}(t'_2)\hat{H}(t'_1)$$

$$+ \left(\frac{-i}{\hbar}\right)^3 \int_{t_a}^{t_b} dt'_3 \int_{t_a}^{t_3} dt'_2 \int_{t_a}^{t_2} dt'_1\, \hat{H}(t'_3)\hat{H}(t'_2)\hat{H}(t'_1) + \ldots, \tag{1.239}$$

known as the *Neumann-Liouville expansion* or *Dyson series*. An interesting modification of this is the so-called *Magnus expansion* to be derived in Eq. (2A.25).

Note that each integral has the time arguments in the Hamilton operators ordered *causally*: Operators with later times stand to left of those with earlier times. It is useful to introduce a *time-ordering operator* which, when applied to an arbitrary product of operators,

$$\hat{O}_n(t_n) \cdots \hat{O}_1(t_1), \tag{1.240}$$

reorders the times successively. More explicitly we define

$$\hat{T}(\hat{O}_n(t_n) \cdots \hat{O}_1(t_1)) \equiv \hat{O}_{i_n}(t_{i_n}) \cdots \hat{O}_{i_1}(t_{i_1}), \tag{1.241}$$

where $t_{i_n}, \ldots, t_{i_1}$ are the times $t_n, \ldots, t_1$ relabeled in the causal order, so that

$$t_{i_n} > t_{i_{n-1}} > \ldots > t_{i_1}. \tag{1.242}$$

Any c-number factors in (1.241) can be pulled out in front of the $\hat{T}$ operator. With this formal operator, the Neumann-Liouville expansion can be rewritten in a more compact way. Take, for instance, the third term in (1.239)

$$\int_{t_a}^{t_b} dt_2 \int_{t_a}^{t_2} dt_1\, \hat{H}(t_2)\hat{H}(t_1). \tag{1.243}$$

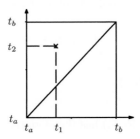

Figure 1.3 Illustration of time-ordering procedure in Eq. (1.243).

The integration covers the triangle above the diagonal in the square $t_1, t_2 \in [t_a, t_b]$ in the (t_1, t_2) plane (see Fig. 1.2). By comparing this with the missing integral over the lower triangle

$$\int_{t_a}^{t_b} dt_2 \int_{t_2}^{t_b} dt_1 \, \hat{H}(t_2)\hat{H}(t_1) \tag{1.244}$$

we see that the two expressions coincide except for the order of the operators. This can be corrected with the use of a time-ordering operator $\hat{T}$. The expression

$$\hat{T} \int_{t_a}^{t_b} dt_2 \int_{t_2}^{t_b} dt_1 \, \hat{H}(t_2)\hat{H}(t_1) \tag{1.245}$$

is equal to (1.243) since it may be rewritten as

$$\int_{t_a}^{t_b} dt_2 \int_{t_2}^{t_b} dt_1 \, \hat{H}(t_1)\hat{H}(t_2) \tag{1.246}$$

or, after interchanging the order of integration, as

$$\int_{t_a}^{t_b} dt_1 \int_{t_a}^{t_1} dt_2 \, \hat{H}(t_1)\hat{H}(t_2). \tag{1.247}$$

Apart from the dummy integration variables $t_2 \leftrightarrow t_1$, this double integral coincides with (1.243). Since the time arguments are properly ordered, (1.243) can trivially be multiplied with the time-ordering operator. The conclusion of this discussion is that (1.243) can alternatively be written as

$$\frac{1}{2}\hat{T} \int_{t_a}^{t_b} dt_2 \int_{t_a}^{t_b} dt_1 \, \hat{H}(t_2)\hat{H}(t_1). \tag{1.248}$$

On the right-hand side, the integrations now run over the full square in the t_1, t_2-plane so that the two integrals can be factorized into

$$\frac{1}{2}\hat{T} \left(\int_{t_a}^{t_b} dt \, \hat{H}(t) \right)^2 . \tag{1.249}$$

Similarly, we may rewrite the nth-order term of (1.239) as

$$\frac{1}{n!}\hat{T}\int_{t_a}^{t_b} dt_n \int_{t_a}^{t_b} dt_{n-1} \cdots \int_{t_a}^{t_b} dt_1\, \hat{H}(t_n)\hat{H}(t_{n-1}) \cdots \hat{H}(t_1)$$

$$= \frac{1}{n!}\hat{T}\left[\int_{t_a}^{t_b} dt\, \hat{H}(t)\right]^n. \tag{1.250}$$

The time evolution operator $\hat{U}(t_b, t_a)$ has therefore the series expansion

$$\hat{U}(t_b, t_a) = 1 - \frac{i}{\hbar}\hat{T}\int_{t_a}^{t_b} dt\, \hat{H}(t) + \frac{1}{2!}\left(\frac{-i}{\hbar}\right)^2 \hat{T}\left(\int_{t_a}^{t_b} dt\, \hat{H}(t)\right)^2$$

$$+ \ldots + \frac{1}{n!}\left(\frac{-i}{\hbar}\right)^n \hat{T}\left(\int_{t_a}^{t_b} dt\, \hat{H}(t)\right)^n + \ldots . \tag{1.251}$$

The right-hand side of $\hat{T}$ contains simply the power series expansion of the exponential so that we can write

$$\hat{U}(t_b, t_a) = \hat{T}\exp\left\{-\frac{i}{\hbar}\int_{t_a}^{t_b} dt\, \hat{H}(t)\right\}. \tag{1.252}$$

If $\hat{H}$ does not depend on the time, the time-ordering operation is superfluous, the integral can be done trivially, and we recover the previous result (1.232).

Note that a small variation $\delta\hat{H}(t)$ of $\hat{H}(t)$ changes $\hat{U}(t_b, t_a)$ by

$$\delta\hat{U}(t_b, t_a) = -\frac{i}{\hbar}\int_{t_a}^{t_b} dt'\, \hat{T}\exp\left\{-\frac{i}{\hbar}\int_{t'}^{t_b} dt\, \hat{H}(t)\right\} \delta\hat{H}(t')\, \hat{T}\exp\left\{-\frac{i}{\hbar}\int_{t_a}^{t'} dt\, \hat{H}(t)\right\}$$

$$= -\frac{i}{\hbar}\int_{t_a}^{t_b} dt'\, \hat{U}(t_b, t')\, \delta\hat{H}(t')\, \hat{U}(t', t_a). \tag{1.253}$$

A simple application for this relation is given in Appendix 1A.

1.7 Properties of the Time Evolution Operator

By construction, $\hat{U}(t_b, t_a)$ has some important properties:

a) Fundamental composition law
If two time translations are performed successively, the corresponding operators $\hat{U}$ are related by

$$\hat{U}(t_b, t_a) = \hat{U}(t_b, t')\hat{U}(t', t_a), \qquad t' \in (t_a, t_b). \tag{1.254}$$

This composition law makes the operators $\hat{U}$ a representation of the abelian group of time translations. For time-independent Hamiltonians with $\hat{U}(t_b, t_a)$ given by

(1.232), the proof of (1.254) is trivial. In the general case (1.252), it follows from the simple manipulation valid for $t_b > t_a$:

$$\hat{T} \exp\left(-\frac{i}{\hbar} \int_{t'}^{t_b} \hat{H}(t)\, dt\right) \hat{T} \exp\left(-\frac{i}{\hbar} \int_{t_a}^{t'} \hat{H}(t)\, dt\right)$$

$$= \hat{T} \left[\exp\left(-\frac{i}{\hbar} \int_{t'}^{t_b} \hat{H}(t)\, dt\right) \exp\left(-\frac{i}{\hbar} \int_{t_a}^{t'} \hat{H}(t)\, dt\right)\right] \qquad (1.255)$$

$$= \hat{T} \exp\left(-\frac{i}{\hbar} \int_{t_a}^{t_b} \hat{H}(t)\, dt\right).$$

b) Unitarity

The expression (1.252) for the time evolution operator $\hat{U}(t_b, t_a)$ was derived only for the *causal* (or *retarded*) time arguments, i.e., for t_b later than t_a. We may, however, define $\hat{U}(t_b, t_a)$ also for the *anticausal* (or *advanced*) case where t_b lies before t_a. To be consistent with the above composition law (1.254), we must have

$$\hat{U}(t_b, t_a) \equiv \hat{U}(t_a, t_b)^{-1}. \qquad (1.256)$$

Indeed, when considering two states at successive times

$$|\Psi(t_a)\rangle = \hat{U}(t_a, t_b)|\Psi(t_b)\rangle, \qquad (1.257)$$

the order of succession is inverted by multiplying both sides by $\hat{U}^{-1}(t_a, t_b)$:

$$|\Psi(t_b)\rangle = \hat{U}(t_a, t_b)^{-1}|\Psi(t_a)\rangle, \quad t_b < t_a. \qquad (1.258)$$

The operator on the right-hand side is defined to be the time evolution operator $\hat{U}(t_b, t_a)$ from the later time t_a to the earlier time t_b.

If the Hamiltonian is independent of time, with the time evolution operator being

$$\hat{U}(t_a, t_b) = e^{-i(t_a - t_b)\hat{H}/\hbar}, \qquad t_a > t_b, \qquad (1.259)$$

the unitarity of the operator $\hat{U}(t_b, t_a)$ is obvious:

$$\hat{U}^\dagger(t_b, t_a) = \hat{U}(t_b, t_a)^{-1}, \qquad t_b < t_a. \qquad (1.260)$$

Let us verify this property for a general time-dependent Hamiltonian. There, a direct solution of the Schrödinger equation (1.163) for the state vector shows that the operator $\hat{U}(t_b, t_a)$ for $t_b < t_a$ has a representation just like (1.252), except for a reversed time order of its arguments. One writes this in the form [compare (1.252)]

$$\hat{U}(t_b, t_a) = \hat{\bar{T}} \exp\left\{\frac{i}{\hbar} \int_{t_a}^{t_b} \hat{H}(t)\, dt\right\}, \qquad (1.261)$$

where $\bar{T}$ denotes the time-antiordering operator, with an obvious definition analogous to (1.241), (1.242). This operator satisfies the relation

$$\left[\hat{T}\left(\hat{O}_1(t_1)\hat{O}_2(t_2)\right)\right]^\dagger = \hat{\bar{T}}\left(\hat{O}_2^\dagger(t_2)\hat{O}_1^\dagger(t_1)\right), \qquad (1.262)$$

with an obvious generalization to the product of n operators. We can therefore conclude right away that

$$\hat{U}^\dagger(t_b, t_a) = \hat{U}(t_a, t_b), \quad t_b > t_a. \tag{1.263}$$

With $\hat{U}(t_a, t_b) \equiv \hat{U}(t_b, t_a)^{-1}$, this proves the unitarity relation (1.260), in general.

c) Schrödinger equation for $\hat{U}(t_b, t_a)$
Since the operator $\hat{U}(t_b, t_a)$ rules the relation between arbitrary wave functions at different times,

$$|\Psi(t_b)\rangle = \hat{U}(t_b, t_a)|\Psi(t_a)\rangle, \tag{1.264}$$

the Schrödinger equation (1.228) implies that the operator $\hat{U}(t_b, t_a)$ satisfies the corresponding equations

$$i\hbar\partial_t \hat{U}(t, t_a) = \hat{H}\hat{U}(t, t_a), \tag{1.265}$$

$$i\hbar\partial_t \hat{U}(t, t_a)^{-1} = -\hat{U}(t, t_a)^{-1}\hat{H}, \tag{1.266}$$

with the initial condition

$$\hat{U}(t_a, t_a) = 1. \tag{1.267}$$

1.8 Heisenberg Picture of Quantum Mechanics

The unitary time evolution operator $\hat{U}(t, t_a)$ may be used to give a different formulation of quantum mechanics bearing the closest resemblance to classical mechanics. This formulation, called the *Heisenberg picture* of quantum mechanics, is in a way more closely related to classical mechanics than the Schrödinger formulation. Many classical equations remain valid by simply replacing the canonical variables $p_i(t)$ and $q_i(t)$ in phase space by Heisenberg *operators*, to be denoted by $p_{Hi}(t)$, $q_{Hi}(t)$. Originally, Heisenberg postulated that they are matrices, but later it became clear that these matrices had to be functional matrix elements of operators, whose indices can be partly continuous. The classical equations hold for the Heisenberg operators and as long as the canonical commutation rules (1.93) are respected at any given time. In addition, $q_i(t)$ must be Cartesian coordinates. In this case we shall always use the letter x for the position variable, as in Section 1.4, rather than q, the corresponding Heisenberg operators being $x_{Hi}(t)$. Suppressing the subscripts i, the canonical equal-time commutation rules are

$$[p_H(t), x_H(t)] = -i\hbar,$$

$$[p_H(t), p_H(t)] = 0, \tag{1.268}$$

$$[x_H(t), x_H(t)] = 0.$$

According to Heisenberg, classical equations involving Poisson brackets remain valid if the Poisson brackets are replaced by $i/\hbar$ times the matrix commutators at equal times. The canonical commutation relations (1.268) are a special case of this rule, recalling the fundamental Poisson brackets (1.25). The Hamilton equations of motion (1.24) turn into the Heisenberg equations

$$
\begin{aligned}
\dot{p}_H(t) &= \frac{i}{\hbar}\left[H_H, p_H(t)\right], \\
\dot{x}_H(t) &= \frac{i}{\hbar}\left[H_H, x_H(t)\right],
\end{aligned}
\tag{1.269}
$$

where

$$
H_H \equiv H(p_H(t), x_H(t), t) \tag{1.270}
$$

is the Hamiltonian in the Heisenberg picture. Similarly, the equation of motion for an arbitrary observable function $O(p_i(t), x_i(t), t)$ derived in (1.20) goes over into the matrix commutator equation for the Heisenberg observable

$$
O_H(t) \equiv O(p_H(t), x_H(t), t), \tag{1.271}
$$

namely,

$$
\frac{d}{dt}O_H = \frac{i}{\hbar}[H_H, O_H] + \frac{\partial}{\partial t}O_H. \tag{1.272}
$$

These rules are referred to as *Heisenberg's correspondence principle*.

The relation between Schrödinger's and Heisenberg's picture is supplied by the time evolution operator. Let $\hat{O}$ be an arbitrary observable in the Schrödinger description

$$
\hat{O}(t) \equiv O(\hat{p}, \hat{x}, t). \tag{1.273}
$$

If the states $|\Psi_a(t)\rangle$ are an arbitrary complete set of solutions of the Schrödinger equation, where a runs through discrete and continuous indices, the operator $\hat{O}(t)$ can be specified in terms of its functional matrix elements

$$
O_{ab}(t) \equiv \langle \Psi_a(t)|\hat{O}(t)|\Psi_b(t)\rangle. \tag{1.274}
$$

We can now use the unitary operator $\hat{U}(t, 0)$ to go to a new time-independent basis $|\Psi_{Ha}\rangle$, defined by

$$
|\Psi_a(t)\rangle \equiv \hat{U}(t, 0)|\Psi_{Ha}\rangle. \tag{1.275}
$$

Simultaneously, we transform the Schrödinger operators of the canonical coordinates $\hat{p}$ and $\hat{x}$ into the time-dependent canonical *Heisenberg operators* $\hat{p}_H(t)$ and $\hat{x}_H(t)$ via

$$
\begin{aligned}
\hat{p}_H(t) &\equiv \hat{U}(t, 0)^{-1}\, \hat{p}\, \hat{U}(t, 0), &\tag{1.276} \\
\hat{x}_H(t) &\equiv \hat{U}(t, 0)^{-1}\, \hat{x}\, \hat{U}(t, 0). &\tag{1.277}
\end{aligned}
$$

At the time $t = 0$, the Heisenberg operators $\hat{p}_H(t)$ and $\hat{x}_H(t)$ coincide with the time-independent Schrödinger operators $\hat{p}$ and $\hat{x}$, respectively. An arbitrary observable $\hat{O}(t)$ is transformed into the associated Heisenberg operator as

$$
\begin{aligned}
\hat{O}_H(t) &\equiv \hat{U}(t, t_a)^{-1}O(\hat{p}, \hat{x}, t)\hat{U}(t, t_a) \\
&\equiv O\left(\hat{p}_H(t), \hat{x}_H(t), t\right).
\end{aligned}
\tag{1.278}
$$

The Heisenberg matrices $O_H(t)_{ab}$ are then obtained from the Heisenberg operators $\hat{O}_H(t)$ by sandwiching $\hat{O}_H(t)$ between the time-independent basis vectors $|\Psi_{Ha}\rangle$:

$$O_H(t)_{ab} \equiv \langle \Psi_{Ha}|\hat{O}_H(t)|\Psi_{Hb}\rangle. \tag{1.279}$$

Note that the time dependence of these matrix elements is now completely due to the time dependence of the operators,

$$\frac{d}{dt}O_H(t)_{ab} \equiv \langle \Psi_{Ha}|\frac{d}{dt}\hat{O}_H(t)|\Psi_{Hb}\rangle. \tag{1.280}$$

This is in contrast to the Schrödinger representation (1.274), where the right-hand side would have contained two more terms from the time dependence of the wave functions. Due to the absence of such terms in (1.280) it is possible to study the equation of motion of the Heisenberg matrices independently of the basis by considering directly the Heisenberg operators. It is straightforward to verify that they do indeed satisfy the rules of Heisenberg's correspondence principle. Consider the time derivative of an arbitrary observable $\hat{O}_H(t)$,

$$\begin{aligned}
\frac{d}{dt}\hat{O}_H(t) &= \left(\frac{d}{dt}\hat{U}^{-1}(t,t_a)\right)\hat{O}(t)\hat{U}(t,t_a) \\
&+ \hat{U}^{-1}(t,t_a)\left(\frac{\partial}{\partial t}\hat{O}(t)\right)\hat{U}(t,t_a) + \hat{U}^{-1}(t,t_a)\hat{O}(t)\left(\frac{d}{dt}\hat{U}(t,t_a)\right),
\end{aligned}$$

which can be rearranged as

$$\left[\left(\frac{d}{dt}\hat{U}^{-1}(t,t_a)\right)\hat{U}(t,t_a)\right]\hat{U}^{-1}(t,t_a)\hat{O}(t)\hat{U}(t,t_a) \tag{1.281}$$

$$+\left[\hat{U}^{-1}(t,t_a)\hat{O}(t)\hat{U}(t,t_a)\right]\hat{U}^{-1}(t,t_a)\frac{d}{dt}\hat{U}(t,t_a) + \hat{U}^{-1}(t,t_a)\left(\frac{\partial}{\partial t}\hat{O}(t)\right)\hat{U}(t,t_a).$$

Using (1.265), we obtain

$$\frac{d}{dt}\hat{O}_H(t) = \frac{i}{\hbar}\left[\hat{U}^{-1}\hat{H}\hat{U}, \hat{O}_H\right] + \hat{U}^{-1}\left(\frac{\partial}{\partial t}\hat{O}(t)\right)\hat{U}. \tag{1.282}$$

After inserting (1.278), we find the equation of motion for the Heisenberg operator:

$$\frac{d}{dt}\hat{O}_H(t) = \frac{i}{\hbar}\left[\hat{H}_H, \hat{O}_H(t)\right] + \left(\frac{\partial}{\partial t}\hat{O}\right)_H(t). \tag{1.283}$$

By sandwiching this equation between the complete time-independent basis states $|\Psi_a\rangle$ in the Hilbert space, it holds for the matrices and turns into the Heisenberg equation of motion. For the phase space variables $p_H(t)$, $x_H(t)$ themselves, these equations reduce, of course, to the Hamilton equations of motion (1.269).

Thus we have shown that Heisenberg's matrix quantum mechanics is completely equivalent to Schrödinger's quantum mechanics, and that the Heisenberg matrices obey the same Hamilton equations as the classical observables.

1.9   Interaction Picture and Perturbation Expansion

For some physical systems, the Hamiltonian operator can be split into two contributions

$$\hat{H} = \hat{H}_0 + \hat{V}, \tag{1.284}$$

where $\hat{H}_0$ is a so-called free Hamiltonian operator for which the Schrödinger equation $\hat{H}_0|\psi(t)\rangle = i\hbar\partial_t|\psi(t)\rangle$ can be solved, and $\hat{V}$ is an interaction potential which perturbs these solutions slightly. In this case it is useful to describe the system in Dirac's *interaction picture*. We remove the time evolution of the unperturbed Schrödinger solutions and define the states

$$|\psi_I(t)\rangle \equiv e^{i\hat{H}_0 t/\hbar}|\psi(t)\rangle. \tag{1.285}$$

Their time evolution comes entirely from the interaction potential $\hat{V}$. It is governed by the time evolution operator

$$\hat{U}_I(t_b, t_a) \equiv e^{iH_0 t_b/\hbar} e^{-iH t_b/\hbar} e^{iH t_a/\hbar} e^{-iH_0 t_a/\hbar}, \tag{1.286}$$

and reads

$$|\psi_I(t_b)\rangle = \hat{U}_I(t_b, t_a)|\psi_I(t_a)\rangle. \tag{1.287}$$

If $\hat{V} = 0$, the states $|\psi_I(t_b)\rangle$ are time-independent and coincide with the Heisenberg states (1.275) of the operator $\hat{H}_0$.

The operator $\hat{U}_I(t_b, t_a)$ satisfies the equation of motion

$$i\hbar\partial_{t_b}\hat{U}_I(t_b, t_a) = V_I(t_b)\hat{U}_I(t_b, t_a), \tag{1.288}$$

where

$$\hat{V}_I(t) \equiv e^{i\hat{H}_0 t/\hbar}\hat{V}e^{-i\hat{H}_0 t/\hbar} \tag{1.289}$$

is the potential in the interaction picture. This equation of motion can be turned into an integral equation

$$\hat{U}_I(t_b, t_a) = 1 - \frac{i}{\hbar}\int_{t_a}^{t_b} dt V_I(t)\hat{U}_I(t, t_a). \tag{1.290}$$

Inserting Eq. (1.289), this reads

$$\hat{U}_I(t_b, t_a) = 1 - \frac{i}{\hbar}\int_{t_a}^{t_b} dt\, e^{i\hat{H}_0 t/\hbar}Ve^{-i\hat{H}_0 t/\hbar}\hat{U}_I(t, t_a). \tag{1.291}$$

This equation can be iterated to find a perturbation expansion for the operator $\hat{U}_I(t_b, t_a)$ in powers of the interaction potential:

$$\hat{U}_I(t_b, t_a) = 1 - \frac{i}{\hbar}\int_{t_a}^{t_b} dt\, e^{i\hat{H}_0 t/\hbar}Ve^{-i\hat{H}_0 t/\hbar}$$
$$+ \left(-\frac{i}{\hbar}\right)^2 \int_{t_a}^{t_b} dt \int_{t_a}^{t} dt'\, e^{i\hat{H}_0 t/\hbar}Ve^{-i\hat{H}_0(t-t')/\hbar}Ve^{-i\hat{H}_0 t'/\hbar} + \ldots\,. \tag{1.292}$$

Inserting on the left-hand side the operator (1.286), this can also be rewritten as

$$e^{-iH(t_b-t_a)/\hbar} = e^{-iH_0(t_b-t_a)/\hbar} - \frac{i}{\hbar}\int_{t_a}^{t_b} dt\, e^{-i\hat{H}_0(t_b-t)/\hbar}V e^{-i\hat{H}_0(t-t_a)/\hbar}$$

$$+\left(-\frac{i}{\hbar}\right)^2\int_{t_a}^{t_b} dt\int_{t_a}^{t} dt'\, e^{-i\hat{H}_0(t_b-t)/\hbar}V e^{-i\hat{H}_0(t-t')/\hbar}V e^{-i\hat{H}_0(t'-t_a)/\hbar} + \dots. \quad (1.293)$$

This expansion is seen to be the recursive solution of the integral equation

$$e^{-iH(t_b-t_a)/\hbar} = e^{-iH_0(t_b-t_a)/\hbar} - \frac{i}{\hbar}\int_{t_a}^{t_b} dt\, e^{-i\hat{H}_0(t_b-t)/\hbar}V e^{-i\hat{H}(t-t_a)/\hbar}. \quad (1.294)$$

Note that the lowest-order correction agrees with the previous formula (1.253)

1.10 Time Evolution Amplitude

In the subsequent development, an important role will be played by the matrix elements of the time evolution operator in the localized basis states,

$$(\mathbf{x}_b t_b | \mathbf{x}_a t_a) \equiv \langle \mathbf{x}_b | \hat{U}(t_b, t_a) | \mathbf{x}_a \rangle. \quad (1.295)$$

They are referred to as *time evolution* amplitudes. The functional matrix $(\mathbf{x}_b t_b | \mathbf{x}_a t_a)$ is also called the *propagator* of the system. For a system with a time-independent Hamiltonian operator where $\hat{U}(t_b, t_a)$ is given by (1.259), the propagator is simply

$$(\mathbf{x}_b t_b | \mathbf{x}_a t_a) = \langle \mathbf{x}_b | \exp[-i\hat{H}(t_b - t_a)/\hbar] | \mathbf{x}_a \rangle. \quad (1.296)$$

Due to the operator equations (1.265), the propagator satisfies the Schrödinger equation

$$[H(-i\hbar\partial_{\mathbf{x}_b}, \mathbf{x}_b, t_b) - i\hbar\partial_{t_b}]\,(\mathbf{x}_b t_b | \mathbf{x}_a t_a) = 0. \quad (1.297)$$

In the quantum mechanics of nonrelativistic particles, only the propagators from earlier to later times will be relevant. It is therefore customary to introduce the so-called *causal time evolution operator* or *retarded time evolution operator*:[9]

$$\hat{U}^R(t_b, t_a) \equiv \begin{cases} \hat{U}(t_b, t_a), & t_b \geq t_a, \\ 0, & t_b < t_a, \end{cases} \quad (1.298)$$

and the associated *causal time evolution amplitude* or *retarded time evolution amplitude*

$$(\mathbf{x}_b t_b | \mathbf{x}_a t_a)^R \equiv \langle \mathbf{x}_b | \hat{U}^R(t_b, t_a) | \mathbf{x}_a \rangle. \quad (1.299)$$

Since this differs from (1.295) only for $t_b < t_a$, and since all formulas in the subsequent text will be used only for $t_b > t_a$, we shall often omit the superscript R. To abbreviate the case distinction in (1.298), it is convenient to use the *Heaviside function* defined by

$$\Theta(t) \equiv \begin{cases} 1 & \text{for} \quad t > 0, \\ 0 & \text{for} \quad t \leq 0, \end{cases} \quad (1.300)$$

---

[9]Compare this with the retarded *Green functions* to be introduced in Section 18.1

and write

$$U^R(t_b, t_a) \equiv \Theta(t_b - t_a)\hat{U}(t_b, t_a), \qquad (\mathbf{x}_b t_b | \mathbf{x}_a t_a)^R \equiv \Theta(t_b - t_a)(\mathbf{x}_b t_b | \mathbf{x}_a t_a). \quad (1.301)$$

There exists also another Heaviside function which differs from (1.300) only by the value at $t_b = t_a$:

$$\Theta^R(t) \equiv \begin{cases} 1 & \text{for} \quad t \geq 0, \\ 0 & \text{for} \quad t < 0. \end{cases} \quad (1.302)$$

Both Heaviside functions have the property that their derivative yields Dirac's δ-function

$$\partial_t \Theta(t) = \delta(t). \quad (1.303)$$

If it is not important which Θ-function is used we shall ignore the superscript.

The retarded propagator satisfies the Schrödinger equation

$$\left[H(-i\hbar \partial_{\mathbf{x}_b}, \mathbf{x}_b, t_b)^R - i\hbar \partial_{t_b} \right] (\mathbf{x}_b t_b | \mathbf{x}_a t_a)^R = -i\hbar \delta(t_b - t_a) \delta^{(3)}(\mathbf{x}_b - \mathbf{x}_a). \quad (1.304)$$

The nonzero right-hand side arises from the extra term

$$-i\hbar \left[\partial_{t_b} \Theta(t_b - t_a) \right] \langle \mathbf{x}_b t_b | \mathbf{x}_a t_a \rangle = -i\hbar \delta(t_b - t_a) \langle \mathbf{x}_b t_b | \mathbf{x}_a t_a \rangle = -i\hbar \delta(t_b - t_a) \langle \mathbf{x}_b t_a | \mathbf{x}_a t_a \rangle \quad (1.305)$$

and the initial condition $\langle \mathbf{x}_b t_a | \mathbf{x}_a t_a \rangle = \langle \mathbf{x}_b | \mathbf{x}_a \rangle$, due to (1.267).

If the Hamiltonian does not depend on time, the propagator depends only on the time difference $t = t_b - t_a$. The retarded propagator vanishes for $t < 0$. Functions $f(t)$ with this property have a characteristic Fourier transform. The integral

$$\tilde{f}(E) \equiv \int_0^\infty dt \; f(t) e^{iEt/\hbar} \quad (1.306)$$

is an analytic function in the upper half of the complex energy plane. This analyticity property is necessary and sufficient to produce a factor $\Theta(t)$ when inverting the Fourier transform via the energy integral

$$f(t) \equiv \int_{-\infty}^\infty \frac{dE}{2\pi\hbar} \; \tilde{f}(E) e^{-iEt/\hbar}. \quad (1.307)$$

For $t < 0$, the contour of integration may be closed by an infinite semicircle in the upper half-plane at no extra cost. Since the contour encloses no singularities, it can be contracted to a point, yielding $f(t) = 0$.

The Heaviside function $\Theta(t)$ itself is the simplest retarded function, with a Fourier representation containing just a single pole just below the origin of the complex energy plane:

$$\Theta(t) = \int_{-\infty}^\infty \frac{dE}{2\pi} \frac{i}{E + i\eta} e^{-iEt}, \quad (1.308)$$

where η is an infinitesimally small positive number. The integral representation is undefined for $t = 0$ and there are, in fact, infinitely many possible definitions for the Heaviside function depending on the value assigned to the function at the origin. A special role is played by the average of the Heaviside functions (1.302) and (1.300), which is equal to $1/2$ at the origin:

$$\bar{\Theta}(t) \equiv \begin{cases} 1 & \text{for} \quad t > 0, \\ \frac{1}{2} & \text{for} \quad t = 0, \\ 0 & \text{for} \quad t < 0. \end{cases} \tag{1.309}$$

Usually, the difference in the value at the origin does not matter since the Heaviside function appears only in integrals accompanied by some smooth function $f(t)$. This makes the Heaviside function a distribution with respect to smooth test functions $f(t)$ as defined in Eq. (1.162). All three distributions $\Theta_r(t)$, $\Theta^l(t)$, and $\bar{\Theta}(t)$ define the same linear functional of the test functions by the integral

$$\Theta[f] = \int dt \, \Theta(t - t') f(t'), \tag{1.310}$$

and this is an element in the linear space of all distributions.

As announced after Eq. (1.162), path integrals will specify, in addition, integrals over products of distribution and thus give rise to an important extension of the theory of distributions in Chapter 10. In this, the Heaviside function $\bar{\Theta}(t - t')$ plays the main role.

While discussing the concept of distributions let us introduce, for later use, the closely related distribution

$$\epsilon(t - t') \equiv \Theta(t - t') - \Theta(t' - t) = \bar{\Theta}(t - t') - \bar{\Theta}(t' - t), \tag{1.311}$$

which is a step function jumping at the origin from -1 to 1 as follows:

$$\epsilon(t - t') \equiv \begin{cases} 1 & \text{for} \quad t > t', \\ 0 & \text{for} \quad t = t', \\ -1 & \text{for} \quad t < t'. \end{cases} \tag{1.312}$$

1.11 Fixed-Energy Amplitude

The Fourier-transform of the retarded time evolution amplitude (1.299)

$$(\mathbf{x}_b|\mathbf{x}_a)_E = \int_{-\infty}^{\infty} dt_b e^{iE(t_b - t_a)/\hbar} (\mathbf{x}_b t_b|\mathbf{x}_b t_a)^R = \int_{t_a}^{\infty} dt_b e^{iE(t_b - t_a)/\hbar} (\mathbf{x}_b t_b|\mathbf{x}_b t_a) \tag{1.313}$$

is called the *fixed-energy amplitude*.

If the Hamiltonian does not depend on time, we insert here Eq. (1.296) and find that the fixed-energy amplitudes are matrix elements

$$(\mathbf{x}_b|\mathbf{x}_a)_E = \langle \mathbf{x}_b|\hat{R}(E)|\mathbf{x}_a \rangle \tag{1.314}$$

of the so-called of the so-called *resolvent operator*

$$\hat{R}(E) = \frac{i\hbar}{E - \hat{H} + i\eta}, \tag{1.315}$$

which is the Fourier transform of the retarded time evolution operator (1.298):

$$\hat{R}(E) = \int_{-\infty}^{\infty} dt_b\, e^{iE(t_b - t_a)/\hbar} \hat{U}^R(t_b, t_a) = \int_{t_a}^{\infty} dt_b\, e^{iE(t_b - t_a)/\hbar} \hat{U}(t_b, t_a). \tag{1.316}$$

Let us suppose that the time-independent Schrödinger equation is completely solved, i.e., that one knows all solutions $|\psi_n\rangle$ of the equation

$$\hat{H}|\psi_n\rangle = E_n|\psi_n\rangle. \tag{1.317}$$

These satisfy the completeness relation

$$\sum_n |\psi_n\rangle\langle\psi_n| = 1, \tag{1.318}$$

which can be inserted on the right-hand side of (1.296) between the Dirac brackets leading to the *spectral representation*

$$(\mathbf{x}_b t_b | \mathbf{x}_a t_a) = \sum_n \psi_n(\mathbf{x}_b)\psi_n^*(\mathbf{x}_a) \exp\left[-iE_n(t_b - t_a)/\hbar\right], \tag{1.319}$$

with

$$\psi_n(\mathbf{x}) = \langle \mathbf{x}|\psi_n\rangle \tag{1.320}$$

being the wave functions associated with the eigenstates $|\psi_n\rangle$. Applying the Fourier transform (1.313), we obtain

$$(\mathbf{x}_b | \mathbf{x}_a)_E = \sum_n \psi_n(\mathbf{x}_b)\psi_n^*(\mathbf{x}_a) R_n(E) = \sum_n \psi_n(\mathbf{x}_b)\psi_n^*(\mathbf{x}_a)\frac{i\hbar}{E - E_n + i\eta}. \tag{1.321}$$

The fixed-energy amplitude (1.313) contains as much information on the system as the time evolution amplitude, which is recovered from it by the inverse Fourier transformation

$$(\mathbf{x}_b t_a | \mathbf{x}_a t_a) = \int_{-\infty}^{\infty} \frac{dE}{2\pi\hbar} e^{-iE(t_b - t_a)/\hbar} (\mathbf{x}_b | \mathbf{x}_a)_E. \tag{1.322}$$

The small $i\eta$-shift in the energy E in (1.321) may be thought of as being attached to each of the energies E_n, which are thus placed by an infinitesimal piece *below* the real energy axis. Then the exponential behavior of the wave functions is slightly damped, going to zero at infinite time:

$$e^{-i(E_n - i\eta)t/\hbar} \to 0. \tag{1.323}$$

This so-called $i\eta$-prescription ensures the causality of the Fourier representation (1.322). When doing the Fourier integral (1.322), the exponential $e^{iE(t_b - t_a)/\hbar}$ makes it always possible to close the integration contour along the energy axis by an infinite semicircle in the complex energy plane, which lies in the upper half-plane for $t_b < t_a$ and in the lower half-plane for $t_b > t_a$. The $i\eta$-prescription guarantees that for $t_b < t_a$, there is no pole inside the closed contour making the propagator vanish. For $t_b > t_a$, on the other hand, the poles in the lower half-plane give, via Cauchy's residue theorem, the spectral representation (1.319) of the propagator. An $i\eta$-prescription will appear in another context in Section 2.3.

If the eigenstates are nondegenerate, the residues at the poles of (1.321) render directly the products of eigenfunctions (barring degeneracies which must be discussed separately). For a system with a continuum of energy eigenvalues, there is a cut in the complex energy plane which may be thought of as a closely spaced sequence of poles. In general, the wave functions are recovered from the discontinuity of the amplitudes $(\mathbf{x}_b|\mathbf{x}_a)_E$ across the cut, using the formula

$$\text{disc}\left(\frac{i\hbar}{E - E_n}\right) \equiv \frac{i\hbar}{E - E_n + i\eta} - \frac{i\hbar}{E - E_n - i\eta} = 2\pi\hbar\delta(E - E_n). \qquad (1.324)$$

Here we have employed the relation[10], valid inside integrals over E:

$$\frac{1}{E - E_n \pm i\eta} = \frac{\mathcal{P}}{E - E_n} \mp i\pi\delta(E - E_n), \qquad (1.325)$$

where $\mathcal{P}$ indicates that the principal value of the integral has to be taken.

The energy integral over the discontinuity of the fixed-energy amplitude (1.321) $(\mathbf{x}_b|\mathbf{x}_a)_E$ reproduces the completeness relation (1.318) taken between the local states $\langle\mathbf{x}_b|$ and $|\mathbf{x}_a\rangle$,

$$\int_{-\infty}^{\infty} \frac{dE}{2\pi\hbar}\text{disc}\,(\mathbf{x}_b|\mathbf{x}_a)_E = \sum_n \psi_n(\mathbf{x}_b)\psi_n^*(\mathbf{x}_a) = \langle\mathbf{x}_b|\mathbf{x}_a\rangle = \delta^{(D)}(\mathbf{x}_b - \mathbf{x}_a). \qquad (1.326)$$

The completeness relation reflects the following property of the resolvent operator:

$$\int_{-\infty}^{\infty} \frac{dE}{2\pi\hbar}\text{disc}\,\hat{R}(E) = \hat{1}. \qquad (1.327)$$

In general, the system possesses also a continuous spectrum, in which case the completeness relation contains a spectral integral and (1.318) has the form

$$\sum_n |\psi_n\rangle\langle\psi_n| + \int d\nu \, |\psi_\nu\rangle\langle\psi_\nu| = 1. \qquad (1.328)$$

The continuum causes a branch cut along in the complex energy plane, and (1.326) includes an integral over the discontinuity along the cut. The cut will mostly be omitted, for brevity.

1.12   Free-Particle Amplitudes

For a free particle with a Hamiltonian operator $\hat{H} = \hat{\mathbf{p}}^2/2M$, the spectrum is continuous. The eigenfunctions are (1.189) with energies $E(\mathbf{p}) = \mathbf{p}^2/2M$. Inserting

---

[10]This is often referred to as *Sochocki's formula*. It is the beginning of an expansion in powers of $\eta > 0$: $1/(x \pm i\eta) = \mathcal{P}/x \mp i\pi\delta(x) + \eta\left[\pi\delta'(x) \pm id_x\mathcal{P}/x\right] + \mathcal{O}(\eta^2)$.

the completeness relation (1.187) into Eq. (1.296), we obtain for the time evolution amplitude of a free particle the Fourier representation

$$(\mathbf{x}_b t_b | \mathbf{x}_a t_a) = \int \frac{d^D p}{(2\pi\hbar)^D} \exp\left\{ \frac{i}{\hbar} \left[\mathbf{p}(\mathbf{x}_b - \mathbf{x}_a) - \frac{\mathbf{p}^2}{2M}(t_b - t_a) \right] \right\}. \qquad (1.329)$$

The momentum integrals can easily be done. First we perform a quadratic completion in the exponent and rewrite it as

$$\mathbf{p}(\mathbf{x}_b - \mathbf{x}_a) - \frac{1}{2M}\mathbf{p}^2(t_b - t_a) = \frac{1}{2M}\left(\mathbf{p} - \frac{1}{M}\frac{\mathbf{x}_b - \mathbf{x}_a}{t_b - t_a}\right)^2 (t_b - t_a) - \frac{M}{2}\frac{(\mathbf{x}_b - \mathbf{x}_a)^2}{t_b - t_a}. \qquad (1.330)$$

Then we replace the integration variables by the shifted momenta $\mathbf{p}' = \mathbf{p} - (\mathbf{x}_b - \mathbf{x}_a)/(t_b - t_a)M$, and the amplitude (1.329) becomes

$$(\mathbf{x}_b t_b | \mathbf{x}_a t_a) = F(t_b - t_a) \exp\left[\frac{i}{\hbar} \frac{M}{2} \frac{(\mathbf{x}_b - \mathbf{x}_a)^2}{t_b - t_a} \right], \qquad (1.331)$$

where $F(t_b - t_a)$ is the integral over the shifted momenta

$$F(t_b - t_a) \equiv \int \frac{d^D p'}{(2\pi\hbar)^D} \exp\left\{ -\frac{i}{\hbar} \frac{\mathbf{p}'^2}{2M}(t_b - t_a) \right\}. \qquad (1.332)$$

This can be performed using the *Fresnel integral formula*

$$\int_{-\infty}^{\infty} \frac{dp}{\sqrt{2\pi}} \exp\left(i\frac{a}{2}p^2 \right) = \frac{1}{\sqrt{|a|}} \left\{ \begin{array}{ll} \sqrt{i}, & a > 0, \\ 1/\sqrt{i}, & a < 0. \end{array} \right. \qquad (1.333)$$

Here the square root $\sqrt{i}$ denotes the phase factor $e^{i\pi/4}$: This follows from the Gauss formula

$$\int_{-\infty}^{\infty} \frac{dp}{\sqrt{2\pi}} \exp\left(-\frac{\alpha}{2}p^2 \right) = \frac{1}{\sqrt{\alpha}}, \qquad \mathrm{Re}\,\alpha > 0, \qquad (1.334)$$

by continuing α analytically from positive values into the right complex half-plane. As long as $\mathrm{Re}\,\alpha > 0$, this is straightforward. On the boundaries, i.e., on the positive and negative imaginary axes, one has to be careful. At $\alpha = \pm ia + \eta$ with $a \gtrless 0$ and infinitesimal $\eta > 0$, the integral is certainly convergent yielding (1.333). But the integral also converges for $\eta = 0$, as can easily be seen by substituting $x^2 = z$. See Appendix 1B.

Note that differentiation of Eq. (1.334) with respect to α yields the more general Gaussian integral formula

$$\int_{-\infty}^{\infty} \frac{dp}{\sqrt{2\pi}} p^{2n} \exp\left(-\frac{\alpha}{2}p^2 \right) = \frac{1}{\sqrt{\alpha}} \frac{(2n-1)!!}{\alpha^n} \qquad \mathrm{Re}\,\alpha > 0, \qquad (1.335)$$

where $(2n - 1)!!$ is defined as the product $(2n - 1) \cdot (2n - 3) \cdots 1$. For odd powers p^{2n+1}, the integral vanishes. In the Fresnel formula (1.333), an extra integrand p^{2n} produces a factor $(i/a)^n$.

Since the Fresnel formula is a special analytically continued case of the Gauss formula, we shall in the sequel always speak of Gaussian integrations and use Fresnel's name only if the imaginary nature of the quadratic exponent is to be emphasized.

Applying this formula to (1.332), we obtain

$$F(t_b - t_a) = \frac{1}{\sqrt{2\pi i\hbar(t_b - t_a)/M}^D}, \tag{1.336}$$

so that the full time evolution amplitude of a free massive point particle is

$$(\mathbf{x}_b t_b | \mathbf{x}_a t_a) = \frac{1}{\sqrt{2\pi i\hbar(t_b - t_a)/M}^D} \exp\left[\frac{i}{\hbar}\frac{M}{2}\frac{(\mathbf{x}_b - \mathbf{x}_a)^2}{t_b - t_a}\right]. \tag{1.337}$$

In the limit $t_b \to t_a$, the left-hand side becomes the scalar product $\langle \mathbf{x}_b | \mathbf{x}_a \rangle = \delta^{(D)}(\mathbf{x}_b - \mathbf{x}_a)$, implying the following limiting formula for the δ-function

$$\delta^{(D)}(\mathbf{x}_b - \mathbf{x}_a) = \lim_{t_b - t_a \to 0} \frac{1}{\sqrt{2\pi i\hbar(t_b - t_a)/M}^D} \exp\left[\frac{i}{\hbar}\frac{M}{2}\frac{(\mathbf{x}_b - \mathbf{x}_a)^2}{t_b - t_a}\right]. \tag{1.338}$$

Inserting Eq. (1.331) into (1.313), we have for the fixed-energy amplitude the integral representation

$$(\mathbf{x}_b | \mathbf{x}_a)_E = \int_0^\infty d(t_b - t_a) \int \frac{d^D p}{(2\pi\hbar)^D} \exp\left\{\frac{i}{\hbar}\left[\mathbf{p}(\mathbf{x}_b - \mathbf{x}_a) + (t_b - t_a)\left(E - \frac{\mathbf{p}^2}{2M}\right)\right]\right\}. \tag{1.339}$$

Performing the time integration yields

$$(\mathbf{x}_b | \mathbf{x}_a)_E = \int \frac{d^D p}{(2\pi\hbar)^D} \exp\left[i\mathbf{p}(\mathbf{x}_b - \mathbf{x}_a)\right] \frac{i\hbar}{E - \mathbf{p}^2/2M + i\eta}, \tag{1.340}$$

where we have inserted a damping factor $e^{-\eta(t_b - t_a)}$ into the integral to ensure convergence at large $t_b - t_a$. For a more explicit result it is more convenient to calculate the Fourier transform (1.337):

$$(\mathbf{x}_b | \mathbf{x}_a)_E = \int_0^\infty d(t_b - t_a) \frac{1}{\sqrt{2\pi i\hbar(t_b - t_a)/M}^D} \exp\left\{\frac{i}{\hbar}\left[E(t_b - t_a) + \frac{M}{2}\frac{(\mathbf{x}_b - \mathbf{x}_a)^2}{t_b - t_a}\right]\right\}. \tag{1.341}$$

For $E < 0$, we set

$$\kappa \equiv \sqrt{-2ME/\hbar^2}, \tag{1.342}$$

and perform the integral with the help of the formula[11]

$$\int_0^\infty dt\, t^{\nu-1} e^{-i\gamma t + i\beta/t} = 2\left(\frac{\beta}{\gamma}\right)^{\nu/2} e^{-i\nu\pi/2} K_{-\nu}(2\sqrt{\beta\gamma}), \tag{1.343}$$

---

[11]I.S. Gradshteyn and I.M. Ryzhik, *Table of Integrals, Series, and Products*, Academic Press, New York, 1980, Formulas 3.471.10, 3.471.11, and 8.432.6

where $K_\nu(z)$ is the modified Bessel function which satifies $K_\nu(z) = K_{-\nu}(z)$[12] The result is

$$(\mathbf{x}_b|\mathbf{x}_a)_E = -i\frac{2M}{\hbar}\frac{\kappa^{D-2}}{(2\pi)^{D/2}}\frac{K_{D/2-1}(\kappa R)}{(\kappa R)^{D/2-1}}, \tag{1.344}$$

where $R \equiv |\mathbf{x}_b - \mathbf{x}_a|$. The simplest modified Bessel function is[13]

$$K_{1/2}(z) = K_{-1/2}(z) = \sqrt{\frac{\pi}{2z}}e^{-z}, \tag{1.345}$$

so that we find for $D = 1, 2, 3$, the amplitudes

$$-i\frac{M}{\hbar}\frac{1}{\kappa}e^{-\kappa R}, \qquad -i\frac{M}{\hbar}\frac{1}{\pi}K_0(\kappa R), \qquad -i\frac{M}{\hbar}\frac{1}{2\pi R}e^{-\kappa R}. \tag{1.346}$$

At $R = 0$, the amplitude (1.344) is finite for all $D \le 2$, where we can use the small-argument behavior of the associated Bessel function[14]

$$K_\nu(z) = K_{-\nu}(z) \approx \frac{1}{2}\Gamma(\nu)\left(\frac{z}{2}\right)^{-\nu} \quad \text{for } \operatorname{Re}\nu > 0, \tag{1.347}$$

to obtain

$$(\mathbf{x}|\mathbf{x})_E = -i\frac{2M}{\hbar}\frac{\kappa^{D-2}}{(4\pi)^{D/2}}\Gamma(1 - D/2). \tag{1.348}$$

This result can be continued analytically to $D > 2$, which will be needed later (for example in Subsection 4.9.4).

For $E > 0$ we set

$$k \equiv \sqrt{2ME/\hbar^2} \tag{1.349}$$

and use the formula[15]

$$\int_0^\infty dt\, t^{\nu-1}e^{i\gamma t + i\beta/t} = i\pi\left(\frac{\beta}{\gamma}\right)^{\nu/2} e^{-i\nu\pi/2}H_{-\nu}^{(1)}(2\sqrt{\beta\gamma}), \tag{1.350}$$

where $H_\nu^{(1)}(z)$ is the Hankel function, to find

$$(\mathbf{x}_b|\mathbf{x}_a)_E = \frac{M\pi}{\hbar}\frac{k^{D-2}}{(2\pi)^{D/2}}\frac{H_{D/2-1}(kR)}{(kR)^{D/2-1}}. \tag{1.351}$$

The relation[16]

$$K_\nu(-iz) = \frac{\pi}{2}ie^{i\nu\pi/2}H_\nu^{(1)}(z) \tag{1.352}$$

---

[12]*ibid.*, Formula 8.486.16

[13]M. Abramowitz and I. Stegun, *Handbook of Mathematical Functions*, Dover, New York, 1965, Formula 10.2.17.

[14]*ibid.*, Formula 9.6.9.

[15]*ibid.*, Formulas 3.471.11 and 8.421.7.

[16]*ibid.*, Formula 8.407.1.

connects the two formulas with each other when continuing the energy from negative to positive values, which replaces κ by $e^{-i\pi/2}k = -ik$.

For large distances, the asymptotic behavior[17]

$$K_\nu(z) \approx \sqrt{\frac{\pi}{2z}} e^{-z}, \qquad H_\nu^{(1)}(z) \approx \sqrt{\frac{2}{\pi z}} e^{i(z-\nu\pi/2-\pi/4)} \qquad (1.353)$$

shows that the fixed-energy amplitude behaves for $E < 0$ like

$$(\mathbf{x}_b|\mathbf{x}_a)_E \approx -i\frac{M}{\hbar}\kappa^{D-2}\frac{1}{(2\pi)^{(D-1)/2}}\frac{1}{(\kappa R)^{(D-1)/2}}e^{-\kappa R/\hbar}, \qquad (1.354)$$

and for $E > 0$ like

$$(\mathbf{x}_b|\mathbf{x}_a)_E \approx \frac{M}{\hbar}k^{D-2}\frac{1}{(2\pi i)^{(D-1)/2}}\frac{1}{(kR)^{(D-1)/2}}e^{ikR/\hbar}. \qquad (1.355)$$

For $D = 1$ and 3, these asymptotic expressions hold for all R.

1.13   Quantum Mechanics of General Lagrangian Systems

An extension of the quantum-mechanical formalism to systems described by a set of completely general Lagrange coordinates $q_1, \ldots, q_N$ is not straightforward. Only in the special case of q_i $(i = 1, \ldots, N)$ being merely a curvilinear reparametrization of a D-dimensional Euclidean space are the above correspondence rules sufficient to quantize the system. Then $N = D$ and a variable change from x^i to q_j in the Schrödinger equation leads to the correct quantum mechanics. It will be useful to label the curvilinear coordinates by Greek superscripts and write q^μ instead of q_j. This will help when we write all ensuing equations in a form that is manifestly covariant under coordinate transformations. In the original definition of generalized coordinates in Eq. (1.1), this was unnecessary since transformation properties were ignored. For the Cartesian coordinates we shall use Latin indices alternatively as sub- or superscripts. The coordinate transformation $x^i = x^i(q^\mu)$ implies the relation between the derivatives $\partial_\mu \equiv \partial/\partial q^\mu$ and $\partial_i \equiv \partial/\partial x^i$:

$$\partial_\mu = e^i{}_\mu(q)\partial_i, \qquad (1.356)$$

with the transformation matrix

$$e^i{}_\mu(q) \equiv \partial_\mu x^i(q) \qquad (1.357)$$

called *basis D-ad* (in 3 dimensions triad, in 4 dimensions tetrad, etc.). Let $e_i{}^\mu(q) = \partial q^\mu/\partial x^i$ be the inverse matrix (assuming it exists) called the *reciprocal D-ad*, satisfying with $e^i{}_\mu$ the orthogonality and completeness relations

$$e^i{}_\mu\, e_i{}^\nu = \delta_\mu{}^\nu, \qquad e^i{}_\mu\, e_j{}^\mu = \delta^i{}_j. \qquad (1.358)$$

---

[17] *ibid.*, Formulas 8.451.6 and 8.451.3.

Then, (1.356) is inverted to

$$\partial_i = e_i{}^\mu(q)\partial_\mu \qquad (1.359)$$

and yields the curvilinear transform of the Cartesian quantum-mechanical momentum operators

$$\hat{p}_i = -i\hbar\partial_i = -i\hbar e_i{}^\mu(q)\partial_\mu. \qquad (1.360)$$

The free-particle Hamiltonian operator

$$\hat{H}_0 = \hat{T} = \frac{1}{2M}\hat{\mathbf{p}}^2 = -\frac{\hbar^2}{2M}\boldsymbol{\nabla}^2 \qquad (1.361)$$

goes over into

$$\hat{H}_0 = -\frac{\hbar^2}{2M}\Delta, \qquad (1.362)$$

where Δ is the Laplacian expressed in curvilinear coordinates:

$$\begin{aligned}
\Delta &= \partial_i^2 = e^{i\mu}\partial_\mu e_i{}^\nu\partial_\nu \\
&= e^{i\mu}e_i{}^\nu\partial_\mu\partial_\nu + (e^{i\mu}\partial_\mu e_i{}^\nu)\partial_\nu.
\end{aligned} \qquad (1.363)$$

At this point one introduces the *metric tensor*

$$g_{\mu\nu}(q) \equiv e_{i\mu}(q)e^i{}_\nu(q), \qquad (1.364)$$

its inverse

$$g^{\mu\nu}(q) = e^{i\mu}(q)e_i{}^\nu(q), \qquad (1.365)$$

defined by $g^{\mu\nu}g_{\nu\lambda} = \delta^\mu{}_\lambda$, and the so-called *affine connection*

$$\Gamma_{\mu\nu}{}^\lambda(q) = -e^i{}_\nu(q)\partial_\mu e_i{}^\lambda(q) = e_i{}^\lambda(q)\partial_\mu e^i{}_\nu(q). \qquad (1.366)$$

Then the Laplacian takes the form

$$\Delta = g^{\mu\nu}(q)\partial_\mu\partial_\nu - \Gamma_\mu{}^{\mu\nu}(q)\partial_\nu, \qquad (1.367)$$

with $\Gamma_\mu{}^{\lambda\nu}$ being defined as the *contraction*

$$\Gamma_\mu{}^{\lambda\nu} \equiv g^{\lambda\kappa}\Gamma_{\mu\kappa}{}^\nu. \qquad (1.368)$$

The reason why (1.364) is called a metric tensor is obvious: An infinitesimal square distance between two points in the original Cartesian coordinates

$$ds^2 \equiv d\mathbf{x}^2 \qquad (1.369)$$

becomes in curvilinear coordinates

$$ds^2 = \frac{\partial\mathbf{x}}{\partial q^\mu}\frac{\partial\mathbf{x}}{\partial q^\nu}dq^\mu dq^\nu = g_{\mu\nu}(q)dq^\mu dq^\nu. \qquad (1.370)$$

The infinitesimal volume element $d^D x$ is given by

$$d^D x = \sqrt{g}\, d^D q, \tag{1.371}$$

where

$$g(q) \equiv \det\left(g_{\mu\nu}(q)\right) \tag{1.372}$$

is the determinant of the metric tensor. Using this determinant, we form the quantity

$$\Gamma_\mu \equiv g^{-1/2}(\partial_\mu g^{1/2}) = \frac{1}{2} g^{\lambda\kappa}(\partial_\mu g_{\lambda\kappa}) \tag{1.373}$$

and see that it is equal to the once-contracted connection

$$\Gamma_\mu = \Gamma_{\mu\lambda}{}^\lambda. \tag{1.374}$$

With the inverse metric (1.365) we have furthermore

$$\Gamma_\mu{}^{\mu\nu} = -\partial_\mu g^{\mu\nu} - \Gamma_\mu{}^{\nu\mu}. \tag{1.375}$$

We now take advantage of the fact that the derivatives $\partial_\mu, \partial_\nu$ applied to the coordinate transformation $x^i(q)$ commute causing $\Gamma_{\mu\nu}{}^\lambda$ to be symmetric in $\mu\nu$, i.e., $\Gamma_{\mu\nu}{}^\lambda = \Gamma_{\nu\mu}{}^\lambda$ and hence $\Gamma_\mu{}^{\nu\mu} = \Gamma^\nu$. Together with (1.373) we find the rotation

$$\Gamma_\mu{}^{\mu\nu} = -\frac{1}{\sqrt{g}}(\partial_\mu g^{\mu\nu}\sqrt{g}), \tag{1.376}$$

which allows the Laplace operator Δ to be rewritten in the more compact form

$$\Delta = \frac{1}{\sqrt{g}}\partial_\mu g^{\mu\nu}\sqrt{g}\partial_\nu. \tag{1.377}$$

This expression is called the *Laplace-Beltrami operator*.[18]

Thus we have shown that for a Hamiltonian in a Euclidean space

$$H(\hat{\mathbf{p}}, \mathbf{x}) = \frac{1}{2M}\hat{\mathbf{p}}^2 + V(\mathbf{x}), \tag{1.378}$$

the Schrödinger equation in curvilinear coordinates becomes

$$\hat{H}\psi(q,t) \equiv \left[-\frac{\hbar^2}{2M}\Delta + V(q)\right]\psi(q,t) = i\hbar\partial_t\psi(q,t), \tag{1.379}$$

where $V(q)$ is short for $V(\mathbf{x}(q))$. The scalar product of two wave functions $\int d^D x \psi_2^*(\mathbf{x},t)\psi_1(\mathbf{x},t)$, which determines the transition amplitudes of the system, transforms into

$$\int d^D q \sqrt{g}\, \psi_2^*(q,t)\psi_1(q,t). \tag{1.380}$$

---

[18]More details will be given later in Eqs. (11.12)–(11.18).

It is important to realize that this Schrödinger equation would *not* be obtained by a straightforward application of the canonical formalism to the coordinate-transformed version of the Cartesian Lagrangian

$$L(\mathbf{x}, \dot{\mathbf{x}}) = \frac{M}{2}\dot{\mathbf{x}}^2 - V(\mathbf{x}).\tag{1.381}$$

With the velocities transforming as

$$\dot{x}^i = e^i{}_\mu(q)\dot{q}^\mu,\tag{1.382}$$

the Lagrangian becomes

$$L(q, \dot{q}) = \frac{M}{2}g_{\mu\nu}(q)\dot{q}^\mu\dot{q}^\nu - V(q).\tag{1.383}$$

Up to a factor M, the metric is equal to the Hessian metric of the system, which depends here only on q^μ [recall (1.12)]:

$$H_{\mu\nu}(q) = Mg_{\mu\nu}(q).\tag{1.384}$$

The canonical momenta are

$$p_\mu \equiv \frac{\partial L}{\partial \dot{q}^\mu} = Mg_{\mu\nu}\dot{q}^\nu.\tag{1.385}$$

The associated quantum-mechanical momentum operators $\hat{p}_\mu$ have to be Hermitian in the scalar product (1.380) and must satisfy the canonical commutation rules (1.268):

$$\begin{aligned}
[\hat{p}_\mu, \hat{q}^\nu] &= -i\hbar\delta_\mu{}^\nu, \\
[\hat{q}^\mu, \hat{q}^\nu] &= 0, \\
[\hat{p}_\mu, \hat{p}_\nu] &= 0.
\end{aligned}\tag{1.386}$$

An obvious solution is

$$\hat{p}_\mu = -i\hbar g^{-1/4}\partial_\mu g^{1/4}, \quad \hat{q}^\mu = q^\mu.\tag{1.387}$$

The commutation rules are true for $-i\hbar g^{-z}\partial_\mu g^z$ with any power z, but only $z = 1/4$ produces a Hermitian momentum operator:

$$\int d^3q\sqrt{g}\,\Psi_2^*(q, t)[-i\hbar g^{-1/4}\partial_\mu g^{1/4}\Psi_1(q, t)] = \int d^3q\,g^{1/4}\,\Psi_2^*(q, t)[-i\hbar\partial_\mu g^{1/4}\Psi_1(q, t)]$$

$$= \int d^3q\sqrt{g}\,[-i\hbar g^{-1/4}\partial_\mu g^{1/4}\Psi_2(q, t)]^*\Psi_1(q, t),\tag{1.388}$$

as is easily verified by partial integration.

In terms of the quantity (1.373), this can also be rewritten as

$$\hat{p}_\mu = -i\hbar(\partial_\mu + \tfrac{1}{2}\Gamma_\mu).\tag{1.389}$$

Consider now the classical Hamiltonian associated with the Lagrangian (1.383), which by (1.385) is simply

$$H = p_\mu \dot{q}^\mu - L = \frac{1}{2M} g_{\mu\nu}(q) p^\mu p^\nu + V(q). \tag{1.390}$$

When trying to turn this expression into a Hamiltonian operator, we encounter the operator-ordering problem discussed in connection with Eq. (1.101). The correspondence principle requires replacing the momenta p_μ by the momentum operators $\hat{p}_\mu$, but it does not specify the position of these operators with respect to the coordinates q^μ contained in the inverse metric $g^{\mu\nu}(q)$. An important constraint is provided by the required Hermiticity of the Hamiltonian operator, but this is not sufficient for a unique specification. We may, for instance, define the canonical Hamiltonian operator as

$$\hat{H}_{\text{can}} \equiv \frac{1}{2M} \hat{p}^\mu g_{\mu\nu}(q) \hat{p}^\nu + V(q), \tag{1.391}$$

in which the momentum operators have been arranged symmetrically around the inverse metric to achieve Hermiticity. This operator, however, is not equal to the correct Schrödinger operator in (1.379). The kinetic term contains what we may call the *canonical Laplacian*

$$\Delta_{\text{can}} = \left(\partial_\mu + \tfrac{1}{2} \Gamma_\mu \right) g^{\mu\nu}(q) \left(\partial_\nu + \tfrac{1}{2} \Gamma_\nu \right). \tag{1.392}$$

It differs from the Laplace-Beltrami operator (1.377) in (1.379) by

$$\Delta - \Delta_{\text{can}} = -\tfrac{1}{2} \partial_\mu (g^{\mu\nu} \Gamma_\nu) - \tfrac{1}{4} g^{\mu\nu} \Gamma_\nu \Gamma_\mu. \tag{1.393}$$

The correct Hamiltonian operator could be obtained by suitably distributing pairs of dummy factors of $g^{1/4}$ and $g^{-1/4}$ symmetrically between the canonical operators [5]:

$$\hat{H} = \frac{1}{2M} g^{-1/4} \hat{p}_\mu g^{1/4} g^{\mu\nu}(q) g^{1/4} \hat{p}_\nu g^{-1/4} + V(q). \tag{1.394}$$

This operator has the same classical limit (1.390) as (1.391). Unfortunately, the correspondence principle does not specify how the classical factors have to be ordered before being replaced by operators.

The simplest system exhibiting the breakdown of the canonical quantization rules is a free particle in a plane described by radial coordinates $q^1 = r, q^2 = \varphi$:

$$x^1 = r \cos \varphi, \quad x^2 = r \sin \varphi. \tag{1.395}$$

Since the infinitesimal square distance is $ds^2 = dr^2 + r^2 d\varphi^2$, the metric reads

$$g_{\mu\nu} = \begin{pmatrix} 1 & 0 \\ 0 & r^2 \end{pmatrix}_{\mu\nu}. \tag{1.396}$$

It has a determinant

$$g = r^2 \tag{1.397}$$

and an inverse

$$g^{\mu\nu} = \begin{pmatrix} 1 & 0 \\ 0 & r^{-2} \end{pmatrix}^{\mu\nu}. \tag{1.398}$$

The Laplace-Beltrami operator becomes

$$\Delta = \frac{1}{r}\partial_r r \partial_r + \frac{1}{r^2}\partial_\varphi^{\,2}. \tag{1.399}$$

The canonical Laplacian, on the other hand, reads

$$\begin{aligned}\Delta_{\text{can}} &= (\partial_r + 1/2r)^2 + \frac{1}{r^2}\partial_\varphi^{\,2} \\ &= \partial_r^{\,2} + \frac{1}{r}\partial_r - \frac{1}{4r^2} + \frac{1}{r^2}\partial_\varphi^{\,2}. \end{aligned} \tag{1.400}$$

The discrepancy (1.393) is therefore

$$\Delta_{\text{can}} - \Delta = -\frac{1}{4r^2}. \tag{1.401}$$

Note that this discrepancy arises even though there is no apparent ordering problem in the naively quantized canonical expression $\hat{p}^\mu g_{\mu\nu}(q)\,\hat{p}^\nu$ in (1.400). Only the need to introduce dummy $g^{1/4}$- and $g^{-1/4}$-factors creates such problems, and a specification of the order is required to obtain the correct result.

If the Lagrangian coordinates q_i do not merely reparametrize a Euclidean space but specify the points of a general geometry, we cannot proceed as above and derive the Laplace-Beltrami operator by a coordinate transformation of a Cartesian Laplacian. With the canonical quantization rules being unreliable in curvilinear coordinates there are, at first sight, severe difficulties in quantizing such a system. This is why the literature contains many proposals for handling this problem [6]. Fortunately, a large class of non-Cartesian systems allows for a unique quantum-mechanical description on completely different grounds. These systems have the common property that their Hamiltonian can be expressed in terms of the generators of a group of motion in the general coordinate frame. For symmetry reasons, the correspondence principle must then be imposed not on the Poisson brackets of the canonical variables p and q, but on those of the group generators and the coordinates. The brackets containing two group generators specify the structure of the group, those containing a generator and a coordinate specify the defining representation of the group in configuration space. The replacement of these brackets by commutation rules constitutes the proper generalization of the canonical quantization from Cartesian to non-Cartesian coordinates. It is called *group quantization*. The replacement rule will be referred to as the *group correspondence principle*. The canonical commutation rules in Euclidean space may be viewed as a special case of the commutation rules between group generators, i.e., of the *Lie algebra* of the group. In a Cartesian coordinate frame, the group of motion is the Euclidean group containing translations and rotations. The generators of translations and rotations are the momenta and the angular momenta, respectively. According to the group correspondence principle, the Poisson brackets between the generators and the co-ordinates are to be replaced by commutation rules. Thus, in a Euclidean space, the commutation rules between group generators and coordinates lead to the canonical quantization rules, and this appears to be the deeper reason why the canonical rules are correct. In systems whose energy depends on generators of the group of motion other than those of translations, for instance on the angular momenta, the commutators between the generators have to be used for quantization rather than the canonical commutators between positions and momenta.

The prime examples for such systems are a particle on the surface of a sphere or a spinning top whose quantization will now be discussed.

1.14 Particle on the Surface of a Sphere

For a particle moving on the surface of a sphere of radius r with coordinates

$$x^1 = r\sin\theta\cos\varphi, \quad x^2 = r\sin\theta\sin\varphi, \quad x^3 = r\cos\theta, \tag{1.402}$$

the Lagrangian reads

$$L = \frac{Mr^2}{2}(\dot{\theta}^2 + \sin^2\theta\,\dot{\varphi}^2). \tag{1.403}$$

The canonical momenta are

$$p_\theta = Mr^2\dot{\theta}, \quad p_\varphi = Mr^2\sin^2\theta\,\dot{\varphi}, \tag{1.404}$$

and the classical Hamiltonian is given by

$$H = \frac{1}{2Mr^2}\left(p_\theta^2 + \frac{1}{\sin^2\theta}p_\varphi^2\right). \tag{1.405}$$

According to the canonical quantization rules, the momenta should become operators

$$\hat{p}_\theta = -i\hbar\frac{1}{\sin^{1/2}\theta}\partial_\theta\sin^{1/2}\theta, \quad \hat{p}_\varphi = -i\hbar\partial_\varphi. \tag{1.406}$$

But as explained in the previous section, these momentum operators are not expected to give the correct Hamiltonian operator when inserted into the Hamiltonian (1.405). Moreover, there exists no proper coordinate transformation from the surface of the sphere to Cartesian coordinates[19] such that a particle on a sphere cannot be treated via the safe Cartesian quantization rules (1.268):

$$\begin{aligned}
[\hat{p}_i, \hat{x}^j] &= -i\hbar\delta_i{}^j, \\
[\hat{x}^i, \hat{x}^j] &= 0, \\
[\hat{p}_i, \hat{p}_j] &= 0.
\end{aligned} \tag{1.407}$$

---

[19]There exist, however, certain infinitesimal nonholonomic coordinate transformations which are multivalued and can be used to transform infinitesimal distances in a curved space into those in a flat one. They are introduced and applied in Sections 10.2 and Appendix 10A, leading once more to the same quantum mechanics as the one described here.

The only help comes from the group properties of the motion on the surface of the sphere. The angular momentum

$$\mathbf{L} = \mathbf{x} \times \mathbf{p} \tag{1.408}$$

can be quantized uniquely in Cartesian coordinates and becomes an operator

$$\hat{\mathbf{L}} = \hat{\mathbf{x}} \times \hat{\mathbf{p}} \tag{1.409}$$

whose components satisfy the commutation rules of the Lie algebra of the rotation group

$$[\hat{L}_i, \hat{L}_j] = i\hbar \hat{L}_k \quad (i, j, k \text{ cyclic}). \tag{1.410}$$

Note that there is no factor-ordering problem since the $\hat{x}^i$'s and the $\hat{p}_i$'s appear with different indices in each $\hat{L}_k$. An important property of the angular momentum operator is its homogeneity in $\mathbf{x}$. It has the consequence that when going from Cartesian to spherical coordinates

$$x^1 = r \sin\theta \cos\varphi, \quad x^2 = r \sin\theta \sin\varphi, \quad x^3 = r \cos\theta, \tag{1.411}$$

the radial coordinate cancels making the angular momentum a differential operator involving only the angles θ, φ:

$$\begin{aligned} \hat{L}_1 &= i\hbar \left(\sin\varphi \, \partial_\theta + \cot\theta \cos\varphi \, \partial_\varphi \right), \\ \hat{L}_2 &= -i\hbar \left(\cos\varphi \, \partial_\theta - \cot\theta \sin\varphi \, \partial_\varphi \right), \\ \hat{L}_3 &= -i\hbar \partial_\varphi. \end{aligned} \tag{1.412}$$

There is then a natural way of quantizing the system which makes use of these operators $\hat{L}_i$. We re-express the classical Hamiltonian (1.405) in terms of the classical angular momenta

$$\begin{aligned} L_1 &= Mr^2 \left(-\sin\varphi \, \dot{\theta} - \sin\theta \cos\theta \cos\varphi \, \dot{\varphi} \right), \\ L_2 &= Mr^2 \left(\cos\varphi \, \dot{\theta} - \sin\theta \cos\theta \sin\varphi \, \dot{\varphi} \right), \\ L_3 &= Mr^2 \sin^2\theta \, \dot{\varphi} \end{aligned} \tag{1.413}$$

as

$$H = \frac{1}{2Mr^2} \mathbf{L}^2, \tag{1.414}$$

and replace the angular momenta by the operators (1.412). The result is the Hamiltonian operator:

$$\hat{H} = \frac{1}{2Mr^2} \hat{\mathbf{L}}^2 = -\frac{\hbar^2}{2Mr^2} \left[\frac{1}{\sin\theta} \partial_\theta \left(\sin\theta \, \partial_\theta \right) + \frac{1}{\sin^2\theta} \partial_\varphi^2 \right]. \tag{1.415}$$

The eigenfunctions diagonalizing the rotation-invariant operator $\hat{\mathbf{L}}^2$ are well known. They can be chosen to diagonalize simultaneously one component of $\hat{L}_i$, for instance the third one, $\hat{L}_3$, in which case they are equal to the spherical harmonics

$$Y_{lm}(\theta, \varphi) = (-1)^m \left[\frac{2l+1}{4\pi} \frac{(l-m)!}{(l+m)!}\right]^{1/2} P_l^m(\cos\theta) e^{im\varphi}, \tag{1.416}$$

with $P_l^m(z)$ being the associated Legendre polynomials

$$P_l^m(z) = \frac{1}{2^l l!}(1-z^2)^{m/2} \frac{d^{l+m}}{dx^{l+m}}(z^2-1)^l. \tag{1.417}$$

The spherical harmonics are orthonormal with respect to the rotation-invariant scalar product

$$\int_0^\pi d\theta \sin\theta \int_0^{2\pi} d\varphi \, Y_{lm}^*(\theta, \varphi) Y_{l'm'}(\theta, \varphi) = \delta_{ll'}\delta_{mm'}. \tag{1.418}$$

Two important lessons can be learned from this group quantization. First, the correct Hamiltonian operator (1.415) does not agree with the canonically quantized one which would be obtained by inserting Eqs. (1.406) into (1.405). The correct result would, however, arise by distributing dummy factors

$$g^{-1/4} = r^{-1}\sin^{-1/2}\theta, \quad g^{1/4} = r\sin^{1/2}\theta \tag{1.419}$$

between the canonical momentum operators as observed earlier in Eq. (1.394). Second, just as in the case of polar coordinates, the correct Hamiltonian operator is equal to

$$\hat{H} = -\frac{\hbar^2}{2M}\Delta, \tag{1.420}$$

where Δ is the Laplace-Beltrami operator associated with the metric

$$g_{\mu\nu} = r^2 \begin{pmatrix} 1 & 0 \\ 0 & \sin^2\theta \end{pmatrix}, \tag{1.421}$$

i.e.,

$$\Delta = \frac{1}{r^2}\left[\frac{1}{\sin\theta}\partial_\theta(\sin\theta\partial_\theta) + \frac{1}{\sin^2\theta}\partial_\varphi^2\right]. \tag{1.422}$$

1.15 Spinning Top

For a spinning top, the optimal starting point is again not the classical Lagrangian but the Hamiltonian expressed in terms of the classical angular momenta. In the symmetric case in which two moments of inertia coincide, it is written as

$$H = \frac{1}{2I_\xi}(L_\xi{}^2 + L_\eta{}^2) + \frac{1}{2I_\zeta}L_\zeta{}^2, \tag{1.423}$$

where L_ξ, L_η, L_ζ are the components of the orbital angular momentum in the directions of the principal body axes with $I_\xi, I_\eta \equiv I_\xi, I_\zeta$ being the corresponding moments of inertia. The classical angular momentum of an aggregate of mass points is given by

$$\mathbf{L} = \sum_\nu \mathbf{x}_\nu \times \mathbf{p}_\nu, \tag{1.424}$$

where the sum over ν runs over all mass points. The angular momentum possesses a unique operator

$$\hat{\mathbf{L}} = \sum_\nu \hat{\mathbf{x}}_\nu \times \hat{\mathbf{p}}_\nu, \tag{1.425}$$

with the commutation rules (1.410) between the components $\hat{L}_i$. Since rotations do not change the distances between the mass points, they commute with the constraints of the rigid body. If the center of mass of the rigid body is placed at the origin, the only dynamical degrees of freedom are the orientations in space. They can uniquely be specified by the rotation matrix which brings the body from some standard orientation to the actual one. We may choose the standard orientation to have the principal body axes aligned with the x, y, z-directions, respectively. An arbitrary orientation is obtained by applying all finite rotations to each point of the body. They are specified by the 3×3 orthonormal matrices R_{ij}. The space of these matrices has three degrees of freedom. It can be decomposed, omitting the matrix indices as

$$R(\alpha, \beta, \gamma) = R_3(\alpha) R_2(\beta) R_3(\gamma), \tag{1.426}$$

where $R_3(\alpha)$, $R_3(\gamma)$ are rotations around the z-axis by angles α, γ, respectively, and $R_2(\beta)$ is a rotation around the y-axis by β. These rotation matrices can be expressed as exponentials

$$R_i(\delta) \equiv e^{-i\delta L_i/\hbar}, \tag{1.427}$$

where δ is the rotation angle and L_i are the 3×3 matrix generators of the rotations with the elements

$$(L_i)_{jk} = -i\hbar\epsilon_{ijk}. \tag{1.428}$$

It is easy to check that these generators satisfy the commutation rules (1.410) of angular momentum operators. The angles α, β, γ are referred to as *Euler angles*.

The 3×3 rotation matrices make it possible to express the infinitesimal rotations around the three coordinate axes as differential operators of the three Euler angles. Let $\psi(R)$ be the wave function of the spinning top describing the probability amplitude of the different orientations which arise from a standard orientation by the rotation matrix $R = R(\alpha, \beta, \gamma)$. Under a further rotation by $R(\alpha', \beta', \gamma')$, the wave function goes over into $\psi'(R) = \psi(R^{-1}(\alpha', \beta', \gamma')R)$. The transformation may be described by a unitary differential operator

$$\hat{U}(\alpha', \beta', \gamma') \equiv e^{-i\alpha'\hat{L}_3} e^{-i\beta'\hat{L}_2} e^{-i\gamma'\hat{L}_3}, \tag{1.429}$$

where $\hat{L}_i$ is the representation of the generators in terms of differential operators. To calculate these we note that the 3×3 -matrix $R^{-1}(\alpha, \beta, \gamma)$ has the following derivatives

$$
\begin{aligned}
-i\hbar\partial_\alpha R^{-1} &= R^{-1}L_3, \\
-i\hbar\partial_\beta R^{-1} &= R^{-1}(\cos\alpha\, L_2 - \sin\alpha\, L_1), \\
-i\hbar\partial_\gamma R^{-1} &= R^{-1}\left[\cos\beta\, L_3 + \sin\beta(\cos\alpha\, L_1 + \sin\alpha\, L_2)\right].
\end{aligned}
\tag{1.430}
$$

The first relation is trivial, the second follows from the rotation of the generator

$$
e^{-i\alpha L_3/\hbar}L_2 e^{i\alpha L_3/\hbar} = \cos\alpha\, L_2 - \sin\alpha\, L_1,
\tag{1.431}
$$

which is a consequence of *Lie's expansion formula*

$$
e^{-iA}Be^{iA} = 1 - i[A, B] + \frac{i^2}{2!}[A, [A, B]] + \dots,
\tag{1.432}
$$

together with the commutation rules (1.428) of the 3×3 matrices L_i. The third requires, in addition, the rotation

$$
e^{-i\beta L_2/\hbar}L_3 e^{i\beta L_2/\hbar} = \cos\beta L_3 + \sin\beta L_1.
\tag{1.433}
$$

Inverting the relations (1.430), we find the differential operators generating the rotations [7]:

$$
\begin{aligned}
\hat{L}_1 &= i\hbar\left(\cos\alpha\cot\beta\,\partial_\alpha + \sin\alpha\,\partial_\beta - \frac{\cos\alpha}{\sin\beta}\partial_\gamma\right), \\
\hat{L}_2 &= i\hbar\left(\sin\alpha\cot\beta\,\partial_\alpha - \cos\alpha\,\partial_\beta - \frac{\sin\alpha}{\sin\beta}\partial_\gamma\right), \\
\hat{L}_3 &= -i\hbar\partial_\alpha.
\end{aligned}
\tag{1.434}
$$

After exponentiating these differential operators we derive

$$
\begin{aligned}
\hat{U}(\alpha', \beta', \gamma')R^{-1}\hat{U}^{-1}(\alpha', \beta', \gamma')(\alpha, \beta, \gamma) &= R^{-1}(\alpha, \beta, \gamma)R(\alpha', \beta', \gamma'), \\
\hat{U}(\alpha', \beta', \gamma')R(\alpha, \beta, \gamma)\hat{U}^{-1}(\alpha', \beta', \gamma') &= R^{-1}(\alpha', \beta', \gamma')R(\alpha, \beta, \gamma),
\end{aligned}
\tag{1.435}
$$

so that $\hat{U}(\alpha', \beta', \gamma')\psi(R) = \psi'(R)$, as desired.

In the Hamiltonian (1.423), we need the components of $\hat{\mathbf{L}}$ along the body axes. They are obtained by rotating the 3×3 matrices L_i by $R(\alpha, \beta, \gamma)$ into

$$
\begin{aligned}
L_\xi &= RL_1R^{-1} = \cos\gamma\cos\beta(\cos\alpha\, L_1 + \sin\alpha\, L_2) \\
&\quad + \sin\gamma(\cos\alpha\, L_2 - \sin\alpha\, L_1) - \cos\gamma\sin\beta\, L_3, \\
L_\eta &= RL_2R^{-1} = -\sin\gamma\cos\beta(\cos\alpha\, L_1 + \sin\alpha\, L_2) \\
&\quad + \cos\gamma(\cos\alpha\, L_2 - \sin\alpha\, L_1) + \sin\gamma\sin\beta\, L_3, \\
L_\zeta &= RL_3R^{-1} = \cos\beta\, L_3 + \sin\beta(\cos\alpha\, L_1 + \sin\alpha\, L_2),
\end{aligned}
\tag{1.436}
$$

and replacing $L_i \rightarrow \hat{L}_i$ in the final expressions. Inserting (1.434), we find the operators

$$
\begin{aligned}
\hat{L}_\xi &= i\hbar\left(-\cos\gamma\cot\beta\,\partial_\gamma - \sin\gamma\,\partial_\beta + \frac{\cos\gamma}{\sin\beta}\partial_\alpha\right), \\
\hat{L}_\eta &= i\hbar\left(\sin\gamma\cot\beta\,\partial_\gamma - \cos\gamma\,\partial_\beta - \frac{\sin\gamma}{\sin\beta}\partial_\alpha\right\}, \\
\hat{L}_\zeta &= -i\hbar\partial_\gamma.
\end{aligned}
\tag{1.437}
$$

Note that these commutation rules have an opposite sign with respect to those in Eqs. (1.410) of the operators $\hat{L}_i$:[20]

$$
[\hat{L}_\xi, \hat{L}_\eta] = -i\hbar\hat{L}_\zeta, \quad \xi, \eta, \zeta = \text{cyclic}.
\tag{1.438}
$$

The sign is most simply understood by writing

$$
\hat{L}_\xi = a_\xi^i \hat{L}_i, \quad \hat{L}_\eta = a_\eta^i \hat{L}_i, \quad \hat{L}_\zeta = a_\zeta^i \hat{L}_i,
\tag{1.439}
$$

where a_ξ^i, a_η^i, a_ζ^i, are the components of the body axes. Under rotations these behave like $[\hat{L}_i, a_\xi^j] = i\hbar\epsilon_{ijk}a_\xi^k$, i.e., they are vector operators. It is easy to check that this property produces the sign reversal in (1.438) with respect to (1.410).

The correspondence principle is now applied to the Hamiltonian in Eq. (1.423) by placing operator hats on the L_a's. The energy spectrum and the wave functions can then be obtained by using only the group commutators between $\hat{L}_\xi, \hat{L}_\eta, \hat{L}_\zeta$. The spectrum is

$$
E_{L\Lambda} = \hbar^2\left[\frac{1}{2I_\xi}L(L+1) + \left(\frac{1}{2I_\zeta} - \frac{1}{2I_\xi}\right)\Lambda^2\right],
\tag{1.440}
$$

where $L(L+1)$ with $L = 0, 1, 2, \ldots$ are the eigenvalues of $\hat{\mathbf{L}}^2$, and $\Lambda = -L, \ldots, L$ are the eigenvalues of $\hat{L}_\zeta$. The wave functions are the representation functions of the rotation group. If the Euler angles α, β, γ are used to specify the orientation of the body axes, the wave functions are

$$
\psi_{L\Lambda m}(\alpha, \beta, \gamma) = D_{m\Lambda}^L(-\alpha, -\beta, -\gamma).
\tag{1.441}
$$

Here m' are the eigenvalues of $\hat{L}_3$, the magnetic quantum numbers, and $D_{m\Lambda}^L(\alpha, \beta, \gamma)$ are the representation matrices of angular momentum L. In accordance with (1.429), one may decompose

$$
D_{mm'}^L(\alpha, \beta, \gamma) = e^{-i(m\alpha + m'\gamma)}d_{mm'}^L(\beta),
\tag{1.442}
$$

with the matrices

$$
\begin{aligned}
d_{mm'}^L(\beta) &= \left[\frac{(L+m')!(L-m')!}{(L+m)!(L-m)!}\right]^{1/2} \\
&\times \left(\cos\frac{\beta}{2}\right)^{m+m'}\left(-\sin\frac{\beta}{2}\right)^{m-m'} P_{L-m'}^{(m'-m,m'+m)}(\cos\beta).
\end{aligned}
\tag{1.443}
$$

---

[20]When applied to functions not depending on α, then, after replacing $\beta \rightarrow \theta$ and $\gamma \rightarrow \varphi$, the operators agree with those in (1.412), up to the sign of $\hat{L}_1$.

For $j = 1/2$, these form the spinor representation of the rotations around the y-axis

$$d^{1/2}_{m'm}(\beta) = \begin{pmatrix} \cos\beta/2 & -\sin\beta/2 \\ \sin\beta/2 & \cos\beta/2 \end{pmatrix}. \tag{1.444}$$

The indices have the order $+1/2, -1/2$. The full spinor representation function $D^{1/2}(\alpha, \beta, \gamma)$ in (1.442) is most easily obtained by inserting into the general expression (1.429) the representation matrices of spin $1/2$ for the generators $\hat{L}_i$ with the commutation rules (1.410), the famous *Pauli spin matrices*:

$$\sigma^1 = \begin{pmatrix} 0 & 1 \\ 1 & 0 \end{pmatrix}, \quad \sigma^2 = \begin{pmatrix} 0 & -i \\ i & 0 \end{pmatrix}, \quad \sigma^3 = \begin{pmatrix} 1 & 0 \\ 0 & -1 \end{pmatrix}. \tag{1.445}$$

Thus we can write

$$D^{1/2}(\alpha, \beta, \gamma) = e^{-i\alpha\sigma_3/2} e^{-i\beta\sigma_2/2} e^{-i\gamma\sigma_3/2}. \tag{1.446}$$

The first and the third factor yield the pure phase factors in (1.442). The function $d^{1/2}_{m'm}(\beta)$ is obtained by a simple power series expansion of $e^{-i\beta\sigma^2/2}$, using the fact that $(\sigma^2)^{2n} = 1$ and $(\sigma^2)^{2n+1} = \sigma^2$:

$$e^{-i\beta\sigma^2/2} = \cos\beta/2 - i\sin\beta/2\,\sigma^2, \tag{1.447}$$

which is equal to (1.444).

For $j = 1$, the representation functions (1.443) form the vector representation

$$d^1_{m'm}(\beta) = \begin{pmatrix} \frac{1}{2}(1+\cos\beta) & -\frac{1}{\sqrt{2}}\sin\beta & \frac{1}{2}(1-\cos\beta) \\ \frac{1}{\sqrt{2}}\sin\beta & \cos\beta & -\frac{1}{\sqrt{2}}\sin\beta \\ \frac{1}{2}(1-\cos\beta) & \frac{1}{\sqrt{2}}\sin\beta & \frac{1}{2}(1+\cos\beta) \end{pmatrix}. \tag{1.448}$$

where the indices have the order $+1/2, -1/2$. The vector representation goes over into the ordinary rotation matrices $R_{ij}(\beta)$ by mapping the states $|1m\rangle$ onto the spherical unit vectors $\boldsymbol{\epsilon}(0) = \hat{\mathbf{z}}$, $\boldsymbol{\epsilon}(\pm 1) = \mp(\hat{\mathbf{x}} \pm i\hat{\mathbf{y}})/2$ using the matrix elements $\langle i|1m\rangle = \epsilon^i(m)$. Hence $R(\beta)\boldsymbol{\epsilon}(m) = \sum_{m'=-1}^{1} \boldsymbol{\epsilon}(m')d^1_{m'm}(\beta)$.

The representation functions $D^1(\alpha, \beta, \gamma)$ can also be obtained by inserting into the general exponential (1.429) the representation matrices of spin 1 for the generators $\hat{L}_i$ with the commutation rules (1.410). In Cartesian coordinates, these are simply $(\hat{L}_i)_{jk} = -i\epsilon_{ijk}$, where ϵ_{ijk} is the completely antisymmetric tensor with $\epsilon_{123} = 1$. In the spherical basis, these become $(\hat{L}_i)_{mm'} = \langle m|i\rangle(\hat{L}_i)_{ij}\langle j|m'\rangle = \epsilon^*_i(m)(\hat{L}_i)_{ij}\epsilon_j(m')$. The exponential $(e^{-i\beta\hat{L}_2})_{mm'}$ is equal to (1.448).

The functions $P_l^{(\alpha,\beta)}(z)$ are the Jacobi polynomials [8], which can be expressed in terms of hypergeometric functions as

$$P_l^{(\alpha,\beta)} \equiv \frac{(-1)^l}{l!}\frac{\Gamma(l+\beta+1)}{\Gamma(\beta+1)}F(-l, l+1+\alpha+\beta; 1+\beta; (1+z)/2), \tag{1.449}$$

where

$$F(a, b; c; z) \equiv 1 + \frac{ab}{c} z + \frac{a(a + 1) b(b + 1)}{c(c + 1)} \frac{z^2}{2!} + \dots . \tag{1.450}$$

The rotation functions $d^L_{mm'}(\beta)$ satisfy the differential equation

$$\left(-\frac{d^2}{d\beta^2} - \cot \beta \frac{d}{d\beta} + \frac{m^2 + m'^2 - 2mm' \cos \beta}{\sin^2 \beta}\right) d^L_{mm'}(\beta) = L(L + 1)d^L_{mm'}(\beta). \tag{1.451}$$

The scalar products of two wave functions have to be calculated with a measure of integration that is invariant under rotations:

$$\langle \psi_2 | \psi_1 \rangle \equiv \int_0^{2\pi} \int_0^\pi \int_0^{2\pi} d\alpha d\beta \sin \beta d\gamma \; \psi_2^*(\alpha, \beta, \gamma)\psi_1(\alpha, \beta, \gamma). \tag{1.452}$$

The above eigenstates (1.442) satisfy the orthogonality relation

$$\int_0^{2\pi} \int_0^\pi \int_0^{2\pi} d\alpha d\beta \sin \beta d\gamma \; D^{L_1 *}_{m'_1 m_1}(\alpha, \beta, \gamma) D^{L_2}_{m'_2 m_2}(\alpha, \beta, \gamma)$$

$$= \delta_{m'_1 m'_2} \; \delta_{m_1 m_2} \delta_{L_1 L_2} \frac{8\pi^2}{2L_1 + 1}. \tag{1.453}$$

Let us also contrast in this example the correct quantization via the commutation rules between group generators with the canonical approach which would start out with the classical Lagrangian. In terms of Euler angles, the Lagrangian reads

$$L = \frac{1}{2}[I_\xi(\omega_\xi{}^2 + \omega_\eta{}^2) + I_\zeta\omega_\zeta{}^2], \tag{1.454}$$

where $\omega_\xi, \omega_\eta, \omega_\zeta$ are the angular velocities measured along the principal axes of the top. To find these we note that the components in the rest system $\omega_1, \omega_2, \omega_3$ are obtained from the relation

$$\omega_k L_k = i\dot{R}R^{-1} \tag{1.455}$$

as

$$\begin{aligned}
\omega_1 &= -\dot{\beta} \sin \alpha + \dot{\gamma} \sin \beta \cos \alpha, \\
\omega_2 &= \dot{\beta} \cos \alpha + \dot{\gamma} \sin \beta \sin \alpha, \\
\omega_3 &= \dot{\gamma} \cos \beta + \dot{\alpha}.
\end{aligned} \tag{1.456}$$

After the rotation (1.436) into the body-fixed system, these become

$$\begin{aligned}
\omega_\xi &= \dot{\beta} \sin \gamma - \dot{\alpha} \sin \beta \cos \gamma, \\
\omega_\eta &= \dot{\beta} \cos \gamma + \dot{\alpha} \sin \beta \sin \gamma, \\
\omega_\zeta &= \dot{\alpha} \cos \beta + \dot{\gamma}.
\end{aligned} \tag{1.457}$$

Explicitly, the Lagrangian is

$$L = \frac{1}{2}[I_\xi(\dot{\beta}^2 + \dot{\alpha}^2 \sin^2 \beta) + I_\zeta(\dot{\alpha} \cos \beta + \dot{\gamma})^2]. \tag{1.458}$$

Considering α, β, γ as Lagrange coordinates q^μ with $\mu = 1, 2, 3$, this can be written in the form (1.383) with the Hessian metric [recall (1.12) and (1.384)]:

$$g_{\mu\nu} = \begin{pmatrix} I_\xi \sin^2\beta + I_\zeta \cos^2\beta & 0 & I_\zeta \cos\beta \\ 0 & I_\xi & 0 \\ I_\zeta \cos\beta & 0 & I_\zeta \end{pmatrix}, \tag{1.459}$$

whose determinant is

$$g = I_\xi^2 I_\zeta \sin^2\beta. \tag{1.460}$$

Hence the measure $\int d^3q\sqrt{g}$ in the scalar product (1.380) agrees with the rotation-invariant measure (1.452) up to a trivial constant factor. Incidentally, this is also true for the asymmetric top with $I_\xi \neq I_\eta \neq I_\zeta$, where $g = I_\xi^2 I_\zeta \sin^2\beta$, although the metric $g_{\mu\nu}$ is then much more complicated (see Appendix 1C).

The canonical momenta associated with the Lagrangian (1.454) are, according to (1.383),

$$\begin{aligned} p_\alpha &= \partial L/\partial\dot\alpha = I_\xi \dot\alpha \sin^2\beta + I_\zeta \cos\beta(\dot\alpha\cos\beta + \dot\gamma), \\ p_\beta &= \partial L/\partial\dot\beta = I_\xi \dot\beta, \\ p_\gamma &= \partial L/\partial\dot\gamma = I_\zeta (\dot\alpha\cos\beta + \dot\gamma). \end{aligned} \tag{1.461}$$

After inverting the metric to

$$g^{\mu\nu} = \frac{1}{I_\xi \sin^2\beta} \begin{pmatrix} 1 & 0 & -\cos\beta \\ 0 & \sin^2\beta & 0 \\ -\cos\beta & 0 & \cos^2\beta + I_\xi \sin^2\beta/I_\zeta \end{pmatrix}^{\mu\nu}, \tag{1.462}$$

we find the classical Hamiltonian

$$H = \frac{1}{2}\left[\frac{1}{I_\xi}p_\beta{}^2 + \left(\frac{\cos^2\beta}{I_\xi \sin^2\beta} + \frac{1}{I_\zeta}\right)p_\gamma{}^2 + \frac{1}{I_\xi \sin^2\beta}p_\alpha{}^2 - \frac{2\cos\beta}{I_\xi \sin^2\beta}p_\alpha p_\gamma\right]. \tag{1.463}$$

This Hamiltonian has no apparent ordering problem. One is therefore tempted to replace the momenta simply by the corresponding Hermitian operators which are, according to (1.387),

$$\begin{aligned} \hat{p}_\alpha &= -i\hbar\partial_\alpha, \\ \hat{p}_\beta &= -i\hbar(\sin\beta)^{-1/2}\partial_\beta(\sin\beta)^{1/2} = -i\hbar(\partial_\beta + \tfrac{1}{2}\cot\beta), \\ \hat{p}_\gamma &= -i\hbar\partial_\gamma. \end{aligned} \tag{1.464}$$

Inserting these into (1.463) gives the canonical Hamiltonian operator

$$\hat{H}_{\text{can}} = \hat{H} + \hat{H}_{\text{discr}}, \tag{1.465}$$

with

$$\begin{aligned} \hat{H} \equiv -\frac{\hbar^2}{2I_\xi}\Big[&\partial_\beta{}^2 + \cot\beta\,\partial_\beta + \left(\frac{I_\xi}{I_\zeta} + \cot^2\beta\right)\partial_\gamma{}^2 \\ &+ \frac{1}{\sin^2\beta}\partial_\alpha{}^2 - \frac{2\cos\beta}{\sin^2\beta}\partial_\alpha\partial_\gamma\Big] \end{aligned} \tag{1.466}$$

and

$$\hat{H}_{\text{discr}} \equiv \frac{1}{2}(\partial_\beta \cot \beta) + \frac{1}{4}\cot^2 \beta = \frac{1}{4\sin^2 \beta} - \frac{3}{4}. \tag{1.467}$$

The first term $\hat{H}$ agrees with the correct quantum-mechanical operator derived above. Indeed, inserting the differential operators for the body-fixed angular momenta (1.437) into the Hamiltonian (1.423), we find $\hat{H}$. The term $\hat{H}_{\text{discr}}$ is the discrepancy between the canonical and the correct Hamiltonian operator. It exists even though there is no apparent ordering problem, just as in the radial coordinate expression (1.400). The correct Hamiltonian could be obtained by replacing the classical $p_\beta{}^2$ term in H by the operator $g^{-1/4}\hat{p}_\beta g^{1/2}\hat{p}_\beta g^{-1/4}$, by analogy with the treatment of the radial coordinates in $\hat{H}$ of Eq. (1.394).

As another similarity with the two-dimensional system in radial coordinates and the particle on the surface of the sphere, we observe that while the canonical quantization fails, the Hamiltonian operator of the symmetric spinning top is correctly given by the Laplace-Beltrami operator (1.377) after inserting the metric (1.459) and the inverse (1.462). It is straightforward although tedious to verify that this is also true for the completely asymmetric top [which has quite a complicated metric given in Appendix 1C, see Eqs. (1C.2), and (1C.4)]. This is an important nontrivial result, since for a spinning top, the Lagrangian cannot be obtained by reparametrizing a particle in a Euclidean space with curvilinear coordinates. The result suggests that a replacement

$$g_{\mu\nu}(q)p^\mu p^\nu \rightarrow -\hbar^2 \Delta \tag{1.468}$$

produces the correct Hamiltonian operator in any non-Euclidean space.[21]

What is the characteristic non-Euclidean property of the α, β, γ space? As we shall see in detail in Chapter 10, the relevant quantity is the curvature scalar R. The exact definition will be found in Eq. (10.42). For the asymmetric spinning top we find (see Appendix 1C)

$$R = \frac{(I_\xi + I_\eta + I_\zeta)^2 - 2(I_\xi^2 + I_\eta^2 + I_\zeta^2)}{2I_\xi I_\eta I_\zeta}. \tag{1.469}$$

Thus, just like a particle on the surface of a sphere, the spinning top corresponds to a particle moving in a space with constant curvature. In this space, the correct correspondence principle can also be deduced from symmetry arguments. The geometry is most easily understood by observing that the α, β, γ space may be considered as the surface of a sphere in four dimensions, as we shall see in more detail in Chapter 8.

An important non-Euclidean space of physical interest is encountered in the context of general relativity. Originally, gravitating matter was assumed to move in a spacetime with an arbitrary local curvature. In newer developments of the theory one also allows for the presence of a nonvanishing torsion. In such a general situation,

---

[21] If the space has curvature and no torsion, this is the correct answer. If torsion is present, the correct answer will be given in Chapters 10 and 8.

where the group quantization rule is inapplicable, the correspondence principle has always been a matter of controversy [see the references after (1.401)] to be resolved in this text. In Chapters 10 and 8 we shall present a new quantum equivalence principle which is based on an application of simple geometrical principles to path integrals and which will specify a natural and unique passage from classical to quantum mechanics in any coordinate frame.[22] The configuration space may carry curvature and a certain class of torsions (gradient torsion). Several arguments suggest that our principle is correct. For the above systems with a Hamiltonian which can be expressed entirely in terms of generators of a group of motion in the underlying space, the new quantum equivalence principle will give the same results as the group quantization rule.

1.16 Scattering

Most observations of quantum phenomena are obtained from scattering processes of fundamental particles.

1.16.1 Scattering Matrix

Consider a particle impinging with a momentum $\mathbf{p}_a$ and energy $E = E_a = \mathbf{p}_a^2/2M$ upon a nonzero potential concentrated around the origin. After a long time, it will be found far from the potential with some momentum $\mathbf{p}_b$. The energy will be unchanged: $E = E_b = \mathbf{p}_b^2/2M$. The probability amplitude for such a process is given by the time evolution amplitude in the momentum representation

$$(\mathbf{p}_b t_b | \mathbf{p}_a t_a) \equiv \langle \mathbf{p}_b | e^{-i\hat{H}(t_b - t_a)/\hbar} | \mathbf{p}_a \rangle, \tag{1.470}$$

where the limit $t_b \to \infty$ and $t_a \to -\infty$ has to be taken. Long before and after the collision, this amplitude oscillates with a frequency $\omega = E/\hbar$ characteristic for free particles of energy E. In order to have a time-independent limit, we remove these oscillations, from (1.470), and define the scattering matrix (S-matrix) by the limit

$$\langle \mathbf{p}_b | \hat{S} | \mathbf{p}_a \rangle \equiv \lim_{t_b - t_a \to \infty} e^{i(E_b t_b - E_a t_a)/\hbar} \langle \mathbf{p}_b | e^{-i\hat{H}(t_b - t_a)/\hbar} | \mathbf{p}_a \rangle. \tag{1.471}$$

Most of the impinging particles will not scatter at all, so that this amplitude must contain a leading term, which is separated as follows:

$$\langle \mathbf{p}_b | \hat{S} | \mathbf{p}_a \rangle = \langle \mathbf{p}_b | \mathbf{p}_a \rangle + \langle \mathbf{p}_b | \hat{S} | \mathbf{p}_a \rangle', \tag{1.472}$$

where

$$\langle \mathbf{p}_b | \mathbf{p}_a \rangle = \langle \mathbf{p}_b | e^{-i\hat{H}(t_b - t_a)/\hbar} | \mathbf{p}_a \rangle = (2\pi\hbar)^3 \delta^{(3)}(\mathbf{p}_b - \mathbf{p}_a) \tag{1.473}$$

shows the normalization of the states [recall (1.186)]. This leading term is commonly subtracted from (1.471) to find the true scattering amplitude. Moreover,

---

[22]H. Kleinert, Mod. Phys. Lett. A *4*, 2329 (1989) (http://www.physik.fu-berlin.de/~kleinert/199); Phys. Lett. B *236*, 315 (1990) (*ibid*.http/202).

since potential scattering conserves energy, the remaining amplitude contains a δ-function ensuring energy conservation, and it is useful to divide this out, defining the so-called T-matrix by the decomposition

$$\langle \mathbf{p}_b|\hat{S}|\mathbf{p}_a\rangle \equiv (2\pi\hbar)^3\delta^{(3)}(\mathbf{p}_a - \mathbf{p}_a) - 2\pi\hbar i\delta(E_b - E_a)\langle \mathbf{p}_b|\hat{T}|\mathbf{p}_a\rangle. \qquad (1.474)$$

From the definition (1.471) and the hermiticity of $\hat{H}$ it follows that the scattering matrix is a unitary matrix. This expresses the physical fact that the total probability of an incident particle to re-emerge at some time is unity (in quantum field theory the situation is more complicated due to emission and absorption processes).

In the basis states $|\mathbf{p^m}\rangle$ introduced in Eq. (1.180) which satisfy the completeness relation (1.182) and are normalized to unity in a finite volume V, the unitarity is expressed as

$$\sum_{\mathbf{m'}}\langle \mathbf{p^m}|\hat{S}^\dagger|\mathbf{p^{m'}}\rangle\langle \mathbf{p^{m'}}|\hat{S}|\mathbf{p^{m''}}\rangle = \sum_{\mathbf{m'}}\langle \mathbf{p^m}|\hat{S}|\mathbf{p^{m'}}\rangle\langle \mathbf{p^{m'}}|\hat{S}^\dagger|\mathbf{p^{m''}}\rangle = 1. \qquad (1.475)$$

Remembering the relation (1.185) between the discrete states $|\mathbf{p^m}\rangle$ and their continuous limits $|\mathbf{p}\rangle$, we see that

$$\langle \mathbf{p}_b{}^{\mathbf{m'}}|\hat{S}|\mathbf{p}_a{}^{\mathbf{m}}\rangle \approx \frac{1}{L^3}\langle \mathbf{p}_b|\hat{S}|\mathbf{p}_a\rangle, \qquad (1.476)$$

where L^3 is the spatial volume, and $\mathbf{p}_b^{\mathbf{m}}$ and $\mathbf{p}_a^{\mathbf{m}}$ are the discrete momenta closest to $\mathbf{p}_b$ and $\mathbf{p}_a$. In the continuous basis $|\mathbf{p}\rangle$, the unitarity relation reads

$$\int \frac{d^3p}{(2\pi\hbar)^3}\langle \mathbf{p}_b|\hat{S}^\dagger|\mathbf{p}\rangle\langle \mathbf{p}|\hat{S}|\mathbf{p}_a\rangle = \int \frac{d^3p}{(2\pi\hbar)^3}\langle \mathbf{p}_b|\hat{S}|\mathbf{p}\rangle\langle \mathbf{p}|\hat{S}^\dagger|\mathbf{p}_a\rangle = 1. \qquad (1.477)$$

1.16.2 Cross Section

The absolute square of $\langle \mathbf{p}_b|\hat{S}|\mathbf{p}_a\rangle$ gives the probability $P_{\mathbf{p}_b\leftarrow\mathbf{p}_a}$ for the scattering from the initial momentum state $\mathbf{p}_a$ to the final momentum state $\mathbf{p}_b$. Omitting the unscattered particles, we have

$$P_{\mathbf{p}_b\leftarrow\mathbf{p}_a} = \frac{1}{L^6}\,2\pi\hbar\delta(0)\,2\pi\hbar\delta(E_b - E_a)|\langle \mathbf{p}_b|\hat{T}|\mathbf{p}_a\rangle|^2. \qquad (1.478)$$

The factor $\delta(0)$ at zero energy is made finite by imagining the scattering process to take place with an incident time-independent plane wave over a finite total time T. Then $2\pi\hbar\delta(0) = \int dt\, e^{iEt/\hbar}|_{E=0} = T$, and the probability is proportional to the time T:

$$P_{\mathbf{p}_b\leftarrow\mathbf{p}_a} = \frac{1}{L^6}\,T\,2\pi\hbar\delta(E_b - E_a)|\langle \mathbf{p}_b|\hat{T}|\mathbf{p}_a\rangle|^2. \qquad (1.479)$$

By summing this over all discrete final momenta, or equivalently, by integrating this over the phase space of the final momenta [recall (1.184)], we find the total probability per unit time for the scattering to take place

$$\frac{dP}{dt} = \frac{1}{L^6}\int \frac{d^3p_b L^3}{(2\pi\hbar)^3}\,2\pi\hbar\delta(E_b - E_a)|\langle \mathbf{p}_b|\hat{T}|\mathbf{p}_a\rangle|^2. \qquad (1.480)$$

The momentum integral can be split into an integral over the final energy and the final solid angle. For non-relativistic particles, this goes as follows

$$\int \frac{d^3 p_b}{(2\pi\hbar)^3} = \frac{1}{(2\pi\hbar)^3}\frac{M}{(2\pi\hbar)^3}\int d\Omega \int_0^\infty dE_b \, p_b, \tag{1.481}$$

where $d\Omega = d\phi_b d\cos\theta_b$ is the element of solid angle into which the particle is scattered. The energy integral removes the δ-function in (1.480), and makes p_b equal to p_a.

The differential scattering cross section $d\sigma/d\Omega$ is defined as the probability that a single impinging particle ends up in a solid angle $d\Omega$ per unit time and unit current density. From (1.480) we identify

$$\frac{d\sigma}{d\Omega} = \frac{d\dot{P}}{d\Omega}\frac{1}{j} = \frac{1}{L^3}\frac{Mp}{(2\pi\hbar)^3}2\pi\hbar|T_{\mathbf{p}_b\mathbf{p}_a}|^2\frac{1}{j}, \tag{1.482}$$

where we have set

$$\langle \mathbf{p}_b|\hat{T}|\mathbf{p}_a\rangle \equiv T_{\mathbf{p}_b\mathbf{p}_a}, \tag{1.483}$$

for brevity. In a volume L^3, the current density of a single impinging particle is given by the velocity $v = p/M$ as

$$j = \frac{1}{L^3}\frac{p}{M}, \tag{1.484}$$

so that the differential cross section becomes

$$\frac{d\sigma}{d\Omega} = \frac{M^2}{(2\pi\hbar)^2}|T_{\mathbf{p}_b\mathbf{p}_a}|^2. \tag{1.485}$$

If the scattered particle moves relativistically, we have to replace the constant mass M in (1.481) by $E = \sqrt{p^2 + M^2}$ inside the momentum integral, where $p = |\mathbf{p}|$, so that

$$\begin{aligned}\int \frac{d^3 p}{(2\pi\hbar)^3} &= \frac{1}{(2\pi\hbar)^3}\int d\Omega \int_0^\infty dp\, p^2 \\ &= \frac{1}{(2\pi\hbar)^3}\int d\Omega \int_0^\infty dE E\, p. \end{aligned} \tag{1.486}$$

In the relativistic case, the initial current density is not proportional to p/M but to the relativistic velocity $v = p/E$ so that

$$j = \frac{1}{L^3}\frac{p}{E}. \tag{1.487}$$

Hence the cross section becomes

$$\frac{d\sigma}{d\Omega} = \frac{E^2}{(2\pi\hbar)^2}|T_{\mathbf{p}_b\mathbf{p}_a}|^2. \tag{1.488}$$

1.16.3   Born Approximation

To lowest order in the interaction strength, the operator $\hat{S}$ in (1.471) is

$$\hat{S} \approx 1 - i\hat{V}/\hbar. \tag{1.489}$$

For a time-independent scattering potential, this implies

$$T_{\mathbf{p}_b \mathbf{p}_a} \approx V_{\mathbf{p}_b \mathbf{p}_a}/\hbar, \tag{1.490}$$

where

$$V_{\mathbf{p}_b \mathbf{p}_a} \equiv \langle \mathbf{p}_b | \hat{V} | \mathbf{p}_a \rangle = \int d^3x\, e^{i(\mathbf{p}_b - \mathbf{p}_a)\mathbf{x}/\hbar} V(\mathbf{x}) = \tilde{V}(\mathbf{p}_b - \mathbf{p}_a) \tag{1.491}$$

is a function of the *momentum transfer* $\mathbf{q} \equiv \mathbf{p}_b - \mathbf{p}_a$ only. Then (1.488) reduces to the so called *Born approximation* (Born 1926)

$$\frac{d\sigma}{d\Omega} \approx \frac{E^2}{(2\pi\hbar)^2\hbar^2} |V_{\mathbf{p}_b \mathbf{p}_a}|^2. \tag{1.492}$$

The amplitude whose square is equal to the differential cross section is usually denoted by $f_{\mathbf{p}_b \mathbf{p}_a}$, i.e., one writes

$$\frac{d\sigma}{d\Omega} = |f_{\mathbf{p}_b \mathbf{p}_a}|^2. \tag{1.493}$$

By comparison with (1.492) we identify

$$f_{\mathbf{p}_b \mathbf{p}_a} \equiv -\frac{M}{2\pi\hbar} R_{\mathbf{p}_b \mathbf{p}_a}, \tag{1.494}$$

where we have chosen the sign to agree with the convention in the textbook by Landau and Lifshitz [9].

1.16.4   Partial Wave Expansion and Eikonal Approximation

The scattering amplitude is usually expanded in partial waves with the help of Legendre polynomials $P_l(z) \equiv P_l^0(z)$ [see (1.417)] as

$$f_{\mathbf{p}_b \mathbf{p}_a} = \frac{\hbar}{2ip} \sum_{l=0}^{\infty} (2l+1) P_l(\cos\theta) \left(e^{2i\partial_l(p)} - 1 \right) \tag{1.495}$$

where $p \equiv |\mathbf{p}| = |\mathbf{p}_b| = |\mathbf{p}_a|$ and θ is the scattering defined by $\cos\theta \equiv \mathbf{p}_b\mathbf{p}_b/|\mathbf{p}_b||\mathbf{p}_a|$. In terms of θ, the momentum transfer $\mathbf{q} = \mathbf{p}_b - \mathbf{p}_a$ has the size $|\mathbf{q}| = 2p\sin(\theta/2)$.

For small θ, we can use the asymptotic form of the Legendre polynomials[23]

$$P_l^{-m}(\cos\theta) \approx \frac{1}{l^m} J_m(l\theta), \tag{1.496}$$

---

[23]M. Abramowitz and I. Stegun, op. cit., Formula 9.1.71.

to rewrite (1.495) approximately as an integral

$$f^{\mathrm{ei}}_{\mathbf{P}_b\mathbf{P}_a} = \frac{p}{i\hbar} \int db\, b\, J_0(qb) \left\{ \exp\left[2i\delta_{pb/\hbar}(p) \right] - 1 \right\}, \tag{1.497}$$

where $b \equiv l\hbar/p$ is the so called *impact parameter* of the scattering process. This is the *eikonal approximation* to the scattering amplitude. As an example, consider Coulomb scattering where $V(r) = Ze^2/r$ and (2.749) yields

$$\chi^{\mathrm{ei}}_{\mathbf{b},\mathbf{P}}[\mathbf{v}] = -\frac{Ze^2 M}{|\mathbf{P}|}\frac{1}{\hbar} \int_{-\infty}^{\infty} dz\, \frac{1}{\sqrt{b^2 + z^2}}. \tag{1.498}$$

The integral diverges logarithmically, but in a physical sample, the potential is screened at some distance R by opposite charges. Performing the integral up to R yields

$$\begin{aligned}
\chi^{\mathrm{ei}}_{\mathbf{b},\mathbf{P}}[\mathbf{v}] &= -\frac{Ze^2 M}{|\mathbf{P}|}\frac{1}{\hbar} \int_{b}^{R} dr\, \frac{1}{\sqrt{r^2 - b^2}} = -\frac{Ze^2 M}{|\mathbf{P}|}\frac{1}{\hbar} \log \frac{R + \sqrt{R^2 - b^2}}{b} \\
&\approx -2\frac{Ze^2 M}{|\mathbf{P}|}\frac{1}{\hbar} \log \frac{2R}{b}.
\end{aligned} \tag{1.499}$$

This implies

$$\exp\left(\chi^{\mathrm{ei}}_{\mathbf{b},\mathbf{P}}\right) \approx \left(\frac{b}{2R}\right)^{2i\gamma}, \tag{1.500}$$

where

$$\gamma \equiv \frac{Ze^2 M}{|\mathbf{P}|}\frac{1}{\hbar} \tag{1.501}$$

is a dimensionless quantity since $e^2 = \hbar c \alpha$ where α is the dimensionless *fine-structure constant*[24]

$$\alpha = \frac{e^2}{\hbar c} = 1/137.035\,997\,9\ldots\,. \tag{1.502}$$

The integral over the impact parameter in (1.497) can now be performed and yields

$$f^{\mathrm{ei}}_{\mathbf{P}_b\mathbf{P}_a} \approx \frac{\hbar}{2ip}\frac{1}{\sin^{2+2i\gamma}(\theta/2)}\frac{\Gamma(1+i\gamma)}{\Gamma(-i\gamma)} e^{-2i\gamma \log(2pR/\hbar)}. \tag{1.503}$$

Remarkably, this is the exact quantum mechanical amplitude of Coulomb scattering, except for the last phase factor which accounts for a finite screening length. This amplitude contains poles at momentum variables $p = p_n$ whenever

$$i\gamma_n \equiv \frac{Ze^2 M \hbar}{p_n} = -n, \quad n = 1, 2, 3, \ldots\,. \tag{1.504}$$

---

[24]Throughout this book we use electromagnetic units where the electric field $\mathbf{E} = -\boldsymbol{\nabla}\phi$ has the energy density $\mathcal{H} = \mathbf{E}^2/8\pi + \rho\phi$, where ρ is the charge density, so that $\boldsymbol{\nabla} \cdot \mathbf{E} = 4\pi\rho$ and $e^2 = \hbar c\,\alpha$. The fine-structure constant is measured most precisely via the *quantum Hall effect*, see M.E. Cage et al., IEEE Trans. Instrum. Meas. **38**, 284 (1989). The magnetic field satisfies Ampère's law $\boldsymbol{\nabla} \times \mathbf{B} = 4\pi\mathbf{j}$, where $\mathbf{j}$ is the current density.

This corresponds to energies

$$E^{(n)} = -\frac{p_n^2}{2M} = -\frac{MZ^2e^4}{\hbar^2}\frac{1}{2n^2}, \qquad (1.505)$$

which are the well-known energy values of hydrogen-like atoms with nuclear charge Ze. The prefactor $E_H \equiv e^2/a_H = Me^4/\hbar^2 = 4.359 \times 10^{-11}$ erg $= 27.210$ eV, is equal to twice the *Rydberg energy* (see also p. 954).

1.16.5 Scattering Amplitude from Time Evolution Amplitude

There exists a heuristic formula expressing the scattering amplitude as a limit of the time evolution amplitude. For this we express the δ-function in the energy as a large-time limit

$$\delta(E_b - E_a) = \frac{M}{p_b}\delta(p_b - p_a) = \frac{M}{p_b}\lim_{t_b \to \infty}\left(\frac{t_b}{2\pi\hbar M/i}\right)^{1/2}\exp\left[-\frac{i}{\hbar}\frac{t_b}{2M}(p_b - p_a)^2\right], \qquad (1.506)$$

where $p_b = |\mathbf{p}_b|$. Inserting this into Eq. (1.474) and setting sloppily $p_b = p_a$ for elastic scattering, the δ-function is removed and we obtain the following expression for the scattering amplitude

$$f_{\mathbf{p}_b\mathbf{p}_a} = \frac{p_b}{M}\frac{\sqrt{2\pi\hbar M/i}^3}{(2\pi\hbar)^3}\lim_{t_b \to \infty}\frac{1}{t_b^{1/2}}e^{iE_b(t_b-t_a)/\hbar}\left[(\mathbf{p}_b t_b|\mathbf{p}_a t_a) - \langle\mathbf{p}_b|\mathbf{p}_a\rangle\right]. \qquad (1.507)$$

This treatment of a δ-function is certainly unsatisfactory. A satisfactory treatment will be given in the path integral formulation in Section 2.22. At the present stage, we may proceed with more care with the following operator calculation. We rewrite the limit (1.471) with the help of the time evolution operator (2.5) as follows:

$$\begin{aligned}\langle\mathbf{p}_b|\hat{S}|\mathbf{p}_a\rangle &\equiv \lim_{t_b-t_a \to \infty}e^{i(E_b t_b - E_a t_a)/\hbar}(\mathbf{p}_b t_b|\mathbf{p}_a t_a) \\ &= \lim_{t_b,-t_a \to \infty}\langle\mathbf{p}_b|\hat{U}_I(t_b, t_a)|\mathbf{p}_a\rangle,\end{aligned} \qquad (1.508)$$

where $\hat{U}_I(t_b, t_a)$ is the time evolution operator in Dirac's interaction picture (1.286).

1.16.6 Lippmann-Schwinger Equation

From the definition (1.286) it follows that the operator $\hat{U}_I(t_b, t_a)$ satisfies the same composition law (1.254) as the ordinary time evolution operator $\hat{U}(t, t_a)$:

$$\hat{U}_I(t, t_a) = \hat{U}_I(t, t_b)\hat{U}_I(t_b, t_a). \qquad (1.509)$$

Now we observe that

$$e^{-iH_0t/\hbar}\hat{U}_I(t, t_a) = e^{-iHt/\hbar}\hat{U}_I(0, t_a) = \hat{U}_I(0, t_a - t)e^{-iH_0t/\hbar}, \qquad (1.510)$$

so that in the limit $t_a \to -\infty$

$$e^{-iH_0t/\hbar}\hat{U}_I(t, t_a) = e^{-iHt/\hbar}\hat{U}_I(0, t_a) \longrightarrow \hat{U}_I(0, t_a)e^{-iH_0t/\hbar}, \qquad (1.511)$$

and therefore

$$\lim_{t_a \to -\infty} \hat{U}_I(t_b, t_a) = \lim_{t_a \to -\infty} e^{iH_0t_b/\hbar}e^{-iHt_b/\hbar}\hat{U}_I(0, t_a) = \lim_{t_a \to -\infty} e^{iH_0t_b/\hbar}\hat{U}_I(0, t_a)e^{-iH_0t_b/\hbar}, (1.512)$$

which allows us to rewrite the scattering matrix (1.508) as

$$\langle \mathbf{p}_b|\hat{S}|\mathbf{p}_a \rangle \equiv \lim_{t_b, -t_a \to \infty} e^{i(E_b - E_a)t_b/\hbar}\langle \mathbf{p}_b|\hat{U}_I(0, t_a)|\mathbf{p}_a \rangle. \qquad (1.513)$$

Note that in contrast to (1.471), the time evolution of the initial state goes now only over the negative time axis rather than the full one.

Taking the matrix elements of Eq. (1.291) between free-particle states $\langle \mathbf{p}_b|$ and $|\mathbf{p}_b \rangle$, and using Eqs. (1.291) and (1.511), we obtain at $t_b = 0$

$$\langle \mathbf{p}_b|\hat{U}_I(0, t_a)|\mathbf{p}_b \rangle = \langle \mathbf{p}_b|\mathbf{p}_b \rangle - \frac{i}{\hbar} \int_{-\infty}^{0} dt\, e^{i(E_b - E_a - i\eta)t/\hbar}\langle \mathbf{p}_b|\hat{V}\hat{U}_I(0, t_a)|\mathbf{p}_b \rangle. \quad (1.514)$$

A small damping factor $e^{\eta t/\hbar}$ is inserted to ensure convergence at $t = -\infty$. For a time-independent potential, the integral can be done and yields

$$\langle \mathbf{p}_b|\hat{U}_I(0, t_a)|\mathbf{p}_b \rangle = \langle \mathbf{p}_b|\mathbf{p}_b \rangle - \frac{1}{E_b - E_a - i\eta}\langle \mathbf{p}_b|\hat{V}\hat{U}_I(0, t_a)|\mathbf{p}_b \rangle. \qquad (1.515)$$

This is the famous *Lippmann-Schwinger equation*. Inserting this into (1.513), we obtain the equation for the scattering matrix

$$\langle \mathbf{p}_b|\hat{S}|\mathbf{p}_a \rangle = \lim_{t_b, -t_a \to \infty} e^{i(E_b - E_a)t_b} \left[\langle \mathbf{p}_b|\mathbf{p}_a \rangle - \frac{1}{E_b - E_a - i\eta}\langle \mathbf{p}_b|\hat{V}\hat{U}_I(0, t_a)|\mathbf{p}_b \rangle \right]. \quad (1.516)$$

The first term in brackets is nonzero only if the momenta $\mathbf{p}_a$ and $\mathbf{p}_b$ are equal, in which case also the energies are equal, $E_b = E_a$, so that the prefactor can be set equal to one. In front of the second term, the prefactor oscillates rapidly as the time t_b grows large, making any finite function of E_b vanish, as a consequence of the Riemann-Lebesgue lemma. The second term contains, however, a pole at $E_b = E_a$ for which the limit has to be done more carefully. The prefactor has the property

$$\lim_{t_b \to \infty} \frac{e^{i(E_b - E_a)t_b/\hbar}}{E_b - E_a - i\eta} = \begin{cases} 0, & E_b \neq E_a, \\ i/\eta, & E_b = E_a. \end{cases} \qquad (1.517)$$

It is easy to see that this property defines a δ-function in the energy:

$$\lim_{t_b \to \infty} \frac{e^{i(E_b - E_a)t_b/\hbar}}{E_b - E_a - i\eta} = 2\pi i\delta(E_b - E_a). \qquad (1.518)$$

Indeed, let us integrate the left-hand side together with a smooth function $f(E_b)$, and set

$$E_b \equiv E_a + \xi/t_b. \tag{1.519}$$

Then the E_b-integral is rewritten as

$$\int_{-\infty}^{\infty} d\xi \frac{e^{i\xi}}{\xi + i\eta} f\left(E_a + \xi/t_a\right). \tag{1.520}$$

In the limit of large t_a, the function $f(E_a)$ can be taken out of the integral and the contour of integration can then be closed in the upper half of the complex energy plane, yielding $2\pi i$. Thus we obtain from (1.516) the formula (1.474), with the T-matrix

$$\langle \mathbf{p}_b | \hat{T} | \mathbf{p}_a \rangle = \frac{1}{\hbar} \langle \mathbf{p}_b | \hat{V} \hat{U}_I(0, t_a) | \mathbf{p}_b \rangle. \tag{1.521}$$

For a small potential $\hat{V}$, we approximate $\hat{U}_I(0, t_a) \approx 1$, and find the Born approximation (1.490).

The Lippmann-Schwinger equation can be recast as an integral equation for the T-matrix. Multiplying the original equation (1.515) by the matrix $\langle \mathbf{p}_b | \hat{V} | \mathbf{p}_a \rangle = V_{\mathbf{p}_b \mathbf{p}_c}$ from the left, we obtain

$$T_{\mathbf{p}_b \mathbf{p}_a} = V_{\mathbf{p}_b \mathbf{p}_a} - \int \frac{d^3 p_c}{(2\pi\hbar)^3} V_{\mathbf{p}_b \mathbf{p}_c} \frac{1}{E_c - E_a - i\eta} T_{\mathbf{p}_c \mathbf{p}_a}. \tag{1.522}$$

To extract physical information from the T-matrix (1.521) it is useful to analyze the behavior of the interacting state $\hat{U}_I(0, t_a) | \mathbf{p}_a \rangle$ in $\mathbf{x}$-space. From Eq. (1.511), we see that it is an eigenstate of the full Hamiltonian operator $\hat{H}$ with the initial energy E_a. Multiplying this state by $\langle \mathbf{x} |$ from the left, and inserting a complete set of momentum eigenstates, we calculate

$$\langle \mathbf{x} | \hat{U}_I(0, t_a) | \mathbf{p}_a \rangle = \int \frac{d^3 p}{(2\pi\hbar)^3} \langle \mathbf{x} | \mathbf{p} \rangle \langle \mathbf{p} | \hat{U}_I(0, t_a) | \mathbf{p}_a \rangle = \int \frac{d^3 p}{(2\pi\hbar)^3} \langle \mathbf{x} | \mathbf{p} \rangle \langle \mathbf{p} | \hat{U}_I(0, t_a) | \mathbf{p}_a \rangle.$$

Using Eq. (1.515), this becomes

$$\langle \mathbf{x} | \hat{U}_I(0, t_a) | \mathbf{p}_a \rangle = \langle \mathbf{x} | \mathbf{p}_a \rangle + \int d^3 x' \int \frac{d^3 p_b}{(2\pi\hbar)^3} \frac{e^{i\mathbf{p}_b(\mathbf{x} - \mathbf{x}')/\hbar}}{E_a - \mathbf{p}_b^2/2M + i\eta} V(\mathbf{x}') \langle \mathbf{x}' | \hat{U}_I(0, t_a) | \mathbf{p}_a \rangle. \tag{1.523}$$

The function

$$(\mathbf{x} | \mathbf{x}')_{E_a} = \int \frac{d^3 p_b}{(2\pi\hbar)^3} e^{i\mathbf{p}_b(\mathbf{x} - \mathbf{x}')/\hbar} \frac{i\hbar}{E_a - \mathbf{p}^2/2M + i\eta} \tag{1.524}$$

is recognized as the fixed-energy amplitude (1.340) of the free particle. In three dimensions it reads [see (1.355)]

$$(\mathbf{x} | \mathbf{x}')_{E_a} = -\frac{2Mi}{\hbar} \frac{e^{ip_a|\mathbf{x} - \mathbf{x}'|/\hbar}}{4\pi|\mathbf{x} - \mathbf{x}'|}, \qquad p_a = \sqrt{2ME_a}. \tag{1.525}$$

In order to find the scattering amplitude, we consider the wave function (1.523) far away from the scattering center, i.e., at large $|\mathbf{x}|$. Under the assumption that $V(\mathbf{x}')$ is nonzero only for small $\mathbf{x}'$, we approximate $|\mathbf{x} - \mathbf{x}'| \approx r - \hat{\mathbf{x}}\mathbf{x}'$, where $\hat{\mathbf{x}}$ is the unit vector in the direction of $\mathbf{x}$, and (1.523) becomes

$$\langle \mathbf{x}|\hat{U}_I(0,t_a)|\mathbf{p}_a\rangle \approx e^{i\mathbf{p}_a\mathbf{x}/\hbar} - \frac{e^{ip_a r}}{4\pi r}\int d^4x' e^{-ip_a\hat{\mathbf{x}}\mathbf{x}'}\frac{2M}{\hbar^2}V(\mathbf{x}')\langle \mathbf{x}'|\hat{U}_I(0,t_a)|\mathbf{p}_a\rangle. \tag{1.526}$$

In the limit $t_a \to -\infty$, the factor multiplying the spherical wave factor $e^{ip_a r/\hbar}/r$ is the scattering amplitude $f(\hat{\mathbf{x}})_{\mathbf{p}_a}$, whose absolute square gives the cross section. For scattering to a final momentum $\mathbf{p}_b$, the outgoing particles are detected far away from the scattering center in the direction $\hat{\mathbf{x}} = \hat{\mathbf{p}}_b$. Because of energy conservation, we may set $p_a\hat{\mathbf{x}} = \mathbf{p}_b$ and obtain the formula

$$f_{\mathbf{p}_b\mathbf{p}_a} = \lim_{t_a\to-\infty} -\frac{M}{2\pi\hbar^2}\int d^4x_b e^{-i\mathbf{p}_b\mathbf{x}_b}V(\mathbf{x}_b)\langle \mathbf{x}_b|\hat{U}_I(0,t_a)|\mathbf{p}_a\rangle. \tag{1.527}$$

By studying the interacting state $\hat{U}_I(0,t_a)|\mathbf{p}_a\rangle$ in $\mathbf{x}$-space, we have avoided the singular δ-function of energy conservation.

We are now prepared to derive formula (1.507) for the scattering amplitude. We observe that in the limit $t_a \to -\infty$, the amplitude $\langle \mathbf{x}_b|\hat{U}_I(0,t_a)|\mathbf{p}_a\rangle$ can be obtained from the time evolution amplitude $(\mathbf{x}_b t_b|\mathbf{x}_a t_a)$ as follows:

$$\begin{aligned}\langle \mathbf{x}_b|\hat{U}_I(0,t_a)|\mathbf{p}_a\rangle &= \langle \mathbf{x}_b|\hat{U}(0,t_a)|\mathbf{p}_a\rangle e^{-iE_a t_a/\hbar} \tag{1.528}\\ &= \lim_{t_a\to-\infty}\left(\frac{-2\pi i\hbar t_a}{M}\right)^{3/2}(\mathbf{x}_b t_b|\mathbf{x}_a t_a)e^{i(\mathbf{p}_a\mathbf{x}_a - p_a^2 t_a/2M)/\hbar}\Big|_{\mathbf{x}_a=\mathbf{p}_a t_a/M}.\end{aligned}$$

This follows directly from the Fourier transformation

$$\langle \mathbf{x}_b|\hat{U}(0,t_a)|\mathbf{p}_a\rangle e^{-iE_a t_a/\hbar} = \int d^3x_a(\mathbf{x}_b t_b|\mathbf{x}_a t_a)e^{i(\mathbf{p}_a\mathbf{x}_a - p_a^2 t_a/2M)/\hbar}, \tag{1.529}$$

by substituting the dummy integration variable $\mathbf{x}_a$ by $\mathbf{p}t_a/M$. Then the right-hand side becomes

$$\left(\frac{-t_a}{M}\right)^3\int d^3p\,(\mathbf{x}_b 0|\mathbf{p}t_a\ t_a)e^{i(\mathbf{p}_a\mathbf{p} - p_a^2)t_a/2M\hbar}. \tag{1.530}$$

Now, for large $-t_a$, the momentum integration is squeezed to $\mathbf{p} = \mathbf{p}_a$, and we obtain (1.528). The appropriate limiting formula for the δ-function

$$\delta^{(D)}(\mathbf{p}_b - \mathbf{p}_a) = \lim_{t_a\to-\infty}\frac{(-t_a)^{D/2}}{\sqrt{2\pi i\hbar M}^D}\exp\left\{-\frac{i}{\hbar}\frac{t_a}{2M}(\mathbf{p}_b - \mathbf{p}_a)^2\right\} \tag{1.531}$$

is easily obtained from Eq. (1.338) by an obvious substitution of variables. Its complex conjugate for $D = 1$ was written down before in Eq. (1.506) with t_a replaced by $-t_b$. The exponential on the right-hand side can just as well be multiplied by a factor $e^{i(p_b^2-p_a^2)^2/2M\hbar}$ which is unity when both sides are nonzero, so that it becomes $e^{-i(\mathbf{p}_a\mathbf{P}-p_a^2)t_a/2M\hbar}$. In this way we obtain a representation of the δ-function by which the Fourier integral (1.530) goes over into (1.528). The phase factor $e^{i(\mathbf{P}_a\mathbf{x}_a-p_a^2t_a/2M)/\hbar}$ on the right-hand side of Eq. (1.528), which is unity in the limit performed in that equation, is kept in Eq. (4.548) for later convenience.

Formula (1.528) is a reliable starting point for extracting the scattering amplitude $f_{\mathbf{p}_b\mathbf{p}_a}$ from the time evolution amplitude in $\mathbf{x}$-space $(\mathbf{x}_b\,0|\mathbf{x}_a t_a)$ at $\mathbf{x}_a = \mathbf{p}_a t_a/M$ by extracting the coefficient of the outcoming spherical wave $e^{ip_a r/\hbar}/r$.

As a cross check we insert the free-particle amplitude (1.337) into (1.528) and obtain the free undisturbed wave function $e^{i\mathbf{p}_a\mathbf{x}}$, which is the correct first term in Eq. (1.523) associated with unscattered particles.

1.17 Classical and Quantum Statistics

Consider a physical system with a constant number of particles N whose Hamiltonian has no explicit time dependence. If it is brought into contact with a thermal reservoir at a temperature T and has reached equilibrium, its thermodynamic properties can be obtained through the following rules: At the level of classical mechanics, each volume element in phase space

$$\frac{dp\,dq}{h} = \frac{dp\,dq}{2\pi\hbar} \tag{1.532}$$

is occupied with a probability proportional to the *Boltzmann factor*

$$e^{-H(p,q)/k_B T}, \tag{1.533}$$

where k_B is the *Boltzmann constant*,

$$k_B = 1.3806221(59) \times 10^{-16}\,\text{erg/Kelvin}. \tag{1.534}$$

The number in parentheses indicates the experimental uncertainty of the two digits in front of it. The quantity $1/k_B T$ has the dimension of an inverse energy and is commonly denoted by β. It will be called the *inverse temperature*, forgetting about the factor k_B. In fact, we shall sometimes take T to be measured in energy units k_B times Kelvin rather than in Kelvin. Then we may drop k_B in all formulas.

The integral over the Boltzmann factors of all phase space elements,[25]

$$Z_{\text{cl}}(T) \equiv \int \frac{dp\,dq}{2\pi\hbar}\, e^{-H(p,q)/k_B T}, \tag{1.535}$$

is called the *classical partition function*. It contains all classical thermodynamic information of the system. Of course, for a general Hamiltonian system with many

---

[25]In the sequel we shall always work at a fixed volume V and therefore suppress the argument V everywhere.

degrees of freedom, the phase space integral is $\prod_n \int dp_n \, dq_n / 2\pi\hbar$. The reader may wonder why an expression containing Planck's quantum $\hbar$ is called *classical*. The reason is that $\hbar$ can really be omitted in calculating any thermodynamic average. In classical statistics it merely supplies us with an irrelevant normalization factor which makes Z dimensionless.

1.17.1 Canonical Ensemble

In quantum statistics, the Hamiltonian is replaced by the operator $\hat{H}$ and the integral over phase space by the trace in the Hilbert space. This leads to the *quantum-statistical partition function*

$$Z(T) \equiv \text{Tr}\left(e^{-\hat{H}/k_B T}\right) \equiv \text{Tr}\left(e^{-H(\hat{p},\hat{x})/k_B T}\right), \tag{1.536}$$

where $\text{Tr}\,\hat{O}$ denotes the trace of the operator $\hat{O}$. If $\hat{H}$ is an N-particle Schrödinger Hamiltonian, the quantum-statistical system is referred to as a *canonical ensemble*. The right-hand side of (1.536) contains the position operator $\hat{x}$ in Cartesian coordinates rather than $\hat{q}$ to ensure that the system can be quantized canonically. In cases such as the spinning top, the trace formula is also valid but the Hilbert space is spanned by the representation states of the angular momentum operators. In more general Lagrangian systems, the quantization has to be performed differently in the way to be described in Chapters 10 and 8.

At this point we make an important observation: The quantum partition function is related in a very simple way to the quantum-mechanical time evolution operator. To emphasize this relation we shall define the trace of this operator for time-independent Hamiltonians as the *quantum-mechanical partition function*:

$$Z_{\text{QM}}(t_b - t_a) \equiv \text{Tr}\left(\hat{U}(t_b, t_a)\right) = \text{Tr}\left(e^{-i(t_b - t_a)\hat{H}/\hbar}\right). \tag{1.537}$$

Obviously the quantum-statistical partition function $Z(T)$ may be obtained from the quantum-mechanical one by continuing the time interval $t_b - t_a$ to the negative imaginary value

$$t_b - t_a = -\frac{i\hbar}{k_B T} \equiv -i\hbar\beta. \tag{1.538}$$

This simple formal relation shows that the trace of the time evolution operator contains all information on the thermodynamic equilibrium properties of a quantum system.

1.17.2 Grand-Canonical Ensemble

For systems containing many bodies it is often convenient to study their equilibrium properties in contact with a particle reservoir characterized by a chemical potential μ. For this one defines what is called the *grand-canonical quantum-statistical partition function*

$$Z_G(T, \mu) = \text{Tr}\left(e^{-(\hat{H} - \mu\hat{N})/k_B T}\right). \tag{1.539}$$

Here $\hat{N}$ is the operator counting the number of particles in each state of the ensemble. The combination of operators in the exponent,

$$\hat{H}_G = \hat{H} - \mu\hat{N}, \tag{1.540}$$

is called the *grand-canonical Hamiltonian*.

Given a partition function $Z(T)$ at a fixed particle number N, the *free energy* is defined by

$$F(T) = -k_B T \log Z(T). \tag{1.541}$$

Its grand-canonical version at a fixed chemical potential is

$$F_G(T, \mu) = -k_B T \log Z_G(T, \mu). \tag{1.542}$$

The *average energy* or *internal energy* is defined by

$$E = \mathrm{Tr}\left(\hat{H} e^{-\hat{H}/k_B T}\right) \Big/ \mathrm{Tr}\left(e^{-\hat{H}/k_B T}\right). \tag{1.543}$$

It may be obtained from the partition function $Z(T)$ by forming the temperature derivative

$$E = Z^{-1} k_B T^2 \frac{\partial}{\partial T} Z(T) = k_B T^2 \frac{\partial}{\partial T} \log Z(T). \tag{1.544}$$

In terms of the free energy (1.541), this becomes

$$E = T^2 \frac{\partial}{\partial T}\left(-F(T)/T\right) = \left(1 - T\frac{\partial}{\partial T}\right) F(T). \tag{1.545}$$

For a grand-canonical ensemble we may introduce an *average particle number* defined by

$$N = \mathrm{Tr}\left(\hat{N} e^{-(\hat{H}-\mu\hat{N})/k_B T}\right) \Big/ \mathrm{Tr}\left(e^{-(\hat{H}-\mu\hat{N})/k_B T}\right). \tag{1.546}$$

This can be derived from the grand-canonical partition function as

$$N = Z_G^{-1}(T, \mu) k_B T \frac{\partial}{\partial \mu} Z_G(T, \mu) = k_B T \frac{\partial}{\partial \mu} \log Z_G(T, \mu), \tag{1.547}$$

or, using the grand-canonical free energy, as

$$N = -\frac{\partial}{\partial \mu} F_G(T, \mu). \tag{1.548}$$

The average energy in a grand-canonical system,

$$E = \mathrm{Tr}\left(\hat{H} e^{-(\hat{H}-\mu\hat{N})/k_B T}\right) \Big/ \mathrm{Tr}\left(e^{-(\hat{H}-\mu\hat{N})/k_B T}\right), \tag{1.549}$$

can be obtained by forming, by analogy with (1.544) and (1.545), the derivative

$$\begin{aligned}
E - \mu N &= Z_G^{-1}(T, \mu) k_B T^2 \frac{\partial}{\partial T} Z_G(T, \mu) \\
&= \left(1 - T\frac{\partial}{\partial T}\right) F_G(T, \mu).
\end{aligned} \tag{1.550}$$

For a large number of particles, the density is a rapidly growing function of energy. For a system of N free particles, for example, the number of states up to energy E is given by

$$N(E) = \sum_{\mathbf{p}_i} \Theta(E - \sum_{i=1}^{N} \mathbf{p}_i^2/2M), \qquad (1.551)$$

where each of the particle momenta $\mathbf{p}_i$ is summed over all discrete momenta $\mathbf{p^m}$ in (1.179) available to a single particle in a finite box of volume $V = L^3$. For a large V, the sum can be converted into an integral[26]

$$N(E) = V^N \prod_{i=1}^{N} \left[\int \frac{d^3 p_i}{(2\pi\hbar)^3} \right] \Theta(E - \sum_{i=1}^{N} \mathbf{p}_i^2/2M), \qquad (1.552)$$

which is simply $[V/(2\pi\hbar)^3]^N$ times the volume Ω_{3N} of a $3N$-dimensional sphere of radius $\sqrt{2ME}$:

$$
\begin{aligned}
N(E) &= \left[\frac{V}{(2\pi\hbar)^3} \right]^N \Omega_{3N} \\
&\equiv \left[\frac{V}{(2\pi\hbar)^3} \right]^N \frac{(2\pi ME)^{3N/2}}{\Gamma\left(\frac{3}{2}N + 1\right)}.
\end{aligned}
\qquad (1.553)
$$

Recall the well-known formula for the volume of a unit sphere in D dimensions:

$$\Omega_D = \pi^{D/2}/\Gamma(D/2 + 1). \qquad (1.554)$$

The surface is [see Eqs. (8.116) and (8.117) for a derivation]

$$S_D = 2\pi^{D/2}/\Gamma(D/2). \qquad (1.555)$$

This follows directly from the integral[27]

$$
\begin{aligned}
S_D &= \int d^D p\, \delta(p-1) = \int d^D p\, 2\delta(p^2 - 1) = \int d^D p \int_{-\infty}^{\infty} \frac{d\lambda}{\pi} e^{i\lambda(p^2 - 1)} \quad (1.556) \\
&= \int_{-\infty}^{\infty} \frac{d\lambda}{\pi} \left(\frac{\pi}{-i\lambda} \right)^{D/2} e^{-i\lambda} = \frac{2\pi^{D/2}}{\Gamma(D/2)} \quad (1.557)
\end{aligned}
$$

Therefore, the density per energy $\rho = \partial N/\partial E$ is given by

$$\rho(E) = \left[\frac{V}{(2\pi\hbar)^3} \right]^N 2\pi M \frac{(2\pi ME)^{3N/2-1}}{\Gamma(\frac{3}{2}N)}. \qquad (1.558)$$

It grows with the very large power $E^{3N/2}$ in the energy. Nevertheless, the integral for the partition function (1.579) is convergent, due to the overwhelming exponential

---

[26]Remember, however, the exception noted in the footnote to Eq. (1.184) for systems possessing a condensate.

[27]I. S. Gradshteyn and I. M. Ryzhik, op. cit., Formula 3.382.7.

falloff of the Boltzmann factor, e^{-E/k_BT}. As the two functions $\rho(e)$ and e^{-e/k_BT} are multiplied with each other, the result is a function which peaks very sharply at the average energy E of the system. The position of the peak depends on the temperature T. For the free N particle system, for example,

$$\rho(E)e^{-E/k_BT} \sim e^{(3N/2-1)\log E - E/k_BT}. \tag{1.559}$$

This function has a sharp peak at

$$E(T) = k_BT\left(\frac{3N}{2} - 1\right) \approx k_BT\frac{3N}{2}. \tag{1.560}$$

The width of the peak is found by expanding (1.559) in $\delta E = E - E(T)$:

$$\exp\left\{\frac{3N}{2}\log E(T) - \frac{E(T)}{k_BT} - \frac{1}{2E^2(T)}\frac{3N}{2}(\delta E)^2 + \ldots\right\}. \tag{1.561}$$

Thus, as soon as δE gets to be of the order of $E(T)/\sqrt{N}$, the exponential is reduced by a factor of two with respect to $E(T) \approx k_BT\,3N/2$. The deviation is of a relative order $1/\sqrt{N}$, i.e., the peak is very sharp. With N being very large, the peak at $E(T)$ of width $E(T)/\sqrt{N}$ can be idealized by a δ-function, and we may write

$$\rho(E)e^{-E/k_BT} \approx \delta(E - E(T))N(T)e^{-E(T)/k_BT}. \tag{1.562}$$

The quantity $N(T)$ measures the total number of states over which the system is distributed at the temperature T.

The entropy $S(T)$ is now defined in terms of $N(T)$ by

$$N(T) = e^{S(T)/k_B}. \tag{1.563}$$

Inserting this with (1.562) into (1.579), we see that in the limit of a large number of particles N:

$$Z(T) = e^{-[E(T)-TS(T)]/k_BT}. \tag{1.564}$$

Using (1.541), the free energy can thus be expressed in the form

$$F(T) = E(T) - TS(T). \tag{1.565}$$

By comparison with (1.545) we see that the entropy may be obtained from the free energy directly as

$$S(T) = -\frac{\partial}{\partial T}F(T). \tag{1.566}$$

For grand-canonical ensembles we may similarly consider

$$Z_G(T,\mu) = \int dE\,dn\,\rho(E,n)e^{-(E-\mu n)/k_BT}, \tag{1.567}$$

where

$$\rho(E,n)e^{-(E-\mu n)/k_BT} \tag{1.568}$$

is now strongly peaked at $E = E(T, \mu)$, $n = N(T, \mu)$ and can be written approximately as

$$\rho(E, n)e^{-(E-\mu n)/k_B T} \approx \delta\left(E - E(T, \mu)\right)\delta\left(n - N(T, \mu)\right)$$
$$\times \ e^{S(T,\mu)/k_B}e^{-[E(T,\mu)-\mu N(T,\mu)]/k_B T}. \tag{1.569}$$

Inserting this back into (1.567) we find for large N

$$Z_G(T, \mu) = e^{-[E(T,\mu)-\mu N(T,\mu)-TS(T,\mu)]/k_B T}. \tag{1.570}$$

For the grand-canonical free energy (1.542), this implies the relation

$$F_G(T, \mu) = E(T, \mu) - \mu N(T, \mu) - TS(T, \mu). \tag{1.571}$$

By comparison with (1.550) we see that the entropy can be calculated directly from the derivative of the grand-canonical free energy

$$S(T, \mu) = -\frac{\partial}{\partial T}F_G(T, \mu). \tag{1.572}$$

The particle number is, of course, found from the derivative (1.548) with respect to the chemical potential, as follows directly from the definition (1.567).

The canonical free energy and the entropy appearing in the above equations depend on the particle number N and the volume V of the system, i.e., they are more explicitly written as $F(T, N, V)$ and $S(T, N, V)$, respectively.

In the arguments of the grand-canonical quantities, the particle number N is replaced by the chemical potential μ.

Among the arguments of the grand-canonical energy $F_G(T, \mu, V)$, the volume V is the only one which grows with the system. Thus $F_G(T, \mu, V)$ must be directly proportional to V. The proportionality constant defines the *pressure p* of the system:

$$F_G(T, \mu, V) \equiv -p(T, \mu, V)V. \tag{1.573}$$

Under infinitesimal changes of the three variables, $F_G(T, \mu, V)$ changes as follows:

$$dF_G(T, \mu, V) = -SdT + \mu dN - pdV. \tag{1.574}$$

The first two terms on the right-hand side follow from varying Eq. (1.571) at a fixed volume. When varying the volume, the definition (1.573) renders the last term.

Inserting (1.573) into (1.571), we find *Euler's relation*:

$$E = TS + \mu N - pV. \tag{1.575}$$

The energy has S, N, V as natural variables. Equivalently, we may write

$$F = -\mu N - pV, \tag{1.576}$$

where T, N, V are the natural variables.

1.18   Density of States and Tracelog

In many thermodynamic calculations, a quantity of fundamental interest is the density of states. To define it, we express the canonical partition function

$$Z(T) = \text{Tr}\left(e^{-\hat{H}/k_B T}\right) \tag{1.577}$$

as a sum over the Boltzmann factors of all eigenstates $|n\rangle$ of the Hamiltonian:, i.e.

$$Z(T) = \sum_n e^{-E_n/k_B T}. \tag{1.578}$$

This can be rewritten as an integral:

$$Z(T) = \int dE\, \rho(E) e^{-E/k_B T}. \tag{1.579}$$

The quantity

$$\rho(E) = \sum_n \delta(E - E_n) \tag{1.580}$$

specifies the *density of states* of the system in the energy interval $(E, E + dE)$. It may also be written formally as a trace of the density of states operator $\hat{\rho}(E)$:

$$\rho(E) = \text{Tr}\,\hat{\rho}(E) \equiv \text{Tr}\,\delta(E - \hat{H}). \tag{1.581}$$

The density of states is obviously the Fourier transform of the canonical partition function (1.577):

$$\rho(E) = \int_{-i\infty}^{\infty} \frac{d\beta}{2\pi i} e^{\beta E}\, \text{Tr}\left(e^{-\beta \hat{H}}\right) = \int_{-i\infty}^{\infty} \frac{d\beta}{2\pi i} e^{\beta E}\, Z(1/k_B \beta). \tag{1.582}$$

The integral

$$N(E) = \int^{E} dE'\, \rho(E') \tag{1.583}$$

is the number of states up to energy E. The integration may start anywhere below the ground state energy. The function $N(E)$ is a sum of Heaviside step functions (1.309):

$$N(E) = \sum_n \Theta(E - E_n). \tag{1.584}$$

This equation is correct only with the Heaviside function which is equal to $1/2$ at the origin, not with the one-sided version (1.302), as we shall see later. Indeed, if integrated to the energy of a certain level E_n, the result is

$$N(E_n) = (n + 1/2). \tag{1.585}$$

This formula will serve to determine the energies of bound states from approximations to $\omega(E)$ in Section 4.7, for instance from the Bohr-Sommerfeld condition (4.190) via the relation (4.210). In order to apply this relation one must be sure that all levels have different energies. Otherwise $N(E)$ jumps at E_n by half the degeneracy of this level. In Eq. (4A.9) we shall exhibit an example for this situation.

An important quantity related to $\rho(E)$ which will appear frequently in this text is the trace of the logarithm, short *tracelog*, of the operator $\hat{H} - E$.

$$\text{Tr} \log(\hat{H} - E) = \sum_n \log(E_n - E). \tag{1.586}$$

It may be expressed in terms of the density of states (1.581) as

$$\text{Tr} \log(\hat{H} - E) = \text{Tr} \int_{-\infty}^{\infty} dE' \, \delta(E' - \hat{H}) \log(E' - E) = \int_{-\infty}^{\infty} dE' \, \rho(E') \log(E' - E). \tag{1.587}$$

The tracelog of the Hamiltonian operator itself can be viewed as a limit of an *operator zeta function* associated with $\hat{H}$:

$$\hat{\zeta}_{\hat{H}}(\nu) = \text{Tr} \, \hat{H}^{-\nu}, \tag{1.588}$$

whose trace is the generalized zeta-function

$$\zeta_{\hat{H}}(\nu) \equiv \text{Tr} \left[\hat{\zeta}_{\hat{H}}(\nu) \right] = \text{Tr} \left(\hat{H}^{-\nu} \right) = \sum_n E_n^{-\nu}. \tag{1.589}$$

For a linearly spaced spectrum $E_n = n$ with $n = 1, 2, 3 \ldots$, this reduces to Riemann's zeta function (2.519).

From the generalized zeta function we can obtain the tracelog by forming the derivative

$$\text{Tr} \log \hat{H} = -\partial_\nu \, \zeta_{\hat{H}}(\nu)\big|_{\nu=0}. \tag{1.590}$$

By differentiating the tracelog (1.586) with respect to E, we find the trace of the resolvent (1.315):

$$\partial_E \text{Tr} \log(\hat{H} - E) = \text{Tr} \frac{1}{E - \hat{H}} = \sum_n \frac{1}{E - E_n} = \frac{1}{i\hbar} \sum_n R_n(E) = \frac{1}{i\hbar} \text{Tr} \, \hat{R}(E). \tag{1.591}$$

Recalling Eq. (1.325) we see that the imaginary part of this quantity slightly above the real E-axis yields the density of states

$$-\frac{1}{\pi} \text{Im} \, \partial_E \text{Tr} \log(\hat{H} - E - i\eta) = \sum_n \delta(E - E_n) = \rho(E). \tag{1.592}$$

By integrating this over the energy we obtain the number of states function $N(E)$ of Eq. (1.583):

$$-\frac{1}{\pi} \text{Im} \, \text{Tr} \log(E - \hat{H}) = \sum_n \Theta(E - E_n) = N(E). \tag{1.593}$$

Appendix 1A   Simple Time Evolution Operator

Consider the simplest nontrivial time evolution operator of a spin-1/2 particle in a magnetic field $\mathbf{B}$. The reduced Hamiltonian operator is $\hat{H}_0 = -\mathbf{B} \cdot \boldsymbol{\sigma}/2$, so that the time evolution operator reads, in natural units with $\hbar = 1$,

$$e^{-i\hat{H}_0(t_b - t_a)} = e^{i(t_b - t_a)\mathbf{B} \cdot \boldsymbol{\sigma}/2}. \tag{1A.1}$$

Expanding this as in (1.293) and using the fact that $(\mathbf{B} \cdot \boldsymbol{\sigma})^{2n} = B^{2n}$ and $(\mathbf{B} \cdot \boldsymbol{\sigma})^{2n+1} = B^{2n}(\mathbf{B} \cdot \boldsymbol{\sigma})$, we obtain

$$e^{-i\hat{H}_0(t_b - t_a)} = \cos B(t_b - t_a)/2 + i\hat{\mathbf{B}} \cdot \boldsymbol{\sigma} \sin B(t_b - t_a)/2, \tag{1A.2}$$

where $\hat{\mathbf{B}} \equiv \mathbf{B}/|\mathbf{B}|$. Suppose now that the magnetic field is not constant but has a small time-dependent variation $\delta\mathbf{B}(t)$. Then we obtain from (1.253) [or the lowest expansion term in (1.293)]

$$\delta e^{-i\hat{H}_0(t_b - t_a)} = \int_{t_a}^{t_b} dt\, e^{-i\hat{H}_0(t_b - t)}\delta\mathbf{B}(t) \cdot \boldsymbol{\sigma} e^{-i\hat{H}_0(t - t_a)}. \tag{1A.3}$$

Using (1A.2), the integrand on the right-hand side becomes

$$\left[\cos B(t_b - t)/2 + i\hat{\mathbf{B}} \cdot \boldsymbol{\sigma} \sin B(t_b - t)/2\right]\delta\mathbf{B}(t) \cdot \boldsymbol{\sigma}\left[\cos B(t - t_a)/2 + i\hat{\mathbf{B}} \cdot \boldsymbol{\sigma} \sin B(t - t_a)/2\right]. \tag{1A.4}$$

We simplify this with the help of the formula [recall (1.445)]

$$\sigma^i \sigma^j = \delta_{ij} + i\epsilon_{ijk}\sigma^k \tag{1A.5}$$

so that

$$\hat{\mathbf{B}} \cdot \boldsymbol{\sigma}\, \delta\mathbf{B}(t) \cdot \boldsymbol{\sigma} = \hat{\mathbf{B}} \cdot \delta\mathbf{B}(t) + i[\hat{\mathbf{B}} \times \delta\mathbf{B}(t)] \cdot \boldsymbol{\sigma}, \quad \delta\mathbf{B}(t) \cdot \boldsymbol{\sigma}\, \hat{\mathbf{B}} \cdot \boldsymbol{\sigma} = \hat{\mathbf{B}} \cdot \delta\mathbf{B}(t) - i[\hat{\mathbf{B}} \times \delta\mathbf{B}(t)] \cdot \boldsymbol{\sigma}, \tag{1A.6}$$

and

$$\begin{aligned}\hat{\mathbf{B}} \cdot \boldsymbol{\sigma}\, \delta\mathbf{B}(t) \cdot \boldsymbol{\sigma}\, \hat{\mathbf{B}} \cdot \boldsymbol{\sigma} &= \left[\hat{\mathbf{B}} \cdot \delta\mathbf{B}(t)\right]\hat{\mathbf{B}} \cdot \boldsymbol{\sigma} + i[\hat{\mathbf{B}} \times \delta\mathbf{B}(t)] \cdot \boldsymbol{\sigma}\, \hat{\mathbf{B}} \cdot \boldsymbol{\sigma} \\ &= i[\hat{\mathbf{B}} \times \delta\mathbf{B}(t)] \cdot \hat{\mathbf{B}} + \left\{[\hat{\mathbf{B}} \cdot \delta\mathbf{B}(t)]\hat{\mathbf{B}} - [\hat{\mathbf{B}} \times \delta\mathbf{B}(t)] \times \hat{\mathbf{B}}\right\} \cdot \boldsymbol{\sigma}. \tag{1A.7}\end{aligned}$$

The first term on the right-hand side vanishes, the second term is equal to $\delta\mathbf{B}$, since $\hat{\mathbf{B}}^2 = 1$. Thus we find for the integrand in (1A.4):

$$\begin{aligned}&\cos B(t_b - t)/2 \cos B(t - t_a)/2\, \delta\mathbf{B}(t) \cdot \boldsymbol{\sigma} \\ +\,&i \sin B(t_b - t)/2 \cos B(t - t_a)/2\{\hat{\mathbf{B}} \cdot \delta\mathbf{B}(t) + i[\hat{\mathbf{B}} \times \delta\mathbf{B}(t)] \cdot \boldsymbol{\sigma}\} \\ +\,&i \cos B(t_b - t)/2 \sin B(t - t_a)/2\{\hat{\mathbf{B}} \cdot \delta\mathbf{B}(t) - i[\hat{\mathbf{B}} \times \delta\mathbf{B}(t)] \cdot \boldsymbol{\sigma}\} \\ +\,&\sin B(t_b - t)/2 \sin B(t - t_a)/2\, \delta\mathbf{B} \cdot \boldsymbol{\sigma} \tag{1A.8}\end{aligned}$$

which can be combined to give

$$\left\{\cos B[(t_b + t_a)/2 - t]\, \delta\mathbf{B}(t) - \sin B[(t_b + t_a)/2 - t]\, [\hat{\mathbf{B}} \times \delta\mathbf{B}(t)]\right\} \cdot \boldsymbol{\sigma} + i \sin B(t_b - t_a)/2\, \hat{\mathbf{B}} \cdot \delta\mathbf{B}(t). \tag{1A.9}$$

Integrating this from t_a to t_b we obtain the variation (1A.3).

Appendix 1B   Convergence of the Fresnel Integral

Here we prove the convergence of the Fresnel integral (1.333) by relating it to the Gauss integral. According to Cauchy's integral theorem, the sum of the integrals along the three pieces of the

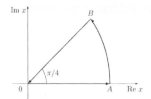

Figure 1.4 Triangular closed contour for Cauchy integral

closed contour shown in Fig. 1.4 vanishes since the integrand e^{-z^2} is analytic in the triangular domain:

$$\oint dz\,e^{-z^2} = \int_0^A dz\,e^{-z^2} + \int_A^B dz\,e^{-z^2} + \int_B^O dz\,e^{-z^2} = 0. \tag{1B.1}$$

Let R be the radius of the arc. Then we substitute in the three integrals the variable z as follows:

$$
\begin{array}{llll}
0\,A: & z = p, & dz = dp, & z^2 = p^2 \\
B\,0: & z = pe^{i\pi/4}, & dz = dp\,e^{i\pi/4}, & z^2 = ip^2 \\
AB: & z = R\,e^{i\varphi}, & dz = i\,R dp, & z^2 = p^2,
\end{array}
$$

and obtain the equation

$$\int_0^R dp\,e^{-p^2} + e^{i\pi/4}\int_R^0 dp\,e^{-ip^2} + \int_0^{\pi/4} d\varphi\,iR\,e^{-R^2(\cos 2\varphi + i\sin 2\varphi)+i\varphi} = 0. \tag{1B.2}$$

The first integral converges rapidly to $\sqrt{\pi}/2$ for $R \to \infty$. The last term goes to zero in this limit. To see this we estimate its absolute value as follows:

$$\left| \int_0^{\pi/4} d\varphi\,iR\,e^{-R^2(\cos 2\varphi + i\sin 2\varphi)+i\varphi} \right| < R \int_0^{\pi/4} d\varphi\,e^{-R^2\cos 2\varphi}. \tag{1B.3}$$

The right-hand side goes to zero exponentially fast, except for angles φ close to $\pi/4$ where the cosine in the exponent vanishes. In the dangerous regime $\alpha \in (\pi/4 - \epsilon, \pi/4)$ with small $\epsilon > 0$, one certainly has $\sin 2\varphi > \sin 2\alpha$, so that

$$R \int_\alpha^{\pi/4} d\varphi\,e^{-R^2\cos 2\varphi} < R \int_\alpha^{\pi/4} d\varphi\,\frac{\sin 2\varphi}{\sin 2\alpha}\,e^{-R^2\cos 2\varphi}. \tag{1B.4}$$

The right-hand integral can be performed by parts and yields

$$\alpha R\,e^{-R^2\cos 2\alpha} + \frac{1}{R\sin 2\alpha}\left[e^{-R^2\cos 2\varphi}\right]_{\varphi=\alpha}^{\varphi=\pi/4}, \tag{1B.5}$$

which goes to zero like $1/R$ for large R. Thus we find from (1B.2) the limiting formula $\int_\infty^0 dp\,e^{-ip^2} = -e^{-i\pi/4}\sqrt{\pi}/2$, or

$$\int_\infty^\infty dp\,e^{-ip^2} = e^{-i\pi/4}\sqrt{\pi}, \tag{1B.6}$$

which goes into Fresnel's integral formula (1.333) by substituting $p \to p\sqrt{a/2}$.

Appendix 1C   The Asymmetric Top

The Lagrangian of the asymmetric top with three different moments of inertia reads

$$L = \frac{1}{2}[I_\xi \omega_\xi{}^2 + I_\eta \omega_\eta{}^2 + I_\zeta \omega_\zeta{}^2]. \tag{1C.1}$$

It has the Hessian metric [recall (1.12) and (1.384)]

$$
\begin{aligned}
g_{11} &= I_\xi \sin^2\beta + I_\zeta \cos^2\beta - (I_\xi - I_\eta)\sin^2\beta\sin^2\gamma, \\
g_{21} &= -(I_\xi - I_\eta)\sin\beta\sin\gamma\cos\gamma, \\
g_{31} &= I_\zeta \cos\beta, \\
g_{22} &= I_\eta + (I_\xi - I_\eta)\sin^2\gamma, \\
g_{32} &= 0, \\
g_{33} &= I_\zeta,
\end{aligned}
\tag{1C.2}
$$

rather than (1.459). The determinant is

$$
g = I_\xi I_\eta I_\zeta \sin^2\beta,
\tag{1C.3}
$$

and the inverse metric has the components

$$
\begin{aligned}
g^{11} &= \frac{1}{g}\{I_\eta + (I_\xi - I_\eta)\sin^2\gamma\}I_\zeta, \\
g^{21} &= \frac{1}{g}\sin\beta\sin\gamma\cos\gamma(I_\xi - I_\eta)I_\zeta, \\
g^{31} &= \frac{1}{g}\{\cos\beta[-\sin^2\gamma(I_\xi - I_\eta) - I_\eta]\}I_\zeta, \\
g^{22} &= \frac{1}{g}\{\sin^2\beta[I_\xi - \sin^2\gamma(I_\xi - I_\eta)]\}I_\zeta, \\
g^{32} &= \frac{1}{g}\{\sin\beta\cos\beta\sin\gamma\cos\gamma(I_\eta - I_\xi)\}I_\zeta, \\
g^{33} &= \frac{1}{g}\{\sin^2\beta I_\xi I_\eta + \cos^2\beta I_\eta I_\zeta + \cos^2\beta\sin^2\gamma\,(I_\xi - I_\eta)I_\zeta\}.
\end{aligned}
\tag{1C.4}
$$

From this we find the components of the Riemann connection, the Christoffel symbol defined in Eq. (1.70):

$$
\begin{aligned}
\bar\Gamma_{11}{}^1 &= [\cos\beta\cos\gamma\sin\gamma(I_\eta^2 - I_\eta I_\zeta - I_\xi^2 + I_\xi I_\zeta)]/I_\xi I_\eta, \\
\bar\Gamma_{21}{}^1 &= \{\cos\beta[\sin^2\gamma(I_\xi^2 - I_\eta^2 - (I_\xi - I_\eta)I_\zeta) \\
&\qquad +I_\eta(I_\xi + I_\eta - I_\zeta)]\}/2\sin\beta I_\xi I_\eta, \\
\bar\Gamma_{31}{}^1 &= \{\cos\gamma\sin\gamma[I_\eta^2 - I_\xi^2 + (I_\xi - I_\eta)I_\zeta]\}/2I_\xi I_\eta, \\
\bar\Gamma_{22}{}^1 &= 0, \\
\bar\Gamma_{32}{}^1 &= [\sin^2\gamma(I_\xi^2 - I_\eta^2 - (I_\xi - I_\eta)I_\zeta) - I_\eta(I_\xi - I_\eta + I_\zeta)]/2\sin\beta I_\xi I_\eta, \\
\bar\Gamma_{33}{}^1 &= 0, \\
\bar\Gamma_{11}{}^2 &= \{\cos\beta\sin\beta[\sin^2\gamma(I_\eta^2 - I_\xi^2 - I_\zeta(I_\xi - I_\eta)) - I_\xi(I_\xi - I_\zeta)]\}/I_\xi I_\eta, \\
\bar\Gamma_{21}{}^2 &= \{\cos\beta\cos\gamma\sin\gamma[I_\xi^2 - I_\eta^2 - I_\zeta(I_\xi - I_\eta)]\}/2I_\xi I_\eta, \\
\bar\Gamma_{31}{}^2 &= \{\sin\beta[\sin^2\gamma(I_\xi^2 - I_\eta^2 - I_\zeta(I_\xi - I_\eta)) - I_\xi(I_\xi - I_\eta - I_\zeta)]\}/2I_\xi I_\eta, \\
\bar\Gamma_{22}{}^2 &= 0, \\
\bar\Gamma_{32}{}^2 &= [\cos\gamma\sin\gamma(I_\xi^2 - I_\eta^2 - I_\zeta(I_\xi - I_\eta))]/2I_\xi I_\eta, \\
\bar\Gamma_{33}{}^2 &= 0, \\
\bar\Gamma_{11}{}^3 &= \{\cos\gamma\sin\gamma[\sin^2\beta(I_\xi I_\eta(I_\xi - I_\eta) - I_\zeta(I_\xi^2 - I_\eta^2) + I_\zeta^2(I_\xi - I_\eta)) \\
&\qquad +(I_\xi^2 - I_\eta^2)I_\zeta - I_\zeta^2(I_\xi - I_\eta)]\}/I_\xi I_\eta I_\zeta, \\
\bar\Gamma_{21}{}^3 &= \{\sin^2\beta[\sin^2\gamma(2I_\xi I_\eta(I_\eta - I_\xi) + I_\zeta(I_\xi^2 - I_\eta^2) - I_\zeta^2(I_\xi - I_\eta)) \\
&\qquad +I_\xi I_\eta(I_\xi - I_\eta) + I_\eta I_\zeta(I_\eta - I_\zeta)] - \sin^2\gamma((I_\xi^2 - I_\eta^2)I_\zeta - I_\zeta^2(I_\xi - I_\eta))
\end{aligned}
$$

$$-I_\eta I_\zeta (I_\xi + I_\eta - I_\zeta)\}/2\sin\beta I_\xi I_\eta I_\zeta,$$

$$\bar\Gamma_{31}{}^3 = [\cos\beta\cos\gamma\sin\gamma(I_\xi^2 - I_\eta^2 - I_\zeta(I_\xi - I_\eta))]/2I_\xi I_\eta,$$

$$\bar\Gamma_{22}{}^3 = \cos\gamma\sin\gamma(I_\eta - I_\xi)/I_\zeta,$$

$$\bar\Gamma_{32}{}^3 = \{\cos\beta[\sin^2\gamma(I_\eta^2 - I_\xi^2 + (I_\xi - I_\eta)I_\zeta) + I_\eta(I_\xi - I_\eta + I_\zeta)]\}/2\sin\beta I_\eta I_\xi,$$

$$\bar\Gamma_{33}{}^3 = 0. \tag{1C.5}$$

The other components follow from the symmetry in the first two indices $\bar\Gamma_{\mu\nu}{}^\lambda = \bar\Gamma_{\nu\mu}^\lambda$. From this Christoffel symbol we calculate the Ricci tensor, to be defined in Eq. (10.41),

$$\bar R_{11} = \{\sin^2\beta[\sin^2\gamma(I_\eta^3 - I_\xi^3 - (I_\xi I_\eta - I_\zeta^2)(I_\xi - I_\eta))$$

$$+((I_\xi + I_\zeta)^2 - I_\eta^2)(I_\xi - I_\zeta)] + I_\zeta^3 - I_\zeta(I_\xi - I_\eta)^2\}/2I_\xi I_\eta I_\zeta,$$

$$\bar R_{21} = \{\sin\beta\sin\gamma\cos\gamma[I_\eta^3 - I_\xi^3 + (I_\xi I_\eta - I_\zeta^2)(I_\eta - I_\xi)]\}/2I_\xi I_\eta I_\zeta,$$

$$\bar R_{31} = -\{\cos\beta[(I_\xi - I_\eta)^2 - I_\zeta^2]\}/2I_\xi I_\eta,$$

$$\bar R_{22} = \{\sin^2\gamma[I_\xi^3 - I_\eta^3 + (I_\xi I_\eta - I_\zeta^2)(I_\xi - I_\eta)] + I_\eta^3 - (I_\xi - I_\zeta)^2 I_\eta\}/2I_\xi I_\eta I_\zeta,$$

$$\bar R_{32} = 0,$$

$$\bar R_{33} = -[(I_\xi - I_\eta)^2 - I_\zeta^2]/2I_\xi I_\eta. \tag{1C.6}$$

Contraction with $g^{\mu\nu}$ gives the curvature scalar

$$\bar R = [2(I_\xi I_\eta + I_\eta I_\zeta + I_\zeta I_\xi) - I_\xi^2 - I_\eta^2 - I_\zeta^2]/2I_\xi I_\eta I_\zeta. \tag{1C.7}$$

Since the space under consideration is free of torsion, the Christoffel symbol $\bar\Gamma_{\mu\nu}{}^\lambda$ is equal to the full affine connection $\Gamma_{\mu\nu}{}^\lambda$. The same thing is true for the curvature scalars $\bar R$ and R calculated from $\bar\Gamma_{\mu\nu}{}^\lambda$ and $\Gamma_{\mu\nu}{}^\lambda$, respectively.

Notes and References

For more details see some standard textbooks:
I. Newton, *Mathematische Prinzipien der Naturlehre*, Wiss. Buchgesellschaft, Darmstadt, 1963;
J.L. Lagrange, *Analytische Mechanik*, Springer, Berlin, 1887;
G. Hamel, *Theoretische Mechanik*, Springer, Berlin, 1949;
A. Sommerfeld, *Mechanik*, Harri Deutsch, Frankfurt, 1977;
W. Weizel, *Lehrbuch der Theoretischen Physik*, Springer, Berlin, 1963;
H. Goldstein, *Classical Mechanics*, Addison-Wesley, Reading, 1950;
L.D. Landau and E.M. Lifshitz, *Mechanics*, Pergamon, London, 1965;
R. Abraham and J.E. Marsden, *Foundations of Mechanics*, Benjamin, New York, 1967;
C.L. Siegel and J.K. Moser, *Lectures on Celestial Mechanics*, Springer, Berlin, 1971;
P.A.M. Dirac, *The Principles of Quantum Mechanics*, Clarendon, Oxford, 1958;
L.D. Landau and E.M. Lifshitz, *Quantum Mechanics*, Pergamon, London, 1965;
A. Messiah, *Quantum Mechanics*, Vols. I and II, North-Holland , Amsterdam, 1961;
L.I. Schiff, *Quantum Mechanics*, 3rd ed., McGraw-Hill, New York, 1968;
E. Merzbacher, *Quantum Mechanics*, 2nd ed, Wiley, New York, 1970;
L.D. Landau and E.M. Lifshitz, *Statistical Physics*, Pergamon, London, 1958;
E.M. Lifshitz and L.P. Pitaevski, *Statistical Physics*, Vol. 2, Pergamon, London, 1987.

The particular citations in this chapter refer to the publications

[1] For an elementary introduction see the book
H.B. Callen, *Classical Thermodynamics*, John Wiley and Sons, New York, 1960. More details are also found later in Eqs. (4.56) and (4.57).

[2] The integrability conditions are named after the mathematician of complex analysis H.A. Schwarz, a student of K. Weierstrass, who taught at the Humboldt-University of Berlin from 1892–1921.

[3] L. Schwartz, *Théorie des distributions*, Vols.I-II, Hermann & Cie, Paris, 1950-51; I.M. Gelfand and G.E. Shilov, *Generalized functions*, Vols.I-II, Academic Press, New York-London, 1964-68.

[4] An exception occurs in the theory of Bose-Einstein condensation where the single state $\mathbf{p} = 0$ requires a separate treatment since it collects a large number of particles in what is called a *Bose-Einstein condensate*. See p. 169 in the above-cited textbook by L.D. Landau and E.M. Lifshitz on *Statistical Mechanics*. Bose-Einstein condensation will be discussed in Sections 7.2.1 and 7.2.4.

[5] This was first observed by
B. Podolsky, Phys. Rev. **32**, 812 (1928).

[6] B.S. DeWitt, Rev. Mod. Phys. **29**, 377 (1957);
K.S. Cheng, J.Math. Phys. **13**, 1723 (1972);
H. Kamo and T. Kawai, Prog. Theor. Phys. **50**, 680, (1973);
T. Kawai, Found. Phys. **5**, 143 (1975);
H. Dekker, Physica A **103**, 586 (1980);
G.M. Gavazzi, Nuovo Cimento **101**A, 241 (1981).
See also the alternative approach by
N.M.J. Woodhouse, *Geometric Quantization*, Oxford University Press, Oxford, 1992.

[7] C. van Winter, Physica **20**, 274 (1954).

[8] For detailed properties of the representation matrices of the rotation group, see
A.R. Edmonds, *Angular Momentum in Quantum Mechanics*, Princeton University Press, 1960.

[9] L.D. Landau and E.M. Lifshitz, *Quantum Mechanics*, Pergamon, London, 1965.

2

Path Integrals — Elementary Properties and Simple Solutions

The operator formalism of quantum mechanics and quantum statistics may not always lead to the most transparent understanding of quantum phenomena. There exists another, equivalent formalism in which operators are avoided by the use of infinite products of integrals, called *path integrals*. In contrast to the Schrödinger equation, which is a differential equation determining the properties of a state at a time from their knowledge at an infinitesimally earlier time, path integrals yield the quantum-mechanical amplitudes in a global approach involving the properties of a system at *all times*.

2.1 Path Integral Representation of Time Evolution Amplitudes

The path integral approach to quantum mechanics was developed by Feynman[1] in 1942. In its original form, it applies to a point particle moving in a Cartesian coordinate system and yields the transition amplitudes of the time evolution operator between the localized states of the particle (recall Section 1.7)

$$(x_b t_b | x_a t_a) = \langle x_b | \hat{U}(t_b, t_a) | x_a \rangle, \quad t_b > t_a. \tag{2.1}$$

For simplicity, we shall at first assume the space to be one-dimensional. The extension to D Cartesian dimensions will be given later. The introduction of curvilinear coordinates will require a little more work. A further generalization to spaces with a nontrivial geometry, in which curvature and torsion are present, will be described in Chapters 10–11.

2.1.1 Sliced Time Evolution Amplitude

We shall be interested mainly in the causal or retarded time evolution amplitudes [see Eq. (1.299)]. These contain all relevant quantum-mechanical information and

---

[1]For the historical development, see Notes and References at the end of this chapter.

possess, in addition, pleasant analytic properties in the complex energy plane [see the remarks after Eq. (1.306)]. This is why we shall always assume, from now on, the causal sequence of time arguments $t_b > t_a$.

Feynman realized that due to the fundamental composition law of the time evolution operator (see Section 1.7), the amplitude (2.1) could be sliced into a large number, say $N + 1$, of time evolution operators, each acting across an infinitesimal time slice of thickness $\epsilon \equiv t_n - t_{n-1} = (t_b - t_a)/(N + 1) > 0$:

$$(x_b t_b | x_a t_a) = \langle x_b | \hat{U}(t_b, t_N) \hat{U}(t_N, t_{N-1}) \cdots \hat{U}(t_n, t_{n-1}) \cdots \hat{U}(t_2, t_1) \hat{U}(t_1, t_a) | x_a \rangle. \quad (2.2)$$

When inserting a complete set of states between each pair of $\hat{U}$'s,

$$\int_{-\infty}^{\infty} dx_n | x_n \rangle \langle x_n | = 1, \quad n = 1, \ldots, N, \quad (2.3)$$

the amplitude becomes a product of N-integrals

$$(x_b t_b | x_a t_a) = \prod_{n=1}^{N} \left[\int_{-\infty}^{\infty} dx_n \right] \prod_{n=1}^{N+1} (x_n t_n | x_{n-1} t_{n-1}), \quad (2.4)$$

where we have set $x_b \equiv x_{N+1}$, $x_a \equiv x_0$, $t_b \equiv t_{N+1}$, $t_a \equiv t_0$. The symbol $\Pi[\cdots]$ denotes the product of the quantities within the brackets. The integrand is the product of the amplitudes for the infinitesimal time intervals

$$(x_n t_n | x_{n-1} t_{n-1}) = \langle x_n | e^{-i\epsilon \hat{H}(t_n)/\hbar} | x_{n-1} \rangle, \quad (2.5)$$

with the Hamiltonian operator

$$\hat{H}(t) \equiv H(\hat{p}, \hat{x}, t). \quad (2.6)$$

The further development becomes simplest under the assumption that the Hamiltonian has the standard form, being the sum of a kinetic and a potential energy:

$$H(p, x, t) = T(p, t) + V(x, t). \quad (2.7)$$

For a sufficiently small slice thickness, the time evolution operator

$$e^{-i\epsilon \hat{H}/\hbar} = e^{-i\epsilon(\hat{T}+\hat{V})/\hbar} \quad (2.8)$$

is factorizable as a consequence of the *Baker-Campbell-Hausdorff formula* (to be proved in Appendix 2A)

$$e^{-i\epsilon(\hat{T}+\hat{V})/\hbar} = e^{-i\epsilon\hat{V}/\hbar} e^{-i\epsilon\hat{T}/\hbar} e^{-i\epsilon^2 \hat{X}/\hbar^2}, \quad (2.9)$$

where the operator $\hat{X}$ has the expansion

$$\hat{X} \equiv \frac{i}{2}[\hat{V}, \hat{T}] - \frac{\epsilon}{\hbar} \left(\frac{1}{6}[\hat{V}, [\hat{V}, \hat{T}]] - \frac{1}{3}[[\hat{V}, \hat{T}], \hat{T}] \right) + \mathcal{O}(\epsilon^2). \quad (2.10)$$

The omitted terms of order $\epsilon^4, \epsilon^5, \ldots$ contain higher commutators of $\hat{V}$ and $\hat{T}$. If we neglect, for the moment, the $\hat{X}$-term which is suppressed by a factor ϵ^2, we calculate for the local matrix elements of $e^{-i\epsilon\hat{H}/\hbar}$ the following simple expression:

$$\langle x_n | e^{-i\epsilon H(\hat{p},\hat{x},t_n)/\hbar} | x_{n-1} \rangle \approx \int_{-\infty}^{\infty} dx \langle x_n | e^{-i\epsilon V(\hat{x},t_n)/\hbar} | x \rangle \langle x | e^{-i\epsilon T(\hat{p},t_n)/\hbar} | x_{n-1} \rangle$$

$$= \int_{-\infty}^{\infty} dx \langle x_n | e^{-i\epsilon V(\hat{x},t_n)/\hbar} | x \rangle \int_{-\infty}^{\infty} \frac{dp_n}{2\pi\hbar} e^{ip_n(x-x_{n-1})/\hbar} e^{-i\epsilon T(p_n,t_n)/\hbar}. \qquad (2.11)$$

Evaluating the local matrix elements,

$$\langle x_n | e^{-i\epsilon V(\hat{x},t_n)/\hbar} | x \rangle = \delta(x_n - x) e^{-i\epsilon V(x_n,t_n)/\hbar}, \qquad (2.12)$$

this becomes

$$\langle x_n | e^{-i\epsilon H(\hat{p},\hat{x},t_n)/\hbar} | x_{n-1} \rangle \approx$$

$$\int_{-\infty}^{\infty} \frac{dp_n}{2\pi\hbar} \exp \left\{ ip_n(x_n - x_{n-1})/\hbar - i\epsilon[T(p_n,t_n) + V(x_n,t_n)]/\hbar \right\}. \qquad (2.13)$$

Inserting this back into (2.4), we obtain *Feynman's path integral formula*, consisting of the multiple integral

$$(x_b t_b | x_a t_a) \approx \prod_{n=1}^{N} \left[\int_{-\infty}^{\infty} dx_n \right] \prod_{n=1}^{N+1} \left[\int_{-\infty}^{\infty} \frac{dp_n}{2\pi\hbar} \right] \exp \left(\frac{i}{\hbar} \mathcal{A}^N \right), \qquad (2.14)$$

where $\mathcal{A}^N$ is the sum

$$\mathcal{A}^N = \sum_{n=1}^{N+1} [p_n(x_n - x_{n-1}) - \epsilon H(p_n, x_n, t_n)]. \qquad (2.15)$$

2.1.2 Zero-Hamiltonian Path Integral

Note that the path integral (2.14) with zero Hamiltonian produces the Hilbert space structure of the theory via a chain of scalar products:

$$(x_b t_b | x_a t_a) \approx \prod_{n=1}^{N} \left[\int_{-\infty}^{\infty} dx_n \right] \prod_{n=1}^{N+1} \left[\int_{-\infty}^{\infty} \frac{dp_n}{2\pi\hbar} \right] e^{i \sum_{n=1}^{N+1} p_n(x_n - x_{n-1})/\hbar}, \qquad (2.16)$$

which is equal to

$$(x_b t_b | x_a t_a) \approx \prod_{n=1}^{N} \left[\int_{-\infty}^{\infty} dx_n \right] \prod_{n=1}^{N+1} \langle x_n | x_{n-1} \rangle = \prod_{n=1}^{N} \left[\int_{-\infty}^{\infty} dx_n \right] \prod_{n=1}^{N+1} \delta(x_n - x_{n-1})$$

$$= \delta(x_b - x_a). \qquad (2.17)$$

whose continuum limit is

$$(x_b t_b | x_a t_a) = \int \mathcal{D}x \int \frac{\mathcal{D}p}{2\pi\hbar} e^{i \int dt p(t)\dot{x}(t)/\hbar} = \langle x_b | x_a \rangle = \delta(x_b - x_a). \qquad (2.18)$$

In the operator expression (2.2), the right-hand side follows from the fact that for zero Hamiltonian the time evolution operators $\hat{U}(t_n, t_{n-1})$ are all equal to unity.

At this point we make the important observation that a momentum variable p_n *inside* the product of momentuma integrations in the expression (2.16) can be generated by a derivative $\hat{p}_n \equiv -i\hbar\partial_{x_n}$ *outside* of it. In Subsection 2.1.4 we shall go to the *continuum limit* of time slicing in which the slice thickness ϵ goes to zero. In this limit, the discrete variables x_n and p_n become functions $x(t)$ and $p(t)$ of the continuous time t, and the momenta p_n become differential operators $p(t) = -i\hbar\partial_{x(t)}$, satisfying the commutation relations with $x(t)$:

$$[\hat{p}(t), x(t)] = -i\hbar. \tag{2.19}$$

These are the canonical *equal-time commutation relations* of Heisenberg.

This observation forms the basis for deriving, from the path integral (2.14), the Schrödinger equation for the time evolution amplitude.

2.1.3 Schrödinger Equation for Time Evolution Amplitude

Let us split from the product of integrals in (2.14) the final time slice as a factor, so that we obtain the recursion relation

$$(x_b t_b | x_a t_a) \approx \int_{-\infty}^{\infty} dx_N (x_b t_b | x_N t_N)\,(x_N t_N | x_a t_a), \tag{2.20}$$

where

$$(x_b t_b | x_N t_N) \approx \int_{-\infty}^{\infty} \frac{dp_b}{2\pi\hbar} e^{(i/\hbar)[p_b(x_b - x_N) - \epsilon H(p_b, x_b, t_b)]}. \tag{2.21}$$

The momentum p_b *inside* the integral can be generated by a differential operator $\hat{p}_b \equiv -i\hbar\partial_{x_b}$ *outside* of it. The same is true for any function of p_b, so that the Hamiltonian can be moved before the momentum integral yielding

$$(x_b t_b | x_N t_N) \approx e^{-i\epsilon H(-i\hbar\partial_{x_b}, x_b, t_b)/\hbar} \int_{-\infty}^{\infty} \frac{dp_b}{2\pi\hbar} e^{ip_b(x_b - x_N)/\hbar} = e^{-i\epsilon H(-i\hbar\partial_{x_b}, x_b, t_b)/\hbar}\delta(x_b - x_N). \tag{2.22}$$

Inserting this back into (2.20) we obtain

$$(x_b t_b | x_a t_a) \approx e^{-i\epsilon H(-i\hbar\partial_{x_b}, x_b, t_b)/\hbar}(x_b\, t_b - \epsilon | x_a t_a), \tag{2.23}$$

or

$$\frac{1}{\epsilon}\left[(x_b\, t_b + \epsilon | x_a t_a) - (x_b t_b | x_a t_a)\right] \approx \frac{1}{\epsilon}\left[e^{-i\epsilon H(-i\partial_{x_b}, x_b, t_b + \epsilon)/\hbar} - 1\right](x_b t_b | x_a t_a). \tag{2.24}$$

In the limit $\epsilon \to 0$, this goes over into the differential equation for the time evolution amplitude

$$i\hbar\partial_{t_b}(x_b t_b | x_a t_a) = H(-i\hbar\partial_{x_b}, x_b, t_b)(x_b\, t_b | x_a t_a), \tag{2.25}$$

which is precisely the Schrödinger equation (1.297) of operator quantum mechanics.

2.1.4 Convergence of of the Time-Sliced Evolution Amplitude

Some remarks are necessary concerning the convergence of the time-sliced expression (2.14) to the quantum-mechanical amplitude in the continuum limit, where the thickness of the time slices $\epsilon = (t_b - t_a)/(N + 1) \to 0$ goes to zero and the number N of slices tends to ∞. This convergence can be proved for the standard kinetic energy $T = p^2/2M$ only if the potential $V(x,t)$ is sufficiently *smooth*. For time-independent potentials this is a consequence of the *Trotter product formula* which reads

$$e^{-i(t_b-t_a)\hat{H}/\hbar} = \lim_{N\to\infty} \left(e^{-i\epsilon\hat{V}/\hbar} e^{-i\epsilon\hat{T}/\hbar} \right)^{N+1}. \tag{2.26}$$

If T and V are c-numbers, this is trivially true. If they are operators, we use Eq. (2.9) to rewrite the left-hand side of (2.26) as

$$e^{-i(t_b-t_a)\hat{H}/\hbar} \equiv \left(e^{-i\epsilon(\hat{T}+\hat{V})/\hbar} \right)^{N+1} \equiv \left(e^{-i\epsilon\hat{V}/\hbar} e^{-i\epsilon\hat{T}/\hbar} e^{-i\epsilon^2\hat{X}/\hbar^2} \right)^{N+1}.$$

The Trotter formula implies that the commutator term $\hat{X}$ proportional to ϵ^2 does not contribute in the limit $N \to \infty$. The mathematical conditions ensuring this require functional analysis too technical to be presented here (for details, see the literature quoted at the end of the chapter). For us it is sufficient to know that the Trotter formula holds for operators which are bounded from below and that for most physically interesting potentials, it cannot be used to derive Feynman's time-sliced path integral representation (2.14), even in systems where the formula is known to be valid. In particular, the short-time amplitude may be different from (2.13). Take, for example, an attractive Coulomb potential $V(x) \propto -1/|x|$ for which the Trotter formula has been proved to be valid. Feynman's time-sliced formula, however, diverges even for two time slices. This will be discussed in detail in Chapter 12. Similar problems will be found for other physically relevant potentials such as $V(x) \propto l(l + D - 2)\hbar^2/|x|^2$ (centrifugal barrier) and $V(\theta) \propto m^2\hbar^2/\sin^2\theta$ (angular barrier near the poles of a sphere). In all these cases, the commutators in the expansion (2.10) of $\hat{X}$ become more and more singular. In fact, as we shall see, the expansion does not even converge, even for an infinitesimally small ϵ. All atomic systems contain such potentials and the Feynman formula (2.14) cannot be used to calculate an approximation for the transition amplitude. A new path integral formula has to be found. This will be done in Chapter 12. Fortunately, it is possible to eventually reduce the more general formula via some transformations back to a Feynman type formula with a bounded potential in an auxiliary space. Thus the above derivation of Feynman's formula for such potentials will be sufficient for the further development in this book. After this it serves as an *independent* starting point for all further quantum-mechanical calculations.

In the sequel, the symbol $\approx$ in all time-sliced formulas such as (2.14) will imply that an equality emerges in the *continuum limit* $N \to \infty$, $\epsilon \to 0$ unless the potential has singularities of the above type. In the action, the continuum limit is without subtleties. The sum $\mathcal{A}^N$ in (2.15) tends towards the integral

$$\mathcal{A}[p, x] = \int_{t_a}^{t_b} dt \, [p(t)\dot{x}(t) - H(p(t), x(t), t)] \tag{2.27}$$

under quite general circumstances. This expression is recognized as the classical canonical action for the path $x(t), p(t)$ in phase space. Since the position variables x_{N+1} and x_0 are fixed at their initial and final values x_b and x_a, the paths satisfy the boundary condition $x(t_b) = x_b$, $x(t_a) = x_a$.

In the same limit, the product of infinitely many integrals in (2.14) will be called a *path integral*. The limiting measure of integration is written as

$$\lim_{N\to\infty} \prod_{n=1}^{N} \left[\int_{-\infty}^{\infty} dx_n\right] \prod_{n=1}^{N+1} \left[\int_{-\infty}^{\infty} \frac{dp_n}{2\pi\hbar}\right] \equiv \int_{x(t_a)=x_a}^{x(t_b)=x_b} \mathcal{D}'x \int \frac{\mathcal{D}p}{2\pi\hbar}. \tag{2.28}$$

By definition, there is always one more p_n-integral than x_n-integrals in this product. While x_0 and x_{N+1} are held fixed and the x_n-integrals are done for $n = 1, \ldots, N$, each pair (x_n, x_{n-1}) is accompanied by one p_n-integral for $n = 1, \ldots, N+1$. The situation is recorded by the prime on the functional integral $\mathcal{D}'x$. With this definition, the amplitude can be written in the short form

$$(x_b t_b | x_a t_a) = \int_{x(t_a)=x_a}^{x(t_b)=x_b} \mathcal{D}'x \int \frac{\mathcal{D}p}{2\pi\hbar} e^{i\mathcal{A}[p,x]/\hbar}. \tag{2.29}$$

The path integral has a simple intuitive interpretation: Integrating over all paths corresponds to summing over all histories along which a physical system can possibly evolve. The exponential $e^{i\mathcal{A}[p,x]/\hbar}$ is the quantum analog of the Boltzmann factor $e^{-E/k_B T}$ in statistical mechanics. Instead of an exponential probability, a pure phase factor is assigned to each possible history: The total amplitude for going from x_a, t_a to x_b, t_b is obtained by adding up the phase factors for all these histories,

$$(x_b t_b | x_a t_a) = \sum_{\substack{\text{all histories} \\ (x_a, t_a) \rightsquigarrow (x_b, t_b)}} e^{i\mathcal{A}[p,x]/\hbar}, \tag{2.30}$$

where the sum comprises all paths in phase space with fixed endpoints x_b, x_a in x-space.

2.1.5 Time Evolution Amplitude in Momentum Space

The above observed asymmetry in the functional integrals over x and p is a consequence of keeping the endpoints fixed in *position space*. There exists the possibility of proceeding in a conjugate way keeping the initial and final *momenta* p_b and p_a fixed. The associated time evolution amplitude can be derived by going through the same steps as before but working in the momentum space representation of the Hilbert space, starting from the matrix elements of the time evolution operator

$$(p_b t_b | p_a t_a) \equiv \langle p_b | \hat{U}(t_b, t_a) | p_a \rangle. \tag{2.31}$$

The time slicing proceeds as in (2.2)–(2.4), with all x's replaced by p's, except in the completeness relation (2.3) which we shall take as

$$\int_{-\infty}^{\infty} \frac{dp}{2\pi\hbar} |p\rangle\langle p| = 1, \qquad (2.32)$$

corresponding to the choice of the normalization of states [compare (1.186)]

$$\langle p_b | p_a \rangle = 2\pi\hbar\delta(p_b - p_a). \qquad (2.33)$$

In the resulting product of integrals, the integration measure has an opposite asymmetry: there is now one more x_n-integral than p_n-integrals. The sliced path integral reads

$$(p_b t_b | p_a t_a) \approx \prod_{n=1}^{N} \left[\int_{-\infty}^{\infty} \frac{dp_n}{2\pi\hbar} \right] \prod_{n=0}^{N} \left[\int_{-\infty}^{\infty} dx_n \right]$$

$$\times \exp\left\{ \frac{i}{\hbar} \sum_{n=0}^{N} [-x_n(p_{n+1} - p_n) - \epsilon H(p_n, x_n, t_n)] \right\}. \qquad (2.34)$$

The relation between this and the x-space amplitude (2.14) is simple: By taking in (2.14) the first and last integrals over p_1 and p_{N+1} out of the product, renaming them as p_a and p_b, and rearranging the sum $\sum_{n=1}^{N+1} p_n(x_n - x_{n-1})$ as follows

$$\sum_{n=1}^{N+1} p_n(x_n - x_{n-1}) = p_{N+1}(x_{N+1} - x_N) + p_N(x_N - x_{N-1}) + \dots$$
$$\dots + p_2(x_2 - x_1) + p_1(x_1 - x_0)$$

$$= p_{N+1}x_{N+1} - p_1 x_0$$
$$- (p_{N+1} - p_N)x_N - (p_N - p_{N-1})x_{N-1} - \dots - (p_2 - p_1)x_1$$

$$= p_{N+1}x_{N+1} - p_1 x_0 - \sum_{n=1}^{N} (p_{n+1} - p_n)x_n, \qquad (2.35)$$

the remaining product of integrals looks as in Eq. (2.34), except that the lowest index n is one unit larger than in the sum in Eq. (2.34). In the limit $N \to \infty$ this does not matter, and we obtain the Fourier transform

$$(x_b t_b | x_a t_a) = \int \frac{dp_b}{2\pi\hbar} e^{ip_b x_b/\hbar} \int \frac{dp_a}{2\pi\hbar} e^{-ip_a x_a/\hbar} (p_b t_b | p_a t_a). \qquad (2.36)$$

The inverse relation is

$$(p_b t_b | p_a t_a) = \int dx_b \, e^{-ip_b x_b/\hbar} \int dx_a \, e^{ip_a x_a/\hbar} (x_b t_b | x_a t_a). \qquad (2.37)$$

In the continuum limit, the amplitude (2.34) can be written as a path integral

$$(p_b t_b | p_a t_a) = \int_{p(t_a)=p_a}^{p(t_b)=p_b} \frac{\mathcal{D}'p}{2\pi\hbar} \int \mathcal{D}x \, e^{i\bar{A}[p,x]/\hbar}, \qquad (2.38)$$

where

$$\bar{A}[p, x] = \int_{t_a}^{t_b} dt \left[-\dot{p}(t)x(t) - H(p(t), x(t), t) \right] = A[p, x] - p_b x_b + p_a x_a. \qquad (2.39)$$

2.1.6   Quantum-Mechanical Partition Function

A path integral symmetric in p and x arises when considering the quantum-mechanical partition function defined by the trace (recall Section 1.17)

$$Z_{\mathrm{QM}}(t_b, t_a) = \mathrm{Tr}\left(e^{-i(t_b-t_a)\hat{H}/\hbar}\right). \tag{2.40}$$

In the local basis, the trace becomes an integral over the amplitude $(x_b t_b | x_a t_a)$ with $x_b = x_a$:

$$Z_{\mathrm{QM}}(t_b, t_a) = \int_{-\infty}^{\infty} dx_a (x_a t_b | x_a t_a). \tag{2.41}$$

The additional trace integral over $x_{N+1} \equiv x_0$ makes the path integral for Z_{QM} symmetric in p_n and x_n:

$$\int_{-\infty}^{\infty} dx_{N+1} \prod_{n=1}^{N} \left[\int_{-\infty}^{\infty} dx_n\right] \prod_{n=1}^{N+1} \left[\int_{-\infty}^{\infty} \frac{dp_n}{2\pi\hbar}\right] = \prod_{n=1}^{N+1} \left[\iint_{-\infty}^{\infty} \frac{dx_n dp_n}{2\pi\hbar}\right]. \tag{2.42}$$

In the continuum limit, the right-hand side is written as

$$\lim_{N\to\infty} \prod_{n=1}^{N+1} \left[\iint_{-\infty}^{\infty} \frac{dx_n dp_n}{2\pi\hbar}\right] \equiv \oint \mathcal{D}x \int \frac{\mathcal{D}p}{2\pi\hbar}, \tag{2.43}$$

and the measures are related by

$$\int_{-\infty}^{\infty} dx_a \int_{x(t_a)=x_a}^{x(t_b)=x_a} \mathcal{D}'x \int \frac{\mathcal{D}p}{2\pi\hbar} \equiv \oint \mathcal{D}x \int \frac{\mathcal{D}p}{2\pi\hbar}. \tag{2.44}$$

The symbol $\oint$ indicates the periodic boundary condition $x(t_a) = x(t_b)$. In the momentum representation we would have similarly

$$\int_{-\infty}^{\infty} \frac{dp_a}{2\pi\hbar} \int_{p(t_a)=p_a}^{p(t_b)=p_a} \frac{\mathcal{D}'p}{2\pi\hbar} \int \mathcal{D}x \equiv \oint \frac{\mathcal{D}p}{2\pi\hbar} \int \mathcal{D}x, \tag{2.45}$$

with the periodic boundary condition $p(t_a) = p(t_b)$, and the same right-hand side. Hence, the quantum-mechanical partition function is given by the path integral

$$Z_{\mathrm{QM}}(t_b, t_a) = \oint \mathcal{D}x \int \frac{\mathcal{D}p}{2\pi\hbar} e^{i\mathcal{A}[p,x]/\hbar} = \oint \frac{\mathcal{D}p}{2\pi\hbar} \int \mathcal{D}x\, e^{i\bar{\mathcal{A}}[p,x]/\hbar}. \tag{2.46}$$

In the right-hand exponential, the action $\bar{\mathcal{A}}[p, x]$ can be replaced by $\mathcal{A}[p, x]$, since the extra terms in (2.39) are removed by the periodic boundary conditions. In the time-sliced expression, the equality is easily derived from the rearrangement of the sum (2.35), which shows that

$$\left.\sum_{n=1}^{N+1} p_n(x_n - x_{n-1})\right|_{x_{N+1}=x_0} = \left.-\sum_{n=0}^{N} (p_{n+1} - p_n)x_n\right|_{p_{N+1}=p_0}. \tag{2.47}$$

In the path integral expression (2.46) for the partition function, the rules of quantum mechanics appear as a natural generalization of the rules of classical statistical mechanics, as formulated by Planck. According to these rules, each volume element in phase space $dx dp/h$ is occupied with the exponential probability $e^{-E/k_B T}$. In the path integral formulation of quantum mechanics, each volume element in the *path phase space* $\prod_n dx(t_n) dp(t_n)/h$ is associated with a pure phase factor $e^{i\mathcal{A}[p,x]/\hbar}$. We see here a manifestation of the correspondence principle which specifies the transition from classical to quantum mechanics. In path integrals, it looks somewhat more natural than in the historic formulation, where it requires the replacement of all classical phase space variables p, x by operators, a rule which was initially hard to comprehend.

2.1.7 Feynman's Configuration Space Path Integral

Actually, in his original paper, Feynman did not give the path integral formula in the above phase space formulation. Since the kinetic energy in (2.7) has usually the form $T(p, t) = p^2/2M$, he focused his attention upon the Hamiltonian

$$H = \frac{p^2}{2M} + V(x, t), \tag{2.48}$$

for which the time-sliced action (2.15) becomes

$$\mathcal{A}^N = \sum_{n=1}^{N+1} \left[p_n(x_n - x_{n-1}) - \epsilon \frac{p_n^2}{2M} - \epsilon V(x_n, t_n) \right]. \tag{2.49}$$

It can be quadratically completed to

$$\mathcal{A}^N = \sum_{n=1}^{N+1} \left[-\frac{\epsilon}{2M} \left(p_n - \frac{x_n - x_{n-1}}{\epsilon} M \right)^2 + \frac{M}{2} \epsilon \left(\frac{x_n - x_{n-1}}{\epsilon} \right)^2 - \epsilon V(x_n, t_n) \right]. \tag{2.50}$$

The momentum integrals in (2.14) may then be performed using the Fresnel integral formula (1.333), yielding

$$\int_{-\infty}^{\infty} \frac{dp_n}{2\pi\hbar} \exp\left[-\frac{i}{\hbar} \frac{\epsilon}{2M} \left(p_n - M \frac{x_n - x_{n-1}}{\epsilon} \right)^2 \right] = \frac{1}{\sqrt{2\pi\hbar i\epsilon/M}}, \tag{2.51}$$

and we arrive at the alternative representation

$$(x_b t_b | x_a t_a) \approx \frac{1}{\sqrt{2\pi\hbar i\epsilon/M}} \prod_{n=1}^{N} \left[\int_{-\infty}^{\infty} \frac{dx_n}{\sqrt{2\pi\hbar i\epsilon/M}} \right] \exp\left(\frac{i}{\hbar} \mathcal{A}^N \right), \tag{2.52}$$

where $\mathcal{A}^N$ is now the sum

$$\mathcal{A}^N = \epsilon \sum_{n=1}^{N+1} \left[\frac{M}{2} \left(\frac{x_n - x_{n-1}}{\epsilon} \right)^2 - V(x_n, t_n) \right], \tag{2.53}$$

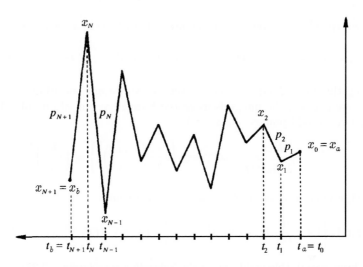

Figure 2.1   Zigzag paths, along which a point particle explores all possible ways of reaching the point x_b at a time t_b, starting from x_a at t_a. The time axis is drawn from right to left to have the same direction as the operator order in Eq. (2.2).

with $x_{N+1} = x_b$ and $x_0 = x_a$. Here the integrals run over all paths in *configuration space* rather than phase space. They account for the fact that a quantum-mechanical particle starting from a given initial point x_a will explore all possible ways of reaching a given final point x_b. The amplitude of each path is $\exp(i\mathcal{A}^N/\hbar)$. See Fig. 2.1 for a geometric illustration of the path integration. In the continuum limit, the sum (2.53) converges towards the action in the Lagrangian form:

$$\mathcal{A}[x] = \int_{t_a}^{t_b} dt\, L(x, \dot{x}) = \int_{t_a}^{t_b} dt\left[\frac{M}{2}\dot{x}^2 - V(x,t)\right]. \tag{2.54}$$

Note that this action is a local functional of $x(t)$ in the temporal sense as defined in Eq. (1.27).[2]

For the time-sliced Feynman path integral, one verifies the Schrödinger equation as follows: As in (2.20), one splits off the last slice as follows:

$$
\begin{aligned}
(x_b t_b | x_a t_a) &\approx \int_{-\infty}^{\infty} dx_N\, (x_b t_b | x_N t_N)\,(x_N t_N | x_a t_a) \\
&= \int_{-\infty}^{\infty} d\Delta x\, (x_b t_b | x_b - \Delta x\, t_b - \epsilon)\,(x_b - \Delta x\, t_b - \epsilon | x_a t_a), \tag{2.55}
\end{aligned}
$$

where

$$(x_b t_b | x_b - \Delta x\, t_b - \epsilon) \approx \frac{1}{\sqrt{2\pi\hbar i\epsilon/M}} \exp\left\{\epsilon\frac{i}{\hbar}\left[\frac{M}{2}\left(\frac{\Delta x}{\epsilon}\right)^2 - V(x_b, t_b)\right]\right\}. \tag{2.56}$$

---

[2]A functional $F[x]$ is called local if it can be written as an integral $\int dt f(x(t), \dot{x}(t))$; it is called *ultra-local* if it has the form $\int dt f(x(t))$.

We now expand the amplitude in the integral of (2.55) in a Taylor series

$$(x_b - \Delta x \, t_b - \epsilon | x_a t_a) = \left[1 - \Delta x \, \partial_{x_b} + \frac{1}{2} (\Delta x)^2 \partial_{x_b}^2 + \ldots \right] (x_b, t_b - \epsilon | x_a t_a). \quad (2.57)$$

Inserting this into (2.55), the odd powers of Δx do not contribute. For the even powers, we perform the integrals using the Fresnel version of formula (1.335), and obtain zero for odd powers of Δx, and

$$\int_{-\infty}^{\infty} \frac{d\Delta x}{\sqrt{2\pi\hbar i\epsilon/M}} (\Delta x)^{2n} \exp\left\{ \epsilon \frac{i}{\hbar} \frac{M}{2} \left(\frac{\Delta x}{\epsilon} \right)^2 \right\} = \left(i \frac{\hbar\epsilon}{M} \right)^n \quad (2.58)$$

for even powers, so that the integral in (2.55) becomes

$$(x_b t_b | x_a t_a) = \left[1 + \epsilon \frac{i\hbar}{2M} \partial_{x_b}^2 + \mathcal{O}(\epsilon^2) \right] \left[1 - \epsilon \frac{i}{\hbar} V(x_b, t_b) + \mathcal{O}(\epsilon^2) \right] (x_b, t_b - \epsilon | x_a t_a). \quad (2.59)$$

In the limit $\epsilon \to 0$, this yields again the Schrödinger equation. (2.23).

In the continuum limit, we write the amplitude (2.52) as a path integral

$$(x_b t_b | x_a t_a) \equiv \int_{x(t_a)=x_a}^{x(t_b)=x_b} \mathcal{D}x \, e^{i\mathcal{A}[x]/\hbar}. \quad (2.60)$$

This is Feynman's original formula for the quantum-mechanical amplitude (2.1). It consists of a sum over all paths in configuration space with a phase factor containing the form of the action $\mathcal{A}[x]$.

We have used the same measure symbol $\mathcal{D}x$ for the paths in configuration space as for the completely different paths in phase space in the expressions (2.29), (2.38), (2.44), (2.45). There should be no danger of confusion. Note that the extra dp_n-integration in the phase space formula (2.14) results now in one extra $1/\sqrt{2\pi\hbar i\epsilon/M}$ factor in (2.52) which is not accompanied by a dx_n-integration.

The Feynman amplitude can be used to calculate the quantum-mechanical partition function (2.41) as a configuration space path integral

$$Z_{\mathrm{QM}} = \oint \mathcal{D}x \, e^{i\mathcal{A}[x]/\hbar}. \quad (2.61)$$

As in (2.43), (2.44), the symbol $\oint \mathcal{D}x$ indicates that the paths have equal endpoints $x(t_a) = x(t_b)$, the path integral being the continuum limit of the product of integrals

$$\oint \mathcal{D}x \approx \prod_{n=1}^{N+1} \int_{-\infty}^{\infty} \frac{dx_n}{\sqrt{2\pi i\hbar\epsilon/M}}. \quad (2.62)$$

There is no extra $1/\sqrt{2\pi i\hbar\epsilon/M}$ factor as in (2.52) and (2.60), due to the integration over the initial (= final) position $x_b = x_a$ representing the quantum-mechanical trace. The use of the same symbol $\oint \mathcal{D}x$ as in (2.46) should not cause any confusion since (2.46) is always accompanied by an integral $\int \mathcal{D}p$.

For the sake of generality we might point out that it is not necessary to slice the time axis in an equidistant way. In the continuum limit $N \to \infty$, the canonical path integral (2.14) is *indifferent* to the choice of the infinitesimal spacings

$$\epsilon_n = t_n - t_{n-1}. \tag{2.63}$$

The configuration space formula contains the different spacings ϵ_n in the following way: When performing the p_n integrations, we obtain a formula of the type (2.52), with each ϵ replaced by ϵ_n, i.e.,

$$(x_b t_b | x_a t_a) \approx \frac{1}{\sqrt{2\pi\hbar i\epsilon_b/M}} \prod_{n=1}^{N} \left[\int_{-\infty}^{\infty} \frac{dx_n}{\sqrt{2\pi i\hbar\epsilon_n/M}} \right]$$
$$\times \exp\left\{ \frac{i}{\hbar} \sum_{n=1}^{N+1} \left[\frac{M}{2} \frac{(x_n - x_{n-1})^2}{\epsilon_n} - \epsilon_n V(x_n, t_n) \right] \right\}. \tag{2.64}$$

To end this section, an important remark is necessary: It would certainly be possible to *define* the path integral for the time evolution amplitude (2.29), without going through Feynman's time-slicing procedure, as the solution of the Schrödinger differential equation [see Eq. (1.304))]:

$$[\hat{H}(-i\hbar\partial_x, x) - i\hbar\partial_t](x\,t|x_a t_a) = -i\hbar\delta(t - t_a)\delta(x - x_a). \tag{2.65}$$

If one possesses an orthonormal and complete set of wave functions $\psi_n(x)$ solving the time-independent Schrödinger equation $\hat{H}\psi_n(x) = E_n\psi_n(x)$, this solution is given by the spectral representation (1.319)

$$(x_b t_b | x_a t_a) = \Theta(t_b - t_a) \sum_n \psi_n(x_b)\psi_n^*(x_a) e^{-iE_n(t_b - t_a)/\hbar}, \tag{2.66}$$

where $\Theta(t)$ is the Heaviside function (1.300). This definition would, however, run contrary to the very purpose of Feynman's path integral approach, which is to understand a quantum system from the global all-time fluctuation point of view. The goal is to find all properties from the globally defined time evolution amplitude, in particular the Schrödinger wave functions.[3] The global approach is usually more complicated than Schrödinger's and, as we shall see in Chapters 8 and 12–14, contains novel subtleties caused by the finite time slicing. Nevertheless, it has at least four important advantages. First, it is conceptually very attractive by formulating a quantum theory without operators which describe quantum fluctuations by close analogy with thermal fluctuations (as will be seen later in this chapter). Second, it links quantum mechanics smoothly with classical mechanics (as will be shown in Chapter 4). Third, it offers new variational procedures for the approximate study of complicated quantum-mechanical and -statistical systems (see Chapter 5). Fourth,

---

[3]Many publications claiming to have solved the path integral of a system have violated this rule by implicitly using the Schrödinger equation, although camouflaged by complicated-looking path integral notation.

it gives a natural geometric access to the dynamics of particles in spaces with curvature and torsion (see Chapters 10–11). This has recently led to results where the operator approach has failed due to operator-ordering problems, giving rise to a unique and correct description of the quantum dynamics of a particle in spaces with curvature and torsion. From this it is possible to derive a unique extension of Schrödinger's theory to such general spaces whose predictions can be tested in future experiments.[4]

2.2 Exact Solution for the Free Particle

In order to develop some experience with Feynman's path integral formula we consider in detail the simplest case of a free particle, which in the canonical form reads

$$(x_b t_b | x_a t_a) = \int_{x(t_a)=x_a}^{x(t_b)=x_b} \mathcal{D}'x \int \frac{\mathcal{D}p}{2\pi\hbar} \exp\left[\frac{i}{\hbar}\int_{t_a}^{t_b} dt \left(p\dot{x} - \frac{p^2}{2M}\right)\right], \qquad (2.67)$$

and in the pure configuration form:

$$(x_b t_b | x_a t_a) = \int_{x(t_a)=x_a}^{x(t_b)=x_b} \mathcal{D}x \exp\left[\frac{i}{\hbar}\int_{t_a}^{t_b} dt \frac{M}{2}\dot{x}^2\right]. \qquad (2.68)$$

Since the integration limits are obvious by looking at the left-hand sides of the equations, they will be omitted from now on, unless clarity requires their specification.

2.2.1 Direct Solution

The problem is solved most easily in the configuration form. The time-sliced expression to be integrated is given by Eqs. (2.52), (2.53) where we have to set $V(x) = 0$. The resulting product of Gaussian integrals can easily be done successively using formula (1.333), from which we derive the simple rule

$$\int dx' \frac{1}{\sqrt{2\pi i\hbar A\epsilon/M}} \exp\left[\frac{i}{\hbar}\frac{M}{2}\frac{(x''-x')^2}{A\epsilon}\right] \frac{1}{\sqrt{2\pi i\hbar B\epsilon/M}} \exp\left[\frac{i}{\hbar}\frac{M}{2}\frac{(x'-x)^2}{B\epsilon}\right]$$

$$= \frac{1}{\sqrt{2\pi i\hbar(A+B)\epsilon/M}} \exp\left[\frac{i}{\hbar}\frac{M}{2}\frac{(x''-x)^2}{(A+B)\epsilon}\right], \qquad (2.69)$$

which leads directly to the free-particle amplitude

$$(x_b t_b | x_a t_a) = \frac{1}{\sqrt{2\pi i\hbar(N+1)\epsilon/M}} \exp\left[\frac{i}{\hbar}\frac{M}{2}\frac{(x_b - x_a)^2}{(N+1)\epsilon}\right]. \qquad (2.70)$$

After inserting $(N+1)\epsilon = t_b - t_a$, this reads

$$(x_b t_b | x_a t_a) = \frac{1}{\sqrt{2\pi i\hbar(t_b - t_a)/M}} \exp\left[\frac{i}{\hbar}\frac{M}{2}\frac{(x_b - x_a)^2}{t_b - t_a}\right]. \qquad (2.71)$$

---

[4]H. Kleinert, Mod. Phys. Lett. A *4*, 2329 (1989) (http://www.physik.fu-berlin.de/~kleinert/199); Phys. Lett. B *236*, 315 (1990) (*ibid.*http/202).

Note that the free-particle amplitude happens to be independent of the number $N + 1$ of time slices. The amplitude (2.71) agrees, of course, with the Schrödinger result (1.337) for $D = 1$.

The calculation shows that the path integrals (2.67) and (2.68) possess a simple solvable generalization to a time-dependent mass $M(t) = Mg(t)$:

$$(x_b t_b | x_a t_a) = \int_{x(t_a)=x_a}^{x(t_b)=x_b} \mathcal{D}'x \int \frac{\mathcal{D}p}{2\pi\hbar} \exp\left[\frac{i}{\hbar} \int_{t_a}^{t_b} dt \left(p\dot{x} - \frac{p^2}{2Mg(t)}\right)\right], \qquad (2.72)$$

and in the pure configuration form:

$$(x_b t_b | x_a t_a) = \int_{x(t_a)=x_a}^{x(t_b)=x_b} \mathcal{D}x\sqrt{g} \exp\left[\frac{i}{\hbar} \int_{t_a}^{t_b} dt \frac{M}{2} g(t)\dot{x}^2(t)\right]. \qquad (2.73)$$

Here the measure of integration $\int \mathcal{D}x\sqrt{g}$ symbolizes the continuum limit of the product [compare (2.52)]:

$$\int_{x(t_a)=x_a}^{x(t_b)=x_b} \mathcal{D}x\sqrt{g} \equiv \frac{1}{\sqrt{2\pi\hbar i\epsilon/Mg(t_b)}} \prod_{n=1}^{N} \left[\int_{-\infty}^{\infty} \frac{dx_n}{\sqrt{2\pi\hbar i\epsilon/Mg(t_n)}}\right]. \qquad (2.74)$$

The factor $g(t_n)$ enters in each of the integrations (2.69), where the previous time slicing parameters ϵ becomes now $\epsilon_n = \epsilon/g(t_n)$, and we find instead of (2.71) the amplitude

$$(x_b t_b | x_a t_a) = \frac{1}{\sqrt{2\pi i\hbar M^{-1} \int_{t_a}^{t_b} g^{-1}(t)}} \exp\left[\frac{i}{\hbar} \frac{M}{2} \frac{(x_b - x_a)^2}{\int_{t_a}^{t_b} g^{-1}(t)}\right]. \qquad (2.75)$$

This has the Fourier representation

$$(x_b t_b | x_a t_a) = \int \frac{dp}{2\pi\hbar} \exp\left\{\frac{i}{\hbar}\left[ip(x_b - x_a) - \frac{p^2}{2M} \int_{t_a}^{t_b} g^{-1}(t)\right]\right\}. \qquad (2.76)$$

2.2.2   Fluctuations around the Classical Path

There exists another method of calculating this amplitude which is somewhat more involved than the simple case at hand, but which turns out to be useful for the treatment of a certain class of nontrivial path integrals, after a suitable generalization. This method is based on all paths with respect to the classical path, i.e., all paths are split into the classical path

$$x_{\rm cl}(t) = x_a + \frac{x_b - x_a}{t_b - t_a}(t - t_a), \qquad (2.77)$$

along which the free particle would run following the equation of motion

$$\ddot{x}_{\rm cl}(t) = 0, \qquad (2.78)$$

plus deviations $\delta x(t)$:

$$x(t) = x_{\text{cl}}(t) + \delta x(t). \tag{2.79}$$

Since the initial and final points are fixed at x_a, x_b, respectively, the deviations vanish at the endpoints:

$$\delta x(t_a) = \delta x(t_b) = 0. \tag{2.80}$$

The deviations $\delta x(t)$ are referred to as the *quantum fluctuations* of the particle orbit. In mathematics, the boundary conditions (2.80) are referred to as *Dirichlet boundary conditions*. When inserting the decomposition (2.79) into the action we observe that due to the equation of motion (2.78) for the classical path, the action separates into the sum of a classical and a purely quadratic fluctuation term

$$
\begin{aligned}
\frac{M}{2} \int_{t_a}^{t_b} dt \, &\left\{ \dot{x}_{\text{cl}}^2(t) + 2\dot{x}_{\text{cl}}(t)\delta\dot{x}(t) + [\delta\dot{x}(t)]^2 \right\} \\
&= \frac{M}{2} \int_{t_a}^{t_b} dt \dot{x}_{\text{cl}}^2 + M\dot{x}\delta x \Big|_{t_a}^{t_b} - M \int_{t_a}^{t_b} dt \ddot{x}_{\text{cl}}\delta x + \frac{M}{2} \int_{t_a}^{t_b} dt(\delta\dot{x})^2 \\
&= \frac{M}{2} \left[\int_{t_a}^{t_b} dt \dot{x}_{\text{cl}}^2 + \int_{t_a}^{t_b} dt(\delta\dot{x})^2 \right].
\end{aligned}
$$

The absence of a mixed term is a general consequence of the extremality property of the classical path,

$$\delta\mathcal{A}\Big|_{x(t)=x_{\text{cl}}(t)} = 0. \tag{2.81}$$

It implies that a quadratic *fluctuation expansion* around the classical action

$$\mathcal{A}_{\text{cl}} \equiv \mathcal{A}[x_{\text{cl}}] \tag{2.82}$$

can have no linear term in $\delta x(t)$, i.e., it must start as

$$\mathcal{A} = \mathcal{A}_{\text{cl}} + \frac{1}{2} \int_{t_a}^{t_b} dt \int_{t_a}^{t_b} dt' \frac{\delta^2\mathcal{A}}{\delta x(t)\delta x(t')} \delta x(t)\delta x(t') \Big|_{x(t)=x_{\text{cl}}(t)} + \dots . \tag{2.83}$$

With the action being a sum of two terms, the amplitude factorizes into the product of a classical amplitude $e^{i\mathcal{A}_{\text{cl}}/\hbar}$ and a fluctuation factor $F_0(t_b - t_a)$,

$$(x_b t_b | x_a t_a) = \int \mathcal{D}x \, e^{i\mathcal{A}[x]/\hbar} = e^{i\mathcal{A}_{\text{cl}}/\hbar} F_0(t_b, t_a). \tag{2.84}$$

For the free particle with the classical action

$$\mathcal{A}_{\text{cl}} = \int_{t_a}^{t_b} dt \frac{M}{2} \dot{x}_{\text{cl}}^2, \tag{2.85}$$

the function factor $F_0(t_b - t_a)$ is given by the path integral

$$F_0(t_b - t_a) = \int \mathcal{D}\delta x(t) \exp\left[\frac{i}{\hbar} \int_{t_a}^{t_b} dt \frac{M}{2} (\delta\dot{x})^2 \right]. \tag{2.86}$$

Due to the vanishing of $\delta x(t)$ at the endpoints, this does not depend on x_b, x_a but only on the initial and final times t_b, t_a. The time translational invariance reduces this dependence further to the time difference $t_b - t_a$. The subscript zero of $F_0(t_b - t_a)$ indicates the free-particle nature of the fluctuation factor. After inserting (2.77) into (2.85), we find immediately

$$
\mathcal{A}_{\text{cl}} = \frac{M}{2} \frac{(x_b - x_a)^2}{t_b - t_a}.
\tag{2.87}
$$

The fluctuation factor, on the other hand, requires the evaluation of the multiple integral

$$
F_0^N(t_b - t_a) = \frac{1}{\sqrt{2\pi\hbar i\epsilon/M}} \prod_{n=1}^{N} \left[\int_{-\infty}^{\infty} \frac{d\delta x_n}{\sqrt{2\pi\hbar i\epsilon/M}} \right] \exp\left(\frac{i}{\hbar} \mathcal{A}_{\text{fl}}^N \right),
\tag{2.88}
$$

where $\mathcal{A}_{\text{fl}}^N$ is the time-sliced fluctuation action

$$
\mathcal{A}_{\text{fl}}^N = \frac{M}{2} \epsilon \sum_{n=1}^{N+1} \left(\frac{\delta x_n - \delta x_{n-1}}{\epsilon} \right)^2.
\tag{2.89}
$$

At the end, we have to take the continuum limit

$$
N \to \infty, \quad \epsilon = (t_b - t_a)/(N + 1) \to 0.
$$

2.2.3 Fluctuation Factor

The remainder of this section will be devoted to calculating the fluctuation factor (2.88). Before doing this, we shall develop a general technique for dealing with such time-sliced expressions. Due to the frequent appearance of the fluctuating δx-variables, we shorten the notation by omitting all δ's and working only with x-variables.

A useful device for manipulating sums on a sliced time axis such as (2.89) is the difference operator ∇ and its conjugate $\overline{\nabla}$, defined by

$$
\nabla x(t) \equiv \frac{1}{\epsilon}[x(t + \epsilon) - x(t)], \qquad \overline{\nabla} x(t) \equiv \frac{1}{\epsilon}[x(t) - x(t - \epsilon)].
\tag{2.90}
$$

They are two different discrete versions of the time derivative ∂_t, to which both reduce in the continuum limit $\epsilon \to 0$:

$$
\nabla, \overline{\nabla} \xrightarrow{\epsilon \to 0} \partial_t,
\tag{2.91}
$$

if they act upon differentiable functions. Since the discretized time axis with $N + 1$ steps constitutes a one-dimensional lattice, the difference operators $\nabla, \overline{\nabla}$ are also called *lattice derivatives*.

For the coordinates $x_n = x(t_n)$ at the discrete times t_n we write

$$
\begin{aligned}
\nabla x_n &= \frac{1}{\epsilon}(x_{n+1} - x_n), && N \geq n \geq 0, \\
\overline{\nabla} x_n &= \frac{1}{\epsilon}(x_n - x_{n-1}), && N+1 \geq n \geq 1.
\end{aligned}
\tag{2.92}
$$

The time-sliced action (2.89) can then be expressed in terms of ∇x_n or $\overline{\nabla} x_n$ as (writing x_n instead of δx_n)

$$
\mathcal{A}_{\mathrm{fl}}^N = \frac{M}{2}\epsilon \sum_{n=0}^{N}(\nabla x_n)^2 = \frac{M}{2}\epsilon \sum_{n=1}^{N+1}(\overline{\nabla} x_n)^2.
\tag{2.93}
$$

In this notation, the limit $\epsilon \to 0$ is most obvious: The sum $\epsilon \sum_n$ goes into the integral $\int_{t_a}^{t_b} dt$, whereas both $(\nabla x_n)^2$ and $(\overline{\nabla} x_n)^2$ tend to $\dot{x}^2$, so that

$$
\mathcal{A}_{\mathrm{fl}}^N \to \int_{t_a}^{t_b} dt \frac{M}{2}\dot{x}^2.
\tag{2.94}
$$

Thus, the time-sliced action becomes the Lagrangian action.

Lattice derivatives have properties quite similar to ordinary derivatives. One only has to be careful in distinguishing ∇ and $\overline{\nabla}$. For example, they allow for the useful operation *summation by parts* which is analogous to integration by parts. Recall the rule for the integration by parts

$$
\int_{t_a}^{t_b} dt\, g(t)\dot{f}(t) = g(t)f(t)\Big|_{t_a}^{t_b} - \int_{t_a}^{t_b} dt\, \dot{g}(t)f(t).
\tag{2.95}
$$

On the lattice, this relation yields for functions $f(t) \to x_n$ and $g(t) \to p_n$:

$$
\epsilon \sum_{n=1}^{N+1} p_n \overline{\nabla} x_n = p_n x_n\big|_0^{N+1} - \epsilon \sum_{n=0}^{N}(\nabla p_n)x_n.
\tag{2.96}
$$

This follows directly by rewriting (2.35).

For functions vanishing at the endpoints, i.e., for $x_{N+1} = x_0 = 0$, we can omit the surface terms and shift the range of the sum on the right-hand side to obtain the simple formula [see also Eq. (2.47)]

$$
\sum_{n=1}^{N+1} p_n \overline{\nabla} x_n = -\sum_{n=0}^{N}(\nabla p_n)x_n = -\sum_{n=1}^{N+1}(\nabla p_n)x_n.
\tag{2.97}
$$

The same thing holds if both $p(t)$ and $x(t)$ are periodic in the interval $t_b - t_a$, so that $p_0 = p_{N+1}$, $x_0 = x_{N+1}$. In this case, it is possible to shift the sum on the right-hand side by one unit arriving at the more symmetric-looking formula

$$
\sum_{n=1}^{N+1} p_n \overline{\nabla} x_n = -\sum_{n=1}^{N+1}(\nabla p_n)x_n.
\tag{2.98}
$$

In the time-sliced action (2.89) the quantum fluctuations x_n ($\hat{=}\delta x_n$) vanish at the ends, so that (2.97) can be used to rewrite

$$\sum_{n=1}^{N+1}(\nabla x_n)^2 = -\sum_{n=1}^{N} x_n \nabla \overline{\nabla} x_n. \tag{2.99}$$

In the ∇x_n -form of the action (2.93), the same expression is obtained by applying formula (2.97) from the right- to the left-hand side and using the vanishing of x_0 and x_{N+1}:

$$\sum_{n=0}^{N}(\nabla x_n)^2 = -\sum_{n=1}^{N+1} x_n \overline{\nabla}\nabla x_n = -\sum_{n=1}^{N} x_n \overline{\nabla}\nabla x_n. \tag{2.100}$$

The right-hand sides in (2.99) and (2.100) can be written in matrix form as

$$-\sum_{n=1}^{N} x_n \nabla \overline{\nabla} x_n \equiv -\sum_{n,n'=1}^{N} x_n (\nabla\overline{\nabla})_{nn'} x_{n'},$$

$$-\sum_{n=1}^{N} x_n \overline{\nabla}\nabla x_n \equiv -\sum_{n,n'=1}^{N} x_n (\overline{\nabla}\nabla)_{nn'} x_{n'}, \tag{2.101}$$

with the same $N \times N$ -matrix

$$\nabla\overline{\nabla} \equiv \overline{\nabla}\nabla \equiv \frac{1}{\epsilon^2}\begin{pmatrix} -2 & 1 & 0 & \cdots & 0 & 0 & 0 \\ 1 & -2 & 1 & \cdots & 0 & 0 & 0 \\ \vdots & & & & & & \vdots \\ 0 & 0 & 0 & \cdots & 1 & -2 & 1 \\ 0 & 0 & 0 & \cdots & 0 & 1 & -2 \end{pmatrix}. \tag{2.102}$$

This is obviously the lattice version of the double time derivative ∂_t^2, to which it reduces in the continuum limit $\epsilon \to 0$. It will therefore be called the *lattice Laplacian*.

A further common property of lattice and ordinary derivatives is that they can both be diagonalized by going to Fourier components. When decomposing

$$x(t) = \int_{-\infty}^{\infty} d\omega\, e^{-i\omega t} x(\omega), \tag{2.103}$$

and applying the lattice derivative ∇, we find

$$\nabla x(t_n) = \int_{-\infty}^{\infty} d\omega\, \frac{1}{\epsilon}\left(e^{-i\omega(t_n+\epsilon)} - e^{-i\omega t_n}\right) x(\omega) \tag{2.104}$$

$$= \int_{-\infty}^{\infty} d\omega\, e^{-i\omega t_n}\frac{1}{\epsilon}(e^{-i\omega\epsilon} - 1)\, x(\omega).$$

Hence, on the Fourier components, ∇ has the eigenvalues

$$\frac{1}{\epsilon}(e^{-i\omega\epsilon} - 1). \tag{2.105}$$

In the continuum limit $\epsilon \to 0$, this becomes the eigenvalue of the ordinary time derivative ∂_t, i.e., $-i$ times the frequency of the Fourier component ω. As a reminder of this we shall denote the eigenvalue of $i\nabla$ by Ω and have

$$(i\nabla x)(\omega) = \Omega\, x(\omega) \equiv \frac{i}{\epsilon}(e^{-i\omega\epsilon} - 1)\, x(\omega). \tag{2.106}$$

For the conjugate lattice derivative we find similarly

$$(i\overline{\nabla} x)(\omega) = \overline{\Omega}\, x(\omega) \equiv -\frac{i}{\epsilon}(e^{i\omega\epsilon} - 1)\, x(\omega), \tag{2.107}$$

where $\overline{\Omega}$ is the complex-conjugate number of Ω, i.e., $\overline{\Omega} \equiv \Omega^*$. As a consequence, the eigenvalues of the negative lattice Laplacian $-\nabla\overline{\nabla} \equiv -\overline{\nabla}\nabla$ are real and nonnegative:

$$\frac{i}{\epsilon}(e^{-i\omega\epsilon} - 1)\frac{i}{\epsilon}(1 - e^{i\omega\epsilon}) = \frac{1}{\epsilon^2}[2 - 2\cos(\omega\epsilon)] \geq 0. \tag{2.108}$$

Of course, Ω and $\overline{\Omega}$ have the same continuum limit ω.

When decomposing the quantum fluctuations $x(t)$ $[\hat{=}\delta x(t)]$ into their Fourier components, not all eigenfunctions occur. Since $x(t)$ vanishes at the initial time $t = t_a$, the decomposition can be restricted to the sine functions and we may expand

$$x(t) = \int_0^\infty d\omega \sin\omega(t - t_a)\, x(\omega). \tag{2.109}$$

The vanishing at the final time $t = t_b$ is enforced by a restriction of the frequencies ω to the discrete values

$$\nu_m = \frac{\pi m}{t_b - t_a} = \frac{\pi m}{(N + 1)\epsilon}. \tag{2.110}$$

Thus we are dealing with the Fourier series

$$x(t) = \sum_{m=1}^{\infty} \sqrt{\frac{2}{(t_b - t_a)}} \sin\nu_m(t - t_a)\, x(\nu_m) \tag{2.111}$$

with real Fourier components $x(\nu_m)$. A further restriction arises from the fact that for finite ϵ, the series has to represent $x(t)$ only at the discrete points $x(t_n)$, $n = 0,\dots,N + 1$. It is therefore sufficient to carry the sum only up to $m = N$ and to expand $x(t_n)$ as

$$x(t_n) = \sum_{m=1}^{N} \sqrt{\frac{2}{N + 1}} \sin\nu_m(t_n - t_a)\, x(\nu_m), \tag{2.112}$$

where a factor $\sqrt{\epsilon}$ has been removed from the Fourier components, for convenience. The expansion functions are orthogonal,

$$\frac{2}{N + 1} \sum_{n=1}^{N} \sin\nu_m(t_n - t_a)\, \sin\nu_{m'}(t_n - t_a) = \delta_{mm'}, \tag{2.113}$$

and complete:

$$\frac{2}{N+1} \sum_{m=1}^{N} \sin \nu_m(t_n - t_a) \; \sin \nu_m(t_{n'} - t_a) = \delta_{nn'} \tag{2.114}$$

(where $0 < m, m' < N + 1$). The orthogonality relation follows by rewriting the left-hand side of (2.113) in the form

$$\frac{2}{N+1}\frac{1}{2} \text{Re} \sum_{n=0}^{N+1} \left\{ \exp\left[\frac{i\pi(m-m')}{N+1}n\right] - \exp\left[\frac{i\pi(m+m')}{N+1}n\right] \right\}, \tag{2.115}$$

with the sum extended without harm by a trivial term at each end. Being of the geometric type, this can be calculated right away. For $m = m'$ the sum adds up to 1, while for $m \neq m'$ it becomes

$$\frac{2}{N+1}\frac{1}{2}\text{Re}\left[\frac{1 - e^{i\pi(m-m')}e^{i\pi(m-m')/(N+1)}}{1 - e^{i\pi(m-m')/(N+1)}} - (m' \to -m')\right]. \tag{2.116}$$

The first expression in the curly brackets is equal to 1 for even $m - m' \neq 0$; while being imaginary for odd $m - m'$ [since $(1 + e^{i\alpha})/(1 - e^{i\alpha})$ is equal to $(1 + e^{i\alpha})(1 - e^{-i\alpha})/|1 - e^{i\alpha}|^2$ with the imaginary numerator $e^{i\alpha} - e^{-i\alpha}$]. For the second term the same thing holds true for even and odd $m + m' \neq 0$, respectively. Since $m - m'$ and $m + m'$ are either both even or both odd, the right-hand side of (2.113) vanishes for $m \neq m'$ [remembering that $m, m' \in [0, N+1]$ in the expansion (2.112), and thus in (2.116)]. The proof of the completeness relation (2.114) can be carried out similarly.

Inserting now the expansion (2.112) into the time-sliced fluctuation action (2.89), the orthogonality relation (2.113) yields

$$\mathcal{A}_{\text{fl}}^N = \frac{M}{2}\epsilon \sum_{n=0}^{N} (\nabla x_n)^2 = \frac{M}{2}\epsilon \sum_{m=1}^{N+1} x(\nu_m)\Omega_m\overline{\Omega}_m x(\nu_m). \tag{2.117}$$

Thus the action decomposes into a sum of independent quadratic terms involving the discrete set of eigenvalues

$$\Omega_m\overline{\Omega}_m = \frac{1}{\epsilon^2}[2 - 2\cos(\nu_m\epsilon)] = \frac{1}{\epsilon^2}\left[2 - 2\cos\left(\frac{\pi m}{N+1}\right)\right], \tag{2.118}$$

and the fluctuation factor (2.88) becomes

$$F_0^N(t_b - t_a) = \frac{1}{\sqrt{2\pi\hbar i\epsilon/M}} \prod_{n=1}^{N} \left[\int_{-\infty}^{\infty} \frac{dx_n}{\sqrt{2\pi\hbar i\epsilon/M}}\right]$$

$$\times \prod_{m=1}^{N} \exp\left\{\frac{i}{\hbar}\frac{M}{2}\epsilon\Omega_m\overline{\Omega}_m \left[x(\nu_m)\right]^2\right\}. \tag{2.119}$$

Before performing the integrals, we must transform the measure of integration from the local variables x_n to the Fourier components $x(\nu_m)$. Due to the orthogonality relation (2.113), the transformation has a unit determinant implying that

$$\prod_{n=1}^{N} dx_n = \prod_{m=1}^{N} dx(\nu_m). \tag{2.120}$$

With this, Eq. (2.119) can be integrated with the help of Fresnel's formula (1.333). The result is

$$F_0^N(t_b - t_a) = \frac{1}{\sqrt{2\pi\hbar i\epsilon/M}} \prod_{m=1}^{N} \frac{1}{\sqrt{\epsilon^2 \Omega_m \overline{\Omega}_m}}. \tag{2.121}$$

To calculate the product we use the formula[5]

$$\prod_{m=1}^{N} \left(1 + x^2 - 2x \cos \frac{m\pi}{N+1}\right) = \frac{x^{2(N+1)} - 1}{x^2 - 1}. \tag{2.122}$$

Taking the limit $x \to 1$ gives

$$\prod_{m=1}^{N} \epsilon^2 \Omega_m \overline{\Omega}_m = \prod_{m=1}^{N} 2\left(1 - \cos \frac{m\pi}{N+1}\right) = N + 1. \tag{2.123}$$

The time-sliced fluctuation factor of a free particle is therefore simply

$$F_0^N(t_b - t_a) = \frac{1}{\sqrt{2\pi i\hbar(N+1)\epsilon/M}}, \tag{2.124}$$

or, expressed in terms of $t_b - t_a$,

$$F_0(t_b - t_a) = \frac{1}{\sqrt{2\pi i\hbar(t_b - t_a)/M}}. \tag{2.125}$$

As in the amplitude (2.71) we have dropped the superscript N since this final result is independent of the number of time slices.

Note that the dimension of the fluctuation factor is 1/length. In fact, one may introduce a length scale associated with the time interval $t_b - t_a$,

$$l(t_b - t_a) \equiv \sqrt{2\pi\hbar(t_b - t_a)/M}, \tag{2.126}$$

and write

$$F_0(t_b - t_a) = \frac{1}{\sqrt{il(t_b - t_a)}}. \tag{2.127}$$

With (2.125) and (2.87), the full time evolution amplitude of a free particle (2.84) is again given by (2.71)

$$(x_b t_b | x_a t_a) = \frac{1}{\sqrt{2\pi i\hbar(t_b - t_a)/M}} \exp\left[\frac{i}{\hbar} \frac{M}{2} \frac{(x_b - x_a)^2}{t_b - t_a}\right]. \tag{2.128}$$

It is straightforward to generalize this result to a point particle moving through any number D of Cartesian space dimensions. If $\mathbf{x} = (x_1, \ldots, x_D)$ denotes the spatial coordinates, the action is

$$\mathcal{A}[\mathbf{x}] = \frac{M}{2} \int_{t_a}^{t_b} dt\, \dot{\mathbf{x}}^2. \tag{2.129}$$

---

[5]I.S. Gradshteyn and I.M. Ryzhik, *op. cit.*, Formula 1.396.2.

Being quadratic in $\mathbf{x}$, the action is the sum of the actions for each component. Hence, the amplitude factorizes and we find

$$(\mathbf{x}_b t_b | \mathbf{x}_a t_a) = \frac{1}{\sqrt{2\pi i \hbar (t_b - t_a)/M}^D} \exp\left[\frac{i}{\hbar} \frac{M}{2} \frac{(\mathbf{x}_b - \mathbf{x}_a)^2}{t_b - t_a}\right], \qquad (2.130)$$

in agreement with the quantum-mechanical result in D dimensions (1.337).

It is instructive to present an alternative calculation of the product of eigenvalues in (2.121) which does not make use of the Fourier decomposition and works entirely in configuration space. We observe that the product

$$\prod_{m=1}^{N} \epsilon^2 \Omega_m \overline{\Omega}_m \qquad (2.131)$$

is the determinant of the diagonalized $N \times N$ -matrix $-\epsilon^2 \nabla \overline{\nabla}$. This follows from the fact that for any matrix, the determinant is the product of its eigenvalues. The product (2.131) is therefore also called the *fluctuation determinant* of the free particle and written

$$\prod_{m=1}^{N} \epsilon^2 \Omega_m \overline{\Omega}_m \equiv \det_N(-\epsilon^2 \nabla \overline{\nabla}). \qquad (2.132)$$

With this notation, the fluctuation factor (2.121) reads

$$F_0^N(t_b - t_b) = \frac{1}{\sqrt{2\pi \hbar i \epsilon/M}} \left[\det_N(-\epsilon^2 \nabla \overline{\nabla})\right]^{-1/2}. \qquad (2.133)$$

Now one realizes that the determinant of $\epsilon^2 \overline{\nabla}\nabla$ can be found very simply from the explicit $N \times N$ matrix (2.102) by induction: For $N = 1$ we see directly that

$$\det_{N=1}(-\epsilon^2 \nabla \overline{\nabla}) = |2| = 2. \qquad (2.134)$$

For $N = 2$, the determinant is

$$\det_{N=2}(-\epsilon^2 \nabla \overline{\nabla}) = \begin{vmatrix} 2 & -1 \\ -1 & 2 \end{vmatrix} = 3. \qquad (2.135)$$

A recursion relation is obtained by developing the determinant twice with respect to the first row:

$$\det_N(-\epsilon^2 \nabla \overline{\nabla}) = 2 \det_{N-1}(-\epsilon^2 \nabla \overline{\nabla}) - \det_{N-2}(-\epsilon^2 \nabla \overline{\nabla}). \qquad (2.136)$$

With the initial condition (2.134), the solution is

$$\det_N(-\epsilon^2 \nabla \overline{\nabla}) = N + 1, \qquad (2.137)$$

in agreement with the previous result (2.123).

Let us also find the time evolution amplitude in momentum space. A simple Fourier transform of the initial and final positions according to the rule (2.37) yields

$$\begin{aligned} (\mathbf{p}_b t_b | \mathbf{p}_a t_a) &= \int d^D x_b \, e^{-i\mathbf{p}_b \mathbf{x}_b/\hbar} \int d^D x_a \, e^{i\mathbf{p}_a \mathbf{x}_a/\hbar} (\mathbf{x}_b t_b | \mathbf{x}_a t_a) \\ &= (2\pi)^D \delta^{(D)}(\mathbf{p}_b - \mathbf{p}_a) e^{-i\mathbf{p}_b^2 (t_b - t_a)/2M\hbar}. \end{aligned} \qquad (2.138)$$

2.2.4 Finite Slicing Properties of Free-Particle Amplitude

The time-sliced free-particle time evolution amplitude (2.70) happens to be independent of the number N of time slices used for their calculation. We have pointed this out earlier for the fluctuation factor (2.124). Let us study the origin of this independence for the classical action in the exponent. The difference equation of motion

$$-\overline{\nabla}\nabla x(t) = 0 \tag{2.139}$$

is solved by the same linear function

$$x(t) = At + B, \tag{2.140}$$

as in the continuum. Imposing the initial conditions gives

$$x_{\rm cl}(t_n) = x_a + (x_b - x_a)\frac{n}{N+1}. \tag{2.141}$$

The time-sliced action of the fluctuations is calculated, via a summation by parts on the lattice [see (2.96)]. Using the difference equation $\overline{\nabla}\nabla x_{\rm cl} = 0$, we find

$$
\begin{aligned}
\mathcal{A}_{\rm cl} &= \epsilon \sum_{n=1}^{N+1} \frac{M}{2}(\nabla x_{\rm cl})^2 \\
&= \frac{M}{2}\left(x_{\rm cl}\nabla x_{\rm cl}\Big|_{n=0}^{N+1} - \epsilon\sum_{n=0}^{N} x_{\rm cl}\nabla\overline{\nabla}x_{\rm cl}\right) \\
&= \frac{M}{2}x_{\rm cl}\nabla x_{\rm cl}\Big|_{n=0}^{N+1} = \frac{M}{2}\frac{(x_b - x_a)^2}{t_b - t_a}.
\end{aligned}
\tag{2.142}
$$

This coincides with the continuum action for any number of time slices.

In the operator formulation of quantum mechanics, the ϵ-independence of the amplitude of the free particle follows from the fact that in the absence of a potential $V(x)$, the two sides of the Trotter formula (2.26) coincide for any N.

2.3 Exact Solution for Harmonic Oscillator

A further problem to be solved along similar lines is the time evolution amplitude of the linear oscillator

$$
\begin{aligned}
(x_b t_b | x_a t_a) &= \int \mathcal{D}'x \int \frac{\mathcal{D}p}{2\pi\hbar} \exp\left\{\frac{i}{\hbar}\mathcal{A}[p, x]\right\} \\
&= \int \mathcal{D}x \exp\left\{\frac{i}{\hbar}\mathcal{A}[x]\right\},
\end{aligned}
\tag{2.143}
$$

with the canonical action

$$\mathcal{A}[p, x] = \int_{t_a}^{t_b} dt\left(p\dot{x} - \frac{1}{2M}p^2 - \frac{M\omega^2}{2}x^2\right), \tag{2.144}$$

and the Lagrangian action

$$\mathcal{A}[x] = \int_{t_a}^{t_b} dt\, \frac{M}{2}(\dot{x}^2 - \omega^2 x^2). \tag{2.145}$$

2.3.1   Fluctuations around the Classical Path

As before, we proceed with the Lagrangian path integral, starting from the time-sliced form of the action

$$
\mathcal{A}^N = \epsilon \frac{M}{2} \sum_{n=1}^{N+1} \left[(\overline{\nabla} x_n)^2 - \omega^2 x_n^2 \right]. \tag{2.146}
$$

The path integral is again a product of Gaussian integrals which can be evaluated successively. In contrast to the free-particle case, however, the direct evaluation is now quite complicated; it will be presented in Appendix 2B. It is far easier to employ the fluctuation expansion, splitting the paths into a classical path $x_{\mathrm{cl}}(t)$ plus fluctuations $\delta x(t)$. The fluctuation expansion makes use of the fact that the action is quadratic in $x = x_{\mathrm{cl}} + \delta x$ and decomposes into the sum of a classical part

$$
\mathcal{A}_{\mathrm{cl}} = \int_{t_a}^{t_b} dt \frac{M}{2} (\dot{x}_{\mathrm{cl}}^2 - \omega^2 x_{\mathrm{cl}}^2), \tag{2.147}
$$

and a fluctuation part

$$
\mathcal{A}_{\mathrm{fl}} = \int_{t_a}^{t_b} dt \frac{M}{2} [(\delta \dot{x})^2 - \omega^2 (\delta x)^2], \tag{2.148}
$$

with the boundary condition

$$
\delta x(t_a) = \delta x(t_b) = 0. \tag{2.149}
$$

There is no mixed term, due to the extremality of the classical action. The equation of motion is

$$
\ddot{x}_{\mathrm{cl}} = -\omega^2 x_{\mathrm{cl}}. \tag{2.150}
$$

Thus, as for a free-particle, the total time evolution amplitude splits into a classical and a fluctuation factor:

$$
(x_b t_b | x_a t_a) = \int \mathcal{D}x \, e^{i\mathcal{A}[x]/\hbar} = e^{i\mathcal{A}_{\mathrm{cl}}/\hbar} F_\omega(t_b - t_a). \tag{2.151}
$$

The subscript of F_ω records the frequency of the oscillator.

The classical orbit connecting initial and final points is obviously

$$
x_{\mathrm{cl}}(t) = \frac{x_b \sin \omega(t - t_a) + x_a \sin \omega(t_b - t)}{\sin \omega(t_b - t_a)}. \tag{2.152}
$$

Note that this equation only makes sense if $t_b - t_a$ is not equal to an integer multiple of π/ω which we shall always assume from now on.[6]

---

[6]For subtleties in the immediate neighborhood of the singularities which are known as *caustic phenomena*, see Notes and References at the end of the chapter, as well as Section 4.8.

After an integration by parts we can rewrite the classical action $\mathcal{A}_{\mathrm{cl}}$ as

$$\mathcal{A}_{\mathrm{cl}} = \int_{t_a}^{t_b} dt \frac{M}{2} \left[x_{\mathrm{cl}}(-\ddot{x}_{\mathrm{cl}} - \omega^2 x_{\mathrm{cl}}) \right] + \frac{M}{2} x_{\mathrm{cl}} \dot{x}_{\mathrm{cl}} \Big|_{t_a}^{t_b} . \qquad (2.153)$$

The first term vanishes due to the equation of motion (2.150), and we obtain the simple expression

$$\mathcal{A}_{\mathrm{cl}} = \frac{M}{2} [x_{\mathrm{cl}}(t_b) \dot{x}_{\mathrm{cl}}(t_b) - x_{\mathrm{cl}}(t_a) \dot{x}_{\mathrm{cl}}(t_a)]. \qquad (2.154)$$

Since

$$\dot{x}_{\mathrm{cl}}(t_a) = \frac{\omega}{\sin \omega(t_b - t_a)} [x_b - x_a \cos \omega(t_b - t_a)], \qquad (2.155)$$

$$\dot{x}_{\mathrm{cl}}(t_b) = \frac{\omega}{\sin \omega(t_b - t_a)} [x_b \cos \omega(t_b - t_a) - x_a], \qquad (2.156)$$

we can rewrite the classical action as

$$\mathcal{A}_{\mathrm{cl}} = \frac{M\omega}{2 \sin \omega(t_b - t_a)} \left[(x_b^2 + x_a^2) \cos \omega(t_b - t_a) - 2 x_b x_a \right]. \qquad (2.157)$$

2.3.2 Fluctuation Factor

We now turn to the fluctuation factor. With the matrix notation for the lattice operator $-\nabla\overline{\nabla} - \omega^2$, we have to solve the multiple integral

$$F_\omega^N(t_b, t_a) = \frac{1}{\sqrt{2\pi\hbar i\epsilon/M}} \prod_{n=1}^{N} \left[\int_{-\infty}^{\infty} \frac{d\delta x_n}{\sqrt{2\pi\hbar i\epsilon/M}} \right]$$

$$\times \exp \left\{ \frac{i}{\hbar} \frac{M}{2} \epsilon \sum_{n,n'=1}^{N} \delta x_n [-\nabla\overline{\nabla} - \omega^2]_{nn'} \delta x_{n'} \right\}. \qquad (2.158)$$

When going to the Fourier components of the paths, the integral factorizes in the same way as for the free-particle expression (2.119). The only difference lies in the eigenvalues of the fluctuation operator which are now

$$\Omega_m \overline{\Omega}_m - \omega^2 = \frac{1}{\epsilon^2} [2 - 2\cos(\nu_m \epsilon)] - \omega^2 \qquad (2.159)$$

instead of $\Omega_m \overline{\Omega}_m$. For times t_b, t_a where all eigenvalues are positive (which will be specified below) we obtain from the upper part of the Fresnel formula (1.333) directly

$$F_\omega^N(t_b, t_a) = \frac{1}{\sqrt{2\pi\hbar i\epsilon/M}} \prod_{m=1}^{N} \frac{1}{\sqrt{\epsilon^2 \Omega_m \overline{\Omega}_m - \epsilon^2 \omega^2}}. \qquad (2.160)$$

The product of these eigenvalues is found by introducing an auxiliary frequency $\tilde{\omega}$ satisfying

$$\sin \frac{\epsilon \tilde{\omega}}{2} \equiv \frac{\epsilon \omega}{2}. \tag{2.161}$$

Then we decompose the product as

$$\prod_{m=1}^{N} [\epsilon^2 \Omega_m \overline{\Omega}_m - \epsilon^2 \omega^2] = \prod_{m=1}^{N} [\epsilon^2 \Omega_m \overline{\Omega}_m] \prod_{m=1}^{N} \left[\frac{\epsilon^2 \Omega_m \overline{\Omega}_m - \epsilon^2 \omega^2}{\epsilon^2 \Omega_m \overline{\Omega}_m} \right]$$

$$= \prod_{m=1}^{N} [\epsilon^2 \Omega_m \overline{\Omega}_m] \left[\prod_{m=1}^{N} \left(1 - \frac{\sin^2 \frac{\epsilon \tilde{\omega}}{2}}{\sin^2 \frac{m\pi}{2(N+1)}} \right) \right]. \tag{2.162}$$

The first factor is equal to $(N + 1)$ by (2.123). The second factor, the product of the ratios of the eigenvalues, is found from the standard formula[7]

$$\prod_{m=1}^{N} \left(1 - \frac{\sin^2 x}{\sin^2 \frac{m\pi}{2(N+1)}} \right) = \frac{1}{\sin 2x} \frac{\sin[2(N+1)x]}{(N+1)}. \tag{2.163}$$

With $x = \tilde{\omega}\epsilon/2$, we arrive at the fluctuation determinant

$$\det_N (-\epsilon^2 \nabla \overline{\nabla} - \epsilon^2 \omega^2) = \prod_{m=1}^{N} [\epsilon^2 \Omega_m \overline{\Omega}_m - \epsilon^2 \omega^2] = \frac{\sin \tilde{\omega}(t_b - t_a)}{\sin \epsilon \tilde{\omega}}, \tag{2.164}$$

and the fluctuation factor is given by

$$F_\omega^N (t_b, t_a) = \frac{1}{\sqrt{2\pi i\hbar/M}} \sqrt{\frac{\sin \tilde{\omega}\epsilon}{\epsilon \sin \tilde{\omega}(t_b - t_a)}}, \quad t_b - t_a < \pi/\tilde{\omega}, \tag{2.165}$$

where, as we have agreed earlier in Eq. (1.333), $\sqrt{i}$ means $e^{i\pi/4}$, and $t_b - t_a$ is always larger than zero.

2.3.3   The $i\eta$-Prescription and Maslov-Morse Index

The result (2.165) is initially valid only for

$$t_b - t_a < \pi/\tilde{\omega}. \tag{2.166}$$

In this time interval, all eigenvalues in the fluctuation determinant (2.164) are positive, and the upper version of the Fresnel formula (1.333) applies to each of the integrals in (2.158) [this was assumed in deriving (2.160)]. If $t_b - t_a$ grows larger than $\pi/\tilde{\omega}$, the smallest eigenvalue $\Omega_1 \overline{\Omega}_1 - \omega^2$ becomes negative and the integration over the associated Fourier component has to be done according to the lower case of the Fresnel formula (1.333). The resulting amplitude carries an extra phase factor

---

[7]I.S. Gradshteyn and I.M. Ryzhik, op. cit., Formula 1.391.1.

$e^{-i\pi/2}$ and remains valid until $t_b - t_a$ becomes larger than $2\pi/\tilde{\omega}$, where the second eigenvalue becomes negative introducing a further phase factor $e^{-i\pi/2}$.

All phase factors emerge naturally if we associate with the oscillator frequency ω an infinitesimal negative imaginary part, replacing everywhere ω by $\omega - i\eta$ with an infinitesimal $\eta > 0$. This is referred to as the *$i\eta$-prescription*. Physically, it amounts to attaching an infinitesimal damping term to the oscillator, so that the amplitude behaves like $e^{-i\omega t - \eta t}$ and dies down to zero after a very long time (as opposed to an unphysical antidamping term which would make it diverge after a long time). Now, each time that $t_b - t_a$ passes an integer multiple of $\pi/\tilde{\omega}$, the square root of $\sin\tilde{\omega}(t_b - t_a)$ in (2.165) passes a singularity in a specific way which ensures the proper phase.[8] With such an $i\eta$-prescription it will be superfluous to restrict $t_b - t_a$ to the range (2.166). Nevertheless it will sometimes be useful to exhibit the phase factor arising in this way in the fluctuation factor (2.165) for $t_b - t_a > \pi/\tilde{\omega}$ by writing

$$F_\omega^N(t_b, t_a) = \frac{1}{\sqrt{2\pi i\hbar/M}} \sqrt{\frac{\sin\tilde{\omega}\epsilon}{\epsilon|\sin\tilde{\omega}(t_b - t_a)|}} e^{-i\nu\pi/2}, \qquad (2.167)$$

where ν is the number of zeros encountered in the denominator along the trajectory. This number is called the *Maslov-Morse index* of the trajectory[9].

2.3.4 Continuum Limit

Let us now go to the continuum limit, $\epsilon \to 0$. Then the auxiliary frequency $\tilde{\omega}$ tends to ω and the fluctuation determinant becomes

$$\det_N(-\epsilon^2\nabla\overline{\nabla} - \epsilon^2\omega^2) \xrightarrow{\epsilon\to 0} \frac{\sin\omega(t_b - t_a)}{\omega\epsilon}. \qquad (2.168)$$

The fluctuation factor $F_\omega^N(t_b - t_a)$ goes over into

$$F_\omega(t_b - t_a) = \frac{1}{\sqrt{2\pi i\hbar/M}} \sqrt{\frac{\omega}{\sin\omega(t_b - t_a)}}, \qquad (2.169)$$

with the phase for $t_b - t_a > \pi/\omega$ determined as above.

In the limit $\omega \to 0$, both fluctuation factors agree, of course, with the free-particle result (2.125).

In the continuum limit, the ratios of eigenvalues in (2.162) can also be calculated in the following simple way. We perform the limit $\epsilon \to 0$ directly in each factor. This gives

$$\frac{\epsilon^2\Omega_m\overline{\Omega}_m - \epsilon^2\omega^2}{\epsilon^2\Omega_m\overline{\Omega}_m} = 1 - \frac{\epsilon^2\omega^2}{2 - 2\cos(\nu_m\epsilon)}$$
$$\xrightarrow{\epsilon\to 0} 1 - \frac{\omega^2(t_b - t_a)^2}{\pi^2 m^2}. \qquad (2.170)$$

---

[8]In the square root, we may equivalently assume $t_b - t_a$ to carry a small negative imaginary part. For a detailed discussion of the phases of the fluctuation factor in the literature, see Notes and References at the end of the chapter.

[9]V.P. Maslov and M.V. Fedoriuk, *Semi-Classical Approximations in Quantum Mechanics*, Reidel, Boston, 1981.

As the number N goes to infinity we wind up with an infinite product of these factors. Using the well-known infinite-product formula for the sine function[10]

$$\sin x = x \prod_{m=1}^{\infty} \left(1 - \frac{x^2}{m^2\pi^2} \right), \tag{2.171}$$

we find, with $x = \omega(t_b - t_a)$,

$$\prod_m \frac{\Omega_m \overline{\Omega}_m}{\Omega_m \overline{\Omega}_m - \omega^2} \xrightarrow{\epsilon \to 0} \prod_{m=1}^{\infty} \frac{\nu_m^2}{\nu_m^2 - \omega^2} = \frac{\omega(t_b - t_a)}{\sin \omega(t_b - t_a)}, \tag{2.172}$$

and obtain once more the fluctuation factor in the continuum (2.169).

Multiplying the fluctuation factor with the classical amplitude, the time evolution amplitude of the linear oscillator in the continuum reads

$$
\begin{aligned}
(x_b t_b | x_a t_a) &= \int \mathcal{D}x(t) \exp\left[\frac{i}{\hbar} \int_{t_a}^{t_b} dt \frac{M}{2} (\dot{x}^2 - \omega^2 x^2) \right] \\
&= \frac{1}{\sqrt{2\pi i \hbar / M}} \sqrt{\frac{\omega}{\sin \omega(t_b - t_a)}} \\
&\quad \times \exp\left\{ \frac{i}{2\hbar} \frac{M\omega}{\sin \omega(t_b - t_a)} [(x_b^2 + x_a^2) \cos \omega(t_b - t_a) - 2x_b x_a] \right\}.
\end{aligned}
\tag{2.173}
$$

The result can easily be extended to any number D of dimensions, where the action is

$$\mathcal{A} = \int_{t_a}^{t_b} dt \frac{M}{2} \left(\dot{\mathbf{x}}^2 - \omega^2 \mathbf{x}^2 \right). \tag{2.174}$$

Being quadratic in $\mathbf{x}$, the action is the sum of the actions of each component leading to the factorized amplitude:

$$
\begin{aligned}
(\mathbf{x}_b t_b | \mathbf{x}_a t_a) &= \prod_{i=1}^{D} \left(x_b^i t_b | x_a^i t_a \right) = \frac{1}{\sqrt{2\pi i \hbar / M}^D} \sqrt{\frac{\omega}{\sin \omega(t_b - t_a)}}^D \\
&\quad \times \exp\left\{ \frac{i}{2\hbar} \frac{M\omega}{\sin \omega(t_b - t_a)} [(\mathbf{x}_b^2 + \mathbf{x}_a^2) \cos \omega(t_b - t_a) - 2\mathbf{x}_b \mathbf{x}_a] \right\},
\end{aligned}
\tag{2.175}
$$

where the phase of the second square root for $t_b - t_a > \pi/\omega$ is determined as in the one-dimensional case [see Eq. (1.543)].

2.3.5 Useful Fluctuation Formulas

It is worth realizing that when performing the continuum limit in the ratio of eigenvalues (2.172), we have actually calculated the ratio of the functional determinants of the *differential* operators

$$\frac{\det(-\partial_t^2 - \omega^2)}{\det(-\partial_t^2)}. \tag{2.176}$$

---

[10]I.S. Gradshteyn and I.M. Ryzhik, op. cit., Formula 1.431.1.

Indeed, the eigenvalues of $-\partial_t^2$ in the space of real fluctuations vanishing at the endpoints are simply

$$\nu_m^2 = \left(\frac{\pi m}{t_b - t_a}\right)^2, \tag{2.177}$$

so that the ratio (2.176) is equal to the product

$$\frac{\det(-\partial_t^2 - \omega^2)}{\det(-\partial_t^2)} = \prod_{m=1}^{\infty} \frac{\nu_m^2 - \omega^2}{\nu_m^2}, \tag{2.178}$$

which is the same as (2.172). This observation should, however, not lead us to believe that the entire fluctuation factor

$$F_\omega(t_b - t_a) = \int \mathcal{D}\delta x \exp\left\{\frac{i}{\hbar}\int_{t_a}^{t_b} dt \frac{M}{2}[(\delta\dot{x})^2 - \omega^2(\delta x)^2]\right\} \tag{2.179}$$

could be calculated via the continuum determinant

$$F_\omega(t_b, t_a) \xrightarrow{\epsilon \to 0} \frac{1}{\sqrt{2\pi\hbar i\epsilon/M}} \frac{1}{\sqrt{\det(-\partial_t^2 - \omega^2)}} \qquad \text{(false)}. \tag{2.180}$$

The product of eigenvalues in $\det(-\partial_t^2 - \omega^2)$ would be a strongly divergent expression

$$\det(-\partial_t^2 - \omega^2) = \prod_{m=1}^{\infty}(\nu_m^2 - \omega^2) \tag{2.181}$$

$$= \prod_{m=1}^{\infty}(\nu_m^2)\prod_{m=1}^{\infty}\left[\frac{\nu_m^2 - \omega^2}{\nu_m^2}\right] = \prod_{m=1}^{\infty}\left[\frac{\pi^2 m^2}{(t_b - t_a)^2}\right] \times \frac{\sin\omega(t_b - t_a)}{\omega(t_b - t_a)}.$$

Only *ratios* of determinants $-\nabla\overline{\nabla} - \omega^2$ with different ω's can be replaced by their differential limits. Then the common divergent factor in (2.181) cancels.

Let us look at the origin of this strong divergence. The eigenvalues on the lattice and their continuum approximation start both out for small m as

$$\Omega_m\overline{\Omega}_m \approx \nu_m^2 \approx \frac{\pi^2 m^2}{(t_b - t_a)^2}. \tag{2.182}$$

For large $m \leq N$, the eigenvalues on the lattice saturate at $\Omega_m\overline{\Omega}_m \to 2/\epsilon^2$, while the ν_m^2's keep growing quadratically in m. This causes the divergence.

The correct time-sliced formulas for the fluctuation factor of a harmonic oscillator is summarized by the following sequence of equations:

$$\begin{aligned}
F_\omega^N(t_b - t_a) &= \frac{1}{\sqrt{2\pi\hbar i\epsilon/M}}\prod_{n=1}^{N}\left[\int\frac{d\delta x_n}{\sqrt{2\pi\hbar i\epsilon/M}}\right]\exp\left[\frac{i}{\hbar}\frac{M}{2\epsilon}\delta x^T(-\epsilon^2\nabla\overline{\nabla} - \epsilon^2\omega^2)\delta x\right] \\
&= \frac{1}{\sqrt{2\pi\hbar i\epsilon/M}}\frac{1}{\sqrt{\det_N(-\epsilon^2\nabla\overline{\nabla} - \epsilon^2\omega^2)}}, \tag{2.183}
\end{aligned}$$

where in the first expression, the exponent is written in matrix notation with x^T denoting the transposed vector x whose components are x_n. Taking out a free-particle determinant $\det_N(-\epsilon^2\overline{\nabla}\nabla)$, formula (2.137), leads to the ratio formula

$$F_\omega^N(t_b - t_a) = \frac{1}{\sqrt{2\pi\hbar i(t_b - t_a)/M}}\left[\frac{\det_N(-\epsilon^2\nabla\overline{\nabla} - \epsilon^2\omega^2)}{\det_N(-\epsilon^2\nabla\overline{\nabla})}\right]^{-1/2}, \tag{2.184}$$

which yields

$$F_\omega^N(t_b - t_a) = \frac{1}{\sqrt{2\pi i\hbar/M}} \sqrt{\frac{\sin \tilde{\omega}\epsilon}{\epsilon \sin \tilde{\omega}(t_b - t_a)}},\tag{2.185}$$

If with $\tilde{\omega}$ of Eq. (2.161). If we are only interested in the continuum limit, we may let ϵ go to zero on the right-hand side of (2.184) and evaluate

$$
\begin{aligned}
F_\omega(t_b - t_a) &= \frac{1}{\sqrt{2\pi\hbar i(t_b - t_a)/M}} \left[\frac{\det(-\partial_t^2 - \omega^2)}{\det(-\partial_t^2)}\right]^{-1/2} \\
&= \frac{1}{\sqrt{2\pi\hbar i(t_b - t_a)/M}} \prod_{m=1}^\infty \left[\frac{\nu_m^2 - \omega^2}{\nu_m^2}\right]^{-1/2} \\
&= \frac{1}{\sqrt{2\pi\hbar i(t_b - t_a)/M}} \sqrt{\frac{\omega(t_b - t_a)}{\sin \omega(t_b - t_a)}}.
\end{aligned}\tag{2.186}
$$

Let us calculate also here the time evolution amplitude in momentum space. The Fourier transform of initial and final positions of (2.175) [as in (2.138)] yields

$$
\begin{aligned}
(\mathbf{p}_b t_b | \mathbf{p}_a t_a) &= \int d^D x_b\, e^{-i\mathbf{p}_b \mathbf{x}_b/\hbar} \int d^D x_a\, e^{i\mathbf{p}_a \mathbf{x}_a/\hbar} (\mathbf{x}_b t_b | \mathbf{x}_a t_a) \\
&= \frac{(2\pi\hbar)^D}{\sqrt{2\pi i\hbar}^D} \frac{1}{\sqrt{M\omega \sin \omega(t_b - t_a)}^D} \\
&\quad \times \exp\left\{\frac{i}{\hbar} \frac{1}{2M\omega \sin \omega(t_b - t_a)} \left[(\mathbf{p}_b^2 + \mathbf{p}_a^2)\cos \omega(t_b - t_a) - 2\mathbf{p}_b \mathbf{p}_a\right]\right\}.
\end{aligned}\tag{2.187}
$$

The limit $\omega \to 0$ reduces to the free-particle expression (2.138), not quite as directly as in the x-space amplitude (2.175). Expanding the exponent

$$
\begin{aligned}
&\frac{1}{2M\omega \sin \omega(t_b - t_a)} \left[(\mathbf{p}_b^2 + \mathbf{p}_a)\cos \omega(t_b - t_a) - 2\mathbf{p}_b \mathbf{p}_a^2\right] \\
&= \frac{1}{2M\omega^2(t_b - t_a)} \left\{(\mathbf{p}_b - \mathbf{p}_a)^2 - \frac{1}{2}(\mathbf{p}_b^2 + \mathbf{p}_a^2)[\omega(t_b - t_a)]^2 + \dots\right\},
\end{aligned}\tag{2.188}
$$

and going to the limit $\omega \to 0$, the leading term in (2.187)

$$\frac{(2\pi)^D}{\sqrt{2\pi i\omega^2(t_b - t_a)\hbar M}^D} \exp\left\{\frac{i}{\hbar} \frac{1}{2M\omega^2(t_b - t_a)} (\mathbf{p}_b - \mathbf{p}_a)^2\right\}\tag{2.189}$$

tends to $(2\pi\hbar)^D \delta^{(D)}(\mathbf{p}_b - \mathbf{p}_a)$ [recall (1.531)], while the second term in (2.188) yields a factor $e^{-i\mathbf{p}^2(t_b - t_a)/2M\hbar}$, so that we recover indeed (2.138).

2.3.6 Oscillator Amplitude on Finite Time Lattice

Let us calculate the exact time evolution amplitude for a finite number of time slices. In contrast to the free-particle case in Section 2.2.4, the oscillator amplitude is no longer equal to its continuum limit but ϵ-dependent. This will allow us to study some typical convergence properties of path integrals in the continuum limit. Since the fluctuation factor was initially calculated at a finite ϵ in (2.167), we only need to find the classical action for finite ϵ. To maintain time reversal invariance at any finite ϵ, we work with a slightly different sliced potential term in the action than before in (2.146), using

$$\mathcal{A}^N = \epsilon \frac{M}{2} \sum_{n=1}^{N+1} \left[(\overline{\nabla} x_n)^2 - \omega^2 (x_n^2 + x_{n-1}^2)/2 \right], \tag{2.190}$$

or, written in another way,

$$\mathcal{A}^N = \epsilon \frac{M}{2} \sum_{n=0}^{N} \left[(\nabla x_n)^2 - \omega^2 (x_{n+1}^2 + x_n^2)/2 \right]. \tag{2.191}$$

This differs from the original time-sliced action (2.146) by having the potential $\omega^2 x_n^2$ replaced by the more symmetric one $\omega^2 (x_n^2 + x_{n-1}^2)/2$. The gradient term is the same in both cases and can be rewritten, after a summation by parts, as

$$\epsilon \sum_{n=1}^{N+1} (\overline{\nabla} x_n)^2 = \epsilon \sum_{n=0}^{N} (\nabla x_n)^2 = \left[x_b \overline{\nabla} x_b - x_a \nabla x_a \right] - \epsilon \sum_{n=1}^{N} x_n \nabla \overline{\nabla} x_n. \tag{2.192}$$

This leads to a time-sliced action

$$\mathcal{A}^N = \frac{M}{2} \left(x_b \overline{\nabla} x_b - x_a \nabla x_a \right) - \epsilon \frac{M}{4} \omega^2 (x_b^2 + x_a^2) - \epsilon \frac{M}{2} \sum_{n=1}^{N} x_n (\nabla \overline{\nabla} + \omega^2) x_n. \tag{2.193}$$

Since the variation of $\mathcal{A}^N$ is performed at fixed endpoints x_a and x_b, the fluctuation factor is the same as in (2.158). The equation of motion on the sliced time axis is

$$(\nabla \overline{\nabla} + \omega^2) x_{\mathrm{cl}}(t) = 0. \tag{2.194}$$

Here it is understood that the time variable takes only the discrete lattice values t_n. The solution of this difference equation with the initial and final values x_a and x_b, respectively, is given by

$$x_{\mathrm{cl}}(t) = \frac{1}{\sin \tilde{\omega}(t_b - t_a)} \left[x_b \sin \tilde{\omega}(t - t_a) + x_a \sin \tilde{\omega}(t_b - t) \right], \tag{2.195}$$

where $\tilde{\omega}$ is the auxiliary frequency introduced in (2.161). To calculate the classical action on the lattice, we insert (2.195) into (2.193). After some trigonometry, and replacing $\epsilon^2 \omega^2$ by $4 \sin^2(\tilde{\omega}\epsilon/2)$, the action resembles closely the continuum expression (2.157):

$$\mathcal{A}_{\mathrm{cl}}^N = \frac{M}{2\epsilon} \frac{\sin \tilde{\omega}\epsilon}{\sin \tilde{\omega}(t_b - t_a)} \left[(x_b^2 + x_a^2) \cos \tilde{\omega}(t_b - t_a) - 2 x_b x_a \right]. \tag{2.196}$$

The total time evolution amplitude on the sliced time axis is

$$(x_b t_b | x_a t_a) = e^{i \mathcal{A}_{\mathrm{cl}}^N / \hbar} F_\omega^N (t_b - t_a), \tag{2.197}$$

with sliced action (2.196) and the sliced fluctuation factor (2.167).

2.4   Gelfand-Yaglom Formula

In many applications one encounters a slight generalization of the oscillator fluctuation problem: The action is harmonic but contains a time-dependent frequency $\Omega^2(t)$ instead of the constant oscillator frequency ω^2. The associated fluctuation factor is

$$F(t_b, t_a) = \int \mathcal{D}\delta x(t) \exp \left(\frac{i}{\hbar} \mathcal{A} \right), \tag{2.198}$$

with the action

$$\mathcal{A} = \int_{t_a}^{t_b} dt \frac{M}{2} [(\delta \dot{x})^2 - \Omega^2(t)(\delta x)^2]. \tag{2.199}$$

Since $\Omega(t)$ may not be translationally invariant in time, the fluctuation factor depends now in general on both the initial and final times. The ratio formula (2.184) holds also in this more general case, i.e.,

$$F^N(t_b, t_a) = \frac{1}{\sqrt{2\pi\hbar i (t_b - t_a)/M}} \left[\frac{\det_N(-\epsilon^2 \nabla \overline{\nabla} - \epsilon^2 \Omega^2)}{\det_N(-\epsilon^2 \nabla \overline{\nabla})} \right]^{-1/2}. \tag{2.200}$$

Here $\Omega^2(t)$ denotes the diagonal matrix

$$\Omega^2(t) = \begin{pmatrix} \Omega_N^2 & & \\ & \ddots & \\ & & \Omega_1^2 \end{pmatrix}, \tag{2.201}$$

with the matrix elements $\Omega_n^2 = \Omega^2(t_n)$.

2.4.1 Recursive Calculation of Fluctuation Determinant

In general, the full set of eigenvalues of the matrix $-\nabla \overline{\nabla} - \Omega^2(t)$ is quite difficult to find, even in the continuum limit. It is, however, possible to derive a powerful difference equation for the fluctuation determinant which can often be used to find its value without knowing all eigenvalues. The method is due to Gelfand and Yaglom.[11]

Let us denote the determinant of the $N \times N$ fluctuation matrix by D_N, i.e.,

$$D_N \equiv \det_N \left(-\epsilon^2 \overline{\nabla} \nabla - \epsilon^2 \Omega^2 \right) \tag{2.202}$$

$$\equiv \begin{vmatrix} 2 - \epsilon^2 \Omega_N^2 & -1 & 0 & \dots & 0 & 0 & 0 \\ -1 & 2 - \epsilon^2 \Omega_{N-1}^2 & -1 & \dots & 0 & 0 & 0 \\ \vdots & & & & & & \vdots \\ 0 & 0 & 0 & \dots & -1 & 2 - \epsilon^2 \Omega_2^2 & -1 \\ 0 & 0 & 0 & \dots & 0 & -1 & 2 - \epsilon^2 \Omega_1^2 \end{vmatrix}.$$

By expanding this along the first column, we obtain the recursion relation

$$D_N = (2 - \epsilon^2 \Omega_N^2) D_{N-1} - D_{N-2}, \tag{2.203}$$

which may be rewritten as

$$\epsilon^2 \left[\frac{1}{\epsilon} \left(\frac{D_N - D_{N-1}}{\epsilon} - \frac{D_{N-1} - D_{N-2}}{\epsilon} \right) + \Omega_N^2 D_{N-1} \right] = 0. \tag{2.204}$$

---

[11]I.M. Gelfand and A.M. Yaglom, J. Math. Phys. *1*, 48 (1960).

Since the equation is valid for all N, it implies that the determinant D_N satisfies the difference equation

$$(\nabla\overline{\nabla} + \Omega_{N+1}^2)D_N = 0. \tag{2.205}$$

In this notation, the operator $-\nabla\overline{\nabla}$ is understood to act on the dimensional label N of the determinant. The determinant D_N may be viewed as the discrete values of a function of $D(t)$ evaluated on the sliced time axis. Equation (2.205) is called the *Gelfand-Yaglom formula*. Thus the determinant as a function of N is the solution of the classical difference equation of motion and the desired result for a given N is obtained from the final value $D_N = D(t_{N+1})$. The initial conditions are

$$\begin{aligned} D_1 &= (2 - \epsilon^2\Omega_1^2), \\ D_2 &= (2 - \epsilon^2\Omega_1^2)(2 - \epsilon^2\Omega_2^2) - 1. \end{aligned} \tag{2.206}$$

2.4.2   Examples

As an illustration of the power of the Gelfand-Yaglom formula, consider the known case of a constant $\Omega^2(t) \equiv \omega^2$ where the Gelfand-Yaglom formula reads

$$(\nabla\overline{\nabla} + \omega^2)D_N = 0. \tag{2.207}$$

This is solved by a linear combination of $\sin(N\tilde{\omega}\epsilon)$ and $\cos(N\tilde{\omega}\epsilon)$, where $\tilde{\omega}$ is given by (2.161). The solution satisfying the correct boundary condition is obviously

$$D_N = \frac{\sin(N+1)\epsilon\tilde{\omega}}{\sin\epsilon\tilde{\omega}}. \tag{2.208}$$

Indeed, the two lowest elements are

$$\begin{aligned} D_1 &= 2\cos\epsilon\tilde{\omega}, \\ D_2 &= 4\cos^2\epsilon\tilde{\omega} - 1, \end{aligned} \tag{2.209}$$

which are the same as (2.206), since $\epsilon^2\Omega^2 \equiv \epsilon^2\omega^2 = 2(1 - \cos\tilde{\omega}\epsilon)$.

The Gelfand-Yaglom formula becomes especially easy to handle in the continuum limit $\epsilon \to 0$. Then, by considering the renormalized function

$$D_{\mathrm{ren}}(t_N) = \epsilon D_N, \tag{2.210}$$

the initial conditions $D_1 = 2$ and $D_2 = 3$ can be re-expressed as

$$(\epsilon D)_1 = D_{\mathrm{ren}}(t_a) = 0, \tag{2.211}$$

$$\frac{\epsilon D_2 - \epsilon D_1}{\epsilon} = (\nabla\epsilon D)_1 \xrightarrow{\epsilon \to 0} \dot{D}_{\mathrm{ren}}(t_a) = 1. \tag{2.212}$$

The difference equation for D_N turns into the differential equation for $D_{\mathrm{ren}}(t)$:

$$[\partial_t^2 + \Omega^2(t)]D_{\mathrm{ren}}(t) = 0. \tag{2.213}$$

The situation is pictured in Fig. 2.2. The determinant D_N is $1/\epsilon$ times the value of the function $D_{\mathrm{ren}}(t)$ at t_b. This value is found by solving the differential equation starting from t_a with zero value and unit slope.

As an example, consider once more the harmonic oscillator with a fixed frequency ω. The equation of motion in the continuum limit is solved by

$$D_{\mathrm{ren}}(t) = \frac{1}{\omega}\sin\omega(t - t_a), \tag{2.214}$$

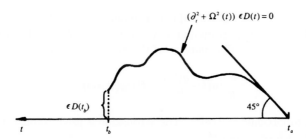

Figure 2.2 Solution of equation of motion with zero initial value and unit initial slope. Its value at the final time is equal to ϵ times the discrete fluctuation determinant $D_N = D(t_b)$.

which satisfies the initial conditions (2.212). Thus we find the fluctuation determinant to become, for small ϵ,

$$\det(-\epsilon^2 \nabla\overline{\nabla} - \epsilon^2 \omega^2) \xrightarrow{\epsilon \to 0} \frac{1}{\epsilon} \frac{\sin \omega(t_b - t_a)}{\omega}, \tag{2.215}$$

in agreement with the earlier result (2.208). For the free particle, the solution is $D_{\mathrm{ren}}(t) = t - t_a$ and we obtain directly the determinant $\det_N(-\epsilon^2 \nabla\overline{\nabla}) = (t_b - t_a)/\epsilon$.

For time-dependent frequencies $\Omega(t)$, an analytic solution of the Gelfand-Yaglom initial-value problem (2.211), (2.212), and (2.213) can be found only for special classes of functions $\Omega(t)$. In fact, (2.213) has the form of a Schrödinger equation of a point particle in a potential $\Omega^2(t)$, and the classes of potentials for which the Schrödinger equation can be solved are well-known.

2.4.3   Calculation on Unsliced Time Axis

In general, the most explicit way of expressing the solution is by linearly combining $D_{\mathrm{ren}} = \epsilon D_N$ from any two independent solutions $\xi(t)$ and $\eta(t)$ of the homogeneous differential equation

$$[\partial_t^2 + \Omega^2(t)]x(t) = 0. \tag{2.216}$$

The solution of (2.213) is found from a linear combination

$$D_{\mathrm{ren}}(t) = \alpha\xi(t) + \beta\eta(t). \tag{2.217}$$

The coefficients are determined from the initial condition (2.212), which imply

$$\begin{aligned}
\alpha\xi(t_a) + \beta\eta(t_a) &= 0, \\
\alpha\dot{\xi}(t_a) + \beta\dot{\eta}(t_a) &= 1,
\end{aligned} \tag{2.218}$$

and thus

$$D_{\mathrm{ren}}(t) = \frac{\xi(t)\eta(t_a) - \xi(t_a)\eta(t)}{\dot{\xi}(t_a)\eta(t_a) - \xi(t_a)\dot{\eta}(t_a)}. \tag{2.219}$$

The denominator is recognized as the time-independent *Wronski determinant* of the two solutions

$$W \equiv \xi(t) \overset{\leftrightarrow}{\partial_t} \eta(t) \equiv \xi(t)\dot{\eta}(t) - \dot{\xi}(t)\eta(t) \tag{2.220}$$

at the initial point t_a. The right-hand side is independent of t.

The Wronskian is an important quantity in the theory of second-order differential equations. It is defined for all equations of the Sturm-Liouville type

$$\frac{d}{dt}\left[a(t)\frac{dy(t)}{dt}\right] + b(t)y(t) = 0, \tag{2.221}$$

for which it is proportional to $1/a(t)$. The Wronskian is used to construct the Green function for all such equations.[12]

In terms of the Wronskian, Eq. (2.219) has the general form

$$D_{\text{ren}}(t) = -\frac{1}{W}\left[\xi(t)\eta(t_a) - \xi(t_a)\eta(t)\right]. \tag{2.222}$$

Inserting $t = t_b$ gives the desired determinant

$$D_{\text{ren}} = -\frac{1}{W}\left[\xi(t_b)\eta(t_a) - \xi(t_a)\eta(t_b)\right]. \tag{2.223}$$

Note that the same functional determinant can be found from by evaluating the function

$$\tilde{D}_{\text{ren}}(t) = -\frac{1}{W}\left[\xi(t_b)\eta(t) - \xi(t)\eta(t_b)\right] \tag{2.224}$$

at t_a. This also satisfies the homogenous differential equation (2.213), but with the initial conditions

$$\tilde{D}_{\text{ren}}(t_b) = 0, \qquad \dot{\tilde{D}}_{\text{ren}}(t_b) = -1. \tag{2.225}$$

It will be useful to emphasize at which ends the Gelfand-Yaglom boundary conditions are satisfied by denoting $D_{\text{ren}}(t)$ and $\tilde{D}_{\text{ren}}(t)$ by $D_a(t)$ and $D_b(t)$, respectively, summarizing their symmetric properties as

$$[\partial_t^2 + \Omega^2(t)]D_a(t) = 0 ; \quad D_a(t_a) = 0, \quad \dot{D}_a(t_a) = 1, \tag{2.226}$$
$$[\partial_t^2 + \Omega^2(t)]D_b(t) = 0 ; \quad D_b(t_b) = 0, \quad \dot{D}_b(t_b) = -1, \tag{2.227}$$

with the determinant being obtained from either function as

$$D_{\text{ren}} = D_a(t_b) = D_b(t_a). \tag{2.228}$$

In contrast to this we see from the explicit equations (2.222) and (2.224) that the time derivatives of two functions at opposite endpoints are in general not related. Only for frequencies $\Omega(t)$ with time reversal invariance, one has

$$\dot{D}_a(t_b) = -\dot{D}_b(t_a), \quad \text{for} \quad \Omega(t) = \Omega(-t). \tag{2.229}$$

---

[12]For its typical use in classical electrodynamics, see J.D. Jackson, *Classical Electrodynamics*, John Wiley & Sons, New York, 1975, Section 3.11.

For arbitrary $\Omega(t)$, one can derive a relation

$$\dot{D}_a(t_b) + \dot{D}_b(t_a) = -2 \int_{t_a}^{t_b} dt\, \Omega(t)\dot{\Omega}(t)D_a(t)D_b(t). \tag{2.230}$$

As an application of these formulas, consider once more the linear oscillator, for which two independent solutions are

$$\xi(t) = \cos\omega t, \quad \eta(t) = \sin\omega t. \tag{2.231}$$

Hence

$$W = \omega, \tag{2.232}$$

and the fluctuation determinant becomes

$$D_{\mathrm{ren}} = -\frac{1}{\omega}(\cos\omega t_b \sin\omega t_a - \cos\omega t_a \sin\omega t_b) = \frac{1}{\omega}\sin\omega(t_b - t_a). \tag{2.233}$$

2.4.4 D'Alembert's Construction

It is important to realize that the construction of the solutions of Eqs. (2.226) and (2.227) requires only the knowledge of one solution of the homogenous differential equation (2.216), say $\xi(t)$. A second linearly independent solution $\eta(t)$ can always be found with the help of a formula due to d'Alembert,

$$\eta(t) = w\,\xi(t) \int^t \frac{dt'}{\xi^2(t')}, \tag{2.234}$$

where w is some constant. Differentiation yields

$$\dot{\eta} = \frac{\dot{\xi}\eta}{\xi} + \frac{w}{\xi}, \quad \ddot{\eta} = \frac{\ddot{\xi}\eta}{\xi}. \tag{2.235}$$

The second equation shows that with $\xi(t)$, also $\eta(t)$ is a solution of the homogenous differential equation (2.216). From the first equation we find that the Wronski determinant of the two functions is equal to w:

$$W = \xi(t)\dot{\eta}(t) - \dot{\xi}(t)\eta(t) = w. \tag{2.236}$$

Inserting the solution (2.234) into the formulas (2.222) and (2.224), we obtain explicit expressions for the Gelfand-Yaglom functions in terms of one arbitrary solution of the homogenous differential equation (2.216):

$$D_{\mathrm{ren}}(t) = D_a(t) = \xi(t)\xi(t_a)\int_{t_a}^t \frac{dt'}{\xi^2(t')}, \quad \tilde{D}_{\mathrm{ren}}(t) = D_b(t) = \xi(t_b)\xi(t)\int_t^{t_b} \frac{dt'}{\xi^2(t')}. \tag{2.237}$$

The desired functional determinant is

$$D_{\mathrm{ren}} = \xi(t_b)\xi(t_a)\int_{t_a}^{t_b} \frac{dt'}{\xi^2(t')}. \tag{2.238}$$

2.4.5   Another Simple Formula

There exists yet another useful formula for the functional determinant. For this we solve the homogenous differential equation (2.216) for an arbitrary initial position x_a and initial velocity $\dot{x}_a$ at the time t_a. The result may be expressed as the following linear combination of $D_a(t)$ and $D_b(t)$:

$$x(x_a, \dot{x}_a; t) = \frac{1}{D_b(t_a)} \left[D_b(t) - D_a(t) \dot{D}_b(t_a) \right] x_a + D_a(t) \dot{x}_a \; . \qquad (2.239)$$

We then see that the Gelfand-Yaglom function $D_{\mathrm{ren}}(t) = D_a(t)$ can be obtained from the partial derivative

$$D_{\mathrm{ren}}(t) = \frac{\partial x(x_a, \dot{x}_a; t)}{\partial \dot{x}_a}. \qquad (2.240)$$

This function obviously satisfies the Gelfand-Yaglom initial conditions $D_{\mathrm{ren}}(t_a) = 0$ and $\dot{D}_{\mathrm{ren}}(t_a) = 1$ of (2.211) and (2.212), which are a direct consequence of the fact that x_a and $\dot{x}_a$ are independent variables in the function $x(x_a, \dot{x}_a; t)$, for which $\partial x_a / \partial \dot{x}_a = 0$ and $\partial \dot{x}_a / \partial \dot{x}_a = 1$.

The fluctuation determinant $D_{\mathrm{ren}} = D_a(t_b)$ is then given by

$$D_{\mathrm{ren}} = \frac{\partial x_b}{\partial \dot{x}_a}, \qquad (2.241)$$

where x_b abbreviates the function $x(x_a, \dot{x}_a; t_b)$. It is now obvious that the analogous equations (2.227) are satisfied by the partial derivative $D_b(t) = -\partial x(t)/\partial x_b$, where $x(t)$ is expressed in terms of the final position x_b and velocity $\dot{x}_b$ as $x(t) = x(x_b, \dot{x}_b; t)$

$$x(x_b, \dot{x}_b; t) = \frac{1}{D_a(t_b)} \left[D_a(t) + D_b(t) \dot{D}_a(t_b) \right] x_b - D_b(t) \dot{x}_b \; , \qquad (2.242)$$

so that we obtain the alternative formula

$$D_{\mathrm{ren}} = -\frac{\partial x_a}{\partial \dot{x}_b}. \qquad (2.243)$$

These results can immediately be generalized to functional determinants of differential operators of the form $-\partial_t^2 \delta_{ij} - \Omega_{ij}^2(t)$ where the time-dependent frequency is a $D \times D$-dimensional matrix $\Omega_{ij}^2(t)$, $(i, j = 1, \ldots, D)$. Then the associated Gelfand-Yaglom function $D_a(t)$ becomes a matrix $D_{ij}(t)$ satisfying the initial conditions $D_{ij}(t_a) = 0$, $\dot{D}_{ij}(t_b) = \delta_{ij}$, and the desired functional determinant D_{ren} is equal to the ordinary determinant of $D_{ij}(t_b)$:

$$D_{\mathrm{ren}} = \mathrm{Det} \left[-\partial_t^2 \delta_{ij} - \Omega_{ij}^2(t) \right] = \det D_{ij}(t_b). \qquad (2.244)$$

The homogeneous differential equation and the initial conditions are obviously satisfied by the partial derivative matrix $D_{ij}(t) = \partial x^i(t)/\partial \dot{x}_a^j$, so that the explicit representations of $D_{ij}(t)$ in terms of the general solution of the classical equations of motion $\left[-\partial_t^2 \delta_{ij} - \Omega_{ij}^2(t) \right] x_j(t) = 0$ become

$$D_{\mathrm{ren}} = \det \frac{\partial x_b^i}{\partial \dot{x}_a^j} = \det \left(-\frac{\partial x_a^i}{\partial \dot{x}_b^j} \right). \qquad (2.245)$$

A further couple of formulas for functional determinants can be found by constructing a solution of the homogeneous differential equation (2.216) which passes through specific initial and final points x_a and x_b at t_a and t_b, respectively:

$$x(x_b, x_a; t) = \frac{D_b(t)}{D_b(t_a)} x_a + \frac{D_a(t)}{D_a(t_b)} x_b. \qquad (2.246)$$

The Gelfand-Yaglom functions $D_a(t)$ and $D_b(t)$ can therefore be obtained from the partial derivatives

$$\frac{D_a(t)}{D_a(t_b)} = \frac{\partial x(x_b, x_a; t)}{\partial x_b}, \qquad \frac{D_b(t)}{D_b(t_a)} = \frac{\partial x(x_b, x_a; t)}{\partial x_a}. \tag{2.247}$$

At the endpoints, Eqs. (2.246) yield

$$\dot{x}_a = \frac{\dot{D}_b(t_a)}{D_b(t_a)} x_a + \frac{1}{D_a(t_b)} x_b, \tag{2.248}$$

$$\dot{x}_b = -\frac{1}{D_b(t_a)} x_a + \frac{\dot{D}_a(t_b)}{D_a(t_b)} x_b, \tag{2.249}$$

so that the fluctuation determinant $D_{\mathrm{ren}} = D_a(t_b) = D_b(t_a)$ is given by the formulas

$$D_{\mathrm{ren}} = \left(\frac{\partial \dot{x}_a}{\partial x_b}\right)^{-1} = -\left(\frac{\partial \dot{x}_b}{\partial x_a}\right)^{-1}, \tag{2.250}$$

where $\dot{x}_a$ and $\dot{x}_b$ are functions of the independent variables x_a and x_b. The equality of these expressions with the previous ones in (2.241) and (2.243) is a direct consequence of the mathematical identity for partial derivatives

$$\left.\frac{\partial x_b}{\partial \dot{x}_a}\right|_{x_a} = \left(\left.\frac{\partial \dot{x}_a}{\partial x_b}\right|_{x_a}\right)^{-1}. \tag{2.251}$$

Let us emphasize that all functional determinants calculated in this Chapter apply to the fluctuation factor of paths with fixed endpoints. In mathematics, this property is referred to as Dirichlet boundary conditions. In the context of quantum statistics, we shall also need such determinants for fluctuations with periodic boundary conditions, for which the Gelfand-Yaglom method must be modified. We shall see in Section 2.11 that this causes considerable complications in the lattice derivation, which will make it desirable to find a simpler derivation of both functional determinants. This will be found in Section 3.27 in a continuum formulation.

In general, the homogenous differential equation (2.216) with time-dependent frequency $\Omega(t)$ cannot be solved analytically. The equation has the same form as a Schrödinger equation for a point particle in one dimension moving in a one dimensional potential $\Omega^2(t)$, and there are only a few classes of potentials for which the solutions are known in closed form. Fortunately, however, the functional determinant will usually arise in the context of quadratic fluctuations around classical solutions in time-independent potentials (see in Section 4.3). If such a classical solution is known analytically, it will provide us automatically with a solution of the homogeneous differential equation (2.216). Some important examples will be discussed in Sections 17.4 and 17.11.

2.4.6   Generalization to D Dimensions

The above formulas have an obvious generalization to a D-dimensional version of the fluctuation action (2.199)

$$\mathcal{A} = \int_{t_a}^{t_b} dt \frac{M}{2} [(\delta \dot{\mathbf{x}})^2 - \delta \mathbf{x}^T \mathbf{\Omega}^2(t) \delta \mathbf{x}], \tag{2.252}$$

where $\mathbf{\Omega}^2(t)$ is a $D \times D$ matrix with elements $\Omega_{ij}^2(t)$. The fluctuation factor (2.200) generalizes to

$$F^N(t_b, t_a) = \frac{1}{\sqrt{2\pi\hbar i(t_b - t_a)/M}^{\,D}} \left[\frac{\det_N(-\epsilon^2 \nabla \overline{\nabla} - \epsilon^2 \mathbf{\Omega}^2)}{\det_N(-\epsilon^2 \nabla \overline{\nabla})}\right]^{-1/2}. \tag{2.253}$$

The fluctuation determinant is found by Gelfand-Yaglom's construction from a formula

$$D_{\text{ren}} = \det \mathbf{D}_a(t_b) = \det \mathbf{D}_b(t_a), \tag{2.254}$$

with the matrices $\mathbf{D}_a(t)$ and $\mathbf{D}_b(t)$ satisfying the classical equations of motion and initial conditions corresponding to (2.226) and (2.227):

$$[\partial_t^2 + \mathbf{\Omega}^2(t)]\mathbf{D}_a(t) = 0 \; ; \quad \mathbf{D}_a(t_a) = 0, \quad \dot{\mathbf{D}}_a(t_a) = \mathbf{1}, \tag{2.255}$$

$$[\partial_t^2 + \mathbf{\Omega}^2(t)]\mathbf{D}_b(t) = 0 \; ; \quad \mathbf{D}_b(t_b) = 0, \quad \dot{\mathbf{D}}_b(t_b) = -\mathbf{1}, \tag{2.256}$$

where $\mathbf{1}$ is the unit matrix in D dimensions. We can then repeat all steps in the last section and find the D-dimensional generalization of formulas (2.250):

$$D_{\text{ren}} = \left(\det \frac{\partial \ddot{x}_a^i}{\partial x_b^j}\right)^{-1} = \left[\det \left(-\frac{\partial \ddot{x}_b^i}{\partial x_a^j}\right)\right]^{-1}. \tag{2.257}$$

2.5　Harmonic Oscillator with Time-Dependent Frequency

The results of the last section put us in a position to solve exactly the path integral of a harmonic oscillator with arbitrary time-dependent frequency $\Omega(t)$. We shall first do this in coordinate space, later in momentum space.

2.5.1　Coordinate Space

Consider the path integral

$$(x_b t_b | x_a t_a) = \int \mathcal{D}x \exp\left\{\frac{i}{\hbar}\mathcal{A}[x]\right\}, \tag{2.258}$$

with the Lagrangian action

$$\mathcal{A}[x] = \frac{M}{2} \int_{t_a}^{t_b} dt \left[\dot{x}^2(t) - \Omega^2(t)x^2(t)\right], \tag{2.259}$$

which is harmonic with a time-dependent frequency. As in Eq. (2.14), the result can be written as a product of a fluctuation factor and an exponential containing the classical action:

$$(x_b t_b | x_a t_a) = \int \mathcal{D}x \, e^{i\mathcal{A}[x]/\hbar} = F_\Omega(t_b, t_a)e^{i\mathcal{A}_{\text{cl}}/\hbar}. \tag{2.260}$$

From the discussion in the last section we know that the fluctuation factor is, by analogy with (2.169), and recalling (2.241),

$$F_\Omega(t_b, t_a) = \frac{1}{\sqrt{2\pi i\hbar/M}} \frac{1}{\sqrt{D_a(t_b)}}. \tag{2.261}$$

The determinant $D_a(t_b) = D_{\text{ren}}$ may be expressed in terms of partial derivatives according to formulas (2.241) and (2.250):

$$F_\Omega(t_b, t_a) = \frac{1}{\sqrt{2\pi i\hbar/M}} \left(\frac{\partial x_b}{\partial \dot{x}_a} \right)^{-1/2} = \frac{1}{\sqrt{2\pi i\hbar/M}} \left(\frac{\partial \dot{x}_a}{\partial x_b} \right)^{1/2}, \tag{2.262}$$

where the first partial derivative is calculated from the function $x(x_a, \dot{x}_a; t)$, the second from $\dot{x}(x_b, x_a; t)$. Equivalently we may use (2.243) and the right-hand part of Eq. (2.250) to write

$$F_\Omega(t_b, t_a) = \frac{1}{\sqrt{2\pi i\hbar/M}} \left(-\frac{\partial x_a}{\partial \dot{x}_b} \right)^{-1/2} = \frac{1}{\sqrt{2\pi i\hbar/M}} \left(-\frac{\partial \dot{x}_b}{\partial x_a} \right)^{1/2}. \tag{2.263}$$

It remains to calculate the classical action $\mathcal{A}_{cl}$. This can be done in the same way as in Eqs. (2.153) to (2.157). After a partial integration, we have as before

$$\mathcal{A}_{cl} = \frac{M}{2}(x_b \dot{x}_b - x_a \dot{x}_a). \tag{2.264}$$

Exploiting the linear dependence of $\dot{x}_b$ and $\dot{x}_a$ on the endpoints x_b and x_a, we may rewrite this as

$$\mathcal{A}_{cl} = \frac{M}{2} \left(x_b \frac{\partial \dot{x}_b}{\partial x_b} x_b - x_a \frac{\partial \dot{x}_a}{\partial x_a} x_a + x_b \frac{\partial \dot{x}_b}{\partial x_a} x_a - x_a \frac{\partial \dot{x}_a}{\partial x_b} x_b \right). \tag{2.265}$$

Inserting the partial derivatives from (2.248) and (2.249) and using the equality of $D_a(t_b)$ and $D_b(t_a)$, we obtain the classical action

$$\mathcal{A}_{cl} = \frac{M}{2D_a(t_b)} \left[x_b^2 \dot{D}_a(t_b) - x_a^2 \dot{D}_b(t_a) - 2x_b x_a \right]. \tag{2.266}$$

Note that there exists another simple formula for the fluctuation determinant D_{ren}:

$$D_{ren} = D_a(t_b) = D_b(t_a) = -M \left(\frac{\partial^2}{\partial x_b \partial x_a} \mathcal{A}_{cl} \right)^{-1}. \tag{2.267}$$

For the harmonic oscillator with time-independent frequency ω, the Gelfand-Yaglom function $D_a(t)$ of Eq. (2.233) has the property (2.229) due to time reversal invariance, and (2.266) reproduces the known result (2.157).

The expressions containing partial derivatives are easily extended to D dimensions: We simply have to replace the partial derivatives $\partial x_b/\partial \dot{x}_a$, $\partial \dot{x}_b/\partial \dot{x}_a, \dots$ by the corresponding $D \times D$ matrices, and write the action as the associated quadratic form.

The D-dimensional versions of the fluctuation factors (2.262) are

$$F_\Omega(t_b, t_a) = \frac{1}{\sqrt{2\pi i\hbar/M}^D} \left[\det \frac{\partial x_b^i}{\partial \dot{x}_a^j} \right]^{-1/2} = \frac{1}{\sqrt{2\pi i\hbar/M}^D} \left[\det \frac{\partial \dot{x}_a^i}{\partial x_b^j} \right]^{1/2}. \tag{2.268}$$

All formulas for fluctuation factors hold initially only for sufficiently short times $t_b - t_a$. For larger times, they carry phase factors determined as before in (2.167). The fully defined expression may be written as

$$
F_\Omega(t_b, t_a) = \frac{1}{\sqrt{2\pi i\hbar/M}^D} \left| \det \frac{\partial x_b^i}{\partial \dot{x}_a^j} \right|^{-1/2} e^{-i\nu\pi/2} = \frac{1}{\sqrt{2\pi i\hbar/M}^D} \left| \det \frac{\partial \dot{x}_a^i}{\partial x_b^j} \right|^{1/2} e^{-i\nu\pi/2},
$$

(2.269)

where ν is the Maslov-Morse index. In the one-dimensional case it counts the turning points of the trajectory, in the multidimensional case the number of zeros in determinant $\det \partial x_b^i / \partial \dot{x}_a^j$ along the trajectory, if the zero is caused by a reduction of the rank of the matrix $\partial x_b^i / \partial \dot{x}_a^j$ by one unit. If it is reduced by more than one unit, ν increases accordingly. In this context, the number ν is also called the *Morse index* of the trajectory.

The zeros of the functional determinant are also called *conjugate points*. They are generalizations of the turning points in one-dimensional systems. The surfaces in **x**-space, on which the determinant vanishes, are called *caustics*. The conjugate points are the places where the orbits touch the caustics.[13]

Note that for infinitesimally short times, all fluctuation factors and classical actions coincide with those of a free particle. This is obvious for the time-independent harmonic oscillator, where the amplitude (2.175) reduces to that of a free particle in Eq. (2.130) in the limit $t_b \to t_a$. Since a time-dependent frequency is constant over an infinitesimal time, this same result holds also here. Expanding the solution of the equations of motion for infinitesimally short times as

$$
\mathbf{x}_b \approx (t_b - t_a)\dot{\mathbf{x}}_a + \mathbf{x}_a, \qquad \mathbf{x}_a \approx -(t_b - t_a)\dot{\mathbf{x}}_b + \mathbf{x}_b,
$$

(2.270)

we have immediately

$$
\frac{\partial x_b^i}{\partial \dot{x}_a^j} = \delta_{ij}(t_b - t_a), \qquad \frac{\partial x_a}{\partial \dot{x}_b^j} = -\delta_{ij}(t_b - t_a).
$$

(2.271)

Similarly, the expansions

$$
\dot{\mathbf{x}}_b \approx \dot{\mathbf{x}}_a \approx \frac{\mathbf{x}_b - \mathbf{x}_a}{t_b - t_a}
$$

(2.272)

lead to

$$
\frac{\partial \dot{x}_b^i}{\partial x_a^j} = -\delta_{ij}\frac{1}{t_b - t_a}, \qquad \frac{\partial \dot{x}_a^i}{\partial x_b^j} = \delta_{ij}\frac{1}{t_b - t_a}.
$$

(2.273)

Inserting the expansions (2.271) or (2.272) into (2.264) (in D dimensions), the action reduces approximately to the free-particle action

$$
\mathcal{A}_{\mathrm{cl}} \approx \frac{M}{2}\frac{(\mathbf{x}_b - \mathbf{x}_a)^2}{t_b - t_a}.
$$

(2.274)

---

[13]See M.C. Gutzwiller, *Chaos in Classical and Quantum Mechanics*, Springer, Berlin, 1990.

2.5.2   Momentum Space

Let us also find the time evolution amplitude in momentum space. For this we write the classical action (2.265) as a quadratic form

$$\mathcal{A}_{\rm cl} = \frac{M}{2}\,(x_b, x_a)\, A \begin{pmatrix} x_b \\ x_a \end{pmatrix} \tag{2.275}$$

with a matrix

$$A = \begin{pmatrix} \dfrac{\partial \dot{x}_b}{\partial x_b} & \dfrac{\partial \dot{x}_b}{\partial x_a} \\[2ex] -\dfrac{\partial \dot{x}_a}{\partial x_b} & -\dfrac{\partial \dot{x}_a}{\partial x_a} \end{pmatrix}. \tag{2.276}$$

The inverse of this matrix is

$$A^{-1} = \begin{pmatrix} \dfrac{\partial x_b}{\partial \dot{x}_b} & -\dfrac{\partial x_b}{\partial \dot{x}_a} \\[2ex] \dfrac{\partial x_a}{\partial \dot{x}_b} & -\dfrac{\partial x_a}{\partial \dot{x}_a} \end{pmatrix}. \tag{2.277}$$

The partial derivatives of x_b and x_a are calculated from the solution of the homogeneous differential equation (2.216) specified in terms of the final and initial velocities $\dot{x}_b$ and $\dot{x}_a$:

$$x(\dot{x}_b, \dot{x}_a; t) = \frac{1}{D_a(t_b)\dot{D}_b(t_a)+1}$$
$$\times \left\{ \left[D_a(t) + D_b(t)\dot{D}_a(t_b) \right] \dot{x}_a + \left[-D_b(t) + D_a(t)\dot{D}_b(t_a) \right] \dot{x}_b \right\}, \tag{2.278}$$

which yields

$$x_a = \frac{1}{D_a(t_b)\dot{D}_b(t_a)+1} \left[D_b(t_a)\dot{D}_a(t_a)\dot{x}_b - D_b(t_a)\dot{x}_b \right], \tag{2.279}$$

$$x_b = \frac{1}{D_a(t_b)\dot{D}_b(t_a)+1} \left[D_a(t_b)\dot{x}_a + D_a(t_b)\dot{D}_b(t_a)\dot{x}_b \right], \tag{2.280}$$

so that

$$A^{-1} = \frac{D_a(t_b)}{D_a(t_b)\dot{D}_b(t_a)+1} \begin{pmatrix} \dot{D}_b(t_a) & -1 \\ -1 & -\dot{D}_a(t_b) \end{pmatrix}. \tag{2.281}$$

The determinant of A is the Jacobian

$$\det A = -\frac{\partial(\dot{x}_b, \dot{x}_a)}{\partial(x_b, x_a)} = -\frac{D_a(t_b)\dot{D}_b(t_a)+1}{D_a(t_b)D_b(t_a)}. \tag{2.282}$$

We can now perform the Fourier transform of the time evolution amplitude and find, via a quadratic completion,

$$(p_b t_b | p_a t_a) = \int dx_b \, e^{-ip_b x_b/\hbar} \int dx_a \, e^{ip_a x_a/\hbar} (x_b t_b | x_a t_a) \tag{2.283}$$

$$= \sqrt{\frac{2\pi\hbar}{iM}} \sqrt{\frac{D_a(t_b)}{D_a(t_b)\dot{D}_b(t_a)+1}}$$

$$\times \exp\left\{ \frac{i}{\hbar}\frac{1}{2M} \frac{D_a(t_b)}{D_a(t_b)\dot{D}_b(t_a)+1} \left[-\dot{D}_b(t_a)p_b^2 + \dot{D}_a(t_b)p_a^2 - 2p_b p_a \right] \right\}.$$

Inserting here $D_a(t_b) = \sin\omega(t_b - t_a)/\omega$ and $\dot{D}_a(t_b) = \cos\omega(t_b - t_a)$, we recover the oscillator result (2.187).

In D dimensions, the classical action has the same quadratic form as in (2.275)

$$\mathcal{A}_{\text{cl}} = \frac{M}{2} (\mathbf{x}_b^T, \mathbf{x}_a^T) \, \mathbf{A} \begin{pmatrix} \mathbf{x}_b \\ \mathbf{x}_a \end{pmatrix} \tag{2.284}$$

with a matrix $\mathbf{A}$ generalizing (2.276) by having the partial derivatives replaced by the corresponding $D \times D$-matrices. The inverse is the $2D \times 2D$-version of (2.277), i.e.

$$\mathbf{A} = \begin{pmatrix} \dfrac{\partial \dot{\mathbf{x}}_b}{\partial \mathbf{x}_b} & \dfrac{\partial \dot{\mathbf{x}}_b}{\partial \mathbf{x}_a} \\[2mm] -\dfrac{\partial \dot{\mathbf{x}}_a}{\partial \mathbf{x}_b} & -\dfrac{\partial \dot{\mathbf{x}}_a}{\partial \mathbf{x}_a} \end{pmatrix}, \qquad \mathbf{A}^{-1} = \begin{pmatrix} \dfrac{\partial \mathbf{x}_b}{\partial \dot{\mathbf{x}}_b} & -\dfrac{\partial \mathbf{x}_b}{\partial \dot{\mathbf{x}}_a} \\[2mm] \dfrac{\partial \mathbf{x}_a}{\partial \dot{\mathbf{x}}_b} & -\dfrac{\partial \mathbf{x}_a}{\partial \dot{\mathbf{x}}_a} \end{pmatrix}. \tag{2.285}$$

The determinant of such a block matrix

$$\mathbf{A} = \begin{pmatrix} a & b \\ c & d \end{pmatrix} \tag{2.286}$$

is calculated after a triangular decomposition

$$\mathbf{A} = \begin{pmatrix} a & b \\ c & d \end{pmatrix} = \begin{pmatrix} a & 0 \\ c & 1 \end{pmatrix} \begin{pmatrix} 1 & a^{-1}b \\ 0 & d - ca^{-1}b \end{pmatrix} = \begin{pmatrix} 1 & b \\ 0 & d \end{pmatrix} \begin{pmatrix} a - bd^{-1}c & 0 \\ d^{-1}c & 1 \end{pmatrix} \tag{2.287}$$

in two possible ways as

$$\det \begin{pmatrix} a & b \\ c & d \end{pmatrix} = \det a \cdot \det (d - ca^{-1}b) = \det (a - bd^{-1}c) \cdot \det d, \tag{2.288}$$

depending whether $\det a$ or $\det b$ is nonzero. The inverse is in the first case

$$\mathbf{A} = \begin{pmatrix} 1 & -a^{-1}bx \\ 0 & x \end{pmatrix} \begin{pmatrix} a^{-1} & 0 \\ -ca^{-1} & 1 \end{pmatrix} = \begin{pmatrix} a^{-1} + a^{-1}bxca^{-1} & -a^{-1}bx \\ -xca^{-1} & x \end{pmatrix}, \quad x \equiv (d - ca^{-1}b)^{-1}. \tag{2.289}$$

The resulting amplitude in momentum space is

$$(\mathbf{p}_b t_b | \mathbf{p}_a t_a) = \int dx_b \, e^{-i\mathbf{p}_b \mathbf{x}_b/\hbar} \int dx_a \, e^{i\mathbf{p}_a \mathbf{x}_a/\hbar} (\mathbf{x}_b t_b | \mathbf{x}_a t_a)$$

$$= \frac{2\pi}{\sqrt{2\pi i\hbar M}} \frac{1}{\sqrt{D_{\text{ren}} \det \mathbf{A}}} \exp \left\{ \frac{i}{\hbar} \frac{1}{2M} \left[(\mathbf{p}_b^T, \mathbf{p}_a^T) \, \mathbf{A}^{-1} \begin{pmatrix} \mathbf{p}_b \\ \mathbf{p}_a \end{pmatrix} \right] \right\}. \tag{2.290}$$

Also in momentum space, the amplitude (2.290) reduces to the free-particle one in Eq. (2.138) in the limit of infinitesimally short time $t_b - t_a$: For the time-independent harmonic oscillator, this was shown in Eq. (2.189), and the time-dependence of $\Omega(t)$ becomes irrelevant in the limit of small $t_b - t_a \to 0$.

2.6 Free-Particle and Oscillator Wave Functions

In Eq. (1.331) we have expressed the time evolution amplitude of the free particle (2.71) as a Fourier integral

$$(x_b t_b | x_a t_a) = \int \frac{dp}{(2\pi\hbar)} e^{ip(x_b - x_a)/\hbar} e^{-ip^2(t_b - t_a)/2M\hbar}. \tag{2.291}$$

This expression contains the information on all stationary states of the system. To find these states we have to perform a spectral analysis of the amplitude. Recall that according to Section 1.7, the amplitude of an arbitrary time-independent system possesses a spectral representation of the form

$$(x_b t_b | x_a t_a) = \sum_{n=0}^{\infty} \psi_n(x_b) \psi_n^*(x_a) e^{-iE_n(t_b - t_a)/\hbar}, \qquad (2.292)$$

where E_n are the eigenvalues and $\psi_n(x)$ the wave functions of the stationary states. In the free-particle case the spectrum is continuous and the spectral sum is an integral. Comparing (2.292) with (2.291) we see that the Fourier decomposition itself happens to be the spectral representation. If the sum over n is written as an integral over the momenta, we can identify the wave functions as

$$\psi_p(x) = \frac{1}{\sqrt{2\pi\hbar}} e^{ipx}. \qquad (2.293)$$

For the time evolution amplitude of the harmonic oscillator

$$(x_b t_b | x_a t_a) = \frac{1}{\sqrt{2\pi i\hbar \sin\left[\omega(t_b - t_a)\right]/M\omega}} \qquad (2.294)$$

$$\times \exp\left\{\frac{iM\omega}{2\hbar \sin\left[\omega(t_b - t_a)\right]}\left[(x_b^2 + x_a^2)\cos\omega(t_b - t_a) - 2x_b x_a\right]\right\},$$

the procedure is not as straight-forward. Here we must make use of a summation formula for *Hermite polynomials* (see Appendix 2C) $H_n(x)$ due to Mehler:[14]

$$\frac{1}{\sqrt{1-a^2}} \exp\left\{-\frac{1}{2(1-a^2)}\left[(x^2 + x'^2)(1 + a^2) - 4xx'a\right]\right\}$$

$$= \exp(-x^2/2 - x'^2/2) \sum_{n=0}^{\infty} \frac{a^n}{2^n n!} H_n(x) H_n(x'), \qquad (2.295)$$

with

$$H_0(x) = 1, \ H_1(x) = 2x, \ H_2(x) = 4x^2 - 2, \ldots, \ H_n(x) = (-1)^n e^{x^2} \frac{d^n}{dx^n} e^{-x^2}. \qquad (2.296)$$

Identifying

$$x \equiv \sqrt{M\omega/\hbar}\, x_b, \quad x' \equiv \sqrt{M\omega/\hbar}\, x_a, \quad a \equiv e^{-i\omega(t_b - t_a)}, \qquad (2.297)$$

so that

$$\frac{a}{1-a^2} = \frac{1}{2i\sin\left[\omega(t_b - t_a)\right]}, \quad \frac{1+a^2}{1-a^2} = \frac{1 + e^{-2i\omega(t_b - t_a)}}{1 - e^{-2i\omega(t_b - t_a)}} = \frac{\cos\left[\omega(t_b - t_a)\right]}{i\sin\left[\omega(t_b - t_a)\right]}$$

---

[14]See P.M. Morse and H. Feshbach, *Methods of Theoretical Physics*, McGraw-Hill, New York, Vol. I, p. 781 (1953).

we arrive at the spectral representation

$$(x_b t_b | x_a t_a) = \sum_{n=0}^{\infty} \psi_n(x_b) \psi_n(x_a) e^{-i(n+1/2)\omega(t_b-t_a)}. \tag{2.298}$$

From this we deduce that the harmonic oscillator has the energy eigenvalues

$$E_n = \hbar\omega(n + 1/2) \tag{2.299}$$

and the wave functions

$$\psi_n(x) = N_n \lambda_\omega^{-1/2} e^{-x^2/2\lambda_\omega^2} H_n(x/\lambda_\omega). \tag{2.300}$$

Here, λ_ω is the natural length scale of the oscillator

$$\lambda_\omega \equiv \sqrt{\frac{\hbar}{M\omega}}, \tag{2.301}$$

and N_n the normalization constant

$$N_n = (1/2^n n! \sqrt{\pi})^{1/2}. \tag{2.302}$$

It is easy to check that the wave functions satisfy the orthonormality relation

$$\int_{-\infty}^{\infty} dx \, \psi_n(x) \psi_{n'}(x)^* = \delta_{nn'}, \tag{2.303}$$

using the well-known orthogonality relation of Hermite polynomials[15]

$$\frac{1}{2^n n! \sqrt{\pi}} \int_{-\infty}^{\infty} dx \, e^{-x^2} H_n(x) H_{n'}(x) = \delta_{n,n'}. \tag{2.304}$$

2.7 General Time-Dependent Harmonic Action

A simple generalization of the harmonic oscillator with time-dependent frequency allows also for a time-dependent mass, so that the action (2.305) becomes

$$\mathcal{A}[x] = \int_{t_a}^{t_b} dt \frac{M}{2} \left[g(t)\dot{x}^2(t) - \Omega^2(t)x^2(t) \right], \tag{2.305}$$

with some dimensionless time-dependent factor $g(t)$. This factor changes the measure of path integration so that the time evolution amplitude can no longer be calculated from (2.258). To find the correct measure we must return to the canonical path integral (2.29) which now reads

$$(x_b t_b | x_a t_a) = \int_{x(t_a)=x_a}^{x(t_b)=x_b} \mathcal{D}'x \int \frac{\mathcal{D}p}{2\pi\hbar} e^{i\mathcal{A}[p,x]/\hbar}, \tag{2.306}$$

with the canonical action

$$\mathcal{A}[p,x] = \int_{t_a}^{t_b} dt \left[p\dot{x} - \frac{p^2}{2Mg(t)} - \frac{M}{2}\Omega^2(t)x^2(t) \right]. \tag{2.307}$$

---

[15]I.S. Gradshteyn and I.M. Ryzhik, *op. cit.*, Formula 7.374.1.

Integrating the momentum variables out in the sliced form of this path integral as in Eqs. (2.49)–(2.51) yields

$$(x_b t_b | x_a t_a) \approx \frac{1}{\sqrt{2\pi\hbar i\epsilon/Mg(t_{N+1})}} \prod_{n=1}^{N} \left[\int_{-\infty}^{\infty} \frac{dx_n}{\sqrt{2\pi\hbar i\epsilon/Mg(t_n)}} \right] \exp\left(\frac{i}{\hbar} \mathcal{A}^N \right). \tag{2.308}$$

The continuum limit of this path integral is written as

$$(x_b t_b | x_a t_a) = \int \mathcal{D}x \sqrt{g} \exp\left\{ \frac{i}{\hbar} \mathcal{A}[x] \right\}, \tag{2.309}$$

with the action (2.305).

The classical orbits solve the equation of motion

$$\left[-\partial_t g(t)\partial_t - \Omega^2(t) \right] x(t) = 0, \tag{2.310}$$

which, by the transformation

$$\tilde{x}(t) = \sqrt{g(t)} x(t), \quad \tilde{\Omega}^2(t) = \frac{1}{g(t)} \left[\Omega^2(t) + \frac{\dot{g}^2(t)}{4g(t)} - \frac{\ddot{g}(t)}{2} \right], \tag{2.311}$$

can be reduced to the previous form

$$\sqrt{g(t)} \left[-\partial_t^2 - \tilde{\Omega}^2(t) \right] \tilde{x}(t) = 0. \tag{2.312}$$

The result of the path integration is therefore

$$(x_b t_b | x_a t_a) = \int \mathcal{D}x \sqrt{g}\, e^{i\mathcal{A}[x]/\hbar} = F(x_b, t_b; x_a, t_a) e^{i\mathcal{A}_{\mathrm{cl}}/\hbar}, \tag{2.313}$$

with a fluctuation factor [compare (2.261)]

$$F(x_b, t_b; x_a, t_a) = \frac{1}{\sqrt{2\pi i\hbar/M}} \frac{1}{\sqrt{D_a(t_b)}}, \tag{2.314}$$

where $D_a(t_b)$ is found from a generalization of the formulas (2.262)–(2.267). The classical action is

$$\mathcal{A}_{\mathrm{cl}} = \frac{M}{2} (g_b x_b \dot{x}_b - g_a x_a \dot{x}_a), \tag{2.315}$$

where $g_b \equiv g(t_b)$, $g_a \equiv g(t_a)$. The solutions of the equation of motion can be expressed in terms of modified Gelfand-Yaglom functions (2.226) and (2.227) with the properties

$$[\partial_t g(t)\partial_t + \Omega^2(t)]D_a(t) = 0 ; \quad D_a(t_a) = 0, \quad \dot{D}_a(t_a) = 1/g_a, \tag{2.316}$$

$$[\partial_t g(t)\partial_t + \Omega^2(t)]D_b(t) = 0 ; \quad D_b(t_b) = 0, \quad \dot{D}_b(t_b) = -1/g_b, \tag{2.317}$$

as in (2.246):

$$x(x_b, x_a; t) = \frac{D_b(t)}{D_b(t_a)} x_a + \frac{D_a(t)}{D_a(t_b)} x_b. \tag{2.318}$$

This allows us to write the classical action (2.315) in the form

$$\mathcal{A}_{\mathrm{cl}} = \frac{M}{2 D_a(t_b)} \left[g_b x_b^2 \dot{D}_a(t_b) - g_a x_a^2 \dot{D}_b(t_a) - 2x_b x_a \right]. \tag{2.319}$$

From this we find, as in (2.267),

$$D_{\text{ren}} = D_a(t_b) = D_b(t_a) = -M \left(\frac{\partial^2 \mathcal{A}_{\text{cl}}}{\partial x_b \partial x_a} \right)^{-1}, \qquad (2.320)$$

so that the fluctuation factor becomes

$$F(x_b, t_b; x_a, t_a) = \frac{1}{\sqrt{2\pi i \hbar}} \sqrt{-\frac{\partial^2 \mathcal{A}_{\text{cl}}}{\partial x_b \partial x_a}}. \qquad (2.321)$$

As an example take a free particle with a time-dependent mass term, where

$$D_a(t) = \int_{t_a}^{t} dt' \, g^{-1}(t'), \quad D_b(t) = \int_{t}^{t_b} dt' \, g^{-1}(t'), \quad D_{\text{ren}} = D_a(t_b) = D_b(t_a) = \int_{t_a}^{t_b} dt' \, g^{-1}(t'), \quad (2.322)$$

and the classical action reads

$$\mathcal{A}_{\text{cl}} = \frac{M}{2} \frac{(x_b - x_a)^2}{D_a(t_b)}. \qquad (2.323)$$

The result can easily be generalized to an arbitrary harmonic action

$$\mathcal{A} = \int_{t_a}^{t_b} dt \, \frac{M}{2} \left[g(t)\dot{x}^2 + 2b(t)x\dot{x} - \Omega^2(t)x^2 \right], \qquad (2.324)$$

which is extremized by the Euler-Lagrange equation [recall (1.8)]

$$\left[\partial_t g(t)\partial_t + \dot{b}(t) + \Omega^2(t) \right] x = 0. \qquad (2.325)$$

The solution of the path integral (2.313) is again given by (2.313), with the fluctuation factor (2.321), where $\mathcal{A}_{\text{cl}}$ is the action (2.324) along the classical path connecting the endpoints.

A further generalization to D dimensions is obvious by adapting the procedure in Subsection 2.4.6, which makes Eqs. (2.316)–(2.318). matrix equations.

2.8 Path Integrals and Quantum Statistics

The path integral approach is useful to also understand the thermal equilibrium properties of a system. We assume the system to have a *time-independent* Hamiltonian and to be in contact with a reservoir of temperature T. As explained in Section 1.7, the bulk thermodynamic quantities can be determined from the quantum-statistical partition function

$$Z = \text{Tr} \left(e^{-\hat{H}/k_B T} \right) = \sum_n e^{-E_n/k_B T}. \qquad (2.326)$$

This, in turn, may be viewed as an analytic continuation of the quantum-mechanical partition function

$$Z_{\text{QM}} = \text{Tr} \left(e^{-i(t_b - t_a)\hat{H}/\hbar} \right) \qquad (2.327)$$

to the imaginary time

$$t_b - t_a = -\frac{i\hbar}{k_B T} \equiv -i\hbar\beta. \qquad (2.328)$$

In the local particle basis $|x\rangle$, the quantum-mechanical trace corresponds to an integral over all positions so that the quantum-statistical partition function can be obtained by integrating the time evolution amplitude over $x_b = x_a$ and evaluating it at the analytically continued time:

$$Z \equiv \int_{-\infty}^{\infty} dx\, z(x) = \int_{-\infty}^{\infty} dx\, \langle x|e^{-\beta \hat{H}}|x\rangle = \int_{-\infty}^{\infty} dx\, (x\, t_b|x\, t_a)|_{t_b-t_a=-i\hbar\beta}. \quad (2.329)$$

The diagonal elements

$$z(x) \equiv \langle x|e^{-\beta \hat{H}}|x\rangle = (x\, t_b|x\, t_a)|_{t_b-t_a=-i\hbar\beta} \quad (2.330)$$

play the role of a *partition function density*. For a harmonic oscillator, this quantity has the explicit form [recall (2.173)]

$$z_\omega(x) = \frac{1}{\sqrt{2\pi\hbar/M}}\sqrt{\frac{\omega}{\sinh \hbar\beta\omega}}\exp\left\{-\frac{M\omega}{\hbar}\tanh\frac{\hbar\beta\omega}{2}\,x^2\right\}. \quad (2.331)$$

By splitting the Boltzmann factor $e^{-\beta \hat{H}}$ into a product of $N+1$ factors $e^{-\epsilon \hat{H}/\hbar}$ with $\epsilon = \hbar/k_B T(N+1)$, we can derive for Z a similar path integral representation just as for the corresponding quantum-mechanical partition function in (2.40), (2.46):

$$Z \equiv \prod_{n=1}^{N+1}\left[\int_{-\infty}^{\infty} dx_n\right] \quad (2.332)$$

$$\times \quad \langle x_{N+1}|e^{-\epsilon \hat{H}/\hbar}|x_N\rangle\langle x_N|e^{-\epsilon \hat{H}/\hbar}|x_{N-1}\rangle \times \ldots \times \langle x_2|e^{-\epsilon \hat{H}/\hbar}|x_1\rangle\langle x_1|e^{-\epsilon \hat{H}/\hbar}|x_{N+1}\rangle.$$

As in the quantum-mechanical case, the matrix elements $\langle x_n|e^{-\epsilon \hat{H}/\hbar}|x_{n-1}\rangle$ are re-expressed in the form

$$\langle x_n|e^{-\epsilon \hat{H}/\hbar}|x_{n-1}\rangle \approx \int_{-\infty}^{\infty}\frac{dp_n}{2\pi\hbar}e^{ip_n(x_n-x_{n-1})/\hbar-\epsilon H(p_n,x_n)/\hbar}, \quad (2.333)$$

with the only difference that there is now no imaginary factor i in front of the Hamiltonian. The product (2.332) can thus be written as

$$Z \approx \prod_{n=1}^{N+1}\left[\int_{-\infty}^{\infty} dx_n \int_{-\infty}^{\infty}\frac{dp_n}{2\pi\hbar}\right]\exp\left(-\frac{1}{\hbar}\mathcal{A}_e^N\right), \quad (2.334)$$

where $\mathcal{A}_e^N$ denotes the sum

$$\mathcal{A}_e^N = \sum_{n=1}^{N+1}\left[-ip_n(x_n-x_{n-1})+\epsilon H(p_n,x_n)\right]. \quad (2.335)$$

In the continuum limit $\epsilon \to 0$, the sum goes over into the integral

$$\mathcal{A}_e[p,x] = \int_0^{\hbar\beta} d\tau[-ip(\tau)\dot{x}(\tau)+H(p(\tau),x(\tau))], \quad (2.336)$$

and the partition function is given by the path integral

$$Z = \int \mathcal{D}x \int \frac{\mathcal{D}p}{2\pi\hbar} e^{-\mathcal{A}_e[p,x]/\hbar}. \tag{2.337}$$

In this expression, $p(\tau), x(\tau)$ may be considered as paths running along an "imaginary time axis" $\tau = it$. The expression $\mathcal{A}_e[p,x]$ is very similar to the mechanical canonical action (2.27). Since it governs the quantum-statistical path integrals it is called *quantum-statistical action* or *Euclidean action*, indicated by the subscript e. The name alludes to the fact that a D-dimensional Euclidean space extended by an imaginary-time axis $\tau = it$ has the same geometric properties as a $D+1$-dimensional Euclidean space. For instance, a four-vector in a Minkowski spacetime has a square length $dx^2 = -(cdt)^2 + (d\mathbf{x})^2$. Continued to an imaginary time, this becomes $dx^2 = (cd\tau)^2 + (d\mathbf{x})^2$ which is the square distance in a Euclidean four-dimensional space with four-vectors $(c\tau, \mathbf{x})$.

The integrand of the Euclidean action (2.337) is the *Euclidean Lagrangian* L_e. It is related to the Hamiltonian by the *Euclidean Legendre transform* [compare (1.9)]

$$H = L_e + i\frac{\partial L_e}{\partial \dot{x}}\dot{x} = L_e + ip\dot{x} \tag{2.338}$$

in which $\dot{x}$ is eliminated in favor of $p = \partial L_e/\partial \dot{x}$ [compare (1.10)].

Just as in the path integral for the quantum-mechanical partition function (2.46), the measure of integration $\oint \mathcal{D}x \int \mathcal{D}p/2\pi\hbar$ in the quantum-statistical expression (2.337) is automatically symmetric in all p's and x's:

$$\oint \mathcal{D}x \int \frac{\mathcal{D}p}{2\pi\hbar} = \oint \frac{\mathcal{D}p}{2\pi\hbar} \int \mathcal{D}x = \prod_{n=1}^{N+1} \iint_{-\infty}^{\infty} \frac{dx_n dp_n}{2\pi\hbar}. \tag{2.339}$$

The symmetry is of course due to the trace integration over all initial $\equiv$ final positions.

Most remarks made in connection with Eq. (2.46) carry over to the present case. The above path integral (2.337) is a natural extension of the rules of classical statistical mechanics. According to these, each cell in phase space $dxdp/h$ is occupied with equal statistical weight, with the probability factor e^{-E/k_BT}. In quantum statistics, the *paths* of all particles fluctuate evenly over the cells in *path phase space* $\prod_n dx(\tau_n)dp(\tau_n)/h$ $(\tau_n \equiv n\epsilon)$, each path carrying a probability factor $e^{-\mathcal{A}_e/\hbar}$ involving the Euclidean action of the system.

2.9 Density Matrix

The partition function does not determine any local thermodynamic quantities. Important local information resides in the thermal analog of the time evolution amplitude $\langle x_b|e^{-\hat{H}/k_BT}|x_a\rangle$. Consider, for instance, the diagonal elements of this amplitude renormalized by a factor Z^{-1}:

$$\rho(x_a) \equiv Z^{-1}\langle x_a|e^{-\hat{H}/k_BT}|x_a\rangle. \tag{2.340}$$

They determine the thermal average of the particle density of a quantum-statistical system. Due to (2.332), the factor Z^{-1} makes the spatial integral over ρ equal to unity:

$$\int_{-\infty}^{\infty} dx\, \rho(x) = 1. \tag{2.341}$$

By inserting into (2.340) a complete set of eigenfunctions $\psi_n(x)$ of the Hamiltonian operator $\hat{H}$, we find the spectral decomposition

$$\rho(x_a) = \sum_n |\psi_n(x_a)|^2 e^{-\beta E_n} \Big/ \sum_n e^{-\beta E_n}. \tag{2.342}$$

Since $|\psi_n(x_a)|^2$ is the probability distribution of the system in the eigenstate $|n\rangle$, while the ratio $e^{-\beta E_n}/\sum_n e^{-\beta E_n}$ is the normalized probability to encounter the system in the state $|n\rangle$, the quantity $\rho(x_a)$ represents the normalized average particle density in space as a function of temperature.

Note the limiting properties of $\rho(x_a)$. In the limit $T \to 0$, only the lowest energy state survives and $\rho(x_a)$ tends towards the particle distribution in the ground state

$$\rho(x_a) \xrightarrow{T \to 0} |\psi_0(x_a)|^2. \tag{2.343}$$

In the opposite limit of high temperatures, quantum effects are expected to become irrelevant and the partition function should converge to the classical expression (1.535) which is the integral over the phase space of the Boltzmann distribution

$$Z \xrightarrow{T \to \infty} Z_{\mathrm{cl}} = \int_{-\infty}^{\infty} dx \int_{-\infty}^{\infty} \frac{dp}{2\pi\hbar} e^{-H(p,x)/k_B T}. \tag{2.344}$$

We therefore expect the large-T limit of $\rho(x)$ to be equal to the *classical particle distribution*

$$\rho(x) \xrightarrow{T \to \infty} \rho_{\mathrm{cl}}(x) = Z_{\mathrm{cl}}^{-1} \int_{-\infty}^{\infty} \frac{dp}{2\pi\hbar} e^{-H(p,x)/k_B T}. \tag{2.345}$$

Within the path integral approach, this limit will be discussed in more detail in Section 2.13. At this place we roughly argue as follows: When going in the original time-sliced path integral (2.332) to large T, i.e., small $\tau_b - \tau_a = \hbar/k_B T$, we may keep only a single time slice and write

$$Z \approx \left[\int_{-\infty}^{\infty} dx \right] \langle x|e^{-\epsilon \hat{H}/\hbar}|x\rangle, \tag{2.346}$$

with

$$\langle x|e^{-\epsilon \hat{H}}|x\rangle \approx \int_{-\infty}^{\infty} \frac{dp_n}{2\pi\hbar} e^{-\epsilon H(p_n,x)/\hbar}. \tag{2.347}$$

After substituting $\epsilon = \tau_b - \tau_a$ this gives directly (2.345). Physically speaking, the path has at high temperatures "no (imaginary) time" to fluctuate, and only one term in the product of integrals needs to be considered.

If $H(p, x)$ has the standard form

$$H(p, x) = \frac{p^2}{2M} + V(x), \qquad (2.348)$$

the momentum integral is Gaussian in p and can be done using the formula

$$\int_{-\infty}^{\infty} \frac{dp}{2\pi\hbar} e^{-ap^2/2\hbar} = \frac{1}{\sqrt{2\pi\hbar a}}. \qquad (2.349)$$

This leads to the pure x-integral for the classical partition function

$$Z_{\text{cl}} = \int_{-\infty}^{\infty} \frac{dx}{\sqrt{2\pi\hbar^2/M k_B T}} e^{-V(x)/k_B T} = \int_{-\infty}^{\infty} \frac{dx}{l_e(\hbar\beta)} e^{-\beta V(x)}. \qquad (2.350)$$

In the second expression we have introduced the length

$$l_e(\hbar\beta) \equiv \sqrt{2\pi\hbar^2\beta/M}. \qquad (2.351)$$

It is the thermal (or Euclidean) analog of the characteristic length $l(t_b - t_a)$ introduced before in (2.126). It is called the *de Broglie wavelength associated with the temperature* $T = 1/k_B\beta$ or, in short, the *thermal de Broglie wavelength*.

Omitting the x-integration in (2.350) renders the large-T limit $\rho(x)$, the classical particle distribution

$$\rho(x) \xrightarrow{T\to\infty} \rho_{\text{cl}}(x) = Z_{\text{cl}}^{-1} \frac{1}{l_e(\hbar\beta)} e^{-\bar{V}(x)}. \qquad (2.352)$$

For a free particle, the integral over x in (2.350) diverges. If we imagine the length of the x-axis to be very large but finite, say equal to L, the partition function is equal to

$$Z_{\text{cl}} = \frac{L}{l_e(\hbar\beta)}. \qquad (2.353)$$

In D dimensions, this becomes

$$Z_{\text{cl}} = \frac{V_D}{l_e^D(\hbar\beta)}, \qquad (2.354)$$

where V_D is the volume of the D-dimensional system. For a harmonic oscillator with potential $M\omega^2 x^2/2$, the integral over x in (2.350) is finite and yields, in the D-dimensional generalization

$$Z_{\text{cl}} = \frac{l_\omega^D}{l^D(\hbar\beta)}, \qquad (2.355)$$

where

$$l_\omega \equiv \sqrt{\frac{2\pi}{\beta M \omega^2}} \qquad (2.356)$$

denotes the classical length scale defined by the frequency of the harmonic oscillator. It is related to the quantum-mechanical one λ_ω of Eq. (2.301) by

$$l_\omega l_e(\hbar\beta) = 2\pi \lambda_\omega^2. \tag{2.357}$$

Thus we obtain the *mnemonic rule* for going over from the partition function of a harmonic oscillator to that of a free particle: we must simply replace

$$l_\omega \xrightarrow[\omega \to 0]{} L, \tag{2.358}$$

or

$$\frac{1}{\omega} \xrightarrow[\omega \to 0]{} \sqrt{\frac{\beta M}{2\pi}} L. \tag{2.359}$$

The real-time version of this is, of course,

$$\frac{1}{\omega} \xrightarrow[\omega \to 0]{} \sqrt{\frac{(t_b - t_a)M}{2\pi\hbar}} L. \tag{2.360}$$

Let us write down a path integral representation for $\rho(x)$. Omitting in (2.337) the final trace integration over $x_b \equiv x_a$ and normalizing the expression by a factor Z^{-1}, we obtain

$$
\begin{aligned}
\rho(x_a) &= Z^{-1} \int_{x(0)=x_a}^{x(\hbar\beta)=x_b} \mathcal{D}'x \int \frac{\mathcal{D}p}{2\pi\hbar} e^{-\mathcal{A}_e[p,x]/\hbar} \\
&= Z^{-1} \int_{x(0)=x_a}^{x(\hbar\beta)=x_b} \mathcal{D}x e^{-\mathcal{A}_e[x]/\hbar}.
\end{aligned}
\tag{2.361}
$$

The thermal equilibrium expectation of an arbitrary Hermitian operator $\hat{O}$ is given by

$$\langle \hat{O} \rangle_T \equiv Z^{-1} \sum_n e^{-\beta E_n} \langle n|\hat{O}|n \rangle. \tag{2.362}$$

In the local basis $|x\rangle$, this becomes

$$\langle \hat{O} \rangle_T = Z^{-1} \iint_{-\infty}^{\infty} dx_b dx_a \langle x_b|e^{-\beta\hat{H}}|x_a \rangle \langle x_a|\hat{O}|x_b \rangle. \tag{2.363}$$

An arbitrary function of the position operator $\hat{x}$ has the expectation

$$\langle f(\hat{x}) \rangle_T = Z^{-1} \iint_{-\infty}^{\infty} dx_b dx_a \langle x_b|e^{-\beta\hat{H}}|x_a \rangle \delta(x_b - x_a) f(x_a) = \int dx \rho(x) f(x). \tag{2.364}$$

The particle density $\rho(x_a)$ determines the thermal averages of local observables.

If f depends also on the momentum operator $\hat{p}$, then the off-diagonal matrix elements $\langle x_b|e^{-\beta\hat{H}}|x_a \rangle$ are also needed. They are contained in the *density matrix* introduced for pure quantum systems in Eq. (1.221), and reads now in a thermal ensemble of temperature T:

$$\rho(x_b, x_a) \equiv Z^{-1} \langle x_b|e^{-\beta\hat{H}}|x_a \rangle, \tag{2.365}$$

whose diagonal values coincide with the above particle density $\rho(x_a)$.

It is useful to keep the analogy between quantum mechanics and quantum statistics as close as possible and to introduce the time translation operator along the imaginary time axis

$$\hat{U}_e(\tau_b, \tau_a) \equiv e^{-(\tau_b - \tau_a)\hat{H}/\hbar}, \qquad \tau_b > \tau_a, \tag{2.366}$$

defining its local matrix elements as *imaginary* or *Euclidean* time evolution amplitudes

$$(x_b \tau_b | x_a \tau_a) \equiv \langle x_b | \hat{U}_e(\tau_b, \tau_a) | x_a \rangle, \qquad \tau_b > \tau_a. \tag{2.367}$$

As in the real-time case, we shall only consider the causal time-ordering $\tau_b > \tau_a$. Otherwise the partition function and the density matrix do not exist in systems with energies up to infinity. Given the imaginary-time amplitudes, the partition function is found by integrating over the diagonal elements

$$Z = \int_{-\infty}^{\infty} dx(x\,\hbar\beta | x\,0), \tag{2.368}$$

and the density matrix

$$\rho(x_b, x_a) = Z^{-1}(x_b\hbar\beta | x_a 0). \tag{2.369}$$

For the sake of generality we may sometimes also consider the imaginary-time evolution operators for time-dependent Hamiltonians and the associated amplitudes. They are obtained by time-slicing the local matrix elements of the operator

$$\hat{U}(\tau_b, \tau_a) = T_\tau \exp\left[-\frac{1}{\hbar} \int_{\tau_a}^{\tau_b} d\tau \hat{H}(-i\tau)\right]. \tag{2.370}$$

Here T_τ is an ordering operator along the imaginary-time axis.

It must be emphasized that the usefulness of the operator (2.370) in describing thermodynamic phenomena is restricted to the Hamiltonian operator $\hat{H}(t)$ depending very weakly on the physical time t. The system has to remain close to equilibrium at all times. This is the range of validity of the so-called *linear response theory* (see Chapter 18 for more details).

The imaginary-time evolution amplitude (2.367) has a path integral representation which is obtained by dropping the final integration in (2.334) and relaxing the condition $x_b = x_a$:

$$(x_b \tau_b | x_a \tau_a) \approx \prod_{n=1}^{N} \left[\int_{-\infty}^{\infty} dx_n\right] \prod_{n=1}^{N+1} \left[\int_{-\infty}^{\infty} \frac{dp_n}{2\pi\hbar}\right] \exp\left(-\mathcal{A}_e^N/\hbar\right). \tag{2.371}$$

The time-sliced Euclidean action is

$$\mathcal{A}_e^N = \sum_{n=1}^{N+1} \left[-ip_n(x_n - x_{n-1}) + \epsilon H(p_n, x_n, \tau_n)\right] \tag{2.372}$$

(we have omitted the factor $-i$ in the τ-argument of H). In the continuum limit this is written as a path integral

$$(x_b \tau_b | x_a \tau_a) = \int \mathcal{D}'x \int \frac{\mathcal{D}p}{2\pi\hbar} \exp\left\{-\frac{1}{\hbar}\mathcal{A}_e[p, x]\right\} \tag{2.373}$$

[by analogy with (2.337)]. For a Hamiltonian of the standard form (2.7),

$$H(p, x, \tau) = \frac{p^2}{2M} + V(x, \tau),$$

with a smooth potential $V(x, \tau)$, the momenta can be integrated out, just as in (2.51), and the Euclidean version of the pure x-space path integral (2.52) leads to (2.53):

$$
\begin{aligned}
(x_b \tau_b | x_a \tau_a) &= \int \mathcal{D}x \exp\left\{-\frac{1}{\hbar}\int_0^{\hbar\beta} d\tau \left[\frac{M}{2}(\partial_\tau x)^2 + V(x, \tau)\right]\right\} \\
&\approx \frac{1}{\sqrt{2\pi\hbar\epsilon/M}} \prod_{n=1}^{N} \left[\int_{-\infty}^{\infty} \frac{dx_n}{\sqrt{2\pi\beta/M}}\right] \\
&\quad \times \exp\left\{-\frac{1}{\hbar}\epsilon \sum_{n=1}^{N+1} \left[\frac{M}{2}\left(\frac{x_n - x_{n-1}}{\epsilon}\right)^2 + V(x_n, \tau_n)\right]\right\}.
\end{aligned}
\tag{2.374}
$$

From this we calculate the quantum-statistical partition function

$$
\begin{aligned}
Z &= \int_{-\infty}^{\infty} dx \, (x\,\hbar\beta | x\,0) \\
&= \int dx \int_{x(0)=x}^{x(\hbar\beta)=x} \mathcal{D}x \, e^{-\mathcal{A}_e[x]/\hbar} = \oint \mathcal{D}x \, e^{-\mathcal{A}_e[x]/\hbar},
\end{aligned}
\tag{2.375}
$$

where $\mathcal{A}_e[x]$ is the Euclidean version of the Lagrangian action

$$\mathcal{A}_e[x] = \int_{\tau_a}^{\tau_b} d\tau \left[\frac{M}{2}x'^2 + V(x, \tau)\right]. \tag{2.376}$$

The prime denotes differentiation with respect to the imaginary time. As in the quantum-mechanical partition function in (2.61), the path integral $\oint \mathcal{D}x$ now stands for

$$\oint \mathcal{D}x \approx \prod_{n=1}^{N+1} \int_{-\infty}^{\infty} \frac{dx_n}{\sqrt{2\pi\hbar\epsilon/M}}. \tag{2.377}$$

It contains no extra $1/\sqrt{2\pi\hbar\epsilon/M}$ factor, as in (2.374), due to the trace integration over the exterior x.

The condition $x(\hbar\beta) = x(0)$ is most easily enforced by expanding $x(\tau)$ into a Fourier series

$$x(\tau) = \sum_{m=-\infty}^{\infty} \frac{1}{\sqrt{N+1}} e^{-i\omega_m \tau} x_m, \tag{2.378}$$

with the Matsubara frequencies

$$\omega_m \equiv 2\pi m k_B T/\hbar = \frac{2\pi m}{\hbar\beta}, \quad m = 0, \pm 1, \pm 2, \ldots . \tag{2.379}$$

When considered as functions on the entire τ-axis, the paths are periodic in $\hbar\beta$ at any τ, i.e.,

$$x(\tau) = x(\tau + \hbar\beta). \tag{2.380}$$

Thus the path integral for the quantum-statistical partition function comprises all periodic paths with a period $\hbar\beta$. In the time-sliced path integral (2.374), the coordinates $x(\tau)$ are needed only at the discrete times $\tau_n = n\epsilon$. Correspondingly, the sum over m in (2.378) can be restricted to run from $m = -N/2$ to $N/2$ for even N and from $-(N-1)/2$ to $(N+1)/2$ for odd N (see Fig. 2.3). In order to have a real $x(\tau_n)$, we must require that

$$x_m = x_{-m}^* \quad (\text{modulo } N+1). \tag{2.381}$$

Note that the Matsubara frequencies in the expansion of the paths $x(\tau)$ are now twice as big as the frequencies ν_m in the quantum fluctuations (2.110) (after analytic continuation of $t_b - t_a$ to $-i\hbar/k_B T$). Still, they have about the same total number, since they run over positive *and* negative integers. An exception is the zero frequency $\omega_m = 0$, which is *included* here, in contrast to the frequencies ν_m in (2.110) which run only over positive $m = 1, 2, 3, \ldots$. This is necessary to describe paths with arbitrary nonzero endpoints $x_b = x_a = x$ (included in the trace).

2.10 Quantum Statistics of the Harmonic Oscillator

The harmonic oscillator is a good example for solving the quantum-statistical path integral. The τ-axis is sliced at $\tau_n = n\epsilon$, with $\epsilon \equiv \hbar\beta/(N+1)$ $(n = 0, \ldots, N+1)$, and the partition function is given by the $N \to \infty$ -limit of the product of integrals

$$Z_\omega^N = \prod_{n=0}^{N} \left[\int_{-\infty}^{\infty} \frac{dx_n}{\sqrt{2\pi\hbar\epsilon/M}} \right] \exp\left(-\mathcal{A}_e^N/\hbar\right), \tag{2.382}$$

where $\mathcal{A}_e^N$ is the time-sliced Euclidean oscillator action

$$\mathcal{A}_e^N = \frac{M}{2\epsilon} \sum_{n=1}^{N+1} x_n(-\epsilon^2 \nabla\overline{\nabla} + \epsilon^2\omega^2)x_n. \tag{2.383}$$

Integrating out the x_n's, we find immediately

$$Z_\omega^N = \frac{1}{\sqrt{\det_{N+1}(-\epsilon^2\nabla\overline{\nabla} + \epsilon^2\omega^2)}}. \tag{2.384}$$

Let us evaluate the fluctuation determinant via the product of eigenvalues which diagonalize the matrix $-\epsilon^2\nabla\overline{\nabla} + \epsilon^2\omega^2$ in the sliced action (2.383). They are

$$\epsilon^2\Omega_m\overline{\Omega}_m + \epsilon^2\omega^2 = 2 - 2\cos\omega_m\epsilon + \epsilon^2\omega^2, \tag{2.385}$$

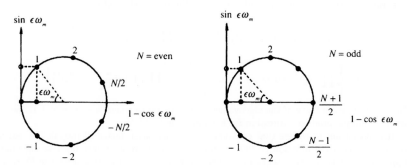

Figure 2.3 Illustration of the eigenvalues (2.385) of the fluctuation matrix in the action (2.383) for even and odd N.

with the Matsubara frequencies ω_m. For $\omega = 0$, the eigenvalues are pictured in Fig. 2.3. The action (2.383) becomes diagonal after going to the Fourier components x_m. To do this we arrange the real and imaginary parts $\operatorname{Re} x_m$ and $\operatorname{Im} x_m$ in a row vector

$$(\operatorname{Re} x_1, \operatorname{Im} x_1; \operatorname{Re} x_2, \operatorname{Im} x_2; \ldots; \operatorname{Re} x_n, \operatorname{Im} x_n; \ldots),$$

and see that it is related to the time-sliced positions $x_n = x(\tau_n)$ by a transformation matrix with the rows

$$
\begin{aligned}
T_{mn}x_n &= (T_m)_n x_n \\
&= \sqrt{\frac{2}{N+1}}\Big(\frac{1}{\sqrt{2}}, \ \cos\frac{m}{N+1}2\pi\cdot 1, \sin\frac{m}{N+1}2\pi\cdot 1, \\
&\qquad \cos\frac{m}{N+1}2\pi\cdot 2, \sin\frac{m}{N+1}2\pi\cdot 2, \ldots \\
&\qquad \ldots, \cos\frac{m}{N+1}2\pi\cdot n, \sin\frac{m}{N+1}2\pi\cdot n, \ldots\Big)_n x_n .
\end{aligned}
\tag{2.386}
$$

For each row index $m = 0, \ldots, N$, the column index n runs from zero to $N/2$ for even N, and to $(N+1)/2$ for odd N. In the odd case, the last column $\sin\frac{m}{N+1}2\pi\cdot n$ with $n = (N+1)/2$ vanishes identically and must be dropped, so that the number of columns in T_{mn} is in both cases $N+1$, as it should be. For odd N, the second-last column of T_{mn} is an alternating sequence ± 1. Thus, for a proper normalization, it has to be multiplied by an extra normalization factor $1/\sqrt{2}$, just as the elements in the first column. An argument similar to (2.115), (2.116) shows that the resulting matrix is orthogonal. Thus, we can diagonalize the sliced action in (2.383) as follows

$$
\mathcal{A}_e^N = \frac{M}{2}\epsilon
\begin{cases}
\Big[\omega^2 x_0^2 + 2\sum_{m=1}^{N/2}(\Omega_m\overline{\Omega}_m + \omega^2)|x_m|^2\Big] & \text{for } N = \text{even}, \\[2mm]
\Big[\omega^2 x_0^2 + (\Omega_{(N+1)/2}\overline{\Omega}_{(N+1)/2} + \omega^2)x_{N+1}^2 \\
\quad + 2\sum_{m=1}^{(N-1)/2}(\Omega_m\overline{\Omega}_m + \omega^2)|x_m|^2\Big] & \text{for } N = \text{odd}.
\end{cases}
\tag{2.387}
$$

Thanks to the orthogonality of T_{mn}, the measure $\prod_n \int_{-\infty}^{\infty} dx(\tau_n)$ transforms simply into

$$\int_{-\infty}^{\infty} dx_0 \prod_{m=1}^{N/2} \int_{-\infty}^{\infty} d\operatorname{Re} x_m \int_{-\infty}^{\infty} d\operatorname{Im} x_m \quad \text{for } N = \text{even},$$

$$\int_{-\infty}^{\infty} dx_0 \int_{-\infty}^{\infty} dx_{(N+1)/2} \prod_{m=1}^{(N-1)/2} \int_{-\infty}^{\infty} d\operatorname{Re} x_m \int_{-\infty}^{\infty} d\operatorname{Im} x_m \quad \text{for } N = \text{odd}.$$

$$\tag{2.388}$$

By performing the Gaussian integrals we obtain the partition function

$$
\begin{aligned}
Z_\omega^N &= \left[\det_{N+1}(-\epsilon^2 \nabla\overline{\nabla} + \epsilon^2\omega^2)\right]^{-1/2} = \left[\prod_{m=0}^{N}(\epsilon^2\Omega_m\overline{\Omega}_m + \epsilon^2\omega^2)\right]^{-1/2} \\
&= \left\{\prod_{m=0}^{N}\left[2(1-\cos\omega_m\epsilon) + \epsilon^2\omega^2\right]\right\}^{-1/2} = \left[\prod_{m=0}^{N}\left(4\sin^2\frac{\omega_m\epsilon}{2} + \epsilon^2\omega^2\right)\right]^{-1/2}.
\end{aligned}
\tag{2.389}
$$

Thanks to the periodicity of the eigenvalues under the replacement $n \to n + N + 1$, the result has become a unique product expression for both even and odd N.

It is important to realize that contrary to the fluctuation factor (2.160) in the real-time amplitude, the partition function (2.389) contains the square root of only positive eigenmodes as a unique result of Gaussian integrations. There are no phase subtleties as in the Fresnel integral (1.333).

To calculate the product, we observe that upon decomposing

$$
\sin^2\frac{\omega_m\epsilon}{2} = \left(1 + \cos\frac{\omega_m\epsilon}{2}\right)\left(1 - \cos\frac{\omega_m\epsilon}{2}\right),
\tag{2.390}
$$

the sequence of first factors

$$
1 + \cos\frac{\omega_m\epsilon}{2} \equiv 1 + \cos\frac{\pi m}{N+1}
\tag{2.391}
$$

runs for $m = 1, \dots N$ through the same values as the sequence of second factors

$$
1 - \cos\frac{\omega_m\epsilon}{2} = 1 - \cos\frac{\pi m}{N+1} \equiv 1 + \cos\pi\frac{N+1-m}{N+1},
\tag{2.392}
$$

except in an opposite order. Thus, separating out the $m = 0$ -term, we rewrite (2.389) in the form

$$
Z_\omega^N = \frac{1}{\epsilon\omega}\left[\prod_{m=1}^{N}2\left(1 - \cos\frac{\omega_m\epsilon}{2}\right)\right]^{-1}\left[\prod_{m=1}^{N}\left(1 + \frac{\epsilon^2\omega^2}{4\sin^2\frac{\omega_m\epsilon}{2}}\right)\right]^{-1/2}.
\tag{2.393}
$$

The first factor on the right-hand side is the quantum-mechanical fluctuation determinant of the free-particle determinant $\det_N(-\epsilon^2\nabla\overline{\nabla}) = N+1$ [see (2.123)], so that we obtain for both even and odd N

$$
Z_\omega^N = \frac{k_B T}{\hbar\omega}\left[\prod_{m=1}^{N}\left(1 + \frac{\epsilon^2\omega^2}{4\sin^2\frac{\omega_m\epsilon}{2}}\right)\right]^{-1/2}.
\tag{2.394}
$$

To evaluate the remaining product, we must distinguish again between even and odd cases of N. For even N, where every eigenvalue occurs twice (see Fig. 2.3), we obtain

$$
Z_\omega^N = \frac{k_B T}{\hbar\omega}\left[\prod_{m=1}^{N/2}\left(1 + \frac{\epsilon^2\omega^2}{4\sin^2\frac{m\pi}{N+1}}\right)\right]^{-1}.
\tag{2.395}
$$

For odd N, the term with $m = (N+1)/2$ occurs only once and must be treated separately so that

$$
Z_\omega^N = \frac{k_B T}{\hbar\omega}\left[\left(1 + \frac{\epsilon^2\omega^2}{4}\right)^{1/2}\prod_{m=1}^{(N-1)/2}\left(1 + \frac{\epsilon^2\omega^2}{4\sin^2\frac{m\pi}{N+1}}\right)\right]^{-1}.
\tag{2.396}
$$

We now introduce the parameter $\tilde{\omega}_e$, the Euclidean analog of (2.161), via the equations

$$
\sin i\frac{\tilde{\omega}_e\epsilon}{2} \equiv i\frac{\omega\epsilon}{2}, \quad \sinh\frac{\tilde{\omega}_e\epsilon}{2} \equiv \frac{\omega\epsilon}{2}.
\tag{2.397}
$$

In the odd case, the product formula[16]

$$\prod_{m=1}^{(N-1)/2} \left[1 - \frac{\sin^2 x}{\sin^2 \frac{m\pi}{(N+1)}} \right] = \frac{2}{\sin 2x} \frac{\sin[(N+1)x]}{(N+1)} \tag{2.398}$$

[similar to (2.163)] yields, with $x = \tilde{\omega}_e \epsilon/2$,

$$Z_\omega^N = \frac{k_B T}{\hbar \omega} \left[\frac{1}{\sinh(\tilde{\omega}_e \epsilon/2)} \frac{\sinh[(N+1)\tilde{\omega}_e \epsilon/2]}{N+1} \right]^{-1}. \tag{2.399}$$

In the even case, the formula[17]

$$\prod_{m=1}^{N/2} \left[1 - \frac{\sin^2 x}{\sin^2 \frac{m\pi}{(N+1)}} \right] = \frac{1}{\sin x} \frac{\sin[(N+1)x]}{(N+1)}, \tag{2.400}$$

produces once more the same result as in Eq. (2.399). Inserting Eq. (2.397) leads to the partition function on the sliced imaginary time axis:

$$Z_\omega^N = \frac{1}{2\sinh(\hbar \tilde{\omega}_e \beta/2)}. \tag{2.401}$$

The partition function can be expanded into the following series

$$Z_\omega^N = e^{-\hbar\tilde{\omega}_e/2k_B T} + e^{-3\hbar\tilde{\omega}_e/2k_B T} + e^{-5\hbar\tilde{\omega}_e/2k_B T} + \dots . \tag{2.402}$$

By comparison with the general spectral expansion (2.326), we display the energy eigenvalues of the system:

$$E_n = \left(n + \frac{1}{2} \right) \hbar \tilde{\omega}_e. \tag{2.403}$$

They show the typical linearly rising oscillator sequence with

$$\tilde{\omega}_e = \frac{2}{\epsilon} \operatorname{arsinh} \frac{\omega \epsilon}{2} \tag{2.404}$$

playing the role of the frequency on the sliced time axis, and $\hbar\tilde{\omega}_e/2$ being the zero-point energy.

In the continuum limit $\epsilon \to 0$, the time-sliced partition function Z_ω^N goes over into the usual oscillator partition function

$$Z_\omega = \frac{1}{2\sinh(\beta\hbar\omega/2)}. \tag{2.405}$$

In D dimensions this becomes, of course, $[2\sinh(\beta\hbar\omega/2)]^{-D}$, due to the additivity of the action in each component of $\mathbf{x}$.

Note that the continuum limit of the product in (2.394) can also be taken factor by factor. Then Z_ω becomes

$$Z_\omega = \frac{k_B T}{\hbar\omega} \left[\prod_{m=1}^{\infty} \left(1 + \frac{\omega^2}{\omega_m^2} \right) \right]^{-1}. \tag{2.406}$$

According to formula (2.171), the product $\prod_{m=1}^{\infty} \left(1 + \frac{x^2}{m^2\pi^2} \right)$ converges rapidly against $\sinh x/x$ and we find with $x = \hbar\omega\beta/2$

$$Z_\omega = \frac{k_B T}{\hbar\omega} \frac{\hbar\omega/2k_B T}{\sinh(\hbar\omega/2k_B T)} = \frac{1}{2\sinh(\beta\hbar\omega/2)}. \tag{2.407}$$

---

[16] I.S. Gradshteyn and I.M. Ryzhik, op. cit., Formula 1.391.1.
[17] *ibid.*, formula 1.391.3.

As discussed after Eq. (2.181), the continuum limit can be taken in each factor since the product in (2.394) contains only ratios of frequencies.

Just as in the quantum-mechanical case, this procedure of obtaining the continuum limit can be summarized in the sequence of equations arriving at a ratio of differential operators

$$
\begin{aligned}
Z_\omega^N &= \left[\det_{N+1}(-\epsilon^2 \nabla \overline{\nabla} + \epsilon^2 \omega^2) \right]^{-1/2} \\
&= \left[\det'_{N+1}(-\epsilon^2 \nabla \overline{\nabla}) \right]^{-1/2} \left[\frac{\det_{N+1}(-\epsilon^2 \nabla \overline{\nabla} + \epsilon^2 \omega^2)}{\det'_{N+1}(-\epsilon^2 \nabla \overline{\nabla})} \right]^{-1/2} \\
&\xrightarrow{\epsilon \to 0} \frac{k_B T}{\hbar} \left[\frac{\det(-\partial_\tau^2 + \omega^2)}{\det'(-\partial_\tau^2)} \right]^{-1/2} = \frac{k_B T}{\hbar \omega} \prod_{m=1}^{\infty} \left[\frac{\omega_m^2 + \omega^2}{\omega_m^2} \right]^{-1}.
\end{aligned} \tag{2.408}
$$

In the $\omega = 0$-determinants, the zero Matsubara frequency is excluded to obtain a finite expression. This is indicated by a prime. The differential operator $-\partial_\tau^2$ acts on real functions which are periodic under the replacement $\tau \to \tau + \hbar\beta$. Remember that each eigenvalue ω_m^2 of $-\partial_\tau^2$ occurs twice, except for the zero frequency $\omega_0 = 0$, which appears only once.

Let us finally mention that the results of this section could also have been obtained directly from the quantum-mechanical amplitude (2.173) [or with the discrete times from (2.197)] by an analytic continuation of the time difference $t_b - t_a$ to imaginary values $-i(\tau_b - \tau_a)$:

$$
\begin{aligned}
(x_b \tau_b | x_a \tau_a) &= \frac{1}{\sqrt{2\pi\hbar/M}} \sqrt{\frac{\omega}{\sinh \omega(\tau_b - \tau_a)}} \\
&\times \exp\left\{ -\frac{1}{2\hbar} \frac{M\omega}{\sinh \omega(\tau_b - \tau_a)} [(x_b^2 + x_a^2) \cosh \omega(\tau_b - \tau_a) - 2x_b x_a] \right\}.
\end{aligned} \tag{2.409}
$$

By setting $x = x_b = x_a$ and integrating over x, we obtain [compare (2.331)]

$$
\begin{aligned}
Z_\omega = \int_{-\infty}^{\infty} dx \, (x\,\tau_b | x\,\tau_a) &= \frac{1}{\sqrt{2\pi\hbar(\tau_b - \tau_a)/M}} \sqrt{\frac{\omega(\tau_b - \tau_a)}{\sinh[\omega(\tau_b - \tau_a)]}} \\
&\times \frac{\sqrt{2\pi\hbar \sinh[\omega(\tau_b - \tau_a)]/\omega M}}{2 \sinh[\omega(\tau_b - \tau_a)/2]} = \frac{1}{2 \sinh[\omega(\tau_b - \tau_a)/2]}.
\end{aligned} \tag{2.410}
$$

Upon equating $\tau_b - \tau_a = \hbar\beta$, we retrieve the partition function (2.405). A similar treatment of the discrete-time version (2.197) would have led to (2.401). The main reason for presenting an independent direct evaluation in the space of real periodic functions was to display the frequency structure of periodic paths and to see the difference with respect to the quantum-mechanical paths with fixed ends. We also wanted to show how to handle the ensuing product expressions.

For applications in polymer physics (see Chapter 15) one also needs the partition function of all path fluctuations with open ends

$$
\begin{aligned}
Z_\omega^{\text{open}} &= \int_{-\infty}^{\infty} dx_b \int_{-\infty}^{\infty} dx_a \, (x_b \tau_b | x_a \tau_a) = \frac{1}{\sqrt{2\pi\hbar(\tau_b - \tau_a)/M}} \sqrt{\frac{\omega(\tau_b - \tau_a)}{\sinh[\omega(\tau_b - \tau_a)]}} \frac{2\pi\hbar}{M\omega} \\
&= \sqrt{\frac{2\pi\hbar}{M\omega}} \frac{1}{\sqrt{\sinh[\omega(\tau_b - \tau_a)]}}.
\end{aligned} \tag{2.411}
$$

The prefactor is $\sqrt{2\pi}$ times the length scale λ_ω of Eq. (2.301).

2.11 Time-Dependent Harmonic Potential

It is often necessary to calculate thermal fluctuation determinants for the case of a time-dependent frequency $\Omega(\tau)$ which is periodic under $\tau \to \tau + \hbar\beta$. As in Section 2.3.6, we consider the amplitude

$$
(x_b \tau_b | x_a \tau_a) = \int \mathcal{D}'x \int \frac{\mathcal{D}p}{2\pi\hbar} e^{-\int_{\tau_a}^{\tau_b} d\tau [-ip\dot{x} + p^2/2M + M\Omega^2(\tau)x^2/2]/\hbar}
$$

$$= \int \mathcal{D}x e^{-\int_{\tau_a}^{\tau_b} d\tau [M\dot{x}^2 + \Omega^2(\tau)x^2]/2\hbar}. \tag{2.412}$$

The time-sliced fluctuation factor is [compare (2.200)]

$$F^N(\tau_a - \tau_b) = \det_{N+1}[-\epsilon^2 \nabla \overline{\nabla} + \epsilon \Omega^2(\tau)]^{-1/2}, \tag{2.413}$$

with the continuum limit

$$F(\tau_a - \tau_b) = \frac{k_B T}{\hbar} \left[\frac{\det(-\partial_\tau^2 + \Omega^2(\tau))}{\det'(-\partial_\tau^2)} \right]^{-1/2}. \tag{2.414}$$

Actually, in the thermal case it is preferable to use the oscillator result for normalizing the fluctuation factor, rather than the free-particle result, and to work with the formula

$$F(\tau_b, \tau_a) = \frac{1}{2\sinh(\beta\hbar\omega/2)} \left[\frac{\det(-\partial_\tau^2 + \Omega^2(\tau))}{\det(-\partial_\tau^2 + \omega^2)} \right]^{-1/2}. \tag{2.415}$$

This has the advantage that the determinant in the denominator contains no zero eigenvalue which would require a special treatment as in (2.408); the operator $-\partial_\tau^2 + \omega^2$ is positive.

As in the quantum-mechanical case, the spectrum of eigenvalues is not known for general $\Omega(\tau)$. It is, however, possible to find a differential equation for the entire determinant, analogous to the Gelfand-Yaglom formula (2.207), with the initial condition (2.212), although the derivation is now much more tedious. The origin of the additional difficulties lies in the periodic boundary condition which introduces additional nonvanishing elements -1 in the upper right and lower left corners of the matrix $-\epsilon^2 \nabla \overline{\nabla}$ [compare (2.102)]:

$$-\epsilon^2 \nabla \overline{\nabla} = \begin{pmatrix} 2 & -1 & 0 & \dots & 0 & 0 & -1 \\ -1 & 2 & -1 & \dots & 0 & 0 & 0 \\ \vdots & & & & & & \vdots \\ 0 & 0 & 0 & \dots & -1 & 2 & -1 \\ -1 & 0 & 0 & \dots & 0 & -1 & 2 \end{pmatrix}. \tag{2.416}$$

To better understand the relation with the previous result we shall replace the corner elements -1 by $-\alpha$ which can be set equal to zero at the end, for a comparison. Adding to $-\epsilon^2 \nabla \overline{\nabla}$ a time-dependent frequency matrix we then consider the fluctuation matrix

$$-\epsilon^2 \nabla \overline{\nabla} + \epsilon^2 \Omega^2 = \begin{pmatrix} 2 + \epsilon^2 \Omega_{N+1}^2 & -1 & 0 & \dots & 0 & -\alpha \\ -1 & 2 + \epsilon^2 \Omega_N^2 & -1 & \dots & 0 & 0 \\ \vdots & & & & & \vdots \\ -\alpha & 0 & 0 & \dots & -1 & 2 + \epsilon^2 \Omega_1^2 \end{pmatrix}. \tag{2.417}$$

Let us denote the determinant of this $(N+1) \times (N+1)$ matrix by $\tilde{D}_{N+1}$. Expanding it along the first column, it is found to satisfy the equation

$$\tilde{D}_{N+1} = (2 + \epsilon^2 \Omega_{N+1}^2) \tag{2.418}$$

$$\times \det_N \begin{pmatrix} 2 + \epsilon^2 \Omega_N^2 & -1 & 0 & \dots & 0 & 0 \\ \vdots & & & & & \vdots \\ 0 & 0 & 0 & \dots & -1 & 2 + \epsilon^2 \Omega_1^2 \end{pmatrix}$$

$$+\det_N \begin{pmatrix} -1 & 0 & 0 & 0 & \dots & 0 & -\alpha \\ -1 & 2 + \epsilon^2 \Omega_{N-1}^2 & -1 & 0 & \dots & 0 & 0 \\ 0 & -1 & 2 + \epsilon^2 \Omega_{N-2}^2 & -1 & \dots & 0 & 0 \\ \vdots & & & & & & \vdots \\ 0 & 0 & 0 & 0 & \dots & -1 & 2 + \epsilon^2 \Omega_1^2 \end{pmatrix}$$

$$+(-1)^{N+1}\alpha\det{}_N\begin{pmatrix} -1 & 0 & 0 & \ldots & 0 & -\alpha \\ 2+\epsilon^2\Omega_N^2 & -1 & 0 & \ldots & 0 & 0 \\ -1 & 2+\epsilon^2\Omega_{N-1}^2 & -1 & \ldots & 0 & 0 \\ \vdots & & & & & \vdots \\ 0 & 0 & 0 & \ldots & 2+\epsilon^2\Omega_2^2 & -1 \end{pmatrix}.$$

The first determinant was encountered before in Eq. (2.202) (except that there it appeared with $-\epsilon^2\Omega^2$ instead of $\epsilon^2\Omega^2$). There it was denoted by D_N, satisfying the difference equation

$$\left(-\epsilon^2\overline{\nabla}\nabla+\epsilon^2\Omega_{N+1}^2\right)D_N=0,\tag{2.419}$$

with the initial conditions

$$\begin{aligned} D_1 &= 2+\epsilon^2\Omega_1^2, \\ D_2 &= (2+\epsilon^2\Omega_1^2)(2+\epsilon^2\Omega_2^2)-1. \end{aligned}\tag{2.420}$$

The second determinant in (2.418) can be expanded with respect to its first column yielding

$$-D_{N-1}-\alpha.\tag{2.421}$$

The third determinant is more involved. When expanded along the first column it gives

$$(-1)^N\left[1+(2+\epsilon^2\Omega_N^2)H_{N-1}-H_{N-2}\right],\tag{2.422}$$

with the $(N-1)\times(N-1)$ determinant

$$H_{N-1}\equiv(-1)^{N-1}\tag{2.423}$$

$$\times\det{}_{N-1}\begin{pmatrix} 0 & 0 & 0 & \ldots & 0 & 0 & -\alpha \\ 2+\epsilon^2\Omega_{N-1}^2 & -1 & 0 & \ldots & 0 & 0 & 0 \\ -1 & 2+\epsilon^2\Omega_{N-2}^2 & -1 & \ldots & 0 & 0 & 0 \\ \vdots & & & & & & \vdots \\ 0 & 0 & 0 & \ldots & -1 & 2+\epsilon^2\Omega_2^2 & -1 \end{pmatrix}.$$

By expanding this along the first column, we find that H_N satisfies the same difference equation as D_N:

$$(-\epsilon^2\overline{\nabla}\nabla+\epsilon^2\Omega_{N+1}^2)H_N=0.\tag{2.424}$$

However, the initial conditions for H_N are different:

$$H_2 = \begin{vmatrix} 0 & -\alpha \\ 2+\epsilon^2\Omega_2^2 & -1 \end{vmatrix}=\alpha(2+\epsilon^2\Omega_2^2),\tag{2.425}$$

$$\begin{aligned} H_3 &= -\begin{vmatrix} 0 & 0 & -\alpha \\ 2+\epsilon^2\Omega_3^2 & -1 & 0 \\ -1 & 2+\epsilon^2\Omega_2^2 & -1 \end{vmatrix} \\ &= \alpha\left[(2+\epsilon^2\Omega_2^2)(2+\epsilon^2\Omega_3^2)-1\right]. \end{aligned}\tag{2.426}$$

They show that H_N is in fact equal to αD_{N-1}, provided we shift Ω_N^2 by one lattice unit upwards to Ω_{N+1}^2. Let us indicate this by a superscript +, i.e., we write

$$H_N=\alpha D_{N-1}^+.\tag{2.427}$$

Thus we arrive at the equation

$$\begin{aligned} \tilde{D}_{N+1} &= (2+\epsilon^2\Omega_N^2)D_N-D_{N-1}-\alpha \\ &\quad -\alpha[1+(2+\epsilon^2\Omega_N^2)\alpha D_{N-2}^+-\alpha D_{N-3}^+]. \end{aligned}\tag{2.428}$$

Using the difference equations for D_N and D_N^+, this can be brought to the convenient form

$$\tilde{D}_{N+1} = D_{N+1} - \alpha^2 D_{N-1}^+ - 2\alpha. \tag{2.429}$$

For quantum-mechanical fluctuations with $\alpha = 0$, this reduces to the earlier result in Section 2.3.6. For periodic fluctuations with $\alpha = 1$, the result is

$$\tilde{D}_{N+1} = D_{N+1} - D_{N-1}^+ - 2. \tag{2.430}$$

In the continuum limit, $D_{N+1} - D_{N-1}^+$ tends towards $2\dot{D}_{\mathrm{ren}}$, where $D_{\mathrm{ren}}(\tau) = D_a(t)$ is the imaginary-time version of the Gelfand-Yaglom function in Section 2.4 solving the homogenous differential equation (2.213), with the initial conditions (2.211) and (2.212), or Eqs. (2.226). The corresponding properties are now:

$$\left[-\partial_\tau^2 + \Omega^2(\tau)\right] D_{\mathrm{ren}}(\tau) = 0, \quad D_{\mathrm{ren}}(0) = 0, \quad \dot{D}_{\mathrm{ren}}(0) = 1. \tag{2.431}$$

In terms of $D_{\mathrm{ren}}(\tau)$, the determinant is given by the *Gelfand-Yaglom-like* formula

$$\det(-\epsilon^2 \overline{\nabla}\nabla + \epsilon\Omega^2)_T \xrightarrow{\epsilon \to 0} 2[\dot{D}_{\mathrm{ren}}(\hbar\beta) - 1], \tag{2.432}$$

and the partition function reads

$$Z_\Omega = \frac{1}{\sqrt{2\left[\dot{D}_{\mathrm{ren}}(\hbar\beta) - 1\right]}}. \tag{2.433}$$

The result may be checked by going back to the amplitude $(x_b t_b | x_a t_a)$ of Eq. (2.260), continuing it to imaginary times $t = i\tau$, setting $x_b = x_a = x$, and integrating over all x. The result is

$$Z_\Omega = \frac{1}{2\sqrt{\dot{D}_a(t_b) - 1}}, \quad t_b = i\hbar\beta, \tag{2.434}$$

in agreement with (2.433).

As an example, take the harmonic oscillator for which the solution of (2.431) is

$$D_{\mathrm{ren}}(\tau) = \frac{1}{\omega}\sinh\omega\tau \tag{2.435}$$

[the analytically continued (2.214)]. Then

$$2[\dot{D}_{\mathrm{ren}}(\tau) - 1] = 2(\cosh\beta\hbar\omega - 1) = 4\sinh^2(\beta\hbar\omega/2), \tag{2.436}$$

and we find the correct partition function:

$$\begin{aligned} Z_\omega &= \left\{2[\dot{D}_{\mathrm{ren}}(\tau) - 1]\right\}^{-1/2}\Big|_{\tau=\hbar\beta} \\ &= \frac{1}{2\sinh(\beta\hbar\omega/2)}. \end{aligned} \tag{2.437}$$

On a sliced imaginary-time axis, the case of a constant frequency $\Omega^2 \equiv \omega^2$ is solved as follows. From Eq. (2.208) we take the ordinary Gelfand-Yaglom function D_N, and continue it to Euclidean $\tilde{\omega}_{\mathrm{e}}$, yielding the imaginary-time version

$$D_N = \frac{\sinh(N+1)\tilde{\omega}_{\mathrm{e}}\epsilon}{\sinh\tilde{\omega}_{\mathrm{e}}\epsilon}. \tag{2.438}$$

Then we use formula (2.430), which simplifies for a constant $\Omega^2 \equiv \omega^2$ for which $D_{N-1}^+ = D_{N-1}$, and calculate

$$
\begin{aligned}
\tilde{D}_{N+1} &= \frac{1}{\sinh \tilde{\omega}_e \epsilon} [\sinh(N+2)\tilde{\omega}_e \epsilon - \sinh N\tilde{\omega}_e \epsilon] - 2 \\
&= 2 [\cosh(N+1)\tilde{\omega}_e \epsilon - 1] = 4 \sinh^2[(N+1)\tilde{\omega}_e \epsilon/2].
\end{aligned}
\tag{2.439}
$$

Inserting this into Eq. (2.384) yields the partition function

$$
Z_\omega = \frac{1}{\sqrt{\tilde{D}_{N+1}}} = \frac{1}{2\sinh(\hbar\tilde{\omega}_e\beta/2)},
\tag{2.440}
$$

in agreement with (2.401).

2.12 Functional Measure in Fourier Space

There exists an alternative definition for the quantum-statistical path integral which is useful for some applications (for example in Section 2.13 and in Chapter 5). The limiting product formula (2.408) suggests that instead of summing over all zigzag configurations of paths on a sliced time axis, a path integral may be defined with the help of the Fourier components of the paths on a continuous time axis. As in (2.378), but with a slightly different normalization of the coefficients, we expand these paths here as

$$
x(\tau) = x_0 + \eta(\tau) \equiv x_0 + \sum_{m=1}^{\infty} \left(x_m e^{i\omega_m \tau} + \text{c.c.} \right), \quad x_0 = \text{real}, \quad x_{-m} \equiv x_m^*.
\tag{2.441}
$$

Note that the temporal integral over the time-dependent fluctuations $\eta(\tau)$ is zero, $\int_0^{\hbar/k_B T} d\tau\, \eta(\tau) = 0$, so that the zero-frequency component x_0 is the temporal average of the fluctuating paths:

$$
x_0 = \bar{x} \equiv \frac{k_B T}{\hbar} \int_0^{\hbar/k_B T} d\tau\, x(\tau).
\tag{2.442}
$$

In contrast to (2.378) which was valid on a sliced time axis and was therefore subject to a restriction on the range of the m-sum, the present sum is unrestricted and runs over all Matsubara frequencies $\omega_m = 2\pi m k_B T/\hbar = 2\pi m/\hbar\beta$. In terms of x_m, the Euclidean action of the linear oscillator is

$$
\begin{aligned}
\mathcal{A}_e &= \frac{M}{2} \int_0^{\hbar/k_B T} d\tau \left(\dot{x}^2 + \omega^2 x^2 \right) \\
&= \frac{M\hbar}{k_B T} \left[\frac{\omega^2}{2} x_0^2 + \sum_{m=1}^{\infty} (\omega_m^2 + \omega^2)|x_m|^2 \right].
\end{aligned}
\tag{2.443}
$$

The integration variables of the time-sliced path integral were transformed to the Fourier components x_m in Eq. (2.386). The product of integrals $\prod_n \int_{-\infty}^{\infty} dx(\tau_n)$

turned into the product (2.388) of integrals over real and imaginary parts of x_m. In the continuum limit, the result is

$$\int_{-\infty}^{\infty} dx_0 \prod_{m=1}^{\infty} \int_{-\infty}^{\infty} d\,\mathrm{Re}\,x_m \int_{-\infty}^{\infty} d\,\mathrm{Im}\,x_m. \qquad (2.444)$$

Placing the exponential $e^{-\mathcal{A}_e/\hbar}$ with the frequency sum (2.443) into the integrand, the product of Gaussian integrals renders a product of inverse eigenvalues $(\omega_m^2 + \omega^2)^{-1}$ for $m = 1, \ldots, \infty$, with some infinite factor. This may be determined by comparison with the known continuous result (2.408) for the harmonic partition function. The infinity is of the type encountered in Eq. (2.181), and must be divided out of the measure (2.444). The correct result (2.406) is obtained from the following measure of integration in Fourier space

$$\oint \mathcal{D}x \equiv \int_{-\infty}^{\infty} \frac{dx_0}{l_e(\hbar\beta)} \prod_{m=1}^{\infty} \left[\int_{-\infty}^{\infty} \int_{-\infty}^{\infty} \frac{d\,\mathrm{Re}\,x_m\, d\,\mathrm{Im}\,x_m}{\pi k_B T / M\omega_m^2} \right]. \qquad (2.445)$$

The divergences in the product over the factors $(\omega_m^2 + \omega^2)^{-1}$ discussed after Eq. (2.181) are canceled by the factors ω_m^2 in the measure. It will be convenient to introduce a short-hand notation for the measure on the right-hand side, writing it as

$$\oint \mathcal{D}x \equiv \int_{-\infty}^{\infty} \frac{dx_0}{l_e(\hbar\beta)} \oint \mathcal{D}'x. \qquad (2.446)$$

The denominator of the x_0-integral is the length scale $l_e(\hbar\beta)$ associated with β defined in Eq. (2.351).

Then we calculate

$$Z_\omega^{x_0} \equiv \oint \mathcal{D}'x\, e^{-\mathcal{A}_e/\hbar} = \prod_{m=1}^{\infty} \left[\int_{-\infty}^{\infty} \int_{-\infty}^{\infty} \frac{d\,\mathrm{Re}\,x_m\, d\,\mathrm{Im}\,x_m}{\pi k_B T / M\omega_m^2} \right] e^{-M\hbar\left[\omega^2 x_0^2/2 + \sum_{m=1}^{\infty} (\omega_m^2 + \omega^2)|x_m|^2 \right]/k_B T}$$

$$= e^{-M\omega^2 x_0^2/2k_B T} \prod_{m=1}^{\infty} \left[\frac{\omega_m^2 + \omega^2}{\omega_m^2} \right]^{-1}. \qquad (2.447)$$

The final integral over the zero-frequency component x_0 yields the partition function

$$Z_\omega = \oint \mathcal{D}x\, e^{-\mathcal{A}_e/\hbar} = \int_{-\infty}^{\infty} \frac{dx_0}{l_e(\hbar\beta)} Z_\omega^{x_0} = \frac{k_B T}{\hbar\omega} \prod_{m=1}^{\infty} \left[\frac{\omega_m^2 + \omega^2}{\omega_m^2} \right]^{-1}, \qquad (2.448)$$

as in (2.408).

The same measure can be used for the more general amplitude (2.412), as is obvious from (2.414). With the predominance of the kinetic term in the measure of path integrals [the divergencies discussed after (2.181) stem only from it], it can easily be shown that the same measure is applicable to any system with the standard kinetic term.

It is also possible to find a Fourier decomposition of the paths and an associated integration measure for the open-end partition function in Eq. (2.411). We begin by considering the slightly reduced set of all paths satisfying the Neumann boundary conditions

$$\dot{x}(\tau_a) = v_a = 0, \qquad \dot{x}(\tau_b) = v_b = 0. \tag{2.449}$$

They have the Fourier expansion

$$x(\tau) = x_0 + \eta(\tau) = x_0 + \sum_{n=1}^{\infty} x_n \cos \nu_n (\tau - \tau_a), \quad \nu_n = n\pi/\beta. \tag{2.450}$$

The frequencies ν_n are the Euclidean version of the frequencies (3.64) for Dirichlet boundary conditions. Let us calculate the partition function for such paths by analogy with the above periodic case by a Fourier decomposition of the action

$$\mathcal{A}_e = \frac{M}{2} \int_0^{\hbar/k_B T} d\tau \left(\dot{x}^2 + \omega^2 x^2\right) = \frac{M\hbar}{k_B T} \left[\frac{\omega^2}{2} x_0^2 + \frac{1}{2} \sum_{n=1}^{\infty} (\nu_n^2 + \omega^2) x_n^2 \right], \tag{2.451}$$

and of the measure

$$\oint \mathcal{D}x \equiv \int_{-\infty}^{\infty} \frac{dx_0}{l_e(\hbar\beta)} \prod_{n=1}^{\infty} \left[\int_{-\infty}^{\infty} \int_{-\infty}^{\infty} \frac{d\,x_n}{\pi k_B T/2M\nu_n^2} \right]$$

$$\equiv \int_{-\infty}^{\infty} \frac{dx_0}{l_e(\hbar\beta)} \oint \mathcal{D}'x. \tag{2.452}$$

We now perform the path integral over all fluctuations at fixed x_0 as in (2.447):

$$Z_\omega^{N,x_0} \equiv \oint \mathcal{D}'x \, e^{-\mathcal{A}_e/\hbar} = \prod_{n=1}^{\infty} \left[\int_{-\infty}^{\infty} \int_{-\infty}^{\infty} \frac{d\,x_n}{\pi k_B T/2M\nu_n^2} \right] e^{-M\hbar[\omega^2 x_0^2/2 + \sum_{n=1}^{\infty}(\nu_n^2 + \omega^2)|x_n|^2]/k_B T}$$

$$= e^{-M\omega^2 x_0^2/2k_B T} \prod_{n=1}^{\infty} \left[\frac{\nu_n^2 + \omega^2}{\nu_n^2} \right]^{-1}. \tag{2.453}$$

Using the product formula (2.181), this becomes

$$Z_\omega^{N,x_0} = \sqrt{\frac{\omega\hbar\beta}{\sinh\omega\hbar\beta}} \exp\left(-\beta \frac{M}{2}\omega^2 x_0^2\right). \tag{2.454}$$

The final integral over the zero-frequency component x_0 yields the partition function

$$Z_\omega^N = \frac{1}{l_e(\hbar\beta)} \sqrt{\frac{2\pi\hbar}{M\omega}} \frac{1}{\sqrt{\sinh\omega\hbar\beta}}. \tag{2.455}$$

We have replaced the denominator in the prefactor $1/l_e(\hbar\beta)$ by the length scale $1/l_e(\hbar\beta)$ of Eq. (2.351). Apart from this prefactor, the Neumann partition function coincides precisely with the open-end partition function Z_ω^{open} in Eq. (2.411).

What is the reason for this coincidence up to a trivial factor, even though the paths satisfying Neumann boundary conditions do *not* comprise all paths with open ends? Moreover, the integrals over the endpoints in the defining equation (2.411) do not force the endpoint *velocities*, but rather endpoint *momenta* to vanish. Indeed, recalling Eq. (2.187) for the time evolution amplitude in momentum space we can see immediately that the partition function with open ends Z_ω^{open} in Eq. (2.411) is identical to the imaginary-time amplitude with vanishing endpoint momenta:

$$Z_\omega^{\text{open}} = (p_b\, \hbar\beta | p_a\, 0)|_{p_b=p_a=0}. \tag{2.456}$$

Thus, the sum over all paths with arbitrary open ends is equal to the sum of all paths satisfying Dirichlet boundary conditions in momentum space. Only classically, the vanishing of the endpoint momenta implies the vanishing of the endpoint velocities. From the general discussion of the time-sliced path integral in phase space in Section 2.1 we know that fluctuating paths have $M\dot{x} \neq p$. The fluctuations of the difference are controlled by a Gaussian exponential of the type (2.51). This leads to the explanation of the trivial factor between Z_ω^{open} and Z_ω^{N}. The difference between $M\dot{x}$ and p appears only in the last short-time intervals at the ends. But at short time, the potential does not influence the fluctuations in (2.51). This is the reason why the fluctuations at the endpoints contribute only a trivial overall factor $l_e(\hbar\beta)$ to the partition function Z_ω^{N}.

2.13 Classical Limit

The alternative measure of the last section serves to show, somewhat more convincingly than before, that in the high-temperature limit the path integral representation of any quantum-statistical partition function reduces to the classical partition function as stated in Eq. (2.344). We start out with the Lagrangian formulation (2.374). Inserting the Fourier decomposition (2.441), the kinetic term becomes

$$\int_0^{\hbar\beta} d\tau\, \frac{M}{2}\dot{x}^2 = \frac{M\hbar}{k_B T}\sum_{m=1}^{\infty}\omega_m^2|x_m|^2, \tag{2.457}$$

and the partition function reads

$$Z = \oint \mathcal{D}x\, \exp\left[-\frac{M}{k_B T}\sum_{m=1}^{\infty}\omega_m^2|x_m|^2 - \frac{1}{\hbar}\int_0^{\hbar/k_B T} d\tau\, V\Big(x_0 + \sum_{m=-\infty}^{\infty}{}' x_m e^{-i\omega_m\tau}\Big)\right]. \tag{2.458}$$

The summation symbol with a prime implies the absence of the $m = 0$ -term. The measure is the product (2.445) of integrals of all Fourier components.

We now observe that for large temperatures, the Matsubara frequencies for $m \neq 0$ diverge like $2\pi m k_B T/\hbar$. This has the consequence that the Boltzmann factor for the $x_{m\neq 0}$ fluctuations becomes sharply peaked around $x_m = 0$. The average size of x_m is $\sqrt{k_B T/M}/\omega_m = \hbar/2\pi m\sqrt{M k_B T}$. If the potential $V\left(x_0 + \sum_{m=-\infty}^{\infty}{}' x_m e^{-i\omega_m\tau}\right)$ is a smooth function of its arguments, we can approximate it by $V(x_0)$ plus terms containing higher powers of x_m. For large temperatures, these are small on the average and can be ignored. The leading term $V(x_0)$ is time-independent. Hence we obtain in the high-temperature limit

$$Z \xrightarrow{T\to\infty} \oint \mathcal{D}x \exp\left[-\frac{M}{k_B T}\sum_{m=1}^{\infty}\omega_m^2|x_m|^2 - \frac{1}{k_B T}V(x_0)\right]. \qquad (2.459)$$

The right-hand side is quadratic in the Fourier components x_m. With the measure of integration (2.445), we perform the integrals over x_m and obtain

$$Z \xrightarrow{T\to\infty} Z_{\rm cl} = \int_{-\infty}^{\infty}\frac{dx_0}{l_e(\hbar\beta)}e^{-V(x_0)/k_B T}. \qquad (2.460)$$

This agrees with the classical statistical partition function (2.350).

The derivation reveals an important prerequisite for the validity of the classical limit: It holds only for sufficiently smooth potentials. We shall see in Chapter 8 that for singular potentials such as $-1/|x|$ (Coulomb), $1/|x|^2$ (centrifugal barrier), $1/\sin^2\theta$ (angular barrier), this condition is not fulfilled and the classical limit is no longer given by (2.460). The particle distribution $\rho(x)$ at a fixed x does not have this problem. It always tends towards the naively expected classical limit (2.352):

$$\rho(x) \xrightarrow{T\to\infty} Z_{\rm cl}^{-1}e^{-V(x)/k_B T}. \qquad (2.461)$$

The convergence is nonuniform in x, which is the reason why the limit does not always carry over to the integral (2.460). This will be an important point in deriving in Chapter 12 a new path integral formula valid for singular potentials. At first, we shall ignore such subtleties and continue with the conventional discussion valid for smooth potentials.

2.14 Calculation Techniques on Sliced Time Axis via the Poisson Formula

In the previous sections we have used tabulated product formulas such as (2.122), (2.163), (2.171), (2.398), (2.400) to find fluctuation determinants on a finite sliced time axis. With the recent interest in lattice models of quantum field theories, it is useful to possess an efficient calculational technique to derive such product formulas (and related sums). Consider, as a typical example, the quantum-statistical partition function for a harmonic oscillator of frequency ω on a time axis with $N+1$ slices of thickness ϵ,

$$Z = \prod_{m=0}^{N}[2(1-\cos\omega_m\epsilon)+\epsilon^2\omega^2]^{-1/2}, \qquad (2.462)$$

with the product running over all Matsubara frequencies $\omega_m = 2\pi m k_B T/\hbar$. Instead of dealing with this product it is advantageous to consider the free energy

$$F = -k_B T\log Z = \frac{1}{2}k_B T\sum_{m=0}^{N}\log[2(1-\cos\omega_m\epsilon)+\epsilon^2\omega^2]. \qquad (2.463)$$

We now observe that by virtue of Poisson's summation formula (1.205), the sum can be rewritten as the following combination of a sum and an integral:

$$F = \frac{1}{2}k_B T(N+1) \sum_{n=-\infty}^{\infty} \int_0^{2\pi} \frac{d\lambda}{2\pi} e^{i\lambda n(N+1)} \log[2(1-\cos\lambda) + \epsilon^2\omega^2]. \tag{2.464}$$

The sum over n squeezes λ to integer multiples of $2\pi/(N+1) = \omega_m\epsilon$ which is precisely what we want.

We now calculate the integrals in (2.464):

$$\int_0^{2\pi} \frac{d\lambda}{2\pi} e^{i\lambda n(N+1)} \log[2(1-\cos\lambda) + \epsilon^2\omega^2]. \tag{2.465}$$

For this we rewrite the logarithm of an arbitrary positive argument as the limit

$$\log a = \lim_{\delta \to 0} \left[-\int_\delta^\infty \frac{d\tau}{\tau} e^{-\tau a/2} \right] + \log(2\delta) + \gamma, \tag{2.466}$$

where

$$\gamma \equiv -\Gamma'(1)/\Gamma(1) = \lim_{N \to \infty} \left(\sum_{n=1}^N \frac{1}{n} - \log N \right) \approx 0.5773156649\ldots \tag{2.467}$$

is the Euler-Mascheroni constant. Indeed, the function

$$E_1(x) = \int_x^\infty \frac{dt}{t} e^{-t} \tag{2.468}$$

is known as the *exponential integral* with the small-x expansion[18]

$$E_1(x) = -\gamma - \log x - \sum_{k=1}^\infty \frac{(-x)^k}{kk!}. \tag{2.469}$$

With the representation (2.466) for the logarithm, the free energy can be rewritten as

$$F = \frac{1}{2\epsilon} \sum_{n=-\infty}^{\infty} \lim_{\delta \to 0} \left\{ -\int_\delta^\infty \frac{d\tau}{\tau} \int_0^{2\pi} \frac{d\lambda}{2\pi} e^{i\lambda n(N+1) - \tau[2(1-\cos\lambda) + \epsilon^2\omega^2]/2} - \delta_{n0}\left[\log(2\delta) + \gamma\right] \right\}. \tag{2.470}$$

The integral over λ is now performed[19] giving rise to a modified Bessel function $I_{n(N+1)}(\tau)$:

$$F = \frac{1}{2\epsilon} \sum_{n=-\infty}^{\infty} \lim_{\delta \to 0} \left\{ -\int_\delta^\infty \frac{d\tau}{\tau} I_{n(N+1)}(\tau) e^{-\tau(2+\epsilon^2\omega^2)/2} - \delta_{n0}\left[\log(2\delta) + \gamma\right] \right\}. \tag{2.471}$$

If we differentiate this with respect to $\epsilon^2\omega^2 \equiv m^2$, we obtain

$$\frac{\partial F}{\partial m^2} = \frac{1}{4\epsilon} \sum_{n=-\infty}^{\infty} \int_0^\infty d\tau I_{n(N+1)}(\tau) e^{-\tau(2+m^2)/2} \tag{2.472}$$

and perform the τ-integral, using the formula valid for $\mathrm{Re}\,\nu > -1$, $\mathrm{Re}\,\alpha > \mathrm{Re}\,\mu$

$$\int_0^\infty d\tau\, I_\nu(\mu\tau) e^{-\tau\alpha} = \mu^\nu \frac{(\alpha - \sqrt{\alpha^2 - \mu^2})^{-\nu}}{\sqrt{\alpha^2 - \mu^2}}, \quad = \mu^{-\nu} \frac{(\alpha - \sqrt{\alpha^2 - \mu^2})^\nu}{\sqrt{\alpha^2 - \mu^2}}, \tag{2.473}$$

---

[18]I.S. Gradshteyn and I.M. Ryzhik, op. cit., Formula 8.214.2.
[19]I.S. Gradshteyn and I.M. Ryzhik, *op. cit.*, Formulas 8.411.1 and 8.406.1.

to find

$$\frac{\partial F}{\partial m^2} = \frac{1}{2\epsilon} \sum_{n=-\infty}^{\infty} \frac{1}{\sqrt{(m^2+2)^2-4}} \left[\frac{m^2+2-\sqrt{(m^2+2)^2-4}}{2} \right]^{|n|(N+1)} . \tag{2.474}$$

From this we obtain F by integration over $m^2 + 1$. The $n = 0$-term under the sum gives

$$\log[(m^2 + 2 + \sqrt{(m^2+2)^2-4})/2] + \text{const} \tag{2.475}$$

and the $n \neq 0$-terms:

$$-\frac{1}{|n|(N+1)}[(m^2 + 2 + \sqrt{(m^2+2)^2-4})/2]^{-|n|(N+1)} + \text{const}, \tag{2.476}$$

where the constants of integration can depend on $n(N + 1)$. They are adjusted by going to the limit $m^2 \to \infty$ in (2.471). There the integral is dominated by the small-τ regime of the Bessel functions

$$I_\alpha(z) \sim \frac{1}{|\alpha|!} \left(\frac{z}{2}\right)^\alpha [1 + O(z^2)], \tag{2.477}$$

and the first term in (2.471) becomes

$$-\frac{1}{(|n|(N+1))!} \int_\delta^\infty \frac{d\tau}{\tau} \left(\frac{\tau}{2}\right)^{|n|(N+1)} e^{-\tau m^2/2}$$
$$\approx \left\{ \begin{array}{ll} \log m^2 + \gamma + \log(2\delta) & n = 0 \\ -(m^2)^{-|n|(N+1)}/|n|(N+1) & n \neq 0 \end{array} \right\} . \tag{2.478}$$

The limit $m^2 \to \infty$ in (2.475), (2.476) gives, on the other hand, $\log m^2 + \text{const}$ and $-(m^2)^{-|n|(N+1)}/|n|(N+1) + \text{const}$, respectively. Hence the constants of integration must be zero. We can therefore write down the free energy for $N + 1$ time steps as

$$\begin{aligned} F &= \frac{1}{2\beta} \sum_{m=0}^{N} \log[2(1 - \cos(\omega_m \epsilon)) + \epsilon^2 \omega^2] \\ &= \frac{1}{2\epsilon} \left\{ \log\left[\left(\epsilon^2\omega^2 + 2 + \sqrt{(\epsilon^2\omega^2+2)^2-4}\right)/2 \right] \right. \\ &\quad \left. -\frac{2}{N+1} \sum_{n=1}^{\infty} \frac{1}{n} \left[\left(\epsilon^2\omega^2 + 2 + \sqrt{(\epsilon^2\omega^2+2)^2-4}\right)/2 \right]^{-|n|(N+1)} \right\}. \end{aligned} \tag{2.479}$$

Here it is convenient to introduce the parameter

$$\epsilon\tilde{\omega}_e \equiv \log\left\{ \left[\epsilon^2\omega^2 + 2 + \sqrt{(\epsilon^2\omega^2+2)^2-4} \right]/2 \right\}, \tag{2.480}$$

which satisfies

$$\cosh(\epsilon\tilde{\omega}_e) = (\epsilon^2\omega^2 + 2)/2, \quad \sinh(\epsilon\tilde{\omega}_e) = \sqrt{(\epsilon^2\omega^2+2)^2-4}/2, \tag{2.481}$$

or

$$\sinh(\epsilon\tilde{\omega}_e/2) = \epsilon\omega/2.$$

Thus it coincides with the parameter introduced in (2.397), which brings the free energy (2.479) to the simple form

$$\begin{aligned} F &= \frac{\hbar}{2} \left[\tilde{\omega}_e - \frac{2}{\epsilon(N+1)} \sum_{n=1}^{\infty} \frac{1}{n} e^{-\epsilon\tilde{\omega}_e n(N+1)} \right] \\ &= \frac{1}{2} \left[\hbar\tilde{\omega}_e + 2k_B T \log(1 - e^{-\beta\hbar\tilde{\omega}_e}) \right] \\ &= \frac{1}{\beta} \log\left[2\sinh(\beta\hbar\tilde{\omega}_e/2) \right], \end{aligned} \tag{2.482}$$

whose continuum limit is

$$F \stackrel{\epsilon \to 0}{=} \frac{1}{\beta} \log\left[2 \sinh\left(\beta\hbar\omega/2\right)\right] = \frac{\hbar\omega}{2} + \frac{1}{\beta} \log(1 - e^{-\beta\hbar\omega}). \tag{2.483}$$

2.15   Field-Theoretic Definition of Harmonic Path Integrals by Analytic Regularization

A slight modification of the calculational techniques developed in the last section for the quantum partition function of a harmonic oscillator can be used to define the harmonic path integral in a way which neither requires time slicing, as in the original Feynman expression (2.64), nor a precise specification of the integration measure in terms of Fourier components, as in Section 2.12. The path integral for the partition function

$$Z_\omega = \oint \mathcal{D}x\, e^{-\int_0^{\hbar\beta} M[\dot{x}^2(\tau) + \omega^2 x^2(\tau)]/2} = \oint \mathcal{D}x\, e^{-\int_0^{\hbar\beta} Mx(\tau)[-\partial_\tau^2 + \omega^2]x(\tau)/2} \tag{2.484}$$

is formally evaluated as

$$Z_\omega = \frac{1}{\sqrt{\mathrm{Det}\,(-\partial_\tau^2 + \omega^2)}} = e^{-\frac{1}{2}\mathrm{Tr}\,\log(-\partial_\tau^2 + \omega^2)}. \tag{2.485}$$

Since the determinant of an operator is the product of all its eigenvalues, we may write, again formally,

$$Z_\omega = \prod_{\omega'} \frac{1}{\sqrt{\omega'^2 + \omega^2}}. \tag{2.486}$$

The product runs over an infinite set of quantities which grow with ω'^2, thus being certainly divergent. It may be turned into a divergent sum by rewriting Z_ω as

$$Z_\omega \equiv e^{-F_\omega/k_B T} = e^{-\frac{1}{2}\sum_{\omega'} \log(\omega'^2 + \omega^2)}. \tag{2.487}$$

This expression has two unsatisfactory features. First, it requires a proper definition of the formal sum over a continuous set of frequencies. Second, the logarithm of the dimensionful arguments $\omega_m^2 + \omega^2$ must be turned into a meaningful expression. The latter problem would be removed if we were able to exchange the logarithm by $\log[(\omega'^2 + \omega^2)/\omega^2]$. This would require the formal sum $\sum_{\omega'} \log \omega^2$ to vanish. We shall see below in Eq. (2.512) that this is indeed one of the pleasant properties of analytic regularization.

At finite temperatures, the periodic boundary conditions along the imaginary-time axis make the frequencies ω' in the spectrum of the differential operator $-\partial_\tau^2 + \omega^2$ discrete, and the sum in the exponent of (2.487) becomes a sum over all Matsubara frequencies $\omega_m = 2\pi k_B T/\hbar$ $(m = 0, \pm 1, \pm 2, \ldots)$:

$$Z_\omega = \exp\left[-\frac{1}{2} \sum_{m=-\infty}^{\infty} \log(\omega_m^2 + \omega^2)\right]. \tag{2.488}$$

For the free energy $F_\omega \equiv (1/\beta) \log Z_\omega$, this implies

$$F_\omega = \frac{1}{2\beta} \mathrm{Tr} \log(-\partial_\tau^2 + \omega^2)\Big|_{\mathrm{per}} = \frac{1}{2\beta} \sum_{m=-\infty}^{\infty} \log(\omega_m^2 + \omega^2). \tag{2.489}$$

where the subscript per emphasizes the periodic boundary conditions in the τ-interval $(0, \hbar\beta)$.

2.15.1 Zero-Temperature Evaluation of the Frequency Sum

In the limit $T \to 0$, the sum in (2.489) goes over into an integral, and the free energy becomes

$$F_\omega \equiv \frac{1}{2\beta} \mathrm{Tr} \log(-\partial_\tau^2 + \omega^2)\Big|_{\pm\infty} = \frac{\hbar}{2} \int_{-\infty}^{\infty} \frac{d\omega'}{2\pi} \log(\omega'^2 + \omega^2), \tag{2.490}$$

where the subscript $\pm\infty$ indicates the vanishing boundary conditions of the eigenfunctions at $\tau = \pm\infty$. Thus, at low temperature, we can replace the frequency sum in the exponent of (2.487) by

$$\sum_{\omega'} \xrightarrow[T \to 0]{} \hbar\beta \int_{-\infty}^{\infty} \frac{d\omega'}{2\pi}. \tag{2.491}$$

This could have been expected on the basis of Planck's rules for the phase space invoked earlier on p. 97 to explain the measure of path integration. According to these rules, the volume element in the phase space of energy and time has the measure $\int dt\, dE/h = \int dt\, d\omega/2\pi$. If the integrand is independent of time, the temporal integral produces an overall factor , which for the imaginary-time interval $(0, \hbar\beta)$ of statistical mechanics is equal to $\hbar\beta = \hbar/k_B T$, thus explaining the integral version of the sum (2.491).

The integral on the right-hand side of (2.490) diverges at large ω'. This is called an *ultraviolet divergence* (UV-divergence), alluding to the fact that the ultraviolet regime of light waves contains the high frequencies of the spectrum.

The important observation is now that the divergent integral (2.490) can be made finite by a mathematical technique called *analytic regularization*.[20] This is based on rewriting the logarithm $\log(\omega'^2 + \omega^2)$ in the derivative form:

$$\log(\omega'^2 + \omega^2) = -\frac{d}{d\epsilon}(\omega'^2 + \omega^2)^{-\epsilon}\Big|_{\epsilon=0}. \tag{2.492}$$

Equivalently, we may obtain the logarithm from an $\epsilon \to 0$ -limit of the function

$$l_{\mathrm{MS}}(\epsilon) = -\frac{1}{\epsilon}(\omega'^2 + \omega^2)^{-\epsilon} + \frac{1}{\epsilon}. \tag{2.493}$$

---

[20]G. 't Hooft and M. Veltman, Nucl. Phys. B **44**, 189 (1972). Analytic regularization is at present the only method that allows to renormalize nonabelian gauge theories without destroying gauge invariance. See also the review by G. Leibbrandt, Rev. Mod. Phys. **74**, 843 (1975).

The subtraction of the pole term $1/\epsilon$ is commonly referred to a *minimal subtraction*. Indicating this process by a subscript MS, we may write

$$l_{\mathrm{MS}}(\epsilon) = -\frac{1}{\epsilon}(\omega'^2 + \omega^2)^{-\epsilon}\Big|_{\mathrm{MS},\ \epsilon \to 0}. \tag{2.494}$$

Using the derivative formula (2.492), the trace of the logarithm in the free energy (2.490) takes the form

$$\frac{1}{\hbar\beta}\mathrm{Tr}\log(-\partial_\tau^2 + \omega^2) = -\frac{d}{d\epsilon}\int_{-\infty}^{\infty}\frac{d\omega'}{2\pi}(\omega'^2 + \omega^2)^{-\epsilon}\Big|_{\epsilon=0}. \tag{2.495}$$

We now set up a useful integral representation, due to Schwinger, for a power $a^{-\epsilon}$ generalizing (2.466). Using the defining integral representation for the Gamma function

$$\int_0^{\infty}\frac{d\tau}{\tau}\tau^\mu e^{-\tau\omega^2} = \omega^{-\mu/2}\Gamma(\mu), \tag{2.496}$$

the desired generalization is

$$a^{-\epsilon} = \frac{1}{\Gamma(\epsilon)}\int_0^{\infty}\frac{d\tau}{\tau}\tau^\epsilon e^{-\tau a}. \tag{2.497}$$

This allows us to re-express (2.495) as

$$\frac{1}{\hbar\beta}\mathrm{Tr}\log(-\partial_\tau^2 + \omega^2) = -\frac{d}{d\epsilon}\frac{1}{\Gamma(\epsilon)}\int_{-\infty}^{\infty}\frac{d\omega'}{2\pi}\int_0^{\infty}\frac{d\tau}{\tau}\tau^\epsilon e^{-\tau(\omega'^2+\omega^2)}\Big|_{\epsilon=0}. \tag{2.498}$$

As long as ϵ is larger than zero, the τ-integral converges absolutely, so that we can interchange the τ- and ω'-integrations, and obtain

$$\frac{1}{\hbar\beta}\mathrm{Tr}\log(-\partial_\tau^2 + \omega^2) = -\frac{d}{d\epsilon}\frac{1}{\Gamma(\epsilon)}\int_0^{\infty}\frac{d\tau}{\tau}\tau^\epsilon\int_{-\infty}^{\infty}\frac{d\omega'}{2\pi}e^{-\tau(\omega'^2+\omega^2)}\Big|_{\epsilon=0}. \tag{2.499}$$

At this point we can perform the Gaussian integral over ω' using formula (1.334), and find

$$\frac{1}{\hbar\beta}\mathrm{Tr}\log(-\partial_\tau^2 + \omega^2) = -\frac{d}{d\epsilon}\frac{1}{\Gamma(\epsilon)}\int_0^{\infty}\frac{d\tau}{\tau}\tau^\epsilon\frac{1}{2\sqrt{\tau\pi}}e^{-\tau\omega^2}\Big|_{\epsilon=0}. \tag{2.500}$$

For small ϵ, the τ-integral is divergent at the origin. It can, however, be defined by an analytic continuation of the integral starting from the regime $\epsilon > 1/2$, where it converges absolutely, to $\epsilon = 0$. The continuation must avoid the pole at $\epsilon = 1/2$. Fortunately, this continuation is trivial since the integral can be expressed in terms of the Gamma function, whose analytic properties are well-known. Using the integral formula (2.496), we obtain

$$\frac{1}{\hbar\beta}\mathrm{Tr}\log(-\partial_\tau^2 + \omega^2) = -\frac{1}{2\sqrt{\pi}}\omega^{1-2\epsilon}\frac{d}{d\epsilon}\frac{1}{\Gamma(\epsilon)}\Gamma(\epsilon - 1/2)\Big|_{\epsilon=0}. \tag{2.501}$$

The right-hand side has to be continued analytically from $\epsilon > 1/2$ to $\epsilon = 0$. This is easily done using the defining property of the Gamma function $\Gamma(x) = \Gamma(1+x)/x$, from which we find $\Gamma(-1/2) = -2\Gamma(1/2) = -2\sqrt{\pi}$, and $1/\Gamma(\epsilon) \approx \epsilon/\Gamma(1+\epsilon) \approx \epsilon$. The derivative with respect to ϵ leads to the free energy of the harmonic oscillator at low temperature via analytic regularization:

$$\frac{1}{\hbar\beta}\mathrm{Tr}\log(-\partial_\tau^2 + \omega^2) = \int_{-\infty}^{\infty}\frac{d\omega'}{2\pi}\log(\omega'^2 + \omega^2) = \omega, \qquad (2.502)$$

so that the free energy of the oscillator at zero-temperature becomes

$$F_\omega = \frac{\hbar\omega}{2}. \qquad (2.503)$$

This agrees precisely with the result obtained from the lattice definition of the path integral in Eq. (2.405), or from the path integral (3.805) with the Fourier measure (2.445).

With the above procedure in mind, we shall often use the sloppy formula expressing the derivative of Eq. (2.497) at $\epsilon = 0$:

$$\log a = -\int_0^{\infty}\frac{d\tau}{\tau}e^{-\tau a}. \qquad (2.504)$$

This formula differs from the correct one by a minimal subtraction and can be used in all calculations with analytic regularization. Its applicability is based on the possibility of dropping the frequency integral over $1/\epsilon$ in the alternative correct expression

$$\frac{1}{\hbar\beta}\mathrm{Tr}\log(-\partial_\tau^2 + \omega^2) = -\frac{1}{\epsilon}\int_{-\infty}^{\infty}\frac{d\omega'}{2\pi}\left[\frac{1}{\epsilon}(\omega'^2 + \omega^2)^{-\epsilon} - \frac{1}{\epsilon}\right]_{\epsilon\to0}. \qquad (2.505)$$

In fact, within analytic regularization one may set all integrals over arbitrary pure powers of the frequency equal to zero:

$$\int_0^{\infty}d\omega'\,(\omega')^\alpha = 0 \quad \text{for all } \alpha. \qquad (2.506)$$

This is known as *Veltman's rule*.[21] It is a special limit of a frequency integral which is a generalization of the integral in (2.495):

$$\int_{-\infty}^{\infty}\frac{d\omega'}{2\pi}\frac{(\omega'^2)^\gamma}{(\omega'^2 + \omega^2)^\epsilon} = \frac{\Gamma(\gamma+1/2)}{2\pi\Gamma(\epsilon)}(\omega^2)^{\gamma+1/2-\epsilon}. \qquad (2.507)$$

This equation may be derived by rewriting the left-hand side as

$$\frac{1}{\Gamma(\epsilon)}\int_{-\infty}^{\infty}\frac{d\omega'}{2\pi}(\omega'^2)^\gamma\int_0^{\infty}\frac{d\tau}{\tau}\tau^\epsilon e^{-\tau(\omega'^2 + \omega^2)}. \qquad (2.508)$$

---

[21]See the textbook H. Kleinert and V. Schulte-Frohlinde, *Critical Properties of ϕ^4-Theories*, World Scientific, Singapore, 2001 (http://www.physik.fu-berlin.de/~kleinert/b8).

The integral over ω' is performed as follows:

$$\int_{-\infty}^{\infty} \frac{d\omega'}{2\pi} (\omega'^2)^\gamma e^{-\tau(\omega'^2+\omega^2)} = \frac{1}{2\pi} \int_0^\infty \frac{d\omega'^2}{\omega'^2} (\omega'^2)^{\gamma+1/2} e^{-\tau(\omega'^2+\omega^2)} = \frac{\tau^{-\gamma-1/2}}{2\pi} \Gamma(\gamma+1/2),$$
(2.509)

leading to a τ-integral in (2.508)

$$\int_0^\infty \frac{d\tau}{\tau} \tau^{\epsilon-\gamma-1/2} e^{-\tau\omega^2} = (\omega^2)^{\gamma+1/2+\epsilon},$$
(2.510)

and thus to the formula (2.507). The Veltman rule (2.506) follows from this directly in the limit $\epsilon \to 0$, since $1/\Gamma(\epsilon) \to 0$ on the right-hand side. This implies that the subtracted $1/\epsilon$ term in (2.505) gives no contribution.

The vanishing of all integrals over pure powers by Veltman's rule (2.506) was initially postulated in the process of developing a finite quantum field theory of weak and electromagnetic interactions. It has turned out to be extremely useful for the calculation of critical exponents of second-order phase transitions from field theories.[21]

An important consequence of Veltman's rule is to make the logarithms of dimensionful arguments in the partition functions (2.487) and the free energy (2.489) meaningful quantities. First, since $\int d(\omega'/2\pi) \log \omega^2 = 0$, we can divide the argument of the logarithm in (2.490) by ω^2 without harm, and make them dimensionless. At finite temperatures, we use the equality of the sum and the integral over an ω_m-independent quantity c

$$k_B T \sum_{m=-\infty}^{\infty} c = \int_{-\infty}^{\infty} \frac{d\omega_m}{2\pi} c$$
(2.511)

to show that also

$$k_B T \sum_{m=-\infty}^{\infty} \log \omega^2 = 0,$$
(2.512)

so that we have, as a consequence of Veltman's rule, that the Matsubara frequency sum over the constant $\log \omega^2$ vanishes for all temperatures. For this reason, also the argument of the logarithm in the free energy (2.489) can be divided by ω^2 without change, thus becoming dimensionless.

2.15.2   Finite-Temperature Evaluation of the Frequency Sum

At finite temperature, the free energy contains an additional term consisting of the difference between the Matsubara sum and the frequency integral

$$\Delta F_\omega = \frac{k_B T}{2} \sum_{m=-\infty}^{\infty} \log\left(\frac{\omega_m^2}{\omega^2}+1\right) - \frac{\hbar}{2} \int_{-\infty}^{\infty} \frac{d\omega_m}{2\pi} \log\left(\frac{\omega_m^2}{\omega^2}+1\right),$$
(2.513)

where we have used dimensionless logarithms as discussed at the end of the last subsection. The sum is conveniently split into a subtracted, manifestly convergent expression

$$\Delta_1 F_\omega = k_B T \sum_{m=1}^\infty \left[\log\left(\frac{\omega_m^2}{\omega^2}+1\right) - \log\frac{\omega_m^2}{\omega^2}\right] = k_B T \sum_{m=1}^\infty \log\left(1 + \frac{\omega^2}{\omega_m^2}\right), \quad (2.514)$$

and a divergent sum

$$\Delta_2 F_\omega = k_B T \sum_{m=1}^\infty \log\frac{\omega_m^2}{\omega^2}. \quad (2.515)$$

The convergent part is most easily evaluated. Taking the logarithm of the product in Eq. (2.406) and recalling (2.407), we find

$$\prod_{m=1}^\infty \left(1 + \frac{\omega^2}{\omega_m^2}\right) = \frac{\sinh(\beta\hbar\omega/2)}{\beta\hbar\omega/2}, \quad (2.516)$$

and therefore

$$\Delta F_1 = \frac{1}{\beta}\log\frac{\sinh(\beta\hbar\omega/2)}{\beta\hbar\omega/2}. \quad (2.517)$$

The divergent sum (2.515) is calculated by analytic regularization as follows: We rewrite

$$\sum_{m=1}^\infty \log\frac{\omega_m^2}{\omega^2} = -\left[2\frac{d}{d\epsilon}\sum_{m=1}^\infty \left(\frac{\omega_m}{\omega}\right)^{-\epsilon}\right]_{\epsilon\to0} = -\left[2\frac{d}{d\epsilon}\left(\frac{2\pi}{\beta\hbar\omega}\right)^{-\epsilon}\sum_{m=1}^\infty m^{-\epsilon}\right]_{\epsilon\to0}, \quad (2.518)$$

and express the sum over $m^{-\epsilon}$ in terms of Riemann's zeta function

$$\zeta(z) = \sum_{m=1}^\infty m^{-z}. \quad (2.519)$$

This sum is well defined for $z > 1$, and can be continued analytically into the entire complex z-plane. The only singularity of this function lies at $z = 1$, where in the neighborhood $\zeta(z) \approx 1/z$. At the origin, $\zeta(z)$ is regular, and satisfies[22]

$$\zeta(0) = -1/2, \qquad \zeta'(0) = -\frac{1}{2}\log 2\pi, \quad (2.520)$$

such that we may approximate

$$\zeta(z) \approx -\frac{1}{2}(2\pi)^z, \qquad z \approx 0. \quad (2.521)$$

Hence we find

$$\sum_{m=1}^\infty \log\frac{\omega_m^2}{\omega^2} = -\left[2\frac{d}{d\epsilon}\left(\frac{2\pi}{\beta\hbar\omega}\right)^{-\epsilon}\zeta(\epsilon)\right]_{\epsilon\to0} = \frac{d}{d\epsilon}(\beta\hbar\omega)^\epsilon\bigg|_{\epsilon\to0} = \log\hbar\omega\beta. \quad (2.522)$$

---

[22]I.S. Gradshteyn and I.M. Ryzhik, *op. cit.*, Formula 9.541.4.

thus determining $\Delta_2 F_\omega$ in Eq. (2.515).

By combining this with (2.517) and the contribution $-\hbar\omega/2$ from the integral (2.513), the finite-temperature part (2.489) of the free energy becomes

$$\Delta F_\omega = \frac{1}{\beta} \log(1 - e^{-\hbar\beta\omega}). \tag{2.523}$$

Together with the zero-temperature free energy (2.503), this yields the dimensionally regularized sum formula

$$
\begin{aligned}
F_\omega &= \frac{1}{2\beta} \operatorname{Tr} \log(-\partial_\tau^2 + \omega^2) = \frac{1}{2\beta} \sum_{m=-\infty}^{\infty} \log(\omega_m^2 + \omega^2) = \frac{\hbar\omega}{2} + \frac{1}{\beta} \log(1 - e^{-\hbar\omega/k_B T}) \\
&= \frac{1}{\beta} \log\left(2 \sinh \frac{\hbar\omega\beta}{2}\right),
\end{aligned}
\tag{2.524}
$$

in agreement with the properly normalized free energy (2.483) at all temperatures.

Note that the property of the zeta function $\zeta(0) = -1/2$ in Eq. (2.520) leads once more to our earlier result (2.512) that the Matsubara sum of a constant c vanishes:

$$\sum_{m=-\infty}^{\infty} c = \sum_{m=-\infty}^{-1} c + c + \sum_{m=1}^{\infty} c = 0, \tag{2.525}$$

since

$$\sum_{m=1}^{\infty} 1 = \sum_{m=-1}^{-\infty} 1 = \zeta(0) = -1/2. \tag{2.526}$$

As mentioned before, this allows us to divide ω^2 out of the logarithms in the sum in Eq. (2.524) and rewrite this sum as

$$\frac{1}{2\beta} \sum_{m=-\infty}^{\infty} \log\left(\frac{\omega_m^2}{\omega^2} + 1\right) = \frac{1}{\beta} \sum_{m=1}^{\infty} \log\left(\frac{\omega_m^2}{\omega^2} + 1\right) \tag{2.527}$$

2.15.3 Quantum-Mechanical Harmonic Oscillator

This observation leads us directly to the analogous quantum-mechanical discussion. Starting from the fluctuation factor (2.86) of the free particle which can formally be written as

$$F_0(\Delta t) = \int \mathcal{D}\delta x(t) \exp\left[\frac{i}{\hbar} \int_{t_a}^{t_b} dt \frac{M}{2} \delta x(-\partial_t^2)\delta x\right] = \frac{1}{\sqrt{2\pi\hbar i \Delta t/M}}, \tag{2.528}$$

where $\Delta t \equiv t_b - t_a$ [recall (2.125)]. The path integral of the harmonic oscillator has the fluctuation factor [compare (2.186)]

$$F_\omega(\Delta t) = F_0(\Delta t) \left[\frac{\operatorname{Det}(-\partial_t^2 - \omega^2)}{\operatorname{Det}(-\partial_t^2)}\right]^{-1/2}. \tag{2.529}$$

The ratio of the determinants has the Fourier decomposition

$$\frac{\mathrm{Det}\,(-\partial_t^2 - \omega^2)}{\mathrm{Det}\,(-\partial_t^2)} = \exp\left\{\sum_{n=1}^{\infty}\left[\log(\nu_n^2 - \omega^2) - \log \nu_n^2\right]\right\}, \tag{2.530}$$

where $\nu_n = n\pi/\Delta t$ [recall (2.110)], and was calculated in Eq. (2.186) to be

$$\frac{\mathrm{Det}\,(-\partial_t^2 - \omega^2)}{\mathrm{Det}\,(-\partial_t^2)} = \frac{\sin \omega \Delta t}{\omega \Delta t}. \tag{2.531}$$

This result can be reproduced with the help of formulas (2.527) and (2.524). We raplace β by $2\Delta t$, and use again $\Sigma_n 1 = \zeta(0) = -1/2$ to obtain

$$\begin{aligned}
\mathrm{Det}\,(-\partial_t^2 - \omega^2) &= \sum_{n=1}^{\infty} \log(\nu_n^2 + \omega^2)\Big|_{\omega \to i\omega} = \sum_{n=1}^{\infty}\left[\log\left(\frac{\nu_n^2}{\omega^2} + 1\right) + \log \omega^2\right]_{\omega \to i\omega} \\
&= \left[\sum_{n=1}^{\infty} \log\left(\frac{\nu_n^2}{\omega^2} + 1\right) - \frac{1}{2}\log \omega^2\right]_{\omega \to i\omega} = \log\left(2\frac{\sin \omega \Delta t}{\omega}\right). \tag{2.532}
\end{aligned}$$

For $\omega = 0$ this reproduces Formula (2.522). Inserting this and (2.532) into (2.530), we recover the result (2.531). Thus we find the amplitude

$$(x_b t_b | x_a t_a) = \frac{1}{\sqrt{\pi i/M}}\mathrm{Det}^{-1/2}(-\partial_t^2 - \omega^2)e^{i\mathcal{A}_{\mathrm{cl}}/\hbar} = \frac{1}{\sqrt{\pi i/M}}\sqrt{\frac{\omega}{2\sin \omega \Delta t}}e^{i\mathcal{A}_{\mathrm{cl}}/\hbar}, \tag{2.533}$$

in agreement with (2.173).

2.15.4 Tracelog of the First-Order Differential Operator

The trace of the logarithm in the free energy (2.490) can obviously be split into two terms

$$\mathrm{Tr}\log(-\partial_\tau^2 + \omega^2) = \mathrm{Tr}\log(\partial_\tau + \omega) + \mathrm{Tr}\log(-\partial_\tau + \omega). \tag{2.534}$$

Since the left-hand side is equal to $\beta\hbar\omega$ by (2.502), and the two integrals must be the same, we obtain the low-temperature result

$$\begin{aligned}
\mathrm{Tr}\log(\partial_\tau + \omega) = \mathrm{Tr}\log(-\partial_\tau + \omega) &= \hbar\beta\int_{-\infty}^{\infty}\frac{d\omega}{2\pi}\log(-i\omega' + \omega) = \hbar\beta\int_{-\infty}^{\infty}\frac{d\omega}{2\pi}\log(i\omega' + \omega) \\
&= \frac{\hbar\beta\omega}{2}. \tag{2.535}
\end{aligned}$$

The same result could be obtained from analytic continuation of the integrals over $\partial_\epsilon(\pm i\omega' + \omega)^\epsilon$ to $\epsilon = 0$.

For a finite temperature, we may use Eq. (2.524) to find

$$\mathrm{Tr}\log(\partial_\tau + \omega) = \mathrm{Tr}\log(-\partial_\tau + \omega) = \frac{1}{2}\mathrm{Tr}\log(-\partial_\tau^2 + \omega^2) = \log\left(2\sinh\frac{\beta\hbar\omega}{2}\right), \tag{2.536}$$

which reduces to (2.535) for $T \to 0$.

The result is also the same if there is an extra factor i in the argument of the tracelog. To see this we consider the case of time-independent frequency where Veltman's rule (2.506) tells us that it does not matter whether one evaluates integrals over $\log(i\omega' \mp \omega)$ or over $\log(\omega' \pm i\omega)$.

Let us also replace ω' by $i\omega'$ in the zero-temperature tracelog (2.502) of the second-order differential operator $(-\partial_\tau^2 + \omega^2)$. Then we rotate the contour of integration clockwise in the complex plane to find

$$\int_{-\infty}^{\infty} \frac{d\omega'}{2\pi} \log(-\omega'^2 + \omega^2 - i\eta) = \omega, \quad \omega \geq 0, \tag{2.537}$$

where an infinitesimal positive η prescribes how to bypass the singularities at $\omega' = \pm\omega \mp i\eta$ along the rotated contour of integration. Recall the discussion of the $i\eta$-prescription in Section 3.3. The integral (2.537) can be split into the integrals

$$\int_{-\infty}^{\infty} \frac{d\omega'}{2\pi} \log[\omega' \pm (\omega - i\eta)] = i\frac{\omega}{2}, \quad \omega \geq 0. \tag{2.538}$$

Hence formula (2.535) can be generalized to arbitrary complex frequencies $\omega = \omega_R + i\omega_I$ as follows:

$$\int_{-\infty}^{\infty} \frac{d\omega'}{2\pi} \log(\omega' \pm i\omega) = \mp\epsilon(\omega_R)\frac{\omega}{2}, \tag{2.539}$$

and

$$\int_{-\infty}^{\infty} \frac{d\omega'}{2\pi} \log(\omega' \pm \omega) = -i\epsilon(\omega_I)\frac{\omega}{2}, \tag{2.540}$$

where $\epsilon(x) = \Theta(x) - \Theta(-x) = x/|x|$ is the antisymmetric Heaviside function (1.311), which yields the sign of its argument. The formulas (2.539) and (2.540) are the large-time limit of the more complicated sums

$$\frac{k_B T}{\hbar} \sum_{m=-\infty}^{\infty} \log(\omega_m \pm i\omega) = \frac{k_B T}{\hbar} \log\left[2\epsilon(\omega_R)\sinh\frac{\hbar\omega}{2k_B T}\right], \tag{2.541}$$

and

$$\frac{k_B T}{\hbar} \sum_{m=-\infty}^{\infty} \log(\omega_m \pm \omega) = \frac{k_B T}{\hbar} \log\left[-2i\epsilon(\omega_I)\sin\frac{\hbar\omega}{2k_B T}\right]. \tag{2.542}$$

The first expression is periodic in the imaginary part of ω, with period $2\pi k_B T$, the second in the real part. The determinants possess a meaningful large-time limit only if the periodic parts of ω vanish. In many applications, however, the fluctuations will involve sums of logarithms (2.542) and (2.541) with different complex frequencies ω, and only the sum of the imaginary or real parts will have to vanish to obtain a meaningful large-time limit. On these occasions we may use the simplified formulas (2.539) and (2.540). Important examples will be encountered in Section 18.9.2.

In Subsection 3.3.2, Formula (2.536) will be generalized to arbitrary positive time-dependent frequencies $\Omega(\tau)$, where it reads [see (3.133)]

$$
\begin{aligned}
\text{Tr} \log \left[\pm \partial_\tau + \Omega(\tau) \right] &= \log \left\{ 2 \sinh \left[\frac{1}{2} \int_0^{\hbar\beta} d\tau'' \, \Omega(\tau'') \right] \right\} \\
&= \frac{1}{2} \int_0^{\hbar\beta} d\tau'' \, \Omega(\tau'') + \log \left[1 - e^{-\int_0^{\hbar\beta} d\tau'' \, \Omega(\tau'')} \right]. \quad (2.543)
\end{aligned}
$$

2.15.5 Gradient Expansion of the One-Dimensional Tracelog

Formula (2.543) may be used to calculate the trace of the logarithm of a second-order differential equation with arbitrary frequency as a semiclassical expansion. We introduce the Planck constant $\hbar$ and the potential $w(\tau) \equiv \hbar\Omega(\tau)$, and factorize as in (2.534):

$$
\text{Det} \left[-\hbar^2 \partial_\tau^2 + w^2(\tau) \right] = \text{Det} \left[-\hbar \partial_\tau - \bar{w}(\tau) \right] \times \text{Det} \left[\hbar \partial_\tau - \bar{w}(\tau) \right], \quad (2.544)
$$

where the function $\bar{w}(\tau)$ satisfies the *Riccati differential equation*:[23]

$$
\hbar \partial_\tau \bar{w}(\tau) + \bar{w}^2(\tau) = w^2(\tau). \quad (2.545)
$$

By solving this we obtain the trace of the logarithm from (2.543):

$$
\text{Tr} \log \left[-\hbar^2 \partial_\tau^2 + w^2(\tau) \right] = \log \left\{ 4 \sinh^2 \left[\frac{1}{2\hbar} \int_0^{\hbar\beta} d\tau' \, \bar{w}(\tau') \right] \right\}. \quad (2.546)
$$

The exponential of this yields the functional determinant. For constant $\bar{w}(\tau) = \omega$ this agrees with the result (2.431) of the Gelfand-Yaglom formula for periodic boundary conditions.

This agreement is no coincidence. We can find the solution of any Riccati differential equation if we know how to solve the second-order differential equation (2.431). Imposing the Gelfand-Yaglom boundary conditions in (2.431), we find $D_{\text{ren}}(\tau)$ and from this the functional determinant $2[\dot{D}_{\text{ren}}(\hbar\beta) - 1]$. Comparison with (2.546) shows that the solution of the Riccati differential equation (2.545) is given by

$$
\bar{w}(\tau) = 2\hbar \partial_\tau \, \text{arsinh} \sqrt{[\dot{D}_{\text{ren}}(\tau) - 1]/2}. \quad (2.547)
$$

For the harmonic oscillator where $\dot{D}_{\text{ren}}(\tau)$ is equal to (2.435), this leads to the constant $\bar{w}(\tau) = \hbar\omega$, as it should.

If we cannot solve the second-order differential equation (2.431), a solution to the Riccati equation (2.545) can still be found as a power series in $\hbar$:

$$
\bar{w}(\tau) = \sum_{n=0}^{\infty} \bar{w}_n(\tau) \hbar^n, \quad (2.548)
$$

---

[23]Recall the general form of the Riccati differential equation $y' = f(\tau)y + g(\tau)y^2 + h(y)$, which is an inhomogeneous version of the Bernoulli differential equation $y' = f(\tau)y + g(\tau)y^n$ for $n = 2$.

which provides us with a so-called *gradient expansion* of the trace of the logarithm. The lowest-order coefficient function $\bar{w}_0(\tau)$ is obviously equal to $w(\tau)$. The higher ones obey the recursion relation

$$\bar{w}_n(\tau) = -\frac{1}{2w(\tau)} \left(\dot{\bar{w}}_{n-1}(\tau) + \sum_{k=1}^{n-1} \bar{w}_{n-k}(\tau)\bar{w}_k(\tau) \right), \qquad n \geq 1. \tag{2.549}$$

These are solved for $n = 0, 1, 2, 3$ by

$$\left\{ \sqrt{v(\tau)}, -\frac{v'(\tau)}{4\,v(\tau)}, \quad -\frac{5\,v'(\tau)^2}{32\,v(\tau)^{5/2}} + \frac{v''(\tau)}{8\,v(\tau)^{3/2}}, \quad -\frac{15\,v'(\tau)^3}{64\,v(\tau)^4} + \frac{9\,v'(\tau)\,v''(\tau)}{32\,v(\tau)^3} - \frac{v^{(3)}(\tau)}{16\,v(\tau)^2}, \right.$$
$$\left. -\frac{1105\,v'(\tau)^4}{2048\,v(\tau)^{11/2}} + \frac{221\,v'(\tau)^2\,v''(\tau)}{256\,v(\tau)^{9/2}} - \frac{19\,v''(\tau)^2}{128\,v(\tau)^{7/2}} - \frac{7\,v'(\tau)\,v^{(3)}(\tau)}{32\,v(\tau)^{7/2}} + \frac{v^{(4)}(\tau)}{32\,v(\tau)^{5/2}} \right\}, (2.550)$$

where $v(\tau) \equiv w^2(\tau)$. The series can, of course, be trivially extended to any desired orders.

2.15.6 Duality Transformation and Low-Temperature Expansion

There exists another method of calculating the finite-temperature part of the free energy (2.489) which is worth presenting at this place, due to its broad applicability in statistical mechanics. For this we rewrite (2.513) in the form

$$\Delta F_\omega = \frac{k_B T}{2} \left(\sum_{m=-\infty}^{\infty} -\frac{\hbar}{k_B T} \int_{-\infty}^{\infty} \frac{d\omega_m}{2\pi} \right) \log(\omega_m^2 + \omega^2). \tag{2.551}$$

Changing the integration variable to m, this becomes

$$\Delta F_\omega = \frac{k_B T}{2} \left(\sum_{m=-\infty}^{\infty} - \int_{-\infty}^{\infty} dm \right) \log\left[\left(\frac{2\pi k_B T}{\hbar} \right)^2 m^2 + \omega^2 \right]. \tag{2.552}$$

Within analytic regularization, this expression is rewritten with the help of formula (2.504) as

$$\Delta F_\omega = -\frac{k_B T}{2} \int_0^\infty \frac{d\tau}{\tau} \left(\sum_{m=-\infty}^{\infty} - \int_{-\infty}^{\infty} dm \right) e^{-\tau\left[(2\pi k_B T/\hbar)^2 m^2 + \omega^2 \right]}. \tag{2.553}$$

The duality transformation proceeds by performing the sum over the Matsubara frequencies with the help of Poisson's formula (1.205) as an integral $\int d\mu$ plus by an extra sum over integer numbers n. This brings (2.553) to the form (expressing the temperature in terms of β),

$$\Delta F_\omega = -\frac{1}{2\beta} \int_0^\infty \frac{d\tau}{\tau} \int_{-\infty}^{\infty} d\mu \left(\sum_{n=-\infty}^{\infty} e^{2\pi\mu n i} - 1 \right) e^{-\tau\left[(2\pi/\hbar\beta)^2 \mu^2 + \omega^2 \right]}. \tag{2.554}$$

The parentheses contain the sum $2 \sum_{n=1}^{\infty} e^{2\pi \mu n i}$. After a quadratic completion of the exponent

$$2\pi \mu n i - \tau \left(\frac{2\pi}{\hbar \beta}\right)^2 \mu^2 = -\tau \left(\frac{2\pi}{\hbar \beta}\right)^2 \left[\mu - i\frac{n\hbar^2 \beta^2}{4\pi \tau}\right]^2 - \frac{1}{4\tau}(\hbar \beta n)^2, \qquad (2.555)$$

the integral over μ can be performed, with the result

$$\Delta F_\omega = -\frac{\hbar}{2\sqrt{\pi}} \int_0^\infty \frac{d\tau}{\tau} \tau^{-1/2} \sum_{n=1}^\infty e^{-(n\hbar \beta)^2/4\tau - \tau \omega^2}. \qquad (2.556)$$

Now we may use the integral formula [compare (1.343)][24]

$$\int_0^\infty \frac{d\tau}{\tau} \tau^\nu e^{-a^2/\tau - b^2 \tau} = 2\left(\frac{a}{b}\right)^\nu K_\nu(2ab), \qquad K_\nu(2ab) = K_{-\nu}(2ab), \qquad (2.557)$$

to obtain the sum over modified Bessel functions

$$\Delta F_\omega = -\frac{\hbar \omega}{2\sqrt{\pi}} \sum_{n=1}^\infty 2\,(n\beta \hbar \omega)^{-1/2}\,\sqrt{2} K_{1/2}(n\beta \hbar \omega). \qquad (2.558)$$

The modified Bessel functions with index $1/2$ are particularly simple:

$$K_{1/2}(z) = \sqrt{\frac{\pi}{2z}} e^{-z}. \qquad (2.559)$$

Inserting this into (2.558), the sum is a simple geometric one, and may be performed as follows:

$$\Delta F_\omega = -\frac{1}{\beta} \sum_{n=1}^\infty \frac{1}{n} e^{-\beta \hbar \omega n} = \frac{1}{\beta} \log\left(1 - e^{-\beta \hbar \omega}\right), \qquad (2.560)$$

in agreement with the previous result (2.523).

The effect of the duality tranformation may be rephrased in another way. It converts the sum over Matsubara frequencies ω_m in (2.514):

$$S(\beta \hbar \omega) = k_B T \sum_{m=1}^\infty \log\left(1 + \frac{\omega^2}{\omega_m^2}\right) \qquad (2.561)$$

into a sum over the quantum numbers n of the harmonic oscillator:

$$S(\beta \hbar \omega) = \frac{\beta \hbar \omega}{2} - \log \beta \hbar \omega - \sum_{n=1}^\infty \frac{1}{n} e^{-n\beta \hbar \omega}. \qquad (2.562)$$

The sum (2.561) converges fast at high temperatures, where it can be expanded in powers of ω^2:

$$S(\beta \hbar \omega) = -\sum_{k=1}^\infty \frac{(-1)^k}{k}\left(\sum_{m=1}^\infty \frac{1}{m^{2k}}\right)\left[\left(\frac{\beta \hbar \omega}{2\pi}\right)^2\right]^k. \qquad (2.563)$$

---

[24]I.S. Gradshteyn and I.M. Ryzhik, *op. cit.*, Formulas 3.471.9 and 8.486.16.

The expansion coefficients are equal to Riemann's zeta function $\zeta(z)$ of Eq. (2.519) at even arguments $z = 2k$, so that we may write

$$S(\beta\hbar\omega) = -\sum_{m=1}^{\infty} \frac{(-1)^k}{k}\zeta(2k)\left(\frac{\beta\hbar\omega}{2\pi}\right)^{2k}. \tag{2.564}$$

At even positive arguments, the values of the zeta function are related to the Bernoulli numbers by[25]

$$\zeta(2n) = \frac{(2\pi)^{2n}}{2(2n)!}|B_{2n}|. \tag{2.565}$$

The Bernoulli numbers are defined by the expansion

$$\frac{t}{e^t - 1} = \sum_{n=0}^{\infty} B_n \frac{t^n}{n!}. \tag{2.566}$$

The lowest nonzero Bernoulli numbers are $B_0 = 1$, $B_1 = -1/2$, $B_2 = 1/2$, $B_4 = -1/30, \dots$. The Bernoulli numbers determine also the values of the zeta functions at negative odd arguments:

$$\zeta(1 - 2n) = -\frac{B_{2n}}{2n}, \tag{2.567}$$

this being a consequence of the general identity[26]

$$\zeta(z) = 2^z \pi^{z-1} \sin(\pi z/2)\Gamma(1-z)\zeta(1-z) = 2^{z-1}\pi^z\zeta(1-z)/\Gamma(z)\cos\frac{z\pi}{2}. \tag{2.568}$$

Typical values of $\zeta(z)$ which will be needed here are[27]

$$\zeta(2) = \frac{\pi^2}{6}, \quad \zeta(4) = \frac{\pi^4}{90}, \quad \zeta(5) = \frac{\pi^6}{945}, \quad \dots, \quad \zeta(\infty) = 1. \tag{2.569}$$

In contrast to the Matsubara frequency sum (2.561) and its expansion (2.564), the dually transformed sum over the quantum numbers n in (2.562) converges rapidly for low temperatures. It converges everywhere except at very large temperatures, where it diverges logarithmically. The precise behavior can be calculated as follows: For large T there exists a large number N which is still much smaller than $1/\beta\hbar\omega$, such that $e^{-\beta\hbar\omega N}$ is close to unity. Then we split the sum as

$$\sum_{n=1}^{\infty}\frac{1}{n}e^{-n\beta\hbar\omega} \approx \sum_{n=1}^{N-1}\frac{1}{n} + \sum_{n=N}^{\infty}\frac{1}{n}e^{-n\beta\hbar\omega}. \tag{2.570}$$

---

[25] *ibid.*, Formulas 9.542 and 9.535.

[26] *ibid.*, Formula 9.535.2.

[27] Other often-needed values are $\zeta(0) = -1/2$, $\zeta'(0) = -\log(2\pi)/2$, $\zeta(-2n) = 0$, $\zeta(3) \approx 1.202057$, $\zeta(5) \approx 1.036928, \dots$.

Since N is large, the second sum can be approximated by an integral

$$\int_N^\infty \frac{dn}{n} e^{-n\beta\hbar\omega} = \int_{N\beta\hbar\omega}^\infty \frac{dx}{x} e^{-x},$$

which is an exponential integral $E_1(N\beta\hbar\omega)$ of Eq. (2.468) with the large-argument expansion $-\gamma - \log(N\beta\hbar\omega)$ of Eq. (2.469).

The first sum in (2.570) is calculated with the help of the Digamma function

$$\psi(z) \equiv \frac{\Gamma'(z)}{\Gamma(z)}. \tag{2.571}$$

This has an expansion[28]

$$\psi(z) = -\gamma - \sum_{n=0}^\infty \left(\frac{1}{n+z} - \frac{1}{n+1} \right), \tag{2.572}$$

which reduces for integer arguments to

$$\psi(N) = -\gamma + \sum_{n=1}^{N-1} \frac{1}{n}, \tag{2.573}$$

and has the large-z expansion

$$\psi(z) \approx \log z - \frac{1}{2z} - \sum_{n=1}^\infty \frac{B_{2n}}{2n z^{2n}}. \tag{2.574}$$

Combining this with (2.469), the logarithm of N cancels, and we find for the sum in (2.570) the large-T behavior

$$\sum_{n=1}^\infty \frac{1}{n} e^{-n\beta\hbar\omega} \underset{T\to\infty}{\approx} -\log\beta\hbar\omega + \mathcal{O}(\beta). \tag{2.575}$$

This cancels the logarithm in (2.562).

The low-temperature series (2.562) can be used to illustrate the power of analytic regularization. Suppose we want to extract from it the large-T behavior, where the sum

$$g(\beta\hbar\omega) \equiv \sum_{n=1}^\infty \frac{1}{n} e^{-n\beta\hbar\omega} \tag{2.576}$$

converges slowly. We would like to expand the exponentials in the sum into powers of ω, but this gives rise to sums over positive powers of n. It is possible to make sense of these sums by analytic continuation. For this we introduce a generalization of (2.576):

$$\zeta_\nu(e^{\beta\hbar\omega}) \equiv \sum_{n=1}^\infty \frac{1}{n^\nu} e^{-n\beta\hbar\omega}, \tag{2.577}$$

---

[28]I.S. Gradshteyn and I.M. Ryzhik, *op. cit.*, Formula 1.362.1.

which reduces to $g(\beta\hbar\omega)$ for $\nu = 1$. This sum is evaluated by splitting it into an integral over n and a difference between sum and integral:

$$\zeta_\nu(e^{\beta\hbar\omega}) = \int_0^\infty dn\,\frac{1}{n^\nu}e^{-n\beta\hbar\omega} + \left(\sum_{n=1}^\infty - \int_0^\infty\right)\frac{1}{n^\nu}e^{-n\beta\hbar\omega}. \tag{2.578}$$

The integral is convergent for $\nu < 1$ and yields $\Gamma(1-\nu)(\beta\hbar\omega)^\nu$ via the integral formula (2.496). For other ν's it is defined by analytic continuation. The remainder may be expanded sloppily in powers of ω and yields

$$\zeta_\nu(e^{\beta\hbar\omega}) = \int_0^\infty dn\,\frac{1}{n^\nu}e^{-n\beta\hbar\omega} + \left(\sum_{n=1}^\infty - \int_0^\infty\right)\frac{1}{n^\nu} + \sum_{k=1}^\infty\left[\left(\sum_{n=1}^\infty - \int_0^\infty\right)n^{k-\nu}\right]\frac{(-1)^k}{k!}(\beta\hbar\omega)^k. \tag{2.579}$$

The second term is simply the Riemann zeta function $\zeta(\nu)$ [recall (2.519)]. Since the additional integral vanishes due to Veltman' rule (2.506), the zeta function may also be defined by

$$\left(\sum_{n=1}^\infty - \int_0^\infty\right)\frac{1}{\nu^k} = \zeta(\nu), \tag{2.580}$$

If this formula is applied to the last term in (2.579), we obtain the so-called *Robinson expansion*[29]

$$\zeta_\nu(e^{\beta\hbar\omega}) = \Gamma(1-\nu)(\beta\hbar\omega)^{\nu-1} + \zeta(\nu) + \sum_{k=1}^\infty\frac{1}{k!}(-\beta\hbar\omega)^k\zeta(\nu-k). \tag{2.581}$$

This expansion will later play an important role in the discussion of Bose-Einstein condensation [see Eq. (7.38)].

For various applications it is useful to record also the auxiliary formula

$$\left(\sum_{n=1}^\infty - \int_0^\infty\right)\frac{e^{n\beta\hbar\omega}}{n^\nu} = \sum_{k=1}^\infty\frac{1}{k!}(-\beta\hbar\omega)^k\zeta(\nu-k) \equiv \bar\zeta_\nu(e^{\beta\hbar\omega}), \tag{2.582}$$

since in the sum minus the integral, the first Robinson terms are absent and the result can be obtained from a naive Taylor expansion of the exponents $e^{n\beta\hbar\omega}$ and the summation formula (2.580).

From (2.581) we can extract the desired sum (2.576) by going to the limit $\nu \to 1$. Close to the limit, the Gamma function has a pole $\Gamma(1-\nu) = 1/(1-\nu)-\gamma+\mathcal{O}(\nu-1)$. From the identity

$$2^z\Gamma(1-z)\zeta(1-z)\sin\frac{\pi z}{2} = \pi^{1-z}\zeta(z) \tag{2.583}$$

and (2.520) we see that $\zeta(\nu)$ behaves near $\nu = 1$ like

$$\zeta(\nu) = \frac{1}{\nu-1} + \gamma + \mathcal{O}(\nu-1) = -\Gamma(1-\nu) + \mathcal{O}(\nu-1). \tag{2.584}$$

---

[29]J.E. Robinson, Phys. Rev. *83*, 678 (1951).

Hence the first two terms in (2.581) can be combined to yield for $\nu \to 1$ the finite result $\lim_{\nu \to 1} \Gamma(1 - \nu)\left[(\beta\hbar\omega)^{\nu-1} - 1\right] = -\log\beta\hbar\omega$. The remaining terms contain in the limit the values $\zeta(0) = -1/2$, $\zeta(-1)$, $\zeta(-2)$, etc. Here we use the property of the zeta function that it vanishes at even negative arguments, and that the function at arbitrary negative argument is related to one at positive argument by the identity (2.583). This implies for the expansion coefficients in (2.581) with $k = 1, 2, 3, \ldots$ in the limit $\nu \to 1$:

$$\zeta(-2p) = 0, \qquad \zeta(1 - 2p) = \frac{1}{p}(-1)^p \frac{(2p)!}{(2\pi)^{2p}}\zeta(2p), \quad p = 1, 2, 3, \ldots . \qquad (2.585)$$

Hence we obtain for the expansion (2.581) in the limit $\nu \to 1$:

$$g(\beta\hbar\omega) = \zeta_1(e^{\beta\hbar\omega}) = -\log\beta\hbar\omega + \frac{\beta\hbar\omega}{2} + \sum_{k=1}^{\infty} \zeta(2k)\frac{(-1)^k}{k!}(\beta\hbar\omega)^{2k}. \qquad (2.586)$$

This can now be inserted into Eq. (2.562) and we recover the previous expansion (2.564) for $S(\beta\hbar\omega)$ which was derived there by a proper duality transformation.

It is interesting to observe what goes wrong if we forget the separation (2.578) of the sum into integral plus sum-minus-integral and its regularization. For this we re-expand (2.576) directly, and illegally, in powers of ω. Then we obtain for $\nu = 1$ the formal expansion

$$\zeta_1(e^{\beta\hbar\omega}) = \sum_{p=0}^{\infty}\left(\sum_{n=1}^{\infty} n^{p-1}\right)\frac{(-1)^p}{p!}(\beta\hbar\omega)^p = -\zeta(1) + \sum_{p=1}^{\infty}\zeta(1-p)\frac{(-1)^p}{p!}(\beta\hbar\omega)^p,$$
$$(2.587)$$

which contains the infinite quantity $\zeta(1)$. The correct result (2.586) is obtained from this by replacing the infinite quantity $\zeta(1)$ by $-\log\beta\hbar\omega$, which may be viewed as a regularized $\zeta_{\text{reg}}(1)$:

$$\zeta(1) \to \zeta_{\text{reg}}(1) = -\log\beta\hbar\omega. \qquad (2.588)$$

The above derivation of the Robinson expansion can be supplemented by a dual version as follows. With the help of Poisson's formula (1.197) we rewrite the sum (2.577) as an integral over n and an auxiliary sum over integer numbers m, after which the integral over n can be performed yielding

$$\zeta_\nu(e^{\beta\hbar\omega}) \equiv \sum_{m=-\infty}^{\infty} \int_0^{\infty} dn\, e^{(2\pi im + \beta\hbar\omega)n}\frac{1}{n^\nu} = \Gamma(1-\nu)(-\beta\hbar\omega)^{\nu-1}$$

$$+ \Gamma(1-\nu)2\,\mathrm{Re}\sum_{m=1}^{\infty}(-\beta\hbar\omega - 2\pi im)^{\nu-1}. \qquad (2.589)$$

The sum can again be expanded in powers of ω

$$2\,\mathrm{Re}\sum_{m=1}^{\infty}(-2\pi im)^{\nu-1}\left(1 + \frac{\beta\hbar\omega}{2\pi im}\right)^{\nu-1}$$

$$= 2\sum_{k=0}^{\infty}\binom{\nu-1}{k}\cos[(1-\nu-k)\pi/2]\,(2\pi)^{\nu-1-k}\zeta(1-\nu+k)\,(\beta\hbar\omega)^k . \quad (2.590)$$

Using the relation (2.583) for zeta-functions, the expansion (2.589) is seen to coincide with (2.577).

Note that the representation (2.589) of $\zeta_\nu(e^{\beta\hbar\omega})$ is a sum over Matsubara frequencies $\omega_m = 2\pi m/\beta$ [recall Eq. (2.379)]:

$$
\begin{aligned}
\zeta_\nu(e^{\beta\hbar\omega}) &\equiv \sum_{m=-\infty}^{\infty} \int_0^\infty dn\, e^{(i\omega_m + \hbar\omega)\beta n} \frac{1}{n^\nu} \\
&= \Gamma(1-\nu)(-\beta\hbar\omega)^{\nu-1}\left[1 + 2\,\mathrm{Re}\sum_{m=1}^{\infty}(1 + i\omega_m/\hbar\omega)^{\nu-1}\right]. \quad (2.591)
\end{aligned}
$$

The first term coming from the integral over n in (2.578) is associated with the zero Matsubara frequency. This term represents the high-temperature or classical limit of the expansion. The remainder contains the sum over all nonzero Matsubara frequencies, and thus the effect of quantum fluctuations.

It should be mentioned that the first two terms in the low-temperature expansion (2.562) can also be found from the sum (2.561) with the help of the Euler-Maclaurin formula[30] for a sum over discrete points $t = a + (k+\kappa)\Delta$ of a function $F(t)$ from $k = 0$ to $K \equiv (b-a)/\Delta$:

$$
\begin{aligned}
\sum_{k=0}^{K} F(a+k\Delta) &= \frac{1}{\Delta}\int_a^b dt\, F(t) + \frac{1}{2}[F(a) + F(b)] \\
&+ \sum_{p=1}^{\infty} \frac{\Delta^{2p-1}}{(2p)!} B_{2p}\left[F^{(2p-1)}(b) - F^{(2p-1)}(a)\right], \quad (2.592)
\end{aligned}
$$

or, more generally for $t = a + (k+\kappa)\Delta$,

$$
\sum_{k=0}^{K-1} F(a+(k+\kappa)\Delta) = \frac{1}{\Delta}\int_a^b dt\, F(t) + \sum_{p=1}^{\infty} \frac{\Delta^{p-1}}{p!} B_p(\kappa)\left[F^{(p-1)}(b) - F^{(p-1)}(a)\right],
$$
$$(2.593)$$

where $B_n(\kappa)$ are the Bernoulli functions defined by a generalization of the expansion (2.566):

$$
\frac{te^{\kappa t}}{e^t - 1} = \sum_{n=0}^{\infty} B_n(\kappa)\frac{t^n}{n!}. \quad (2.594)
$$

At $\kappa = 0$, the Bernoulli functions start out with the Bernoulli numbers: $B_n(0) = B_n$. The function $B_0(\kappa)$ is equal to 1 everywhere.

Another way of writing formula (2.595) is

$$
\sum_{k=0}^{K-1} F(a+(k+\kappa)\Delta) = \frac{1}{\Delta}\int_a^b dt\left[1 + \sum_{p=0}^{\infty}\frac{\Delta^p}{p!}B_p(\kappa)\partial_t^p\right]F(t). \quad (2.595)
$$

This implies that a sum over discrete values of a function can be replaced by an integral over a gradient expansion of the function.

---

[30]M. Abramowitz and I. Stegun, op. cit., Formulas 23.1.30 and 23.1.32.

Using the first Euler-Maclaurin formula (2.592) with $a = \omega_1^2$, $b = \omega_M^2$, and $\Delta = \omega_1$, we find

$$\sum_{m=0}^{M} \left[\log(\omega_m^2 + \omega^2) - \log(\omega_m^2) \right] = \left\{ \pi \frac{\omega}{\omega_1} + \frac{\omega_m}{\omega_1} \left[\log(\omega_m^2 + \omega^2) - 2 \right] \right\} \Big|_{m=1}^{m=M} - \left\{ \omega = 0 \right\} \Big|_{m=1}^{m=M}$$
$$+ \frac{1}{2} \left\{ \log(\omega_1^2 + \omega^2) + \log(\omega_M^2 + \omega^2) \right\} - \left\{ \omega = 0 \right\}. \quad (2.596)$$

For small T, the leading two terms on the right-hand side are

$$\pi \frac{\omega}{\omega_1} - \frac{1}{2} \log \frac{\omega^2}{\omega_1^2}, \quad (2.597)$$

in agreement with the first two terms in the low-temperature series (2.562). Note that the Euler-Maclaurin formula is unable to recover the exponentially small terms in (2.562), since they are not expandable in powers of T.

The transformation of high- into low-temperature expansions is an important tool for analyzing phase transitions in models of statistical mechanics.[31]

2.16 Finite-N Behavior of Thermodynamic Quantities

Thermodynamic fluctuations in Euclidean path integrals are often imitated in computer simulations. These are forced to employ a sliced time axis. It is then important to know in which way the time-sliced thermodynamic quantities converge to their continuum limit. Let us calculate the internal energy E and the specific heat at constant volume C for finite N from (2.482). Using (2.482) we have

$$\frac{\partial(\beta \tilde{\omega}_e)}{\partial \beta} = \frac{\omega}{\cosh(\epsilon \tilde{\omega}_e/2)},$$
$$\frac{\partial(\epsilon \tilde{\omega}_e)}{\partial \beta} = \frac{2}{\beta} \tanh(\epsilon \tilde{\omega}_e/2), \quad (2.598)$$

and find the internal energy

$$E = \frac{\partial}{\partial \beta}(\beta F) = \frac{\hbar}{2} \coth(\beta \hbar \tilde{\omega}_e/2) \frac{\partial(\beta \tilde{\omega}_e)}{\partial \beta}$$
$$= \frac{\hbar \omega}{2} \frac{\coth(\beta \hbar \tilde{\omega}_e/2)}{\cosh(\epsilon \tilde{\omega}_e/2)}. \quad (2.599)$$

The specific heat at constant volume is given by

$$\frac{1}{k_B} C = -\beta^2 \frac{\partial^2}{\partial \beta^2}(\beta F) = -\beta^2 \frac{\partial}{\partial \beta} E \quad (2.600)$$
$$= \frac{1}{4} \beta^2 \hbar^2 \omega^2 \left[\frac{1}{\sinh^2(\beta \hbar \tilde{\omega}_e/2)} + \coth(\beta \hbar \tilde{\omega}_e/2) \tanh(\epsilon \tilde{\omega}_e/2) \frac{\epsilon}{\hbar \beta} \right] \frac{1}{\cosh^2(\epsilon \tilde{\omega}_e/2)}.$$

Plots are shown in Fig. 2.5 for various N using natural units with $\hbar = 1$, $k_B = 1$. At high

---

[31]See H. Kleinert, *Gauge Fields in Condensed Matter*, Vol. I Superflow and Vortex Lines, World Scientific, Singapore, 1989, pp. 1–742 (http://www.physik.fu-berlin.de/~kleinert/b1).

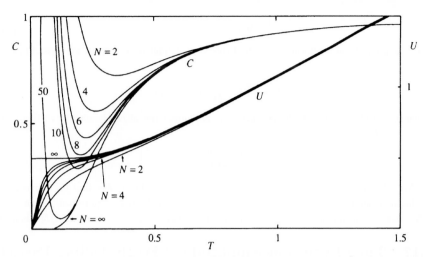

Figure 2.4   Finite-lattice effects in internal energy E and specific heat C at constant volume, as a function of the temperature for various numbers $N+1$ of time slices. Note the nonuniform way in which the exponential small-T behavior of $C \propto e^{-\omega/T}$ is approached in the limit $N \to \infty$.

temperatures, F, E, and C are independent of N:

$$F \;\to\; \frac{1}{\beta}\log \beta, \tag{2.601}$$

$$E \;\to\; \frac{1}{\beta} = T, \tag{2.602}$$

$$C \;\to\; 1. \tag{2.603}$$

These limits are a manifestation of the *Dulong-Petit law*: An oscillator has one kinetic and one potential degree of freedom, each carrying an internal energy $T/2$ and a specific heat $1/2$. At low temperatures, on the other hand, E and C are strongly N-dependent (note that since F and E are different at $T = 0$, the entropy of the lattice approximation does not vanish as it must in the continuum limit). Thus, the convergence $N \to \infty$ is highly nonuniform. After reaching the limit, the specific heat goes to zero for $T \to 0$ exponentially fast, like $e^{-\omega/T}$. The quantity ω is called *activation energy*.[32] It is the energy difference between the ground state and the first excited state of the harmonic oscillator. For large but finite N, on the other hand, the specific heat has the large value $N+1$ at $T = 0$. This is due to $\tilde{\omega}_e$ and $\cosh^2(\epsilon\tilde{\omega}_e/2)$ behaving, for a finite N and $T \to 0$ (where ϵ becomes large) like

$$\tilde{\omega}_e \;\to\; \frac{1}{\epsilon}\log(\epsilon^2\omega^2),$$
$$\cosh(\epsilon\tilde{\omega}_e/2) \;\to\; \epsilon\omega/2. \tag{2.604}$$

Hence

$$E \;\xrightarrow{\;T\to 0\;}\; \frac{1}{\beta}\coth[(N+1)\log(\epsilon\omega)] \;\xrightarrow{\;T\to 0\;}\; 0, \tag{2.605}$$

---

[32]Note that in a D-dimensional solid the lattice vibrations can be considered as an ensemble of harmonic oscillators with energies ω ranging from zero to the Debye frequency. Integrating over the corresponding specific heats with the appropriate density of states, $\int d\omega\,\omega^{D-1}e^{-\omega/k_B T}$, gives the well-known power law at low temperatures $C \propto T^D$.

$$C \xrightarrow{T \to 0} N + 1. \tag{2.606}$$

The reason for the nonuniform approach of the $N \to \infty$ limit is obvious: If we expand (2.482) in powers of ϵ, we find

$$\tilde{\omega}_e = \omega \left(1 - \frac{1}{24} \epsilon^2 \omega^2 + \dots \right). \tag{2.607}$$

When going to low T at finite N the corrections are quite large, as can be seen by writing (2.607), with $\epsilon = \hbar \beta / (N + 1)$, as

$$\tilde{\omega}_e = \omega \left[1 - \frac{1}{24} \frac{\hbar^2 \omega^2}{k_b^2 T^2 (N + 1)^2} + \dots \right]. \tag{2.608}$$

Note that (2.607) contains no corrections of the order ϵ. This implies that the convergence of all thermodynamic quantities in the limit $N \to \infty, \epsilon \to 0$ at fixed T is quite fast — one order in $1/N$ faster than we might at first expect [the Trotter formula (2.26) also shows the $1/N^2$-behavior].

2.17 Time Evolution Amplitude of Freely Falling Particle

The gravitational potential of a particle on the surface of the earth is

$$V(\mathbf{x}) = V_0 + M \mathbf{g} \cdot \mathbf{x}, \tag{2.609}$$

where $-\mathbf{g}$ is the earth's acceleration vector pointing towards the ground, and V_0 some constant. The equation of motion reads

$$\ddot{\mathbf{x}} = -\mathbf{g}, \tag{2.610}$$

which is solved by

$$\mathbf{x} = \mathbf{x}_a + \mathbf{v}_a(t - t_a) + \frac{\mathbf{g}}{2}(t - t_a)^2, \tag{2.611}$$

with the initial velocity

$$\mathbf{v}_a = \frac{\mathbf{x}_b - \mathbf{x}_a}{t_b - t_a} - \frac{\mathbf{g}}{2}(t_b - t_a). \tag{2.612}$$

Inserting this into the action

$$\mathcal{A} = \int_{t_a}^{t_b} dt \left(\frac{M}{2} \dot{\mathbf{x}}^2 - V_0 - \mathbf{g} \cdot \mathbf{x} \right), \tag{2.613}$$

we obtain the classical action

$$\mathcal{A}_{\mathrm{cl}} = -V_0(t_b - t_a) + \frac{M}{2} \frac{(\mathbf{x}_b - \mathbf{x}_a)^2}{t_b - t_a} - \frac{1}{2}(t_b - t_a)\mathbf{g} \cdot (\mathbf{x}_b + \mathbf{x}_a) - \frac{1}{24}(t_b - t_a)^3 \mathbf{g}^2. \tag{2.614}$$

Since the quadratic part of (2.613) is the same as for a free particle, also the fluctuation factor is the same [see (2.125)], and we find the time evolution amplitude

$$(\mathbf{x}_b t_b | \mathbf{x}_a t_a) = \frac{1}{\sqrt{2\pi i \hbar \omega (t_b - t_a)/M}^3} e^{-\frac{i}{\hbar} V_0 (t_b - t_a)}$$

$$\times \exp \left\{ \frac{iM}{2\hbar} \left[\frac{(\mathbf{x}_b - \mathbf{x}_a)^2}{t_b - t_a} - (t_b - t_a)\mathbf{g} \cdot (\mathbf{x}_b + \mathbf{x}_a) - \frac{1}{12}(t_b - t_a)^3 \mathbf{g}^2 \right] \right\}. \tag{2.615}$$

The potential (2.609) can be considered as a limit of a harmonic potential

$$V(\mathbf{x}) = V_0 + \frac{M}{2}\omega^2(\mathbf{x} - \mathbf{x}_0)^2 \qquad (2.616)$$

for

$$\omega \to 0, \quad \mathbf{x}_0 = -\mathbf{g}/\omega^2 \to -\infty \approx \hat{\mathbf{g}}, \quad V_0 = -M\mathbf{x}_0^2/2 = -Mg^2/2\omega^4 \to -\infty, \quad (2.617)$$

keeping

$$\mathbf{g} = -M\omega^2\mathbf{x}_0, \qquad (2.618)$$

and

$$v_0 = V_0 + \frac{M}{2}\omega^2\mathbf{x}_0^2 \qquad (2.619)$$

fixed. If we perform this limit in the amplitude (2.175), we find of course (2.615).

The wave functions can be obtained most easily by performing this limiting procedure on the wave functions of the harmonic oscillator. In one dimension, we set $n = E/\omega$ and find that the spectral representation (2.292) goes over into

$$(x_b t_b | x_a t_a) = \int dE\, A_E(x_b) A_E^*(x_a) e^{-iE(t_b - t_a)/\hbar}, \qquad (2.620)$$

with the wave functions

$$A_E(x) = \frac{1}{\sqrt{l\varepsilon}} \operatorname{Ai}\left(\frac{x}{l} - \frac{E}{\varepsilon}\right). \qquad (2.621)$$

Here $\varepsilon \equiv (\hbar^2 g^2 M/2)^{1/3}$ and $l \equiv (\hbar^2/2M^2g)^{1/3} = \varepsilon/Mg$ are the natural units of energy and length, respectively, and $\operatorname{Ai}(z)$ is the *Airy function* solving the differential equation

$$\operatorname{Ai}''(z) = z\operatorname{Ai}(z), \qquad (2.622)$$

For positive z, the Airy function can be expressed in terms of modified Bessel functions $I_\nu(\xi)$ and $K_\nu(\xi)$:[33]

$$\operatorname{Ai}(z) = \frac{\sqrt{z}}{2}[I_{-1/3}(2z^{3/2}/3) - I_{1/3}(2z^{3/2}/3)] = \frac{1}{\pi}\sqrt{\frac{z}{3}}K_{1/3}\left(2z^{3/2}/3\right). \qquad (2.623)$$

For large z, this falls off exponentially:

$$\operatorname{Ai}(z) \to \frac{1}{2\sqrt{\pi}z^{1/4}}e^{-2z^{3/2}/3}, \quad z \to \infty. \qquad (2.624)$$

For negative z, an analytic continuation[34]

$$\begin{aligned}
I_\nu(\xi) &= e^{-\pi\nu i/2}J(e^{\pi i/2}\xi), \quad -\pi < \arg\xi \le \pi/2, \\
I_\nu(\xi) &= e^{-\pi\nu i/2}J(e^{\pi i/2}\xi), \quad \pi/2 < \arg\xi \le \pi,
\end{aligned} \qquad (2.625)$$

---

[33] A compact description of the properties of Bessel functions is found in M. Abramowitz and I. Stegun, op. cit., Chapter 10. The Airy function is expressed in Formulas 10.4.14.
[34] I.S. Gradshteyn and I.M. Ryzhik, *op. cit.*, Formulas 8.406.

leads to

$$\mathrm{Ai}(z) = \frac{1}{3}\sqrt{z}\left[J_{-1/3}(2(-z)^{3/2}/3) + J_{1/3}(2(-z)^{3/2}/3)\right],\qquad(2.626)$$

where $J_{1/3}(\xi)$ are ordinary Bessel functions. For large arguments, these oscillate like

$$J_\nu(\xi) \to \sqrt{\frac{2}{\pi\xi}}\cos(\xi - \pi\nu/2 - \pi/4) + \mathcal{O}(\xi^{-1}),\qquad(2.627)$$

from which we obtain the oscillating part of the Airy function

$$\mathrm{Ai}(z) \to \frac{1}{\sqrt{\pi}z^{1/4}}\sin\left[2(-z)^{3/2}/3 + \pi/4\right],\quad z \to -\infty.\qquad(2.628)$$

The Airy function has the simple Fourier representation

$$\mathrm{Ai}(x) = \int_{-\infty}^{\infty}\frac{dk}{2\pi}e^{i(xk+k^3/3)}.\qquad(2.629)$$

In fact, the momentum space wave functions of energy E are

$$\langle p|E\rangle = \sqrt{\frac{l}{\varepsilon}}e^{-i(pE-p^3/6M)l/\varepsilon\hbar}\qquad(2.630)$$

fulfilling the orthogonality and completeness relations

$$\int\frac{dp}{2\pi\hbar}\langle E'|p\rangle\langle p|E\rangle = \delta(E'-E),\quad \int dE\,\langle p'|E\rangle\langle E|p\rangle = 2\pi\hbar\delta(p'-p).\qquad(2.631)$$

The Fourier transform of (2.630) is equal to (2.621), due to (2.629).

2.18 Charged Particle in Magnetic Field

Having learned how to solve the path integral of the harmonic oscillator we are ready to study also a more involved harmonic system of physical importance: a charged particle in a magnetic field. This problem was first solved by L.D. Landau in 1930 in Schrödinger theory.[35]

2.18.1 Action

The magnetic interaction of a particle of charge e is given by

$$\mathcal{A}_{\mathrm{mag}} = \frac{e}{c}\int_{t_a}^{t_b} dt\,\dot{\mathbf{x}}(t)\cdot\mathbf{A}(\mathbf{x}(t)),\qquad(2.632)$$

where $\mathbf{A}(\mathbf{x})$ is the vector potential of the magnetic field. The total action is

$$A[\mathbf{x}] = \int_{t_a}^{t_b} dt\left[\frac{M}{2}\dot{\mathbf{x}}^2(t) + \frac{e}{c}\dot{\mathbf{x}}(t)\cdot\mathbf{A}(\mathbf{x}(t))\right].\qquad(2.633)$$

---

[35]L.D. Landau and E.M. Lifshitz, *Quantum Mechanics*, Pergamon, London, 1965.

Suppose now that the particle moves in a homogeneous magnetic field $\mathbf{B}$ pointing along the z-direction. Such a field can be described by a vector potential

$$\mathbf{A}(\mathbf{x}) = (0, Bx, 0). \tag{2.634}$$

But there are other possibilities. The magnetic field

$$\mathbf{B}(\mathbf{x}) = \boldsymbol{\nabla} \times \mathbf{A}(\mathbf{x}) \tag{2.635}$$

as well as the magnetic interaction (2.632) are invariant under gauge transformations

$$\mathbf{A}(\mathbf{x}) \to \mathbf{A}(\mathbf{x}) + \boldsymbol{\nabla}\Lambda(\mathbf{x}), \tag{2.636}$$

where $\Lambda(\mathbf{x})$ are arbitrary single-valued functions of $\mathbf{x}$. As such they satisfy the Schwarz integrability condition [compare (1.41)–(1.42)]

$$(\partial_i \partial_j - \partial_j \partial_i)\Lambda(\mathbf{x}) = 0. \tag{2.637}$$

For instance, the axially symmetric vector potential

$$\tilde{\mathbf{A}}(\mathbf{x}) = \frac{1}{2}\mathbf{B} \times \mathbf{x} \tag{2.638}$$

gives the same magnetic field; it differs from (2.634) by a gauge transformation

$$\tilde{\mathbf{A}}(\mathbf{x}) = \mathbf{A}(\mathbf{x}) + \boldsymbol{\nabla}\Lambda(\mathbf{x}), \tag{2.639}$$

with the gauge function

$$\Lambda(\mathbf{x}) = -\frac{1}{2}B\,xy. \tag{2.640}$$

In the canonical form, the action reads

$$\mathcal{A}[\mathbf{p}, \mathbf{x}] = \int_{t_a}^{t_b} dt \left\{ \mathbf{p} \cdot \dot{\mathbf{x}} - \frac{1}{2M} \left[\mathbf{p} - \frac{e}{c}\mathbf{A}(\mathbf{x}) \right]^2 \right\}. \tag{2.641}$$

The magnetic interaction of a point particle is thus included in the path integral by the so-called *minimal substitution* of the momentum variable:

$$\mathbf{p} \longrightarrow \mathbf{P} \equiv \mathbf{p} - \frac{e}{c}\mathbf{A}(\mathbf{x}). \tag{2.642}$$

For the vector potential (2.634), the action (2.641) becomes

$$\mathcal{A}[\mathbf{p}, \mathbf{x}] = \int_{t_a}^{t_b} dt \left[\mathbf{p} \cdot \dot{\mathbf{x}} - H(\mathbf{p}, \mathbf{x}) \right], \tag{2.643}$$

with the Hamiltonian

$$H(\mathbf{p}, \mathbf{x}) = \frac{\mathbf{p}^2}{2M} + \frac{1}{8}M\omega_L^2\mathbf{x}^2 - \frac{1}{2}\omega_L l_z(\mathbf{p}, \mathbf{x}), \tag{2.644}$$

where $\mathbf{x} = (x, y)$ and $\mathbf{p} = (p_x, p_y)$ and

$$l_z(\mathbf{p}, \mathbf{x}) = (\mathbf{x} \times \mathbf{p})_z = xp_y - yp_x \tag{2.645}$$

is the z-component of the orbital angular momentum. In a Schrödinger equation, the last term in $H(\mathbf{p}, \mathbf{x})$ is diagonal on states with a given angular momentum around the z-axis. We have introduced the field-dependent frequency

$$\omega_L = \frac{e}{Mc}B, \tag{2.646}$$

called *Landau frequency* or *cyclotron frequency*. This can also be written in terms of the *Bohr magneton*

$$\mu_B \equiv \frac{\hbar e}{Mc}, \tag{2.647}$$

as

$$\omega_L = \mu_B B/\hbar. \tag{2.648}$$

The first two terms in (2.644) describe a harmonic oscillator in the xy-plane with a field-dependent *magnetic frequency*

$$\omega_B \equiv \frac{\omega_L}{2}. \tag{2.649}$$

Note that in the gauge gauge (2.634), the Hamiltonian would have the rotationally noninvariant form

$$H(\mathbf{p}, \mathbf{x}) = \frac{\mathbf{p}^2}{2M} + \frac{1}{2}M\omega_L^2 x^2 - \omega_L x p_y \tag{2.650}$$

rather than (2.644), implying oscillations of frequency ω_L in the x-direction and a free motion in the y-direction.

The time-sliced form of the canonical action (2.641) reads

$$\mathcal{A}_e^N = \sum_{n=1}^{N+1} \left\{ \mathbf{p}_n(\mathbf{x}_n - \mathbf{x}_{n-1}) - \frac{1}{2M}\left[p_{xn}^2 + (p_{yn} - Bx_n)^2 + p_{zn}^2 \right] \right\}, \tag{2.651}$$

and the associated time-evolution amplitude for the particle to run from $\mathbf{x}_a$ to $\mathbf{x}_b$ is given by

$$(\mathbf{x}_b t_b | \mathbf{x}_a t_a) = \prod_{n=1}^{N} \left[\int d^3 x_n \right] \prod_{n=1}^{N+1} \left[\int \frac{d^3 p_n}{(2\pi\hbar)^3} \right] \exp\left(\frac{i}{\hbar}\mathcal{A}_e^N \right), \tag{2.652}$$

with the time-sliced action

$$\mathcal{A}^N = \sum_{n=1}^{N+1} \left\{ \mathbf{p}_n(\mathbf{x}_n - \mathbf{x}_{n-1}) - \frac{1}{2M}\left[p_{xn}^2 + (p_{yn} - Bx_n)^2 + p_{zn}^2 \right] \right\}. \tag{2.653}$$

2.18.2   Gauge Properties

Note that the time evolution amplitude is not gauge-invariant. If we use the vector potential in some other gauge

$$\mathbf{A}'(\mathbf{x}) = \mathbf{A}(\mathbf{x}) + \mathbf{\nabla}\Lambda(\mathbf{x}), \tag{2.654}$$

the action changes by a surface term

$$\Delta\mathcal{A} = \frac{e}{c}\int_{t_a}^{t_b} dt\,\dot{\mathbf{x}}\cdot\mathbf{\nabla}\Lambda(\mathbf{x}) = \frac{e}{c}[\Lambda(\mathbf{x}_b) - \Lambda(\mathbf{x}_a)]. \tag{2.655}$$

The amplitude is therefore multiplied by a phase factor on both ends

$$(\mathbf{x}_b t_b|\mathbf{x}_a t_a)_A \to (\mathbf{x}_b t_b|\mathbf{x}_a t_a)_{A'} = e^{ie\Lambda(\mathbf{x}_b)/c\hbar}(\mathbf{x}_b t_b|\mathbf{x}_a t_a)_A e^{-ie\Lambda(\mathbf{x}_a)/c\hbar}. \tag{2.656}$$

For the observable particle distribution $(\mathbf{x}\,t_b|\mathbf{x}\,t_a)$, the phase factors are obviously irrelevant. But all other observables of the system must also be independent of the phases $\Lambda(\mathbf{x})$ which is assured if they correspond to gauge-invariant operators.

2.18.3   Time-Sliced Path Integration

Since the action $\mathcal{A}^N$ contains the variables y_n and z_n only in the first term $\sum_{n=1}^{N+1} i\mathbf{p}_n\mathbf{x}_n$, we can perform the y_n, z_n integrations and find a product of N Δ-functions in the y- and z-components of the momenta $\mathbf{p}_n$. If the projections of $\mathbf{p}$ to the yz-plane are denoted by $\mathbf{p}'$, the product is

$$(2\pi\hbar)^2\delta^{(2)}\left(\mathbf{p}'_{N+1} - \mathbf{p}'_N\right)\cdots(2\pi\hbar)^2\delta^{(2)}\left(\mathbf{p}'_2 - \mathbf{p}'_1\right). \tag{2.657}$$

These allow performing all p_{yn}, p_{zn}-integrals, except for one overall p_y, p_z. The path integral reduces therefore to

$$(\mathbf{x}_b t_b|\mathbf{x}_a t_a) = \int_{-\infty}^{\infty}\frac{dp_y dp_z}{(2\pi\hbar)^2}\prod_{n=1}^{N}\left[\int_{-\infty}^{\infty} dx_n\right]\prod_{n=1}^{N+1}\left[\int\frac{dp_{xn}}{2\pi\hbar}\right] \tag{2.658}$$

$$\times\ \exp\left\{\frac{i}{\hbar}\left[p_y(y_b - y_a) + p_z(z_b - z_a) - (t_b - t_a)\frac{p_z^2}{2M}\right]\right\}\exp\left(\frac{i}{\hbar}\mathcal{A}_x^N\right),$$

where $\mathcal{A}_x^N$ is the time-sliced action involving only a one-dimensional path integral over the x-component of the path, $x(t)$, with the sliced action

$$\mathcal{A}_x^N = \sum_{n=1}^{N+1}\left[p_{xn}(x_n - x_{n-1}) - \frac{p_{xn}^2}{2M} - \frac{1}{2M}\left(p_y - \frac{e}{c}Bx_n\right)^2\right]. \tag{2.659}$$

This is the action of a one-dimensional harmonic oscillator with field-dependent frequency ω_B whose center of oscillation depends on p_y and lies at

$$x_0 = p_y/M\omega_L. \tag{2.660}$$

The path integral over $x(t)$ is harmonic and known from (2.173):

$$(x_b t_b | x_a t_a)_{x_0} = \sqrt{\frac{M\omega_L}{2\pi i\hbar \sin \omega_L(t_b - t_a)}}$$

$$\times \exp\left(\frac{i}{\hbar} \frac{M\omega_L}{2\sin \omega_L(t_b - t_a)} \left\{ \left[(x_b - x_0)^2 + (x_a - x_0)^2\right] \cos \omega_L(t_b - t_a) \right.\right.$$

$$\left.\left. -2(x_b - x_0)(x_a - x_0) \right\} \right). \tag{2.661}$$

Doing the p_z-integral in (2.658), we arrive at the formula

$$(\mathbf{x}_b t_b | \mathbf{x}_a t_a) = \frac{1}{\sqrt{2\pi i\hbar(t_b - t_a)/M}} e^{i\frac{M}{2\hbar}\frac{(z_b - z_a)^2}{t_b - t_a}} (\mathbf{x}_b^\perp t_b | \mathbf{x}_a^\perp t_a), \tag{2.662}$$

with the amplitude orthogonal to the magnetic field

$$(\mathbf{x}_b^\perp t_b | \mathbf{x}_a^\perp t_a) \equiv \frac{M\omega_L}{2\pi\hbar} \int_{-\infty}^\infty dx_0 \, e^{iM\omega_L x_0(y_b - y_a)/\hbar}(x_b t_b | x_a t_a)_{x_0}. \tag{2.663}$$

After a quadratic completion in x_0, the total exponent in (2.663) reads

$$\frac{iM\omega_L}{2\hbar}\left[-(x_b^2 + x_a^2)\tan[\omega_L(t_b - t_a)/2] + (x_b - x_a)^2\frac{1}{\sin \omega_L(t_b - t_a)}\right]$$

$$-i\frac{M\omega_L}{\hbar}\tan[\omega_L(t_b - t_a)/2]\left(x_0 - \frac{x_b + x_a}{2} - \frac{y_b - y_a}{2\tan[\omega_L(t_b - t_a)/2]}\right)^2$$

$$+i\frac{M\omega_L}{2\hbar}\left[\frac{(x_b + x_a)^2}{2}\tan[\omega_L(t_b - t_a)/2] + \frac{(y_b - y_a)^2}{2\tan[\omega_L(t_b - t_a)/2]}\right]$$

$$+i\frac{M\omega_L}{2\hbar}(x_b + x_a)(y_b - y_a). \tag{2.664}$$

The integration $M\omega_L \int_{-\infty}^\infty dx_0/2\pi\hbar$ removes the second term and results in a factor

$$\frac{M\omega_L}{2\pi\hbar}\sqrt{\frac{\pi\hbar}{iM\omega_L \tan[\omega_L(t_b - t_a)/2]}}. \tag{2.665}$$

By rearranging the remaining terms, we arrive at the amplitude

$$(\mathbf{x}_b t_b | \mathbf{x}_a t_a) = \sqrt{\frac{M}{2\pi i\hbar(t_b - t_a)}}^3 \frac{\omega_L(t_b - t_a)/2}{\sin[\omega_L(t_b - t_a)/2]}\exp\left[\frac{i}{\hbar}(\mathcal{A}_{\mathrm{cl}} + \mathcal{A}_{\mathrm{sf}})\right], \tag{2.666}$$

with an action

$$\mathcal{A}_{\mathrm{cl}} = \frac{M}{2}\left\{\frac{(z_b - z_a)^2}{t_b - t_a} + \frac{\omega_L}{2}\cot[\omega_L(t_b - t_a)/2]\left[(x_b - x_a)^2 + (y_b - y_a)^2\right]\right.$$

$$\left. + \quad \omega_L(x_a y_b - x_b y_a)\right\} \tag{2.667}$$

and the surface term

$$\mathcal{A}_{\mathrm{sf}} = \frac{M\omega_L}{2}(x_b y_b - x_a y_a) = \frac{e}{2c}B\,xy\Big|_a^b. \tag{2.668}$$

2.18.4   Classical Action

Since the action is harmonic, the amplitude is again a product of a phase $e^{i\mathcal{A}_{\text{cl}}}$ and a fluctuation factor. A comparison with (2.639) and (2.656) shows that the surface term would be absent if the amplitude $(\mathbf{x}_b t_b | \mathbf{x}_a t_a)_{\bar{A}}$ were calculated with the vector potential in the axially symmetric gauge (2.638). Thus $\mathcal{A}_{\text{cl}}$ must be equal to the classical action in this gauge. Indeed, the orthogonal part can be rewritten as

$$\mathcal{A}_{\text{cl}}^{\perp} = \int_{t_a}^{t_b} dt \left\{ \frac{M}{2} \frac{d}{dt} (x\dot{x} + y\dot{y}) + \frac{M}{2} \left[x(-\ddot{x} + \omega_L \dot{y}) + y(-\ddot{y} - \omega_L \dot{x}) \right] \right\}. \qquad (2.669)$$

The equations of motion are

$$\ddot{x} = \omega_L \dot{y}, \qquad \ddot{y} = -\omega_L \dot{x}, \qquad (2.670)$$

reducing the action of a classical orbit to

$$\mathcal{A}_{\text{cl}}^{\perp} = \frac{M}{2} (x\dot{x} + y\dot{y}) \Big|_{t_a}^{t_b} = \frac{M}{2} \left([x_b\dot{x}_b - x_a\dot{x}_a] + [y_b\dot{y}_b - y_a\dot{y}_a] \right). \qquad (2.671)$$

The orbits are easily determined. By inserting the two equations in (2.670) into each other we see that $\dot{x}$ and $\dot{y}$ perform independent oscillations:

$$\ddot{\dot{x}} + \omega_L^2 \dot{x} = 0, \qquad \ddot{\dot{y}} + \omega_L^2 \dot{y} = 0. \qquad (2.672)$$

The general solution of these equations is

$$x = \frac{1}{\sin \omega_L (t_b - t_a)} \left[(x_b - x_0) \sin \omega_L (t - t_a) - (x_a - x_0) \sin \omega_L (t - t_b) \right] + x_0, \quad (2.673)$$

$$y = \frac{1}{\sin \omega_L (t_b - t_a)} \left[(y_b - y_0) \sin \omega_L (t - t_a) - (y_a - y_0) \sin \omega_L (t - t_b) \right] + y_0, \quad (2.674)$$

where we have incorporated the boundary condition $x(t_{a,b}) = x_{a,b}$, $y(t_{a,b}) = y_{a,b}$. The constants x_0, y_0 are fixed by satisfying (2.670). This gives

$$x_0 = \frac{1}{2} \left[(x_b + x_a) + (y_b - y_a) \cot \frac{\omega_L}{2} (t_b - t_a) \right], \qquad (2.675)$$

$$y_0 = \frac{1}{2} \left[(y_b + y_a) - (x_b - x_a) \cot \frac{\omega_L}{2} (t_b - t_a) \right]. \qquad (2.676)$$

Now we calculate

$$x_b\dot{x}_b = \frac{\omega_L}{\sin \omega_L (t_b - t_a)} x_b \left[(x_0 - x_a) + (x_b - x_0) \cos \omega_L (t_b - t_a) \right], \qquad (2.677)$$

$$x_a\dot{x}_a = \frac{\omega_L}{\sin \omega_L (t_b - t_a)} x_a \left[(x_0 - x_a) \cos \omega_L (t_b - t_a) + (x_b - x_0) \right], \qquad (2.678)$$

and hence

$$\begin{aligned} x_b\dot{x}_b - x_a\dot{x}_a &= \omega_L x_0 (x_b + x_a) \tan \frac{\omega_L}{2} (t_b - t_a) \\ &\quad + \frac{\omega_L}{\sin \omega_L (t_b - t_a)} \left[(x_b^2 + x_a^2) \cos \omega_L (t_b - t_a) - 2x_b x_a \right] \\ &= \frac{\omega_L}{2} \left[(x_b - x_a)^2 \cot \frac{\omega_L}{2} (t_b - t_a) + (x_b + x_a)(y_b - y_a) \right]. \end{aligned} \qquad (2.679)$$

Similarly we find

$$
\begin{aligned}
y_b\dot{y}_b - y_a\dot{y}_a &= \omega_L y_0 (y_b + y_a) \tan \frac{\omega_L}{2}(t_b - t_a) \\
&\quad + \frac{\omega_L}{\sin\omega_L(t_b - t_a)}[(y_b^2 + y_a^2)\cos\omega_L(t_b - t_a) - 2y_b y_a] \\
&= \frac{\omega_L}{2}\left[(y_b - y_a)^2 \cot \frac{\omega_L}{2}(t_b - t_a) - (x_b - x_a)(y_b + y_a)\right].
\end{aligned}
\tag{2.680}
$$

Inserted into (2.671), this yields the classical action for the orthogonal motion

$$
\mathcal{A}_{\mathrm{cl}}^{\perp} = \frac{M}{2}\left\{\frac{\omega_L}{2}\cot[\omega_L(t_b - t_a)/2]\left[(x_b - x_a)^2 + (y_b - y_a)^2\right] + \omega_L(x_a y_b - x_b y_a)\right\},
\tag{2.681}
$$

which is indeed the orthogonal part of the action (2.667).

2.18.5 Translational Invariance

It is interesting to see how the amplitude ensures the translational invariance of all physical observables. The first term in the classical action is trivially invariant. The last term reads

$$
\Delta\mathcal{A} = \frac{M\omega_L}{2}(x_a y_b - x_b y_a).
\tag{2.682}
$$

Under a translation by a distance $\mathbf{d}$,

$$
\mathbf{x} \to \mathbf{x} + \mathbf{d},
\tag{2.683}
$$

this term changes by

$$
\frac{M\omega_L}{2}[d_x(y_b - y_a) + d_y(x_a - x_b)] = \frac{M\omega_L}{2}[(\mathbf{d}\times\mathbf{x})_b - (\mathbf{d}\times\mathbf{x})_a]_z
\tag{2.684}
$$

causing the amplitude to change by a pure gauge transformation as

$$
(\mathbf{x}_b t_b|\mathbf{x}_a t_a) \to e^{ie\Lambda(\mathbf{x}_b)/c\hbar}(\mathbf{x}_b t_b|\mathbf{x}_a t_a)e^{-ie\Lambda(\mathbf{x}_a)/c\hbar},
\tag{2.685}
$$

with the phase

$$
\Lambda(\mathbf{x}) = -\frac{M\omega_L\hbar c}{2e}[\mathbf{d}\times\mathbf{x}]_z.
\tag{2.686}
$$

Since observables involve only gauge-invariant quantities, such transformations are irrelevant.

This will be done in 2.23.3.

2.19   Charged Particle in Magnetic Field plus Harmonic Potential

For application in Chapter 5 we generalize the above magnetic system by adding a harmonic oscillator potential, thus leaving the path integral solvable. For simplicity, we consider only the orthogonal part of the resulting system with respect to the magnetic field. Omitting orthogonality symbols, the Hamiltonian is the same as in (2.644). Without much more work we may solve the path integral of a more general system in which the harmonic potential in (2.644) has a different frequency $\omega \neq \omega_L$, and thus a Hamiltonian

$$H(\mathbf{p}, \mathbf{x}) = \frac{\mathbf{p}^2}{2M} + \frac{1}{2} M \omega^2 \mathbf{x}^2 - \omega_B l_z(\mathbf{p}, \mathbf{x}). \tag{2.687}$$

The associated Euclidean action

$$\mathcal{A}_e[\mathbf{p}, \mathbf{x}] = \int_{\tau_a}^{\tau_b} d\tau \left[-i\mathbf{p} \cdot \dot{\mathbf{x}} + H(\mathbf{p}, \mathbf{x}) \right] \tag{2.688}$$

has the Lagrangian form

$$\mathcal{A}_e[\mathbf{x}] = \int_0^{\hbar\beta} d\tau \left\{ \frac{M}{2} \dot{\mathbf{x}}^2(\tau) + \frac{1}{2} M(\omega^2 - \omega_B^2) \mathbf{x}^2(\tau) - i M \omega_B [\mathbf{x}(\tau) \times \dot{\mathbf{x}}(\tau)]_z \right\}. \tag{2.689}$$

At this point we observe that the system is stable only for $\omega \geq \omega_B$. The action (2.689) can be written in matrix notation as

$$\mathcal{A}_{\rm cl} = \int_0^{\hbar\beta} d\tau \left[\frac{M}{2} \frac{d}{d\tau} (\mathbf{x}\dot{\mathbf{x}}) + \frac{M}{2} \mathbf{x}^T \mathbf{D}_{\omega^2,B} \mathbf{x} \right], \tag{2.690}$$

where $\mathbf{D}_{\omega^2,B}$ is the 2×2 -matrix

$$\mathbf{D}_{\omega^2,B}(\tau, \tau') \equiv \begin{pmatrix} -\partial_\tau^2 + \omega^2 - \omega_B^2 & -2i\omega_B \partial_\tau \\ 2i\omega_B \partial_\tau & -\partial_\tau^2 + \omega^2 - \omega_B^2 \end{pmatrix} \delta(\tau - \tau'). \tag{2.691}$$

Since the path integral is Gaussian, we can immediately calculate the partition function

$$Z = \frac{1}{(2\pi\hbar/M)^2} \det \mathbf{D}_{\omega^2,B}^{-1/2}. \tag{2.692}$$

By expanding $\mathbf{D}_{\omega^2,B}(\tau, \tau')$ in a Fourier series

$$\mathbf{D}_{\omega^2,B}(\tau, \tau') = \frac{1}{\hbar\beta} \sum_{m=-\infty}^{\infty} \tilde{\mathbf{D}}_{\omega^2,B}(\omega_m) e^{-i\omega_m(\tau - \tau')}, \tag{2.693}$$

we find the Fourier components

$$\tilde{\mathbf{D}}_{\omega^2,B}(\omega_m) = \begin{pmatrix} \omega_m^2 + \omega^2 - \omega_B^2 & -2\omega_B\omega_m \\ 2\omega_B\omega_m & \omega_m^2 + \omega^2 - \omega_B^2 \end{pmatrix}, \tag{2.694}$$

with the determinants

$$\det \tilde{\mathbf{D}}_{\omega^2,B}(\omega_m) = (\omega_m^2 + \omega^2 - \omega_B^2)^2 + 4\omega_B^2\omega_m^2. \tag{2.695}$$

These can be factorized as

$$\det \tilde{\mathbf{D}}_{\omega^2,B}(\omega_m) = (\omega_m^2 + \omega_+^2)(\omega_m^2 + \omega_-^2), \tag{2.696}$$

with

$$\omega_\pm \equiv \omega \pm \omega_B. \tag{2.697}$$

The eigenvectors of $\tilde{\mathbf{D}}_{\omega^2,B}(\omega_m)$ are

$$\mathbf{e}_+ = \frac{1}{\sqrt{2}}\begin{pmatrix} 1 \\ i \end{pmatrix}, \qquad \mathbf{e}_- = -\frac{1}{\sqrt{2}}\begin{pmatrix} 1 \\ -i \end{pmatrix}, \tag{2.698}$$

with eigenvalues

$$d_\pm = \omega_m^2 + \omega^2 \pm 2i\omega_m\omega_B = (\omega_m + i\omega_\pm)(\omega_m - i\omega_\mp). \tag{2.699}$$

Thus the right- and left-circular combinations $x_\pm = \pm(x \pm iy)/\sqrt{2}$ diagonalize the Lagrangian (2.690) to

$$\mathcal{A}_{\mathrm{cl}} = \int_0^{\hbar\beta} d\tau \left\{ \frac{M}{2}\frac{d}{d\tau}\left(x_+^*\dot{x}_+ + x_-^*\dot{x}_-\right)\right.$$
$$\left. + \frac{M}{2}\left[x_+^*(-\partial_\tau - \omega_+)(-\partial_\tau + \omega_-)x_+ + x_-^*(-\partial_\tau - \omega_-)(-\partial_\tau + \omega_+)x_-\right]\right\}. \tag{2.700}$$

Continued back to real times, the components $\mathrm{x}_\pm(t)$ are seen to oscillate independently with the frequencies $\omega_\pm$.

The factorization (2.696) makes (2.700) an action of two independent harmonic oscillators of frequencies $\omega_\pm$. The associated partition function has therefore the product form

$$Z = \frac{1}{2\sinh(\hbar\beta\omega_+/2)}\frac{1}{2\sinh(\hbar\beta\omega_-/2)}. \tag{2.701}$$

For the original system of a charged particle in a magnetic field discussed in Section 2.18, the partition function is obtained by going to the limit $\omega \to \omega_B$ in the Hamiltonian (2.687). Then $\omega_- \to 0$ and the partition function (2.701) diverges, since the system becomes translationally invariant in space. From the mnemonic replacement rule (2.359) we see that in this limit we must replace the relevant vanishing inverse frequency by an expression proportional to the volume of the system. The role of ω^2 in (2.359) is played here by the frequency in front of the x^2-term of the Lagrangian (2.689). Since there are two dimensions, we must replace

$$\frac{1}{\omega^2 - \omega_B^2} \xrightarrow[\omega \to \omega_B]{} \frac{1}{2\omega\omega_-} \xrightarrow[\omega_- \to 0]{} \frac{\beta}{2\pi/M}V_2, \tag{2.702}$$

and thus

$$Z \xrightarrow[\omega_- \to 0]{} \frac{1}{2\sinh(\hbar\beta\omega)}\frac{V_2}{\lambda_\omega^2}, \tag{2.703}$$

where λ_ω is the quantum-mechanical length scale in Eqs. (2.301) and (2.357) of the harmonic oscillator.

2.20 Gauge Invariance and Alternative Path Integral Representation

The action (2.633) of a particle in an external ordinary potential $V(\mathbf{x}, t)$ and a vector potential $\mathbf{A}(\mathbf{x}, t)$ can be rewritten with the help of an arbitrary space- and time-dependent gauge function $\Lambda(\mathbf{x}, t)$ in the following form:

$$\mathcal{A}[\mathbf{x}] = \int_{t_a}^{t_b} dt \left\{ \frac{M}{2} \dot{\mathbf{x}}^2 + \frac{e}{c} \dot{\mathbf{x}}(t)[\mathbf{A}(\mathbf{x}, t) + \boldsymbol{\nabla}\Lambda(\mathbf{x}, t)] - V(\mathbf{x}, t) + \frac{e}{c} \partial_t \Lambda(\mathbf{x}, t) \right\}$$
$$- \frac{e}{c} \left[\Lambda(\mathbf{x}_b, t_b) - \Lambda(\mathbf{x}_a, t_a) \right]. \tag{2.704}$$

The $\Lambda(\mathbf{x}, t)$-terms inside the integral are canceled by the last two surface terms making the action independent of $\Lambda(\mathbf{x}, t)$.

We may now choose a particular function $\Lambda(\mathbf{x}, t)$ equal to c/e-times the classical action $A(\mathbf{x}, t)$ which solves the Hamilton-Jacobi equation (1.65), i.e.,

$$\frac{1}{2M} \left[\boldsymbol{\nabla}A(\mathbf{x}, t) - \frac{e}{c}\mathbf{A}(\mathbf{x}, t) \right]^2 + \partial_t A(\mathbf{x}, t) + V(\mathbf{x}, t) = 0. \tag{2.705}$$

Then we obtain the following alternative expression for the action:

$$\mathcal{A}[\mathbf{x}] = \int_{t_a}^{t_b} dt \frac{1}{2M} \left[M\dot{\mathbf{x}} - \boldsymbol{\nabla}A(\mathbf{x}, t) + \frac{e}{c}\mathbf{A}(\mathbf{x}, t) \right]^2$$
$$+ A(\mathbf{x}_b, t_b) - A(\mathbf{x}_a, t_a). \tag{2.706}$$

For two infinitesimally different solutions of the Hamilton-Jacobi equation, the difference between the associated action functions δA satisfies the differential equation

$$\mathbf{v} \cdot \boldsymbol{\nabla}\delta A + \partial_t \delta A = 0, \tag{2.707}$$

where $\mathbf{v}$ is the classical velocity field

$$\mathbf{v}(\mathbf{x}, t) \equiv (1/M) \left[\boldsymbol{\nabla}A(\mathbf{x}, t) - \frac{e}{c}\mathbf{A}(\mathbf{x}, t) \right]. \tag{2.708}$$

The differential equation (2.707) expresses the fact that two solutions $A(\mathbf{x}, t)$ for which the particle energy and momenta at $\mathbf{x}$ and t differ by δE and $\delta\mathbf{p}$, respectively, satisfy the kinematic relation $\delta E = \mathbf{p} \cdot \delta\mathbf{p}/M = \dot{\mathbf{x}}_{\rm cl} \cdot \boldsymbol{\nabla}\delta A$. This follows directly from $E = \mathbf{p}^2/2M$. The so-constrained variations δE and $\delta\mathbf{p}$ leave the action (2.706) invariant.

A sequence of changes δA of this type can be used to make the function $A(\mathbf{x}, t)$ coincide with the action $A(\mathbf{x}, t; \mathbf{x}_a, t_a)$ of paths which start out from $\mathbf{x}_a, t_a$ and arrive at $\mathbf{x}, t$. In terms of this action function, the path integral representation of the time evolution amplitude takes the form

$$(\mathbf{x}_b t_b | \mathbf{x}_a t_a) = e^{iA(\mathbf{x}_b, t_b; \mathbf{x}_a, t_a)/\hbar} \int_{\mathbf{x}(t_a)=\mathbf{x}_a}^{\mathbf{x}(t_b)=\mathbf{x}_b} \mathcal{D}\mathbf{x} \tag{2.709}$$
$$\times \exp\left\{ \frac{i}{\hbar} \int_{t_a}^{t_b} dt \frac{1}{2M} \left[M\dot{\mathbf{x}} - \boldsymbol{\nabla}A(\mathbf{x}, t; \mathbf{x}_a, t_a) + \frac{e}{c}\mathbf{A}(\mathbf{x}, t) \right]^2 \right\}$$

or, using $\mathbf{v}(\mathbf{x}(t), t)$,

$$(\mathbf{x}_b t_b | \mathbf{x}_a t_a) = e^{iA(\mathbf{x}_b, t_b; \mathbf{x}_a, t_a)/\hbar} \int_{\mathbf{x}(t_a)=\mathbf{x}_a}^{\mathbf{x}(t_b)=\mathbf{x}_b} \mathcal{D}\mathbf{x} \exp\left[\frac{i}{\hbar} \int_{t_a}^{t_b} dt \frac{M}{2} (\dot{\mathbf{x}} - \mathbf{v})^2 \right].$$
$$\tag{2.710}$$

The fluctuations are now controlled by the deviations of the instantaneous velocity $\dot{x}(t)$ from local value of the classical velocity field $\mathbf{v}(\mathbf{x}, t)$. Since the path integral attempts to keep the deviations as small as possible, we call $\mathbf{v}(\mathbf{x}, t)$ the *desired velocity* of the particle at $\mathbf{x}$ and t. Introducing momentum variables $\mathbf{p}(t)$, the amplitude may be written as a phase space path integral

$$(\mathbf{x}_b t_b | \mathbf{x}_a t_a) = e^{iA(\mathbf{x}_b, t_b; \mathbf{x}_a, t_a)/\hbar} \int_{\mathbf{x}(t_a)=\mathbf{x}_a}^{\mathbf{x}(t_b)=\mathbf{x}_b} \mathcal{D}'\mathbf{x} \int \frac{\mathcal{D}\mathbf{p}}{2\pi\hbar}$$

$$\times \; \exp\left(\frac{i}{\hbar} \int_{t_a}^{t_b} dt \left\{ \mathbf{p}(t) \left[\dot{\mathbf{x}}(t) - \mathbf{v}(\mathbf{x}(t), t) \right] - \frac{1}{2M} \mathbf{p}^2(t) \right\} \right), \qquad (2.711)$$

which will be used in Section 18.15 to give a stochastic interpretation of quantum processes.

2.21 Velocity Path Integral

There exists yet another form of writing the path integral in which the fluctuating velocities play a fundamental role and which will later be seen to be closely related to path integrals in the so-called stochastic calculus to be introduced in Sections 18.13 and 18.640. We observe that by rewriting the path integral as

$$(\mathbf{x}_b t_b | \mathbf{x}_a t_a) = \int \mathcal{D}^3 x \, \delta\left(\mathbf{x}_b - \mathbf{x}_a - \int_{t_a}^{t_b} dt \, \dot{\mathbf{x}}(t) \right) \exp\left\{ \frac{i}{\hbar} \int_{t_a}^{t_b} dt \left[\frac{M}{2} \dot{\mathbf{x}}^2 - V(\mathbf{x}) \right] \right\}, \qquad (2.712)$$

the δ-function allows us to include the last variable $\mathbf{x}_n$ in the integration measure of the time-sliced version of the path integral. Thus all time-sliced time derivatives $(\mathbf{x}_{n+1} - \mathbf{x}_n)/\epsilon$ for $n = 0$ to N are integrated over implying that they can be considered as independent fluctuating variables $\mathbf{v}_n$. In the potential, the dependence on the velocities can be made explicit by inserting

$$\mathbf{x}(t) = \mathbf{x}_b - \int_t^{t_b} dt \, \mathbf{v}(t), \qquad (2.713)$$

$$\mathbf{x}(t) = \mathbf{x}_a + \int_{t_a}^t dt \, \mathbf{v}(t), \qquad (2.714)$$

$$\mathbf{x}(t) = \mathbf{X} + \frac{1}{2} \int_{t_a}^{t_b} dt' \, \mathbf{v}(t') \epsilon(t' - t), \qquad (2.715)$$

where

$$\mathbf{X} \equiv \frac{\mathbf{x}_b + \mathbf{x}_a}{2} \qquad (2.716)$$

is the average position of the endpoints and $\epsilon(t - t')$ is the antisymmetric combination of Heaviside functions introduced in Eq. (1.311).

In the first replacement, we obtain the *velocity path integral*

$$(\mathbf{x}_b t_b | \mathbf{x}_a t_a) = \int \mathcal{D}^3 v \, \delta\left(\mathbf{x}_b - \mathbf{x}_a - \int_{t_a}^{t_b} dt \, \mathbf{v}(t) \right) \exp\left\{ \frac{i}{\hbar} \int_{t_a}^{t_b} dt \left[\frac{M}{2} \mathbf{v}^2 - V\left(\mathbf{x}_b - \int_t^{t_b} dt \, \mathbf{v}(t) \right) \right] \right\}. \qquad (2.717)$$

The measure of integration is normalized to make

$$\int \mathcal{D}^3 v \, \exp\left\{ \frac{i}{\hbar} \int_{t_a}^{t_b} dt \left[\frac{M}{2} \mathbf{v}^2 \right] \right\} = 1. \qquad (2.718)$$

The correctness of this normalization can be verified by evaluating (2.717) for a free particle. Inserting the Fourier representation for the δ-function

$$\delta\left(\mathbf{x}_b - \mathbf{x}_a - \int_{t_a}^{t_b} dt \, \mathbf{v}(t) \right) = \int \frac{d^3 p}{(2\pi i)^3} \exp\left[\frac{i}{\hbar} \mathbf{p} \left(\mathbf{x}_b - \mathbf{x}_a - \int_{t_a}^{t_b} dt \, \mathbf{v}(t) \right) \right], \qquad (2.719)$$

we can complete the square in the exponent and integrate out the $\mathbf{v}$-fluctuations using (2.718) to obtain

$$(\mathbf{x}_b t_b | \mathbf{x}_a t_a) = \int \frac{d^3 p}{(2\pi i)^3} \exp\left\{ \frac{i}{\hbar} \left[\mathbf{p}\,(\mathbf{x}_b - \mathbf{x}_a) - \frac{\mathbf{p}^2}{2M}(t_b - t_a) \right] \right\}. \tag{2.720}$$

This is precisely the spectral representation (1.329) of the free-particle time evolution amplitude (1.331) [see also Eq. (2.51)].

A more symmetric velocity path integral is obtained by choosing the third replacement (2.715). This leads to the expression

$$
\begin{aligned}
(\mathbf{x}_b t_b | \mathbf{x}_a t_a) &= \int \mathcal{D}^3 v\, \delta\left(\Delta \mathbf{x} - \int_{t_a}^{t_b} dt\, \mathbf{v}(t) \right) \exp\left\{ \frac{i}{\hbar} \int_{t_a}^{t_b} dt\, \frac{M}{2}\, \mathbf{v}^2 \right\} \\
&\quad \times \exp\left\{ -\frac{i}{\hbar} \int_{t_a}^{t_b} dt\, V\left(\mathbf{X} + \frac{1}{2} \int_{t_a}^{t_b} dt'\, \mathbf{v}(t') \epsilon(t' - t) \right) \right\}.
\end{aligned}
\tag{2.721}
$$

The velocity representations are particularly useful if we want to know integrated amplitudes such as

$$
\begin{aligned}
\int d^3 x_a\, (\mathbf{x}_b t_b | \mathbf{x}_a t_a) &= \int \mathcal{D}^3 v \exp\left\{ \frac{i}{\hbar} \int_{t_a}^{t_b} dt\, \frac{M}{2}\, \mathbf{v}^2 \right\}. \\
&\quad \times \exp\left\{ -\frac{i}{\hbar} \int_{t_a}^{t_b} dt\, V\left(\mathbf{x}_b - \frac{1}{2} \int_t^{t_b} dt'\, \mathbf{v}(t') \right) \right\},
\end{aligned}
\tag{2.722}
$$

which will be of use in the next section.

2.22 Path Integral Representation of the Scattering Matrix

In Section 1.16 we have seen that the description of scattering processes requires several nontrivial limiting procedures on the time evolution amplitude. Let us see what these procedures yield when applied to the path integral representation of this amplitude.

2.22.1 General Development

Formula (1.471) for the scattering matrix expressed in terms of the time evolution operator in momentum space has the following path integral representation:

$$\langle \mathbf{p}_b | \hat{S} | \mathbf{p}_a \rangle \equiv \lim_{t_b - t_a \to \infty} e^{i(E_b t_b - E_a t_a)/\hbar} \int d^3 x_b \int d^3 x_a\, e^{-i(\mathbf{P}_b \mathbf{x}_b - \mathbf{P}_a \mathbf{x}_a)/\hbar} (\mathbf{x}_b t_b | \mathbf{x}_a t_a). \tag{2.723}$$

Introducing the momentum transfer $\mathbf{q} \equiv (\mathbf{p}_b - \mathbf{p}_a)$, we rewrite $e^{-i(\mathbf{P}_b \mathbf{x}_b - \mathbf{P}_a \mathbf{x}_a)/\hbar}$ as $e^{-i\mathbf{q}\mathbf{x}_b/\hbar} e^{-i\mathbf{P}_a(\mathbf{x}_b - \mathbf{x}_a)/\hbar}$, and observe that the amplitude including the exponential prefactor $e^{-i\mathbf{P}_a(\mathbf{x}_b - \mathbf{x}_a)/\hbar}$ has the path integral representation:

$$e^{-i\mathbf{P}_a(\mathbf{x}_b - \mathbf{x}_a)/\hbar} (\mathbf{x}_b t_b | \mathbf{x}_a t_a) = \int \mathcal{D}^3 x \exp\left\{ \frac{i}{\hbar} \int_{t_a}^{t_b} dt \left[\frac{M}{2}\, \dot{\mathbf{x}}^2 - \mathbf{p}_a \dot{\mathbf{x}} - V(\mathbf{x}) \right] \right\}. \tag{2.724}$$

The linear term in $\dot{\mathbf{x}}$ is eliminated by shifting the path from $\mathbf{x}(t)$ to

$$\mathbf{y}(t) = \mathbf{x}(t) - \frac{\mathbf{p}_a}{M}\, t \tag{2.725}$$

leading to

$$e^{-i\mathbf{P}_a(\mathbf{x}_b - \mathbf{x}_b)/\hbar}(\mathbf{x}_b t_b | \mathbf{x}_a t_a) = e^{-i\mathbf{p}_a^2(t_b - t_a)/2M\hbar} \int \mathcal{D}^3 y \exp\left\{ \frac{i}{\hbar} \int_{t_a}^{t_b} dt \left[\frac{M}{2}\, \dot{\mathbf{y}}^2 - V\left(\mathbf{y} + \frac{\mathbf{P}}{M}\, t \right) \right] \right\}. \tag{2.726}$$

Inserting everything into (2.723) we obtain

$$\langle \mathbf{p}_b|\hat{S}|\mathbf{p}_a\rangle \equiv \lim_{t_b-t_a\to\infty} e^{i\mathbf{q}^2 t_b/2M\hbar} \int d^3 y_b\, e^{-i\mathbf{q}\mathbf{y}_b/\hbar} \int d^3 y_a$$
$$\times \int \mathcal{D}^3 y \exp\left\{\frac{i}{\hbar}\int_{t_a}^{t_b} dt\left[\frac{M}{2}\dot{\mathbf{y}}^2 - V\left(\mathbf{y}+\frac{\mathbf{p}_a}{M}t\right)\right]\right\}. \tag{2.727}$$

In the absence of an interaction, the path integral over $\mathbf{y}(t)$ gives simply

$$\int d^3 y_a \frac{1}{\sqrt{2\pi\hbar i(t_b-t_a)/M}} \exp\left[\frac{i}{\hbar}\frac{M}{2}\frac{(\mathbf{y}_b-\mathbf{y}_a)^2}{t_b-t_a}\right] = 1, \tag{2.728}$$

and the integral over $\mathbf{y}_a$ yields

$$\langle \mathbf{p}_b|\hat{S}|\mathbf{p}_a\rangle|_{V\equiv 0} = \lim_{t_b-t_a\to\infty} e^{i\mathbf{q}^2(t_b-t_a)/8M\hbar}(2\pi\hbar)^3\delta^{(3)}(\mathbf{q}) = (2\pi\hbar)^3\delta^{(3)}(\mathbf{p}_b-\mathbf{p}_a), \tag{2.729}$$

which is the contribution from the unscattered beam to the scattering matrix in Eq. (1.474).

The first-order contribution from the interaction reads, after a Fourier decomposition of the potential,

$$\langle \mathbf{p}_b|\hat{S}_1|\mathbf{p}_a\rangle = -\frac{i}{\hbar}\lim_{t_b-t_a\to\infty} e^{i\mathbf{q}^2 t_b/2M\hbar} \int d^3 y_b\, e^{-i\mathbf{q}\mathbf{y}_b/\hbar} \int \frac{d^3 Q}{(2\pi\hbar)^3} V(\mathbf{Q}) \int d^3 y_a$$
$$\times \int_{t_a}^{t_b} dt' \exp\left(\frac{i}{\hbar}\frac{\mathbf{p}_a\mathbf{Q}}{M}t'\right)\int \mathcal{D}^3 y \exp\left\{\frac{i}{\hbar}\int_{t_a}^{t_b} dt\left[\frac{M}{2}\dot{\mathbf{y}}^2 + \delta(t'-t)\mathbf{Q}\mathbf{y}\right]\right\}. \tag{2.730}$$

The harmonic path integral was solved in one dimension for an arbitrary source $j(t)$ in Eq. (3.168). For $\omega = 0$ and the particular source $j(t) = \delta(t'-t)\mathbf{Q}$ the result reads, in three dimensions,

$$\frac{1}{\sqrt{2\pi i\hbar(t_b-t_a)/M}^3} \exp\left[\frac{i}{\hbar}\frac{M}{2}\frac{(\mathbf{y}_b-\mathbf{y}_a)^2}{t_b-t_a}\right]$$
$$\times \exp\left(\frac{i}{\hbar}\frac{1}{t_b-t_a}\left\{[\mathbf{y}_b(t'-t_a)+\mathbf{y}_a(t_b-t')]\mathbf{Q} - \frac{1}{2M}(t_b-t')(t'-t_a)\mathbf{Q}^2\right\}\right). \tag{2.731}$$

Performing here the integral over $\mathbf{y}_a$ yields

$$\exp\left\{\frac{i}{\hbar}\mathbf{Q}\mathbf{y}_b\right\}\exp\left\{-\frac{i}{\hbar}\frac{1}{2M}(t_b-t')\mathbf{Q}^2\right\}. \tag{2.732}$$

The integral over $\mathbf{y}_b$ in (2.730) leads now to a δ-function $(2\pi\hbar)^3\delta^{(3)}(\mathbf{Q}-\mathbf{q})$, such that the exponential prefactor in (2.730) is canceled by part of the second factor in (2.732).

In the limit $t_b-t_a \to \infty$, the integral over t' produces a δ-function $2\pi\hbar\delta(\mathbf{p}_b\mathbf{Q}/M + \mathbf{Q}^2/2M) = 2\pi\hbar\delta(E_b - E_a)$ which enforces the conservation of energy. Thus we find the well-known Born approximation

$$\langle \mathbf{p}_b|\hat{S}_1|\mathbf{p}_a\rangle = -2\pi i\delta(E_b - E_a)V(\mathbf{q}). \tag{2.733}$$

In general, we subtract the unscattered particle term (2.729) from (2.727), to obtain a path integral representation for the T-matrix [for the definition recall (1.474)]:

$$2\pi\hbar i\delta(E_b - E_a)\langle \mathbf{p}_b|\hat{T}|\mathbf{p}_a\rangle \equiv -\lim_{t_b-t_a\to\infty} e^{i\mathbf{q}^2 t_b/2M\hbar}\int d^3 y_b\, e^{-i\mathbf{q}\mathbf{y}_b/\hbar}\int d^3 y_a$$
$$\times \int \mathcal{D}^3 y \exp\left(\frac{i}{\hbar}\int_{t_a}^{t_b} dt\frac{M}{2}\dot{\mathbf{y}}^2\right)\left\{\exp\left[-\frac{i}{\hbar}\int_{t_a}^{t_b} dt\, V\left(\mathbf{y}+\frac{\mathbf{p}_a}{M}t\right)\right] - 1\right\}. \tag{2.734}$$

It is preferable to find a formula which does not contain the δ-function of energy conservation as a factor on the left-hand side. In order to remove this we observe that its origin lies in the time-translational invariance of the path integral in the limit $t_b - t_a \to \infty$. If we go over to a shifted time variable $t \to t + t_0$, and change simultaneously $\mathbf{y} \to \mathbf{y} - \mathbf{p}_a t_0 / M$, then the path integral remains the same except for shifted initial and final times $t_b + t_0$ and $t_a + t_0$. In the limit $t_b - t_a \to \infty$, the integrals $\int_{t_a+t_0}^{t_b+t_0} dt$ can be replaced again by $\int_{t_a}^{t_b} dt$. The only place where a t_0-dependence remains is in the prefactor $e^{-i\mathbf{q}\mathbf{y}_b/\hbar}$ which changes to $e^{-i\mathbf{q}\mathbf{y}_b/\hbar} e^{i\mathbf{q}\mathbf{p}_a t_0/M\hbar}$. Among all path fluctuations, there exists one degree of freedom which is equivalent to a temporal shift of the path. This is equivalent to an integral over t_0 which yields a δ-function $2\pi\hbar\delta\left(\mathbf{q}\mathbf{p}_a/M\right) = 2\pi\hbar\delta\left(E_b - E_a\right)$. We only must make sure to find the relation between this temporal shift and the corresponding measure in the path integral. This is obviously a shift of the path as a whole in the direction $\hat{\mathbf{p}}_a \equiv \mathbf{p}_a/|\mathbf{p}_a|$. The formal way of isolating this degree of freedom proceeds according to a method developed by Faddeev and Popov[36] by inserting into the path integral (2.727) the following integral representation of unity:

$$1 = \frac{|\mathbf{p}_a|}{M} \int_{-\infty}^{\infty} dt_0 \, \delta\left(\hat{\mathbf{p}}_a(\mathbf{y}_b + \mathbf{p}_a t_0/M)\right). \tag{2.735}$$

In the following, we shall drop the subscript a of the incoming beam, writing

$$\mathbf{p} \equiv \mathbf{p}_a, \qquad p \equiv |\mathbf{p}_a| = |\mathbf{p}_b|. \tag{2.736}$$

After the above shift in the path integral, the δ-function in (2.735) becomes $\delta\left(\hat{\mathbf{p}}_a\mathbf{y}_b\right)$ inside the path integral, with no t_0-dependence. The integral over t_0 can now be performed yielding the δ-function in the energy. Removing this from the equation we obtain the path integral representation of the T-matrix

$$\langle \mathbf{p}_b | \hat{T} | \mathbf{p}_a \rangle \equiv i\frac{p}{M} \lim_{t_b-t_a \to \infty} e^{i\mathbf{q}^2(t_b-t_a)/8M\hbar} \int d^3y_b \, \delta\left(\hat{\mathbf{p}}_a\mathbf{y}_b\right) e^{-i\mathbf{q}\mathbf{y}_b/\hbar} \int d^3y_a$$
$$\times \int \mathcal{D}^3y \exp\left(\frac{i}{\hbar}\int_{t_a}^{t_b} dt \, \frac{M}{2}\dot{\mathbf{y}}^2\right) \left\{\exp\left[-\frac{i}{\hbar}\int_{t_a}^{t_b} dt \, V\left(\mathbf{y} + \frac{\mathbf{P}}{M}t\right)\right] - 1\right\}. \tag{2.737}$$

At this point it is convenient to go over to the velocity representation of the path integral (2.721). This enables us to perform trivially the integral over $\mathbf{y}_b$, and we obtain the $\mathbf{y}$ version of (2.722). The δ-function enforces a vanishing longitudinal component of $\mathbf{y}_b$. The transverse component of $\mathbf{y}_b$ will be denoted by $\mathbf{b}$:

$$\mathbf{b} \equiv \mathbf{y}_b - (\hat{\mathbf{p}}_a\mathbf{y}_b)\hat{\mathbf{p}}/a. \tag{2.738}$$

Hence we find the path integral representation

$$\langle \mathbf{p}_b | \hat{T} | \mathbf{p}_a \rangle \equiv i\frac{p}{M} \lim_{t_b-t_a \to \infty} e^{i\mathbf{q}^2 t_b/2M\hbar} \int d^2b \, e^{-i\mathbf{q}\mathbf{b}/\hbar}$$
$$\times \int \mathcal{D}^3v \exp\left(\frac{i}{\hbar}\int_{t_a}^{t_b} dt \, \frac{M}{2}\mathbf{v}^2\right) \left[e^{i\chi_{\mathbf{b},\mathbf{p}}[\mathbf{v}]} - 1\right], \tag{2.739}$$

where the effect of the interaction is contained in the scattering phase

$$\chi_{\mathbf{b},\mathbf{p}}[\mathbf{v}] \equiv -\frac{1}{\hbar}\int_{t_a}^{t_b} dt \, V\left(\mathbf{b} + \frac{\mathbf{P}}{M}t - \int_{t}^{t_b} dt' \, \mathbf{v}(t')\right). \tag{2.740}$$

We can go back to a more conventional path integral by replacing the velocity paths $\mathbf{v}(t)$ by $\dot{\mathbf{y}}(t) = -\int_{t}^{t_b} \mathbf{v}(t)$. This vanishes at $t = t_b$. Equivalently, we can use paths $\mathbf{z}(t)$ with periodic boundary conditions and subtract from these $\mathbf{z}(t_b) = \mathbf{z}_b$.

From $\langle \mathbf{p}_b | \hat{T} | \mathbf{p}_a \rangle$ we obtain the scattering amplitude $f_{\mathbf{p}_b,\mathbf{p}_a}$, whose square gives the differential cross section, by multiplying it with a factor $-M/2\pi\hbar$ [see Eq. (1.494)].

---

[36]L.D. Faddeev and V.N. Popov, Phys. Lett. B *25*, 29 (1967).

Note that in the velocity representation, the evaluation of the harmonic path integral integrated over y_a in (2.730) is much simpler than before where we needed the steps (2.731), (2.732). After the Fourier decomposition of $V(\mathbf{x})$ in (2.740), the relevant integral is

$$\int \mathcal{D}^3 v \exp\left\{\frac{i}{\hbar}\int_{t_a}^{t_b} dt\left[\frac{M}{2}\mathbf{v}^2 - \Theta(t_b - t')\mathbf{Q}\,\mathbf{v}\right]\right\} = e^{-\frac{i}{2M\hbar}\int_{t_a}^{t_b} dt\,\Theta^2(t_b-t')\mathbf{Q}^2} = e^{-\frac{i}{2M\hbar}(t_b-t')\mathbf{Q}^2}. \qquad (2.741)$$

The first factor in (2.732) comes directly from the argument $\mathbf{Y}$ in the Fourier representation of the potential

$$V\left(\mathbf{y}_b + \frac{\mathbf{p}}{M}t - \int_t^{t_b} dt'\,\mathbf{v}(t')\right)$$

in the velocity representation of the S-matrix (2.727).

2.22.2 Improved Formulation

The prefactor $e^{i\mathbf{q}^2 t_b/2M\hbar}$ in Formula (2.739) is an obstacle to taking a more explicit limit $t_b - t_a \to \infty$ on the right-hand side. To overcome this, we represent this factor by an auxiliary path integral[37] over some vector field $\mathbf{w}(t)$:

$$e^{i\mathbf{q}^2 t_b/2M\hbar} = \int \mathcal{D}^3 w \exp\left[-\frac{i}{\hbar}\int_{t_a}^{t_b} dt\,\frac{M}{2}\mathbf{w}^2(t)\right] e^{i\int_{t_a}^{t_b} dt\,\Theta(t)\mathbf{w}(t)\mathbf{q}/\hbar}. \qquad (2.742)$$

The last factor changes the exponential $e^{-i\mathbf{q}\mathbf{b}/\hbar}$ in (2.739) into $e^{-i\mathbf{q}\left[\mathbf{b}+\int_{t_a}^{t_b} dt\,\Theta(t)\mathbf{w}(t)\right]/\hbar}$. Since $\mathbf{b}$ is a dummy variable of integration, we can equivalently replace $\mathbf{b} \to \mathbf{b_w} \equiv \mathbf{b} - \int_{t_a}^{t_b} dt\,\Theta(t)\mathbf{w}(t)$ in the scattering phase $\chi_{\mathbf{b},\mathbf{p}}[\mathbf{v}]$ and remain with

$$
\begin{aligned}
f_{\mathbf{p}_b\mathbf{p}_a} &= \lim_{t_b - t_a \to \infty}\frac{p}{2\pi i\hbar}\int d^2 b\,e^{-i\mathbf{q}\mathbf{b}/\hbar}\int \mathcal{D}^3 w \\
&\times \int \mathcal{D}^3 v \exp\left[\frac{i}{\hbar}\int_{-\infty}^{\infty} dt\,\frac{M}{2}\left(\mathbf{v}^2 - \mathbf{w}^2\right)\right]\left[\exp\left(i\chi_{\mathbf{b_w},\mathbf{p}}\right) - 1\right].
\end{aligned}
\qquad (2.743)
$$

The scattering phase in this expression can be calculated from formula (2.740) with the integral taken over the entire t-axis:

$$\chi_{\mathbf{b_w},\mathbf{p}}[\mathbf{v},\mathbf{w}] = -\frac{1}{\hbar}\int_{-\infty}^{\infty} dt\,V\left(\mathbf{b} + \frac{\mathbf{p}}{M}t - \int_{t_a}^{t_b} dt'\,\left[\Theta(t'-t)\mathbf{v}(t') - \Theta(t')\mathbf{w}(t')\right]\right). \qquad (2.744)$$

The fluctuations of $\mathbf{w}(t)$ are necessary to correct for the fact that the outgoing particle does not run, on the average, with the velocity $\mathbf{p}/M = \mathbf{p}_a/M$ but with velocity $\mathbf{p}_b/M = (\mathbf{p}+\mathbf{q})/M$.

We may also go back to a more conventional path integral by inserting $\mathbf{y}(t) = -\int_t^{t_b}\mathbf{v}(t)$ and setting similarly $\mathbf{z}(t) = -\int_t^{t_b}\mathbf{w}(t)$. Then we obtain the alternative representation

$$
\begin{aligned}
f_{\mathbf{p}_b\mathbf{p}_a} &= \lim_{t_b - t_a \to \infty}\frac{p}{2\pi i\hbar}\int d^2 b\,e^{-i\mathbf{q}\mathbf{b}/\hbar}\int d^3 y_a\int d^3 z_a \\
&\times \int \mathcal{D}^3 y\int \mathcal{D}^3 z \exp\left[\frac{i}{\hbar}\int_{t_a}^{t_b} dt\,\frac{M}{2}\left(\dot{\mathbf{y}}^2 - \dot{\mathbf{z}}^2\right)\right]\left[e^{i\chi_{\mathbf{b_z},\mathbf{p}}[\mathbf{y}]} - 1\right],
\end{aligned}
\qquad (2.745)
$$

with

$$\chi_{\mathbf{b_z},\mathbf{p}}[\mathbf{y}] \equiv -\frac{1}{\hbar}\int_{t_a}^{t_b} dt\,V\left(\mathbf{b} + \frac{\mathbf{p}}{M}t + \mathbf{y}(t) - \mathbf{z}(0)\right), \qquad (2.746)$$

where the path integrals run over all paths with $\mathbf{y}_b = 0$ and $\mathbf{z}_b = 0$. In Section 3.26 this path integral will be evaluated perturbatively.

---

[37]See R. Rosenfelder, notes of a lecture held at the ETH Zürich in 1979: *Pfadintegrale in der Quantenphysik*, 126 p., PSI Report 97-12, ISSN 1019-0643, and Lecture held at the 7th Int. Conf. on Path Integrals in Antwerpen, *Path Integrals from Quarks to Galaxies*, 2002.

2.22.3   Eikonal Approximation to the Scattering Amplitude

To lowest approximation, we neglect the fluctuating variables $\mathbf{y}(t)$ and $\mathbf{z}(t)$ in (2.746). Since the integral

$$\int d^3 y_a \int d^3 z_a \int \mathcal{D}^3 y \int \mathcal{D}^3 z \exp\left[\frac{i}{\hbar}\int_{t_a}^{t_b} dt\, \frac{M}{2}\left(\dot{\mathbf{y}}^2 - \dot{\mathbf{z}}^2\right)\right] \tag{2.747}$$

in (2.745) has unit normalization [recall the calculation of (2.732)], we obtain directly the eikonal approximation to the scattering amplitude

$$f_{\mathbf{p}_b \mathbf{p}_a}^{\text{ei}} \equiv \frac{p}{2\pi i\hbar}\int d^2 b\, e^{-i\mathbf{q}\mathbf{b}/\hbar}\left[\exp\left(i\chi_{\mathbf{b},\mathbf{p}}^{\text{ei}}\right) - 1\right], \tag{2.748}$$

with

$$\chi_{\mathbf{b},\mathbf{p}}^{\text{ei}} \equiv -\frac{1}{\hbar}\int_{-\infty}^{\infty} dt\, V\left(\mathbf{b} + \frac{\mathbf{p}}{M}t\right). \tag{2.749}$$

The time integration can be converted into a line integration along the direction of the incoming particles by introducing a variable $z \equiv pt/M$. Then we can write

$$\chi_{\mathbf{b},\mathbf{p}}^{\text{ei}} \equiv -\frac{M}{p}\frac{1}{\hbar}\int_{-\infty}^{\infty} dz\, V\left(\mathbf{b} + \hat{\mathbf{p}}z\right). \tag{2.750}$$

If $V(\mathbf{x})$ is rotationally symmetric, it depends only on $r \equiv |\mathbf{x}|$. Then we shall write the potential as $V(r)$ and calculate (2.750) as the integral

$$\chi_{\mathbf{b},\mathbf{p}}^{\text{ei}} \equiv -\frac{M}{p}\frac{1}{\hbar}\int_{-\infty}^{\infty} dz\, V\left(\sqrt{b^2 + z^2}\right). \tag{2.751}$$

Inserting this into (2.748), we can perform the integral over all angles between $\mathbf{q}$ and $\mathbf{b}$ using the formula

$$\frac{1}{2\pi}\int_{-\pi}^{\pi} d\theta\, \exp\left(\frac{i}{\hbar}qb\cos\theta\right) = J_0(qb), \tag{2.752}$$

where $J_0(\xi)$ is the Bessel function, and find

$$f_{\mathbf{p}_b \mathbf{p}_a}^{\text{ei}} = \frac{p}{i\hbar}\int db\, b\, J_0(qb)\left[\exp\left(i\chi_{\mathbf{b},\mathbf{p}}^{\text{ei}}\right) - 1\right]. \tag{2.753}$$

The variable of integration b coincides with the impact parameter b introduced in Eq. (1.497). The result (2.753) is precisely the eikonal approximation (1.497) with $\chi_{\mathbf{b},\mathbf{p}}^{\text{ei}}/2$ playing the role of the scattering phases $\delta_l(p)$ of angular momentum $l = pb/\hbar$:

$$\chi_{\mathbf{b},\mathbf{p}}^{\text{ei}} = 2i\delta_{pb/\hbar}(p). \tag{2.754}$$

2.23   Heisenberg Operator Approach to Time Evolution Amplitude

An interesting alternative to the path integral derivation of the time evolution amplitudes of harmonic systems is based on quantum mechanics in the Heisenberg picture. It bears a close similarity with the path integral derivation in that it requires solving the classical equations of motion with given initial and final positions to obtain the exponential of the classical action $e^{i\mathcal{A}/\hbar}$. The fluctuation factor, however, which accompanies this exponential is obtained quite differently from commutation rules of the operatorial orbits at different times as we shall now demonstrate.

2.23.1   Free Particle

We want to calculate the matrix element of the time evolution operator

$$(\mathbf{x}\,t|\mathbf{x}'\,0) = \langle \mathbf{x}|e^{-i\hat{H}t/\hbar}|\mathbf{x}'\rangle, \tag{2.755}$$

where $\hat{H}$ is the Hamiltonian operator

$$\hat{H} = H(\hat{\mathbf{p}}) = \frac{\hat{\mathbf{p}}^2}{2M}. \tag{2.756}$$

We shall calculate the time evolution amplitude (2.755) by solving the differential equation

$$
\begin{aligned}
i\hbar\partial_t\langle \mathbf{x}\,t|\mathbf{x}'\,0\rangle &\equiv \langle \mathbf{x}|\hat{H}\,e^{-i\hat{H}t/\hbar}|\mathbf{x}'\rangle = \langle \mathbf{x}|e^{-i\hat{H}t/\hbar}\left[e^{i\hat{H}t/\hbar}\hat{H}\,e^{-i\hat{H}t/\hbar}\right]|\mathbf{x}'\rangle \\
&= \langle \mathbf{x}\,t|H(\hat{\mathbf{p}}(t))|\mathbf{x}'\,0\rangle.
\end{aligned}
\tag{2.757}
$$

The argument contains now the time-dependent Heisenberg picture of the operator $\hat{\mathbf{p}}$. The evaluation of the right-hand side will be based on re-expressing the operator $H(\hat{\mathbf{p}}(t))$ as a function of initial and final position operators in such a way that all final position operators stand to the left of all initial ones:

$$\hat{H} = H(\hat{\mathbf{x}}(t), \hat{\mathbf{x}}(0); t). \tag{2.758}$$

Then the matrix elements on the right-hand side can immediately be evaluated using the eigenvalue equations

$$\langle \mathbf{x}\,t|\hat{\mathbf{x}}(t) = \mathbf{x}\langle \mathbf{x}\,t|, \qquad \hat{\mathbf{x}}(0)|\mathbf{x}'\,0\rangle = \mathbf{x}'|\mathbf{x}'\,0\rangle, \tag{2.759}$$

as being

$$\langle \mathbf{x}\,t|H(\hat{\mathbf{x}}(t), \hat{\mathbf{x}}(0); t)|\hat{\mathbf{x}}\,0\rangle = H(\mathbf{x}, \mathbf{x}'; t)\langle \mathbf{x}\,t|\mathbf{x}'\,0\rangle, \tag{2.760}$$

and the differential equation (2.757) becomes

$$i\hbar\partial_t\langle \mathbf{x}\,t|\mathbf{x}'\,0\rangle \equiv H(\mathbf{x}, \mathbf{x}'; t)\langle \mathbf{x}\,t|\mathbf{x}'\,0\rangle, \tag{2.761}$$

or

$$\langle \mathbf{x}\,t|\mathbf{x}'\,0\rangle = C(\mathbf{x}, \mathbf{x}')E(\mathbf{x}, \mathbf{x}'; t) \equiv C(\mathbf{x}, \mathbf{x}')e^{-i\int_t dt'\,H(\mathbf{x},\mathbf{x}';t')/\hbar}. \tag{2.762}$$

The prefactor $C(\mathbf{x}, \mathbf{x}')$ contains a possible constant of integration resulting from the time integral in the exponent.

The Hamiltonian operator is brought to the time-ordered form (2.758) by solving the Heisenberg equations of motion

$$\frac{d\hat{\mathbf{x}}(t)}{dt} = \frac{i}{\hbar}\left[\hat{H}, \hat{\mathbf{x}}(t)\right] = \frac{\hat{\mathbf{p}}(t)}{M}, \tag{2.763}$$

$$\frac{d\hat{\mathbf{p}}(t)}{dt} = \frac{i}{\hbar}\left[\hat{H}, \hat{\mathbf{p}}(t)\right] = 0. \tag{2.764}$$

The second equation shows that the momentum is time-independent:

$$\hat{\mathbf{p}}(t) = \hat{\mathbf{p}}(0), \tag{2.765}$$

so that the first equation is solved by

$$\hat{\mathbf{x}}(t) - \hat{\mathbf{x}}(0) = t\,\frac{\hat{\mathbf{p}}(t)}{M}, \tag{2.766}$$

which brings (2.756) to

$$\hat{H} = \frac{M}{2t^2}\left[\hat{\mathbf{x}}(t) - \hat{\mathbf{x}}(0)\right]^2. \tag{2.767}$$

This is not yet the desired form (2.758) since there is one factor which is not time-ordered. The proper order is achieved by rewriting $\hat{H}$ as

$$\hat{H} = \frac{M}{2t^2} \left\{ \hat{\mathbf{x}}^2(t) - 2\hat{\mathbf{x}}(t)\hat{\mathbf{x}}(0) + \hat{\mathbf{x}}^2(0) + [\hat{\mathbf{x}}(t), \hat{\mathbf{x}}(0)] \right\}, \tag{2.768}$$

and calculating the commutator from Eq. (2.766) and the canonical commutation rule $[\hat{p}_i, \hat{x}_j] = -i\hbar\delta_{ij}$ as

$$[\hat{\mathbf{x}}(t), \hat{\mathbf{x}}(0)] = -\frac{i\hbar}{M}Dt, \tag{2.769}$$

so that we find the desired expression

$$\hat{H} = H(\hat{\mathbf{x}}(t), \hat{\mathbf{x}}(0); t) = \frac{M}{2t^2} \left[\hat{\mathbf{x}}^2(t) - 2\hat{\mathbf{x}}(t)\hat{\mathbf{x}}(0) + \hat{\mathbf{x}}^2(0) \right] - i\hbar\frac{D}{2t}. \tag{2.770}$$

Its matrix elements (2.760) can now immediately be written down:

$$H(\mathbf{x}, \mathbf{x}'; t) = \frac{M}{2t^2} (\mathbf{x} - \mathbf{x}')^2 - i\hbar\frac{D}{2t}. \tag{2.771}$$

From this we find directly the exponential factor in (2.762)

$$E(\mathbf{x}, \mathbf{x}'; t) = e^{-i \int dt\, H(\mathbf{x},\mathbf{x}';t)/\hbar} = \exp\left[\frac{i}{\hbar} \frac{M}{2t} (\mathbf{x} - \mathbf{x}')^2 - \frac{D}{2} \log t \right]. \tag{2.772}$$

Inserting (2.772) into Eq. (2.762), we obtain

$$\langle \mathbf{x}\, t | \mathbf{x}'\, 0 \rangle = C(\mathbf{x}, \mathbf{x}') \frac{1}{t^{D/2}} \exp\left[\frac{i}{\hbar} \frac{M}{2t} (\mathbf{x} - \mathbf{x}')^2 \right]. \tag{2.773}$$

A possible constant of integration in (2.772) depending on $\mathbf{x}, \mathbf{x}'$ is absorbed in the prefactor $C(\mathbf{x}, \mathbf{x}')$. This is fixed by differential equations involving $\mathbf{x}$:

$$-i\hbar\boldsymbol{\nabla}\langle \mathbf{x}\, t | \mathbf{x}'\, 0 \rangle = \langle \mathbf{x} | \hat{\mathbf{p}} e^{-i\hat{H}t/\hbar} | \mathbf{x}' \rangle = \langle \mathbf{x} | e^{-i\hat{H}t} \left[e^{i\hat{H}t/\hbar} \hat{\mathbf{p}} e^{-i\hat{H}t/\hbar} \right] | \mathbf{x}' \rangle = \langle \mathbf{x}\, t | \hat{\mathbf{p}}(t) | \mathbf{x}'\, 0 \rangle.$$

$$i\hbar\boldsymbol{\nabla}'\langle \mathbf{x}\, t | \mathbf{x}'\, 0 \rangle = \langle \mathbf{x} | e^{-i\hat{H}t/\hbar} \hat{\mathbf{p}} | \mathbf{x}' \rangle \langle \mathbf{x}\, t | \hat{\mathbf{p}}(0) | \mathbf{x}'\, 0 \rangle. \tag{2.774}$$

Inserting (2.766) and using the momentum conservation (2.765), these become

$$-i\hbar\boldsymbol{\nabla}\langle \mathbf{x}\, t | \mathbf{x}'\, 0 \rangle = \frac{M}{t}(\mathbf{x} - \mathbf{x}') \langle \mathbf{x}\, t | \mathbf{x}'\, 0 \rangle,$$

$$i\hbar\boldsymbol{\nabla}'\langle \mathbf{x}\, t | \mathbf{x}'\, 0 \rangle = \frac{M}{t}(\mathbf{x} - \mathbf{x}') \langle \mathbf{x}\, t | \mathbf{x}'\, 0 \rangle. \tag{2.775}$$

Inserting here the previous result (2.773), we obtain the conditions

$$-i\boldsymbol{\nabla} C(\mathbf{x}, \mathbf{x}') = 0, \quad i\boldsymbol{\nabla}' C(\mathbf{x}, \mathbf{x}') = 0, \tag{2.776}$$

which are solved only by a constant C. The constant, in turn, is fixed by the initial condition

$$\lim_{t \to 0} \langle \mathbf{x}\, t | \mathbf{x}'\, 0 \rangle = \delta^{(D)}(\mathbf{x} - \mathbf{x}'), \tag{2.777}$$

to be

$$C = \sqrt{\frac{M}{2\pi i\hbar}}^D, \tag{2.778}$$

so that we find the correct free-particle amplitude (2.130)

$$\langle \mathbf{x}\, t | \mathbf{x}'\, 0 \rangle \equiv \sqrt{\frac{M}{2\pi i\hbar t}}^D \exp\left[\frac{i}{\hbar} \frac{M}{2t} (\mathbf{x} - \mathbf{x}')^2 \right]. \tag{2.779}$$

Note that the fluctuation factor $1/t^{D/2}$ emerges in this approach as a consequence of the commutation relation (2.769).

2.23.2   Harmonic Oscillator

Here we are dealing with the Hamiltonian operator

$$\hat{H} = H(\hat{\mathbf{p}}, \hat{\mathbf{x}}) = \frac{\hat{\mathbf{p}}^2}{2M} + \frac{M\omega^2}{2}\mathbf{x}^2, \tag{2.780}$$

which has to be brought again to the time-ordered form (2.758). We must now solve the Heisenberg equations of motion

$$\frac{d\hat{\mathbf{x}}(t)}{dt} = \frac{i}{\hbar}\left[\hat{H}, \hat{\mathbf{x}}(t)\right] = \frac{\hat{\mathbf{p}}(t)}{M}, \tag{2.781}$$

$$\frac{d\hat{\mathbf{p}}(t)}{dt} = \frac{i}{\hbar}\left[\hat{H}, \hat{\mathbf{p}}(t)\right] = -M\omega^2\hat{\mathbf{x}}(t). \tag{2.782}$$

By solving these equations we obtain [compare (2.156)]

$$\hat{\mathbf{p}}(t) = M\frac{\omega}{\sin\omega t}\left[\hat{\mathbf{x}}(t)\cos\omega t - \hat{\mathbf{x}}(0)\right]. \tag{2.783}$$

Inserting this into (2.780), we obtain

$$\hat{H} = \frac{M\omega^2}{2\sin^2\omega t}\left\{\left[\hat{\mathbf{x}}(t)\cos\omega t - \hat{\mathbf{x}}(0)\right]^2 + \sin^2\omega t\,\hat{\mathbf{x}}^2(t)\right\}, \tag{2.784}$$

which is equal to

$$\hat{H} = \frac{M\omega^2}{2\sin^2\omega t}\left\{\hat{\mathbf{x}}^2(t) + \hat{\mathbf{x}}^2(0) - 2\cos\omega t\,\hat{\mathbf{x}}(t)\hat{\mathbf{x}}(0) + \cos\omega t\,[\hat{\mathbf{x}}(t), \hat{\mathbf{x}}(0)]\right\}. \tag{2.785}$$

By commuting Eq. (2.783) with $\hat{\mathbf{x}}(t)$, we find the commutator [compare (2.769)]

$$[\hat{\mathbf{x}}(t), \hat{\mathbf{x}}(0)] = -\frac{i\hbar}{M}D\frac{\sin\omega t}{\omega}, \tag{2.786}$$

so that we find the matrix elements of the Hamiltonian operator in the form (2.760) [compare (2.771)]

$$H(\mathbf{x}, \mathbf{x}'; t) = \frac{M\omega^2}{2\sin^2\omega t}\left(\mathbf{x}^2 + \mathbf{x}'^2 - 2\cos\omega t\,\mathbf{x}\mathbf{x}'\right) - i\hbar\frac{D}{2}\omega\cot\omega t. \tag{2.787}$$

This has the integral [compare (2.772)]

$$\int dt\, H(\mathbf{x}, \mathbf{x}'; t) = -\frac{M\omega}{2\sin\omega t}\left[\left(\mathbf{x}^2 + \mathbf{x}'^2\right)\cos\omega t - 2\mathbf{x}\mathbf{x}'\right] - i\hbar\frac{D}{2}\log\frac{\sin\omega t}{\omega}. \tag{2.788}$$

Inserting this into Eq. (2.762), we find precisely the harmonic oscillator amplitude (2.175), apart from the factor $C(\mathbf{x}, \mathbf{x}')$. This is again determined by the differential equations (2.774), leaving only a simple normalization factor fixed by the initial condition (2.777) with the result (2.778).

Again, the fluctuation factor has its origin in the commutator (2.786).

2.23.3   Charged Particle in Magnetic Field

We now turn to a charged particle in three dimensions in a magnetic field treated in Section 2.18, where the Hamiltonian operator is most conveniently expressed in terms of the operator of the covariant momentum (2.642),

$$\hat{\mathbf{P}} \equiv \hat{\mathbf{p}} - \frac{e}{c}\mathbf{A}(\hat{\mathbf{x}}), \tag{2.789}$$

as [compare (2.641)]

$$\hat{H} = H(\hat{\mathbf{p}}, \hat{\mathbf{x}}) = \frac{\hat{\mathbf{P}}^2}{2M}. \tag{2.790}$$

In the presence of a magnetic field, its components do not commute but satisfy the commutation rules:

$$[\hat{P}_i, \hat{P}_j] = -\frac{e}{c}[\hat{p}_i, \hat{A}_j] - \frac{e}{c}[\hat{A}_i, \hat{p}_j] = i\frac{e\hbar}{c}(\nabla_i A_j - \nabla_j A_i) = i\frac{e\hbar}{c}B_{ij}, \tag{2.791}$$

where $B_{ij} = \epsilon_{ijk}B_K$ is the usual antisymmetric tensor representation of the magnetic field.

We now have to solve the Heisenberg equations of motion

$$\frac{d\hat{\mathbf{x}}(t)}{dt} = \frac{i}{\hbar}\left[\hat{H}, \hat{\mathbf{x}}(t)\right] = \frac{\hat{\mathbf{P}}(t)}{M} \tag{2.792}$$

$$\frac{d\hat{\mathbf{P}}(t)}{dt} = \frac{i}{\hbar}\left[\hat{H}, \hat{\mathbf{P}}(t)\right] = \frac{e}{Mc}B(\hat{\mathbf{x}}(t))\hat{\mathbf{P}}(t) + i\frac{e\hbar}{Mc}\nabla_j B_{ji}(\hat{\mathbf{x}}(t)), \tag{2.793}$$

where $B(\hat{\mathbf{x}}(t))\hat{\mathbf{P}}(t)$ is understood as the product of the matrix $B_{ij}(\hat{\mathbf{x}}(t))$ with the vector $\hat{\mathbf{P}}$. In a *constant* field, where $B_{ij}(\hat{\mathbf{x}}(t))$ is a constant matrix B_{ij}, the last term in the second equation is absent and we find directly the solution

$$\hat{\mathbf{P}}(t) = e^{\Omega_L t}\hat{\mathbf{P}}(0), \tag{2.794}$$

where Ω_L is a matrix version of the Landau frequency (2.646)

$$\Omega_{L\,ij} \equiv \frac{e}{Mc}B_{ij}, \tag{2.795}$$

which can also be rewritten with the help of the Landau frequency vector

$$\boldsymbol{\omega}_L \equiv \frac{e}{Mc}\mathbf{B} \tag{2.796}$$

and the 3×3-generators of the rotation group

$$(L_k)_{ij} \equiv -i\epsilon_{kij} \tag{2.797}$$

as

$$\Omega_L = i\,\mathbf{L} \cdot \boldsymbol{\omega}_L. \tag{2.798}$$

Inserting this into Eq. (2.792), we find

$$\hat{\mathbf{x}}(t) = \hat{\mathbf{x}}(0) + \frac{e^{\Omega_L t} - 1}{\Omega_L}\frac{\hat{\mathbf{P}}(0)}{M}, \tag{2.799}$$

where the matrix on the right-hand side is again defined by a power series expansion

$$\frac{e^{\Omega_L t} - 1}{\Omega_L} = t + \Omega_L \frac{t^2}{2} + \Omega_L^2 \frac{t^3}{3!} + \dots . \tag{2.800}$$

We can invert (2.799) to find

$$\frac{\hat{\mathbf{P}}(0)}{M} = \frac{\Omega_L/2}{\sinh \Omega_L t/2}e^{-\Omega_L t/2}\left[\hat{\mathbf{x}}(t) - \hat{\mathbf{x}}(0)\right]. \tag{2.801}$$

Using (2.794), this implies

$$\hat{\mathbf{P}}(t) = MN(\Omega_L t)\left[\hat{\mathbf{x}}(t) - \hat{\mathbf{x}}(0)\right], \tag{2.802}$$

with the matrix

$$N(\Omega_L t) \equiv \frac{\Omega_L/2}{\sinh \Omega_L t/2} e^{\Omega_L t/2}. \tag{2.803}$$

By squaring (2.802) we obtain

$$\frac{\hat{\mathbf{P}}^2(t)}{2M} = \frac{M}{2} \left[\hat{\mathbf{x}}(t) - \hat{\mathbf{x}}(0)\right]^T K(\Omega_L t) \left[\hat{\mathbf{x}}(t) - \hat{\mathbf{x}}(0)\right], \tag{2.804}$$

where

$$K(\Omega_L t) = N^T(\Omega_L t)N(\Omega_L t). \tag{2.805}$$

Using the antisymmetry of the matrix Ω_L, we can rewrite this as

$$K(\Omega_L t) = N(-\Omega_L t)N(\Omega_L t) = \frac{\Omega_L^2/4}{\sinh^2 \Omega_L t/2}. \tag{2.806}$$

The commutator between two operators $\hat{\mathbf{x}}(t)$ at different times is, due to Eq. (2.799),

$$[\hat{\mathbf{x}}_i(t), \hat{\mathbf{x}}_j(0)] = -\frac{i}{M} \left(\frac{e^{\Omega_L t} - 1}{\Omega_L}\right)_{ij}, \tag{2.807}$$

and

$$[\hat{\mathbf{x}}_i(t), \hat{\mathbf{x}}_j(0)] + [\hat{\mathbf{x}}_j(t), \hat{\mathbf{x}}_i(0)] = -\frac{i}{M} \left(\frac{e^{\Omega_L t} - 1}{\Omega_L} + \frac{e^{\Omega_L^T t} - 1}{\Omega_L^T}\right)_{ij}$$

$$= -\frac{i}{M} \left(\frac{e^{\Omega_L t} - e^{-\Omega_L t}}{\Omega_L}\right)_{ij} = -2\frac{i}{M} \left[\frac{\sinh \Omega_L t}{\Omega_L}\right]_{ij}. \tag{2.808}$$

Respecting this, we can expand (2.804) in powers of operators $\hat{\mathbf{x}}(t)$ and $\hat{\mathbf{x}}(0)$, thereby time-ordering the later operators to the left of the earlier ones as follows:

$$H(\hat{\mathbf{x}}(t), \hat{\mathbf{x}}(0)) = \frac{M}{2} \left[\hat{\mathbf{x}}^T(t)K(\Omega_L t)\hat{\mathbf{x}}(t) - 2\hat{\mathbf{x}}^T K(\Omega_L t)\hat{\mathbf{x}}(0) + \hat{\mathbf{x}}^T K(\Omega_L t)\hat{\mathbf{x}}(0)\right]$$

$$- \frac{i\hbar}{2}\text{tr} \left[\frac{\Omega_L}{2}\coth \frac{\Omega_L t}{2}\right]. \tag{2.809}$$

This has to be integrated in t, for which we use the formulas

$$\int dt\, K(\Omega_L t) = \int dt\, \frac{\Omega_L^2/2}{\sinh^2 \Omega_L t/2} = -\frac{\Omega_L}{2}\coth \frac{\Omega_L t}{2}, \tag{2.810}$$

and

$$\int dt\, \frac{1}{2}\text{tr}\left[\frac{\Omega_L}{2}\coth\frac{\Omega_L t}{2}\right] = \text{tr}\log\frac{\sinh \Omega_L t/2}{\Omega_L/2} = \text{tr}\log\frac{\sinh \Omega_L t/2}{\Omega_L t/2} + 3\log t, \tag{2.811}$$

these results following again from a Taylor expansion of both sides. The factor 3 in the last term is due to the three-dimensional trace. We can then immediately write down the exponential factor $E(\mathbf{x}, \mathbf{x}'; t)$ in (2.762):

$$E(\mathbf{x}, \mathbf{x}'; t) = \frac{1}{t^{3/2}} \exp\left\{\frac{i}{\hbar}\frac{M}{2}(\mathbf{x}-\mathbf{x}')^T \left(\frac{\Omega_L}{2}\coth\frac{\Omega_L t}{2}\right)(\mathbf{x}-\mathbf{x}') - \frac{1}{2}\text{tr}\log\frac{\sinh \Omega_L t/2}{\Omega_L t/2}\right\}. \tag{2.812}$$

The last term gives rise to a prefactor

$$\left[\det \frac{\sinh \Omega_L t/2}{\Omega_L t/2}\right]^{-1/2}. \tag{2.813}$$

As before, the time-independent integration factor $C(\mathbf{x}, \mathbf{x}')$ in (2.762) is fixed by differential equations in $\mathbf{x}$ and $\mathbf{x}'$, which involve here the covariant derivatives:

$$
\begin{aligned}
\left[-i\hbar\boldsymbol{\nabla} - \frac{e}{c}\mathbf{A}(\mathbf{x})\right] \langle \mathbf{x}\, t|\mathbf{x}'\, 0\rangle &= \langle\mathbf{x}|\hat{\mathbf{P}}e^{-i\hat{H}t/\hbar}|\mathbf{x}'\rangle = \langle\mathbf{x}|e^{-i\hat{H}t/\hbar}\left[e^{i\hat{H}t/\hbar}\hat{\mathbf{P}}e^{-i\hat{H}t/\hbar}\right]|\mathbf{x}'\rangle \\
&= \langle\mathbf{x}\, t|\hat{\mathbf{P}}(t)|\mathbf{x}'\, 0\rangle = L(\Omega_L t)(\mathbf{x} - \mathbf{x}')\langle\mathbf{x}\, t|\mathbf{x}'\, 0\rangle, \quad (2.814)
\end{aligned}
$$

$$
\begin{aligned}
\left[\ i\hbar\boldsymbol{\nabla}' - \frac{e}{c}\mathbf{A}(\mathbf{x})\right] \langle \mathbf{x}\, t|\mathbf{x}'\, 0\rangle &= \langle\mathbf{x}|e^{-i\hat{H}t/\hbar}\hat{\mathbf{P}}|\mathbf{x}'\rangle \\
&= \langle\mathbf{x}\, t|\hat{\mathbf{P}}(0)|\mathbf{x}'\, 0\rangle = L(\Omega_L t)(\mathbf{x} - \mathbf{x}')\langle\mathbf{x}\, t|\mathbf{x}'\, 0\rangle. \quad (2.815)
\end{aligned}
$$

Calculating the partial derivative we find

$$
\begin{aligned}
-i\hbar\boldsymbol{\nabla}\langle\mathbf{x}\, t|\mathbf{x}'\, 0\rangle &= [-i\hbar\boldsymbol{\nabla}C(\mathbf{x}, \mathbf{x}')]E(\mathbf{x}, \mathbf{x}'; t) + C(\mathbf{x}, \mathbf{x}')[-i\hbar\boldsymbol{\nabla}E(\mathbf{x}, \mathbf{x}'; t)] \\
&= [-i\hbar\boldsymbol{\nabla}C(\mathbf{x}, \mathbf{x}')]E(\mathbf{x}, \mathbf{x}'; t) + C(\mathbf{x}, \mathbf{x}')M\left(\frac{\Omega_L}{2}\coth\frac{\Omega_L t}{2}\right)(\mathbf{x} - \mathbf{x}')E(\mathbf{x}, \mathbf{x}'; t).
\end{aligned}
$$

Subtracting the right-hand side of (2.814) leads to

$$
M\left(\frac{\Omega_L}{2}\coth\frac{\Omega_L t}{2}\right)(\mathbf{x} - \mathbf{x}') - ML(\Omega_L t)(\mathbf{x} - \mathbf{x}') = -\frac{M}{2}\Omega_L(\mathbf{x} - \mathbf{x}'), \quad (2.816)
$$

so that $C(\mathbf{x}, \mathbf{x}')$ satisfies the time-independent differential equation

$$
\left[-i\hbar\boldsymbol{\nabla} - \frac{e}{c}A(\mathbf{x}) - \frac{M}{2}\Omega_L(\mathbf{x} - \mathbf{x}')\right]C(\mathbf{x}, \mathbf{x}') = 0. \quad (2.817)
$$

A similar equation is found from the second equation (2.815):

$$
\left[\ i\hbar\boldsymbol{\nabla}' - \frac{e}{c}A(\mathbf{x}) - \frac{M}{2}\Omega_L(\mathbf{x} - \mathbf{x}')\right]C(\mathbf{x}, \mathbf{x}') = 0. \quad (2.818)
$$

These equations are solved by

$$
C(\mathbf{x}, \mathbf{x}') = C \exp\left\{\frac{i}{\hbar}\int_{\mathbf{x}'}^{\mathbf{x}} d\boldsymbol{\xi}\left[\frac{e}{c}\mathbf{A}(\boldsymbol{\xi}) + \frac{M}{2}\Omega_L(\boldsymbol{\xi} - \mathbf{x}')\right]\right\}. \quad (2.819)
$$

The contour of integration is arbitrary since the vector field in brackets,

$$
\frac{e}{c}\mathbf{A}'(\boldsymbol{\xi}) \equiv \frac{e}{c}\mathbf{A}(\boldsymbol{\xi}) + \frac{\Omega_L}{2}(\boldsymbol{\xi} - \mathbf{x}') = \frac{e}{c}\left[\mathbf{A}(\boldsymbol{\xi}) - \frac{1}{2}\mathbf{B}\times(\boldsymbol{\xi} - \mathbf{x}')\right] \quad (2.820)
$$

has a vanishing curl, $\boldsymbol{\nabla}\times\mathbf{A}'(\mathbf{x}) = 0$. We can therefore choose the contour to be a straight line connecting $\mathbf{x}'$ and $\mathbf{x}$, in which case $d\boldsymbol{\xi}$ points in the same direction of $\mathbf{x} - \mathbf{x}'$ as $\boldsymbol{\xi} - \mathbf{x}'$ so that the cross product vanishes. Hence we may write for a straight-line connection the Ω_L-term

$$
C(\mathbf{x}, \mathbf{x}') = C \exp\left\{i\frac{e}{c}\int_{\mathbf{x}'}^{\mathbf{x}} d\boldsymbol{\xi}\, \mathbf{A}(\boldsymbol{\xi})\right\}. \quad (2.821)
$$

Finally, the normalization constant C is fixed by the initial condition (2.777) to have the value (2.778).

Collecting all terms, the amplitude is

$$
\begin{aligned}
\langle\mathbf{x}\, t|\mathbf{x}'\, 0\rangle &= \frac{1}{\sqrt{2\pi i\hbar^2 t/M}^3}\left[\det\frac{\sinh\Omega_L t/2}{\Omega_L t/2}\right]^{-1/2}\exp\left\{i\frac{e}{c}\int_{\mathbf{x}'}^{\mathbf{x}} d\boldsymbol{\xi}\, \mathbf{A}(\boldsymbol{\xi})\right\} \\
&\quad \times \exp\left\{\frac{i}{\hbar}\frac{M}{2}(\mathbf{x} - \mathbf{x}')^T\left(\frac{\Omega_L}{2}\coth\frac{\Omega_L t}{2}\right)(\mathbf{x} - \mathbf{x}')\right\}. \quad (2.822)
\end{aligned}
$$

All expressions simplify if we assume the magnetic field to point in the z-direction, in which case the frequency matrix becomes

$$\Omega_L = \begin{pmatrix} 0 & \omega_L & 0 \\ -\omega_L & 0 & 0 \\ 0 & 0 & 0 \end{pmatrix}, \tag{2.823}$$

so that

$$\cos \frac{\Omega_L t}{2} = \begin{pmatrix} \cos \omega_L t/2 & 0 & 0 \\ 0 & \cos \omega_L t/2 & 0 \\ 0 & 0 & 1 \end{pmatrix}, \tag{2.824}$$

and

$$\frac{\sinh \Omega_L t/2}{\Omega_L t/2} = \begin{pmatrix} 0 & \sin \omega_L t/2 & 0 \\ -\sin \omega_L t/2 & 0 & 0 \\ 0 & 0 & 1 \end{pmatrix}, \tag{2.825}$$

whose determinant is

$$\det \frac{\sinh \Omega_L t/2}{\Omega_L t/2} = \left(\frac{\sinh \omega_L t/2}{\omega_L t/2} \right)^2. \tag{2.826}$$

Let us calculate the exponential involving the vector potential in (2.822) explicitly. We choose the gauge in which the vector potential points in the y-direction [recall (2.634)], and parametrize the straight line between $\mathbf{x}'$ and $\mathbf{x}$ as

$$\boldsymbol{\xi} = \mathbf{x}' + s(\mathbf{x} - \mathbf{x}'), \quad s \in [0, 1]. \tag{2.827}$$

Then we find

$$\begin{aligned} \int_{\mathbf{x}'}^{\mathbf{x}} d\boldsymbol{\xi} \, \mathbf{A}(\boldsymbol{\xi}) &= B(y - y') \int_0^1 ds \, [x' + s(x - x')] = B(y - y')(x + x') \\ &= B(xy - x'y') + B(x'y - xy'). \end{aligned} \tag{2.828}$$

Inserting this and (2.826) into (2.762), we recover the earlier result (2.666).

Appendix 2A   Baker-Campbell-Hausdorff Formula and Magnus Expansion

The standard Baker-Campbell-Hausdorff formula, from which our formula (2.9) can be derived, reads

$$e^{\hat{A}} e^{\hat{B}} = e^{\hat{C}}, \tag{2A.1}$$

where

$$\hat{C} = \hat{B} + \int_0^1 dt \, g(e^{\operatorname{ad} A \, t} e^{\operatorname{ad} B})[\hat{A}], \tag{2A.2}$$

and $g(z)$ is the function

$$g(z) \equiv \frac{\log z}{z - 1} = \sum_{n=0}^{\infty} \frac{(1 - z)^n}{n + 1}, \tag{2A.3}$$

and $\operatorname{ad} B$ is the operator associated with $\hat{B}$ in the so-called *adjoint representation*, which is defined by

$$\operatorname{ad} B[\hat{A}] \equiv [\hat{B}, \hat{A}]. \tag{2A.4}$$

One also defines the trivial adjoint operator $(\operatorname{ad}B)^0[\hat{A}] = 1[\hat{A}] \equiv \hat{A}$. By expanding the exponentials in Eq. (2A.2) and using the power series (2A.3), one finds the explicit formula

$$
\hat{C} = \hat{B} + \hat{A} + \sum_{n=1}^{\infty} \frac{(-1)^n}{n+1} \sum_{p_i,q_i;p_i+q_i \geq 1} \frac{1}{1 + \sum_{i=1}^{n} p_i}
$$
$$
\times \frac{(\operatorname{ad}A)^{p_1}}{p_1!} \frac{(\operatorname{ad}B)^{q_1}}{q_1!} \cdots \frac{(\operatorname{ad}A)^{p_n}}{p_n!} \frac{(\operatorname{ad}B)^{q_n}}{q_n!}[\hat{A}].
\tag{2A.5}
$$

The lowest expansion terms are

$$
\begin{aligned}
\hat{C} &= \hat{B} + \hat{A} - \frac{1}{2} \left[\tfrac{1}{2}\operatorname{ad}A + \operatorname{ad}B + \tfrac{1}{6}(\operatorname{ad}A)^2 + \tfrac{1}{2}\operatorname{ad}A \operatorname{ad}B + \tfrac{1}{2}(\operatorname{ad}B)^2 + \ldots \right][\hat{A}] \\
&\quad + \frac{1}{3}\left[\tfrac{1}{3}(\operatorname{ad}A)^2 + \tfrac{1}{2}\operatorname{ad}A \operatorname{ad}B + \tfrac{1}{2}\operatorname{ad}B \operatorname{ad}A + (\operatorname{ad}B)^2 + \ldots \right][\hat{A}] \\
&= \hat{A} + \hat{B} + \frac{1}{2}[\hat{A},\hat{B}] + \frac{1}{12}([\hat{A},[\hat{A},\hat{B}]] + [\hat{B},[\hat{B},\hat{A}]]) + \frac{1}{24}[\hat{A},[[\hat{A},\hat{B}],\hat{B}]] \ldots .
\end{aligned}
\tag{2A.6}
$$

The result can be rearranged to the closely related *Zassenhaus formula*

$$
e^{\hat{A}+\hat{B}} = e^{\hat{A}} e^{\hat{B}} \, e^{\hat{Z}_2} e^{\hat{Z}_3} e^{\hat{Z}_4} \cdots ,
\tag{2A.7}
$$

where

$$
\hat{Z}_2 = \frac{1}{2}[\hat{B},\hat{A}]
\tag{2A.8}
$$

$$
\hat{Z}_3 = -\frac{1}{3}[\hat{B},[\hat{B},\hat{A}]] - \frac{1}{6}[\hat{A},[\hat{B},\hat{A}]])
\tag{2A.9}
$$

$$
\hat{Z}_4 = \frac{1}{8}\left([[[\hat{B},\hat{A}],\hat{B}],\hat{B}] + [[[\hat{B},\hat{A}],\hat{A}],\hat{B}] \right) + \frac{1}{24}[[[\hat{B},\hat{A}],\hat{A}],\hat{A}]
\tag{2A.10}
$$
$$
\vdots
$$

To prove the expansion (2A.6), we derive and solve a differential equation for the operator function

$$
\hat{C}(t) = \log(e^{\hat{A}t}e^{\hat{B}}).
\tag{2A.11}
$$

Its value at $t = 1$ will supply us with the desired result $\hat{C}$ in (2A.5). The starting point is the observation that for any operator $\hat{M}$,

$$
e^{\hat{C}(t)}\hat{M}e^{-\hat{C}(t)} = e^{\operatorname{ad}C(t)}[\hat{M}],
\tag{2A.12}
$$

by definition of $\operatorname{ad}C$. Inserting (2A.11), the left-hand side can also be rewritten as $e^{\hat{A}t}e^{\hat{B}}\hat{M}e^{-\hat{B}}e^{-\hat{A}t}$, which in turn is equal to $e^{\operatorname{ad}At}e^{\operatorname{ad}B}[\hat{M}]$, by definition (2A.4). Hence we have

$$
e^{\operatorname{ad}C(t)} = e^{\operatorname{ad}At}e^{\operatorname{ad}B}.
\tag{2A.13}
$$

Differentiation of (2A.11) yields

$$
e^{\hat{C}(t)}\frac{d}{dt}e^{-\hat{C}(t)} = -\hat{A}.
\tag{2A.14}
$$

The left-hand side, on the other hand, can be rewritten in general as

$$
e^{\hat{C}(t)}\frac{d}{dt}e^{-\hat{C}(t)} = -f(\operatorname{ad}C(t))[\dot{\hat{C}}(t)],
\tag{2A.15}
$$

where

$$
f(z) \equiv \frac{e^z - 1}{z}.
\tag{2A.16}
$$

This will be verified below. It implies that

$$f(\operatorname{ad} C(t))[\dot{\hat{C}}(t)] = \hat{A}. \tag{2A.17}$$

We now define the function $g(z)$ as in (2A.3) and see that it satisfies

$$g(e^z)f(z) \equiv 1. \tag{2A.18}$$

We therefore have the trivial identity

$$\dot{\hat{C}}(t) = g(e^{\operatorname{ad} C(t)})f(\operatorname{ad} C(t))[\dot{\hat{C}}(t)]. \tag{2A.19}$$

Using (2A.17) and (2A.13), this turns into the differential equation

$$\dot{\hat{C}}(t) = g(e^{\operatorname{ad} C(t)})[\hat{A}] = e^{\operatorname{ad} A t}e^{\operatorname{ad} B}[\hat{A}], \tag{2A.20}$$

from which we find directly the result (2A.2).

To complete the proof we must verify (2A.15). The expression is not simply equal to $-e^{\hat{C}(t)}\dot{\hat{C}}(t)Me^{-\hat{C}(t)}$ since $\dot{\hat{C}}(t)$ does not, in general, commute with $\hat{C}(t)$. To account for this consider the operator

$$\hat{O}(s,t) \equiv e^{\hat{C}(t)s}\frac{d}{dt}e^{-\hat{C}(t)s}. \tag{2A.21}$$

Differentiating this with respect to s gives

$$\begin{aligned}
\partial_s \hat{O}(s,t) &= e^{\hat{C}(t)s}\hat{C}(t)\frac{d}{dt}\left(e^{-\hat{C}(t)s}\right) - e^{\hat{C}(t)s}\frac{d}{dt}\left(\hat{C}(t)e^{-\hat{C}(t)s}\right) \\
&= -e^{\hat{C}(t)s}\dot{\hat{C}}(t)e^{-\hat{C}(t)s} \\
&= -e^{\operatorname{ad} C(t)s}[\dot{\hat{C}}(t)].
\end{aligned} \tag{2A.22}$$

Hence

$$\begin{aligned}
\hat{O}(s,t) - \hat{O}(0,t) &= \int_0^s ds' \partial_{s'}\hat{O}(s',t) \\
&= -\sum_{n=0}^{\infty}\frac{s^{n+1}}{(n+1)!}\left(\operatorname{ad} C(t)\right)^n [\dot{\hat{C}}(t)],
\end{aligned} \tag{2A.23}$$

from which we obtain

$$\hat{O}(1,t) = e^{\hat{C}(t)}\frac{d}{dt}e^{-\hat{C}(t)} = -f(\operatorname{ad} C(t))[\dot{\hat{C}}(t)], \tag{2A.24}$$

which is what we wanted to prove.

Note that the final form of the series for $\hat{C}$ in (2A.6) can be rearranged in many different ways, using the Jacobi identity for the commutators. It is a nontrivial task to find a form involving the smallest number of terms.[38]

The same mathematical technique can be used to derive a useful modification of the Neumann-Liouville expansion or Dyson series (1.239) and (1.251). This is the so-called *Magnus expansion*[39], in which one writes $\hat{U}(t_b, t_a) = e^{\hat{E}}$, and expands the exponent $\hat{E}$ as

$$\hat{E} = -\frac{i}{\hbar}\int_{t_a}^{t_b}dt_1\,\hat{H}(t_1) + \frac{1}{2}\left(\frac{-i}{\hbar}\right)^2\int_{t_a}^{t_b}dt_2\int_{t_a}^{t_2}dt_1\left[\hat{H}(t_2),\hat{H}(t_1)\right]$$

---

[38]For a discussion see J.A. Oteo, J. Math. Phys. *32*, 419 (1991).

[39]See A. Iserles, A. Marthinsen, and S.P. Nørsett, *On the implementation of the method of Magnus series for linear differential equations*, BIT **39**, 281 (1999) (http://www.damtp.cam.ac.uk/user/ai/Publications).

$$+ \frac{1}{4} \left(\frac{-i}{\hbar} \right)^3 \left\{ \int_{t_a}^{t_b} dt_3 \int_{t_a}^{t_3} dt_2 \int_{t_a}^{t_2} dt_1 \left[\hat{H}(t_3), \left[\hat{H}(t_2), \hat{H}(t_1) \right] \right] \right.$$

$$\left. + \frac{1}{3} \int_{t_a}^{t_b} dt_3 \int_{t_a}^{t_b} dt_2 \int_{t_a}^{t_b} dt_1 \left[\left[\hat{H}(t_3), \hat{H}(t_2) \right], \hat{H}(t_1) \right] \right\} + \dots, \qquad (2A.25)$$

which converges faster than the Neumann-Liouville expansion.

Appendix 2B   Direct Calculation of the Time-Sliced Oscillator Amplitude

After time-slicing, the amplitude (2.143) becomes a multiple integral over short-time amplitudes [using the action (2.190)]

$$(x_n \epsilon | x_{n-1} 0) = \frac{1}{\sqrt{2\pi\hbar i\epsilon/M}} \exp\left\{ \frac{i}{\hbar} \frac{M}{2} \left[\frac{(x_n - x_{n-1})^2}{\epsilon} - \epsilon\omega^2 \frac{1}{2} (x_n^2 + x_{n-1}^2) \right] \right\}. \qquad (2B.26)$$

We shall write this as

$$(x_n \epsilon | x_{n-1} 0) = \mathcal{N}_1 \exp\left\{ \frac{i}{\hbar} \left[a_1 (x_n^2 + x_{n-1}^2) - 2 b_1 x_n x_{n-1} \right] \right\}, \qquad (2B.27)$$

with

$$a_1 = \frac{M}{2\epsilon} \left[1 - 2 \left(\frac{\omega\epsilon}{2} \right)^2 \right], \quad b_1 = \frac{M}{2\epsilon},$$

$$\mathcal{N}_1 = \frac{1}{\sqrt{2\pi\hbar i\epsilon/M}}. \qquad (2B.28)$$

When performing the intermediate integrations in a product of N such amplitudes, the result must have the same general form

$$(x_N \epsilon | x_{N-1} 0) = \mathcal{N}_N \exp\left\{ \frac{i}{\hbar} \left[a_N (x_N^2 + x_0^2) - 2 b_N x_N x_0 \right] \right\}. \qquad (2B.29)$$

Multiplying this by a further short-time amplitude and integrating over the intermediate position gives the recursion relations

$$\mathcal{N}_{N+1} = \mathcal{N}_1 \mathcal{N}_N \sqrt{\frac{i\pi\hbar}{a_N + a_1}}, \qquad (2B.30)$$

$$a_{N+1} = \frac{a_N^2 - b_N^2 + a_1 a_N}{a_1 + a_N} = \frac{a_1^2 - b_1^2 + a_1 a_N}{a_1 + a_N}, \qquad (2B.31)$$

$$b_{N+1} = \frac{b_1 b_N}{a_1 + a_N}. \qquad (2B.32)$$

From (2B.31) we find

$$a_N^2 = b_N^2 + a_1^2 - b_1^2, \qquad (2B.33)$$

and the only nontrivial recursion relation to be solved is that for b_N. With (2B.32) it becomes

$$b_{N+1} = \frac{b_1 b_N}{a_1 + \sqrt{b_N^2 - (b_1^2 - a_1^2)}}, \qquad (2B.34)$$

or

$$\frac{1}{b_{N+1}} = \frac{1}{b_1} \left(\frac{a_1}{b_N} + \sqrt{1 - \frac{b_1^2 - a_1^2}{b_N^2}} \right). \qquad (2B.35)$$

We now introduce the auxiliary frequency $\tilde{\omega}$ of Eq. (2.161). Then

$$a_1 = \frac{M}{2\epsilon}\cos\tilde{\omega},\tag{2B.36}$$

and the recursion for b_{N+1} reads

$$\frac{1}{b_{N+1}} = \frac{\cos\tilde{\omega}\epsilon}{b_N} + \frac{2\epsilon}{M}\sqrt{1 - \frac{M^2\sin^2\tilde{\omega}\epsilon}{4\epsilon^2\,b_N^2}}.\tag{2B.37}$$

By introducing the reduced quantities

$$\beta_N \equiv \frac{2\epsilon}{M}b_N,\tag{2B.38}$$

with

$$\beta_1 = 1,\tag{2B.39}$$

the recursion becomes

$$\frac{1}{\beta_{N+1}} = \frac{\cos\tilde{\omega}\epsilon}{\beta_N} + \sqrt{1 - \frac{\sin^2\tilde{\omega}\epsilon}{\beta_N^2}}.\tag{2B.40}$$

For $N = 1, 2$, this determines

$$\frac{1}{\beta_2} = \cos\tilde{\omega}\epsilon + \sqrt{1 - \sin^2\tilde{\omega}\epsilon} = \frac{\sin 2\tilde{\omega}\epsilon}{\sin\tilde{\omega}\epsilon},$$

$$\frac{1}{\beta_3} = \cos\tilde{\omega}\epsilon\frac{\sin 2\tilde{\omega}\epsilon}{\sin\tilde{\omega}\epsilon} + \sqrt{1 - \sin^2\tilde{\omega}\epsilon\frac{\sin^2 2\tilde{\omega}\epsilon}{\sin^2\tilde{\omega}\epsilon}} = \frac{\sin 3\tilde{\omega}\epsilon}{\sin\tilde{\omega}\epsilon}.\tag{2B.41}$$

We therefore expect the general result

$$\frac{1}{\beta_{N+1}} = \frac{\sin\tilde{\omega}(N+1)\epsilon}{\sin\tilde{\omega}\epsilon}.\tag{2B.42}$$

It is easy to verify that this solves the recursion relation (2B.40). From (2B.38) we thus obtain

$$b_{N+1} = \frac{M}{2\epsilon}\frac{\sin\tilde{\omega}\epsilon}{\sin\tilde{\omega}(N+1)\epsilon}.\tag{2B.43}$$

Inserting this into (2B.30) and (2B.33) yields

$$a_{N+1} = \frac{M}{2\epsilon}\sin\tilde{\omega}\epsilon\frac{\cos\tilde{\omega}(N+1)\epsilon}{\sin\tilde{\omega}(N+1)\epsilon},\tag{2B.44}$$

$$\mathcal{N}_{N+1} = \mathcal{N}_1\sqrt{\frac{\sin\tilde{\omega}\epsilon}{\sin\tilde{\omega}(N+1)\epsilon}},\tag{2B.45}$$

such that (2B.29) becomes the time-sliced amplitude (2.197).

Appendix 2C   Derivation of Mehler Formula

Here we briefly sketch the derivation of Mehler's formula.[40] It is based on the observation that the left-hand side of Eq. (2.295), let us call it $F(x, x')$, is the Fourier transform of the function

$$\tilde{F}(k, k') = \pi\, e^{-(k^2 + k'^2 + akk')/2},\tag{2C.46}$$

---

[40]See P.M. Morse and H. Feshbach, *Methods of Theoretical Physics*, McGraw-Hill, New York, Vol. I, p. 781 (1953).

as can easily be verified by performing the two Gaussian integrals in the Fourier representation

$$F(x; x') = \int_{-\infty}^{\infty} \int_{-\infty}^{\infty} \frac{dk}{2\pi} \frac{dk'}{2\pi} e^{ikx+ik'x} \tilde{F}(k, k'). \qquad (2C.47)$$

We now consider the right-hand side of (2.295) and form the Fourier transform by recognizing the exponential $e^{k^2/2-ikx}$ as the generating function of the Hermite polynomials[41]

$$e^{k^2/2-ikx} = \sum_{n=0}^{\infty} \frac{(-ik/2)^n}{n!} H_n(x). \qquad (2C.48)$$

This leads to

$$\tilde{F}(k, k') = \int_{-\infty}^{\infty} \int_{-\infty}^{\infty} dx \, dx' F(x, x') e^{-ikx-ik'x} = e^{-(k^2+k'^2)/2}$$

$$\times \int_{-\infty}^{\infty} \int_{-\infty}^{\infty} dx \, dx' F(x, x') \sum_{n=0}^{\infty} \sum_{n'=0}^{\infty} \frac{(-ik/2)^n}{n!} \frac{(-ik'/2)^{n'}}{n'!} H_n(x) H_{n'}(x). \qquad (2C.49)$$

Inserting here the expansion on the right-hand side of (2.295) and using the orthogonality relation of Hermite polynomials (2.304), we obtain once more (2C.47).

Notes and References

The basic observation underlying path integrals for time evolution amplitudes goes back to the historic article
P.A.M. Dirac, Physikalische Zeitschrift der Sowjetunion **3**, 64 (1933).
He observed that the short-time propagator is the exponential of $i/\hbar$ times the classical action.
See also
P.A.M. Dirac, *The Principles of Quantum Mechanics*, Oxford University Press, Oxford, 1947;
E.T. Whittaker, Proc. Roy. Soc. Edinb. **61**, 1 (1940).

Path integrals in configuration space were invented by R. P. Feynman in his 1942 Princeton thesis.
The theory was published in 1948 in
R.P. Feynman, Rev. Mod. Phys. **20**, 367 (1948).

The mathematics of path integration had previously been developed by
N. Wiener, J. Math. Phys. **2**, 131 (1923); Proc. London Math. Soc. **22**, 454 (1924); Acta Math. **55**, 117 (1930);
N. Wiener, *Generalized Harmonic Analysis and Tauberian Theorems*, MIT Press, Cambridge, Mass., 1964,
after some earlier attempts by
P.J. Daniell, Ann. Math. **19**, 279; **20**, 1 (1918); **20**, 281 (1919); **21**, 203 (1920);
discussed in
M. Kac, Bull. Am. Math. Soc. **72**, Part II, 52 (1966).
Note that even the name *path integral* appears in Wiener's 1923 paper.

Further important papers are
I.M. Gelfand and A.M. Yaglom, J. Math. Phys. **1**, 48 (1960);
S.G. Brush, Rev. Mod. Phys. **33**, 79 (1961);
E. Nelson, J. Math. Phys. **5**, 332 (1964);
A.M. Arthurs, ed., *Functional Integration and Its Applications*, Clarendon Press, Oxford, 1975,

---

[41]I.S. Gradshteyn and I.M. Ryzhik, *op. cit.*, Formula 8.957.1.

C. DeWitt-Morette, A. Maheshwari, and B.L. Nelson, Phys. Rep. **50**, 255 (1979);
D.C. Khandekar and S.V. Lawande, Phys. Rep. **137**, 115 (1986).

The general harmonic path integral is derived in
M.J. Goovaerts, Physica **77**, 379 (1974); C.C. Grosjean and M.J. Goovaerts, J. Comput. Appl.
Math. **21**, 311 (1988); G. Junker and A. Inomata, Phys. Lett. A **110**, 195 (1985).

The Feynman path integral was applied to thermodynamics by
M. Kac, Trans. Am. Math. Soc. **65**, 1 (1949);
M. Kac, *Probability and Related Topics in Physical Science*, Interscience, New York, 1959, Chapter IV.

A good selection of earlier textbooks on path integrals is
R.P. Feynman, A.R. Hibbs, *Quantum Mechanics and Path Integrals*, McGraw Hill, New York 1965,
L.S. Schulman, *Techniques and Applications of Path Integration*, Wiley-Interscience, New York, 1981,
F.W. Wiegel, *Introduction to Path-Integral Methods in Physics and Polymer Science*, World Scientific, Singapore, 1986.
G. Roepstorff, *Path Integral Approach to Quantum Physics*, Springer, Berlin, 1994.

The path integral in phase space is reviewed by
C. Garrod, Rev. Mod. Phys. **38**, 483 (1966).

The path integral for the most general quadratic action has been studied in various ways by
D.C. Khandekar and S.V. Lawande, J. Math. Phys. **16**, 384 (1975); **20**, 1870 (1979);
V.V. Dodonov and V.I. Manko, Nuovo Cimento **44B**, 265 (1978);
A.D. Janussis, G.N. Brodimas, and A. Streclas, Phys. Lett. A **74**, 6 (1979);
C.C. Gerry, J. Math. Phys. **25**, 1820 (1984);
B.K. Cheng, J. Phys. A **17**, 2475 (1984);
G. Junker and A. Inomata, Phys. Lett. A **110**, 195 (1985);
H. Kleinert, J. Math. Phys. **27**, 3003 (1986) (http://www.physik.fu-berlin.de/~kleinert/144).

The caustic phenomena near the singularities of the harmonic oscillator amplitude at $t_b - t_a =$ integer multiples of π/ω, in particular the phase of the fluctuation factor (2.167), have been discussed by
J.M. Souriau, in *Group Theoretical Methods in Physics, IVth International Colloquium*, Nijmegen, 1975, ed. by A. Janner, *Springer Lecture Notes in Physics*, **50**;
P.A. Horvathy, Int. J. Theor. Phys. **18**, 245 (1979).
See in particular the references therein.

The amplitude for the freely falling particle is discussed in
G.P. Arrighini, N.L. Durante, C. Guidotti, Am. J. Phys. **64**, 1036 (1996);
B.R. Holstein, Am. J. Phys. **69**, 414 (1997).

For the Baker-Campbell-Hausdorff formula see
J.E. Campbell, Proc. London Math. Soc. **28**, 381 (1897); **29**, 14 (1898);
H.F. Baker, *ibid.*, **34**, 347 (1902); **3**, 24 (1905);
F. Hausdorff, Berichte Verhandl. Sächs. Akad. Wiss. Leipzig, Math. Naturw. Kl. **58**, 19 (1906);
W. Magnus, Comm. Pure and Applied Math **7**, 649 (1954), Chapter IV;
J.A. Oteo, J. Math. Phys. **32**, 419 (1991);
See also the internet address
E.W. Weisstein, http://mathworld.wolfram.com/baker-hausdorffseries.html.

The Zassenhaus formula is derived in

W. Magnus, Comm. Pure and Appl. Mathematics, **7**, 649 (1954); C. Quesne, *Disentangling q-Exponentials*, (math-ph/0310038).

For Trotter's formula see the original paper:
E. Trotter, Proc. Am. Math. Soc. **10**, 545 (1958).
The mathematical conditions for its validity are discussed by
E. Nelson, J. Math. Phys. **5**, 332 (1964);
T. Kato, in *Topics in Functional Analysis*, ed. by I. Gohberg and M. Kac, Academic Press, New York 1987.
Faster convergent formulas:
M. Suzuki, Comm. Math. Phys. **51**, 183 (1976); Physica A **191**, 501 (1992);
H. De Raedt and B. De Raedt, Phys. Rev. A **28**, 3575 (1983);
W. Janke and T. Sauer, Phys. Lett. A **165**, 199 (1992).
See also
M. Suzuki, Physica A **191**, 501 (1992).

The path integral representation of the scattering amplitude is developed in
W.B. Campbell, P. Finkler, C.E. Jones, and M.N. Misheloff, Phys. Rev. D **12**, 12, 2363 (1975).
See also:
H.D.I. Abarbanel and C. Itzykson, Phys. Rev. Lett. **23**, 53 (1969);
R. Rosenfelder, see Footnote 37.

The alternative path integral representation in Section 2.18 is due to
M. Roncadelli, Europhys. Lett. **16**, 609 (1991); J. Phys. A **25**, L997 (1992);
A. Defendi and M. Roncadelli, Europhys. Lett. **21**, 127 (1993).

3

External Sources, Correlations, and Perturbation Theory

Important information on every quantum-mechanical system is carried by the correlation functions of the path $x(t)$. They are defined as the expectation values of products of path positions at different times, $x(t_1)\cdots x(t_n)$, to be calculated as functional averages. Quantities of this type are observable in simple scattering experiments. The most efficient extraction of correlation functions from a path integral proceeds by adding to the Lagrangian an external time-dependent mechanical force term disturbing the system linearly, and by studying the response to the disturbance. A similar linear term is used extensively in quantum field theory, for instance in quantum electrodynamics where it is no longer a mechanical force, but a source of fields, i.e., a charge or a current density. For this reason we shall call this term generically source or current term.

In this chapter, the procedure is developed for the harmonic action, where a linear source term does not destroy the solvability of the path integral. The resulting amplitude is a simple functional of the current. Its functional derivatives will supply all correlation functions of the system, and for this reason it is called the *generating functional* of the theory. It serves to derive the celebrated *Wick rule* for calculating the correlation functions of an arbitrary number of $x(t)$. This forms the basis for perturbation expansions of anharmonic theories.

3.1 External Sources

Consider a harmonic oscillator with an action

$$\mathcal{A}_\omega = \int_{t_a}^{t_b} dt\, \frac{M}{2}(\dot{x}^2 - \omega^2 x^2).\qquad(3.1)$$

Let it be disturbed by an external *source* or *current* $j(t)$ coupled linearly to the particle coordinate $x(t)$. The source action is

$$\mathcal{A}_j = \int_{t_a}^{t_b} dt\, x(t)j(t).\qquad(3.2)$$

209

The total action

$$\mathcal{A} = \mathcal{A}_\omega + \mathcal{A}_j \tag{3.3}$$

is still harmonic in x and $\dot{x}$, which makes it is easy to solve the path integral in the presence of a source term. In particular, the source term does not destroy the factorization property (2.151) of the time evolution amplitude into a classical amplitude $e^{i\mathcal{A}_{j,\mathrm{cl}}/\hbar}$ and a fluctuation factor $F_{\omega,j}(t_b, t_a)$,

$$(x_b t_b | x_a t_a)_\omega^j = e^{(i/\hbar)\mathcal{A}_{j,\mathrm{cl}}} F_{\omega,j}(t_b, t_a). \tag{3.4}$$

Here $\mathcal{A}_{j,\mathrm{cl}}$ is the action for the classical orbit $x_{j,\mathrm{cl}}(t)$ which minimizes the total action $\mathcal{A}$ in the presence of the source term and which obeys the equation of motion

$$\ddot{x}_{j,\mathrm{cl}}(t) + \omega^2 x_{j,\mathrm{cl}}(t) = j(t). \tag{3.5}$$

In the sequel, we shall first work with the classical orbit $x_{\mathrm{cl}}(t)$ extremizing the action *without* the source term:

$$x_{\mathrm{cl}}(t) = \frac{x_b \sin\omega(t - t_a) + x_a \sin\omega(t_b - t)}{\sin\omega(t_b - t_a)}. \tag{3.6}$$

All paths will be written as a sum of the classical orbit $x_{\mathrm{cl}}(t)$ and a fluctuation $\delta x(t)$:

$$x(t) = x_{\mathrm{cl}}(t) + \delta x(t). \tag{3.7}$$

Then the action separates into a classical and a fluctuating part, each of which contains a source-free and a source term:

$$
\begin{aligned}
\mathcal{A} = \mathcal{A}_\omega + \mathcal{A}_j &\equiv \mathcal{A}_{\mathrm{cl}} + \mathcal{A}_{\mathrm{fl}} \\
&= (\mathcal{A}_{\omega,\mathrm{cl}} + \mathcal{A}_{j,\mathrm{cl}}) + (\mathcal{A}_{\omega,\mathrm{fl}} + \mathcal{A}_{j,\mathrm{fl}}).
\end{aligned} \tag{3.8}
$$

The time evolution amplitude can be expressed as

$$
\begin{aligned}
(x_b t_b | x_a t_a)_\omega^j &= e^{(i/\hbar)\mathcal{A}_{\mathrm{cl}}} \int \mathcal{D}x \exp\left(\frac{i}{\hbar}\mathcal{A}_{\mathrm{fl}}\right) \\
&= e^{(i/\hbar)(\mathcal{A}_{\omega,\mathrm{cl}} + \mathcal{A}_{j,\mathrm{cl}})} \int \mathcal{D}x \exp\left[\frac{i}{\hbar}(\mathcal{A}_{\omega,\mathrm{fl}} + \mathcal{A}_{j,\mathrm{fl}})\right].
\end{aligned} \tag{3.9}
$$

The classical action $\mathcal{A}_{\omega,\mathrm{cl}}$ is known from Eq. (2.157):

$$\mathcal{A}_{\omega,\mathrm{cl}} = \frac{M\omega}{2\sin\omega(t_b - t_a)}\left[(x_b^2 + x_a^2)\cos\omega(t_b - t_a) - 2x_b x_a\right]. \tag{3.10}$$

The classical source term is known from (3.6):

$$
\begin{aligned}
\mathcal{A}_{j,\mathrm{cl}} &= \int_{t_a}^{t_b} dt\, x_{\mathrm{cl}}(t) j(t) \\
&= \frac{1}{\sin\omega(t_b - t_a)} \int_{t_a}^{t_b} dt [x_a \sin\omega(t_b - t) + x_b \sin\omega(t - t_a)] j(t).
\end{aligned} \tag{3.11}
$$

Consider now the fluctuating part of the action, $\mathcal{A}_{\mathrm{fl}} = \mathcal{A}_{\omega,\mathrm{fl}} + \mathcal{A}_{j,\mathrm{fl}}$. Since $x_{\mathrm{cl}}(t)$ extremizes the action without the source, $\mathcal{A}_{\mathrm{fl}}$ contains a term linear in $\delta x(t)$. After a partial integration [making use of the vanishing of $\delta x(t)$ at the ends] it can be written as

$$\mathcal{A}_{\mathrm{fl}} = \frac{M}{2} \int_{t_a}^{t_b} dt\,dt'\, \delta x(t) D_{\omega^2}(t,t')\delta x(t') + \int_{t_a}^{t_b} dt\, \delta x(t) j(t), \tag{3.12}$$

where $D_{\omega^2}(t,t')$ is the differential operator

$$D_{\omega^2}(t,t') = (-\partial_t^2 - \omega^2)\delta(t-t') = \delta(t-t')(-\partial_{t'}^2 - \omega^2), \quad t,t' \in (t_a,t_b). \tag{3.13}$$

It may be considered as a functional matrix in the space of the t-dependent functions vanishing at t_a, t_b. The equality of the two expressions is seen as follows. By partial integrations one has

$$\int_{t_a}^{t_b} dt\, f(t)\partial_t^2 g(t) = \int_{t_a}^{t_b} dt\, \partial_t^2 f(t)g(t), \tag{3.14}$$

for any $f(t)$ and $g(t)$ vanishing at the boundaries (or for periodic functions in the interval). The left-hand side can directly be rewritten as $\int_{t_a}^{t_b} dt\,dt'\, f(t)\delta(t-t')\partial_{t'}^2 g(t')$, the right-hand side as $\int_{t_a}^{t_b} dt\,dt'\, \partial_t^2 f(t)\delta(t-t')g(t')$, and after further partial integrations, as $\int dt\,dt'\, f(t)\partial_t^2\delta(t-t')g(t)$.

The inverse $D_{\omega^2}^{-1}(t,t')$ of the functional matrix (3.13) is formally defined by the relation

$$\int_{t_a}^{t_b} dt'\, D_{\omega^2}(t'',t')D_{\omega^2}^{-1}(t',t) = \delta(t''-t), \quad t'',t \in (t_a,t_b), \tag{3.15}$$

which shows that it is the standard classical Green function of the harmonic oscillator of frequency ω:

$$G_{\omega^2}(t,t') \equiv D_{\omega^2}^{-1}(t,t') = (-\partial_t^2 - \omega^2)^{-1}\delta(t-t'), \quad t,t' \in (t_a,t_b). \tag{3.16}$$

This definition is not unique since it leaves room for an additional arbitrary solution $H(t,t')$ of the homogeneous equation $\int_{t_a}^{t_b} dt'\, D_{\omega^2}(t'',t')H(t',t) = 0$. This freedom will be removed below by imposing appropriate boundary conditions.

In the fluctuation action (3.12), we now perform a *quadratic completion* by a shift of $\delta x(t)$ to

$$\delta\tilde{x}(t) \equiv \delta x(t) + \frac{1}{M}\int_{t_a}^{t_b} dt'\, G_{\omega^2}(t,t')j(t'). \tag{3.17}$$

Then the action becomes quadratic in both $\delta\tilde{x}$ and j:

$$\mathcal{A}_{\mathrm{fl}} = \int_{t_a}^{t_b} dt \int_{t_a}^{t_b} dt' \left[\frac{M}{2}\delta\tilde{x}(t) D_{\omega^2}(t,t')\delta\tilde{x}(t') - \frac{1}{2M}j(t)G_{\omega^2}(t,t')j(t') \right]. \tag{3.18}$$

The Green function obeys the same boundary condition as the fluctuations $\delta x(t)$:

$$G_{\omega^2}(t,t') = 0 \quad \text{for} \quad \begin{cases} t = t_b, & t' \text{ arbitrary}, \\ t \text{ arbitrary}, & t' = t_a. \end{cases} \tag{3.19}$$

Thus, the shifted fluctuations $\delta\tilde{x}(t)$ of (3.17) also vanish at the ends and run through the same functional space as the original $\delta x(t)$. The measure of path integration $\int \mathcal{D}\delta x(t)$ is obviously unchanged by the simple shift (3.17). Hence the path integral $\int \mathcal{D}\delta\tilde{x}$ over $e^{i\mathcal{A}_{fl}/\hbar}$ with the action (3.18) gives, via the first term in $\mathcal{A}_{fl}$, the harmonic fluctuation factor $F_\omega(t_b - t_a)$ calculated in (2.169):

$$F_\omega(t_b - t_a) = \frac{1}{\sqrt{2\pi i\hbar/M}} \sqrt{\frac{\omega}{\sin\omega(t_b - t_a)}}. \tag{3.20}$$

The source part in (3.18) contributes only a trivial exponential factor

$$F_{j,fl} = \exp\left\{\frac{i}{\hbar}\mathcal{A}_{j,fl}\right\}, \tag{3.21}$$

whose exponent is quadratic in $j(t)$:

$$\mathcal{A}_{j,fl} = -\frac{1}{2M}\int_{t_a}^{t_b} dt \int_{t_a}^{t_b} dt'\, j(t) G_{\omega^2}(t,t') j(t'). \tag{3.22}$$

The total time evolution amplitude in the presence of a source term can therefore be written as the product

$$(x_b t_b | x_a t_a)_\omega^j = (x_b t_b | x_a t_a)_\omega F_{j,cl} F_{j,fl}, \tag{3.23}$$

where $(x_b t_b | x_a t)_\omega$ is the source-free time evolution amplitude

$$(x_b t_b | x_a t_a)_\omega = e^{(i/\hbar)\mathcal{A}_{\omega,cl}} F_\omega(t_b - t_a) = \frac{1}{\sqrt{2\pi i\hbar/M}}\sqrt{\frac{\omega}{\sin\omega(t_b - t_a)}}$$

$$\times \exp\left\{\frac{i}{2\hbar}\frac{M\omega}{\sin\omega(t_b - t_a)}[(x_b^2 + x_a^2)\cos\omega(t_b - t_a) - 2x_b x_a]\right\}, \tag{3.24}$$

and $F_{j,cl}$ is an amplitude containing the classical action (3.11):

$$F_{j,cl} = e^{(i/\hbar)\mathcal{A}_{j,cl}}$$

$$= \exp\left\{\frac{i}{\hbar}\frac{1}{\sin\omega(t_b - t_a)}\int_{t_a}^{t_b} dt[x_a \sin\omega(t_b - t) + x_b \sin\omega(t - t_a)]\, j(t)\right\}. \tag{3.25}$$

To complete the result we need to know the Green function $G_{\omega^2}(t, t')$ explicitly, which will be calculated in the next section.

3.2   Green Function of Harmonic Oscillator

According to Eq. (3.16), the Green function in Eq. (3.22) is obtained by inverting the second-order differential operator $-\partial_t^2 - \omega^2$:

$$G_{\omega^2}(t, t') = (-\partial_t^2 - \omega^2)^{-1}\delta(t - t'), \quad t, t' \in (t_a, t_b). \tag{3.26}$$

As remarked above, this function is defined only up to solutions of the homogeneous differential equation associated with the operator $-\partial_t^2 - \omega^2$. The boundary conditions removing this ambiguity are the same as for the fluctuations $\delta x(t)$, i.e., $G_{\omega^2}(t, t')$ vanishes if either t or t' or both hit an endpoint t_a or t_b (Dirichlet boundary condition). The Green function is symmetric in t and t'. For the sake of generality, we shall find the Green function also for the more general differential equation with time-dependent frequency,

$$[-\partial_t^2 - \Omega^2(t)]G_{\Omega^2}(t, t') = \delta(t - t'), \tag{3.27}$$

with the same boundary conditions.

There are several ways of calculating this explicitly.

3.2.1 Wronski Construction

The simplest way proceeds via the so-called *Wronski construction*, which is based on the following observation. For different time arguments, $t > t'$ or $t < t'$, the Green function $G_{\Omega^2}(t, t')$ has to solve the homogeneous differential equations

$$(-\partial_t^2 - \omega^2)G_{\Omega^2}(t, t') = 0, \quad (-\partial_{t'}^2 - \omega^2)G_{\Omega^2}(t, t') = 0. \tag{3.28}$$

It must therefore be a linear combination of two independent solutions of the homogeneous differential equation in t as well as in t', and it must satisfy the Dirichlet boundary condition of vanishing at the respective endpoints.

Constant Frequency

If $\Omega^2(t) \equiv \omega^2$, this implies that for $t > t'$, $G_{\omega^2}(t, t')$ must be proportional to $\sin\omega(t_b - t)$ as well as to $\sin\omega(t' - t_a)$, leaving only the solution

$$G_{\omega^2}(t, t') = C\sin\omega(t_b - t)\sin\omega(t' - t_a), \quad t > t'. \tag{3.29}$$

For $t < t'$, we obtain similarly

$$G_{\omega^2}(t, t') = C\sin\omega(t_b - t')\sin\omega(t - t_a), \quad t < t'. \tag{3.30}$$

The two cases can be written as a single expression

$$G_{\omega^2}(t, t') = C\sin\omega(t_b - t_>)\sin\omega(t_< - t_a), \tag{3.31}$$

where the symbols $t_>$ and $t_<$ denote the larger and the smaller of the times t and t', respectively. The unknown constant C is fixed by considering coincident times $t = t'$. There, the time derivative of $G_{\omega^2}(t, t')$ must have a discontinuity which gives rise to the δ-function in (3.15). For $t > t'$, the derivative of (3.29) is

$$\partial_t G_{\omega^2}(t, t') = -C\omega\cos\omega(t_b - t)\sin\omega(t' - t_a), \tag{3.32}$$

whereas for $t < t'$

$$\partial_t G_{\omega^2}(t, t') = C\omega \sin \omega(t_b - t') \cos \omega(t - t_a). \tag{3.33}$$

At $t = t'$ we find the discontinuity

$$\partial_t G_{\omega^2}(t, t')|_{t=t'+\epsilon} - \partial_t G_{\omega^2}(t, t')|_{t=t'-\epsilon} = -C\omega \sin \omega(t_b - t_a). \tag{3.34}$$

Hence $-\partial_t^2 G_{\omega^2}(t, t')$ is proportional to a δ-function:

$$- \partial_t^2 G_{\omega^2}(t, t') = C\omega \sin \omega(t_b - t_a)\delta(t - t'). \tag{3.35}$$

By normalizing the prefactor to unity, we fix C and find the desired Green function:

$$G_{\omega^2}(t, t') = \frac{\sin \omega(t_b - t_>) \sin \omega(t_< - t_a)}{\omega \sin \omega(t_b - t_a)}. \tag{3.36}$$

It exists only if $t_b - t_a$ is not equal to an integer multiple of π/ω. This restriction was encountered before in the amplitude without external sources; its meaning was discussed in the two paragraphs following Eq. (2.166).

The constant in the denominator of (3.36) is the Wronski determinant (or *Wronskian*) of the two solutions $\xi(t) = \sin \omega(t_b - t)$ and $\eta(t) = \sin \omega(t - t_a)$ which was introduced in (2.220):

$$W[\xi(t), \eta(t)] \equiv \xi(t)\dot{\eta}(t) - \dot{\xi}(t)\eta(t). \tag{3.37}$$

An alternative expression for (3.36) is

$$G_{\omega^2}(t, t') = \frac{-\cos \omega(t_b - t_a - |t - t'|) + \cos \omega(t_b + t_a - t - t')}{2\omega \sin \omega(t_b - t_a)}. \tag{3.38}$$

In the limit $\omega \to 0$ we obtain the free-particle Green function

$$
\begin{aligned}
G_0(t, t') &= \frac{1}{(t_b - t_a)}(t_b - t_>)(t_< - t_a) \\
&= \frac{1}{t_b - t_a}\left[-tt' - \frac{1}{2}(t_b - t_a)|t - t'| + \frac{1}{2}(t_a + t_b)(t + t') - t_a t_b\right]. \tag{3.39}
\end{aligned}
$$

Time-Dependent Frequency

It is just as easy to find the Green functions of the more general differential equation (3.27) with a time-dependent oscillator frequency $\Omega(t)$. We construct first a so-called *retarded Green function* (compare page 38) as a product of a Heaviside function with a smooth function

$$G_{\Omega^2}(t, t') = \Theta(t - t')\Delta(t, t'). \tag{3.40}$$

Inserting this into the differential equation (3.27) we find

$$
\begin{aligned}
[-\partial_t^2 - \Omega^2(t)]G_{\Omega^2}(t,t') &= \Theta(t-t')\left[-\partial_t^2 - \Omega^2(t)\right]\Delta(t,t') \\
&- \dot{\delta}(t-t') - 2\partial_t\Delta(t,t')\delta(t-t').
\end{aligned}
\tag{3.41}
$$

Expanding

$$
\Delta(t,t') = \Delta(t,t) + [\partial_t\Delta(t,t')]_{t=t'}(t-t') + \frac{1}{2}[\partial_t^2\Delta(t,t')]_{t=t'}(t-t')^2 + \dots , \tag{3.42}
$$

and using the fact that

$$
(t-t')\dot{\delta}(t-t') = -\delta(t-t'), \quad (t-t')^n\dot{\delta}(t-t') = 0 \quad \text{for} \quad n > 1, \tag{3.43}
$$

the second line in (3.41) can be rewritten as

$$
-\dot{\delta}(t-t')\Delta(t,t') - \delta(t-t')\partial_t\Delta(t,t'). \tag{3.44}
$$

By choosing the initial conditions

$$
\Delta(t,t) = 0, \quad \dot{\Delta}(t,t')|_{t'=t} = -1, \tag{3.45}
$$

we satisfy the inhomogeneous differential equation (3.27) provided $\Delta(t,t')$ obeys the homogeneous differential equation

$$
[-\partial_t^2 - \Omega^2(t)]\Delta(t,t') = 0, \quad \text{for} \quad t > t' . \tag{3.46}
$$

This equation is solved by a linear combination

$$
\Delta(t,t') = \alpha(t')\xi(t) + \beta(t')\eta(t) \tag{3.47}
$$

of any two independent solutions $\eta(t)$ and $\xi(t)$ of the homogeneous equation

$$
[-\partial_t^2 - \Omega^2(t)]\xi(t) = 0, \quad [-\partial_t^2 - \Omega^2(t)]\eta(t) = 0. \tag{3.48}
$$

Their Wronski determinant $W = \xi(t)\dot{\eta}(t) - \dot{\xi}(t)\eta(t)$ is nonzero and, of course, time-independent, so that we can determine the coefficients in the linear combination (3.47) from (3.45) and find

$$
\Delta(t,t') = \frac{1}{W}\left[\xi(t)\eta(t') - \xi(t')\eta(t)\right]. \tag{3.49}
$$

The right-hand side contains the so-called Jacobi commutator of the two functions $\xi(t)$ and $\eta(t)$. Here we list a few useful algebraic properties of $\Delta(t,t')$:

$$
\Delta(t,t') = \frac{\Delta(t_b,t)\Delta(t',t_a) - \Delta(t,t_a)\Delta(t_b,t')}{\Delta(t_b,t_a)}, \tag{3.50}
$$

$$
\Delta(t_b,t)\partial_{t_b}\,\Delta(t_b,t_a) - \Delta(t,t_a) = \Delta(t_b,t_a)\partial_t\Delta(t_b,t), \tag{3.51}
$$

$$\Delta(t, t_a)\partial_{t_b}\Delta(t_b, t_a) - \Delta(t_b, t) = \Delta(t_b, t_a)\partial_t\Delta(t, t_a). \tag{3.52}$$

The retarded Green function (3.40) is so far not the unique solution of the differential equation (3.27), since one may always add a general solution of the homogeneous differential equation (3.48):

$$G_{\Omega^2}(t, t') = \Theta(t - t')\Delta(t, t') + a(t')\xi(t) + b(t')\eta(t), \tag{3.53}$$

with arbitrary coefficients $a(t')$ and $b(t')$. This ambiguity is removed by the Dirichlet boundary conditions

$$\begin{aligned} G_{\Omega^2}(t_b, t) &= 0, \quad t_b \neq t, \\ G_{\Omega^2}(t, t_a) &= 0, \quad t \neq t_a. \end{aligned} \tag{3.54}$$

Imposing these upon (3.53) leads to a simple algebraic pair of equations

$$\begin{aligned} a(t)\xi(t_a) + b(t)\eta(t_a) &= 0, \tag{3.55} \\ a(t)\xi(t_b) + b(t)\eta(t_b) &= \Delta(t, t_b). \tag{3.56} \end{aligned}$$

Denoting the 2×2 -coefficient matrix by

$$\Lambda = \begin{pmatrix} \xi(t_a) & \eta(t_a) \\ \xi(t_b) & \eta(t_b) \end{pmatrix}, \tag{3.57}$$

we observe that under the condition

$$\det \Lambda = W\Delta(t_a, t_b) \neq 0, \tag{3.58}$$

the system (3.56) has a unique solution for the coefficients $a(t)$ and $b(t)$ in the Green function (3.53). Inserting this into (3.54) and using the identity (3.50), we obtain from this Wronski's general formula corresponding to (3.36)

$$G_{\Omega^2}(t, t') = \frac{\Theta(t - t')\Delta(t_b, t)\Delta(t', t_a) + \Theta(t' - t)\Delta(t, t_a)\Delta(t_b, t')}{\Delta(t_a, t_b)}. \tag{3.59}$$

At this point it is useful to realize that the functions in the numerator coincide with the two specific linearly independent solutions $D_a(t)$ and $D_b(t)$ of the homogenous differential equations (3.48) which were introduced in Eqs. (2.226) and (2.227). Comparing the initial conditions of $D_a(t)$ and $D_b(t)$ with that of the function $\Delta(t, t')$ in Eq. (3.45), we readily identify

$$D_a(t) \equiv \Delta(t, t_a), \qquad D_b(t) \equiv \Delta(t_b, t), \tag{3.60}$$

and formula (3.59) can be rewritten as

$$G_{\Omega^2}(t, t') = \frac{\Theta(t - t')D_b(t)D_a(t') + \Theta(t' - t)D_a(t)D_b(t')}{D_a(t_b)}. \tag{3.61}$$

It should be pointed out that this equation renders a unique and well-defined Green function if the differential equation $[-\partial_t^2 - \Omega^2(t)]y(t) = 0$ has no solutions with Dirichlet boundary conditions $y(t_a) = y(t_b) = 0$, generally called *zero-modes*. A zero mode would cause problems since it would certainly be one of the independent solutions of (3.49), say $\eta(t)$. Due to the property $\eta(t_a) = \eta(t_b) = 0$, however, the determinant of Λ would vanish, thus destroying the condition (3.58) which was necessary to find (3.59). Indeed, the function $\Delta(t, t')$ in (3.49) would remain undetermined since the boundary condition $\eta(t_a) = 0$ together with (3.55) implies that also $\xi(t_a) = 0$, making $W = \xi(t)\dot\eta(t) - \dot\xi(t)\eta(t)$ vanish at the initial time t_a, and thus for all times.

3.2.2 Spectral Representation

A second way of specifying the Green function explicitly is via its *spectral representation*.

Constant Frequency

For constant frequency $\Omega(t) \equiv \omega$, the fluctuations $\delta x(t)$ which satisfy the differential equation

$$(-\partial_t^2 - \omega^2)\, \delta x(t) = 0, \tag{3.62}$$

and vanish at the ends $t = t_a$ and $t = t_b$, are expanded into a complete set of orthonormal functions:

$$x_n(t) = \sqrt{\frac{2}{t_b - t_a}} \sin \nu_n(t - t_a), \tag{3.63}$$

with the frequencies [compare (2.110)]

$$\nu_n = \frac{\pi n}{t_b - t_a}. \tag{3.64}$$

These functions satisfy the orthonormality relations

$$\int_{t_a}^{t_b} dt\, x_n(t)x_{n'}(t) = \delta_{nn'}. \tag{3.65}$$

Since the operator $-\partial_t^2 - \omega^2$ is diagonal on $x_n(t)$, this is also true for the Green function $G_{\omega^2}(t, t') = (-\partial_t^2 - \omega^2)^{-1}\delta(t - t')$. Let G_n be its eigenvalues defined by

$$\int_{t_a}^{t_b} dt\, G_{\omega^2}(t, t')x_n(t') = G_n x_n(t). \tag{3.66}$$

Then we expand $G_{\omega^2}(t, t')$ as follows:

$$G_{\omega^2}(t, t') = \sum_{n=1}^{\infty} G_n x_n(t)x_n(t'). \tag{3.67}$$

By definition, the eigenvalues of $G_{\omega^2}(t, t')$ are the inverse eigenvalues of the differential operator $(-\partial_t^2 - \omega^2)$, which are $\nu_n^2 - \omega^2$. Thus

$$G_n = (\nu_n^2 - \omega^2)^{-1}, \tag{3.68}$$

and we arrive at the spectral representation of $G_{\omega^2}(t, t')$:

$$G_{\omega^2}(t, t') = \frac{2}{t_b - t_a} \sum_{n=1}^{\infty} \frac{\sin \nu_n(t - t_a) \sin \nu_n(t' - t_a)}{\nu_n^2 - \omega^2}. \tag{3.69}$$

We may use the trigonometric relation

$$\sin \nu_n(t_b - t) = -\sin \nu_n[(t - t_a) - (t_b - t_a)] = -(-1)^n \sin \nu_n(t - t_a)$$

to rewrite (3.69) as

$$G_{\omega^2}(t, t') = \frac{2}{t_b - t_a} \sum_{n=1}^{\infty} (-1)^{n+1} \frac{\sin \nu_n(t_b - t) \sin \nu_n(t' - t_a)}{\nu_n^2 - \omega^2}. \tag{3.70}$$

These expressions make sense only if $t_b - t_a$ is not equal to an integer multiple of π/ω, where one of the denominators in the sums vanishes. This is the same range of $t_b - t_a$ as in the Wronski expression (3.36).

Time-Dependent Frequency

The spectral representation can also be written down for the more general Green function with a time-dependent frequency defined by the differential equation (3.27). If $y_n(t)$ are the eigenfunctions solving the differential equation with eigenvalue λ_n

$$K(t)y_n(t) = \lambda_n y_n(t), \tag{3.71}$$

and if these eigenfunctions satisfy the orthogonality and completeness relations

$$\int_{t_a}^{t_b} dt \, y_n(t) y_{n'}(t) = \delta_{nn'}, \tag{3.72}$$

$$\sum_n y_n(t) y_n(t') = \delta(t - t'), \tag{3.73}$$

and if, moreover, there exists no zero-mode for which $\lambda_n = 0$, then $G_{\Omega^2}(t, t')$ has the spectral representation

$$G_{\Omega^2}(t, t') = \sum_n \frac{y_n(t) y_n(t')}{\lambda_n}. \tag{3.74}$$

This is easily verified by multiplication with $K(t)$ using (3.71) and (3.73).

It is instructive to prove the equality between the Wronskian construction and the spectral representations (3.36) and (3.70). It will be useful to do this in several steps. In the present context, some of these may appear redundant. They will, however, yield intermediate results which will be needed in Chapters 7 and 18 when discussing path integrals occurring in quantum field theories.

3.3   Green Functions of First-Order Differential Equation

An important quantity of statistical mechanics are the Green functions $G_\Omega^{\mathrm{p}}(t,t')$ which solve the *first-order* differential equation

$$[i\partial_t - \Omega(t)]\, G_\Omega(t,t') = i\delta(t-t'), \quad t-t' \in [0, t_b - t_a), \tag{3.75}$$

or its Euclidean version $G_{\Omega,\mathrm{e}}^{\mathrm{p}}(\tau,\tau'')$ which solves the differential equation, obtained from (3.75) for $t = -i\tau$:

$$[\partial_\tau - \Omega(\tau)]\, G_{\Omega,\mathrm{e}}(\tau,\tau') = \delta(\tau-\tau'), \quad \tau - \tau' \in [0, \hbar\beta). \tag{3.76}$$

These can be calculated for an arbitrary function $\Omega(t)$.

3.3.1   Time-Independent Frequency

Consider first the simplest case of a Green function $G_\omega^{\mathrm{p}}(t,t')$ with fixed frequency ω which solves the *first-order* differential equation

$$(i\partial_t - \omega)G_\omega^{\mathrm{p}}(t,t') = i\delta(t-t'), \quad t-t' \in [0, t_b - t_a). \tag{3.77}$$

The equation determines $G_\omega^{\mathrm{p}}(t,t')$ only up to a solution $H(t,t')$ of the homogeneous differential equation $(i\partial_t - \omega)H(t,t') = 0$. The ambiguity is removed by imposing the *periodic* boundary condition

$$G_\omega^{\mathrm{p}}(t,t') \equiv G_\omega^{\mathrm{p}}(t-t') = G_\omega^{\mathrm{p}}(t-t'+t_b-t_a), \tag{3.78}$$

indicated by the superscript p. With this boundary condition, the Green function $G_\omega^{\mathrm{p}}(t,t')$ is translationally invariant in time. It depends only on the difference between t and t' and is periodic in it.

The spectral representation of $G_\omega^{\mathrm{p}}(t,t')$ can immediately be written down, assuming that $t_b - t_a$ does not coincide with an even multiple of π/ω:

$$G_\omega^{\mathrm{p}}(t-t') = \frac{1}{t_b - t_a} \sum_{m=-\infty}^{\infty} e^{-i\omega_m(t-t')} \frac{i}{\omega_m - \omega}. \tag{3.79}$$

The frequencies ω_m are twice as large as the previous ν_m's in (3.64):

$$\omega_m \equiv \frac{2\pi m}{t_b - t_a}, \quad m = 0, \pm 1, \pm 2, \pm 3, \ldots . \tag{3.80}$$

As for the periodic orbits in Section 2.9, there are "about as many" ω_m as ν_m, since there is an ω_m for each positive *and* negative integer m, whereas the ν_m are all positive (see the last paragraph in that section). The frequencies (3.80) are the real-time analogs of the Matsubara frequencies (2.379) of quantum statistics with the usual correspondence $t_b - t_a = -i\hbar/k_B T$ of Eq. (2.328).

To calculate the spectral sum, we use the Poisson summation formula in the form (1.197):

$$\sum_{m=-\infty}^{\infty} f(m) = \int_{-\infty}^{\infty} d\mu \sum_{n=-\infty}^{\infty} e^{2\pi i \mu n} f(\mu). \tag{3.81}$$

Accordingly, we rewrite the sum over ω_m as an integral over ω', followed by an auxiliary sum over n which squeezes the variable ω' onto the proper discrete values $\omega_m = 2\pi m/(t_b - t_a)$:

$$G_\omega^{\mathrm{p}}(t) = \sum_{n=-\infty}^{\infty} \int_{-\infty}^{\infty} \frac{d\omega'}{2\pi} e^{-i\omega'[t-(t_b-t_a)n]} \frac{i}{\omega' - \omega}. \tag{3.82}$$

At this point it is useful to introduce another Green function $G_\omega(t-t')$ associated with the first-order differential equation (3.77) on an *infinite* time interval:

$$G_\omega(t) = \int_{-\infty}^{\infty} \frac{d\omega'}{2\pi} e^{-i\omega't} \frac{i}{\omega' - \omega}. \tag{3.83}$$

In terms of this function, the periodic Green function (3.82) can be written as a sum which exhibits in a most obvious way the periodicity under $t \to t + (t_b - t_a)$:

$$G_\omega^{\mathrm{p}}(t) = \sum_{n=-\infty}^{\infty} G_\omega(t - (t_b - t_a)n). \tag{3.84}$$

The advantage of using $G_\omega(t - t')$ is that the integral over ω' in (3.83) can easily be done. We merely have to prescribe how to treat the singularity at $\omega' = \omega$. This also removes the freedom of adding a homogeneous solution $H(t, t')$. To make the integral unique, we replace ω by $\omega - i\eta$ where η is a very small positive number, i.e., by the $i\eta$-prescription introduced after Eq. (2.166). This moves the pole in the integrand of (3.83) into the lower half of the complex ω'-plane, making the integral over ω' in $G_\omega(t)$ fundamentally different for $t < 0$ and for $t > 0$. For $t < 0$, the contour of integration can be closed in the complex ω'-plane by a semicircle in the *upper* half-plane at no extra cost, since $e^{-i\omega't}$ is exponentially small there (see Fig. 3.1). With the integrand being analytic in the upper half-plane we can contract the contour to zero and find that the integral vanishes. For $t > 0$, on the other hand, the contour is closed in the lower half-plane containing a pole at $\omega' = \omega - i\eta$. When contracting the contour to zero, the integral picks up the residue at this pole and yields a factor $-2\pi i$. At the point $t = 0$, finally, we can close the contour either way. The integral over the semicircles is now nonzero, $\mp 1/2$, which has to be subtracted from the residues 0 and 1, respectively, yielding $1/2$. Hence we find

$$\begin{aligned}
G_\omega(t) &= \int_{-\infty}^{\infty} \frac{d\omega'}{2\pi} e^{-i\omega't} \frac{i}{\omega' - \omega + i\eta} \\
&= e^{-i\omega t} \times \begin{cases} 1 & \text{for} \quad t > 0, \\ \frac{1}{2} & \text{for} \quad t = 0, \\ 0 & \text{for} \quad t < 0. \end{cases}
\end{aligned} \tag{3.85}$$

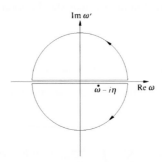

Figure 3.1 Pole in Fourier transform of Green functions $G_\omega^{\mathrm{p,a}}(t)$, and infinite semicircles in the upper (lower) half-plane which extend the integrals to a closed contour for $t < 0$ ($t > 0$).

The vanishing of the Green function for $t < 0$ is the *causality property* of $G_\omega(t)$ discussed in (1.306) and (1.307). It is a general property of functions whose Fourier transforms are analytic in the upper half-plane.

The three cases in (3.85) can be collected into a single formula using the Heaviside function $\Theta(t)$ of Eq. (1.309):

$$G_\omega(t) = e^{-i\omega t}\bar{\Theta}(t). \qquad (3.86)$$

The periodic Green function (3.84) can then be written as

$$G_\omega^{\mathrm{p}}(t) = \sum_{n=-\infty}^{\infty} e^{-i\omega[t-(t_b-t_a)n]}\bar{\Theta}(t - (t_b - t_a)n). \qquad (3.87)$$

Being periodic in $t_b - t_a$, its explicit evaluation can be restricted to the basic interval

$$t \in [0, t_b - t_a). \qquad (3.88)$$

Inside the interval $(0, t_b - t_a)$, the sum can be performed as follows:

$$
\begin{aligned}
G_\omega^{\mathrm{p}}(t) &= \sum_{n=-\infty}^{0} e^{-i\omega[t-(t_b-t_a)n]} = \frac{e^{-i\omega t}}{1 - e^{-i\omega(t_b-t_a)}} \\
&= -i\frac{e^{-i\omega[t-(t_b-t_a)/2]}}{2\sin[\omega(t_b-t_a)/2]}, \qquad t \in (0, t_b - t_a).
\end{aligned}
\qquad (3.89)
$$

At the point $t = 0$, the initial term with $\bar{\Theta}(0)$ contributes only $1/2$ so that

$$G_\omega^{\mathrm{p}}(0) = G_\omega^{\mathrm{p}}(0+) - \frac{1}{2}. \qquad (3.90)$$

Outside the basic interval (3.88), the Green function is determined by its periodicity. For instance,

$$G_\omega^{\mathrm{p}}(t) = -i\frac{e^{-i\omega[t+(t_b-t_a)/2]}}{2\sin[\omega(t_b-t_a)/2]}, \qquad t \in (-(t_b - t_a), 0). \qquad (3.91)$$

Note that as t crosses the upper end of the interval $[0, t_b - t_a)$, the sum in (3.87) picks up an additional term (the term with $n = 1$). This causes a jump in $G_\omega^{\mathrm{p}}(t)$ which enforces the periodicity. At the upper point $t = t_b - t_a$, there is again a reduction by $1/2$ so that $G_\omega^{\mathrm{p}}(t_b - t_a)$ lies in the middle of the jump, just as the value $1/2$ lies in the middle of the jump of the Heaviside function $\bar{\Theta}(t)$.

The periodic Green function is of great importance in the quantum statistics of Bose particles (see Chapter 7). After a continuation of the time to imaginary values, $t \to -i\tau$, $t_b - t_a \to -i\hbar/k_B T$, it takes the form

$$G_{\omega,\mathrm{e}}^{\mathrm{p}}(\tau) = \frac{1}{1 - e^{-\hbar\omega/k_B T}} e^{-\omega\tau}, \qquad \tau \in (0, \hbar\beta), \tag{3.92}$$

where the subscript e records the Euclidean character of the time. The prefactor is related to the average boson occupation number of a particle state of energy $\hbar\omega$, given by the *Bose-Einstein distribution function*

$$n_\omega^{\mathrm{b}} = \frac{1}{e^{\hbar\omega/k_B T} - 1}. \tag{3.93}$$

In terms of it,

$$G_{\omega,\mathrm{e}}^{\mathrm{p}}(\tau) = (1 + n_\omega^{\mathrm{b}})e^{-\omega\tau}, \qquad \tau \in (0, \hbar\beta). \tag{3.94}$$

The τ-behavior of the subtracted periodic Green function $G_{\omega,\mathrm{e}}^{\mathrm{p}\prime}(\tau) \equiv G_{\omega,\mathrm{e}}^{\mathrm{p}}(\tau) - 1/\hbar\beta\omega$ is shown in Fig. 3.2.

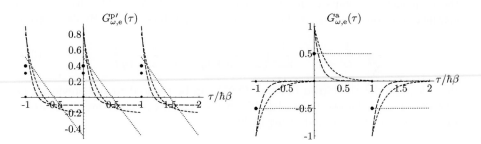

Figure 3.2 Subtracted periodic Green function $G_{\omega,\mathrm{e}}^{\mathrm{p}\prime} \equiv G_{\omega,\mathrm{e}}^{\mathrm{p}}(\tau) - 1/\hbar\beta\omega$ and antiperiodic Green function $G_{\omega,\mathrm{e}}^{\mathrm{a}}(\tau)$ for frequencies $\omega = (0, 5, 10)/\hbar\beta$ (with increasing dash length). The points show the values at the jumps of the three functions (with increasing point size) corresponding to the relation (3.90).

As a next step, we consider a Green function $G_{\omega^2}^{\mathrm{p}}(t)$ associated with the second-order differential operator $-\partial_t^2 - \omega^2$,

$$G_{\omega^2}^{\mathrm{p}}(t, t') = (-\partial_t^2 - \omega^2)^{-1}\delta(t - t'), \qquad t - t' \in [t_a, t_b), \tag{3.95}$$

which satisfies the periodic boundary condition:

$$G_{\omega^2}^{\mathrm{p}}(t, t') \equiv G_{\omega^2}^{\mathrm{p}}(t - t') = G_{\omega^2}^{\mathrm{p}}(t - t' + t_b - t_a). \tag{3.96}$$

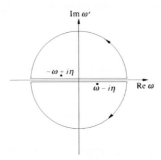

Figure 3.3   Two poles in Fourier transform of Green function $G_{\omega^2}^{\mathrm{P,a}}(t)$.

Just like $G_\omega^{\mathrm{P}}(t,t')$, this periodic Green function depends only on the time difference $t-t'$. It obviously has the spectral representation

$$G_{\omega^2}^{\mathrm{P}}(t) = \frac{1}{t_b - t_a} \sum_{m=-\infty}^{\infty} e^{-i\omega_m t} \frac{1}{\omega_m^2 - \omega^2}, \tag{3.97}$$

which makes sense as long as $t_b - t_a$ is not equal to an even multiple of π/ω. At infinite $t_b - t_a$, the sum becomes an integral over ω_m with singularities at $\pm\omega$ which must be avoided by an $i\eta$-prescription, which adds a negative imaginary part to the frequency ω [compare the discussion after Eq. (2.166)]. This fixes also the continuation from small $t_b - t_a$ beyond the multiple values of π/ω. By decomposing

$$\frac{1}{\omega'^2 - \omega^2 + i\eta} = \frac{1}{2i\omega} \left(\frac{i}{\omega' - \omega + i\eta} - \frac{i}{\omega' + \omega - i\eta} \right), \tag{3.98}$$

the calculation of the Green function (3.97) can be reduced to the previous case. The positions of the two poles of (3.98) in the complex ω'-plane are illustrated in Fig. 3.3. In this way we find, using (3.89),

$$\begin{aligned} G_{\omega^2}^{\mathrm{P}}(t) &= \frac{1}{2\omega i} [G_\omega^{\mathrm{P}}(t) - G_{-\omega}^{\mathrm{P}}(t)] \\ &= -\frac{1}{2\omega} \frac{\cos\omega[t - (t_b - t_a)/2]}{\sin[\omega(t_b - t_a)/2]}, \qquad t \in [0, t_b - t_a). \end{aligned} \tag{3.99}$$

In $G_{-\omega}^{\mathrm{P}}(t)$ one must keep the small negative imaginary part attached to the frequency ω. For an infinite time interval $t_b - t_a$, this leads to a Green function $G_{\omega^2}^{\mathrm{P}}(t - t')$: also

$$G_{-\omega}(t) = -e^{-i\omega t}\bar{\Theta}(-t). \tag{3.100}$$

The directional change in encircling the pole in the ω'-integral leads to the exchange $\bar{\Theta}(t) \to -\bar{\Theta}(-t)$.

Outside the basic interval $t \in [0, t_b - t_a)$, the function is determined by its periodicity. For $t \in [-(t_b - t_a), 0)$, we may simply replace t by $|t|$.

As a further step we consider another Green function $G^{\mathrm{a}}_\omega(t, t')$. It fulfills the same first-order differential equation $i\partial_t - \omega$ as $G^{\mathrm{p}}_\omega(t, t')$:

$$(i\partial_t - \omega)G^{\mathrm{a}}_\omega(t, t') = i\delta(t - t'), \quad t - t' \in [0, t_b - t_a), \tag{3.101}$$

but in contrast to $G^{\mathrm{p}}_\omega(t, t')$ it satisfies the antiperiodic boundary condition

$$G^{\mathrm{a}}_\omega(t, t') \equiv G^{\mathrm{a}}_\omega(t - t') = -G^{\mathrm{a}}_\omega(t - t' + t_b - t_a). \tag{3.102}$$

As for periodic boundary conditions, the Green function $G^{\mathrm{a}}_\omega(t, t')$ depends only on the time difference $t - t'$. In contrast to $G^{\mathrm{p}}_\omega(t, t')$, however, $G^{\mathrm{a}}_\omega(t, t')$ changes sign under a shift $t \to t + (t_b - t_a)$. The Fourier expansion of $G^{\mathrm{a}}_\omega(t - t')$ is

$$G^{\mathrm{a}}_\omega(t) = \frac{1}{t_b - t_a} \sum_{m=-\infty}^{\infty} e^{-i\omega^{\mathrm{f}}_m t} \frac{i}{\omega^{\mathrm{f}}_m - \omega}, \tag{3.103}$$

where the frequency sum covers the *odd* Matsubara-like frequencies

$$\omega^{\mathrm{f}}_m = \frac{\pi(2m + 1)}{t_b - t_a}. \tag{3.104}$$

The superscript f stands for *fermionic* since these frequencies play an important role in the statistical mechanics of particles with Fermi statistics to be explained in Section 7.10 [see Eq. (7.414)].

The antiperiodic Green functions are obtained from a sum similar to (3.82), but modified by an additional phase factor $e^{i\pi n} = (-)^n$. When inserted into the Poisson summation formula (3.81), such a phase is seen to select the half-integer numbers in the integral instead of the integer ones:

$$\sum_{m=-\infty}^{\infty} f(m + 1/2) = \int_{-\infty}^{\infty} d\mu \sum_{n=-\infty}^{\infty} (-)^n e^{2\pi i \mu n} f(\mu). \tag{3.105}$$

Using this formula, we can expand

$$\begin{aligned}
G^{\mathrm{a}}_\omega(t) &= \sum_{n=-\infty}^{\infty} \int_{-\infty}^{\infty} \frac{d\omega'}{2\pi} (-)^n e^{-i\omega'[t-(t_b-t_a)n]} \frac{i}{\omega' - \omega + i\eta} \\
&= \sum_{n=-\infty}^{\infty} (-)^n G_\omega(t - (t_b - t_a)n),
\end{aligned} \tag{3.106}$$

or, more explicitly,

$$G^{\mathrm{a}}_\omega(t) = \sum_{n=-\infty}^{\infty} e^{-i\omega[t-(t_b-t_a)n]} (-)^n \bar{\Theta}(t - (t_b - t_a)n). \tag{3.107}$$

For $t \in [0, t_b - t_a)$, this gives

$$\begin{aligned}
G^{\mathrm{a}}_\omega(t) &= \sum_{n=-\infty}^{0} e^{-i\omega[t-(t_b-t_a)n]} (-)^n = \frac{e^{i\omega t}}{1 + e^{-i\omega(t_b-t_a)}} \\
&= \frac{e^{-i\omega[t-(t_b-t_a)/2]}}{2\cos[\omega(t_b-t_a)/2]}, \quad t \in [0, t_b - t_a).
\end{aligned} \tag{3.108}$$

Outside the interval $t \in [0, t_b - t_a)$, the function is defined by its antiperiodicity. The τ-behavior of the antiperiodic Green function $G^a_{\omega,e}(\tau)$ is also shown in Fig. 3.2.

In the limit $\omega \to 0$, the right-hand side of (3.108) is equal to $1/2$, and the antiperiodicity implies that

$$G^a_0(t) = \frac{1}{2} \epsilon(t), \quad t \in [-(t_b - t_a), (t_b - t_a)]. \tag{3.109}$$

Antiperiodic Green functions play an important role in the quantum statistics of Fermi particles. After analytically continuing t to the imaginary time $-i\tau$ with $t_b - t_a \to -i\hbar/k_B T$, the expression (3.108) takes the form

$$G^a_{\omega,e}(\tau) = \frac{1}{1 + e^{-\hbar\omega/k_B T}} e^{-\omega\tau}, \quad \tau \in [0, \hbar\beta). \tag{3.110}$$

The prefactor is related to the average Fermi occupation number of a state of energy $\hbar\omega$, given by the *Fermi-Dirac distribution function*

$$n^f_\omega = \frac{1}{e^{\hbar\omega/k_B T} + 1}. \tag{3.111}$$

In terms of it,

$$G^a_{\omega,e}(\tau) = (1 - n^f_\omega)e^{-\omega\tau}, \quad \tau \in [0, \hbar\beta). \tag{3.112}$$

With the help of $G^a_\omega(t)$, we form the antiperiodic analog of (3.97), (3.99), i.e., the antiperiodic Green function associated with the second-order differential operator $-\partial^2_t - \omega^2$:

$$\begin{aligned} G^a_{\omega^2}(t) &= \frac{1}{t_b - t_a} \sum_{m=0}^\infty e^{-i\omega^f_m t} \frac{1}{\omega^{f\,2}_m - \omega^2} \\ &= \frac{1}{2\omega i}[G^a_\omega(t) - G^a_{-\omega}(t)] \\ &= -\frac{1}{2\omega} \frac{\sin\omega[t - (t_b - t_a)/2]}{\cos[\omega(t_b - t_a)/2]}, \quad t \in [0, t_b - t_a]. \end{aligned} \tag{3.113}$$

Outside the basic interval $t \in [0, t_b - t_a]$, the Green function is determined by its antiperiodicity. If, for example, $t \in [-(t_b - t_a), 0]$, one merely has to replace t by $|t|$.

Note that the Matsubara sums

$$G^p_{\omega^2,e}(0) = \frac{1}{\hbar\beta} \sum_{m=-\infty}^\infty \frac{1}{\omega^2_m + \omega^2}, \quad G^p_{\omega,e}(0) = \frac{1}{\hbar\beta} \sum_{m=-\infty}^\infty \frac{1}{\omega^{f\,2}_m + \omega^2}, \tag{3.114}$$

can also be calculated from the combinations of the simple Green functions (3.79) and (3.103):

$$\frac{1}{2\omega}\left[G^p_{\omega,e}(\eta) + G^p_{\omega,e}(-\eta)\right] = \frac{1}{2\omega}\left[G^p_{\omega,e}(\eta) + G^p_{\omega,e}(\hbar\beta - \eta)\right] = \frac{1}{2\omega}\left(1 + n^b_\omega\right)\left(1 + e^{-\beta\omega}\right)$$

$$= \frac{1}{2\omega} \coth \frac{\hbar\omega\beta}{2}, \tag{3.115}$$

$$\frac{1}{2\omega} \left[G^{\mathrm{a}}_{\omega,\mathrm{e}}(\eta) + G^{\mathrm{a}}_{\omega,\mathrm{e}}(-\eta) \right] = \frac{1}{2\omega} \left[G^{\mathrm{a}}_{\omega,\mathrm{e}}(\eta) - G^{\mathrm{a}}_{\omega,\mathrm{e}}(\hbar\beta - \eta) \right] = \frac{1}{2\omega} \left(1 - n^{\mathrm{f}}_\omega \right) \left(1 - e^{-\beta\omega} \right)$$

$$= \frac{1}{2\omega} \tanh \frac{\hbar\omega\beta}{2}, \tag{3.116}$$

where η is an infinitesimal positive number needed to specify on which side of the jump the Green functions $G^{\mathrm{p,a}}_{\omega,\mathrm{e}}(\tau)$ at $\tau = 0$ have to be evaluated (see Fig. 3.2).

3.3.2 Time-Dependent Frequency

The above results (3.89) and (3.108) for the periodic and antiperiodic Green functions of the first-order differential operator $(i\partial_t - \omega)$ can easily be found also for arbitrary time-dependent frequencies $\Omega(t)$, thus solving (3.75). We shall look for the retarded version which vanishes for $t < t'$. This property is guaranteed by the ansatz containing the Heaviside function (1.309):

$$G_\Omega(t,t') = \bar{\Theta}(t-t')g(t,t'). \tag{3.117}$$

Using the property (1.303) of the Heaviside function, that its time derivative yields the δ-function, and normalizing $g(t,t)$ to be equal to 1, we see that $g(t,t')$ must solve the homogenous differential equation

$$[i\partial_t - \Omega(t)] \, g(t,t') = 0. \tag{3.118}$$

The solution is

$$g(t,t') = K(t')e^{-i\int_c^t dt'' \, \Omega(t'')}. \tag{3.119}$$

The condition $g(t,t) = 1$ fixes $K(t) = e^{i\int_c^t dt'' \, \Omega(t'')}$, so that we obtain

$$G_\Omega(t,t') = \bar{\Theta}(t-t')e^{-i\int_{t'}^t dt'' \, \Omega(t'')}. \tag{3.120}$$

The most general Green function is a sum of this and an arbitrary solution of the homogeneous equation (3.118):

$$G_\Omega(t,t') = \left[\bar{\Theta}(t-t') + C(t') \right] e^{-i\int_{t'}^t dt'' \, \Omega(t'')}. \tag{3.121}$$

For a periodic frequency $\Omega(t)$ we impose periodic boundary conditions upon the Green function, setting $G_\Omega(t_a,t') = G_\Omega(t_b,t')$. This is ensured if for $t_b > t > t' > t_a$:

$$C(t')e^{-i\int_{t'}^{t_a} dt'' \, \Omega(t'')} = [1 + C(t')] \, e^{-i\int_{t'}^{t_b} dt'' \, \Omega(t'')}. \tag{3.122}$$

This equation is solved by a t'-independent $C(t')$:

$$C = n^{\mathrm{p}}_\Omega \equiv \frac{1}{e^{i\int_{t_a}^{t_b} dt'' \, \Omega(t'')} - 1}. \tag{3.123}$$

Hence we obtain the periodic Green function

$$G_\Omega^{\mathrm{p}}(t,t') = \left[\bar{\Theta}(t-t') + n_\Omega^{\mathrm{p}}\right] e^{-i\int_{t'}^{t} dt''\, \Omega(t'')}. \tag{3.124}$$

For antiperiodic boundary conditions we obtain the same equation with n_Ω^{p} replaced by $-n_\Omega^{\mathrm{a}}$ where

$$n_\Omega^{\mathrm{a}} \equiv \frac{1}{e^{i\int_{t_a}^{t_b} dt\, \Omega(t)} + 1}. \tag{3.125}$$

Note that a sign change in the time derivative of the first-order differential equation (3.75) to

$$[-i\partial_t - \Omega(t)]\, G_\Omega(t,t') = i\delta(t-t') \tag{3.126}$$

has the effect of interchanging in the time variable t and t' of the Green function Eq. (3.120).

If the frequency $\Omega(t)$ is a matrix, all exponentials have to be replaced by time-ordered exponentials [recall (1.252)]

$$e^{i\int_{t_a}^{t_b} dt\, \Omega(t)} \;\rightarrow\; \hat{T} e^{i\int_{t_a}^{t_b} dt\, \Omega(t)}. \tag{3.127}$$

As remarked in Subsection 2.15.4, this integral cannot, in general, be calculated explicitly. A simple formula is obtained only if the matrix $\Omega(t)$ varies only little around a fixed matrix Ω_0.

For imaginary times $\tau = it$ we generalize the results (3.92) and (3.110) for the periodic and antiperiodic imaginary-time Green functions of the first-order differential equation (3.76) to time-dependent periodic frequencies $\Omega(\tau)$. Here the Green function (3.120) becomes

$$G_\Omega(\tau,\tau') = \bar{\Theta}(\tau - \tau') e^{-\int_{\tau'}^{\tau} d\tau''\, \Omega(\tau'')}, \tag{3.128}$$

and the periodic Green function (3.124):

$$G_\Omega(\tau,\tau') = \left[\bar{\Theta}(\tau - \tau') + n^{\mathrm{b}}\right] e^{-\int_{\tau'}^{\tau} d\tau''\, \Omega(\tau'')}, \tag{3.129}$$

where

$$n^{\mathrm{b}} \equiv \frac{1}{e^{\int_0^{\hbar\beta} d\tau''\, \Omega(\tau'')} - 1} \tag{3.130}$$

is the generalization of the Bose distribution function in Eq. (3.93). For antiperiodic boundary conditions we obtain the same equation, except that the generalized Bose distribution function is replaced by the negative of the generalized Fermi distribution function in Eq. (3.111):

$$n^{\mathrm{f}} \equiv \frac{1}{e^{\int_0^{\hbar\beta} d\tau''\, \Omega(\tau'')} + 1}. \tag{3.131}$$

For the opposite sign of the time derivative in (3.128), the arguments τ and τ' are interchanged.

From the Green functions (3.124) or (3.128) we may find directly the trace of the logarithm of the operators $[-i\partial_t + \Omega(t)]$ or $[\partial_\tau + \Omega(\tau)]$. At imaginary time, we multiply $\Omega(\tau)$ with a strength parameter g, and use the formula

$$\mathrm{Tr}\log\left[\partial_\tau + g\Omega(\tau)\right] = \int_0^g dg'\, G_{g'\Omega}(\tau, \tau). \qquad (3.132)$$

Inserting on the right-hand side of Eq. (3.129), we find for $g = 1$:

$$\begin{aligned}
\mathrm{Tr}\log\left[\partial_\tau + \Omega(\tau)\right] &= \log\left\{2\sinh\left[\frac{1}{2}\int_0^{\hbar\beta} d\tau''\, \Omega(\tau'')\right]\right\} \\
&= \frac{1}{2}\int_0^{\hbar\beta} d\tau''\, \Omega(\tau'') + \log\left[1 - e^{-\int_0^{\hbar\beta} d\tau''\, \Omega(\tau'')}\right], \quad (3.133)
\end{aligned}$$

which reduces at low temperature to

$$\mathrm{Tr}\log\left[\partial_\tau + \Omega(\tau)\right] = \frac{1}{2}\int_0^{\hbar\beta} d\tau''\, \Omega(\tau''). \qquad (3.134)$$

The result is the same for the opposite sign of the time derivative and the trace of the logarithm is sensitive only to $\bar{\Theta}(\tau - \tau')$ at $\tau = \tau'$, where it is equal to $1/2$.

As an exercise for dealing with distributions it is instructive to rederive this result in the following perturbative way. For a positive $\Omega(\tau)$, we introduce an infinitesimal positive quantity η and decompose

$$\begin{aligned}
\mathrm{Tr}\log\left[\pm\partial_\tau + \Omega(\tau)\right] &= \mathrm{Tr}\log\left[\pm\partial_\tau + \eta\right] + \mathrm{Tr}\log\left[1 + (\pm\partial_\tau + \eta)^{-1}\Omega(\tau)\right] \quad (3.135) \\
&= \mathrm{Tr}\log\left[\pm\partial_\tau + \eta\right] + \mathrm{Tr}\log\left[1 + (\pm\partial_\tau + \eta)^{-1}\Omega(\tau)\right].
\end{aligned}$$

The first term $\mathrm{Tr}\log\left[\pm\partial_\tau + \eta\right] = \mathrm{Tr}\log\left[\pm\partial_\tau + \eta\right] = \int_{-\infty}^{\infty} d\omega \log\omega$ vanishes since $\int_{-\infty}^{\infty} d\omega \log\omega = 0$ in dimensional regularization by Veltman's rule [see (2.506)]. Using the Green functions

$$\left[\pm\partial_\tau + \eta\right]^{-1}(\tau, \tau') = \left\{\begin{matrix} \bar{\Theta}(\tau - \tau') \\ \bar{\Theta}(\tau' - \tau) \end{matrix}\right\}, \qquad (3.136)$$

the second term can be expanded in a Taylor series

$$\sum_{n=1}^{\infty} \frac{(-1)^{n+1}}{n}\int d\tau_1\cdots d\tau_n\, \Omega(\tau_1)\bar{\Theta}(\tau_1 - \tau_2)\Omega(\tau_2)\bar{\Theta}(\tau_2 - \tau_3)\cdots\Omega(\tau_n)\bar{\Theta}(\tau_n - \tau_1). \quad (3.137)$$

For the lower sign of $\pm\partial_\tau$, the Heaviside functions have reversed arguments $\tau_2 - \tau_1, \tau_3 - \tau_2, \ldots, \tau_1 - \tau_n$. The integrals over a cyclic product of Heaviside functions in (3.137) are zero since the arguments $\tau_1, \ldots, \tau_n$ are time-ordered which makes the argument of the last factor $\bar{\Theta}(\tau_n - \tau_1)$ [or $\bar{\Theta}(\tau_1 - \tau_n)$] negative and thus $\bar{\Theta}(\tau_n - \tau_1) = 0$ [or $\bar{\Theta}(\tau_1 - \tau_n)$]. Only the first term survives yielding

$$\int d\tau_1\, \Omega(\tau_1)\bar{\Theta}(\tau_1 - \tau_1) = \frac{1}{2}\int d\tau\, \Omega(\tau), \qquad (3.138)$$

such that we re-obtain the result (3.134).

This expansion (3.133) can easily be generalized to an arbitrary matrix $\Omega(\tau)$ or a time-dependent operator, $\hat{H}(\tau)$. Since $\hat{H}(\tau)$ and $\hat{H}(\tau')$ do not necessarily commute, the generalization is

$$\text{Tr}\log[\hbar\partial_\tau + \hat{H}(\tau)] = \frac{1}{2\hbar}\text{Tr}\left[\int_0^{\hbar\beta} d\tau\,\hat{H}(\tau)\right] - \sum_{n=1}^{\infty}\frac{1}{n}\text{Tr}\left[\hat{T}e^{-n\int_0^{\hbar\beta} d\tau''\,\hat{H}(\tau'')/\hbar}\right], \quad (3.139)$$

where $\hat{T}$ is the time ordering operator (1.241). Each term in the sum contains a power of the time evolution operator (1.255).

3.4 Summing Spectral Representation of Green Function

After these preparations we are ready to perform the spectral sum (3.70) for the Green function of the differential equation of second order with Dirichlet boundary conditions. Setting $t_2 \equiv t_b - t$, $t_1 \equiv t' - t_a$, we rewrite (3.70) as

$$
\begin{aligned}
G_{\omega^2}(t,t') &= \frac{2}{t_b - t_a}\sum_{n=1}^{\infty}\frac{(-1)^{n+1}}{(2i)^2}\frac{(e^{i\nu_n t_2} - e^{-i\nu_n t_2})(e^{i\nu_n t_1} - e^{-i\nu_n t_1})}{\nu_n^2 - \omega^2} \\
&= \frac{1}{2}\frac{1}{t_b - t_a}\sum_{n=1}^{\infty}(-1)^n\frac{\left[(e^{-i\nu_n(t_2+t_1)} - e^{-i\nu_n(t_2-t_1)}) + \text{c.c.}\right]}{\nu_n^2 - \omega^2} \\
&= \frac{1}{2}\frac{1}{t_b - t_a}\sum_{n=-\infty}^{\infty}(-1)^n\frac{e^{-i\nu_n(t_2+t_1)} - e^{-i\nu_n(t_2-t_1)}}{\nu_n^2 - \omega^2}.
\end{aligned}
\quad (3.140)
$$

We now separate even and odd frequencies ν_n and write these as bosonic and fermionic Matsubara frequencies $\omega_m = \nu_{2m}$ and $\omega_m^{\text{f}} = \nu_{2m+1}$, respectively, recalling the definitions (3.80) and (3.104). In this way we obtain

$$
\begin{aligned}
G_{\omega^2}(t,t') = \frac{1}{2}\Bigg\{ &\frac{1}{t_b - t_a}\sum_{m=-\infty}^{\infty}\frac{e^{-i\omega_m(t_2+t_1)}}{\omega_m^2 - \omega^2} - \frac{1}{t_b - t_a}\sum_{m=-\infty}^{\infty}\frac{e^{-i\omega_m^{\text{f}}(t_2+t_1)}}{\omega_m^{\text{f}\,2} - \omega^2} \\
&- \frac{1}{t_b - t_a}\sum_{m=-\infty}^{\infty}\frac{e^{-i\omega_m(t_2-t_1)}}{\omega_m^2 - \omega^2} + \frac{1}{t_b - t_a}\sum_{m=-\infty}^{\infty}\frac{e^{-i\omega_m^{\text{f}}(t_2-t_1)}}{\omega_m^{\text{f}\,2} - \omega^2}\Bigg\}.
\end{aligned}
\quad (3.141)
$$

Inserting on the right-hand side the periodic and antiperiodic Green functions (3.99) and (3.108), we obtain the decomposition

$$G_{\omega^2}(t,t') = \frac{1}{2}\left[G_\omega^{\text{p}}(t_2+t_1) - G_\omega^{\text{a}}(t_2+t_1) - G_\omega^{\text{p}}(t_2-t_1) + G_\omega^{\text{a}}(t_2-t_1)\right]. \quad (3.142)$$

Using (3.99) and (3.113) we find that

$$G_\omega^{\text{p}}(t_2+t_1) - G_\omega^{\text{p}}(t_2-t_1) = \frac{\sin\omega[t_2 - (t_b - t_a)/2]\sin\omega t_1}{\omega\sin[\omega(t_b - t_a)/2]}, \quad (3.143)$$

$$G_\omega^{\text{a}}(t_2+t_1) - G_\omega^{\text{a}}(t_2-t_1) = -\frac{\cos\omega[t_2 - (t_b - t_a)/2]\sin\omega t_1}{\omega\cos[\omega(t_b - t_a)/2]}, \quad (3.144)$$

such that (3.142) becomes

$$G_{\omega^2}(t,t') = \frac{1}{\omega \sin \omega(t_b - t_a)} \sin \omega t_2 \sin \omega t_1, \tag{3.145}$$

in agreement with the earlier result (3.36).

An important limiting case is

$$t_a \to -\infty, \quad t_b \to \infty. \tag{3.146}$$

Then the boundary conditions become irrelevant and the Green function reduces to

$$G_{\omega^2}(t,t') = -\frac{i}{2\omega} e^{-i\omega|t-t'|}, \tag{3.147}$$

which obviously satisfies the second-order differential equation

$$(-\partial_t^2 - \omega^2)G_{\omega^2}(t,t') = \delta(t - t'). \tag{3.148}$$

The periodic and antiperiodic Green functions $G_{\omega^2}^{\mathrm{p}}(t,t')$ and $G_{\omega^2}^{\mathrm{a}}(t,t')$ at finite $t_b - t_a$ in Eqs. (3.99) and (3.113) are obtained from $G_{\omega^2}(t,t')$ by summing over all periodic repetitions [compare (3.106)]

$$
\begin{aligned}
G_{\omega^2}^{\mathrm{p}}(t,t') &= \sum_{n=-\infty}^{\infty} G(t + n(t_b - t_a), t'), \\
G_{\omega^2}^{\mathrm{a}}(t,t') &= \sum_{n=-\infty}^{\infty} (-1)^n G_{\omega^2}(t + n(t_b - t_a), t').
\end{aligned}
\tag{3.149}
$$

For completeness let us also sum the spectral representation with the normalized wave functions [compare (3.98)–(3.69)]

$$x_0(t) = \sqrt{\frac{1}{t_b - t_a}}, \qquad x_n(t) = \sqrt{\frac{2}{t_b - t_a}} \cos \nu_n(t - t_a), \tag{3.150}$$

which reads:

$$G_{\omega^2}^{\mathrm{N}}(t,t') = \frac{2}{t_b - t_a} \left[-\frac{1}{2\omega^2} + \sum_{n=1}^{\infty} \frac{\cos \nu_n(t - t_a) \cos \nu_n(t' - t_a)}{\nu_n^2 - \omega^2} \right]. \tag{3.151}$$

It satisfies the *Neumann boundary conditions*

$$\partial_t G_{\omega^2}^{\mathrm{N}}(t,t')\big|_{t=t_b} = 0, \qquad \partial_{t'} G_{\omega^2}^{\mathrm{N}}(t,t')\big|_{t'=t_a} = 0. \tag{3.152}$$

The spectral representation (3.151) can be summed by a decomposition (3.140), if that the lowest line has a plus sign between the exponentials, and (3.142) becomes

$$G_{\omega^2}^{\mathrm{N}}(t,t') = \frac{1}{2} \left[G_\omega^{\mathrm{p}}(t_2 + t_1) - G_\omega^{\mathrm{a}}(t_2 + t_1) + G_\omega^{\mathrm{p}}(t_2 - t_1) - G_\omega^{\mathrm{a}}(t_2 - t_1) \right]. \tag{3.153}$$

Using now (3.99) and (3.113) we find that

$$G_\omega^p(t_2 + t_1) + G_\omega^p(t_2 - t_1) = -\frac{\cos\omega[t_2 - (t_b - t_a)/2]\cos\omega t_1}{\omega\sin[\omega(t_b - t_a)/2]}, \tag{3.154}$$

$$G_\omega^a(t_2 + t_1) + G_\omega^a(t_2 - t_1) = -\frac{\sin\omega[t_2 - (t_b - t_a)/2]\cos\omega t_1}{\omega\cos[\omega(t_b - t_a)/2]}, \tag{3.155}$$

and we obtain instead of (3.145):

$$G_{\omega^2}^N(t, t') = -\frac{1}{\omega\sin\omega(t_b - t_a)}\cos\omega(t_b - t_>)\,\cos\omega(t_< - t_a), \tag{3.156}$$

which has the small-ω expansion

$$G_{\omega^2}^N(t, t') \underset{\omega^2 \approx 0}{\approx} -\frac{1}{(t_b - t_a)\omega^2} + \frac{t_b - t_a}{3} - \frac{1}{2}|t - t'| - \frac{1}{2}(t + t') + \frac{1}{2(t_b - t_a)}\left(t^2 + t'^2\right). \tag{3.157}$$

3.5 Wronski Construction for Periodic and Antiperiodic Green Functions

The Wronski construction in Subsection 3.2.1 of Green functions with time-dependent frequency $\Omega(t)$ satisfying the differential equation (3.27)

$$[-\partial_t^2 - \Omega^2(t)]G_{\Omega^2}(t, t') = \delta(t - t') \tag{3.158}$$

can easily be carried over to the Green functions $G_{\Omega^2}^{p,a}(t, t')$ with periodic and antiperiodic boundary conditions. As in Eq. (3.53) we decompose

$$G_{\Omega^2}^{p,a}(t, t') = \bar{\Theta}(t - t')\Delta(t, t') + a(t')\xi(t) + b(t')\eta(t), \tag{3.159}$$

with independent solutions of the homogenous equations $\xi(t)$ and $\eta(t)$, and insert this into (3.27), where $\delta^{p,a}(t - t')$ is the periodic version of the δ-function

$$\delta^{p,a}(t - t') \equiv \sum_{n=-\infty}^{\infty} \delta(t - t' - n\hbar\beta)\left\{\begin{matrix} 1 \\ (-1)^n \end{matrix}\right\}, \tag{3.160}$$

and $\Omega(t)$ is assumed to be periodic or antiperiodic in $t_b - t_a$. This yields again for $\Delta(t, t')$ the homogeneous initial-value problem (3.46), (3.45),

$$[-\partial_t^2 - \Omega^2(t)]\Delta(t, t') = 0; \quad \Delta(t, t) = 0, \quad \partial_t\Delta(t, t')|_{t'=t} = -1. \tag{3.161}$$

The periodic boundary conditions lead to the system of equations

$$\begin{aligned} a(t)[\xi(t_b) \mp \xi(t_a)] + b(t)[\eta(t_b) \mp \eta(t_a)] &= -\Delta(t_b, t), \\ a(t)[\dot{\xi}(t_b) \mp \dot{\xi}(t_a)] + b(t)[\dot{\eta}(t_b) \mp \dot{\eta}(t_a)] &= -\partial_t\Delta(t_b, t). \end{aligned} \tag{3.162}$$

Defining now the constant 2×2 -matrices

$$\bar{\Lambda}^{\mathrm{p,a}}(t_a, t_b) = \begin{pmatrix} \xi(t_b) \mp \xi(t_a) & \eta(t_b) \mp \eta(t_a) \\ \dot{\xi}(t_b) \mp \dot{\xi}(t_a) & \dot{\eta}(t_b) \mp \dot{\eta}(t_a) \end{pmatrix}, \tag{3.163}$$

the condition analogous to (3.58),

$$\det \bar{\Lambda}^{\mathrm{p,a}}(t_a, t_b) = W \, \bar{\Delta}^{\mathrm{p,a}}(t_a, t_b) \neq 0, \tag{3.164}$$

with

$$\bar{\Delta}^{\mathrm{p,a}}(t_a, t_b) = 2 \pm \partial_t \Delta(t_a, t_b) \pm \partial_t \Delta(t_b, t_a), \tag{3.165}$$

enables us to obtain the unique solution to Eqs. (3.162). After some algebra using the identities (3.51) and (3.52), the expression (3.159) for Green functions with periodic and antiperiodic boundary conditions can be cast into the form

$$G_{\Omega^2}^{\mathrm{p,a}}(t, t') = G_{\Omega^2}(t, t') \mp \frac{[\Delta(t, t_a) \pm \Delta(t_b, t)][\Delta(t', t_a) \pm \Delta(t_b, t')]}{\bar{\Delta}^{\mathrm{p,a}}(t_a, t_b)\Delta(t_a, t_b)}, \tag{3.166}$$

where $G_{\Omega^2}(t, t')$ is the Green function (3.59) with Dirichlet boundary conditions. As in (3.59) we may replace the functions on the right-hand side by the solutions $D_a(t)$ and $D_b(t)$ defined in Eqs. (2.226) and (2.227) with the help of (3.60).

The right-hand side of (3.166) is well-defined unless the operator $K(t) = -\partial_t^2 - \Omega^2(t)$ has a zero-mode, say $\eta(t)$, with periodic or antiperiodic boundary conditions $\eta(t_b) = \pm\eta(t_a)$, $\dot{\eta}(t_b) = \pm\dot{\eta}(t_a)$, which would make the determinant of the 2×2 -matrix $\bar{\Lambda}^{\mathrm{p,a}}$ vanish.

3.6   Time Evolution Amplitude in Presence of Source Term

Given the Green function $G_{\omega^2}(t, t')$, we can write down an explicit expression for the time evolution amplitude. The quadratic source contribution to the fluctuation factor (3.21) is given explicitly by

$$\mathcal{A}_{j,\mathrm{fl}} = -\frac{1}{2M} \int_{t_a}^{t_b} dt \int_{t_a}^{t_b} dt' \, G_{\omega^2}(t, t') \, j(t) j(t') \tag{3.167}$$

$$= -\frac{1}{M} \frac{1}{\omega \sin \omega(t_b - t_a)} \int_{t_a}^{t_b} dt \int_{t_a}^{t} dt' \, \sin \omega(t_b - t) \sin \omega(t' - t_a) j(t) j(t').$$

Altogether, the path integral in the presence of an external source $j(t)$ reads

$$(x_b t_b | x_a t_a)_\omega^j = \int \mathcal{D}x \, \exp\left\{ \frac{i}{\hbar} \int_{t_a}^{t_b} dt \left[\frac{M}{2} (\dot{x}^2 - \omega^2 x^2) + jx \right] \right\} = e^{(i/\hbar)\mathcal{A}_{j,\mathrm{cl}}} F_{\omega,j}(t_b, t_a), \tag{3.168}$$

with a total classical action

$$\mathcal{A}_{j,\mathrm{cl}} = \frac{1}{2} \frac{M\omega}{\sin \omega(t_b - t_a)} \left[(x_b^2 + x_a^2) \cos \omega(t_b - t_a) - 2x_b x_a \right]$$

$$+ \frac{1}{\sin \omega(t_b - t_a)} \int_{t_a}^{t_b} dt [x_a \sin \omega(t_b - t) + x_b \sin \omega(t - t_a)] j(t), \tag{3.169}$$

and the fluctuation factor composed of (2.169) and a contribution from the current term $e^{i\mathcal{A}_{j,\mathrm{fl}}/\hbar}$:

$$F_{\omega,j}(t_b,t_a) = F_\omega(t_b,t_a)e^{i\mathcal{A}_{j,\mathrm{fl}}/\hbar} = \frac{1}{\sqrt{2\pi i\hbar/M}}\sqrt{\frac{\omega}{\sin\omega(t_b-t_a)}}$$

$$\times \exp\left\{-\frac{i}{\hbar M\omega\sin\omega(t_b-t_a)}\int_{t_a}^{t_b}dt\int_{t_a}^t dt'\sin\omega(t_b-t)\sin\omega(t'-t_a)j(t)j(t')\right\}. \quad (3.170)$$

This expression is easily generalized to arbitrary time-dependent frequencies. Using the two independent solutions $D_a(t)$ and $D_b(t)$ of the homogenous differential equations (3.48), which were introduced in Eqs. (2.226) and (2.227), we find for the action (3.169) the general expression, composed of the harmonic action (2.266) and the current term $\int_{t_a}^{t_b}dt x_{\mathrm{cl}}(t)j(t)$ with the classical solution (2.246):

$$\mathcal{A}_{j,\mathrm{cl}} = \frac{M}{2D_a(t_b)}\left[x_b^2\dot{D}_a(t_b)-x_a^2\dot{D}_b(t_a)-2x_bx_a\right] + \frac{1}{D_a(t_b)}\int_{t_a}^{t_b}dt\,[x_bD_a(t)+x_aD_b(t)]j(t). \quad (3.171)$$

The fluctuation factor is composed of the expression (2.261) for the current-free action, and the generalization of (3.167) with the Green function (3.61):

$$F_{\omega,j}(t_b,t_a) = F_\omega(t_b,t_a)e^{i\mathcal{A}_{j,\mathrm{fl}}/\hbar} = \frac{1}{\sqrt{2\pi i\hbar/M}}\frac{1}{\sqrt{D_a(t_b)}}\exp\left\{-\frac{i}{2\hbar MD_a(t_b)}\right.$$

$$\times \left. \int_{t_a}^{t_b}dt\int_{t_a}^t dt'\,j(t)\left[\bar{\Theta}(t-t')D_b(t)D_a(t')+\bar{\Theta}(t'-t)D_a(t)D_b(t')\right]j(t')\right\}. \quad (3.172)$$

For applications to statistical mechanics which becomes possible after an analytic continuation to imaginary times, it is useful to write (3.169) and (3.170) in another form. We introduce the Fourier transforms of the current

$$A(\omega) \equiv \frac{1}{M\omega}\int_{t_a}^{t_b}dt\,e^{-i\omega(t-t_a)}j(t), \quad (3.173)$$

$$B(\omega) \equiv \frac{1}{M\omega}\int_{t_a}^{t_b}dt\,e^{-i\omega(t_b-t)}j(t) = -e^{-i\omega(t_b-t_a)}A(-\omega), \quad (3.174)$$

and see that the classical source term in the exponent of (3.168) can be written as

$$\mathcal{A}_{j,\mathrm{cl}} = -i\frac{M\omega}{\sin\omega(t_b-t_a)}\left\{\left[x_b(e^{i\omega(t_b-t_a)}A-B)\right]+x_a(e^{i\omega(t_b-t_a)}B-A)\right\}. \quad (3.175)$$

The source contribution to the quadratic fluctuations in Eq. (3.167), on the other hand, can be rearranged to yield

$$\mathcal{A}_{j,\mathrm{fl}} = \frac{i}{4M\omega}\int_{t_a}^{t_b}dt\int_{t_b}^t dt'\,e^{-i\omega|t-t'|}j(t)j(t')-\frac{M\omega}{2\sin\omega(t_b-t_a)}\left[e^{i\omega(t_b-t_a)}(A^2+B^2)-2AB\right]. \quad (3.176)$$

This is seen as follows: We write the Green function between $j(t), j(t')$ in (3.168) as

$$- \left[\sin \omega(t_b - t) \sin \omega(t' - t_a) \bar{\Theta}(t - t') + \sin \omega(t_b - t') \sin \omega(t - t_a) \bar{\Theta}(t' - t) \right]$$
$$= \frac{1}{4} \left[\left(e^{i\omega(t_b - t_a)} e^{-i\omega(t - t')} + \text{c.c.} \right) - \left(e^{i\omega(t_b + t_a)} e^{-i\omega(t + t')} + \text{c.c.} \right) \right] \bar{\Theta}(t - t')$$
$$+ \left\{ t \leftrightarrow t' \right\}. \tag{3.177}$$

Using $\bar{\Theta}(t - t') + \bar{\Theta}(t' - t) = 1$, this becomes

$$\frac{1}{4} \left\{ - \left(e^{i\omega(t_b + t_a)} e^{-i\omega(t' + t)} + \text{c.c.} \right) \right.$$
$$+ e^{i\omega(t_b - t_a)} \left(e^{-i\omega(t - t')} \bar{\Theta}(t - t') + e^{-i\omega(t' - t)} \bar{\Theta}(t' - t) \right) \tag{3.178}$$
$$\left. + e^{-i\omega(t_b - t_a)} \left[e^{i\omega(t - t')} (1 - \bar{\Theta}(t' - t)) + e^{i\omega(t' - t)} (1 - \bar{\Theta}(t - t')) \right] \right\}.$$

A multiplication by $j(t), j(t')$ and an integration over the times t, t' yield

$$\frac{1}{4} \left[- e^{i\omega(t_b - t_a)} 4M^2 \omega^2 (B^2 + A^2) \right. \tag{3.179}$$
$$\left. + \left(e^{i\omega(t_b - t_a)} - e^{-i\omega(t_b - t_a)} \right) \int_{t_a}^{t_b} dt \int_{t_b}^{t_b} dt' e^{-i\omega|t - t'|} j(t) j(t') + 4M^2 \omega^2 2AB \right],$$

thus leading to (3.176).

If the source $j(t)$ is time-independent, the integrals in the current terms of the exponential of (3.169) and (3.170) can be done, yielding the j-dependent exponent

$$\frac{i}{\hbar} \mathcal{A}_j = \frac{i}{\hbar} \left(\mathcal{A}_{j,\text{cl}} + \mathcal{A}_{j,\text{fl}} \right) = \frac{i}{\hbar} \left\{ \frac{1}{\omega \sin \omega(t_b - t_a)} [1 - \cos \omega(t_b - t_a)](x_b + x_a) j \right.$$
$$\left. + \frac{1}{2M\omega^3} \left[\omega(t_b - t_a) + 2 \frac{\cos \omega(t_b - t_a) - 1}{\sin \omega(t_b - t_a)} \right] j^2 \right\}. \tag{3.180}$$

Substituting $(1 - \cos \alpha)$ by $\sin \alpha \tan(\alpha/2)$, this yields the total source action becomes

$$\mathcal{A}_j = \frac{1}{\omega} \tan \frac{\omega(t_b - t_a)}{2} (x_b + x_a) j + \frac{1}{2M\omega^3} \left[\omega (t_b - t_a) - 2 \tan \frac{\omega(t_b - t_a)}{2} \right] j^2. \tag{3.181}$$

This result could also have been obtained more directly by taking the potential plus a constant-current term in the action

$$- \int_{t_a}^{t_b} dt \left(\frac{M}{2} \omega^2 x^2 - xj \right), \tag{3.182}$$

and by completing it quadratically to the form

$$- \int_{t_a}^{t_b} dt \frac{M}{2} \omega^2 \left(x - \frac{j}{M\omega^2} \right)^2 + \frac{t_b - t_a}{2M\omega^2} j^2. \tag{3.183}$$

This is a harmonic potential shifted in x by $-j/M\omega^2$. The time evolution amplitude can thus immediately be written down as

$$(x_b t_b | x_a t_a)_\omega^{j=\text{const}} = \sqrt{\frac{M\omega}{2\pi i\hbar \sin \omega(t_b - t_a)}} \exp\left(\frac{i}{2\hbar} \frac{M\omega}{\sin \omega(t_b - t_a)}\right.$$

$$\times \left\{\left[\left(x_b - \frac{j}{M\omega^2}\right)^2 + \left(x_a - \frac{j}{M\omega^2}\right)^2\right] \cos \omega(t_b - t_a)\right. \qquad (3.184)$$

$$\left.\left. -2\left(x_b - \frac{j}{M\omega^2}\right)\left(x_a - \frac{j}{M\omega^2}\right)\right\} + \frac{i}{\hbar} \frac{t_b - t_a}{2M\omega^2} j^2\right).$$

In the free-particle limit $\omega \to 0$, the result becomes particularly simple:

$$(x_b t_a | x_a t_a)_0^{j=\text{const}} = \frac{1}{\sqrt{2\pi i\hbar(t_b - t_a)/M}} \exp\left[\frac{i}{\hbar} \frac{M}{2} \frac{(x_b - x_a)^2}{t_b - t_a}\right]$$

$$\times \exp\left\{\frac{i}{\hbar}\left[\frac{1}{2}(x_b + x_a)(t_b - t_a)j - \frac{1}{24M}(t_b - t_a)^3 j^2\right]\right\}. \quad (3.185)$$

As a cross check, we verify that the total exponent is equal to $i/\hbar$ times the classical action

$$\mathcal{A}_{j,\text{cl}} = \int_{t_a}^{t_b} dt \left(\frac{M}{2}\dot{x}_{j,\text{cl}}^2 + jx_{j,\text{cl}}\right), \qquad (3.186)$$

calculated for the classical orbit $x_{j,\text{cl}}(t)$ connecting x_a and x_b in the presence of the constant current j. This satisfies the Euler-Lagrange equation

$$\ddot{x}_{j,\text{cl}} = j/M, \qquad (3.187)$$

which is solved by

$$x_{j,\text{cl}}(t) = x_a + \left[x_b - x_a - \frac{j}{2M}(t_b - t_a)^2\right]\frac{t - t_a}{t_b - t_a} + \frac{j}{2M}(t - t_a)^2. \qquad (3.188)$$

Inserting this into the action yields

$$\mathcal{A}_{j,\text{cl}} = \frac{M}{2}\frac{(x_b - x_a)^2}{t_b - t_a} + \frac{1}{2}(x_b + x_a)(t_b - t_a)j - \frac{(t_b - t_a)^3}{24}\frac{j^2}{M}, \qquad (3.189)$$

just as in the exponent of (3.185).

Let us remark that the calculation of the oscillator amplitude $(x_a t_b | x_a t)_\omega^j$ in (3.168) could have proceeded alternatively by using the orbital separation

$$x(t) = x_{j,\text{cl}}(t) + \delta x(t), \qquad (3.190)$$

where $x_{j,\text{cl}}(t)$ satisfies the Euler-Lagrange equations with the time-dependent source term

$$\ddot{x}_{j,\text{cl}}(t) + \omega^2 x_{j,\text{cl}}(t) = j(t)/M, \qquad (3.191)$$

rather than the orbital separation of Eq. (3.7),

$$x(t) = x_{\rm cl}(t) + \delta x(t),$$

where $x_{\rm cl}(t)$ satisfied the Euler-Lagrange equation with no source. For this *inhomogeneous* differential equation we would have found the following solution passing through x_a at $t = t_a$ and x_b at $t = t_b$:

$$x_{j,\rm cl}(t) = x_a \frac{\sin \omega(t_b - t)}{\sin \omega(t_b - t_a)} + x_b \frac{\sin \omega(t - t_a)}{\sin \omega(t_b - t_a)} + \frac{1}{M} \int_{t_a}^{t_b} dt' G_{\omega^2}(t, t') j(t'). \quad (3.192)$$

The Green function $G_{\omega^2}(t, t')$ appears now at the classical level. The separation (3.190) in the total action would have had the advantage over (3.7) that the source causes no linear term in $\delta x(t)$. Thus, there would be no need for a quadratic completion; the classical action would be found from a pure surface term plus one half of the source part of the action

$$
\begin{aligned}
\mathcal{A}_{\rm cl} &= \int_{t_a}^{t_b} dt \left[\frac{M}{2}(\dot{x}_{\rm cl,j}^2 - \omega^2 x_{j,\rm cl}^2) + j x_{j,\rm cl} \right] = \frac{M}{2} x_{j,\rm cl} \dot{x}_{j,\rm cl} \Big|_{t_a}^{t_b} \\
&+ \int_{t_a}^{t_b} dt \left[\frac{M}{2} x_{j,\rm cl} \left(-\ddot{x}_{j,\rm cl} - \omega^2 x_{j,\rm cl} + \frac{j}{M} \right) \right] + \frac{1}{2} \int_{t_a}^{t_b} dt x_{j,\rm cl} j \\
&= \frac{M}{2}(x_b \dot{x}_b - x_a \dot{x}_a) \Big|_{x = x_{j,\rm cl}} + \frac{1}{2} \int_{t_a}^{t_b} dt x_{j,\rm cl}(t) j(t). \quad (3.193)
\end{aligned}
$$

Inserting $x_{j,\rm cl}$ from (3.192) and $G_{\omega^2}(t, t')$ from (3.36) leads once more to the exponent in (3.168). The fluctuating action quadratic in $\delta x(t)$ would have given the same fluctuation factor as in the $j = 0$-case, i.e., the prefactor in (3.168) with no further j^2 (due to the absence of a quadratic completion).

3.7 Time Evolution Amplitude at Fixed Path Average

Another interesting quantity to be needed in Chapter 15 is the Fourier transform of the amplitude (3.184):

$$(x_b t_b | x_a t_a)_\omega^{x_0} = (t_b - t_a) \int_{-\infty}^{\infty} \frac{dj}{2\pi\hbar} e^{-ij(t_b - t_a)x_0/\hbar} (x_b t_b | x_a t_a)_\omega^j. \quad (3.194)$$

This is the amplitude for a particle to run from x_a to x_b along restricted paths whose temporal average $\bar{x} \equiv (t_b - t_a)^{-1} \int_{t_a}^{t_b} dt\, x(t)$ is held fixed at x_0:

$$(x_b t_b | x_a t_a)_\omega^{x_0} = \int \mathcal{D}x\, \delta(x_0 - \bar{x}) \exp \left\{ \frac{i}{\hbar} \int_{t_a}^{t_b} dt \frac{M}{2}(\dot{x}^2 - \omega^2 x^2) \right\}. \quad (3.195)$$

This property of the paths follows directly from the fact that the integral over the time-independent source j (3.194) produces a δ-function $\delta((t_b - t_a)x_0 - \int_{t_a}^{t_b} dt\, x(t))$. Restricted amplitudes of this type will turn out to have important applications later in Subsection 3.25.1 and in Chapters 5, 10, and 15.

The integral over j in (3.194) is done after a quadratic completion in $\mathcal{A}_j - j(t_b - t_a)x_0$ with $\mathcal{A}_j$ of (3.181):

$$\mathcal{A}_j - j(t_b - t_a)x_0 = \frac{1}{2M\omega^3}\left[\omega\,(t_b - t_a) - 2\tan\frac{\omega(t_b - t_a)}{2}\right](j - j_0)^2 + \mathcal{A}^{x_0}, \quad (3.196)$$

with

$$j_0 = \frac{M\omega^2}{\omega(t_b - t_a) - 2\tan\frac{\omega(t_b - t_a)}{2}}\left[\omega(t_b - t_a)\,x_0 - \tan\frac{\omega(t_b - t_a)}{2}(x_b + x_a)\right],$$

and

$$\mathcal{A}^{x_0} = -\frac{M\omega}{2\left[\omega(t_b - t_a) - 2\tan\frac{\omega(t_b - t_a)}{2}\right]}\left[\omega(t_b - t_a)x_0 - \tan\frac{\omega(t_b - t_a)}{2}(x_a + x_b)\right]^2.$$

With the completed quadratic exponent (3.196), the Gaussian integral over j in (3.194) can immediately be done, yielding

$$(x_b t_b | x_a t_a)^{x_0}_\omega = (x_b t_b | x_a t_a)\sqrt{\frac{iM\omega^3/2\pi\hbar}{\omega(t_b - t_a) - 2\tan\frac{\omega(t_b - t_a)}{2}}}\,\exp\left(\frac{i}{\hbar}\mathcal{A}^{x_0}\right). \quad (3.197)$$

If we set $x_b = x_a$ and integrate over $x_b = x_a$, we find the quantum-mechanical version of the partition function at fixed x_0:

$$Z^{x_0}_\omega = \frac{1}{\sqrt{2\pi\hbar(t_b - t_a)/Mi}}\frac{\omega(t_b - t_a)/2}{\sin[\omega(t_b - t_a)/2]}\exp\left[-\frac{i}{2\hbar}(t_b - t_a)M\omega^2 x_0^2\right]. \quad (3.198)$$

As a check we integrate this over x_0 and recover the correct Z_ω of Eq. (2.410).

We may also integrate over both ends independently to obtain the partition function

$$Z^{\text{open},x_0}_\omega = \sqrt{\frac{\omega(t_b - t_a)}{\sin\omega(t_b - t_a)}}\exp\left[-\frac{i}{2\hbar}(t_b - t_a)M\omega^2 x_0^2\right]. \quad (3.199)$$

Integrating this over x_0 and going to imaginary times leads back to the partition function Z^{open}_ω of Eq. (2.411).

3.8 External Source in Quantum-Statistical Path Integral

In the last section we have found the quantum-mechanical time evolution amplitude in the presence of an external source term. Let us now do the same thing for the quantum-statistical case and calculate the path integral

$$(x_b\hbar\beta | x_a 0)^j_\omega = \int \mathcal{D}x(\tau)\exp\left\{-\frac{1}{\hbar}\int_0^{\hbar\beta} d\tau\left[\frac{M}{2}(\dot{x}^2 + \omega^2 x^2) - j(\tau)x(\tau)\right]\right\}. \quad (3.200)$$

This will be done in two ways.

3.8.1   Continuation of Real-Time Result

The desired result is obtained most easily by an analytic continuation of the quantum-mechanical results (3.23), (3.168) in the time difference $t_b - t_a$ to an imaginary time $-i\hbar(\tau_b - \tau_a) = -i\hbar\beta$. This gives immediately

$$(x_b\hbar\beta|x_a0)^j_\omega = \sqrt{\frac{M}{2\pi\hbar^2\beta}}\sqrt{\frac{\omega\hbar\beta}{\sinh\omega\hbar\beta}}\exp\left\{-\frac{1}{\hbar}\mathcal{A}^{\mathrm{ext}}_{\mathrm{e}}[j]\right\}, \qquad (3.201)$$

with the extended classical Euclidean oscillator action

$$\mathcal{A}^{\mathrm{ext}}_{\mathrm{e}}[j] = \mathcal{A}_{\mathrm{e}} + \mathcal{A}^j_{\mathrm{e}} = \mathcal{A}_{\mathrm{e}} + \mathcal{A}^j_{1,\mathrm{e}} + \mathcal{A}^j_{2,\mathrm{e}}, \qquad (3.202)$$

where $\mathcal{A}_{\mathrm{e}}$ is the Euclidean action

$$\mathcal{A}_{\mathrm{e}} = \frac{M\omega}{2\sinh\beta\hbar\omega}\left[(x_b^2 + x_a^2)\cosh\omega\hbar\beta - 2x_bx_a\right], \qquad (3.203)$$

while the linear and quadratic Euclidean source terms are

$$\mathcal{A}^j_{1,\mathrm{e}} = -\frac{1}{\sinh\omega\hbar\beta}\int_{\tau_a}^{\tau_b}d\tau[x_a\sinh\omega(\hbar\beta - \tau) + x_b\ \sinh\omega\tau]j(\tau), \qquad (3.204)$$

and

$$\mathcal{A}^j_{2,\mathrm{e}} = -\frac{1}{M}\int_0^{\hbar\beta}d\tau\int_0^\tau d\tau'\ j(\tau)\,G_{\omega^2,\mathrm{e}}(\tau,\tau')j(\tau'), \qquad (3.205)$$

where $G_{\omega^2,\mathrm{e}}(\tau,\tau')$ is the Euclidean version of the Green function (3.36) with Dirichlet boundary conditions:

$$\begin{aligned} G_{\omega^2,\mathrm{e}}(\tau,\tau') &= \frac{\sinh\omega(\hbar\beta - \tau_>)\sinh\omega\tau_<}{\omega\sinh\omega\hbar\beta} \\ &= \frac{\cosh\omega(\hbar\beta - |\tau - \tau'|) - \cosh\omega(\hbar\beta - \tau - \tau')}{2\omega\sinh\omega\hbar\beta}, \end{aligned} \qquad (3.206)$$

satisfying the differential equation

$$(-\partial_\tau^2 + \omega^2)\,G_{\omega^2,\mathrm{e}}(\tau,\tau') = \delta(\tau - \tau'). \qquad (3.207)$$

It is related to the real-time Green function (3.36) by

$$G_{\omega^2,\mathrm{e}}(\tau,\tau') = i\,G_{\omega^2}(-i\tau, -i\tau'), \qquad (3.208)$$

the overall factor i accounting for the replacement $\delta(t - t') \to i\delta(\tau - \tau')$ on the right-hand side of (3.148) in going to (3.207) when going from the real time t to the Euclidean time $-i\tau$. The symbols $\tau_>$ and $\tau_<$ in the first line (3.206) denote the larger and the smaller of the Euclidean times τ and τ', respectively.

The source terms (3.204) and (3.205) can be rewritten as follows:

$$\mathcal{A}_{1,e}^{j} = -\frac{M\omega}{\sinh \omega \hbar \beta} \left\{ \left[x_b (e^{-\beta \hbar \omega} A_e - B_e) \right] x_a (e^{-\beta \hbar \omega} B_e - A_e) \right\}, \qquad (3.209)$$

and

$$\mathcal{A}_{2,e}^{j} = -\frac{1}{4M\omega} \int_0^{\hbar \beta} d\tau \int_0^{\hbar \beta} d\tau' e^{-\omega|\tau - \tau'|} j(\tau) j(\tau')$$
$$+ \frac{M\omega}{2 \sinh \omega \hbar \beta} \left[e^{\beta \hbar \omega} (A_e^2 + B_e^2) - 2 A_e B_e \right]. \qquad (3.210)$$

We have introduced the Euclidean versions of the functions $A(\omega)$ and $B(\omega)$ in Eqs. (3.173) and (3.174) as

$$A_e(\omega) \equiv iA(\omega)|_{t_b - t_a = -i\hbar\beta} = \frac{1}{M\omega} \int_0^{\hbar\beta} d\tau e^{-\omega\tau} j(\tau), \qquad (3.211)$$

$$B_e(\omega) \equiv iB(\omega)|_{t_b - t_a = -i\hbar\beta} = \frac{1}{M\omega} \int_0^{\hbar\beta} d\tau e^{-\omega(\hbar\beta - \tau)} j(\tau) = -e^{-\beta\hbar\omega} A_e(-\omega). \quad (3.212)$$

From (3.201) we now calculate the quantum-statistical partition function. Setting $x_b = x_a = x$, the first term in the action (3.202) becomes

$$\mathcal{A}_e = \frac{M\omega}{\sinh \beta \hbar \omega} 2 \sinh^2(\omega \hbar \beta / 2) x^2. \qquad (3.213)$$

If we ignore the second and third action terms in (3.202) and integrate (3.201) over x, we obtain, of course, the free partition function

$$Z_\omega = \frac{1}{2 \sinh(\beta \hbar \omega / 2)}. \qquad (3.214)$$

In the presence of j, we perform a quadratic completion in x and obtain a source-dependent part of the action (3.202):

$$\mathcal{A}_e^j = \mathcal{A}_{\mathrm{fl},e}^j + \mathcal{A}_{\mathrm{r},e}^j, \qquad (3.215)$$

where the additional term $\mathcal{A}_{\mathrm{r},e}^j$ is the remainder left by a quadratic completion. It reads

$$\mathcal{A}_{\mathrm{r},e}^j = -\frac{M\omega}{2 \sinh \omega \beta} e^{\beta \hbar \omega} (A_e + B_e)^2. \qquad (3.216)$$

Combining this with $\mathcal{A}_{\mathrm{fl},e}^j$ of (3.210) gives

$$\mathcal{A}_{\mathrm{fl},e}^j + \mathcal{A}_{\mathrm{r},e}^j = -\frac{1}{4M\omega} \int_0^{\hbar\beta} d\tau \int_0^{\hbar\beta} d\tau' e^{-\omega|\tau - \tau'|} j(\tau) j(\tau') - \frac{M\omega}{\sinh(\beta\hbar\omega/2)} e^{\beta\hbar\omega/2} A_e B_e.$$
$$(3.217)$$

This can be rearranged to the total source term

$$\mathcal{A}_e^j = -\frac{1}{4M\omega} \int_0^{\hbar\beta} d\tau \int_0^{\hbar\beta} d\tau' \frac{\cosh\omega(|\tau - \tau'| - \hbar\beta/2)}{\sinh(\beta\hbar\omega/2)} j(\tau)j(\tau'). \tag{3.218}$$

This is proved by rewriting the latter integrand as

$$\frac{1}{2\sinh(\beta\hbar\omega/2)} \left\{ \left[e^{\omega(\tau-\tau')}e^{-\beta\hbar\omega/2} + (\omega \to -\omega) \right] \bar{\Theta}(\tau - \tau') \right.$$
$$\left. + \left[e^{\omega(\tau'-\tau)}e^{-\beta\hbar\omega/2} + (\omega \to -\omega) \right] \bar{\Theta}(\tau' - \tau) \right\} j(\tau)j(\tau').$$

In the second and fourth terms we replace $e^{\beta\hbar\omega/2}$ by $e^{-\beta\hbar\omega/2} + 2\sinh(\beta\hbar\omega/2)$ and integrate over τ, τ', with the result (3.217).

The expression between the currents in (3.218) is recognized as the Euclidean version of the periodic Green function $G_{\omega^2}^{\mathrm{P}}(\tau)$ in (3.99):

$$G_{\omega^2,e}^{\mathrm{P}}(\tau) \equiv iG_{\omega^2}^{\mathrm{P}}(-i\tau)|_{t_b-t_a=-i\hbar\beta}$$
$$= \frac{1}{2\omega} \frac{\cosh\omega(\tau - \hbar\beta/2)}{\sinh(\beta\hbar\omega/2)}, \quad \tau \in [0, \hbar\beta]. \tag{3.219}$$

In terms of (3.218), the partition function of an oscillator in the presence of the source term is

$$Z_\omega[j] = Z_\omega \exp\left(-\frac{1}{\hbar}\mathcal{A}_e^j\right). \tag{3.220}$$

For completeness, let us also calculate the partition function of all paths with open ends in the presence of the source $j(t)$, thus generalizing the result (2.411). Integrating (3.201) over initial and final positions x_a and x_b we obtain

$$Z_\omega^{\mathrm{open}}[j] = \sqrt{\frac{2\pi\hbar}{M\omega}} \frac{1}{\sqrt{\sinh[\omega(\tau_b - \tau_a)]}} e^{-(\mathcal{A}_{2,e}^j + \tilde{\mathcal{A}}_{2,e}^j)/\hbar}, \tag{3.221}$$

where

$$\tilde{\mathcal{A}}_{2,e}^j = -\frac{1}{M} \int_0^{\hbar\beta} d\tau \int_0^\tau d\tau' j(\tau)\tilde{G}_{\omega^2}(\tau, \tau')j(\tau'), \tag{3.222}$$

with

$$\tilde{G}_{\omega^2}(\tau, \tau') = \frac{1}{2\omega\sinh^3\omega\hbar\beta}\{\cosh\omega\hbar\beta[\sinh\omega(\hbar\beta-\tau)\sinh\omega(\hbar\beta-\tau') + \sinh\omega\tau\sinh\omega\tau']$$
$$+ \sinh\omega(\hbar\beta-\tau)\sinh\omega\tau' + \sinh\omega(\hbar\beta-\tau')\sinh\omega\tau\}. \tag{3.223}$$

By some trigonometric identities, this can be simplified to

$$\tilde{G}_{\omega^2}(\tau, \tau') = \frac{1}{\omega} \frac{\cosh\omega(\hbar\beta - \tau - \tau')}{\sinh\omega\hbar\beta}. \tag{3.224}$$

The first step is to rewrite the curly brackets in (3.223) as

$$\sinh \omega\tau \left[\cosh \omega\hbar\beta \sinh \omega\tau' + \sinh \omega(\hbar\beta - \tau') \right]$$
$$+ \sinh \omega(\hbar\beta - \tau') \left[\cosh \omega\hbar\beta \sinh \omega(\hbar\beta - \tau) + \sinh \omega(\hbar\beta - ((\hbar\beta - \tau)) \right]. \quad (3.225)$$

The first bracket is equal to $\sinh \beta\hbar\omega \cosh \omega\tau$, the second to $\sinh \beta\hbar\omega \cosh \omega(\hbar\beta - \tau')$, so that we arrive at

$$\sinh \omega\hbar\beta \left[\sinh \omega\tau \cosh \omega\tau' + \sinh \omega(\hbar\beta - \tau) \cosh \omega(\hbar\beta - \tau') \right]. \quad (3.226)$$

The bracket is now rewritten as

$$\frac{1}{2} \left[\sinh \omega(\tau + \tau') + \sinh \omega(\tau - \tau') + \sinh \omega(2\hbar\beta - \tau - \tau') + \sinh \omega(\tau' - \tau) \right], \quad (3.227)$$

which is equal to

$$\frac{1}{2} \left[\sinh \omega(\hbar\beta + \tau + \tau' - \hbar\beta) + \sinh \omega(\hbar\beta + \hbar\beta - \tau - \tau') \right], \quad (3.228)$$

and thus to

$$\frac{1}{2} \left[2 \sinh \omega\hbar\beta \cosh \omega(\hbar\beta - \tau - \tau') \right], \quad (3.229)$$

such that we arrive indeed at (3.224). The source action in the exponent in (3.221) is therefore:

$$(\mathcal{A}_{2,\mathrm{e}}^j + \tilde{\mathcal{A}}_{2,\mathrm{e}}^j) = -\frac{1}{M} \int_0^{\hbar\beta} d\tau \int_0^{\tau} d\tau' \, j(\tau) G_{\omega^2,\mathrm{e}}^{\mathrm{open}}(\tau, \tau') j(\tau'), \quad (3.230)$$

with (3.205)

$$G_{\omega^2,\mathrm{e}}^{\mathrm{open}}(\tau, \tau') = \frac{\cosh \omega(\hbar\beta - |\tau - \tau'|) + \cosh \omega(\hbar\beta - \tau - \tau')}{2\omega \sinh \omega\hbar\beta}$$
$$= \frac{\cosh \omega(\hbar\beta - \tau_>) \cosh \omega\tau_<}{\omega \sinh \omega\hbar\beta}. \quad (3.231)$$

This Green function coincides precisely with the Euclidean version of Green function $G_{\omega^2}^{\mathrm{N}}(t, t')$ in Eq. (3.151) using the relation (3.208). This coincidence should have been expected after having seen in Section 2.12 that the partition function of all paths with open ends can be calculated, up to a trivial factor $l_\mathrm{e}(\hbar\beta)$ of Eq. (2.351), as a sum over all paths satisfying Neumann boundary conditions (2.449), which is calculated using the measure (2.452) for the Fourier components.

In the limit of small-ω, the Green function (3.231) reduces to

$$G_{\omega^2,\mathrm{e}}^{\mathrm{open}}(\tau, \tau') \underset{\omega^2 \approx 0}{\approx} \frac{1}{\beta\omega^2} + \frac{\beta}{3} - \frac{1}{2}|\tau - \tau'| - \frac{1}{2}(\tau + \tau') + \frac{1}{2\beta}\left(\tau^2 + \tau'^2\right), \quad (3.232)$$

which is the imaginary-time version of (3.157).

3.8.2 Calculation at Imaginary Time

Let us now see how the partition function with a source term is calculated directly in the imaginary-time formulation, where the periodic boundary condition is used from the outset. Thus we consider

$$Z_\omega[j] = \int \mathcal{D}x(\tau)\, e^{-\mathcal{A}_e[j]/\hbar}, \tag{3.233}$$

with the Euclidean action

$$\mathcal{A}_e[j] = \int_0^{\hbar\beta} d\tau \left[\frac{M}{2}(\dot{x}^2 + \omega^2 x^2) - j(\tau)x(\tau) \right]. \tag{3.234}$$

Since $x(\tau)$ satisfies the periodic boundary condition, we can perform a partial integration of the kinetic term without picking up a boundary term $x\dot{x}|_{t_a}^{t_b}$. The action becomes

$$\mathcal{A}_e[j] = \int_0^{\hbar\beta} d\tau \left[\frac{M}{2}x(\tau)(-\partial_\tau^2 + \omega^2)x(\tau) - j(\tau)x(\tau) \right]. \tag{3.235}$$

Let $D_e(\tau, \tau')$ be the functional matrix

$$D_{\omega^2,e}(\tau, \tau') \equiv (-\partial_\tau^2 + \omega^2)\delta(\tau - \tau'), \quad \tau - \tau' \in [0, \hbar\beta]. \tag{3.236}$$

Its functional inverse is the Euclidean Green function,

$$G_{\omega^2,e}^{\mathrm{p}}(\tau, \tau') = G_{\omega^2,e}^{\mathrm{p}}(\tau - \tau') = D_{\omega^2,e}^{-1}(\tau, \tau') = (-\partial_\tau^2 + \omega^2)^{-1}\delta(\tau - \tau'), \tag{3.237}$$

with the periodic boundary condition.

Next we perform a quadratic completion by shifting the path:

$$x \to x' = x - \frac{1}{M}G_{\omega^2,e}^P\, j. \tag{3.238}$$

This brings the Euclidean action to the form

$$\mathcal{A}_e[j] = \int_0^{\hbar\beta} d\tau \frac{M}{2}x'(-\partial_\tau^2 + \omega^2)x' - \frac{1}{2M}\int_0^{\hbar\beta} d\tau \int_0^{\hbar\beta} d\tau'\, j(\tau)G_{\omega^2,e}^{\mathrm{p}}(\tau - \tau')j(\tau'). \tag{3.239}$$

The fluctuations over the periodic paths $x'(\tau)$ can now be integrated out and yield for $j(\tau) \equiv 0$

$$Z_\omega = \mathrm{Det}\, D_{\omega^2,e}^{-1/2}. \tag{3.240}$$

As in Subsection 2.15.2, we find the functional determinant by rewriting the product of eigenvalues as

$$\mathrm{Det}\, D_{\omega^2,e} = \prod_{m=-\infty}^{\infty} (\omega_m^2 + \omega^2) = \exp\left[\sum_{m=-\infty}^{\infty} \log(\omega_m^2 + \omega^2) \right], \tag{3.241}$$

and evaluating the sum in the exponent according to the rules of analytic regularization. This leads directly to the partition function of the harmonic oscillator as in Eq. (2.407):

$$Z_\omega = \frac{1}{2\sinh(\beta\hbar\omega/2)}. \tag{3.242}$$

The generating functional for $j(\tau) \neq 0$ is therefore

$$Z[j] = Z_\omega \exp\left\{-\frac{1}{\hbar}\mathcal{A}_e^j[j]\right\}, \tag{3.243}$$

with the source term:

$$\mathcal{A}_e^j[j] = -\frac{1}{2M}\int_0^{\hbar\beta}d\tau\int_0^{\hbar\beta}d\tau'\, j(\tau)G_{\omega^2,e}^p(\tau-\tau')j(\tau'). \tag{3.244}$$

The Green function of imaginary time is calculated as follows. The eigenfunctions of the differential operator $-\partial_\tau^2$ are $e^{-i\omega_m\tau}$ with eigenvalues ω_m^2, and the periodic boundary condition forces ω_m to be equal to the thermal Matsubara frequencies $\omega_m = 2\pi m/\hbar\beta$ with $m = 0, \pm1, \pm2, \dots$. Hence we have the Fourier expansion

$$G_{\omega^2,e}^p(\tau) = \frac{1}{\hbar\beta}\sum_{m=-\infty}^{\infty}\frac{1}{\omega_m^2+\omega^2}e^{-i\omega_m\tau}. \tag{3.245}$$

In the zero-temperature limit, the Matsubara sum becomes an integral, yielding

$$G_{\omega^2,e}^p(\tau)\underset{T=0}{=}\int\frac{d\omega_m}{2\pi}\frac{1}{\omega_m^2+\omega^2}e^{-i\omega_m\tau} = \frac{1}{2\omega}e^{-\omega|\tau|}. \tag{3.246}$$

The frequency sum in (3.245) may be written as such an integral over ω_m, provided the integrand contains an additional Poisson sum (3.81):

$$\sum_{\bar{m}=-\infty}^{\infty}\delta(m-\bar{m}) = \sum_{n=-\infty}^{\infty}e^{i2\pi nm} = \sum_{n=-\infty}^{\infty}e^{in\omega_m\hbar\beta}. \tag{3.247}$$

This implies that the finite-temperature Green function (3.245) is obtained from (3.246) by a periodic repetition:

$$\begin{aligned}
G_{\omega^2,e}^p(\tau) &= \sum_{n=-\infty}^{\infty}\frac{1}{2\omega}e^{-\omega|\tau+n\hbar\beta|}\\
&= \frac{1}{2\omega}\frac{\cosh\omega(\tau-\hbar\beta/2)}{\sinh(\beta\hbar\omega/2)}, \quad \tau\in[0,\hbar\beta].
\end{aligned} \tag{3.248}$$

A comparison with (3.97), (3.99) shows that $G_{\omega^2,e}^p(\tau)$ coincides with $G_{\omega^2}^p(t)$ at imaginary times, as it should.

Note that for small ω, the Green function has the expansion

$$G_{\omega^2,e}^p(\tau) = \frac{1}{\hbar\beta\omega^2} + \frac{\tau^2}{2\hbar\beta} - \frac{\tau}{2} + \frac{\hbar\beta}{12} + \dots. \tag{3.249}$$

The first term diverges in the limit $\omega \to 0$. Comparison with the spectral representation (3.245) shows that it stems from the zero Matsubara frequency contribution to the sum. If this term is omitted, the subtracted Green function

$$G^{p\prime}_{\omega^2,e}(\tau) \equiv G^{p}_{\omega^2,e}(\tau) - \frac{1}{\hbar\beta\omega^2} \tag{3.250}$$

has a well-defined $\omega \to 0$ limit

$$G^{p\prime}_{0,e}(\tau) = \frac{1}{\hbar\beta} \sum_{m=\pm1,\pm2,\ldots} \frac{1}{\omega_m^2} e^{-i\omega_m\tau} = \frac{\tau^2}{2\hbar\beta} - \frac{\tau}{2} + \frac{\hbar\beta}{12}, \tag{3.251}$$

the right-hand side being correct only for $\tau \in [0, \hbar\beta]$. Outside this interval it must be continued periodically. The subtracted Green function $G^{p\prime}_{\omega^2,e}(\tau)$ is plotted for different frequencies ω in Fig. 3.4.

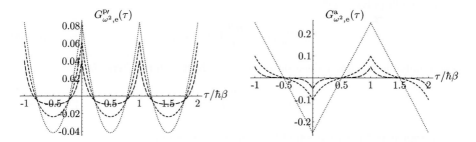

Figure 3.4 Subtracted periodic Green function $G^{p\prime}_{\omega^2,e}(\tau) \equiv G^{p}_{\omega^2,e}(\tau) - 1/\hbar\beta\omega^2$ and antiperiodic Green function $G^{a}_{\omega^2,e}(\tau)$ for frequencies $\omega = (0,\ 5,\ 10)/\hbar\beta$ (with increasing dash length). Compare Fig. 3.2.

The limiting expression (3.251) can, incidentally, be derived using the methods developed in Subsection 2.15.6. We rewrite the sum as

$$\frac{1}{\hbar\beta} \sum_{m=\pm1,\pm2,\ldots} \frac{(-1)^m}{\omega_m^2} e^{-i\omega_m(\tau-\hbar\beta/2)} \tag{3.252}$$

and expand

$$-\frac{2}{\hbar\beta}\left(\frac{\hbar\beta}{2\pi}\right)^2 \sum_{n=0,2,4,\ldots} \frac{1}{n!}\left[-i\frac{2\pi}{\hbar\beta}(\tau-\hbar\beta/2)\right]^n \sum_{m=1}^{\infty} \frac{(-1)^{m-1}}{m^{2-n}}. \tag{3.253}$$

The sum over m on the right-hand side is Riemann's eta function[1]

$$\eta(z) \equiv \sum_{m=1}^{\infty} \frac{(-1)^{m-1}}{m^z}, \tag{3.254}$$

---

[1] M. Abramowitz and I. Stegun, op. cit., Formula 23.2.19.

which is related to the zeta function (2.519) by

$$\eta(z) = (1 - 2^{1-z})\zeta(z). \tag{3.255}$$

Since the zeta functions of negative integers are all zero [recall (2.585)], only the terms with $n = 0$ and 2 contribute in (3.253). Inserting

$$\eta(0) = -\zeta(0) = 1/2, \quad \eta(2) = \zeta(2)/2 = \pi^2/12, \tag{3.256}$$

we obtain

$$-\frac{2}{\hbar\beta}\left(\frac{\hbar\beta}{2\pi}\right)^2\left[\frac{\pi^2}{12} - \frac{1}{4}\left(\frac{2\pi}{\hbar\beta}\right)^2(\tau - \hbar\beta/2)^2\right] = \frac{\tau^2}{2\hbar\beta} - \frac{\tau}{2} + \frac{\hbar\beta}{12}, \tag{3.257}$$

in agreement with (3.251).

It is worth remarking that the Green function (3.251) is directly proportional to the Bernoulli polynomial $B_2(z)$:

$$G^{\text{p}'}_{0,\text{e}}(\tau) = \frac{\hbar\beta}{2}B_2(\tau/\hbar\beta). \tag{3.258}$$

These polynomials are defined in terms of the Bernoulli numbers B_k as[2]

$$B_n(x) = \sum_{k=0}^n \binom{n}{k} B_k z^{n-k}. \tag{3.259}$$

They appear in the expansion of the generating function[3]

$$\frac{e^{zt}}{e^t - 1} = \sum_{n=0}^\infty B_n(z)\frac{t^{n-1}}{n!}, \tag{3.260}$$

and have the expansion

$$B_{2n}(z) = (-1)^{n-1}\frac{2(2n)!}{(2\pi)^{2n}}\sum_{k=0}^\infty \frac{\cos(2\pi kz)}{k^{2n}}, \tag{3.261}$$

with the special cases

$$B_1(z) = z - 1/2, \quad B_2(z) = z^2 - z + 1/6, \ldots. \tag{3.262}$$

By analogy with (3.248), the antiperiodic Green function can be obtained from an antiperiodic repetition

$$\begin{aligned}
G^{\text{a}}_{\omega^2,\text{e}}(\tau) &= \sum_{n=-\infty}^\infty \frac{(-1)^n}{2\omega}e^{-\omega|\tau+n\hbar\beta|} \\
&= \frac{1}{2\omega}\frac{\sinh\omega(\tau - \hbar\beta/2)}{\cosh(\beta\hbar\omega/2)}, \quad \tau \in [0, \hbar\beta],
\end{aligned} \tag{3.263}$$

---

[2]I.S. Gradshteyn and I.M. Ryzhik, *op. cit.*, Formula 9.620.
[3]ibid. Formula 9.621.

which is an analytic continuation of (3.113) to imaginary times. In contrast to (3.249), this has a finite $\omega \to 0$ limit

$$G^{\mathrm{a}}_{\omega^2,\mathrm{e}}(\tau) = \frac{\tau}{2} - \frac{\hbar\beta}{4}, \quad \tau \in [0, \hbar\beta]. \tag{3.264}$$

For a plot of the antiperiodic Green function for different frequencies ω see again Fig. 3.4.

The limiting expression (3.264) can again be derived using an expansion of the type (3.253). The spectral representation in terms of odd Matsubara frequencies (3.104)

$$G^{\mathrm{a}}_{\omega^2,\mathrm{e}}(\tau) \equiv \frac{1}{\hbar\beta} \sum_{m=-\infty}^{\infty} \frac{1}{\omega_m^{\mathrm{f}\,2}} e^{-i\omega_m^{\mathrm{f}}\tau} \tag{3.265}$$

is rewritten as

$$\frac{1}{\hbar\beta} \sum_{m=-\infty}^{\infty} \frac{1}{\omega_m^{\mathrm{f}\,2}} \cos(\omega_m^{\mathrm{f}}\tau) = \frac{1}{\hbar\beta} \sum_{m=-\infty}^{\infty} \frac{(-1)^m}{\omega_m^{\mathrm{f}\,2}} \sin[\omega_m^{\mathrm{f}}(\tau - \hbar\beta/2)]. \tag{3.266}$$

Expanding the sin function yields

$$\frac{2}{\hbar\beta} \left(\frac{\hbar\beta}{2\pi}\right)^2 \sum_{n=1,3,5,\ldots} \frac{(-1)^{(n-1)/2}}{n!} \left[\frac{2\pi}{\hbar\beta}(\tau - \hbar\beta/2)\right]^n \sum_{m=0}^{\infty} \frac{(-1)^m}{\left(m+\frac{1}{2}\right)^{2-n}}. \tag{3.267}$$

The sum over m at the end is 2^{2-n} times Riemann's beta function[4] $\beta(2-n)$, which is defined as

$$\beta(z) \equiv \frac{1}{2^z} \sum_{m=0}^{\infty} \frac{(-1)^m}{\left(m+\frac{1}{2}\right)^z}, \tag{3.268}$$

and is related to Riemann's zeta function

$$\zeta(z, q) \equiv \sum_{m=0}^{\infty} \frac{1}{(m+q)^z}. \tag{3.269}$$

Indeed, we see immediately that

$$\sum_{m=0}^{\infty} \frac{(-1)^m}{(m+q)^z} = \zeta(z, q) - 2^{2-z}\zeta(z, (q+1)/2), \tag{3.270}$$

so that

$$\beta(z) \equiv \frac{1}{2^z} \left[\zeta(z, 1/2) - 2^{2-z}\zeta(z, 3/4)\right]. \tag{3.271}$$

Near $z = 1$, the function $\zeta(z, q)$ behaves like[5]

$$\zeta(z, q) = \frac{1}{z-1} - \psi(q) + \mathcal{O}(z-1), \tag{3.272}$$

---

[4] M. Abramowitz and I. Stegun, *op. cit.*, Formula 23.2.21.
[5] I.S. Gradshteyn and I.M. Ryzhik, *op. cit.*, Formula 9.533.2.

where $\psi(z)$ is the Digamma function (2.571). Thus we obtain in the limit $z \to 1$:

$$\beta(1) = \lim_{z \to 1} \frac{1}{2} \left[\zeta(z, 1/2) - 2^{1-z}\zeta(z, 3/4) \right] = \frac{1}{2} \left[-\psi(1/2) + \psi(3/4) + \log 2 \right] = \frac{\pi}{4}.$$

(3.273)

The last result follows from the specific values [compare (2.573)]:

$$\psi(1/2) = -\gamma - 2\log 2, \quad \psi(3/4) = -\gamma - 3\log 2 + \frac{\pi}{2}. \tag{3.274}$$

For negative odd arguments, the beta function (3.268) vanishes, so that there are no further contributions. Inserting this into (3.267) the only surviving e $n = 1$ -term yields once more (3.264).

Note that the relation (3.273) could also have been found directly from the expansion (2.572) of the Digamma function, which yields

$$\beta(1) = \frac{1}{4} \left[\psi(3/2) - \psi(1/4) \right], \tag{3.275}$$

and is equal to (3.273) due to $\psi(1/4) = -\gamma - 3\log 2 - \pi/2$.

For currents $j(\tau)$, which are periodic in $\hbar\beta$, the source term (3.244) can also be written more simply:

$$\mathcal{A}_e^j[j] = -\frac{1}{4M\omega} \int_0^{\hbar\beta} d\tau \int_{-\infty}^{\infty} d\tau' e^{-\omega|\tau-\tau'|} j(\tau)j(\tau'). \tag{3.276}$$

This follows directly by rewriting (3.276), by analogy with (3.149), as a sum over all periodic repetitions of the zero-temperature Green function (3.246):

$$G_{\omega^2, e}^p(\tau) = \frac{1}{2\omega} \sum_{n=-\infty}^{\infty} e^{-\omega|\tau + n\hbar\beta|}. \tag{3.277}$$

When inserted into (3.244), the factors $e^{-n\beta\hbar\omega}$ can be removed by an irrelevant periodic temporal shift in the current $j(\tau') \to j(\tau' - n\hbar\beta)$ leading to (3.276).

For a time-dependent periodic or antiperiodic potential $\Omega(\tau)$, the Green function $G_{\omega^2, e}^{p,a}(\tau)$ solving the differential equation

$$[\partial_\tau^2 - \Omega^2(\tau)]G_{\omega^2, e}^{p,a}(\tau, \tau') = \delta^{p,a}(\tau - \tau'), \tag{3.278}$$

with the periodic or antiperiodic δ-function

$$\delta^{p,a}(\tau - \tau') = \sum_{n=-\infty}^{\infty} \delta(\tau - \tau' - n\hbar\beta) \left\{ \begin{array}{c} 1 \\ (-1)^n \end{array} \right\}, \tag{3.279}$$

can be expressed[6] in terms of two arbitrary solutions $\xi(\tau)$ and $\eta(\tau)$ of the homogenous differential equation in the same way as the real-time Green functions in Section 3.5:

$$G_{\omega^2, e}^{p,a}(\tau, \tau') = G_{\omega^2, e}(\tau, \tau') \mp \frac{[\Delta(\tau, \tau_a) \pm \Delta(\tau_b, \tau)][\Delta(\tau', \tau_a) \pm \Delta(\tau_b, \tau')]}{\bar{\Delta}^{p,a}(\tau_a, \tau_b)\Delta(\tau_a, \tau_b)}, \tag{3.280}$$

---

[6]See H. Kleinert and A. Chervyakov, Phys. Lett. A *245*, 345 (1998) (quant-ph/9803016); J. Math. Phys. B *40*, 6044 (1999) (physics/9712048).

where $G_{\omega^2,\mathrm{e}}(\tau,\tau')$ is the imaginary-time Green function with Dirichlet boundary conditions corresponding to (3.206):

$$G_{\omega^2,\mathrm{e}}(\tau,\tau') = \frac{\bar{\Theta}(\tau-\tau')\Delta(\tau_b,\tau)\Delta(\tau',\tau_a) + \bar{\Theta}(\tau-\tau')\Delta(\tau_b,\tau')\Delta(\tau,\tau_a)}{\Delta(\tau_a,\tau_b)}, \qquad (3.281)$$

with

$$\Delta(\tau,\tau') = \frac{1}{W}\left[\xi(\tau)\eta(\tau') - \xi(\tau')\eta(\tau)\right], \qquad W = \xi(\tau)\dot{\eta}(\tau) - \dot{\xi}(\tau)\eta(\tau), \qquad (3.282)$$

and

$$\bar{\Delta}^{\mathrm{p,a}}(\tau_a,\tau_b) = 2 \pm \partial_\tau\Delta(\tau_a,\tau_b) \pm \partial_\tau\Delta(\tau_b,\tau_a). \qquad (3.283)$$

Let also write down the imaginary-time versions of the periodic or antiperiodic Green functions for time-dependent frequencies. Recall the expressions for constant frequency $G_\omega^{\mathrm{p}}(t)$ and $G_\omega^{\mathrm{a}}(t)$ of Eqs. (3.94) and (3.112) for $\tau \in (0,\hbar\beta)$:

$$\begin{aligned}
G_{\omega,\mathrm{e}}^{\mathrm{p}}(\tau) &= \frac{1}{\hbar\beta}\sum_m e^{-i\omega_m\tau}\frac{-1}{i\omega_m - \omega} = e^{-\omega(\tau-\hbar\beta/2)}\frac{1}{2\sinh(\beta\hbar\omega/2)} \\
&= (1 + n_\omega^{\mathrm{b}})e^{-\omega\tau}, \qquad (3.284)
\end{aligned}$$

and

$$\begin{aligned}
G_{\omega,\mathrm{e}}^{\mathrm{a}}(\tau) &= \frac{1}{\hbar\beta}\sum_m e^{-i\omega_m^{\mathrm{f}}\tau}\frac{-1}{i\omega_m^{\mathrm{f}} - \omega} = e^{-\omega(\tau-\hbar\beta/2)}\frac{1}{2\cosh(\beta\hbar\omega/2)} \\
&= (1 - n_\omega^{\mathrm{f}})e^{-\omega\tau}, \qquad (3.285)
\end{aligned}$$

the first sum extending over the even Matsubara frequencies, the second over the odd ones. The Bose and Fermi distribution functions $n_\omega^{\mathrm{b,f}}$ were defined in Eqs. (3.93) and (3.111).

For $\tau < 0$, periodicity or antiperiodicity determine

$$G_{\omega,\mathrm{e}}^{\mathrm{p,a}}(\tau) = \pm G_{\omega,\mathrm{e}}^{\mathrm{p,a}}(\tau + \hbar\beta). \qquad (3.286)$$

The generalization of these expressions to time-dependent periodic and antiperiodic frequencies $\Omega(\tau)$ satisfying the differential equations

$$[-\partial_\tau - \Omega(\tau)]G_{\Omega,\mathrm{e}}^{\mathrm{p,a}}(\tau,\tau') = \delta^{\mathrm{p,a}}(\tau - \tau') \qquad (3.287)$$

has for $\beta \to \infty$ the form

$$G_{\Omega,\mathrm{e}}^{\mathrm{p,a}}(\tau,\tau') = \bar{\Theta}(\tau-\tau')e^{-\int_0^\tau d\tau'\,\Omega(\tau')}. \qquad (3.288)$$

Its periodic superposition yields for finite β a sum analogous to (3.277):

$$G_{\Omega,\mathrm{e}}^{\mathrm{p,a}}(\tau,\tau') = \sum_{n=0}^\infty e^{-\int_0^{\tau+n\hbar\beta} d\tau'\,\Omega(\tau')}\left\{\begin{array}{c} 1 \\ (-1)^n \end{array}\right\}, \qquad \hbar\beta > \tau > \tau' > 0, \qquad (3.289)$$

which reduces to (3.284), (3.285) for a constant frequency $\Omega(\tau) \equiv \omega$.

3.9 Lattice Green Function

As in Chapter 2, it is easy to calculate the above results also on a sliced time axis. This is useful when it comes to comparing analytic results with Monte Carlo lattice simulations. We consider here only the Euclidean versions; the quantum-mechanical ones can be obtained by analytic continuation to real times.

The Green function $G_{\omega^2}(\tau, \tau')$ on an imaginary-time lattice with infinitely many lattice points of spacing ϵ reads [instead of the Euclidean version of (3.147)]:

$$G_{\omega^2}(\tau, \tau') = \frac{\epsilon}{2 \sinh \epsilon \tilde{\omega}_e} e^{-\tilde{\omega}_e|\tau - \tau'|} = \frac{1}{2\omega} \frac{1}{\cosh(\epsilon \tilde{\omega}_e/2)} e^{-\tilde{\omega}_e|\tau - \tau'|}, \tag{3.290}$$

where $\tilde{\omega}_e$ is given, as in (2.404), by

$$\tilde{\omega}_e = \frac{2}{\epsilon} \text{arsinh} \frac{\epsilon \omega}{2}. \tag{3.291}$$

This is derived from the spectral representation

$$G_{\omega^2}(\tau, \tau') = G_{\omega^2}(\tau - \tau') = \int \frac{d\omega'}{2\pi} e^{-i\omega'(\tau - \tau')} \frac{\epsilon^2}{2(1 - \cos \epsilon \omega') + \epsilon^2 \omega^2} \tag{3.292}$$

by rewriting it as

$$G_{\omega^2}(\tau, \tau') = \int_0^\infty ds \int \frac{d\omega'}{2\pi} e^{-i\omega' \epsilon n} e^{-s[2(1 - \cos \epsilon \omega') + \epsilon^2 \omega^2]/\epsilon^2}, \tag{3.293}$$

with $n \equiv (\tau' - \tau)/\epsilon$, performing the ω'-integral which produces a Bessel function $I_{(\tau - \tau')/\epsilon}(2s/\epsilon^2)$, and subsequently the integral over s with the help of formula (2.473). The Green function (3.290) is defined only at discrete $\tau_n = n\hbar\beta/(N+1)$. If it is summed over all periodic repetitions $n \to n + k(N+1)$ with $k = 0, \pm 1, \pm 2, \dots$, one obtains the lattice analog of the periodic Green function (3.248):

$$\begin{aligned} G_e^p(\tau) &= \frac{1}{\hbar\beta} \sum_{m=-\infty}^{\infty} \frac{\epsilon^2}{2(1 - \cos \epsilon \omega_m) + \epsilon^2 \omega^2} e^{-i\omega_m \tau} \\ &= \frac{1}{2\omega} \frac{1}{\cosh(\epsilon \tilde{\omega}/2)} \frac{\cosh \tilde{\omega}(\tau - \hbar\beta/2)}{\sinh(\hbar \tilde{\omega} \beta/2)}, \quad \tau \in [0, \hbar\beta]. \end{aligned} \tag{3.294}$$

3.10 Correlation Functions, Generating Functional, and Wick Expansion

Equipped with the path integral of the harmonic oscillator in the presence of an external source it is easy to calculate the correlation functions of any number of position variables $x(\tau)$. We consider here only a system in thermal equilibrium and study the behavior at imaginary times. The real-time correlation functions can be discussed similarly. The precise relation between them will be worked out in Chapter 18.

In general, i.e., also for nonharmonic actions, the thermal correlation functions of n-variables $x(\tau)$ are defined as the functional averages

$$\begin{aligned} G_{\omega^2}^{(n)}(\tau_1, \dots, \tau_n) &\equiv \langle x(\tau_1) x(\tau_2) \cdots x(\tau_n) \rangle \\ &\equiv Z^{-1} \int \mathcal{D}x \, x(\tau_1) x(\tau_2) \cdots x(\tau_n) \exp\left(-\frac{1}{\hbar} \mathcal{A}_e\right). \end{aligned} \tag{3.295}$$

They are also referred to as *n-point functions*. In operator quantum mechanics, the same quantities are obtained from the thermal expectation values of time-ordered products of Heisenberg position operators $\hat{x}_H(\tau)$:

$$G^{(n)}_{\omega^2}(\tau_1, \ldots, \tau_n) = Z^{-1}\mathrm{Tr}\left\{\hat{T}_\tau\left[\hat{x}_H(\tau_1)\hat{x}_H(\tau_2)\cdots\hat{x}_H(\tau_n)e^{-\hat{H}/k_B T}\right]\right\}, \qquad (3.296)$$

where Z is the partition function

$$Z = e^{-F/k_B T} = \mathrm{Tr}(e^{-\hat{H}/k_B T}) \qquad (3.297)$$

and $\hat{T}_\tau$ is the time-ordering operator. Indeed, by slicing the imaginary-time evolution operator $e^{-\hat{H}\tau/\hbar}$ at discrete times in such a way that the times τ_i of the n position operators $x(\tau_i)$ are among them, we find that $G^{(n)}_{\omega^2}(\tau_1, \ldots, \tau_n)$ has precisely the path integral representation (3.295).

By definition, the path integral with the product of $x(\tau_i)$ in the integrand is calculated as follows. First we sort the times τ_i according to their time order, denoting the reordered times by $\tau_{t(i)}$. We also set $\tau_b \equiv \tau_{t(n+1)}$ and $\tau_a \equiv \tau_{t(0)}$. Assuming that the times $\tau_{t(i)}$ are different from one another, we slice the time axis $\tau \in [\tau_a, \tau_b]$ into the intervals $[\tau_b, \tau_{t(n)}], [\tau_{t(n)}, \tau_{t(n-1)}], [\tau_{t(n-2)}, \tau_{t(n-3)}], \ldots, [\tau_{t(4)}, \tau_{t(3)}], [\tau_{t(2)}, \tau_{t(1)}], [\tau_{t(1)}, \tau_a]$. For each of these intervals we calculate the time evolution amplitude $(x_{t(i+1)}\tau_{t(i+1)}|x_{t(i)}\tau_{t(i)})$ as usual. Finally, we recombine the amplitudes by performing the intermediate $x(\tau_{t(i)})$-integrations, with an extra factor $x(\tau_i)$ at each τ_i, i.e.,

$$\begin{aligned}G^{(n)}_{\omega^2}(\tau_1, \ldots, \tau_n) = \prod_{i=1}^{n+1}&\left[\int_{-\infty}^{\infty}dx_{\tau_{t(i)}}\right](x_{t(n+1)}\tau_b|x_{t(n)}\tau_{t(n)})\cdot x(\tau_{t(n)})\cdot\ldots\\
&\cdot(x_{t(i+1)}\tau_{t(i+1)}|x_{t(i)}\tau_{t(i)})\cdot x(\tau_{t(i)})\cdot(x_{t(i)}\tau_{t(i)}|x_{t(i-1)}\tau_{t(i-1)})\cdot x(\tau_{t(i-1)})\\
&\cdot\ldots\cdot(x_{t(2)}\tau_{t(2)}|x_{t(1)}\tau_{t(1)})\cdot x(\tau_{t(1)})\cdot(x_{t(1)}\tau_{t(1)}|x_{t(0)}\tau_a).\end{aligned} \qquad (3.298)$$

We have set $x_{t(n+1)} \equiv x_b = x_a \equiv x_{t(0)}$, in accordance with the periodic boundary condition. If two or more of the times τ_i are equal, the intermediate integrals are accompanied by the corresponding power of $x(\tau_i)$.

Fortunately, this rather complicated-looking expression can be replaced by a much simpler one involving functional derivatives of the thermal partition function $Z[j]$ in the presence of an external current j. From the definition of $Z[j]$ in (3.233) it is easy to see that all correlation functions of the system are obtained by the functional formula

$$G^{(n)}_{\omega^2}(\tau_1, \ldots, \tau_n) = \left[Z[j]^{-1}\hbar\frac{\delta}{\delta j(\tau_1)}\cdots\hbar\frac{\delta}{\delta j(\tau_n)}Z[j]\right]_{j=0}. \qquad (3.299)$$

This is why $Z[j]$ is called the *generating functional* of the theory.

In the present case of a harmonic action, $Z[j]$ has the simple form (3.243), (3.244), and we can write

$$\begin{aligned}G^{(n)}_{\omega^2}(\tau_1, \ldots, \tau_n) = &\left[\hbar\frac{\delta}{\delta j(\tau_1)}\cdots\hbar\frac{\delta}{\delta j(\tau_n)}\right. \qquad (3.300)\\
&\left.\times\exp\left\{\frac{1}{2\hbar M}\int_0^{\hbar\beta}d\tau\int_0^{\hbar\beta}d\tau'j(\tau)G^p_{\omega^2,e}(\tau-\tau')j(\tau')\right\}\right]_{j=0},\end{aligned}$$

where $G^{\mathrm{p}}_{\omega^2,\mathrm{e}}(\tau - \tau')$ is the Euclidean Green function (3.248). Expanding the exponential into a Taylor series, the differentiations are easy to perform. Obviously, any odd number of derivatives vanishes. Differentiating (3.243) twice yields the two-point function [recall (3.248)]

$$G^{(2)}_{\omega^2}(\tau, \tau') = \langle x(\tau) x(\tau') \rangle = \frac{\hbar}{M} G^{\mathrm{p}}_{\omega^2,\mathrm{e}}(\tau - \tau'). \tag{3.301}$$

Thus, up to the constant prefactor, the two-point function coincides with the Euclidean Green function (3.248). Inserting (3.301) into (3.300), all n-point functions are expressed in terms of the two-point function $G^{(2)}_{\omega^2}(\tau, \tau')$: Expanding the exponential into a power series, the expansion term of order $n/2$ carries the numeric prefactors $1/(n/2)! \cdot 1/2^{n/2}$ and consists of a product of $n/2$ factors $\int_0^{\hbar\beta} d\tau' j(\tau) G^{(2)}_{\omega^2}(\tau, \tau') j(\tau')/\hbar^2$. The n-point function is obtained by functionally differentiating this term n times. The result is a sum over products of $n/2$ factors $G^{(2)}_{\omega^2}(\tau, \tau')$ with $n!$ permutations of the n time arguments. Most of these products coincide, for symmetry reasons. First, $G^{(2)}_{\omega^2}(\tau, \tau')$ is symmetric in its arguments. Hence $2^{n/2}$ of the permutations correspond to identical terms, their number canceling one of the prefactors. Second, the $n/2$ Green functions $G^{(2)}_{\omega^2}(\tau, \tau')$ in the product are identical. Of the $n!$ permutations, subsets of $(n/2)!$ permutations produce identical terms, their number canceling the other prefactor. Only $n!/[(n/2)!2^{n/2}] = (n - 1) \cdot (n - 3) \cdots 1 = (n - 1)!!$ terms are different. They all carry a unit prefactor and their sum is given by the so-called *Wick rule* or *Wick expansion*:

$$G^{(n)}_{\omega^2}(\tau_1, \ldots, \tau_n) = \sum_{\text{pairs}} G^{(2)}_{\omega^2}(\tau_{p(1)}, \tau_{p(2)}) \cdots G^{(2)}_{\omega^2}(\tau_{p(n-1)}, \tau_{p(n)}). \tag{3.302}$$

Each term is characterized by a different pair configurations of the time arguments in the Green functions. These pair configurations are found most simply by the following rule: Write down all time arguments in the n-point function $\tau_1 \tau_2 \tau_3 \tau_4 \ldots \tau_n$. Indicate a pair by a common symbol, say $\dot\tau_{p(i)} \dot\tau_{p(i+1)}$, and call it a *pair contraction* to symbolize a Green function $G^{(2)}_{\omega^2}(\tau_{p(i)}, \tau_{p(i+1)})$. The desired $(n - 1)!!$ pair configurations in the Wick expansion (3.302) are then found iteratively by forming $n - 1$ single contractions

$$\dot\tau_1 \dot\tau_2 \tau_3 \tau_4 \ldots \tau_n + \dot\tau_1 \tau_2 \dot\tau_3 \tau_4 \ldots \tau_n + \dot\tau_1 \tau_2 \tau_3 \dot\tau_4 \ldots \tau_n + \ldots + \dot\tau_1 \tau_2 \tau_3 \tau_4 \ldots \dot\tau_n, \tag{3.303}$$

and by treating the remaining $n - 2$ uncontracted variables in each of these terms likewise, using a different contraction symbol. The procedure is continued until all variables are contracted.

In the literature, one sometimes another shorter formula under the name of Wick's rule, stating that a single harmonically fluctuating variable satisfies the equality of expectations:

$$\left\langle e^{Kx} \right\rangle = e^{K^2 \langle x^2 \rangle /2}. \tag{3.304}$$

This follows from the observation that the generating functional (3.233) may also be viewed as Z_ω times the expectation value of the source exponential

$$Z_\omega[j] = Z_\omega \times \left\langle e^{\int d\tau j(\tau)x(\tau)/\hbar} \right\rangle. \tag{3.305}$$

Thus we can express the result (3.243) also as

$$\left\langle e^{\int d\tau j(\tau)x(\tau)/\hbar} \right\rangle = e^{(1/2M\hbar)\int d\tau \int d\tau' j(\tau)G^{\mathrm{P}}_{\omega^2,\mathrm{e}}(\tau,\tau')j(\tau')}. \tag{3.306}$$

Since $(\hbar/M)G^{\mathrm{P}}_{\omega^2,\mathrm{e}}(\tau,\tau')$ in the exponent is equal to the correlation function $G^{(2)}_{\omega^2}(\tau,\tau') = \langle x(\tau)x(\tau')\rangle$ by Eq. (3.301), we may also write

$$\left\langle e^{\int d\tau j(\tau)x(\tau)/\hbar} \right\rangle = e^{\int d\tau \int d\tau' j(\tau)\langle x(\tau)x(\tau')\rangle j(\tau')/2\hbar^2}. \tag{3.307}$$

Considering now a discrete time axis sliced at $t = t_n$, and inserting the special source current $j(\tau_n) = K\delta_{n,0}$, for instance, we find directly (3.304).

The Wick theorem in this form has an important physical application. The intensity of the sharp diffraction peaks observed in *Bragg scattering* of X-rays on crystal planes is reduced by thermal fluctuations of the atoms in the periodic lattice. The reduction factor is usually written as e^{-2W} and called the *Debye-Waller factor*. In the Gaussian approximation it is given by

$$e^{-W} \equiv \left\langle e^{-\nabla \cdot \mathbf{u}(\mathbf{x})} \right\rangle = e^{-\Sigma_\mathbf{k}\langle |\mathbf{k}\cdot\mathbf{u}(\mathbf{k})|^2\rangle/2}, \tag{3.308}$$

where $\mathbf{u}(\mathbf{x})$ is the atomic displacement field.

If the fluctuations take place around $\langle x(\tau)\rangle \neq 0$, then (3.304) goes obviously over into

$$\left\langle e^{Px} \right\rangle = e^{P\langle x(\tau)\rangle + P^2\langle x - \langle x(\tau)\rangle\rangle^2/2}. \tag{3.309}$$

3.10.1   Real-Time Correlation Functions

The translation of these results to real times is simple. Consider, for example, the harmonic fluctuations $\delta x(t)$ with Dirichlet boundary conditions, which vanish at t_b and t_a. Their correlation functions can be found by using the amplitude (3.23) as a generating functional, if we replace $x(t) \to \delta x(t)$ and $x_b = x_a \to 0$. Differentiating twice with respect to the external currents $j(t)$ we obtain

$$G^{(2)}_{\omega^2}(t,t') = \langle x(t)x(t')\rangle = i\frac{\hbar}{M}G_{\omega^2}(t-t'), \tag{3.310}$$

with the Green function $G_{\omega^2}(t-t')$ of Eq. (3.36), which vanishes if $t = t_b$ or $t = t_a$. The correlation function of $\dot{x}(t)$ is

$$\langle \dot{x}(t)\dot{x}(t')\rangle = i\frac{\hbar}{M}\frac{\cos\omega(t_b - t_>) \cos\omega(t_< - t_a)}{\omega \sin\omega(t_b - t_a)}, \tag{3.311}$$

and has the value

$$\langle \dot{x}(t_b)\dot{x}(t_b)\rangle = i\frac{\hbar}{M}\cot\omega(t_b - t_a).$$ (3.312)

As an application, we use this result to calculate once more the time evolution amplitude $(\mathbf{x}_b t_b|\mathbf{x}_a t_a)$ in a way closely related to the operator method in Section 2.23. We observe that the time derivative of this amplitude has the path integral representation [compare (2.761)]

$$i\hbar\partial_{t_b}(\mathbf{x}_b t_b|\mathbf{x}_a t_a) = -\int \mathcal{D}^D x\, L(\mathbf{x}_b, \dot{\mathbf{x}}_b)e^{i\int_{t_a}^{t_b} dt\, L(\mathbf{x},\dot{\mathbf{x}})/\hbar} = -\langle L(\mathbf{x}_b, \dot{\mathbf{x}}_b)\rangle\, (\mathbf{x}_b t_b|\mathbf{x}_a t_a),$$ (3.313)

and calculate the expectation value $\langle L(\mathbf{x}_b, \dot{\mathbf{x}}_b)\rangle$ as a sum of the classical Lagrangian $L(\mathbf{x}_{\mathrm{cl}}(t_b), \dot{\mathbf{x}}_{\mathrm{cl}}(t_b))$ and the expectation value of the fluctuating part of the Lagrangian $\langle L_{\mathrm{fl}}(\mathbf{x}_b, \dot{\mathbf{x}}_b)\rangle \equiv \langle\, [L(\mathbf{x}_b, \dot{\mathbf{x}}_b) - L(\mathbf{x}_{\mathrm{cl}}(t_b), \dot{\mathbf{x}}_{\mathrm{cl}}(t_b))]\,\rangle$. If the Lagrangian has the standard form $L = M\dot{\mathbf{x}}^2/2 - V(\mathbf{x})$, then only the kinetic term contributes to $\langle L_{\mathrm{fl}}(\mathbf{x}_b, \dot{\mathbf{x}}_b)\rangle$, so that

$$\langle L_{\mathrm{fl}}(\mathbf{x}_b, \dot{\mathbf{x}}_b)\rangle = \frac{M}{2}\langle\delta\dot{\mathbf{x}}_b^2\rangle.$$ (3.314)

There is no contribution from $\langle V(\mathbf{x}_b) - V(\mathbf{x}_{\mathrm{cl}}(t_b))\rangle$, due to the Dirichlet boundary conditions.

The temporal integral over $-[L(\mathbf{x}_{\mathrm{cl}}(t_b), \dot{\mathbf{x}}_{\mathrm{cl}}(t_b)) - \langle L_{\mathrm{fl}}(\mathbf{x}_b, \dot{\mathbf{x}}_b)\rangle]$ agrees with the operator result (2.788), and we obtain the time evolution amplitude from the formula

$$(\mathbf{x}_b t_b|\mathbf{x}_a t_a) = C(\mathbf{x}_b, \mathbf{x}_a)e^{iA(\mathbf{x}_b,\mathbf{x}_a;t_b-t_a)/\hbar}\exp\left(\frac{i}{\hbar}\int_{t_a}^{t_b} dt_{b'}\frac{M}{2}\langle\delta\dot{\mathbf{x}}_{b'}^2\rangle\right),$$ (3.315)

where $A(\mathbf{x}_b, \mathbf{x}_a; t_b - t_a)$ is the classical action $\mathcal{A}[\mathbf{x}_{\mathrm{cl}}]$ expressed as a function of the endpoints [recall (4.87)]. The constant of integration $C(\mathbf{x}_b, \mathbf{x}_a)$ is fixed as in (2.774) by solving the differential equation

$$-i\hbar\nabla_b(\mathbf{x}_b t_b|\mathbf{x}_a t_a) = \langle \mathbf{p}_b\rangle(\mathbf{x}_b t_b|\mathbf{x}_a t_a) = \mathbf{p}_{\mathrm{cl}}(t_b)(\mathbf{x}_b t_b|\mathbf{x}_a t_a),$$ (3.316)

and a similar equation for $\mathbf{x}_a$ [compare (2.775)]. Since the prefactor $\mathbf{p}_{\mathrm{cl}}(t_b)$ on the right-hand side is obtained from the derivative of the exponential $e^{iA(\mathbf{x}_b,\mathbf{x}_a;t_b-t_a)/\hbar}$ in (3.315), due to the general relation (4.88), the constant of integration $C(\mathbf{x}_b, \mathbf{x}_a)$ is actually independent of $\mathbf{x}_b$ and $\mathbf{x}_a$. Thus we obtain from (3.315) once more the known result (3.315).

As an example, take the harmonic oscillator. The terms linear in $\delta\mathbf{x}(t) = \mathbf{x}(t) - \mathbf{x}_{\mathrm{cl}}(t)$ vanish since they are they are odd in $\delta\mathbf{x}(t)$ while the exponent in (3.313) is even. Inserting on the right-hand side of (3.314) the correlation function (3.311), we obtain in D dimensions

$$\langle L_{\mathrm{fl}}(\mathbf{x}_b, \dot{\mathbf{x}}_b)\rangle = \frac{M}{2}\langle\delta\dot{\mathbf{x}}_b^2\rangle = i\frac{\hbar\omega}{2}D\cot\omega(t_b - t_a),$$ (3.317)

which is precisely the second term in Eq. (2.787), with the appropriate opposite sign.

3.11 Correlation Functions of Charged Particle in Magnetic Field and Harmonic Potential

It is straightforward to find the correlation functions of a charged particle in a magnetic and an extra harmonic potential discussed in Section 2.19. They are obtained by inverting the functional matrix (2.691):

$$\mathbf{G}^{(2)}_{\omega^2,B}(\tau,\tau') = \frac{\hbar}{M}\,\mathbf{D}^{-1}_{\omega^2,B}(\tau,\tau'). \tag{3.318}$$

By an ordinary matrix inversion of (2.694), we obtain the Fourier expansion

$$\mathbf{G}^{(2)}_B(\tau,\tau') = \frac{1}{\hbar\beta}\sum_{m=-\infty}^{\infty}\tilde{\mathbf{G}}_{\omega^2,B}(\omega_m)e^{-i\omega_m(\tau-\tau')}, \tag{3.319}$$

with

$$\tilde{\mathbf{G}}^{(2)}_{\omega^2,B}(\omega_m) = \frac{\hbar}{M}\frac{1}{(\omega_m^2+\omega_+^2)(\omega_m^2+\omega_-^2)}\begin{pmatrix} \omega_m^2+\omega^2-\omega_B^2 & 2\omega_B\omega_m \\ -2\omega_B\omega_m & \omega_m^2+\omega^2-\omega_B^2, \end{pmatrix}. \tag{3.320}$$

Since $\omega_+^2+\omega_-^2 = 2(\omega^2+\omega_B^2)$ and $\omega_+^2-\omega_-^2 = 4\omega\omega_B$, the diagonal elements can be written as

$$\frac{1}{2(\omega_m^2+\omega_+^2)(\omega_m^2+\omega_-^2)}\left[(\omega_m^2+\omega_+^2)+(\omega_m^2+\omega_-^2)-4\omega_B^2\right]$$
$$= \frac{1}{2}\left\{\left[\frac{1}{\omega_m^2+\omega_+^2}+\frac{1}{\omega_m^2+\omega_-^2}\right]+\frac{\omega_B}{\omega}\left[\frac{1}{\omega_m^2+\omega_+^2}-\frac{1}{\omega_m^2+\omega_-^2}\right]\right\}. \tag{3.321}$$

Recalling the Fourier expansion (3.245), we obtain directly the diagonal periodic correlation function

$$G^{(2)}_{\omega^2,B,xx} = \frac{\hbar}{4M\omega}\left[\frac{\cosh\omega_+(|\tau-\tau'|-\hbar\beta/2)}{\sinh(\omega_+\hbar\beta/2)}+\frac{\cosh\omega_-(|\tau-\tau'|-\hbar\beta/2)}{\sinh(\omega_-\hbar\beta/2)}\right], \tag{3.322}$$

which is equal to $G^{(2)}_{\omega^2,B,yy}$. The off-diagonal correlation functions have the Fourier components

$$\frac{2\omega_B\omega_m}{(\omega_m^2+\omega_+^2)(\omega_m^2+\omega_-^2)} = \frac{\omega_m}{2\omega}\left[\frac{1}{\omega_m^2+\omega_+^2}-\frac{1}{\omega_m^2+\omega_-^2}\right]. \tag{3.323}$$

Since ω_m are the Fourier components of the derivative $i\partial_\tau$, we can write

$$G^{(2)}_{\omega^2,B,xy}(\tau,\tau') = -G^{(2)}_{\omega^2,B,yx}(\tau,\tau') = \frac{\hbar}{2M}\,i\partial_\tau\left[\frac{1}{2\omega_+}\frac{\cosh\omega_+(|\tau-\tau'|-\hbar\beta/2)}{\sinh(\omega_+\hbar\beta/2)}\right.$$
$$\left.-\frac{1}{2\omega_-}\frac{\cosh\omega_-(|\tau-\tau'|-\hbar\beta/2)}{\sinh(\omega_-\hbar\beta/2)}\right]. \tag{3.324}$$

Performing the derivatives yields

$$G^{(2)}_{\omega^2,B,xy}(\tau,\tau') = G^{(2)}_{\omega^2,B,yx}(\tau,\tau') = \frac{\hbar\epsilon(\tau-\tau')}{2Mi}\left[\frac{1}{2\omega_+}\frac{\sinh\omega_+(|\tau-\tau'|-\hbar\beta/2)}{\sinh(\omega_+\hbar\beta/2)}\right.$$
$$\left.-\frac{1}{2\omega_-}\frac{\sinh\omega_-(|\tau-\tau'|-\hbar\beta/2)}{\sinh(\omega_-\hbar\beta/2)}\right], \quad (3.325)$$

where $\epsilon(\tau-\tau')$ is the step function (1.311).

For a charged particle in a magnetic field without an extra harmonic oscillator we have to take the limit $\omega \to \omega_B$ in these equations. Due to translational invariance of the limiting system, this exists only after removing the zero-mode in the Matsubara sum. This is done most simply in the final expressions by subtracting the high-temperature limits at $\tau = \tau'$. In the diagonal correlation functions (3.322) this yields

$$G^{(2)\prime}_{\omega^2,B,xx}(\tau,\tau') = G^{(2)\prime}_{\omega^2,B,yy}(\tau,\tau') = G^{(2)}_{\omega^2,B,xx} - \frac{1}{\beta M\omega_+\omega_-}, \quad (3.326)$$

where the prime indicates the subtraction. Now one can easily go to the limit $\omega \to \omega_B$ with the result

$$G^{(2)\prime}_{\omega^2,B,xx}(\tau,\tau') = G^{(2)\prime}_{\omega^2,B,yy}(\tau,\tau') = \frac{\hbar}{4M\omega}\left[\frac{\cosh 2\omega(|\tau-\tau'|-\hbar\beta/2)}{\sinh(\beta\hbar\omega)} - \frac{1}{\omega\hbar\beta}\right]. \quad (3.327)$$

For the subtracted off-diagonal correlation functions (3.325) we find

$$G^{(2)\prime}_{\omega^2,B,xy}(\tau,\tau') = -G^{(2)\prime}_{\omega^2,B,yx}(\tau,\tau') = G^{(2)}_{\omega^2,B,xy} + \frac{\hbar\omega_B}{2Mi\omega_+\omega_-}\epsilon(\tau-\tau'). \quad (3.328)$$

For more details see the literature.[7]

3.12 Correlation Functions in Canonical Path Integral

Sometimes it is desirable to know the correlation functions of position and momentum variables

$$G^{(m,n)}_{\omega^2}(\tau_1,\ldots,\tau_m;\tau_1,\ldots,\tau_n) \equiv \langle x(\tau_1)x(\tau_2)\cdots x(\tau_m)p(\tau_1)p(\tau_2)\cdots p(\tau_n)\rangle \quad (3.329)$$
$$\equiv Z^{-1}\int \mathcal{D}x(\tau)\int\frac{\mathcal{D}p(\tau)}{2\pi}x(\tau_1)x(\tau_2)\cdots x(\tau_m)p(\tau)\,p(\tau_1)p(\tau_2)\cdots p(\tau_n)\exp\left(-\frac{1}{\hbar}\mathcal{A}_e\right).$$

These can be obtained from a direct extension of the generating functional (3.233) by another source $k(\tau)$ coupled linearly to the momentum variable $p(\tau)$:

$$Z[j,k] = \int \mathcal{D}x(\tau)\,e^{-\mathcal{A}_e[j,k]/\hbar}. \quad (3.330)$$

---

[7]M. Bachmann, H. Kleinert, and A. Pelster, Phys. Rev. A 62, 52509 (2000) (quant-ph/0005074); Phys. Lett. A 279, 23 (2001) (quant-ph/0005100).

3.12.1   Harmonic Correlation Functions

For the harmonic oscillator, the generating functional (3.330) is denoted by $Z_\omega[j,k]$ and its Euclidean action reads

$$\mathcal{A}_e[j,k] = \int_0^{\hbar\beta} d\tau \left[-ip(\tau)\dot{x}(\tau) + \frac{1}{2M}p^2 + \frac{M}{2}\omega^2 x^2 - j(\tau)x(\tau) - k(\tau)p(\tau) \right], \tag{3.331}$$

the partition function is denoted by $Z_\omega[j,k]$. Introducing the vectors in phase space $\mathbf{V}(\tau) = (p(\tau), x(\tau))$ and $\mathbf{J}(\tau) = (j(\tau), k(\tau))$, this can be written in matrix form as

$$\mathcal{A}_e[\mathbf{J}] = \int_0^{\hbar\beta} d\tau \left(\frac{1}{2}\mathbf{V}^T \mathbf{D}_{\omega^2,e}\mathbf{V} - \mathbf{V}^T\mathbf{J} \right), \tag{3.332}$$

where $\mathbf{D}_{\omega^2,e}(\tau,\tau')$ is the functional matrix

$$\mathbf{D}_{\omega^2,e}(\tau,\tau') \equiv \begin{pmatrix} M\omega^2 & i\partial_\tau \\ -i\partial_\tau & M^{-1} \end{pmatrix} \delta(\tau - \tau'), \quad \tau - \tau' \in [0, \hbar\beta]. \tag{3.333}$$

Its functional inverse is the Euclidean Green function,

$$\begin{aligned}
\mathbf{G}^P_{\omega^2,e}(\tau,\tau') &= \mathbf{G}^P_{\omega^2,e}(\tau-\tau') = \mathbf{D}^{-1}_{\omega^2,e}(\tau,\tau') \\
&= \begin{pmatrix} M^{-1} & -i\partial_\tau \\ i\partial_\tau & M\omega^2 \end{pmatrix} (-\partial_\tau^2 + \omega^2)^{-1}\delta(\tau - \tau'),
\end{aligned} \tag{3.334}$$

with the periodic boundary condition. After performing a quadratic completion as in (3.238) by shifting the path:

$$\mathbf{V} \to \mathbf{V}' = \mathbf{V} + \mathbf{G}^P_{\omega^2,e}\mathbf{J}, \tag{3.335}$$

the Euclidean action takes the form

$$\mathcal{A}_e[\mathbf{J}] = \int_0^{\hbar\beta} d\tau \frac{1}{2}\mathbf{V}'^T \mathbf{D}_{\omega^2,e}\mathbf{V}' - \frac{1}{2}\int_0^{\hbar\beta} d\tau \int_0^{\hbar\beta} d\tau' \mathbf{J}^T(\tau')\mathbf{G}^P_{\omega^2,e}(\tau-\tau')\mathbf{J}(\tau'). \tag{3.336}$$

The fluctuations over the periodic paths $\mathbf{V}'(\tau)$ can now be integrated out and yield for $\mathbf{J}(\tau) \equiv 0$ the oscillator partition function

$$Z_\omega = \mathrm{Det}\, \mathbf{D}^{-1/2}_{\omega^2,e}. \tag{3.337}$$

A Fourier decomposition into Matsubara frequencies

$$\mathbf{D}_{\omega^2,e}(\tau,\tau') = \frac{1}{\hbar\beta}\sum_{m=-\infty}^{\infty} \mathbf{D}^P_{\omega^2,e}(\omega_m)e^{-i\omega_m(\tau-\tau')}, \tag{3.338}$$

has the components

$$\mathbf{D}^P_{\omega^2,e}(\omega_m) = \begin{pmatrix} M^{-1} & \omega_m \\ -\omega_m & M\omega^2 \end{pmatrix}, \tag{3.339}$$

with the determinants

$$\det \mathbf{D}^P_{\omega^2,e}(\omega_m) = \omega_m^2 + \omega^2, \tag{3.340}$$

and the inverses

$$\mathbf{G}^P_e(\omega_m) = [\mathbf{D}^P_{\omega^2,e}(\omega_m)]^{-1} = \begin{pmatrix} M\omega^2 & -\omega_m \\ \omega_m & M^{-1} \end{pmatrix} \frac{1}{\omega_m^2 + \omega^2}. \tag{3.341}$$

The product of determinants (3.340) for all ω_m required in the functional determinant of Eq. (3.337) is calculated with the rules of analytic regularization in Section 2.15, and yields the same partition function as in (3.241), and thus the same partition function (3.242):

$$Z_\omega = \frac{1}{\prod_{m=1}^\infty \sqrt{\omega_m^2 + \omega^2}} = \frac{1}{2\sinh(\beta\hbar\omega/2)}. \tag{3.342}$$

We therefore obtain for arbitrary sources $\mathbf{J}(\tau) = (j(\tau), k(\tau)) \neq 0$ the generating functional

$$Z[\mathbf{J}] = Z_\omega \exp\left\{-\frac{1}{\hbar}\mathcal{A}_e^{\mathbf{J}}[\mathbf{J}]\right\}, \tag{3.343}$$

with the source term

$$\mathcal{A}_e^{\mathbf{J}}[\mathbf{J}] = -\frac{1}{2}\int_0^{\hbar\beta} d\tau \int_0^{\hbar\beta} d\tau' \, \mathbf{J}^T(\tau)\mathbf{G}_{\omega^2,e}^{\mathrm{P}}(\tau,\tau')\mathbf{J}(\tau'). \tag{3.344}$$

The Green function $\mathbf{G}_{\omega^2,e}^{\mathrm{P}}(\tau,\tau')$ follows immediately from Eq. (3.334) and (3.237):

$$\mathbf{G}_{\omega^2,e}^{\mathrm{P}}(\tau,\tau') = \mathbf{G}_{\omega^2,e}^{\mathrm{P}}(\tau-\tau') = \mathbf{D}_{\omega^2,e}^{-1}(\tau,\tau') = \begin{pmatrix} M^{-1} & -i\partial_\tau \\ i\partial_\tau & M\omega^2 \end{pmatrix} G_{\omega^2,e}^{\mathrm{P}}(\tau-\tau'), \tag{3.345}$$

where $G_{\omega^2,e}^{\mathrm{P}}(\tau-\tau')$ is the simple periodic Green function (3.248). From the functional derivatives of (3.343) with respect to $j(\tau)/\hbar$ and $k(\tau)/\hbar$ as in (3.299), we now find the correlation functions

$$G_{\omega^2,e,xx}^{(2)}(\tau,\tau') \equiv \langle x(\tau)x(\tau')\rangle = \frac{\hbar}{M}G_{\omega^2,e}^{\mathrm{P}}(\tau-\tau'), \tag{3.346}$$

$$G_{\omega^2,e,xp}^{(2)}(\tau,\tau') \equiv \langle x(\tau)p(\tau')\rangle = -i\hbar\dot{G}_{\omega^2,e}^{\mathrm{P}}(\tau-\tau'), \tag{3.347}$$

$$G_{\omega^2,e,px}^{(2)}(\tau,\tau') \equiv \langle p(\tau)x(\tau')\rangle = i\hbar\dot{G}_{\omega^2,e}^{\mathrm{P}}(\tau-\tau'), \tag{3.348}$$

$$G_{\omega^2,e,pp}^{(2)}(\tau,\tau') \equiv \langle p(\tau)p(\tau')\rangle = \hbar M\omega^2 G_{\omega^2,e}^{\mathrm{P}}(\tau-\tau'). \tag{3.349}$$

The correlation function $\langle x(\tau)x(\tau')\rangle$ is the same as in the pure configuration space formulation (3.301). The mixed correlation function $\langle p(\tau)x(\tau')\rangle$ is understood immediately by rewriting the current-free part of the action (3.331) as

$$\mathcal{A}_e[0,0] = \int_0^{\hbar\beta} d\tau \left[\frac{1}{2M}(p - iM\dot{x})^2 + \frac{M}{2}\left(\dot{x}^2 + \omega^2 x^2\right)\right], \tag{3.350}$$

which shows that $p(\tau)$ fluctuates harmonically around the classical momentum for imaginary time $iM\dot{x}(\tau)$. It is therefore not surprising that the correlation function $\langle p(\tau)x(\tau')\rangle$ comes out to be the same as that of $iM\langle\dot{x}(\tau)x(\tau')\rangle$. Such an analogy is no longer true for the correlation function $\langle p(\tau)p(\tau')\rangle$. In fact, the correlation function $\langle\dot{x}(\tau)\dot{x}(\tau')\rangle$ is equal to

$$\langle\dot{x}(\tau)\dot{x}(\tau')\rangle = -\hbar M\partial_\tau^2 G_{\omega^2,e}^{\mathrm{P}}(\tau-\tau'). \tag{3.351}$$

Comparison with (3.349) reveals the relation

$$\begin{aligned} \langle p(\tau)p(\tau')\rangle &= \langle\dot{x}(\tau)\dot{x}(\tau')\rangle + \frac{\hbar}{M}\left(-\partial_\tau^2 + \omega^2\right) G_{\omega^2,e}^{\mathrm{P}}(\tau-\tau') \\ &= \langle\dot{x}(\tau)\dot{x}(\tau')\rangle + \frac{\hbar}{M}\delta(\tau-\tau'). \end{aligned} \tag{3.352}$$

The additional δ-function on the right-hand side is the consequence of the fact that $p(\tau)$ is not equal to $iM\dot{x}$, but fluctuates around it harmonically.

For the canonical path integral of a particle in a uniform magnetic field solved in Section 2.18, there are analogous relations. Here we write the canonical action (2.641) with a vector potential (2.638) in the Euclidean form as

$$
\mathcal{A}_e[\mathbf{p}, \mathbf{x}] = \int_0^{\hbar\beta} d\tau \left\{ \frac{1}{2M} \left[\mathbf{p} - \frac{e}{c}\mathbf{B} \times \mathbf{x} - iM\dot{\mathbf{x}} \right]^2 + \frac{M}{2}\omega^2\mathbf{x}^2 \right\},
\tag{3.353}
$$

showing that $\mathbf{p}(\tau)$ fluctuates harmonically around the classical momentum $\mathbf{p}_{cl}(\tau) = (e/c)\mathbf{B} \times \mathbf{x} - iM\dot{\mathbf{x}}$. For a magnetic field pointing in the z-direction we obtain, with the frequency $\omega_B = \omega_L/2$ of Eq. (2.646), the following relations between the correlation functions involving momenta and those involving only coordinates given in (3.322), (3.324), (3.324):

$$
G^{(2)}_{\omega^2,B,xp_x}(\tau, \tau') \equiv \langle x(\tau)p_x(\tau')\rangle = iM\partial_{\tau'}G^{(2)}_{\omega^2,B,xx}(\tau, \tau') - M\omega_B G^{(2)}_{\omega^2,B,xy}(\tau, \tau'),
\tag{3.354}
$$

$$
G^{(2)}_{\omega^2,B,xp_y}(\tau, \tau') \equiv \langle x(\tau)p_y(\tau')\rangle = iM\partial_{\tau'}G^{(2)}_{\omega^2,B,xy}(\tau, \tau') + M\omega_B G^{(2)}_{\omega^2,B,xx}(\tau, \tau'),
\tag{3.355}
$$

$$
G^{(2)}_{\omega^2,B,zp_z}(\tau, \tau') \equiv \langle z(\tau)p_z(\tau')\rangle = iM\partial_{\tau'}G^{(2)}_{\omega^2,B,zz}(\tau, \tau'),
\tag{3.356}
$$

$$
G^{(2)}_{\omega^2,B,p_xp_x}(\tau, \tau') \equiv \langle p_x(\tau)p_x(\tau')\rangle = -M^2\partial_\tau\partial_{\tau'}G^{(2)}_{\omega^2,B,xx}(\tau, \tau') - 2iM^2\omega_B\partial_\tau G^{(2)}_{\omega^2,B,xy}(\tau, \tau')
$$
$$
+ M^2\omega_B^2\, G^{(2)}_{\omega^2,B,xx}(\tau, \tau') + \hbar M\delta(\tau - \tau'),
\tag{3.357}
$$

$$
G^{(2)}_{\omega^2,B,p_xp_y}(\tau, \tau') \equiv \langle p_x(\tau)p_y(\tau')\rangle = -M^2\partial_\tau\partial_{\tau'}G^{(2)}_{\omega^2,B,xy}(\tau, \tau') + iM^2\partial_\tau G^{(2)}_{\omega^2,B,xx}(\tau, \tau')
$$
$$
+ M^2\omega_B^2 G^{(2)}_{\omega^2,B,xy}(\tau, \tau'),
\tag{3.358}
$$

$$
G^{(2)}_{\omega^2,B,p_zp_z}(\tau, \tau') \equiv \langle p_z(\tau)p_z(\tau')\rangle = -M^2\partial_\tau\partial_{\tau'}G^{(2)}_{\omega^2,B,zz}(\tau, \tau') + \hbar M\delta(\tau - \tau').
\tag{3.359}
$$

Only diagonal correlations between momenta contain the extra δ-function on the right-hand side according to the rule (3.352). Note that $\partial_\tau\partial_{\tau'}G^{(2)}_{\omega^2,B,ab}(\tau, \tau') = -\partial_\tau^2 G^{(2)}_{\omega^2,B,ab}(\tau, \tau')$. Each correlation function is, of course, invariant under time translations, depending only on the time difference $\tau - \tau'$.

The correlation functions $\langle x(\tau)x(\tau')\rangle$ and $\langle x(\tau)y(\tau')\rangle$ are the same as before in Eqs. (3.324) and (3.325).

3.12.2　Relations between Various Amplitudes

A slight generalization of the generating functional (3.330) contains paths with fixed endpoints rather than all periodic paths. If the endpoints are held fixed in configuration space, one defines

$$
(x_b\,\hbar\beta|x_a\,0)[j, k] = \int_{x(0)=x_a}^{x(\hbar\beta)=x_b} Dx\frac{Dp}{2\pi\hbar} \exp\left\{ -\frac{1}{\hbar}\mathcal{A}_e[j, k] \right\}.
\tag{3.360}
$$

If the endpoints are held fixed in momentum space, one defines

$$
(p_b\,\hbar\beta|p_a\,0)[j, k] = \int_{p(0)=p_a}^{p(\hbar\beta)=p_b} Dx\frac{Dp}{2\pi\hbar} \exp\left\{ -\frac{1}{\hbar}\mathcal{A}_e[j, k] \right\}.
\tag{3.361}
$$

The two are related by a Fourier transformation

$$
(p_b\,\hbar\beta|p_a\,0)[j, k] = \int_{-\infty}^{+\infty} dx_a \int_{-\infty}^{+\infty} dx_b e^{-i(p_bx_b - p_ax_a)/\hbar}(x_b\,\hbar\beta|x_a\,0)[j, k].
\tag{3.362}
$$

We now observe that in the canonical path integral, the amplitudes (3.360) and (3.361) with fixed endpoints can be reduced to those with vanishing endpoints with modified sources. The modification consists in shifting the current $k(\tau)$ in the action by the source term $ix_b\delta(\tau_b - \tau) -$

$ix_a\delta(\tau - \tau_a)$ and observe that this produces in (3.361) an overall phase factor in the limit $\tau_b \uparrow \hbar\beta$ and $\tau_a \downarrow 0$:

$$\lim_{\tau_b \uparrow \hbar\beta} \lim_{\tau_a \downarrow 0} (p_b \, \hbar\beta|p_a \, 0)[j(\tau), k(\tau) + ix_b\delta(\tau_b - \tau) - ix_a\delta(\tau - \tau_a)]$$

$$= \exp\left\{\frac{i}{\hbar}(p_b x_b - p_a x_a)\right\} (p_b \, \hbar\beta|p_a \, 0)[j(\tau), k(\tau)]. \tag{3.363}$$

By inserting (3.363) into the inverse of the Fourier transformation (3.362),

$$(x_b \, \hbar\beta|x_a \, 0)[j, k] = \int_{-\infty}^{+\infty} \frac{dp_a}{2\pi\hbar} \int_{-\infty}^{+\infty} \frac{dp_b}{2\pi\hbar} e^{i(p_b x_b - p_a x_a)/\hbar} (p_b \, \hbar\beta|p_a \, 0)[j, k], \tag{3.364}$$

we obtain

$$(x_b \, \hbar\beta|x_a \, 0)[j, k] = \lim_{\tau_b \uparrow \hbar\beta} \lim_{\tau_a \downarrow 0} (0 \, \hbar\beta|0 \, 0)[j(\tau), k(\tau) + ix_b\delta(\tau_b - \tau) - ix_a\delta(\tau - \tau_a)]. \tag{3.365}$$

In this way, the fixed-endpoint path integral (3.360) can be reduced to a path integral with vanishing endpoints but additional δ-terms in the current $k(\tau)$ coupled to the momentum $p(\tau)$.

There is also a simple relation between path integrals with fixed equal endpoints and periodic path integrals. The measures of integration are related by

$$\int_{x(0)=x}^{x(\hbar\beta)=x} \frac{\mathcal{D}x\mathcal{D}p}{2\pi\hbar} = \oint \frac{\mathcal{D}x\mathcal{D}p}{2\pi\hbar} \delta(x(0) - x). \tag{3.366}$$

Using the Fourier decomposition of the delta function, we rewrite (3.366) as

$$\int_{x(0)=x}^{x(\hbar\beta)=x} \frac{\mathcal{D}x\mathcal{D}p}{2\pi\hbar} = \lim_{\tau_a' \downarrow 0} \int_{-\infty}^{+\infty} \frac{dp_a}{2\pi\hbar} e^{ip_a x/\hbar} \oint \frac{\mathcal{D}x\mathcal{D}p}{2\pi\hbar} e^{-i\int_0^{\hbar\beta} d\tau \, p_a \delta(\tau - \tau_a')x(\tau)/\hbar}. \tag{3.367}$$

Inserting now (3.367) into (3.365) leads to the announced desired relation

$$(x_b \, \hbar\beta|x_a 0)[k, j] = \lim_{\tau_b \uparrow \hbar\beta} \lim_{\tau_a \downarrow 0} \lim_{\tau_a' \downarrow 0} \int_{-\infty}^{+\infty} \frac{dp_a}{2\pi\hbar}$$

$$\times Z[j(\tau) - ip_a\delta(\tau - \tau_a'), k(\tau) + ix_b\delta(\tau_b - \tau) - ix_a\delta(\tau - \tau_a)], \tag{3.368}$$

where $Z[j, k]$ is the thermodynamic partition function (3.330) summing all periodic paths. When using (3.368) we must be careful in evaluating the three limits. The limit $\tau_a' \downarrow 0$ has to be evaluated prior to the other limits $\tau_b \uparrow \hbar\beta$ and $\tau_a \downarrow 0$.

3.12.3 Harmonic Generating Functionals

Here we write down explicitly the harmonic generating functionals with the above shifted source terms:

$$\tilde{k}(\tau) = k(\tau) + ix_b\delta(\tau_b - \tau) - ix_a\delta(\tau - \tau_a), \quad \tilde{j}(\tau) = j(\tau) - ip\delta(\tau - \tau_a'), \tag{3.369}$$

leading to the factorized generating functional

$$Z_\omega[\tilde{k}, \tilde{j}] = Z_\omega^{(0)}[0, 0]Z_\omega^{(1)}[k, j]Z_\omega^{\mathrm{p}}[k, j]. \tag{3.370}$$

The respective terms on the right-hand side of (3.370) read in detail

$$Z_\omega^{(0)}[0, 0] = Z_\omega \exp\left(\frac{1}{2\hbar^2}\left\{-p^2 G_{xx}^{\mathrm{p}}(\tau_a', \tau_a') - 2p\left[x_a G_{xp}^{\mathrm{p}}(\tau_a', \tau_a) + x_b G_{xp}^{\mathrm{p}}(\tau_a', \tau_b)\right]\right.\right.$$

$$-x_a^2 G_{pp}^{\mathrm{p}}(\tau_a, \tau_a) - x_b^2 G_{pp}^{\mathrm{p}}(\tau_b, \tau_b) + 2x_a x_b G_{pp}^{\mathrm{p}}(\tau_a, \tau_b)\}\big) , \tag{3.371}$$

$$Z_\omega^{(1)}[k,j] = \exp\left(\frac{1}{\hbar^2}\int_0^{\hbar\beta} d\tau \left\{ j(\tau)[-ipG_{xx}^{\mathrm{p}}(\tau,\tau_a') + ix_b G_{xp}^{\mathrm{p}}(\tau,\tau_b) - ix_a G_{xp}^{\mathrm{p}}(\tau,\tau_a)] \right.\right.$$

$$\left.\left. + k(\tau)[-ipG_{xp}^{\mathrm{p}}(\tau,\tau_a') + ix_b G_{pp}^{\mathrm{p}}(\tau,\tau_b) - ix_a G_{pp}^{\mathrm{p}}(\tau,\tau_a)]\right\}\right), \tag{3.372}$$

$$Z_\omega^{\mathrm{p}}[k,j] = \exp\left\{\frac{1}{2\hbar^2}\int_0^{\hbar\beta}d\tau_1 \int_0^{\hbar\beta}d\tau_2 \left[(j(\tau_1), k(\tau_2))\right.\right.$$

$$\left.\left. \times \begin{pmatrix} G_{xx}^{\mathrm{p}}(\tau_1,\tau_1) & G_{xp}^{\mathrm{p}}(\tau_1,\tau_2) \\ G_{px}^{\mathrm{p}}(\tau_1,\tau_2) & G_{pp}^{\mathrm{p}}(\tau_1,\tau_2) \end{pmatrix}\begin{pmatrix} j(\tau_2) \\ k(\tau_2) \end{pmatrix}\right]\right\}, \tag{3.373}$$

where Z_ω is given by (3.342) and $G_{xp}^{\mathrm{p}}(\tau_1,\tau_2)$ etc. are the periodic Euclidean Green functions $G_{\omega^2,e,ab}^{(2)}(\tau_1,\tau_2)$ defined in Eqs. (3.346)–(3.349) in an abbreviated notation. Inserting (3.370) into (3.368) and performing the Gaussian momentum integration, over the exponentials in $Z_\omega^{(0)}[0,0]$ and $Z_\omega^{(1)}[k,j]$, the result is

$$(x_b\,\hbar\beta|x_a\,0)[k,j] = (x_b\,\hbar\beta|x_a\,0)[0,0] \times \exp\left\{\frac{1}{\hbar}\int_0^{\hbar\beta}d\tau\,[x_{\mathrm{cl}}(\tau)j(\tau) + p_{\mathrm{cl}}(\tau)k(\tau)]\right\}$$

$$\times \exp\left\{\frac{1}{2\hbar^2}\int_0^{\hbar\beta}d\tau_1 \int_0^{\hbar\beta}d\tau_2\,[(j(\tau_1),k(\tau_2))\begin{pmatrix} G_{xx}^{(\mathrm{D})}(\tau_1,\tau_2) & G_{xp}^{(\mathrm{D})}(\tau_1,\tau_2) \\ G_{px}^{(\mathrm{D})}(\tau_1,\tau_2) & G_{pp}^{(\mathrm{D})}(\tau_1,\tau_2) \end{pmatrix}\begin{pmatrix} j(\tau_2) \\ k(\tau_2) \end{pmatrix}]\right\}, \tag{3.374}$$

where the Green functions $G_{ab}^{(\mathrm{D})}(\tau_1,\tau_2)$ have now Dirichlet boundary conditions. In particular, the Green function $G_{ab}^{(\mathrm{D})}(\tau_1,\tau_2)$ is equal to (3.36) continued to imaginary time. The Green functions $G_{xp}^{(\mathrm{D})}(\tau_1,\tau_2)$ and $G_{pp}^{(\mathrm{D})}(\tau_1,\tau_2)$ are Dirichlet versions of Eqs. (3.346)–(3.349) which arise from the above Gaussian momentum integrals.

After performing the integrals, the first factor without currents is

$$(x_b\,\hbar\beta|x_a 0)[0,0] = \lim_{\tau_b\uparrow\hbar\beta}\lim_{\tau_a\downarrow 0}\lim_{\tau_a'\downarrow 0}\frac{Z_\omega}{2\pi\hbar}\sqrt{\frac{2\pi\hbar^2}{G_{xx}^{\mathrm{p}}(\tau_a',\tau_a')}}$$

$$\times \exp\left[\frac{1}{2\hbar^2}\left(x_a^2\left\{\frac{G_{xp}^{\mathrm{p}}{}^2(\tau_a',\tau_a)}{G_{xx}^{\mathrm{p}}(\tau_a',\tau_a')} - G_{pp}^{\mathrm{p}}(\tau_a,\tau_a)\right\} + x_b^2\left\{\frac{G_{xp}^{\mathrm{p}}{}^2(\tau_a',\tau_b)}{G_{xx}^{\mathrm{p}}(\tau_a',\tau_a')} - G_{pp}^{\mathrm{p}}(\tau_b,\tau_b)\right\}\right.\right.$$

$$\left.\left. - 2x_a x_b\left\{\frac{G_{xp}^{\mathrm{p}}(\tau_a',\tau_a)G_{xp}^{\mathrm{p}}(\tau_a',\tau_b)}{G_{xx}^{\mathrm{p}}(\tau_a',\tau_a')} - G_{pp}^{\mathrm{p}}(\tau_a,\tau_b)\right\}\right)\right]. \tag{3.375}$$

Performing the limits using

$$\lim_{\tau_a\downarrow 0}\lim_{\tau_a'\downarrow 0} G_{xp}^{\mathrm{p}}(\tau_a',\tau_a) = -i\frac{\hbar}{2}, \tag{3.376}$$

where the order of the respective limits turns out to be important, we obtain the amplitude (2.409):

$$(x_b\,\hbar\beta|x_a 0)[0,0] = \sqrt{\frac{M\omega}{2\pi\hbar\sinh\hbar\beta\omega}}$$

$$\times \exp\left\{-\frac{M\omega}{2\hbar\sinh\hbar\beta\omega}\left[(x_a^2 + x_b^2)\cosh\hbar\beta\omega - 2x_a x_b\right]\right\}. \tag{3.377}$$

The first exponential in (3.374) contains a complicated representation of the classical path

$$x_{\mathrm{cl}}(\tau) = \lim_{\tau_b\uparrow\hbar\beta}\lim_{\tau_a\downarrow 0}\lim_{\tau_a'\downarrow 0}\frac{i}{\hbar}\left\{x_a\left[\frac{G_{xp}^{\mathrm{p}}(\tau_a',\tau_a)G_{xx}^{\mathrm{p}}(\tau,\tau_a')}{G_{xx}^{\mathrm{p}}(\tau_a',\tau_a')} + G_{xp}^{\mathrm{p}}(\tau_a,\tau)\right]\right.$$

$$-x_b\left[\frac{G_{xp}^{\mathrm{p}}(\tau_a',\tau_b)G_{xx}^{\mathrm{p}}(\tau,\tau_a')}{G_{xx}^{\mathrm{p}}(\tau_a',\tau_a')}+G_{xp}^{\mathrm{p}}(\tau_b,\tau)\right]\right\}, \tag{3.378}$$

and of the classical momentum

$$p_{\mathrm{cl}}(\tau) = \lim_{\tau_b\uparrow\hbar\beta}\lim_{\tau_a\downarrow 0}\lim_{\tau_a'\downarrow 0}\frac{i}{\hbar}\left\{x_a\left[\frac{G_{xp}^{\mathrm{p}}(\tau_a',\tau_a)G_{xp}^{\mathrm{p}}(\tau_a',\tau)}{G_{xx}^{\mathrm{p}}(\tau_a',\tau_a')}-G_{pp}^{\mathrm{p}}(\tau_a,\tau)\right]\right.$$
$$\left.-x_b\left[\frac{G_{xp}^{\mathrm{p}}(\tau_a',\tau_b)G_{xp}^{\mathrm{p}}(\tau_a',\tau)}{G_{xx}^{\mathrm{p}}(\tau_a',\tau_a')}-G_{pp}^{\mathrm{p}}(\tau_b,\tau)\right]\right\}. \tag{3.379}$$

Indeed, inserting the explicit periodic Green functions (3.346)–(3.349) and going to the limits we obtain

$$x_{\mathrm{cl}}(\tau) = \frac{x_a\sinh\omega(\hbar\beta-\tau)+x_b\sinh\omega\tau}{\sinh\hbar\beta\omega} \tag{3.380}$$

and

$$p_{\mathrm{cl}}(\tau) = iM\omega\frac{-x_a\cosh\omega(\hbar\beta-\tau)+x_b\cosh\omega\tau}{\sinh\hbar\beta\omega}, \tag{3.381}$$

the first being the imaginary-time version of the classical path (3.6), the second being related to it by the classical relation $p_{\mathrm{cl}}(\tau)=iM\,dx_{\mathrm{cl}}(\tau)/d\tau$.

The second exponential in (3.374) quadratic in the currents contains the Green functions with Dirichlet boundary conditions

$$G_{xx}^{(\mathrm{D})}(\tau_1,\tau_2) = G_{xx}^{\mathrm{p}}(\tau_1,\tau_2)-\frac{G_{xx}^{\mathrm{p}}(\tau_1,0)G_{xx}^{\mathrm{p}}(\tau_2,0)}{G_{xx}^{\mathrm{p}}(\tau_1,\tau_1)}, \tag{3.382}$$

$$G_{xp}^{(\mathrm{D})}(\tau_1,\tau_2) = G_{xp}^{\mathrm{p}}(\tau_1,\tau_2)+\frac{G_{xx}^{\mathrm{p}}(\tau_1,0)G_{xp}^{\mathrm{p}}(\tau_2,0)}{G_{xx}^{\mathrm{p}}(\tau_1,\tau_1)}, \tag{3.383}$$

$$G_{px}^{(\mathrm{D})}(\tau_1,\tau_2) = G_{px}^{\mathrm{p}}(\tau_1,\tau_2)+\frac{G_{xp}^{\mathrm{p}}(\tau_1,0)G_{xx}^{\mathrm{p}}(\tau_2,0)}{G_{xx}^{\mathrm{p}}(\tau_1,\tau_1)}, \tag{3.384}$$

$$G_{pp}^{(\mathrm{D})}(\tau_1,\tau_2) = G_{pp}^{\mathrm{p}}(\tau_1,\tau_2)-\frac{G_{xp}^{\mathrm{p}}(\tau_1,0)G_{xp}^{\mathrm{p}}(\tau_2,0)}{G_{xx}^{\mathrm{p}}(\tau_1,\tau_1)}. \tag{3.385}$$

After applying some trigonometric identities, these take the form

$$G_{xx}^{(\mathrm{D})}(\tau_1,\tau_2) = \frac{\hbar}{2M\omega\sinh\hbar\beta\omega}\left[\cosh\omega(\hbar\beta-|\tau_1-\tau_2|)-\cosh\omega(\hbar\beta-\tau_1-\tau_2)\right], \tag{3.386}$$

$$G_{xp}^{(\mathrm{D})}(\tau_1,\tau_2) = \frac{i\hbar}{2\sinh\hbar\beta\omega}\left\{\theta(\tau_1-\tau_2)\sinh\omega(\hbar\beta-|\tau_1-\tau_2|)\right.$$
$$\left.-\theta(\tau_2-\tau_1)\sinh\omega(\hbar\beta-|\tau_2-\tau_1|)+\sinh\omega(\hbar\beta-\tau_1-\tau_2)\right\}, \tag{3.387}$$

$$G_{px}^{(\mathrm{D})}(\tau_1,\tau_2) = -\frac{i\hbar}{2\sinh\hbar\beta\omega}\left\{\theta(\tau_1-\tau_2)\sinh\omega(\hbar\beta-|\tau_1-\tau_2|)\right.$$
$$\left.-\theta(\tau_2-\tau_1)\sinh\omega(\hbar\beta-|\tau_2-\tau_1|)-\sinh\omega(\hbar\beta-\tau_1-\tau_2)\right\}, \tag{3.388}$$

$$G_{pp}^{(\mathrm{D})}(\tau_1,\tau_2) = \frac{M\hbar\omega}{2\sinh\hbar\beta\omega}\left[\cosh\omega(\hbar\beta-|\tau_1-\tau_2|)+\cosh\omega(\hbar\beta-\tau_1-\tau_2)\right]. \tag{3.389}$$

The first correlation function is, of course, the imaginary-time version of the Green function (3.206). Observe the symmetry properties under interchange of the time arguments:

$$G_{xx}^{(\mathrm{D})}(\tau_1,\tau_2) = G_{xx}^{(\mathrm{D})}(\tau_2,\tau_1), \qquad G_{xp}^{(\mathrm{D})}(\tau_1,\tau_2) = -G_{xp}^{(\mathrm{D})}(\tau_2,\tau_1), \tag{3.390}$$

$$G_{px}^{(\mathrm{D})}(\tau_1,\tau_2) = -G_{px}^{(\mathrm{D})}(\tau_2,\tau_1), \qquad G_{pp}^{(\mathrm{D})}(\tau_1,\tau_2) = G_{pp}^{(\mathrm{D})}(\tau_2,\tau_1), \tag{3.391}$$

and the identity

$$G_{xp}^{(\mathrm{D})}(\tau_1, \tau_2) = G_{px}^{(\mathrm{D})}(\tau_2, \tau_1). \tag{3.392}$$

In addition, there are the following derivative relations between the Green functions with Dirichlet boundary conditions:

$$G_{xp}^{(\mathrm{D})}(\tau_1, \tau_2) = -iM\frac{\partial}{\partial \tau_1}G_{xx}^{(\mathrm{D})}(\tau_1, \tau_2) = iM\frac{\partial}{\partial \tau_2}G_{xx}^{(\mathrm{D})}(\tau_1, \tau_2), \tag{3.393}$$

$$G_{px}^{(\mathrm{D})}(\tau_1, \tau_2) = iM\frac{\partial}{\partial \tau_1}G_{xx}^{(\mathrm{D})}(\tau_1, \tau_2) = -iM\frac{\partial}{\partial \tau_2}G_{xx}^{(\mathrm{D})}(\tau_1, \tau_2), \tag{3.394}$$

$$G_{pp}^{(\mathrm{D})}(\tau_1, \tau_2) = \hbar M\delta(\tau_1-\tau_2) - M^2\frac{\partial^2}{\partial \tau_1\partial \tau_2}G_{xx}^{(\mathrm{D})}(\tau_1-\tau_2). \tag{3.395}$$

Note that Eq. (3.382) is a nonlinear alternative to the additive decomposition (3.142) of a Green function with Dirichlet boundary conditions: into Green functions with periodic boundary conditions.

3.13　Particle in Heat Bath

The results of Section 3.8 are the key to understanding the behavior of a quantum-mechanical particle moving through a dissipative medium at a fixed temperature T. We imagine the coordinate $x(t)$ a particle of mass M to be coupled linearly to a *heat bath* consisting of a great number of harmonic oscillators $X_i(\tau)$ ($i = 1, 2, 3, \ldots$) with various masses M_i and frequencies Ω_i. The imaginary-time path integral in this heat bath is given by

$$
\begin{aligned}
(x_b\hbar\beta|x_a0) = {}& \prod_i \oint \mathcal{D}X_i(\tau)\int_{x(0)=x_a}^{x(\hbar\beta)=x_b}\mathcal{D}x(\tau) \\
& \times \exp\left\{-\frac{1}{\hbar}\int_0^{\hbar\beta}d\tau\sum_i\left[\frac{M_i}{2}(\dot{X}_i^2 + \Omega_i^2 X_i^2)\right]\right\} \\
& \times \exp\left\{-\frac{1}{\hbar}\int_0^{\hbar\beta}d\tau\left[\frac{M}{2}\dot{x}^2 + V(x(\tau)) - \sum_i c_i X_i(\tau)x(\tau)\right]\right\} \times \frac{1}{\prod_i Z_i},
\end{aligned}
\tag{3.396}
$$

where we have allowed for an arbitrary potential $V(x)$. The partition functions of the individual bath oscillators

$$
\begin{aligned}
Z_i & \equiv \oint \mathcal{D}X_i(\tau)\exp\left\{-\frac{1}{\hbar}\int_0^{\hbar\beta}d\tau\left[\frac{M_i}{2}(\dot{X}_i^2 + \Omega_i^2 X_i^2)\right]\right\} \\
& = \frac{1}{2\sinh(\hbar\beta\Omega_i/2)}
\end{aligned}
\tag{3.397}
$$

have been divided out, since their thermal behavior is trivial and will be of no interest in the sequel. The path integrals over $X_i(\tau)$ can be performed as in Section 3.1 leading for each oscillator label i to a source expression like (3.243), in which $c_i x(\tau)$ plays the role of a current $j(\tau)$. The result can be written as

$$
(x_b\hbar\beta|x_a0) = \int_{x(0)=x_a}^{x(\hbar\beta)=x_b}\mathcal{D}x(\tau)\exp\left\{-\frac{1}{\hbar}\int_0^{\hbar\beta}d\tau\left[\frac{M}{2}\dot{x}^2 + V(x(\tau))\right] - \frac{1}{\hbar}\mathcal{A}_{\mathrm{bath}}[x]\right\},
\tag{3.398}
$$

where $\mathcal{A}_{\text{bath}}[x]$ is a *nonlocal* action for the particle motion generated by the bath

$$\mathcal{A}_{\text{bath}}[x] = -\frac{1}{2} \int_0^{\hbar\beta} d\tau \int_0^{\hbar\beta} d\tau' x(\tau)\alpha(\tau - \tau')x(\tau'). \qquad (3.399)$$

The function $\alpha(\tau - \tau')$ is the weighted periodic correlation function (3.248):

$$\begin{aligned}
\alpha(\tau - \tau') &= \sum_i c_i^2 \frac{1}{M_i} G_{\Omega_i^2,\text{e}}^{\text{p}}(\tau - \tau') \\
&= \sum_i \frac{c_i^2}{2M_i\Omega_i} \frac{\cosh\Omega_i(|\tau - \tau'| - \hbar\beta/2)}{\sinh(\Omega_i\hbar\beta/2)}.
\end{aligned} \qquad (3.400)$$

Its Fourier expansion has the Matsubara frequencies $\omega_m = 2\pi k_B T/\hbar$

$$\alpha(\tau - \tau') = \frac{1}{\hbar\beta} \sum_{m=-\infty}^{\infty} \alpha_m e^{-i\omega_m(\tau - \tau')}, \qquad (3.401)$$

with the coefficients

$$\alpha_m = \sum_i \frac{c_i^2}{M_i} \frac{1}{\omega_m^2 + \omega_i^2}. \qquad (3.402)$$

Alternatively, we can write the bath action in the form corresponding to (3.276) as

$$\mathcal{A}_{\text{bath}}[x] = -\frac{1}{2} \int_0^{\hbar\beta} d\tau \int_{-\infty}^{\infty} d\tau' x(\tau)\alpha_0(\tau - \tau')x(\tau'), \qquad (3.403)$$

with the weighted nonperiodic correlation function [recall (3.277)]

$$\alpha_0(\tau - \tau') = \sum_i \frac{c_i^2}{2M_i\Omega_i} e^{-\Omega_i|\tau - \tau'|}. \qquad (3.404)$$

The bath properties are conveniently summarized by the *spectral density of the bath*

$$\rho_{\text{b}}(\omega') \equiv 2\pi \sum_i \frac{c_i^2}{2M_i\Omega_i} \delta(\omega' - \Omega_i). \qquad (3.405)$$

The frequencies Ω_i are by definition positive numbers. The spectral density allows us to express $\alpha_0(\tau - \tau')$ as the spectral integral

$$\alpha_0(\tau - \tau') = \int_0^{\infty} \frac{d\omega'}{2\pi} \rho_{\text{b}}(\omega') e^{-\omega'|\tau - \tau'|}, \qquad (3.406)$$

and similarly

$$\alpha(\tau - \tau') = \int_0^{\infty} \frac{d\omega'}{2\pi} \rho_{\text{b}}(\omega') \frac{\cosh\omega'(|\tau - \tau'| - \hbar\beta/2)}{\sinh(\omega'\hbar\beta/2)}. \qquad (3.407)$$

For the Fourier coefficients (3.402), the spectral integral reads

$$\alpha_m = \int_0^\infty \frac{d\omega'}{2\pi} \rho_b(\omega') \frac{2\omega'}{\omega_m^2 + \omega'^2}. \tag{3.408}$$

It is useful to subtract from these coefficients the first term α_0, and to invert the sign of the remainder making it positive definite. Thus we split

$$\alpha_m = 2 \int_0^\infty \frac{d\omega'}{2\pi} \frac{\rho_b(\omega')}{\omega'} \left(1 - \frac{\omega_m^2}{\omega_m^2 + \omega'^2} \right) = \alpha_0 - g_m. \tag{3.409}$$

Then the Fourier expansion (3.401) separates as

$$\alpha(\tau - \tau') = \alpha_0 \delta^P(\tau - \tau') - g(\tau - \tau'), \tag{3.410}$$

where $\delta^P(\tau - \tau')$ is the periodic δ-function (3.279):

$$\delta^P(\tau - \tau') = \frac{1}{\hbar\beta} \sum_{m=-\infty}^{\infty} e^{-i\omega_m(\tau-\tau')} = \sum_{n=-\infty}^{\infty} \delta(\tau - \tau' - n\hbar\beta), \tag{3.411}$$

the right-hand sum following from Poisson's summation formula (1.197). The subtracted correlation function

$$g(\tau - \tau') = \frac{1}{\hbar\beta} \sum_{m=-\infty}^{\infty} g(\omega_m) e^{-i\omega_m(\tau-\tau')}, \tag{3.412}$$

has the coefficients

$$g_m = \sum_i \frac{c_i^2}{M_i} \frac{\omega_m^2}{\omega_m^2 + \Omega_i^2} = \int_0^\infty \frac{d\omega'}{2\pi} \frac{\rho_b(\omega')}{\omega'} \frac{2\omega_m^2}{\omega_m^2 + \omega'^2}. \tag{3.413}$$

The corresponding decomposition of the bath action (3.399) is

$$\mathcal{A}_{\text{bath}}[x] = \mathcal{A}_{\text{loc}} + \mathcal{A}'_{\text{bath}}[x], \tag{3.414}$$

where

$$\mathcal{A}'_{\text{bath}}[x] = \frac{1}{2} \int_0^{\hbar\beta} d\tau \int_0^{\hbar\beta} d\tau' x(\tau) g(\tau - \tau') x(\tau'), \tag{3.415}$$

and

$$\mathcal{A}_{\text{loc}} = -\frac{\alpha_0}{2} \int_0^{\hbar\beta} d\tau \, x^2(\tau), \tag{3.416}$$

is a local action which can be added to the original action in Eq. (3.398), changing merely the curvature of the potential $V(x)$. Because of this effect, it is useful to introduce a *frequency shift* $\Delta\omega^2$ via the equation

$$M\Delta\omega^2 \equiv -\alpha_0 = -2 \int_0^\infty \frac{d\omega'}{2\pi} \frac{\rho_b(\omega')}{\omega'} = -\sum_i \frac{c_i^2}{M_i \Omega_i^2}. \tag{3.417}$$

Then the local action (3.416) becomes

$$\mathcal{A}_{\text{loc}} = \frac{M}{2}\Delta\omega^2 \int_0^{\hbar\beta} d\tau\, x^2(\tau). \tag{3.418}$$

This can be absorbed into the potential of the path integral (3.398), yielding a *renormalized potential*

$$V_{\text{ren}}(x) = V(x) + \frac{M}{2}\Delta\omega^2\, x^2. \tag{3.419}$$

With the decomposition (3.414), the path integral (3.398) acquires the form

$$(x_b\hbar\beta|x_a 0) = \int_{x(0)=x_a}^{x(\hbar\beta)=x_b} \mathcal{D}x(\tau)\exp\left\{-\frac{1}{\hbar}\int_0^{\hbar\beta} d\tau\left[\frac{M}{2}\dot{x}^2 + V_{\text{ren}}(x(\tau))\right] - \frac{1}{\hbar}\mathcal{A}'_{\text{bath}}[x]\right\}. \tag{3.420}$$

The subtracted correlation function (3.412) has the property

$$\int_0^{\hbar\beta} d\tau\, g(\tau - \tau') = 0. \tag{3.421}$$

Thus, if we rewrite in (3.415)

$$x(\tau)x(\tau') = \frac{1}{2}\{x^2(\tau) + x^2(\tau') - [x(\tau) - x(\tau')]^2\}, \tag{3.422}$$

the first two terms do not contribute, and we remain with

$$\mathcal{A}'_{\text{bath}}[x] = -\frac{1}{4}\int_0^{\hbar\beta} d\tau \int_0^{\hbar\beta} d\tau'\, g(\tau - \tau')[x(\tau) - x(\tau')]^2. \tag{3.423}$$

If the oscillator frequencies Ω_i are densely distributed, the function $\rho_b(\omega')$ is continuous. As will be shown later in Eqs. (18.208) and (18.317), an oscillator bath introduces in general a *friction force* into classical equations of motion. If this is to have the usual form $-M\gamma\dot{x}(t)$, the spectral density of the bath must have the approximation

$$\rho_b(\omega') \approx 2M\gamma\omega' \tag{3.424}$$

[see Eqs. (18.208), (18.317)]. This approximation is characteristic for *Ohmic dissipation*. In general, a typical friction force increases with ω only for small frequencies; for larger ω, it decreases again. An often applicable phenomenological approximation is the so-called *Drude form*

$$\rho_b(\omega') \approx 2M\gamma\omega'\frac{\omega_D^2}{\omega_D^2 + \omega'^2}, \tag{3.425}$$

where $1/\omega_D \equiv \tau_D$ is *Drude's relaxation time*. For times much shorter than the Drude time τ_D, there is no dissipation. In the limit of large ω_D, the Drude form describes again Ohmic dissipation.

Inserting (3.425) into (3.413), we obtain the Fourier coefficients for Drude dissipation

$$g_m = 2M\gamma\omega_D^2 \int_0^\infty \frac{d\omega}{2\pi} \frac{1}{\omega_D^2 + \omega^2} \frac{2\omega_m^2}{\omega_m^2 + \omega^2} = M|\omega_m|\gamma \frac{\omega_D}{|\omega_m| + \omega_D}. \qquad (3.426)$$

It is customary, to factorize

$$g_m \equiv M|\omega_m|\gamma_m, \qquad (3.427)$$

so that Drude dissipation corresponds to

$$\gamma_m = \gamma \frac{\omega_D}{|\omega_m| + \omega_D}, \qquad (3.428)$$

and Ohmic dissipation to $\gamma_m \equiv \gamma$.

The Drude form of the spectral density gives rise to a frequency shift (3.417)

$$\Delta\omega^2 = -\gamma\omega_D, \qquad (3.429)$$

which goes to infinity in the Ohmic limit $\omega_D \to \infty$.

3.14 Heat Bath of Photons

The heat bath in the last section was a convenient phenomenological tool to reproduce the Ohmic friction observed in many physical systems. In nature, there can be various different sources of dissipation. The most elementary of these is the deexcitation of atoms by radiation, which at zero temperature gives rise to the natural line width of atoms. The photons may form a thermally equilibrated gas, the most famous example being the cosmic black-body radiation which is a gas of the photons of 3 K left over from the big bang 15 billion years ago (and which create a sizable fraction of the blips on our television screens).

The theoretical description is quite simple. We decompose the vector potential $\mathbf{A}(\mathbf{x}, t)$ of electromagnetism into Fourier components of wave vector $\mathbf{k}$

$$\mathbf{A}(\mathbf{x}, t) = \sum_{\mathbf{k}} c_{\mathbf{k}}(\mathbf{x})\mathbf{X}_{\mathbf{k}}(t), \qquad c_{\mathbf{k}} = \frac{e^{i\mathbf{k}\mathbf{x}}}{\sqrt{2\Omega_{\mathbf{k}}V}}, \qquad \sum_{\mathbf{k}} = \int \frac{d^3kV}{(2\pi)^3}. \qquad (3.430)$$

The Fourier components $\mathbf{X}_{\mathbf{k}}(t)$ can be considered as a sum of harmonic oscillators of frequency $\Omega_{\mathbf{k}} = c|\mathbf{k}|$, where c is the light velocity. A photon of wave vector $\mathbf{k}$ is a quantum of $\mathbf{X}_{\mathbf{k}}(t)$. A certain number N of photons with the same wave vector can be described as the Nth excited state of the oscillator $\mathbf{X}_{\mathbf{k}}(t)$. The statistical sum of these harmonic oscillators led Planck to his famous formula for the energy of black-body radiation for photons in an otherwise empty cavity whose walls have a temperature T. These will form the bath, and we shall now study its effect on the quantum mechanics of a charged point particle. Its coupling to the vector potential is given by the interaction (2.632). Comparison with the coupling to the heat bath in Eq. (3.396) shows that we simply have to replace $-\sum_i c_i X_i(\tau) x(\tau)$ by $-\sum_{\mathbf{k}} c_{\mathbf{k}} \mathbf{X}_{\mathbf{k}}(\tau) \dot{x}(\tau)$. The bath action (3.399) takes then the form

$$A_{\text{bath}}[x] = -\frac{1}{2} \int_0^{\hbar\beta} d\tau \int_0^{\hbar\beta} d\tau' \dot{x}^i(\tau) \alpha^{ij}(\mathbf{x}(\tau), \tau; \mathbf{x}(\tau'), \tau') \dot{x}^j(\tau'), \qquad (3.431)$$

where $\alpha^{ij}(\mathbf{x}, \tau; \mathbf{x}', \tau')$ is a 3×3 matrix generalization of the correlation function (3.400):

$$\alpha^{ij}(\mathbf{x}, \tau; \mathbf{x}', \tau') = \frac{e^2}{\hbar c^2} \sum_{\mathbf{k}} c_{-\mathbf{k}}(\mathbf{x}) c_{\mathbf{k}}(\mathbf{x}') \langle X^i_{-\mathbf{k}}(\tau) X^j_{\mathbf{k}}(\tau') \rangle. \qquad (3.432)$$

We now have to account for the fact that there are two polarization states for each photon, which are transverse to the momentum direction. We therefore introduce a transverse Kronecker symbol

$$^T\delta^{ij}_{\mathbf{k}} \equiv (\delta^{ij} - k^i k^j / \mathbf{k}^2) \qquad (3.433)$$

and write the correlation function of a single oscillator $X^i_{-\mathbf{k}}(\tau)$ as

$$G^{ij}_{-\mathbf{k}'\mathbf{k}}(\tau - \tau') = \langle \hat{X}^i_{-\mathbf{k}'}(\tau) \hat{X}^j_{\mathbf{k}'}(\tau') \rangle = \hbar \,^T\delta^{ij}_{\mathbf{k}} \, \delta_{\mathbf{k}\mathbf{k}'} \, G^{\text{p}}_{\omega^2, e\,\mathbf{k}}(\tau - \tau'), \qquad (3.434)$$

with

$$G^{\text{p}}_{\omega^2, e\,\mathbf{k}}(\tau - \tau') \equiv \frac{1}{2\Omega_{\mathbf{k}}} \frac{\cosh \Omega_{\mathbf{k}}(|\tau - \tau'| - \hbar\beta/2)}{\sinh(\Omega_{\mathbf{k}}\hbar\beta/2)}. \qquad (3.435)$$

Thus we find

$$\alpha^{ij}(\mathbf{x}, \tau; \mathbf{x}', \tau') = \frac{e^2}{c^2} \int \frac{d^3 k}{(2\pi)^3} \,^T\delta^{ij}_{\mathbf{k}} \frac{e^{i\mathbf{k}(\mathbf{x}-\mathbf{x}')}}{2\Omega_{\mathbf{k}}} \frac{\cosh \Omega_{\mathbf{k}}(|\tau - \tau'| - \hbar\beta/2)}{\sinh(\Omega_{\mathbf{k}}\hbar\beta/2)}. \qquad (3.436)$$

At zero temperature, and expressing $\Omega_{\mathbf{k}} = c|\mathbf{k}|$, this simplifies to

$$\alpha^{ij}(\mathbf{x}, \tau; \mathbf{x}', \tau') = \frac{e^2}{c^3} \int \frac{d^3 k}{(2\pi)^3} \,^T\delta^{ij}_{\mathbf{k}} \frac{e^{i\mathbf{k}(\mathbf{x}-\mathbf{x}')-c|\mathbf{k}||\tau-\tau'|}}{2|\mathbf{k}|}. \qquad (3.437)$$

Forgetting for a moment the transverse Kronecker symbol and the prefactor e^2/c^2, the integral yields

$$G^R_e(\mathbf{x}, \tau; \mathbf{x}', \tau') = \frac{1}{4\pi^2 c^2} \frac{1}{(\tau - \tau')^2 + (\mathbf{x} - \mathbf{x}')^2 / c^2}, \qquad (3.438)$$

which is the imaginary-time version of the well-known retarded Green function used in electromagnetism. If the system is small compared to the average wavelengths in the bath we can neglect the retardation and omit the term $(\mathbf{x} - \mathbf{x}')^2 / c^2$. In the finite-temperature expression (3.437) this amounts to neglecting the $\mathbf{x}$-dependence.

The transverse Kronecker symbol can then be averaged over all directions of the wave vector and yields simply $2\delta^{ij}/3$, and we obtain the approximate function

$$\alpha^{ij}(\mathbf{x}, \tau; \mathbf{x}', \tau') = \frac{2e^2}{3c^2} \delta^{ij} \frac{1}{2\pi c^2} \int \frac{d\omega}{2\pi} \omega \frac{\cosh \omega(|\tau - \tau'| - \hbar\beta/2)}{\sinh(\omega\hbar\beta/2)}. \tag{3.439}$$

This has the generic form (3.407) with the spectral function of the photon bath

$$\rho_{\mathrm{pb}}(\omega') = \frac{e^2}{3c^2\pi}\omega'. \tag{3.440}$$

This has precisely the Ohmic form (3.424), but there is now an important difference: the bath action (3.431) contains now the time derivatives of the paths $\mathbf{x}(\tau)$. This gives rise to an extra factor ω'^2 in (3.424), so that we may define a spectral density for the photon bath:

$$\rho_{\mathrm{pb}}(\omega') \approx 2M\gamma\omega'^3, \qquad \gamma = \frac{e^2}{6c^2\pi M}. \tag{3.441}$$

In contrast to the usual friction constant γ in the previous section, this has the dimension 1/frequency.

3.15   Harmonic Oscillator in Ohmic Heat Bath

For a harmonic oscillator in an Ohmic heat bath, the partition function can be calculated as follows. Setting

$$V_{\mathrm{ren}}(x) = \frac{M}{2}\omega^2 x^2, \tag{3.442}$$

the Fourier decomposition of the action (3.420) reads

$$\mathcal{A}_e = \frac{M\hbar}{k_B T} \left\{ \frac{\omega^2}{2}x_0^2 + \sum_{m=1}^{\infty} \left[\omega_m^2 + \omega^2 + \omega_m\gamma_m\right] |x_m|^2 \right\}. \tag{3.443}$$

The harmonic potential is the full renormalized potential (3.419). Performing the Gaussian integrals using the measure (2.445), we obtain the partition function for the damped harmonic oscillator of frequency ω [compare (2.406)]

$$Z_\omega^{\mathrm{damp}} = \frac{k_B T}{\hbar\omega} \left\{ \prod_{m=1}^{\infty} \left[\frac{\omega_m^2 + \omega^2 + \omega_m\gamma_m}{\omega_m^2}\right] \right\}^{-1}. \tag{3.444}$$

For the Drude dissipation (3.426), this can be written as

$$Z_\omega^{\mathrm{damp}} = \frac{k_B T}{\hbar\omega} \prod_{m=1}^{\infty} \frac{\omega_m^2(\omega_m + \omega_D)}{\omega_m^3 + \omega_m^2\omega_D + \omega_m(\omega^2 + \gamma\omega_D) + \omega_D\omega^2}. \tag{3.445}$$

Let w_1, w_2, w_3 be the roots of the cubic equation

$$w^3 - w^2\omega_D + w(\omega^2 + \gamma\omega_D) - \omega^2\omega_D = 0. \tag{3.446}$$

Then we can rewrite (3.445) as

$$Z_\omega^{\text{damp}} = \frac{k_B T}{\hbar \omega} \prod_{m=1}^\infty \frac{\omega_m}{\omega_m + w_1} \frac{\omega_m}{\omega_m + w_2} \frac{\omega_m}{\omega_m + w_3} \frac{\omega_m + \omega_D}{\omega_m}. \tag{3.447}$$

Using the product representation of the Gamma function[8]

$$\Gamma(z) = \lim_{n\to\infty} \frac{n^z}{z} \prod_{m=1}^n \frac{m}{m+z} \tag{3.448}$$

and the fact that

$$w_1 + w_2 + w_3 - \omega_D = 0, \qquad w_1 w_2 w_3 = \omega^2 \omega_D, \tag{3.449}$$

the partition function (3.447) becomes

$$Z_\omega^{\text{damp}} = \frac{1}{2\pi} \frac{\omega}{\omega_1} \frac{\Gamma(w_1/\omega_1)\Gamma(w_2/\omega_1)\Gamma(w_3/\omega_1)}{\Gamma(\omega_D/\omega_1)}, \tag{3.450}$$

where $\omega_1 = 2\pi k_B T/\hbar$ is the first Matsubara frequency, such that $w_i/\omega_1 = w_i\beta/2\pi$. In the Ohmic limit $\omega_D \to \infty$, the roots w_1, w_2, w_3 reduce to

$$w_1 = \gamma/2 + i\delta, \qquad w_1 = \gamma/2 - i\delta, \qquad w_3 = \omega_D - \gamma, \tag{3.451}$$

with

$$\delta \equiv \sqrt{\omega^2 - \gamma^2/4}, \tag{3.452}$$

and (3.450) simplifies further to

$$Z_\omega^{\text{damp}} = \frac{1}{2\pi} \frac{\omega}{\omega_1} \Gamma(w_1/\omega_1)\Gamma(w_2/\omega_1). \tag{3.453}$$

For vanishing friction, the roots w_1 and w_2 become simply $w_1 = i\omega$, $w_2 = -i\omega$, and the formula[9]

$$\Gamma(1-z)\Gamma(z) = \frac{\pi}{\sin \pi z} \tag{3.454}$$

can be used to calculate

$$\Gamma(i\omega/\omega_1)\Gamma(-i\omega/\omega_1) = \frac{\omega_1}{\omega} \frac{\pi}{\sinh(\pi\omega/\omega_1)} = \frac{\omega_1}{\omega} \frac{\pi}{\sinh(\omega\hbar/2k_B T)}, \tag{3.455}$$

showing that (3.450) goes properly over into the partition function (3.214) of the undamped harmonic oscillator.

The free energy of the system is

$$\begin{aligned} F(T) = \; & -k_B T \left[\log(\omega/2\pi\omega_1) - \log\Gamma(\omega_D/\omega_1) \right. \\ & \left. + \log\Gamma(w_1/\omega_1) + \log\Gamma(w_2/\omega_1) + \log\Gamma(w_3/\omega_1)\right]. \end{aligned} \tag{3.456}$$

---

[8]I.S. Gradshteyn and I.M. Ryzhik, *op. cit.*, Formula 8.322.
[9]*ibid.*, Formula 8.334.3.

Using the large-z behavior of $\log \Gamma(z)$[10]

$$\log \Gamma(z) = \left(z - \frac{1}{2}\right) \log z - z + \frac{1}{2} \log 2\pi + \frac{1}{12z} - \frac{1}{360z^3} - \mathcal{O}(1/z^5), \qquad (3.457)$$

we find the free energy at low temperature

$$F(T) \sim E_0 - \left(\frac{1}{w_1} + \frac{1}{w_2} + \frac{1}{w_1} - \frac{w^2}{w_1 w_2 w_3}\right) \frac{\pi}{6\hbar}(k_B T)^2 = E_0 - \frac{\gamma\pi}{6\omega^2\hbar}(k_B T)^2, \quad (3.458)$$

where

$$E_0 = -\frac{\hbar}{2\pi}\left[w_1 \log(w_1/\omega_D) + w_2 \log(w_2/\omega_D) + w_3 \log(w_3/\omega_D)\right] \qquad (3.459)$$

is the ground state energy.

For small friction, this reduces to

$$E_0 = \frac{\hbar\omega}{2} + \frac{\gamma}{2\pi} \log \frac{\omega_D}{\omega} - \frac{\gamma^2}{16\omega}\left(1 + \frac{4\omega}{\pi\omega_D}\right) + \mathcal{O}(\gamma^3). \qquad (3.460)$$

The T^2-behavior of $F(T)$ in Eq. (3.458) is typical for Ohmic dissipation.

At zero temperature, the Matsubara frequencies $\omega_m = 2\pi m k_B T/\hbar$ move arbitrarily close together, so that Matsubara sums become integrals according to the rule

$$\frac{1}{\hbar\beta} \sum_m \xrightarrow[T\to 0]{} \int_0^\infty \frac{d\omega_m}{2\pi}. \qquad (3.461)$$

Applying this limiting procedure to the logarithm of the product formula (3.445), the ground state energy can also be written as an integral

$$E_0 = \frac{\hbar}{2\pi} \int_0^\infty d\omega_m \log\left[\frac{\omega_m^3 + \omega_m^2 \omega_D + \omega_m(\omega^2 + \gamma\omega_D) + \omega_D\omega^2}{\omega_m^2(\omega_m + \omega_D)}\right], \qquad (3.462)$$

which shows that the energy E_0 increases with the friction coefficient γ.

It is instructive to calculate the density of states defined in (1.580). Inverting the Laplace transform (1.579), we have to evaluate

$$\rho(\varepsilon) = \frac{1}{2\pi i} \int_{\eta - i\infty}^{\eta + i\infty} d\beta\, e^{i\varepsilon\beta}\, Z_\omega^{\mathrm{damp}}(\beta), \qquad (3.463)$$

where η is an infinitesimally small positive number. In the absence of friction, the integral over $Z_\omega(\beta) = \sum_{n=0}^\infty e^{-\beta\hbar\omega(n+1/2)}$ yields

$$\rho(\varepsilon) = \sum_{n=0}^\infty \delta(\varepsilon - (n + 1/2)\hbar\omega). \qquad (3.464)$$

In the presence of friction, we expect the sharp δ-function spikes to be broadened. The calculation is done as follows: The vertical line of integration in the complex β-plane in (3.463) is moved all the way to the left, thereby picking up the poles

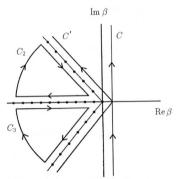

Figure 3.5 Poles in complex β-plane of Fourier integral (3.463) coming from the Gamma functions of (3.450)

of the Gamma functions which lie at negative integer values of $w_i\beta/2\pi$. From the representation of the Gamma function[11]

$$\Gamma(z) = \int_1^\infty dt\, t^{z-1} e^{-t} + \sum_{n=0}^\infty \frac{(-1)^n}{n!(z+n)} \tag{3.465}$$

we see the size of the residues. Thus we obtain the sum

$$\rho(\varepsilon) = \frac{1}{\omega} \sum_{n=1}^\infty \sum_{i=1}^3 R_{n,i} e^{-2\pi n\varepsilon/w_i}, \tag{3.466}$$

$$R_{n,1} = \frac{\omega}{w_1^2} \frac{(-1)^{n-1}}{(n-1)!} \frac{\Gamma(-nw_2/w_1)\Gamma(-nw_3/w_1)}{\Gamma(-nw_D/w_1)}, \tag{3.467}$$

with analogous expressions for $R_{n,2}$ and $R_{n,3}$. The sum can be done numerically and yields the curves shown in Fig. 3.6 for typical underdamped and overdamped situations. There is an isolated δ-function at the ground state energy E_0 of (3.459) which is not widened by the friction. Right above E_0, the curve continues from a finite value $\rho(E_0 + 0) = \gamma\pi/6\omega^2$ determined by the first expansion term in (3.458).

3.16   Harmonic Oscillator in Photon Heat Bath

It is straightforward to extend this result to a photon bath where the spectral density is given by (3.441) and (3.468) becomes

$$Z_\omega^{\text{damp}} = \frac{k_B T}{\hbar\omega} \left\{ \prod_{m=1}^\infty \left[\frac{\omega_m^2 + \omega^2 + \omega_m^3 \gamma}{\omega_m^2} \right] \right\}^{-1}, \tag{3.468}$$

---

[10] *ibid.*, Formula 8.327.
[11] *ibid.*, Formula 8.314.

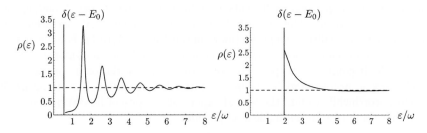

Figure 3.6 Density of states for weak and strong damping in natural units. On the left, the parameters are $\gamma/\omega = 0.2$, $\omega_D/\omega = 10$, on the right $\gamma/\omega = 5$, $\omega_D/\omega = 10$. For more details see Hanke and Zwerger in Notes and References.

with $\gamma = e^2/6c^2\pi M$. The power of ω_m accompanying the friction constant is increased by two units. Adding a Drude correction for the high-frequency behavior we replace γ by $\omega_m/(\omega_m + \omega_D)$ and obtain instead of (3.445)

$$Z_\omega^{\text{damp}} = \frac{k_B T}{\hbar\omega} \prod_{m=1}^{\infty} \frac{\omega_m^2 (\omega_m + \omega_D)(1 + \gamma\omega_D)}{\omega_m^3 (1 + \gamma\omega_D) + \omega_m^2 \omega_D + \omega_m\omega^2 + \omega_D\omega^2}. \tag{3.469}$$

The resulting partition function has again the form (3.450), except that w_{123} are the solutions of the cubic equation

$$w^3(1 + \gamma\omega_D) - w^2\omega_D + w\omega^2 - \omega^2\omega_D = 0. \tag{3.470}$$

Since the electromagnetic coupling is small, we can solve this equation to lowest order in γ. If we also assume ω_D to be large compared to ω, we find the roots

$$w_1 \approx \gamma_{\text{pb}}^{\text{eff}}/2 + i\omega, \qquad w_1 \approx \gamma_{\text{pb}}^{\text{eff}}/2 - i\omega, \qquad w_3 \approx \omega_D/(1 + \gamma_{\text{pb}}^{\text{eff}}\omega_D/\omega^2), \tag{3.471}$$

where we have introduced an effective friction constant of the photon bath

$$\gamma_{\text{pb}}^{\text{eff}} = \frac{e^2}{6c^2\pi M}\omega^2, \tag{3.472}$$

which has the dimension of a frequency, just as the usual friction constant γ in the previous heat bath equations (3.451).

3.17 Perturbation Expansion of Anharmonic Systems

If a harmonic system is disturbed by an additional anharmonic potential $V(x)$, to be called *interaction*, the path integral can be solved exactly only in exceptional cases. These will be treated in Chapters 8, 13, and 14. For sufficiently smooth and small potentials $V(x)$, it is always possible to expand the full partition in powers of the interaction strength. The result is the so-called *perturbation series*. Unfortunately, it only renders reliable numerical results for very small $V(x)$ since, as we shall prove in Chapter 17, the expansion coefficients grow for large orders k like $k!$, making the series strongly divergent. The can only be used for extremely small perturbations. Such expansions are called *asymptotic* (more in Subsection 17.10.1). For this reason we are forced to develop a more powerful technique of studying anharmonic systems in Chapter 5. It combines the perturbation series with a variational approach and will yield very accurate energy levels up to arbitrarily large interaction strengths. It is therefore worthwhile to find the formal expansion in spite of its divergence.

Consider the quantum-mechanical amplitude

$$(x_b t_b | x_a t_a) = \int_{x(t_a)=x_a}^{x(t_b)=x_b} \mathcal{D}x \exp\left\{ \frac{i}{\hbar} \int_{t_a}^{t_b} dt \left[\frac{M}{2}\dot{x}^2 - M\frac{\omega^2}{2}x^2 - V(x) \right] \right\}, \quad (3.473)$$

and expand the integrand in powers of $V(x)$, which leads to the series

$$(x_b t_b | x_a t_a) = \int_{x(t_a)=x_a}^{x(t_b)=x_b} \mathcal{D}x \left[1 - \frac{i}{\hbar} \int_{t_a}^{t_b} dt V(x(t)) \right.$$
$$- \frac{1}{2!\hbar^2} \int_{t_a}^{t_b} dt_2 V(x(t_2)) \int_{t_a}^{t_b} dt_1 V(x(t_1))$$
$$\left. + \frac{i}{3!\hbar^3} \int_{t_a}^{t_b} dt_3 V(x(t_3)) \int_{t_a}^{t_b} dt_2 V(x(t_2)) \int_{t_a}^{t_b} dt_1 V(x(t_1)) + \dots \right]$$
$$\times \exp\left[\frac{i}{\hbar} \int_{t_a}^{t_b} dt \frac{M}{2} \left(\dot{x}^2 - \omega^2 x^2 \right) \right]. \quad (3.474)$$

If we decompose the path integral in the nth term into a product (2.4), the expansion can be rewritten as

$$(x_b t_b | x_a t_a) = (x_b t_b | x_a t_a) - \frac{i}{\hbar} \int_{t_a}^{t_b} dt_1 \int dx_1 (x_b t_b | x_1 t_1) V(x_1)(x_b t_b | x_a t_a) \quad (3.475)$$
$$- \frac{1}{2!\hbar^2} \int_{t_a}^{t_b} dt_2 \int_{t_a}^{t_b} dt_1 \int dx_1 dx_2 (x_b t_b | x_2 t_2) V(x_2)(x_2 t_2 | x_1 t_1) V(x_1)(x_1 t_1 | x_a t_a) + \dots.$$

A similar expansion can be given for the Euclidean path integral of a partition function

$$Z = \oint \mathcal{D}x \exp\left\{ -\frac{1}{\hbar} \int_0^{\hbar\beta} d\tau \left[\frac{M}{2}(\dot{x}^2 + \omega^2 x^2) + V(x) \right] \right\}, \quad (3.476)$$

where we obtain

$$Z = \int \mathcal{D}x \left[1 - \frac{1}{\hbar} \int_0^{\hbar\beta} d\tau V(x(\tau)) + \frac{1}{2!\hbar^2} \int_0^{\hbar\beta} d\tau_2 V(x(\tau_2)) \int_0^{\hbar\beta} d\tau_1 V(x(\tau_1)) \right.$$
$$\left. - \frac{1}{3!\hbar^3} \int_0^{\hbar\beta} d\tau_3 V(x(\tau_3)) \int_0^{\hbar\beta} d\tau_2 V(x(\tau_2)) \int_0^{\hbar\beta} d\tau_1 V(x(\tau_1)) + \dots \right]$$
$$\times \exp\left\{ -\frac{1}{\hbar} \int_0^{\hbar\beta} dt \left[\frac{M}{2}\dot{x}^2 + M\frac{\omega^2}{2}x^2 \right] \right\}. \quad (3.477)$$

The individual terms are obviously expectation values of powers of the Euclidean interaction

$$\mathcal{A}_{\text{int,e}} \equiv \int_0^{\hbar\beta} d\tau V(x(\tau)), \quad (3.478)$$

calculated within the harmonic-oscillator partition function Z_ω. The expectation values are defined by

$$\langle \ldots \rangle_\omega \equiv Z_\omega^{-1} \int \mathcal{D}x \ldots \exp\left\{ -\frac{1}{\hbar} \int_0^{\hbar\beta} d\tau \left[\frac{M}{2} (\dot{x}^2 + \omega^2 x^2) \right] \right\}. \qquad (3.479)$$

With these, the perturbation series can be written in the form

$$Z = \left(1 - \frac{1}{\hbar} \langle \mathcal{A}_{\text{int,e}} \rangle_\omega + \frac{1}{2!\hbar^2} \langle \mathcal{A}_{\text{int,e}}^2 \rangle_\omega - \frac{1}{3!\hbar^3} \langle \mathcal{A}_{\text{int,e}}^3 \rangle_\omega + \ldots \right) Z_\omega. \qquad (3.480)$$

As we shall see immediately, it is preferable to resum the prefactor into an exponential of a series

$$1 - \frac{1}{\hbar} \langle \mathcal{A}_{\text{int,e}} \rangle_\omega + \frac{1}{2!\hbar^2} \langle \mathcal{A}_{\text{int,e}}^2 \rangle_\omega - \frac{1}{3!\hbar^3} \langle \mathcal{A}_{\text{int,e}}^3 \rangle_\omega + \ldots$$

$$= \exp\left\{ -\frac{1}{\hbar} \langle \mathcal{A}_{\text{int,e}} \rangle_\omega + \frac{1}{2!\hbar^2} \langle \mathcal{A}_{\text{int,e}}^2 \rangle_{\omega,\text{c}} - \frac{1}{3!\hbar^3} \langle \mathcal{A}_{\text{int,e}}^3 \rangle_{\omega,\text{c}} + \ldots \right\}. \qquad (3.481)$$

The expectation values $\langle \rangle_{\omega,\text{c}}$ are called *cumulants*. They are related to the original expectation values by the *cumulant expansion*:[12]

$$\langle \mathcal{A}_{\text{int,e}}^2 \rangle_{\omega,\text{c}} \equiv \langle \mathcal{A}_{\text{int,e}}^2 \rangle_\omega - \langle \mathcal{A}_{\text{int,e}} \rangle_\omega^2 \qquad (3.482)$$

$$= \langle [\mathcal{A}_{\text{int,e}} - \langle \mathcal{A}_{\text{int,e}} \rangle_\omega]^2 \rangle_\omega,$$

$$\langle \mathcal{A}_{\text{int,e}}^3 \rangle_{\omega,\text{c}} \equiv \langle \mathcal{A}_{\text{int,e}}^3 \rangle_\omega - 3\langle \mathcal{A}_{\text{int,e}}^2 \rangle_\omega \langle \mathcal{A}_{\text{int,e}} \rangle_\omega + 2\langle \mathcal{A}_{\text{int,e}} \rangle_\omega^3 \qquad (3.483)$$

$$= \langle [\mathcal{A}_{\text{int,e}} - \langle \mathcal{A}_{\text{int,e}} \rangle_\omega]^3 \rangle_\omega,$$

$$\vdots$$

The cumulants contribute directly to the free energy $F = -(1/\beta)\log Z$. From (3.481) and (3.480) we conclude that the anharmonic potential $V(x)$ shifts the free energy of the harmonic oscillator $F_\omega = (1/\beta)\log[2\sinh(\hbar\beta\omega/2)]$ by

$$\Delta F = \frac{1}{\beta} \left(\frac{1}{\hbar} \langle \mathcal{A}_{\text{int,e}} \rangle_\omega - \frac{1}{2!\hbar^2} \langle \mathcal{A}_{\text{int,e}}^2 \rangle_{\omega,\text{c}} + \frac{1}{3!\hbar^3} \langle \mathcal{A}_{\text{int,e}}^3 \rangle_{\omega,\text{c}} + \ldots \right). \qquad (3.484)$$

Whereas the original expectation values $\left\langle \mathcal{A}_{\text{int,e}}^n \right\rangle_\omega$ grow for large β with the nth power of β, due to contributions of n disconnected diagrams of first order in g which are integrated independently over τ from 0 to $\hbar\beta$, the cumulants $\left\langle \mathcal{A}_{\text{int,,e}}^n \right\rangle_\omega$ are proportional to β, thus ensuring that the free energy F has a finite limit, the ground state energy E_0. In comparison with the ground state energy of the unperturbed harmonic system, the energy E_0 is shifted by

$$\Delta E_0 = \lim_{\beta\to\infty} \frac{1}{\beta} \left(\frac{1}{\hbar} \langle \mathcal{A}_{\text{int,e}} \rangle_\omega - \frac{1}{2!\hbar^2} \langle \mathcal{A}_{\text{int,e}}^2 \rangle_{\omega,\text{c}} + \frac{1}{3!\hbar^3} \langle \mathcal{A}_{\text{int,e}}^3 \rangle_{\omega,\text{c}} + \ldots \right). \qquad (3.485)$$

---

[12]Note that the subtracted expressions in the second lines of these equations are particularly simple only for the lowest two cumulants given here.

There exists a simple functional formula for the perturbation expansion of the partition function in terms of the generating functional $Z_\omega[j]$ of the unperturbed harmonic system. Adding a source term into the action of the path integral (3.476), we define the generating functional of the interacting theory:

$$Z[j] = \oint \mathcal{D}x \exp\left\{-\frac{1}{\hbar}\int_0^{\hbar\beta} d\tau \left[\frac{M}{2}\left(\dot{x}^2 + \omega^2 x^2\right) + V(x) - jx\right]\right\}. \qquad (3.486)$$

The interaction can be brought outside the path integral in the form

$$Z[j] = e^{-\frac{1}{\hbar}\int_0^{\hbar\beta} d\tau V(\delta/\delta j(\tau))} Z_\omega[j] \ . \qquad (3.487)$$

The interacting partition function is obviously

$$Z = Z[0]. \qquad (3.488)$$

Indeed, after inserting on the right-hand side the explicit path integral expression for $Z[j]$ from (3.233):

$$Z_\omega[j] = \int \mathcal{D}x \exp\left\{-\frac{1}{\hbar}\int_0^{\hbar\beta} d\tau \left[\frac{M}{2}(\dot{x}^2 + \omega^2 x^2) - jx\right]\right\}, \qquad (3.489)$$

and expanding the exponential in the prefactor

$$e^{-\frac{1}{\hbar}\int_0^{\hbar\beta} d\tau V(\delta/\delta j(\tau))} = 1 - \frac{1}{\hbar}\int_0^{\hbar\beta} d\tau V(\delta/\delta j(\tau))$$

$$+\frac{1}{2!\hbar^2}\int_0^{\hbar\beta} d\tau_2 V(\delta/\delta j(\tau_2))\int_0^{\hbar\beta} d\tau_1 V(\delta/\delta j(\tau_1)) \qquad (3.490)$$

$$-\frac{1}{3!\hbar^3}\int_0^{\hbar\beta} d\tau_3 V(\delta/\delta j(\tau_3))\int_0^{\hbar\beta} d\tau_2 V(\delta/\delta(\tau_2))\int_0^{\hbar\beta} d\tau_1 V(\delta/\delta(\tau_1)) + \ldots ,$$

the functional derivatives of $Z[j]$ with respect to the source $j(\tau)$ generate inside the path integral precisely the expansion (3.480), whose cumulants lead to formula (3.484) for the shift in the free energy.

Before continuing, let us mention that the partition function (3.476) can, of course, be viewed as a generating functional for the calculation of the expectation values of the action and its powers. We simply have to form the derivatives with respect to $\hbar^{-1}$:

$$\langle \mathcal{A}^n \rangle = Z^{-1}\frac{\partial^n}{\partial \hbar^{-1n}} Z_\omega[j]\bigg|_{\hbar^{-1}=0}. \qquad (3.491)$$

For a harmonic oscillator where Z is given by (3.242), this yields

$$\langle \mathcal{A} \rangle = \lim_{\hbar\to\infty} Z_\omega^{-1} \hbar^2 \frac{\partial}{\partial \hbar} Z_\omega = \lim_{\hbar\to\infty} \hbar\frac{\hbar\omega\beta}{2\sinh\hbar\omega\beta/2} = 0. \qquad (3.492)$$

The same result is, incidentally, obtained by calculating the expectation value of the action with analytic regularization:

$$\left\langle \dot{x}^2(\tau) \right\rangle_\omega + \omega^2 \left\langle x^2(\tau) \right\rangle_\omega = \int \frac{d\omega'}{2\pi}\frac{\omega'^2}{\omega'^2 + \omega^2} + \int \frac{d\omega'}{2\pi}\frac{\omega^2}{\omega'^2 + \omega^2} = \int \frac{d\omega'}{2\pi} = 0. \,(3.493)$$

The integral vanishes by Veltman's rule (2.506).

3.18　Rayleigh-Schrödinger and Brillouin-Wigner Perturbation Expansion

The expectation values in formula (3.484) can be evaluated by means of the so-called *Rayleigh-Schrödinger perturbation expansion*, also referred to as *old-fashioned perturbation expansion*. This expansion is particularly useful if the potential $V(x)$ is not a polynomial in x. Examples are $V(x) = \delta(x)$ and $V(x) = 1/x$. In these two cases the perturbation expansions can be summed to all orders, as will be shown for the first example in Section 9.5. For the second example the reader is referred to the literature.[13] We shall explicitly demonstrate the procedure for the ground state and the excited energies of an anharmonic oscillator. Later we shall also give expansions for scattering amplitudes.

To calculate the free-energy shift ΔF in Eq. (3.484) to first order in $V(x)$, we need the expectation

$$\langle \mathcal{A}_{\mathrm{int,e}}\rangle_\omega \equiv Z_\omega^{-1} \int_0^{\hbar\beta} d\tau_1 \int dx dx_1 (x\, \hbar\beta | x_1\, \tau_1)_\omega V(x_1)(x_1\tau_1 | x\, 0)_\omega. \tag{3.494}$$

The time evolution amplitude on the right describes the temporal development of the harmonic oscillator located initially at the point x, from the imaginary time 0 up to τ_1. At the time τ_1, the state is subject to the interaction depending on its position $x_1 = x(\tau_1)$ with the amplitude $V(x_1)$. After that, the state is carried to the final state at the point x by the other time evolution amplitude.

To second order we have to calculate the expectation in $V(x)$:

$$\frac{1}{2}\langle \mathcal{A}_{\mathrm{int,e}}^2\rangle_\omega \equiv Z_\omega^{-1} \int_0^{\hbar\beta} d\tau_2 \int_0^{\hbar\beta} d\tau_1 \int dx dx_2 dx_1 (x\, \hbar\beta | x_2\tau_2)_\omega V(x_2)$$
$$\times (x_2\tau_2 | x_1\tau_1)_\omega V(x_1)(x_1\tau_1 | x\, 0)_\omega. \tag{3.495}$$

The integration over τ_1 is taken only up to τ_2 since the contribution with $\tau_1 > \tau_2$ would merely render a factor 2.

The explicit evaluation of the integrals is facilitated by the spectral expansion (2.298). The time evolution amplitude at imaginary times is given in terms of the eigenstates $\psi_n(x)$ of the harmonic oscillator with the energy $E_n = \hbar\omega(n + 1/2)$:

$$(x_b\tau_b | x_a\tau_a)_\omega = \sum_{n=0}^\infty \psi_n(x_b)\psi_n^*(x_a)e^{-E_n(\tau_b-\tau_a)/\hbar}. \tag{3.496}$$

The same type of expansion exists also for the real-time evolution amplitude. This leads to the Rayleigh-Schrödinger perturbation expansion for the energy shifts of all excited states, as we now show.

The amplitude can be projected onto the eigenstates of the harmonic oscillator. For this, the two sides are multiplied by the harmonic wave functions $\psi_n^*(x_b)$ and

---

[13]M.J. Goovaerts and J.T. Devreese, J. Math. Phys. *13*, 1070 (1972).

$\psi_n(x_a)$ of quantum number n and integrated over x_b and x_a, respectively, resulting in the expansion

$$\int dx_b dx_a \psi_n^*(x_b)(x_b t_b | x_a t_a) \psi_n(x_a) = \int dx_b dx_a \psi_n^*(x_b)(x_b t_b | x_a t_a)_\omega \psi_n(x_a)$$
$$\times \left(1 + \frac{i}{\hbar} \langle n | \mathcal{A}_{\text{int}} | n \rangle_\omega - \frac{1}{2!\hbar^2} \langle n | \mathcal{A}_{\text{int}}^2 | n \rangle_\omega - \frac{i}{3!\hbar^3} \langle n | \mathcal{A}_{\text{int}}^3 | n \rangle_\omega + \dots \right), \quad (3.497)$$

with the interaction

$$\mathcal{A}_{\text{int}} \equiv - \int_{t_a}^{t_b} dt \, V(x(t)). \quad (3.498)$$

The expectation values are defined by

$$\langle n | \dots | n \rangle_\omega \equiv Z_{\text{QM},\omega,n}^{-1} \int dx_b dx_a \psi_n^*(x_b) \left(\int_{x(t_a)=x_a}^{x(t_b)=x_b} \mathcal{D}x \dots e^{i\mathcal{A}_\omega/\hbar} \right) \psi_n(x_a), \quad (3.499)$$

where

$$Z_{\text{QM},\omega,n} \equiv e^{-i\omega(n+1/2)(t_b - t_a)} \quad (3.500)$$

is the projection of the quantum-mechanical partition function of the harmonic oscillator

$$Z_{\text{QM},\omega} = \sum_{n=0}^{\infty} e^{-i\omega(n+1/2)(t_b - t_a)}$$

[see (2.40)] onto the nth excited state.

The expectation values are calculated as in (3.494), (3.495). To first order in $V(x)$, one has

$$\langle n | \mathcal{A}_{\text{int}} | n \rangle_\omega \equiv -Z_{\text{QM},\omega,n}^{-1} \int_{t_a}^{t_b} dt_1 \int dx_b dx_a dx_1 \psi_n^*(x_b)(x_b t_b | x_1 t_1)_\omega$$
$$\times V(x_1)(x_1 t_1 | x_a t_a)_\omega \psi_n(x_a). \quad (3.501)$$

The time evolution amplitude on the right-hand side describes the temporal development of the initial state $\psi_n(x_a)$ from the time t_a to the time t_1, where the interaction takes place with an amplitude $-V(x_1)$. After that, the time evolution amplitude on the left-hand side carries the state to $\psi_n^*(x_b)$.

To second order in $V(x)$, the expectation value is given by the double integral

$$\frac{1}{2} \langle n | \mathcal{A}_{\text{int}}^2 | n \rangle_\omega \equiv Z_{\text{QM},\omega,n}^{-1} \int_{t_a}^{t_b} dt_2 \int_{t_a}^{t_2} dt_1 \int dx_b dx_a dx_2 dx_1$$
$$\times \psi_n^*(x_b)(x_b t_b | x_2 t_2)_\omega V(x_2)(x_2 t_2 | x_1 t_1)_\omega V(x_1)(x_1 t_1 | x_a t_a)_\omega \psi_n(x_a). \quad (3.502)$$

As in (3.495), the integral over t_1 ends at t_2.

By analogy with (3.481), we resum the corrections in (3.497) to bring them into the exponent:

$$1 + \frac{i}{\hbar} \langle n | \mathcal{A}_{\text{int}} | n \rangle_\omega - \frac{1}{2!\hbar^2} \langle n | \mathcal{A}_{\text{int}}^2 | n \rangle_\omega - \frac{i}{3!\hbar^3} \langle n | \mathcal{A}_{\text{int}}^3 | n \rangle_\omega + \dots \quad (3.503)$$

$$= \exp \left\{ \frac{i}{\hbar} \langle n | \mathcal{A}_{\text{int}} | n \rangle_\omega - \frac{1}{2!\hbar^2} \langle n | \mathcal{A}_{\text{int}}^2 | n \rangle_{\omega,c} - \frac{i}{3!\hbar^3} \langle n | \mathcal{A}_{\text{int}}^3 | n \rangle_{\omega,c} + \dots \right\}.$$

The cumulants in the exponent are

$$
\begin{aligned}
\langle n|\mathcal{A}_{\mathrm{int}}^2|n\rangle_{\omega,c} &\equiv \langle n|\mathcal{A}_{\mathrm{int}}^2|n\rangle_\omega - \langle n|\mathcal{A}_{\mathrm{int}}|n\rangle_\omega^2 \\
&= \langle n|[\mathcal{A}_{\mathrm{int}} - \langle n|\mathcal{A}_{\mathrm{int}}|n\rangle_\omega]^2|n\rangle_\omega, &\qquad(3.504)\\
\langle n|\mathcal{A}_{\mathrm{int}}^3|n\rangle_{\omega,c} &\equiv \langle n|\mathcal{A}_{\mathrm{int}}^3|n\rangle_\omega - 3\langle n|\mathcal{A}_{\mathrm{int}}^2|n\rangle_\omega\langle n|\mathcal{A}_{\mathrm{int}}|n\rangle_\omega + 2\langle n|\mathcal{A}_{\mathrm{int}}|n\rangle_\omega^3 \\
&= \langle n|[\mathcal{A}_{\mathrm{int}} - \langle n|\mathcal{A}_{\mathrm{int}}|n\rangle_\omega]^3|n\rangle_\omega, &\qquad(3.505)
\end{aligned}
$$

$$\vdots$$

From (3.503), we obtain the energy shift of the nth oscillator energy

$$
\Delta E_n = \lim_{t_b - t_a \to \infty} \frac{i\hbar}{t_b - t_a} \left\{ \frac{i}{\hbar}\langle n|\mathcal{A}_{\mathrm{int}}|n\rangle_\omega - \frac{1}{2!\hbar^2}\langle n|\mathcal{A}_{\mathrm{int}}^2|n\rangle_{\omega,c} \right.
$$
$$
\left. - \frac{i}{3!\hbar^3}\langle n|\mathcal{A}_{\mathrm{int}}^3|n\rangle_{\omega,c} + \dots \right\}, \qquad(3.506)
$$

which is a generalization of formula (3.485) which was valid only for the ground state energy. At $n = 0$, the new formula goes over into (3.485), after the usual analytic continuation of the time variable.

The cumulants can be evaluated further with the help of the real-time version of the spectral expansion (3.496):

$$
(x_b t_b|x_a t_a)_\omega = \sum_{n=0}^\infty \psi_n(x_b)\psi_n^*(x_a)e^{-iE_n(t_b - t_a)/\hbar}. \qquad(3.507)
$$

To first order in $V(x)$, it leads to

$$
\langle n|\mathcal{A}_{\mathrm{int}}|n\rangle_\omega \equiv -\int_{t_a}^{t_b} dt \int dx\,\psi_n^*(x)V(x)\psi_n(x) \equiv -(t_b - t_a)V_{nn}. \qquad(3.508)
$$

To second order in $V(x)$, it yields

$$
\frac{1}{2}\langle n|\mathcal{A}_{\mathrm{int}}^2|n\rangle_\omega \equiv Z_{\mathrm{QM},\omega,n}^{-1} \int_{t_a}^{t_b} dt_2 \int_{t_a}^{t_2} dt_1 \qquad(3.509)
$$
$$
\times \sum_k e^{-iE_n(t_b - t_2)/\hbar - iE_k(t_2 - t_1)/\hbar - iE_n(t_1 - t_a)/\hbar}V_{nk}V_{kn}.
$$

The right-hand side can also be written as

$$
\int_{t_a}^{t_b} dt_2 \int_{t_a}^{t_2} dt_1 \sum_k e^{i(E_n - E_k)t_2/\hbar + i(E_k - E_n)t_1/\hbar}V_{nk}V_{kn} \qquad(3.510)
$$

and becomes, after the time integrations,

$$
-\sum_k \frac{V_{nk}V_{kn}}{E_k - E_n}\left\{ i\hbar(t_b - t_a) - \frac{\hbar^2}{E_n - E_k}\left[e^{i(E_n - E_k)(t_b - t_a)/\hbar} - 1 \right] \right\}.
$$
$$\qquad(3.511)$$

As it stands, the sum makes sense only for the $E_k \neq E_n$ -terms. In these, the second term in the curly brackets can be neglected in the limit of large time differences $t_b - t_a$. The term with $E_k = E_n$ must be treated separately by doing the integral directly in (3.510). This yields

$$V_{nn}V_{nn}\frac{(t_b - t_a)^2}{2}, \qquad (3.512)$$

so that

$$\frac{1}{2}\langle n|\mathcal{A}_{\text{int}}^2|n\rangle_\omega = -\sum_{m \neq n}\frac{V_{nm}V_{mn}}{E_m - E_n}i\hbar(t_b - t_a) + V_{nn}V_{nn}\frac{(t_b - t_a)^2}{2}. \qquad (3.513)$$

The same result could have been obtained without the special treatment of the $E_k = E_n$ -term by introducing artificially an infinitesimal energy difference $E_k - E_n = \epsilon$ in (3.511), and by expanding the curly brackets in powers of $t_b - t_a$.

When going over to the cumulants $\frac{1}{2}\langle n|\mathcal{A}_{\text{int}}^2|n\rangle_{\omega,\text{c}}$ according to (3.504), the $k = n$ - term is eliminated and we obtain

$$\frac{1}{2}\langle n|\mathcal{A}_{\text{int}}^2|n\rangle_{\omega,\text{c}} = -\sum_{k \neq n}\frac{V_{nk}V_{kn}}{E_k - E_n}i\hbar(t_b - t_a). \qquad (3.514)$$

For the energy shifts up to second order in $V(x)$, we thus arrive at the simple formula

$$\Delta_1 E_n + \Delta_2 E_n = V_{nn} - \sum_{k \neq n}\frac{V_{nk}V_{kn}}{E_k - E_n}. \qquad (3.515)$$

The higher expansion coefficients become rapidly complicated. The correction of third order in $V(x)$, for example, is

$$\Delta_3 E_n = \sum_{k \neq n}\sum_{l \neq n}\frac{V_{nk}V_{kl}V_{ln}}{(E_k - E_n)(E_l - E_n)} - V_{nn}\sum_{k \neq n}\frac{V_{nk}V_{kn}}{(E_k - E_n)^2}. \qquad (3.516)$$

For comparison, we recall the well-known formula of *Brillouin-Wigner equation*[14]

$$\Delta E_n = \bar{R}_{nn}(E_n + \Delta E_n), \qquad (3.517)$$

where $\bar{R}_{nn}(E)$ are the diagonal matrix elements $\langle n|\hat{\bar{R}}(E)|n\rangle$ of the *level shift operator* $\hat{\bar{R}}(E)$ which solves the integral equation

$$\hat{\bar{R}}(E) = \hat{V} + \hat{V}\frac{1 - \hat{P}_n}{E - \hat{H}_\omega}\hat{\bar{R}}(E). \qquad (3.518)$$

The operator $\hat{P}_n \equiv |n\rangle\langle n|$ is the projection operator onto the state $|n\rangle$. The factors $1 - \hat{P}_n$ ensure that the sums over the intermediate states exclude the quantum

---

[14]L. Brillouin and E.P. Wigner, J. Phys. Radium **4**, 1 (1933); M.L. Goldberger and K.M. Watson, *Collision Theory*, John Wiley & Sons, New York, 1964, pp. 425–430.

number n of the state under consideration. The integral equation is solved by the series expansion in powers of $\hat{V}$:

$$\hat{\bar{R}}(E) = \hat{V} + \hat{V}\frac{1 - \hat{P}_n}{E - \hat{H}_\omega}\hat{V} + \hat{V}\frac{1 - \hat{P}_n}{E - \hat{H}_\omega}\hat{V}\frac{1 - \hat{P}_n}{E - \hat{H}_\omega}\hat{V} + \ldots . \tag{3.519}$$

Up to the third order in $\hat{V}$, Eq. (3.517) leads to the *Brillouin-Wigner perturbation expansion*

$$E - E_n = R_{nn}(E) = V_{nn} + \sum_{k\neq n}\frac{V_{nk}V_{kn}}{E - E_k} + \sum_{k\neq n}\sum_{l\neq n}\frac{V_{nk}V_{kl}V_{ln}}{(E - E_k)(E - E_l)} + \ldots , \tag{3.520}$$

which is an implicit equation for $\Delta E_n = E - E_n$. The Brillouin-Wigner equation (3.517) may be converted into an explicit equation for the level shift ΔE_n:

$$\Delta E_n = R_{nn}(E_n) + R_{nn}(E_n)R'_{nn}(E_n) + [R_{nn}(E_n)R'_{nn}(E_n)^2 + \tfrac{1}{2}R^2_{nn}(E_n)R''_{nn}(E_n)]$$
$$+ [R_{nn}(E_n)R'_{nn}(E_n)^3 + \tfrac{3}{2}R^2_{nn}(E_n)R'_{nn}(E_n)R''_{nn}(E_n) + \tfrac{1}{6}R^3_{nn}(E_n)R'''_{nn}(E_n)] + \ldots . \tag{3.521}$$

Inserting (3.520) on the right-hand side, we recover the standard *Rayleigh-Schrödinger perturbation expansion* of quantum mechanics, which coincides precisely with the above perturbation expansion of the path integral whose first three terms were given in (3.515) and (3.516). Note that starting from the third order, the explicit solution (3.521) for the level shift introduces more and more extra disconnected terms with respect to the simple systematics in the Brillouin-Wigner expansion (3.520).

For arbitrary potentials, the calculation of the matrix elements V_{nk} can become quite tedious. A simple technique to find them is presented in Appendix 3A.

The calculation of the energy shifts for the particular interaction $V(x) = gx^4/4$ is described in Appendix 3B. Up to order g^3, the result is

$$\begin{aligned}
\Delta E_n &= \frac{\hbar\omega}{2}(2n + 1) + \frac{g}{4}3(2n^2 + 2n + 1)a^4 \\
&- \left(\frac{g}{4}\right)^2 2(34n^3 + 51n^2 + 59n + 21)a^8\frac{1}{\hbar\omega} \\
&+ \left(\frac{g}{4}\right)^3 4\cdot 3(125n^4 + 250n^3 + 472n^2 + 347n + 111)a^{12}\frac{1}{\hbar^2\omega^2} .
\end{aligned} \tag{3.522}$$

The perturbation series for this as well as arbitrary polynomial potentials can be carried out to high orders via recursion relations for the expansion coefficients. This is done in Appendix 3C.

3.19   Level-Shifts and Perturbed Wave Functions from Schrödinger Equation

It is instructive to rederive the perturbation expansion from ordinary operator Schrödinger theory. This derivation provides us also with the perturbed eigenstates to any desired order.

The Hamiltonian operator $\hat{H}$ is split into a free and an interacting part

$$\hat{H} = \hat{H}_0 + \hat{V}. \tag{3.523}$$

Let $|n\rangle$ be the eigenstates of $\hat{H}_0$ and $|\psi^{(n)}\rangle$ those of $\hat{H}$:

$$\hat{H}_0|n\rangle = E_0^{(n)}|n\rangle, \qquad \hat{H}|\psi^{(n)}\rangle = E^{(n)}|\psi^{(n)}\rangle. \tag{3.524}$$

We shall assume that the two sets of states $|n\rangle$ and $|\psi^{(n)}\rangle$ are orthogonal sets, the first with unit norm, the latter normalized by scalar products

$$a_n^{(n)} \equiv \langle n|\psi^{(n)}\rangle = 1. \tag{3.525}$$

Due to the completeness of the states $|n\rangle$, the states $|\psi^{(n)}\rangle$ can be expanded as

$$|\psi^{(n)}\rangle = |n\rangle + \sum_{m \neq n} a_m^{(n)}|m\rangle, \tag{3.526}$$

where

$$a_m^{(n)} \equiv \langle m|\psi^{(n)}\rangle \tag{3.527}$$

are the components of the interacting states in the free basis. Projecting the right-hand Schrödinger equation in (3.524) onto $\langle m|$ and using (3.527), we obtain

$$E_0^{(m)}a_m^{(n)} + \langle m|\hat{V}|\psi^{(n)}\rangle = E^{(n)}a_m^{(n)}. \tag{3.528}$$

Inserting here (3.526), this becomes

$$E_0^{(m)}a_m^{(n)} + \langle m|\hat{V}|n\rangle + \sum_{k \neq n} a_k^{(n)}\langle m|\hat{V}|k\rangle = E^{(n)}a_m^{(n)}, \tag{3.529}$$

and for $m = n$, due to the special normalization (3.525),

$$E_0^{(n)} + \langle n|\hat{V}|n\rangle + \sum_{k \neq n} a_k^{(n)}\langle n|\hat{V}|k\rangle = E^{(n)}. \tag{3.530}$$

Multiplying this equation with $a_m^{(n)}$ and subtracting it from (3.529), we eliminate the unknown exact energy $E^{(n)}$, and obtain a set of coupled algebraic equations for $a_m^{(n)}$:

$$a_m^{(n)} = \frac{1}{E_0^{(n)} - E_0^{(m)}}\left[\langle m - a_m^{(n)}n|\hat{V}|n\rangle + \sum_{k \neq n} a_k^{(n)}\langle m - a_m^{(n)}n|\hat{V}|k\rangle\right], \tag{3.531}$$

where we have introduced the notation $\langle m - a_m^{(n)}n|$ for the combination of states $\langle m| - a_m^{(n)}\langle n|$, for brevity.

   This equation can now easily be solved perturbatively order by order in powers of the interaction strength. To count these, we replace $\hat{V}$ by $g\hat{V}$ and expand $a_m^{(n)}$ as well as the energies $E^{(n)}$ in powers of g as:

$$a_m^{(n)}(g) = \sum_{l=1}^{\infty} a_{m,l}^{(n)}(-g)^l \qquad (m \neq n), \tag{3.532}$$

and

$$E^{(n)} = E_0^{(n)} - \sum_{l=1}^{\infty}(-g)^l E_l^{(n)}. \tag{3.533}$$

Inserting these expansions into (3.530), and equating the coefficients of g, we immediately find the perturbation expansion of the energy of the nth level

$$E_1^{(n)} = \langle n|\hat{V}|n\rangle, \tag{3.534}$$

$$E_l^{(n)} = \sum_{k\neq n} a_{k,l-1}^{(n)} \langle n|\hat{V}|k\rangle \quad l > 1. \tag{3.535}$$

The expansion coefficients $a_{m,l}^{(n)}$ are now determined by inserting the ansatz (3.532) into (3.531). This yields

$$a_{m,1}^{(n)} = \frac{\langle m|\hat{V}|n\rangle}{E_0^{(m)} - E_0^{(n)}}, \tag{3.536}$$

and for $l > 1$:

$$a_{m,l}^{(n)} = \frac{1}{E_0^{(m)} - E_0^{(n)}} \left[-a_{m,l-1}^{(n)} \langle n|\hat{V}|n\rangle + \sum_{k\neq n} a_{k,l-1}^{(n)} \langle m|\hat{V}|k\rangle - \sum_{l'=1}^{l-2} a_{m,l'}^{(n)} \sum_{k\neq n} a_{k,l-1-l'}^{(n)} \langle n|\hat{V}|k\rangle \right]. \tag{3.537}$$

Using (3.534) and (3.535), this can be simplified to

$$a_{m,l}^{(n)} = \frac{1}{E_0^{(m)} - E_0^{(n)}} \left[\sum_{k\neq n} a_{k,l-1}^{(n)} \langle m|\hat{V}|k\rangle - \sum_{l'=1}^{l-1} a_{m,l'}^{(n)} E_{l-l'}^{(n)} \right]. \tag{3.538}$$

Together with (3.534), (3.535), and (3.536), this is a set of recursion relations for the coefficients $a_{m,l}^{(n)}$ and $E_l^{(n)}$.

The recursion relations allow us to recover the perturbation expansions (3.515) and (3.516) for the energy shift. The second-order result (3.515), for example, follows directly from (3.537) and (3.538), the latter giving

$$E_2^{(n)} = \sum_{k\neq n} a_{k,1}^{(n)} \langle n|\hat{V}|k\rangle = \sum_{k\neq n} \frac{\langle k|\hat{V}|n\rangle \langle n|\hat{V}|k\rangle}{E_0^{(k)} - E_0^{(n)}}. \tag{3.539}$$

If the potential $\hat{V} = V(\hat{x})$ is a polynomial in $\hat{x}$, its matrix elements $\langle n|\hat{V}|k\rangle$ are nonzero only for n in a finite neighborhood of k, and the recursion relations consist of finite sums which can be solved exactly.

3.20   Calculation of Perturbation Series via Feynman Diagrams

The expectation values in formula (3.484) can be evaluated also in another way which can be applied to all potentials which are simple polynomials pf x. Then the partition function can be expanded into a sum of integrals associated with certain *Feynman diagrams*.

The procedure is rooted in the Wick expansion of correlation functions in Section 3.10. To be specific, we assume the anharmonic potential to have the form

$$V(x) = \frac{g}{4} x^4. \tag{3.540}$$

The graphical expansion terms to be found will be typical for all so-called φ^4-theories of quantum field theory.

To calculate the free energy shift (3.484) to first order in g, we have to evaluate the harmonic expectation of $\mathcal{A}_{\text{int,e}}$. This is written as

$$\langle \mathcal{A}_{\text{int,e}} \rangle_\omega = \frac{g}{4} \int_0^{\hbar\beta} d\tau \langle x^4(\tau) \rangle_\omega. \tag{3.541}$$

The integrand contains the correlation function

$$\langle x(\tau_1)x(\tau_2)x(\tau_3)x(\tau_4) \rangle_\omega = G_{\omega^2}^{(4)}(\tau_1, \tau_2, \tau_3, \tau_4)$$

at identical time arguments. According to the Wick rule (3.302), this can be expanded into the sum of three pair terms

$$G_{\omega^2}^{(2)}(\tau_1, \tau_2)G_{\omega^2}^{(2)}(\tau_3, \tau_4) + G_{\omega^2}^{(2)}(\tau_1, \tau_3)G_{\omega^2}^{(2)}(\tau_2, \tau_4) + G_{\omega^2}^{(2)}(\tau_1, \tau_4)G_{\omega^2}^{(2)}(\tau_2, \tau_3),$$

where $G_{\omega^2}^{(2)}(\tau, \tau')$ are the periodic Euclidean Green functions of the harmonic oscillator [see (3.301) and (3.248)]. The expectation (3.541) is therefore equal to the integral

$$\langle \mathcal{A}_{\text{int,e}} \rangle_\omega = 3\frac{g}{4} \int_0^{\hbar\beta} d\tau G_{\omega^2}^{(2)}(\tau, \tau)^2. \tag{3.542}$$

The right-hand side is pictured by the Feynman diagram

3

Because of its shape this is called a two-loop diagram. In general, a Feynman diagram consists of lines meeting at points called *vertices*. A line connecting two points represents the Green function $G_{\omega^2}^{(2)}(\tau_1, \tau_2)$. A vertex indicates a factor $g/4\hbar$ and a variable τ to be integrated over the interval $(0, \hbar\beta)$. The present simple diagram has only one point, and the τ-arguments of the Green functions coincide. The number underneath counts how often the integral occurs. It is called the *multiplicity* of the diagram.

To second order in $V(x)$, the harmonic expectation to be evaluated is

$$\langle \mathcal{A}_{\text{int,e}}^2 \rangle_\omega = \left(\frac{g}{4}\right)^2 \int_0^{\hbar\beta} d\tau_2 \int_0^{\hbar\beta} d\tau_1 \langle x^4(\tau_2)x^4(\tau_1) \rangle_\omega. \tag{3.543}$$

The integral now contains the correlation function $G_{\omega^2}^{(8)}(\tau_1, \ldots, \tau_8)$ with eight time arguments. According to the Wick rule, it decomposes into a sum of $7!! = 105$ products of four Green functions $G_{\omega^2}^{(2)}(\tau, \tau')$. Due to the coincidence of the time arguments, there are only three different types of contributions to the integral (3.543):

$$\langle \mathcal{A}_{\text{int,e}}^2 \rangle_\omega = \left(\frac{g}{4}\right)^2 \int_0^{\hbar\beta} d\tau_2 \int_0^{\hbar\beta} d\tau_1 \Big[72 G_{\omega^2}^{(2)}(\tau_2, \tau_2)G_{\omega^2}^{(2)}(\tau_2, \tau_1)^2 G_{\omega^2}^{(2)}(\tau_1, \tau_1)$$
$$+ 24 G_{\omega^2}^{(2)}(\tau_2, \tau_1)^4 + 9 G_{\omega^2}^{(2)}(\tau_2, \tau_2)^2 G_{\omega^2}^{(2)}(\tau_1, \tau_1)^2 \Big]. \tag{3.544}$$

The integrals are pictured by the following Feynman diagrams composed of three loops:

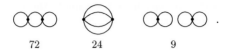

$$72 \qquad\qquad 24 \qquad\qquad 9$$

They contain two vertices indicating two integration variables τ_1, τ_2. The first two diagrams with the shape of three bubbles in a chain and of a watermelon, respectively, are *connected* diagrams, the third is *disconnected*. When going over to the cumulant $\langle \mathcal{A}^2_{\text{int,e}} \rangle_{\omega,\text{c}}$, the disconnected diagram is eliminated.

To higher orders, the counting becomes increasingly tedious and it is worth developing computer-algebraic techniques for this purpose. Figure 3.7 shows the diagrams for the free-energy shift up to four loops. The cumulants eliminate precisely all disconnected diagrams. This diagram-rearranging property of the logarithm is very general and happens to every order in g, as can be shown with the help of functional differential equations.

$$\beta F = \beta F_\omega + \underset{3}{\bigcirc\!\bigcirc} - \frac{1}{2!}\left(\underset{72}{\bigcirc\!\bigcirc\!\bigcirc} + \underset{24}{\ominus} \right)$$

$$+ \frac{1}{3!}\left(\underset{2592}{\bigcirc\!\bigcirc\!\bigcirc\!\bigcirc} + \underset{1728}{\text{Y}} + \underset{3456}{\ominus} + \underset{1728}{\triangle} \right) + \ldots$$

Figure 3.7 Perturbation expansion of free energy up to order g^3 (four loops).

The lowest-order term βF_ω containing the free energy of the harmonic oscillator [recall Eqs. (3.242) and (2.524)]

$$F_\omega = \frac{1}{\beta} \log\left(2 \sinh \frac{\beta \hbar \omega}{2} \right) \tag{3.545}$$

is often represented by the one-loop diagram

$$\beta F_\omega = -\frac{1}{2} \text{Tr} \log G^{(2)}_{\omega^2} = -\frac{1}{2\hbar\beta} \int_0^{\hbar\beta} d\tau \left[\log G^{(2)}_{\omega^2} \right](\tau,\tau) = -\frac{1}{2}\,\bigcirc . \tag{3.546}$$

With it, the graphical expansion in Fig. 3.7 starts more systematically with one loop rather than two. The systematics is, however, not perfect since the line in the one-loop diagram does not show that integrand contains a logarithm. In addition, the line is not connected to any vertex.

All τ-variables in the diagrams are integrated out. The diagrams have no open lines and are called *vacuum diagrams*.

The calculation of the diagrams in Fig. 3.7 is simplified with the help of a factorization property: If a diagram consists of two subdiagrams touching each other at a single vertex, its Feynman integral factorizes into those of the subdiagrams. Thanks to this property, we only have to evaluate the following integrals (omitting the factors $g/4\hbar$ for each vertex)

$$
\bigcirc = \int_0^{\hbar\beta} d\tau\, G^{(2)}_{\omega^2}(\tau, \tau) = \hbar\beta a^2,
$$

$$
\infty = \int_0^{\hbar\beta}\int_0^{\hbar\beta} d\tau_1 d\tau_2\; G^{(2)}_{\omega^2}(\tau_1, \tau_2)^2 \equiv \hbar\beta \frac{1}{\omega} a_2^4,
$$

$$
\bigcirc = \int_0^{\hbar\beta}\int_0^{\hbar\beta}\int_0^{\hbar\beta} d\tau_1 d\tau_2 d\tau_3\; G^{(2)}_{\omega^2}(\tau_1, \tau_2) G^{(2)}_{\omega^2}(\tau_2, \tau_3) G^{(2)}_{\omega^2}(\tau_3, \tau_1)
$$

$$
\equiv \hbar\beta \left(\frac{1}{\omega}\right)^2 a_3^6,
$$

$$
\ominus = \int_0^{\hbar\beta}\int_0^{\hbar\beta} d\tau_1 d\tau_2\; G^{(2)}_{\omega^2}(\tau_1, \tau_2)^4 \equiv \hbar\beta \frac{1}{\omega} a_2^8,
$$

$$
\ominus = \int_0^{\hbar\beta}\int_0^{\hbar\beta}\int_0^{\hbar\beta} d\tau_1 d\tau_2 d\tau_3\; G^{(2)}_{\omega^2}(\tau_1, \tau_2) G^{(2)}_{\omega^2}(\tau_2, \tau_3) G^{(2)}_{\omega^2}(\tau_3, \tau_1)^3
$$

$$
\equiv \hbar\beta \left(\frac{1}{\omega}\right)^2 a_3^{10},
$$

$$
\triangle = \int_0^{\hbar\beta}\int_0^{\hbar\beta}\int_0^{\hbar\beta} d\tau_1 d\tau_2 d\tau_3\; G^{(2)}_{\omega^2}(\tau_1, \tau_2)^2 G^{(2)}_{\omega^2}(\tau_2, \tau_3)^2 G^{(2)}_{\omega^2}(\tau_3, \tau_1)^2
$$

$$
\equiv \hbar\beta \left(\frac{1}{\omega}\right)^2 a_3^{12}. \tag{3.547}
$$

Note that in each expression, the last τ-integral yields an overall factor $\hbar\beta$, due to the translational invariance along the τ-axis. The others give rise to a factor $1/\omega$, for dimensional reasons. The temperature-dependent quantities a_V^{2L} are labeled by the number of vertices V and lines L of the associated diagrams. Their dimension is length to the nth power [corresponding to the dimension of the n $x(\tau)$-variables in the diagram]. For more than four loops, there can be more than one diagram for each V and L, such that one needs an additional label in a_V^{2L} to specify the diagram uniquely. Each a_V^{2L} may be written as a product of the basic length scale $(\hbar/M\omega)^L$ multiplied by a function of the dimensionless variable $x \equiv \beta\hbar\omega$:

$$
a_V^{2L} = \left(\frac{\hbar}{M\omega}\right)^L \alpha_V^{2L}(x). \tag{3.548}
$$

The functions $\alpha_V^{2L}(x)$ are listed in Appendix 3D.

As an example for the application of the factorization property, take the Feynman integral of the second third-order diagram in Fig. 3.7 (called a "daisy" diagram because of its shape):

$$
\text{\small(diagram)} = \int_0^{\hbar\beta}\int_0^{\hbar\beta}\int_0^{\hbar\beta} d\tau_1 d\tau_2 d\tau_3 G^{(2)}_{\omega^2}(\tau_1,\tau_2) G^{(2)}_{\omega^2}(\tau_2,\tau_3) G^{(2)}_{\omega^2}(\tau_3,\tau_1)
$$

$$
\times G^{(2)}_{\omega^2}(\tau_1,\tau_1) G^{(2)}_{\omega^2}(\tau_2,\tau_2) G^{(2)}_{\omega^2}(\tau_3,\tau_3).
$$

It decomposes into a product between the third integral in (3.547) and three powers of the first integral:

$$
\text{\small(diagram)} \;\rightarrow\; \text{\small(circle)} \times \text{\small(circle)}^{\,3}.
$$

Thus we can immediately write

$$
\text{\small(diagram)} = \hbar\beta \left(\frac{1}{\omega}\right)^2 a_3^6 (a^2)^3.
$$

In terms of a_V^{2L}, the free energy becomes

$$
F = F_\omega + \frac{g}{4} 3a^4 - \frac{1}{2!\hbar\omega}\left(\frac{g}{4}\right)^2 \left(72a^2 a_2^4 a^2 + 24 a_2^8\right) \tag{3.549}
$$

$$
+ \frac{1}{3!\hbar^2\omega^2}\left(\frac{g}{4}\right)^3 \left[2592a^2 (a_2^4)^2 a^2 + 1728 a_3^6 (a^2)^3 + 3456 a_3^{10} a^2 + 1728 a_3^{12}\right] + \dots .
$$

In the limit $T \to 0$, the integrals (3.547) behave like

$$
a_2^4 \;\rightarrow\; a^4, \qquad a_3^6 \;\rightarrow\; \frac{3}{2}a^6,
$$

$$
a_2^8 \;\rightarrow\; \frac{1}{2}a^8, \qquad a_3^{10} \;\rightarrow\; \frac{5}{8}a^{10}, \tag{3.550}
$$

$$
a_3^{12} \;\rightarrow\; \frac{3}{8}a^{12},
$$

and the free energy reduces to

$$
F = \frac{\hbar\omega}{2} + \frac{g}{4} 3a^4 - \left(\frac{g}{4}\right)^2 42 a^8 \frac{1}{\hbar\omega} + \left(\frac{g}{4}\right)^3 4\cdot 333 a^{12}\left(\frac{1}{\hbar\omega}\right)^2 + \dots . \tag{3.551}
$$

In this limit, it is simpler to calculate the integrals (3.547) directly with the zero-temperature limit of the Green function (3.301), which is $G^{(2)}_{\omega^2}(\tau,\tau') = a^2 e^{-\omega|\tau-\tau'|}$ with $a^2 = \hbar/2\omega M$ [see (3.248)]. The limits of integration must, however, be shifted by half a period to $\int_{-\hbar\beta/2}^{\hbar\beta/2} d\tau$ before going to the limit, so that one evaluates $\int_{-\infty}^{\infty} d\tau$ rather than $\int_0^{\infty} d\tau$ (the latter would give the wrong limit since it misses the left-hand side of the peak at $\tau = 0$). Before integration, the integrals are conveniently split as in Eq. (3D.1).

3.21 Perturbative Definition of Interacting Path Integrals

In Section 2.15 we have seen that it is possible to define a harmonic path integral without time slicing by dimensional regularization. With the techniques developed so far, this definition can trivially be extended to path integrals with interactions, if these can be treated perturbatively. We recall that in Eq. (3.480), the partition function of an interacting system can be expanded in a series of harmonic expectation values of powers of the interaction. The procedure is formulated most conveniently in terms of the generating functional (3.486) using formula (3.487) for the generating functional with interactions and Eq. (3.488) for the associated partition. The harmonic generating functional on the right-hand side of (3.487),

$$Z_\omega[j] = \oint \mathcal{D}x \exp\left\{-\frac{1}{\hbar}\int_0^{\hbar\beta} d\tau\left[\frac{M}{2}(\dot{x}^2 + \omega^2 x^2) - jx\right]\right\}, \qquad (3.552)$$

can be evaluated with analytic regularization as described in Section (2.15) and yields, after a quadratic completion [recall (3.243), (3.244)]:

$$Z_\omega[j] = \frac{1}{2\sin(\omega\hbar\beta/2)}\exp\left\{\frac{1}{2M\hbar}\int_0^{\hbar\beta} d\tau\int_0^{\hbar\beta} d\tau'\, j(\tau)G^{\mathrm{p}}_{\omega^2,\mathrm{e}}(\tau-\tau')j(\tau')\right\}, \quad (3.553)$$

where $G^{\mathrm{p}}_{\omega^2,\mathrm{e}}(\tau)$ is the periodic Green function (3.248)

$$G^{\mathrm{p}}_{\omega^2,\mathrm{e}}(\tau) = \frac{1}{2\omega}\frac{\cosh\omega(\tau-\hbar\beta/2)}{\sinh(\beta\hbar\omega/2)}, \quad \tau\in[0,\hbar\beta]. \qquad (3.554)$$

As a consequence, Formula (3.487) for the generating functional of an interacting theory

$$Z[j] = e^{-\frac{1}{\hbar}\int_0^{\hbar\beta} d\tau V(\hbar\delta/\delta j(\tau))}\, Z_\omega[j]\,, \qquad (3.555)$$

is completely defined by analytic regularization. By expanding the exponential prefactor as in Eq. (3.490), the full generating functional is obtained from the harmonic one without any further path integration. Only functional differentiations are required to find the generating functional of all interacting interacting correlation functions $Z[j]$ from the harmonic one $Z_\omega[j]$.

This procedure yields the perturbative definition of arbitrary path integrals. It is widely used in the quantum field theory of particle physics[15] and critical phenomena[16] It is also the basis for an important extension of the theory of distributions to be discussed in detail in Sections 10.6–10.11.

It must be realized, however, that the perturbative definition is not a complete definition. Important contributions to the path integral may be missing: all those

---

[15]C. Itzykson and J.-B. Zuber, *Quantum Field Theory*, McGraw-Hill (1985).

[16]H. Kleinert and V. Schulte-Frohlinde, *Critical Properties of ϕ^4-Theories*, World Scientific, Singapore 2001, pp. 1–487 (www.physik.fu-berlin.de/~kleinert/re.html#b8).

which are not expandable in powers of the interaction strength g. Such contributions are essential in understanding many physical phenomena, for example, tunneling, to be discussed in Chapter 17. Interestingly, however, information on such phenomena can, with appropriate resummation techniques to be developed in Chapter 5, also be extracted from the large-order behavior of the perturbation expansions, as will be shown in Subsection 17.10.4.

3.22 Generating Functional of Connected Correlation Functions

In Section 3.10 we have seen that the correlation functions obtained from the functional derivatives of $Z[j]$ via relation (3.295) contain many disconnected parts. The physically relevant free energy $F[j] = -k_B T \log Z[j]$, on the other hand, contains only in the connected parts of $Z[j]$. In fact, from statistical mechanics we know that meaningful description of a very large thermodynamic system can only be given in terms of the free energy which is directly proportional to the total volume V. The partition function $Z = e^{-F/k_B T}$ has no meaningful infinite-volume limit, also called the *thermodynamic limit*, since it contains a power series in V. Only the free energy density $f \equiv F/V$ has an infinite-volume limit. The expansion of $Z[j]$ diverges therefore for $V \to \infty$. This is why in thermodynamics we always go over to the free energy density by taking the logarithm of the partition function. This is calculated entirely from the connected diagrams.

Due to this thermodynamic experience we expect the logarithm of $Z[j]$ to provide us with a generating functional for all connected correlation functions. To avoid factors $k_B T$ we define this functional as

$$W[j] = \log Z[j], \tag{3.556}$$

and shall now prove that the functional derivatives of $W[j]$ produce precisely the connected parts of the Feynman diagrams for each correlation function.

Consider the connected correlation functions $G_c^{(n)}(\tau_1, \ldots, \tau_n)$ defined by the functional derivatives

$$G_c^{(n)}(\tau_1, \ldots, \tau_n) = \frac{\delta}{\delta j(\tau_1)} \cdots \frac{\delta}{\delta j(\tau_n)} W[j] . \tag{3.557}$$

Ultimately, we shall be interested only in these functions with zero external current, where they reduce to the physically relevant connected correlation functions. For the general development in this section, however, we shall consider them as functionals of $j(\tau)$, and set $j = 0$ only at the end.

Of course, given all connected correlation functions $G_c^{(n)}(\tau_1, \ldots, \tau_n)$, the full correlation functions $G^{(n)}(\tau_1, \ldots, \tau_n)$ in Eq. (3.295) can be recovered via simple composition laws from the connected ones. In order to see this clearly, we shall derive the general relationship between the two types of correlation functions in Section 3.22.2. First, we shall prove the connectedness property of the derivatives (3.557).

3.22.1   Connectedness Structure of Correlation Functions

We first prove that the generating functional $W[j]$ collects *only* connected diagrams in its Taylor coefficients $\delta^n W / \delta j(\tau_1) \ldots \delta j(\tau_n)$. Later, after Eq. (3.585), we shall see that these functional derivatives comprise *all* connected diagrams in $G^{(n)}(\tau_1, \ldots, \tau_n)$.

Let us write the path integral for the generating functional $Z[j]$ as follows (here we use natural units with $\hbar = 1$):

$$Z[j] = \int \mathcal{D}x \, e^{-\mathcal{A}_e[x,j]/\hbar}, \tag{3.558}$$

with the action

$$\mathcal{A}_e[x, j] = \int_0^{\hbar\beta} d\tau \left[\frac{M}{2} \left(\dot{x}^2 + \omega^2 x^2 \right) + V(x) - j(\tau)x(\tau) \right]. \tag{3.559}$$

In the following structural considerations we shall use natural physical units in which $\hbar = 1$, for simplicity of the formulas. By analogy with the integral identity

$$\int dx \, \frac{d}{dx} \, e^{-F(x)} = 0,$$

which holds by partial integration for any function $F(x)$ which goes to infinity for $x \to \pm\infty$, the functional integral satisfies the identity

$$\int \mathcal{D}x \, \frac{\delta}{\delta x(\tau)} e^{-\mathcal{A}_e[x,j]} = 0, \tag{3.560}$$

since the action $\mathcal{A}_e[x, j]$ goes to infinity for $x \to \pm\infty$. Performing the functional derivative, we obtain

$$\int \mathcal{D}x \, \frac{\delta \mathcal{A}_e[x, j]}{\delta x(\tau)} e^{-\mathcal{A}_e[x,j]} = 0. \tag{3.561}$$

To be specific, let us consider the anharmonic oscillator with potential $V(x) = \lambda x^4/4!$. We have chosen a coupling constant $\lambda/4!$ instead of the previous g in (3.540) since this will lead to more systematic numeric factors. The functional derivative of the action yields the classical equation of motion

$$\frac{\delta \mathcal{A}_e[x, j]}{\delta x(\tau)} = M(-\ddot{x} + \omega^2 x) + \frac{\lambda}{3!}x^3 - j = 0, \tag{3.562}$$

which we shall write as

$$\frac{\delta \mathcal{A}_e[x, j]}{\delta x(\tau)} = G_0^{-1}x + \frac{\lambda}{3!}x^3 - j = 0, \tag{3.563}$$

where we have set $G_0(\tau, \tau') \equiv G^{(2)}$ to get free space for upper indices. With this notation, Eq. (3.561) becomes

$$\int \mathcal{D}x \, \left\{ G_0^{-1}x(\tau) + \frac{\lambda}{3!}x^3(\tau) - j(\tau) \right\} e^{-\mathcal{A}_e[x,j]} = 0. \tag{3.564}$$

We now express the paths $x(\tau)$ as functional derivatives with respect to the source current $j(\tau)$, such that we can pull the curly brackets in front of the integral. This leads to the functional differential equation for the generating functional $Z[j]$:

$$\left\{ G_0^{-1} \frac{\delta}{\delta j(\tau)} + \frac{\lambda}{3!} \left[\frac{\delta}{\delta j(\tau)} \right]^3 - j(\tau) \right\} Z[j] = 0. \qquad (3.565)$$

With the short-hand notation

$$Z_{j(\tau_1)j(\tau_2)...j(\tau_n)}[j] \equiv \frac{\delta}{\delta j(\tau_1)} \frac{\delta}{\delta j(\tau_2)} \cdots \frac{\delta}{\delta j(\tau_n)} Z[j], \qquad (3.566)$$

where the arguments of the currents will eventually be suppressed, this can be written as

$$G_0^{-1} Z_{j(\tau)} + \frac{\lambda}{3!} Z_{j(\tau)j(\tau)j(\tau)} - j(\tau) = 0. \qquad (3.567)$$

Inserting here (3.556), we obtain a functional differential equation for $W[j]$:

$$G_0^{-1} W_j + \frac{\lambda}{3!} \left(W_{jjj} + 3W_{jj}W_j + W_j^3 \right) - j = 0. \qquad (3.568)$$

We have employed the same short-hand notation for the functional derivatives of $W[j]$ as in (3.566) for $Z[j]$,

$$W_{j(\tau_1)j(\tau_2)...j(\tau_n)}[j] \equiv \frac{\delta}{\delta j(\tau_1)} \frac{\delta}{\delta j(\tau_2)} \cdots \frac{\delta}{\delta j(\tau_n)} W[j], \qquad (3.569)$$

suppressing the arguments $\tau_1, \ldots, \tau_n$ of the currents, for brevity. Multiplying (3.568) functionally by G_0 gives

$$W_j = -\frac{\lambda}{3!} G_0 \left(W_{jjj} + 3W_{jj}W_j + W_j^3 \right) + G_0 j. \qquad (3.570)$$

We have omitted the integral over the intermediate τ's, for brevity. More specifically, we have written $G_0 j$ for $\int d\tau' \, G_0(\tau, \tau')j(\tau')$. Similar expressions abbreviate all functional products. This corresponds to a functional version of *Einstein's summation convention*.

Equation (3.570) may now be expressed in terms of the one-point correlation function

$$G_c^{(1)} = W_j, \qquad (3.571)$$

defined in (3.557), as

$$G_c^{(1)} = -\frac{\lambda}{3!} G_0 \left\{ G_{cjj}^{(1)} + 3G_{cj}^{(1)} G_c^{(1)} + \left[G_c^{(1)} \right]^3 \right\} + G_0 j. \qquad (3.572)$$

The solution to this equation is conveniently found by a diagrammatic procedure displayed in Fig. 3.8. To lowest, zeroth, order in λ we have

$$G_c^{(1)} = G_0\, j. \tag{3.573}$$

From this we find by functional integration the zeroth order generating functional $W_0[j]$

$$W_0[j] = \int \mathcal{D}j\, G_c^{(1)} = \frac{1}{2} j G_0 j, \tag{3.574}$$

up to a j-independent constant. Subscripts of $W[j]$ indicate the order in the interaction strength λ.

Reinserting (3.573) on the right-hand side of (3.572) gives the first-order expression

$$G_c^{(1)} = -G_0 \frac{\lambda}{3!} \left[3G_0 G_0 j + (G_0 j)^3 \right] + G_0 j, \tag{3.575}$$

represented diagrammatically in the second line of Fig. 3.8. Equation (3.575) can be integrated functionally in j to obtain $W[j]$ up to first order in λ. Diagrammatically, this process amounts to multiplying each open lines in a diagram by a current j, and dividing the arising j^ns by n. Thus we arrive at

$$W_0[j] + W_1[j] = \frac{1}{2} j G_0 j - \frac{\lambda}{4} G_0 (G_0 j)^2 - \frac{\lambda}{24} (G_0 j)^4, \tag{3.576}$$

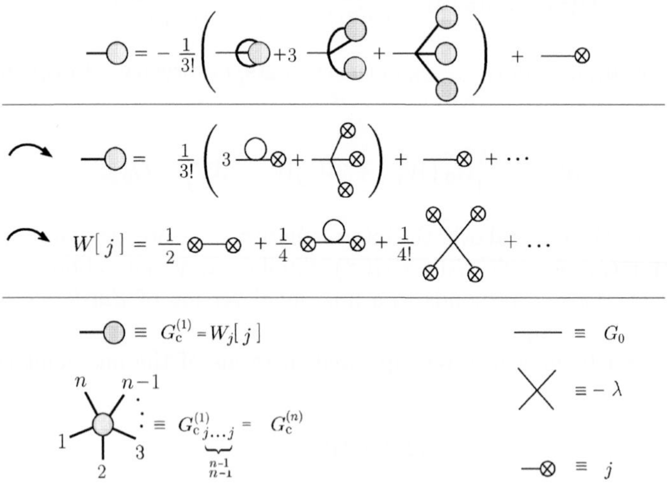

Figure 3.8 Diagrammatic solution of recursion relation (3.570) for the generating functional $W[j]$ of all connected correlation functions. First line represents Eq. (3.572), second (3.575), third (3.576). The remaining lines define the diagrammatic symbols.

as illustrated in the third line of Fig. 3.8. This procedure can be continued to any order in λ.

The same procedure allows us to prove that the generating functional $W[j]$ collects *only* connected diagrams in its Taylor coefficients $\delta^n W/\delta j(x_1)\ldots\delta j(x_n)$. For the lowest two orders we can verify the connectedness by inspecting the third line in Fig. 3.8. The diagrammatic form of the recursion relation shows that this topological property remains true for all orders in λ, by induction. Indeed, if we suppose it to be true for some n, then all $G_c^{(1)}$ inserted on the right-hand side are connected, and so are the diagrams constructed from these when forming $G_c^{(1)}$ to the next, $(n+1)$st, order.

Note that this calculation is unable to recover the value of $W[j]$ at $j=0$ which is an unknown integration constant of the functional differential equation. For the purpose of generating correlation functions, this constant is irrelevant. We have seen in Fig. 3.7 that $W[0]$, which is equal to $-F/k_B T$, consists of the sum of all connected vacuum diagrams contained in $Z[0]$.

3.22.2 Correlation Functions versus Connected Correlation Functions

Using the logarithmic relation (3.556) between $W[j]$ and $Z[j]$ we can now derive general relations between the n-point functions and their connected parts. For the one-point function we find

$$G^{(1)}(\tau) = Z^{-1}[j]\frac{\delta}{\delta j(\tau)}Z[j] = \frac{\delta}{\delta j(\tau)}W[j] = G_c^{(1)}(\tau). \tag{3.577}$$

This equation implies that the one-point function representing the ground state expectation value of the path $x(\tau)$ is always connected:

$$\langle x(\tau)\rangle \equiv G^{(1)}(\tau) = G_c^{(1)}(\tau) = X. \tag{3.578}$$

Consider now the two-point function, which decomposes as follows:

$$\begin{aligned}
G^{(2)}(\tau_1,\tau_2) &= Z^{-1}[j]\frac{\delta}{\delta j(\tau_1)}\frac{\delta}{\delta j(\tau_2)}Z[j] \\
&= Z^{-1}[j]\frac{\delta}{\delta j(\tau_1)}\left\{\left(\frac{\delta}{\delta j(\tau_2)}W[j]\right)Z[j]\right\} \\
&= Z^{-1}[j]\left\{W_{j(\tau_1)j(\tau_2)} + W_{j(\tau_1)}W_{j(\tau_2)}\right\}Z[j] \\
&= G_c^{(2)}(\tau_1,\tau_2) + G_c^{(1)}(\tau_1)G_c^{(1)}(\tau_2).
\end{aligned} \tag{3.579}$$

In addition to the connected diagrams with two ends there are two connected diagrams ending in a single line. These are absent in a x^4-theory at $j=0$ because of the symmetry of the potential, which makes all odd correlation functions vanish. In that case, the two-point function is automatically connected.

For the three-point function we find

$$
\begin{aligned}
G^{(3)}\left(\tau_1, \tau_2, \tau_3\right) &= Z^{-1}[j] \frac{\delta}{\delta j(\tau_1)} \frac{\delta}{\delta j(\tau_2)} \frac{\delta}{\delta j(\tau_3)} Z[j] \\
&= Z^{-1}[j] \frac{\delta}{\delta j(\tau_1)} \frac{\delta}{\delta j(\tau_2)} \left\{ \left[\frac{\delta}{\delta j(\tau_3)} W[j] \right] Z[j] \right\} \\
&= Z^{-1}[j] \frac{\delta}{\delta j(\tau_1)} \left\{ \left[W_{j(\tau_3)j(\tau_2)} + W_{j(\tau_2)}W_{j(\tau_3)} \right] Z[j] \right\} \\
&= Z^{-1}[j] \left\{ W_{j(\tau_1)j(\tau_2)j(\tau_3)} + \left(W_{j(\tau_1)}W_{j(\tau_2)j(\tau_3)} + W_{j(\tau_2)}W_{j(\tau_1)j(\tau_3)} \right. \right. \\
&\qquad \left. \left. + W_{j(\tau_3)}W_{j(\tau_1)j(\tau_2)} \right) + W_{j(\tau_1)}W_{j(\tau_2)}W_{j(\tau_3)} \right\} Z[j] \\
&= G_{\mathrm{c}}^{(3)}\left(\tau_1, \tau_2, \tau_3\right) + \left[G_{\mathrm{c}}^{(1)}(\tau_1) G_{\mathrm{c}}^{(2)}(\tau_2, \tau_3) + 2 \text{ perm} \right] + G_{\mathrm{c}}^{(1)}(\tau_1) G_{\mathrm{c}}^{(1)}(\tau_2) G_{\mathrm{c}}^{(1)}(\tau_3),
\end{aligned}
\tag{3.580}
$$

and for the four-point function

$$
\begin{aligned}
G^{(4)}\left(\tau_1, \ldots, \tau_4\right) &= G_{\mathrm{c}}^{(4)}\left(\tau_1, \ldots, \tau_4\right) + \left[G_{\mathrm{c}}^{(3)}\left(\tau_1, \tau_2, \tau_3\right) G_{\mathrm{c}}^{(1)}(\tau_4) + 3 \text{ perm} \right] \\
&\quad + \left[G_{\mathrm{c}}^{(2)}\left(\tau_1, \tau_2\right) G_{\mathrm{c}}^{(2)}\left(\tau_3, \tau_4\right) + 2 \text{ perm} \right] \\
&\quad + \left[G_{\mathrm{c}}^{(2)}\left(\tau_1, \tau_2\right) G_{\mathrm{c}}^{(1)}(\tau_3) G_{\mathrm{c}}^{(1)}(\tau_4) + 5 \text{ perm} \right] \\
&\quad + G_{\mathrm{c}}^{(1)}(\tau_1) \cdots G_{\mathrm{c}}^{(1)}(\tau_4).
\end{aligned}
\tag{3.581}
$$

In the pure x^4-theory there are no odd correlation functions, because of the symmetry of the potential.

For the general correlation function $G^{(n)}$, the total number of terms is most easily retrieved by dropping all indices and differentiating with respect to j (the arguments $\tau_1, \ldots, \tau_n$ of the currents are again suppressed):

$$
\begin{aligned}
G^{(1)} &= e^{-W} \left(e^W \right)_j = W_j = G_{\mathrm{c}}^{(1)} \\
G^{(2)} &= e^{-W} \left(e^W \right)_{jj} = W_{jj} + W_j{}^2 = G_{\mathrm{c}}^{(2)} + G_{\mathrm{c}}^{(1)2} \\
G^{(3)} &= e^{-W} \left(e^W \right)_{jjj} = W_{jjj} + 3W_{jj}W_j + W_j{}^3 = G_{\mathrm{c}}^{(3)} + 3G_{\mathrm{c}}^{(2)}G_{\mathrm{c}}^{(1)} + G_{\mathrm{c}}^{(1)3} \\
G^{(4)} &= e^{-W} \left(e^W \right)_{jjjj} = W_{jjjj} + 4W_{jjj}W_j + 3W_{jj}{}^2 + 6W_{jj}W_j{}^2 + W_j{}^4 \\
&= G_{\mathrm{c}}^{(4)} + 4G_{\mathrm{c}}^{(3)}G_{\mathrm{c}}^{(1)} + 3G_{\mathrm{c}}^{(2)2} + 6G_{\mathrm{c}}^{(2)}G_{\mathrm{c}}^{(1)2} + G_{\mathrm{c}}^{(1)4}.
\end{aligned}
\tag{3.582}
$$

All equations follow from the recursion relation

$$
G^{(n)} = G_j^{(n-1)} + G^{(n-1)} G_{\mathrm{c}}^{(1)}, \quad n \geq 2,
\tag{3.583}
$$

if one uses $G_j^{(n-1)} = G_{\mathrm{c}}^{(n)}$ and the initial relation $G^{(1)} = G_{\mathrm{c}}^{(1)}$. By comparing the first four relations with the explicit expressions (3.579)–(3.581) we see that the numerical factors on the right-hand side of (3.582) refer to the permutations of the arguments $\tau_1, \tau_2, \tau_3, \ldots$ of otherwise equal expressions. Since there is no problem in reconstructing the explicit permutations we shall henceforth write all composition laws in the short-hand notation (3.582).

The formula (3.582) and its generalization is often referred to as *cluster decomposition*, or also as the *cumulant expansion*, of the correlation functions.

We can now prove that the connected correlation functions collect precisely all connected diagrams in the n-point functions. For this we observe that the decomposition rules can be inverted by repeatedly differentiating both sides of the equation $W[j] = \log Z[j]$ functionally with respect to the current j:

$$
\begin{aligned}
G_{\mathrm{c}}^{(1)} &= G^{(1)} \\
G_{\mathrm{c}}^{(2)} &= G^{(2)} - G^{(1)}G^{(1)} \\
G_{\mathrm{c}}^{(3)} &= G^{(3)} - 3G^{(2)}G^{(1)} + 2G^{(1)3} \\
G_{\mathrm{c}}^{(4)} &= G^{(4)} - 4G^{(3)}G^{(1)} + 12G^{(2)}G^{(1)2} - 3G^{(2)2} - 6G^{(1)4}.
\end{aligned}
\tag{3.584}
$$

Each equation follows from the previous one by one more derivative with respect to j, and by replacing the derivatives on the right-hand side according to the rule

$$
G_j^{(n)} = G^{(n+1)} - G^{(n)}G^{(1)}.
\tag{3.585}
$$

Again the numerical factors imply different permutations of the arguments and the subscript j denotes functional differentiations with respect to j.

Note that Eqs. (3.584) for the connected correlation functions are valid for symmetric as well as asymmetric potentials $V(x)$. For symmetric potentials, the equations simplify, since all terms involving $G^{(1)} = X = \langle x \rangle$ vanish.

It is obvious that any connected diagram contained in $G^{(n)}$ must also be contained in $G_{\mathrm{c}}^{(n)}$, since all the terms added or subtracted in (3.584) are products of $G_j^{(n)}$'s, and thus necessarily disconnected. Together with the proof in Section 3.22.1 that the correlation functions $G_{\mathrm{c}}^{(n)}$ contain *only* the connected parts of $G^{(n)}$, we can now be sure that $G_{\mathrm{c}}^{(n)}$ contains precisely the connected parts of $G^{(n)}$.

3.22.3 Functional Generation of Vacuum Diagrams

The functional differential equation (3.570) for $W[j]$ contains all information on the connected correlation functions of the system. However, it does not tell us anything about the vacuum diagrams of the theory. These are contained in $W[0]$, which remains an undetermined constant of functional integration of these equations.

In order to gain information on the vacuum diagrams, we consider a modification of the generating functional (3.558), in which we set the external source j equal to zero, but generalize the source $j(\tau)$ in (3.558) coupled linearly to $x(\tau)$ to a bilocal form $K(\tau, \tau')$ coupled linearly to $x(\tau)x(\tau')$:

$$
Z[K] = \int \mathcal{D}x(\tau)\, e^{-\mathcal{A}_{\mathrm{e}}[x,K]},
\tag{3.586}
$$

where $\mathcal{A}_{\mathrm{e}}[x, K]$ is the Euclidean action

$$
\mathcal{A}_{\mathrm{e}}[x, K] \equiv \mathcal{A}_0[x] + \mathcal{A}^{\mathrm{int}}[x] + \frac{1}{2}\int d\tau \int d\tau'\, x(\tau)K(\tau, \tau')x(\tau').
\tag{3.587}
$$

When forming the functional derivative with respect to $K(\tau, \tau')$ we obtain the correlation function in the presence of $K(\tau, \tau')$:

$$G^{(2)}(\tau, \tau') = -2Z^{-1}[K]\frac{\delta Z}{\delta K(\tau, \tau')}. \qquad (3.588)$$

At the end we shall set $K(\tau, \tau') = 0$, just as previously the source j. When differentiating $Z[K]$ twice, we obtain the four-point function

$$G^{(4)}(\tau_1, \tau_2, \tau_3, \tau_4) = 4Z^{-1}[K]\frac{\delta^2 Z}{\delta K(\tau_1, \tau_2)\delta K(\tau_3, \tau_4)}. \qquad (3.589)$$

As before, we introduce the functional $W[K] \equiv \log Z[K]$. Inserting this into (3.588) and (3.589), we find

$$G^{(2)}(\tau, \tau') = 2\frac{\delta W}{\delta K(\tau, \tau')}, \qquad (3.590)$$

$$G^{(4)}(\tau_1, \tau_2, \tau_3, \tau_4) = 4\left[\frac{\delta^2 W}{\delta K(\tau_1, \tau_2)\delta K(\tau_3, \tau_4)} + \frac{\delta W}{\delta K(\tau_1, \tau_2)}\frac{\delta W}{\delta K(\tau_3, \tau_4)}\right]. \qquad (3.591)$$

With the same short notation as before, we shall use again a subscript K to denote functional differentiation with respect to K, and write

$$G^{(2)} = 2W_K, \qquad G^{(4)} = 4\left[W_{KK} + W_K W_K\right] = 4W_{KK} + G^{(2)}G^{(2)}. \qquad (3.592)$$

From Eq. (3.582) we know that in the absence of a source j and for a symmetric potential, $G^{(4)}$ has the connectedness structure

$$G^{(4)} = G_c^{(4)} + 3G_c^{(2)}G_c^{(2)}. \qquad (3.593)$$

This shows that in contrast to W_{jjjj}, the derivative W_{KK} does not directly yield a connected four-point function, but two disconnected parts:

$$4W_{KK} = G_c^{(4)} + 2G_c^{(2)}G_c^{(2)}, \qquad (3.594)$$

the two-point functions being automatically connected for a symmetric potential. More explicitly, (3.594) reads

$$\frac{4\delta^2 W}{\delta K(\tau_1, \tau_2)\delta K(\tau_3, \tau_4)}$$
$$= G_c^{(4)}(\tau_1, \tau_2, \tau_3, \tau_4) + G_c^{(2)}(\tau_1, \tau_3)G_c^{(2)}(\tau_2, \tau_4) + G_c^{(2)}(\tau_1, \tau_4)G_c^{(2)}(\tau_2, \tau_3). \quad (3.595)$$

Let us derive functional differential equations for $Z[K]$ and $W[K]$. By analogy with (3.560) we start out with the trivial functional differential equation

$$\int \mathcal{D}x\, x(\tau)\frac{\delta}{\delta x(\tau')}e^{-\mathcal{A}_e[x, K]} = -\delta(\tau - \tau')Z[K], \qquad (3.596)$$

which is immediately verified by a functional integration by parts. Performing the functional derivative yields

$$\int \mathcal{D}x \, x(\tau) \frac{\delta \mathcal{A}_e[x, K]}{\delta x(\tau')} e^{-\mathcal{A}_e[x, K]} = \delta(\tau - \tau') Z[K], \tag{3.597}$$

or

$$\int \mathcal{D}x \int d\tau \int d\tau' \left\{ x(\tau) G_0^{-1}(\tau, \tau') x(\tau') + \frac{\lambda}{3!} x(\tau) x^3(\tau') \right\} e^{-\mathcal{A}_e[x, K]} = \delta(\tau - \tau') Z[K]. \tag{3.598}$$

For brevity, we have absorbed the source in the free-field correlation function G_0:

$$G_0 \to [G_0^{-1} - K]^{-1}. \tag{3.599}$$

The left-hand side of (3.598) can obviously be expressed in terms of functional derivatives of $Z[K]$, and we obtain the functional differential equation whose short form reads

$$G_0^{-1} Z_K + \frac{\lambda}{3} Z_{KK} = \frac{1}{2} Z. \tag{3.600}$$

Inserting $Z[K] = e^{W[K]}$, this becomes

$$G_0^{-1} W_K + \frac{\lambda}{3} (W_{KK} + W_K W_K) = \frac{1}{2}. \tag{3.601}$$

It is useful to reconsider the functional $W[K]$ as a functional $W[G_0]$. Then $\delta G_0 / \delta K = G_0^2$, and the derivatives of $W[K]$ become

$$W_K = G_0^2 W_{G_0}, \quad W_{KK} = 2 G_0^3 W_{G_0} + G_0^4 W_{G_0 G_0}, \tag{3.602}$$

and (3.601) takes the form

$$G_0 W_{G_0} + \frac{\lambda}{3} (G_0^4 W_{G_0 G_0} + 2 G_0^3 W_{G_0} + G_0^4 W_{G_0} W_{G_0}) = \frac{1}{2}. \tag{3.603}$$

This equation is represented diagrammatically in Fig. 3.9. The zeroth-order solution

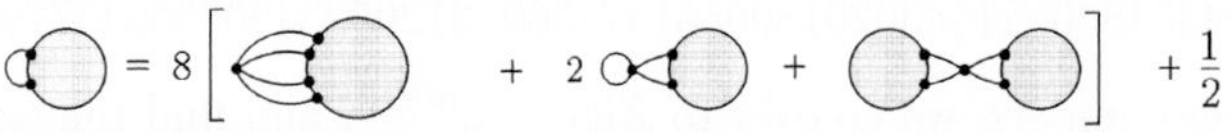

$$G_0 W_{G_0} = 8 \frac{-1}{4!} \left[\lambda G_0^4 W_{G_0 G_0} + 2 G_0 \lambda G_0^2 W_{G_0} + W_{G_0} G_0^2 \lambda G_0^2 W_{G_0} \right] + \frac{1}{2}$$

Figure 3.9 Diagrammatic representation of functional differential equation (3.603). For the purpose of finding the multiplicities of the diagrams, it is convenient to represent here by a vertex the coupling strength $-\lambda/4!$ rather than $g/4$ in Section 3.20.

to this equation is obtained by setting $\lambda = 0$:

$$W^{(0)}[G_0] = \frac{1}{2} \text{Tr} \log(G_0). \qquad (3.604)$$

Explicitly, the right-hand side is equal to the one-loop contribution to the free energy in Eq. (3.546), apart from a factor $-\beta$.

The corrections are found by iteration. For systematic treatment, we write $W[G_0]$ as a sum of a free and an interacting part,

$$W[G_0] = W^{(0)}[G_0] + W^{\text{int}}[G_0], \qquad (3.605)$$

insert this into Eq. (3.603), and find the differential equation for the interacting part:

$$G_0 W_{G_0}^{\text{int}} + \frac{\lambda}{3}(G_0^4 W_{G_0 G_0}^{\text{int}} + 3G_0^3 W_{G_0}^{\text{int}} + G_0^4 W_{G_0}^{\text{int}} W_{G_0}^{\text{int}}) = 6\frac{-\lambda}{4!} G_0^2. \qquad (3.606)$$

This equation is solved iteratively. Setting $W^{\text{int}}[G_0] = 0$ in all terms proportional to λ, we obtain the first-order contribution to $W^{\text{int}}[G_0]$:

$$W^{\text{int}}[G_0] = 3\frac{-\lambda}{4!} G_0^2. \qquad (3.607)$$

This is precisely the contribution (3.542) of the two-loop Feynman diagram (apart from the different normalization of g).

In order to see how the iteration of Eq. (3.606) may be solved systematically, let us ignore for the moment the functional nature of Eq. (3.606), and treat G_0 as an ordinary real variable rather than a functional matrix. We expand $W[G_0]$ in a Taylor series:

$$W^{\text{int}}[G_0] = \sum_{p=1}^{\infty} \frac{1}{p!} W_p \left(\frac{-\lambda}{4!}\right)^p (G_0)^{2p}, \qquad (3.608)$$

and find for the expansion coefficients the recursion relation

$$W_{p+1} = 4\left\{ [2p(2p-1) + 3(2p)] W_p + \sum_{q=1}^{p-1} \binom{p}{q} 2q\, W_q \times 2(p-q) W_{p-q} \right\}. \qquad (3.609)$$

Solving this with the initial number $W_1 = 3$, we obtain the multiplicities of the connected vacuum diagrams of pth order:

3, 96, 9504, 1880064, 616108032, 301093355520, 205062331760640, 185587468924354560, 215430701800551874560, 312052349085504377978880. (3.610)

To check these numbers, we go over to $Z[G] = e^{W[G_0]}$, and find the expansion:

$$\begin{aligned} Z[G_0] &= \exp\left[\frac{1}{2}\text{Tr}\log G_0 + \sum_{p=1}^{\infty} \frac{1}{p!} W_p \left(\frac{-\lambda}{4!}\right)^p (G_0)^{2p}\right] \\ &= \text{Det}^{1/2}[G_0]\left[1 + \sum_{p=1}^{\infty} \frac{1}{p!} z_p \left(\frac{-\lambda}{4!}\right)^p (G_0)^{2p}\right]. \qquad (3.611) \end{aligned}$$

The expansion coefficients z_p count the total number of vacuum diagrams of order p. The exponentiation (3.611) yields $z_p = (4p - 1)!!$, which is the correct number of Wick contractions of p interactions x^4.

In fact, by comparing coefficients in the two expansions in (3.611), we may derive another recursion relation for W_p:

$$W_p + 3 \binom{p-1}{1} W_{p-1} + 7 \cdot 5 \cdot 3 \binom{p-1}{2} + \ldots + (4p - 5)!! \binom{p-1}{p-1} = (4p - 1)!!, \quad (3.612)$$

which is fulfilled by the solutions of (3.609).

In order to find the associated Feynman diagrams, we must perform the differentiations in Eq. (3.606) functionally. The numbers W_p become then a sum of diagrams, for which the recursion relation (3.609) reads

$$W_{p+1} = 4 \left[G_0^4 \frac{d^2}{d\cap^2} W_p + 3 \cdot G_0^3 \frac{d}{d\cap} W_p + \sum_{q=1}^{p-1} \binom{p}{q} \left(\frac{d}{d\cap} W_q \right) G_0^2 \cdot G_0^2 \left(\frac{d}{d\cap} W_{p-q} \right) \right],$$

$$(3.613)$$

where the differentiation $d/d\cap$ removes one line connecting two vertices in all possible ways. This equation is solved diagrammatically, as shown in Fig. 3.10.

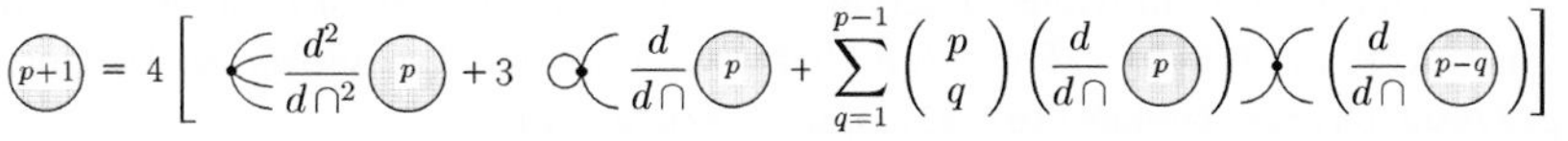

$$W_{p+1} = 4 \left[G_0^4 \frac{d^2}{d\cap^2} W_p + 3 \cdot G_0^3 \frac{d}{d\cap} W_p + \sum_{q=1}^{p-1} \binom{p}{q} \left(\frac{d}{d\cap} W_q \right) G_0^2 \cdot G_0^2 \left(\frac{d}{d\cap} W_{p-q} \right) \right]$$

Figure 3.10 Diagrammatic representation of recursion relation (3.609). A vertex represents the coupling strength $-\lambda$.

Starting the iteration with $W_1 = 3\,\infty$, we have $dW_p/d\cap = 6\,Q$ and $d^2 W_p/d\cap^2 = 6\,\times$. Proceeding to order five loops and going back to the usual vertex notation $-\lambda$, we find the vacuum diagrams with their weight factors as shown in Fig. 3.11. For more than five loops, the reader is referred to the paper quoted in Notes and References, and to the internet address from which Mathematica programs can be downloaded which solve the recursion relations and plot all diagrams of $W[0]$ and the resulting two- and four-point functions.

3.22.4   Correlation Functions from Vacuum Diagrams

The vacuum diagrams contain information on all correlation functions of the theory. One may rightly say that the vacuum is the world. The two- and four-point functions are given by the functional derivatives (3.592) of the vacuum functional $W[K]$. Diagrammatically, a derivative with respect to K corresponds to cutting one line of a vacuum diagram in all possible ways. Thus, all diagrams of the two-point function

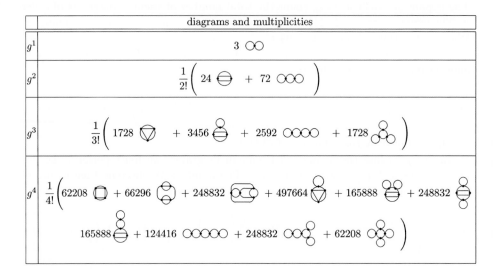

diagrams and multiplicities

Figure 3.11 Vacuum diagrams up to five loops and their multiplicities. The total numbers to orders g^n are 3, 96, 9504, 1880064, respectively. In contrast to Fig. 3.10, and to the previous diagrammatic notation in Fig. 3.7, a vertex stands here for $-\lambda/4!$ for brevity. For more than five loops see the tables on the internet (`http://www.physik.fu-berlin/~kleinert/b3/programs`).

$G^{(2)}$ can be derived from such cuts, multiplied by a factor 2. As an example, consider the first-order vacuum diagram of $W[K]$ in Fig. 3.11. Cutting one line, which is possible in two ways, and recalling that in Fig. 3.11 a vertex stands for $-\lambda/4!$ rather than $-\lambda$, as in the other diagrams, we find

$$W_1[0] = \frac{1}{8} \; \text{⚭} \quad \longrightarrow \quad G_1^{(2)}(\tau_1, \tau_2) = 2 \times \frac{1}{8} \, 2 \; \underset{\tau_1 \quad \tau_2}{\text{◯}} \; . \tag{3.614}$$

The second equation in (3.592) tells us that all connected contributions to the four-point function $G^{(4)}$ may be obtained by cutting two lines in all combinations, and multiplying the result by a factor 4. As an example, take the second-order vacuum diagrams of $W[0]$ with the proper translation of vertices by a factor 4!, which are

$$W_2[0] = \frac{1}{16} \; \text{◯◯◯} \quad + \frac{1}{48} \; \text{⊖} \; . \tag{3.615}$$

Cutting two lines in all possible ways yields the following contributions to the connected diagrams of the two-point function:

$$G^{(4)} = 4 \times \left(2 \cdot 1 \cdot \frac{1}{16} + 4 \cdot 3 \cdot \frac{1}{48} \right) \; \text{✕✕} \; . \tag{3.616}$$

It is also possible to find all diagrams of the four-point function from the vacuum diagrams by forming a derivative of $W[0]$ with respect to the coupling constant $-\lambda$, and multiplying the result by a factor $4!$. This follows directly from the fact that this differentiation applied to $Z[0]$ yields the correlation function $\int d\tau \langle x^4 \rangle$. As an example, take the first diagram of order g^3 in Table 3.11 (with the same vertex convention as in Fig. 3.11):

$$W_2[0] = \frac{1}{48} \; \bigcirc \!\!\!\!\Delta \; . \tag{3.617}$$

Removing one vertex in the three possible ways and multiplying by a factor $4!$ yields

$$G^{(4)} = 4! \times \frac{1}{48} \; 3 \; \asymp \; . \tag{3.618}$$

3.22.5 Generating Functional for Vertex Functions. Effective Action

Apart from the connectedness structure, the most important step in economizing the calculation of Feynman diagrams consists in the decomposition of higher connected correlation functions into *one-particle irreducible* vertex functions and one-particle irreducible two-particle correlation functions, from which the full amplitudes can easily be reconstructed. A diagram is called one-particle irreducible if it cannot be decomposed into two disconnected pieces by cutting a single line.

There is, in fact, a simple algorithm which supplies us in general with such a decomposition. For this purpose let us introduce a new generating functional $\Gamma[X]$, to be called the *effective action* of the theory. It is defined via a Legendre transformation of $W[j]$:

$$- \Gamma[X] \equiv W[j] - W_j \, j. \tag{3.619}$$

Here and in the following, we use a short-hand notation for the functional multiplication, $W_j \, j = \int d\tau \, W_j(\tau) j(\tau)$, which considers fields as vectors with a continuous index τ. The new variable X is the functional derivative of $W[j]$ with respect to $j(\tau)$ [recall (3.569)]:

$$X(\tau) \equiv \frac{\delta W[j]}{\delta j(\tau)} \equiv W_{j(\tau)} = \langle x \rangle_{j(\tau)}, \tag{3.620}$$

and thus gives the ground state expectation of the field operator in the presence of the current j. When rewriting (3.619) as

$$- \Gamma[X] \equiv W[j] - X \, j, \tag{3.621}$$

and functionally differentiating this with respect to X, we obtain the equation

$$\Gamma_X[X] = j. \tag{3.622}$$

This equation shows that the physical path expectation $X(\tau) = \langle x(\tau) \rangle$, where the external current is zero, extremizes the effective action:

$$\Gamma_X[X] = 0. \tag{3.623}$$

We shall study here only physical systems for which the path expectation value is a constant $X(\tau) \equiv X_0$. Thus we shall not consider systems which possess a time-dependent $X_0(\tau)$, although such systems can also be described by x^4-theories by admitting more general types of gradient terms, for instance $x(\partial^2 - k_0^2)^2 x$. The ensuing τ-dependence of $X_0(\tau)$ may be oscillatory.[17] Thus we shall assume a constant

$$X_0 = \langle x \rangle|_{j=0}, \tag{3.624}$$

which may be zero or non-zero, depending on the phase of the system.

Let us now demonstrate that the effective action contains all the information on the proper vertex functions of the theory. These can be found directly from the functional derivatives:

$$\Gamma^{(n)}(\tau_1, \ldots, \tau_n) \equiv \frac{\delta}{\delta X(\tau_1)} \cdots \frac{\delta}{\delta X(\tau_n)} \Gamma[X]. \tag{3.625}$$

We shall see that the proper vertex functions are obtained from these functions by a Fourier transform and a simple removal of an overall factor $(2\pi)^D \delta(\sum_{i=1}^n \omega_i)$ to ensure momentum conservation. The functions $\Gamma^{(n)}(\tau_1, \ldots, \tau_n)$ will therefore be called *vertex functions*, without the adjective *proper* which indicates the absence of the δ-function. In particular, the Fourier transforms of the vertex functions $\Gamma^{(2)}(\tau_1, \tau_2)$ and $\Gamma^{(4)}(\tau_1, \tau_2, \tau_3, \tau_4)$ are related to their proper versions by

$$\Gamma^{(2)}(\omega_1, \omega_2) = 2\pi\delta(\omega_1 + \omega_2)\,\bar{\Gamma}^{(2)}(\omega_1), \tag{3.626}$$

$$\Gamma^{(4)}(\omega_1, \omega_2, \omega_3, \omega_4) = 2\pi\delta\left(\sum_{i=1}^4 \omega_i\right)\bar{\Gamma}^{(4)}(\omega_1, \omega_2, \omega_3, \omega_4). \tag{3.627}$$

For the functional derivatives (3.625) we shall use the same short-hand notation as for the functional derivatives (3.569) of $W[j]$, setting

$$\Gamma_{X(\tau_1)\ldots X(\tau_n)} \equiv \frac{\delta}{\delta X(\tau_1)} \cdots \frac{\delta}{\delta X(\tau_n)} \Gamma[X]. \tag{3.628}$$

The arguments $\tau_1, \ldots, \tau_n$ will usually be suppressed.

In order to derive relations between the derivatives of the effective action and the connected correlation functions, we first observe that the connected one-point function $G_c^{(1)}$ at a nonzero source j is simply the path expectation X [recall (3.578)]:

$$G_c^{(1)} = X. \tag{3.629}$$

---

[17]In higher dimensions there can be crystal- or quasicrystal-like modulations. See, for example, H. Kleinert and K. Maki, Fortschr. Phys. **29**, 1 (1981) (http://www.physik.fu-berlin.de/~kleinert/75). This paper was the first to investigate in detail icosahedral quasicrystalline structures discovered later in aluminum.

Second, we see that the connected two-point function at a nonzero source j is given by

$$G_c^{(2)} = G_j^{(1)} = W_{jj} = \frac{\delta X}{\delta j} = \left(\frac{\delta j}{\delta X}\right)^{-1} = \Gamma_{XX}^{-1}. \tag{3.630}$$

The inverse symbols on the right-hand side are to be understood in the functional sense, i.e., Γ_{XX}^{-1} denotes the functional matrix:

$$\Gamma_{X(\tau)X(y)}^{-1} \equiv \left[\frac{\delta^2\Gamma}{\delta X(\tau)\delta X(\tau')}\right]^{-1}, \tag{3.631}$$

which satisfies

$$\int d\tau' \, \Gamma_{X(\tau)X(\tau')}^{-1}\Gamma_{X(\tau')X(\tau'')} = \delta(\tau - \tau''). \tag{3.632}$$

Relation (3.630) states that the second derivative of the effective action determines directly the connected correlation function $G_c^{(2)}(\omega)$ of the interacting theory in the presence of the external source j. Since j is an auxiliary quantity, which eventually be set equal to zero thus making X equal to X_0, the actual physical propagator is given by

$$G_c^{(2)}\Big|_{j=0} = \Gamma_{XX}^{-1}\Big|_{X=X_0}. \tag{3.633}$$

By Fourier-transforming this relation and removing a δ-function for the overall momentum conservation, the full propagator $G_{\omega^2}(\omega)$ is related to the vertex function $\Gamma^{(2)}(\omega)$, defined in (3.626) by

$$G_{\omega^2}(\omega) \equiv \bar{G}^{(2)}(k) = \frac{1}{\Gamma^{(2)}(\omega)}. \tag{3.634}$$

The third derivative of the generating functional $W[j]$ is obtained by functionally differentiating W_{jj} in Eq. (3.630) once more with respect to j, and applying the chain rule:

$$W_{jjj} = -\Gamma_{XX}^{-2}\Gamma_{XXX}\frac{\delta X}{\delta j} = -\Gamma_{XX}^{-3}\Gamma_{XXX} = -G_c^{(2)\,3}\Gamma_{XXX}. \tag{3.635}$$

This equation has a simple physical meaning. The third derivative of $W[j]$ on the left-hand side is the full three-point function at a nonzero source j, so that

$$G_c^{(3)} = W_{jjj} = -G_c^{(2)\,3}\Gamma_{XXX}. \tag{3.636}$$

This equation states that the full three-point function arises from a third derivative of $\Gamma[X]$ by attaching to each derivation a full propagator, apart from a minus sign. We shall express Eq. (3.636) diagrammatically as follows:

where

denotes the connected n-point function, and

the negative n-point vertex function.

For the general analysis of the diagrammatic content of the effective action, we observe that according to Eq. (3.635), the functional derivative of the correlation function G with respect to the current j satisfies

$$G_c^{(2)}{}_j = W_{jjj} = G_c^{(3)} = -G_c^{(2)^3} \Gamma_{XXX}. \tag{3.637}$$

This is pictured diagrammatically as follows:

$$\tag{3.638}$$

This equation may be differentiated further with respect to j in a diagrammatic way. From the definition (3.557) we deduce the trivial recursion relation

$$G_c^{(n)}(\tau_1, \ldots, \tau_n) = \frac{\delta}{\delta j(\tau_n)} G_c^{(n-1)}(\tau_1, \ldots, \tau_{n-1}), \tag{3.639}$$

which is represented diagrammatically as

$$n > 2.$$

By applying $\delta/\delta j$ repeatedly to the left-hand side of Eq. (3.637), we generate all higher connected correlation functions. On the right-hand side of (3.637), the chain rule leads to a derivative of all correlation functions $G = G_c^{(2)}$ with respect to j, thereby changing a line into a line with an extra three-point vertex as indicated in the diagrammatic equation (3.638). On the other hand, the vertex function Γ_{XXX} must be differentiated with respect to j. Using the chain rule, we obtain for any n-point vertex function:

$$\Gamma_{X \ldots Xj} = \Gamma_{X \ldots XX} \frac{\delta X}{\delta j} = \Gamma_{X \ldots XX} G_c^{(2)}, \tag{3.640}$$

which may be represented diagrammatically as

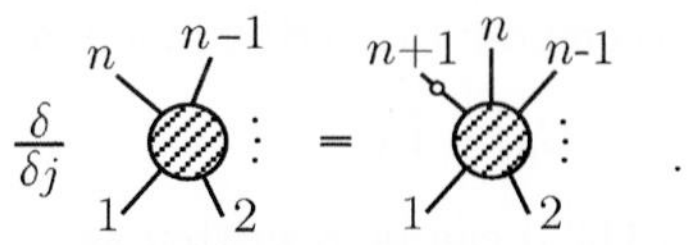

With these diagrammatic rules, we can differentiate (3.635) any number of times, and derive the diagrammatic structure of the connected correlation functions with an arbitrary number of external legs. The result up to $n = 5$ is shown in Fig. 3.12.

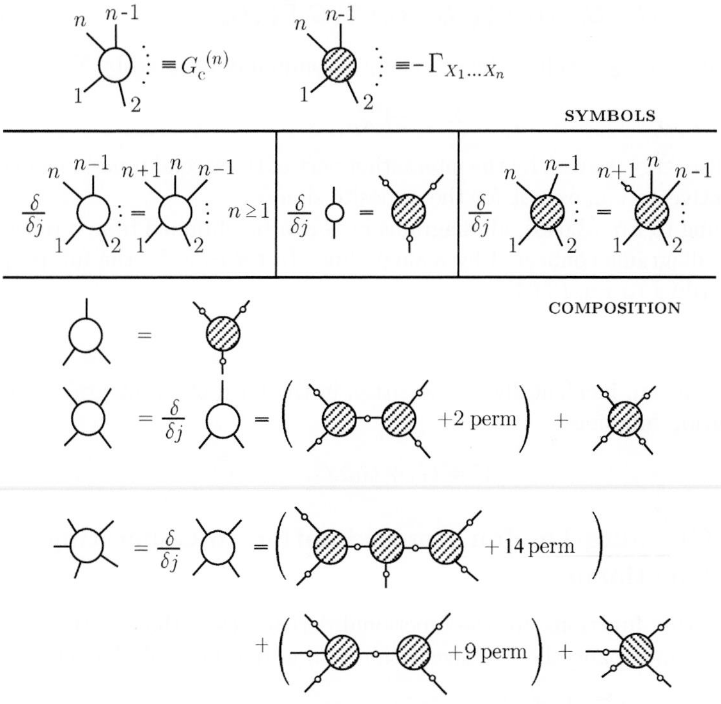

Figure 3.12 Diagrammatic differentiations for deriving tree decomposition of connected correlation functions. The last term in each decomposition yields, after amputation and removal of an overall δ-function of momentum conservation, precisely all one-particle irreducible diagrams.

The diagrams generated in this way have a tree-like structure, and for this reason they are called *tree diagrams*. The tree decomposition reduces all diagrams to their one-particle irreducible contents.

The effective action $\Gamma[X]$ can be used to prove an important composition theorem: The full propagator G can be expressed as a geometric series involving the so-called *self-energy*. Let us decompose the vertex function as

$$\bar{\Gamma}^{(2)} = G_0^{-1} + \bar{\Gamma}^{\text{int}}_{XX}, \tag{3.641}$$

such that the full propagator (3.633) can be rewritten as

$$G = \left(1 + G_0\bar{\Gamma}^{\text{int}}_{XX}\right)^{-1} G_0. \tag{3.642}$$

Expanding the denominator, this can also be expressed in the form of an integral equation:

$$G = G_0 - G_0\bar{\Gamma}^{\text{int}}_{XX}G_0 + G_0\bar{\Gamma}^{\text{int}}_{XX}G_0\bar{\Gamma}^{\text{int}}_{XX}G_0 - \dots \; . \tag{3.643}$$

The quantity $-\bar{\Gamma}^{\text{int}}_{XX}$ is called the self-energy, commonly denoted by Σ:

$$\Sigma \equiv -\bar{\Gamma}^{\text{int}}_{XX}, \tag{3.644}$$

i.e., the self-energy is given by the interacting part of the second functional derivative of the effective action, except for the opposite sign.

According to Eq. (3.643), all diagrams in G can be obtained from a repetition of self-energy diagrams connected by a single line. In terms of Σ, the full propagator reads, according to Eq. (3.642):

$$G \equiv [G_0^{-1} - \Sigma]^{-1}. \tag{3.645}$$

This equation can, incidentally, be rewritten in the form of an integral equation for the correlation function G:

$$G = G_0 + G_0\Sigma G. \tag{3.646}$$

3.22.6 Ginzburg-Landau Approximation to Generating Functional

Since the vertex functions are the functional derivatives of the effective action [see (3.625)], we can expand the effective action into a functional Taylor series

$$\Gamma[X] = \sum_{n=0}^{\infty} \frac{1}{n!} \int d\tau_1 \dots d\tau_n \Gamma^{(n)}(\tau_1, \dots, \tau_n) X(\tau_1) \dots X(\tau_n). \tag{3.647}$$

The expansion in the number of loops of the generating functional $\Gamma[X]$ collects systematically the contributions of fluctuations. To zeroth order, all fluctuations are neglected, and the effective action reduces to the initial action, which is the mean-field approximation to the effective action. In fact, in the absence of loop diagrams, the vertex functions contain only the lowest-order terms in $\Gamma^{(2)}$ and $\Gamma^{(4)}$:

$$\Gamma_0^{(2)}(\tau_1, \tau_2) = M\left(-\partial_{\tau_1}^2 + \omega^2\right)\delta(\tau_1 - \tau_2), \tag{3.648}$$

$$\Gamma_0^{(4)}(\tau_1, \tau_2, \tau_3, \tau_4) = \lambda\,\delta(\tau_1 - \tau_2)\delta(\tau_1 - \tau_3)\delta(\tau_1 - \tau_4). \tag{3.649}$$

Inserted into (3.647), this yields the zero-loop approximation to $\Gamma[X]$:

$$\Gamma_0[X] = \frac{M}{2!} \int d\tau \, [(\partial_\tau X)^2 + \omega^2 X^2] + \frac{\lambda}{4!} \int d\tau \, X^4. \qquad (3.650)$$

This is precisely the original action functional (3.559). By generalizing $X(\tau)$ to be a magnetization vector field, $X(\tau) \to \mathbf{M}(\mathbf{x})$, which depends on the three-dimensional space variables $\mathbf{x}$ rather than the Euclidean time, the functional (3.650) coincides with the phenomenological energy functional set up by Ginzburg and Landau to describe the behavior of magnetic materials near the Curie point, which they wrote as[18]

$$\Gamma[\mathbf{M}] = \int d^3x \left[\frac{1}{2} \sum_{i=1}^{3} (\partial_i \mathbf{M})^2 + \frac{m^2}{2!} \mathbf{M}^2 + \frac{\lambda}{4!} \mathbf{M}^4 \right]. \qquad (3.651)$$

The use of this functional is also referred to as *mean-field theory* or *mean-field approximation* to the full theory.

3.22.7 Composite Fields

Sometimes it is of interest to study also correlation functions in which two fields coincide at one point, for instance

$$G^{(1,n)}(\tau, \tau_1, \ldots, \tau_n) = \frac{1}{2} \langle x^2(\tau) x(\tau_1) \cdots x(\tau_n) \rangle. \qquad (3.652)$$

If multiplied by a factor $M\omega^2$, the composite operator $M\omega^2 x^2(\tau)/2$ is precisely the frequency term in the action energy functional (3.559). For this reason one speaks of a *frequency insertion*, or, since in the Ginzburg-Landau action (3.651) the frequency ω is denoted by the mass symbol m, one speaks of a *mass insertion* into the correlation function $G^{(n)}(\tau_1, \ldots, \tau_n)$.

Actually, we shall never make use of the full correlation function (3.652), but only of the integral over τ in (3.652). This can be obtained directly from the generating functional $Z[j]$ of all correlation functions by differentiation with respect to the square mass in addition to the source terms

$$\int d\tau \, G^{(1,n)}(\tau, \tau_1, \ldots, \tau_n) = -Z^{-1} \frac{\partial}{M\partial\omega^2} \frac{\delta}{\delta j(\tau_1)} \cdots \frac{\delta}{\delta j(\tau_n)} Z[j] \bigg|_{j=0}. \qquad (3.653)$$

By going over to the generating functional $W[j]$, we obtain in a similar way the connected parts:

$$\int d\tau \, G_c^{(1,n)}(\tau, \tau_1, \ldots, \tau_n) = -\frac{\partial}{M\partial\omega^2} \frac{\delta}{\delta j(\tau_1)} \cdots \frac{\delta}{\delta j(\tau_n)} W[j] \bigg|_{j=0}. \qquad (3.654)$$

---

[18]L.D. Landau, J.E.T.P. **7**, 627 (1937).

The right-hand side can be rewritten as

$$\int d\tau\, G_c^{(1,n)}(\tau, \tau_1, \ldots, \tau_n) = -\frac{\partial}{M\partial\omega^2} G_c^{(n)}(\tau_1, \ldots, \tau_n). \tag{3.655}$$

The connected correlation functions $G_c^{(1,n)}(\tau, \tau_1, \ldots, \tau_n)$ can be decomposed into tree diagrams consisting of lines and one-particle irreducible vertex functions $\Gamma^{(1,n)}(\tau, \tau_1, \ldots, \tau_n)$. If integrated over τ, these are defined from Legendre transform (3.619) by a further differentiation with respect to $M\omega^2$:

$$\int d\tau\, \Gamma^{(1,n)}(\tau, \tau_1, \ldots, \tau_n) = -\frac{\partial}{M\partial\omega^2} \frac{\delta}{\delta X(\tau_1)} \cdots \frac{\delta}{\delta X(\tau_n)} \Gamma[X]\Big|_{X_0}, \tag{3.656}$$

implying the relation

$$\int d\tau\, \Gamma^{(1,n)}(\tau, \tau_1, \ldots, \tau_n) = -\frac{\partial}{M\partial\omega^2} \Gamma^{(n)}(\tau_1, \ldots, \tau_n). \tag{3.657}$$

3.23 Path Integral Calculation of Effective Action by Loop Expansion

Path integrals give the most direct access to the effective action of a theory avoiding the cumbersome Legendre transforms. The derivation will proceed diagrammatically loop by loop, which will turn out to be organized by the powers of the Planck constant $\hbar$. This will now be kept explicit in all formulas. For later applications to quantum mechanics we shall work with real time.

3.23.1 General Formalism

Consider the generating functional of all Green functions

$$Z[j] = e^{iW[j]/\hbar}, \tag{3.658}$$

where $W[j]$ is the generating functional of all *connected* Green functions. The vacuum expectation of the field, the average

$$X(t) \equiv \langle x(t) \rangle, \tag{3.659}$$

is given by the first functional derivative

$$X(t) = \delta W[j]/\delta j(t). \tag{3.660}$$

This can be inverted to yield $j(t)$ as a functional of $X(t)$:

$$j(t) = j[X](t), \tag{3.661}$$

which leads to the Legendre transform of $W[j]$:

$$\Gamma[X] \equiv W[j] - \int dt\, j(t) X(t), \tag{3.662}$$

where the right-hand side is replaced by (3.661). This is the *effective action* of the theory. The effective action for time independent $\mathbf{X}(t) \equiv \mathbf{X}$ defines the *effective potential*

$$V^{\text{eff}}(X) \equiv -\frac{1}{t_b - t_a}\Gamma[X].\tag{3.663}$$

The first functional derivative of the effective action gives back the current

$$\frac{\delta\Gamma[X]}{\delta X(t)} = -j(t).\tag{3.664}$$

The generating functional of all connected Green functions can be recovered from the effective action by the inverse Legendre transform

$$W[j] = \Gamma[X] + \int dt\, j(t)X(t).\tag{3.665}$$

We now calculate these quantities from the path integral formula (3.558) for the generating functional $Z[j]$:

$$Z[j] = \int \mathcal{D}x(t)e^{(i/\hbar)\left\{\mathcal{A}[x]+\int dt\, j(t)x(t)\right\}}.\tag{3.666}$$

With (3.658), this amounts to the path integral formula for $\Gamma[X]$:

$$e^{\frac{i}{\hbar}\left\{\Gamma[X]+\int dt\, j(t)X(t)\right\}} = \int \mathcal{D}x(t)e^{(i/\hbar)\left\{\mathcal{A}[x]+\int dt\, j(t)x(t)\right\}}.\tag{3.667}$$

The action quantum $\hbar$ is a measure for the size of quantum fluctuations. Under many physical circumstances, quantum fluctuations are small, which makes it desirable to develop a method of evaluating (3.667) as an expansion in powers of $\hbar$.

3.23.2   Mean-Field Approximation

For $\hbar \to 0$, the path integral over the path $x(t)$ in (3.666) is dominated by the classical solution $x_{\text{cl}}(t)$ which extremizes the exponent

$$\left.\frac{\delta\mathcal{A}[x]}{\delta x(t)}\right|_{x=x_{\text{cl}}(t)} = -j(t),\tag{3.668}$$

and is a functional of $j(t)$ which may be written, more explicitly, as $x_{\text{cl}}(t)[j]$. At this level we can identify

$$W[j] = \Gamma[X] + \int dt\, j(t)X(t) \approx \mathcal{A}[x_{\text{cl}}[j]] + \int dt\, j(t)x_{\text{cl}}(t)[j].\tag{3.669}$$

By differentiating $W[j]$ with respect to j, we have from the general first part of Eq. (3.659):

$$X = \frac{\delta W}{\delta j} = \frac{\delta\Gamma}{\delta X}\frac{\delta X}{\delta j} + X + j\frac{\delta X}{\delta j}.\tag{3.670}$$

Inserting the classical equation of motion (3.668), this becomes

$$X = \frac{\delta \mathcal{A}}{\delta x_{\mathrm{cl}}} \frac{\delta x_{\mathrm{cl}}}{\delta j} + x_{\mathrm{cl}} + j \frac{\delta x_{\mathrm{cl}}}{\delta j} = x_{\mathrm{cl}}. \tag{3.671}$$

Thus, to this approximation, $X(t)$ coincides with the classical path $x_{\mathrm{cl}}(t)$. Replacing $x_{\mathrm{cl}}(t) \to X(t)$ on the right-hand side of Eq. (3.669), we obtain the lowest-order result, which is of zeroth order in $\hbar$, the classical approximation to the effective action:

$$\Gamma_0[X] = \mathcal{A}[X]. \tag{3.672}$$

For an anharmonic oscillator in N dimensions with unit mass and an interaction $\mathbf{x}^4$, where $\mathbf{x} = (x_1, \ldots, x_N)$, which is symmetric under N-dimensional rotations $O(N)$, the lowest-order effective action reads

$$\Gamma_0[\mathbf{X}] = \int dt \left[\frac{1}{2} \left(\dot{X}_a^2 - \omega^2 X_a^2 \right) - \frac{g}{4!} \left(X_a^2 \right)^2 \right], \tag{3.673}$$

where repeated indices $a, b, \ldots$ are summed from 1 to N following Einstein's summation convention. The effective potential (3.663) is simply the initial potential

$$V_0^{\mathrm{eff}}(\mathbf{X}) = V(\mathbf{X}) = \frac{\omega^2}{2} X_a^2 + \frac{g}{4!} \left(X_a^2 \right)^2. \tag{3.674}$$

For $\omega^2 > 0$, this has a minimum at $\mathbf{X} \equiv \mathbf{0}$, and there are only two non-vanishing vertex functions $\Gamma^{(n)}(t_1, \ldots, t_n)$:

For $n = 2$:

$$\begin{aligned}
\Gamma^{(2)}(t_1, t_2)_{ab} &\equiv \left. \frac{\delta^2 \Gamma}{\delta X_a(t_1) \delta X_b(t_2)} \right|_{X_a = 0} = \left. \frac{\delta^2 \mathcal{A}}{x_a(t_1) x_b(t_2)} \right|_{x_a = X_a = 0} \\
&= (-\partial_t^2 - \omega^2) \delta_{ab} \delta(t_1 - t_2).
\end{aligned} \tag{3.675}$$

This determines the inverse of the propagator:

$$\Gamma^{(2)}(t_1, t_2)_{ab} = [i\hbar G^{-1}]_{ab}(t_1, t_2). \tag{3.676}$$

Thus we find to this zeroth-order approximation that $G_{ab}(t_1, t_2)$ is equal to the free propagator:

$$G_{ab}(t_1, t_2) = G_{0ab}(t_1, t_2). \tag{3.677}$$

For $n = 4$:

$$\Gamma^{(4)}(t_1, t_2, t_3, t_4)_{abcd} \equiv \frac{\delta^4 \Gamma}{\delta X_a(t_1) \delta X_b(t_2) \delta X_c(t_3) \delta X_d(t_4)} = g T_{abcd}, \tag{3.678}$$

with

$$T_{abcd} = \frac{1}{3} (\delta_{ab} \delta_{cd} + \delta_{ac} \delta_{bd} + \delta_{ad} \delta_{bc}). \tag{3.679}$$

According to the definition of the effective action, all diagrams of the theory can be composed from the propagator $G_{ab}(t_1, t_2)$ and this vertex via tree diagrams. Thus we see that in this lowest approximation, we recover precisely the subset of all original Feynman diagrams with a tree-like topology. These are all diagrams which do not involve any loops. Since the limit $\hbar \to 0$ corresponds to the classical equations of motion with no quantum fluctuations we conclude: Classical theory corresponds to tree diagrams.

For $\omega^2 < 0$ the discussion is more involved since the minimum of the potential (3.674) lies no longer at $\mathbf{X} = \mathbf{0}$, but at a nonzero vector $\mathbf{X}_0$ with an arbitrary direction, and a length

$$|\mathbf{X}_0| = \sqrt{-6\omega^2/g}. \tag{3.680}$$

The second functional derivative (3.675) at $\mathbf{X}$ is anisotropic and reads

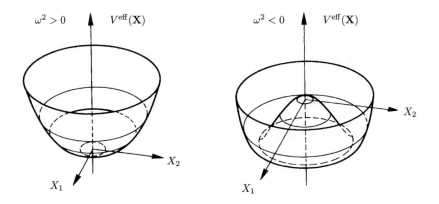

Figure 3.13 Effective potential for $\omega^2 > 0$ and $\omega^2 < 0$ in mean-field approximation, pictured for the case of two components X_1, X_2. The right-hand figure looks like a Mexican hat or the bottom of a champaign bottle..

$$
\begin{aligned}
\Gamma^{(2)}(t_1, t_2)_{ab} &\equiv \left. \frac{\delta^2 \Gamma}{\delta X_a(t_1) \delta X_b(t_2)} \right|_{X_a \neq 0} = \left. \frac{\delta^2 \mathcal{A}}{x_a(t_1) x_b(t_2)} \right|_{x_a = X_a \neq 0} \\
&= \left[-\partial_t^2 - \omega^2 - \frac{g}{6} \left(\delta_{ab} X_c^2 + 2 X_a X_b \right) \right] \delta_{ab} \delta(t_1 - t_2).
\end{aligned} \tag{3.681}
$$

This is conveniently separated into longitudinal and transversal derivatives with respect to the direction $\hat{\mathbf{X}} = \mathbf{X}/|\mathbf{X}|$. We introduce associated projection matrices:

$$P_{Lab}(\hat{\mathbf{X}}) = \hat{X}_a \hat{X}_b, \qquad P_{Tab}(\hat{\mathbf{X}}) = \delta_{ab} - \hat{X}_a \hat{X}_b, \tag{3.682}$$

and decompose

$$\Gamma^{(2)}(t_1, t_2)_{ab} = \Gamma_L^{(2)}(t_1, t_2)_{ab} P_{Lab}(\hat{\mathbf{X}}) + \Gamma_T^{(2)}(t_1, t_2)_{ab} P_{Tab}(\hat{\mathbf{X}}), \tag{3.683}$$

where

$$\Gamma_L^{(2)}(t_1, t_2)_{ab} = \left[-\partial_t^2 - \left(\omega^2 + \frac{g}{6} \mathbf{X}^2 \right) \right] \delta(t_1 - t_2), \qquad (3.684)$$

and

$$\Gamma_L^{(2)}(t_1, t_2)_{ab} = \left[-\partial_t^2 - \left(\omega^2 + 3\frac{g}{6} \mathbf{X}^2 \right) \right] \delta(t_1 - t_2). \qquad (3.685)$$

This can easily be inverted to find the propagator

$$\mathcal{G}(t_1, t_2)_{ab} = i\hbar \left[\Gamma^{(2)}(t_1, t_2) \right]_{ab}^{-1} = \mathcal{G}_L(t_1, t_2)_{ab} P_{Lab}(\hat{\mathbf{X}}) + \mathcal{G}_T(t_1, t_2)_{ab} P_{Tab}(\hat{\mathbf{X}}), \quad (3.686)$$

where

$$\mathcal{G}_L(t_1, t_2)_{ab} = \frac{i\hbar}{\Gamma_L(t_1, t_2)} = \frac{i\hbar}{-\partial_t^2 - \omega_L^2(\mathbf{X})}, \qquad (3.687)$$

$$\mathcal{G}_T(t_1, t_2)_{ab} = \frac{i\hbar}{\Gamma_T^{(2)}(t_1, t_2)} = \frac{i\hbar}{-\partial_t^2 - \omega_T^2(\mathbf{X})} \qquad (3.688)$$

are the longitudinal and transversal parts of the Green function. For convenience, we have introduced the $\mathbf{X}$-dependent frequencies of the longitudinal and transversal Green functions:

$$\omega_L^2(\mathbf{X}) \equiv \omega^2 + 3\frac{g}{6} \mathbf{X}^2, \quad \omega_T^2(\mathbf{X}) \equiv \omega^2 + \frac{g}{6} \mathbf{X}^2. \qquad (3.689)$$

To emphasize the fact that this propagator is a functional of $\mathbf{X}$ we represent it by the calligraphic letter $\mathcal{G}$. For $\omega^2 > 0$, we perform the fluctuation expansion around the minimum of the potential (3.663) at $\mathbf{X} = 0$, where the two Green functions coincide, both having the same frequency ω:

$$\mathcal{G}_L(t_1, t_2)_{ab} |_{\mathbf{X}=0} = \mathcal{G}_T(t_1, t_2)_{ab} |_{\mathbf{X}=0} = \mathcal{G}(t_1, t_2)_{ab} |_{\mathbf{X}=0} = \frac{i\hbar}{-\partial_t^2 - \omega^2}, \qquad (3.690)$$

For $\omega^2 < 0$, however, where the minimum lies at the vector $\mathbf{X}_0$ of length (3.680), they are different:

$$\mathcal{G}_L(t_1, t_2)_{ab} |_{\mathbf{X}=\mathbf{X}_0} = \frac{i\hbar}{-\partial_t^2 + 2\omega^2}, \quad \mathcal{G}_T(t_1, t_2)_{ab} |_{\mathbf{X}=\mathbf{X}_0} = \frac{i\hbar}{-\partial_t^2}. \qquad (3.691)$$

Since the curvature of the potential at the minimum in radial direction of $\mathbf{X}$ is positive at the minimum, the longitudinal part has now the positive frequency $-2\omega^2$. The movement along the valley of the minimum, on the other hand, does not increase the energy. For this reason, the transverse part has zero frequency. This feature, observed here in lowest order of the fluctuation expansion, is a very general one, and can be found in the effective action to any loop order. In quantum field theory, there exists a theorem asserting this called *Nambu-Goldstone theorem*. It states that if a quantum field theory without long-range interactions has a continuous symmetry which is broken by a nonzero expectation value of the field corresponding to the present $\mathbf{X}$ [recall (3.659)], then the fluctuations transverse to it have a zero mass. They are called *Nambu-Goldstone modes* or, because of their bosonic nature, *Nambu-Goldstone bosons*. The exclusion of long-range interactions is necessary, since these can mix with the zero-mass modes and make it massive. This happens, for example, in a superconductor where they make the magnetic field massive, giving it a finite penetration depth, the famous *Meissner effect*. One expresses this pictorially by saying that the long-range mode can eat up the Nambu-Goldstone modes and become massive. The same mechanism is used in elementary particle physics to explain the mass of the $W^\pm$ and Z^0 vector bosons as a consequence of having eaten up a would be Nambu-Goldstone boson of an auxiliary Higgs-field theory.

In quantum-mechanical systems, however, a nonzero expectation value with the associated zero frequency mode in the transverse direction is found only as an artifact of perturbation theory. If all fluctuation corrections are summed, the minimum of the effective potential lies always at the origin. For example, it is well known, that the ground state wave functions of a particle in a double-well potential is symmetric, implying a zero expectation value of the particle position. This symmetry is caused by quantum-mechanical tunneling, a phenomenon which will be discussed in detail in Chapter 17. This phenomenon is of a nonperturbative nature which cannot be described by an effective potential calculated order by order in the fluctuation expansion. Such a potential does, in general, posses a nonzero minimum at some $\mathbf{X}_0$ somewhere near the zero-order minimum (3.680). Due to this shortcoming, it is possible to derive the Nambu-Goldstone theorem from the quantum-mechanical effective action in the loop expansion, even though the nonzero expectation value $\mathbf{X}_0$ assumed in the derivation of the zero-frequency mode does not really exist in quantum mechanics. The derivation will be given in Section 3.24.

The use of the initial action to approximate the effective action neglecting corrections caused by the fluctuations is referred to as *mean-field approximation*.

3.23.3   Corrections from Quadratic Fluctuations

In order to find the first $\hbar$-correction to the mean-field approximation we expand the action in powers of the fluctuations of the paths around the classical solution

$$\delta x(t) \equiv x(t) - x_{\mathrm{cl}}(t), \qquad (3.692)$$

and perform a perturbation expansion. The quadratic term in $\delta x(t)$ is taken to be the free-particle action, the higher powers in $\delta x(t)$ are the interactions. Up to second order in the fluctuations $\delta x(t)$, the action is expanded as follows:

$$\mathcal{A}[x_{\mathrm{cl}} + \delta x] + \int dt\, j(t)\, [x_{\mathrm{cl}}(t) + \delta x(t)]$$

$$= \mathcal{A}[x_{\mathrm{cl}}] + \int dt\, j(t)\, x_{\mathrm{cl}}(t) + \int dt \left\{ j(t) + \left.\frac{\delta \mathcal{A}}{\delta x(t)}\right|_{x=x_{\mathrm{cl}}} \right\} \delta x(t)$$

$$+ \frac{1}{2} \int dt\, dt'\ \delta x(t) \left.\frac{\delta^2 \mathcal{A}}{\delta x(t)\delta x(t')}\right|_{x=x_{\mathrm{cl}}} \delta x(t') + \mathcal{O}\left((\delta x)^3\right). \qquad (3.693)$$

The curly bracket multiplying the linear terms in the variation $\delta x(t)$ vanish due to the extremality property of the classical path x_{cl} expressed by the equation of motion (3.668). Inserting this expansion into (3.667), we obtain the approximate expression

$$Z[j] \approx e^{(i/\hbar)\left\{A[x_{\mathrm{cl}}] + \int dt\, j(t)x_{\mathrm{cl}}(t)\right\}} \int \mathcal{D}\delta x \exp\left\{\frac{i}{\hbar}\int dt\, dt'\; \delta x(t)\frac{\delta^2 A}{\delta x(t)\delta x(t')}\bigg|_{x=x_{\mathrm{cl}}} \delta x(t')\right\}.$$
(3.694)

We now observe that the fluctuations $\delta x(t)$ will be of average size $\sqrt{\hbar}$ due to the $\hbar$-denominator in the Fresnel exponent. Thus the fluctuations $(\delta x)^n$ are of average size $\sqrt{\hbar}^n$. The approximate path integral (3.694) is of the Fresnel type and my be integrated to yield

$$e^{(i/\hbar)\left\{A[x_{\mathrm{cl}}] + \int dt\, j(t)x_{\mathrm{cl}}(t)\right\}} \left[\det \frac{\delta^2 A}{\delta x(t)\delta x(t')}\right]^{-1/2}_{x=x_{\mathrm{cl}}}$$
(3.695)

$$= e^{(i/\hbar)\left\{A[x_{\mathrm{cl}}] + \int dt\, j(t)x_{\mathrm{cl}}(t) + i(\hbar/2)\mathrm{Tr}\log[\delta^2 A/\delta x(t)\delta x(t')|_{x=x_{\mathrm{cl}}}]\right\}}.$$

Comparing this with the left-hand side of (3.667), we find that to first order in $\hbar$, the effective action may be recovered by equating

$$\Gamma[X] + \int dt\, j(t)X(t) = A[x_{\mathrm{cl}}[j]] + \int dt\, j(t)\, x_{\mathrm{cl}}(t)[j] + \frac{i\hbar}{2}\mathrm{Tr}\log\frac{\delta^2 A\,[x_{\mathrm{cl}}[j]]}{\delta x(t)\delta x(t')}.$$
(3.696)

In the limit $\hbar \to 0$, the tracelog term disappears and (3.696) reduces to the classical expression (3.669).

To include the $\hbar$-correction into $\Gamma[X]$, we expand $W[j]$ as

$$W[j] = W_0[j] + \hbar W_1[j] + \mathcal{O}(\hbar^2).$$
(3.697)

Correspondingly, the path X differs from X_{cl} by a correction term of order $\hbar$:

$$X = x_{\mathrm{cl}} + \hbar\, X_1 + \mathcal{O}(\hbar^2).$$
(3.698)

Inserting this into (3.696), we find

$$\Gamma[X] + \int dt\, jX = A\,[X - \hbar X_1] + \int dt\, jX - \hbar \int dt\, jX_1$$
$$+ \frac{i}{2}\hbar\,\mathrm{Tr}\log\frac{\delta^2 A}{\delta x_a \delta x_b}\bigg|_{x=X-\hbar X_1} + \mathcal{O}\left(\hbar^2\right).$$
(3.699)

Expanding the action up to the same order in $\hbar$ gives

$$\Gamma[X] = A[X] - \hbar \int dt \left\{\frac{\delta A[X]}{\delta X} + j\right\} X_1 + \frac{i}{2}\hbar\,\mathrm{Tr}\log\frac{\delta^2 A}{\delta x_a \delta x_b}\bigg|_{x=X} + \mathcal{O}\left(\hbar^2\right).$$
(3.700)

Due to (3.668), the curly-bracket term is only of order $\hbar^2$, so that we find the one-loop form of the effective action

$$\Gamma[X] = \Gamma_0[X] + \hbar\Gamma_1[X] = \int dt \left[\frac{1}{2}\dot{X}^2 - \frac{\omega^2}{2}X_a^2 - \frac{g}{4!}\left(X_a^2\right)^2\right]$$
$$+ \frac{i}{2}\hbar \operatorname{Tr}\log\left[-\partial_t^2 - \omega^2 - \frac{g}{6}\left(\delta_{ab}X_c^2 + 2X_aX_b\right)\right]. \quad (3.701)$$

Using the decomposition (3.683), the tracelog term can be written as a sum of transversal and longitudinal parts

$$\hbar\Gamma_1[X] = \frac{i}{2}\hbar \operatorname{Tr}\log\Gamma_L^{(2)}(t_1, t_2)_{ab} + \frac{i}{2}(N-1)\hbar \operatorname{Tr}\log\Gamma_T^{(2)}(t_1, t_2)_{ab} \quad (3.702)$$
$$= \frac{i}{2}\hbar \operatorname{Tr}\log\left(-\partial_t^2 - \omega_L^2(\mathbf{X})\right) + \frac{i}{2}(N-1)\hbar \operatorname{Tr}\log\left(-\partial_t^2 - \omega_T^2(\mathbf{X})\right).$$

What is the graphical content in the Green functions at this level of approximation? Assuming $\omega^2 > 0$, we find for $\mathbf{j} = 0$ that the minimum lies at $\bar{\mathbf{X}} = \mathbf{0}$, as in the mean-field approximation. Around this minimum, we may expand the tracelog in powers of $\mathbf{X}$. For the simplest case of a single X-variable, we obtain

$$\frac{i}{2}\hbar \operatorname{Tr}\log\left(-\partial_t^2 - \omega^2 - \frac{g}{2}X^2\right) = \frac{i}{2}\hbar \operatorname{Tr}\log\left(-\partial_t^2 - \omega^2\right) + \frac{i}{2}\hbar \operatorname{Tr}\log\left(1 + \frac{i}{-\partial_t^2 - \omega^2}ig\frac{X^2}{2}\right)$$
$$= i\frac{\hbar}{2}\operatorname{Tr}\log\left(-\partial_t^2 - \omega^2\right) - i\frac{\hbar}{2}\sum_{n=1}^{\infty}\left(-i\frac{g}{2}\right)^n \frac{1}{n}\operatorname{Tr}\left(\frac{i}{-\partial_t^2 - \omega^2}X^2\right)^n. \quad (3.703)$$

If we insert

$$G_0 = \frac{i}{-\partial_t^2 - \omega^2}, \quad (3.704)$$

this can be written as

$$i\frac{\hbar}{2}\operatorname{Tr}\log\left(-\partial_t^2 - \omega^2\right) - i\frac{\hbar}{2}\sum_{n=1}^{\infty}\left(-i\frac{g}{2}\right)^n \frac{1}{n}\operatorname{Tr}\left(G_0X^2\right)^n. \quad (3.705)$$

More explicitly, the terms with $n = 1$ and $n = 2$ read:

$$-\frac{\hbar}{4}g\int dt\,dt'\,\delta(t-t')G_0(t,t')X^2(t')$$
$$+i\hbar\frac{g^2}{16}\int dt\,dt'\,dt''\,\delta^4(t-t'')G_0(t,t')X^2(t')G_0(t',t'')X^2(t'') + \cdots. \quad (3.706)$$

The expansion terms of (3.705) for $n \geq 1$ correspond obviously to the Feynman diagrams (omitting multiplicity factors)

$$\mathcal{A}[x_{cl}] = \quad \bigcirc \quad + \quad \bigcirc \quad + \quad \bigcirc \quad + \quad \times\!\!\bigcirc\!\!\times \quad + \quad \cdots. \quad (3.707)$$

The series (3.705) is therefore a sum of all diagrams with one loop and any number of fundamental X^4-vertices

To systematize the entire expansion (3.705), the tracelog term is [compare (3.546)] pictured by a single-loop diagram

$$i\frac{\hbar}{2}\text{Tr}\log\left(-\partial_t^2 - \omega^2\right) = \frac{1}{2}\;\bigcirc\;. \tag{3.708}$$

The first two diagrams in (3.707) contribute corrections to the vertices $\Gamma^{(2)}$ and $\Gamma^{(4)}$. The remaining diagrams produce higher vertex functions and lead to more involved tree diagrams. In Fourier space we find from (3.706)

$$\Gamma^{(2)}(q) = q^2 - \omega^2 - \hbar\frac{g}{2}\int\frac{dk}{2\pi}\frac{i}{k^2 - \omega^2 + i\eta} \tag{3.709}$$

$$\Gamma^{(4)}(q_i) = g - i\frac{g^2}{2}\left[\int\frac{dk}{2\pi}\frac{i}{k^2 - \omega^2 + i\eta}\frac{i}{(q_1 + q_2 - k)^2 - \omega^2 + i\eta} + 2\text{ perm}\right]. \tag{3.710}$$

We may write (3.709) in Euclidean form as

$$\Gamma^{(2)}(q) = -q^2 - \omega^2 - \hbar\frac{g}{2}\int\frac{dk}{2\pi}\frac{1}{k^2 + \omega^2}$$

$$= -\left(q^2 + \omega^2 + \hbar\frac{g}{2}\frac{1}{2\omega}\right), \tag{3.711}$$

$$\Gamma^{(4)}(q_i) = g - \hbar\frac{g^2}{2}\left[I\left(q_1 + q_2\right) + 2\text{ perm}\right], \tag{3.712}$$

with the Euclidean two-loop integral

$$I(q_1 + q_2) = \int\frac{dk}{2\pi}\frac{1}{k^2 + \omega^2}\frac{i}{\left(q_1 + q_2 - k\right)^2 + \omega^2}, \tag{3.713}$$

to be calculated explicitly in Chapter 10. It is equal to $J((q_1 + q_2)^2)/2\pi$ with the functions $J(z)$ of Eq. (10.259).

For $\omega^2 < 0$ where the minimum of the effective action lies at $\bar{\mathbf{X}} \neq \mathbf{0}$, the expansion of the trace of the logarithm in (3.701) must distinguish longitudinal and transverse parts.

3.23.4 Effective Action to Order $\hbar^2$

Let us now find the next correction to the effective action.[19] Instead of truncating the expansion (3.693), we keep all terms, reorganizing only the linear and quadratic terms as in (3.694). This yields

$$e^{(i/\hbar)\{\Gamma[X]+jX\}} = e^{i(\hbar/2)W[j]} = e^{(i/\hbar)\{(A[x_{\text{cl}}]+jx_{\text{cl}})+(i\hbar/2)\text{Tr}\log A_{xx}[x_{\text{cl}}]\}}\, e^{(i/\hbar)\hbar^2 W_2[x_{\text{cl}}]}. \tag{3.714}$$

---

[19]R. Jackiw, Phys. Rev. D *9*, 1687 (1976)

The functional $W_2[x_{\rm cl}]$ is defined by the path integral over the fluctuations

$$e^{(i/\hbar)\hbar^2 W_2[x_{\rm cl}]} = \frac{\int \mathcal{D}x \exp \frac{i}{\hbar} \left\{ \frac{1}{2}\delta x \mathcal{D}[x_{\rm cl}]\delta x + \mathcal{R}[x_{\rm cl}, \delta x] \right\}}{\int \mathcal{D}\delta x \exp \frac{i}{\hbar} \left\{ \frac{1}{2}\delta x \mathcal{A}_{xx}[x_{\rm cl}]\delta x \right\}}, \tag{3.715}$$

where $\mathcal{D}[x_{\rm cl}] \equiv \mathcal{A}_{xx}[x_{\rm cl}]$ is the second functional derivative of the action at $x = x_{\rm cl}$. The subscripts x of $\mathcal{A}_{xx}$ denote functional differentiation. For the anharmonic oscillator:

$$\mathcal{D}[x_{\rm cl}] \equiv \mathcal{A}_{xx}[x_{\rm cl}] = -\partial_t^2 - \omega^2 - \frac{g}{2}x_{\rm cl}^2. \tag{3.716}$$

The functional $\mathcal{R}$ collects all unharmonic terms:

$$\mathcal{R}[x_{\rm cl}, \delta x] = \mathcal{A}[x_{\rm cl} + \delta x] - \mathcal{A}[x_{\rm cl}] - \int dt\, \mathcal{A}_x[x_{\rm cl}](t)\delta x(t)$$
$$- \frac{1}{2} \int dt dt'\, \delta x(t)\mathcal{A}_{xx}[x_{\rm cl}](t, t')\delta x(t'). \tag{3.717}$$

In condensed functional vector notation, we shall write expressions like the last term as

$$\frac{1}{2} \int dt dt'\, \delta x(t)\mathcal{A}_{xx}[x_{\rm cl}](t, t')\delta x(t') \to \frac{1}{2}\delta x \mathcal{A}_{xx}[x_{\rm cl}]\delta x. \tag{3.718}$$

By construction, $\mathcal{R}$ is at least cubic in δx. The path integral (3.715) may thus be considered as the generating functional $Z^{\rm fl}$ of a fluctuating variable $\delta x(\tau)$ with a propagator

$$\mathcal{G}[x_{\rm cl}] = i\hbar\{\mathcal{A}_{xx}[x_{\rm cl}]\}^{-1} \equiv i\hbar\mathcal{D}^{-1}[x_{\rm cl}],$$

and an interaction $\mathcal{R}[x_{\rm cl}, \dot{x}]$, both depending on j via $x_{\rm cl}$. We know from the previous sections, and will immediately see this explicitly, that $\hbar^2 W_2[x_{\rm cl}]$ is of order $\hbar^2$. Let us write the full generating functional $W[j]$ in the form

$$W[j] = \mathcal{A}[x_{\rm cl}] + x_{\rm cl}j + \hbar\Delta_1[x_{\rm cl}], \tag{3.719}$$

where the last term collects one- and two-loop corrections (in higher-order calculations, of course, also higher loops):

$$\Delta_1[x_{\rm cl}] = \frac{i}{2}\mathrm{Tr}\log \mathcal{D}[x_{\rm cl}] + \hbar W_2[x_{\rm cl}]. \tag{3.720}$$

From (3.719) we find the vacuum expectation value $X = \langle x \rangle$ as the functional derivative

$$X = \frac{\delta W[j]}{\delta j} = x_{\rm cl} + \hbar\Delta_{1x_{\rm cl}}[x_{\rm cl}]\frac{\delta x_{\rm cl}}{\delta j}, \tag{3.721}$$

implying the correction term X_1:

$$X_1 = \Delta_{1x_{\rm cl}}[x_{\rm cl}]\frac{\delta x_{\rm cl}}{\delta j}. \tag{3.722}$$

The only explicit dependence of $W[j]$ on j comes from the second term in (3.719). In all others, the j-dependence is due to $x_{\rm cl}[j]$. We may use this fact to express j as a function of $x_{\rm cl}$. For this we consider $W[j]$ for a moment as a functional of $x_{\rm cl}$:

$$W[x_{\rm cl}] = \mathcal{A}[x_{\rm cl}] + x_{\rm cl}\, j[x_{\rm cl}] + \hbar\Delta_1[x_{\rm cl}]. \tag{3.723}$$

The combination $W[x_{\rm cl}] - jX$ gives us the effective action $\Gamma[X]$ [recall (3.662)]. We therefore express $x_{\rm cl}$ in (3.723) as $X - \hbar X_1 - \mathcal{O}(\hbar^2)$ from (3.698), and re-expand everything around X rather than $x_{\rm cl}$, yields

$$\begin{aligned}
\Gamma[X] &= \mathcal{A}[X] - \hbar\mathcal{A}_X[X]X_1 - \hbar X_1\, j[X] + \hbar^2 X_1\, j_X[X]X_1 + \frac{1}{2}\hbar^2 X_1\, \mathcal{D}[X]X_1 \\
&\quad + \hbar\Delta_1[X] - \hbar^2\Delta_{1X}[X]X_1 + \mathcal{O}(\hbar^3).
\end{aligned} \tag{3.724}$$

Since the action is extremal at $x_{\rm cl}$, we have

$$\mathcal{A}_X[X - \hbar X_1] = -j[X] + \mathcal{O}(\hbar^2), \tag{3.725}$$

and thus

$$\mathcal{A}_X[X] = -j[X] + \hbar\mathcal{A}_{XX}[X]X_1 + \mathcal{O}(\hbar^2) = -j[X] + \hbar\mathcal{D}[X]X_1 + \mathcal{O}(\hbar^2), \tag{3.726}$$

and therefore:

$$\Gamma[X] = \mathcal{A}[X] + \hbar\Delta_1[X] + \hbar^2\left\{-\frac{1}{2}X_1\mathcal{D}[X]X_1 + X_1 j_X[X]X_1 - \Delta_{1X}X_1\right\}. \tag{3.727}$$

From (3.722) we see that

$$\frac{\delta j}{\delta x_{\rm cl}}X_1 = \Delta_{1x_{\rm cl}}[x_{\rm cl}]. \tag{3.728}$$

Replacing $x_{\rm cl} \to X$ with an error of order $\hbar$, this implies

$$\frac{\delta j}{\delta X}X = \Delta_{1X}[X] + \mathcal{O}(\hbar). \tag{3.729}$$

Inserting this into (3.727), the last two terms in the curly brackets cancel, and the only remaining $\hbar^2$-terms are

$$-\frac{\hbar^2}{2}X_1\mathcal{D}[X]X_1 + \hbar^2 W_2[X] + \mathcal{O}(\hbar^3). \tag{3.730}$$

From the classical equation of motion (3.668) one has a further equation for $\delta j/\delta x_{\rm cl}$:

$$\frac{\delta j}{\delta x_{\rm cl}} = -\mathcal{A}_{xx}[x_{\rm cl}] = -\mathcal{D}[x_{\rm cl}]. \tag{3.731}$$

Inserting this into (3.722) and replacing again $x_{\rm cl} \to X$, we find

$$X_1 = -\mathcal{D}^{-1}[X]\Delta_{1X}[X] + \mathcal{O}(\hbar). \tag{3.732}$$

We now express $\Delta_{1X}[X]$ via (3.720). This yields

$$\Delta_{1X}[X] = \frac{i}{2}\mathrm{Tr}\left(\mathcal{D}^{-1}[X]\frac{\delta}{\delta X}\mathcal{D}[X]\right) + \hbar W_{2X}[X] + \mathcal{O}(\hbar^2). \tag{3.733}$$

Inserting this into (3.732) and further into (3.727), we find for the effective action the expansion up to the order $\hbar^2$:

$$\begin{aligned}
\Gamma[X] &= \mathcal{A}[X] + \hbar\Gamma_1[X] + \hbar^2\Gamma_2[X] \\
&= \mathcal{A}[X] + i\frac{\hbar}{2}\mathrm{Tr}\log\mathcal{D}[X] + \hbar^2 W_2[X] \\
&\quad + \frac{\hbar^2}{2}\frac{1}{2}\mathrm{Tr}\left(\mathcal{D}^{-1}[X]\frac{\delta}{\delta X}\mathcal{D}[X]\right)\mathcal{D}^{-1}[X]\frac{1}{2}\mathrm{Tr}\left(\mathcal{D}^{-1}[X]\frac{\delta}{\delta X}\mathcal{D}[X]\right). \tag{3.734}
\end{aligned}$$

We now calculate $W_2[X]$ to lowest order in $\hbar$. The remainder $\mathcal{R}[X;x]$ in (3.717) has the expansion

$$\mathcal{R}[X;\delta x] = \frac{1}{3!}\mathcal{A}_{XXX}[X]\delta x\,\delta x\,\delta x + \frac{1}{4!}\mathcal{A}_{XXXX}[X]\delta x\,\delta x\,\delta x\,\delta x + \dots\ . \tag{3.735}$$

Being interested only in the $\hbar^2$-corrections, we have simply replaced x_{cl} by X. In order to obtain $W_2[X]$, we have to calculate all connected vacuum diagrams for the interaction terms in $\mathcal{R}[X;\delta x]$ with a $\delta x(t)$-propagator

$$\mathcal{G}[X] = i\hbar\{\mathcal{A}_{XX}[X]\}^{-1} \equiv i\hbar\mathcal{D}^{-1}[X].$$

Since every contraction brings in a factor $\hbar$, we can truncate the expansion (3.735) after δx^4. Thus, the only contributions to $i\hbar W_2[X]$ come from the connected vacuum diagrams

$$\tag{3.736}$$

where a line stands now for $\mathcal{G}[X]$, a four-vertex for

$$(i/\hbar)\mathcal{A}_{XXXX}[X] = (i/\hbar)\mathcal{D}_{XX}[X], \tag{3.737}$$

and a three-vertex for

$$(i/\hbar)\mathcal{A}_{XXX}[X] = (i/\hbar)\mathcal{D}_X[X]. \tag{3.738}$$

Only the first two diagrams are one-particle irreducible. As a pleasant result, the third diagram which is one-particle reducible cancels with the last term in (3.734). To see this we write that term more explicitly as

$$\frac{\hbar^2}{8}\mathcal{D}^{-1}_{X_1X_2}\mathcal{A}_{X_1X_2X_3}\mathcal{D}^{-1}_{X_3X_{3'}}\mathcal{A}_{X_{3'}X_{1'}X_{2'}}\mathcal{D}^{-1}_{X_{1'}X_{2'}}, \tag{3.739}$$

which corresponds precisely to the third diagram in $\Gamma_2[X]$, except for an opposite sign. Note that the diagram has a multiplicity 9.

Thus, at the end, only the one-particle irreducible vacuum diagrams contribute to the $\hbar^2$-correction to $\Gamma[X]$:

$$i\Gamma_2[X] = i\frac{3}{4!}\mathcal{D}_{12}^{-1}\mathcal{A}_{X_1X_2X_3X_4}\mathcal{D}_{34}^{-1} + i\frac{1}{4!^2}\mathcal{A}_{X_1X_2X_3}\mathcal{D}_{X_1X_{1'}}^{-1}\mathcal{D}_{X_2X_{2'}}^{-1}\mathcal{D}_{X_3X_{3'}}^{-1}\mathcal{A}_{X_1X_2X_3}.$$

(3.740)

Their diagrammatic representation is

$$\frac{i}{\hbar}\hbar^2\Gamma_2[X] = \tfrac{1}{8}\underset{\hbar^2}{\bigcirc\!\bigcirc} \ + \tfrac{1}{12}\underset{\hbar^2}{\ominus} \ .$$

(3.741)

The one-particle irreducible nature of the diagrams is found to all orders in $\hbar$.

3.23.5 Finite-Temperature Two-Loop Effective Action

At finite temperature, and in D dimensions, the expansion proceeds with the imaginary-time versions of the **X**-dependent Green functions (3.687) and (3.688)

$$\mathcal{G}_L(\tau_1, \tau_2) = \frac{\hbar}{2M\omega_L}\frac{\cosh(\omega_L|\tau_1 - \tau_2| - \hbar\beta\omega_L/2)}{\sinh(\hbar\beta\omega_L/2)},$$

(3.742)

and

$$\mathcal{G}_T(\tau_1, \tau_2) = \frac{\hbar}{2M\omega_T}\frac{\cosh(\omega_T|\tau_1 - \tau_2| - \hbar\beta\omega_T/2)}{\sinh(\hbar\beta\omega_T/2)},$$

(3.743)

where we have omitted the argument **X** in $\omega_L(\mathbf{X})$ and $\omega_T(\mathbf{X})$. Treating here the general rotationally symmetric potential $V(\mathbf{x}) = v(x)$, $x = \sqrt{\mathbf{x}^2}$, the two frequencies are

$$\omega_L^2(\mathbf{X}) \equiv \frac{1}{M}v''(X), \quad \omega_T^2(\mathbf{X}) \equiv \frac{1}{MX}v'(X).$$

(3.744)

We also decompose the vertex functions into longitudinal and transverse parts. The three-point vertex is a sum

$$\frac{\partial^3 v(X)}{\partial X_i\partial X_j\partial X_k} = P_{ijk}^L v'''(X) + P_{ijk}^T\left[\frac{v''(X)}{X} - \frac{v'(X)}{X^2}\right],$$

(3.745)

with the symmetric tensors

$$P_{ijk}^L \equiv \frac{X_iX_jX_k}{X^3} \quad \text{and} \quad P_{ijk}^T \equiv \delta_{ij}\frac{X_k}{X} + \delta_{ik}\frac{X_j}{X} + \delta_{jk}\frac{X_i}{X} - 3P_{ijk}^L.$$

(3.746)

The four-point vertex reads

$$\frac{\partial^4 v(X)}{\partial X_i\partial X_j\partial X_k\partial X_l} = P_{ijkl}^L v^{(4)}(X) + P_{ijkl}^T\frac{v'''(X)}{X} + P_{ijkl}^S\left[\frac{v''(X)}{X^2} - \frac{v'(X)}{X^3}\right],$$

(3.747)

with the symmetric tensors

$$P_{ijkl}^L = \frac{X_iX_jX_kX_l}{X^4},$$

(3.748)

$$P_{ijkl}^T = \delta_{ij}\frac{X_kX_l}{X^2} + \delta_{ik}\frac{X_jX_l}{X^2} + \delta_{il}\frac{X_jX_k}{X^2} + \delta_{jk}\frac{X_iX_l}{X^2} + \delta_{jl}\frac{X_iX_k}{X^2} + \delta_{kl}\frac{X_iX_k}{X^2} - 6P_{ijkl}^L,$$

(3.749)

$$P_{ijkl}^S = \delta_{ij}\delta_{kl} + \delta_{ik}\delta_{jl} + \delta_{il}\delta_{jk} - 3P_{ijkl}^L - 3P_{ijkl}^T.$$

(3.750)

The tensors obey the following relations:

$$\frac{X_i}{X}P_{ijk}^L = P_{jk}^L\,, \qquad \frac{X_i}{X}P_{ijk}^T = P_{jk}^T\,, \tag{3.751}$$

$$P_{ij}^L P_{ikl}^L = P_{jkl}^L\,, \quad P_{ij}^T P_{ikl}^T = \frac{X_k}{X}P_{jl}^T + \frac{X_l}{X}P_{jk}^T\,, \quad P_{ij}^L P_{ikl}^T = \frac{X_j}{X}P_{kl}^T\,, \quad P_{ij}^T P_{ikl}^L = 0\,, \tag{3.752}$$

$$P_{hij}^L P_{hkl}^L = P_{ijkl}^L\,, \qquad P_{hij}^T P_{hkl}^T = P_{ij}^T P_{kl}^T + P_{ik}^T P_{jl}^T + P_{il}^T P_{jk}^T + P_{jk}^T P_{il}^T + P_{jl}^T P_{ik}^T\,, \tag{3.753}$$

$$P_{hij}^L P_{hkl}^T = P_{ij}^L P_{kl}^T\,, \quad P_{hij}^T P_{hkl}^L = P_{ij}^T P_{kl}^L\,, \quad P_{ij}^L P_{ijkl}^L = P_{kl}^L\,, \quad P_{ij}^T P_{ijkl}^T = (D-1)P_{kl}^L\,, \tag{3.754}$$

$$P_{ij}^L P_{ijkl}^T = P_{kl}^T\,, \qquad P_{ij}^T P_{ijkl}^L = 0\,, \tag{3.755}$$

$$P_{ij}^L P_{ijkl}^S = -2P_{kl}^T\,, \quad P_{ij}^T P_{ijkl}^S = (D+1)P_{kl}^T - 2(D-1)P_{kl}^L\,. \tag{3.756}$$

Instead of the effective action, the diagrammatic expansion (3.741) yields now the free energy

$$(i/\hbar)\Gamma[\mathbf{X}] \rightarrow -\beta F(\mathbf{X})\,. \tag{3.757}$$

Using the above formulas we obtain immediately the mean field contribution to the free energy

$$-\beta F_{\mathrm{MF}} = -\int_0^{\hbar\beta} d\tau \left[\frac{M}{2}\dot{\mathbf{X}}^2 + v(X)\right]\,, \tag{3.758}$$

and the one-loop contribution [from the trace-log term in Eq. (3.734)]:

$$-\beta F_{1-\mathrm{loop}} = -\log\left[2\sinh(\hbar\beta\omega_L/2)\right] - (D-1)\log\left[2\sinh(\hbar\beta\omega_T/2)\right]\,. \tag{3.759}$$

The first of the two-loop diagrams in (3.741) yields the contribution to the free energy

$$-\beta\Delta_1 F_{2-\mathrm{loop}} = -\beta\left\{\mathcal{G}_L^2(\tau,\tau)v^{(4)}(X) + (D^2-1)\mathcal{G}_T^2(\tau,\tau)\left[\frac{v''(X)}{X^2} - \frac{v'(X)}{X^3}\right]\right.$$
$$\left. + 2(D-1)\mathcal{G}_L(\tau,\tau)\mathcal{G}_T(\tau,\tau)\left[\frac{v'''(X)}{X} - \frac{2v''(X)}{X^2} + \frac{2v'(X)}{X^3}\right]\right\}\,. \tag{3.760}$$

From the second diagram we obtain the contribution

$$-\beta\Delta_2 F_{2-\mathrm{loop}} = \frac{1}{\hbar^2}\int_0^{\hbar\beta}d\tau_1\int_0^{\hbar\beta}d\tau_2\left\{\mathcal{G}_L^3(\tau_1,\tau_2)\left[v'''(X)\right]^2\right.$$
$$\left. + 3(D-1)\mathcal{G}_L(\tau_1,\tau_2)\mathcal{G}_T^2(\tau_1,\tau_2)\left[\frac{v''(X)}{X} - \frac{v'(X)}{X^2}\right]^2\right\}\,. \tag{3.761}$$

The explicit evaluation yields

$$-\beta\Delta_1 F_{2-\mathrm{loop}} = -\frac{\hbar^2\beta}{(2M)^2}\left\{\frac{1}{\omega_L^2}\coth^2(\hbar\beta\omega_L/2)v^{(4)}(X)\right. \tag{3.762}$$
$$+ \frac{D^2-1}{\omega_T^2}\coth^2(\hbar\beta\omega_T/2)\left[\frac{v''(X)}{X^2} - \frac{v'(X)}{X^3}\right]$$
$$\left. + \frac{2(D-1)}{\omega_L\omega_T}\coth(\hbar\beta\omega_L/2)\coth(\hbar\beta\omega_T/2)\left[\frac{v'''(X)}{X} - \frac{2v''(X)}{X^2} + \frac{2v'(X)}{X^3}\right]\right\}\,.$$

and

$$-\beta\Delta_2 F_{2-\mathrm{loop}} = \frac{2\hbar^2\beta}{\omega_L}\frac{1}{(2M\omega_L)^3}[v'''(X)]^2\left[\frac{1}{3} + \frac{1}{\sinh^2(\hbar\beta\omega_L/2)}\right]$$
$$+ \frac{6\hbar^2\beta(D-1)}{2\omega_T+\omega_L}\frac{1}{2M\omega_L}\frac{1}{(2M\omega_T)^2}\left[\frac{v''(X)}{X} - \frac{v'(X)}{X^2}\right]^2 \tag{3.763}$$
$$\times\left[\coth^2(\hbar\beta\omega_T/2) + \frac{\omega_T}{\omega_L}\frac{1}{\sinh^2(\hbar\beta\omega_T/2)} + \frac{\omega_T}{2\omega_T-\omega_L}\frac{\sinh[\hbar\beta(2\omega_T-\omega_L)/2]}{\sinh(\hbar\beta\omega_L/2)\sinh^2(\hbar\beta\omega_T/2)}\right]\,.$$

In the limit of zero temperature, the effective potential in the free energy becomes

$$
V_{\text{eff}}(X) \underset{T \to 0}{=} v(X) + \frac{\hbar \omega_L}{2} + (D-1)\frac{\hbar \omega_T}{2} + \frac{\hbar^2}{8(2M)^2} \left\{ \frac{1}{\omega_L^2} v^{(4)}(X) \right.
$$
$$
+ \frac{D^2-1}{\omega_T^2} \left[\frac{v''(X)}{X^2} - \frac{v'(X)}{X^3} \right] + \frac{2(D-1)}{\omega_L \omega_T} \left[\frac{v'''(X)}{X} - \frac{2v''(X)}{X^2} + \frac{2v'(X)}{X^3} \right] \right\}
$$
$$
- \frac{\hbar^2}{6(2M)^3} \left\{ \frac{1}{3\omega_L^4}[v'''(X)]^2 + \frac{3(D-1)}{2\omega_T + \omega_L} \frac{1}{\omega_L \omega_T^2} \left[\frac{v''(X)}{X} - \frac{v'(X)}{X^2} \right]^2 \right\} + \mathcal{O}(\hbar^3). \tag{3.764}
$$

For the one-dimensional potential

$$
V(x) = \frac{M}{2} \omega^2 x^2 + \frac{g_3}{3!} x^3 + \frac{g_4}{4!} x^4, \tag{3.765}
$$

the effective potential becomes, up to two loops,

$$
V_{\text{eff}}(X) = \frac{M}{2} \omega^2 X^2 + g_3 X^3 + g_4 X^4 + \frac{1}{\beta} \log \left(2 \sinh \hbar \beta \omega/2 \right) + \hbar^2 \frac{g_4}{8(2M\omega)^2} \frac{1}{\tanh^2(\hbar \beta \omega/2)}
$$
$$
- \frac{\hbar^2}{6\omega} \frac{(g_3 + g_4 X)^2}{(2M\omega)^3} \left[\frac{1}{3} + \frac{1}{\sinh^2(\hbar \beta \omega/2)} \right] + \mathcal{O}(\hbar^3), \tag{3.766}
$$

whose $T \to 0$ limit is

$$
V_{\text{eff}}(X) \underset{T \to 0}{=} \frac{M}{2} \omega^2 X^2 + \frac{g_3}{3!} X^3 + \frac{g_4}{4!} X^4 + \frac{\hbar \omega}{2} + \hbar^2 \frac{g_4}{8(2M\omega)^2}
$$
$$
- \frac{\hbar^2}{18\omega} \frac{(g_3 + g_4 X)^2}{(2M\omega)^3} + \mathcal{O}(\hbar^3). \tag{3.767}
$$

If the potential is a polynomial in $\mathbf{X}$, the effective potential at zero temperature can be solved more efficiently than here and to much higher loop orders with the help of recursion relations. This will be shown in Appendix 3C.5.

3.23.6 Background Field Method for Effective Action

In order to find the rules for the loop expansion to any order, let us separate the total effective action into a sum of the classical action $\mathcal{A}[\mathbf{X}]$ and a term $\Gamma^{\text{fl}}[\mathbf{X}]$ which collects the contribution of all quantum fluctuations:

$$
\Gamma[\mathbf{X}] = \mathcal{A}[\mathbf{X}] + \Gamma^{\text{fl}}[\mathbf{X}]. \tag{3.768}
$$

To calculate the fluctuation part $\Gamma^{\text{fl}}[\mathbf{X}]$, we expand the paths $\mathbf{x}(t)$ around some arbitrarily chosen background path $\mathbf{X}(t)$:[20]

$$
\mathbf{x}(t) = \mathbf{X}(t) + \delta \mathbf{x}(t), \tag{3.769}
$$

and calculate the generating functional $W[\mathbf{j}]$ by performing the path integral over the fluctuations:

$$
\exp \left\{ \frac{i}{\hbar} W[\mathbf{j}] \right\} = \int \mathcal{D}\delta \mathbf{x} \exp \left\{ \frac{i}{\hbar} \left(\mathcal{A}[\mathbf{X} + \delta \mathbf{x}] + \mathbf{j}[\mathbf{X}](\mathbf{X} + \delta \mathbf{x}) \right) \right\}. \tag{3.770}
$$

---

[20]In the theory of fluctuating fields, this is replaced by a more general *background field* which explains the name of the method.

From $W[\mathbf{j}]$ we find a $\mathbf{j}$-dependent expectation value $\mathbf{X^j} = \langle \mathbf{x} \rangle^{\mathbf{j}}$ as $\mathbf{X^j} = \delta W[\mathbf{j}]/\delta \mathbf{j}$, and the Legendre transform $\Gamma[\mathbf{X}] = W[\mathbf{j}] - \mathbf{j}\mathbf{X^j}$. In terms of $\mathbf{X^j}$, Eq. (3.770) can be rewritten as

$$\exp\left\{ \frac{i}{\hbar} \left(\Gamma[\mathbf{X^j}] + \mathbf{j}[\mathbf{X^j}]\,\mathbf{X^j} \right) \right\} = \int \mathcal{D}\delta\mathbf{x} \, \exp\left\{ \frac{i}{\hbar} \left(\mathcal{A}[\mathbf{X} + \delta\mathbf{x}] + \mathbf{j}[\mathbf{X}](\mathbf{X} + \delta\mathbf{x}) \right) \right\}. \quad (3.771)$$

The expectation value $\mathbf{X^j}$ has the property of extremizing $\Gamma[\mathbf{X}]$, i.e., it satisfies the equation

$$\mathbf{j} = -\left. \frac{\delta\Gamma[\mathbf{X}]}{\partial\mathbf{X}} \right|_{\mathbf{X}=\mathbf{X^j}} = -\Gamma_{\mathbf{X}}[\mathbf{X^j}]. \quad (3.772)$$

We now choose $\mathbf{j}$ in such a way that $\mathbf{X^j}$ equals the initially chosen $\mathbf{X}$, and find

$$\exp\left\{ \frac{i}{\hbar}\Gamma[\mathbf{X}] \right\} = \int \mathcal{D}\delta\mathbf{x} \, \exp\left(\frac{i}{\hbar} \left\{ \mathcal{A}[\mathbf{X} + \delta\mathbf{x}] - \Gamma_{\mathbf{X}}[\mathbf{X}]\delta\mathbf{x} \right\} \right). \quad (3.773)$$

This is a functional integro-differential equation for the effective action $\Gamma[\mathbf{X}]$ which we can solve perturbatively order by order in $\hbar$. This is done diagrammatically. The diagrammatic elements are lines representing the propagator (3.686)

$$\underline{\hspace{2cm}} = \mathcal{G}_{ab}[\mathbf{X}] \equiv i\hbar \left[\frac{\delta^2 \mathcal{A}[\mathbf{X}]}{\delta X_a \delta X_b} \right]^{-1}_{ab}, \quad (3.774)$$

and vertices

$$= \frac{\delta^n \mathcal{A}[\mathbf{X}]}{\delta X_{a_1} \delta X_{a_2} \ldots \delta X_{a_n}}. \quad (3.775)$$

From the explicit calculations in the last two subsections we expect the effective action to be the sum of all one-particle irreducible vacuum diagrams formed with these propagators and vertices. This will now be proved to all orders in perturbation theory.

We introduce an auxiliary generating functional $\tilde{W}\left[\mathbf{X},\tilde{\mathbf{j}}\right]$ which governs the correlation functions of the fluctuations $\delta\mathbf{x}$ around the above fixed backgound $\mathbf{X}$:

$$\exp\left\{ i\tilde{W}\left[\mathbf{X},\tilde{\mathbf{j}}\right] /\hbar \right\} \equiv \int \mathcal{D}\delta\mathbf{x} \, \exp\left(\frac{i}{\hbar} \left\{ \tilde{\mathcal{A}}\left[\mathbf{X},\delta\mathbf{x}\right] + \int dt\,\tilde{\mathbf{j}}(t)\,\delta\mathbf{x}(t) \right\} \right), \quad (3.776)$$

with the action of fluctuations

$$\tilde{\mathcal{A}}[\mathbf{X},\delta\mathbf{x}] = \mathcal{A}[\mathbf{X} + \delta\mathbf{x}] - \mathcal{A}[\mathbf{X}] - \mathcal{A}_{\mathbf{X}}[\mathbf{X}]\delta\mathbf{x}, \quad (3.777)$$

whose expansion in powers of $\delta\mathbf{x}(t)$ starts out with a quadratic term. A source $\tilde{\mathbf{j}}(t)$ is coupled to the fluctuations $\delta\mathbf{x}(t)$. By comparing (3.776) with (3.773) we see that for the special choice of the current

$$\tilde{\mathbf{j}} = -\Gamma_{\mathbf{X}}[\mathbf{X}] + \mathcal{A}_{\mathbf{X}}[\mathbf{X}] = -\tilde{\Gamma}_{\mathbf{X}}[\mathbf{X}], \qquad (3.778)$$

the right-hand sides coincide, such that the auxiliary functional $\tilde{W}[\mathbf{X}, \tilde{\mathbf{j}}]$ contains precisely the diagrams in $\Gamma^{\mathrm{fl}}[\mathbf{X}]$ which we want to calculate. We now form the Legendre transform of $\tilde{W}[\mathbf{X}, \tilde{\mathbf{j}}]$, which is an auxiliary effective action with two arguments:

$$\tilde{\Gamma}\left[\mathbf{X}, \tilde{\mathbf{X}}\right] \equiv \tilde{W}[\mathbf{X}, \tilde{\mathbf{j}}] - \int dt\, \tilde{\mathbf{j}}\, \tilde{\mathbf{X}}, \qquad (3.779)$$

with the auxiliary conjugate variable

$$\tilde{\mathbf{X}} = \frac{\delta\tilde{W}[\mathbf{X}, \tilde{\mathbf{j}}]}{\delta\tilde{\mathbf{j}}} = \tilde{\mathbf{X}}[\mathbf{X}, \tilde{\mathbf{j}}]. \qquad (3.780)$$

This is the expectation value of the fluctuations $\langle\delta\mathbf{x}\rangle$ in the path integral (3.776). If $\tilde{\mathbf{j}}$ has the value (3.778), this expectation vanishes, i.e. $\tilde{\mathbf{X}} = 0$. The auxiliary action $\tilde{\Gamma}[\mathbf{X}, 0]$ coincides with the fluctuating part $\Gamma^{\mathrm{fl}}[\mathbf{X}]$ of the effective action which we want to calculate.

The functional derivatives of $\tilde{W}[\mathbf{X}, \tilde{\mathbf{j}}]$ with respect to $\tilde{\mathbf{j}}$ yield all connected correlation functions of the fluctuating variables $\delta\mathbf{x}(t)$. The functional derivatives of $\tilde{\Gamma}\left[\mathbf{X}, \tilde{\mathbf{X}}\right]$ with respect to $\tilde{\mathbf{X}}$ select from these the one-particle irreducible correlation functions. For $\tilde{\mathbf{X}} = 0$, only vacuum diagrams survive.

Thus we have proved that the full effective action is obtained from the sum of the classical action $\Gamma_0[\mathbf{X}] = \mathcal{A}[\mathbf{X}]$, the one-loop contribution $\Gamma_1[\mathbf{X}]$ given by the trace of the logarithm in Eq. (3.702), the two-loop contribution $\Gamma_2[\mathbf{X}]$ in (3.741), and the sum of all connected one-particle irreducible vacuum diagrams with more than two loops

$$\frac{i}{\hbar}\sum_{n\geq 3} i\hbar^n \Gamma_n[\mathbf{X}] = \frac{1}{8}\;\text{⬡}\; + \frac{1}{12}\;\text{⬡}\; + \frac{1}{48}\;\text{⬡}\; + \frac{1}{16}\;\text{⬡⬡}$$

$$+ \frac{1}{8}\;\text{⬡}\; + \frac{1}{8}\;\text{⬡}\; + \frac{1}{24}\;\text{⊗}\; + \frac{1}{16}\;\text{⬡}\; . \qquad (3.781)$$

Observe that in the expansion of $\Gamma[X]/\hbar$, each line carries a factor $\hbar$, whereas each n-point vertex contributes a factor $\hbar^{-1}$. The contribution of an n-loop diagram to $\Gamma[X]$ is therefore of order $\hbar^n$. The higher-loop diagrams are most easily generated by a recursive treatment of the type developed in Subsection 3.22.3.

For a harmonic oscillator, the expansion stops after the trace of the logarithm (3.702), and reads simply, in one dimension:

$$\begin{aligned}
\Gamma[X] &= \mathcal{A}[X] + \frac{i}{2}\hbar\,\text{Tr}\log\Gamma^{(2)}(t_b, t_a) \\
&= \int_{t_a}^{t_b} dt\left[\frac{M}{2}\dot{X}^2 - \frac{M\omega^2}{2}X^2\right] + \frac{i}{2}\hbar\,\text{Tr}\log\left(-\partial_t^2 - \omega^2\right). \qquad (3.782)
\end{aligned}$$

Evaluating the trace of the logarithm we find for a constant X the effective potential (3.663):

$$V^{\text{eff}}(X) = V(X) - \frac{i}{2(t_b - t_a)} \log\{2\pi i \sin[\omega(t_b - t_a)]/M\omega\}. \tag{3.783}$$

If the boundary conditions are periodic, so that the analytic continuation of the result can be used for quantum statistical calculations, the result is

$$V^{\text{eff}}(X) = V(X) - \frac{i}{(t_b - t_a)} \log\{2i \sin[\omega(t_b - t_a)/2]\}. \tag{3.784}$$

It is important to keep in mind that a line in the above diagrams contains an infinite series of fundamental Feynman diagrams of the original perturbation expansion, as can be seen by expanding the denominators in the propagator $\mathcal{G}_{ab}$ in Eqs. (3.686)–(3.688) in powers of $\mathbf{X}^2$. This expansion produces a sum of diagrams which can be obtained from the loop diagrams in the expansion of the trace of the logarithm in (3.707) by cutting the loop.

If the potential is a polynomial in $\mathbf{X}$, the effective potential at zero temperature can be solved most efficiently to high loop orders with the help of recursion relations. This is shown in detail in Appendix 3C.5.

3.24　Nambu-Goldstone Theorem

The appearance of a zero-frequency mode as a consequence of a nonzero expectation value $\mathbf{X}$ can easily be proved for any continuous symmetry and to all orders in perturbation theory by using the full effective action. To be more specific we consider as before the case of $O(N)$-symmetry, and perform infinitesimal symmetry transformations on the currents $\mathbf{j}$ in the generating functional $W[\mathbf{j}]$:

$$j_a \to j_a - i\epsilon_{cd}(L_{cd})_{ab} j_b, \tag{3.785}$$

where L_{cd} are the $N(N-1)/2$ generators of $O(N)$-rotations with the matrix elements

$$(L_{cd})_{ab} = i(\delta_{ca}\delta_{db} - \delta_{da}\delta_{cb}), \tag{3.786}$$

and ϵ_{ab} are the infinitesimal angles of the rotations. Under these, the generating functional is assumed to be invariant:

$$\delta W[\mathbf{j}] = 0 = \int dt \frac{\delta W[\mathbf{j}]}{\delta j_a(x)} i(L_{cd})_{ab} j_b \epsilon_{cd} = 0. \tag{3.787}$$

Expressing the integrand in terms of Legendre-transformed quantities via Eqs. (3.620) and (3.622), we obtain

$$\int dt X_a(t) i(L_{cd})_{ab} \frac{\delta\Gamma[\mathbf{X}]}{\delta X_b(t)} \epsilon_{cd} = 0. \tag{3.788}$$

This expresses the infinitesimal invariance of the effective action $\Gamma[\mathbf{X}]$ under infinitesimal rotations

$$X_a \to X_a - i\epsilon_{cd}\left(L_{cd}\right)_{ab} X_b.$$

The invariance property (3.788) is called the *Ward-Takakashi identity* for the functional $\Gamma[\mathbf{X}]$. It can be used to find an infinite set of equally named identities for all vertex functions by forming all $\Gamma[\mathbf{X}]$ functional derivatives of $\Gamma[\mathbf{X}]$ and setting $\mathbf{X}$ equal to the expectation value at the minimum of $\Gamma[\mathbf{X}]$. The first derivative of $\Gamma[\mathbf{X}]$ gives directly from (3.788) (dropping the infinitesimal parameter ϵ_{cd})

$$
\begin{aligned}
\left(L_{cd}\right)_{ab} j_b(t) &= \left(L_{cd}\right)_{ab} \frac{\delta\Gamma[\mathbf{X}]}{\delta X(t)_b} \\
&= -\int dt' X_{a'}(t') \left(L_{cd}\right)_{a'b} \frac{\delta^2\Gamma[\mathbf{X}]}{\delta X_b(t')\delta X_n(t)}.
\end{aligned}
\tag{3.789}
$$

Denoting the expectation value at the minimum of the effective potential by $\bar{\mathbf{X}}$, this yields

$$
\int dt'\, \bar{X}_{a'}(t') \left(L_{cd}\right)_{a'b} \frac{\delta^2\Gamma[\mathbf{X}]}{\delta X_b(t')\delta X_a(t)}\bigg|_{\mathbf{X}(t)=\bar{\mathbf{X}}} = 0.
\tag{3.790}
$$

Now the second derivative is simply the vertex function $\Gamma^{(2)}(t',t)$ which is the functional inverse of the correlation function $G^{(2)}(t',t)$. The integral over t selects the zero-frequency component of the Fourier transform

$$
\tilde{\Gamma}^{(2)}(\omega') \equiv \int dt'\, e^{i\omega't}\Gamma^{(2)}(t',t).
\tag{3.791}
$$

If we define the Fourier components of $\Gamma^{(2)}(t',t)$ accordingly, we can write (3.790) in Fourier space as

$$
X_{a'}^0 \left(L_{cd}\right)_{a'b} \tilde{G}_{ba}^{-1}(\omega'=0) = 0.
\tag{3.792}
$$

Inserting the matrix elements (3.786) of the generators of the rotations, this equation shows that for $\bar{\mathbf{X}} \neq 0$, the fully interacting transverse propagator has to possess a singularity at $\omega' = 0$. In quantum field theory, this implies the existence of $N-1$ massless particles, the Nambu-Goldstone boson. The conclusion may be drawn only if there are no massless particles in the theory from the outset, which may be "eaten up" by the Nambu-Goldstone boson, as explained earlier in the context of Eq. (3.688).

As mentioned before at the end of Subsection 3.23.1, the Nambu-Goldstone theorem does not have any consequences for quantum mechanics since fluctuations are too violent to allow for the existence of a nonzero expectation value $\mathbf{X}$. The effective action calculated to any finite order in perturbation theory, however, is incapable of reproducing this physical property and does have a nonzero extremum and ensuing transverse zero-frequency modes.

3.25   Effective Classical Potential

The loop expansion of the effective action $\Gamma[X]$ in (3.768), consisting of the trace of the logarithm (3.702) and the one-particle irreducible diagrams (3.741), (3.781) and the associated effective potential $V(X)$ in Eq. (3.663), can be continued in a straightforward way to imaginary times setting $t_b - t_a \to -i\hbar\beta$ to form the Euclidean effective potential $\Gamma_e[X]$. For the harmonic oscillator, where the expansion stops after the trace of the logarithm and the effective potential reduces to the simple expression (3.782), we find the imaginary-time version

$$V^{\text{eff}}(X) = V(X) + \frac{1}{\beta}\log\left(2\sinh\frac{\beta\hbar\omega}{2}\right). \tag{3.793}$$

Since the effective action contains the effect of all fluctuations, the minimum of the effective potential $V(X)$ should yield directly the full quantum statistical partition function of a system:

$$Z = \exp[-\beta V(X)\big|_{\text{min}}]. \tag{3.794}$$

Inserting the harmonic oscillator expression (3.793) we find indeed the correct result (2.405).

For anharmonic systems, we expect the loop expansion to be able to approximate $V(X)$ rather well to yield a good approximation for the partition function via Eq. (3.794). It is easy to realize that this cannot be true. We have shown in Section 2.9 that for high temperatures, the partition function is given by the integral [recall (2.351)]

$$Z_{\text{cl}} = \int_{-\infty}^{\infty}\frac{dx}{l_e(\hbar\beta)}e^{-V(x)/k_BT}. \tag{3.795}$$

This integral can in principle be treated by the same background field method as the path integral, albeit in a much simpler way. We may write $x = X + \delta x$ and find a loop expansion for an effective potential. This expansion evaluated at the extremum will yield a good approximation to the integral (3.795) only if the potential is very close to a harmonic one. For any more complicated shape, the integral at small β will cover the entire range of x and can therefore only be evaluated numerically. Thus we can never expect a good result for the partition function of anharmonic systems at high temperatures, if it is calculated from Eq. (3.794).

It is easy to find the culprit for this problem. In a one-dimensional system, the correlation functions of the fluctuations around X are given by the correlation function [compare (3.301), (3.248), and (3.687)]

$$\begin{aligned}
\langle \delta x(\tau)\delta x(\tau')\rangle &= G^{(2)}_{\Omega^2(X)}(\tau,\tau') = \frac{\hbar}{M}G^{\text{p}}_{\Omega^2(X),\text{e}}(\tau-\tau') \\
&= \frac{\hbar}{M}\frac{1}{2\Omega(X)}\frac{\cosh\Omega(X)(|\tau-\tau'|-\hbar\beta/2)}{\sinh[\Omega(X)\hbar\beta/2]}, \quad |\tau-\tau'|\in[0,\hbar\beta], \tag{3.796}
\end{aligned}$$

with the X-dependent frequency given by

$$\Omega^2(X) = \omega^2 + 3\frac{g}{6}X^2. \tag{3.797}$$

At equal times $\tau = \tau'$, this specifies the square width of the fluctuations $\delta x(\tau)$:

$$\left\langle [\delta x(\tau)]^2 \right\rangle = \frac{\hbar}{M} \frac{1}{2\Omega(X)} \coth \frac{\Omega(X)\hbar\beta}{2}. \tag{3.798}$$

The point is now that for large temperatures T, this width grows linearly in T

$$\left\langle [\delta x(\tau)]^2 \right\rangle \xrightarrow{T\to\infty} \frac{k_B T}{M\Omega^2}. \tag{3.799}$$

The linear behavior follows the historic *Dulong-Petit law* for the classical fluctuation width of a harmonic oscillator [compare with the Dulong-Petit law (2.601) for the thermodynamic quantities]. It is a direct consequence of the *equipartition theorem* for purely thermal fluctuations, according to which the potential energy has an average $k_B T/2$:

$$\frac{M\Omega^2}{2}\left\langle x^2 \right\rangle = \frac{k_B T}{2}. \tag{3.800}$$

If we consider the spectral representation (3.245) of the correlation function,

$$G^{\mathrm{p}}_{\Omega^2,\mathrm{e}}(\tau - \tau') = \frac{1}{\hbar\beta} \sum_{m=-\infty}^{\infty} \frac{1}{\omega_m^2 + \Omega^2} e^{-i\omega_m(\tau-\tau')}, \tag{3.801}$$

we see that the linear growth is entirely due to term with zero Matsubara frequency.

The important observation is now that if we remove this zero frequency term from the correlation function and form the *subtracted correlation function* [recall (3.250)]

$$G^{\mathrm{p}'}_{\Omega^2,\mathrm{e}}(\tau) \equiv G^{\mathrm{p}}_{\Omega^2,\mathrm{e}}(\tau) - \frac{1}{\hbar\beta\Omega^2} = \frac{1}{2\Omega}\frac{\cosh\Omega(|\tau|-\hbar\beta/2)}{\sinh[\Omega\hbar\beta/2]} - \frac{1}{\hbar\beta\Omega^2}, \tag{3.802}$$

we see that the subtracted square width

$$a_\Omega^2 \equiv G^{\mathrm{p}'}_{\Omega^2,\mathrm{e}}(0) = \frac{1}{2\Omega}\coth\frac{\Omega\hbar\beta}{2} - \frac{1}{\hbar\beta\Omega^2} \tag{3.803}$$

decrease for large T. This is shown in Fig. 3.14. Due to this decrease, there exists a method to substantially improve perturbation expansions with the help of the so-called effective classical potential.

3.25.1 Effective Classical Boltzmann Factor

The above considerations lead us to the conclusion that a useful approximation for partition function can be obtained only by expanding the path integral in powers of

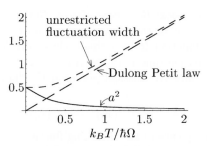

Figure 3.14 Local fluctuation width compared with the unrestricted fluctuation width of harmonic oscillator and its linear Dulong-Petit approximation. The vertical axis shows units of $\hbar/M\Omega$, a quantity of dimension length^2.

the subtracted fluctuations $\delta' x(\tau)$ which possess no zero Matsubara frequency. The quantity which is closely related to the effective potential $V^{\text{eff}}(X)$ in Eq. (3.663) but allows for a more accurate evaluation of the partition function is the *effective classical potential* $V^{\text{eff cl}}(x_0)$. Just as $V^{\text{eff}}(X)$, it contains the effects of *all* quantum fluctuations, but it keeps separate track of the thermal fluctuations which makes it a convenient tool for numerical treatment of the partition function. The definition starts out similar to the background method in Subsection 3.23.6 in Eq. (3.769). We split the paths as in Eq. (2.441) into a time-independent constant background x_0 and a fluctuation $\eta(\tau)$ with zero temporal average $\bar{\eta} = 0$:

$$x(\tau) = x_0 + \eta(\tau) \equiv x_0 + \sum_{m=1}^{\infty} \left(x_m e^{i\omega_m \tau} + \text{c.c.} \right), \quad x_0 = \text{real}, \quad x_{-m} \equiv x_m^*, \quad (3.804)$$

and write the partition function using the measure (2.446) as

$$Z = \oint \mathcal{D}x \, e^{-\mathcal{A}_e/\hbar} = \int_{-\infty}^{\infty} \frac{dx_0}{l_e(\hbar\beta)} \oint \mathcal{D}'x \, e^{-\mathcal{A}_e/\hbar}, \quad (3.805)$$

where

$$\oint \mathcal{D}'x \, e^{-\mathcal{A}_e/\hbar} = \prod_{m=1}^{\infty} \left[\int_{-\infty}^{\infty} \int_{-\infty}^{\infty} \frac{d \operatorname{Re} x_m \, d \operatorname{Im} x_m}{\pi k_B T/M\omega_m^2} \right] e^{-\mathcal{A}_e/\hbar}. \quad (3.806)$$

Comparison of (2.445) with the integral expression (2.350) for the classical partition function Z_{cl} suggests writing the path integral over the components with nonzero Matsubara frequencies as a Boltzmann factor

$$B(x_0) \equiv e^{-V^{\text{eff cl}}(x_0)/k_B T} \quad (3.807)$$

and defined the quantity $V^{\text{eff cl}}(x_0)$ as the effective classical potential. The full partition function is then given by the integral

$$Z = \int_{-\infty}^{\infty} \frac{dx_0}{l_e(\hbar\beta)} e^{-V^{\text{eff cl}}(x_0)/k_B T}, \quad (3.808)$$

where the *effective classical Boltzmann factor* $B(x_0)$ contains all information on the quantum fluctuations of the system and allows to calculate the full quantum statistical partition function from a single classically looking integral. At high-temperature, the partition function (3.808) takes the classical limit (2.460). Thus, by construction, the effective classical potential $V^{\text{eff cl}}(x_0)$ will approach the initial potential $V(x_0)$:

$$V^{\text{eff cl}}(x_0) \xrightarrow{T \to \infty} V(x_0). \tag{3.809}$$

This is a direct consequence of the shrinking fluctuation width (3.803) for growing temperature.

The path integral representation of the effective classical Boltzmann factor

$$B(x_0) \equiv \oint \mathcal{D}'x \, e^{-\mathcal{A}_e/\hbar} \tag{3.810}$$

can also be written as a path integral in which one has inserted a δ-function to ensure the path average

$$\bar{x} \equiv \frac{1}{\hbar\beta} \int_0^{\hbar\beta} d\tau \, x(\tau). \tag{3.811}$$

Let us introduce the slightly modified δ-function [recall (2.351)]

$$\tilde{\delta}(\bar{x} - x_0) \equiv l_e(\hbar\beta)\delta(\bar{x} - x_0) = \sqrt{\frac{2\pi\hbar^2\beta}{M}} \, \delta(\bar{x} - x_0). \tag{3.812}$$

Then we can write

$$B(x_0) \equiv e^{-V^{\text{eff cl}}(x_0)/k_BT} = \oint \mathcal{D}'x \, e^{-\mathcal{A}_e/\hbar} = \oint \mathcal{D}x \, \tilde{\delta}(\bar{x} - x_0) \, e^{-\mathcal{A}_e/\hbar}$$

$$= \oint \mathcal{D}\eta \, \tilde{\delta}(\bar{\eta}) \, e^{-\mathcal{A}_e/\hbar}. \tag{3.813}$$

As a check we evaluate the effective classical Boltzmann factor for the harmonic action (2.443). With the path splitting (3.804), it reads

$$\mathcal{A}_e[x_0 + \eta] = \hbar\beta \frac{M\omega^2}{2}x_0^2 + \frac{M}{2}\int_0^{\hbar\beta} d\tau \left[\dot{\eta}^2(\tau) + \omega^2\eta^2(\tau)\right]. \tag{3.814}$$

After representing the δ function by a Fourier integral

$$\tilde{\delta}(\bar{\eta}) = l_e(\hbar\beta) \int_{-i\infty}^{i\infty} \frac{d\lambda}{2\pi i} \exp\left(\lambda \frac{1}{\hbar\beta} \int d\tau \, \eta(\tau)\right), \tag{3.815}$$

we find the path integral

$$B_\omega(x_0) = \oint \mathcal{D}\eta \, \tilde{\delta}(\bar{\eta}) \, e^{-\mathcal{A}_e/\hbar} = e^{-\beta M\omega^2 x_0^2/2} l_e(\hbar\beta) \int_{-i\infty}^{i\infty} \frac{d\lambda}{2\pi i}$$

$$\times \oint \mathcal{D}\eta \exp\left\{-\frac{1}{\hbar}\int_0^{\hbar\beta} d\tau \left[\frac{M}{2}\dot{\eta}^2(\tau) - \frac{\lambda}{\beta}\eta(\tau)\right]\right\}. \tag{3.816}$$

The path integral over $\eta(\tau)$ in the second line can now be performed without the restriction $\hbar\beta = 0$ and yields, recalling (3.552), (3.553), and inserting there $j(\tau) = \lambda/\beta$, we obtain for the path integral over $\eta(\tau)$ in the second line of (3.816):

$$\frac{1}{2\sinh(\beta\hbar\omega/2)} \exp\left\{\frac{\lambda^2}{2M\hbar\beta^2}\int_0^{\hbar\beta}d\tau\int_0^{\hbar\beta}d\tau'\, G^{\mathrm{p}}_{\omega^2,\mathrm{e}}(\tau - \tau')\right\}. \tag{3.817}$$

The integrals over τ, τ' are most easily performed on the spectral representation (3.245) of the correlation function:

$$\int_0^{\hbar\beta}d\tau\int_0^{\hbar\beta}d\tau'\, G^{\mathrm{p}}_{\omega^2,\mathrm{e}}(\tau - \tau') = \int_0^{\hbar\beta}d\tau\int_0^{\hbar\beta}d\tau'\,\frac{1}{\hbar\beta}\sum_{m=-\infty}^{\infty}\frac{1}{\omega_m^2 + \omega^2}e^{-i\omega_m(\tau-\tau')} = \frac{\hbar\beta}{\omega^2}. \tag{3.818}$$

The expression (3.817) has to be integrated over λ and yields

$$\frac{1}{2\sinh(\beta\hbar\omega/2)}\int_{-i\infty}^{i\infty}\frac{d\lambda}{2\pi i}\exp\left(\frac{\lambda^2}{2M\omega^2\beta}\right) = \frac{1}{2\sinh(\beta\hbar\omega/2)}\frac{1}{l_{\mathrm{e}}(\hbar\beta)}\,\omega\hbar\beta. \tag{3.819}$$

Inserting this into (3.816) we obtain the local Boltzmann factor

$$B_\omega(x_0) \equiv e^{-V_\omega^{\mathrm{eff\,cl}}(x_0)/k_BT} = \oint \mathcal{D}\eta\,\tilde{\delta}(\bar{\eta})\,e^{-\mathcal{A}_{\mathrm{e}}/\hbar} = \frac{\beta\hbar\omega/2}{\sinh(\beta\hbar\omega/2)}e^{-\beta M\omega^2 x_0^2}. \tag{3.820}$$

The final integral over x_0 in (3.805) reproduces the correct partition function (2.407) of the harmonic oscillator.

3.25.2 Effective Classical Hamiltonian

It is easy to generalize the expression (3.813) to phase space, where we define the *effective classical Hamiltonian* $H^{\mathrm{eff\,cl}}(p_0, x_0)$ and the associated Boltzmann factor $B(p_0, x_0)$ by the path integral

$$B(p_0, x_0) \equiv \exp\left[-\beta H^{\mathrm{eff\,cl}}(p_0, x_0)\right] \equiv \oint \mathcal{D}x\oint\frac{\mathcal{D}p}{2\pi\hbar}\,\delta(x_0 - \bar{x})2\pi\hbar\delta(p_0 - \bar{p})\,e^{-\mathcal{A}_{\mathrm{e}}[p,x]/\hbar}, \tag{3.821}$$

where $\bar{x} = \int_0^{\hbar\beta}d\tau\,x(\tau)/\hbar\beta$ and $\bar{p} = \int_0^{\hbar\beta}d\tau\,p(\tau)/\hbar\beta$ are the temporal averages of position and momentum, and $\mathcal{A}_{\mathrm{e}}[p,x]$ is the Euclidean action in phase space

$$\mathcal{A}_{\mathrm{e}}[p, x] = \int_0^{\hbar\beta}d\tau\,[-ip(\tau)\dot{x}(\tau) + H(p(\tau), x(\tau))]. \tag{3.822}$$

The full quantum-mechanical partition function is obtained from the classical-looking expression [recall (2.344)]

$$Z = \int_{-\infty}^{\infty}dx_0\int_{-\infty}^{\infty}\frac{dp_0}{2\pi\hbar}e^{-\beta H^{\mathrm{eff\,cl}}(p_0,x_0)}. \tag{3.823}$$

The definition is such that in the classical limit, $H^{\text{eff cl}}(p_0, x_0))$ becomes the ordinary Hamiltonian $H(p_0, x_0)$.

For a harmonic oscillator, the effective classical Hamiltonian can be directly deduced from Eq. (3.820) by "undoing" the p_0-integration:

$$B_\omega(p_0, x_0) \equiv e^{-H_\omega^{\text{eff cl}}(p_0, x_0)/k_B T} = l_{\text{e}}(\hbar\beta) \frac{\beta\hbar\omega/2}{\sinh(\beta\hbar\omega/2)} e^{-\beta(p_0^2/2M + M\omega^2 x_0^2)}. \qquad (3.824)$$

Indeed, inserting this into (3.823), we recover the harmonic partition function (2.407).

Consider a particle in three dimensions moving in a constant magnetic field B along the z-axis. For the sake of generality, we allow for an additional harmonic oscillator centered at the origin with frequencies $\omega_\parallel$ in z-direction and $\omega_\perp$ in the xy-plane (as in Section 2.19). It is then easy to calculate the effective classical Boltzmann factor for the Hamiltonian [recall (2.687)]

$$H(\mathbf{p}, \mathbf{x}) = \frac{1}{2M}\mathbf{p}^2 + \frac{M}{2}\omega_\perp^2 \mathbf{x}_\perp^2(\tau) + \frac{M}{2}\omega_\parallel^2 z^2(\tau) + \omega_B l_z(\mathbf{p}(\tau), \mathbf{x}(\tau)), \qquad (3.825)$$

where $l_z(\mathbf{p}, \mathbf{x})$ is the z-component of the angular momentum defined in Eq. (2.645). We have shifted the center of momentum integration to $\mathbf{p}_0$, for later convenience (see Subsection 5.11.2). The vector $\mathbf{x}^\perp = (x, y)$ denotes the orthogonal part of $\mathbf{x}$. As in the generalized magnetic field action (2.687), we have chosen different frequencies in front of the harmonic oscillator potential and of the term proportional to l_z, for generality. The effective classical Boltzmann factor follows immediately from (2.701) by "undoing" the momentum integrations in p_x, p_y, and using (3.824) for the motion in the z-direction:

$$B(\mathbf{p}_0, \mathbf{x}_0) = e^{-\beta H^{\text{eff cl}}(\mathbf{p}_0, \mathbf{x}_0)} = l_{\text{e}}^3(\hbar\beta) \frac{\hbar\beta\omega_+/2}{\sinh\hbar\beta\omega_+/2} \frac{\hbar\beta\omega_-/2}{\sinh\hbar\beta\omega_-/2} \frac{\hbar\beta\omega_\parallel/2}{\sinh\hbar\beta\omega_\parallel/2} e^{-\beta H(\mathbf{p}_0, \mathbf{x}_0)},$$
$$(3.826)$$

where $\omega_\pm \equiv \omega_B \pm \omega_\perp$, as in (2.697). As in Eq. (3.820), the restrictions of the path integrals over $\mathbf{x}$ and $\mathbf{p}$ to the fixed averages $\mathbf{x}_0 = \bar{\mathbf{x}}$ and $\mathbf{p}_0 = \bar{\mathbf{p}}$ give rise to the extra numerators in comparison to (2.701).

3.25.3 High- and Low-Temperature Behavior

We have remarked before in Eq. (3.809) that in the limit $T \to \infty$, the effective classical potential $V^{\text{eff cl}}(x_0)$ converges by construction against the initial potential $V(x_0)$. There exists, in fact, a well-defined power series in $\hbar\omega/k_B T$ which describes this approach. Let us study this limit explicitly for the effective classical potential of the harmonic oscillator calculated in (3.820), after rewriting it as

$$\begin{aligned} V_\omega^{\text{eff cl}}(x_0) &= k_B T \log\frac{\sinh(\hbar\omega/2k_B T)}{\hbar\omega/2k_B T} + \frac{M}{2}\omega^2 x_0^2 \qquad (3.827) \\ &= \frac{M}{2}\omega^2 x_0^2 + \frac{\hbar\omega}{2} + k_B T \left[\log(1 - e^{-\hbar\omega/k_B T}) - \log\frac{\hbar\omega}{k_B T}\right]. \end{aligned}$$

Due to the subtracted logarithm of ω in the brackets, the effective classical potential has a power series

$$V_\omega^{\text{eff cl}}(x_0) = \frac{M}{2}\omega^2 x_0^2 + \hbar\omega \left[\frac{1}{24}\frac{\hbar\omega}{k_B T} - \frac{1}{2880}\left(\frac{\hbar\omega}{k_B T}\right)^3 + \dots\right]. \qquad (3.828)$$

This pleasant high-temperature behavior is in contrast to that of the effective potential which reads for the harmonic oscillator

$$\begin{aligned}
V_\omega^{\text{eff}}(x_0) &= k_B T \log\left[2\sinh(\hbar\omega/2k_B T)\right] + \frac{M}{2}\omega^2 x_0^2 \\
&= \frac{M}{2}\omega^2 x_0^2 + \frac{\hbar\omega}{2} + k_B T \log(1 - e^{-\hbar\omega/k_B T}),
\end{aligned} \qquad (3.829)$$

as we can see from (3.793). The logarithm of ω prevents this from having a power series expansion in $\hbar\omega/k_B T$, reflecting the increasing width of the unsubtracted fluctuations.

Consider now the opposite limit $T \to 0$, where the final integral over the Boltzmann factor $B(x_0)$ can be calculated exactly by the saddle-point method. In this limit, the effective classical potential $V^{\text{eff cl}}(x_0)$ coincides with the Euclidean version of the effective potential:

$$V^{\text{eff cl}}(x_0) \xrightarrow[T \to 0]{} V^{\text{eff}}(x_0) \equiv \Gamma_e[X]/\beta\Big|_{X=x_0}, \qquad (3.830)$$

whose real-time definition was given in Eq. (3.663).

Let us study this limit again explicitly for the harmonic oscillator, where it becomes

$$V_\omega^{\text{eff cl}}(x_0) \xrightarrow[]{T \to 0} \frac{\hbar\omega}{2} + \frac{M}{2}\omega^2 x_0^2 - k_B T \log\frac{\hbar\omega}{k_B T}, \qquad (3.831)$$

i.e., the additional constant tends to $\hbar\omega/2$. This is just the quantum-mechanical zero-point energy which guarantees the correct low-temperature limit

$$\begin{aligned}
Z_\omega &\xrightarrow[]{T \to 0} e^{-\hbar\omega/2k_B T}\frac{\hbar\omega}{k_B T}\int_{-\infty}^{\infty}\frac{dx_0}{l_e(\hbar\beta)}e^{-M\omega^2 x_0^2/2k_B T} \\
&= e^{-\hbar\omega/2k_B T}.
\end{aligned} \qquad (3.832)$$

The limiting partition function is equal to the Boltzmann factor with the zero-point energy $\hbar\omega/2$.

3.25.4　Alternative Candidate for Effective Classical Potential

It is instructive to compare this potential with a related expression which can be defined in terms of the partition function density defined in Eq. (2.330):

$$\tilde{V}_\omega^{\text{eff cl}}(x) \equiv k_B T \log\left[l_e(\hbar\beta)\, z(x)\right]. \qquad (3.833)$$

This quantity shares with $V_\omega^{\text{eff cl}}(x_0)$ the property that it also yields the partition function by forming the integral [compare (2.329)]:

$$Z = \int_{-\infty}^{\infty} \frac{dx_0}{l_e(\hbar\beta)} e^{-\tilde{V}^{\text{eff cl}}(x_0)/k_B T}. \tag{3.834}$$

It may therefore be considered as an alternative candidate for an effective classical potential.

For the harmonic oscillator, we find from Eq. (2.331) the explicit form

$$\tilde{V}_\omega^{\text{eff cl}}(x) = -\frac{k_B T}{2} \log \frac{2\hbar\omega}{k_B T} + \frac{\hbar\omega}{2} + k_B T \left[\log \left(1 - e^{-2\hbar\omega/k_B T} \right) + \frac{M\omega}{\hbar} \tanh \frac{\hbar\omega}{k_B T} x^2 \right]. \tag{3.835}$$

This shares with the effective potential $V^{\text{eff}}(X)$ in Eq. (3.829) the unpleasant property of possessing no power series representation in the high-temperature limit.

The low-temperature limit of $\tilde{V}_\omega^{\text{eff cl}}(x)$ looks at first sight quite similar to (3.831):

$$\tilde{V}^{\text{eff cl}}(x_0) \xrightarrow{T \to 0} \frac{\hbar\omega}{2} + k_B T \frac{M\omega}{\hbar} x^2 - \frac{k_B T}{2} \log \frac{2\hbar\omega}{k_B T}, \tag{3.836}$$

and the integration leads to the same result (3.832) in only a slightly different way:

$$Z_\omega \xrightarrow{T \to 0} e^{-\hbar\omega/2k_B T} \sqrt{\frac{2\hbar\omega}{k_B T}} \int_{-\infty}^{\infty} \frac{dx}{l_e(\hbar\beta)} e^{-M\omega x^2/\hbar}$$

$$= e^{-\hbar\omega/2k_B T}. \tag{3.837}$$

There is, however, an important difference of (3.836) with respect to (3.831). The width of a local Boltzmann factor formed from the partition function density (2.330):

$$\tilde{B}(x) \equiv l_e(\hbar\beta) z(x) = e^{-\tilde{V}^{\text{eff cl}}(x)/k_B T} \tag{3.838}$$

is much wider than that of the effective classical Boltzmann factor $B(x_0) = e^{-V^{\text{eff cl}}(x_0)/k_B T}$. Whereas $B(x_0)$ has a finite width for $T \to 0$, the Boltzmann factor $\tilde{B}(x)$ has a width growing to infinity in this limit. Thus the integral over x in (3.837) converges much more slowly than that over x_0 in (3.832). This is the principal reason for introducing $V^{\text{eff cl}}(x_0)$ as an effective classical potential rather than $\tilde{V}^{\text{eff cl}}(x_0)$.

3.25.5 Harmonic Correlation Function without Zero Mode

By construction, the correlation functions of $\eta(\tau)$ have the desired subtracted form (3.802):

$$\langle \eta(\tau)\eta(\tau') \rangle_\omega = \frac{\hbar}{M} G_{\omega^2,e}^{p\prime}(\tau - \tau') = \frac{\hbar}{2M\omega} \frac{\cosh \omega(|\tau - \tau'| - \hbar\beta/2)}{\sinh(\beta\hbar\omega/2)} - \frac{1}{\hbar\beta\omega^2}, \tag{3.839}$$

with the square width as in (3.803):

$$\langle \eta^2(\tau) \rangle_\omega \equiv a_\omega^2 = G_{\omega^2,e}^{p\prime}(0) = \frac{1}{2\omega} \coth \frac{\beta\hbar\omega}{2} - \frac{1}{\hbar\beta\omega^2}, \tag{3.840}$$

which decreases with increasing temperature. This can be seen explicitly by adding a current term $-\int d\tau\, j(\tau)\eta(\tau)$ to the action (3.814) which winds up in the exponent of (3.816), replacing λ/β by $j(\tau) + \lambda/\beta$ and multiplies the exponential in (3.817) by a factor

$$\frac{1}{2M\hbar\beta^2} \left\{ \int_0^{\hbar\beta} d\tau \int_0^{\hbar\beta} d\tau' \left[\lambda^2 + \lambda\beta j(\tau) + \lambda\beta j(\tau') \right] G_{\omega^2,e}^{p}(\tau - \tau') \right\}$$

$$\times \exp \left\{ \frac{1}{2M\hbar} \int_0^{\hbar\beta} d\tau \int_0^{\hbar\beta} d\tau'\, j(\tau) G_{\omega^2,e}^{p}(\tau - \tau') j(\tau') \right\}. \tag{3.841}$$

In the first exponent, one of the τ-integrals over $G_{\omega^2,e}^{p}(\tau - \tau')$, say τ', produces a factor $1/\omega^2$ as in (3.818), so that the first exponent becomes

$$\frac{1}{2M\hbar\beta^2} \left\{ \lambda^2 \frac{\hbar\beta}{\omega^2} + 2\frac{\lambda\beta}{\omega^2} \int_0^{\hbar\beta} d\tau\, j(\tau) \right\}. \tag{3.842}$$

If we now perform the integral over λ, the linear term in λ yields, after a quadratic completion, a factor

$$\exp \left\{ -\frac{1}{2M\beta\hbar^2\omega^2} \int_0^{\hbar\beta} d\tau \int_0^{\hbar\beta} d\tau' j(\tau) j(\tau') \right\}. \tag{3.843}$$

Combined with the second exponential in (3.841) this leads to a generating functional for the subtracted correlation functions (3.839):

$$Z_\omega^{x_0}[j] = \frac{\beta\hbar\omega/2}{\sin(\beta\hbar\omega/2)} e^{-\beta M\omega^2 x_0^2/2} \exp \left\{ \frac{1}{2M\hbar} \int_0^{\hbar\beta} d\tau \int_0^{\hbar\beta} d\tau'\, j(\tau) G_{\omega^2,e}^{p\prime}(\tau - \tau') j(\tau') \right\}. \tag{3.844}$$

For $j(\tau) \equiv 0$, this reduces to the local Boltzmann factor (3.820).

3.25.6 Perturbation Expansion

We can now apply the perturbation expansion (3.480) to the path integral over $\eta(\tau)$ in Eq. (3.813) for the effective classical Boltzmann factor $B(x_0)$. We take the action

$$\mathcal{A}_e[x] = \int_0^{\hbar\beta} d\tau \left[\frac{M}{2} \dot{x}^2 + V(x) \right], \tag{3.845}$$

and rewrite it as

$$\mathcal{A}_e = \hbar\beta V(x_0) + \mathcal{A}_e^{(0)}[\eta] + \mathcal{A}_{\text{int},e}[x_0; \eta], \tag{3.846}$$

with an unperturbed action

$$\mathcal{A}_{\mathrm{e}}^{(0)}[\eta] = \int_0^{\hbar\beta} d\tau \left[\frac{M}{2}\dot{\eta}^2(\tau) + \frac{M}{2}\Omega^2(x_0)\eta^2(\tau) \right], \quad \Omega^2(x_0) \equiv V''(x_0)/M, \quad (3.847)$$

and an interaction

$$\mathcal{A}_{\mathrm{int,e}}[x_0;\eta] = \int_0^{\hbar\beta} d\tau \, V^{\mathrm{int}}(x_0;\eta(\tau)), \quad (3.848)$$

containing the subtracted potential

$$V^{\mathrm{int}}(x_0;\eta(\tau)) = V(x_0 + \eta(\tau)) - V(x_0) - V'(x_0)\eta(\tau) - \frac{1}{2}V''(x_0)\eta^2(\tau). \quad (3.849)$$

This has a Taylor expansion starting with the cubic term

$$V^{\mathrm{int}}(x_0;\eta) = \frac{1}{3!}V'''(x_0)\eta^3 + \frac{1}{4!}V^{(4)}(x_0)\eta^4 + \dots. \quad (3.850)$$

Since $\eta(\tau)$ has a zero temporal average, the linear term $\int_0^{\hbar\beta} d\tau \, V'(x_0)\eta(\tau)$ is absent in (3.847). The effective classical Boltzmann factor $B(x_0)$ in (3.813) has then the perturbation expansion [compare (3.480)]

$$B(x_0) = \left(1 - \frac{1}{\hbar}\left\langle \mathcal{A}_{\mathrm{int,e}} \right\rangle_\Omega^{x_0} + \frac{1}{2!\hbar^2}\left\langle \mathcal{A}_{\mathrm{int,e}}^2 \right\rangle_\Omega^{x_0} - \frac{1}{3!\hbar^3}\left\langle \mathcal{A}_{\mathrm{int,e}}^3 \right\rangle_\Omega^{x_0} + \dots \right) B_\Omega(x_0). \quad (3.851)$$

The harmonic expectation values are defined with respect to the harmonic path integral

$$B_\Omega(x_0) = \int \mathcal{D}\eta \, \tilde{\delta}(\bar{\eta}) \, e^{-\mathcal{A}_{\mathrm{e}}^{(0)}[\eta]//\hbar}. \quad (3.852)$$

For an arbitrary functional $F[x]$ one has to calculate

$$\langle F[x]\rangle_\Omega^{x_0} = B_\Omega^{-1}(x_0) \int \mathcal{D}\eta \, \tilde{\delta}(\bar{\eta}) \, F[x] \, e^{-\mathcal{A}_{\mathrm{e}}^{(0)}[\eta]/\hbar}. \quad (3.853)$$

Some calculations of local expectation values are conveniently done with the explicit Fourier components of the path integral. Recalling (3.806) and expanding the action (3.814) in its Fourier components using (3.804), they are given by the product of integrals

$$\langle F[x]\rangle_\Omega^{x_0} = [Z_\Omega^{x_0}]^{-1} \prod_{m=1}^{\infty} \left[\int \frac{dx_m^{\mathrm{re}} dx_m^{\mathrm{im}}}{\pi k_B T/M\omega_m^2} \right] e^{-\frac{M}{k_B T}\Sigma_{m=1}^{\infty}[\omega_m^2 + \Omega^2(x_0)]|x_m|^2} F[x]. \quad (3.854)$$

This implies the correlation functions for the Fourier components

$$\left\langle x_m x_{m'}^* \right\rangle_\Omega^{x_0} = \delta_{mm'} \frac{k_B T}{M} \frac{1}{\omega_m^2 + \Omega^2(x_0)}. \quad (3.855)$$

From these we can calculate once more the correlation functions of the fluctuations $\eta(\tau)$ as follows:

$$\langle \eta(\tau)\eta(\tau')\rangle_\Omega^{x_0} = \left\langle \sum_{m,m'\neq 0}^{\infty} x_m x_{m'}^* e^{-i(\omega_m-\omega_{m'})\tau} \right\rangle_\Omega^{x_0} = 2\frac{1}{M\beta}\sum_{m=1}^{\infty}\frac{1}{\omega_m^2+\Omega^2(x_0)}. \quad (3.856)$$

Performing the sum gives once more the subtracted correlation function Eq. (3.839), whose generating functional was calculated in (3.844).

The calculation of the harmonic averages in (3.851) leads to a similar loop expansion as for the effective potential in Subsection 3.23.6 using the background field method. The path average x_0 takes over the role of the background X and the non-zero Matsubara frequency part of the paths $\eta(\tau)$ corresponds to the fluctuations. The only difference with respect to the earlier calculations is that the correlation functions of $\eta(\tau)$ contain no zero-frequency contribution. Thus they are obtained from the subtracted Green functions $G_{\Omega^2(x_0),\mathrm{e}}^{\mathrm{p}\prime}(\tau)$ defined in Eq. (3.802).

All Feynman diagrams in the loop expansion are one-particle irreducible, just as in the loop expansion of the effective potential. The reducible diagrams are absent since there is no linear term in the interaction (3.850). This trivial absence is an advantage with respect to the somewhat involved proof required for the effective action in Subsection 3.23.6. The diagrams in the two expansions are therefore precisely the same and can be read off from Eqs. (3.741) and (3.781). The only difference lies in the replacement $X \to x_0$ in the analytic expressions for the lines and vertices. In addition, there is the final integral over x_0 to obtain the partition function Z in Eq. (3.808). This is in contrast to the partition function expressed in terms of the effective potential $V^{\mathrm{eff}}(X)$, where only the extremum has to be taken.

3.25.7 Effective Potential and Magnetization Curves

The effective classical potential $V^{\mathrm{eff\,cl}}(x_0)$ in the Boltzmann factor (3.807) allows us to estimate the *effective potential* defined in Eq. (3.663). It can be derived from the generating functional $Z[j]$ restricted to time-independent external source $j(\tau) \equiv j$, in which case $Z[j]$ reduces to a mere function of j:

$$Z(j) = \int \mathcal{D}x(\tau) \exp\left\{-\int_0^\beta d\tau\left[\frac{1}{2}\dot{x}^2 + V(x(\tau))\right] + \beta j\bar{x}\right\}, \quad (3.857)$$

where $\bar{x}$ is the path average of $x(\tau)$. The function $Z(j)$ is obtained from the effective classical potential by a simple integral over x_0:

$$Z(j) = \int_{-\infty}^{\infty} \frac{dx_0}{\sqrt{2\pi\beta}} e^{-\beta[V^{\mathrm{eff\,cl}}(x_0)-jx_0]}. \quad (3.858)$$

The effective potential $V^{\mathrm{eff}}(X)$ is equal to the Legendre transform of $W(j) = \log Z(j)$:

$$V^{\mathrm{eff}}(X) = -\frac{1}{\beta}W(j) + Xj, \quad (3.859)$$

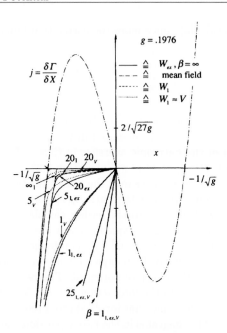

Figure 3.15 Magnetization curves in double-well potential $V(x) = -x^2/2 + gx^4/4$ with $g = 0.4$, at various inverse temperatures β. The integral over these curves returns the effective potential $V^{\text{eff}}(X)$. The curves arising from the approximate effective potential $W_1(x_0)$ are labeled by β_1 (- - -) and the exact curves (found by solving the Schrödinger equation numerically) by β_{ex} (——). For comparison we have also drawn the classical curves ($\cdots$) obtained by using the potential $V(x_0)$ in Eqs. (3.861) and (3.858) rather than $W_1(x_0)$. They are labeled by β_V. Our approximation $W_1(x_0)$ is seen to render good magnetization curves for all temperatures above $T = 1/\beta \sim 1/10$. The label β carries several subscripts if the corresponding curves are indistinguishable on the plot. Note that all approximations are monotonous, as they should be (except for the mean field, of course).

where the right-hand side is to be expressed in terms of X using

$$X = X(j) = \frac{1}{\beta}\frac{d}{dj}W(j). \tag{3.860}$$

To picture the effective potential, we calculate the average value of $x(\tau)$ from the integral

$$X = Z(j)^{-1}\int_{-\infty}^{\infty}\frac{dx_0}{\sqrt{2\pi\beta}}\, x_0\, \exp\left\{-\beta[V^{\text{eff cl}}(x_0) - jx_0]\right\} \tag{3.861}$$

and plot $X = X(j)$. By exchanging the axes we display the inverse $j = j(X)$ which is the slope of the effective potential:

$$j(X) = \frac{dV^{\text{eff}}(X)}{dX}. \tag{3.862}$$

The curves $j(X)$ are shown in Fig. 3.15 for the double-well potential with a coupling strength $g = 0.4$ at various temperatures.

Note that the x_0-integration makes $j(X)$ necessarily a monotonous function of X. The effective potential is therefore always a convex function of X, no matter what the classical potential looks like. This is in contrast to $j(X)$ *before* fluctuations are taken into account, the *mean-field approximation* to (3.862) [recall the discussion in Subsection 3.23.1], which is given by

$$j = dV(X)/dX. \tag{3.863}$$

For the double-well potential, this becomes

$$j = -X + gX^3. \tag{3.864}$$

Thus, the mean-field effective potential coincides with the classical potential $V(X)$, which is obviously not convex.

In magnetic systems, j is a constant magnetic field and X its associated magnetization. For this reason, plots of $j(X)$ are referred to as *magnetization curves*.

3.25.8 First-Order Perturbative Result

To first order in the interaction $V^{\text{int}}(x_0; \eta)$, the perturbation expansion (3.851) becomes

$$B(x_0) = \left(1 - \frac{1}{\hbar} \langle \mathcal{A}_{\text{int,e}} \rangle_\Omega^{x_0} + \dots \right) B_\Omega(x_0), \tag{3.865}$$

and we have to calculate the harmonic expectation value of $\mathcal{A}_{\text{int,e}}$. Let us assume that the interaction potential possesses a Fourier transform

$$V^{\text{int}}(x_0; \eta(\tau)) = \int_{-\infty}^{\infty} \frac{dk}{2\pi} e^{ik(x_0 + \eta(\tau))} \tilde{V}^{\text{int}}(k). \tag{3.866}$$

Then we can write the expectation of (3.848) as

$$\langle \mathcal{A}_{\text{int,e}}[x_0; \eta] \rangle_\Omega^{x_0} = \int_0^{\hbar\beta} d\tau \int_{-\infty}^{\infty} \frac{dk}{2\pi} \tilde{V}^{\text{int}}(k) e^{ikx_0} \left\langle e^{ik\eta(\tau)} \right\rangle_\Omega^{x_0}. \tag{3.867}$$

We now use Wick's rule in the form (3.304) to calculate

$$\left\langle e^{ik\eta(\tau)} \right\rangle_\Omega^{x_0} = e^{-k^2 \langle \eta^2(\tau) \rangle_\Omega^{x_0}/2}. \tag{3.868}$$

We now use Eq. (3.840) to write this as

$$\left\langle e^{ik\eta(\tau)} \right\rangle_\Omega^{x_0} = e^{-k^2 a_{\Omega(x_0)}^2/2}. \tag{3.869}$$

Thus we find for the expectation value (3.867):

$$\langle \mathcal{A}_{\text{int,e}}[x_0; \eta] \rangle_\Omega^{x_0} = \int_0^{\hbar\beta} d\tau \int_{-\infty}^{\infty} \frac{dk}{2\pi} \tilde{V}^{\text{int}}(k) e^{ikx_0 - k^2 a_{\Omega(x_0)}^2/2}. \tag{3.870}$$

Due to the periodic boundary conditions satisfied by the correlation function and the associated invariance under time translations, this result is independent of τ, so that the τ-integral can be performed trivially, yielding simply a factor $\hbar\beta$. We now reinsert the Fourier coefficients of the potential

$$\tilde{V}^{\text{int}}(k) = \int_{-\infty}^{\infty} dx\, V^{\text{int}}(x_0; \eta)\, e^{-ik(x_0+\eta)}, \tag{3.871}$$

perform the integral over k via a quadratic completion, and obtain

$$\left\langle V^{\text{int}}(x(\tau)) \right\rangle_{\Omega}^{x_0} \equiv V_{a_{\Omega}^2}^{\text{int}}(x_0) = \int_{-\infty}^{\infty} \frac{dx_0'}{\sqrt{2\pi a_{\Omega(x_0)}^2}} e^{-\eta^2/2a_{\Omega(x_0)}^2}\, V^{\text{int}}(x_0; \eta). \tag{3.872}$$

The expectation $\left\langle V^{\text{int}}(x(\tau)) \right\rangle_{\Omega}^{x_0} \equiv V_{a_{\Omega}^2}^{\text{int}}(x_0)$ of the potential arises therefore from a convolution integral of the original potential with a Gaussian distribution of square width $a_{\Omega(x_0)}^2$. The convolution integral smears the original interaction potential $V_{a_{\Omega}^2}^{\text{int}}(x_0)$ out over a length scale $a_{\Omega(x_0)}$. In this way, the approximation accounts for the quantum-statistical path fluctuations of the particle.

As a result, we can write the first-order Boltzmann factor (3.865) as follows:

$$B(x_0) \approx \frac{\Omega(x_0)\hbar\beta}{2\sin[\Omega(x_0)\hbar\beta/2]} \exp\left\{-\beta M\Omega(x_0)^2 x_0^2/2 - \beta V_{a_{\Omega}^2}^{\text{int}}(x_0)\right\}. \tag{3.873}$$

Recalling the harmonic effective classical potential (3.831), this may be written as a Boltzmann factor associated with the first-order effective classical potential

$$V^{\text{eff cl}}(x_0) \approx V_{\Omega(x_0)}^{\text{eff cl}}(x_0) + V_{a_{\Omega}^2}^{\text{int}}(x_0). \tag{3.874}$$

Given the power series expansion (3.850) of the interaction potential

$$V^{\text{int}}(x_0; \eta) = \sum_{k=3}^{\infty} \frac{1}{k!} V^{(k)}(x_0)\hbar^k, \tag{3.875}$$

we may use the integral formula

$$\int_{-\infty}^{\infty} \frac{d\eta}{\sqrt{2\pi a^2}} e^{-\eta^2/2a^2} \eta^k = \left\{ \begin{array}{c} (k-1)!!\, a^k \\ 0 \end{array} \right\} \quad \text{for } k = \left\{ \begin{array}{c} \text{even} \\ \text{odd} \end{array} \right\}, \tag{3.876}$$

we find the explicit smeared potential

$$V_{a^2}^{\text{int}}(x_0) = \sum_{k=4,6,\ldots}^{\infty} \frac{(k-1)!!}{k!} V^{(k)}(x_0) a^k(x_0). \tag{3.877}$$

3.26 Perturbative Approach to Scattering Amplitude

In Eq. (2.745) we have derived a path integral representation for the scattering amplitude. It involves calculating a path integral of the general form

$$\int d^3 y_a \int d^3 z_a \int \mathcal{D}^3 y \int \mathcal{D}^3 z \exp\left[\frac{i}{\hbar}\int_{t_a}^{t_b} dt\, \frac{M}{2}\left(\dot{\mathbf{y}}^2 - \dot{\mathbf{z}}^2\right)\right] F[\mathbf{y}(t) - \mathbf{z}(0)], \qquad (3.878)$$

where the paths $\mathbf{y}(t)$ and $\mathbf{z}(t)$ vanish at the final time $t = t_b$ whereas the initial positions are integrated out. In lowest approximation, we may neglect the fluctuations in $\mathbf{y}(t)$ and $\mathbf{z}(0)$ and obtain the eikonal approximation (2.748). In order to calculate higher-order corrections to path integrals of the form (3.878) we find the generating functional of all correlation functions of $\mathbf{y}(t) - \mathbf{z}(0)$.

3.26.1 Generating Functional

For the sake of generality we calculate the harmonic path integral over $\mathbf{y}$:

$$Z[\mathbf{j_y}] \equiv \int d^3 y_a \int \mathcal{D}^3 y \exp\left\{\frac{i}{\hbar}\int_{t_a}^{t_b} dt\,\left[\frac{M}{2}\left(\dot{\mathbf{y}}^2 - \omega^2 \mathbf{y}^2\right) - \mathbf{j_y}\mathbf{y}\right]\right\}. \qquad (3.879)$$

This differs from the amplitude calculated in (3.168) only by an extra Fresnel integral over the initial point and a trivial extension to three dimensions. This yields

$$\begin{aligned}
Z[\mathbf{j_y}] &= \int d^3 y_a (\mathbf{y}_b t_b | \mathbf{y}_a t_a)_\omega^{\mathbf{j_y}} \\
&= \exp\left\{\frac{i}{\hbar}\int_{t_a}^{t_b} dt\,\frac{1}{\sin\omega(t_b - t_a)}\left[\mathbf{y}_b\left(\sin\left[\omega(t - t_a)\right] + \sin\left[\omega(t_b - t)\right]\right)\mathbf{j_y}\right]\right\} \\
&\quad \times \exp\left\{-\frac{i}{\hbar^2}\frac{\hbar}{M}\int_{t_a}^{t_b} dt \int_{t_a}^{t} dt'\, \mathbf{j_y}(t)\bar{G}_{\omega^2}(t,t')\mathbf{j_y}(t')\right\},
\end{aligned} \qquad (3.880)$$

where $\bar{G}_{\omega^2}(t,t')$ is obtained from the Green function (3.36) with Dirichlet boundary conditions by adding the result of the quadratic completion in the variable $\mathbf{y}_b - \mathbf{y}_a$ preceding the evaluation of the integral over $d^3 y_a$:

$$\bar{G}_{\omega^2}(t,t') = \frac{1}{\omega \sin\omega(t_b - t_a)}\sin\omega(t_b - t_>)\left[\sin\omega(t_< - t_a) + \sin\omega(t_b - t_<)\right]. \qquad (3.881)$$

We need the special case $\omega = 0$ where

$$\bar{G}_{\omega^2}(t,t') = t_b - t_>. \qquad (3.882)$$

In contrast to $G_{\omega^2}(t,t')$ of (3.36), this Green function vanishes only at the final time. This reflects the fact that the path integral (3.878) is evaluated for paths $\mathbf{y}(t)$ which vanish at the final time $t = t_b$.

A similar generating functional for $\mathbf{z}(t)$ leads to the same result with opposite sign in the exponent. Since the variable $\mathbf{z}(t)$ appears only with time argument zero in (3.878), the relevant generating functional is

$$\begin{aligned}
Z[\mathbf{j}] &\equiv \int d^3 y_a \int d^3 z_a \int \mathcal{D}^3 y \int \mathcal{D}^3 z \\
&\quad \times \exp\left(\frac{i}{\hbar}\int_{t_a}^{t_b} dt\,\left\{\frac{M}{2}\left[\dot{\mathbf{y}}^2 - \dot{\mathbf{z}}^2 - \omega^2\left(\mathbf{y}^2 - \mathbf{z}^2\right)\right] - \mathbf{j}\mathbf{y}_\mathbf{z}\right\}\right),
\end{aligned} \qquad (3.883)$$

with $\mathbf{y}_b = \mathbf{z}_b = 0$, where we have introduced the subtracted variable

$$\mathbf{y}_\mathbf{z}(t) \equiv \mathbf{y}(t) - \mathbf{z}(0), \qquad (3.884)$$

for brevity. From the above calculations we can immediately write down the result

$$Z[\mathbf{j}] = \frac{1}{D_\omega} \exp\left\{-\frac{i}{\hbar^2}\frac{\hbar}{2M}\int_{t_a}^{t_b} dt \int_{t_a}^{t_b} dt'\, \mathbf{j}(t)\bar{G}'_{\omega^2}(t,t')\mathbf{j}(t')\right\}, \tag{3.885}$$

where D_ω is the functional determinant associated with the Green function (3.881) which is obtained by integrating (3.881) over $t \in (t_b, t_a)$ and over ω^2:

$$D_\omega = \frac{1}{\cos^2[\omega(t_b - t_a)]}\exp\left[\int_0^{t_b-t_a}\frac{dt}{t}\,(\cos\omega t - 1)\right], \tag{3.886}$$

and $\bar{G}'_{\omega^2}(t,t')$ is the subtracted Green function (3.881):

$$\bar{G}'_{\omega^2}(t,t') \equiv \bar{G}_{\omega^2}(t,t') - \bar{G}_{\omega^2}(0,0). \tag{3.887}$$

For $\omega = 0$ where $D_\omega = 1$, this is simply

$$\bar{G}'_0(t,t') \equiv -t_>, \tag{3.888}$$

where $t_>$ denotes the larger of the times t and t'. It is important to realize that thanks to the subtraction in the Green function (3.882) caused by the $\mathbf{z}(0)$-fluctuations, the limits $t_a \to -\infty$ and $t_b \to \infty$ can be taken in (3.885) without any problems.

3.26.2   Application to Scattering Amplitude

We can now apply this result to the path integral (2.745). With the abbreviation (3.884) we write it as

$$\begin{aligned}
f_{\mathbf{p}_b\mathbf{p}_a} &= \frac{p}{2\pi i\hbar}\int d^2b\, e^{-i\mathbf{q}\mathbf{b}/\hbar} \\
&\times \int \mathcal{D}^3 y_\mathbf{z}\exp\left\{\frac{i}{\hbar}\int_{-\infty}^{\infty} dt\,\frac{M}{2}\,\mathbf{y}_\mathbf{z}[\bar{G}'_0(t,t')]^{-1}\mathbf{y}_\mathbf{z}\right\}\left[e^{i\chi_{\mathbf{b},\mathbf{p}}[\mathbf{y}_\mathbf{z}]} - 1\right], \tag{3.889}
\end{aligned}$$

where $[\bar{G}'_0(t,t')]^{-1}$ is the functional inverse of the subtracted Green function (3.888), and $\chi_{\mathbf{b},\mathbf{p}}[\mathbf{y}_\mathbf{z}]$ the integral over the interaction potential $V(\mathbf{x})$:

$$\chi_{\mathbf{b},\mathbf{p}}[\mathbf{y}_\mathbf{z}] \equiv -\frac{1}{\hbar}\int_{-\infty}^{\infty} dt\, V\left(\mathbf{b} + \frac{\mathbf{p}}{M}t + \mathbf{y}_\mathbf{z}(t)\right). \tag{3.890}$$

3.26.3   First Correction to Eikonal Approximation

The first correction to the eikonal approximation (2.748) is obtained by expanding (3.890) to first order in $\mathbf{y}_\mathbf{z}(t)$. This yields

$$\chi_{\mathbf{b},\mathbf{p}}[\mathbf{y}] = \chi_{\mathbf{b},\mathbf{p}}^{\text{ei}} - \frac{1}{\hbar}\int_{-\infty}^{\infty} dt\, \boldsymbol{\nabla}V\left(\mathbf{b} + \frac{\mathbf{p}}{M}t\right)\mathbf{y}_\mathbf{z}(t). \tag{3.891}$$

The additional terms can be considered as an interaction

$$-\frac{1}{\hbar}\int_{-\infty}^{\infty} dt\, \mathbf{y}_\mathbf{z}(t)\,\mathbf{j}(t), \tag{3.892}$$

with the current

$$\mathbf{j}(t) = \boldsymbol{\nabla}V\left(\mathbf{b} + \frac{\mathbf{p}}{M}t\right). \tag{3.893}$$

Using the generating functional (3.885), this is seen to yield an additional scattering phase

$$
\Delta_1 \chi_{\mathbf{b},\mathbf{p}}^{\text{ei}} = \frac{1}{2M\hbar} \int_{-\infty}^{\infty} dt_1 \int_{-\infty}^{\infty} dt_2 \, \nabla V \left(\mathbf{b} + \frac{\mathbf{p}}{M} t_1 \right) \nabla V \left(\mathbf{b} + \frac{\mathbf{p}}{M} t_2 \right) t_>. \tag{3.894}
$$

To evaluate this we shall always change, as in (2.750), the time variables $t_{1,2}$ to length variables $z_{1,2} \equiv p_{1,2} t/M$ along the direction of $\mathbf{p}$.

For spherically symmetric potentials $V(r)$ with $r \equiv |\mathbf{x}| = \sqrt{b^2 + z^2}$, we may express the derivatives parallel and orthogonal to the incoming particle momentum $\mathbf{p}$ as follows:

$$
\nabla_\| V = z \, V'/r, \quad \nabla_\perp V = \mathbf{b} \, V'/r. \tag{3.895}
$$

Then (3.894) reduces to

$$
\Delta_1 \chi_{\mathbf{b},\mathbf{p}}^{\text{ei}} = \frac{M^2}{2\hbar p^3} \int_{-\infty}^{\infty} dz_1 \int_{-\infty}^{\infty} dz_2 \frac{V'(r_1)}{r_1} \frac{V'(r_2)}{r_2} \left(b^2 + z_1 z_2 \right) z_1. \tag{3.896}
$$

The part of the integrand before the bracket is obviously symmetric under $z \to -z$ and under the exchange $z_1 \leftrightarrow z_2$. For this reason we can rewrite

$$
\Delta_1 \chi_{\mathbf{b},\mathbf{p}}^{\text{ei}} = \frac{M^2}{\hbar p^3} \int_{-\infty}^{\infty} dz_1 \, z_1 \frac{V'(r_1)}{r_1} \int_{-\infty}^{\infty} dz_2 \frac{V'(r_2)}{r_2} \left(b^2 - z_2^2 \right). \tag{3.897}
$$

Now we use the relations (3.895) in the opposite direction as

$$
z V'/r = \partial_z V, \quad b V'/r = \partial_b V, \tag{3.898}
$$

and performing a partial integration in z_1 to obtain[21]

$$
\Delta_1 \chi_{\mathbf{b},\mathbf{p}}^{\text{ei}} = -\frac{M^2}{\hbar p^3} (1 + b\partial_b) \int_{-\infty}^{\infty} dz \, V^2 \left(\sqrt{b^2 + z^2} \right). \tag{3.899}
$$

Compared to the leading eikonal phase (2.751), this is suppressed by a factor $V(0)M/p^2$.

Note that for the Coulomb potential where $V^2(\sqrt{b^2 + z^2}) \propto 1/(b^2 + z^2)$, the integral is proportional to $1/b$ which is annihilated by the factor $1 + b\partial_b$. Thus there is no first correction to the eikonal approximation (1.503).

3.26.4   Rayleigh-Schrödinger Expansion of Scattering Amplitude

In Section 1.16 we have introduced the scattering amplitude as the limiting matrix element [see (1.513)]

$$
\langle \mathbf{p}_b | \hat{S} | \mathbf{p}_a \rangle \equiv \lim_{t_b - t_a \to \infty} e^{i(E_b - E_a) t_b / \hbar} (\mathbf{p}_b 0 | \mathbf{p}_a t_a) e^{-i E_a t_a / \hbar}. \tag{3.900}
$$

A perturbation expansion for these quantities can be found via a Fourier transformation of the expansion (3.474). We only have to set the oscillator frequency of the harmonic part of the action equal to zero, since the particles in a scattering process are free far away from the scattering center. Since scattering takes usually place in three dimensions, all formulas will be written down in such a space.

---

[21] This agrees with results from Schrödinger theory by S.J. Wallace, Ann. Phys. *78*, 190 (1973); S. Sarkar, Phys. Rev. D *21*, 3437 (1980). It differs from R. Rosenfelder's result (see Footnote 37 on p. 193) who derives a prefactor $p \cos(\theta/2)$ instead of the incoming momentum p.

We shall thus consider the perturbation expansion of the amplitude

$$(\mathbf{p}_b 0 | \mathbf{p}_a t_a) = \int d^3 x_b d^3 x_a e^{-i\mathbf{p}_b \mathbf{x}_b} (\mathbf{x}_b 0 | \mathbf{x}_a t_a) e^{i\mathbf{p}_a \mathbf{x}_a}, \tag{3.901}$$

where $(\mathbf{x}_b 0 | \mathbf{x}_a t_a)$ is expanded as in (3.474). The immediate result looks as in the expansion (3.497), if we replace the external oscillator wave functions $\psi_n(x_b)$ and $\psi_a(x_b)$ by free-particle plane waves $e^{-i\mathbf{p}_b \mathbf{x}_b}$ and $e^{i\mathbf{p}_a \mathbf{x}_a}$:

$$
\begin{aligned}
(\mathbf{p}_b 0 | \mathbf{p}_a t_a) = {} & (\mathbf{p}_b 0 | \mathbf{p}_a t_a)_0 \\
& + \frac{i}{\hbar} \langle \mathbf{p}_b | \mathcal{A}_{\text{int}} | \mathbf{p}_a \rangle_0 - \frac{1}{2!\hbar^2} \langle \mathbf{p}_b | \mathcal{A}_{\text{int}}^2 | \mathbf{p}_a \rangle_0 - \frac{i}{3!\hbar^3} \langle \mathbf{p}_b | \mathcal{A}_{\text{int}}^3 | \mathbf{p}_a \rangle_0 + \ldots .
\end{aligned} \tag{3.902}
$$

Here

$$(\mathbf{p}_b 0 | \mathbf{p}_a t_a)_0 = (2\pi\hbar)^3 \delta^{(3)}(\mathbf{p}_b - \mathbf{p}_a) e^{i\mathbf{p}_b^2 t_a / 2M\hbar} \tag{3.903}$$

is the free-particle time evolution amplitude in momentum space [recall (2.138)] and the matrix elements are defined by

$$\langle \mathbf{p}_b | \ldots | \mathbf{p}_a \rangle_0 \equiv \int d^3 x_b d^3 x_a e^{-i\mathbf{p}_b \mathbf{x}_b} \left(\int \mathcal{D}^3 x \ldots e^{i\mathcal{A}_0/\hbar} \right) e^{i\mathbf{p}_a \mathbf{x}_a}. \tag{3.904}$$

In contrast to (3.497) we have not divided out the free-particle amplitude (3.903) in this definition since it is too singular. Let us calculate the successive terms in the expansion (3.902). First

$$
\begin{aligned}
\langle \mathbf{p}_b | \mathcal{A}_{\text{int}} | \mathbf{p}_a \rangle_0 = {} & - \int_{t_a}^{0} dt_1 \int d^3 x_b d^3 x_a d^3 x_1 e^{-i\mathbf{p}_b \mathbf{x}_b} (\mathbf{x}_b 0 | \mathbf{x}_1 t_1)_0 \\
& \times V(\mathbf{x}_1)(\mathbf{x}_1 t_1 | \mathbf{x}_a t_a)_0 e^{i\mathbf{p}_a \mathbf{x}_a}.
\end{aligned} \tag{3.905}
$$

Since

$$
\begin{aligned}
\int d^3 x_b e^{-i\mathbf{p}_b \mathbf{x}_b} (\mathbf{x}_b t_b | \mathbf{x}_1 t_1)_0 &= e^{-i\mathbf{p}_b \mathbf{x}_1} e^{-i\mathbf{p}_b^2 (t_b - t_1)/2M\hbar}, \\
\int d^3 x_a (\mathbf{x}_1 t_1 | \mathbf{x}_a t_a)_0 e^{-i\mathbf{p}_b \mathbf{x}_b} &= e^{-i\mathbf{p}_a \mathbf{x}_1} e^{i\mathbf{p}_a^2 (t_1 - t_a)/2M\hbar},
\end{aligned} \tag{3.906}
$$

this becomes

$$\langle \mathbf{p}_b | \mathcal{A}_{\text{int}} | \mathbf{p}_a \rangle_0 = - \int_{t_a}^{0} dt_1 e^{i(\mathbf{p}_b^2 - \mathbf{p}_a^2) t_1 / 2M\hbar} V_{\mathbf{p}_b \mathbf{p}_a} e^{i\mathbf{p}_a^2 t_a / 2M\hbar}, \tag{3.907}$$

where

$$V_{\mathbf{p}_b \mathbf{p}_a} \equiv \langle \mathbf{p}_b | \hat{V} | \mathbf{p}_a \rangle = \int d^3 x e^{i(\mathbf{p}_b - \mathbf{p}_a) \mathbf{x}/\hbar} V(\mathbf{x}) = \tilde{V}(\mathbf{p}_b - \mathbf{p}_a) \tag{3.908}$$

[recall (1.491)]. Inserting a damping factor $e^{\eta t_1}$ into the time integral, and replacing $\mathbf{p}^2/2M$ by the corresponding energy E, we obtain

$$\frac{i}{\hbar} \langle \mathbf{p}_b | \mathcal{A}_{\text{int}} | \mathbf{p}_a \rangle_0 = - \frac{1}{E_b - E_a - i\eta} V_{\mathbf{p}_b \mathbf{p}_a} e^{iE_a t_a}. \tag{3.909}$$

Inserting this together with (3.903) into the expansion (3.902), we find for the scattering amplitude (3.900) the first-order approximation

$$\langle \mathbf{p}_b | \hat{S} | \mathbf{p}_a \rangle \equiv \lim_{t_b - t_a \to \infty} e^{i(E_b - E_a)t_b/\hbar} \left[(2\pi\hbar)^3 \delta^{(3)}(\mathbf{p}_b - \mathbf{p}_a) - \frac{1}{E_b - E_a - i\eta} V_{\mathbf{p}_b \mathbf{p}_a} \right]$$
(3.910)

corresponding precisely to the first-order approximation of the operator expression (1.516), the Born approximation.

Continuing the evaluation of the expansion (3.902) we find that $V_{\mathbf{p}_b \mathbf{p}_a}$ in (3.910) is replaced by the T-matrix [recall (1.474)]

$$T_{\mathbf{p}_b \mathbf{p}_a} = V_{\mathbf{p}_b \mathbf{p}_a} - \int \frac{d^3 p_c}{(2\pi\hbar)^3} V_{\mathbf{p}_b \mathbf{p}_c} \frac{1}{E_c - E_a - i\eta} V_{\mathbf{p}_c \mathbf{p}_a}$$
(3.911)
$$+ \int \frac{d^3 p_c}{(2\pi\hbar)^3} \int \frac{d^3 p_d}{(2\pi\hbar)^3} V_{\mathbf{p}_b \mathbf{p}_c} \frac{1}{E_c - E_a - i\eta} V_{\mathbf{p}_c \mathbf{p}_d} \frac{1}{E_d - E_a - i\eta} V_{\mathbf{p}_d \mathbf{p}_a} + \cdots .$$

This amounts to an integral equation

$$T_{\mathbf{p}_b \mathbf{p}_a} = V_{\mathbf{p}_b \mathbf{p}_a} - \int \frac{d^3 p_c}{(2\pi\hbar)^3} V_{\mathbf{p}_b \mathbf{p}_c} \frac{1}{E_c - E_a - i\eta} T_{\mathbf{p}_c \mathbf{p}_a},$$
(3.912)

which is recognized as the Lippmann-Schwinger equation (1.522) for the T-matrix.

3.27 Functional Determinants from Green Functions

In Subsection 3.2.1 we have seen that there exists a simple method, due to Wronski, for constructing Green functions of the differential equation (3.27),

$$\mathcal{O}(t) G_{\omega^2}(t, t') \equiv [-\partial_t^2 - \Omega^2(t)] G_{\omega^2}(t, t') = \delta(t - t'),$$
(3.913)

with Dirichlet boundary conditions. That method did not require any knowledge of the spectrum and the eigenstates of the differential operator $\mathcal{O}(t)$, except for the condition that zero-modes are absent. The question arises whether this method can be used to find also functional determinants.[22] The answer is positive, and we shall now demonstrate that Gelfand and Yaglom's initial-value problem (2.211), (2.212), (2.213) with the Wronski construction (2.223) for its solution represents the most concise formula for the functional determinant of the operator $\mathcal{O}(t)$. Starting point is the observation that a functional determinant of an operator $\mathcal{O}$ can be written as

$$\text{Det}\, \mathcal{O} = e^{\text{Tr}\,\log \mathcal{O}},$$
(3.914)

and that a Green function of a harmonic oscillator with an arbitrary time-dependent frequency has the integral

$$\text{Tr} \left\{ \int_0^1 dg\, \Omega^2(t) [-\partial_t^2 - g\Omega^2(t)]^{-1} \delta(t - t') \right\} = -\text{Tr}\left\{ \log[-\partial_t^2 - \Omega^2(t)]\delta(t - t') \right\}$$
$$+ \text{Tr}\left\{ \log[-\partial_t^2]\delta(t - t') \right\}.$$
(3.915)

---

[22]See the reference in Footnote 6 on p. 247.

If we therefore introduce a strength parameter $g \in [0,1]$ and an auxiliary Green function $G_g(t, t')$ satisfying the differential equation

$$\mathcal{O}_g(t) G_g(t, t') \equiv [-\partial_t^2 - g\Omega^2(t)] G_g(t, t') = \delta(t - t'), \qquad (3.916)$$

we can express the ratio of functional determinants $\mathrm{Det}\, \mathcal{O}_1 / \mathrm{Det}\, \mathcal{O}_0$ as

$$\mathrm{Det}\, (\mathcal{O}_0^{-1} \mathcal{O}_1) = e^{-\int_0^1 dg\, \mathrm{Tr}\, [\Omega^2(t) G_g(t, t')]}. \qquad (3.917)$$

Knowing of the existence of Gelfand-Yaglom's elegant method for calculating functional determinants in Section 2.4, we now try to relate the right-hand side in (3.917) to the solution of the Gelfand-Yaglom's equations (2.213), (2.211), and (2.212):

$$\mathcal{O}_g(t) D_g(t) = 0; \quad D_g(t_a) = 0, \quad \dot{D}_g(t_a) = 1. \qquad (3.918)$$

By differentiating these equations with respect to the parameter g, we obtain for the g-derivative $D'_g(t) \equiv \partial_g D_g(t)$ the inhomogeneous initial-value problem

$$\mathcal{O}_g(t) D'_g(t) = \Omega^2(t) D_g(t); \quad D'_g(t_a) = 0, \quad \dot{D}'_g(t_a) = 0. \qquad (3.919)$$

The unique solution of equations (3.918) can be expressed as in Eq. (2.219) in terms of an arbitrary set of solutions $\eta_g(t)$ and $\xi_g(t)$ as follows

$$D_g(t) = \frac{\xi_g(t_a)\eta_g(t) - \xi_g(t)\eta_g(t_a)}{W_g} = \Delta_g(t, t_a), \qquad (3.920)$$

where W_g is the constant Wronski determinant

$$W_g = \xi_g(t)\dot{\eta}_g(t) - \eta_g(t)\dot{\xi}_g(t). \qquad (3.921)$$

We may also write

$$D_g(t_b) = \frac{\mathrm{Det}\Lambda_g}{W_g} = \Delta_g(t_b, t_a), \qquad (3.922)$$

where Λ_g is the constant 2×2 -matrix

$$\Lambda_g = \begin{pmatrix} \xi_g(t_a) & \eta_g(t_a) \\ \xi_g(t_b) & \eta_g(t_b) \end{pmatrix}. \qquad (3.923)$$

With the help of the solution $\Delta_g(t, t')$ of the homogenous initial-value problem (3.918) we can easily construct a solution of the inhomogeneous initial-value problem (3.919) by superposition:

$$D'_g(t) = \int_{t_a}^{t} dt' \Omega^2(t') \Delta_g(t, t') \Delta_g(t', t_a). \qquad (3.924)$$

Comparison with (3.59) shows that at the final point $t = t_b$

$$D'_g(t_b) = \Delta_g(t_b, t_a) \int_{t_a}^{t_b} dt'\, \Omega^2(t') G_g(t', t'). \qquad (3.925)$$

Together with (3.922), this implies the following equation for the integral over the Green function which solves (3.913) with Dirichlet's boundary conditions:

$$\mathrm{Tr}\left[\Omega^2(t)G_g(t,t')\right] = -\partial_g \log\left(\frac{\det \Lambda_g}{W_g}\right) = -\partial_g \log D_g(t_b). \tag{3.926}$$

Inserting this into (3.915), we find for the ratio of functional determinants the simple formula

$$\mathrm{Det}\,(\mathcal{O}_0^{-1}\mathcal{O}_g) = C(t_b,t_a)D_g(t_b). \tag{3.927}$$

The constant of g-integration, which still depends in general on initial and final times, is fixed by applying (3.927) to the trivial case $g = 0$, where $\mathcal{O}_0 = -\partial_t^2$ and the solution to the initial-value problem (3.918) is

$$D_0(t) = t - t_a. \tag{3.928}$$

At $g = 0$, the left-hand side of (3.927) is unity, determining $C(t_b,t_a) = (t_b - t_a)^{-1}$ and the final result for $g = 1$:

$$\mathrm{Det}\,(\mathcal{O}_0^{-1}\mathcal{O}_1) = \frac{\det \Lambda_1}{W_1}\bigg/\frac{\mathrm{Det}\Lambda_0}{W_0} = \frac{D_1(t_b)}{t_b - t_a}, \tag{3.929}$$

in agreement with the result of Section 2.7.

The same method permits us to find the Green function $G_{\omega^2}(\tau,\tau')$ governing quantum statistical harmonic fluctuations which satisfies the differential equation

$$\mathcal{O}_g(\tau)G_g^{\mathrm{p,a}}(\tau,\tau') \equiv [\partial_\tau^2 - g\Omega^2(\tau)]G_g^{\mathrm{p,a}}(\tau,\tau') = \delta^{\mathrm{p,a}}(\tau - \tau'), \tag{3.930}$$

with periodic and antiperiodic boundary conditions, frequency $\Omega(\tau)$, and δ-function. The imaginary-time analog of (3.915) for the ratio of functional determinants reads

$$\mathrm{Det}\,(\mathcal{O}_0^{-1}\mathcal{O}_1) = e^{-\int_0^1 dg \mathrm{Tr}\,[\Omega^2(\tau)G_g(\tau,\tau')]}. \tag{3.931}$$

The boundary conditions satisfied by the Green function $G_g^{\mathrm{p,a}}(\tau,\tau')$ are

$$\begin{aligned}
G_g^{\mathrm{p,a}}(\tau_b,\tau') &= \pm G_g^{\mathrm{p,a}}(\tau_a,\tau'), \\
\dot{G}_g^{\mathrm{p,a}}(\tau_b,\tau') &= \pm \dot{G}_g^{\mathrm{p,a}}(\tau_a,\tau').
\end{aligned} \tag{3.932}$$

According to Eq. (3.166), the Green functions are given by

$$G_g^{\mathrm{p,a}}(\tau,\tau') = G_g(\tau,\tau') \mp \frac{[\Delta_g(\tau,\tau_a) \pm \Delta_g(\tau_b,\tau)][\Delta_g(\tau',\tau_a) \pm \Delta_g(\tau_b,\tau')]}{\bar{\Delta}_g^{\mathrm{p,a}}(\tau_a,\tau_b)\cdot\Delta_g(\tau_a,\tau_b)}, \tag{3.933}$$

where [compare (3.49)]

$$\Delta(\tau,\tau') = \frac{1}{W}\left[\xi(\tau)\eta(\tau') - \xi(\tau')\eta(\tau)\right], \tag{3.934}$$

with the Wronski determinant $W = \xi(\tau)\dot{\eta}(\tau) - \dot{\xi}(\tau)\eta(\tau)$, and [compare (3.165)]

$$\bar{\Delta}_g^{\text{p,a}}(\tau_a, \tau_b) = 2 \pm \partial_\tau \Delta_g(\tau_a, \tau_b) \pm \partial_\tau \Delta_g(\tau_b, \tau_a). \tag{3.935}$$

The solution is unique provided that

$$\det \bar{\Lambda}_g^{\text{p,a}} = W_g \bar{\Delta}_g^{\text{p,a}}(\tau_a, \tau_b) \neq 0. \tag{3.936}$$

The right-hand side is well-defined unless the operator $\mathcal{O}_g(t)$ has a zero-mode with $\eta_g(t_b) = \pm\eta_g(t_a)$, $\dot{\eta}_g(t_b) = \pm\dot{\eta}_g(t_a)$, which would make the determinant of the 2×2 -matrix $\bar{\Lambda}_g^{\text{p,a}}$ vanish.

We are now in a position to rederive the functional determinant of the operator $\mathcal{O}(\tau) = \partial_\tau^2 - \Omega^2(\tau)$ with periodic or antiperiodic boundary conditions more elegantly than in Section 2.11. For this we formulate again a homogeneous initial-value problem, but with boundary conditions dual to Gelfand and Yaglom's in Eq. (3.918):

$$\mathcal{O}_g(\tau)\bar{D}_g(\tau) = 0; \quad \bar{D}_g(\tau_a) = 1, \quad \dot{\bar{D}}_g(\tau_a) = 0. \tag{3.937}$$

In terms of the previous arbitrary set $\eta_g(t)$ and $\xi_g(t)$ of solutions of the homogeneous differential equation, the unique solution of (3.937) reads

$$\bar{D}_g(\tau) = \frac{\xi_g(\tau)\dot{\eta}_g(\tau_a) - \dot{\xi}_g(\tau_a)\eta_g(\tau)}{W_g}. \tag{3.938}$$

This can be combined with the time derivative of (3.920) at $\tau = \tau_b$ to yield

$$\dot{D}_g(\tau_b) + \bar{D}_g(\tau_b) = \pm[2 - \bar{\Delta}_g^{\text{p,a}}(\tau_a, \tau_b)]. \tag{3.939}$$

By differentiating Eqs. (3.937) with respect to g, we obtain the following inhomogeneous initial-value problem for $\bar{D}'_g(\tau) = \partial_g \bar{D}_g(\tau)$:

$$\mathcal{O}_g(\tau)\bar{D}'_g(\tau) = \Omega^2(\tau)\bar{D}'_g(\tau); \quad \bar{D}'_g(\tau_a) = 1, \quad \dot{\bar{D}}'_g(\tau_a) = 0, \tag{3.940}$$

whose general solution reads by analogy with (3.924)

$$\bar{D}'_g(\tau) = -\int_{\tau_a}^{\tau} d\tau' \, \Omega^2(\tau')\Delta_g(\tau, \tau')\dot{\Delta}_g(\tau_a, \tau'), \tag{3.941}$$

where the dot on $\dot{\Delta}_g(\tau_a, \tau')$ acts on the first imaginary-time argument. With the help of identities (3.939) and (3.940), the combination $\dot{D}'(\tau) + \bar{D}'_g(\tau)$ at $\tau = \tau_b$ can now be expressed in terms of the periodic and antiperiodic Green functions (3.166), by analogy with (3.925),

$$\dot{D}'_g(\tau_b) + \bar{D}'_g(\tau_b) = \pm\bar{\Delta}_g^{\text{p,a}}(\tau_a, \tau_b)\int_{\tau_a}^{\tau_b} d\tau \, \Omega^2(\tau)G_g^{\text{p,a}}(\tau, \tau). \tag{3.942}$$

Together with (3.939), this gives for the temporal integral on the right-hand side of (3.917) the simple expression analogous to (3.926)

$$
\begin{aligned}
\mathrm{Tr}\,[\Omega^2(\tau)G_g^{\mathrm{p,a}}(\tau,\tau')] &= -\partial_g \log\left(\frac{\det \bar{\Lambda}_g^{\mathrm{p,a}}}{Wg}\right) \\
&= -\partial_g \log\left[2 \mp \dot{D}_g(\tau_b) \mp \bar{D}_g(\tau_b)\right],
\end{aligned}
\tag{3.943}
$$

so that we obtain the ratio of functional determinants with periodic and antiperiodic boundary conditions

$$
\mathrm{Det}\,(\tilde{\mathcal{O}}^{-1}\mathcal{O}_g) = C(t_b,t_a)\left[2 \mp \dot{D}_g(\tau_b) \mp \bar{D}_g(\tau_b)\right],
\tag{3.944}
$$

where $\tilde{\mathcal{O}} = \mathcal{O}_0 - \omega^2 = \partial_\tau^2 - \omega^2$. The constant of integration $C(t_b,t_a)$ is fixed in the way described after Eq. (3.915). We go to $g = 1$ and set $\Omega^2(\tau) \equiv \omega^2$. For the operator $\mathcal{O}_1^\omega \equiv -\partial_\tau^2 - \omega^2$, we can easily solve the Gelfand-Yaglom initial-value problem (3.918) as well as the dual one (3.937) by

$$
D_1^\omega(\tau) = \frac{1}{\omega}\sin\omega(\tau - \tau_a), \quad \bar{D}_1^\omega(\tau) = \cos\omega(\tau - \tau_a),
\tag{3.945}
$$

so that (3.944) determines $C(t_b,t_a)$ by

$$
1 = C(t_b,t_a)\begin{cases} 4\sin^2[\omega(\tau_b - \tau_a)/2] & \text{periodic case,} \\ 4\cos^2[\omega(\tau_b - \tau_a)/2] & \text{antiperiodic case.} \end{cases}
\tag{3.946}
$$

Hence we find the final results for periodic boundary conditions

$$
\mathrm{Det}\,(\tilde{\mathcal{O}}^{-1}\mathcal{O}_1) = \frac{\det \bar{\Lambda}_1^{\mathrm{P}}}{W_1}\bigg/\frac{\mathrm{Det}\bar{\Lambda}_1^{\omega\,\mathrm{P}}}{W_1^\omega} = \frac{2 - \dot{D}_1(\tau_b) - \bar{D}_1(\tau_b)}{4\sin^2[\omega(\tau_b - \tau_a)/2]},
\tag{3.947}
$$

and for antiperiodic boundary conditions

$$
\mathrm{Det}\,(\tilde{\mathcal{O}}^{-1}\mathcal{O}_1) = \frac{\det \bar{\Lambda}_1^{\mathrm{a}}}{W_1}\bigg/\frac{\mathrm{Det}\bar{\Lambda}_1^{\omega\,\mathrm{a}}}{W_1^\omega} = \frac{2 + \dot{D}_1(\tau_b) + \bar{D}_1(\tau_b)}{4\cos^2[\omega(\tau_b - \tau_a)/2]}.
\tag{3.948}
$$

The intermediate expressions in (3.929), (3.947), and (3.948) show that the ratios of functional determinants are ordinary determinants of two arbitrary independent solutions ξ and η of the homogeneous differential equation $\mathcal{O}_1(t)y(t) = 0$ or $\mathcal{O}_1(\tau)y(\tau) = 0$. As such, the results are manifestly invariant under arbitrary linear transformations of these functions $(\xi, \eta) \to (\xi', \eta')$.

It is useful to express the above formulas for the ratio of functional determinants (3.929), (3.947), and (3.948) in yet another form. We rewrite the two independent solutions of the homogenous differential equation $[-\partial_t^2 - \Omega^2(t)]y(t) = 0$ as follows

$$
\xi(t) = q(t)\cos\phi(t), \quad \eta(t) = q(t)\sin\phi(t).
\tag{3.949}
$$

The two functions $q(t)$ and $\phi(t)$ parametrizing $\xi(t)$ and $\eta(t)$ satisfy the constraint

$$\dot{\phi}(t)q^2(t) = W, \tag{3.950}$$

where W is the constant Wronski determinant. The function $q(t)$ is a soliton of the Ermankov-Pinney equation[23]

$$\ddot{q} + \Omega^2(t)q - W^2 q^{-3} = 0. \tag{3.951}$$

For Dirichlet boundary conditions we insert (3.949) into (3.929), and obtain the ratio of fluctuation determinants in the form

$$\mathrm{Det}\,(\mathcal{O}_0^{-1}\mathcal{O}_1) = \frac{1}{W}\frac{q(t_a)q(t_b)\sin[\phi(t_b) - \phi(t_a)]}{t_b - t_a}. \tag{3.952}$$

For periodic or antiperiodic boundary conditions with a corresponding frequency $\Omega(t)$, the functions $q(t)$ and $\phi(t)$ in Eq. (3.949) have the same periodicity. The initial value $\phi(t_a)$ may always be assumed to vanish, since otherwise $\xi(t)$ and $\eta(t)$ could be combined linearly to that effect. Substituting (3.949) into (3.947) and (3.948), the function $q(t)$ drops out, and we obtain the ratios of functional determinants for periodic boundary conditions

$$\mathrm{Det}\,(\tilde{\mathcal{O}}^{-1}\mathcal{O}_1) = 4\sin^2\frac{\phi(t_b)}{2}\Big/4\sin^2\frac{\omega(t_b - t_a)}{2}, \tag{3.953}$$

and for antiperiodic boundary conditions

$$\mathrm{Det}\,(\tilde{\mathcal{O}}^{-1}\mathcal{O}_1) = 4\cos^2\frac{\phi(t_b)}{2}\Big/4\cos^2\frac{\omega(t_b - t_a)}{2}. \tag{3.954}$$

For a harmonic oscillator with $\Omega(t) \equiv \omega$, Eq. (3.951) is solved by

$$q(t) \equiv \sqrt{\frac{W}{\omega}}, \tag{3.955}$$

and Eq. (3.950) yields

$$\phi(t) = \omega(t - t_a). \tag{3.956}$$

Inserted into (3.952), (3.953), and (3.954) we reproduce the known results:

$$\mathrm{Det}\,(\mathcal{O}_0^{-1}\mathcal{O}_1) = \frac{\sin\omega(t_b - t_a)}{\omega(t_b - t_a)}, \qquad \mathrm{Det}\,(\tilde{\mathcal{O}}^{-1}\mathcal{O}_1) = 1.$$

---

[23]For more details see J. Rezende, J. Math. Phys. *25*, 3264 (1984).

Appendix 3A   Matrix Elements for General Potential

The matrix elements $\langle n|\hat{V}|m\rangle$ can be calculated for an arbitrary potential $\hat{V} = V(\hat{x})$ as follows: We represent $V(\hat{x})$ by a Fourier integral as a superposition of exponentials

$$V(\hat{x}) = \int_{-i\infty}^{i\infty} \frac{dk}{2\pi i} V(k) \exp(k\hat{x}), \tag{3A.1}$$

and express $\exp(k\hat{x})$ in terms of creation and annihilation operators as $\exp(k\hat{x}) = \exp[k(\hat{a}+\hat{a}^\dagger)/\sqrt{2}]$, set $k \equiv \sqrt{2}\epsilon$, and write down the obvious equation

$$\langle n|e^{\epsilon\sqrt{2}\hat{x}}|m\rangle = \frac{1}{\sqrt{n!m!}} \frac{\partial^n}{\partial\alpha^n} \frac{\partial^m}{\partial\beta^m} \langle 0|e^{\alpha\hat{a}}e^{\epsilon(\hat{a}+\hat{a}^\dagger)}e^{\beta\hat{a}^\dagger}|0\rangle\bigg|_{\alpha=\beta=0}. \tag{3A.2}$$

We now make use of the Baker-Campbell-Hausdorff Formula (2A.1) with (2A.6), and rewrite

$$e^{\hat{A}}e^{\hat{B}} = e^{\hat{A}+\hat{B}+\frac{1}{2}[\hat{A},\hat{B}]+\frac{1}{12}([\hat{A},[\hat{A},\hat{B}]]+[\hat{B},[\hat{B},\hat{A}]])+\cdots}. \tag{3A.3}$$

Identifying $\hat{A}$ and $\hat{B}$ with $\hat{a}$ and $\hat{a}^\dagger$, the property $[\hat{a}, \hat{a}^\dagger] = 1$ makes this relation very simple:

$$e^{\epsilon(\hat{a}+\hat{a}^\dagger)} = e^{\epsilon\hat{a}}e^{\epsilon\hat{a}^\dagger}e^{-\epsilon^2/2}, \tag{3A.4}$$

and the matrix elements (3A.2) become

$$\langle 0|e^{\alpha\hat{a}}e^{\epsilon(\hat{a}+\hat{a}^\dagger)}e^{\beta\hat{a}^\dagger}|0\rangle = \langle 0|e^{(\alpha+\epsilon)\hat{a}}e^{(\beta+\epsilon)\hat{a}^\dagger}|0\rangle e^{-\epsilon^2/2}. \tag{3A.5}$$

The bra and ket states on the right-hand side are now eigenstates of the annihilation operator $\hat{a}$ with eigenvalues $\alpha + \epsilon$ and $\beta + \epsilon$, respectively. Such states are known as *coherent states*. Using once more (3A.3), we obtain

$$\langle 0|e^{(\alpha+\epsilon)\hat{a}}e^{(\beta+\epsilon)\hat{a}^\dagger}|0\rangle = e^{(\epsilon+\alpha)(\epsilon+\beta)}, \tag{3A.6}$$

and (3A.2) becomes simply

$$\langle n|e^{\epsilon\sqrt{2}\hat{x}}|m\rangle = \frac{1}{\sqrt{n!m!}} \frac{\partial^n}{\partial\alpha^n} \frac{\partial^m}{\partial\beta^m} e^{(\alpha+\epsilon)(\beta+\epsilon)}e^{-\epsilon^2/2}\bigg|_{\alpha=\beta=0}. \tag{3A.7}$$

We now calculate the derivatives

$$\frac{\partial^n}{\partial\alpha^n} \frac{\partial^m}{\partial\beta^m} e^{(\epsilon+\alpha)(\epsilon+\beta)}\bigg|_{\alpha=\beta=0} = \frac{\partial^n}{\partial\alpha^n} (\epsilon+\alpha)^m e^{\epsilon(\epsilon+\alpha)}\bigg|_{\alpha=0}. \tag{3A.8}$$

Using the chain rule of differentiation for products $f(x) = g(x)\,h(x)$:

$$f^{(n)}(x) = \sum_{l=0}^{n} \binom{n}{l} g^{(l)}(x)h^{(n-l)}(x), \tag{3A.9}$$

the right-hand side becomes

$$\frac{\partial^n}{\partial\alpha^n} (\epsilon+\alpha)^m e^{\epsilon(\epsilon+\alpha)}\bigg|_{\alpha=0} = \sum_{l=0}^{n} \binom{n}{l} \frac{\partial^l}{\partial\alpha^l}(\epsilon+\alpha)^m \frac{\partial^{n-l}}{\partial\alpha^{n-l}} e^{\epsilon(\epsilon+\alpha)}\bigg|_{\alpha=0}$$

$$= \sum_{l=0}^{n} \binom{n}{l} m(m-1)\cdots(m-l+1)\epsilon^{n+m-2l}e^{\epsilon^2}. \tag{3A.10}$$

Hence we find

$$\langle n|e^{\epsilon\sqrt{2}\hat{x}}|m\rangle = \frac{1}{\sqrt{n!m!}}\sum_{l=0}^{n}\binom{n}{l}\binom{m}{l}l!\,\epsilon^{n+m-2l}e^{\epsilon^2/2}. \tag{3A.11}$$

From this we obtain the matrix elements of single powers $\hat{x}^p$ by forming, with the help of (3A.9) and $(\partial^q/\partial\epsilon^q)e^{\epsilon^2/2}|_{\epsilon=0} = q!!$, the derivatives

$$\frac{\partial^p}{\partial\epsilon^p}\epsilon^{n+m-2l}e^{\epsilon^2/2}\bigg|_{\epsilon=0} = \binom{p}{n+m-2l}[2l-(n+m-p)]!! = \frac{p!}{2^{l-(n+m-p)/2}[l-p-(n+m-p)/2]!}. \tag{3A.12}$$

The result is

$$\langle n|\hat{x}^p|m\rangle = \frac{1}{\sqrt{n!m!}}\sum_{l=(n+m-p)/2}^{\min(n,m)}\binom{n}{l}\binom{m}{l}l!\,\frac{p!}{2^{l+p-(n+m)/2}[l-(n+m-p)/2]!}. \tag{3A.13}$$

For the special case of a pure fourth-order interaction, this becomes

$$\begin{aligned}
\langle n|\hat{x}^4|n-4\rangle &= \sqrt{n-3}\sqrt{n-2}\sqrt{n-1}\sqrt{n},\\
\langle n|\hat{x}^4|n-2\rangle &= (4n-2)\sqrt{n-1}\sqrt{n},\\
\langle n|\hat{x}^4|n\rangle &= 6n^2+6n+3,\\
\langle n|\hat{x}^4|n+2\rangle &= (4n+6)\sqrt{n+1}\sqrt{n+2},\\
\langle n|\hat{x}^4|n+4\rangle &= \sqrt{n+1}\sqrt{n+2}\sqrt{n+3}\sqrt{n+4}.
\end{aligned} \tag{3A.14}$$

For a general potential (3A.1) we find

$$\langle n|V(\hat{x})|m\rangle = \frac{1}{\sqrt{n!m!}}\sum_{l=0}^{n}\binom{n}{l}\binom{m}{l}l!\,\frac{1}{2^{l-(n+m)/2}}\int_{-i\infty}^{i\infty}\frac{dk}{2\pi i}V(k)k^{n+m-2l}e^{k^2/4}. \tag{3A.15}$$

Appendix 3B Energy Shifts for $gx^4/4$-Interaction

For the specific polynomial interaction $V(x) = gx^4/4$, the shift of the energy $E^{(n)}$ to any desired order is calculated most simply as follows. Consider the expectations of powers $\hat{x}^4(z_1)\hat{x}^4(z_2)\cdots\hat{x}^4(z_n)$ of the operator $\hat{x}(z) = (\hat{a}^\dagger z + \hat{a}z^{-1})$ between the excited oscillator states $\langle n|$ and $|n\rangle$. Here $\hat{a}$ and $\hat{a}^\dagger$ are the usual creation and annihilation operators of the harmonic oscillator, and $|n\rangle = (a^\dagger)^n|0\rangle/\sqrt{n!}$. To evaluate these expectations, we make repeated use of the commutation rules $[\hat{a},\hat{a}^\dagger] = 1$ and of the ground state property $\hat{a}|0\rangle = 0$. For $n = 0$ this gives

$$\begin{aligned}
\langle x^4(z)\rangle_\omega &= 3,\\
\langle x^4(z_1)x^4(z_2)\rangle_\omega &= 72z_1^{-2}z_2^2 + 24z_1^{-4}z_2^4 + 9,\\
\langle x^4(z_1)x^4(z_2)x^4(z_3)\rangle_\omega &= 27\cdot 8z_1^{-2}z_2^2 + 63\cdot 32z_1^{-2}z_2^{-2}z_3^4\\
&\quad +351\cdot 8z_1^{-2}z_3^4 + 9\cdot 8z_1^{-4}z_2^4 + 63\cdot 32z_1^{-4}z_2^2z_3^2 + 369\cdot 8z_1^{-4}z_3^4\\
&\quad +27\cdot 8z_2^{-2}z_3^2 + 9\cdot 8z_2^{-4}z_3^4 + 27.
\end{aligned} \tag{3B.1}$$

The cumulants are

$$\begin{aligned}
\langle x^4(z_1)x^4(z_2)\rangle_{\omega,c} &= 72z_1^{-2}z_2^2 + 24z_1^{-4}z_2^4,\\
\langle x^4(z_1)x^4(z_2)x^4(z_3)\rangle_{\omega,c} &= 288(7z_1^{-2}z_2^{-2}z_3^4 + 9z_1^{-2}z_3^4 + 7z_1^{-4}z_2^2z_3^2 + 10z_1^{-4}z_3^4).
\end{aligned} \tag{3B.2}$$

The powers of z show by how many steps the intermediate states have been excited. They determine the energy denominators in the formulas (3.515) and (3.516). Apart from a factor $(g/4)^n$ and a factor $1/(2\omega)^{2n}$ which carries the correct length scale of $x(z)$, the energy shifts $\Delta E = \Delta_1 E_0 + \Delta_2 E_0 + \Delta_3 E_0$ are thus found to be given by

$$\Delta_1 E_0 = 3,$$

$$\Delta_2 E_0 = -\left(72 \cdot \frac{1}{2} + 24 \cdot \frac{1}{4}\right), \tag{3B.3}$$

$$\Delta_3 E_0 = 288\left(7 \cdot \frac{1}{2} \cdot \frac{1}{4} + 9 \cdot \frac{1}{2} \cdot \frac{1}{2} + 7 \cdot \frac{1}{4} \cdot \frac{1}{2} + 10 \cdot \frac{1}{4} \cdot \frac{1}{4}\right) = 333 \cdot 4.$$

Between excited states, the calculation is somewhat more tedious and yields

$$\langle x^4(z)\rangle_\omega = 6n^2 + 6n + 3, \tag{3B.4}$$

$$\begin{aligned}
\langle x^4(z_1)x^4(z_2)\rangle_{\omega,c} &= (16n^4 + 96n^3 + 212n^2 + 204n + 72)z_1^{-2}z_2^2 \\
&+ (n^4 + 10n^3 + 35n^2 + 50n + 24)z_1^{-4}z_2^4 \\
&+ (n^4 - 6n^3 + 11n^2 - 6n)z_1^4 z_2^{-4} \\
&+ (16n^4 - 32n^3 + 20n^2 - 4n)z_1^2 z_2^{-2}, \tag{3B.5}
\end{aligned}$$

$$\begin{aligned}
\langle x^4(z_1)x^4(z_2)x^4(z_3)\rangle_{\omega,c} &= [(16n^6 + 240n^5 + 1444n^4 + 4440n^3 + 7324n^2 + 6120n + 2016) \\
&\qquad \times (z_1^{-2}z_2^{-2}z_3^4 + z_1^{-4}z_2^2 z_3^2) \\
&+ (384n^5 + 2880n^4 + 8544n^3 + 12528n^2 + 9072n + 2592)z_1^{-2}z_3^2 \\
&+ (48n^5 + 600n^4 + 2880n^3 + 6600n^2 + 7152n + 2880)z_1^{-4}z_3^4 \\
&+ (16n^6 - 144n^5 + 484n^4 - 744n^3 + 508n^2 - 120n)z_1^4 z_2^{-2} z_3^{-2} \\
&+ (-48n^5 + 360n^4 - 960n^3 + 1080n^2 - 432n)z_1^4 z_3^{-4} \\
&+ (16n^6 + 48n^5 + 4n^4 - 72n^3 - 20n^2 + 24n)z_1^2 z_2^{-4} z_3^2 \\
&+ (-384n^5 + 960n^4 - 864n^3 + 336n^2 - 48n)z_1^2 z_3^{-2} \\
&+ (16n^6 - 144n^5 + 484n^4 - 744n^3 + 508n^2 - 120n)z_1^2 z_2^2 z_3^{-4} \\
&+ (16n^6 + 48n^5 + 4n^4 - 72n^3 - 20n^2 + 24n)z_1^{-2}z_2^4 z_3^{-2}]. \tag{3B.6}
\end{aligned}$$

From these we obtain the reduced energy shifts:

$$\Delta_1 E_0 = 6n^2 + 6n + 3, \tag{3B.7}$$

$$\begin{aligned}
\Delta_2 E_0 &= -(16n^4 + 96n^3 + 212n^2 + 204n + 72) \cdot \tfrac{1}{2} \\
&\quad -(n^4 + 10n^3 + 35n^2 + 50n + 24) \cdot \tfrac{1}{4} \\
&\quad -(n^4 - 6n^3 + 11n^2 - 6n) \cdot \tfrac{-1}{4} \\
&\quad -(16n^4 - 32n^3 + 20n^2 - 4n) \cdot \tfrac{-1}{2} \\
&= 2 \cdot (34n^3 + 51n^2 + 59n + 21), \tag{3B.8}
\end{aligned}$$

$$\begin{aligned}
\Delta_3 E_0 &= [(16n^6 + 240n^5 + 1444n^4 + 4440n^3 + 7324n^2 + 6120n + 2016) \cdot (\tfrac{1}{2} \cdot \tfrac{1}{4} + \tfrac{1}{4} \cdot \tfrac{1}{2}) \\
&\quad +(384n^5 + 2880n^4 + 8544n^3 + 12528n^2 + 9072n + 2592) \cdot \tfrac{1}{2} \cdot \tfrac{1}{2} \\
&\quad +(48n^5 + 600n^4 + 2880n^3 + 6600n^2 + 7152n + 2880) \cdot \tfrac{1}{4} \cdot \tfrac{1}{4} \\
&\quad +(16n^6 - 144n^5 + 484n^4 - 744n^3 + 508n^2 - 120n) \cdot \tfrac{1}{4} \cdot \tfrac{1}{2} \\
&\quad +(-48n^5 + 360n^4 - 960n^3 + 1080n^2 - 432n) \cdot \tfrac{1}{4} \cdot \tfrac{1}{4} \\
&\quad +(16n^6 + 48n^5 + 4n^4 - 72n^3 - 20n^2 + 24n) \cdot \tfrac{1}{2} \cdot \tfrac{1}{4} \\
&\quad +(-384n^5 + 960n^4 - 864n^3 + 336n^2 - 48n) \cdot \tfrac{1}{2} \cdot \tfrac{1}{2} \\
&\quad +(16n^6 - 144n^5 + 484n^4 - 744n^3 + 508n^2 - 120n) \cdot \tfrac{1}{2} \cdot \tfrac{1}{4} \\
&\quad +(16n^6 + 48n^5 + 4n^4 - 72n^3 - 20n^2 + 24n) \cdot \tfrac{1}{2} \cdot \tfrac{-1}{2}] \\
&= 4 \cdot 3 \cdot (125n^4 + 250n^3 + 472n^2 + 347n + 111). \tag{3B.9}
\end{aligned}$$

Appendix 3C   Recursion Relations for Perturbation Coefficients of Anharmonic Oscillator

Bender and Wu[24] were the first to solve to high orders recursion relations for the perturbation coefficients of the ground state energy of an anharmonic oscillator with a potential $x^2/2 + gx^4/4$. Their relations are similar to Eqs. (3.534), (3.535), and (3.536), but not the same. Extending their method, we derive here a recursion relation for the perturbation coefficients of all energy levels of the anharmonic oscillator in any number of dimensions D, where the radial potential is $l(l + D - 2)/2r^2 + r^2/2 + (g/2)(a_4 r^4 + a_6 r^6 + \ldots + a_{2q} x^{2q})$, where the first term is the centrifugal barrier of angular momentum l in D dimensions. We shall do this in several steps.

3C.1   One-Dimensional Interaction x^4

In natural physical units with $\hbar = 1, \omega = 1, M = 1$, the Schrödinger equation to be solved reads

$$\left(-\frac{1}{2}\frac{d^2}{dx^2} + \frac{1}{2}x^2 + gx^4 \right) \psi^{(n)}(x) = E^{(n)}\psi^{(n)}(x). \tag{3C.1}$$

At $g = 0$, this is solved by the harmonic oscillator wave functions

$$\psi^{(n)}(x, g = 0) = N^n e^{-x^2/2} H_n(x), \tag{3C.2}$$

with proper normalization constant N^n, where $H_n(x)$ are the Hermite polynomial of nth degree

$$H_n(x) = \sum_{p=0}^{n} h_n^p x^p. \tag{3C.3}$$

Generalizing this to the anharmonic case, we solve the Schrödinger equation (3C.1) with the power series ansatz

$$\psi^{(n)}(x) = e^{-x^2/2} \sum_{k=0}^{\infty} (-g)^k \, \Phi_k^{(n)}(x), \tag{3C.4}$$

$$E^{(n)} = \sum_{k=0}^{\infty} g^k E_k^{(n)}. \tag{3C.5}$$

To make room for derivative symbols, the superscript of $\Phi_k^{(n)}(x)$ is now dropped. Inserting (3C.4) and (3C.5) into (3C.1) and equating the coefficients of equal powers of g, we obtain the equations

$$x\Phi_k'(x) - n\Phi_k(x) = \frac{1}{2}\Phi_k''(x) - x^4\Phi_{k-1}(x) + \sum_{k'=1}^{k}(-1)^{k'} E_{k'}^{(n)}\Phi_{k-k'}(x), \tag{3C.6}$$

where we have inserted the unperturbed energy

$$E_0^{(n)} = n + 1/2, \tag{3C.7}$$

and defined $\Phi_k(x) \equiv 0$ for $k < 0$. The functions $\Phi_k(x)$ are anharmonic versions of the Hermite polynomials. They turn out to be polynomials of $(4k + n)$th degree:

$$\Phi_k(x) = \sum_{p=0}^{4k+n} A_k^p x^p. \tag{3C.8}$$

---

[24]C.M. Bender and T.T. Wu, Phys. Rev. *184*, 1231 (1969); Phys. Rev. D *7*, 1620 (1973).

In a more explicit notation, the expansion coefficients A_k^p would of course carry the dropped superscript of $\Phi_k^{(n)}$. All higher coefficients vanish:

$$A_k^p \equiv 0 \quad \text{for} \quad p \geq 4k + n + 1. \tag{3C.9}$$

From the harmonic wave functions (3C.2),

$$\Phi_0(x) = N^n H_n(x) = N^n \sum_{p=0}^{n} h_n^p x^p, \tag{3C.10}$$

we see that the recursion starts with

$$A_0^p = h_n^p N^n. \tag{3C.11}$$

For levels with an even principal quantum number n, the functions $\Phi_k(x)$ are symmetric. It is convenient to choose the normalization $\psi^{(n)}(0) = 1$, such that $N^n = 1/h_n^0$ and

$$A_k^0 = \delta_{0k}. \tag{3C.12}$$

For odd values of n, the wave functions $\Phi_k(x)$ are antisymmetric. Here we choose the normalization $\psi^{(n)\prime}(0) = 3$, so that $N^n = 3/h_n^1$ and

$$A_k^1 = 3\delta_{0k}. \tag{3C.13}$$

Defining

$$A_k^p \equiv 0 \quad \text{for} \quad p < 0 \quad \text{or} \quad k < 0, \tag{3C.14}$$

we find from (3C.6), by comparing coefficients of x^p,

$$(p - n)A_k^p = \frac{1}{2}(p + 2)(p + 1)A_k^{p+2} + A_{k-1}^{p-4} + \sum_{k'=1}^{k} (-1)^{k'} E_{k'}^{(n)} A_{k-k'}^p. \tag{3C.15}$$

The last term on the right-hand side arises after exchanging the order of summation as follows:

$$\sum_{k'=1}^{k} (-1)^{k'} E_{k'}^{(n)} \sum_{p=0}^{4(k-k')+n} A_{k-k'}^p x^p = \sum_{p=0}^{4k+n} x^p \sum_{k'=1}^{k} (-1)^{k'} E_{k'}^{(n)} A_{k-k'}^p. \tag{3C.16}$$

For even n, Eq. (3C.15) with $p = 0$ and $k > 0$ yields [using (3C.14) and (3C.12)] the desired expansion coefficients of the energies

$$E_k^{(n)} = -(-1)^k A_k^2. \tag{3C.17}$$

For odd n, we take Eq. (3C.15) with $p = 1$ and odd $k > 0$ and find [using (3C.13) and (3C.14)] the expansion coefficients of the energies:

$$E_k^{(n)} = -(-1)^k A_k^3. \tag{3C.18}$$

For even n, the recursion relations (3C.15) obviously relate only coefficients carrying even indices with each other. It is therefore useful to set

$$n = 2n' , \quad p = 2p' , \quad A_k^{2p'} = C_k^{p'} , \tag{3C.19}$$

leading to

$$2(p' - n')C_k^{p'} = (2p' + 1)(p' + 1)C_k^{p'+1} + C_{k-1}^{p'-2} - \sum_{k'=1}^{k} C_{k'}^1 C_{k-k'}^{p'}. \tag{3C.20}$$

For odd n, the substitution

$$n = 2n' + 1 \ , \quad p = 2p' + 1 \ , \quad A_k^{2p'+1} = C_k^{p'} \ , \tag{3C.21}$$

leads to

$$2(p' - n')C_k^{p'} = (2p' + 3)(p' + 1)C_k^{p'+1} + C_{k-1}^{p'-2} - \sum_{k'=1}^{k} C_{k'}^1 C_{k-k'}^{p'}. \tag{3C.22}$$

The rewritten recursion relations (3C.20) and (3C.22) are the same for even and odd n, except for the prefactor of the coefficient $C_k^{p'+1}$. The common initial values are

$$C_0^{p'} = \begin{cases} h_n^{2p'}/h_n^0 & \text{for} \quad 0 \le p' \le n', \\ 0 & \text{otherwise.} \end{cases} \tag{3C.23}$$

The energy expansion coefficients are given in either case by

$$E_k^{(n)} = -(-1)^k C_k^1. \tag{3C.24}$$

The solution of the recursion relations proceeds in three steps as follows. Suppose we have calculated for some value of k all coefficients $C_{k-1}^{p'}$ for an upper index in the range $1 \le p' \le 2(k-1) + n'$.

In a first step, we find $C_k^{p'}$ for $1 \le p' \le 2k + n'$ by solving Eq. (3C.20) or (3C.22), starting with $p' = 2k + n'$ and lowering p' down to $p' = n' + 1$. Note that the knowledge of the coefficients C_k^1 (which determine the yet unknown energies and are contained in the last term of the recursion relations) is not required for $p' > n'$, since they are accompanied by factors $C_0^{p'}$ which vanish due to (3C.23).

Next we use the recursion relation with $p' = n'$ to find equations for the coefficients C_k^1 contained in the last term. The result is, for even k,

$$C_k^1 = \left[(2n' + 1)(n' + 1)C_k^{n'+1} + C_{k-1}^{n'-2} - \sum_{k'=1}^{k-1} C_{k'}^1 C_{k-k'}^{n'} \right] \frac{1}{C_0^{n'}} \ . \tag{3C.25}$$

For odd k, the factor $(2n' + 1)$ is replaced by $(2n' + 3)$. These equations contain once more the coefficients $C_k^{n'}$.

Finally, we take the recursion relations for $p' < n'$, and relate the coefficients $C_k^{n'-1}, \ldots, C_k^1$ to $C_k^{n'}$. Combining the results we determine from Eq. (3C.24) all expansion coefficients $E_k^{(n)}$.

The relations can easily be extended to interactions which are an arbitrary linear combination

$$V(x) = \sum_{n=2}^{\infty} a_{2n} \epsilon^n x^{2n}. \tag{3C.26}$$

A short Mathematica program solving the relations can be downloaded from the internet.[25]

The expansion coefficients have the remarkable property of growing, for large order k, like

$$E_k^{(n)} \longrightarrow -\frac{1}{\pi} \sqrt{\frac{6}{\pi}} \frac{12}{n!} (-3)^k \Gamma(k + n + 1/2). \tag{3C.27}$$

This will be shown in Eq. (17.323). Such a factorial growth implies the perturbation expansion to have a zero radius of convergence. The reason for this will be explained in Section 17.10. At the expansion point $g = 0$, the energies possess an essential singularity. In order to extract meaningful numbers from a Taylor series expansion around such a singularity, it will be necessary to find a convergent resummation method. This will be provided by the variational perturbation theory to be developed in Section 5.14.

---

[25]See http://www.physik.fu-berlin/~kleinert/b3/programs.

3C.2   General One-Dimensional Interaction

Consider now an arbirary interaction which is expandable in a power series

$$v(x) = \sum_{k=1}^{\infty} g^k v_{k+2} x^{k+2}. \tag{3C.28}$$

Note that the coupling constant corresponds now to the square root of the previous one, the lowest interaction terms being $g v_3 x^3 + g^2 v_4 x^4 + \dots$. The powers of g count the number of loops of the associated Feynman diagrams. Then Eqs. (3C.6) and (3C.15) become

$$x\Phi'_k(x) - n\Phi_k(x) = \frac{1}{2}\Phi''_k(x) - \sum_{k'=1}^{k} (-1)^{k'} v_{k'+2} x^{k'+2} \Phi_{k-k'} + \sum_{k'=1}^{k} (-1)^{k'} E_{k'}^{(n)} \Phi_{k-k'}(x), \tag{3C.29}$$

and

$$(p-n)A_k^p = \frac{1}{2}(p+2)(p+1)A_k^{p+2} - \sum_{k'=1}^{k} (-1)^{k'} v_{k'+2} A_{k-k'}^{p-j-2} + \sum_{k'=1}^{k} (-1)^{k'} E_{k'}^{(n)} A_{k-k'}^p. \tag{3C.30}$$

The expansion coefficients of the energies are, as before, given by (3C.17) and (3C.18) for even and odd n, respectively, but the recursion relation (3C.30) has to be solved now in full.

3C.3   Cumulative Treatment of Interactions x^4 and x^3

There exists a slightly different recursive treatment which we shall illustrate for the simplest mixed interaction potential

$$V(x) = \frac{M}{2}\omega^2 x^2 + g v_3 x^3 + g^2 v_4 x^4. \tag{3C.31}$$

Instead of the ansatz (3C.4) we shall now factorize the wave function of the ground state as follows:

$$\psi^{(n)}(x) = \left(\frac{M\omega}{\pi\hbar}\right)^{1/4} \exp\left[-\frac{M\omega}{2\hbar}x^2 + \phi^{(n)}(x)\right], \tag{3C.32}$$

i.e., we allow for powers series expansion in the exponent:

$$\phi^{(n)}(x) = \sum_{k=1}^{\infty} g^k \phi_k^{(n)}(x). \tag{3C.33}$$

We shall find that this expansion contains fewer terms than in the Bender-Wu expansion of the correction factor in Eq. (3C.4). For completeness, we keep here physical dimensions with explicit constants $\hbar, \omega, M$.

Inserting (3C.32) into the Schrödinger equation

$$\left[-\frac{\hbar^2}{2M}\frac{d^2}{dx^2} + \left(\frac{M}{2}\omega^2 x^2 + g v_3 x^3 + g^2 v_4 x^4\right) - E^{(n)}\right]\psi^{(n)}(x) = 0, \tag{3C.34}$$

we obtain, after dropping everywhere the superscript (n), the differential equation for $\phi^{(n)}(x)$:

$$-\frac{\hbar^2}{2M}\phi''(x) + \hbar\omega x\,\phi'(x) - \frac{\hbar^2}{2M}\left[\phi'(x)\right]^2 + g v_3 x^3 + g^2 v_4 x^4 = n\hbar\omega + \epsilon, \tag{3C.35}$$

where ϵ denotes the correction to the harmonic energy

$$E = \hbar\omega\left(n + \frac{1}{2}\right) + \epsilon. \tag{3C.36}$$

We shall calculate ϵ as a power series in g:

$$\epsilon = \sum_{k=1}^{\infty} g^k \, \epsilon_k \, . \tag{3C.37}$$

From now on we shall consuder only the ground state with $n = 0$. Inserting expansion (3C.33) into (3C.35), and comparing coefficients, we obtain the infinite set of differential equations for $\phi_k(x)$:

$$-\frac{\hbar^2}{2M}\phi_k''(x) + \hbar\omega\, x\, \phi_k'(x) - \frac{\hbar^2}{2M}\sum_{l=1}^{k-1}\phi_{k-l}'(x)\,\phi_l'(x) + \delta_{k,1}v_3 x^3 + \delta_{k,2}v_4 x^4 = \epsilon_k. \tag{3C.38}$$

Assuming that $\phi_k(x)$ is a polynomial, we can show by induction that its degree cannot be greater than $k + 2$, i.e.,

$$\phi_k(x) = \sum_{m=1}^{\infty} c_m^{(k)}\, x^m \, , \qquad \text{with } c_m^{(k)} \equiv 0 \quad \text{for } m > k+2 \, , \tag{3C.39}$$

The lowest terms $c_0^{(k)}$ have been omitted since they will be determined at the end the normalization of the wave function $\psi(x)$. Inserting (3C.39) into (3C.38) for $k = 1$, we find

$$c_1^{(1)} = -\frac{v_3}{M\omega^2}, \quad c_2^{(1)} = 0, \quad c_3^{(1)} = -\frac{v_3}{3\hbar\omega}, \quad \epsilon_1 = 0 \, . \tag{3C.40}$$

For $k = 2$, we obtain

$$c_1^{(2)} = 0, \quad c_2^{(2)} = \frac{7v_3^2}{8M^2\omega^4} - \frac{3v_4}{4M\omega^2}, \quad c_3^{(2)} = 0, \quad c_4^{(2)} = \frac{v_3^2}{8M\hbar\omega^3} - \frac{v_4}{4\hbar\omega}, \tag{3C.41}$$

$$\epsilon_2 = -\frac{11v_3^2\hbar^2}{8M^3\omega^4} + \frac{3v_4\hbar^2}{4M^2\omega^2} \, . \tag{3C.42}$$

For the higher-order terms we must solve the recursion relations

$$c_m^{(k)} = \frac{(m+2)(m+1)\hbar}{2mM\omega}c_{m+2}^{(k)} + \frac{\hbar}{2mM\omega}\sum_{l=1}^{k-1}\sum_{n=1}^{m+1} n(m+2-n)\,c_n^{(l)}\,c_{m+2-n}^{(k-l)}, \tag{3C.43}$$

$$\epsilon_k = -\frac{\hbar^2}{M}c_2^{(k)} - \frac{\hbar^2}{2M}\sum_{l=1}^{k-1}c_1^{(l)}\,c_1^{(k-l)} \, . \tag{3C.44}$$

Evaluating this for $k = 3$ yields

$$c_1^{(3)} = -\frac{5v_3^3\hbar}{M^4\omega^7} + \frac{6v_3v_4\hbar}{M^3\omega^5}, \quad c_2^{(3)} = 0, \quad c_3^{(3)} = -\frac{13v_3^3}{12M^3\omega^6} + \frac{3v_3v_4}{2M^2\omega^4},$$

$$c_4^{(3)} = 0, \quad c_5^{(3)} = -\frac{v_3^3}{10M^2\hbar\omega^5} + \frac{v_3v_4}{5M\hbar\omega^3}, \quad \epsilon_3 = 0, \tag{3C.45}$$

and for $k = 4$:

$$c_1^{(4)} = 0, \quad c_2^{(4)} = \frac{305v_3^4\hbar}{32M^5\omega^9} - \frac{123v_3^2v_4\hbar}{8M^4\omega^7} + \frac{21v_4^2\hbar}{8M^3\omega^5}, \quad c_3^{(4)} = 0, \quad c_4^{(4)} = \frac{99v_3^4}{64M^4\omega^8} - \frac{47v_3^2v_4}{16M\omega^6} + \frac{11v_4^2}{16M^2\omega^4},$$

$$c_5^{(4)} = 0, \quad c_6^{(4)} = \frac{5v_3^4}{48M^3\hbar\omega^7} - \frac{v_3^2v_4}{4M^2\hbar\omega^5} + \frac{v_4^2}{12M\hbar\omega^3}, \tag{3C.46}$$

$$\epsilon_4 = -\frac{465v_3^4\hbar^3}{32M^6\omega^9} + \frac{171v_3^2v_4\hbar^3}{8M^5\omega^7} - \frac{21v_4^2\hbar^3}{8M^4\omega^5}. \tag{3C.47}$$

The general form of the coefficients is, now in natural units with $\hbar = 1$, $M = 1$,:

$$c_m^{(k)} = \sum_{\lambda=0}^{\lfloor k/2 \rfloor} \frac{v_3^{k-2\lambda} v_4^{\lambda}}{\omega^{5k/2-m/2-2\lambda}} c_{m,\lambda}^{(k)}, \quad \text{with } c_{m,\lambda}^{(k)} \equiv 0 \text{ for } m > k+2, \text{ or } \lambda > \left\lfloor \frac{k}{2} \right\rfloor, \quad (3C.48)$$

$$\epsilon_k = \sum_{\lambda=0}^{\lfloor k/2 \rfloor} \frac{v_3^{k-2\lambda} v_4^{\lambda}}{\omega^{5k/2-1-2\lambda}} \epsilon_{k,\lambda}. \quad (3C.49)$$

This leads to the recursion relations

$$c_{m,\lambda}^{(k)} = \frac{(m+2)(m+1)}{2m} c_{m+2,\lambda}^{(k)} + \frac{1}{2m} \sum_{l=1}^{k-1} \sum_{n=1}^{m+1} \sum_{\lambda'=0}^{\lambda} n(m+2-n) c_{n,\lambda-\lambda'}^{(l)} c_{m+2-n,\lambda'}^{(k-l)}, \quad (3C.50)$$

with $c_{m,\lambda}^{(k)} \equiv 0$ for $m > k+2$ or $\lambda > \lfloor k/2 \rfloor$. The starting values follow by comparing (3C.40) and (3C.41) with (3C.48):

$$c_{1,0}^{(1)} = -1, \quad c_{2,0}^{(1)} = 0, \quad c_{3,0}^{(1)} = -\frac{1}{3}, \quad (3C.51)$$

$$c_{1,0}^{(2)} = 0, \quad c_{1,1}^{(2)} = 0, \quad c_{2,0}^{(2)} = \frac{7}{8}, \quad c_{2,1}^{(2)} = -\frac{3}{4},$$

$$c_{3,0}^{(2)} = 0, \quad c_{3,1}^{(2)} = 0, \quad c_{4,0}^{(2)} = \frac{1}{8}, \quad c_{4,1}^{(2)} = -\frac{1}{4}. \quad (3C.52)$$

The expansion coefficients $\epsilon_{k,\lambda}$ for the energy corrections ϵ_k are obtained by inserting (3C.49) and (3C.48) into (3C.44) and going to natural units:

$$\epsilon_{k,\lambda} = -c_{2,\lambda}^{(k)} - \frac{1}{2} \sum_{l=1}^{k-1} \sum_{\lambda'=0}^{\lambda} c_{1,\lambda-\lambda'}^{(l)} c_{1,\lambda'}^{(k-l)}. \quad (3C.53)$$

Table 3.1 shows the nonzero even energy corrections ϵ_k up to the tenth order.

3C.4 Ground-State Energy with External Current

In the presence of a constant external current j, the time-independent Schrödinger reads

$$-\frac{\hbar^2}{2M} \psi''(x) + \left(\frac{M}{2} \omega^2 x^2 + g v_3 x^3 + g^2 v_4 x^4 - jx \right) \psi(x) = E \psi(x). \quad (3C.54)$$

For zero coupling constant $g = 0$, we may simply introduce the new variables x' and E':

$$x' = x - \frac{j}{M\omega^2} \quad \text{and} \quad E' = E + \frac{j^2}{2M\omega^2}, \quad (3C.55)$$

and the system becomes a harmonic oscillator in x' with energy $E' = \hbar\omega/2$. Thus we make the ansatz for the wave function

$$\psi(x) \propto e^{\phi(x)}, \quad \text{with} \quad \phi(x) = \frac{j}{\hbar\omega} x - \frac{M\omega}{2\hbar} x^2 + \sum_{k=1}^{\infty} g^k \phi_k(x), \quad (3C.56)$$

and for the energy

$$E(j) = \frac{\hbar\omega}{2} - \frac{j^2}{2M\omega^2} + \sum_{k=1}^{\infty} g^k \epsilon_k. \quad (3C.57)$$

k	ϵ_k
2	$\dfrac{-11v_3^2 + 6v_4\omega^2}{8\omega^4}$
4	$-\dfrac{465v_3^4 - 684v_3^2v_4\omega^2 + 84v_4^2\omega^4}{32\omega^9}$
6	$\dfrac{-39709v_3^6 + 91014v_3^4v_4\omega^2 - 47308v_3^2v_4^2\omega^4 + 2664v_4^3\omega^6}{128\omega^{14}}$
8	$-3(6416935v_3^8 - 19945048v_3^6v_4\omega^2 + 18373480v_3^4v_4^2\omega^4 \\ + 4962400v_3^2v_4^3\omega^6 164720v_4^4\omega^8)/(2048\omega^{19})$
10	$(-2944491879v_3^{10} + 11565716526v_3^8v_4\omega^2 - 15341262168v_3^6v_4^2\omega^4 \\ + 7905514480v_3^4v_4^3\omega^6 - 1320414512v_3^2v_4^4\omega^8 + 29335392v_4^5\omega^{10})/(8192\omega^{24})$

Table 3.1 Expansion coefficients for the ground-state energy of the anharmonic oscillator (3C.31) up to the 10th order.

The equations (3C.38) become now

$$-\frac{\hbar^2}{2M}\phi_k''(x) - \frac{\hbar^2}{2M}\sum_{l=1}^{k-1}\phi_{k-l}'(x)\phi_l'(x) + \left(\hbar\omega x - \frac{j\hbar}{M\omega}\right)\phi_k'(x) + \delta_{k,1}v_3x^3 + \delta_{k,2}v_4x^4 = \epsilon_k.$$

(3C.58)

The results are now for $k = 1$

$$c_1^{(1)} = -\frac{v_3}{M\omega^2} - \frac{j^2v_3}{M^2\hbar\omega^5}, \quad c_2^{(1)} = -\frac{jv_3}{2M\hbar\omega^3}, \quad c_3^{(1)} = -\frac{v_3}{3\hbar\omega}, \quad \epsilon_1 = \frac{3\hbar jv_3}{2M\omega^3} + \frac{j^3v_3}{M^3\omega^6},$$

(3C.59)

and for $k = 2$:

$$c_1^{(2)} = \frac{17jv_3^2}{4M^3\omega^6} + \frac{4j^3v_3^2}{M^4\hbar\omega^9} - \frac{5jv_4}{2M^2\omega^4} - \frac{j^3v_4}{M^3\hbar\omega^7},$$

$$c_2^{(2)} = \frac{7v_3^2}{8M^2\omega^4} + \frac{3j^2v_3^2}{2M^3\hbar\omega^7} - \frac{3v_4}{4M\omega^2} - \frac{j^2v_4}{2M^2\hbar\omega^5},$$

$$c_3^{(2)} = \frac{jv_3^2}{2M^2\hbar\omega^5} - \frac{jv_4}{3M\hbar\omega^3}, \quad c_4^{(2)} = \frac{v_3^2}{8M\hbar\omega^3} - \frac{v_4}{4\hbar\omega},$$

$$\epsilon_2 = -\frac{11\hbar^2v_3^2}{8M^3\omega^4} - \frac{27\hbar j^2v_3^2}{4M^4\omega^7} - \frac{9j^4v_3^2}{2M^5\omega^{10}} + \frac{3\hbar^2v_4}{4M^2\omega^2} + \frac{3\hbar j^2v_4}{M^3\omega^5} + \frac{j^4v_4}{M^4\omega^8}.$$

(3C.60)

The recursive equations (3C.43) and (3C.44) become

$$c_m^{(k)} = \frac{(m+2)(m+1)\hbar}{2mM\omega}c_{m+2}^{(k)} + \frac{\hbar}{2mM\omega}\sum_{l=1}^{k-1}\sum_{n=1}^{m+1}n(m+2-n)c_n^{(l)}c_{m+2-n}^{(k-l)}$$

$$+ \frac{j(m+1)}{Mm\omega^2}c_{m+1}^{(k)},$$

(3C.61)

$$\epsilon_k = -\frac{j\hbar}{M\omega}c_1^{(k)} - \frac{\hbar^2}{M}c_2^{(k)} - \frac{\hbar^2}{2M}\sum_{l=1}^{k-1}c_1^{(l)}c_1^{(k-l)}. \tag{3C.62}$$

Table 3.2 shows the energy corrections ϵ_k in the presence of an external current up to the sixth order using natural units, $\hbar = 1$, $M = 1$.

k	ϵ_k
1	$\dfrac{v_3 j(2j^2 + 3\omega^3)}{2\omega^6}$
2	$\dfrac{2v_4\omega^2(4j^4 + 12j^2\omega^3 + 3\omega^6) - v_3^2(36j^4 + 54j^2\omega^3 + 11\omega^6)}{8\omega^{10}}$
3	$\dfrac{v_3 j[3v_3^2(36j^4 + 63j^2\omega^3 + 22\omega^6) - 2v_4\omega^2(24j^4 + 66j^2\omega^3 + 31\omega^6)]}{4\omega^{14}}$
4	$[36v_3^2v_4\omega^2(112j^6 + 324j^4\omega^3 + 212j^2\omega^6 + 19\omega^9)$ $-4v_4^2\omega^4(64j^6 + 264j^4\omega^3 + 248j^2\omega^6 + 21\omega^9)$ $-3v_3^4(2016j^6 + 4158j^4\omega^3 + 2112j^2\omega^6 + 155\omega^9)]/(32\omega^{10})$
5	$v_3 j[27v_3^4(1728j^6 + 4158j^4\omega^3 + 2816j^2\omega^6 + 465\omega^9)$ $+4v_4^2\omega^4(1536j^6 + 6408j^4\omega^3 + 7072j^2\omega^6 + 1683\omega^9)$ $-12v_3^2v_4\omega^2(3456j^6 + 10908j^4\omega^3 + 9176j^2\omega^6 + 1817\omega^9)]/(32\omega^{22})$
6	$[8v_4^3\omega^6(1536j^8 + 8544j^6\omega^3 + 14144j^4\omega^6 + 6732j^2\omega^9 + 333\omega^{12})$ $-4v_3^2v_4^2\omega^4(103680j^8 + 454032j^6\omega^3 + 584928j^4\omega^6 + 221706j^2\omega^9 + 11827\omega^{12})$ $+6v_3^4v_4\omega^2(285120j^8 + 991224j^6\omega^3 + 1024224j^4\omega^6 + 323544j^2\omega^9 + 15169\omega^{12})$ $-v_3^6(1539648j^8 + 4266108j^6\omega^3 + 3649536j^4\omega^6 + 979290j^2\omega^9 + 39709\omega^{12})]/(128\omega^{26})$

Table 3.2 Expansion coefficients for the ground-state energy of the anharmonic oscillator (3C.31) in the presence of an external current up to the 6th order.

3C.5 Recursion Relation for Effective Potential

It is possible to derive a recursion relation directly for the zero-temperature effective potential (5.259). To this we observe that according to Eq. (3.771), the fluctuating part of the effective potential is given by the Euclidean path integral

$$e^{-\beta V_{\text{eff}}^{\text{fl}}(X)} = \int \mathcal{D}\delta x \exp\left(-\frac{1}{\hbar}\left\{\mathcal{A}[X+\delta x] - \mathcal{A}[X] - \mathcal{A}_X[X]\delta x - V_{\text{eff}\,X}^{\text{fl}}(X)\delta x\right\}\right). \tag{3C.63}$$

This can be rewritten as [recall (3.768)]

$$e^{-\beta[V_{\text{eff}}(X) - V(X)]} = \oint \mathcal{D}\delta x \exp\left\{-\frac{1}{\hbar}\int_0^{\hbar\beta}d\tau\right.$$
$$\left. \times \left[\frac{M}{2}\delta\dot{x}^2(\tau) + V(X+\delta x) - V(X) - V'_{\text{eff}}(X)\delta x\right]\right\}. \tag{3C.64}$$

Going back to the integration variable $x = X + \delta x$, and taking all terms depending only on X to the left-hand side, this becomes

$$e^{-\beta[V_{\text{eff}}(X) - V'_{\text{eff}}(X)X]} = \oint \mathcal{D}x \, \exp\left\{-\frac{1}{\hbar}\int_0^{\hbar\beta} d\tau\left[\frac{M}{2}\dot{x}^2(\tau) + V(x) - V'_{\text{eff}}(X)x\right]\right\}. \tag{3C.65}$$

In the limit of zero temperature, the right-hand side is equal to $e^{-\beta E^{(0)}(X)}$, where $E^{(0)}(X)$ is the ground state of the Schrödinger equation associated with the path integral. Hence we obtain

$$-\frac{\hbar^2}{2M}\psi''(x) + [V(x) - V'_{\text{eff}}(X)x]\psi(x) = [V_{\text{eff}}(X) - V'_{\text{eff}}(X)X]\psi(x). \tag{3C.66}$$

For the mixed interaction of the previous subsection, this reads

$$-\frac{\hbar^2}{2M}\psi''(x) + \left[\frac{M}{2}\omega^2 x^2 + gv_3 x^3 + g^2 v_4 x^4 - V'_{\text{eff}}(X)x\right]\psi(x)$$
$$= [V_{\text{eff}}(X) - V'_{\text{eff}}(X)X]\psi(x), \tag{3C.67}$$

and may be solved recursively. We expand the effective potential in powers of is expanded in the coupling constant g:

$$V_{\text{eff}}(X) = \sum_{k=0}^{\infty} g^k V_k(X), \tag{3C.68}$$

and assume $V_k(X)$ to be a polynomial in X:

$$V_k(X) = \sum_{m=0}^{k+2} C_m^{(k)} X^m. \tag{3C.69}$$

Comparison with Eq. (3C.54) shows that we may set $j = V'_{\text{eff}}(X)$ and calculate $V_{\text{eff}}(X) - V'_{\text{eff}}(X)X$ by analogy to the energy in (3C.57). Inserting the ansatz (3C.68), (3C.69) into (3C.67) we find all equations for $V_k(X)$ by comparing coefficients of g^k and X^m. It turns out that for even or odd k, also $V_k(X)$ is even or odd in X, respectively. Table 3.3 shows the first six orders of the effective potential, which have been obtained in this way.

The equations for $V_k(X)$ are obtained as follows. We insert into (3C.67) the ansatz for the wave function $\psi(x) \propto e^{\phi(x)}$ with

$$\phi(x) = \frac{M\omega X}{\hbar}x - \frac{M\omega}{2\hbar}x^2 + \sum_{k=1}^{\infty} g^k \phi_k(x), \tag{3C.70}$$

and expand

$$V_{\text{eff}}(X) = \frac{\hbar\omega}{2} + \frac{M}{2}\omega^2 X^2 + \sum_{k=1}^{\infty} g^k V_k(X), \tag{3C.71}$$

to obtain the set of equations

$$-\frac{\hbar^2}{2M}\phi_k''(x) - \frac{\hbar^2}{2M}\sum_{l=1}^{k-1}\phi_{k-l}'(x)\phi_l'(x) + \hbar\omega(x-X)\phi_k'(x) - xV_k'(X) + \delta_{k,1}v_3 x^3 + \delta_{k,2}v_4 x^4$$
$$= V_k(X) - V_k'(X)X. \tag{3C.72}$$

From these we find for $k = 1$:

$$c_1^{(1)} = \frac{v_3}{2M\omega^2} + \frac{2v_3 X^2}{\hbar\omega}, \quad c_2^{(1)} = -\frac{v_3 X}{2\hbar\omega}, \quad c_3^{(1)} = -\frac{v_3}{3\hbar\omega}, \quad V_1(X) = \frac{3v_3\hbar}{2M\omega} + v_3 X^3, \tag{3C.73}$$

k	$V_k(X)$
0	$\dfrac{\omega}{2} + \dfrac{\omega^2}{2} X^2$
1	$v_3 X^3 + \dfrac{3v_3}{2\omega} X$
2	$v_4 X^4 - \dfrac{v_3^2(1 + 9\omega X^2) + v_4\omega^2(3 + 12\omega X^2)}{4\omega^4}$
3	$\dfrac{v_3 X[3v_3^2(4 + 9\omega X^2) - 2v_4\omega^2(13 + 18\omega X^2)]}{4\omega^6}$
4	$-[4v_4^2\omega^4(21 + 104\omega X^2 + 72\omega^2 X^4) - 12v_3^2 v_4\omega^2(13 + 152\omega X^2 + 108\omega^2 X^4)$ $+v_3^4(51 + 864\omega X^2 + 810\omega^2 X^4)]/(32\omega^9)$
5	$3v_3 X[9v_3^4(51 + 256\omega X^2 + 126\omega^2 X^4) + 4v_4^2\omega^4(209 + 544\omega X^2 + 216\omega^2 X^4)$ $-4v_3^2 v_4\omega^2(341 + 1296\omega X^2 + 540\omega^2 X^4)]/(32\omega^{11})$
6	$[24v_4^3\omega^6(111 + 836\omega X^2 + 1088\omega^2 X^4 + 288\omega^3 X^6)$ $-36v_3^2 v_4^2\omega^4(365 + 5654\omega X^2 + 8448\omega^2 X^4 + 2160\omega^3 X^6)$ $+6v_3^4 v_4\omega^2(2129 + 46008\omega X^2 + 85248\omega^2 X^4 + 22680\omega^3 X^6)$ $-v_3^6(3331 + 90882\omega X^2 + 207360\omega^2 X^4 + 61236\omega^3 X^6)]/(128\omega^{14})$

Table 3.3   Effective potential of the anharmonic oscillator (3C.31) up to the 6th order, expanded in the coupling constant g (in natural units with $\hbar = 1$ and $M = 1$). The lowest terms agree, of course, with the two-loop result (3.767).

and for $k = 2$:

$$c_1^{(2)} = -\frac{13v_3^2 X}{4M^2\omega^4} - \frac{2v_3^2 X^3}{M\hbar\omega^3} + \frac{7v_4 X}{2M\omega^2} + \frac{3v_4 X^3}{\hbar\omega}, \quad c_2^{(2)} = \frac{v_3^2}{8M^2\omega^4} - \frac{3v_4}{4M\omega^2} - \frac{v_4 X^2}{2\hbar\omega},$$

$$c_3^{(2)} = \frac{v_3^2 X}{2M\hbar\omega^3} - \frac{v_4 X}{3\hbar\omega} \quad c_4^{(2)} = \frac{v_3^2}{8M\hbar\omega^3} - \frac{v_4}{4\hbar\omega}, \tag{3C.74}$$

$$V_2(X) = -\frac{\hbar^2 v_3^2}{4M^3\omega^4} - \frac{9\hbar v_3^2 X^2}{4M^2\omega^3} + \frac{3\hbar^2 v_4}{4M^2\omega^2} + \frac{3\hbar v_4 X^2}{M\omega} + v_4 X^4. \tag{3C.75}$$

For $k \geq 3$, we must solve recursively

$$c_m^{(k)} = \frac{(m+2)(m+1)\hbar}{2mM\omega} c_{m+2}^{(k)} + \frac{\hbar}{2mM\omega} \sum_{l=1}^{k-1}\sum_{n=1}^{m+1} n(m+2-n)c_n^{(l)} c_{m+2-n}^{(k-l)}$$

$$+ \frac{X(m+1)}{m} c_{m+1}^{(k)} \text{ for } m \geq 2 \text{ and with } c_m^{(k)} \equiv 0 \text{ for } m > k+2, \tag{3C.76}$$

$$c_1^{(k)} = \frac{3\hbar}{M\omega} c_3^{(k)} + 2X c_2^{(k)} + \frac{\hbar}{M\omega} \sum_{l=1}^{k-1} \left(c_2^{(k-l)} c_1^{(l)} + c_1^{(k-l)} c_2^{(l)} \right) + \frac{1}{\hbar\omega} V_k'(X), \tag{3C.77}$$

$$V_k(X) = -\frac{\hbar^2}{M}c_2^{(k)} - \frac{3\hbar^2}{M}Xc_3^{(k)} - 2\hbar\omega X^2 c_2^{(k)} - \frac{\hbar^2}{M}X\sum_{l=1}^{k-1}\left(c_2^{(k-l)}c_1^{(l)} + c_1^{(k-l)}c_2^{(l)}\right)$$
$$-\frac{\hbar^2}{2M}\sum_{l=1}^{k-1}c_1^{(l)}c_1^{(k-l)}. \tag{3C.78}$$

The results are listed in Table 3.3.

3C.6 Interaction r^4 in D-Dimensional Radial Oscillator

It is easy to generalize these relations further to find the perturbation expansions for the eigenvalues of the radial Schrödinger equation of an anharmonic oscillator in D dimensions

$$\left[-\frac{1}{2}\frac{d^2}{dr^2} - \frac{1}{2}\frac{D-1}{r}\frac{d}{dr} + \frac{l(l+D-2)}{2r^2} + \frac{1}{2}r^2 + \frac{g}{4}r^4\right]R_n(r) = E^{(n)}R_n(r). \tag{3C.79}$$

The case $g = 0$ will be solved in Section 9.2, with the energy eigenvalues

$$E^{(n)} = 2n' + l + D/2 = n + D/2, \qquad n = 0,1,2,3,\ldots \quad, l = 0,1,2,3,\ldots. \tag{3C.80}$$

For a fixed principal quantum number $n = 2n_r + l$, the angular momentum runs through $l = 0,2,\ldots,n$ for even, and $l = 1,3,\ldots,n$ for odd n. There are $(n+1)(n+2)/2$ degenerate levels. Removing a factor r^l from $R_n(r)$, and defining $R_n(r) = r^l w_n(r)$, the Schrödinger equation becomes

$$\left(-\frac{1}{2}\frac{d^2}{dr^2} - \frac{1}{2}\frac{2l+D-1}{r}\frac{d}{dr} + \frac{1}{2}r^2 + \frac{g}{4}r^4\right)w_n(r) = E^{(n)}w_n(r). \tag{3C.81}$$

The second term modifies the differential equation (3C.6) to

$$r\Phi_k'(r) - 2n'\Phi_k(r) = \frac{1}{2}\Phi_k''(r) + \frac{(2l+D-1)}{2r}\Phi_k'(r) + r^4\Phi_{k-1}(r) + \sum_{k'=1}^{k}(-1)^{k'}E_{k'}^{(n)}\Phi_{k-k'}(r). \tag{3C.82}$$

The extra terms change the recursion relation (3C.15) into

$$(p-2n')A_k^p = \frac{1}{2}[(p+2)(p+1) + (p+2)(2l+D-1)]A_k^{p+2} + A_{k-1}^{p-4} + \sum_{k'=1}^{k}(-1)^{k'}E_{k'}^{(n)}A_{k-k'}^p. \tag{3C.83}$$

For even $n = 2n' + l$ with $l = 0,2,4,\ldots,n$, we normalize the wave functions by setting

$$C_k^0 = (2l+D)\delta_{0k}, \tag{3C.84}$$

rather than (3C.12), and obtain

$$2(p'-n')C_k^{p'} = [(2p'+1)(p'+1) + (p'+1)(l+D/2-1/2)]C_k^{p'+1} + C_{k-1}^{p'-2} - \sum_{k'=1}^{k}C_{k'}^1 C_{k-k'}^{p'}, \tag{3C.85}$$

instead of (3C.20).

For odd $n = 2n' + l$ with $l = 1,3,5,\ldots,n$, the equations analogous to (3C.13) and (3C.22) are

$$C_k^1 = 3(2l+D)\delta_{0k} \tag{3C.86}$$

and

$$2(p'-n')C_k^{p'} = [(2p'+3)(p'+1) + (p'+3/2)(l+D/2-1/2)]C_k^{p'+1} + C_{k-1}^{p'-2} - \sum_{k'=1}^{k}C_{k'}^1 C_{k-k'}^{p'}. \tag{3C.87}$$

In either case, the expansion coefficients of the energy are given by

$$E_k^{(n)} = -\frac{(-1)^k}{2}\frac{2l+D+1}{2l+D}C_k^1. \tag{3C.88}$$

3C.7   Interaction r^{2q} in D Dimensions

A further extension of the recursion relation applies to interactions $gx^{2q}/4$. Then Eqs. (3C.20) and (3C.22) are changed in the second terms on the right-hand side which become $C_k^{p'-q}$. In a first step, these equations are now solved for $C_k^{p'}$ for $1 \le p \le qk + n$, starting with $p' = qk + n'$ and lowering p' down to $p' = n' + 1$. As before, the knowledge of the coefficients C_k^1 (which determine the yet unknown energies and are contained in the last term of the recursion relations) is not required for $p' > n'$. The second and third steps are completely analogous to the case $q = 2$.

The same generalization applies to the D-dimensional case.

3C.8   Polynomial Interaction in D Dimensions

If the Schrödinger equation has the general form

$$
\left[-\frac{1}{2}\frac{d^2}{dr^2} - \frac{1}{2}\frac{D-1}{r}\frac{d}{dr} + \frac{l(l+D-2)}{2r^2} + \frac{1}{2}r^2 \right.
$$
$$
\left. + \frac{g}{4}(a_4 r^4 + a_6 r^6 + \ldots + a_{2q} x^{2q}) \right] R_n(r) = E^{(n)} R_n(r), \tag{3C.89}
$$

we simply have to replace in the recursion relations (3C.85) and (3C.87) the second term on the right-hand side as follows

$$
C_{k-1}^{p'-2} \to a_4 C_{k-1}^{p'-2} + a_6 C_{k-1}^{p'-3} + \ldots + a_{2q} C_{k-1}^{p'-q}, \tag{3C.90}
$$

and perform otherwise the same steps as for the potential $gr^{2q}/4$ alone.

Appendix 3D   Feynman Integrals for $T \neq 0$

The calculation of the Feynman integrals (3.547) can be done straightforwardly with the help of the symbolic program Mathematica. The first integral in Eqs. (3.547) is trivial. The second and forth integrals are simple, since one overall integration over, say, τ_3 yields merely a factor $\hbar\beta$, due to translational invariance of the integrand along the τ-axis. The triple integrals can then be split as

$$
\int_0^{\hbar\beta}\int_0^{\hbar\beta}\int_0^{\hbar\beta} d\tau_1 d\tau_2 d\tau_3 \; f(|\tau_1 - \tau_2|, |\tau_2 - \tau_3|, |\tau_3 - \tau_1|)
$$
$$
= \hbar\beta \int_0^{\hbar\beta}\int_0^{\hbar\beta} d\tau_1 d\tau_2 \; f(|\tau_1 - \tau_2|, |\tau_2|, |\tau_1|) \tag{3D.1}
$$
$$
= \hbar\beta \left(\int_0^{\hbar\beta} d\tau_2 \int_0^{\tau_2} d\tau_1 f(\tau_2 - \tau_1, \tau_2, \tau_1) + \int_0^{\hbar\beta} d\tau_2 \int_{\tau_2}^{\hbar\beta} d\tau_1 f(\tau_1 - \tau_2, \tau_2, \tau_1) \right),
$$

to ensure that the arguments of the Green function have the same sign in each term. The lines represent the thermal correlation function

$$
G^{(2)}(\tau, \tau') = \frac{\hbar}{2M\omega} \frac{\cosh\omega[|\tau - \tau'| - \hbar\beta/2]}{\sinh(\omega\hbar\beta/2)}. \tag{3D.2}
$$

With the dimensionless variable $x \equiv \omega\hbar\beta$, the result for the quantities α_V^{2L} defined in (3.547) in the Feynman diagrams with L lines and V vertices is

$$
a^2 = \frac{1}{2}\coth\frac{x}{2}, \tag{3D.3}
$$
$$
\alpha_2^4 = \frac{1}{8}\frac{1}{\sinh^2\frac{x}{2}}(x + \sinh x), \tag{3D.4}
$$

$$\alpha_3^6 = \frac{1}{64} \frac{1}{\sinh^3 \frac{x}{2}} \left(-3 \cosh \frac{x}{2} + 2 x^2 \cosh \frac{x}{2} + 3 \cosh \frac{3x}{2} + 6 x \sinh \frac{x}{2} \right), \tag{3D.5}$$

$$\alpha_2^8 = \frac{1}{256} \frac{1}{\sinh^4 \frac{x}{2}} \left(6 x + 8 \sinh x + \sinh 2x \right), \tag{3D.6}$$

$$\alpha_3^{10} = \frac{1}{4096} \frac{1}{\sinh^5 \frac{x}{2}} \left(-40 \cosh \frac{x}{2} + 24 x^2 \cosh \frac{x}{2} + 35 \cosh \frac{3x}{2} \right.$$
$$\left. +5 \cosh \frac{5x}{2} + 72 x \sinh \frac{x}{2} + 12 x \sinh \frac{3x}{2} \right), \tag{3D.7}$$

$$\alpha_3^{12} = \frac{1}{16384} \frac{1}{\sinh^6 \frac{x}{2}} \left(-48 + 32 x^2 - 3 \cosh x + 8 x^2 \cosh x \right.$$
$$\left. +48 \cosh 2x + 3 \cosh 3x + 108 x \sinh x \right), \tag{3D.8}$$

$$\alpha_2^6 = \frac{1}{24} \frac{1}{\sinh^2 \frac{x}{2}} \left(5 + 24 \cosh x \right), \tag{3D.9}$$

$$\alpha_3^8 = \frac{1}{72} \frac{1}{\sinh^3 \frac{x}{2}} \left(3 x \cosh \frac{x}{2} + 9 \sinh \frac{x}{2} + \sinh \frac{3x}{2} \right), \tag{3D.10}$$

$$\alpha_{3'}^{10} = \frac{1}{2304} \frac{1}{\sinh^4 \frac{x}{2}} \left(30 x + 104 \sinh x + 5 \sinh 2x \right). \tag{3D.11}$$

For completeness, we have also listed the integrals α_2^6, α_3^8, and $\alpha_{3'}^{10}$, corresponding to the three diagrams

$$\tag{3D.12}$$

respectively, which occur in perturbation expansions with a cubic interaction potential x^3. These will appear in a modified version in Chapter 5.

In the low-temperature limit where $x = \omega \hbar \beta \to \infty$, the x-dependent factors α_V^{2L} in Eqs. (3D.3)–(3D.11) converge towards the constants

$$1/2, \quad 1/4, \quad 3/16, \quad 1/32, \quad 5/(8 \cdot 2^5), \quad 3/(8 \cdot 2^6), \quad 1/12, \quad 1/18, 5/(9 \cdot 2^5), \tag{3D.13}$$

respectively. From these numbers we deduce the relations (3.550) and, in addition,

$$a_2^6 \to \frac{2}{3} a^6, \quad a_3^8 \to \frac{8}{9} a^8, \quad a_{3'}^{10} \to \frac{5}{9} a^{10}. \tag{3D.14}$$

In the high-temperature limit $x \to 0$, the Feynman integrals $\hbar \beta (1/\omega)^{V-1} a_V^{2L}$ with L lines and V vertices diverge like $\beta^V (1/\beta)^L$. The first V factors are due to the V-integrals over τ, the second are the consequence of the product of $n/2$ factors a^2. Thus, a_V^{2L} behaves for $x \to 0$ like

$$a_V^{2L} \propto \left(\frac{\hbar}{M\omega} \right)^L x^{V-1-L}. \tag{3D.15}$$

Indeed, the x-dependent factors α_V^{2L} in (3D.3)–(3D.11) grow like

$$\alpha^2 \approx 1/x + x/12 + \dots,$$
$$\alpha_2^4 \approx 1/x + x^3/720,$$
$$\alpha_2^6 \approx 1/x + x^5/30240 + \dots,$$
$$\alpha_2^8 \approx 1/x^3 + x/120 - x^3/3780 + x^5/80640 + \dots,$$
$$\alpha_3^{10} \approx 1/x^3 + x/240 - x^3/15120 + x^7/6652800 + \dots,$$

$$\alpha_3^{12} \approx 1/x^4 + 1/240 + x^2/15120 - x^6/4989600 + 701\,x^8/34871316480 + \dots,$$
$$\alpha_2^6 \approx 1/x^2 + x^2/240 - x^4/6048 + \dots,$$
$$\alpha_3^8 \approx 1/x^2 + x^2/720 - x^6/518400 + \dots,$$
$$\alpha_{3'}^{10} \approx 1/x^3 + x/360 - x^5/1209600 + 629\,x^9/261534873600 + \dots . \tag{3D.16}$$

For the temperature behavior of these Feynman integrals see Fig. 3.16. We have plotted the reduced Feynman integrals $\hat{a}_V^{2L}(x)$ in which the low-temperature behaviors (3.550) and (3D.14) have been divided out of a_V^{2L}.

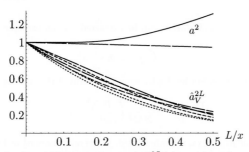

Figure 3.16 Plot of reduced Feynman integrals $\hat{a}_V^{2L}(x)$ as a function of $L/x = Lk_BT/\hbar\omega$. The integrals (3D.4)–(3D.11) are indicated by decreasing dash-lengths.

The integrals (3D.4) and (3D.5) for a_2^4 and a_3^6 can be obtained from the integral (3D.3) for a^2 by the operation

$$\frac{\hbar^n}{n!M^n}\left(-\frac{\partial}{\partial\omega^2}\right)^n = \frac{\hbar^n}{n!M^n}\left(-\frac{1}{2\omega}\frac{\partial}{\partial\omega}\right)^n, \tag{3D.17}$$

with $n = 1$ and $n = 2$, respectively. This follows immediately from the fact that the Green function

$$G_\omega^{(2)}(\tau,\tau') = \frac{1}{\hbar M\beta}\sum_{m=-\infty}^{\infty} e^{-i\omega_m(\tau-\tau')}\frac{\hbar}{\omega_m^2+\omega^2}, \tag{3D.18}$$

with ω^2 shifted to $\omega^2 + \delta\omega^2$ can be expanded into a geometric series

$$pace-1.9cm G_{\sqrt{\omega^2+\delta\omega^2}}^{(2)}(\tau,\tau') = \frac{1}{\hbar M\beta}\sum_{m=-\infty}^{\infty} e^{-i\omega_m(\tau-\tau')}\left[\frac{\hbar}{\omega_m^2+\omega^2} - \frac{\delta\omega^2}{\hbar}\frac{\hbar^2}{(\omega_m^2+\omega^2)^2}\right.$$
$$\left. pace1.0cm + \left(\frac{\delta\omega^2}{\hbar}\right)^2\frac{\hbar^3}{(\omega_m^2+\omega^2)^3} + \dots\right], \tag{3D.19}$$

which corresponds to a series of convoluted τ-integrals

$$G_{\sqrt{\omega^2+\delta\omega^2}}^{(2)}(\tau,\tau') = G_\omega^{(2)}(\tau,\tau') - \frac{M\delta\omega^2}{\hbar}\int_0^{\hbar\beta} d\tau_1 G_\omega^{(2)}(\tau,\tau_1)G_\omega^{(2)}(\tau_1,\tau') \tag{3D.20}$$
$$+ \left(\frac{M\delta\omega^2}{\hbar}\right)^2\int_0^{\hbar\beta}\int_0^{\hbar\beta} d\tau_1 d\tau_2 G_\omega^{(2)}(\tau,\tau_1)G_\omega^{(2)}(\tau_1,\tau_2)G_\omega^{(2)}(\tau_2,\tau') + \dots .$$

In the diagrammatic representation, the derivatives (3D.17) insert n points into a line. In quantum field theory, this operation is called a *mass insertion*. Similarly, the Feynman integral (3D.7) is obtained from (3D.6) via a differentiation (3D.17) with $n = 1$ [see the corresponding diagrams in (3.547)]. A factor 4 must be removed, since the differentiation inserts a point into each of the four lines which are indistinguishable. Note that from these rules, we obtain directly the relations 1, 2, and 4 of (3.550).

Note that the same type of expansion allows us to derive the three integrals from the one-loop diagram (3.546). After inserting (3D.20) into (3.546) and re-expanding the logarithm we find the series of Feynman integrals

$$
\overset{\omega^2+\delta\omega^2}{\bigcirc} \longrightarrow \bigcirc + \frac{M\delta\omega^2}{\hbar}\,\bigcirc - \left(\frac{M\delta\omega^2}{\hbar}\right)^2\frac{1}{2}\,\bigcirc + \left(\frac{M\delta\omega^2}{\hbar}\right)^3\frac{1}{3}\,\bigcirc - \dots ,
$$

from which the integrals (3D.3)–(3D.5) can be extracted. As an example, consider the Feynman integral

$$
\bigcirc = \hbar\beta\frac{1}{\omega}\,a_2^4.
$$

It is obtained from the second-order Taylor expansion term of the tracelog as follows:

$$
-\frac{1}{2}\hbar\beta\frac{1}{\omega}a_1^4 = \frac{\hbar^2}{2!M^2}\left(\frac{\partial}{\partial\omega^2}\right)^2 [-2\beta V_\omega]. \tag{3D.21}
$$

A straightforward calculation, on the other hand, yields once more a_2^4 of Eq. (3D.5).

Notes and References

The theory of generating functionals in quantum field theory is elaborated by
J. Rzewuski, *Field Theory*, Hafner, New York, 1969.
For the usual operator derivation of the Wick expansion, see
S.S. Schweber, *An Introduction to Relativistic Quantum Field Theory*, Harper and Row, New York, 1962, p. 435.

The derivation of the recursion relation in Fig. 3.10 was given in
H. Kleinert, Fortschr. Physik. **30**, 187 (1986) (http://www.physik.fu-berlin.de/~kleinert/82), Fortschr. Physik. **30**, 351 (1986) (*ibid*.http/84).
See in particular Eqs. (51)–(61).
Its efficient graphical evaluation is given in
H. Kleinert, A. Pelster, B. Kastening, M. Bachmann, *Recursive Graphical Construction of Feynman Diagrams and Their Multiplicities in x^4- and in x^2A-Theory*, Phys. Rev. D **61**, 085017 (2000) (hep-th/9907044).
This paper develops a Mathematica program for a fast calculation of diagrams beyond five loops, which can be downloaded from the internet at *ibid*.http/b3/programs.

The Mathematica program solving the Bender-Wu-like recursion relations for the general anharmonic potential (3C.26) is found in the same directory. This program was written in collaboration with W. Janke.

The path integral calculation of the effective action in Section 3.23 can be found in
R. Jackiw, Phys. Rev. D **9**, 1686 (1974).
See also
C. De Dominicis, J. Math. Phys. **3**, 983 (1962),
C. De Dominicis and P.C. Martin, *ibid*. **5**, 16, 31 (1964),
B.S. DeWitt, in *Dynamical Theory of Groups and Fields*, Gordon and Breach, N.Y., 1965,
A.N. Vassiliev and A.K. Kazanskii, Teor. Math. Phys. **12**, 875 (1972),
J.M. Cornwall, R. Jackiw, and E. Tomboulis, Phys. Rev. D **10**, 1428 (1974),
and the above papers by the author in Fortschr. Physik **30**.

The path integral of a particle in a dissipative medium is discussed in A.O. Caldeira and A.J. Leggett, Ann. Phys. **149**, 374 (1983), **153**; 445(E) (1984).
See also
A.J. Leggett, Phys. Rev. B **30**, 1208 (1984);
A.I. Larkin, and Y.N. Ovchinnikov, Zh. Eksp. Teor. Fiz. **86**, 719 (1984) [Sov. Phys. JETP **59**, 420 (1984)]; J. Stat. Phys. **41**, 425 (1985);
H. Grabert and U. Weiss, Z. Phys. B **56**, 171 (1984);
L.-D. Chang and S. Chakravarty, Phys. Rev. B **29**, 130 (1984);
D. Waxman and A.J. Leggett, Phys. Rev. B **32**, 4450 (1985);
P. Hänggi, H. Grabert, G.-L. Ingold, and U. Weiss, Phys. Rev. Lett. **55**, 761 (1985);
D. Esteve, M.H. Devoret, and J.M. Martinis, Phys. Rev. B **34**, 158 (1986);
E. Freidkin, P. Riseborough, and P. Hänggi, Phys. Rev. B **34**, 1952 (1986);
H. Grabert, P. Olschowski and U. Weiss, Phys. Rev. B **36**, 1931 (1987),
and in the textbook
U. Weiss, *Quantum Dissipative Systems*, World Scientific, Singapore, 1993.
See also Notes and References in Chapter 18.

For alternative approaches to the damped oscillator see
F. Haake and R. Reibold, Phys. Rev. A **32**, 2462 (1985),
A. Hanke and W. Zwerger, Phys. Rev. E **52**, 6875 (1995);
S. Kehrein and A. Mielke, Ann. Phys. (Leipzig) **6**, 90 (1997) (cond-mat/9701123).
X.L. Li, G.W. Ford, and R.F. O'Connell, Phys. Rev. A **42**, 4519 (1990).

The effective potential (5.259) was derived in D dimensions by
H. Kleinert and B. Van den Bossche, Nucl. Phys. B **632**, 51 (2002) (http://arxiv.org/abs/cond-mat/0104102">cond-mat/0104102).
By inserting $D = 1$ and changing the notation appropriately, one finds (5.259).
The finite-temperature expressions (3.760)–(3.763) are taken from S.F. Brandt, *Beyond Effective Potential via Variational Perturbation Theory*, M.S. thesis, FU-Berlin 2004 (http://hbar.wustl.edu/~sbrandt/diplomarbeit.pdf). See also
S.F. Brandt, H. Kleinert, and A. Pelster, J. Math. Phys. **46**, 032101 (2005) (quant-ph/0406206) .

4

Semiclassical Time Evolution Amplitude

The path integral approach renders a clear intuitive understanding of quantum-mechanical effects in terms of quantum fluctuations, exhibiting precisely how the laws of classical mechanics are modified by these fluctuations. In some limiting situations, the modifications may be small, for instance, if an electron in an atom is highly excited. Its wave packet encircles the nucleus in almost the same way as a point particle in classical mechanics. Then it is relatively easy to calculate quite accurate quantum-mechanical amplitudes by expanding them around classical expressions in powers of the fluctuation width.

4.1 Wentzel-Kramers-Brillouin (WKB) Approximation

In Schrödinger's theory, an important step towards understanding the relation between classical and quantum mechanics consists in proving that the center of a Schrödinger wave packet moves like a classical particle. The approach to the classical limit is described by the so-called *eikonal approximation*, or the *Wentzel-Kramers-Brillouin approximation* (short: *WKB approximation*), which proceeds as follows:

First, one rewrites the time-independent Schrödinger equation of a point particle

$$\left\{ -\frac{\hbar^2}{2M}\boldsymbol{\nabla}^2 - [E - V(\mathbf{x})] \right\} \psi(\mathbf{x}) = 0 \tag{4.1}$$

in the form

$$\left[-\hbar^2\boldsymbol{\nabla}^2 - p^2(\mathbf{x}) \right] \psi(\mathbf{x}) = 0, \tag{4.2}$$

where

$$p(\mathbf{x}) \equiv \sqrt{2M[E - V(\mathbf{x})]} \tag{4.3}$$

is the *local classical momentum* of the particle.

In a second step, one re-expresses the wave function as an exponential

$$\psi(\mathbf{x}) = e^{iS(\mathbf{x})/\hbar}. \tag{4.4}$$

369

For the exponent $S(\mathbf{x})$, called the *eikonal*, the Schrödinger equation amounts to the a differential equation :

$$- i\hbar \boldsymbol{\nabla}^2 S(\mathbf{x}) + [\boldsymbol{\nabla} S(\mathbf{x})]^2 - p^2(\mathbf{x}) = 0. \tag{4.5}$$

To solve this equation approximately, one assumes that the function $p(\mathbf{x})$ shows little relative change over the de Broglie wavelength

$$\lambda \equiv \frac{2\pi}{k(\mathbf{x})} \equiv \frac{2\pi\hbar}{p(\mathbf{x})}, \tag{4.6}$$

i.e.,

$$\varepsilon \equiv \frac{2\pi\hbar}{p(\mathbf{x})} \left| \frac{\boldsymbol{\nabla} p(\mathbf{x})}{p(\mathbf{x})} \right| \ll 1. \tag{4.7}$$

This condition is called the *WKB condition*. In the extreme case of $p(\mathbf{x})$ being a constant the condition is certainly fulfilled and the Riccati equation is solved by the trivial eikonal

$$S = \mathbf{p}\mathbf{x} + \text{const}, \tag{4.8}$$

which makes (4.4) a plain wave. For slow variations, the first term in the Riccati equation is much smaller than the others and can be treated systematically in a "smoothness expansion". Since the small ratio (4.7) carries a prefactor $\hbar$, the Planck constant may be used to count the powers of the smallness parameter ε, i.e., it may formally be considered as a small expansion parameter.

The limit $\hbar \to 0$ of the equation determines the lowest-order approximation to the eikonal, $S_0(\mathbf{x})$, by

$$[\boldsymbol{\nabla} S_0(\mathbf{x})]^2 - p^2(\mathbf{x}) = 0. \tag{4.9}$$

Being independent of $\hbar$, this is a classical equation. Indeed, it is equivalent to the Hamilton-Jacobi differential equation of classical mechanics: For time-independent systems, the action can be written as

$$A(\mathbf{x}, t) = S_0(\mathbf{x}) - tE, \tag{4.10}$$

where $S_0(\mathbf{x})$ is defined by

$$S_0(\mathbf{x}) \equiv \int dt\, \mathbf{p}(t)\dot{\mathbf{x}}(t), \tag{4.11}$$

and E is the constant energy of the orbit under consideration. The action solves the Hamilton-Jacobi equation (1.64). In three dimensions, it is a function $A(\mathbf{x}, t)$ of the orbital endpoints. According to Eq. (1.62), the derivative of $A(\mathbf{x}, t)$ is equal to the momentum $\mathbf{p}$. Since E is a constant, the same thing holds for $S_0(\mathbf{x}, t)$. Hence

$$\mathbf{p} \equiv \boldsymbol{\nabla} A(\mathbf{x}) = \boldsymbol{\nabla} S_0(\mathbf{x}). \tag{4.12}$$

In terms of the action A, the Hamilton-Jacobi equation reads

$$\frac{1}{2M}(\boldsymbol{\nabla} A)^2 + V(\mathbf{x}) = -\partial_t A. \tag{4.13}$$

By inserting Eqs. (4.10) and (4.12), we recover Eq. (4.9). This is why $S_0(\mathbf{x})$ is also called the *classical eikonal*.

The corrections to the classical eikonal are calculated most systematically by imagining $\hbar \neq 0$ to be a small quantity and expanding the eikonal around $S_0(x)$ in a power series in $\hbar$:

$$S = S_0 - i\hbar S_1 + (-i\hbar)^2 S_2 + (-i\hbar)^3 S_3 + \dots \ . \tag{4.14}$$

This is called the *semiclassical expansion* of the eikonal. Inserting it into the Riccati equation, we find the sequence of *WKB equations*

$$
\begin{aligned}
(\boldsymbol{\nabla} S_0)^2 - p^2 &= 0, \\
\boldsymbol{\nabla}^2 S_0 + 2\boldsymbol{\nabla} S_0 \cdot \boldsymbol{\nabla} S_1 &= 0, \\
\boldsymbol{\nabla}^2 S_1 + (\boldsymbol{\nabla} S_1)^2 + 2\boldsymbol{\nabla} S_0 \cdot \boldsymbol{\nabla} S_2 &= 0, \\
\boldsymbol{\nabla}^2 S_2 + 2\boldsymbol{\nabla} S_1 \cdot \boldsymbol{\nabla} S_2 + 2\boldsymbol{\nabla} S_0 \cdot \boldsymbol{\nabla} S_3 &= 0, \\
&\ \vdots \\
\boldsymbol{\nabla}^2 S_n + \sum_{m=0}^{n+1} \boldsymbol{\nabla} S_m \cdot \boldsymbol{\nabla} S_{n+1-m} &= 0, \\
&\ \vdots
\end{aligned}
\tag{4.15}
$$

Note that these equations involve only the vectors

$$\mathbf{q}_n = \boldsymbol{\nabla} S_n \tag{4.16}$$

and allow for a successive determination of $S_0, S_1, S_2, \dots$, giving higher and higher corrections to the eikonal $S(\mathbf{x})$.

In one dimension we recognize in (4.5) the Riccati differential equation (2.545) fulfilled by $\partial_\tau S(x)$, if we identify x with $\tau\sqrt{2M}$, and $v(\tau)$ with $E - V(\tau)$, where (4.5) reads

$$-i\hbar\partial_\tau[\partial_\tau S(\tau)] + [\partial_\tau S(\tau)]^2 = p^2(\tau) = v(\tau). \tag{4.17}$$

If we re-express the expansion terms (2.550) in terms of $w(\tau) = \sqrt{v(\tau)}$, we may replace $w(\tau), w'(\tau), \dots$ by $p(x), p'(x), \dots$, and find directly

$$
\begin{aligned}
q_0(x) &= \pm p(x), \quad q_1(x) = -\frac{1}{2}\frac{p'(x)}{p(x)} = V'/4(E-V), \\
q_2(x) &= \pm\left[\frac{1}{4}\frac{p''(x)}{p^2(x)} - \frac{3}{8}\frac{p'^2(x)}{p^3(x)}\right] = \mp\frac{4V''(E-V) + 5V'^2}{32\sqrt{2M}(E-V)^{5/2}} \equiv \mp p(x)g(x), \\
q_3(x) &= \frac{1}{2}g'(x) = \frac{15V'^3 + 18V'V''(E-V) + 4V'''(E-V)^2}{128M(E-V)^4}.
\end{aligned}
\tag{4.18}
$$

Note that $q_2(x)$ can also be written as

$$q_2(x) = \frac{1}{4S'(x)}\{S', x\},$$

where $\{h, x\}$ denotes the so-called *Schwartz derivative* [1]

$$\{h, x\} \equiv \frac{h'''}{h'} - \frac{3}{2}\left(\frac{h''}{h'}\right)^2 \tag{4.19}$$

of the function $h(x)$. The equation for $q_2(x)$ defines also the quantity $g(x) \equiv q_2(x)/p(x)$ appearing in the equation for $q_3(x)$.

It is useful to introduce also the expansion

$$q(x) \equiv S'(x) = q_0(x) + \hbar q_1(x) + \hbar^2 q_2(x) + \ldots . \tag{4.20}$$

Then the eikonal has the expansion

$$S(x) = \int dx\, q(x) = \pm \int dx\, p(x)[1 + \hbar^2 g(x)] + \hbar\frac{i}{2}[\log p(x) + \hbar^2 g(x)] \pm \ldots . \tag{4.21}$$

Keeping only terms up to the order $\hbar$, which is possible if $\hbar^2|\epsilon(x)| \ll 1$, we find the (as yet unnormalized) *WKB wave function*

$$\psi_{\text{WKB}}(x) = e^{(i/\hbar)\int^x dx' q(x')} = \frac{1}{\sqrt{p(x)}} e^{\pm(i/\hbar)\int^x dx' p(x')}. \tag{4.22}$$

In the classically accessible regime $V(x) \leq E$, this is an oscillating wave function; in the inaccessible regime $V(x) \geq E$, it decreases or increases exponentially. The transition from one to the other is nontrivial since for $V(x) \approx E$, the WKB approximation breaks down. After some analytic work[1], however, it is possible to derive simple *connection rules* for the linearly independent solutions. Let $k(x) \equiv p(x)/\hbar$ in the oscillating regime and $\kappa(x) \equiv |p(x)|/\hbar$ in the exponential regime. Suppose that there is a crossover at $x = a$ connecting an inaccessible regime on the left of $x = a$ with an accessible one on the right. Then the connection rules are

$$
\begin{array}{ccc}
V(x) > E & & V(x) < E \\[4pt]
\dfrac{1}{\sqrt{\kappa}} e^{-\int_x^a dx'\kappa} & \longleftrightarrow & \dfrac{2}{\sqrt{k}} \cos\left(\int_a^x dx'k - \dfrac{\pi}{4}\right),
\end{array} \tag{4.23}
$$

$$
\dfrac{1}{\sqrt{\kappa}} e^{\int_x^a dx'\kappa} \quad \longleftarrow\!\!\longrightarrow \quad -\dfrac{1}{\sqrt{k}} \sin\left(\int_a^x dx'k - \dfrac{\pi}{4}\right). \tag{4.24}
$$

In the opposite situation at the point $x = b$, they turn into

$$
\begin{array}{ccc}
V(x) < E & & V(x) > E \\[4pt]
\dfrac{2}{\sqrt{k}} \cos\left(\int_x^b dx'k - \dfrac{\pi}{4}\right) & \longleftarrow\!\!\longrightarrow & \dfrac{1}{\sqrt{\kappa}} e^{-\int_b^x dx'\kappa},
\end{array} \tag{4.25}
$$

$$
-\dfrac{1}{\sqrt{k}} \sin\left(\int_x^b dx'k - \dfrac{\pi}{4}\right) \quad \longleftrightarrow \quad \dfrac{2}{\sqrt{\kappa}} e^{\int_b^x dx'\kappa}. \tag{4.26}
$$

---

[1] R.E. Langer, Phys. Rev. *51*, 669 (1937). See also W.H. Furry, Phys. Rev. *71*, 360 (1947), and the textbooks S. Flügge, *Practical Quantum Mechanics*, Springer, Berlin, 1974; L.I. Schiff, *Quantum Mechanics*, McGraw-Hill, New York, 1955; N. Fröman and P.O. Fröman, *JWKB-Approximation*, North-Holland, Amsterdam, 1965.

The connection rules can be used safely only along the direction of the double arrows.

For their derivation one solves the Schrödinger equation exactly in the neighborhood of the turning points where the potential rises or falls approximately linearly. These solutions are connected with adjacent WKB wave functions. The connection formulas can also be found directly by a formal trick: When approaching the dangerous turning points, one escapes into the complex x-plane and passes around the singularities at a finite distance. This has to be sufficiently large to preserve the WKB condition (4.7), but small enough to allow for the linear approximation of the potential near the turning point. Take for example the connection rule at $x = b$. When approaching the turning point at $x = b$ from the right, the function $\kappa(x)$ is approximately proportional to $\sqrt{x - b}$. Going around this zero in the upper complex half-plane takes $\kappa(x)$ into $ik(x)$ and the wave function $\sqrt{\kappa}^{-1} e^{-\int_b^x dx' \kappa}$ becomes $e^{-i\pi/4} \sqrt{k}^{-1} e^{-i \int_b^x dx' k}$. Going around the turning point in the lower half-plane produces $e^{i\pi/4} \sqrt{k}^{-1} e^{i \int_b^x dx' k}$. The sum of the two terms is $2\sqrt{k}^{-1} \cos(\int_x^b dx' k - \pi/4)$. The argument does not show why one should use the sum rather than the average. This becomes clear only after a more detailed discussion found in quantum-mechanical textbooks[2]. The simplest derivation of the connection formulas is based on the large-distance behaviors (2.623) and (2.628) of wave the function to the right and left of a linearly rising potential and applying this to the linearly rising section of the general potential near the turning point.

In a simple potential well, the function $p(x)$ has two zeros, say one at $x = a$ and one at $x = b$. The bound-state wave functions must satisfy the boundary condition to vanish exponentially fast at $x = \pm\infty$. Imposing these, the connection formulas lead to the *semiclassical* or *Bohr-Sommerfeld quantization rule*

$$\int_a^b dx\, k(x) = (n + 1/2)\pi, \quad n = 0, \pm 1, \pm 2, \ldots . \tag{4.27}$$

It was pointed out by Dunham[3] that due to the square-root nature of the cut between a and b, this condition implies that the anticlockwise contour integral around the cut over the eikonal expansion (4.21) up to order $\hbar$, where $q(x) \approx \pm p(x) + i(\hbar/2) \log p(x)$, is an integer multiple of $\hbar$:

$$\oint dx\, q(x) = 2\pi n\hbar. \tag{4.28}$$

The contour encloses the classical turning points and no other singularities of $q(x)$. This property ensures the single-valuedness of the wave function $\psi(x) = \exp[iS(x)/\hbar] = \exp[i \int dx\, q(x)/\hbar]$ at the semiclassical level.

Moreover, the single-valuedness is a necessary property to *all* orders in the expansion (4.21). This makes (4.28) an exact quantization rule.[4] In fact, in Sub-

---

[2]See for example E. Merzbacher, *Quantum Theory*, John Wiley & Sons, New York, 1970, p. 122; M.L. Goldberger and K.M. Watson, *Collision Theory*, John Wiley & Sons, New York, 1964, p. 324. The analytic argument is given in L.D. Landau and E.M. Lifshitz, *Quantum Mechanics*, Pergamon, London, 1965, p. 158.

[3]J.L. Dunham, Phys. Rev. *41*, 21 (1932)

[4]J.R. Krieger and C. Rosenzweig, Phys. Rev. *164*, 171 (1967).

section 4.9.6 we shall arrive at the quantization condition (4.28) from a different starting point and calculate vary accurate energy levels for the quartic potential $gx^4/4$ by evaluating (4.28) to high orders in $\hbar$.

Note that the first term $-i\hbar q_1(x)$ in the semiclassical expansion (4.20) of $q(x) = S'(x)$ contributes, via Cauchy's residue theorem, a value $-\pi\hbar$ to the contour integral (4.28). This expansion term can therefore be moved from the left- to the right-hand side of (4.28) by simply changing the right-hand side from $2\pi n\hbar$ to $2\pi(n+\frac{1}{2})\hbar$. This is the origin of the difference between the Bohr-Sommerfeld quantization condition (4.27) and the old the Bohr quantization condition which has only $n\pi$ on the right-hand side of (4.27). Remarkably, the next odd expansion terms $q_3(x)$ is a pure derivative which does not contribute to the contour integral in (4.28) at all. Using all terms up to $q_4(x)$, and applying partial integrations, the quantization condition (4.28) can be re-written as

$$\oint \left\{ (E-V)^{1/2} - \tilde{\hbar}^2 \frac{V'^2}{32(E-V)^{5/2}} - \tilde{\hbar}^4 \frac{49V'^4 - 16(E-V)^2 V'V''''}{2048(E-V)^{11/2}} + \ldots \right\} = 2\pi(n+\tfrac{1}{2})\tilde{\hbar},$$

where $\tilde{\hbar} \equiv \hbar/\sqrt{2M}$. In terms of $\tilde{q}^2(x) \equiv 2Mq^2(x) = E - V(x)$, this becomes

$$\oint \left\{ \tilde{q} - \frac{\tilde{\hbar}^2}{32} \frac{(\tilde{q}^2)'^2}{\tilde{q}^5} - \frac{\tilde{\hbar}^4}{2048} \left[\frac{49(\tilde{q}^2)'^4}{\tilde{q}^{11}} - \frac{16(\tilde{q}^2)'(\tilde{q}^2)'''}{\tilde{q}^7} \right] + \ldots \right\} = 2\pi(n+\tfrac{1}{2})\tilde{\hbar}. \quad (4.29)$$

For the harmonic oscillator, the semiclassical quantization rule (4.27) gives the exact energy levels. Indeed, for an energy E, the classical crossover points with $V(x_E) = E$ are

$$x_c = \pm\sqrt{\frac{2E}{M\omega^2}}, \quad (4.30)$$

to be identified in (4.27) with a and b, respectively. Inserting further

$$k(x) = \frac{p(x)}{\hbar} = \sqrt{\frac{2M}{\hbar^2}\left(E - \frac{1}{2}M\omega^2 x^2\right)}, \quad (4.31)$$

we obtain the WKB approximation for the energy levels

$$\int_{-x_E}^{x_E} dx' k(x) = \frac{E}{\hbar\omega}\pi = (n+\tfrac{1}{2})\pi, \quad (4.32)$$

which indeed coincides with the exact ones. Only nonnegative values $n = 0, 1, 2, \ldots$ lead to oscillatory waves.

As an example consider the quartic potential $V(x) = gx^4/2$ for which the Schrödinger equation cannot be solved exactly. Inserting this into equation (4.32), we obtain

$$\frac{1}{\hbar} \int_{-x_E}^{x_E} dx \sqrt{2M(E - gx^4/4)} \equiv \nu(E)\pi = (n+\tfrac{1}{2})\pi, \quad (4.33)$$

with the turning points $\pm x_E = \pm(4E/g)^{1/4}$. The integral is done using the formula[5]

$$\int_0^1 dt\, t^{\mu-1}(1-t^\lambda)^{\nu-1} = \frac{1}{\lambda} B(\mu/\lambda, \nu) \tag{4.34}$$

which for $\mu = 1$, $\lambda = 4$, and $\nu = 3/2$ yields $(1/4)B(1/4, 3/2)$, so that the left-hand side of Eq.(4.33) can be written with $\Gamma(1/4) = 2\pi/\Gamma(3/4)\sqrt{2}$ as $\nu(E)\pi = E^{3/4}2M^{1/2}\pi^{3/2}/3\hbar g^{1/4}\Gamma^2(3/4)$, and (4.33) determines the energy with principal quantum numebr n in the Bohr-Sommerfeld approximation by the consition $\nu(E) = n + \frac{1}{2}$ resulting in

$$E_{\mathrm{BS}}^{(n)} = \hbar\omega\kappa_{\mathrm{BS}}^{(n)}\left(\frac{g\hbar}{4M^2\omega^3}\right)^{1/3}, \tag{4.35}$$

with

$$\kappa_{\mathrm{BS}}^{(n)} = \frac{1}{\pi^{2/3}}\left(\frac{3}{2}\right)^{4/3}\Gamma^{8/3}(3/4)\left(n+\tfrac{1}{2}\right)^{4/3} \approx 0.688\,253\,702 \times 2(n+\tfrac{1}{2})^{4/3}. \tag{4.36}$$

This large-n result may be compared with the precise of $\kappa^{(0)} = 0.667986\ldots$ to be derived in Section 5.19 (see Table 5.9).

If the potential contains a centrifugal term $V_{\mathrm{cf}}(r) = l(l+1)\hbar^2/2Mr^2$ in addition to the potential $V(\mathbf{x})$, the singularity at $r = 0$ is too strong to apply the WKB approximation. In addition, the factor $\hbar^2$ in front of this term ruins the systematics of the expansion terms (4.14) if one starts out with $(\nabla S_0)^2 = p^2 = 2M[E - V - l(l+1)\hbar^2/2Mr^2]$. In this situation, the semiclassical treatment needs a modification sketched.

For a rotationally symmetric potential, the wave function in the Schrödinger equation (4.1) can be factorized into a radial part $R(r)/r$ and spherical harmonic $Y_{lm}(\theta, \phi)$ [see (1.416)]. The radial wave function satisfies the radial Schrödinger equation

$$\left\{-\frac{\hbar^2}{2M}\partial_r^2 - [E - V(r) - V_{\mathrm{cf}}(r)]\right\}R(r) = 0. \tag{4.37}$$

In this equation, one moves the singularity at $r = 0$ to $\xi = -\infty$ by a coordinate transformation $r = e^\xi$, leading to the following nonsingular equation $\xi(\xi) \equiv e^{-\xi/2}R(e^\xi)$:

$$\left[-\frac{\hbar^2}{2M}\partial_\xi^2 - q^2(\xi)\right]\chi(\xi) = 0. \tag{4.38}$$

with

$$\tilde{q}^2(\xi) \equiv e^{2\xi}\left[E - V(e^\xi) - V_{\mathrm{cf}}(e^\xi) - \frac{\hbar^2 e^{-2\xi}}{8M}\right] = e^{2\xi}\left[E - V(e^\xi) - \frac{(l+\frac{1}{2})^2\hbar^2 e^{-2\xi}}{2M}\right]. \tag{4.39}$$

Inserting this into the quantization condition (4.29), only the first semiclassical term contributes, and all the higher terms vaish, leading to the *exact* energies for the

---

[5]I.S. Gradshteyn and I.M. Ryzhik, op. cit., Formula 3.251.1.

Coulomb potential $V(r) = -e^2/r$ with an additional centrifugal barrier $V_{cf}(x)$. They are the same as the energies obtained from the initial Bohr-Sommerfeld quantization condition (4.27) if we exchange the factor $l(l+1)$ in the centrifugal barrier by $(l+\frac{1}{2})^2$. This exchange is called the *Langer correction*. For a one-dimensional hydrogen atom, the singularity of the Coulomb potential at $r = 0$ is also so strong that one must change the variable r to ξ. Inserting $\tilde{q}^2(\xi) = e^{2\xi}\left[E + e^2 e^{-\xi} - \frac{1}{8}e^{-2\pi}\right]$ into the Bohr-Sommerfeld quantization condition (4.27), we obtain the equation, re-expressed in terms of r,

$$\int_{r_a}^{r_b} (dr/r)\sqrt{2(-1/8 + e^2 r + Er^2)} = \frac{\pi}{2} + \frac{\pi}{\sqrt{-2E}} = (n+\tfrac{1}{2})\pi,$$

where $r_{a,b}$ are the turning points $r_{a,b} = \left(e^2 \mp \sqrt{2e^2 + E}\right)/(-4E)$. For simplicity, we have gone to natural units with $\hbar = 1$, $M = 1$. Thus we find the semiclassical energies $E = -e^4/2n^2$ $(n = 1, 2, 3, \ldots)$, which are exact. Without the $1/8$-term from the variable change $r = e^\xi$, the integral would have been equal to $\pi/\sqrt{-2E}$ yielding the approximate energies $E = -e^4/2(n+\frac{1}{2})^2$. These will be derived once more from another semiclassical expansion at the end of Appendix 4A.

4.2   Saddle Point Approximation

Let us now look at the semiclassical expansion within the path integral approach to quantum mechanics. Consider the time evolution amplitude

$$(x_b t_b | x_a t_a) = \int \mathcal{D}x \, e^{i\mathcal{A}[x]/\hbar}, \tag{4.40}$$

imagining Planck's constant $\hbar$ to be again a free parameter which is very small compared to the typical fluctuations of the action. With $\hbar$ appearing in the denominator of an imaginary exponent we see that in the limit $\hbar \to 0$, the path integral becomes a sum of rapidly oscillating terms which will approximately cancel each other. This phenomenon is known from ordinary integrals

$$\int \frac{dx}{\sqrt{2\pi i\hbar}} e^{ia(x)/\hbar}, \tag{4.41}$$

which converge to zero for $\hbar \to 0$ according to the Riemann-Lebesgue lemma. The precise behavior is given by the saddle point expansion of integrals which we shall first recapitulate.

4.2.1   Ordinary Integrals

The evaluation of an integral of the type (4.41) proceeds for small $\hbar$ via the so-called *saddle point approximation*. In the limit $\hbar \to 0$, the integral is dominated by the extremum of the function $a(x)$ with the smallest absolute value, call it x_{cl} (assuming it to be unique, for simplicity), where

$$a'(x_{cl}) = 0. \tag{4.42}$$

In the path integral, the point x_{cl} in this example corresponds to the classical orbit for which the functional derivative vanishes. This is the reason for using the subscript cl. For x near the extremum, the oscillations of the integrand are weakest. The leading oscillatory behavior of the integral is given by

$$\int_{-\infty}^{\infty} \frac{dx}{\sqrt{2\pi i\hbar}} e^{ia(x)/\hbar} \xrightarrow{\hbar \to 0} \text{const} \times e^{ia(x_{\text{cl}})/\hbar}, \qquad (4.43)$$

with a constant proportionality factor independent of $\hbar$. This can be calculated by expanding $a(x)$ around its extremum as

$$a(x) = a(x_{\text{cl}}) + \frac{1}{2} a''(x_{\text{cl}})(\delta x)^2 + \frac{1}{3!} a^{(3)}(x_{\text{cl}})(\delta x)^3 + \dots , \qquad (4.44)$$

where $\delta x \equiv x - x_{\text{cl}}$ is the deviation from x_{cl}. It is the analog of the *quantum fluctuation* introduced in Section 2.2. Due to (4.42), the linear term in δx is absent. If $a''(x_{\text{cl}}) \neq 0$ and the higher derivatives are neglected, the integral is of the Fresnel type and can be done, yielding

$$\int_{-\infty}^{\infty} \frac{dx}{\sqrt{2\pi i\hbar}} e^{ia(x)/\hbar} \to e^{ia(x_{\text{cl}})/\hbar} \int_{-\infty}^{\infty} \frac{d\delta x}{\sqrt{2\pi i\hbar}} e^{ia''(x_{\text{cl}})(\delta x)^2/2\hbar} = \frac{e^{ia(x_{\text{cl}})/\hbar}}{\sqrt{a''(x_{\text{cl}})}}. \qquad (4.45)$$

The right-hand side is the *saddle point approximation* to the integral (4.41).

The saddle point approximation may be viewed as the consequence of the classical limit of the exponential function:

$$e^{ia(x)/\hbar} \xrightarrow{\hbar \to 0} \frac{\sqrt{2\pi i\hbar}}{\sqrt{a''(x_{\text{cl}})}} \delta(x - x_{\text{cl}}) . \qquad (4.46)$$

Corrections can be calculated perturbatively by expanding the integral in powers of $\hbar$, leading to what is called the *saddle point expansion*. For this we expand the remaining exponent in powers of δx:

$$\exp\left\{ \frac{i}{\hbar} \left[\frac{1}{3!} a^{(3)}(x_{\text{cl}})(\delta x)^3 + \frac{1}{4!} a^{(4)}(x_{\text{cl}})(\delta x)^4 + \dots \right] \right\}$$

$$= 1 + \frac{i}{\hbar} \left[\frac{1}{3!} a^{(3)}(x_{\text{cl}})(\delta x)^3 + \frac{1}{4!} a^{(4)}(x_{\text{cl}})(\delta x)^4 + \dots \right]$$

$$- \frac{1}{\hbar^2} \left[\frac{1}{72} a^{(3)}(x_{\text{cl}})^2 (\delta x)^6 + \dots \right] + \dots \qquad (4.47)$$

and perform the resulting integrals of the type

$$\int_{-\infty}^{\infty} \frac{d\delta x}{\sqrt{2\pi i\hbar}} e^{ia''(x_{\text{cl}})(\delta x)^2/2\hbar} (\delta x)^n = \begin{cases} \dfrac{(n-1)!!}{[a''(x_{\text{cl}})]^{(1+n)/2}} (i\hbar)^{n/2}, & n = \text{even}, \\ 0, & n = \text{odd}. \end{cases} \qquad (4.48)$$

Each factor δx in (4.47) introduces a power $\sqrt{\hbar/a''(x_{\text{cl}})}$. This is the average relative size of the quantum fluctuations. The increasing powers of $\hbar$ ensure the decreasing importance of the higher terms for small $\hbar$. For instance, the fourth-order term $a^{(4)}(x_{cl})(\delta x)^4/4!$ is accompanied by $\hbar$, and the lowest correction amounts to a factor

$$1 - ia^{(4)}(x_{cl})\frac{3!!}{4!}\frac{\hbar}{[a''(x_{cl})]^2}. \tag{4.49}$$

The cubic term $a^{(3)}(x_{cl})(\delta x)^3/3!$ yields a factor

$$1 + i[a^{(3)}(x_{cl})]^2\frac{5!!}{72}\frac{\hbar}{[a''(x_{cl})]^3}. \tag{4.50}$$

Thus we obtain the saddle point expansion to the integral is

$$\int_{-\infty}^{\infty}\frac{dx}{\sqrt{2\pi i\hbar}}e^{ia(x)/\hbar} = \frac{e^{ia(x_{cl})/\hbar}}{\sqrt{a''(x_{cl})}}\left\{1 - i\hbar\left[\frac{1}{8}\frac{a^{(4)}(x_{cl})}{[a''(x_{cl})]^2} - \frac{5}{24}\frac{[a^{(3)}(x_{cl})]^2}{[a''(x_{cl})]^3}\right] + \mathcal{O}(\hbar^2)\right\}. \tag{4.51}$$

Expectation values in this integral can also be expanded in powers of $\hbar$, for instance $\langle x\rangle = x_{cl} + \langle\delta x\rangle$ where

$$\langle\delta x\rangle = -i\hbar\frac{1}{2}\frac{a^{(3)}(x_{cl})}{[a''(x_{cl})]^2}$$
$$- \hbar^2\left[\frac{2}{3}\frac{a^{(3)}(x_{cl})a^{(4)}(x_{cl})}{[a''(x_{cl})]^4} - \frac{5}{8}\frac{[a^{(3)}(x_{cl})]^3}{[a''(x_{cl})]^5} - \frac{1}{8}\frac{a^{(5)}(x_{cl})}{[a''(x_{cl})]^3}\right] + \mathcal{O}(\hbar^3). \tag{4.52}$$

Since the saddle point expansion is organized in powers of $\hbar$, it corresponds precisely to the semiclassical expansion of the eikonal in the previous section.

The saddle point expansion can be used for very small $\hbar$ to calculate an integral with increasing accuracy. It is impossible, however, to achieve arbitrary accuracy since the resulting series is divergent for all physically interesting systems. It is merely an *asymptotic series* whose usefulness decreases rapidly with an increasing size of the expansion parameter. A variational expansion must be used to achieve convergence. For more details, see Sections 5.15 and 17.9.

An important property of the semiclassical approximation is that Fourier transformations become very simple. Consider the Fourier integral

$$\int_{-\infty}^{\infty}dx\, e^{-ipx/\hbar}e^{ia(x)/\hbar}. \tag{4.53}$$

For small $\hbar$, this can be done in the saddle point approximation according to the rule (4.46), and obtain

$$\int_{-\infty}^{\infty}dx\, e^{-ipx/\hbar}e^{ia(x)/\hbar} \rightarrow \sqrt{2\pi i\hbar}\frac{e^{i[a(x_{cl})-px_{cl}]/\hbar}}{\sqrt{a''(x_{cl})}}, \tag{4.54}$$

where x_{cl} is now the extremum of the action with a source term p, i.e., it is determined by the equation $p = a'(x_{cl})$. Note that the formula holds also if the exponential carries an x-dependent prefactor, since the x-dependence gives only corrections of the order of $\hbar$ in the exponent:

$$\int_{-\infty}^{\infty}dx\, e^{-ipx/\hbar}c(x)e^{ia(x)/\hbar} \rightarrow \sqrt{2\pi i\hbar}\, c(x_{cl})\frac{e^{i[a(x_{cl})-px_{cl}]/\hbar}}{\sqrt{a''(x_{cl})}}. \tag{4.55}$$

If the equation $p = a'(x_{cl})$ is inverted to find x_{cl} as a function $x_{cl}(p)$, the exponent $a(x_{cl}) - px_{cl}$ may be considered as a function of p:

$$b(p) = a(x_{cl}) - px_{cl}, \quad p = a'(x_{cl}). \tag{4.56}$$

This function is recognized as being the Legendre transform of the function $a(x)$ [recall (1.9)].

The original function $a(x)$ can be recovered from $b(p)$ via an inverse Legendre transformation

$$a(x) = b(p_{cl}) + xp_{cl}, \quad x = -b'(p_{cl}). \tag{4.57}$$

This formalism is the basis for many thermodynamic calculations. For large statistical systems, fluctuations of global properties such as the volume and the total internal energy are very small so that the saddle point approximation is very good. In this chapter, the formalism will be applied on many occasions.

4.2.2 Path Integrals

A similar saddle point expansion exists for the path integral (4.40). For small $\hbar$, the amplitudes $e^{iA/\hbar}$ from the various paths will mostly cancel each other by interference. The dominant contribution comes from the functional regime where the oscillations are weakest, which is from extremum of the action

$$\delta A[x] = 0. \tag{4.58}$$

This gives the classical Euler-Lagrange equation of motion. For a point particle with the action

$$A[x] = \int_{t_a}^{t_b} dt \left[\frac{M}{2} \dot{x}^2 - V(x) \right], \tag{4.59}$$

it reads

$$M\ddot{x} = -V'(x). \tag{4.60}$$

Let $x_{cl}(t)$ denote the classical orbit. After multiplying (4.60) by $\dot{x}$, an integration in t yields the law of energy conservation

$$E = \frac{M}{2}\dot{x}_{cl}^2 + V(x_{cl}) = \text{const}. \tag{4.61}$$

This implies that the classical momentum

$$p_{cl}(t) \equiv M\dot{x}_{cl}(t) \tag{4.62}$$

can be written as

$$p_{cl}(t) = p(x_{cl}(t)), \tag{4.63}$$

where $p(x)$ is the local classical momentum defined in (4.3). From (4.61), the time dependence of the classical orbit $x_{cl}(t)$ is given by

$$t - t_0 = \int_0^{x_{cl}} dx \frac{M}{p(x)} = \int_0^{x_{cl}} dx \frac{M}{\sqrt{2M[E - V(x)]}}. \tag{4.64}$$

When solving the integral on the right-hand side we find for a given time interval $t = t_b - t_a$ the energy for which a pair of positions x_a, x_b can be connected by a classical orbit:

$$E = E(x_b, x_a; t_b - t_a). \tag{4.65}$$

The classical action is given by

$$\begin{aligned}
\mathcal{A}[x_{cl}] &= \int_{t_a}^{t_b} dt \left[\frac{M}{2} \dot{x}_{cl}^2 - V(x_{cl}) \right] \\
&= \int_{t_a}^{t_b} dt \left[p_{cl}(t)\dot{x}_{cl} - H(p_{cl}, x_{cl}) \right] \\
&= \int_{x_a}^{x_b} dx p(x) - (t_b - t_a)E.
\end{aligned} \tag{4.66}$$

Just like E, the classical action is a function of x_b, x_a and $t_b - t_a$, to be denoted by $A(x_b, x_a; t_b - t_a)$, for which (4.66) reads more explicitly

$$A(x_b, x_a; t_b - t_a) \equiv \int_{x_a}^{x_b} dx\, p(x) - (t_b - t_a)E(x_b, x_a; t_b - t_a). \tag{4.67}$$

Recalling (4.11), the first term on the right-hand side is seen to be the classical eikonal

$$S(x_b, x_a; E) = \int_{x_a}^{x_b} dx\, p(x), \tag{4.68}$$

where E is the energy function (4.65) and $p(x)$ is given by (4.3),

The eikonal may be viewed as a functional $S_E[x]$ of paths $x(t)$ of a fixed energy, in which case it is extremal on the classical orbits. This was observed as early as 1744 by Maupertius. The proof for this is quite simple: We insert the classical momentum (4.3) into $S_E[x]$ and write

$$S_E[x] \equiv \int p(x)dx = \int dt\, p(x)\dot{x} = \int dt\, L_E(x, \dot{x}) = \int dt \sqrt{2M[E - V(x)]}\dot{x}, \tag{4.69}$$

thus introducing a Lagrangian $L_E(x, \dot{x})$ for this problem. The associated Euler-Lagrange equation reads

$$\frac{d}{dt} \frac{\partial L_E}{\partial \dot{x}} = \frac{\partial L_E}{\partial x}. \tag{4.70}$$

Inserting $L_E(x, \dot{x}) = p(x)\dot{x}$ we find the correct equation of motion $\dot{p} = -V'(x)$.

There is an interesting geometrical aspect to this variational procedure. In order to see this let us go to D dimensions and write the eikonal (4.69) as

$$S_E[\mathbf{x}] = \int dt\, L_E(\mathbf{x}, \dot{\mathbf{x}}) = \int dt \sqrt{g_{ij}(\mathbf{x})\dot{x}^i(t)\dot{x}^j(t)}, \tag{4.71}$$

with an energy-dependent metric

$$g_{ij}(\mathbf{x}) = p_E^2(\mathbf{x})\delta_{ij}. \tag{4.72}$$

Then the Euler-Lagrange equations for $\mathbf{x}(t)$ coincides with the equation (1.72) for the geodesics in a Riemannian space with a metric $g_{ij}(\mathbf{x})$. In this way, the dynamical problem has been reduced to a geometric problem. The metric $g_{ij}(\mathbf{x})$ may be called *dynamical metric* of the space with respect to the potential $V(\mathbf{x})$. This geometric view is further enhanced by the fact that the eikonal (4.69) is, in fact, independent of the parametrization of the trajectory. Instead of the time t we could have used any parameter τ to describe $\mathbf{x}(\tau)$ and write the eikonal (4.69) as

$$S_E[\mathbf{x}] = \int d\tau \, \sqrt{g_{ij}(\mathbf{x})\dot{x}^i(\tau)\dot{x}^j(\tau)}. \tag{4.73}$$

Einstein has certainly been inspired by this ancient description of classical trajectories when geometrizing the relativistic Kepler motion by attributing a dynamical Riemannian geometry to spacetime.

It is worth pointing out a subtlety in this variational principle, in view of a closely related situation to be encountered later in Chapter 10. The variations are supposed to be carried out at a fixed energy

$$E = \frac{M}{2}\dot{\mathbf{x}}^2 + V(\mathbf{x}). \tag{4.74}$$

This is a nonholonomic constraint which destroys the independence of the variation $\delta x(t)$ and $\delta \dot{x}$. They are related by

$$\dot{\mathbf{x}}\,\delta\dot{\mathbf{x}} = -\frac{1}{M}\boldsymbol{\nabla}V(\mathbf{x})\delta\mathbf{x}. \tag{4.75}$$

It is, however, possible to regain the independence by allowing for a simultaneous variation of the time argument in $\mathbf{x}(t)$ when varying $\mathbf{x(t)}$. As a consequence, we can no longer employ the standard equality $\delta\dot{x} = d\delta x/dt$ which is necessary for the derivation of the Euler-Lagrange equation (4.70). Instead, we calculate

$$\delta\dot{\mathbf{x}} = \frac{d\mathbf{x} + d\delta\mathbf{x}}{dt + d\delta t} - \dot{\mathbf{x}} = \frac{d}{dt}\delta\mathbf{x} - \dot{\mathbf{x}}\frac{d}{dt}\delta t, \tag{4.76}$$

which shows that variation and time derivatives no longer commute with each other. Combining this with the relation (4.75) we see that the variations of $\mathbf{x}$ and $\dot{\mathbf{x}}$ can be made independent if we vary t along the orbit according to the relation

$$\dot{\mathbf{x}}^2\frac{d}{dt}\delta t = \dot{\mathbf{x}}\frac{d}{dt}\delta\mathbf{x} + \frac{1}{M}\boldsymbol{\nabla}V(\mathbf{x})\delta\mathbf{x}. \tag{4.77}$$

With (4.76), the variations of the eikonal (4.71) are

$$\delta S_E[\mathbf{x}] = \int dt \, \left(\frac{\partial L_E}{\partial\dot{\mathbf{x}}}\frac{d}{dt}\delta\mathbf{x} + \frac{\partial L_E}{\partial\mathbf{x}}\delta\mathbf{x}\right) + \int dt \, \left(L_E - \frac{\partial L_E}{\partial\dot{\mathbf{x}}}\dot{\mathbf{x}}\right)\frac{d}{dt}\delta t, \tag{4.78}$$

where we have kept the usual commutativity of variation and time derivative of the time itself. In the second integral, we may set

$$-L_E + \frac{\partial L_E}{\partial\dot{\mathbf{x}}}\dot{\mathbf{x}} \equiv H_E. \tag{4.79}$$

The function H_E arises from L_E by the same combination of $L_E(\mathbf{x}, \dot{\mathbf{x}})$ and $\partial L_E / \partial \dot{\mathbf{x}}(\mathbf{x}, \dot{\mathbf{x}})$ as in a Legendre transformation which brings a Lagrangian to the associated Hamiltonian [recall (1.13)]. But in contrast to the usual procedure we do not eliminate $\dot{\mathbf{x}}$ in favor of a canonical momentum variable $\partial L_E / \partial \dot{\mathbf{x}}$ [recall (1.14)], i.e., the H_E is a function $H_E(\mathbf{x}, \dot{\mathbf{x}})$. Note that it is *not* equal to the energy.

The variation (4.78) shows that the extra variation δt of the time does not change the Euler-Lagrange equations for the above Lagrangian in Eq. (4.71), $L_E = \sqrt{2M[E - V(\mathbf{x})]\dot{\mathbf{x}}^2}$. Being linear in $\dot{\mathbf{x}}$, the associated H_E vanishes identically, so that the second term disappears and we recover the ordinary equation of motion

$$\frac{d}{dt}\frac{\partial L_E}{\partial \dot{\mathbf{x}}} = \frac{\partial L_E}{\partial \mathbf{x}}. \tag{4.80}$$

In general, however, we must keep the second term. Expressing $d\delta t/dt$ via (4.77), we find

$$\begin{aligned}
\delta S_E[\mathbf{x}] &= \int dt \left[\frac{\partial L_E}{\partial \dot{\mathbf{x}}} - H_E \frac{\dot{\mathbf{x}}}{\dot{\mathbf{x}}^2} \right] \frac{d}{dt} \delta \mathbf{x} \\
&+ \int dt \left[\frac{\partial L_E}{\partial \mathbf{x}} - H_E \frac{1}{\dot{\mathbf{x}}^2} \frac{1}{M} \boldsymbol{\nabla} V(\mathbf{x}) \right] \delta \mathbf{x},
\end{aligned} \tag{4.81}$$

and the general equation of motion becomes

$$\frac{d}{dt} \left[\frac{\partial L_E}{\partial \dot{\mathbf{x}}} - H_E \frac{\dot{\mathbf{x}}}{\dot{\mathbf{x}}^2} \right] = \frac{\partial L_E}{\partial \mathbf{x}} - H_E \frac{1}{\dot{\mathbf{x}}^2} \frac{1}{M} \boldsymbol{\nabla} V(\mathbf{x}), \tag{4.82}$$

rather than (4.80). Let us illustrate this by rewriting the eikonal as a functional

$$S_E[\mathbf{x}] = \int dt \, L'_E(\mathbf{x}, \dot{\mathbf{x}}) = M \int dt \, \dot{\mathbf{x}}^2(t), \tag{4.83}$$

which is the same functional as (4.69) as long as the energy E is kept fixed. If we insert the new Lagrangian L'_E into (4.82), we obtain the correct equation of motion

$$M\ddot{\mathbf{x}} = -\boldsymbol{\nabla} V(\mathbf{x}). \tag{4.84}$$

In this case, the equation of motion can actually be found more directly. We vary the eikonal (4.83) as follows:

$$\delta S_E[\mathbf{x}] = M \int \delta dt \, \dot{\mathbf{x}}^2 + M \int dt \, \dot{\mathbf{x}} \, \delta \dot{\mathbf{x}} + M \int dt \, \dot{\mathbf{x}} \, \delta \dot{\mathbf{x}}. \tag{4.85}$$

In the last term we insert the relation (4.76) and write

$$\delta S_E[\mathbf{x}] = M \left[\int \delta dt \, \dot{\mathbf{x}}^2 + \int dt \, \dot{\mathbf{x}} \, \delta \dot{\mathbf{x}} + \int dt \, \dot{\mathbf{x}} \frac{d}{dt} \delta \mathbf{x} - \int dt \, \dot{\mathbf{x}}^2 \frac{d}{dt} \delta t \right]. \tag{4.86}$$

The two terms containing δt cancel each other, so that relation (4.77) is no longer needed. Using now (4.75), we obtain directly the equation of motion (4.84).

With the help of the eikonal (4.68), we write the classical action (4.66) as

$$A(x_b, x_a; t_b - t_a) \equiv S(x_b, x_a; E) - (t_b - t_a)E, \tag{4.87}$$

where E is given by (4.65).

The action has the property that its derivatives with respect to the endpoints x_b, x_a at a fixed $t_b - t_a$ yield the initial and final classical momenta:

$$\frac{\partial}{\partial x_{b,a}} A(x_b, x_a; t_b - t_a) = \pm p(x_{b,a}). \tag{4.88}$$

Indeed, the differentiation gives

$$\frac{\partial A}{\partial x_b} = p(x_b) + \left[\int_{x_a}^{x_b} dx \frac{\partial p(x)}{\partial E} - (t_b - t_a) \right] \frac{\partial E}{\partial x_b}, \tag{4.89}$$

and using

$$\frac{\partial p(x)}{\partial E} = \frac{M}{p(x)} = \frac{1}{\dot{x}}, \tag{4.90}$$

we see that

$$\int_{x_a}^{x_b} dx \frac{\partial p(x)}{\partial E} = \int_{t_a}^{t_b} dt = t_b - t_a, \tag{4.91}$$

so that the bracket in (4.89) vanishes, and (4.88) is indeed fulfilled [compare also (4.12)]. The relation (4.91) implies that the eikonal (4.68) has the energy derivative

$$\frac{\partial}{\partial E} S(x_b, x_a; E) = t_b - t_a. \tag{4.92}$$

As a conjugate relation, the derivative of the action with respect to the time t_b at fixed x_b gives the energy with a minus sign [compare (4.10)]:

$$\frac{\partial}{\partial t_b} A(x_b, x_a; t_b - t_a) = -E(x_b, x_a; t_b - t_a). \tag{4.93}$$

This is easily verified:

$$\frac{\partial}{\partial t_b} A = \left[\int_{x_a}^{x_b} dx \frac{\partial p}{\partial E} - (t_b - t_a) \right] \frac{\partial E}{\partial t_b} - E = -E. \tag{4.94}$$

Thus, the classical action function $A(x_b, x_a; t_b - t_a)$ and the eikonal $S(x_b, x_a; E)$ are Legendre transforms of each other.

The equation

$$\frac{1}{2M} (\partial_x A)^2 + V(x) = \partial_t A \tag{4.95}$$

is, of course, the Hamilton-Jacobi equation (4.13) of classical mechanics.

We have therefore found the leading term in the semiclassical approximation to the amplitude [corresponding to the approximation (4.43)]:

$$(x_b t_b | x_a t_a) \xrightarrow{\hbar \to 0} \text{const} \times e^{iA(x_b, x_a; t_b - t_a)/\hbar}. \tag{4.96}$$

In general, this leading term will be multiplied by a fluctuation factor

$$(x_b t_b | x_a t_a) = e^{iA(x_b, x_a; t_b - t_a)/\hbar} F(x_b, x_a; t_b - t_a). \tag{4.97}$$

In contrast to the purely harmonic case in Eq. (2.151) this will depend on the initial and final coordinates x_a and x_b.

The calculation of the leading contribution to the fluctuation factor is the next step in the saddle point expansion of the path integral (4.40). For this we expand the action (4.59) in the neighborhood of the classical orbit in powers of the fluctuations

$$\delta x(t) = x(t) - x_{cl}(t). \tag{4.98}$$

This yields the fluctuation expansion

$$
\begin{aligned}
\mathcal{A}[x, \dot{x}] = {} & \mathcal{A}[x_{cl}] + \int_{t_a}^{t_b} dt \, \frac{\delta \mathcal{A}}{\delta x(t)} \delta x(t) \\
& + \frac{1}{2} \int_{t_a}^{t_b} dt dt' \frac{\delta^2 \mathcal{A}}{\delta x(t) \delta x(t')} \delta x(t) \delta x(t') \\
& + \frac{1}{3!} \int_{t_a}^{t_b} dt dt' dt'' \frac{\delta^3 \mathcal{A}}{\delta x(t) \delta x(t') \delta x(t'')} \delta x(t) \delta x(t') \delta x(t'') + \ldots,
\end{aligned} \tag{4.99}
$$

where all functional derivatives on the right-hand side are evaluated along the classical orbit $x(t) = x_{cl}(t)$. The linear term in the quantum fluctuation $\delta x(t)$ is absent since $\mathcal{A}[x, \dot{x}]$ is extremal at $x_{cl}(t)$. For a point particle, the quadratic term is

$$\frac{1}{2} \int_{t_a}^{t_b} dt dt' \frac{\delta^2 \mathcal{A}}{\delta x(t) \delta x(t')} \delta x(t) \delta x(t') = \int_{t_a}^{t_b} dt \left[\frac{M}{2} (\delta \dot{x})^2 + \frac{1}{2} V''(x_{cl}(t))(\delta x)^2 \right]. \tag{4.100}$$

Thus the fluctuations behave like those of a harmonic oscillator with a time-dependent frequency

$$\Omega^2(t) = \frac{1}{M} V''(x_{cl}(t)). \tag{4.101}$$

By definition, the fluctuations vanish at the endpoints:

$$\delta x(t_a) = 0, \quad \delta x(t_b) = 0. \tag{4.102}$$

If we include only the quadratic terms in the fluctuation expansion (4.99), we can integrate out the fluctuations in the path integral (4.40). Since $x(t)$ and $\delta x(t)$

differ only by a fixed additive function $x_{\rm cl}(t)$, the measure of the path integral over $x(t)$ transforms trivially into that over $\delta x(t)$. Thus we conclude that the leading semiclassical limit of the amplitude is given by the product

$$(x_b t_b | x_a t_a)_{\rm sc} = e^{iA(x_b, x_a; t_b - t_a)/\hbar} F_{\rm sc}(x_b, x_a; t_b - t_a), \qquad (4.103)$$

with the semiclassical fluctuation factor [compare (2.198)]

$$
\begin{aligned}
F_{\rm sc}(x_b, x_a; t_b - t_a) &= \int \mathcal{D}\delta x(t) \exp\left\{ \frac{i}{\hbar} \int_{t_b}^{t_a} dt \frac{M}{2} [\delta \dot{x}^2 - \Omega^2(t)\delta x^2] \right\} \\
&= \frac{1}{\sqrt{2\pi i \epsilon \hbar / M}} \det\left(-\overline{\nabla}\nabla - \Omega^2(t) \right)^{-1/2} \\
&= \frac{1}{\sqrt{2\pi i \hbar (t_b - t_a)/M}} \sqrt{\frac{\det(-\partial_t^2)}{\det(-\partial_t^2 - \Omega^2(t))}}. \qquad (4.104)
\end{aligned}
$$

In principle, we would now have to solve the differential equation

$$[-\partial_t^2 - \Omega^2(t)]y_n(t) = [-\partial_t^2 - V''(x_{\rm cl}(t))/M]y_n(t) = \lambda_n y_n(t), \qquad (4.105)$$

and find the energies of the eigenmodes $y_n(t)$ of the fluctuations. The ratio of fluctuation determinants

$$\frac{D^0}{D} = \frac{\det(-\partial_t^2)}{\det(-\partial_t^2 - \Omega^2(t))} \qquad (4.106)$$

in the second line of (4.104) would then be found from the product of ratios of eigenvalues, λ_n/λ_n^0, where λ_n^0 are the eigenvalues of the differential equation

$$-\partial_t^2 y_n(t) = \lambda_n^0 y_n(t). \qquad (4.107)$$

Fortunately, we can save ourselves all this work using the Gelfand-Yaglom method of Section 2.4 which provides a much simpler and more direct way of calculating fluctuation determinants with a time-dependent frequency without the knowledge of the eigenvalues λ_n.

4.3   Van Vleck-Pauli-Morette Determinant

According to the Gelfand-Yaglom method of Section 2.4, a functional determinant of the form

$$\det(-\partial_t^2 - \Omega^2(t))$$

is found by solving the differential equation (4.105) at zero eigenvalue

$$[-\partial_t^2 - \Omega^2(t)]D_a(t) = 0,$$

with the initial conditions

$$D_a(t_a) = 0, \qquad \dot{D}_a(t_a) = 1. \qquad (4.108)$$

Then $D_a(t_b)$ is the desired fluctuation determinant. In Eq. (2.238), we have constructed the solution to these equations in terms of an arbitrary solution $\xi(t)$ of the homogenous equation

$$[-\partial_t^2 - \Omega^2(t)]\xi(t) = 0 \tag{4.109}$$

as

$$D_{\text{ren}} = \xi(t)\xi(t_a) \int_{t_a}^{t_b} \frac{dt'}{\xi^2(t')}. \tag{4.110}$$

In general, it is difficult to find an analytic solution to Eq. (4.109). In the present fluctuation problem, however, the time-dependent frequency $\Omega(t)$ has a special form $\Omega^2(t) = V''(x_{\text{cl}}(t))/M$ of (4.101). We shall now prove that, just as in the purely harmonic action in Section 2.5, all information on the fluctuation determinant is contained in the classical orbit $x_{\text{cl}}(t)$, and ultimately in the mixed spatial derivatives $\partial_{x_b}\partial_{x_a}$ of the classical action $A(x_b, x_a; t_b - t_a)$. In fact, the solution $\xi(t)$ is simply equal to the velocity

$$\xi(t) = \dot{x}_{\text{cl}}(t). \tag{4.111}$$

This is seen directly by differentiating the equation of motion (4.60) with respect to t, yielding

$$\partial_t[M\ddot{x}_{\text{cl}} + V'(x_{\text{cl}}(t))] = [M\partial_t^2 + V''(x_{\text{cl}}(t))]\dot{x}_{\text{cl}}(t) = 0, \tag{4.112}$$

which is precisely the homogenous differential equation (4.109) for $\dot{x}_{\text{cl}}(t)$.

There is a simple symmetry argument to understand (4.111) as a completely general consequence of the time translation invariance of the system. The fluctuation $\delta x(t) \propto \dot{x}_{\text{cl}}(t)$ describes an infinitesimal translation of the classical solution $x_{\text{cl}}(t)$ in time, $x_{\text{cl}}(t) \to x_{\text{cl}}(t+\epsilon) = x_{\text{cl}}+\epsilon\dot{x}_{\text{cl}}+\dots$. Interpreted as a *translational fluctuation* of the solution $x_{\text{cl}}(t)$ along the time axis it cannot carry any energy λ_n and $y_0(t) \propto \dot{x}_{\text{cl}}(t)$ must therefore solve Eq. (4.105) with $\lambda_0 = 0$.

With the special solution (4.111), the functional determinant (4.110) becomes

$$D_{\text{ren}} = \dot{x}_{\text{cl}}(t_b)\dot{x}_{\text{cl}}(t_a) \int_{t_a}^{t_b} \frac{dt}{\dot{x}_{\text{cl}}^2(t)}. \tag{4.113}$$

Note that also the Green-function of the quadratic fluctuations associated with Eq. (4.109) can be given explicitly in terms of the classical solution $x_{\text{cl}}(t)$. For Dirichlet boundary conditions, it is equal to the combination (3.61) of the solutions $D_a(t)$ and $D_b(t)$ of the homogeneous differential equation (4.109) satisfying the boundary conditions (2.226) and (2.227), whose d'Alembert construction (2.237) becomes here

$$D_a(t) = \dot{x}_{\text{cl}}(t)\dot{x}_{\text{cl}}(t_a) \int_{t_a}^{t} \frac{dt}{\dot{x}_{\text{cl}}^2(t)}, \quad D_b(t) = \dot{x}_{\text{cl}}(t_b)\dot{x}_{\text{cl}}(t) \int_{t}^{t_b} \frac{dt}{\dot{x}_{\text{cl}}^2(t)}. \tag{4.114}$$

In Eqs. (2.250) and (2.267) we have found two simple expressions for the fluctuation determinant in terms of the classical action

$$D_{\text{ren}} = -\left(\frac{\partial \dot{x}_b}{\partial x_a}\right)^{-1} = -M\left[\frac{\partial^2}{\partial x_b \partial x_a}\mathcal{A}_{\text{cl}}\right]^{-1}. \tag{4.115}$$

These were derived for purely quadratic actions with an arbitrary time-dependent frequency $\Omega^2(t)$. But they hold for any action. First, the equality between the second and third expression is a consequence of the general relation (4.88). Second, we may consider the semiclassical approximation to the path integral as an *exact* path integral associated with the lowest quadratic approximation to the action in (4.99), (4.100):

$$\mathcal{A}_{\mathrm{qu}}[x, \dot{x}] = \mathcal{A}[x_{\mathrm{cl}}] + \int_{t_a}^{t_b} dt \left[\frac{M}{2}(\delta\dot{x})^2 + \Omega^2(t)(\delta x)^2 \right], \qquad (4.116)$$

with $\Omega^2(t) = V''(x_{\mathrm{cl}}(t))/M$ of (4.101). Then, since the classical orbit running from x_a to x_b satisfies the equation of motion (4.112), also a slightly different orbit $(x_{\mathrm{cl}} + \delta x_{\mathrm{cl}})(t)$ from $x'_a = x_a + \delta x_a$ to $x'_b = x_a + \delta x_b$ satisfies (4.112). Although the small change of the classical orbit gives rise to a slightly different frequency $\Omega^2(t) = V''((x_{\mathrm{cl}} + \delta x_{\mathrm{cl}})(t))/M$, this contributes only to second order in δx_a and δx_b. As a consequence, the derivative $D_a(t) = -\partial \dot{x}_b(t)/\partial x_a$ satisfies Eq. (4.112) as well. Also the boundary conditions of $D_a(t)$ are the same as those of $D_a(t)$ in Eqs. (2.226). Hence the quantity $D_a(t_b)$ is the correct fluctuation determinant also for the general action in the semiclassical approximation under study.

Another way to derive this formula makes use of the general relation (4.88), from which we find

$$\frac{\partial}{\partial x_b}\frac{\partial}{\partial x_a} A(x_b, x_a; t_b - t_a) = \frac{\partial}{\partial x_a}p(x_b) = \frac{M}{p(x_b)}\frac{\partial E}{\partial x_a}. \qquad (4.117)$$

On the right-hand side we have suppressed the arguments of the function $E(x_b, x_a; t_b - t_a)$. After rewriting

$$\begin{aligned} \frac{\partial E}{\partial x_a} &= -\frac{\partial}{\partial x_a}\frac{\partial A}{\partial t_b} = -\frac{\partial}{\partial t_b}\frac{\partial A}{\partial x_a} \\ &= \frac{\partial}{\partial t_b}p(x_a) = \frac{M}{p(x_a)}\frac{\partial E}{\partial t_b}, \end{aligned} \qquad (4.118)$$

we see that

$$\frac{\partial}{\partial x_b}\frac{\partial}{\partial x_a} A(x_b, x_a; t_b - t_a) = \frac{1}{\dot{x}(t_b)\dot{x}(t_a)}\frac{\partial E}{\partial t_b}. \qquad (4.119)$$

From (4.64) we calculate

$$\begin{aligned} \frac{\partial E}{\partial t_b} &= \left(\frac{\partial t_b}{\partial E}\right)^{-1} = \left[-\int_{x_a}^{x_b} dx \frac{M}{p^2}\frac{\partial p}{\partial E}\right]^{-1} \\ &= \left[-\int_{x_a}^{x_b} dx \frac{M^2}{p^3}\right]^{-1} = \left[-M\int_{t_a}^{t_b}\frac{dt}{p^2}\right]^{-1} = \left[-\frac{1}{M}\int_{t_a}^{t_b}\frac{dt}{\dot{x}_{\mathrm{cl}}^2(t)}\right]^{-1}. \end{aligned} \qquad (4.120)$$

Inserting this into (4.119), we obtain once more formula (4.115) for the fluctuation determinant.

A relation following from (4.92):

$$\frac{\partial E}{\partial t_b} = \left(\frac{\partial^2 S}{\partial E^2}\right)^{-1}, \tag{4.121}$$

leads to an alternative expression

$$D_{\mathrm{ren}} = -M\dot{x}_{\mathrm{cl}}(t_b)\dot{x}_{\mathrm{cl}}(t_a)\frac{\partial^2 S}{\partial E^2}. \tag{4.122}$$

The fluctuation factor is therefore also here [recall the normalization from Eqs. (2.200), (2.202), and (2.210)]

$$F(x_b, t_a; t_b - t_a) = \frac{1}{\sqrt{2\pi i\hbar/M}}\left[\frac{\partial \dot{x}_b}{\partial x_a}\right]^{1/2} = \frac{1}{\sqrt{2\pi i\hbar}}[-\partial_{x_b}\partial_{x_a}A(x_b, x_a; t_b - t_a)]^{1/2}. \tag{4.123}$$

Its D-dimensional generalization of (4.123) is

$$F(\mathbf{x}_b, \mathbf{x}_a; t_b - t_a) = \frac{1}{\sqrt{2\pi i\hbar}^D}\left\{\det_D[-\partial_{x_b^i}\partial_{x_a^j}A(\mathbf{x}_b, \mathbf{x}_a; t_b - t_a)]\right\}^{1/2}, \tag{4.124}$$

and the semiclassical time evolution amplitude reads

$$(\mathbf{x}_b t_b | \mathbf{x}_a t_a) = \frac{1}{\sqrt{2\pi i\hbar}^D}\left\{\det_D[-\partial_{x_b^i}\partial_{x_a^j}A(\mathbf{x}_b, \mathbf{x}_a; t_b - t_a)]\right\}^{1/2}e^{iA(\mathbf{x}_b, \mathbf{x}_a; t_b - t_a)/\hbar}. \tag{4.125}$$

The $D \times D$-determinant in the curly brackets is the so-called *Van Vleck-Pauli-Morette determinant*.[6] It is the analog of the determinant in the right-hand part of Eq. (2.268). As discussed there, the result is initially valid only as long as the fluctuation determinant is regular. Otherwise we must replace the determinant by its absolute value, and multiply the fluctuation factor by the phase factor $e^{-i\nu/2}$ with the Maslov-Morse index ν [see Eq. (2.269)]. Using the relation (4.88) in D dimensions

$$\partial_{x_b^i}\partial_{x_a^j}A(\mathbf{x}_b, \mathbf{x}_a; t_b - t_a) = \frac{\partial p_b^i}{\partial x_a^j}, \tag{4.126}$$

we shall often write (4.125) as

$$(\mathbf{x}_b t_b | \mathbf{x}_a t_a) = \frac{1}{\sqrt{2\pi i\hbar}^D}\left[\det_D\left(-\frac{\partial \mathbf{p}_b}{\partial \mathbf{x}_a}\right)\right]^{1/2}e^{iA(\mathbf{x}_b, \mathbf{x}_a; t_b - t_a)/\hbar}, \tag{4.127}$$

where the subscripts a and b can be interchanged in the determinant, if the sign is changed [recall (2.250)]. This concludes the calculation of the semiclassical approximation to the time evolution amplitude.

---

[6]J.H. Van Vleck, Proc. Nat. Acad. Sci. (USA) *14*, 178 (1928); W. Pauli, *Selected Topics in Field Quantization*, MIT Press, Cambridge, Mass. (1973); C. DeWitt-Morette, Phys. Rev. *81*, 848 (1951).

As a simple application, we use this formula to write down the semiclassical amplitude for a free particle and a harmonic oscillator. The first has the classical action

$$A(x_b, x_a; t_b - t_a) = \frac{M}{2} \frac{(x_b - x_a)^2}{t_b - t_a}, \qquad (4.128)$$

and Eq. (4.115) gives

$$D_{\text{ren}} = t_b - t_a, \qquad (4.129)$$

as it should. The harmonic-oscillator action is

$$A(x_b, x_a; t_b - t_a) = \frac{M\omega}{2 \sin \omega(t_b - t_a)} \left[(x_b^2 + x_a^2) \cos \omega(t_b - t_a) - 2x_b x_a \right], \quad (4.130)$$

and has the second derivative

$$- \partial_{x_b} \partial_{x_a} A = \frac{M\omega}{\sin \omega(t_b - t_a)}, \qquad (4.131)$$

so that (4.123) coincides with fluctuation factor (2.214).

4.4   Fundamental Composition Law for Semiclassical Time Evolution Amplitude

The determinant ensures that the semiclassical approximation for the time evolution amplitude satisfies the fundamental composition law (2.4) in D dimensions

$$(x_b t_b | x_a t_a) = \prod_{n=1}^{N} \left[\int_{-\infty}^{\infty} d^D x_n \right] \prod_{n=1}^{N+1} (x_n t_n | x_{n-1} t_{n-1}), \qquad (4.132)$$

if the intermediate x-integrals are evaluated in the saddle point approximation. To leading order in $\hbar$, only those intermediate x-values contribute which lie on the classical trajectory determined by the endpoints of the combined amplitude. To next order in $\hbar$, the quadratic correction to the intermediate integrals renders an inverse square root of the fluctuation determinant. If two such amplitudes are connected with each other by an intermediate integration according to the composition law (4.132), the product of the two fluctuation factors turns into the correct fluctuation factor of the combined time interval. This is seen after rewriting the matrix $\partial_{x_b^i} \partial_{x_a^j} A(x_b, x_a; t_b - t_a)$ with the help of (4.12) as $\partial p_b / \partial x_a$. The intermediate integral over x in the product of two amplitudes receives a contribution only from continuous paths since, at the saddle point, the adjacent momenta have to be equal:

$$\frac{\partial}{\partial x_b} A(x_b, x; t_b - t) + \frac{\partial}{\partial x} A(x, x_a; t - t_a) = -p'(x_b, x; t_b - t) + p(x, x_a; t - t_a) = 0.$$
$$(4.133)$$

To obtain the combined amplitude, we obviously need the relation

$$\det{}_D\left(-\left.\frac{\partial \mathbf{p}_b}{\partial \mathbf{x}}\right|_{\mathbf{x}_b}\right)\left\{\det{}_D\left(-\left.\frac{\partial \mathbf{p}'}{\partial \mathbf{x}}\right|_{\mathbf{x}_b}+\left.\frac{\partial \mathbf{p}}{\partial \mathbf{x}}\right|_{\mathbf{x}_a}\right)_{\mathbf{p}'=\mathbf{p}}\right\}^{-1}\det{}_D\left(-\left.\frac{\partial \mathbf{p}}{\partial \mathbf{x}_a}\right|_{\mathbf{x}}\right)$$
$$= \det{}_D\left(-\left.\frac{\partial \mathbf{p}_b}{\partial \mathbf{x}_a}\right|_{\mathbf{x}_b}\right), \tag{4.134}$$

where we have indicated explicitly the variables kept fixed in $\mathbf{p}'(\mathbf{x}_b, \mathbf{x}; t_b - t)$ and $\mathbf{p}(\mathbf{x}, \mathbf{x}_a; t - t_a)$ when forming the partial derivatives. To prove (4.134), we use the product rule for determinants

$$\det{}_D^{-1}\left(-\left.\frac{\partial \mathbf{p}_b}{\partial \mathbf{x}}\right|_{\mathbf{x}_b}\right)\det{}_D\left(-\left.\frac{\partial \mathbf{p}_b}{\partial \mathbf{x}_a}\right|_{\mathbf{x}_b}\right)=\det{}_D\left(\left.\frac{\partial \mathbf{x}}{\partial \mathbf{x}_a}\right|_{\mathbf{x}_b}\right) \tag{4.135}$$

to rewrite (4.134) as

$$\det{}_D\left(-\left.\frac{\partial \mathbf{p}}{\partial \mathbf{x}_a}\right|_{\mathbf{x}}\right)=\det{}_D\left(-\left.\frac{\partial \mathbf{p}'}{\partial \mathbf{x}}\right|_{\mathbf{x}_b}+\left.\frac{\partial \mathbf{p}}{\partial \mathbf{x}}\right|_{\mathbf{x}_a}\right)_{\mathbf{p}'=\mathbf{p}}\det{}_D\left(\left.\frac{\partial \mathbf{x}}{\partial \mathbf{x}_a}\right|_{\mathbf{x}_b}\right). \tag{4.136}$$

This equation is true due to the chain rule of differentiation applied to the momentum $\mathbf{p}'(\mathbf{x}_b, \mathbf{x}; t_b - t) = \mathbf{p}(\mathbf{x}, \mathbf{x}_a; t - t_a)$, after expressing $\mathbf{p}(\mathbf{x}, \mathbf{x}_a; t - t_a)$ explicitly in terms of the variables $\mathbf{x}_b$ and $\mathbf{x}_a$ as $\mathbf{p}(\mathbf{x}(\mathbf{x}_b, \mathbf{x}_a; t_b - t_a), \mathbf{x}_a; t - t_a)$, to enable us to hold $\mathbf{x}_b$ fixed in the second partial derivative:

$$\left.\frac{\partial \mathbf{p}'}{\partial \mathbf{x}}\right|_{\mathbf{x}_b}=\left.\frac{\partial \mathbf{p}}{\partial \mathbf{x}}\right|_{\mathbf{x}_b}=\left.\frac{\partial \mathbf{p}(\mathbf{x}(\mathbf{x}_b, \mathbf{x}_a; t_b - t_a), \mathbf{x}_a; t - t_a)}{\partial \mathbf{x}}\right|_{\mathbf{x}_b}=\left.\frac{\partial \mathbf{p}}{\partial \mathbf{x}}\right|_{\mathbf{x}}\left.\frac{\partial \mathbf{x}_a}{\partial \mathbf{x}}\right|_{\mathbf{x}_b}+\left.\frac{\partial \mathbf{p}}{\partial \mathbf{x}}\right|_{\mathbf{x}_a}.$$
$$\tag{4.137}$$

It may be expected, and can indeed be proved, that it is possible to proceed in the opposite direction and derive the semiclassical expressions (4.125) and (4.127) with the Van Vleck-Pauli-Morette determinant from the fundamental composition law (4.132).[7]

In the semiclassical approximation, the composition law (4.132) can also be written as a temporal integral (in D dimensions)

$$(\mathbf{x}_b t_b | \mathbf{x}_a t_a) = \int dt (\mathbf{x}_b t_b | \mathbf{x}_{\mathrm{cl}}(t) t)\,\dot{\mathbf{x}}_{\mathrm{cl}}(t)\,(\mathbf{x}_{\mathrm{cl}}(t) t | \mathbf{x}_a t_a) \tag{4.138}$$

over a classical orbit $\mathbf{x}_{\mathrm{cl}}(t)$, where the t-integration is done in the saddle point approximation, assuming that the fluctuation determinant does not happen to be degenerate.

Just as in the saddle point expansion of ordinary integrals, it is possible to calculate higher corrections in $\hbar$. The result is a saddle point expansion of the path

---

[7]H. Kleinert and B. Van den Bossche, Berlin preprint 2000 (http://www.physik.fu-berlin.de/~kleinert/301).

integral which is again a semiclassical expansion. The counting of the $\hbar$-powers is the same as for the integral. The lowest approximation is of the exponential form $e^{i\mathcal{A}_{\text{cl}}/\hbar}$. Thus, in the exponent, the leading term is of order $1/\hbar$.[8] The fluctuation factor F contributes to this an additive term $\log F$, which is of order $\hbar^0$. To first order in $\hbar$, one finds expressions containing the third and fourth functional derivative of the action in the expansion (4.99), corresponding to the expressions (4.49) and (4.50) in the integral. Unfortunately, the functional case offers little opportunity for further analytic corrections, so we shall not dwell on this more academic possibility.

4.5 Semiclassical Fixed-Energy Amplitude

As pointed out at the end of Subsection 4.2.1, we have observed that the semiclassical approximation allows for a simple evaluation of Fourier integrals. As an application of the rules presented there, let us evaluate the Fourier transform of the time evolution amplitude, the fixed-energy amplitude introduced in (1.307). It is given by the temporal integral

$$(x_b|x_a)_E = \frac{1}{\sqrt{2\pi i\hbar}} \int_{t_a}^{\infty} dt_b [-\partial_{x_b}\partial_{x_a} A(x_b, x_a; t_b - t_a)]^{1/2}$$
$$\times e^{i[A(x_b,x_a;t_b-t_a)+(t_b-t_a)E]/\hbar}, \tag{4.139}$$

which may be evaluated in the same saddle point approximation as the path integral. The extremum lies at

$$\frac{\partial}{\partial t} A(x_b, x_a; t_b - t_a) = -E. \tag{4.140}$$

Because of (4.93), the left-hand side is the function $-E(x_b, x_a; t_b - t_a)$. At the extremum, the time interval $t_b - t_a$ is some function of the endpoints and the energy E:

$$t_b - t_a = t(x_b, x_a; E). \tag{4.141}$$

The exponent is equal to the eikonal function $S(x_b, x_a; E)$ of Eq. (4.87), whose derivative with respect to the energy gives [recalling (4.92)]

$$\frac{\partial}{\partial E} S(x_b, x_a; E) = t(x_b, x_a; E). \tag{4.142}$$

The expansion of the exponent around the extremum has the quadratic term

$$\frac{i}{\hbar} \frac{\partial^2 A(x_b, x_a; t_b - t_a)}{\partial t_b^2} [t_b - t_a - t(x_b, x_a; E)]^2. \tag{4.143}$$

The time integral over t_b yields a factor

$$\sqrt{2\pi i\hbar} \left[\frac{\partial^2 A(x_b, x_a; t_b - t_a)}{\partial t_b^2} \right]^{-1/2}. \tag{4.144}$$

---

[8]Since $\hbar$ has the dimension of an action, the dimensionless number $\hbar/\mathcal{A}_{\text{cl}}$ should really be used as an appropriate dimensionless expansion parameter, but it has become customary to count directly the orders in $\hbar$.

With this, the fixed-energy amplitude has precisely the form (4.55):

$$(x_b|x_a)_E = \left[-\partial_{x_b}\partial_{x_a} A(x_b, x_a; t)/\partial_t^2 A(x_b, x_a; t)\right]^{1/2} e^{iS(x_b, x_a; E)/\hbar}. \tag{4.145}$$

Since the fluctuation factor has to be evaluated at a fixed energy E, it is advantageous to express it in terms of $S(x_b, x_a; E)$. For $\partial_t^2 A$, the evaluation is simple since

$$\frac{\partial^2 A}{\partial t^2} = -\frac{\partial E}{\partial t} = -\left(\frac{\partial t}{\partial E}\right)^{-1} = -\left(\frac{\partial^2 S}{\partial E^2}\right)^{-1}. \tag{4.146}$$

For $\partial_{x_b}\partial_{x_a} A$, we observe that the spatial derivatives of the action must be performed at a fixed time, so that a variation of x_b implies also a change of the energy $E(x_b, x_a; t)$. This is found from the condition

$$\frac{\partial t}{\partial x_b} = 0, \tag{4.147}$$

which after inserting (4.142), goes over into

$$\left.\frac{\partial^2 S}{\partial E \partial x_b}\right|_t = \frac{\partial^2 S}{\partial E \partial x_b} + \frac{\partial^2 S}{\partial E^2}\frac{\partial E}{\partial x_b} = 0. \tag{4.148}$$

We now use the relation

$$\left.\frac{\partial A}{\partial x_b}\right|_t = \left.\frac{\partial S}{\partial x_b}\right|_t - \left.\frac{\partial E}{\partial x_b}\right|_t t = \frac{\partial S}{\partial x_b} + \frac{\partial S}{\partial E}\left.\frac{\partial E}{\partial x_b}\right|_t - \left.\frac{\partial E}{\partial x_b}\right|_t t = \left.\frac{\partial S}{\partial x_b}\right|_E \tag{4.149}$$

and find from it

$$\begin{aligned}
\left.\frac{\partial^2 A}{\partial x_b \partial x_a}\right|_t &= \frac{\partial^2 S}{\partial x_b \partial x_a} + \frac{\partial^2 S}{\partial x_b \partial E}\frac{\partial E}{\partial x_a} \\
&= \frac{\partial^2 S}{\partial x_b \partial x_a} - \frac{\partial^2 S}{\partial x_a \partial E}\frac{\partial^2 S}{\partial x_b \partial E}\bigg/\frac{\partial^2 S}{\partial E^2}.
\end{aligned} \tag{4.150}$$

Thus the fixed-energy amplitude (4.139) takes the simple form

$$(x_b|x_a)_E = D_S^{1/2} e^{iS(x_b, x_a; E)/\hbar}, \tag{4.151}$$

with the 2×2-determinant

$$D_S = \begin{vmatrix} \dfrac{\partial^2 S}{\partial x_b \partial x_a} & \dfrac{\partial^2 S}{\partial E \partial x_a} \\[2ex] \dfrac{\partial^2 S}{\partial x_b \partial E} & \dfrac{\partial^2 S}{\partial E^2} \end{vmatrix}. \tag{4.152}$$

The determinant can be simplified by the fact that a differentiation of the Hamilton-Jacobi equation

$$H\left(\frac{\partial S}{\partial x_b}, x_b\right) = E \tag{4.153}$$

with respect to x_a leads to the equation

$$\frac{\partial H}{\partial p_b} \frac{\partial^2 S}{\partial x_b \partial x_a} = \dot{x}_b \frac{\partial^2 S}{\partial x_b \partial x_a} = 0. \tag{4.154}$$

It implies the vanishing of the upper left element in (4.152), reducing D_S to

$$D_S = -\frac{\partial^2 S}{\partial x_b \partial E} \frac{\partial^2 S}{\partial x_a \partial E}. \tag{4.155}$$

Since $\partial S/\partial x_{b,a} = \pm p_{b,a}$ and $\partial p/\partial E = 1/\dot{x}$, one arrives at

$$D_S = \frac{1}{\dot{x}_b \dot{x}_a}. \tag{4.156}$$

Let us calculate the semiclassical fixed-energy amplitude for a free particle. The classical action function is

$$A(x_b, x_a; t_b - t_a) = \frac{M}{2} \frac{(x_b - x_a)^2}{t_b - t_a}, \tag{4.157}$$

so that the function $E(x_b, x_a; t_b - t_a)$ is given by

$$E(x_b, x_a; t_b - t_a) = -\frac{\partial}{\partial t_b} \frac{M}{2} \frac{(x_b - x_a)^2}{t_b - t_a} = \frac{M}{2} \frac{(x_b - x_a)^2}{(t_b - t_a)^2}. \tag{4.158}$$

By a Legendre transformation, or directly from the defining equation (4.68), we calculate

$$S(x_b, x_a; E) = \sqrt{2ME}|x_b - x_a|. \tag{4.159}$$

From this we calculate the determinant (4.156) as

$$D_s = \frac{M}{2E}, \tag{4.160}$$

and the fixed-energy amplitude (4.151) becomes

$$(x_b|x_a)_E = \sqrt{\frac{M}{2E}} e^{i\sqrt{2ME}|x_b - x_a|/\hbar}. \tag{4.161}$$

4.6 Semiclassical Amplitude in Momentum Space

The simple way of finding Fourier transforms in the semiclassical approximation can be used to derive easily amplitudes in momentum space. Consider first the time evolution amplitude $(x_b t_b | x_a t_a)_{\text{sc}}$. The momentum space version is given by the two-dimensional Fourier integral [recall (2.37) and insert (4.103)]

$$(p_b t_b | p_a t_a)_{\text{sc}} = \int dx_b\, dx_a\, e^{-i(p_b x_b - p_a x_a)/\hbar} e^{iA(x_b, x_a; t_b - t_a)/\hbar} F(x_b, x_a; t_b - t_a). \tag{4.162}$$

The semiclassical evaluation according to the general rule (4.55) yields

$$(p_b t_b | p_a t_a)_{\mathrm{sc}} = \frac{\sqrt{2\pi i\hbar}}{\sqrt{\det H}} [-\partial_{x_b}\partial_{x_a} A(x_b, x_a; t_b - t_a)]^{1/2} e^{i[A(x_b, x_a; t_b - t_a) - p_b x_b + p_a x_a]/\hbar}, \quad (4.163)$$

where H is the matrix

$$H = \begin{pmatrix} \partial_{x_b}^2 A(x_b, x_a; t_b - t_a) & \partial_{x_b}\partial_{x_a} A(x_b, x_a; t_b - t_a) \\ \partial_{x_a}\partial_{x_b} A(x_b, x_a; t_b - t_a) & \partial_{x_a}^2 A(x_b, x_a; t_b - t_a) \end{pmatrix}. \quad (4.164)$$

The exponent must be evaluated at the extremum with respect to x_b and x_a, which lies at

$$p_b = \partial_{x_b} A(x_b, x_a; t_b - t_a), \qquad p_a = -\partial_{x_b} A(x_b, x_a; t_b - t_a). \quad (4.165)$$

The exponent contains then the Legendre transform of the action $A(x_b, x_a; t_b - t_a)$ which depends naturally on p_b and p_a:

$$A(p_b, p_a; t_b - t_a) = A(x_b, x_a; t_b - t_a) - p_b x_b + p_a x_a. \quad (4.166)$$

The inverse Legendre transformation to (4.165) is

$$x_b = -\partial_{p_b} A(x_b, x_a; t_b - t_a), \qquad x_a = \partial_{x_b} A(x_b, x_a; t_b - t_a). \quad (4.167)$$

The important observation which greatly simplifies the result is that for a 2×2 matrix H_{ab} with $(a, b = 1, 2)$, the matrix element $-H_{12}/\det H$ is equal to H_{12}. By writing the matrix H and its inverse as

$$H = \begin{pmatrix} \dfrac{\partial p_b}{\partial x_b} & \dfrac{\partial p_b}{\partial x_a} \\ -\dfrac{\partial p_a}{\partial x_b} & -\dfrac{\partial p_a}{\partial x_a} \end{pmatrix}, \qquad H^{-1} = \begin{pmatrix} \dfrac{\partial x_b}{\partial p_b} & -\dfrac{\partial x_b}{\partial p_a} \\ \dfrac{\partial x_a}{\partial p_b} & -\dfrac{\partial x_a}{\partial p_a} \end{pmatrix}, \quad (4.168)$$

we see that, just as in the Eqs. (2.276) and (2.277):

$$H_{12}^{-1} = \frac{\partial x_a}{\partial p_b} = \frac{\partial^2 A(p_b, p_a; t_b - t_a)}{\partial p_b \partial p_a}. \quad (4.169)$$

As a result, the semiclassical time evolution amplitude in momentum space (4.163) takes the simple form

$$(p_b t_b | p_a t_a)_{\mathrm{sc}} = \frac{2\pi\hbar}{\sqrt{2\pi i\hbar}} [-\partial_{p_b}\partial_{p_a} A(p_b, p_a; t_b - t_a)]^{1/2} e^{iA(p_b, p_a; t_b - t_a)/\hbar}. \quad (4.170)$$

In D dimensions, this becomes

$$(\mathbf{p}_b t_b | \mathbf{p}_a t_a) = \frac{1}{\sqrt{2\pi i\hbar}^D} \left\{ \det{}_D[-\partial_{p_b^i}\partial_{p_a^j} A(\mathbf{p}_b, \mathbf{p}_a; t_b - t_a)] \right\}^{1/2} e^{iA(\mathbf{p}_b, \mathbf{p}_a; t_b - t_a)/\hbar}, \quad (4.171)$$

or

$$(\mathbf{p}_b t_b | \mathbf{p}_p t_a) = \frac{1}{\sqrt{2\pi i\hbar}^D} \left\{ \det_D \left[-\frac{\partial \mathbf{p}_b}{\partial \mathbf{x}_a} \right] \right\}^{1/2} e^{iA(\mathbf{p}_b, \mathbf{p}_a; t_b - t_a)/\hbar}, \qquad (4.172)$$

these results being completely analogous to the x-space expression (4.125) and (4.127), respectively. As before, the subscripts a and b can be interchanged in the determinant.

If we apply these formulas to the harmonic oscillator with a time-dependent frequency, we obtain precisely the amplitude (2.283). Thus in this case, the semiclassical time evolution amplitude $(p_b t_b | p_a t_a)_{sc}$ happens to coincide with the exact one.

For a free particle with the action $A(x_b, x_a; t_b - t_a) = M(x_b - x_a)^2/2(t_b - t_a)$, the formula (4.163) cannot be applied since determinant of H vanishes, so that the saddle point approximation is inapplicable. The formal infinity one obtains when trying to apply Eq. (4.163) is a reflection of the δ-function in the exact expression (2.138), which has no semiclassical approximation. The Legendre transform of the action can, however, be calculated correctly and yields via the derivatives $p_a = p_b \equiv p = A(x_b, x_a; t_b - t_a) = M(x_b - x_a)/2(t_b - t_a)$ the expression

$$A(p_b, p_a; t_b - t_a) = -\frac{p^2}{2}(t_b - t_a), \qquad (4.173)$$

which agrees with the exponent of (2.138).

4.7 Semiclassical Quantum-Mechanical Partition Function

From the result (4.103) we can easily derive the quantum-mechanical partition function (1.537) in semiclassical approximation:

$$Z_{QM}^{sc}(t_b - t_a) = \int dx_a (x_a t_b | x_a t_a)_{sc} = \int dx_a F(x_a, x_a; t_b - t_a) e^{iA(x_a, x_a; t_b - t_a)/\hbar}. \qquad (4.174)$$

Within the semiclassical approximation the path integral, as the final trace integral may be performed using the saddle point approximation. At the saddle point one has [as in (4.133)]

$$\frac{\partial}{\partial x_a} A(x_a, x_a; t_b - t_a) = \frac{\partial}{\partial x_b} A(x_b, x_a; t_b - t_a) \Big|_{x_b = x_a} + \frac{\partial}{\partial x_a} A(x_b, x_a; t_b - t_a) \Big|_{x_b = x_a}$$
$$= p_b - p_a = 0, \qquad (4.175)$$

i.e., only classical orbits contribute whose momenta are equal at the coinciding endpoints. This restricts the orbits to periodic solutions of the equations of motion. The semiclassical limit selects, among all paths with $x_a = x_b$, the paths solving the equation of motion, ensuring the continuity of the internal momenta along these paths. The integration in (4.174) enforces the equality of the initial and final momenta on these paths and permits a continuation of the equations of motion beyond the final time t_b in a periodic fashion, leading to periodic orbits. Along each of these orbits, the energy $E(x_a, x_a, t_b - t_a)$ and the action $A(x_a, x_a, t_b - t_a)$ do not depend on the choice of x_a. The phase factor $e^{iA/\hbar}$ in the integral (4.174) is therefore a constant. The integral must be performed over a full period between the turning points of each orbit in the forward and backward direction. It contains a nontrivial x_a-dependence only in the fluctuation factor. Thus, (4.174) can be written as

$$Z_{\rm QM}^{\rm sc}(t_b - t_a) = \left[\int dx_a F(x_a, x_a; t_b - t_a) \right] e^{iA(x_a, x_a; t_b - t_a)/\hbar}. \qquad (4.176)$$

For the integration over the fluctuation factor we use the expression (4.123) and the equation

$$\frac{\partial}{\partial x_b} \frac{\partial}{\partial x_a} A(x_b, x_a; t_b - t_a) = -\frac{1}{\dot{x}_b \dot{x}_a} \frac{\partial^2 A}{\partial t_b^2}, \qquad (4.177)$$

following from (4.119) and (4.93), and have

$$F(x_b, x_a; t_b - t_a) = \frac{1}{\sqrt{2\pi i\hbar}} \left[\frac{1}{\dot{x}(t_b)\dot{x}(t_a)} \frac{\partial^2 A}{\partial t_b^2} \right]^{1/2}. \qquad (4.178)$$

Inserting $x_a = x_b$ leads to

$$F(x_a, x_a; t_b - t_a) = \frac{1}{\sqrt{2\pi i\hbar}} \frac{1}{\dot{x}_a} \left[\frac{\partial^2 A}{\partial t_b^2} \right]^{1/2}. \qquad (4.179)$$

The action of a periodic path does not depend on x_a, so that the x_b-integration in (4.174) requires only integrating $1/\dot{x}_a$ forward and back, which produces the total period:

$$t_b - t_a = 2 \int_{x_-}^{x_+} dx_a \frac{1}{\dot{x}_a} = 2 \int_{x_-}^{x_+} dx \frac{M}{\sqrt{2M[E - V(x)]}}. \qquad (4.180)$$

Hence we obtain from (4.174):

$$Z_{\rm QM}^{\rm sc}(t_b - t_a) = \frac{t_b - t_a}{\sqrt{2\pi i\hbar}} \left| \frac{\partial^2 A}{\partial t_b^2} \right|^{1/2} e^{iA(t_b - t_a)/\hbar - i\pi}. \qquad (4.181)$$

There is a phase factor $e^{-i\pi}$ associated with a Maslov-Morse index $\nu = 2$, first introduced in the fluctuation factor (2.269). In the present context, this phase factor arises from the fact that when doing the integral (4.176), the periodic orbit passes through the turning points x_- and x_+ where the integrand of (4.180) becomes singular, even though the integral remains finite. Near the turning points, the semiclassical approximation breaks down, as discussed in Section 4.1 in the context of the WKB approximation to the Schrödinger equation. This breakdown required special attention in the derivation of the connection formulas relating the wave functions on one side of the turning points to those on the other side. There, the breakdown was circumvented by escaping into the complex x-plane. When going around the singularity in the clockwise sense, the prefactor $1/p(x) = 1/\sqrt{2M(E - V(x))}^{-1/2}$ acquired a phase factor $e^{-i\pi/2}$. For a periodic orbit, both turning points had to be encircled producing twice this phase factor, which is precisely the phase $e^{-i\pi}$ given in (4.181).

The result (4.181) takes an especially simple form after a Fourier transform action:

$$
\begin{aligned}
\tilde{Z}_{\text{QM}}^{\text{sc}}(E) &= \int_{t_a}^{\infty} dt_b e^{iE(t_b - t_a)/\hbar} Z_{\text{QM}}^{\text{sc}}(t_b - t_a) \\
&= \frac{1}{\sqrt{2\pi i\hbar}} \int_{t_a}^{\infty} dt_b (t_b - t_a) \left| \frac{\partial^2 A}{\partial t_b^2} \right|^{1/2} e^{i[A(t_b - t_a) + (t_b - t_a)E]/\hbar - i\pi}.
\end{aligned}
\tag{4.182}
$$

In the semiclassical approximation, the main contribution to the integral at a given energy E comes from the time where $t_b - t_a$ is equal to the period of the particle orbit with this energy. It is determined as in (4.139) by the extremum of

$$
A(t_b - t_a) + (t_b - t_a)E. \tag{4.183}
$$

Thus it satisfies

$$
-\frac{\partial}{\partial t_b} A(t_b - t_a) = E. \tag{4.184}
$$

As in (4.140), the extremum determines the period $t_b - t_a$ of the orbit with an energy E. It will be denoted by $t(E)$. The second derivative of the exponent is $(i/\hbar)\partial^2 A(t_b - t_a)/\partial t_b^2$. For this reason, the quadratic correction in the saddle point approximation to the integral over t_b cancels the corresponding prefactor in (4.182) and leads to the simple expression

$$
\tilde{Z}_{\text{QM}}^{\text{sc}}(E) = t(E)e^{i[A(t) + t(E)E]/\hbar - i\pi}. \tag{4.185}
$$

The exponent contains again the eikonal $S(E) = A(t) + t(E)E$, the Legendre transform of the action $A(t)$ defined by

$$
S(E) = A(t) - t\frac{\partial A(t)}{\partial t}, \tag{4.186}
$$

where the variable t has to be replaced by $E(t) = -\partial A(t)/\partial t$. Via the inverse Legendre transformation, the derivative $\partial S(E)/\partial E = t$ leads back to

$$
A(t) = S(E) - \frac{\partial S(E)}{\partial E}E. \tag{4.187}
$$

Explicitly, $S(E)$ is given by the integral (4.68):

$$
S(E) = 2\int_{x_-}^{x_+} dx\, p(x) = 2\int_{x_-}^{x_+} dx\, \sqrt{2M[E - V(x)]}. \tag{4.188}
$$

Finally, we have to take into account that the periodic orbit is repeatedly traversed for an arbitrary number of times. Each period yields a phase factor $e^{iS(E)/\hbar - i\pi}$. The sum is

$$\tilde{Z}_{\text{QM}}^{\text{sc}}(E) = \sum_{n=1}^{\infty} t(E) e^{in[S(E)/\hbar - \pi]} = -t(E) \frac{e^{iS(E)/\hbar}}{1 + e^{iS(E)/\hbar}}. \tag{4.189}$$

This expression possesses poles in the complex energy plane at points where the eikonal satisfies the condition

$$S(E_n) = 2\pi\hbar(n + 1/2), \quad n = 0, \pm 1, \pm 2, \dots . \tag{4.190}$$

This condition agrees precisely with the Bohr-Sommerfeld rule (4.27) for semiclassical quantization. At the poles, one has

$$\tilde{Z}_{\text{QM}}^{\text{sc}}(E) \approx t(E) \frac{i\hbar}{S'(E_n)(E - E_n)}. \tag{4.191}$$

Due to (4.92), the pole terms acquire the simple form

$$\tilde{Z}_{\text{QM}}^{\text{sc}}(E) \approx \frac{i\hbar}{E - E_n}. \tag{4.192}$$

From (4.189) we derive the density of states defined in (1.580). For this we use the general formula

$$\rho(E) = \frac{1}{2\pi\hbar} \text{disc}\, \tilde{Z}_{\text{QM}}(E), \tag{4.193}$$

where disc $Z_{\text{QM}}(E)$ is the discontinuity $Z_{\text{QM}}(E + i\eta) - Z_{\text{QM}}(E - i\eta)$ across the singularities defined in Eq. (1.324). If we equip the energies E_n in (4.192) with the usual small imaginary part $-i\eta$, we can also write (4.193) as

$$\rho(E) = \frac{1}{\pi\hbar} \text{Re}\, \tilde{Z}_{\text{QM}}(E). \tag{4.194}$$

Inserting here the sum (4.189), we obtain the semiclassical approximation

$$\bar{\rho}_{\text{sc}}(E) = \frac{t(E)}{\pi\hbar} \sum_{n=1}^{\infty} \cos\{n[S(E)/\hbar - \pi]\} \tag{4.195}$$

or

$$\Delta\rho_{\text{sc}}(E) = \frac{t(E)}{2\pi\hbar} \left(-1 + \sum_{n=-\infty}^{\infty} e^{in[S(E)/\hbar - \pi]} \right). \tag{4.196}$$

We have added a Δ-symbol to this quantity since it is really the semiclasscial correction to the classical density of states $\rho(E)$, as we sall see in a moment. With the help of Poisson's summation formula (1.197), this goes over into

$$\Delta\rho_{\text{sc}}(E) = -\frac{t(E)}{2\pi\hbar} + \frac{t(E)}{\hbar} \sum_{n=-\infty}^{\infty} \delta[S(E)/\hbar - 2\pi(n + 1/2)]. \tag{4.197}$$

The right-hand side contains δ-functions which are singular at the semiclassical energy values (4.190). Using once more the relation (4.92), the formula $\delta(ax) = a^{-1}\delta(x)$ leads to the simple expression

$$\Delta\rho_{\rm sc}(E) = -\frac{t(E)}{2\pi\hbar} + \sum_{n=-\infty}^{\infty} \delta(E - E_n). \tag{4.198}$$

This result has a surprising property: Consider the spacing between the energy levels

$$\Delta E_n = E_n - E_{n-1} = 2\pi\hbar\frac{\Delta E_n}{\Delta S_n} \tag{4.199}$$

and average the sum in (4.198) over a small energy interval ΔE containing several energy levels. Then we obtain an average density of states:

$$\rho_{\rm av}(E) = \frac{S'(E)}{2\pi\hbar} = \frac{t(E)}{2\pi\hbar}. \tag{4.200}$$

It cancels precisely the first term in (4.198). Thus, the semiclassical formula (4.189) possesses a vanishing average density of states. This cannot be correct and we conclude that in the derivation of the formula, a contribution must have been overlooked. This contribution comes from the classical partition function. Within the above analysis of periodic orbits, there are also those which return to the point of departure after an infinitesimally small time (which leaves them with no time to fluctuate). The expansion (4.189) does not contain them, since the saddle point approximation to the time integration (4.182) used for its derivation fails at short times. The reason for this failure is the singular behavior of the fluctuation factor $\propto 1/(t_b - t_a)^{1/2}$ in (4.103).

In order to recover the classical contribution, one simply uses the short-time amplitude in the form (2.347) to calculate the purely classical contribution to $Z(E)$:

$$Z_{\rm cl}(E) \equiv \int dx \int \frac{dp}{2\pi\hbar}\frac{i\hbar}{E - H(p,x)}. \tag{4.201}$$

This implies a classical contribution to the density of states

$$\rho_{\rm cl}(E) \equiv \int dx\, \rho_{\rm cl}(E;x), \tag{4.202}$$

which is a spatial integral over the *classical local density of states*

$$\rho_{\rm cl}(E;x) \equiv \int \frac{dp}{2\pi\hbar}\delta[E - H(p,x)]. \tag{4.203}$$

The δ-function in the integrand can be rewritten as

$$\delta(E - H(p,x)) = \frac{M}{p(E;x)}\left[\delta(p - p(E;x)) + \delta(p + p(E;x))\right], \tag{4.204}$$

where $p(E; x)$ is the local momentum associated with the energy E

$$p(E; x) = \sqrt{2M[E - V(x)]}, \tag{4.205}$$

which was defined in (4.3), except that we have now added the energy to the argument, to have a more explicit notation. It is then trivial to evaluate the integral (4.203) and (4.202) yielding the classical local density of states

$$\rho_{\text{cl}}(E; x) = \frac{1}{\pi\hbar} \frac{M}{p(E; x)}, \tag{4.206}$$

and its integral

$$\rho_{\text{cl}}(E) = \int dx \, \frac{1}{\pi\hbar} \frac{M}{p(E; x)} = \frac{1}{2\pi\hbar} t(E) = \rho_{\text{av}}(E), \tag{4.207}$$

which coincides with the average classical density of states in (4.200).

Thus the full semiclassical density of states consists of the sum of (4.198) and (4.207):

$$\rho_{\text{sc}}(E) = \rho_{\text{cl}}(E) + \Delta\rho_{\text{sc}}(E). \tag{4.208}$$

This has, on the average, the correct classical value.

Note that by Eq. (4.200), the eikonal $S(E)$ is related to the integral over the classical density of states $\rho_{\text{cl}}(E)$ by a factor $2\pi\hbar$:

$$S(E) = 2\pi\hbar \int_{-\infty}^{E} dE \, \rho_{\text{cl}}(E). \tag{4.209}$$

Recalling the definition (1.584) of the number of states up to the energy E we see that

$$S(E) = 2\pi\hbar N(E), \tag{4.210}$$

which shows that the Bohr-Sommerfeld quantization condition (4.190) is the semiclassical version of the completely general equation (1.585).

4.8   Multi-Dimensional Systems

The D-dimensional generalization of the classical partition function (4.201) reads

$$Z_{\text{cl}}(E) \equiv \int d^D x \int \frac{d^D p}{2\pi\hbar} \frac{i\hbar}{E - H(\mathbf{p}, \mathbf{x})}, \tag{4.211}$$

and of the density of states (4.203):

$$\rho_{\text{cl}}(E; \mathbf{x}) \equiv \int \frac{d^D p}{(2\pi\hbar)^D} \delta[E - H(\mathbf{p}, \mathbf{x})]. \tag{4.212}$$

The Hamiltonian of the standard form $H(\mathbf{p}, \mathbf{x}) = \mathbf{p}^2/2M + V(\mathbf{x})$ allows us to perform the momentum integration by separating it into radial and angular parts,

$$\int \frac{d^D p}{(2\pi\hbar)^D} = \int dp \, p^{D-1} \int d\hat{\mathbf{p}}. \tag{4.213}$$

The angular integral yields the surface of a unit sphere in D dimensions:

$$\int d\hat{\mathbf{p}} = S_D = \frac{2\pi^{D/2}}{\Gamma(D/2)}. \tag{4.214}$$

The δ-function $\delta(E - H(\mathbf{p}, \mathbf{x}))$ can again be rewritten as in (4.204), which selects the momenta of magnitude

$$p(E; \mathbf{x}) = \sqrt{2M[E - V(\mathbf{x})]}. \tag{4.215}$$

Thus we find

$$\rho_{\mathrm{cl}}(E) \equiv \int d^D x \, \rho_{\mathrm{cl}}(E; \mathbf{x}), \tag{4.216}$$

where $\rho_{\mathrm{cl}}(E; \mathbf{x})$ is the *classical local density of states*.

$$\rho_{\mathrm{cl}}(E; \mathbf{x}) = S_D \frac{M}{p^2(E; \mathbf{x})} \frac{p^D(E; \mathbf{x})}{(2\pi\hbar)^D} = \frac{1}{(4\pi\hbar^2)^{D/2}} \frac{2M}{\Gamma(D/2)} \{2M[E - V(\mathbf{x})]\}^{D/2-1}, \tag{4.217}$$

generalizing expression (4.207). The number of states with energies between E and $E + dE$ in the volume element $d^D x$ is $dE d^3 x \rho_{\mathrm{cl}}(E; \mathbf{x})$.

For completeness we state some features of the semiclassical results which appear when generalizing the theory to D dimensions. For a detailed derivation see the rich literature on this subject quoted at the end of the chapter.

For an arbitrary number D of dimensions. the Van Vleck-Pauli-Morette determinant (4.124) takes the form

$$F(\mathbf{x}_b, \mathbf{x}_a; t_b - t_a) = \frac{1}{\sqrt{2\pi i\hbar}^D} \left| \det_D[-\partial_{x_b^i} \partial_{x_a^j} A(\mathbf{x}_b, \mathbf{x}_a; t_b - t_a)] \right|^{1/2} e^{-i\pi\nu/2}, \tag{4.218}$$

where ν is the Maslov-Morse index.

The fixed-energy amplitude becomes the sum over all periodic orbits:[9]

$$(\mathbf{x}_b|\mathbf{x}_a)_E = \frac{1}{\sqrt{2\pi i\hbar}^{D-1}} \sum_p |D_S|^{1/2} e^{iS(\mathbf{x}_b, \mathbf{x}_a; E)/\hbar - i\pi\nu'/2}, \tag{4.219}$$

where $S(\mathbf{x}_b, \mathbf{x}_a; E)$ is the D-dimensional generalization of (4.68) and D_S the $(D + 1) \times (D + 1)$-determinant:

$$D_S = (-1)^{D+1} \det \begin{pmatrix} \dfrac{\partial^2 S}{\partial \mathbf{x}_b \partial \mathbf{x}_a} & \dfrac{\partial^2 S}{\partial E \partial \mathbf{x}_a} \\ \dfrac{\partial^2 S}{\partial \mathbf{x}_b \partial E} & \dfrac{\partial^2 S}{\partial E \partial E} \end{pmatrix}. \tag{4.220}$$

---

[9]M.C. Gutzwiller, J. Math. Phys. *8*, 1979 (1967); *11*, 1791 (1970); *12*, 343 (1971).

The factor $(-1)^{D+1}$ makes the determinant positive for short trajectories. The index ν' differs from ν by one unit if $\partial^2 S/\partial E^2 = \partial t(E)/\partial E$ is negative.

In D dimensions, the Hamilton-Jacobi equation leads to

$$\frac{\partial H}{\partial \mathbf{p}_b} \cdot \frac{\partial^2 S}{\partial \mathbf{x}_b \partial \mathbf{x}_a} = \dot{\mathbf{x}}_b \cdot \frac{\partial^2 S}{\partial \mathbf{x}_b \partial \mathbf{x}_a} = 0, \tag{4.221}$$

instead of (4.154). Only the longitudinal projection of the $D \times D$-matrix $\partial^2 S/\partial \mathbf{x}_b \partial \mathbf{x}_a$ along the direction of motion vanishes now. In this direction

$$\dot{\mathbf{x}}_b \cdot \frac{\partial^2 S}{\partial \mathbf{x}_b \partial E} = 1, \tag{4.222}$$

so that the determinant (4.220) can be reduced to

$$D_S = \frac{1}{|\dot{\mathbf{x}}_b||\dot{\mathbf{x}}_a|} \det \left(-\frac{\partial^2 S}{\partial \mathbf{x}_b^\perp \partial \mathbf{x}_a^\perp} \right), \tag{4.223}$$

instead of (4.156). Here $\mathbf{x}_{b,a}^\perp$ denotes the deviations from the orbit orthogonal to $\dot{\mathbf{x}}_{b,a}$, and we have used (2.288) to arrive at (4.223).

As an example, let us write down the D-dimensional generalization of the free-particle amplitude (4.161). The eikonal is obviously

$$S(\mathbf{x}_a, \mathbf{x}_b; E) = \sqrt{2ME}|\mathbf{x}_b - \mathbf{x}_a|, \tag{4.224}$$

and the determinant (4.223) becomes

$$D_S = \frac{M}{2E} \frac{(2ME)^{(D-1)/2}}{|\mathbf{x}_a - \mathbf{x}_b|^{D-1}}. \tag{4.225}$$

Thus we find

$$(\mathbf{x}_b|\mathbf{x}_a)_E = \sqrt{\frac{M}{2E}} \frac{1}{(2\pi i\hbar)^{(D-1)/2}} \frac{(2ME)^{(D-1)/4}}{|\mathbf{x}_a - \mathbf{x}_b|^{(D-1)/2}} e^{i\sqrt{2ME}|\mathbf{x}_b - \mathbf{x}_a|/\hbar}. \tag{4.226}$$

For $D = 1$, this reduces to (4.161).

Note that the semiclassical result coincides with the large-distance behavior (1.355) of the exact result (1.351), since the semiclassical limit implies a large momenta k in the Bessel function (1.351).

When calculating the partition function, one has to perform a D-dimensional integral over all $\mathbf{x}_b = \mathbf{x}_a$. This is best decomposed into a one-dimensional integral along the orbit and a $D-1$-dimensional one orthogonal to it. The eikonal function $S(\mathbf{x}_a, \mathbf{x}_a; E)$ is constant along the orbit, as in the one-dimensional case. When leaving the orbit, however, this is no longer true. The quadratic deviation of S orthogonal to the orbit is

$$\frac{1}{2}(\mathbf{x} - \mathbf{x}^\perp)^T \frac{\partial^2 S(\mathbf{x}, \mathbf{x}; E)}{\partial \mathbf{x}^\perp \partial \mathbf{x}^\perp}(\mathbf{x} - \mathbf{x}^\perp), \tag{4.227}$$

where the superscript T denotes the transposed vector to be multiplied from the left with the matrix in the middle. After the exact trace integration along the orbit and a quadratic approximation in the transversal direction for each primitive orbit, which is not repeated, we obtain the contribution to the partition function

$$Z_{sc} = t(E) \frac{\left| \dfrac{\partial^2 S(\mathbf{x}_b, \mathbf{x}_a; E)}{\partial \mathbf{x}_b^\perp \partial \mathbf{x}_a^\perp} \right|_{\mathbf{x}_b = \mathbf{x}_a = \mathbf{x}}^{1/2}}{\left| \dfrac{\partial^2 S(\mathbf{x}, \mathbf{x}; E)}{\partial \mathbf{x}^\perp \partial \mathbf{x}^\perp} \right|^{1/2}} \sum_{n=1}^{\infty} e^{in[S(E)/\hbar - i\pi\nu/2]}, \qquad (4.228)$$

where ν is the Maslov-Morse index of the orbit. The ratio of the determinants is conveniently expressed in terms of the determinant of the so-called *stability matrix* M in phase space, which is introduced in classical mechanics as follows:

Consider a classical orbit in phase space and vary slightly the initial point, moving it orthogonally away from the orbit by $\delta \mathbf{x}_a^\perp, \delta \mathbf{p}_a^\perp$. This produces variations at the final point $\delta \mathbf{x}_b^\perp, \delta \mathbf{p}_b^\perp$, related to those at the initial point by the linear equation

$$\begin{pmatrix} \delta \mathbf{x}_b^\perp \\ \delta \mathbf{p}_b^\perp \end{pmatrix} = \begin{pmatrix} A & B \\ C & D \end{pmatrix} \begin{pmatrix} \delta \mathbf{x}_a^\perp \\ \delta \mathbf{p}_a^\perp \end{pmatrix} \equiv M \begin{pmatrix} \delta \mathbf{x}_a^\perp \\ \delta \mathbf{p}_a^\perp \end{pmatrix}. \qquad (4.229)$$

The $2(D-1) \times 2(D-1)$-dimensional matrix is the stability matrix M. It can be expressed in terms of the second derivatives of $S(\mathbf{x}_b, \mathbf{x}_a; E)$. These appear in the relation

$$\begin{pmatrix} \delta \mathbf{p}_a^\perp \\ \delta \mathbf{p}_b^\perp \end{pmatrix} = \begin{pmatrix} -a & -b \\ b^T & c \end{pmatrix} \begin{pmatrix} \delta \mathbf{x}_a^\perp \\ \delta \mathbf{x}_b^\perp \end{pmatrix}, \qquad (4.230)$$

where a, b, and c are the $(D-1) \times (D-1)$-dimensional matrices

$$a = \frac{\partial^2 S}{\partial \mathbf{x}_a^\perp \partial \mathbf{x}_a^\perp}, \quad b = \frac{\partial^2 S}{\partial \mathbf{x}_a^\perp \partial \mathbf{x}_b^\perp}, \quad c = \frac{\partial^2 S}{\partial \mathbf{x}_b^\perp \partial \mathbf{x}_b^\perp}. \qquad (4.231)$$

From this one calculates the matrix elements of the stability matrix (4.229):

$$A = -b^{-1}a, \quad B = -b^{-1}, \quad C = b^T - cb^{-1}a, \quad D = -cb^{-1}. \qquad (4.232)$$

The stability properties of the classical orbits are classified by the eigenvalues of the stability matrix (4.229). In three dimensions, the eigenvalues are given by the zeros of the characteristic polynomial of the 4×4 -matrix M:

$$P(\lambda) = |M - \lambda| = \begin{vmatrix} A - \lambda & B \\ C & D - \lambda \end{vmatrix} = \begin{vmatrix} -b^{-1}a - \lambda & -b^{-1} \\ b^T - cb^{-1}a & -cb^{-1} - \lambda \end{vmatrix}. \qquad (4.233)$$

The usual manipulations bring this to the form

$$\begin{aligned} P(\lambda) &= \begin{vmatrix} -b^{-1}a - \lambda & -b^{-1} \\ b^T + \lambda & -\lambda \end{vmatrix} = \frac{1}{|b|} \begin{vmatrix} -a - \lambda b & -1 \\ b^T + (a+c)\lambda + \lambda^2 b & 0 \end{vmatrix} \\ &= \frac{1}{|b|} \left| b^T + (a+c)\lambda + \lambda^2 b \right|. \end{aligned} \qquad (4.234)$$

Precisely this expression appears, with $\mathbf{x}_b = \mathbf{x}_a$, in the prefactor of (4.228) if this is rewritten as

$$
\frac{\left|\dfrac{\partial^2 S}{\partial \mathbf{x}_b^\perp \partial \mathbf{x}_a^\perp}\right|^{1/2}_{\mathbf{x}_b \approx \mathbf{x}_a = \mathbf{x}}}{\left|\dfrac{\partial^2 S}{\partial \mathbf{x}_b^\perp \partial \mathbf{x}_b^\perp} + 2\dfrac{\partial S}{\partial \mathbf{x}_b^\perp}\dfrac{\partial S}{\partial \mathbf{x}_a^\perp} + \dfrac{\partial^2 S}{\partial \mathbf{x}_a^\perp \partial \mathbf{x}_a^\perp}\right|^{1/2}_{\mathbf{x}_b = \mathbf{x}_a = \mathbf{x}}}.
\tag{4.235}
$$

Due to (4.232), this coincides with $P(1)^{-1/2}$. The semiclassical limit to the quantum-mechanical partition function takes therefore the simple form referred to as *Gutzwiller's trace formula*

$$
Z_{\mathrm{sc}}(E) = t(E)\frac{1}{P(1)^{1/2}}\frac{e^{iS(E)-i\pi\nu/2}}{1 - e^{iS(E)-i\pi\nu/2}}.
\tag{4.236}
$$

The energy eigenvalues lie at the poles and satisfy the quantization rules [compare (4.27), (4.190)]

$$
S(E_n) = 2\pi\hbar(n + \nu/4).
\tag{4.237}
$$

The eigenvalues of the stability matrix come always in pairs $\lambda, 1/\lambda$, as is obvious from (4.234). For this reason, one has to classify only two eigenvalues. These must be either both real or mutually complex-conjugate. One distinguishes the following cases:

1. *elliptic,*   if $\lambda = e^{i\chi}, e^{-i\chi}$,   with a real phase $\chi \neq 0$,

2. *direct parabolic,*   if $\lambda = 1$,
   inverse parabolic, $\lambda = -1$,

3. *direct hyperbolic,*   if $\lambda = e^{\pm\chi}$,
   inverse hyperbolic, $\lambda = -e^{\pm\chi}$,

4. *loxodromic,*   if $\lambda = e^{u\pm v}$.

In these cases,

$$
P(1) = \prod_{i=1}^{2}(\lambda_i - 1)(1/\lambda_i - 1)
\tag{4.238}
$$

has the values

1. $4\sin^2(\chi/2)$,

2. 0 or 4,

3. $-4\sinh^2(\chi/2)$ or $4\cosh^2(\chi/2)$,

4. $4\sin[(u + v)/2]\sin[(u - v)/2]$.

Only in the parabolic case are the equations of motion integrable, this being obviously an exception rather than a rule, since it requires the fulfillment of the equation $a + c = \pm 2b$. Actually, since the transverse part of the trace integration in the partition function results in a singular determinant in the denominator of (4.236), this case requires a careful treatment to arrive at the correct result.[10] In general, a system will show a mixture of elliptic and hyperbolic behavior, and the particle orbits exhibit what is called a *smooth chaos*. In the case of a purely hyperbolic behavior one speaks of a *hard chaos*, which is simpler to understand. The semiclassical approximation is based precisely on those orbits of a system which are exceptional in a chaotic system, namely, the periodic orbits.

The expression (4.236) also serves to obtain the semiclassical density of states in D-dimensional systems via Eq. (4.193). In D dimensions the paths, with vanishing length contribute to the partition function the classical expression [compare (4.211)]. Application of semiclassical formulas has led to surprisingly simple explanations of extremely complex experimental data on highly excited atomic spectra which classically behave in a chaotic manner.

For completeness, let us also state the momentum space representation of the semiclassical fixed-energy amplitude (4.145). It is given by the momentum space analog of (4.219):

$$(\mathbf{p}_b|\mathbf{p}_a)_E = \frac{(2\pi\hbar)^D}{\sqrt{2\pi i\hbar}^{D-1}} \sum_p |\tilde{D}_S|^{1/2} e^{iS(\mathbf{p}_b, \mathbf{p}_a; E)/\hbar - i\pi\nu'/2}, \qquad (4.239)$$

where $S(\mathbf{p}_b, \mathbf{p}_a; E)$ is the Legendre transform of the eikonal

$$S(\mathbf{p}_b, \mathbf{p}_a; E) = S(\mathbf{p}_b, \mathbf{p}_a; E) - \mathbf{p}_b\mathbf{x}_b + \mathbf{p}_a\mathbf{x}_a, \qquad (4.240)$$

evaluated at the classical momenta $\mathbf{p}_b = \partial_{\mathbf{p}_b} S(\mathbf{p}_b, \mathbf{p}_a; E)$ and $\mathbf{p}_a = \partial_{\mathbf{p}_a} S(\mathbf{p}_b, \mathbf{p}_a; E)$. The determinant can be brought to the form:

$$D_S = \frac{1}{|\dot{\mathbf{p}}_b||\dot{\mathbf{p}}_a|} \det\left(-\frac{\partial^2 S}{\partial \mathbf{p}_b^\perp \partial \mathbf{p}_a^\perp}\right), \qquad (4.241)$$

where $\mathbf{p}_a^\perp$ is the momentum orthogonal to $\dot{\mathbf{p}}_a$.

This formula cannot be applied to the free particle fixed-energy amplitude (3.216) for the same degeneracy reason as before.

Higher $\hbar$-corrections to the trace formula (4.236) have also been derived, but the resulting expressions are very complicated to handle. See the citations at the end of this chapter.

4.9 Quantum Corrections to Classical Density of States

There exists a simple way of calculating quantum corrections to the semiclassical expressions (4.207) and its D-dimensional generalization (4.216) for the density of

---

[10]M.V. Berry and M. Tabor, J. Phys. A *10*, 371 (1977), Proc. Roy. Soc. A *356*, 375 (1977).

states. To derive them we introduce an operator δ-function $\delta(E - \hat{H})$ via the spectral representation

$$\delta(E - \hat{H}) \equiv \sum_n \delta(E - E_n)|n\rangle\langle n|, \tag{4.242}$$

where $|n\rangle$ are the eigenstates of the Hamiltonian operator $\hat{H}$. The δ-function (4.242) has the Fourier representation [recall (1.193)]

$$\delta(E - \hat{H}) = \int_{-\infty}^{\infty} \frac{dt}{2\pi\hbar} e^{-i(\hat{H}-E)t/\hbar}. \tag{4.243}$$

Its matrix elements between eigenstates $|\mathbf{x}\rangle$ of the position operator,

$$\rho(E; \mathbf{x}) = \langle \mathbf{x}|\delta(E - \hat{H})|\mathbf{x}\rangle = \int_{-\infty}^{\infty} \frac{dt}{2\pi\hbar} e^{iEt/\hbar}\langle \mathbf{x}|e^{-i\hat{H}t/\hbar}|\mathbf{x}\rangle, \tag{4.244}$$

define the *quantum-mechanical local density of states*. The amplitude on the right-hand side is the time evolution amplitude

$$\langle \mathbf{x}|e^{-i\hat{H}t/\hbar}|\mathbf{x}\rangle = (\mathbf{x}\,t|\mathbf{x}\,0), \tag{4.245}$$

which can be represented by a path integral as described in Chapter 2. In the semiclassical limit, only the short-time behavior of $(\mathbf{x}\,t|\mathbf{x}\,0)$ is relevant.

4.9.1 One-Dimensional Case

For a one-dimensional harmonic oscillator, the short-time expansion of (4.245) can easily be written down. For short times $t \equiv t_b - t_a$ compared to the period $1/\omega$, we expand the amplitude (2.173) at equal initial and final space points $x = x_a = x_b$ as follows in a power series of t:

$$(x\,t_b|x\,t_a) = \frac{1}{\sqrt{2\pi i\hbar t/M}} e^{-i\frac{M}{2}\omega^2 x^2 t/\hbar}\left\{1 + \frac{t^2}{12}\omega^2 - \frac{i}{\hbar}\frac{t^3}{24}M\omega^4 x^2 + \ldots\right\}, \tag{4.246}$$

This expansion is valid for an arbitrary smooth potential $V(x)$ if the exponential prefactor containing the harmonic potential is replaced by

$$e^{-iV(x)t/\hbar}, \tag{4.247}$$

whereas ω^2 and $M\omega^4 x^2$ are substituted as follows:

$$\omega^2 \rightarrow \frac{1}{M}V''(x), \tag{4.248}$$

$$M\omega^4 x^2 \rightarrow \frac{1}{M}[V'(x)]^2. \tag{4.249}$$

Hence:

$$(x\,t_b|x\,t_a) = \frac{1}{\sqrt{2\pi i\hbar t/M}} e^{-iV(x)t/\hbar}\left\{1 + \frac{t^2}{12M}V''(x) - \frac{i}{\hbar}\frac{t^3}{24M}[V'(x)]^2 + \ldots\right\}. \tag{4.250}$$

Inserting this into (4.244) yields the local density of states

$$\rho(E;x) = \int_{-\infty}^{\infty} \frac{dt}{2\pi\hbar} \frac{1}{\sqrt{2\pi i\hbar t/M}} e^{-i[V(x)-E]t/\hbar}$$

$$\times \left\{1 + \frac{t^2}{12M}V''(x) - \frac{i}{\hbar}\frac{t^3}{24M}[V'(x)]^2 + \dots\right\}. \qquad (4.251)$$

For positive $E - V(x)$, the integration along the real axis can be deformed into the upper complex plane to enclose the square-root cut along the positive imaginary t-axis in the anti-clockwise sense. Setting $t = i\tau$ and using the fact that the discontinuity across a square root cut produces a factor two, we have

$$\rho(E;x) = 2\int_0^{\infty} \frac{d\tau}{2\pi\hbar} \frac{1}{\sqrt{2\pi\hbar\tau/M}} e^{-[E-V(x)]\tau/\hbar} \left\{1 - \frac{\tau^2}{12M}V''(x) - \frac{1}{\hbar}\frac{\tau^3}{24M}[V'(x)]^2 + \dots\right\}. \qquad (4.252)$$

The first term can easily be integrated for $E > V(x)$, and yields the classical local density of states (4.207):

$$\rho_{\rm cl}(E;x) = \frac{1}{\pi\hbar} \frac{M}{\sqrt{2M[E-V(x)]}} = \frac{M}{\pi\hbar} \frac{1}{p(E;x)}. \qquad (4.253)$$

In order to calculate the effect of the correction terms in the expansion (4.252), we observe that a factor τ in the integrand is the same as a derivative $\hbar d/dV$ applied to the exponential. Thus we find directly the semiclassical expansion for the density of states (4.252), valid for $E > V(x)$:

$$\rho(E;x) = \left\{1 - \frac{\hbar^2}{12M}V''(x)\frac{d^2}{dV^2} - \frac{\hbar^2}{24M}[V'(x)]^2\frac{d^3}{dV^3} + \dots\right\}\rho_{\rm cl}(E;x). \qquad (4.254)$$

Inserting (4.253) and performing the differentiations with respect to V we obtain

$$\rho(E;x) = \frac{1}{\pi\hbar}\sqrt{\frac{M}{2}}\left\{\frac{1}{[E-V(x)]^{1/2}} - \frac{\hbar^2}{12M}V''(x)\frac{3}{4}\frac{1}{[E-V(x)]^{5/2}}\right.$$

$$\left. - \frac{\hbar^2}{24M}[V'(x)]^2\frac{15}{8}\frac{1}{[E-V(x)]^{7/2}} + \dots\right\}. \qquad (4.255)$$

Note that the proceeds in powers of higher gradients of the potential; it is a *gradient expansion*.

The integral over (4.255) yields a gradient expansion for $\rho(E)$ [2]. The second term can be integrated by parts which, under the assumption that $V(x)$ vanishes at the boundaries, simply changes the sign of the third term, so that we find

$$\rho(E) = \frac{1}{\pi\hbar}\sqrt{\frac{M}{2}}\int dx\left\{\frac{1}{[E-V(x)]^{1/2}} + \frac{\hbar^2}{24M}[V'(x)]^2\frac{15}{8}\frac{1}{[E-V(x)]^{7/2}} + \dots\right\}. \qquad (4.256)$$

4.9.2 Arbitrary Dimensions

In D dimensions, the short-time expansion of the time evolution amplitude (4.250) takes the form

$$(\mathbf{x}\,t_b|\mathbf{x}\,t_a) = \frac{1}{\sqrt{2\pi i\hbar t/M}^D}e^{-iV(\mathbf{x})t/\hbar}\left\{1 + \frac{t^2}{12M}\boldsymbol{\nabla}^2 V(\mathbf{x}) - \frac{i}{\hbar}\frac{t^3}{24M}[\boldsymbol{\nabla}V(\mathbf{x})]^2 + \dots\right\}.$$

(4.257)

Recalling the $i\eta$-prescription on page 115, according to which the singularity at $t = 0$ has to be shifted slightly into the upper half plane by replacing $t \to t - i\eta$, we use the formula[11]

$$\int_{-\infty}^{\infty}\frac{dt}{2\pi}\frac{1}{(it+\eta)^{\nu}}e^{ita} = \Theta(a)\frac{a^{\nu-1}}{\Gamma(\nu)}e^{-a\eta},$$

and obtain the obvious generalization of (4.254):

$$\rho(E;\mathbf{x}) = \left\{1 - \frac{\hbar^2}{12M}\boldsymbol{\nabla}^2 V(\mathbf{x})\frac{d^2}{dV^2} - \frac{\hbar^2}{24M}[\boldsymbol{\nabla}V(\mathbf{x})]^2\frac{d^3}{dV^3} + \dots\right\}\rho_{\mathrm{cl}}(E;\mathbf{x}), \quad (4.258)$$

where $\rho_{\mathrm{cl}}(E;\mathbf{x})$ is the classical D-dimensional local density of states (4.217). The way this appears here is quite different from that in the earlier classical calculation (4.212), which may be expressed with the help of the local momentum (4.215) as an integral

$$\rho_{\mathrm{cl}}(E;\mathbf{x}) = \int\frac{d^D p}{(2\pi\hbar)^D}\delta[E - H(\mathbf{p},\mathbf{x})] = \int\frac{d^D p}{(2\pi\hbar)^D}\frac{M}{p(E;\mathbf{x})}\delta[p - p(E;\mathbf{x})]. \quad (4.259)$$

In order to see the relation to the appearance in (4.258) we insert the Fourier decomposition of the leading term of the short-time expansion of the time evolution amplitude

$$(\mathbf{x}\,t|\mathbf{x}\,0)_{\mathrm{cl}} = \int\frac{d^D p}{(2\pi\hbar)^D}e^{-i[\mathbf{p}^2/2M+V(\mathbf{x})]t/\hbar} \quad (4.260)$$

into the integral representation (4.244) which takes the form

$$\rho(E;\mathbf{x}) = \int_{-\infty}^{\infty}\frac{dt}{2\pi\hbar}\int\frac{d^D p}{(2\pi\hbar)^D}e^{-i[p^2-p^2(E;\mathbf{x})]t/2M\hbar}. \quad (4.261)$$

By doing the integral over the time first, the size of the momentum is fixed to the local momentum $p^2(E;\mathbf{x})$ resulting in the original representation (4.259). The expression (4.217) for the density of states, on the other hand, corresponds to first integrating over *all* momenta. The time integration selects from the result of this the correct local momenta $p^2(E;\mathbf{x})$.

---

[11]I.S. Gradshteyn and I.M. Ryzhik, *op. cit.*, Formula 3.382.6. The formula is easily derived by expressing $(it + \eta)^{-\nu} = \Gamma^{-1}(\nu)\int_0^{\infty}d\tau\,\tau^{\nu-1}e^{-\tau(it+\eta)}$.

This generalizes (4.255) to

$$
\rho(E;\mathbf{x}) = \left(\frac{M}{2\pi\hbar^2}\right)^{D/2} \left\{ \frac{1}{\Gamma(D/2)}[E - V(\mathbf{x})]^{D/2-1} \right.
$$
$$
- \frac{\hbar^2}{12M}\left[\boldsymbol{\nabla}^2 V(\mathbf{x})\right] \frac{1}{\Gamma(D/2-2)}[E - V(\mathbf{x})]^{D/2-3}
$$
$$
\left. + \frac{\hbar^2}{24M}\left[\boldsymbol{\nabla} V(\mathbf{x})\right]^2 \frac{1}{\Gamma(D/2-3)}[E - V(\mathbf{x})]^{D/2-4} + \ldots \right\}. \tag{4.262}
$$

When integrating the density (4.262) over all $\mathbf{x}$, the second term in the curly brackets can again be converted into the third term changing its sign, as in (4.256). The right-hand side can easily be integrated for all pure power potentials. This will be done in Appendix 4A.

4.9.3   Bilocal Density of States

It is useful to generalize the local density of states (4.244) and introduce a *bilocal density of states*:

$$
\rho(E;\mathbf{x}_b,\mathbf{x}_a) = \langle \mathbf{x}_b|\delta(E - \hat{H})|\mathbf{x}_a\rangle = \int_{-\infty}^{\infty} \frac{dt}{2\pi\hbar} e^{iEt/\hbar}\,\langle \mathbf{x}_b|e^{-i\hat{H}t/\hbar}|\mathbf{x}_a\rangle
$$
$$
= \int_{-\infty}^{\infty} \frac{dt}{2\pi\hbar} e^{iEt/\hbar}\,(\mathbf{x}_b\, t|\mathbf{x}_a\, 0). \tag{4.263}
$$

The semiclassical expansion requires now the nondiagonal version of the short-time expansions (4.257). For the one-dimensional harmonic oscillator, the expansion (4.246) is generalized to

$$
(x_b t_b|x_a t_a) = \frac{1}{\sqrt{2\pi i\hbar t/M}} e^{iM(x_b-x_a)^2/2t\hbar} e^{-iM\omega^2\bar{x}^2 t/2\hbar}
$$
$$
\times \left\{ 1 + \frac{t^2}{12}\omega^2 - \frac{i}{\hbar}\frac{t^3}{24}M\omega^4\bar{x}^2 - \frac{i}{\hbar}\frac{t}{24}(x_b - x_a)^2 M\omega^2 + \ldots \right\}, \tag{4.264}
$$

where $\bar{x} = (x_b+x_a)/2$ is the mean position of the two endpoints. In this expansion we have included all terms whose size is of the order t^3, keeping in mind that $(x_b - x_a)^2$ is of the order $\hbar$ in a finite amplitude. Going to D dimensions and performing the substitutions (4.248) and (4.249), this expansion is generalized to

$$
(\mathbf{x}_b t_b|\mathbf{x}_a t_a) = \frac{1}{\sqrt{2\pi i\hbar t/M}^D} e^{iM(\mathbf{x}_b-\mathbf{x}_a)^2/2t\hbar - iV(\bar{\mathbf{x}})t/\hbar} \tag{4.265}
$$
$$
\times \left\{ 1 + \frac{t^2}{12M}\boldsymbol{\nabla}^2 V(\bar{\mathbf{x}}) - \frac{i}{\hbar}\frac{t^3}{24M}[\boldsymbol{\nabla} V(\bar{\mathbf{x}})]^2 - \frac{i}{\hbar}\frac{t}{24}[(\mathbf{x}_b-\mathbf{x}_a)\boldsymbol{\nabla}]^2 V(\bar{\mathbf{x}}) + \ldots \right\}.
$$

For a derivation without substitution trick in (4.248) and (4.249) see Appendix 4B.

Inserting this amplitude into the integral in Eq. (4.263), we obtain the bilocal density of states

$$\rho(E; \mathbf{x}_b, \mathbf{x}_a) = \int_{-\infty}^{\infty} \frac{dt}{2\pi\hbar} \frac{1}{\sqrt{2\pi i\hbar t/M}^D} e^{iM(\mathbf{x}_b - \mathbf{x}_a)^2/2t\hbar} e^{-i[V(\bar{\mathbf{x}}) - E]t/\hbar}$$

$$\times \left\{ 1 + \frac{t^2}{12M} \boldsymbol{\nabla}^2 V(\bar{\mathbf{x}}) - \frac{i}{\hbar} \frac{t^3}{24M} [\boldsymbol{\nabla} V(\bar{\mathbf{x}})]^2 - \frac{i}{\hbar} \frac{t}{24} [(\mathbf{x}_b - \mathbf{x}_a) \boldsymbol{\nabla}]^2 V(\bar{\mathbf{x}}) + \ldots \right\}. \quad (4.266)$$

The first term in the integrand is simply the time evolution amplitude of the free-particle in a constant potential $V(\bar{\mathbf{x}})$ which has the Fourier decomposition [recall (1.331)]:

$$(\mathbf{x}_b t_b | \mathbf{x}_a t_a)_{\text{cl}} = \int \frac{d^D p}{(2\pi\hbar)^D} e^{i\mathbf{p}(\mathbf{x}_b - \mathbf{x}_a)/\hbar} e^{-iH(\mathbf{p}, \bar{\mathbf{x}})t/\hbar}. \quad (4.267)$$

Indeed, inserting this into (4.263), and performing the integration over time, we find

$$\rho_{\text{cl}}(E; \mathbf{x}_b, \mathbf{x}_a) = \int \frac{d^D p}{(2\pi\hbar)^D} \delta(E - H(\mathbf{p}, \bar{\mathbf{x}})) e^{i\mathbf{p}(\mathbf{x}_b - \mathbf{x}_a)/\hbar}. \quad (4.268)$$

Decomposing the momentum integral into radial and angular parts as in (4.213), we can integrate out the radial part as in (4.203), whereas the angular integral yields the following function of $R = |\mathbf{x}_b - \mathbf{x}_a|$:

$$\int d\hat{\mathbf{p}}\, e^{i\mathbf{p}(\mathbf{x}_b - \mathbf{x}_a)/\hbar} = S_D(pR/\hbar), \quad (4.269)$$

which is a direct generalization of the surface of a sphere in D dimensions (4.214). It reduces to it for $p = 0$. This integral will be calculated in Section 9.1. The result is

$$S_D(z) = (2\pi)^{D/2} J_{D/2-1}(z)/z^{D/2-1}, \quad (4.270)$$

where $J_\nu(z)$ are Bessel functions. For small z, these behave like[12]

$$J_\nu(z) \approx \frac{(z/2)^\nu}{\Gamma(\nu + 1)}, \quad (4.271)$$

thus ensuring that $S_D(kR)$ is indeed equal to S_D at $R = 0$.

Altogether, the classical limit of the bilocal density of states is

$$\rho_{\text{cl}}(E; \mathbf{x}_b, \mathbf{x}_a) = \left(\frac{1}{2\pi\hbar^2}\right)^{D/2} M \frac{J_{D/2-1}(p(E; \bar{\mathbf{x}})R/\hbar)}{(R/\hbar)^{D/2-1}}. \quad (4.272)$$

At $\mathbf{x}_b = \mathbf{x}_a$, this reduces to the density (4.212).

---

[12] M. Abramowitz and I. Stegun, *op. cit.*, Formula 9.1.7.

In three dimensions, the Bessel function becomes

$$J_{1/2}(z) = \sqrt{\frac{2}{\pi z}} \sin z, \tag{4.273}$$

and (4.272) yields

$$\rho_{\rm cl}(E; \mathbf{x}_b, \mathbf{x}_a) = \left(\frac{1}{2\pi\hbar^2}\right)^{3/2} \frac{1}{\Gamma(3/2)} \frac{M}{\sqrt{2}} \frac{\sin[p(E;\bar{\mathbf{x}})R/\hbar]}{R/\hbar}. \tag{4.274}$$

From the D-dimensional version of the short-time expansion (4.266) we obtain, after using once more the equivalence of t and $i\hbar d/dV$,

$$\rho(E; \mathbf{x}_b, \mathbf{x}_a) = \left\{1 - \frac{\hbar^2}{12M}\left[\boldsymbol{\nabla}^2 V(\bar{\mathbf{x}})\right]\frac{d^2}{dV^2} - \frac{\hbar^2}{24M}[\boldsymbol{\nabla} V(\bar{\mathbf{x}})]^2 \frac{d^3}{dV^3}\right.$$
$$\left. + \frac{1}{24}[(\mathbf{x}_b - \mathbf{x}_a)\boldsymbol{\nabla}]^2 V(\bar{\mathbf{x}})\frac{d}{dV} + \ldots\right\}\rho_{\rm cl}(E; \mathbf{x}_b, \mathbf{x}_a). \tag{4.275}$$

4.9.4 Gradient Expansion of Tracelog of Hamiltonian Operator

Starting point is formula (1.587) for the tracelog of the Hamiltonian operator. By performing the trace in the local basis $|\mathbf{x}\rangle$, we arrive at the useful formula involving the density of states (4.244)

$$\mathrm{Tr} \log \hat{H} = \int d^D x \int_{-\infty}^{\infty} dE\, \rho(E; \mathbf{x}) \log E. \tag{4.276}$$

Inserting here the classical density of states (4.217), and integrating over the classical spectrum $E \in (E_0, \infty)$, where E_0 is the bottom of the potential $V(\mathbf{x})$, we obtain the classical limit of the tracelog:

$$[\mathrm{Tr} \log \hat{H}]_{\rm cl} = \int d^D x \int_{E_0}^{\infty} dE\, \rho_{\rm cl}(E; \mathbf{x}) \log E = \left(\frac{M}{2\pi\hbar^2}\right)^{D/2} \int d^D x\, I_{D/2}(V(\mathbf{x})), \tag{4.277}$$

where

$$I_\alpha(V) \equiv \frac{1}{\Gamma(\alpha)} \int_V^{\infty} dE\, (E - V)^{\alpha-1} \log E. \tag{4.278}$$

The integrals $I_{D/2}(V(\mathbf{x}))$ diverge, but can be calculated with the techniques explained in Section 2.15 from the analytically regularized integrals[13]

$$I_\alpha^\eta(V) \equiv \frac{1}{\Gamma(\alpha)} \int_V^{\infty} dE\, (E - V)^{\alpha-1} E^{-\eta} = V^{\alpha-\eta} \frac{\Gamma(-\alpha+\eta)}{\Gamma(\eta)}. \tag{4.279}$$

Since $E^{-\eta} = 1 - \eta \log E + \mathcal{O}(\eta^2)$, the coefficient of $-\eta$ in the Taylor series of $I_\alpha^\eta(V)$ will yield the desired integral. Since $1/\Gamma(\eta) \approx \eta$, we obtain directly

$$I_\alpha(V) = -\Gamma(-\alpha)V^\alpha, \tag{4.280}$$

so that (4.277) becomes

$$[\mathrm{Tr} \log \hat{H}]_{\rm cl} = -\left(\frac{M}{2\pi\hbar^2}\right)^{D/2} \Gamma(-D/2) \int d^D x\, [V(\mathbf{x})]^{D/2}. \tag{4.281}$$

---

[13]I.S. Gradshteyn and I.M. Ryzhik, *op. cit.*, Formula 3.196.2.

The same result can be obtained with the help of formulas (4.244), (4.257), and (2.504) as

$$[\text{Tr}\log\hat{H}]_{\text{cl}} = -\int d^D x \int_{-\infty}^{\infty} dE \int_{-\infty}^{\infty} \frac{dt}{2\pi\hbar} \frac{e^{it[E-V(\mathbf{x})]/\hbar}}{(2\pi i\hbar t/M)^{D/2}} \int_0^{\infty} \frac{dt'}{t'} e^{-iEt'/\hbar}. \tag{4.282}$$

Integrating over the energy yields

$$[\text{Tr}\log\hat{H}]_{\text{cl}} = -\int d^D x \int_0^{\infty} \frac{dt}{t} \frac{1}{(2\pi i\hbar t/M)^{D/2}} e^{-itV(\mathbf{x})/\hbar}. \tag{4.283}$$

Deforming the contour of integration by the substitution $t = -i\tau$, we arrive at the integral representation of the Gamma function (2.496) which reproduces immediately the result (4.281).

The quantum corrections are obtained by multiplying this with the prefactor in curly brackets in the expansion (4.258):

$$\begin{aligned}
\text{Tr}\log\hat{H} &= -\left(\frac{M}{2\pi\hbar^2}\right)^{D/2} \Gamma(-D/2) \\
&\times \int d^D x \left\{1 - \frac{\hbar^2}{12M}\boldsymbol{\nabla}^2 V(\mathbf{x})\frac{d^2}{dV^2} - \frac{\hbar^2}{24M}[\boldsymbol{\nabla}V(\mathbf{x})]^2\frac{d^3}{dV^3} + \dots\right\}[V(\mathbf{x})]^{D/2}. \tag{4.284}
\end{aligned}$$

The second term can be integrated by parts, which replaces $\boldsymbol{\nabla}^2 V(\mathbf{x}) \to -[\boldsymbol{\nabla}V(\mathbf{x})]^2 d/dV$, so that we obtain the gradient expansion

$$\text{Tr}\log\hat{H} = -\left(\frac{M}{2\pi\hbar^2}\right)^{D/2} \Gamma(-D/2) \int d^D x \left\{1 + \frac{\hbar^2}{24M}[\boldsymbol{\nabla}V(\mathbf{x})]^2\frac{d^3}{dV^3} + \dots\right\}[V(\mathbf{x})]^{D/2}. \tag{4.285}$$

The curly brackets can obviously be replaced by

$$\left\{1 - \frac{\hbar^2}{24M}\frac{\Gamma(3-D/2)}{\Gamma(-D/2)}\frac{[\boldsymbol{\nabla}V(\mathbf{x})]^2}{[V(\mathbf{x})]^3} + \dots\right\}. \tag{4.286}$$

In one dimension and with $M = 1/2$, this amounts to the formula

$$\text{Tr}\log[-\hbar^2\partial_x^2 + V(x)] = \frac{1}{\hbar}\int dx \sqrt{V(x)}\left\{1 + \frac{\hbar^2}{32}\frac{[V'(x)]^2}{V^3(x)} + \dots\right\}. \tag{4.287}$$

It is a useful exercise to rederive this with the help of the Gelfand-Yaglom method in Section 2.4.

There exists another method for deriving the gradient expansion (4.284). We split $V(\mathbf{x})$ into a constant term V and a small $\mathbf{x}$-dependent term $\delta V(\mathbf{x})$, and rewrite

$$\begin{aligned}
\text{Tr}\log\left[-\frac{\hbar^2}{2M}\boldsymbol{\nabla}^2 + V(\mathbf{x})\right] &= \text{Tr}\log\left[-\frac{\hbar^2}{2M}\boldsymbol{\nabla}^2 + V + \delta V(\mathbf{x})\right] \\
&= \text{Tr}\log\left(-\frac{\hbar^2}{2M}\boldsymbol{\nabla}^2 + V\right) + \text{Tr}\log(1 + \Delta_V\,\delta V), \tag{4.288}
\end{aligned}$$

where Δ_V denotes the functional matrix

$$\Delta_V(\mathbf{x}, \mathbf{x}') = \left(-\frac{\hbar^2}{2M}\boldsymbol{\nabla}^2 + V\right)^{-1} = \int \frac{d^D p}{(2\pi\hbar)^D}\frac{e^{i\mathbf{p}(\mathbf{x}-\mathbf{x}')/\hbar}}{\mathbf{p}^2/2M + V} \equiv \Delta_V(\mathbf{x}-\mathbf{x}'). \tag{4.289}$$

This coincides with the fixed-energy amplitude $(i/\hbar)(\mathbf{x}|\mathbf{x}')_E$ at $E = -V$ [recall Eq. (1.344)].

The first term in (4.288) is equal to (4.281) if we replace $V(\mathbf{x})$ in that expression by the constant V, so that we may write

$$\text{Tr}\log\hat{H} = [\text{Tr}\log\hat{H}]_{\text{cl}}\Big|_{V(\mathbf{x})\to V} + \text{Tr}\log(1 + \Delta_V\,\delta V). \tag{4.290}$$

We now expand the remainder

$$\text{Tr} \log\left(1 + \Delta_V \, \delta V\right) = \text{Tr} \, \Delta_V \, \delta V - \frac{1}{2} \text{Tr}\left(\Delta_V \, \delta V\right)^2 + \dots \,, \tag{4.291}$$

and evaluate the expansion terms. The first term is simply

$$\text{Tr} \, \Delta_V \, \delta V = \int d^D x \, \Delta_V(\mathbf{x}, \mathbf{x}) \delta V(\mathbf{x}) = \Delta_V(\mathbf{0}) \int d^D x \, \delta V(\mathbf{x}). \tag{4.292}$$

where

$$\Delta_V(\mathbf{0}) = \int \frac{d^D p}{(2\pi\hbar)^D} \frac{1}{\mathbf{p}^2/2M + V} = \partial_V \left[\text{Tr} \log \hat{H}\right]_{\text{cl}} \big|_{V(\mathbf{x}) \to V}. \tag{4.293}$$

The result of the integration was given in Eq. (1.348).

The second term in the remainder (4.291) reads explicitly

$$-\frac{1}{2} \text{Tr} \left(\Delta_V \, \delta V\right)^2 = -\frac{1}{2} \int d^D x \int d^D x' \, \Delta_V(\mathbf{x}, \mathbf{x}') \delta V(\mathbf{x}') \Delta_V(\mathbf{x}', \mathbf{x}) \delta V(\mathbf{x}). \tag{4.294}$$

We now make use of the operator relation

$$[f(\hat{A}), \hat{B}] = f'(\hat{A})[\hat{A}, \hat{B}] - \frac{1}{2} f''(\hat{A})[\hat{A}, [\hat{A}, \hat{B}]] + \dots \,, \tag{4.295}$$

to expand

$$\delta V \, \Delta_V = \Delta_V \, \delta V + \Delta_V^2 \left[\hat{T}, \delta V\right] + \Delta_V^3 \left[\hat{T}, [\hat{T}, \delta V]\right] + \dots \,, \tag{4.296}$$

where $\hat{T}$ is the operator of the kinetic energy $\hat{\mathbf{p}}^2/2M$. It commutes with any function $f(\mathbf{x})$ as follows:

$$[\hat{T}, f] = -\frac{\hbar^2}{2M} \left[\left(\boldsymbol{\nabla}^2 f\right) + 2\left(\boldsymbol{\nabla} f\right) \cdot \boldsymbol{\nabla}\right], \tag{4.297}$$

$$[\hat{T}, [\hat{T}, f]] = \frac{\hbar^4}{4M^2} \left\{\left[(\boldsymbol{\nabla}^2)^2 f\right] + 4[\boldsymbol{\nabla}\boldsymbol{\nabla}^2 f] \cdot \boldsymbol{\nabla} + 4[\nabla_i \nabla_i f]\nabla_i \nabla_j\right\}, \tag{4.298}$$

$$\vdots$$

Inserting this into (4.294), we obtain a first contribution

$$-\frac{1}{2} \int d^D x \int d^D x' \, \Delta_V(\mathbf{x}, \mathbf{x}') \Delta_V(\mathbf{x}', \mathbf{x})[\delta V(\mathbf{x})]^2. \tag{4.299}$$

The spatial integrals are performed by going to momentum space, where we derive the general formula

$$\int d^D x \int d^D x_1 \cdots \int d^D x_n \, \Delta_V(\mathbf{x}, \mathbf{x}_1) \Delta_V(\mathbf{x}_1, \mathbf{x}_2) \cdots \Delta_V(\mathbf{x}_{n-1}, \mathbf{x}_n) \Delta_V(\mathbf{x}_n, \mathbf{x})$$

$$= \int \frac{d^D p}{(2\pi\hbar)^D} \frac{1}{(\mathbf{p}^2/2M + V)^{n+1}} = \frac{(-1)^n}{n!} \partial_V^n \, \Delta_V(\mathbf{0}). \tag{4.300}$$

This simplifies (4.299) to

$$\frac{1}{2} \partial_V \Delta_V(\mathbf{0}) \int d^D x \, [\delta V(\mathbf{x})]^2. \tag{4.301}$$

We may now combine the non-gradient terms of $\delta V(\mathbf{x})$ consisting of the first term in (4.290), of (4.292), and of (4.301), and replace in the latter $\Delta_V(\mathbf{0})$ according to (4.293), to obtain the first three expansion terms of $[\text{Tr} \log \hat{H}]_{\text{cl}}$ with the full $\mathbf{x}$-dependent $V(\mathbf{x})$ in Eq. (4.281).

The next contribution to (4.294) coming from (4.297) is

$$\frac{\hbar^2}{4M} \int d^D x \int d^D x_1 \int d^D x_2 \, \Delta_V(\mathbf{x}, \mathbf{x}_1) \Delta_V(\mathbf{x}_1, \mathbf{x}_2) \Delta_V(\mathbf{x}_2, \mathbf{x})$$
$$\times \left\{ \left[\nabla^2 \delta V(\mathbf{x}) \right] \delta V(\mathbf{x}) + 2 \left[\nabla \delta V(\mathbf{x}) \right]^2 + 2 [\nabla \delta V(\mathbf{x})] \delta V(\mathbf{x}) \nabla \right\}, \qquad (4.302)$$

where the last ∇ acts on the first $\mathbf{x}$ in $\Delta_V(\mathbf{x}, \mathbf{x}_1)$, due to the trace. It does not contribute to the integral since it is odd in $\mathbf{x} - \mathbf{x}_1$.

We now perform the integrals over $\mathbf{x}_1$ and $\mathbf{x}_2$ using formula (4.300) and find

$$\frac{\hbar^2}{8M} \int d^D x \left\{ \left[\nabla^2 \delta V(\mathbf{x}) \right] [\delta V(\mathbf{x})] + 2 \left[\nabla \delta V(\mathbf{x}) \right]^2 \right\} \partial_V^2 \Delta_V(0). \qquad (4.303)$$

The first term can be integrated by parts, after which it removes half of the second term.

A third contribution to (4.294) which contains only the lowest gradients of $\delta V(\mathbf{x})$ comes from the third term in (4.298):

$$-\frac{\hbar^4}{8M^2} \int d^D x \int d^D x_1 \int d^D x_2 \int d^D x_3 \Delta_V(\mathbf{x}, \mathbf{x}_1) \Delta_V(\mathbf{x}_1, \mathbf{x}_2) \Delta_V(\mathbf{x}_2, \mathbf{x}_3) \Delta_V(\mathbf{x}_3, \mathbf{x})$$
$$\times 4 \left[\nabla_i \nabla_j \delta V(\mathbf{x}) \right] \delta V(\mathbf{x}) \nabla_i \nabla_j, \qquad (4.304)$$

where the last $\nabla_i \nabla_j$ acts again on we men the first $\mathbf{x}$ in $\Delta_V(\mathbf{x}, \mathbf{x}_1)$, as a consequence of the trace. In momentum space, we encounter the integral

$$\int \frac{d^D p}{(2\pi\hbar)^D} \frac{-4 p_i p_j / \hbar^2}{(\mathbf{p}^2/2M + V)^4} = -\frac{8M}{\hbar^2} \frac{\delta_{ij}}{D} \int \frac{d^D p}{(2\pi\hbar)^D} \left[\frac{1}{(\mathbf{p}^2/2M + V)^3} - \frac{V}{(\mathbf{p}^2/2M + V)^4} \right]$$
$$= -\frac{8M}{\hbar^2} \frac{\delta_{ij}}{D} \left(\frac{1}{2} \partial_V^2 + \frac{1}{6} V \partial_V^3 \right) \Delta_V(0). \qquad (4.305)$$

so that the third contribution to (4.294) reads, after an integration by parts,

$$-\frac{\hbar^2}{M} \int d^D x \left[\nabla \delta V(\mathbf{x}) \right]^2 \frac{\delta_{ij}}{D} \left(\frac{1}{2} \partial_V^2 + \frac{1}{6} V \partial_V^3 \right) \Delta_V(0). \qquad (4.306)$$

Combining all gradient terms in $[\nabla V(\mathbf{x})]^2$ and replacing $\Delta_V(0)$ according to (4.293), we recover the previous result (4.285) with the curly brackets (4.286).

For the one-dimensional tracelog, this leads to the formula

$$\mathrm{Tr} \log[-\hbar^2 \partial_x^2 + V(x)] = \frac{1}{\hbar} \int dx \, \sqrt{V(x)} \left\{ 1 + \frac{\hbar^2}{32} \frac{[V'(x)]^2}{V^3(x)} + \dots \right\}. \qquad (4.307)$$

It is a useful exercise to rederive this with the help of the Gelfand-Yaglom method in Section 2.4.

This expansion can actually be deduced, and carried to much higher order, with the help of the gradient expansion of the trace of the logarithm of the operator $-\hbar^2 \partial_\tau^2 + w^2(\tau)$ derived in Subsection 2.15.4. If we replace τ by $x/\hbar$ and $v(\tau)$ by $V(x)$, we obtain from (2.548)–(2.550):

$$\frac{1}{\hbar} \int dx \, \sqrt{V(x)} \left\{ 1 - \hbar \frac{V'}{4V^{3/2}} - \hbar^2 \left(\frac{5V'^2}{32V^3} - \frac{V''}{8V^2} \right) - \hbar^3 \left(\frac{15V'^3}{64V^{9/2}} - \frac{9V'V''}{32V^{7/2}} + \frac{V^{(3)}}{16V^{5/2}} \right) \right.$$
$$\left. - \hbar^4 \left(\frac{1105V'^4}{2048V^6} - \frac{221V'^2V''}{256V^5} + \frac{19V''^2}{128V^4} + \frac{7V'V^{(3)}}{32V^4} - \frac{V^{(4)}}{32V^3} \right) \right\}. \qquad (4.308)$$

The $\hbar$-term in the curly brackets vanishes if $V(x)$ is the same at the boundaries, and the $\hbar^2$ term goes over into the $\hbar^2$-term in (4.307).

4.9.5 Local Density of States on Circle

For future use, let us also calculate this determinant for x on a circle $x = (0, b)$, so that, as a side result, we obtain also the gradient expansion of the tracelog of the operator $(-\partial_\tau^2 + \omega^2(\tau))$ at a finite temperature. For this we recall that for a τ-independent frequency, the starting point is Eq. (2.556), according to which the tracelog of the operator $(-\partial_\tau^2 + \omega^2)$ with periodic boundary conditions in $\tau \in (0, \hbar\beta)$ is given by

$$F_\omega = -\frac{\hbar}{\sqrt{\pi}} \int_0^\infty \frac{d\tau}{\tau} \tau^{-1/2} \left[1 + 2 \sum_{n=1}^\infty e^{-(n\hbar\beta)^2/4\tau} \right] e^{-\tau\omega^2}. \tag{4.309}$$

The first term is the zero-temperature expression, the second comes from the Poisson summation formula and gives the finite-temperature effects. In the first (classical) term of the density (4.252), the factor $1/\sqrt{2\pi\hbar\tau/M}$ came from the integral over the Boltzmann factor involving the kinetic energy $\int_{-\infty}^\infty (dk/2\pi) e^{-\tau\hbar k^2/2M}$. For periodic boundary conditions in $x \in (0, b)$, this is changed to $(1/b) \sum_m e^{-\tau\hbar k_m^2/2M}$, where $k_m = 2\pi m/b$. By Poisson's formula (1.205), this can be replaced by the integral and an auxiliary sum

$$\frac{1}{b} \sum_m e^{-\tau\hbar k_m^2/2M} = \sum_{n=-\infty}^\infty \int_{-\infty}^\infty \frac{dk}{2\pi} e^{-\tau\hbar k^2/2M + ibkn} = \frac{1}{\sqrt{2\pi\hbar\tau/M}} \sum_{n=-\infty}^\infty e^{-n^2 M b^2/2\hbar\tau}. \tag{4.310}$$

If the sum is inserted into the integral (4.252), we obtain the density $\rho(E; x)$ on a circle of circumference b, with the classical contribution

$$\rho_{\rm cl}(b, E; x) = 2 \int_0^\infty \frac{d\tau}{2\pi\hbar} \sum_{n=-\infty}^\infty \frac{e^{-n^2 M b^2/2\hbar\tau - \tau[E - V(x)]/\hbar}}{(2\pi\hbar\tau/M)^{1/2}}. \tag{4.311}$$

The $n = 0$-term in the sum leads back to the original expression (4.252) on an infinite x-axis. The τ-integrals are now done with the help of formula (2.557) which yields, due to $K_\nu(z) = K_{-\nu}(z)$,

$$\int_0^\infty \frac{d\tau}{\tau} \tau^\nu e^{-n^2 M b^2/2\hbar\tau - [E - V(x)]\tau/\hbar} = 2 \left[\frac{nMb}{p(E; x)} \right]^\nu K_\nu(np(E; x)b/\hbar), \tag{4.312}$$

and we obtain, instead of (4.253),

$$\rho_{\rm cl}(b, E; x) = \frac{1}{\pi\hbar} \frac{1}{\sqrt{2\pi\hbar/M}} 2 \sum_{n=0}^\infty \left[\frac{nMb}{p(E; x)} \right]^{1/2} K_{1/2}(np(E; x)b/\hbar). \tag{4.313}$$

Inserting $K_{1/2}(z) = \sqrt{\pi/2z}\, e^{-z}$ [recall (2.559)], this becomes

$$\rho_{\rm cl}(b, E; x) = \frac{M}{\pi\hbar} \frac{1}{p(E; x)} \left(1 + 2 \sum_{n=1}^\infty e^{-np(E;x)b/\hbar} \right). \tag{4.314}$$

The sum $\sum_{n=1}^\infty \alpha^n$ is equal to $\alpha/(1 - \alpha)$, so that we obtain

$$\rho_{\rm cl}(b, E; x) = \frac{M}{\pi\hbar} \frac{\coth[p(E; x)b/2\hbar]}{p(E; x)} = \frac{M}{\pi\hbar} \frac{\coth \sqrt{2M[E - V(x)]}\, b/2\hbar}{\sqrt{2M[E - V(x)]}}. \tag{4.315}$$

For $b \to \infty$, this reduces to the previous density (4.253).

If we include the higher powers of τ in (4.252), we obtain the generalization of expression (4.254):

$$\rho(b, E; x) = \left\{ 1 - \frac{\hbar^2}{12M} V''(x) \frac{d^2}{dV^2} - \frac{\hbar^2}{24M} [V'(x)]^2 \frac{d^3}{dV^3} + \ldots \right\} \rho_{\rm cl}(b, E; x). \tag{4.316}$$

The tracelog is obtained by integrating this over $dE \log E$ from $V(x)$ to infinity. The integral diverges, and we must employ analytic regularization. We proceed as in (4.282), by using the real-time version of (4.311) and rewriting $\log E$ as an integral $-\int_0^\infty (dt'/t')e^{-iEt'/\hbar}$, so that the leading term in (4.316) is given by

$$[\operatorname{Tr} \log \hat{H}]_{\mathrm{cl}} = -\int_0^b dx \int_{-\infty}^\infty dE \int_{-\infty}^\infty \frac{dt}{2\pi\hbar} \sum_{n=-\infty}^\infty \frac{e^{-in^2 Mb^2/2\hbar t + it[E-V(x)]/\hbar}}{(2\pi i\hbar t/M)^{1/2}} \int_0^\infty \frac{dt'}{t'} e^{-iEt'/\hbar}. \quad (4.317)$$

The integral over E leads now to

$$[\operatorname{Tr} \log \hat{H}]_{\mathrm{cl}} = -\sum_{n=-\infty}^\infty \int_0^b dx \int_0^\infty \frac{dt}{t} \frac{1}{(2\pi i\hbar t/M)^{1/2}} e^{-in^2 Mb^2/2\hbar t - itV(x)/\hbar}. \quad (4.318)$$

Deforming again the contour of integration by the substitution $t = -i\tau$, creating the τ^{-1} in the denominator by an integration over $V(x)/\hbar$, we see that

$$
\begin{aligned}
[\operatorname{Tr} \log \hat{H}]_{\mathrm{cl}} &= -\pi \int_0^b dx \int_{V(x)}^\infty dV \rho_{\mathrm{cl}}(b, E; x)\Big|_{E-V(x)\to V} \\
&= \pi \int_0^b dx \int_{V(x)}^\infty \frac{dV}{E_b} \frac{1}{4\pi b} \frac{\coth \frac{1}{2}\sqrt{V/E_b}}{\frac{1}{2}\sqrt{V/E_b}}, \quad (4.319)
\end{aligned}
$$

where $E_b \equiv \hbar/2Mb^2$ is the energy associated with the length b. The integration over V produces a factor $\hbar/\tau$ in the integrand.

Thus we obtain

$$\operatorname{Tr} \log \hat{H} = \int_0^b dx \left\{ 1 + \frac{\hbar^2}{24M} [V'(x)]^2 \frac{d^3}{dV^3} + \dots \right\} \frac{2}{b} \log \left\{ 2 \sinh \left[\frac{1}{2} \sqrt{\frac{V(x)}{E_b}} \right] \right\}. \quad (4.320)$$

For $M = \hbar^2/2$, this give us the finite-b correction to formula (4.307). Replacing x by the Euclidean time τ, b by $\hbar\beta$, and $V(x)$ by the time-dependent square frequency $\omega^2(\tau)$, we obtain the gradient expansion

$$\operatorname{Tr} \log[-\partial_\tau^2 + \omega^2(\tau)] = \int_0^{\hbar\beta} d\tau \left\{ 1 + \frac{[\partial_\tau \omega^2(\tau)]^2}{12} \partial_{\omega^2}^3 + \dots \right\} \frac{2}{\hbar\beta} \log \left[2 \sinh \frac{\hbar\beta\omega(\tau)}{2} \right]. \quad (4.321)$$

4.9.6 Quantum Corrections to Bohr-Sommerfeld Approximation

The expansion (4.308) can be used to obtain a higher-order expansion of the density of states $\rho(E)$, thereby extending Eq. (4.256). For this we recall Eq. (1.592) according to which we can calculate the exact density of states from the formula

$$\rho(E) = -\frac{1}{\pi} \partial_E \operatorname{Im} \operatorname{Tr} \log \left\{ -\partial_x^2 + [V(x) - E] \right\}. \quad (4.322)$$

Integrating this over the energy yields, according to Eq. (1.593), the number of states times π, and thus the simple *exact* quantization condition for a nondegenerate one-dimensional system:

$$-\operatorname{Im} \operatorname{Tr} \log \left\{ -\partial_x^2 + [V(x) - E] \right\} = \pi(n + 1/2). \quad (4.323)$$

By comparison with Eq. (4.209) we may define a fullly quantum corrected version of the classical eikonal:

$$S_{\mathrm{qc}}(E) = -2\hbar \operatorname{Im} \operatorname{Tr} \log \left\{ -\partial_x^2 + [V(x) - E] \right\}. \quad (4.324)$$

The semiclassical expansion of this can be obtained from our earlier result (4.308) by replacing $V(x) \to V(x) - E$, so that $\sqrt{V(x)} \to -i\sqrt{E - V(x)}$, yielding

$$S_{\rm qc}(E) = 2 \int dx \sqrt{E - V(x)} \left\{ 1 + \hbar^2 \left[\frac{5V'^2}{32(E-V)^3} + \frac{V''}{8(E-V)^2} \right] \right.$$

$$\left. -\hbar^4 \left[\frac{1105V'^4}{2048(E-V)^6} + \frac{221V'^2 V''}{256(E-V)^5} + \frac{19V''^2}{128(E-V)^4} + \frac{7V'V^{(3)}}{32(E-V)^4} + \frac{V^{(4)}}{32(E-V)^3} + \cdots \right] \right\}. \quad (4.325)$$

The first term in the expansion corresponds to the Bohr-Sommerfeld approximation (4.27), the remaining ones yield the quantum corrections. The integrand agrees, of course, with the WKB expansion of the eikonal (4.14) after inserting the expansion terms $S_0, S_1, \ldots$, obtained by integrating the relations (4.20) over the functions $q_0(x), q_1(x), \ldots$ of Eq. (4.18).

Using Eq. (4.322) we obtain from $S_{\rm qc}(E)$ the density of states

$$\rho(E) = \frac{1}{2\pi\hbar} S_{\rm qc}(E) = \frac{1}{2\pi\hbar} \int dx \frac{1}{\sqrt{E-V}} \left\{ 1 - \hbar^2 \left[\frac{25V'^2}{32(E-V)^3} + \frac{3V''}{8(E-V)^2} \right] \right.$$

$$\left. +\hbar^3 \left[\frac{12155V'^4}{2048(E-V)^6} + \frac{1989V'^2 V''}{256(E-V)^5} + \frac{133V''^2}{128(E-V)^4} + \frac{49V'V^{(3)}}{32(E-V)^4} + \frac{5V^{(4)}}{32(E-V)^3} \right] \right\}. \quad (4.326)$$

Let us calculate the quantum corrections to the semiclassical energies for a purely quartic potential $V(x) = gx^4/4$, where the integral over the first term in (4.325) between the turning points $\pm x_E = \pm (4E/g)^{1/4}$ gave the Bohr-Sommerfeld approximation (4.33). The integrals of the higher terms in (4.325) are divergent, but can be calculated in analytically regularized form using once more the integral formula (4.34). This extends the Bohr-Sommerfeld equation $\nu(E) = n + 1/2$ to the exact equation $N(E) \equiv S_{\rm qc}/2\pi\hbar = n + 1/2$ [recall (1.593)]. If we express $N(E)$ in terms of $\nu(E)$ defined in Eq. (4.33) rather than E, we obtain the expansion

$$N(\nu) = \nu - \frac{1}{12\pi\nu} + \frac{11\pi^2}{10368\Gamma^8(\frac{3}{4})\nu^3} + \frac{4697\pi}{1866240\Gamma^8(\frac{3}{4})\nu^5} - \frac{390065\pi^4}{501645312\Gamma^{16}(\frac{3}{4})\nu^7}$$

$$-\frac{53352893\pi^3}{7739670528\Gamma^{16}(\frac{3}{4})\nu^9} + \cdots = n + 1/2. \quad (4.327)$$

The function is plotted in Fig. 4.327 for increasing orders in y. Given a solution $\nu^{(n)}$ of this equation, we obtain the energy $E^{(n)}$ from Eq. (4.35) with

$$\kappa^{(n)} = [\nu^{(n)} 3\Gamma(3/2)^2/2\sqrt{\pi}]^{4/3}. \quad (4.328)$$

For large n, where $\nu^{(n)} \to n$, we recover the Bohr-Sommerfeld result (4.36). We can invert the series (4.327) and obtain

$$\nu^{(n)} = (n + \tfrac{1}{2}) \left[1 + \frac{0.026525823}{(n+\frac{1}{2})^2} - \frac{0.002762954}{(n+\frac{1}{2})^4} - \frac{0.001299177}{(n+\frac{1}{2})^6} \right.$$

$$\left. + \frac{0.003140091}{(n+\frac{1}{2})^8} + \frac{0.007594497}{(n+\frac{1}{2})^{10}} + \cdots \right]. \quad (4.329)$$

The results are compared with the exact ones in Table 4.1, which approach rapidly the Bohr-Sommerfeld limit $0.688\,253\,702\ldots$. The approach is illustrated in the right-hand part of Fig. 4.1 where $\log[\nu^{(n)}/(n+\frac{1}{2})^{4/3} - 1] = \log[E^{(n)}/E^{(n)}_{BS} - 1]$ is plotted once for the exact values and once for the semiclassical expansion in Fig. 4.1. The second excited states is very well represented by the series. Although the series is only asymptotic, the results for higher n are excellent. For a detailed study of the convergence properties of the semiclassical expansion see Ref. [3].

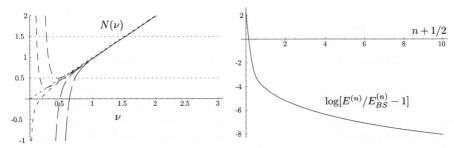

Figure 4.1 Determination of energy eigenvalues $E^{(n)}$ for purely quartic potential $gx^4/4$ in semiclassical expansion. The intersections of $N(\nu)$ with the horizontal lines yield $\nu^{(n)}$, from which $E^{(n)}$ is obtained via Eqs. (4.328) and (4.35). The increasing dash lengths show the expansions of $N(\nu)$ to increasing orders in ν. For the ground state with $n = 0$, the expansion is too divergent to give improvements to the lowest approximation without resumming the series; Right: Comparison between exact and semiclassical energies. The plot is for $\log[E^{(n)}/E_{BS}^{(n)} - 1]$.

Table 4.1 Particle energies in purely anharmonic potential $gx^4/4$ for $n = 0, 2, 4, 6, 8, 10$.

n	$E^{(n)}/(g\hbar/4M^2)^{1/3}$	$\kappa^{(n)}/2(n+1/2)^{4/3}$
0	0.667 986 259 155 777 108 3	0.841 609 948 950 895 526
2	4.696 795 386 863 646 196 2	0.692 125 685 914 981 314
4	10.244 308 455 438 771 076 0	0.689 449 772 359 340 765
6	16.711 890 073 897 950 947 1	0.688 828 486 600 234 466
8	23.889 993 634 572 505 935 5	0.688 590 146 947 993 676
10	31.659 456 477 221 552 442 8	0.688 474 290 179 981 433

Having obtained the quantum-corrected eikonal $S_{\mathrm{qc}}(E)$ we can write down a quantum-corrected partition function replacing the classical eikonal $S(E)$ in Eq. (4.236):

$$Z_{\mathrm{qc}}(E) = t_{\mathrm{qc}}(E) \frac{1}{P_{\mathrm{qc}}(1)^{1/2}} \frac{e^{iS_{\mathrm{qc}}(E) - i\pi\nu/2}}{1 - e^{iS_{\mathrm{qc}}(E) - i\pi\nu/2}}. \tag{4.330}$$

where $t_{\mathrm{qc}}(E) \equiv S_{\mathrm{qc}}'(E)$ [compare (4.198)], and $P_{\mathrm{qc}}(1)^{1/2}$ is the quantum-corrected determinant (4.234) whose calculation will require extra work.

4.10 Thomas-Fermi Model of Neutral Atoms

The density of states calculated in the last section forms the basis for the *Thomas-Fermi model* of neutral atoms. If an atom has a large nuclear charge Z, most of the electrons move in orbits with large quantum numbers. For $Z \to \infty$, we expect them to be described by semiclassical limiting formulas, which for decreasing values of Z require quantum corrections. The largest quantum correction is expected for electrons near the nucleus which must be calculated separately.

4.10.1 Semiclassical Limit

Filling up all negative energy states with electrons of both spin directions produces some local particle density $n(\mathbf{x})$, which is easily calculated from the classical local density (4.217) over all negative energies, yielding the *Thomas-Fermi density of states*

$$\rho_{\mathrm{cl}}^{(-)}(\mathbf{x}) = \int_{V(\mathbf{x})}^{0} dE\, \rho_{\mathrm{cl}}(E; \mathbf{x}) = \left(\frac{M}{2\pi\hbar^2}\right)^{D/2} \frac{1}{\Gamma(D/2 + 1)} [-V(\mathbf{x})]^{D/2}. \tag{4.331}$$

This expression can also be obtained directly from the phase space integral over the accessible free-particle energies. At each point $\mathbf{x}$, the electrons occupy all levels up to a *Fermi energy*

$$E_F = \frac{p_F(\mathbf{x})^2}{2M} + V(\mathbf{x}). \tag{4.332}$$

The associated local *Fermi momentum* is equal to the local momentum function (4.215) at $E = E_F$:

$$p_F(\mathbf{x}) = p(E_F; \mathbf{x}) = \sqrt{2M[E_F - V(\mathbf{x})]}. \tag{4.333}$$

The electrons fill up the entire Fermi sphere $|\mathbf{p}| \le p_F(\mathbf{x})$:

$$\rho_{\mathrm{cl}}^{(-)}(\mathbf{x}) = \int_{|\mathbf{p}| \le p_F(\mathbf{x})} \frac{d^D p}{(2\pi\hbar)^D} = \frac{1}{(2\pi\hbar)^D} S_D \int_0^{p_F(\mathbf{x})} dp\, p^{D-1} = \frac{1}{(2\pi\hbar)^D} \frac{2\pi^{D/2}}{\Gamma(D/2)} \frac{p_F^D(\mathbf{x})}{D}. \tag{4.334}$$

For neutral atoms, the Fermi energy is zero and we recover the density (4.331).

By occupying each state of negative energy twice, we find the classical electron density

$$n(\mathbf{x}) = 2\rho_{\mathrm{cl}}^{(-)}(\mathbf{x}). \tag{4.335}$$

The potential energy density associated with the levels of negative energy is obviously

$$E^{(-)}_{\text{pot TF}}(\mathbf{x}) = V(\mathbf{x})\rho^{(-)}(\mathbf{x}) = -\left(\frac{M}{2\pi\hbar^2}\right)^{D/2}\frac{1}{\Gamma(D/2+1)}[-V(\mathbf{x})]^{D/2+1}. \quad (4.336)$$

To find the kinetic energy density we integrate

$$\begin{aligned}
E^{(-)}_{\text{kin TF}}(\mathbf{x}) &= \int_{V(\mathbf{x})}^{0} dE\,[E - V(\mathbf{x})]\rho_{\text{cl}}(E;\mathbf{x}) \\
&= \frac{D/2}{D/2+1}\left(\frac{M}{2\pi\hbar^2}\right)^{D/2}\frac{1}{\Gamma(D/2+1)}[-V(\mathbf{x})]^{D/2+1}. \quad (4.337)
\end{aligned}$$

As in the case of the density of states (4.334), this expression can be obtained directly from the phase space integral over the free-particle energies

$$E^{(-)}_{\text{kin TF}}(\mathbf{x}) = \int_{|\mathbf{p}|\leq p_F(\mathbf{x})}\frac{d^D p}{(2\pi\hbar)^D}\frac{p^2}{2M}. \quad (4.338)$$

Performing the momentum integral on the right-hand side yields the energy density

$$E^{(-)}_{\text{kin TF}}(\mathbf{x}) = \frac{1}{(2\pi\hbar)^D}S_D\frac{1}{2M}\int_0^{p_F(\mathbf{x})}dp\,p^{D+1} = \frac{1}{(2\pi\hbar)^D}\frac{S_D}{D+2}\frac{p_F^{D+2}(\mathbf{x})}{2M}, \quad (4.339)$$

in agreement with (4.337). The sum of the two is the Thomas-Fermi energy density

$$\begin{aligned}
E^{(-)}_{\text{TF}}(\mathbf{x}) &= \int_{V(\mathbf{x})}^{0}dE\,E\,\rho_{\text{cl}}(E;\mathbf{x}) \\
&= -\frac{1}{D/2+1}\left(\frac{M}{2\pi\hbar^2}\right)^{D/2}\frac{1}{\Gamma(D/2+1)}[-V(\mathbf{x})]^{D/2+1}. \quad (4.340)
\end{aligned}$$

The three energies are related by

$$E^{(-)}_{\text{TF}}(\mathbf{x}) = -\frac{1}{D/2}E^{(-)}_{\text{kin TF}}(\mathbf{x}) = \frac{1}{D/2+1}E^{(-)}_{\text{pot TF}}(\mathbf{x}). \quad (4.341)$$

Note that the Thomas-Fermi model can also be applied to ions.[14] Then the energy levels are filled up to a nonzero Fermi energy E_F, so that the density of states (4.331) and the kinetic energy (4.337) have $-V$ replaced by $E_F - V$. This follows immediately from the representations (4.334) and (4.338) where the right-hand sides depend only on $p_F(\mathbf{x}) = \sqrt{2M[E_F - V(\mathbf{x})]}$. In the potential energy (4.336), the expression $(-V)^{D/2+1}$ is replaced by $(-V)(E_F - V)^{D/2}$, whereas in the Thomas-Fermi energy density (4.340) it becomes $(1 - E_F\partial/\partial E_F)(E_F - V)^{D/2+1}$.

---

[14]H.J. Brudner and S. Borowitz, Phys. Rev. *120*, 2054 (1960).

4.10.2 Self-Consistent Field Equation

The total electrostatic potential energy $V(\mathbf{x})$ caused by the combined charges of the nucleus and the electron cloud is found by solving the Poisson equation

$$\boldsymbol{\nabla}^2 V(\mathbf{x}) = 4\pi e^2 [Z\delta^{(3)}(\mathbf{x}) - n(\mathbf{x})] \equiv 4\pi e^2 [n_C(\mathbf{x}) - n(\mathbf{x})]. \tag{4.342}$$

The nucleus is treated as a point charge which by itself gives rise to the Coulomb potential

$$V_C(\mathbf{x}) = -\frac{Ze^2}{r}. \tag{4.343}$$

Recall that in these units $e^2 = \alpha\hbar c$, where α is the dimensionless fine-structure constant (1.502). A single electron near the ground state of this potential has orbits with diameters of the order na_H/Z, where n is the principal quantum number and a_H the *Bohr radius* of the hydrogen atom, which will be discussed in detail in Chapter 13. The latter is expressed in terms of the electron charge e and mass M as

$$a_H = \frac{\hbar^2}{Me^2} = \frac{1}{\alpha}\lambda_M^C. \tag{4.344}$$

This equation implies that a_H is about 137 times larger than the *Compton wavelength* of the electron

$$\lambda_M^C \equiv \hbar/Mc \approx 3.861\,593\,23 \times 10^{-13}\,\text{cm}. \tag{4.345}$$

It is convenient to describe the screening effect of the electron cloud upon the Coulomb potential (4.343) by a multiplicative dimensionless function $f(\mathbf{x})$. Restricting our attention to the ground state, which is rotationally symmetric, we shall write the solution of the Poisson equation (4.342) as

$$V(\mathbf{x}) = -\frac{Ze^2}{r}f(r). \tag{4.346}$$

At the origin the function $f(r)$ is normalized to unity,

$$f(0) = 1, \tag{4.347}$$

to ensure that the nuclear charge is not changed by the electrons.

It is useful to introduce a length scale of the electron cloud

$$a_{TF} = \frac{1}{e^2 Z^{1/3}}\frac{2\pi\hbar^2}{M}\left[\frac{\Gamma(5/2)}{2\cdot 4\pi}\right]^{2/3} = \frac{1}{2}\left(\frac{3\pi}{4}\right)^{2/3}\frac{a_H}{Z^{1/3}} \approx 0.8853\frac{a_H}{Z^{1/3}}, \tag{4.348}$$

which is larger than the smallest orbit a_H/Z by roughly a factor $Z^{2/3}$. All length scales will now be specified in units of a_{TF}, i.e., we set

$$r = a_{TF}\,\xi. \tag{4.349}$$

In these units, the electron density (4.335) becomes simply

$$n(\mathbf{x}) = -\frac{(2Ze^2M)^{3/2}}{3\pi^2\hbar^3}\left[\frac{f(\xi)}{a_{\mathrm{TF}}\,\xi}\right]^{3/2} = \frac{Z}{4\pi a_{\mathrm{TF}}^3}a_{\mathrm{TF}}^3\left[\frac{f(\xi)}{\xi}\right]^{3/2}. \tag{4.350}$$

The left-hand side of the Poisson equation (4.342) reads

$$\boldsymbol{\nabla}^2 V(\mathbf{x}) = \frac{1}{r}\frac{d}{dr^2}rV(\mathbf{x}) = -\frac{Ze^2}{a_{\mathrm{TF}}^3}\frac{1}{\xi}f''(\xi), \tag{4.351}$$

so that we obtain the self-consistent Thomas-Fermi equation

$$f''(\xi) = \frac{1}{\sqrt{\xi}}f^{3/2}(\xi), \qquad \xi > 0. \tag{4.352}$$

This equation ensures that the volume integral over the above electron density equal to Z. The condition $\xi > 0$ excludes the nuclear charge from the equation, whose correct magnitude is incorporated by the initial condition (4.347).

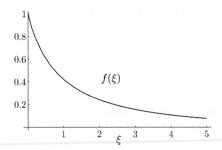

Figure 4.2   Solution for screening function $f(\xi)$ in Thomas-Fermi model.

The differential equation (4.352) is solved by the function shown in Fig. 4.2. Near the origin, it starts out like

$$f(\xi) = 1 - s\xi + \dots\,, \tag{4.353}$$

with a slope

$$s \approx 1.58807. \tag{4.354}$$

For large ξ, it goes to zero like

$$f(\xi) \approx \frac{144}{\xi^3}. \tag{4.355}$$

This power falloff is a weakness of the model since the true screened potential should fall off exponentially fast. The right-hand side by itself happens to be an exact solution of (4.352), but does not satisfy the desired boundary condition $f(0) = 1$.

4.10.3 Energy Functional of Thomas-Fermi Atom

Let us derive an energy functional whose functional extremization yields the Thomas-Fermi equation (4.352). First, there is the kinetic energy of the spin-up and spin-down electrons in a potential $V(\mathbf{x})$. It is given by the volume integral over twice the Thomas-Fermi expression (4.337):

$$E_{\text{kin}}^{(-)} = 2\frac{3}{5}\left(\frac{M}{2\pi\hbar^2}\right)^{3/2}\frac{1}{\Gamma(5/2)}\int d^3x\,[-V(\mathbf{x})]^{5/2}. \tag{4.356}$$

This can be expressed in terms of the electron density (4.335) as

$$E_{\text{kin}}^{(-)} = \frac{3}{5}\kappa\int d^3x\,n^{5/3}(\mathbf{x}), \tag{4.357}$$

where

$$\kappa \equiv \frac{\hbar}{2M}\left(3\pi^2\right)^{2/3}. \tag{4.358}$$

The potential energy

$$E_{\text{pot}}^{(-)} = \int d^3x\,V(\mathbf{x})n(\mathbf{x}) \tag{4.359}$$

is related to $E_{\text{kin}}^{(-)}$ via relation (4.341) as

$$E_{\text{pot}}^{(-)} = -\frac{5}{3}E_{\text{kin}}^{(-)}, \tag{4.360}$$

and the total electron energy in the potential $V(\mathbf{x})$ is

$$E_{\text{e}}^{(-)} = E_{\text{kin}}^{(-)} + E_{\text{pot}}^{(-)} = \frac{2}{5}E_{\text{pot}}^{(-)}. \tag{4.361}$$

We now observe that if we consider the energy as a functional of an *arbitrary* density $n(\mathbf{x})$,

$$E_{\text{e}}^{(-)} = \mathcal{E}_{\text{e}}[n] \equiv \frac{3}{5}\kappa\int d^3x\,n^{5/3}(\mathbf{x}) + \int d^3x\,V(\mathbf{x})n(\mathbf{x}), \tag{4.362}$$

the physical particle density (4.335) constitutes a minimum of the functional, which satisfies $\kappa n^{2/3}(\mathbf{x}) = -V(\mathbf{x})$. In the Thomas-Fermi atom, $V(\mathbf{x})$ on the right-hand side of (4.362) is, of course, the nuclear Coulomb potential, i.e.,

$$E_{\text{pot}}^{(-)} = E_{\text{C}}^{(-)} \equiv \int d^3x\,V_{\text{C}}(\mathbf{x})\,n(\mathbf{x}). \tag{4.363}$$

The energy $E_{\text{e}}^{(-)}$ has to be supplemented by the energy due to the Coulomb repulsion between the electrons

$$E_{\text{ee}}^{(-)} = \mathcal{E}_{\text{ee}}[n] = \frac{e^2}{2}\int d^3x\,d^3x'\,n(\mathbf{x})\frac{1}{|\mathbf{x}-\mathbf{x}'|}n(\mathbf{x}'). \tag{4.364}$$

The physical energy density should now be obtained from the minimum of the combined energy functional

$$\mathcal{E}_{\rm tot}[n] = \frac{3}{5}\kappa \int d^3x\, n^{5/3}(\mathbf{x}) + \int d^3x\, V_{\rm C}(\mathbf{x})\, n(\mathbf{x}) + \frac{e^2}{2}\int d^3x d^3x'\, n(\mathbf{x})\frac{1}{|\mathbf{x}-\mathbf{x}'|}n(\mathbf{x}').$$

(4.365)

Since we are not very familiar with extremizing nonlocal functionals, it will be convenient to turn this into a local functional. This is done as follows. We introduce an auxiliary local field $\varphi(\mathbf{x})$ and rewrite the interaction term as

$$\mathcal{E}_{\rm ee}[n,\varphi] = \mathcal{E}_{\varphi}[n,\varphi] - \mathcal{E}_{\varphi\varphi}[\varphi] \equiv \int d^3x\, \varphi(\mathbf{x})n(\mathbf{x}) - \frac{1}{8\pi e^2}\int d^3x\, \boldsymbol{\nabla}\varphi(\mathbf{x})\boldsymbol{\nabla}\varphi(\mathbf{x}).$$

(4.366)

Extremizing this in $\varphi(\mathbf{x})$, under the assumption of a vanishing $\varphi(\mathbf{x})$ at spatial infinity, yields the electric potential of the electron cloud

$$\boldsymbol{\nabla}^2\varphi(\mathbf{x}) = -4\pi e^2 n(\mathbf{x}),$$

(4.367)

which is the same as (4.342), but without the nuclear point charge at the origin. Inserting this into (4.366) we reobtain precisely to the repulsive electron-electron interaction energy (4.364).

Replacing the last term in (4.365) by the functional (4.366), we obtain a total energy functional $\mathcal{E}_{\rm tot}[n,\varphi]$, for which it is easy to find the extremum with respect to $n(\mathbf{x})$. This lies at

$$\kappa\, n^{2/3}(\mathbf{x}) = -V(\mathbf{x}),$$

(4.368)

where

$$V(\mathbf{x}) = V_{\rm C}(\mathbf{x}) + \varphi(\mathbf{x})$$

(4.369)

is the combined Thomas-Fermi potential of the nucleus and the electron cloud solving the Poisson equation (4.342).

If the extremal density (4.368) is inserted into the total energy functional $\mathcal{E}_{\rm tot}[n,\varphi]$, we may use the relation (4.360) to derive the following functional of $\varphi(\mathbf{x})$:

$$\mathcal{E}_{\rm tot}[\varphi] = -\frac{2}{5}\int d^3x\, V(\mathbf{x})n(\mathbf{x}) + \frac{1}{8\pi e^2}\int d^3x\, \varphi(\mathbf{x})\boldsymbol{\nabla}^2\varphi(\mathbf{x}).$$

(4.370)

When extremizing this expression with respect to $\varphi(\mathbf{x})$ we must remember that $V(\mathbf{x}) = V_{\rm C}(\mathbf{x}) + \varphi(\mathbf{x})$ is also present in $n(\mathbf{x})$ with a power $3/2$. The extremum lies therefore again at a field satisfying the Poisson equation (4.367).

4.10.4　Calculation of Energies

We now proceed to calculate explicitly the energies occuring in Eq. (4.370). They turn out to depend only on the slope of the screening function $f(\xi)$ at the origin. Consider first $E_{\rm pot}^{(-)}$. The common prefactor appearing in all energy expressions can be expressed in terms of the Thomas-Fermi length scale $a_{\rm TF}$ of Eq. (4.348) as

$$2\frac{M^{3/2}}{(2\pi\hbar)^{3/2}}\frac{4\pi}{\Gamma(5/2)}e^3 = \frac{Z^{-1/2}}{a_{\rm TF}^{3/2}}.$$

(4.371)

We therefore obtain the simple energy integral involving the screening function $f(\xi)$:

$$E_{\text{pot}}^{(-)} = -\frac{Z^2 e^2}{a} \int_0^\infty d\xi \frac{1}{\sqrt{\xi}} f^{5/2}(\xi). \tag{4.372}$$

The interaction energy between the electrons at the extremal $\varphi(\mathbf{x})$ satisfying (4.367) becomes simply

$$E_{\text{ee}}^{(-)} = \frac{1}{2} \int d^3x \, n(\mathbf{x}) \varphi(\mathbf{x}), \tag{4.373}$$

which can be rewritten as

$$E_{\text{ee}}^{(-)} = \frac{1}{2} \int d^3x \, n(\mathbf{x}) [V(\mathbf{x}) - V_{\text{C}}(\mathbf{x})] = \frac{1}{2} E_{\text{pot}}^{(-)} - \frac{1}{2} E_{\text{C}}^{(-)}. \tag{4.374}$$

Inserting this into (4.370) we find the alternative expression for the total energy

$$E_{\text{tot}}^{(-)} = \frac{2}{5} E_{\text{pot}}^{(-)} - E_{\text{ee}}^{(-)} = -\frac{1}{10} E_{\text{pot}}^{(-)} + \frac{1}{2} E_{\text{C}}^{(-)}. \tag{4.375}$$

The energy $E_{\text{C}}^{(-)}$ of the electrons in the Coulomb potential is evaluated as follows. Replacing $n(\mathbf{x})$ by $-\nabla^2 \varphi(\mathbf{x})/4\pi e^2$, we have, after two partial integrations with vanishing boundary terms and recalling (4.342),

$$E_{\text{C}}^{(-)} = -\frac{1}{4\pi e^2} \int d^3x \, \varphi(\mathbf{x}) \nabla^2 V_{\text{C}}(\mathbf{x}) = -\int d^3x \, \varphi(\mathbf{x}) n_{\text{C}}(\mathbf{x}). \tag{4.376}$$

Now, since

$$\varphi(\mathbf{x}) = V(\mathbf{x}) - V_{\text{C}}(\mathbf{x}) = -\frac{Ze^2}{r} [f(\xi) - 1], \qquad n_{\text{C}} = Z\delta^{(3)}(\mathbf{x}), \tag{4.377}$$

we see that the Coulomb energy $E_{\text{C}}^{(-)}$ depends only on $\varphi(0)$, which can be expressed in terms of the negative slope (4.354) of the function $f(\xi)$ as:

$$\varphi(0) = \frac{Ze^2}{a} s. \tag{4.378}$$

Thus we obtain

$$E_{\text{C}}^{(-)} = -\frac{Z^2 e^2}{a} s. \tag{4.379}$$

We now turn to the integral associated with the potential energy in Eq. (4.372):

$$I[f] = \int_0^\infty d\xi \frac{1}{\sqrt{\xi}} f^{5/2}(\xi). \tag{4.380}$$

By a trick it can again be expressed in terms of the slope parameter s. We express the energy functional (4.370) in terms of the screening function $f(\xi)$ as

$$\mathcal{E}_{\text{tot}}[\varphi] = -\frac{Z^2 e^2}{a_{\text{TF}}} \varepsilon[f], \tag{4.381}$$

with the dimensionless functional

$$\varepsilon[f] \equiv \frac{2}{5}I[f] + \frac{1}{2}J[f] = \int_0^\infty d\xi \left\{ \frac{2}{5}\frac{1}{\sqrt{\xi}}f^{5/2}(\xi) - \frac{1}{2}[f(\xi)-1]f''(\xi) \right\}. \tag{4.382}$$

The second integral can also be rewritten as

$$J[f] = \int_0^\infty d\xi \, [f'(\xi)]^2. \tag{4.383}$$

This follows from a partial integration

$$J[f] = -\int_0^\infty d\xi \, [f(\xi)-1]f''(\xi) = \int_0^\infty d\xi \, [f'(\xi)]^2 - [f(\xi)-1]f'(\xi)|_0^\infty, \tag{4.384}$$

inserting the boundary condition $f(\xi) - 1 = 0$ at $\xi = 0$ and $f'(\xi) = 0$ at $\xi = \infty$.

We easily verify that the Euler-Lagrange equation following from $\varepsilon[f]$ is the Thomas-Fermi differential equation (4.352).

As a next step in calculating the integrals I and J, we make use of the fact that under a scaling transformation

$$f(\xi) \to \bar{f}(\xi) = f(\lambda\xi), \tag{4.385}$$

the functional $\varepsilon[f]$ goes over into

$$\varepsilon_\lambda[f] = \frac{2}{5}\lambda^{-1/2}I[f] + \frac{1}{2}\lambda J[f]. \tag{4.386}$$

This must be extremal at $\lambda = 1$, from which we deduce that for $f(\xi)$ satisfying the differential equation (4.352):

$$I[f] = \frac{5}{2}J[f]. \tag{4.387}$$

This relation permits us to express the integral J in terms of the slope of $f(\xi)$ at the origin. For this we separate the two terms in (4.384), and replace $f''(\xi)$ via the Thomas-Fermi differential equation (4.352) to obtain

$$J = -\int_0^\infty d\xi \, f(\xi)f''(\xi) + \int_0^\infty d\xi \, f''(\xi) = -\int_0^\infty d\xi \, \frac{1}{\sqrt{\xi}}f^{5/2}(\xi) - f'(0) = -I + s. \tag{4.388}$$

Together with (4.387), this implies

$$I = \frac{5}{7}s, \qquad J = \frac{2}{7}s. \tag{4.389}$$

Thus we obtain for the various energies:

$$\begin{aligned}
&E_{\text{kin}}^{(-)} = \frac{3}{5}\frac{Z^2e^2}{a_{\text{TF}}}\frac{5}{7}s, \qquad
E_{\text{pot}}^{(-)} = -\frac{Z^2e^2}{a_{\text{TF}}}\frac{5}{7}s, \qquad
E_{\text{e}}^{(-)} = -\frac{2}{5}\frac{Z^2e^2}{a_{\text{TF}}}\frac{5}{7}s, \\
&E_{\text{C}}^{(-)} = -\frac{Z^2e^2}{a_{\text{TF}}}s, \qquad
E_{\text{ee}}^{(-)} = \frac{Z^2e^2}{a_{\text{TF}}}\frac{1}{7}s,
\end{aligned} \tag{4.390}$$

and the total energy is

$$E_{\text{tot}}^{(-)} = \frac{2}{5} E_{\text{pot}}^{(-)} - E_{\text{ee}}^{(-)} = -\frac{1}{10} E_{\text{pot}}^{(-)} + \frac{1}{2} E_{\text{C}}^{(-)} = -\frac{3}{7} \frac{Z^2 e^2}{a_{\text{TF}}} s \approx -0.7687 \, Z^{7/3} \frac{e^2}{a_H}. \quad (4.391)$$

The energy increases with the nuclear charge Z like $Z^2 / a_{\text{TF}} \propto Z^{7/3}$.

At the extremum, we may express the energy functional $\varepsilon[f]$ with the help of (4.388) as

$$\bar{\varepsilon}[f] = -\frac{1}{10} \int_0^\infty d\xi \frac{1}{\sqrt{\xi}} f^{5/2}(\xi) - \frac{1}{2} f(0) f'(0), \quad (4.392)$$

or in a form corresponding to (4.375):

$$\bar{\bar{\varepsilon}}[f] \equiv -\frac{1}{10} I[f] + \frac{1}{2} J_{\text{C}}[f] = -\frac{1}{10} \int_0^\infty d\xi \frac{1}{\sqrt{\xi}} f^{5/2}(\xi) + \frac{1}{2} \int_0^\infty d\xi \frac{1}{\sqrt{\xi}} f^{3/2}(\xi). \quad (4.393)$$

Using the Thomas-Fermi equation (4.352), the second integral corresponding to the Coulomb energy can be reduced to a surface term yielding $J_{\text{C}}[f] = s$, so that (4.393) gives the same total energy as (4.382).

4.10.5 Virial Theorem

Note that the total energy is equal in magnitude and opposite in sign to the kinetic energy. This is a general consequence of the so-called *viral theorem* for Coulomb systems. The kinetic energy of the many-electron Schrödinger equation contains the Laplace differential operator proportional to $\boldsymbol{\nabla}^2$, whereas the Coulomb potentials are proportional to $1/r$. For this reason, a rescaling $x \to \lambda x$ changes the sum of kinetic and total potential energies

$$E_{\text{tot}} = E_{\text{kin}} + E_{\text{pot}} \quad (4.394)$$

into

$$\lambda^2 E_{\text{kin}} + \lambda E_{\text{pot}}. \quad (4.395)$$

Since this must be extremal at $\lambda = 1$, one has the relation

$$2 E_{\text{kin}} + E_{\text{pot}} = 0, \quad (4.396)$$

which proves the virial theorem

$$E_{\text{tot}} = -E_{\text{kin}}. \quad (4.397)$$

In the Thomas-Fermi model, the role of total potential energy is played by the combination $E_{\text{pot}}^{(-)} - E_{\text{ee}}^{(-)}$, and Eq. (4.390) shows that the theorem is satisfied.

4.10.6 Exchange Energy

In many-body theory it is shown that due to the Fermi statistics of the electronic wave functions, there exists an additional electron-electron *exchange interaction* which we shall now take into account. For this purpose we introduce the bilocal density of all states of negative energy by analogy with (4.331):

$$\rho_{\text{cl}}^{(-)}(\mathbf{x}_b, \mathbf{x}_a) = \int_{V(\bar{\mathbf{x}})}^{0} dE \rho_{\text{cl}}^{(-)}(E; \mathbf{x}_b, \mathbf{x}_a). \tag{4.398}$$

In three dimensions we insert (4.274) and rewrite the energy integral as

$$\int_{V(\bar{\mathbf{x}})}^{0} dE = \frac{1}{M} \int_{0}^{p_F(\bar{\mathbf{x}})} dp\, p, \tag{4.399}$$

with the Fermi momentum $p_F(\bar{\mathbf{x}})$ of the neutral atom at the point $\bar{\mathbf{x}}$ [see (4.333)]. In this way we find

$$\rho_{\text{cl}}^{(-)}(\mathbf{x}_b, \mathbf{x}_a) = \frac{p_F^3(\bar{\mathbf{x}})}{2\pi^2 \hbar^3} \frac{1}{z^3} (\sin z - z \cos z), \tag{4.400}$$

where

$$z \equiv p_F(\bar{\mathbf{x}}) R/\hbar. \tag{4.401}$$

This expression can, incidentally, be obtained alternatively by analogy with the local expression (4.334) from a momentum integral over free wavefunctions

$$\rho_{\text{cl}}^{(-)}(\mathbf{x}_b, \mathbf{x}_a) = \int_{|\mathbf{p}| \leq p_F(\bar{\mathbf{x}})} \frac{d^3 p}{(2\pi\hbar)^D} e^{i\mathbf{p}(\mathbf{x}_b - \mathbf{x}_a)/\hbar}. \tag{4.402}$$

The simplest way to derive the exchange energy is to re-express the density of states $\rho^{(-)}(E; \mathbf{x})$ as the diagonal elements of the bilocal density

$$\rho^{(-)}(\mathbf{x}) = \rho^{(-)}(\mathbf{x}_b, \mathbf{x}_a) \tag{4.403}$$

and rewrite the electron-electron energy (4.364) as

$$E_{\text{ee}}^{(-)} = 4 \times \frac{e^2}{2} \int d^3 x d^3 x' \rho^{(-)}(\mathbf{x}, \mathbf{x}) \frac{1}{4\pi|\mathbf{x} - \mathbf{x}'|} \rho^{(-)}(\mathbf{x}', \mathbf{x}'). \tag{4.404}$$

The factor 4 accounts for the four different spin pairs in the first and the second bilocal density.

$$\uparrow \uparrow; \uparrow \uparrow; \quad \uparrow \uparrow; \downarrow \downarrow; \quad \downarrow \downarrow; \uparrow \uparrow; \quad \downarrow \downarrow; \downarrow \downarrow.$$

In the first and last case, there exists an exchange interaction which is obtained by interchanging the second arguments of the bilocal densities and changing the sign. This yields

$$E_{\text{exch}}^{(-)} = -2 \times \frac{e^2}{2} \int d^3 x d^3 x' \rho^{(-)}(\mathbf{x}, \mathbf{x}') \frac{1}{4\pi|\mathbf{x} - \mathbf{x}'|} \rho^{(-)}(\mathbf{x}', \mathbf{x}). \tag{4.405}$$

The integral over $\mathbf{x} - \mathbf{x}'$ may be performed using the formula

$$\int_0^\infty dz z^2 \frac{1}{z} \left[\frac{1}{z^3} (\sin z - z \cos z) \right]^2 = \frac{1}{4}, \tag{4.406}$$

and we obtain the exchange energy

$$E_{\text{exch}}^{(-)} = -\frac{e^2}{4\pi^3} \int d^3\bar{\mathbf{x}} \left[\frac{p_F(\bar{\mathbf{x}})}{\hbar} \right]^4. \tag{4.407}$$

Inserting

$$p_F(\mathbf{x}) = \sqrt{\frac{2Z}{a_H a_{\text{TF}}} \frac{f(\xi)}{\xi}}, \tag{4.408}$$

the exchange energy becomes

$$E_{\text{exch}}^{(-)} = -\frac{4}{\pi^2} \frac{a_{\text{TF}}}{a_H} I_2 \frac{Z^2}{a_H} \approx -0.3588\, Z^{5/3} \frac{e^2}{a_H} I_2, \tag{4.409}$$

where I_2 is the integral

$$I_2 \equiv \int_0^\infty d\xi\, f^2(\xi) \approx 0.6154. \tag{4.410}$$

Hence we obtain

$$E_{\text{exch}}^{(-)} \approx -0.2208 Z^{5/3} \frac{e^2}{a_H}, \tag{4.411}$$

giving rise to a correction factor

$$C_{\text{exch}}(Z) = 1 + 0.2872\, Z^{-2/3} \tag{4.412}$$

to the Thomas-Fermi energy (4.391).

4.10.7   Quantum Correction Near Origin

The Thomas-Fermi energy with exchange corrections calculated so far would be reliable for large-Z only if the potential was smooth so that the semiclassical approximation is applicable. Near the origin, however, the Coulomb potential is singular and this condition is no longer satisfied. Some more calculational effort is necessary to account for the quantum effects near the singularity, based on the following observation [4]. For levels with an energy smaller than some value $\varepsilon < 0$, which is large compared to the ground state energy $Z^2 e^2/a_H$, but much smaller than the average Thomas Fermi energy per particle $Z^2 e^2/aZ \sim Z^2 e^2/a_H Z^{2/3}$, i.e., for

$$\frac{Z^2 e^2}{a_H} \frac{1}{Z^{2/3}} \ll -\varepsilon \ll \frac{Z^2 e^2}{a_H}, \tag{4.413}$$

we have to recalculate the energy. Let us define a parameter ν by

$$-\varepsilon \equiv \frac{Z^2}{2a_H \nu^2}, \tag{4.414}$$

which satisfies

$$1 \ll \nu^2 \ll Z^{2/3}. \tag{4.415}$$

The contribution of the levels with energy

$$\frac{p^2}{2M} - \frac{Ze^2}{r} < -\varepsilon \tag{4.416}$$

to $E_{\text{kin,TF}}^{(-)}$ is given by an integral like (4.339), where the momentum runs from 0 to $p_{-\varepsilon}(\mathbf{x}) = \sqrt{2M[-\varepsilon - V(\mathbf{x})]}$. In the kinetic energy (4.340), the potential $V(\mathbf{x})$ is simply replaced by $-\varepsilon - V(\mathbf{x})$, and the spatial integral covers the small sphere of radius $r_{\max}$, where $-\varepsilon - V(\mathbf{x}) > 0$. For the screening function $f(\xi) = V(r)r/Ze^2$ this implies the replacement

$$f(\xi) \to [-\varepsilon - V(r)]\frac{r}{Ze^2} = f(\xi) - \xi/\xi_{\text{m}}, \tag{4.417}$$

where

$$\xi_{\text{m}} = \frac{2\nu^2}{Z}\frac{a_H}{a} \tag{4.418}$$

is small of the order $Z^{-2/3}$. Using relation (4.397) between total and kinetic energies we find the additional total energy

$$\Delta E_{\text{tot}}^{(-)} = -\frac{3}{5}\frac{Z^2e^2}{a}\int_0^{\xi_{\max}} d\xi \frac{1}{\sqrt{\xi}}\left[f(\xi) - \xi/\xi_{\text{m}}\right]^{5/2}, \tag{4.419}$$

where $\xi_{\max} \equiv r_{\max}/a$ is the place at which the integrand vanishes, i.e., where

$$\xi_{\max} = Ze^2\varepsilon f(\xi_{\max})/a, \tag{4.420}$$

this being the dimensionless version of

$$-\varepsilon - V(r_{\max}) = 0. \tag{4.421}$$

Under the condition (4.415), the slope of $f(\xi)$ may be ignored and we can use the approximation

$$\xi_{\max} \approx \xi_{\text{m}} \tag{4.422}$$

corresponding to $r_{\max} = Ze^2/\varepsilon$, with an error of relative order $Z^{-2/3}$. After this, the integral

$$\int_0^{1/c} d\lambda \frac{1}{\sqrt{\lambda}}(1 - c\lambda)^{5/2} = \frac{1}{\sqrt{c}}B\left(1/2, 7/2\right) = \frac{1}{\sqrt{c}}\frac{5}{8}\pi, \tag{4.423}$$

yielding a Beta function $B(x, y) \equiv \Gamma(x)\Gamma(y)/\Gamma(x + y)$, leads to an energy

$$\Delta E_{\text{tot}}^{(-)} = -\frac{3}{5}\frac{Z^2e^2}{a}\frac{5}{8}\frac{\pi}{M}\sqrt{\frac{a_H}{2a}}\frac{\nu}{Z^{1/3}}, \tag{4.424}$$

showing that the correction to the energy will be of relative order $1/Z^{1/3}$. Expressing a in terms of a_H via (4.348), we find

$$\Delta E_{\text{tot}}^{(-)} = -\frac{Z^2 e^2}{a_H} \nu. \tag{4.425}$$

The point is now that this energy can easily be calculated more precisely. Since the slope of the screening function can be ignored in the small selected radius, the potential is Coulomb-like and we may simply sum all occupied exact quantum-mechanical energies E_n in a Coulomb potential $-Ze^2/r$ which lie below the total energy $-\varepsilon$. They depend on the principal quantum number n in the well-known way:

$$E_n = -\frac{Z^2 e^2}{a_H} \frac{1}{2n^2}. \tag{4.426}$$

Each level occurs with angular momentum $l = 0, \dots, n-1$, and with two spin directions so that the total degeneracy is $2n^2$. By Eq. (4.414), the maximal energy $-\varepsilon$ corresponds to a maximal quantum number $n_{\max} = \nu$. The sum of all energies E_n up to the energy ε is therefore given by

$$\begin{aligned}
\Delta_{\text{QM}} E_{\text{tot}}^{(-)} &= -2 \frac{Z^2 e^2}{a_H} \frac{1}{2} \sum_{n=0}^{\nu} 1 \\
&= -\frac{Z^2 e^2}{a_H} [\nu],
\end{aligned} \tag{4.427}$$

where $[\nu]$ is the largest integer number smaller than ν. The difference between the semiclassical energy (4.424) and the true quantum-mechanical one (4.427) yields the desired quantum correction

$$\Delta E_{\text{corr}}^{(-)} = -\frac{Z^2 e^2}{a_H} ([\nu] - \nu). \tag{4.428}$$

For large ν, we must average over the step function $[\nu]$, and find

$$\langle [\nu] \rangle = \nu - \frac{1}{2}, \tag{4.429}$$

and therefore

$$\Delta E_{\text{corr}}^{(-)} = \frac{Z^2 e^2}{a_H} \frac{1}{2}. \tag{4.430}$$

This is the correction to the energy of the atom due to the failure of the quasi-classical expansion near the singularity of the Coulomb potential. With respect to the Thomas-Fermi energy (4.391) which grows with increasing nuclear charge Z like $-0.7687\, Z^{7/3} e^2/a_H$, this produces a correction factor

$$C_{\text{sing}}(Z) = 1 - \frac{7 a_{\text{TF}}}{6 a_H s} \approx 1 - 0.6504\, Z^{-1/3} \tag{4.431}$$

to the Thomas-Fermi energy (4.391).

4.10.8   Systematic Quantum Corrections to Thomas-Fermi Energies

Just as for the density of states in Section 4.9, we can derive the quantum corrections to the energies in the Thomas-Fermi atom. The electrons fill up all negative-energy levels in the combined potential $V(\mathbf{x})$. The density of states in these levels can be selected by a Heaviside function of the negative Hamiltonian operator as follows:

$$\rho^{(-)}(\mathbf{x}) = \langle \mathbf{x}|\Theta(-\hat{H})|\mathbf{x}\rangle. \tag{4.432}$$

Using the Fourier representation (1.308) for the Heaviside function we write

$$\Theta(-\hat{H}) = \int_{-\infty}^{\infty} \frac{dt}{2\pi i(t - i\eta)} e^{-i\hat{H}t/\hbar}, \tag{4.433}$$

and obtain the integral representation

$$\rho^{(-)}(\mathbf{x}) = \int_{-\infty}^{\infty} \frac{dt}{2\pi i(t - i\eta)} (\mathbf{x}\, t|\mathbf{x}\, 0). \tag{4.434}$$

Inserting the short-time expansion (4.257), and the correspondence $t \to i\hbar d/dV$ in the time integral, we find

$$\rho(\mathbf{x}) = \left\{ 1 - \frac{\hbar^2}{12M}\boldsymbol{\nabla}^2 V(\mathbf{x})\frac{d^2}{dV^2} - \frac{\hbar^2}{24M}[\boldsymbol{\nabla}V(\mathbf{x})]^2\frac{d^3}{dV^3} + \dots \right\} \rho_{\mathrm{TF}}(\mathbf{x}). \tag{4.435}$$

The potential energy is simply given by

$$E_{\mathrm{pot}}^{(-)}(\mathbf{x}) = \langle \mathbf{x}|V(\mathbf{x})\Theta(-\hat{H})|\mathbf{x}\rangle = \int d^3x V(\mathbf{x})\rho^{(-)}(\mathbf{x}). \tag{4.436}$$

With (4.435), this becomes

$$E_{\mathrm{pot}}^{(-)}(\mathbf{x}) = V(\mathbf{x})\left\{ 1 - \frac{\hbar^2}{12M}\boldsymbol{\nabla}^2 V(\mathbf{x})\frac{d^2}{dV^2} - \frac{\hbar^2}{24M}[\boldsymbol{\nabla}V(\mathbf{x})]^2\frac{d^3}{dV^3} + \dots \right\} \rho_{\mathrm{TF}}(\mathbf{x}). \tag{4.437}$$

For the energy of all negative-energy states we may introduce a density function

$$E^{(-)}(\mathbf{x}) = \langle \mathbf{x}|\hat{H}\Theta(-\hat{H})|\mathbf{x}\rangle. \tag{4.438}$$

The derivative of this with respect to $V(\mathbf{x})$ is equal to the density of states (4.432):

$$\frac{\partial}{\partial V(\mathbf{x})} E^{(-)}(\mathbf{x}) = \rho^{(-)}(\mathbf{x}). \tag{4.439}$$

This follows right-away from $\partial\hat{H}/\partial V(\mathbf{x}) = 1$ and $\partial[x\Theta(x)]/\delta x = \Theta(x)$.

Inserting the representation (4.433), the factor $\hat{H}$ can be obtained by applying the differential operator $i\hbar\partial_t$ to the exponential function. After a partial integration, we arrive at the integral representation

$$E^{(-)}(\mathbf{x}) = i\hbar \int_{-\infty}^{\infty} \frac{dt}{2\pi i(t - i\eta)^2} (\mathbf{x}\, t|\mathbf{x}\, 0). \tag{4.440}$$

Inserting on the right-hand side the expansion (4.257), the leading term produces the local *Thomas-Fermi energy* density (4.340):

$$E_{\mathrm{TF}}^{(-)}(\mathbf{x}) = -\left(\frac{M}{2\pi\hbar}\right)^{D/2} \frac{1}{\Gamma(D/2 + 2)}[-V(\mathbf{x})]^{D/2+1}. \tag{4.441}$$

The short-time expansion terms yield, with the correspondence $t \to i\hbar d/dV$, the energy including the quantum corrections

$$E^{(-)}(\mathbf{x}) = \left\{ 1 - \frac{\hbar^2}{12M}\boldsymbol{\nabla}^2 V(\mathbf{x})\frac{d^2}{dV^2} - \frac{\hbar^2}{24M}[\boldsymbol{\nabla}V(\mathbf{x})]^2\frac{d^3}{dV^3} + \dots \right\} E_{\text{TF}}^{(-)}(\mathbf{x}). \tag{4.442}$$

One may also calculate selectively the kinetic energy density from the expression

$$E^{(-)}(\mathbf{x}) = \frac{1}{2M}\langle \mathbf{x}|\hat{\mathbf{p}}^2 \Theta(-\hat{H})|\mathbf{x}\rangle. \tag{4.443}$$

This can obviously be extracted from (4.440) by a differentiation with respect to the mass:

$$\begin{aligned}
E_{\text{kin}}^{(-)}(\mathbf{x}) &= -M\frac{\partial}{\partial M}i\hbar\int_{-\infty}^{\infty}\frac{dt}{2\pi i(t-i\eta)^2}\langle \mathbf{x}|e^{-i\hat{H}t/\hbar}|\mathbf{x}\rangle \\
&= -M\frac{\partial}{\partial M}E^{(-)}(\mathbf{x}).
\end{aligned} \tag{4.444}$$

According to Eq. (4.442), the first quantum correction to the energy $E_{\text{e}}^{(-)}$ is

$$\begin{aligned}
\Delta E_{\text{e}}^{(-)} &= -\int d^3x\left[-\frac{\hbar^2}{12M}\boldsymbol{\nabla}^2 V\frac{d^2}{dV^2} - \frac{\hbar^2}{24M}(\boldsymbol{\nabla}V)^2\frac{d^3}{dV^3}\right] \\
&\qquad\qquad \times 2\,\frac{2}{5}\left(\frac{M}{2\pi\hbar}\right)^{3/2}\frac{1}{\Gamma(5/2)}(-V)^{5/2} \\
&= \frac{\sqrt{2M}}{12\hbar\pi^2}\int d^3x\left[(-V)^{1/2}\boldsymbol{\nabla}^2 V - \frac{1}{4}(-V)^{-3/2}(\boldsymbol{\nabla}V)^2\right].
\end{aligned} \tag{4.445}$$

It is useful to bring the second term to a more convenient form. For this we note that by the chain rule of differentiation

$$\boldsymbol{\nabla}^2 V^{3/2} = \frac{3}{2}\boldsymbol{\nabla}\left[V^{1/2}\boldsymbol{\nabla}V\right] = \frac{3}{4}\left[V^{-1/2}(\boldsymbol{\nabla}V)^2 + 2V^{1/2}\boldsymbol{\nabla}^2 V\right]. \tag{4.446}$$

As a consequence we find

$$\Delta E_{\text{e}}^{(-)} = \frac{\sqrt{2M}}{24\hbar\pi^2}\int d^3x\left[(-V)^{1/2}\boldsymbol{\nabla}^2 V - \frac{2}{3}\boldsymbol{\nabla}^2(-V)^{3/2}\right]. \tag{4.447}$$

This energy evaluated with the potential V determined above describes directly the lowest correction to the total energy. To prove this, consider the new total energy [recall (4.366)]

$$E_{\text{tot}}^{(-)} = E_{\text{e}}^{(-)} + \Delta E_{\text{e}}^{(-)} - \mathcal{E}_{\varphi\varphi}[\varphi]. \tag{4.448}$$

Extremizing this in the field $\varphi(\mathbf{x})$ and denoting the new extremal field by $\varphi(\mathbf{x})+\Delta\varphi(\mathbf{x})$, we obtain for $\Delta\varphi(\mathbf{x})$ the field equation:

$$\frac{\delta}{\delta V(\mathbf{x})}[E_{\text{e}}^{(-)} + \Delta E_{\text{e}}^{(-)}] + \frac{1}{e^2}\boldsymbol{\nabla}^2[\varphi(\mathbf{x}) + \Delta\varphi(\mathbf{x})] = 0. \tag{4.449}$$

Taking advantage of the initial extremality condition (4.367), we derive the field equation

$$\frac{1}{e^2}\boldsymbol{\nabla}^2\Delta\varphi(\mathbf{x}) = -\frac{\delta\Delta E_{\text{e}}^{(-)}}{\delta V(\mathbf{x})}. \tag{4.450}$$

If we now expand the corrected energy up to first order in $\Delta E_{\text{e}}^{(-)}$ and $\Delta\varphi(\mathbf{x})$, we obtain

$$E_{\text{tot}}^{(-)} = E_{\text{e}}^{(-)} + \Delta E_{\text{e}}^{(-)} + \int d^3x\frac{\delta E_{\text{e}}^{(-)}}{\delta V(\mathbf{x})}\Delta\varphi(\mathbf{x}) - \mathcal{E}_{\varphi\varphi}[\varphi] - \frac{1}{4\pi e^2}\int d^3x\,\boldsymbol{\nabla}\varphi(\mathbf{x})\boldsymbol{\nabla}\Delta\varphi(\mathbf{x}). \tag{4.451}$$

Due to the extremality property (4.367) of the uncorrected energy at the original field $\varphi(\mathbf{x})$, the second and fourth terms cancel each other, and the correction to the total energy is indeed given by (4.447).

Actually, the statement that the energy (4.447) is the next quantum correction is not quite true. When calculating the first quantum correction in Subsection 4.10.7, we subtracted the contribution of all orbits with total energies

$$E < -\varepsilon. \tag{4.452}$$

After that we calculated in (4.427) the *exact* quantum corrections coming from the neighborhood

$$r < r_{\max} = Ze^2\varepsilon \tag{4.453}$$

of the origin. Thus we have to omit this neighborhood from all successive terms in the semiclassical expansion, in particular from (4.447). According to the remarks after Eq. (4.340), the energy $E_e^{(-)}$ of electrons filling all levels up to a total energy E_F is found from (4.336) by replacing $(-V)^{D/2+1}$ by $(1 - E_F\partial/E_F)(E_F - V)^{D/2+1}$. The energy level satisfying (4.452) correspond to a Fermi level E_F, so that the energy of the electrons in these levels is

$$E_e^{(-)} = 2\frac{2}{5}E_{\text{pot TF}}^{(-)} = -2\frac{2}{5}\left(\frac{M}{2\pi\hbar}\right)^{3/2}\frac{1}{\Gamma(5/2)}(1 - E_F\partial_{E_F})\int d^3x[-E_F - V(\mathbf{x})]^{5/2}. \tag{4.454}$$

We therefore have to subtract from the correction (4.445) a term

$$\Delta_{\text{sub}}E_e^{(-)} = (1 - E_F\partial_{E_F})\frac{\sqrt{2M}}{12\hbar\pi^2}\int d^3x\left[(-E_F - V)^{1/2}\nabla^2 V - \frac{1}{4}(-E_F - V)^{-3/2}(\nabla V)^2\right]. \tag{4.455}$$

The true correction can then be decomposed into a contribution from the finite region outside the small sphere

$$\Delta E_{\text{outside}}^{(-)} = \frac{\sqrt{2M}}{24\hbar\pi^2}\int_{r \geq r_{\max}} d^3x(-V)^{1/2}\nabla^2 V, \tag{4.456}$$

plus a subtracted contribution from the inside

$$\Delta E_{\text{inside}}^{(-)} = \frac{\sqrt{2M}}{24\hbar\pi^2}\int_{r < r_{\max}} d^3x\left[(-V)^{1/2} - (1 - E_F\partial_{E_F})(-E_F - V)^{1/2}\right]\nabla^2 V, \tag{4.457}$$

plus a pure gradient term

$$\Delta E_{\text{grad}}^{(-)} = -\frac{\sqrt{2M}}{24\hbar\pi^2}\frac{2}{3}\left[\int d^3x\nabla^2(-V)^{3/2} - \int_{r < r_{\max}} d^3x\nabla^2(1 - E_F\partial_{E_F})(-E_F - V)^{3/2}\right]. \tag{4.458}$$

The last two volume integrals can be converted into surface integrals. Either integrand vanishes on its outer surface [recall (4.420)]. At the inner surface, an infinitesimal sphere around the origin, the integrands coincide so that the energy (4.458) vanishes.

The energy (4.457) does not vanish but can be ignored in the present approximation. At the δ-function at the nuclear charge in $\nabla^2 V$, the difference $(-V)^{1/2} - (1 - E_F\partial_{E_F})(-E_F - V)^{1/2}$ vanishes. In the integral, we may therefore replace $\nabla^2 V(\mathbf{x})$ by $-4\pi e^2 n(\mathbf{x})$ [dropping the δ-function in (4.342)]. In the small neighborhood of the origin, the integral is suppressed by a power of $Z^{-4/3}$ as will be seen below.

Thus, only the outside energy (4.456) needs to be evaluated, where the small sphere excludes the nuclear charge, so that we may replace $\nabla^2 V(\mathbf{x})$ as in the last integral. Thus we obtain

$$\Delta E_{\text{outside}}^{(-)} = -\frac{2e^2M^2M}{9\pi^3\hbar^4}\int_{r < r_{\max}} d^3x[-V(\mathbf{x})]^2. \tag{4.459}$$

Expressing V in terms of the screening function and going to reduced variables, the quantum correction takes the final form

$$\Delta E_{\text{outside}}^{(-)} = -\frac{8}{9\pi^2} \frac{a}{a_H} \left[\int d\xi f^2(\xi) \right] Z^{5/3}$$

$$\approx -0.07971 \frac{e^2}{a_H} I_2 Z^{5/3} \approx -0.04905 Z^{5/3} \frac{e^2}{a_H}, \tag{4.460}$$

where I_2 is the integral over $f^2(\xi)$ calculated in Eq. (4.410). We have re-extended the integration over the entire space with a relative error of order $Z^{-2/3}$, due to the smallness of the sphere. The correction factor to the leading Thomas-Fermi energy caused by this is

$$C_{\text{QM}} = 1 + 0.06381 \, Z^{-2/3}. \tag{4.461}$$

In the reduced variables, the order of magnitude of the ignored energy (4.457) can most easily be estimated. It reads

$$\Delta E_{\text{inside}}^{(-)} = -\frac{8}{9\pi^2} \frac{a}{a_H} \int_0^{\xi_{\text{max}}} d\xi \left[f^2(\xi) - \sqrt{f(\xi) - \xi/\xi_{\text{m}}} f^{3/2}(\xi) \right] Z^{5/3}. \tag{4.462}$$

Since ξ_{max} and ξ_{m} are of the order $Z^{-2/3}$, this energy is of the relative order $Z^{-4/3}$ and thus negligible since we want to find here only corrections up to $Z^{-2/3}$.

Observe that the quantum correction (4.463) is of relative order $Z^{-2/3}$ and precisely a fraction $2/9$ of the exchange energy (4.409). Both energies together are therefore

$$\Delta E_{\text{inside}}^{(-)} + E_{\text{exch}}^{(-)} = \frac{11}{9} E_{\text{exch}}^{(-)} \approx -0.2699 \, Z^{5/3} \frac{e^2}{a_H}. \tag{4.463}$$

The corrections of order $Z^{-2/3}$ can be collected in the expression

$$\Delta E_{\text{inside}}^{(-)} + E_{\text{exch}}^{(-)} = -0.7687 \, Z^{7/3} \frac{e^2}{a_H} \times C_2(Z), \tag{4.464}$$

where $C_2(Z)$ is the correction factor

$$C_2(Z) = 1 + 0.3510 \, Z^{-2/3} + \dots \, . \tag{4.465}$$

Including also the $Z^{-1/3}$-correction (4.431) from the origin we obtain the total energy

$$E_{\text{tot}}^{(-)} = -0.7687 \, Z^{7/3} \frac{e^2}{a_H} \times C_{\text{tot}}(Z), \tag{4.466}$$

with the total correction factor

$$C_{\text{tot}}(Z) = 1 - 0.6504 \, Z^{-1/3} + 0.3510 \, Z^{-2/3} + \dots \, . \tag{4.467}$$

This large-Z approximation is surprisingly accurate. The experimental binding energy of mercury with $Z = 80$ is

$$E_{\text{Hg}}^{\text{exp}} \approx -18130, \tag{4.468}$$

in units of $e^2/a_H = 2 \, \text{Ry}$, whereas the large-Z formula (4.466) with the correction factor (4.467) yields the successive approximations including the first, second, and third term in (4.467):

$$E_{\text{Hg}} \approx -(21200, 18000, 18312) \quad \text{for} \quad C(80) = (1, 0.849, 0.868). \tag{4.469}$$

Even at the lowest value $Z = 1$, the binding energy of the hydrogen atom

$$E_{\text{H}}^{\text{exp}} \approx -\frac{1}{2} \tag{4.470}$$

is quite rapidly approached by the successive approximations

$$E_{\text{H}} \approx -(0.7687, 0.2687, 0.5386) \quad \text{for} \quad C(1) = (1, 0.350, 0.701). \tag{4.471}$$

4.11   Classical Action of Coulomb System

Consider an electron of mass M in an attractive Coulomb potential of a proton at the coordinate origin

$$V_C(\mathbf{x}) = -\frac{e^2}{r}. \tag{4.472}$$

If the proton is substituted by a heavier nucleus, e^2 has to be multiplied by the charge Z of that nucleon. The Lagrangian of this system is

$$L = \frac{M}{2}\dot{\mathbf{x}}^2 - \frac{e^2}{r}. \tag{4.473}$$

Because of rotational invariance, the orbital angular momentum is conserved and the motion is restricted to a plane, say $x - y$. If ϕ denotes the azimuthal angle in this plane, the constant orbital momentum is

$$l = Mr^2\dot{\phi}. \tag{4.474}$$

The conserved energy is

$$E = \frac{M}{2}(\dot{r}^2 + r^2\dot{\phi}^2) - \frac{e^2}{r}. \tag{4.475}$$

Together with (4.474) we find

$$\dot{r} = \frac{1}{M}p_E(r); \qquad p_E(r) = \sqrt{2M[E - V_{\text{eff}}(r)]}, \tag{4.476}$$

where $V_{\text{eff}}(r) = V_C(r) + V_l(r)$ is the sum of the Coulomb potential and the *angular barrier* potential

$$V_l(r) = \frac{l^2}{2Mr^2}. \tag{4.477}$$

The differential equation (4.476) is solved by the integral relation

$$t = \int dr \, \frac{M}{p_E(r)}. \tag{4.478}$$

For the angle ϕ, Eq. (4.474) implies

$$\frac{d\phi}{dr} = \frac{l}{r^2 p_E(r)}, \tag{4.479}$$

which is solved by the integral

$$\phi = l \int dr \frac{1}{r^2 p_E(r)}. \tag{4.480}$$

Inserting $p_E(r)$ from (4.476) this becomes explicitly

$$\phi = l \int \frac{dr}{r} \left[2ME \left(r^2 + \frac{e^2 r}{E} - \frac{l^2}{2ME} \right) \right]^{-1/2}, \tag{4.481}$$

while (4.478) reads

$$t = M \int dr\, r \left[2ME \left(r^2 + \frac{e^2 r}{E} - \frac{l^2}{2ME} \right) \right]^{-1/2}. \qquad (4.482)$$

Consider now the motion for negative energies. Defining

$$\bar{p}_E \equiv p_{-E}(\infty) = \sqrt{-2ME}, \qquad (4.483)$$

and introducing the parameters

$$a \equiv \frac{e^2}{2|E|} = \frac{Me^2}{\bar{p}_E^2}, \quad \epsilon^2 \equiv 1 - \frac{l^2 \bar{p}_E^2}{M^2 e^4} = 1 - \frac{l^2}{aMe^2} = 1 - \frac{l^2 v_a^2}{e^4}, \quad v_a = \sqrt{\frac{e^2}{aM}} = \frac{\bar{p}_E}{M}, \qquad (4.484)$$

we obtain

$$\phi = \frac{l}{Mv_a} \int \frac{dr}{r} \frac{1}{\sqrt{a^2\epsilon^2 - (r - a)^2}}, \qquad (4.485)$$

$$t = \frac{1}{v_a} \int dr \frac{r}{\sqrt{a^2\epsilon^2 - (r - a)^2}}, \qquad (4.486)$$

where v_a is the velocity associated with the momentum $\bar{p}_E$. The ratio

$$\omega = \frac{v_a}{a} \qquad (4.487)$$

is the inverse period of the orbit, also called mean motion, which satisfies $\omega^2 a^3 = e^2/M$, the third *Kepler law*. In the limit $E \to 0$, the major semiaxis a becomes infinite, and so does ω. The eccentricity vanishes and the orbit is parabolic.

Introducing the variable

$$h \equiv a(1 - \epsilon^2) = \frac{l^2}{Me^2} = \frac{l^2}{\bar{p}_E^2 a}, \qquad (4.488)$$

and observing that

$$\frac{l}{Mv_a} = a\sqrt{1 - \epsilon^2}, \qquad (4.489)$$

the first equation is solved by

$$\frac{h}{r} = 1 + \epsilon \cos(\phi - \phi_0). \qquad (4.490)$$

This follows immediately from the fact that

$$\sin(\phi - \phi_0) = \frac{\sqrt{1 - \epsilon^2}}{\epsilon r} \sqrt{a^2\epsilon^2 - (r - a)^2}. \qquad (4.491)$$

The relation (4.490) describes an ellipse with principal axes $a = h/(1 - \epsilon^2)$, $b = h/\sqrt{1 - \epsilon^2}$, and an eccentricity ϵ (see Fig. 4.3). In the orbital plane, the Cartesian

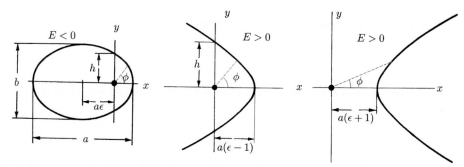

Figure 4.3 Orbits in Coulomb potential showing the parameter h and the eccentricity ϵ of ellipse ($E < 0$) and hyperbola ($E > 0$) in attractive and repulsive cases.

coordinates of the motion are

$$x = h\frac{\cos(\phi - \phi_0)}{1 + \epsilon \sin(\phi - \phi_0)}, \qquad y = h\frac{\sin(\phi - \phi_0)}{1 + \epsilon \sin(\phi - \phi_0)}. \tag{4.492}$$

For positive energy, we define a momentum

$$p_E = p_E(\infty) = \sqrt{2ME}, \tag{4.493}$$

and the parameters

$$a \equiv \frac{e^2}{2|E|} = \frac{Me^2}{p_E^2}, \quad \epsilon^2 \equiv 1 + \frac{l^2 p_E^2}{M^2 e^4} = 1 + \frac{l^2}{aMe^2} = 1 + \frac{l^2 v_a^2}{e^4}, \quad v_a = \sqrt{\frac{e^2}{aM}} = \frac{p_E}{M}. \tag{4.494}$$

The eccentricity ϵ is now larger than unity. Apart from this, the solutions to the equations of motion are the same as before, and the orbits are hyperbolas as shown in Fig. 4.3. The y-coordinate above the focus is now

$$h = a(\epsilon^2 - 1) = \frac{l^2}{Me^2} = \frac{l^2}{p_E^2 a}. \tag{4.495}$$

For a repulsive interaction, we change the sign of e^2 in the above equations. The equation (4.490) for r becomes now

$$\frac{h}{r} = -1 + \epsilon \cos(\phi - \phi_0), \tag{4.496}$$

and yields the right-hand hyperbola shown in Fig. 4.3.

For later discussions in Chapter 13, where we shall solve the path integral of the Coulomb system exactly, we also note that by introducing a new variable in terms of the variable ξ, to so-called *eccentric anomaly*

$$r = a(1 - \epsilon \cos \xi), \tag{4.497}$$

we can immediately perform the integral (4.486) to find

$$t = \frac{a}{v_a} \int d\xi \, (1 - \epsilon \cos \xi) = \frac{1}{\omega}(\xi - \epsilon \sin \xi), \tag{4.498}$$

where we have chosen the integration constant to zero. Using Eq. (4.496) we see that

$$x = r \cos \phi = \frac{h - r}{\epsilon} = a(\cos \xi - \epsilon), \tag{4.499}$$

$$y = r \sin \phi = \sqrt{r^2 - x^2} = a\sqrt{1 - \epsilon^2} \sin \xi = b \sin \xi. \tag{4.500}$$

Equations (4.497) and (4.498) represent a parametric representation of the orbit. From Eqs. (4.498) and (4.497) we see that

$$\frac{dt}{r} = \frac{1}{a}d(\xi/\omega), \tag{4.501}$$

exhibiting ξ/ω is a path-dependent *pseudotime*. As a function of this pseudotime, the coordinates x and y oscillate harmonically.

The pseudotime facilitates the calculation of the classical action $A(\mathbf{x}_b, \mathbf{x}_a; t_b - t_a)$, as done first in the eighteenth century by Lambert.[15] If we denote the derivative with respect to ξ by a prime, the classical action reads

$$A = \int_{\xi_a}^{\xi_b} d\xi \left[\frac{M}{2r} a\omega \left(x'^2 + y'^2 \right) + \frac{e^2}{a\omega} \right]. \tag{4.502}$$

For an elliptic orbit with principal axes a, b, this becomes

$$A = \int_{\xi_a}^{\xi_b} d\xi \left[\frac{M}{2} \omega \frac{a^2 \sin^2 \xi + b^2 \cos^2 \xi}{1 - \epsilon \cos \xi} \right] + M\omega a^2 (\xi_b - \xi_a). \tag{4.503}$$

After performing the integral using the formula

$$\int \frac{d\xi}{1 - \epsilon \cos \xi} = \frac{2}{1 - \epsilon^2} \arctan \frac{\sqrt{1 - \epsilon^2} \tan(\xi/2)}{1 - \epsilon}, \tag{4.504}$$

we find

$$A = \frac{M}{2} a^2 \omega \left[3(\xi_b - \xi_a) + \epsilon(\sin \xi_b - \sin \xi_a) \right]. \tag{4.505}$$

Introducing the parameters $\alpha, \beta, \gamma,$ and δ by the relations

$$\cos \alpha \equiv \epsilon \cos[(\xi_b + \xi_a)/2], \quad \beta \equiv (\xi_b - \xi_a)/2, \quad \gamma \equiv \alpha + \beta, \quad \delta \equiv \alpha - \beta, \tag{4.506}$$

---

[15] Johann Heinrich Lambert (1728–1777) was an ingenious autodidactic taylor's son who with 16 years found *Lambert's law* for the apparent motion of comets (and planets) on the sky: If the sun lies on the concave (convex) side of the apparent orbit, comet is closer to (farther from) the sun than the earth. In addition, he laid the foundations to photometry.

the action becomes

$$A = \frac{M}{2}a^2\omega\left[(3\gamma + \sin\gamma) - (3\delta + \sin\delta)\right]. \tag{4.507}$$

Using (4.498), we find for the elapsed time $t_b - t_a$ the relation

$$t_b - t_a = \frac{1}{\omega}\left[(\gamma - \sin\gamma) \mp (\delta - \sin\delta)\right]. \tag{4.508}$$

The $\mp$-signs apply to an ellipse whose short or long arc connects the two endpoints, respectively. The parameters γ and δ in the action and in the elapsed time are related to the endpoints $\mathbf{x}_b$ and $\mathbf{x}_a$ by

$$r_b + r_a + R = 4a\sin^2(\gamma/2) \equiv 4a\rho_+, \qquad r_b + r_a - R = 4a\sin^2(\delta/2) \equiv 4a\rho_-, \tag{4.509}$$

where $r_b \equiv |\mathbf{x}_b|$, $r_a \equiv |\mathbf{x}_a|$, $R \equiv |\mathbf{x}_b - \mathbf{x}_a|$, and $\rho_\pm \in [0,1]$. Expressing the semimajor axis in (4.507) in terms of ω as $a = (e^2/M\omega^2)^{1/3}$, and ω in terms of $t_b - t_a$ we obtain the desired classical action $A(\mathbf{x}_b, \mathbf{x}_a; t_b - t_a)$.

The elapsed time depends on the endpoints via a transcendental equation which can only be solved by a convergent power series (Lambert's series)

$$t_b - t_a = \frac{1}{\sqrt{2}\omega}\sum_{j=1}^{\infty}\frac{(2j)!}{2^{2j}j!^2}\frac{1}{j^2 - 1/4}\epsilon^{j-1}\left(\rho_+^{j+1/2} \pm \rho_-^{j+1/2}\right). \tag{4.510}$$

In the limit of a parabolic orbit the series has only the first term, yielding

$$t_b - t_a = \frac{\sqrt{2}}{3\omega}(\rho_+^{3/2} \pm \rho_-^{3/2}). \tag{4.511}$$

We can also express a and ω in terms of the energy E as $e^2/(-2E)$ and $\sqrt{-2E/Ma^2}$, respectively, and go over to the eikonal $S(\mathbf{x}_b, \mathbf{x}_a; E)$ via the Legendre transformation (4.87). Substituting for $t_b - t_a$ the relation (4.508), we obtain for the short arc of the ellipse

$$S(\mathbf{x}_b, \mathbf{x}_a; E) = A(\mathbf{x}_b, \mathbf{x}_a; E) + (t_b - t_a)E = \frac{M}{2}a^2\omega\,4(\gamma - \delta). \tag{4.512}$$

For a complete orbit, the expression (4.505) yields a total action

$$A = \frac{M}{2}a^2\omega\,2\pi(1+2) = \frac{M}{2}v_a^2\frac{2\pi}{\omega}(1+2) = \frac{\bar{p}_E}{2M}\frac{2\pi}{\omega}(1+2), \tag{4.513}$$

where the numbers 1 and 2 indicate the source of the contributions from the kinetic and potential parts of the action (4.502), respectively. Since the action is the difference between kinetic and potential energy, the average potential energy is minus twice as big as the kinetic energy, which is the single-particle version of the virial theorem observed in the Thomas-Fermi approximation (4.396).

For positive energies E, the eccentricity is $\epsilon > 1$ and the orbit is a hyperbola (see Fig. 4.3). Then equations (4.497)–(4.500) become

$$r = a(\epsilon \cosh \xi - 1), \qquad t = \frac{a}{v_a}(\xi - \epsilon \sin \xi), \tag{4.514}$$

$$x = -a\epsilon(\cosh \xi - \epsilon), \qquad y = a\sqrt{\epsilon^2 - 1}\sinh \xi. \tag{4.515}$$

The orbits take a simple form in momentum space. Using Eq. (4.474), we find from (4.492):

$$p_x = -\frac{l}{h}\left[\epsilon + \sin(\phi - \phi_0)\right], \qquad p_y = \frac{l}{h}\cos(\phi - \phi_0). \tag{4.516}$$

As a function of time, the momenta describe a circle of radius

$$p_0 = \frac{l}{h} = \frac{Me^2}{l} = \frac{\bar{p}_E}{\sqrt{1 - \epsilon^2}} \tag{4.517}$$

around a center on the p_y-axis with (see Fig. 4.4)

$$p_x^c = -\frac{l}{h}\epsilon = -\frac{\epsilon}{\sqrt{1 - \epsilon^2}}. \tag{4.518}$$

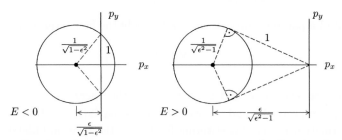

Figure 4.4 Circular orbits in momentum space in units $\bar{p}_E$ for $E < 0$ and p_E for $E > 0$.

For positive energies, the above solutions can be used to describe the scattering of electrons or ions on a central atom. For helium nuclei obtained by α-decay of radioactive atoms, this is the famous *Rutherford scattering* process. The potential is then repulsive, and e^2 in the potential (4.472) must be replaced by $-2Ze^2$, where $2e$ is the charge of the projectiles and Ze the charge of the central atom. As we can see on Fig. 4.5, the trajectories in an attractive potential are simply related to those in a repulsive potential. The momentum $\bar{p}_E$ is the asymptotic momentum of the projectile, and may be called p_∞. The impact parameter b of the projectile fixes the angular momentum via

$$l = bp_\infty = b\bar{p}_E. \tag{4.519}$$

Inserting l and $\bar{p}_E$ we see that b coincides with the previous parameter b.

The relation of the impact parameter b with the scattering angle θ may be taken from Fig. 4.5, which shows that (see also Fig. 4.4)

$$\tan\frac{\theta}{2} = \frac{p_0}{p_\infty} = \frac{1}{\epsilon^2 - 1}. \tag{4.520}$$

Thus we have

$$b = a\cot\frac{\theta}{2}. \tag{4.521}$$

The particles impinging into a circular annulus of radii b and $b + db$ come out between the angles θ and $\theta + d\theta$, with $db/d\theta = -a/2\sin^2(\theta/2)$. The area of the annulus $d\sigma = 2\pi b\,db$ is the differential cross section for this scattering process. The absolute ratio with respect to the associated solid angle $d\Omega \equiv 2\pi\sin\theta\,d\theta$ is then

$$\frac{d\sigma}{d\Omega} = \frac{b}{\sin\theta}\left|\frac{db}{d\theta}\right| = \frac{a^2}{4\sin^4(\theta/2)} = \frac{Z^2\alpha^2 M^2}{4p_\infty^4\sin^4(\theta/2)}. \tag{4.522}$$

The right-hand side is the famous *Rutherford formula*, which arises after express-ing a in terms of the incoming momentum p_∞ [recall (4.494)] as $a = Ze^2/2E = Z\alpha\hbar c M/p_\infty^2$.

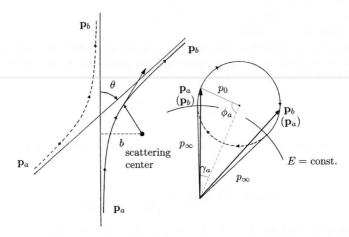

Figure 4.5 Geometry of scattering in momentum space. The solid curves are for at-tractive Coulomb potential, the dashed curves for repulsive (Rutherford scattering). The right-hand part of the figure shows the circle on which the momentum moves from $\mathbf{p}_a$ to $\mathbf{p}_b$ as the angle ϕ runs from ϕ_a to ϕ_b. The distance b is the impact parameter, and θ is the scattering angle.

Let us also calculate the classical eikonal in momentum space. We shall do this for the attractive interaction at positive energy. Inserting (4.492) and (4.516) we have

$$S(\mathbf{p}_b, \mathbf{p}_a; E) = -l \int_{\phi_a}^{\phi_b} \frac{d\phi}{1 + \epsilon(\phi - \phi_0)}. \tag{4.523}$$

Using the formula[16]

$$\int \frac{d\xi}{1 + \epsilon \cos \xi} = \frac{2}{\sqrt{\epsilon^2 - 1}} \log \frac{1 + \epsilon + \sqrt{\epsilon^2 - 1} \tan(\xi/2)}{1 - \epsilon - \sqrt{\epsilon^2 - 1} \tan(\xi/2)}. \tag{4.524}$$

Inserting $l/\sqrt{1 - \epsilon^2} = M v_a a = M e^2/p_\infty$, we find after some algebra

$$S(\mathbf{p}_b, \mathbf{p}_a; E) = -\frac{M e^2}{p_\infty} \log \frac{\zeta + 1}{\zeta - 1}, \qquad p_\infty = \sqrt{2ME}, \tag{4.525}$$

with

$$\zeta \equiv \sqrt{1 + \frac{(p_b^2 - p_\infty^2)(p_b^2 - p_\infty^2)}{p_\infty^2 |\mathbf{p}_b - \mathbf{p}_a|^2}}. \tag{4.526}$$

The expression (4.525) has no definite limit if the impinging particle comes in from spatial infinity where p_a becomes equal to p_∞. There is a logarithmic divergence which is due to the infinite range of the Coulomb potential. In nature, the charges are always screened at some finite radius R, after which the logarithmic divergence disappears. This was discussed before when deriving the eikonal approximation (1.499) to Coulomb scattering.

There is a simple geometric meaning to the quantity ζ. Since the force is central, the change in momentum along a classical orbit is always in the direction of the center, so that (4.523) can also be written as

$$S(\mathbf{p}_b, \mathbf{p}_a; E) = -\int_{\mathbf{P}_a}^{\mathbf{P}_b} r \, dp. \tag{4.527}$$

Expressing r in terms of momentum and total energy, this becomes for an attractive (repulsive) potential

$$S(\mathbf{p}_b, \mathbf{p}_a; E) = \pm 2M e^2 \int_{\mathbf{P}_a}^{\mathbf{P}_b} \frac{dp}{p^2 - 2ME}. \tag{4.528}$$

Now we observe, that for $E < 0$, the integrand is the arc length on a sphere of radius $1/\bar{p}_E$ in a four-dimensional momentum space. Indeed, the three-dimensional momentum space can be mapped onto the surface of a four-dimensional unit sphere by the following transformation

$$n_4 \equiv \frac{p^2 - \bar{p}_E^2}{p^2 + \bar{p}_E^2}, \qquad \mathbf{n} \equiv \frac{2\bar{p}_E \mathbf{p}}{p^2 + \bar{p}_E^2}. \tag{4.529}$$

---

[16]See the previous footnote.

Then we find that

$$\frac{dp}{p^2 + \bar{p}_E^2} = \frac{d\vartheta}{2\bar{p}_E}, \tag{4.530}$$

where $d\vartheta$ is the infinitesimal arc length on the unit sphere. But then the eikonal becomes simply

$$S(\mathbf{p}_b, \mathbf{p}_a; E) = \pm \frac{2Me^2}{\bar{p}_E} \vartheta_{ba}, \tag{4.531}$$

where ϑ is the angular difference between the images of the momenta $\mathbf{p}_b$ and $\mathbf{p}_a$. This is easily calculated. From (4.529) we find directly

$$\cos \vartheta_{ba} = \frac{4\bar{p}_E^2 \, \mathbf{p}_b \cdot \mathbf{p}_a + (p_b^2 - \bar{p}_E^2)(p_a^2 - \bar{p}_E^2)}{(p_b^2 + \bar{p}_E^2)(p_a^2 + \bar{p}_E^2)} = 1 - 2\frac{\bar{p}_E^2(\mathbf{p}_b - \mathbf{p}_a)^2}{(p_b^2 + \bar{p}_E^2)(p_a^2 + \bar{p}_E^2)}. \tag{4.532}$$

Continuing E analytically to positive energies, we may replace $\bar{p}_E$ by $ip_\infty = 2ME$, and obtain

$$\cos \vartheta_{ba} = 1 + 2\frac{p_\infty^2(\mathbf{p}_b - \mathbf{p}_a)^2}{(p_b^2 - p_\infty^2)(p_a^2 - p_\infty^2)}. \tag{4.533}$$

Hence ϑ becomes imaginary, $\vartheta = i\bar{\vartheta}$, with

$$\sin \frac{\bar{\vartheta}}{2} = \frac{p_\infty^2(\mathbf{p}_b - \mathbf{p}_a)^2}{(p_b^2 - p_\infty^2)(p_a^2 - p_\infty^2)}, \tag{4.534}$$

and the eikonal function (4.535) takes the form

$$S(\mathbf{p}_b, \mathbf{p}_a; E) = \pm \frac{2Me^2}{p_\infty} \bar{\vartheta}_{ba}. \tag{4.535}$$

This is precisely the expression (4.525) with $\zeta = 1/\tanh(\bar{\vartheta}_{ab}/2)$.

4.12   Semiclassical Scattering

Let us also derive the semiclassical limit for the scattering amplitude.

4.12.1   General Formulation

Consider a particle impinging with a momentum $\mathbf{p}_a$ and energy $E = E_a = \mathbf{p}_a^2/2M$ upon a nonzero potential concentrated around the origin. After a long time, it will be found far from the potential with some momentum $\mathbf{p}_b$ and the same energy $E = E_b = \mathbf{p}_b^2/2M$. Let us derive the scattering amplitude for such a process from the heuristic formula (1.507):

$$f_{\mathbf{p}_b \mathbf{p}_a} = \frac{p_b}{M} \frac{\sqrt{2\pi\hbar M/i}^3}{(2\pi\hbar)^3} \lim_{t_b \to \infty} \frac{1}{t_b^{1/2}} e^{iE_b(t_b - t_a)/\hbar} \left[(\mathbf{p}_b t_b | \mathbf{p}_a t_a) - \langle \mathbf{p}_b | \mathbf{p}_a \rangle \right]. \tag{4.536}$$

In the semiclassical approximation we replace the exact propagator in the momentum representation by a sum over all classical trajectories and associated phases, connecting $\mathbf{p}_a$ to $\mathbf{p}_b$ in the time $t_b - t_a$. According to formula (4.172) we have in three dimensions

$$(\mathbf{p}_b t_b | \mathbf{p}_a t_a) - \langle \mathbf{p}_b | \mathbf{p}_a \rangle = \frac{(2\pi\hbar)^3}{(2\pi\hbar/i)^{3/2}} \sum_{\text{class. traj.}}' \left| \det\left(-\frac{\partial \mathbf{x}_a}{\partial \mathbf{p}_b} \right) \right|^{1/2} e^{iA(\mathbf{p}_b,\mathbf{p}_a;t_b-t_a)/\hbar - i\nu\pi\hbar/2}.$$

$$(4.537)$$

The sum carries a prime to indicate that unscattered trajectories are omitted. The classical action in momentum space is

$$A(\mathbf{p}_b, \mathbf{p}_a; t_b - t_a) = \int_{\mathbf{p}_b}^{\mathbf{p}_a} \mathbf{x} \cdot \dot{\mathbf{p}} - \int_{t_a}^{t_b} H \, dt = S(\mathbf{p}_b, \mathbf{p}_a; E) - E(t_b - t_a), \quad (4.538)$$

where $S(\mathbf{p}_b, \mathbf{p}_a; E)$ is the eikonal function introduced in Eqs. (4.240) and (4.68).

Inserting (4.537) into (1.507) we obtain the semiclassical scattering transition amplitude

$$f_{\mathbf{p}_b\mathbf{p}_a} = \lim_{t_b \to \infty} \sum_{\text{class. traj.}}' \frac{p_b}{\sqrt{t_b M}} \left| \det \frac{\partial \mathbf{x}_a}{\partial \mathbf{p}_b} \right|^{1/2} e^{iS(\mathbf{p}_b,\mathbf{p}_a;E)/\hbar - i\nu\pi/2}. \quad (4.539)$$

The determinant has a simple physical meaning. To see this we rewrite

$$\left. \frac{\partial \mathbf{x}_a}{\partial \mathbf{p}_b} \right|_{\mathbf{p}_a} = \left. \frac{\partial \mathbf{p}_b}{\partial \mathbf{x}_a} \right|_{\mathbf{p}_a}^{-1}, \quad (4.540)$$

so that (4.541) becomes

$$f_{\mathbf{p}_b\mathbf{p}_a} = \lim_{t_b \to \infty} \sum_{\text{class. traj.}}' \frac{p}{\sqrt{t_b M}} \left| \det \frac{\partial \mathbf{p}_b}{\partial \mathbf{x}_a} \right|_{\mathbf{p}_a}^{-1/2} e^{iS(\mathbf{p}_b,\mathbf{p}_a;E)/\hbar - i\nu\pi/2}. \quad (4.541)$$

We now note that for large t_b

$$\mathbf{p}_b = \mathbf{p}(t_b) = M\mathbf{x}_b(t_b)/t_b \quad (4.542)$$

along any trajectory. Thus we find

$$f_{\mathbf{p}_b\mathbf{p}_a} = \lim_{t_b \to \infty} \sum_{\text{class. traj.}}' r_b \left| \det \frac{\partial \mathbf{x}_b}{\partial \mathbf{x}_a} \right|_{\mathbf{p}_b}^{-1/2} e^{iS(\mathbf{p}_b,\mathbf{p}_a;E)/\hbar - i\nu\pi/2}, \quad (4.543)$$

where $r_b = |\mathbf{x}_b|$.

From the definition of the scattering amplitude (1.494) we expect the prefactor of the exponential to be equal to the square root of the classical differential cross section $d\sigma_{\text{cl}}/d\Omega$. Let us choose convenient coordinates in which the particle trajectories start out at a point with cartesian coordinates $\mathbf{x}_a = (x_a, y_a, z_a)$ with a large negative z_a and a momentum $\mathbf{p}_a \approx p_a\hat{\mathbf{z}}$, where $\hat{\mathbf{z}}$ is the direction of the z-axis. The final points $\mathbf{x}_b$ of the trajectories will be described in spherical coordinates. If $\hat{\mathbf{p}}_b = (\sin\theta_b \cos\phi_b, \sin\theta \sin\phi_b, \cos\theta_b)$ denotes the direction of the final momentum $\mathbf{p}_b =$

$p_b \hat{\mathbf{p}}_b$, then $\mathbf{x}_b = r_b \hat{\mathbf{p}}$. Let us introduce an auxiliary triplet of spherical coordinates $\mathbf{s}_b \equiv (r_b, \theta_b, \phi_b)$. Then we factorize the determinant in (4.543) as

$$\det \frac{\partial \mathbf{x}_b}{\partial \mathbf{x}_a} = \det \frac{\partial \mathbf{x}_b}{\partial \mathbf{s}_a} \times \det \frac{\partial \mathbf{s}_b}{\partial \mathbf{x}_a} = r_b^2 \det \frac{\partial \mathbf{s}_b}{\partial \mathbf{x}_a}.$$

We further calculate

$$\det \frac{\partial \mathbf{s}_b}{\partial \mathbf{x}_a} = \begin{pmatrix} \dfrac{\partial r_b}{\partial x_a} & \dfrac{\partial r_b}{\partial y_a} & \dfrac{\partial r_b}{\partial z_a} \\ \dfrac{\partial \theta_b}{\partial x_a} & \dfrac{\partial \theta_b}{\partial y_a} & \dfrac{\partial \theta_b}{\partial z_a} \\ \dfrac{\partial \phi_b}{\partial x_a} & \dfrac{\partial \phi_b}{\partial y_a} & \dfrac{\partial \phi_b}{\partial z_a} \end{pmatrix}. \tag{4.544}$$

Long after the collision, for $t_b \to \infty$, a small change of the starting point *along* the trajectory dz_a will not affect the scattering angle. Thus we may approximate the matrix elements in the third column by $\partial z_a \approx \partial \phi / \partial z_a \approx 0$. After the same amount of time the particle will only wind up at a slightly more distant r_b, where $dr_b \approx dz_b$. Thus we may replace the matrix element in the right upper corner by 1, so that the determinant (4.544) becomes in the limit

$$\lim_{t_b \to \infty} \det \frac{\partial \mathbf{s}_b}{\partial \mathbf{x}_a} \approx \det \begin{pmatrix} \dfrac{\partial \theta_b}{\partial x_a} & \dfrac{\partial \theta_b}{\partial y_a} \\ \dfrac{\partial \phi_b}{\partial x_a} & \dfrac{\partial \phi_b}{\partial y_a} \end{pmatrix} = \frac{d\theta_b d\phi_b}{dx_b dy_b} = \frac{d\Omega}{d\sigma}, \tag{4.545}$$

where $d\Omega = \sin \theta_b d\theta d\phi_b$ is the element of the solid angle of the emerging trajectories, and $d\sigma$ the area element in the $x - y$ -plane, for which the trajectories arrive in an element of the final solid angle $d\Omega$. Thus we obtain

$$\left[\det \frac{\partial \mathbf{x}_b}{\partial \mathbf{x}_a} \right]^{-1} \approx \frac{1}{r_b^2} \frac{d\sigma}{d\Omega}. \tag{4.546}$$

The ratio $d\sigma/d\Omega$ is precisely the classical differential cross section of the scattering process.

Combining (4.543) and (4.545), we see that the contribution of an individual trajectory to the semiclassical amplitude is of the expected form [5]

$$f_{\mathbf{p}_b \mathbf{p}_a} = \sqrt{\frac{d\sigma_{\text{cl}}}{d\Omega}} \times \sum_{\text{class. traj.}}' e^{iS(\mathbf{p}_b, \mathbf{p}_a; E)/\hbar - i\nu\pi/2}. \tag{4.547}$$

Note that this equation is also valid for some potentials which are not restricted to a finite regime around the origin, such as the Coulomb potentials. In the operator theory of quantum-mechanical scattering processes, such potentials always cause considerable problems since the outgoing wave functions remain distorted even at large distances from the scattering center.

Usually, there are only a few trajectories contributing to a process with a given scattering angle. If the actions of these trajectories differ by less than $\hbar$, the semiclassical approximation fails since the fluctuation integrals overlap. Examples are the light scattering causing the ordinary rainbow in nature, and glory effects seen at night around the moonlight.

We now turn to a derivation of the amplitude (4.547) from the more reliable formula (1.528) for the interacting wave function

$$\langle \mathbf{x}_b | \hat{U}_I(0, t_a) | \mathbf{p}_a \rangle = \lim_{t_a \to -\infty} \left(\frac{-2\pi i \hbar t_a}{M} \right)^{3/2} (\mathbf{x}_b t_b | \mathbf{x}_a t_a) e^{i(\mathbf{p}_a \mathbf{x}_a - p_a^2 t_a / 2M)/\hbar} \bigg|_{\mathbf{x}_a = \mathbf{p}_a t_a / M},$$

by isolating the factor of $e^{i p_a r_b}/r_b$ for large r_b, as discussed at the end of Section 1.16. On the right-hand side we now insert the x-space form (4.127) of the semiclassical amplitude, and use (4.87) to write

$$\langle \mathbf{x}_b | \hat{U}_I(0, t_a) | \mathbf{p}_a \rangle = \lim_{t_a \to -\infty} \left(\frac{-t_a}{M} \right)^{3/2} \left[\det{}_3 \left(-\frac{\partial \mathbf{p}_a}{\partial \mathbf{x}_b} \right) \right]^{1/2}$$
$$\times \; e^{i[S(\mathbf{x}_b, \mathbf{x}_a; E_a) + i\mathbf{p}_a \mathbf{x}_a - i\nu\pi/2]/\hbar} \bigg|_{\mathbf{x}_a = \mathbf{p}_a t_a / M}. \tag{4.548}$$

Now we observe that

$$\left(\frac{-t_a}{M} \right)^{3/2} \left[\det{}_3 \left(-\frac{\partial \mathbf{p}_a}{\partial \mathbf{x}_b} \right) \right]^{1/2} = \left[\det{}_3 \left(\frac{\partial \mathbf{x}_a}{\partial \mathbf{x}_b} \right) \right]^{1/2}. \tag{4.549}$$

In Eq. (4.546) we have found that this determinant is equal to $\sqrt{d\sigma/d\Omega}/r_b$, bringing Eq. (4.548) to the form

$$\langle \mathbf{x}_b | \hat{U}_I(0, t_a) | \mathbf{p}_a \rangle = \lim_{t_a \to -\infty} \frac{1}{r_b} \sqrt{\frac{d\sigma_{\rm cl}}{d\Omega}} e^{i[S(\mathbf{x}_b, \mathbf{x}_a; E_a) + i\mathbf{p}_a \mathbf{x}_a - i\nu\pi/2]/\hbar} \bigg|_{\mathbf{x}_a = \mathbf{p}_a t_a / M}. \tag{4.550}$$

For large $\mathbf{x}_b$ in the direction of the final momentum $\mathbf{p}_b$, we can rewrite the exponent as [recalling (4.240)]

$$S(\mathbf{x}_b, \mathbf{x}_a; E_a) + i\mathbf{p}_a \mathbf{x}_a = p_b r_b + S(\mathbf{p}_b, \mathbf{p}_a; E_a) \tag{4.551}$$

so that (4.550) consists of an outgoing spherical wave function $e^{i p_b r_b/\hbar}/r_b$ multiplied by the scattering amplitude

$$f_{\mathbf{p}_b \mathbf{p}_a} = \sqrt{\frac{d\sigma_{\rm cl}}{d\Omega}} \times \sum_{\rm class.\ traj.}{}' e^{iS(\mathbf{p}_b, \mathbf{p}_a; E)/\hbar - i\nu\pi/2}, \tag{4.552}$$

the same as in (4.547).

4.12.2   Semiclassical Cross Section of Mott Scattering

If the scattering particle is distinguishable from the target particles, the extra phase in the semiclassical formula (4.552) does not change the classical result (4.522). A quantum-mechanical effect becomes visible only if we consider electron-electron scattering, also referred to as *Mott scattering*. The potential is repulsive, and the above Coulomb potential holds for the relative motion of the two identical particles in their center-of-mass frame. Moreover, the identity of particles requires us to add the amplitudes for the trajectories going to $\mathbf{p}_b$ and to $-\mathbf{p}_b$ [see Fig. 4.6], so that the differential cross section is

$$\frac{d\sigma_{\mathrm{sc}}}{d\Omega} = |f_{\mathbf{p}_b\mathbf{p}_a} - f_{\mathbf{p}_b,-\mathbf{p}_a}|^2. \tag{4.553}$$

The minus sign accounts for the Fermi statistics of the two electrons. For two identical bosons, we have to use a plus sign instead. Now the eikonal function $S(\mathbf{p}, \mathbf{p}, E)$ enters into the result. According to Eq. (4.525), this is given by

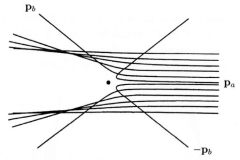

Figure 4.6   Classical trajectories in Coulomb potential plotted in the center-of-mass frame. For identical particles, trajectories which merge with a scattering angle θ and $\pi - \theta$ are indistinguishable. Their amplitudes must be subtracted from each other, yielding the differential cross section (4.553).

$$S(\mathbf{p}_b, \mathbf{p}_a; E) = -\frac{M\hbar c\alpha}{p_\infty} \log \frac{\sqrt{1+\Delta}+1}{\sqrt{1+\Delta}-1}, \tag{4.554}$$

where $p_\infty = \sqrt{2ME}$ is the impinging momentum at infinite distance, and

$$\Delta \equiv \frac{(p_b^2 - p_\infty^2)(p_a{}^2 - p_\infty^2)}{p_\infty^2 |\mathbf{p}_b - \mathbf{p}_a|^2}. \tag{4.555}$$

The eikonal function is needed only for momenta $\mathbf{p}_b$, $\mathbf{p}_a$ in the asymptotic regime where $p_b, p_a \approx p_\infty$, so that Δ is small and

$$S(\mathbf{p}_b, \mathbf{p}_a; E) \approx \frac{M\hbar c\alpha}{p_\infty} \log \Delta, \tag{4.556}$$

which may be rewritten as

$$S(\mathbf{p}, \mathbf{p}_a E) \approx 2\sigma_0 - \frac{M\hbar c\alpha}{p_\infty} \log(\sin^2 \theta/2), \tag{4.557}$$

with

$$\sigma_0 = \frac{M\hbar c\alpha}{2p_\infty} \log\left[\frac{(p_b^2 - p_\infty^2)(p_a{}^2 - p_\infty^2)}{p_\infty^4}\right], \tag{4.558}$$

and the scattering angle determined by $\cos\theta = [\mathbf{p}_b \cdot \mathbf{p}_a/p_b p_a]$. The logarithmically diverging constant σ_0 for $p_a = p_b p_\infty$ does, fortunately, not depend on the scattering angle, and is therefore the semiclassical approximation for the phase shift at angular momentum $l = 0$. It therefore drops, fortunately, out of the difference of the amplitudes in Eq. (4.553). Inserting (4.557) with (4.558) into (4.552) and (4.553), we obtain the differential cross section for Mott scattering (see Fig. 4.7 for a plot)

$$\frac{d\sigma}{d\Omega} = \left(\frac{\hbar c\alpha}{4E}\right)^2 \left\{\frac{1}{\sin^4\theta/2} + \frac{1}{\cos^4\theta/2} \pm \frac{1}{\sin^2\theta/2\cos^2\theta/2} 2\cos\left[\frac{2\alpha Mc}{p_\infty}\log(\cot\theta/2)\right]\right\}. \tag{4.559}$$

This semiclassical result happens to be identical to the exact result. The exactness is caused by two properties of the Coulomb motion: First there is only one trajectory for each scattering angle, second the motion can be mapped onto that of a harmonic oscillator in four dimensions, as we shall see in Chapter 13.

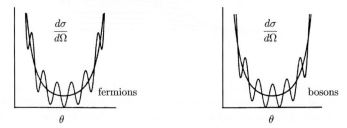

Figure 4.7 Oscillations in differential Mott scattering cross section caused by statistics. For scattering angle $\theta = 90^0$, the cross section vanishes due to the Pauli exclusion principle. The right-hand plot shows the situation for identical bosons.

Appendix 4A Semiclassical Quantization for Pure Power Potentials

Let us calculate the local density of states (4.262) for the general pure power potential $V(\mathbf{x}) = gx^p/p$ in D dimensions. For $D = 1$ and $p = 2$, we shall recover the exact spectrum of the harmonic oscillator. For $p = 4$, we shall find the energies of the purely quartic potential which can be compared with the strong-coupling limit of the anharmonic oscillator with $V(x) = \omega^2 x^2/2 + gx^4/4$ to be calculated in Section 5.16. The integrals on the right-hand side of Eq. (4.262) can be calculated using the formula

$$\int d^D x \, r^\mu [E - V(\mathbf{x})]^\nu = S_D \int_0^{r_E} dr \, r^{D-1+\mu} \left(E - \frac{g}{p} r^p\right)^\nu$$

$$= S_D \frac{\Gamma(1+\nu)\Gamma((D+\mu)/p)}{p\,\Gamma(1+\nu+(D+\mu)/p)}\left(\frac{g}{p}\right)^{-(D+\mu)/p} E^{\nu+(D+\mu)/p}, \quad p>0, \quad (4A.1)$$

where $r_E = (pE/g)^{1/p}$. For $p<0$, the result is

$$\int d^D x\, r^\mu [E-V(\mathbf{x})]^\nu = S_D \frac{\Gamma(1+\nu)\Gamma(-\nu-(D+\mu)/p)}{p\,\Gamma(1-(D+\mu)/p)}\left(-\frac{g}{p}\right)^{-(D+\mu)/p} (-E)^{\nu+(D+\mu)/p}, \quad p<0. \quad (4A.2)$$

Recalling (4.214), we find the total density of states for $p>0$:

$$\rho(E) = \frac{2}{\Gamma\left(\frac{D}{2}\right)p}\left(\frac{M}{2\hbar^2}\right)^{D/2}\left(\frac{g}{p}\right)^{-D/p}\frac{\Gamma\left(\frac{D}{p}\right)}{\Gamma\left(\frac{D}{2}+\frac{D}{p}\right)}\left\{1-\frac{\hbar^2}{24M}\left(\frac{g}{p}\right)^{2/p}\right.$$

$$\times \left.\frac{p^2\Gamma\left(\frac{D-2}{2}+2\right)\Gamma\left(\frac{D}{2}+\frac{D}{p}\right)}{\Gamma\left(\frac{D-2}{2}(1+\frac{2}{p})\right)\Gamma\left(\frac{D}{p}\right)}E^{-1-2/p}+\dots\right\}E^{(D/2)(1+2/p)-1}, \quad p>0, \quad (4A.3)$$

and for $p<0$:

$$\rho(E) = -\frac{2}{\Gamma(\frac{D}{2})p}\left(\frac{M}{2\hbar^2}\right)^{D/2}\left(-\frac{g}{p}\right)^{-D/p}\frac{\Gamma\left(1-\frac{D}{2}-\frac{D}{p}\right)}{\Gamma\left(1-\frac{D}{p}\right)}\left\{1+\frac{\hbar^2}{24M}\left(-\frac{g}{p}\right)^{2/p}\right.$$

$$\times \left.\frac{p^2\Gamma\left(1-\frac{D-2}{2}(1+\frac{2}{p})\right)\Gamma\left(1-\frac{D}{p}\right)}{\Gamma\left(-1-\frac{D-2}{p}\right)\Gamma\left(1-\frac{D}{2}-\frac{D}{p}\right)}(-E)^{-1-2/p}+\dots\right\}(-E)^{(D/2)(1+2/p)-1}, \quad p<0. \quad (4A.4)$$

For a harmonic oscillator with $p=2$ and $g=M\omega^2$, we obtain

$$\rho(E)=\frac{1}{(\hbar\omega)^D}\left\{\frac{1}{\Gamma(D)}E^{D-1}-\frac{\hbar^2\omega^2}{24}\frac{D}{\Gamma(D-2)}E^{D-3}+\dots\right\}. \quad (4A.5)$$

In one and two dimension, only the first term survives and $\rho(E) = 1/\hbar\omega$ or $\rho(E) = E/(\hbar\omega)^2$. Inserting this into Eq. (1.583), we find the number of states $N(E) = \int_0^E dE'\,\rho(E') = E/\hbar\omega$ or $E^2/2(\hbar\omega)^2$. According to the exact quantization condition (1.585), we set $N(E) = n+1/2$ the exact energies $E_n = (n+1/2)\hbar\omega$. Since the semiclassical expansion (4A.3) contains only the first term, the exact quantization condition (1.585) agrees with the Bohr-Sommerfeld quantization condition (4.190).

In two dimensions we obtain $\rho(E) = E/(\hbar\omega)^2$ and $N(E) = E^2/2(\hbar\omega)^2$. Here the exact quantization condition $N(E) = n+1/2$ cannot be used to find the energies E_n, due to the degeneracies of the energy eigenvalues E_n. In order to see that it is nevertheless a true equation, let us expand the known partition function (2.405) of the two-dimensional oscillator as

$$Z = \frac{1}{[2\sinh(\hbar\beta\omega/2)]^2} = e^{-\hbar\beta\omega}\frac{1}{(1-e^{-\hbar\beta\omega})^2} = \sum_{n=0}^{\infty}(n+1)e^{-(n+1)\hbar\beta\omega}. \quad (4A.6)$$

This shows that the energies are $E_n = (n+1)\hbar\omega$ withy $n+1$ -fold degeneracy. The density of states is found from this by the Fourier transform (1.582):

$$\rho(E) = \sum_{n'=0}^{\infty}(n'+1)\delta(E-\hbar\omega(n'+1)). \quad (4A.7)$$

Integrating this over E yields the number of states

$$N(E) = \int_0^E dE'\,\rho(E') = \sum_{n'=0}^{\infty}(n'+1)\Theta(E-E_n). \quad (4A.8)$$

For $E = E_n$ this becomes [recall (1.309)]

$$N(E_n) = \sum_{n'=0}^{n-1} (n'+1) + (n+1)\frac{1}{2} = \frac{1}{2}(n+1)^2. \tag{4A.9}$$

This shows that the exact energies $E_n = (n+1)\hbar\omega$ of the two-dimensional oscillator satisfy the quantization condition $N(E_n) = (n+1)^2/2$ rather than (1.585).

For a quartic potential $gx^4/4$, Eq. (4A.3) becomes

$$\rho(E) = \frac{1}{2\Gamma(\frac{D}{2})} \frac{\Gamma(\frac{D}{4})}{\Gamma(\frac{3D}{4})} \left(\frac{g\hbar^4}{M^2}\right)^{-\frac{D}{4}} \left\{ 1 - \left(\frac{g\hbar^4}{M^2}\right)^{\frac{1}{2}} \frac{\Gamma(\frac{3}{2}+\frac{D}{4})\Gamma(\frac{3D}{4})E^{-\frac{3}{2}}}{3\Gamma(\frac{D}{4})\Gamma\left(\frac{3}{4}(D-2)\right)} + \dots \right\} E^{\frac{3D}{4}-1}. \tag{4A.10}$$

Integrating this over E yields $N(E)$. Setting $N(E) = n + 1/2$ in one dimension, we obtain the Bohr-Sommerfeld energies (4.35) plus a first quantum correction. Since we have studied these corrections to high order in Subsection 4.9.6 (see Fig. 4.1), we do not write the result down here.

A physically important case is $p = -1$, $g = \hbar c\alpha$, with α of Eq. (1.502), where $V(x) = -\hbar c\alpha/r$ becomes the Coulomb potential. Here we obtain from (4A.10):

$$\rho(E) = \frac{2}{\Gamma(\frac{D}{2})} \left(\frac{Mg^2}{2\pi\hbar^2}\right)^{D/2} \frac{\Gamma\left(1+\frac{D}{2}\right)}{\Gamma(1+D)}$$
$$\times \left\{ 1 - \frac{\hbar^2}{24Mg^{-2}} \frac{p^2\,\Gamma\left(\frac{D-2}{2}\right)\Gamma(1+D)}{\Gamma(D-3)\,\Gamma\left(1+\frac{D}{2}\right)} E + \dots \right\} (-E)^{(D/2)(1+2/p)-1}. \tag{4A.11}$$

For $D = 1$, only the leading term survives and

$$\rho(E) = \sqrt{\frac{Mg^2}{2\hbar^2}} (-E)^{-3/2}, \tag{4A.12}$$

implying that

$$N(E) = 2\sqrt{\frac{Mg^2}{2\hbar^2}} (-E)^{-1/2}. \tag{4A.13}$$

In order to find the bound-state energies, we must watch out for a subtlety in one dimension: only the positive half-space is accessible to the particle in a Coulomb potential, due to the strong singularity at the origin. For this reason, the "surface of a sphere" S_D for $D = 1$, which is equal to 2, must be replaced by 1, so that we must equate $N(E)/2$ to $n + 1/2$. This yields the spectrum $E_n = -\alpha^2 Mc^2/2(n+1/2)^2$. The exact energies follow the same formula with n^2 instead of $(n+1/2)^2$ [7]. In contrast to the harmonic oscillator, the Bohr-Sommerfeld approximation yields here the correct energies only for large n. We have seen at the end of Section 4.1 how to correct this defect: we must go to a new variable ξ by the coordinate transformation $r = e^\xi$. This moves the singularity at $r = 0$ to $\xi = -\infty$ and the semiclassical result becomes *exact*.

Appendix 4B Derivation of Semiclassical Time Evolution Amplitude

Here we derive the semiclassical approximation to the time evolution amplitude (4.265). We shall do this for imaginary times $\tau = it$. Decomposing the path $\mathbf{x}(\tau)$ into path average of the ends points $\bar{\mathbf{x}} = (\mathbf{x}_b + \mathbf{x}_a)/2$ and fluctuations $\boldsymbol{\eta}(\tau)$, we calculate the imaginary-time amplitude

$$(\mathbf{x}_b\tau_b|\mathbf{x}_a\tau_a) = \int_{\boldsymbol{\eta}(\tau_a)=-\Delta\mathbf{x}/2}^{\boldsymbol{\eta}(\tau_b)=\Delta\mathbf{x}/2} \mathcal{D}\boldsymbol{\eta} \exp\left\{ -\frac{1}{\hbar} \int_{\tau_a}^{\tau_b} d\tau \left[\frac{M}{2}\dot{\boldsymbol{\eta}}^2(\tau) + V\left(\bar{\mathbf{x}} + \boldsymbol{\eta}(\tau)\right) \right] \right\}, \tag{4B.1}$$

where $\Delta \mathbf{x} \equiv \mathbf{x}_b - \mathbf{x}_a$. For smooth potentials we expand

$$V\left(\overline{\mathbf{x}} + \boldsymbol{\eta}(\tau)\right) = V(\overline{\mathbf{x}}) + \partial_i V(\overline{\mathbf{x}})\, \eta_i(\tau) + \frac{1}{2}\, \partial_i \partial_j V(\overline{\mathbf{x}})\, \eta_i(\tau)\, \eta_j(\tau) + \dots, \tag{4B.2}$$

where $V_{ij\dots}(\overline{\mathbf{x}}) \equiv \partial_i \partial_j \cdots V(\overline{\mathbf{x}})$, and rewrite the path integral (4B.1) as

$$
\begin{aligned}
(\mathbf{x}_b \tau_b | \mathbf{x}_a \tau_a) =\ & e^{-\beta V(\overline{\mathbf{x}})} \int_{\boldsymbol{\eta}(\tau_a)=-\Delta\mathbf{x}/2}^{\boldsymbol{\eta}(\tau_b)=\Delta\mathbf{x}/2} \mathcal{D}\boldsymbol{\eta}\ \exp\left\{ -\frac{1}{\hbar} \int_{\tau_a}^{\tau_b} d\tau\, \frac{M}{2}\, \dot{\boldsymbol{\eta}}^2(\tau) \right\} \\
& \times \left\{ 1 - \frac{1}{\hbar} \int_{\tau_a}^{\tau_b} d\tau \left[V_i(\overline{\mathbf{x}})\, \eta_i(\tau) + \frac{1}{2}\, V_{ij}(\overline{\mathbf{x}})\, \eta_i(\tau)\, \eta_j(\tau) + \dots \right] \right. \\
& \left. + \frac{1}{2\hbar^2} \int_{\tau_a}^{\tau_b} d\tau \int_{\tau_a}^{\tau_b} d\tau' \left[V_i(\overline{\mathbf{x}}) V_j(\overline{\mathbf{x}})\, \eta_i(\tau) \eta_j(\tau') + \dots \right] + \dots \right\}.
\end{aligned}
\tag{4B.3}
$$

At this point it is useful to introduce an auxiliary harmonic imaginary-time amplitude

$$(\Delta\mathbf{x}/2\,\tau_b | -\Delta\mathbf{x}/2\,\tau_a) = \int_{\boldsymbol{\eta}(\tau_a)=-\Delta\mathbf{x}/2}^{\boldsymbol{\eta}(\tau_b)=\Delta\mathbf{x}/2} \mathcal{D}\boldsymbol{\eta}\ \exp\left\{ -\frac{1}{\hbar} \int_{\tau_a}^{\tau_b} d\tau\, \frac{M}{2}\, \dot{\boldsymbol{\eta}}^2(\tau) \right\} \tag{4B.4}$$

and the harmonic expectation values

$$\langle F[\boldsymbol{\eta}] \rangle \equiv \frac{1}{(\Delta\mathbf{x}/2\,\tau_b | -\Delta\mathbf{x}/2\,\tau_a)} \int_{\boldsymbol{\eta}(\tau_a)=-\Delta\mathbf{x}/2}^{\boldsymbol{\eta}(\tau_b)=\Delta\mathbf{x}/2} \mathcal{D}\boldsymbol{\eta}\, F[\boldsymbol{\eta}] \exp\left\{ -\frac{1}{\hbar} \int_{\tau_a}^{\tau_b} d\tau\, \frac{M}{2}\, \dot{\boldsymbol{\eta}}^2(\tau) \right\}, \tag{4B.5}$$

which allows us to rewrite (4B.3) more concisely as

$$
\begin{aligned}
(\mathbf{x}_b \tau_b | \mathbf{x}_a \tau_a) =\ & e^{-\beta V(\overline{\mathbf{x}})} (\Delta\mathbf{x}/2\,\tau_b | -\Delta\mathbf{x}/2\,\tau_a) \left\{ 1 - \frac{1}{\hbar} \int_{\tau_a}^{\tau_b} d\tau\, V_i(\overline{\mathbf{x}})\, \langle \eta_i(\tau) \rangle \right. \\
& \left. - \frac{1}{2\hbar}\, V_{ij}(\overline{\mathbf{x}}) \int_{\tau_a}^{\tau_b} d\tau\, \langle \eta_i(\tau)\eta_j(\tau) \rangle + \frac{1}{2\hbar^2}\, V_i(\overline{\mathbf{x}}) V_j(\overline{\mathbf{x}}) \int_{\tau_a}^{\tau_b} d\tau \int_{\tau_a}^{\tau_b} d\tau'\, \langle \eta_i(\tau)\eta_j(\tau') \rangle + \dots \right\}.
\end{aligned}
\tag{4B.6}
$$

The amplitudes (4B.4) reads explicitly, with $\Delta\tau \equiv \tau_b - \tau_a$:

$$(\Delta\mathbf{x}/2\,\tau_b | -\Delta\mathbf{x}/2\,\tau_a) = \left(\frac{M}{2\pi\hbar\Delta\tau} \right)^{D/2} \exp\left\{ -\frac{M}{2\hbar\Delta\tau}\, (\Delta\mathbf{x})^2 \right\}, \tag{4B.7}$$

and (4B.5) can be calculated from the generating functional

$$(\Delta\mathbf{x}/2\,\tau_b | -\Delta\mathbf{x}/2\,\tau_a)[\mathbf{j}] = \int_{\boldsymbol{\eta}(\tau_a)=-\Delta\mathbf{x}/2}^{\boldsymbol{\eta}(\tau_b)=\Delta\mathbf{x}/2} \mathcal{D}\boldsymbol{\eta}\ \left\{ -\frac{1}{\hbar} \int_{\tau_a}^{\tau_b} d\tau \left[\frac{M}{2}\, \dot{\boldsymbol{\eta}}^2(\tau) - \mathbf{j}(\tau)\, \boldsymbol{\eta}(\tau) \right] \right\}, \tag{4B.8}$$

whose explicit solution is

$$
\begin{aligned}
(\Delta\mathbf{x}/2\,\tau_b | -\Delta\mathbf{x}/2\,\tau_a)[\mathbf{j}] =\ & \left(\frac{M}{2\pi\hbar\Delta\tau} \right)^{D/2} \exp\left\{ -\frac{M}{2\hbar\Delta\tau}\, (\Delta\mathbf{x})^2 + \frac{1}{\hbar\Delta\tau} \int_{\tau_a}^{\tau_b} d\tau (\tau - \overline{\tau})\, \Delta\mathbf{x}\, \mathbf{j}(\tau) \right. \\
& \left. + \frac{1}{2\hbar} \int_{\tau_a}^{\tau_b} d\tau \int_{\tau_a}^{\tau_b} d\tau'\, \frac{\Theta(\tau-\tau')(\Delta\tau - \tau)\tau' + \Theta(\tau'-\tau)(\Delta\tau - \tau')\tau}{M\Delta\tau}\, \mathbf{j}(\tau)\, \mathbf{j}(\tau') \right\}.
\end{aligned}
$$

The expectation values in (4B.6) and (4B.5) are obtained from the functional derivatives

$$\langle \eta_i(\tau) \rangle = \frac{1}{(\Delta\mathbf{x}/2\,\tau_b | -\Delta\mathbf{x}/2\,\tau_a)} \frac{\hbar\delta}{\delta j_i(\tau)} (\Delta\mathbf{x}/2\,\tau_b | -\Delta\mathbf{x}/2\,\tau_a)[\mathbf{j}] \Big|_{\mathbf{j}=0}, \tag{4B.9}$$

$$\langle \eta_i(\tau)\eta_j(\tau') \rangle = \frac{1}{(\Delta\mathbf{x}/2\,\tau_b | -\Delta\mathbf{x}/2\,\tau_a)} \frac{\hbar\delta}{\delta j_i(\tau)} \frac{\hbar\delta}{\delta j_j(\tau')} (\Delta\mathbf{x}/2\,\tau_b | -\Delta\mathbf{x}/2\,\tau_a)[\mathbf{j}] \Big|_{\mathbf{j}=0}. \tag{4B.10}$$

This yields the $\Delta \mathbf{x}$-dependent expectation value

$$\langle \eta_i(\tau) \rangle = (\tau - \bar{\tau}) \frac{\Delta x_i}{\Delta \tau}, \tag{4B.11}$$

and the $\Delta \mathbf{x}$-dependent correlation function

$$
\begin{aligned}
\langle \eta_i(\tau) \eta_j(\tau') \rangle &= \frac{\hbar}{M \Delta \tau} \left[\Theta(\tau - \tau')(\Delta \tau - \tau)\tau' + \Theta(\tau' - \tau)(\Delta \tau - \tau')\tau \right] \delta_{ij} \\
&+ (\tau - \bar{\tau})(\tau' - \bar{\tau}) \frac{\Delta x_i}{\Delta \tau} \frac{\Delta x_j}{\Delta \tau} \\
&\equiv A^2(\tau, \tau') \delta_{ij} + B^2(\tau, \tau') \Delta x_i \Delta x_j \equiv G_{ij}(\tau, \tau'),
\end{aligned}
\tag{4B.12}
$$

suppressing the argument $\Delta \mathbf{x}$ in $G_{ij}(\tau, \tau')$, for brevity. Note that

$$\int_{\tau_a}^{\tau_b} d\tau \, \langle \eta_i(\tau) \rangle = 0, \tag{4B.13}$$

and

$$\int_{\tau_a}^{\tau_b} d\tau \int_{\tau_a}^{\tau_b} d\tau' \, \langle \eta_i(\tau) \eta_j(\tau') \rangle = \int_{\tau_a}^{\tau_b} d\tau \int_{\tau_a}^{\tau_b} d\tau' G_{ij}(\tau, \tau') = \frac{\hbar}{12M} \Delta \tau (\Delta \tau^2 - \tau_a^2) \delta_{ij}, \tag{4B.14}$$

$$\int_{\tau_a}^{\tau_b} d\tau \, \langle \eta_i(\tau) \eta_j(\tau) \rangle = \int_{\tau_a}^{\tau_b} d\tau G_{ij}(\tau, \tau) = \frac{\hbar}{M} \left[\frac{\Delta \tau^2}{6} - \tau_a^2 \right] \delta_{ij} + \frac{\Delta \tau}{12} \Delta x_i \Delta x_j. \tag{4B.15}$$

Thus we obtain the semiclassical imaginary-time amplitude

$$
\begin{aligned}
(\mathbf{x}_b \tau_b | \mathbf{x}_a \tau_a) &= \left(\frac{M}{2\pi\hbar\Delta\tau} \right)^{D/2} \exp \left\{ -\frac{M}{2\hbar\Delta\tau} \Delta \mathbf{x}^2 - \frac{\Delta\tau}{\hbar} V(\bar{\mathbf{x}}) \right\} \\
&\times \left\{ 1 - \frac{\Delta\tau^2}{12M} \boldsymbol{\nabla}^2 V(\bar{\mathbf{x}}) - \frac{\Delta\tau}{24\hbar}(\Delta\mathbf{x}\boldsymbol{\nabla})^2 V(\bar{\mathbf{x}}) + \frac{\Delta\tau^3}{24M\hbar} [\boldsymbol{\nabla} V(\bar{\mathbf{x}})]^2 + \ldots \right\}.
\end{aligned}
\tag{4B.16}
$$

This agrees precisely with the real-time amplitude (4.265).

For the partition function at inverse temperature $\beta = (\tau_b - \tau_a)/\hbar$, this implies the semiclassical approximation

$$
\begin{aligned}
Z &= \int d^D x \, (\mathbf{x}\,\hbar\beta|\mathbf{x}\,0) \\
&\approx \left(\frac{M}{2\pi\hbar^2\beta} \right)^{D/2} \int d^D \bar{x} \left(1 - \frac{\hbar^2\beta^2}{12M} \boldsymbol{\nabla}^2 V(\bar{\mathbf{x}}) + \frac{\hbar^2\beta^3}{24M} [\boldsymbol{\nabla} V(\bar{\mathbf{x}})]^2 \right) e^{-\beta V(\bar{\mathbf{x}})}.
\end{aligned}
\tag{4B.17}
$$

A partial integration simplifies this to

$$
\begin{aligned}
Z &\approx \left(\frac{M}{2\pi\hbar^2\beta} \right)^{D/2} \int d^D \bar{x} \left(1 - \frac{\hbar^2\beta^2}{24M} \boldsymbol{\nabla}^2 V(\bar{\mathbf{x}}) \right) e^{-\beta V(\bar{\mathbf{x}})} \\
&\approx \int d^D \bar{x} \left(\frac{M}{2\pi\hbar^2\beta} \right)^{D/2} \exp \left(-\beta V(\bar{\mathbf{x}}) - \frac{\hbar^2\beta^2}{24M} \boldsymbol{\nabla}^2 V(\bar{\mathbf{x}}) \right).
\end{aligned}
\tag{4B.18}
$$

Actually, it is easy to calculate all terms in (4B.16) proportional to $V(\bar{\mathbf{x}})$ and its derivatives. Instead of the expansion (4B.6), we evaluate

$$(\mathbf{x}_b\tau_b | \mathbf{x}_a\tau_a) = e^{-\beta V(\bar{\mathbf{x}})} (\Delta\mathbf{x}/2\,\tau_b | -\Delta\mathbf{x}/2\,\tau_a) \left\{ 1 - \frac{1}{\hbar} \int_{\tau_a}^{\tau_b} d\tau \, \langle V(\bar{\mathbf{x}} + \boldsymbol{\eta}(\tau)) - V(\bar{\mathbf{x}}) \rangle \right\}. \tag{4B.19}$$

By rewriting $V(\overline{\mathbf{x}} + \boldsymbol{\eta})$ as a Fourier integral

$$V(\overline{\mathbf{x}} + \boldsymbol{\eta}) = \int \frac{d^D k}{(2\pi)^D} \tilde{V}(\mathbf{k}) \exp\left[i\mathbf{k}\left(\overline{\mathbf{x}} + \boldsymbol{\eta}\right)\right], \tag{4B.20}$$

we obtain

$$
\begin{aligned}
(\mathbf{x}_b \tau_b | \mathbf{x}_a \tau_a) &= (\Delta\mathbf{x}/2\,\tau_b | -\Delta\mathbf{x}/2\,\tau_a) \\
&\times e^{-\beta V(\overline{\mathbf{x}})} \left\{ 1 - \frac{1}{\hbar} \int_{\tau_a}^{\tau_b} d\tau \int \frac{d^D k}{(2\pi)^D} \tilde{V}(\mathbf{k}) e^{i\mathbf{k}\overline{\mathbf{x}}} \left\langle e^{i\mathbf{k}\boldsymbol{\eta}(\tau)} - 1 \right\rangle \right\}.
\end{aligned} \tag{4B.21}
$$

The expectation value can be calculated using Wick's theorem (3.307) as

$$\int_{\tau_a}^{\tau_b} d\tau \left\langle e^{i\mathbf{k}\boldsymbol{\eta}(\tau)} \right\rangle = \int_{\tau_a}^{\tau_b} d\tau e^{-k_i k_j \langle \eta_i(\tau)\eta_j(\tau)\rangle/2} = \int_{\tau_a}^{\tau_b} d\tau e^{-k_i k_j [A^2(\tau,\tau)\delta_{ij} + B^2(\tau,\tau)\Delta x_i \Delta x_j]/2}. \tag{4B.22}$$

where $A^2(\tau,\tau)$, $B^2(\tau,\tau)$ are from (4B.12):

$$A^2(\tau,\tau) = \frac{\hbar}{M\Delta\tau}(\Delta\tau - \tau)\tau, \qquad B^2(\tau,\tau) = \frac{(\tau - \bar{\tau})^2}{\Delta\tau^2}. \tag{4B.23}$$

Inserting the inverse of the Fourier decomposition (4B.20),

$$\tilde{V}(\mathbf{k}) = \int d^D \eta\, V(\overline{\mathbf{x}} + \boldsymbol{\eta}) \exp\left[-i\mathbf{k}\left(\overline{\mathbf{x}} + \boldsymbol{\eta}\right)\right], \tag{4B.24}$$

where $\boldsymbol{\eta}$ is now a time-independent variable of integration, we find

$$\int_{\tau_a}^{\tau_b} d\tau \left\langle e^{i\mathbf{k}\boldsymbol{\eta}(\tau)} \right\rangle = \int_{\tau_a}^{\tau_b} d\tau \int d^D \eta\, V(\overline{\mathbf{x}} + \boldsymbol{\eta}) \int \frac{d^D k}{(2\pi)^D} e^{-(1/2)k_i G_{ij}(\tau,\tau)k_j - ik_i \eta_i(\tau)}. \tag{4B.25}$$

After a quadratic completion of the exponent, the momentum integral can be performed and yields

$$\int_{\tau_a}^{\tau_b} d\tau \left\langle e^{i\mathbf{k}\boldsymbol{\eta}(\tau)} \right\rangle = \int d^D \eta\, V(\overline{\mathbf{x}} + \boldsymbol{\eta}) \int_{\tau_a}^{\tau_b} d\tau \,[\det G(\tau,\tau)]^{-1/2}\, e^{-(1/2)\eta_i G_{ij}^{-1}(\tau,\tau)\eta_j}. \tag{4B.26}$$

Using the transverse and longitudinal projection matrices

$$P_{ij}^T = \delta_{ij} - \frac{\Delta x_i \Delta x_j}{(\Delta x)^2}, \qquad P_{ij}^L = \frac{\Delta x_i \Delta x_j}{(\Delta x)^2}, \tag{4B.27}$$

satisfying $P^{T2} = P^T$, $P^{L2} = P^L$, we can decompose $G_{ij}(\tau,\tau')$ as

$$G_{ij} \equiv A^2(\tau,\tau') P_{ij}^T + \left[A^2(\tau,\tau') + B^2(\tau,\tau')(\Delta x)^2\right] P_{ij}^L. \tag{4B.28}$$

It is then easy to find the determinant

$$\det G(\tau,\tau') = [A^2(\tau,\tau')]^{D-1}[A^2(\tau,\tau') + B^2(\tau,\tau')(\Delta x)^2], \tag{4B.29}$$

and the inverse matrix

$$G_{ij}^{-1}(\tau) = \frac{1}{A^2(\tau,\tau')}\left[\delta_{ij} - \frac{\Delta x_i \Delta x_j}{(\Delta x)^2}\right] + \frac{1}{A^2(\tau,\tau') + B^2(\tau,\tau')(\Delta x)^2}\frac{\Delta x_i \Delta x_j}{(\Delta x)^2}. \tag{4B.30}$$

Inserting (4B.26) back into (4B.21), and taking the correction into the exponent, we arrive at

$$(\mathbf{x}_b \tau_b | \mathbf{x}_a \tau_a) = (\Delta\mathbf{x}/2\,\tau_b | -\Delta\mathbf{x}/2\,\tau_a) e^{-(1/\hbar)\int_{\tau_a}^{\tau_b} d\tau V_{\rm sm}(\overline{\mathbf{x}},\tau)}, \tag{4B.31}$$

where $V_{\rm sm}(\overline{\bf x}, \tau)$ is the harmonically smeared potential

$$V_{\rm sm}(\overline{\bf x}, \tau) \equiv [\det G(\tau, \tau)]^{-1/2} \int d^D \eta\, V(\overline{\bf x} + {\boldsymbol\eta}) e^{-(1/2)\eta_i G_{ij}^{-1}(\tau)\eta_j}. \tag{4B.32}$$

By expanding $V(\overline{\bf x} + {\boldsymbol\eta})$ to second order in ${\boldsymbol\eta}$, the exponent in (4B.31) becomes

$$-\beta V(\overline{\bf x}) - \frac{1}{2\hbar} V_{ij}(\overline{\bf x}) \int_{\tau_a}^{\tau_b} d\tau\, G_{ij}(\tau, \tau) + \dots \,. \tag{4B.33}$$

According to Eq. (4B.15), we have

$$\int_{\tau_a}^{\tau_b} d\tau\, G_{ij}(\tau, \tau) = \frac{\hbar}{M} \frac{\Delta\tau^2}{6} \delta_{ij} + \frac{\Delta\tau}{12} \Delta x_i\, \Delta x_j, \tag{4B.34}$$

so that we reobtain the first two correction terms in the curly brackets of (4B.16)

The calculation of the higher-order corrections becomes quite tedious. One rewrites the expansion (4B.2) as $V\left(\overline{\bf x} + {\boldsymbol\eta}(\tau)\right) = e^{\eta_i(\tau)\partial_i} V(\overline{\bf x})$ and (4B.19) as

$$(\mathbf{x}_b \tau_b | \mathbf{x}_a \tau_a) = (\Delta\mathbf{x}/2\, \tau_b | -\Delta\mathbf{x}/2\, \tau_a) \times \sum_{n=0}^{\infty} \frac{(-1)^n}{n!} \left\langle \prod_{n=0}^{\infty} \int_{\tau_a}^{\tau_b} d\tau_n\, e^{\eta_i(\tau_n)\partial_i} V(\overline{\bf x}) \right\rangle. \tag{4B.35}$$

Now we apply Wick's rule (3.307) for harmonically fluctuating variables, to re-express

$$\left\langle e^{\eta(\tau)\partial_i} \right\rangle = e^{\langle \eta_i(\tau)\eta_j(\tau)\rangle/2} = e^{G_{ij}(\tau,\tau)\partial_i\partial_j/2},$$

$$\left\langle e^{\eta_i(\tau)\partial_i} e^{\eta_i(\tau')\partial_i} \right\rangle = e^{[\langle\eta_i(\tau)\eta_j(\tau)\rangle\partial_i\partial_j + \langle\eta_i(\tau')\eta_j(\tau')\rangle\partial_i\partial_j + 2\langle\eta_i(\tau)\eta_j(\tau')\rangle\partial_i\partial_j]/2}$$

$$= e^{[G_{ij}(\tau,\tau)\partial_i\partial_j + G_{ij}(\tau',\tau')\partial_i\partial_j + 2G_{ij}(\tau,\tau')\partial_i\partial_j]/2}$$

$$\vdots \qquad . \tag{4B.36}$$

Expanding the exponentials and performing the τ-integrals in (4B.35) yields all desired higher-order corrections to (4B.16).

For $\Delta\mathbf{x} = 0$, the expansion has been driven to high orders in Ref. [6] (including a minimal interaction with a vector potential).

Notes and References

For the eikonal expansion, see the original works by
G. Wentzel, Z. Physik **38**, 518 (1926);
H.A. Kramers, Z. Physik **39**, 828 (1926);
L. Brillouin, C. R. Acad. Sci. Paris **183**, 24 (1926);
V.P. Maslov and M.V. Fedoriuk, *Semiclassical Approximation in Quantum Mechanics*, Reidel, Dordrecht, 1982;
J.B. Delos, *Semiclassical Calculation of Quantum Mechanical Wave Functions*, Adv. Chem. Phys. **65**, 161 (1986);
M.V. Berry and K.E. Mount, *Semiclassical Wave Mechanics*, Rep. Prog. Phys. **35**, 315 (1972);
and the references quoted in the footnotes.

For the semiclassical expansion of path integrals see
R. Dashen, B. Hasslacher and A. Neveu, Phys. Rev. D **10**, 4114, 4130 (1974),
R. Rajaraman, Phys. Rep. **21C**, 227 (1975);
S. Coleman, Phys. Rev. D **15**, 2929 (1977); and in *The Whys of Subnuclear Physics*, Erice Lectures 1977, Plenum Press, 1979, ed. by A. Zichichi.

Recent semiclassical treatments of atomic systems are given in
R.S. Manning and G.S. Ezra, Phys. Rev. **50**, 954 (1994).
Chaos **2**, 19 (1992).
Semiclassical scattering is treated in
J.M. Rost and E.J. Heller, J. Phys. B **27**, 1387 (1994).
For the semiclassical approach to chaotic systems see the textbook
M.C. Gutzwiller, *Chaos in Classical and Quantum Mechanics*, Springer, Berlin, 1990,
where the trace formula (4.236) is derived. In Section 12.4 of that book, the action (4.507) and the
eikonal (4.525) of the Coulomb system are calculated. Sections 6.3 and 6.4 discuss the properties
of the stability matrix in (4.229).
Applications to complex highly excited atomic spectra are described by
H. Friedrich and D. Wintgen, Phys. Rep. **183**, 37 (1989);
P. Cvitanović and B. Eckhardt, Phys. Rev. Lett. **63**, 823 (1991);
G. Tanner, P. Scherer, E.B. Bogomolny, B. Eckhardt, and D. Wintgen, Phys. Rev. Lett. **67**, 2410
(1991);
G.S. Ezra, K. Richter, G. Tanner, and D. Wintgen, J. Phys. B **24**, L413 (1991);
B. Eckhardt and D. Wintgen, J. Phys. A **24**, 4335 (1991);
D. Wintgen, K. Richter, and G. Tanner, Chaos **2**, 19 (1992).

P. Gaspard, D. Alonso, and I. Burghardt, Adv. Chem. Phys. XC **105** (1995);
B. Grémaud, Phys. Rev. E **65**, 056207 (2002); E **72**, 046208 (2005).

The individual citations refer to the following works:

[1] The derivative due to E.L. Schwartz is a differential invariant of the *conformal transforma-
    tion* $h = (aq+b)/(cq+d)$ in the complex plane. It is a relation between h and q which does
    not depend on a, b, c, d. To derove it one may assume, for a moment, an artificial dependence
    of h and q on an auxiliary parameter x, and calculate three derivatives with respect to x of
    $f(x) = ch(x)q(x) + dh(x) - aq(x) - b$. Since $f(x)$ does not really depend on x, the four linear
    equations f, f', f'', f''' of a, b, c, d are zero, so that the determinant must vanish, which is
    equal to $-2(h'q')^2$ times the Schwartz derivative $\{h, q\}$.

[2] C.M. Fraser, Z. Phys. C **28**, 101 (1985);
    J.Iliopoulos, C. Itzykson, A. Martin, Rev. Mod. Phys. **47**, 165 (1975);
    K. Kikkawa, Prog. Theor. Phys. **56**, 947 (1976);
    H. Kleinert, Fortschr. Phys. **26**, 565 (1978);
    R. MacKenzie, F. Wilczek, and A. Zee, Phys. Rev. Lett. **53**, 2203 (1984);
    I.J.R. Aitchison and C.M. Fraser, Phys. Lett. B **146**, 63 (1984).

[3] C.M. Bender, K. Olaussen, and P.S. Wang, Phys. Rev. D **16**, 1740 (1977).

[4] J. Schwinger, Phys. Phys. A **22**, 1827 (1980), A**24**, 2353 (1981).

[5] The form (4.547) of the scattering amplitude was first derived by
    P. Pechukas, Phys. Rev. **181**, 166 (1969).
    See also
    J.M. Rost and E.J. Heller, J. Phys. B **27**, 1387 (1994).
    For rainbow and glory scattering see the paper by Pechukas and by
    K.W. Ford and J.A. Wheeler, Ann. Phys. **7**, 529 (1959).

For the semiclassical treatment of the Coulomb problem see
A. Northcliffe and I.C. Percival, J. Phys. B **1**, 774, 784 (1968); A. Northcliffe, I.C. Percival,
and M.J. Roberts, J. Phys. B **2**, 590, 578 (1968).

For an alternative path integral formula for the scattering matrix see
W.B. Campbell, P. Finkler, C.E. Jones, and M.N. Misheloff, Phys. Rev. D **12**, 2363 (1975).

[6] D. Fliegner, M.G. Schmidt, and C. Schubert, Z. Phys. C**64**, 111 (1994) (hep-ph/9401221);
D. Fliegner, P. Haberl, M.G. Schmidt, and C. Schubert, Ann. Phys. (N.Y.) **264**, 51 (1998)
(hep-th/9707189).

[7] S.P. Alliluev, Zh. Eksp. Teor. Fiz. **33**, 200 (1957) [Sov. Phys.–JETP **6**, 156 (1958)].

Who can believe what varies every day,
Nor ever was, nor will be at a stay?
JOHN DRYDEN (1631-1700), Hind and the Panther (1687)

5

Variational Perturbation Theory

Most path integrals cannot be performed exactly. It is therefore necessary to develop approximation procedures which allow us to approach the exact result with any desired accuracy, at least in principle. The perturbation expansion of Chapter 3 does not serve this purpose since it diverges for any coupling strength. Similar divergencies appear in the semiclassical expansion of Chapter 4.

The present chapter develops a convergent approximation procedure to calculate Euclidean path integrals at a finite temperature. The basis for this procedure is a variational approach due to Feynman and Kleinert, which was recently extended to a systematic and uniformly convergent *variational perturbation expansion* [1].

5.1 Variational Approach to Effective Classical Partition Function

Starting point for the variational approach will be the path integral representation (3.813) for the effective classical potential introduced in Section 3.25. Explicitly, the effective classical Boltzmann factor $B(x_0)$ for a quantum system with an action

$$\mathcal{A}_e = \int_0^{\hbar\beta} d\tau \left[\frac{M}{2}\dot{x}^2(\tau) + V(x(\tau)) \right] \tag{5.1}$$

has the path integral representation [recall (3.813) and (3.806)]

$$B(x_0) \equiv e^{-V^{\text{eff cl}}(x_0)/k_BT} = \oint \mathcal{D}'x\, e^{-\mathcal{A}_e/\hbar} = \prod_{m=1}^{\infty} \left[\int \frac{dx_m^{\text{re}} dx_m^{\text{im}}}{\pi k_BT/M\omega_m^2} \right] \times \tag{5.2}$$

$$\times \exp\left[-\frac{M}{k_BT} \sum_{m=1}^{\infty} \omega_m^2 |x_m|^2 - \frac{1}{\hbar} \int_0^{\hbar/k_BT} d\tau V\left(x_0 + \sum_{m=1}^{\infty} (x_m e^{-i\omega_m\tau} + \text{c.c.}) \right) \right],$$

with the notation $x_m^{\text{re}} = \text{Re}\, x_m$, $x_m^{\text{im}} = \text{Im}\, x_m$. To make room for later subscripts, we shall in this chapter write $\mathcal{A}$ instead of $\mathcal{A}_e$. In Section 3.25 we have derived a perturbation expansion for $B(x_0) = e^{-\beta V^{\text{eff cl}}(x_0)}$. Here we shall find a simple but quite accurate approximation for $B(x_0)$ whose effective classical potential $V^{\text{eff cl}}(x_0)$ approaches the exact expression always from above.

5.2 Local Harmonic Trial Partition Function

The desired approximation is obtained by comparing the path integral in question with a solvable trial path integral. The trial path integral consists of a suitable superposition of local harmonic oscillator path integrals centered at arbitrary average positions x_0, each with an own frequency $\Omega^2(x_0)$. The coefficients of the superposition and the frequencies are chosen in such a way that the effective classical potential of the trial system is an optimal upper bound to the true effective classical potential. In systems with a smooth or at least not too singular potential, the accuracy of the approximation will be very good. In Section 5.13 we show how to use this approximation as a starting point of a systematic variational perturbation expansion which permits improving the result to any desired accuracy.

As a local trial action we shall take the harmonic action (3.847), which may also be considered as the action of a harmonic oscillator centered around some point x_0:

$$
\mathcal{A}_\Omega^{x_0} = \int_0^{\hbar/k_BT} d\tau M \left[\frac{\dot{x}^2}{2} + \Omega^2(x_0) \frac{(x - x_0)^2}{2} \right]. \tag{5.3}
$$

However, instead of using the specific frequency $\Omega^2(x_0) \equiv V''(x_0)/M$ in (3.847), we shall choose $\Omega(x_0)$ to be an as yet unknown local trial frequency. The effective classical Boltzmann factor $B(x_0)$ associated with this trial action can be taken directly from (3.820). We simply replace the harmonic potential $M\omega^2x^2/2$ in the defining expression (3.813) by the local trial potential $\Omega(x_0)(x - x_0)^2/2$. Then the first term in the fluctuation expansion (3.814) of the action vanishes, and we obtain, instead of (3.820), the local Boltzmann factor

$$
B_\Omega(x_0) \equiv e^{-V^{\text{eff cl}}(x_0)/k_BT} = \frac{\hbar\Omega(x_0)/2k_BT}{\sinh[\hbar\Omega(x_0)/2k_BT]} \equiv Z_\Omega^{x_0}. \tag{5.4}
$$

The exponential $\exp(-\beta M\omega^2x_0^2/2)$ in (3.820) is absent since the local trial potential vanishes at $x = x_0$. The local Boltzmann factor $B_\Omega(x_0)$ is a local partition function of paths whose temporal average is restricted to x_0, and this fact will be emphasized by the alternative notation $Z_\Omega^{x_0}$ which we shall now find convenient to use. The effective classical potential of the harmonic oscillators may also be viewed as a local free energy associated with the local partition function $Z_\Omega^{x_0}$ which we defined as

$$
F_\Omega^{x_0} \equiv -k_BT \log Z_\Omega^{x_0}, \tag{5.5}
$$

such that we may identify

$$
V_\Omega^{\text{eff cl}}(x_0) = F_\Omega^{x_0} = k_BT \log \left\{ \frac{\sinh[\hbar\Omega(x_0)/2k_BT]}{\hbar\Omega(x_0)/2k_BT} \right\}. \tag{5.6}
$$

The harmonic path integral associated with the local partition function $Z_\Omega^{x_0}$ is

$$
\begin{aligned}
Z_\Omega^{x_0} &= \int \mathcal{D}x(\tau)\tilde{\delta}(\bar{x} - x_0)e^{-\mathcal{A}_\Omega^{x_0}/\hbar} \\
&= \prod_{m=1}^{\infty} \left[\int \frac{dx_m^{\text{re}}dx_m^{\text{im}}}{\pi k_BT/M\omega_m^2} \right] e^{-\frac{M}{k_BT}\sum_{m=1}^{\infty}[\omega_m^2 + \Omega^2(x_0)]|x_m|^2},
\end{aligned} \tag{5.7}
$$

where $\tilde{\delta}(\bar{x} - x_0)$ is the slightly modified δ-function introduced in Eq. (3.812).

We now define the *local expectation values* of an arbitrary functional $F[x(\tau)]$ within the harmonic path integral (5.7):

$$\langle F[x(\tau)] \rangle_\Omega^{x_0} \equiv [Z_\Omega^{x_0}]^{-1} \int \mathcal{D}x \, \tilde{\delta}(\bar{x} - x_0) e^{-A_\Omega^{x_0}/\hbar} \, F[x(\tau)]. \tag{5.8}$$

The effective classical potential can then be re-expressed as a path integral

$$
\begin{aligned}
e^{-V^{\text{eff cl}}(x_0)/k_B T} &= Z^{x_0} = \int \mathcal{D}x \, \tilde{\delta}(\bar{x} - x_0) e^{-A/\hbar} \\
&\equiv \int \mathcal{D}x \, \tilde{\delta}(\bar{x} - x_0) e^{-A_\Omega^{x_0}/\hbar} e^{-(A - A_\Omega^{x_0})/\hbar} \\
&= Z_\Omega^{x_0} \left\langle e^{-(A - A_\Omega^{x_0})/\hbar} \right\rangle_\Omega^{x_0}.
\end{aligned}
\tag{5.9}
$$

We now take advantage of the fact that the expectation value on the right-hand side possesses an easily calculable bound given by the *Jensen-Peierls inequality*:

$$\left\langle e^{-(A - A_\Omega^{x_0})/\hbar} \right\rangle_\Omega^{x_0} \geq e^{-\left\langle A/\hbar - A_\Omega^{x_0}/\hbar \right\rangle_\Omega^{x_0}}. \tag{5.10}$$

This implies that the effective classical potential has an upper bound

$$V^{\text{eff cl}}(x_0) \leq F_\Omega^{x_0}(x_0) + k_B T \left\langle A/\hbar - A_\Omega^{x_0}/\hbar \right\rangle_\Omega^{x_0}. \tag{5.11}$$

The Jensen-Peierls inequality is a consequence of the *convexity* of the exponential function: The average of two exponentials is always larger than the exponential at the average point (see Fig. 5.1):

$$\frac{e^{-x_1} + e^{-x_2}}{2} \geq e^{-\frac{x_1 + x_2}{2}}. \tag{5.12}$$

This convexity property of the exponential function can be generalized to an exponential functional. Let $\mathcal{O}[x]$ be an arbitrary functional in the space of paths $x(\tau)$, and

$$\langle \mathcal{O}[x] \rangle \equiv \int \mathcal{D}\mu[x] \mathcal{O}[x] \tag{5.13}$$

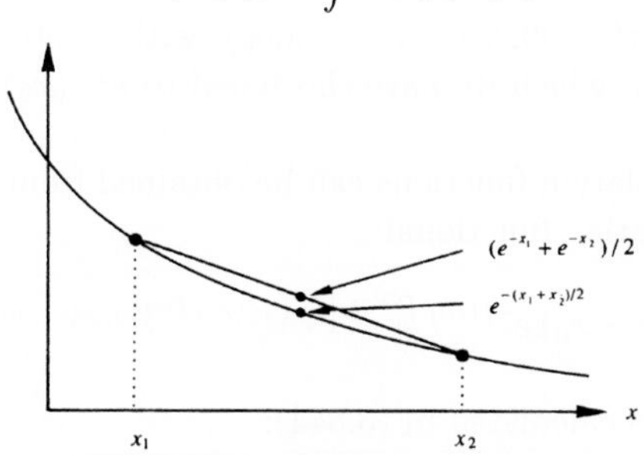

Figure 5.1 Illustration of convexity of exponential function e^{-x}, satisfying $\langle e^{-x} \rangle \geq e^{-\langle x \rangle}$ everywhere.

an expectation value in this space. The measure of integration $\mu[x]$ is supposed to be normalized so that $\langle 1 \rangle = 1$. Then (5.12) generalizes to

$$\left\langle e^{-\mathcal{O}} \right\rangle \geq e^{-\langle \mathcal{O} \rangle}. \tag{5.14}$$

To prove this we first observe that the inequality (5.12) remains valid if x_1, x_2 are replaced by the values of an arbitrary function $\mathcal{O}(x)$:

$$\frac{e^{-\mathcal{O}(x_1)} + e^{-\mathcal{O}(x_2)}}{2} \geq e^{-\frac{\mathcal{O}(x_1)+\mathcal{O}(x_2)}{2}}. \tag{5.15}$$

This inequality is then generalized with the help of any positive measure $\mu(x)$, with unit normalization, $\int d\mu(x) = 1$, to

$$\int d\mu(x) e^{-\mathcal{O}(x)} \geq e^{\int d\mu(x) e^{-\mathcal{O}(x)}}, \tag{5.16}$$

and further to

$$\int \mathcal{D}\mu[x] e^{-\mathcal{O}[x]} \geq e^{\int \mathcal{D}\mu[x] e^{-\mathcal{O}[x]}}, \tag{5.17}$$

where $\mu[x]$ is any positive functional measure with the normalization $\int \mathcal{D}\mu[x] = 1$. This shows that Eq. (5.14) is true, and hence also the Jensen-Peierls inequality (5.10).

Since the kinetic energies in the two actions $\mathcal{A}$ and $\mathcal{A}_\Omega^{x_0}$ in (5.11) are equal, the inequality (5.11) can also be written as

$$V^{\text{eff cl}}(x_0) \leq F_\Omega^{x_0} + \frac{k_B T}{\hbar} \int_0^{\hbar/k_B T} d\tau \left\langle \left[V(x(\tau)) - M \frac{\Omega^2(x_0)}{2}(x(\tau) - x_0)^2 \right] \right\rangle_\Omega^{x_0}. \tag{5.18}$$

The local expectation value on the right-hand side is easily calculated. Recalling the definition (3.853) we have to use the correlation functions of $\eta(\tau)$ without the zero frequency in Eq. (3.839):

$$\langle \eta(\tau)\eta(\tau') \rangle_\Omega^{x_0} \equiv a_{\tau\tau'}^2(x_0) = \frac{\hbar}{M} \left[\frac{1}{2\Omega(x_0)} \frac{\cosh \Omega(x_0)(|\tau - \tau'| - \hbar\beta/2)}{\sinh(\Omega(x_0)\hbar\beta/2)} - \frac{1}{\hbar\beta\Omega^2} \right], \tag{5.19}$$

valid for arguments $\tau, \tau' \in [0, \hbar\beta]$. By analogy with (3.803) we may denote this quantity by $a_{\Omega(x_0)}^2(\tau, \tau')$, which we have shortened to $a_{\tau\tau'}^2(x_0)$, to avoid a pile-up of indices.

All subtracted correlation functions can be obtained from the functional derivatives of the local generating functional

$$Z_\Omega^{x_0}[j] \equiv \int \mathcal{D}x(\tau)\tilde{\delta}(\bar{x} - x_0) e^{-(1/\hbar)\int_0^{\hbar\beta} d\tau \left[\frac{M}{2}\left\{ \dot{x}^2(\tau) + \Omega^2(x_0)[x(\tau)-x_0]^2 \right\} - j(\tau)[x(\tau)-x_0] \right]}, \tag{5.20}$$

whose explicit form was calculated in (3.844):

$$Z_\Omega^{x_0}[j] = Z_\Omega^{x_0} \exp\left\{ \frac{1}{2M\hbar} \int_0^{\hbar\beta} d\tau \int_0^{\hbar\beta} d\tau' j(\tau) G_{\Omega^2(x_0),\text{e}}^{\text{p}\prime}(\tau - \tau')j(\tau') \right\}. \tag{5.21}$$

If desired, the Green function can be continued analytically to the real-time retarded Green function

$$G_{\Omega^2}^R(t, t') = \frac{M}{\hbar} \Theta(t - t') \langle x(t)x(t') \rangle \tag{5.22}$$

in a way to be explained later in Section 18.2.

With the help of the spectral decomposition of these correlation functions to be derived in Section 18.2, it is possible to show that the lowest frequency content of the zero-temperature Green function (5.22) contains the information on the lowest excitation energy of a physical system. For the frequncy ω of an oscillator Green function this is trivially true. It is then obvious that in the above approximation where the Green function is a superposition of oscillator Green functions of frequencies $\Omega(x_0)$, the minimal zero-temperature value of $\hbar\Omega(x_0)$ gives an approximation to the energy difference between ground and first excited states. In Table 5.1 we see that for the anharmonic oscillator, this approximation is quite good.

As shown in Fig. 3.14, the *local fluctuation square width*

$$\left\langle \eta^2(\tau) \right\rangle_{\Omega}^{x_0} = \left\langle (x(\tau) - x_0)^2 \right\rangle_{\Omega}^{x_0} = a_{\tau\tau}^2(x_0) \equiv a_{\Omega(x_0)}^2, \tag{5.23}$$

with the explicit form (3.803),

$$a_{\Omega(x_0)}^2 = \frac{2}{M\beta} \sum_{m=1}^{\infty} \frac{1}{\omega_m^2 + \Omega^2(x_0)} = \frac{1}{M\beta\Omega^2(x_0)} \left[\frac{\hbar\beta\Omega(x_0)}{2} \coth \frac{\hbar\beta\Omega(x_0)}{2} - 1 \right], \tag{5.24}$$

goes to zero for high temperature like

$$a_{\Omega(x_0)}^2 \xrightarrow[T \to \infty]{} \hbar^2 / 12 M k_B T. \tag{5.25}$$

This is in contrast to the unrestricted expectation $\langle (x(\tau) - x_0)^2 \rangle_{\Omega(x_0)}$ of the harmonic oscillator which includes the $\omega_m = 0$ -term in the spectral decomposition (3.800):

$$a_{\text{tot}}^2 \equiv \left\langle (x(\tau) - x_0)^2 \right\rangle_{\Omega(x_0)} = a_{\Omega(x_0)}^2 + \frac{k_B T}{M\Omega^2(x_0)}. \tag{5.26}$$

This grows linearly with T following the equipartition theorem (3.800).

As discussed in Section 3.25, this difference is essential for the reliability of a perturbation expansion of the effective classical potential. It will also be essential for the quality of the variational approach in this chapter.

The local fluctuation square width $a_{\Omega(x_0)}^2$ measures the importance of quantum fluctuations at nonzero temperatures. These decrease with increasing temperatures. In contrast, the square width of the $\omega_0 = 0$ -term grows with the temperature showing the growing importance of classical fluctuations.

This behavior of the fluctuation width is in accordance with our previous observation after Eq. (3.837) on the finite width of the Boltzmann factor $B(x_0) = e^{-V^{\text{eff cl}}(x_0)/k_B T}$ for low temperatures in comparison to the diverging width of the alternative Boltzmann factor (3.838) formed from the partition function density $\tilde{B}(x_0) \equiv l_e(\hbar\beta)\,z(x) = e^{-\tilde{V}^{\text{eff cl}}(x_0)/k_B T}$.

Since $a^2_{\Omega(x_0)}$ is finite at *all* temperatures, the quantum fluctuations can be treated approximately. The approximation improves with growing temperatures where $a^2_{\Omega(x_0)}$ tends to zero. The thermal fluctuations, on the other hand, diverge at high temperatures. Their evaluation requires a numeric integration over x_0 in the final effective classical partition function (3.808).

Having determined $a^2_{\Omega(x_0)}$, the calculation of the local expectation value $\langle V(x(\tau))\rangle^{x_0}_\Omega$ is quite easy following the steps in Subsection 3.25.8. The result is the smearing formula analogous to (3.872): We write $V(x(\tau))$ as a Fourier integral

$$V(x(\tau)) = \int_{-\infty}^{\infty} \frac{dk}{2\pi} e^{ikx(\tau)} \tilde{V}(k), \tag{5.27}$$

and obtain with the help of Wick's rule (3.304) the expectation value

$$\langle V(x(\tau))\rangle^{x_0}_\Omega = V_{a^2}(x_0) \equiv \int_{-\infty}^{\infty} \frac{dk}{2\pi} \tilde{V}(k) e^{ikx_0 - a^2(x_0)k^2/2}. \tag{5.28}$$

For brevity, we have used the shorter notation $a^2(x_0)$ for $a^2_{\Omega(x_0)}$, and shall do so in the remainder of this chapter. Reinsert the Fourier coefficients of the potential

$$\tilde{V}(k) = \int_{-\infty}^{\infty} dx\, V(x) e^{-ikx}, \tag{5.29}$$

we may perform the integral over k and obtain the convolution integral

$$\langle V(x(\tau))\rangle^{x_0}_\Omega = V_{a^2}(x_0) \equiv \int_{-\infty}^{\infty} \frac{dx'_0}{\sqrt{2\pi a^2(x_0)}} e^{-(x'_0-x_0)^2/2a^2(x_0)}\, V(x'_0). \tag{5.30}$$

As in (3.872), the convolution integral smears the original potential $V(x_0)$ out over a length scale $a(x_0)$, thus accounting for the effects of quantum-statistical path fluctuations.

The expectation value $\langle (x(\tau)-x_0)^2\rangle^{x_0}_\Omega$ in Eq. (5.26) is, of course, a special case of this general *smearing rule*:

$$(x-x_0)^2_{a^2} = \int_{-\infty}^{\infty} \frac{dx'}{\sqrt{2\pi a^2}} e^{-(1/2a^2)(x'-x_0)^2} (x'-x_0)^2 = a^2(x_0). \tag{5.31}$$

Hence we obtain for the effective classical potential the approximation

$$W_1^\Omega(x_0) \equiv F_\Omega^{x_0} + V_{a^2}(x_0) - \frac{M}{2}\Omega^2(x_0)a^2(x_0), \tag{5.32}$$

which by the Jensen-Peierls inequality lies always above the true result:

$$W_1^\Omega(x_0) \geq V^{\text{eff cl}}(x_0). \tag{5.33}$$

A minimization of $W_1^\Omega(x_0)$ in $\Omega(x_0)$ produces an optimal variational approximation to be denoted by $W_1(x_0)$.

For the harmonic potential $V(x) = M\omega^2 x^2/2$, the smearing process leads to $V_{a^2}(x_0) = M\omega^2(x_0^2 + a^2)/2$. The extremum of $W_1^\Omega(x_0)$ lies at $\Omega(x_0) \equiv \omega$, so that the optimal upper bound is

$$W_1(x_0) = F_\omega^{x_0} + M\omega^2 \frac{x_0^2}{2}. \tag{5.34}$$

Thus, for the harmonic oscillator, $W_1^\Omega(x_0)$ happens to coincide with the exact effective classical potential $V_\omega^{\text{eff cl}}(x_0)$ found in (3.827).

5.3   The Optimal Upper Bound

We now determine the frequency $\Omega(x_0)$ of the local trial oscillator which optimizes the upper bound in Eq. (5.33). The derivative of $W_1^\Omega(x_0)$ with respect to $\Omega^2(x_0)$ has two terms:

$$\frac{dW_1^\Omega(x_0)}{d\Omega^2(x_0)} = \frac{\partial W_1^\Omega(x_0)}{\partial\Omega^2(x_0)} + \frac{\partial W_1^\Omega(x_0)}{\partial a^2(x_0)}\bigg|_{\Omega(x_0)} \frac{\partial a^2(x_0)}{\partial\Omega^2(x_0)}.$$

The first term is

$$\frac{\partial W_1(x_0)}{\partial\Omega^2(x_0)} = \frac{M}{2}\left\{ \frac{k_B T}{M\Omega^2(x_0)} \left[\frac{\hbar\Omega}{2k_B T} \coth\left(\frac{\hbar\Omega}{2k_B T} \right) - 1 \right] - a^2(x_0) \right\}. \tag{5.35}$$

It vanishes automatically due to (5.24). Thus we only have to minimize $W_1^\Omega(x_0)$ with respect to $a^2(x_0)$ by satisfying the condition

$$\frac{\partial W_1^\Omega(x_0)}{\partial a^2(x_0)} = 0. \tag{5.36}$$

Inserting (5.32), this determines the trial frequency

$$\Omega^2(x_0) = \frac{2}{M} \frac{\partial V_{a^2}(x_0)}{\partial a^2(x_0)}. \tag{5.37}$$

In the Fourier integral (5.28) for $V_{a^2}(x_0)$, the derivative $2(\partial/\partial a^2)V_{a^2}$ is represented by a factor $-k^2$ which, in turn, is equivalent to $\partial/\partial x_0^2$. This leads to the alternative equation:

$$\Omega^2(x_0) = \frac{1}{M} \left[\frac{\partial^2}{\partial x_0^2} V_{a^2}(x_0) \right]_{a^2 = a^2(x_0)}. \tag{5.38}$$

Note that the partial derivatives must be taken at *fixed* a^2 which is to be set equal to $a^2(x_0)$ at the end.

The potential $W_1^\Omega(x_0)$ with the extremal $\Omega^2(x_0)$ and the associated $a^2(x_0)$ of (5.24) constitutes the *Feynman-Kleinert approximation* $W_1(x_0)$ to the effective classical potential $V^{\text{eff cl}}(x_0)$.

It is worth noting that due to the vanishing of the partial derivative $\partial W_1^\Omega(x_0)/\partial \Omega^2(x_0)$ in (5.35) we may consider $\Omega^2(x_0)$ and $a^2(x_0)$ as arbitrary variational parameters in the expression (5.32) for $W_1^\Omega(x_0)$. Then the *independent* variation of $W_1^\Omega(x_0)$ with respect to these two parameters yields both (5.24) *and* the minimization condition (5.37) for $\Omega^2(x_0)$.

From the extremal $W_1^\Omega(x_0)$ we obtain the approximation for the partition function and the free energy [recall (3.834)]

$$Z_1 = e^{-F_1/k_B T} = \int_{-\infty}^\infty \frac{dx_0}{l_e(\hbar\beta)} e^{-W_1(x_0)/k_B T} \le Z, \tag{5.39}$$

where $l_e(\hbar\beta)$ is the thermal de Broglie length defined in Eq. (2.351). We leave it to the reader to calculate the second derivative of $W_1^\Omega(x_0)$ with respect to $\Omega^2(x_0)$ and to prove that it is nonnegative, implying that the above extremal solution is a local *minimum*.

5.4 Accuracy of Variational Approximation

The accuracy of the approximate effective classical potential $W_1(x_0)$ can be estimated by the following observation: In the limit of high temperatures, the approximation is perfect by construction, due the shrinking width (5.25) of the nonzero frequency fluctuations. This makes $W_1(x_0)$ in (5.31) converge against $V(x_0)$, just as the exact effective classical potential in Eq. (3.809).

In the opposite limit of low temperatures, the integral over x_0 in the general expression (3.808) is dominated by the minimum of the effective classical potential. If its position is denoted by x_m, we have the saddle point approximation (see Section 4.2)

$$Z \xrightarrow[T\to 0]{} e^{-V^{\text{eff cl}}(x_m)/k_B T} \int \frac{dx_0}{l_e(\hbar\beta)} e^{-[V^{\text{eff cl}}(x_m)]''(x_0-x_m)^2/2k_B T}. \tag{5.40}$$

The exponential of the prefactor yields the leading low-temperature behavior of the free energy:

$$F \xrightarrow[T\to 0]{} V^{\text{eff cl}}(x_m). \tag{5.41}$$

The Gaussian integral over x_0 contributes a term

$$\Delta F = k_B T \log \left\{ \frac{\hbar}{k_B T} \sqrt{\frac{[V^{\text{eff cl}}(x_m)]''}{M}} \right\}, \tag{5.42}$$

which accounts for the entropy of x_0 fluctuations around x_m [recall Eq. (1.565)]. Moreover, at zero temperature, the free energy F converges against the ground state energy $E^{(0)}$ of the system, so that

$$E^{(0)} = V^{\text{eff cl}}(x_m). \tag{5.43}$$

The minimum of the approximate effective classical potential, $W_1^\Omega(x_0)$ with respect to $\Omega(x_0)$ supplies us with a variational approximation to the free energy F_1, which in the limit $T \to 0$ yields a variational approximation to the ground state energy

$$E_1^{(0)} = F_1|_{T=0} \equiv W_1(x_m)|_{T=0}. \tag{5.44}$$

By taking the $T \to 0$ limit in (5.32) we see that

$$\lim_{T \to 0} W_1^\Omega(x_0) = \frac{1}{2}\left[\hbar\Omega(x_0) - M\Omega^2(x_0)a^2(x_0)\right] + V_{a^2}(x_0). \tag{5.45}$$

In the same limit, Eq. (5.24) gives

$$\lim_{T \to 0} a^2(x_0) = \frac{\hbar}{2M\Omega(x_0)}, \tag{5.46}$$

so that

$$\lim_{T \to 0} W_1^\Omega(x_0) = \frac{1}{4}\hbar\Omega(x_0) + V_{a^2}(x_0) = \frac{1}{8}\frac{\hbar^2}{Ma^2(x_0)} + V_{a^2}(x_0). \tag{5.47}$$

The right-hand side is recognized as the expectation value of the Hamiltonian operator

$$\hat{H} = \frac{\hat{p}^2}{2M} + V(x) \tag{5.48}$$

in a normalized Gaussian wave packet of width a centered at x_0:

$$\psi(x) = \frac{1}{(2\pi a^2)^{1/4}} \exp\left[-\frac{1}{4a^2}(x - x_0)^2\right]. \tag{5.49}$$

Indeed,

$$\left\langle \hat{H} \right\rangle_\psi \equiv \int_{-\infty}^{\infty} dx\,\psi^*(x)\hat{H}\psi(x) = \frac{1}{8}\frac{\hbar^2}{Ma^2} + V_{a^2}(x_0). \tag{5.50}$$

Let E_1 be the minimum of this expectation under the variation of x_0 and a^2:

$$E_1 = \min_{x_0, a^2} \left\langle \hat{H} \right\rangle_\psi. \tag{5.51}$$

This is the variational approximation to the ground state energy provided by the *Rayleigh-Ritz method*.

In the low temperature limit, the approximation F_1 to the free energy converges toward E_1:

$$\lim_{T \to 0} F_1 = E_1. \tag{5.52}$$

The approximate effective classical potential $W_1(x_0)$ is for all temperatures and x_0 more accurate than the estimate of the ground state energy E_0 by the minimal expectation value (5.51) of the Hamiltonian operator in a Gaussian wave packet. For potentials with a pronounced unique minimum of quadratic shape, this estimate is known to be excellent.

Table 5.1 Comparison of variational energy $E_1 = \lim_{T \to 0} F_1$, obtained from Gaussian trial wave function, with exact ground state energy $E_{\text{ex}}^{(0)}$. The energies of the first two excited states $E_{\text{ex}}^{(1)}$ and $E_{\text{ex}}^{(2)}$ are listed as well. The level splitting $\Delta E_{\text{ex}}^{(0)} = E_{\text{ex}}^{(1)} - E_{\text{ex}}^{(0)}$ to the first excited state is shown in column 6. We see that it is well approximated by the value of $\Omega(0)$, as it should (see the discussion after Eq. (5.21)).

$g/4$	E_1	$E_{\text{ex}}^{(0)}$	$E_{\text{ex}}^{(1)}$	$E_{\text{ex}}^{(2)}$	$\Delta E_{\text{ex}}^{(0)}$	$\Omega(0)$	$a^2(0)$
0.1	0.5603	0.559146	1.76950	3.13862	1.21035	1.222	0.4094
0.2	0.6049	0.602405	1.95054	3.53630	1.34810	1.370	0.3650
0.3	0.6416	0.637992	2.09464	3.84478	1.45665	1.487	0.3363
0.4	0.6734	0.668773	2.21693	4.10284	1.54816	1.585	0.3154
0.5	0.7017	0.696176	2.32441	4.32752	1.62823	1.627	0.2991
0.6	0.7273	0.721039	2.42102	4.52812	1.69998	1.749	0.2859
0.7	0.7509	0.743904	2.50923	4.71033	1.76533	1.819	0.2749
0.8	0.7721	0.765144	2.59070	4.87793	1.82556	1.884	0.2654
0.9	0.7932	0.785032	2.66663	5.03360	1.86286	1.944	0.2572
1.0	0.8125	0.803771	2.73789	5.17929	1.93412	2.000	0.2500
10	1.5313	1.50497	5.32161	10.3471	3.81694	4.000	0.1250
50	2.5476	2.49971	8.91510	17.4370	6.41339	6.744	0.0741
100	3.1924	3.13138	11.1873	21.9069	8.05590	8.474	0.0590
500	5.4258	5.31989	19.0434	37.3407	13.7235	14.446	0.0346
1000	6.8279	6.69422	23.9722	47.0173	17.2780	18.190	0.0275

In Table 5.1 we list the energies $E_1 = W_1(0)$ for a particle in an anharmonic oscillator potential. Its action will be specified in Section 5.7, where the approximation $W_1(x_0)$ will be calculated and discussed in detail. The table shows that this approximation promises to be quite good [2].

With the effective classical potential having good high- and low-temperature limits, it is no surprise that the approximation is quite reliable at all temperatures.

Even if the potential minimum is not smooth, the low-temperature limit can be of acceptable accuracy. An example is the three-dimensional Coulomb system for which the limit (5.51) becomes (with the obvious optimal choice $x_0 = 0$)

$$E_1 = \min_a \left(\frac{3}{8} \frac{\hbar^2}{Ma^2} - \frac{2}{\sqrt{\pi}} \frac{e^2}{\sqrt{2a^2}} \right) = -\frac{3}{8} \frac{\hbar^2}{Ma_{\min}^2}. \qquad (5.53)$$

The minimal value of a is $a_{\min} = \sqrt{9\pi/32} a_H$ where $a_H = \hbar^2/Me^2$ is the Bohr radius (4.344) of the hydrogen atom. In terms of it, the minimal energy has the value $E_1 = -(4/3\pi)e^2/a_H$. This is only 15% percent different from the true ground state energy of the Coulomb system $E_{\text{ex}}^{(0)} = -(1/2)e^2/a_H$. Such a high degree of accuracy may seem somewhat surprising since the exact Coulomb wave function $\psi(\mathbf{x}) = (\pi a_H^3)^{-1/2} \exp(-r/a_H)$ is far from being a Gaussian.

The partition function of the Coulomb system can be calculated only after subtracting the free-particle partition function and screening the $1/r$-behavior down to a finite range. The effective classical free energy F_1 of the Coulomb potential obtained by this method is, at any temperature, more accurate than the difference between E_1 and $E_{\text{ex}}^{(0)}$. More details will be given in Section 5.10.

5.5   Weakly Bound Ground State Energy in Finite-Range Potential Well

The variational approach allows us to derive a simple approximation for the bound-state energy in an arbitrarily shaped potential of finite range, for which the binding energy is very weak. Precisely speaking, the falloff of the ground state wave function has to lie outside the range of the potential. A typical example for this situation is the binding of electrons to Cooper pairs in a superconductor. The attractive force comes from the electron-phonon interaction which is weakened by the Coulomb repulsion. The potential has a complicated shape, but the binding energy is so weak that the wave function of a Cooper pair reaches out to several thousand lattice spacings, which is much larger than the range of the potential, which extends only over maximally a hundred lattice spacings. In this case one may practically replace the potential by an equivalent δ-function potential.

The present considerations apply to this situation. Let us assume the absolute minimum of the potential to lie at the origin. The first-order variational energy at the origin is given by

$$W_1(0) = \frac{\Omega}{2} + V_{a^2}(0), \tag{5.54}$$

where by (5.30)

$$V_{a^2}(0) = \int_{-\infty}^{\infty} \frac{dx_0'}{\sqrt{2\pi a^2}} e^{-x_0'^2/2a^2}\, V(x_0'). \tag{5.55}$$

By assumption, the binding energy is so small that the ground state wave function does not fall off within the range of $V(x_0)$. Hence we can approximate

$$V_{a^2}(0) \approx \sqrt{\frac{\Omega}{\pi}} \int_{-\infty}^{\infty} dx_0'\, V(x_0'), \tag{5.56}$$

where we have inserted $a^2 = 1/2\Omega$. Extremizing this in Ω yields the approximate ground state energy

$$E_1^{(0)} \approx -\frac{1}{2\pi}\left[-\int_{-\infty}^{\infty} dx_0'\, V(x_0')\right]^2. \tag{5.57}$$

By applying this result to a simple δ-function potential at the origin,

$$V(x) = -g\delta(x), \qquad g > 0, \tag{5.58}$$

we find an approximate ground state energy

$$E_1^{(0)} = -\frac{1}{2\pi}g^2. \tag{5.59}$$

The exact value is

$$E^{(0)} = -\frac{1}{2}g^2. \tag{5.60}$$

The failure of the variational approximation is due to the fact that outside the range of the potential, the wave function is a simple exponential $e^{-k|x|}$ with $k = \sqrt{-2E^{(0)}}$, and not a Gaussian. In fact, if we consider the expectation value of the Hamiltonian operator

$$H = -\frac{1}{2}\frac{d^2}{dx^2} - g\delta(x) \tag{5.61}$$

for a normalized trial wave function

$$\psi(x) = \sqrt{K}e^{-K|x|}, \tag{5.62}$$

we obtain a variational energy

$$W_1 = \frac{K^2}{2} - gK, \tag{5.63}$$

whose minimum gives the exact ground state energy (5.60). Thus, problems of the present type call for the development of a variational perturbation theory for which Eq. (5.54) and (5.55) read

$$W(0) = \frac{K^2}{2} + V_K(0), \tag{5.64}$$

where

$$V_K(0) = K \int_{-\infty}^{\infty} \frac{dx_0'}{a} e^{-K|x_0|} V(x_0). \tag{5.65}$$

For an arbitrary attractive potential whose range is much shorter than a, this leads to the correct energy for a weakly bound ground state

$$E_1^{(0)} \approx -\frac{1}{2}\left[-\int_{-\infty}^{\infty} dx_0' \, V(x_0)\right]^2. \tag{5.66}$$

5.6 Possible Direct Generalizations

Let us remark that there is a possible immediate generalization of the above variational procedure. One may treat higher components x_m with $m > 0$ accurately, say up to $m = \bar{m} - 1$, where $\bar{m}$ is some integer > 1, using the ansatz

$$\begin{aligned}
Z_{\bar{m}} &\equiv \int \mathcal{D}x(\tau) \exp\left\{ -\frac{1}{\hbar} \int_0^{\hbar/k_B T} \frac{M}{2}\left[\frac{\dot{x}^2(\tau)}{2} + \Omega^2(x_0,\dots,x_{\bar{m}}) \right.\right. \\
&\quad \times \left.\left. \left(x(\tau) - x_0 - \sum_{m=1}^{\bar{m}-1}(x_m e^{-i\omega_m \tau} + \text{c.c.}) \right)^2 \right]\right\} e^{-(1/k_B T)L_{\bar{m}}(x_0,\dots,x_{\bar{m}})} \\
&= \int_{-\infty}^{\infty} \frac{dx_0}{l_e(\hbar\beta)} \prod_{n=1}^{\bar{m}-1}\left[\int \frac{dx_m^{\text{re}}\, dx_m^{\text{im}}}{\pi k_B T/M\omega_m^2} \frac{\omega_m^2 + \Omega^2(x_0)}{\omega_m^2} \right] \\
&\quad \times \frac{\hbar\Omega(x_0)/2k_B T}{\sinh(\hbar\Omega(x_0)/2k_B T)} e^{-L_{\bar{m}}(x_0,\dots,x_{\bar{m}})/k_B T}, \tag{5.67}
\end{aligned}$$

with the trial function $L_{\bar{m}}$:

$$L_{\bar{m}}(x_0,\ldots,x_{\bar{m}}) = \frac{k_BT}{\hbar}\int_0^{\hbar/k_BT} d\tau\, V_{a_{\bar{m}}^2}\left(x_0 + \sum_{m=1}^{\bar{m}-1}(x_m e^{-i\omega_m\tau} + \text{c.c.})\right) - \frac{M}{2}\Omega^2(x_0,\ldots,x_{\bar{m}})a_{\bar{m}}^2,$$

(5.68)

and a smearing square width of the potential

$$a_{\bar{m}}^2 = \frac{2k_BT}{M}\sum_{m=\bar{m}}^{\infty}\frac{1}{\omega_m^2+\Omega^2}$$

(5.69)

$$= \frac{k_BT}{M\Omega^2}\left(\frac{\hbar\Omega}{2k_BT}\coth\frac{\hbar\Omega}{2k_BT}-1\right) - \frac{2k_BT}{M}\sum_{m=1}^{\bar{m}-1}\frac{1}{\omega_m^2+\Omega^2}.$$

For the partition function alone the additional work turns out to be not very rewarding since it renders only small improvements. It turns out that in the low-temperature limit $T \to 0$, the free energy is still equal to the optimal expectation of the Hamiltonian operator in the Gaussian wave packet (5.49).

Note that the ansatz (5.7) [as well as (5.67)] cannot be improved by allowing the trial frequency $\Omega(x_0)$ to be a matrix $\Omega_{mm'}(x_0)$ in the space of Fourier components x_m [i.e., by using $\sum_{m,m'}\Omega_{mm'}(x_0)x_m^*x_{m'}$ instead of $\Omega(x_0)\sum_m|x_m|^2$]. This would also lead to an exactly integrable trial partition function. However, after going through the minimization procedure one would fall back to the diagonal solution $\Omega_{mm'}(x_0) = \delta_{mm'}\Omega(x_0)$.

5.7 Effective Classical Potential for Anharmonic Oscillator and Double-Well Potential

For a typical application of the approximation method consider the Euclidean action

$$\mathcal{A}[x] = \int_0^{\hbar/k_BT} d\tau\left[\frac{M}{2}\left(\dot{x}^2+\omega^2 x^2\right) + \frac{g}{4}x^4\right].$$

(5.70)

Let us write $1/k_BT$ as β and use natural units with $M = 1$, $\hbar = k_B = 1$. We have to distinguish two cases:

a) Case $\omega^2 > 0$, Anharmonic Oscillator

Setting $\omega^2 = 1$, the smeared potential (5.30) is according to formula (3.876):

$$V_{a^2}(x_0) = \frac{x_0^2}{2} + \frac{g}{4}x_0^4 + \frac{a^2}{2} + \frac{3}{2}gx_0^2a^2 + \frac{3g}{4}a^4.$$

(5.71)

Differentiating this with respect to $a^2/2$ gives, via (5.37),

$$\Omega^2(x_0) = [1 + 3gx_0^2 + 3ga^2(x_0)].$$

(5.72)

This equation is solved at each x_0 by iteration together with (5.24),

$$a^2(x_0) = \frac{1}{\beta\Omega^2(x_0)}\left[\frac{\beta\Omega(x_0)}{2}\coth\frac{\beta\Omega(x_0)}{2}-1\right].$$

(5.73)

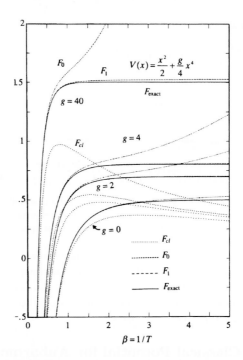

Figure 5.2 Approximate free energy F_1 of anharmonic oscillator as compared with the exact energy F_{ex}, the classical limit $F_{cl} = -(1/\beta)\log\int_{-\infty}^{\infty}(dx/\sqrt{2\pi\beta})e^{-\beta V(x)}$, as well as an earlier approximation $F_0 = -(1/\beta)\log Z_0$ of Feynman's corresponding to F_1 for the nonoptimal choice $\Omega = 0$, $a^2 = \beta/12$. Note that F_0, F_1 satisfy the inequality $F_{0,1} \geq F$, while F_{cl} does not.

An initial approximation such as $\Omega(x_0) = 0$ is inserted into (5.73) to find $a^2(x_0) \equiv \beta/12$, which serves to calculate from (5.72) an improved $\Omega^2(x_0)$, and so on. The iteration converges rapidly. Inserting the final $a^2(x_0), \Omega^2(x_0)$ into (5.71) and (5.32), we obtain the desired approximation $W_1(x_0)$ to the effective classical potential $V^{\text{eff cl}}(x_0)$. By performing the integral (5.39) in x_0 we find the approximate free energy F_1 plotted as a function of β in Fig. 5.2. The exact free-energy values are obtained from the known energy eigenvalues of the anharmonic oscillator. They are seen to lie closely below the approximate F_1 curve. For comparison, we have also plotted the classical approximation $F_{cl} = -(1/\beta)\log Z_{cl}$ which does not satisfy the Jensen-Peierls inequality and lies *below* the exact curve.

In his book on statistical mechanics [3], Feynman gives another approximation, called here F_0, which can be obtained from the present $W_1(x_0)$ by ending the iteration of (5.72), (5.73) after the first step, i.e., by using the constant nonminimal variational parameters $\Omega(x_0) \equiv 0$, $a^2(x_0) \equiv \hbar^2\beta/12M$. This leads to the approximation

$$V_{a^2}(x_0) \approx V(x_0) + \frac{1}{2}\frac{\hbar^2\beta}{12M}V''(x_0) + \frac{1}{8}\left(\frac{\hbar^2\beta}{12M}\right)^2 V^{(4)}(x_0) + \dots \, , \tag{5.74}$$

referred to as Wigner's expansion [4]. The approximation F_0 is good only at higher temperatures, as seen in Fig. 5.2. Just like F_1, the curve F_0 lies also above the exact curve since it is subject to the Jensen-Peierls inequality. Indeed, the inequality holds for the potential $W_1(x_0)$ in the general form (5.32), i.e., irrespective of the minimization in $a^2(x_0)$. Thus it is valid for arbitrary $\Omega^2(x_0)$, in particular for $\Omega^2(x_0) \equiv 0$.

In the limit $T \to 0$, the free energy F_1 yields the following approximation for the ground state energy $E^{(0)}$ of the anharmonic oscillator:

$$E_1^{(0)} = \frac{1}{4\Omega} + \frac{\Omega}{4} + \frac{3}{4}\frac{g}{4\Omega^2}. \tag{5.75}$$

This approximation is very good for all coupling strengths, including the strong-coupling limit. In this limit, the optimal frequency and energy have the expansions

$$\Omega_1 = \left(\frac{g}{4}\right)^{1/3}\left[6^{1/3} + \frac{1}{2^{1/3}3^{4/3}}\frac{1}{(g/4)^{2/3}} + \dots\right], \tag{5.76}$$

and

$$E_1^{(0)}(g) \approx \left(\frac{g}{4}\right)^{1/3}\left[\left(\frac{3}{4}\right)^{4/3} + \frac{1}{2^{7/3}3^{1/3}}\frac{1}{(g/4)^{2/3}} + \dots\right]$$

$$\approx \left(\frac{g}{4}\right)^{1/3}\left[0.681420 + 0.13758\frac{1}{(g/4)^{2/3}} + \dots\right]. \tag{5.77}$$

The coefficients are quite close to the precise limiting expression to be calculated in Section 5.15 (listed in Table 5.9).

b) Case $\omega^2 < 0$: The Double-Well Potential

For $\omega^2 = -1$, we slightly modify the potential by adding a constant $1/4g$, so that it becomes

$$V(x) = -\frac{x^2}{2} + \frac{g}{4}x^4 + \frac{1}{4g}. \tag{5.78}$$

The additional constant ensures a smooth behavior of the energies in the limit $g \to 0$. Since the potential possesses now two symmetric minima, it is called the *double-well potential*. Its smeared version $V_{a^2}(x_0)$ can be taken from (5.71), after a sign change in the first and third terms (and after adding the constant $1/4g$).

Now the trial frequency

$$\Omega^2(x_0) = -1 + 3gx_0^2 + 3ga^2(x_0) \tag{5.79}$$

can become negative, although it turns out to remain always larger than $-4\pi^2/\beta^2$, since the solution is incapable of crossing the first singularity in the sum (5.24) from the right. Hence the smearing square width $a^2(x_0)$ is always positive. For $\Omega^2 \in (-4\pi^2/\beta^2, 0)$, the sum (5.24) gives

$$
\begin{aligned}
a^2(x_0) &= \frac{2}{\beta} \sum_{m=1}^{\infty} \frac{1}{\omega_m^2 + \Omega^2(x_0)} \\
&= \frac{1}{\beta\Omega^2(x_0)} \left(\frac{\beta|\Omega(x_0)|}{2} \cot \frac{\beta|\Omega(x_0)|}{2} - 1 \right),
\end{aligned}
\tag{5.80}
$$

which is the expression (5.73), continued analytically to imaginary $\Omega(x_0)$. The above procedure for finding $a^2(x_0)$ and $\Omega^2(x_0)$ by iteration of (5.79) and (5.80) is not applicable near the central peak of the double well, where it does not converge. There one finds the solution by searching for the zero of the function of $\Omega^2(x_0)$

$$
f(\Omega^2(x_0)) \equiv a^2(x_0) - \frac{1}{3g}[1 + \Omega^2(x_0) - 3gx_0^2],
\tag{5.81}
$$

with $a^2(x_0)$ calculated from (5.80) or (5.73). At $T = 0$, the curves have for $g \leq g_c$ two symmetric nontrivial minima at $\pm x_m$ with

$$
x_m = \sqrt{\frac{1 - 3ga^2}{g}},
\tag{5.82}
$$

where Eq. (5.79) becomes

$$
\Omega^2(x_m) = 2 - 6ga^2(x_m).
\tag{5.83}
$$

These disappear for

$$
g > g_c = \frac{4}{9}\sqrt{\frac{2}{3}} \approx 0.3629 \ .
\tag{5.84}
$$

The resulting effective classical potentials and the free energies are plotted in Figs. 5.3 and 5.4.

It is useful to compare the approximate effective classical potential $W_1(x)$ with the true one $V^{\mathrm{eff\ cl}}(x)$ in Fig. 5.5. The latter was obtained by Monte Carlo simulations of the path integral of the double-well potential, holding the path average $\bar{x} = (1/\beta)\int_0^\beta d\tau\, x(\tau)$ fixed at x_0. The coupling strength is chosen as $g = 0.4$, where the worst agreement is expected.

In the limit $T \to 0$, the approximation F_1 yields an approximation $E_1^{(0)}$ for the ground state energy. In the strong-coupling limit, the leading behavior is the same as in Eq. (5.77) for the anharmonic oscillator.

Let us end this section with the following remark. The entire approximation procedure can certainly also be applied to a time-sliced path integral in which the time axis contains $N + 1$ discrete points $\tau_n = n\epsilon$, $n = 0, 1, \ldots N$. The only change in the above treatment consists in the replacement

$$
\omega_m^2 \to \Omega_m\bar{\Omega}_m = \frac{1}{\epsilon^2}[2 - 2\cos(\epsilon\omega_m)].
\tag{5.85}
$$

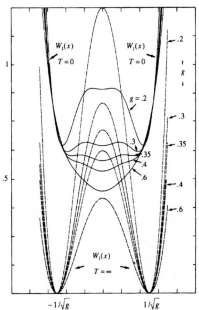

Figure 5.3 Effective classical potential of double well $V(x) = -x^2/2 + gx^4/4 + 1/4g$ at various g for $T = 0$ and $T = \infty$ [where it is equal to the potential $V(x)$ itself]. The quantum fluctuations at $T = 0$ smear out the double well completely if $g \gtrsim 0.4$, but not if $g = 0.2$.

Hence the expression for the smearing square width parameter $a^2(x_0)$ of (5.24) is replaced by

$$
\begin{aligned}
a^2(x_0) &= \frac{2k_BT}{M}\frac{\partial}{\partial\Omega^2(x_0)}\log\prod_{m=1}^{m_{\max}}\left[\Omega_m\overline{\Omega}_m + \Omega^2(x_0)\right] \\
&= \frac{k_BT}{M}\frac{1}{\Omega}\frac{\partial}{\partial\Omega}\log\frac{\sinh(\hbar\Omega_N(x_0)/2k_BT)}{\hbar\Omega(x_0)/2k_BT} \\
&= \frac{k_BT}{M\Omega^2(x_0)}\left[\frac{\hbar\Omega(x_0)}{2k_BT}\coth\frac{\hbar\Omega_N(x_0)}{2k_BT}\frac{1}{\cosh(\epsilon\Omega_N(x_0)/2)} - 1\right],
\end{aligned}
\tag{5.86}
$$

where $m_{\max} = N/2$ for even and $(N-1)/2$ for odd N [recall (2.389)], and $\Omega_N(x_0)$ is defined by

$$
\sinh[\epsilon\Omega_N(x_0)/2] \equiv \epsilon\Omega(x_0)/2
\tag{5.87}
$$

[see Eq. (2.397)]. The trial potential $W_1(x_0)$ now reads

$$
W_1(x_0) \equiv k_BT\log\frac{\sinh\hbar\Omega_N(x_0)/2k_BT}{\hbar\Omega(x_0)/2k_BT} + V_{a^2(x_0)}(x_0) - \frac{M}{2}\Omega^2(x_0)a^2(x_0),
\tag{5.88}
$$

rather than (5.32). Minimizing this in $a^2(x_0)$ gives again (5.37) and (5.38) for $\Omega^2(x_0)$.

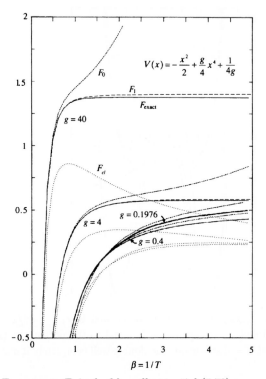

Figure 5.4 Free energy F_1 in double-well potential (5.78), compared with the exact free energy F_{ex}, the classical limit F_{cl}, and Feynman's approximation F_0 (which coincides with F_1 for the nonminimal values $\Omega = 0$, $a^2 = \beta/12$).

In Fig. 5.6 we have plotted the resulting approximate effective classical potential $W_1(x_0)$ of the double-well potential (5.78) with $g = 0.4$ at a fixed large value $\beta = 20$ for various numbers of lattice points $N + 1$. It is interesting to compare these plots with the exact curves, obtained again from Monte Carlo simulations. For $N = 1$, the agreement is exact. For small N, the agreement is good near and outside the potential minima. For larger N, the exact effective classical potential has oscillations which are not reproduced by the approximation.

5.8 Particle Densities

It is possible to find approximate particle densities from the optimal effective classical potential $W_1(x_0)$ [5, 6]. Certainly, the results cannot be as accurate as those for the free energies. In Schrödinger quantum mechanics, it is well known that variational methods can give quite accurate energies even if the trial wave functions are only of moderate quality. This has also been seen in the Eq. (5.53) estimate to the ground state energy of the Coulomb system by a Gaussian wave packet. The energy

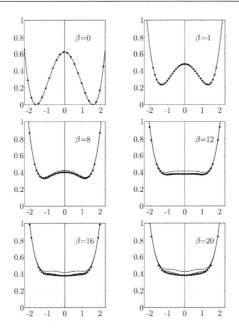

Figure 5.5   Comparison of approximate effective classical potential $W_1(x_0)$ (dashed curves) and $W_3(x_0)$ (solid curves) with exact $V^{\mathrm{eff\,cl}}(x_0)$ (dots) at various inverse temperatures $\beta = 1/T$. The data dots are obtained from Monte Carlo simulations using 10^5 configurations [W. Janke and H. Kleinert, Chem. Phys. Lett. *137*, 162 (1987) (http://www.physik.fu-berlin.de/~kleinert/154)]. We have picked the worst case, $g = 0.4$. The solid lines represent the higher approximation $W_3(x_0)$, to be calculated in Section 5.13.

is a rather global property of the system. For physical quantities such as particle densities which contain local information on the wave functions, the approximation is expected to be much worse. Let us nevertheless calculate particle densities of a quantum-mechanical system. For this we tie down the periodic particle orbit in the trial partition function Z_1 for an arbitrary time at a particular position, say x_a. Mathematically, this is enforced with the help of a δ-function:

$$\delta(x_a - x(\tau)) = \delta\left(x_a - x_0 - \sum_{m=1}^{\infty}(x_m e^{-i\omega_m \tau} + \mathrm{c.c.})\right)$$

$$= \int_{-\infty}^{\infty}\frac{dk}{2\pi}\exp\left\{ik\left[x_a - x_0 - \sum_{m=1}^{\infty}(x_m e^{-i\omega_m \tau} + \mathrm{c.c.})\right]\right\}. \qquad (5.89)$$

With this, we write the path integral for the particle density [compare (2.351)]

$$\rho(x_a) = Z^{-1}\oint \mathcal{D}x\,\delta(x_a - x(\tau))e^{-\mathcal{A}/\hbar} \qquad (5.90)$$

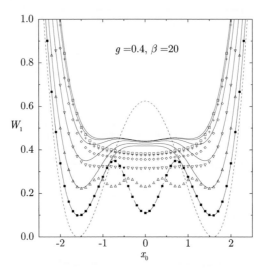

Figure 5.6 Effective classical potential $W_1(x_0)$ for double-well potential (5.78) with $g = 0.4$ at fixed low temperature $T = 1/\beta = 1/20$, for various numbers of time slices $N+1 = 2$ (■), 4 (△), 8 (▽), 16 (◇), 32 (+), 64 (□). The dashed line represents the original potential $V(x_0)$. For the source of the data points, see the previous figure caption.

and decompose

$$\rho(x_a) = Z^{-1} \int_{-\infty}^{\infty} \frac{dx_0}{l_e(\hbar\beta)} \int \frac{dk}{2\pi} e^{ik(x_a - x_0)}$$
$$\times \int \mathcal{D}x\,\tilde{\delta}(\bar{x} - x_0) e^{-ik\Sigma_{m=1}^{\infty}(x_m + \text{c.c.})} e^{-\mathcal{A}/\hbar}. \qquad (5.91)$$

The approximation $W_1(x_0)$ is based on a quasiharmonic treatment of the x_m-fluctuations for $m > 0$. For harmonic fluctuations we use Wick's rule of Section 3.10 to evaluate

$$\left\langle e^{-ik\Sigma_{m=1}^{\infty}(x_m e^{-i\omega_m\tau} + \text{c.c.})} \right\rangle_\Omega^{x_0} \approx e^{-k^2\Sigma_{m=1}^{\infty}\langle|x_m|^2\rangle_\Omega^{x_0}} \equiv e^{-k^2 a^2/2}, \qquad (5.92)$$

which is true for any τ. Thus we could have chosen any τ in the δ-function (5.89) to find the distribution function. Inserting (5.92) into (5.91) we can integrate out k and find the approximation to the particle density

$$\rho(x_a) \approx Z^{-1} \int_{-\infty}^{\infty} \frac{dx_0}{l_e(\hbar\beta)} \frac{e^{-(x_a - x_0)^2/2a^2(x_0)}}{\sqrt{2\pi a^2(x_0)}} e^{-V^{\text{eff cl}}(x_0)/k_B T}. \qquad (5.93)$$

By inserting for $V^{\text{eff cl}}(x_0)$ the approximation $W_1(x_0)$, which for Z yields the approximation Z_1, we arrive at the corresponding approximation for the particle distribution function:

$$\rho_1(x_a) = Z_1^{-1} \int_{-\infty}^{\infty} \frac{dx_0}{l_e(\hbar\beta)} \frac{e^{-(x_a - x_0)^2/2a^2(x_0)}}{\sqrt{2\pi a^2(x_0)}} e^{-W_1(x_0)/k_B T}. \qquad (5.94)$$

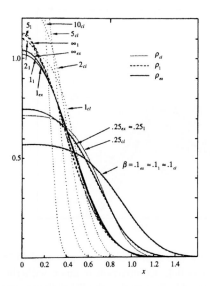

Figure 5.7 Approximate particle density (5.94) of anharmonic oscillator for $g = 40$, as compared with the exact density $\rho(x) = Z^{-1} \sum_n |\psi_n(x)|^2 e^{-\beta E_n}$, obtained by integrating the Schrödinger equation numerically. The curves are labeled by their β values with the subscripts 1, ex, cl indicating the approximation.

This has obviously the correct normalization $\int_{-\infty}^{\infty} dx_a \rho_1(x_a) = 1$. Figure 5.7 shows a comparison of the approximate particle distribution functions of the anharmonic oscillator with the exact ones. Both agree reasonably well with each other. In Fig. 5.8, the same plot is given for the double-well potential at a coupling $g = 0.4$. Here the agreement at very low temperature is not as good as in Fig. 5.7. Compare, for example, the zero-temperature curve ∞_1 with the exact curve ∞_{ex}. The first has only a single central peak, the second a double peak. The reason for this discrepancy is the correspondence of the approximate distribution to an optimal Gaussian wave function which happens to be centered at the origin, in spite of the double-well shape of the potential. In Fig. 5.3 we see the reason for this: The approximate effective classical potential $W_1(x_0)$ has, at small temperatures up to $T \sim 1/10$, only one minimum at the origin, and this becomes the center of the optimal Gaussian wave function. For larger temperatures, there are two minima and the approximate distribution function $\rho_1(x)$ corresponds roughly to two Gaussian wave packets centered around these minima. Then, the agreement with the exact distribution becomes better. We have intentionally chosen the coupling $g = 0.4$, where the result would be about the worst. For $g \gg 0.4$, both the true and the approximate distributions have a single central peak. For $g \ll 0.4$, both have two peaks at small temperatures. In both limits, the approximation is acceptable.

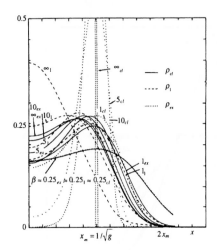

Figure 5.8 Particle density (5.94) in double-well potential (5.78) for the worst choice of the coupling constant, $g = 0.4$. Comparison is made with the exact density $\rho(x) = Z^{-1} \sum_n |\psi_n(x)|^2 e^{-\beta E_n}$ obtained by integrating the Schrödinger equation numerically. The curves are labeled by their β values with the subscripts 1, ex, cl indicating the approximation. For $\beta \to \infty$, the distribution tends to the Gaussian $e^{-x^2/2a^2}/\sqrt{2\pi a^2}$ with $a^2 = 1.030$ (see Table 5.1).

5.9 Extension to D Dimensions

The method can easily be extended to approximate the path integral of a particle moving in a D-dimensional $\mathbf{x}$-space. Let x_i be the D components of $\mathbf{x}$. Then the trial frequency $\Omega_{ij}^2(\mathbf{x}_0)$ in (5.7) must be taken as a $D \times D$ -matrix. In the special case of $V(\mathbf{x})$ being rotationally symmetric and depending only on $r = \sqrt{\mathbf{x}^2}$, we may introduce, as in the discussion of the effective action in Eqs. (3.681)–(3.686), longitudinal and transverse parts of $\Omega_{ij}^2(\mathbf{x}_0)$ via the decomposition

$$\Omega_{ij}^2(\mathbf{x}_0) = \Omega_L^2(r_0) P_{Lij}(\mathbf{x}_0) + \Omega_T^2(r_0) P_{Tij}(\mathbf{x}_0), \tag{5.95}$$

where

$$P_{Lij}(\hat{\mathbf{x}}_0) = x_{0i} x_{0j}/r_0^2, \qquad P_{Tij}(\hat{\mathbf{x}}_0) = \delta_{ij} - x_{0i} x_{0j}/r_0^2, \tag{5.96}$$

are projection matrices into the longitudinal and transverse directions of $\mathbf{x}_0$. Analogous projections of a vector are defined by $\mathbf{x}_L \equiv P_L(\mathbf{x}_0)\mathbf{x}$, $\mathbf{x}_T \equiv P_T(\mathbf{x}_0)\mathbf{x}$. Then the anisotropic generalization of $W_1^\Omega(\mathbf{x}_0)$ becomes

$$\begin{aligned}
W_1^\Omega(r_0) &= k_B T \left[\log \frac{\sinh \hbar\Omega_L/2k_B T}{\hbar\Omega_L/2k_B T} + (D-1) \log \frac{\sinh \hbar\Omega_T/2k_B T}{\hbar\Omega_T/2k_B T} \right] \\
&\quad - \frac{M}{2} \left[\Omega_L^2 a_L^2 + (D-1)\Omega_T^2 a_T^2 \right] + V_{a_L, a_T},
\end{aligned} \tag{5.97}$$

with all functions on the right-hand side depending only on r_0. The bold-face superscript indicates the presence of two variational frequencies $\mathbf{\Omega} \equiv (\Omega_L, \Omega_T)$. The smeared potential is now given by

$$V_{a_L,a_T}(r_0) = \frac{1}{\sqrt{2\pi a_L^2}}\frac{1}{\sqrt{2\pi a_T^2}^{D-1}}\prod_{i=1}^{D}\left[\int_{-\infty}^{\infty}d\delta x_i\right]$$

$$\times\ \exp\left[-\frac{1}{2}\left(a_L^{-2}\delta x_L^2 + a_T^{-2}\delta \mathbf{x}_T^2\right)\right]V(\mathbf{x}_0 + \delta\mathbf{x}),\qquad(5.98)$$

which can also be written as

$$V_{a_L,a_T}(r_0) = \frac{1}{\sqrt{(2\pi)^D a_L^2 a_T^{2D-2}}}\int d\delta x_L\int d^{D-1}\delta x_T \exp\left[-\frac{\delta x_L^2}{2a_L^2} - \frac{\delta x_T^2}{2a_T^2}\right]V(\mathbf{x}_0 + \delta\mathbf{x}). \ (5.99)$$

For higher temperatures where the smearing widths a_L^2, a_T^2 are small, we set $V(\mathbf{x}) \equiv v(r^2)$, so that

$$V(\mathbf{x}) = v(r_0^2 + 2r_0\delta x_L + \delta x_L^2 + \delta\mathbf{x}_T^2) = \sum_{n=0}^{\infty}\frac{1}{n!}v^{(n)}(r_0^2)\left(2r_0\delta x_L + \delta x_L^2 + \delta\mathbf{x}_T^2\right)^n$$

$$= \left[v(r_0^2 + \partial_\lambda)e^{\lambda\left(2r_0\delta x_L + \delta x_L^2 + \delta\mathbf{x}_T^2\right)}\right]_{\lambda=0}.\qquad(5.100)$$

Inserting this into the right-hand side of (5.98), we find

$$V_{a_L,a_T}(r_0) = \left[v(r_0^2 + \partial_\lambda)\frac{1}{\sqrt{1 - 2a_L^2\lambda}}\frac{1}{\sqrt{1 - 2a_T^2\lambda}^{D-1}}e^{2r_0^2\lambda^2 a_L^2/(1-2a_L^2\lambda)}\right]_{\lambda=0},\qquad(5.101)$$

which has the expansion

$$V_{a_L,a_T}(r_0) = v(r_0^2) + v'(r_0^2)[a_L^2 + (D - 1)a_T^2]$$

$$+ \frac{1}{2}v''(r_0^2)[3a_L^4 + 2(D - 1)a_L^2 a_T^2 + (D^2 - 1)a_T^4 + 4r_0^2 a_L^2] + \dots.\qquad(5.102)$$

The prime abbreviates the derivative with respect to r_0^2.

In general it is useful to insert into (5.98) the Fourier representation for the potential

$$V(\mathbf{x}) = \int\frac{d^D k}{(2\pi)^D}e^{i\mathbf{k}\mathbf{x}}V(\mathbf{k}),\qquad(5.103)$$

which makes the $\mathbf{x}$-integration Gaussian, so that (5.99) becomes

$$V_{a_L^2,a_T^2}(r_0) = \int\frac{d^D k}{(2\pi)^D}V(\mathbf{k})\exp\left(-\frac{a_L^2}{2}k_L^2 - \frac{a_T^2}{2}k_T^2 - ir_0 k_L\right).\qquad(5.104)$$

Exploiting the rotational symmetry of the potential by writing $V(\mathbf{k}) \equiv v(k^2)$, we decompose the measure of integration as

$$\int\frac{d^D k}{(2\pi)^D} = \frac{S_{D-1}}{(2\pi)^D}\int_{-\infty}^{\infty}dk_L\int_{0}^{\infty}dk_T\,k_T^{D-2},\qquad(5.105)$$

where

$$S_D = \frac{2\pi^{D/2}}{\Gamma(D/2)} \qquad (5.106)$$

is the surface of a sphere in D dimensions, and further with $k_L = k\cos\phi$, $k_T = k\sin\phi$:

$$\int \frac{d^D k}{(2\pi)^D} = \frac{S_{D-1}}{(2\pi)^D} \int_{-1}^{1} d\cos\phi \int_{0}^{\infty} dk\, k^{D-1}. \qquad (5.107)$$

This brings (5.104) to the form

$$\begin{aligned}
V_{a_L^2, a_T^2}(r_0) &= \frac{S_{D-1}}{(2\pi)^D} \int_{-1}^{1} du \int_{-1}^{\infty} dk\, k^{D-1} v(k^2) \\
&\times \exp\left\{ -\frac{1}{2} \left[\left(a_L^2 u^2 + a_T^2 (1 - u^2)\right) k^2 - i r_0 u \right] \right\}.
\end{aligned} \qquad (5.108)$$

The final effective classical potential is found by minimizing $W_1(r_0)$ at each r_0 in $a_L, a_T, \Omega_L, \Omega_T$. To gain a rough idea about the solution, it is usually of advantage to study first the *isotropic approximation* obtained by assuming $a_T^2(r_0) = a_L^2(r_0)$, and to proceed later to the anisotropic approximation.

5.10 Application to Coulomb and Yukawa Potentials

The effective classical potential can be useful also for singular potentials as long as the smearing procedure makes sense. An example is the Yukawa potential

$$V(r) = -(e^2/r)e^{-mr}, \qquad (5.109)$$

which reduces to the Coulomb potential for $m \equiv 0$. Using the Fourier representation

$$V(r) = 4\pi \int \frac{d^3 k}{(2\pi)^3} \frac{e^{i\mathbf{k}\mathbf{x}}}{k^2 + m^2} = 4\pi \int_{0}^{\infty} d\tau \int \frac{d^3 k}{(2\pi)^3} e^{-\tau(k^2 + m^2)}, \qquad (5.110)$$

we easily calculate the isotropically smeared potential

$$\begin{aligned}
V_{a^2}(r_0) &= -e^2 \int \frac{d^3 x}{\sqrt{2\pi a^2}^3} \exp\left[-\frac{1}{2a^2} (\mathbf{x}_0 - \mathbf{x})^2 \right] V(r) \\
&= -e^2\, 2\pi e^{m^2 a^2/2} \int_{a^2}^{\infty} da'^2 \frac{1}{\sqrt{2\pi a'^2}^3} \exp\left(-r_0^2/2a'^2 - m^2 a'^2/2 \right) \\
&= -e^2 \frac{2e^{m^2 a^2/2}}{\sqrt{\pi}} \frac{1}{r_0} \int_{0}^{r_0/\sqrt{2a^2}} dt\, e^{-(t^2 + m^2 r_0^2/4t^2)}.
\end{aligned} \qquad (5.111)$$

In the Coulomb limit $m \to 0$, the smeared potential becomes equal to the Coulomb potential multiplied by an error function,

$$V_{a^2}(r_0) = -\frac{e^2}{r_0} \mathrm{erf}(r_0/\sqrt{2a^2}), \qquad (5.112)$$

where the error function is defined by

$$\text{erf}(z) \equiv \frac{2}{\sqrt{\pi}} \int_0^z dx e^{-x^2}. \tag{5.113}$$

The smeared potential is no longer singular at the origin,

$$V_{a^2}(0) = -\frac{e^2}{a}\sqrt{\frac{2}{\pi}} e^{m^2 a^2/2}. \tag{5.114}$$

The singularity has been removed by quantum fluctuations. In this way the effective classical potential explains the stability of matter in quantum physics, i.e., the fact that atomic electrons do not fall into the origin. The effective classical potential of the Coulomb system is then by the isotropic version of (5.97)

$$W_1^{\Omega}(\mathbf{x}_0) = \frac{3}{\beta} \ln \frac{\sinh[\hbar\beta\Omega(\mathbf{x}_0)/2]}{\hbar\beta\Omega(\mathbf{x}_0)/2} - \frac{e^2}{|\mathbf{x}_0|}\text{erf}\left[\frac{|\mathbf{x}_0|}{\sqrt{2a^2(\mathbf{x}_0)}}\right] - \frac{3}{2}M\Omega^2(\mathbf{x}_0)a^2(\mathbf{x}_0). \tag{5.115}$$

Minimizing $W_1(r_0)$ with respect to $a^2(r_0)$ gives an equation analogous to Eq. (5.37) determining the frequency $\Omega^2(r_0)$ to be

$$\Omega^2(r_0) = e^2\frac{2}{3}\frac{1}{\sqrt{2\pi}}\frac{1}{(a^2)^{3/2}}e^{-r_0^2/2a^2}. \tag{5.116}$$

We have gone to atomic units in which $e = M = \hbar = k_B = 1$, so that energies and temperatures are measured in units of $E_0 = Me^4/M\hbar^2 \approx 4.36 \times 10^{-11}\text{erg} \approx 27.21\text{eV}$ and $T_0 = E_0/k_B \approx 31575\text{K}$, respectively. Solving (5.116) together with (5.24), we find $a^2(r_0)$ and the approximate effective classical potential (5.32). The result is shown in Fig. 5.9 as a dashed curve. The above approximation may now be improved by treating the fluctuations anisotropically, as described in the previous section, with different $\Omega^2(\mathbf{x}_0)$ for radial and tangential fluctuations $\delta\mathbf{x}_L$ and $\delta\mathbf{x}_T$, and the effective potential following Eqs. (5.97) and (5.98). For the anisotropically smeared Coulomb potential we calculate from (5.108):

$$V_{a_L^2,a_T^2}(r_0) = -\sqrt{\frac{1}{2\pi}}\int_{-1}^1 du \frac{e^2}{\sqrt{a_L^2 u^2 + a_T^2(1-u^2)}}\exp\left\{-\frac{r_0^2 u^2}{2\left[a_L^2 u^2 + a_T^2(1-u^2)\right]}\right\}. \tag{5.117}$$

Introducing the variable $\lambda = \sqrt{a_L^2}u/\sqrt{a_L^2\lambda^2 + a_T^2(1-\lambda^2)}$, which runs through the same interval $[-1,1]$ as u, we rewrite this as

$$V_{a_L^2,a_T^2}(r_0) = -e^2\sqrt{\frac{a_L^2}{2\pi}}\int_{-1}^1 d\lambda \frac{\exp[-(r_0^2/2a_L^2)\lambda^2]}{a_L^2(1-\lambda^2) + a_T^2\lambda^2}. \tag{5.118}$$

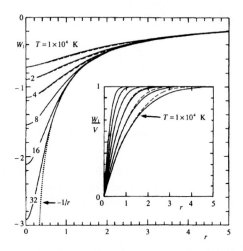

Figure 5.9 Approximate effective classical potential $W_1(r_0)$ of Coulomb system at various temperatures (in multiples of 10^4K). It is calculated once in the isotropic (dashed curves) and once in the anisotropic approximation. The improvement is visible in the insert which shows W_1/V. The inverse temperature values of the different curves are $\beta = 31.58, 15.78, 7.89, 3.945, 1.9725, 0.9863, 0$ atomic units, respectively.

Extremization of $W_1(r_0)$ in Eq. (5.97) yields the equations for the trial frequencies

$$\Omega_T^2(r_0) \equiv \frac{\partial}{\partial a_T^2} V_{a_T^2, a_L^2} = e^2 \sqrt{\frac{a_L^2}{2\pi}} \int_{-1}^{1} d\lambda \frac{\lambda^2 \exp[-(r_0^2/2a_L^2)\lambda^2]}{[a_L^2(1-\lambda^2) + a_T^2\lambda^2]^2}, \tag{5.119}$$

$$\Omega^2(r_0) \equiv \frac{1}{3}[\Omega_L^2 + 2\Omega_T^2] = \frac{1}{3}\left[\frac{\partial}{\partial a_L^2} + 2\frac{\partial}{\partial a_T^2}\right] V_{a_T^2, a_L^2}$$

$$= e^2 \frac{2}{3} \frac{1}{\sqrt{2\pi}} \frac{1}{\sqrt{a_L^2 a_T^4}} \exp(-r_0^2/2a_L^2). \tag{5.120}$$

These equations have to be solved together with

$$a_{L,T}^2(r_0) = \frac{1}{\beta \Omega_{L,T}^2(r_0)} \left[\frac{\beta \Omega_{L,T}(r_0)}{2} \coth \frac{\beta \Omega_{L,T}(r_0)}{2} - 1\right]. \tag{5.121}$$

Upon inserting the solutions into (5.97), we find the approximate effective classical potential plotted in Fig. 5.9 as a solid curve. Let us calculate the approximate particle distribution functions using a three-dimensional anisotropic version of Eq. (5.94). With the potential $W_1(r_0)$, we arrive at the integral [6]

$$\rho_1(r) = \int d^3x_0 \frac{e^{-(z_0-r)^2/2a_L^2}}{\sqrt{2\pi a_L^2(r_0)}} \frac{e^{-(x_0^2+y_0^2)/2a_T^2}}{2\pi a_T^2(r_0)} \frac{e^{-\beta W_1(r_0)}}{(2\pi\beta)^{3/2}}$$

$$= 2\pi \int_0^\infty dr_0 \int_{-1}^1 d\lambda \frac{e^{-(\lambda r_0-r)^2/2a_L^2}}{\sqrt{2\pi a_L^2(r_0)}} \frac{e^{-r_0^2(1-\lambda^2)/2a_T^2}}{2\pi a_T^2(r_0)} \frac{e^{-\beta W_1(r_0)}}{(2\pi\beta)^{3/2}}. \tag{5.122}$$

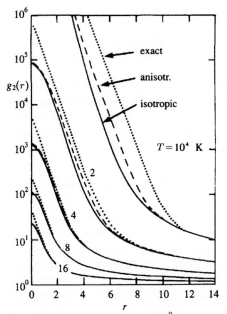

Figure 5.10   Particle distribution $g(r) \equiv \sqrt{2\pi\beta^3}\rho(r)$ in Coulomb potential at different temperatures T (the same as in Fig. 5.9), calculated once in the isotropic and once in the anisotropic approximation. The dotted curves show the classical distribution. For low and intermediate temperatures the exact distributions of R.G. Storer, J. Math. Phys. *9*, 964 (1968) are well represented by the two lowest energy levels for which $\rho(r)=\pi^{-1}e^{-2r}e^{\beta/2}+(1/8\pi)(1-r+r^2/2)e^{-r}e^{\beta/8}+(1/3^8\pi)(243-324r+216r^2-48r^3+4r^4)e^{-2r/3}e^{\beta/18}$.

The resulting curves to different temperatures are plotted in Fig. 5.10 and compared with the exact distribution as given by Storer. The distribution obtained from the earlier isotropic approximation (5.116) to the trial frequency $\Omega^2(r_0)$ is also shown.

5.11   Hydrogen Atom in Strong Magnetic Field

The recent discovery of magnetars [7] has renewed interest in the behavior of charged particle systems in the presence of extremely strong external magnetic fields. In this new type of neutron stars, electrons and protons from decaying neutrons produce magnetic fields B reaching up to 10^{15} G, much larger than those in neutron stars and white dwarfs, where B is of order $10^{10} - 10^{12}$ G and $10^6 - 10^8$ G, respectively.

Analytic treatments of the strong-field properties of an atomic system are difficult, even in the zero-temperature limit. The reason is a logarithmic asymptotic behavior of the ground state energy to be derived in Eq. (5.132). In the weak-field limit, on the other hand, perturbative approaches yield well-known series expansions in powers of B^2 up to B^{60} [8]. These are useful, however, only for $B \ll B_0$, where B_0 is the atomic magnetic field strength $B_0 = e^3 M^2/\hbar^3 \approx 2.35 \times 10^5\,\text{T} = 2.35 \times 10^9\,\text{G}$.

So far, the most reliable values for strong uniform fields were obtained by numerical calculations [9].

The variational approach can be used to derive a single analytic expression for the effective classical potential applicable to all field strengths and temperatures [10]. The Hamiltonian of the electron in a hydrogen atom in a uniform external magnetic field pointing along the positive z-axis is the obvious extension of the expression in Eq. (2.644) by a Coulomb potential:

$$H(\mathbf{p}, \mathbf{x}) = \frac{1}{2M}\mathbf{p}^2 + \frac{M}{2}\omega_B^2\mathbf{x}^2 - \omega_B l_z(\mathbf{p}, \mathbf{x}) - \frac{e^2}{|\mathbf{x}|}, \qquad (5.123)$$

where ω_B denotes the B-dependent magnetic frequency $\omega_L/2 = eB/2Mc$ of Eq. (2.649), i.e., half the Landau or cyclotron frequency. The magnetic vector potential has been chosen in the symmetric gauge (2.638). Recall that l_z is the z-component of the orbital angular momentum $l_z(\mathbf{p}, \mathbf{x}) = (\mathbf{x} \times \mathbf{p})_z$ [see (2.645)].

At first, we restrict ourselves here to zero temperature. From the imaginary-time version of the classical action (2.638) we see that the particle distribution function in the orthogonal direction of the magnetic field is, for $x_b = 0, y_b = 0$, proportional to

$$\exp\left(-\frac{M}{2}\omega_B\mathbf{x}_a^2\right). \qquad (5.124)$$

This is the same distribution as for a transverse harmonic oscillator with frequency ω_B. Being at zero temperature, the first-order variational energy requires knowing the smeared potential at the origin. Allowing for a different smearing width $a_\parallel^2$ and $a_\perp^2$ along an orthogonal to the magnetic field, we may use Eq. (5.118) to write

$$V_{a_\parallel^2, a_\perp^2}(0) = -e^2\sqrt{\frac{1}{2\pi a_\parallel^2}} \int_{-1}^{1} d\lambda \frac{1}{(1 - \lambda^2) + \lambda^2 a_\perp^2/a_\parallel^2}. \qquad (5.125)$$

Performing the integral yields

$$V_{a_\parallel^2, a_\perp^2}(0) = -e^2\sqrt{\frac{1}{2\pi(a_\parallel^2 - a_\perp^2)}}\, 2\,\text{arccosh}\frac{a_\parallel}{a_\perp}. \qquad (5.126)$$

Since the ground state energies of the parallel and orthogonal oscillators are $\Omega_\parallel/2$ and $2 \times \Omega_\perp/2$, we obtain immediately the first-order variational energy

$$W_1(0) = \Omega_\perp + \frac{\Omega_\parallel}{2} + V_{a_\parallel^2, a_\perp^2}(0) + \frac{M}{2}(\omega_\parallel^2 - \Omega_\parallel^2)a_\parallel^2 + M(\omega_B^2 - \Omega_\perp^2)a_\perp^2, \qquad (5.127)$$

with $\omega_\parallel = 0$ and $a_{\parallel,\perp}^2 = 1/2\Omega_{\parallel,\perp}$. In this expression we have ignored the second term in the Hamiltonian (5.123), since the angular momentum l_z of the ground state must have a zero expectation value.

For very strong magnetic fields, the transverse variational frequency Ω_T will become equal to ω_B, such that in this limit

$$W(0) \approx \omega_B + \frac{\Omega_\parallel}{4} - e^2 \sqrt{\frac{\Omega_\parallel}{\pi}} \log \frac{8\omega_B}{\Omega_\parallel}. \qquad (5.128)$$

Extremizing this in $\Omega_\parallel$ yields

$$\Omega_\parallel \approx \frac{4e^4}{\pi} \log^2 \frac{2\pi\omega_B}{e^4}, \qquad (5.129)$$

and thus an approximate ground state energy

$$E_1^{(0)} \approx \omega_B - \frac{e^4}{\pi} \log^2 \frac{2\pi\omega_B}{e^4}. \qquad (5.130)$$

The approach to very strong fields can be found by extremizing the energy (5.127) also in $\Omega_\perp$. Going over to atomic units with $e = 1$ and $m = 1$, where energies are measured in units of $\epsilon_0 = Me^4/\hbar^2 \equiv 2\,\mathrm{Ryd} \approx 27.21\,\mathrm{eV}$, temperatures in $\epsilon_0/k_B \approx 3.16 \times 10^5\,\mathrm{K}$, distances in Bohr radii $a_B = \hbar^2/Me^2 \approx 0.53 \times 10^{-8}\,\mathrm{cm}$, and magnetic field strengths in $B_0 = e^3 M^2/\hbar^3 \approx 2.35 \times 10^5\,\mathrm{T} = 2.35 \times 10^9\,\mathrm{G}$, the extremization yields

$$\Omega_\parallel \approx \frac{B}{2}, \qquad \sqrt{\Omega_\perp} = \frac{2}{\sqrt{\pi}} \left(\ln B - 2\ln\ln B + \frac{2a}{\ln B} + \frac{a^2}{\ln^2 B} + b \right) + \mathcal{O}(\ln^{-3} B), \qquad (5.131)$$

with abbreviations $a = 2 - \ln 2 \approx 1.307$ and $b = \ln(\pi/2) - 2 \approx -1.548$. The associated optimized ground state energy is, up to terms of order $\ln^{-2} B$,

$$E^{(0)}(B) = \frac{B}{2} - \frac{1}{\pi} \left\{ \ln^2 B - 4\ln B\,\ln\ln B + 4\ln^2\ln B - 4b\ln\ln B + 2(b+2)\ln B + b^2 \right.$$
$$\left. - \frac{1}{\ln B} \left[8\ln^2\ln B - 8b\ln\ln B + 2b^2 \right] \right\} + \mathcal{O}(\ln^{-2} B). \qquad (5.132)$$

The prefactor $1/\pi$ of the leading $\ln^2 B$-term using a variational ansatz of the type (5.64), (5.65) for the transverse degree of freedom is in contrast to the value $1/2$ calculated in the textbook by Landau and Lifshitz [11]. The calculation of higher orders in variational perturbation theory would drive our value towards $1/2$.

The convergence of the expansion (5.132) is quite slow. At a magnetic field strength $B = 10^5 B_0$, which corresponds to $2.35 \times 10^{10}\,\mathrm{T} = 2.35 \times 10^{14}\,\mathrm{G}$, the contribution from the first six terms is $22.87\,[2\,\mathrm{Ryd}]$. The next three terms suppressed by a factor $\ln^{-1} B$ contribute $-2.29\,[2\,\mathrm{Ryd}]$, while an estimate for the $\ln^{-2} B$-terms yields nearly $-0.3\,[2\,\mathrm{Ryd}]$. Thus we find $\varepsilon^{(1)}(10^5) = 20.58 \pm 0.3\,[2\,\mathrm{Ryd}]$.

Table 5.2 lists the values of the first six terms of Eq. (5.132). This shows in particular the significance of the second term in (5.132), which is of the same order of the leading first term, but with an opposite sign.

Table 5.2   Example for competing leading six terms in large-B expansion (5.132) at $B = 10^5 B_0 \approx 2.35 \times 10^{14}$ G.

$(1/\pi)\ln^2 B$	$-(4/\pi)\ln B \ln\ln B$	$(4/\pi)\ln^2\ln B$	$-(4b/\pi)\ln\ln B$	$[2(b+2)/\pi]\ln B$	b^2/π
42.1912	-35.8181	7.6019	4.8173	3.3098	0.7632

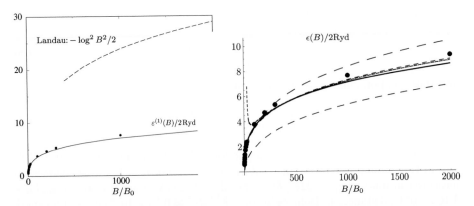

Figure 5.11   First-order variational result for binding energy (5.133) as a function of the strength of the magnetic field. The dots indicate the values derived in the reference given in Ref. [12]. The long-dashed curve on the left-hand side shows the simple estimate $0.5\ln^2 B$ of the textbook by Landau and Lifshitz [11]. The right-hand side shows the successive approximations from the strong-field expansion (5.134) for $N = 0, 1, 2, 3, 4$, with decreasing dash length. Fat curve is our variational approximation.

The field dependence of the binding energy

$$\varepsilon(B) \equiv \frac{B}{2} - E^{(0)} \tag{5.133}$$

is plotted in Fig. 5.11, where it is compared with the results of other authors who used completely different methods, with satisfactory agreement [12]. On the strong-coupling side we have plotted successive orders of a strong-field expansion [24]. The curves result from an iterative solution of the sequence of implicit equations for the quantity $w(B) = \sqrt{\varepsilon(4B)/2}$ for $N = 1, 2, 3, 4$:

$$w = \frac{1}{2}\log\frac{B}{w^2} + \sum_{n=1}^{N} a_n(B, w), \tag{5.134}$$

where

$$a_1 \equiv -\frac{1}{2}(\gamma + \log 2), \quad a_2 \equiv \frac{\pi^2}{12w}, \quad a_3 \equiv -\sqrt{\frac{2}{B}}\left(\log\frac{B}{2w^2} - \sqrt{\pi}w\right), \quad a_4 = \sqrt{\frac{2}{B}}\frac{\sqrt{\pi}}{2}D, \tag{5.135}$$

and D denotes the integral $D \equiv \gamma - 2 \int_0^\infty dy \, (y/\sqrt{y^2 + 1} - 1) \log y \approx -0.03648$, where $\gamma \approx 0.5773$ is the Euler-Mascheroni constant (2.467).

Our results are of similar accuracy as those of other first-order calculations based on an operator optimization method [25]. The advantage of our variational approach is that it yields good results for all magnetic field strengths and temperatures, and that it can be improved systematically by methods to be developed in Section 5.13, with rapid convergence. The figure shows also the energy of Landau and Lifshitz which grossly overestimates the binding energies even at very large magnetic fields, such as $2000B_0 \propto 10^{12}\,\text{G}$. Obviously, the nonleading terms in Eq. (5.132) give important contributions to the asymptotic behavior even at such large magnetic fields. As an peculiar property of the asymptotic behavior, the absolute value of the difference between the Landau-Lifshitz result and our approximation (5.132) diverges with increasing magnetic field strengths B. Only the relative difference decreases.

5.11.1 Weak-Field Behavior

Let us also calculate the weak-field behavior of the variational energy (5.127). Setting $\eta \equiv \Omega_\parallel / \Omega_\perp$, we rewrite $W_1(0)$ as

$$W_1(0) = \frac{\Omega_\perp}{2}\left(1 + \frac{\eta}{2}\right) + \frac{B^2}{8\Omega_\perp} + \sqrt{\frac{\eta\Omega_\perp}{\pi}} \frac{1}{\sqrt{1-\eta}} \ln \frac{1 - \sqrt{1-\eta}}{1 + \sqrt{1-\eta}}. \qquad (5.136)$$

This is minimized in η and $\Omega_\perp$ by expanding $\eta(B)$ and $\Omega(B)$ in powers of B^2 with unknown coefficients, and inserting these expansions into extremality equations. The expansion coefficients are then determined order by order. The optimal expansions are inserted into (5.136), yielding the optimized binding energy $\varepsilon^{(1)}(B)$ as a power series

$$W_1(0) = \sum_{n=0}^{\infty} \varepsilon_n B^{2n}. \qquad (5.137)$$

The coefficients ε_n are listed in Table 5.3 and compared with the exact ones. Of course, the higher-order coefficients of this first-order variational approximation become rapidly inaccurate, but the results can be improved, if desired, by going to higher orders in variational perturbation theory of Section 5.13.

5.11.2 Effective Classical Hamiltonian

The quantum statistical properties of the system at an arbitrary temperature are contained in the effective classical potential $H^{\text{eff cl}}(\mathbf{p}_0, \mathbf{x}_0)$ defined by the three-dimensional version of Eq. (3.821):

$$B(\mathbf{p}_0, \mathbf{x}_0) \equiv e^{-\beta H^{\text{eff cl}}(\mathbf{p}_0, \mathbf{x}_0)} \equiv \oint \mathcal{D}^3 x \oint \frac{\mathcal{D}^3 p}{(2\pi\hbar)^3} \, \delta^{(3)}(\mathbf{x}_0 - \bar{\mathbf{x}})(2\pi\hbar)^3 \delta(\mathbf{p}_0 - \bar{\mathbf{p}}) \, e^{-\mathcal{A}[\mathbf{p},\mathbf{x}]/\hbar},$$
$$(5.138)$$

Table 5.3 Perturbation coefficients up to order B^6 in weak-field expansions of variational parameters, and binding energy in comparison to exact ones (from J.E. Avron et al. and B.G. Adams et al. quoted in Notes and References).

n	0	1	2	3
η_n	1.0	$-\dfrac{405\pi^2}{7168} \approx -0.5576$	$\dfrac{16828965\pi^4}{1258815488} \approx 1.3023$	$\dfrac{3886999332075\pi^6}{884272562962432} \approx -4.2260$
Ω_n	$\dfrac{32}{9\pi} \approx 1.1318$	$\dfrac{99\pi}{224} \approx 1.3885$	$-\dfrac{1293975\pi^3}{19668992} \approx -2.03982$	$\dfrac{524431667187\pi^5}{27633517592576} \approx 5.8077$
ε_n	$-\dfrac{4}{3\pi} \approx -0.4244$	$\dfrac{9\pi}{128} \approx 0.2209$	$-\dfrac{8019\pi^3}{1835008} \approx -0.1355$	$\dfrac{256449807\pi^5}{322256764928} \approx 0.2435$
$\varepsilon_n^{\text{ex}}$	-0.5	0.25	$-\dfrac{53}{192} \approx -0.2760$	$\dfrac{5581}{4608} \approx 1.2112$

where $\mathcal{A}_e[\mathbf{p}, \mathbf{x}]$ is the Euclidean action

$$\mathcal{A}_e[\mathbf{p}, \mathbf{x}] = \int_0^{\hbar\beta} d\tau\, [-i\mathbf{p}(\tau)\dot{\mathbf{x}}(\tau) + H(\mathbf{p}(\tau), \mathbf{x}(\tau))], \tag{5.139}$$

and $\overline{\mathbf{x}} = \int_0^{\hbar\beta} d\tau\, \mathbf{x}(\tau)/\hbar\beta$ and $\overline{\mathbf{p}} = \int_0^{\hbar\beta} d\tau\, \mathbf{p}(\tau)/\hbar\beta$ are the temporal averages of position and momentum. Note that the deviations of $\mathbf{p}(\tau)$ from the average $\mathbf{p}_0$ share with $\mathbf{x}(\tau) - \mathbf{x}_0$ the property that the averages of the squares go to zero with increasing temperatures like $1/T$, and remains finite for $T \to 0$. while the expectation of $\mathbf{p}^2$ grows linearly with T (Dulong-Petit law). For $T \to 0$, the averages of the squares of $\mathbf{p}(\tau)$ remain finite. This property is the basis for a reliable accuracy of the variational treatment.

Thus we separate the action (5.139) (omitting the subscript e) as

$$\mathcal{A}_e[\mathbf{p}, \mathbf{x}] = \beta H(\mathbf{p}_0, \mathbf{x}_0) + \mathcal{A}_\Omega^{\mathbf{p}_0, \mathbf{x}_0}[\mathbf{p}, \mathbf{x}] + \mathcal{A}_{\text{int}}[\mathbf{p}, \mathbf{x}], \tag{5.140}$$

where $\mathcal{A}_\Omega^{\mathbf{p}_0, \mathbf{x}_0}[\mathbf{p}, \mathbf{x}]$ is the most general harmonic trial action containing the magnetic field. It has the form (3.825), except that we use capital frequencies to emphasize that they are now variational parameters:

$$\mathcal{A}_\Omega^{\mathbf{p}_0, \mathbf{x}_0}[\mathbf{p}, \mathbf{x}] = \int_0^{\hbar\beta} d\tau\, \Big\{ -i[\mathbf{p}(\tau) - \mathbf{p}_0] \cdot \dot{\mathbf{x}}(\tau) + \frac{1}{2M}[\mathbf{p}(\tau) - \mathbf{p}_0]^2$$
$$+ \Omega_B l_z(\mathbf{p}(\tau) - \mathbf{p}_0, \mathbf{x}(\tau) - \mathbf{x}_0) + \frac{M}{2}\Omega_\perp^2 [\mathbf{x}_\perp(\tau) - \mathbf{x}_{0\perp}]^2 + \frac{M}{2}\Omega_\parallel^2 [z(\tau) - z_0]^2 \Big\}. \tag{5.141}$$

The vector $\mathbf{x}^\perp = (x, y)$ is the projection of $\mathbf{x}$ orthogonal to $\mathbf{B}$.

The trial frequencies $\Omega = (\Omega_B, \Omega_\perp, \Omega_\parallel)$ are arbitrary functions of $\mathbf{p}_0$, $\mathbf{x}_0$, and $\mathbf{B}$. Inserting the decomposition (5.140) into (5.138), we expand the exponential of the interaction, $\exp\{-\mathcal{A}_{\text{int}}[\mathbf{p}, \mathbf{x}]/\hbar\}$, and obtain a series of expectation values of powers of the interaction $\langle \mathcal{A}_{\text{int}}^n[\mathbf{p}, \mathbf{x}] \rangle_\Omega^{\mathbf{p}_0, \mathbf{x}_0}$, defined in general by the path integral

$$\langle \mathcal{O}[\mathbf{p}, \mathbf{x}] \rangle_\Omega^{\mathbf{p}_0, \mathbf{x}_0} = \frac{1}{Z_\Omega^{\mathbf{p}_0, \mathbf{x}_0}} \oint \mathcal{D}^3 x \frac{\mathcal{D}^3 p}{(2\pi\hbar)^3} \mathcal{O}[\mathbf{p}, \mathbf{x}]\, \delta^{(3)}(\mathbf{x}_0 - \overline{\mathbf{x}})(2\pi\hbar)^3 \delta(\mathbf{p}_0 - \overline{\mathbf{p}})$$
$$\times\, \exp\Big\{ -\frac{1}{\hbar}\mathcal{A}_\Omega^{\mathbf{p}_0, \mathbf{x}_0}[\mathbf{p}, \mathbf{x}] \Big\}, \tag{5.142}$$

where the local partition function in phase space $Z_\Omega^{\mathbf{p}_0,\mathbf{x}_0}$ is the normalization factor which ensures that $\langle 1 \rangle_\Omega^{\mathbf{p}_0,\mathbf{x}_0} = 1$. From Eq. (3.826) we know that

$$Z_\Omega^{\mathbf{p}_0,\mathbf{x}_0} \equiv e^{-\beta F_\Omega^{\mathbf{p}_0,\mathbf{x}_0}} = l_e^3(\hbar\beta) \frac{\hbar\beta\Omega_+/2}{\sinh\hbar\beta\Omega_+/2} \frac{\hbar\beta\Omega_-/2}{\sinh\hbar\beta\Omega_-/2} \frac{\hbar\beta\Omega_\parallel/2}{\sinh\hbar\beta\Omega_\parallel/2}, \qquad (5.143)$$

where $\Omega_\pm \equiv \Omega_B \pm \Omega_\perp$. In comparison to (3.826), the classical Boltzmann factor $e^{-\beta H(\mathbf{p}_0,\mathbf{x}_0)}$ is absent due to the shift of the integration variables in the action (5.141). Note that the fluctuations $\mathbf{p}(\tau) - \mathbf{p}_0$ decouple from $\mathbf{p}_0$ just as $\mathbf{x}(\tau) - \mathbf{x}_0$ decoupled from $\mathbf{x}_0$ due to the absence of zero frequencies in the fluctuations.

Rewriting the perturbation series as a cumulant expansion, evaluating the expectation values, and integrating out the momenta on the right-hand side of Eq. (5.138) leads to a series representation for the effective classical potential $V_{\text{eff}}(\mathbf{x}_0)$. Since it is impossible to sum up the series, the perturbation expansion must be truncated, leading to an Nth-order approximation $W_\Omega^{(N)}(\mathbf{x}_0)$ for the effective classical potential. Since the parameters $\boldsymbol{\Omega}$ are arbitrary, $W_\Omega^{(N)}(\mathbf{x}_0)$ should depend *minimally* on $\boldsymbol{\Omega}$. This determines the optimal values of $\boldsymbol{\Omega}$ to be equal to $\boldsymbol{\Omega}^{(N)}(\mathbf{x}_0) = (\Omega_B^{(N)}(\mathbf{x}_0), \Omega_\perp^{(N)}(\mathbf{x}_0), \Omega_\parallel^{(N)}(\mathbf{x}_0))$ of Nth order. Reinserting these into $W_\Omega^{(N)}(\mathbf{x}_0)$ yields the optimal approximation $W^{(N)}(\mathbf{x}_0) \equiv W_{\boldsymbol{\Omega}^{(N)}}^{(N)}(\mathbf{x}_0)$.

The first-order approximation to the effective classical potential is then, with $\omega_\parallel = 0, \omega_\perp = \omega_B$,

$$\begin{aligned} W_\Omega^{(1)}(\mathbf{x}_0) &= F_\Omega^{\mathbf{p}_0,\mathbf{x}_0} - \frac{M}{2}\Omega_B(\mathbf{x}_0)[\omega_B - \Omega_B(\mathbf{x}_0)]\, b_\perp^2(\mathbf{x}_0)\, a_\perp^2(\mathbf{x}_0) \\ &+ \frac{M}{2}\left[\omega_B^2 - \Omega_\perp^2(\mathbf{x}_0)\right] - \frac{1}{2}\Omega_\parallel^2 a_\parallel^2(\mathbf{x}_0) - \left\langle \frac{e^2}{|\mathbf{x}|} \right\rangle_\Omega^{\mathbf{p}_0,\mathbf{x}_0}. \end{aligned} \qquad (5.144)$$

The smearing of the Coulomb potential is performed as in Section 5.10. This yields the result (5.117). with the longitudinal width

$$a_\parallel^2(\mathbf{x}_0) = G_{zz}^{(2)\,\mathbf{x}_0}(\tau,\tau) = \frac{1}{\beta M\Omega_\parallel^2(\mathbf{x}_0)}\left[\frac{\hbar\beta\Omega_\parallel(\mathbf{x}_0)}{2}\coth\frac{\hbar\beta\Omega_\parallel(\mathbf{x}_0)}{2} - 1\right], \qquad (5.145)$$

and an analog transverse width.

The quantity $b_\perp^2(\mathbf{x}_0)$ is new in this discussion based on the canonical path integral. It denotes the expectation value associated with the z-component of the angular momentum

$$b_\perp^2(\mathbf{x}_0) \equiv \frac{1}{M\Omega_B}\langle l_z \rangle_\Omega^{\mathbf{p}_0,\mathbf{x}_0}, \qquad (5.146)$$

which can also be written as

$$b_\perp^2(\mathbf{x}_0) = \frac{2}{M\Omega_{T1}}\langle x(\tau)p_y(\tau) \rangle_\Omega^{\mathbf{p}_0,\mathbf{x}_0}. \qquad (5.147)$$

According to Eq. (3.355), the correlation function $\langle x(\tau)p_y(\tau) \rangle_\Omega^{\mathbf{p}_0,\mathbf{x}_0}$ is given by

$$\langle x(\tau)p_y(\tau') \rangle_\Omega^{\mathbf{p}_0,\mathbf{x}_0} = iM\partial_{\tau'}G_{\omega^2,B,xx}^{(2)}(\tau,\tau') - M\omega_B G_{\omega^2,B,xy}^{(2)}(\tau,\tau'), \qquad (5.148)$$

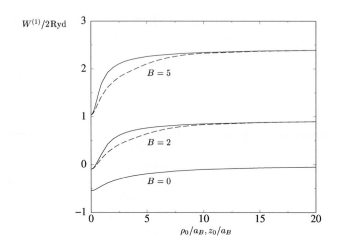

Figure 5.12 Effective classical potential of atom in strong magnetic field plotted along two directions: once as a function of the coordinate $\rho_0 = \sqrt{x_0^2 + y_0^2}$ perpendicular to the field lines at $z_0 = 0$ (solid curves), and once parallel to the magnetic field as a function of z_0 at $\rho_0 = 0$ (dashed curves). The inverse temperature is fixed at $\beta = 100$, and the strengths of the magnetic field B are varied (all in natural units).

where the expressions on the right-hand side are those of Eqs. (3.326) and (3.328), with ω replaced by Ω.

The variational energy (5.144) is minimized at each $\mathbf{x}_0$, and the resulting $W^{(N)}(\mathbf{x}_0)$ is displayed for a low temperature and different magnetic fields in Fig. 5.12. The plots show the anisotropy with respect to the magnetic field direction. The anisotropy grows when lowering the temperature and increasing the field strength. Far away from the proton at the origin, the potential becomes isotropic, due to the decreasing influence of the Coulomb interaction. Analytically, this is seen by going to the limits $\rho_0 \to \infty$ or $z_0 \to \infty$, where the expectation value of the Coulomb potential tends to zero, leaving an effective classical potential

$$W_\Omega^{(1)}(\mathbf{x}_0) \to F_\Omega^{\mathbf{p}_0,\mathbf{x}_0} - M\Omega_B(\omega_\perp - \Omega_B)\, b_\perp^2 + M\left(\omega_\perp^2 - \Omega_\perp^2\right)a_\perp^2 - \frac{M}{2}\Omega_\parallel^2 a_\parallel^2. \quad (5.149)$$

This is $\mathbf{x}_0$-independent, and optimization yields the constants $\Omega_B^{(1)} = \Omega_\perp^{(1)} = \omega_B$ and $\Omega_\parallel^{(1)} = 0$, with the asymptotic energy

$$W^{(1)}(\mathbf{x}_0) \to -\frac{1}{\beta}\log\frac{\beta\hbar\omega_B}{\sinh\beta\hbar\omega_B}. \quad (5.150)$$

The $B = 0$-curves agree, of course, with those obtained from the previous variational perturbation theory of the hydrogen atom [26].

For large temperatures, the anisotropy decreases since the violent thermal fluctuations have a smaller preference of the z-direction.

5.12   Variational Approach to Excitation Energies

As explained in Section 5.4, the success of the above variational treatment is rooted in the fact that for smooth potentials, the ground state energy can be approximated quite well by the optimal expectation value of the Hamiltonian operators in a Gaussian wave packet. The question arises as to whether the energies of excited states can also be obtained by calculating an optimized expectation value between excited oscillator wave functions. If the potential shape has only a rough similarity with that of a harmonic oscillator, when there are no multiple minima, the answer is positive. Consider again the anharmonic oscillator with the action

$$\mathcal{A}[x] = \int_0^{\hbar\beta} d\tau \left[\frac{M}{2} \left(\dot{x}^2 + \omega^2 x^2 \right) + \frac{1}{4} g x^4 \right]. \tag{5.151}$$

As for the ground state, we replace the action $\mathcal{A}$ by $\mathcal{A}_\Omega^{x_0} + \mathcal{A}_{\text{int}}^{x_0}$ with the trial oscillator action centered around the arbitrary point x_0

$$\mathcal{A}_\Omega^{x_0} = \int_0^{\hbar\beta} d\tau M \left[\frac{1}{2} \dot{x}^2 + \frac{\Omega^2}{2} (x - x_0)^2 \right], \tag{5.152}$$

and a remainder

$$\mathcal{A}_{\text{int}}^{x_0} = \int_0^\beta d\tau \left[M \frac{\omega^2 - \Omega^2}{2} (x - x_0)^2 + \frac{g}{4} x^4 \right] \tag{5.153}$$

to be treated as an interaction. Let $\psi_\Omega^{(n)}(x - x_0)$ be the wave functions of the trial oscillator.[1] With these, we form the projections

$$Z_\Omega^{(n)}(x_0) \equiv \int dx_b dx_a \psi_\Omega^{(n)*}(x_b)(x_b \tau_b | x_a \tau_a) \psi_\Omega^{(n)}(x_a)$$

$$= \int dx_b dx_a \psi_\Omega^{(n)*}(x_b) \left(\int_{(x_a,0) \rightsquigarrow (x_b,\hbar\beta)} \mathcal{D}x e^{-\mathcal{A}/\hbar} \right) \psi_\Omega^{(n)}(x_a). \tag{5.154}$$

If the temperature tends to zero, an optimization in the parameters x_0 and $\Omega(x_0)$ should yield information on the energy $E^{(n)}$ of this state by containing an exponentially decreasing function

$$Z^{(n)} \approx e^{-\beta E^{(n)}}. \tag{5.155}$$

Since the trial wave functions $\psi_\Omega^{(n)}(x - x_0)$ are not the true ones, the behavior (5.155) contains an admixture of Boltzmann factors $e^{-\beta E^{(n')}}$ with $n' \neq n$, which have to be eliminated. They are easily recognized by the powers of g which they carry.

The calculation of (5.154) is done in the same approximation that rendered good results for the ground state, i.e., we approximate the part of $Z^{(n)}$ behaving like (5.155) as follows:

$$Z^{(n)} \approx Z_\Omega^{(n)} e^{-\left\langle \mathcal{A}_{\text{int}}^{x_0}/\hbar \right\rangle_\Omega^{(n)}}, \tag{5.156}$$

---

[1] In contrast to the earlier notation, we now use superscripts in parentheses to indicate the principal quantum numbers. The subscripts specify the level of approximation.

where $Z_\Omega^{(n)}$ denotes the contribution of the nth excited state to the oscillator partition function

$$Z_\Omega^{(n)} = e^{-\beta\hbar\Omega(n+1/2)}, \tag{5.157}$$

and $\langle \mathcal{A}_{int}^{x_0}/\hbar \rangle_\Omega^{(n)}$ stands for the expectation of $\mathcal{A}_{int}^{x_0}/\hbar$ in the state $\psi_\Omega^{(n)}(x-x_0)$. This approximation corresponds precisely to the first term in the perturbation expansion (3.506) (after continuing to imaginary times).

Note that the approximation on the right-hand side of (5.156) is not necessarily smaller than the left-hand side, as in the Jensen-Peierls inequality (5.10), since the measure of integration in (5.154) is no longer positive.

For the action (5.151), the best value of x_0 lies at the coordinate origin. This simplifies the calculation of the expectation $\langle \mathcal{A}_{int}^{x_0}/\hbar \rangle_\Omega^{(n)}$. The expectation of x^{2k} in the state $\psi_\Omega^{(n)}(x)$ is given by

$$\left\langle x^{2k} \right\rangle_\Omega^{(n)} = \frac{\hbar^k}{(M\Omega)^k} n_{2k}, \tag{5.158}$$

where

$$n_2 = (n+1/2), \quad n_4 = \frac{3}{2}(n^2+n+1/2), \quad n_6 = \frac{5}{4}(2n^3+3n^2+4n+3/2),$$

$$n_8 = \frac{1}{16}(70n^4+140n^3+344n^2+280n+105), \quad \dots . \tag{5.159}$$

After inserting (5.153) into (5.156), the expectations (5.158) yield the approximation

$$Z^{(n)} \approx \exp\left\{ -\beta \left[\frac{1}{2}\left(\Omega^2+\omega^2\right)\frac{\hbar n_2}{\Omega} + \frac{g}{4}\frac{\hbar^2 n_4}{M^2\Omega^2} \right] \right\}. \tag{5.160}$$

With the dimensionless coupling constant $g' \equiv g\hbar/M^2\omega^3$, this corresponds to the variational energies

$$W^{(n)} = \hbar\left[\frac{1}{2}\left(\Omega+\frac{\omega^2}{\Omega}\right)n_2 + \frac{g'}{4}\frac{\omega^3}{\Omega^2}n_4 \right]. \tag{5.161}$$

They are optimized by the extremal Ω-values

$$\Omega = \Omega_1^{(n)} = \begin{cases} \frac{2}{\sqrt{3}}\omega \cosh\left[\frac{1}{3}\text{arcosh}\left(g/g^{(n)}\right)\right] \\ \frac{2}{\sqrt{3}}\omega \cos\left[\frac{1}{3}\arccos(g/g^{(n)})\right] \end{cases} \text{for} \quad \begin{matrix} g > g^{(n)}, \\ g < g^{(n)}, \end{matrix} \tag{5.162}$$

where

$$g^{(n)} \equiv \frac{2}{3\sqrt{3}}\frac{n_2}{n_4}\frac{M^2\omega^3}{\hbar}. \tag{5.163}$$

The optimized $W^{(n)}$ yield the desired approximations $E_{app}^{(n)}$ for the excited energies of the anharmonic oscillator.

For large g, the trial frequency grows like

$$\Omega^{(n)} \equiv 6^{1/3} \left(\frac{g\hbar}{4M^2\omega^3} \right)^{1/3} \tilde{n}, \tag{5.164}$$

where

$$\tilde{n} \equiv \frac{4n_4/3}{2n_2} = \frac{n^2 + n + 1/2}{n + 1/2}, \tag{5.165}$$

making the energy $E_{\text{app}}^{(n)}$ grow like

$$E_{\text{BS}}^{(n)} \to \hbar\omega\kappa^{(n)} \left(\frac{g\hbar}{4M^2\omega^3} \right)^{1/3}, \quad \text{for large } g, \tag{5.166}$$

with

$$\kappa^{(n)} = \frac{3 \cdot 6^{1/3}}{8} (2n + 1)\, \tilde{n}^{1/3} \approx 0.68142 \times (2n + 1)\, \tilde{n}^{1/3}. \tag{5.167}$$

For $n = 0$, this is in good agreement with the precise growth behavior to be calculated in Section 5.16 where we shall find $\kappa_{\text{ex}}^{(0)} = 0.667\,986\ldots$ (see Table 5.9).

In the limit of large g *and* n, this can be compared with the exact behavior obtained from the semiclassical approximation of Bohr and Sommerfeld, which gives the same leading powers in g and n as in (5.166), but a 3% larger prefactor $0.688\,253\,702 \times 2$ [recall (4.36)]. The exact values of $E^{(n)}/(g\hbar/4M^2)$ and $\frac{1}{2}\kappa^{(n)}/(n + 1/2)^{4/3}$ are for $n = 0, 2, 4, 6, 8, 10$ are shown in Table 5.4.

A comparison of the approximate energies $E_{\text{app}}^{(n)}$ with the precise numerical solutions of the Schrödinger equation in natural units $\hbar = 1$, $M = 1$ in Table 5.5 shows an excellent agreement for all coupling strengths.

Near the strong-coupling limit, the optimal frequencies $\Omega_1^{(n)}$ and the approximate energies $E_1^{(n)}$ behave as follows [in natural units; compare (5.76) and (5.77)]:

$$\begin{aligned}
\Omega_1^{(n)} &= 6^{1/3} \left(\frac{\tilde{n}g}{4} \right)^{1/3} \left[1 + \frac{1}{9} \left(\frac{3}{4} \right)^{1/3} \frac{1}{(\tilde{n}g/4)^{2/3}} + \ldots \right], \\
E_1^{(n)} &= \left(\frac{3}{4} \right)^{4/3} (2n + 1) \left(\frac{\tilde{n}g}{4} \right)^{1/3} \left[1 + \frac{6^{1/3}}{9} \frac{1}{(\tilde{n}g/4)^{2/3}} + \ldots \right].
\end{aligned} \tag{5.168}$$

Table 5.4 Approach of variational energies of nth excited state to Bohr-Sommerfeld approximation with increasing n. Values in the last column converge rapidly towards the Bohr-Sommerfeld value $0.688\,253\,702\ldots$ in Eq. (4.36).

n	$E^{(n)}/(g\hbar/4M^2)$	$\frac{1}{2}\kappa^{(n)}/(n + 1/2)^{4/3}$
0	0.667 986 259 155 777 108 3	0.841 609 948 112 105 001
2	4.696 795 386 863 646 196 2	0.692 125 685 914 981 314
4	10.244 308 455 438 771 076 0	0.689 449 772 359 340 765
6	16.711 890 073 897 950 947 1	0.688 828 486 600 234 466
8	23.889 993 634 572 505 935 5	0.688 590 146 947 993 676
10	31.659 456 477 221 552 442 8	0.688 474 290 179 981 433

Table 5.5 Energies of the nth excited states of anharmonic oscillator $\omega^2 x^2/2 + g x^4/4$ for various coupling strengths g (in natural units). In each entry, the upper number shows the energies obtained from a numerical integration of the Schrödinger equation, whereas the lower number is our variational result.

$g/4$	$E^{(0)}$	$E^{(1)}$	$E^{(2)}$	$E^{(3)}$	$E^{(4)}$	$E^{(5)}$	$E^{(6)}$	$E^{(7)}$	$E^{(8)}$
0.1	0.559 146	1.769 50	3.138 62	4.628 88	6.220 30	7.899 77	9.657 84	11.4873	13.3790
	0.560 307	1.773 39	3.138 24	4.621 93	6.205 19	7.875 22	9.622 76	11.4407	13.3235
0.2	0.602 405	1.950 54	3.536 30	5.291 27	7.184 46	9.196 34	11.313 2	13.5249	15.8222
	0.604 901	1.958 04	3.534 89	5.278 55	7.158 70	9.156 13	11.257 3	13.4522	15.7328
0.3	0.637 992	2.094 64	3.844 78	5.796 57	7.911 75	10.1665	12.5443	15.0328	17.6224
	0.641 630	2.104 98	3.842 40	5.779 48	7.878 23	10.1151	12.4736	14.9417	17.5099
0.4	0.668 773	2.216 93	4.102 84	6.215 59	8.511 41	10.9631	13.5520	16.2642	19.0889
	0.673 394	2.229 62	4.099 59	6.194 95	8.471 69	10.9028	13.4698	16.1588	18.9591
0.5	0.696 176	2.324 41	4.327 52	6.578 40	9.028 78	11.6487	14.4177	17.3220	20.3452
	0.701 667	2.339 19	4.323 52	6.554 75	8.983 83	11.5809	14.3257	17.2029	20.2009
0.6	0.721 039	2.421 02	4.528 12	6.901 05	9.487 73	12.2557	15.1832	18.2535	21.4542
	0.727 296	2.437 50	4.523 43	6.874 77	9.438 25	12.1816	15.0828	18.1256	21.2974
0.7	0.743 904	2.509 23	4.710 33	7.193 27	9.902 61	12.8039	15.8737	19.0945	22.4530
	0.750 859	2.527 29	4.705 01	7.164 64	9.849 11	12.7240	15.7658	18.9573	22.2852
0.8	0.765 144	2.590 70	4.877 93	7.461 45	10.2828	13.3057	16.5053	19.8634	23.3658
	0.772 736	2.610 21	4.872 04	7.430 71	10.2257	13.2206	16.3907	19.7179	23.1880
0.9	0.785 032	2.666 63	5.033 60	7.710 07	10.6349	13.7700	17.0894	20.5740	24.2091
	0.793 213	2.687 45	5.027 18	7.677 39	10.5744	13.6801	16.9687	20.4209	24.0221
1	0.803 771	2.737 89	5.179 29	7.942 40	10.9636	14.2031	17.6340	21.2364	24.9950
	0.812 500	2.759 94	5.172 37	7.907 93	10.9000	14.1090	17.5076	21.0763	24.7996
10	1.504 97	5.321 61	10.3471	16.0901	22.4088	29.2115	36.4369	44.0401	51.9865
	1.531 25	5.382 13	10.3244	15.9993	22.2484	28.9793	36.1301	43.6559	51.5221
50	2.499 71	8.915 10	17.4370	27.1926	37.9385	49.5164	61.8203	74.7728	88.3143
	2.547 58	9.023 38	17.3952	27.0314	37.6562	49.1094	61.2842	74.1029	87.5059
100	3.131 38	11.1873	21.9069	34.1825	47.7072	62.2812	77.7708	94.0780	111.128
	3.192 44	11.3249	21.8535	33.9779	47.3495	61.7660	77.0924	93.2307	110.106
500	5.319 89	19.0434	37.3407	58.3016	81.4012	106.297	132.760	160.622	189.756
	5.425 76	19.2811	37.2477	57.9489	80.7856	105.411	131.595	159.167	188.001
1000	6.694 22	23.9722	47.0173	73.4191	102.516	133.877	167.212	202.311	239.012
	6.827 95	24.2721	46.9000	72.9741	101.740	132.760	165.743	200.476	236.799

The tabulated energies can be used to calculate an approximate partition function at all temperatures:

$$Z \approx \sum_{n=0}^{\infty} e^{-\beta E_{\mathrm{app}}^{(n)}}. \qquad (5.169)$$

The resulting free energies $F_1' = -\log Z/\beta$ agree well with the previous variational results of the Feynman-Kleinert approximation — in the plots of Fig. 5.2, the curves are indistinguishable. This is not astonishing since both approximations are dominated near zero temperature by the same optimal energy $E_{\mathrm{app}}^{(0)}$, while approaching the semiclassical behavior at high temperatures. The previous free energy F_1 does so exactly, the free energy F_1' to a very good approximation.

By combining the Boltzmann factors with the oscillator wave functions $\psi_\Omega^{(n)}(x)$, we also calculate the density matrix $(x_b \tau_b | x_a \tau_a)$ and the distribution functions $\rho(x) = (x \, \hbar\beta | x \, 0)$ in this approximation:

$$(x_b \tau_b | x_a \tau_a) \approx \sum_{n=0}^{\infty} \psi^{(n)}(x_b) \psi^{(n)*}(x_a) e^{-\beta E_{\text{app}}^{(n)}}. \tag{5.170}$$

They are in general less accurate than the earlier calculated particle density $\rho_1(x_0)$ of (5.94).

5.13　Systematic Improvement of Feynman-Kleinert Approximation. Variational Perturbation Theory

A systematic improvement of the variational approach leads to a convergent *variational perturbation expansion* for the effective classical potential of a quantum-mechanical system [27]. To derive it, we expand the action in powers of the deviations of the path from its average $x_0 = \bar{x}$:

$$\delta x(\tau) \equiv x(\tau) - x_0. \tag{5.171}$$

The expansion reads

$$\mathcal{A} = V(x_0) + \mathcal{A}_\Omega^{x_0} + \mathcal{A}_{\text{int}}'^{x_0}, \tag{5.172}$$

where $\mathcal{A}_\Omega^{x_0}$ is the quadratic action of the deviations $\delta x(\tau)$

$$\mathcal{A}_\Omega^{x_0} = \int_0^{\hbar/k_B T} d\tau \frac{M}{2} \left\{ [\delta\dot{x}(\tau)]^2 + \Omega^2(x_0)[\delta x(\tau)]^2 \right\}, \tag{5.173}$$

and $\mathcal{A}_{\text{int}}'^{x_0}$ contains all higher powers in $\delta x(\tau)$:

$$\mathcal{A}_{\text{int}}'^{x_0} = \int_0^{\hbar\beta} d\tau \left\{ \frac{g_2}{2!} [\delta x(\tau)]^2 + \frac{g_3}{3!} [\delta x(\tau)]^3 + \frac{g_4}{4!} [\delta x(\tau)]^4 + \dots \right\}. \tag{5.174}$$

The coupling constants are, in general, x_0-dependent:

$$g_i(x_0) = V^{(i)}(x_0) - \Omega^2 \delta_{i2}, \qquad V^{(i)}(x_0) \equiv \frac{d^i V(x_0)}{dx_0^i}. \tag{5.175}$$

For the anharmonic oscillator, they take the values

$$\begin{aligned} g_2(x_0) &= M[\omega^2 - \Omega^2(x_0)] + 3g x_0^2, \\ g_3(x_0) &= 6g x_0, \\ g_4(x_0) &= 6g. \end{aligned} \tag{5.176}$$

Introducing the parameter

$$r^2 = 2M(\omega^2 - \Omega^2)/g, \tag{5.177}$$

which has the dimension length square, we write $g_2(x_0)$ as

$$g_2(x_0) = g(r^2/2 + 3x_0^2). \tag{5.178}$$

With the decomposition (5.172), the Feynman-Kleinert approximation (5.32) to the effective classical potential can be written as

$$W_1^{\Omega}(x_0) = V(x_0) + F_{\Omega}^{x_0} + \frac{1}{\hbar\beta} \langle \mathcal{A}_{\text{int}}^{x_0} \rangle_{\Omega}^{x_0}. \tag{5.179}$$

To generalize this, we replace the local free energy $F_{\Omega}^{x_0}$ by $F_{\Omega}^{x_0} + \Delta F^{x_0}$, where ΔF^{x_0} denotes the local analog of the cumulant expansion (3.484). This leads to the variational perturbation expansion for the effective classical potential

$$
\begin{aligned}
V^{\text{eff cl}}(x_0) &= V(x_0) + F_{\Omega}^{x_0} + \frac{1}{\hbar\beta} \langle \mathcal{A}_{\text{int}}^{x_0} \rangle_{\Omega}^{x_0} \\
&\quad - \frac{1}{2!\hbar^2\beta} \langle \mathcal{A}_{\text{int}}^{x_0\,2} \rangle_{\Omega,c}^{x_0} + \frac{1}{3!\hbar^3\beta} \langle \mathcal{A}_{\text{int}}^{x_0\,3} \rangle_{\Omega,c}^{x_0} + \cdots
\end{aligned}
\tag{5.180}
$$

in terms of the connected expectation values of powers of the interaction:

$$\langle \mathcal{A}_{\text{int}}^{x_0\,2} \rangle_{\Omega,c}^{x_0} \equiv \langle \mathcal{A}_{\text{int}}^{x_0\,2} \rangle_{\Omega}^{x_0} - \langle \mathcal{A}_{\text{int}}^{x_0} \rangle_{\Omega}^{x_0\,2}, \tag{5.181}$$

$$\langle \mathcal{A}_{\text{int}}^{x_0\,3} \rangle_{\Omega,c}^{x_0} \equiv \langle \mathcal{A}_{\text{int}}^{x_0\,3} \rangle_{\Omega}^{x_0} - 3 \langle \mathcal{A}_{\text{int}}^{x_0\,2} \rangle_{\Omega}^{x_0} \langle \mathcal{A}_{\text{int}}^{x_0} \rangle_{\Omega}^{x_0} + 2 \langle \mathcal{A}_{\text{int}}^{x_0} \rangle_{\Omega}^{x_0\,3}, \tag{5.182}$$

$$\vdots$$

By construction, the infinite sum (5.180) is independent of the choice of the trial frequency $\Omega(x_0)$.

When truncating (5.180) after the Nth order, we obtain the approximation $W_N^{\Omega}(x_0)$ to the effective classical potential $V^{\text{eff cl}}(x_0)$. In many applications, it is sufficient to work with the third-order approximation

$$
\begin{aligned}
W_3^{\Omega}(x_0) &= V(x_0) + F_{\Omega}^{x_0} + \frac{1}{\hbar\beta} \langle \mathcal{A}_{\text{int}}'^{x_0} \rangle_{\Omega}^{x_0} \\
&\quad - \frac{1}{2\hbar^2\beta} \langle \mathcal{A}_{\text{int}}'^{x_0\,2} \rangle_{\Omega,c}^{x_0} + \frac{1}{6\hbar^3\beta} \langle \mathcal{A}_{\text{int}}'^{x_0\,3} \rangle_{\Omega,c}^{x_0}.
\end{aligned}
\tag{5.183}
$$

In contrast to (5.180), the truncated sums $W_N(x_0)$ *do* depend on $\Omega(x_0)$. Since the infinite sum is $\Omega(x_0)$-independent, the best truncated sum $W_N(x_0)$ should lie at the frequency $\Omega_N(x_0)$ where $W_N^{\Omega}(x_0)$ depends minimally on it. The optimal $\Omega_N(x_0)$ is called the *frequency of least dependence*. Thus we require for (5.183)

$$\frac{\partial W_3^{\Omega}(x_0)}{\partial \Omega(x_0)} = 0. \tag{5.184}$$

At the frequency $\Omega_3(x_0)$ fixed by this condition, Eq. (5.183) yields the desired third-order approximation $W_3(x_0)$ to the effective classical potential.

The explicit calculation of the expectation values on the right-hand side of (5.183) proceeds according to the rules of Section 3.18. We have to evaluate the Feynman integrals associated with all vacuum diagrams. These are composed of p-particle vertices carrying the coupling constant $g_p/p!\hbar$, and of lines representing the correlation function of the fluctuations introduced in (5.23):

$$\langle \delta x(\tau)\delta x(\tau')\rangle_\Omega^{x_0} = \frac{\hbar}{M}G'^{x_0}_\Omega(\tau - \tau')$$
$$= a^2(\tau - \tau', x_0). \tag{5.185}$$

Since this correlation function contains no zero frequency, diagrams of the type shown in Fig. 5.13 do not contribute. Their characteristic property is to fall apart when cutting a single line. They are called *one-particle reducible diagrams*. A vacuum subdiagram connected with the remainder by a single line is called a *tadpole diagram*, alluding to its biological shape. Tadpole diagrams do not contribute to the variational perturbation expansion since they vanish as a consequence of energy conservation: the connecting line ending in the vacuum, must have a vanishing frequency where the spectral representation of the correlation function $\langle \delta x(\tau)\delta x(\tau')\rangle_\Omega^{x_0}$ has no support, by construction.

The number of Feynman diagrams to be evaluated is reduced by ignoring at first all diagrams containing the vertices g_2. The omitted diagrams can be recovered from the diagrams without g_2-vertices by calculating the latter at the initial frequency ω, and by replacing ω by a modified local trial frequency,

$$\omega \to \tilde{\Omega}(x_0) \equiv \sqrt{\Omega^2(x_0) + g_2(x_0)/M}. \tag{5.186}$$

After this replacement, which will be referred to as the *square-root trick*, all diagrams are re-expanded in powers of g [remembering that $g_2(x_0)$ is by (5.178) proportional to g] up to the maximal power g^3.

5.14 Applications of Variational Perturbation Expansion

The third-order approximation $W_3(x_0)$ is far more accurate than $W_1(x_0)$. This will now be illustrated by performing the variational perturbation expansion for the an-

Figure 5.13 Structure of a one-particle reducible vacuum diagram. The dashed box encloses a so-called tadpole diagram. Such diagrams vanish in the present expansion since an ending line cannot carry any energy and since the correlation function $\langle \delta x(\tau)\delta x(\tau')\rangle$ contains no zero frequency.

harmonic oscillator and the double-well potential. The reason for the great increase in accuracy will become clear in Section 5.15, where we shall demonstrate that this expansion *converges* rapidly towards the exact result at *all* coupling strengths, in contrast to ordinary perturbation expansions which diverges even for arbitrarily small values of g.

5.14.1 Anharmonic Oscillator at $T = 0$

Consider first the case of zero temperature, where the calculation is simplest and the approximation should be the worst. At $T = 0$, only the point $x_0 = 0$ contributes, and $\delta x(\tau)$ coincides with the path itself $\equiv x(\tau)$. Thus we may omit the superscript x_0 in all equations, so that the interaction in (5.182) becomes writing it as

$$\mathcal{A}_{\text{int}} = \frac{g}{4} \int_0^{\hbar\beta} d\tau (r^2 x^2 + x^4). \tag{5.187}$$

The effect of the r^2-term is found by replacing the frequency ω in the original perturbation expansion for the anharmonic oscillator according to the square-root trick (5.186), which for $x_0 = 0$ is simply

$$\omega \to \tilde{\Omega} \equiv \sqrt{\Omega^2 + (\omega^2 - \Omega^2)} = \sqrt{\Omega^2 + gr^2/2M}. \tag{5.188}$$

After this replacement, all terms are re-expanded in powers of g. Finally, r^2 is again replaced by

$$r^2 \to \frac{2M}{g}(\omega^2 - \Omega^2). \tag{5.189}$$

Since the interaction is even in x, the zero-temperature expansion is automatically free of tadpole diagrams.

The perturbation expansion to third order was given in Eq. (3.551). With the above replacement it leads to the free energy

$$F = \frac{\hbar\tilde{\Omega}}{2} + \frac{g}{4} 3a^4 + \left(\frac{g}{4}\right)^2 42a^8 \frac{1}{\hbar\tilde{\Omega}} + \left(\frac{g}{4}\right)^3 4 \cdot 333a^{12} \left(\frac{1}{\hbar\tilde{\Omega}}\right)^2, \tag{5.190}$$

with

$$a^2 = \frac{\hbar}{2M\tilde{\Omega}}. \tag{5.191}$$

The higher orders can most easily be calculated with the help of the Bender-Wu recursion relations derived in Appendix 3C.[2]

By expanding F in powers of g up to g^3 we obtain

$$W_3^\Omega = \frac{\hbar\Omega}{2} + g(3a^4 - r^2 a^2)/4 - g^2(21a^8/8 + 3a^6 r^2/4 + a^4 r^4/16)/\hbar\Omega$$
$$+ g^3(333a^{12}/16 + 105a^{10}r^2/16 + 3a^8 r^4/4 + a^6 r^6/32)/(\hbar\Omega)^2. \tag{5.192}$$

After the replacement (5.189), we minimize W_3 in Ω, and obtain the third-order

---

[2] The Mathematica program is available on the internet under `http://www.physik.fu-berlin/.de~kleinert/294/programs`.

Table 5.6 Second- and third-order approximations to ground state energy, in units of $\hbar\omega$, of anharmonic oscillator at various coupling constants g in comparison with exact values $E_{\mathrm{ex}}^{(0)}(g)$ and the Feynman-Kleinert approximation $E_1^{(0)}(g)$ of previous section.

$g/4$	$E_{\mathrm{ex}}^{(0)}(g)$	$E_1^{(0)}(g)$	$E_2^{(0)}(g)$	$E_3^{(0)}(g)$
0.1	0.559146	0.560307371	0.559152139	0.559154219
0.2	0.602405	0.604900748	0.602450713	0.602430621
0.3	0.637992	0.641629862	0.638088735	0.638035760
0.4	0.668773	0.673394715	0.668922455	0.668834137
0.5	0.696176	0.701661643	0.696376950	0.696253632
0.6	0.721039	0.727295668	0.721288789	0.721131776
0.7	0.743904	0.750857818	0.744199436	0.744010317
0.8	0.765144	0.772736359	0.765483301	0.765263697
0.9	0.785032	0.793213066	0.785412037	0.785163494
1	0.803771	0.812500000	0.804190095	0.803914053
10	1.50497	1.53125000	1.50674000	1.50549750
50	2.49971	2.54758040	2.50312133	2.50069963
100	3.13138	3.19244404	3.13578530	3.13265656
500	5.31989	5.42575605	5.32761969	5.32211709
1000	6.69422	6.82795331	6.70400326	6.69703286

approximation $E_3^{(0)}(g)$ to the ground state energy. Its accuracy at various coupling strengths is seen in Table 5.6 where it is compared with the exact values obtained from numerical solutions of the Schrödinger equation. The improvement with respect to the earlier approximation W_1 is roughly a factor 50. The maximal error is now smaller than 0.05%.

We shall see in the last subsection that up to rather high orders N, the minimum happens to be unique.

Observe that when truncating the expansion (5.183) after the second order and working with the approximation $W_2^\Omega(x_0)$, there exists no minimum in Ω, as can be seen in Fig. 5.14. The reason for this is the alternating sign of the cumulants in (5.183). This gives an alternating sign to the highest power a and thus to the highest power of $1/\Omega$ in the g^n-terms of Eq. (5.192), causing the trial energy of order N to diverge for $\Omega \to 0$ like $(-1)^{N-1} g^N \times (1/\Omega)^{3N-1}$. Since the trial energy goes for large Ω to positive infinity, only the odd approximations are guaranteed to possess a minimum.

The second-order approximation W_2 can nevertheless be used to find an improved energy value. As shown by Fig. 5.14, the frequency of least dependence Ω_2 is well defined. It is the frequency where the Ω-dependence of W_2 has its minimal absolute value. Thus we optimize Ω with the condition

$$\frac{\partial^2 W_2^\Omega}{\partial \Omega^2} = 0. \tag{5.193}$$

This leads to the energy values $E_2^{(0)}(g)$ listed in Table 5.6. They are more accurate than the values $E_1^{(0)}(g)$ by an order of magnitude.

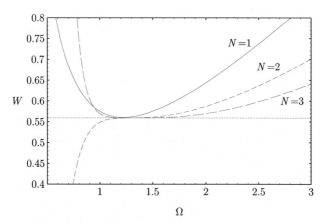

Figure 5.14 Typical Ω-dependence of approximations $W_{1,2,3}$ at $T = 0$. The coupling constant has the value $g = 0.4$. The second-order approximation W_2^Ω has no extremum. Here the minimal Ω-dependence lies at the turning point, and the condition $\partial^2 W_2^\Omega/\partial\Omega^2$ renders the best approximation to the energy (short dashes).

5.14.2 Anharmonic Oscillator for $T > 0$

Consider now the anharmonic oscillator at a finite temperature, where the expansion (5.183) consists of the sum of one-particle irreducible vacuum diagrams

$$W_3^{\tilde{\Omega}} = \frac{1}{\beta}\left\{ -\frac{1}{2} \overset{\tilde{\Omega}}{\bigcirc} + \underset{3}{\overset{\tilde{\Omega}}{\bigcirc\bigcirc}} \right.$$

$$-\frac{1}{2!}\left(\underset{72}{\overset{\tilde{\Omega}}{\bigcirc\bigcirc\bigcirc}} + \underset{24}{\overset{\tilde{\Omega}}{\ominus}} + \underset{6}{\overset{\tilde{\Omega}}{\ominus}} \right) \tag{5.194}$$

$$+\frac{1}{3!}\left(\underset{2592}{\overset{\tilde{\Omega}}{\bigcirc\bigcirc\bigcirc\bigcirc}} + \underset{1728}{\overset{\tilde{\Omega}}{\bigcirc}} + \underset{3456}{\overset{\tilde{\Omega}}{\ominus}} + \underset{1728}{\overset{\tilde{\Omega}}{\triangle}} + \underset{648}{\overset{\tilde{\Omega}}{\triangle}} + \underset{648}{\overset{\tilde{\Omega}}{\ominus}} \right) \left. \right\}.$$

The vertices represent the couplings $g_3(x_0)/3!\hbar$, $g_4(x_0)/4!\hbar$, whereas the lines stand for the correlation function $G''^{x_0}_{\tilde{\Omega}}(\tau, \tau')$. The numbers under each diagram are their multiplicities acting as factors.

Only the five integrals associated with the diagrams

$$\bigcirc \quad ; \quad \ominus \quad ; \quad \ominus \quad ; \quad \triangle \quad ; \quad \triangle$$

need to be evaluated explicitly; all others arise by the expansion of $\tilde{\Omega}$ or by factorization. The explicit form of three of these integrals can be found in the first, forth, and sixth of Eqs. (3.547). The results of the integrations are listed in Appendix 5A.

Only the second and fourth diagrams are new since they involve vertices with three legs. They can be found in Eqs. (3D.7) and (3D.8).

In quantum field theory one usually calculates Feynman integrals in momentum space. At finite temperatures, this requires the evaluation of multiple sums over Matsubara frequencies. The present quantum-mechanical example corresponds to a $D = 1$ -dimensional quantum field theory. Here it is more convenient to evaluate the integrals in τ-space. The diagrams

for example, are found by performing the integrals

$$\hbar\beta a^2 \equiv \int_0^{\hbar\beta} d\tau\, G^2(\tau, \tau),$$

$$\hbar\beta \left(\frac{1}{\omega}\right)^2 a_3^{12} \equiv \int_0^{\hbar\beta}\int_0^{\hbar\beta}\int_0^{\hbar\beta} G^2(\tau_1, \tau_2)G^2(\tau_2, \tau_3)G^2(\tau_3, \tau_1)\, d\tau_1 d\tau_2 d\tau_3.$$

The factor $\hbar\beta$ on the left-hand side is due to an overall τ-integral and reflects the temporal translation symmetry of the system; the factors $1/\omega$ arise from the remaining τ-integrations whose range is limited by the correlation time $1/\omega$.

In general, the β- and x_0-dependent parameters a_V^{2L} have the dimension of a length to the nth power and the associated diagrams consist of m vertices and $n/2$ lines (defining $a_1^2 \equiv a^2$).

We now use the rule (5.188) to replace ω by $\tilde{\Omega}$ and expand everything in powers of g_2 up to the third order. The expansion can be performed diagrammatically in each Feynman diagram. Letting a dot on a line indicate the coupling $g_2/2\hbar$, the one-loop diagram is expanded as follows:

$$\overset{\tilde{\Omega}}{\bigcirc} = \bigcirc - 2\,\bigcirc + \bigcirc - \frac{1}{3}\,\bigcirc. \tag{5.195}$$

The other diagrams are expanded likewise:

$$\overset{\tilde{\Omega}}{\infty} = \infty - \infty + \frac{1}{6}\,\infty + \frac{1}{6}\,\infty,$$

$$\frac{1}{2}\overset{\tilde{\Omega}}{\ominus} = \frac{1}{2}\ominus - \frac{1}{6}\ominus,$$

$$\frac{1}{2}\overset{\tilde{\Omega}}{\ominus} = \frac{1}{2}\ominus - \frac{1}{6}\ominus,$$

$$\frac{1}{2}\ \overset{\tilde{\Omega}}{\bigcirc\!\!\bigcirc\!\!\bigcirc}\ =\ \frac{1}{2}\ \bigcirc\!\!\bigcirc\!\!\bigcirc\ -\frac{1}{6}\ \bigcirc\!\!\bigcirc\!\!\bigcirc\ -\frac{1}{6}\ \bigcirc\!\!\bigcirc\!\!\bigcirc\ .$$

In this way, we obtain from (5.194) the complete graphical expansion for $W_3^\Omega(x_0)$ including all vertices associated with the coupling $g_2(x_0)$:

$$
\begin{aligned}
\beta W_3^\Omega(x_0) = &-\frac{1}{2}\,\bigcirc + \left(\frac{1}{2}\,\bigcirc + \frac{1}{8}\,\bigcirc\!\!\bigcirc\right)\\
&-\frac{1}{2!}\left[\frac{1}{2}\,\bigcirc + \frac{1}{2}\,\bigcirc\!\!\bigcirc + \frac{1}{8}\,\bigcirc\!\!\bigcirc\!\!\bigcirc + \frac{1}{24}\,\bigoplus + \frac{1}{6}\,\ominus\right]\\
&+\frac{1}{3!}\left[\bigcirc + 3\left(\frac{1}{4}\,\bigcirc\!\!\bigcirc + \frac{1}{2}\,\bigcirc\!\!\bigcirc\right)\right.\\
&\qquad +3\left(\frac{1}{4}\,\bigcirc\!\!\bigcirc\!\!\bigcirc + \frac{1}{4}\,\bigcirc\!\!\bigcirc\!\!\bigcirc + \frac{1}{6}\,\bigoplus + \frac{1}{2}\,\ominus\right)\\
&\qquad +\left(\frac{3}{16}\,\bigcirc\!\!\bigcirc\!\!\bigcirc\!\!\bigcirc + \frac{1}{8}\,Y + \frac{1}{4}\,\ominus + \frac{1}{8}\,\triangle\right.\\
&\qquad\qquad\left.\left. +\frac{3}{4}\,\diamond + \frac{3}{4}\,\ominus\right)\right].
\end{aligned}
\tag{5.196}
$$

In the latter diagrams, the vertices represent directly the couplings $g_n/\hbar$. The denominators $n!$ of the previous vertices $g_n/n!\hbar$ have been combined with the multiplicities of the diagrams yielding the indicated prefactors. The corresponding analytic expression for $W_3^\Omega(x_0)$ is

$$
\begin{aligned}
W_3^\Omega(x_0) = {}& V(x_0) + F_\Omega^{x_0} + \left(\frac{g_2}{2}a^2 + \frac{g_4}{8}a^4\right)\\
&-\frac{1}{2!\hbar\Omega}\left[\frac{g_2^2}{2}a_2^4 + \frac{g_2 g_4}{2}a_2^4 a^2 + \frac{g_4^2}{8}a_2^4 a^4 + \frac{g_4^2}{24}a_2^8 + \frac{g_3^2}{6}a_2^6\right]\\
&+\frac{1}{3!\hbar^2\Omega^2}\left[g_2^3 a_3^6 + 3\left(\frac{g_2^2 g_4}{4}(a_2^4)^2 + \frac{g_2^2 g_4}{2}a_3^6 a^2\right)\right.\\
&\qquad +3\left(\frac{g_2 g_4^2}{4}(a_2^4)^2 a^2 + \frac{g_2 g_4^2}{4}a_3^6 a^4 + \frac{g_2 g_4^2}{6}a_3^{10} + \frac{g_2 g_3^2}{2}a_3^8\right)\\
&\qquad +\left(\frac{3g_4^3}{16}(a_2^4 a^2)^2 + \frac{g_4^3}{8}a_3^6 a^6 + \frac{g_4^3}{4}a_3^{10}a^2 + \frac{g_4^3}{8}a_3^{12} + \frac{3g_3^2 g_4}{4}a_{3'}^{10} + \frac{3g_3^2 g_4}{4}a_3^8 a^2\right)\biggr].
\end{aligned}
\tag{5.197}
$$

The quantities a_V^{2L} are ordered in the same way as the associated diagrams in (5.196). As before, we have omitted the variable x_0 in all but the first three terms, for brevity.

The optimal trial frequency $\Omega(x_0)$ is found numerically by searching, at each value of x_0, for the real roots of the first derivate of $W_3^\Omega(x_0)$ with respect to $\Omega(x_0)$. Just as for zero temperature, the solution happens to be unique.

By calculating the integral

$$Z_3 = \int_{-\infty}^{\infty} \frac{dx_0}{l_e(\hbar\beta)} e^{-W_3(x_0)/k_B T}, \tag{5.198}$$

we obtain the approximate free energy

$$F_3 = -\frac{1}{\beta} \ln Z_3. \tag{5.199}$$

The results are listed in Table 5.7 for various coupling constants g and temperatures. They are compared with the exact free energy

$$F_{\text{ex}} = -\frac{1}{\beta} \log \sum_n \exp(-\beta E_n), \tag{5.200}$$

whose energies E_n were obtained by numerically solving the Schrödinger equation.

g	β	F_1	F_3	F_{ex}
0.002	2.0	0.427937	0.427937	0.427741
0.4	1.0	0.226084	0.226075	0.226074
	5.0	0.559155	0.558678	0.558675
2.0	1.0	0.492685	0.492578	0.492579
	5.0	0.699431	0.696180	0.696118
	10.0	0.700934	0.696285	0.696176
4.0	1.0	0.657396	0.6571051	0.6571049
	5.0	0.809835	0.803911	0.803758
20	1.0	1.18102	1.17864	1.17863
	5.0	1.24158	1.22516	1.22459
	10.0	1.24353	1.22515	1.22459
200	5.0	2.54587	2.50117	2.49971
2000	0.1	2.6997	2.69834	2.69834
	1.0	5.40827	5.32319	5.31989
	10.0	5.4525	5.3225	5.3199
80000	0.1	18.1517	18.0470	18.0451
	3.0	18.501	18.146	18.137

Table 5.7 Free energy of anharmonic oscillator with potential $V(x) = x^2/2 + gx^4/4$ for various coupling strengths g and $\beta = 1/kT$.

We see that to third order, the new approximation yields energies which are better than those of F_1 by a factor of 30 to 50. The remaining difference with respect to the exact energies lies in the fourth digit.

In the high-temperature limit, all approximations $W_N(x_0)$ tend to the classical result $V(x_0)$, as they should. Thus, for small β, the approximations $W_3(x_0)$ and $W_1(x_0)$ are practically indistinguishable.

The accuracy is worst at zero temperature. Using the $T \to 0$ -limits of the Feynman integrals a_V^{2L} given in (3.550) and (3D.14), the approximation W_3^{Ω} takes the simple form

$$
\begin{aligned}
W_3^{\Omega}(x_0) &= V(x_0) + \frac{\hbar\Omega(x_0)}{2} + \frac{g_2}{2}a^2 + \frac{g_4}{8}a^4 \\
&\quad - \frac{1}{2\hbar\Omega}\left[\frac{g_2^2}{2}a^4 + \frac{g_2 g_4}{2}a^6 + \frac{g_3^2}{6}\frac{2}{3}a^6 + \frac{g_4^2}{24}\frac{7}{2}a^8\right] \\
&\quad + \frac{1}{6\hbar^2\Omega^2}\left[g_2^3\frac{3}{2}a^6 + g_2 g_3^2\frac{4}{3}a^8 + g_2{}^2 g_4 3a^8 + g_3^2 g_4\frac{13}{3}a^{10} + g_2 g_4^2\frac{35}{16}a^{10} + g_4^3\frac{37}{64}a^{12}\right],
\end{aligned}
\tag{5.201}
$$

where $a^2 = \hbar/2M\Omega$. As in (5.197), we have omitted the arguments x_0 in Ω and g_i, for brevity. At zero temperature, the remaining integral over x_0 in the partition function (5.198) receives its only contribution from the point $x_0 = 0$, where $W_3(0)$ is minimal. There it reduces to the energy W_3 of Eq. (5.192).

5.15 Convergence of Variational Perturbation Expansion

For a single interaction x^p, the approximation W_N at zero temperature can easily be carried to high orders [13, 14]. The perturbation coefficients are available exactly from recursion relations, which were derived for the anharmonic oscillator with $p = 4$ in Appendix 3C. The starting point is the ordinary perturbation expansion for the energy levels of the anharmonic oscillator

$$
E^{(n)} = \omega \sum_{k=0}^{\infty} E_k^{(n)}\left(\frac{g}{4\omega^3}\right)^k.
\tag{5.202}
$$

It was remarked in (3C.27) and will be proved in Section 17.10 [see Eq. (17.323)] that the coefficients $E_k^{(n)}$ grow for large k like

$$
E_k^{(n)} \longrightarrow -\frac{1}{\pi}\sqrt{\frac{6}{\pi}}\frac{12^n}{n!}(-3)^k\Gamma(k+n+1/2).
\tag{5.203}
$$

Using Stirling's formula[3]

$$
n! \approx (2\pi)^{1/2}n^{n-1/2}e^{-n},
\tag{5.204}
$$

this amounts to

$$
E_k^{(n)} \longrightarrow -\frac{2\sqrt{3}}{\pi}\frac{(-4)^n}{n!}\left[\frac{-3(k+n)}{e}\right]^{n+k}.
\tag{5.205}
$$

Thus, $E_k^{(n)}$ grows faster than any power in k. Such a strong growth implies that the expansion has a zero radius of convergence. It is a manifestation of the fact that the energy possesses an essential singularity in the complex g-plane at the expansion point $g = 0$. The series is a so-called asymptotic series. The precise form of the

---

[3]M. Abramowitz and I. Stegun, op. cit., Formulas 6.1.37 and 6.1.38.

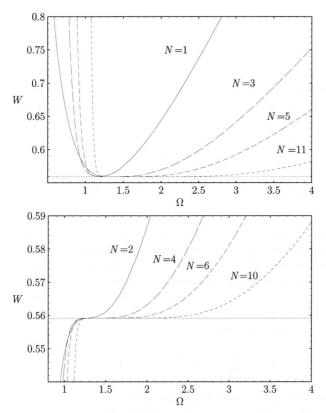

Figure 5.15 Typical Ω-dependence of Nth approximations W_N at $T = 0$ for increasing orders N. The coupling constant has the value $g/4 = 0.1$. The dashed horizontal line indicates the exact energy.

singularity will be calculated in Section 17.10 with the help of the semiclassical approximation.

If we want to extract meaningful numbers from a divergent perturbation series such as (5.202), it is necessary to find a convergent resummation procedure. Such a procedure is supplied by the variational perturbation expansion, as we now demonstrate for the ground state energy of the anharmonic oscillator.

Truncating the infinite sum (5.202) after the Nth term, the replacement (5.188) followed by a re-expansion in powers of g up to order N leads to the approximation W_N at zero temperature:

$$W_N^\Omega = \Omega \sum_{l=0}^{N} \varepsilon_l^{(0)} \left(\frac{g}{4\Omega^3}\right)^l, \tag{5.206}$$

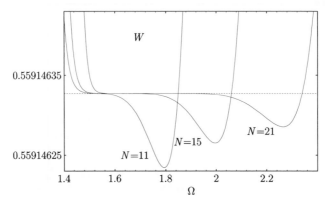

Figure 5.16 New plateaus in W_N developing for higher orders $N \geq 15$ in addition to the minimum which now gives worse results. For $N = 11$ the new plateau is not yet extremal, but it is the proper region of least Ω-dependence yielding the best approximation to the exact energy indicated by the dashed horizontal line. The minimum has fallen far below this value and is no longer useful. The figure looks similar for all couplings (in the plot, $g = 0.4$). The reason is the scaling property (5.215) proved in Appendix 5B.

with the re-expansion coefficients

$$\varepsilon_l^{(0)} = \sum_{j=0}^{l} E_j^{(0)} \binom{(1 - 3j)/2}{l - j} (-4\sigma)^{l-j}. \tag{5.207}$$

Here σ denotes the dimensionless function of Ω

$$\sigma \equiv -\frac{1}{2}\Omega r^2 = \frac{\Omega(\Omega^2 - 1)}{g}. \tag{5.208}$$

In Fig. 5.15 we have plotted the Ω-dependence of W_N for increasing N at the coupling constant $g/4 = 0.1$. For odd and even N, an increasingly flat plateau develops at the optimal energy.

At larger orders $N \geq 15$, the initially flat plateau is deformed into a minimum with a larger curvature and is no longer a good approximation. However, a new plateau has developed yielding the best energy. This is seen on the high-resolution plot in Fig. 5.16. At $N = 11$, the new plateau is not yet extremal but close to the correct energy.

The worsening extrema in Fig. 5.16 correspond here to points leaving the optimal dashed into the upward direction. The newly forming plateaus lie always on the dashed curve.

The set of all extremal Ω_N-values for odd N up to $N = 91$ is shown in Fig. 5.17. The optimal frequencies with smallest curvature are marked by a fat dot. In Subsection 17.10.5. we shall derive that

$$\sigma_N = \frac{\Omega_N(\Omega_N^2 - 1)}{g} \tag{5.209}$$

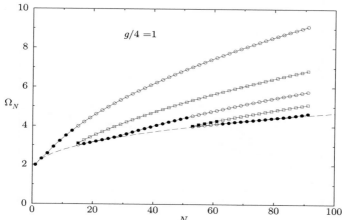

Figure 5.17 Trial frequencies Ω_N extremizing the variational approximation W_N at $T = 0$ for odd $N \leq 91$. The coupling is $g/4 = 1$. The dashed curve corresponds to the approximation (5.211) [related to Ω_N via (5.208)]. The frequencies on this curve produce the fastest convergence. The worsening extrema in Fig. 5.16 correspond here to points leaving the optimal dashed into the upward direction. The newly forming plateaus lie always on the dashed curve.

grows for large N like

$$\sigma_N \approx cN, \quad c = 0.186047\ldots\,, \tag{5.210}$$

so that Ω_N grows like $\Omega_N \approx (cNg)^{1/3}$. For smaller N, the best Ω_N-values in Fig. 5.17 can be fitted with the help of the corrected formula (5.210):

$$\sigma_N \approx cN \left(1 + \frac{6.85}{N^{2/3}}\right). \tag{5.211}$$

The associated Ω_N-curve is shown as a dashed line. It is the lower envelope of the extremal frequencies.

The set of extremal and turning point frequencies Ω_N is shown in Fig. 5.18 for even and odd N up to $N = 30$. The optimal extrema with smallest curvature are again marked by a fat dot. The theoretical curve for an optimal convergence calculated from (5.211) and (5.209) is again plotted as a dashed line.

In Table 5.8, we illustrate the precision reached for large orders N at various coupling constants g by a comparison with accurate energies derived from numerical solutions of the Schrödinger equation.

The approach to the exact energy values is illustrated in Fig. 5.19 which shows that a good convergence is achieved by using the lowest of all extremal frequencies, which lie roughly on the dashed theoretical curve in Fig. 5.17 and specify the position of the plateaus. The frequencies Ω_N on the higher branches leaving the dashed curve in that figure, on the other hand, do not yield converging energy values. The

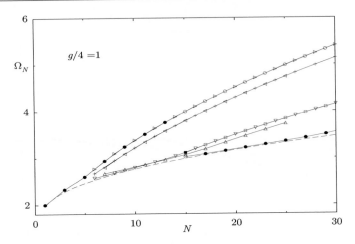

Figure 5.18 Extremal and turning point frequencies Ω_N in variational approximation W_N at $T = 0$ for even and odd $N \leq 30$. The coupling is $g/4 = 1$. The dashed curve corresponds to the approximation (5.211) [related to Ω_N via (5.208)].

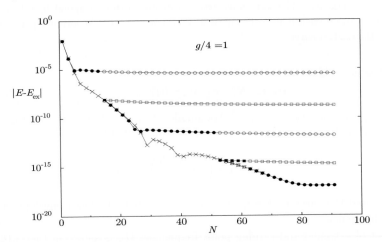

Figure 5.19 Difference between approximate ground state energies $E = W_N$ and exact energies E_{ex} for odd N corresponding to the Ω_N-values shown in Fig. 5.17. The coupling is $g/4 = 1$. The lower curve follows roughly the error estimate to be derived in Eq. (17.409). The extrema in Fig. (5.17) which move away from the dashed curve lie here on horizontal curves whose accuracy does not increase.

sharper minima in Fig. 5.16 correspond precisely to those branches which no longer determine the region of weakest Ω-dependence.

The anharmonic oscillator has the remarkable property that a plot of the Ω_N-values in the N, σ_N-plane is universal in the coupling strengths g; the plots do not depend on g. To see the reason for this, we reinsert explicitly the frequency ω (which

N	$g/4 =0.1$	$g/4 = 0.3$	$g/4 =0.5$	$g/4 =1.0$	$g/4 =2.0$
1	0.5603073711	0.6416298621	0.7016616429	0.8125000000	0.9644035598
2	0.5591521393	0.6380887347	0.6963769499	0.8041900946	0.9522936298
3	0.5591542188	0.6380357598	0.6962536326	0.8039140528	0.9517997694
4	0.5591457408	0.6379878713	0.6961684978	0.8037563457	0.9515444198
5	0.5591461596	0.6379899084	0.6961717475	0.8037615232	0.9515517450
10	0.5591463266	0.6379917677	0.6961757782	0.8037705329	0.9515682249
15	0.5591463272	0.6379917838	0.6961758231	0.8037706596	0.9515684933
20	0.5591463272	0.6379917836	0.6961743059	0.8037706575	0.9515684887
25	0.5591463272	0.6379917832	0.6961758208	0.8037706513	0.9515584121
exact	0.5591463272	0.6379917832	0.6961758208	0.8037706514	0.9515684727

N	$g/4 =50$	$g/4 =200$	$g/4 =1000$	$g/4 =8000$	$g/4 =20000$
1	2.5475803996	4.0084608812	6.8279533136	13.635282593	18.501658712
2	2.5031213253	3.9365586048	6.7040032606	13.386598486	18.163979967
3	2.5006996279	3.9325538203	6.6970328638	13.372561189	18.144908389
4	2.4995980125	3.9307488127	6.6930036178	13.366269038	18.136361642
5	2.4996213227	3.9307857892	6.6939667971	13.366395347	18.136533060
10	2.4997071960	3.9309286743	6.6942161680	13.366898079	18.137216200
15	2.4997089403	3.9309316283	6.6942213631	13.366908583	18.137230481
20	2.4997089079	3.9309315732	6.6942212659	13.366908387	18.137230214
25	2.4997087731	3.9309313396	6.6942208522	13.366907551	18.137230022
exact	2.4997087726	3.9309313391	6.6942208505	13.366907544	18.137229073

Table 5.8 Comparison of the variational approximations W_N at $T = 0$ for increasing N with the exact ground state energy at various coupling constants g.

was earlier set equal to unity). Then the re-expanded energy W_N in Eq. (5.206) has the general scaling form

$$W_N^\Omega = \Omega w_N(\hat{g}, \hat{\omega}^2),\qquad(5.212)$$

where w_N is a dimensionless function of the reduced coupling constant and frequency

$$\hat{g} \equiv \frac{g}{\Omega^3},\qquad \hat{\omega} \equiv \frac{\omega}{\Omega},\qquad(5.213)$$

respectively. When differentiating (5.212),

$$\frac{d}{d\Omega}W_N = \left[1 - 3\hat{g}\frac{d}{d\hat{g}} - 2\hat{\omega}^2\frac{d}{d\hat{\omega}^2}\right] w_N(\hat{g}, \hat{\omega}^2),\qquad(5.214)$$

we discover that the right-hand side can be written as a product of $\hat{g}^N$ and a dimensionless polynomial of order N depending only on $\sigma = \Omega(\Omega^2 - \omega^2)/g$:

$$\frac{d}{d\Omega}W_N^\Omega = \hat{g}^N p_N(\sigma).\qquad(5.215)$$

A proof of this will be given in Appendix 5B for any interaction x^p. The universal optimal σ_N-values are obtained from the zeros of $p_N(\sigma)$.

It is possible to achieve the same universality for the optimal frequencies of the even approximations W_N by determining them from the extrema of $p_N(\sigma)$ rather than from the turning points of W_N as a function of Ω.

The universal functions $p_N(\sigma)$ are found most easily by replacing the variable σ in the coefficients $\varepsilon_l^{(0)}$ of the re-expansion (5.206) by its $\omega = 0$ -limit $\sigma|_{\omega=0} = \Omega^3/g = 1/\hat{g}$. This yields the simpler expression

$$W_N^\Omega = \Omega w_N(\hat{g},0) = \Omega\sum_{l=0}^{N}\varepsilon_l^{(0)}\left(\frac{\hat{g}}{4}\right)^l,\qquad(5.216)$$

with

$$\varepsilon_l^{(0)} = \sum_{j=0}^{l} E_j^{(0)} \left(\begin{array}{c} (1 - 3j)/2 \\ l - j \end{array} \right) (-4/\hat{g})^{l-j}. \tag{5.217}$$

The derivative of W_N with respect to Ω yields

$$p_N(\sigma) = \hat{g}^{-N} \left[1 - 3\hat{g} \frac{d}{d\hat{g}} \right] w_N(\hat{g}, 0) \Big|_{\hat{g}=1/\sigma}. \tag{5.218}$$

In Section 17.10, we show the re-expansion coefficients $\varepsilon_k^{(0)}$ in (5.207) to be for large k proportional to $E_k^{(0)}$:

$$\varepsilon_k^{(0)} \approx e^{-2\sigma_N} E_k^{(0)}, \qquad \sigma_N = \frac{\Omega_N(\Omega_N^2 - 1)}{g} \tag{5.219}$$

[see Eq. (17.396)]. Thus, at any fixed Ω, the re-expanded series has the same asymptotic growth as the original series with the same vanishing radius of convergence. The behavior (5.219) can be seen in Fig. 5.20(a) where we have plotted the logarithm of the absolute value of the kth term

$$S_k = \varepsilon_k^{(0)} \left(\frac{\hat{g}}{4} \right)^k \tag{5.220}$$

of the re-expanded perturbation series (5.202) for various optimal values Ω_N and $g = 40$. All curves show a growth $\propto k^k$. The terms in the original series start growing immediately (precocious growth). Those in the re-expanded series, on the other hand, decrease initially and go through a minimum before they start growing (retarded growth). The dashed curves indicate the analytically calculated asymptotic behavior (5.219).

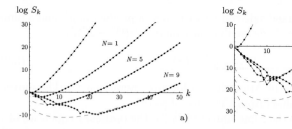

Figure 5.20   Logarithmic plot of kth terms in re-expanded perturbation series at a coupling constant $g/4 = 1$:
(a) Frequencies Ω_N extremizing the approximation W_N. The dashed curves indicate the theoretical asymptotic behavior (5.219).
(b) Frequencies Ω_N corresponding to the dashed curve in Fig. 5.17. The minima lie for each N precisely at $k = N$, producing the fastest convergence. The curves labeled $\Omega = \omega$ indicate the kth term in the original perturbation series.

The increasingly retarded growth is the reason why energies obtained from the variational expansion converge towards the exact result. Consider the terms S_k of the resummed series with frequencies Ω_N taken from the theoretical curve of optimal convergence in Fig. 5.17 (or 5.18). In Fig. 5.20(b) we see that the terms S_k are minimal at $k = N$, i.e., at the last term contained in the approximation W_N. In general, a divergent series yields an optimal result if it is truncated after the smallest term S_k. The size of the last term gives the order of magnitude of the error in the truncated evaluation. The re-expansion makes it possible to find, for every N, a frequency Ω_N which makes the truncation optimal in this sense.

5.16   Variational Perturbation Theory for Strong-Coupling Expansion

From the $\omega \to 0$ -limit of (5.206), we obtain directly the strong-coupling behavior of W_N. Since $\Omega = (g/\hat{g})^{1/3}$, we can write

$$W_N = (g/\hat{g})^{1/3} w_N(\hat{g}, 0), \qquad (5.221)$$

and evaluate this at the optimal value $\hat{g} = 1/\sigma_N$. The large-g behavior of W_N is therefore

$$W_N \xrightarrow[g\to\infty]{} (g/4)^{1/3} b_0, \qquad (5.222)$$

with the coefficient

$$b_0 = (4/\hat{g})^{1/3} w_N(\hat{g}, 0)|_{\hat{g}=1/\sigma_N}. \qquad (5.223)$$

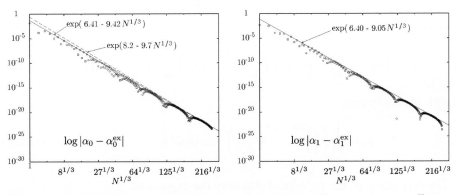

Figure 5.21   Logarithmic plot of N-behavior of strong-coupling expansion coefficients b_0 and b_1.

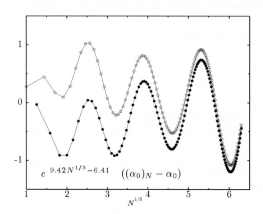

Figure 5.22 Oscillations of approximate strong-coupling expansion coefficient b_0 as a function of N when approaching exponentially fast the exact limit. The exponential behavior has been factored out. The upper and lower points show the odd-N and even-N approximations, respectively.

The higher corrections to the leading behavior (5.222) are found just as easily. By expanding $w_N(\hat{g}, \hat{\omega}^2)$ in powers of $\hat{\omega}^2$,

$$
W_N = \left(\frac{g}{\hat{g}} \right)^{1/3} \left[1 + \hat{\omega}^2 \frac{d}{d\hat{\omega}^2} + \frac{\hat{\omega}^4}{2!} \left(\frac{d}{d\hat{\omega}^2} \right)^2 + \ldots \right] w_N(\hat{g}, \hat{\omega}^2) \Bigg|_{\hat{\omega}^2 = 0}, \tag{5.224}
$$

and inserting

$$
\hat{\omega}^2 = \frac{\hat{g}^{2/3}}{(g/\omega^3)^{2/3}}, \tag{5.225}
$$

we obtain the expansion

$$
W_N = \left(\frac{g}{4} \right)^{1/3} \left[b_0 + b_1 \left(\frac{g}{4\omega^3} \right)^{-2/3} + b_2 \left(\frac{g}{4\omega^3} \right)^{-4/3} + \ldots \right], \tag{5.226}
$$

with the coefficients

$$
b_n = \frac{1}{n!} \left(\frac{\hat{g}}{4} \right)^{(2n-1)/3} \left(\frac{d}{d\hat{\omega}^2} \right)^n w_N(\hat{g}, \hat{\omega}^2) \Bigg|_{\hat{g}=1/\sigma_N, \hat{\omega}^2=0}. \tag{5.227}
$$

The derivatives on the right-hand side have the expansions

$$
\left(\frac{d}{d\hat{\omega}^2} \right)^n w_N(\hat{g}, \hat{\omega}^2) = \sum_{l=0}^{N} \varepsilon_{nl}^{(0)} \left(\frac{\hat{g}}{4} \right)^l, \tag{5.228}
$$

Table 5.9 Coefficients b_n of strong-coupling expansion of ground state energy of anharmonic oscillator obtained from a perturbation expansion of order 251. An extremely precise value for b_0 was given by F. Vinette and J. Čížek, J. Math. Phys. *32*, 3392 (1991): $b_0 = 0.667\,986\,259\,155\,777\,108\,270\,962\,016\,198\,601\,994\,304\,049\,36\ldots$.

n	b_n
0	$0.667\,986\,259\,155\,777\,108\,270\,96$
1	$0.143\,668\,783\,380\,864\,910\,020\,3$
2	$-\,0.008\,627\,565\,680\,802\,279\,128$
3	$0.000\,818\,208\,905\,756\,349\,543$
4	$-\,0.000\,082\,429\,217\,130\,077\,221$
5	$0.000\,008\,069\,494\,235\,040\,966$
6	$-\,0.000\,000\,727\,977\,005\,945\,775$
7	$0.000\,000\,056\,145\,997\,222\,354$
8	$-\,0.000\,000\,002\,949\,562\,732\,712$
9	$-\,0.000\,000\,000\,064\,215\,331\,954$
10	$0.000\,000\,000\,048\,214\,263\,787$
11	$-\,0.000\,000\,000\,008\,940\,319\,867$
12	$0.000\,000\,000\,001\,205\,637\,215$
13	$-\,0.000\,000\,000\,000\,130\,347\,650$
14	$0.000\,000\,000\,000\,010\,760\,089$
15	$-\,0.000\,000\,000\,000\,000\,445\,890\,1$
16	$-\,0.000\,000\,000\,000\,000\,058\,989\,8$
17	$0.000\,000\,000\,000\,000\,019\,196\,00$
18	$-\,0.000\,000\,000\,000\,000\,003\,288\,13$
19	$0.000\,000\,000\,000\,000\,000\,429\,62$
20	$-\,0.000\,000\,000\,000\,000\,000\,044\,438$
21	$0.000\,000\,000\,000\,000\,000\,003\,230\,5$
22	$-\,0.000\,000\,000\,000\,000\,000\,000\,031\,4$

with the re-expansion coefficients

$$\frac{1}{n!}\varepsilon_{nl}^{(0)} = \sum_{j=0}^{l-n} E_j^{(0)} \binom{(1-3j)/2}{l-j} \binom{l-j}{n} (-1)^{l-j-n}(4/\hat{g})^{l-j}. \tag{5.229}$$

For increasing N, the coefficients $b_0, b_1, \ldots$ converge rapidly against the values shown in Table 5.9. From the logarithmic plot in Fig. 5.21 we extract a convergence

$$|b_0 - b_0^{\mathrm{ex}}| \sim e^{8.2-9.7N^{1/3}}, \qquad |b_1 - b_1^{\mathrm{ex}}| \sim e^{6.4-9.1N^{1/3}}. \tag{5.230}$$

This behavior will be derived in Subsection 17.10.5, where we shall find that for any $g > 0$, the error decreases at large N roughly like $e^{-[9.7+(cg)^{-2/3}]\,N^{1/3}}$ [see Eq. (17.409)].

The approach to this limiting behavior is oscillatory, as seen in Fig. 5.22 where we have removed the exponential falloff and plotted $e^{-6.5+9.42N^{1/3}}(b_0 - b_0^{\mathrm{ex}})$ against N [16].

For the proof of the convergence of the variational perturbation expansion in Subsection 17.10.5. it will be important to know that the strong-coupling expansion for the ground state energy

$$E^{(0)} = \left(\frac{g}{4}\right)^{1/3} \left[b_0 + b_1 \left(\frac{g}{4\omega^3}\right)^{-2/3} + b_2 \left(\frac{g}{4\omega^3}\right)^{-4/3} + \ldots \right], \tag{5.231}$$

converges for large enough

$$g > g_{\mathrm{s}}. \tag{5.232}$$

The same is true for the excited energies.

5.17 General Strong-Coupling Expansions

The coefficients of the strong-coupling expansion can be derived for any divergent perturbation series

$$E_N(g) = \sum_{n=0}^{N} a_n g^n, \tag{5.233}$$

for which we know that it behaves at large couplings g like

$$E(g) = g^{p/q} \sum_{m=0}^{M} b_m (g^{-2/q})^m. \tag{5.234}$$

The series (5.233) can trivially be rewritten as

$$E_N(g) = \omega^p \sum_{n=0}^{N} a_n \left(\frac{g}{\omega^q}\right)^n, \tag{5.235}$$

with $\omega = 1$. We now apply the square-root trick (5.188) and replace ω by the identical expression

$$\omega \equiv \sqrt{\Omega^2 + (\omega^2 - \Omega^2)}, \tag{5.236}$$

containing a dummy scaling parameter Ω. The series (5.235) is then re-expanded in powers of g up to the order N, thereby treating $\omega^2 - \Omega^2$ as a quantity of order g. The result is most conveniently expressed in terms of dimensionless parameters $\hat{g} \equiv g/\Omega^q$ and $\sigma \equiv (1 - \hat{\omega}^2)/\hat{g}$, where $\hat{\omega} \equiv \omega/\Omega$. Then the replacement (5.236) amounts to

$$\omega \longrightarrow \Omega(1 - \sigma\hat{g})^{1/2}, \tag{5.237}$$

so that the re-expanded series reads explicitly

$$W_N(\hat{g}, \sigma) = \Omega^p \sum_{n=0}^{N} \varepsilon_n(\sigma) (\hat{g})^n, \tag{5.238}$$

with the coefficients:

$$\varepsilon_n(\sigma) = \sum_{j=0}^{n} a_j \binom{(p - qj)/2}{n - j} (-\sigma)^{n-j}. \tag{5.239}$$

For any fixed g, we form the first and second derivatives of $W_N^\Omega(g)$ with respect to Ω, calculate the Ω-values of the extrema and the turning points, and select the smallest of these as the optimal scaling parameter Ω_N. The function $W_N(g) \equiv W_N(g, \Omega_N)$ constitutes the Nth variational approximation $E_N(g)$ to the function $E(g)$.

We now take this approximation to the strong-coupling limit $g \to \infty$. For this we observe that (5.238) has the general scaling form

$$W_N^\Omega(g) = \Omega^p w_N(\hat{g}, \hat{\omega}^2). \tag{5.240}$$

For dimensional reasons, the optimal Ω_N increases with g for large g like $\Omega_N \approx g^{1/q} c_N$, so that $\hat{g} = c_N^{-q}$ and $\sigma = 1/\hat{g} = c_N^q$ remain finite in the strong-coupling limit, whereas $\hat{\omega}^2$ goes to zero like $1/[c_N(g/\omega^q)^{1/q}]^2$. Hence

$$W_N^{\Omega_N}(g) \approx g^{p/q} c_N^p w_N(c_N^{-q}, 0). \tag{5.241}$$

Here c_N plays the role of the variational parameter to be determined by the lowest extremum or turning point of $c_N^p w_N(c_N^{-q}, 0)$.

The full strong-coupling expansion is obtained by expanding $w_N(\hat{g}, \hat{\omega}^2)$ in powers of $\hat{\omega}^2 = (g/\omega^q \hat{g})^{-2/q}$ at a fixed $\hat{g}$. The result is

$$W_N(g) = g^{p/q} \left[\bar{b}_0(\hat{g}) + \bar{b}_1(\hat{g}) \left(\frac{g}{\omega^q} \right)^{-2/q} + \bar{b}_2(\hat{g}) \left(\frac{g}{\omega^q} \right)^{-4/q} + \ldots \right], \tag{5.242}$$

with

$$\bar{b}_n(\hat{g}) = \frac{1}{n!} \left(\frac{d}{d\hat{\omega}^2} \right)^n w_N^{(n)}(\hat{g}, \hat{\omega}^2) \Big|_{\hat{\omega}^2 = 0} \hat{g}^{(2n-p)/q}, \tag{5.243}$$

with respect to $\hat{\omega}^2$. Explicitly:

$$\frac{1}{n!} w_N^{(n)}(\hat{g}, 0) = \sum_{l=0}^{N} (-1)^{l+n} \sum_{j=0}^{l-n} a_j \binom{(p-qj)/2}{l-j} \binom{l-j}{n} (-\hat{g})^j. \tag{5.244}$$

Since $\hat{g} = c_N^{-q}$, the coefficients $\bar{b}_n(\hat{g})$ may be written as functions of the parameter c:

$$\bar{b}_n(c) = \sum_{l=0}^{N} a_l \sum_{j=0}^{N-l} \binom{(p-lq)/2}{j} \binom{j}{n} (-1)^{j-n} c^{p-lq-2n}. \tag{5.245}$$

The values of c which optimize $W_N(g)$ for fixed g yield the desired values of c_N. The optimization may be performed stepwise using directly the expansion coefficients $\bar{b}_n(c)$. First we optimize the leading coefficient $\bar{b}_0(c)$ as a function of c and identifying the smallest of them as c_N. Next we have to take into account that for large but finite α, the trial frequency Ω has corrections to the behavior $\hat{g}^{1/q} c$. The coefficient c will depend on $\hat{g}$ like

$$c(\hat{g}) = c + \gamma_1 \left(\frac{\hat{g}}{\omega^q}\right)^{-2/q} + \gamma_2 \left(\frac{\hat{g}}{\omega^q}\right)^{-4/q} + \dots, \qquad (5.246)$$

requiring a re-expansion of c-dependent coefficients $\bar{b}_n$ in (5.242). The expansion coefficients c and γ_n for $n = 1, 2, \dots$ are determined by extremizing $\bar{b}_{2n}(c)$. The final result can again be written in the form (5.242) with $\bar{b}(c)_n$ replaced by the final b_n:

$$W_N(g) = g^{p/q} \left[b_0 + b_1 \left(\frac{g}{\omega^q}\right)^{-2/q} + b_2 \left(\frac{g}{\omega^q}\right)^{-4/q} + \dots\right]. \qquad (5.247)$$

The final b_n are determined by the equations shown in Table 5.10. The two leading coefficients receive no correction and are omitted.

The extremal values of $\hat{g}$ will have a strong-coupling expansion corresponding to (5.246):

$$\hat{g} = c_N^{-q} \left[1 + \delta_1 \left(\frac{g}{\omega^q}\right)^{-2/q} + \delta_2 \left(\frac{g}{\omega^q}\right)^{-4/q} + \cdots\right]. \qquad (5.248)$$

Table 5.10 Equations determining coefficients b_n in strong-coupling expansion (5.247) from the functions $\bar{b}_n(c)$ in (5.245) and their derivatives. For brevity, we have suppressed the argument c in the entries.

n	b_n	$-\gamma_{n-1}$
2	$\bar{b}_2 + \gamma_1 \bar{b}_1' + \frac{1}{2}\gamma_1^2 \bar{b}_0''$	$\bar{b}_1'/\bar{b}_0''$
3	$\bar{b}_3 + \gamma_2 \bar{b}_1' + \gamma_1 \bar{b}_2' + \gamma_1\gamma_2 \bar{b}_0'' + \frac{1}{2}\gamma_1^2 \bar{b}_1'' + \frac{1}{6}\gamma_1^3 \bar{b}_0^{(3)}$	$(\bar{b}_2' + \gamma_1 \bar{b}_1'' + \frac{1}{2}\gamma_1^2 \bar{b}_0^{(3)})/\bar{b}_0''$
4	$\bar{b}_4 + \gamma_3 \bar{b}_1' + \gamma_2 \bar{b}_2' + \gamma_1 \bar{b}_3' + (\frac{1}{2}\gamma_2^2 + \gamma_1\gamma_3)\bar{b}_0''$	$(\bar{b}_3' + \gamma_2 \bar{b}_1'' + \gamma_1 \bar{b}_2'' + \gamma_1\gamma_2 \bar{b}_0^{(3)}$
	$+\gamma_1\gamma_2 \bar{b}_1'' + \frac{1}{2}\gamma_1^2 \bar{b}_2'' + \frac{1}{2}\gamma_1^2\gamma_2 \bar{b}_0^{(3)} + \frac{1}{6}\gamma_1^3 \bar{b}_1^{(3)} + \frac{1}{24}\gamma_1^4 \bar{b}_0^{(4)}$	$+\frac{1}{2}\gamma_1^2 \bar{b}_1^{(3)} + \frac{1}{6}\gamma_1^3 \bar{b}_0^{(4)})/\bar{b}_0''$

The convergence of the general strong-coupling expansion is similar to the one observed for the anharmonic oscillator. This will be seen in Subsection 17.10.5.

The general strong-coupling expansion has important applications in the theory of critical phenomena. This theory renders expansions of the above type for the so-called critical exponents, which have to be evaluated at infinitely strong (bare) couplings of scalar field theories with $g\phi^4$ interactions. The results of these applications are better than those obtained previously with a much more involved theory based on a combination of renormalization group equations and Padé-Borel resummation techniques [28]. The critical exponents have power series expansions in powers of g/ω in the physically most interesting three-dimensional systems, where $\omega^2/2$ is the factor in front of the quadratic field term ϕ^2. The important phenomenon observed in such systems is the appearance of *anomalous dimensions*. These imply that the expansion terms $(g/\omega)^n$ cannot simply be treated with the square-root trick (5.236). The anomalous dimension requires that $(g/\omega)^n$ must be treated as if it were $(g/\omega^q)^n$

when applying the square-root trick. Thus we must use the *anomalous square-root trick*

$$\omega \to \Omega \left[1 + \left(\frac{\omega}{\Omega} \right)^{2/q} \right]^{q/2} . \tag{5.249}$$

The power $2/q$ appearing in the strong-coupling expansion (5.234) is experimentally observable since it governs the approach of the system to the scaling limit. This exponent is usually denoted by the letter ω, and is referred to as the *Wegner exponent* [29]. This exponent ω is not to be confused with the frequency ω in the present discussion. In superfluid helium, for example, this critical exponent is very close to the value $4/5$, implying $q \approx 5/2$. The Wegner exponent of fluctuating quantum fields cannot be deduced, as in quantum mechanics, from simple scaling analyses of the action. It is, however, calculable by applying variational perturbation theory to the logarithmic derivative of the power series of the other critical exponents. These are called β-functions and have to vanish for the correct ω. This procedure is referred to as *dynamical determination* of ω and has led to values in excellent agreement with experiment [17].

In the above variational procedure, the existence of anomalous dimensions is signalized by the appearance of a nonzero slope in the plateaus of Figs. 5.14. If this happens, the power q in the replacement (5.249) must be modified until the plateaus are flat. This method is a practical alternative to determining the Wegner exponent $\omega = 2/q$ via the β-function.

5.18 Variational Interpolation between Weak and Strong-Coupling Expansions

The possibility of calculating the strong-coupling coefficients from the perturbation coefficients can be used to find a *variational interpolation* of a function with known weak- and strong-coupling coefficients [18]. Such pairs of expansions are known for many other physical systems, for example most lattice models of statistical mechanics [19]. If applied to the ground state energy of the anharmonic oscillator, this method converges exponentially fast [15]. The weak-coupling expansion of the ground state energy of the anharmonic oscillator has the form (5.233). In natural units with $\hbar = M = \omega = 1$, the lowest coefficient a_0 is trivially determined to be $a_0 = 1/2$ by the ground state energy of the harmonic oscillator. If we identify $\alpha = g/4$ with the coupling constant in (5.233), to save factors $1/4$, the first coefficient is $a_1 = 3/4$ [see (5.190)]. We have seen before in Section 5.12 that even the lowest order variational perturbation theory yields leading strong-coupling coefficient in excellent agreement with the exact one [with a maximal error of $\approx 2\%$, see Eq. (5.167)]. In Fig. 5.23 we have plotted the relative deviation of the variational approximation from the exact one in percent.

The strong-coupling behavior is known from (5.226). It starts out like $g^{1/3}$, followed by powers of $g^{-1/3}, g^{-1}, g^{-5/3}$. Comparison with (5.234) shows that this

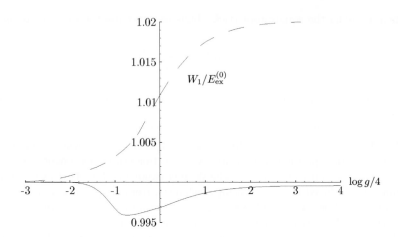

Figure 5.23 Ratio of approximate and exact ground state energy of anharmonic oscillator from lowest-order variational interpolation. Dashed curve shows first-order Feynman-Kleinert approximation $W_1(g)$. The accuracy is everywhere better than 99.5 %. For comparison, we also display the much worse (although quite good) variational perturbation result using the exact $a_1^{ex} = 3/4$.

corresponds to $p = 1$ and $q = 3$. The leading coefficient is given in Table 5.9 with extreme accuracy: $b_0 = 0.667\,986\,259\,155\,777\,108\,270\,962\,016\,919\,860\ldots$.

In a variational interpolation, this value is used to determine an approximate a_1 (forgetting that we know the exact value $a_1^{ex} = 3/4$). The energy (5.238) reads for $N = 1$ (with $\alpha = g/4$ instead of g):

$$W_1(\alpha, \Omega) = \left(\frac{\Omega}{2} + \frac{1}{2\Omega}\right) a_0 + \frac{a_1}{\Omega^2}\alpha. \tag{5.250}$$

Equation (5.245) yields, for $n = 0$:

$$b_0 = \frac{c}{2}a_0 + \frac{a_1}{c^2}. \tag{5.251}$$

Minimizing b_0 with respect to c we find $c = c_1 \equiv 2(a_1/2a_0)^{1/3}$ with $b_0 = 3a_0c_1/4 = 3(a_0^2 a_1/2)^{1/3}/2$. Inserting this into (5.251) fixes $a_1 = 2(2/3b_0)^3/a_0^2 = 0.773\,970\ldots$, quite close to the exact value $3/4$. With our approximate a_1 we calculate $W_1(\alpha, \Omega)$ at its minimum, where

$$\Omega_1 = \begin{cases} \frac{2}{\sqrt{3}}\omega \cosh\left[\frac{1}{3}\mathrm{acosh}(g/g^{(0)})\right] \\ \frac{2}{\sqrt{3}}\omega \cos\left[\frac{1}{3}\arccos(g/g^{(0)})\right] \end{cases} \text{for} \quad \begin{matrix} g > g^{(0)}, \\ g < g^{(0)}, \end{matrix} \tag{5.252}$$

with $g^{(0)} \equiv 2\omega^3 a_0/3\sqrt{3}a_1$. The result is shown in Fig. 5.23. Since the difference with respect to the exact solution would be too small to be visible on a direct plot of the energy, we display the ratio with respect to the exact energy $W_1(g)/E_{ex}^0$. The accuracy is everywhere better than 99.5 %.

5.19   Systematic Improvement of Excited Energies

The variational method for the energies of excited states developed in Section 5.12 can also be improved systematically. Recall the n-dependent level shift formulas (3.515) and (3.516), according to which

$$
\begin{aligned}
\Delta E^{(n)} &= \Delta_1 E^{(n)} + \Delta_2 E^{(n)} + \Delta_3 E^{(n)} = \Delta V_{nn} - \sum_{k \neq n} \frac{\Delta V_{nk} \Delta V_{kn}}{E_k - E_n} \\
&+ \sum_{k \neq n} \sum_{l \neq n} \frac{\Delta V_{nk} \Delta V_{kl} \Delta V_{ln}}{(E_k - E_n)(E_l - E_n)} - \Delta V_{nn} \sum_{k \neq n} \frac{\Delta V_{nk} \Delta V_{kn}}{(E_k - E_n)^2}.
\end{aligned}
\tag{5.253}
$$

By applying the substitution rule (5.188) to the total energies

$$
E^{(n)} = \hbar\Omega(n + 1/2) + \Delta E^{(n)},
$$

and expanding each term in powers of g up to g^3, we find the contributions to the level shift

$$
\Delta_1 E^{(n)} = \frac{g}{4} \left[3(2n^2 + 2n + 1)a^4 + (2n + 1)a^2 r^2 \right],
$$

$$
\begin{aligned}
\Delta_2 E^{(n)} = -\left(\frac{g}{4}\right)^2 &\left[2(34n^3 + 51n^2 + 59n + 21)a^8 \right. \\
&\left. + 4 \cdot 3(2n^2 + 2n + 1)a^6 r^2 + (2n + 1)a^4 r^4 \right] \frac{1}{\hbar\Omega},
\end{aligned}
$$

$$
\begin{aligned}
\Delta_3 E^{(n)} = \left(\frac{g}{4}\right)^3 &\left[4 \cdot 3(125n^4 + 250n^3 + 472n^2 + 347n + 111)a^{12} \right. \\
&+ 4 \cdot 5(34n^3 + 51n^2 + 59n + 21)a^{10} r^2 \\
&\left. + 16 \cdot 3(2n^2 + 2n + 1)a^8 r^4 + 2 \cdot (2n + 1)a^6 r^6 \right] \frac{1}{\hbar^2 \Omega^2},
\end{aligned}
\tag{5.254}
$$

which for $n = 0$ reduce to the corresponding terms in (5.192). The extremization in Ω leads to energies which lie only very little above the exact values for all n. This is illustrated in Table 5.11 for $n = 8$ (compare with the energies in Table 5.5). A sum over the Boltzmann factors $e^{-\beta E_3^{(n)}}$ produces an approximate partition function Z_3 which deviates from the exact one by less than 50.1%.

It will be interesting to use the improved variational approach for the calculation of density matrices, particle distributions, and magnetization curves.

Table 5.11 Higher approximations to excited energy with $n = 8$ of anharmonic oscillator at various coupling constants g. The third-order approximation $E_3^{(8)}(g)$ is compared with the exact values $E_{\text{ex}}^{(8)}(g)$, with the approximation $E_1^{(8)}(g)$ of the last section, and with the lower approximation of even order $E_2^{(8)}(g)$ (all in units of $\hbar\omega$).

$g/4$	$E_{\text{ex}}^{(8)}(g)$	$E_1^{(8)}(g)$	$E_2^{(8)}(g)$	$E_3^{(8)}(g)$
0.1	13.3790	13.3235257	13.3766211	13.3847643
0.2	15.8222	15.7327929	15.8135994	15.8275802
0.3	17.6224	17.5099190	17.6099785	17.6281810
0.4	19.0889	18.9591071	19.0742800	19.0958388
0.5	20.3452	20.2009502	20.3287326	20.3531080
0.6	21.4542	21.2974258	21.4361207	21.4629384
0.7	22.4530	22.2851972	22.4335694	22.4625543
0.8	23.3658	23.1879959	23.3451009	23.3760415
0.9	24.2091	24.0221820	24.1872711	24.2199988
1	24.9950	24.7995745	24.9720376	25.0064145
10	51.9865	51.5221384	51.9301030	51.9986710
50	88.3143	87.5058600	88.2154879	88.3500454
100	111.128	110.105819	111.002842	111.173183
500	189.756	188.001018	189.540577	189.833415
1000	239.012	236.799221	238.740320	239.109584

5.20   Variational Treatment of Double-Well Potential

Let us also calculate the approximate effective classical potential of third order $W_3(x_0)$ for the double-well potential

$$V(x) = M\frac{\omega^2}{2}x^2 + \frac{g}{4}x^4 + \frac{M^2\omega^4}{4g}, \qquad \omega^2 = -1. \qquad (5.255)$$

In the expression (5.197), the sign change of ω^2 affects only the coupling $g_2(x_0)$, which becomes

$$g_2(x_0) = M[-1 - \Omega^2(x_0)] + 3gx_0^2 = gr^2/2 + 3gx_0^2 \qquad (5.256)$$

[recall (5.176)]. Note the constant energy $M^2\omega^4/4g$ in $V(x)$ which shifts the minima of the potential to zero [compare (5.78)].

To see the improved accuracy of W_3 with respect to the first approximation $W_1(x_0)$ discussed in Section 5.7 [corresponding to the first line of (5.197)], we study the limit of zero temperature where the accuracy is expected to be the worst. In this limit, $W_3(x_0)$ reduces to (5.201) and is easily minimized in x_0 and Ω.

At larger coupling constants $g > g_c \approx 0.3$, the energy has a minimum at $x_0 = 0$. For $g \le g_c$, there is an additional symmetric pair of minima at $x_0 = \pm x_m \ne 0$ (recall Figs. 5.5 and 5.6). The resulting $W_3(0)$ is plotted in Fig. 5.24 together with $W_1(0)$. The figure also contains the first excited energy which is obtained by setting $\omega^2 = -1$ in $r^2 = 2M(\omega^2 - \Omega^2)/g$ of Eqs. (5.253)–(5.255).

For small couplings g, the energies $W_1(0)$, $W_3(0)$, ... diverge and the minima at $x = \pm x_m$ of Eq. (5.82) become relevant. Moreover, there is quantum tunneling across the central barrier from one minimum to the other which takes place for $g \le g_c \approx 0.3$ and is unaccounted for by $W_3(0)$ and $W_3(x_m)$. Tunneling leads to a level splitting to be calculated in Chapter 17. In this chapter, we test the accuracy of $W_1(x_m)$ and $W_3(x_m)$ by comparing them with the averages of the two lowest energies. Figure 5.24 shows that the accuracy of the approximation $W_3(x_m)$ is quite good.

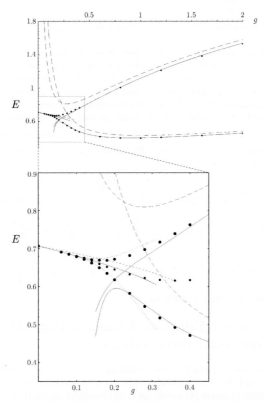

Figure 5.24 Lowest two energies in double-well potential as function of coupling strength g. The approximations are $W_1(0)$ (dashed line) and $W_3(0)$ (solid line). The dots indicate numeric results of the Schrödinger equation. The lower part of the figure shows $W_1(x_m)$ and $W_3(0)$ in comparison with the average of the Schrödinger energies (small dots). Note that W_1 misses the slope by 25%. Tunneling causes a level splitting to be calculated in Chapter 17 (dotted curves).

Note that the approximation $W_1(x_m)$ does not possess the correct slope in g, which is missed by 25%. In fact, a Taylor expansion of $W_1(x_m)$ reads

$$W_1(x_m) = \frac{\sqrt{2}}{2} - \frac{3}{16}g - \frac{9\sqrt{2}}{128}g^2 - \frac{27}{256}g^3 + \dots , \tag{5.257}$$

whereas the true expansion starts out with

$$E^{(0)} = \frac{\sqrt{2}}{2} - \frac{1}{4}g + \dots . \tag{5.258}$$

The optimal frequency associated with (5.257) has the expansion

$$\Omega_1(x_m) = \sqrt{2} - \frac{3}{4}g - \frac{27\sqrt{2}}{64}g^2 - \frac{27}{32}g^3 + \dots .$$

Let us also compare the x_0-behavior of $W_3(x_0)$ with that of the true effective classical potential calculated numerically by Monte Carlo simulations. The curves are plotted in Fig. 5.5, and the agreement is seen to be excellent. There are significant deviations only for low temperatures with $\beta \gtrsim 20$.

At zero temperature, there exists a simple way of recovering the effective classical potential from the classical potential calculated up to two loops in Eq. (3.767). As we learned in Eq. (3.830), x_0 coincides at zero temperature. with X in Eq. (3.767) Thus we merely have to employ the square-root trick (5.186) to the effective potential (3.767) and interchange X by x_0 to obtain the variational approximation $W_1(x_0)$ to the effective classical potantial to be varied. Explicitly, we replace, as in (5.188), the one-loop contribtions, ground state energies $\hbar\omega_{T,L}$ in (3.767) by $\sqrt{\Omega_{T,L}^2 + (\omega_{T,L}^2(X) - \Omega_{T,L}^2)} \approx \Omega_{T,L} + (\omega_{T,L}^2(X) - \Omega_{T,L}^2)/2\Omega_{T,L}$, and exchange $\omega_{L,T}(X)$ of the remaining two-loop terms by $\Omega_{T,L}$. This yields for the D-dimensional rotated double-well potential (the Mexical hat potential in Fig. 3.13]:

$$
\begin{aligned}
W(x_0) \underset{T\to 0}{=} &\left\{ v(X) + \hbar\left[\frac{\Omega_L}{4} + \frac{\omega_L^2(X)}{4\Omega_L}\right] + (D-1)\hbar\left[\frac{\Omega_T}{4} + \frac{\omega_T^2(X)}{4\Omega_T}\right] + \frac{\hbar^2}{8(2M)^2}\left\{\frac{1}{\Omega_L^2}v^{(4)}(X)\right.\right. \\
&+ \frac{D^2-1}{\Omega_T^2}\left[\frac{v''(X)}{X^2} - \frac{v'(X)}{X^3}\right] + \frac{2(D-1)}{\Omega_L\Omega_T}\left[\frac{v'''(X)}{X} - \frac{2v''(X)}{X^2} + \frac{2v'(X)}{X^3}\right]\right\} \\
&- \frac{\hbar^2}{6(2M)^3}\left\{\frac{1}{3\Omega_L^4}[v'''(X)]^2 + \frac{3(D-1)}{2\Omega_T+\Omega_L}\frac{1}{\Omega_L\Omega_T^2}\left[\frac{v''(X)}{X} - \frac{v'(X)}{X^2}\right]^2\right\} + \mathcal{O}(\hbar^3)\right\}_{X\to x_0}. \quad (5.259)
\end{aligned}
$$

This has to be extremized in $\Omega_{T,L}$ at fixed x_0.

5.21 Higher-Order Effective Classical Potential for Nonpolynomial Interactions

The systematic improvement of the Feynman-Kleinert approximation in Section 5.13 was based on Feynman diagrams and therefore applicable only to polynomial potentials. If we want to calculate higher-order effective classical potentials for nonpolynomial interactions such as the Coulomb interaction, we need a generalization of the smearing rule (5.30) to the correlation functions of interaction potentials which occur in the expansion (5.180). The second-order term, for example, requires the calculation of

$$
\langle \mathcal{A}_{\text{int}}^{x_0\,2} \rangle_{\Omega,c}^{x_0} = \int_0^{\hbar\beta} d\tau \int_0^{\hbar\beta} d\tau' \, \langle V_{\text{int}}^{x_0}(x(\tau)) V_{\text{int}}^{x_0}(x(\tau')) \rangle_{\Omega,c}^{x_0}, \quad (5.260)
$$

where

$$
V_{\text{int}}^{x_0}(x) = V(x) - \frac{1}{2}M\Omega^2(x_0)(x-x_0)^2. \quad (5.261)
$$

Thus we need an efficient smearing formula for local expectations of the form

$$
\begin{aligned}
\langle F_1(x(\tau_1))\ldots F_n(x(\tau_n)) \rangle_{\Omega}^{x_0} = &\frac{1}{Z_{\Omega}^{x_0}} \\
&\times \oint \mathcal{D}x(\tau) F_1(x(\tau_1)) \cdots F_n(x(\tau_n)) \delta(\bar{x}-x_0)\exp\left\{-\frac{1}{\hbar}\mathcal{A}_{\Omega}^{x_0}[x(\tau)]\right\}, \quad (5.262)
\end{aligned}
$$

where $\mathcal{A}_{\Omega}^{x_0}[x(\tau)]$ and $Z_{\Omega}^{x_0}$ are the local action and partition function of Eqs. (5.3) and (5.4). After rearranging the correlation functions to connected ones according to Eqs. (5.182) we find the cumulant expansion for the effective classical potential [see (5.180)]

$$
V^{\text{eff,cl}}(x_0) = F_{\Omega}^{x_0} + \frac{1}{\hbar\beta}\int_0^{\hbar\beta} d\tau_1 \, \langle V_{\text{int}}^{x_0}(x(\tau_1)) \rangle_{\Omega,c}^{x_0}
$$

$$-\frac{1}{2\hbar^2\beta}\int_0^{\hbar\beta}d\tau_1\int_0^{\hbar\beta}d\tau_2\,\langle V_{\text{int}}^{x_0}(x(\tau_1))V_{\text{int}}^{x_0}(x(\tau_2))\rangle_{\Omega,c}^{x_0} \tag{5.263}$$

$$+\frac{1}{6\hbar^3\beta}\int_0^{\hbar\beta}d\tau_1\int_0^{\hbar\beta}d\tau_2\int_0^{\hbar\beta}d\tau_3\,\langle V_{\text{int}}^{x_0}(x(\tau_1))V_{\text{int}}^{x_0}(x(\tau_2))V_{\text{int}}^{x_0}(x(\tau_3))\rangle_{\Omega,c}^{x_0}+\cdots.$$

It differs from the previous expansion (5.180) for polynomial interactions by the potential $V(x)$ not being expanded around x_0. The first term on the right-hand side is the local free energy (5.6).

5.21.1   Evaluation of Path Integrals

The local pair correlation function was given in Eq. (5.19):

$$\langle\delta x(\tau)\delta x(\tau')\rangle_\Omega^{x_0}\equiv\langle[x(\tau)-x_0][x(\tau')-x_0]\rangle_\Omega^{x_0}=\frac{\hbar}{M}G_\Omega^{(2)x_0}(\tau,\tau')=a_{\tau\tau'}^2(x_0), \tag{5.264}$$

with [recall (5.19)–(5.24)]

$$a_{\tau\tau'}^2(x_0)=\frac{\hbar}{2M\Omega(x_0)}\frac{\cosh[\Omega(x_0)(\tau-\tau'-\hbar\beta/2)]}{\sinh[\Omega(x_0)\hbar\beta/2]}-\frac{1}{M\beta\Omega^2(x_0)},\qquad\tau\in(0,\hbar\beta). \tag{5.265}$$

Higher correlation functions are expanded in products of these according to Wick's rule (3.302). For an even number of $\delta x(\tau)$'s one has

$$\langle\delta x(\tau_1)\delta x(\tau_2)\cdots\delta x(\tau_n)\rangle_\Omega^{x_0}=\sum_{\text{pairs}}a_{\tau_{p(1)}\tau_{p(2)}}^2(x_0)\cdots a_{\tau_{p(n-1)}\tau_{p(n)}}^2(x_0), \tag{5.266}$$

where the sum runs over all $(n-1)!!$ pair contractions. For an exponential, Wick's rule implies

$$\left\langle\exp\left[i\int_0^{\hbar\beta}d\tau j(\tau)\,\delta x(\tau)\right]\right\rangle_\Omega^{x_0}=\exp\left[-\frac{1}{2}\int_0^{\hbar\beta}d\tau\int_0^{\hbar\beta}d\tau'j(\tau)a_{\tau\tau'}^2(x_0)j(\tau')\right]. \tag{5.267}$$

Inserting $j(\tau)=\sum_{i=1}^n k_i\delta(\tau-\tau_i)$, this gives for the expectation value of a sum of exponentials

$$\left\langle\exp\left[i\sum_{i=1}^n k_i\,\delta x(\tau_i)\right]\right\rangle_\Omega^{x_0}=\exp\left[-\frac{1}{2}\sum_{i=1}^n\sum_{j=1}^n a_{\tau_i\tau_j}^2(x_0)k_ik_j\right]. \tag{5.268}$$

By Fourier-decomposing the functions $F(x(\tau))=\int(dk/2\pi)F(k)\exp ik\,[x_0+\delta x(\tau)]$ in (5.262), we obtain from (5.268) the new smearing formula

$$\langle F_1(x(\tau_1))\cdots F_n(x(\tau_n))\rangle_\Omega^{x_0}=\left\{\prod_{k=1}^n\int_{-\infty}^{+\infty}d\,\delta x_k F_k(x_0+\delta x_k)\right\}$$

$$\times\frac{1}{\sqrt{(2\pi)^n\mathrm{Det}\left[a_{\tau_k\tau_{k'}}^2(x_0)\right]}}\exp\left\{-\frac{1}{2}\sum_{k=1}^n\sum_{k'=1}^n\delta x_k\,a_{\tau_k\tau_{k'}}^{-2}(x_0)\,\delta x_{k'}\right\}, \tag{5.269}$$

where $a_{\tau_i\tau_j}^{-2}(x_0)$ is the inverse of the $n\times n$ -matrix $a_{\tau_i\tau_j}^2(x_0)$. This smearing formula determines the harmonic expectation values in the variational perturbation expansion (5.263) as convolutions with Gaussian functions.

For $n=1$ and only the diagonal elements $a^2(x_0)=a_{\tau\tau}^2(x_0)$ appear in the smearing formula (5.269), which reduces to the previous one in Eq. (5.30) $[F(x(\tau))=V(x(\tau))]$.

For polynomials $F(x(\tau))$, we set $x(\tau) = x_0 + \delta x(\tau)$ and expand in powers of $\delta x(\tau)$, and see that the smearing formula (5.269) reproduces the Wick expansion (5.266).

For two functions, the smearing formula (5.269) reads explicitly

$$
\langle F_1(x(\tau_1))F_2(x(\tau_2))\rangle_\Omega^{x_0} = \int\limits_{-\infty}^{+\infty} dx_1 \int\limits_{-\infty}^{+\infty} dx_2\, F_1(x_1)F_2(x_2)\, \frac{1}{\sqrt{(2\pi)^2[a^4(x_0) - a^4_{\tau_1\tau_2}(x_0)]}}
$$

$$
\times \exp\left\{-\frac{a^2(x_0)(x_1-x_0)^2 - 2a^2_{\tau_1\tau_2}(x_0)(x_1-x_0)(x_2-x_0) + a^2(x_0)(x_2-x_0)^2}{2[a^4(x_0) - a^4_{\tau_1\tau_2}(x_0)]}\right\}. \tag{5.270}
$$

Specializing $F_2(x(\tau_2))$ to quadratic functions in $x(\tau)$, we obtain from this

$$
\left\langle F_1(x(\tau_1))\,[x(\tau_2) - x_0]^2\right\rangle_\Omega^{x_0} = \langle F_1(x(\tau_1))\rangle_\Omega^{x_0}\, a^2(x_0)\left[1 - \frac{a^4_{\tau_1\tau_2}(x_0)}{a^4(x_0)}\right]
$$

$$
+ \left\langle F_1(x(\tau_1))\,[x(\tau_1) - x_0]^2\right\rangle_\Omega^{x_0}\, \frac{a^4_{\tau_1\tau_2}(x_0)}{a^4(x_0)}, \tag{5.271}
$$

and

$$
\left\langle [x(\tau_1) - x_0]^2\,[x(\tau_2) - x_0]^2\right\rangle_\Omega^{x_0} = a^4(x_0) + 2a^4_{\tau_1\tau_2}(x_0). \tag{5.272}
$$

5.21.2 Higher-Order Smearing Formula in D Dimensions

The smearing formula can easily be generalized to D-dimensional systems, where the local pair correlation function (5.264) becomes a $D \times D$-dimensional matrix:

$$
\langle \delta x_i(\tau)\delta x_j(\tau')\rangle_\Omega^{x_0} = a^2_{ij;\tau\tau'}(\mathbf{x}_0). \tag{5.273}
$$

For rotationally-invariant systems, the matrix can be decomposed in the same way as the trial frequency $\Omega^2_{ij}(\mathbf{x}_0)$ in (5.95) into longitudinal and transversal components with respect to $\mathbf{x}_0$:

$$
a^2_{ij;\tau\tau'}(\mathbf{x}_0) = a^2_{L;\tau\tau'}(r_0)P_{L;ij}(\hat{\mathbf{x}}_0) + a^2_{T;\tau\tau'}(r_0)P_{T;ij}(\hat{\mathbf{x}}_0), \tag{5.274}
$$

where $P_{L;ij}(\hat{\mathbf{x}}_0)$ and $P_{T;ij}(\hat{\mathbf{x}}_0)$ are longitudinal and transversal projection matrices introduced in (5.96). Denoting the matrix (5.274) by $\mathbf{a}^2_{\tau,\tau'}(\mathbf{x}_0)$, we can write the D-dimensional generalization of the smearing formula (5.275) as

$$
\langle F_1(\mathbf{x}(\tau_1))\cdots F_n(\mathbf{x}(\tau_n))\rangle_\Omega^{\mathbf{x}_0} = \left\{\prod_{k=1}^n \int\limits_{-\infty}^{+\infty} d\,\delta x_k F_k(\mathbf{x}_0 + \delta\mathbf{x}_k)\right\}
$$

$$
\times \frac{1}{\sqrt{(2\pi)^n \mathrm{Det}\left[\mathbf{a}^2_{\tau_k\tau_{k'}}(\mathbf{x}_0)\right]}} \exp\left\{-\frac{1}{2}\sum_{k=1}^n\sum_{k'=1}^n \delta\mathbf{x}_k\, \mathbf{a}^{-2}_{\tau_k\tau_{k'}}(\mathbf{x}_0)\,\delta\mathbf{x}_{k'}\right\}. \tag{5.275}
$$

The inverse $D \times D$-matrix $\mathbf{a}^{-2}_{\tau_k\tau_{k'}}(\mathbf{x}_0)$ is formed by simply inverting the $n \times n$-matrices $\mathbf{a}^{-2}_{L;\tau_k\tau_{k'}}(r_0), \mathbf{a}^{-2}_{T;\tau_k\tau_{k'}}(r_0)$ in the projection formula (5.274) with projection matrices $\mathbf{P}_L(\hat{\mathbf{x}}_0)$ and $\mathbf{P}_T(\hat{\mathbf{x}}_0)$:

$$
\mathbf{a}^{-2}_{\tau_k\tau_{k'}}(\mathbf{x}_0) = \mathbf{a}^{-2}_{L;\tau_k\tau_{k'}}(r_0)\mathbf{P}_L(\hat{\mathbf{x}}_0) + \mathbf{a}^{-2}_{T;\tau_k\tau_{k'}}(r_0)\mathbf{P}_T(\hat{\mathbf{x}}_0). \tag{5.276}
$$

In D dimensions, the trial potential contains a $D \times D$ frequency matrix and reads

$$
\frac{M}{2}\Omega^2_{ij}(\mathbf{x}_0)(x_i - x_{0i})(x_j - x_{0j}),
$$

with the analogous decomposition

$$\Omega_{ij}^2(\mathbf{x}_0) = \Omega_L^2(\mathbf{x}_0) \frac{x_{0i}x_{0j}}{r_0^2} + \Omega_T^2(\mathbf{x}_0) \left(\delta_{ij} - \frac{x_{0i}x_{0j}}{r_0^2} \right). \tag{5.277}$$

The interaction potential (5.261) becomes

$$V_{\text{int}}^{\mathbf{x}_0}(\mathbf{x}) = V(\mathbf{x}) - \frac{M}{2}\Omega_{ij}^2(\mathbf{x}_0)(x_i - x_{0i})(x_j - x_{0j}). \tag{5.278}$$

To first order, the anisotropic smearing formula (5.275) reads

$$\langle F_1(\mathbf{x}(\tau_1))\rangle_{\Omega_T,\Omega_L}^{r_0} = \int\limits_{-\infty}^{+\infty} d^3x_1 F_1(\mathbf{x}_1) \frac{1}{\sqrt{(2\pi)^3 a_T^4 a_L^2}} \exp\left\{ -\frac{(x_{1L} - r_0)^2}{2a_L^2} - \frac{\mathbf{x}_{1T}^2}{2a_T^2} \right\}, \tag{5.279}$$

with the special cases

$$\left\langle [\delta\mathbf{x}(\tau_1)]_T^2 \right\rangle_{\Omega_T,\Omega_L}^{r_0} = 2a_T^2, \qquad \left\langle [\delta\mathbf{x}(\tau_1)]_L^2 \right\rangle_{\Omega_T,\Omega_L}^{r_0} = a_L^2. \tag{5.280}$$

Inserting this into formula (5.263) we obtain the first-order approximation for the effective classical potential

$$W_1(\mathbf{x}_0) = F_\Omega^{\mathbf{x}_0} + \frac{1}{\hbar\beta} \int\limits_0^{\hbar\beta} d\tau_1 \left\langle V_{\text{int}}^{\mathbf{x}_0}(\mathbf{x}(\tau_1)) \right\rangle_{\Omega,c}^{\mathbf{x}_0}, \tag{5.281}$$

in agreement with the earlier result (5.97).

To second-order, the smearing formula (5.275) yields [33]

$$\langle F_1(\mathbf{x}(\tau_1)) F_2(\mathbf{x}(\tau_2))\rangle_{\Omega_T,\Omega_L}^{r_0} = \int\limits_{-\infty}^{+\infty} d^3x_1 \int\limits_{-\infty}^{+\infty} d^3x_2 F_1(\mathbf{x}_1) F_2(\mathbf{x}_2)$$

$$\times \frac{1}{(2\pi)^3 (a_T^4 - a_{T\tau_1\tau_2}^4)\sqrt{a_L^4 - a_{L\tau_1\tau_2}^4}} \exp\left\{ -\frac{a_T^2\mathbf{x}_{1T}^2 - 2a_{T\tau_1\tau_2}^2\mathbf{x}_{1T}\mathbf{x}_{2T} + a_T^2\mathbf{x}_{2T}^2}{2(a_T^4 - a_{T\tau_1\tau_2}^4)} \right\}$$

$$\times \exp\left\{ -\frac{a_L^2(x_{1L} - r_0)^2 - 2a_{L\tau_1\tau_2}^2(x_{1L} - r_0)(x_{2L} - r_0) + a_L^2(x_{2L} - r_0)^2}{2(a_L^4 - a_{L\tau_1\tau_2}^4)} \right\}, \tag{5.282}$$

so that rule (5.271) for expectation values generalizes to

$$\left\langle F_1(\mathbf{x}(\tau_1)) [\delta\mathbf{x}(\tau_2)]_T^2 \right\rangle_{\Omega_T,\Omega_L}^{r_0} = 2a_T^2 \left[1 - \frac{a_{T\tau_1\tau_2}^4}{a_T^4} \right] \langle F_1(\mathbf{x}(\tau_1))\rangle_{\Omega_T,\Omega_L}^{r_0}$$

$$+ \frac{a_{T\tau_1\tau_2}^4}{a_T^4} \left\langle F_1(\mathbf{x}(\tau_1)) [\delta\mathbf{x}(\tau_1)]_T^2 \right\rangle_{\Omega_T,\Omega_L}^{r_0}, \tag{5.283}$$

$$\left\langle F_1(\mathbf{x}(\tau_1)) [\delta\mathbf{x}(\tau_2)]_L^2 \right\rangle_{\Omega_T,\Omega_L}^{r_0} = a_L^2 \left[1 - \frac{a_{L\tau_1\tau_2}^4}{a_L^4} \right] \langle F_1(\mathbf{x}(\tau_1))\rangle_{\Omega_T,\Omega_L}^{r_0}$$

$$+ \frac{a_{L\tau_1\tau_2}^4}{a_L^4} \left\langle F_1(\mathbf{x}(\tau_1))[\delta\mathbf{x}(\tau_1)]_L^2 \right\rangle_{\Omega_T,\Omega_L}^{r_0}. \tag{5.284}$$

Specializing $F(\mathbf{x})$ to quadratic function, we obtain the generalizations of (5.272)

$$\left\langle [\delta\mathbf{x}(\tau_1)]_T^2 [\delta\mathbf{x}(\tau_2)]_T^2 \right\rangle_{\Omega_T,\Omega_L}^{r_0} = 4a_T^4 + 4a_{T\tau_1\tau_2}^4, \tag{5.285}$$

$$\left\langle [\delta\mathbf{x}(\tau_1)]_T^2 [\delta\mathbf{x}(\tau_2)]_L^2 \right\rangle_{\Omega_T,\Omega_L}^{r_0} = 2a_T^2 a_L^2, \tag{5.286}$$

$$\left\langle [\delta\mathbf{x}(\tau_1)]_L^2 [\delta\mathbf{x}(\tau_2)]_L^2 \right\rangle_{\Omega_T,\Omega_L}^{r_0} = a_L^4 + 2a_{L\tau_1\tau_2}^4. \tag{5.287}$$

5.21.3   Isotropic Second-Order Approximation to Coulomb Problem

To demonstrate the use of the higher-order smearing formula (5.275), we calculate the effective classical potential of the three-dimensional Coulomb potential

$$V(\mathbf{x}) = -\frac{e^2}{|\mathbf{x}|} \tag{5.288}$$

to second order in variational perturbation theory, thus going beyond the earlier results in Eq. (5.53) and Section 5.10. The interaction potential corresponding to (5.278) is

$$V_{\text{int}}^{\mathbf{x}_0}(\mathbf{x}) = -\frac{e^2}{|\mathbf{x}|} - \frac{M}{2}\Omega^2(\mathbf{x}_0)(\mathbf{x} - \mathbf{x}_0)^2 . \tag{5.289}$$

For simplicity, we consider only the isotropic approximation with only a single trial frequency. Then all formulas derived in the beginning of this section have a trivial extension to three dimensions. Better results will, of course be obtained with two trial frequencies $\Omega_L^2(r_0)$ and $\Omega_T^2(r_0)$ of Section 5.9.

The Fourier transform $4\pi e^2/|\mathbf{k}|^2$ of the Coulomb potential $e^2/|\mathbf{x}|$ is most conveniently written in a proper-time type of representation as

$$V(\mathbf{k}) = \frac{4\pi e^2}{2}\int_0^\infty d\sigma e^{-\sigma \mathbf{k}^2/2 - i\mathbf{k}\mathbf{x}}, \tag{5.290}$$

where σ has the dimension length square. The lowest-order smeared potentials were calculated before in Section 5.10. For brevity, we consider here only the isotropic approximation in which longitudinal and transverse trial frequencies are identified [compare (5.112)]:

$$\left\langle [\mathbf{x}(\tau_1) - \mathbf{x}_0]^2 \right\rangle_\Omega^{\mathbf{x}_0} = 3a^2(\mathbf{x}_0) , \qquad \left\langle \frac{1}{|\mathbf{x}(\tau_1)|} \right\rangle_\Omega^{\mathbf{x}_0} = \frac{1}{|\mathbf{x}_0|}\,\text{erf}\left[\frac{|\mathbf{x}_0|}{\sqrt{2a^2(\mathbf{x}_0)}}\right] . \tag{5.291}$$

The first-order variational approximation to the effective classical potential (5.281) is then given by the earlier-calculated expression (5.115).

To second order in variational perturbation theory we calculate expectation values

$$\left\langle [\mathbf{x}(\tau_1) - \mathbf{x}_0]^2 [\mathbf{x}(\tau_2) - \mathbf{x}_0]^2 \right\rangle_\Omega^{\mathbf{x}_0} = 9a^4(\mathbf{x}_0) + 6a_{\tau_1 \tau_2}^4(\mathbf{x}_0) , \tag{5.292}$$

$$\left\langle [\mathbf{x}(\tau_1) - \mathbf{x}_0]^2 \frac{1}{|\mathbf{x}(\tau_2)|} \right\rangle_\Omega^{\mathbf{x}_0} = \frac{2[3a^4(\mathbf{x}_0) - a_{\tau_1 \tau_2}^4(\mathbf{x}_0)]}{\sqrt{\pi a^6(\mathbf{x}_0)}} , \tag{5.293}$$

which follow from the obvious generalization of (5.271), (5.272) to three dimensions. More involved is the Coulomb-Coulomb correlation function

$$\left\langle \frac{1}{|\mathbf{x}(\tau_1)|} \frac{1}{|\mathbf{x}(\tau_2)|} \right\rangle_\Omega^{\mathbf{x}_0} = \frac{1}{2\pi}\int_0^{+\infty} d\sigma_1 \int_0^{+\infty} d\sigma_2 \frac{1}{\sqrt{[a^2(\mathbf{x}_0) + \sigma_1][a^2(\mathbf{x}_0) + \sigma_2] - a_{\tau_1 \tau_2}^4(\mathbf{x}_0)}^3}$$

$$\times \exp\left\{ -\mathbf{x}_0^2 \frac{a^2(\mathbf{x}_0) + \sigma_1/2 + \sigma_2/2 - a_{\tau_1 \tau_2}^2(\mathbf{x}_0)}{[a^2(\mathbf{x}_0) + \sigma_1][a^2(\mathbf{x}_0) + \sigma_2] - a_{\tau_1 \tau_2}^4(\mathbf{x}_0)} \right\} . \tag{5.294}$$

Using these smearing results we calculate the second connected correlation functions of the interaction potential (5.289) appearing in (5.263) and find the effective classical potential to second order in variational perturbation theory

$$W_2(\mathbf{x}_0) = W_1(\mathbf{x}_0) + \left[\frac{Me^2\Omega(\mathbf{x}_0)}{\hbar\sqrt{2\pi a^6(\mathbf{x}_0)}} - \frac{3M^2\Omega^3(\mathbf{x}_0)}{4\hbar} \right] l^4(\mathbf{x}_0)$$

$$- \frac{e^4}{2\hbar}\int_0^{\hbar\beta} d\tau \left\langle \frac{1}{|\mathbf{x}(\tau)|} \frac{1}{|\mathbf{x}(0)|} \right\rangle_\Omega^{\mathbf{x}_0} , \tag{5.295}$$

with the abbreviation

$$l^4(\mathbf{x}_0) \equiv \frac{\hbar \left[4 + \hbar^2 \beta^2 \Omega^2(\mathbf{x}_0) - 4\cosh \hbar\beta\Omega(\mathbf{x}_0) + \hbar\beta\Omega(\mathbf{x}_0)\sinh \hbar\beta\Omega(\mathbf{x}_0)\right]}{8\beta M^2 \Omega^3(\mathbf{x}_0)\sinh[\hbar\beta\Omega(\mathbf{x}_0)/2]}, \tag{5.296}$$

the symbol indicating that this is a quantity of dimension length to the forth power. After an extremization of (5.295) with respect to the trial frequency $\Omega(\mathbf{x}_0)$, which has to be done numerically, we obtain the second-order approximation for the effective classical potential of the Coulomb system plotted in Fig. 5.25 for various temperatures. The curves lie all below the first-order ones, and the difference between the two decreases with increasing temperature and increasing distance from the origin.

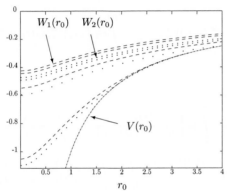

Figure 5.25 Isotropic approximation to effective classical potential of Coulomb system in the first (lines) and second order (dots). The temperatures are $10^{-4}, 10^{-3}, 10^{-2}, 10^{-1}$, and ∞ from top to bottom in atomic units. Compare also Fig. 5.9.

5.21.4 Anisotropic Second-Order Approximation to Coulomb Problem

The first-order effective classical potential $W_1(\mathbf{x}_0)$ was derived in Eqs. (5.97) and (5.117)–(5.121). To obtain the second-order approximation $W_2(\mathbf{x}_0)$, we insert the Coulomb potential in the representation (5.290) into the second-order smearing formula (5.282), and find

$$\left\langle \frac{1}{|\mathbf{x}(\tau_1)|}\left[\delta\mathbf{x}(\tau_2)\right]_T^2 \right\rangle_{\Omega_T,\Omega_L}^{r_0} = \sqrt{\frac{2a_L^2}{\pi}} \int_0^1 d\lambda\, e^{-\frac{r_0^2}{2a_L^2}\lambda^2} \left\{ \frac{2a_T^2}{(a_T^2 - a_L^2)\lambda^2 + a_L^2} - \frac{2a_{T\tau_1\tau_2}^4 \lambda^2}{[(a_T^2 - a_L^2)\lambda^2 + a_L^2]^2} \right\},$$

$$\left\langle \frac{1}{|\mathbf{x}(\tau_1)|}\left[\delta\mathbf{x}(\tau_2)\right]_L^2 \right\rangle_{\Omega_T,\Omega_L}^{r_0} = \sqrt{\frac{2a_L^2}{\pi}} \int_0^1 d\lambda\, e^{-\frac{r_0^2}{2a_L^2}\lambda^2} \frac{a_L^6 + a_{L\tau_1\tau_2}^4[r_0^2\lambda^4 - a_L^2\lambda^2]}{a_L^4[(a_T^2 - a_L^2)\lambda^2 + a_L^2]}. \tag{5.297}$$

These are special cases of the more general expectation value

$$\left\langle \frac{1}{|\mathbf{x}(\tau_1)|} F(\mathbf{x}(\tau_2)) \right\rangle_{\Omega_T,\Omega_L}^{r_0} = \frac{1}{2\pi^2} \int_0^{+\infty} d\sigma \frac{\exp\left\{-\dfrac{a_L^2 r_0^2}{2[a_L^4 - a_{L\tau_1\tau_2}^4 + 2a_L^2\sigma]}\right\}}{[a_T^4 - a_{T\tau_1\tau_2}^4 + 2a_T^2\sigma]\sqrt{a_L^4 - a_{L\tau_1\tau_2}^4 + 2a_L^2\sigma}}$$

$$\times \int d^3x\, F(\mathbf{x}) \exp\left\{-\frac{(a_T^2 + 2\sigma)\mathbf{x}_T^2}{2[a_T^4 - a_{T\tau_1\tau_2}^4 + 2a_T^2\sigma]} - \frac{(a_L^2 + 2\sigma)(x_L - r_0)^2 + 2a_{L\tau_1\tau_2}^2 r_0(x_L - r_0)}{2[a_L^4 - a_{L\tau_1\tau_2}^4 + 2a_L^2\sigma]}\right\}, \tag{5.298}$$

which furthermore leads to

$$
\left\langle \frac{1}{|\mathbf{x}(\tau_1)|} \frac{1}{|\mathbf{x}(\tau_2)|} \right\rangle^{r_0}_{\Omega_T, \Omega_L} = \frac{2}{\pi} \int\limits_{0}^{+\infty} d\sigma_1 \int\limits_{0}^{+\infty} d\sigma_2 \, \frac{1}{[a_T^2 + 2\sigma_1][a_T^2 + 2\sigma_2] - a_{T\tau_1\tau_2}^4}
$$
$$
\times \frac{1}{\sqrt{[a_L^2 + 2\sigma_1][a_L^2 + 2\sigma_2] - a_{L\tau_1\tau_2}^4}} \exp\left\{ -\frac{r_0^2[a_L^2 + \sigma_1 + \sigma_2 - a_{L\tau_1\tau_2}^2]}{[a_L^2 + 2\sigma_1][a_L^2 + 2\sigma_2] - a_{L\tau_1\tau_2}^4} \right\}. \quad (5.299)
$$

From these smearing results we calculate the second-order approximation to the effective classical potential

$$
W_2(\mathbf{x}_0) = F_\Omega^{\mathbf{x}_0} + \frac{1}{\hbar\beta} \int\limits_{0}^{\hbar\beta} d\tau_1 \, \langle V_{\text{int}}^{\mathbf{x}_0}(\mathbf{x}(\tau_1)) \rangle^{\mathbf{x}_0}_{\Omega,c} - \frac{1}{2\hbar^2\beta} \int\limits_{0}^{\hbar\beta} d\tau_1 \int\limits_{0}^{\hbar\beta} d\tau_2 \, \langle V_{\text{int}}^{\mathbf{x}_0}(\mathbf{x}(\tau_1)) V_{\text{int}}^{\mathbf{x}_0}(\mathbf{x}(\tau_2)) \rangle^{\mathbf{x}_0}_{\Omega,c}. \quad (5.300)
$$

The result is

$$
W_2^{\Omega_T, \Omega_L}(r_0) = W_1^{\Omega_T, \Omega_L}(r_0)
$$
$$
+ \frac{e^2 M}{2\hbar} \sqrt{\frac{2a_L^2}{\pi}} \int\limits_{0}^{1} d\lambda \left\{ \frac{2\Omega_T l_T^4 \lambda^2}{[(a_T^2 - a_L^2)\lambda^2 + a_L^2]^2} - \frac{\Omega_L l_L^4 [r_0^2 \lambda^4 - a_L^2 \lambda^2]}{a_L^4 [(a_T^2 - a_L^2)\lambda^2 + a_L^2]} \right\} e^{-r_0^2 \lambda^2 / 2a_L^2}
$$
$$
- \frac{M^2 [2\Omega_T^3 l_T^4 + \Omega_L^3 l_L^4]}{4\hbar} - \frac{e^4}{2\hbar^2\beta} \int\limits_{0}^{\hbar\beta} d\tau_1 \int\limits_{0}^{\hbar\beta} d\tau_2 \left\langle \frac{1}{|\mathbf{x}(\tau_1)|} \frac{1}{|\mathbf{x}(\tau_2)|} \right\rangle^{r_0}_{\Omega_T, \Omega_L, c}, \quad (5.301)
$$

with the abbreviation

$$
l_{T,L}^4 = \frac{\hbar \left[4 + \hbar^2\beta^2\Omega_{T,L}^2 - 4\cosh\hbar\beta\Omega_{T,L} + \hbar\beta\Omega_{T,L}\sinh\hbar\beta\Omega_{T,L} \right]}{8\beta M^2\Omega_{T,L}^3 \sinh[\hbar\beta\Omega_{T,L}/2]}, \quad (5.302)
$$

which is a quantity of dimension (length)^4. After an extremization of (5.115) and (5.301) with respect to the trial frequencies Ω_T, Ω_L which has to be done numerically, we obtain the second-order approximation for the effective classical potential of the Coulomb system plotted in Fig. 5.26 for various temperatures. The second order curves lie all below the first-order ones, and the difference between the two decreases with increasing temperature and increasing distance from the origin.

5.21.5 Zero-Temperature Limit

As a cross check of our result we take (5.301) to the limit $T \to 0$. Just as in the lowest-order discussion in Sect. (5.4), the x_0-integral can be evaluated in the saddle-point approximation which becomes exact in this limit, so that the minimum of $W_N(\mathbf{x}_0)$ in x_0 yields the nth approximation to the free energy at $T = 0$ and thus the nth approximations $E_N^{(0)}$ the ground state energy $E^{(0)}$ of the Coulomb system. In this limit, the results should coincide with those derived from a direct variational treatment of the Rayleigh-Schrödinger perturbation expansion in Section 3.18. With the help of such a treatment, we shall also carry the approximation to the next order, thereby illustrating the convergence of the variational perturbation expansions. For symmetry reasons, the minimum of the effective classical potential occurs for all temperatures at the origin, such that we may restrict (5.115) and (5.301) to this point. Recalling the zero-temperature limit of the two-point correlations (5.19) from (3.246),

$$
\lim_{\beta\to\infty} a_{\tau\tau'}^2(\mathbf{x}_0) = \frac{\hbar}{2M\Omega(\mathbf{x}_0)} \exp\left\{ -\Omega(\mathbf{x}_0) |\tau - \tau'| \right\}, \quad (5.303)
$$

we immediately deduce for the first order approximation (5.115) with $\Omega = \Omega(0)$ the limit

$$E_1^{(0)}(\Omega) = \lim_{\beta \to \infty} W_1^\Omega(0) = \frac{3}{4}\hbar\Omega - \frac{2}{\sqrt{\pi}}\sqrt{\frac{M\Omega}{\hbar}}e^2 . \qquad (5.304)$$

In the second-order expression (5.301), the zero-temperature limit is more tedious to take. Performing the integrals over σ_1 and σ_2, we obtain the connected correlation function

$$\left\langle \frac{1}{|\mathbf{x}(\tau_1)|}\frac{1}{|\mathbf{x}(\tau_2)|} \right\rangle_{\Omega,c}^{\mathbf{x}_0} = \frac{1}{a_{\tau_1\tau_2}^4(0)} - \frac{2}{\pi a_{\tau_1\tau_2}^4(0)}\arctan\sqrt{\frac{a_{\tau_1\tau_2}^2(0)}{a_{\tau_1\tau_2}(0)}-1} - \frac{2}{\pi a_{\tau_1\tau_2}^2(0)} . \qquad (5.305)$$

Inserting (5.303), setting $\tau_1 = 0$ and integrating over the imaginary times $\tau = \tau_2 \in [0, \hbar\beta]$, we find

$$\int_0^{\hbar\beta} d\tau \left\langle \frac{1}{|\mathbf{x}(\tau)|}\frac{1}{|\mathbf{x}(0)|} \right\rangle_{\Omega,c}^{\mathbf{x}_0} \underset{\beta\to\infty}{\approx} \frac{4M}{\hbar^2\beta\Omega}\left\{ e^{\hbar\beta\Omega} - 1 - \hbar\beta\Omega - \frac{\hbar^2\beta^2\Omega^2}{\pi} - \frac{2}{\pi} \right.$$

$$\left. \times \left[e^{\hbar\beta\Omega}\arcsin\sqrt{1-e^{-2\hbar\beta\Omega}} + \frac{1}{2}\ln\alpha(\beta) - \frac{1}{8}[\ln\alpha(\beta)]^2 - \frac{1}{2}\int_{\alpha(\beta)}^1 du\,\frac{\ln u}{1+u} \right] \right\}, \qquad (5.306)$$

with the abbreviation

$$\alpha(\beta) = \frac{1-\sqrt{1-e^{-2\hbar\beta\Omega}}}{1+\sqrt{1-e^{-2\hbar\beta\Omega}}} . \qquad (5.307)$$

Inserting this into (5.301) and going to the limit $\beta \to \infty$ we obtain

$$E_2^{(0)}(\Omega) = \lim_{\beta\to\infty} W_2^\Omega(0) = \frac{9}{16}\hbar\Omega - \frac{3}{2\sqrt{\pi}}\sqrt{\frac{M\Omega}{\hbar}}e^2 - \frac{4}{\pi}\left(1+\ln 2 - \frac{\pi}{2}\right)\frac{M}{\hbar^2}e^4 . \qquad (5.308)$$

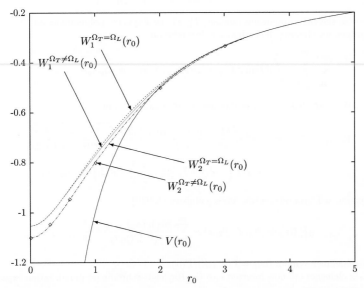

Figure 5.26 Isotropic and anisotropic approximations to effective classical potential of Coulomb system in first and second order at temperature 0.1 in atomic units. The lowest line represents the high temperature limit in which all isotropic and anisotropic approximations coincide.

Postponing for a moment the extremization of (5.304) and (5.308) with respect to the trial frequency Ω, let us first rederive this result from a variational treatment of the ordinary Rayleigh-Schrödinger perturbation expansion for the ground state energy in Section 3.18. According to the replacement rule (5.186), we must first calculate the ground state energy for a Coulomb potential in the presence of a harmonic potential of frequency ω:

$$V(\mathbf{x}) = \frac{M}{2}\omega^2\mathbf{x}^2 - \frac{e^2}{|\mathbf{x}|}. \tag{5.309}$$

After this, we make the trivial replacement $\omega \to \sqrt{\Omega^2 + \omega^2 - \Omega^2}$ and re-expand the energy in powers of $\omega^2 - \Omega^2$, considering this quantity as being of the order e^2 and truncating the re-expansion accordingly. At the end we go to $\omega = 0$, since the original Coulomb system contains no oscillator potential. The result of this treatment will be precisely the expansions (5.304) and (5.308).

The Rayleigh-Schrödinger perturbation expansion of the ground state energy $E_N^{(0)}(\omega)$ for the potential (5.309) in Section 3.18 requires knowledge of the matrix elements of the Coulomb potential (5.288) with respect to the eigenfunctions of the harmonic oscillator with the frequency ω:

$$V_{n,l,m;n',l',m'} = \int_0^{2\pi} d\varphi \int_0^\pi d\vartheta \sin\vartheta \int_0^\infty dr\, r^2\, \psi_{n,l,m}^*(r,\vartheta,\varphi) \frac{-e^2}{r} \psi_{n',l',m'}(r,\vartheta,\varphi), \tag{5.310}$$

where [see (9.67), (9.68), and (9.53)]

$$\psi_{n,l,m}(r,\vartheta,\varphi) = \sqrt{\frac{2n!}{\Gamma(n+l+3/2)}} \sqrt[4]{\frac{M\omega}{\hbar}} \left(\frac{M\omega}{\hbar}r^2\right)^{(l+1)/2}$$
$$\times L_n^{l+1/2}\left(\frac{M\omega}{\hbar}r^2\right) \exp\left\{-\frac{M\omega}{2\hbar}r^2\right\} Y_{l,m}(\vartheta,\varphi). \tag{5.311}$$

Here n denotes the radial quantum number, $L_n^\alpha(x)$ the Laguerre polynomials and $Y_{l,m}(\vartheta,\varphi)$ the spherical harmonics obeying the orthonormality relation

$$\int_0^{2\pi} d\varphi \int_0^\pi d\vartheta \sin\vartheta\, Y_{l,m}^*(\vartheta,\varphi) Y_{l',m'}(\vartheta,\varphi) = \delta_{l,l'}\delta_{m,m'}. \tag{5.312}$$

Inserting (5.311) into (5.310), and evaluating the integrals, we find

$$V_{n,l,m;n',l',m'} = -e^2 \sqrt{\frac{M\omega}{\pi\hbar}} \frac{\Gamma(l+1)\Gamma(n+1/2)}{\Gamma(l+3/2)} \sqrt{\frac{\Gamma(n'+l+3/2)}{n!n'!\Gamma(n+l+3/2)}}$$
$$\times {}_3F_2\left(-n', l+1, \frac{1}{2}; l+\frac{3}{2}, \frac{1}{2} - n; 1\right) \delta_{l,l'}\,\delta_{m,m'}, \tag{5.313}$$

with the generalized hypergeometric series [compare (1.450)]

$$ {}_3F_2(\alpha_1,\alpha_2,\alpha_3;\beta_1,\beta_2;x) = \sum_{k=0}^\infty \frac{(\alpha_1)_k(\alpha_2)_k(\alpha_3)_k}{(\beta_1)_k(\beta_2)_k} \frac{x^k}{k!} \tag{5.314}$$

and the Pochhammer symbol $(\alpha)_k = \Gamma(\alpha+k)/\Gamma(\alpha)$.

These matrix elements are now inserted into the Rayleigh-Schrödinger perturbation expansion for the ground state energy

$$E^{(0)}(\omega) = E_{0,0,0} + V_{0,0,0;0,0,0} + \sum_{n,l,m}' \frac{V_{0,0,0;n,l,m} V_{n,l,m;0,0,0}}{E_{0,0,0} - E_{n,l,m}}$$

$$- \sum_{n,l,m}' V_{0,0,0;0,0,0} \frac{V_{0,0,0;n,l,m} V_{n,l,m;0,0,0}}{[E_{0,0} - E_{n,l,m}]^2}$$

$$+ \sum_{n,l,m}' \sum_{n',l',m'}' \frac{V_{0,0,0;n,l,m} V_{n,l,m;n',l',m'} V_{n',l',m';0,0,0}}{[E_{0,0} - E_{n,l,m}][E_{0,0} - E_{n',l',m'}]} + \dots , \tag{5.315}$$

the denominators containing the energy eigenvalues of the harmonic oscillator

$$E_{n,l,m} = \hbar\omega \left(2n + l + \frac{3}{2} \right). \tag{5.316}$$

The primed sums in (5.315) run over all values of the quantum numbers $n, l = -\infty, \dots, +\infty$ and $m = -l, \dots, +l$, excluding those for which the denominators vanish. For the first three orders we obtain from (5.313)–(5.316)

$$E^{(0)}(\omega) = \frac{3}{2}\hbar\omega - \frac{2}{\sqrt{\pi}} \sqrt{\frac{M\omega}{\hbar}} e^2 - \frac{4}{\pi}\left(1 + \ln 2 - \frac{\pi}{2}\right) \frac{M}{\hbar^2} e^4 - c \sqrt{\frac{M^3}{\hbar^7\omega}} e^6 + \dots , \tag{5.317}$$

with the constant

$$c = \frac{1}{\pi^{3/2}} \left\{ \sum_{n=1}^{\infty} \frac{1 \cdot 3 \cdots (2n-1)}{2 \cdot 4 \cdots 2n} \frac{1}{n^2(n+1/2)} - \sum_{n=1}^{\infty} \sum_{n'=1}^{\infty} \frac{1 \cdot 3 \cdots (2n-1)}{2 \cdot 4 \cdots 2n} \right.$$

$$\left. \times \frac{1 \cdot 3 \cdots (2n'-1)}{2 \cdot 4 \cdots 2n'} \frac{{}_3F_2\left(-n', l+1, \frac{1}{2}; l+\frac{3}{2}, \frac{1}{2} - n; 1\right)}{n\,n'(n+1/2)} \right\} \approx 0.031801. \tag{5.318}$$

Since we are interested only in the energies in the pure Coulomb system with $\omega = 0$, the variational re-expansion procedure described after (5.309) becomes particularly simple: We simply have to replace ω by $\sqrt{\Omega^2 - \Omega^2}$ which is appropriately re-expanded in the second Ω^2, thereby considering Ω^2 as a quantity of order e^2. For the first term in the energy (5.317) which is proportional to Ω itself this amounts to a multiplication by a factor $(1-1)^{1/2}$ which is re-expanded in the second "1" up to the third order as $1 - \frac{1}{2} - \frac{1}{8} - \frac{1}{16} = \frac{5}{16}$. The term $3\omega/2$ in (5.317) becomes therefore $15/32\omega$. By the same rule, the factor $\omega^{1/2}$ in the second term of the energy (5.317) goes over into $\Omega^{1/2}(1-1)^{1/4}$, re-expanded to second order in the second "1", i.e., into $\Omega^{1/4}(1 - \frac{1}{4} - \frac{3}{32}) = \frac{21}{32}$. The next term in (5.317) happens to be independent of ω and needs no re-expansion, whereas the last term remains unchanged since it is already of highest order in e^2. In this way we obtain from (5.317) the third-order variational perturbation expansion

$$E_3^{(0)}(\Omega) = \frac{15}{32}\hbar\Omega - \frac{21}{16\sqrt{\pi}} \sqrt{\frac{M\Omega}{\hbar}} e^2 - \frac{4}{\pi}\left(1 + \ln 2 - \frac{\pi}{2}\right) \frac{M}{\hbar^2} e^4 - c \sqrt{\frac{M^3}{\hbar^7\Omega}} e^6. \tag{5.319}$$

Extremizing (5.304), (5.308), and (5.317) successively with respect to the trial frequency Ω we find to orders $1, 2$, and 3 the optimal values

$$\Omega_1 = \Omega_2 = \frac{16}{9\pi} \frac{Me^4}{\hbar^3}, \qquad \Omega_3 = c' \frac{Me^4}{\hbar^3}, \tag{5.320}$$

with $c' \approx 0.52621$. The corresponding approximations to the ground state energy are

$$E_N^{(0)}(\Omega^N) = -\gamma_N \frac{Me^4}{\hbar^2}, \tag{5.321}$$

with the constants

$$\gamma_1 = \frac{4}{3\pi} \approx 0.42441, \qquad \gamma_2 = \frac{5 + 4\ln 2}{\pi} - 2 \approx 0.47409, \qquad \gamma_3 \approx 0.49012, \tag{5.322}$$

approaching exponentially fast the exact value $\gamma = 0.5$, as shown in Fig. 5.27.

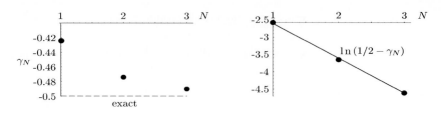

Figure 5.27 Approach of the variational approximations of first, second, and third order to the correct ground state energy -0.5, in atomic units.

5.22 Polarons

An important role in the development of variational methods for the approximate solution of path integrals was played by the *polaron* problem [34]. Polarons arise when electrons travel through ionic crystals thereby producing an electrostatic deformation in their neighborhood. If $\mathbf{P}^i(\mathbf{x}, t)$ denotes electric polarization density caused by the displacement of the positive against the negative ions, an electron sees a local ionic charge distribution

$$\rho^i(\mathbf{x}, t) = \boldsymbol{\nabla} \cdot \mathbf{P}^i(\mathbf{x}, t), \tag{5.323}$$

which gives rise to an electric potential satisfying

$$\boldsymbol{\nabla}^2 A^0(\mathbf{x}, t) = 4\pi \boldsymbol{\nabla} \cdot \mathbf{P}^i(\mathbf{x}, t). \tag{5.324}$$

The Fourier transform of this,

$$A_{\mathbf{k}}^0(t) \equiv \int_{-\infty}^{\infty} d^3x \, A^0(\mathbf{x}, t) e^{-i\mathbf{k}\mathbf{p}}, \tag{5.325}$$

and that of $\mathbf{P}^i(\mathbf{x}, t)$,

$$\mathbf{P}_{\mathbf{k}}^i(t) = \int_{-\infty}^{\infty} d^3x \, \mathbf{P}^i(\mathbf{x}, t) e^{-i\mathbf{k}\mathbf{x}}, \tag{5.326}$$

are related by

$$A_{\mathbf{k}}^0(t) = -\frac{4\pi}{\mathbf{k}^2} i\mathbf{k} \cdot \mathbf{P}_{\mathbf{k}}^i(t). \tag{5.327}$$

Only longitudinal phonons which have $\mathbf{P}_{\mathbf{k}}^i(t) \propto \mathbf{k}$ and correspond to density fluctuations in the crystal contribute. For these, an electron at position $\mathbf{x}(t)$ experiences an electric potential

$$A^0(\mathbf{x}, t) = -\sum_{\mathbf{k}} A_{\mathbf{k}}^0(t) \frac{e^{i\mathbf{k}\mathbf{x}}}{\sqrt{V}} = i \sum_{\mathbf{k}} \frac{4\pi}{|\mathbf{k}|} \mathbf{P}_{\mathbf{k}}(\tau) \frac{e^{i\mathbf{k}\mathbf{x}}}{\sqrt{V}}. \tag{5.328}$$

In the regime of optical phonons, each Fourier component oscillates with approximately the same frequency ω, the frequency of longitudinal optical phonons. The variables $\mathbf{P}_{\mathbf{k}}^i(t)$ have therefore a Lagrangian

$$L(t) = \frac{1}{2\mu} \sum_{\mathbf{k}} \left[\dot{\mathbf{P}}_{-\mathbf{k}}^i(t) \dot{\mathbf{P}}_{\mathbf{k}}^i(t) - \omega^2 \mathbf{P}_{-\mathbf{k}}^i(t) \mathbf{P}_{\mathbf{k}}^i(t) \right], \tag{5.329}$$

with some material constant μ and $\mathbf{P}^{i}_{-\mathbf{k}}(t) = \mathbf{P}^{i}_{\mathbf{k}}{}^{*}(t)$, since the polarization is a real field. This can be expressed in terms of measurable properties of the crystal. For this we note that the interaction of the polarization field with a given total charge distribution $\rho(\mathbf{x}, t)$ is described by a Lagrangian

$$L_{\text{int}}(t) = -\int d^3x \, \rho(\mathbf{x}, t) V(\mathbf{x}, t). \tag{5.330}$$

Inserting (5.325) and performing a partial

$$L_{\text{int}}(t) = 4\pi \int d^3x \, \frac{1}{\boldsymbol{\nabla}^2} \boldsymbol{\nabla}\rho(\mathbf{x}, t) \cdot \mathbf{P}^{i}(\mathbf{x}', t). \tag{5.331}$$

Recalling the Gauss law $\boldsymbol{\nabla} \cdot \mathbf{D}(\mathbf{x}, t) = 4\pi\rho(\mathbf{x}, t)$ we identify the factor of $\mathbf{P}^{i}(\mathbf{x}, t)$ with the total electric displacement field and write

$$L_{\text{int}}(t) = 4\pi \int d^3x \, \mathbf{D}(\mathbf{x}, t) \cdot \mathbf{P}^{i}(\mathbf{x}, t). \tag{5.332}$$

In combination with (5.329) this leads to an equation of motion

$$\frac{1}{\mu} \left[\ddot{\mathbf{P}}^{i}_{\mathbf{k}}(t) + \omega^2 \mathbf{P}^{i}_{\mathbf{k}}(t) \right] = \mathbf{D}_{\mathbf{k}}(t). \tag{5.333}$$

If we go over to the temporal Fourier components $\mathbf{P}^{i}_{\omega',\mathbf{k}}$ of the ionic polarization, we find the relation

$$\frac{1}{\mu} \left(\omega^2 - \omega'^2 \right) \mathbf{P}^{i}_{\omega',\mathbf{k}} = \mathbf{D}_{\omega',\mathbf{k}}. \tag{5.334}$$

For very slow deformations, this becomes

$$\frac{\omega^2}{\mu} \mathbf{P}^{i}_{\omega',\mathbf{k}} \approx \mathbf{D}_{0,\mathbf{k}}. \tag{5.335}$$

Using the general relation

$$\mathbf{D}_{\omega',\mathbf{k}} = \mathbf{E}_{\omega',\mathbf{k}} + 4\pi\mathbf{P}_{\omega',\mathbf{k}}, \tag{5.336}$$

where $4\pi\mathbf{P}_{\omega',\mathbf{k}}$ contains both ionic *and* electronic polarizations, we obtain

$$4\pi\mathbf{P}_{\omega',\mathbf{k}} = \left(1 - \frac{1}{\epsilon_{\omega'}} \right) \mathbf{D}_{\omega',\mathbf{k}}, \tag{5.337}$$

with $\epsilon_{\omega'}$ being the dielectric constant at frequency ω'. For a slowly moving electron, the lattice deformations have small frequencies, and we can write the time-dependent equation

$$4\pi\mathbf{P}_{\mathbf{k}}(t) \approx \left(1 - \frac{1}{\epsilon_0} \right) \mathbf{D}_{\mathbf{k}}(t). \tag{5.338}$$

By comparison with Eq. (5.335) we determine the parameter μ. Before we can do so, however, we must subtract from (5.338) the contribution of the electrons. These fulfill the approximate time-dependent equation

$$4\pi \mathbf{P}_{\mathbf{k}}^{\text{el}}(t) \approx \left(1 - \frac{1}{\epsilon_\infty}\right) \mathbf{D}_{\mathbf{k}}(t), \tag{5.339}$$

where ϵ_∞ is the dielectric constant at high frequency where only electrons can follow the field oscillations. The purely ionic polarization field is therefore given by

$$4\pi \mathbf{P}_{\mathbf{k}}^{\text{i}}(t) \approx \left(\frac{1}{\epsilon_\infty} - \frac{1}{\epsilon_0}\right) \mathbf{D}_{\mathbf{k}}(t). \tag{5.340}$$

By comparison with (5.334) we identify

$$\mu \approx \frac{\omega^2}{4\pi}\left(\frac{1}{\epsilon_\infty} - \frac{1}{\epsilon_0}\right). \tag{5.341}$$

5.22.1 Partition Function

The partition function of the combined system of an electron and the oscillating polarization is therefore described by the imaginary-time path integral

$$Z = \int \mathcal{D}^3 x \prod_{\mathbf{k}} \int \mathcal{D}^3 \mathbf{P}_{\mathbf{k}} \exp\left(-\frac{1}{\hbar}\int_0^{\hbar\beta} d\tau \left\{\frac{M}{2}\dot{\mathbf{x}}^2(\tau)\right.\right.$$
$$\left.\left. + \sum_{\mathbf{k}} \frac{1}{2\mu}\left[\dot{\mathbf{P}}_{-\mathbf{k}}^{\text{i}}(t)\dot{\mathbf{P}}_{\mathbf{k}}^{\text{i}}(t) - \omega^2 \mathbf{P}_{-\mathbf{k}}^{\text{i}}(t)\mathbf{P}_{\mathbf{k}}^{\text{i}}(t)\right] + ie\sum_{\mathbf{k}} \frac{4\pi}{|\mathbf{k}|}\mathbf{P}_{\mathbf{k}}(\tau)\frac{e^{i\mathbf{k}\mathbf{x}(\tau)}}{\sqrt{V}}\right\}\right), \tag{5.342}$$

where V is the volume of the system. The path integral is Gaussian in the Fourier components $\mathbf{P}_{\mathbf{k}}(\tau)$. These can therefore be integrated out with the rules of Subsection 3.8.2. For the correlation function of the polarizations we shall use the representation of the Green function (3.248) as a sum of periodic repetitions of the zero-temperature Green function

$$\langle \mathbf{P}_{\mathbf{k}}(\tau)\mathbf{P}_{-\mathbf{k}}(\tau)\rangle = \frac{\hbar}{2\omega}\sum_{n=-\infty}^{\infty} e^{-\omega|\tau-\tau'+n\hbar\beta|}. \tag{5.343}$$

Abbreviating

$$\sum_{n=-\infty}^{\infty} e^{-\omega|\tau-\tau'+n\hbar\beta|} \equiv e_{\text{per}}^{-\omega|\tau-\tau'|} = \frac{1}{2\omega}\left(e^{-\omega|\tau-\tau'|} - e^{-\omega(\hbar\beta-|\tau-\tau'|)}\right)\frac{1}{1-e^{-\hbar\beta\omega}}, \tag{5.344}$$

the right-hand side being valid for $\hbar\beta > \tau, \tau' > 0$, we find

$$Z = \int \mathcal{D}^3 x \exp\left\{-\frac{1}{\hbar}\int_0^{\hbar\beta} d\tau \frac{M}{2}\dot{\mathbf{x}}^2(\tau)\right.$$
$$\left. + \frac{\mu}{2\hbar^2} 4\pi e^2 \frac{\hbar}{2\omega}\frac{1}{V}\sum_{\mathbf{k}}\frac{4\pi}{\mathbf{k}^2}\int_0^{\hbar\beta} d\tau \int_0^{\hbar\beta} d\tau' e^{i\mathbf{k}[\mathbf{x}(\tau)-\mathbf{x}(\tau')]}e_{\text{per}}^{-\omega|\tau-\tau'|}\right\}. \tag{5.345}$$

Performing the sum over all wave vectors $\mathbf{k}$ using the formula

$$\frac{1}{V} \sum_{\mathbf{k}} \frac{4\pi}{\mathbf{k}^2} e^{i\mathbf{k}\mathbf{x}} = \frac{1}{|\mathbf{x}|}, \tag{5.346}$$

we obtain the path integral

$$Z = \int \mathcal{D}^3 x \, \exp\left\{-\frac{1}{\hbar}\left[\int_0^{\hbar\beta} d\tau \frac{M}{2}\dot{\mathbf{x}}^2(\tau) - \frac{a}{2\sqrt{2}}\int_0^{\hbar\beta} d\tau \int_0^{\hbar\beta} d\tau' \frac{e_{\text{per}}^{-\omega|\tau-\tau'|}}{|\mathbf{x}(\tau)-\mathbf{x}(\tau')|}\right]\right\}, \tag{5.347}$$

where

$$a \equiv \frac{\mu}{\hbar} 4\pi e^2 \frac{\hbar}{2\omega}\sqrt{2}. \tag{5.348}$$

The factor $\sqrt{2}$ is a matter of historic convention. Staying with this convention, we use the characteristic length scale (9.73) associated with the mass M and the frequency ω:

$$\lambda_\omega \equiv \sqrt{\frac{\hbar}{M\omega}}. \tag{5.349}$$

This length scale will appear in the wave functions of the harmonic oscillator in Eq. (9.73). Using this we introduce a dimensionless coupling constant α defined by

$$\alpha \equiv \frac{1}{\sqrt{2}}\left(\frac{1}{\epsilon_\omega} - \frac{1}{\epsilon_0}\right)\frac{e^2}{\hbar\omega\lambda_\omega}. \tag{5.350}$$

A typical value of α is 5 for sodium chloride. In different crystals it varies between 1 and 20, thus requiring a strong-coupling treatment. In terms of α, one has

$$a = \hbar\omega^2\lambda_\omega\alpha. \tag{5.351}$$

The expression (5.348) is the famous path integral of the polaron problem written down in 1955 by Feynman [20] and solved approximately by a variational perturbation approach.

In order to allow for later calculations of a particle density in an external potential, we decompose the paths in a Fourier series with fixed endpoints $\mathbf{x}(\beta) \equiv \mathbf{x}_b = \mathbf{x}_a$:

$$\mathbf{x}(\tau) = \mathbf{x}_a + \sum_{n=1}^\infty \mathbf{x}_n \sin\nu_n\tau, \quad \nu_n \equiv n\pi/\hbar\beta. \tag{5.352}$$

The path integral is then the limit $N \to \infty$ of the product of integrals

$$\int \mathcal{D}^3 x \equiv \int \frac{d^2 x_a}{\sqrt{2\pi\hbar^2\beta/M}^3} \prod_{n=1}^\infty \left[\int \frac{d^3 x_n}{\sqrt{4\pi/M\nu_n^2\beta}^3}\right]. \tag{5.353}$$

The correctness of this measure is verified by considering the free particle in which case the action is

$$\mathcal{A}_0 = \frac{1}{2}\sum_{n=1}^\infty A_n^0 \mathbf{x}_n^2, \quad \text{with} \quad A_n^0 \equiv \frac{M}{2}\hbar\beta\nu_n^2. \tag{5.354}$$

The Fourier components $\mathbf{x}_n$ can be integrated leaving a final integral

$$Z = \int \frac{d^2 x_a}{\sqrt{2\pi\hbar^2\beta/M}^3}, \qquad (5.355)$$

which is the correct partition function of a free particle [compare with the one-dimensional expression (3.808)].

The endpoints $\mathbf{x}_b = \mathbf{x}_a$ do not appear in the integrand of (5.347) as a manifestation of translational invariance. The integral over the endpoints produces therefore a total volume factor V. We may imagine performing the path integral with fixed endpoints which produces the particle density.

5.22.2 Harmonic Trial System

The harmonic trial system used by Feynman as a starting point of his variational treatment has the generating functional

$$Z^{\Omega,C}[\mathbf{j}] = \int \mathcal{D}^3 x \, \exp\left\{ -\frac{1}{\hbar} \int_0^{\hbar\beta} d\tau \left[\frac{M}{2}\dot{\mathbf{x}}^2(\tau) - \mathbf{j}(\tau)\mathbf{x}(\tau) \right] \right.$$
$$\left. -\frac{C}{2}\frac{M}{\hbar} \int_0^{\hbar\beta} d\tau \int_0^{\hbar\beta} d\tau' \, [\mathbf{x}(\tau) - \mathbf{x}(\tau')]^2 \, e_{\mathrm{per}}^{-\Omega|\tau-\tau'|} \right\}. \qquad (5.356)$$

The external current is Fourier-decomposed in the same way as $\mathbf{x}(\tau)$ in (5.352). To preserve translational invariance, we assume the current to vanish at the endpoints: $\mathbf{j}_a = 0$. Then the first two terms in the action in (5.356) are

$$\mathcal{A}[\mathbf{j}] = \frac{1}{2} \sum_{n=1}^{\infty} \left(A_n^0 \mathbf{x}_n^2 - \beta \, \mathbf{j}_n \mathbf{x}_n \right). \qquad (5.357)$$

The Fourier decomposition of the double integral in (5.356) reads

$$\int_0^{\hbar\beta} d\tau \int_0^{\hbar\beta} d\tau' \left[\sum_{n=1}^{\infty} (\sin\nu_n\tau - \sin\nu_n\tau') \, \mathbf{x}_n \right]^2 e_{\mathrm{per}}^{-\Omega|\tau-\tau'|}. \qquad (5.358)$$

With the help of trigonometric identities and a change of variables to $\sigma = (\tau+\tau')/2$ and $\Delta\tau = (\tau-\tau')/2$, this becomes

$$2\int_0^{\hbar\beta} d\sigma \int_0^{2\hbar\beta} d\Delta\tau \left[\sum_{n=1}^{\infty} \cos\nu_n\sigma \, \sin(\nu_n\Delta\tau/2) \, \mathbf{x}_n \right]^2 e_{\mathrm{per}}^{-\Omega|\Delta\tau|}. \qquad (5.359)$$

Integrating out σ leaves

$$\hbar\beta \int_0^{2\hbar\beta} d\Delta\tau \sum_{n=1}^{\infty} \sin^2(\nu_n\Delta\tau/2) \, \mathbf{x}_n^2 \, e_{\mathrm{per}}^{-\Omega|\Delta\tau|}, \qquad (5.360)$$

and performing the integral over $\delta\tau$ gives for (5.358) the result

$$\frac{\hbar\beta}{2}\left(1-e^{-2\hbar\beta\Omega}\right)\left(1+2\sum_{n'=1}^{\infty}e^{-n'\hbar\beta\Omega}\right)\sum_{n=1}^{\infty}\mathbf{x}_n^2\frac{\nu_n^2}{\nu_n^2+\Omega^2}. \tag{5.361}$$

Hence we can write the interaction term in (5.356) as

$$-\frac{C_\beta M\beta}{4\Omega}\sum_{n=1}^{\infty}\mathbf{x}_n^2\frac{\nu_n^2}{\nu_n^2+\Omega^2}, \qquad C_\beta\equiv C\left(1-e^{-2\hbar\beta\Omega}\right)\coth\frac{\hbar\beta\Omega}{2}. \tag{5.362}$$

This changes A_n^0 in (5.357) into

$$A_n\equiv\frac{M\hbar\beta\nu_n^2}{2}\left(1+\frac{C_\beta}{\Omega}\frac{1}{\nu_n^2+\Omega^2}\right)=A_n^0\left(1+\frac{C_\beta}{\Omega}\frac{1}{\nu_n^2+\Omega^2}\right). \tag{5.363}$$

The trial partition function without external source is then approximately equal to

$$Z^{\Omega,C}\underset{T\approx 0}{\equiv}\int\mathcal{D}^3x\int\frac{d^2x_a}{\sqrt{2\pi\hbar^2\beta/M}^3}\prod_{n=1}^{\infty}\sqrt{\frac{A_n^0}{A_n}}^3. \tag{5.364}$$

The product is calculated as follows:

$$\prod_{n=1}^{\infty}\sqrt{\frac{A_n^0}{A_n}}^3=\prod_{n=1}^{\infty}\sqrt{1+\frac{C_\beta}{\Omega}\frac{1}{\nu_n^2+\Omega^2}}^{-3}=e^{-\beta F^{\Omega,C}}, \tag{5.365}$$

resulting in the approximate free energy

$$F^{\Omega,C}\underset{T\approx 0}{=}\frac{1}{\beta}\sum_{n=1}^{\infty}\log\frac{\nu_n^2+\Gamma_\beta^2}{\nu_n^2+\Omega^2}, \tag{5.366}$$

where we have introduced the function of the trial frequency Ω:

$$\Gamma_\beta^2(\Omega)\equiv\Omega^2+C_\beta/\Omega. \tag{5.367}$$

With the help of formula (2.171) we find therefore

$$F^{\Omega,C}=\frac{3}{2\beta}\log\frac{\sinh\hbar\beta\Gamma_\beta}{\sinh\hbar\beta\Omega}. \tag{5.368}$$

For simplicity, we shall from now on consider only the low-temperature regime where $C_\beta-> C$, $\Gamma_\beta\to\Omega^2+C/\Omega$, and the free energy (5.368) becomes approximately

$$F^{\Omega,C}\underset{T\approx 0}{=}\frac{3\hbar}{2}\left(\Gamma-\Omega\right)\equiv E_0^{\Omega,C}. \tag{5.369}$$

The right-hand side is the ground state energy of the harmonic trial system (5.356).

In Feynman's variational approach, the ground state energy of the polaron is smaller than this given by the minimum of [compare (5.18), (5.32) and (5.45)]

$$E_0 \leq E_0^{\Omega,C} + \Delta E_{\text{int}}^{\Omega,C} - \Delta E_{\text{int,harm}}^{\Omega,C}, \tag{5.370}$$

where the two additional terms are the limits $\beta \to \infty$ of the harmonic expectation values

$$\Delta E_{\text{int}}^{\Omega,C} = -\frac{1}{\hbar\beta} \langle \mathcal{A}_{\text{int}} \rangle^{\Omega,C} \equiv \frac{1}{\hbar\beta} \left\langle -\frac{a}{2\sqrt{2}} \int_0^{\hbar\beta} d\tau \int_0^{\hbar\beta} d\tau' \frac{e^{-\omega|\tau-\tau'|}}{|\mathbf{x}(\tau) - \mathbf{x}(\tau')|} \right\rangle^{\Omega,C}, \tag{5.371}$$

and

$$\Delta E_{\text{int,harm}}^{\Omega,C} = -\frac{1}{\hbar\beta} \langle \mathcal{A}_{\text{int,harm}} \rangle^{\Omega,C} \equiv \frac{1}{\hbar\beta} \left\langle \frac{C}{2} \int_0^{\hbar\beta} d\tau \int_0^{\hbar\beta} d\tau' \, [\mathbf{x}(\tau) - \mathbf{x}(\tau')]^2 \, e^{-\Omega|\tau-\tau'|} \right\rangle^{\Omega,C}. \tag{5.372}$$

The calculation of the first expectation value is most easily done using the Fourier decomposition (5.345), where we must find the expectation value

$$\left\langle e^{i\mathbf{k}[\mathbf{x}(\tau) - \mathbf{x}(\tau')]} \right\rangle^{\Omega,C}. \tag{5.373}$$

In the trial path integral (5.356), the exponential corresponds to a source

$$\mathbf{j}(\tau'') = \hbar\mathbf{k} \left[\delta^{(3)}(\tau - \tau'') - \delta^{(3)}(\tau' - \tau'') \right], \tag{5.374}$$

in terms of which (5.373) reads

$$\left\langle e^{i \int d\tau \mathbf{j}(\tau)\mathbf{x}(\tau)/\hbar} \right\rangle^{\Omega,C}. \tag{5.375}$$

Introducing the correlation function

$$\langle x_i(\tau) x_j(\tau') \rangle^{\Omega,C} \equiv \delta_{ij} G^{\Omega,\Gamma}(\tau, \tau'), \tag{5.376}$$

and using Wick's rule (3.306) for harmonically fluctuating paths, the expectation value (5.375) is equal to

$$\left\langle e^{i \int d\tau \mathbf{j}(\tau)\mathbf{x}(\tau)/\hbar} \right\rangle^{\Omega,C} = \exp\left\{ -\frac{1}{2\hbar^2} \int_0^{\hbar\beta} d\tau \int_0^{\hbar\beta} d\tau' \, \mathbf{j}(\tau) \, G^{\Omega,\Gamma}(\tau, \tau') \, \mathbf{j}(\tau') \right\}. \tag{5.377}$$

Inserting the special source (5.373), we obtain

$$\left\langle e^{i\mathbf{k}[\mathbf{x}(\tau) - \mathbf{x}(\tau')]} \right\rangle^{\Omega,C} = I^{\Omega,C}(\mathbf{k}, \tau, \tau') \equiv \exp\left[\mathbf{k}^2 \, \bar{G}^{\Omega,\Gamma}(\tau, \tau') \right], \tag{5.378}$$

where the exponent contains the subtracted Green function

$$\bar{G}^{\Omega,\Gamma}(\tau, \tau') \equiv G^{\Omega,\Gamma}(\tau, \tau') - \frac{1}{2} G^{\Omega,\Gamma}(\tau, \tau) - \frac{1}{2} G^{\Omega,\Gamma}(\tau', \tau'). \tag{5.379}$$

The Green function $G^{\Omega,\Gamma}(\tau, \tau')$ itself has the Fourier expansion

$$G^{\Omega,\Gamma}(\tau, \tau') = \hbar \sum_{n=1}^{\infty} \frac{\sin \nu_n \tau \sin \nu_n \tau'}{2A_n/\hbar\beta} = \frac{\hbar}{M} \sum_{n=1}^{\infty} \frac{\nu_n^2 + \Omega^2}{\nu_n^2(\nu_n^2 + \Gamma_\beta^2)} \sin \nu_n \tau \sin \nu_n \tau'. \quad (5.380)$$

It solves the Euler-Lagrange equation which extremizes the action in (5.356) for a source $\mathbf{j}(\tau) = M\boldsymbol{\delta}(\tau - \tau')$:

$$\left\{ -\partial_\tau^2 + 2C \int_0^{\hbar\beta} d\tau' \left[x_i(\tau) - x_i(\tau') \right] e_{\text{per}}^{-\Omega|\tau-\tau'|} \right\} G^{\Omega,\Gamma}(\tau, \tau') = \delta(\tau - \tau'). \quad (5.381)$$

Decomposing

$$\frac{\nu_n^2 + \Omega^2}{\nu_n^2(\nu_n^2 + \Gamma_\beta^2)} = \frac{\Omega^2}{\Gamma_\beta^2} \times \frac{1}{\nu_n^2} + \frac{\Gamma_\beta^2 - \Omega^2}{\Gamma_\beta^2} \times \frac{1}{\nu_n^2 + \Gamma_\beta^2}, \quad (5.382)$$

we obtain a combination of ordinary Green functions of the second-order operator differential equation (3.236), but with Dirichlet boundary conditions. For such Green functions, the spectral sum over n was calculated in Section 3.4 for real time [see (3.36) and (3.145)]. The imaginary-time result is

$$G_{\omega^2}(\tau, \tau') = \sum_{n=1}^{\infty} \frac{\sin \nu_n \tau \sin \nu_n \tau'}{\nu_n^2 + \omega^2} = \frac{\sinh \omega(\hbar\beta - \tau) \sinh \omega\tau'}{\omega \sinh \omega\hbar\beta}, \quad \text{for } \tau > \tau' > 0. \quad (5.383)$$

In the low-temperature limit, this becomes

$$G_{\omega^2}(\tau, \tau') = \frac{1}{2\omega} \left(e^{-\omega(\tau-\tau')} - e^{-\omega(\tau+\tau')} \right), \quad \text{for } \tau > \tau' > 0, \quad (5.384)$$

such that

$$\bar{G}_{\Gamma^2}(\tau, \tau') = \frac{1}{2\Gamma} \left(e^{-\Gamma(\tau-\tau')} - 1 - e^{-\Gamma(\tau+\tau')} + \frac{1}{2}e^{-2\Gamma\tau} + \frac{1}{2}e^{-2\Gamma\tau'} \right), \quad \text{for } \tau > \tau' > 0. \quad (5.385)$$

In the limit $\Gamma \to 0$, this becomes $-\frac{1}{2}|\tau - \tau'|$. We therefore obtain at zero temperature

$$\bar{G}^{\Omega,\Gamma}(\tau, \tau') \underset{T=0}{=} \frac{\hbar}{2M} \left[\frac{\Omega^2}{\Gamma^2} \bar{G}_0(\tau, \tau') + \frac{\Gamma^2 - \Omega^2}{\Gamma^2} \bar{G}_0(\tau, \tau') \right] \quad (5.386)$$

$$= -\frac{\hbar}{2M} \left\{ \frac{\Omega^2}{\Gamma^2} |\tau - \tau'| + \frac{\Gamma^2 - \Omega^2}{\Gamma^3} \left(1 - e^{-\Gamma|\tau-\tau'|} + e^{-\Gamma(\tau+\tau')} - \frac{1}{2}e^{-2\Gamma\tau} - \frac{1}{2}e^{-2\Gamma\tau'} \right) \right\},$$

to be inserted into (5.378) to get the expectation value $\left\langle e^{i\mathbf{k}[\mathbf{x}(\tau)-\mathbf{x}(\tau')]} \right\rangle$. The last three terms can be avoided by shifting the time interval under consideration and thus the Fourier expansion (5.352) from $(0, \hbar\beta)$ to $(-\hbar\beta/2, \hbar\beta/2)$, which changes Green function (5.387) to

$$G_{\omega^2}(\tau, \tau') = \sum_{n=1}^{\infty} \frac{\sin \nu_n(\tau + \hbar\beta/2) \sin \nu_n(\tau' + \hbar\beta/2)}{\nu_n^2 + \omega^2}$$

$$= \frac{\sinh \omega(\hbar\beta/2 - \tau) \sinh \omega(\tau' + \hbar\beta/2)}{\omega \sinh \omega\hbar\beta}, \quad \text{for } \tau > \tau' > 0. \quad (5.387)$$

We have seen before at the end of Section 3.20 that such a shift is important when discussing the limit $T \to 0$ which we want to do in the sequel. With the symmetric limits of integration, the Green function (5.384) looses its last term [compare with (3.147) for real times] and (5.386) simplifies to

$$\bar{G}^{\Omega,\Gamma}(\tau,\tau') \underset{T=0}{\approx} -\frac{\hbar}{2M}\left\{\frac{\Omega^2}{\Gamma^2}|\tau-\tau'| + \frac{\Gamma^2-\Omega^2}{\Gamma^3}\left(1-e^{-\Gamma|\tau-\tau'|}\right)\right\}. \quad (5.388)$$

At any temperature, we have the complicated expression for $\tau > \tau'$:

$$\bar{G}^{\Omega,\Gamma}(\tau,\tau') = -\frac{\hbar}{2M}\left\{\frac{\Omega^2}{\Gamma_\beta^2}\left[\tau-\tau'-\frac{1}{\hbar\beta}(\tau-\tau')^2\right]\right. \quad (5.389)$$
$$\left. -\frac{\Gamma_\beta^2-\Omega^2}{\Gamma_\beta^2\sinh\hbar\beta\Gamma_\beta}[\sinh\Gamma_\beta(\hbar\beta/2-\tau)\,\sinh\Gamma_\beta(\tau'+\hbar\beta/2)-(\tau'\to\tau)-(\tau\to\tau')]\right\}.$$

With the help of the Fourier integral (5.346) we find from this the expectation value of the interaction in (5.347):

$$\left\langle\frac{1}{|\mathbf{x}(\tau)-\mathbf{x}(\tau')|}\right\rangle = \int\frac{d^3k}{(2\pi)^3}\frac{4\pi}{\mathbf{k}^2}\left\langle e^{i\mathbf{k}[\mathbf{x}(\tau)-\mathbf{x}(\tau')]}\right\rangle = \int\frac{d^3k}{(2\pi)^3}\frac{4\pi}{\mathbf{k}^2}I^{\Omega,C}(\mathbf{k},\tau,\tau'). \quad (5.390)$$

For zero temperature, this leads directly to the expectation value of the interaction in (5.347):

$$\int_0^{\hbar\beta}d\tau\int_0^{\hbar\beta}d\tau'\left\langle\frac{e^{-\omega|\tau-\tau'|}}{|\mathbf{x}(\tau)-\mathbf{x}(\tau')|}\right\rangle \approx 2\hbar\beta\int_0^{\hbar\beta/2}d\Delta\tau\,e^{-\omega\Delta\tau}\int\frac{d^3k}{(2\pi)^3}\frac{4\pi}{\mathbf{k}^2}e^{\mathbf{k}^2\bar{G}^{\Omega,\Gamma}(\Delta\tau,0)}$$
$$\approx 4\frac{\hbar\beta}{\sqrt{2\pi\omega\lambda_\omega}}\int_0^\infty d\Delta\tau\frac{e^{-\omega\Delta\tau}}{\sqrt{-2\bar{G}^{\Omega,\Gamma}(\Delta\tau,0)}}. \quad (5.391)$$

The expectation value of the harmonic trial interaction in (5.356), on the other hand, is simply found from the correlation function (5.376) [or equivalently from the second derivative of $I^{\Omega,C}(\mathbf{k},\tau,\tau')$ with respect to the momenta]:

$$\left\langle[\mathbf{x}(\tau)-\mathbf{x}(\tau')]^2\right\rangle^{\Omega,C} = -6\bar{G}^{\Omega,\Gamma}(\tau,\tau'). \quad (5.392)$$

At low temperatures, this leads to an integral

$$\frac{CM}{2\hbar}\int_0^{\hbar\beta}d\tau\int_0^{\hbar\beta}d\tau'\left\langle[\mathbf{x}(\tau)-\mathbf{x}(\tau')]^2\right\rangle^{\Omega,C}e_{\text{per}}^{-\Omega|\tau-\tau'|}\underset{T=0}{=}\hbar\beta\frac{3C}{4\Omega\Gamma}. \quad (5.393)$$

This expectation value contributes to the ground state energy a term

$$\Delta E_{\text{int, var}}^{\Omega,C} = \frac{3\hbar C}{4\Omega\Gamma}. \quad (5.394)$$

Note that this term can be derived from the derivative of the ground state energy (5.368) as $C\partial_C E_0^{\Omega,C}$. Together with $-a/2\sqrt{2}$ times the result of (5.391), the inequality (5.370) for the ground state energy becomes

$$E_0 \leq \frac{3\hbar}{4\Gamma}(\Gamma - \Omega)^2 - \hbar\omega \frac{\alpha\omega}{\sqrt{\pi\omega}} \int_0^\infty d\Delta\tau \frac{e^{-\omega\Delta\tau}}{\sqrt{-2\bar{G}^{\Omega,\Gamma}(\Delta\tau,0)}}. \tag{5.395}$$

This has to be minimized in Ω and C, or equivalently, in Ω and Γ. Considering the low-temperature limit, we have taken the upper limit of integration to infinity (the frequency ω corresponds usually to temperatures of the order of 1000 K).

For small α, the optimal parameters Ω and Γ differ by terms of order α. We can therefore expand the integral in (5.395) and find that the minimum lies, in natural units with $\hbar = \omega = 1$, at $\Omega = 3$ and $\Gamma = 3[1 + 2\alpha(1 - P)/3\Gamma$, where $P = 2[(1 - \Gamma)^{1/2} - 1]$. From this we obtain the upper bound

$$E_0 \leq -\alpha - \frac{\alpha^2}{81} + \ldots \approx -\alpha - 0.0123\alpha^2 + \ldots . \tag{5.396}$$

This agrees well with the perturbative result [21]

$$E_w^{ex} = -\alpha - 0.0159196220\alpha^2 - 0.000806070048\alpha^3 - O(\alpha^4). \tag{5.397}$$

The second term has the exact value

$$\left[\frac{1}{\sqrt{2}} - \log\left(1 + 3\sqrt{2}/4\right)\right]\alpha^2. \tag{5.398}$$

In the strong-coupling region, the best parameters are $\Omega = 1$, $\Gamma = 4\alpha^2/9\pi - [4(\log 2 + \gamma/2) - 1]$, where $\gamma \approx 0.5773156649$ is the Euler-Mascheroni constant (2.467). At these values, we obtain the upper bound

$$E_0 \leq -\frac{\alpha^2}{3\pi} - 3\left(\frac{1}{4} + \log 2\right) + \mathcal{O}(\alpha^{-2}) \approx -0.1061\alpha - 2.8294 + \mathcal{O}(\alpha^{-2}). \tag{5.399}$$

This agrees reasonably well with the precise strong-coupling expansion [23].

$$E_s^{ex} = -0.108513\alpha^2 - 2.836 - O(\alpha^{-2}). \tag{5.400}$$

The numerical results for variational parameters and energy are shown in Table 5.12.

5.22.3 Effective Mass

By performing a shift in the velocity of the path integral (5.347), Feynman calculated also an effective mass for the polaron. The result is

$$M^{eff} = M\left[1 + \frac{\alpha}{3}\frac{\omega}{\sqrt{\pi\omega}}\Gamma^2 \int_0^\infty d\Delta\tau \frac{(\Delta\tau)^2 e^{-\omega\Delta\tau}}{\sqrt{-2\bar{G}^{\Omega,\Gamma}(\Delta\tau,0)}}\right]. \tag{5.401}$$

Table 5.12 Numerical results for variational parameters and energy.

α	Γ	Ω	E_0	$\Delta E_0^{(2)}$	E_{tot}	correction
1	3.110	2.871	-1.01	-0.0035	-1.02	0.35 %
3	3.421	2.560	-3.13	-0.031	-3.16	1.0 %
5	4.034	2.140	-5.44	-0.083	-5.52	1.5 %
7	5.810	1.604	-8.11	-0.13	-8.24	1.6 %
9	9.850	1.282	-11.5	-0.17	-11.7	1.4 %
11	15.41	1.162	-15.7	-0.22	-15.9	1.4 %
15	30.08	1.076	-26.7	-0.39	-27.1	1.5 %

The reduced effective mass $m \equiv M^{\text{eff}}/M$ has the weak-coupling expansion

$$m_{\text{w}} = 1 + \frac{\alpha}{6} + 2.469136 \times 10^{-2}\alpha^2 + 3.566719 \times 10^{-3}\alpha^3 + \dots \quad (5.402)$$

and behaves for strong couplings like

$$
\begin{aligned}
m_{\text{s}} &\approx \frac{16}{81\pi^2}\alpha^4 - \frac{4}{3\pi}(1+\log 4)\,\alpha^2 + 11.85579 + \dots \\
&\approx 0.020141\alpha^4 - 1.012775\alpha^2 + 11.85579 + \dots .
\end{aligned} \quad (5.403)
$$

The exact expansions are [31]

$$m_{\text{w}}^{\text{ex}} = 1 + \frac{\alpha}{6} + 2.362763 \times 10^{-2}\alpha^2 + O(\alpha^4), \quad (5.404)$$

$$m_{\text{s}}^{\text{ex}} = 0.0227019\alpha^4 + O(\alpha^2). \quad (5.405)$$

5.22.4 Second-Order Correction

With some effort, also the second-order contribution to the variational energy has been calculated at zero temperature [32]. It gives a contribution to the ground state energy

$$\Delta E_0^{(2)} = -\frac{1}{2\hbar\beta}\frac{1}{\hbar^2}\left\langle(\mathcal{A}_{\text{int}} - \mathcal{A}_{\text{int,harm}})^2\right\rangle_c^{\Omega,C}. \quad (5.406)$$

Recall the definitions of the interactions in Eqs. (5.371) and (5.372). There are three terms

$$\Delta E_0^{(2,1)} = -\frac{1}{2\hbar\beta}\frac{1}{\hbar^2}\left\{\left\langle\mathcal{A}_{\text{int}}^2\right\rangle^{\Omega,C} - \left[\langle\mathcal{A}_{\text{int}}\rangle^{\Omega,C}\right]^2\right\}, \quad (5.407)$$

$$\Delta E_0^{(2,2)} = 2\frac{1}{2\hbar\beta}\frac{1}{\hbar^2}\left\{\left\langle\mathcal{A}_{\text{int}}\mathcal{A}_{\text{int,harm}}\right\rangle^{\Omega,C} - \langle\mathcal{A}_{\text{int}}\rangle^{\Omega,C}\langle\mathcal{A}_{\text{int,harm}}\rangle^{\Omega,C}\right\}, \quad (5.408)$$

and

$$\Delta E_0^{(2,3)} = -\frac{1}{2\hbar\beta}\frac{1}{\hbar^2}\left\{\left\langle \mathcal{A}_{\text{int,harm}}^2\right\rangle^{\Omega,C} - \left[\left\langle \mathcal{A}_{\text{int,harm}}\right\rangle^{\Omega,C}\right]^2\right\}. \tag{5.409}$$

The second term can be written as

$$\Delta E_0^{(2,2)} = -\frac{1}{2\hbar\beta}\frac{1}{\hbar^2}\left\{-2C\partial_C\left\langle \mathcal{A}_{\text{int}}\right\rangle^{\Omega,C}\right\}, \tag{5.410}$$

the third as

$$\Delta E_0^{(2,3)} = -\frac{1}{2\hbar\beta}\frac{1}{\hbar^2}\left\{\left[1 - C\partial_C\right]\left\langle \mathcal{A}_{\text{int,harm}}\right\rangle^{\Omega,C}\right\}. \tag{5.411}$$

The final expression is rather involved and given in Appendix 5C. The second-order correction leads to the second term (5.398) found in perturbation theory. In the strong coupling limit, it changes the leading term $-\alpha^2/3\pi \approx -0.1061$ in (5.399) into

$$-\frac{1}{4\pi} - \frac{2}{\pi}\sum_{n=1}^{\infty}\frac{(2n)!}{2^{4n}(n!)^2 n(2n+1)} = \frac{-17 + 64\arcsin(\frac{\sqrt{2-\sqrt{3}}}{2}) - 32\log(4\frac{\sqrt{2-\sqrt{3}}}{2})}{4\pi}, \tag{5.412}$$

which is approximately equal to -0.1078. The corrections are shown numerically in the previous Table 5.12.

5.22.5   Polaron in Magnetic Field, Bipolarons, Small Polarons, Polaronic Excitons, and More

Feynman's solution of the polaron problem has instigated a great deal of research on this subject [34]. There are many publications dealing with a polaron in a magnetic field. In particular, there was considerable discussion on the validity of the Jensen-Peierls inequality (5.10) in the presence of a magnetic field until it was shown by Larsen in 1985 that the variational energy does indeed lie below the exact energy for sufficiently strong magnetic fields. On the basis of this result he criticized the entire approach. The problem was, however, solved by Devreese and collaborators who determined the range of variational parameters for which the inequality remained valid.

In the light of the systematic higher-order variational perturbation theory developed in this chapter we do not consider problems with the inequality any more as an obstacle to variational procedures. The optimization procedure introduced in Section 5.13 for even and odd approximations does not require an inequality. We have seen that for higher orders, the exact result will be approached rapidly with exponential convergence. The inequality is useful only in Feynman's original lowest-order variational approach where it is important to know the direction of the error. For higher orders, the importance of this information decreases rapidly since the convergence behavior allows us to estimate the limiting value quantitatively, whereas the inequality tells us merely the sign of the error which is often quite large in the lowest-order variational approach, for instance in the Coulomb system.

There is also considerable interest in bound states of two polarons called bipolarons. Such investigations have become popular since the discovery of high-temperature superconductivity.[3]

5.22.6 Variational Interpolation for Polaron Energy and Mass

Let us apply the method of variational interpolation developed in Section 5.18 to the polaron. Starting from the presently known weak-coupling expansions (5.397) and (5.404) we fix a few more expansion coefficients such that the curves fit also the strong-coupling expansions (5.400) and (5.405). We find it convenient to make the series start out with α^0 by removing an overall factor $-\alpha$ from E and deal with the quantity $-E_w^{ex}/\alpha$. Then we see from (5.400) that the correct leading power in the strong-coupling expansion requires taking $p = 1, q = 1$. The knowledge of b_0 and b_1 allows us to extend the known weak coupling expansion (5.397) by two further expansion terms. Their coefficients a_3, a_4 are solutions of the equations [recall (5.245)]

$$b_0 = \frac{35}{128}a_0 c + a_1 + \frac{15}{8}\frac{a_2}{c} + \frac{2a_3}{c^2} + \frac{a_4}{c^3}, \tag{5.413}$$

$$b_1 = \frac{35}{32}\frac{a_0}{c} - \frac{5}{4}\frac{a_2}{c^3} - \frac{a_3}{c^3}. \tag{5.414}$$

The constant c governing the growth of Ω_N for $\alpha \to \infty$ is obtained by extremizing b_0 in c, which yields the equation

$$\frac{35}{128}a_0 - \frac{15}{8}\frac{a_2}{c^2} - \frac{4a_3}{c^3} - \frac{4a_4}{c^5} = 0. \tag{5.415}$$

The simultaneous solution of (5.413)–(5.415) renders

$$c_4 = 0.09819868,$$
$$a_3 = 6.43047343 \times 10^{-4}, \tag{5.416}$$
$$a_4 = -8.4505836 \times 10^{-5}.$$

The re-expanded energy (5.238) reads explicitly as a function of α and Ω (for E including the earlier-removed factor $-\alpha$)

$$W_4(\alpha, \Omega) = a_0\alpha\left(-\frac{35}{128}\Omega - \frac{35}{32\Omega} + \frac{35}{64\Omega^3} - \frac{7}{32\Omega^5} + \frac{5}{128\Omega^7}\right) - a_1\alpha^2$$
$$+ a_2\alpha^3\left(-\frac{15}{8\Omega} + \frac{5}{4\Omega^3} - \frac{3}{8\Omega^5}\right) + a_3\alpha^4\left(-\frac{2}{\Omega^2} + \frac{1}{\Omega^4}\right) - a_4\alpha^5\frac{1}{\Omega^3}. \tag{5.417}$$

Extremizing this we find Ω_4 as a function of α [it turns out to be quite well approximated by the simple function $\Omega_4 \approx c_4\alpha + 1/(1+0.07\alpha)$]. This is to be compared with the optimal frequency obtained from minimizing the lower approximation $W_2(\alpha, \Omega)$:

$$\Omega_2^2 = 1 + \frac{4a_2}{3a_0}x^2 + \sqrt{\left(1 + \frac{4a_2}{3a_0}x^2\right)^2 - 1}, \tag{5.418}$$

which behaves like $c_2\alpha + 1 + \dots$ with $c_2 = \sqrt{8a_2/3a_0} \approx 0.120154$. The resulting energy is shown in Fig. 5.28, where it is compared with the Feynman variational energy. For completeness, we have also plotted the weak-coupling expansion, the strong-coupling expansion, the lower approximation $W_2(\alpha)$, and two Padé approximants given in Ref. [22] as upper and lower bounds to the energy.

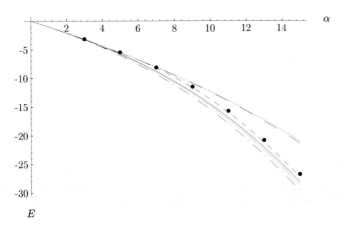

Figure 5.28 Variational interpolation of polaron energy (solid line) between the weak-coupling expansion (dashed) and the strong-coupling expansion (short-dashed) shown in comparison with Feynman's variational approximation (fat dots), which is an upper bound to the energy. The dotted curves are upper and lower bounds coming from Padé approximants [22]. The dot-dashed curve shows the variational perturbation theory $W_2(\alpha)$ which does not make use of the strong-coupling information.

Consider now the effective mass of the polaron, where the strong-coupling behavior (5.405) fixes $p = 4, q = 1$. The coefficient b_0 allows us to determine an approximate coefficient a_3 and to calculate the variational perturbation expansion $W_3(\alpha)$. From (5.245) we find the equation

$$b_0 = -a_1 c^3/8 + a_3 c, \qquad (5.419)$$

whose minimum lies at $c_3 = \sqrt{8a_2/3a_0}$ where $b_0 = \sqrt{32a_3^3/27a_1}$. Equating b_0 of Eq. (5.419) with the leading coefficient in the strong-coupling expansion (5.405), we obtain $a_3 = [27a_1 b_0^2/32]^{1/3} \approx 0.0416929$.

The variational expression for the polaron mass is from (5.238)

$$W_3(\alpha, \omega) = a_0 + a_1\alpha\left(-\frac{\Omega^3}{8} + \frac{3\Omega}{4} + \frac{3}{8\Omega}\right) + a_2\alpha^2 + a_3\alpha^3\Omega. \qquad (5.420)$$

This is extremal at

$$\Omega_3^2 = 1 + \frac{4a_3}{3a_1}x^2 + \sqrt{\left(1 + \frac{4a_3}{3a_1}x^2\right)^2 - 1}. \qquad (5.421)$$

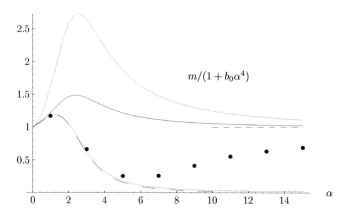

Figure 5.29 Variational interpolation of polaron effective mass between the weak-(dashed) and strong-coupling expansions (short-dashed). To see better the differences between the strongly rising functions, we have divided out the asymptotic behavior $m_{as} = 1 + b_0\alpha^4$ before plotting the curves. The fat dots show Feynman's variational approximation. The dotted curves are upper and lower bounds coming from Padé approximants [22].

From this we may find once more $c_3 = \sqrt{8a_2/3a_0}$. The approximation $W_3(\alpha) = W_3(\alpha, \Omega_3)$ for the polaron mass is shown in Fig. 5.29, where it is compared with the weak and strong-coupling expansions and with Feynman's variational result. To see better the differences between the curves which all grow fast with α, we have divided out the asymptotic behavior $m_{as} = 1 + b_0\alpha^4$ before plotting the data. As for the energy, we have again displayed two Padé approximants given in Ref. [22] as upper and lower bounds to the energy. Note that our interpolation differs considerably from Feynman's and higher order expansion coefficients in the weak- or the strong-coupling expansions will be necessary to find out which is the true behavior of the model.

Our curve has, incidentally, the strong-coupling expansion

$$m^s = 0.0227019\alpha^4 + 0.125722\alpha^2 + 1.15304 + O(\alpha^{-2}), \tag{5.422}$$

the second term $\propto \alpha^2$-term being in sharp contrast with Feynman's expression (5.403). On the weak-coupling side, a comparison of our expansion with Feynman's in Eq. (5.402) shows that our coefficient $a_3 \approx 0.0416929$ is about 10 times larger than his.

Both differences are the reason for our curve forming a positive arch in Fig. 2, whereas Feynman's has a valley. It will be interesting to find out how the polaron mass really behaves. This would be possible by calculating a few more terms in either the weak- or the strong-coupling expansion.

Note that our interpolation algorithm is much more powerful than Padé's. First, we can account for an arbitrary fractional leading power behavior α^p as $\alpha \to \infty$. Second, the successive lower powers in the strong-coupling expansion can be spaced by an arbitrary $2/q$. Third, our functions have in general a cut in the complex α-plane approximating the cuts in the function to be interpolated (see the discussion in Subsection 17.10.4). Padé approximants, in contrast, have always an integer power behavior in the strong-coupling limit, a unit spacing in the strong-coupling expansion, and poles to approximate cuts.

5.23   Density Matrices

In path integrals with fixed end points, the separate treatment of the path average $x_0 \equiv (1/\hbar\beta) \int_0^{\hbar\beta} d\tau\, x(\tau)$ looses its special virtues. Recall that the success of this separation in the variational approach was based on the fact that for fixed x_0, the fluctuation square width $a^2(x_0)$ shrinks to zero for large temperatures like $\hbar^2/12Mk_BT$ [recall (5.25)]. A similar shrinking occurs for paths whose endpoints held fixed, which is the case in path integrals for the density. Thus there is no need for a separate treatment of x_0, and one may develop a variational perturbation theory for fixed endpoints instead. These may, moreover, be taken to be different from one another $x_b \neq x_a$, thus allowing us to calculate directly density matrices.[4]

The density matrix is defined by the normalized expression

$$\rho(x_b, x_a) = \frac{1}{Z}\tilde{\rho}(x_b, x_a)\,, \tag{5.423}$$

where $\tilde{\rho}(x_b, x_a)$ is the unnormalized transition amplitude given by the path integral

$$\tilde{\rho}(x_b, x_a) = (x_b\,\hbar\beta|x_a 0) = \int_{(x_a,0)\rightsquigarrow(x_b,\hbar/k_BT)} \mathcal{D}x\, \exp\left\{-\mathcal{A}[x]/\hbar\right\}, \tag{5.424}$$

summing all paths with the fixed endpoints $x(0) = x_a$ and $x(\hbar/k_BT) = x_b$. The diagonal matrix elements of the density matrix in the integrand yield, of course, the particle density (5.90). The diagonal elements coincide with the partition function density $z(x)$ introduced in Eq. (2.331).

The partition function divided out in (5.423) is found from the trace

$$Z = \int_{-\infty}^{\infty} dx\, \tilde{\rho}(x, x). \tag{5.425}$$

5.23.1   Harmonic Oscillator

As usual in the variational approach, we shall base the approximations to be developed on the exactly solvable density matrix of the harmonic oscillator. For the sake of generality, this will be assumed to be centered around x_m, with an action

$$\mathcal{A}_{\Omega,x_m}[x] = \int_0^{\hbar/k_BT} d\tau \left\{ \frac{1}{2}M\dot{x}^2(\tau) + \frac{1}{2}M\Omega^2[x(\tau) - x_m]^2 \right\}. \tag{5.426}$$

---

[4]H. Kleinert, M. Bachmann, and A. Pelster, Phys. Rev. A *60*, 3429 (1999) (quant-ph/9812063).

Its unnormalized density matrix is [see (2.409)]

$$\tilde{\rho}_0^{\,\Omega,x_m}(x_b, x_a) = \sqrt{\frac{M\Omega}{2\pi\hbar \sinh \hbar\Omega/k_B T}}$$

$$\times \exp\left\{-\frac{M\Omega}{2\hbar \sinh \hbar\Omega/k_B T}\left[(\tilde{x}_b^2 + \tilde{x}_a^2)\cosh \hbar\Omega/k_B T - 2\tilde{x}_b\tilde{x}_a\right]\right\}, \quad (5.427)$$

with the abbreviation

$$\tilde{x}(\tau) \equiv x(\tau) - x_m. \quad (5.428)$$

At fixed endpoints x_b, x_a and oscillation center x_m, the quantum-mechanical correlation functions are given by the path integral

$$\langle O_1(x(\tau_1))\, O_2(x(\tau_2))\cdots\rangle_{x_b,x_a}^{\Omega,x_m} = \frac{1}{\tilde{\rho}_0^{\,\Omega,x_m}(x_b, x_a)}$$

$$\times \int_{(x_a,0)\rightsquigarrow(x_b,\hbar/k_B T)} \mathcal{D}x\, O_1(x(\tau_1))\, O_2(x(\tau_2))\cdots \exp\left\{-\mathcal{A}_{\Omega,x_m}[x]/\hbar\right\}. \quad (5.429)$$

The path $x(\tau)$ at a fixed imaginary time τ has a distribution

$$p(x,\tau) \equiv \langle \delta(x - x(\tau))\rangle_{x_b,x_a}^{\Omega,x_m} = \frac{1}{\sqrt{2\pi b^2(\tau)}}\exp\left[-\frac{(\tilde{x} - x_{\rm cl}(\tau))^2}{2b^2(\tau)}\right], \quad (5.430)$$

where $x_{\rm cl}(\tau)$ is the classical path of a particle in the harmonic potential

$$x_{\rm cl}(\tau) = \frac{\tilde{x}_b \sinh \Omega\tau + \tilde{x}_a \sinh \Omega(\hbar/k_B T - \tau)}{\sinh \hbar\Omega/k_B T}, \quad (5.431)$$

and $b^2(\tau)$ is the square width

$$b^2(\tau) = \frac{\hbar}{2M\Omega}\left\{\coth \frac{\hbar\Omega}{k_B T} - \frac{\cosh[\Omega(2\tau - \hbar/k_B T)]}{\sinh \hbar\Omega/k_B T}\right\}. \quad (5.432)$$

In contrast to the square width $a^2(x_0)$ in Eq. (5.24) this depends on the Euclidean time τ, which makes calculations more cumbersome than before. Since the τ lies in the interval $0 \le \tau \le \hbar/k_B T$, the width (5.432) is bounded by

$$b^2(\tau) \le \frac{\hbar}{2M\Omega}\tanh\frac{\hbar\Omega}{2k_B T}, \quad (5.433)$$

thus sharing with $a^2(x_0)$ the property of remaining finite at all temperatures. The temporal average of (5.432) is

$$\overline{b^2} = \frac{k_B T}{\hbar}\int_0^{\hbar/k_B T} d\tau\, b^2(\tau) = \frac{\hbar}{2M\Omega}\left(\coth\frac{\hbar\Omega}{k_B T} - \frac{k_B T}{\hbar\Omega}\right). \quad (5.434)$$

Just as $a^2(x_0)$, this goes to zero for $T \to \infty$. Note however, that the asymptotic behavior is

$$\overline{b^2} \xrightarrow[T\to\infty]{} \hbar\Omega/6k_B T, \quad (5.435)$$

which is twice as big as that of $a^2(x_0)$ in Eq. (5.25) (see Fig. 5.30).

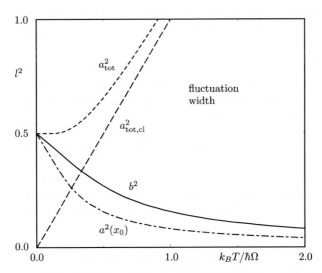

Figure 5.30 Temperature dependence of fluctuation widths of any point $x(\tau)$ on the path in a harmonic oscillator (l^2 is the generic square length in units of $\hbar/M\Omega$). The quantity a^2 (dashed) is the quantum-mechanical width, whereas $a^2(x_0)$ (dash-dotted) shares the width after separating out the fluctuations around the path average x_0. The quantity a_{cl}^2 (long-dashed) is the width of the classical distribution, and b^2 (solid curve) is the fluctuation width at fixed ends which is relevant for the calculation of the density matrix by variational perturbation theory (compare Fig. 3.14).

5.23.2 Variational Perturbation Theory for Density Matrices

To obtain a variational approximation for the density matrix, we separate the full action into the harmonic trial action and a remainder

$$\mathcal{A}[x] = \mathcal{A}_{\Omega,x_m}[x] + \mathcal{A}_{\mathrm{int}}[x], \tag{5.436}$$

with an interaction

$$\mathcal{A}_{\mathrm{int}}[x(\tau)] = \int_0^{\hbar\beta} d\tau \, V_{\mathrm{int}}(x(\tau)), \tag{5.437}$$

where the interaction potential is the difference between the original one $V(x)$ and the inserted displaced harmonic oscillator:

$$V_{\mathrm{int}}(x(\tau)) = V(x(\tau)) - \frac{1}{2}M\Omega^2[x(\tau) - x_m]^2. \tag{5.438}$$

The path integral (5.424) is then expanded perturbatively around the harmonic expression (5.427) as

$$\tilde{\rho}(x_b, x_a) = \tilde{\rho}_0^{\,\Omega,x_m}(x_b, x_a)\left[1 - \frac{1}{\hbar}\big\langle \mathcal{A}_{\mathrm{int}}[x]\big\rangle_{x_b,x_a}^{\Omega,x_m} + \frac{1}{2\hbar^2}\Big\langle \mathcal{A}_{\mathrm{int}}^2[x]\Big\rangle_{x_b,x_a}^{\Omega,x_m} - \dots\right], \tag{5.439}$$

with the harmonic expectation values defined in (5.429). The sum can be evaluated as an exponential of its connected parts, going over to the cumulant expansion:

$$\tilde{\rho}(x_b, x_a) = \tilde{\rho}_0^{\,\Omega, x_m}(x_b, x_a) \exp\left[-\frac{1}{\hbar}\langle \mathcal{A}_{\text{int}}[x]\rangle_{x_b, x_a, c}^{\Omega, x_m} + \frac{1}{2\hbar^2}\langle \mathcal{A}_{\text{int}}^2[x]\rangle_{x_b, x_a, c}^{\Omega, x_m} - \dots\right], \quad (5.440)$$

where the cumulants are defined as usual [see (3.482), (3.483)]. The series (5.440) is truncated after the N-th term, resulting in the N-th order approximant for the quantum statistical density matrix

$$\tilde{\rho}_N^{\,\Omega, x_m}(x_b, x_a) = \tilde{\rho}_0^{\,\Omega, x_m}(x_b, x_a) \exp\left[\sum_{n=1}^N \frac{(-1)^n}{n!\hbar^n}\langle \mathcal{A}_{\text{int}}^n[x]\rangle_{x_b, x_a, c}^{\Omega, x_m}\right], \quad (5.441)$$

which explicitly depends on the two variational parameters Ω and x_m.

By analogy with classical statistics, where the Boltzmann distribution in configuration space is controlled by the classical potential $V(x)$ according to [recall (2.352)]

$$\tilde{\rho}_{\text{cl}}(x) = \sqrt{\frac{M}{2\pi\hbar^2\beta}}\exp\left[-\beta V(x)\right], \quad (5.442)$$

we shall now work with the alternative type of *effective classical potential* $\tilde{V}^{\text{eff,cl}}(x_a, x_b)$ introduced in Subsection 3.25.4. It governs the unnormalized density matrix [see Eq. (3.834)]

$$\tilde{\rho}(x_b, x_a) = \sqrt{\frac{M}{2\pi\hbar^2\beta}}\exp\left[-\beta\tilde{V}^{\text{eff,cl}}(x_b, x_a)\right]. \quad (5.443)$$

Variational approximations to $\tilde{V}^{\text{eff,cl}}(x_b, x_a)$ of Nth order are obtained from (5.427), (5.441), and (5.443) as a cumulant expansion

$$\begin{aligned}
\tilde{W}_N^{\Omega, x_m}(x_b, x_a) &= \frac{1}{2\beta}\ln\frac{\sinh\hbar\beta\Omega}{\hbar\beta\Omega} + \frac{M\Omega}{2\hbar\beta\sinh\hbar\beta\Omega}\left\{(\tilde{x}_b^2 + \tilde{x}_a^2)\cosh\hbar\beta\Omega - 2\tilde{x}_b\tilde{x}_a\right\} \\
&\quad - \frac{1}{\beta}\sum_{n=1}^N \frac{(-1)^n}{n!\hbar^n}\langle \mathcal{A}_{\text{int}}^n[x]\rangle_{x_b, x_a, c}^{\Omega, x_m}.
\end{aligned} \quad (5.444)$$

They have to be optimized in the variational parameters Ω and x_m for a pair of endpoints x_b, x_a. The result is denoted by $\tilde{W}_N(x_b, x_a)$. The optimal values $\Omega(x_a, x_b)$ and $x_m(x_a, x_b)$ are denoted by $\Omega_N(x_a, x_b), x_m^N(x_a, x_b)$. The Nth-order approximation for the normalized density matrix is then given by

$$\rho_N(x_b, x_a) = Z_N^{-1}\tilde{\rho}_N^{\,\Omega_N^2, x_m^N}(x_b, x_a), \quad (5.445)$$

where the corresponding partition function reads

$$Z_N = \int_{-\infty}^{\infty} dx_a\, \tilde{\rho}_N^{\,\Omega_N^2, x_m^N}(x_a, x_a). \quad (5.446)$$

In principle, one could also optimize the entire ratio (5.445), but this would be harder to do in practice. Moreover, the optimization of the unnormalized density matrix is the only option, if the normalization diverges due to singularities of the potential, for example in the hydrogen atom

5.23.3   Smearing Formula for Density Matrices

In order to calculate the connected correlation functions in the variational perturbation expansion (5.441), we must find efficient formulas for evaluating expectation values (5.429) of any power of the interaction (5.437)

$$
\langle \, \mathcal{A}_{\mathrm{int}}^n[x] \, \rangle_{x_b,x_a}^{\Omega,x_m} \;=\; \frac{1}{\tilde{\rho}_0^{\,\Omega,x_m}(x_b,x_a)} \int_{\tilde{x}_a,0}^{\tilde{x}_b,\hbar\beta} \mathcal{D}\tilde{x} \prod_{l=1}^{n} \left[\int_0^{\hbar\beta} d\tau_l \, V_{\mathrm{int}}(\tilde{x}(\tau_l) + x_m) \right]
$$

$$
\times \; \exp\left\{ -\frac{1}{\hbar} \mathcal{A}_{\Omega,x_m}[\tilde{x} + x_m] \right\}. \tag{5.447}
$$

This can be done by an extension of the smearing formula (5.30). For this we rewrite the interaction potential as

$$
V_{\mathrm{int}}(\tilde{x}(\tau_l) + x_m) = \int_{-\infty}^{\infty} dz_l \, V_{\mathrm{int}}(z_l + x_m) \int_{-\infty}^{\infty} \frac{d\lambda_l}{2\pi} e^{i\lambda_l z_l} \exp\left[-\int_0^{\hbar\beta} d\tau \, i\lambda_l \delta(\tau - \tau_l)\tilde{x}(\tau) \right],
$$

$$
\tag{5.448}
$$

and introduce a current

$$
J(\tau) = \sum_{l=1}^{n} i\hbar\lambda_l \delta(\tau - \tau_l), \tag{5.449}
$$

so that (5.447) becomes

$$
\langle \, \mathcal{A}_{\mathrm{int}}^n[x] \, \rangle_{x_b,x_a}^{\Omega,x_m} = \frac{1}{\tilde{\rho}_0^{\,\Omega,x_m}(x_b,x_a)} \prod_{l=1}^{n} \left[\int_0^{\hbar\beta} d\tau_l \int_{-\infty}^{\infty} dz_l V_{\mathrm{int}}(z_l + x_{\min}) \int_{-\infty}^{\infty} \frac{d\lambda_l}{2\pi} e^{i\lambda_l z_l} \right] K^{\Omega,x_m}[j].
$$

$$
\tag{5.450}
$$

The kernel $K^{\Omega,x_m}[j]$ represents the generating functional for all correlation functions of the displaced harmonic oscillator

$$
K^{\Omega,x_m}[j] = \int_{\tilde{x}_a,0}^{\tilde{x}_b,\hbar\beta} \mathcal{D}\tilde{x} \, \exp\left\{ -\frac{1}{\hbar} \int_0^{\hbar\beta} d\tau \left[\frac{m}{2}\dot{\tilde{x}}^2(\tau) + \frac{1}{2}M\Omega^2\tilde{x}^2(\tau) + j(\tau)\tilde{x}(\tau) \right] \right\}. \tag{5.451}
$$

For zero current j, this generating functional reduces to the Euclidean harmonic propagator (5.427):

$$
K^{\Omega,x_m}[j = 0] = \tilde{\rho}_0^{\,\Omega,x_m}(x_b,x_a), \tag{5.452}
$$

and the solution of the functional integral (5.451) is given by (recall Section 3.1)

$$
K^{\Omega,x_m}[j] \;=\; \tilde{\rho}_0^{\,\Omega,x_m}(x_b,x_a) \exp\left[-\frac{1}{\hbar} \int_0^{\hbar\beta} d\tau \, j(\tau) \, x_{\mathrm{cl}}(\tau) \right.
$$

$$
\left. +\frac{1}{2\hbar^2} \int_0^{\hbar\beta} d\tau \int_0^{\hbar\beta} d\tau' \, j(\tau) \, G_{\Omega^2}^{(2)}(\tau,\tau') \, j(\tau') \right], \tag{5.453}
$$

where $x_{cl}(\tau)$ denotes the classical path (5.431) and $G_{\Omega^2}^{(2)}(\tau, \tau')$ the harmonic Green function with Dirichlet boundary conditions (3.386), to be written here as

$$G_{\Omega^2}^{(2)}(\tau, \tau') = \frac{\hbar}{2M\Omega} \frac{\cosh \Omega(|\tau - \tau'| - \hbar\beta) - \cosh \Omega(\tau + \tau' - \hbar\beta)}{\sinh \hbar\beta\Omega}. \tag{5.454}$$

The expression (5.453) can be simplified by using the explicit expression (5.449) for the current j. This leads to a generating functional

$$K^{\Omega, x_m}[j] = \tilde{\rho}_0^{\,\Omega, x_m}(x_b, x_a) \, \exp\left(-i\boldsymbol{\lambda}^T \mathbf{x}_{cl} - \frac{1}{2}\boldsymbol{\lambda}^T G \boldsymbol{\lambda}\right), \tag{5.455}$$

where we have introduced the n-dimensional vectors $\boldsymbol{\lambda} = (\lambda_1, \ldots, \lambda_n)$ and $\mathbf{x}_{cl} = (x_{cl}(\tau_1), \ldots, x_{cl}(\tau_n))^T$ with the superscript T denoting transposition, and the symmetric $n \times n$-matrix G whose elements are $G_{kl} = G_{\Omega^2}^{(2)}(\tau_k, \tau_l)$. Inserting (5.455) into (5.450), and performing the integrals with respect to $\lambda_1, \ldots, \lambda_n$, we obtain the n-th order smearing formula for the density matrix

$$\langle \mathcal{A}_{int}^n[x] \rangle_{x_b, x_a}^{\Omega, x_m} = \prod_{l=1}^{n} \left[\int_0^{\hbar\beta} d\tau_l \int_{-\infty}^{\infty} dz_l \, V_{int}(z_l + x_m)\right]$$

$$\times \frac{1}{\sqrt{(2\pi)^n \det G}} \exp\left\{-\frac{1}{2} \sum_{k,l=1}^{n} [z_k - x_{cl}(\tau_k)] \, G_{kl}^{-1} \, [z_l - x_{cl}(\tau_l)]\right\}. \tag{5.456}$$

The integrand contains an n-dimensional Gaussian distribution describing both thermal and quantum fluctuations around the harmonic classical path $x_{cl}(\tau)$ of Eq. (5.431) in a trial oscillator centered at x_m, whose width is governed by the Green functions (5.454).

For closed paths with coinciding endpoints ($x_b = x_a$), formula (5.456) leads to the n-th order smearing formula for particle densities

$$\rho(x_a) = \frac{1}{Z}\tilde{\rho}(x_a, x_a) = \frac{1}{Z} \oint \mathcal{D}x \, \delta(x(\tau = 0) - x_a) \, \exp\{-\mathcal{A}[x]/\hbar\}, \tag{5.457}$$

which can be written as

$$\langle \mathcal{A}_{int}^n[x] \rangle_{x_a, x_a}^{\Omega, x_m} = \frac{1}{\rho_0^{\Omega, x_m}(x_a)} \prod_{l=1}^{n} \left[\int_0^{\hbar\beta} d\tau_l \int_{-\infty}^{\infty} dz_l \, V_{int}(z_l + x_m)\right]$$

$$\times \frac{1}{\sqrt{(2\pi)^{n+1} \det a^2}} \exp\left(-\frac{1}{2} \sum_{k,l=0}^{n} z_k \, a_{kl}^{-2} \, z_l\right), \tag{5.458}$$

with $z_0 = \tilde{x}_a$. Here a denotes a symmetric $(n+1) \times (n+1)$-matrix whose elements $a_{kl}^2 = a^2(\tau_k, \tau_l)$ are obtained from the harmonic Green function for periodic paths $G_{\Omega^2}^{(2)}(\tau, \tau')$ of Eq. (3.301):

$$a^2(\tau, \tau') \equiv \frac{\hbar}{M} G_{\Omega^2}^{(2)}(\tau, \tau') = \frac{\hbar}{2M\Omega} \frac{\cosh \Omega(|\tau - \tau'| - \hbar\beta/2)}{\sinh \hbar\beta\Omega/2}. \tag{5.459}$$

The diagonal elements $a^2 = a(\tau, \tau)$ are all equal to the fluctuation square width (5.24).

Both smearing formulas (5.456) and (5.458) allow us to calculate all harmonic expectation values for the variational perturbation theory of density matrices and particle densities in terms of ordinary Gaussian integrals. Unfortunately, in many applications containing nonpolynomial potentials, it is impossible to solve neither the spatial nor the temporal integrals analytically. This circumstance drastically increases the numerical effort in higher-order calculations.

5.23.4   First-Order Variational Approximation

The first-order variational approximation gives usually a reasonable estimate for any desired quantity. Let us investigate the classical and the quantum-mechanical limit of this approximation. To facilitate the discussion, we first derive a new representation for the first-order smearing formula (5.458) which allows a direct evaluation of the imaginary time integral. The resulting expression will depend only on temperature, whose low- and high-temperature limits can easily be extracted.

Alternative First-Order Smearing Formula

For simplicity, we restrict ourselves to the case of particle densities and allow only symmetric potentials $V(x)$ centered at the origin. If $V(x)$ has only one minimum at the origin, then also x_m will be zero. If $V(x)$ has several symmetric minima, then x_m goes to zero only at sufficiently high temperatures as in Section 5.7.

To first order, the smearing formula (5.458) reads

$$
\langle \mathcal{A}_{\mathrm{int}}[x] \rangle_{x_a,x_a}^{\Omega} = \frac{1}{\rho_0^{\Omega}(x_a)} \int\limits_0^{\hbar\beta} d\tau \int\limits_{-\infty}^{\infty} \frac{dz}{2\pi} V_{\mathrm{int}}(z) \frac{1}{\sqrt{a_{00}^2 - a_{01}^2}} \exp\left\{ -\frac{1}{2} \frac{(z^2 + x_a^2)a_{00} - 2z x_a a_{01}}{a_{00}^2 - a_{01}^2} \right\}.
$$

$$(5.460)$$

Expanding the exponential with the help of Mehler's formula (2.295), we obtain the following expansion in terms of Hermite polynomials $H_n(x)$:

$$
\langle \mathcal{A}_{\mathrm{int}}[x] \rangle_{x_a,x_a}^{\Omega} = \sum_{n=0}^{\infty} \frac{\hbar\beta}{2^n n!} C_{\beta}^{(n)} H_n\left(\frac{z}{\sqrt{2a_{00}^2}} \right) \int\limits_{-\infty}^{\infty} \frac{dz}{\sqrt{2\pi a_{00}^2}} V_{\mathrm{int}}(z) e^{-z^2/2a_{00}^2} H_n\left(\frac{z}{\sqrt{2a_{00}^2}} \right).
$$

$$(5.461)$$

Its temperature dependence stems from the diagonal elements of the harmonic Green function (5.459). The dimensionless functions $C_{\beta}^{(n)}$ are defined by

$$
C_{\beta}^{(n)} = \frac{1}{\hbar\beta} \int\limits_0^{\hbar\beta} d\tau \left(\frac{a_{01}^2}{a_{00}^2} \right)^n.
$$

$$(5.462)$$

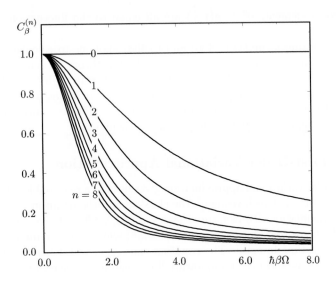

Figure 5.31 Temperature-dependence of first 9 functions $C_\beta^{(n)}$, where $\beta = 1/k_B T$.

Inserting (5.459) and performing the integral over τ, we obtain

$$C_\beta^{(n)} = \frac{1}{2^n \cosh^n \hbar\beta\Omega/2} \sum_{k=0}^{n} \binom{n}{k} \frac{\sinh \hbar\beta\Omega(n/2-k)}{\hbar\beta\Omega(n/2-k)}. \qquad (5.463)$$

At high temperatures, these functions of β go to unity:

$$\lim_{\beta\to 0} C_\beta^{(n)} = 1. \qquad (5.464)$$

Their zero-temperature limits are

$$\lim_{\beta\to\infty} C_\beta^{(n)} = \begin{cases} 1, & n = 0, \\ \dfrac{2}{\hbar\beta\Omega n}, & n > 0. \end{cases} \qquad (5.465)$$

According to (5.444), the first-order approximation to the new effective potential is given by

$$\tilde{W}_1^\Omega(x_a) = \frac{1}{2\beta} \ln \frac{\sinh \hbar\beta\Omega}{\hbar\beta\Omega} + \frac{M\Omega}{\hbar\beta} x_a^2 \tanh \frac{\hbar\beta\Omega}{2} + V_{a^2}^\Omega(x_a), \qquad (5.466)$$

with the smeared interaction potential

$$V_{a^2}^\Omega(x_a) = \frac{1}{\hbar\beta} \langle \mathcal{A}_{\text{int}}[x] \rangle_{x_a,x_a}^\Omega. \qquad (5.467)$$

It is instructive to discuss separately the limits $\beta \to 0$ and $\beta \to \infty$ to see the effects of pure classical and pure quantum fluctuations.

a) Classical Limit of Effective Classical Potential

In the classical limit $\beta \to 0$, the first-order effective classical potential (5.466) reduces to

$$\tilde{W}_1^{\Omega,\mathrm{cl}}(x_a) = \frac{1}{2}M\Omega^2 x_a^2 + \lim_{\beta \to 0} V_{a^2}^{\Omega}(x_a). \tag{5.468}$$

The second term is determined by inserting the high-temperature limit of the total fluctuation width (5.26):

$$a_{\mathrm{tot,cl}}^2 = \frac{k_B T}{M\Omega^2}, \tag{5.469}$$

and of the polynomials (5.464) into the expansion (5.461), leading to

$$V_{a^2}^{\Omega}(x_a) \underset{T \to \infty}{\approx} \sum_{n=0}^{\infty} \frac{1}{2^n n!} H_n\left(\sqrt{\frac{M\Omega^2\beta}{2}}x_a\right) \int_{-\infty}^{\infty} \frac{dz}{\sqrt{2\pi/M\Omega^2\beta}} V_{\mathrm{int}}(z) e^{-M\Omega^2\beta z^2/2} H_n\left(\frac{M\Omega^2\beta}{2}z\right). \tag{5.470}$$

Then we make use of the completeness relation for Hermite polynomials

$$\frac{1}{\sqrt{\pi}} e^{-x^2} \sum_{n=0}^{\infty} \frac{1}{2^n n!} H_n(x) H_n(x') = \delta(x - x'), \tag{5.471}$$

which may be derived from Mehler's formula (2.295) in the limit $b \to 1^-$, to reduce the smeared interaction potential $V_{a^2}^{\Omega}(x_a)$ to the pure interaction potential (5.438):

$$\lim_{\beta \to 0} V_{a^2}^{\Omega}(x_a) = V_{\mathrm{int}}(x_a). \tag{5.472}$$

Recalling (5.438) we see that the first-order effective classical potential (5.468) approaches the classical one:

$$\lim_{\beta \to 0} \tilde{W}_1^{\Omega,\mathrm{cl}}(x_a) = V(x_a). \tag{5.473}$$

This is a consequence of the vanishing fluctuation width b^2 of the paths around the classical orbits. This property is universal to all higher-order approximations to the effective classical potential (5.444). Thus all correction terms with $n > 1$ must disappear in the limit $\beta \to 0$,

$$\lim_{\beta \to 0} \frac{-1}{\beta} \sum_{n=2}^{\infty} \frac{(-1)^n}{n!\hbar^n} \langle \mathcal{A}_{\mathrm{int}}^n[x] \rangle_{x_a,x_a,c}^{\Omega} = 0. \tag{5.474}$$

b) Zero-Temperature Limit

At low temperatures, the first-order effective classical potential (5.466) becomes

$$\tilde{W}_1^{\Omega,\mathrm{qm}}(x_a) = \frac{\hbar\Omega}{2} + \lim_{\beta \to \infty} V_{a^2}^{\Omega}(x_a). \tag{5.475}$$

The zero-temperature limit of the smeared potential in the second term defined in (5.467) follows from Eq. (5.461) by taking into account the limiting procedure for the polynomials $C_\beta^{(n)}$ in (5.465) and the zero-temperature limit of the total fluctuation width (5.26), which is equal to the zero-temperature limit of $a^2(x_0)$: $a_{\mathrm{tot},0}^2 \underset{T=0}{=} a_{\mathrm{tot}}^2 = \hbar/2M\Omega$. Thus we obtain with $H_0(x) = 1$ and the inverse length $\kappa \equiv 1/\lambda_\Omega = \sqrt{M\Omega/\hbar}$ [recall (2.301)]:

$$\lim_{\beta\to\infty} V_{a^2}^\Omega(x_a) = \int_{-\infty}^{\infty} dz \sqrt{\frac{\kappa^2}{\pi}} H_0(\kappa z)^2 \exp\{-\kappa^2 z^2\} V_{\mathrm{int}}(z). \tag{5.476}$$

Introducing the harmonic eigenvalues

$$E_n^\Omega = \hbar\Omega\left(n + \frac{1}{2}\right), \tag{5.477}$$

and the harmonic eigenfunctions [recall (2.299) and (2.300)]

$$\psi_n^\Omega(x) = \frac{1}{\sqrt{n!2^n}}\left(\frac{\kappa^2}{\pi}\right)^{1/4} e^{-\frac{1}{2}\kappa^2 x^2} H_n(\kappa x), \tag{5.478}$$

we can re-express the zero-temperature limit of the first-order effective classical potential (5.475) with (5.476) by

$$\tilde{W}_1^{\Omega,\mathrm{qm}}(x_a) = E_0^\Omega + \langle\,\psi_0^\Omega\,|\,V_{\mathrm{int}}\,|\,\psi_0^\Omega\,\rangle. \tag{5.479}$$

This is recognized as the first-order Rayleigh-Schrödinger perturbative result for the ground state energy.

For the discussion of the quantum-mechanical limit of the first-order normalized density,

$$\rho_1^\Omega(x_a) = \frac{\tilde\rho_1^\Omega(x_a)}{Z} = \rho_0^\Omega(x_a)\,\frac{\exp\left\{-\frac{1}{\hbar}\langle\,\mathcal{A}_{\mathrm{int}}[x]\,\rangle_{x_a,x_a}^\Omega\right\}}{\int_{-\infty}^{\infty} dx_a\,\rho_0^\Omega(x_a)\exp\left\{-\frac{1}{\hbar}\langle\,\mathcal{A}_{\mathrm{int}}[x]\,\rangle_{x_a,x_a}^\Omega\right\}}, \tag{5.480}$$

we proceed as follows. First we expand (5.480) up to first order in the interaction, leading to

$$\rho_1^\Omega(x_a) = \rho_0^\Omega(x_a)\left[1 - \frac{1}{\hbar}\left(\langle\,\mathcal{A}_{\mathrm{int}}[x]\,\rangle_{x_a,x_a}^\Omega - \int_{-\infty}^{\infty} dx_a\,\rho_0^\Omega(x_a)\,\langle\,\mathcal{A}_{\mathrm{int}}[x]\,\rangle_{x_a,x_a}^\Omega\right)\right]. \tag{5.481}$$

Inserting (5.428) and (5.461) into the third term in (5.481), and assuming Ω not to depend explicitly on x_a, the x_a-integral reduces to the orthonormality relation for Hermite polynomials

$$\frac{1}{2^n n!\sqrt{\pi}}\int_{-\infty}^{\infty} dx_a H_n(x_a)H_0(x_a)e^{-x_a^2} = \delta_{n0}, \tag{5.482}$$

so that the third term in (5.481) eventually becomes

$$
-\int_{-\infty}^{\infty} dx_a\, \rho_0^{\Omega}(x_a)\, \langle \mathcal{A}_{\mathrm{int}}[x]\rangle_{x_a,x_a}^{\Omega} = -\beta \int_{-\infty}^{\infty} dz\, \sqrt{\frac{\kappa^2}{\pi}}\, V_{\mathrm{int}}(z)\, \exp\{-\kappa^2 z^2\}\, H_0(\kappa z). \quad (5.483)
$$

But this is just the $n = 0$ -term of (5.461) with an opposite sign, thus canceling the zeroth component of the second term in (5.481), which would have been divergent for $\beta \to \infty$.

The resulting expression for the first-order normalized density is

$$
\rho_1^{\Omega}(x_a) = \rho_0^{\Omega}(x_a) \left[1 - \sum_{n=1}^{\infty} \frac{\beta}{2^n n!} C_{\beta}^{(n)} H_n(\kappa x_a) \int_{-\infty}^{\infty} dz\, \sqrt{\frac{\kappa^2}{\pi}}\, V_{\mathrm{int}}(z)\, \exp(-\kappa^2 z^2)\, H_n(\kappa z) \right].
$$
$$(5.484)$$

The zero-temperature limit of $C_{\beta}^{(n)}$ is from (5.465) and (5.477)

$$
\lim_{\beta \to \infty} \beta C_{\beta}^{(n)} = \frac{2}{E_n^{\Omega} - E_0^{\Omega}}, \quad (5.485)
$$

so that we obtain from (5.484) the limit

$$
\begin{aligned}
\rho_1^{\Omega}(x_a) &= \rho_0^{\Omega}(x_a) \left[1 - 2 \sum_{n=1}^{\infty} \frac{1}{2^n n!} \frac{1}{E_n^{\Omega} - E_0^{\Omega}} H_n(\kappa x_a) \right. \\
&\quad \left. \times \int_{-\infty}^{\infty} dz\, \sqrt{\frac{\kappa^2}{\pi}}\, V_{\mathrm{int}}(z)\, \exp\{-\kappa^2 z^2\} H_n(\kappa z) H_0(\kappa z) \right].
\end{aligned} \quad (5.486)
$$

Taking into account the harmonic eigenfunctions (5.478), we can rewrite (5.486) as

$$
\rho_1^{\Omega}(x_a) = |\psi_0(x_a)|^2 = [\psi_0^{\Omega}(x_a)]^2 - 2\psi_0^{\Omega}(x_a) \sum_{n>0} \psi_n^{\Omega}(x_a) \frac{\langle \psi_n^{\Omega} | V_{\mathrm{int}} | \psi_0^{\Omega}\rangle}{E_n^{\Omega} - E_0^{\Omega}}, \quad (5.487)
$$

which is just equivalent to the harmonic first-order Rayleigh-Schrödinger result for particle densities.

Summarizing the results of this section, we have shown that our method has properly reproduced the high- and low-temperature limits. Due to relation (5.487), the variational approach for particle densities can be used to determine approximately the ground state wave function $\psi_0(x_a)$ for the system of interest.

5.23.5 Smearing Formula in Higher Spatial Dimensions

Most physical systems possess many degrees of freedom. This requires an extension of our method to higher spatial dimensions. In general, we must consider anisotropic harmonic trial systems, where the previous variational parameter Ω^2 becomes a $D \times D$ -matrix $\Omega_{\mu\nu}^2$ with $\mu, \nu = 1, 2, \ldots, D$.

a) Isotropic Approximation

An isotropic trial ansatz

$$\Omega_{\mu\nu}^2 = \Omega^2 \delta_{\mu\nu} \tag{5.488}$$

can give rough initial estimates for the properties of the system. In this case, the n-th order smearing formula (5.458) generalizes directly to

$$
\langle \mathcal{A}_{\text{int}}^n[\mathbf{r}] \rangle_{\mathbf{r}_a,\mathbf{r}_a}^{\Omega} = \frac{1}{\rho_0^{\Omega}(\mathbf{r}_a)} \prod_{l=1}^{n} \left[\int_0^{\hbar\beta} d\tau_l \int d^D z_l \, V_{\text{int}}(\mathbf{z}_l) \right]
$$
$$
\times \frac{1}{\sqrt{(2\pi)^{n+1} \det a^2}^D} \exp\left[-\frac{1}{2} \sum_{k,l=0}^{n} \mathbf{z}_k \, a_{kl}^{-2} \, \mathbf{z}_l \right], \tag{5.489}
$$

with the D-dimensional vectors $\mathbf{z}_l = (z_{1l}, z_{2l}, \ldots, z_{Dl})^T$. Note, that Greek labels $\mu, \nu, \ldots = 1, 2, \ldots, D$ specify spatial indices and Latin labels $k, l, \ldots = 0, 1, 2, \ldots, n$ refer to the different imaginary times. The vector $\mathbf{z}_0$ denotes $\mathbf{r}_a$, the matrix a^2 is the same as in Subsection 5.23.3. The harmonic density reads

$$\rho_0^{\Omega}(\mathbf{r}) = \sqrt{\frac{1}{2\pi a_{00}^2}}^D \exp\left[-\frac{1}{2 a_{00}^2} \sum_{\mu=1}^{D} x_{\mu}^2 \right]. \tag{5.490}$$

b) Anisotropic Approximation

In the discussion of the anisotropic approximation, we shall only consider radially-symmetric potentials $V(\mathbf{r}) = V(|\mathbf{r}|)$ because of their simplicity and their major occurrence in physics. The trial frequencies decompose naturally into a radial frequency Ω_L and a transverse one Ω_T as in (5.95):

$$\Omega_{\mu\nu}^2 = \Omega_L^2 \frac{x_{a\mu} x_{a\nu}}{r_a^2} + \Omega_T^2 \left(\delta_{\mu\nu} - \frac{x_{a\mu} x_{a\nu}}{r_a^2} \right), \tag{5.491}$$

with $r_a = |\mathbf{r}_a|$. For practical reasons we rotate the coordinate system by $\bar{\mathbf{x}}_n = U \mathbf{x}_n$ so that $\bar{\mathbf{r}}_a$ points along the first coordinate axis,

$$(\bar{\mathbf{r}}_a)_\mu \equiv \bar{z}_{\mu 0} = \begin{cases} r_a, & \mu = 1, \\ 0, & 2 \leq \mu \leq D, \end{cases} \tag{5.492}$$

and Ω^2-matrix is diagonal:

$$\overline{\Omega^2} = \begin{pmatrix} \Omega_L^2 & 0 & 0 & \cdots & 0 \\ 0 & \Omega_T^2 & 0 & \cdots & 0 \\ 0 & 0 & \Omega_T^2 & \cdots & 0 \\ \vdots & \vdots & \vdots & \ddots & \vdots \\ 0 & 0 & 0 & \cdots & \Omega_T^2 \end{pmatrix} = U \, \Omega^2 \, U^{-1}. \tag{5.493}$$

After this rotation, the *anisotropic* n-th order smearing formula in D dimensions reads

$$\langle \mathcal{A}_{\text{int}}^n[\mathbf{r}] \rangle_{\bar{\mathbf{r}}_a,\bar{\mathbf{r}}_a}^{\Omega_{L,T}} = \frac{1}{\rho_0^{\Omega_{L,T}}(\bar{\mathbf{r}}_a)} \prod_{l=1}^{n} \left[\int_0^{\hbar\beta} d\tau_l \int d^D \bar{z}_l \, V_{\text{int}}(|\bar{\mathbf{z}}_l|) \right] (2\pi)^{-D(n+1)/2} \tag{5.494}$$

$$\times (\det a_L^2)^{-1/2} (\det a_T^2)^{-(D-1)/2} e^{-\frac{1}{2} \sum_{k,l=0}^{n} \bar{z}_{1k} a_{L\,kl}^{-2} \bar{z}_{1l}} e^{-\frac{1}{2} \sum_{\mu=2}^{D} \sum_{k,l=1}^{n} \bar{z}_{\mu k} a_{T\,kl}^{-2} \bar{z}_{\mu l}}.$$

The components of the longitudinal and transversal matrices a_L^2 and a_T^2 are

$$a_{L\,kl}^2 = a_L^2(\tau_k, \tau_l), \quad a_{T\,kl}^2 = a_T^2(\tau_k, \tau_l),$$ (5.495)

where the frequency Ω in (5.459) must be substituted by the new variational parameters Ω_L, Ω_T, respectively. For the harmonic density in the rotated system $\rho_0^{\Omega_{L,T}}(\bar{\mathbf{r}})$ which is used to normalize (5.494), we find

$$\rho_0^{\Omega_{L,T}}(\bar{\mathbf{r}}) = \sqrt{\frac{1}{2\pi a_{L00}^2}} \sqrt{\frac{1}{2\pi a_{T00}^2}}^{D-1} \exp\left[-\frac{1}{2\,a_{L00}^2}\bar{x}_1^2 - \frac{1}{2\,a_{T00}^2}\sum_{\mu=2}^{D}\bar{x}_\mu^2 \right].$$ (5.496)

Appendix 5A   Feynman Integrals for $T \neq 0$ without Zero Frequency

The Feynman integrals needed in variational perturbation theory of the anharmonic oscillator at nonzero temperature can be calculated in close analogy to those of ordinary perturbation theory in Section 3.20. The calculation proceeds as explained in Appendix 3D, except that the lines represent now the thermal correlation function (5.19) with the zero-frequency subtracted from the spectral decomposition:

$$G_{\tilde{\Omega}}^{(2)x_0}(\tau, \tau') = \frac{1}{2\tilde{\Omega}}\left[\frac{\cosh\tilde{\Omega}(|\tau - \tau'| - \hbar\beta/2)}{\sinh(\tilde{\Omega}\hbar\beta/2)} \right] - \frac{1}{\hbar\beta\tilde{\Omega}^2}.$$

With the dimensionless variable $x \equiv \hbar\beta\tilde{\omega}$, the results for the quantities a_V^{2L} defined of each Feynman diagram with L lines and V vertices as in (3.547), but now without the zero Matsubara frequency, are [compare with the results (3D.3)–(3D.11)]

$$a^2 = \frac{\hbar}{\tilde{\omega}}\frac{1}{x}\left(\frac{x}{2}\coth\frac{x}{2} - 1 \right),$$ (5A.1)

$$a_2^4 = \left(\frac{\hbar}{\tilde{\omega}}\right)^2 \frac{1}{8x}\frac{1}{\sinh^2\frac{x}{2}}\left(4 + x^2 - 4\,\cosh x + x\sinh x \right),$$ (5A.2)

$$a_3^6 = \left(\frac{\hbar}{\tilde{\omega}}\right)^3 \frac{1}{64x}\frac{1}{\sinh^3\frac{x}{2}}\left(-3\,x\cosh\frac{x}{2} + 2\,x^3\cosh\frac{x}{2} + 3\,x\cosh\frac{3x}{2} \right.$$

$$\left. +48\,\sinh\frac{x}{2} + 6\,x^2\sinh\frac{x}{2} - 16\,\sinh\frac{3x}{2} \right),$$ (5A.3)

$$a_2^8 = \left(\frac{\hbar}{\omega}\right)^4 \frac{1}{768x^3}\frac{1}{\sinh^4\frac{x}{2}}\left(-864 + 18\,x^4 + 1152\,\cosh x + 32\,x^2\cosh x \right.$$

$$-288\,\cosh 2x - 32\,x^2\cosh 2x - 288\,x\sinh x + 24\,x^3\sinh x$$

$$\left. +144\,x\sinh 2x + 3\,x^3\sinh 2x \right),$$ (5A.4)

$$a_3^{10} = \left(\frac{\hbar}{\tilde{\omega}}\right)^5 \frac{1}{4096x^3}\frac{1}{\sinh^5\frac{x}{2}}\left(672\,x\cosh\frac{x}{2} - 8\,x^3\cosh\frac{x}{2} + 24\,x^5\cosh\frac{x}{2} \right.$$

$$-1008\,x\cosh\frac{3x}{2} + 3\,x^3\cosh\frac{3x}{2} + 336\,x\cosh\frac{5x}{2} + 5\,x^3\cosh\frac{5x}{2}$$

$$-7680\,\sinh\frac{x}{2} - 352\,x^2\sinh\frac{x}{2} + 72\,x^4\sinh\frac{x}{2} + 3840\,\sinh\frac{3x}{2}$$

$$\left. +224\,x^2\sinh\frac{3x}{2} + 12\,x^4\sinh\frac{3x}{2} - 768\,\sinh\frac{5x}{2} - 64\,x^2\sinh\frac{5x}{2} \right),$$ (5A.5)

$$a_3^{12} = \left(\frac{\hbar}{\tilde{\omega}}\right)^6 \frac{1}{49152 x^4} \frac{1}{\sinh^6 \frac{x}{2}} \left(-107520 - 7360\, x^2 + 624\, x^4 + 96\, x^6\right.$$

$$+161280 \cosh x + 12000\, x^2 \cosh x - 777\, x^4 \cosh x + 24\, x^6 \cosh x$$
$$-64512 \cosh 2x - 5952\, x^2 \cosh 2x + 144\, x^4 \cosh 2x + 10752 \cosh 3x$$
$$-28800\, x \sinh x + 1312\, x^2 \cosh 3x + 9\, x^4 \cosh 3x + 1120\, x^3 \sinh x$$
$$+324\, x^5 \sinh x + 23040\, x \sinh 2x - 320\, x^3 \sinh 2x - 5760\, x \sinh 3x$$
$$\left. -160\, x^3 \sinh 3x\right), \tag{5A.6}$$

$$a_2^6 = \left(\frac{\hbar}{\tilde{\omega}}\right)^3 \frac{1}{24 x^2} \frac{1}{\sinh^2 \frac{x}{2}} \left(-24 - 4\, x^2 + 24 \cosh x + x^2 \cosh x - 9\, x \sinh x\right),$$

$$a_3^8 = \left(\frac{\hbar}{\tilde{\omega}}\right)^4 \frac{1}{288 x^2} \frac{1}{\sinh^3 \frac{x}{2}} \left(45\, x \cosh \frac{x}{2} - 6\, x^3 \cosh \frac{x}{2} - 45\, x \cosh \frac{3x}{2}\right.$$

$$\left. -432 \sinh \frac{x}{2} - 54\, x^2 \sinh \frac{x}{2} + 144 \sinh \frac{3x}{2} + 4\, x^2 \sinh \frac{3x}{2}\right), \tag{5A.7}$$

$$a_{3'}^{10} = \left(\frac{\hbar}{\tilde{\omega}}\right)^5 \frac{1}{2304 x^3} \frac{1}{\sinh^4 \frac{x}{2}} \left(-3456 - 414\, x^2 - 6\, x^4 + 4608 \cosh x +\right.$$

$$496\, x^2 \cosh x - 1152 \cosh 2x - 82\, x^2 \cosh 2x - 1008\, x \sinh x -$$
$$\left. 16\, x^3 \sinh x + 504\, x \sinh 2x + 5\, x^3 \sinh 2x\right). \tag{5A.8}$$

Six of these integrals are the analogs of those in Eqs. (3.547). In addition there are the three integrals a_2^6, a_3^8, and $a_{3'}^{10}$, corresponding to the three diagrams

$$\tag{5A.9}$$

respectively, which are needed in Subsection 5.14.2. They have been calculated with zero Matsubara frequency in Eqs. (3D.8)–(3D.11).

In the low-temperature limit where $x = \Omega\hbar\beta \to \infty$, the x-dependent factors in Eqs. (5A.1)–(5A.8) converge towards the same constants (3D.13) as those with zero Matsubara frequency, and the same limiting relations hold as in Eqs. (3.550) and (3D.14).

The high-temperature limits $x \to 0$, however, are quite different from those in Eq. (3D.16). The present Feynman integrals all vanish rapidly for increasing temperatures. For L lines and V vertices, $\hbar\beta(1/\tilde{\omega})^{V-1} a_V^{2L}$ goes to zero like $\beta^V (\beta/12)^L$. The first V factors are due to the V-integrals over τ, the second are the consequence of the product of $n/2$ factors a^2. Thus a_V^{2L} behaves like

$$a_V^{2L} \propto \left(\frac{\hbar}{\tilde{\omega}}\right)^L x^{V-1+L}. \tag{5A.10}$$

Indeed, the x-dependent factors in (5A.1)–(5A.8) vanish now like

$$x/12, \quad x^3/720, \quad x^5/30240,$$
$$x^5/241920, \quad x^7/11404800, \quad 193 x^8/47551795200,$$
$$x^4/30240, \quad x^6/1814400, \quad x^7/59875200, \tag{5A.11}$$

respectively. When expanding (5A.1)–(5A.8) into a power series, the lowest powers cancel each other. For the temperature behavior of these Feynman integrals see Fig. 5.32. We have plotted the reduced Feynman integrals $\hat{a}_V^{2L}(x)$ in which the low-temperature behaviors (3.550) and (3D.14) have been divided out of a_V^{2L}.

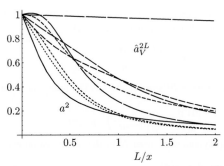

Figure 5.32 Plot of the reduced Feynman integrals $\hat{a}_V^{2L}(x)$ as functions of $L/x = Lk_BT/\hbar\omega$. The integrals (3D.4)–(3D.11) are indicated by decreasing dash-lengths. Compare Fig. 3.16.

The integrals (5A.2) and (5A.3) for a_2^4 and a_3^6 can be obtained from the integral (5A.1) for a^2 via the operation

$$\frac{\hbar^n}{n!}\left(-\frac{\partial}{\partial\tilde{\omega}^2}\right)^n = \frac{\hbar^n}{n!}\left(-\frac{1}{2\tilde{\omega}}\frac{\partial}{\partial\tilde{\omega}}\right)^n, \tag{5A.12}$$

with $n=1$ and $n=2$, respectively. This is derived following the same steps as in Eqs. (3D.18)–(3D.20). The absence of the zero Matsubara frequency does not change the argument.

Also, as in Eqs. (5.195)–(3D.21), the same type of expansion allows us to derive the three integrals from the one-loop diagram (3.546).

Appendix 5B Proof of Scaling Relation for Extrema of W_N

Here we prove the scaling relation (5.215), according to which the derivative of the Nth approximation W_N to the ground state energy can be written as [14]

$$\frac{d}{d\Omega}W_N^\Omega = \left(\frac{g}{\Omega^3}\right)^N p_N(\sigma), \tag{5B.1}$$

where $p_N(\sigma)$ is a polynomial of order N in the scaling variable $\sigma = \Omega(\Omega^2-1)/g$.

For the sake of generality, we consider an anharmonic oscillator with a potential gx^P whose power P is arbitrary. The ubiquitous factor $1/4$ accompanying g is omitted, for convenience. The energy eigenvalue of the ground state (or any excited state) has an Nth order perturbation expansion

$$E_N(g) = \omega\sum_{l=0}^N E_l\left(\frac{g}{\omega^{(P+2)/2}}\right)^l, \tag{5B.2}$$

where E_l are rational numbers. After the replacement (5.188), the series is re-expanded at fixed r in powers of g up to order N, and we obtain

$$W_N^\Omega = \Omega\sum_{l=0}^N \varepsilon_l(\sigma)\left(\frac{g}{\Omega^{(P+2)/2}}\right)^l, \tag{5B.3}$$

with the re-expansion coefficients [compare (5.207)]

$$\varepsilon_l(\sigma) = \sum_{j=0}^l E_j\binom{(1-\frac{P+2}{2}j)/2}{l-j}(-\sigma)^{l-j}. \tag{5B.4}$$

Here σ is a scaling variable for the potential gx^P generalizing (5.208) (note that it is four times as big as the previous σ, due to the different normalization of g):

$$\sigma \equiv \frac{\Omega^{(P-2)/2}(\Omega^2 - \omega^2)}{g}. \tag{5B.5}$$

We now show that the derivative $dW_N(g, \Omega)/d\Omega$ has the following scaling form generalizing (5B.1):

$$\frac{dW_N^\Omega}{d\Omega} = \left(\frac{g}{\Omega^{(P+2)/2}}\right)^N p_N(\sigma), \tag{5B.6}$$

where $p_N(\sigma)$ is the following polynomial of order N in the scaling variable σ:

$$
\begin{aligned}
p_N(\sigma) &= -2\frac{d\varepsilon_{N+1}(\sigma)}{d\sigma} \\
&= 2\sum_{j=0}^{N} E_j \left(\frac{(1 - \frac{P+2}{2}j)/2}{N+1-j}\right)(N+1-j)(-\sigma)^{N-j}.
\end{aligned} \tag{5B.7}
$$

The proof starts by differentiating (5B.3) with respect to Ω, yielding

$$\frac{dW_N^\Omega}{d\Omega} = \sum_{l=0}^{N}\left[\varepsilon_l(\sigma) - \frac{P+2}{2}l\varepsilon_l(\sigma) + \Omega\frac{d\varepsilon_l}{d\Omega}\right]\left(\frac{g}{\Omega^{(P+2)/2}}\right)^l. \tag{5B.8}$$

Using the chain rule of differentiation we see from (5B.4) that

$$\Omega\frac{d\varepsilon_l}{d\Omega} = \left[2\frac{\Omega^{(P+2)/2}}{g} + \frac{P-2}{2}\sigma\right]\frac{d\varepsilon_l}{d\sigma}, \tag{5B.9}$$

and (5B.8) can be rewritten as

$$\frac{dW_N^\Omega}{d\Omega} = \sum_{l=0}^{N}\left[\left(1 - \frac{P+2}{2}l\right)\varepsilon_l(\sigma) + \left(2\frac{\Omega^{(P+2)/2}}{g} + \frac{P-2}{2}\sigma\right)\frac{d\varepsilon_l}{d\sigma}\right]\left(\frac{g}{\Omega^{(P+2)/2}}\right)^l. \tag{5B.10}$$

After rearranging the sum, this becomes

$$
\begin{aligned}
\frac{dW_N^\Omega}{d\Omega} &= 2\frac{d\varepsilon_0}{d\sigma}\left(\frac{g}{\Omega^{(P+2)/2}}\right)^{-1} \\
&+ \sum_{l=0}^{N-1}\left[\left(1 - \frac{P+2}{2}l\right)\varepsilon_l + \frac{P-2}{2}\sigma\frac{d\varepsilon_l}{d\sigma} + 2\frac{d\varepsilon_{l+1}}{d\sigma}\right]\left(\frac{g}{\Omega^{(P+2)/2}}\right)^l \\
&+ \left[\left(1 - \frac{P+2}{2}N\right)\varepsilon_N + \frac{P-2}{2}\sigma\frac{d\varepsilon_N}{d\sigma}\right]\left(\frac{g}{\Omega^{(P+2)/2}}\right)^N.
\end{aligned} \tag{5B.11}
$$

The first term vanishes trivially since ε_0 happens to be independent of σ. The sum in the second line vanishes term by term:

$$\left(1 - \frac{P+2}{2}l\right)\varepsilon_l + \frac{P-2}{2}\sigma\frac{d\varepsilon_l}{d\sigma} + 2\frac{d\varepsilon_{l+1}}{d\sigma} = 0, \quad l = 1 \ldots N-1. \tag{5B.12}$$

To see this we form the derivative

$$2\frac{d\varepsilon_{l+1}}{d\sigma} = 2\sum_{j=0}^{l} E_j \left(\frac{(1 - \frac{P+2}{2}j)/2}{l+1-j}\right)(j-l-1)(-\sigma)^{l-j}, \tag{5B.13}$$

and use the identity

$$2 \left(\begin{array}{c} (1 - \frac{P+2}{2}j)/2 \\ l+1-j \end{array} \right) = \frac{\frac{P-2}{2}j + 2l - 1}{j - l - 1} \left(\begin{array}{c} (1 - \frac{P+2}{2}j)/2 \\ l - j \end{array} \right) \tag{5B.14}$$

to rewrite (5B.13) as

$$2\frac{d\varepsilon_{l+1}}{d\sigma} = \sum_{j=0}^{l} E_j \left(\begin{array}{c} (1 - \frac{P+2}{2}j)/2 \\ l - j \end{array} \right) \left(\frac{P-2}{2}j + 2l - 1 \right) (-\sigma)^{l-j}, \tag{5B.15}$$

implying

$$\frac{P-2}{2}\sigma\frac{d\varepsilon_l}{d\sigma} = \sum_{j=0}^{l} E_j \left(\begin{array}{c} (1 - \frac{P+2}{2}j)/2 \\ l - j \end{array} \right) \frac{P-2}{2}(l-j)(-\sigma)^{l-j}. \tag{5B.16}$$

By combining this with (5B.4), (5B.13), we obtain Eq. (5B.12) which proves that the second line in (5B.11) vanishes.

Thus we are left with the last term on the right-hand side of Eq. (5B.11). Using (5B.12) for $l = N$ leads to

$$\frac{dW_N^{\Omega}}{d\Omega} = -2 \left(\frac{g}{\Omega^{(P+2)/2}} \right)^N \frac{d\varepsilon_{N+1}(\sigma)}{d\sigma}. \tag{5B.17}$$

When expressing $d\varepsilon_{N+1}(\sigma)/d\sigma$ with the help of (5B.4), we arrive at

$$\frac{dW_N^{\Omega}}{d\Omega} = 2 \left(\frac{g}{\Omega^{(P+2)/2}} \right)^N \sum_{j=0}^{N} E_j \left(\begin{array}{c} (1 - \frac{P+2}{2}j)/2 \\ N+1-j \end{array} \right) (N+1-j)(-\sigma)^{N-j}. \tag{5B.18}$$

This proves the scaling relation (5B.6) with the polynomial (5B.7).

The proof can easily be extended to physical quantities $Q_N(g)$ with a different physical dimension α, which have an expansion

$$Q_N(g) = \omega^\alpha \sum_{l=0}^{N} E_l \left(\frac{g}{\omega^{(P+2)/2}} \right)^l, \qquad \alpha \neq 1, \tag{5B.19}$$

rather than (5B.2). In this case the quantity $[Q_N(g)]^{1/\alpha}$ has again an expansion like (5B.2). By rewriting $Q_N(g)$ as $\{[Q_N(g)]^{1/\alpha}\}^\alpha$ and forming the derivative using the chain rule we see that the derivative vanishes whenever the polynomial $p_N(\sigma)$ vanishes, which is formed from $[Q_N(g)]^{1/\alpha}$ as in Eq. (5B.7).

Appendix 5C   Second-Order Shift of Polaron Energy

For brevity, we introduce the dimensionless variable $\rho \equiv \omega\Delta\tau$ and

$$F[\rho] \equiv -2\Gamma^2 \bar{G}^{\Omega,\Gamma}(\rho, 0). \tag{5C.1}$$

Going to natural units with $\hbar = M = \omega = 1$, Feynman's variational energy (5.395) takes the form

$$E_0 = \frac{3}{4\Gamma}(\Gamma - \Omega)^2 - \frac{1}{\sqrt{\pi}}\Gamma \int_0^\infty d\rho\, e^{-\rho} F^{-1/2}(\rho). \tag{5C.2}$$

The second-order correction (5.406) reads

$$\Delta E_0^{(2)} = -\frac{4\Gamma^2}{\pi}\Gamma^2 I + \frac{1}{2\Gamma\sqrt{\pi}}(\Gamma^2 - \Omega^2) \int_0^\infty d\rho\, e^{-\rho} F^{-1/2}(\rho) - \frac{3}{16\Gamma^3}(\Gamma^2 - \Omega^2)^2$$
$$-\frac{1}{4\sqrt{\pi}}(\Gamma^2 - \Omega^2) \int_0^\infty d\rho\, e^{-\rho} F^{-3/2}(\rho) \left\{ \left(1 + \frac{\Omega^2}{\Gamma}\right)(1 - e^{-\Gamma\rho}) + \frac{\Gamma^2 - \Omega^2}{\Gamma}\rho e^{-\Gamma\rho} \right\}, \tag{5C.3}$$

where I denotes the integral

$$I = \frac{1}{4} \int_0^\infty \int_0^\infty \int_0^\infty d\rho_1 d\rho_2 d\rho_3 e^{-\rho_1} F^{-1/2}(\rho_1) e^{-\rho_2} F^{-1/2}(\rho_2) \left(\frac{\arcsin Q}{Q} - 1 \right), \qquad (5\text{C}.4)$$

with

$$\begin{aligned} Q = Q_1 \quad &\text{for} \quad \rho_3 - \rho_2 + \rho_1 \geq 0 \quad \text{and} \quad \rho_3 - \rho_2 \geq 0, \\ Q = Q_2 \quad &\text{for} \quad \rho_3 - \rho_2 + \rho_1 \geq 0 \quad \text{and} \quad \rho_3 - \rho_2 < 0, \qquad (5\text{C}.5) \\ Q = Q_3 \quad &\text{for} \quad \rho_3 - \rho_2 + \rho_1 < 0 \quad \text{and} \quad \rho_3 - \rho_2 < 0, \end{aligned}$$

and

$$Q_1 = \frac{1}{2} F^{-1/2}(\rho_1^2) F^{-1/2}(\rho_2^2) \frac{\Gamma^2 - \Omega^2}{\Gamma} e^{-\Gamma \rho_3} (1 - e^{-\Gamma \rho_1})(e^{\Gamma \rho} - 1), \qquad (5\text{C}.6)$$

$$Q_2 = \frac{1}{2} F^{-1/2}(\rho_1) F^{-1/2}(\rho_2)$$
$$\times \left\{ \frac{\Gamma^2 - \Omega^2}{\Gamma} \left[e^{-\Gamma(\rho_2 - \rho_3)} - e^{-\Gamma \rho_3} (1 + e^{-\Gamma(\rho_1 - \rho_2)} - e^{-\Gamma \rho_1}) \right] - 2\Omega^2 (\rho_2 - \rho_3) \right\}, \qquad (5\text{C}.7)$$

$$Q_3 = \frac{1}{2} F^{-1/2}(\rho_1) F^{-1/2}(\rho_2)$$
$$\times \left\{ \frac{\Gamma^2 - \Omega^2}{\Gamma} \left[-e^{-\Gamma \rho_3} (1 - e^{-\Gamma \rho_1}) - e^{-\Gamma(\rho_2 - \rho_3)} (e^{\Gamma \rho_1} - 1) \right] - 2\Omega^2 \rho_1 \right\}. \qquad (5\text{C}.8)$$

Notes and References

The first-order variational approximation to the effective classical partition function $V^{\text{eff cl}}(x_0)$ presented in this chapter was developed in 1983 by
R.P. Feynman and H. Kleinert, Phys. Rev. A **34**, 5080 (1986) (http://www.physik.fu-berlin.de/~kleinert/159).
For further development see:
H. Kleinert, Phys. Lett. A **118**, 195 (1986) (*ibid*.http/148); B **181**, 324 (1986) (*ibid*.http/151);
W. Janke and B.K. Chang, Phys. Lett. B **129**, 140 (1988);
W. Janke, in *Path Integrals from meV to MeV*, ed. by V. Sa-yakanit et al., World Scientific, Singapore, 1990.
A detailed discussion of the accuracy of the approach in comparison with several other approximation schemes is given by
S. Srivastava and Vishwamittar, Phys. Rev. A **44**, 8006 (1991).

For a similar, independent development containing applications to simple quantum field theories, see
R. Giachetti and V. Tognetti, Phys. Rev. Lett. **55**, 912 (1985); Int. J. Magn. Mater.
54-57, 861 (1986);
R. Giachetti, V. Tognetti, and R. Vaia, Phys. Rev. B **33**, 7647 (1986); Phys. Rev. A **37**, 2165 (1988); Phys. Rev. A **38**, 1521, 1638 (1988);
Physica Scripta **40**, 451 (1989). R. Giachetti, V. Tognetti, A. Cuccoli, and R. Vaia, lecture presented at the XXVI Karpacz School of Theoretical Physics, Karpacz, Poland, 1990.
See also
R. Vaia and V. Tognetti, Int. J. Mod. Phys. B **4**, 2005 (1990);
A. Cuccoli, V. Tognetti, and R. Vaia, Phys. Rev. B **41**, 9588 (1990); A **44**, 2743 (1991);
A. Cuccoli, A. Maradudin, A.R. McGurn, V. Tognetti, and R. Vaia, Phys. Rev. D **46**, 8839 (1992).

The variational approach has solved some old problems in quantum crystals by extending in a simple way the classical methods into the quantum regime. See
V.I. Yukalov, Mosc. Univ. Phys. Bull. **31**, 10-15 (1976);
S. Liu, G.K. Horton, and E.R. Cowley, Phys. Lett. A **152**, 79 (1991);
A. Cuccoli, A. Macchi, M. Neumann, V. Tognetti, and R. Vaia, Phys. Rev. B **45**, 2088 (1992).

The systematic extension of the variational approach was developed by
H. Kleinert, Phys. Lett. A **173**, 332 (1993) (quant-ph/9511020).
See also
J. Jaenicke and H. Kleinert, Phys. Lett. A **176**, 409 (1993) (*ibid*.http/217);
H. Kleinert and H. Meyer, Phys. Lett. A **184**, 319 (1994) (hep-th/9504048).

A similar convergence mechanism was first observed within an order-dependent mapping technique in the seminal paper by
R. Seznec and J. Zinn-Justin, J. Math. Phys. **20**, 1398 (1979).
For an introduction into various resummation procedures see
C.M. Bender and S.A. Orszag, *Advanced Mathematical Methods for Scientists and Engineers*, McGraw-Hill, New York, 1978.

The proof of the convergence of the variational perturbation expansion to be given in Subsection 17.10.5 went through the following stages: First a weak estimate was found for the anharmonic integral:
I.R.C. Buckley, A. Duncan, H.F. Jones, Phys. Rev. D **47**, 2554 (1993);
C.M. Bender, A. Duncan, H.F. Jones, Phys. Rev. D **49**, 4219 (1994).
This was followed by a similar extension to the quantum-mechanical case:
A. Duncan and H.F. Jones, Phys. Rev. D **47**, 2560 (1993);
C. Arvanitis, H.F. Jones, and C.S. Parker, Phys.Rev. D **52**, 3704 (1995) (hep-ph/9502386);
R. Guida, K. Konishi, and H. Suzuki, Ann. Phys. 241, 152 (1995) (hep-th/9407027).

The exponentially fast convergence observed in the calculation of the strong-coupling coefficients of Table 5.9 was, however, not explained. The accuracy in the table was reached by working up to the order 251 with 200 digits in Ref. [13].

The analytic properties of the strong-coupling expansion were studied by
C.M. Bender and T.T. Wu, Phys. Rev. **184**, 1231 (1969); Phys. Rev. Lett. **27**, 461 (1971); Phys. Rev. D **7**, 1620 (1973); *ibid.* D **7**, 1620 (1973);
C.M. Bender, J. Math. Phys. **11**, 796 (1970);
T. Banks and C.M. Bender, J. Math. Phys. **13**, 1320 (1972);
J.J. Loeffel and A. Martin, Cargèse Lectures on Physics (1970);
D. Bessis ed., Gordon and Breach, New York 1972, Vol. 5, p.415;
B. Simon, Ann. Phys. (N.Y.) **58**, 76 (1970); Cargèse Lectures on Physics (1970), D. Bessis ed., Gordon and Breach, New York 1972, Vol. 5, p. 383.

The problem of tunneling at low barriers (*sliding*) was solved by
H. Kleinert, Phys. Lett. B **300**, 261 (1993) (*ibid*.http/214).
See also Chapter 17. Some of the present results are contained in
H. Kleinert, *Pfadintegrale in Quantenmechanik, Statistik und Polymerphysik*, B.-I. Wissenschaftsverlag, Mannheim, 1993.

A variational approach to tunneling is also used in chemical physics:
M.J. Gillan, J. Phys. C **20**, 362 (1987);
G.A. Voth, D. Chandler, and W.H. Miller, J. Chem. Phys. **91**, 7749 (1990);
G.A. Voth and E.V. OGorman, J. Chem. Phys. **94**, 7342 (1991);
G.A. Voth, Phys. Rev. A **44**, 5302 (1991).

Variational approaches without the separate treatment of x_0 have been around in the literature for some time:

T. Barnes and G.I. Ghandour, Phys. Rev. D **22**, 924 (1980);
B.S. Shaverdyan and A.G. Usherveridze, Phys. Lett. B **123**, 316 (1983);
K. Yamazaki, J. Phys. A **17**, 345 (1984);
H. Mitter and K. Yamazaki, J. Phys. A **17**, 1215 (1984);
P.M. Stevenson, Phys. Rev. D **30**, 1712 (1985); D **32**, 1389 (1985);
P.M. Stevenson and R. Tarrach, Phys. Lett. B **176**, 436 (1986);
A. Okopinska, Phys. Rev. D **35**, 1835 (1987); D **36**, 2415 (1987);
W. Namgung, P.M. Stevenson, and J.F. Reed, Z. Phys. C **45**, 47 (1989);
U. Ritschel, Phys. Lett. B **227**, 44 (1989); Z. Phys. C **51**, 469 (1991);
M.H. Thoma, Z. Phys. C **44**, 343 (1991);
I. Stancu and P.M. Stevenson, Phys. Rev. D **42**, 2710 (1991);
R. Tarrach, Phys. Lett. B **262**, 294 (1991);
H. Haugerud and F. Raunda, Phys. Rev. D **43**, 2736 (1991);
A.N. Sissakian, I.L. Solovtsov, and O.Y. Shevchenko, Phys. Lett. B **313**, 367 (1993).

Different applications of variational methods to density matrices are given in
V.B. Magalinsky, M. Hayashi, and H.V. Mendoza, J. Phys. Soc. Jap. **63**, 2930 (1994);
V.B. Magalinsky, M. Hayashi, G.M. Martinez Peña, and R. Reyes Sánchez, Nuovo Cimento B **109**, 1049 (1994).

The particular citations in this chapter refer to the publications

[1] H. Kleinert, Phys. Lett. A **173**, 332 (1993) (quant-ph/9511020).

[2] The energy eigenvalues of the anharmonic oscillator are taken from
F.T. Hioe, D. MacMillan, and E.W. Montroll, Phys. Rep. **43**, 305 (1978); W. Caswell, Ann. Phys. (N.Y.) **123**, 153 (1979); R.L. Somorjai, and D.F. Hornig, J. Chem. Phys. **36**, 1980 (1962).
See also
K. Banerjee, Proc. Roy. Soc. A **364**, 265 (1978);
R. Balsa, M. Plo, J.G. Esteve, A.F. Pacheco, Phys. Rev. D **28**, 1945 (1983);
and most accurately
F. Vinette and J. Čížek, J. Math. Phys. **32**, 3392 (1991);
E.J. Weniger, J. Čížek, J. Math. Phys. **34**, 571 (1993).

[3] R.P. Feynman, *Statistical Mechanics*, Benjamin, Reading, 1972, Section 3.5.

[4] M. Hillary, R.F. O'Connell, M.O. Scully, and E.P. Wigner, Phys. Rep. **106**, 122 (1984).

[5] H. Kleinert, Phys. Lett. A **118**, 267 (1986) (*ibid*.http/145).

[6] For a detailed discussion of the effective classical potential of the Coulomb system see
W. Janke and H. Kleinert, Phys. Lett. A **118**, 371 (1986) (*ibid*.http/153).

[7] C. Kouveliotou et al., Nature **393**, 235 (1998); Astroph. J. **510**, L115 (1999);
K. Hurley et al., Astroph. J. **510**, L111 (1999);
V.M. Kaspi, D. Chakrabarty, and J. Steinberger, Astroph. J. **525**, L33 (1999);
B. Zhang and A.K. Harding, (astro-ph/0004067).

[8] The perturbation expansion of the ground state energy in powers of the magnetic field B was driven to high orders in
J.E. Avron, B.G. Adams, J. Čížek, M. Clay, M.L. Glasser, P. Otto, J. Paldus, and E. Vrscay, Phys. Rev. Lett. **43**, 691 (1979);
B.G. Adams, J.E. Avron, J. Čížek, P. Otto, J. Paldus, R.K. Moats, and H.J. Silverstone, Phys. Rev. A **21**, 1914 (1980).
This was possible on the basis of the dynamical group O(4,1) and the tilting operator (13.181) found by the author in his Ph.D. thesis. See
H. Kleinert, *Group Dynamics of Elementary Particles*, Fortschr. Physik **6**, 1 (1968)

(*ibid*.http/1);
H. Kleinert, *Group Dynamics of the Hydrogen Atom*, Lectures in Theoretical Physics, edited by W.E. Brittin and A.O. Barut, Gordon and Breach, N.Y. 1968, pp. 427-482 (*ibid*.http/4).

[9] Precise numeric calculations of the ground state energy of the hydrogen atom in a magnetic field were made by
H. Ruder, G. Wunner, H. Herold, and F. Geyer, *Atoms in Strong Magnetic Fields* (Springer-Verlag, Berlin, 1994).

[10] M. Bachmann, H. Kleinert, and A. Pelster, Phys. Rev. A **62**, 52509 (2000) (quant-ph/0005074), Phys. Lett. A **279**, 23 (2001) (quant-ph/000510).

[11] L.D. Landau and E.M. Lifshitz, *Quantum Mechanics*, Pergamon, London, 1965.

[12] J.C. LeGuillou and J. Zinn-Justin, Ann. Phys. (N.Y.) **147**, 57 (1983).

[13] W. Janke and H. Kleinert, Phys. Rev. Lett. **75**, 2787 (1995) (quant-ph/9502019).

[14] The high accuracy became possible due to a scaling relation found in
W. Janke and H. Kleinert, Phys. Lett. A **199**, 287 (1995) (quant-ph/9502018).

[15] For the proof of the exponentially fast convergence see
H. Kleinert and W. Janke, Phys. Lett. A **206**, 283 (1995) (quant-ph/9502019);
R. Guida, K. Konishi, and H. Suzuki, Ann. Phys. **249**, 109 (1996) (hep-th/9505084).
The proof will be given in Subsection 17.10.5.

[16] The oscillatory behavior around the exponential convergence shown in Fig. 5.22 was explained in
H. Kleinert and W. Janke, Phys. Lett. A **206**, 283 (1995) (quant-ph/9502019)
in terms of the convergence radius of the strong-coupling expansion (see Section 5.15).

[17] H. Kleinert, Phys. Rev. D **60** , 085001 (1999) (hep-th/9812197); Phys. Lett. B **463**, 69 (1999) (cond-mat/9906359).
See also Chapters 19–20 in the textbook
H. Kleinert and V. Schulte-Frohlinde, *Critical Properties of Φ^4-Theories*, World Scientific, Singapore 2001 (*ibid*.http/b8)

[18] H. Kleinert, Phys. Lett. A **207**, 133 (1995).

[19] See for example the textbooks
H. Kleinert, *Gauge Fields in Condensed Matter*, Vol. I Superflow and Vortex Lines, Vol. II Stresses and Defects, World Scientific, Singapore, 1989 (*ibid*.http/b1).

[20] R.P. Feynman, Phys. Rev. **97**, 660 (1955).

[21] S. Höhler and A. Müllensiefen, Z. Phys. *157*, 159 (1959); M.A. Smondyrev, Theor. Math. Fiz. *68*, 29 (1986); O.V. Selyugin and M.A. Smondyrev, Phys. Stat. Sol. (b) *155*, 155 (1989).

[22] N.N. Bogoliubov (jun) and V.N. Plechko, Teor. Mat. Fiz. [Sov. Phys.-Theor. Math. Phys.], **65**, 423 (1985); Riv. Nuovo Cimento *11*, 1 (1988).

[23] S.J. Miyake, J. Phys. Soc. Japan, **38**, 81 (1975).

[24] J.E. Avron, I.W. Herbst, B. Simon, Phys. Rev. A **20**, 2287 (1979).

[25] I.D. Feranshuk and L.I. Komarov, J. Phys. A: Math. Gen. **17**, 3111 (1984).

[26] H. Kleinert, W. Kürzinger, and A. Pelster, J. Phys. A: Math. Gen. **31**, 8307 (1998) (quant-ph/9806016).

[27] H. Kleinert, Phys. Lett. A **173**, 332 (1993) (quant-ph/9511020).

[28] H. Kleinert, Phys. Rev. D **57**, 2264 (1998) and Addendum: Phys. Rev. D **58**, 107702 (1998).

[29] F.J. Wegner, Phys. Rev. B **5**, 4529 (1972); B *6*, 1891 (1972).

[30] H. Kleinert, Phys. Lett. A **207**, 133 (1995) (quant-ph/9507005).

[31] J. Rössler, J. Phys. Stat. Sol. **25**, 311 (1968).

[32] J.T. Marshall and L.R. Mills, Phys. Rev. B *2*, 3143 (1970).

[33] Higher-order smearing formulas for nonpolynomial interactions were derived in
H. Kleinert, W. Kürzinger and A. Pelster, J. Phys. A **31**, 8307 (1998) (quant-ph/9806016).

[34] The polaron problem is solved in detail in the textbook
R.P. Feynman, *Statistical Mechanics*, Benjamin, New York, 1972, Chapter 8.
Extensive numerical evaluations are found in
T.D. Schultz, Phys. Rev. **116**, 526 (1959);
See also
M. Dineykhan, G.V. Efimov, G. Ganbold, and S.N. Nedelko, *Oscillator Representation in Quantum Physics*, Springer, Berlin, 1995.
An excellent review article is
J.T. Devreese, *Polarons*, Review article in Encyclopedia of Applied Physics, **14**, 383 (1996) (cond-mat/0004497).
This article contains ample references on work concerning polarons in magnetic fields, for instance
F.M. Peeters, J.T. Devreese, Phys. Stat. Sol. B **110**, 631 (1982); Phys. Rev. B **25**, 7281, 7302 (1982);
Xiaoguang Wu, F.M. Peeters, J.T. Devreese, Phys. Rev. B **32**, 7964 (1985);
F. Brosens and J.T. Devreese, Phys. Stat. Sol. B **145**, 517 (1988).
For discussion of the validity of the Jensen-Peierls inequality (5.10) in the presence of a magnetic field, see
J.T. Devreese and F. Brosens, Solid State Communs. **79**, 819 (1991); Phys. Rev. B **45**, 6459 (1992); Solid State Communs. **87**, 593 (1993);
D. Larsen in *Landau Level Spectroscopy*, Vol. 1, G. Landwehr and E. Rashba (eds.), North Holland, Amsterdam, 1991, p. 109.
The paper
D. Larsen, Phys. Rev. B **32**, 2657 (1985)
shows that the variational energy can lie lower than the exact energy.
The review article by Devreese contains numerous references on bipolarons, small polarons, and polaronic excitations. For instance:
J.T. Devreese, J. De Sitter, M.J. Goovaerts, Phys. Rev. B **5**, 2367 (1972);
L.F. Lemmens, J. De Sitter, J.T. Devreese, Phys. Rev. B **8**, 2717 (1973);
J.T. Devreese, L.F. Lemmens, J. Van Royen, Phys. Rev. B **15**, 1212 (1977);
J. Thomchick, L.F. Lemmens, J.T. Devreese, Phys. Rev. B **14**, 1777 (1976);
F.M. Peeters, Xiaoguang Wu, J.T. Devreese, Phys. Rev. B **34**, 1160 (1986);
F.M. Peeters, J.T. Devreese, Phys. Rev. B **34**, 7246 (1986); B **35**, 3745 (1987);
J.T. Devreese, S.N. Klimin, V.M. Fomin, F. Brosens, Solid State Communs. **114**, 305 (2000).
There exists also a broad collection of articles in
E.K.H. Salje, A.S. Alexandrov, W.Y. Liang (eds.), *Polarons and Bipolarons in High-T_c Superconductors and Related Materials*, Cambridge University Press, Cambridge, 1995.
A generalization of the harmonic trial path integral (5.356), in which the exponential function $e^{-\Omega|\tau-\tau'|}$ at zero temperature is replaced by $f(|\tau - \tau'|)$, has been proposed by
M. Saitoh, J. Phys. Soc. Japan. **49**, 878 (1980),
and further studied by
R. Rosenfelder and A.W. Schreiber, Phys. Lett. A **284**, 63 (2001) (cond-mat/0011332).

In spite of a much higher numerical effort, this generalization improves the ground state energy only by at most 0.1 % (the weak-coupling expansion coefficient -0.012346 in (5.396) is changed to -0.012598, while the strong-coupling coefficients in (5.399) remain unchanged at this level of accuracy. For the effective mass, the lowest nontrivial weak-coupling coefficient of 1^2 in (5.402) is changed by 0.0252 % while the strong-coupling coefficients in (5.403) remain unchanged at this level of accuracy.

Aevo rarissima nostro, simplicitas.
Simplicity, a very rare thing in our age.
OVID, Ars Amatoria, Book 1, 241

6

Path Integrals with Topological Constraints

The path integral representations of the time evolution amplitudes considered so far were derived for orbits $x(t)$ fluctuating in Euclidean space with Cartesian coordinates. Each coordinate runs from minus infinity to plus infinity. In many physical systems, however, orbits are confined to a topologically restricted part of a Cartesian coordinate system. This changes the quantum-mechanical completeness relation and with it the derivation of the path integral from the time-sliced time evolution operator in Section 2.1. We shall consider here only a point particle moving on a circle, in a half-space, or in a box. The path integral treatment of these systems is the prototype for any extension to more general topologies.

6.1 Point Particle on Circle

For a point particle on a circle, the orbits are specified in terms of an angular variable $\varphi(t) \in [0, 2\pi]$ subject to the topological constraint that $\varphi = 0$ and $\varphi = 2\pi$ be identical points.

The initial step in the derivation of the path integral for such a system is the same as before: The time evolution operator is decomposed into a product

$$\langle \varphi_b t_b | \varphi_a t_a \rangle = \langle \varphi_b | \exp\left[-\frac{i}{\hbar}(t_b - t_a)\hat{H} \right] |\varphi_a\rangle \equiv \langle \varphi_b | \prod_{n=1}^{N+1} \exp\left(-\frac{i}{\hbar}\epsilon\hat{H} \right) |\varphi_a\rangle. \quad (6.1)$$

The restricted geometry shows up in the completeness relations to be inserted between the factors on the right-hand side for $n = 1, \ldots, N$:

$$\int_0^{2\pi} d\varphi_n |\varphi_n\rangle\langle\varphi_n| = 1. \quad (6.2)$$

If the integrand is singular at $\varphi = 0$, the integrations must end at an infinitesimal piece below 2π. Otherwise there is the danger of double-counting the contributions from the identical points $\varphi = 0$ and $\varphi = 2\pi$. The orthogonality relations on these intervals are

$$\langle \varphi_n | \varphi_{n-1} \rangle = \delta(\varphi_n - \varphi_{n-1}), \quad \varphi_n \in [0, 2\pi). \quad (6.3)$$

The δ-function can be expanded into a complete set of periodic functions on the circle:

$$\delta\left(\varphi_n - \varphi_{n-1}\right) = \sum_{m_n=-\infty}^{\infty} \frac{1}{2\pi} \exp[im_n(\varphi_n - \varphi_{n-1})]. \tag{6.4}$$

For a trivial system with no Hamiltonian, the scalar products (6.4) lead to the following representation of the transition amplitude:

$$(\varphi_b t_b|\varphi_a t_a)_0 = \prod_{n=1}^{N}\left[\int_0^{2\pi} d\varphi_n\right] \prod_{n=1}^{N+1}\left[\sum_{m_n} \frac{1}{2\pi}\right] \exp\left[i\sum_{n=1}^{N+1} m_n(\varphi_n - \varphi_{n-1})\right]. \tag{6.5}$$

We now introduce a Hamiltonian $H(p, \varphi)$. At each small time step, we calculate

$$\begin{aligned}
(\varphi_n t_n|\varphi_{n-1} t_{n-1}) &= \langle\varphi_n|\exp\left[-\frac{i}{\hbar}\epsilon\hat{H}(p, \varphi)\right]|\varphi_{n-1}\rangle \\
&= \exp\left[-\frac{i}{\hbar}\epsilon\hat{H}(-i\hbar\partial_{\varphi_n}, \varphi_n)\right]\langle\varphi_n|\varphi_{n-1}\rangle.
\end{aligned}$$

Replacing the scalar products by their spectral representation (6.4), this becomes

$$\begin{aligned}
(\varphi_n t_n|\varphi_{n-1} t_{n-1}) &= \langle\varphi_n|\exp\left[-\frac{i}{\hbar}\epsilon\hat{H}(p, \varphi)\right]|\varphi_{n-1}\rangle \\
&= \exp\left[-\frac{i}{\hbar}\epsilon\hat{H}(-i\hbar\partial_{\varphi_n}, \varphi_n)\right]\sum_{m_n=-\infty}^{\infty} \frac{1}{2\pi} \exp[im_n(\varphi_n - \varphi_{n-1})]. \tag{6.6}
\end{aligned}$$

By applying the operator in front of the sum to each term, we obtain

$$(\varphi_n\, t_n|\varphi_{n-1}\, t_{n-1}) = \sum_{m_n=-\infty}^{\infty} \frac{1}{2\pi} \exp\left[im_n(\varphi_n - \varphi_{n-1}) - \frac{i\epsilon}{\hbar}H(\hbar m_n, \varphi_n)\right]. \tag{6.7}$$

The total amplitude can therefore be written as

$$\begin{aligned}
(\varphi_b t_b|\varphi_a t_a) &\approx \prod_{n=1}^{N}\left[\int_0^{2\pi} d\varphi_n\right] \prod_{n=1}^{N+1}\left[\sum_{m_n=-\infty}^{\infty} \frac{1}{2\pi}\right] \\
&\quad \times \exp\left\{i\sum_{n=1}^{N+1}\left[m_n(\varphi_n - \varphi_{n-1}) - \frac{1}{\hbar}\epsilon H(\hbar m_n, \varphi_n)\right]\right\}.
\end{aligned} \tag{6.8}$$

This is the desired generalization of the original path integral from Cartesian to cyclic coordinates. As a consequence of the indistinguishability of $\varphi(t)$ and $\varphi(t) + 2\pi n$, the momentum integrations have turned into sums over integer numbers. The sums reflect the fact that the quantum-mechanical wave functions $(1/\sqrt{2\pi})\exp(ip_\varphi\varphi/\hbar)$ are single-valued.

The discrete momenta enter into (6.8) via a "momentum step sum" rather than a proper path integral. At first sight, such an expression looks somewhat hard to deal with in practical calculations. Fortunately, it can be turned into a more comfortable equivalent form, involving a proper continuous path integral. This is possible at the expense of a single additional infinite sum which guarantees the cyclic invariance in the variable φ. To find the equivalent form, we recall Poisson's formula (1.197),

$$\sum_{l=-\infty}^{\infty} e^{2\pi i k l} = \sum_{m=-\infty}^{\infty} \delta(k-m), \tag{6.9}$$

to make the right-hand side of (6.4) a periodic sum of δ-functions, so that (6.3) becomes

$$\langle \varphi_n | \varphi_{n-1} \rangle = \sum_{l=-\infty}^{\infty} \delta(\varphi_n - \varphi_{n-1} + 2\pi l). \tag{6.10}$$

A Fourier decomposition of the δ-functions yields

$$\langle \varphi_n | \varphi_{n-1} \rangle = \sum_{l=-\infty}^{\infty} \int_{-\infty}^{\infty} \frac{dk_n}{2\pi} \exp[ik_n(\varphi_n - \varphi_{n-1}) + 2\pi i k_n l]. \tag{6.11}$$

Note that the right-hand side reduces to (6.4) when applying Poisson's summation formula (6.9) to the l-sum, which produces a sum of δ-functions for the integer values of $k_n = m_n = 0, \pm 1, \pm 2, \ldots$. Using this expansion rather than (6.4), the amplitude (6.5) with no Hamiltonian takes the form

$$(\varphi_b t_b | \varphi_a t_a)_0 = \prod_{n=1}^{N} \left[\int_0^{2\pi} d\varphi_n \right] \prod_{n=1}^{N+1} \left[\int_{-\infty}^{\infty} \frac{dk_n}{2\pi} \sum_{l_n=-\infty}^{\infty} \right] e^{i \sum_{n=1}^{N+1} [k_n(\varphi_n - \varphi_{n-1}) + 2\pi k_n l_n]}. \tag{6.12}$$

In this expression, we observe that the sums over l_n can be absorbed into the variables φ_n by extending their range of integration from $[0, 2\pi)$ to $(-\infty, \infty)$. Only in the last sum $\sum_{l_{N+1}}$, this is impossible, and we arrive at

$$(\varphi_b t_b | \varphi_a t_a)_0 = \sum_{l=-\infty}^{\infty} \prod_{n=1}^{N} \left[\int_{-\infty}^{\infty} d\varphi_n \right] \prod_{n=1}^{N+1} \left[\int_{-\infty}^{\infty} \frac{dk_n}{2\pi} \right] e^{i \sum_{n=1}^{N+1} k_n(\varphi_n - \varphi_{n-1} + 2\pi l \delta_{n,N+1})}. \tag{6.13}$$

The right-hand side looks just like an $\hat{H} \equiv 0$ -amplitude of an ordinary particle which would read

$$(\varphi_b t_b | \varphi_a t_a)_{0,\text{noncyclic}} = \prod_{n=1}^{N} \left[\int_{-\infty}^{\infty} d\varphi_n \right] \prod_{n=1}^{N+1} \left[\int_{-\infty}^{\infty} \frac{dk_n}{2\pi} \right] e^{i \sum_{n=1}^{N+1} k_n(\varphi_n - \varphi_{n-1})}. \tag{6.14}$$

The amplitude (6.13) differs from this by the sum over paths running over all periodic repetitions of the final point $\varphi_b + 2\pi n, t_b$. The amplitude (6.13) may therefore be written as a sum over all periodically repeated final points of the amplitude (6.14):

$$(\varphi_b t_b | \varphi_a t_a)_0 = \sum_{l=-\infty}^{\infty} (\varphi_b + 2\pi l, t_b | \varphi_a t_a)_{0,\text{noncyclic}} . \tag{6.15}$$

In each term on the right-hand side, the Hamiltonian can be inserted as usual, and we arrive at the time-sliced formula

$$(\varphi_b t_b | \varphi_a t_a) \approx \sum_{l=-\infty}^{\infty} \prod_{n=1}^{N} \left[\int_{-\infty}^{\infty} d\varphi_n \right] \prod_{n=1}^{N+1} \left[\int_{-\infty}^{\infty} \frac{dp_n}{2\pi\hbar} \right]$$

$$\times \exp \left\{ \frac{i}{\hbar} \sum_{n=1}^{N+1} [p_n(\varphi_n - \varphi_{n-1} + 2\pi l \delta_{n,N+1}) - \epsilon H(p_n, \varphi_n)] \right\}. \tag{6.16}$$

In the continuum limit, this tends to the path integral

$$(\varphi_b t_b | \varphi_a t_a) \xrightarrow{\epsilon \to 0} \sum_{l=-\infty}^{\infty} \int_{\varphi_a \rightsquigarrow \varphi_b + 2\pi l} \mathcal{D}\varphi(t) \int \frac{\mathcal{D}p(t)}{2\pi\hbar} \exp \left\{ \frac{i}{\hbar} \int_{t_a}^{t_b} dt[p\dot\varphi - H(p, \varphi)] \right\}. \tag{6.17}$$

The way in which this path integral has replaced the sum over all paths on the circle $\varphi \in [0, 2\pi)$ by the sum over all paths with the same action on the entire φ-axis is illustrated in Fig. 6.1.

As an example, consider a free particle moving on a circle with a Hamiltonian

$$H(p, \varphi) = \frac{p^2}{2M}. \tag{6.18}$$

The ordinary noncyclic path integral is

$$(\varphi_b t_b | \varphi_a t_a)_{\text{noncyclic}} = \frac{1}{\sqrt{2\pi\hbar i(t_b - t_a)/M}} \exp \left[\frac{i}{\hbar} \frac{M}{2} \frac{(\varphi_b - \varphi_a)^2}{t_b - t_a} \right]. \tag{6.19}$$

Using Eq. (6.15), the cyclic amplitude is given by the periodic Gaussian

$$(\varphi_b t_b | \varphi_a t_a) = \frac{1}{\sqrt{2\pi\hbar i(t_b - t_a)/M}} \sum_{l=-\infty}^{\infty} \exp \left[\frac{i}{\hbar} \frac{M}{2} \frac{(\varphi_b - \varphi_a + 2\pi l)^2}{t_b - t_a} \right]. \tag{6.20}$$

The same amplitude could, of course, have been obtained by a direct quantum-mechanical calculation based on the wave functions

$$\psi_m(\varphi) = \frac{1}{\sqrt{2\pi}} e^{im\varphi} \tag{6.21}$$

and the energy eigenvalues

$$H = \frac{\hbar^2}{2M} m^2. \tag{6.22}$$

Within operator quantum mechanics, we find

$$(\varphi_b t_b | \varphi_a t_a) = \langle \varphi_b | \exp \left[-\frac{i}{\hbar}(t_b - t_a)\hat{H} \right] | \varphi_a \rangle$$

$$= \sum_{m=-\infty}^{\infty} \psi_m(\varphi_b)\psi_m^*(\varphi_a) \exp \left[-\frac{i}{\hbar} \frac{\hbar^2 m^2}{2M}(t_b - t_a) \right]$$

$$= \sum_{m=-\infty}^{\infty} \frac{1}{2\pi} \exp \left[im(\varphi_b - \varphi_a) - i \frac{\hbar m^2}{2M}(t_b - t_a) \right]. \tag{6.23}$$

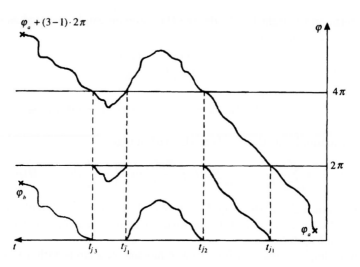

Figure 6.1 Path with 3 jumps from 2π to 0 at $t_{j_1}, t_{j_2}, t_{j_3}$, and with one jump from 0 to 2π at $t_{\bar{j}_1}$. It can be drawn as a smooth path in the *extended zone scheme*, arriving at $\varphi^{(n,\bar{n})} = \varphi_b + (n - \bar{n})2\pi$, where n and $\bar{n}$ count the number of jumps of the first and the second type, respectively.

If the sum over m is converted into an integral over p and a dual l-sum via Poisson's formula (6.9), this coincides with the previous result:

$$
\begin{aligned}
(\varphi_b t_b | \varphi_a t_a) &= \sum_{l=-\infty}^{\infty} \int_{-\infty}^{\infty} \frac{dp}{2\pi\hbar} \exp\left\{ \frac{i}{\hbar} \left[p(\varphi_b - \varphi_a + 2\pi l) - \frac{p^2}{2M}(t_b - t_a) \right] \right\} \\
&= \frac{1}{\sqrt{2\pi\hbar i(t_b - t_a)/M}} \sum_{l=-\infty}^{\infty} \exp\left[\frac{i}{\hbar} \frac{M}{2} \frac{(\varphi_b - \varphi_a + 2\pi l)^2}{t_b - t_a} \right].
\end{aligned}
\tag{6.24}
$$

6.2 Infinite Wall

In the case of an infinite wall, only a half-space, say $x = r > 0$, is accessible to the particle, and the completeness relation reads

$$
\int_0^\infty dr |r\rangle\langle r| = 1.
\tag{6.25}
$$

For singular integrands, the origin has to be omitted from the integration. The orthogonality relation is

$$
\langle r | r' \rangle = \delta(r - r'); \qquad r, r' > 0.
\tag{6.26}
$$

Given a free particle moving in such a geometry, we want to calculate

$$
(r_b t_b | r_a t_a) = \langle r_b | \prod_{n=1}^{N+1} \exp\left(-\frac{i}{\hbar} \hat{H}\epsilon \right) | r_a \rangle.
\tag{6.27}
$$

As usual, we insert N completeness relations between the $N+1$ factors. In the case of a vanishing Hamiltonian, the amplitude (6.27) becomes

$$(r_b t_b | r_a t_a)_0 = \prod_{n=1}^{N} \left[\int_0^\infty dr_n \right] \prod_{n=1}^{N+1} \langle r_n | r_{n-1} \rangle = \langle r_b | r_a \rangle. \qquad (6.28)$$

For each scalar product $\langle r_n | r_{n-1} \rangle = \delta(r_n - r_{n-1})$, we substitute its spectral representation appropriate to the infinite-wall boundary at $r = 0$. It consists of a superposition of the free-particle wave functions vanishing at $r = 0$:

$$\langle r | r' \rangle = 2 \int_0^\infty \frac{dk}{\pi} \sin kr \sin kr' \qquad (6.29)$$

$$= \int_{-\infty}^\infty \frac{dk}{2\pi} [\exp ik(r - r') - \exp ik(r + r')] = \delta(r - r') - \delta(r + r').$$

This Fourier representation does a bit more than what we need. In addition to the δ-function at $r = r'$, there is also a δ-function at the unphysical reflected point $r = -r'$. The reflected point plays a similar role as the periodically repeated points in the representation (6.11). For the same reason as before, we retain the reflected points in the formula as though r' were permitted to become zero or negative. Thus we rewrite the Fourier representation (6.29) as

$$\langle r | r' \rangle = \sum_{x=\pm r} \int_{-\infty}^\infty \frac{dp}{2\pi\hbar} \exp \left[\frac{i}{\hbar} p(x - x') + i\pi(\sigma(x) - \sigma(x')) \right]_{x'=r'}, \qquad (6.30)$$

where

$$\sigma(x) \equiv \Theta(-x) \qquad (6.31)$$

with the Heaviside function $\Theta(x)$ of Eq. (1.309). For symmetry reasons, it is convenient to liberate both the initial and final positions r and r' from their physical half-space and to introduce the localized states $|x\rangle$ whose scalar product exists on the entire x-axis:

$$\langle x | x' \rangle = \sum_{x''=\pm x} \int_{-\infty}^\infty \frac{dp}{2\pi\hbar} \exp \left[\frac{i}{\hbar} p(x'' - x') + i\pi(\sigma(x'') - \sigma(x')) \right]$$

$$= \delta(x - x') - \delta(x + x'). \qquad (6.32)$$

With these states, we write

$$\langle r | r' \rangle = \langle x | x' \rangle |_{x=r, x'=r'}. \qquad (6.33)$$

We now take the trivial transition amplitude with zero Hamiltonian

$$(r_b t_b | r_a t_a)_0 = \delta(r_b - r_a), \qquad (6.34)$$

extend it with no harm by the reflected δ-function

$$(r_b t_b | r_a t_a)_0 = \delta(r_b - r_a) - \delta(r_b + r_a), \qquad (6.35)$$

and factorize it into many time slices:

$$(r_b t_b | r_a t_a)_0 = \prod_{n=1}^{N} \left[\int_0^\infty dr_n \right] \prod_{n=1}^{N+1} \left[\sum_{x_n = \pm r_n} (x_n \epsilon | x_{n-1} 0)_0 \right] \tag{6.36}$$

$(r_b = r_{N+1}, r_a = r_0)$, where the trivial amplitude of a single slice is

$$(x_n \epsilon | x_{n-1} 0)_0 = \langle x_n | x_{n-1} \rangle, \qquad x \in (-\infty, \infty). \tag{6.37}$$

With the help of (6.32), this can be written as

$$\begin{aligned}
(r_b t_b | r_a t_a)_0 &= \prod_{n=1}^{N} \left[\int_0^\infty dr_n \right] \prod_{n=1}^{N+1} \left[\sum_{x_n = \pm r_n} \int_{-\infty}^\infty \frac{dp_n}{2\pi\hbar} \right] \\
&\times \exp \left\{ \sum_{n=1}^{N+1} \left[\frac{i}{\hbar} p(x_n - x_{n-1}) + i\pi(\sigma(x_n) - \sigma(x_{n-1})) \right] \right\}.
\end{aligned} \tag{6.38}$$

The sum over the reflected points $x_n = \pm r_n$ is now combined, at each n, with the integral $\int_0^\infty dr_n$ to form an integral over the entire x-axis, including the unphysical half-space $x < 0$. Only the last sum cannot be accommodated in this way, so that we obtain the path integral representation for the trivial amplitude

$$\begin{aligned}
(r_b t_b | r_a t_a)_0 &= \sum_{x_b = \pm r_b} \prod_{n=1}^{N} \left[\int_{-\infty}^\infty dx_n \right] \prod_{n=1}^{N+1} \left[\int_{-\infty}^\infty \frac{dp_n}{2\pi\hbar} \right] \\
&\times \exp \left\{ \sum_{n=1}^{N+1} \left[\frac{i}{\hbar} p(x_n - x_{n-1}) + i\pi(\sigma(x_n) - \sigma(x_{n-1})) \right] \right\}.
\end{aligned} \tag{6.39}$$

The measure of this path integral is now of the conventional type, integrating over all paths which fluctuate through the entire space. The only special feature is the final symmetrization in $x_b = \pm r_b$.

It is instructive to see in which way the final symmetrization together with the phase factor $\exp[i\pi\sigma(x)] = \pm 1$ eliminates all the wrong paths in the extended space, i.e., those which cross the origin into the unphysical subspace. This is illustrated in Fig. 6.2. Note that having assumed $x_a = r_a > 0$, the initial phase $\sigma(x_a)$ can be omitted. We have kept it merely for symmetry reasons.

In the continuum limit, the exponent corresponds to an action

$$\mathcal{A}_0^\sigma[p, x] = \int_{t_a}^{t_b} dt[p\dot{x} + \hbar\pi\partial_t\sigma(x)] \equiv \mathcal{A}_0[p, x] + \mathcal{A}_{\text{topol}}^\sigma. \tag{6.40}$$

The first term is the usual canonical expression in the absence of a Hamiltonian. The second term is new. It is a pure boundary term:

$$\mathcal{A}_{\text{topol}}^\sigma[x] = \hbar\pi(\sigma(x_b) - \sigma(x_a)), \tag{6.41}$$

which keeps track of the topology of the half space $x > 0$ embedded in the full space $x \in (-\infty, \infty)$. This is why the action carries the subscript "topol".

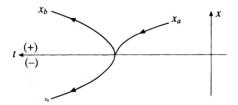

Figure 6.2 Illustration of path counting near reflecting wall. Each path touching the wall once is canceled by a corresponding path of equal action crossing the wall once into the unphysical regime (the path is mirror-reflected after the crossing). The phase factor $\exp[i\pi\sigma(x_b)]$ provides for the opposite sign in the path integral. Only paths not touching the wall at all cannot be canceled in the path integral.

The topological action (6.41) can be written formally as a local coupling of the velocity at the origin:

$$A_{\text{topol}}^{\sigma}[x] = -\pi\hbar \int_{t_a}^{t_b} dt\,\dot{x}(t)\delta(x(t)). \tag{6.42}$$

This follows directly from

$$\sigma(x_b) - \sigma(x_a) = \int_{x_a}^{x_b} dx\,\sigma'(x) = \int_{x_a}^{x_b} dx\,\Theta'(-x) = -\int_{x_a}^{x_b} dx\,\delta(x). \tag{6.43}$$

Consider now a free point particle in the right half-space with the usual Hamiltonian

$$H = \frac{p^2}{2M}. \tag{6.44}$$

The action reads

$$A[p, x] = \int_{t_a}^{t_b} dt[p\dot{x} - p^2/2M - \hbar\pi\dot{x}(t)\delta(x(t))], \tag{6.45}$$

and the time-sliced path integral looks like (6.39), except for additional energy terms $-p_n^2/2M$ in the action. Since the new topological term is a pure boundary term, all the extended integrals in (6.39) can be evaluated right away in the same way as for a free particle in the absence of an infinite wall. The result is

$$
\begin{aligned}
(r_b t_b | r_a t_a) &= \sum_{x_b = \pm r_b} \frac{1}{\sqrt{2\pi\hbar i(t_b - t_a)/M}} \\
&\quad \times \exp\left[\frac{i}{\hbar}\frac{M}{2}\frac{(x_b - x_a)^2}{t_b - t_a} + i\pi(\sigma(x_b) - \sigma(x_a))\right] \\
&= \frac{1}{\sqrt{2\pi\hbar i(t_b - t_a)/M}}\left\{\exp\left[\frac{i}{\hbar}\frac{M}{2}\frac{(r_b - r_a)^2}{t_b - t_a}\right] - (r_b \to -r_b)\right\}
\end{aligned}
\tag{6.46}
$$

with $x_a = r_a$.

This is indeed the correct result: Inserting the Fourier transform of the Gaussian (Fresnel) distribution we see that

$$
\begin{aligned}
(r_b t_b | r_a t_a) &= \int_{-\infty}^{\infty} \frac{dp}{2\pi\hbar} \left\{ \exp\left[\frac{i}{\hbar} p(r_b - r_a)\right] - (r_b \to -r_b) \right\} e^{-ip^2(t_b - t_a)/2M\hbar} \\
&= 2 \int_0^{\infty} \frac{dp}{2\pi\hbar} \sin(pr_b/\hbar) \sin(pr_a/\hbar) \exp\left[-\frac{i}{\hbar}\frac{p^2}{2M}(t_b - t_a)\right], \quad (6.47)
\end{aligned}
$$

which is the usual spectral representation of the time evolution amplitude.

Note that the first part of (6.46) may be written more symmetrically as

$$
(r_b t_b | r_a t_a) = \frac{1}{\sqrt{2\pi\hbar i(t_b - t_a)/M}} \frac{1}{2} \sum_{\substack{x_a = \pm r_a \\ x_b = \pm r_b}} \exp\left[\frac{i}{\hbar}\frac{M}{2}\frac{(x_b - x_a)^2}{t_b - t_a} + i\pi(\sigma(x_b) - \sigma(x_a))\right].
$$
(6.48)

In this form, the phase factors $e^{i\pi\sigma(x)}$ are related to what may be considered as even and odd "spherical harmonics" in one dimension [more after (9.60)]

$$
Y_{e,\o}(\hat{x}) = \frac{1}{\sqrt{2}}(\Theta(x) \pm \Theta(-x)),
$$

namely,

$$
Y_e(\hat{x}) = \frac{1}{\sqrt{2}}, \quad Y_\o(\hat{x}) = \frac{1}{\sqrt{2}} e^{i\pi\sigma(x)}. \quad (6.49)
$$

The amplitude (6.48) is therefore simply the odd "partial wave" of the free-particle amplitude

$$
(r_b t_b | r_a t_a) = \sum_{\substack{\hat{x}_b, \hat{x}_a \\ |x_b| = r_b, |x_a| = r_a}} Y_\o^*(\hat{x}_b) \langle x_b t_b | x_a t_a \rangle Y_\o(\hat{x}_a), \quad (6.50)
$$

which is what we would also have obtained from Schrödinger quantum mechanics.

6.3 Point Particle in Box

If a point particle is confined between two infinitely high walls in the interval $x \in (0, d)$, we speak of a *particle in a box*.[1] The box is a geometric constraint. Since the wave functions vanish at the walls, the scalar product between localized states is given by the quantum-mechanical orthogonality relation for $r \in (0, d)$:

$$
\langle r | r' \rangle = \frac{2}{d} \sum_{k_\nu > 0} \sin k_\nu r \sin k_\nu r', \quad (6.51)
$$

---

[1]See W. Janke and H. Kleinert, Lett. Nuovo Cimento **25**, 297 (1979) (http://www.physik.fu-berlin.de/~kleinert/64).

where k_ν runs over the discrete positive momenta

$$k_\nu = \frac{\pi}{d}\nu, \quad \nu = 1, 2, 3, \ldots . \tag{6.52}$$

We can write the restricted sum in (6.51) also as a sum over all momenta k_ν with $\nu = 0, \pm 1, \pm 2, \ldots$:

$$\langle r|r' \rangle = \frac{1}{2d} \sum_{k_\nu} \left[e^{ik_\nu(r-r')} - e^{ik_\nu(r+r')} \right]. \tag{6.53}$$

With the help of the Poisson summation formula (6.9), the right-hand side is converted into an integral and an auxiliary sum:

$$\langle r|r' \rangle = \sum_{l=-\infty}^{\infty} \int_{-\infty}^{\infty} \frac{dk}{2\pi} \left[e^{ik(r-r'+2dl)} - e^{ik(r+r'+2dl)} \right]. \tag{6.54}$$

Using the potential $\sigma(x)$ of (6.31), this can be re-expressed as

$$\langle r|r' \rangle = \sum_{x=\pm r} \sum_{l=-\infty}^{\infty} \int_{-\infty}^{\infty} \frac{dk}{2\pi} e^{ik(x-x'+2dl)+i\pi(\sigma(x)-\sigma(x'))}. \tag{6.55}$$

The trivial path integral for the time evolution amplitude with a zero Hamiltonian is again obtained by combining a sequence of scalar products (6.51):

$$(r_b t_b | r_a t_a)_0 = \prod_{n=1}^{N} \left[\int_0^d dr_n \right] \langle r_n | r_{n-1} \rangle \tag{6.56}$$

$$= \prod_{n=1}^{N} \left[\int_0^d dr_n \right] \prod_{n=1}^{N+1} \frac{2}{d} \sum_{k_\nu} \sin k_{\nu_n} r_n \sin k_{\nu_{n-1}} r_{n-1}.$$

The alternative spectral representation (6.55) allows us to extend the restricted integrals over x_n and sums over k_ν to complete phase space integrals, and we may write

$$(r_b t_b | r_a t_a)_0 = \sum_{x_b=\pm r_b} \sum_{l=-\infty}^{\infty} \prod_{n=1}^{N} \left[\int_{-\infty}^{\infty} dx_n \right] \prod_{n=1}^{N+1} \left[\int_{-\infty}^{\infty} \frac{dp_n}{2\pi\hbar} \right] \exp\left(\frac{i}{\hbar} \mathcal{A}_0^N \right), \tag{6.57}$$

with the time-sliced $H \equiv 0$ -action:

$$\mathcal{A}_0^N = \sum_{n=1}^{N+1} \left[p_n(x_n - x_{n-1}) + \hbar\pi(\sigma(x_n) - \sigma(x_{n-1})) \right]. \tag{6.58}$$

The final x_b is summed over all periodically repeated endpoints $r_b + 2dl$ and their reflections $-r_b + 2dl$.

We now add dynamics to the above path integral by introducing some Hamiltonian $H(p, x)$, so that the action reads

$$\mathcal{A} = \int_{t_a}^{t_b} dt[p\dot{x} - H(p, x) - \hbar\pi\dot{x}\delta(x)]. \tag{6.59}$$

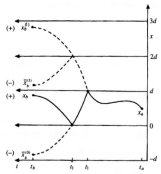

Figure 6.3 Illustration of path counting in a box. A path reflected once on the upper and once on the lower wall of the box is eliminated by a path with the same action running to $x_b^{(1)}$ and to $\bar{x}_b^{(0)}, \bar{x}_b^{(1)}$. The latter receive a negative sign in the path integral from the phase factor $\exp[i\pi\sigma(x_b)]$. Only paths remaining completely within the walls have no partner for cancellation.

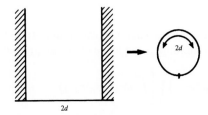

Figure 6.4 A particle in a box is topologically equivalent to a particle on a circle with an infinite wall at one point.

The amplitude is written formally as the path integral

$$(r_b t_b | r_a t_a) = \sum_{l=-\infty}^{\infty} \sum_{x_b = \pm r_b + 2dl} \int \mathcal{D}x \int \frac{\mathcal{D}p}{2\pi\hbar} \exp\left(\frac{i}{\hbar}\mathcal{A}\right). \quad (6.60)$$

In the time-sliced version, the action is

$$\mathcal{A}^N = \mathcal{A}_0^N - \epsilon \sum_{n=1}^{N+1} H(p_n, x_n). \quad (6.61)$$

The way in which the sum over the final positions $x_b = \pm r_b + 2dl$ together with the phase factor $\exp[i\pi\sigma(x_b)]$ eliminates the unphysical paths is illustrated in Fig. 6.3. The mechanism is obviously a combination of the previous two. A particle in a box of length d behaves like a particle on a circle of circumference $2d$ with a periodic boundary condition, containing an infinite wall at one point. This is illustrated in Fig. 6.4. The periodicity in $2d$ selects the momenta

$$k_\nu = (\pi/d)\nu, \quad \nu = 1, 2, 3, \ldots,$$

as it should.

For a free particle with $H = p^2/2M$, the integrations over x_n, p_n can be done as usual and we obtain the amplitude $(x_a = r_a)$

$$(r_b t_b | r_a t_a) = \sum_{l=-\infty}^{\infty} \sum_{x_b = \pm r_b + 2dl} \frac{1}{\sqrt{2\pi\hbar i (t_b - t_a)/M}} \left[e^{\frac{i}{\hbar} \frac{M}{2} \frac{(x_b - x_a + 2dl)^2}{t_b - t_a}} - (x_b \to -x_b) \right].$$

$$(6.62)$$

A Fourier transform and an application of Poisson's formula (6.9) shows that this is, of course, equal to the quantum-mechanical expression

$$
\begin{aligned}
(r_b t_b | r_a t_a) &= \langle r_b | \exp\left[-\frac{i}{\hbar}(t_b - t_a)\hat{H} \right] | r_a \rangle \\
&= \frac{2}{d} \sum_{\nu=1}^{\infty} \sin k_\nu r_b \sin k_\nu r_a \exp\left[-i\hbar \frac{k_\nu^2}{2M}(t_b - t_a) \right].
\end{aligned}
$$

$$(6.63)$$

In analogy with the discussion in Section 2.6, we identify in the exponentials the eigenvalues of the energy levels labeled by $\nu - 1 = 0, 1, 2, \ldots$:

$$E^{(\nu-1)} = \frac{\hbar^2 k_\nu^2}{2M}, \qquad \nu = 1, 2, 3 \ldots .$$

$$(6.64)$$

The factors in front determine the wave functions associated with these energies:

$$\psi^{(\nu-1)}(x) = \frac{2}{d} \sin k_\nu x.$$

$$(6.65)$$

6.4 Strong-Coupling Theory for Particle in Box

The strong-coupling theory developed in Chapter 5 open up the possibility of treating quantum-mechanical systems with hard-wall potentials via perturbation theory. After converting divergent weak-coupling expansions into convergent strong-coupling expansions, the strong-coupling limit of a function can be evaluated from its weak-coupling expansion with any desired accuracy. Due to the combination with the variational procedure, new classes of physical systems become accessible to perturbation theory. For instance, the important problem of the pressure exerted by a stack of membranes upon enclosing walls has been solved by this method.[2]

Here we illustrate the working of that theory for the system treated in the previous section, the point particle in a one-dimensional box.

This is just a quantum-mechanical exercise for the treatment of physically more interesting problems. The ground state energy of this system has, according to Eq. (6.64), the value $E^{(0)} = \pi^2/2d^2$. For simplicity, we shall now use natural units in which we can omit Planck and Boltzmann constants everywhere, setting them equal to unity: $\hbar = 1, k_B = 1$. We shall now demonstrate how this result is found via strong-coupling theory from a perturbation expansion.

---

[2]See Notes and References.

6.4.1 Partition Function

The discussion becomes simplest by considering the quantum statistical partition function of the particle. It is given by the Euclidean path integral (always in natural units)

$$Z = \int \mathcal{D}u(\tau) e^{\frac{1}{2}\int_0^{\hbar\beta} d\tau (\partial u)^2} , \qquad (6.66)$$

where the shifted particle coordinate $u(\tau) \equiv x(\tau) - d/2$ is restricted to the symmetric interval $-d/2 \leq u(\tau) \leq d/2$. Since such a hard-wall restriction is hard to treat analytically in (6.66), we make the hard-walls soft by adding to the Euclidean action E in the exponent of (6.66) a potential term diverging near the walls. Thus we consider the auxiliary Euclidean action

$$\mathcal{A}_e = \frac{1}{2} \int_0^{\hbar\beta} d\tau \left\{ [\partial u(\tau)]^2 + V(u(\tau)) \right\}, \qquad (6.67)$$

where $V(u)$ is given by

$$V(u) = \frac{\omega^2}{2} \left(\frac{d}{\pi} \tan \frac{\pi u}{d} \right)^2 = \frac{\omega^2}{2} \left(u^2 + \frac{2}{3} g u^4 + \dots \right). \qquad (6.68)$$

On the right-hand side we have introduced a parameter $g \equiv \pi^2/d^2$.

6.4.2 Perturbation Expansion

The expansion of the potential in powers of g can now be treated perturbatively, leading to an expansion of Z around the harmonic part of the partition function. In this, the integrations over $u(\tau)$ run over the entire u-axis, and can be integrated out as described in Section 2.15. The result is [see Eq. (2.487)]

$$Z_\omega = e^{-(1/2)\mathrm{Tr}\log(\partial^2 + \omega^2)}. \qquad (6.69)$$

For $\beta \to \infty$, the exponent gives a free energy density $F = -\beta^{-1}\log Z$ equal to the ground state energy of the harmonic oscillator

$$F_\omega = \frac{\omega}{2}. \qquad (6.70)$$

The treatment of the interaction terms can be organized in powers of g, and give rise to an expansion of the free energy with the generic form

$$F = F_\omega + \omega \sum_{k=1}^{\infty} a_k \left(\frac{g}{\omega} \right)^k. \qquad (6.71)$$

The calculation of the coefficients a_k in this expansion proceeds as follows. First we expand the potential in (6.67) to identify the power series for the interaction energy

$$\begin{aligned}
\mathcal{A}_e^{\text{int}} &= \frac{\omega^2}{2} \int d\tau \left(g v_4 u^4 + g^2 v_6 u^6 + g^3 v_8 u^8 + \dots \right) \\
&= \frac{\omega^2}{2} \sum_{k=1}^{\infty} \int d\tau \, g^k v_{2k+2} [u^2(\tau)]^{k+1} ,
\end{aligned} \qquad (6.72)$$

with coefficients

$$v_4 = \frac{2}{3}, \quad v_6 = \frac{17}{45}, \quad v_8 = \frac{62}{315}, \quad v_{10} = \frac{1382}{14175}, \quad v_{12} = \frac{21844}{467775}, \quad v_{14} = \frac{929569}{42567525},$$

$$v_{16} = \frac{6404582}{638512875}, \quad v_{18} = \frac{443861162}{97692469875}, \quad v_{20} = \frac{18888466084}{9280784638125}, \quad v_{22} = \frac{113927491862}{126109485376875},$$

$$v_{24} = \frac{58870668456604}{147926426347074375}, \quad v_{26} = \frac{8374643517010684}{48076088562799171875}, \quad v_{28} = \frac{689005380505609448}{9086380738369043484375},$$

$$v_{30} = \frac{129848163681107301953}{3952575621190533915703125}, \quad v_{32} = \frac{1736640792209901647222}{1225298442569066551386796875},$$

$$v_{34} = \frac{41878123149529303891392 2}{6873924262812457532799304687 5}, \quad \cdots . \tag{6.73}$$

The interaction terms $\int d\tau \, [u^2(\tau)]^{k+1}$ and their products are expanded according to Wick's rule in Section 3.10 into sums of products of harmonic two-point correlation functions

$$\langle u(\tau_1)u(\tau_2)\rangle = \int \frac{dk}{2\pi} \frac{e^{ik(\tau_1-\tau_2)}}{k^2+\omega^2} = \frac{e^{-\omega|\tau_1-\tau_2|}}{2\omega}. \tag{6.74}$$

Associated local expectation values are $\langle u^2\rangle = 1/2\omega$, and

$$\langle u\partial u\rangle = \int \frac{dk}{2\pi} \frac{k}{k^2+\omega^2} = 0$$

$$\langle \partial u\partial u\rangle = \int \frac{dk}{2\pi} \frac{k^2}{k^2+\omega^2} = -\frac{\omega}{2}, \tag{6.75}$$

where the last integral is calculated using dimensional regularization in which $\int dk \, k^\alpha = 0$ for all α. The Wick contractions are organized with the help of the Feynman diagrams as explained in Section 3.20. Only the connected diagrams contribute to the free energy density. The graphical expansion of free energy up to four loops is

$$F = \frac{\omega}{2} + \left(\frac{\omega^2}{2}\right) \left\{ gv_4 \, 3 \, \infty + g^2 v_6 15 \, \text{⊳○} + g^3 v_8 105 \, \text{✕} \right\}$$

$$- \frac{1}{2!}\left(\frac{\omega^2}{2}\right)^2 \left\{ g^2 v_4^2 \, [72 \, \text{○○○} + 24 \, \ominus] + g^3 \, 2v_4 v_6 \, \left[540 \, \text{⊳○○} + 360 \, \ominus\!\!\circ\right] \right\}$$

$$+ \frac{1}{3!}\left(\frac{\omega^2}{2}\right)^3 g^3 v_4^3 \left\{ 2592 \, \text{○○○○} + 1728 \, \text{⊶} + 3456 \, \ominus\!\!\circ + 1728 \, \text{⊘} \right\}. \tag{6.76}$$

Note different numbers of loops contribute to the terms of order g^n. The calculation of the diagrams in Eq. (6.76) is simplified by the factorization property: If a diagram consists of two subdiagrams touching each other at a single vertex, the associated Feynman integral factorizes into those of the subdiagrams. In each diagram, the last t-integral yields an overall factor β, due to translational invariance along the t-axis, the others produce a factor $1/\omega$. Using the explicit expression (6.75) for the lines in the diagrams, we find the following values for the Feynman integrals:

$$\text{○○○} = \beta \frac{1}{16\omega^5}, \qquad \text{○○○○} = \beta \frac{1}{64\omega^8},$$

$$\ominus \;=\; \beta\,\frac{1}{32\omega^5}\,, \qquad\qquad \eightfigure \;=\; \beta\,\frac{3}{128\omega^8}\,,$$

$$\dumbbell \;=\; \beta\,\frac{1}{32\omega^6}\,, \qquad\qquad \lollipop \;=\; \beta\,\frac{5}{8\cdot 64\omega^8}\,, \tag{6.77}$$

$$\circledcirc \;=\; \beta\,\frac{1}{32\omega^6}\,, \qquad\qquad \triangledot \;=\; \beta\,\frac{3}{8\cdot 64\omega^8}\,.$$

Adding all contributions in (6.76), we obtain up to the order g^3:

$$F_3 \;=\; \omega\left\{\frac{1}{2}+\frac{3}{8}v_4\left(\frac{g}{\omega}\right)+\left[\frac{15}{16}v_6-\frac{21}{32}v_4^2\right]\left(\frac{g}{\omega}\right)^2+\left[\frac{105}{32}v_8-\frac{45}{8}v_4\,v_6+\frac{333}{128}v_4^3\right]\left(\frac{g}{\omega}\right)^3\right\}, \tag{6.78}$$

which has the generic form (6.71).

We can go to higher orders by extending the Bender-Wu recursion relation (3C.20) for the ground state energy of the quartic anharmonic oscillator as follows:

$$2p'C_n^{p'} = (p'+1)(2p'+1)C_n^{p'} + \frac{1}{2}\sum_{k=1}^{n}v_{2k+2}C_{n-k}^{p'-k-1} - \sum_{k=1}^{n-1}C_k^1 C_{n-k}^{p'},\; 1\le p'\le 2n,$$

$$C_0^0 = 1,\;\; C_n^{p'} = 0 \quad (n\ge 1, p'<1). \tag{6.79}$$

After solving these recursion relations, the coefficients a_k in (6.71) are given by $a_k = (-1)^{k+1}C_{k,1}$. For brevity, we list here the first sixteen expansion coefficients for F, calculated with the help of MATHEMATICA of REDUCE programs:[3]

$$a_0 = \frac{1}{2},\, a_1 = \frac{1}{4},\, a_2 = \frac{1}{16},\, a_3 = 0,\, a_4 = -\frac{1}{256},\, a_5 = 0,$$

$$a_6 = \frac{1}{2048},\, a_7 = 0,\, a_8 = -\frac{5}{65536},\, a_9 = 0,$$

$$a_{10} = \frac{7}{524288},\, a_{11} = 0,\, a_{12} = -\frac{21}{8388608},\, a_{13} = 0,$$

$$a_{14} = \frac{33}{67108864},\, a_{15} = 0,\, a_{16} = -\frac{429}{4294967296},\dots\;. \tag{6.80}$$

6.4.3 Variational Strong-Coupling Approximations

We are now ready to calculate successive strong-coupling approximations to the function $F(g)$. It will be convenient to remove the expected correct d dependence π^2/d^2 from $F(g)$, and study the function $\tilde{F}(\bar{g}) \equiv F(g)/g$ which depends only on the dimensionless reduced coupling constant $\bar{g} = g/\omega$. The limit $\omega \to 0$ corresponds to a strong-coupling limit in the reduced coupling constant $\bar{g}$. According to the general theory of variational perturbation theory and its strong-coupling limit in Sections 5.14 and 5.17, the Nth order approximation to the strong-coupling limit of $\tilde{F}(\bar{g})$, to be denoted by $\tilde{F}^*$, is found by replacing, in the series truncated after the

---

[3]The programs can be downloaded from www.physik.fu-berlin.de/~kleinert/b5/programs

Nth term, $\tilde{F}_N(g/\omega)$, the frequency ω by the identical expression $\sqrt{\Omega^2 - gr^2/2M}$, where

$$r^2 \equiv 2M(\Omega^2 - \omega^2)/g. \qquad (6.81)$$

For a moment, this is treated as an independent variable, whereas Ω is a dummy parameter. Then the square root is expanded binomially in powers of g, and $\tilde{F}_N(g/\sqrt{\Omega^2 - gr^2/2M})$ is re-expanded up to order g^N. After that, r is replaced by its proper value. In this way we obtain a function $\tilde{F}_N(g, \Omega)$ which depends on Ω, which thus becomes a variational parameter. The best approximation is obtained by extremizing $\tilde{F}_N(g, \Omega)$ with respect to ω. Setting $\omega = 0$, we go to the strong-coupling limit $g \to \infty$. There the optimal Ω grows proportionally to g, so that $g/\Omega = c^{-1}$ is finite, and the variational expression $\tilde{F}_N(g, \Omega)$ becomes a function of $f_N(c)$. In this limit, the above re-expansion amounts simply to replacing each power ω^n in each expansion terms of $\tilde{F}_N(\bar{g})$ by the binomial expansion of $(1 - 1)^{-n/2}$ truncated after the $(N - n)$th term, and replacing $\bar{g}$ by c^{-1}. The first nine variational functions $f_N(c)$ are listed in Table 6.1. The functions $f_N(c)$ are minimized starting from $f_2(c)$ and searching the minimum of each successive $f_3(c)$, $f_3(c)$, ... nearest to the previous one. The functions $f_N(c)$ together with their minima are plotted in Fig. 6.5. The minima lie at

Table 6.1 First eight variational functions $f_N(c)$.

$$f_2(c) = \tfrac{1}{4} + \tfrac{1}{16\,c} + \tfrac{3\,c}{16}$$
$$f_3(c) = \tfrac{1}{4} + \tfrac{3}{32\,c} + \tfrac{5\,c}{32}$$
$$f_4(c) = \tfrac{1}{4} - \tfrac{1}{256\,c^3} + \tfrac{15}{128\,c} + \tfrac{35\,c}{256}$$
$$f_5(c) = \tfrac{1}{4} - \tfrac{5}{512\,c^3} + \tfrac{35}{256\,c} + \tfrac{63\,c}{512}$$
$$f_6(c) = \tfrac{1}{4} + \tfrac{1}{2048\,c^5} - \tfrac{35}{2048\,c^3} + \tfrac{315}{2048\,c} + \tfrac{231\,c}{2048}$$
$$f_7(c) = \tfrac{1}{4} + \tfrac{7}{4096\,c^5} - \tfrac{105}{4096\,c^3} + \tfrac{693}{4096\,c} + \tfrac{429\,c}{4096}$$
$$f_8(c) = \tfrac{1}{4} - \tfrac{5}{65536\,c^7} + \tfrac{63}{16384\,c^5} - \tfrac{1155}{32768\,c^3} + \tfrac{3003}{16384\,c} + \tfrac{6435\,c}{65536}$$
$$f_9(c) = \tfrac{1}{4} - \tfrac{45}{131072\,c^7} + \tfrac{231}{32768\,c^5} - \tfrac{3003}{65536\,c^3} + \tfrac{6435}{32768\,c} + \tfrac{12155\,c}{131072}$$

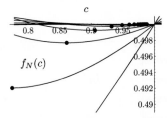

Figure 6.5 Variational functions $f_N(c)$ for particle between walls up to $N = 16$ are shown together with their minima whose y-coordinates approach rapidly the correct limiting value $1/2$.

$$
\begin{aligned}
(N, f_N^{\min}) \ = \ & (2, 0.466506), \ (3, 0.492061), \ (4, 0.497701), \\
& (5, 0.499253), \ (6, 0.499738), \ (7, 0.499903), \\
& (8, 0.499963), \ (9, 0.499985), \ (10, 0.499994), \\
& (11, 0.499998), \ (12, 0.499999), \ (13, 0.5000), \\
& (14, 0.50000), \ (15, 0.50000), \ (16, 0.5000).
\end{aligned} \tag{6.82}
$$

They converge exponentially fast against the known result $1/2$, as shown in Fig. 6.6.

6.4.4 Special Properties of Expansion

The alert reader will have noted that the expansion coefficients (6.80) possess two special properties: First, they lack the factorial growth at large orders which would be found for a single power $[u^2(\tau)]^{k+1}$ of the interaction potential, as mentioned in Eq.(3C.27) and will be proved in Eq. (17.323). The factorial growth is canceled by the specific combination of the different powers in the interaction (6.72), making the series (6.71) convergent inside a certain circle. Still, since this circle has a finite radius (the ratio test shows that it is unity), this convergent series cannot be evaluated in the limit of large g which we want to do, so that variational strong-coupling theory is not superfluous. However, there is a second remarkable property of the coefficients (6.80): They contain an infinite number of zeros in the sequence of coefficients for each odd number, except for the first one. We may take advantage of this property by separating off the irregular term $a_1 g = g/4 = \pi^2/4d^2$, setting $\alpha = g^2/4\omega^2$, and rewriting $\tilde{F}(\bar{g})$ as

$$
\tilde{F}(\alpha) = \frac{1}{4}\left[1 + \frac{1}{\sqrt{\alpha}}h(\alpha)\right], \qquad h(\alpha) \equiv \sum_{n=0}^{N} 2^{2n+1} a_{2n}\alpha^n. \tag{6.83}
$$

Inserting the numbers (6.80), the expansion of $h(\alpha)$ reads

$$
h(\alpha) = 1 + \frac{\alpha}{2} - \frac{\alpha^2}{8} + \frac{\alpha^3}{16} - \frac{5}{128}\alpha^4 + \frac{7}{256}\alpha^5 - \frac{21}{1024}\alpha^6 + \frac{33}{2048}\alpha^7 - \frac{429}{32768}\alpha^8 + \dots. \tag{6.84}
$$

We now realize that this is the binomial power series expansion of $\sqrt{1+\alpha}$. Substituting this into (6.83), we find the exact ground state energy for the Euclidean action (6.67)

$$
E^{(0)} = \frac{\pi^2}{4d^2}\left(1 + \sqrt{1 + \frac{1}{\alpha}}\right) = \frac{\pi^2}{4d^2}\left(1 + \sqrt{1 + 4\omega^2\frac{d^4}{\pi^4}}\right). \tag{6.85}
$$

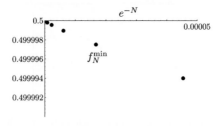

Figure 6.6 Exponentially fast convergence of strong-coupling approximations towards exact value.

Here we can go directly to the strong-coupling limit $\alpha \to \infty$ to recover the exact ground state energy $E^{(0)} = \pi^2/2d^2$.

The energy (6.85) can of course be obtained directly by solving the Schrödinger equation associated with the potential (6.72),

$$\frac{1}{2}\left\{-\frac{\partial^2}{\partial x^2} + \left[\frac{\lambda(1-\lambda)}{\cos^2 x} - 1\right]\right\}\psi(x) = \frac{d^2}{\pi^2}E\psi(x), \tag{6.86}$$

where we have replaced $u \to dx/\pi$ and set $\omega^2 d^4/\pi^4 \equiv \lambda(\lambda - 1)$, so that

$$\lambda = \frac{1}{2}\left(1 + \sqrt{1 + 4\omega^2\frac{d^4}{\pi^4}}\right). \tag{6.87}$$

Equation (6.86) is of the Pöschl-Teller type [see Subsection 14.4.5], and has the ground state wave function, to be derived in Eq. (14.163),

$$\psi_0(x) = \text{const} \times \cos^\lambda x, \tag{6.88}$$

with the eigenvalue $\pi^2 E^{(0)}/d^2 = (\lambda^2 - 1)/2$, which agrees of course with Eq. (6.85).

If we were to apply the variational procedure to the series $h(\alpha)/\sqrt{\alpha}$ in F of Eq. (6.85), by replacing the factor $1/\omega^{2n}$ contained in each power α^n by $\Omega = \sqrt{\Omega^2 - r\alpha}$ and re-expanding now in powers of α rather than g, we would find that all approximation $h_N(c)$ would possess a minimum with unit value, such that the corresponding extremal functions $f_N(c)$ yield the correct final energy in *each* order N.

6.4.5 Exponentially Fast Convergence

With the exact result being known, let us calculate the exponential approach of the variational approximations observed in Fig. (6.6). Let us write the exact energy (6.85) as

$$E^{(0)} = \frac{1}{4}(g + \sqrt{g^2 + 4\omega^2}). \tag{6.89}$$

After the replacement $\omega \to \sqrt{\Omega^2 - \rho g}$, this becomes

$$E^{(0)} = \frac{\Omega}{4}\left(\hat{g} + \sqrt{\hat{g}^2 - 4\rho\hat{g} + 4}\right), \tag{6.90}$$

where $\hat{g} \equiv g/\Omega^2$. The Nth-order approximant $f_N(g)$ of $E^{(0)}$ is obtained by expanding (6.91) in powers of $\hat{g}$ up to order N,

$$f_N(g) = \Omega \sum_0^N h_k(\rho)\hat{g}^k, \tag{6.91}$$

and substituting ρ by $2Mr^2 = (1 - \hat{\omega}^2)/\hat{g}$ [compare (6.81)], with $\hat{\omega}^2 \equiv \omega^2/\Omega^2$. The resulting function of $\hat{g}$ is then optimized.

It is straightforward to find an integral representation for $F_N(g)$. Setting $r\hat{g} \equiv z$, we have

$$F_N = \frac{1}{2\pi i}\oint_{C_0}\frac{dz}{z^{N+1}}\frac{1 - z^{N+1}}{1 - z}f(z), \tag{6.92}$$

where the contour C_0 refers to small circle around the origin and

$$\begin{aligned}F(z) &= \frac{\Omega}{4}\left(\frac{z}{r} + \sqrt{\frac{z^2}{r^2} - 4z + 4}\right)\\ &= \frac{1}{4r}\left(z + \sqrt{(z - z_1)(z - z_2)}\right),\end{aligned} \tag{6.93}$$

with branch points at $z_{1,2} = 2r^2 \left(1 \pm \sqrt{1 - 1/r^2}\right)$. For $z < 1$, we rewrite

$$
\begin{aligned}
1 - z^{N+1} &= (1 - z)(1 + z + \ldots + z^N) = (1 - z)(N + 1) \\
&\quad -(1 - z)^2 \left[N + (N - 1)z + \ldots + z^{N-1}\right]
\end{aligned}
\tag{6.94}
$$

and estimate this for $z \approx 1$ as

$$
1 - z^{N+1} = (1 - z)(N + 1) + \mathcal{O}(|1 - z|^2 N^2).
\tag{6.95}
$$

Dividing the approximant (6.92) by Ω, and indicating this by a hat, we use (6.94) to write $\hat{F}_N$ as a sum over the discontinuities across the two branch cuts:

$$
\begin{aligned}
\hat{F}_N &= \frac{(N+1)}{2\pi i} \oint_{C_0} \frac{dz \hat{F}(z)}{z^{N+1}} \hat{F}(z) = \frac{(N+1)}{N!} \hat{F}^{(N)}(0) \\
&= (N+1) \sum_{i=1}^{2} \int_{z_i}^{\infty} \frac{dz}{z^{N+1}} \hat{F}(z).
\end{aligned}
\tag{6.96}
$$

The integrals yield a constant plus a product

$$
\Delta \hat{F}_N \approx \frac{(N+1)(N - \frac{3}{2})!}{N!} \frac{1}{(r^2)^N} \frac{1}{(1 + r^2)^N},
\tag{6.97}
$$

which for large N can be approximated using Stirling's formula (5.204) by

$$
\Delta \hat{F}_N \approx \frac{A}{(r^2)^N \sqrt{N}} e^{-r^2 N}.
\tag{6.98}
$$

In the strong-coupling limit of interest here, $\hat{\omega}^2 = 0$, and $r = 1/\hat{g} = \Omega/g = c$. In Fig. 6.5 we see that the optimal c-values tend to unity for $N \to \infty$, so that $\Delta \hat{f}_N$ goes to zero like e^{-N}, as observed in Fig. 6.6.

Notes and References

There exists a large body of literature on this subject, for example
L.S. Schulman, J. Math. Phys. **12**, 304 (1971);
M.G.G. Laidlaw and C. DeWitt-Morette, Phys. Rev. D **3**, 1375 (1971);
J.S. Dowker, J. Phys. A **5**, 936 (1972);
P.A. Horvathy, Phys. Lett. A **76**, 11 (1980) and in *Differential Geometric Methods in Math. Phys.*, Lecture Notes in Mathematics **905**, Springer, Berlin, 1982;
J.J. Leinaas and J. Myrheim, Nuovo Cimento **37**, 1, (1977).
The latter paper is reprinted in the textbook
F. Wilczek, *Fractional Statistics and Anyon Superconductivity*, World Scientific, 1990.
See further
P.A. Horvathy, G. Morandi, and E.C.G. Sudarshan, Nuovo Cimento D **11**, 201 (1989),
and the textbook
L.S. Schulman, *Techniques and Applications of Path Integration*, Wiley, New York, 1981.

It is possible to account for the presence of hard walls using infinitely high δ-functions:
C. Grosche, Phys. Rev. Lett. **71**, 1 (1993); Ann. Phys. **2**, 557 (1993); (hep-th/9308081); (hep-th/9308082); (hep-th/9402110);
M.J. Goovaerts, A. Babcenco, and J.T. Devreese, J. Math. Phys. **14**, 554 (1973);
C. Grosche, J. Phys. A Math. Gen. **17**, 375 (1984).

The physically important problem of membranes between walls has been discussed in
W. Helfrich, Z. Naturforsch. A **33**, 305 (1978);
W. Helfrich and R.M. Servuss, Nuovo Cimento D **3**, 137 (1984);
W. Janke and H. Kleinert, Phys. Lett. **58**, 144 (1987) (http://www.physik.fu-berlin.de/
~kleinert/143);
W. Janke, H. Kleinert, and H. Meinhardt, Phys. Lett. B **217**, 525 (1989) (*ibid.*http/184);
G. Gompper and D.M. Kroll, Europhys. Lett. **9**, 58 (1989);
R.R. Netz and R. Lipowski, Europhys. Lett. **29**. 345 (1995);
F. David, J. de Phys. **51**, C7-115 (1990);
H. Kleinert, Phys. Lett. A **257**, 269 (1999) (cond-mat/9811308);
M. Bachmann, H. Kleinert, A. Pelster, Phys. Lett. A **261**, 127 (1999) (cond-mat/9905397).

The problem has been solved with the help of the strong-coupling variational perturbation theory
developed in Chapter 5 by
H. Kleinert, Phys. Lett. A **257**, 269 (1999) (cond-mat/9811308);
M. Bachmann, H. Kleinert, and A. Pelster, Phys. Lett. A **261**, 127 (1999) (cond-mat/9905397).

The quantum-mechanical calculation presented in Section 6.4 is taken from
H. Kleinert, A. Chervyakov, and B. Hamprecht, Phys. Lett. A **260**, 182 (1999) (cond-mat/9906241).

Mirum, quod divina natura dedit agros.
It's wonderful that divine nature has given us fields.
VARRO, 82 B.C.

7

Many Particle Orbits – Statistics and Second Quantization

Realistic physical systems usually contain groups of identical particles such as specific atoms or electrons. Focusing on a single group, we shall label their orbits by $\mathbf{x}^{(\nu)}(t)$ with $\nu = 1, 2, 3, \ldots, N$. Their Hamiltonian is invariant under the group of all $N!$ permutations of the orbital indices ν. Their Schrödinger wave functions can then be classified according to the irreducible representations of the permutation group.

Not all possible representations occur in nature. In more than two space dimensions, there exists a *superselection rule*, whose origin is yet to be explained, which eliminates all complicated representations and allows only for the two simplest ones to be realized: those with complete symmetry and those with complete antisymmetry. Particles which appear always with symmetric wave functions are called *bosons*. They all carry an integer-valued spin. Particles with antisymmetric wave functions are called *fermions*[1] and carry a spin whose value is half-integer.

The symmetric and antisymmetric wave functions give rise to the characteristic statistical behavior of fermions and bosons. Electrons, for example, being spin-1/2 particles, appear only in antisymmetric wave functions. The antisymmetry is the origin of the famous *Pauli exclusion principle*, allowing only a single particle of a definite spin orientation in a quantum state, which is the principal reason for the existence of the *periodic system* of elements, and thus of matter in general. The atoms in a gas of helium, on the other hand, have zero spin and are described by symmetric wave functions. These can accommodate an infinite number of particles in a single quantum state giving rise to the famous phenomenon of Bose-Einstein condensation. This phenomenon is observable in its purest form in the absence of interactions, where at zero temperature all particles condense in the ground state. In interacting systems, Bose-Einstein statistics can lead to the stunning quantum state of superfluidity.

The particular association of symmetry and spin can be explained within relativistic quantum field theories in spaces with more than two dimensions where it is shown to be intimately linked with the *locality* and *causality* of the theory.

---

[1] Had M. Born as editor of Zeitschrift für Physik not kept a paper by P. Jordan in his suitcase for half a year in 1925, they would be called *jordanons*. See the bibliographical notes by B. Schroer (hep-th/0303241).

In two dimensions there can be particles with an exceptional statistical behavior. Their properties will be discussed in Section 7.5. In Chapter 16, such particles will serve to explain the fractional quantum Hall effect.

The problem to be solved in this chapter is how to incorporate the statistical properties into a path integral description of the orbits of a many-particle system. Afterwards we describe the formalism of *second quantization* or *field quantization* in which the path integral of *many* identical particle orbits is abandoned in favor of a path integral over a *single fluctuating field* which is able to account for the statistical properties in a most natural way.

7.1　Ensembles of Bose and Fermi Particle Orbits

For bosons, the incorporation of the statistical properties into the orbital path integrals is quite easy. Consider, for the moment, distinguishable particles. Their many-particle time evolution amplitude is given by the path integral

$$(\mathbf{x}_b^{(1)}, \ldots, \mathbf{x}_b^{(N)}; t_b | \mathbf{x}_a^{(1)}, \ldots, \mathbf{x}_a^{(N)}; t_a) = \prod_{\nu=1}^{N} \left[\int \mathcal{D}^D x^{(\nu)} \right] e^{i\mathcal{A}^{(N)}/\hbar}, \tag{7.1}$$

with an action of the typical form

$$\mathcal{A}^{(N)} = \int_{t_a}^{t_b} dt \left\{ \sum_{\nu=1}^{N} \left[\frac{M^{(\nu)}}{2} \dot{\mathbf{x}}^{(\nu)2} - V(\mathbf{x}^{(\nu)}) \right] - \frac{1}{2} \sum_{\nu \neq \nu'=1}^{N} V_{\text{int}}(\mathbf{x}^{(\nu)} - \mathbf{x}^{(\nu')}) \right\}, \tag{7.2}$$

where $V(\mathbf{x}^{(\nu)})$ is some common background potential for all particles interacting via the pair potential $V_{\text{int}}(\mathbf{x}^{(\nu)} - \mathbf{x}^{(\nu')})$. We shall ignore interactions involving more than two particles at the same time, for simplicity.

If we want to apply the path integral (7.1) to indistinguishable particles of spin zero, we merely have to *add* to the sum over all paths $\mathbf{x}^{(\nu)}(t)$ running to the final positions $\mathbf{x}_b^{(\nu)}$ the sum of all paths running to the indistinguishable permuted final positions $\mathbf{x}_b^{(p(\nu))}$. The amplitude for n bosons reads therefore

$$(\mathbf{x}_b^{(1)}, \ldots, \mathbf{x}_b^{(N)}; t_b | \mathbf{x}_a^{(1)}, \ldots, \mathbf{x}_a^{(N)}; t_a) = \sum_{p(\nu)} (\mathbf{x}_b^{(p(1))}, \ldots, \mathbf{x}_b^{(p(N))}; t_b | \mathbf{x}_a^{(1)}, \ldots, \mathbf{x}_a^{(N)}; t_a), \tag{7.3}$$

where $p(\nu)$ denotes the $N!$ permutations of the indices ν. For bosons of higher spin, the same procedure applies to each subset of particles with equal spin orientation.

A similar discussion holds for fermions. Their Schrödinger wave function requires complete antisymmetrization in the final positions. Correspondingly, the amplitude (7.1) has to be summed over all permuted final positions $\mathbf{x}_b^{(p(\nu))}$, with an extra minus sign for each odd permutation $p(\nu)$. Thus, the path integral involves both *sums and differences* of paths. So far, the measure of path integration has always been a true *sum* over paths. For this reason it will be preferable to attribute the alternating sign to an interaction between the orbits, to be called a *statistics interaction*. This interaction will be derived in Section 7.4.

For the statistical mechanics of Bose- and Fermi systems consider the imaginary-time version of the amplitude (7.3):

$$(\mathbf{x}_b^{(1)}, \ldots, \mathbf{x}_b^{(N)}; \hbar\beta | \mathbf{x}_a^{(1)}, \ldots, \mathbf{x}_a^{(N)}; 0) = \sum_{p(\nu)} \epsilon_{p(\nu)} (\mathbf{x}_b^{(p(1))}, \ldots, \mathbf{x}_b^{(p(N))}; \hbar\beta | \mathbf{x}_a^{(1)}, \ldots, \mathbf{x}_a^{(N)}; 0),$$
(7.4)

where $\epsilon_{p(\nu)} = \pm 1$ is the parity of even and odd permutations $p(\nu)$, to be used for Bosons and Fermions, respectively. Its spatial trace integral yields the partition function of N-particle orbits:

$$Z^{(N)} = \frac{1}{N!} \int d^D x^{(1)} \cdots d^D x^{(N)} (\mathbf{x}^{(1)}, \ldots, \mathbf{x}^{(N)}; \hbar\beta | \mathbf{x}^{(1)}, \ldots, \mathbf{x}^{(N)}; 0).$$
(7.5)

A factor $1/N!$ accounts for the indistinguishability of the permuted final configurations.

For free particles, each term in the sum (7.4) factorizes:

$$(\mathbf{x}_b^{(p(1))}, \ldots, \mathbf{x}_b^{(p(N))}; \hbar\beta | \mathbf{x}_a^{(1)}, \ldots, \mathbf{x}_a^{(N)}; 0)_0 = (\mathbf{x}_b^{(p(1))} \hbar\beta | \mathbf{x}_a^{(1)} 0)_0 \cdots (\mathbf{x}_b^{(p(N))} \hbar\beta | \mathbf{x}_a^{(N)} 0)_0, (7.6)$$

where each factor has a path integral representation

$$(\mathbf{x}_b^{(p(\nu))} \hbar\beta | \mathbf{x}_a^{(\nu)} 0)_0 = \int_{\mathbf{x}^{(\nu)}(0) = \mathbf{x}_a^{(\nu)}}^{\mathbf{x}^{(\nu)}(\hbar\beta) = \mathbf{x}_b^{(p(\nu))}} \mathcal{D}^D x^{(\nu)} \exp\left[-\frac{1}{\hbar} \int_0^{\hbar\beta} d\tau \frac{M}{2} \dot{\mathbf{x}}^{(\nu)2}(\tau) \right],$$
(7.7)

which is solved by the imaginary-time version of (2.130):

$$(\mathbf{x}_b^{(p(\nu))} \hbar\beta | \mathbf{x}_a^{(\nu)} 0)_0 = \frac{1}{\sqrt{2\pi\hbar^2\beta/M}^D} \exp\left\{ -\frac{1}{\hbar} \frac{M}{2} \frac{\left[\mathbf{x}_b^{(p(\nu))} - \mathbf{x}_a^{(\nu)} \right]^2}{\hbar\beta} \right\}.$$
(7.8)

The partition function can therefore be rewritten in the form

$$Z_0^{(N)} = \frac{1}{\sqrt{2\pi\hbar^2\beta/M}^{ND}} \frac{1}{N!} \int d^D x^{(1)} \cdots d^D x^{(N)} \sum_{p(\nu)} \epsilon_{p(\nu)} \prod_{\nu=1}^{N} \exp\left\{ -\frac{1}{\hbar} \frac{M}{2} \frac{\left[\mathbf{x}^{(p(\nu))} - \mathbf{x}^{(\nu)} \right]^2}{\hbar\beta} \right\}.$$
(7.9)

This is a product of Gaussian convolution integrals which can easily be performed as before when deriving the time evolution amplitude (2.70) for free particles with the help of Formula (2.69). Each convolution integral simply extends the temporal length in the fluctuation factor by $\hbar\beta$. Due to the indistinguishability of the particles, only a few paths will have their end points connected to their own initial points, i.e., they satisfy periodic boundary conditions in the interval $(0, \hbar\beta)$. The sum over permutations connects the final point of some paths to the initial point of a different path, as illustrated in Fig. 7.1. Such paths satisfy periodic boundary conditions on an interval $(0, w\hbar\beta)$, where w is some integer number. This is seen most clearly by drawing the paths in Fig. 7.1 in an extended zone scheme shown in Fig. 7.2, which is reminiscent of Fig. 6.1. The extended zone scheme can, moreover, be placed on a

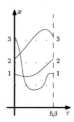

Figure 7.1 Paths summed in partition function (7.9). Due to indistinguishability of particles, final points of one path may connect to initial points of another.

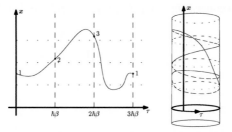

Figure 7.2 Periodic representation of paths summed in partition function (7.9), once in extended zone scheme, and once on D-dimensional hypercylinder embedded in $D+1$ dimensions. The paths are shown in Fig. 7.1. There is now only one closed path on the cylinder. In general there are various disconnected parts of closed paths.

hypercylinder, illustrated in the right-hand part of Fig. 7.2. In this way, all paths decompose into mutually disconnected groups of closed paths winding around the cylinder, each with a different *winding number* w [1]. An example for a connected path which winds three times 3 around the D-dimensional cylinder contributes to the partition function a factor [using Formula (2.69)]:

$$\Delta Z_0^{(N)3} = \frac{1}{\sqrt{2\pi\hbar^2\beta/M}^{3D}} \int d^D x^{(1)} d^D x^{(2)} d^D x^{(3)} \exp\left\{-\frac{1}{\hbar}\frac{M}{2}\frac{\left[\mathbf{x}^{(3)}-\mathbf{x}^{(2)}\right]^2}{\hbar\beta}\right\}$$

$$\times \exp\left\{-\frac{1}{\hbar}\frac{M}{2}\frac{\left[\mathbf{x}^{(2)}-\mathbf{x}^{(1)}\right]^2}{\hbar\beta}\right\} \exp\left\{-\frac{1}{\hbar}\frac{M}{2}\frac{\left[\mathbf{x}^{(1)}-\mathbf{x}^{(3)}\right]^2}{\hbar\beta}\right\} = \frac{V_D}{\sqrt{2\pi\hbar^2 3\beta/M}^D}. \quad (7.10)$$

For cycles of length w the contribution is

$$\Delta Z_0^{(N)\,w} = Z_0(w\beta), \quad (7.11)$$

where $Z_0(w\beta)$ is the partition function of a free particle in a D-dimensional volume V_D for an imaginary-time interval $w\hbar\beta$:

$$Z_0(w\beta) = \frac{V_D}{\sqrt{2\pi\hbar^2 w\beta/M}^D}. \quad (7.12)$$

In terms of the thermal de Broglie length $l_e(\hbar\beta) \equiv \sqrt{2\pi\hbar^2\beta/M}$ associated with the temperature $T = 1/k_B\beta$ [recall (2.351)], this can be written as

$$Z_0(w\beta) = \frac{V_D}{l_e^D(w\hbar\beta)}. \tag{7.13}$$

There is an additional factor $1/w$ in Eq. (7.11), since the number of connected windings of the total $w!$ closed paths is $(w-1)!$. In group theoretic language, it is the number of *cycles* of length w, usually denoted by $(1, 2, 3, \ldots, w)$, plus the $(w-1)!$ permutations of the numbers $2, 3, \ldots, w$. They are illustrated in Fig. 7.3 for $w = 2, 3, 4$. In a decomposition of all $N!$ permutations as products of cycles, the number of elements consisting of C_1, C_2, C_3, $\ldots$ cycles of length 1, 2, 3, $\ldots$ contains

$$M(C_1, C_2, \ldots, C_N) = \frac{N!}{\prod_{w=1}^{N} C_w! w^{C_w}} \tag{7.14}$$

elements [2].

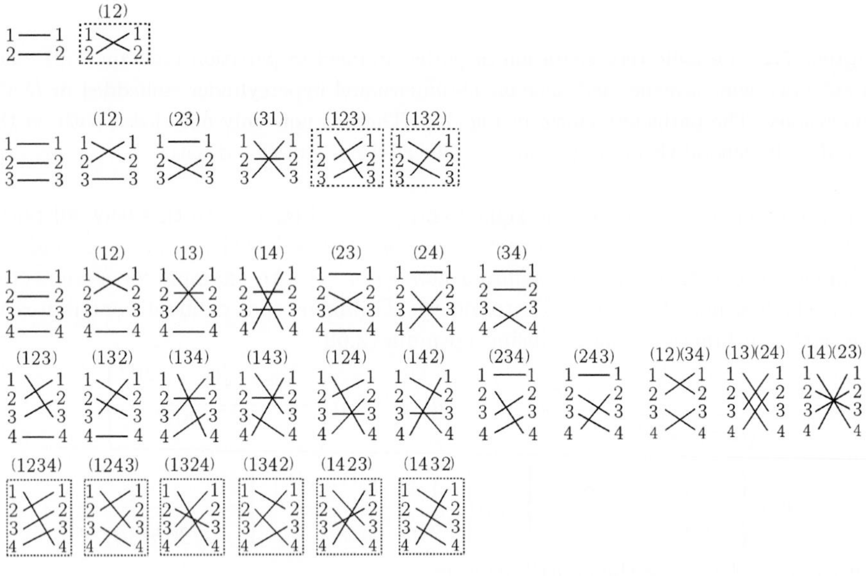

Figure 7.3 Among the $w!$ permutations of the different windings around the cylinder, $(w-1)!$ are connected. They are marked by dotted frames. In the cycle notation for permutation group elements, these are (12) for two elements, (123), (132) for three elements, (1234), (1243), (1324), (1342), (1423), (1432) for four elements. The cycles are shown on top of each graph, with trivial cycles of unit length omitted. The graphs are ordered according to a decreasing number of cycles.

With the knowledge of these combinatorial factors we can immediately write down the canonical partition function (7.9) of N bosons or fermions as the sum of all orbits around the cylinder, decomposed into cycles:

$$Z_0^{(N)}(\beta) = \frac{1}{N!} \sum_{p(\nu)} \epsilon_{p(\nu)} M(C_1, \ldots, C_N) \prod_{\substack{w=1 \\ N = \Sigma_w w C_w}}^{N} \left[Z_0(w\beta) \right]^{C_w} . \tag{7.15}$$

The sum can be reordered as follows:

$$Z_0^{(N)} = \frac{1}{N!} \sum_{\substack{C_1,\ldots,C_N \\ N = \Sigma_w w C_w}} M(C_1, \ldots, C_N) \, \epsilon_{w,C_1,\ldots,C_n} \prod_{w=1}^{N} \left[Z_0(w\beta) \right]^{C_w} . \tag{7.16}$$

The parity $\epsilon_{w,C_1,\ldots,C_n}$ of permutations is equal to $(\pm 1)^{\Sigma_w (w+1)C_w}$. Inserting (7.14), the sum (7.16) can further be regrouped to

$$\begin{aligned} Z_0^{(N)}(\beta) &= \sum_{\substack{C_1,\ldots,C_N \\ N = \Sigma_w w C_w}} \frac{1}{\prod_{w=1}^{N} C_w! \, w^{C_w}} (\pm 1)^{\Sigma_w (w+1)C_w} \prod_{w=1}^{N} \left[Z_0(w\beta) \right]^{C_w} \\ &= \sum_{\substack{C_1,\ldots,C_N \\ N = \Sigma_w w C_w}} \prod_{w=1}^{N} \frac{1}{C_w!} \left[(\pm 1)^{w-1} \frac{Z_0(w\beta)}{w} \right]^{C_w} . \end{aligned} \tag{7.17}$$

For $N = 0$, this formula yields the trivial partition function $Z_0^{(0)}(\beta) = 1$ of the no-particle state, the *vacuum*. For $N = 1$, i.e., a single particle, we find $Z_0^{(1)}(\beta) = Z_0(\beta)$. The higher $Z_0^{(N)}$ can be written down most efficiently if we introduce a characteristic temperature

$$T_c^{(0)} \equiv \frac{2\pi\hbar^2}{k_B M} \left[\frac{N}{V_D \zeta(D/2)} \right]^{2/D} , \tag{7.18}$$

and measure the temperature T in units of $T_c^{(0)}$, defining a reduced temperature $t \equiv T/T_c^{(0)}$. Then we can rewrite $Z_0^{(1)}(\beta)$ as $t^{D/2} V_D$. Introducing further the N-dependent variable

$$\tau_N \equiv \left[\frac{N}{\zeta(D/2)} \right]^{2/D} t, \tag{7.19}$$

we find $Z_0^{(1)} = \tau_1^{D/2}$. A few low-N examples are for bosons and fermions:

$$\begin{aligned} Z_0^{(2)} &= \pm 2^{-1-D/2} \tau_2^{D/2} + \tau_2^D , \\ Z_0^{(3)} &= \pm 3^{-1-D/2} \tau_3^{D/2} + 2^{-1-D/2} \tau_3^D \pm 3^{-1} 2^{-1} \tau_3^{3D/2} , \\ Z_0^{(4)} &= \pm 2^{-2-D} \tau_4^{D/2} + \left(2^{-3-D} + 3^{-1-D/2} \right) \tau_4^D \pm 2^{-2-\frac{D}{2}} \tau_4^{3D/2} + 3^{-1} 2^{-3} \tau_4^{2D} . \end{aligned} \tag{7.20}$$

From $Z_0^{(N)}(\beta)$ we calculate the specific heat [recall (2.600)] of the free canonical ensemble:

$$C_0^{(N)} = T \frac{d^2}{dT^2} [T \log Z_0^{(N)}] = \tau_N \frac{d^2}{d\tau_N^2} [\tau_N \log Z_0^{(N)}], \tag{7.21}$$

and plot it [3] in Fig. 7.4 against t for increasing particle number N. In the limit $N \to \infty$, the curves approach a limiting form with a phase transition at $T = T_c^{(0)}$, which will be derived from a grand-canonical ensemble in Eqs. (7.67) and (7.70).

The partition functions can most easily be calculated with the help of a recursion relation [4, 5], starting from $Z_0^{(0)} \equiv 1$:

$$Z_0^{(N)}(\beta) = \frac{1}{N} \sum_{n=1}^{N} (\pm 1)^{n-1} Z_0(n\beta) Z_0^{(N-n)}(\beta). \tag{7.22}$$

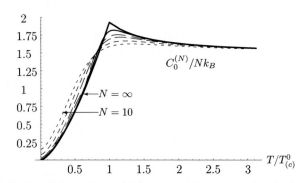

Figure 7.4 Plot of the specific heat of free Bose gas with $N = 10, 20, 50, 100, 500, \infty$ particles. The curve approaches for large T the Dulong-Petit limit $3k_B N/2$ corresponding to the three harmonic kinetic degrees of freedom in the classical Hamiltonian $\mathbf{p}^2/2M$. There are no harmonic potential degrees of freedom.

This relation is proved with the help of the grand-canonical partition function which is obtained by forming the sum over all canonical partition functions $Z_0^{(N)}(\beta)$ with a weight factor z^N:

$$Z_{G0}(\beta) \equiv \sum_{N=0}^{\infty} Z_0^{(N)}(\beta)\, z^N. \tag{7.23}$$

The parameter z is the Boltzmann factor of one particle with the chemical potential μ:

$$z = z(\beta) \equiv e^{\beta\mu}. \tag{7.24}$$

It is called the *fugacity* of the ensemble. Inserting the cycle decompositions (7.17), the sum becomes

$$Z_{G0}(\beta) = \sum_{\substack{C_1,\ldots,C_N \\ N = \Sigma_w wC_w}} \prod_{w=1}^{N} \frac{1}{C_w!} \left[(\pm 1)^{w-1} \frac{Z_0(w\beta)e^{w\beta\mu}}{w} \right]^{C_w}. \tag{7.25}$$

The right-hand side may be rearranged to

$$Z_{G0}(\beta) = \prod_{w=1}^{\infty} \sum_{C_w=0}^{\infty} \frac{1}{C_w!} \left[(\pm 1)^{w-1} \frac{Z_0(w\beta)e^{w\beta\mu}}{w} \right]^{C_w}$$

$$= \exp\left[\sum_{w=1}^{\infty}(\pm 1)^{w-1}\frac{Z_0(w\beta)}{w}e^{w\beta\mu}\right]. \tag{7.26}$$

From this we read off the grand-canonical free energy [recall (1.542)] of noninteracting identical particles

$$F_G(\beta) \equiv -\frac{1}{\beta}\log Z_{G0}(\beta) = -\frac{1}{\beta}\sum_{w=1}^{\infty}(\pm 1)^{w-1}\frac{Z_0(w\beta)}{w}e^{w\beta\mu}. \tag{7.27}$$

This is simply the sum of the contributions (7.11) of connected paths to the canonical partition function which wind $w = 1, 2, 3, \ldots$ times around the cylinder [1, 6]. Thus we encounter the same situation as observed before in Section 3.20: the free energy of any quantum-mechanical system can be obtained from the perturbation expansion of the partition function by keeping only the connected diagrams.

The canonical partition function is obviously obtained from (7.27) by forming the derivative:

$$Z_0^{(N)}(\beta) = \frac{1}{N!}\frac{\partial^N}{\partial z^N}Z_{G0}(\beta)\Big|_{z=0}. \tag{7.28}$$

It is now easy to derive the recursion relation (7.22). From the explicit form (7.27), we see that

$$\frac{\partial}{\partial z}Z_{G0} = -\left(\frac{\partial}{\partial z}\beta F_G\right)Z_{G0}. \tag{7.29}$$

Applying to this $N-1$ more derivatives yields

$$\frac{\partial^{N-1}}{\partial z^{N-1}}\left[\frac{\partial}{\partial z}Z_{G0}\right] = -\sum_{l=0}^{N-1}\frac{(N-1)!}{l!(N-l-1)!}\left(\frac{\partial^{l+1}}{\partial z^{l+1}}\beta F_G\right)\frac{\partial^{N-l-1}}{\partial z^{N-l-1}}Z_{G0}.$$

To obtain from this $Z_0^{(N)}$ we must divide this equation by $N!$ and evaluate the derivatives at $z = 0$. From (7.27) we see that the $l + 1$st derivative of the grand-canonical free energy is

$$\frac{\partial^{l+1}}{\partial z^{l+1}}\beta F_G\Big|_{z=0} = -(\pm 1)^l\, l!\, Z_0((l+1)\beta). \tag{7.30}$$

Thus we obtain

$$\frac{1}{N!}\frac{\partial^N}{\partial z^N}Z_{G0}\Big|_{z=0} = \frac{1}{N}\sum_{l=0}^{N-1}(\pm 1)^l Z_0((l+1)\beta)\frac{1}{(N-l-1)!}\frac{\partial^{N-l-1}}{\partial z^{N-l-1}}Z_{G0}\Big|_{z=0}.$$

Inserting here (7.28) and replacing $l \to n-1$ we obtain directly the recursion relation (7.22).

The grand-canonical free energy (7.27) may be simplified by using the property

$$Z_0(w\beta) = Z_0(\beta)\frac{1}{w^{D/2}} \tag{7.31}$$

of the free-particle partition function (7.12), to remove a factor $1/\sqrt{w}^D$ from $Z_0(w\beta)$. This brings (7.27) to the form

$$F_G = -\frac{1}{\beta}Z_0(\beta)\sum_{w=1}^{\infty}(\pm 1)^{w-1}\frac{e^{w\beta\mu}}{w^{D/2+1}}. \tag{7.32}$$

The average number of particles is found from the derivative with respect to the chemical potential[2]

$$N = -\frac{\partial}{\partial\mu}F_G = Z_0(\beta)\sum_{w=1}^{\infty}(\pm 1)^{w-1}\frac{e^{w\beta\mu}}{w^{D/2}}. \tag{7.33}$$

The sums over w converge certainly for negative or vanishing chemical potential μ, i.e., for fugacities smaller than unity. In Section 7.3 we shall see that for fermions, the convergence extends also to positive μ.

If the particles have a nonzero spin S, the above expressions carry a multiplicity factor $g_S = 2S + 1$, which has the value 2 for electrons.

The grand-canonical free energy (7.32) will now be studied in detail thereby revealing the interesting properties of many-boson and many-fermion orbits, the ability of the former to undergo *Bose-Einstein condensation*, and of the latter to form a *Fermi sphere* in momentum space.

7.2 Bose-Einstein Condensation

We shall now discuss the most interesting phenomenon observable in systems containing a large number of bosons, the Bose-Einstein condensation process.

7.2.1 Free Bose Gas

For bosons, the above thermodynamic functions (7.32) and (7.33) contain the functions

$$\zeta_\nu(z) \equiv \sum_{w=1}^{\infty}\frac{z^w}{w^\nu}. \tag{7.34}$$

These start out for small z like z, and increase for $z \to 1$ to $\zeta(\nu)$, where $\zeta(z)$ is Riemann's zeta function (2.519). The functions $\zeta_\nu(z)$ are called *Polylogarithmic functions* in the mathematical literature [7], where they are denoted by $\mathrm{Li}_\nu(z)$. They are related to the *Hurwitz zeta function* $\zeta(\nu, a, z) \equiv \sum_{w=0}^{\infty} z^w/(w + a)^\nu$ as $\zeta_\nu(z) = z\zeta(\nu, 1, z)$. The functions $\phi(z, \nu, a) = \zeta(\nu, a, z)$ are also known as *Lerch functions*.

In terms of the functions $\zeta_\nu(z)$, and the explicit form (7.12) of $Z_0(\beta)$, we may write F_G and N of Eqs. (7.32) and (7.33) simply as

---

[2]In grand-canonical ensembles, one always deals with the average particle number $\langle N \rangle$ for which one writes N in all thermodynamic equations [recall (1.548)]. This should be no lead to confusion.

$$F_G = -\frac{1}{\beta}Z_0(\beta)\zeta_{D/2+1}(z) = -\frac{1}{\beta}\frac{V_D}{l_e^D(\hbar\beta)}\zeta_{D/2+1}(z), \qquad (7.35)$$

$$N = Z_0(\beta)\zeta_{D/2}(z) = \frac{V_D}{l_e^D(\hbar\beta)}\zeta_{D/2}(z). \qquad (7.36)$$

The most interesting range where we want to know the functions $\zeta_\nu(z)$ is for negative small chemical potential μ. There the convergence is very slow and it is useful to find a faster-convergent representation. As in Subsection 2.15.6 we rewrite the sum over w for $z = e^{\beta\mu}$ as an integral plus a difference between sum and integral

$$\zeta_\nu(e^{\beta\mu}) \equiv \int_0^\infty dw\,\frac{e^{w\beta\mu}}{w^\nu} + \left(\sum_{w=1}^\infty - \int_0^\infty dw\right)\frac{e^{w\beta\mu}}{w^\nu}. \qquad (7.37)$$

The integral yields $\Gamma(1-\nu)(-\beta\mu)^{\nu-1}$, and the remainder may be expanded sloppily in powers of μ to yield the Robinson expansion (2.579):

$$\zeta_\nu(e^{\beta\mu}) = \Gamma(1-\nu)(-\beta\mu)^{\nu-1} + \sum_{k=0}^\infty \frac{1}{k!}(\beta\mu)^k\zeta(\nu-k). \qquad (7.38)$$

There exists a useful integral representation for the functions $\zeta_\nu(z)$:

$$\zeta_\nu(z) \equiv \frac{1}{\Gamma(\nu)}i_\nu(\beta\mu), \qquad (7.39)$$

where $i_\nu(\alpha)$ denotes the integral

$$i_\nu(\alpha) \equiv \int_0^\infty d\varepsilon\,\frac{\varepsilon^{\nu-1}}{e^{\varepsilon-\alpha}-1}, \qquad (7.40)$$

containing the Bose distribution function (3.93):

$$n_\varepsilon^b = \frac{1}{e^{\varepsilon-\alpha}-1}. \qquad (7.41)$$

Indeed, by expanding the denominator in the integrand in a power series

$$\frac{1}{e^{\varepsilon-\alpha}-1} = \sum_{w=1}^\infty e^{-w\varepsilon}e^{w\alpha}, \qquad (7.42)$$

and performing the integrals over ε, we obtain directly the series (7.34).

It is instructive to express the grand-canonical free energy F_G in terms of the functions $i_\nu(\alpha)$. Combining Eqs. (7.35) with (7.39) and (7.40), we obtain

$$F_G = -\frac{1}{\beta}Z_0(\beta)\frac{i_{D/2+1}(\beta\mu)}{\Gamma(D/2+1)} = -\frac{1}{\beta\Gamma(D/2+1)}\frac{V_D}{\sqrt{2\pi\hbar^2\beta/M}^D}\int_0^\infty d\varepsilon\,\frac{\varepsilon^{D/2}}{e^{\varepsilon-\beta\mu}-1}. \qquad (7.43)$$

The integral can be brought to another form by partial integration, using the fact that

$$\frac{1}{e^{\varepsilon - \beta\mu} - 1} = \frac{\partial}{\partial\varepsilon} \log(1 - e^{-\varepsilon + \beta\mu}). \tag{7.44}$$

The boundary terms vanish, and we find immediately:

$$F_G = \frac{1}{\beta\Gamma(D/2)} \frac{V_D}{\sqrt{2\pi\hbar^2\beta/M}^D} \int_0^\infty d\varepsilon\, \varepsilon^{D/2-1} \log(1 - e^{-\varepsilon + \beta\mu}). \tag{7.45}$$

This expression is obviously equal to the sum over momentum states of oscillators with energy $\hbar\omega_\mathbf{p} \equiv \mathbf{p}^2/2M$, evaluated in the thermodynamic limit $N \to \infty$ with fixed particle density N/V, where the momentum states become continuous:

$$F_G = \frac{1}{\beta} \sum_\mathbf{p} \log(1 - e^{-\beta\hbar\omega_\mathbf{p} + \beta\mu}). \tag{7.46}$$

This is easily verified if we rewrite the sum with the help of formula (1.555) for the surface of a unit sphere in D dimensions and a change of variables to the reduced particle energy $\varepsilon = \beta\mathbf{p}^2/2M$ as an integral

$$\begin{aligned}
\sum_\mathbf{p} &\to V_D \int \frac{d^D p}{(2\pi\hbar)^D} = V_D S_D \frac{1}{(2\pi\hbar)^D} \int dp\, p^{D-1} = V_D S_D \frac{(2M/\beta)^{D/2}}{(2\pi\hbar)^D} \frac{1}{2} \int d\varepsilon\, \varepsilon^{D/2-1} \\
&= \frac{1}{\Gamma(D/2)} \frac{V_D}{\sqrt{2\pi\hbar^2\beta/M}^D} \int_0^\infty d\varepsilon\, \varepsilon^{D/2-1}.
\end{aligned} \tag{7.47}$$

Another way of expressing this limit is

$$\sum_\mathbf{p} \to \int_0^\infty d\varepsilon\, N_\varepsilon, \tag{7.48}$$

where N_ε is the reduced *density of states* per unit energy interval:

$$N_\varepsilon \equiv \frac{1}{\Gamma(D/2)} \frac{V_D}{\sqrt{2\pi\hbar^2\beta/M}^D} \varepsilon^{D/2-1}. \tag{7.49}$$

The free energy of each oscillator (7.46) differs from the usual harmonic oscillator expression (2.483) by a missing ground-state energy $\hbar\omega_\mathbf{p}/2$. The origin of this difference will be explained in Sections 7.7 and 7.14.

The particle number corresponding to the integral representations (7.45) and (7.46) is

$$N = -\frac{\partial}{\partial\mu} F_G = \frac{1}{\Gamma(D/2)} \frac{V_D}{\sqrt{2\pi\hbar^2\beta/M}^D} \int_0^\infty d\varepsilon \frac{\varepsilon^{D/2-1}}{e^{\varepsilon - \beta\mu} - 1} \approx \sum_\mathbf{p} \frac{1}{e^{\beta\hbar\omega_\mathbf{p} - \beta\mu} - 1}. \tag{7.50}$$

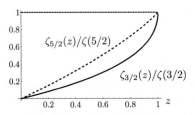

Figure 7.5   Plot of functions $\zeta_\nu(z)$ for $\nu = 3/2$ and $5/2$ appearing in Bose-Einstein thermodynamics.

For a given particle number N, Eq. (7.36) allows us to calculate the fugacity as a function of the inverse temperature, $z(\beta)$, and from this the chemical potential $\mu(\beta) = \beta^{-1} \log z(\beta)$. This is most simply done by solving Eq. (7.36) for β as a function of z, and inverting the resulting function $\beta(z)$. The required functions $\zeta_\nu(z)$ are shown in Fig. 7.5.

There exists a solution $z(\beta)$ only if the total particle number N is smaller than the characteristic function defined by the right-hand side of (7.36) at unit fugacity $z(\beta) = 1$, or zero chemical potential $\mu = 0$:

$$N < \frac{V_D}{l_e^D(\hbar\beta)} \zeta(D/2). \qquad (7.51)$$

Since $l_e(\hbar\beta)$ decreases with increasing temperature, this condition certainly holds at sufficiently high T. For decreasing temperature, the solution exists only as long as the temperature is higher than the critical temperature $T_c = 1/k_B \beta_c$, determined by

$$N = \frac{V_D}{l_e^D(\hbar\beta_c)} \zeta(D/2). \qquad (7.52)$$

This determines the critical density of the atoms. The de Broglie length at the critical temperature will appear so frequently that we shall abbreviate it by ℓ_c:

$$l_e(\hbar\beta_c) = \ell_c \equiv \left[\frac{N}{V_D \zeta(D/2)} \right]^{-1/D} . \qquad (7.53)$$

The critical density is reached at the characteristic temperature $T_c^{(0)}$ introduced in Eq. (7.18). Note that for a two-dimensional system, Eq. (7.18) yields $T_c^{(0)} = 0$, due to $\zeta(1) = \infty$, implying the nonexistence of a condensate. One can observe, however, definite experimental signals for the vicinity of a transition. In fact, we have neglected so far the interaction between the atoms, which is usually repulsive. This will give rise to a special type of phase transition called Kosterlitz-Thouless transition. For a discussion of this transition see other textbooks [8].

By combining (7.18) with (7.36) we obtain an equation for the temperature dependence of z above T_c:

$$1 = \left(\frac{T}{T_c}\right)^{D/2} \frac{\zeta_{D/2}(z(T))}{\zeta(D/2)}, \quad T > T_c. \tag{7.54}$$

This is solved most easily by calculating T/T_c as a function of $z = e^{\beta\mu}$.

If the temperature drops below T_c, the system can no longer accommodate all particles N in a normal state. A certain fraction of them, say $N_{\text{cond}}(T)$, is forced to condense in the ground state of zero momentum, forming the so-called *Bose-Einstein condensate*. The condensate acts like a particle reservoir with a chemical potential zero.

Both phases can be described by the single equation for the number of normal particles, i.e., those outside the condensate:

$$N_{\text{n}}(T) = \frac{V_D}{l_e^D(\hbar\beta)} \zeta_{D/2}(z(\beta)). \tag{7.55}$$

For $T > T_c$, all particles are normal and the relation between μ and the temperature is found from the equation $N_{\text{n}}(T) = N$, where (7.55) reduces to (7.36). For $T < T_c$, however, the chemical potential vanishes so that $z = 1$ and (7.55) reduces to

$$N_{\text{n}}(T) = \frac{V_D}{l_e^D(\hbar\beta)} \zeta_{D/2}(1), \tag{7.56}$$

which yields the temperature dependence of the number of normal particles:

$$\frac{N_{\text{n}}(T)}{N} = \left(\frac{T}{T_c}\right)^{D/2}, \quad T < T_c. \tag{7.57}$$

The density of particles in the condensate is therefore given by

$$\frac{N_{\text{cond}}(T)}{N} = 1 - \frac{N_{\text{n}}(T)}{N} = 1 - \left(\frac{T}{T_c}\right)^{D/2}. \tag{7.58}$$

We now calculate the internal energy which is, according to the general thermodynamic relation (1.550), given by

$$E = F_G + TS + \mu N = F_G - T\partial_T F_G + \mu N = \partial_\beta(\beta F_G) + \mu N. \tag{7.59}$$

Expressing N as $-\partial F_G/\partial\mu$, we can also write

$$E = F_G + (\beta\partial_\beta - \mu\partial_\mu)F_G. \tag{7.60}$$

Inserting (7.35) we see that only the β-derivative of the prefactor contributes since $(\beta\partial_\beta - \mu\partial_\mu)$ applied to any function of $z = e^{\beta\mu}$ vanishes. Thus we obtain directly

$$E = -\frac{D}{2} F_G, \tag{7.61}$$

which becomes with (7.35) and (7.36):

$$E = -\frac{D}{2\beta}Z_0(\beta)\zeta_{D/2+1}(z) = -\frac{D}{2}\frac{\zeta_{D/2+1}(z)}{\zeta_{D/2}(z)}Nk_BT. \tag{7.62}$$

The entropy is found using the thermodynamic relation (1.571):

$$S = \frac{1}{T}(E - \mu N - F_G) = \frac{1}{T}\left(-\frac{D+2}{2}F_G - \mu N\right), \tag{7.63}$$

or, more explicitly,

$$S = k_B\left[\frac{D+2}{2}Z_0(\beta)\zeta_{D/2+1}(z) - \beta\mu N\right] = k_BN\left[\frac{D+2}{2}\frac{\zeta_{D/2+1}(z)}{\zeta_{D/2}(z)} - \beta\mu\right]. \tag{7.64}$$

For $T < T_c$, the entropy is given by (7.64) with $\mu = 0$, $z = 1$ and N replaced by the number N_n of normal particles of Eq. (7.57):

$$S_< = k_BN\left(\frac{T}{T_c}\right)^{D/2}\frac{(D+2)}{2}\frac{\zeta_{D/2+1}(1)}{\zeta_{D/2}(1)}, \quad T < T_c. \tag{7.65}$$

The particles in the condensate do not contribute since they are in a unique state. They do not contribute to E and F_G either since they have zero energy and $\mu = 0$. Similarly we find from (7.62):

$$E_< = \frac{D}{2}Nk_BT\left(\frac{T}{T_c}\right)^{D/2}\frac{\zeta_{D/2+1}(1)}{\zeta_{D/2}(1)} = \frac{D}{D+2}TS_<, \quad T < T_c. \tag{7.66}$$

The specific heat C at a constant volume in units of k_B is found for $T < T_c$ from (7.65) via the relation $C = T\partial_T S|_N$ [recall (2.600)]:

$$C = k_BN\left(\frac{T}{T_c}\right)^{D/2}\frac{(D+2)D}{4}\frac{\zeta_{D/2+1}(1)}{\zeta_{D/2}(1)}, \quad T < T_c. \tag{7.67}$$

For $T > T_c$, the chemical potential at fixed N satisfies the equation

$$\beta\partial_\beta(\beta\mu) = \frac{D}{2}\frac{\zeta_{D/2}(z)}{\zeta_{D/2-1}(z)}. \tag{7.68}$$

This follows directly from the vanishing derivative $\beta\partial_\beta N = 0$ implied by the fixed particle number N. Applying the derivative to Eq. (7.36) and using the relation $z\partial_z\zeta_\nu(z) = \zeta_{\nu-1}(z)$, as well as $\beta\partial_\beta f(z) = z\partial_z f(z)\beta\partial_\beta(\beta\mu)$, we obtain

$$\begin{aligned}\beta\partial_\beta N &= [\beta\partial_\beta Z_0(\beta)]\zeta_{D/2}(z) + Z_0(\beta)\beta\partial_\beta\zeta_{D/2}(z)\\ &= -\frac{D}{2}Z_0(\beta)\zeta_{D/2}(z) + Z_0(\beta)\,\zeta_{D/2-1}(z)\beta\partial_\beta(\beta\mu) = 0,\end{aligned} \tag{7.69}$$

thus proving (7.68).

The specific heat C at a constant volume in units of k_B is found from the derivative $C = T\partial_T S|_N = -\beta^2 \partial_\beta E|_N$, using once more (7.68):

$$C = k_B N \left[\frac{(D+2)D}{4} \frac{\zeta_{D/2+1}(z)}{\zeta_{D/2}(z)} - \frac{D^2}{4} \frac{\zeta_{D/2}(z)}{\zeta_{D/2-1}(z)} \right], \quad T > T_c. \tag{7.70}$$

At high temperatures, C tends to the Dulong-Petit limit $Dk_B N/2$ since for small z all $\zeta_\nu(z)$ behave like z.

Consider now the physical case $D = 3$, where the second denominator in (7.70) contains $\zeta_{1/2}(z)$. As the temperature approaches the critical point from above, z tends to unity from below and $\zeta_{1/2}(z)$ diverges. Thus $1/\zeta_{1/2}(1) = 0$ and the second term in (7.70) disappears, yielding a maximal value in three dimensions

$$C_{\text{max}} = k_B N \frac{15}{4} \frac{\zeta_{5/2}(1)}{\zeta_{3/2}(1)} \approx k_B N\, 1.92567. \tag{7.71}$$

This value is the same as the critical value of Eq. (7.67) below T_c. The specific heat is therefore continuous at T_c. It shows, however, a marked kink. To calculate the jump in the slope we calculate the behavior of the thermodynamic quantities for $T \gtrsim T_c$. As T passes T_c from below, the chemical potential starts becoming smaller than zero, and we can expand Eq. (7.54)

$$1 = \left(\frac{T}{T_c} \right)^{3/2} \left[1 + \frac{\Delta \zeta_{3/2}(z)}{\zeta_{3/2}(1)} \right], \tag{7.72}$$

where the symbol Δ in front of a quantity indicates that the same quantity at zero chemical potential is subtracted. Near T_c, we can approximate

$$\left(\frac{T}{T_c} \right)^{3/2} - 1 \approx -\frac{\Delta \zeta_{3/2}(z)}{\zeta_{3/2}(1)}. \tag{7.73}$$

We now use the Robinson expansion (7.38) to approximate for small negative μ:

$$\zeta_{3/2}(e^{\beta\mu}) = \Gamma(-1/2)(-\beta\mu)^{1/2} + \zeta(3/2) + \beta\mu\zeta(1/2) + \dots, \tag{7.74}$$

with $\Gamma(-1/2) = -2\sqrt{\pi}$. The right-hand side of (7.73) becomes therefore $-\Delta\zeta_{3/2}(z)/\zeta_{3/2}(1) = -2\sqrt{\pi}/\zeta(3/2)(-\beta\mu)^{1/2}$. Inserting this into Eq. (7.73), we obtain the temperature dependence of $-\mu$ for $T \gtrsim T_c$:

$$-\mu \approx \frac{1}{4\pi} k_B T_c\, \zeta^2(3/2) \left[\left(\frac{T}{T_c} \right)^{3/2} - 1 \right]^2. \tag{7.75}$$

The leading square-root term on the right-hand side of (7.74) can also be derived from the integral representation (7.40) which receives for small α its main contribution from $\alpha \approx 0$, where $\Delta i_{3/2}(\alpha)$ can be approximated by

$$\Delta i_{3/2}(\alpha) = \int_0^\infty dz\, z^{1/2} \left(\frac{1}{e^{z-\alpha}-1} - \frac{1}{e^z-1} \right) \approx \alpha \int_0^\infty dz \frac{1}{z^{1/2}(z-\alpha)} = -\pi\sqrt{-\alpha}. \tag{7.76}$$

Using the relation (7.61) for $D = 3$, we calculate the derivative of the energy with respect to the chemical potential from

$$\frac{\partial E}{\partial \mu}\bigg|_{T,V} = -\frac{3}{2}\frac{\partial F_G}{\partial \mu}\bigg|_{T,V}. \tag{7.77}$$

This allows us to find the internal energy slightly above the critical temperature T_c, where $-\mu$ is small, as

$$E \approx E_< + \frac{3}{2}N\mu = E_< - \frac{3}{8\pi}Nk_BT_c\,\zeta^2(3/2)\left[\left(\frac{T}{T_c}\right)^{3/2} - 1\right]^2. \tag{7.78}$$

Forming the derivative of this with respect to the temperature we find that the slope of the specific heat below and above T_c jumps at T_c by

$$\Delta\left(\frac{\partial C}{\partial T}\right) \approx \frac{27}{16\pi}\zeta^2(3/2)\,N\frac{k_B}{T_c} \equiv 3.6658\,N\frac{k_B}{T_c}, \tag{7.79}$$

the individual slopes being from (7.66) with $D = 3$ (using $\partial C/\partial T = \partial^2 E/\partial T^2$):

$$\left(\frac{\partial C_<}{\partial T}\right) = \frac{15}{4}\frac{3}{2}\frac{\zeta(5/2)}{\zeta(3/2)}\frac{Nk_B}{T_c} \approx 2.8885\frac{Nk_B}{T_c}, \quad T < T_c, \tag{7.80}$$

and from (7.78):

$$\left(\frac{\partial C_>}{\partial T}\right) = \left(\frac{\partial C_<}{\partial T}\right) - \Delta\left(\frac{\partial C}{\partial T}\right) \approx -0.7715\frac{Nk_B}{T_c}, \quad T > T_c. \tag{7.81}$$

The specific heat of the three-dimensional Bose gas is plotted in Fig. 7.6, where it is compared with the specific heat of superfluid helium for the appropriate atomic parameters $n = 22.22$ nm^{-3}, $M = 6.695 \times 10^{-27}$ kg, where the critical temperature is $T_c \approx 3.145$K, which is somewhat larger than $T_c \approx 2.17$ K of helium.

There are two major disagreements due to the strong interactions in helium. First, the small-T behavior is $(T/T_c)^{3/2}$ rather than the physical $(T/T_c)^3$ due to phonons. Second, the Dulong-Petit limit of an interacting system is closer to that of harmonic oscillators which is twice as big as the free-particle case. Recall that according to the Dulong-Petit law (2.601), C receives a contribution $Nk_BT/2$ per *harmonic* degree of freedom, potential as well as kinetic. Free particle energies are only harmonic in the momentum.

In 1995, Bose-Einstein condensation was observed in a dilute gas in a way that fits the above simple theoretical description [9]. When ^{87}Rb atoms were cooled down in a magnetic trap to temperatures less than 170 nK, about 50 000 atoms were observed to form a condensate, a kind of "superatom". Recently, such condensates have been set into rotation and shown to become perforated by vortex lines [10] just like rotating superfluid helium II.

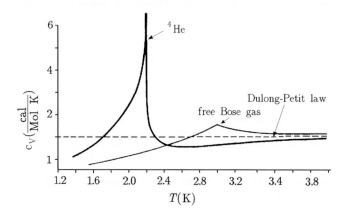

Figure 7.6 Specific heat of ideal Bose gas with phase transition at T_c. For comparison, we have also plotted the specific heat of superfluid helium for the same atomic parameters. The experimental curve is reduced by a factor 2 to have the same Dulong-Petit limit as free bosons.

7.2.2 Bose Gas in Finite Box

The condensation process can be understood better by studying a system in which there is a large but finite number of bosons enclosed in a large cubic box of size L. Then the sum on the right-hand side of Eq. (7.50) gives the contribution of the discrete momentum states to the total particle number. This implies that the function $\zeta_{D/2}(z)$ in Eq. (7.36) has to be replaced by a function $\zeta_{D/2}^{\text{box}}(z)$ defined by the sum over the discrete momentum vectors $\mathbf{p_n} = \hbar\pi(n_1, n_2, \ldots, n_D)/L$ with $n_i = 1, 2, 3, \ldots$:

$$N = \frac{V_D}{l_e^D(\hbar\beta)} \zeta_{D/2}^{\text{box}}(z) = \sum_{\mathbf{P_n}} \frac{1}{e^{\beta \mathbf{p_n}^2/2M - \beta\mu} - 1}. \qquad (7.82)$$

This can be expressed in terms of the one-dimensional auxiliary partition function of a particle in a one-dimensional "box":

$$Z_1(b) \equiv \sum_{n=1}^{\infty} e^{-bn^2/2}, \quad b \equiv \beta\hbar^2\pi^2/ML^2 = \pi l_e^2(\hbar\beta)/2L^2, \qquad (7.83)$$

as

$$N = \frac{V_D}{l_e^D(\hbar\beta)} \zeta_{D/2}^{\text{box}}(z) = \sum_w Z_1^D(wb) z^w. \qquad (7.84)$$

The function $Z_1(b)$ is related to the *elliptic theta function*

$$\vartheta_3(u, z) \equiv 1 + 2\sum_{n=1}^{\infty} z^{n^2} \cos 2nu \qquad (7.85)$$

by $Z_1(b) = [\vartheta_3(0, e^{-b/2}) - 1]/2$. The small-$b$ behavior of this function is easily calculated following the technique of Subsection 2.15.6. We rewrite the sum with the help of Poisson's summation formula (1.205) as a sum over integrals

$$
\vartheta_3(0, e^{-b/2}) = \sum_{k=-\infty}^{\infty} e^{-k^2 b/2} = \sum_{m=-\infty}^{\infty} \int_{-\infty}^{\infty} dk\, e^{-k^2 b/2 + 2\pi i k m}
$$
$$
= \sqrt{\frac{2\pi}{b}} \left(1 + 2 \sum_{m=1}^{\infty} e^{-2\pi^2 m^2/b} \right). \tag{7.86}
$$

Thus, up to exponentially small corrections, we may replace $\vartheta_3(0, e^{-b/2})$ by $\sqrt{2\pi/b}$, so that for small b (i.e., large $L/\sqrt{\beta}$):

$$
Z_1(b) = \sqrt{\frac{\pi}{2b}} - \frac{1}{2} + \mathcal{O}(e^{-2\pi^2/b}). \tag{7.87}
$$

For large b, $Z_1(b)$ falls exponentially fast to zero.

In the sum (7.82), the lowest energy level with $\mathbf{p}_{1,\ldots,1} = \hbar\pi(1,\ldots,1)/L$ plays a special role. Its contribution to the total particle number is the number of particles in the condensate:

$$
N_{\mathrm{cond}}(T) = \frac{1}{e^{Db/2 - \beta\mu} - 1} = \frac{z_D}{1 - z_D}, \qquad z_D \equiv e^{\beta\mu - Db/2}. \tag{7.88}
$$

This number diverges for $z_D \to 1$, where the box function $\zeta_{D/2}^{\mathrm{box}}(z)$ has a pole $1/(Db/2 - \beta\mu)$. This pole prevents $\beta\mu$ from becoming exactly equal to $Db/2$ when solving the equation (7.82) for the particle number in the box.

For a large but finite system near $T = 0$, almost all particles will go into the condensate, so that $Db/2 - \beta\mu$ will be very small, of the order $1/N$, but not zero. The thermodynamic limit can be performed smoothly by defining a regularized function $\bar\zeta_{D/2}^{\mathrm{box}}(z)$ in which the lowest, singular term in the sum (7.82) is omitted. Let us define the number of *normal* particles which have not condensed into the state of zero momentum as $N_{\mathrm{n}}(T) = N - N_{\mathrm{cond}}(T)$. Then we can rewrite Eq. (7.82) as an equation for the number of normal particles

$$
N_{\mathrm{n}}(T) = \frac{V_D}{l_{\mathrm{e}}^D(\hbar\beta)} \bar\zeta_{D/2}^{\mathrm{box}}(z(\beta)), \tag{7.89}
$$

which reads more explicitly

$$
N_{\mathrm{n}}(T) = S_D(z_D) \equiv \sum_{w=1}^{\infty} [Z_1^D(wb) e^{wDb/2} - 1] z_D^w. \tag{7.90}
$$

A would-be critical point may now be determined by setting here $z_D = 1$ and equating the resulting N_{n} with the total particle number N. If N is sufficiently large, we need only the small-b limit of $S_D(1)$ which is calculated in Appendix 7A [see Eq. (7A.12)], so that the associated temperature $T_c^{(1)}$ is determined from the equation

$$N = \sqrt{\frac{\pi}{2b_c}}^{-3} \zeta(3/2) + \frac{3\pi}{4b_c^{(1)}} \log C_3 b_c + \dots , \qquad (7.91)$$

where $C_3 \approx 0.0186$. In the thermodynamic limit, the critical temperature $T_c^{(0)}$ is obtained by ignoring the second term, yielding

$$N = \sqrt{\frac{\pi}{2b_c^{(0)}}}^{-3} \zeta(3/2), \qquad (7.92)$$

in agreement with Eq. (7.52), if we recall b from (7.83). Using this we rewrite (7.91) as

$$1 \equiv \left(\frac{T_c^{(1)}}{T_c^{(0)}}\right)^{3/2} + \frac{3}{2N} \frac{\pi}{2b_c^{(0)}} \log C_3 b_c^{(0)}. \qquad (7.93)$$

Expressing $b_c^{(0)}$ in terms of N from (7.92), this implies

$$\frac{\delta T_c^{(1)}}{T_c^{(0)}} \approx \frac{1}{\zeta^{2/3}(3/2)N^{1/3}} \log \frac{2}{\pi C_3} \frac{N^{2/3}}{\zeta^{2/3}(3/2)}. \qquad (7.94)$$

Experimentally, the temperature $T_c^{(1)}$ is not immediatly accessible. What is easy to find is the place where the condensate density has the largest curvature, i.e., where $d^3 N_{\text{cond}}/dT^3 = 0$. The associated temperature T_c^{exp} is larger than $T_c^{(1)}$ by a factor $1 + \mathcal{O}(1/N)$, so that it does not modify the leading finite-size correction to order in $1/N^{1/3}$. The proof is given in Appendix 7B.

7.2.3 Effect of Interactions

Superfluid helium has a lower transition temperature than the ideal Bose gas. This should be expected since the atomic repulsion impedes the condensation process of the atoms in the zero-momentum state. Indeed, a simple perturbative calculation based on a potential whose Fourier transform behaves for small momenta like

$$\tilde{V}(\mathbf{k}) = g \left[1 - r_{\text{eff}}^2 \mathbf{k}^2/6 + \dots\right] \qquad (7.95)$$

gives a negative shift $\Delta T_c \equiv T_c - T_c^{(0)}$ proportional to the particle density [11]:

$$\frac{\Delta T_c}{T_c^{(0)}} = -\frac{1}{3\hbar^2} M r_{\text{eff}}^2 g \frac{N}{V}, \qquad (7.96)$$

where V is the three-dimensional volume. When discussing the interacting system we shall refer to the previously calculated critical temperature of the free system as $T_c^{(0)}$. This result follows from the fact that for small g and r_{eff}, the free-particle energies $\epsilon_0(\mathbf{k}) = \hbar^2 \mathbf{k}^2/2M$ are changed to $\epsilon(\mathbf{k}) = \epsilon(0) + \hbar^2 \mathbf{k}^2/2M^*$ with a renormalized inverse effective mass $1/M^* = [1 - M r_{\text{eff}}^2 g N/3\hbar^2 V]/M$. Inserting this into Eq. (7.18), from which we may extract the equation for the temperature shift $\Delta T_c/T_c^{(0)} = M/M^* - 1$, we obtain indeed the result (7.96). The parameter $r_{\rm eff}$ is called the *effective range* of the potential. The parameter g in (7.95) can be determined by measuring the s-wave *scattering length* a in a two-body scattering experiment. The relation is obtained from a solution of the Lippmann-Schwinger equation (1.522) for the T-matrix. In a dilute gas, this yields in three dimensions

$$g = \frac{2\pi\hbar^2}{M/2}\, a. \tag{7.97}$$

The denominator $M/2$ is the reduced mass of the two identical bosons. In $D = 2$ dimensions where g has the dimension energy×length^D, Eq. (7.97) is replaced by [12, 13, 14, 15, 17]

$$g = \frac{\pi^{D/2-1}}{2^{2-D}\Gamma(1 - D/2)(na^D)^{D-2} + \Gamma(D/2 - 1)} \frac{2\pi\hbar^2}{M/2} a^{D-2} \equiv \gamma(D, na^D) \frac{2\pi\hbar^2}{M/2} a^{D-2}. \tag{7.98}$$

In the limit $D \to 2$, this becomes

$$g = -\frac{2\pi\hbar^2/M}{\ln(e^\gamma na^2/2)}, \tag{7.99}$$

where γ is the Euler-Mascheroni constant. The logarithm in the denominator implies that the effective repulsion decreases only very slowly with decreasing density [16].

For a low particle density N/V, the effective range $r_{\rm eff}$ becomes irrelevant and the shift ΔT_c depends on the density with a lower power $(N/V)^{\kappa/3}$, $\kappa < 3$. The low-density limit can be treated by keeping in (7.95) only the first term corresponding to a pure δ-function repulsion

$$V(\mathbf{x} - \mathbf{x}') = g\, \delta^{(3)}(\mathbf{x} - \mathbf{x}'). \tag{7.100}$$

For this interaction, the lowest-order correction to the energy is in D dimensions

$$\Delta E = g \int d^D x\, n^2(\mathbf{x}), \tag{7.101}$$

where $n(\mathbf{x})$ is the local particle density. For a homogeneous gas, this changes the grand-canonical free energy from (7.35) to

$$F_G = -\frac{1}{\beta}\frac{V_D}{l_e^D(\hbar\beta)}\zeta_{D/2+1}(z) + g\, V_D \left[\frac{\zeta_{D/2}(z)}{l_e^D(\hbar\beta)}\right]^2 + \mathcal{O}(g^2), \tag{7.102}$$

where we have substituted $n(\mathbf{x})$ by the constant density N/V_D of Eq. (7.36). We now introduce a length parameter α proportional to the coupling constant g:

$$g \equiv \frac{2}{\beta}\frac{\alpha}{l_e(\hbar\beta)} l_e^D(\hbar\beta). \tag{7.103}$$

In three dimensions, α coincides in the dilute limit (small na^D) with the s-wave scattering length a of Eq. (7.97). In in D dimensions, the relation is

$$\alpha = \gamma(D, na^D) \left[\frac{a}{l_e(\hbar\beta)}\right]^{D-3} a. \qquad (7.104)$$

In two dimensions, this has the limit

$$\alpha = -\frac{l_e(\hbar\beta)}{\ln(e^\gamma na^2/2)}. \qquad (7.105)$$

We further introduce the reduced dimensionless coupling parameter $\hat{\alpha} \equiv \alpha/l_e(\hbar\beta) = \gamma(D, na^D)\,[a/l_e(\hbar\beta)]^{D-2}$, for brevity [recall (7.104)]. In terms of $\hat{\alpha}$, the grand-canonical free energy (7.102) takes the form

$$F_G = -\frac{1}{\beta}\frac{V_D}{l_e^D(\hbar\beta)}\left\{\zeta_{D/2+1}(z) - 2\hat{\alpha}[\zeta_{D/2}(z)]^2\right\} + \mathcal{O}(\alpha^2). \qquad (7.106)$$

A second-order perturbation calculation extends this by a term [18, 19]

$$\Delta F_G = -\frac{1}{\beta}\frac{V_D}{l_e^D(\hbar\beta)}\left\{8\hat{\alpha}^2 h_D(z)\right\}, \qquad (7.107)$$

where $h_D(z) = h_D^{(1)}(z) + h_D^{(2)}(z)$ is the sum of two terms. The first is simply

$$h_D^{(1)}(z) \equiv [\zeta_{D/2}(z)]^2 \zeta_{D/2-1}(z), \qquad (7.108)$$

the second has been calculated only for $D = 3$:

$$h_3^{(2)}(z) \equiv \sum_{n_1,n_2,n_3=1}^{\infty} \frac{z^{n_1+n_2+n_3}}{\sqrt{n_1 n_2 n_3}(n_1+n_2)(n_1+n_3)}. \qquad (7.109)$$

The associated particle number is

$$N = \frac{V_D}{l_e^D(\hbar\beta)}\left\{\zeta_{D/2}(z) - 4\hat{\alpha}\zeta_{D/2-1}(z)\zeta_{D/2}(z) + 8\hat{\alpha}^2 z\frac{dh_D(z)}{dz}\right\} + \mathcal{O}(\alpha^3). \qquad (7.110)$$

For small α, we may combine the equations for F_G and N and derive the following simple relation between the shift in the critical temperature and the change in the particle density caused by the interaction

$$\frac{\Delta T_c}{T_c^{(0)}} \approx -\frac{2}{D}\frac{\Delta n}{n}, \qquad (7.111)$$

where Δn is the change in the density at the critical point caused by the interaction. The equation is correct to lowest order in the interaction strength.

   The calculation of $\Delta T_c/T_c^{(0)}$ in three dimensions has turned out to be a difficult problem. The reason is that the perturbation series for the right-hand side of (7.111) is found to be an expansion in powers $(T - T_c^{(0)})^{-n}$ which needs a strong-coupling evaluation for $T \to T_c^{(0)}$. Many theoretical papers have given completely different result, even in sign, and Monte Carlo data to indicate sign and order of magnitude. The ideal tool for such a calculation, field theoretic variational perturbation theory, has only recently been developed [20], and led to a result in rough agreement with Monte Carlo data [21]. Let us briefly review the history.

All theoretical results obtained in the literature have the generic form

$$\frac{\Delta T_c}{T_c^{(0)}} = c[\zeta(3/2)]^{\kappa/3} \left[\frac{a}{\ell_c}\right]^{\kappa} = c\, a^{\kappa} \left(\frac{N}{V}\right)^{\kappa/3}, \qquad (7.112)$$

where the right-hand part of the equation follows from the middle part via Eq. (7.53). In an early calculation [22] based on the δ-function potential (7.100), κ was found to be $1/2$, with a downward shift of T_c. More recent studies, however, have lead to the opposite sign [23]–[38]. The exponents κ found by different authors range from $\kappa = 1/2$ [23, 18, 31] to $\kappa = 3/2$ [19]. The most recent calculations yield $\kappa = 1$ [24]–[28], i.e., a direct proportionality of $\Delta T_c/T_c^{(0)}$ to the s-wave scattering length a, a result also found by Monte Carlo simulations [29, 30, 38], and by an extrapolation of experimental data measured in the strongly interacting superfluid ^{4}He after diluting it with the help of Vycor glass [40]. As far as the proportionality constant c is concerned, the literature offers various values which range for $\kappa = 1$ from $c = 0.34 \pm 0.03$ [29] to $c = 5.1$ [40]. A recent negative value $c \approx -0.93$ [41] has been shown to arise from a false assumption on the relation between canonical and grand-canonical partition function [42]. An older Monte Carlo result found $c \approx 2.3$ [30] lies close to the theoretical results of Refs. [23, 27, 28], who calculated $c_1 \approx -8\zeta(1/2)/3\zeta^{1/3}(3/2) \approx 2.83$, $8\pi/3\zeta^{4/3}(3/2) \approx 2.33$, 1.9, respectively, while the extrapolation of the experimental data on ^{4}He in Vycor glass favored $c = 5.1$, near the theoretical estimate $c = 4.66$ of Stoof [24]. The latest Monte Carlo data, however, point towards a smaller value $c \approx 1.32 \pm 0.02$, close to theoretical numbers 1.48 and 1.14 in Refs. [35, 39] and 1.14 ± 0.11 in Ref. [21].

There is no space here to discuss in detail how the evaluation is done. We only want to point out an initially surprising result that in contrast to a potential with a finite effective range in (7.96), the δ-function repulsion (7.100) with a pure phase shift causes no change of T_c linear $an^{1/D}$ if only the first perturbative correction to the grand-canonical free energy in Eq. (7.106) is taken into account. A simple nonperturbative approach which shows this goes as follows: We observe that the one-loop expression for the free energy may be considered as the extremum with respect to σ of the variational expression

$$F_G^{\sigma} = -\frac{1}{\beta} \frac{V_D}{l_e^D(\hbar\beta)} \left[\zeta_{D/2+1}(ze^{\sigma}) - \frac{\sigma^2}{8\hat{\alpha}}\right], \qquad (7.113)$$

which is certainly correct to first order in a. We have introduced the reduced dimensionless coupling parameter $\hat{\alpha} \equiv \alpha/l_e(\hbar\beta) = \gamma(D, na^D)[a/l_e(\hbar\beta)]^{D-2}$, for brevity [recall (7.104)]. The extremum lies at $\sigma = \Sigma$ which solves the implicit equation

$$\Sigma \equiv -4\hat{\alpha}\zeta_{D/2}(ze^{\Sigma}). \qquad (7.114)$$

The extremal free energy is

$$F_G^\Sigma = -\frac{1}{\beta} \frac{V_D}{l_e^D(\hbar\beta)} \left\{ \zeta_{D/2+1}(Z) - 2\hat{a}[\zeta_{D/2}(Z)]^2 \right\}, \quad Z \equiv z e^\Sigma. \tag{7.115}$$

The equation for the particle number, obtained from the derivative of the free energy (7.113) with respect to $-\mu$, reads

$$N = \frac{V_D}{l_e^D(\hbar\beta)} \zeta_{D/2}(Z), \tag{7.116}$$

and fixes Z. This equation has the same form as in the free case (7.36), so that the phase transition takes place at $Z = 1$, implying the same transition temperature as in the free system to this order. As discussed above, a nonzero shift (7.112) can only be found after summing up many higher-order Feynman diagrams which all diverge at the critical temperature.

Let us also point out that the path integral approach is not an adequate tool for discussing the condensed phase below the critical temperature, for which infinitely many particle orbits are needed. In the condensed phase, a quantum field theoretic formulation is more appropriate (see Section 7.6). The result of such a discussion is that due to the positive value of c in (7.112), the phase diagram has an unusual shape shown in Fig. 7.7. The nose in the phase transition curve implies that the system can undergo Bose-Einstein condensation slightly *above* $T_c^{(0)}$ if the interaction is *increased* (see Ref. [43]).

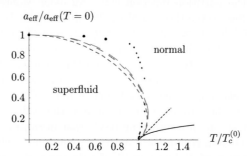

Figure 7.7 Reentrant transition in phase diagram of Bose-Einstein condensation for different interaction strengths. Curves were obtained in Ref. [43] from a variationally improved one-loop approximation to field-theoretic description with properly imposed positive slope at $T_c^{(0)}$ (dash length increasing with order of variational perturbation theory). Short solid curve and dashed straight line starting at $T_c^{(0)}$ are due to Ref. [38] and [21]. Diamonds correspond to the Monte-Carlo data of Ref. [29] and dots stem from Ref. [44], both scaled to their critical value $a_{\text{eff}}(T = 0) \approx 0.63$.

An important consequence of a repulsive short-range interaction is a change in the particle excitation energies below T_c. It was shown by Bogoliubov [46] that this changes the energy from the quadratic form $\epsilon(\mathbf{p}) = \mathbf{p}^2/2M$ to

$$\epsilon(\mathbf{p}) = \sqrt{\epsilon^2(\mathbf{p}) + 2gn\epsilon(\mathbf{p})}, \tag{7.117}$$

which starts linearly like $\epsilon(\mathbf{p}) = \sqrt{gn/M}\,|\mathbf{p}| = \hbar\sqrt{4\pi an/M}\,|\mathbf{p}|$ for small $\mathbf{p}$, the slope defining the second sound velocity. In the strongly interacting superfluid helium, the momentum dependence has the form shown in Fig. 7.8. It was shown by Bogoliubov

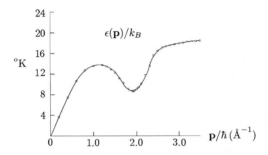

Figure 7.8   Energies of elementary excitations of superfluid ^{4}He measured by neutron scattering showing excitation energy of NG bosons [after R.A. Cowley and A.D. Woods, Can. J. Phys *49*, 177 (1971)].

[46] that a system in which $\epsilon(\mathbf{p}) > v_c\,|\mathbf{p}|$ for some finite critical velocity v_c will display superfluidity as long as it moves with velocity $\mathbf{v} = \partial\epsilon(\mathbf{p})/\partial\mathbf{p}$ smaller than v_c. This follows by forming the free energy in a frame moving with velocity $\mathbf{v}$. It is given by the Legendre transform:

$$f \equiv \epsilon(\mathbf{p}) - \mathbf{v}\cdot\mathbf{p}, \qquad (7.118)$$

which has a minimum at $\mathbf{p} = 0$ if as long as $|\mathbf{v}| < v_c$, implying that the particles do not move. Seen from the moving frame, they keep moving with a constant velocity $-\mathbf{v}$, without slowing down. This is in contrast to particles with the free spectrum $\epsilon(\mathbf{p}) = \mathbf{p}^2/2M$ for which f has a minimum at $\mathbf{p} = M\mathbf{v}$, implying that these particles always move with the same velocity as the moving frame. Note that the second sound waves of long wavelength have an energy $\epsilon(\mathbf{p} = c|\mathbf{p}|$ just like light waves in the vacuum, with light velocity exchanged by the sound velocity. Even though the superfluid is nonrelativistic, the sound waves behave like relativistic particles. This phenomenon is known from the sound waves in crystals which behave similar to relativistic massless particles. In fact, the Debye theory of specific heat is very similar to Planck's black-body theory, except that the lattice size appears explicitly as a short-wavelength cutoff. It has recently been speculated that the relativistically invariant world we observe is merely the long-wavelength limit of a world crystal whose lattice constant is very small [47], of the order of the Planck length $\ell_P \approx 8.08 \times 10^{-33}$ cm.

7.2.4   Bose-Einstein Condensation in Harmonic Trap

In a harmonic magnetic trap, the path integral (7.7) of the individual orbits becomes

$$(\mathbf{x}_b^{(p(\nu))}\hbar\beta|\mathbf{x}_a^{(\nu)}\,0)_\omega = \int_{\mathbf{x}^{(\nu)}(\tau_a)=\mathbf{x}_a^{(\nu)}}^{\mathbf{x}^{(p(\nu))}(\tau_b)=\mathbf{x}_b^{(\nu)}} \mathcal{D}^D x^{(\nu)}\exp\left\{-\frac{1}{\hbar}\int_0^{\hbar\beta}d\tau\frac{M}{2}\left[\dot{\mathbf{x}}^{(\nu)2}+\omega^2\mathbf{x}^{(\nu)2}\right]\right\}, \quad (7.119)$$

and is solved by [recall (2.409)]

$$(\mathbf{x}_b^{(p(\nu))}\hbar\beta|\mathbf{x}_a^{(\nu)}\,0)_\omega = \frac{1}{\sqrt{2\pi\hbar/M}^D}\sqrt{\frac{\omega}{\sinh\beta\hbar\omega}}^D \qquad\qquad (7.120)$$

$$\times\exp\left\{-\frac{1}{2\hbar}\frac{M\omega}{\sinh\beta\hbar\omega}[(\mathbf{x}_b^{(p(\nu))})^2+\mathbf{x}_a^{(\nu)\,2})\cosh\beta\hbar\omega-2\mathbf{x}_b^{(p(\nu))}\mathbf{x}_a^{(\nu)}]\right\}.$$

The partition function (7.5) can therefore be rewritten in the form

$$Z_\omega^{(N)} = \sqrt{\frac{M\omega}{2\pi\hbar\sinh\beta\hbar\omega}}^{ND}\frac{1}{N!}\int d^D x^{(1)}\cdots d^D x^{(N)}$$

$$\times\sum_{p(\nu)}\exp\left\{-\frac{1}{2\hbar}\frac{M\omega}{\sinh\beta\hbar\omega}[(\mathbf{x}^{(p(\nu))\,2}+\mathbf{x}^{(\nu)\,2})\cosh\beta\hbar\omega-2\mathbf{x}^{(p(\nu))}\mathbf{x}^{(\nu)}]\right\}. \quad (7.121)$$

7.2.5   Thermodynamic Functions

With the same counting arguments as before we now obtain from the connected paths, which wind some number $w = 1, 2, 3, \ldots$ times around the cylinder, a contribution to the partition function

$$\Delta Z_\omega^{(N)\,w} = Z_\omega(w\beta)\frac{1}{w} = \int d^D x\, z_\omega(w\beta;\mathbf{x})\frac{1}{w}, \qquad (7.122)$$

where

$$Z_\omega(\beta) = \frac{1}{[2\sinh(\beta\hbar\omega/2)]^D} \qquad\qquad (7.123)$$

is the D-dimensional harmonic partition function (2.405), and $z_\omega(w\beta;\mathbf{x})$ the associated density [compare (2.331) and (2.410)]:

$$z_\omega(\beta;\mathbf{x}) = \sqrt{\frac{\omega M}{2\pi\hbar\sinh\beta\hbar\omega}}^D\exp\left[-\frac{M\omega}{\hbar}\tanh\frac{\beta\hbar\omega}{2}\,\mathbf{x}^2\right]. \qquad (7.124)$$

Its spatial integral is the partition function of a free particle at an imaginary-time interval β [compare (2.410)]. The sum over all connected contributions (7.122) yields the grand-canonical free energy

$$F_G = -\frac{1}{\beta}\sum_{w=1}^\infty Z_\omega(w\beta)\frac{e^{w\beta\mu}}{w} = -\frac{1}{\beta}\sum_{w=1}^\infty\int d^D x\, z_\omega(w\beta;\mathbf{x})\frac{e^{w\beta\mu}}{w}. \qquad (7.125)$$

Note the important difference between this and the free-boson expression (7.27). Whereas in (7.27), the winding number appeared as a factor $w^{-D/2}$ which was removed from $Z_0(w\beta)$ by writing $Z_0(w\beta) = Z_0(\beta)/w^{D/2}$ which lead to (7.32), this is no longer possible here. The average number of particles is thus given by

$$N = -\frac{\partial}{\partial\mu}F_G = \sum_{w=1}^{\infty} Z_w(w\beta)e^{w\beta\mu} = \int d^D x \sum_{w=1}^{\infty} z_w(w\beta; \mathbf{x})e^{w\beta\mu}. \qquad (7.126)$$

Since $Z_w(w\beta) \approx e^{-wD\hbar\omega\beta/2}$ for large w, the sum over w converges only for $\mu < D\hbar\omega/2$. Introducing the fugacity associated with the ground-state energy

$$z_D(\beta) = e^{-(D\hbar\omega/2-\mu)\beta} = e^{-D\beta\hbar\omega/2}z, \qquad (7.127)$$

by analogy with the fugacity (7.24) of the zero-momentum state, we may rewrite (7.126) as

$$N = -\frac{\partial}{\partial\mu}F_G = Z_w(\beta)\zeta_D(\beta\hbar\omega; z_D). \qquad (7.128)$$

Here $\zeta_D(\beta\hbar\omega; z_D)$ are generalizations of the functions (7.34):

$$\zeta_D(\beta\hbar\omega; z_D) \equiv \sum_{w=1}^{\infty}\left[\frac{\sinh(\omega\hbar\beta/2)}{\sinh(w\omega\hbar\beta/2)}\right]^D e^{w\beta\mu} = Z_w^{-1}(\beta)\sum_{w=1}^{\infty}\frac{1}{(1-e^{-2w\beta\hbar\omega})^{D/2}}z_D^w, \quad (7.129)$$

which reduce to $\zeta_D(z)$ in the trapless limit $\omega \to 0$, where $z_D \to z$. Expression (7.128) is the closest we can get to the free-boson formula (7.36).

We may define local versions of the functions $\zeta_D(\beta\hbar\omega; z_D)$ as in Eq.(7.124):

$$\zeta_D(\beta\hbar\omega; z_D; \mathbf{x}) \equiv Z_w^{-1}(\beta)\sum_{w=1}^{\infty}\frac{1}{(1-e^{-2w\beta\hbar\omega})^{D/2}}\left(\frac{\omega M}{\pi\hbar}\right)^{D/2}e^{-M\omega\tanh(w\beta\hbar\omega/2)\mathbf{x}^2/\hbar}z_D^w, (7.130)$$

in terms of which the particle number (7.128) reads

$$N = -\frac{\partial}{\partial\mu}F_G = \int d^D x\, n_w(\mathbf{x}) \equiv Z_w(\beta)\int d^D x\, \zeta_D(\beta\hbar\omega; z_D; \mathbf{x}), \qquad (7.131)$$

and the free energy (7.125) becomes

$$F_G = \int d^D x\, f(\mathbf{x}) \equiv -\frac{1}{\beta}Z_w(\beta)\int d^D x\int_0^{z_D}\frac{dz}{z}\,\zeta_D(\beta\hbar\omega; z; \mathbf{x}). \qquad (7.132)$$

For small trap frequency ω, the function (7.130) has a simple limiting form:

$$\zeta_D(\beta\hbar\omega; z_D; \mathbf{x}) \overset{\omega\approx 0}{\approx} \left(\frac{\beta\hbar\omega}{2\pi}\right)^{D/2}\frac{1}{\lambda_\omega^D}\sum_{w=1}^{\infty}\frac{1}{w^{D/2}}e^{-w\beta(M\omega^2\mathbf{x}^2/2-\mu)}, \qquad (7.133)$$

where λ_ω is the oscillator length scale $\sqrt{\hbar/M\omega}$ of Eq. (2.301). Together with the prefactor $Z_w(\hbar\beta)$ in (7.131), which for small ω becomes $Z_w(\hbar\beta) \approx 1/(\beta\hbar\omega)^D$, this yields the particle density

$$n_0(\beta; z; \mathbf{x}) = \frac{1}{l_e^D(\hbar\beta)}\sum_{w=1}^{\infty}\frac{1}{w^{D/2}}e^{-w\beta[V(\mathbf{x})-\mu]} = \frac{1}{l_e^D(\hbar\beta)}\zeta_{D/2}(e^{-\beta[V(\mathbf{x})-\mu]}), \qquad (7.134)$$

where $V(\mathbf{x}) = M\omega^2\mathbf{x}^2/2$ is the oscillator potential and $l_e(\hbar\beta)$ the thermal length scale (2.351). There is only one change with respect to the corresponding expression for the density in the homogeneous gas [compare (7.36)]: the fugacity $z = e^{\beta\mu}$ in the argument by the *local fugacity*

$$z(\mathbf{x}) \equiv e^{-\beta[V(\mathbf{x})-\mu]}. \tag{7.135}$$

The function $\zeta_D(\beta\hbar\omega; z_D)$ starts out like z_D, and diverges for $z_D \to 1$ like $[2\sinh(\beta\hbar\omega/2)]^D/(z_D-1)$. This divergence is the analog of the divergence of the box function (7.82) which reflects the formation of a condensate in the discrete ground state with particle number [compare (7.88)]

$$N_{\mathrm{cond}} = \frac{1}{e^{D\beta\hbar\omega/2 - \beta\mu} - 1}. \tag{7.136}$$

This number diverges for $\mu \to D\hbar\omega/2$ or $z_D \to 1$.

In a box with a finite the number of particles, Eq. (7.36) for the particle number was replaced by the equation for the normal particles (7.89) containing the regularized box functions $\bar\zeta_{D/2}^{\mathrm{box}}(z)$. In the thermodynamic limit, this turned into the function $\zeta_{D/2}(z)$ in which the momentum sum was evaluated as an integral. The present functions $\zeta_D(\beta\hbar\omega; z_D)$ governing the particle number in the harmonic trap play precisely the role of the previous box functions, with a corresponding singular term which has to be subtracted, thus defining a regularized function

$$\bar\zeta_D(\beta\hbar\omega; z_D) \equiv \zeta_D(\beta\hbar\omega; z_D) - Z_\omega^{-1}(\beta)\frac{z_D}{1-z_D} = Z_\omega^{-1}(\beta)S_D(\beta\hbar\omega; z_D), \tag{7.137}$$

where $S(\beta\hbar\omega, z_D)$ is the sum

$$S(\beta\hbar\omega, z_D) \equiv \sum_{w=1}^{\infty}\left[\frac{1}{(1-e^{-w\beta\hbar\omega})^D} - 1\right]z_D^w \tag{7.138}$$

The local function (7.130) has a corresponding divergence and can be regularized by a similar subtraction.

The sum (7.138) governs directly the number of *normal* particles

$$N_{\mathrm{n}}(T) = Z_\omega(\beta)\bar\zeta_D(\beta\hbar\omega; z_D) \equiv S_D(\beta\hbar\omega, z_D). \tag{7.139}$$

The replacement of the singular equation (7.128) by the regular (7.139) of completely analogous to the replacement of the singular (7.82) by the regular (7.89) in the box.

7.2.6 Critical Temperature

Bose-Einstein condensation can be observed as a proper phase transition in the thermodynamic limit, in which N goes to infinity, at a constant average particle density in the trap defined by $N/\lambda_\omega^D \propto N\omega^D$. In this limit, ω goes to zero and the sum (7.139) becomes $\zeta_D(z_D)/\beta\hbar\omega^D$. The associated particle equation

$$N_{\mathrm{n}} = \frac{1}{(\beta\hbar\omega)^D}\zeta_D(z_D) \tag{7.140}$$

can be solved only as long as $z_D < 1$. Above T_c, this equation determines z_D as a function of T from the condition that all particles are normal, $N_n = N$. Below T_c, it determines the temperature dependence of the number of normal particles by inserting $z_D = 1$, where

$$N_n = \frac{1}{(\beta \hbar \omega)^D} \zeta(D). \tag{7.141}$$

The particles in the ground state from a condensate, whose fraction is given by $N_{cond}(T)/N \equiv 1 - N_n(T)/N$. This is plotted in Fig. 7.9 as a function of the temperature for a total particle number $N = 40\,000$.

The critical point with $T = T_c$, $\mu = \mu_c = D\omega/2$ is reached if N_n is equal to the total particle number N where

$$k_B T_c^{(0)} = \hbar \omega \left[\frac{N}{\zeta(D)} \right]^{1/D}. \tag{7.142}$$

This formula has a solution only for $D > 1$.

Inserting (7.142) back into (7.162), we may re-express the normal fraction as a function of the temperature as follows:

$$\frac{N_n^{(0)}}{N} \approx \left(\frac{T}{T_c^{(0)}} \right)^D, \tag{7.143}$$

and the condensate fraction as

$$\frac{N_{cond}^{(0)}}{N} \approx 1 - \left(\frac{T}{T_c^{(0)}} \right)^D. \tag{7.144}$$

Including the next term in (7.162), the condensate fraction becomes

$$\frac{N_{cond}^{(0)}}{N} \approx 1 - \left(\frac{T}{T_c^{(0)} + \delta T_c} \right)^D. \tag{7.145}$$

It is interesting to re-express the critical temperature (7.142) in terms of the particle density at the origin which is at the critical point, according to Eq. (7.134),

$$n_0(\mathbf{0}) = \frac{1}{l_e^D(\hbar \beta_c)} \zeta(D/2), \tag{7.146}$$

where we have shortened the notation on an obvious way. From this we obtain

$$k_B T_c^{(0)} = \frac{2\pi \hbar^2}{M} \left[\frac{n_0(\mathbf{0})}{\zeta(D/2)} \right]^{2/D}. \tag{7.147}$$

Comparing this with the critical temperature *without a trap* in Eq. (7.18) we see that both expressions agree if we replace N/V by the uniform density $n_0(\mathbf{0})$. As an

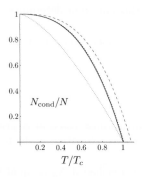
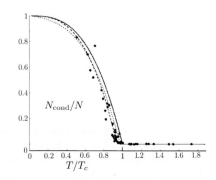

Figure 7.9 Condensate fraction $N_{\rm cond}/N \equiv 1 - N_{\rm n}/N$ as function of temperature for total number of N particles. The long- and short-dashed curves on the left-hand show the zeroth and first-order approximations (7.144) and (7.163). The dotted curve displays the free-boson behavior (7.58). The right-hand figure shows experimental data for 40 000 ^{87}Rb atoms near T_c by Ensher *et. al.*, Phys. Rev. Lett. **77**, 4984 (1996). The solid curve describes noninteracting Bose gas in a harmonic trap [cf. (7.144)]. The dotted curve corrects for finite-N effects [cf. (7.163)]. The dashed curve is the best fit to the data. The transition lies at $T_c \approx 280\,{\rm nK}$, about 3% below the dotted curve.

obvious generalization, we conclude that Bose condensation in *any* trap will set in when the density at the lowest point reaches the critical value determined by (7.147) [and (7.18)].

Note the different power D and argument in $\zeta(D)$ in Eq. (7.142) in comparison with the free-boson argument $D/2$ in Eq. (7.18). This has the consequence that in contrast to the free Bose gas, the gas in a trap can form a condensate in two dimensions. There is now, however, a problem in $D = 1$ dimension where (7.142) gives a vanishing transition temperature. The leading-order expression (7.142) for critical temperature can also be calculated from a simple statistical consideration. For small ω, the density of states available to the bosons is given by the classical expression (4.216):

$$\rho_{\rm cl}(E) = \left(\frac{M}{2\pi\hbar^2}\right)^{D/2} \frac{1}{\Gamma(D/2)} \int d^D x \, [E - V(\mathbf{x})]^{D/2-1}. \qquad (7.148)$$

The number of normal particles is given by the equation

$$N_{\rm n} = \int_{E_{\rm min}}^{\infty} dE \, \frac{\rho_{\rm cl}(E)}{e^{E/k_B T} - 1}, \qquad (7.149)$$

where $E_{\rm min}$ is the classical ground state energy. For a harmonic trap, the spatial integral in (7.148) can be done and is proportional to E^D, after which the integral over E in (7.149) yields $[k_B T/\hbar\omega]^D \zeta(D) = (T/T_c^{(0)})^D N$, in agreement with (7.143).

Alternatively we may use the phase space formula

$$
\begin{aligned}
N_n &= \int d^D x \frac{d^D p}{(2\pi\hbar)^D} \frac{1}{e^{\beta[p^2/2M+V(\mathbf{x})]} - 1} = \sum_{n=1}^{\infty} \int d^D x \frac{d^D p}{(2\pi\hbar)^D} e^{-n\beta[p^2/2M+V(\mathbf{x})]} \\
&= \sum_{n=1}^{\infty} \frac{1}{\sqrt{2\pi\hbar^2 n\beta/M}^D} \int d^D x \, e^{-n\beta V(\mathbf{x})},
\end{aligned}
\tag{7.150}
$$

where the spatial integration produces a factor $\sqrt{2\pi/M\omega^2 n\beta}$ so that the right-hand side becomes again $(T/T_c^{(0)})^D N$.

7.2.7　More General Anisotropic Trap

The equation (7.150) for the particle number can be easily calculated for a more general trap where the potential has the anisotropic power behavior

$$
V(\mathbf{x}) = \frac{M}{2}\tilde{\omega}^2 \tilde{a}^2 \sum_{i=1}^{D} \left(\frac{|x_i|}{a_i}\right)^{p_i},
\tag{7.151}
$$

where $\tilde{\omega}$ is some frequency parameter and $\tilde{a}$ is the geometric average $\tilde{a} \equiv \left[\Pi_{i=1}^{D} a_i\right]^{1/D}$. Inserting (7.151) into (7.150) we encounter a product of integrals

$$
\prod_{i=1}^{D} \int_{-\infty}^{\infty} dx \, e^{-n\beta M\tilde{\omega}^2 \tilde{a}^2 (|x_i|/a_i)^{p_i}/2} = \prod_{i=1}^{D} \frac{a_i}{(\beta M\tilde{\omega}^2 \tilde{a}^2/2)^{1/p_i}} \Gamma(1 + 1/p_i),
\tag{7.152}
$$

so that the right-hand side of (7.150) becomes $(T/T_c^{(0)})^{\tilde{D}} N$, with the critical temperature

$$
k_B T_c^{(0)} = \frac{M\tilde{a}^2 \tilde{\omega}^2}{2} \left(\frac{\hbar\tilde{\omega}}{M\tilde{a}^2\tilde{\omega}^2}\right)^{D/\tilde{D}} \left[\frac{N\pi^{D/2}}{\zeta(\tilde{D}) \prod_{i=1}^{D} \Gamma(1 + 1/p_i)}\right]^{1/\tilde{D}},
\tag{7.153}
$$

where $\tilde{D}$ is the dimensionless parameter

$$
\tilde{D} \equiv \frac{D}{2} + \sum_{i=1}^{D} \frac{1}{p_i}.
\tag{7.154}
$$

which takes over the role of D in the harmonic formula (7.142). A harmonic trap with different oscillator frequencies $\omega_1, \ldots, \omega_D$ along the D Cartesian axes, is a special case of (7.151) with $p_i \equiv 2$, $\omega_i^2 = \tilde{\omega}^2\tilde{a}^2/a_i^2$ and $\tilde{D} = D$, and formula (7.153) reduces to (7.142) with ω replaced by the geometric average of the frequencies $\tilde{\omega} \equiv (\omega_1 \cdots \omega_D)^{1/D}$. The parameter $\tilde{a}$ disappears from the formula. A free Bose gas in a box of size $V_D = \prod_{i=1}^{D}(2a_i) = 2^D\tilde{a}^D$ is described by (7.151) in the limit $p_i \to \infty$ where $\tilde{D} = D/2$. Then Eq. (7.153) reduces to

$$
k_B T_c^{(0)} = \frac{\pi\hbar^2}{2M\tilde{a}^2} \left[\frac{N}{\zeta(D/2)}\right]^{2/D} = \frac{2\pi\hbar^2}{M} \left[\frac{N}{V_D\zeta(D/2)}\right]^{2/D},
\tag{7.155}
$$

in agreement with (7.53) and (7.18).

Another interesting limiting case is that of a box of length $L = 2a_1$ in the x-direction with $p_1 = \infty$, and two different oscillators of frequency ω_2 and ω_3 in the other two directions. To find $T_c^{(0)}$ for such a Bose gas we identify $\tilde{\omega}^2 \tilde{a}^2/a_{2,3}^2 = \omega_{2,3}^2$ in the potential (7.151), so that $\tilde{\omega}^4/\tilde{a}^2 = \omega_2^2 \omega_3^2/a_1^2$, and obtain

$$
k_B T_c^{(0)} = \hbar \tilde{\omega} \left(\frac{\pi \hbar}{2M\tilde{\omega}} \right)^{1/5} \left[\frac{N}{a_1 \zeta(5/2)} \right]^{2/5} = \hbar \tilde{\omega} \left(\frac{2\pi \lambda_{\omega_1} \lambda_{\omega_2}}{L^2} \right)^{1/5} \left[\frac{N}{\zeta(5/2)} \right]^{2/5}. \quad (7.156)
$$

7.2.8 Rotating Bose-Einstein Gas

Another interesting potential can be prepared in the laboratory by rotating a Bose condensate [48] with an angular velocity Ω around the z-axis. The vertical trapping frequencies is $\omega_z \approx 2\pi \times 11.0 \, \text{Hz} \approx 0.58 \times \text{nK}$, the horizontal one is $\omega_\perp \approx 6 \times \omega_z$. The centrifugal forces create an additional repulsive harmonic potential, bringing the rotating potential to the form

$$
V(\mathbf{x}) = \frac{M\omega_z^2}{2} \left(z^2 + 36\eta r_\perp^2 + \frac{\kappa}{2} \frac{r_\perp^4}{\lambda_{\omega_z}^2} \right) = \frac{\hbar \omega_z}{2} \left(\frac{z^2}{\lambda_{\omega_z}^2} + 36\eta \frac{r_\perp^2}{\lambda_{\omega_z}^2} + \frac{\kappa}{2} \frac{r_\perp^4}{\lambda_{\omega_z}^4} \right), \quad (7.157)
$$

where $r_\perp^2 = x^2 + y^2$, $\eta \equiv 1 - \Omega^2/\omega_\perp^2$, $\kappa \approx 0.4$, and $\lambda_{\omega_z} \equiv 3.245 \, \mu\text{m} \approx 1.42 \times 10^{-3} \, \text{K}$. For $\Omega > \omega_\perp$, η turns negative and the potential takes the form of a Mexican hat as shown in Fig. 3.13, with a circular minimum at $r_m^2 = -36\eta \lambda_{\omega_z}^2/\kappa$. For large rotation speed, the potential may be approximated by a circular harmonic well, so that we may apply formula (7.156) with $a_1 = 2\pi r_m$, to obtain the η-independent critical temperature

$$
k_B T_c^{(0)} \approx \hbar \omega_z \left(\frac{\kappa}{\pi} \right)^{1/5} \left[\frac{N}{\zeta(5/2)} \right]^{2/5}. \quad (7.158)
$$

For $\kappa = 0.4$ and $N = 300\,000$, this yields $T_c \approx 53\text{nK}$.

At the critical rotation speed $\Omega = \omega_\perp$, the potential is purely quartic $r_\perp = \sqrt{(x^2 + y^2)}$. To estimate $T_c^{(0)}$ we approximate it for a moment by the slightly different potential (7.151) with the powers $p_1 = 2, p_2 = 4, p_3 = 4$, $a_1 = \lambda_{\omega_z}, a_2 = a_3 = \lambda_{\omega_z}(\kappa/2)^{1/4}$, so that formula (7.153) becomes

$$
k_B T_c^{(0)} = \hbar \omega_z \left[\frac{\pi^2 \kappa}{16\Gamma^4(5/4)} \right]^{1/5} \left[\frac{N}{\zeta(5/2)} \right]^{2/5}. \quad (7.159)
$$

It is easy to change this result so that it holds for the potential $\propto r^4 = (x + y)^4$ rather than $x^4 + y^4$: we multiply the right-hand side of equation (7.149) for N by a factor

$$
\frac{2\pi \int r \, dr dx dy \, e^{-r^4}}{\int dx dy \, e^{-x^4 - y^4}} = \frac{\pi^{3/2}}{\Gamma[5/4]^2}. \quad (7.160)
$$

This factor arrives inversely in front of N in Eq. (7.161), so that we obtain the critical temperature in the critically rotating Bose gas

$$k_B T_c^{(0)} = \hbar \omega_z \left(\frac{4\kappa}{\pi} \right)^{1/5} \left[\frac{N}{\zeta(5/2)} \right]^{2/5}. \qquad (7.161)$$

The critical temperature at $\Omega = \omega_\perp$ is therefore by a factor $4^{1/5} \approx 1.32$ larger than at infinite Ω. Actually, this limit is somewhat academic in a semiclassical approximation since the quantum nature of the oscillator should be accounted for.

7.2.9 Finite-Size Corrections

Experiments never take place in the thermodynamic limit. The particle number is finite and for comparison with the data we must calculate finite-size corrections coming from finite N where ll $\omega \approx 1/N^{1/D}$ is small but nonzero. The transition is no longer sharp and the definition of the critical temperature is not precise. As in the thermodynamic limit, we shall identify it by the place where $z_D = 1$ in Eq. (7.139) for $N_n = N$. For $D > 3$, the corrections are obtained by expanding the first term in the sum (7.138) in powers of ω and performing the sums over w and subtracting $\zeta(0)$ for the sum $\sum_{w=1}^{\infty} 1$:

$$N_n(T_c) = \frac{1}{(\beta \hbar \omega)^D} \left[\zeta(D) + \frac{\beta \hbar \omega}{2} D \zeta(D-1) + \frac{(\omega \hbar \beta)^2}{24} D(3D-1) \zeta(D-2) + \ldots \right]$$
$$- \zeta(0). \qquad (7.162)$$

The higher expansion terms contain logarithmically divergent expressions, for instance in one dimension the first term $\zeta(1)$, and in three dimensions $\zeta(D-2) = \zeta(1)$. These indicate that the expansion powers of $\beta \hbar \omega$ is has been done improperly at a singular point. Only the terms whose ζ-function have a positive argument can be trusted. A careful discussion along the lines of Subsection 2.15.6 reveals that $\zeta(1)$ must be replaced by $\zeta_{\text{reg}}(1) = -\log(\beta \hbar \omega) + \text{const.}$, similar to the replacement (2.588). The expansion is derived in Appendix 7A.

For $D > 1$, the expansion (7.162) can be used up to the $\zeta(D-1)$ terms and yields the finite-size correction to the number of normal particles

$$\frac{N_n}{N} = \left(\frac{T}{T_c^{(0)}} \right)^D + \frac{D}{2} \frac{\zeta(D-1)}{\zeta^{1-1/D}(D)} \frac{1}{N^{1/D}}. \qquad (7.163)$$

Setting $N_n = N$, we obtain a shifted critical temperature by a relative amount

$$\frac{\delta T_c}{T_c^{(0)}} = -\frac{1}{2} \frac{\zeta(D-1)}{\zeta^{(D-1)/D}(D)} \frac{1}{N^{1/D}} + \cdots. \qquad (7.164)$$

In three dimensions, the first correction shifts the critical temperature $T_c^{(0)}$ downwards by 2% for 40 000 atoms.

Note that correction (7.164) has no direct ω-dependence whose size enters only implicitly via $T_c^{(0)}$ of Eq. (7.142). In an anisotropic harmonic trap, the temperature shift would carry a dimensionless factor $\tilde{\omega}/\bar{\omega}$ where $\bar{\omega}$ is the arithmetic mean $\bar{\omega} \equiv \left(\sum_{i=1}^{D} \omega_i\right)/D$.

The higher finite-size corrections for smaller particle numbers are all calculated in Appendix 7A. The result can quite simply be deduced by recalling that according to the Robinson expansion (2.581), the first term in the naive, wrong power series expansion of $\zeta_\nu(e^{-b}) = \sum_{w=1}^{\infty} e^{-wb}/w^\nu = \sum_{k=0}^{\infty}(-b)^k \zeta(\nu - k)/k!$ is corrected by changing the leading term $\zeta(\nu)$ to $\Gamma(1 - \nu)(-b)^{\nu-1} + \zeta(\nu)$ which remains finite for all positive integer ν. Hence we may expect that the correct equation for the critical temperature is obtained by performing this change in Eq. (7.162) on each $\zeta(\nu)$. This expectation is confirmed in Appendix 7A. It yields for $D = 3, 2, 1$ the equations for the number of particles in the excited states

$$N_n = \frac{1}{\beta\hbar\omega}\left[-(\log\beta\hbar\omega - \gamma) - \frac{\beta\hbar\omega}{2}\zeta(0) + \frac{(\beta\hbar\omega)^2}{12}\zeta(-1) + \ldots\right], \qquad (7.165)$$

$$N_n = \frac{1}{(\beta\hbar\omega)^2}\left[\zeta(2) - \beta\hbar\omega\left(\log\beta\hbar\omega - \gamma + \frac{1}{2}\right) - \frac{7(\beta\hbar\omega)^2}{12}\zeta(0) + \ldots\right], \quad (7.166)$$

$$N_n = \frac{1}{(\beta\hbar\omega)^3}\left[\zeta(3) + \frac{3\beta\hbar\omega}{2}\zeta(2) - (\beta\hbar\omega)^2\left(\log\beta\hbar\omega - \gamma + \frac{19}{24}\right) + \ldots\right], \quad (7.167)$$

where $\gamma = 0.5772\ldots$ is the Euler-Mascheroni number (2.467). Note that all nonlogarithmic expansion terms coincide with those of the naive expansion (7.162).

These equations may be solved for β at $N_n = N$ to obtain the critical temperature of the would-be phase transition.

Once we study the position of would-be transitions at finite size, it makes sense to include also the case $D = 1$ where the thermodynamic limit has no transition at all. There is a strong increase of number of particles in the ground state at a "critical temperature" determined by equating Eq. (7.165) with the total particle number N, which yields

$$k_B T_c^{(0)} = \hbar\omega N \frac{1}{(-\log\beta\hbar\omega + \gamma)} \approx \hbar\omega N \frac{1}{\log N}, \qquad D = 1. \qquad (7.168)$$

Note that this result can also be found also from the divergent naive expansion (7.162) by inserting for the divergent quantity $\zeta(1)$ the dimensionally regularized expression $\zeta_{\text{reg}}(1) = -\log(\beta\hbar\omega)$ of Eq. (2.588).

7.2.10 Entropy and Specific Heat

By comparing (7.126) with (7.125) we see that the grand-canonical free energy can be obtained from N_n of Eq. (7.162) by a simple multiplication with $-1/\beta$ and an increase of the arguments of the zeta-functions $\zeta(\nu)$ by one unit. Hence we have, up to first order corrections in ω

$$F_G(\beta, \mu_c) = -\frac{1}{\beta(\beta\hbar\omega)^D}\left[\zeta(D+1) + \beta\hbar\omega\frac{D}{2}\zeta(D) + \ldots\right]. \qquad (7.169)$$

From this we calculate immediately the entropy $S = -\partial_T F_G = k_B \beta^2 \partial_\beta F_G$ as

$$S = -k_B(D+1)\beta F_G = k_B(D+1)\frac{1}{(\beta\hbar\omega)^D}\left[\zeta(D+1) + \beta\hbar\omega\frac{D}{2}\zeta(D) + \ldots\right]. \quad (7.170)$$

In terms of the lowest-order critical temperature (7.142), this becomes

$$S = k_B N(D+1)\left\{\left(\frac{T}{T_c^{(0)}}\right)^D \frac{\zeta(D+1)}{\zeta(D)} + \frac{D}{2}\left[\frac{\zeta(D)}{N}\right]^{1/D}\left(\frac{T}{T_c^{(0)}}\right)^{D-1} + \ldots\right\}. \quad (7.171)$$

From this we obtain the specific heat $C = T\partial_T S$ below T_c:

$$C = k_B N(D+1)\left\{D\left(\frac{T}{T_c^{(0)}}\right)^D \frac{\zeta(D+1)}{\zeta(D)} + \frac{D(D-1)}{2}\left[\frac{\zeta(D)}{N}\right]^{1/D}\left(\frac{T}{T_c^{(0)}}\right)^{D-1} + \ldots\right\}. \quad (7.172)$$

At the critical temperature, this has the maximal value

$$C_{\max} \approx k_B N(D+1)D\frac{\zeta(D+1)}{\zeta(D)}\left\{1 + \left[\frac{D-1}{2}\frac{\zeta^{1+1/D}(D)}{\zeta(D+1)} - \frac{D}{2}\frac{\zeta(D-1)}{\zeta^{1-1/D}(D)}\right]\frac{1}{N^{1/D}}\right\}. \quad (7.173)$$

In three dimensions, the lowest two approximations have their maximum at

$$C_{\max}^{(0)} \approx k_B N\, 10.805, \qquad C_{\max}^{(1)} \approx k_B N\, 9.556. \quad (7.174)$$

Above T_c, we expand the total particle number (7.126) in powers of ω as in (7.162). The fugacity of the ground state is now different from unity:

$$N(\beta,\mu) = \frac{1}{(\beta\hbar\omega)^D}\left[\zeta_D(z_D) + \beta\hbar\omega\frac{D}{2}\zeta_{D-1}(z_D) + \ldots\right]. \quad (7.175)$$

The grand-canonical free energy is

$$F_G(\beta,\mu) = -\frac{1}{\beta(\beta\hbar\omega)^D}\left[\zeta_{D+1}(z_D) + \beta\hbar\omega\frac{D}{2}\zeta_D(z_D) + \ldots\right], \quad (7.176)$$

and the entropy

$$S(\beta,\mu) = k_B\frac{1}{(\beta\hbar\omega)^D}\left\{(D+1)\zeta_{D+1}(z_D) + \frac{1}{2}\left(\beta\hbar\omega D^2 - 2\beta\mu\right)\zeta_D(z_D) + \ldots\right\}. \quad (7.177)$$

The specific heat C is found from the derivative $-\beta\partial_\beta S|_N$ as

$$C(\beta,\mu) = k_B\frac{1}{(\beta\hbar\omega)^D}\left\{(D+1)D\,\zeta_{D+1}(z_D)\right.$$
$$\left. + \frac{1}{2}\left[D\left(D^2+1\right)\beta\hbar\omega - D\beta\mu - \beta\partial_\beta(\beta\mu)\right]\zeta_D(z_D) + \ldots\right\}. \quad (7.178)$$

The derivative $\beta\partial_\beta(\beta\mu)$ is found as before from the condition: $\partial N(\beta,\mu(\beta))/\partial\beta = 0$, implying

$$0 = \frac{1}{(\beta\hbar\omega)^D}\left\{-\left[D\zeta_D(z_D) + \beta\hbar\omega\frac{D}{2}(D-1)\zeta_{D-1}(z_D)\right]\right.$$
$$\left.+ \left[\zeta_{D-1}(z_D) + \beta\hbar\omega\frac{D}{2}\zeta_{D-2}(z_D)\right]\left[\beta\partial_\beta(\beta\mu)-\beta\hbar\omega\frac{D}{2}\right] + \ldots\right\}, \quad (7.179)$$

so that we obtain

$$\beta\partial_\beta(\beta\mu) = D\frac{\zeta_D(z_D) + \beta\hbar\omega\dfrac{D}{2}\zeta_{D-2}(z_D) + \ldots}{\zeta_{D-1}(z_D) + \beta\hbar\dfrac{D}{2}\omega\zeta_{D-2}(z_D) + \ldots}. \quad (7.180)$$

Let us first consider the lowest approximation, where

$$N(\beta,\mu) \approx \frac{1}{(\beta\hbar\omega)^D}\zeta_D(z), \quad (7.181)$$

$$F_G(\beta,\mu) \approx -\frac{1}{\beta}\frac{1}{(\beta\hbar\omega)^D}\zeta_{D+1}(z) = -N\frac{1}{\beta}\frac{\zeta_{D+1}(z)}{\zeta_D(z)}, \quad (7.182)$$

$$S(\beta,\mu) \approx \frac{1}{T}\left[\frac{D+1}{\beta}\frac{1}{(\beta\hbar\omega)^D}\zeta_{D+1}(z) - \mu N\right] = Nk_B\left[\frac{D+1}{(\beta\hbar\omega)^D}\frac{\zeta_{D+1}(z)}{\zeta_D(z)} - \beta\mu\right], \quad (7.183)$$

so that from (7.60)

$$E(\beta,\mu) \approx \frac{D}{\beta}\frac{1}{(\beta\hbar\omega)^D}\zeta_{D+1}(z) = N\frac{D}{\beta}\frac{\zeta_{D+1}(z)}{\zeta_D(z)}. \quad (7.184)$$

The chemical potential at fixed N satisfies the equation

$$\beta\partial_\beta(\beta\mu) = D\frac{\zeta_D(z)}{\zeta_{D-1}(z)}. \quad (7.185)$$

The specific heat at a constant volume is

$$C = k_B N\left[(D+1)D\frac{\zeta_{D+1}(z)}{\zeta_D(z)} - D^2\frac{\zeta_D(z)}{\zeta_{D-1}(z)}\right]. \quad (7.186)$$

At high temperatures, C tends to the Dulong-Petit limit DNk_B since for small z all $\zeta_\nu(z)$ behave like z. This is twice a big as the Dulong-Petit limit of the free Bose gas since there are twice as many harmonic modes.

As the temperature approaches the critical point from above, z tends to unity from below and we obtain a maximal value in three dimensions

$$C_{\max}^{(0)} = k_B N\left[12\frac{\zeta_4(1)}{\zeta_3(1)} - 9\frac{\zeta_3(1)}{\zeta_2(1)}\right] \approx k_B N\,4.22785. \quad (7.187)$$

The specific heat for a fixed large number N of particles in a trap has a much sharper peak than for the free Bose gas. The two curves are compared in Fig. 7.10, where we also show how the peak is rounded for different finite numbers N.[3]

---

[3]P.W. Courteille, V.S. Bagnato, and V.I. Yukalov, Laser Physics **2**, 659 (2001).

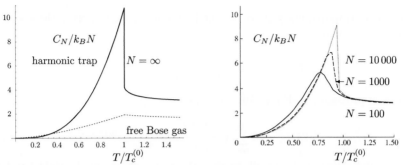

Figure 7.10   Peak of specific heat for infinite (left-hand plot) and various finite numbers 100, 1000, 10 000 of particles N (right-hand plots) in harmonic trap. The large-N curve is compared with that of a free Bose gas.

7.2.11   Interactions in Harmonic Trap

Let us now study the effect of interactions on a Bose gas in an isotropic harmonic trap. This is most easily done by adding to the free part (7.132) the interaction (7.101) with $n(\mathbf{x})$ taken from (7.131), to express the grand-canonical free energy by analogy with (7.102) as

$$F_G = \int d^D x \left\{ -\frac{1}{\beta} Z_\omega(\beta) \int_0^{z_D} \frac{dz}{z} \zeta_D(\beta\hbar\omega; z; \mathbf{x}) + g \left[Z_\omega(\beta)\zeta_D(\beta\hbar\omega; z_D; \mathbf{x}) \right]^2 \right\}. \tag{7.188}$$

Using the relation (7.103), this takes a form more similar to (7.106):

$$F_G = -\frac{1}{\beta} Z_\omega(\beta) \int d^D x \left\{ \int_0^{z_D} \frac{dz}{z} \zeta_D(\beta\hbar\omega; z; \mathbf{x}) - 2\hat{a}\, l_e^D(\hbar\beta) Z_\omega(\beta) \left[\zeta_D(\beta\hbar\omega; z_D; \mathbf{x}) \right]^2 \right\}. \tag{7.189}$$

As in Eq. (7.113), we now construct the variational free energy to be extremized with respect to the local parameter $\sigma(\mathbf{x})$. Moreover, we shall find it convenient to express $\zeta(z)$ as $\int_0^z (dz'/z')\zeta(z')$. This leads to the variational expression

$$F_G^\sigma = -\frac{1}{\beta} Z_\omega(\beta) \int d^3 x \left[\int_0^{z_D e^{\sigma(\mathbf{x})}} \frac{dz}{z} \zeta_D(\beta\hbar\omega; z; \mathbf{x}) - \frac{\sigma^2(\mathbf{x})}{8\tilde{a}} \right], \tag{7.190}$$

where $z_D = e^{-(D\hbar\omega/2 - \mu)\beta} = e^{-D\beta\hbar\omega/2} z$ and

$$\tilde{a} \equiv \hat{a}\, l_e^D(\hbar\beta) Z_\omega(\beta) = \frac{a}{l_e(\hbar\beta)}\, l_e^D(\hbar\beta) Z_\omega(\beta). \tag{7.191}$$

The extremum lies at $\sigma(\mathbf{x}) = \Sigma(\mathbf{x})$ where by analogy with (7.114):

$$\Sigma(\mathbf{x}) \equiv -4\tilde{a}\zeta_D(\beta\hbar\omega; z_D e^{\Sigma(\mathbf{x})}; \mathbf{x}). \tag{7.192}$$

For a small trap frequency ω, we use the function $\zeta_D(\beta\hbar\omega; z_D; \mathbf{x})$ in the approximate form (7.133), written as

$$\zeta_D(\beta\hbar\omega; z_D; \mathbf{x}) \equiv \sum_{w=1}^{\infty} \sqrt{\frac{\omega^2 M \beta}{2\pi w}}^D z_D^w e^{-Mw\beta\omega^2\mathbf{x}^2/2}. \tag{7.193}$$

In this approximation, Eq. (7.192) becomes

$$\Sigma(\mathbf{x}) \overset{\omega\approx 0}{\approx} -4\tilde{a} \sum_{w=1}^{\infty} \sqrt{\frac{\omega^2 M \beta}{2\pi w}}^D z_D^w e^{w\Sigma(\mathbf{x})} e^{-Mw\beta\omega^2\mathbf{x}^2/2}. \tag{7.194}$$

For small $\tilde{a}$, $\Sigma(\mathbf{x})$ is also small, so that the factor $e^{w\Sigma(\mathbf{x})}$ on the right-hand side is close to unity and can be omitted. This will be inserted into the equation for the particle number above T_c:

$$N = \int d^D x\, n(\mathbf{x}) = Z_\omega(\beta) \int d^D x\, \bar{\zeta}_D(\beta\hbar\omega; z_D e^{\Sigma(\mathbf{x})}; \mathbf{x}). \tag{7.195}$$

Recall that in the thermodynamic limit for $D > 1$ where the phase transition properly exists, $\zeta_D(\beta\hbar\omega; z_D; \mathbf{x})$ and $\bar{\zeta}_D(\beta\hbar\omega; z_D; \mathbf{x})$ coincide, due to (7.139) and (7.141).

From (7.195) we may derive the following equation for the critical temperature as a function of z_D:

$$1 = \frac{Z_\omega(\beta)}{Z_\omega(\beta_c^{(0)})} \int d^D x \frac{\bar{\zeta}_D(\beta\hbar\omega; z_D e^{\Sigma(\mathbf{x})}; \mathbf{x})}{\bar{\zeta}_D(\beta_c^{(0)}\hbar\omega; 1)}, \tag{7.196}$$

where $T_c^{(0)}$ is the critical temperature in the trap *without* the repulsive interaction. The critical temperature T_c of the interacting system is reached if the second argument of $\bar{\zeta}_D(\beta\hbar\omega; z_D e^{\Sigma(\mathbf{x})}; \mathbf{x})$ hits the boundary of the unit convergence radius of the expansion (7.130) for $\mathbf{x} = 0$, i.e., if $z_D e^{\Sigma(0)} = 1$. Thus we find the equation for T_c:

$$1 = \frac{Z_\omega(\beta_c)}{Z_\omega(\beta_c^{(0)})} \int d^D x \frac{\bar{\zeta}_D(\beta_c\hbar\omega; e^{\Sigma_c(\mathbf{x})-\Sigma_c(0)}; \mathbf{x})}{\bar{\zeta}_D(\beta_c^{(0)}\hbar\omega; 1)}, \tag{7.197}$$

where the subscript of $\Sigma_c(\mathbf{x})$ indicates that β in (7.194) has been set equal to β_c. In particular, $\tilde{a}$ contained in $\Sigma_c(\mathbf{x})$ is equal to $\tilde{a}_c \equiv a/\ell_c$. Since this is small by assumption, we expand the numerator of (7.197) as

$$1 \approx \frac{Z_\omega(\beta_c)}{Z_\omega(\beta_c^{(0)})} \left\{ 1 + \frac{1}{\bar{\zeta}_D(\beta_c^{(0)}\hbar\omega; 1)} \int d^D x\, \Delta\bar{\zeta}_D(\beta_c^{(0)}\hbar\omega; 1; \mathbf{x}) \right\}, \tag{7.198}$$

where the integral has the explicit small-ω form

$$\int d^D x \sum_{w=1}^{\infty} \sqrt{\frac{\omega^2 M \beta_c^{(0)}}{2\pi w}}^D e^{-Mw\beta_c^{(0)}\omega^2\mathbf{x}^2/2} \left(e^{w[\Sigma_c(\mathbf{x})-\Sigma_c(0)]} - 1 \right). \tag{7.199}$$

In the subtracted term we have used the fact that for small ω, $\bar{\zeta}_D(\beta_c^{(0)}\hbar\omega; 1) \approx \bar{\zeta}_D(\beta_c\hbar\omega; 1) \approx \zeta(D)$ is independent of β [see (7.139) and (7.141)].

Next we approximate near $T_c^{(0)}$:

$$\frac{Z_\omega(\beta_c)}{Z_\omega(\beta_c^{(0)})} \equiv 1 + D\frac{\Delta T_c}{T_c^{(0)}}\frac{\beta_c^{(0)}\hbar\omega/2}{\tanh(\beta_c^{(0)}\hbar\omega/2)} \overset{\omega\approx0}{\approx} 1 + D\frac{\Delta T_c}{T_c^{(0)}}, \qquad (7.200)$$

such that Eq. (7.198) can be solved for $\Delta T_c/T_c^{(0)}$:

$$\frac{\Delta T_c}{T_c^{(0)}} \approx -\frac{1}{D}\frac{1}{\zeta(D)}\int d^D x\,\Delta\bar\zeta_D(\beta_c^{(0)}\hbar\omega;1;\mathbf{x}). \qquad (7.201)$$

On the right-hand side, we now insert (7.199) with the small quantity $\Sigma_c(\mathbf{x})$ approximated by Eq. (7.194) at $\beta_c^{(0)}$, in which the factor $e^{w\Sigma_c(\mathbf{x})}$ on the right-hand side is replaced by 1, and find for the integral (7.199):

$$-4\tilde{a}_c\int d^D x \sum_{w=1}^{\infty}\sqrt{\frac{\omega^2 M\beta_c^{(0)}}{2\pi w}}^{\,D} e^{-Mw\beta_c^{(0)}\omega^2\mathbf{x}^2/2}\; w\sum_{w'=1}^{\infty}\sqrt{\frac{\omega^2 M\beta_c^{(0)}}{2\pi w'}}^{\,D}\left(e^{Mw'\beta_c^{(0)}\omega^2\mathbf{x}^2/2}-1\right).$$

The integral leads to

$$-4\tilde{a}_c\sqrt{\frac{\omega^2 M\beta_c^{(0)}}{2\pi}}^{\,D}\sum_{w,w'=1}^{\infty}\frac{1}{w^{D/2-1}w'^{D/2}}\left[\frac{1}{(w+w')^{D/2}}-\frac{1}{w^{D/2}}\right] \equiv -4\hat{a}_c S(D), \quad (7.202)$$

where $S(D)$ abbreviates the double sum, whose prefactor has been simplifies to $-4\hat{a}_c$ using (7.191) and the fact that for small ω, $Z_\omega(\beta_c^{(0)}) \approx (\omega\hbar\beta_c^{(0)})^{-D}$. For $D = 3$, the double sum has the value $S(3) \approx 1.2076-\zeta(2)\zeta(3/2) \approx -3.089$. Inserting everything into (7.201), we obtain for small a and small ω the shift in the critical temperature

$$\frac{\Delta T_c}{T_c^{(0)}} \approx \frac{4\hat{a}_c}{D}\frac{S(D)}{\zeta(D)} \overset{D=3}{\approx} -3.427\frac{a}{\ell_c}. \qquad (7.203)$$

In contrast to the free Bose gas, where a small δ-function repulsion does not produce any shift using the same approximation as here [recall (7.116)], and only a high-loop calculation leads to an upwards shift proportional to a, the critical temperature of the trapped Bose gas is shifted *downwards*. We can express ℓ_c in terms of the length scale $\lambda_\omega \equiv \sqrt{\hbar/M\omega}$ associated with the harmonic oscillator [recall Eq. (2.301)] and rewrite $\ell_c = \lambda_\omega\sqrt{2\pi\beta_c\hbar\omega}$. Together with the relation (7.142), we find

$$\frac{\Delta T_c}{T_c^{(0)}} \approx -3.427\frac{1}{\sqrt{2\pi}[\zeta(3)]^{1/6}}\frac{a}{\lambda_\omega}N^{1/6} \approx -1.326\frac{a}{\lambda_\omega}N^{1/6}. \qquad (7.204)$$

Note that since ω is small, the temperature shift formula (7.201) can also be expressed in terms of the zero-ω density (7.134) as

$$\Delta T_c \overset{\omega,\tilde{a}\approx0}{\approx} 2g\frac{\int d^3x\,[\partial_\mu n_0(\mathbf{x})]\,[n_0(\mathbf{x})-n_0(\mathbf{0})]}{\int d^3x\,\partial_T n_0(\mathbf{x})}, \qquad (7.205)$$

where we have omitted the other arguments β and z of $n_0(\beta; z; \mathbf{x})$, for brevity. To derive this formula we rewrite the grand-canonical free energy (7.190) as

$$F_G = -\frac{1}{\beta} \int d^3x \left[\int_0^{ze^{\beta\nu(\mathbf{x})}} \frac{dz'}{z'} n_0(\beta; z'; \mathbf{x}) - \beta \frac{\nu^2(\mathbf{x})}{4g} \right], \tag{7.206}$$

so that the particle number equation Eq. (7.195) takes the form

$$N = \int d^3x \, n_0(\beta; ze^{\beta\nu(\mathbf{x})}; \mathbf{x}). \tag{7.207}$$

Extremizing F_G in $\nu(\mathbf{x})$ yields the self-consistent equation

$$\nu(\mathbf{x}) = -2gn_0(\beta; ze^{\beta\nu(\mathbf{x})}; \mathbf{x}). \tag{7.208}$$

As before, the critical temperature is reached for $ze^{\beta\nu(0)} = 1$, implying that

$$N = \int d^3x \, n_0(\beta_c; ze^{-2\beta g[n_0(\mathbf{x})-n_0(0)]}; \mathbf{x}). \tag{7.209}$$

In the exponent we have omitted again the arguments β and z of $n_0(\beta; z; \mathbf{x})$. If we now impose the condition of constant N, $\Delta N = (\Delta T_c \partial_T + \Delta \mu \partial_\mu) N = 0$, and insert $\Delta \mu = -2g[n_0(\mathbf{x}) - n_0(\mathbf{0})]$, we find (7.205). Inserting into (7.205) the density (7.134) for the general trap (7.151), we find the generalization of (7.203):

$$\frac{\Delta T_c}{T_c^{(0)}} \approx \frac{4\hat{a}_c}{\tilde{D}} \frac{1}{\zeta(\tilde{D})} \sum_{w,w'=1}^{\infty} \frac{1}{w^{D/2-1}w'^{D/2}} \left[\frac{1}{(w+w')^{\tilde{D}-D/2}} - \frac{1}{w^{\tilde{D}-D/2}} \right], \tag{7.210}$$

which vanishes for the homogeneous gas, as concluded before on the basis of Eq. (7.116).

Let us compare the result (7.204) with the experimental temperature shift for ^{87}Rb in a trap with a critical temperature $T_c \approx 280\,\mathrm{nK}$ which lies about 3% below the noninteracting Bose gas temperature (see Fig. 7.9). Its thermal de Broglie length is calculated best in atomic units. Then the fundamental length scale is the Bohr radius $a_H = \hbar/M_p c \alpha$, where M_p is the proton mass and $\alpha \approx 1/137.035$ the fine-structure constant. The fundamental energy scale is $E_H = M_e c^2 \alpha^2$. Writing now the thermal de Broglie length at the critical temperature as

$$\frac{a}{\ell_c} = \frac{2\pi\hbar^2}{Mk_B T_c} = \sqrt{2\pi}\sqrt{\frac{E_H}{k_B T_c}}\sqrt{\frac{M_e}{87M_p}}\,a_H, \tag{7.211}$$

we estimate with $\sqrt{E_H/k_B T_c} \approx \sqrt{27.21\,\mathrm{eV}/(280 \times 10^{-9}/11\,604.447\,\mathrm{eV})} \approx 1.06 \times 10^6$ and $\sqrt{M_e/87M_p} \approx \sqrt{0.511\mathrm{eV}/(87 \times 938.27\mathrm{eV})} \approx 0.002502$ such that $\ell_c \approx 6646 a_H$.

The triplet s-wave scattering length of ^{87}Rb is $a \approx (106 \pm 4)\,a_H$ such that we find from (7.203)

$$\frac{\Delta T_c}{T_c^{(0)}} \approx -5.4\%, \tag{7.212}$$

which is compatible with the experimentally data of the trap in Fig. 7.9.

Let us finally mention recent studies of more realistic systems in which bosons in a trap interact with longer-range interactions [45].

7.3   Gas of Free Fermions

For fermions, the thermodynamic functions (7.32) and (7.33) contain the functions

$$\zeta_\nu^{\mathrm{f}}(z) \equiv \sum_{w=1}^{\infty} (-1)^{w-1} \frac{z^w}{w^\nu}, \tag{7.213}$$

which starts out for small z like z. For $z = 1$, this becomes

$$\zeta_\nu^{\mathrm{f}}(1) = \frac{1}{1} - \frac{1}{2^\nu} + \frac{1}{3^\nu} - \frac{1}{4^\nu} + \ldots = \sum_{k=0}^{\infty} \frac{1}{k^\nu} - 2^{1-\nu} \sum_{k=0}^{\infty} \frac{1}{k^\nu} = \left(1 - 2^{1-\nu}\right) \zeta(\nu). \tag{7.214}$$

In contrast to $\zeta_\nu(z)$ in Eq. (7.34) this function is perfectly well-defined for all chemical potentials by analytic continuation. The reason is the alternating sign in the series (7.213). The analytic continuation is achieved by expressing $\zeta_\nu^{\mathrm{f}}(z)$ as an integral by analogy with (7.39), (7.40):

$$\zeta_n^{\mathrm{f}}(z) \equiv \frac{1}{\Gamma(n)} i_n^{\mathrm{f}}(\alpha), \tag{7.215}$$

where $i_n^{\mathrm{f}}(\alpha)$ are the integrals

$$i_n^{\mathrm{f}}(\alpha) \equiv \int_0^\infty d\varepsilon \frac{\varepsilon^{n-1}}{e^{\varepsilon - \alpha} + 1}. \tag{7.216}$$

In the integrand we recognize the Fermi distribution function of Eq. (3.111), in which $\omega\hbar\beta$ is replaced by $\varepsilon - \alpha$:

$$n_\varepsilon^{\mathrm{f}} = \frac{1}{e^{\varepsilon - \alpha} + 1}. \tag{7.217}$$

The quantity ε plays the role of a reduced energy $\varepsilon = E/k_B T$, and α is a reduced chemical potential $\alpha = \mu/k_B T$.

Let us also here express the grand-canonical free energy F_G in terms of the functions $i_n^{\mathrm{f}}(\alpha)$. Combining Eqs. (7.32), (7.213), and (7.216) we obtain for fermions with $g_S = 2S + 1$ spin orientations:

$$F_G = -\frac{1}{\beta} Z_0(\beta) \frac{g_S}{\Gamma(D/2+1)} i_{D/2+1}^{\mathrm{f}}(\beta\mu) = -\frac{1}{\beta\Gamma(D/2+1)} \frac{g_S V_D}{\sqrt{2\pi\hbar^2\beta/M}^D} \int_0^\infty d\varepsilon \frac{\varepsilon^{D/2}}{e^{\varepsilon - \beta\mu} + 1}, \tag{7.218}$$

and the integral can be brought by partial integration to the form

$$F_G = -\frac{1}{\beta\Gamma(D/2)} \frac{g_S V_D}{\sqrt{2\pi\hbar^2\beta/M}^D} \int_0^\infty d\varepsilon \, \varepsilon^{D/2-1} \log(1 + e^{-\varepsilon + \beta\mu}). \tag{7.219}$$

Recalling Eq. (7.47), this can be rewritten as a sum over momenta of oscillators with energy $\hbar\omega_{\mathbf{p}} \equiv \mathbf{p}^2/2M$:

$$F_G = -\frac{g_S}{\beta} \sum_{\mathbf{p}} \log(1 + e^{-\beta\hbar\omega_{\mathbf{p}} + \beta\mu}). \tag{7.220}$$

This free energy will be studied in detail in Section 7.14.

The particle number corresponding to the integral representations (7.219) and (7.220) is

$$N = -\frac{\partial}{\partial \mu} F_G = \frac{1}{\Gamma(D/2)} \frac{g_S V_D}{\sqrt{2\pi\hbar^2\beta/M}^D} \int_0^\infty d\varepsilon \frac{\varepsilon^{D/2-1}}{e^{\varepsilon+\beta\mu}-1} = g_S \sum_{\mathbf{p}} \frac{1}{e^{\beta\hbar\omega_{\mathbf{p}}+\beta\mu}-1}. \quad (7.221)$$

Recalling the reduced density of states, this may be written with the help of the Fermi distribution function n_ω^{f} of Eq. (7.217) as

$$N = g_S V_D \int_0^\infty d\varepsilon \, N_\varepsilon \, n_\varepsilon^{\mathrm{f}}. \quad (7.222)$$

The Bose function contains a pole at $\alpha = 0$ which prevents the existence of a solution for positive α. In the analytically continued fermionic function (7.215), on the other hand, the point $\alpha = 0$ is completely regular.

Consider now a Fermi gas close to zero temperature which is called the degenerate limit. Then the reduced variables $\varepsilon = E/k_B T$ and $\alpha = \mu/k_B T$ become very large and the distribution function (7.217) reduces to

$$n_\varepsilon^{\mathrm{f}} = \left\{ \begin{array}{ll} 1 & \\ 0 \end{array} \text{ for } \begin{array}{l} \varepsilon < \alpha \\ \varepsilon > \alpha \end{array} \right\} = \Theta(\varepsilon - \alpha). \quad (7.223)$$

All states with energy E lower than the chemical potential μ are filled, all higher states are empty. The chemical potential μ at zero temperature is called *Fermi energy* E_{F}:

$$\mu\Big|_{T=0} \equiv E_{\mathrm{F}}. \quad (7.224)$$

The Fermi energy for a given particle number N in a volume V_D is found by performing the integral (7.222) at $T = 0$:

$$N = g_S V_D \int_0^{E_{\mathrm{F}}} d\varepsilon N_\varepsilon = \frac{1}{\Gamma(D/2+1)} \frac{g_S V_D E_{\mathrm{F}}^{D/2}}{\sqrt{2\pi\hbar^2/M}^D} = \frac{1}{\Gamma(D/2+1)} \frac{g_S V_D}{\sqrt{4\pi}^D} \left(\frac{p_{\mathrm{F}}}{\hbar}\right)^D, \quad (7.225)$$

where

$$p_{\mathrm{F}} \equiv \sqrt{2M E_{\mathrm{F}}} \quad (7.226)$$

is the *Fermi momentum* associated with the Fermi energy. Equation (7.225) is solved for E_{F} by

$$E_{\mathrm{F}} = \frac{2\pi\hbar^2}{M} \left[\frac{\Gamma(D/2+1)}{g_S}\right]^{2/D} \left(\frac{N}{V_D}\right)^{2/D}, \quad (7.227)$$

and for the Fermi momentum by

$$p_{\mathrm{F}} = 2\sqrt{\pi}\hbar \left[\frac{\Gamma(D/2+1)}{g_S}\right]^{1/D} \left(\frac{N}{V_D}\right)^{1/D} \hbar. \quad (7.228)$$

Note that in terms of the particle number N, the density of states per unit energy interval and volume can be written as

$$N_\varepsilon \equiv \frac{d}{2}\frac{N}{V_D}\sqrt{\frac{k_B T}{E_F}}^{-D/2}\varepsilon^{D/2-1}. \tag{7.229}$$

As the gas is heated slightly, the degeneracy in the particle distribution function in Eq. (7.223) softens. The degree to which this happens is governed by the size of the ratio kT/E_F. It is useful to define a characteristic temperature, the so-called *Fermi temperature*

$$T_F \equiv \frac{E_F}{k_B} = \frac{1}{k_B}\frac{p_F{}^2}{2M}. \tag{7.230}$$

For electrons in a metal, p_F is of the order of $\hbar/1\text{Å}$. Inserting $M = m_e = 9.109558 \times 10^{-28}$ g, further $k_B = 1.380622 \times 10^{-16}$ erg/K and $\hbar = 6.0545919 \times 10^{-27}$ erg sec, we see that T_F has the order of magnitude

$$T_F \approx 44\,000\text{K}. \tag{7.231}$$

Hence, even far above room temperatures the relation $T/T_F \ll 1$ is quite well fulfilled and can be used as an expansion parameter in evaluating the thermodynamic properties of the electron gas at nonzero temperature.

Let us calculate the finite-T effects in $D = 3$ dimensions. From Eq. (7.225) we obtain

$$N = N(T,\mu) \equiv \frac{g_S V}{l_e^3(\hbar\beta)}\frac{2}{\sqrt{\pi}}i_{3/2}^f\left(\frac{\mu}{k_B T}\right). \tag{7.232}$$

Expressing the particle number in terms of the Fermi energy with the help of Eq. (7.225), we obtain for the temperature dependence of the chemical potential an equation analogous to (7.54):

$$1 = \left(\frac{k_B T}{E_F}\right)^{3/2}\frac{3}{2}i_{3/2}^f\left(\frac{\mu}{k_B T}\right). \tag{7.233}$$

To evaluate this equation, we write the integral representation (7.216) as

$$
\begin{aligned}
i_n^f(\alpha) &= \int_{-\alpha}^{\infty} dx\,\frac{(\alpha+x)^{n-1}}{e^x+1} \\
&= \int_0^{\alpha} dx\,\frac{(\alpha-x)^{n-1}}{e^{-x}+1} + \int_0^{\infty} dx\,\frac{(\alpha+x)^{n-1}}{e^x+1}. \tag{7.234}
\end{aligned}
$$

where $\varepsilon = x + \alpha$. In the first integral we substitute $1/(e^{-x}+1) = 1 - 1/(e^x+1)$, and obtain

$$i_n^f(\alpha) = \int_0^{\alpha} dx\,x^{n-1} + \int_0^{\infty} dx\,\frac{(\alpha+x)^{n-1}-(\alpha-x)^{n-1}}{e^x+1} + \int_\alpha^{\infty} dx\,\frac{(\alpha-x)^{n-1}}{e^x+1}. \tag{7.235}$$

In the limit $\alpha \to \infty$, only the first term survives, whereas the last term is exponentially small, so that it can be ignored in a series expansion in powers of $1/\alpha$. The second term is expanded in such a series:

$$2 \sum_{k=\text{odd}}^{\infty} \binom{n-1}{k} \alpha^{n-1-k} \int_0^{\infty} dx \frac{x^k}{e^x + 1} = 2 \sum_{k=\text{odd}} \frac{(n-1)!}{(n-1-k)!} \alpha^{n-1-k} (1 - 2^{-k}) \zeta(k+1).$$

(7.236)

In the last equation we have used the integral formula for Riemann's ζ-function[4]

$$\int_0^{\infty} dx \frac{x^{\nu-1}}{e^{\mu x} + 1} = \mu^{-\nu}(1 - 2^{1-\nu}) \zeta(\nu).$$

(7.237)

At even positive and odd negative integer arguments, the zeta function is related to the Bernoulli numbers by Eq. (2.567). The lowest values of $\zeta(x)$ occurring in the expansion (7.236) are $\zeta(2)$, $\zeta(4), \ldots$, whose values were given in (2.569), so that the expansion of $i_n^f(\alpha)$ starts out like

$$i_n^f(\alpha) = \frac{1}{n} \alpha^n + 2(n-1) \frac{1}{2} \zeta(2) \alpha^{n-2} + 2(n-1)(n-2)(n-3) \frac{7}{8} \zeta(4) \alpha^{n-4} + \ldots .$$

(7.238)

Inserting this into Eq. (7.233) where $n = 3/2$, we find the low-temperature expansion

$$1 = \left(\frac{k_B T}{E_F}\right)^{3/2} \frac{3}{2} \left[\frac{2}{3} \left(\frac{\mu}{k_B T}\right)^{3/2} + \frac{\pi^2}{12} \left(\frac{\mu}{k_B T}\right)^{-1/2} + \frac{7\pi^4}{3 \cdot 320} \left(\frac{\mu}{k_B T}\right)^{-5/2} \ldots \right],$$

(7.239)

implying for μ the expansion

$$\mu = E_F \left[1 - \frac{\pi^2}{12} \left(\frac{k_B T}{E_F}\right)^2 + \frac{7\pi^4}{720} \left(\frac{k_B T}{E_F}\right)^4 + \ldots \right].$$

(7.240)

These expansions are asymptotic. They have a zero radius of convergence, diverging for any T. They can, however, be used for calculations if T is sufficiently small or at all T after a variational resummation à la Section 5.18.

We now turn to the grand-canonical free energy F_G. In terms of the function (7.216), this reads

$$F_G = -\frac{1}{\beta} \frac{g_S V}{l_e^3(\hbar\beta)} \frac{1}{\Gamma(5/2)} i_{5/2}^f(\alpha).$$

(7.241)

Using again (7.238), this has the expansion

$$F_G(T, \mu, V) = F_G(0, \mu, V) \left[1 + \frac{5\pi^2}{8} \left(\frac{k_B T}{\mu}\right)^2 - \frac{7\pi^4}{384} \left(\frac{k_B T}{\mu}\right)^4 + \ldots \right],$$

(7.242)

where

$$F_G(0, \mu, V) \equiv -\frac{2}{5} g_S V \frac{\sqrt{2} M^{3/2}}{3\pi^2 \hbar^3} (\mu)^{3/2} \mu = -\frac{2}{5} N \left(\frac{\mu}{E_F}\right)^{3/2} \mu.$$

---

[4] I.S. Gradshteyn and I.M. Ryzhik, *op. cit.*, Formula 3.411.3.

By differentiating F_G with respect to the temperature at fixed μ, we obtain the low-temperature behavior of the entropy

$$S \;=\; k_B \frac{\pi^2}{2} \frac{k_B T}{E_F} N + \dots \;. \tag{7.243}$$

From this we find a specific heat at constant volume

$$C \;=\; T \frac{\partial S}{\partial T}\bigg|_{V,N} \;\overset{T\approx 0}{\approx}\; S = k_B \frac{\pi^2}{2} \frac{k_B T}{E_F} N + \dots \;. \tag{7.244}$$

This grows linearly with increasing temperature and saturates at the constant value $3k_B N/2$ which obeys the Dulong-Petit law of Section 2.12 corresponding to three kinetic and no potential harmonic degrees of freedom in the classical Hamiltonian $\mathbf{p}^2/2M$. See Fig. 7.11 for the full temperature behavior. The linear behavior is

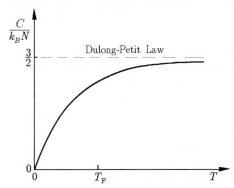

Figure 7.11 Temperature behavior of specific heat of free Fermi gas. As in the free Bose gas, the Dulong-Petit rule gives a high-temperature limit $3k_B N/2$ for the three harmonic kinetic degrees of freedom in the classical Hamiltonian $\mathbf{p}^2/2M$. There are no harmonic potential degrees of freedom.

due to the progressive softening near the surface of the Fermi distribution which makes more and more electrons thermally excitable. It is detected experimentally in metals at low temperature where the contribution of lattice vibrations freezes out as $(T/T_D)^3$. Here T_D is the Debye temperature which characterizes the elastic stiffness of the crystal and ranges from $T_D \approx 90$K in soft metals like lead over $T_D \approx 389$K for aluminum to $T_D \approx 1890$K for diamond. The experimental size of the slope is usually larger than the free electron gas value in (7.244). This can be explained mainly by the interactions with the lattice which result in a large effective mass $M_{\mathrm{eff}} > M$.

Note that the quantity $F_G(0, \mu, V)$ is temperature dependent via the chemical potential μ. Inserting (7.240) into (7.242) we find the complete T-dependence

$$F_G(T, \mu, V) = F_G(0, E_F, V) \left[1 + \frac{5\pi^2}{12} \left(\frac{k_B T}{E_F} \right)^2 - \frac{\pi^4}{16} \left(\frac{k_B T}{E_F} \right)^4 + \dots \right], \tag{7.245}$$

where

$$F_G(0, E_F, V) = -\frac{2}{5} N E_F. \tag{7.246}$$

As in the boson gas, we have a relation (7.61) between energy and grand-canonical free energy:

$$E = -\frac{3}{2} F_G, \tag{7.247}$$

such that equation (7.245) supplies us with the low-temperature behavior of the internal energy:

$$E = \frac{3}{5} N E_F \left[1 + \frac{5\pi^2}{12} \left(\frac{k_B T}{E_F} \right)^2 - \frac{\pi^4}{16} \left(\frac{k_B T}{E_F} \right)^4 + \dots \right]. \tag{7.248}$$

The first term is the energy of the zero-temperature Fermi sphere. Using the relation $c_V = \partial E / V \partial T$, the second term yields once more the leading $T \to 0$ -behavior (7.244) of specific heat.

This behavior of the specific heat can be observed in metals where the conduction electrons behave like a free electron gas. Due to *Bloch's theorem*, a single electron in a perfect lattice behaves just like a free particle. For many electrons, this is still approximately true, if the mass of the electrons is replaced by an effective mass.

Another important macroscopic system where (7.244) can be observed is a liquid consisting of the fermionic isotope ^{3}He. There are two electron spins and an odd number of nucleon spins which make this atom a fermion. Also there the strong interactions in the liquid produce a screening effect which raises to an effective value of the mass to 8 times that of the atom.

7.4 Statistics Interaction

First, we consider only two identical particles; the generalization to n particles will be obvious. For simplicity, we ignore the one-body potentials $V(\mathbf{x}^{(\nu)})$ in (7.2) since they cause only inessential complications. The total orbital action is then

$$\mathcal{A} = \int_{t_a}^{t_b} dt \left[\frac{M^{(1)}}{2} \dot{\mathbf{x}}^{(1)2} + \frac{M^{(2)}}{2} \dot{\mathbf{x}}^{(2)2} - V_{\text{int}}(\mathbf{x}^{(1)} - \mathbf{x}^{(2)}) \right]. \tag{7.249}$$

The standard change of variables to center-of-mass and relative coordinates

$$\mathbf{X} = (M^{(1)}\mathbf{x}^{(1)} + M^{(2)}\mathbf{x}^{(2)})/(M^{(1)} + M^{(2)}), \qquad \mathbf{x} = (\mathbf{x}^{(1)} - \mathbf{x}^{(2)}), \tag{7.250}$$

respectively, separates the action into a free center-of-mass and a relative action

$$\mathcal{A} = \mathcal{A}_{\text{CM}} + \mathcal{A}_{\text{rel}} = \int_{t_a}^{t_b} dt \frac{M}{2} \dot{\mathbf{X}}^2 + \int_{t_a}^{t_b} dt \left[\frac{\mu}{2} \dot{\mathbf{x}}^2 - V_{\text{int}}(\mathbf{x}) \right], \tag{7.251}$$

with a total mass $M = M^{(1)} + M^{(2)}$ and a reduced mass $\mu = M^{(1)}M^{(2)}/(M^{(1)} + M^{(2)})$. Correspondingly, the time evolution amplitude of the two-body system factorizes into that of an ordinary free particle of mass M, $(\mathbf{X}_b t_b | \mathbf{X}_a t_a)$, and a relative amplitude $(\mathbf{x}_b t_b | \mathbf{x}_a t_a)$. The path integral for the center-of-mass motion is solved as in Chapter 2. Only the relative amplitude is influenced by the particle statistics and needs a separate treatment for bosons and fermions.

First we work in one dimension only. Many of the formulas arising in this case are the same as those of Section 6.2, where we derived the path integral for a particle moving in a half-space $x = r > 0$; only the interpretation is different. We take care of the indistinguishability of the particles by restricting x to the positive semiaxis $x = r \geq 0$; the opposite vector $-x$ describes an identical configuration. The completeness relation of local states reads therefore

$$\int_0^\infty dr |r\rangle\langle r| = 1. \tag{7.252}$$

To write down the orthogonality relation, we must specify the bosonic or fermionic nature of the wave functions. Since these are symmetric or antisymmetric, respectively, we express $\langle r_b | r_a \rangle$ in terms of the complete set of functions with these symmetry properties:

$$\langle r_b | r_a \rangle = 2 \int_0^\infty \frac{dp}{\pi\hbar} \left\{ \begin{array}{c} \cos pr_b/\hbar \ \cos pr_a/\hbar \\ \sin pr_b/\hbar \ \sin pr_a/\hbar \end{array} \right\}. \tag{7.253}$$

This may be rewritten as

$$\langle r_b | r_a \rangle = \int_{-\infty}^\infty \frac{dp}{2\pi\hbar} \left(e^{ip(r_b - r_a)/\hbar} \pm e^{ip(r_b + r_a)/\hbar} \right) = \delta(r_b - r_a) \pm \delta(r_b + r_a). \tag{7.254}$$

The infinitesimal time evolution amplitude of relative motion is then, in the canonical formulation,

$$(r_n \epsilon | r_{n-1} 0) = \langle r_n | e^{-i\epsilon \hat{H}_{\rm rel}/\hbar} | r_{n-1} \rangle$$
$$= \int_{-\infty}^\infty \frac{dp}{2\pi\hbar} \left(e^{ip(r_n - r_{n-1})/\hbar} \pm e^{ip(r_n + r_{n-1})/\hbar} \right) e^{-i\epsilon H_{\rm rel}(p, r_n)/\hbar}, \tag{7.255}$$

where $H_{\rm rel}(p, x)$ is the Hamiltonian of relative motion associated with the action $\mathcal{A}_{\rm rel}$ in Eq. (7.251). By combining $N + 1$ factors, we find the time-sliced amplitude

$$(r_b t_b | r_a t_a) = \prod_{n=1}^N \left[\int_0^\infty dr_n \right] \prod_{n=1}^{N+1} \left[\int_{-\infty}^\infty \frac{dp_n}{2\pi\hbar} \right] \tag{7.256}$$
$$\times \left\{ \exp\left[\frac{i}{\hbar} \sum_{n=1}^{N+1} p_n (r_n - r_{n-1}) \right] \pm \exp\left[\frac{i}{\hbar} \sum_{n=1}^{N+1} p_n (r_n + r_{n-1}) \right] \right\} e^{-\frac{i}{\hbar}\epsilon \sum_{n=1}^{N+1} H_{\rm rel}(p_n, r_n)},$$

valid for bosons and fermions, respectively. By extending the radial integral over the entire space it is possible to remove the term after the $\pm$sign by writing

$$(r_b t_b | r_a t_a) = \sum_{x_b = \pm r_b} \prod_{n=1}^N \left[\int_{-\infty}^\infty dx_n \right] \prod_{n=1}^{N+1} \left[\int_{-\infty}^\infty \frac{dp_n}{2\pi\hbar} \right] \tag{7.257}$$
$$\times \exp\left\{ \frac{i}{\hbar} \sum_{a=1}^{N+1} [p_n(x_n - x_{n-1}) - \epsilon H_{\rm rel}(p_n, x_n) + \hbar\pi(\sigma(x_n) - \sigma(x_{n-1}))] \right\},$$

where the function $\sigma(x)$ vanishes identically for bosons while being equal to

$$\sigma(x) = \Theta(-x) \qquad (7.258)$$

for fermions, where $\Theta(x)$ is the Heaviside function (1.309). As usual, we have identified $x_b \equiv x_{N+1}$ and $x_a \equiv x_0$ which is equal to r_a. The final sum over $x_b = \pm r_b$ accounts for the indistinguishability of the two orbits. The phase factors $e^{i\pi\sigma(x_n)}$ give the necessary minus signs when exchanging two fermion positions.

Let us use this formula to calculate explicitly the path integral for a free two-particle relative amplitude. In the bosonic case with a vanishing σ-term, we simply obtain the free-particle amplitude summed over the final positions $\pm r_b$:

$$(r_b t_b | r_a t_a) = \frac{1}{\sqrt{2\pi\hbar i(t_b - t_a)/\mu}} \left\{ \exp\left[\frac{i}{\hbar}\frac{\mu}{2}\frac{(r_b - r_a)^2}{t_b - t_a}\right] + (r_b \to -r_b) \right\}. \qquad (7.259)$$

For fermions, the phases $\sigma(x_n)$ in (7.257) cancel each other successively, except for the boundary term

$$e^{i\pi(\sigma(x_b)-\sigma(x_a))}. \qquad (7.260)$$

When summing over $x_b = \pm r_b$ in (7.257), this causes a sign change of the term with $x_b = -r_b$ and leads to the antisymmetric amplitude

$$(r_b t_b | r_a t_a) = \frac{1}{\sqrt{2\pi\hbar i(t_b - t_a)/\mu}} \left\{ \exp\left[\frac{i}{\hbar}\frac{\mu}{2}\frac{(r_b - r_a)^2}{(t_b - t_a)}\right] - (r_b \to -r_b) \right\}. \qquad (7.261)$$

Let us also write down the continuum limit of the time-sliced action (7.257). It reads

$$\mathcal{A} = \mathcal{A}_{\rm rel} + \mathcal{A}_{\rm f} = \int_{t_a}^{t_b} dt\, [p\dot{x} - H_{\rm rel}(p, x) + \hbar\pi\dot{x}(t)\partial_x\sigma(x(t))]. \qquad (7.262)$$

The last term is the desired Fermi statistics interaction. It can also be written as

$$\mathcal{A}_{\rm f} = -\hbar\pi \int_{t_a}^{t_b} dt\dot{x}(t)\delta(x(t)) = \hbar\pi \int_{t_a}^{t_b} dt\partial_t\Theta(-x(t)). \qquad (7.263)$$

The right-hand expression shows clearly the pure boundary character of $\mathcal{A}_{\rm f}$, which does not change the equations of motion. Such an interaction is called a *topological interaction*.

Since the integrals in (7.257) over x and p now cover the *entire* phase space and $\sigma(x)$ enters only at the boundaries of the time axis, it is possible to add to the action any potential $V_{\rm int}(r)$. As long as the ordinary path integral can be performed, also the path integral with the additional σ-terms in (7.257) can be done immediately.

It is easy to generalize this result to any number of fermion orbits $x^{(\nu)}(t)$, $\nu = 1, \ldots, n$. The statistics interaction is then $\sum_{\nu<\nu'} \mathcal{A}_{\rm f}[x^{(\nu,\nu')}]$ with the distance vectors $x^{(\nu,\nu')} \equiv x^{(\nu)} - x^{(\nu')}$. When summing over all permuted final positions, the many-fermion wave functions become antisymmetric. The amplitude is given by the generalization of Eq. (7.257):

$$
(x_b^{(\nu_b)}; t_b | x_a^{(\nu_a)}; t_a) = \sum_{p(\nu_b)} \prod_{\nu=1}^{n} \left\{ \prod_{n=1}^{N} \left[\int_{-\infty}^{\infty} dx_n^{(\nu)} \right] \prod_{n=1}^{N+1} \left[\int_{-\infty}^{\infty} \frac{dp_n^{(\nu)}}{2\pi\hbar} \right] \right\}
$$

$$
\times \exp \left(\frac{i}{\hbar} \sum_{n=1}^{N+1} \left\{ \sum_{\nu=1}^{n} \left[p_n^{(\nu)}(x_n^{(\nu)} - x_{n-1}^{(\nu)}) - \epsilon H_{\text{rel}}(p_n^{(\nu)}, x_n^{(\nu)}) \right] \right. \right.
$$

$$
\left. \left. + \hbar\pi \sum_{\nu < \nu'} \left[\sigma(x_n^{(\nu,\nu')}) - \sigma(x_{n-1}^{(\nu,\nu')}) \right] \right\} \right), \quad (7.264)
$$

where $\Sigma_{p(\nu_b)}$ denotes the sum over all permutations of the final positions. The phases $\exp[i\pi\sigma(x)]$ produce the complete antisymmetry for fermions.

Consider now two particles moving in a two-dimensional space. Let the relative motion be described in terms of polar coordinates. For distinguishable particles, the scalar product of localized states is

$$
\langle r_b \varphi_b | r_a \varphi_a \rangle = \int_0^{\infty} dk\, k \sum_{m=-\infty}^{\infty} i_m(kr_b) i_m(kr_a) \frac{1}{2\pi} e^{im(\varphi_b - \varphi_a)}
$$

$$
= \frac{1}{\sqrt{r_b r_a}} \delta(r_b - r_a)\delta(\varphi_b - \varphi_a). \quad (7.265)
$$

This follows straightforwardly by expanding the exponentials $e^{i\mathbf{kx}} = e^{ikr\cos\varphi}$ in the scalar product

$$
\langle \mathbf{x}_b | \mathbf{x}_a \rangle = \int \frac{d^2 k}{(2\pi)^2} e^{i\mathbf{kx}_b} e^{-i\mathbf{kx}_a} = \delta^{(2)}(\mathbf{x}_b - \mathbf{x}_a) \quad (7.266)
$$

into Bessel functions, according to the well-known formula[5]

$$
e^{a\cos\varphi} = \sum_{m=-\infty}^{\infty} i_m(a) e^{im\varphi}, \quad (7.267)
$$

and by rewriting $\delta^{(2)}(\mathbf{x}_b - \mathbf{x}_a)$ as $(r_b r_a)^{-1/2}\delta(r_b - r_a)\delta(\varphi_b - \varphi_a)$. For indistinguishable particles, the angle φ is restricted to a half-space, say $\varphi \in [0, \pi)$. When considering bosons or fermions, the phase factor $e^{im(\varphi_b - \varphi_a)}$ must be replaced by $e^{im(\varphi_b - \varphi_a)} \pm e^{im(\varphi_b + \pi - \varphi_a)}$, respectively. In the product of such amplitudes in a time-sliced path integral, the $\pm$-terms in (7.256) can again be accounted for by completing the half-space in φ to the full space $[-\pi, \pi)$ and introducing the field $\sigma(\varphi)$. By including a Hamiltonian and returning to Euclidean coordinates x_1, x_2, we arrive at the relative amplitude

$$
(\mathbf{x}_b t_b | \mathbf{x}_a t) = \int \mathcal{D}^2 x \int \frac{\mathcal{D}^2 p}{2\pi} \left\{ \exp\left[\frac{i}{\hbar}\mathcal{A}_{\text{rel}} + \frac{i}{\hbar}\mathcal{A}_{\text{f}} \right] + (\mathbf{x}_b \to -\mathbf{x}_b) \right\}, \quad (7.268)
$$

with an obvious time slicing as in (7.257).

---

[5]I.S. Gradshteyn and I.M. Ryzhik, op. cit., Formula 6.633.1.

The Fermi statistics interaction $\mathcal{A}_{\mathrm{f}}$ looks in polar coordinates just like (7.263), but with x replaced by φ:

$$\mathcal{A}_{\mathrm{f}} = \hbar\pi \int_{t_a}^{t_b} dt\,\dot{\varphi}(t)\partial_{\varphi}\sigma(\varphi(t)). \tag{7.269}$$

Adapting the step function $\sigma(\varphi)$ to the periodic nature of the variable φ, we continue this function periodically in φ. Equivalently, we replace it by a step function $\sigma_{\mathrm{p}}(\varphi)$ which jumps by one unit at every integer multiple of π and write

$$\mathcal{A}_{\mathrm{f}} = \hbar \int_{t_a}^{t_b} dt\,\dot{\mathbf{x}}(t) \cdot \mathbf{a}(\mathbf{x}(t)), \tag{7.270}$$

with a vector potential

$$\mathbf{a}(\mathbf{x}) \equiv \pi\boldsymbol{\nabla}\sigma_{\mathrm{p}}(\varphi). \tag{7.271}$$

When calculating particle distributions or partition functions which satisfy periodic boundary conditions, this coupling is invariant under local gauge transformations of the vector potential

$$\mathbf{a}(\mathbf{x}) \to \mathbf{a}(\mathbf{x}) + \boldsymbol{\nabla}\Lambda(\mathbf{x}), \tag{7.272}$$

with smooth and single-valued functions $\Lambda(\mathbf{x})$, i.e., with $\Lambda(\mathbf{x})$ satisfying the integrability condition of Schwarz:

$$(\partial_i\partial_j - \partial_j\partial_i)\Lambda(\mathbf{x}) = 0. \tag{7.273}$$

Taking advantage of gauge invariance, we can in (7.271) replace $\sigma_{\mathrm{p}}(\varphi)$ by *any* function of $\mathbf{x}$ as long as it changes by one unit when going from φ_b to $\varphi_b + \pi$. A convenient choice is

$$\sigma_{\mathrm{p}}(\mathbf{x}) = \frac{1}{\pi}\varphi(\mathbf{x}) \equiv \frac{1}{\pi}\arctan\frac{x_2}{x_1}. \tag{7.274}$$

With this, the statistics interaction (7.270) becomes

$$\mathcal{A}_{\mathrm{f}} = \hbar \int_{t_a}^{t_b} dt\,\dot{\mathbf{x}}\partial_{\mathbf{x}}\varphi(\mathbf{x}) = \hbar \int_{t_a}^{t_b} dt\,\epsilon_{ij}\frac{x_i\dot{x}_j}{\mathbf{x}^2}, \tag{7.275}$$

where ϵ_{ij} is the antisymmetric unit tensor of Levi-Civita in two dimensions.

Just like the expression (7.263), this is a purely topological interaction. By comparison with (7.270), we identify the vector potential of the statistics interaction as

$$a_i(\mathbf{x}) = \partial_i\varphi = -\epsilon_{ij}\frac{x_j}{\mathbf{x}^2}. \tag{7.276}$$

The Fermi statistics remains obviously in operation if we choose, instead of the vector potential (7.276), an arbitrary odd multiple of it:

$$a_i(\mathbf{x}) = \partial_i\varphi = -(2n+1)\epsilon_{ij}\frac{x_j}{\mathbf{x}^2}, \qquad n = 0, \pm 1, \pm 2, \ldots . \tag{7.277}$$

The even multiples

$$a_i(\mathbf{x}) = \partial_i\varphi = -2n\epsilon_{ij}\frac{x_j}{\mathbf{x}^2}, \qquad n = 0, \pm 1, \pm 2, \ldots, \qquad (7.278)$$

on the other hand, give rise to Bose statistics.

For more than two particles, the amplitude (7.268) is generalized to the two-dimensional analog of Eq. (7.264). In one and two space dimensions we have thus succeeded in taking care of the indistinguishability of the particles and the fermionic nature by the simple statistics interaction terms (7.263) and (7.275). The indistinguishability of the particles requires that the path integral over all paths from the initial point $\mathbf{x}_a$ to the final point $\mathbf{x}_b$ has to be extended by those paths which run to the reflected point $-\mathbf{x}_b$. The statistics interaction guarantees the antisymmetry of the resulting amplitude.

7.5   Fractional Statistics

The above considerations raise an important question. Is it possible that particles with an arbitrary real multiple of the statistical gauge interaction (7.275) exist in nature? Such particles would show an unusual statistical behavior. If the prefactor is denoted by μ_0 and the statistics interaction reads

$$\mathcal{A}_{\mathrm{f}} = \hbar\mu_0 \int_{t_a}^{t_b} dt\, \dot{\mathbf{x}} \cdot \boldsymbol{\nabla}\varphi(\mathbf{x}) = \hbar\mu_0 \int_{t_a}^{t_b} dt \epsilon_{ij}\frac{x_i\dot{x}_j}{\mathbf{x}^2}, \qquad (7.279)$$

an interchange of the orbital endpoints in the path integral gives rise to a phase factor $e^{i\pi\mu_0}$. If μ_0 is even or odd, the amplitude describes bosons or fermions, respectively. For rational values of μ_0, however, the particles are neither one nor the other. They are called *anyons*. The phase of the amplitude returns to its initial value only after the particles have been rotated around each other several times. The statistical behavior of such particles will be studied in detail in Section 16.2. There we shall see that for two ordinary particles, an anyonic statistical behavior can be generated by a magnetic interaction. An interaction of the form (7.279) arises from an infinitesimally thin magnetic flux tube of strength $\Phi = \mu_0\Phi_0$ with $\Phi_0 = 2\pi\hbar c/e$. Indeed, the magnetic interaction is given by the gauge-invariant expression

$$\mathcal{A}_{\mathrm{mg}} = \frac{e}{c} \int_{t_a}^{t_b} dt\, \dot{\mathbf{x}}(t)\mathbf{A}(\mathbf{x}(t)), \qquad (7.280)$$

and the vector potential of a thin magnetic flux tube of flux Φ reads

$$A_i(\mathbf{x}) = \frac{\Phi}{2\pi}\partial_i\varphi = -\Phi\epsilon_{ij}\frac{x_j}{\mathbf{x}^2}. \qquad (7.281)$$

For the flux $\Phi_0 = 2\pi\hbar c/e$ or an odd multiple thereof, the magnetic interaction coincides with the statistics interaction of two fermions (7.275). Bose statistics holds if Φ is zero or an even multiple of Φ_0. The magnetic field can be chosen to produce any value of μ_0. This analogy will permit us to calculate the second virial coefficient of a gas of anyons in Section 16.3. There we shall also see that the statistical parameter μ_0 determines the behavior of the wave functions near the origin. While the wave functions of bosons and fermions carry either even or odd azimuthal angular momenta m, respectively, and vanish like $|\mathbf{x}|^m$ for $|\mathbf{x}| \to 0$, those of anyons can carry any integer m, behaving like $|\mathbf{x}|^{|m+\mu_0|}$ with a noninteger exponent.

We shall demonstrate in Section 16.2 that flux tubes whose flux Φ is an integer multiple of Φ_0, i.e., those with a flux corresponding to Fermi or Bose statistics, have a vanishing scattering amplitude with respect to particles of charge e (*Aharonov-Bohm effect*).

Such flux tubes can be used as a theoretical artifact to construct the vector potential of a magnetic monopole. Although magnetic fields can have no sources, a monopole can be brought in from infinity inside an infinitely thin tube of flux $\Phi = n\Phi_0$ (n = integer), called a *Dirac string*. Since this cannot be detected by any electromagnetic scattering experiment the endpoint of the string behaves like a magnetic monopole.[6] In an important aspect, the analogy between the magnetic and statistics interaction is not perfect and the present path integral is different from the one governing the magnetic scattering amplitude: The magnetic scattering amplitude deals with two different particles, one with an electric and the other with a magnetic charge. The paths are therefore summed with a fixed endpoint. In the statistics case, on the other hand, the sum includes the final point $\mathbf{x}_b$ and the reflected point $-\mathbf{x}_b$.

For this reason, the magnetic analogy can be used to impose arbitrary statistics only upon two particles and not upon an ensemble of many identical particles. The analogy has nevertheless been useful to guide recent theoretical developments, in particular the explanation of the fractional quantum Hall effect (to be discussed in Sections 16.13–16.12).

Particles in two dimensions with fractional statistics have recently become a source of inspiration in field theory, leading to many new and interesting insights.

7.6 Second-Quantized Bose Fields

We have seen above that the path integral of a system with many identical particles can become quite cumbersome to handle. Fortunately, there exists a much simpler and more efficient path integral description of many-particle systems. In the Schrödinger formulation of quantum mechanics, it is possible to generalize the single-particle Schrödinger equation to an arbitrary and *variable* number of particles by letting the complex Schrödinger fields $\psi(\mathbf{x}, t)$ be *field operators* rather than complex c-numbers. These are denoted by $\hat{\psi}(\mathbf{x}, t)$ and postulated to satisfy the harmonic-oscillator commutation relations at each point $\mathbf{x}$ in space. To impose properly such local quantization rules, space is discretized into little cubes of volume ϵ^3, centered

---

[6] See also the discussion in H. Kleinert, Int. J. Mod. Phys. A *7*, 4693 (1992) (http://www.physik.fu-berlin.de/~kleinert/203); Phys. Lett. B *246*, 127 (1990) (*ibid.*http/205).

around the points $\mathbf{x_n} = \epsilon(n_1, n_2, n_3)$, with $n_{1,2,3}$ running through all integers. If we omit the subscripts $\mathbf{n}$, for brevity, the quantization rules are

$$\begin{aligned}
[\hat{\psi}(\mathbf{x}, t), \hat{\psi}^\dagger(\mathbf{x}', t)] &= \delta_{\mathbf{xx}'}, \\
[\hat{\psi}^\dagger(\mathbf{x}, t), \hat{\psi}^\dagger(\mathbf{x}', t)] &= 0, \\
[\hat{\psi}(\mathbf{x}, t), \hat{\psi}(\mathbf{x}', t)] &= 0.
\end{aligned} \tag{7.282}$$

The commutativity of the operators at different places ensures the independence of the associated oscillators. Imposing the conditions (7.282) is referred to as *second quantization* or *field quantization*. One also speaks of a *quantization of particle number*. The commutation relations generate an infinite-dimensional Hilbert space at each space point $\mathbf{x}$. Applying the operator $\hat{\psi}^\dagger(\mathbf{x}, t)$ n times to the ground state of the harmonic oscillator $|0\rangle_\mathbf{x}$ at $\mathbf{x}$ creates states with n excitations at $\mathbf{x}$:

$$|n, \mathbf{x}\rangle = \frac{1}{\sqrt{n!}}[\hat{\psi}^\dagger(\mathbf{x}, 0)]^n |0\rangle. \tag{7.283}$$

These states are interpreted as states describing n particles at the point $\mathbf{x}$. The ground state of all oscillators is

$$|0\rangle \equiv \prod_\mathbf{x} |0\rangle_\mathbf{x}. \tag{7.284}$$

It is called the *vacuum state* of the system. The total number of particles at each time is measured by the operator

$$\hat{N}(t) = \sum_\mathbf{x} \hat{\psi}^\dagger(\mathbf{x}, t)\hat{\psi}(\mathbf{x}, t). \tag{7.285}$$

The simplest classical action, whose quantum theory has the above structure, describes an ensemble of free bosons with a chemical potential μ:

$$\mathcal{A}[\psi^*, \psi] = \sum_\mathbf{x} \int_{t_a}^{t_b} dt \left[\psi^*(i\hbar\partial_t + \mu)\psi(\mathbf{x}, t) - \frac{\hbar^2}{2M}\psi^*\nabla_\mathbf{x}\overline{\nabla}_\mathbf{x}\psi(\mathbf{x}, t) \right]. \tag{7.286}$$

The symbols $\nabla_\mathbf{x}, \overline{\nabla}_\mathbf{x}$ denote the difference operators on the discretized three-dimensional space, each component $\nabla_i, \overline{\nabla}_i$ being defined in the same way as the difference operators $\nabla, \overline{\nabla}$ on the sliced time axis in Eqs. (2.92). The eigenvalues on a plane wave of momentum $\mathbf{p}$ are

$$-i\hbar\nabla_i e^{i\mathbf{px}/\hbar} = P_i e^{i\mathbf{px}/\hbar}, \quad -i\hbar\overline{\nabla}_i e^{i\mathbf{px}/\hbar} = \bar{P}_i e^{i\mathbf{px}/\hbar}, \tag{7.287}$$

with

$$P_i = -i\frac{\hbar}{\epsilon}\left[e^{i\epsilon p_i/\hbar} - 1 \right], \quad \bar{P}_i = P_i^*. \tag{7.288}$$

By Fourier decomposing the field

$$\psi(\mathbf{x}, t) = \sqrt{\frac{\epsilon^3}{V}} \sum_\mathbf{p} e^{i\mathbf{px}/\hbar} a_\mathbf{p}(t), \tag{7.289}$$

the difference operators $\nabla_{\mathbf{x}}$, $\overline{\nabla}_{\mathbf{x}}$ are diagonalized and the action is decomposed into a direct sum of fields $a_{\mathbf{p}}^*(t)$, $a_{\mathbf{p}}(t)$ of a fixed momentum $\mathbf{p}$,

$$\mathcal{A}[a^*, a] = \hbar \sum_{\mathbf{p}} \int_{t_a}^{t_b} dt \left[a_{\mathbf{p}}^*(t) i \partial_t a_{\mathbf{p}}(t) - \omega(\mathbf{p}) a_{\mathbf{p}}^*(t) a_{\mathbf{p}}(t) \right], \qquad (7.290)$$

where $\omega(\mathbf{p})$ denotes the single-particle frequencies

$$\omega(\mathbf{p}) \equiv \frac{1}{\hbar} \left[\frac{|\mathbf{P}|^2}{2M} - \mu \right], \qquad (7.291)$$

with

$$|\mathbf{P}|^2 \equiv \frac{\hbar^2}{\epsilon^2} \sum_i 2 \left[1 - \cos \left(\frac{\epsilon p_i}{\hbar} \right) \right]. \qquad (7.292)$$

The extremization of (7.286) gives the field equation

$$\left(i\hbar \partial_t + \mu + \frac{\hbar^2}{2M} \nabla_{\mathbf{x}} \overline{\nabla}_{\mathbf{x}} \right) \psi(\mathbf{x}, t) = 0. \qquad (7.293)$$

This is the ordinary free-particle Schrödinger equation (the *first-quantized* field equation), apart from a constant shift in the energy by the chemical potential μ. Recall that the chemical potential guarantees a fixed average particle number which, in experiments, is enforced by contact with an appropriate particle reservoir (see Section 1.17). In momentum space, the field equation reads

$$[i\partial_t - \omega(\mathbf{p})] a_{\mathbf{p}}(t) = 0.$$

Knowing the general relation between the operator and the path integral description of quantum mechanics, we expect that the above rules of second quantization of operators can be accounted for by assuming the field variables $a^*_{\mathbf{p}}(t)$ and $a_{\mathbf{p}}(t)$ in the action to be fluctuating c-number variables and summing over all their configurations with an amplitude $\exp\{(i/\hbar)\mathcal{A}[a^*, a]\}$. The precise form of this path integral can be inferred from the oscillator nature of the commutation relations (7.282). After the Fourier transform (7.289), the components $\hat{a}_{\mathbf{p}}(t)$, $\hat{a}_{\mathbf{p}}^\dagger(t)$ satisfy

$$\begin{aligned}
[\hat{a}_{\mathbf{p}}(t), \hat{a}_{\mathbf{p}'}^\dagger(t)] &= \delta_{\mathbf{p}\mathbf{p}'}, \\
[\hat{a}_{\mathbf{p}}^\dagger(t), \hat{a}_{\mathbf{p}'}^\dagger(t)] &= 0, \\
[\hat{a}_{\mathbf{p}}(t), \hat{a}_{\mathbf{p}'}(t)] &= 0.
\end{aligned} \qquad (7.294)$$

Since the oscillators at different momenta $\mathbf{p}$ are independent of each other and since the action is a direct sum, we may drop the subscript $\mathbf{p}$ in the sequel and consider fields of a single momentum only.

The commutators (7.294) are the same as those of a harmonic oscillator, of course, obtained from the usual canonical commutators

$$[\hat{p}, \hat{x}] = -i\hbar \qquad (7.295)$$

by the canonical transformation

$$\hat{a}^\dagger = \sqrt{M/2\hbar\omega}(\omega\hat{x} - i\hat{p}/M), \quad \hat{a} = \sqrt{M/2\hbar\omega}(\omega\hat{x} + i\hat{p}/M). \tag{7.296}$$

Note that within the present context, the oscillator momentum $\hat{p}$ is the conjugate momentum of the field operator and has no relation to the particle momentum $\mathbf{p}$ (there exists a field operator for each particle momentum $\mathbf{p}$). The transformation (7.296) changes the Hamiltonian of the harmonic oscillator

$$\hat{H}_\omega = \frac{1}{2M}\hat{p}^2 + \frac{M\omega^2}{2}\hat{x}^2 \tag{7.297}$$

into the creation and annihilation operator form

$$\hat{H}_\omega = \frac{\hbar\omega}{2}(\hat{a}^\dagger\hat{a} + \hat{a}\,\hat{a}^\dagger). \tag{7.298}$$

The classical action in the canonical form

$$\mathcal{A}[p, q] = \int_{t_a}^{t_b} dt \, [p\dot{q} - H_\omega(p, q)] \tag{7.299}$$

turns into

$$\mathcal{A}[a^*, a] = \hbar \int_{t_a}^{t_b} dt \, (a^* i\partial_t a - \omega a^* a). \tag{7.300}$$

If one wants to describe quantum statistics, one has to replace $t \to -i\tau$ and use the Euclidean action (with $\beta = 1/k_B T$)

$$\mathcal{A}_e[a^*, a] = \hbar \int_0^{\hbar\beta} d\tau \, (a^* \partial_\tau a + \omega a^* a), \tag{7.301}$$

which coincides precisely with the action (7.290) for particles of a single momentum.

7.7 Fluctuating Bose Fields

We set up a path integral formulation which replaces this second-quantized operator structure. Since we have studied the harmonic oscillator extensively in real and imaginary time and since we know how to go back and forth between quantum-mechanical and -statistical expressions, we consider here only the case of imaginary time with the Euclidean action (7.301). For simplicity, we calculate only the partition function. The extension to density matrices is straightforward. Correlation functions will be discussed in detail in Chapter 18.

Since the action (7.300) of the harmonic oscillator is merely a rewritten canonical action (7.299), the partition function of the harmonic oscillator is given by the path integral [see (2.337)]

$$Z_\omega = \oint \mathcal{D}x(\tau) \int \frac{\mathcal{D}p(\tau)}{2\pi\hbar} \exp\left[-\int_0^{\hbar\beta} d\tau \, (a^* \partial_\tau a + \omega a^* a)\right], \tag{7.302}$$

where the quantum-mechanical trace requires the orbits $x(\tau)$ to be periodic in $\tau \to \tau + \hbar\beta$, with a Fourier expansion

$$x(\tau) = \frac{1}{\sqrt{\hbar\beta}} \sum_{m=-\infty}^{\infty} x_m e^{-i\omega_m \tau}, \quad \omega_m = 2\pi m/\hbar\beta. \tag{7.303}$$

The momentum integrations are unrestricted.

If the momentum states were used as the diagonal basis for the derivation of the path integral, the measure would be $\int \mathcal{D}x \oint (\mathcal{D}p/2\pi\hbar)$. Then $p(\tau)$ is periodic under $\tau \to \tau + \hbar\beta$ and the $x(\tau)$-integrations are unrestricted. This would give a different expression at the time-sliced level; the continuum limit $\epsilon \to 0$, however, would be the same.

Since the explicit conjugate variables in the action are now a and a^*, it is customary to express the measure of the path integral in terms of these variables and write

$$Z_\omega \equiv \oint \frac{\mathcal{D}a^*(\tau)\mathcal{D}a(\tau)}{\pi} \exp\left\{ -\int_0^{\hbar\beta} d\tau \left(a^* \partial_\tau a + \omega a^* a \right) \right\}, \tag{7.304}$$

where $\oint \mathcal{D}a^* \mathcal{D}a$ stands for the measure

$$\oint \mathcal{D}a^* \mathcal{D}a = \oint_{-\infty}^{\infty} \mathcal{D}\,\mathrm{Re}\,a \int_{-\infty}^{\infty} \mathcal{D}\,\mathrm{Im}\,a\,. \tag{7.305}$$

With the action being the time-sliced oscillator action, the result of the path integration in the continuum limit is known from (2.407) to be

$$Z_\omega = \frac{1}{2\sinh(\hbar\omega\beta/2)}. \tag{7.306}$$

In the context of second quantization, this is not really the desired result. For large β, the partition function (7.306) behaves like

$$Z_\omega \to e^{-\hbar\omega\beta/2}, \tag{7.307}$$

exhibiting in the exponent the oscillator ground-state energy $E_0 = \hbar\omega/2$. In the second-quantized interpretation, however, the ground state is the no-particle state. Hence its energy should be zero. In the operator formulation, this can be achieved by an appropriate operator ordering, choosing the Hamiltonian operator to be

$$\hat{H} = \hbar\omega\hat{a}^\dagger\hat{a}, \tag{7.308}$$

rather than the oscillator expression (7.298). In the path integral, the same goal is achieved by suitably time-slicing the path integral (7.304) and writing

$$Z_\omega^N = \prod_{n=0}^{N} \left[\int \frac{da_n^* da_n}{\pi} \right] \exp\left\{ -\frac{1}{\hbar} \mathcal{A}_\omega^N \right\}, \tag{7.309}$$

with the sliced action

$$A_\omega^N = \hbar \sum_{n=1}^{N} [a_n^*(a_n - a_{n-1}) + \epsilon\omega a_n^* a_{n-1}].$$

(7.310)

Expressed in terms of the difference operator, it reads

$$A_\omega^N = \hbar\epsilon \sum_{n=1}^{N} a_n^* \left[(1 - \epsilon\omega)\overline{\nabla} + \omega\right] a_n.$$

(7.311)

The $a(\tau)$-orbits are taken to be periodic functions of τ, with a Fourier expansion

$$a(\tau) = \frac{1}{\sqrt{\hbar\beta}} \sum_{m=-\infty}^{\infty} a_m e^{-i\omega_m \tau}, \quad \omega_m = 2\pi m/\hbar\beta.$$

(7.312)

Note that in contrast to the coefficients x_m in expansion (7.303), a_m and a_{-m} are *independent* of each other, since $a(\tau)$ is complex. The periodicity of $a(\tau)$ arises as follows: In the time-sliced path integral derived in the x-basis with integration variables $x_0, \ldots, x_{N+1}$ and $p_1, \ldots, p_{N+1}$, we introduce a fictitious momentum variable p_0 which is set identically equal to p_{N+1}. Then the time-sliced $\int_0^{\hbar\beta} d\tau p\dot{x}$ term, $\sum_{n=1}^{N+1} p_n \overline{\nabla} x_n$, can be replaced by $-\sum_{n=1}^{N+1} x_n \nabla p_n$ [see the rule of partial integration on the lattice, Eq. (2.98)] or by $-\sum_{n=1}^{N+1} x_n \overline{\nabla} p_n$. The first term in the time-sliced action (7.310) arises by symmetrizing the above two lattice sums.

In order to perform the integrals in (7.309), we make use of the Gaussian formula valid for $\text{Re}\, A > 0$,

$$\int \frac{da_n^* \, da_n}{\pi} e^{-a_n^* A_n a_n} = \frac{1}{A_n}, \quad \text{Re}\, A_n > 0.$$

(7.313)

By taking a product of N of these, we have

$$\prod_{n=0}^{N} \left[\int \frac{da_n^* \, da_n}{\pi}\right] e^{-\sum_n a_n^* A_n a_n} = \prod_{n=1}^{N+1} \frac{1}{A_n}, \quad \text{Re}\, A_n > 0.$$

(7.314)

This is obviously a special case of the matrix formula

$$Z = \prod_{n=0}^{N} \left[\int \frac{da_n^* \, da_n}{\pi}\right] e^{-\sum_{n,m} a_n^* A_{nm} a_m} = \frac{1}{\det A},$$

(7.315)

in which the matrix $A = A^d$ has only diagonal elements with a positive real part. Now we observe that the measure of integration is certainly invariant under any unitary transformation of the components a_n:

$$a_n \to \sum_{n'} U_{n,n'} a_{n'}.$$

(7.316)

So is the determinant of A:

$$\det A \to \det\left(U A^d U^\dagger\right) = \det A^d.$$

(7.317)

But then formula (7.315) holds for any matrix A which can be diagonalized by a unitary transformation and has only eigenvalues with a positive real part. In the present case, the possibility of diagonalizing A is guaranteed by the fact that A satisfies $AA^\dagger = A^\dagger A$, i.e., it is a *normal* matrix. This property makes the Hermitian and anti-Hermitian parts of A commute with each other, allowing them to be diagonalized simultaneously.

In the partition function (7.309), the $(N+1) \times (N+1)$ matrix A has the form

$$
A = \epsilon(1 - \epsilon\omega)\overline{\nabla} + \epsilon\omega = \begin{pmatrix} 1 & 0 & 0 & \cdots & 0 & -1+\epsilon\omega \\ -1+\epsilon\omega & 1 & 0 & \cdots & 0 & 0 \\ 0 & -1+\epsilon\omega & 1 & \cdots & 0 & 0 \\ 0 & 0 & -1+\epsilon\omega & \cdots & 0 & 0 \\ \vdots & & & & & \vdots \\ 0 & 0 & 0 & \cdots & -1+\epsilon\omega & 1 \end{pmatrix}.
$$

$$(7.318)$$

This matrix acts on a complex vector space. Its determinant can immediately be calculated by a repeated expansion along the first row, giving

$$
\det{}_{N+1}A = 1 - (1 - \epsilon\omega)^{N+1}.
\tag{7.319}
$$

Hence we obtain the time-sliced partition function

$$
Z_\omega^N = \frac{1}{\det_{N+1}[\epsilon(1-\epsilon\omega)\overline{\nabla} + \epsilon\omega]} = \frac{1}{1 - (1 - \epsilon\omega)^{N+1}}.
\tag{7.320}
$$

It is useful to introduce the auxiliary frequency

$$
\bar\omega_{\rm e} \equiv -\frac{1}{\epsilon}\log(1 - \epsilon\omega).
\tag{7.321}
$$

The subscript e records the Euclidean nature of the time [in analogy with the frequencies $\tilde\omega_{\rm e}$ of Eq. (2.397)]. In terms of $\bar\omega_{\rm e}$, Z_ω^N takes the form

$$
Z_\omega^N = \frac{1}{1 - e^{-\beta\hbar\bar\omega_{\rm e}}}.
\tag{7.322}
$$

This is the well-known partition function of Bose particles for a single state of energy $\bar\omega_{\rm e}$. It has the expansion

$$
Z_\omega = 1 + e^{-\beta\hbar\bar\omega_{\rm e}} + e^{-2\beta\hbar\bar\omega_{\rm e}} + \dots \, ,
\tag{7.323}
$$

in which the nth term exhibits the Boltzmann factor for an occupation of a particle state by n particles, in accordance with the Hamiltonian operator

$$
\hat H_\omega = \hbar\bar\omega_{\rm e}\hat N = \hbar\bar\omega_{\rm e}a^\dagger a.
\tag{7.324}
$$

In the continuum limit $\epsilon \to 0$, the auxiliary frequency tends to ω,

$$\bar{\omega}_e \xrightarrow{\epsilon \to 0} \omega, \tag{7.325}$$

and Z_ω^N reduces to

$$Z_\omega = \frac{1}{1 - e^{-\beta\hbar\omega}}. \tag{7.326}$$

The generalization of the partition function to a system with a time-dependent frequency $\Omega(\tau)$ reads

$$Z_\omega^N = \prod_{n=0}^{N} \left[\int \frac{da_n^\dagger da_n}{\pi} \right] \exp\left\{ -\frac{1}{\hbar} \mathcal{A}^N \right\}, \tag{7.327}$$

with the sliced action

$$\mathcal{A}_\omega^N = \hbar \sum_{n=1}^{N} \left[a_n^*(a_n - a_{n-1}) + \epsilon \Omega_n a_n^* a_{n-1} \right], \tag{7.328}$$

or, expressed in terms of the difference operator $\overline{\nabla}$,

$$\mathcal{A}_\omega^N = \hbar\epsilon \sum_{n=1}^{N} a_n^* \left[(1 - \epsilon\Omega_n)\overline{\nabla} + \Omega_n \right] a_n. \tag{7.329}$$

The result is

$$Z_\omega^N = \frac{1}{\det_{N+1}[\epsilon(1 - \epsilon\Omega)\overline{\nabla} + \epsilon\Omega]} = \frac{1}{1 - \prod_{n=0}^{N}(1 - \epsilon\Omega_n)}. \tag{7.330}$$

Here we introduce the auxiliary frequency

$$\bar{\Omega}_e \equiv -\frac{1}{(N+1)\epsilon} \sum_{n=0}^{N} \log(1 - \epsilon\Omega_n), \tag{7.331}$$

which brings Z_ω^N to the form

$$Z_\omega^N = \frac{1}{1 - e^{-\beta\hbar\bar{\Omega}_e}}. \tag{7.332}$$

For comparison, let us also evaluate the path integral directly in the continuum limit. Then the difference operator (7.318) becomes the differential operator

$$(1 - \epsilon\omega)\overline{\nabla} + \omega \to \partial_\tau + \omega, \tag{7.333}$$

acting on periodic complex functions $e^{-i\omega_m \tau}$ with the Matsubara frequencies ω_m. Hence the continuum partition function of a harmonic oscillator could be written as

$$\begin{aligned} Z_\omega &= \oint \frac{\mathcal{D}a^*\mathcal{D}a}{\pi} \exp\left[-\int_0^{\hbar\beta} d\tau \, (a^*\partial_\tau a + \omega a^* a) \right] \\ &= \mathcal{N}_\omega \frac{1}{\det(\partial_\tau + \omega)}. \end{aligned} \tag{7.334}$$

The normalization constant is fixed by comparison with the time-sliced result. The operator $\partial_\tau + \omega$ has the eigenvalues $-i\omega_m + \omega$. The product of these is calculated by considering the ratios with respect to the $\omega = 0$ -values

$$\prod_{m=-\infty,\neq 0}^{\infty} \frac{-i\omega_m + \omega}{-i\omega_m} = \frac{\sinh(\hbar\omega\beta/2)}{\hbar\omega\beta/2}. \tag{7.335}$$

This product is the ratio of functional determinants

$$\frac{\det(\partial_\tau + \omega)}{\det'(\partial_\tau)} = \omega \frac{\sinh(\hbar\omega\beta/2)}{\hbar\omega\beta/2}, \tag{7.336}$$

where the prime on the determinant with $\omega = 0$ denotes the omission of the zero frequency $\omega_0 = 0$ in the product of eigenvalues; the prefactor ω accounts for this.

Note that this ratio formula of continuum fluctuation determinants gives naturally only the harmonic oscillator partition function (7.306), not the second-quantized one (7.322). Indeed, after fixing the normalization factor $\mathcal{N}_\omega$ in (7.334), the path integral in the continuum formulation can be written as

$$
\begin{aligned}
Z_\omega &= \oint \frac{\mathcal{D}a^*\mathcal{D}a}{\pi} \exp\left[-\int_0^{\hbar\beta} d\tau\,(a^*\partial_\tau a + \omega a^* a)\right] \\
&= \frac{k_B T}{\hbar} \frac{\det'(\partial_\tau)}{\det(\partial_\tau + \omega)} = \frac{1}{2\sinh(\hbar\omega\beta/2)}.
\end{aligned} \tag{7.337}
$$

In the continuum, the relation with the oscillator fluctuation factor can be established most directly by observing that in the determinant, the operator $\partial_\tau + \omega$ can be replaced by the conjugate operator $-\partial_\tau + \omega$, since all eigenvalues come in complex-conjugate pairs, except for the $m = 0$ -value, which is real. Hence the determinant of $\partial_\tau + \omega$ can be substituted everywhere by

$$\det(\partial_\tau + \omega) = \det(-\partial_\tau + \omega) = \sqrt{\det(-\partial_\tau^2 + \omega^2)}, \tag{7.338}$$

rewriting the partition function (7.337) as

$$
\begin{aligned}
Z_\omega &= \frac{k_B T}{\hbar} \frac{\det'(\partial_\tau)}{\det(\partial_\tau + \omega)} \\
&= \frac{k_B T}{\hbar} \left[\frac{\det'(-\partial_\tau^2)}{\det(-\partial_\tau^2 + \omega^2)}\right]^{1/2} = \frac{1}{2\sinh(\hbar\omega\beta/2)},
\end{aligned} \tag{7.339}
$$

where the second line contains precisely the oscillator expressions (2.394).

A similar situation holds for an arbitrary time-dependent frequency where the partition function is

$$
\begin{aligned}
Z_\omega &= \oint \frac{\mathcal{D}a^*(\tau)\mathcal{D}a(\tau)}{\pi} \exp\left\{-\int_0^{\hbar\beta} d\tau\,\left[a^\dagger\partial_\tau a + \Omega(\tau)a^\dagger a\right]\right\} \\
&= \frac{k_B T}{\hbar} \left[\frac{\det'(-\partial_\tau^2)}{\det(-\partial_\tau^2 + \Omega^2(\tau))}\right]^{1/2} = \frac{1}{2\sinh(\hbar\omega\beta/2)} \left[\frac{\det(-\partial_\tau^2 + \omega^2)}{\det(-\partial_\tau^2 + \Omega^2(\tau))}\right]^{1/2}. \tag{7.340}
\end{aligned}
$$

While the oscillator partition function can be calculated right-away in the continuum limit after forming ratios of eigenvalues, the second-quantized path integral depends sensitively on the choice $a_n^\dagger a_{n-1}$ in the action (7.310). It is easy to verify that the alternative slicings $a_{n+1}^\dagger a_n$ and $a_n^\dagger a_n$ would have led to the partition functions $[e^{\beta\hbar\omega} - 1]^{-1}$ and $[2\sinh(\hbar\omega\beta/2)]^{-1}$, respectively. The different time slicings produce obviously the same physics as the corresponding time-ordered Hamiltonian operators $\hat{H} = \hat{T}\hat{a}^\dagger(t)\hat{a}(t')$ in which t' approaches t once from the right, once from the left, and once symmetrically from both sides.

It is easy to decide which of these mathematically possible approaches is the physically correct one. Classical mechanics is invariant under canonical transformations. Thus we require that path integrals have the same invariance. Since the classical actions (7.300) and (7.301) arise from oscillator actions by the canonical transformation (7.296), the associated partition functions must be the same. This fixes the time-slicing to the symmetric one.

Another argument in favor of this symmetric ordering was given in Subsection 2.15.4. We shall see that in order to ensure invariance of path integrals under coordinate transformations, which is guaranteed in Schrödinger theory, path integrals should be defined by dimensional regularization. In this framework, the symmetric fixing emerges automatically.

It must, however, be pointed out that the symmetric fixing gives rise to an important and poorly understood physical problem in many-body theory. Since each harmonic oscillator in the world has a ground-state energy ω, each momentum state of each particle field in the world possesses a nonzero vacuum energy $\hbar\omega$ (thus for each element in the periodic system). This would lead to a divergence in the cosmological constant, and thus to a catastrophic universe. So far, the only idea to escape this is to imagine that the universe contains for each Bose field a Fermi field which, as we shall see in Eq. (7.427), contributes a negative vacuum energy to the ground state. Some people have therefore proposed that the world is described by a theory with a broken supersymmetry, where an underlying supersymmetric action contains fermions and bosons completely symmetrically. Unfortunately, all theories proposed so far possess completely unphysical particle spectra.

7.8   Coherent States

As long as we calculate the partition function of the harmonic oscillator in the variables $a^*(\tau)$ and $a(\tau)$, the path integrals do not differ from those of the harmonic-oscillator (except for the possibly absent ground-state energy). The situation changes if we want to calculate the path integral (7.334) for specific initial and final values $a_a = a(\tau_a)$ and $a_b = a(\tau_b)$, implying also $a_a^* = a^*(\tau_a)$ and $a_b^* = a^*(\tau_b)$ by complex conjugation. In the definition of the canonical path integral in Section 2.1 we had to choose between measures (2.44) and (2.45), depending on which of the two completeness relations

$$\int dx \, |x\rangle\langle x| = 1, \qquad \int \frac{dp}{2\pi} \, |p\rangle\langle p| = 1 \qquad (7.341)$$

we wanted to insert into the factorized operator version of the Boltzmann factor $e^{-\beta\hat{H}}$ into products of $e^{-\epsilon\hat{H}}$. The time-sliced path integral (7.309), on the other hand, runs over $a^*(\tau)$ and $a(\tau)$ corresponding to an apparent completeness relation

$$\int dx \frac{dp}{2\pi} |x\,p\rangle\langle x\,p| = 1. \tag{7.342}$$

This resolution of the identity is at first sight surprising, since in a quantum-mechanical system *either* x *or* p can be specified, but not both. Thus we expect (7.342) to be structurally different from the completeness relations in (7.341). In fact, (7.342) may be called an *overcompleteness relation*.

In order to understand this, we form coherent states [49] similar to those used earlier in Eq. (3A.5) [49]:

$$|z\rangle \equiv e^{z\hat{a}^\dagger - z^*\hat{a}}|0\rangle, \qquad \langle z| \equiv \langle 0|e^{-z\hat{a}^\dagger + z^*\hat{a}}. \tag{7.343}$$

The Baker-Campbell-Hausdorff formula (2.9) allows us to rewrite

$$e^{z\hat{a}^\dagger - z^*\hat{a}} = e^{z^*z[\hat{a}^\dagger,\hat{a}]/2}e^{z\hat{a}^\dagger}e^{-z^*\hat{a}} = e^{-z^*z/2}e^{z\hat{a}^\dagger}e^{-z^*\hat{a}}. \tag{7.344}$$

Since $\hat{a}$ annihilates the vacuum state, we may expand

$$|z\rangle = e^{-z^*z/2}e^{z\hat{a}^\dagger}|0\rangle = e^{-z^*z/2}\sum_{n=0}^{\infty}\frac{z^n}{\sqrt{n!}}|n\rangle. \tag{7.345}$$

The states $|n\rangle$ and $\langle n|$ can be recovered from the coherent states $|z\rangle$ and $\langle z|$ by the operations:

$$|n\rangle = \left[|z\rangle e^{z^*z/2}\overleftarrow{\partial}_z^n\right]_{z=0}\frac{1}{\sqrt{n!}}, \qquad \langle n| = \frac{1}{\sqrt{n!}}\left[\partial_{z^*}^n e^{z^*z/2}\langle z|\right]_{z=0}. \tag{7.346}$$

For an operator $\hat{O}$, the trace can be calculated from the integral over the diagonal elements

$$\text{tr}\,\hat{O} = \int \frac{dz^*dz}{\pi}\langle z|\hat{O}|z\rangle = \int \frac{dz^*dz}{\pi}e^{-z^*z}\sum_{m,n=0}^{\infty}\frac{z^{*m}}{\sqrt{m!}}\frac{z^n}{\sqrt{n!}}\langle m|\hat{O}|n\rangle. \tag{7.347}$$

Setting $z = re^{i\phi}$, this becomes

$$\text{tr}\,\hat{O} = \int dr^2 \left[\frac{d\phi}{2\pi}e^{-i(m-n)\phi}\right]e^{-r^2}\sum_{m,n=0}^{\infty}\left(r^2\right)^{(m+n)/2}\frac{1}{\sqrt{m!}}\frac{1}{\sqrt{n!}}\langle m|\hat{O}|n\rangle. \tag{7.348}$$

The integral over ϕ gives a Kronecker symbol $\delta_{m,n}$ and the integral over r^2 cancels the factorials, so that we remain with the diagonal sum

$$\text{tr}\,\hat{O} = \int dr^2 e^{-r^2}\sum_{n=0}^{\infty}\left(r^2\right)^n\frac{1}{n!}\langle n|\hat{O}|n\rangle = \sum_{n=0}^{\infty}\langle n|\hat{O}|n\rangle. \tag{7.349}$$

The sum on the right-hand side of (7.345) allows us to calculate immediately the scalar product of two such states:

$$\langle z_1 | z_2 \rangle = e^{-z_1^* z_1/2 - z_2^* z_2/2 + z_1^* z_2}. \tag{7.350}$$

We identify the states in formula (7.342) with these coherent states:

$$|x\,p\rangle \equiv |z\rangle, \quad \text{where } z \equiv (x + ip)/\sqrt{2}. \tag{7.351}$$

Then (7.345) can be written as

$$|x\,p\rangle = e^{-(x^2 + p^2)/4} \sum_{n=0}^{\infty} \frac{(x + ip)^n}{\sqrt{2^n n!}} |0\rangle, \tag{7.352}$$

and

$$\int dx \frac{dp}{2\pi} |x\,p\rangle\langle x\,p| = \int dx \frac{dp}{2\pi} |x\,p\rangle\langle x\,p| e^{-(x^2 + p^2)/2} \sum_{m,n=0}^{\infty} \frac{(x - ip)^m}{\sqrt{2^m m!}} \frac{(x + ip)^n}{\sqrt{2^n n!}} |m\rangle\langle n|. \tag{7.353}$$

Setting $(x - ip)/\sqrt{2} \equiv re^{i\phi}$, this can be rewritten as

$$\int dx \frac{dp}{2\pi} |x\,p\rangle\langle x\,p| = \int dr^2 \left[\frac{d\phi}{2\pi} e^{-i(m-n)\phi} \right] e^{-r^2} \sum_{m,n=0}^{\infty} \left(r^2 \right)^{(m+n)/2} \frac{1}{\sqrt{m!}} \frac{1}{\sqrt{n!}} |m\rangle\langle n|. \tag{7.354}$$

The angular integration enforces $m = n$, and the integrals over r^2 cancel the factorials, as in (7.348), thus proving the resolution of the identity (7.342), which can also be written as

$$\int \frac{dz^* dz}{\pi} |z\rangle\langle z| = 1. \tag{7.355}$$

This resolution of the identity can now be inserted into a product decomposition of a Boltzmann operator

$$\langle z_b | e^{-\beta \hat{H}_\omega} | z_a \rangle = \langle z_b | e^{-\beta \hat{H}_\omega/(N+1)} e^{-\beta \hat{H}_\omega/(N+1)} \cdots e^{-\beta \hat{H}_\omega/(N+1)} | z_a \rangle, \tag{7.356}$$

to arrive at a sliced path integral [compare (2.2)–(2.4)]

$$\langle z_b | e^{-\beta \hat{H}_\omega} | z_a \rangle = \prod_{n=1}^{N} \left[\int \frac{dz_n^* dz_n}{\pi} \right] \prod_{n=1}^{N+1} \langle z_n | e^{-\epsilon \hat{H}_\omega} | z_{n-1} \rangle, \quad z_0 = z_a, \; z_{N+1} = z_b, \; \epsilon \equiv \beta/(N+1). \tag{7.357}$$

We now calculate the matrix elements $\langle z_n | e^{-\epsilon \hat{H}_\omega} | z_{n-1} \rangle$ and find

$$\langle z_n | e^{-\epsilon \hat{H}_\omega} | z_{n-1} \rangle \approx \langle z_n | 1 - \epsilon \hat{H}_\omega | z_{n-1} \rangle = \langle z_n | z_{n-1} \rangle - \epsilon \langle z_n | \hat{H}_\omega | z_{n-1} \rangle. \tag{7.358}$$

Using (7.350) we find

$$\langle z_n | z_{n-1} \rangle = e^{-z_n^* z_n/2 - z_{n-1}^* z_{n-1}/2 + z_n^* z_{n-1}} = e^{-(1/2) \left[z_n^* (z_n - z_{n-1}) - \left(z_n^* - z_{n-1}^* \right) z_{n-1} \right]}. \tag{7.359}$$

The matrix elements of the operator Hamiltonian (7.298) is easily found. The coherent states (7.345) are eigenstates of the annihilation operator $\hat{a}$ with eigenvalue z:

$$\hat{a}|z\rangle = e^{-z^*z/2} \sum_{n=0}^{\infty} \frac{z^n}{\sqrt{n!}} \hat{a}|n\rangle = e^{-z^*z/2} \sum_{n=1}^{\infty} \frac{z^n}{\sqrt{(n-1)!}} |n-1\rangle = z|z\rangle. \qquad (7.360)$$

Thus we find immediately

$$\langle z_n|\hat{H}_\omega|z_{n-1}\rangle = \hbar\omega\langle z_n|(\hat{a}^\dagger\hat{a} + \hat{a}\,\hat{a}^\dagger)|z_{n-1}\rangle = \hbar\omega\left(z_n^\dagger z_{n-1} + \frac{1}{2}\right). \qquad (7.361)$$

Inserting this together with (7.359) into (7.358), we obtain for small ϵ the path integral

$$\langle z_b|e^{-\beta\hat{H}_\omega}|z_a\rangle = \prod_{n=1}^{N}\left[\int \frac{dz_n^* dz_n}{\pi}\right] e^{-\mathcal{A}_\omega^N[z^*,z]/\hbar}, \qquad (7.362)$$

with the time-sliced action

$$\mathcal{A}_\omega^N[z^*,z] = \hbar\epsilon \sum_{n=1}^{N+1}\left\{\frac{1}{2}\left[z_n^*\overline{\nabla}z_n - (\overline{\nabla}z_n^*)z_{n-1}\right] + \omega\left(z_n^*z_{n-1} + \frac{1}{2}\right)\right\}. \qquad (7.363)$$

The gradient terms can be regrouped using formula (2.35), and rewriting its right-hand side as $p_{N+1}x_{N+1} - p_0 x_0 + \sum_{n=1}^{N+1}(p_n - p_{n-1})x_{n-1}$. This leads to

$$\mathcal{A}_\omega^N[z^*,z] = \frac{\hbar}{2}(-z_b^* z_b + z_a^* z_a) + \hbar\epsilon \sum_{n=1}^{N+1}\left\{z_n^*\overline{\nabla}z_n + \omega\left(z_n^*z_{n-1} + \frac{1}{2}\right)\right\}. \qquad (7.364)$$

Except for the surface terms which disappear for periodic paths, this action agrees with the time-sliced Euclidean action (7.310), except for a trivial change of variables $a \rightarrow z$.

As a brief check of formula (7.362) we set $N = 0$ and find

$$\mathcal{A}_\omega^0[z^*,z] = \frac{\hbar}{2}(-z_b^* z_b + z_a^* z_a) + \hbar z_b^*(z_b - z_a) + \omega\left(z_b^* z_a + \frac{1}{2}\right), \qquad (7.365)$$

and the short-time amplitude (7.364) becomes

$$\langle z_b|e^{-\epsilon\hat{H}_\omega}|z_a\rangle = \exp\left[-\frac{1}{2}(z_b^* z_b + z_a^* z_a) + z_b^* z_a - \epsilon\hbar\omega\left(z_b^* z_a + \frac{1}{2}\right)\right]. \qquad (7.366)$$

Applying the recovery operations (7.346) we find

$$\langle 0|e^{-\epsilon\hat{H}_\omega}|0\rangle = \left[e^{(z_b^* z_b + z_a^* z_a)/2}\langle z_b|e^{-\epsilon\hat{H}_\omega}|z_a\rangle\right]_{z^*=0,z=0} = e^{-\epsilon\hbar\omega/2}, \qquad (7.367)$$

$$\langle 1|e^{-\epsilon\hat{H}_\omega}|1\rangle = \left\{\partial_z\left[e^{(z_b^* z_b + z_a^* z_a)/2}\langle z_b|e^{-\epsilon\hat{H}_\omega}|z_a\rangle\right]\overleftarrow{\partial}_z^*\right\}_{z^*=0,z=0} = e^{-\epsilon 3\hbar\omega/2}, \qquad (7.368)$$

$$\langle 0|e^{-\epsilon\hat{H}_\omega}|1\rangle = \langle 1|e^{-\epsilon\hat{H}_\omega}|0\rangle = 0. \qquad (7.369)$$

Thus we have shown that for fixed ends, the path integral gives the amplitude for an initial coherent state $|z_a\rangle$ to go over to a final coherent state $|z_b\rangle$. The partition function (7.337) is obtained from this amplitude by forming the diagonal integral

$$Z_\omega = \int \frac{dz^* dz}{\pi}\langle z|e^{-\beta\hat{H}_\omega}|z\rangle. \qquad (7.370)$$

7.9   Second-Quantized Fermi Fields

The existence of the periodic system of elements is based on the fact that electrons can occupy each orbital state only once (counting spin-up and -down states separately). Particles with this statistics are called *fermions*. In the above Hilbert space in which n-particle states at a point $\mathbf{x}$ are represented by oscillator states $|n, \mathbf{x}\rangle$, this implies that the particle occupation number n can take only the values

$$n = 0 \quad \text{(no electron)},$$
$$n = 1 \quad \text{(one electron)}.$$

It is possible to construct such a restricted many-particle Hilbert space explicitly by subjecting the quantized fields $\hat{\psi}^\dagger(\mathbf{x})$, $\hat{\psi}(\mathbf{x})$ or their Fourier components $\hat{a}_{\mathbf{p}}^\dagger$, $\hat{a}_{\mathbf{p}}$ to *anticommutation relations*, instead of the commutation relations (7.282), i.e., by postulating

$$[\hat{\psi}(\mathbf{x}, t), \hat{\psi}^\dagger(\mathbf{x}', t)]_+ = \delta_{\mathbf{x}\mathbf{x}'},$$
$$[\hat{\psi}^\dagger(\mathbf{x}, t), \hat{\psi}^\dagger(\mathbf{x}', t)]_+ = 0,$$
$$[\hat{\psi}(\mathbf{x}, t), \hat{\psi}(\mathbf{x}', t)]_+ = 0, \qquad (7.371)$$

or for the Fourier components

$$[\hat{a}_{\mathbf{p}}(t), \hat{a}_{\mathbf{p}'}^\dagger(t)]_+ = \delta_{\mathbf{p}\mathbf{p}'},$$
$$[\hat{a}_{\mathbf{p}}^\dagger(t), \hat{a}_{\mathbf{p}'}^\dagger(t)]_+ = 0,$$
$$[\hat{a}_{\mathbf{p}}(t), \hat{a}_{\mathbf{p}'}(t)]_+ = 0. \qquad (7.372)$$

Here $[\hat{A}, \hat{B}]_+$ denotes the anticommutator of the operators $\hat{A}$ and $\hat{B}$

$$[\hat{A}, \hat{B}]_+ \equiv \hat{A}\hat{B} + \hat{B}\hat{A}. \qquad (7.373)$$

Apart from the anticommutation relations, the second-quantized description of Fermi fields is completely analogous to that of Bose fields in Section 7.6.

7.10   Fluctuating Fermi Fields

The question arises as to whether it is possible to find a path integral formulation which replaces the anticommuting operator structure. The answer is affirmative, but at the expense of a somewhat unconventional algebraic structure. The fluctuating paths can no longer be taken as c-numbers. Instead, they must be described by anticommuting variables.

7.10.1   Grassmann Variables

Mathematically, such objects are known under the name of *Grassmann variables*. They are defined by the algebraic property

$$\theta_1 \theta_2 = -\theta_2 \theta_1, \qquad (7.374)$$

which makes them *nilpotent*:

$$\theta^2 = 0. \tag{7.375}$$

These variables have the curious consequence that an arbitrary function of them possesses only two Taylor coefficients, F_0 and F_1,

$$F(\theta) = F_0 + F_1\theta. \tag{7.376}$$

They are obtained from $F(\theta)$ as follows:

$$F_0 = F(0), \tag{7.377}$$

$$F_1 = F' \equiv \frac{\partial}{\partial\theta}F.$$

The existence of only two parameters in $F(\theta)$ is the reason why such functions naturally collect amplitudes of two local fermion states, F_0 for zero occupation, F_1 for a single occupation.

It is now possible to define integrals over functions of these variables in such a way that the previous path integral formalism remains applicable without a change in the notation, leading to the same results as the second-quantized theory with anticommutators. Recall that for ordinary real functions, integrals are linear functionals. We postulate this property also for integrals with Grassmann variables. Since an arbitrary function of a Grassmann variable $F(\theta)$ is at most linear in θ, its integral is completely determined by specifying only the two fundamental integrals $\int d\theta$ and $\int d\theta\,\theta$. The values which render the correct physics with a conventional path integral notation are

$$\int \frac{d\theta}{\sqrt{2\pi}} = 0, \tag{7.378}$$

$$\int \frac{d\theta}{\sqrt{2\pi}}\theta = 1. \tag{7.379}$$

Using the linearity property, an arbitrary function $F(\theta)$ is found to have the integral

$$\int \frac{d\theta}{\sqrt{2\pi}}F(\theta) = F_1 = F'. \tag{7.380}$$

Thus, integration of $F(\theta)$ coincides with differentiation. This must be remembered whenever Grassmann integration variables are to be changed: The integral is transformed with the *inverse* of the usual Jacobian. The obvious equation

$$\int \frac{d\theta}{\sqrt{2\pi}}F(c \cdot \theta) = c \cdot F' = c \cdot \int \frac{d\theta'}{\sqrt{2\pi}}F(\theta') \tag{7.381}$$

for any complex number c implies the relation

$$\int \frac{d\theta}{\sqrt{2\pi}}F(\theta'(\theta)) = \int \frac{d\theta'}{\sqrt{2\pi}}\left[\frac{d\theta}{d\theta'}\right]^{-1}F(\theta'). \tag{7.382}$$

For ordinary integration variables, the Jacobian $d\theta/d\theta'$ would appear without the power -1.

When integrating over a product of two functions $F(\theta)$ and $G(\theta)$, the rule of integration by parts holds with the opposite sign with respect to that for ordinary integrals:

$$\int \frac{d\theta}{\sqrt{2\pi}} G(\theta) \frac{\partial}{\partial\theta} F(\theta) = \int \frac{d\theta}{\sqrt{2\pi}} \left[\frac{\partial}{\partial\theta} G(\theta)\right] F(\theta). \tag{7.383}$$

There exists a simple generalization of the Dirac δ-function to Grassmann variables. We shall define this function by the integral identity

$$\int \frac{d\theta'}{\sqrt{2\pi}} \delta(\theta - \theta') F(\theta') \equiv F(\theta). \tag{7.384}$$

Inserting the general form (7.376) for $F(\theta)$, we see that the function

$$\delta(\theta - \theta') = \theta' - \theta \tag{7.385}$$

satisfies (7.384). Note that the δ-function is a Grassmann variable and, in contrast to Dirac's δ-function, antisymmetric. Its derivative has the property

$$\delta'(\theta - \theta') \equiv \partial_\theta \delta(\theta - \theta') = -1. \tag{7.386}$$

It is interesting to see that δ' shares with Dirac's δ' the following property:

$$\int \frac{d\theta'}{\sqrt{2\pi}} \delta'(\theta - \theta') F(\theta') = -F'(\theta), \tag{7.387}$$

with the opposite sign of the Dirac case. This follows from the above rule of partial integration, or simpler, by inserting (7.386) and the explicit decomposition (7.376) for $F(\theta)$.

The integration may be extended to complex Grassmann variables which are combinations of two real Grassmann variables θ_1, θ_2:

$$a^* = \frac{1}{\sqrt{2}}(\theta_1 - i\theta_2), \qquad a = \frac{1}{\sqrt{2}}(\theta_1 + i\theta_2). \tag{7.388}$$

The measure of integration is defined by

$$\int \frac{da^* da}{\pi} \equiv \int \frac{d\theta_2 d\theta_1}{2\pi i} \equiv -\int \frac{da\, da^*}{\pi}. \tag{7.389}$$

Using (7.378) and (7.379) we see that the integration rules for complex Grassmann variables are

$$\int \frac{da^* da}{\pi} = 0, \quad \int \frac{da^* da}{\pi} a = 0, \quad \int \frac{da^* da}{\pi} a^* = 0, \tag{7.390}$$

$$\int \frac{da^* da}{\pi} a^* a = \int \frac{d\theta_2 d\theta_1}{2\pi i} i\theta_1 \theta_2 = 1. \tag{7.391}$$

Every function of a^*a has at most two terms:

$$F(a^*a) = F_0 + F_1 \, a^*a. \tag{7.392}$$

In particular, the exponential $\exp\{-a^*Aa\}$ with a complex number A has the Taylor series expansion

$$e^{-a^*Aa} = 1 - a^*Aa. \tag{7.393}$$

Thus we find the following formula for the Gaussian integral:

$$\int \frac{da^* \, da}{\pi} e^{-a^*Aa} = A. \tag{7.394}$$

The integration rule (7.390) can be used directly to calculate the Grassmann version of the product of integrals (7.315). For a matrix A which can be diagonalized by a unitary transformation, we obtain directly

$$Z^{\mathrm{f}} = \prod_n \left[\int \frac{da_n^* da_n}{\pi} \right] e^{i\Sigma_{n,n'} a_n^* A_{n,n'} a_{n'}} = \det A. \tag{7.395}$$

Remarkably, the fermion integration yields precisely the inverse of the boson result (7.315).

7.10.2 Fermionic Functional Determinant

Consider now the time-sliced path integral of the partition function written like (7.309) but with fermionic anticommuting variables. In order to find the same results as in operator quantum mechanics it is necessary to require the anticommuting Grassmann fields $a(\tau), a^*(\tau)$ to be *antiperiodic* on the interval $\tau \in (0, \hbar\beta)$, i.e.,

$$a(\hbar\beta) = -a(0), \tag{7.396}$$

or in the sliced form

$$a_{N+1} = -a_0. \tag{7.397}$$

Then the exponent of (7.395) has the same form as in (7.315), except that the matrix A of Eq. (7.398) is replaced by

$$A^{\mathrm{f}} = \epsilon(1 - \epsilon\omega)\overline{\nabla}_\tau + \epsilon\omega = \begin{pmatrix} 1 & 0 & 0 & \dots & 0 & 1 - \epsilon\omega \\ -1 + \epsilon\omega & 1 & 0 & \dots & 0 & 0 \\ 0 & -1 + \epsilon\omega & 1 & \dots & 0 & 0 \\ 0 & 0 & -1 + \epsilon\omega & \dots & 0 & 0 \\ \vdots & & & & & \vdots \\ 0 & 0 & 0 & \dots & -1 + \epsilon\omega & 1 \end{pmatrix}, \tag{7.398}$$

where the rows and columns are counted from 1 to $N + 1$. The element in the upper right corner is positive and thus has the opposite sign of the bosonic matrix in (7.318). This makes an important difference: While for $\omega = 0$ the bosonic matrix gave

$$\det\left(-\epsilon\overline{\nabla}\right)_{\omega=0} = 0, \tag{7.399}$$

due to translational invariance in τ, we now have

$$\det\left(-\epsilon\overline{\nabla}\right)_{\omega=0} = 2. \tag{7.400}$$

The determinant of the fermionic matrix (7.398) can be calculated by a repeated expansion along the first row and is found to be

$$\det{}_{N+1}A = 1 + (1 - \epsilon\omega)^{N+1}. \tag{7.401}$$

Hence we obtain the time-sliced fermion partition function

$$Z_{\omega}^{\mathrm{f},N} = \det{}_{N+1}[\epsilon(1 - \epsilon\omega)\overline{\nabla} + \epsilon\omega] = 1 + (1 - \epsilon\omega)^{N+1}. \tag{7.402}$$

As in the boson case, we introduce the auxiliary frequency

$$\bar{\omega}_e \equiv -\frac{1}{\epsilon}\log(1 - \epsilon\omega) \tag{7.403}$$

and write $Z_{\omega}^{\mathrm{f},N}$ in the form

$$Z_{\omega}^{N} = 1 + e^{-\beta\hbar\bar{\omega}_e}. \tag{7.404}$$

This partition function displays the typical property of Fermi particles. There are only two terms, one for the zero-particle and one for the one-particle state at a point. Their energies are 0 and $\hbar\bar{\omega}_e$, corresponding to the Hamiltonian operator

$$\hat{H}_{\omega} = \hbar\bar{\omega}_e\hat{N} = \hbar\bar{\omega}_e a^{\dagger}a. \tag{7.405}$$

In the continuum limit $\epsilon \to 0$, where $\bar{\omega}_e \to \omega$, the partition function $Z{\omega}^{N}$ goes over into

$$Z_{\omega} = 1 + e^{-\beta\hbar\omega}. \tag{7.406}$$

Let us generalize also the fermion partition function to a system with a time-dependent frequency $\Omega(\tau)$, where it reads

$$Z_{\omega}^{\mathrm{f},N} = \prod_{n=0}^{N}\left[\int \frac{da_n^* da_n}{\pi}\right]\exp\left(-\frac{1}{\hbar}\mathcal{A}_{\omega}^{N}\right), \tag{7.407}$$

with the sliced action

$$\mathcal{A}_{\omega}^{N} = \hbar\sum_{n=1}^{N}\left[a_n^*(a_n - a_{n-1}) + \epsilon\Omega_n a_n^* a_{n-1}\right], \tag{7.408}$$

or, expressed in terms of the difference operator $\overline{\nabla}$,

$$\mathcal{A}_{\omega}^{N} = \hbar\epsilon\sum_{n=1}^{N} a_n^*\left[(1 - \epsilon\Omega_n)\overline{\nabla} + \Omega_n\right]a_n. \tag{7.409}$$

The result is

$$Z_\omega^{\mathrm{f},N} = \det_{N+1}[\epsilon(1 - \epsilon\Omega)\overline{\nabla} + \epsilon\omega] = 1 - \prod_{n=0}^{N}(1 - \epsilon\Omega_n). \qquad (7.410)$$

As in the bosonic case, it is useful to introduce the auxiliary frequency

$$\bar{\Omega}_e \equiv -\frac{1}{(N+1)\epsilon}\sum_{n=0}^{N}\log(1 - \epsilon\Omega_n), \qquad (7.411)$$

and write $Z_\omega^{\mathrm{f},N}$ in the form

$$Z_\omega^{\mathrm{f},N} = 1 + e^{-\beta\hbar\bar{\Omega}_e}. \qquad (7.412)$$

If we attempt to write down a path integral formula for fermions directly in the continuum limit, we meet the same phenomenon as in the bosonic case. The difference operator (7.398) turns into the corresponding differential operator

$$(1 - \epsilon\omega)\overline{\nabla} + \omega \rightarrow \partial_\tau + \omega, \qquad (7.413)$$

which now acts upon periodic complex functions $e^{-i\omega_m^{\mathrm{f}}\tau}$ with the *odd* Matsubara frequencies

$$\omega_m^{\mathrm{f}} = \pi(2m + 1)k_B T/\hbar, \quad m = 0, \pm1, \pm2, \ldots. \qquad (7.414)$$

The continuum partition function can be written as a path integral

$$\begin{aligned} Z_\omega^{\mathrm{f}} &= \oint \frac{\mathcal{D}a^*\mathcal{D}a}{\pi} \exp\left[-\int_0^{\hbar\beta} d\tau\,(a^*\partial_\tau a + \omega a^* a)\right] \\ &= \mathcal{N}_\omega \det(\partial_\tau + \omega), \end{aligned} \qquad (7.415)$$

with some normalization constant $\mathcal{N}_\omega$ determined by comparison with the time-sliced result. To calculate Z_ω^{f}, we take the eigenvalues of the operator $\partial_\tau + \omega$, which are now $-i\omega_m^{\mathrm{f}} + \omega$, and evaluate the product of ratios

$$\prod_{m=-\infty}^{\infty} \frac{-i\omega_m^{\mathrm{f}} + \omega}{-i\omega_m^{\mathrm{f}}} = \cosh(\hbar\omega\beta/2). \qquad (7.416)$$

This corresponds to the ratio of functional determinants

$$\frac{\det(\partial_\tau + \omega)}{\det(\partial_\tau)} = \cosh(\hbar\omega\beta/2). \qquad (7.417)$$

In contrast to the boson case (7.336), no prime is necessary on the determinant of ∂_τ since there is no zero frequency in the product of eigenvalues (7.416). Setting $\mathcal{N}_\omega = 1/2\det(\partial_\tau)$, the ratio formula produces the correct partition function

$$Z_\omega^{\mathrm{f}} = 2\cosh(\hbar\omega\beta/2). \qquad (7.418)$$

Thus we may write the free-fermion path integral in the continuum form explicitly as follows:

$$Z_\omega^{\mathrm{f}} = \oint \frac{\mathcal{D}a^*\mathcal{D}a}{\pi} \exp\left[-\int_0^{\hbar\beta} d\tau\,(a^*\partial_\tau a + \omega a^* a)\right] = 2\frac{\det(\partial_\tau + \omega)}{\det(\partial_\tau)}$$
$$= 2\cosh(\hbar\omega\beta/2). \tag{7.419}$$

The determinant of the operator $\partial_\tau + \omega$ can again be replaced by

$$\det(\partial_\tau + \omega) = \det(-\partial_\tau + \omega) = \sqrt{\det(-\partial_\tau^2 + \omega^2)}. \tag{7.420}$$

As in the bosonic case, this Fermi analog of the harmonic oscillator partition function agrees with the results of dimensional regularization in Subsection 2.15.4 which will ensure invariance of path integrals under a change of variables, as will be seen in Section 10.6. The proper fermionic time-sliced partition function corresponding to the dimensional regularization in Subsection 2.15.4 is obtained from a fermionic version of the time-sliced oscillator partition function by evaluating

$$Z_\omega^{\mathrm{f},N} = \left[\det_{N+1}(-\epsilon^2\nabla\bar{\nabla} + \epsilon^2\omega^2)\right]^{1/2} = \prod_{m=0}^{N}\left[\epsilon^2\Omega_m^{\mathrm{f}}\bar{\Omega}_m^{\mathrm{f}} + \epsilon^2\omega^2\right]^{1/2}$$
$$\equiv \prod_{m=0}^{N}\left[2(1 - \cos\omega_m^{\mathrm{f}}\epsilon) + \epsilon^2\omega^2\right]^{1/2} = \prod_{m=0}^{N}\left[2\sin^2\frac{\epsilon\omega_m^{\mathrm{f}}}{2} + \epsilon^2\omega^2\right]^{1/2}, \tag{7.421}$$

with a product over the odd Matsubara frequencies ω_m^{f}. The result is

$$Z_\omega^{\mathrm{f},N} = 2\cosh(\hbar\tilde{\omega}_{\mathrm{e}}\beta), \tag{7.422}$$

with $\tilde{\omega}_{\mathrm{e}}$ given by

$$\sinh(\tilde{\omega}_{\mathrm{e}}/2) = \epsilon\omega/2. \tag{7.423}$$

This follows from the Fermi analogs of the product formulas (2.398), (2.400):[7]

$$\prod_{m=0}^{N/2-1}\left(1 - \frac{\sin^2 x}{\sin^2\frac{(2m+1)\pi}{2(N+1)}}\right) = \frac{\cos(N+1)x}{\cos x}, \quad N = \text{even}, \tag{7.424}$$

$$\prod_{m=0}^{(N-1)/2}\left(1 - \frac{\sin^2 x}{\sin^2\frac{(2m+1)\pi}{2(N+1)}}\right) = \cos(N+1)x, \quad N = \text{odd}. \tag{7.425}$$

For odd N, where all frequencies occur twice, we find from (7.425) that

$$\prod_{m=0}^{N}\left(1 - \frac{\sin^2 x}{\sin^2\frac{(2m+1)\pi}{2(N+1)}}\right)^{1/2} = \cos(N+1)x, \tag{7.426}$$

---

[7]I.S. Gradshteyn and I.M. Ryzhik, op. cit., Formulas 1.391.2, 1.391.4.

and thus, with (7.423), directly (7.422). For even N, where the frequency with $m = N/2$ occurs only once, formula (7.424) gives once more the same answer, thus proving (7.426) for even *and* odd N.

There exists no real fermionic oscillator action since x^2 and $\dot{x}^2$ would vanish identically for fermions, due to the nilpotency (7.375) of Grassmann variables. The product of eigenvalues in Eq. (7.421) emerges naturally from a path integral in which the action (7.408) is replaced by a symmetrically sliced action.

An important property of the partition function (7.418) of (7.422) is that the ground-state energy is *negative*:

$$E^{(0)} = -\frac{\hbar\omega}{2}. \tag{7.427}$$

As discussed at the end of Section 7.7, such a fermionic vacuum energy is required for each bosonic vacuum energy to avoid an infinite vacuum energy of the world, which would produce an infinite cosmological constant, whose experimentally observed value is extremely small.

7.10.3 Coherent States for Fermions

For the bosonic path integral (7.304) we have studied in Section 7.8, the case that the endpoint values $a_a = a(\tau_a)$ and $a_b = a(\tau_b)$ of the paths $a(\tau)$ are held fixed. The result was found to be the matrix element of the Boltzmann operator $e^{-\beta\hat{H}_\omega}$ between coherent states $|a\rangle = e^{-a^*a/2}e^{a\hat{a}^\dagger}|0\rangle$ [recall (7.345)]. There exists a similar interpretation for the fermion path integral (7.415) if we hold the endpoint values $a_a = a(\tau_a)$ and $a_b = a(\tau_b)$ of the Grassmann paths fixed. By analogy with Eq. (7.345) we introduce coherent states [50]

$$|\zeta\rangle \equiv e^{-\zeta^*\zeta/2}e^{a^\dagger\zeta}|0\rangle = e^{-\zeta^*\zeta/2}\Big(|0\rangle - \zeta|1\rangle\Big). \tag{7.428}$$

The corresponding adjoint states read

$$\langle\zeta| \equiv e^{-\zeta^*\zeta/2}\langle 0|e^{\zeta^*a} = e^{-\zeta^*\zeta/2}\Big(\langle 0| + \zeta^*\langle 1|\Big). \tag{7.429}$$

Note that for consistency of the formalism, the Grassmann elements ζ anticommute with the fermionic operators. The states $|0\rangle$ and $\langle 1|$ and their conjugates $\langle 0|$ and $\langle 1|$ can be recovered from the coherent states $|\zeta\rangle$ and $\langle\zeta|$ by the operations:

$$|n\rangle = \left[|\zeta\rangle e^{\zeta^*\zeta/2}\overleftarrow{\partial}_\zeta^n\right]_{\zeta=0}\frac{1}{\sqrt{n!}}, \qquad \langle n| = \frac{1}{\sqrt{n!}}\left[\partial_{\zeta^*}^n e^{\zeta^*\zeta/2}\langle\zeta|\right]_{\zeta=0}. \tag{7.430}$$

These formula simplify here to

$$|0\rangle = \left[|\zeta\rangle e^{\zeta^*\zeta/2}\right]_{\zeta=0}, \qquad \langle 0| = \left[e^{\zeta^*\zeta/2}\langle\zeta|\right]_{\zeta=0}, \tag{7.431}$$

$$|1\rangle = \left[|\zeta\rangle e^{\zeta^*\zeta/2}\overleftarrow{\partial}_\zeta\right]_{\zeta=0}, \qquad \langle 1| = \left[\partial_{\zeta^*}e^{\zeta^*\zeta/2}\langle\zeta|\right]_{\zeta=0}. \tag{7.432}$$

For an operator $\hat{\mathcal{O}}$, the trace can be calculated from the integral over the *antidiagonal* elements

$$\text{tr}\,\hat{\mathcal{O}} = \int \frac{d\zeta^* d\zeta}{\pi}\,\langle -\zeta|\hat{\mathcal{O}}|\zeta\rangle = \int \frac{d\zeta^* d\zeta}{\pi}\,e^{-\zeta^*\zeta}\big(\langle 0| - \zeta^*\langle 1|\big)\hat{\mathcal{O}}\big(|0\rangle - \zeta|1\rangle\big). \quad (7.433)$$

Using the integration rules (7.390) and (7.391), this becomes

$$\text{tr}\,\hat{\mathcal{O}} = \langle 0|\hat{\mathcal{O}}|0\rangle + \langle 1|\hat{\mathcal{O}}|1\rangle. \quad (7.434)$$

The states $|\zeta\rangle$ form an overcomplete set in the one-fermion Hilbert space. The scalar products are [compare (7.350)]:

$$\begin{aligned}
\langle \zeta_1|\zeta_2\rangle &= e^{-\zeta_1^*\zeta_1/2 - \zeta_2^*\zeta_2/2 + \zeta_1^*\zeta_2} \\
&= e^{-\zeta_1^*(\zeta_1-\zeta_2)/2 + (\zeta_1^* - \zeta_2^*)\zeta_2/2}.
\end{aligned} \quad (7.435)$$

The resolution of the identity (7.355) is now found as follows [recall (7.390)]:

$$\begin{aligned}
\int \frac{d\zeta^* d\zeta}{\pi}\,|\zeta\rangle\langle\zeta| &= \int \frac{d\zeta^* d\zeta}{\pi} e^{-\zeta^*\zeta}\Big[|0\rangle\langle 0| - \zeta|1\rangle\langle 0| + \zeta^*|0\rangle\langle 1|\Big] \\
&= \int \frac{d\zeta^* d\zeta}{\pi}\Big[|0\rangle\langle 0| + |1\rangle\langle 0|\zeta + \zeta^*|0\rangle\langle 1| + \zeta\zeta^*\big(|0\rangle\langle 0| + |1\rangle\langle 0|\big)\Big] = 1. \quad (7.436)
\end{aligned}$$

We now insert this resolution of the identity into the product of Boltzmann factors

$$\langle \zeta_b|e^{-\beta\hat{H}_\omega}|\zeta_a\rangle = \langle \zeta_b|e^{-\epsilon\hat{H}_\omega}e^{-\epsilon\hat{H}_\omega}\cdots e^{-\epsilon\hat{H}_\omega}|\zeta_a\rangle \quad (7.437)$$

where $\epsilon \equiv \beta/(N+1)$, and obtain by analogy with (7.362) the time-sliced path integral

$$\langle z_b|e^{-\beta\hat{H}_\omega}|z_a\rangle = \prod_{n=1}^{N}\left[\int \frac{dz_n^* dz_n}{\pi}\right] e^{-\mathcal{A}_e[z^*,z]/\hbar}, \quad (7.438)$$

with the a time-sliced action similar to the bosonic one in (7.364):

$$\mathcal{A}_\omega^N[\zeta^*,\zeta] = \frac{\hbar}{2}\big(-\zeta_b^*\zeta_b + \zeta_a^*\zeta_a\big) + \hbar\epsilon\sum_{n=1}^{N+1}\left\{\zeta_n^*\overline{\nabla}\zeta_n + \omega\left(\zeta_n^*\zeta_{n-1} + \frac{1}{2}\right)\right\}. \quad (7.439)$$

Except for the surface term which disappears for antiperiodic paths, this agrees with the time-sliced Euclidean action (7.310), except for a trivial change of variables $a \to \zeta$.

We have shown that as in the Bose case the path integral with fixed ends gives the amplitude for an initial coherent state $|\zeta_a\rangle$ to go over to a final coherent state $|\zeta_b\rangle$. The fermion partition function (7.419) is obtained from this amplitude by forming the trace of the operator $e^{-\beta\hat{H}_\omega}$, which by formula (7.434) is given by the integral over the *antidiagonal* matrix elements

$$Z_\omega^{\text{f}} = \int \frac{d\zeta^* d\zeta}{\pi}\,\langle -\zeta|e^{-\beta\hat{H}_\omega}|\zeta\rangle. \quad (7.440)$$

The antidiagonal matrix elements lead to antiperiodic boundary conditions of the fermionic path integral.

7.11   Hilbert Space of Quantized Grassmann Variable

To understand the Hilbert space associated with a path integral over a Grassmann variable we recall that a path integral with zero Hamiltonian serves to define the Hilbert space via all its scalar products as shown in Eq. (2.18):

$$(x_b t_b | x_a t_a) = \int \mathcal{D}x \int \frac{\mathcal{D}p}{2\pi\hbar} \exp\left[i \int dt\, p(t)\dot{x}(t)\right] = \langle x_b | x_a \rangle = \delta(x_b - x_a). \qquad (7.441)$$

A momentum variable inside the integral corresponds to a derivative operator $\hat{p} \equiv -i\hbar\partial_x$ outside the amplitude, and this operator satisfies with $\hat{x} = x$ the canonical commutation relation $[\hat{p}, \hat{x}] = -i\hbar$ [see (2.19)].

By complete analogy with this it is possible to create the Hilbert space of spinor indices with the help of a path integral over anticommuting Grassmann variables. In order to understand the Hilbert space, we shall consider three different cases.

7.11.1   Single Real Grassmann Variable

First we consider the path integral of a real Grassmann field with zero Hamiltonian

$$\int \frac{\mathcal{D}\theta}{2\pi} \exp\left[\frac{i}{\hbar} \int dt\, \frac{i\hbar}{2}\theta(t)\dot{\theta}(t)\right]. \qquad (7.442)$$

From the Lagrangian

$$\mathcal{L}(t) = \frac{i}{2}\hbar\theta(t)\dot{\theta}(t) \qquad (7.443)$$

we obtain a canonical momentum

$$p_\theta = \frac{\partial \mathcal{L}}{\partial \dot{\theta}} = -\frac{i\hbar}{2}\theta. \qquad (7.444)$$

Note the minus sign in (7.444) arising from the fact that the derivative with resect to $\dot{\theta}$ anticommutes with the variable θ on its left.

The canonical momentum is proportional to the dynamical variable. The system is therefore subject to a constraint

$$\chi = p_\theta + \frac{i\hbar}{2}\theta = 0. \qquad (7.445)$$

In the Dirac classification this is a second-class constraint, in which case the quantization proceeds by forming the classical *Dirac brackets* rather than the Poisson brackets (1.21), and replacing them by $\pm i/\hbar$ times commutation or anticommutation relations, respectively. For n dynamical variables q_i and m constraints χ_p the Dirac brackets are defined by

$$\{A, B\}_D = \{A, B\} - \{A, \chi_p\}C^{pq}\{\chi_q, B\}, \qquad (7.446)$$

where C^{pq} is the inverse of the matrix

$$C^{pq} = \{\chi_p, \chi_q\}. \qquad (7.447)$$

For Grassmann variables p_i, q_i, the Poisson bracket (1.21) carries by definition an overall minus sign if A contains an odd product of Grassmann variables. Applying this rule to the present system we insert $A = p_\theta$ and $B = \theta$ into the Poisson bracket (1.21) we see that it vanishes. The constraint (7.445), on the other hand, satisfies

$$\{\chi, \chi\} = \left\{ p_\theta + \frac{i\hbar}{2}\theta, p_\theta + \frac{i\hbar}{2}\theta \right\} = -i\hbar\{p_\theta, \theta\} = -i\hbar. \tag{7.448}$$

Hence $C = -i\hbar$ with an inverse $i/\hbar$. The Dirac bracket is therefore

$$\{p_\theta, \theta\}_D = \{p_\theta, \theta\} - \frac{i}{\hbar}\{p_\theta, \chi\}\{\chi, \theta\} = 0 - \frac{-i}{\hbar}\left(\frac{i\hbar}{2}\right)(\hbar) = -\frac{\hbar}{2}. \tag{7.449}$$

With the substitution rule $\{A, B\}_D \to (-i/\hbar)[\hat{A}, \hat{B}]_+$, we therefore obtain the canonical equal-time anticommutation relation for this constrained system:

$$[\hat{p}(t), \hat{\theta}(t)]_+ = -\frac{i\hbar}{2}, \tag{7.450}$$

or, because of (7.444),

$$[\hat{\theta}(t), \hat{\theta}(t)]_+ = 1. \tag{7.451}$$

The proportionality of p_θ and θ has led to a factor $1/2$ on the right-hand side with respect to the usual canonical anticommutation relation.

Let $\psi(\theta)$ be an arbitrary wave function of the general form (7.376):

$$\psi(\theta) = \psi_0 + \psi_1\theta. \tag{7.452}$$

The scalar product in the space of all wave functions is defined by the integral

$$\langle\psi'|\psi\rangle \equiv \int \frac{d\theta}{2\pi}\psi'^*(\theta)\psi(\theta) = \psi_0'^*\psi_1 + \psi_1'^*\psi_0 . \tag{7.453}$$

In the so-defined Hilbert space, the operator $\hat{\theta}$ is diagonal, while the operator $\hat{p}$ is given by the differential operator

$$\hat{p} = i\hbar\partial_\theta, \tag{7.454}$$

to satisfy (7.450).

The matrix elements of the operator $\hat{p}$ are

$$\langle\psi'|\hat{p}|\psi\rangle \equiv \int \frac{d\theta}{2\pi}\psi'^*(\theta)i\hbar\frac{\partial}{\partial\theta}\psi(\theta) = i\hbar\psi_1'^*\psi_1. \tag{7.455}$$

By calculating

$$\langle\hat{p}\psi'|\psi\rangle \equiv \int \frac{d\theta}{2\pi}\left[i\hbar\frac{\partial}{\partial\theta}\psi'(\theta)\right]^*\psi(\theta) = -i\hbar\psi_1'^*\psi_1, \tag{7.456}$$

we see that the operator $\hat{p}$ is anti-Hermitian, this being in accordance with the opposite sign in the rule (7.383) of integration by parts.

Let $|\theta\rangle$ be the local eigenstates of which the operator $\hat{\theta}$ is diagonal:

$$\hat{\theta}|\theta\rangle = \theta|\theta\rangle. \tag{7.457}$$

The operator $\hat{\theta}$ is Hermitian, such that

$$\langle\theta|\hat{\theta} = \langle\theta|\theta = \theta\langle\theta|. \tag{7.458}$$

The scalar products satisfy therefore the usual relation

$$(\theta' - \theta)\langle\theta'|\theta\rangle = 0. \tag{7.459}$$

On the other hand, the general expansion rule (7.376) tells us that the scalar product $S = \langle\theta'|\theta\rangle$ must be a linear combination of $S_0 + S_1\theta + S_1'\theta' + S_2\theta\theta'$. Inserting this into (7.459), we find

$$\langle\theta'|\theta\rangle = -\theta' + \theta + S_2\theta\theta', \tag{7.460}$$

where the proportionality constants S_0 and S_1 are fixed by the property

$$\langle\theta'|\theta\rangle = \int \frac{d\theta''}{2\pi}\langle\theta'|\theta''\rangle\langle\theta''|\theta\rangle. \tag{7.461}$$

The constant S_2 is an arbitrary real number. Recalling (7.385) we see that Eq. (7.460) implies that the scalar product $\langle\theta'|\theta\rangle$ is equal to a δ-function:

$$\langle\theta'|\theta\rangle = \delta(\theta - \theta'), \tag{7.462}$$

just as in ordinary quantum mechanics. Note the property

$$\langle\theta'|\theta\rangle^* = -\langle\theta|\theta'\rangle = \langle-\theta| - \theta'\rangle, \tag{7.463}$$

and the fact that since the scalar product $\langle\theta'|\theta\rangle$ is a Grassmann object, a Grassmann variable anticommutes with the scalar product. Having assumed in (7.458) that the Grassmann variable θ can be taken to the left of the bra-vector $\langle\theta|$, the ket-vector $|\theta\rangle$ must be treated like a Grassmann variable, i.e.,

$$\hat{\theta}|\theta\rangle = \theta|\theta\rangle = -|\theta\rangle\theta. \tag{7.464}$$

The momentum operator has the following matrix elements

$$\langle\theta'|\hat{p}|\theta\rangle = -i\hbar\partial_{\theta'}\langle\theta'|\theta\rangle = i\hbar. \tag{7.465}$$

Let $|p\rangle$ be an eigenstate of $\hat{p}$ with eigenvalue ip, then its scalar product with $|\theta\rangle$ satisfies

$$\langle\theta|\hat{p}|p\rangle = \int \frac{d\theta'}{2\pi}\langle\theta|\hat{p}|\theta'\rangle\langle\theta'|p\rangle = i\hbar\int \frac{d\theta'}{2\pi}\langle\theta'|p\rangle = i\hbar\partial_{\theta'}\langle\theta'|p\rangle, \tag{7.466}$$

the last step following from the rule (7.380). Solving (7.466) we find

$$\langle\theta|p\rangle = e^{i\theta p/\hbar},\tag{7.467}$$

the right-hand side being of course equal to $1 + i\theta p/\hbar$.

It is easy to find an orthonormal set of basis vectors in the space of wave functions (7.452):

$$\psi_+(\theta) \equiv \frac{1}{\sqrt{2}}(1+\theta)\,, \quad \psi_-(\theta) \equiv \frac{1}{\sqrt{2}}(1-\theta)\,.\tag{7.468}$$

We can easily check that these are orthogonal to each other and that they have the scalar products

$$\int \frac{d\theta}{2\pi} \psi_\pm^*(\theta)\psi_\pm(\theta) = \pm 1.\tag{7.469}$$

The Hilbert space contains states of negative norm which are referred to as *ghosts*. Because of the constraint, only half of the Hilbert space is physical. For more details on these problems see the literature on supersymmetric quantum mechanics.

7.11.2 Quantizing Harmonic Oscillator with Grassmann Variables

Let us now turn to the more important physical system containing two Grassmann variables θ_1 and θ_2, combined to complex Grassmann variables (7.388). The Lagrangian is assumed to have the same form as that of an ordinary harmonic oscillator:

$$\mathcal{L}(t) = \hbar\left[a^*(t)i\partial_t a(t) - \omega a^*(t)a(t)\right].\tag{7.470}$$

We may treat $a(t)$ and $a^*(t)$ as independent variables, such that there is no constraint in the system. The classical equation of motion

$$i\dot{a}(t) = \omega a(t)\tag{7.471}$$

is solved by

$$a(t) = e^{-i\omega t}a(0)\,, \quad a^\dagger(t) = e^{-i\omega t}a^\dagger(0).\tag{7.472}$$

The canonical momentum reads

$$p_a(t) = \frac{\partial\mathcal{L}(t)}{\partial\dot{a}(t)} = -i\hbar a(t),\tag{7.473}$$

and the system is quantized by the equal-time anticommutation relation

$$[\hat{p}_a(t), \hat{a}(t)]_+ = -i\hbar,\tag{7.474}$$

or

$$[\hat{a}^\dagger(t), \hat{a}(t)]_+ = 1.\tag{7.475}$$

In addition we have

$$[\hat{a}(t), \hat{a}(t)]_+ = 0\,, \quad [\hat{a}^\dagger(t), \hat{a}^\dagger(t)]_+ = 0.\tag{7.476}$$

Due to these anticommutation relations, the time-independent number operator

$$\hat{N} \equiv a^\dagger(t)a(t) \tag{7.477}$$

satisfies the commutation relations

$$[\hat{N}, a^\dagger(t)] = a^\dagger(t), \qquad [\hat{N}, a(t)] = -a(t). \tag{7.478}$$

We can solve the algebra defined by (7.475), (7.476), and (7.478) for any time, say $t = 0$, in the usual way, defining a ground state $|0\rangle$ by the condition

$$a|0\rangle = 0, \tag{7.479}$$

and an excited state $|1\rangle$ as

$$|1\rangle \equiv a^\dagger|0\rangle. \tag{7.480}$$

These are the only states, and the Hamiltonian operator $\hat{H} = \omega\hat{N}$ possesses the eigenvalues 0 and ω on them. Let $\psi(a)$ be wave functions in the representation where the operator a is diagonal. The canonically conjugate operator $\hat{p}_a = i\hbar\hat{a}^\dagger$ has then the form

$$\hat{p}_a \equiv -i\hbar\partial_a. \tag{7.481}$$

7.11.3 Spin System with Grassmann Variables

For the purpose of constructing path integrals of relativistic electrons later in Chapter 19 we discuss here another system with Grassmann variables.

Pauli Algebra

First we introduce three real Grassmann fields θ^i, $i = 1, 2, 3$, and consider the path integral

$$\prod_{i=1}^{3}\left[\int \mathcal{D}\theta^i\right]\exp\left[\frac{i}{\hbar}\int dt\,\frac{i\hbar}{4}\theta^i(t)\dot{\theta}^i(t)\right]. \tag{7.482}$$

The equation of motion is

$$\dot{\theta}^i(t) = 0, \tag{7.483}$$

so that $\theta^i(t)$ are time independent variables. The three momentum operators lead now to the three-dimensional version of the equal-time anticommutation relation (7.451)

$$[\hat{\theta}^i(t), \hat{\theta}^j(t)]_+ = 2\delta^{ij}, \tag{7.484}$$

where the time arguments can be omitted due to (7.483). The algebra is solved with the help of the Pauli spin matrices (1.445). The solution of (7.484) is obviously

$$\langle B|\hat{\theta}^i|A\rangle = \sigma^i_{BA}, \qquad A, B = 1, 2. \tag{7.485}$$

Let us now add in the exponent of the trivial path integral a Hamiltonian

$$H_B = -\mathbf{S} \cdot \mathbf{B}(t), \tag{7.486}$$

where

$$S^i \equiv -\frac{i}{4}\epsilon^{ijk}\theta^j\theta^k \qquad (7.487)$$

plays the role of a spin vector. This can be verified by calculating the canonical commutation relations between the operators

$$[\hat{S}^i, \hat{S}^j] = i\epsilon_{ijk}\hat{S}^k, \qquad (7.488)$$

and

$$[\hat{S}^i, \hat{\theta}^j] = i\epsilon_{ijk}\hat{\theta}^k. \qquad (7.489)$$

Thus, the operator $\hat{H}$ describes the coupling of a spin vector to a magnetic field $\mathbf{B}(t)$. Using the commutation relation (7.488), we find the Heisenberg equation (1.283) for the Grassmann variables:

$$\dot{\boldsymbol{\theta}} = \mathbf{B} \times \boldsymbol{\theta}, \qquad (7.490)$$

which goes over into a similar equation for the spin vector:

$$\dot{\mathbf{S}} = \mathbf{B} \times \mathbf{S}. \qquad (7.491)$$

The important observation is now that the path integral (7.492) with the magnetic Hamiltonian (7.486) with fixed ends $\theta_b^i = \theta^i(\tau_b)$ and $\theta_a^i = \theta^i(\tau_a)$ written as

$$\int_{\theta_a^i=\theta^i(\tau_a)}^{\theta_b^i=\theta^i(\tau_b)} \mathcal{D}^3\theta \exp\left[\frac{i}{\hbar}\int dt\left(\frac{i\hbar}{4}\theta^i\dot{\theta}^i + B^i\frac{i}{4}\epsilon^{ijk}\theta^j\theta^k\right)\right], \qquad (7.492)$$

represents the matrix (recall Appendix 1A)

$$\hat{T}\exp\left(\frac{i}{\hbar}\int dt\,\mathbf{B}(t)\cdot\frac{\boldsymbol{\sigma}}{2}\right), \qquad (7.493)$$

where $\hat{T}$ is the time-ordering operator defined in Eq. (1.241). This operator is necessary for a time-dependent $\mathbf{B}(t)$ field since the matrices $\mathbf{B}(t)\boldsymbol{\sigma}/2$ do not in general commute with each other. Whenever we encounter time ordered exponentials of integrals over matrices, these can be transformed into a fluctuating path integral over Grassmann variables.

In the applications to come, we need only the trace of the matrix (7.493). According to Eq. (7.440), this is found from the integral over (7.492) with $\theta_b^i = -\theta_a^i$, which means performing the integral over all antiperiodic Grassmann paths $\theta^i(\tau)$. If we want to find individual matrix elements, we have to make use of suitably extended recovery formulas (7.431) and (7.432).

The result may be expressed in a slightly different notation using the spin tensors

$$S^{ij} \equiv \epsilon^{ijk}S^k = \frac{1}{2i}\theta^i\theta^j, \qquad (7.494)$$

whose matrix elements satisfy the rotation algebra

$$[\hat{S}^{ij}, \hat{S}^{kl}] = i\left(\delta^{ik}\hat{S}^{jl} - \delta^{il}\hat{S}^{jk} + \delta^{jl}\hat{S}^{ik} - \delta^{jk}\hat{S}^{il}\right), \qquad (7.495)$$

and which have the matrix representation

$$\langle B|\hat{S}^{ij}|A\rangle = \frac{1}{2}\sigma_{BA}^{ij} \equiv \frac{1}{4i}[\sigma^i,\sigma^j]_{BA} = \epsilon^{ijk}\frac{\sigma^k}{2}. \tag{7.496}$$

Note the normalization

$$\sigma^{12} = \sigma^3. \tag{7.497}$$

Introducing the analogous magnetic field tensor

$$F^{ij} \equiv \epsilon^{ijk}B^k, \tag{7.498}$$

we can write the final result also in the tensorial form

$$\hat{T}\exp\left(\frac{i}{4\hbar}\int dt\, F^{ij}(t)\sigma^{ij}\right) = \int_{\theta_a^i=\theta^i(\tau_a)}^{\theta_b^i=\theta^i(\tau_b)}\mathcal{D}^3\theta\,\exp\left[\frac{i}{\hbar}\int dt\left(\frac{i\hbar}{4}\theta^i\dot{\theta}^i + \frac{i}{4}F^{jk}\theta^j\theta^k\right)\right]. \tag{7.499}$$

The trace of the left-hand side is, of course, given by the path integral over all antiperiodic Grassmann paths on the right-hand side.

Dirac Algebra

There exists a similar path integral suitable for describing relativistic spin systems. If we introduce four Grassmann variables θ^μ, $\mu = 0, 1, 2, 3$, the path integral

$$\prod_{\mu=0}^{3}\left[\int \mathcal{D}\theta^\mu\right]\exp\left\{\frac{i}{\hbar}\int dt\left[-\frac{i}{4}\theta_\mu(t)\dot{\theta}^\mu(t)\right]\right\}, \tag{7.500}$$

leads to an equation of motion

$$\dot{\theta}^\mu(t) = 0, \tag{7.501}$$

and an operator algebra at equal times

$$\{\hat{\theta}^\mu(t),\hat{\theta}^\nu(t)\} = 2g^{\mu\nu}, \tag{7.502}$$

where $g_{\mu\nu}$ is the metric in Minkowski space

$$g^{\mu\nu} = \begin{pmatrix} 1 & 0 & 0 & 0 \\ 0 & -1 & 0 & 0 \\ 0 & 0 & -1 & 0 \\ 0 & 0 & 0 & -1 \end{pmatrix}. \tag{7.503}$$

The time argument in (7.502) can again be dropped due to (7.501). The algebra (7.502) is solved by the matrix elements [recall (7.485)]

$$\langle\beta|\hat{\theta}^\mu(t)|\alpha\rangle = (\gamma^\mu)_{\beta\alpha}, \quad \beta,\alpha = 1,2,3,4, \tag{7.504}$$

where γ^μ, γ_5 are composed of 2×2-matrices 0, 1, and σ^i as follows:

$$\gamma^0 \equiv \begin{pmatrix} 0 & 1 \\ -1 & 0 \end{pmatrix}, \quad \gamma^i \equiv \begin{pmatrix} 0 & \sigma^i \\ -\sigma^i & 0 \end{pmatrix}, \quad \gamma_5 \equiv i\gamma^0\gamma^1\gamma^2\gamma^3 = \begin{pmatrix} -1 & 0 \\ 0 & 1 \end{pmatrix} \equiv \gamma^5. \tag{7.505}$$

One may also introduce an additional Grassmann variable θ_5 such that the path integral

$$\prod_{\mu=0}^{3}\left[\int \mathcal{D}\theta_5\right]\exp\left[\frac{i}{\hbar}\int dt\,\frac{i}{4}\theta_5(t)\dot{\theta}_5(t)\right], \tag{7.506}$$

produces the matrix elements of $\gamma_5(t)$:

$$\langle\beta|\hat{\theta}_5(t)|\alpha\rangle = (\gamma_5)_{\beta\alpha}\,, \qquad \beta, \alpha = 1, 2, 3, 4. \tag{7.507}$$

The Grassmann variables θ^μ and θ_5 anticommute with each other, and so do the matrices $\gamma_5\gamma^\mu$ and γ_5.

In terms of the Grassmann variables θ^μ it is possible to rewrite a four-dimensional version of the time ordered 2×2 matrix integral (7.493) as a path integral without time ordering:

$$\hat{T}\exp\left(\frac{i}{2\hbar}\int dt\, F^{\mu\nu}\Sigma_{\mu\nu}\right) = \int \mathcal{D}^4\theta\,\exp\left\{\frac{i}{\hbar}\int dt\left[-\frac{i\hbar}{4}\theta_\mu\dot{\theta}^\mu + \frac{i}{4}F_{\mu\nu}\theta^\mu\theta^\nu\right]\right\}, \tag{7.508}$$

where $\Sigma^{\mu\nu}$ is the Minkowski space generalization of the spin tensor matrix $\sigma^{ij}/2$ in (7.496):

$$\Sigma^{\mu\nu} \equiv \frac{i}{4}[\gamma^\mu, \gamma^\nu] = -\Sigma^{\nu\mu}. \tag{7.509}$$

As a check we use (7.505) and find

$$\Sigma^{12} = \frac{1}{2}\left(\begin{array}{cc} \sigma^{12} & 0 \\ 0 & \sigma^{12} \end{array}\right) = \frac{1}{2}\left(\begin{array}{cc} \sigma^3 & 0 \\ 0 & \sigma^3 \end{array}\right), \tag{7.510}$$

in agreement with (7.497).

As remarked before, we shall need in the applications to come only the trace of the matrix (7.493), which is found, according to Eq. (7.440), from the integral over (7.492) with $\theta_b^i = -\theta_a^i$, i.e., by performing the integral over all antiperiodic Grassmann paths $\theta^i(\tau)$. If we want to find individual matrix elements, we have to make use of suitably extended recovery formulas (7.431) and (7.432).

The path integral over all antiperiodic paths for $F_{\mu\nu} = 0$ fixes the normalization. The left-hand side is equal to 4, so that we have

$$\int \mathcal{D}^4\theta\,\exp\left\{\frac{i}{\hbar}\int dt\left[-\frac{i\hbar}{4}\theta_\mu\dot{\theta}^\mu\right]\right\} = 4. \tag{7.511}$$

This normalization factor agrees with Eqs. (7.415) and (7.418), where we found the path integral of single complex fermion field to carry a normalization factor 2. For four real fields this corresponds to a factor 4.

In the presence of a nonzero field tensor, the result of the path integral (7.508) is therefore

$$\int \mathcal{D}x\, e^{(i/4\hbar)\int dt\{[-i\hbar\theta_\mu\dot{\theta}^\mu + iF_{\mu\nu}\theta^\mu\theta^\nu]\}} = 4\,\mathrm{Det}^{1/2}\left[\delta_{\mu\nu}\,\partial_t - \frac{1}{\hbar}F_{\mu\nu}(x(\tau))\right]. \tag{7.512}$$

Relation between Harmonic Oscillator, Pauli and Dirac Algebra

There exists a simple relation between the path integrals of the previous three paragraphs. We simply observe that the combinations

$$\hat{a}^\dagger = \frac{1}{2}\left(\sigma^1 + i\sigma^2\right) = \frac{1}{2}\sigma^+ = \begin{pmatrix} 0 & 0 \\ 1 & 0 \end{pmatrix}, \quad \hat{a} = \frac{1}{2}\left(\sigma^1 - i\sigma^2\right) = \frac{1}{2}\sigma^- = \begin{pmatrix} 0 & 1 \\ 0 & 0 \end{pmatrix}, \quad (7.513)$$

satisfy the anticommutation rules

$$[\hat{a}, \hat{a}^\dagger]_+ = 1, \quad [\hat{a}^\dagger, \hat{a}^\dagger]_+ = 1, \quad [\hat{a}, \hat{a}]_+ = 0. \quad (7.514)$$

The vacuum state annihilated by a is the spin-down spinor, and the one-particle state is the spin-up spinor:

$$|0\rangle = \begin{pmatrix} 0 \\ 1 \end{pmatrix}, \quad |1\rangle = \begin{pmatrix} 1 \\ 0 \end{pmatrix}. \quad (7.515)$$

A similar construction can be found for the Dirac algebra. There are now two types of creation and annihilation operators

$$\hat{a}^\dagger = \frac{1}{2}\left(\gamma^1 + i\gamma^2\right) = \frac{1}{2}\begin{pmatrix} 0 & \sigma^+ \\ -\sigma^+ & 0 \end{pmatrix}, \quad \hat{a} = \frac{1}{2}\left(-\gamma^1 + i\gamma^2\right) = \frac{1}{2}\begin{pmatrix} 0 & -\sigma^- \\ \sigma^- & 0 \end{pmatrix}, \quad (7.516)$$

and

$$\hat{b}^\dagger = \frac{1}{2}\left(\gamma^0 + \gamma^3\right) = \frac{1}{2}\begin{pmatrix} 0 & i\sigma^2\sigma^+ \\ -i\sigma^2\sigma^+ & 0 \end{pmatrix}, \quad \hat{b} = \frac{1}{2}\left(\gamma^1 - \gamma^3\right) = \frac{1}{2}\begin{pmatrix} 0 & i\sigma^2\sigma^- \\ -i\sigma^2\sigma^- & 0 \end{pmatrix}. \quad (7.517)$$

The states $|n_a, a_b\rangle$ with n_a quanta a and n_b quanta b are the following:

$$|0,0\rangle = \frac{1}{\sqrt{2}}\begin{pmatrix} 0 \\ 1 \\ 0 \\ 1 \end{pmatrix}, \quad |1,0\rangle = a^\dagger|0,0\rangle = \frac{1}{\sqrt{2}}\begin{pmatrix} 1 \\ 0 \\ -1 \\ 0 \end{pmatrix}, \quad (7.518)$$

$$|0,1\rangle = b^\dagger|0,0\rangle = \frac{1}{\sqrt{2}}\begin{pmatrix} 0 \\ -1 \\ 0 \\ 1 \end{pmatrix}, \quad |1,1\rangle = a^\dagger b^\dagger|0,0\rangle = \frac{1}{\sqrt{2}}\begin{pmatrix} 1 \\ 0 \\ 1 \\ 0 \end{pmatrix}. \quad (7.519)$$

From these relations we can easily deduce the proper recovery formulas generalizing (7.431) and (7.432).

7.12   External Sources in a^*, a -Path Integral

In Chapter 3, the path integral of the harmonic oscillator was solved in the presence of an arbitrary external current $j(\tau)$. This yielded the generating functional $Z[j]$ for the calculation of all correlation functions of $x(\tau)$. In the present context we are interested in the generating functional of correlations of $a(\tau)$ and $a^*(\tau)$. Thus we also need the path integrals quadratic in $a(\tau)$ and $a^*(\tau)$ coupled to external currents. Consider the Euclidean action

$$\mathcal{A}[a^*, a] + \mathcal{A}^{\text{source}} = \hbar \int_0^{\hbar\beta} d\tau [a^*(\tau)\partial_\tau a(\tau) + \omega a^*a(\tau) - (\eta^*(\tau)a(\tau) + \text{c.c.})], \quad (7.520)$$

with periodic boundary conditions for $a(\tau)$ and $a^*(\tau)$. For simplicity, we shall use the continuum formulation of the partition function, so that our results will correspond to the harmonic oscillator time slicing, with the energies $E_n = (n + 1/2)\hbar\omega$. There is no problem in going over to the second-quantized formulation with the energies $E_n = n\hbar\omega$. The partition function is given by the path integral

$$Z_\omega[\eta^*, \eta] = \oint \frac{\mathcal{D}a^*\mathcal{D}a}{\pi} \exp\left\{-\frac{1}{\hbar}\left(\mathcal{A}[a^*, a] + \mathcal{A}^{\text{source}}\right)\right\}. \quad (7.521)$$

As in Chapter 3, we define the functional matrix between $a^*(\tau)$, $a(\tau')$ as

$$D_{\omega,e}(\tau, \tau') \equiv (\partial_\tau + \omega)\delta(\tau - \tau'), \quad \tau, \tau' \in (0, \hbar\beta). \quad (7.522)$$

The inverse is the Euclidean Green function

$$G^{\text{p}}_{\omega,e}(\tau, \tau') = D^{-1}_{\omega,e}(\tau, \tau'), \quad (7.523)$$

satisfying the periodic boundary condition. We now complete the square and rewrite the action (7.520), using the shifted fields

$$a'(\tau) = a(\tau) - \int_0^{\hbar\beta} d\tau' \, G^{\text{p}}_{\omega,e}(\tau, \tau')\eta(\tau') \quad (7.524)$$

as

$$\mathcal{A}_e = \hbar \int_0^{\hbar\beta} d\tau \int_0^{\hbar\beta} d\tau' [a'^*(\tau)D_{\omega,e}(\tau, \tau')a'(\tau') - \eta^*(\tau)G^{\text{p}}_{\omega,e}(\tau, \tau')\eta(\tau')]. \quad (7.525)$$

On an infinite β-interval the Green function can easily be written down in terms of the Heaviside function (1.309):

$$G_{\omega,e}(\tau, \tau') = G_{\omega,e}(\tau - \tau') = e^{-\omega(\tau-\tau')}\Theta(\tau - \tau'). \quad (7.526)$$

As we have learned in Section 3.3, the periodic Green function is obtained from this by forming a periodic sum

$$G^{\text{p}}_{\omega,e}(\tau, \tau') = G^{\text{p}}_{\omega,e}(\tau - \tau') = \sum_{n=-\infty}^{\infty} e^{-\omega(\tau-\tau'-n\hbar\beta)}\Theta(\tau - \tau' - n\hbar\beta), \quad (7.527)$$

which is equal to

$$G^{\mathrm{p}}_{\omega,\mathrm{e}}(\tau) = \frac{e^{-\omega(\tau-\hbar\beta/2)}}{2\sinh(\omega\hbar\beta/2)} = (1+n^{\mathrm{b}}_{\omega})e^{-\omega\tau}, \qquad (7.528)$$

where n_{ω} is the Bose-Einstein distribution function

$$n_{\omega} = \frac{1}{e^{\beta\hbar\omega}-1} \qquad (7.529)$$

[compare (3.92) and (3.93)].

The same considerations hold for anticommuting variables with antiperiodic boundary conditions, in which case we find once more the action (7.525) with the antiperiodic Green function [compare (3.112)]

$$G^{a}_{\omega,\mathrm{e}}(\tau) = \frac{e^{-\omega(\tau-\hbar\beta/2)}}{2\cosh(\omega\hbar\beta/2)} = (1-n^{\mathrm{f}}_{\omega})e^{-\omega\tau}, \qquad (7.530)$$

where n_{ω} is the Fermi-Dirac distribution function [see (3.111)]:

$$n^{\mathrm{f}}_{\omega} = \frac{1}{e^{\beta\hbar\omega}+1}. \qquad (7.531)$$

In either case, we may decompose the currents in (7.525) into real and imaginary parts j and k via

$$\eta = \sqrt{\omega/2M\hbar}(j-i\omega Mk), \qquad (7.532)$$
$$\eta^* = \sqrt{\omega/2M\hbar}(j+i\omega Mk),$$

and write the source part of the original action (7.520) in the form

$$\mathcal{A}^{\mathrm{source}} = \hbar \int_0^{\hbar\beta} d\tau(a^*\eta+\eta^*a) = \int_0^{\hbar\beta} d\tau(jx+kp). \qquad (7.533)$$

Hence, the real current $\eta = \eta^* = \sqrt{\omega/2M\hbar}\, j$ corresponds to the earlier source term (3.2). Inserting this current into the action (7.525), it yields the quadratic source term

$$\mathcal{A}^{\mathrm{s}}_{\mathrm{e}} = -\frac{1}{2M\omega} \int_0^{\hbar\beta} d\tau \int_0^{\tau} d\tau' \frac{e^{-\omega(\tau-\tau')}}{1-e^{-\beta\hbar\omega}} j(\tau)j(\tau'). \qquad (7.534)$$

This can also be rewritten as

$$\mathcal{A}^{\mathrm{s}}_{\mathrm{e}} = -\frac{1}{2M\omega} \sum_{n=0}^{\infty} \int_0^{\hbar\omega} d\tau \int_0^{\tau} d\tau' e^{-\omega(\tau-\tau'+n\hbar\beta)} j(\tau)j(\tau'), \qquad (7.535)$$

which becomes for a current periodic in $\hbar\beta$:

$$\mathcal{A}^{\mathrm{s}}_{\mathrm{e}} = -\frac{1}{2M\omega} \int_0^{\hbar\omega} d\tau \int_{-\infty}^{\tau} d\tau'\ e^{-\omega(\tau-\tau')} j(\tau)j(\tau'). \qquad (7.536)$$

Interchanging τ and τ', this is also equal to

$$\mathcal{A}^{\mathrm{s}}_{\mathrm{e}} = -\frac{1}{4M\omega} \int_0^{\hbar\omega} d\tau \int_{-\infty}^{\infty} d\tau' e^{-\omega|\tau-\tau'|} j(\tau)j(\tau'), \qquad (7.537)$$

in agreement with (3.276).

7.13 Generalization to Pair Terms

There exists an important generalization of these considerations to the case of the quadratic frequency term $\hbar \int d\tau \, \Omega^2(\tau) a^*(\tau) a(\tau)$ being extended to the more general quadratic form

$$\hbar \int d\tau \left[\Omega^2(\tau) a^*(\tau) a(\tau) + \frac{1}{2} \Delta^*(\tau) a^2(\tau) + \frac{1}{2} \Delta(\tau) a^{*2}(\tau) \right]. \tag{7.538}$$

The additional off-diagonal terms in a^*, a are called *pair terms*. They play an important role in the theory of superconductivity. The basic physical mechanism for this phenomenon will be explained in Section 17.10. Here we just mention that the lattice vibrations give rise to the formation of bound states between pairs of electrons, called *Cooper pairs*. By certain manipulations of the path integral of the electron field in the second-quantized interpretation, it is possible to introduce a complex pair field $\Delta_{\mathbf{x}}(t)$ at each space point which is coupled to the electron field in an action of the type (7.538). The partition function to be studied is then of the generic form

$$Z = \oint \frac{\mathcal{D}a^* \, \mathcal{D}a}{\pi} \exp\left[-\int_0^{\hbar\beta} d\tau \left(a^* \partial_\tau a + \Omega^2 a^* a + \tfrac{1}{2} \Delta^* a^2 + \tfrac{1}{2} \Delta \, a^{*2} \right) \right]. \tag{7.539}$$

It is easy to calculate this partition function on the basis of the previous formulas. To this end we rewrite the action in the matrix form, using the field doublets

$$f(\tau) \equiv \begin{pmatrix} a(\tau) \\ a^*(\tau) \end{pmatrix}, \tag{7.540}$$

as

$$\mathcal{A}_e = \frac{\hbar}{2} \int_0^{\hbar\beta} d\tau \, f^{*T}(\tau) \begin{pmatrix} \partial_\tau + \Omega(\tau) & \Delta(\tau) \\ \Delta^*(\tau) & \mp(\partial_\tau \pm \Omega(\tau)) \end{pmatrix} f(\tau), \tag{7.541}$$

where the derivative terms require a partial integration to obtain this form. The partition function can be written as

$$Z = \int \frac{\mathcal{D}f^* \, \mathcal{D}f}{\pi} e^{-\frac{1}{2} f^* M f}, \tag{7.542}$$

with the matrix

$$M = \begin{pmatrix} \partial_\tau + \Omega(\tau) & \Delta(\tau) \\ \Delta^*(\tau) & \mp(\partial_\tau \pm \Omega(\tau)) \end{pmatrix}. \tag{7.543}$$

The fields f^* and f are not independent of each other, since

$$f^* = \begin{pmatrix} 0 & 1 \\ 1 & 0 \end{pmatrix} f. \tag{7.544}$$

Thus, there are only half as many independent integrations as for a usual complex field. This has the consequence that the functional integration in (7.542) gives only the square root of the determinant of M,

$$Z = \mathcal{N}^{b,f} \det \begin{pmatrix} \partial_\tau + \Omega(\tau) & \Delta(\tau) \\ \Delta^*(\tau) & \mp(\partial_\tau + \Omega(\tau)) \end{pmatrix}^{\mp 1/2}, \tag{7.545}$$

with the normalization factors $\mathcal{N}^{b,f}$ being fixed by comparison with (7.337) and (7.419).

A fluctuation determinant of this type occurs in the theory of superconductivity [32], with constant parameters ω, Δ. When applied to functions oscillating like $e^{-i\omega_m^f \tau}$, the matrix M becomes

$$M = \begin{pmatrix} -i\omega_m^f + \omega & \Delta \\ \Delta^* & \mp(-i\omega_m^f \pm \omega) \end{pmatrix}. \tag{7.546}$$

It is brought to diagonal form by what is called, in this context, a *Bogoliubov transformation* [33]

$$M \to M^d = \begin{pmatrix} -i\omega_m^f + \omega_\Delta & 0 \\ 0 & \mp(-i\omega_m^f \pm \omega_\Delta) \end{pmatrix}, \tag{7.547}$$

where ω_Δ is the frequency

$$\omega_\Delta = \sqrt{\omega^2 + \Delta^2}. \tag{7.548}$$

In a superconductor, these frequencies correspond to the energies of the quasi-particles associated with the electrons. They generalize the quasi-particles introduced in Landau's theory of Fermi liquids. The partition function (7.539) is then given by

$$
\begin{aligned}
Z_{\omega,\Delta} &= \mathcal{N}^{b,f} \det \begin{pmatrix} \partial_\tau + \omega & \Delta \\ \Delta^* & \mp\partial_\tau \pm \omega \end{pmatrix}^{\mp 1/2} \\
&= N^{b,f} \left[\det\left(-\partial_\tau^2 + \omega_\Delta^2\right)\right]^{\mp 1/2} = \begin{cases} [2\sinh(\hbar\beta\omega_\Delta/2)]^{-1}, \\ 2\cosh(\hbar\beta\omega_\Delta/2), \end{cases}
\end{aligned}
\tag{7.549}
$$

for bosons and fermions, respectively. This is equal to the partition function of a symmetrized Hamilton operator

$$\hat{H} = \frac{\omega_\Delta}{2}\Delta\left(\hat{a}^\dagger\hat{a} + \hat{a}\hat{a}^\dagger\right) = \omega_\Delta\left(\hat{a}^\dagger\hat{a} \pm \frac{1}{2}\right), \tag{7.550}$$

with the eigenvalue spectrum $\omega_\Delta\left(n \pm \frac{1}{2}\right)$, where $n = 0, 1, 2, 3\ldots$ for bosons and $n = 0, 1$ for fermions:

$$Z_{\omega,\Delta} = \sum_{n}^{\infty,1} e^{-\hbar\omega_\Delta(n\pm 1/2)\beta}. \tag{7.551}$$

In the second-quantized interpretation where zero-point energies are omitted, this becomes

$$Z_{\omega,\Delta} = \begin{cases} (1 - e^{-\hbar\beta\omega_\Delta})^{-1}, \\ (1 + e^{-\hbar\beta\omega_\Delta}). \end{cases} \tag{7.552}$$

7.14   Spatial Degrees of Freedom

In the path integral treatment of the last sections, the particles have been restricted to a particular momentum state $\mathbf{p}$. In a three-dimensional volume, the fields $a^*(\tau)$, $\Delta(\tau)$ and their frequency ω depend on $\mathbf{p}$. With this trivial extension one obtains a free *quantum field theory*.

7.14.1   Grand-Canonical Ensemble of Particle Orbits from Free Fluctuating Field

The free particle action becomes a sum over momentum states

$$\mathcal{A}_e[a^*, a] = \hbar \int_0^{\hbar\beta} d\tau \sum_{\mathbf{p}} \left[a^*_{\mathbf{p}} \partial_\tau a_{\mathbf{p}} + \omega(\mathbf{p}) a^*_{\mathbf{p}} a_{\mathbf{p}} \right]. \tag{7.553}$$

The time-sliced partition function is given by the product

$$Z = \prod_{\mathbf{p}} \left\{ \det \left[\epsilon(1 - \epsilon\omega)\overline{\nabla} + \epsilon\omega(\mathbf{p}) \right] \right\}^{\mp 1} \tag{7.554}$$

$$= \begin{cases} \exp\left[-\sum_{\mathbf{p}} \log\left(1 - e^{-\hbar\beta\bar{\omega}_e(\mathbf{p})} \right) \right] & \text{for bosons,} \\ \exp\left[\sum_{\mathbf{p}} \log\left(1 + e^{-\hbar\beta\bar{\omega}_e(\mathbf{p})} \right) \right] & \text{for fermions.} \end{cases}$$

The free energy

$$F = -k_B T \log Z \tag{7.555}$$

is for bosons

$$F = k_B T \sum_{\mathbf{p}} \log\left(1 - e^{-\hbar\beta\bar{\omega}_e(\mathbf{p})} \right), \tag{7.556}$$

and for fermions

$$F = -k_B T \sum_{\mathbf{p}} \log\left(1 + e^{-\hbar\beta\bar{\omega}_e(\mathbf{p})} \right). \tag{7.557}$$

In a large volume, the momentum sum may be replaced by the integral

$$\sum_{\mathbf{p}} \to \int \frac{d^D p\, V}{(2\pi\hbar)^D}. \tag{7.558}$$

For infinitely thin time slices, $\epsilon \to 0$ and $\bar{\omega}_e(\mathbf{p})$ reduces to $\omega(\mathbf{p})$ and the expressions (7.556), (7.557) turn into the usual free energies of bosons and fermions. They agree completely with the expressions (7.46) and (7.220) derived from the sum over orbits.

Next we may introduce a fluctuating field in space and imaginary time

$$\psi(\mathbf{x}, \tau) = \frac{1}{\sqrt{V}} \sum_{\mathbf{p}} e^{i\mathbf{p}\mathbf{x}} a_{\mathbf{p}}(\tau), \tag{7.559}$$

and rewrite the action (7.553) in the local form

$$\mathcal{A}_e[\psi^*, \psi] = \int_0^{\hbar\beta} d\tau \int d^D x \left[\psi^*(\mathbf{x}, \tau)\hbar\partial_\tau \psi(\mathbf{x}, \tau) + \frac{\hbar^2}{2M} \boldsymbol{\nabla}\psi^*(\mathbf{x}, \tau)\boldsymbol{\nabla}\psi(\mathbf{x}, \tau) \right]. \tag{7.560}$$

The partition function is given by the functional integral

$$Z = \oint \mathcal{D}\psi \mathcal{D}\psi^* e^{-\mathcal{A}_e[\psi^*,\psi]/\hbar}. \tag{7.561}$$

Thus we see that the functional integral over a fluctuating field yields precisely the same partition function as the sum over a grand-canonical ensemble of fluctuating orbits. Bose or Fermi statistics are naturally accounted for by using complex or Grassmann field variables with periodic or antiperiodic boundary conditions, respectively. The theory based on the action (7.560) is completely equivalent to the second-quantized theory of field operators.

In order to distinguish the second-quantized or quantum field description of many particle systems from the former path integral description of many particle orbits, the former is referred to as the *first-quantized* approach, or also the *world-line* approach.

The action (7.560) can be generalized further to include an external potential $V(\mathbf{x}, \tau)$, i.e., it may contain a general Schrödinger operator $\hat{H}(\tau) = \hat{\mathbf{p}}^2 + V(\mathbf{x}, \tau)$ instead of the gradient term:

$$\mathcal{A}_e[\psi^*, \psi] = \int_0^{\hbar\beta} d\tau \int d^D x \left\{ \psi^*(\mathbf{x}, \tau)\hbar\partial_\tau\psi(\mathbf{x}, \tau) + \psi^*(\mathbf{x}, \tau)\left[\hat{H}(\tau) - \mu\right]\psi(\mathbf{x}, \tau)\right\}. \tag{7.562}$$

For the sake of generality we have also added a chemical potential to enable the study of grand-canonical ensembles. This action can be used for a second-quantized description of the free Bose gas in an external magnetic trap potential $V(\mathbf{x})$, which would, of course, lead to the same results as the first-quantized approach in Section 7.2.4. The free energy associated with this action

$$F = \frac{1}{\beta}\text{Tr}\log[\hbar\partial_\tau + \hat{H}(\tau) - \mu] \tag{7.563}$$

was calculated in Eq. (3.139) as an expansion

$$F = \frac{1}{2\hbar\beta}\text{Tr}\left[\int_0^{\hbar\beta} d\tau \, \hat{H}(\tau)\right] - \frac{1}{\beta}\sum_{n=1}^{\infty}\frac{1}{n}\text{Tr}\left\{\hat{T}e^{-n\int_0^{\hbar\beta} d\tau'' [\hat{H}(\tau'') - \mu]/\hbar}\right\}, \tag{7.564}$$

The sum can be evaluated in the semiclassical expansion developed in Section 4.9. For simplicity, we consider here only time-independent external potentials, where we must calculate $\text{Tr}\left[e^{-n\beta(\hat{H}-\mu)}\right]$. Its semiclassical limit was given in Eq. (4.257) continued to imaginary time. From this we obtain

$$F = \frac{1}{2}\text{Tr}\,\hat{H} - \frac{1}{\beta}\frac{1}{\sqrt{2\pi\hbar^2\beta/M}^D}\sum_{n=1}^{\infty}\int d^D x \frac{[z(\mathbf{x})]^n}{n^{D/2+1}}, \tag{7.565}$$

where $z(\mathbf{x}) \equiv e^{-\beta[V(\mathbf{x})-\mu]}$ is the local fugacity (7.24). The sum agrees with the previous first-quantized result in (7.132) and (7.133). The first term is due to the symmetric treatment of the fields in the action (7.562) [recall the discussion after Eq. (7.340)].

One may calculate quantum corrections to this expansion by including the higher gradient terms of the semiclassical expansion (4.257). If the potential is time-dependent, the expansion (4.257) must be generalized accordingly.

7.14.2   First versus Second Quantization

There exists a simple set of formulas which illustrates nicely the difference between first and second quantization, i.e., between path and field quantization. Both may be though of as being based on two different representations of the Dirac δ-function. The first-quantized representation is

$$\delta^{(D)}(\mathbf{x}_b - \mathbf{x}_a) = \int_{\mathbf{x}(t_a)=\mathbf{x}_a}^{\mathbf{x}(t_b)=\mathbf{x}_b} \mathcal{D}^D x \oint \frac{\mathcal{D}^D p}{(2\pi\hbar)^D} e^{(i/\hbar)\int_{t_a}^{t_b} dt\, \mathbf{p}\dot{\mathbf{x}}}, \tag{7.566}$$

the second-quantized representation

$$i\hbar\delta^{(D)}(\mathbf{x}_b - \mathbf{x}_a)\delta(t_b - t_a) = \oint \mathcal{D}\psi\mathcal{D}\psi^* \,\psi(\mathbf{x}_b, t_b)\psi^*(\mathbf{x}_a, t_a)e^{(i/\hbar)\int d^D x \int_{-\infty}^{\infty} dt\psi^*(\mathbf{x},t)\psi(\mathbf{x},t)}. \tag{7.567}$$

The first representation is turned into a transition amplitude by acting upon it with the time-evolution operator $e^{-i\hat{H}(t_b-t_a)}$, which yields

$$(\mathbf{x}_b t_b | \mathbf{x}_a t_a) = e^{i\hat{H}(t_b-t_a)}\delta^{(D)}(\mathbf{x}_b - \mathbf{x}_a) = \int \mathcal{D}^D x \oint \frac{\mathcal{D}^D p}{(2\pi\hbar)^D} e^{(i/\hbar)\int_{t_a}^{t_b} dt(\mathbf{p}\dot{\mathbf{x}}-H)}. \tag{7.568}$$

By multiplying this with the Heaviside function $\Theta(t_b - t_a)$, we obtain the solution of the inhomogeneous Schrödinger equation

$$(i\hbar\partial_t - \hat{H})\Theta(t_b - t_a)(\mathbf{x}_b t_b | \mathbf{x}_a t_a) = i\hbar\delta^{(D)}(\mathbf{x}_b - \mathbf{x}_a)\delta(t_b - t_a). \tag{7.569}$$

This may be expressed as a path integral representation for the resolvent

$$\left\langle \mathbf{x}_b t_b \left| \frac{i\hbar}{i\hbar\partial_t - \hat{H}} \right| \mathbf{x}_a t_a \right\rangle = \Theta(t_b - t_a)\int_{\mathbf{x}(t_a)=\mathbf{x}_a}^{\mathbf{x}(t_b)=\mathbf{x}_b} \mathcal{D}^D x \oint \frac{\mathcal{D}^D p}{(2\pi\hbar)^D} e^{(i/\hbar)\int_{t_a}^{t_b} dt(\mathbf{p}\dot{\mathbf{x}}-H)}. \tag{7.570}$$

The same quantity is obtained from the second representation (7.567) by changing the integrand in the exponent from $\psi^*(\mathbf{x},t)\psi(\mathbf{x},t)$ to $\psi^*(\mathbf{x},t)(i\hbar\partial_t - \hat{H})\psi(\mathbf{x},t)$:

$$\left\langle \mathbf{x}_b t_b \left| \frac{i\hbar}{i\hbar\partial_t - \hat{H}} \right| \mathbf{x}_a t_a \right\rangle$$
$$= \oint \mathcal{D}\psi\mathcal{D}\psi^* \,\psi(\mathbf{x}_b, t_b)\psi^*(\mathbf{x}_a, t_a)e^{(i/\hbar)\int d^D x \int_{-\infty}^{\infty} dt\psi^*(\mathbf{x},t)(i\hbar\partial_t - \hat{H})\psi(\mathbf{x},t)}. \tag{7.571}$$

This is the second-quantized functional integral representation of the resolvent.

7.14.3   Interacting Fields

The interaction between particle orbits in a grand-canonical ensemble can be accounted for by anharmonic terms in the particle fields. A pair interaction between orbits, for example, corresponds to a fourth-order self interaction. An example is the interaction in the Bose-Einstein condensate corresponding to the energy in Eq. (7.101). Expressed in terms of the fields it reads

$$\mathcal{A}_{\mathrm{e}}^{\mathrm{int}}[\psi^*, \psi] = -\int_0^{\hbar\beta} d\tau\, \Delta E = -\frac{g}{2} \int_0^{\hbar\beta} d\tau \int d^3x\, \psi^*(\mathbf{x}, \tau + \eta)\psi^*(\mathbf{x}, \tau + \eta)\psi(\mathbf{x}, \tau)\psi(\mathbf{x}, \tau),$$
(7.572)

where $\eta > 0$ is an infinitesimal time shift. It is then possible to develop a perturbation theory in terms of Feynman diagrams by complete analogy with the treatment in Section 3.20 of the anharmonic oscillator with a fourth-order self interaction. The free correlation function is the momentum sum of oscillator correlation functions:

$$\langle \psi(\mathbf{x}, \tau)\psi^*(\mathbf{x}', \tau') \rangle = \sum_{\mathbf{p}, \mathbf{p}'} \langle a_{\mathbf{p}}(\tau) a_{\mathbf{p}'}^*(\tau') \rangle e^{i(\mathbf{p}\mathbf{x} - \mathbf{p}'\mathbf{x}')} = \sum_{\mathbf{p}} \langle a_{\mathbf{p}}(\tau) a_{\mathbf{p}}^*(\tau') \rangle e^{i\mathbf{p}(\mathbf{x} - \mathbf{x}')}. \quad (7.573)$$

The small $\eta > 0$ in (7.572) is necessary to specify the side of the jump of the correlation functions (recall Fig. 3.2). The expectation value of ΔE is given by (7.101), with a prefactor g rather than $g/2$ due to the two possible Wick contractions. Inserting the periodic correlation function (7.528), we obtain the Fourier integral

$$\langle \psi(\mathbf{x}, \tau)\psi^*(\mathbf{x}', \tau') \rangle = \sum_{\mathbf{p}} (1 + n_{\omega_{\mathbf{p}}}) e^{-\omega_{\mathbf{p}}(\tau - \tau') + i\mathbf{p}(\mathbf{x} - \mathbf{x}')}. \quad (7.574)$$

Recalling the representation (3.284) of the periodic Green function in terms of a sum over Matsubara frequencies, this can also be written as

$$\langle \psi(\mathbf{x}, \tau)\psi^*(\mathbf{x}', \tau') \rangle = \frac{1}{\hbar\beta} \sum_{\omega_m, \mathbf{p}} \frac{-1}{i\omega_m - \omega_{\mathbf{p}}} e^{-i\omega_m(\tau - \tau') + i\mathbf{p}(\mathbf{x} - \mathbf{x}')}. \quad (7.575)$$

The terms in the free energies (7.102) and (7.107) with the two parts (7.108) and (7.109) can then be shown to arise from the Feynman diagrams in the first line of Fig. 3.7.

In a grand-canonical ensemble, the energy $\hbar\omega_{\mathbf{p}}$ in (7.575) is replaced by $\hbar\omega_{\mathbf{p}} - \mu$. The same replacement appears in $\omega_{\mathbf{p}}$ of Eq. (7.574) which brings the distribution function $n_{\omega_{\mathbf{p}}}$ to [recall (7.529)]

$$n_{\omega_{\mathbf{p}} - \mu/\hbar} = \frac{1}{z^{-1} e^{\beta \hbar \omega_{\mathbf{p}}} - 1}, \quad (7.576)$$

where z is the fugacity $z = e^{\beta\mu}$. The expansion of the Feynman integrals in powers of z yields directly the expressions (7.102) and (7.107).

7.14.4 Effective Classical Field Theory

For the purpose of studying phase transitions, a functional integral over fields $\psi(\mathbf{x}, \tau)$ with an interaction (7.572) must usually be performed at a finite temperature. Then is often advisable to introduce a direct three-dimensional extension of the effective classical potential $V^{\mathrm{eff\,cl}}(x_0)$ introduced in Section 3.25 and used efficiently in Chapter 5. In a field theory we can set up, by analogy, an *effective classical action* which is a functional of the three-dimensional field with zero Matsubara frequency $\phi(\mathbf{x}) \equiv \psi_0(\mathbf{x})$. The advantages come from the reasons discussed in Section 3.25, that the zero-frequency fluctuations have a linearly diverging fluctuation width at high temperature, following the Dulong-Petit law. Thus only the nonzero-modes can be treated efficiently by the perturbative methods explained in Subsection 3.25.6.

By analogy with the splitting of the measure of path integration in Eq. (3.805), we may factorize the functional integral (7.561) into zero- and nonzero-Matsubara frequency parts as follows:

$$Z = \oint \mathcal{D}\psi \mathcal{D}\psi^* e^{-\mathcal{A}_e[\psi^*,\psi]} = \oint \mathcal{D}\psi_0 \mathcal{D}\psi_0^* \oint \mathcal{D}'\psi \mathcal{D}'\psi^* e^{-\mathcal{A}_e[\psi^*,\psi]}, \qquad (7.577)$$

and introduce the Boltzmann factor [compare (3.810)] contain the effective classical action

$$B[\psi_0^*, \psi_0] \equiv e^{-\mathcal{A}^{\text{eff cl}}[\psi_0^*,\psi_0]} \equiv \oint \mathcal{D}'\psi \mathcal{D}'\psi^* \, e^{-\mathcal{A}_e[\psi^*,\psi]}, \qquad (7.578)$$

to express the partition function as a *functional integral* over time-independent fields in three dimensions as:

$$Z = \oint \mathcal{D}\psi_0 \mathcal{D}\psi_0^* e^{-\mathcal{A}^{\text{eff cl}}[\psi_0^*,\psi_0]}. \qquad (7.579)$$

In Subsection 3.25.1 we have seen that the full effective classical potential $V^{\text{eff cl}}(x_0)$ in Eq. (3.809) reduces in the high-temperature limit to the initial potential $V(x_0)$. For the same reason, the full effective classical action in the functional integral (7.577) can be approximated at high temperature by the *bare effective classical action*, which is simply the zero-frequency part of the initial action:

$$\mathcal{A}_b^{\text{eff cl}}[\psi_0^*, \psi_0] = \beta \int d^3x \left\{ \psi_0^*(\mathbf{x}) \left(-\frac{1}{2m} \nabla^2 - \mu \right) \psi_0(\mathbf{x}) + \frac{2\pi a}{m} \left[\psi_0^*(\mathbf{x}) \psi_0(\mathbf{x}) \right]^2 \right\}. \qquad (7.580)$$

This follows directly from the fact that, at high temperature, the fluctuations in the functional integral (7.578) are strongly suppressed by the large Matsubara frequencies in the kinetic terms.

Remarkably, the absence of a shift in the critical temperature in the first-order energy (7.113) deduced from Eq. (7.116) implies that the chemical potential in the effective classical action does not change at this order [34]. For this reason, the lowest-order shift in the critical temperature of a weakly interacting Bose-Einstein condensate can be calculated entirely from the three-dimensional effective classical field theory (7.577) with the bare effective classical action (7.580) in the Boltzmann factor.

The action (7.580) may be brought to a more conventional form by introducing the differently normalized two-component fields $\phi = (\phi_1, \phi_2)$ related to the original complex field ψ by $\psi(\mathbf{x}) = \sqrt{MT}[\phi_1(\mathbf{x}) + i\phi_2(\mathbf{x})]$. If we also define a square mass $m^2 \equiv -2M\mu$ and a quartic coupling $u = 48\pi a MT$, the bare effective classical action reads

$$\mathcal{A}_b^{\text{eff cl}}[\psi_0^*, \psi_0] = \mathcal{A}[\phi] = \int d^3x \left[\frac{1}{2}|\nabla\phi|^2 + \frac{1}{2}m^2\phi^2 + \frac{u}{4!}(\phi^2)^2 \right]. \qquad (7.581)$$

In the field theory governed by the action (7.581), the relation (7.111) for the shift of the critical temperature to lowest order in the coupling constant becomes

$$
\frac{\Delta T_c}{T_c^{(0)}} \approx -\frac{2}{3}\frac{mT_c^{(0)}}{n}\left\langle \Delta\phi^2 \right\rangle = -\frac{4\pi}{3}\frac{(mT_c^{(0)})^2}{n}4!\left\langle \frac{\Delta\phi^2}{u} \right\rangle a
$$
$$
= -\frac{4\pi}{3}\frac{(2\pi)^2}{[\zeta(3/2)]^{4/3}}4!\left\langle \frac{\Delta\phi^2}{u} \right\rangle an^{1/3}, \tag{7.582}
$$

where $\left\langle \Delta\phi^2 \right\rangle$ is the shift in the expectation value of ϕ^2 caused by the interaction. Since a repulsive interaction pushes particles apart, $\left\langle \Delta\phi^2 \right\rangle$ is negative, thus explaining the positive shift in the critical temperature. The evaluation of the expectation value $\left\langle \Delta\phi^2 \right\rangle$ from the path integral (7.577) with the bare three-dimensional effective classical action (7.581) in the exponent can now proceed within one of the best-studied field theories in the literature [51]. The theoretical tools for calculating strong-coupling results in this theory are well developed, and this has made it possible to drive the calculations of the shift to the five-loop order [21].

Some low-order corrections to the effective classical action (7.580) have been calculated from the path integral (7.578) in Ref. [34].

7.15 Bosonization

The path integral formulation of quantum-mechanical systems is very flexible. Just as integrals can be performed in different variables of integration, so can path integrals in different path variables. This has important applications in many-body systems, which show a rich variety of so-called collective phenomena. Practically all fermion systems show collective excitations such as sound, second sound, and spin waves. These are described phenomenologically by bosonic fields. In addition, there are phase transitions whose description requires a bosonic order parameter. Superconductivity of electron systems is a famous example where a bosonic order parameter appears in a fermion system—the energy gap. In all these cases it is useful to transform the initial path variables to so-called collective path variables, in higher dimensions these become *collective fields* [32].

Let us illustrate the technique with simple a model defined by the Lagrangian

$$
L(t) = a^*(t)i\partial_t a(t) - \frac{\varepsilon}{2}\left[a^*(t)a(t)\right]^2, \tag{7.583}
$$

where a^*, a are commuting or anticommuting variables. All Green functions can be calculated from the generating functional

$$
Z[\eta^*,\eta] = N\int \mathcal{D}a^*\mathcal{D}a\, \exp\left[i\int dt\,(L + \eta^*a + a^*\eta)\right], \tag{7.584}
$$

where we have omitted an irrelevant overall factor.

In operator language, the model is defined by the Hamiltonian operator

$$
\hat{H} = \varepsilon(\hat{a}^\dagger\hat{a})^2/2, \tag{7.585}
$$

where $\hat{a}^\dagger$, $\hat{a}$ are creation and annihilation operator of either a boson or a fermion at a point. In the boson case, the eigenstates are

$$|n\rangle = \frac{1}{\sqrt{n!}}(\hat{a}^\dagger)^n|0\rangle, \quad n = 0, 1, 2, \ldots , \tag{7.586}$$

with an energy spectrum

$$E_n = \varepsilon\frac{n^2}{2}. \tag{7.587}$$

In the fermion case, there are only two solutions

$$|0\rangle \qquad \text{with} \quad E_0 = 0, \tag{7.588}$$

$$|1\rangle = a^\dagger|0\rangle \quad \text{with} \quad E_1 = \frac{\varepsilon}{2}. \tag{7.589}$$

Here the Green functions are obtained from the generating functional

$$Z[\eta^\dagger, \eta] = \langle 0|\hat{T} \exp\left[i \int dt(\eta^*\hat{a} + \hat{a}^\dagger\eta)\right]|0\rangle, \tag{7.590}$$

where $\hat{T}$ is the time ordering operator (1.241). The functional derivatives with respect to the sources η^*, η generate all Green functions of the type (3.296) at $T = 0$.

7.15.1 Collective Field

At this point we introduce an additional *collective field* via the Hubbard-Stratonovich transformation [52]

$$\exp\left\{-i\frac{\varepsilon}{2} \int dt[a^*(t)a(t)]^2\right\} = N \int \mathcal{D}\rho(t) \exp\left\{i \int dt \left[\frac{\rho^2(t)}{2\varepsilon} - \rho(t)a^*(t)a(t)\right]\right\}, \tag{7.591}$$

where N is some irrelevant factor. Equivalently we multiply the partition function (7.584) with the trivial Gaussian path integral

$$1 = N \int \mathcal{D}\rho(t) \exp\left\{\int dt \frac{1}{2\varepsilon} \left[\rho(t) - \varepsilon a^\dagger(t)a(t)\right]^2\right\}, \tag{7.592}$$

to obtain the generating functional

$$Z[\eta^\dagger, \eta] = N \int \mathcal{D}a^*\mathcal{D}a\mathcal{D}\rho$$
$$\times \exp\left\{\int dt \left[a^*(t)i\partial_t a(t) - \varepsilon\rho(t)a^*(t)a(t) + \frac{\rho^2(t)}{2\varepsilon} + \eta^*(t)a(t) + a^*(t)\eta(t)\right]\right\}. \tag{7.593}$$

From (7.591) we see that the collective field $\rho(t)$ fluctuates harmonically around ε times particle density. By extremizing the action in (7.593) with respect to $\rho(t)$ we obtain the classical equality:

$$\rho(t) = \varepsilon a^*(t)a(t). \tag{7.594}$$

The virtue of the Hubbard-Stratonovich transformation is that the fundamental variables a^*, a appear now quadratically in the action and can be integrated out to yield a path integral involving only the collective variable $\rho(t)$:

$$Z[\eta^*, \eta] = N \int \mathcal{D}\rho \exp \left\{ i\mathcal{A}[\rho] - \int dt dt' \eta^*(t) G_\rho(t, t') \eta(t') \right\}, \qquad (7.595)$$

with the collective action

$$\mathcal{A}[\rho] = \pm i \mathrm{Tr} \log \left(iG_\rho^{-1} \right) + \int dt \frac{\rho^2(t)}{2}, \qquad (7.596)$$

where G_ρ denotes the Green function of the fundamental path variables in an external $\rho(t)$ background potential, which satisfies [compare (3.75)]

$$[i\partial_t - \rho(t)] G_\rho(t, t') = i\delta(t - t'). \qquad (7.597)$$

The Green function was found in Eq. (3.124). Here we find the solution once more in a different way which will be useful in the sequel. We introduce an auxiliary field

$$\varphi(t) = \int^t \rho(t') dt', \qquad (7.598)$$

in terms of which $G_\rho(t, t')$ is simply

$$G_\rho(t, t') = e^{-i\varphi(t)} e^{i\varphi(t')} G_0(t - t'), \qquad (7.599)$$

with G_0 being the free-field propagator of the fundamental particles. At this point one has to specify the boundary condition on $G_0(t - t')$. They have to be adapted to the physical situation of the system. Suppose the time interval is infinite. Then G_0 is given by Eq. (3.100), so that we obtain, as in (3.124),

$$G_\rho(t, t') = e^{-i\varphi(t)} e^{i\varphi(t')} \bar{\Theta}(t - t'). \qquad (7.600)$$

The Heaviside function $\bar{\Theta}$ defined in Eq. (1.309) corresponds to $\rho(t)$ coupling to the symmetric operator combination $(\hat{a}^\dagger \hat{a} + \hat{a} \hat{a}^\dagger)/2$. The products $\hat{a}^\dagger \hat{a}$ in the Hamiltonian (7.585), however, have the creation operators left of the annihilation operators. In this case we must use the Heaviside function $\Theta(t - t')$ of Eq. (1.300). If this function is inserted into (7.600), the right-hand side of (7.600) vanishes at $t = t'$.

Using this property we see that the functional derivative of the tracelog in (7.596) vanishes:

$$\frac{\delta}{\delta\rho(t)} \left\{ \pm i \mathrm{Tr} \log(iG_\rho^{-1}) \right\} = \frac{\delta}{\delta\rho(t)} \left\{ \pm i \mathrm{Tr} \log \left[i\partial_t - \rho(t) \right] \right\} = \mp G_\rho(t, t')|_{t'=t} = 0. \quad (7.601)$$

This makes the tracelog an irrelevant constant. The generating functional is then simply

$$Z[\eta^\dagger, \eta] = N \int \mathcal{D}\varphi(t) \exp \left\{ \frac{i}{2\varepsilon} \int dt \, \dot{\varphi}(t)^2 - \int dt dt' \eta^\dagger(t) \eta(t') e^{-i\varphi(t)} e^{-i\varphi(t')} \Theta(t - t') \right\},$$

$$\qquad (7.602)$$

where we have used the Jacobian

$$\mathcal{D}\rho = \mathcal{D}\varphi \, \text{Det}(\partial_t) = \mathcal{D}\varphi, \tag{7.603}$$

since $\text{Det}(\partial_t) = 1$ according to (2.535). Observe that the transformation (7.598) provides $\varphi(t)$ with a standard kinetic term $\dot{\varphi}^2(t)$.

The original theory has been transformed into a new theory involving the collective field φ. In a three-dimensional generalization of this theory to electron gases, the field φ describes density waves [32].

7.15.2 Bosonized versus Original Theory

The first term in the bosonized action in (7.602) is due to a Lagrangian

$$L_0(t) = \frac{1}{2\varepsilon}\dot{\varphi}(t)^2 \tag{7.604}$$

describing free particles on the infinite φ-axis. Its correlation function is divergent and needs a small frequency, say κ, to exist. Its small-κ expansion reads

$$
\begin{aligned}
\langle \varphi(t)\varphi(t') \rangle &= \varepsilon \int \frac{d\omega}{2\pi} \frac{i}{\omega^2 - \kappa^2 + i\epsilon} e^{-i\omega(t-t')} \\
&= \frac{\varepsilon}{2\kappa} e^{-\kappa|t-t'|} = \frac{\varepsilon}{2\kappa} - \varepsilon\frac{i}{2}|t-t'| + \mathcal{O}(\kappa).
\end{aligned}
\tag{7.605}
$$

The first term plays an important role in calculating the bosonized correlation functions generated by $Z[\eta^*, \eta]$ in Eq. (7.602). Differentiating this n times with respect to η and η^* and setting them equal to zero we obtain

$$
\begin{aligned}
G^{(n)}(t_n, \dots, t_1; t'_n, \dots, t'_1) &= (-1)^n \langle e^{-i\varphi(t_n)} \cdots e^{-i\varphi(t_1)} \, e^{i\varphi(t'_n)} \cdots e^{i\varphi(t'_1)} \rangle \\
&\times \sum_p \epsilon_p \Theta(t_{p_1} - t_1) \cdots \Theta(t_{p_n} - t_1),
\end{aligned}
\tag{7.606}
$$

where the sum runs over all permutations p of the time labels of $t_1 \dots t_n$ making sure that these times are all later than the times $t'_1 \dots t'_n$. The factor ϵ_p reflects the commutation properties of the sources η, η^*, being equal to 1 for all permutations if η and η^* commute. For Grassmann variables η, η^*, the factor ϵ_p is equal to -1 if the permutation p is odd. The expectation value is defined by the path integral

$$\langle \, \dots \, \rangle \equiv \frac{1}{Z[0,0]} \int \mathcal{D}\varphi(t) \dots e^{(i/2\varepsilon) \int dt \, \dot{\varphi}(t)^2}. \tag{7.607}$$

Let us calculate the expectation value of the exponentials in (7.606), writing it as

$$\left\langle \exp\left[i\sum_{i=1}^{2n} q_i \varphi(t_i) \right] \right\rangle = \left\langle \exp\left[\sum_i \int dt \, \delta(t - t_i) q_i \varphi(t) \right] \right\rangle, \tag{7.608}$$

where we have numbered the times as $t_1, t_2, \ldots t_{2n}$ rather than $t_1, t_1', t_2, t_2', \ldots t_n, t_n'$. Half of the $2n$ "charges" q_i are 1, the other half are -1. Using Wick's theorem in the form (3.307) we find

$$
\left\langle e^{i \sum_{i=1}^{2n} q_i \varphi(t_i)} \right\rangle = \exp\left[-\frac{1}{2} \int dt dt' \sum_{i=1}^{2n} q_i \delta(t - t_i) \langle \varphi(t)\varphi(t') \rangle (t') \sum_j q_j \delta(t - t_j) \right]
$$

$$
= \exp\left[-\frac{1}{2} \sum_{i,j=1}^{2n} q_i q_j \langle \varphi(t_i)\varphi(t_j) \rangle \right]. \tag{7.609}
$$

Inserting here the small-κ expansion (7.605), we see that the $1/\kappa$-term gives a prefactor

$$
\exp\left[-\frac{\varepsilon}{4\kappa} \left(\sum_i q_i \right)^2 \right], \tag{7.610}
$$

which vanishes since the sum of all "charges" q_i is zero. Hence we can go to the limit $\kappa \to 0$ and obtain

$$
\left\langle \exp\left[i \sum_{i=1}^{2n} q_i \varphi(t_i) \right] \right\rangle = \exp\left[\varepsilon \frac{i}{2} \sum_{i>j} q_i q_j |t_i - t_j| \right]. \tag{7.611}
$$

Note that the limit $\kappa \to 0$ makes all correlation functions vanish which do not contain an equal number of $a^*(t)$ and $a(t)$ variables. It ensures "charge conservation".

For one positive and one negative charge we obtain, after a multiplication by a Heaviside function $\Theta(t_1 - t_1')$, the two-point function

$$
G^{(2)}(t, t') = e^{-i\varepsilon(t-t')/2} \Theta(t - t'). \tag{7.612}
$$

This agrees with the operator result derived from the generating functional (7.590), where

$$
G^{(2)}(t, t') = \langle 0|\hat{T} \hat{a}(t) \hat{a}^\dagger(t')|0 \rangle = \Theta(t - t')\langle 0|\hat{a}(t)\hat{a}^\dagger(t')|0 \rangle e^{-i\varepsilon(t-t')/2} \Theta(t - t'). \tag{7.613}
$$

Inserting the Heisenberg equation [recall (1.277)]

$$
a(t) = e^{it\hat{H}} a e^{-it\hat{H}}, \qquad a^\dagger(t') = e^{it'\hat{H}} a^\dagger e^{-it'\hat{H}}. \tag{7.614}
$$

we find

$$
\langle 0|\hat{T}\hat{a}(t)\hat{a}^\dagger(t')|0\rangle = \Theta(t - t')\langle 0|e^{i\varepsilon(\hat{a}^\dagger \hat{a})^2 t/2} a e^{-\frac{i}{2}(\hat{a}^\dagger \hat{a})^2 (t-t')} \hat{a}^\dagger e^{-i(\hat{a}^\dagger \hat{a})^2 t'/2}|0\rangle
$$

$$
= \Theta(t - t')\langle 1|e^{-\frac{i}{2}(\hat{a}^\dagger \hat{a})^2 (t-t')}|1\rangle = \Theta(t - t') e^{-i\varepsilon(t-t')/2}. \tag{7.615}
$$

Let us compare the above calculations with an operator evaluation of the bosonized theory. The Hamiltonian operator associated with the Lagrangian (7.604) is $\hat{H} = \varepsilon \hat{p}^2/2$. The states in the Hilbert space are eigenstates of the momentum operator $\hat{p} = -i\partial/\partial\varphi$:

$$
\{\varphi|p\} = \frac{1}{\sqrt{2\pi}} e^{ip\varphi}. \tag{7.616}
$$

Here, curly brackets are a modified Dirac notation for states of the bosonized theory, which distinguishes them from the states of the original theory created by products of operators $\hat{a}^\dagger(t)$ acting on $|0\rangle$. In operator language, the generating functional of the theory (7.602) reads

$$Z[\eta^\dagger, \eta] = \frac{1}{\{0|0\}}\{0|\hat{T}\exp\left[-\int dt dt' \eta^\dagger(t)\eta(t')e^{-i\hat{\varphi}(t)}e^{i\hat{\varphi}(t')}\Theta(t-t')\right]|0\}, \quad (7.617)$$

where $\hat{\varphi}(t)$ are free-particle operators.

We obtain all Green functions of the initial operators $\hat{a}(t)$, $\hat{a}^\dagger(t)$ by forming functional derivatives of the generating functional $Z[\eta^\dagger, \eta]$ with respect to $\eta^\dagger, \eta$. Take, for instance, the two-point function

$$\langle 0|\hat{T}\hat{a}(t)\hat{a}^\dagger(t')|0\rangle = -\left.\frac{\delta^{(2)}Z}{\delta\eta^\dagger(t)\delta\eta(t')}\right|_{\eta^\dagger,\eta=0} = \frac{1}{\{0|0\}}\{0|e^{-i\hat{\varphi}(t)}e^{i\hat{\varphi}(t')}|0\}\Theta(t-t'). \quad (7.618)$$

Inserting here the Heisenberg equation (1.277),

$$\hat{\varphi}(t) = e^{i\varepsilon\frac{\hat{p}^2}{2}t}\,\hat{\varphi}(0)\,e^{-i\varepsilon\frac{\hat{p}^2}{2}t}, \quad (7.619)$$

we can evaluate the matrix element in (7.618) as follows:

$$\{0|e^{i\varepsilon\frac{\hat{p}^2}{2}t}2e^{-i\hat{\varphi}(0)}e^{-i\varepsilon\frac{\hat{p}^2}{2}(t-t')}e^{i\hat{\varphi}(0)}e^{-i\varepsilon\frac{\hat{p}^2}{2}t'}|0\} = \{0|e^{-i\hat{\varphi}(0)}e^{-i\varepsilon\frac{\hat{p}^2}{2}(t-t')}e^{i\hat{\varphi}(0)}|0\}. \quad (7.620)$$

The state $e^{i\hat{\varphi}(0)}|0\}$ is an eigenstate of $\hat{p}$ with unit momentum $|1\}$ and the same norm as $|0\}$, so that (7.620) equals

$$\frac{1}{\{0|0\}}\{1|1\}\,e^{-i\varepsilon(t-t')/2} = e^{-i\varepsilon(t-t')/2} \quad (7.621)$$

and the Green function (7.618) becomes, as in (7.613) and (7.615):

$$\langle 0|\hat{T}\hat{a}(t)\hat{a}^\dagger(t')|0\rangle = e^{-i\varepsilon(t-t')/2}\Theta(t-t'). \quad (7.622)$$

Observe that the Fermi and Bose statistics of the original operators $\hat{a}, \hat{a}^\dagger$ enters to result only via the commutation properties of the sources.

There exists also the opposite phenomenon that a bosonic theory possesses solutions which behave like fermions. This will be discussed in Section 8.12.

Appendix 7A   Treatment of Singularities in Zeta-Function

Here we show how to evaluate the sums which determine the would-be critical temperatures of a Bose gas in a box and in a harmonic trap.

7A.1   Finite Box

According to Eqs. (7.83), (7.88), and (7.90), the relation between temperature $T = \hbar^2\pi^2/bML^2k_B$ and the fugacity z_D at a fixed particle number N in a finite D-dimensional box is determined by the equation

$$N = N_{\mathrm{n}}(T) + N_{\mathrm{cond}}(T) = S_D(z_D) + \frac{z_D}{1 - z_D}, \tag{7A.1}$$

where $S_D(z_D)$ is the subtracted infinite sum

$$S_D(z_D) \equiv \sum_{w=1}^{\infty} [Z_1^D(wb)e^{wDb/2} - 1]z_D^w, \tag{7A.2}$$

containing the Dth power of one-particle partition function in the box $Z_1(b) = \sum_{k=1}^{\infty} e^{-bk^2/2}$. The would-be critical temperature is found by equating this sum at $z_D = 1$ with the total particle number N. We shall rewrite $Z_1(b)$ as

$$Z_1(b) = e^{-b/2}\left[1 + e^{-3b/2}\sigma_1(b)\right], \tag{7A.3}$$

where $\sigma_1(b)$ is related to the elliptic theta function (7.85) by

$$\sigma_1(b) \equiv \sum_{k=2}^{\infty} e^{-(k^2-4)b/2} = \frac{e^{2b}}{2}\left[\vartheta_3(0, e^{-b/2}) - 1 - 2e^{-b/2}\right]. \tag{7A.4}$$

According to Eq. (7.87), this has the small-b behavior

$$\sigma_1(b) = \sqrt{\frac{\pi}{2b}}e^{2b} - e^{3b/2} - \frac{1}{2}e^{2b} + \ldots. \tag{7A.5}$$

The omitted terms are exponentially small as long as $b < 1$ [see the sum over m in Eq. (7.86)]. For large b, these terms become important to ensure an exponentially fast falloff like $e^{-3b/2}$. Inserting (7A.3) into (7A.2), we find

$$S_D(1) \equiv D\sum_{w=1}^{\infty}\left[\sigma_1(wb)e^{-3wb/2} + \frac{D-1}{2}\sigma_1^2(wb)e^{-6wb/2} + \frac{(D-1)(D-2)}{6}\sigma_1^3(wb)e^{-9wb/2}\right]. \tag{7A.6}$$

Inserting here the small-b expression (7A.5), we obtain

$$S_2(1) \equiv \sum_{w=1}^{\infty}\left(\frac{\pi}{2wb}e^{wb} - \sqrt{\frac{\pi}{2wb}}e^{wb} + \frac{1}{4}e^{wb} - 1\right) + \ldots, \tag{7A.7}$$

$$S_3(1) \equiv \sum_{w=1}^{\infty}\left(\sqrt{\frac{\pi}{2wb}}^3 e^{3wb/2} - \frac{3}{2}\frac{\pi}{2wb}e^{3wb/2} + \frac{3}{4}\sqrt{\frac{\pi}{2wb}}e^{3wb/2} - \frac{1}{8}e^{3wb/2} - 1\right) + \ldots, \tag{7A.8}$$

the dots indicating again exponentially small terms. The sums are convergent only for negative b, this being a consequence of the approximate nature of these expressions. If we evaluate them in this regime, the sums produce Polylogarithmic functions (7.34), and we find, using also $\sum_{w=1}^{\infty} 1 = \zeta(0) = -1/2$ from (2.520),

$$S_2(1) = \zeta_1(e^b) - \sqrt{\frac{\pi}{2b}}\zeta_{1/2}(e^b) + \frac{1}{4}\zeta_0(e^b) - \zeta(0) + \ldots. \tag{7A.9}$$

$$S_3(1) = \sqrt{\frac{\pi}{2b}}^3 \zeta_{3/2}(e^{3b/2}) - \frac{3}{2}\frac{\pi}{2b}\zeta_1(e^{3b/2}) + \frac{3}{4}\sqrt{\frac{\pi}{2b}}\zeta_1(e^{3b/2}) - \frac{1}{8}\zeta_0(e^{3b/2}) - \zeta(0) + \ldots. \tag{7A.10}$$

These expressions can now be expanded in powers of b with the help of the Robinson expansion (2.579). Afterwards, b is continued analytically to positive values and we obtain

$$S_2(1) = -\frac{\pi}{2b}\log(C_2 b) - \sqrt{\frac{\pi}{2b}}\zeta(1/2) + \frac{1}{8}(3 - 2\pi) + \mathcal{O}(b^{1/2}), \tag{7A.11}$$

$$S_3(1) = \sqrt{\frac{\pi}{2b}}^3 \zeta(3/2) + \frac{3\pi}{4b}\log(C_3 b) + \frac{3}{4}\sqrt{\frac{\pi}{2b}}\zeta(1/2)(1 + \pi) + \frac{9}{16}(1 + \pi) + \mathcal{O}(b^{1/2}). \tag{7A.12}$$

The constants $C_{2,3}, C'_{2,3}$ inside the logarithms turn out to be complex, implying that the limiting expressions (7A.7) and (7A.8) cannot be used reliably. A proper way to proceed gors as follows. We subtract from $S_D(1)$ terms which remove the small-b singularties by means of modifications of (7A.7) and (7A.8) which have the same small-b expansion up to b^0:

$$\tilde{S}_2(1) \equiv \sum_{w=1}^{\infty}\left(\frac{\pi}{2wb} - \sqrt{\frac{\pi}{2wb}} + \frac{4\pi - 3}{4}\right)e^{-wb} + \dots, \tag{7A.13}$$

$$\tilde{S}_3(1) \equiv \sum_{w=1}^{\infty}\left[\sqrt{\frac{\pi}{2wb}}^3 - \frac{3}{2}\frac{\pi}{2wb} + \frac{3}{4}(1 + 2\pi)\sqrt{\frac{\pi}{2wb}} - \frac{9}{8}(1 + 2\pi)\right]e^{-3wb/2} + \dots. \tag{7A.14}$$

In these expressions, the sums over w can be performed for positive b yielding

$$\tilde{S}_2(1) \equiv \frac{\pi}{2b}\zeta_1(e^{-b}) - \sqrt{\frac{\pi}{2b}}\zeta_{1/2}(e^{-b}) + \frac{4\pi - 3}{4}\zeta_0(e^{-b}) + \dots, \tag{7A.15}$$

$$\tilde{S}_3(1) \equiv \sqrt{\frac{\pi}{2b}}^3 \zeta_{3/2}(e^{-3b/2}) - \frac{3}{2}\frac{\pi}{2b}\zeta_1(e^{-3b/2}) + \frac{3}{4}(1 + 2\pi)\sqrt{\frac{\pi}{2b}}\zeta_{1/2}(e^{-3b/2}) - \frac{9}{8}(1 + 2\pi)\zeta_0(e^{-3b/2}) + \dots. \tag{7A.16}$$

Inserting again the Robinson expansion (2.579), we obtain once more the above expansions (7A.11) and (7A.12), but now with the well-determined real constants

$$\tilde{C}_2 = e^{3/2\pi - 2 + \sqrt{2}} \approx 0.8973, \qquad \tilde{C}_3 = \frac{3}{2}e^{-2 + 1/\sqrt{3} - 1/\pi} \approx 0.2630. \tag{7A.17}$$

The subtracted expressions $S_D(1) - \tilde{S}_D(1)$ are smooth near the origin, so that the leading small-b behavior of the sums over these can simply be obtained from a numeric integral over w:

$$\int_0^{\infty} dw[S_2(1) - \tilde{S}_2(1)] = -\frac{1.1050938}{b}, \qquad \int_0^{\infty} dw[S_3(1) - \tilde{S}_3(1)] = 3.0441. \tag{7A.18}$$

These modify the constants $\tilde{C}_{2,3}$ to

$$C_2 = 1.8134, \qquad C_3 = 0.9574. \tag{7A.19}$$

The corrections to the sums over $S_D(1) - \tilde{S}_D(1)$ are of order b^0 an higher and already included in the expansions (7A.11) and (7A.12), which were only unreliable as far as $C_{2,3}$ os concerned.

Let us calculate from (7A.12) the finite-size correction to the critical temperature by equating $S_3(1)$ with N. Expressing this in terms of $b_c^{(0)}$ via (7.92), and introducing the ratio $\hat{b}_c \equiv b_c/b_c^{(0)}$ which is close to unity, we obtain the expansion in powers of the small quantity $2b_c^{(0)}/\pi = [\zeta(3/2)/N]^{2/3}$:

$$\hat{b}_c^{3/2} = 1 + \sqrt{\frac{2b_c^{(0)}}{\pi}}\frac{3\sqrt{\hat{b}_c}}{2\zeta(3/2)}\log(C_3 b_c^{(0)}\hat{b}_c) + \frac{2b_c^{(0)}}{\pi}\frac{3}{4\zeta(3/2)}\zeta(1/2)(1 + \pi)\hat{b}_c + \dots. \tag{7A.20}$$

To lowest order, the solution is simply

$$\hat{b}_c = 1 + \sqrt{\frac{2b_c^{(0)}}{\pi}} \frac{1}{\zeta(3/2)} \log(C_3 b_c^{(0)}) + \dots \,, \tag{7A.21}$$

yielding the would-be critical temperature to first in $1/N^{1/3}$ as stated in (7.94). To next order, we insert into the last term the zero-order solution $\hat{b}_c \approx 1$, and in the second term the first-order solution (7A.21) to find

$$
\begin{aligned}
\hat{b}_c^{3/2} =\ & 1 + \frac{3}{2}\sqrt{\frac{2b_c^{(0)}}{\pi}} \frac{1}{\zeta(3/2)} \log(C_3 b_c^{(0)}) \\
& + \frac{2b_c^{(0)}}{\pi} \frac{3}{4\zeta(3/2)} \left\{ \zeta(1/2)(1+\pi) + \frac{1}{\zeta(3/2)} \left[2\log(C_3 b_c^{(0)}) + \log^2(C_3 b_c^{(0)}) \right] \right\} + \dots \,. \tag{7A.}
\end{aligned}
$$

Replacing $b_c^{(0)}$ by $(2/\pi)\,[\zeta(3/2)/N]^{2/3}$, this gives us the ratio $(T_c^{(0)}/T_c)^{3/2}$ between finite- and infinite size critical temperatures T_c and $T_c^{(0)}$. The first and second-order corrections are plotted in Fig. 7.12, together with precise results from numeric solution of the equation $N = S_3(1)$.

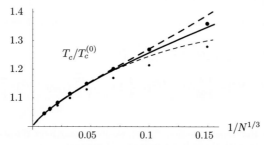

Figure 7.12 Finite-size corrections to the critical temperature for $N = 300$ to infinity calculated once from the formula $N = S_3(1)$ (solid curve) and once from the expansion (7A.22) (short-dashed up to order $[b_c^{(0)}]^{1/2} \propto 1/N^{1/3}$, long-dashed up to the order $b_c^{(0)} \propto 1/N^{2/3}$). The fat dots show the peaks in the second derivative $d^2 N_{\text{cond}}(T)/dT^2$. The small dots show the corresponding values for canonical ensembles, for comparison.

7A.2 Harmonic Trap

The sum relevant for the would-be phase transition in a harmonic trap is (7.138),

$$S_D(b, z_D) = \sum_{w=1}^{\infty} \left[\frac{1}{(1-e^{-wb})^D} - 1 \right] z_D^w, \tag{7A.23}$$

which determines the number of normal particles in the harmonic trap via Eq. (7.139). We consider only the point $z_D = 1$ which determines the critical temperature by the condition $N_{\text{n}} = N$. Restricting ourselves to the physical cases $D = 1, 2, 3$, we rewrite the sum as

$$S_D(b,1) = \sum_{w=1}^{\infty} D\left[e^{-wb} - \frac{(D-1)}{2} e^{-2wb} + \frac{(D-1)(D-2)}{6} e^{-3wb} \right] \frac{1}{(1-e^{-wb})^D}. \tag{7A.24}$$

According to the method developed in the evaluation of Eq. (2.579) we obtain such a sum in two steps. First we go to small b where the sum reduces to an integral over w. After this we calculate the difference between sum and integral by a naive power series expansion.

As it stands, the sum (7A.24) cannot be converted into an integral due to singularities at $w = 0$. These have to be first removed by subtractions. Thus we decompose $S_D(b, 1)$ into a subtracted sum plus a remainder as

$$S_D(b,1) = \bar{S}_D(b,1) + \Delta_D S_D(b,1) + b\frac{D}{2}\Delta_{D-1}S_D(b,1) + b^2\frac{D(3D-1)}{24}\Delta_{D-2}S_D(b,1), \quad (7A.25)$$

where

$$\begin{aligned}
\bar{S}_D(b,1) &= \sum_{w=1}^{\infty} D\left[e^{-wb} - \frac{D-1}{2}e^{-2wb} + \frac{(D-1)(D-2)}{6}e^{-3wb}\right] \\
&\times \left[\frac{1}{(1-e^{-wb})^D} - \frac{1}{w^D b^D} - \frac{D}{2w^{D-1}b^{D-1}} - \frac{D(3D-1)}{24w^{D-2}b^{D-2}}\right]
\end{aligned} \quad (7A.26)$$

is the subtracted sum and

$$\Delta_{D'}S_D(b,1) \equiv \frac{D}{b^D}\left[\zeta_{D'}(e^{-b}) - \frac{D-1}{2}\zeta_{D'}(e^{-2b}) + \frac{(D-1)(D-2)}{6}\zeta_{D'}(e^{-3b})\right] \quad (7A.27)$$

collects the remainders. The subtracted sum can now be done in the limit of small b as an integral over w, using the well-known integral formula for the Beta function:

$$\int_0^{\infty} dx\,\frac{e^{-ax}}{(1-e^{-x})^b} = B(a, 1-b) = \frac{\Gamma(a)\Gamma(1-b)}{\Gamma(1+a-b)}. \quad (7A.28)$$

This yields the small-b contributions to the subtracted sums

$$\begin{aligned}
\bar{S}_1(b,1) &\underset{b\to 0}{\to} \frac{1}{b}\left(\gamma - \frac{7}{12}\right) \equiv s_1, \\
\bar{S}_2(b,1) &\underset{b\to 0}{\to} \frac{1}{b}\left(\gamma + \log 2 - \frac{9}{8}\right) \equiv s_2, \\
\bar{S}_3(b,1) &\underset{b\to 0}{\to} \frac{1}{b}\left(\gamma + \log 3 - \frac{19}{24}\right) \equiv s_3,
\end{aligned} \quad (7A.29)$$

where $\gamma = 0.5772\ldots$ is the Euler-Mascheroni number (2.467). The remaining sum-minus-integral is obtained by a series expansion of $1/(1 - e^{-wb})^D$ in powers of b and performing the sums over w using formula (2.582). However, due to the subtractions, the corrections are all small of order $(1/b^D)\mathcal{O}(b^3)$, and will be ignored here. Thus we obtain

$$S_D(b,1) = \frac{s_D}{b^D} + \Delta S_D(b,1) + \frac{1}{b^D}\mathcal{O}(b^3). \quad (7A.30)$$

We now expand $\Delta_{D'}S_D(b,1)$ using Robinson's formula (2.581) up to b^2/b^D and find

$$\begin{aligned}
\Delta_{D'}S_1(b,1) &= \frac{1}{b}\zeta_{D'}(e^{-b}), \\
\Delta_{D'}S_2(b,1) &= \frac{1}{b^2}\left[2\zeta_{D'}(e^{-b}) - \zeta_{D'}(e^{-2b})\right], \quad (7A.31) \\
\Delta_{D'}S_3(b,1) &= \frac{1}{b^3}\left[3\zeta_{D'}(e^{-b}) - 3\zeta_{D'}(e^{-2b}) + \zeta_{D'}(e^{-3b})\right], \quad (7A.32)
\end{aligned}$$

where

$$\begin{aligned}
\zeta_1(e^{-b}) &= -\log\left(1 - e^{-b}\right) = -\log b + \frac{b}{2} - \frac{b^2}{24} + \ldots, \\
\zeta_2(e^{-b}) &= \zeta(2) + b(\log b - 1) - \frac{b^2}{4} + \ldots, \quad (7A.33) \\
\zeta_3(e^{-b}) &= \zeta(3) - \frac{b}{6}\zeta(2) - \frac{b^2}{2}\left(\log b - \frac{3}{2}\right) + \ldots, \quad (7A.34)
\end{aligned}$$

The results are

$$S_1(b,1) = \frac{1}{b}\left[(-\log b + \gamma) + \frac{b}{4} - \frac{b^2}{144} + \cdots\right],$$

$$S_2(b,1) = \frac{1}{b^2}\left[\zeta(2) - b\left(\log b - \gamma + \frac{1}{2}\right) + \frac{7b^2}{24} + \cdots\right],\tag{7A.35}$$

$$S_3(b,1) = \frac{1}{b^3}\left[\zeta(3) + \frac{3b}{2}\zeta(2) - b^2\left(\log b - \gamma + \frac{19}{24}\right) + \cdots\right],$$

as stated in Eqs. (7.165)–(7.167).

Note that the calculation cannot be shortened by simply expanding the factor $1/(1 - e^{-wb})^D$ in the unsubtracted sum (7A.24) in powers of w, which would yield the result (7A.25) without the first term $\bar{S}_1(b,1)$, and thus without the integrals (7A.29).

Appendix 7B   Experimental versus Theoretical Would-be Critical Temperature

In Fig. 7.12 we have seen that there is only a small difference between the theoretical would-be critical temperature T_c of a finite system calculated from the equation $N = S_2(1)$ and and the experimental T_c^{exp} determined from the maximum of the second derivative of the condensate fraction $\rho \equiv N_{\mathrm{cond}}/N$. Let us extimate the difference. The temperature T_c is found by solving the equation [recall (7.90)]

$$N = S_D(1) \equiv \sum_{w=1}^{\infty}[Z_1^D(wb_c)e^{wDb/2} - 1].\tag{7B.1}$$

To find the latter we must solve the full equation (7A.1) and search for the maximum of $d^2 N_{\mathrm{cond}}(T)/dT^2$. The last term in (7A.1) can be expressed in terms of the condensate fraction $\rho(T)$ as $N_{\mathrm{cond}}(T) = \rho(T)N$. Near the critical point, we set $\delta z_D \equiv 1 - z_D$ and see that it is equal to $\delta z_D = 1/(1 + N\rho)$

In the critical regime, ρ and $1/N\rho$ are both of the same order $1/\sqrt{N}$. Thus we expand $S_D(z_D)$ to lowest order in $\delta z_D = 1/N\rho + \dots$, and replace (7A.1) by

$$N \approx [S_D(1) + \partial_b S_D(1)\delta b - \partial_{z_D} S_D(1)\,(1/N\rho - 1/N + \dots)] + \rho N,\tag{7B.2}$$

or

$$\frac{b\partial_b S_D(1)}{S_D(1)}\frac{\delta b}{b} \approx \frac{\partial_{z_D} S_D(1)}{S_D(1)}\frac{1}{N\rho} - \rho + \cdots,\tag{7B.3}$$

In this leading-order discussion we may ignore the finite-size correction to $S_D(1)$, i.e., the difference between $T_c^{(0)}$ and $T_c^{(1)}$, so that $S_D(1) = (\pi/2b_c^{(0)})^{D/2}$, and we may approximate the left-hand side of (7B.3) by $(D/2)\delta T/T_c^{(0)}$. Abbreviating $\partial_{z_D} S_D(1)/S_D(1)$ by A which is of order unity, we must find $\rho(T)$ from the the the equation

$$\frac{D}{2}t \approx \frac{A}{N\rho} - \rho - \frac{A}{N},\tag{7B.4}$$

where $t \equiv \delta T/T_c^{(0)}$. This yields

$$\rho(t) = \sqrt{\frac{A}{N}} - \frac{D}{4}t + \frac{D^2}{32}\sqrt{\frac{N}{A}}t^2 - \frac{D^4}{2048}\sqrt{\frac{N}{A}}^3 t^4 + \cdots.\tag{7B.5}$$

The maximum of $d^2\rho(t)/dt^2$ lies at $t = 0$, so that the experimental would-be critical coincides with the theoretical $T_c^{(1)}$ to this order.

If we carry the expansion of (7B.2) to the second order in $\rho \approx 1/N\rho \approx 1/\sqrt{N}$, we find a shift of the order $1/N$. As an example, take $D = 3$, assume $A = 1$, and use the full T-dependence rather than the lowest-order expansion in Eq. (7B.4):

$$\left(\frac{T}{T_c^{(0)}}\right)^{3/2} = (1+t)^{3/2} = \frac{1}{N\rho} - \rho. \tag{7B.6}$$

For simplicity, we ignore the $1/N$ terms in (7B.3). The resulting $\rho(t)$ and $d^2\rho/dt^2$ are plotted in Fig. 7.13. The maximum of $d^2\rho/dt^2$ lies at $t \approx 8/9N$, which can be ignored if we know T_c only up to order $\mathcal{O}(N^{-1/3})$ or $\mathcal{O}(N^{-2/3})$, as in Eqs. (7.91) and (7A.22).

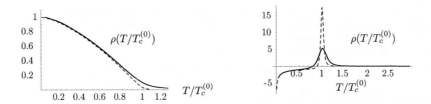

Figure 7.13   Plots of condensate fraction and its second derivative for Bose Gas in a finite box following Eq. (7B.6) for $N = 100$ and 1000.

Note that a convenient way to determine the would-be critical temperature from numerical data is via the maximal curvature of the chemical potential $\mu(T)$, i.e., from the maximum of $-d^2\mu(T)/dT^2$. Since $z_D = e^{\mu - Db/2} = 1 - \delta z_D \approx 1 + \mu - Db/2$, the second derivative of μ is related to that of δz_D by

$$-\frac{d^2\mu}{dt^2} \approx \frac{d^2\delta z_D}{dt^2} - Db_c^{(0)}. \tag{7B.7}$$

The second term is of the order $1/N^{2/D}$ and can be ignored as we shall see immediately. The equation shows that the second derivative of $-\mu$ coincides with that of δz_D, which is related to $d^2\rho/dt^2$ by

$$\frac{d^2\delta z_D}{dt^2} \approx \frac{d^2}{dt^2}\frac{1}{N\rho} = \frac{1}{N}\left[\frac{2}{\rho^3}\left(\frac{d\rho}{dt}\right)^2 - \frac{1}{\rho^2}\frac{d^2\rho}{dt^2}\right] = \frac{D^2}{16}\sqrt{\frac{N}{A^3}}\left[1 - \frac{3D^2}{32}\frac{N}{A}t^2 + \mathcal{O}(t^4)\right]. \tag{7B.8}$$

This is again maximal at $t = 0$, implying that the determination of $T_c^{(1)}$ by this procedure coincides with the previous one. The neglected term of order $1/N^{2/D}$ has a relative suppression of $1/N^{2/D+1/2}$, which can be ignored for larger N.

Notes and References

Path integrals of many identical particles are discussed in the textbook
R.P. Feynman, *Statistical Mechanics*, Benjamin, Reading, 1972.
See also the papers by
L.S. Schulman, Phys. Rev. **176**, 1558 (1968);
M.G.G. Laidlaw and C. DeWitt-Morette, Phys. Rev. D **3**, 5 (1971);
F.J. Bloore, in *Differential Geometric Methods in Math. Phys.*, Lecture Notes in Mathematics **905**, Springer, Berlin, 1982;
P.A. Horvathy, G. Morandi, and E.C.G. Sudarshan, Nuovo Cimento D **11**, 201 (1989).

Path Integrals over Grassmann variables are discussed in detail in
F.A. Berezin, *The Method of Second Quantization*, Nauka, Moscow, 1965;
J. Rzewuski, *Field Theory*, Hafner, New York, 1969;
L.P. Singh and F. Steiner, Phys. Lett. B **166**, 155 (1986);
M. Henneaux and C. Teitelboim, *Quantization of Gauge Systems*, Princeton Univ. Press, 1992.

For fractional statistics see the papers by
F. Wilczek and A. Zee, Phys. Rev. Lett. **51**, 2250 (1983);
Y.S. Wu, *ibid.*, **53**, 111 (1984);
and the reprint collection by
A. Shapere and F. Wilczek and *Geometric Phases in Physics*, World Scientific, Singapore, 1989, p. 284.
The simple derivation of the statistics interaction given in this text is taken from
H. Kleinert, Phys. Lett. B **235**, 381 (1989) (wwwK/104, where (wwwK is short for http://www.physik.fu-berlin.de/~kleinert).
See also the notes and references in Chapter 16.

I thank Prof. Carl Wieman for his permission to publish Fig. 7.9 of the JILA BEC group.

The particular citations in this chapter refer to the publications

[1] R.P. Feynman, *Statistical Mechanics*, Benjamin, Reading, 1972.

[2] J. Flachsmeyer, *Kombinatorik*, Dtsch. Verl. der Wiss., Berlin, 1972. p. 128, Theorem 11.3. The number n_z^n of permutations of n elements consisting of z cycles satisfies the recursion relation $n_z^n = n_{z-1}^{n-1} + (n-1) n_z^{n-1}$, with the initial conditions $n_n^n = 1$ and $n_1^n = (n-1)!$.

[3] The Mathematica program for this can be downloaded from the internet (wwwK/b5/pgm7).

[4] F. Brosens, J.T. Devreese, L.F. Lemmens, Phys. Rev. E *55*, 227 (1997). See also K. Glaum, Berlin Ph.D. thesis (in preparation).

[5] P. Bormann and G. Franke, J. Chem. Phys. **98**, 2484 (1984).

[6] S. Bund and A. Schakel, Mod. Phys. Lett. B **13**, 349 (1999).

[7] See E.W. Weisstein's internet page mathworld.wolfram.com.

[8] See, for example, Chapter 11 in H. Kleinert, *Gauge Fields in Condensed Matter*, op. cit., Vol. I, World Scientific, Singapore, 1989 (wwwK/b1).

[9] The first observation was made at JILA with ^{87}Ru:
M.H. Anderson, J.R. Ensher, M.R. Matthews, C.E. Wieman, and E.A. Cornell, Science **269**, 198 (1995).
It was followed by a condensate of ^{7}Li at Rice University:
C.C. Bradley, C.A. Sackett, J.J. Tollet, and R.G. Hulet, Phys. Rev. Lett. **75**, 1687 (1995);
and in ^{30}Na at MIT:
K.B. Davis, M.-O. Mewes, M.R. Andrews, and N.J. van Druten, , D.S. Durfee, D.M. Kurn, W. Ketterle, Phys. Rev. Lett. **75**, 3969 (1995).

[10] J.R. Abo-Shaeer, C. Raman, J.M. Vogels, and W. Ketterle, Science **292**, 476 (2001).

[11] A.L. Fetter and J.D. Walecka, *Quantum Theory of Many-Particle Systems* McGraw–Hill, New York, 1971, Sec. 28.

[12] V.N. Popov, Theor. Math. Phys. **11**, 565 (1972); *Functional Integrals in Quantum Field Theory and Statistical Physics* (Reidel, Dordrecht, 1983).

[13] M. Schick, Phys. Rev. A **3**, 1067 (1971).

[14] J.O. Andersen, U. Al Khawaja, and H.T.C. Stoof, Phys. Rev. Lett. **88**, 070407 (2002).

[15] H.T.C. Stoof and M. Bijlsma, Phys. Rev. E **47**, 939 (1993); U. Al Khawaja, J.O. Andersen, N.P. Proukakis, and H.T.C. Stoof, Phys. Rev. A **A**, 013615 (2002).

[16] D.S. Fisher and P.C. Hohenberg, Phys. Rev. B **37**, 4936 (1988).

[17] F. Nogueira and H. Kleinert, Phys. Rev. B **73**, 104515 (2006) (cond-mat/0503523).

[18] K. Huang, Phys. Rev. Lett. **83**, 3770 (1999).

[19] K. Huang, C.N. Yang, Phys. Rev. **105**, 767 (1957); K. Huang, C.N. Yang, J.M. Luttinger, Phys. Rev. **105**, 776 (1957); T.D. Lee, K. Huang, C.N. Yang, Phys. Rev. **106**, 1135 (1957); K. Huang, in *Studies in Statistical Mechanics*, North-Holland, Amsterdam, 1964, Vol. II.

[20] H. Kleinert, *Strong-Coupling Behavior of Phi⁴-Theories and Critical Exponents*, Phys. Rev. D **57** , 2264 (1998); Addendum: Phys. Rev. D **58**, 107702 (1998) (cond-mat/9803268); *Seven Loop Critical Exponents from Strong-Coupling ϕ^4-Theory in Three Dimensions*, Phys. Rev. D **60** , 085001 (1999) (hep-th/9812197); *Theory and Satellite Experiment on Critical Exponent alpha of Specific Heat in Superfluid Helium*, Phys. Lett. A **277**, 205 (2000) (cond-mat/9906107); *Strong-Coupling ϕ^4-Theory in $4 - \epsilon$ Dimensions, and Critical Exponent*, Phys. Lett. B **434** , 74 (1998) (cond-mat/9801167); *Critical Exponents without beta-Function*, Phys. Lett. B **463**, 69 (1999) (cond-mat/9906359).
See also the textbook:
H. Kleinert and V. Schulte-Frohlinde, *Critical Properties of ϕ^4-Theories*, World Scientific, Singapore, 2001 (wwwK/b8).

[21] H. Kleinert, *Five-Loop Critical Temperature Shift in Weakly Interacting Homogeneous Bose-Einstein Condensate* (cond-mat/0210162).

[22] T. Toyoda, Ann. Phys. (N.Y.) **141**, 154 (1982).

[23] A. Schakel, Int. J. Mod. Phys. B **8**, 2021 (1994); *Boulevard of Broken Symmetries*, (cond-mat/9805152). The upward shift $\Delta T_c/T_c$ by a weak δ-function repulsion reported in this paper is, however, due to neglecting a two-loop contribution to the free energy. If this is added, the shift disappears. See the remarks below in [34].

[24] H.T.C. Stoof, Phys. Rev. A **45**, 8398 (1992).

[25] M. Holzmann, P. Grueter, and F. Laloe, Eur. Phys. J. B **10**, 239 (1999).

[26] G. Baym, J.-P. Blaizot, M. Holzmann, F. Laloe, and D. Vautherin, Phys. Rev. Lett. **83**, 1703 (1999).

[27] G. Baym, J.-P. Blaizot, and J. Zinn-Justin, Europhys. Lett. **49**, 150 (2000) (cond-mat/9907241)

[28] M. Holzmann, G. Baym, J.-P. Blaizot, and F. Laloe, Phys. Rev. Lett. **87**, 120403 (2001) (cond-mat/0107129).

[29] P. Grueter, D. Ceperley, and F. Laloe, Phys. Rev. Lett. **79**, 3549 (1997).

[30] M. Holzmann and W. Krauth, Phys. Rev. Lett. **83**, 2687 (1999).

[31] M. Bijlsma and H.T.C. Stoof, Phys. Rev. A **54**, 5085 (1996).

[32] H. Kleinert, *Collective Quantum Fields*, Fortschr. Phys. **26**, 565 (1978) http://www.physik.fu-berlin.de/~kleinert/55).

[33] N.N. Bogoliubov, Zh. Eksp. Teor. Fiz. **34**, 58 (1958).

[34] P. Arnold and B. Tomášik, Phys. Rev. A **62**, 063604 (2000).
This paper starts out from the 3+1-dimensional initial theory and derives from it the three-dimensional effective classical field theory as defined in Subsection 7.14.4. This program was initiated much earlier by
A. Schakel, Int. J. Mod. Phys. B **8**, 2021 (1994);
and explained in detail in Ref. [23]. Unfortunately, the author did not go beyond the one-loop level so that he found a positive shift $\Delta T_c/T_c$, and did not observe its cancellation at the two-loop level. See the recent paper
A. Schakel, *Zeta Function Regularization of Infrared Divergences in Bose-Einstein Condensation*, (cond-mat/0301050).

[35] F.F. de Souza Cruz, M.B. Pinto, R.O. Ramos, Phys. Rev. B **64**, 014515 (2001);
F.F. de Souza Cruz, M.B. Pinto, R.O. Ramos, P. Sena, Phys. Rev. A **65**, 053613 (2002) (cond-mat/0112306).

[36] P. Arnold, G. Moore, and B. Tomášik, Phys. Rev. A **65**, 013606 (2002) (cond-mat/0107124).

[37] G. Baym, J.-P. Blaizot, M. Holzmann, F. Laloe, and D. Vautherin (cond-mat/0107129).

[38] P. Arnold and G. Moore, Phys. Rev. Lett. **87**, 120401 (2001); V.A. Kashurnikov, N.V. Prokof´ev, and B.V. Svistunov, Phys. Rev. Lett. **87**, 120402 (2001).

[39] J.L. Kneur, M.B. Pinto, R.O. Ramos, Phys. Rev. A **68**, 43615 (2003) (cond-mat/0207295).

[40] J.D. Reppy, B.C. Crooker, B. Hebral, A.D. Corwin, J. He, and G.M. Zassenhaus, cond-mat/9910013 (1999).

[41] M. Wilkens, F. Illuminati, and M. Krämer, J. Phys. B **33**, L779 (2000) (cond-mat/0001422).

[42] E.J. Mueller, G. Baym, and M. Holzmann, J. Phys. B **34**, 4561 (2001) (cond-mat/0105359).

[43] H. Kleinert, S. Schmidt, and A. Pelster, Phys. Rev. Lett. **93**, 160402 (2004) (cond-mat/0307412).

[44] E.L. Pollock and K.J. Runge, Phys. Rev. B **46**, 3535 (1992).

[45] J. Tempere, F. Brosens, L.F. Lemmens, and J.T. Devreese, Phys. Rev. A **61**, 043605 (2000) (cond-mat/9911210).

[46] N.N. Bogoliubov, *On the Theory of Superfluidity* , Izv. Akad. Nauk SSSR (Ser. Fiz.) **11**, 77 (1947); Sov.Phys.-JETP **7**, 41(1958).
See also
J.G. Valatin, Nuovo Cimento **7**, 843 (1958);
K. Huang, C.N. Yang, J.M. Luttinger, Phys. Rev. **105**, 767, 776 (1957).

[47] See pp. 1356–1369 in
H. Kleinert, *Gauge Fields in Condensed Matter*, op. cit., Vol. II, World Scientific, Singapore, 1989 (wwwK/b2); H. Kleinert, *Gravity as Theory of Defects in a Crystal with Only Second-Gradient Elasticity*, Ann. d. Physik, **44**, 117 (1987) (wwwK/172);
H. Kleinert and J. Zaanen, (gr-qc/0307033).

[48] V. Bretin, V.S. Stock, Y. Seurin, F. Chevy, and J. Dalibard, *Fast Rotation of a Bose-Einstein Condensate*, Phys. Rev. Lett. **92**, 050403 (2004).

[49] For the beginnings see
J.R. Klauder, Ann. of Physics **11**, 123 (1960).
The present status is reviewed in
J.R. Klauder, *The Current State of Coherent States*, (quant-ph/0110108).

[50] G. Junker and J.R. Klauder, Eur. Phys. J. C **4**, 173 (1998) (quant-phys/9708027).

[51] See the textbook cited at the end of Ref. [20].

[52] R. Stratonovich Sov. Phys. Dokl. **2**, 416 (1958):
J. Hubbard, Phys. Rev. Lett. **3**, 77 (1959);
B. Mühlschlegel, J. Math. Phys. , **3**, 522 (1962);
J. Langer, Phys. Rev. A **134**, 553 (1964); T.M. Rice, Phys. Rev. A **140** 1889 (1965); J. Math. Phys. **8**, 1581 (1967); A.V. Svidzinskij, Teor. Mat. Fiz. **9**, 273 (1971); D. Sherrington, J. Phys. C **4**, 401 (1971).

8

Path Integrals in Polar and Spherical Coordinates

Many physical systems possess rotational symmetry. In operator quantum mechanics, this property is of great help in finding wave functions and energies of a system. If a rotationally symmetric Schrödinger equation is transformed to spherical coordinates, it separates into a radial and several angular differential equations. The latter are universal and have well-known solutions. Only the radial equation contains specific information on the dynamics of the system. Being an ordinary one-dimensional Schrödinger equation, it can be solved with the usual techniques.

In the path integral approach, a similar coordinate transformation is possible, although it makes things initially more complicated rather than simpler. First, the use of non-Cartesian coordinates causes nontrivial problems of the kind observed in Chapter 6, where the configuration space was topologically constrained. Such problems can be solved as in Chapter 6 using the knowledge of the correct procedure in Cartesian coordinates. A second complication is more severe: When studying a system at a given angular momentum, the presence of a centrifugal barrier destroys the possibility of setting up a time-sliced path integral of the Feynman type as in Chapter 2. The recent solution of the latter problem has paved the way for two major advances in path integration which will be presented in Chapters 10, 11, and 12.

8.1 Angular Decomposition in Two Dimensions

Consider a two-dimensional quantum-mechanical system with rotational invariance. In Schrödinger quantum mechanics, it is convenient to introduce polar coordinates

$$\mathbf{x} = r(\cos\varphi, \sin\varphi), \tag{8.1}$$

and to split the differential equation into a radial and an azimuthal one which are solved separately. Let us try to follow the same approach in path integrals. To avoid the complications associated with path integrals in the canonical formulation [1],all calculations will be done in the Lagrange formulation. It will, moreover, be advantageous to work with the imaginary-time amplitude (the thermal density matrix) to avoid carrying around factors of i. Thus we start out with the path integral

$$(\mathbf{x}_b T_b | \mathbf{x}_a T_a) = \int \mathcal{D}^2 x(\tau) \exp \left\{ -\frac{1}{\hbar} \int_{T_a}^{T_b} d\tau \left[\frac{M}{2} \dot{\mathbf{x}}^2 + V(r) \right] \right\}. \tag{8.2}$$

It is time-sliced in the standard way into a product of integrals

$$(\mathbf{x}_b T_b | \mathbf{x}_a T_a) \approx \prod_{n=1}^{N} \left[\int \frac{d^2 x_n}{2\pi\hbar\epsilon/M} \right] \exp \left\{ -\frac{\epsilon}{\hbar} \sum_{n=1}^{N+1} \left[\frac{M}{2\epsilon^2} (\mathbf{x}_n - \mathbf{x}_{n-1})^2 + V(r_n) \right] \right\}. \tag{8.3}$$

When going over to polar coordinates, the measure of integration changes to

$$\prod_{n=1}^{N} \int_0^\infty dr_n r_n \int_0^{2\pi} \frac{d\varphi_n}{2\pi\hbar\epsilon/M}$$

and the kinetic term becomes

$$\exp \left[-\frac{1}{\hbar} \frac{M}{2\epsilon} (\mathbf{x}_n - \mathbf{x}_{n-1})^2 \right] = \exp \left\{ -\frac{1}{\hbar} \frac{M}{2\epsilon} \left[r_n^2 + r_{n-1}^2 - 2 r_n r_{n-1} \cos(\varphi_n - \varphi_{n-1}) \right] \right\}. \tag{8.4}$$

To do the φ_n-integrals, it is useful to expand (8.4) into a factorized series using the formula

$$e^{a\cos\varphi} = \sum_{m=-\infty}^{\infty} I_m(a) e^{im\varphi}, \tag{8.5}$$

where $I_m(z)$ are the modified Bessel functions. Then (8.4) becomes

$$\exp \left[-\frac{1}{\hbar} \frac{M}{2\epsilon} (\mathbf{x}_n - \mathbf{x}_{n-1})^2 \right]$$

$$= \exp \left[-\frac{1}{\hbar} \frac{M}{2\epsilon} (r_n^2 + r_{n-1}^2) \right] \sum_{m=-\infty}^{\infty} I_m \left(\frac{M}{\hbar\epsilon} r_n r_{n-1} \right) e^{im(\varphi_n - \varphi_{n-1})}. \tag{8.6}$$

In the discretized path integral (8.3), there are $N+1$ factors of this type. The N-integrations over the φ_n-variables can now be performed and produce N Kronecker δ's:

$$\prod_{n=1}^{N} 2\pi \delta_{m_n, m_{n-1}}. \tag{8.7}$$

These can be used to eliminate all but one of the sums over m, so that we arrive at the amplitude

$$(\mathbf{x}_b T_b | \mathbf{x}_a T_a) \approx \frac{2\pi}{2\pi\hbar\epsilon/M} \prod_{n=1}^{N} \left[\int_0^\infty \frac{dr_n r_n 2\pi}{2\pi\hbar\epsilon/M} \right] \sum_{m=-\infty}^{\infty} \frac{1}{2\pi} e^{im(\varphi_b - \varphi_a)}$$

$$\times \prod_{n=1}^{N+1} \left\{ \exp \left[-\frac{1}{\hbar} \frac{M}{2\epsilon} (r_n^2 + r_{n-1}^2) \right] I_m \left(\frac{M}{\hbar\epsilon} r_n r_{n-1} \right) \right\} \exp \left[-\frac{\epsilon}{\hbar} \sum_{n=1}^{N+1} V(r_n) \right]. \tag{8.8}$$

We now define the *radial time evolution amplitudes* by the following expansion with respect to the azimuthal quantum numbers m:

$$(\mathbf{x}_b T_b | \mathbf{x}_a T_a) = \sum_{m=-\infty}^{\infty} \frac{1}{\sqrt{r_b r_a}} (r_b T_b | r_a T_a)_m \frac{1}{2\pi} e^{im(\varphi_b - \varphi_a)}. \tag{8.9}$$

The amplitudes $(r_b T_b | r_a T_a)_m$ are obviously given by the *radial path integral*

$$(r_b T_b | r_a T_a)_m \approx \frac{1}{\sqrt{2\pi\hbar\epsilon/M}} \prod_{n=1}^{N} \left[\int_0^{\infty} \frac{dr_n}{\sqrt{2\pi\hbar\epsilon/M}} \right]$$

$$\times \prod_{n=1}^{N+1} \left[\exp\left\{ -\frac{M}{2\epsilon\hbar}(r_n - r_{n-1})^2 \right\} \tilde{I}_m \left(\frac{M r_n r_{n-1}}{\hbar\epsilon} \right) \right] \exp\left\{ -\frac{\epsilon}{\hbar} \sum_{n=1}^{N+1} V(r_n) \right\}. \tag{8.10}$$

Here we have introduced slightly different modified Bessel functions

$$\tilde{I}_m(z) \equiv \sqrt{2\pi z} e^{-z} I_m(z). \tag{8.11}$$

They will also be called "Bessel functions", for short. They have the asymptotic behavior

$$\tilde{I}_m(z) \xrightarrow{z \to \infty} 1 - \frac{m^2 - 1/4}{2z} + \ldots = e^{-\frac{m^2 - 1/4}{2z}} + \ldots , \tag{8.12}$$

$$\tilde{I}_m(z) \xrightarrow{z \to 0} 2\sqrt{\pi}(z/2)^{m+1/2} + \ldots . \tag{8.13}$$

In the case of a free particle with $V(r) = 0$, it is easy to perform all the intermediate integrals over r_n in (8.10). Two neighboring figures in the product require the integral

$$\int_0^{\infty} dr' \exp\left(-\frac{r''^2}{2\epsilon_2} - \frac{r^2}{2\epsilon_1} \right) \exp\left[-r'^2 \left(\frac{1}{2\epsilon_2} + \frac{1}{2\epsilon_1} \right) \right] \frac{\sqrt{r''r'}}{\epsilon_2} I_m \left(\frac{r''r'}{\epsilon_2} \right) \frac{\sqrt{r'r}}{\epsilon_1} I_m \left(\frac{r'r}{\epsilon_1} \right)$$

$$= \exp\left[-\frac{r''^2 + r^2}{2(\epsilon_1 + \epsilon_2)} \right] \frac{\sqrt{r''r}}{\epsilon_1 + \epsilon_2} I_m \left(\frac{r''r}{\epsilon_1 + \epsilon_2} \right). \tag{8.14}$$

For simplicity, the units in this formula are $M = 1, \hbar = 1$. The right-hand side of (8.14) follows directly from the formula

$$\int_0^{\infty} dr r e^{-r^2/\epsilon} I_\nu(\beta r) I_\nu(\alpha r) = \frac{\epsilon}{2} e^{(\alpha^2 + \beta^2)\epsilon/4} I_\nu(\epsilon\alpha\beta/2), \tag{8.15}$$

after identifying ϵ, α, β as

$$\begin{aligned} \epsilon &= 2\epsilon_1\epsilon_2/(\epsilon_1 + \epsilon_2), \\ \alpha &= r/\epsilon_1, \\ \beta &= r''/\epsilon_2. \end{aligned} \tag{8.16}$$

Thus, the integrals in (8.10) with $V(r) = 0$ can successively be performed yielding the thermal amplitude for $\tau_b > \tau_a$:

$$(r_b \tau_b | r_a \tau_a)_m = \frac{M}{\hbar} \frac{\sqrt{r_b r_a}}{\tau_b - \tau_a} \exp \left[-\frac{M}{2\hbar} \frac{r_b^2 + r_a^2}{(\tau_b - \tau_a)} \right] I_m \left(\frac{M}{\hbar} \frac{r_b r_a}{\tau_b - \tau_a} \right). \tag{8.17}$$

Note that the same result could have been obtained more directly from the imaginary-time amplitude of a free particle in two dimensions,

$$(\mathbf{x}_b \tau_b | \mathbf{x}_a \tau_a) = \frac{M}{2\pi\hbar(\tau_b - \tau_a)} \exp \left[-\frac{M}{2\hbar} \frac{(\mathbf{x}_b - \mathbf{x}_a)^2}{\tau_b - \tau_a} \right], \tag{8.18}$$

by rewriting the right-hand side as

$$\frac{M}{2\pi\hbar (\tau_b - \tau_a)} \exp \left(-\frac{M}{2\hbar} \frac{r_b^2 + r_a^2}{\tau_b - \tau_a} \right) \exp \left[\frac{M}{2\hbar} \frac{r_b r_a}{\tau_b - \tau_a} \cos (\varphi_b - \varphi_a) \right],$$

and expanding the second exponential according to (8.5) into the series

$$\sum_{m=-\infty}^{\infty} I_m \left(\frac{M}{\hbar} \frac{r_b r_a}{\tau_b - \tau_a} \right) e^{im(\varphi_b - \varphi_a)}. \tag{8.19}$$

A comparison of the coefficients with those in (8.9) gives the radial amplitudes (8.17).

Due to (8.14), the radial amplitude satisfies a fundamental composition law corresponding to (2.4), which reads for $\tau_b > \tau > \tau_a$

$$\int_0^\infty dr\, r\, (r_b \tau_b | r\, \tau)_m (r\, \tau | r_a \tau_a)_m = (r_b \tau_b | r_a \tau_a)_m. \tag{8.20}$$

8.2 Trouble with Feynman's Path Integral Formula in Radial Coordinates

In the above calculation we have shown that the expression (8.10) is certainly the correct radial path integral. It is, however, not of the Feynman type. In operator quantum mechanics we learn that the action of a particle moving in a potential $V(r)$ at a fixed angular momentum $L_3 = m\hbar$ contains a centrifugal barrier $\hbar^2(m^2 - 1/4)/2Mr^2$ and reads

$$\mathcal{A}_m = \int_{\tau_a}^{\tau_b} d\tau \left(\frac{M}{2} \dot{r}^2 + \frac{\hbar^2}{2M} \frac{m^2 - 1/4}{r^2} + V(r) \right). \tag{8.21}$$

This is shown by separating the Hamiltonian operator into radial and azimuthal coordinates, over fixing the azimuthal angular momentum L_3, and choosing for it the quantum-mechanical value $\hbar m$. According to Feynman's rules, the radial amplitude therefore should simply be given by the path integral

$$(r_b \tau_b | r_a \tau_a)_m = \int_0^\infty \mathcal{D}r \exp \left(-\frac{1}{\hbar} \mathcal{A}_m \right). \tag{8.22}$$

The reader may object to using the word "classical" in the presence of a term proportional to $\hbar^2$ in the action. In this section, however, $\hbar m$ is merely meant to be a parameter specifying the azimuthal momentum $p_\varphi \equiv \hbar m$ in the classical centrifugal barrier $p_\varphi^2/2Mr^2$. It is parametrized in terms of a dimensionless number m which does not necessarily have the integer values required by the quantization of the azimuthal motion.

By naively time-slicing (8.22) according to Feynman's rules of Section 2.1 we would have defined it by the finite-N expression

$$
(r_b \tau_b | r_a \tau_a)_m \approx \frac{1}{\sqrt{2\pi\hbar\epsilon/M}} \prod_{n=1}^{N} \left[\int_0^\infty \frac{dr_n}{\sqrt{2\pi\hbar\epsilon/M}} \right]
$$

$$
\times \exp\left\{ -\frac{1}{\hbar} \sum_{n=1}^{N+1} \left[\frac{M}{2\epsilon}(r_n - r_{n-1})^2 + \epsilon\frac{\hbar^2}{2M}\frac{m^2 - 1/4}{r_n r_{n-1}} + \epsilon V(r_n) \right] \right\}. \quad (8.23)
$$

Actually, the denominators in the centrifugal barrier could have been chosen to be r_n^2. This would make a negligible difference for small ϵ. Note that in contrast to a standard Feynman path integral in one dimension, the integrations over r cover only the semi-axis $r \geq 0$ rather than the complete r-axis. This represents no problem since we have learned in Chapter 6 how to treat such half-spaces.

The expression (8.23) is now a place for an unpleasant surprise: For $m = 0$, a time-sliced Feynman path integral formula cannot possibly exist since the potential has an abyss at small r. This leads to a phenomenon which will be referred to as the *path collapse*, to be understood physically and resolved later in Chapter 12. At this place we merely point out the mathematical origin of the problem, by comparing the naively time-sliced expression (8.23) with the certainly correct one (8.10). The singularity would be of no consequence if the two expressions were to converge towards each other in the continuum limit $\epsilon \to 0$. At first sight, this seems to be the case. After all, ϵ is assumed to be infinitesimally small so that we may replace the "Bessel function" $\tilde{I}_m(Mr_n r_{n-1}/\hbar\epsilon)$ by its asymptotic form (8.12),

$$
\tilde{I}_m\left(\frac{Mr_n r_{n-1}}{\hbar\epsilon}\right) \xrightarrow{\epsilon \to 0} \exp\left(-\epsilon\frac{\hbar}{2M}\frac{m^2 - 1/4}{r_n r_{n-1}}\right). \quad (8.24)
$$

For a fixed set of r_n, i.e., for a given path, the continuum limit $\epsilon \to 0$ makes the integrands (8.10) and (8.23) coincide, the difference being of the order ϵ^2. Unfortunately, the path integral requires the limit to be taken *after* the integrations over the dr_n. The integrals, however, do not exist at $m = 0$. For paths moving very close to the singularity at $r = 0$, the approximation (8.24) breaks down. In fact, the large-z expansion

$$
\tilde{I}_m(z) = 1 - \frac{m^2 - 1/4}{2z} + \frac{(m^2 - 1/4)(m^2 - 1/9)}{2!(2z)^2} - \cdots \quad (8.25)
$$

with $z = Mr_n r_{n-1}/\hbar\epsilon$ is never convergent even for a very small ϵ. The series shows only an asymptotic convergence (more on this subject in Section 17.9). If we want to evaluate $\tilde{I}_m(z) = \sqrt{2\pi z}e^{-z}I_m(z)$ for all z we have to use the convergent power series expansion of $I_m(z)$ around $z = 0$:

$$I_m(z) = \left(\frac{z}{2}\right)^m \left[\frac{1}{0!m!} + \frac{1}{1!(m+1)!}\left(\frac{z}{2}\right)^2 + \frac{1}{2!(m+2)!}\left(\frac{z}{2}\right)^4 + \ldots\right].$$

It is known from the Schrödinger theory that the leading power z^m determines the threshold behavior of the quantum-mechanical particle distribution near the origin. This is qualitatively different from the exponentially small distribution $\exp\left[-\epsilon\hbar(m^2 - 1/4)/2Mr^2\right]$ contained in each time slice of the Feynman formula (8.23) for $|m| > 1/2$.

The root of these troubles is an anomalous behavior in the high-temperature limit of the partition function. In this limit, the imaginary time difference $\tau_b - \tau_a = \hbar/k_BT$ is very small and it is usually sufficient to keep only a single slice in a time-sliced path integral (see Sections 2.9, 2.13). If this were true also here, the formula (8.23) would lead, in the absence of a potential $V(r)$, to the classical particle distribution [compare (2.347)]

$$(r_a\,\tau_a + \epsilon|r_a\tau_a) = \frac{1}{\sqrt{2\pi\hbar\epsilon/M}}\exp\left(-\epsilon\frac{\hbar}{2M}\frac{m^2 - 1/4}{r_a^2}\right). \tag{8.26}$$

If we subtract the barrier-free distribution, this amounts to the classical partition function[1]

$$Z_{\rm cl} = \int_0^\infty \frac{dr}{\sqrt{2\pi a}}\left[\exp\left(-a\frac{m^2 - 1/4}{2r^2}\right) - 1\right] = -\frac{1}{2}\sqrt{m^2 - \frac{1}{4}}, \tag{8.27}$$

where we have abbreviated the factor $\epsilon\hbar/M$ by a. The integral is temperature-independent and converges only for $|m| > 1/2$.

Compare this result with the proper high-temperature limit of the exact partition function calculated with the use of (8.17) and with the same subtraction as before. It reads for *all* T

$$Z = \frac{1}{2}\int_0^\infty dz\,e^{-z}[I_m(z) - I_{1/2}(z)]. \tag{8.28}$$

As in the classical expression, there is no temperature dependence. The integral

$$\int_0^\infty dz\,e^{-\alpha z}I_\mu(z) = (\alpha^2 - 1)^{-1/2}(\alpha + \sqrt{\alpha^2 - 1})^{-\mu} \tag{8.29}$$

[see formula (2.473)] converges for arbitrary real ν and $\alpha > 1$, and gives in the limit $\alpha \to 1$

$$Z = -\frac{1}{2}(m - 1/2). \tag{8.30}$$

---

[1]This follows by expanding the formula $\int_0^\infty dx\,e^{-a/x^2 - bx^2} = \sqrt{\pi/4b}\,e^{-2\sqrt{ab}}$ in powers of $\sqrt{b}$, subtracting the $a = 0$ -term, and taking the limit $b \to 0$.

This is different from the classical result (8.27) and agrees with it only in the limit of large m.[2]

Thus we conclude that a time-sliced path integral containing a centrifugal barrier can only give the correct amplitude when using the Euclidean action

$$\tilde{\mathcal{A}}_m^N = \sum_{n=1}^{N+1} \left[\frac{M}{2\epsilon}(r_n - r_{n-1})^2 - \hbar \log \tilde{I}_m \left(\frac{M}{\hbar\epsilon} r_n r_{n-1} \right) \right], \tag{8.31}$$

in which the neighborhood of the singularity is treated quantum-mechanically. The naively time-sliced classical action in (8.23) is of no use. The centrifugal barrier renders therefore a counterexample to Feynman rules of path integration according to which quantum-mechanical amplitudes should be obtainable from a sum over all histories of exponentials which involve only the classical expression for the short-time actions.

It is easy to see where the derivation of the time-sliced path integral in Section 2.1 breaks down. There, the basic ingredient was the Trotter product formula (2.26), which for imaginary time reads

$$e￿^{-\beta(\hat{T}+\hat{V})} = \lim_{N \to \infty} \left(e^{-\epsilon\hat{T}} e^{-\epsilon\hat{V}} \right)^{N+1}, \quad \epsilon \equiv \beta/(N+1). \tag{8.32}$$

An exact identity is $e^{-\beta(\hat{T}+\hat{V})} \equiv \left(e^{-\epsilon\hat{T}} e^{-\epsilon\hat{V}} e^{i\epsilon^2 X} \right)^{N+1}$ with X given in Eq. (2.10) consisting of a sum of higher and higher commutators between $\hat{V}$ and $\hat{T}$. The Trotter formula neglects these commutators. In the presence of a centrifugal barrier, however, this is not permitted. Although the neglected commutators carry increasing powers of the small quantity ϵ, they are more and more divergent at $r = 0$, like ϵ^n/r^{2n}. The same terms occur in the asymptotic expansion (8.25) of the "Bessel function". In the proper action (8.31), these terms are present.

It should be noted that for a Hamiltonian possessing a centrifugal barrier V_{cb} in addition to an arbitrary smooth potential V, i.e., for a Hamiltonian operator of the form $\hat{H} = \hat{T} + \hat{V}_{cf} + \hat{V}$, the Trotter formula *is* applicable in the form

$$e^{-\beta H} = \lim_{N \to \infty} \left(e^{-\epsilon(\hat{T}+\hat{V}_{cb})} e^{-\epsilon\hat{V}} \right)^{N+1}, \quad \epsilon \equiv \beta/(N+1). \tag{8.33}$$

It leads to a valid time-sliced path integral formula

$$(r_b \tau_b | r_a \tau_a)_m \approx \prod_{n=1}^{N} \left[\int_0^\infty dr_n \right] \prod_{n=1}^{N+1} \left[\int \frac{dp_n}{2\pi\hbar} \right] \exp\left\{ -\frac{1}{\hbar} \tilde{\mathcal{A}}_m^N[p, r] \right\}, \tag{8.34}$$

---

[2]If m were not merely a dimensionless number parametrizing an arbitrary centrifugal barrier with fixed $p_\varphi^2/2Mr^2$, as it is in this section, the classical limit at a fixed p_φ would eliminate the problem since for $\hbar \to 0$, the number m would become infinitely large, leading to the correct high-temperature limit of the partition function. For a fixed finite m, however, the discrepancy is unavoidable.

with the sliced action

$$
\tilde{A}_m^N[p, r] = \sum_{n=1}^{N+1} \left[-ip_n(r_n - r_{n-1}) + \epsilon \frac{p_n^2}{2M} \right.
$$
$$
\left. -\hbar \log \tilde{I}_m \left(\frac{M}{\hbar\epsilon} r_n r_{n-1} \right) + \epsilon V(r_n) \right]. \tag{8.35}
$$

After integrating out the momenta, this becomes

$$
(r_b \tau_b | r_a \tau_a)_m \approx \frac{1}{\sqrt{2\pi\hbar\epsilon/M}} \prod_{n=1}^N \left[\int_0^\infty \frac{dr_n}{\sqrt{2\pi\hbar\epsilon/M}} \right] \exp \left\{ -\frac{1}{\hbar} \tilde{A}_m^N[r] \right\}, \tag{8.36}
$$

with the action

$$
\tilde{A}_m^N[r, \dot{r}] = \sum_{n=1}^{N+1} \left[\frac{M}{2\epsilon}(r_n - r_{n-1})^2 - \hbar \log \tilde{I}_m \left(\frac{M r_n r_{n-1}}{\hbar\epsilon} \right) + \epsilon V(r_n) \right]. \tag{8.37}
$$

The path integral formula (8.36) can in principle be used to find the amplitude for a fixed angular momentum of some solvable systems.

An example is the radial harmonic oscillator at an angular momentum m, although it should be noted that this particularly simple example does not really require calculating the integrals in (8.36). The result can be found much more simply from a direct angular momentum decomposition of the amplitude (2.175). After a continuation to imaginary times $t = -i\tau$ and an expansion of part of the exponent with the help of (8.5), it reads for $D = 2$

$$
(\mathbf{x}_b \tau_b | \mathbf{x}_a \tau_a) = \frac{1}{2\pi} \frac{M\omega}{\hbar \sinh[\omega(\tau_b - \tau_a)]} e^{-\frac{M\omega}{2\hbar} \coth[\omega(\tau_b - \tau_a)](r_b^2 + r_a^2)}
$$
$$
\times \sum_{m=-\infty}^{\infty} I_m \left(\frac{M\omega r_b r_a}{\hbar \sinh[\omega(\tau_b - \tau_a)]} \right) e^{im(\varphi_b - \varphi_a)}. \tag{8.38}
$$

By comparison with (8.9), we extract the radial amplitude

$$
(r_b \tau_b | r_a \tau_a)_m = \frac{M}{\hbar} \frac{\omega \sqrt{r_b r_a}}{\coth[\omega(\tau_b - \tau_a)]} I_m \left(\frac{M\omega r_b r_a}{\hbar \sinh[\omega(\tau_b - \tau_a)]} \right). \tag{8.39}
$$

The limit $\omega \to 0$ gives the free-particle result

$$
(r_b \tau_b | r_a \tau_a)_m = \frac{M}{\hbar} \frac{\sqrt{r_b r_a}}{\tau_b - \tau_a} I_m \left(\frac{M}{\hbar} \frac{r_b r_a}{\tau_b - \tau_a} \right). \tag{8.40}
$$

8.3　Cautionary Remarks

It is important to emphasize that we obtained the correct amplitudes by performing the time slicing in Cartesian coordinates followed by the transformation to the polar coordinates in the time-sliced expression. Otherwise we would have easily missed the factor $-1/4$ in the centrifugal barrier. To see what can go wrong let us proceed illegally and do the change of variables in the initial continuous action. Thus we try to calculate the path integral

$$
(\mathbf{x}_b \tau_b | \mathbf{x}_a \tau_a) = \sum_{l=-\infty,\infty} \int_0^\infty \mathcal{D}r \, r \int_{-\infty}^\infty \mathcal{D}\varphi
$$
$$
\times \exp \left\{ \frac{1}{\hbar} \int_{\tau_a}^{\tau_b} d\tau \left[\frac{M}{2} (\dot{r}^2 + r^2 \dot{\varphi}^2) + V(r) \right] \right\} \Bigg|_{\varphi(\tau_b)=\varphi_b+2\pi l}. \tag{8.41}
$$

The summation over all periodic repetitions of the final azimuthal angle φ accounts for its multivaluedness according to the rules of Section 6.1. If the expression (8.41) is time-sliced straightforwardly it reads

$$
\sum_{\varphi_{N+1}=\varphi_b+2\pi l, l=-\infty,\infty} \frac{1}{2\pi\hbar\epsilon/M} \prod_{n=1}^N \left[\int_0^\infty \frac{dr_n r_n}{\sqrt{2\pi\hbar\epsilon/M}} \int_{-\infty}^\infty \frac{d\varphi_n}{\sqrt{2\pi\hbar\epsilon/M}} \right]
$$
$$
\times \exp \left\{ -\frac{1}{\hbar} \epsilon \sum_{n=1}^{N+1} \left[\frac{M}{2\epsilon^2} [(r_n - r_{n-1})^2 + r_n^2 (\varphi_n - \varphi_{n-1})^2] + V(r_n) \right] \right\}, \tag{8.42}
$$

where $r_b = r_{N+1}, \varphi_a = \varphi_0, r_a = r_0$. The integrals can be treated as follows. We introduce the momentum integrals over $(p_\varphi)_n$ conjugate to φ_n, writing for each n [with the short notation $(p_\varphi)_n \equiv p_n$]

$$
\frac{r_n}{\sqrt{2\pi\hbar\epsilon/M}} e^{-(M/2\hbar\epsilon)r_n^2(\varphi_n - \varphi_{n-1})^2} \stackrel{\cdot}{=} \int_{-\infty}^\infty \frac{dp_n}{2\pi\hbar} \exp \left\{ -\frac{1}{\hbar} \frac{\epsilon}{2M} \frac{p_n^2}{r_n^2} + \frac{i}{\hbar} p_n (\varphi_n - \varphi_{n-1}) \right\}. \tag{8.43}
$$

After this, the integrals over φ_n $(n = 1 \ldots N)$ in (8.42) enforce all p_n to be equal to each other, i.e., $p_n \equiv p$ for $n = 1, \ldots, N+1$, and we remain with

$$
(\mathbf{x}_b \tau_b | \mathbf{x}_a \tau_a) \approx \frac{1}{2\pi\hbar\epsilon/M} \frac{1}{r_b} \sum_{l-\infty}^\infty \int_{-\infty}^\infty \frac{dp}{2\pi\hbar} e^{(i/\hbar)p(\varphi_b - \varphi_a + 2\pi l)} \prod_{n=1}^N \left[\int_0^\infty \frac{dr_n}{2\pi\hbar\epsilon/M} \right]
$$
$$
\times \exp \left\{ -\frac{\epsilon}{\hbar} \sum_{n=1}^{N+1} \left[\frac{M}{2} \frac{(r_n - r_{n-1})^2}{\epsilon^2} + \frac{p^2}{2Mr_n^2} + V(r_n) \right] \right\}. \tag{8.44}
$$

Performing the sum over l with the help of Poisson's formula (6.9) changes the integral over $dp/2\pi\hbar$ into a sum over integers m and yields the angular momentum decomposition (8.9) with the partial-wave amplitudes

$$
(r_b \tau_b | r_a \tau_a)_m \approx \sqrt{r_b r_a} \frac{1}{2\pi\hbar\epsilon/M} \frac{1}{r_b} \prod_{n=1}^N \left[\int_0^\infty \frac{dr_n}{2\pi\hbar\epsilon/M} \right]
$$
$$
\times \exp \left\{ -\frac{\epsilon}{\hbar} \sum_{n=1}^{N+1} \left[\frac{M}{2} \frac{(r_n - r_{n-1})^2}{\epsilon^2} + \frac{\hbar^2}{2M} \frac{m^2}{r_n^2} + V(r_n) \right] \right\}. \tag{8.45}
$$

This wrong result differs from the correct one in Eq. (8.10) in three respect. First, it does not possess the proper centrifugal term $-\hbar \log \tilde{I}_m(M r_b r_a/\hbar\epsilon)$. Second, there is a spurious overall factor $\sqrt{r_a/r_b}$. Third, in comparison with the limiting expression (8.24), the centrifugal barrier lacks the term $1/4$.

It is possible to restore the term $1/4$ by observing that in the time-sliced expression, the factor $\sqrt{r_b/r_a}$ is equal to

$$
\prod_{n=1}^{N+1} \sqrt{\frac{r_b}{r_a}} = \prod_{n=1}^{N+1} \left(1 + \frac{r_n - r_{n-1}}{r_{n-1}}\right)^{-1/2} = \prod_{n=1}^{N+1} \left(1 - \frac{1}{2}\frac{r_n - r_{n-1}}{\sqrt{r_n r_{n-1}}} + \frac{1}{8}\frac{(r_n - r_{n-1})^2}{r_n r_{n-1}} - \dots\right)
$$

$$
= \exp\left(-\frac{1}{2}\sum_{n=1}^{N+1} \frac{r_n - r_{n-1}}{\sqrt{r_n r_{n-1}}} + \dots\right). \tag{8.46}
$$

This, in turn, can be incorporated into the kinetic term of (8.45) via a quadratic completion leading to

$$
\exp\left\{-\frac{\epsilon}{\hbar}\sum_{n=1}^{N+1}\left[\frac{M}{2}\left(\frac{r_n - r_{n-1}}{\epsilon} + i\frac{\hbar}{2M\sqrt{r_n r_{n-1}}}\right)^2 + \frac{\hbar^2}{2M}\frac{m^2 - 1/4}{r_n^2}\right]\right\}. \tag{8.47}
$$

The centrifugal barrier is now correct, but the kinetic term is wrong. In fact, it does not even correspond to a Hermitian Hamiltonian operator, as can be seen by introducing momentum integrations and completing the square to

$$
\exp\left\{-ip_n(r_n - r_{n-1}) + \frac{\epsilon}{\hbar}\left[\frac{p_n^2}{2M} - i\hbar\frac{p_n}{2M r_n}\right]\right\}. \tag{8.48}
$$

The last term is an imaginary energy. Only by dropping it artificially would the time-sliced action acquire the Feynman form (8.23), while still being beset with the problem of nonexistence for $m = 0$ (path collapse) and the nonuniform convergence of the path integrations to be solved in Chapter 12.

The lesson of this is the following: A naive time slicing *cannot* be performed in curvilinear coordinates. It can safely be done in the Cartesian formulation.

Fortunately, a systematic modification of the naive slicing rules has recently been found which makes them applicable to non-Cartesian systems. This will be shown in Chapters 10 and 11.

In the sequel it is useful to maintain, as far as possible, the naive notation for the radial path integral (8.23) and the continuum limit of the action (8.21). The places where care has to be taken in the time-slicing process will be emphasized by setting the centrifugal barrier in quotation marks and defining

$$
{}^{\text{“}}\epsilon\frac{\hbar^2}{2M}\frac{m^2 - 1/4}{r_n r_{n-1}}{}^{\text{”}} \equiv -\hbar \log \tilde{I}_m\left(\frac{M r_n r_{n-1}}{\hbar\epsilon}\right). \tag{8.49}
$$

Thus we shall write the properly sliced action (8.37) as

$$
\tilde{A}_m^N[r, \dot{r}] = \sum_{n=1}^{N+1}\left[\frac{M}{2\epsilon}(r_n - r_{n-1})^2 + {}^{\text{“}}\epsilon\frac{\hbar^2}{2\mu}\frac{m^2 - 1/4}{r_n r_{n-1}}{}^{\text{”}} + \epsilon V(r_n)\right],
$$

$$
\tag{8.50}
$$

and emphasize the need for the non-naive time slicing of the continuum action correspondingly:

$$\mathcal{A}_m = \int_{\tau_a}^{\tau_b} d\tau \left[\frac{M}{2}\dot{r}^2 + \text{``} \frac{\hbar^2}{2M} \frac{m^2 - 1/4}{r^2} \text{''} + V(r) \right]. \tag{8.51}$$

8.4 Time Slicing Corrections

It is interesting to find the origin of the above difficulties. For this purpose, we take the Cartesian kinetic terms expressed in terms of polar coordinates

$$(\mathbf{x}_n - \mathbf{x}_{n-1})^2 = (r_n - r_{n-1})^2 + 2r_n r_{n-1}[1 - \cos(\varphi_n - \varphi_{n-1})], \tag{8.52}$$

and treat it perturbatively in the coordinate differences [2]. Expanding the cosine into a power series we obtain the time-sliced action

$$\mathcal{A}^N = \frac{M}{2\epsilon} \sum_{n=1}^{N+1} (\mathbf{x}_n - \mathbf{x}_{n-1})^2 = \frac{M}{2\epsilon} \sum_{n=1}^{N+1} \left\{ (r_n - r_{n-1})^2 \right.$$
$$\left. + 2r_n r_{n-1} \left[\frac{1}{2!}(\varphi_n - \varphi_{n-1})^2 - \frac{1}{4!}(\varphi_n - \varphi_{n-1})^4 + \dots \right] \right\}. \tag{8.53}$$

In contrast to the naively time-sliced expression (8.42), we now keep the quartic term $(\varphi_n - \varphi_{n-1})^4$. To see how it contributes, consider a single intermediate integral

$$\int \frac{d\varphi_{n-1}}{\sqrt{2\pi\epsilon/a}} e^{-(a/2\epsilon)\left[(\varphi_n - \varphi_{n-1})^2 + a_4(\varphi_n - \varphi_{n-1})^4 + \dots\right]}. \tag{8.54}$$

The first term in the exponent restricts the width of the fluctuations of the difference $\varphi_n - \varphi_{n-1}$ to

$$\langle (\varphi_n - \varphi_{n-1})^2 \rangle_0 = \frac{\epsilon}{a}. \tag{8.55}$$

If we rescale the arguments, $\varphi_n \to \sqrt{\epsilon}u_n$, the integral takes the form

$$\int \frac{du_{n-1}}{\sqrt{2\pi/a}} e^{-(a/2)\left[(u_n - u_{n-1})^2 + \epsilon a_4(u_n - u_{n-1})^4 + \dots\right]}. \tag{8.56}$$

This shows that each higher power in the difference $u_n - u_{n-1}$ is suppressed by an additional factor $\sqrt{\epsilon}$. We now expand the integrand in powers of $\sqrt{\epsilon}$ and use the integrals

$$\int \frac{du}{\sqrt{2\pi/a}} e^{-(a/2)u^2} \begin{Bmatrix} u^2 \\ u^4 \\ \vdots \\ u^{2n} \end{Bmatrix} = \begin{Bmatrix} a^{-1} \\ 3a^{-2} \\ \vdots \\ (2n-1)!!a^{-n} \end{Bmatrix}, \tag{8.57}$$

with odd powers of u giving trivially 0, to find an expansion of the integral (8.56). It begins as follows:

$$1 - \epsilon \frac{a}{2} a_4 \, 3a^{-2} + \mathcal{O}(\epsilon^2). \tag{8.58}$$

This can be thought of as coming from the equivalent integral

$$\int \frac{d\varphi_{n-1}}{\sqrt{2\pi\epsilon/a}} e^{-(a/2\epsilon)(\varphi_n-\varphi_{n-1})^2-3a_4\epsilon/2a+\dots} . \tag{8.59}$$

The quartic term in (8.54),

$$\Delta A = \frac{a}{2\epsilon}a_4(\varphi_n-\varphi_{n-1})^4, \tag{8.60}$$

has generated an effective action-like term in the exponent:

$$A_{\text{eff}} = \epsilon\frac{3a_4}{2a}. \tag{8.61}$$

This is obviously due to the expectation value $\langle\Delta A\rangle_0$ of the quartic term and we can record, for later use, the perturbative formula

$$A_{\text{eff}} = \langle\Delta A\rangle_0. \tag{8.62}$$

If u is a vector in D dimensions to be denoted by $\mathbf{u}$, with the quadratic term being $(a/2\epsilon)(\mathbf{u}_n - \mathbf{u}_{n-1})^2$, the integrals (8.57) are replaced by

$$\int \frac{d^D u}{\sqrt{2\pi/a}^D} e^{-(a/2)\mathbf{u}^2} \left\{ \begin{array}{c} u_i u_j \\ u_i u_j u_k u_l \\ \vdots \\ u_{i_1}\cdot\ldots\cdot u_{i_{2n}} \end{array} \right\} = \left\{ \begin{array}{c} a^{-1}\delta_{ij} \\ a^{-2}(\delta_{ij}\delta_{kl} + \delta_{ik}\delta_{jl} + \delta_{il}\delta_{jk}) \\ \vdots \\ a^{-n}\delta_{i_1\dots i_{2n}} \end{array} \right\}, \tag{8.63}$$

where $\delta_{i_1\dots i_{2n}}$ will be referred to as *contraction tensors*, defined iteratively by the recursion relation

$$\delta_{i_1\dots i_{2n}} = \delta_{i_1 i_2}\delta_{i_3 i_4\dots i_{2n}} + \delta_{i_1 i_3}\delta_{i_2 i_4\dots i_{2n}} + \dots + \delta_{i_1 i_{2n}}\delta_{i_2 i_3\dots i_{2n-1}}. \tag{8.64}$$

A comparison with the Wick expansion (3.303) shows that this recursion relation amounts to $\delta_{i_1\dots i_{2n}}$ possessing a Wick-like expansion into the sum of products of Kronecker δ's, each representing a pair contraction. Indeed, the integral formulas (8.63) can be derived by adding a source term $\mathbf{j}\cdot\mathbf{u}$ to the exponent in the integrand, completing the square, and differentiating the resulting $e^{(1/2a)\mathbf{j}^2}$ with respect to the "current" components j_i. For vectors φ_i, a possible quartic term in the exponent of (8.54) may have the form

$$\Delta A = \frac{a}{2\epsilon}(a_4)_{ijkl}(\varphi_n-\varphi_{n-1})_i(\varphi_n-\varphi_{n-1})_j(\varphi_n-\varphi_{n-1})_k(\varphi_n-\varphi_{n-1})_l. \tag{8.65}$$

Then the factor $3a_4$ in A_{eff} of Eq. (8.61) is replaced by the three contractions

$$(a_4)_{ijkl}(\delta_{ij}\delta_{kl} + 2 \text{ more pair terms}).$$

Applying the simple result (8.61) to the action (8.53), where $a = (M/\hbar)2r_n r_{n-1}$ and $a_4 = -1/4!$, we find that the naively time-sliced kinetic term of the φ field

$$\frac{M}{2\epsilon} r_n r_{n-1} (\varphi_n - \varphi_{n-1})^2$$

is extended to

$$\frac{M}{2\epsilon} r_n r_{n-1} (\varphi_n - \varphi_{n-1})^2 - \epsilon \hbar^2 \frac{1/4}{2M r_n r_{n-1}} + \dots .$$

Thus, the lowest perturbative correction due to the fourth-order expansion term of $\cos(\varphi_n - \varphi_{n-1})$ supplies precisely the 1/4-term in the centrifugal barrier which was missing in (8.45). Proceeding in this fashion, the higher powers in the expansion of $\cos(\varphi_n - \varphi_{n-1})$ give higher and higher contributions $(\epsilon/r_n r_{n-1})^n$. Eventually, they would of course produce the entire asymptotic expansion of the "Bessel function" in the correct time-sliced action (8.37).

Note that the failure of this series to converge destroys the justification for truncating the perturbation series after any finite number of terms. In particular, the knowledge of the large-order behavior (the "tail end" of the series) [3] is needed to recover the correct threshold behavior $\propto r^m$ in the amplitudes observed in (8.26).

The reader may rightfully object that the integral (8.56) should really contain an exponential factor $\exp\left[-(a/2)(u_{n-1} - u_{n-2})^2\right]$ from the adjacent time slice which also contains the variable u_{n-1} and which has been ignored in the integral (8.56) over u_{n-1}. In fact, with the abbreviations $\bar{u}_{n-1} \equiv (u_n + u_{n-2})/2$, $\delta \equiv u_{n-1} - \bar{u}_{n-1}$, $\Delta \equiv u_n - u_{n-2}$, the complete integrand containing the variable u_{n-1} can be written as

$$\int \frac{d\delta}{\sqrt{2\pi/a}} e^{-a\delta^2} e^{-a\Delta^2/4} \left\{ 1 - \frac{a}{2}\epsilon a_4 [(-\delta + \Delta/2)^4 + (\delta + \Delta/2)^4] + \dots \right\}. \qquad (8.66)$$

When doing the integral over δ, each even power δ^{2n} gives a factor $(1/2a)(2n - 1)!!$ and we observe that the mean value of the fluctuating u_{n-1} is different from what it was above, when we singled out the expression (8.56) and ignored the u_{n-1} dependence of the adjacent integral. Instead of u_n in (8.56), the mean value of u_{n-1} is now the average position of the neighbors, $\bar{u}_{n-1} = (u_n + u_{n-2})/2$. Moreover, instead of the width of the u_{n-1} fluctuations being $\langle (u_n - \bar{u}_{n-1})^2 \rangle_0 \sim 1/a$, as in (8.55), it is now given by half this value:

$$\langle (u_{n-1} - \bar{u}_{n-1})^2 \rangle_0 = \frac{1}{2a}. \qquad (8.67)$$

At first, these observations seem to invalidate the above perturbative evaluation of (8.56). Fortunately, this objection ignores an important fact which cancels the apparent mistake, and the result (8.62) of the sloppy derivation is correct after all. The argument goes as follows: The integrand of a single time slice is a sharply peaked function of the coordinate difference whose width is of the order ϵ and goes to zero in the continuum limit. If such a function is integrated together with some *smooth* amplitude, it is sensitive only to a small neighborhood of a point in the amplitude. The sharply peaked function is a would-be δ-function that can be effectively replaced by a δ-function plus correction terms which contain increasing derivatives of δ-functions multiplied by corresponding powers of $\sqrt{\epsilon}$. Indeed, let us take the integrand of the model integral (8.54),

$$\frac{1}{\sqrt{2\pi\epsilon/a}}e^{-(a/2\epsilon)\left[(\varphi_n-\varphi_{n-1})^2+a_4(\varphi_n-\varphi_{n-1})^4+\dots\right]}, \tag{8.68}$$

and integrate it over φ_{n-1} together with a smooth amplitude $\psi(\varphi_{n-1})$ which plays the same role of a test function in mathematics [recall Eq. (1.162)]:

$$\int_{-\infty}^{\infty}\frac{d\varphi_{n-1}}{\sqrt{2\pi\epsilon/a}}e^{-(a/2\epsilon)\left[(\varphi_n-\varphi_{n-1})^2+a_4(\varphi_n-\varphi_{n-1})^4+\dots\right]}\psi(\varphi_{n-1}). \tag{8.69}$$

For small ϵ, we expand $\psi(\varphi_{n-1})$ around φ_n,

$$\psi(\varphi_{n-1}) = \psi(\varphi_n) - (\varphi_n - \varphi_{n-1})\psi'(\varphi_n) + \frac{1}{2}(\varphi_n - \varphi_{n-1})^2\psi''(\varphi_n) + \dots \; , \tag{8.70}$$

and (8.69) becomes

$$(1 - A_{\text{eff}})\psi(\varphi_n) + \frac{\epsilon}{2a}\psi''(\varphi_n) + \dots \; . \tag{8.71}$$

This shows that the amplitude for a single time slice, when integrated together with a smooth amplitude, can be expanded into a series consisting of a δ-function and its derivatives:

$$\frac{1}{\sqrt{2\pi\epsilon/a}}e^{-(a/2\epsilon)\left[(\varphi_n-\varphi_{n-1})^2+a_4(\varphi_n-\varphi_{n-1})^4+\dots\right]}$$
$$= (1 - A_{\text{eff}})\delta(\varphi_n - \varphi_{n-1}) + \frac{\epsilon}{2a}\delta''(\varphi_n - \varphi_{n-1}) + \dots \; . \tag{8.72}$$

The right-hand side may be viewed as the result of a simpler would-be δ-function

$$\frac{1}{\sqrt{2\pi\epsilon/a}}e^{-(a/2\epsilon)(\varphi_n-\varphi_{n-1})^2}e^{-A_{\text{eff}}}, \tag{8.73}$$

correct up to terms of order ϵ. This is precisely what we found in (8.59).

The problems observed above arise only if the would-be δ-function in (8.72) is integrated together with another sharply peaked neighbor function which is itself a would-be δ-function. Indeed, in the theory of distributions, it is strictly forbidden to form integrals over products of two proper distributions. For the would-be distributions at hand the rule is not quite as strict and integrals over products can be formed. The crucial expansion (8.72), however, is no longer applicable if the accompanying function is a would-be δ-function, and a more careful treatment is required.

The correctness of formula (8.62) derives from the fact that each time slice has, for sufficiently small ϵ, a large number of neighbors at earlier and later times. If the integrals are done for all these neighbors, they render a smooth amplitude before and a smooth amplitude after the slice under consideration. Thus, each intermediate integral in the time-sliced product contains a would-be δ-function multiplied on the right- and left-hand side with a *smooth* amplitude. In each such integral, the replacement (8.59) and thus formula (8.62) is correct. The only exceptions are time slices near the endpoints. Their integrals possess the above subtleties. The relative number of these, however, goes to zero in the continuum limit $\epsilon \to 0$. Hence they do not change the final result (8.62).

For completeness, let us state that the presence of a cubic term in the single-sliced action (8.68) has the following δ-function expansion

$$\frac{1}{\sqrt{2\pi\epsilon/a}} e^{-(a/2\epsilon)\left[(\varphi_n-\varphi_{n-1})^2+a_3(\varphi_n-\varphi_{n-1})^3+a_4(\varphi_n-\varphi_{n-1})^4+\ldots\right]} \tag{8.74}$$

$$= (1 - A_{\text{eff}})\delta(\varphi_n - \varphi_{n-1}) + 3a_3\frac{\epsilon}{2a}\delta'(\varphi_n - \varphi_{n-1}) + \frac{\epsilon}{2a}\delta''(\varphi_n - \varphi_{n-1}) + \ldots$$

corresponding to an "effective action"

$$A_{\text{eff}} = \frac{\epsilon}{2a}\left[3a_4 - \frac{15}{4}a_3^2\right]. \tag{8.75}$$

Using (8.75), the left-hand side of (8.74) can also be replaced by the would-be δ-function

$$\frac{1}{\sqrt{2\pi\epsilon/a}} e^{-(a/2\epsilon)(\varphi_n-\varphi_{n-1})^2} e^{-A_{\text{eff}}}\left[1 - \frac{3a_3}{2}(\varphi_n - \varphi_{n-1}) + \ldots\right], \tag{8.76}$$

which has the same leading terms in the δ-function expansion.

In D dimensions, the term $3a_3(\varphi_n - \varphi_{n-1})$ has the general form

$$[(a_3)_{ijj} + (a_3)_{jij} + (a_3)_{jji}](\varphi_n - \varphi_{n-1})_i,$$

and the term $15a_3^2$ in A_{eff} becomes

$$(a_3)_{ijk}(a_3)_{i'j'k'}(\delta_{ii'}\delta_{jj'}\delta_{kk'} + 14 \text{ more pair terms}).$$

8.5 Angular Decomposition in Three and More Dimensions

Let us now extend the two-dimensional development of Section 8.2 and study the radial path integrals of particles moving in three and more dimensions. Consider the amplitude for a rotationally invariant action in D dimensions

$$(\mathbf{x}_b\tau_b|\mathbf{x}_a\tau_a) = \int \mathcal{D}^D x(\tau) \exp\left\{-\frac{1}{\hbar}\int_{\tau_a}^{\tau_b}\left[\frac{M}{2}\dot{\mathbf{x}}^2 + V(r)\right]\right\}. \tag{8.77}$$

By time-slicing this in Cartesian coordinates, the kinetic term gives an integrand

$$\exp\left[-\frac{1}{\hbar}\frac{M}{2\epsilon}\sum_{n=1}^{N+1}(r_n^2 + r_{n-1}^2 - 2r_n r_{n-1}\cos\Delta\vartheta_n)\right],\tag{8.78}$$

where $\Delta\vartheta_n$ is the relative angle between the vectors $\mathbf{x}_n$ and $\mathbf{x}_{n-1}$.

8.5.1 Three Dimensions

In three dimensions, we go over to the spherical coordinates

$$\mathbf{x} = r(\cos\theta\cos\varphi, \cos\theta\sin\varphi, \sin\theta)\tag{8.79}$$

and write

$$\cos\Delta\vartheta_n = \cos\theta_n\cos\theta_{n-1} + \sin\theta_n\sin\theta_{n-1}\cos(\varphi_n - \varphi_{n-1}).\tag{8.80}$$

The integration measure in the time-sliced version of (8.77),

$$\frac{1}{\sqrt{2\pi\hbar\epsilon/M}^3}\prod_{n=1}^{N}\int\frac{d^3x_n}{\sqrt{2\pi\hbar\epsilon/M}^3},\tag{8.81}$$

becomes

$$\frac{1}{\sqrt{2\pi\hbar\epsilon/M}^3}\prod_{n=1}^{N}\int\frac{dr_n r_n^2 d\cos\theta_n d\varphi_n}{\sqrt{2\pi\hbar\epsilon/M}^3}.\tag{8.82}$$

To perform the integrals, we use the spherical analog of the expansion (8.5)

$$e^{h\cos\Delta\vartheta_n} = \sqrt{\frac{\pi}{2h}}\sum_{l=0}^{\infty}I_{l+1/2}(h)(2l+1)P_l(\cos\Delta\vartheta_n),\tag{8.83}$$

where $P_l(z)$ are the Legendre polynomials. These, in turn, can be decomposed into spherical harmonics

$$Y_{lm}(\theta,\varphi) = (-1)^m\left[\frac{2l+1}{4\pi}\frac{(l-m)!}{(l+m)!}\right]^{1/2}P_l^m(\cos\theta)e^{im\varphi},\tag{8.84}$$

with the help of the addition theorem

$$\frac{2l+1}{4\pi}P_l(\cos\Delta\vartheta_n) = \sum_{m=-l}^{l}Y_{lm}(\theta_n,\varphi_n)Y_{lm}^*(\theta_{n-1},\varphi_{n-1}),\tag{8.85}$$

the sum running over all azimuthal (magnetic) quantum numbers m. The right-hand side of $P_l^m(z)$ contains the associated Legendre polynomials

$$P_l^m(z) = \frac{(1-z^2)^{m/2}}{2^l l!}\frac{(l-m)!}{(l+m)!}\frac{d^{l+m}}{dz^{l+m}}(z^2-1)^l,\tag{8.86}$$

which are solutions of the differential equation[3]

$$\left[-\frac{1}{\sin\theta}\frac{d}{d\theta}\left(\sin\theta\frac{d}{d\theta}\right)+\frac{m^2}{\sin^2\theta}\right]P_l^m(\cos\theta)=l(l+1)P_l^m(\cos\theta). \tag{8.87}$$

Thus, the expansion (8.83) becomes

$$e^{h\cos\Delta\vartheta_n}=\sqrt{\frac{\pi}{2h}}4\pi\sum_{l=0}^{\infty}I_{l+1/2}(h)\sum_{m=-l}^{l}Y_{lm}(\theta_n,\varphi_n)Y_{lm}^*(\theta_{n-1},\varphi_{n-1}). \tag{8.88}$$

Inserted into (8.78), it leads to the time-sliced path integral

$$(\mathbf{x}_b\tau_b|\mathbf{x}_a\tau_a)\approx\frac{4\pi}{\sqrt{2\pi\hbar\epsilon/M}^3}\prod_{n=1}^{N}\left[\int_0^{\infty}\frac{dr_n r_n^2 d\cos\theta_n d\varphi_n 4\pi}{\sqrt{2\pi\hbar\epsilon/M}^3}\right]$$

$$\times\prod_{n=1}^{N+1}\left[\left(\frac{\hbar\epsilon\pi}{2Mr_n r_{n-1}}\right)^{1/2}\sum_{l_n=0}^{\infty}\sum_{m_n=-l_n}^{l_n}I_{l_n+1/2}\left(\frac{M}{\hbar\epsilon}r_n r_{n-1}\right)\right.$$

$$\times\left.Y_{l_n m_n}(\theta_n,\varphi_n)Y_{l_n m_n}^*(\theta_{n-1},\varphi_{n-1})\right]\exp\left\{-\frac{\epsilon}{\hbar}\sum_{n=1}^{N+1}\left[\frac{M}{2}\frac{r_n^2+r_{n-1}^2}{\epsilon^2}+V(r_n)\right]\right\}. \tag{8.89}$$

The intermediate φ_n- and $\cos\theta_n$-integrals can now all be done using the orthogonality relation

$$\int_{-1}^{1}d\cos\theta\int_{-\pi}^{\pi}d\varphi\,Y_{lm}^*(\theta,\varphi)Y_{l'm'}(\theta,\varphi)=\delta_{ll'}\delta_{mm'}. \tag{8.90}$$

Each φ_n-integral yields a product of Kronecker symbols $\delta_{l_n l_{n-1}}\delta_{m_n m_{n-1}}$. Only the initial and the final spherical harmonics survive, $Y_{l_{N+1}m_{N+1}}$ and $Y_{l_0 m_0}^*$, since they are not subject to integration. Thus we arrive at the angular momentum decomposition

$$(\mathbf{x}_b\tau_b|\mathbf{x}_a\tau_a)=\sum_{l=0}^{\infty}\sum_{m=-l}^{l}\frac{1}{r_b r_a}(r_b\tau_b|r_a\tau_a)_l Y_{lm}(\theta_b,\varphi_b)Y_{lm}^*(\theta_a,\varphi_a), \tag{8.91}$$

with the radial amplitude

$$(r_b\tau_b|r_a\tau_a)_l\approx\frac{4\pi r_b r_a}{\sqrt{2\pi\hbar\epsilon/M}^3}\prod_{n=1}^{N}\left[\int\frac{dr_n r_n^2 4\pi}{\sqrt{2\pi\hbar\epsilon/M}^3}\right]\prod_{n=1}^{N+1}\left[\frac{\hbar\epsilon}{2Mr_n r_{n-1}}\right] \tag{8.92}$$

$$\times\prod_{n=1}^{N+1}\left[\tilde{I}_{l+1/2}(\frac{M}{\hbar\epsilon}r_n r_{n-1})\right]\exp\left\{-\frac{\epsilon}{\hbar}\sum_{n=1}^{N+1}\left[\frac{M}{2\epsilon^2}(r_n-r_{n-1})^2+V(r_n)\right]\right\}.$$

---

[3]Note that $y_l^m(\cos\theta)=\sqrt{\sin\theta}P_l^m(\cos\theta)$ satisfies $\left[-\frac{d}{d\theta^2}-\frac{1}{4}+\frac{m^2-1/4}{\sin^2\theta}\right]y_l^m=l(l+1)y_l^m$. This differential equation will be used later in Eq. (8.196).

The factors $\prod_a^N r_n^2 / \prod_a^{N+1} r_n r_{n-1}$ pile up to $1/r_b r_a$ and cancel the prefactor $r_b r_a$. Together with the remaining product, the integration measure takes the usual one-dimensional form

$$\frac{1}{\sqrt{2\pi\hbar\epsilon/M}} \prod_{n=1}^{N} \left[\int_0^\infty \frac{dr_n}{\sqrt{2\pi\hbar\epsilon/M}} \right]. \tag{8.93}$$

If we were to use here the large-argument limit (8.24) of the Bessel function, the integrand would become $\exp(-\mathcal{A}_l^N/\hbar)$, with the time-sliced radial action

$$\mathcal{A}_l^N = \epsilon \sum_{n=1}^{N+1} \left[\frac{M}{2\epsilon^2} (r_n - r_{n-1})^2 + \frac{\hbar^2}{2M} \frac{l(l+1)}{r_n r_{n-1}} + V(r_n) \right]. \tag{8.94}$$

The associated radial path integral

$$(r_b \tau_b | r_a \tau_a)_l \approx \frac{1}{\sqrt{2\pi\epsilon\hbar/M}} \prod_{n=1}^{N} \left[\int_0^\infty \frac{dr_n}{\sqrt{2\pi\hbar\epsilon/M}} \right] \exp\left(-\frac{1}{\hbar} \mathcal{A}_l^N \right) \tag{8.95}$$

agrees precisely with what would have been obtained by naively time-slicing the continuum path integral

$$(r_b \tau_b | r_a \tau_a)_l = \int_0^\infty \mathcal{D}r(\tau) e^{-\frac{1}{\hbar}\mathcal{A}_l[r]}, \tag{8.96}$$

with the radial action

$$\mathcal{A}_l[r] = \int_{\tau_a}^{\tau_b} d\tau \left[\frac{M}{2} \dot{r}^2 + \frac{\hbar^2}{2M} \frac{l(l+1)}{r^2} + V(r) \right]. \tag{8.97}$$

In particular, this would contain the correct centrifugal barrier

$$V_{\text{cf}} = \frac{\hbar^2}{2M} \frac{l(l+1)}{r^2}. \tag{8.98}$$

However, as we know from the discussion in Section 8.2, Eq. (8.95) is incorrect and must be replaced by (8.92), due to the non uniformity of the continuum limit $\epsilon \to 0$ in the integrand of (8.92).

8.5.2 D Dimensions

The generalization to D dimensions is straightforward. The main place where the dimension enters is the expansion of

$$e^{h \cos \Delta\vartheta_n} = e^{-\frac{M}{\hbar\epsilon} r_n r_{n-1} \cos \Delta\vartheta_n}, \tag{8.99}$$

in which $\Delta\vartheta_n$ is the relative angle between D-dimensional vectors $\mathbf{x}_n$ and $\mathbf{x}_{n-1}$. The expansion reads [compare with (8.5) and (8.83)]

$$e^{h \cos \Delta\vartheta_n} = \sum_{l=0}^{\infty} a_l(h) \frac{l + D/2 - 1}{D/2 - 1} \frac{1}{S_D} C_l^{(D/2-1)}(\cos \Delta\vartheta_n), \tag{8.100}$$

where S_D is the surface of a unit sphere in D dimensions (1.555), and

$$a_l(h) \equiv (2\pi)^{D/2} h^{1-D/2} I_{l+D/2-1}(h)$$

$$\equiv e^h \tilde{a}_l(h) = e^h \left(\frac{2\pi}{h}\right)^{(D-1)/2} \tilde{I}_{l+D/2-1}(h). \qquad (8.101)$$

The functions $C_l^{(\alpha)}(\cos\vartheta)$ are the ultra-spherical Gegenbauer polynomials. The expansion (8.100) follows from the completeness of the polynomials $C_l^{(\nu)}(\cos\vartheta)$ at fixed ν, using the integration formulas[4]

$$\int_0^\pi d\vartheta \sin^\nu \vartheta e^{h\cos\vartheta} C_l^{(\nu)}(\cos\vartheta) = \pi \frac{2^{1-\nu}\Gamma(2\nu+l)}{l!\Gamma(\nu)} h^{-\nu} I_{\nu+l}(h), \qquad (8.102)$$

$$\int_0^\pi d\vartheta \sin^\nu \vartheta C_l^{(\nu)}(\cos\vartheta) C_{l'}^{(\nu)}(\cos\vartheta) = \pi \frac{2^{1-2\nu}\Gamma(2\nu+l)}{l!(l+\nu)\Gamma(\nu)^2} \delta_{ll'}. \qquad (8.103)$$

The Gegenbauer polynomials are related to Jacobi polynomials, which are defined in terms of hypergeometric functions (1.450) by[5]

$$P_l^{(\alpha,\beta)}(z) \equiv \frac{1}{l!} \frac{\Gamma(l+1+\beta)}{\Gamma(1+\beta)} F(-l, l+1+\alpha+\beta; 1+\beta; (1+z)/2). \qquad (8.104)$$

The relation is

$$C_l^{(\nu)}(z) = \frac{\Gamma(2\nu+l)\Gamma(\nu+1/2)}{\Gamma(2\nu)\Gamma(\nu+l+1/2)} P_l^{(\nu-1/2,\nu-1/2)}(z). \qquad (8.105)$$

This follows from the defining equation[6]

$$C_l^{(\nu)}(z) = \frac{1}{l!} \frac{\Gamma(l+2\nu)}{\Gamma(2\nu)} F(-l, l+2\nu; 1/2+\nu; (1+z)/2). \qquad (8.106)$$

For $D=2$ and 3, one has[7]

$$\lim_{\nu\to 0} \frac{1}{\nu} C_l^{(\nu)}(\cos\vartheta) = \frac{1}{2l}\cos l\vartheta, \qquad (8.107)$$

$$C_l^{(1/2)}(\cos\vartheta) = P_l^{(0,0)}(\cos\vartheta) = P_l(\cos\vartheta), \qquad (8.108)$$

and the expansion (8.100) reduces to (8.5) and (8.7), respectively. For $D=4$

$$C_l^{(1)}(\cos\vartheta) = \frac{\sin(l+1)\beta}{\sin\beta}. \qquad (8.109)$$

---

[4]I. S. Gradshteyn and I. M. Ryzhik, op. cit., Formulas 7.321 and 7.313.
[5]M. Abramowitz and I. Stegun, *op. cit.*, Formula 15.4.6.
[6]ibid., Formula 15.4.5.
[7]I.S. Gradshteyn and I.M. Ryzhik, op. cit.,*ibid.*, Formula 8.934.4.

According to an addition theorem, the Gegenbauer polynomials can be decomposed into a sum of pairs of D-dimensional ultra-spherical harmonics $Y_{l\mathbf{m}}(\hat{\mathbf{x}})$.[8] The label $\mathbf{m}$ stands collectively for the set of *magnetic* quantum numbers $m_1, m_2, m_3, ..., m_{D-1}$ with $1 \le m_1 \le m_2 \le ... \le |m_{D-2}|$. The direction $\hat{\mathbf{x}}$ of a vector $\mathbf{x}$ is specified by $D-1$ polar angles

$$
\begin{aligned}
\hat{x}^1 &= \sin\varphi_{D-1}\cdots\sin\varphi_1, \\
\hat{x}^2 &= \sin\varphi_{D-1}\cdots\cos\varphi_1, \\
&\vdots \\
\hat{x}^D &= \cos\varphi_{D-1},
\end{aligned}
\tag{8.110}
$$

with the ranges

$$
0 \le \varphi_1 < 2\pi, \tag{8.111}
$$
$$
0 \le \varphi_i < \pi, \quad i \ne 1. \tag{8.112}
$$

The ultra-spherical harmonics $Y_{l\mathbf{m}}(\hat{\mathbf{x}})$ form an orthonormal and complete set of functions on the D-dimensional unit sphere. For a fixed quantum number l of total angular momentum, the label $\mathbf{m}$ can take

$$
d_l = \frac{(2l + D - 2)(l + D - 3)!}{l!(D-2)!} \tag{8.113}
$$

different values. The functions are orthonormal,

$$
\int d\hat{\mathbf{x}}\, Y_{l\mathbf{m}}^*(\hat{\mathbf{x}}) Y_{l'\mathbf{m}'}(\hat{\mathbf{x}}) = \delta_{ll'}\delta_{\mathbf{m}\mathbf{m}'}, \tag{8.114}
$$

with $\int d\hat{\mathbf{x}}$ denoting the integral over the surface of the unit sphere:

$$
\int d\hat{\mathbf{x}} = \int d\varphi_{D-1}\sin^{D-2}\varphi_{D-1}\int d\varphi_{D-2}\sin^{D-3}\varphi_{D-2}\cdots\int d\varphi_2\sin\varphi_2\int d\varphi_1 . \tag{8.115}
$$

By evaluating this integral over a unit integrand[9] we find $S_D = 2\pi^{D/2}/\Gamma(D/2)$ as anticipated in Eq. (1.555). Since $Y_{00}(\hat{\mathbf{x}})$ is independent of $\hat{\mathbf{x}}$, the integral (8.114) implies that $Y_{00}(\hat{\mathbf{x}}) = 1/\sqrt{S_D}$.

Note that the integral over the unit sphere in D-dimensions can be decomposed recursively into an angular integration with respect to any selected direction, say $\hat{\mathbf{u}}$, in the space followed by an integral over a sphere of radius $\sin\varphi_{D-1}$ in the remaining $D-1$-dimensional space to $\hat{\mathbf{u}}$. If $\hat{\mathbf{x}}^\perp$ denotes the unit vector covering the directions

---

[8] See H. Bateman, *Higher Transcendental Functions*, McGraw-Hill, New York, 1953, Vol. II, Ch. XI; N.H. Vilenkin, *Special Functions and the Theory of Group Representations*, Am. Math. Soc., Providence, RI, 1968.

[9] With the help of the integral formula $\int_0^\pi d\varphi \sin^k\varphi = \sqrt{\pi}\Gamma((k+1)/2)/\Gamma((k+2)/2)$ we find $S_D = \prod_{k=0}^{D-1}\int_0^\pi d\varphi_k \sin^k\varphi_k = 2\pi^{D/2}\prod_{k=1}^{D-2}\Gamma((k+1)/2)/\prod_{k=1}^{D-2}\Gamma((k+2)/2) = 2\pi^{D/2}/\Gamma(D/2)$.

in this remaining space, one decomposes $\hat{\mathbf{x}} = (\cos\varphi_D\hat{\mathbf{u}} + \sin\varphi_D\hat{\mathbf{x}})$, and can factorize the integral measure as

$$\int d^{D-1}\hat{\mathbf{x}} = \int d\varphi_{D-1}\sin^{D-2}\varphi_{D-1}\int d^{D-2}\hat{\mathbf{x}}_\perp . \tag{8.116}$$

For clarity, the dimensionalities of initial and remaining surfaces are marked as superscripts on the measure symbols $d^{D-1}\hat{\mathbf{x}}$ and $d^{D-1}\hat{\mathbf{x}}^\perp$.

For the surface of the sphere, this corresponds to the recursion relation

$$S_D = \frac{\sqrt{\pi}\Gamma((D-1)/2)}{\Gamma(D/2)} \times S_{D-1}, \tag{8.117}$$

which is solved by $S_D = 2\pi^{D/2}/\Gamma(D)$.

In four dimensions, the unit vectors $\hat{\mathbf{x}}$ have a parametrization in terms of polar angles

$$\hat{\mathbf{x}} = (\cos\theta, \sin\theta\cos\psi, \sin\theta\sin\psi\cos\varphi, \sin\theta\sin\psi\sin\varphi), \tag{8.118}$$

with the integration measure

$$d\hat{\mathbf{x}} = d\theta\sin^2\theta d\psi\sin\psi\, d\varphi. \tag{8.119}$$

It is, however, more convenient to go over to another parametrization in terms of the three Euler angles which are normally used in the kinematic description of the spinning top. In terms of these, the unit vectors have the components

$$\begin{aligned}
\hat{x}^1 &= \cos(\theta/2)\cos[(\varphi+\gamma)/2], \\
\hat{x}^2 &= -\cos(\theta/2)\sin[(\varphi+\gamma)/2], \\
\hat{x}^3 &= \sin(\theta/2)\cos[(\varphi-\gamma)/2], \\
\hat{x}^4 &= \sin(\theta/2)\sin[(\varphi-\gamma)/2],
\end{aligned} \tag{8.120}$$

with the angles covering the intervals

$$\theta \in [0,\pi), \quad \varphi \in [0,2\pi), \quad \gamma \in [-2\pi,2\pi). \tag{8.121}$$

We have renamed the usual Euler angles α,β,γ introduced in Section 1.15 calling them φ,θ,γ, since the formulas to be derived for them will be used in a later application in Chapter 13 [see Eq. (13.97)]. There the first two Euler angles coincide with the polar angles φ,θ of a position vector in a three-dimensional space. It is important to note that for a description of the entire surface of the sphere, the range of the angle γ must be twice as large as for the classical spinning top. The associated group space belongs to the covering group, of the rotation group which is equivalent to the group of unimodular matrices in two dimensions called SU(2). It is defined by the matrices

$$g(\varphi,\theta,\gamma) = \exp(i\varphi\sigma_3/2)\exp(i\theta\sigma_2/2)\exp(i\gamma\sigma_3/2), \tag{8.122}$$

where σ_i are the Pauli spin matrices (1.445). In this parametrization, the integration measure reads

$$d\hat{\mathbf{x}} = \frac{1}{8} d\theta \sin\theta \, d\varphi \, d\gamma. \tag{8.123}$$

When integrated over the surface, the two measures give the same result $S_4 = 2\pi^2$. The Euler parametrization has the advantage of allowing the spherical harmonics in four dimensions to be expressed in terms of the well-known representation functions of the rotation group introduced in (1.442), (1.443):

$$Y_{l,m_1,m_2}(\hat{\mathbf{x}}) = \sqrt{\frac{l+1}{2\pi^2}} \mathcal{D}^{l/2}_{m_1 m_2}(\varphi,\theta,\gamma) = \sqrt{\frac{l+1}{2\pi^2}} d^{l/2}_{m_1 m_2}(\theta) e^{i(m_1\varphi + m_2\gamma)}. \tag{8.124}$$

For even and odd l, the numbers m_1, m_2 are both integer or half-integer, respectively.

In arbitrary dimensions $D > 2$, the ultra-spherical Gegenbauer polynomials satisfy the following addition theorem

$$\frac{2l+D-2}{D-2} \frac{1}{S_D} C^{(D/2-1)}_l(\cos\Delta\vartheta_n) = \sum_{\mathbf{m}} Y_{l\mathbf{m}}(\hat{\mathbf{x}}_n) Y^*_{l\mathbf{m}}(\hat{\mathbf{x}}_{n-1}). \tag{8.125}$$

For $D = 3$, this reduces properly to the well-known addition theorem for the spherical harmonics

$$\frac{1}{4\pi}(2l+1) P_l(\cos\Delta\vartheta_n) = \sum_{m=-l}^{l} Y_{lm}(\hat{\mathbf{x}}_n) Y^*_{lm}(\hat{\mathbf{x}}_{n-1}). \tag{8.126}$$

For $D = 4$, it becomes[10]

$$\frac{l+1}{2\pi^2} C^{(1)}_l(\cos\Delta\vartheta_n) = \frac{l+1}{2\pi^2} \sum_{m_1,m_2=-l/2}^{l/2} \mathcal{D}^{l/2}_{m_1 m_2}(\varphi_n,\theta_n,\gamma_n) \mathcal{D}^{l/2\,*}_{m_1 m_2}(\varphi_{n-1},\theta_{n-1},\gamma_{n-1}), \tag{8.127}$$

where the angle $\Delta\vartheta_n$ is related to the Euler angles of the vectors $\mathbf{x}_n$, $\mathbf{x}_{n-1}$ by

$$\begin{aligned}
\cos\Delta\vartheta_n = {}& \cos(\theta_n/2)\cos(\theta_{n-1}/2)\cos[(\varphi_n - \varphi_{n-1} + \gamma_n - \gamma_{n-1})/2] \\
& + \sin(\theta_n/2)\sin(\theta_{n-1}/2)\cos[(\varphi_n - \varphi_{n-1} - \gamma_n + \gamma_{n-1})/2].
\end{aligned} \tag{8.128}$$

Using (8.125), we can rewrite the expansion (8.100) in the form

$$e^{h(\cos\Delta\vartheta_n - 1)} = \sum_{l=0}^{\infty} \tilde{a}_l(h) \sum_{\mathbf{m}} Y_{l\mathbf{m}}(\hat{\mathbf{x}}_n) Y^*_{l\mathbf{m}}(\hat{\mathbf{x}}_{n-1}). \tag{8.129}$$

This is now valid for any dimension D, including the case $D = 2$ where the left-hand side of (8.125) involves the limiting procedure (8.107). We shall see in Chapter 9

---

[10]Note that $C^{(1)}_l(\cos\Delta\vartheta_n)$ coincides with the trace over the representation functions (1.443) of the rotation group, i.e., it is equal to $\sum_{m=-l/2}^{l/2} d^{l/2}_{m,m}(\Delta\vartheta_n)$.

in connection with Eq. (9.61) that it also makes sense to apply this expansion to the case $D = 1$ where the "partial-wave expansion" degenerates into a separation of even and odd wave functions. In four dimensions, we shall mostly prefer the expansion

$$
e^{h(\cos \Delta \vartheta_n - 1)} = \sum_{l=0}^{\infty} \tilde{a}_l(h) \frac{l+1}{2\pi^2} \sum_{m_1, m_2 = -l/2}^{l/2} \mathcal{D}_{m_1 m_2}^{l/2}(\varphi_n, \theta_n, \gamma_n) \mathcal{D}_{m_1 m_2}^{l/2 \, *}(\varphi_{n-1}, \theta_{n-1}, \gamma_{n-1}),
$$

(8.130)

where the sum over m_1, m_2 runs for even and odd l over integer and half-integer numbers, respectively.

The reduction of the time evolution amplitude in D dimensions to a radial path integral proceeds from here on in the same way as in two and three dimensions. The generalization of (8.89) reads

$$
(\mathbf{x}_b \tau_b | \mathbf{x}_a \tau_a) \approx \frac{1}{\sqrt{2\pi\hbar\epsilon/M}^D} \prod_{n=1}^{N} \left[\int_0^\infty \frac{dr_n r_n^{D-1} d\hat{\mathbf{x}}_n}{\sqrt{2\pi\hbar\epsilon/M}^D} \right]
$$
$$
\times \prod_{n=1}^{N+1} \left[\left(\frac{2\pi\hbar\epsilon}{Mr_n r_{n-1}} \right)^{(D-1)/2} \sum_{l_n=0}^{\infty} \tilde{I}_{D/2-1+l_n} \left(\frac{M}{\hbar\epsilon} r_n r_{n-1} \right) \right.
$$
$$
\left. \times \sum_{\mathbf{m}_n} Y_{l_n \mathbf{m}_n}(\hat{\mathbf{x}}_n) Y_{l_n \mathbf{m}_n}^*(\hat{\mathbf{x}}_{n-1}) \right] \exp \left\{ -\frac{\epsilon}{\hbar} \sum_{n=1}^{N+1} \left[\frac{M}{2\epsilon^2} (r_n - r_{n-1})^2 + V(r_n) \right] \right\}.
$$

(8.131)

By performing the angular integrals and using the orthogonality relations (8.114), the product of sums over $l_n, \mathbf{m}_n$ reduces to a single sum over $l, \mathbf{m}$, just as in the three-dimensional amplitude (8.91). The result is the spherical decomposition

$$
(\mathbf{x}_b \tau_b | \mathbf{x}_a \tau_a) = \frac{1}{(r_b r_a)^{(D-1)/2}} \sum_{l=0}^{\infty} (r_b \tau_b | r_a \tau_a)_l \sum_{\mathbf{m}} Y_{l\mathbf{m}}(\hat{\mathbf{x}}_b) Y_{l\mathbf{m}}^*(\hat{\mathbf{x}}_a),
$$

(8.132)

where $(r_b \tau_b | r_a \tau_a)_l$ is the purely radial amplitude

$$
(r_b \tau_b | r_a \tau_a)_l \approx \frac{1}{\sqrt{2\pi\hbar\epsilon/M}} \prod_{n=1}^{N} \left[\int_0^\infty \frac{dr_n}{\sqrt{2\pi\hbar\epsilon/M}} \right] \exp \left\{ -\frac{1}{\hbar} \mathcal{A}_l^N[r] \right\},
$$

(8.133)

with the time-sliced action

$$
\mathcal{A}_l^N[r] = \epsilon \sum_{n=1}^{N+1} \left[\frac{M}{2\epsilon^2} (r_n - r_{n-1})^2 - \frac{\hbar}{\epsilon} \log \tilde{I}_{l+D/2-1} \left(\frac{M}{\hbar\epsilon} r_n r_{n-1} \right) + V(r_n) \right]. \quad (8.134)
$$

As before, the product $\prod_{n=1}^{N+1} 1/(r_n r_{n-1})^{(D-1)/2}$ has removed the product $\prod_{n=1}^{N} r_n^{D-1}$ in the measure as well as the factor $(r_b r_a)^{(D-1)/2}$ in front of it, leaving only the standard one-dimensional measure of integration.

In the continuum limit $\epsilon \to 0$, the asymptotic expression (8.24) for the Bessel function brings the action to the form

$$\mathcal{A}_l^N[r,\bar{r}] \approx \epsilon \sum_{n=1}^{N+1} \left[\frac{M}{2\epsilon^2}(r_n - r_{n-1})^2 + \frac{\hbar^2}{2M} \frac{(l+D/2-1)^2 - 1/4}{r_n r_{n-1}} + V(r_n) \right]. \quad (8.135)$$

This looks again like the time-sliced version of the radial path integral in D dimensions

$$(r_b \tau_b | r_a \tau_a)_l = \int \mathcal{D}r(\tau) \exp \left\{ -\frac{1}{\hbar} \mathcal{A}_l[r] \right\}, \quad (8.136)$$

with the continuum action

$$\mathcal{A}_l[r] = \int_{\tau_a}^{\tau_b} d\tau \left[\frac{M}{2}\dot{r}^2 + {}^{``}\frac{\hbar^2}{2M} \frac{(l+D/2-1)^2 - 1/4}{r^2}{}^{"} + V(r) \right]. \quad (8.137)$$

As in Eq. (8.50), we have written the centrifugal barrier as

$$ {}^{``}\frac{\hbar^2}{2Mr^2}[(l+D/2-1)^2 - 1/4]{}^{"}, \quad (8.138)$$

to emphasize the subtleties of the time-sliced radial path integral, with the understanding that the time-sliced barrier reads [as in (8.51)]

$$ {}^{``}\frac{\epsilon\hbar^2}{2Mr_n r_{n-1}}[(l+D/2-1)^2 - 1/4)]{}^{"} \equiv -\hbar\log\tilde{I}_{l+D/2-1}(\frac{M}{\hbar\epsilon}r_n r_{n-1}). \quad (8.139)$$

8.6   Radial Path Integral for Harmonic Oscillator and Free Particle in D Dimensions

For the harmonic oscillator and the free particle, there is no need to perform the radial path integral (8.133) with the action (8.134). As in (8.38), we simply take the known amplitude in D dimensions, (2.175), continue it to imaginary times $t = -i\tau$, and expand it with the help of (8.129):

$$(\mathbf{x}_b \tau_b | \mathbf{x}_a \tau_a) = \frac{1}{\sqrt{2\pi\hbar/M}^D} \sqrt{\frac{\omega}{\sinh[\omega(\tau_b - \tau_a)]}}^D \quad (8.140)$$

$$\times \exp\left\{ -\frac{1}{\hbar} \frac{M\omega}{\sinh[\omega(\tau_b - \tau_a)]}(r_b^2 + r_a^2)\cosh[\omega(\tau_b - \tau_a)] \right\}$$

$$\times \sum_{l=0}^{\infty} a_l \left(\frac{M\omega r_b r_a}{\hbar \sinh[\omega(\tau_b - \tau_a)]} \right) \sum_{\mathbf{m}} Y_{l\mathbf{m}}(\hat{\mathbf{x}}_b) Y_{l\mathbf{m}}^*(\hat{\mathbf{x}}_a).$$

Comparing this with Eq. (8.132) and remembering (8.101), we identify the radial amplitude as

$$(r_b \tau_b | r_a \tau_a)_l = \frac{M}{\hbar} \frac{\omega\sqrt{r_b r_a}^{D-1}}{\sinh[\omega(\tau_b - \tau_a)]}$$

$$\times e^{-(M\omega/2\hbar)\coth[\omega(\tau_b - \tau_a)](r_b^2 + r_a^2)} I_{l+D/2-1}\left(\frac{M\omega r_b r_a}{\hbar \sinh[\omega(\tau_b - \tau_a)]} \right), \quad (8.141)$$

generalizing (8.39). The limit $\omega \to 0$ yields the amplitude for a free particle

$$(r_b \tau_b | r_a \tau_a)_l = \frac{M}{\hbar} \frac{\sqrt{r_b r_a}^{D-1}}{(\tau_b - \tau_a)} e^{-M(r_b^2 + r_a^2)/2\hbar(\tau_b - \tau_a)} I_{l+D/2-1} \left(\frac{M r_b r_a}{\hbar(\tau_b - \tau_a)} \right). \tag{8.142}$$

Comparing this with (8.40) on the one hand and Eqs. (8.139), (8.137) with (8.49), (8.51) on the other hand, we conclude: An analytical continuation in D yields the path integral for a linear oscillator in the presence of an arbitrary $1/r^2$-potential as follows:

$$(r_b \tau_b | r_a \tau_a)_l = \int_0^\infty \mathcal{D}r(\tau) \exp\left[-\frac{1}{\hbar} \int_{\tau_a}^{\tau_b} d\tau \left(\frac{M}{2}\dot{r}^2 + \text{``}\frac{\hbar^2}{2M}\frac{\mu^2 - 1/4}{r^2}\text{''} + \frac{M}{2}\omega^2 r^2 \right) \right]$$

$$= \frac{M}{\hbar} \frac{\omega\sqrt{r_b r_a}^{D-1}}{\sinh[\omega(\tau_b - \tau_a)]} e^{-(M\omega/2\hbar)\coth[\omega(\tau_b - \tau_a)](r_b^2 + r_a^2)} I_\mu \left(\frac{M\omega r_b r_a}{\hbar \sinh[\omega(\tau_b - \tau_a)]} \right). \tag{8.143}$$

Here μ is some strength parameter which initially takes the values $\mu = l + D/2 - 1$ with integer l and D. By analytic continuation, the range of validity is extended to all real $\mu > 0$. The justification for the continuation procedure follows from the fact that the integral formula (8.14) holds for arbitrary $m = \mu \geq 0$. The amplitude (8.143) satisfies therefore the fundamental composition law (8.20) for all real $m = \mu \geq 0$. The harmonic oscillator with an arbitrary extra centrifugal barrier potential

$$V_{\text{extra}}(r) = \hbar^2 \frac{l_{\text{extra}}^2}{2Mr^2} \tag{8.144}$$

has therefore the radial amplitude (8.143) with

$$\mu = \sqrt{(l + D/2 - 1)^2 + l_{\text{extra}}^2}. \tag{8.145}$$

For a finite number $N + 1$ of time slices, the radial amplitude is known from the angular momentum expansion of the finite-N oscillator amplitude (2.197) in its obvious extension to D dimensions. It can also be calculated directly as in Appendix 2B by a successive integration of (8.131), using formula (8.14). The iteration formulas are the Euclidean analogs of those derived in Appendix 2B, with the prefactor of the amplitude being $2\pi \mathcal{N}_1^2 \mathcal{N}_{N+1}^2 \sqrt{r_b r_a}$, with the exponent $-a_{N+1}(r_b^2 + r_a^2)/\hbar$, and with the argument of the Bessel function $2b_{N+1}r_b r_a/\hbar$. In this way we obtain precisely the expression (8.143), except that $\sinh[\omega(\tau_b - \tau_a)]$ is replaced by $\sinh[\tilde{\omega}(N+1)\epsilon]\epsilon/\sinh\tilde{\omega}\epsilon$ and $\cosh[\omega(\tau_b - \tau_a)]$ by $\cosh[\tilde{\omega}(N+1)\epsilon]$.

8.7 Particle *near* the Surface of a Sphere in D Dimensions

With the insight gained in the previous sections, it is straightforward to calculate exactly a certain class of auxiliary path integrals. They involve only angular variables and will be called path integrals of a point particle moving *near* the surface of a sphere in D dimensions. The resulting amplitudes lead eventually to the physically more relevant amplitudes describing the behavior of a particle *on* the surface of a sphere.

On the surface of a sphere of radius r, the position of the particle as a function of time is specified by a unit vector $\mathbf{u}(t)$. The Euclidean action is

$$\mathcal{A} = \frac{M}{2}r^2 \int_{\tau_a}^{\tau_b} d\tau \, \dot{\mathbf{u}}^2(\tau). \tag{8.146}$$

The precise way of time-slicing this action is not known from previous discussions. It cannot be *deduced* from the time-sliced action in Cartesian coordinates, nor from its angular momentum decomposition. A new geometric feature makes the previous procedures inapplicable: The surface of a sphere is a Riemannian space with nonzero intrinsic curvature. Sections 1.13 to 1.15 have shown that the motion in a curved space does not follow the canonical quantization rules of operator quantum mechanics. The same problem is encountered here in another form: Right in the beginning, we are not allowed to time-slice the action (8.146) in a straightforward way. The correct slicing is found in two steps. First we use the experience gained with the angular momentum decomposition of time-sliced amplitudes in a Euclidean space to introduce and solve the earlier mentioned auxiliary time-sliced path integral *near* the surface of the sphere. In a second step we shall implement certain corrections to properly describe the action *on* the sphere. At the end, we have to construct the correct measure of path integration which will not be what one naively expects. To set up the auxiliary path integral *near* the surface of a sphere we observe that the kinetic term of a time slice in D dimensions

$$\frac{M}{2\epsilon} \sum_{n=1}^{N+1} (r_n^2 + r_{n-1}^2 - 2r_n r_{n-1} \cos \Delta\vartheta_n) \tag{8.147}$$

decomposes into radial and angular parts as

$$-\frac{M}{2\epsilon} \sum_{n=1}^{N+1} (r_n^2 + r_{n-1}^2 - 2r_n r_{n-1}) + \frac{M}{2\epsilon} \sum_{n=1}^{N+1} 2r_n r_{n-1}(1 - \cos \Delta\vartheta_n). \tag{8.148}$$

The angular factor can be written as

$$-\frac{M}{2\epsilon} \sum_{n=1}^{N+1} r_n r_{n-1}(\hat{\mathbf{x}}_n - \hat{\mathbf{x}}_{n-1})^2, \tag{8.149}$$

where $\hat{\mathbf{x}}_n$, $\hat{\mathbf{x}}_{n-1}$ are the unit vectors pointing in the directions of $\mathbf{x}_n$, $\mathbf{x}_{n-1}$ [recall (8.110)]. Restricting all radial variables r_n to the surface of a sphere of a fixed radius r and identifying $\hat{\mathbf{x}}$ with $\mathbf{u}$ leads us directly to the time-sliced path integral *near* the surface of the sphere in D dimensions:

$$(\mathbf{u}_b \tau_b | \mathbf{u}_a \tau_a) \approx \frac{1}{\sqrt{2\pi\hbar\epsilon/Mr^2}^{D-1}} \prod_{n=1}^{N} \left[\int \frac{d\mathbf{u}_n}{\sqrt{2\pi\hbar\epsilon/Mr^2}^{D-1}} \right] \exp\left(-\frac{1}{\hbar}\mathcal{A}^N\right), \tag{8.150}$$

with the sliced action

$$\mathcal{A}^N = \frac{M}{2\epsilon} r^2 \sum_{n=1}^{N+1} (\mathbf{u}_n - \mathbf{u}_{n-1})^2. \tag{8.151}$$

The measure $d\mathbf{u}_n$ denotes infinitesimal surface elements on the sphere in D dimensions [recall (8.115)]. Note that although the endpoints $\mathbf{u}_n$ lie all on the sphere, the paths remain only *near* the sphere since the path sections between the points leave the surface and traverse the embedding space along a straight line. This will be studied further in Section 8.8.

As mentioned above, this amplitude can be solved exactly. In fact, for each time interval ϵ, the exponential

$$\exp\left[-\frac{Mr^2}{2\hbar\epsilon}(\mathbf{u}_n - \mathbf{u}_{n-1})^2\right] = \exp\left[-\frac{Mr^2}{\hbar\epsilon}(1 - \cos\Delta\vartheta_n)\right] \tag{8.152}$$

can be expanded into spherical harmonics according to formulas (8.100)–(8.101),

$$\exp\left[-\frac{Mr^2}{2\hbar\epsilon}(\mathbf{u}_n - \mathbf{u}_{n-1})^2\right] = \sum_{l=0}^{\infty} \tilde{a}_l(h) \frac{l + D/2 - 1}{D/2 - 1} \frac{1}{S_D} C_l^{(D/2-1)}(\cos\Delta\vartheta_n)$$

$$= \sum_{l=0}^{\infty} \tilde{a}_l(h) \sum_{\mathbf{m}} Y_{l\mathbf{m}}(\mathbf{u}_n) Y_{l\mathbf{m}}^*(\mathbf{u}_{n-1}), \tag{8.153}$$

where

$$\tilde{a}_l(h) = \left(\frac{2\pi}{h}\right)^{(D-1)/2} \tilde{I}_{l+D/2-1}(h), \quad h = \frac{Mr^2}{\hbar\epsilon}. \tag{8.154}$$

For each adjacent pair $(n+1, n), (n, n-1)$ of such factors in the sliced path integral, the integration over the intermediate $\mathbf{u}_n$ variable can be done using the orthogonality relation (8.114). In this way, (8.150) produces the time-sliced amplitude

$$(\mathbf{u}_b\tau_b|\mathbf{u}_a\tau_a) = \left(\frac{h}{2\pi}\right)^{(N+1)(D-1)/2} \sum_{l=0}^{\infty} \tilde{a}_l(h)^{N+1} \sum_{\mathbf{m}} Y_{l\mathbf{m}}(\mathbf{u}_b) Y_{l\mathbf{m}}^*(\mathbf{u}_a). \tag{8.155}$$

We now go to the continuum limit $N \to \infty$, $\epsilon = (\tau_b - \tau_a)/(N+1) \to 0$, where [recall (8.11)]

$$\left(\frac{h}{2\pi}\right)^{(N+1)(D-1)/2} \tilde{a}_l(h)^{N+1} = \left[\tilde{I}_{l+D/2-1}\left(\frac{Mr^2}{\hbar\epsilon}\right)\right]^{N+1}$$

$$\xrightarrow{\epsilon\to 0} \exp\left\{-(\tau_b - \tau_a)\hbar\frac{(l + D/2 - 1)^2 - 1/4}{2Mr^2}\right\}. \tag{8.156}$$

Thus, the final time evolution amplitude for the motion *near* the surface of the sphere is

$$(\mathbf{u}_b\tau_b|\mathbf{u}_a\tau_a) = \sum_{l=0}^{\infty} \exp\left[-\frac{\hbar L_2}{2Mr^2}(\tau_b - \tau_a)\right] \sum_{\mathbf{m}} Y_{l\mathbf{m}}(\mathbf{u}_b) Y_{l\mathbf{m}}^*(\mathbf{u}_a), \tag{8.157}$$

with

$$L_2 \equiv (l + D/2 - 1)^2 - 1/4. \qquad (8.158)$$

For $D = 3$, this amounts to an expansion in terms of associated Legendre polynomials

$$
(\mathbf{u}_b \tau_b | \mathbf{u}_a \tau_a) = \sum_{l=0}^{\infty} \frac{2l+1}{4\pi} \exp\left\{ -\frac{\hbar L_2}{2Mr^2}(\tau_b - \tau_a) \right\}
$$
$$
\times \sum_{m=-l}^{l} \frac{(l-m)!}{(l+m)!} P_l^m(\cos\theta_b) P_l^m(\cos\theta_a) e^{im(\varphi_b - \varphi_a)}. \qquad (8.159)
$$

If the initial point lies at the north pole of the sphere, this simplifies to

$$
(\mathbf{u}_b \tau_b | \hat{\mathbf{z}}_a \tau_a) = \sum_{l=0}^{\infty} \frac{2l+1}{4\pi} \exp\left[-\frac{\hbar L_2}{2Mr^2}(\tau_b - \tau_a) \right] P_l(\cos\theta_b) P_l(1), \qquad (8.160)
$$

where $P_l(1) = 1$. By rotational invariance the same result holds for arbitrary directions of $\mathbf{u}_a$, if θ_b is replaced by the difference angle ϑ between $\mathbf{u}_b$ and $\mathbf{u}_a$.

In four dimensions, the most convenient expansion uses again the representation functions of the rotation group, so that (8.157) reads

$$
(\mathbf{u}_b \tau_b | \mathbf{u}_a \tau_a) = \sum_{l=0}^{\infty} \exp\left[-\frac{\hbar L_2}{2Mr^2}(\tau_b - \tau_a) \right] \qquad (8.161)
$$
$$
\times \frac{l+1}{2\pi^2} \sum_{m_1, m_2 = -l/2}^{l/2} \mathcal{D}_{m_1 m_2}^{l/2}(\varphi_b, \theta_b, \gamma_b) \mathcal{D}_{m_1 m_2}^{l/2\,*}(\varphi_a, \theta_a, \gamma_a).
$$

These results will be needed in Sections 8.9 and 10.4 to calculate the amplitudes *on* the surface of a sphere. First, however, we extract some more information from the amplitudes *near* the surface of the sphere.

8.8 Angular Barriers *near* the Surface of a Sphere

In Section 8.5 we have projected the path integral of a free particle in three dimensions into a state of fixed angular momentum l finding a radial path integral containing a singular potential, the centrifugal barrier. This could not be treated via the standard time-slicing formalism. The projection of the path integral, however, supplied us with a valid time-sliced action and yielded the correct amplitude. A similar situation occurs if we project the path integral near the surface of a sphere into a fixed azimuthal quantum number m. The physics very near the poles of a sphere is almost the same as that on the tangential surfaces at the poles. Thus, at a fixed two-dimensional angular momentum, the tangential surfaces contain centrifugal barriers. We expect analogous centrifugal barriers at a fixed azimuthal quantum number m near the poles of a sphere at a fixed azimuthal quantum number m. These will be called *angular barriers*.

8.8.1 Angular Barriers in Three Dimensions

Consider first the case $D = 3$ where the azimuthal decomposition is

$$(\mathbf{u}_b \tau_b | \mathbf{u}_a \tau_a) = \sum_m (\sin \theta_b \tau_b | \sin \theta_a \tau_a)_m \frac{1}{2\pi} e^{im(\varphi_b - \varphi_a)}. \tag{8.162}$$

It is convenient to introduce also the differently normalized amplitude

$$(\theta_b \tau_b | \theta_a \tau_a)_m \equiv \sqrt{\sin \theta_b \sin \theta_a} (\sin \theta_b \tau_b | \sin \theta_a \tau_a)_m, \tag{8.163}$$

in terms of which the expansion reads

$$(\mathbf{u}_b \tau_b | \mathbf{u}_a \tau_a) = \sum_m \frac{1}{\sqrt{\sin \theta_b \sin \theta_a}} (\theta_b \tau_b | \theta_a \tau_a)_m \frac{1}{2\pi} e^{im(\varphi_b - \varphi_a)}. \tag{8.164}$$

While the amplitude $(\sin \theta_b \tau_b | \sin \theta_a \tau_a)_m$ has the equal-time limit

$$(\sin \theta_b \tau | \sin \theta_a \tau)_m = \frac{1}{\sin \theta_a} \delta(\theta_b - \theta_a) \tag{8.165}$$

corresponding to the invariant measure of the θ-integration on the surface of the sphere $\int d\theta \sin \theta$, the new amplitude $(\theta_b \tau_b | \theta_a \tau)_m$ has the limit

$$(\theta_b \tau | \theta_a \tau)_m = \delta(\theta_b - \theta_a) \tag{8.166}$$

with a simple δ-function, just as for a particle moving on the coordinate interval $\theta \in (0, 2\pi)$ with an integration measure $\int d\theta$. The renormalization is analogous to that of the radial amplitudes in (8.9).

The projected amplitude can immediately be read off from Eq. (8.157):

$$\begin{aligned}
(\theta_b \tau_b | \theta_a \tau_a)_m &= \sqrt{\sin \theta_b \sin \theta_a} \\
&\times \sum_{l=m}^{\infty} \exp\left[-\frac{\hbar l(l+1)}{2Mr^2}(\tau_b - \tau_a) \right] 2\pi Y_{lm}(\theta_b, 0) Y_{lm}^*(\theta_a, 0).
\end{aligned} \tag{8.167}$$

In terms of associated Legendre polynomials [recall (8.84)], this reads

$$\begin{aligned}
(\theta_b \tau_b | \theta_a \tau_a)_m &= \sqrt{\sin \theta_b \sin \theta_a} \sum_{l=m}^{\infty} \exp\left\{ -\frac{\hbar l(l+1)}{2Mr^2}(\tau_b - \tau_a) \right\} \\
&\times \frac{(2l+1)}{2} \frac{(l-m)!}{(l+m)!} P_l^m(\cos \theta_b) P_l^m(\cos \theta_a).
\end{aligned} \tag{8.168}$$

Let us look at the time-sliced path integral associated with this amplitude. We start from Eq. (8.150) for $D = 3$,

$$(\mathbf{u}_b \tau_b | \mathbf{u}_a \tau_a) \approx \frac{1}{2\pi \hbar \epsilon / Mr^2} \prod_{n=1}^{N} \left[\int \frac{d\cos \theta_n d\varphi_n}{2\pi \hbar \epsilon / Mr^2} \right] \exp\left(-\frac{1}{\hbar} \mathcal{A}^N \right), \tag{8.169}$$

and use the addition theorem

$$\cos \Delta \vartheta_n = \cos \theta_n \cos \theta_{n-1} + \sin \theta_n \sin \theta_{n-1} \cos(\varphi_n - \varphi_{n-1}) \tag{8.170}$$

to expand the exponent as

$$
\begin{aligned}
\exp\left[-\frac{Mr^2}{2\hbar\epsilon}(\mathbf{u}_n - \mathbf{u}_{n-1})^2\right] &= \exp\left[-\frac{Mr^2}{\hbar\epsilon}(1 - \cos\Delta\vartheta_n)\right] \\
&= \exp\left[-\frac{Mr^2}{\hbar\epsilon}(1 - \cos\theta_n \cos\theta_{n-1} - \sin\theta_n \sin\theta_{n-1})\right] \\
&\times \frac{1}{\sqrt{2\pi h_n}} \sum_{m=-\infty}^{\infty} \tilde{I}_m(h_n) e^{im(\varphi_n - \varphi_{n-1})},
\end{aligned}
\tag{8.171}
$$

where h_n is defined as

$$h_n \equiv \frac{Mr^2}{\hbar\epsilon} \sin\theta_n \sin\theta_{n-1}. \tag{8.172}$$

By doing successively the φ_n-integrations, we wind up with the path integral for the projected amplitude

$$(\theta_b \tau_b | \theta_a \tau_a)_m \approx \frac{1}{\sqrt{2\pi\epsilon\hbar/Mr^2}} \prod_{n=1}^{N}\left[\int_0^\pi \frac{d\theta_n}{\sqrt{2\pi\epsilon\hbar/Mr^2}}\right] \exp\left(-\frac{1}{\hbar}\mathcal{A}_m^N\right), \tag{8.173}$$

where $\mathcal{A}_m^N$ is the sliced action

$$\mathcal{A}_m^N = \sum_{n=1}^{N+1}\left\{\frac{Mr^2}{\epsilon}[1 - \cos(\theta_n - \theta_{n-1})] - \hbar \log \tilde{I}_m(h_n)\right\}. \tag{8.174}$$

For small ϵ, this can be approximated (setting $\Delta\theta_n \equiv \theta_n - \theta_{n-1}$) by

$$\mathcal{A}_m^N \approx \epsilon \sum_{n=1}^{N+1}\left\{\frac{Mr^2}{2\epsilon^2}\left[(\Delta\theta_n)^2 - \frac{1}{12}(\Delta\theta_n)^4 + \ldots\right] + \frac{\hbar^2}{2Mr^2}\frac{m^2 - 1/4}{\sin\theta_n \sin\theta_{n-1}}\right\}, \tag{8.175}$$

with the continuum limit

$$\mathcal{A}_m = \int_{\tau_a}^{\tau_b} d\tau \left(\frac{Mr^2}{2}\dot\theta^2 - \frac{\hbar^2}{8Mr^2} + \frac{\hbar^2}{2Mr^2}\frac{m^2 - 1/4}{\sin^2\theta}\right). \tag{8.176}$$

This action has a $1/\sin^2\theta$ -singularity at $\theta = 0$ and $\theta = \pi$, i.e., at the north and south poles of the sphere, whose similarity with the $1/r^2$-singularity of the centrifugal barrier justifies the name "angular barriers".

By analogy with the problems discussed in Section 8.2, the amplitude (8.173) with the naively time-sliced action (8.175) does not exist for $m = 0$, this being the path collapse problem to be solved in Chapter 12. With the full time-sliced action (8.174), however, the path integral is stable for all m. In this stable expression, the successive integration of the intermediate variables using formula (8.14) gives certainly the correct result (8.168).

To do such a calculation, we start out from the product of integrals (8.173) and expand in each factor $I_m(h_n)$ with the help of the addition theorem

$$\sqrt{\frac{2\zeta}{\pi}} e^{\zeta \cos\theta_n \cos\theta_{n-1}} I_m(\zeta \sin\theta_n \sin\theta_{n-1})$$
$$= \sum_{l=m}^{\infty} I_{l+1/2}(\zeta)(2l+1)\frac{(l-m)!}{(l+m)!}P_l^m(\cos\theta_n)P_l^m(\cos\theta_{n-1}), \qquad (8.177)$$

where $\zeta \equiv Mr^2/\hbar\epsilon$. This theorem follows immediately from a comparison of two expansions

$$e^{-\zeta(1-\cos\Delta\theta_n)} = e^{-\zeta[1-\cos\theta_n\cos\theta_{n-1}-\sin\theta_n\sin\theta_{n-1}\cos(\varphi_n-\varphi_{n-1})]} \qquad (8.178)$$
$$\times \frac{1}{\sqrt{2\pi\zeta\sin\theta_n\sin\theta_{n-1}}}\sum_{m=-\infty}^{\infty}\tilde{I}_m(\zeta\sin\theta_n\sin\theta_{n-1})e^{im(\varphi_n-\varphi_{n-1})},$$

$$e^{-\zeta(1-\cos\Delta\theta_n)} = e^{-\zeta}\sqrt{\frac{\pi}{2\zeta}}\sum_{l=0}^{\infty}(2l+1)I_{l+1/2}(\zeta)P_l(\cos\Delta\theta_n). \qquad (8.179)$$

The former is obtained with the help (8.5), the second is taken from (8.83). After the comparison, the Legendre polynomialis expanded via the addition theorem (8.85), which we rewrite with (8.84) as

$$P_l(\cos\Delta\theta_n) = \sum_{m=-l}^{l}\frac{(l-m)!}{(l+m)!}P_l^m(\theta_n)P_l^m(\theta_{n-1})e^{im(\varphi_n-\varphi_{n-1})}. \qquad (8.180)$$

We now recall the orthogonality relation (8.50), rewritten as

$$\int_{-1}^{1}\frac{d\cos\theta}{\sin^2\theta}P_l^m(\cos\theta)P_{l'}^m(\cos\theta) = \frac{(l+m)!}{(l-m)!}\frac{2}{2l+1}\delta_{ll'}. \qquad (8.181)$$

This allows us to do all angular integrations in (8.174). The result

$$(\theta_b\tau_b|\theta_a\tau_a)_m = \sqrt{\sin\theta_b\sin\theta_a}\sum_{l=m}^{\infty}[\tilde{I}_{m+l+1/2}(\zeta)]^{N+1}$$
$$\times\frac{(2l+1)}{2}\frac{(l-m)!}{(l+m)!}P_l^m(\cos\theta_b)P_l^m(\cos\theta_a) \qquad (8.182)$$

is the solution of the time-sliced path integral (8.173).

In the continuum limit, $[\tilde{I}_{m+l+1/2}(\zeta)]^{N+1}$ is dominated by the leading asymptotic term of (8.12) so that

$$[\tilde{I}_{m+l+1/2}(\zeta)]^{N+1} \approx \exp\left[-\frac{\hbar}{2Mr^2}L_2(\tau_b - \tau_a)\right], \qquad (8.183)$$

leading to the previously found expression (8.168).

We have gone through this calculation in detail for the following purpose. Later applications will require an analytic continuation of the path integral from integer values of m to arbitrary real values $\mu \geq 0$. With the present calculation, such a continuation is immediately possible by rewriting (8.182) with the help of the relation

$$P_l^m(z) = (-)^m P_l^{-m} \frac{(l+m)!}{(l-m)!} \tag{8.184}$$

as

$$
(\theta_b \tau_b | \theta_a \tau_a)_\mu = \sqrt{\sin \theta_b \sin \theta_a} \sum_{n=0}^{\infty} [\tilde{I}_{n+\mu+1/2}(\zeta)]^{N+1}
$$
$$
\times \frac{(2n+2\mu+1)}{2} \frac{(n+2\mu)!}{n!} P_{n+\mu}^{-\mu}(\cos \theta_b) P_{n+\mu}^{-\mu}(\cos \theta_a). \tag{8.185}
$$

Here, μ can be an arbitrary real number if the factorials $(n+2\mu)!$ and $n!$ are defined as $\Gamma(n+2\mu+1)$ and $\Gamma(n+1)$. In the continuum limit, (8.185) becomes

$$
(\theta_b \tau_b | \theta_a \tau_a)_\mu = \sqrt{\sin \theta_b \sin \theta_a} \sum_{n=0}^{\infty} \exp\left[-\frac{\hbar(n+\mu)(n+\mu+1)}{2Mr^2}(\tau_b - \tau_a)\right]
$$
$$
\times \frac{(2n+2\mu+1)}{2} \frac{(n+2\mu)!}{n!} P_{n+\mu}^{-\mu}(\cos \theta_b) P_{n+\mu}^{-\mu}(\cos \theta_a). \tag{8.186}
$$

We prove this to solve the time-sliced path integral (8.173) for arbitrary real values of $m = \mu$ [4] by using the addition theorem[11]

$$
(\sin \alpha \sin \beta)^{-\mu} J_\mu(z \sin \alpha \sin \beta) e^{iz \cos \alpha \cos \beta} = \frac{2^{2\mu+1} \Gamma^2(\mu+1/2)}{\sqrt{2\pi z}}
$$
$$
\times \sum_{n=0}^{\infty} \frac{i^n n!(n+\mu+1/2)}{\Gamma(n+2\mu+1)} J_{n+\mu+1/2}(z) C_n^{(\mu+1/2)}(\cos \alpha) C_n^{(\mu+1/2)}(\cos \beta). \tag{8.187}
$$

After substituting z by $\zeta e^{-i\pi/2}$ this turns into

$$
(\sin \alpha \sin \beta)^{-\mu} I_\mu(\zeta \sin \alpha \sin \beta) \exp(\zeta \cos \alpha \cos \beta) = \frac{2^{2\mu+1} \Gamma^2(\mu+1/2)}{\sqrt{2\pi \zeta}}
$$
$$
\times \sum_{n=0}^{\infty} \frac{n!(n+\mu+1/2)}{\Gamma(n+2\mu+1)} I_{n+\mu+1/2}(\zeta) C_n^{(\mu+1/2)}(\cos \alpha) C_n^{(\mu+1/2)}(\cos \beta). \tag{8.188}
$$

The Gegenbauer polynomials $C_n^{(\mu+1/2)}(z)$ can be expressed, for arbitrary μ, by means of Eq. (8.105) in terms of Jacobi polynomials $P_n^{(\mu,\mu)}$, and these further in terms of Legendre functions $P_{n+\mu}^{-\mu}$, using the formula

$$
P_n^{(\mu,\mu)}(z) = (-2)^\mu \frac{(n+\mu)!}{n!} (1-z^2)^{-\mu/2} P_{n+\mu}^{-\mu}(z). \tag{8.189}
$$

---

[11]G.N. Watson, *Theory of Bessel Functions*, Cambridge University Press, 1952, Ch. 11.6, Eq. (11.9).

Thus[12]

$$C_n^{(\mu+1/2)}(z) = \frac{\Gamma(n + 2\mu + 1)\Gamma(\mu + 1)}{\Gamma(2\mu + 1)n!} \left[\frac{1 - z^2}{4}\right]^{-\mu/2} P_{n+\mu}^{-\mu}(z). \tag{8.190}$$

We can now perform the integrations in the time-sliced path integral by means of the known continuation of the orthogonality relation (8.181) to arbitrary real values of μ:

$$\int_{-1}^{1} \frac{d\cos\theta}{\sin^2\theta} P_{n+\mu}^{-\mu}(\cos\theta) P_{n'+\mu}^{-\mu}(\cos\theta) = \frac{n!}{(n + 2\mu)!} \frac{2}{2n + 1\mu + 1} \delta_{nn'}. \tag{8.191}$$

Note that for noninteger μ, the Legendre functions $P_{n+m}^{-m}(\cos\theta)$ are no longer polynomials as in (1.417). Instead, they are defined in terms of the hypergeometric function as follows:

$$P_\nu^\mu(z) = \frac{1}{\Gamma(1 - \mu)} \left(\frac{1 + z}{1 - z}\right)^{\mu/2} F(-\nu, \nu + 1; 1 - \mu; (1 - z)/2). \tag{8.192}$$

The integral formula (8.191) is a consequence of the orthogonality of the Gegenbauer polynomials (8.103), which is applied here in the form

$$\int_{-1}^{1} dz \, (1 - z^2)^\mu C_n^{(\mu+1/2)}(z) C_{n'}^{(\mu+1/2)}(z) = \delta_{nn'} \frac{\pi 2^{-2\mu} \Gamma(2\mu + 2 + n)}{n!(n + \mu)[\Gamma(\mu + 1/2)]^2}. \tag{8.193}$$

Using (8.188), (8.190) and (8.191), the integrals in the product (8.206) can all be performed as before, resulting in the amplitude (8.185) with the continuum limit (8.186), both valid for arbitrary real values of $m = \mu \geq 0$.

The continuation to arbitrary real values of μ has an important application: The action (8.176) of the projected motion of a particle near the surface of the sphere coincides with the action of a particle moving in the so-called Pöschl-Teller potential [5]:

$$V(\theta) = \frac{\hbar^2}{2Mr^2} \frac{s(s + 1)}{\sin^2\theta} \tag{8.194}$$

with the strength parameter $s = m - 1/2$. After the continuation of arbitrary real $m = \mu \geq 0$, the amplitude (8.186) describes this system for any potential strength. This fact will be discussed further in Chapter 14 where we develop a general method for solving a variety of nontrivial path integrals.

Note that the amplitude $(\sin\theta_b\tau_b|\sin\theta_a\tau_a)_m$ satisfies the Schrödinger equation

$$\left[\frac{\hbar^2}{Mr^2}\left(-\frac{1}{2}\frac{1}{\sin\theta}\frac{d}{d\theta}\sin\theta\frac{d}{d\theta} + \frac{m^2}{2\sin^2\theta}\right) + \hbar\partial_\tau\right](\sin\theta \, \tau|\sin\theta_a\tau_a)_m$$
$$= \hbar\delta(\tau - \tau_a)\delta(\cos\theta - \cos\theta_a). \tag{8.195}$$

---

[12]I.S. Gradshteyn and I.M. Ryzhik, op. cit., Formula 8.936.

This follows from the differential equation obeyed by the Legendre polynomials $P_l^m(\cos\theta)$ in (8.87). The new amplitude $(\theta\,\tau|\theta_a\tau_a)_m$, on the other hand, satisfies the equation [corresponding to that of $\sqrt{\sin\theta}P_l^m(\cos\theta)$ in the footnote to Eq. (8.87)]

$$\left[\frac{\hbar^2}{Mr^2}\left(-\frac{1}{2}\frac{d}{d\theta^2}-\frac{1}{8}+\frac{m^2-1/4}{2\sin^2\theta}\right)+\hbar\partial_\tau\right](\theta\,\tau|\theta_a\tau_a)_m=\hbar\delta(\tau-\tau_a)\delta(\theta-\theta_a). \quad (8.196)$$

8.8.2　Angular Barriers in Four Dimensions

In four dimensions, the angular momentum decomposition reads in terms of Euler angles [see (8.161)]

$$(\mathbf{u}_b\tau_b|\mathbf{u}_a\tau_a)=\sum_{l=0}^\infty\exp\left\{-\frac{\hbar L_2}{2Mr^2}(\tau_b-\tau_a)\right\}$$
$$\times\frac{l+1}{2\pi^2}\sum_{m_1m_2=-l/2}^{l/2}d_{m_1m_2}^{l/2}(\theta_b)d_{m_1m_2}^{l/2}(\theta_a)e^{im_1(\varphi_b-\varphi_a)+im_2(\gamma_b-\gamma_a)}, \quad (8.197)$$

with

$$L_2\equiv(l+1)^2-1/4=4(l/2)(l/2+1)+3/4 \quad (8.198)$$

and m_1,m_2 running over integers or half-integers depending on $l/2$. We now define the projected amplitudes by the expansion

$$(\mathbf{u}_b\tau_b|\mathbf{u}_a\tau_a)=8\sum_{m_1m_2}(\sin\theta_b\tau_b|\sin\theta_a\tau_a)_{m_1m_2}\frac{1}{2\pi}e^{im_1(\varphi_b-\varphi_a)}\frac{1}{4\pi}e^{im_2(\gamma_b-\gamma_a)}. \quad (8.199)$$

As in (8.163), it is again convenient to introduce the differently normalized amplitude $(\theta_b\tau_b|\theta_a\tau_a)_m$ defined by

$$(\theta_b\tau_b|\theta_a\tau_a)_{m_1m_2}\equiv\sqrt{\sin\theta_b\sin\theta_a}(\sin\theta_b\tau_b|\sin\theta_a\tau_a)_{m_1m_2}, \quad (8.200)$$

in terms of which the expansion becomes [compare (8.162)]

$$(\mathbf{u}_b\tau_b|\mathbf{u}_a\tau_a)=\sum_m\frac{8}{\sqrt{\sin\theta_b\sin\theta_a}}(\theta_b\tau_b|\theta_a\tau_a)_{m_1m_2}\frac{1}{2\pi}e^{im(\varphi_b-\varphi_a)}\frac{1}{4\pi}e^{im(\gamma_b-\gamma_a)}. \quad (8.201)$$

A comparison with (8.197) gives immediately the projected amplitude

$$(\theta_b\tau_b|\theta_a\tau_a)_{m_1m_2}=\sqrt{\sin\theta_b\sin\theta_a} \quad (8.202)$$
$$\times\sum_l^\infty\exp\left\{-\frac{\hbar[(l+1)^2-1/4]}{2Mr^2}(\tau_b-\tau_a)\right\}\frac{l+1}{2}d_{m_1m_2}^{l/2}(\theta_b)d_{m_1m_2}^{l/2}(\theta_a),$$

in which l is summed in even steps from the larger value of $|2m_1|,|2m_2|$ to infinity.

Let us write down the time-sliced path integral leading to this amplitude. According to (8.150)–(8.152), it is given by

$$
(\mathbf{u}_b T_b | \mathbf{u}_a T_a) \approx \frac{1}{\sqrt{2\pi\hbar\epsilon/Mr^2}^3} \prod_{n=1}^{N} \left[\int_0^\pi \int_0^{2\pi} \int_0^{4\pi} \frac{d\theta_n \sin\theta_n \, d\varphi_n \, d\gamma_n}{8\sqrt{2\pi\hbar\epsilon/Mr^2}^3} \right] \exp\left(-\frac{1}{\hbar}\mathcal{A}^N\right).
$$

(8.203)

In each time slice we make use of the addition theorem (8.128) and expand the exponent with (8.6) as

$$
\exp\left[-\frac{Mr^2}{2\hbar\epsilon}(\mathbf{u}_n - \mathbf{u}_{n-1})^2\right] = \exp\left[-\frac{Mr^2}{\hbar\epsilon}(1 - \cos\Delta\vartheta_n)\right]
$$

$$
= \exp\left\{-\frac{Mr^2}{2\hbar\epsilon}[1 - \cos(\theta_n/2)\cos(\theta_{n-1}/2) - \sin(\theta_n/2)\sin(\theta_{n-1}/2)]\right\}
$$

$$
\times \frac{1}{\sqrt{2\pi h_n^c}} \frac{1}{\sqrt{4\pi h_n^s}} \sum_{m_1,m_2=-\infty}^{\infty} \tilde{I}_{|m_1+m_2|}(h_n^c)\tilde{I}_{|m_1-m_2|}(h_n^s)
$$

(8.204)

$$
\times \exp\{im_1(\varphi_n - \varphi_{n-1}) + im_2(\gamma_n - \gamma_{n-1})\},
$$

where h_n^c and h_n^s are given by

$$
h_n^c = \frac{Mr^2}{\hbar\epsilon}\cos(\theta_n/2)\cos(\theta_{n-1}/2), \qquad h_n^s = \frac{Mr^2}{\hbar\epsilon}\sin(\theta_n/2)\sin(\theta_{n-1}/2). \quad (8.205)
$$

By doing successively the φ_n- and γ_n-integrations, we wind up with the path integral for the projected amplitude

$$
(\theta_b T_b | \theta_a T_a)_{m_1 m_2} \approx \frac{1}{\sqrt{2\pi\epsilon\hbar/4Mr^2}} \prod_{n=1}^{N} \left[\int_0^\pi \frac{d\theta_n}{\sqrt{2\pi\epsilon\hbar/4Mr^2}} \right] \exp\left\{-\frac{1}{\hbar}\mathcal{A}_{m_1 m_2}^N\right\}, \quad (8.206)
$$

where $\mathcal{A}_{m_1 m_2}^N$ is the sliced action

$$
\mathcal{A}_{m_1 m_2}^N = \sum_{n=1}^{N+1} \left\{ \frac{Mr^2}{\epsilon}[1 - \cos[(\theta_n - \theta_{n-1})/2]] \right.
$$

$$
\left. -\hbar \log \tilde{I}_{|m_1+m_2|}(h_n^c) - \hbar \log \tilde{I}_{|m_1-m_2|}(h_n^s) \right\}.
$$

(8.207)

For small ϵ, this can be approximated (setting $\Delta\theta_n \equiv \theta_n - \theta_{n-1}$) by

$$
\mathcal{A}_{m_1 m_2}^N \to \epsilon \sum_{n=1}^{N+1} \left\{ \frac{Mr^2}{2\epsilon^2}[(\Delta\theta_n/2)^2 - \frac{1}{12}(\Delta\theta_n/2)^4 + \ldots] \right.
$$

$$
\left. + \frac{\hbar^2}{2Mr^2}\frac{(m_1+m_2)^2 - 1/4}{\cos(\theta_n/2)\cos(\theta_{n-1}/2)} + \frac{\hbar^2}{2Mr^2}\frac{(m_1-m_2)^2 - 1/4}{\sin(\theta_n/2)\sin(\theta_{n-1}/2)} \right\}, \quad (8.208)
$$

with the continuum limit

$$\mathcal{A}_{m_1 m_2} = \int_{\tau_a}^{\tau_b} d\tau \left(\frac{Mr^2}{8} \dot{\theta}^2 - \frac{\hbar^2}{8Mr^2} + \frac{\hbar^2}{2Mr^2} \frac{|m_1 + m_2|^2 - 1/4}{\cos^2(\theta/2)} \right.$$
$$\left. + \frac{\hbar^2}{2Mr^2} \frac{|m_1 - m_2|^2 - 1/4}{\sin^2(\theta/2)} \right). \tag{8.209}$$

After introducing the auxiliary mass

$$\mu = M/4 \tag{8.210}$$

and rearranging the potential terms, we can write the action equivalently as

$$\mathcal{A}_{m_1 m_2} = \int_{\tau_a}^{\tau_b} d\tau \left(\frac{\mu r^2}{2} \dot{\theta}^2 - \frac{\hbar^2}{32\mu r^2} + \frac{\hbar^2}{2\mu r^2} \frac{m_1^2 + m_2^2 - 1/4 - 2m_1 m_2 \cos\theta}{\sin^2\theta} \right). \tag{8.211}$$

Just as in the previous system, this action contains an angular barrier $1/\sin^2\theta$ at $\theta = 0$, and $\theta = \pi$, so that the amplitude (8.206) with the naively time-sliced action (8.175) does not exist for $m_1 = m_2$ or $m_1 = -m_2$, due to path collapse. Only with the properly time-sliced action (8.207) is the path integral stable and solvable by successive integrations with the result (8.202).

As before, the path integral (8.206) is initially only defined and solved by (8.202) if both m_1 and m_2 have integer or half-integer values. The path integral and its solution can, however, be continued to arbitrary real values of $m_1 = \mu_1 \geq 0$ and its $m_2 = \mu_2 \geq 0$. For this we rewrite (8.202) in the form [4]

$$(\theta_b \tau_b | \theta_a \tau_a)_{\mu_1 \mu_2} = \sqrt{\sin\theta_b \sin\theta_a} \tag{8.212}$$
$$\times \sum_{n=0}^{\infty} \exp\left\{ \frac{\hbar[(n + \mu_1 + 1)^2 - 1/4]}{2Mr^2} (\tau_b - \tau_a) \right\} \frac{n + \mu_1 + 1}{2} d_{\mu_1 \mu_2}^{n+\mu_1}(\theta_b) d_{\mu_1 \mu_2}^{n+\mu_1}(\theta_a),$$

assuming that $\mu_1 \geq \mu_2$. The products of the rotation functions $d_{\mu_1 \mu_2}^{\lambda}(\theta)$ have a well-defined analytic continuation to arbitrary real values of the indices μ_1, μ_2, λ, as can be seen by expressing them in terms of Jacobi polynomials via formula (1.443).

To perform the path integral in the analytically continued case, we use the expansion valid for all μ_+, μ_-,[13]

$$\frac{z}{2} J_{\mu_+}(z \cos\alpha \cos\beta) J_{\mu_-}(z \sin\alpha \sin\beta) = \cos^{\mu_+}\alpha \cos^{\mu_+}\beta \sin^{\mu_-}\alpha \sin^{\mu_-}\beta$$
$$\times \sum_{n=0}^{\infty} (-1)^n (\mu_+ + \mu_- + 2n + 1) J_{\mu_+ + \mu_- + 2n + 1}(z)$$
$$\times \frac{\Gamma(n + \mu_+ + \mu_- + 1)\Gamma(n + \mu_- + 1)}{n! \Gamma(n + \mu_+ + 1)[\Gamma(\mu_- + 1)]^2}$$
$$\times F(-n, n + \mu_+ + \mu_- + 1; \mu_- + 1; \sin^2\alpha)$$
$$\times F(-n, n + \mu_+ + \mu_- + 1; \mu_- + 1; \sin^2\beta) \tag{8.213}$$

---

[13]G.N. Watson, op. cit., Chapter 11.6, Gl. (11.6), (1).

with $\zeta \equiv Mr^2/\hbar\epsilon$. The hypergeometric functions appearing on the right-hand side have a first argument with a negative integer value. They are therefore proportional to the Jacobi polynomials $P_n^{(\mu_-,\mu_+)}$:

$$P_n^{(\mu_-,\mu_+)}(x) = \frac{1}{n!}\frac{\Gamma(n+\mu_-+1)}{\Gamma(\mu_-+1)}F\left(-n,n+\mu_++\mu_-+1;1+\mu_-;(1-x)/2\right) \quad (8.214)$$

[recall (1.443) and the identity $P_n^{(\mu_-,\mu_+)}(x) = (-)^n P_n^{(\mu_-,\mu_+)}(-x)$]. Inserting $z = i\zeta$, $\alpha = \theta_n/2$, $\beta = \theta_{n-1}/2$, and expressing the Jacobi polynomials in terms of rotation functions continued to real-valued μ_1, μ_2, we obtain from (8.213) for $\mu_1 \geq \mu_2$

$$I_{\mu_+}\left(\zeta \cos\frac{\theta_n}{2}\cos\frac{\theta_{n-1}}{2}\right)I_{\mu_-}\left(\zeta \sin\frac{\theta_n}{2}\sin\frac{\theta_{n-1}}{2}\right)$$
$$= \frac{4}{\zeta}\sum_{n=0}^{\infty}I_{2n+\mu_++\mu_-+1}(\zeta)\frac{(n+\mu_1+\mu_2)!(n+\mu_1-\mu_2)!}{(n+2\mu_1)!n!}d_{\mu_1\mu_2}^{n+\mu_1}(\theta_n)d_{\mu_1\mu_2}^{n+\mu_1}(\theta_{n-1}). \quad (8.215)$$

Now we make use of the orthogonality relation [compare (1.452)]

$$\int_{-1}^{1}d\cos\theta\, d_{\mu_1\mu_2}^{n+\mu_1}(\theta)d_{\mu_1\mu_2}^{n'+\mu_1}(\theta) = \delta_{nn'}\frac{2}{2n+1}, \quad (8.216)$$

which for real μ_1, μ_2 follows from the corresponding relation for Jacobi polynomials[14]

$$\int_{-1}^{1}dx\,(1-x)^{\mu_-}(1+x)^{\mu_+}P_n^{(\mu_-,\mu_+)}(x)P_{n'}^{(\mu_-,\mu_+)}(x)$$
$$= \delta_{nn'}\frac{2^{\mu_++\mu_-+1}\Gamma(\mu_++n+1)\Gamma(\mu_-+n+1)}{n!(\mu_++\mu_-+1+2n)\Gamma(\mu_++\mu_-+n+1)}, \quad (8.217)$$

valid for $\mathrm{Re}\,\mu_+ > -1$, $\mathrm{Re}\,\mu_- > -1$. Performing all θ_n-integrations in (8.206) yields the time-sliced amplitude

$$(\theta_b\,\tau_b|\theta_a\,\tau_a)_{\mu_1\mu_2} = \sqrt{\sin\theta_b\sin\theta_a}\sum_{n=0}^{\infty}\left[\tilde{I}_{2n+\mu_++\mu_-+1}(\zeta)\right]^{N+1}d_{\mu_1\mu_2}^{n+\mu_1}(\theta_b)d_{\mu_1\mu_2}^{n+\mu_1}(\theta_a), \quad (8.218)$$

valid for all real $\mu_1 \geq \mu_2 \geq 0$. In the continuum limit, this becomes

$$(\theta_b\,\tau_b|\theta_a\,\tau_a)_{\mu_1\mu_2} = \sqrt{\sin\theta_b\sin\theta_a}\sum_{n=0}^{\infty}e^{-E_n(\tau_b-\tau_a)/\hbar}d_{\mu_1\mu_2}^{n+\mu_1}(\theta_b)d_{\mu_1\mu_2}^{n+\mu_1}(\theta_a), \quad (8.219)$$

with

$$E_n = \frac{\hbar}{2Mr^2}[(2n+\mu_++\mu_-+1)^2-1/4], \quad (8.220)$$

which proves (8.212).

---

[14]I.S. Gradshteyn and I.M. Ryzhik, op. cit., Formula 7.391.

Apart from the projected motion of a particle near the surface of the sphere, the amplitude (8.212) describes also a particle moving in the general Pöschl-Teller potential[15]

$$V_{\mathcal{PT}'}(\theta) = \frac{\hbar^2}{2Mr^2}\left[\frac{s_1(s_1+1)}{\sin^2(\theta/2)} + \frac{s_2(s_2+1)}{\cos^2(\theta/2)}\right]. \tag{8.221}$$

Due to the analytic continuation to arbitrary real m_1, m_2 the parameters s_1 and s_2 are arbitrary with the potential strength parameters $s_1 = m_1 + m_2 - 1/2$ and $s_2 = m_1 - m_2 - 1/2$. This will be discussed further in Chapter 14.

Recalling the differential equation (1.451) satisfied by the rotation functions $d^{l/2}_{m_1m_2}(\theta)$, we see that the original projected amplitude (8.206) obeys the Schrödinger equation

$$\left[\frac{\hbar^2}{2\mu r^2}\left(-\frac{1}{\sin\theta}\frac{d}{d\theta}\sin\theta\frac{d}{d\theta} + \frac{3}{16} + \frac{m_1^2 + m_2^2 - 2m_1m_2\cos\theta}{\sin^2\theta}\right) + \hbar\partial_\tau\right]$$
$$\times(\sin\theta\,\tau|\sin\theta_a\tau_a)_{m_1m_2} = \hbar\delta(\tau - \tau_a)\delta(\cos\theta - \cos\theta_a). \tag{8.222}$$

The extra term $3/16$ is necessary to account for the energy difference between the motion near the surface of a sphere in four dimensions, whose energy is $(\hbar^2/2\mu r^2)[(l/2)(l/2 + 1) + 3/16]$ [see (8.157)], and that of a symmetric spinning top with angular momentum $L = l/2$ in three dimensions, whose energy is $(\hbar^2/2\mu r^2)(l/2)(l/2 + 1)$, as shown in the next section in detail.

The amplitude $(\theta_b\tau_b|\theta_a\tau_a)_{m_1m_2}$ defined in (8.200) satisfies the differential equation

$$\left[\frac{\hbar^2}{2\mu r^2}\left(-\frac{d^2}{d\theta^2} - \frac{1}{16} + \frac{m_1^2 + m_2^2 - 1/4 - 2m_1m_2\cos\theta}{\sin^2\theta}\right) + \hbar\partial_\tau\right]$$
$$\times(\theta\,\tau|\theta_a\tau_a)_{m_1m_2} = \hbar\delta(\tau - \tau_a)\delta(\theta - \theta_a). \tag{8.223}$$

This is, of course, precisely the Schrödinger equation associated with the action (8.211).

8.9 Motion on a Sphere in D Dimensions

The wave functions in the time evolution amplitude *near* the surface of a sphere are also correct for the motion *on* a sphere. This is not true for the energies, for which the amplitude (8.157) gives

$$E_l = \frac{\hbar^2}{2Mr^2}(L_2^2)_l, \tag{8.224}$$

with

$$(L_2^2)_l = (l + D/2 - 1)^2 - 1/4, \quad l = 0, 1, 2, \ldots . \tag{8.225}$$

As we know from Section 1.14, the energies should be equal to

$$E_l = \frac{\hbar^2}{2Mr^2}(\hat{L}^2)_l, \tag{8.226}$$

---

[15]See Footnote 15.

where $(\hat{L}^2)_l$ denotes the eigenvalues of the square of the angular momentum operator. In D dimensions, the eigenvalues are known from the Schrödinger theory to be

$$(\hat{L}^2)_l = l(l + D - 2), \quad l = 0, 1, 2, \ldots . \tag{8.227}$$

Apart from the trivial case $D = 1$, the two energies are equal only for $D = 3$, where $(L_2^2)_l \equiv (\hat{L}^2)_l = l(l + 1)$. For all other dimensions, we shall have to remove the difference

$$\Delta(L_2^2)_l \equiv \hat{L}^2 - L_2^2 = \frac{1}{4} - \left(\frac{D}{2} - 1\right)^2 = -\frac{(D-1)(D-3)}{4}. \tag{8.228}$$

The simplest nontrivial case where the difference appears is for $D = 2$ where the role of l is played by the magnetic quantum number m and $(L_2^2)_m = m^2 - 1/4$, whereas the correct energies should be proportional to $(\hat{L}^2)_m = m^2$.

Two changes are necessary in the time-sliced path integral to find the correct energies. First, the time-sliced action (8.151) must be modified to measure the proper distance on the surface rather than the Euclidean distance in the embedding space. Second, we will have to correct the measure of path integration. The modification of the action is simply

$$\mathcal{A}^N_{\text{on sphere}} = \frac{M}{\epsilon} r^2 \sum_{n=1}^{N+1} \frac{(\Delta\vartheta_n)^2}{2}, \tag{8.229}$$

in addition to

$$\mathcal{A}^N = \frac{M}{\epsilon} r^2 \sum_{n=1}^{N+1} (1 - \cos\Delta\vartheta_n). \tag{8.230}$$

Since the time-sliced path integral was solved exactly with the latter action, it is convenient to expand the true action around the solvable one as follows:

$$\mathcal{A}^N_{\text{on sphere}} = \frac{M}{\epsilon} r^2 \sum_{n=1}^{N+1} \left[(1 - \cos\Delta\vartheta_n) + \frac{1}{24}(\Delta\vartheta_n)^4 - \ldots\right]. \tag{8.231}$$

There is no need to go to higher than the fourth order in $\Delta\vartheta_n$, since these do not contribute to the relevant order ϵ. For $D = 2$, the correction of the action is sufficient to transform the path integral *near* the surface of the sphere into one *on* the sphere, which in this reduced dimension is merely a circle. On a circle, $\Delta\vartheta_n = \varphi_n - \varphi_{n-1}$ and the measure of path integration becomes

$$\frac{1}{\sqrt{2\pi\hbar\epsilon/Mr^2}} \prod_{n=1}^{N+1} \int_{-\pi/2}^{\pi/2} \frac{d\varphi_n}{\sqrt{2\pi\hbar\epsilon/Mr^2}}. \tag{8.232}$$

The quartic term $(\Delta\vartheta_n)^4 = (\varphi_n - \varphi_{n-1})^4$ can be replaced according to the rules of perturbation theory by its expectation [see (8.62)]

$$\langle(\Delta\vartheta_n)^4\rangle_0 = 3\frac{\epsilon\hbar}{Mr^2}. \tag{8.233}$$

The correction term in the action

$$\Delta_{\mathrm{qu}}\mathcal{A}^N = \frac{M}{\epsilon}r^2 \sum_{n=1}^{N+1} \frac{1}{24}(\Delta\vartheta_n)^4 \tag{8.234}$$

has, therefore, the expectation

$$\langle\Delta_{\mathrm{qu}}\mathcal{A}^N\rangle_0 = (N+1)\epsilon\frac{\hbar^2/4}{2Mr^2}. \tag{8.235}$$

This supplies precisely the missing term which raises the energy from the *near*-the-surface value $E_m = \hbar^2(m^2 - 1/4)/2Mr^2$ to the proper *on*-the-sphere value $E_m = \hbar^2 m^2/2Mr^2$.

   In higher dimensions, the path integral near the surface of a sphere requires a second correction. The difference (8.228) between $\hat{L}^2$ and L_2^2 is negative. Since the expectation of the quartic correction term alone is always positive, it can certainly not explain the difference.[16] Let us calculate first its contribution at arbitrary D. For very small ϵ, the fluctuations near the surface of the sphere lie close to the $D-1$ -dimensional tangent space. Let $\Delta\mathbf{x}_n$ be the coordinates in this space. Then we can write the quartic correction term as

$$\Delta_{\mathrm{qu}}\mathcal{A}^N \approx \frac{M}{\epsilon}\sum_{n=1}^{N+1}\frac{1}{24r^2}(\Delta\mathbf{x}_n)^4, \tag{8.236}$$

where the components $(\Delta\mathbf{x}_n)_i$ have the correlations

$$\langle(\Delta\mathbf{x}_n)_i(\Delta\mathbf{x}_n)_j\rangle_0 = \frac{\hbar\epsilon}{M}\delta_{ij}. \tag{8.237}$$

Thus, according to the rule (8.62), $\Delta_{\mathrm{qu}}\mathcal{A}^N$ has the expectation

$$\langle\Delta\mathcal{A}^N\rangle_0 = (N+1)\epsilon\frac{\hbar^2}{2Mr^2}\Delta_{\mathrm{qu}}L_2^2, \tag{8.238}$$

where $\Delta_{\mathrm{qu}}L_2^2$ is the contribution of the quartic term to the value L_2^2:

$$\Delta_{\mathrm{qu}}L_2^2 = \frac{D^2-1}{12}. \tag{8.239}$$

This result is obtained using the contraction rules for the tensor

$$\langle\Delta x_i\Delta x_j\Delta x_k\Delta x_l\rangle_0 = \left(\frac{\epsilon\hbar}{M}\right)^2(\delta_{ij}\delta_{kl} + \delta_{ik}\delta_{jl} + \delta_{il}\delta_{jk}), \tag{8.240}$$

which follow from the integrals (8.63).

---

[16]This was claimed by G. Junker and A. Inomata, in *Path Integrals from* meV *to* MeV, edited by M.C. Gutzwiller, A. Inomata, J.R. Klauder, and L. Streit (World Scientific, Singapore, 1986), p.333.

Incidentally, the same result can also be derived in a more pedestrian way: The term $(\Delta \mathbf{x}_n)^4$ can be decomposed into $D - 1$ quartic terms of the individual components Δx_{ni}, and $(D-1)(D-2)$ mixed quadratic terms $(\Delta x_{ni})^2 (\Delta x_{nj})^2$ with $i \neq j$. The former have an expectation $(D-1) \cdot 3(\epsilon\hbar/Mr)^2$, the latter $(D-1)(D-2) \cdot (\epsilon\hbar/Mr)^2$. When inserted into (8.236), they lead to (8.238).

Thus we remain with a final difference in D dimensions:

$$\Delta_{\mathrm{f}} L_2^2 = \Delta L_2^2 - \Delta_{\mathrm{qu}} L_2^2 = -\frac{1}{3}(D-1)(D-2). \tag{8.241}$$

This difference can be removed only by the measure of the path integral. Near the sphere we have used the measure

$$\prod_{n=1}^{N} \left[\int \frac{d^{D-1}\mathbf{u}_n}{\sqrt{2\pi\hbar\epsilon/Mr^2}^{D-1}} \right]. \tag{8.242}$$

In Chapter 10 we shall argue that this measure is incorrect. We shall find that the measure (8.242) receives a correction factor

$$\prod_{n=1}^{N} \left[1 + \frac{D-2}{6}(\Delta\vartheta_n)^2 \right] \tag{8.243}$$

[see the factor $(1 + i\Delta\mathcal{A}_J^\epsilon)$ of Eq. (10.151)]. Setting $(\Delta\vartheta_n)^2 = (\Delta\mathbf{x}_n/r)^2$, the expectation of this factor becomes

$$\prod_{n=1}^{N} \left[1 + \frac{(D-2)(D-1)}{6r^2}\frac{\epsilon\hbar}{M} \right] \tag{8.244}$$

corresponding to a correction term in the action

$$\langle \Delta\mathcal{A}_f^N \rangle_0 = (N+1)\epsilon\frac{\hbar^2}{2Mr^2}\Delta_f L_2^2, \tag{8.245}$$

with $\Delta_f L_2^2$ given by (8.241). This explains the remaining difference between the eigenvalues $(L_2)_l$ and $(\hat{L})_l^2$.

In summary, the time evolution amplitude on the D-dimensional sphere reads [6]

$$(\mathbf{u}_b\tau_b|\mathbf{u}_a\tau_a) = \sum_{l=0}^{\infty} \exp\left[-\frac{\hbar\hat{L}^2}{2Mr^2}(\tau_b - \tau_a) \right] \sum_{\mathbf{m}} Y_{l\mathbf{m}}(\mathbf{u}_b)Y_{l\mathbf{m}}^*(\mathbf{u}_a), \tag{8.246}$$

with

$$\hat{L}^2 = l(l + D - 2), \tag{8.247}$$

which are precisely the eigenvalues of the squared angular momentum operator of Schrödinger quantum mechanics. For $D = 3$ and $D = 4$, the amplitude (8.246) coincides with the more specific representations (8.160) and (8.161), if L_2^2 is replaced by $\hat{L}^2$.

Finally, let us emphasize that in contrast to the amplitude (8.157) *near* the surface of the sphere, the normalization of the amplitude (8.246) *on* the sphere is

$$\int d^{D-1}\mathbf{u}_b\,(\mathbf{u}_b\tau_b|\mathbf{u}_a\tau_a) = 1. \qquad (8.248)$$

This follows from the integral

$$\int d^{D-1}\mathbf{u}_b \sum_m Y_{lm}(\mathbf{u}_b)Y_{lm}^*(\mathbf{u}_a) = \delta_{l0}\int d^{D-1}\mathbf{u}_b\,Y_{00}(\mathbf{u}_b)Y_{00}^*(\mathbf{u}_a)$$

$$= \delta_{l0}\int d^{D-1}\mathbf{u}_b\,1/S_D = \delta_{l0}. \qquad (8.249)$$

This is in contrast to the amplitude near the surface which satisfies

$$\int d^{D-1}\mathbf{u}_b\,(\mathbf{u}_b\tau_b|\mathbf{u}_a\tau_a) = \exp\left[-\frac{(D/2-1)^2 - 1/4}{2\mu r^2}(\tau_b - \tau_a)\right]. \qquad (8.250)$$

We end this section with the following observation. In the continuum, the Euclidean path integral on the surface of a sphere can be rewritten as a path integral in flat space with an auxiliary path integral over a Lagrange multiplier $\lambda(\tau)$ in the form[17]

$$(\mathbf{x}_b\tau_b|\mathbf{x}_a\tau_a) = \int_{-i\infty}^{i\infty}\int \mathcal{D}^2x(\tau)\frac{\mathcal{D}\lambda(\tau)}{2\pi i\hbar}\exp\left(-\frac{1}{\hbar}\int_{\tau_a}^{\tau_b}d\tau\left\{\frac{M}{2}\dot{\mathbf{x}}^2 + \frac{\lambda(\tau)}{2r}[x^2(\tau) - r^2]\right\}\right). \qquad (8.251)$$

A naive time slicing of this expression would *not* yield the correct energy spectrum on the sphere. The slicing would lead to the product of integrals

$$(\mathbf{u}_b t_b|\mathbf{u}_a t_a) \approx \prod_{n=1}^{N}\left[\int d^D x_n\right]\prod_{n=1}^{N}\left[\int \frac{d\lambda_n}{2\pi i\hbar/\epsilon}\right]\exp\left(\frac{i}{\hbar}\mathcal{A}^N\right), \qquad (8.252)$$

with $\mathbf{u} \equiv \mathbf{x}/|\mathbf{x}|$ and the time-sliced action

$$\mathcal{A}^N = \sum_{n=1}^{N+1}\left[\frac{M}{2\epsilon}(\mathbf{x}_n - \mathbf{x}_{n-1})^2 + \epsilon\frac{\lambda_n}{2r}(\mathbf{x}_n^2 - r^2)\right]. \qquad (8.253)$$

Integrating out the λ_n's would produce precisely the expression (8.150) with the action (8.151) *near* the surface of the sphere. The δ-functions arising from the λ_n-integrations would force only the intermediate positions $\mathbf{x}_n$ to lie on the sphere; the sliced kinetic terms, however, would not correspond to the geodesic distance. Also, the measure of path integration would be wrong.

---

[17]The field-theoretic generalization of this path integral, in which τ is replaced by a d-dimensional spatial vector $\mathbf{x}$, is known as the $O(D)$-symmetric *nonlinear σ-model* in d dimensions. In statistical mechanics it corresponds to the well-studied classical $O(D)$ Heisenberg model in d dimensions.

8.10   Path Integrals on Group Spaces

In Section 8.3, we have observed that the surface of a sphere in four dimensions is equivalent to the covering group of rotations in three dimensions, i.e., with the group SU(2). Since we have learned how to write down an exactly solvable time-sliced path integral *near* and *on* the surface of the sphere, the equivalence opens up the possibility of performing path integrals for the motion of a mechanical system *near* and *on* the group space of SU(2). The most important system to which the path integral on the group space of SU(2) can be applied is the spinning top, whose Schrödinger quantum mechanics was discussed in Section 1.15. Exploiting the above equivalence we are able to describe the same quantum mechanics in terms of path integrals. The theory to be developed for this particular system will, after a suitable generalization, be applicable to systems whose dynamics evolves on any group space.

First, we discuss the path integral *near* the group space using the exact result of the path integral *near* the surface of the sphere in four dimensions. The crucial observation is the following: The time-sliced action near the surface

$$A^N = \frac{M}{2\epsilon} r^2 \sum_{n=1}^{N+1} (\mathbf{u}_n - \mathbf{u}_{n-1})^2 = \frac{Mr^2}{\epsilon} \sum_{n=1}^{N+1} (1 - \cos \Delta \vartheta_n) \qquad (8.254)$$

can be rewritten in terms of the group elements $g(\varphi, \theta, \gamma)$ defined in Eq. (8.122) as

$$A^N = \frac{M}{\epsilon} r^2 \sum_{n=1}^{N+1} \left[1 - \frac{1}{2} \mathrm{tr}(g_n g_{n-1}^{-1}) \right], \qquad (8.255)$$

with the obvious notation

$$g_n = g(\varphi_n, \theta_n, \gamma_n). \qquad (8.256)$$

This follows after using the explicit matrix form for g, which reads

$$g(\varphi, \theta, \gamma) = \exp(i\varphi\sigma_3/2) \exp(i\theta\sigma_2/2) \exp(i\gamma\sigma_3/2) \qquad (8.257)$$
$$= \begin{pmatrix} e^{i\varphi/2} & 0 \\ 0 & e^{-i\varphi/2} \end{pmatrix} \begin{pmatrix} \cos(\theta/2) & \sin(\theta/2) \\ -\sin(\theta/2) & \cos(\theta/2) \end{pmatrix} \begin{pmatrix} e^{i\gamma/2} & 0 \\ 0 & e^{-i\gamma/2} \end{pmatrix}.$$

After a little algebra we find

$$\frac{1}{2} \mathrm{tr}(g_n g_{n-1}^{-1}) = \cos(\theta_n/2) \cos(\theta_{n-1}/2) \cos[(\varphi_n - \varphi_{n-1} + \gamma_n - \gamma_{n-1})/2]$$
$$+ \sin(\theta_n/2) \sin(\theta_{n-1}/2) \cos[(\varphi_n - \varphi_{n-1} - \gamma_n + \gamma_{n-1})/2],$$
$$(8.258)$$

just as in (8.128). The invariant group integration measure is usually defined to be normalized to unity, i.e.,

$$\int dg \equiv \frac{1}{16\pi^2} \int_0^\pi \int_0^{2\pi} \int_0^{4\pi} d\theta \sin\theta d\varphi d\gamma = \frac{1}{2\pi^2} \int d^3\mathbf{u} = 1. \qquad (8.259)$$

We shall renormalize the time evolution amplitude $(\mathbf{u}_b \tau_b | \mathbf{u}_a \tau_a)$ near the surface of the four-dimensional sphere accordingly, making it a properly normalized amplitude for the corresponding group elements $(g_b \tau_b | g_a \tau_a)$. Thus we define

$$(\mathbf{u}_b \tau_b | \mathbf{u}_a \tau_a) \equiv \frac{1}{2\pi^2}(g_b \tau_b | g_a \tau_a). \tag{8.260}$$

The path integral (8.150) then turns into the following path integral for the motion near the group space [compare also (8.203)]:

$$(g_b \tau_b | g_a \tau_a) \approx \frac{2\pi^2}{\sqrt{2\pi\hbar\epsilon/Mr^2}^3} \prod_{n=1}^{N} \left[\int \frac{2\pi^2 dg_n}{\sqrt{2\pi\hbar\epsilon/Mr^2}^3} \right] \exp\left(-\frac{1}{\hbar}\mathcal{A}^N\right). \tag{8.261}$$

Let us integrate this expression within the group space language. For this we expand the exponential as in (8.130):

$$\exp\left\{-\frac{Mr^2}{\hbar\epsilon}\left[1 - \frac{1}{2}\mathrm{tr}(g_n g_{n-1}^{-1})\right]\right\} = \sum_{l=0}^{\infty} \tilde{a}_l(h)\frac{l+1}{2\pi^2}C_l^{(1)}(\cos\Delta\vartheta_n) \tag{8.262}$$

$$= \sum_{l=0}^{\infty} \tilde{a}_l(h)\frac{l+1}{2\pi^2} \sum_{m_1,m_2=-l/2}^{l/2} \mathcal{D}_{m_1 m_2}^{l/2}(\varphi_n,\theta_n,\gamma_n)\mathcal{D}_{m_1 m_2}^{l/2\,*}(\varphi_{n-1},\theta_{n-1},\gamma_{n-1}).$$

In general terms, the right-hand side corresponds to the well-known *character expansion* for the group SU(2):

$$\exp\left[\frac{h}{2}\mathrm{tr}(g_n g_{n-1}^{-1})\right] = \frac{1}{h}\sum_{l=0}^{\infty}(l+1)I_{l+1}(h)\chi^{(l/2)}(g_n g_{n-1}^{-1}). \tag{8.263}$$

Here $\chi^{l/2}(g)$ are the so-called *characters*, the traces of the representation matrices of the group element g, i.e.,

$$\chi^{(l/2)}(g) = \mathcal{D}_{mm}^{l/2}(g). \tag{8.264}$$

The relation between the two expansions is obvious if we use the representation properties of the $\mathcal{D}_{m_1 m_2}^{l/2}$ functions and their unitarity to write

$$\chi^{(l/2)}(g_n g_{n-1}^{-1}) = \mathcal{D}_{mm'}^{l/2}(g_n)\mathcal{D}_{mm'}^{l/2*}(g_{n-1}). \tag{8.265}$$

This leads directly to (8.262) [see also the footnote to (8.127)]. Having done the character expansion in each time slice, the intermediate group integrations can all be performed using the orthogonality relations of group characters

$$\int dg \chi^{(L)}(g_1 g^{-1})\chi^{(L')}(g g_2^{-1}) = \delta_{LL'}\frac{1}{d_L}\chi^{(L)}(g_1 g_2^{-1}). \tag{8.266}$$

The result of the integrations is, of course, the same amplitude as before in (8.161):

$$(g_b\tau_b|g_a\tau_a) = \sum_{l=0}^{\infty} \exp\left[-\frac{\hbar L_2}{2Mr^2}(\tau_b - \tau_a)\right] \tag{8.267}$$

$$\times (l+1) \sum_{m_1,m_2=-l/2}^{l/2} \mathcal{D}_{m_1m_2}^{l/2}(\varphi_n, \theta_n, \gamma_n)\mathcal{D}_{m_1m_2}^{l/2\,*}(\varphi_{n-1}, \theta_{n-1}, \gamma_{n-1}).$$

Given this amplitude *near* the group space we can find the amplitude for the motion *on* the group space, by adding to the energy near the sphere $E = \hbar^2[(l/2 + 1)^2 - 1/4]/2Mr^2$ the correction $\Delta E = \hbar^2\Delta L_2^2/2Mr^2$ associated with Eq. (8.228). For $D = 4$, $L_2^2 = (l/2)(l/2+1)+3/4$ has to be replaced by $\hat{L}^2 = L_2^2+\Delta L_2^2 = (l/2)(l/2+1)$, and the energy changes by

$$\Delta E = -\frac{3\hbar^2}{8M}. \tag{8.268}$$

Otherwise the amplitude is the same as in (8.267) [6].

Character expansions of the exponential of the type (8.263) and the orthogonality relation (8.266) are general properties of group representations. The above time-sliced path integral can therefore serve as a prototype for the quantum mechanics of other systems moving *near* or *on* more general group spaces than SU(2).

Note that there is no problem in proceeding similarly with noncompact groups [7]. In this case we would start out with a treatment of the path integral *near* and *on* the surface of a hyperboloid rather than a sphere in four dimensions. The solution would correspond to the path integral near and on the group space of the covering group SU(1,1) of the Lorentz group O(2,1). The main difference with respect to the above treatment would be the appearance of hyperbolic functions of the second Euler angle θ rather than trigonometric functions.

An important family of noncompact groups whose path integral can be obtained in this way are the Euclidean groups [8] consisting of rotations and translations. Their *Lie algebra* comprises the momentum operators $\hat{\mathbf{p}}$, whose representation on the spatial wave functions has the Schrödinger form $\hat{\mathbf{p}} = -i\hbar\boldsymbol{\nabla}$. Thus, the canonical commutation rules in a Euclidean space form part of the representation algebra of these groups. Within a Euclidean group, the separation of the path integral into a radial and an azimuthal part is an important tool in obtaining all group representations.

8.11 Path Integral of Spinning Top

We are now also in a position to solve the time-sliced path integral of a spinning top by reducing it to the previous case of a particle moving on the group space SU(2). Only in one respect is the spinning top different: the equivalent "particle" does not move on the covering space SU(2) of the rotation group, but on the rotation group O(3) itself. The angular configurations with Euler angles γ and $\gamma+2\pi$ are physically indistinguishable. The physical states form a representation space of O(3) and the time evolution amplitude must reflect this. The simplest possibility to incorporate the O(3) topology is to add the two amplitudes leading from the initial configuration $\varphi_a, \theta_a, \gamma_a$ to the two identical final ones $\varphi_b, \theta_b, \gamma_b$ and $\varphi_b, \theta_b, \gamma_b + 2\pi$. This yields the amplitude of the spinning top:

$$(\varphi_b, \theta_b, \gamma_b \; \tau_b | \varphi_b, \theta_b, \gamma_b \; \tau_a)_{\text{top}} \tag{8.269}$$
$$= (\varphi_b, \theta_b, \gamma_b \; \tau_b | \varphi_b, \theta_b, \gamma_b \; \tau_a) + (\varphi_b, \theta_b, \gamma_b + 2\pi \; \tau_b | \varphi_b, \theta_b, \gamma_b \; \tau_a).$$

The sum eliminates all half-integer representation functions $D_{mm'}^{l/2}(\theta)$ in the expansion (8.267) of the amplitude.

Instead of the sum we could have also formed another representation of the operation $\gamma \to \gamma + 2\pi$, the antisymmetric combination

$$(\varphi_b, \theta_b, \gamma_b \; \tau_b | \varphi_b, \theta_b, \gamma_b \; \tau_a)_{\text{fermionic}} \tag{8.270}$$
$$= (\varphi_b, \theta_b, \gamma_b \; \tau_b | \varphi_b, \theta_b, \gamma_b \; \tau_a) - (\varphi_b, \theta_b, \gamma_b + 2\pi \; \tau_b | \varphi_b, \theta_b, \gamma_b \; \tau_a).$$

Here the expansion (8.267) retains only the half-integer angular momenta $l/2$. As discussed in Chapter 7, half-integer angular momenta are associated with fermions such as electrons, protons, muons, and neutrinos. This is indicated by the subscript "fermionic". In spite of this, the above amplitude cannot be used to describe a single fermion since this has only one fixed spin $1/2$, while (8.270) contains all possible fermionic spins at the same time.

In principle, there is no problem in also treating the non-spherical top. In the formulation near the group space, the gradient term in the action,

$$\frac{1}{\epsilon^2} \left[1 - \frac{1}{2} \mathrm{tr}(g_n g_{n-1}^{-1}) \right], \tag{8.271}$$

has to be separated into time-sliced versions of the different angular velocities. In the continuum these are defined by

$$\omega_a = -i\mathrm{tr}\,(\dot{\sigma}_a g^{-1}), \quad a = \xi, \eta, \zeta. \tag{8.272}$$

The gradient term (8.271) has the symmetric continuum limit $\dot{\omega}_a^2$. With the different moments of inertia I_ξ, I_η, I_ζ, the asymmetric sliced gradient term reads

$$\frac{1}{\epsilon^2} \left\{ I_\xi \left[1 - \frac{1}{2} \mathrm{tr}(g_n \sigma_\xi g_{n-1}^{-1}) \right] + I_\eta \left[1 - \frac{1}{2} \mathrm{tr}(g_n \sigma_\eta g_{n-1}^{-1}) \right] \right.$$
$$\left. + I_\zeta \left[1 - \frac{1}{2} \mathrm{tr}(g_n \sigma_\zeta g_{n-1}^{-1}) \right] \right\}, \tag{8.273}$$

rather than (8.271). The amplitude near the top is an appropriate generalization of (8.267). The calculation of the correction term ΔE, however, is more complicated than before and remains to be done, following the rules explained above.

8.12 Path Integral of Spinning Particle

The path integral of a particle on the surface of a sphere contains states of all integer angular momenta $l = 1, 2, 3, \ldots$. The path integral on the group space $SU(2)$ contains also all half-integer spins $s = \frac{1}{2}, \frac{3}{2}, \frac{5}{2}, \ldots$.

The question arises whether it is possible to set up a path integral which contains only a single spinning particle, for instance of spin $s = 1/2$. Thus we need a path integral which for each time slice spans precisely one irreducible representation space of the rotation group, consisting of the $2s + 1$ states $|s\, s_3\rangle$ for $s_3 = -s, \ldots, s$. In order to sum over paths, we must parametrize this space in terms of a continuous variable. This is possible by selecting a particular spin state, for example the state $|ss\rangle$ pointing in the z-direction, and rotating it into an arbitrary direction $\mathbf{u} = (\sin\theta\cos\varphi, \sin\theta\sin\varphi, \cos\theta)$ with the help of some rotation, for instance

$$|\theta\,\varphi\rangle \equiv R^s(\theta, \varphi)|ss\rangle \equiv e^{-iS_3\varphi}e^{-iS_2\theta}|ss\rangle, \tag{8.274}$$

where S_i are matrix generators of the rotation group of spin s, which satisfy the commutation rules of the generators $\hat{L}_i$ in (1.410). The states (8.274) are nonabelian versions of the coherent states (7.343). They can be expanded into the $2s + 1$ spin states $|s\, s_3\rangle$ as follows:

$$|\theta\,\varphi\rangle = \sum_{s_3=-s}^{s} |s\, s_3\rangle\langle s\, s_3|\hat{R}(\theta, \varphi)|ss\rangle = \sum_{s_3=-s}^{s} |s\, s_3\rangle e^{-is_3\varphi}d_{s_3\,s}^s(\theta), \tag{8.275}$$

where $d_{mm'}^j(\theta)$ are the representation matrices of $e^{-iS_2\theta}$ with angular momentum j given in Eq. (1.443). For $s = 1/2$, where the matrix $d_{mm'}^j$ has the form (1.444), the states (8.275) are

$$|\theta\varphi\rangle = e^{i\varphi/2}\cos\theta/2|\tfrac{1}{2}\,\tfrac{1}{2}\rangle - e^{-i\varphi/2}\sin\theta/2|\tfrac{1}{2}\,-\tfrac{1}{2}\rangle. \tag{8.276}$$

At this point it is useful to introduce the so-called *monopole spherical harmonics* defined by

$$Y_{m\,q}^j(\theta, \varphi) \equiv \sqrt{\frac{2j+1}{4\pi}}\,e^{im\varphi}d_{m\,q}^j(\theta). \tag{8.277}$$

Comparison with Eq. (8.124) shows that these are simply the ultra-spherical harmonics $Y_{j\,m\,q}(\hat{\mathbf{x}})$ with vanishing third Euler angle γ. They satisfy the orthogonality relation

$$\int_{-1}^{1} d\cos\theta \int_{0}^{2\pi} d\varphi\, Y_{m\,q}^{j\,*}(\theta, \varphi)Y_{m'\,q}^{j'}(\theta, \varphi) = \delta_{jj'}\delta_{mm'}, \tag{8.278}$$

and the completeness relation

$$\sum_{j,m} Y_{m\,q}^j(\theta, \varphi)Y_{m\,q}^{j\,*}(\theta', \varphi') = \delta(\cos\theta - \cos\theta')\delta(\varphi - \varphi'). \tag{8.279}$$

We define now the covariant looking states

$$|\mathbf{u}\rangle \equiv \sqrt{\frac{2j+1}{4\pi}}|\theta\,\varphi\rangle = \sum_{s_3=-s}^{s} |s\,s_3\rangle\,Y_{s_3\,s}^{s\,*}(\theta,\varphi), \qquad (8.280)$$

and write the angular integral as an integral over the surface of the unit sphere:

$$\int_{-1}^{1} d\cos\theta \int_{0}^{2\pi} d\varphi = \int d^3u\,\delta(\mathbf{u}^2 - 1) \equiv \int d\mathbf{u}. \qquad (8.281)$$

From (8.278) we deduce that the states $|\mathbf{u}\rangle$ are complete in the space of spin-s states:

$$\int d\mathbf{u}\,|\mathbf{u}\rangle\langle\mathbf{u}| = \sum_{s_3=-s}^{s}\sum_{s_3'=-s}^{s}|s\,s_3\rangle \int_{-1}^{1} d\cos\theta \int_{0}^{2\pi} d\varphi Y_{s_3\,s}^{s\,*}(\theta,\varphi)Y_{s_3'\,s}^{s}(\theta,\varphi)\langle s s_3'|$$

$$= \sum_{s_3=-s}^{s}|s\,s_3\rangle\,\langle s\,s_3| = 1^s. \qquad (8.282)$$

The states are not orthogonal, however. Writing $Y_{s_3\,s}^{s}(\theta,\varphi)$ as $Y_{s_3\,s}^{s}(\mathbf{u})$, we see that

$$\langle\mathbf{u}|\mathbf{u}'\rangle = \sum_{s_3=-s}^{s}\sum_{s_3'=-s}^{s}Y_{s_3\,s}^{s}(\mathbf{u})\langle s\,s_3|ss_3'\rangle Y_{s_3'\,s}^{s}(\mathbf{u}') = \sum_{s_3=-s}^{s}Y_{s_3\,s}^{s}(\mathbf{u})Y_{s_3\,s}^{s\,*}(\mathbf{u}'). \qquad (8.283)$$

The right-hand side can be calculated as follows:

$$\frac{2j+1}{4\pi}\langle ss|e^{i\theta S_2}e^{i\varphi S_3}e^{-i\varphi' S_3}e^{-i\theta' S_2}|ss\rangle = \frac{2j+1}{4\pi}\langle ss|e^{-isAS_3}e^{-i\beta S_2}|ss\rangle$$

$$= \frac{2j+1}{4\pi}e^{-isAS_3}d_{ss}^{s}(\beta) = \frac{2j+1}{4\pi}e^{-isAS_3}\left(\frac{1+\mathbf{u}\cdot\mathbf{u}'}{2}\right)^{s}, \qquad (8.284)$$

where β is the angle between $\mathbf{u}$ and $\mathbf{u}'$, and $A(\mathbf{u},\mathbf{u}',\hat{\mathbf{z}})$ is the area of the spherical triangle on the unit sphere formed by the three points in the argument. For a radius R, the area is equal to R^2 time the angular excess E of the triangle, defined as the amount by which the sum of the angles in the triangle is larger that π. An explicit formula for E is the spherical generalization of *Heron's formula* for the area A of a triangle [9]:

$$A = \sqrt{s(s-a)(s-b)(s-c)}, \qquad s = (a+b+c)/2 = \text{semiperimeter}. \qquad (8.285)$$

The angular excess on a sphere is

$$\tan\frac{E}{4} = \sqrt{\tan\frac{\phi_s}{2}\tan\frac{\phi_s-\phi_a}{2}\tan\frac{\phi_s-\phi_b}{2}\tan\frac{\phi_s-\phi_c}{2}}, \qquad (8.286)$$

where ϕ_a, ϕ_b, ϕ_c are the *angular lengths* of the sides of the triangle and

$$\phi_s = (\phi_a + \phi_b + \phi_c)/2 \qquad (8.287)$$

is the angular semiperimeter on the sphere.

We can now set up a path integral for the scalar product (8.284):

$$\langle \mathbf{u}_b | \mathbf{u}_a \rangle = \left[\prod_{n=1}^{N} \int d\mathbf{u}_n \right] \langle \mathbf{u}_b | \mathbf{u}_N \rangle \langle \mathbf{u}_N | \mathbf{u}_{N-1} \rangle \langle \mathbf{u}_{N-1} | \cdots | \mathbf{u}_1 \rangle \langle \mathbf{u}_1 | \mathbf{u}_a \rangle. \quad (8.288)$$

For large N, the intermediate $\mathbf{u}_n$-vectors will all lie close to their neighbors and we can write approximately

$$\langle \mathbf{u}_n | \mathbf{u}_{n-1} \rangle \approx \frac{2s+1}{4\pi} e^{i\Delta A_n} \left[1 + \tfrac{1}{2} \mathbf{u}_n (\mathbf{u}_n - \mathbf{u}_{n-1}) \right]^s \approx \frac{2s+1}{4\pi} e^{i\Delta A_n + \frac{s}{2} \mathbf{u}_n (\mathbf{u}_n - \mathbf{u}_{n-1})}. \quad (8.289)$$

Let us take this expression to the continuum limit. We introduce a time parameter labeling the chain of $\mathbf{u}_n$-vectors by $t_n = n\epsilon$, and find the small-ϵ approximation

$$\frac{2s+1}{4\pi} \exp\left[i\epsilon\, s\, \cos\theta\, \dot{\varphi} - \frac{s}{4}\epsilon^2 \left(\dot{\theta}^2 + \sin^2\theta\, \dot{\varphi}^2 \right) + \dots \right]. \quad (8.290)$$

The first term is obtained from the scalar product

$$\begin{aligned}
\langle \theta\, \varphi | i\partial_t | \theta\, \varphi \rangle &= \langle ss | e^{iS_2\theta} e^{iS_3\varphi} i\partial_t e^{-iS_3\varphi} e^{-iS_2\theta} | ss \rangle \\
&= \langle ss | e^{iS_2\theta} e^{iS_3\varphi} \left(\dot{\varphi} S_3 e^{-iS_3\varphi} e^{-iS_2\theta} + e^{-iS_3\varphi} \dot{\theta} S_2 e^{-iS_2\theta} \right) | ss \rangle \\
&= \langle ss | (\cos\theta S_3 - \sin\theta S_1)\dot{\varphi} + \dot{\theta} S_2 | ss \rangle = s \cos\theta\, \dot{\varphi}. \quad (8.291)
\end{aligned}$$

This result is actually not completely correct. The reason is that the angular variables in the states (8.274) are cyclic variables. For integer spins, θ and φ are cyclic in 2π, for half-integer spins in 4π. Thus there can be jumps by 2π or 4π in these angles which do not change the states (8.274). In writing down the approximation (8.290) we must assume that we are at a safe distances from such singularities. If we get close to them, we must change the direction of the quantization axis.

Keeping this in mind we can express the scalar product in the limit $\epsilon \to 0$ by the path integral

$$\langle \mathbf{u}_b | \mathbf{u}_a \rangle = \int \mathcal{D}\mathbf{u}\, e^{\frac{i}{\hbar} \int_{t_a}^{t_b} dt\, \hbar s \cos\theta\, \dot{\varphi}}, \quad (8.292)$$

where $\int \mathcal{D}\mathbf{u}$ is defined by the limit $N \to \infty$ of the product of integrals

$$\int \mathcal{D}\mathbf{u} \equiv \lim_{N\to\infty} \left[\frac{2s+1}{4\pi} \prod_{n=1}^{N} \int d\mathbf{u}_n \right]. \quad (8.293)$$

The path integral fixes the Hilbert space of the spin theory. It is the analog of the zero-Hamiltonian path integral in Eqs. (2.17) and (2.18). Comparing (8.292) with (2.18), we see that $s\hbar \cos\theta$ plays the role of a canonically conjugate momentum of the variable φ.

For a specific spin dynamics we must add, as in (2.15), a Hamiltonian $H(\cos\theta, \varphi)$, and arrive at the general path integral representation for the time evolution amplitude of a spinning particle [10]

$$(\mathbf{u}_b t_b | \mathbf{u}_a t_a) = \int \mathcal{D}\mathbf{u}\, e^{i \int_{t_a}^{t_b} dt\, [\hbar s \cos\theta\, \dot{\varphi} - H(\theta, \varphi)]/\hbar}. \quad (8.294)$$

The above path integral has a remarkable property which is worth emphasizing. For half-integer spins $s = \frac{1}{2}, \frac{3}{2}, \frac{5}{2}, \ldots$ it is able to describe the physics of a fermion in terms of a field theory involving a unit vector field $\mathbf{u}$ which describes the direction of the spin state:

$$\langle \mathbf{u}|\hat{\mathbf{S}}|\mathbf{u}\rangle = \langle ss|\hat{R}^{-1}(\theta, \varphi)\hat{\mathbf{S}}\hat{R}(\theta, \varphi)|ss\rangle = R_{ij}(\theta, \varphi)\langle ss|S_j|ss\rangle = s\mathbf{u}. \tag{8.295}$$

This follows from the vector property of the spin matrices $\hat{\mathbf{S}}$. The matrices $R_{ij}(\theta, \varphi)$ are the defining 3×3 matrices of the rotation group (the so-called *adjoint representation*). Thus we describe a fermion in terms of a Bose field.

The above path integral is only the simplest illustration for a more general phenomenon. In 1961, Skyrme pointed out that a certain field configuration of pions is capable of behaving in many respects like a nucleon [11], in particular its fermionic properties.

In two dimensions, Bose field theories are even more powerful and can describe particles with any commutation rule, called anyons in Section 7.5. This will be shown in Chapter 16.

In Chapter 10 we shall see that the action in (8.292) can be interpreted as the action of a particle of charge e on the surface of the unit sphere whose center contains a fictitious magnetic monopole of charge $g = -4\pi\hbar cs/e$. The associated vector potential $\mathbf{A}^{(g)}(\mathbf{u})$ will be given in Eq. (10A.59). Coupling this minimally to a particle of charge e as in Eq. (2.633) on the surface of the sphere yields the action

$$\mathcal{A}_0 = \frac{e}{c} \int_{t_a}^{t_b} \mathbf{A}^{(g)}(\mathbf{u}) \cdot \dot{\mathbf{u}} = \hbar s \int_{t_a}^{t_b} dt \, \cos\theta \, \dot{\varphi}, \tag{8.296}$$

where the magnetic flux is supplied to the monopole by two infinitesimally thin flux tubes, the famous *Dirac strings*, one from below and one from above. The field $\mathbf{A}^{(-s)}$ in (8.296) is the average of the two expressions in (10A.61) for $g = -4\pi\hbar cs/e$. We can easily change the supply line to a single string from above, by choosing the states

$$|\theta \, \varphi\rangle' \equiv \hat{R}(\theta, \varphi)|ss\rangle = e^{-iS_3\varphi}e^{-iS_2\theta}e^{iS_3\varphi}|ss\rangle = |\theta \, \varphi\rangle e^{is\varphi}, \tag{8.297}$$

rather than $|\theta \, \varphi\rangle$ of Eq. (8.274) for the construction of the path integral. The physics is the same since the string is an artifact of the choice of the quantization axis.

In terms of Cartesian coordinates, the action (8.296) with a flux supplied from the north pole can also be expressed in terms of the vector $\mathbf{u}(t)$ as [compare (10A.59)]

$$\mathcal{A}_0 = \hbar s \int dt \, \frac{\hat{\mathbf{z}} \times \mathbf{u(t)}}{1 - u_z(t)} \cdot \dot{\mathbf{u}}(t). \tag{8.298}$$

This expression is singular on the north pole of the unit sphere. The singularity can be rotated into an arbitrary direction $\mathbf{n}$, leading to

$$\mathcal{A}_0 = \hbar s \int dt \, \frac{\mathbf{n} \times \mathbf{u(t)}}{1 - \mathbf{n} \cdot \mathbf{u}(t)} \cdot \dot{\mathbf{u}}(t). \tag{8.299}$$

If $\mathbf{u}$ gets close to $\mathbf{n}$ we must change the direction of $\mathbf{n}$. The action (8.299) is referred to as *Wess-Zumino action*.

In Chapter 10 we shall also calculate the curl of the vector potential $\mathbf{A}^{(g)}$ of a monopole of magnetic charge g and find the radial magnetic field accompanied by a singular string contribution along the direction $\mathbf{n}$ of flux supply [compare (10A.54)]:

$$\mathbf{B}^{(g)} = \boldsymbol{\nabla} \times \mathbf{A}^{(g)} = g\,\frac{\mathbf{u}}{|\mathbf{u}|} - 4\pi g \int_0^\infty ds\,\hat{\mathbf{n}}\,\delta^{(3)}(\mathbf{u} - s\,\hat{\mathbf{n}}), \qquad (8.300)$$

The singular contribution is an artifact of the description of the magnetic field. The line from zero to infinity is called a *Dirac string*. Since the magnetic field has no divergence, the magnetic flux emerging at the origin of the sphere must be imported from somewhere at infinity. In the field (8.301) the field is imported along the straight line in the direction $\mathbf{u}$. Indeed, we can easily check that the divergence of (8.301) is zero.

For a closed orbit, the interaction (8.296) can be rewritten by Stokes' theorem as

$$\mathcal{A}_0 = \frac{e}{c}\int dt\,\mathbf{A}^{(g)}(t)\cdot\mathbf{u}(t) = \frac{e}{c}\int d\mathbf{S}\cdot\left[\boldsymbol{\nabla}\times\mathbf{A}^{(g)}\right] = \frac{e}{c}\int d\mathbf{S}\cdot\mathbf{B}^{(g)}, \qquad (8.301)$$

where $\int d\mathbf{S}$ runs over the surface enclosed by the orbit. This surface may or may not contain the Dirac string of the monopole, in which case $\mathcal{A}_0$ differ by $4\pi ge/c$. A path integral over closed orbits of the spinning particle

$$Z_{\mathrm{QM}} = \oint d\mathbf{u}\,e^{i(\mathcal{A}_0 + \mathcal{A})/\hbar}, \qquad (8.302)$$

is therefore invariant under changes of the position of the Dirac string if the monopole charge g satisfies the *Dirac charge quantization condition*

$$\frac{ge}{\hbar c} = s, \qquad (8.303)$$

with $s = $ half-integer or integer.

Dirac was the first to realize that as a consequence of quantum mechanics, an electrically charged particle whose charge satisfies the quantization condition (8.303) sees only the radial monopole field in (8.301), not the field in the string. The string can run along *any line* L without being detectable. This led him to conjecture that there could exist magnetic monopoles of a specific g, which would explain that all charges in nature are integer multiples of the electron charge [16]. More on this subject will be discussed in Section 16.2.

In Chapter 10 we shall learn how to define a monopole field $\mathbf{A}^{(g)}$ which is free of the artificial string singularity [see Eq. (10A.58)]. With the new definition, the divergence of $\mathbf{B}$ is a δ-function at the origin:

$$\boldsymbol{\nabla}\times\mathbf{B}(\mathbf{u}) = 4\pi g\,\delta^{(3)}(\mathbf{u}). \qquad (8.304)$$

8.13 Berry Phase

This phenomenon has a simple physical basis which can be explained most clearly by means of the following gedanken experiment. Consider a thin rod whose dynamics is described by a unit vector field $\mathbf{u}(t)$ with an action

$$\mathcal{A} = \frac{M}{2} \int dt \left[\dot{\mathbf{u}}^2(t) - V(\mathbf{u}^2(t)) \right], \quad \mathbf{u}^2(t) \equiv 1, \tag{8.305}$$

where $\mathbf{u}(t)$ is a unit vector along the rod. This is the same Lagrangian as for a particle on a sphere as in (8.146) (recall p. 738).

Let us suppose that the thin rod is a solenoid carrying a strong magnetic field, and containing at its center a particle of spin $s = 1/2$. Then (8.305) is extended by the action [13]

$$\mathcal{A}_0 = \int dt \, \psi^*(t) \left[i\hbar \partial_t + \gamma \mathbf{u}(t) \cdot \boldsymbol{\sigma} \right] \psi(t), \tag{8.306}$$

where σ^i are the Pauli spin matrices (1.445). For large coupling strength γ and sufficiently slow rotations of the solenoid, the direction of the fermion spin will always be in the ground state of the magnetic field, i.e., its direction will follow the direction of the solenoid adiabatically, pointing always along $\mathbf{u}(t)$. If we parametrize $\mathbf{u}(t)$ in terms of spherical angles $\theta(t), \phi(t)$ as $\mathbf{n} = (\sin\theta \cos\phi, \sin\theta \sin\phi, \cos\theta)$, the associated wave function satisfies $\mathbf{u}(t) \cdot \boldsymbol{\sigma} \, \psi(\mathbf{u}(t)) = (1/2)\psi(\mathbf{u}(t))$, and reads

$$\psi(\mathbf{u}(t)) = \begin{pmatrix} e^{-i\phi(t)/2} \cos\theta(t)/2 \\ e^{i\phi(t)/2} \sin\theta(t)/2 \end{pmatrix}. \tag{8.307}$$

Inserting this into (8.306) we obtain [compare (8.291)]

$$\mathcal{A}_0 = \int dt \, \psi(\mathbf{u}(t)) \, i\hbar \partial_t \, \psi(\mathbf{u}(t)) = \hbar \frac{1}{2} \cos\theta(t) \, \dot{\phi}(t) \equiv \hbar\beta(t). \tag{8.308}$$

The action coincides with the previous expression (8.296). The angle $\beta(t)$ is called *Berry phase* [14].

In this simple model it is obvious why the bosonic theory of the solenoid behaves like a spin-1/2 particle: It simply inherits the physical properties of the enslaved spinor.

The reason why it is a monopole field that causes the spin-1/2 behavior will become clear in another way in Section 14.6. There we shall solve the path integral of a charged particle in a monopole field and show that it behaves like a fermion if its charge e and the monopole charge g have half-integer products $q \equiv eg/\hbar c$.

8.14 Spin Precession

The Wess-Zumino action $\mathcal{A}_0$ adds an interesting kinetic term to the equation of motion of a solenoid. Extremizing $\mathcal{A}_0 + \mathcal{A}$ yields

$$M \left(\ddot{\mathbf{u}} + \dot{\mathbf{u}}^2 \, \mathbf{u} \right) = -\partial_{\mathbf{u}} V(\mathbf{u}) - \frac{\delta}{\delta \mathbf{u}(t)} \mathcal{A}_0. \tag{8.309}$$

The functional derivative of $\mathcal{A}_0$ is most easily calculated starting from the general expression in (8.296)

$$\frac{\delta}{\delta u_i(t)} \hbar \int_{t_a}^{t_b} dt \, \mathbf{A}^{(g)} \cdot \dot{\mathbf{u}} = \hbar \left(\partial_{u_i} A_j^{(g)} \dot{u}_j - \partial_t A_i^{(g)} \right) = \hbar \left[\left(\partial_{u_i} A_j^{(g)} - \partial_{u_j} A_i^{(g)} \right) \dot{u}_j \right]$$

$$= \hbar \left[\left(\boldsymbol{\nabla} \times \mathbf{A}^{(g)} \right) \times \mathbf{u} \right]_i. \tag{8.310}$$

Inserting here the curl of Eq. (8.301), while staying safely away from the singularity, the last term in (8.309) becomes $\hbar g \mathbf{u} \times \dot{\mathbf{u}}$, and the equation of motion (8.309) turns into

$$M \left(\ddot{\mathbf{u}} + \dot{\mathbf{u}}^2 \, \mathbf{u} \right) = -\partial_{\mathbf{u}} V(\mathbf{u}) - \hbar g \, \mathbf{u} \times \dot{\mathbf{u}}. \tag{8.311}$$

Multiplying this vectorially by $\mathbf{u}(t)$ we find [15, 16]

$$M \mathbf{u} \times \ddot{\mathbf{u}} = -\mathbf{u} \times \partial_{\mathbf{u}} V(\mathbf{u}) - \hbar g \left[\mathbf{u} \left(\mathbf{u} \cdot \dot{\mathbf{u}} \right) - \dot{\mathbf{u}} \, \mathbf{u}^2 \right]. \tag{8.312}$$

Since $\mathbf{u}^2 = 1$, this reduces to

$$M \mathbf{u} \times \ddot{\mathbf{u}} - \hbar g \dot{\mathbf{u}} = -\mathbf{u} \times \partial_{\mathbf{u}} V(\mathbf{u}(t)). \tag{8.313}$$

If M is small we obtain for a particle of spin s, where $g = -s$, the so-called *Landau-Lifshitz equation* [17, 18]

$$\hbar s \, \dot{\mathbf{u}} = -\mathbf{u} \times \partial_{\mathbf{u}} V(\mathbf{u}). \tag{8.314}$$

This is a useful equation for studying magnetization fields in ferromagnetic materials.

The interaction energy of a spinning particle with an external magnetic field has the general form

$$H_{\text{int}} = -\boldsymbol{\mu} \cdot \mathbf{B}, \qquad \boldsymbol{\mu} = -\gamma \mathbf{S}, \tag{8.315}$$

where $\boldsymbol{\mu}$ is the magnetic moment, $\mathbf{S}$ the vector of spin matrices, and γ the *gyromagnetic ratio*.

Since $\mathbf{u}$ is the direction vector of the spin, we may identify the magnetic moment of the spin s as

$$\boldsymbol{\mu} = \gamma s \hbar \mathbf{u}, \tag{8.316}$$

so that the interaction energy becomes

$$H_{\text{int}} = -\gamma s \hbar \, \mathbf{u} \cdot \mathbf{B}. \tag{8.317}$$

Inserting this for $V(\mathbf{u})$ in (8.314) yields the equation of motion

$$\dot{\mathbf{u}} = -\gamma \, \mathbf{B} \times \mathbf{u}, \tag{8.318}$$

showing a rate of precession $\boldsymbol{\Omega} = -\gamma \mathbf{B}$.

For comparison we recall the derivation of this result in the conventional way from the Heisenberg equation of motion (1.272):

$$\hbar \, \dot{\mathbf{S}} = i[\hat{H}, \mathbf{S}]. \tag{8.319}$$

Inserting for $\hat{H}$ the interaction energy (8.315) and using the commutation relations (1.410) of the rotation group for the spin matrices $\mathbf{S}$, this yields the Heisenberg equation for spin precession

$$\dot{\mathbf{S}} = -\gamma\, \mathbf{B} \times \mathbf{S}, \tag{8.320}$$

in agreement with the Landau-Lifshitz equation (8.314). This shows that the Wess-Zumino term in the action $\mathcal{A}_0 + \mathcal{A}$ has the ability to render quantum equations of motion from a classical action. This allows us, in particular, to mimic systems of half-integer spins, which are fermions, with a theory containing only a bosonic directional vector field $\mathbf{u}(t)$. This has important applications in statistical mechanics where models of interacting quantum spins for ferro- and antiferromagnets à la Heisenberg can be studied by applying field theoretic methods to vector field theories.

Notes and References

The path integral in radial coordinates was advanced by
S.F. Edwards and Y.V. Gulyaev, Proc. Roy. Soc. London, A **279**, 229 (1964);
D. Peak and A. Inomata, J. Math. Phys. **19**, 1422 (1968);
A. Inomata and V.A. Singh, J. Math. Phys. **19**, 2318 (1978);
C.C. Gerry and V.A. Singh, Phys. Rev. D **20**, 2550 (1979); Nuovo Cimento **73**B, 161 (1983).
The path integral with a Bessel function in the measure has a predecessor in the theory of stochastic differential equations:
J. Pitman and M. Yor, *Bessel Processes and Infinitely Divisible Laws*, in *Stochastic Integrals*, Springer Lecture Notes in Mathematics 851, ed. by D. Williams, 1981, p. 285.
Difficulties with radial path integrals have been noted before by
W. Langguth and A. Inomata, J. Math. Phys. **20**, 499 (1979);
F. Steiner, in *Path Integrals from* meV *to* MeV, edited by M. C. Gutzwiller, A. Inomata, J.R. Klauder, and L. Streit (World Scientific, Singapore, 1986), p. 335.
The solution to the problem was given by
H. Kleinert, Phys. Lett. B **224**, 313 (1989) (http://www.physik.fu-berlin.de/~kleinert/195).
It will be presented in Chapters 12 and 14.

The path integral on a sphere and on group spaces was found by
H. Kleinert, Phys. Lett. B **236**, 315 (1990) (*ibid*.http/202).
Compare also with earlier attempts by
L.S. Schulman, Phys. Rev. **174**, 1558 (1968),
who specified the correct short-time amplitude but did not solve the path integral.

The individual citations refer to

[1] C.C. Gerry, J. Math. Phys. **24**, 874 (1983);
    C. Garrod, Rev. Mod. Phys. **38**, 483 (1966).

[2] These observations are due to
    S.F. Edwards and Y.V. Gulyaev, Proc. Roy. Soc. London, A **279**, 229 (1964).
    See also
    D.W. McLaughlin and L.S. Schulman, J. Math. Phys. **12**, 2520 (1971).

[3] H. Kleinert and V. Schulte-Frohlinde, *Critical Properties of Φ^4-Theories*, World Scientific, Singapore 2001, pp. 1–487 (http://www.physik.fu-berlin.de/~kleinert/b8).

[4] H. Kleinert and I. Mustapic, J. Math. Phys. **33**, 643 (1992) (http://www.physik.fu-berlin.de/~kleinert/207).

[5] G. Pöschl and E. Teller, Z. Phys. **83**, 143 (1933).
See also
S. Flügge, *Practical Quantum Mechanics*, Springer, Berlin, 1974, p. 89.

[6] H. Kleinert, Phys. Lett. B **236**, 315 (1990) (*ibid*.http/202).

[7] M. Böhm and G. Junker, J. Math. Phys. **28**, 1978 (1987).
Note, however, that these authors do not really solve the path integral *on* the group space as they claim but only *near* the group space. Also, many expressions are meaningless due to path collapse.

[8] M. Böhm and G. Junker, J. Math. Phys. **30**, 1195 (1989). See also remarks in [7].

[9] D.D. Ballow and F.H. Steen, *Plane and Spherical Trigonometry with Tables*, Ginn, New York , 1943.

[10] J.R. Klauder, Phys. Rev. D **19**, 2349 (1979);
A. Jevicki and N. Papanicolaou, Nucl. Phys. B **171**, 382 (1980).

[11] T.H.R. Skyrme, Proc. R. Soc. London A **260**, 127 (1961).

[12] P.A.M. Dirac, Proc. Roy. Soc. A **133**, 60 (1931); Phys. Rev. **74**, 817 (1948), Phys. Rev. **74**, 817 (1948).

[13] M. Stone, Phys. Rev. D **33**, 1191 (1986).

[14] For the Berry phase and quantum spin systems see the reprint collection
A. Shapere and F. Wilczek, *Geometric Phases in Physics*, World Scientific, Singapore, 1989, and the textbook
E. Fradkin, *Field Theories of Condensed Matter Systems*, Addison-Wesley, Reading, MA, 1991.

[15] H. Poincaré, *Remarques sur une expérience de Birkeland*, Rendues, **123**, 530 (1896).

[16] E. Witten, Nucl. Phys. B **223**, 422 (1983).

[17] L.D. Landau and E.M. Lifshitz, *Statistical Physics*, Pergamon, London, 1975, Chapter 7.

[18] The ability of the monopole action to generate the Landau-Lifshitz equation in a classical equation of motion was apparently first recognized by
C.F. Valenti and M. Lax, Phys. Rev. B **16**, 4936-4944 (1977).

9

Wave Functions

The fundamental quantity obtained by solving a path integral is the time evolution amplitude or propagator of a system $(\mathbf{x}_b t_b|\mathbf{x}_a t_a)$. In Schrödinger quantum mechanics, on the other hand, one has direct access on the energy spectrum and the wave functions of a system [see (1.94)]. This chapter will explain how to extract this information from the time evolution amplitude $(\mathbf{x}_b t_b|\mathbf{x}_a t_a)$. The crucial quantity for this purpose the Fourier transform of $(\mathbf{x}_b t_b|\mathbf{x}_a t_a)$, the fixed-energy amplitude introduced in Eq. (1.313):

$$(\mathbf{x}_b|\mathbf{x}_a)_E = \int_{t_a}^{\infty} dt_b \exp\{iE(t_b - t_a)/\hbar\}(\mathbf{x}_b t_b|\mathbf{x}_b t_a), \qquad (9.1)$$

which contains as much information on the system as $(\mathbf{x}_b t_b|\mathbf{x}_a t_a)$, and gives, in particular, a direct access to the energy spectrum and the wave functions of the system. This is done via the the spectral decomposition (1.321).

Alternatively, we can work with the causal propagator at imaginary time,

$$(\mathbf{x}_b \tau_b|\mathbf{x}_a \tau_a) = \int \frac{d^D p}{(2\pi\hbar)^D} \exp\left[\frac{i}{\hbar}\mathbf{p}(\mathbf{x}_b - \mathbf{x}_a) - \frac{\mathbf{p}^2}{2M\hbar}(\tau_b - \tau_a)\right], \qquad (9.2)$$

and calculate the fixed-energy amplitude by the Laplace transformation

$$(\mathbf{x}_b|\mathbf{x}_a)_E = -i\int_{\tau_a}^{\infty} d\tau_b \exp\{E(\tau_b - \tau_a)/\hbar\}(\mathbf{x}_b \tau_b|\mathbf{x}_b \tau_a). \qquad (9.3)$$

9.1 Free Particle in D Dimensions

For a free particle in D dimensions, the fixed-energy amplitude was calculated in Eqs. (1.344) and (1.351). It will be instructive to rederive the same result once more using the development in Section 8.5.1. Here we start directly from the spectral representation (1.321), which for a free particle takes the explicit form Eq. (1.340):

$$(\mathbf{x}_b|\mathbf{x}_a)_E = \int \frac{d^D k}{(2\pi)^D} e^{i\mathbf{k}(\mathbf{x}_b - \mathbf{x}_a)} \frac{i\hbar}{E - \hbar^2 \mathbf{k}^2/2M + i\eta}. \qquad (9.4)$$

The momentum integral can now be done as follows. The exponential function $\exp\,(i\mathbf{k}\mathbf{R})$ is written as $\exp\,(ikR\cos\vartheta)$, where $\mathbf{R}$ is the distance vector $\mathbf{x}_b - \mathbf{x}_a$ and ϑ the angle between $\mathbf{k}$ and $\mathbf{R}$. Then we use formula (8.100) with the coefficients (8.101) and the hyperspherical harmonics $Y_{l\mathbf{m}}(\hat{\mathbf{x}})$ of Eq. (8.125) and expand

$$e^{i\mathbf{k}\mathbf{R}} = \sum_{l=0} a_l(ikR) \sum_{\mathbf{m}} Y_{l\mathbf{m}}(\hat{\mathbf{k}}) Y_{l\mathbf{m}}^*(\hat{\mathbf{R}}). \qquad (9.5)$$

The integral over $\mathbf{k}$ follows now directly from the decomposition into size and direction $d^D k = dk d\hat{\mathbf{k}}$, and the orthogonality property (8.114) of the hyperspherical harmonics, according to which

$$\int d\hat{\mathbf{k}}\, Y_{l\mathbf{m}}(\hat{\mathbf{k}}) = \delta_{l0}\delta_{\mathbf{m}0}\sqrt{S_D}, \qquad (9.6)$$

with S_D of Eq. (1.555). Since $Y_{00}(\hat{\mathbf{x}}) = 1/\sqrt{S_D}$, we obtain

$$(\mathbf{x}_b|\mathbf{x}_a)_E = -\frac{2Mi}{(2\pi)^D} \int_0^\infty dk k^{D-1} \frac{1}{k^2 + \kappa^2} a_0(ikR), \qquad (9.7)$$

where

$$\kappa \equiv \sqrt{-2ME/\hbar^2}, \qquad (9.8)$$

as in (1.342). Inserting $a_0(ikR)$ from (8.101),

$$a_0(ikR) = (2\pi)^{D/2} J_{D/2-1}(kR)/(kR)^{D/2-1}, \qquad (9.9)$$

we find

$$(\mathbf{x}_b|\mathbf{x}_a)_E = -\frac{2Mi}{(2\pi)^{D/2}} R^{1-D/2} \int_0^\infty dk k^{D/2} \frac{1}{k^2 + \kappa^2} J_{D/2-1}(kR). \qquad (9.10)$$

The integral

$$\int_0^\infty dk \frac{k^{\nu+1}}{(k^2 + a^2)^{\mu+1}} J_\nu(kb) = \frac{a^{\nu-\mu}b^\mu}{2^\mu \Gamma(\mu+1)} K_{\nu-\mu}(ab) \qquad (9.11)$$

yields once more the fixed-energy amplitude (1.343).

In two dimensions, the amplitude (1.343) becomes [recall (1.346)]

$$(\mathbf{x}_b|\mathbf{x}_a)_E = -\frac{iM}{\pi\hbar} K_0(\kappa|\mathbf{x}_b - \mathbf{x}_a|). \qquad (9.12)$$

It can be decomposed into partial waves by inserting

$$|\mathbf{x}_b - \mathbf{x}_a| = \sqrt{r_b^2 + r_a^2 - 2r_a r_b \cos(\varphi_b - \varphi_a)}. \qquad (9.13)$$

Then a well-known addition theorem for Bessel functions yields the expansion

$$K_0(\kappa|\mathbf{x}_b - \mathbf{x}_a|) = \sum_{m=-\infty}^{\infty} I_m(\kappa r_<) K_m(\kappa r_>) e^{im(\varphi_b - \varphi_a)}, \qquad (9.14)$$

where $r_<$ and $r_>$ are the smaller and larger values of r_a and r_b, respectively. Hence the fixed-energy amplitude turns into

$$(\mathbf{x}_b|\mathbf{x}_a)_E = -\frac{2iM}{\hbar} \sum_m I_m(\kappa r_<) K_m(\kappa r_>) \frac{1}{2\pi} e^{im(\varphi_b-\varphi_a)}. \tag{9.15}$$

This is an analytic function in the complex E-plane. The parameter κ is real for $E < 0$. For $E > 0$, the square root (9.8) allows for two imaginary solutions, $\kappa^\pm \equiv e^{\mp i\pi/2} k \equiv e^{\mp i\pi/2}\sqrt{2ME}/\hbar$, so that the amplitude has a right-hand cut. Its discontinuity specifies the continuum of free-particle states. On top of the cut, we use the analytic continuation formulas (valid for $-\pi/2 < \arg z \le \pi$)[1]

$$\begin{aligned} I_\mu(e^{-i\pi/2}z) &= e^{-i\pi\mu/2} J_\mu(z), \\ K_\mu(e^{-i\pi/2}z) &= i\frac{\pi}{2} e^{i\pi\mu/2} H_\mu^{(1)}(z), \end{aligned} \tag{9.16}$$

to find the fixed-energy amplitude above the cut

$$(\mathbf{x}_b|\mathbf{x}_a)_{E+i\eta} = \frac{\pi M}{\hbar} \sum_m J_m(kr_<) H_m^{(1)}(kr_>) \frac{1}{2\pi} e^{im(\varphi_b-\varphi_a)}. \tag{9.17}$$

The reflection properties[2]

$$\begin{aligned} H_\mu^{(1)}(e^{i\pi}z) &= -H_{-\mu}^{(2)}(z) \equiv -e^{-i\pi\mu} H_\mu^{(2)}(z), \\ J_\mu(e^{i\pi}z) &= e^{i\pi\mu} J_\mu(z) \end{aligned} \tag{9.18}$$

yield the amplitude below the cut

$$(\mathbf{x}_b|\mathbf{x}_a)_{E-i\eta} = -\frac{\pi M}{\hbar} \sum_m J_m(kr_<) H_m^{(2)}(kr_>) \frac{1}{2\pi} e^{im(\varphi_b-\varphi_a)}. \tag{9.19}$$

The discontinuity across the cut follows from the relation

$$J_\mu(z) = \frac{1}{2}\left[H_\mu^{(1)}(z) + H_\mu^{(2)}(z) \right] \tag{9.20}$$

and reads

$$\text{disc}\,(\mathbf{x}_b|\mathbf{x}_a)_E = \frac{2\pi M}{\hbar} \sum_{m=-\infty}^{\infty} J_m(kr_b) J_m(kr_a) \frac{1}{2\pi} e^{im(\varphi_b-\varphi_b)}. \tag{9.21}$$

According to (1.326), the integral over the discontinuity yields the completeness relation

$$\begin{aligned} \int_{-\infty}^{\infty} \frac{dE}{2\pi\hbar} &\text{disc}\,(\mathbf{x}_b|\mathbf{x}_a)_E \\ &= \int_{-\infty}^{\infty} \frac{dE}{2\pi\hbar} \frac{2\pi M}{\hbar} \sum_{m=-\infty}^{\infty} J_m(kr_b) J_m(kr_a) \frac{1}{2\pi} e^{im(\varphi_b-\varphi_b)} = \delta(\mathbf{x}_b - \mathbf{x}_a). \end{aligned} \tag{9.22}$$

---

[1] I.S. Gradshteyn and I.M. Ryzhik, *op. cit.*, Formulas 8.406.1 and 8.407.1.
[2] ibid., Formulas 8.476.1 and 8.476.8.

After replacing the energy integral by a k-integral,

$$\int_{-\infty}^{\infty} \frac{dE}{2\pi\hbar} \frac{2\pi M}{\hbar} = \int_0^{\infty} dk\, k, \qquad (9.23)$$

we can also write

$$
\begin{aligned}
\int_{-\infty}^{\infty} \frac{dE}{2\pi\hbar} \operatorname{disc}(\mathbf{x}_b|\mathbf{x}_a)_E &= \int \frac{d^2k}{(2\pi)^2} e^{i\mathbf{k}(\mathbf{x}_b-\mathbf{x}_a)} = \frac{1}{2\pi} \int_0^{\infty} dk\, k\, J_0(k|\mathbf{x}_b-\mathbf{x}_a|) \\
&= \sum_{m=-\infty}^{\infty} \int_0^{\infty} dk\, k\, J_m(kr_b) J_m(kr_a) \frac{1}{2\pi} e^{im(\varphi_b-\varphi_a)} \\
&= \frac{1}{\sqrt{r_b r_a}} \delta(r_b-r_a)\delta(\varphi_b-\varphi_a).
\end{aligned}
\qquad (9.24)
$$

The last two lines exhibit the well-known completeness relation of the radial wave functions of a free particle.[3]

9.2 Harmonic Oscillator in D Dimensions

The wave functions of the one-dimensional harmonic oscillator have already been derived in Section 2.6 from a spectral decomposition of the time evolution amplitude. This was possible with the help of Mehler's formula. In D dimensions, the fixed-energy amplitude is the best starting point for determining the wave functions. We take the radial propagator (8.141) obtained from the angular momentum decomposition (8.140) or, for the sake of greater generality, the radial amplitude (8.143) with an additional centrifugal barrier, continue it to imaginary time $\tau = it$, and go over to its Laplace transform

$$
\begin{aligned}
(r_b|r_a)_{E,l} = -i\sqrt{M\omega/\hbar}\sqrt{M\omega r_b r_a/\hbar} \int_{\tau_a}^{\infty} d\tau_b\; e^{E(\tau_b-\tau_a)/\hbar} \frac{1}{\sin[\omega(\tau_b-\tau_a)]} \\
\times e^{-\frac{1}{\hbar}\frac{M\omega}{2} \coth[\omega(\tau_b-\tau_a)](r_b^2+r_a^2)} I_\mu\left(\frac{M\omega r_b r_a}{\hbar\sinh[\omega(\tau_b-\tau_a)]}\right).
\end{aligned}
\qquad (9.25)
$$

To evaluate the τ-integral we make use of a standard integral formula for Bessel functions[4]

$$
\begin{aligned}
\int_0^{\infty} dx\, [\coth(x/2)]^{2\nu}\, e^{-\beta\cosh x} J_\mu(\alpha\sinh x) \\
= \frac{\Gamma((1+\mu)/2-\nu)}{\alpha\Gamma(\mu+1)} W_{\nu,\mu/2}\left(\sqrt{\alpha^2+\beta^2}+\beta\right) M_{-\nu,\mu/2}\left(\sqrt{\alpha^2+\beta^2}-\beta\right),
\end{aligned}
\qquad (9.26)
$$

where $W_{\nu,\mu/2}(z)$, $M_{-\nu,\mu/2}(z)$ are the Whittaker functions. The formula is valid for

$$\operatorname{Re}\beta > |\operatorname{Re}\alpha|, \qquad \operatorname{Re}(\mu/2-\nu) > -\frac{1}{2}.$$

---

[3]Compare with Eqs. (3.112) and (3.139) in J.D. Jackson, *Classical Electrodynamics*, John Wiley & Sons, New York, 1975.

[4]I.S. Gradshteyn and I.M. Ryzhik, *op. cit.*, Formula 6.669.1.

By a change of variables
$$\sqrt{\alpha^2 + \beta^2} \pm \beta = t\alpha_{b,a},$$

and
$$\sinh x = (\sinh y)^{-1}, \quad \cosh x = \coth y, \quad \coth(x/2) = e^y, \quad \coth x = \cosh y,$$

with
$$dx = dy/\sinh y,$$

Eq. (9.26) goes over into
$$\int_0^\infty \frac{dy}{\sinh y} e^{2\nu y} \exp\left[-\tfrac{1}{2}t(\alpha_b - \alpha_a)\coth y\right] J_\mu\left(t\frac{\sqrt{\alpha_b \alpha_a}}{\sinh y}\right)$$
$$= \frac{\Gamma\left((1+\mu)/2 - \nu\right)}{t\sqrt{\alpha_b \alpha_a}\,\Gamma(\mu+1)} W_{\nu,\mu/2}(t\alpha_b) M_{-\nu,\mu/2}(-t\alpha_a). \tag{9.27}$$

Using the identity
$$M_{-\nu,\mu/2}(z) \equiv e^{-i(\mu+1)\pi/2} M_{\nu,\mu/2}(-z) \tag{9.28}$$

and changing the sign of α_a in (9.27), this can be turned into
$$\int_0^\infty \frac{dy}{\sinh y} e^{2\nu y} \exp\left[-\tfrac{1}{2}t(\alpha_b + \alpha_a)\coth y\right] I_\mu\left(\frac{t\sqrt{\alpha_b \alpha_a}}{\sinh y}\right)$$
$$= \frac{\Gamma\left((1+\mu)/2 - \nu\right)}{t\sqrt{\alpha_b \alpha_a}\,\Gamma(\mu+1)} W_{\nu,\mu/2}(t\alpha_b) M_{\nu,\mu/2}(t\alpha_a), \tag{9.29}$$

with the range of validity
$$\alpha_b > \alpha_a > 0, \quad \mathrm{Re}\left[(1+\mu)/2 - \nu\right] > 0,$$
$$\mathrm{Re}\,t > 0, \quad |\arg t| < \pi. \tag{9.30}$$

Setting
$$y = \omega(t_b - t_a), \quad \alpha_b = \frac{M}{\hbar}\omega r_b^2, \quad \alpha_a = \frac{M\omega}{\hbar} r_a^2, \quad \nu = E/2\omega\hbar \tag{9.31}$$

in (9.29) brings the radial amplitude (9.25) to the form (valid for $r_b > r_a$)
$$(r_b|r_a)_{E,l} = -i\frac{1}{\omega}\frac{1}{\sqrt{r_b r_a}}\frac{\Gamma((1+\mu)/2 - \nu)}{\Gamma(\mu+1)} W_{\nu,\mu/2}\left(\frac{M\omega}{\hbar} r_b^2\right) M_{\nu,\mu/2}\left(\frac{M\omega}{\hbar} r_a^2\right). \tag{9.32}$$

The Gamma function has poles at
$$\nu = \nu_r \equiv (1+\mu)/2 + n_r \tag{9.33}$$

for integer values of the so-called radial quantum number of the system $n_r = 0, 1, 2, \ldots$. The poles have the form
$$\Gamma\left((1+\mu)/2 - \nu\right) \overset{\nu \sim \nu_r}{\sim} -\frac{(-1)^{n_r}}{n_r!}\frac{1}{\nu - \nu_r}. \tag{9.34}$$

Inserting here the particular value of the parameter μ for the D-dimensional oscillator which is $\mu = D/2 + l - 1$, and remembering that $\nu = E/2\omega\hbar$, we find the energy spectrum

$$E = \hbar\omega \left(2n_r + l + D/2\right).$$ (9.35)

The principal quantum number is defined by

$$n \equiv 2n_r + l$$ (9.36)

and the energy depends on it as follows:

$$E_n = \hbar\omega(n + D/2).$$ (9.37)

For a fixed principal quantum number $n = 2n_r + l$, the angular momentum runs through $l = 0, 2, \ldots, n$ for even, and $l = 1, 3, \ldots, n$ for odd n. There are $d_n = (n + D - 1)!/(D - 1)!n!$ degenerate levels. From the residues $1/(\nu - \nu_r) \sim 2\hbar\omega/(E - E_n)$, we extract the product of radial wave functions at given n_r, l:

$$R_{n_r l}(r_b)\, R_{n_r l}(r_a) = \frac{1}{\sqrt{r_b r_a}} \frac{2(-1)^{n_r}}{\Gamma(\mu + 1)n_r!}$$ (9.38)
$$\times W_{(1+\mu)/2+n_r, \frac{\mu}{2}}(M\omega r_b{}^2/\hbar) M_{(1+\mu)/2+n_r, \frac{\mu}{2}}(M\omega r_a{}^2/\hbar).$$

It is now convenient to express the Whittaker functions in terms of the confluent hypergeometric or Kummer functions:[5]

$$W_{(1+\mu)/2+n_r, \frac{\mu}{2}}(z) = e^{-z/2} z^{(1+\mu)/2} U(-n_r, 1 + \mu, z),$$ (9.39)
$$M_{(1+\mu)/2+n_r, \frac{\mu}{2}}(z) = e^{-z/2} z^{(1+\mu)/2} M(-n_r, 1 + \mu, z).$$ (9.40)

The latter equation follows from the relation

$$M_{-(1+\mu)/2-n_r, \frac{\mu}{2}}(z) = e^{-z/2} z^{(1+\mu)/2} M(1 + \mu + n_r, 1 + \mu, z),$$ (9.41)

after replacing $n_r \to -n_r - \mu - 1$.

For completeness, we also mention the identity

$$M(a, b, z) = e^z M(b - a, b, -z),$$ (9.42)

so that

$$M(1 + \mu + n_r, 1 + \mu, z) = e^z M(-n_r, 1 + \mu, -z).$$ (9.43)

This permits us to rewrite (9.41) as

$$M_{-(1+\mu)/2-n_r, \frac{\mu}{2}}(z) = e^{z/2} z^{(1+\mu)/2} M(-n_r, 1 + \mu, -z),$$ (9.44)

which turns into (9.40) by using (9.28) and appropriately changing the indices.

---

[5] I.S. Gradshteyn and I.M. Ryzhik, *op. cit.*, Formula 9.220.2.

The Kummer function $M(a, b, z)$ has the power series

$$M(a, b, z) \equiv {}_1F_1(a; b; z) = 1 + \frac{a}{b}z + \frac{a(a+1)}{b(b+1)} \frac{z}{z!} + \dots, \tag{9.45}$$

showing that $M_{(1+\mu)/2+n_r, \frac{\mu}{2}}(M\omega r_a^2/2\hbar)$ is an exponential $e^{-M\omega r_a^2/\hbar}$ times a polynomial in r_a of order $2n_r$. A similar expression is obtained for the other factor $W_{(1+\mu)/2+n_r, \frac{\mu}{2}}(M\omega r_b^2/\hbar)$ of Eq. (9.39). Indeed, the Kummer function $U(a, b, z)$ is related to $M(a, b, z)$ by[6]

$$U(a, b, z) = \frac{\pi}{\sin \pi b} \left[\frac{M(a, b, z)}{\Gamma(1+a-b)\Gamma(b)} - z^{1-b} \frac{M(1+a-b, 2-b, z)}{\Gamma(a)\Gamma(2-b)} \right]. \tag{9.46}$$

Since $a = -n_r$ with integer n_r and $1/\Gamma(a) = 0$, we see that only the first term in the brackets is present. Then the identity

$$\Gamma(-\mu)\Gamma(1+\mu) = \pi / \sin[\pi(1+\mu)]$$

leads to the relation

$$U(-n_r, 1+\mu, z) = \frac{\Gamma(-\mu)}{\Gamma(-n_r-\mu)} M(-n_r, 1+\mu, z), \tag{9.47}$$

which is a polynomial in z of order n_r. Thus we have the useful formula

$$W_{(1+\mu)/2+n_r, \frac{\mu}{2}}(z_b) M_{(1+\mu)/2+n_r, \frac{\mu}{2}}(z_a) = \frac{\Gamma(-\mu)}{\Gamma(-n_r-\mu)} e^{-(z_b+z_a)/2}$$
$$\times (z_b z_a)^{(1+\mu)/2} M(-n_r, 1+\mu, z_b) M(-n_r, 1+\mu, z_a). \tag{9.48}$$

We can therefore re-express Eq. (9.38) as

$$R_{n_r l}(r_b) R_{n_r l}(r_a) = \sqrt{\frac{M\omega}{\hbar}} \frac{2(-)^{n_r}\Gamma(-\mu)}{\Gamma(-n_r-\mu)\Gamma(1+\mu)\, n_r!}$$
$$\times e^{-M\omega(r_b^2+r_a^2)/2\hbar} \left(M\omega r_b r_a/\hbar\right)^{1/2+\mu}$$
$$\times M(-n_r, 1+\mu, M\omega r_b^2/\hbar) M(-n_r, 1+\mu, M\omega r_a^2/\hbar). \tag{9.49}$$

We now insert

$$\frac{(-)^{n_r}\Gamma(-\mu)}{\Gamma(-n_r-\mu)} = \frac{\Gamma(n_r+1+\mu)}{\Gamma(1+\mu)}, \tag{9.50}$$

setting $\mu = D/2 + l - 1$, and identify the wave functions as

$$R_{n_r l}(r) = C_{n_r l} \left(M\omega/\hbar\right)^{1/4} (M\omega r^2/\hbar)^{l/2+(D-1)/4}$$
$$\times e^{-M\omega r^2/2\hbar} M(-n_r, l+D/2, M\omega r^2/\hbar), \tag{9.51}$$

---

[6]M. Abramowitz and I. Stegun, *Handbook of Mathematical Functions*, Dover, New York, 1965, Formula 13.1.3.

with the normalization factor

$$C_{n_r l} = \frac{\sqrt{2}}{\Gamma(1+\mu)}\sqrt{\frac{(n_r+\mu)!}{n_r!}}. \tag{9.52}$$

By introducing the Laguerre polynomials [7]

$$L_n^\mu(z) \equiv \frac{(n+\mu)!}{n!\mu!}M(-n,\mu+1,z), \tag{9.53}$$

and using the integral formula[8]

$$\int_0^\infty dz e^{-z} z^\mu L_n^\mu(z) L_{n'}^\mu(z) = \delta_{nn'}\frac{(n+\mu)!}{n!}, \tag{9.54}$$

we find that the radial wave functions satisfy the orthonormality relation

$$\int_0^\infty dr R_{n_r l}(r) R_{n'_r l}(r) = \delta_{n_r n'_r}. \tag{9.55}$$

The radial imaginary-time evolution amplitude has now the spectral representation

$$(r_b \tau_b | r_a \tau_a)_l = \sum_{n_r=0}^\infty R_{n_r l}(r_b) R_{n_r l}(r_a) e^{-E_n(\tau_b-\tau_a)/\hbar}, \tag{9.56}$$

with the energies

$$E_n = \hbar\omega(n+D/2) = \hbar\omega\left(2n_r + l + D/2\right). \tag{9.57}$$

The full causal propagator is given, as in (8.91), by

$$(\mathbf{x}_b\tau_b|\mathbf{x}_a\tau_a) = \frac{1}{(r_b r_a)^{(D-1)/2}}\sum_{l=0}^\infty (r_b\tau_b|r_a\tau_a)_l \sum_m Y_{lm}(\hat{\mathbf{x}}_b)Y_{lm}^*(\hat{\mathbf{x}}_a). \tag{9.58}$$

From this, we extract the wave functions

$$\psi_{n_r l m}(\mathbf{x}) = \frac{1}{r^{(D-1)/2}}R_{n_r l}(r)Y_{lm}(\hat{\mathbf{x}}). \tag{9.59}$$

They have the threshold behavior r^l near the origin.

The one-dimensional oscillator may be viewed as a special case of these formulas. For $D=1$, the partial wave expansion amounts to a separation into even and odd wave functions. There are two "spherical harmonics",

$$Y_{e,\o}(\hat{x}) = \frac{1}{\sqrt{2}}(\Theta(x) \pm \Theta(-x)), \tag{9.60}$$

---

[7]I.S. Gradshteyn and I.M. Ryzhik, op. cit., Formula 8.970 (our definition differs from that in L.D. Landau and E.M. Lifshitz, *Quantum Mechanics*, Pergamon, London, 1965, Eq. (d.13). The relation is $L_n^\mu = (-)^\mu/(n+\mu)!L_{n+\mu}^{\text{L.L.},\mu})$.

[8]ibid., Formula 7.414.3.

and the amplitude has the decomposition

$$(x_b \tau_b | x_a \tau_a) = (r_b \tau_b | r_a \tau_a)_e Y_e(\hat{x}_b) Y_e(\hat{x}_a) + (r_b \tau_b | r_a \tau_a)_\o Y_\o(\hat{x}_b) Y_\o(\hat{x}_a), \tag{9.61}$$

with the "radial" amplitudes

$$(r_b \tau_b | r_a \tau_a)_{e,\o} = (x_b \tau_b | x_a \tau_a) \pm (-x_b \tau_b | x_a \tau_a). \tag{9.62}$$

These are known from Eq. (2.175) to be

$$(r_b \tau_b | r_a \tau_a)_{e,\o} = \frac{1}{\sqrt{2\pi}} \sqrt{\frac{M\omega/\hbar}{\sinh[\omega(\tau_b - \tau_a)]}} \tag{9.63}$$

$$\times \quad \exp\left[-\frac{M\omega}{2\hbar}(r_b^2 + r_a^2)\cot\omega(\tau_b - \tau_a)\right] 2 \begin{Bmatrix} \cosh \\ \sinh \end{Bmatrix} (M\omega r_b r_a/\hbar).$$

The two cases coincide with the integrand of (9.25) for $l = 0$ and 1, respectively, since $\mu = l + D/2 - 1$ takes the values $\pm 1/2$ and

$$\sqrt{z} I_{\mp\frac{1}{2}}(z) = \frac{2}{2\pi} \begin{Bmatrix} \cosh z \, , \\ \sinh z \, . \end{Bmatrix} \tag{9.64}$$

The associated energy spectrum (9.35) is

$$E = \begin{cases} \hbar\omega(2n_r + \frac{1}{2}) & \text{even,} \\ \hbar\omega(2n_r + \frac{3}{2}) & \text{odd,} \end{cases} \tag{9.65}$$

with the radial quantum number $n_r = 0, 1, 2, \ldots$. The two cases follow the single formula

$$E = \hbar\omega(n + \tfrac{1}{2}), \tag{9.66}$$

where the principal quantum number $n = 0, 1, 2, \ldots$ is related to n_r by $n = 2n_r$ and $n = 2n_r + 1$, respectively. The radial wave functions (9.51) become

$$R_{n_r,e}(r) = (M\omega/\hbar)^{1/4} \sqrt{\frac{2\Gamma(n_r + \frac{1}{2})}{\pi n_r!}} M(-n_r, \tfrac{1}{2}, M\omega r^2/\hbar), \tag{9.67}$$

$$R_{n_r,\o}(r) = (M\omega/\hbar)^{1/4} \sqrt{\frac{2\Gamma(n_r + \frac{3}{2})}{(\pi/4)n_r!}} \sqrt{\frac{M\omega r^2}{\hbar}} M(-n_r, \tfrac{3}{2}, M\omega r^2/\hbar). \tag{9.68}$$

The special Kummer functions appearing here are Hermite polynomials

$$M(-n, \tfrac{1}{2}, x^2) = \frac{n!}{(2n)!}(-)^n H_{2n}(x), \tag{9.69}$$

$$M(-n, \tfrac{3}{2}, x^2) = \frac{n!}{(2n+1)!}(-)^n H_{2n+1}(x)/2\sqrt{x}. \tag{9.70}$$

Using the identity

$$\Gamma(z)\Gamma(z + \tfrac{1}{2}) = (2\pi)^{1/2} 2^{-2z+1/2} \Gamma(2z), \tag{9.71}$$

we obtain in either case the radial wave functions [to be compared with the one-dimensional wave functions (2.300)]

$$R_n(r) = N_n \sqrt{2} \lambda_\omega^{-1/2} e^{-r^2/2\lambda_\omega^2} H_n(r/\lambda_\omega), \quad n = 0, 1, 2, \ldots \tag{9.72}$$

with

$$\lambda_\omega \equiv \sqrt{\frac{\hbar}{M\omega}}, \qquad N_n = \frac{1}{\sqrt{2^n n! \sqrt{\pi}}}. \tag{9.73}$$

This formula holds for both even and odd wave functions with $n_r = 2n$ and $n_r = 2n+1$, respectively. It is easy to check that they possess the correct normalization $\int_0^\infty dr R_n^2(r) = 1$. Note that the "spherical harmonics" (9.60) remove a factor $\sqrt{2}$ in (9.72), but compensate for this by extending the $x > 0$ integration to the entire x-axis by the "one-dimensional angular integration".

9.3 Free Particle from $\omega \to 0$ -Limit of Oscillator

The results obtained for the D-dimensional harmonic oscillator in the last section can be used to find the amplitude and wave functions of a free particle in D dimensions in radial coordinates. This is done by taking the limit $\omega \to 0$ at fixed energy E. In the amplitude (9.32) with $W_{\nu,\mu/2}(z), M_{\nu,\mu/2}(z)$ substituted according to (9.39), we rewrite n_r as $(E/\omega\hbar - l - 1)/2$ and go to the limit $\omega \to 0$ at a fixed energy E. Replacing $M\omega r^2/\hbar$ by $k^2 r^2/2n_r \equiv z/n_r$ (where $z = k^2 r^2/2$, and using $E = p^2/2M = \hbar^2 k^2/2M$), we apply the limiting formulas[9]

$$\lim_{n_r \to \infty} \{\Gamma(1 - n_r - b) U(-a, b, \mp z/n_r)\}$$

$$= z^{-\frac{1}{2}(b-1)} \begin{cases} 2K_{b-1}(2\sqrt{z}) \\ -i\pi e^{i\pi b} H_{b-1}^{(1)}(2\sqrt{z}) \quad (\operatorname{Im} z > 0), \end{cases} \tag{9.74}$$

$$\lim_{n_r \to \infty} M(-a, b, \mp z/n_r) / \Gamma(b) = z^{-\frac{1}{2}(b-1)} \begin{cases} I_{b-1}(2\sqrt{z}) \\ J_{b-1}(2\sqrt{z}), \end{cases} \tag{9.75}$$

and obtain the radial wave functions directly from (9.51) and (9.75):

$$R_{n_r l}(r) \xrightarrow{n_r \to \infty} C_{n_r l} (M\omega r^2/\hbar)^{(\mu/2+1/2)} \left(k^2 r^2/2\right)^{-\mu/2} \Gamma(1 + \mu) J_\mu(kr), \tag{9.76}$$

where

$$C_{n_r l} \xrightarrow{n_r \to \infty} \sqrt{2} \left(\frac{E}{2\hbar\omega}\right)^{\mu/2} \frac{1}{\Gamma(1 + \mu)}. \tag{9.77}$$

Hence

$$R_{n_r l}(r) \xrightarrow{n_r \to \infty} r^{1/2} \sqrt{M\omega/\hbar} \sqrt{2} J_\mu(kr). \tag{9.78}$$

---

[9]M. Abramowitz and I. Stegun, *op. cit.*, Formulas 13.3.1–13.3.4.

Inserting these wave functions into the radial time evolution amplitude

$$(r_b\tau_b|r_a\tau_a)_l = \sum_{n_r} R_{n_r l}(r_b) R_{n_r l}(r_b) e^{-E_n(\tau_b-\tau_a)/\hbar}, \tag{9.79}$$

and replacing the sum over n_r by the integral $\int_0^\infty dk\ \hbar k/M\omega$ [in accordance with the $n_r \to \infty$ limit of $E_{n_r} = \omega\hbar(2n_r + l + D/2) \to \hbar^2 k^2/2M$], we obtain the spectral representation of the free-particle propagator

$$(r_b\tau_b|r_a\tau_a)_\mu = \sqrt{r_b r_a} \int_0^\infty dk\ k J_\mu(kr_b) J_\mu(kr_a) e^{-\frac{\hbar k^2}{2M}(\tau_b-\tau_a)}. \tag{9.80}$$

For comparison, we derive the same results directly from the initial spectral representation (9.2) in one dimension

$$(x_b\tau_b|x_a\tau_a) = \int_{-\infty}^\infty \frac{dk}{2\pi} \exp\left[ik(x_b - x_a) - \frac{\hbar k^2}{2M}(\tau_b - \tau_a)\right]. \tag{9.81}$$

Its "angular decomposition" is a decomposition with respect to even and odd wave functions

$$\begin{aligned}
(r_b\tau_b|r_a\tau_a)_{e,\o} &= \int_0^\infty \frac{dk}{\pi}[\cos k(r_b - r_a) \pm \cos k(r_b + r_a)]e^{-\frac{\hbar k^2}{2M}(\tau_b-\tau_a)} \\
&= 2\int_0^\infty \frac{dk}{\pi} \left\{ \begin{matrix} \cos kr_b & \cos kr_a \\ \sin kr_b & \sin kr_a \end{matrix} \right\} e^{-\frac{\hbar k^2}{2M}(\tau_b-\tau_a)}.
\end{aligned} \tag{9.82}$$

In D dimensions we use the expansion (8.100) for $e^{i\mathbf{kx}}$ to calculate the amplitude in the radial form

$$\begin{aligned}
(\mathbf{x}_b\tau_b|\mathbf{x}_a\tau_a) &= \int \frac{d^D k}{(2\pi)^D} e^{i\mathbf{k}(\mathbf{x}_b-\mathbf{x}_a)} e^{-\frac{\hbar k^2}{2M}(\tau_b-\tau_a)} \\
&= \frac{1}{(2\pi)^{D/2}} \int_0^\infty dk k^{2\nu} \frac{1}{(kR)^\nu} J_\nu(kR) e^{-\frac{\hbar k^2}{2M}(\tau_b-\tau_a)},
\end{aligned} \tag{9.83}$$

with $\nu \equiv D/2 - 1$. With the help of the addition theorem for Bessel functions[10] (8.187) we rewrite

$$\frac{1}{(kR)^\nu} J_\nu(kR) = \frac{2^\nu \Gamma(\nu)}{(k^2 r_b r_a)^\nu} \sum_{l=0}^\infty (\nu + l) J_{\nu+l}(kr_b) J_{\nu+l}(kr_a) C_l^{(\nu)}(\Delta\vartheta) \tag{9.84}$$

and expand further according to

$$\frac{1}{(kR)^\nu} J_\nu(kR) = \frac{(2\pi)^{D/2}}{(k^2 r_b r_a)^\nu} \sum_{l=0}^\infty J_{\nu+l}(kr_b) J_{\nu+l}(kr_a) \sum_{\mathbf{m}} Y_{l\mathbf{m}}(\hat{\mathbf{x}}_b) Y_{l\mathbf{m}}^*(\hat{\mathbf{x}}_a), \tag{9.85}$$

to obtain the radial amplitude

$$(r_b\tau_b|r_a\tau_a)_l = \sqrt{r_b r_a} \int_0^\infty dk k J_{\nu+l}(kr_b) J_{\nu+l}(kr_a) e^{-\frac{\hbar k^2}{2M}(\tau_b-\tau_a)}, \tag{9.86}$$

---

[10] I.S. Gradshteyn and I.M. Ryzhik, op. cit., Formula 8.532.

just as in (9.80).

For $D = 1$, this reduces to (9.82) using the particular Bessel functions

$$\sqrt{z} J_{\mp 1/2}(z) = \frac{2}{\sqrt{2\pi}} \left\{ \begin{array}{c} \cos z \\ \sin z \end{array} \right\}. \tag{9.87}$$

9.4 Charged Particle in Uniform Magnetic Field

Let us also find the wave functions of a charged particle in a magnetic field. The amplitude was calculated in Section 2.18. Again we work with the imaginary-time version. Factorizing out the free motion along the direction of the magnetic field, we write

$$(\mathbf{x}_b \tau_b | \mathbf{x}_a \tau_a) = (z_b \tau_b | z_a \tau_a)(\mathbf{x}_b^\perp \tau_b | \mathbf{x}_a^\perp \tau_a), \tag{9.88}$$

with

$$(z_b \tau_b | z_a \tau_a) = \frac{1}{\sqrt{2\pi\hbar(\tau_b - \tau_a)/M}} \exp\left\{ -\frac{M}{2\hbar} \frac{(z_b - z_a)^2}{\tau_b - \tau_a} \right\}, \tag{9.89}$$

and have for the amplitude in the transverse direction

$$(\mathbf{x}_b^\perp \tau_b | \mathbf{x}_a^\perp \tau_a) = \frac{M}{2\pi\hbar(\tau_b - \tau_a)} \frac{\omega(\tau_b - \tau_a)/2}{\sinh\left[\omega(\tau_b - \tau_a)/2\right]} \exp\left[-\mathcal{A}_l^\perp/\hbar \right], \tag{9.90}$$

with the classical transverse action

$$\mathcal{A}_l^\perp = \frac{M\omega}{2} \left\{ \tfrac{1}{2} \coth\left[\omega(\tau_b - \tau_a)/2\right] (\mathbf{x}_b^\perp - \mathbf{x}_a^\perp)^2 + \mathbf{x}_a^\perp \times \mathbf{x}_b^\perp \right\}. \tag{9.91}$$

This result is valid if the vector potential is chosen as

$$\mathbf{A} = \frac{1}{2}\mathbf{B} \times \mathbf{x}. \tag{9.92}$$

In the other gauge with

$$\mathbf{A} = (0, Bx, 0), \tag{9.93}$$

there is an extra surface term, and $\mathcal{A}_{\rm cl}^\perp$ is replaced by

$$\tilde{\mathcal{A}}_{\rm cl}^\perp = \mathcal{A}_{\rm cl}^\perp + \frac{M\omega}{2}(x_b y_b - x_a y_a). \tag{9.94}$$

The calculation of the wave functions is quite different in these two gauges. In the gauge (9.93) we merely recall the expressions (2.661) and (2.663) and write down the integral representation

$$(\mathbf{x}_b^\perp \tau_b | \mathbf{x}_a^\perp \tau_a) = \int \frac{dp_y}{2\pi\hbar} e^{ip_y(y_b - y_a)/\hbar} (x_b \tau_b | x_a \tau_a)_{x_0 = p_y/M\omega}, \tag{9.95}$$

with the oscillator amplitude in the x-direction

$$(x_b \tau_b | x_a \tau_a)_{x_0} = \sqrt{\frac{M\omega}{2\pi\hbar \sinh[\omega(\tau_b - \tau_a)]}} \exp\left(-\frac{1}{\hbar}\mathcal{A}_{\rm cl}^{\rm os} \right), \tag{9.96}$$

and the classical oscillator action centered around x_0

$$
\mathcal{A}_{\text{cl}}^{\text{os}} = \frac{M\omega}{2\sinh[\omega(\tau_b - \tau_a)]} \{[(x_b - x_0)^2 + (x_a - x_0)^2]\cosh[\omega(\tau_b - \tau_b)]
$$
$$
-2(x_b - x_0)(x_a - x_0)\}. \tag{9.97}
$$

The spectral representation of the amplitude (9.96) is then

$$
(x_a\tau_b|x_a\tau_a)_{x_0} = \sum_{n=0}^{\infty} \psi_n(x_b - x_0)\psi_n(x_a - x_0)e^{-\left(n+\frac{1}{2}\right)\omega(\tau_b - \tau_a)}, \tag{9.98}
$$

where $\psi_n(x)$ are the oscillator wave functions (2.300). This leads to the spectral representation of the full amplitude (9.88)

$$
(\mathbf{x}_b\tau_b|\mathbf{x}_a\tau_a) = \int \frac{dp_z}{2\pi\hbar} \int \frac{dp_y}{2\pi\hbar} e^{ip_z(z_b - z_a)/\hbar} \tag{9.99}
$$
$$
\times \sum_{n=0}^{\infty} \psi_n(x_b - p_y/M\omega)\psi_n(x_a - p_y/M\omega)e^{-(n+\frac{1}{2})\omega(\tau_b - \tau_a)}.
$$

The combination of a sum and two integrals exhibits the complete set of wave functions of a particle in a uniform magnetic field. Note that the energy

$$
E_n = (n + \tfrac{1}{2})\hbar\omega \tag{9.100}
$$

is highly degenerate; it does not depend on p_y.

In the gauge $\mathbf{A} = \frac{1}{2}\mathbf{B} \times \mathbf{x}$, the spectral decomposition looks quite different. To derive it, the transverse Euclidean action is written down in radial coordinates [compare Eq. (2.667)] as

$$
\mathcal{A}_{\text{cl}}^{\perp} = \frac{M}{2}\left\{\frac{\omega}{2}\coth\left[\omega(\tau_b - \tau_a)/2\right]\left[r_b^2 + r_a^2 - 2r_br_a\cos(\varphi_b - \varphi_a)\right]\right.
$$
$$
\left. - i\omega r_br_a\sin(\varphi_b - \varphi_a)\right\}. \tag{9.101}
$$

This can be rearranged to

$$
\mathcal{A}_{\text{cl}}^{\perp} = \frac{M}{2}\frac{\omega}{2}\coth\left[\omega(\tau_b - \tau_a)/2\right](r_b^2 + r_a^2)
$$
$$
-\frac{M}{2}\frac{\omega}{\sinh\left[\omega(\tau_b - \tau_a)/2\right]}\cos\left[\varphi_b - \varphi_a - i\omega(\tau_b - \tau_a)/2\right]. \tag{9.102}
$$

We now expand $e^{-\mathcal{A}_{\text{cl}}^{\perp}/\hbar}$ into a series of Bessel functions using (8.5)

$$
e^{-\mathcal{A}_{\text{cl}}^{\perp}/\hbar} = \exp\left\{-\frac{M}{2\hbar}\frac{\omega}{2}\coth\left[\omega(\tau_b - \tau_a)/2\right](r_b^2 + r_a^2)\right\} \tag{9.103}
$$
$$
\times \sum_{m=-\infty}^{\infty} I_m\left(\frac{M\omega}{2\hbar}\frac{r_br_a}{\sinh\left[\omega(\tau_b - \tau_a)/2\right]}\right)e^{m\omega(\tau_b - \tau_a)/2}e^{im(\varphi_b - \varphi_a)}.
$$

The fluctuation factor is the same as before. Hence we obtain the angular decomposition of the transverse amplitude

$$(\mathbf{x}_b^\perp \tau_b | \mathbf{x}_a^\perp \tau_a) = \frac{1}{\sqrt{r_b r_a}} \sum_m (r_b \tau_b | r_a \tau_a)_m \frac{1}{2\pi} e^{im(\varphi_b - \varphi_a)}, \tag{9.104}$$

where

$$(r_b \tau_b | r_a \tau_a)_m = \sqrt{r_b r_a} \frac{M\omega}{2\hbar\eta} \frac{\eta}{\sinh\eta} \exp\left[-\frac{M}{2\hbar} \frac{\omega}{2} \coth\eta (r_b^2 + r_a^2)\right] I_m\left(\frac{M\omega r_b r_a}{2\hbar \sinh\eta}\right) e^{m\eta}, \tag{9.105}$$

with

$$\eta \equiv \omega(\tau_b - \tau_a)/2. \tag{9.106}$$

To find the spectral representation we go to the fixed-energy amplitude

$$(r_b | r_a)_{m,E} = -i \int_{\tau_a}^{\infty} d\tau_b e^{E(\tau_b - \tau_a)/\hbar} (r_b \tau_b | r_a \tau_a)_m \tag{9.107}$$

$$= -i\sqrt{r_b r_a} \frac{M}{\hbar} \int_0^\infty d\eta \; e^{2\nu\eta} \frac{1}{\sinh\eta} e^{-(M\omega/4\hbar)\coth\eta (r_b^2 + r_a^2)} I_m\left(\frac{M\omega r_b r_a}{2\hbar \sinh\eta}\right).$$

The integral is done with the help of formula (9.29) and yields

$$(r_b | r_a)_{m,E} = -i\sqrt{r_b r_a} \frac{M}{\hbar} \frac{\Gamma\left(\frac{1}{2} - \nu + \frac{|m|}{2}\right)}{(M\omega/2\hbar) r_b r_a \Gamma(|m| + 1)}$$

$$\times W_{\nu,|m|/2}\left(\frac{M\omega}{2\hbar} r_b^2\right) M_{\nu,|m|/2}\left(\frac{M\omega}{2\hbar} r_a^2\right), \tag{9.108}$$

with

$$\nu \equiv \frac{E}{\omega\hbar} + \frac{m}{2}. \tag{9.109}$$

The Gamma function $\Gamma(1/2 - \nu - |m|/2)$ has poles at

$$\nu = \nu_r \equiv n_r + \frac{1}{2} + \frac{|m|}{2} \tag{9.110}$$

of the form

$$\Gamma(1/2 - \nu - |m|/2) \approx -\frac{1}{n_r!} \frac{(-1)^{n_r}}{\nu - \nu_r} \sim -\frac{(-1)^{n_r}}{n_r!} \frac{\omega\hbar}{E - E_{n_r m}}. \tag{9.111}$$

The poles lie at the energies

$$E_{n_r m} = \hbar\omega\left(n_r + \frac{1}{2} + \frac{|m|}{2} - \frac{m}{2}\right). \tag{9.112}$$

These are the well-known Landau levels of a particle in a uniform magnetic field. The Whittaker functions at the poles are (for $m > 0$)

$$M_{-\nu,m/2}(z) = e^{z/2}z^{\frac{1+m}{2}}M(-n_r, 1+m, -z), \tag{9.113}$$

$$W_{\nu,m/2}(z) = e^{-z/2}z^{\frac{1+m}{2}}(-)^{n_r}\frac{(n_r+m)!}{m!}M(-n_r, 1+m, z). \tag{9.114}$$

The fixed-energy amplitude near the poles is therefore

$$(r_b|r_a)_{m,E} \sim \frac{i\hbar}{E - E_{n_r m}}R_{n_r m}(r_b)R_{n_r m}(r_a), \tag{9.115}$$

with the radial wave functions[11]

$$R_{n_r m}(r) = \sqrt{r}\left(\frac{M\omega}{\hbar}\right)^{1/2}\sqrt{\frac{(n_r+|m|)!}{n_r!}}\frac{1}{|m|!} \tag{9.116}$$

$$\times \exp\left(-\frac{M\omega}{4\hbar}r^2\right)\left(\frac{M\omega}{2\hbar}r^2\right)^{|m|/2}M\left(-n_r, 1+|m|, \frac{M\omega}{2\hbar}r^2\right).$$

Using Eq. (9.53), they can be expressed in terms of Laguerre polynomials $L_n^\alpha(z)$:

$$R_{n_r m} = \sqrt{r}\left(\frac{M\omega}{\hbar}\right)^{1/2}\sqrt{\frac{n_r!}{(n_r+|m|)!}}\exp\left(-\frac{M\omega}{4\hbar}r^2\right)$$

$$\times \left(\frac{M\omega}{2\hbar}r^2\right)^{|m|/2}L_{n_r}^{|m|}\left(\frac{M\omega}{2\hbar}r^2\right). \tag{9.117}$$

The integral (9.54) ensures the orthonormality of the radial wave functions

$$\int_0^\infty dr\, R_{n_r m}(r)R_{n_r' m}(r) = \delta_{n_r n_r'}. \tag{9.118}$$

A Laplace transformation of the fixed-energy amplitude (9.108) gives, via the residue theorem, the spectral representation of the radial time evolution amplitude

$$(r_b\tau_b|r_a\tau_a) = \sum_{n_r m}R_{n_r m}(r_b)R_{n_r m}(r_a)e^{-E_{n_r m}(\tau_b-\tau_a)/\hbar}, \tag{9.119}$$

with the energies (9.112). The full wave functions in the transverse subspace are, of course,

$$\psi_{n_r m}(\mathbf{x}) = \frac{1}{\sqrt{r}}R_{n_r m}(r)\frac{e^{im\varphi}}{\sqrt{2\pi}}. \tag{9.120}$$

Comparing the energies (9.112) with (9.100), we identify the principal quantum number n as

$$n \equiv n_r + \frac{|m|}{2} - \frac{m}{2}. \tag{9.121}$$

---

[11]Compare with L.D. Landau and E.M. Lifshitz, *Quantum Mechanics*, Pergamon, London, 1965, p. 427.

Note that the infinite degeneracy of the energy levels observed in (9.100) with respect to p_y is now present with respect to m. This energy does not depend on m for $m \geq 0$. The somewhat awkward m-dependence of the energy can be avoided by introducing, instead of m, another quantum number n' related to n, m by

$$m = n' - n. \tag{9.122}$$

The states are then labeled by n, n' with both n and n' taking the values $0, 1, 2, 3, \ldots$. For $n' < n$, one has $n' = n_r$ and $m = n' - n < 0$, whereas for $n' \geq n$ one has $n = n_r$ and $m = n' - n \geq 0$. There exists a natural way of generating the wave functions $\psi_{n_r m}(\mathbf{x})$ such that they appear immediately with the quantum numbers n, n'. For this we introduce the Landau radius

$$a = \sqrt{\frac{2\hbar}{M\omega}} = \sqrt{\frac{2\hbar c}{eB}} \tag{9.123}$$

as a length parameter and define the dimensionless transverse coordinates

$$z = (x + iy)/\sqrt{2}a, \quad z^* = (x - iy)/\sqrt{2}a. \tag{9.124}$$

It is then possible to prove that the $\psi_{n_r m}$'s coincide with the wave functions

$$\psi_{n,n'}(z, z^*) = N_{n,n'} e^{z^* z} \left(-\frac{1}{\sqrt{2}} \partial_{z*} \right)^n \left(-\frac{1}{\sqrt{2}} \partial_z \right)^{n'} e^{-2z^* z}. \tag{9.125}$$

The normalization constants are obtained by observing that the differential operators

$$e^{z^* z} \left(-\frac{1}{\sqrt{2}} \partial_{z*} \right) e^{-z^* z} = \frac{1}{\sqrt{2}} (-\partial_{z*} + z),$$

$$e^{z^* z} \left(-\frac{1}{\sqrt{2}} \partial_z \right) e^{-z^* z} = \frac{1}{\sqrt{2}} (-\partial_z + z^*) \tag{9.126}$$

behave algebraically like two independent creation operators

$$\hat{a}^\dagger = \frac{1}{\sqrt{2}} (-\partial_{z*} + z),$$

$$\hat{b}^\dagger = \frac{1}{\sqrt{2}} (-\partial_z + z^*), \tag{9.127}$$

whose conjugate annihilation operators are

$$\hat{a} = \frac{1}{\sqrt{2}} (\partial_z + z^*),$$

$$\hat{b} = \frac{1}{\sqrt{2}} (\partial_{z*} + z). \tag{9.128}$$

The ground state wave function annihilated by these is

$$\psi_{0,0}(z, z^*) = \langle z, z^*|0\rangle \propto e^{-z^*z}. \tag{9.129}$$

We can therefore write the complete set of wave functions as

$$\psi_{n,n'}(z, z^*) = N_{nn'}\hat{a}^{\dagger n}\hat{b}^{\dagger n'}\psi_{0,0}(z, z^*). \tag{9.130}$$

Using the fact that $\hat{a}^{\dagger*} = \hat{b}^\dagger$, $\hat{b}^{\dagger*} = \hat{a}^\dagger$, and that partial integrations turn $\hat{b}^\dagger, \hat{a}^\dagger$ into $\hat{a}, \hat{b}$, respectively, the normalization integral can be rewritten as

$$\int dx\, dy\ \psi_{n_1,n'_1}(z, z^*)\psi_{n_2,n'_2}(z, z^*)$$
$$= N_{n_1 n'_1}N_{n_2 n'_2}\int dxdy \left[(a^\dagger)^{n_1}(b^\dagger)^{n'_1}e^{-z^*z}\right]\left[(a^\dagger)^{n_2}(b^\dagger)^{n'_2}e^{-z^*z}\right]$$
$$= N_{n_1 n'_1}N_{n_2 n'_2}\int dxdy e^{-2z^*z}\left(a^{n_1}b^{n'_1}a^{\dagger n_2}b^{\dagger n'_2}\right). \tag{9.131}$$

Here the commutation relations between $\hat{a}^\dagger, \hat{b}^\dagger, \hat{a}, \hat{b}$ serve to reduce the parentheses in the last line to

$$n_1!n_2!\ \delta_{n_1 n'_1}\delta_{n_2 n'_2}. \tag{9.132}$$

The trivial integral

$$\int dxdy\ e^{-2z^*z} = \pi\int dr^2\ e^{-r^2/a^2} = \pi a^2 \tag{9.133}$$

shows that the normalization constants are

$$N_{n,n'} = \frac{1}{\sqrt{\pi a^2 n!n'!}}. \tag{9.134}$$

Let us prove the equality of $\psi_{n_r m}$ and $\psi_{n,n'}$ up to a possible overall phase. For this we first observe that z, ∂_{z^*} and z^*, ∂_z carry phase factors $e^{i\varphi}$ and $e^{-i\varphi}$, respectively, so that the two wave functions have obviously the azimuthal quantum number $m = n - n'$. Second, we make sure that the energies coincide by considering the Schrödinger equation corresponding to the action (2.641)

$$\frac{1}{2M}\left(-i\hbar\boldsymbol{\nabla} - \frac{e}{c}\mathbf{A}\right)^2\psi = E\psi. \tag{9.135}$$

In the gauge where

$$\mathbf{A} = (0, Bx, 0),$$

it reads

$$-\frac{\hbar^2}{2M}\left[\partial_x{}^2 + (\partial_y - i\frac{eB}{c}x)^2 + \partial_z{}^2\right]\psi = E\psi, \tag{9.136}$$

and the wave functions can be taken from Eq. (9.99). In the gauge where

$$\mathbf{A} = (-By/2, Bx/2, 0), \tag{9.137}$$

on the other hand, the Schrödinger equation becomes, in cylindrical coordinates,

$$\left[-\frac{\hbar^2}{2M} \left(\partial_r^2 + \frac{1}{r}\partial_r + \frac{1}{r^2}\partial_\varphi^2 + \partial_z^2 \right) - \frac{ie\hbar B}{2Mc}\partial_\varphi + \frac{e^2 B^2}{8Mc^2}r^2 \right] \psi(r,z,\varphi)$$
$$= E\psi(r,z,\varphi). \tag{9.138}$$

Employing a reduced radial coordinate $\rho = r/a$ and factorizing out a plane wave in the z-direction, $e^{ip_z z/\hbar}$, this takes the form

$$\left[\partial_\rho^2 + \frac{1}{\rho}\partial_\rho + \frac{a^2}{\hbar^2}(2ME - p_z^2) - \rho^2 - 2i\partial_\varphi - \frac{1}{\rho^2}\partial_\varphi^2 \right] \psi(r,\varphi) = 0. \tag{9.139}$$

The solutions are

$$\psi_{n_r m}(r,\varphi) \propto e^{im\varphi}e^{-\rho^2/2}\rho^{|m|/2}M\left(-n_r, |m|+\frac{1}{2}, \rho\right), \tag{9.140}$$

where the confluent hypergeometric functions $M\left(-n_r, |m|+\frac{1}{2}, \rho\right)$ are polynomials for integer values of the radial quantum number

$$n_r = n + \frac{1}{2}m - \frac{1}{2}|m| - \frac{1}{2}, \tag{9.141}$$

as in (9.116). The energy is related to the principal quantum number by

$$n + \frac{1}{2} \equiv \frac{a^2}{\hbar^2}\left(2ME - p_z^2\right). \tag{9.142}$$

Since

$$\frac{2Ma^2}{\hbar^2} = \frac{1}{\hbar\omega}, \tag{9.143}$$

the energy is

$$E = \left(n + \frac{1}{2}\right)\hbar\omega + \frac{p_z^2}{2M}. \tag{9.144}$$

We now observe that the Schrödinger equation (9.139) can be expressed in terms of the creation and annihilation operators (9.127), (9.128) as

$$4\left[-(a^\dagger a + 1/2) + \frac{1}{\hbar\omega}\left(E - \frac{p_z^2}{2M}\right) \right]\psi(z,z^*) = 0. \tag{9.145}$$

This proves that the algebraically constructed wave functions $\psi_{n,n'}$ in (9.130) coincide with the wave functions $\psi_{n_r m}$ of (9.116) and (9.140), up to an irrelevant phase. Note that the energy depends only on the number of a-quanta; it is independent of the number of b-quanta.

9.5   Dirac δ-Function Potential

For a particle in a Dirac δ-function potential, the fixed-energy amplitudes $(\mathbf{x}_b|\mathbf{x}_a)_E$ can be calculated by performing a perturbation expansion around a free-particle amplitude and summing it up exactly. For any time-independent potential $V(\mathbf{x})$, in addition to a harmonic potential $M\omega^2\mathbf{x}^2/2$, the perturbation expansion in Eq. (3.475) can be Laplace-transformed in the imaginary time via (9.3) to find

$$
\begin{aligned}
(\mathbf{x}_b|\mathbf{x}_a)_E \; = \; & (\mathbf{x}_b|\mathbf{x}_a)_{\omega,E} - \frac{i}{\hbar}\int d^D x_1 (\mathbf{x}_b|\mathbf{x}_1)_{\omega,E} V(\mathbf{x}_1)(\mathbf{x}_1|\mathbf{x}_a)_{\omega,E} \\
& + \; -\frac{1}{\hbar^2}\int d^D x_2 \int d^D x_1 (\mathbf{x}_b|\mathbf{x}_2)_{\omega,E} V(\mathbf{x}_2)(\mathbf{x}_2|\mathbf{x}_1)_{\omega,E} V(\mathbf{x}_1)(\mathbf{x}_1|\mathbf{x}_a)_{\omega,E} \\
& + \; \dots \; .
\end{aligned}
\tag{9.146}
$$

If the potential is a Dirac δ-function centered around $\mathbf{X}$,

$$
V(\mathbf{x}) = g\,\delta^{(D)}(\mathbf{x} - \mathbf{X}), \qquad g \equiv \frac{\hbar^2}{Ml^{2-D}},
\tag{9.147}
$$

this series simplifies to

$$
(\mathbf{x}_b|\mathbf{x}_a)_E = (\mathbf{x}_b|\mathbf{x}_a)_{\omega,E} - \frac{ig}{\hbar} g(\mathbf{x}_b|\mathbf{X})_{\omega,E}(\mathbf{X}|\mathbf{x}_a)_{\omega,E} - \frac{g^2}{\hbar^2}(\mathbf{x}_b|\mathbf{X})_{\omega,E}(\mathbf{X}|\mathbf{X})_{\omega,E}(\mathbf{X}|\mathbf{x}_a)_{\omega,E} + \dots ,
\tag{9.148}
$$

and can be summed up to

$$
(\mathbf{x}_b|\mathbf{x}_a)_E = (\mathbf{x}_b|\mathbf{x}_a)_{\omega,E} - i\frac{g}{\hbar}\,\frac{(\mathbf{x}_b|\mathbf{X})_{\omega,E}(\mathbf{X}|\mathbf{x}_a)_{\omega,E}}{1 + i\dfrac{g}{\hbar}(\mathbf{X}|\mathbf{X})_{\omega,E}}.
\tag{9.149}
$$

This is, incidentally, true if a δ-function potential is added to an arbitrary solvable fixed-energy amplitude, not just the harmonic one.

If the δ-function is the only potential, we use formula (9.149) with $\omega = 0$, so that $(\mathbf{x}_b|\mathbf{x}_a)_{0,E}$ reduces to the fixed-energy amplitude (9.12) of a free particle, and obtain directly

$$
\begin{aligned}
(\mathbf{x}_b|\mathbf{x}_a)_E \; = \; & -2i\frac{M}{\hbar}\frac{\kappa^{D-2}}{(2\pi)^{D/2}}\frac{K_{D/2-1}(\kappa R)}{(\kappa R)^{D/2-1}} \\
& - \frac{ig}{\hbar}\frac{i\dfrac{2M}{\hbar}\dfrac{\kappa^{D-2}}{(2\pi)^{D/2}}\dfrac{K_{D/2-1}(\kappa R_b)}{(\kappa R_b)^{D/2-1}} \times i\dfrac{2M}{\hbar}\dfrac{\kappa^{D-2}}{(2\pi)^{D/2}}\dfrac{K_{D/2-1}(\kappa R_a)}{(\kappa R_a)^{D/2-1}}}{1 - \dfrac{g}{\hbar}\dfrac{2M}{\hbar}\dfrac{\kappa^{D-2}}{\pi^{D/2}}\dfrac{K_{D/2-1}(\kappa\delta)}{(\kappa\delta)^{D/2-1}}},
\end{aligned}
\tag{9.150}
$$

where $R \equiv |\mathbf{x}_b - \mathbf{x}_a|$ and $R_{a,b} \equiv |\mathbf{x}_{a,b} - \mathbf{X}|$, and δ is an infinitesimal distance regularizing a possible singularity at zero-distance. In $D = 1$ dimension, this reduces to

$$
(x_b|x_a)_E = -i\frac{M}{\hbar}\frac{1}{\kappa}e^{-\kappa R} + i\frac{M}{\hbar\kappa}e^{\kappa(R_b+R_a)}\frac{1}{l\kappa+1},
\tag{9.151}
$$

For an attractive potential with $l < 0$, the second term can be written as

$$- i\frac{1}{\kappa l^2} \frac{\hbar}{E + \hbar^2/2Ml^2} e^{\kappa(R_b + R_a)}, \tag{9.152}$$

exhibiting a pole at the bound-state energy $E_B = -\hbar/2Ml^2$. In its neighborhood, the pole contribution reads

$$\frac{2}{l} e^{-(R_b + R_a)/l} \frac{i\hbar}{E + \hbar^2/2Ml^2}. \tag{9.153}$$

This has precisely the spectral form (1.321) with the normalized bound-state wave function

$$\psi_B(x) = \sqrt{\frac{2}{l}} e^{-|x - X|/l}. \tag{9.154}$$

In $D = 3$ dimensions, the amplitude (9.150) becomes

$$(\mathbf{x}_b|\mathbf{x}_a)_E = -i\frac{M}{\hbar}\frac{1}{2\pi R} e^{-\kappa R} + i\frac{M}{\hbar}\frac{e^{\kappa R_b}}{2\pi R_b}\frac{e^{\kappa R_a}}{2\pi R_a}\frac{1}{1/l + e^{-\kappa\delta}/2\pi\delta}. \tag{9.155}$$

In the limit $\delta \to 0$, the denominator requires renormalization. We introduce a renormalized coupling length scale

$$\frac{1}{l_r} \equiv 1 + \frac{1}{2\pi\delta}, \tag{9.156}$$

and rewrite the last factor in (9.155) as

$$\frac{1}{1/l_r - \kappa/2\pi}. \tag{9.157}$$

For $l_r < 0$, this has a pole at the bound-state energy $E_B = -4\pi^2\hbar^2/2Ml_R^2$ of the form

$$- l_r E_B \frac{1}{E - E_B}. \tag{9.158}$$

The total pole term in (9.155) can therefore be written as

$$\psi_B(\mathbf{x}_b)\psi_B^*(\mathbf{x}_a)\frac{i\hbar}{E - E_B}, \tag{9.159}$$

with $\kappa_B = \sqrt{2ME_B}/\hbar = 2\pi/l_r$ and the normalized bound-state wave functions

$$\psi_B(\mathbf{x}) = \left(\frac{\kappa_B^2}{4\pi^2}\right)^{1/4} \frac{e^{-\kappa_B|\mathbf{x} - \mathbf{X}|}}{r}. \tag{9.160}$$

In $D = 2$ dimensions, the situation is more subtle. It is useful to consider the amplitude (9.150) in $D = 2 + \epsilon$ dimensions where one has

$$
\begin{aligned}
(\mathbf{x}_b|\mathbf{x}_a)_E &= -i\frac{M}{\hbar}\frac{1}{\pi}\frac{K_{\epsilon/2}(\kappa R)}{(2\pi\kappa R)^{\epsilon/2}} \\
&+ i\frac{M^2}{\hbar^2\pi^2}\frac{1}{(2\pi\kappa R_b)^{\epsilon/2}}\frac{1}{(2\pi\kappa R_a)^{\epsilon/2}}\frac{K_{\epsilon/2}(\kappa R_b)K_{\epsilon/2}(\kappa R_a)}{\dfrac{\hbar}{g}+\dfrac{M}{\hbar\pi}\dfrac{1}{(2\pi\kappa\delta)^{\epsilon/2}}K_{\epsilon/2}(\kappa\epsilon)}.
\end{aligned}
\tag{9.161}
$$

Inserting here $K_{\epsilon/2}(\kappa\delta) \approx (1/2)\Gamma(\epsilon/2)(\kappa\delta/2)^{-\epsilon/2}$, the denominator becomes

$$
\frac{\hbar}{g} + \frac{M}{2\hbar\pi}\frac{\Gamma(\epsilon/2)}{(\pi\kappa\delta)^{\epsilon/2}} \approx \frac{\hbar}{g} + \frac{M}{2\hbar\pi}\frac{2}{\epsilon}\left[1 - \frac{\epsilon}{2}\log(\pi\kappa\delta)\right].
\tag{9.162}
$$

Here we introduce a renormalized coupling constant

$$
\frac{1}{g_r} = \frac{1}{g} + \frac{M}{\hbar^2}\frac{1}{\epsilon},
\tag{9.163}
$$

and rewrite the right-hand side as

$$
\frac{\hbar}{g_r} - \frac{M}{2\hbar\pi}\log\pi\kappa\delta.
\tag{9.164}
$$

This has a pole at

$$
\kappa_B = \frac{1}{\pi\delta}e^{2\hbar^2\pi/Mg_r},
\tag{9.165}
$$

indicating a bound-state pole of energy $E_B = -\hbar^2\kappa_B^2/2M$.

We can now go to the limit of $D = 2$ dimensions and find that the pole term in (9.161) has the form

$$
\psi_B(\mathbf{x}_b)\psi_B^*(\mathbf{x}_a)\frac{i\hbar}{E - E_B},
\tag{9.166}
$$

with the normalized bound-state wave function

$$
\psi_B(\mathbf{x}) = \frac{\kappa_B}{\sqrt{\pi}}K_0(\kappa_B|\mathbf{x} - \mathbf{X}|).
\tag{9.167}
$$

Notes and References

The wave functions derived in this chapter from the time evolution amplitude should be compared with those given in standard textbooks on quantum mechanics, such as
L.D. Landau and E.M. Lifshitz, *Quantum Mechanics*, Pergamon, London, 1965. The charged particle in a magnetic field is treated in §111.
The δ-function potential was studied via path integrals by
C. Grosche, Phys. Rev. Letters, **71**, 1 (1993).

10

Spaces with Curvature and Torsion

The path integral of a free particle in spherical coordinates has taught us an important lesson: In a Euclidean space, we were able to obtain the correct time-sliced amplitude in curvilinear coordinates by setting up the sliced action in Cartesian coordinates x^i and transforming them to the spherical coordinates $q^\mu = (r, \theta, \phi)$. It was crucial to do the transformation at the level of the *finite* coordinate differences, $\Delta x^i \to \Delta q^\mu$. This produced higher-order terms in the differences Δq^μ which had to be included up to the order $(\Delta q)^4/\epsilon$. They all contributed to the relevant order ϵ. It is obvious that as long as the space is Euclidean, the same procedure can be used to find the path integral in an arbitrary curvilinear coordinate system q^μ, if we ignore subtleties arising near coordinate singularities which are present in centrifugal barriers, angular barriers, or Coulomb potentials. For these, a special treatment will be developed in Chapters 12–14.

We are now going to develop an entirely nontrivial but quite natural extension of this procedure and define a path integral in an arbitrary metric-affine space with curvature and torsion. It must be emphasized that the quantum theory in such spaces is not uniquely defined by the formalism developed so far. The reason is that also the original Schrödinger theory which was used in Chapter 2 to justify the introduction of path integrals is not uniquely defined in such spaces. In classical physics, the equivalence principle postulated by Einstein is a powerful tool for deducing equations of motion in curved space from those in flat space. At the quantum level, this principle becomes insufficient since it does not forbid the appearance of arbitrary coordinate-independent terms proportional to Planck's quantum $\hbar^2$ and the scalar curvature R to appear in the Schödinger equation. We shall set up a simple extension of Einstein's equivalence principle which will allow us to carry quantum theories from flat to curved spaces which are, moreover, permitted to carry certain classes of torsion.

In such spaces, not only the time-sliced action but also the measure of path integration requires a special treatment. To be valid in general it will be necessary to find construction rules for the time evolution amplitude which do not involve the crutch of Cartesian coordinates. The final formula will be purely intrinsic to the general metric-affine space [1].

A crucial test of the validity of the resulting path integral formula will come from applications to systems whose correct operator quantum mechanics is known on the basis of symmetries and group commutation rules rather than canonical commutation rules. In contrast to earlier approaches, our path integral formula will always yield the same quantum mechanics as operator quantum mechanics quantized via group commutation rules.

Our formula can, of course, also be used for an alternative approach to the path integrals solved before in Chapter 8, where a Euclidean space was parametrized in terms of curvilinear coordinates. There it gives rise to a more satisfactory treatment than before, since it involves only the intrinsic variables of the coordinate systems.

10.1 Einstein's Equivalence Principle

To motivate the present study we invoke *Einstein's equivalence principle*, according to which gravitational forces upon a spinless mass point are indistinguishable from those felt in an accelerating local reference.[1] They are independent of the atomic composition of the particle and strictly proportional to the value of the mass, the same mass that appears in the relation between force and acceleration, in Newton's first law. The strict equality between the two masses, gravitational and inertial, is fundamental to Einstein's equivalence principle. Experimentally, the equality holds to an extremely high degree of accuracy. Any possible small deviation can presently be attributed to extra non-gravitational forces. Einstein realized that as a consequence of this equality, all spinless point particles move in a gravitational field along the same orbits which are independent of their composition and mass. This *universality* of orbital motion permits the gravitational field to be attributed to geometric properties of spacetime.

In Newton's theory of gravity, the gravitational forces between mass points are inversely proportional to their distances in a Euclidean space. In Einstein's geometric theory the forces are explained entirely by a curvature of spacetime. In general the spacetime of general relativity may also carry another geometric property, called torsion. Torsion is supposed to be generated by the spin densities of material bodies. Quantitatively, this may have only extremely small effects, too small to be detected by present-day experiments. But this is only due to the small intrinsic spin of ordinary gravitational matter. In exceptional states of matter such as polarized neutron stars or black holes, torsion can become relevant. It is now generally accepted that spacetime should carry a nonvanishing torsion at least locally at those points which are occupied by spinning elementary particles [55]. This follows from rather general symmetry considerations. The precise equations of motion for the torsion field, on the other hand, are still a matter of speculation. Thus it is an open question whether or not the torsion field is able to propagate into the empty space away from spinning matter.

---

[1]Quotation from his original paper *Über das Relativitätsprinzip und die aus demselben gezogenen Folgerungen*, Jahrbuch der Relativität und Elektonik *4*, 411 (1907): "Wir ... wollen daher im folgenden die völlige physikalische Gleichwertigkeit von Gravitationsfeld und entsprechender Beschleunigung des Bezugssystems annehmen".

Even though the effects of torsion are small we shall keep the discussion as general as possible and study the motion of a particle in a metric-affine space with both curvature and torsion. To prepare the grounds let us first recapitulate a few basic facts about classical orbits of particles in a gravitational field. For simplicity, we assume here only the three-dimensional space to have a nontrivial geometry.[2] Then there is a natural choice of a time variable t which is conveniently used to parametrize the particle orbits.

Starting from the free-particle action we shall then introduce a path integral for the time evolution amplitude in any metric-affine space which determines the quantum mechanics via the quantum fluctuations of the particle orbits.

10.2 Classical Motion of Mass Point in General Metric-Affine Space

On the basis of the equivalence principle, Einstein formulated the rules for finding the classical laws of motion in a gravitational field as a consequence of the geometry of spacetime. Let us recapitulate his reasoning adapted to the present problem of a nonrelativistic point particle in a non-Euclidean geometry.

10.2.1 Equations of Motion

Consider first the action of the particle along the orbit $\mathbf{x}(t)$ in a flat space parametrized with rectilinear, Cartesian coordinates:

$$\mathcal{A} = \int_{t_a}^{t_b} dt \frac{M}{2} (\dot{x}^i)^2, \quad i = 1, 2, 3. \tag{10.1}$$

It is transformed to curvilinear coordinates q^μ, $\mu = 1, 2, 3$, via some functions

$$x^i = x^i(q), \tag{10.2}$$

leading to

$$\mathcal{A} = \int_{t_a}^{t_b} dt \frac{M}{2} g_{\mu\nu}(q) \dot{q}^\mu \dot{q}^\nu, \tag{10.3}$$

where

$$g_{\mu\nu}(q) = \partial_\mu x^i(q) \partial_\nu x^i(q) \tag{10.4}$$

is the *induced metric* for the curvilinear coordinates. Repeated indices are understood to be summed over, as usual.

The length of the orbit in the flat space is given by

$$l = \int_{t_a}^{t_b} dt \sqrt{g_{\mu\nu}(q) \dot{q}^\mu \dot{q}^\nu}. \tag{10.5}$$

---

[2]The generalization to non-Euclidean spacetime will be obvious after the development in Chapter 19.

Both the action (10.3) and the length (10.5) are invariant under arbitrary *reparametrizations of space* $q^\mu \to q'^\mu$.

Einstein's equivalence principle amounts to the postulate that the transformed action (10.3) describes directly the motion of the particle in the presence of a gravitational field caused by other masses. The forces caused by the field are all a result of the geometric properties of the metric tensor.

The equations of motion are obtained by extremizing the action in Eq. (10.3) with the result

$$\partial_t(g_{\mu\nu}\dot{q}^\nu) - \frac{1}{2}\partial_\mu g_{\lambda\nu}\dot{q}^\lambda\dot{q}^\nu = g_{\mu\nu}\ddot{q}^\nu + \bar{\Gamma}_{\lambda\nu\mu}\dot{q}^\lambda\dot{q}^\nu = 0. \qquad (10.6)$$

Here

$$\bar{\Gamma}_{\lambda\nu\mu} \equiv \frac{1}{2}(\partial_\lambda g_{\nu\mu} + \partial_\nu g_{\lambda\mu} - \partial_\mu g_{\lambda\nu}) \qquad (10.7)$$

is the Riemann connection or Christoffel symbol of the first kind [recall (1.70)]. With the help of the Christoffel symbol of the second kind [recall (1.71)]

$$\bar{\Gamma}_{\lambda\nu}{}^\mu \equiv g^{\mu\sigma}\bar{\Gamma}_{\lambda\nu\sigma}, \qquad (10.8)$$

we can write

$$\ddot{q}^\mu + \bar{\Gamma}_{\lambda\nu}{}^\mu\dot{q}^\lambda\dot{q}^\nu = 0. \qquad (10.9)$$

The solutions of these equations are the classical orbits. They coincide with the extrema of the length of a line l in (10.5). Thus, in a curved space, classical orbits are the shortest lines, the *geodesics* [recall (1.72)].

The same equations can also be obtained directly by transforming the equation of motion from

$$\ddot{x}^i = 0 \qquad (10.10)$$

to curvilinear coordinates q^μ, which gives

$$\ddot{x}^i = \frac{\partial x^i}{\partial q^\mu}\ddot{q}^\mu + \frac{\partial^2 x^i}{\partial q^\lambda \partial q^\nu}\dot{q}^\lambda\dot{q}^\nu = 0. \qquad (10.11)$$

At this place it is again useful to employ the quantities defined in Eq. (1.357), the basis triads and their reciprocals

$$e^i{}_\mu(q) \equiv \frac{\partial x^i}{\partial q^\mu}, \quad e_i{}^\mu(q) \equiv \frac{\partial q^\mu}{\partial x^i}, \qquad (10.12)$$

which satisfy the orthogonality and completeness relations (1.358):

$$e_i{}^\mu e^i{}_\nu = \delta^\mu{}_\nu, \quad e_i{}^\mu e^j{}_\mu = \delta_i{}^j. \qquad (10.13)$$

The induced metric can then be written as

$$g_{\mu\nu}(q) = e^i{}_\mu(q)e^i{}_\nu(q). \tag{10.14}$$

Labeling Cartesian coordinates, upper and lower indices i are the same. The indices μ, ν of the curvilinear coordinates, on the other hand, can be lowered only by contraction with the metric $g_{\mu\nu}$ or raised with the inverse metric $g^{\mu\nu} \equiv (g_{\mu\nu})^{-1}$. Using the basis triads, Eq. (10.11) can be rewritten as

$$\frac{d}{dt}(e^i{}_\mu \dot{q}^\mu) = e^i{}_\mu \ddot{q}^\mu + \partial_\nu e^i{}_\mu \dot{q}^\mu \dot{q}^\nu = 0,$$

or as

$$\ddot{q}^\mu + e_i{}^\mu \partial_\lambda e^i{}_\kappa \dot{q}^\kappa \dot{q}^\lambda = 0. \tag{10.15}$$

The quantity in front of $\dot{q}^\kappa \dot{q}^\lambda$ is called the *affine connection*:

$$\Gamma_{\lambda\kappa}{}^\mu = e_i{}^\mu \partial_\lambda e^i{}_\kappa. \tag{10.16}$$

Due to (10.13), it can also be written as [compare (1.366)]

$$\Gamma_{\lambda\kappa}{}^\mu = -e^i{}_\kappa \partial_\lambda e_i{}^\mu. \tag{10.17}$$

Thus we arrive at the transformed flat-space equation of motion

$$\ddot{q}^\mu + \Gamma_{\kappa\lambda}{}^\mu \dot{q}^\kappa \dot{q}^\lambda = 0. \tag{10.18}$$

The solutions of this equation are called the *straightest lines* or *autoparallels*.

If the coordinate transformation functions $x^i(q)$ are smooth and single-valued, their derivatives commute as required by Schwarz's integrability condition

$$(\partial_\lambda \partial_\kappa - \partial_\kappa \partial_\lambda)x^i(q) = 0. \tag{10.19}$$

Then the triads satisfy the identity

$$\partial_\lambda e^i_\kappa - \partial_\kappa e^i_\lambda = 0, \tag{10.20}$$

implying that the connection $\Gamma_{\lambda\kappa}{}^\mu$ is symmetric in the lower indices. In fact, it coincides with the Riemann connection, the Christoffel symbol $\bar{\Gamma}_{\lambda\kappa}{}^\mu$. This follows immediately after inserting $g_{\mu\nu}(q) = e^i{}_\mu(q)e^i{}_\nu(q)$ into (10.7) and working out all derivatives using (10.20). Thus, for a space with curvilinear coordinates q^μ which can be reached by an integrable coordinate transformation from a flat space, the autoparallels coincide with the geodesics.

10.2.2 Nonholonomic Mapping to Spaces with Torsion

It is possible to map the x-space locally into a q-space with torsion via an infinitesimal transformation

$$dx^i = e^i{}_\mu(q)dq^\mu. \tag{10.21}$$

We merely have to assume that the coefficient functions $e^i{}_\mu(q)$ do not satisfy the property (10.20) which follows from the Schwarz integrability condition (10.19):

$$\partial_\lambda e^i{}_\kappa(q) - \partial_\kappa e^i{}_\lambda(q) \neq 0, \tag{10.22}$$

implying that second derivatives in front of $x^i(q)$ do not commute as in Eq. (10.19):

$$(\partial_\lambda\partial_\kappa - \partial_\kappa\partial_\lambda)x^i(q) \neq 0. \tag{10.23}$$

In this case we shall call the differential mapping (10.21) *nonholonomic*, in analogy with the nomenclature for nonintegrable constraints in classical mechanics. The property (10.23) implies that $x^i(q)$ is a multivalued function $x^i(q)$, of which we shall give typical examples below in Eqs. (10.44) and (10.55).

Educated readers in mathematics have been wondering whether such nonholonomic coordinate transformations make any sense. They will understand this concept better if they compare the situation with the quite similar but much simpler creation of magnetic field in a field-free space by *nonholonomic gauge transformations*. More details are explained in Appendix 10A.

From Eq. (10.22) we see that the image space of a nonholonomic mapping carries torsion. The connection $\Gamma_{\lambda\kappa}{}^\mu = e_i{}^\mu e^i{}_{\kappa,\lambda}$ has a nonzero antisymmetric part, called the *torsion tensor*:[3]

$$S_{\lambda\kappa}{}^\mu = \frac{1}{2}(\Gamma_{\lambda\kappa}{}^\mu - \Gamma_{\kappa\lambda}{}^\mu) = \frac{1}{2}e_i{}^\mu\left(\partial_\lambda e^i{}_\kappa - \partial_\kappa e^i{}_\lambda\right). \tag{10.24}$$

In contrast to $\Gamma_{\lambda\kappa}{}^\mu$, the antisymmetric part $S_{\lambda\kappa}{}^\mu$ is a proper tensor under general coordinate transformations. The contracted tensor

$$S_\mu \equiv S_{\mu\lambda}{}^\lambda \tag{10.25}$$

transforms like a vector, whereas the contracted connection $\Gamma_\mu \equiv \Gamma_{\mu\nu}{}^\nu$ does not. Even though $\Gamma_{\mu\nu}{}^\lambda$ is not a tensor, we shall freely lower and raise its indices using contractions with the metric or the inverse metric, respectively: $\Gamma^\mu{}_\nu{}^\lambda \equiv g^{\mu\kappa}\Gamma_{\kappa\nu}{}^\lambda$, $\Gamma_\mu{}^{\nu\lambda} \equiv g^{\nu\kappa}\Gamma_{\mu\kappa}{}^\lambda$, $\Gamma_{\mu\nu\lambda} \equiv g_{\lambda\kappa}\Gamma_{\mu\nu}{}^\kappa$. The same thing will be done with $\bar\Gamma_{\mu\nu}{}^\lambda$.

In the presence of torsion, the affine connection (10.16) is no longer equal to the Christoffel symbol. In fact, by rewriting $\Gamma_{\mu\nu\lambda} = e_{i\lambda}\partial_\mu e^i{}_\nu$ trivially as

$$\Gamma_{\mu\nu\lambda} = \frac{1}{2}\left\{\left[e_{i\lambda}\partial_\mu e^i{}_\nu + \partial_\mu e_{i\lambda}e^i{}_\nu\right] + \left[e_{i\mu}\partial_\nu e^i{}_\lambda + \partial_\nu e_{i\mu}e^i{}_\lambda\right] - \left[e_{i\mu}\partial_\lambda e^i{}_\nu + \partial_\lambda e_{i\mu}e^i{}_\nu\right]\right\}$$
$$+ \frac{1}{2}\left\{\left[e_{i\lambda}\partial_\mu e^i{}_\nu - e_{i\lambda}\partial_\nu e^i{}_\mu\right] - \left[e_{i\mu}\partial_\nu e^i{}_\lambda - e_{i\mu}\partial_\lambda e^i{}_\nu\right] + \left[e_{i\nu}\partial_\lambda e^i{}_\mu - e_{i\nu}\partial_\mu e^i{}_\lambda\right]\right\} \tag{10.26}$$

---

[3]Our notation for the geometric quantities in spaces with curvature and torsion is the same as in J.A. Schouten, *Ricci Calculus*, Springer, Berlin, 1954.

and using $e^i{}_\mu(q)e^i{}_\nu(q) = g_{\mu\nu}(q)$, we find the decomposition

$$\Gamma_{\mu\nu}{}^\lambda = \bar{\Gamma}_{\mu\nu}{}^\lambda + K_{\mu\nu}{}^\lambda, \tag{10.27}$$

where the combination of torsion tensors

$$K_{\mu\nu\lambda} \equiv S_{\mu\nu\lambda} - S_{\nu\lambda\mu} + S_{\lambda\mu\nu} \tag{10.28}$$

is called the *contortion tensor*. It is antisymmetric in the last two indices so that

$$\Gamma_{\mu\nu}{}^\nu = \bar{\Gamma}_{\mu\nu}{}^\nu. \tag{10.29}$$

In the presence of torsion, the shortest and straightest lines are no longer equal. Since the two types of lines play geometrically an equally favored role, the question arises as to which of them describes the correct classical particle orbits. Intuitively, we expect the straightest lines to be the correct trajectories since massive particles possess *inertia* which tend to minimize their deviations from a straight line in spacetime. It is hard to conceive how a particle should know which path to take at each instant in time in order to minimize the path length to a distant point. This would contradict the principle of locality which pervades all laws of physics. Only in a spacetime without torsion is this possible, since there the shortest lines happen to coincide with straightest ones for purely mathematical reasons. In Subsection 10.2.3, the straightest lines will be derived from an action principle.

In Einstein's theory of gravitation, matter produces curvature in four-dimensional Minkowski spacetime, thereby explaining the universal nature of gravitational forces. The flat spacetime metric is

$$\eta_{ab} = \begin{pmatrix} 1 & & & \\ & -1 & & \\ & & -1 & \\ & & & -1 \end{pmatrix}_{ab}, \qquad a, b = 0,\ 1,\ 2,\ 3. \tag{10.30}$$

The *Riemann-Cartan curvature tensor* is defined as the covariant curl of the affine connection:

$$R_{\mu\nu\lambda}{}^\kappa = \partial_\mu \Gamma_{\nu\lambda}{}^\kappa - \partial_\nu \Gamma_{\mu\lambda}{}^\kappa - [\Gamma_\mu, \Gamma_\nu]_\lambda{}^\kappa, \qquad \mu, \nu, \ldots = 0,\ 1,\ 2,\ 3. \tag{10.31}$$

The last term is written in a matrix notation for the connection, in which the tensor components $\Gamma_{\mu\lambda}{}^\kappa$ are viewed as matrix elements $(\Gamma_\mu)_\lambda{}^\kappa$. The matrix commutator in (10.31) is then equal to

$$[\Gamma_\mu, \Gamma_\nu]_\lambda{}^\kappa \equiv (\Gamma_\mu \Gamma_\nu - \Gamma_\nu \Gamma_\mu)_\lambda{}^\kappa = \Gamma_{\mu\lambda}{}^\sigma \Gamma_{\nu\sigma}{}^\kappa - \Gamma_{\nu\lambda}{}^\sigma \Gamma_{\mu\sigma}{}^\kappa. \tag{10.32}$$

Expressing the affine connection (10.16) in (10.31) with the help of Eqs. (10.16) in terms of the four-dimensional generalization of the triads (10.12) and their reciprocals (10.12), the tetrads $e^a{}_\mu$ and their reciprocals $e_a{}^\mu$, we obtain the compact formula

$$R_{\mu\nu\lambda}{}^\kappa = e_a{}^\kappa(\partial_\mu\partial_\nu - \partial_\nu\partial_\mu)e^a{}_\lambda. \tag{10.33}$$

For the mapping (10.21), this implies that not only the coordinate transformation $x^a(q)$, but also its first derivatives fail to satisfy Schwarz's integrability condition:

$$(\partial_\mu \partial_\nu - \partial_\nu \partial_\mu)\partial_\lambda x^a(q) \neq 0. \tag{10.34}$$

Such general transformation matrices $e^a{}_\mu(q)$ will be referred to as *multivalued basis tetrads*.

A transformation for which $x^a(q)$ have commuting derivatives, while the first derivatives $\partial_\mu x^a(q) = e^i{}_\mu(q)$ do not, carries a flat-space region into a purely curved one.

Einstein's original theory of gravity assumes the absence of torsion. The space properties are completely specified by the *Riemann curvature tensor* formed from the Riemann connection (the Christoffel symbol)

$$\bar{R}_{\mu\nu\lambda}{}^\kappa = \partial_\mu \bar{\Gamma}_{\nu\lambda}{}^\kappa - \partial_\nu \bar{\Gamma}_{\mu\lambda}{}^\kappa - [\bar{\Gamma}_\mu, \bar{\Gamma}_\nu]_\lambda{}^\kappa. \tag{10.35}$$

The relation between the two curvature tensors is

$$R_{\mu\nu\lambda}{}^\kappa = \bar{R}_{\mu\nu\lambda}{}^\kappa + \bar{D}_\mu K_{\nu\lambda}{}^\kappa - \bar{D}_\nu K_{\mu\lambda}{}^\kappa - [K_\mu, K_\nu]_\lambda{}^\kappa. \tag{10.36}$$

In the last term, the $K_{\mu\lambda}{}^\kappa$'s are viewed as matrices $(K_\mu)_\lambda{}^\kappa$. The symbols $\bar{D}_\mu$ denote the *covariant derivatives* formed with the Christoffel symbol. Covariant derivatives act like ordinary derivatives if they are applied to a scalar field. When applied to a vector field, they act as follows:

$$\begin{aligned}
\bar{D}_\mu v_\nu &\equiv \partial_\mu v_\nu - \bar{\Gamma}_{\mu\nu}{}^\lambda v_\lambda, \\
\bar{D}_\mu v^\nu &\equiv \partial_\mu v^\nu + \bar{\Gamma}_{\mu\lambda}{}^\nu v^\lambda.
\end{aligned} \tag{10.37}$$

The effect upon a tensor field is the generalization of this; every index receives a corresponding additive $\bar{\Gamma}$ contribution.

Note that the Laplace-Beltrami operator (1.367) applied to a scalar field $\sigma(q)$ can be written as

$$\Delta\,\sigma = g^{\mu\nu} \bar{D}_\mu \bar{D}_\nu \sigma. \tag{10.38}$$

In the presence of torsion, there exists another covariant derivative formed with the affine connection $\Gamma_{\mu\nu}{}^\lambda$ rather than the Christoffel symbol which acts upon a vector field as

$$\begin{aligned}
D_\mu v_\nu &\equiv \partial_\mu v_\nu - \Gamma_{\mu\nu}{}^\lambda v_\lambda, \\
D_\mu v^\nu &\equiv \partial_\mu v^\nu + \Gamma_{\mu\lambda}{}^\nu v^\lambda.
\end{aligned} \tag{10.39}$$

Note by definition of $\Gamma_{\lambda\kappa}{}^\mu$ in (10.16) and (10.17), the covariant derivatives of $e^i{}_\mu$ and $e_i{}^\mu$ vanish:

$$D_\mu e^i{}_\nu \equiv \partial_\mu e^i{}_\nu - \Gamma_{\mu\nu}{}^\lambda e^i{}_\lambda = 0, \qquad D_\mu e_i{}^\nu \equiv \partial_\mu e_i{}^\nu + \Gamma_{\mu\lambda}{}^\nu e_i{}^\lambda = 0. \tag{10.40}$$

This will be of use later.

From either of the two curvature tensors, $R_{\mu\nu\lambda}{}^{\kappa}$ and $\bar{R}_{\mu\nu\lambda}{}^{\kappa}$, one can form the once-contracted tensors of rank two, the *Ricci tensor*

$$R_{\nu\lambda} = R_{\mu\nu\lambda}{}^{\mu}, \tag{10.41}$$

and the *curvature scalar*

$$R = g^{\nu\lambda} R_{\nu\lambda}. \tag{10.42}$$

The celebrated Einstein equation for the gravitational field postulates that the tensor

$$G_{\mu\nu} \equiv R_{\mu\nu} - \frac{1}{2} g_{\mu\nu} R, \tag{10.43}$$

the so-called *Einstein tensor*, is proportional to the symmetric energy-momentum tensor of all matter fields. This postulate was made only for spaces with no torsion, in which case $R_{\mu\nu} = \bar{R}_{\mu\nu}$ and $R_{\mu\nu}$, $G_{\mu\nu}$ are both symmetric. As mentioned before, it is not yet clear how Einstein's field equations should be generalized in the presence of torsion since the experimental consequences are as yet too small to be observed. In this text, we are not concerned with the generation of curvature and torsion but only with their consequences upon the motion of point particles.

It is useful to set up two simple examples for nonholonomic mappings which illustrate the way in which these are capable of generating curvature and torsion from a Euclidean space. The reader not familiar with this subject is advised to consult a textbook on the physics of defects [2]. where such mappings are standard and of great practical importance; every plastic deformation of a material can only be described in terms of such mappings.

As a first example consider the transformation in two dimensions

$$dx^i = \begin{cases} dq^1 & \text{for } i = 1, \\ dq^2 + \epsilon \partial_\mu \phi(q) dq^\mu & \text{for } i = 2, \end{cases} \tag{10.44}$$

with an infinitesimal parameter ϵ and the multi-valued function

$$\phi(q) \equiv \arctan(q^2/q^1). \tag{10.45}$$

The triads reduce to dyads, with the components

$$\begin{aligned} e^1{}_\mu &= \delta^1{}_\mu \ , \\ e^2{}_\mu &= \delta^2{}_\mu + \epsilon \partial_\mu \phi(q) \ , \end{aligned} \tag{10.46}$$

and the torsion tensor has the components

$$e^1{}_\lambda S_{\mu\nu}{}^\lambda = 0, \qquad e^2{}_\lambda S_{\mu\nu}{}^\lambda = \frac{\epsilon}{2}(\partial_\mu \partial_\nu - \partial_\nu \partial_\mu)\phi. \tag{10.47}$$

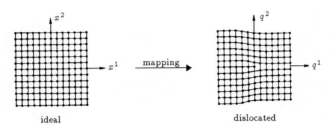

Figure 10.1   Edge dislocation in crystal associated with missing semi-infinite plane of atoms. The nonholonomic mapping from the ideal crystal to the crystal with the dislocation introduces a δ-function type torsion in the image space.

If we differentiate (10.45) formally, we find $(\partial_\mu \partial_\nu - \partial_\nu \partial_\mu)\phi \equiv 0$. This, however, is incorrect at the origin. Using Stokes' theorem we see that

$$\int d^2q(\partial_1\partial_2 - \partial_2\partial_1)\phi = \oint dq^\mu \partial_\mu \phi = \oint d\phi = 2\pi \tag{10.48}$$

for any closed circuit around the origin, implying that there is a δ-function singularity at the origin with

$$e^2{}_\lambda S_{12}{}^\lambda = \frac{\epsilon}{2} 2\pi \delta^{(2)}(q). \tag{10.49}$$

By a linear superposition of such mappings we can generate an arbitrary torsion in the q-space. The mapping introduces no curvature.

In defect physics, the mapping (10.46) is associated with a dislocation caused by a missing or additional layer of atoms (see Fig. 10.1). When encircling a dislocation along a closed path C, its counter image C' in the ideal crystal does not form a closed path. The closure failure is called the *Burgers vector*

$$b^i \equiv \oint_{C'} dx^i = \oint_C dq^\mu e^i{}_\mu. \tag{10.50}$$

It specifies the direction and thickness of the layer of additional atoms. With the help of Stokes' theorem, it is seen to measure the torsion contained in any surface S spanned by C:

$$b^i = \oint_S d^2s^{\mu\nu} \partial_\mu e^i{}_\nu = \oint_S d^2s^{\mu\nu} e^i{}_\lambda S_{\mu\nu}{}^\lambda, \tag{10.51}$$

where $d^2s^{\mu\nu} = -d^2s^{\nu\mu}$ is the projection of an oriented infinitesimal area element onto the plane $\mu\nu$. The above example has the Burgers vector

$$b^i = (0, \epsilon). \tag{10.52}$$

A corresponding closure failure appears when mapping a closed contour C in the ideal crystal into a crystal containing a dislocation. This defines a Burgers vector:

$$b^\mu \equiv \oint_{C'} dq^\mu = \oint_C dx^i e_i{}^\mu. \tag{10.53}$$

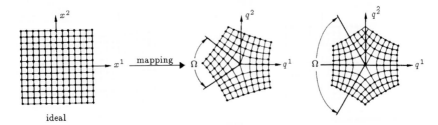

Figure 10.2 Edge disclination in crystal associated with missing semi-infinite section of atoms of angle Ω. The nonholonomic mapping from the ideal crystal to the crystal with the disclination introduces a δ-function type curvature in the image space.

By Stokes' theorem, this becomes a surface integral

$$b^\mu = \oint_S d^2 s^{ij} \partial_i e_j{}^\mu = \oint_S d^2 s^{ij} e_i{}^\nu \partial_\nu e_j{}^\mu$$
$$= -\oint_S d^2 s^{ij} e_i{}^\nu e_j{}^\lambda S_{\nu\lambda}{}^\mu, \tag{10.54}$$

the last step following from (10.17).

As a second example for a nonholonomic mapping, we generate curvature by the transformation

$$x^i = \delta^i{}_\mu [q^\mu + \Omega \epsilon^\mu{}_\nu q^\nu \phi(q)], \tag{10.55}$$

with the multi-valued function (10.45). The symbol $\epsilon_{\mu\nu}$ denotes the antisymmetric Levi-Civita tensor. The transformed metric

$$g_{\mu\nu} = \delta_{\mu\nu} - \frac{2\Omega}{q^\sigma q_\sigma} \epsilon_{\mu\lambda} \epsilon_{\nu\kappa} q^\lambda q^\kappa \tag{10.56}$$

is single-valued and has commuting derivatives. The torsion tensor vanishes since $(\partial_1 \partial_2 - \partial_2 \partial_1) x^{1,2}$ are both proportional to $q^{2,1} \delta^{(2)}(q)$, a distribution identical to zero. The local rotation field $\omega(q) \equiv \frac{1}{2}(\partial_1 x^2 - \partial_2 x^1)$, on the other hand, is equal to the multi-valued function $-\Omega\phi(q)$, thus having the noncommuting derivatives:

$$(\partial_1 \partial_2 - \partial_2 \partial_1) \omega(q) = -2\pi \Omega \delta^{(2)}(q). \tag{10.57}$$

To lowest order in Ω, this determines the curvature tensor, which in two dimensions possesses only one independent component, for instance R_{1212}. Using the fact that $g_{\mu\nu}$ has commuting derivatives, R_{1212} can be written as

$$R_{1212} = (\partial_1 \partial_2 - \partial_2 \partial_1) \omega(q). \tag{10.58}$$

In defect physics, the mapping (10.55) is associated with a disclination which corresponds to an entire section of angle Ω missing in an ideal atomic array (see Fig. 10.2).

It is important to emphasize that our multivalued basis tetrads $e^a{}_\mu(q)$ are *not* related to the standard tetrads or *vierbein fields* $h^a{}_\mu(q)$ used in the theory of gravitation with spinning particles. The difference is explained in Appendix 10B.

10.2.3   New Equivalence Principle

In classical mechanics, many dynamical problems are solved with the help of non-holonomic transformations. Equations of motion are differential equations which remain valid if transformed differentially to new coordinates, even if the transformation is not integrable in the Schwarz sense. Thus we *postulate that the correct equations of motion of point particles in a space with curvature and torsion are the images of the equation of motion in a flat space.* The equations (10.18) for the autoparallels yield therefore the correct trajectories of spinless point particles in a space with curvature and torsion.

This postulate is based on our knowledge of the motion of many physical systems. Important examples are the Coulomb system which will be discussed in detail in Chapter 13, and the spinning top in the body-fixed reference system [3]. Thus the postulate has a good chance of being true, and will henceforth be referred to as a *new equivalence principle*.

10.2.4   Classical Action Principle for Spaces with Curvature and Torsion

Before setting up a path integral for the time evolution amplitude we must find an action principle for the classical motion of a spinless point particle in a space with curvature and torsion, i.e., the movement along autoparallel trajectories. This is a nontrivial task since autoparallels must emerge as the extremals of an action (10.3) involving only the metric tensor $g_{\mu\nu}$. The action is independent of the torsion and carries only information on the Riemann part of the space geometry. Torsion can therefore enter the equations of motion only via some novel feature of the variation procedure. Since we know how to perform variations of an action in the Euclidean x-space, we deduce the correct procedure in the general metric-affine space by transferring the variations $\delta x^i(t)$ under the nonholonomic mapping

$$\dot{q}^\mu = e_i{}^\mu(q)\dot{x}^i \qquad (10.59)$$

into the q^μ-space. Their images are quite different from ordinary variations as illustrated in Fig. 10.3(a). The variations of the Cartesian coordinates $\delta x^i(t)$ are done at fixed endpoints of the paths. Thus they form *closed paths* in the x-space. Their images, however, lie in a space with defects and thus possess a closure failure indicating the amount of torsion introduced by the mapping. This property will be emphasized by writing the images $\delta^S q^\mu(t)$ and calling them *nonholonomic variations*. The superscript indicates the special feature caused by torsion.

Let us calculate them explicitly. The paths in the two spaces are related by the integral equation

$$q^\mu(t) = q^\mu(t_a) + \int_{t_a}^{t} dt'\, e_i{}^\mu(q(t'))\dot{x}^i(t'). \qquad (10.60)$$

For two neighboring paths in x-space differing from each other by a variation $\delta x^i(t)$, equation (10.60) determines the nonholonomic variation $\delta^S q^\mu(t)$:

$$\delta^S q^\mu(t) = \int_{t_a}^t dt' \delta^S [e_i{}^\mu(q(t'))\dot{x}^i(t')]. \tag{10.61}$$

A comparison with (10.59) shows that the variation δ^S and the time derivatives d/dt of $q^\mu(t)$ commute with each other:

$$\delta^S \dot{q}^\mu(t) = \frac{d}{dt}\delta^S q^\mu(t), \tag{10.62}$$

just as for ordinary variations δx^i:

$$\delta \dot{x}^i(t) = \frac{d}{dt}\delta x^i(t). \tag{10.63}$$

Let us also introduce *auxiliary nonholonomic variations* in q-space:

$$\bar{\delta} q^\mu \equiv e_i{}^\mu(q)\delta x^i. \tag{10.64}$$

In contrast to $\delta^S q^\mu(t)$, these vanish at the endpoints,

$$\bar{\delta} q(t_a) = \bar{\delta} q(t_b) = 0, \tag{10.65}$$

just as the usual variations $\delta x^i(t)$, i.e., they form *closed* paths with the unvaried orbits.

Using (10.62), (10.63), and the fact that $\delta^S x^i(t) \equiv \delta x^i(t)$, by definition, we derive from (10.61) the relation

$$\begin{aligned}
\frac{d}{dt}\delta^S q^\mu(t) &= \delta^S e_i{}^\mu(q(t))\dot{x}^i(t) + e_i{}^\mu(q(t))\frac{d}{dt}\delta x^i(t) \\
&= \delta^S e_i{}^\mu(q(t))\dot{x}^i(t) + e_i{}^\mu(q(t))\frac{d}{dt}[e^i{}_\nu(t)\bar{\delta} q^\nu(t)].
\end{aligned} \tag{10.66}$$

After inserting

$$\delta^S e_i{}^\mu(q) = -\Gamma_{\lambda\nu}{}^\mu \delta^S q^\lambda e_i{}^\nu, \qquad \frac{d}{dt}e^i{}_\nu(q) = \Gamma_{\lambda\nu}{}^\mu \dot{q}^\lambda e^i{}_\mu, \tag{10.67}$$

this becomes

$$\frac{d}{dt}\delta^S q^\mu(t) = -\Gamma_{\lambda\nu}{}^\mu \delta^S q^\lambda \dot{q}^\nu + \Gamma_{\lambda\nu}{}^\mu \dot{q}^\lambda \bar{\delta} q^\nu + \frac{d}{dt}\bar{\delta} q^\mu. \tag{10.68}$$

It is useful to introduce the difference between the nonholonomic variation $\delta^S q^\mu$ and an auxiliary closed nonholonomic variation $\bar{\delta} q^\mu$:

$$\delta^S b^\mu \equiv \delta^S q^\mu - \bar{\delta} q^\mu. \tag{10.69}$$

Then we can rewrite (10.68) as a first-order differential equation for $\delta^S b^\mu$:

$$\frac{d}{dt}\delta^S b^\mu = -\Gamma_{\lambda\nu}{}^\mu \delta^S b^\lambda \dot{q}^\nu + 2 S_{\lambda\nu}{}^\mu \dot{q}^\lambda \delta q^\nu. \tag{10.70}$$

After introducing the matrices

$$G^\mu{}_\lambda(t) \equiv \Gamma_{\lambda\nu}{}^\mu(q(t))\dot{q}^\nu(t) \tag{10.71}$$

and

$$\Sigma^\mu{}_\nu(t) \equiv 2 S_{\lambda\nu}{}^\mu(q(t))\dot{q}^\lambda(t), \tag{10.72}$$

equation (10.70) can be written as a vector differential equation:

$$\frac{d}{dt}\delta^S b = -G\delta^S b + \Sigma(t)\,\delta q^\nu(t). \tag{10.73}$$

Although not necessary for the further development, we solve this equation by

$$\delta^S b(t) = \int_{t_a}^{t} dt' U(t, t')\,\Sigma(t')\,\delta q(t'), \tag{10.74}$$

with the matrix

$$U(t, t') = T \exp\left[-\int_{t'}^{t} dt'' G(t'')\right]. \tag{10.75}$$

In the absence of torsion, $\Sigma(t)$ vanishes identically and $\delta^S b(t) \equiv 0$, and the variations $\delta^S q^\mu(t)$ coincide with the auxiliary closed nonholonomic variations $\delta q^\mu(t)$ [see Fig. 10.3(b)]. In a space with torsion, the variations $\delta^S q^\mu(t)$ and $\delta q^\mu(t)$ are different from each other [see Fig. 10.3(c)].

Under an arbitrary nonholonomic variation $\delta^S q^\mu(t) = \delta q^\mu + \delta^S b^\mu$, the action (10.3) changes by

$$\delta^S \mathcal{A} = M \int_{t_a}^{t_b} dt \left(g_{\mu\nu}\dot{q}^\nu \delta^S \dot{q}^\mu + \frac{1}{2}\partial_\mu g_{\lambda\kappa}\delta^S q^\mu \dot{q}^\lambda \dot{q}^\kappa\right). \tag{10.76}$$

After a partial integration of the $\delta\dot{q}$-term we use (10.65), (10.62), and the identity $\partial_\mu g_{\nu\lambda} \equiv \Gamma_{\mu\nu\lambda} + \Gamma_{\mu\lambda\nu}$, which follows directly form the definitions $g_{\mu\nu} \equiv e^i{}_\mu e^i{}_\nu$ and $\Gamma_{\mu\nu}{}^\lambda \equiv e_i{}^\lambda \partial_\mu e^i{}_\nu$, and obtain

$$\delta^S \mathcal{A} = M \int_{t_a}^{t_b} dt \left[-g_{\mu\nu}\left(\ddot{q}^\nu + \bar{\Gamma}_{\lambda\kappa}{}^\nu \dot{q}^\lambda \dot{q}^\kappa\right)\delta q^\mu + \left(g_{\mu\nu}\dot{q}^\nu \frac{d}{dt}\delta^S b^\mu + \Gamma_{\mu\lambda\kappa}\delta^S b^\mu \dot{q}^\lambda \dot{q}^\kappa\right)\right]. \tag{10.77}$$

To derive the equation of motion we first vary the action in a space without torsion. Then $\delta^S b^\mu(t) \equiv 0$, and (10.77) becomes

$$\delta^S \mathcal{A} = -M \int_{t_a}^{t_b} dt\, g_{\mu\nu}(\ddot{q}^\nu + \bar{\Gamma}_{\lambda\kappa}{}^\nu \dot{q}^\lambda \dot{q}^\kappa)\delta q^\nu. \tag{10.78}$$

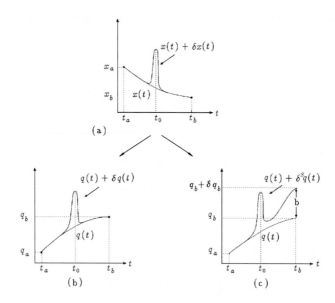

Figure 10.3 Images under holonomic and nonholonomic mapping of fundamental δ-function path variation. In the holonomic case, the paths $x(t)$ and $x(t) + \delta x(t)$ in (a) turn into the paths $q(t)$ and $q(t) + \delta q(t)$ in (b). In the nonholonomic case with $S^\lambda_{\mu\nu} \neq 0$, they go over into $q(t)$ and $q(t) + \delta^S q(t)$ shown in (c) with a closure failure b^μ at t_b analogous to the Burgers vector b^μ in a solid with dislocations.

Thus, the action principle $\delta^S \mathcal{A} = 0$ produces the equation for the geodesics (10.9), which are the correct particle trajectories in the absence of torsion.

In the presence of torsion, $\delta^S b^\mu$ is nonzero, and the equation of motion receives a contribution from the second parentheses in (10.77). After inserting (10.70), the nonlocal terms proportional to $\delta^S b^\mu$ cancel and the total nonholonomic variation of the action becomes

$$
\begin{aligned}
\delta^S \mathcal{A} &= -M \int_{t_a}^{t_b} dt\, g_{\mu\nu} \left[\ddot{q}^\nu + \left(\bar{\Gamma}_{\lambda\kappa}{}^\nu + 2 S^\nu{}_{\lambda\kappa} \right) \dot{q}^\lambda \dot{q}^\kappa \right] \delta q^\mu \\
&= -M \int_{t_a}^{t_b} dt\, g_{\mu\nu} \left(\ddot{q}^\nu + \Gamma_{\lambda\kappa}{}^\nu \dot{q}^\lambda \dot{q}^\kappa \right) \delta q^\mu.
\end{aligned}
\tag{10.79}
$$

The second line follows from the first after using the identity $\Gamma_{\lambda\kappa}{}^\nu = \bar{\Gamma}_{\{\lambda\kappa\}}{}^\nu + 2 S^\nu{}_{\{\lambda\kappa\}}$. The curly brackets indicate the symmetrization of the enclosed indices. Setting $\delta^S \mathcal{A} = 0$ and inserting for $\delta q(t)$ the image under (10.64) of an arbitrary δ-function variation $\delta x^i(t) \propto \epsilon^i \delta(t - t_0)$ gives the autoparallel equations of motions (10.18), which is what we wanted to show.

The above variational treatment of the action is still somewhat complicated and calls for a simpler procedure [4]. The extra term arising from the second parenthesis in the variation (10.77) can be traced to a simple property of the auxiliary closed nonholonomic variations (10.64). To find this we form the time derivative $d_t \equiv d/dt$ of the defining equation (10.64) and find

$$d_t \delta q^\mu(t) = \partial_\nu e_i{}^\mu(q(t))\,\dot{q}^\nu(t)\delta x^i(t) + e_i{}^\mu(q(t))d_t\delta x^i(t). \qquad (10.80)$$

Let us now perform variation δ and t-derivative in the opposite order and calculate $d_t\delta q^\mu(t)$. From (10.59) and (10.13) we have the relation

$$d_t q^\lambda(t) = e_i{}^\lambda(q(t))\,d_t x^i(t)\,. \qquad (10.81)$$

Varying this gives

$$\delta d_t q^\mu(t) = \partial_\nu e_i{}^\mu(q(t))\,\delta q^\nu d_t x^i(t) + e_i{}^\mu(q(t))\delta d_t x^i. \qquad (10.82)$$

Since the variation in x^i-space commute with the t-derivatives [recall (10.63)], we obtain

$$\delta d_t q^\mu(t) - d_t \delta q^\mu(t) = \partial_\nu e_i{}^\mu(q(t))\,\delta q^\nu d_t x^i(t) - \partial_\nu e_i{}^\mu(q(t))\,\dot{q}^\nu(t)\delta x^i(t). \qquad (10.83)$$

After re-expressing $\delta x^i(t)$ and $d_t x^i(t)$ back in terms of $\delta q^\mu(t)$ and $d_t q^\mu(t) = \dot{q}^\mu(t)$, and using (10.17), (10.24), this becomes

$$\delta d_t q^\mu(t) - d_t \delta q^\mu(t) = 2S_{\nu\lambda}{}^\mu \dot{q}^\nu(t)\delta q^\lambda(t). \qquad (10.84)$$

Thus, due to the closure failure in spaces with torsion, the operations d_t and δ do not commute in front of the path $q^\mu(t)$. In other words, in contrast to the open variations δ^S (and of course the usual ones δ), the auxiliary closed nonholonomic variations δ of velocities $\dot{q}^\mu(t)$ no longer coincide with the velocities of variations. This property is responsible for shifting the trajectory from geodesics to autoparallels.

Indeed, let us vary an action

$$\mathcal{A} = \int\limits_{t_1}^{t_2} dt\,L\left(q^\mu(t), \dot{q}^\mu(t)\right) \qquad (10.85)$$

directly by $\delta q^\mu(t)$ and impose (10.84), we find

$$\delta\mathcal{A} = \int\limits_{t_1}^{t_2} dt \left\{ \frac{\partial L}{\partial q^\mu}\delta q^\mu + \frac{\partial L}{\partial \dot{q}^\mu}\frac{d}{dt}\delta q^\mu + 2\,S^\mu{}_{\nu\lambda}\frac{\partial L}{\partial \dot{q}^\mu}\dot{q}^\nu \delta q^\lambda \right\}. \qquad (10.86)$$

After a partial integration of the second term using the vanishing $\delta q^\mu(t)$ at th~ endpoints, we obtain the Euler-Lagrange equation

$$\frac{\partial L}{\partial q^\mu} - \frac{d}{dt}\frac{\partial L}{\partial \dot{q}^\mu} = -2S_{\mu\nu}{}^\lambda \dot{q}^\nu \frac{\partial L}{\partial \ldots}$$

This differs from the standard Euler-Lagrange ~tion due to the torsion tensor. For the action (1 motion

$$M\left[\ddot{q}^\mu + g^{\mu\kappa}\left(\partial_\nu g_{\lambda\kappa} - \frac{1}{2}\partial_\kappa g_{\nu\lambda}\right) - 2S\right.$$

which is once more Eq. (10.18) for autoparallels.

10.3 Path Integral in Spaces with Curvature and Torsion

We now turn to the quantum mechanics of a point particle in a general metric-affine space. Proceeding in analogy with the earlier treatment in spherical coordinates, we first consider the path integral in a flat space with Cartesian coordinates

$$(\mathbf{x}\,t|\mathbf{x}'t') = \frac{1}{\sqrt{2\pi i\epsilon\hbar/M}^D} \prod_{n=1}^{N}\left[\int_{-\infty}^{\infty} dx_n\right] \prod_{n=1}^{N+1} K_0^\epsilon(\Delta\mathbf{x}_n), \tag{10.89}$$

where $K_0^\epsilon(\Delta\mathbf{x}_n)$ is an abbreviation for the short-time amplitude

$$K_0^\epsilon(\Delta\mathbf{x}_n) \equiv \langle\mathbf{x}_n|\exp\left(-\frac{i}{\hbar}\epsilon\hat{H}\right)|\mathbf{x}_{n-1}\rangle = \frac{1}{\sqrt{2\pi i\epsilon\hbar/M}^D}\exp\left[\frac{i}{\hbar}\frac{M}{2}\frac{(\Delta\mathbf{x}_n)^2}{\epsilon}\right], \tag{10.90}$$

with $\Delta\mathbf{x}_n \equiv \mathbf{x}_n - \mathbf{x}_{n-1}$, $\mathbf{x} \equiv \mathbf{x}_{N+1}$, $\mathbf{x}' \equiv \mathbf{x}_0$. A possible external potential has been omitted since this would contribute in an additive way, uninfluenced by the space geometry.

Our basic postulate is that the path integral in a general metric-affine space should be obtained by an appropriate nonholonomic transformation of the amplitude (10.89) to a space with curvature and torsion.

10.3.1 Nonholonomic Transformation of Action

The short-time action contains the square distance $(\Delta\mathbf{x}_n)^2$ which we have to transform to q-space. For an infinitesimal coordinate difference $\Delta\mathbf{x}_n \approx d\mathbf{x}_n$, the square distance is obviously given by $(d\mathbf{x})^2 = g_{\mu\nu}dq^\mu dq^\nu$. For a finite $\Delta\mathbf{x}_n$, however, we know from Chapter 8 that we must expand $(\Delta\mathbf{x}_n)^2$ up to the fourth order in $\Delta q_n{}^\mu = q_n{}^\mu - q_{n-1}{}^\mu$ to find all terms contributing to the relevant order ϵ.

It is important to realize that with the mapping from dx^i to dq^μ not being holonomic, the finite quantity Δq^μ is not uniquely determined by Δx^i. A unique relation can only be obtained by integrating the functional relation (10.60) along a specific path. The preferred path is the classical orbit, i.e., the autoparallel in the q-space. It is characterized by being the image of a straight line in x-space. There the velocity $\dot{x}^i(t)$ is constant, and the orbit has the linear time dependence

$$\Delta x^i(t) = \dot{x}^i(t_0)\Delta t, \tag{10.91}$$

where the time t_0 can lie anywhere on the t-axis. Let us choose for t_0 the final time in each interval (t_n, t_{n-1}). At that time, $\dot{x}_n^i \equiv \dot{x}^i(t_n)$ is related to $\dot{q}_n^\mu \equiv \dot{q}^\mu(t_n)$ by

$$\dot{x}_n^i = e^i{}_\mu(q_n)\dot{q}_n^\mu. \tag{10.92}$$

easy to express $\dot{q}_n^\mu$ in terms of $\Delta q_n^\mu = q_n^\mu - q_{n-1}^\mu$ along the classical orbit. First and $q^\mu(t_{n-1})$ into a Taylor series around t_n. Dropping the time arguments, ty, we have

$$\Delta q \equiv q^\lambda - q'^\lambda = \epsilon\dot{q}^\lambda - \frac{\epsilon^2}{2!}\ddot{q}^\lambda + \frac{\epsilon^3}{3!}\dddot{q}^\lambda + \dots, \tag{10.93}$$

where $\epsilon = t_n - t_{n-1}$ and $\dot{q}^\lambda, \ddot{q}^\lambda, \ldots$ are the time derivatives at the final time t_n. An expansion of this type is referred to as a *postpoint expansion*. Due to the arbitrariness of the choice of the time t_0 in Eq. (10.92), the expansion can be performed around any other point just as well, such as t_{n-1} and $\bar{t}_n = (t_n + t_{n-1})/2$, giving rise to the so-called *prepoint* or *midpoint* expansions of Δq.

Now, the term $\ddot{q}^\lambda$ in (10.93) is given by the equation of motion (10.18) for the autoparallel

$$\ddot{q}^\lambda = -\Gamma_{\mu\nu}{}^\lambda \dot{q}^\mu \dot{q}^\nu. \tag{10.94}$$

A further time derivative determines

$$\dddot{q}^\lambda = -(\partial_\sigma \Gamma_{\mu\nu}{}^\lambda - 2\Gamma_{\mu\nu}{}^\tau \Gamma_{\{\sigma\tau\}}{}^\lambda) \dot{q}^\mu \dot{q}^\nu \dot{q}^\sigma. \tag{10.95}$$

Inserting these expressions into (10.93) and inverting the expansion, we obtain $\dot{q}^\lambda$ at the final time t_n expanded in powers of Δq. Using (10.91) and (10.92) we arrive at the mapping of the finite coordinate differences:

$$\begin{aligned} \Delta x^i &= e^i{}_\lambda \dot{q}^\lambda \Delta t \tag{10.96}\\ &= e^i{}_\lambda \left[\Delta q^\lambda - \frac{1}{2!}\Gamma_{\mu\nu}{}^\lambda \Delta q^\mu \Delta q^\nu + \frac{1}{3!}\left(\partial_\sigma \Gamma_{\mu\nu}{}^\lambda + \Gamma_{\mu\nu}{}^\tau \Gamma_{\{\sigma\tau\}}{}^\lambda \right) \Delta q^\mu \Delta q^\nu \Delta q^\sigma + \ldots \right], \end{aligned}$$

where $e^i{}_\lambda$ and $\Gamma_{\mu\nu}{}^\lambda$ are evaluated at the postpoint.

It is useful to introduce

$$\Delta \xi^\mu \equiv e_i{}^\mu \Delta x^i \tag{10.97}$$

as *autoparallel coordinates* or *normal coordinates* to parametrize the neighborhood of a point q. If the space has no torsion, they are also called *Riemann normal coordinates* or *geodesic coordinates*.

The normal coordinates (10.97) are expanded in (10.150) in powers of Δq^μ around the postpoint. There exists also a prepoint version of $\Delta \xi$ in which all signs of Δq are simply the opposite. The prepoint version, for instance, has the expansion:

$$\Delta \xi^\lambda = \Delta q^\lambda + \frac{1}{2!}\Gamma_{\mu\nu}{}^\lambda \Delta q^\mu \Delta q^\nu + \frac{1}{3!}\left(\partial_\sigma \Gamma_{\mu\nu}{}^\lambda + \Gamma_{\mu\nu}{}^\tau \Gamma_{\{\sigma\tau\}}{}^\lambda \right) \Delta q^\mu \Delta q^\nu \Delta q^\sigma + \ldots. \tag{10.98}$$

In contrast to the finite differences Δq^μ, the normal coordinates $\Delta \xi^\mu$ in the neighborhood of a point are vectors and thus allow for a *covariant Taylor expansion* of a function $f(q^\mu + \Delta q^\mu)$. Its form is found by performing an ordinary Taylor expansion of a function $F(x)$ in Cartesian coordinates

$$F(x + \Delta x) = F(x) + \partial_i F(x)\Delta x^i + \frac{1}{2!}\partial_i \partial_j F(x)\Delta x^i \Delta x^j + \ldots, \tag{10.99}$$

and transforming this to coordinates q^μ. The function $F(x)$ becomes $f(q) = F(x(q))$, and the derivatives $\partial_{i_1} \partial_{i_2} \cdots \partial_{i_n} f(x)$ go over into covariant derivatives:

$e_{i_1}{}^\mu e_{i_2}{}^\mu \cdots e_{i_n}{}^\mu D_{\mu_1} D_{\mu_2} \cdots D_{\mu_n} f(q)$. For instance, $\partial_i F(x) = e_i{}^\mu \partial_\mu f(q) = e_i{}^\mu D_\mu f(q)$, and

$$\partial_i \partial_j f(q) = e_i{}^\mu \partial_\mu e_j{}^\nu \partial_\nu f(q) = \left[e_i{}^\mu e_j{}^\nu \partial_\mu \partial_\nu + e_i{}^\mu (\partial_\mu e_j{}^\nu) \partial_\nu \right] f(q)$$
$$= e_i{}^\mu e_j{}^\nu \left[\partial_\mu \partial_\nu - \Gamma_{\mu\nu}{}^\lambda \partial_\lambda \right] f(q) = e_i{}^\mu e_j{}^\nu D_\mu \partial_\nu f(q) = e_i{}^\mu e_j{}^\nu D_\mu D_\nu f(q), \quad (10.100)$$

where we have used (10.17) to express $\partial_\mu e_j{}^\nu = -\Gamma_{\mu\sigma}{}^\nu e_j{}^\sigma$, and changed dummy indices. The differences Δx^i in (10.99) are replaced by $e^i{}_\mu \Delta \xi^\mu$ with the prepoint expansion (10.98). In this way we arrive at the covariant Taylor expansion

$$f(q + \Delta q) = F(x) + D_\mu f(q) \Delta \xi^\mu + \frac{1}{2!} D_\mu D_\nu f(q) \Delta \xi^\mu \Delta \xi^\nu + \dots . \quad (10.101)$$

Indeed, re-expanding the right-hand side in powers of Δq^μ via (10.98) we may verify that the affine connections cancel against those in the covariant derivatives of $f(q)$, so that (10.101) reduces to the ordinary Taylor expansion of $f(q + \Delta q)$ in powers of Δq.

Note that the expansion (10.96) differs only slightly from a naive Taylor expansion of the difference around the postpoint:

$$\Delta x^i = x^i(q) - x^i(q - \Delta q) = e^i{}_\lambda \Delta q^\lambda - \frac{1}{2} e^i{}_{\nu,\mu} \Delta q^\mu \Delta q^\nu + \frac{1}{3!} e^i{}_{\nu,\mu\sigma} \Delta q^\mu \Delta q^\nu \Delta q^\sigma + \dots , \quad (10.102)$$

where a subscript λ separated by a comma denotes the partial derivative $\partial_\lambda = \partial/\partial q^\lambda$, i.e., $f_{,\lambda} \equiv \partial_\lambda f$. The right-hand side can be rewritten with the help of the completeness relation (10.13) as

$$\Delta x^i = e^i{}_\lambda \left[\Delta q^\lambda - \frac{1}{2} e_j{}^\lambda e^j{}_{\nu,\mu} \Delta q^\mu \Delta q^\nu + \frac{1}{3!} e_j{}^\lambda e^j{}_{\nu,\mu\sigma} \Delta q^\mu \Delta q^\nu \Delta q^\sigma + \dots \right]. \quad (10.103)$$

The expansion coefficients can be expressed in terms of the affine connection (10.16), using the derived relation

$$e_i{}^\lambda e^i{}_{\nu,\mu\sigma} = \partial_\sigma (e_i{}^\lambda e^i{}_{\nu,\mu}) - e^{i\tau} e^i{}_{\nu,\mu} e^j{}_\tau e^{j\lambda}{}_{,\sigma} = \partial_\sigma \Gamma_{\mu\nu}{}^\lambda + \Gamma_{\mu\nu}{}^\tau \Gamma_{\sigma\tau}{}^\lambda. \quad (10.104)$$

Thus we obtain

$$\Delta x^i = e^i{}_\lambda \left[\Delta q^\lambda - \frac{1}{2!} \Gamma_{\mu\nu}{}^\lambda \Delta q^\mu \Delta q^\nu + \frac{1}{3!} \left(\partial_\sigma \Gamma_{\mu\nu}{}^\lambda + \Gamma_{\mu\nu}{}^\tau \Gamma_{\sigma\tau}{}^\lambda \right) \Delta q^\mu \Delta q^\nu \Delta q^\sigma + \dots \right]. \quad (10.105)$$

This differs from the true expansion (10.150) only by the absence of the symmetrization of the indices in the last affine connection.

Inserting (10.96) into the short-time amplitude (10.90), we obtain

$$K_0^\epsilon(\Delta x) = \langle x | \left(-\frac{i}{\hbar} \epsilon \hat{H} \right) | x - \Delta x \rangle = \frac{1}{\sqrt{2\pi i \epsilon \hbar / M}^D} e^{i \mathcal{A}_>^\epsilon (q, q - \Delta q)/\hbar}, \quad (10.106)$$

with the short-time postpoint action

$$
\begin{aligned}
\mathcal{A}_>^\epsilon(q, q - \Delta q) = (\Delta x^i)^2 &= \epsilon \frac{M}{2} g_{\mu\nu} \dot{q}^\mu \dot{q}^\nu \\
&= \frac{M}{2\epsilon} \Big\{ g_{\mu\nu} \Delta q^\mu \Delta q^\nu - \Gamma_{\mu\nu\lambda} \Delta q^\mu \Delta q^\nu \Delta q^\lambda \\
&\quad + \Big[\frac{1}{3} g_{\mu\tau} \left(\partial_\kappa \Gamma_{\lambda\nu}{}^\tau + \Gamma_{\lambda\nu}{}^\delta \Gamma_{\{\kappa\delta\}}{}^\tau \right) + \frac{1}{4} \Gamma_{\lambda\kappa}{}^\sigma \Gamma_{\mu\nu\sigma} \Big] \Delta q^\mu \Delta q^\nu \Delta q^\lambda \Delta q^\kappa + \dots \Big\}.
\end{aligned}
\tag{10.107}
$$

Separating the affine connection into Christoffel symbol and torsion, this can also be written as

$$
\begin{aligned}
\mathcal{A}_>^\epsilon(q, q - \Delta q) = \frac{M}{2\epsilon} \Big\{ &g_{\mu\nu} \Delta q^\mu \Delta q^\nu - \bar{\Gamma}_{\mu\nu\lambda} \Delta q^\mu \Delta q^\nu \Delta q^\lambda \\
&+ \Big[\frac{1}{3} g_{\mu\tau} \left(\partial_\kappa \bar{\Gamma}_{\lambda\nu}{}^\tau + \bar{\Gamma}_{\lambda\nu}{}^\delta \bar{\Gamma}_{\delta\kappa}{}^\tau \right) + \frac{1}{4} \bar{\Gamma}_{\lambda\kappa}{}^\sigma \bar{\Gamma}_{\mu\nu\sigma} + \frac{1}{3} S^\sigma{}_{\lambda\kappa} S_{\sigma\mu\nu} \Big] \Delta q^\mu \Delta q^\nu \Delta q^\lambda \Delta q^\kappa + \dots \Big\}.
\end{aligned}
\tag{10.108}
$$

Note that in contrast to the formulas for the short-time action derived in Chapter 8, the right-hand side contains only *intrinsic* quantities of q-space. For the systems treated there (which all lived in a Euclidean space parametrized with curvilinear coordinates), the present intrinsic result reduces to the previous one.

Take, for example, a two-dimensional Euclidean space parametrized by radial coordinates treated in Section 8.1. The postpoint expansion (10.96) reads for the components r, ϕ of $\dot{q}^\lambda$

$$
\dot{r} = \frac{\Delta r}{\epsilon} + \frac{r(\Delta\phi)^2}{2\epsilon} - \frac{\Delta r(\Delta\phi)^2}{\epsilon} + \dots,
\tag{10.109}
$$

$$
\dot{\phi} = \frac{\Delta\phi}{\epsilon} - \frac{\Delta r \Delta\phi}{\epsilon r} - \frac{(\Delta\phi)^3}{6\epsilon} + \dots.
\tag{10.110}
$$

Inserting these into the short-time action which is here simply

$$
\mathcal{A}^\epsilon = \frac{M}{2} \epsilon (\dot{r}^2 + r^2 \dot{\phi}^2),
\tag{10.111}
$$

we find the time-sliced action

$$
\mathcal{A}^\epsilon = \frac{M}{2\epsilon} \Big[\Delta r^2 + r^2 (\Delta\phi)^2 - r \Delta r (\Delta\phi)^2 - \frac{1}{12} r^2 (\Delta\phi)^4 + \dots \Big].
\tag{10.112}
$$

A symmetrization of the postpoint expressions using the fact that r^2 stands for

$$
r_n^2 = r_n(r_{n-1} + \Delta r_n),
\tag{10.113}
$$

leads to the short-time action displaying the subscripts n

$$
\mathcal{A}^\epsilon = \frac{M}{2\epsilon} \Big[\Delta r_n^2 + r_n r_{n-1} (\Delta\phi_n)^2 - \frac{1}{12} r_n r_{n-1} (\Delta\phi_n)^4 + \dots \Big].
\tag{10.114}
$$

This agrees with the previous expansion of the time-sliced action in Eq. (8.53). While the previous result was obtained from a transformation of the time-sliced

Euclidean action to radial coordinates, the short-time action here is found from a purely intrinsic formulation. The intrinsic method has the obvious advantage of not being restricted to a Euclidean initial space and therefore has the chance of being true in an arbitrary metric-affine space.

At this point we observe that the final short-time action (10.107) could also have been introduced without any reference to the flat reference coordinates x^i. Indeed, the same action is obtained by evaluating the continuous action (10.3) for the small time interval $\Delta t = \epsilon$ along the classical orbit between the points q_{n-1} and q_n. Due to the equations of motion (10.18), the Lagrangian

$$L(q, \dot{q}) = \frac{M}{2} g_{\mu\nu}(q(t)) \, \dot{q}^\mu(t) \dot{q}^\nu(t) \qquad (10.115)$$

is independent of time (this is true for autoparallels as well as geodesics). The short-time action

$$\mathcal{A}^\epsilon(q, q') = \frac{M}{2} \int_{t-\epsilon}^t dt \, g_{\mu\nu}(q(t)) \dot{q}^\mu(t) \dot{q}^\nu(t) \qquad (10.116)$$

can therefore be written in either of the three forms

$$\mathcal{A}^\epsilon = \frac{M}{2} \epsilon g_{\mu\nu}(q) \dot{q}^\mu \dot{q}^\nu = \frac{M}{2} \epsilon g_{\mu\nu}(q') \dot{q}'^\mu \dot{q}'^\nu = \frac{M}{2} \epsilon g_{\mu\nu}(\bar{q}) \dot{\bar{q}}^\mu \dot{\bar{q}}^\nu, \qquad (10.117)$$

where $q^\mu, q'^\mu, \bar{q}^\mu$ are the coordinates at the final time t_n, the initial time t_{n-1}, and the average time $(t_n + t_{n-1})/2$, respectively. The first expression obviously coincides with (10.107). The others can be used as a starting point for deriving equivalent prepoint or midpoint actions. The prepoint action $\mathcal{A}^\epsilon_<$ arises from the postpoint one $\mathcal{A}^\epsilon_>$ by exchanging Δq by $-\Delta q$ and the postpoint coefficients by the prepoint ones. The midpoint action has the most simple-looking appearance:

$$\bar{\mathcal{A}}^\epsilon(\bar{q} + \frac{\Delta q}{2}, \bar{q} - \frac{\Delta q}{2}) = \qquad (10.118)$$
$$\frac{M}{2\epsilon} \left[g_{\mu\nu}(\bar{q}) \Delta q^\mu \Delta q^\nu + \frac{1}{12} g_{\kappa\tau} \left(\partial_\lambda \Gamma_{\mu\nu}{}^\tau + \Gamma_{\mu\nu}{}^\delta \Gamma_{\{\lambda\delta\}}{}^\tau \right) \Delta q^\mu \Delta q^\nu \Delta q^\lambda \Delta q^\kappa + \ldots \right],$$

where the affine connection can be evaluated at any point in the interval (t_{n-1}, t_n). The precise position is irrelevant to the amplitude, producing changes only in higher than the relevant orders of ϵ.

We have found the postpoint action most useful since it gives ready access to the time evolution of amplitudes, as will be seen below. The prepoint action is completely equivalent to it and useful if one wants to describe the time evolution backwards. Some authors favor the midpoint action because of its symmetry and the absence of cubic terms in Δq^μ in the expression (10.118).

The different completely equivalent "anypoint" formulations of the same short-time action, which is *universally defined* by the nonholonomic mapping procedure, must be distinguished from various so-called time-slicing "prescriptions" found in the literature when setting up a lattice approximation to the Lagrangian (10.115). There, a *midpoint prescription* is often favored, in which one approximates L by

$$L(q, \dot{q}) \rightarrow L^\epsilon(q, \Delta q/\epsilon) = \frac{M}{2\epsilon^2} g_{\mu\nu}(\bar{q}) \, \Delta q^\mu(t) \Delta q^\nu(t), \qquad (10.119)$$

and uses the associated short-time action

$$\bar{\mathcal{A}}^\epsilon_{\mathrm{mpp}} = \epsilon L^\epsilon(q, \Delta q/\epsilon) \qquad (10.120)$$

in the exponent of the path integrand. The motivation for this prescription lies in the popularity of H. Weyl's ordering prescription for products of position and momenta in operator quantum mechanics. From the discussion in Section 1.13 we know, however, that the Weyl prescription for the operator order in the kinetic energy $g^{\mu\nu}(\hat{q})\hat{p}_\mu\hat{p}_\nu/2M$ does not lead to the correct Laplace-Beltrami operator in general coordinates. The discussion in this section, on the other hand, will show that the Weyl-ordered action (10.120) differs from the correct midpoint form (10.118) of the action by an additional forth-order term in Δq^μ, implying that the short-time action $\bar{\mathcal{A}}^\epsilon_{\mathrm{mpp}}$ does not lead to the correct physics. Worse shortcomings are found when slicing the short-time action following a pre- or postpoint prescription. There is, in fact, no freedom of choice of different slicing prescriptions, in contrast to ubiquitous statements in the literature. The short-time action is completely fixed as being the unique nonholonomic image of the Euclidean time-sliced action. This also solves uniquely the operator-ordering problem which has plagued theorists for many decades.

In the following, the action $\mathcal{A}^\epsilon$ without subscript will always denote the preferred postpoint expression (10.107):

$$\mathcal{A}^\epsilon \equiv \mathcal{A}^\epsilon_{>}(q, q - \Delta q). \qquad (10.121)$$

10.3.2   Measure of Path Integration

We now turn to the integration measure in the Cartesian path integral (10.89)

$$\frac{1}{\sqrt{2\pi i \epsilon \hbar/M}^D} \prod_{n=1}^{N} d^D x_n.$$

This has to be transformed to the general metric-affine space. We imagine evaluating the path integral starting out from the latest time and performing successively the integrations over $x_N, x_{N-1}, \ldots$, i.e., in each short-time amplitude we integrate over the earlier position coordinate, the prepoint coordinate. For the purpose of this discussion, we relabel the product $\prod_{n=1}^{N} d^D x_n$ by $\prod_{n=2}^{N+1} d x_{n-1}^i$, so that the integration in each time slice (t_n, t_{n-1}) with $n = N+1, N, \ldots$ runs over dx_{n-1}^i.

In a flat space parametrized with curvilinear coordinates, the transformation of the integrals over $d^D x_{n-1}^i$ into those over $d^D q_{n-1}^\mu$ is obvious:

$$\prod_{n=2}^{N+1} \int d^D x_{n-1}^i = \prod_{n=2}^{N+1} \left\{ \int d^D q_{n-1}^\mu \, \det \left[e^i_\mu(q_{n-1}) \right] \right\}. \qquad (10.122)$$

The determinant of $e^i{}_\mu$ is the square root of the determinant of the metric $g_{\mu\nu}$:

$$\det(e^i{}_\mu) = \sqrt{\det g_{\mu\nu}(q)} \equiv \sqrt{g(q)}, \qquad (10.123)$$

and the measure may be rewritten as

$$\prod_{n=2}^{N+1} \int d^D x^i_{n-1} = \prod_{n=2}^{N+1} \left[\int d^D q^\mu_{n-1} \sqrt{g(q_{n-1})} \right]. \qquad (10.124)$$

This expression is not directly applicable. When trying to do the $d^D q^\mu_{n-1}$-integrations successively, starting from the final integration over dq^μ_N, the integration variable q_{n-1} appears for each n in the argument of $\det\left[e^i_\mu(q_{n-1})\right]$ or $g_{\mu\nu}(q_{n-1})$. To make this q_{n-1}-dependence explicit, we expand in the measure (10.122) $e^i_\mu(q_{n-1}) = e^i{}_\mu(q_n - \Delta q_n)$ around the postpoint q_n into powers of Δq_n. This gives

$$dx^i = e^i_\mu(q - \Delta q)dq^\mu = e^i_\mu dq^\mu - e^i{}_{\mu,\nu}dq^\mu \Delta q^\nu + \frac{1}{2}e^i{}_{\mu,\nu\lambda}dq^\mu \Delta q^\nu \Delta q^\lambda + \dots \quad (10.125)$$

omitting, as before, the subscripts of q_n and Δq_n. Thus the Jacobian of the coordinate transformation from dx^i to dq^μ is

$$J_0 = \det(e^i{}_\kappa)\det\left[\delta^\kappa{}_\mu - e_i{}^\kappa e^i{}_{\mu,\nu}\Delta q^\nu + \frac{1}{2}e_i{}^\kappa e^i{}_{\mu,\nu\lambda}\Delta q^\nu \Delta q^\lambda\right], \qquad (10.126)$$

giving the relation between the infinitesimal integration volumes $d^D x^i$ and $d^D q^\mu$:

$$\prod_{n=2}^{N+1} \int d^D x^i_{n-1} = \prod_{n=2}^{N+1} \left\{ \int d^D q^\mu_{n-1} J_{0n} \right\}. \qquad (10.127)$$

The well-known expansion formula

$$\det(1 + B) = \exp\operatorname{tr}\log(1 + B) = \exp\operatorname{tr}(B - B^2/2 + B^3/3 - \dots) \qquad (10.128)$$

allows us now to rewrite J_0 as

$$J_0 = \det(e^i{}_\kappa)\exp\left(\frac{i}{\hbar}\mathcal{A}^\epsilon_{J_0}\right), \qquad (10.129)$$

with the determinant $\det(e^i_\mu) = \sqrt{g(q)}$ evaluated at the postpoint. This equation defines an effective action associated with the Jacobian, for which we obtain the expansion

$$\frac{i}{\hbar}\mathcal{A}^\epsilon_{J_0} = -e_i{}^\kappa e^i{}_{\kappa,\mu}\Delta q^\mu + \frac{1}{2}\left[e_i{}^\mu e^i{}_{\mu,\nu\lambda} - e_i{}^\mu e^i{}_{\kappa,\nu}e_j{}^\kappa e^j{}_{\mu,\lambda}\right]\Delta q^\nu \Delta q^\lambda + \dots . \quad (10.130)$$

The expansion coefficients are expressed in terms of the affine connection (10.16) using the relations:

$$e_{i\nu,\mu}e^i{}_{\kappa,\lambda} = e_i{}^\sigma e^i{}_{\nu,\mu}e_{j\sigma}e^j{}_{\kappa,\lambda} = \Gamma_{\mu\nu}{}^\sigma \Gamma_{\lambda\kappa\sigma} \qquad (10.131)$$

$$e_{i\mu}e^i{}_{\nu,\lambda\kappa} = g_{\mu\tau}[\partial_\kappa(e_i{}^\tau e^i{}_{\nu,\lambda}) - e^{i\sigma}e^i{}_{\nu,\lambda}e^j{}_\sigma e^{j\tau}{}_{,\kappa}]$$

$$= g_{\mu\tau}(\partial_\kappa\Gamma_{\lambda\nu}{}^\tau + \Gamma_{\lambda\nu}{}^\sigma \Gamma_{\kappa\sigma}{}^\tau). \qquad (10.132)$$

The Jacobian action becomes therefore:

$$\frac{i}{\hbar}\mathcal{A}_{J_0}^{\epsilon} = -\Gamma_{\mu\nu}{}^{\nu}\Delta q^{\mu} + \frac{1}{2}\partial_{\mu}\Gamma_{\nu\kappa}{}^{\kappa}\Delta q^{\nu}\Delta q^{\mu} + \dots . \tag{10.133}$$

The same result would, incidentally, be obtained by writing the Jacobian in accordance with (10.124) as

$$J_0 = \sqrt{g(q - \Delta q)}, \tag{10.134}$$

which leads to the alternative formula for the Jacobian action

$$\exp\left(\frac{i}{\hbar}\mathcal{A}_{J_0}^{\epsilon}\right) = \frac{\sqrt{g(q - \Delta q)}}{\sqrt{g(q)}}. \tag{10.135}$$

An expansion in powers of Δq gives

$$\exp\left(\frac{i}{\hbar}\mathcal{A}_{J_0}^{\epsilon}\right) = 1 - \frac{1}{\sqrt{g(q)}}\sqrt{g(q)}_{,\mu}\Delta q^{\mu} + \frac{1}{2\sqrt{g(q)}}\sqrt{g(q)}_{,\mu\nu}\Delta q^{\mu}\Delta q^{\nu} + \dots . \tag{10.136}$$

Using the formula

$$\frac{1}{\sqrt{g}}\partial_{\mu}\sqrt{g} = \frac{1}{2}g^{\sigma\tau}\partial_{\mu}g_{\sigma\tau} = \bar{\Gamma}_{\mu\nu}{}^{\nu}, \tag{10.137}$$

this becomes

$$\exp\left(\frac{i}{\hbar}\mathcal{A}_{J_0}^{\epsilon}\right) = 1 - \bar{\Gamma}_{\mu\nu}{}^{\nu}\Delta q^{\mu} + \frac{1}{2}(\partial_{\mu}\bar{\Gamma}_{\nu\lambda}{}^{\lambda} + \bar{\Gamma}_{\mu\sigma}{}^{\sigma}\bar{\Gamma}_{\nu\lambda}{}^{\lambda})\Delta q^{\mu}\Delta q^{\nu} + \dots , \tag{10.138}$$

so that

$$\frac{i}{\hbar}\mathcal{A}_{J_0}^{\epsilon} = -\bar{\Gamma}_{\mu\nu}{}^{\nu}\Delta q^{\mu} + \frac{1}{2}\partial_{\mu}\bar{\Gamma}_{\nu\lambda}{}^{\lambda}\Delta q^{\mu}\Delta q^{\nu} + \dots . \tag{10.139}$$

In a space without torsion where $\bar{\Gamma}_{\mu\nu}^{\lambda} \equiv \Gamma_{\mu\nu}{}^{\lambda}$, the Jacobian actions (10.133) and (10.139) are trivially equal to each other. But the equality holds also in the presence of torsion. Indeed, when inserting the decomposition (10.27), $\Gamma_{\mu\nu}{}^{\lambda} = \bar{\Gamma}_{\mu\nu}{}^{\lambda} + K_{\mu\nu}{}^{\lambda}$, into (10.133), the contortion tensor drops out since it is antisymmetric in the last two indices and these are contracted in both expressions.

In terms of $\mathcal{A}_{J_{0n}}^{\epsilon}$, we can rewrite the transformed measure (10.122) in the more useful form

$$\prod_{n=2}^{N+1}\int d^{D}x_{n-1}^{i} = \prod_{n=2}^{N+1}\left\{\int d^{D}q_{n-1}^{\mu}\det\left[e_{\mu}^{i}(q_n)\right]\exp\left(\frac{i}{\hbar}\mathcal{A}_{J_{0n}}^{\epsilon}\right)\right\}. \tag{10.140}$$

In a flat space parametrized in terms of curvilinear coordinates, the right-hand sides of (10.122) and (10.140) are related by an ordinary coordinate transformation, and both give the correct measure for a time-sliced path integral. In a general metric-affine space, however, this is no longer true. Since the mapping $dx^i \to dq^\mu$ is nonholonomic, there are in principle infinitely many ways of transforming the path integral measure from Cartesian coordinates to a non-Euclidean space. Among these, there exists a preferred mapping which leads to the correct quantum-mechanical amplitude in all known physical systems. This will serve to solve the path integral of the Coulomb system in Chapter 13.

The clue for finding the correct mapping is offered by an unaesthetic feature of Eq. (10.125): The expansion contains both differentials dq^μ and differences Δq^μ. This is somehow inconsistent. When time-slicing the path integral, the differentials dq^μ in the action are increased to finite differences Δq^μ. Consequently, the differentials in the measure should also become differences. A relation such as (10.125) containing simultaneously differences and differentials should not occur.

It is easy to achieve this goal by changing the starting point of the nonholonomic mapping and rewriting the initial flat space path integral (10.89) as

$$(\mathbf{x}\,t|\mathbf{x}'t') = \frac{1}{\sqrt{2\pi i \epsilon \hbar/M}^D} \prod_{n=1}^{N} \left[\int_{-\infty}^{\infty} d\Delta x_n \right] \prod_{n=1}^{N+1} K_0^\epsilon(\Delta \mathbf{x}_n). \qquad (10.141)$$

Since x_n are Cartesian coordinates, the measures of integration in the time-sliced expressions (10.89) and (10.141) are certainly identical:

$$\prod_{n=1}^{N} \int d^D x_n \equiv \prod_{n=2}^{N+1} \int d^D \Delta x_n. \qquad (10.142)$$

Their images under a nonholonomic mapping, however, are different so that the initial form of the time-sliced path integral is a matter of choice. The initial form (10.141) has the obvious advantage that the integration variables are precisely the quantities Δx_n^i which occur in the short-time amplitude $K_0^\epsilon(\Delta x_n)$.

Under a nonholonomic transformation, the right-hand side of Eq. (10.142) leads to the integral measure in a general metric-affine space

$$\prod_{n=2}^{N+1} \int d^D \Delta x_n \to \prod_{n=2}^{N+1} \left[\int d^D \Delta q_n \, J_n \right], \qquad (10.143)$$

with the Jacobian following from (10.96) (omitting n)

$$J = \frac{\partial(\Delta x)}{\partial(\Delta q)} = \det(e^i{}_\kappa) \det\left[\delta_\mu{}^\lambda - \Gamma_{\{\mu\nu\}}{}^\lambda \Delta q^\nu + \frac{1}{2}\left(\partial_{\{\sigma}\Gamma_{\mu\nu\}}{}^\lambda + \Gamma_{\{\mu\nu}{}^\tau \Gamma_{\{\tau|\sigma\}}{}^\lambda \right) \Delta q^\nu \Delta q^\sigma + \ldots \right]. \qquad (10.144)$$

In a space with curvature and torsion, the measure on the right-hand side of (10.143) replaces the flat-space measure on the right-hand side of (10.124). The curly double brackets around the indices ν, κ, σ, μ indicate a symmetrization in τ and σ followed by a symmetrization in μ, ν, and σ. With the help of formula (10.128) we now calculate the Jacobian action

$$\frac{i}{\hbar} \mathcal{A}_J^\epsilon = -\Gamma_{\{\mu\nu\}}{}^\mu \Delta q^\nu + \frac{1}{2} \left[\partial_{\{\mu}\Gamma_{\nu\kappa\}}{}^\kappa + \Gamma_{\{\nu\kappa}{}^\sigma \Gamma_{\{\sigma|\mu\}\}}{}^\kappa - \Gamma_{\{\nu\kappa\}}{}^\sigma \Gamma_{\{\sigma\mu\}}{}^\kappa \right] \Delta q^\nu \Delta q^\mu + \ldots . \qquad (10.145)$$

This expression differs from the earlier Jacobian action (10.133) by the symmetrization symbols. Dropping them, the two expressions coincide. This is allowed if q^μ are curvilinear coordinates in a flat space. Since then the transformation functions $x^i(q)$ and their first derivatives $\partial_\mu x^i(q)$ are integrable and possess commuting derivatives, the two Jacobian actions (10.133) and (10.145) are identical.

There is a further good reason for choosing (10.142) as a starting point for the nonholonomic transformation of the measure. According to Huygens' principle of wave optics, each point of a wave front is a center of a new spherical wave propagating from that point. Therefore, in a time-sliced path integral, the differences Δx_n^i play a more fundamental role than the coordinates themselves. Intimately related to this is the observation that in the canonical form, a short-time piece of the action reads

$$\int \frac{dp_n}{2\pi\hbar} \exp\left[\frac{i}{\hbar}p_n(x_n - x_{n-1}) - \frac{ip_n^2}{2M\hbar}t\right].$$

Each momentum is associated with a coordinate difference $\Delta x_n \equiv x_n - x_{n-1}$. Thus, we should expect the spatial integrations conjugate to p_n to run over the coordinate differences $\Delta x_n = x_n - x_{n-1}$ rather than the coordinates x_n themselves, which makes the important difference in the subsequent nonholonomic coordinate transformation.

We are thus led to postulate the following time-sliced path integral in q-space:

$$\langle q|e^{-i(t-t')\hat{H}/\hbar}|q'\rangle = \frac{1}{\sqrt{2\pi i\hbar\epsilon/M}^D} \prod_{n=2}^{N+1}\left[\int d^D\Delta q_n \frac{\sqrt{g(q_n)}}{\sqrt{2\pi i\epsilon\hbar/M}^D}\right] e^{i\sum_{n=1}^{N+1}(\mathcal{A}^\epsilon + \mathcal{A}_j^\epsilon)/\hbar},$$

(10.146)

where the integrals over Δq_n may be performed successively from $n = N$ down to $n = 1$.

Let us emphasize that this expression has not been *derived* from the flat space path integral. It is the result of a specific new *quantum equivalence principle* which rules how a flat space path integral behaves under nonholonomic coordinate transformations.

It is useful to re-express our result in a different form which clarifies best the relation with the naively expected measure of path integration (10.124), the product of integrals

$$\prod_{n=1}^{N}\int d^D x_n = \prod_{n=1}^{N}\left[\int d^D q_n \sqrt{g(q_n)}\right].$$

(10.147)

The measure in (10.146) can be expressed in terms of (10.147) as

$$\prod_{n=2}^{N+1}\left[\int d^D\Delta q_n \sqrt{g(q_n)}\right] = \prod_{n=1}^{N}\left[\int d^D q_n \sqrt{g(q_n)}e^{-i\mathcal{A}_{J_0}^\epsilon/\hbar}\right].$$

(10.148)

The corresponding expression for the entire time-sliced path integral (10.146) in the metric-affine space reads

$$\langle q|e^{-i(t-t')\hat{H}/\hbar}|q'\rangle = \frac{1}{\sqrt{2\pi i\hbar\epsilon/M}^D} \prod_{n=1}^{N} \left[\int d^D q_n \frac{\sqrt{g(q_n)}}{\sqrt{2\pi i\hbar\epsilon/M}^D} \right] e^{i\sum_{n=1}^{N+1}(\mathcal{A}^\epsilon + \Delta\mathcal{A}_J^\epsilon)/\hbar},$$

(10.149)

where $\Delta\mathcal{A}_J^\epsilon$ is the difference between the correct and the wrong Jacobian actions in Eqs. (10.133) and (10.145):

$$\Delta\mathcal{A}_J^\epsilon \equiv \mathcal{A}_J^\epsilon - \mathcal{A}_{J_0}^\epsilon.$$

(10.150)

In the absence of torsion where $\Gamma_{\{\mu\nu\}}{}^\lambda = \bar{\Gamma}_{\mu\nu}{}^\lambda$, this simplifies to

$$\frac{i}{\hbar}\Delta\mathcal{A}_J^\epsilon = \frac{1}{6}\bar{R}_{\mu\nu}\Delta q^\mu \Delta q^\nu,$$

(10.151)

where $\bar{R}_{\mu\nu}$ is the Ricci tensor associated with the Riemann curvature tensor, i.e., the contraction (10.41) of the Riemann curvature tensor associated with the Christoffel symbol $\bar{\Gamma}_{\mu\nu}{}^\lambda$.

Being quadratic in Δq, the effect of the additional action can easily be evaluated perturbatively using the methods explained in Chapter 8, according to which $\Delta q^\mu \Delta q^\nu$ may be replaced by its lowest order expectation

$$\langle \Delta q^\mu \Delta q^\nu \rangle_0 = i\epsilon\hbar g^{\mu\nu}(q)/M.$$

Then $\Delta\mathcal{A}_J^\epsilon$ yields the additional effective potential

$$V_{\text{eff}}(q) = -\frac{\hbar^2}{6M}\bar{R}(q),$$

(10.152)

where $\bar{R}$ is the Riemann curvature scalar. By including this potential in the action, the path integral in a curved space can be written down in the naive form (10.147) as follows:

$$\langle q|e^{-i(t-t')\hat{H}/\hbar}|q'\rangle = \frac{1}{\sqrt{2\pi i\hbar\epsilon/M}^D} \prod_{n=1}^{N} \left[\int d^D q_n \frac{\sqrt{g(q_n)}}{\sqrt{2\pi i\epsilon\hbar/M}^D} e^{i\hbar R(q_n)/6M} \right] e^{i\sum_{n=1}^{N+1}\mathcal{A}^\epsilon[q_n]/\hbar}.$$

(10.153)

This time-sliced expression will from now on be the definition of a path integral in curved space written in the continuum notation as

$$\langle q|e^{-i(t-t')\hat{H}/\hbar}|q'\rangle = \int \mathcal{D}^D q \sqrt{g(q)}\, e^{i\int_{t_a}^{t_b} dt\, \mathcal{A}[q]/\hbar}.$$

(10.154)

The integrals over q_n in (10.153) are conveniently performed successively downwards over $\Delta q_{n+1} = q_{n+1} - q_n$ at fixed q_{n+1}. The weights $\sqrt{g(q_n)} = \sqrt{g(q_{n+1} - \Delta q_{n+1})}$ require a postpoint expansion leading to the naive Jacobian J_0 of (10.126) and the Jacobian action $\mathcal{A}_{J_0}^\epsilon$ of Eq. (10.133).

It is important to observe that the above time-sliced definition is automatically invariant under coordinate transformations. This is an immediate consequence of the definition via the nonholonomic mapping from a flat-space path integral.

It goes without saying that the path integral (10.153) also has a phase space version. It is obtained by omitting all $(M/2\epsilon)(\Delta q_n)^2$ terms in the short-time actions $\mathcal{A}^\epsilon$ and extending the multiple integral by the product of momentum integrals

$$\prod_{n=1}^{N+1}\left[\frac{dp_n}{2\pi\hbar\sqrt{g(q_n)}}\right] e^{(i/\hbar)\sum_{n=1}^{N+1}\left[p_{n\mu}\Delta q^\mu-\epsilon\frac{1}{2M}g^{\mu\nu}(q_n)p_{n\mu}p_{n\nu}\right]}. \tag{10.155}$$

When using this expression, all problems which were encountered in the literature with canonical transformations of path integrals disappear.

An important property of the definition of the path integral in spaces with curvature and torsion as a nonholonomic image of a Euclidean path integral is that this image is automatically invariant under ordinary holonomic coordinate transformations.

10.4 Completing the Solution of Path Integral on Surface of Sphere in D Dimensions

The measure of path integration in Eq. (10.146) allows us to finally complete the calculation, initiated in Sections 8.7–8.9, of the path integrals of a point particle on the surface of a sphere on group spaces in any number of dimensions. Indeed, using the result (10.152) we are now able to solve the problems discussed in Section 8.7 in conjunction with the energy formula (8.224). Thus we are finally in a position to find the correct energies and amplitudes of these systems.

A sphere of radius r embedded in D dimensions has an intrinsic dimension $D' \equiv D - 1$ and a curvature scalar

$$\bar{R} = \frac{(D' - 1)D'}{r^2}. \tag{10.156}$$

This is most easily derived as follows. Consider a line element in D dimensions

$$(d\mathbf{x})^2 = (dx^1)^2 + (dx^2)^2 + \ldots + (dx^D)^2 \tag{10.157}$$

and restrict the motion to a spherical surface

$$(x^1)^2 + (x^2)^2 + \ldots + (x^D)^2 = r^2, \tag{10.158}$$

by eliminating x^D. This brings (10.157) to the form

$$(d\mathbf{x})^2 = (dx^1)^2 + (dx^2)^2 + \ldots + (dx^{D'})^2 + \frac{(x^1dx^1 + dx^2 + \ldots + x^{D'}dx^{D'})^2}{r^2 - r'^2}, \tag{10.159}$$

where $r'^2 \equiv (x^1)^2 + (x^2)^2 + \ldots + (x^{D'})^2$. The metric on the D'-dimensional surface is therefore

$$g_{\mu\nu}(x) = \delta_{\mu\nu} + \frac{x^\mu x^\nu}{r^2 - r'^2}. \tag{10.160}$$

Since $\bar{R}$ will be constant on the spherical surface, we may evaluate it for small x^μ ($\mu = 1, \ldots, D'$) where $g_{\mu\nu}(x) \approx \delta_{\mu\nu} + x^\mu x^\nu / r^2$ and the Christoffel symbols (10.7) are $\Gamma_{\mu\nu}{}^\lambda \approx \bar{\Gamma}_{\mu\nu\lambda} \approx \delta_{\mu\nu} x_\lambda / r^2$. Inserting this into (10.35) we obtain the curvature tensor for small x^μ:

$$\bar{R}_{\mu\nu\lambda\kappa} \approx \frac{1}{r^2} \left(\delta_{\mu\kappa} \delta_{\nu\lambda} - \delta_{\mu\lambda} \delta_{\nu\kappa} \right). \tag{10.161}$$

This can be extended covariantly to the full surface of the sphere by replacing $\delta_{\mu\lambda}$ by the metric $g_{\mu\lambda}(x)$:

$$\bar{R}_{\mu\nu\lambda\kappa}(x) = \frac{1}{r^2} \left[g_{\mu\kappa}(x) g_{\nu\lambda}(x) - g_{\mu\lambda}(x) g_{\nu\kappa}(x) \right], \tag{10.162}$$

so that Ricci tensor is [recall (10.41)]

$$\bar{R}_{\nu\kappa}(x) = \bar{R}_{\mu\nu\kappa}{}^\mu(x) = \frac{D' - 1}{r^2} g_{\nu\kappa}(x). \tag{10.163}$$

Contracting this with $g^{\nu\kappa}$ [recall (10.42)] yields indeed the curvature scalar (10.156). The effective potential (10.152) is therefore

$$V_{\text{eff}} = -\frac{\hbar^2}{6Mr^2}(D - 2)(D - 1). \tag{10.164}$$

It supplies precisely the missing energy which changes the energy (8.224) *near* the sphere, corrected by the expectation of the quartic term ϑ_n^4 in the action, to the proper value

$$E_l = \frac{\hbar^2}{2Mr^2} l(l + D - 2). \tag{10.165}$$

Astonishingly, this elementary result of Schrödinger quantum mechanics was found only a decade ago by path integration [5]. Other time-slicing procedures yield extra terms proportional to the scalar curvature $\bar{R}$, which are absent in our theory. Here the scalar curvature is a trivial constant, so it would remain undetectable in atomic experiments which measure only energy differences. The same result will be derived in general in Eqs. (11.25) and in Section 11.3.

An important property of this spectrum is that the ground state energy vanished for all dimensions D. This property would not have been found in the naive measure of path integration on the right-hand side of Eq. (10.147) which is used in most works on this subject. The correction term (10.151) coming from the nonholonomic mapping of the flat-space measure is essential for the correct result.

More evidence for the correctness of the measure in (10.146) will be supplied in Chapter 13 where we solve the path integrals of the most important atomic system, the hydrogen atom.

We remark that for $t \to t'$, the amplitude (10.153) shows the states $|q\rangle$ to obey the covariant orthonormality relation

$$\langle q|q'\rangle = \sqrt{g(q)}^{-1}\delta^{(D)}(q - q'). \tag{10.166}$$

The completeness relation reads

$$\int d^D q \sqrt{g(q)}|q\rangle\langle q| = 1. \tag{10.167}$$

10.5   External Potentials and Vector Potentials

An important generalization of the above path integral formulas (10.146), (10.149), (10.153) of a point particle in a space with curvature and torsion includes the presence of an external potential and a vector potential. These allow us to describe, for instance, a particle in external electric and magnetic fields. The classical action is then

$$\mathcal{A}_{\rm em} = \int_{t_a}^{t_b} dt \left[\frac{e}{c}A_\mu(q(t))\dot{q}^\mu - V(q(t)) \right]. \tag{10.168}$$

To find the time-sliced action we proceed as follows. First we set up the correct time-sliced expression in Euclidean space and Cartesian coordinates. For a single slice it reads, in the postpoint form,

$$\mathcal{A}^\epsilon = \frac{M}{2\epsilon}(\Delta x^i)^2 + \frac{e}{c}A_i(\mathbf{x})\Delta x^i - \frac{e}{2c}A_{i,j}(\mathbf{x})\Delta x^i \Delta x^j - \epsilon V(\mathbf{x}) + \dots . \tag{10.169}$$

As usual, we have neglected terms which do not contribute in the continuum limit. The derivation of this time-sliced expression proceeds by calculating, as in (10.116), the action

$$\mathcal{A}^\epsilon = \int_{t-\epsilon}^t dt L(t) \tag{10.170}$$

along the classical trajectory in Euclidean space, where

$$L(t) = \frac{M}{2}\dot{\mathbf{x}}^2(t) + \frac{e}{c}\mathbf{A}(\mathbf{x}(t))\dot{\mathbf{x}}(t) - V(\mathbf{x}(t)) \tag{10.171}$$

is the classical Lagrangian. In contrast to (10.116), however, the Lagrangian has now a nonzero time derivative (omitting the time arguments):

$$\frac{d}{dt}L = M\dot{\mathbf{x}}\ddot{\mathbf{x}} + \frac{e}{c}\mathbf{A}(\mathbf{x})\ddot{\mathbf{x}} + \frac{e}{c}A_{i,j}(\mathbf{x})\dot{x}^i\dot{x}^j - V_i(\mathbf{x})\dot{x}^i. \tag{10.172}$$

For this reason we cannot simply write down an expression such as (10.117) but we have to expand the Lagrangian around the postpoint leading to the series

$$\mathcal{A}^{\epsilon} = \int_{t-\epsilon}^{t} dt\, L(t) = \epsilon L(t) - \frac{1}{2}\epsilon^2 \frac{d}{dt}L(t) + \dots . \tag{10.173}$$

The evaluation makes use of the equation of motion

$$M\ddot{x}^i = -\frac{e}{c}(A_{i,j}(\mathbf{x}) - A_{j,i}(\mathbf{x}))\dot{x}^j - V_i(\mathbf{x}), \tag{10.174}$$

from which we derive the analog of Eq. (10.96): First we have the postpoint expansion

$$\begin{aligned} \Delta x^i &= -\epsilon\dot{x}^i + \frac{1}{2}\epsilon^2\ddot{x}^i + \dots \\ &= -\epsilon\dot{x}^i - \frac{e}{2Mc}\epsilon^2\left[(A_{i,j} - A_{j,i})\dot{x}^j + V_i(\mathbf{x})\right] + \dots . \end{aligned} \tag{10.175}$$

Inverting this gives

$$\dot{x}^i = -\frac{\Delta x^i}{\epsilon} - \frac{e}{2Mc}(A_{i,j} - A_{j,i})\Delta x^j + \dots . \tag{10.176}$$

When inserted into (10.173), this yields indeed the time-sliced short-time action (10.169).

The quadratic term $\Delta x^i \Delta x^j$ in the action (10.169) can be replaced by the perturbative expectation value

$$\Delta x^i \Delta x^j \rightarrow \langle \Delta x^i \Delta x^j \rangle = \delta_{ij}\, i\frac{\hbar\epsilon}{M}, \tag{10.177}$$

so that $\mathcal{A}^{\epsilon}$ becomes

$$\mathcal{A}^{\epsilon} = \frac{M}{2\epsilon}(\Delta x^i)^2 + \frac{e}{c}A_i(\mathbf{x})\Delta x^i - i\epsilon\frac{\hbar e}{2Mc}A_{i,i}(\mathbf{x}) - \epsilon V(\mathbf{x}) + \dots . \tag{10.178}$$

Incidentally, the action (10.169) could also have been written as

$$\mathcal{A}^{\epsilon} = \frac{M}{2\epsilon}(\Delta x^i)^2 + \frac{e}{c}A_i(\bar{\mathbf{x}})\Delta x^i - \epsilon V(\mathbf{x}) + \dots , \tag{10.179}$$

where $\bar{\mathbf{x}}$ is the midpoint value of the slice coordinates

$$\bar{\mathbf{x}} = \mathbf{x} - \frac{1}{2}\Delta\mathbf{x}, \tag{10.180}$$

i.e., more explicitly,

$$\bar{\mathbf{x}}(t_n) \equiv \frac{1}{2}[\mathbf{x}(t_n) + \mathbf{x}(t_{n-1})]. \tag{10.181}$$

Thus, with an external vector potential in Cartesian coordinates, a midpoint "prescription" for $\mathcal{A}^\epsilon$ happens to yield the correct expression (10.179).

Having found the time-sliced action in Cartesian coordinates, it is easy to go over to spaces with curvature and torsion. We simply insert the nonholonomic transformation (10.96) for the differentials Δx^i. This gives again the short-time action (10.107), extended by the interaction due to the potentials

$$\mathcal{A}^\epsilon_{\text{em}} = \frac{e}{c}A_\mu\Delta q^\mu - \frac{e}{2c}\partial_\nu A_\mu\Delta q^\mu\Delta q^\nu - \epsilon V(q) + \dots \ . \tag{10.182}$$

The second term can be evaluated perturbatively leading to

$$\mathcal{A}^\epsilon_{\text{em}} = \frac{e}{c}A_\mu\Delta q^\mu - i\epsilon\frac{\hbar e}{2Mc}\partial_\mu A^\mu - \epsilon V(q) + \dots \ . \tag{10.183}$$

The sum over all slices,

$$\mathcal{A}^N_{\text{em}} = \sum_{n=1}^{N+1} \mathcal{A}^\epsilon_{\text{em}} , \tag{10.184}$$

has to be added to the action in each time-sliced expression (10.146), (10.149), and (10.153).

10.6 Perturbative Calculation of Path Integrals in Curved Space

In Sections 2.15 and 3.21 we have given a perturbative definition of path integrals which does not require the rather cumbersome time slicing but deals directly with a continuous time. We shall now extend this definition to curved space in such a way that it leads to the same result as the time-sliced definition given in Section 10.3. In particular, we want to ensure that this definition preserves the fundamental property of coordinate independence achieved in the time-sliced definition via the nonholonomic mapping principle, as observed at the end of Subsection 10.3.2. In a perturbative calculation, this property will turn out to be highly nontrivial. In addition, we want to be sure that the ground state energy of a particle on a sphere is zero in any number of dimensions, just as in the time-sliced calculation leading to Eq. (10.165). This implies that also in the perturbative definition of path integral, the operator-ordering problem will be completely solved.

10.6.1 Free and Interacting Parts of Action

The partition function of a point particle in a curved space with an intrinsic dimension D is given by the path integral over all periodic paths on the imaginary-time axis τ:

$$Z = \int \mathcal{D}^D q\sqrt{g}\,e^{-\mathcal{A}[q]}, \tag{10.185}$$

where $\mathcal{A}[q]$ is the Euclidean action

$$\mathcal{A}[q] = \int_0^\beta d\tau \left[\frac{1}{2} g_{\mu\nu}(q(\tau)) \dot{q}^\mu(\tau) \dot{q}^\nu(\tau) + V(q(\tau)) \right]. \tag{10.186}$$

We have set $\hbar$ and the particle mass M equal to unity. For a space with constant curvature, this is a generalization of the action for a particle on a sphere (8.146), also called a *nonlinear σ-model* (see p. 738). The perturbative definition of Sections 2.15 and 3.21 amounts to the following calculation rules. Expand the metric $g_{\mu\nu}(q)$ and the potential $V(q)$ around some point q_a^μ in powers of $\delta q^\mu \equiv q^\mu - q_a^\mu$. After this, separate the action $\mathcal{A}[q]$ into a harmonically fluctuating part

$$\mathcal{A}^{(0)}[q_a; \delta q] \equiv \frac{1}{2} \int_0^\beta d\tau \, g_{\mu\nu}(q_a) \left[\delta \dot{q}^\mu(\tau) \delta \dot{q}^\nu(\tau) + \omega^2 \delta q^\mu(\tau) \delta q^\nu(\tau) \right], \tag{10.187}$$

and an interacting part

$$\mathcal{A}^{\text{int}}[q_a; \delta q] \equiv \mathcal{A}[q] - \mathcal{A}^{(0)}[q_a; \delta q]. \tag{10.188}$$

The second term in (10.187) is called *frequency term* or *mass term*. It is not invariant under coordinate transformations. The implications of this will be seen later. When studying the partition function in the limit of large β, the frequency ω cannot be set equal to zero since this would lead to infinities in the perturbation expansion, as we shall see below.

A delicate problem is posed by the square root of the determinant of the metric in the functional integration measure in (10.185). In a purely formal continuous definition of the measure, we would write it as

$$\int \mathcal{D}^D q \sqrt{g} \equiv \prod_\tau \int d^D q(\tau) \sqrt{g(\tau)} = \left[\prod_\tau \int d^D q(\tau) \sqrt{g(q_a)} \right] \exp \left[\frac{1}{2} \sum_\tau \log \frac{g(q(\tau))}{g(q_a)} \right]. \tag{10.189}$$

The formal sum over all continuous times τ in the exponent corresponds to an integral $\int d\tau$ divided by the spacing of the points, which on a sliced time axis would be the slicing parameter ϵ. Here it is $d\tau$. The ratio $1/d\tau$ may formally be identified with $\delta(0)$, in accordance with the defining integral $\int d\tau \, \delta(\tau) = 1$. The infinity of $\delta(0)$ may be regularized in some way, for instance by a cutoff in the Fourier representation $\delta(0) \equiv \int d\omega/(2\pi)$ at large frequencies ω, a so-called *UV-cutoff*. Leaving the regularization unspecified, we rewrite the measure (10.189) formally as

$$\int \mathcal{D}^D q \sqrt{g} \equiv \left[\prod_\tau \int d^D q(\tau) \sqrt{g(q_a)} \right] \exp \left[\frac{1}{2} \delta(0) \int_0^\beta d\tau \log \frac{g(q(\tau))}{g(q_a)} \right], \tag{10.190}$$

and further as

$$\int \mathcal{D}^D q \sqrt{g(q_a)} \, e^{-\mathcal{A}^g[q]}, \tag{10.191}$$

where we have introduced an effective action associated with the measure:

$$A_g[q] = -\frac{1}{2}\delta(0) \int_0^\beta d\tau \, \log \frac{g(q(\tau))}{g(q_a)}. \tag{10.192}$$

For a perturbative treatment, this action is expanded in powers of $\delta q(\tau)$ and is a functional of this variable:

$$A_g[q_a, \delta q] = -\frac{1}{2}\delta(0) \int_0^\beta d\tau \, [\log g(q_a + \delta q(\tau)) - \log g(q_a)]. \tag{10.193}$$

This is added to (10.188) to yield the total interaction

$$A_{\text{tot}}^{\text{int}}[q_a, \delta q] = A^{\text{int}}[q_a, \delta q] + A_g[q_a, \delta q]. \tag{10.194}$$

The path integral for the partition function is now written as

$$Z = \int \mathcal{D}^D q \, \sqrt{g(q_a)} \, e^{-A^{(0)}[q]} \, e^{-A_{\text{tot}}^{\text{int}}[q]}. \tag{10.195}$$

According to the rules of perturbation theory, we expand the factor $e^{-A_{\text{tot}}^{\text{int}}}$ in powers of the total interaction, and obtain the perturbation series

$$\begin{aligned} Z &= \int \mathcal{D}^D q \, \sqrt{g(q_a)} \left(1 - A_{\text{tot}}^{\text{int}} + \frac{1}{2}\langle A_{\text{tot}}^{\text{int}\,2}\rangle - \ldots\right) e^{-A^{(0)}[q]} \\ &= Z_\omega \left[1 - \langle A_{\text{tot}}^{\text{int}}\rangle + \frac{1}{2!}\langle A_{\text{tot}}^{\text{int}\,2}\rangle - \ldots\right], \end{aligned} \tag{10.196}$$

where

$$Z_\omega \equiv e^{-\beta F_\omega} = \int \mathcal{D}^D q \sqrt{g(q_a)} \, e^{-A^{(0)}[q]} \tag{10.197}$$

is the path integral over the free part, and the symbol $\langle \ldots \rangle$ denotes the expectation values in this path integral

$$\langle \ldots \rangle = Z_\omega^{-1} \int \mathcal{D}q \, \sqrt{g(q_a)} \, (\ldots) \, e^{-A^{(0)}[q]}. \tag{10.198}$$

With the usual definition of the cumulants $\langle A_{\text{tot}}^{\text{int}}\rangle_c = \langle A_{\text{tot}}^{\text{int}}\rangle$, $\langle A_{\text{tot}}^{\text{int}\,2}\rangle_c = \langle A_{\text{tot}}^{\text{int}\,2}\rangle - \langle A_{\text{tot}}^{\text{int}}\rangle^2, \ldots$ [recall (3.482), (3.483)], this can be written as

$$Z \equiv e^{-\beta F} = \exp\left[-\beta F_\omega - \langle A_{\text{tot}}^{\text{int}}\rangle_c + \frac{1}{2!}\langle A_{\text{tot}}^{\text{int}\,2}\rangle_c - \ldots\right], \tag{10.199}$$

where $F_\omega \equiv -\beta^{-1}\log Z$ the free energy associated with Z_ω.

The cumulants are now calculated according to Wick's rule order by order in $\hbar$, treating the δ-function at the origin $\delta(0)$ as if it were finite. The perturbation series will contain factors of $\delta(0)$ and its higher powers. Fortunately, these unpleasant terms will turn out to cancel each other at each order in a suitably defined expansion parameter. On account of these cancellations, we may ultimately discard all terms containing $\delta(0)$, or set $\delta(0)$ equal to zero, in accordance with Veltman's rule (2.506).

The harmonic path integral (10.197) is performed using formulas (2.487) and (2.503). Assuming for a moment what we shall prove below that we may choose coordinates in which $g_{\mu\nu}(q_a) = \delta_{\mu\nu}$, we obtain directly in D dimensions

$$Z_\omega = \int \mathcal{D}q\, e^{-\mathcal{A}_\omega[q]} = \exp\left[-\frac{D}{2}\mathrm{Tr}\log(-\partial^2 + \omega^2)\right] \equiv e^{-\beta F_\omega}. \tag{10.200}$$

The expression in brackets specifies the free energy F_ω of the harmonic oscillator at the inverse temperature β.

10.6.2 Zero Temperature

For simplicity, we fist consider the limit of zero temperature or $\beta \to \infty$. Then F_ω becomes equal to the sum of D ground state energies $\omega/2$ of the oscillator, one for each dimension:

$$F_\omega = \frac{1}{\beta}\frac{D}{2}\mathrm{Tr}\log(-\partial^2 + \omega^2) \underset{\beta\to\infty}{\longrightarrow} \frac{D}{2}\int_{-\infty}^{\infty}\frac{dk}{2\pi}\log(k^2 + \omega^2) = \frac{D}{2}\omega. \tag{10.201}$$

The Wick contractions in the cumulants $\left\langle \mathcal{A}_{\mathrm{tot}}^{\mathrm{int}\,2}\right\rangle_c$ of the expansion (10.198) contain only connected diagrams. They contain temporal integrals which, after suitable partial integrations, become products of the following basic correlation functions

$$G_{\mu\nu}^{(2)}(\tau,\tau') \equiv \langle q_\mu(\tau)q_\nu(\tau')\rangle = \underline{\qquad}\,, \tag{10.202}$$

$$\partial_\tau G_{\mu\nu}^{(2)}(\tau,\tau') \equiv \langle \dot{q}_\mu(\tau)q_\nu(\tau')\rangle = \underline{\qquad}\cdots\,, \tag{10.203}$$

$$\partial_{\tau'} G_{\mu\nu}^{(2)}(\tau,\tau') \equiv \langle q_\mu(\tau)\dot{q}_\nu(\tau')\rangle = \cdots\underline{\qquad}\,, \tag{10.204}$$

$$\partial_\tau\partial_{\tau'} G_{\mu\nu}^{(2)}(\tau,\tau') \equiv \langle \dot{q}_\mu(\tau)\dot{q}_\nu(\tau')\rangle = \cdots\cdots\,. \tag{10.205}$$

The right-hand sides define line symbols to be used for drawing Feynman diagrams for the interaction terms.

Under the assumption $g_{\mu\nu}(q_a) = \delta_{\mu\nu}$, the correlation function $G_{\mu\nu}^{(2)}(\tau,\tau')$ factorizes as

$$G_{\mu\nu}^{(2)}(\tau,\tau') = \delta_{\mu\nu}\,\Delta(\tau - \tau'), \tag{10.206}$$

with $\Delta(\tau - \tau')$ abbreviating the correlation the zero-temperature Green function $G_{\omega^2,e}^{\mathrm{p}}(\tau)$ of Eq. (3.246) (remember the present units with $M = \hbar = 1$):

$$\Delta(\tau - \tau') = \int_{-\infty}^{\infty}\frac{dk}{2\pi}\frac{e^{ik(\tau-\tau')}}{k^2 + \omega^2} = \frac{1}{2\omega}e^{-\omega|\tau-\tau'|}. \tag{10.207}$$

As a consequence, the second correlation function (10.203) has a discontinuity

$$\partial_\tau G_{\mu\nu}^{(2)}(\tau,\tau') = \delta_{\mu\nu}\,i\int_{-\infty}^{\infty}\frac{dk}{2\pi}\frac{e^{ik(\tau-\tau')}k}{k^2 + \omega^2} = \delta_{\mu\nu}\dot{\Delta}(\tau - \tau') \equiv -\frac{1}{2}\delta_{\mu\nu}\epsilon(\tau - \tau')e^{-\omega|\tau-\tau'|}, \tag{10.208}$$

where $\epsilon(\tau - \tau')$ is the distribution defined in Eq. (1.311) which has a jump at $\tau = \tau'$ from -1 to 1. It can be written as an integral over a δ-function:

$$\epsilon(\tau - \tau') \equiv -1 + 2 \int_{-\infty}^{\tau} d\tau'' \, \delta(\tau'' - \tau'). \tag{10.209}$$

The third correlation function (10.204) is simply the negative of (10.203):

$$\partial_{\tau'} G_{\mu\nu}^{(2)}(\tau, \tau') = -\partial_\tau G_{\mu\nu}^{(2)}(\tau, \tau') = -\delta_{\mu\nu} \dot{\Delta}(\tau - \tau'). \tag{10.210}$$

At the point $\tau = \tau'$, the momentum integral (10.208) vanishes by antisymmetry:

$$\partial_\tau G_{\mu\nu}^{(2)}(\tau, \tau')\big|_{\tau=\tau'} = -\partial_{\tau'} G_{\mu\nu}^{(2)}(\tau, \tau')\big|_{\tau=\tau'} = \delta_{\mu\nu} \, i \int_{-\infty}^{\infty} \frac{dk}{2\pi} \frac{k}{k^2 + \omega^2} = \delta_{\mu\nu} \dot{\Delta}(0) = 0. \tag{10.211}$$

The fourth correlation function (10.205) contains a δ-function:

$$\partial_\tau \partial_{\tau'} G_{\mu\nu}^{(2)}(\tau, \tau') = -\partial_\tau^2 G_{\mu\nu}^{(2)}(\tau, \tau') = \delta_{\mu\nu} \int_{-\infty}^{\infty} \frac{dk}{2\pi} \frac{e^{ik(\tau-\tau')} k^2}{k^2 + \omega^2} = -\delta_{\mu\nu} \ddot{\Delta}(\tau - \tau') \tag{10.212}$$

$$= \delta_{\mu\nu} \int_{-\infty}^{\infty} \frac{dk}{2\pi} e^{ik(\tau-\tau')} \left(1 - \frac{\omega^2}{k^2 + \omega^2}\right) = \delta_{\mu\nu} \delta(\tau - \tau') - \omega^2 G_{\mu\nu}^{(2)}(\tau, \tau').$$

The Green functions for $\mu = \nu$ are plotted in Fig. 10.4.

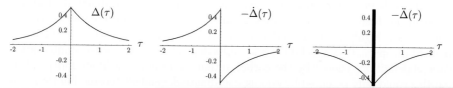

Figure 10.4 Green functions for perturbation expansions in curvilinear coordinates in natural units with $\omega = 1$. The third contains a δ-function at the origin.

The last equation is actually the defining equation for the Green function, which is always the solution of the inhomogeneous equation of motion associated with the harmonic action (10.187), which under the assumption $g_{\mu\nu}(q_a) = \delta_{\mu\nu}$ reads for each component:

$$-\ddot{q}(\tau) + \omega^2 q(\tau) = \delta(\tau - \tau'). \tag{10.213}$$

The Green function $\Delta(\tau - \tau')$ solves this equation, satisfying

$$\ddot{\Delta}(\tau) = \omega^2 \Delta(\tau) - \delta(\tau). \tag{10.214}$$

When trying to evaluate the different terms produced by the Wick contractions, we run into a serious problem. The different terms containing products of time derivatives of Green functions contain effectively products of δ-functions and Heaviside functions. In the mathematical theory of distributions, such integrals are undefined. We shall offer two ways to resolve this problem. One is based on extending the integrals over the time axis to integrals over a d-dimensional *time space*, and continuing the results at the end back to $d = 1$. The extension makes the path integral a *functional integral* of the type used in quantum field theories. It will turn out that this procedure leads to well-defined finite results, also for the initially divergent terms coming from the effective action of the measure (10.193). In addition, and very importantly, it guarantees that the perturbatively defined path integral is invariant under coordinate transformations.

For the time-sliced definition in Section 10.3, coordinate independence was an automatic consequence of the nonholonomic mapping from a flat-space path integral. In the perturbative definition, the coordinate independence has been an outstanding problem for many years, and was only solved recently in Refs. [23]–[25].

In d-dimensional quantum field theory, path integrals between two and four spacetime dimensions have been defined by perturbation expansions for a long time. Initial difficulties in guaranteeing coordinate independence were solved by 't Hooft and Veltman [29] using dimensional regularization with minimal subtractions. For a detailed description of this method see the textbook [30]. Coordinate independence emerges after calculating all Feynman integrals in an arbitrary number of dimensions d, and continuing the results to the desired physical integer value. Infinities occuring in the limit are absorbed into parameters of the action. In contrast, and surprisingly, numerous attempts [31]–[36] to define the simpler *quantum-mechanical* path integrals in curved space by perturbation expansions encountered problems. Although all final results are finite and unique, the Feynman integrals in the expansions are highly singular and mathematically undefined. When evaluated in momentum space, they yield different results depending on the order of integration. Various definitions chosen by the earlier authors were not coordinate-independent, and this could only be cured by adding coordinate-dependent "correction terms" to the classical action — a highly unsatisfactory procedure violating the basic Feynman postulate that physical amplitudes should consist of a sum over all paths with phase factors $e^{i\mathcal{A}/\hbar}$ containing only the classical actions along the paths.

The calculations in d spacetime dimensions and the continuation to $d = 1$ will turn out to be somewhat tedious. We shall therefore find in Subsection 10.11.4 a method of doing the calculations directly for $d = 1$.

10.7 Model Study of Coordinate Invariance

Let us consider first a simple model which exhibits typical singular Feynman integrals encountered in curvilinear coordinates and see how these can be turned into a finite perturbation expansion which is invariant under coordinate transformations. For simplicity, we consider an ordinary harmonic oscillator in one dimension, with the action

$$\mathcal{A}_\omega = \frac{1}{2} \int_0^\beta d\tau \left[\dot{x}^2(\tau) + \omega^2 x^2(\tau) \right]. \tag{10.215}$$

The partition function of this system is exactly given by (10.201):

$$Z_\omega = \int \mathcal{D}x \, e^{-\mathcal{A}_\omega[x]} = \exp\left[-\frac{D}{2}\text{Tr}\log(-\partial^2 + \omega^2)\right] \equiv e^{-\beta F_\omega}. \qquad (10.216)$$

A nonlinear transformation of $x(\tau)$ to some other coordinate $q(\tau)$ turns (10.216) into a path integral of the type (10.185) which has a singular perturbation expansion. For simplicity we assume a specific simple coordinate transformation preserving the reflection symmetry $x \longleftrightarrow -x$ of the initial oscillator, whose power series expansion starts out like

$$x(\tau) = f_\eta(\eta q(\tau)) \equiv \frac{1}{\eta}f(\eta q(\tau)) = q - \frac{\eta}{3}q^3 + a\frac{\eta^2}{5}q^5 - \cdots, \qquad (10.217)$$

where η is an expansion parameter which will play the role of a coupling constant counting the orders of the perturbation expansion. An extra parameter a is introduced for the sake of generality. We shall see that it does not influence the conclusions.

The transformation changes the partition function (10.216) into

$$Z = \int \mathcal{D}q(\tau) \, e^{-\mathcal{A}_J[q]} e^{-\mathcal{A}[q]}, \qquad (10.218)$$

where is $\mathcal{A}[q]$ is the transformed action, whereas

$$\mathcal{A}_J[q] = -\delta(0)\int d\tau \, \log\frac{\partial f(q(\tau))}{\partial q(\tau)} \qquad (10.219)$$

is an effective action coming from the Jacobian of the coordinate transformation

$$J = \prod_\tau \sqrt{\frac{\partial f(q(\tau))}{\partial q(\tau)}}. \qquad (10.220)$$

The Jacobian plays the role of the square root of the determinant of the metric in (10.185), and $\mathcal{A}_J[q]$ corresponds to the effective action $\mathcal{A}_g[\delta q]$ in Eq. (10.193).

The transformed action is decomposed into a free part

$$\mathcal{A}_\omega[q] = \frac{1}{2}\int_0^\beta d\tau \, [\dot{q}^2(\tau) + \omega^2 q^2(\tau)], \qquad (10.221)$$

and an interacting part corresponding to (10.188), which reads to second order in η:

$$\mathcal{A}^{\text{int}}[q] = \int_0^\beta d\tau \left\{-\eta\left[q^2(\tau)\dot{q}^2(\tau) + \frac{\omega^2}{3}q^4(\tau)\right] \right.$$
$$\left. +\eta^2\left[\left(\frac{1}{2}+a\right)q^4(\tau)\dot{q}^2(\tau) + \omega^2\left(\frac{1}{18}+\frac{2a}{5}\right)q^6(\tau)\right]\right\}. \qquad (10.222)$$

This is found from (10.188) by inserting the one-dimensional metric

$$g_{00}(q) = g(q) = [f'(\eta q)]^2 = 1 - 2\eta q^2 + (1+2a)\eta^2 q^4 + \cdots. \qquad (10.223)$$

To the same order in η, the Jacobian action (10.219) is

$$\mathcal{A}_J[q] = -\delta(0) \int_0^\beta d\tau \left[-\eta q^2(\tau) + \eta^2 \left(a - \frac{1}{2} \right) q^4(\tau) \right], \qquad (10.224)$$

and the perturbation expansion (10.199) is to be performed with the total interaction

$$\mathcal{A}_{\text{tot}}^{\text{int}}[q] = \mathcal{A}^{\text{int}}[q] + \mathcal{A}_J[q]. \qquad (10.225)$$

For $\eta = 0$, the transformed partition function (10.218) coincides trivially with (10.216). When expanding Z of Eq. (10.218) in powers of η, we obtain sums of Feynman diagrams contributing to each order η^n. This sum must vanish to ensure coordinate independence of the path integral. From the connected diagrams in the cumulants in (10.199) we obtain the free energy

$$\beta F = \beta F_\omega + \beta \sum_{n=1} \eta^n F_n = \beta F_\omega + \left\langle \mathcal{A}_{\text{tot}}^{\text{int}} \right\rangle_c - \frac{1}{2!} \left\langle \mathcal{A}_{\text{tot}}^{\text{int}\,2} \right\rangle_c + \dots . \qquad (10.226)$$

The perturbative treatment is coordinate-independent if F does not depend on the parameters η and a of the coordinate transformation (10.217). Hence all expansion terms F_n must vanish. This will indeed happen, albeit in a quite nontrivial way.

10.7.1 Diagrammatic Expansion

The graphical expansion for the ground state energy will be carried here only up to three loops. At any order η^n, there exist different types of Feynman diagrams with $L = n+1, n$, and $n-1$ number of loops coming from the interaction terms (10.222) and (10.224), respectively. The diagrams are composed of the three types of lines in (10.202)–(10.205), and new interaction vertices for each power of η. The diagrams coming from the Jacobian action (10.224) are easily recognized by an accompanying power of $\delta(0)$.

First we calculate the contribution to the free energy of the first cumulant $\left\langle \mathcal{A}_{\text{tot}}^{\text{int}} \right\rangle_c$ in the expansion (10.226). The associated diagrams contain only lines whose end points have equal times. Such diagrams will be called *local*.

To lowest order in η, the cumulant contains the terms

$$\beta F_1 = \eta \int_0^\beta d\tau \left\langle - \left[q^2(\tau) \dot{q}^2(\tau) + \frac{\omega^2}{3} q^4(\tau) \right] + \delta(0) q^2(\tau) \right\rangle_c .$$

There are two diagrams originating from the interaction, one from the Jacobian action:

$$\beta F_1 = -\eta \; \text{⬡} \; -\eta \omega^2 \; \text{⬡} \; +\eta \delta(0) \; \text{⬡} . \qquad (10.227)$$

The first cumulant contains also terms of order η^2:

$$\eta^2 \int_0^\beta d\tau \left\langle \left[\left(\frac{1}{2} + a \right) q^4(\tau)\dot{q}^2(\tau) + \omega^2 \left(\frac{1}{18} + \frac{2a}{5} \right) q^6(\tau) \right] - \delta(0) \left(a - \frac{1}{2} \right) q^4(\tau) \right\rangle_c.$$

The interaction gives rise to two three-loop diagrams, the Jacobian action to a single two-loop diagram:

$$\beta F_2^{(1)} = \eta^2 \left[3 \left(\frac{1}{2} + a \right) \; \text{} \; + 15\,\omega^2 \left(\frac{1}{18} + \frac{a}{5} \right) \; \text{} \; - 3 \left(a - \frac{1}{2} \right) \delta(0) \; \text{} \right]. \quad (10.228)$$

We now come to the contribution of the second cumulant $\left\langle \mathcal{A}_{\text{tot}}^{\text{int}\,2} \right\rangle_c$ in the expansion (10.226). Keeping only terms contributing to order η^2 we have to calculate the expectation value

$$-\frac{1}{2!}\eta^2 \int_0^\beta d\tau \int_0^\beta d\tau' \left\langle \left[-q^2(\tau)\dot{q}^2(\tau) - \frac{\omega^2}{3}q^4(\tau) + \delta(0)q^2(\tau) \right] \right.$$
$$\left. \times \left[-q^2(\tau')\dot{q}^2(\tau') - \frac{\omega^2}{3}q^4(\tau') + \delta(0)q^2(\tau') \right] \right\rangle_c. \quad (10.229)$$

Only the connected diagrams contribute to the cumulant, and these are necessarily nonlocal. The simplest diagrams are those containing factors of $\delta(0)$:

$$\beta F_2^{(2)} = -\frac{\eta^2}{2!} \left\{ 2\delta^2(0) \; \text{} \; - 4\delta(0) \left[\; \text{} \; + \; \text{} + 2\,\omega^2 \; \text{} \; \right] \right\}. \quad (10.230)$$

The remaining diagrams have either the form of three bubble in a chain, or of a watermelon, each with all possible combinations of the three line types (10.202)–(10.205). The sum of the three-bubbles diagrams is

$$\beta F_2^{(3)} = -\frac{\eta^2}{2!} \left[4 \; \text{} + 2 \; \text{} + 2 \; \text{} + 8\,\omega^2 \; \text{} + 8\omega^2 \; \text{} + 8\omega^4 \; \text{} \right], \quad (10.231)$$

while the watermelon-like diagrams contribute

$$\beta F_2^{(4)} = -\frac{\eta^2}{2!} 4 \left[\frac{2}{3}\omega^4 \; \text{} + \; \text{} + 4 \; \text{} + \; \text{} + 4\omega^2 \; \text{} \right]. \quad (10.232)$$

Since the equal-time expectation value $\langle \dot{q}(\tau)\,q(\tau) \rangle$ vanishes according to Eq. (10.211), diagrams with a local contraction of a mixed line (10.203) vanish trivially, and have been omitted.

We now show that if we calculate all Feynman integrals in $d = 1 - \varepsilon$ time dimensions and take the limit $\varepsilon \to 0$ at the end, we obtain unique finite results. These have the desired property that the sum of all Feynman diagrams contributing to each order η^n vanishes, thus ensuring invariance of the perturbative expressions (10.196) and (10.199) under coordinate transformations.

10.7.2 Diagrammatic Expansion in d Time Dimensions

As a first step, we extend the dimension of the τ-axis to d, assuming τ to be a vector $\tau \equiv (\tau^0, \dots, \tau^d)$, in which the zeroth component is the physical imaginary time, the others are auxiliary coordinates to be eliminated at the end. Then we replace the harmonic action (10.215) by

$$\mathcal{A}_\omega = \frac{1}{2} \int d^d\tau \left[\partial_\alpha x(\tau) \partial_\alpha x(\tau) + \omega^2 x^2(\tau) \right], \tag{10.233}$$

and the terms $\dot{q}^2$ in the transformed action (10.222) accordingly by $\partial_a q(\tau) \partial_a q(\tau)$.

The correlation functions (10.206), (10.208), and (10.212) are replaced by two-point functions

$$G^{(2)}(\tau, \tau') = \langle q(\tau) q(\tau') \rangle \quad = \Delta(\tau - \tau') \quad = \int \frac{d^d k}{(2\pi)^d} \frac{e^{ik(\tau - \tau')}}{k^2 + \omega^2}, \tag{10.234}$$

and its derivatives

$$G^{(2)}_\alpha(\tau, \tau') = \langle \partial_\alpha q(\tau) q(\tau') \rangle \quad = \Delta_\alpha(\tau - \tau') \quad = \int \frac{d^d k}{(2\pi)^d} \frac{ik_\alpha}{k^2 + \omega^2} e^{ik(\tau - \tau')}, \tag{10.235}$$

$$G^{(2)}_{\alpha\beta}(\tau, \tau') = \langle \partial_\alpha q(\tau) \partial_\beta q(\tau') \rangle = \Delta_{\alpha\beta}(\tau - \tau') = \int \frac{d^d k}{(2\pi)^d} \frac{k_\alpha k_\beta}{k^2 + \omega^2} e^{ik(\tau - \tau')}. \tag{10.236}$$

The configuration space is still one-dimensional so that the indices μ, ν and the corresponding tensors in Eqs. (10.206), (10.208), and (10.212) are absent.

The analytic continuation to $d = 1 - \varepsilon$ time dimensions is most easily performed if the Feynman diagrams are calculated in momentum space. The three types of lines represent the analytic expressions

$$\underline{\qquad} = \frac{1}{p^2 + \omega^2}, \qquad \underline{\cdots} = i \frac{p_\alpha}{p^2 + \omega^2}, \qquad \cdots\cdots = \frac{p_\alpha p_\beta}{p^2 + \omega^2}. \tag{10.237}$$

Most diagrams in the last section converge in one-dimensional momentum space, thus requiring no regularization to make them finite, as we would expect for a quantum-mechanical system. Trouble arises, however, in some multiple momentum integrals, which yield different results depending on the order of their evaluation.

As a typical example, take the Feynman integral

$$\bigodot = -\int d^d \tau_1 \Delta(\tau_1 - \tau_2) \Delta_\alpha(\tau_1 - \tau_2) \Delta_\beta(\tau_1 - \tau_2) \Delta_{\alpha\beta}(\tau_1 - \tau_2). \tag{10.238}$$

For the ordinary one-dimensional Euclidean time, a Fourier transformation yields the triple momentum space integral

$$X = \int \frac{dk}{2\pi} \frac{dp_1}{2\pi} \frac{dp_2}{2\pi} \frac{k^2 (p_1 p_2)}{(k^2 + \omega^2)(p_1^2 + \omega^2)(p_2^2 + \omega^2)[(k + p_1 + p_2)^2 + \omega^2]}. \tag{10.239}$$

Integrating this first over k, then over p_1 and p_2 yields $1/32\omega$. In the order first p_1, then p_2 and k, we find $-3/32\omega$, whereas the order first p_1, then k and p_2, gives again $1/32\omega$. As we shall see below in Eq. (10.284), the correct result is $1/32\omega$.

The unique correct evaluation will be possible by extending the momentum space to d dimensions and taking the limit $d \to 1$ at the end. The way in which the ambiguity will be resolved may be illustrated by a typical Feynman integral

$$Y_d = \int \frac{d^d k}{(2\pi)^d} \frac{d^d p_1}{(2\pi)^d} \frac{d^d p_2}{(2\pi)^d} \frac{k^2 (p_1 p_2) - (k p_1)(k p_2)}{(k^2 + \omega^2)(p_1^2 + \omega^2)(p_2^2 + \omega^2)[(k + p_1 + p_2)^2 + \omega^2]}, (10.240)$$

whose numerator vanished trivially in $d = 1$ dimensions. Due to the different contractions in d dimensions, however, Y_0 will be seen to have the nonzero value $Y_0 = 1/32\omega - (-1/32\omega)$ in the limit $d \to 1$, the result being split according to the two terms in the numerator [to appear in the Feynman integrals (10.282) and (10.284); see also Eq. (10.355)].

The diagrams which need a careful treatment are easily recognized in configuration space, where the one-dimensional correlation function (10.234) is the continuous function (10.207). Its first derivative (10.208) which has a jump at equal arguments is a rather unproblematic distribution, as long as the remaining integrand does not contain δ-functions or their derivatives. These appear with second derivatives of $\Delta(\tau, \tau')$, where the d-dimensional evaluation must be invoked to obtain a unique result.

10.8   Calculating Loop Diagrams

The loop integrals encountered in d dimensions are based on the basic one-loop integral

$$I \equiv \int \frac{d^d k}{k^2 + \omega^2} = \frac{\omega^{d-2}}{(4\pi)^{d/2}} \Gamma\left(1 - d/2\right) \underset{d=1}{=} \frac{1}{2\omega}, \qquad (10.241)$$

where we have abbreviated $d^d k \equiv d^d k/(2\pi)^d$ by analogy with $\hbar \equiv h/2\pi$. The integral exists only for $\omega \neq 0$ since it is otherwise divergent at small k. Such a divergence is called infrared divergence (IR-divergence) and ω plays the role of an infrared (IR) cutoff.

By differentiation with respect to ω^2 we can easily generalize (10.241) to

$$I_\alpha^\beta \equiv \int \frac{d^d k \, (k^2)^\beta}{(k^2 + \omega^2)^\alpha} = \frac{\omega^{d+2\beta-2\alpha}}{(4\pi)^{d/2}} \frac{\Gamma\left(d/2 + \beta\right) \Gamma\left(\alpha - \beta - d/2\right)}{\Gamma\left(d/2\right) \Gamma(\alpha)}. \qquad (10.242)$$

Note that for consistency of dimensional regularization, all integrals over a pure power of the momentum must vanish:

$$I_0^\beta = \int d^d k \, (k^2)^\beta = 0. \qquad (10.243)$$

We recognize *Veltman's rule* of Eq. (2.506).

With the help of Eqs. (10.241) and (10.242) we calculate immediately the local expectation values (10.234) and (10.236) and thus the local diagrams in (10.227) and (10.228):

$$\bigcirc \quad = \quad \langle q^2 \rangle \qquad = \quad \int \frac{d^d k}{k^2 + \omega^2} \underset{d=1}{=} \frac{1}{2\omega}, \tag{10.244}$$

$$\infty \quad = \quad \langle q^2 \rangle^2 \qquad = \quad \left(\int \frac{d^d k}{k^2 + \omega^2} \right)^2 \underset{d=1}{=} \frac{1}{4\omega^2}, \tag{10.245}$$

$$\diagup\!\!\!\!\!\bigcirc \quad = \quad \langle q^2 \rangle^3 \qquad = \quad \left(\int \frac{d^d k}{k^2 + \omega^2} \right)^3 \underset{d=1}{=} \frac{1}{8\omega^3}, \tag{10.246}$$

$$\bigcirc\!\!\cdot\!\!\bigcirc \quad = \quad \langle q^2 \rangle \langle \partial q \, \partial q \rangle = \int \frac{d^d k}{k^2 + \omega^2} \int \frac{d^d p \, p^2}{p^2 + \omega^2} \underset{d=1}{=} -\frac{1}{4}, \tag{10.247}$$

$$\diagdown\!\!\!\bigcirc \quad = \quad \langle qq \rangle^2 \langle \partial q \partial q \rangle = \left(\int \frac{d^d k}{k^2 + \omega^2} \right)^2 \int \frac{d^d p \, p^2}{p^2 + \omega^2} \underset{d=1}{=} -\frac{1}{8\omega}. \tag{10.248}$$

The two-bubble diagrams in Eq. (10.230) can also be easily computed

$$\bigcirc \quad = \int d^d \tau_1 \, \Delta^2(\tau_1 - \tau_2) \int \frac{d^d p}{(p^2 + \omega^2)^2} \underset{d=1}{=} \frac{1}{4\omega^3}, \tag{10.249}$$

$$\infty \quad = \int d^d \tau_1 \, \Delta(\tau_1 - \tau_1) \Delta_\alpha^2(\tau_1 - \tau_2) \int \frac{d^d k}{k^2 + \omega^2} \int \frac{d^d p \, p^2}{(p^2 + \omega^2)^2} \underset{d=1}{=} \frac{1}{8\omega^2}, \tag{10.250}$$

$$\cdot\!\!\infty \quad = \int d^d \tau_1 \, \Delta_{\alpha\alpha}(\tau_1 - \tau_1) \Delta^2(\tau_1 - \tau_2) \int \frac{d^d k \, k^2}{k^2 + \omega^2} \int \frac{d^d p}{(p^2 + \omega^2)^2} \underset{d=1}{=} -\frac{1}{8\omega^2}, \tag{10.251}$$

$$\infty \quad = \int d^d \tau_1 \, \Delta(\tau_1 - \tau_1) \Delta^2(\tau_1 - \tau_2) \int \frac{d^d k}{k^2 + \omega^2} \int \frac{d^d p}{(p^2 + \omega^2)^2} \underset{d=1}{=} \frac{1}{8\omega^4}. \tag{10.252}$$

For the three-bubble diagrams in Eq. (10.231) we find

$$\bigcirc\!\!\cdot\!\!\bigcirc \quad = \int d^d \tau_1 \, \Delta(\tau_1 - \tau_1) \Delta_{\alpha\alpha}^2(\tau_1 - \tau_2) \Delta(\tau_2 - \tau_2)$$
$$= \left(\int \frac{d^d q}{q^2 + \omega^2} \right)^2 \int \frac{d^d p \, (p^2)^2}{(p^2 + \omega^2)^2} \underset{d=1}{=} -\frac{3}{16\omega}, \tag{10.253}$$

$$\cdot\!\!\bigcirc\!\!\bigcirc\!\!\cdot \quad = \int d^d \tau_1 \, \Delta_{\alpha\alpha}(\tau_1 - \tau_1) \Delta^2(\tau_1 - \tau_2) \Delta_{\beta\beta}(\tau_2 - \tau_2)$$
$$= \left[\int \frac{d^d q \, (q^2)^2}{q^2 + \omega^2} \right]^2 \int \frac{d^d k}{(k^2 + \omega^2)^2} \underset{d=1}{=} \frac{1}{16\omega}, \tag{10.254}$$

$$\bigcirc\!\!\cdot\!\!\bigcirc \quad = \int d^d \tau_1 \, \Delta(\tau_1 - \tau_1) \Delta_\alpha^2(\tau_1 - \tau_2) \Delta_{\beta\beta}(\tau_2 - \tau_2)$$
$$= \int \frac{d^d k}{k^2 + \omega^2} \int \frac{d^d p \, p^2}{(p^2 + \omega^2)^2} \int \frac{d^d q \, q^2}{q^2 + \omega^2} \underset{d=1}{=} -\frac{1}{16\omega}, \tag{10.255}$$

$$\cdot\!\!\bigcirc\!\!\bigcirc \quad = \int d^d \tau_1 \, \Delta_{\alpha\alpha}(\tau_1 - \tau_1) \Delta^2(\tau_1 - \tau_2) \Delta(\tau_2 - \tau_2)$$

$$= \int \frac{đ^d k \, k^2}{k^2 + \omega^2} \int \frac{đ^d p}{(p^2 + \omega^2)^2} \int \frac{đ^d q}{q^2 + \omega^2} \underset{d=1}{=} -\frac{1}{16\omega^3}, \qquad (10.256)$$

$$\text{⊂⊃⊂⊃} = \int d^d \tau_1 \, \Delta(\tau_1 - \tau_1) \Delta_\alpha^2(\tau_1 - \tau_2) \Delta(\tau_2 - \tau_2)$$

$$= \int \frac{đ^d k}{k^2 + \omega^2} \int \frac{đ^d p \, p^2}{(p^2 + \omega^2)^2} \int \frac{đ^d q}{q^2 + \omega^2} \underset{d=1}{=} \frac{1}{16\omega^3}, \qquad (10.257)$$

$$\text{⊂⊃⊂⊃⊂⊃} = \int d^d \tau_1 \, \Delta(\tau_1 - \tau_1) \Delta^2(\tau_1 - \tau_2) \Delta(\tau_2 - \tau_2)$$

$$= \int \frac{đ^d k}{k^2 + \omega^2} \int \frac{đ^d p}{(p^2 + \omega^2)^2} \int \frac{đ^d q}{q^2 + \omega^2} \underset{d=1}{=} \frac{1}{16\omega^5}. \qquad (10.258)$$

In these diagrams, it does not make any difference if we replace $\Delta_{\alpha\alpha}^2$ by $\Delta_{\alpha\beta}^2$.

We now turn to the watermelon-like diagrams in Eq. (10.232) which are more tedious to calculate. They require a further basic integral [26]:

$$J(p^2) = \int \frac{đ^d k}{(k^2 + \omega^2)[(k + p)^2 + \omega^2]} = \int_0^1 dx \int \frac{đ^d k}{[k^2 + p^2 x(1 - x) + \omega^2]^2}$$

$$= \frac{\Gamma(2 - d/2)}{(4\pi)^{d/2}} \left(\frac{p^2 + 4\omega^2}{4}\right)^{d/2 - 2} F\left(2 - \frac{d}{2}, \frac{1}{2}; \frac{3}{2}; \frac{p^2}{p^2 + 4\omega^2}\right), \qquad (10.259)$$

where $F(a, b; c; z)$ is the hypergeometric function (1.450). For $d = 1$, the result is simply

$$J(p^2) = \frac{1}{\omega(p^2 + 4\omega^2)}. \qquad (10.260)$$

We also define the more general integrals

$$J_{\alpha_1 \dots \alpha_n}(p) = \int \frac{đ^d k \, k_{\alpha_1} \cdots k_{\alpha_n}}{(k^2 + \omega^2)[(k + p)^2 + \omega^2]}, \qquad (10.261)$$

and further

$$J_{\alpha_1 \dots \alpha_n, \beta_1 \dots \beta_m}(p) = \int \frac{đ^d k \, k_{\alpha_1} \cdots k_{\alpha_n} (k + p)_{\beta_1} \cdots (k + p)_{\beta_m}}{(k^2 + \omega^2)[(k + p)^2 + \omega^2]}. \qquad (10.262)$$

The latter are linear combinations of momenta and the former, for instance

$$J_{\alpha, \beta}(p) = J_\alpha(p) \, p_\beta + J_{\alpha\beta}(p). \qquad (10.263)$$

Using Veltman's rule (10.243), all integrals (10.262) can be reduced to combinations of $p, I, J(p^2)$. Relevant examples for our discussion are

$$J_\alpha(p) = \int \frac{đ^d k \, k_\alpha}{(k^2 + \omega^2)[(k + p)^2 + \omega^2]} = -\frac{1}{2} p_\alpha \, J(p^2), \qquad (10.264)$$

and

$$J_{\alpha\beta}(p) = \int \frac{d^d k \; k_\alpha k_\beta}{(k^2 + \omega^2)[(k + p)^2 + \omega^2]} = \left[\delta_{\alpha\beta} + (d - 2)\frac{p_\alpha p_\beta}{p^2} \right] \frac{I}{2(d - 1)}$$
$$+ \left[-\delta_{\alpha\beta}(p^2 + 4\omega^2) + \frac{p_\alpha p_\beta}{p^2}\left(d\, p^2 + 4\omega^2 \right) \right] \frac{J(p^2)}{4(d - 1)}, \qquad (10.265)$$

whose trace is

$$J_{\alpha\alpha}(p) = \int \frac{d^d k \; k^2}{(k^2 + \omega^2)[(k + p)^2 + \omega^2]} = I - \omega^2\, J(p^2). \qquad (10.266)$$

Similarly we expand

$$J_{\alpha\alpha\beta}(p) = \int \frac{d^d k \; k^2 k_\beta}{(k^2 + \omega^2)[(k + p)^2 + \omega^2]} = \frac{1}{2} p_\beta[-I + \omega^2 J(p^2)]. \qquad (10.267)$$

The integrals appear in the following subdiagrams

$$
\begin{array}{lll}
\raisebox{-0.5em}{$\underset{k+p}{\overset{k}{\bigcirc}}$} = J(p^2), &
\raisebox{-0.5em}{$\underset{k+p,\alpha}{\overset{k}{\bigcirc}}$} = J_{,\alpha}(p), &
\raisebox{-0.5em}{$\underset{k+p,\alpha\beta}{\overset{k}{\bigcirc}}$} = J_{,\alpha\beta}(p), \\[1.5em]
\raisebox{-0.5em}{$\underset{k+p}{\overset{k,\alpha}{\bigcirc}}$} = J_{\alpha}(p), &
\raisebox{-0.5em}{$\underset{k+p,\beta}{\overset{k,\alpha}{\bigcirc}}$} = J_{\alpha,\beta}(p), &
\raisebox{-0.5em}{$\underset{k+p,\beta\gamma}{\overset{k,\alpha}{\bigcirc}}$} = J_{\alpha,\beta\gamma}(p), \\[1.5em]
\raisebox{-0.5em}{$\underset{k+p}{\overset{k,\alpha}{\bigcirc}}$} = J_{\alpha\beta}(p), &
\raisebox{-0.5em}{$\underset{k+p,\gamma}{\overset{k,\alpha\beta}{\bigcirc}}$} = J_{\alpha\beta,\gamma}(p), &
\raisebox{-0.5em}{$\underset{k+p,\gamma\delta}{\overset{k,\alpha\beta}{\bigcirc}}$} = J_{\alpha\beta,\gamma\delta}(p).
\end{array}
\qquad (10.268)
$$

All two- and three-loop integrals needed for the calculation can be brought to the generic form

$$K(a, b) = \int d^d p \; (p^2)^a J^b(p^2), \quad a \geq 0, \quad b \geq 1, \quad a \leq b, \qquad (10.269)$$

and evaluated recursively as follows [27]. From the Feynman parametrization of the first line of Eq. (10.259) we observe that the two basic integrals (10.241) and (10.259) satisfy the differential equation

$$J(p^2) = -\frac{\partial I}{\partial \omega^2} + \frac{1}{2}p^2 \frac{\partial J(p^2)}{\partial \omega^2} - 2p^2 \frac{\partial J(p^2)}{\partial p^2}. \qquad (10.270)$$

Differentiating $K(a + 1, b)$ from Eq. (10.269) with respect to ω^2, and using Eq. (10.270), we find the recursion relation

$$K(a, b) = \frac{2b(d/2 - 1)\, I\, K(a - 1, b - 1) - 2\omega^2(2a - 2 - b + d)K(a - 1, b)}{(b + 1)d/2 - 2b + a}, \quad (10.271)$$

which may be solved for increasing a starting with

$$K(0, 0) = 0, \quad K(0, 1) = \int d^d p\, J(p^2) = I^2,$$
$$K(0, 2) = \int d^d p\, J^2(p^2) = A, \; \ldots, \qquad (10.272)$$

where A is the integral

$$A \equiv \int \frac{d^d p_1 d^d p_2 d^d k}{(p_1^2 + \omega^2)(p_2^2 + \omega^2)(k^2 + \omega^2)[(p_1 + p_2 + k)^2 + \omega^2]}. \tag{10.273}$$

This integral will be needed only in $d = 1$ dimensions where it can be calculated directly from the configuration space version of this integral. For this we observe that the first watermelon-like diagram in (10.232) corresponds to an integral over the product of two diagrams $J(p^2)$ in (10.268):

$$\ominus = \int d^d \tau_1 \Delta(\tau_1 - \tau_2)\Delta(\tau_1 - \tau_2)\Delta(\tau_1 - \tau_2)\Delta(\tau_1 - \tau_2) = \int d^d k \, J^2(k) = A. \tag{10.274}$$

Thus we find A in $d = 1$ dimensions from the simple τ-integral

$$A = \int_{-\infty}^{\infty} d\tau \, \Delta^4(\tau, 0) = \int_{-\infty}^{\infty} dx \left(\frac{1}{2\omega} e^{-\omega|x|} \right)^4 = \frac{1}{32\omega^5}. \tag{10.275}$$

Since this configuration space integral contains no δ-functions, the calculation in $d = 1$ dimension is without subtlety.

With the help of Eqs. (10.271), (10.272), and Veltman's rule (10.243), according to which

$$K(a, 0) \equiv 0, \tag{10.276}$$

we find further the integrals

$$\int d^d p \, p^2 J(p^2) = K(1, 1) = -2\omega^2 I^2, \tag{10.277}$$

$$\int d^d p \, p^2 \, J^2(p^2) = K(1, 2) = \frac{4}{3}(I^3 - \omega^2 A), \tag{10.278}$$

$$\int d^d p \, (p^2)^2 J^2(p^2) = K(2, 2) = -8\omega^2 \frac{(6 - 5d)I^3 + 2d\omega^2 A}{3(4 - 3d)}. \tag{10.279}$$

We are thus prepared to calculate all remaining three-loop contributions from the watermelon-like diagrams in Eq. (10.232). The second is an integral over the product of subdiagrams $J_{\alpha\beta}$ in (10.268) and yields

$$\begin{aligned}
\bigoplus &= \int d^d \tau_1 \, \Delta^2(\tau_1 - \tau_2)\Delta_{\alpha\beta}^2(\tau_1 - \tau_2) \\
&= \int d^d p \, d^d k \, d^d q \frac{(pk)^2}{(p^2 + \omega^2)(k^2 + \omega^2)(q^2 + \omega^2)[(p + k + q)^2 + \omega^2]} \\
&\underset{q \to q - p}{=} \int d^d q \, J_{\alpha\beta}(q)J_{\alpha\beta}(q) = \int d^d k \frac{1}{16}(k^2)^2 J^2(k) \\
&\quad + \int d^d k \frac{1}{4(d-1)} \left\{ dI^2 + \left[(d-2)k^2 - 4\omega^2 \right] I \, J(k) + \frac{1}{4}(k^2 + 4\omega^2)^2 J^2(k) \right\} \\
&= -\frac{\omega^2}{2} \frac{(6 - 5d)I^3 + 2d\omega^2 A}{3(4 - 3d)} - \frac{\omega^2}{6(4 - 3d)} \left[(6 - 5d)I^3 + 2d\omega^2 A \right] \tag{10.280} \\
&= -\frac{\omega^2}{3(4 - 3d)} \left[(8 - 7d)I^3 + (d + 4)\omega^2 A \right] \underset{d=1}{=} -\frac{\omega^2}{3}(I^3 + 5\omega^2 A) = -\frac{3}{32\omega}.
\end{aligned}$$

The third diagram contains two mixed lines. It is an integral over a product of the diagrams $J_\alpha(p)$ and $J_{\beta,\alpha\beta}$ in (10.268) and gives

$$
\begin{aligned}
\ominus \ &= -\int d^d\tau_1 \Delta(\tau_1 - \tau_2)\Delta_\alpha(\tau_1 - \tau_2)\Delta_\beta(\tau_1 - \tau_2)\Delta_{\alpha\beta}(\tau_1 - \tau_2) \\
&= \int d^d k\, d^d p_1 d^d p_2 \frac{(kp_1)(kp_2)}{(k^2+\omega^2)(p_1^2+\omega^2)(p_2^2+\omega^2)\left[(k+p_1+p_2)^2+\omega^2\right]} \\
&\underset{p_2 \to p_2 - k}{=} \int d^d p\, [p_\beta J_\alpha(p)J_{\alpha\beta}(p) + J_\alpha(p)J_{\beta\alpha\beta}(p)] \\
&= -\frac{1}{8}\int d^d p\, p^2 J(p^2)\left[(p^2 + 2\omega^2)J(p^2) - 2I\right] \\
&= -\frac{\omega^2}{6(4-3d)}\left[(8-5d)\,I^3 - 2(4-d)\omega^2 A\right] \underset{d=1}{=} -\frac{\omega^2}{2}(I^3 - 2\omega^2 A) = -\frac{1}{32\omega}.
\end{aligned}
\tag{10.281}
$$

The fourth diagram contains four mixed lines and is evaluated as follows:

$$
\ominus \ = \ -\int d^d\tau_1 \Delta_\alpha(\tau_1 - \tau_2)\Delta_\alpha(\tau_1 - \tau_2)\Delta_\beta(\tau_1 - \tau_2)\Delta_\beta(\tau_1 - \tau_2).
\tag{10.282}
$$

Since the integrand is regular and vanishes at infinity, we can perform a partial integration and rewrite the configuration space integral as

$$
\begin{aligned}
\ominus \ &= \ \int d^d\tau_1 \Delta(\tau_1 - \tau_2)\Delta_{\alpha\alpha}(\tau_1 - \tau_2)\Delta_\beta(\tau_1 - \tau_2)\Delta_\beta(\tau_1 - \tau_2) \\
&+ \ 2\int d^d\tau_1 \Delta(\tau_1 - \tau_2)\Delta_\alpha(\tau_1 - \tau_2)\Delta_\beta(\tau_1 - \tau_2)\Delta_{\alpha\beta}(\tau_1 - \tau_2).
\end{aligned}
\tag{10.283}
$$

The second integral has just been evaluated in (10.282). The first is precisely the integral Eq. (10.239) discussed above. It is calculated as follows:

$$
\begin{aligned}
&\int d^d\tau_1 \Delta(\tau_1 - \tau_2)\Delta_{\alpha\alpha}(\tau_1 - \tau_2)\Delta_\beta(\tau_1 - \tau_2)\Delta_\beta(\tau_1 - \tau_2) \\
&= \ \int d^d k\, d^d p_1 d^d p_2 \frac{k^2\,(p_1 p_2)}{(k^2+\omega^2)(p_1^2+\omega^2)(p_2^2+\omega^2)[(k+p_1+p_2)^2+\omega^2]} \\
&= \ \int d^d p\, [p_\alpha J_\alpha(p)J_{\beta\beta} + J_\alpha(p)J_{\beta\alpha\beta}(p)] = \frac{\omega^2}{4}\int d^d p\, p^2 J^2(p^2) \\
&= \ -\frac{\omega^2}{3}(I^3 - \omega^2 A) \underset{d=1}{=} \frac{1}{32\omega}.
\end{aligned}
\tag{10.284}
$$

Hence we obtain

$$
\ominus \ = \ \frac{1}{32\omega}.
\tag{10.285}
$$

The fifth diagram in (10.232) is an integral of the product of two subdiagrams $J_\alpha(p)$ in (10.268) and yields

$$
\ominus \ = \ \int d^d\tau_1 \Delta(\tau_1 - \tau_2)\Delta_\alpha^2(\tau_1 - \tau_2)\Delta(\tau_1 - \tau_2)
$$

$$= -\int d^dk d^d p_1 d^d p_2 \frac{p_1 p_2}{(k^2+\omega^2)(p_1^2+\omega^2)(p_2^2+\omega^2)\left[(k+p_1+p_2)^2+\omega^2\right]}$$

$$\underset{k\to k-p_2}{=} -\int d^dk d^d p_1 d^d p_2 \frac{p_1 p_2}{\left[(k-p_2)^2+\omega^2\right](p_1^2+\omega^2)(p_2^2+\omega^2)\left[(k+p_1)^2+\omega^2\right]}$$

$$\underset{p_2\to -p_2}{=} \int d^dk d^d p_1 \frac{p_{1\alpha}}{(p_1^2+\omega^2)\left[(p_1+k)^2+\omega^2\right]} \int d^d p_2 \frac{p_{2\alpha}}{(p_2^2+\omega^2)\left[(p_2+k)^2+\omega^2\right]}$$

$$= \int d^dk \, J_\alpha^2(k) = \frac{1}{4}\int d^dk \, k^2 J^2(k^2)$$

$$= \frac{1}{4}\frac{4}{3}(I^3 - \omega^2 A) \underset{d=1}{=} \frac{1}{32\omega^3}. \tag{10.286}$$

We can now sum up all contributions to the free energy in Eqs. (10.227)–(10.232). An immediate simplification arises from the Veltman's rule (10.243). This implies that all δ-functions at the origin are zero in dimensional regularization:

$$\delta^{(d)}(0) = \int \frac{d^dk}{(2\pi)^d} = 0. \tag{10.287}$$

The first-order contribution (10.227) to the free energy is obviously zero by Eqs. (10.245) and (10.247).

The first second-order contribution $\beta F_2^{(1)}$ becomes, from (10.246) and (10.248):

$$F_2^{(1)} = \eta^2 \left[3\left(\frac{1}{2}+a\right)\left(-\frac{1}{8\omega}\right) + 15\omega^2\left(\frac{1}{18}+\frac{a}{5}\right)\right] = -\frac{\eta^2}{12\omega}. \tag{10.288}$$

The parameter a has disappeared from this equation.

The second second-order contribution $\beta F_2^{(2)}$ vanishes trivially, by Veltman's rule (10.287).

The third second-order contribution $\beta F_2^{(3)}$ in (10.231) vanishes nontrivially using (10.253)–(10.258):

$$F_2^{(3)} = -\frac{\eta^2}{2!}\left[\ 4\ \left(-\frac{1}{16\omega}\right) + 2\left(-\frac{3}{16\omega}\right) + 2\left(\frac{1}{16\omega}\right)\right.$$
$$\left. +8\omega^2\left(-\frac{1}{16\omega^3}\right) + 8\omega^2\left(\frac{1}{16\omega^3}\right) + 8\omega^4\left(\frac{1}{16\omega^5}\right)\right] = 0. \tag{10.289}$$

The fourth second-order contribution, finally, associated with the watermelon-like diagrams in (10.232) yield via (10.281), (10.282), (10.285), (10.286), and (10.274):

$$\beta F_2^{(4)} = -\frac{\eta^2}{2!}\left[-\frac{3}{32\omega} + 4\left(-\frac{1}{32\omega}\right) + \frac{1}{32\omega} + 4\omega^2\left(\frac{1}{32\omega^3}\right) + \frac{2}{3}\omega^4\left(\frac{1}{32\omega^5}\right)\ \right] = \frac{\eta^2}{12\omega}, \tag{10.290}$$

canceling (10.288), and thus the entire free energy. This proves the invariance of the perturbatively defined path integral under coordinate transformations.

10.8.1   Reformulation in Configuration Space

The Feynman integrals in momentum space in the last section corresponds in τ-space to integrals over products of distributions. For many applications it is desirable to do the calculations directly in τ-space. This will lead to an extension of distribution theory which allows us to do precisely that.

In dimensional regularization, an important simplification came from Veltman's rule (10.287), according to which the delta function at the origin vanishes. In the more general calculations to come, we shall encounter generalized δ-functions, which are multiple derivatives of the ordinary δ-function:

$$\delta^{(d)}_{\alpha_1\ldots\alpha_n}(\tau) \equiv \partial_{\alpha_1\ldots\alpha_n}\delta^{(d)}(\tau) = \int d^d k (ik)_{\alpha_1}\ldots(ik)_{\alpha_n}e^{ikx}, \tag{10.291}$$

with $\partial_{\alpha_1\ldots\alpha_n} \equiv \partial_{\alpha_1}\ldots\partial_{\alpha_n}$ and $d^d k \equiv d^d k/(2\pi)^d$. By Veltman's rule (10.243), all these vanish at the origin:

$$\delta^{(d)}_{\alpha_1\ldots\alpha_n}(0) = \int d^d k (ik)_{\alpha_1}\ldots(ik)_{\alpha_n} = 0. \tag{10.292}$$

In the extended coordinate space, the correlation function $\Delta(\tau, \tau')$ in (10.234), which we shall also write as $\Delta(\tau - \tau')$, is at equal times given by the integral [compare (10.241)]

$$\Delta(0) = \int \frac{d^d k}{k^2 + \omega^2} = \frac{\omega^{d-2}}{(4\pi)^{d/2}}\Gamma\left(1 - \frac{d}{2}\right) = I \underset{d=1}{=} \frac{1}{2\omega}. \tag{10.293}$$

The extension (10.235) of the time derivative (10.208),

$$\Delta_\alpha(\tau) = \int d^d k \frac{ik_\alpha}{k^2 + \omega^2}e^{ik\tau} \tag{10.294}$$

vanishes at equal times, just like (10.211):

$$\Delta_\alpha(0) = 0. \tag{10.295}$$

This follows directly from a Taylor series expansion of $1/(k^2 + \omega^2)$ in powers of k^2, after imposing (10.292).

The second derivative of $\Delta(\tau)$ has the Fourier representation (10.236). Contracting the indices yields

$$\Delta_{\alpha\alpha}(\tau) = -\int d^d k \frac{k^2}{k^2 + \omega^2}e^{ikx} = -\delta^{(d)}(\tau) + \omega^2\Delta(\tau). \tag{10.296}$$

This equation is a direct consequence of the definition of the correlation function as a solution to the inhomogeneous field equation

$$(-\partial_\alpha^2 + \omega^2)q(\tau) = \delta^{(d)}(\tau). \tag{10.297}$$

Inserting Veltman's rule (10.287) into (10.296), we obtain

$$\Delta_{\alpha\alpha}(0) = \omega^2\,\Delta(0) \underset{d=1}{=} \frac{\omega}{2}. \tag{10.298}$$

This ensures the vanishing of the first-order contribution (10.227) to the free energy

$$F_1 = -g\,\eta\left[-\Delta_{\alpha\alpha}(0) + \omega^2\Delta(0)\right]\Delta(0) = 0. \tag{10.299}$$

The same equation (10.296) allows us to calculate immediately the second-order contribution (10.228) from the local diagrams

$$
\begin{aligned}
F_2^{(1)} &= -\eta^2\, 3g^2 \left[\left(\frac{1}{2} + a \right) \Delta_{\alpha\alpha}(0) - 5 \left(\frac{1}{18} + \frac{a}{5} \right) \omega^2 \Delta(0) \right] \Delta^2(0) \\
&= -\eta^2 \frac{2}{3}\, \omega^2 \Delta^3(0) \underset{d=1}{=} -\frac{\eta^2}{12\omega}.
\end{aligned}
\tag{10.300}
$$

The other contributions to the free energy in the expansion (10.226) require rules for calculating products of two and four distributions, which we are now going to develop.

10.8.2 Integrals over Products of Two Distributions

The simplest integrals are

$$
\begin{aligned}
\int d^d\tau\, \Delta^2(\tau) &= \int d^d p\, d^d k\, \frac{\delta^{(d)}(k+p)}{(p^2+\omega^2)(k^2+\omega^2)} \\
&= \int \frac{d^d k}{(k^2+\omega^2)^2} = \frac{\omega^{d-4}}{(4\pi)^{d/2}} \Gamma\left(2 - \frac{d}{2} \right) = \frac{(2-d)}{2\omega^2} \Delta(0),
\end{aligned}
\tag{10.301}
$$

and

$$
\begin{aligned}
\int d^d\tau\, \Delta_\alpha^2(\tau) &= -\int d^d\tau\, \Delta(\tau) \left[-\delta^{(d)}(\tau) + \omega^2 \Delta(\tau) \right] = \Delta(0) - \omega^2 \int d^d\tau\, \Delta^2(\tau) \\
&= \frac{d}{2} \Delta(0).
\end{aligned}
\tag{10.302}
$$

To obtain the second result we have performed a partial integration and used (10.296).

In contrast to (10.301) and (10.302), the integral

$$
\begin{aligned}
\int d^d\tau\, \Delta_{\alpha\beta}^2(\tau) &= \int d^d p\, d^d k\, \frac{(kp)^2\, \delta^{(d)}(k+p)}{(k^2+\omega^2)(p^2+\omega^2)} \\
&= \int d^d k\, \frac{(k^2)^2}{(k^2+\omega^2)^2} = \int d^d\tau\, \Delta_{\alpha\alpha}^2(\tau)
\end{aligned}
\tag{10.303}
$$

diverges formally in $d=1$ dimension. In dimensional regularization, however, we may decompose $(k^2)^2 = (k^2+\omega^2)^2 - 2\omega^2(k^2+\omega^2) + \omega^4$, and use (10.292) to evaluate

$$
\begin{aligned}
\int d^d\tau\, \Delta_{\alpha\alpha}^2(\tau) &= \int d^d k\, \frac{(k^2)^2}{(k^2+\omega^2)^2} = -2\omega^2 \int \frac{d^d k}{(k^2+\omega^2)} + \omega^4 \int \frac{d^d k}{(k^2+\omega^2)^2} \\
&= -2\omega^2 \Delta(0) + \omega^4 \int d^d\tau\, \Delta^2(\tau).
\end{aligned}
\tag{10.304}
$$

Together with (10.301), we obtain the relation between integrals of products of two distributions

$$
\begin{aligned}
\int d^d\tau\, \Delta_{\alpha\beta}^2(\tau) &= \int d^d\tau\, \Delta_{\alpha\alpha}^2(\tau) = -2\omega^2 \Delta(0) + \omega^4 \int d^d\tau\, \Delta^2(\tau) \\
&= -(1+d/2)\, \omega^2 \Delta(0).
\end{aligned}
\tag{10.305}
$$

An alternative way of deriving the equality (10.303) is to use partial integrations and the identity

$$\partial_\alpha \Delta_{\alpha\beta}(\tau) = \partial_\beta \Delta_{\alpha\alpha}(\tau), \tag{10.306}$$

which follows directly from the Fourier representation (10.294).

Finally, from Eqs. (10.301), (10.302), and (10.305), we observe the useful identity

$$\int d^d\tau \left[\Delta_{\alpha\beta}^2(\tau) + 2\omega^2 \Delta_\alpha^2(\tau) + \omega^4 \Delta^2(\tau) \right] = 0, \tag{10.307}$$

which together with the inhomogeneous field equation (10.296) reduces the calculation of the second-order contribution of all three-bubble diagrams (10.231) to zero:

$$F_2^{(3)} = -g^2 \Delta^2(0) \int d^d\tau \left[\Delta_{\alpha\beta}^2(\tau) + 2\omega^2 \Delta_\alpha^2(\tau) + \omega^4 \Delta^2(\tau) \right] = 0. \tag{10.308}$$

10.8.3 Integrals over Products of Four Distributions

Consider now the more delicate integrals arising from watermelon-like diagrams in (10.232) which contain products of four distributions, a nontrivial tensorial structure, and overlapping divergences. We start from the second to fourth diagrams:

$$\bigcirc = \int d^d\tau \, \Delta^2(\tau) \Delta_{\alpha\beta}^2(\tau), \tag{10.309}$$

$$4 \bigcirc = 4 \int d^d\tau \Delta(\tau) \Delta_\alpha(\tau) \Delta_\beta(\tau) \Delta_{\alpha\beta}(\tau), \tag{10.310}$$

$$\bigcirc = \int d^d\tau \Delta_\alpha(\tau) \Delta_\alpha(\tau) \Delta_\beta(\tau) \Delta_\beta(\tau). \tag{10.311}$$

To isolate the subtleties with the tensorial structure exhibited in Eq. (10.240), we introduce the integral

$$Y_d = \int d^d\tau \, \Delta^2(\tau) \left[\Delta_{\alpha\beta}^2(\tau) - \Delta_{\alpha\alpha}^2(\tau) \right]. \tag{10.312}$$

In $d = 1$ dimension, the bracket vanishes formally, but the limit $d \to 1$ of the integral is nevertheless finite. We now decompose the Feynman diagram (10.309), into the sum

$$\int d^d\tau \, \Delta^2(\tau) \Delta_{\alpha\beta}^2(\tau) = \int d^d\tau \, \Delta^2(\tau) \Delta_{\alpha\alpha}^2(\tau) + Y_d. \tag{10.313}$$

To obtain an analogous decomposition for the other two diagrams (10.310) and (10.311), we derive a few useful relations using the inhomogeneous field equation (10.296), partial integrations, and Veltman's rules (10.287) or (10.292). From the inhomogeneous field equation, there is the relation

$$-\int d^d\tau \, \Delta_{\alpha\alpha}(\tau) \Delta^3(\tau) = \Delta^3(0) - \omega^2 \int d^d\tau \, \Delta^4(\tau). \tag{10.314}$$

By a partial integration, the left-hand side becomes

$$\int d^d\tau \, \Delta_{\alpha\alpha}(\tau)\Delta^3(\tau) = -3 \int d^d\tau \, \Delta_\alpha^2(\tau)\Delta^2(\tau), \tag{10.315}$$

leading to

$$\int d^d\tau \, \Delta_\alpha^2(\tau)\Delta^2(\tau) = \frac{1}{3}\Delta^3(0) - \frac{1}{3}\omega^2 \int d^d\tau \, \Delta^4(\tau). \tag{10.316}$$

Invoking once more the inhomogeneous field equation (10.296) and Veltman's rule (10.287), we obtain the integrals

$$\int d^d\tau \, \Delta_{\alpha\alpha}^2(\tau)\Delta^2(\tau) - \omega^4 \int d^d\tau \, \Delta^4(\tau) + 2\omega^2\Delta^3(0) = 0, \tag{10.317}$$

and

$$\int d^d\tau \, \Delta_{\alpha\alpha}(\tau)\Delta_\beta^2(\tau)\Delta(\tau) = \omega^2 \int d^d\tau \, \Delta_\beta^2(\tau)\Delta^2(\tau). \tag{10.318}$$

Using (10.316), the integral (10.318) takes the form

$$\int d^d\tau \, \Delta_{\alpha\alpha}(\tau)\Delta_\beta^2(\tau)\Delta(\tau) = \frac{1}{3}\omega^2\Delta^3(0) - \frac{1}{3}\omega^4 \int d^d\tau \, \Delta^4(\tau). \tag{10.319}$$

Partial integration, together with Eqs. (10.317) and (10.319), leads to

$$\int d^d\tau \, \partial_\beta \Delta_{\alpha\alpha}(\tau)\Delta_\beta(\tau)\Delta^2(\tau) = -\int d^d\tau \, \Delta_{\alpha\alpha}^2(\tau)\Delta^2(\tau) - 2\int d^d\tau \, \Delta_{\alpha\alpha}(\tau)\Delta_\beta^2(\tau)\Delta(\tau)$$
$$= \frac{4}{3}\omega^2\Delta^3(0) - \frac{1}{3}\omega^4 \int d^d\tau \, \Delta^4(\tau). \tag{10.320}$$

A further partial integration, and use of Eqs. (10.306), (10.318), and (10.320) produces the decompositions of the second and third Feynman diagrams (10.310) and (10.311):

$$4 \int d^d\tau \, \Delta(\tau)\Delta_\alpha(\tau)\Delta_\beta(\tau)\Delta_{\alpha\beta}(\tau) = 4\omega^2 \int d^d\tau \, \Delta^2(\tau)\Delta_\alpha^2(\tau) - 2Y_d, \tag{10.321}$$

and

$$\int d^d\tau \, \Delta_\alpha^2(\tau)\Delta_\beta^2(\tau) = -3\omega^2 \int d^d\tau \, \Delta^2(\tau)\Delta_\alpha^2(\tau) + Y_d. \tag{10.322}$$

We now make the important observation that the subtle integral Y_d of Eq. (10.312) appears in Eqs. (10.313), (10.321), and (10.322) in such a way that it drops out from the sum of the watermelon-like diagrams in (10.232):

$$\ominus + 4 \ominus + \ominus = \int d^d\tau \, \Delta^2(\tau)\Delta_{\alpha\alpha}^2(\tau) + \omega^2 \int d^d\tau \, \Delta^2(\tau)\Delta_\alpha^2(\tau). \tag{10.323}$$

Using now the relations (10.316) and (10.317), the right-hand side becomes a sum of completely regular integrals involving only products of propagators $\Delta(\tau)$.

We now add to this sum the first and last watermelon-like diagrams in Eq. (10.232)

$$\frac{2}{3}\omega^4 \;\;\ominus\;\; = \frac{2}{3}\omega^4 \int d^d\tau\, \Delta^4(\tau), \tag{10.324}$$

and

$$4\omega^2 \;\;\ominus\;\; = 4\omega^2 \int d^d\tau\, \Delta^2(\tau)\Delta_\alpha^2(\tau), \tag{10.325}$$

and obtain for the total contribution of all watermelon-like diagrams in (10.232) the simple expression for $\eta = 1$:

$$
\begin{aligned}
F_2^{(4)} &= -2\eta^2\, g^2 \int d^d\tau\, \Delta^2(\tau) \left[\frac{2}{3}\omega^4\, \Delta^2(\tau) + \Delta_{\alpha\alpha}^2(\tau) + 5\omega^2\, \Delta_\alpha^2(\tau) \right] \\
&= \eta^2 \frac{2}{3}\,\omega^2 \Delta^3(0) \underset{d=1}{=} \frac{\eta^2}{12\omega}.
\end{aligned}
\tag{10.326}
$$

This cancels the finite contribution (10.300), thus making also the second-order free energy in (10.222) vanish, and confirming the invariance of the perturbatively defined path integral under coordinate transformations up to this order.

Thus we have been able to relate all diagrams involving singular time derivatives of correlation functions to integrals over products of the regular correlation function (10.234), where they can be replaced directly by their $d = 1$-version (10.207). The disappearance of the ambiguous integral Y_d in the combination of watermelon-like diagrams (10.323) has the pleasant consequence that ultimately all calculations can be done in $d = 1$ dimensions after all. This leads us to expect that the dimensional regularization may be made superfluous by a more comfortable calculation procedure. This is indeed so and the rules will be developed in Section 10.11. Before we come to this it is useful, however, to point out a pure x-space way of finding the previous results.

10.9   Distributions as Limits of Bessel Function

In dimensional regularization it is, of course, possible to perform the above configuration space integrals over products of distributions without any reference to momentum space integrals. For this we express all distributions explicitly in terms of modified Bessel functions $K_\alpha(y)$.

10.9.1   Correlation Function and Derivatives

The basic correlation function in d-dimension is obtained from the integral in Eq. (10.234), as

$$\Delta(\tau) = c_d\, y^{1-d/2}\, K_{1-d/2}(y), \tag{10.327}$$

where $y \equiv m\,|\tau|$ is reduced length of τ_α, with the usual Euclidean norm $|x| = \sqrt{\tau_1^2 + \ldots + \tau_d^2}$, and $K_{1-d/2}(y)$ is the modified Bessel function. The constant factor in front is

$$c_d = \frac{\omega^{d-2}}{(2\pi)^{d/2}}. \tag{10.328}$$

In one dimension, the correlation function (10.327) reduces to (10.202). The short-distance properties of the correlation functions is governed by the small-y behavior of Bessel function at origin[4]

$$K_\beta(y) \underset{y \approx 0}{\approx} \frac{1}{2}\Gamma(\beta)(y/2)^{\mp\beta}, \quad \text{Re } \beta \gtrless 0. \tag{10.329}$$

In the application to path integrals, we set the dimension equal to $d = 1 - \varepsilon$ with a small positive ε, whose limit $\varepsilon \to 0$ will yield the desired results in $d = 1$ dimension. In this regime, Eq. (10.329) shows that the correlation function (10.327) is regular at the origin, yielding once more (10.293). For $d = 1$, the result is $\Delta(0) = 1/2\omega$, as stated in Eq. (10.298).

The first derivative of the correlation function (10.327), which is the d-dimensional extension of time derivative (10.203), reads

$$\Delta_\alpha(\tau) = -c_d y^{1-d/2} K_{d/2}(y) \partial_\alpha y, \tag{10.330}$$

where $\partial_\alpha y = m\tau_\alpha/|x|$. By Eq. (10.329), this is regular at the origin for $\varepsilon > 0$, such that the antisymmetry $\Delta_\alpha(-x) = -\Delta_\alpha(\tau)$ makes $\Delta_\alpha(0) = 0$, as observed after Eq. (10.294).

Explicitly, the small-τ behavior of the correlation function and its derivative is

$$\Delta(\tau) \propto \text{const.}, \quad \Delta_\alpha(\tau) \propto |\tau|^\varepsilon \partial_\alpha|\tau|. \tag{10.331}$$

In contrast to these two correlation functions, the second derivative

$$\Delta_{\alpha\beta}(\tau) = \Delta(\tau)(\partial_\alpha y)(\partial_\beta y) + \frac{c_d}{(d-2)} y^{d/2} K_{d/2}(y) \partial_{\alpha\beta}y^{2-d}, \tag{10.332}$$

is singular at short distance. The singularity comes from the second term in (10.332):

$$\partial_{\alpha\beta}y^{2-d} = (2-d)\frac{\omega^{2-d}}{|y|^d}\left(\delta_{\alpha\beta} - d\frac{y_\alpha y_\beta}{y^2}\right), \tag{10.333}$$

which is a distribution that is ambiguous at origin, and defined up to the addition of a $\delta^{(d)}(\tau)$-function. It is regularized in the same way as the divergence in the Fourier representation (10.292). Contracting the indices α and β in Eq. (10.333), we obtain

$$\partial^2 y^{2-d} = (2-d)\omega^{2-d} S_d \delta^{(d)}(\tau), \tag{10.334}$$

where $S_d = 2\pi^{d/2}/\Gamma(d/2)$ is the surface of a unit sphere in d dimensions [recall Eq. (1.555)]. As a check, we take the trace of $\Delta_{\alpha\beta}(\tau)$ in Eq. (10.332), and reproduce the inhomogeneous field equation (10.296):

$$\begin{aligned}\Delta_{\alpha\alpha}(\tau) &= \omega^2\,\Delta(\tau) - c_d\,m^{2-d}\,S_d\,\frac{1}{2}\Gamma(d/2)\,2^{d/2}\,\delta^{(d)}(\tau) \\ &= \omega^2\,\Delta(\tau) - \delta^{(d)}(\tau).\end{aligned} \tag{10.335}$$

Since $\delta^{(d)}(\tau)$ vanishes at the origin by (10.292), we find once more Eq. (10.298).

A further relation between distributions is found from the derivative

$$\partial_\alpha \Delta_{\alpha\beta}(\tau) = \partial_\beta\left[-\delta^{(d)}(\tau) + \omega^2\,\Delta(\tau)\right] + \omega S_d\left[\Delta(\tau)|y|^{d-1}(\partial_\beta y)\right]\delta^{(d)}(\tau) = \partial_\beta\Delta_{\lambda\lambda}(\tau). \tag{10.336}$$

---

[4]M. Abramowitz and I. Stegun, op. cit., Formula 9.6.9.

10.9.2 Integrals over Products of Two Distributions

Consider now the integrals over products of such distributions. If an integrand $f(|x|)$ depends only on $|x|$, we may perform the integrals over the directions of the vectors

$$\int d^d\tau\, f(\tau) = S_d \int_0^\infty dr\, r^{d-1} f(r), \qquad r \equiv |x|. \tag{10.337}$$

Using the integral formula[5]

$$\int_0^\infty dy\, y\, K_\beta^2(y) = \frac{1}{2}\frac{\pi\beta}{\sin \pi\beta} = \frac{1}{2}\Gamma(1+\beta)\Gamma(1-\beta), \tag{10.338}$$

we can calculate directly:

$$
\begin{aligned}
\int d^d\tau\, \Delta^2(\tau) &= \omega^{-d} c_d^2\, S_d \int_0^\infty dy\, y\, K_{1-d/2}^2(y) \\
&= \omega^{-d} c_d^2\, S_d \frac{1}{2}\,(1-d/2)\,\Gamma(1-d/2)\,\Gamma(d/2) = \frac{2-d}{2\omega^2}\Delta(0),
\end{aligned}
\tag{10.339}$$

and

$$
\begin{aligned}
\int d^d\tau\, \Delta_\alpha^2(\tau) &= \omega^{2-d} c_d^2\, S_d \int_0^\infty dy\, y\, K_{d/2}^2(y) \\
&= \omega^{2-d} c_d^2\, S_d \frac{1}{2}\Gamma(1+d/2)\,\Gamma(1-d/2) = \frac{d}{2}\Delta(0),
\end{aligned}
\tag{10.340}$$

in agreement with Eqs. (10.301) and (10.302). Inserting $\Delta(0) = 1/2\omega$ from (10.293), these integrals give us once more the values of the Feynman diagrams (10.249), (10.252), (10.253), (10.256), and (10.258).

Note that due to the relation[6]

$$K_{d/2}(y) = -y^{d/2-1}\frac{d}{dy}\left[y^{1-d/2}\,K_{1-d/2}(y)\right], \tag{10.341}$$

the integral over y in Eq. (10.340) can also be performed by parts, yielding

$$
\begin{aligned}
\int d^d\tau\, \Delta_\alpha^2(\tau) &= -\omega^{2-d} c_d^2\, S_d \left(y^{d/2}K_{d/2}\right)\left(y^{1-d/2}K_{1-d/2}\right)\Big|_0^\infty - \omega^2 \int d^d\tau\, \Delta^2(\tau) \\
&= \Delta(0) - \omega^2 \int d^d\tau \Delta^2(\tau).
\end{aligned}
\tag{10.342}$$

The upper limit on the right-hand side gives zero because of the exponentially fast decrease of the Bessel function at infinity. This was obtained before in Eq. (10.302) from a partial integration and the inhomogeneous field equation (10.296).

Using the explicit representations (10.327) and (10.332), we calculate similarly the integral

$$
\begin{aligned}
\int d^d\tau\, \Delta_{\alpha\alpha}^2(\tau) &= \int d^d\tau\, \Delta_{\alpha\beta}^2(\tau) = \omega^4 \int d^d\tau\, \Delta^2(\tau) - \omega^{4-d} c_d^2\, \Gamma(d/2)\,\Gamma(1-d/2)\, S_d \\
&= \omega^4 \int d^d\tau\, \Delta^2(\tau) - 2\omega^2\, \Delta(0) = -(1+d/2)\,\omega^2\Delta(0).
\end{aligned}
\tag{10.343}$$

---

[5]I.S. Gradshteyn and I.M. Ryzhik, *op. cit.*, Formula 6.521.3
[6]ibid., Formulas 8.485 and 8.486.12

The first equality follows from partial integrations. In the last equality we have used (10.339). We have omitted the integral containing the modified Bessel functions

$$(d-1)\left[\int_0^\infty dz K_{d/2}(z)K_{1-d/2}(z) + \frac{d}{2}\int_0^\infty dz\, z^{-1} K_{d/2}^2(z)\right],\qquad(10.344)$$

since this vanishes in one dimension as follows:

$$-\frac{\pi}{4}\Gamma(1-\varepsilon/2)\left[\Gamma(\varepsilon/2) + \Gamma(-\varepsilon/2)\right]\varepsilon^2\,\Gamma(\varepsilon)\underset{\varepsilon\to0}{=}0.$$

Inserting into (10.343) $\Delta(0) = 1/2\omega$ from (10.293), we find once more the value of the right-hand Feynman integral (10.250) and the middle one in (10.253).

By combining the result (10.303) with (10.339) and (10.340), we can derive by proper integrations the fundamental rule in this generalized distribution calculus that the integral over the square of the δ-function vanishes. Indeed, solving the inhomogeneous field equation (10.296) for $\delta^{(d)}(\tau)$, and squaring it, we obtain

$$\int d^d\tau\left[\delta^{(d)}(\tau)\right]^2 = \omega^4\int d^d\tau\,\Delta^2(\tau) + 2\omega^2\int d^d\tau\,\Delta_\alpha^2(\tau) + \int d^d\tau\,\Delta_{\beta\beta}^2(\tau) = 0.\qquad(10.345)$$

Thus we may formally calculate

$$\int d^d\tau\,\delta^{(d)}(\tau)\delta^{(d)}(\tau) = \delta^{(d)}(0) = 0,\qquad(10.346)$$

pretending that one of the two δ-functions is an admissible smooth test function $f(\tau)$ of ordinary distribution theory, where

$$\int d^d\tau\,\delta^{(d)}(\tau)f(\tau) = f(0).\qquad(10.347)$$

10.9.3   Integrals over Products of Four Distributions

The calculation of the configuration space integrals over products of four distributions in $d = 1$ dimension is straightforward as long as they are unique. Only if they are ambiguous, they require a calculation in $d = 1 - \varepsilon$ dimension, with the limit $\varepsilon \to 0$ taken at the end.

A unique case is

$$\begin{aligned}\int d^d\tau\,\Delta^4(\tau) &= c_d^4\,\omega^{-d}S_d\int_0^\infty dy\, y^{3-d} K_{1-d/2}^4(y)\\&\underset{d=1}{=} c_1^4\,\omega^{-1} S_1\frac{\pi^2}{2^4}\Gamma^4\left(\frac{3}{2}-\frac{d}{2}\right)\Gamma(d) = \frac{1}{32\omega^5},\end{aligned}\qquad(10.348)$$

where we have set

$$y \equiv \omega\tau.\qquad(10.349)$$

Similarly, we derive by partial integration

$$\begin{aligned}\int d^d\tau\,\Delta^2(\tau)\Delta_\alpha^2(\tau) &= \omega^{2-d} c_d^4 S_d\int_0^\infty dy y^{3-d} K_{d/2}^2(y)K_{1-d/2}^2(y)\\&= \frac{1}{3}\omega^{2-d} c_d^4 S_d\left[2^{-d-1}\Gamma(d/2)\Gamma^3(1-d/2)\right.\\&\quad\left.+ \int_0^\infty dy\left(y^{1-d/2} K_{1-d/2}\right)^3\frac{d}{dy}\left(y^{d/2} K_{d/2}\right)\right]\\&= \frac{1}{3}\left[\Delta^3(0) - \omega^2\int d^d\tau\,\Delta^4(\tau)\right]\underset{d=1}{=}\frac{1}{32\omega}.\end{aligned}\qquad(10.350)$$

Using (10.327), (10.330), and (10.332), we find for the integral in $d = 1 - \varepsilon$ dimensions

$$\int d^d\tau \, \Delta(\tau)\Delta_\alpha(\tau)\Delta_\beta(\tau)\Delta_{\alpha\beta}(\tau) = \omega^2 \int d^d\tau \, \Delta^2(\tau)\Delta_\alpha^2(\tau) - \frac{1}{2}Y_d, \tag{10.351}$$

where Y_d is the integral

$$Y_d = -2(d-1)\omega^{4-d} c_d^4 S_d \int_0^\infty dy y^{2-d} K_{1-d/2}(y) K_{d/2}^3(y). \tag{10.352}$$

In spite of the prefactor $d - 1$, this has a nontrivial limit for $d \to 1$, the zero being compensated by a pole from the small-y part of the integral at $y = 0$. In order to see this we use the integral representation of the Bessel function [28]:

$$K_\beta(y) = \pi^{-1/2}(y/2)^{-\beta}\Gamma\left(\frac{1}{2}+\beta\right)\int_0^\infty dt(\cosh t)^{-2\beta}\cos(y\sinh t). \tag{10.353}$$

In one dimension where $\beta = 1/2$, this becomes simply $K_{1/2}(y) = \sqrt{\pi/2y}e^{-y}$. For $\beta = d/2$ and $\beta = 1 - d/2$ written as $\beta = (1 \mp \varepsilon)/2$, it is approximately equal to

$$K_{(1\mp\varepsilon)/2}(y) = \pi^{-1/2}(y/2)^{-(1\mp\varepsilon)/2}\Gamma\left(1\mp\frac{\varepsilon}{2}\right)$$
$$\times \left[\frac{\pi}{2}e^{-y} \pm \varepsilon\int_0^\infty dt(\cosh t)^{-1}\ln(\cosh t)\cos(y\sinh t)\right], \tag{10.354}$$

where the t-integral is regular at $y = 0$.[7] After substituting (10.354) into (10.352), we obtain the finite value

$$Y_d \underset{\varepsilon\approx 0}{\approx} 2\left(\omega^{4-d}c_d^4 S_d\right)\varepsilon\frac{\pi^2}{4}\Gamma\left(1+\varepsilon/2\right)\Gamma^3\left(1-\varepsilon/2\right)\times 2^{-5\varepsilon}\Gamma(2\varepsilon)$$
$$\underset{\varepsilon\to 0}{=} \left(\frac{1}{2\omega\pi^2}\right)\frac{\pi^2}{4} = \frac{1}{8\omega}. \tag{10.355}$$

The prefactor $d - 1 = -\varepsilon$ in (10.352) has been canceled by the pole in $\Gamma(2\varepsilon)$.

This integral coincides with the integral (10.312) whose subtle nature was discussed in the momentum space formulation (10.240). Indeed, inserting the Bessel expressions (10.327) and (10.332) into (10.312), we find

$$\int d^d\tau \, \Delta^2(\tau)\left[\Delta_{\alpha\beta}^2(\tau) - \Delta_{\alpha\alpha}^2(\tau)\right]$$
$$= -(d-1)\omega^{4-d}c_d^4 S_d \int_0^\infty dy \left[y^{1-d/2}K_{1-d/2}(y)\right]^2\frac{d}{dy}K_{d/2}^2(y), \tag{10.356}$$

and a partial integration

$$\int_0^\infty dy \left[y^{1-d/2}(y)K_{1-d/2}(y)\right]^2\frac{d}{dy}K_{d/2}^2(y) = 2\int_0^\infty dy \, y^{2-d}K_{1-d/2}(y)K_{d/2}^3(y) \tag{10.357}$$

establishes contact with the integral (10.352) for Y_d. Thus Eq. (10.351) is the same as (10.321).

Knowing Y_d, we also determine, after integrations by parts, the integral

$$\int d^d\tau \, \Delta_\alpha^2(\tau)\Delta_\beta^2(\tau) = -3\omega^2 \int d^d\tau \, \Delta^2(\tau)\Delta_\alpha^2(\tau) + Y_d, \tag{10.358}$$

---

[7]I.S. Gradshteyn and I.M. Ryzhik, *op. cit.*, Formulas 3.511.1 and 3.521.2.

which is the same as (10.322). It remains to calculate one more unproblematic integral over four distributions:

$$\int d^d\tau\, \Delta^2(\tau)\Delta^2_{\lambda\lambda}(\tau) = \left[-2\omega^2\,\Delta^3(0) + \omega^4 \int d^d\tau\, \Delta^4(\tau)\right]_{d=1} = -\frac{7}{32\omega}. \qquad (10.359)$$

Combining this with (10.355) and (10.358) we find the Feynman diagram (10.281). The combination of (10.351) and (10.358) with (10.355) and (10.350), finally, yields the diagrams (10.325) and (10.324), respectively.

Thus we see that there is no problem in calculating integrals over products of distributions in configuration space which produce the same results as dimensional regularization in momentum space.

10.10 Simple Rules for Calculating Singular Integrals

The above methods of calculating the Feynman integrals in d time dimensions with a subsequent limit $d \to 1$ are obviously quite cumbersome. It is preferable to develop a simple procedure of finding the same results directly working with a one-dimensional time. This is possible if we only keep track of some basic aspects of the d-dimensional formulation [37].

Consider once more the ambiguous integrals coming from the first two watermelon diagrams in Eq. (10.232), which in the one-dimensional formulation represent the integrals

$$I_1 = \int_{-\infty}^{\infty} d\tau\, \ddot{\Delta}^2(\tau)\Delta^2(\tau), \qquad (10.360)$$

$$I_2 = \int_{-\infty}^{\infty} d\tau\, \ddot{\Delta}(\tau)\dot{\Delta}^2(\tau)\Delta(\tau), \qquad (10.361)$$

evaluated before in the d-dimensional equations (10.284) and (10.282). Consider first the integral (10.360) which contains a square of a δ-function. We separate this out by writing

$$I_1 = \int_{-\infty}^{\infty} d\tau\, \ddot{\Delta}^2(\tau)\Delta^2(\tau) = I_1^{\text{div}} + I_1^R, \qquad (10.362)$$

with a divergent and a regular part

$$I_1^{\text{div}} = \Delta^2(0) \int_{-\infty}^{\infty} d\tau\, \delta^2(\tau), \qquad I_1^R = \int_{-\infty}^{\infty} d\tau\, \Delta^2(\tau)\left[\ddot{\Delta}^2(\tau) - \delta^2(\tau)\right]. \qquad (10.363)$$

All other watermelon diagrams (10.232) lead to the well-defined integrals

$$\int_{-\infty}^{\infty} d\tau\, \Delta^4(\tau) = \frac{1}{4\omega^2}\Delta^3(0), \qquad (10.364)$$

$$\int_{-\infty}^{\infty} d\tau\, \dot{\Delta}^2(\tau)\Delta^2(\tau) = \frac{1}{4}\Delta^3(0), \qquad (10.365)$$

$$\int_{-\infty}^{\infty} d\tau\, \dot{\Delta}^4(\tau) = \frac{1}{4}\omega^2\,\Delta^3(0), \qquad (10.366)$$

whose D-dimensional versions are (10.274), (10.286), and (10.282). Substituting these and (10.361), (10.362) into (10.232) yields the sum of all watermelon diagrams

$$-\frac{4}{2!}\int_{-\infty}^{\infty} d\tau \left[\Delta^2(\tau)\ddot{\Delta}^2(\tau) + 4\Delta(\tau)\dot{\Delta}^2(\tau)\ddot{\Delta}(\tau) + \dot{\Delta}^4(\tau) + 4\omega^2\Delta^2(\tau)\dot{\Delta}^2(\tau) + \frac{2}{3}\omega^4\Delta^4(\tau)\right]$$

$$= -2\Delta^2(0)\int_{-\infty}^{\infty} d\tau\, \delta^2(\tau) - 2\left(I_1^R + 4I_2\right) - \frac{17}{6}\omega^2\Delta^3(0). \tag{10.367}$$

Adding these to (10.230), (10.231), we obtain the sum of all second-order connected diagrams

$$\Sigma\,(\text{all}) = 3\left[\delta(0) - \int_{-\infty}^{\infty} d\tau\,\delta^2(\tau)\right]\Delta^2(0) - 2\left(I_1^R + 4I_2\right) - \frac{7}{2}\omega^2\Delta^3(0), \tag{10.368}$$

where the integrals I_1^R and I_2 are undefined. The sum has to vanish to guarantee coordinate independence. We therefore equate to zero both the singular and finite contributions in Eq. (10.368). The first yields the rule for the product of two δ-functions: $\delta^2(\tau) = \delta(0)\,\delta(\tau)$. This equality should of course be understood in the distributional sense: it holds after multiplying it with an arbitrary test function and integrating over τ.

$$\int d\tau\, \delta^2(\tau)f(\tau) \equiv \delta(0)f(0). \tag{10.369}$$

The equation leads to a perfect cancellation of all powers of $\delta(0)$ arising from the expansion of the Jacobian action, which is the fundamental reason why the heuristic *Veltman rule* of setting $\delta(0) = 0$ is applicable everywhere without problems.

The vanishing of the regular parts of (10.368) requires the integrals (10.361) and (10.362) to satisfy

$$I_1^R + 4I_2 = -\frac{7}{4}\omega^2\Delta^3(0) = -\frac{7}{32\omega}. \tag{10.370}$$

At this point we run into two difficulties. First, this single equation (10.370) for the two undefined integrals I_1^R and I_2 is insufficient to specify both integrals, so that the requirement of reparametrization invariance alone is not enough to fix all ambiguous temporal integrals over products of distributions. Second, and more seriously, Eq. (10.370) leads to conflicts with standard integration rules based on the use of partial integration and equation of motion, and the independence of the order in which these operations are performed. Indeed, let us apply these rules to the calculation of the integrals I_1^R and I_2 in different orders. Inserting the equation of motion (10.214) into the finite part of the integral (10.362) and making use of the regular integral (10.364), we find immediately

$$I_1^R = \int_{-\infty}^{\infty} d\tau\, \Delta^2(\tau)\left[\ddot{\Delta}^2(\tau) - \delta^2(\tau)\right]$$

$$= -2\omega^2\,\Delta^3(0) + \omega^4\int_{-\infty}^{\infty} d\tau\, \Delta^4(\tau) = -\frac{7}{4}\omega^2\Delta^3(0) = -\frac{7}{32\omega}. \tag{10.371}$$

The same substitution of the equation of motion (10.214) into the other ambiguous integral I_2 of (10.361) leads, after performing the regular integral (10.365), to

$$
\begin{aligned}
I_2 &= -\int_{-\infty}^{\infty} d\tau \, \dot{\Delta}^2(\tau) \, \Delta(\tau) \, \delta(\tau) + \omega^2 \int_{-\infty}^{\infty} d\tau \, \dot{\Delta}^2(\tau) \, \Delta^2(\tau) \\
&= -\frac{1}{8\omega} \int_{-\infty}^{\infty} d\tau \, \epsilon^2(\tau) \, \delta(\tau) + \frac{1}{4}\omega^2 \Delta^3(0) = \frac{1}{8\omega}\left(-I_{\epsilon^2\delta} + \frac{1}{4}\right), \quad (10.372)
\end{aligned}
$$

where $I_{\epsilon^2\delta}$ denotes the undefined integral over a product of distributions

$$
I_{\epsilon^2\delta} = \int_{-\infty}^{\infty} d\tau \, \epsilon^2(\tau)\delta(\tau). \tag{10.373}
$$

The integral I_2 can apparently be fixed by applying partial integration to the integral (10.361) which reduces it to the completely regular form (10.366):

$$
I_2 = \frac{1}{3}\int_{-\infty}^{\infty} d\tau \, \Delta(\tau)\frac{d}{d\tau}\left[\dot{\Delta}^3(\tau)\right] = -\frac{1}{3}\int_{-\infty}^{\infty} d\tau \, \dot{\Delta}^4(\tau) = -\frac{1}{12}\omega^2\Delta^3(0) = -\frac{1}{96\omega}. \tag{10.374}
$$

There are no boundary terms due to the exponential vanishing at infinity of all functions involved. From (10.372) and (10.374) we conclude that $I_{\epsilon^2\delta} = 1/3$. This, however, cannot be correct since the results (10.374) and (10.371) do not obey Eq. (10.370) which is needed for coordinate independence of the path integral. This was the reason why previous authors [32, 35] added a noncovariant correction term $\Delta V = -g^2(q^2/6)$ to the classical action (10.186), which is proportional to $\hbar$ and thus violates Feynman's basic postulate that the phase factors $e^{i\mathcal{A}/\hbar}$ in a path integral should contain only the classical action along the paths.

We shall see below that the correct value of the singular integral I in (10.373) is

$$
I_{\epsilon^2\delta} = \int_{-\infty}^{\infty} d\tau \, \epsilon^2(\tau)\delta(\tau) = 0. \tag{10.375}
$$

From the perspective of the previous sections where all integrals were defined in $d = 1 - \epsilon$ dimensions and continued to $\epsilon \to 0$ at the end, the inconsistency of $I_{\epsilon^2\delta} = 1/3$ is obvious: Arbitrary application of partial integration and equation of motion to one-dimensional integrals is forbidden, and this is the case in the calculation (10.374). Problems arise whenever several dots can correspond to different contractions of partial derivatives $\partial_\alpha, \partial_\beta, \ldots$, from which they arise in the limit $d \to 1$. The different contractions may lead to different integrals.

In the pure one-dimensional calculation of the integrals I_1^R and I_2, all ambiguities can be accounted for by using partial integration and equation of motion (10.214) only according to the following integration rules:

RULE 1. We perform a partial integration which allows us to apply subsequently the equation of motion (10.214).

RULE 2. If the equation of motion (10.214) leads to integrals of the type (10.373), they must be performed using naively the Dirac rule for the δ-function and the property $\epsilon(0) = 0$. Examples are (10.375) and the trivially vanishing integrals for all odd powers of $\epsilon(\tau)$:

$$\int d\tau \, \epsilon^{2n+1}(\tau) \, \delta(\tau) = 0, \quad n = \text{integer}, \tag{10.376}$$

which follow directly from the antisymmetry of $\epsilon^{2n+1}(\tau)$ and the symmetry of $\delta(\tau)$ contained in the regularized expressions (10.330) and (10.332).

RULE 3. The above procedure leaves in general singular integrals, which must be treated once more with the same rules.

Let us show that calculating the integrals I_1^R and I_2 with these rules is consistent with the coordinate independence condition (10.370). In the integral I_2 of (10.361) we first apply partial integration to find

$$
\begin{aligned}
I_2 &= \frac{1}{2} \int_{-\infty}^{\infty} d\tau \, \Delta(\tau) \, \dot\Delta(\tau) \frac{d}{d\tau} \left[\dot\Delta^2(\tau) \right] \\
&= -\frac{1}{2} \int_{-\infty}^{\infty} d\tau \, \dot\Delta^4(\tau) - \frac{1}{2} \int_{-\infty}^{\infty} d\tau \, \Delta(\tau) \, \dot\Delta^2(\tau) \, \ddot\Delta(\tau),
\end{aligned}
\tag{10.377}
$$

with no contributions from the boundary terms. Note that the partial integration (10.374) is forbidden since it does not allow for a subsequent application of the equation of motion (10.214). On the right-hand side of (10.377) it can be applied. This leads to a combination of two regular integrals (10.365) and (10.366) and the singular integral I, which we evaluate with the naive Dirac rule to $I = 0$, resulting in

$$
\begin{aligned}
I_2 &= -\frac{1}{2} \int_{-\infty}^{\infty} d\tau \, \dot\Delta^4(\tau) + \frac{1}{2} \int_{-\infty}^{\infty} d\tau \, \dot\Delta^2(\tau) \, \Delta(\tau) \, \delta(\tau) - \frac{1}{2}\omega^2 \int_{-\infty}^{\infty} d\tau \, \dot\Delta^2(\tau) \, \Delta^2(\tau) \\
&= \frac{1}{16\omega} I - \frac{1}{4}\omega^2 \Delta^3(0) = -\frac{1}{32\omega}.
\end{aligned}
\tag{10.378}
$$

If we calculate the finite part I_1^R of the integral (10.362) with the new rules we obtain a result different from (10.371). Integrating the first term in brackets by parts and using the equation of motion (10.214), we obtain

$$
\begin{aligned}
I_1^R &= \int_{-\infty}^{\infty} d\tau \Delta^2(\tau) \left[\ddot\Delta^2(\tau) - \delta^2(\tau) \right] \\
&= \int_{-\infty}^{\infty} d\tau \left[- \dddot\Delta(\tau) \, \dot\Delta(\tau) \, \Delta^2(\tau) - 2\ddot\Delta(\tau) \, \dot\Delta^2(\tau) \, \Delta(\tau) - \Delta^2(\tau) \, \delta^2(\tau) \right] \\
&= \int_{-\infty}^{\infty} d\tau \left[\dot\Delta(\tau) \, \Delta^2(\tau) \, \dot\delta(\tau) - \Delta^2(\tau) \, \delta^2(\tau) \right] - 2I_2 - \omega^2 \int_{-\infty}^{\infty} d\tau \dot\Delta^2(\tau) \, \Delta^2(\tau) .
\end{aligned}
\tag{10.379}
$$

The last two terms are already known, while the remaining singular integral in brackets must be subjected once more to the same treatment. It is integrated by parts so that the equation of motion (10.214) can be applied to obtain

$$
\begin{aligned}
\int_{-\infty}^{\infty} d\tau \left[\dot\Delta(\tau) \, \Delta^2(\tau) \, \dot\delta(\tau) - \Delta^2(\tau) \, \delta^2(\tau) \right] &= -\int_{-\infty}^{\infty} d\tau \left[\ddot\Delta(\tau) \, \Delta^2(\tau) + 2\dot\Delta^2(\tau) \, \Delta(\tau) \right] \delta(\tau) \\
&\quad - \int_{-\infty}^{\infty} d\tau \Delta^2(\tau) \, \delta^2(\tau) = -\omega^2 \, \Delta^3(0) - \frac{1}{4\omega} I .
\end{aligned}
\tag{10.380}
$$

Inserting this into Eq. (10.379) yields

$$I_1^R = \int_{-\infty}^{\infty} d\tau \Delta^2(\tau) \left[\ddot{\Delta}^2(\tau) - \delta^2(\tau) \right] = -2I_2 - \frac{5}{4}\omega^2 \Delta^3(0) - \frac{1}{4\omega} I = -\frac{3}{32\omega}, \quad (10.381)$$

the right-hand side following from $I = 0$, which is a consequence of Rule 3.

We see now that the integrals (10.378) and (10.381) calculated with the new rules obey Eq. (10.370) which guarantees coordinate independence of the path integral.

The applicability of Rules 1–3 follows immediately from the previously established dimensional continuation [23, 24]. It avoids completely the cumbersome calculations in $1 - \varepsilon$-dimension with the subsequent limit $\varepsilon \to 0$. Only some intermediate steps of the derivation require keeping track of the d-dimensional origin of the rules. For this, we continue the imaginary time coordinate τ to a d-dimensional spacetime vector $\tau \to \tau^\mu = (\tau^0, \tau^1, \ldots, \tau^{d-1})$, and note that the equation of motion (10.214) becomes a scalar field equation of the Klein-Gordon type

$$\left(-\partial_\alpha^2 + \omega^2 \right) \Delta(\tau) = \delta^{(d)}(\tau). \quad (10.382)$$

In d dimensions, the relevant second-order diagrams are obtained by decomposing the harmonic expectation value

$$\int d^d\tau \left\langle q_\alpha^2(\tau) q^2(\tau) q_\beta^2(0) q^2(0) \right\rangle \quad (10.383)$$

into a sum of products of four two-point correlation functions according to the Wick rule. The fields $q_\alpha(\tau)$ are the d-dimensional extensions $q_\alpha(\tau) \equiv \partial_\alpha q(\tau)$ of $\dot{q}(\tau)$. Now the d-dimensional integrals, corresponding to the integrals (10.360) and (10.361), are defined uniquely by the contractions

$$I_1^d = \int d^d\tau \left\langle \overline{q_\alpha(\tau) q_\alpha(\tau) q(\tau) q(\tau) q_\beta(0) q_\beta(0) q(0) q(0)} \right\rangle$$
$$= \int d^d\tau \, \Delta^2(\tau) \, \Delta_{\alpha\beta}^2(\tau), \quad (10.384)$$

$$I_2^d = \int d^d\tau \left\langle \overline{q_\alpha(\tau) q_\alpha(\tau) q(\tau) q(\tau) q_\beta(0) q_\beta(0) q(0) q(0)} \right\rangle$$
$$= \int d^d\tau \, \Delta(\tau) \, \Delta_\alpha(\tau) \, \Delta_\beta(\tau) \, \Delta_{\alpha\beta}(\tau). \quad (10.385)$$

The different derivatives $\partial_\alpha \partial_\beta$ acting on $\Delta(\tau)$ prevent us from applying the field equation (10.382). This obstacle was hidden in the one-dimensional formulation. It can be overcome by a partial integration. Starting with I_2^d, we obtain

$$I_2^d = -\frac{1}{2} \int d^d\tau \, \Delta_\beta^2(\tau) \left[\Delta_\alpha^2(\tau) + \Delta(\tau) \Delta_{\alpha\alpha}(\tau) \right]. \quad (10.386)$$

Treating I_1^d likewise we find

$$I_1^d = -2I_2^d + \int d^d\tau \, \Delta^2(\tau) \, \Delta_{\alpha\alpha}^2(\tau) + 2 \int d^d\tau \, \Delta(\tau) \Delta_\beta^2(\tau) \Delta_{\alpha\alpha}(\tau). \quad (10.387)$$

In the second equation we have used the fact that $\partial_\alpha \Delta_{\alpha\beta} = \partial_\beta \Delta_{\lambda\lambda}$. The right-hand sides of (10.386) and (10.387) contain now the contracted derivatives ∂_α^2 such that we can apply the field equation (10.382). This mechanism works to all orders in the perturbation expansion which is the reason for the applicability of Rules 1 and 2 which led to the results (10.378) and (10.381) ensuring coordinate independence.

The value $I_{\epsilon^2\delta} = 0$ according to the Rule 2 can be deduced from the regularized equation (10.386) in $d = 1 - \varepsilon$ dimensions by using the field equation (10.335) to rewrite I_2^d as

$$
\begin{aligned}
I_2^d &= -\frac{1}{2} \int d^d\tau \, \Delta_\beta^2(\tau) \left[\Delta_\alpha^2(\tau) + \omega^2 \Delta^2(\tau) - \Delta(\tau)\delta^{(d)}(\tau) \right] \\
&\underset{d\approx 1}{\approx} -\frac{1}{32\omega} + \frac{1}{2} \int d^d\tau \, \Delta_\beta^2(\tau) \Delta_\alpha^2(\tau) \Delta(\tau)\delta^{(d)}(\tau) .
\end{aligned}
$$

Comparison with (10.372) yields the regularized expression for $I_{\epsilon^2\delta}$

$$
I_{\epsilon^2\delta}^R = \left[\int_{-\infty}^{\infty} d\tau \, \epsilon^2(\tau)\delta(\tau) \right]^R = 8\omega \int d^d\tau \, \Delta_\beta^2(\tau) \, \Delta(\tau)\delta^{(d)}(\tau) = 0, \tag{10.388}
$$

the vanishing for all $\varepsilon > 0$ being a consequence of the small-τ behavior $\Delta(\tau) \Delta_\alpha^2(\tau) \propto |\tau|^{2\varepsilon}$, which follows directly from (10.331).

Let us briefly discuss an alternative possibility of giving up partial integration completely in ambiguous integrals containing ϵ- and δ-function, or their time derivatives, which makes unnecessary to satisfy Eq. (10.374). This yields a freedom in the definition of integral over product of distribution (10.373) which can be used to fix $I_{\epsilon^2\delta} = 1/4$ from the requirement of coordinate independence [25]. Indeed, this value of I would make the integral (10.372) equal to $I_2 = 0$, such that (10.370) would be satisfied and coordinate independence ensured. In contrast, giving up partial integration, the authors of Refs. [31, 33] have assumed the vanishing $\epsilon^2(\tau)$ at $\tau = 0$ so that the integral $I_{\epsilon^2\delta}$ should vanish as well: $I_{\epsilon^2\delta} = 0$. Then Eq. (10.372) yields $I_2 = 1/32\omega$ which together with (10.371) does not obey the coordinate independence condition (10.370), making yet an another noncovariant quantum correction $\Delta V = g^2(q^2/2)$ necessary in the action, which we reject since it contradicts Feynman's original rules of path integration. We do not consider giving up partial integration as an attractive option since it is an important tool for calculating higher-loop diagrams.

10.11 Perturbative Calculation on Finite Time Intervals

The above calculation rules can be extended with little effort to path integrals of time evolution amplitudes on finite time intervals. We shall use an imaginary time interval with $\tau_a = 0$ and $\tau_b = \beta$ to have the closest connection to statistical mechanics. The ends of the paths will be fixed at τ_a and τ_b to be able to extract quantum-mechanical time evolution amplitudes by a mere replacement $\tau \to -it$. The extension to a finite time interval is nontrivial since the Feynman integrals in frequency space become sums over discrete frequencies whose d-dimensional generalizations can usually not be evaluated with standard formulas. The above ambiguities of the integrals, however, will appear in the sums in precisely the same way as before. The reason is that they stem from ordering ambiguities between q and $\dot{q}$ in the perturbation expansions. These are properties of small time intervals and thus of high frequencies, where the sums can be approximated by integrals. In fact, we have seen in the last section, that all ambiguities can be resolved by a careful treatment of the singularities of the correlation functions at small temporal spacings. For integrals on a time axis it is thus completely irrelevant whether the total time interval is finite or infinite, and the ambiguities can be resolved in the same way as before [38].

This can also be seen technically by calculating the frequency sums in the Feynman integrals of finite-time path integrals with the help of the Euler-Maclaurin formula (2.592) or the equivalent ζ-function methods described in Subsection 2.15.6. The lowest approximation involves the pure frequency integrals whose ambiguities have been resolved in the preceeding sections. The remaining correction terms in powers of the temperature $T = 1/\beta$ are all unique and finite [see Eq. (2.596) or (2.556)].

The calculations of the Feynman integrals will most efficiently proceed in configuration space as described in Subsection 10.8.1. Keeping track of certain minimal features of the unique definition of all singular integrals in d dimensions, we shall develop reduction rules based on the equation of motion and partial integration. These will allow us to bring all singular Feynman integrals to a regular form in which the integrations can be done directly in one dimension. The integration rules will be in complete agreement with much more cumbersome calculations in d dimensions with the limit $d \to 1$ taken at the end.

10.11.1 Diagrammatic Elements

The perturbation expansion for an evolution amplitude over a finite imaginary time proceeds as described in Section 10.6, except that the free energy in Eq. (10.201) becomes [recall (2.524)]

$$
\beta F_\omega = \frac{D}{2\beta}\mathrm{Tr}\log(-\partial^2 + \omega^2) = \frac{D}{2\beta}\sum_n \log(\omega_n^2 + \omega^2)
$$
$$
= \frac{D}{\beta}\log\left[2\sinh\left(\hbar\beta\omega/2\right)\right]. \tag{10.389}
$$

As before, the diagrams contain four types of lines representing the correlation functions (10.202)–(10.191). Their explicit forms are, however, different. It will be convenient to let the frequency ω in the free part of the action (10.187) go to zero. Then the free energy (10.389) diverges logarithmically in ω. This divergence is, however, trivial. As explained in Section 2.9, the divergence is removed by replacing ω by the length of the q-axis according to the rule (2.359). For finite time intervals, the correlation functions are no longer given by (10.207) which would not have a finite limit for $\omega \to 0$. Instead, they satisfy Dirichlet boundary conditions, where we can go to $\omega = 0$ without problem. The finiteness of the time interval removes a possible infrared divergence for $\omega \to 0$. The Dirichlet boundary conditions fix the paths at the ends of the time interval $(0, \beta)$ making the fluctuations vanish, and thus also their correlation functions:

$$
G_{\mu\nu}^{(2)}(0, \tau') = G_{\mu\nu}^{(2)}(\beta, \tau') = 0, \qquad G_{\mu\nu}^{(2)}(\tau, 0) = G_{\mu\nu}^{(2)}(\tau, \beta) = 0. \tag{10.390}
$$

The first correlation function corresponding to (10.206) is now

$$
G_{\mu\nu}^{(2)}(\tau, \tau') = \delta_{\mu\nu}\Delta(\tau, \tau') = \quad\underline{\qquad}\quad, \tag{10.391}
$$

where

$$
\Delta(\tau, \tau') = \Delta(\tau', \tau) = \frac{1}{\beta}(\beta - \tau_>)\tau_< = \frac{1}{2}\left[-\epsilon(\tau - \tau')(\tau - \tau') + \tau + \tau'\right] - \frac{\tau\tau'}{\beta}, \tag{10.392}
$$

abbreviates the Euclidean version of $G_0(t, t')$ in Eq. (3.39). Being a Green function of the free equation of motion (10.213) for $\omega = 0$, this satisfies the inhomogeneous differential equations

$$
\ddot{\Delta}(\tau, \tau') = \Delta''(\tau, \tau') = -\delta(\tau - \tau'), \tag{10.393}
$$

by analogy with Eq. (10.214) for $\omega = 0$. In addition, there is now an independent equation in which the two derivatives act on the different time arguments:

$$\overset{\centerdot}{\Delta}{}^{\centerdot}(\tau,\tau') = \delta(\tau - \tau') - 1/\beta. \tag{10.394}$$

For a finite time interval, the correlation functions (10.203) (10.204) differ by more than just a sign [recall (10.210)]. We therefore must distinguish the derivatives depending on whether the left or the right argument are differentiated. In the following, we shall denote the derivatives with respect to τ or τ' by a dot on the left or right, respectively, writing

$$\overset{\centerdot}{\Delta}(\tau,\tau') \equiv \frac{d}{d\tau}\Delta(\tau,\tau'), \qquad \Delta^{\centerdot}(\tau,\tau') \equiv \frac{d}{d\tau'}\Delta(\tau,\tau'). \tag{10.395}$$

Differentiating (10.392) we obtain explicitly

$$\overset{\centerdot}{\Delta}(\tau,\tau') = -\frac{1}{2}\epsilon(\tau - \tau') + \frac{1}{2} - \frac{\tau'}{\beta}, \quad \Delta^{\centerdot}(\tau,\tau') = \frac{1}{2}\epsilon(\tau - \tau') + \frac{1}{2} - \frac{\tau}{\beta} = \overset{\centerdot}{\Delta}(\tau',\tau). \tag{10.396}$$

The discontinuity at $\tau = \tau'$ which does not depend on the boundary condition is of course the same as before, The two correlation functions (10.208) and (10.210) and their diagrammatic symbols are now

$$\partial_\tau G^{(2)}_{\mu\nu}(\tau,\tau') \equiv \langle \dot{q}_\mu(\tau) q_\nu(\tau') \rangle = \delta_{\mu\nu} \overset{\centerdot}{\Delta}(\tau,\tau') \;\; = \;\; \cdots\!-\!\!-\!\! , \tag{10.397}$$

$$\partial_{\tau'} G^{(2)}_{\mu\nu}(\tau,\tau') \equiv \langle q_\mu(\tau) \dot{q}_\nu(\tau') \rangle = \delta_{\mu\nu} \Delta^{\centerdot}(\tau,\tau') \;\; = \;\; -\!\!-\!\!-\!\cdots\!\! . \tag{10.398}$$

The fourth correlation function (10.212) is now

$$\partial_\tau \partial_{\tau'} G^{(2)}_{\mu\nu}(\tau,\tau') = \delta_{\mu\nu} \overset{\centerdot}{\Delta}{}^{\centerdot}(\tau,\tau') = \;\; \cdots\cdots\cdots\!\! , \tag{10.399}$$

with $\overset{\centerdot}{\Delta}{}^{\centerdot}(\tau,\tau')$ being given by (10.394). Note the close similarity but also the difference of this with respect to the equation of motion (10.393).

10.11.2 Cumulant Expansion of D-Dimensional Free-Particle Amplitude in Curvilinear Coordinates

We shall now calculate the partition function of a point particle in curved space for a finite time interval. Starting point is the integral over the diagonal amplitude of a free point particle of unit mass $(\mathbf{x}_a\beta|\mathbf{x}_a 0)$ in flat D-dimensional space

$$Z = \int d^D x_a \, (\mathbf{x}_a\beta|\mathbf{x}_a 0), \tag{10.400}$$

with the path integral representation

$$(\mathbf{x}_a\beta|\mathbf{x}_a 0)_0 = \int \mathcal{D}^D x \, e^{-\mathcal{A}^{(0)}[\mathbf{x}]}, \tag{10.401}$$

where $\mathcal{A}^{(0)}[\mathbf{x}]$ is the free-particle action

$$\mathcal{A}^{(0)}[x] = \frac{1}{2}\int_0^\beta d\tau \, \dot{\mathbf{x}}^2(\tau). \tag{10.402}$$

Performing the Gaussian path integral leads to

$$(\mathbf{x}_a\beta|\mathbf{x}_a 0)_0 = e^{-(D/2)\mathrm{Tr}\log(-\partial^2)} = [2\pi\beta]^{-D/2}, \tag{10.403}$$

where the trace of the logarithm is evaluated with Dirichlet boundary conditions. The result is of course the D-dimensional imaginary-time version of the fluctuation factor (2.125) in natural units.

A coordinate transformation $x^i(\tau) = x^i(q^\mu(\tau))$ mapping $\mathbf{x}_a$ to q_a^μ brings the action (10.402) to the form (10.186) with $V(q(\tau)) = 0$:

$$\mathcal{A}[q] = \frac{1}{2} \int_0^\beta d\tau \, g_{\mu\nu}(q(\tau)) \dot{q}^\mu(\tau) \dot{q}^\nu(\tau), \quad \text{with} \quad g_{\mu\nu}(q) \equiv \frac{\partial x^i(q)}{\partial q^\mu} \frac{\partial x^i(q)}{\partial q^\nu}. \tag{10.404}$$

In the formal notation (10.189), the measure transforms as follows:

$$\int \mathcal{D}^D x(\tau) \equiv \prod_\tau \int d^D x(\tau) = J \prod_\tau \int d^D q(\tau) \equiv J \int \mathcal{D}^D q \sqrt{g(q_a)}, \tag{10.405}$$

where $g(q) \equiv \det g_{\mu\nu}(q)$ and J is the Jacobian of the coordinate transformation generalizing (10.220) and (10.219)

$$J = \prod_\tau \left[\sqrt{\frac{\partial x^i(q(\tau))}{\delta q^\mu(\tau)}} \Bigg/ \sqrt{\frac{\partial x^i(q_a)}{\delta q_0^\mu}} \right] = \exp\left[\frac{1}{2}\delta(0) \int_0^\beta d\tau \log \frac{g(q(\tau))}{g(q_a)} \right]. \tag{10.406}$$

Thus we may write the transformed path integral (10.401) in the form

$$(\mathbf{x}_a \beta | \mathbf{x}_a 0)_0 \equiv (q_a \beta | q_a 0)_0 = \int \mathcal{D}^D q \, e^{-\mathcal{A}_{\text{tot}}[x]}, \tag{10.407}$$

with the total action in the exponent

$$\mathcal{A}_{\text{tot}}[q] = \int_0^\beta d\tau \left[\frac{1}{2} g_{\mu\nu}(q(\tau)) \dot{q}^\mu(\tau) \dot{q}^\nu(\tau) - \frac{1}{2}\delta(0) \log \frac{g(q(\tau))}{g(q_a)} \right]. \tag{10.408}$$

Following the rules described in Subsection 10.6.1 we expand the action in powers of $\delta q^\mu(\tau) = q^\mu(\tau) - q_a^\mu$. The action can then be decomposed into a free part

$$\mathcal{A}^{(0)}[q_a, \delta q] = \frac{1}{2} \int_0^\beta d\tau \, g_{\mu\nu}(q_a) \delta \dot{q}^\mu(\tau) \delta \dot{q}^\nu(\tau) \tag{10.409}$$

and an interacting part written somewhat more explicitly than in (10.194) with (10.188) and (10.193):

$$\mathcal{A}_{\text{tot}}^{\text{int}}[q_a, \delta q] = \int_0^\beta d\tau \frac{1}{2} [g_{\mu\nu}(q) - g_{\mu\nu}(q_a)] \delta \dot{q}^\mu \delta \dot{q}^\nu$$
$$- \int_0^\beta d\tau \frac{1}{2}\delta(0) \left\{ \left[\frac{g(q_a + \delta q)}{g(q_a)} - 1 \right] - \frac{1}{2} \left[\frac{g(q_a + \delta q)}{g(q_a)} - 1 \right]^2 + \dots \right\}. \tag{10.410}$$

For simplicity, we assume the coordinates to be orthonormal at q_a^μ, i.e., $g_{\mu\nu}(q_a) = \delta_{\mu\nu}$. The path integral (10.407) is now formally defined by a perturbation expansion similar to (10.199):

$$(q_a \beta | q_a 0) = \int \mathcal{D}^D q \, e^{\mathcal{A}^{(0)}[q] - \mathcal{A}_{\text{tot}}^{\text{int}}[q]} = \int \mathcal{D}^D q \, e^{-\mathcal{A}^{(0)}[q]} \left(1 - \mathcal{A}_{\text{tot}}^{\text{int}} + \frac{1}{2} \mathcal{A}_{\text{tot}}^{\text{int}\,2} - \dots \right)$$
$$= (2\pi\beta)^{-D/2} \left[1 - \langle \mathcal{A}_{\text{tot}}^{\text{int}} \rangle + \frac{1}{2} \langle \mathcal{A}_{\text{tot}}^{\text{int}\,2} \rangle - \dots \right],$$
$$= (2\pi\beta)^{-D/2} \exp\left\{ - \langle \mathcal{A}_{\text{tot}}^{\text{int}} \rangle_c + \frac{1}{2} \langle \mathcal{A}_{\text{tot}}^{\text{int}\,2} \rangle_c - \dots \right\} \equiv e^{-\beta f(q)}, \tag{10.411}$$

with the harmonic expectation values

$$\langle \dots \rangle = (2\pi\beta)^{D/2} \int \mathcal{D}^D q(\tau) (\dots) e^{-\mathcal{A}^{(0)}[q]} \tag{10.412}$$

and their cumulants $\left\langle \mathcal{A}_{\rm tot}^{\rm int\, 2} \right\rangle_c = \left\langle \mathcal{A}_{\rm tot}^{\rm int\, 2} \right\rangle - \left\langle \mathcal{A}_{\rm tot}^{\rm int} \right\rangle^2, \ldots$ [recall (3.482), (3.483)], containing only connected diagrams. To emphasize the analogy with the cumulant expansion of the free energy in (10.199), we have defined the exponent in (10.411) as $-\beta f(q)$. This q-dependent quantity $f(q)$ is closely related to the alternative effective classical potential discussed in Subsection 3.25.4, apart from a normalization factor:

$$
e^{-\beta f(q)} = \frac{1}{\sqrt{2\pi\hbar^2/Mk_BT}} e^{-\beta \tilde{V}_\omega^{\rm eff\, cl}(q)}. \tag{10.413}
$$

If our calculation procedure respects coordinate independence, all expansion terms of $\beta f(q)$ must vanish to yield the trivial exact results (10.401).

10.11.3 Propagator in $1-\varepsilon$ Time Dimensions

In the dimensional regularization of the Feynman integrals on an infinite time interval in Subsection 10.7.2 we have continued all Feynman diagrams in momentum space to $d = 1 - \varepsilon$ time dimensions. For the present Dirichlet boundary conditions, this standard continuation of quantum field theory is not directly applicable since the integrals in momentum space become sums over discrete frequencies $\nu_n = \pi n/\beta$ [compare (3.64)]. For such sums one has to set up completely new rules for a continuation, and there are many possibilities for doing this. Fortunately, it will not be necessary to make a choice since we can use the method developed in Subsection 10.10 to avoid continuations altogether. and work in a single physical time dimension. For a better understanding of the final procedure it is, however, useful to see how a dimensional continuation could proceed. We extend the imaginary time coordinate τ to a d-dimensional spacetime vector whose zeroth component is τ: $z^\mu = (\tau, z^1, \ldots, z^{d-1})$. In $d = 1 - \varepsilon$ dimensions, the extended correlation function reads

$$
\Delta(\tau, \mathbf{z}; \tau', \mathbf{z}') = \int \frac{d^\varepsilon k}{(2\pi)^\varepsilon} e^{i\mathbf{k}(\mathbf{z}-\mathbf{z}')} \Delta_\omega(\tau, \tau'), \qquad \text{where} \quad \omega \equiv |\mathbf{k}|. \tag{10.414}
$$

Here the extra ε-dimensional space coordinates $\mathbf{z}$ are assumed to live on infinite axes with translational invariance along all directions. Only the original τ-coordinate lies in a finite interval $0 \leq \tau \leq \beta$, with Dirichlet boundary conditions. The Fourier component in the integrand $\Delta_\omega(\tau, \tau')$ is the usual one-dimensional correlation function of a harmonic oscillator with the $\mathbf{k}$-dependent frequency $\omega = |\mathbf{k}|$. It is the Green function which satisfies on the finite τ-interval the equation of motion

$$
-\ddot{\Delta}_\omega(\tau, \tau') + \omega^2 \Delta_\omega(\tau, \tau') = \delta(\tau - \tau'), \tag{10.415}
$$

with Dirichlet boundary conditions

$$
\Delta_\omega(0, \tau) = \Delta_\omega(\beta, \tau) = 0. \tag{10.416}
$$

The explicit form was given in Eq. (3.36) for real times. Its obvious continuation to imaginary-time is

$$
\Delta_\omega(\tau, \tau') = \frac{\sinh \omega(\beta - \tau_>) \sinh \omega\tau_<}{\omega \sinh \omega\beta}, \tag{10.417}
$$

where $\tau_>$ and $\tau_<$ denote the larger and smaller of the imaginary times τ and τ', respectively.

In d time dimensions, the equation of motion (10.393) becomes a scalar field equation of the Klein-Gordon type. Using Eq. (10.415) we obtain

$$
\begin{aligned}
{}_{\mu\mu}\Delta(\tau, \mathbf{z}; \tau', \mathbf{z}') &= \Delta_{\mu\mu}(\tau, \mathbf{z}; \tau', \mathbf{z}') = \ddot{\Delta}(\tau, \mathbf{z}; \tau', \mathbf{z}') + {}_{\mathbf{zz}}\Delta(\tau, \mathbf{z}; \tau', \mathbf{z}') \\
&= \int \frac{d^\varepsilon k}{(2\pi)^\varepsilon} e^{i\mathbf{k}(\mathbf{z}-\mathbf{z}')} \left[\ddot{\Delta}_\omega(\tau, \tau') - \omega^2 \Delta_\omega(\tau, \tau') \right] = \\
&= -\delta(\tau - \tau')\, \delta^{(\varepsilon)}(\mathbf{z} - \mathbf{z}') \equiv -\delta^{(d)}(z - z').
\end{aligned} \tag{10.418}
$$

The important observation is now that for d spacetime dimensions, perturbation expansion of the path integral yields for the second correlation function $\overset{..}{\Delta}(\tau, \tau')$ in Eqs. (10.515) and (10.516) the extension $_\mu\Delta_\nu(z, z')$. This function differs from the contracted function $_\mu\Delta_\mu(z, z')$, and from $_{\mu\mu}\Delta(z, z')$ which satisfies the field equation (10.418). In fact, all correlation functions $\overset{..}{\Delta}(\tau, \tau')$ encountered in the diagrammatic expansion which have different time arguments always turn out to have the d-dimensional extension $_\mu\Delta_\nu(z, z')$. An important exception is the correlation function at *equal* times $\overset{..}{\Delta}(\tau, \tau)$ whose d-dimensional extension is always $_\mu\Delta_\mu(z, z)$, which satisfies the right-hand equation (10.393) in the $\varepsilon \to 0$ -limit. Indeed, it follows from Eq. (10.414) that

$$_\mu\Delta_\mu(z, z) = \int \frac{d^\varepsilon k}{(2\pi)^\varepsilon} \left[\overset{..}{\Delta}_\omega(\tau, \tau) + \omega^2 \Delta_\omega(\tau, \tau) \right]. \tag{10.419}$$

With the help of Eq. (10.417), the integrand in Eq. (10.419) can be brought to

$$\overset{..}{\Delta}_\omega(\tau, \tau) + \omega^2 \Delta_\omega(\tau, \tau) = \delta(0) - \frac{\omega \cosh \omega(2\tau - \beta)}{\sinh \omega\beta}. \tag{10.420}$$

Substituting this into Eq. (10.419), we obtain

$$_\mu\Delta_\mu(z, z) = \delta^{(d)}(z, z) - I^\varepsilon. \tag{10.421}$$

The integral I^ε is calculated as follows

$$
\begin{aligned}
I^\varepsilon &= \int \frac{d^\varepsilon k}{(2\pi)^\varepsilon} \frac{\omega \cosh \omega(2\tau - \beta)}{\sinh \omega\beta} = \frac{1}{\beta} \frac{S_\varepsilon}{(2\pi\beta)^\varepsilon} \int_0^\infty dz z^\varepsilon \frac{\cosh z(1 - 2\tau/\beta)}{\sinh z} \\
&= \frac{1}{\beta} \frac{S_\varepsilon}{(2\pi\beta)^\varepsilon} \frac{\Gamma(\varepsilon + 1)}{2^{\varepsilon+1}} \left[\zeta\left(\varepsilon + 1, 1 - \tau/\beta\right) + \zeta\left(\varepsilon + 1, \tau/\beta\right) \right],
\end{aligned}
\tag{10.422}
$$

where $S_\varepsilon = 2\pi^{\varepsilon/2}/\Gamma(\varepsilon/2)$ is the surface of a unit sphere in ε dimension [recall Eq. (1.555)], and $\Gamma(z)$ and $\zeta(z, q)$ are gamma and zeta functions, respectively. For small $\varepsilon \to 0$, they have the limits $\zeta(\varepsilon + 1, q) \to 1/\varepsilon - \psi(q)$, and $\Gamma(\varepsilon/2) \to 2/\varepsilon$, so that $I^\varepsilon \to 1/\beta$, proving that the d-dimensional equation (10.421) at coinciding arguments reduces indeed to the one-dimensional equation (10.393). The explicit d-dimensional form will never be needed, since we can always treat $_\mu\Delta_\mu(z, z)$ as one-dimensional functions $\overset{..}{\Delta}(\tau, \tau)$, which can in turn be replaced everywhere by the right-hand side $\delta(0) - 1/\beta$ of (10.394).

10.11.4   Coordinate Independence for Dirichlet Boundary Conditions

Before calculating the path integral (10.411) in curved space with Dirichlet boundary conditions, let us first verify its coordinate independence following the procedure in Section 10.7. Thus we consider the perturbation expansion of the short-time amplitude of a free particle in one general coordinate. The free action is (10.221), and the interactions (10.222) and (10.224), all with $\omega = 0$. Taking the parameter $a = 1$, the actions are

$$\mathcal{A}^{(0)}[q] = \frac{1}{2} \int_0^\beta d\tau \, \dot{q}^2(\tau), \tag{10.423}$$
$$\tag{10.424}$$

and

$$\mathcal{A}_{\text{tot}}^{\text{int}} = \int_0^\beta d\tau \left\{ \left[-\eta q^2(\tau) + \frac{3}{2}\eta^2 q^4(\tau) \right] \dot{q}^2(\tau) - \delta(0) \left[-\eta q^2(\tau) + \frac{1}{2}\eta^2 q^4(\tau) \right] \right\}. \tag{10.425}$$

We calculate the cumulants $\left\langle \mathcal{A}_{\text{tot}}^{\text{int}} \right\rangle_c = \left\langle \mathcal{A}_{\text{tot}}^{\text{int}} \right\rangle$, $\left\langle \mathcal{A}_{\text{tot}}^{\text{int}\,2} \right\rangle_c = \left\langle \mathcal{A}_{\text{tot}}^{\text{int}\,2} \right\rangle - \left\langle \mathcal{A}_{\text{tot}}^{\text{int}} \right\rangle^2, \dots$ [recall (3.482), (3.483)] contributing to th quantity βf in Eq. (10.411) order by order in η. For a better comparison with the previous expansion in Subsection 10.7.1 we shall denote the diagrammatic contributions which are analogous to the different free energy terms $\beta F_n^{(m)}$ of order n by corresponding symbols $\beta f_n^{(m)}$. There are two main differences with respect to Subsection 10.7.1: All diagrams with a prefactor ω are absent, and there are new diagrams involving the correlation functions at equal times $\langle \dot{q}^\mu(\tau) q^\nu(\tau) \rangle$ and $\langle q^\mu(\tau) \dot{q}^\nu(\tau) \rangle$ which previously vanished because of (10.211). Here they have the nonzero value $\dot{\Delta}(\tau,\tau) = \Delta^{\cdot}(\tau,\tau) = 1/2 - \tau/\beta$ by Eq. (10.396). To first order in η, the quantity $f(q)$ in Eq. (10.411) receives a contribution from the first cumulant of the linear terms in η of the interaction (10.425):

$$\beta f_1 = \left\langle \mathcal{A}_{\text{tot}}^{\text{int}} \right\rangle_c = \eta \int_0^\beta d\tau \left\langle -q^2(\tau)\dot{q}^2(\tau) + \delta(0)q^2(\tau) \right\rangle + \mathcal{O}(\eta^2). \tag{10.426}$$

There exists only three diagrams, two originating from the kinetic term and one from the Jacobian action:

$$\beta f_1 = -\eta \; \text{⬭} \; - 2\eta \; \text{⬭} \; + \eta \, \delta(0) \; \text{◯}. \tag{10.427}$$

Note the difference with respect to the diagrams (10.227) for infinite time interval with ω^2-term in the action.

The omitted η^2-terms in (10.426) yield the second-order contribution

$$\beta f_2^{(1)} = \eta^2 \int_0^\beta d\tau \left\langle \frac{3}{2} q^4(\tau)\dot{q}^2(\tau) - \delta(0)\frac{1}{2} q^4(\tau) \right\rangle_c. \tag{10.428}$$

The associated local diagrams are [compare (10.228)]:

$$\beta f_2^{(1)} = \eta^2 \left[\frac{9}{2} \; \text{⬀} \; + 18 \; \text{⬭} \; - \frac{3}{2}\delta(0) \; \text{∞} \right]. \tag{10.429}$$

The second cumulant to order η^2 reads

$$-\frac{1}{2!}\eta^2 \int_0^\beta d\tau \int_0^\beta d\tau' \left\langle \left[-q^2(\tau)\dot{q}^2(\tau) + \delta(0)q^2(\tau) \right] \left[-q^2(\tau')\dot{q}^2(\tau') + \delta(0)q^2(\tau') \right] \right\rangle_c,$$

leading to diagrams containing $\delta(0)$:

$$\beta f_2^{(2)} = -\frac{\eta^2}{2!} \left\{ 2\,\delta^2(0) \; \text{◯} \; - 4\,\delta(0) \left[\; \text{⬭} \; + 4 \; \text{⬭} \; + \; \text{⬭} \; \right] \right\}. \tag{10.430}$$

The remaining diagrams are either of the three-bubble type, or of the watermelon type, each with all possible combinations of the four line types (10.391) and (10.397)–(10.399). The three-bubbles diagrams yield [compare (10.231)]

$$\beta f_2^{(3)} = -\frac{\eta^2}{2!} \left[4 \; \text{⬭} + 2 \; \text{⬭} - 8 \; \text{⬭} + 4 \; \text{⬭} + 4 \; \text{⬭} + 2 \; \text{⬭} - 8 \; \text{⬭} \right]. \tag{10.431}$$

The watermelon-type diagrams contribute the same diagrams as in (10.232) for $\omega = 0$:

$$\beta f_2^{(4)} = -\frac{\eta^2}{2!} 4 \left[\; \text{⬭} \; + 4 \; \text{⬭} \; + \; \text{⬭} \; \right]. \tag{10.432}$$

For coordinate independence, the sum of the first-order diagrams (10.427) has to vanish. Analytically, this amounts to the equation

$$\beta f_1 = -\eta \int_0^\beta d\tau \left[\Delta(\tau,\tau)\dot{\Delta}^{\cdot}(\tau,\tau) + 2\dot{\Delta}^2(\tau,\tau) - \delta(0)\Delta(\tau,\tau) \right] = 0. \tag{10.433}$$

In the d-dimensional extension, the correlation function $\overset{\centerdot}{\Delta}{}^{\centerdot}(\tau,\tau)$ at equal times is the limit $d \to 1$ of the contracted correlation function ${}_\mu\Delta_\mu(x,x)$ which satisfies the d-dimensional field equation (10.418). Thus we can use Eq. (10.394) to replace $\overset{\centerdot}{\Delta}{}^{\centerdot}(\tau,\tau)$ by $\delta(0) - 1/\beta$. This removes the infinite factor $\delta(0)$ in Eq. (10.433) coming from the measure. The remainder is calculated directly:

$$\int_0^\beta d\tau \left[-\frac{1}{\beta}\Delta(\tau,\tau) + 2\,\overset{\centerdot}{\Delta}{}^2(\tau,\tau) \right] = 0. \tag{10.434}$$

This result is obtained without subtleties, since by Eqs. (10.392) and (10.396)

$$\Delta(\tau,\tau) = \tau - \frac{\tau^2}{\beta}, \qquad \overset{\centerdot}{\Delta}{}^2(\tau,\tau) = \frac{1}{4} - \frac{\Delta(\tau,\tau)}{\beta}, \tag{10.435}$$

whose integrals yield

$$\frac{1}{2\beta}\int_0^\beta d\tau\,\Delta(\tau,\tau) = \int_0^\beta d\tau\,\overset{\centerdot}{\Delta}{}^2(\tau,\tau) = \frac{\beta}{12}. \tag{10.436}$$

Let us evaluate the second-order diagrams in $\beta f_2^{(i)}$, $i = 1,2,3,4$. The sum of the local diagrams in (10.429) consists of the integrals by

$$\beta f_2^{(1)} = \frac{3}{2}\eta^2 \int_0^\beta d\tau \left[3\Delta^2(\tau,\tau)\overset{\centerdot\centerdot}{\Delta}{}^{}(\tau,\tau) + 12\Delta(\tau,\tau)\overset{\centerdot}{\Delta}{}^2(\tau,\tau) - \delta(0)\Delta^2(\tau,\tau) \right]. \tag{10.437}$$

Replacing $\overset{\centerdot\centerdot}{\Delta}{}^{}(\tau,\tau)$ in Eq. (10.437) again by $\delta(0) - 1/\beta$, on account of the equation of motion (10.394), and taking into account the right-hand equation (10.435),

$$\beta f_2^{(1)} = \eta^2 \left[3\delta(0)\int_0^\beta d\tau \Delta^2(\tau,\tau) \right] = \eta^2 \frac{\beta^3}{10}\,\delta(0). \tag{10.438}$$

We now calculate the sum of bubble diagrams (10.430)–(10.432), beginning with (10.430) whose analytic form is

$$\beta f_2^{(2)} = -\frac{\eta^2}{2}\int_0^\beta\int_0^\beta d\tau\,d\tau' \left\{ 2\delta^2(0)\Delta^2(\tau,\tau') \right. \tag{10.439}$$
$$\left. -4\,\delta(0)\left[\Delta(\tau,\tau)\overset{\centerdot}{\Delta}{}^2(\tau,\tau') + 4\overset{\centerdot}{\Delta}{}^{}(\tau,\tau)\Delta(\tau,\tau')\overset{\centerdot}{\Delta}{}^{}(\tau,\tau') + \Delta^2(\tau,\tau')\overset{\centerdot\centerdot}{\Delta}{}^{}(\tau,\tau) \right] \right\}.$$

Inserting Eq. (10.394) into the last equal-time term, we obtain

$$\beta f_2^{(2)} = -\frac{\eta^2}{2}\int_0^\beta\int_0^\beta d\tau\,d\tau' \left\{ -2\delta^2(0)\Delta^2(\tau,\tau') \right. \tag{10.440}$$
$$\left. -4\delta(0)\left[\Delta(\tau,\tau)\overset{\centerdot}{\Delta}{}^2(\tau,\tau') + 4\overset{\centerdot}{\Delta}{}^{}(\tau,\tau)\Delta(\tau,\tau')\overset{\centerdot}{\Delta}{}^{}(\tau,\tau') - \Delta^2(\tau,\tau')/\beta \right] \right\}.$$

As we shall see below, the explicit evaluation of the integrals in this sum is not necessary. Just for completeness, we give the result:

$$\begin{aligned}
\beta f_2^{(2)} &= \frac{\eta^2}{2}\left\{ 2\delta^2(0)\frac{\beta^4}{90} + 4\delta(0)\left[\frac{\beta^3}{45} + 4\frac{\beta^3}{180} - \frac{\beta^3}{90} \right] \right\} \\
&= \eta^2 \left\{ \frac{\beta^4}{90}\delta^2(0) + \frac{\beta^3}{15}\delta(0) \right\}.
\end{aligned} \tag{10.441}$$

We now turn to the three-bubbles diagrams (10.432). Only three of these contain the correlation function ${}_\mu\Delta_\nu(x,x') \to \overset{\centerdot}{\Delta}{}^{\centerdot}(\tau,\tau')$ for which Eq. (10.394) is not applicable: the second, fourth, and sixth diagram. The other three-bubble diagrams in (10.432) containing the generalization $_\mu\Delta_\mu(x,x)$ of the equal-time propagator $\dot{\Delta}(\tau,\tau)$ can be calculated using Eq. (10.394).

Consider first a partial sum consisting of the first three three-bubble diagrams in the sum (10.432). This has the analytic form

$$\beta f_2^{(3)}\Big|_{1,2,3} = -\frac{\eta^2}{2} \int_0^\beta \int_0^\beta d\tau \, d\tau' \left\{ 4\,\Delta(\tau,\tau)\,\dot{\Delta}^2(\tau,\tau')\,\Delta(\tau',\tau') \right. \tag{10.442}$$
$$\left. + 2\,\dot{\Delta}(\tau,\tau)\Delta^2(\tau,\tau')\,\dot{\Delta}(\tau',\tau') + 16\,\dot{\Delta}(\tau,\tau)\Delta(\tau,\tau')\,\dot{\Delta}(\tau,\tau')\,\dot{\Delta}(\tau',\tau') \right\}.$$

Replacing $\dot{\Delta}(\tau,\tau)$ and $\dot{\Delta}(\tau',\tau')$ by $\delta(0) - 1/\beta$, according to of (10.394), we see that Eq. (10.442) contains, with opposite sign, precisely the previous sum (10.439) of all one- and two-bubble diagrams. Together they give

$$\beta f_2^{(2)} + \beta f_2^{(3)}\Big|_{1,2,3} = -\frac{\eta^2}{2} \int_0^\beta \int_0^\beta d\tau \, d\tau' \left\{ -\frac{4}{\beta}\Delta(\tau,\tau)\,\dot{\Delta}^2(\tau,\tau') \right.$$
$$\left. + \frac{2}{\beta^2}\Delta^2(\tau,\tau') - \frac{16}{\beta}\,\dot{\Delta}(\tau,\tau)\Delta(\tau,\tau')\,\dot{\Delta}(\tau,\tau') \right\}, \tag{10.443}$$

and can be evaluated directly to

$$\beta f_2^{(2)} + \beta f_2^{(3)}\Big|_{1,2,3} = \frac{\eta^2}{2}\left(\frac{4}{\beta}\frac{\beta^2}{45} - \frac{2}{\beta^2}\frac{\beta^4}{90} + \frac{16}{\beta}\frac{\beta^3}{180}\right) = \frac{\eta^2}{2}\frac{7}{45}\beta^2. \tag{10.444}$$

By the same direct calculation, the Feynman integral in the fifth three-bubble diagram in (10.432) yields

$$\text{⦶⦶} : \quad I_5 = \int_0^\beta \int_0^\beta d\tau \, d\tau' \,\dot{\Delta}(\tau,\tau)\,\dot{\Delta}(\tau,\tau')\,\Delta(\tau,\tau')\,\dot{\Delta}(\tau',\tau') = -\frac{\beta^2}{720}. \tag{10.445}$$

The explicit results (10.444) and (10.445) are again not needed, since the last term in Eq. (10.443) is equal, with opposite sign, to the partial sum of the fourth and fifth three-bubble diagrams in Eq. (10.432). To see this, consider the Feynman integral associated with the sixth three-bubble diagram in Eq. (10.432):

$$\text{⦶⦶} : \quad I_4 = \int_0^\beta \int_0^\beta d\tau \, d\tau' \,\dot{\Delta}(\tau,\tau)\Delta(\tau,\tau')\,\dot{\Delta}(\tau,\tau')\,\dot{\Delta}(\tau',\tau'), \tag{10.446}$$

whose d-dimensional extension is

$$I_4^d = \int_0^\beta \int_0^\beta d^d\tau \, d^d\tau' \,_\alpha\Delta(\tau,\tau)\Delta(\tau,\tau')\,_\alpha\Delta_\beta(\tau,\tau')\Delta_\beta(\tau',\tau'). \tag{10.447}$$

Adding this to the fifth Feynman integral (10.445) and performing a partial integration, we find in one dimension

$$\beta f_2^{(3)}\Big|_{4,5} = -\frac{\eta^2}{2} 16\,(I_4 + I_5) = -\frac{\eta^2}{2} \int_0^\beta \int_0^\beta d\tau \, d\tau'\frac{16}{\beta}\,\dot{\Delta}(\tau,\tau)\,\dot{\Delta}(\tau,\tau')\Delta(\tau,\tau')$$
$$= -\eta^2 \frac{4}{45}\beta^2, \tag{10.448}$$

where we have used $d\left[\dot{\Delta}(\tau,\tau)\right]/d\tau = -1/\beta$ obtained by differentiating (10.435). Comparing (10.448) with (10.443), we find the sum of all bubbles diagrams, except for the sixth and seventh three-bubble diagrams in Eq. (10.432), to be given by

$$\beta f_2^{(2)} + \beta f_2^{(3)}\Big|_{6,7}' = \frac{\eta^2}{2}\frac{2}{15}\beta^2. \tag{10.449}$$

The prime on the sum denotes the exclusion of the diagrams indicated by subscripts. The correlation function $\dot{\Delta}(\tau, \tau')$ in the two remaining diagrams of Eq. (10.432), whose d-dimensional extension is $_\alpha\Delta_\beta(x, x')$, cannot be replaced via Eq. (10.394), and the expression can only be simplified by applying partial integration to the seventh diagram in Eq. (10.432), yielding

$$\bigcirc\!\!\!\!\bigcirc : \quad I_7 = \int_0^\beta \int_0^\beta d\tau\, d\tau'\, \Delta(\tau, \tau)\,\dot{\Delta}\,(\tau, \tau')\,\dot{\Delta}(\tau, \tau')\,\dot{\Delta}'(\tau', \tau')$$

$$\rightarrow \int_0^\beta \int_0^\beta d^d\tau\, d^d\tau'\, \Delta(\tau, \tau)\,_\alpha\Delta(\tau, \tau')\,_\alpha\Delta_\beta(\tau, \tau')\Delta_\beta(\tau', \tau')$$

$$= \frac{1}{2} \int_0^\beta \int_0^\beta d^d\tau\, d^d\tau'\, \Delta(\tau, \tau)\Delta_\beta(\tau', \tau')\partial'_\beta \left[_\alpha\Delta(\tau, \tau')\right]^2$$

$$\rightarrow \frac{1}{2} \int_0^\beta \int_0^\beta d\tau\, d\tau'\, \Delta(\tau, \tau)\,\dot{\Delta}\,(\tau', \tau')\frac{d}{d\tau'}\left[\dot{\Delta}^2(\tau, \tau')\right]$$

$$= \frac{1}{2\beta} \int_0^\beta \int_0^\beta d\tau\, d\tau' \Delta(\tau, \tau)\,\dot{\Delta}^2(\tau, \tau') = \frac{\beta^2}{90}. \tag{10.450}$$

The sixth diagram in the sum (10.432) diverges linearly. As before, we add and subtract the divergence

$$\bigcirc\!\!\!\!\bigcirc : \quad I_6 = \int_0^\beta \int_0^\beta d\tau\, d\tau'\, \Delta(\tau, \tau)\,\dot{\Delta}^2(\tau, \tau')\Delta(\tau', \tau')$$

$$= \int_0^\beta \int_0^\beta d\tau\, d\tau'\, \Delta(\tau, \tau) \left[\dot{\Delta}^2(\tau, \tau') - \delta^2(\tau - \tau')\right]\Delta(\tau', \tau')$$

$$+ \int_0^\beta \int_0^\beta d\tau\, d\tau'\, \Delta^2(\tau, \tau)\,\delta^2(\tau - \tau'). \tag{10.451}$$

In the first, finite term we go to d dimensions and replace $\delta(\tau - \tau') \rightarrow \delta^{(d)}(\tau - \tau') = -\Delta_{\beta\beta}(\tau, \tau')$ using the field equation (10.418). After this, we apply partial integration and find

$$I_6^R \rightarrow \int_0^\beta \int_0^\beta d^d\tau\, d^d\tau'\, \Delta(\tau, \tau) \left[_\alpha\Delta_\beta^2(\tau, \tau') - \Delta_{\gamma\gamma}^2(\tau, \tau')\right]\Delta(\tau', \tau')$$

$$= \int_0^\beta \int_0^\beta d^d\tau\, d^d\tau'\, \{-\partial_\alpha \left[\Delta(\tau, \tau)\right]\Delta_\beta(\tau, \tau')\,_\alpha\Delta_\beta(\tau, \tau')\Delta(\tau', \tau')$$

$$+ \Delta(\tau, \tau)\Delta_\beta(\tau, \tau')\Delta_{\gamma\gamma}(\tau, \tau')\partial'_\beta \left[\Delta(\tau', \tau')\right]\}$$

$$\rightarrow \int_0^\beta \int_0^\beta d\tau\, d\tau'\, 2 \{-\dot{\Delta}\,(\tau, \tau)\,\dot{\Delta}\,(\tau, \tau')\,\dot{\Delta}\,(\tau, \tau')\Delta(\tau', \tau') +$$

$$\Delta(\tau, \tau)\,\dot{\Delta}\,(\tau, \tau')\,\dot{\Delta}\,(\tau', \tau')\,\dot{\Delta}\,(\tau, \tau')\}. \tag{10.452}$$

In going to the last line we have used $d[\Delta(\tau, \tau)]/d\tau = 2\,\dot{\Delta}\,(\tau, \tau)$ following from (10.435). By interchanging the order of integration $\tau \leftrightarrow \tau'$, the first term in Eq. (10.452) reduces to the integral (10.450). In the last term we replace $\dot{\Delta}(\tau, \tau')$ using the field equation (10.393) and the trivial equation (10.376). Thus we obtain

$$I_6 = I_6^R + I_6^{\text{div}} \tag{10.453}$$

with

$$I_6^R = 2 \left(-\frac{\beta^2}{90} - \frac{\beta^2}{120}\right) = \frac{1}{2} \left(-\frac{7\beta^2}{90}\right), \tag{10.454}$$

$$I_6^{\text{div}} = \int_0^\beta \int_0^\beta d\tau\, d\tau'\, \Delta^2(\tau, \tau)\,\delta^2(\tau - \tau'). \tag{10.455}$$

With the help of the identity for distributions (10.369), the divergent part is calculated to be

$$I_6^{\text{div}} = \delta(0) \int_0^\beta d\tau \, \Delta^2(\tau, \tau) = \delta(0) \frac{\beta^3}{30}. \tag{10.456}$$

Using Eqs. (10.450) and (10.453) yields the sum of the sixth and seventh three-bubble diagrams in Eq. (10.432):

$$\beta f_2^{(3)}\Big|_{6,7} = -\frac{\eta^2}{2} (2I_6 + 16I_7) = -\frac{\eta^2}{2} \left[2\delta(0) \frac{\beta^3}{30} + \frac{\beta^2}{10} \right]. \tag{10.457}$$

Adding this to (10.449), we obtain the sum of all bubble diagrams

$$\beta f_2^{(2)} + \beta f_2^{(3)} = -\frac{\eta^2}{2} \left[2\delta(0) \frac{\beta^3}{30} + \frac{\beta^2}{30} \right]. \tag{10.458}$$

The contributions of the watermelon diagrams (10.432) correspond to the Feynman integrals

$$\beta f_2^{(4)} = -2\eta^2 \int_0^\beta \int_0^\beta d\tau d\tau' \left[\Delta^2(\tau, \tau') \dot{\Delta}^2(\tau, \tau') \right. $$
$$\left. + 4\Delta(\tau, \tau') \, \dot{\Delta}(\tau, \tau') \, \Delta^{\cdot}(\tau, \tau') \dot{\Delta}^{\cdot}(\tau, \tau') + \dot{\Delta}^{\cdot 2}(\tau, \tau') \Delta^2(\tau, \tau') \right]. \tag{10.459}$$

The third integral is unique and can be calculated directly:

$$\overset{\cdots}{\bigominus} : \quad I_{10} = \int_0^\beta d\tau \int_0^\beta d\tau' \, \dot{\Delta}^{\cdot 2}(\tau, \tau') \, \Delta^2(\tau, \tau') = \frac{\beta^2}{90}. \tag{10.460}$$

The second integral reads in d dimensions

$$\overset{\cdots}{\bigominus} : \quad I_9 = \int \int d^d\tau \, d^d\tau' \, \Delta(\tau, \tau') \,_\alpha \Delta(\tau, \tau') \Delta_\beta(\tau, \tau') \,_\alpha \Delta_\beta(\tau, \tau'). \tag{10.461}$$

This is integrated partially to yield, in one dimension,

$$I_9 = -\frac{1}{2} I_{10} + I_{9'} \equiv -\frac{1}{2} I_{10} - \frac{1}{2} \int \int d\tau \, d\tau' \Delta(\tau, \tau') \, \Delta^2(\tau, \tau') \, \ddot{\Delta}(\tau, \tau'). \tag{10.462}$$

The integral on the right-hand side is the one-dimensional version of

$$I_{9'} = -\frac{1}{2} \int \int d^d\tau \, d^d\tau' \, \Delta(\tau, \tau') \Delta_\beta^2(\tau, \tau') \,_{\alpha\alpha} \Delta(\tau, \tau'). \tag{10.463}$$

Using the field equation (10.418), going back to one dimension, and inserting $\Delta(\tau, \tau')$, $\Delta^{\cdot}(\tau, \tau')$, and $\ddot{\Delta}(\tau, \tau')$ from (10.392), (10.396), and (10.393), we perform all unique integrals and obtain

$$I_{9'} = \beta^2 \left\{ \frac{1}{48} \int d\tau \, \epsilon^2(\tau) \, \delta(\tau) + \frac{1}{240} \right\}. \tag{10.464}$$

According to Eq. (10.375), the integral over the product of distributions vanishes. Inserting the remainder and (10.460) into Eq. (10.462) gives:

$$I_9 = -\frac{\beta^2}{720}. \tag{10.465}$$

We now evaluate the first integral in Eq. (10.459). Adding and subtracting the linear divergence yields

$$\overset{\cdots}{\bigominus} : \quad I_8 = \int_0^\beta \int_0^\beta d\tau \, d\tau' \, \Delta^2(\tau, \tau') \dot{\Delta}^2(\tau, \tau') = \int_0^\beta \int_0^\beta d\tau d\tau' \Delta^2(\tau, \tau) \delta^2(\tau - \tau')$$
$$+ \int_0^\beta \int_0^\beta d\tau d\tau' \Delta^2(\tau, \tau') \left[\dot{\Delta}^2(\tau, \tau') - \delta^2(\tau - \tau') \right]. \tag{10.466}$$

The finite second part of the integral (10.466) has the d-dimensional extension

$$I_8^R = \int \int d^d\tau\, d^d\tau'\, \Delta^2(\tau, \tau') \left[{}_\alpha\Delta_\beta^2(\tau, \tau') - \Delta_{\gamma\gamma}^2(\tau, \tau') \right], \tag{10.467}$$

which after partial integration and going back to one dimension reduces to a combination of integrals Eqs. (10.465) and (10.464):

$$I_8^R = -2I_9 + 2I_{9'} = -\frac{\beta^2}{72}. \tag{10.468}$$

The divergent part of I_8 coincides with $I_6^{\rm div}$ in Eq. (10.455):

$$I_8^{\rm div} = \int_0^\beta \int_0^\beta d\tau d\tau' \Delta^2(\tau, \tau)\delta^2(\tau - \tau') = I_6^{\rm div} = \delta(0)\frac{\beta^3}{30}. \tag{10.469}$$

Inserting this together with (10.460) and (10.465) into Eq. (10.459), we obtain the sum of watermelon diagrams

$$\beta f_2^{(4)} = -2\eta^2(I_8 + 4I_9 + I_{10}) = -\frac{\eta^2}{2}\left\{ 4\delta(0)\frac{\beta^2}{30} - \frac{\beta^2}{30} \right\}. \tag{10.470}$$

For a flat space in curvilinear coordinates, the sum of the first-order diagrams vanish. To second order, the requirement of coordinate independence implies a vanishing sum of all connected diagrams (10.429)–(10.432). By adding the sum of terms in Eqs. (10.438), (10.458), and (10.470), we find indeed zero, thus confirming coordinate independence. It is not surprising that the integration rules for products of distributions derived in an infinite time interval $\tau \in [0, \infty)$ are applicable for finite time intervals. The singularities in the distributions come in only at a single point of the time axis, so that its total length is irrelevant.

The procedure can easily be continued to higher-loop diagrams to define integrals over higher singular products of ϵ- and δ-functions. At the one-loop level, the cancellation of $\delta(0)$s requires

$$\int d\tau\, \Delta(\tau, \tau)\, \delta(0) = \delta(0) \int d\tau\, \Delta(\tau, \tau). \tag{10.471}$$

The second-order gave, in addition, the rule

$$\int_0^\beta \int_0^\beta d\tau\, d\tau'\, \Delta^2(\tau, \tau)\, \delta^2(\tau - \tau') = \delta(0) \int d\tau\, \Delta^2(\tau, \tau), \tag{10.472}$$

To n-order we can derive the equation

$$\int d\tau_1 \ldots d\tau_n\, \Delta(\tau_1, \tau_2)\, \delta(\tau_1, \tau_2) \cdots \Delta(\tau_n, \tau_1)\, \delta(\tau_n, \tau_1) = \delta(0) \int d\tau\, \Delta^n(\tau, \tau), \tag{10.473}$$

which reduces to

$$\int \int d\tau_1\, d\tau_n\, \Delta^n(\tau_1, \tau_1)\, \delta^2(\tau_1 - \tau_n) = \delta(0) \int d\tau\, \Delta^n(\tau, \tau), \tag{10.474}$$

which is satisfied due to the integration rule (10.369). See Appendix 10C for a general derivation of (10.473).

10.11.5 Time Evolution Amplitude in Curved Space

The same Feynman diagrams which we calculated to verify coordinate independence appear also in the perturbation expansion of the time evolution amplitude in curved space if this is performed in normal or geodesic coordinates.

The path integral in curved space is derived by making the mapping from x^i to q_a^μ in Subsection 10.11.2 nonholonomic, so that it can no longer be written as $x^i(\tau) = x^i(q^\mu(\tau))$ but only as $dx^i(\tau) = e^i{}_\mu(q)dq^\mu(\tau)$. Then the q-space may contain curvature and torsion, and the result of the path integral will not longer be trivial one in Eq. (10.403) but depend on $R_{\mu\nu\lambda}{}^\kappa(q_a)$ and $S_{\mu\nu}{}^\lambda(q_a)$. For simplicity, we shall ignore torsion.

Then the action becomes (10.404) with the metric $g_{\mu\nu}(q) = e^i{}_\mu(q)e^i{}_\nu(q)$. It was shown in Subsection 10.3.2 that under nonholonomic coordinate transformations, the measure of a time-sliced path integral transforms from the flat-space form $\prod_n d^D x_n$ to $\prod_n d^D q\sqrt{g_n}\exp(\Delta t \bar{R}_n/6)$. This had the consequence, in Section 10.4, that the time evolution amplitude for a particle on the surface of a sphere has an energy (10.165) corresponding to the Hamiltonian (1.414) which governs the Schrödinger equation (1.420). It contains a pure Laplace-Beltrami operator in the kinetic part. There is no extra R-term, which would be allowed if only covariance under ordinary coordinate transformations is required. This issue will be discussed in more detail in Subsection 11.1.1.

Below we shall see that for perturbatively defined path integrals, the nonholonomic transformation must carry the flat-space measure into curved space as follows:

$$\mathcal{D}^D x \to \mathcal{D}^D q \sqrt{g} \exp\left(\int_0^\beta d\tau \, \bar{R}/8\right). \tag{10.475}$$

For a D-dimensional space with a general metric $g_{\mu\nu}(q)$ we can make use of the above proven coordinate invariance to bring the metric to the most convenient normal or geodesic coordinates (10.98) around some point q_a. The advantage of these coordinates is that the derivatives and thus the affine connection vanish at this point. Its derivatives can directly be expressed in terms of the curvature tensor:

$$\partial_\kappa \bar{\Gamma}_{\tau\kappa}{}^\mu(q_a) = -\frac{1}{3}\left[\bar{R}_{\tau\kappa\sigma}{}^\mu(q_a) + \bar{R}_{\sigma\kappa\tau}{}^\mu(q_a)\right], \quad \text{for normal coordinates.} \tag{10.476}$$

Assuming q_a to lie at the origin, we expand the metric and its determinant in powers of normal coordinates $\Delta\xi^\mu$ around the origin and find, dropping the smallness symbols Δ in front of q and ξ in the transformation (10.98):

$$g_{\mu\nu}(\xi) = \delta_{\mu\nu} + \eta\,\frac{1}{3}\,\bar{R}_{\mu\lambda\nu\kappa}\,\xi^\lambda\xi^\kappa + \eta^2\,\frac{2}{45}\,\bar{R}_{\lambda\nu\kappa}{}^\delta\,\bar{R}_{\sigma\mu\tau\delta}\,\xi^\lambda\xi^\kappa\xi^\sigma\xi^\tau + \ldots, \tag{10.477}$$

$$g(\xi) = 1 - \eta\,\frac{1}{3}\,\bar{R}_{\mu\nu}\,\xi^\mu\xi^\nu + \eta^2\,\frac{1}{18}\left(\bar{R}_{\mu\nu}\,\bar{R}_{\lambda\kappa} + \frac{1}{5}\,\bar{R}_{\mu\sigma\nu}{}^\tau\,\bar{R}_{\lambda\tau\kappa}{}^\sigma\right)\xi^\mu\xi^\nu\xi^\lambda\xi^\kappa + \ldots. \tag{10.478}$$

These expansions have obviously the same power content in ξ^μ as the previous one-dimensional expansions (10.223) had in q. The interaction (10.410) becomes in normal coordinates, up to order η^2:

$$\begin{aligned}
\mathcal{A}_{\text{tot}}^{\text{int}}[\xi] &= \int_0^\beta d\tau \left\{\left[\eta\,\frac{1}{6}\bar{R}_{\mu\lambda\nu\kappa}\,\xi^\lambda\xi^\kappa + \eta^2\frac{1}{45}\bar{R}_{\lambda\mu\kappa}{}^\delta\,\bar{R}_{\sigma\nu\tau\delta}\,\xi^\lambda\xi^\kappa\xi^\sigma\xi^\tau\right]\dot{\xi}^\mu\dot{\xi}^\nu \right. \\
&\quad \left. + \eta\,\frac{1}{6}\delta(0)\bar{R}_{\mu\nu}\,\xi^\mu\xi^\nu + \eta^2\frac{1}{180}\delta(0)\bar{R}_{\mu\delta\nu}{}^\sigma\,\bar{R}_{\lambda\sigma\kappa}{}^\delta\,\xi^\mu\xi^\nu\xi^\lambda\xi^\kappa\right\}.
\end{aligned} \tag{10.479}$$

This has again the same powers in ξ^μ as the one-dimensional interaction (10.425), leading to the same Feynman diagrams, differing only by the factors associated with the vertices. In one dimension, with the trivial vertices of the interaction (10.425), the sum of all diagrams vanishes. In curved space with the more complicated vertices proportional to $\bar{R}_{\mu\nu\kappa\lambda}$ and $\bar{R}_{\mu\nu}$, the result is nonzero but depends on contractions of the curvature tensor $\bar{R}_{\mu\nu\kappa\lambda}$. The dependence is easily identified for each diagram. All bubble diagrams in (10.430)–(10.432) yield results proportional to $\bar{R}_{\mu\nu}^2$, while the watermelon-like diagrams (10.432) carry a factor $\bar{R}_{\mu\nu\kappa\lambda}^2$.

When calculating the contributions of the first expectation value $\langle\mathcal{A}_{\text{tot}}^{\text{int}}[\xi]\rangle$ to the time evolution amplitude it is useful to reduce the D-dimensional expectation values of (10.479) to one-dimensional ones of (10.425) as follows using the contraction rules (8.63) and (8.64):

$$\langle\xi^\mu\xi^\nu\rangle = \delta^{\mu\nu}\langle\xi\xi\rangle, \tag{10.480}$$

$$\langle\xi^\lambda\xi^\kappa\xi^\mu\xi^\nu\rangle = \left(\delta^{\lambda\kappa}\delta^{\mu\nu} + \delta^{\lambda\mu}\delta^{\nu\kappa} + \delta^{\lambda\nu}\delta^{\kappa\mu}\right)\langle\xi\xi\xi\xi\rangle, \tag{10.481}$$

$$\left\langle\xi^\lambda\xi^\kappa\dot\xi^\mu\dot\xi^\nu\right\rangle = \delta^{\lambda\kappa}\delta^{\mu\nu}\langle\xi\xi\rangle\left\langle\dot\xi\dot\xi\right\rangle + \left(\delta^{\lambda\mu}\delta^{\kappa\nu} + \delta^{\lambda\nu}\delta^{\kappa\mu}\right)\left\langle\xi\dot\xi\right\rangle\left\langle\xi\dot\xi\right\rangle, \tag{10.482}$$

$$\left\langle\xi^\lambda\xi^\kappa\xi^\sigma\xi^\tau\dot\xi^\mu\dot\xi^\nu\right\rangle = \left(\delta^{\lambda\kappa}\delta^{\sigma\tau} + \delta^{\lambda\sigma}\delta^{\kappa\tau} + \delta^{\lambda\tau}\delta^{\kappa\sigma}\right)\delta^{\mu\nu}\langle\xi\xi\rangle\langle\xi\xi\rangle\left\langle\dot\xi\dot\xi\right\rangle$$

$$+\; \left[\delta^{\lambda\mu}\left(\delta^{\kappa\nu}\delta^{\sigma\tau} + \delta^{\sigma\nu}\delta^{\tau\kappa} + \delta^{\tau\nu}\delta^{\kappa\sigma}\right) + \delta^{\kappa\mu}\left(\delta^{\sigma\nu}\delta^{\lambda\tau} + \delta^{\lambda\nu}\delta^{\sigma\tau} + \delta^{\tau\nu}\delta^{\sigma\lambda}\right)\right.$$

$$+\; \left.\delta^{\sigma\mu}\left(\delta^{\tau\nu}\delta^{\lambda\kappa} + \delta^{\lambda\nu}\delta^{\kappa\tau} + \delta^{\kappa\nu}\delta^{\tau\lambda}\right) + \delta^{\tau\mu}\left(\delta^{\lambda\nu}\delta^{\kappa\sigma} + \delta^{\kappa\nu}\delta^{\sigma\lambda} + \delta^{\sigma\nu}\delta^{\lambda\kappa}\right)\right]$$

$$\times\; \langle\xi\xi\rangle\left\langle\xi\dot\xi\right\rangle\left\langle\xi\dot\xi\right\rangle. \tag{10.483}$$

Inserting these into the expectation value of (10.479) and performing the tensor contractions, we obtain

$$\langle\mathcal{A}_{\text{tot}}^{\text{int}}[\xi]\rangle = \int_0^\beta d\tau\left\{\eta\frac{1}{6}\bar R\left[-\langle\xi\xi\rangle\left\langle\dot\xi\dot\xi\right\rangle + \left\langle\xi\dot\xi\right\rangle\left\langle\xi\dot\xi\right\rangle + \eta\,\delta(0)\langle\xi\xi\rangle\right]\right.$$

$$+\; \eta^2\frac{1}{45}\left[\left(\bar R_{\mu\nu}^2 + \bar R_{\mu\nu\lambda\kappa}\bar R^{\mu\nu\lambda\kappa} + \bar R_{\mu\nu\lambda\kappa}\bar R^{\lambda\nu\mu\kappa}\right)\left(\langle\xi\xi\rangle^2\left\langle\dot\xi\dot\xi\right\rangle - \langle\xi\xi\rangle\left\langle\xi\dot\xi\right\rangle^2\right)\right]$$

$$+\; \left.\eta^2\frac{1}{180}\delta(0)\left(\bar R_{\mu\nu}^2 + \bar R_{\mu\nu\lambda\kappa}\bar R^{\mu\nu\lambda\kappa} + \bar R_{\mu\nu\lambda\kappa}\bar R^{\lambda\nu\mu\kappa}\right)\langle\xi\xi\rangle\langle\xi\xi\rangle\right\}. \tag{10.484}$$

Individually, the four tensors in the brackets of (10.483) contribute the tensor contractions, using the antisymmetry of $\bar R_{\mu\nu\lambda\kappa}$ in $\mu\nu$ and the contraction to the Ricci tensor $\bar R_{\lambda\mu\kappa}{}^\delta\,\delta^{\mu\kappa} = \bar R_\lambda{}^\delta$:

$$\bar R_{\lambda\mu\kappa}{}^\delta\bar R_{\sigma\nu\tau\delta}\,\delta^{\lambda\mu}\left(\delta^{\kappa\nu}\delta^{\sigma\tau} + \delta^{\sigma\nu}\delta^{\tau\kappa} + \delta^{\tau\nu}\delta^{\kappa\sigma}\right) = 0,$$

$$\bar R_{\lambda\mu\kappa}{}^\delta\bar R_{\sigma\nu\tau\delta}\,\delta^{\kappa\mu}\left(\delta^{\sigma\nu}\delta^{\lambda\tau} + \delta^{\lambda\nu}\delta^{\sigma\tau} + \delta^{\tau\nu}\delta^{\sigma\lambda}\right) = \bar R_\lambda{}^\delta\left(-\bar R^\lambda{}_\delta + \bar R^\lambda{}_\delta\right) = 0,$$

$$\bar R_{\lambda\mu\kappa}{}^\delta\bar R_{\sigma\nu\tau\delta}\,\delta^{\sigma\mu}\left(\delta^{\tau\nu}\delta^{\lambda\kappa} + \delta^{\lambda\nu}\delta^{\kappa\tau} + \delta^{\kappa\nu}\delta^{\tau\lambda}\right) = \bar R_{\lambda\mu\kappa}{}^\delta\left(\bar R^\sigma{}_\delta\,\delta^{\lambda\kappa} + \bar R_{\sigma\lambda\kappa\delta} + \bar R_{\sigma\kappa\lambda\delta}\right)$$

$$= -\left(\bar R_{\mu\nu}^2 + \bar R_{\mu\nu\lambda\kappa}\bar R^{\mu\nu\lambda\kappa} + \bar R_{\mu\nu\lambda\kappa}\bar R^{\lambda\nu\mu\kappa}\right)$$

$$\bar R_{\lambda\mu\kappa}{}^\delta\bar R_{\sigma\nu\tau\delta}\,\delta^{\tau\mu}\left(\delta^{\lambda\nu}\delta^{\kappa\sigma} + \delta^{\kappa\nu}\delta^{\sigma\lambda} + \delta^{\sigma\nu}\delta^{\lambda\kappa}\right) = \bar R_{\lambda\tau\kappa}{}^\delta\left(R_{\kappa\lambda\tau\delta} + R_{\lambda\kappa\tau\delta}\right) = 0. \tag{10.485}$$

We now use the *fundamental identity* of Riemannian spaces

$$\bar R_{\mu\nu\lambda\kappa} + \bar R_{\mu\lambda\kappa\nu} + \bar R_{\mu\kappa\nu\lambda} = 0. \tag{10.486}$$

By expressing the curvature tensor (10.31) in Riemannian space in terms of the Christoffel symbol (1.70) as

$$R_{\mu\nu\lambda\kappa} = \frac{1}{2}\left(\partial_\mu\partial_\lambda g_{\nu\kappa} - \partial_\mu\partial_\kappa g_{\nu\lambda} - \partial_\nu\partial_\lambda g_{\mu\kappa} + \partial_\nu\partial_\kappa g_{\mu\lambda}\right) - [\bar\Gamma_\mu, \bar\Gamma_\nu]_{\lambda\kappa}, \tag{10.487}$$

we see that the identity (10.486) is a consequence of the symmetry of the metric and the single-valuedness of the metric expressed by the integrability condition[8] $(\partial_\lambda\partial_\kappa - \partial_\kappa\partial_\lambda)g_{\mu\nu} = 0$. Indeed, due to the symmetry of $g_{\mu\nu}$ we find

$$\bar R_{\mu\nu\lambda\kappa} + \bar R_{\mu\lambda\kappa\nu} + \bar R_{\mu\kappa\nu\lambda} = \frac{1}{2}[(\partial_\nu\partial_\kappa - \partial_\kappa\partial_\nu)\,g_{\mu\lambda} - (\partial_\nu\partial_\lambda - \partial_\lambda\partial_\nu)\,g_{\mu\kappa} - (\partial_\lambda\partial_\kappa - \partial_\kappa\partial_\lambda)\,g_{\mu\nu}] = 0.$$

The integrability has also the consequence that

$$R_{\mu\nu\lambda\kappa} = -R_{\mu\nu\kappa\lambda}, \quad R_{\mu\nu\lambda\kappa} = R_{\lambda\kappa\mu\nu}. \tag{10.488}$$

---

[8]For the derivation see p. 1353 in the textbook [2].

Using (10.486) and (10.488) we find that

$$\bar{R}_{\mu\nu\lambda\kappa}\bar{R}^{\lambda\nu\mu\kappa} = \frac{1}{2}\bar{R}_{\mu\nu\lambda\kappa}\bar{R}^{\mu\nu\lambda\kappa}, \tag{10.489}$$

so that the contracted curvature tensors in the parentheses of (10.484) can be replaced by $\bar{R}^2_{\mu\nu} + \frac{3}{2}\bar{R}_{\mu\nu\lambda\kappa}\bar{R}^{\mu\nu\lambda\kappa}$.

We now calculate explicitly the contribution of the first-order diagrams in (10.484) [compare (10.427)]:

$$\beta f_1 = \frac{1}{6}\bar{R}\left[-\eta \;\; \text{⬭} \;\; +\eta \;\; \text{⬭} \;\; +\eta\,\delta(0) \;\; \text{⬭}\;\right]. \tag{10.490}$$

corresponding to the analytic expression [compare (10.433)]:

$$\beta f_1 = -\eta\frac{1}{6}\bar{R}\int_0^\beta d\tau \left[\Delta(\tau,\tau)\dot{}\Delta\dot{}(\tau,\tau) - \dot{}\Delta^2(\tau,\tau) - \delta(0)\Delta(\tau,\tau)\right]. \tag{10.491}$$

Note that the combination of propagators in the brackets is different from the previous one in (10.433). Using the integrals (10.436) we find, setting $\eta = 1$:

$$\beta f_1 = -\frac{1}{6}\bar{R}\int_0^\beta d\tau \left[-\frac{1}{\beta}\Delta(\tau,\tau) - \dot{}\Delta^2(\tau,\tau)\right]. \tag{10.492}$$

Using Eq. (10.436), this becomes

$$\beta f_1 = \frac{1}{6}\bar{R}\int_0^\beta d\tau \frac{3}{2\beta}\Delta(\tau,\tau) = \frac{\beta}{24}\bar{R}. \tag{10.493}$$

Adding to this the similar contribution coming from the nonholonomically transformed measure (10.475), we obtain the first-order expansion of the imaginary-time evolution amplitude

$$(q_a\,\beta|q_a\,0) = \frac{1}{\sqrt{2\pi\beta}^D}\,\exp\left[\frac{\beta}{12}\bar{R}(q_a) + \ldots\right]. \tag{10.494}$$

We now turn to the second-order contributions in η. The sum of the local diagrams (10.429) reads now

$$\beta f_2^{(1)} = \eta^2\frac{1}{45}\left(\bar{R}_{\mu\nu}\bar{R}^{\mu\nu} + \frac{3}{2}\bar{R}_{\mu\nu\lambda\kappa}\bar{R}^{\mu\nu\lambda\kappa}\right)\left[\;\text{⬭} - \text{⬭} + \frac{1}{4}\delta(0)\;\text{⬭}\;\right]. \tag{10.495}$$

In terms of the Feynman integrals, the brackets are equal to [compare (10.437)]

$$\int_0^\beta d\tau \left[\Delta^2(\tau,\tau)\dot{}\Delta\dot{}(\tau,\tau) - \Delta(\tau,\tau)\dot{}\Delta^2(\tau,\tau) + \frac{1}{4}\delta(0)\Delta^2(\tau,\tau)\right]. \tag{10.496}$$

Inserting the equation of motion (10.394) and the right-hand equation (10.435), this becomes

$$\int_0^\beta d\tau \left[-\frac{5}{4\beta}\Delta^2(\tau,\tau) + \frac{5}{4}\delta(0)\Delta^2(\tau,\tau)\right] = \frac{5}{4}\frac{1}{30}\left[1 - \delta(0)\right]. \tag{10.497}$$

Thus we find

$$\beta f_2^{(1)} = -\eta^2\frac{\beta^2}{1080}\left(\bar{R}_{\mu\nu}^2 + \frac{3}{2}\bar{R}_{\mu\nu\kappa\lambda}^2\right)\left[1 - \delta(0)\right]. \tag{10.498}$$

Next we calculate the nonlocal contributions of order $\hbar^2$ to βf coming from the cumulant

$$-\frac{1}{2}\left\langle \mathcal{A}_{\mathrm{tot}}^{\mathrm{int}\,2}\right\rangle_c = -\frac{\eta^2}{2}\frac{1}{36}\int_0^\beta d\tau \int_0^\beta d\tau' \left\langle \left[\delta^2(0)\bar{R}_{\mu\nu}\,\bar{R}_{\mu'\nu'}\,\xi^\mu(\tau)\xi^\nu(\tau)\,\xi^{\mu'}(\tau')\xi^{\nu'}(\tau')\right.\right.$$

$$+2\,\delta(0)\bar{R}_{\mu\lambda\nu\kappa}\,\bar{R}_{\mu'\nu'}\,\xi^\lambda(\tau)\xi^\kappa(\tau)\dot{\xi}^\mu(\tau)\dot{\xi}^\nu(\tau)\,\xi^{\mu'}(\tau')\xi^{\nu'}(\tau') \qquad (10.499)$$

$$\left.\left.+\bar{R}_{\mu\lambda\nu\kappa}\,\bar{R}_{\mu'\lambda'\nu'\kappa'}\,\xi^\lambda(\tau)\xi^\kappa(\tau)\dot{\xi}^\mu(\tau)\dot{\xi}^\nu(\tau)\,\xi^{\lambda'}(\tau')\xi^{\kappa'}(\tau')\dot{\xi}^{\mu'}(\tau')\dot{\xi}^{\nu'}(\tau')\right]\right\rangle_c.$$

The first two terms yield the connected diagrams [compare (10.430)]

$$\beta f_2^{(2)} = -\frac{\eta^2}{2}\frac{\bar{R}_{\mu\nu}\bar{R}^{\mu\nu}}{36}\left\{2\,\delta^2(0)\,\bigcirc - 4\,\delta(0)\left[\,\bigcirc\!\!\bigcirc - 2\;\bigcirc\!\!\bigcirc + \bigcirc\!\!\bullet\,\right]\right\}, \qquad (10.500)$$

and the analytic expression diagrams [compare (10.439)]

$$\beta f_2^{(2)} = -\frac{\eta^2}{2}\frac{\bar{R}_{\mu\nu}\bar{R}^{\mu\nu}}{36}\int_0^\beta\int_0^\beta d\tau\,d\tau'\,\left\{2\delta^2(0)\Delta^2(\tau,\tau')\right. \qquad (10.501)$$

$$\left.-4\,\delta(0)\left[\Delta(\tau,\tau){}^{\boldsymbol{\cdot\cdot}}\!\Delta^{\,2}(\tau,\tau') - 2{}^{\boldsymbol{\cdot}}\!\Delta(\tau,\tau)\Delta(\tau,\tau'){}^{\boldsymbol{\cdot}}\!\Delta(\tau,\tau') + \Delta^2(\tau,\tau'){}^{\boldsymbol{\cdot\cdot}}\!\Delta(\tau,\tau)\right]\right\}.$$

The third term in (10.499) leads to the three-bubble diagrams [compare (10.439)]

$$\beta f_2^{(3)} = -\frac{\eta^2}{2}\frac{\bar{R}_{\mu\nu}\bar{R}^{\mu\nu}}{36}\left[4\,\bigcirc\!\!\bigcirc\!\!\bigcirc + 2\,\bigcirc\!\!\bigcirc\!\!\bigcirc - 8\,\bigcirc\!\!\bigcirc\!\!\bigcirc + 4\,\bigcirc\!\!\bigcirc\!\!\bigcirc + 4\,\bigcirc\!\!\bigcirc\!\!\bigcirc + 2\,\bigcirc\!\!\bigcirc\!\!\bigcirc - 8\,\bigcirc\!\!\bigcirc\!\!\bigcirc\right].$$

$$(10.502)$$

The analytic expression for the diagrams 1,2,3 is [compare (10.442)]

$$\beta f_2^{(3)}\Big|_{1,2,3} = -\frac{\eta^2}{2}\frac{\bar{R}_{\mu\nu}\bar{R}^{\mu\nu}}{36}\int_0^\beta\int_0^\beta d\tau\,d\tau'\,\left\{4\,\Delta(\tau,\tau){}^{\boldsymbol{\cdot\cdot}}\!\Delta^{\,2}(\tau,\tau'){}^{\boldsymbol{\cdot\cdot}}\!\Delta(\tau',\tau')\right. \qquad (10.503)$$

$$\left.+2\,{}^{\boldsymbol{\cdot}}\!\Delta(\tau,\tau)\Delta^2(\tau,\tau'){}^{\boldsymbol{\cdot}}\!\Delta(\tau',\tau') - 8\,{}^{\boldsymbol{\cdot}}\!\Delta(\tau,\tau)\Delta(\tau,\tau'){}^{\boldsymbol{\cdot}}\!\Delta(\tau,\tau'){}^{\boldsymbol{\cdot}}\!\Delta(\tau',\tau')\right\}.$$

and for 4 and 5 [compare (10.448)] :

$$\beta f_2^{(3)}\Big|_{4,5} = -\frac{\eta^2}{2}\frac{\bar{R}_{\mu\nu}\bar{R}^{\mu\nu}}{36}(I_4+I_5) = -\frac{\eta^2}{2}\frac{\bar{R}_{\mu\nu}\bar{R}^{\mu\nu}}{36}\int_0^\beta\int_0^\beta d\tau\,d\tau'\frac{4}{\beta}{}^{\boldsymbol{\cdot}}\!\Delta(\tau,\tau){}^{\boldsymbol{\cdot}}\!\Delta(\tau,\tau')\Delta(\tau,\tau')$$

$$= -\frac{\eta^2}{2}\frac{\bar{R}_{\mu\nu}\bar{R}^{\mu\nu}}{36}\frac{1}{45}\beta^2. \qquad (10.504)$$

For the diagrams 6 and 7, finally, we obtain [compare (10.458)]

$$\beta f_2^{(3)}\Big|_{6,7} = -\frac{\eta^2}{2}\frac{\bar{R}_{\mu\nu}\bar{R}^{\mu\nu}}{36}(2I_6 - 8I_7) = -\frac{\eta^2}{2}\frac{\bar{R}_{\mu\nu}\bar{R}^{\mu\nu}}{36}\left[2\delta(0)\frac{\beta^3}{30} - \frac{\beta^2}{6}\right]. \qquad (10.505)$$

The sum of all bubbles diagrams (10.501) and (10.503) is therefore

$$\beta f_2^{(2)} + \beta f_2^{(3)} = \eta^2\frac{\beta^2}{432}\,\bar{R}_{\mu\nu}^2 - \eta^2\,\delta(0)\frac{\beta^3}{1080}\,\bar{R}_{\mu\nu}^2. \qquad (10.506)$$

This compensates exactly the $\delta(0)$-term proportional to $\bar{R}_{\mu\nu}^2$ in Eq. (10.498), leaving only a finite second-order term

$$\beta f_1^{(2)} + \beta f_2^{(2)} + \beta f_2^{(3)} = \eta^2\frac{\beta^2}{720}\left(\bar{R}_{\mu\nu}^2 - \bar{R}_{\mu\nu\kappa\lambda}^2\right) + \eta^2\delta(0)\frac{\beta^3}{1080}\bar{R}_{\mu\nu\kappa\lambda}^2. \qquad (10.507)$$

Finally we calculate the second-order watermelon diagrams (10.432) which contain the initially ambiguous Feynman integrals we make the following observation. Their sum is [compare (10.432)]

$$\beta f_2^{(4)} = -\frac{\eta^2}{2} 2 \frac{1}{36} \left(\bar{R}_{\mu\nu\lambda\kappa} \bar{R}^{\mu\nu\lambda\kappa} + \bar{R}_{\mu\nu\lambda\kappa} \bar{R}^{\mu\lambda\nu\kappa} \right) \left[\ominus - 2 \ominus + \ominus \right], \tag{10.508}$$

corresponding to the analytic expression

$$\beta f_2^{(3)} = -\frac{\eta^2}{2} 2 \frac{3}{2} \frac{1}{36} \bar{R}^2_{\mu\nu\kappa\lambda} \int_0^\beta \int_0^\beta d\tau d\tau' \left[\Delta^2(\tau,\tau') \dot{\Delta}^2(\tau,\tau') \right.$$
$$\left. - 2 \,{}^{\cdot}\!\Delta(\tau,\tau') \,{}^{\cdot}\!\Delta(\tau,\tau') \,\Delta^{\cdot}(\tau,\tau') \,\Delta^{\cdot}(\tau,\tau') + \,{}^{\cdot}\!\Delta^2(\tau,\tau') \,\Delta^{\cdot 2}(\tau,\tau') \right]$$

$$= -\frac{\eta^2}{24} \bar{R}^2_{\mu\nu\kappa\lambda} \left(I_8 - 2I_9 + I_{10} \right), \tag{10.509}$$

where the integrals I_8, I_9, and I_{10} were evaluated before in Eqs. (10.469), (10.468), (10.465), and (10.460). Substituting the results into Eq. (10.509) and using the rules (10.369) and (10.375), we obtain

$$\beta f_2^{(3)} = -\frac{\eta^2}{24} \bar{R}^2_{\mu\nu\kappa\lambda} \int_0^\beta \int_0^\beta d\tau \, d\tau' \Delta^2(\tau,\tau) \delta^2(\tau - \tau') = -\eta^2 \frac{\beta^3}{720} \bar{R}^2_{\mu\nu\kappa\lambda} \, \delta(0). \tag{10.510}$$

Thus the only role of the watermelon diagrams is to cancel the remaining $\delta(0)$-term proportional to $\bar{R}^2_{\mu\nu\kappa\lambda}$ in Eq. (10.507). It gives no finite contribution.

The remaining total sum of all second-order contribution in Eq. (10.507). changes the diagonal time evolution amplitude (10.494) to

$$(q_a \, \beta | q_a \, 0) = \frac{1}{\sqrt{2\pi\beta}^D} \exp\left[\frac{\beta}{12} \bar{R}(q_a) + \frac{\beta^2}{720} \left(\bar{R}^2_{\mu\nu\kappa\lambda} - \bar{R}^2_{\mu\nu} \right) + \ldots \right]. \tag{10.511}$$

In Chapter 11 we shall see that this expression agrees with what has been derived in Schrödinger quantum mechanics from a Hamiltonian operator $\hat{H} = -\Delta/2$ which contains only is the Laplace-Beltrami operator $\Delta = g^{-1/2} \partial_\mu g^{1/2} g^{\mu\nu}(q) \partial_\nu$ of Eq. (1.377) and no extra R-term:

$$(q_a \, \beta \mid q_a \, 0) \equiv \left\langle e^{\beta\Delta/2} \right\rangle = (q_a \mid e^{\beta\Delta/2} \mid q_a) \tag{10.512}$$

$$= \frac{1}{\sqrt{2\pi\beta}^D} \left\{ 1 + \frac{\beta}{12} \bar{R} + \frac{\beta^2}{2} \left[\frac{1}{144} \bar{R}^2 + \frac{1}{360} \left(\bar{R}^{\mu\nu\kappa\lambda} \bar{R}_{\mu\nu\kappa\lambda} - \bar{R}^{\mu\nu} \bar{R}_{\mu\nu} \right) \right] + \ldots \right\}.$$

This expansion due to DeWitt and Seeley will be derived in Section 11.6, the relevant equation being (11.110).

Summarizing the results we have found that for one-dimensional q-space as well as for a D-dimensional curved space in normal coordinates, our calculation procedure on a one-dimensional τ-axis yields unique results. The procedure uses only the essence of the d-dimensional extension, together with the rules (10.369) and (10.375). The results guarantee the coordinate independence of path integrals. They also agree with the DeWitt-Seeley expansion of the short-time amplitude to be derived in Eq. (11.110). The agreement is ensured by the initially ambiguous integrals I_8 and I_9 satisfying the equations

$$I_8^R + 4I_9 + I_{10} = -\frac{\beta^2}{120}, \tag{10.513}$$

$$I_8^R - 2I_9 + I_{10} = 0, \tag{10.514}$$

as we can see from Eqs. (10.470) and (10.509). Since the integral $I_{10} = \beta^2/90$ is unique, we must have $I_9 = -\beta^2/720$ and $I_8^R = -\beta^2/72$, and this is indeed what we found from our integration rules.

The main role of the d-dimensional extension of the τ-axis is, in this context, to forbid the application of the equation of motion (10.394) to correlation functions $\dot{\Delta}^{\cdot}(\tau,\tau')$. This would fix immediately the finite part of the integral I_8 to the wrong value $I_8^R = -\beta^2/18$, leaving the integral I_9 which fixes the integral over distributions (10.375). In this way, however, we could only satisfy one of the equations (10.513) and (10.514), the other would always be violated. Thus, any regularization different from ours will ruin immediately coordinate independence.

10.11.6   Covariant Results for Arbitrary Coordinates

It must be noted that if we were to use arbitrary rather than Riemann normal coordinates, we would find ambiguous integrals already at the two-loop level:

$$\text{:} \qquad I_{14} = \int_0^\beta \int_0^\beta d\tau\, d\tau'\, \ddot{\Delta}(\tau,\tau')\, \Delta^{\cdot}(\tau,\tau')\, {}^{\cdot}\Delta(\tau,\tau'), \qquad (10.515)$$

$$\text{:} \qquad I_{15} = \int_0^\beta \int_0^\beta d\tau\, d\tau'\, \Delta(\tau,\tau')\, \Delta^{\cdot 2}(\tau,\tau'). \qquad (10.516)$$

Let us show that coordinate independence requires these integrals to have the values

$$I_{14} = \beta/24, \qquad I_{15}^R = -\beta/8, \qquad (10.517)$$

where the superscript R denotes the finite part of an integral. We study first the ambiguities arising in one dimension. Without dimensional extension, the values (10.517) would be incompatible with partial integration and the equation of motion (10.393). In the integral (10.515), we use the symmetry $\ddot{\Delta}(\tau,\tau') = \Delta^{\cdot}(\tau,\tau')$, apply partial integration twice taking care of nonzero boundary terms, and obtain on the one hand

$$I_{14} = \frac{1}{2} \int_0^\beta \int_0^\beta d\tau\, d\tau'\, {}^{\cdot}\Delta(\tau,\tau') \frac{d}{d\tau}\left[\Delta^{\cdot 2}(\tau,\tau')\right] = -\frac{1}{2}\int_0^\beta\int_0^\beta d\tau\, d\tau'\, \Delta^{\cdot 2}(\tau,\tau')\, {}^{\cdot\cdot}\Delta(\tau,\tau')$$

$$= -\frac{1}{6}\int_0^\beta\int_0^\beta d\tau\, d\tau' \frac{d}{d\tau'}\left[\Delta^{\cdot 3}(\tau,\tau')\right] = \frac{1}{6}\int_0^\beta d\tau\left[\Delta^{\cdot 3}(\tau,0) - \Delta^{\cdot 3}(\tau,\beta)\right] = \frac{\beta}{12}. \qquad (10.518)$$

On the other hand, we apply Eq. (10.394) and perform two regular integrals, reducing I_{14} to a form containing an undefined integral over a product of distributions:

$$I_{14} = \int_0^\beta\int_0^\beta d\tau\, d\tau'\, {}^{\cdot}\Delta(\tau,\tau')\,\Delta^{\cdot}(\tau,\tau')\delta(\tau-\tau') - \frac{1}{\beta}\int_0^\beta\int_0^\beta d\tau\, d\tau'\, {}^{\cdot}\Delta(\tau,\tau')\,\Delta^{\cdot}(\tau,\tau')$$

$$= \int_0^\beta\int_0^\beta d\tau\, d\tau'\left[-\frac{1}{4}\epsilon^2(\tau-\tau')\delta(\tau-\tau')\right] + \int_0^\beta d\tau\, \Delta^{\cdot 2}(\tau,\tau) + \frac{\beta}{12}$$

$$= \beta\left[-\frac{1}{4}\int d\tau\, \epsilon^2(\tau)\delta(\tau) + \frac{1}{6}\right]. \qquad (10.519)$$

A third, mixed way of evaluating I_{14} employs one partial integration as in the first line of Eq. (10.518), then the equation of motion (10.393) to reduce I_{14} to yet another form

$$I_{14} = \frac{1}{2}\int_0^\beta\int_0^\beta d\tau\, d\tau'\, \Delta^{\cdot 2}(\tau,\tau')\delta(\tau-\tau') =$$

$$= \frac{1}{8}\int_0^\beta\int_0^\beta d\tau\, d\tau'\epsilon^2(\tau-\tau')\delta(\tau-\tau') + \frac{1}{2}\int_0^\beta d\tau\, \Delta^{\cdot 2}(\tau,\tau)$$

$$= \beta\left[\frac{1}{8}\int d\tau\, \epsilon^2(\tau)\delta(\tau) + \frac{1}{24}\right]. \qquad (10.520)$$

We now see that if we set [compare with the correct equation (10.375)]

$$\int d\tau[\epsilon(\tau)]^2\, \delta(\tau) \equiv \frac{1}{3}, \qquad \text{(false)}, \qquad (10.521)$$

the last two results (10.520) and (10.519) coincide with the first in Eq. (10.518). The definition (10.521) would obviously be consistent with partial integration if we insert $\delta(\tau) = \dot{\epsilon}(\tau)/2$:

$$\int d\tau [\epsilon(\tau)]^2 \, \delta(\tau) = \frac{1}{2} \int d\tau [\epsilon(\tau)]^2 \, \dot{\epsilon}(\tau) = \frac{1}{6} \int d\tau \frac{d}{d\tau} [\epsilon(\tau)]^3 = \frac{1}{3}. \tag{10.522}$$

In spite of this consistency with partial integration and the equation of motion, Eq. (10.521) is incompatible with the requirement of coordinate independence. This can be seen from the discrepancy between the resulting value $I_{14} = \beta/12$ and the necessary (10.517). In earlier work on the subject by other authors [31]– [36], this discrepancy was compensated by adding the above-mentioned (on p. 809) noncovariant term to the classical action, in violation of Feynman's construction rules for path integrals.

A similar problem appears with the other Feynman integral (10.516). Applying first Eq. (10.394) we obtain

$$I_{15} = \int_0^\beta \int_0^\beta d\tau \, d\tau' \Delta(\tau, \tau') \delta^2(\tau - \tau') - \frac{2}{\beta} \int_0^\beta d\tau \Delta(\tau, \tau) + \frac{1}{\beta^2} \int_0^\beta \int_0^\beta d\tau d\tau' \Delta(\tau, \tau'). \tag{10.523}$$

For the integral containing the square of the δ-function we must postulate the integration rule (10.369) to obtain a divergent term

$$I_{15}^{\mathrm{div}} = \delta(0) \int_0^\beta d\tau \, \Delta(\tau, \tau) = \delta(0) \frac{\beta^2}{6}, \tag{10.524}$$

which is proportional to $\delta(0)$, and compensates a similar term from the measure. The remaining integrals in (10.523) are finite and yield the regular part $I_{15}^R = -\beta/4$, which we shall see to be inconsistent with coordinate invariance. In another calculation of I_{15}, we first add and subtract the UV-divergent term, writing

$$I_{15} = \int_0^\beta \int_0^\beta d\tau \, d\tau' \Delta(\tau, \tau') \left[\Delta^{\cdot 2}(\tau, \tau') - \delta^2(\tau - \tau') \right] + \delta(0) \frac{\beta^2}{6}. \tag{10.525}$$

Replacing $\delta^2(\tau - \tau')$ by the square of the left-hand side of the equation of motion (10.393), and integrating the terms in brackets by parts, we obtain

$$
\begin{aligned}
I_{15}^R &= \int_0^\beta \int_0^\beta d\tau \, d\tau' \Delta(\tau, \tau') \left[\Delta^{\cdot 2}(\tau, \tau') - \Delta^{\cdot\cdot 2}(\tau, \tau') \right] \\
&= \int_0^\beta \int_0^\beta d\tau \, d\tau' \left[-\dot{\Delta}(\tau, \tau') \Delta^{\cdot}(\tau, \tau') \dot{\Delta}^{\cdot}(\tau, \tau') - \Delta(\tau, \tau') \Delta^{\cdot}(\tau, \tau') \ddot{\Delta}^{\cdot} d(\tau, \tau') \right] \\
&\quad - \int_0^\beta \int_0^\beta d\tau \, d\tau' \left[-\Delta^{\cdot 2}(\tau, \tau') \Delta^{\cdot}(\tau, \tau') - \Delta(\tau, \tau') \Delta^{\cdot}(\tau, \tau') \Delta^{\cdot}(\tau, \tau') \right] \\
&= -I_{14} + \int_0^\beta \int_0^\beta d\tau \, d\tau' \Delta^{\cdot 2}(\tau, \tau') \Delta^{\cdot}(\tau, \tau') = -I_{14} - \beta/6.
\end{aligned}
\tag{10.526}
$$

The value of the last integral follows from partial integration.

For a third evaluation of I_{15} we insert the equation of motion (10.393) and bring the last integral in the fourth line of (10.526) to

$$-\int_0^\beta \int_0^\beta d\tau \, d\tau' \Delta^{\cdot 2}(\tau, \tau') \delta(\tau - \tau') = -\beta \left[\frac{1}{4} \int d\tau \, \epsilon^2(\tau) \delta(\tau) + \frac{1}{12} \right]. \tag{10.527}$$

All three ways of calculation lead, with the assignment (10.521) to the singular integral, to the same result $I_{15}^R = -\beta/4$ using the rule (10.521). This, however, is again in disagreement with the coordinate-independent value in Eq. (10.517). Note that both integrals I_{14} and I_{15}^R are too large by a factor 2 with respect to the necessary (10.517) for coordinate independence.

How can we save coordinate independence while maintaining the equation of motion and partial integration? The direction in which the answer lies is suggested by the last line of Eq. (10.520): we must find a consistent way to have an integral $\int d\tau \, [\epsilon(\tau)]^2 \, \delta(\tau) = 0$, as in Eq. (10.375), instead of the false value (10.521), which means that we need a reason for forbidding the application of partial integration to this singular integral. For the calculation at the infinite time interval, this problem was solved in Refs. [23]–[25] with the help of dimensional regularization.

In dimensional regularization, we would write the Feynman integral (10.515) in d dimensions as

$$I_{14}^d = \int \int d^d x \, d^d x' \,_\mu\Delta(x, x')\Delta_\nu(x, x') \,_\mu\Delta_\nu(x, x'), \qquad (10.528)$$

and see that the different derivatives on $_\mu\Delta_\nu(x, x')$ prevent us from applying the field equation (10.418), in contrast to the one-dimensional calculation. We can, however, apply partial integration as in the first line of Eq. (10.518), and arrive at

$$I_{14}^d = -\frac{1}{2} \int \int d^d x \, d^d x' \Delta_\nu^2(x, x')\Delta_{\mu\mu}(x, x'). \qquad (10.529)$$

In contrast to the one-dimensional expression (10.518), a further partial integration is impossible. Instead, we may apply the field equation (10.418), go back to one dimension, and apply the integration rule (10.375) as in Eq. (10.520) to obtain the correct result $I_{14} = \beta/24$ guaranteeing coordinate independence.

The Feynman integral (10.516) for I_{15} is treated likewise. Its d-dimensional extension is

$$I_{15}^d = \int \int d^d x \, d^d x' \Delta(x, x') \left[_\mu\Delta_\nu(x, x') \right]^2. \qquad (10.530)$$

The different derivatives on $_\mu\Delta_\nu(x, x')$ make it impossible to apply a dimensionally extended version of equation (10.394) as in Eq. (10.523). We can, however, extract the UV-divergence as in Eq. (10.525), and perform a partial integration on the finite part which brings it to a dimensionally extended version of Eq. (10.526):

$$I_{15}^R = -I_{14} + \int d^d x \, d^d x' \Delta_\nu^2(x, x')\Delta_{\mu\mu}(x, x'). \qquad (10.531)$$

On the right-hand side we use the field equation (10.418), as in Eq. (10.527), return to $d = 1$, and use the rule (10.375) to obtain the result $I_{15}^R = -I_{14} - \beta/12 = -\beta/8$, again guaranteeing coordinate independence.

Thus, by keeping only track of a few essential properties of the theory in d dimensions we indeed obtain a simple consistent procedure for calculating singular Feynman integrals. All results obtained in this way ensure coordinate independence. They agree with what we would obtain using the one-dimensional integration rule (10.375) for the product of two ϵ- and one δ-distribution.

Our procedure gives us unique rules telling us where we are allowed to apply partial integration and the equation of motion in one-dimensional expressions. Ultimately, all integrals are brought to a regular form, which can be continued back to one time dimension for a direct evaluation. This procedure is obviously much simpler than the previous explicit calculations in d dimensions with the limit $d \to 1$ taken at the end.

The coordinate independence would require the equations (10.517). Thus, although the calculation in normal coordinates are simpler and can be carried more easily to higher orders, the perturbation in arbitrary coordinates help to fix more ambiguous integrals.

Let us see how the integrals I_{14} and I_{15} arise in the perturbation expansion of the time evolution amplitude in arbitrary coordinates up to the order η, and that the values in (10.517) are necessary to guarantee a covariant result. We use arbitrary coordinates and expand the metric around the origin. Dropping the increment symbol δ in front of δq^μ, we write:

$$g_{\mu\nu}(q) = \delta_{\mu\nu} + \sqrt{\eta}(\partial_\lambda g_{\mu\nu})q^\lambda + \eta \frac{1}{2}(\partial_\lambda \partial_\kappa g_{\mu\nu})q^\lambda q^\kappa, \tag{10.532}$$

with the expansion parameter η keeping track of the orders of the perturbation series. At the end it will be set equal to unity. The determinant has the expansion to order η:

$$\log g(q) = \sqrt{\eta}\, g^{\mu\nu}(\partial_\lambda g_{\mu\nu})q^\lambda + \eta \frac{1}{2}\, g^{\mu\nu}\left[(\partial_\lambda \partial_\kappa g_{\mu\nu}) - g^{\sigma\tau}(\partial_\lambda g_{\mu\sigma})(\partial_\kappa g_{\nu\tau})\right]q^\lambda q^\kappa. \tag{10.533}$$

The total interaction (10.410) becomes

$$\mathcal{A}_{\text{tot}}^{\text{int}}[q] = \int_0^\beta d\tau \Big\{ \Big[\frac{1}{2}\sqrt{\eta}\,(\partial_\kappa g_{\mu\nu})q^\kappa + \frac{1}{4}\eta\,(\partial_\lambda \partial_\kappa g_{\mu\nu})q^\lambda q^\kappa\Big]\dot{q}^\mu \dot{q}^\nu \tag{10.534}$$
$$- \frac{1}{2}\sqrt{\eta}\,\delta(0)g^{\mu\nu}(\partial_\kappa g_{\mu\nu})q^\kappa - \frac{1}{4}\eta\,\delta(0)\,g^{\mu\nu}\Big[(\partial_\lambda \partial_\kappa g_{\mu\nu}) - g^{\sigma\tau}\,(\partial_\lambda g_{\mu\sigma})(\partial_\kappa g_{\nu\tau})\Big]q^\lambda q^\kappa \Big\}.$$

Using the relations following directly from the definition of the Christoffel symbols (1.70) and (1.71),

$$\partial_\kappa g_{\mu\nu} = -g_{\mu\sigma}g_{\nu\tau}\partial_\kappa g^{\sigma\tau} = \Gamma_{\kappa\mu\nu} + \Gamma_{\kappa\nu\mu} = 2\Gamma_{\kappa\{\mu\nu\}}, \quad g^{\mu\nu}(\partial_\kappa g_{\mu\nu}) = 2\Gamma_{\kappa\mu}{}^\mu,$$
$$\partial_\lambda \partial_\kappa g_{\mu\nu} = \partial_\lambda \Gamma_{\kappa\mu\nu} + \partial_\lambda \Gamma_{\kappa\nu\mu} = \partial_\lambda \Gamma_{\kappa\{\mu\nu\}},$$
$$g^{\mu\nu}(\partial_\tau \partial_\lambda g_{\mu\nu}) = 2g^{\mu\nu}\partial_\kappa \Gamma_{\sigma\mu\nu} = 2\partial_\kappa(g^{\mu\nu}\Gamma_{\sigma\mu\nu}) - 2\Gamma_{\sigma\mu\nu}\partial_\kappa g^{\mu\nu}$$
$$= 2\,(\partial_\kappa \Gamma_{\sigma\mu}{}^\mu + g^{\mu\nu}\Gamma_{\sigma\mu\tau}\Gamma_{\lambda\nu}{}^\tau + \Gamma_{\sigma\mu}{}^\nu\Gamma_{\lambda\nu}{}^\mu), \tag{10.535}$$

this becomes

$$\mathcal{A}_{\text{tot}}^{\text{int}}[q] = \int_0^\beta d\tau \Big\{ \Big[\sqrt{\eta}\,\Gamma_{\kappa\mu\nu}q^\kappa + \frac{\eta}{2}\,\partial_\lambda \Gamma_{\kappa\mu\nu}\,q^\lambda q^\kappa\Big]\dot{q}^\mu \dot{q}^\nu$$
$$- \delta(0)\Big[\sqrt{\eta}\,\Gamma_{\mu\kappa}{}^\mu q^\kappa + \frac{\eta}{2}\,\partial_\lambda \Gamma_{\tau\mu}{}^\mu\,q^\lambda q^\tau\Big] \Big\}. \tag{10.536}$$

The derivative of the Christoffel symbol in the last term can also be written differently using the identity $\partial_\lambda g^{\mu\nu} = -g^{\mu\sigma}g^{\nu\tau}\partial_\lambda g_{\sigma\tau}$ as follows:

$$\partial_\lambda \Gamma_{\tau\mu}{}^\mu = \partial_\lambda g^{\mu\nu}\Gamma_{\tau\mu\nu} = g^{\mu\nu}\partial_\lambda \Gamma_{\tau\mu\nu} = g^{\mu\nu}\partial_\lambda \Gamma_{\tau\mu\nu} - g^{\mu\sigma}g^{\nu\tau}(\partial_\lambda g_{\sigma\tau})\Gamma_{\tau\mu\nu}$$
$$= \partial_\lambda \Gamma_{\tau\mu\nu} - (\Gamma_{\lambda\mu\nu} + \Gamma_{\lambda\nu\mu})\Gamma_\tau{}^{\mu\nu}. \tag{10.537}$$

To first order in η, we obtain from the first cumulant $\langle \mathcal{A}_{\text{tot}}^{\text{int}}[q]\rangle_c$:

$$\beta f_1^{(1)} = \eta \int_0^\beta d\tau \Big\langle \frac{1}{2}\partial_\lambda \Gamma_{\kappa\mu\nu}q^\lambda q^\kappa \dot{q}^\mu \dot{q}^\nu - \delta(0)\frac{1}{2}\partial_\lambda \Gamma_{\tau\mu}{}^\mu q^\lambda q^\tau \Big\rangle_c, \tag{10.538}$$

the diagrams (10.490) corresponding to the analytic expression [compare (10.491)]

$$\beta f_1^{(1)} = \frac{\eta}{2}\partial_\lambda \Gamma_{\kappa\{\mu\nu\}}\int_0^\beta d\tau \Big\{ g^{\mu\nu}g^{\kappa\lambda}\,{}^{\cdot}\!\Delta^{\cdot}(\tau,\tau)\Delta(\tau,\tau) + 2g^{\mu\kappa}g^{\nu\lambda}\,{}^{\cdot}\!\Delta^2(\tau,\tau) - \delta(0)g^{\mu\nu}g^{\kappa\lambda}\Delta(\tau,\tau) \Big\}$$
$$- \frac{\eta}{2}g^{\lambda\kappa}(\Gamma_{\lambda\nu}{}^\mu\Gamma_{\mu\kappa}{}^\nu + g^{\tau\mu}\Gamma_{\tau\kappa}{}^\nu\Gamma_{\mu\lambda\nu})\delta(0)\int_0^\beta d\tau\,\Delta(\tau,\tau). \tag{10.539}$$

Replacing $\,{}^{\cdot}\!\Delta^{\cdot}(\tau,\tau)$ by $\delta(0) - 1/\beta$ according to (10.394), and using the integrals (10.436), the $\delta(0)$-terms in the first integral cancel and we obtain

$$\beta f_1^{(1)} = -\beta\frac{\eta}{4}\Big[\frac{g^{\mu\nu}g^{\kappa\lambda}}{6}(\partial_\lambda \Gamma_{\mu\kappa\nu} - \partial_\mu \Gamma_{\kappa\lambda\nu}) - \frac{g^{\lambda\kappa}}{3}(\Gamma_{\lambda\nu}{}^\mu\Gamma_{\mu\kappa}{}^\nu + g^{\tau\mu}\Gamma_{\tau\kappa}{}^\nu\Gamma_{\mu\lambda\nu})\delta(0)\Big]. \tag{10.540}$$

In addition, there are contributions of order η from the second cumulant

$$-\frac{\eta}{2!}\int_0^\beta d\tau \int_0^\beta d\tau' \Big\langle\, [\Gamma_{\kappa\mu\nu}q^\kappa(\tau)\dot q^\mu(\tau)\dot q^\nu(\tau)+\delta(0)\Gamma_{\mu\kappa}{}^\mu q^\kappa(\tau)]$$
$$\times\,[\Gamma_{\kappa\mu\nu}q^\kappa(\tau')\dot q^\mu(\tau')\dot q^\nu(\tau')+\delta(0)\Gamma_{\mu\kappa}{}^\mu q^\kappa(\tau')]\,\Big\rangle_c,\tag{10.541}$$

These add to the free energy

$$\beta f_1^{(2)}=-\frac{\eta}{2}g^{\lambda\kappa}\Gamma_{\lambda\mu}{}^\mu\Gamma_{\kappa\nu}{}^\nu\Big[\ \ -2\delta(0)\ \ +\ \delta^2(0)\ \ \Big],$$
$$\beta f_1^{(3)}=-\eta\Gamma_{\lambda\mu}{}^\mu\,(g^{\lambda\kappa}\Gamma_{\kappa\nu}{}^\nu+g^{\nu\kappa}\Gamma_{\nu\kappa}{}^\lambda)\Big[\ \ -\delta(0)\ \ \Big],$$
$$\beta f_1^{(4)}=-\frac{\eta}{2}(g^{\mu\lambda}g^{\kappa\tau}\Gamma_{\mu\lambda}{}^\nu\Gamma_{\kappa\tau\nu}+g^{\mu\nu}\Gamma_{\mu\kappa}{}^\kappa\Gamma_{\nu\lambda}{}^\lambda+2g^{\mu\nu}\Gamma_{\mu\nu}{}^\kappa\Gamma_{\kappa\lambda}{}^\lambda)\ \ ,$$
$$\beta f_1^{(5)}=-\frac{\eta}{2}(g^{\mu\kappa}g^{\nu\lambda}\Gamma_{\mu\lambda}{}^\tau\Gamma_{\kappa\beta}+3g^{\mu\kappa}\Gamma_{\mu\lambda}{}^\tau\Gamma_{\tau\kappa}{}^\lambda)\ \ ,$$
$$\beta f_1^{(6)}=-\frac{\eta}{2}g^{\lambda\kappa}(\Gamma_{\lambda\nu}{}^\mu\Gamma_{\mu\kappa}{}^\nu+g^{\mu\tau}\Gamma_{\tau\kappa}{}^\nu\Gamma_{\mu\lambda\nu})\ \ .\tag{10.542}$$

The Feynman integrals associated with the diagrams in the first and second lines are

$$I_{11}=\int\int d\tau\,d\tau'\,\{{}^{\textstyle\cdot}\Delta^{\textstyle\cdot}(\tau,\tau)\Delta(\tau,\tau'){}^{\textstyle\cdot}\Delta^{\textstyle\cdot}(\tau',\tau')-2\delta(0)\,{}^{\textstyle\cdot}\Delta^{\textstyle\cdot}(\tau,\tau)\,\Delta(\tau,\tau')+\delta^2(0)\,\Delta(\tau,\tau')\}\tag{10.543}$$

and

$$I_{12}=\int\int d\tau\,d\tau'\,\{{}^{\textstyle\cdot}\Delta(\tau,\tau){}^{\textstyle\cdot}\Delta(\tau,\tau'){}^{\textstyle\cdot}\Delta^{\textstyle\cdot}(\tau',\tau')-\delta(0)\,{}^{\textstyle\cdot}\Delta(\tau,\tau){}^{\textstyle\cdot}\Delta(\tau,\tau')\},\tag{10.544}$$

respectively. Replacing in Eqs. (10.543) and (10.544) ${}^{\textstyle\cdot}\Delta^{\textstyle\cdot}(\tau,\tau)$ and ${}^{\textstyle\cdot}\Delta^{\textstyle\cdot}(\tau',\tau')$ by $\delta(0)-1/\beta$ leads to cancellation of the infinite factors $\delta(0)$ and $\delta^2(0)$ from the measure, such that we are left with

$$I_{11}=\frac{1}{\beta^2}\int_0^\beta d\tau\int_0^\beta d\tau'\,\Delta(\tau,\tau')=\frac{\beta}{12}\tag{10.545}$$

and

$$I_{12}=-\frac{1}{\beta}\int_0^\beta d\tau\int_0^\beta d\tau'\,{}^{\textstyle\cdot}\Delta(\tau,\tau){}^{\textstyle\cdot}\Delta(\tau,\tau')=-\frac{\beta}{12}.\tag{10.546}$$

The Feynman integral of the diagram in the third line of Eq. (10.542) has d-dimensional extension

$$I_{13}=\int\int d\tau\,d\tau'\,{}^{\textstyle\cdot}\Delta(\tau,\tau)\,\Delta(\tau',\tau'){}^{\textstyle\cdot}\Delta(\tau,\tau')$$
$$\to\int\int d^dx\,d^dx'\,{}_\mu\Delta(x,x)\Delta_\nu(x',x')_\mu\Delta_\nu(x,x').\tag{10.547}$$

Integrating this partially yields

$$I_{13}=\frac{1}{\beta}\int\int d\tau\,d\tau'\,\Delta(\tau,\tau'){}^{\textstyle\cdot}\Delta(\tau',\tau')=\frac{1}{\beta}\int_0^\beta d\tau\int_0^\beta d\tau'\,{}^{\textstyle\cdot}\Delta(\tau,\tau){}^{\textstyle\cdot}\Delta(\tau,\tau')=\frac{\beta}{12},\tag{10.548}$$

where we have interchanged the order of integration $\tau \leftrightarrow \tau'$ in the second line of Eq. (10.548) and used $d[\dot\Delta\,(\tau,\tau)]/d\tau = -1/\beta$. Multiplying the integrals (10.545), (10.546), and (10.548) by corresponding vertices in Eq. (10.542) and adding them together, we obtain

$$\beta f_1^{(2)} + \beta f_1^{(3)} + \beta f_1^{(4)} = -\frac{\eta\beta}{24}\, g^{\mu\nu}g^{\kappa\lambda}\,\Gamma_{\mu\nu}{}^\tau\,\Gamma_{\kappa\lambda\tau}. \tag{10.549}$$

The contributions of the last three diagrams in $\beta f_1^{(5)}$ and $\beta f_1^{(6)}$ of (10.542) are determined by the initially ambiguous integrals (10.515) and (10.516) to be equal to $I_{14} = -\beta/24$ and $I_{15} = -\beta/8 + \delta(0)\beta^2/6$, respectively. Moreover, the $\delta(0)$-part in the latter, when inserted into the last line of Eq. (10.542) for $f_1^{(6)}$, is canceled by the contribution of the local diagram with the factor $\delta(0)$ in $f_1^{(1)}$ of (10.540). We see here an example that with general coordinates, the divergences containing powers of $\delta(0)$ no longer cancel order by order in $\hbar$, but do so at the end.

Thus only the finite part $I_{15}^R = -\beta/24$ remains and we find

$$\begin{aligned}
\beta f_1^{(5)} + \beta f_1^{(6)R} &= -\frac{\eta}{2}\left\{ g^{\mu\kappa}g^{\nu\lambda}\Gamma_{\mu\lambda}{}^\tau\Gamma_{\kappa\nu\tau}\left(I_{14}+I_{15}^R\right) + g^{\lambda\kappa}\Gamma_{\lambda\nu}{}^\mu\Gamma_{\mu\kappa}{}^\nu\left(3I_{14}+I_{15}^R\right)\right\} \\
&= \frac{\eta\beta}{24}\, g^{\mu\nu}g^{\kappa\lambda}\,\Gamma_{\mu\kappa}{}^\sigma\,\Gamma_{\nu\lambda\sigma}.
\end{aligned} \tag{10.550}$$

By adding this to (10.549), we find the sum of all diagrams in (10.542) as follows

$$\sum_{i=2}^{6}\beta f_1^{(i)} = -\frac{\eta\beta}{24}\, g^{\mu\nu}g^{\kappa\lambda}\left(\Gamma_{\mu\nu}{}^\sigma\Gamma_{\kappa\lambda\sigma} - \Gamma_{\mu\kappa}{}^\sigma\Gamma_{jl,n}\right). \tag{10.551}$$

Together with the regular part of (10.540) in the first line, this yields the sum of all first-order diagrams

$$\sum_{i=1}^{6}\beta f_1^{(i)} = -\frac{\eta\beta}{24}\, g^{\mu\nu}g^{\kappa\lambda}\, R_{\lambda\mu\kappa\nu} = \frac{\eta\beta}{24}\,\bar{R}. \tag{10.552}$$

The result is covariant and agrees, of course, with Eq. (10.492) derived with normal coordinate. Note that to obtain this covariant result, the initially ambiguous integrals (10.515) and (10.516) over distributions appearing ni Eq. (10.550) mast satisfy

$$\begin{aligned}
I_{14} + I_{15}^R &= -\frac{\beta}{12}, \\
3I_{14} + I_{15}^R &= 0,
\end{aligned} \tag{10.553}$$

which leaves only the values (10.517).

10.12 Effective Classical Potential in Curved Space

In Chapter 5 we have seen that the partition function of a quantum statistical system in flat space can always be written as an integral over a classical Boltzmann factor $\exp[-\beta V^{\mathrm{eff\,cl}}(\mathbf{x}_0)]$, where $B(\mathbf{x}_0) = V^{\mathrm{eff\,cl}}(\mathbf{x}_0)$ is the so-called *effective classical potential* containing the effects of all quantum fluctuations. The variable of integration is the temporal path average $\mathbf{x}_0 \equiv \beta^{-1}\int_0^\beta d\tau\,\mathbf{x}(\tau)$. In this section we generalize this concept to curved space, and show how to calculate perturbatively the high-temperature expansion of $V^{\mathrm{eff\,cl}}(q_0)$. The requirement of independence under coordinate transformations $q^\mu(\tau) \to q'^\mu(\tau)$ introduces subtleties into the definition and treatment of the path average q_0^μ, and covariance is achieved only with the help of a procedure invented by Faddeev and Popov [49] to deal with gauge freedoms in quantum field theory.

In the literature, attempts to introduce an effective classical potential in curved space around a fixed temporal average $q_0 \equiv \bar{q}(\tau) \equiv \beta^{-1} \int_0^\beta d\tau q(\tau)$ have so far failed and produced a two-loop perturbative result for $V^{\text{eff cl}}(q_0)$ which turned out to deviate from the covariant one by a noncovariant total derivative [34], in contrast to the covariant result (10.494) obtained with Dirichlet boundary conditions. For this reason, perturbatively defined path integrals with periodic boundary conditions in curved space have been of limited use in the presently popular first-quantized worldline approach to quantum field theory (also called the string-inspired approach reviewed in Ref. [50]). In particular, is has so far been impossible to calculate with periodic boundary conditions interesting quantities such as curved-space effective actions, gravitational anomalies, and index densities, all results having been reproduced with Dirichlet boundary conditions [46, 52].

The development in this chapter cures the problems by exhibiting a manifestly covariant integration procedure for periodic paths [51]. It is an adaption of similar procedures used before in the effective action formalism of two-dimensional sigma-models [46]. Covariance is achieved by expanding the fluctuations in the neighborhood of any given point in powers of geodesic coordinates, and by a covariant definition of a path average different from the naive temporal average. As a result, we shall find the same locally covariant perturbation expansion of the effective classical potential as in Eq. (10.494) calculated with Dirichlet boundary conditions.

All problems encountered in the literature occur in the first correction terms linear in β in the time evolution amplitude. It will therefore be sufficient to consider only to lowest-order perturbation expansion. For this reason we shall from now on drop the parameter of smallness η used before.

10.12.1 Covariant Fluctuation Expansion

We want to calculate the partition function from the functional integral over all periodic paths

$$Z = \oint \mathcal{D}^D q \sqrt{g(q)} e^{-\mathcal{A}[q]}, \tag{10.554}$$

where the symbol $\oint$ indicates the periodicity of the paths. By analogy with (2.441), we split the paths into a time-independent and a time-dependent part:

$$q^\mu = q_0^\mu + \eta^\mu(\tau), \tag{10.555}$$

with the goal to express the partition function as in Eq. (3.808) by an ordinary integral over an effective classical partition function

$$Z = \int \frac{d^D q_0}{\sqrt{2\pi\beta}^D} \sqrt{g(q_0)} \, e^{-\beta V^{\text{eff cl}}(q_0)}, \tag{10.556}$$

where $V^{\text{eff cl}}(q_0)$ is the curved-space version of the effective classical partition function. For a covariant treatment, we parametrize the small fluctuations $\eta^\mu(\tau)$ in terms of the prepoint normal coordinates $\Delta\xi^\mu(\tau)$ of the point q_0^μ introduced in Eq. (10.98), which are here geodesic due to the absence of torsion. Omitting the smallness symbols Δ, there will be some *nonlinear* decomposition

$$q^\mu(\tau) = q_0^\mu + \eta^\mu(q_0, \xi), \tag{10.557}$$

where $\eta^\mu(q_0, \xi) = 0$ for $\xi^\mu = 0$. Inverting the relation (10.98) we obtain

$$\eta^\mu(q_0, \xi) = \xi^\mu - \frac{1}{2}\bar{\Gamma}_{\sigma\tau}{}^\mu(q_0)\xi^\sigma\xi^\tau - \frac{1}{6}\bar{\Gamma}_{\sigma\tau\kappa}{}^\mu(q_0)\xi^\sigma\xi^\tau\xi^\kappa - \ldots, \tag{10.558}$$

where the coefficients $\bar{\Gamma}_{\sigma\tau\ldots\kappa}{}^\mu(q_0)$ with more than two subscripts are defined similarly to covariant derivatives with respect to lower indices (they are not covariant quantities):

$$\bar{\Gamma}_{\sigma\tau\kappa}{}^\mu(q_0) = \nabla_\kappa\bar{\Gamma}_{\sigma\tau}{}^\mu = \partial_\kappa\bar{\Gamma}_{\sigma\tau}{}^\mu - 2\bar{\Gamma}_{\kappa\sigma}{}^\nu\bar{\Gamma}_{\nu\tau}{}^\mu, \quad \ldots. \tag{10.559}$$

If the initial coordinates q^μ are themselves geodesic at q_0^μ, all coefficients $\bar{\Gamma}_{\sigma\tau\ldots\kappa}{}^\mu(q_0)$ in Eq. (10.558) are zero, so that $\eta^\mu(\tau) = \xi^\mu(\tau)$, and the decomposition (10.557) is linear. In arbitrary coordinates, however, $\eta^\mu(\tau)$ does not transform like a vector under coordinate transformations, and we must use the nonlinear decomposition (10.557).

We now transform the path integral (10.554) to the new coordinates $\xi^\mu(\tau)$ using Eqs. (10.557)–(10.559). The perturbation expansion for the transformed path integral over $\xi^\mu(\tau)$ is constructed for any chosen q_0^μ by expanding the total action (10.408) including the measure factor (10.406) in powers of small linear fluctuations $\xi^\mu(\tau)$. Being interested only in the lowest-order contributions we shall from now on drop the parameter of smallness η counting the orders in the earlier perturbation expansions. This is also useful since the similar symbol $\eta^\mu(\tau)$ is used here to describe the path fluctuations. The action relevant for the terms to be calculated here consists of a free action, which we write after a partial integration as

$$\mathcal{A}^{(0)}[q_0, \xi] = g_{\mu\nu}(q_0)\int_0^\beta d\tau\, \frac{1}{2}\xi^\mu(\tau)(-\partial_\tau^2)\xi^\nu(\tau), \tag{10.560}$$

and an interaction which contains only the leading terms in (10.479):

$$\mathcal{A}^{\text{int}}_{\text{tot}}[q_0; \xi] = \int_0^\beta d\tau\left[\frac{1}{6}\bar{R}_{\mu\lambda\nu\kappa}\xi^\lambda\xi^\kappa\dot{\xi}^\mu\dot{\xi}^\nu + \frac{1}{6}\delta(0)\bar{R}_{\mu\nu}\xi^\mu\xi^\nu\right]. \tag{10.561}$$

The partition function (10.554) in terms of the coordinates $\xi^\mu(\tau)$ is obtained from the perturbation expansion

$$Z = \oint \mathcal{D}^D\xi(\tau)\sqrt{g(q_0)}\, e^{-\mathcal{A}^{(0)}[q_0,\xi] - \mathcal{A}^{\text{int}}_{\text{tot}}[q_0,\xi]}. \tag{10.562}$$

The path integral (10.562) cannot immediately be calculated perturbatively in the standard way, since the quadratic form of the free action (10.560) is degenerate. The spectrum of the operator $-\partial_\tau^2$ in the space of periodic functions $\xi^\mu(\tau)$ has a zero mode. The zero mode is associated with the fluctuations of the temporal average of $\xi^\mu(\tau)$:

$$\xi_0^\mu = \bar\xi^\mu \equiv \beta^{-1} \int_0^\beta d\tau\, \xi^\mu(\tau). \qquad (10.563)$$

Small fluctuations of ξ_0^μ have the effect of moving the path as a whole infinitesimally through the manifold. The same movement can be achieved by changing q_0^μ infinitesimally. Thus we can replace the integral over the path average ξ_0^μ by an integral over q_0^μ, provided that we properly account for the change of measure arising from such a variable transformation.

Anticipating such a change, the path average (10.563) can be set equal to zero eliminating the zero mode in the fluctuation spectrum. The basic free correlation function $\langle \xi^\mu(\tau)\xi^\nu(\tau')\rangle$ can then easily be found from its spectral representation as shown in Eq. (3.251). The result is

$$\langle \xi^\mu(\tau)\xi^\nu(\tau')\rangle^{q_0} = g^{\mu\nu}(q_0)(-\partial_\tau^2)^{-1}\,\delta(\tau-\tau') = g^{\mu\nu}(q_0)\bar\Delta(\tau,\tau'), \qquad (10.564)$$

where $\bar\Delta(\tau,\tau')$ is a short notation for the translationally invariant periodic Green function $G_{0,e}^{p'}(\tau)$ of the operator $-\partial_\tau^2$ without the zero mode in Eq. (3.251) (for a plot see Fig. 3.4):

$$\bar\Delta(\tau,\tau') = \bar\Delta(\tau-\tau') \equiv \frac{(\tau-\tau')^2}{2\beta} - \frac{|\tau-\tau'|}{2} + \frac{\beta}{12}, \quad \tau,\tau' \in [0,\hbar\beta]. \qquad (10.565)$$

This notation is useful since we shall have to calculate Feynman integrals of precisely the same form as previously with the Dirichlet-type correlation function Eq. (10.392). In contrast to (10.393) and (10.394) for $\Delta(\tau,\tau')$, the translational invariance of the periodic correlation function implies that $\bar\Delta(\tau,\tau') = \bar\Delta(\tau-\tau')$, so that the first time derivatives of $\bar\Delta(\tau,\tau')$ have opposite signs:

$$\dot{\bar\Delta}(\tau,\tau') = -\bar\Delta{}'(\tau,\tau') \equiv \frac{\tau-\tau'}{\beta} - \frac{\epsilon(\tau-\tau')}{2}, \quad \tau,\tau' \in [0,\hbar\beta], \qquad (10.566)$$

and the three possible double time derivatives are equal, up tp a sign:

$$-\ddot{\bar\Delta}(\tau,\tau') = -\bar\Delta{}''(\tau,\tau') = \dot{\bar\Delta}{}'(\tau,\tau') = \delta(\tau-\tau') - 1/\beta. \qquad (10.567)$$

The right-hand side contains an extra term on the right-hand side due to the missing zero eigenmode in the spectral representation of the δ-function:

$$\frac{1}{\beta}\sum_{m\neq 0} e^{-i\omega_m(\tau-\tau')} = \delta(\tau-\tau') - \frac{1}{\beta}. \qquad (10.568)$$

The third equation in (10.567) happens to coincide with the differential equation (10.394) in the Dirichlet case. All three equations have the same right-hand side due to the translational invariance of $\bar\Delta(\tau,\tau')$.

10.12.2 Arbitrariness of q_0^μ

We now take advantage of an important property of the perturbation expansion of the partition function (10.562) around $q^\mu(\tau) = q_0^\mu$: the *independence* of the choice of q_0^μ. The separation (10.557) into a constant q_0^μ and a time-dependent $\xi^\mu(\tau)$ paths must lead to the same result for any nearby constant $q_0'^\mu$ on the manifold. The result must therefore be invariant under an arbitrary infinitesimal displacement

$$q_0^\mu \to q_{0\varepsilon}^\mu = q_0^\mu + \varepsilon^\mu, \quad |\varepsilon| \ll 1. \tag{10.569}$$

In the path integral, this will be compensated by some translation of fluctuation coordinates $\xi^\mu(\tau)$, which will have the general nonlinear form

$$\xi^\mu \to \xi_\varepsilon^\mu = \xi^\mu - \varepsilon^\nu Q_\nu^\mu(q_0, \xi). \tag{10.570}$$

The transformation matrix $Q_\nu^\mu(q_0, \xi)$ satisfies the obvious initial condition $Q_\nu^\mu(q_0, 0) = \delta_\nu^\mu$. The path $q^\mu(\tau) = q^\mu(q_0, \xi(\tau))$ must remain invariant under simultaneous transformations (10.569) and (10.570), which implies that

$$\delta q^\mu \equiv q_\varepsilon^\mu - q^\mu = \varepsilon^\nu D_\nu q^\mu(q_0, \xi) = 0, \tag{10.571}$$

where D_μ is the infinitesimal transition operator

$$D_\mu = \frac{\partial}{\partial q_0^\mu} - Q_\mu^\nu(q_0, \xi)\frac{\partial}{\partial \xi^\nu}. \tag{10.572}$$

Geometrically, the matrix $Q_\nu^\mu(q_0, \xi)$ plays the role of a locally flat nonlinear connection [46]. It can be calculated as follows. We express the vector $q^\mu(q_0, \xi)$ in terms of the geodesic coordinates ξ^μ using Eqs. (10.557), (10.558), and (10.559), and substitute this into Eq. (10.571). The coefficients of ε^ν yield the equations

$$\delta_\nu^\mu + \frac{\partial \eta^\mu(q_0, \xi)}{\partial q_0^\nu} - Q_\nu^\kappa(q_0, \xi)\frac{\partial \eta^\mu(q_0, \xi)}{\partial \xi^\kappa} = 0, \tag{10.573}$$

where by Eq. (10.558):

$$\frac{\partial \eta^\mu(q_0, \xi)}{\partial q_0^\nu} = -\frac{1}{2}\partial_\nu \bar\Gamma_{(\sigma\tau)}{}^\mu(q_0)\xi^\sigma\xi^\tau - \dots, \tag{10.574}$$

and

$$\frac{\partial \eta^\mu(q_0, \xi)}{\partial \xi^\nu} = \delta_\nu^\mu - \bar\Gamma_{(\nu\sigma)}{}^\mu(q_0)\xi^\sigma - \frac{1}{2}\bar\Gamma_{(\nu\sigma\tau)}{}^\mu(q_0)\xi^\sigma\xi^\tau - \dots$$

$$= \delta_\nu^\mu - \bar\Gamma_{\nu\sigma}{}^\mu\xi^\sigma - \frac{1}{3}\left(\partial_\sigma\bar\Gamma_{\nu\tau}^\mu + \frac{1}{2}\partial_\nu\bar\Gamma_{\sigma\tau}{}^\mu - 2\bar\Gamma_{\tau\nu}{}^\kappa\bar\Gamma_{\kappa\sigma}{}^\mu - \bar\Gamma_{\tau\sigma}{}^\kappa\bar\Gamma_{\kappa\nu}{}^\mu\right)\xi^\sigma\xi^\tau - \dots. \tag{10.575}$$

To find $Q_\nu^\mu(q_0, \xi)$, we invert the expansion (10.575) to

$$\left[\left(\frac{\partial \eta(q_0, \xi)}{\partial \xi}\right)^{-1}\right]_\nu^\mu = \delta_\nu^\mu + \bar\Gamma_{\nu\sigma}^\mu\xi^\sigma + \frac{1}{3}\left(\partial_\sigma\bar\Gamma_{\nu\tau}^\mu + \frac{1}{2}\partial_\nu\bar\Gamma_{\sigma\tau}^\mu + \bar\Gamma_{\tau\nu}^\kappa\bar\Gamma_{\kappa\sigma}^\mu - \bar\Gamma_{\tau\sigma}^\kappa\bar\Gamma_{\kappa\nu}^\mu\right)\xi^\sigma\xi^\tau + \dots \tag{10.576}$$

$$= \left(\frac{\partial \xi^\mu(q_0, \eta)}{\partial \eta^\nu}\right)_{\eta = \eta(q_0, \xi)},$$

the last equality indicating that the result (10.576) can also be obtained from the original expansion (10.98) in the present notation [compare (10.558)]:

$$\xi^\mu(q_0,\eta) = \eta^\mu + \frac{1}{2}\tilde{\Gamma}_{\sigma\tau}{}^\mu(q_0)\eta^\sigma\eta^\tau + \frac{1}{6}\tilde{\Gamma}_{\sigma\tau\kappa}{}^\mu(q_0)\eta^\sigma\eta^\tau\eta^\kappa + \ldots, \tag{10.577}$$

with coefficients

$$\begin{aligned}
\tilde{\Gamma}_{\sigma\tau}{}^\mu(q_0) &= \bar{\Gamma}_{\sigma\tau}{}^\mu, \\
\tilde{\Gamma}_{\sigma\tau\kappa}{}^\mu(q_0) &= \bar{\Gamma}_{\sigma\tau\kappa}{}^\mu + 3\bar{\Gamma}_{\kappa\sigma}{}^\nu\bar{\Gamma}_{\nu\tau}{}^\mu = \partial_\kappa\bar{\Gamma}_{\sigma\tau}{}^\mu + \bar{\Gamma}_{\kappa\sigma}{}^\nu\bar{\Gamma}_{\nu\tau}{}^\mu, \\
&\ \vdots
\end{aligned} \tag{10.578}$$

Indeed, differentiating (10.577) with respect to η^ν, and re-expressing the result in terms of ξ^μ via Eq. (10.558), we find once more (10.576).

Multiplying both sides of Eq. (10.573) by (10.576), we express the nonlinear connection $Q_\nu^\mu(q_0,\xi)$ by means of geodesic coordinates $\xi^\mu(\tau)$ as

$$Q_\nu^\mu(q_0,\xi) = \delta_\nu^\mu + \bar{\Gamma}_{\nu\sigma}{}^\mu(q_0)\xi^\sigma + \frac{1}{3}\bar{R}_{\sigma\nu\tau}{}^\mu(q_0)\xi^\sigma\xi^\tau + \ldots. \tag{10.579}$$

The effect of simultaneous transformations (10.569), (10.570) upon the fluctuation function $\eta^\mu = \eta^\mu(q_0,\xi)$ in Eq. (10.558) is

$$\eta^\mu \to \eta'^\mu = \eta^\mu - \varepsilon^\nu \bar{Q}_\nu^\mu(q_0,\eta), \qquad \bar{Q}_\nu^\mu(q_0,0) = \delta_\nu^\mu, \tag{10.580}$$

where the matrix $\bar{Q}_\nu^\mu(q_0,\eta)$ is related to $Q_\nu^\mu(q_0,\xi)$ as follows

$$\bar{Q}_\nu^\mu(q_0,\eta) = \left[Q_\nu^\kappa(q_0,\xi)\frac{\partial\eta^\mu(q_0,\xi)}{\partial\xi^\kappa} - \frac{\partial\eta^\mu(q_0,\xi)}{\partial q_0^\nu}\right]_{\xi=\xi(q_0,\eta)}. \tag{10.581}$$

Applying Eq. (10.573) to the right-hand side of Eq. (10.581) yields $\bar{Q}_\nu^\mu(q_0,\eta) = \delta_\nu^\mu$, as it should to compensate the translation (10.569).

The above independence of q_0^μ will be essential for constructing the correct perturbation expansion for the path integral (10.562). For some special cases of the Riemannian manifold, such as a surface of sphere in $D+1$ dimensions which forms a homogeneous space $O(D)/O(D-1)$, all points are equivalent, and the local independence becomes global. This will be discussed further in Section 10.12.6.

10.12.3 Zero-Mode Properties

We are now prepared to eliminate the zero mode by the condition of vanishing average $\bar{\xi}^\mu = 0$. As mentioned before, the vanishing fluctuation $\xi^\mu(\tau) = 0$ is obviously a classical saddle-point for the path integral (10.562). In addition, because of the symmetry (10.570) there exist other equivalent extrema $\xi_\varepsilon^\mu(\tau) = -\varepsilon^\mu = \text{const}$. The D components of ε^μ correspond to D zero modes which we shall eliminate in favor of a change of q_0^μ. The proper way of doing this is provided by the Faddeev-Popov procedure. We insert into the path integral (10.562) the trivial unit integral, rewritten with the help of (10.569):

$$1 = \int d^D q_0 \, \delta^{(D)}(q_{0\varepsilon} - q_0) = \int d^D q_0 \, \delta^{(D)}(\varepsilon), \qquad (10.582)$$

and decompose the measure of path integration over all periodic paths $\xi^\mu(\tau)$ into a product of an ordinary integral over the temporal average $\xi_0^\mu = \bar{\xi}^\mu$, and a remainder containing only nonzero Fourier components [recall (2.446)]

$$\oint \mathcal{D}^D \xi = \int \frac{d^D \xi_0}{\sqrt{2\pi\beta}^D} \oint \mathcal{D}'^D \xi. \qquad (10.583)$$

According to Eq. (10.570), the path average $\bar{\xi}^\mu$ is translated under ε^μ as follows

$$\bar{\xi}^\mu \to \bar{\xi}_\varepsilon^\mu = \bar{\xi}^\mu - \varepsilon^\nu \frac{1}{\beta} \int_0^\beta d\tau \, Q_\nu^\mu(q_0, \xi(\tau)). \qquad (10.584)$$

Thus we can replace

$$\int \frac{d^D \xi_0}{\sqrt{2\pi\beta}^D} \to \int \frac{d^D \varepsilon}{\sqrt{2\pi\beta}^D} \det \left[\frac{1}{\beta} \int_0^\beta d\tau \, Q_\nu^\mu(q_0, \xi(\tau)) \right]. \qquad (10.585)$$

Performing this replacement in (10.583) and performing the integral over ε^μ in the inserted unity (10.582), we obtain the measure of path integration in terms of q_0^μ and geodesic coordinates of zero temporal average

$$\begin{aligned}
\oint \mathcal{D}^D \xi &= \int d^D q_0 \oint \frac{d^D \xi_0}{\sqrt{2\pi\beta}^D} \delta^{(D)}(\varepsilon) \oint \mathcal{D}'^D \xi \\
&= \int \frac{d^D q_0}{\sqrt{2\pi\beta}^D} \oint \mathcal{D}'^D \xi \det \left[\frac{1}{\beta} \int_0^\beta d\tau \, Q_\nu^\mu(q_0, \xi(\tau)) \right]. \qquad (10.586)
\end{aligned}$$

The factor on the right-hand side is the Faddeev-Popov determinant $\Delta[q_0, \xi]$ for the change from ξ_0^μ to q_0^μ. We shall write it as an exponential:

$$\Delta[q_0, \xi] = \det \left[\frac{1}{\beta} \int_0^\beta d\tau \, Q_\nu^\mu(q_0, \xi) \right] = e^{-\mathcal{A}^{\mathrm{FP}}[q_0, \xi]}, \qquad (10.587)$$

where $\mathcal{A}^{\mathrm{FP}}[q_0, \xi]$ is an auxiliary action accounting for the Faddeev-Popov determinant

$$\mathcal{A}^{\mathrm{FP}}[q_0, \xi] \equiv -\mathrm{tr} \log \left[\frac{1}{\beta} \int_0^\beta d\tau \, Q_\nu^\mu(q_0, \xi) \right], \qquad (10.588)$$

which must be included into the interaction (10.561). Inserting (10.579) into Eq. (10.588), we find explictly

$$\begin{aligned}
\mathcal{A}^{\mathrm{FP}}[q_0, \xi] &= -\mathrm{tr} \log \left[\delta_\nu^\mu + (3\beta)^{-1} \int_0^\beta d\tau \, \bar{R}_{\sigma\nu\tau}{}^\mu(q_0) \xi^\sigma(\tau) \xi^\tau(\tau) + \ldots \right] \\
&= \frac{1}{3\beta} \int_0^\beta d\tau \, \bar{R}_{\mu\nu}(q_0) \xi^\mu \xi^\nu + \ldots . \qquad (10.589)
\end{aligned}$$

The contribution of this action will crucial for obtaining the correct perturbation expansion of the path integral (10.562).

With the new interaction

$$\mathcal{A}_{\mathrm{tot,FP}}^{\mathrm{int}}[q_0,\xi] = \mathcal{A}_{\mathrm{tot}}^{\mathrm{int}}[q_0,\xi] + \mathcal{A}^{\mathrm{FP}}[q_0,\xi] \tag{10.590}$$

the partition function (10.562) can be written as a classical partition function

$$Z = \int \frac{d^D q_0}{\sqrt{2\pi\beta}^D} \sqrt{g(q_0)}\, e^{-\beta V^{\mathrm{eff\,cl}}(q_0)}, \tag{10.591}$$

where $V^{\mathrm{eff\,cl}}(q_0)$ is the curved-space version of the effective classical partition function of Ref. [53]. The effective classical Boltzmann factor

$$B(q_0) \equiv e^{-\beta V^{\mathrm{eff\,cl}}(q_0)} \tag{10.592}$$

is given by the path integral

$$B(q_0) = \oint \mathcal{D}'^D \xi\, \sqrt{g(q_0)} e^{-\mathcal{A}^{(0)}[q_0,\xi] - \mathcal{A}_{\mathrm{tot,FP}}^{\mathrm{int}}[q_0,\xi]}. \tag{10.593}$$

Since the zero mode is absent in the fluctuations on the right-hand side, the perturbation expansion is now straightforward. We expand the path integral (10.593) in powers of the interaction (10.590) around the free Boltzmann factor

$$B_0(q_0) = \oint \mathcal{D}'^D \xi\, \sqrt{g(q_0)} e^{-\int_0^\beta d\tau\, \frac{1}{2} g_{\mu\nu}(q_0)\dot{\xi}^\mu \dot{\xi}^\nu} \tag{10.594}$$

as follows:

$$B(q_0) = B_0(q_0)\left[1 - \left\langle \mathcal{A}_{\mathrm{tot,FP}}^{\mathrm{int}}[q_0,\xi] \right\rangle^{q_0} + \frac{1}{2}\left\langle \mathcal{A}_{\mathrm{tot,FP}}^{\mathrm{int}}[q_0,\xi]^2 \right\rangle^{q_0} - \ldots \right], \tag{10.595}$$

where the q_0-dependent correlation functions are defined by the Gaussian path integrals

$$\langle \ldots \rangle^{q_0} = B^{-1}(q_0) \oint \mathcal{D}'^D \xi\, [\ldots]^{q_0}\, e^{-\mathcal{A}^{(0)}[q_0,\xi]}. \tag{10.596}$$

By taking the logarithm of (10.594), we obtain directly a cumulant expansion for the effective classical potential $V^{\mathrm{eff\,cl}}(q_0)$.

For a proper normalization of the Gaussian path integral (10.594) we diagonalize the free action in the exponent by going back to the orthonormal components Δx^i in (10.97). Omitting again the smallness symbols Δ, the measure of the path integral becomes simply:

$$\oint \mathcal{D}'^D \xi^\mu\, \sqrt{g(q_0)} = \oint \mathcal{D}'^D x^i, \tag{10.597}$$

and we find

$$B_0(q_0) = \oint \mathcal{D}'^D x^i\, e^{-\int_0^\beta d\tau\, \frac{1}{2}\dot{\xi}^2}. \tag{10.598}$$

If we expand the fluctuations $x^i(\tau)$ into the eigenfunctions $e^{-i\omega_m \tau}$ of the operator $-\partial_\tau^2$ for periodic boundary conditions $x^i(0) = x^i(\beta)$,

$$x^i(\tau) = \sum_m x_m^i u_m(\tau) = x_0^i + \sum_{m \neq 0} x_m^i u_m(\tau), \quad x_{-m}^i = x_m^{i\,*}, \quad m > 0, \qquad (10.599)$$

and substitute this into the path integral (10.598), the exponent becomes

$$-\frac{1}{2} \int_0^\beta d\tau \left[\dot{x}^i(\tau) \right]^2 = -\frac{\beta}{2} \sum_{m \neq 0} \omega_m^2 \, x_{-m}^i x_m^i = -\beta \sum_{m>0} \omega_m^2 \, x_m^{i\,*} x_m^i. \qquad (10.600)$$

Remembering the explicit form of the measure (2.445) the Gaussian integrals in (10.598) yield the free-particle Boltzmann factor

$$B_0(q_0) = 1, \qquad (10.601)$$

corresponding to a vanishing effective classical potential in Eq. (10.592).

The perturbation expansion (10.595) becomes therefore simply

$$B(q_0) = 1 - \left\langle \mathcal{A}_{\text{tot,FP}}^{\text{int}}[q_0, \xi] \right\rangle^{q_0} + \frac{1}{2} \left\langle \mathcal{A}_{\text{tot,FP}}^{\text{int}}[q_0, \xi]^2 \right\rangle^{q_0} - \dots \, . \qquad (10.602)$$

The expectation values on the right-hand side are to be calculated with the help of Wick contractions involving the basic correlation function of $\xi^a(\tau)$ associated with the unperturbed action in (10.598):

$$\left\langle x^i(\tau) x^j(\tau') \right\rangle^{q_0} = \delta^{ij} \bar{\Delta}(\tau, \tau'), \qquad (10.603)$$

which is, of course, consistent with (10.564) via Eq. (10.97).

10.12.4 Covariant Perturbation Expansion

We now perform all possible Wick contractions of the fluctuations $\xi^\mu(\tau)$ in the expectation values (10.602) using the correlation function (10.564). We restrict our attention to the lowest-order terms only, since all problems of previous treatments arise already there. Making use of Eqs. (10.565) and (10.567), we find for the interaction (10.561):

$$\left\langle \mathcal{A}_{\text{tot}}^{\text{int}}[q_0, \xi] \right\rangle^{q_0} = \int_0^\beta d\tau \frac{1}{6} \left[\bar{R}_{\mu\lambda\nu\kappa}(q_0) \left\langle \xi^\lambda \xi^\kappa \dot{\xi}^\mu \dot{\xi}^\nu \right\rangle^{q_0} + \delta(0) \bar{R}_{\mu\nu}(q_0) \left\langle \xi^\mu \xi^\nu \right\rangle^{q_0} \right]$$

$$= \frac{1}{72} \bar{R}(q_0) \beta, \qquad (10.604)$$

and for (10.589):

$$\left\langle \mathcal{A}^{\text{FP}}[q_0, \xi] \right\rangle^{q_0} = \int_0^\beta d\tau \frac{1}{3\beta} \bar{R}_{\mu\nu}(q_0) \left\langle \xi^\mu \xi^\nu \right\rangle^{q_0} = \frac{1}{36} \bar{R}(q_0) \beta. \qquad (10.605)$$

The sum of the two contributions yields the manifestly covariant high-temperature expansion up to two loops:

$$B(q_0) = 1 - \left\langle \mathcal{A}_{\text{tot,FP}}^{\text{int}}[q_0, \xi] \right\rangle^{q_0} + \ldots = 1 - \frac{1}{24} \bar{R}(q_0)\beta + \ldots \qquad (10.606)$$

in agreement with the partition function density (10.494) calculated from Dirichlet boundary conditions. The associated partition function

$$Z^{\text{P}} = \int \frac{d^D q_0}{\sqrt{2\pi\beta}^D} \sqrt{g(q_0)} \; B(q_0) \qquad (10.607)$$

coincides with the partition function obtained by integrating over the partition function density (10.494). Note the crucial role of the action (10.589) coming from the Faddeev-Popov determinant in obtaining the correct two-loop coefficient in Eq. (10.606) and the normalization in Eq. (10.607).

The intermediate transformation to the geodesic coordinates $\xi^\mu(\tau)$ has made our calculations rather lengthy if the action is given in arbitrary coordinates, but it guarantees complete independence of the coordinates in the result (10.606). The entire derivation simplifies, of course, drastically if we choose from the outset geodesic coordinates to parametrize the curved space.

10.12.5 Covariant Result from Noncovariant Expansion

Having found the proper way of calculating the Boltzmann factor $B(q_0)$ we can easily set up a procedure for calculating the same covariant result without the use of the geodesic fluctuations $\xi^\mu(\tau)$. Thus we would like to evaluate the path integral (10.594) by a direct expansion of the action in powers of the noncovariant fluctuations $\eta^\mu(\tau)$ in Eq. (10.557). In order to make q_0^μ equal to the path average, $\bar{q}(\tau)$, we now require $\eta^\mu(\tau)$ to have a vanishing temporal average $\eta_0^\mu = \bar{\eta}^\mu = 0$.

Instead of (10.560), the free action reads now

$$\mathcal{A}^{(0)}[q_0, \eta] = g_{\mu\nu}(q_0) \int_0^\beta d\tau \, \frac{1}{2} \eta^\mu(\tau)(-\partial_\tau^2)\eta^\nu(\tau) \,, \qquad (10.608)$$

and the small-β behavior of the path integral (10.607) is governed by the interaction $\mathcal{A}_{\text{tot}}^{\text{int}}[q]$ fo Eq. (10.536) with unit smallness parameter η. In a notation as in (10.561), the interaction (10.536) reads

$$\mathcal{A}_{\text{tot}}^{\text{int}}[q_0; \eta] = \int_0^\beta d\tau \left\{ \left[\Gamma_{\kappa\mu\nu}\eta^\kappa + \frac{1}{2} \partial_\lambda \Gamma_{\kappa\mu\nu} \, \eta^\lambda \eta^\kappa \right] \dot{\eta}^\mu \dot{\eta}^\nu \right.$$
$$\left. - \delta(0) \left[\Gamma_{\mu\kappa}{}^\mu \eta^\kappa + \frac{1}{2} \partial_\lambda \Gamma_{\tau\mu}{}^\mu \eta^\lambda \eta^\tau \right] \right\}. \qquad (10.609)$$

We must deduce the measure of functional integration over η-fluctuations without zero mode $\eta_0^\mu = \bar{\eta}^\mu$ from the proper measure in (10.586) of ξ-fluctuations without zero mode:

$$\oint \mathcal{D}'^D \xi \, J(q_0, \xi) \Delta^{\text{FP}}[q_0, \xi] \equiv \oint \mathcal{D}^D \xi(\tau) J(q_0, \xi) \delta^{(D)}(\xi_0) \Delta^{\text{FP}}[q_0, \xi]. \qquad (10.610)$$

This is transformed to coordinates $\eta^\mu(\tau)$ via Eqs. (10.577) and (10.578) yielding

$$\oint \mathcal{D}'^D \xi \, J(q_0, \xi) \Delta^{\text{FP}}[q_0, \xi] = \oint \mathcal{D}'^D \eta \, \bar{\Delta}^{\text{FP}}[q_0, \eta] \,, \qquad (10.611)$$

where $\bar{\Delta}^{\rm FP}[q_0, \eta]$ is obtained from the Faddeev-Popov determinant $\Delta^{\rm FP}[q_0, \xi]$ of (10.587) by expressed the coordinates $\eta^\mu(\tau)$ in terms of $\xi^\mu(\tau)$, and multiplying the result by a Jacobian accounting for the change of the δ-function of ξ_0 to a δ-function of η_0 via the transformation Eq. (10.577):

$$\bar{\Delta}^{\rm FP}[q_0, \eta] = \Delta^{\rm FP}[q_0, \xi(q_0, \eta)] \times \det\left(\frac{\partial \bar{\eta}^\mu(q_0, \xi)}{\partial \xi^\nu}\right)_{\xi=\xi(q_0, \eta)}. \tag{10.612}$$

The last determinant has the exponential form

$$\det\left(\frac{\partial \bar{\eta}^\mu(q_0, \xi)}{\partial \xi^\nu}\right)_{\xi=\xi(q_0, \eta)} = \exp\left\{\operatorname{tr}\log\left[\frac{1}{\beta}\int_0^\beta d\tau \left(\frac{\partial \bar{\eta}^\mu(q_0, \xi)}{\partial \xi^\nu}\right)_{\xi=\xi(q_0, \eta)}\right]\right\}, \tag{10.613}$$

where the matrix in the exponent has small-η expansion

$$\left(\frac{\partial \bar{\eta}^\mu(q_0, \xi)}{\partial \xi^\nu}\right)_{\xi=\xi(q_0, \eta)} = \delta_\nu^\mu - \bar{\Gamma}_{\nu\sigma}{}^\mu \eta^\sigma$$

$$-\frac{1}{3}\left(\partial_\sigma \bar{\Gamma}_{\nu\tau}{}^\mu + \frac{1}{2}\partial_\nu \bar{\Gamma}_{\sigma\tau}{}^\mu - 2\bar{\Gamma}_{\tau\nu}{}^\kappa \bar{\Gamma}_{\kappa\sigma}{}^\mu + \frac{1}{2}\bar{\Gamma}_{\tau\sigma}{}^\kappa \bar{\Gamma}_{\kappa\nu}{}^\mu\right)\eta^\sigma\eta^\tau + \ldots. \tag{10.614}$$

The factor (10.612) in (10.611) leads to a new contribution to the interaction (10.608), if we rewrite it as

$$\bar{\Delta}^{\rm FP}[q_0, \eta] = e^{-\bar{\mathcal{A}}^{\rm FP}[q_0, \eta]}. \tag{10.615}$$

Combining Eqs. (10.587) and (10.613), we find a new Faddeev-Popov type action for η^μ-fluctuations at vanishing η_0^μ:

$$\bar{\mathcal{A}}^{\rm FP}[q_0, \eta] = \mathcal{A}^{\rm FP}[q_0, \xi(q_0, \eta)] - \operatorname{tr}\log\left[\frac{1}{\beta}\int_0^\beta d\tau \left(\frac{\partial \bar{\eta}^\mu(q_0, \xi)}{\partial \xi^\nu}\right)_{\xi=\xi(q_0, \eta)}\right]$$

$$= \frac{1}{2\beta}\int_0^\beta d\tau \, T_{\sigma\tau}(q_0)\eta^\sigma\eta^\tau + \ldots, \tag{10.616}$$

where

$$T_{\sigma\tau}(q_0) = \left(\partial_\mu \bar{\Gamma}_{\sigma\tau}{}^\mu - 2\bar{\Gamma}_{\sigma\kappa}{}^\mu \bar{\Gamma}_{\mu\tau}{}^\kappa + \bar{\Gamma}_{\kappa\mu}{}^\mu \bar{\Gamma}_{\sigma\tau}{}^\kappa\right). \tag{10.617}$$

The unperturbed correlation functions associated with the action (10.608) are:

$$\langle \eta^\mu(\tau)\eta^\nu(\tau')\rangle^{q_0} = g^{\mu\nu}(q_0)\bar{\Delta}(\tau, \tau') \tag{10.618}$$

and the free Boltzmann factor is the same as in Eq. (10.601). The perturbation expansion of the interacting Boltzmann factor is to be calculated from an expansion like (10.602):

$$B(q_0) = 1 - \langle \mathcal{A}^{\rm int}_{\rm tot, FP}[q_0, \eta]\rangle^{q_0} + \frac{1}{2}\left\langle\left(\mathcal{A}^{\rm int}_{\rm tot, FP}[q_0, \eta]\right)^2\right\rangle^{q_0} - \ldots, \tag{10.619}$$

where the interaction is now

$$\mathcal{A}^{\rm int}_{\rm tot, FP}[q_0, \eta] = \mathcal{A}^{\rm int}_{\rm tot}[q_0, \eta] + \bar{\mathcal{A}}^{\rm FP}[q_0, \eta]. \tag{10.620}$$

As before in the Dirichlet case, the divergences containing powers of $\delta(0)$ no longer cancel order by order, but do so at the end.

The calculations proceed as in the Dirichlet case, except that the correlation functions are now given by (10.565) which depend only on the difference of their arguments.

The first expectation value contributing to (10.619) is given again by $f_1^{(1)}$ of Eq. (10.539), except that the integrals have to be evaluated with the periodic correlation function $\bar{\Delta}(\tau, \tau)$ of Eq. (10.565), which has the properties

$$\bar{\Delta}(\tau, \tau) = \frac{1}{12}, \quad {}^{\bullet}\bar{\Delta}(\tau, \tau) = \bar{\Delta}^{\bullet}(\tau, \tau) = 0. \tag{10.621}$$

Using further the common property ${}^{\bullet}\bar{\Delta}^{\bullet}(\tau, \tau') = \delta(\tau - \tau') - 1/\beta$ of Eq. (10.567), we find directly from (10.539):

$$\begin{aligned}
\bar{f}_1^{(1)} = \left\langle \mathcal{A}_{\text{tot}}^{\text{int}}[q_0, \eta] \right\rangle^{q_0} &= \frac{\beta}{24} g^{\sigma\tau} \left(\partial_\sigma \bar{\Gamma}_{\tau\mu}{}^\mu + g^{\mu\nu} \bar{\Gamma}_{\tau\mu\kappa} \bar{\Gamma}_{\sigma\nu}{}^\kappa + \bar{\Gamma}_{\tau\nu}{}^\mu \bar{\Gamma}_{\sigma\mu}{}^\nu \right) \\
&\quad - \frac{\beta^2}{24} \delta(0) \, g^{\sigma\tau} \left(g^{\mu\nu} \bar{\Gamma}_{\tau\mu\kappa} \bar{\Gamma}_{\sigma\nu}{}^\kappa + \bar{\Gamma}_{\tau\mu}{}^\nu \bar{\Gamma}_{\sigma\nu}{}^\mu \right).
\end{aligned} \tag{10.622}$$

To this we must the expectation value of the Faddeev-Popov action:

$$\left\langle \bar{\mathcal{A}}^{\text{FP}}[q_0, \eta] \right\rangle^{q_0} = -\frac{\beta}{24} g^{\sigma\tau} \left(\partial_\mu \bar{\Gamma}_{\sigma\tau}{}^\mu - 2\bar{\Gamma}_{\sigma\nu}{}^\mu \bar{\Gamma}_{\mu\tau}{}^\nu + \bar{\Gamma}_{\mu\kappa}{}^\mu \bar{\Gamma}_{\sigma\tau}{}^\kappa \right). \tag{10.623}$$

The divergent term with the factor $\delta(0)$ in (10.622) is canceled by the same expression in the second-order contribution to (10.602) which we calculate now, evaluating the second cumulant (10.541) with the periodic correlation function $\bar{\Delta}(\tau, \tau)$ of Eq. (10.565). The diagrams are the same as in (10.543), but their evaluation is much simpler. Due to the absence of the zero modes in $\eta^\mu(\tau)$, all one-particle reducible diagrams vanish, so that the analogs of $f_1^{(2)}$, $f_1^{(3)}$, and $f_1^{(4)}$ in (10.542) are all zero. Only those of $f_1^{(5)}$ and $f_1^{(6)}$ survive, which involve now the Feynman integrals I_{14} and I_{15} evaluated with $\bar{\Delta}(\tau, \tau)$ which are

$$\cdots : \quad \bar{I}_{14} = \int_0^\beta \int_0^\beta d\tau \, d\tau' \, {}^{\bullet}\bar{\Delta}(\tau, \tau') \, \bar{\Delta}(\tau, \tau') \, {}^{\bullet}\bar{\Delta}(\tau, \tau') = -\frac{\beta}{24} + \delta(0) \frac{\beta^2}{12}, \tag{10.624}$$

$$\bigcirc : \quad \bar{I}_{15} = \int_0^\beta \int_0^\beta d\tau \, d\tau' \, \bar{\Delta}(\tau, \tau') \, {}^{\bullet}\bar{\Delta}^{\bullet 2}(\tau, \tau') = -\frac{\beta}{24}. \tag{10.625}$$

This leads to the second cumulant

$$\begin{aligned}
\frac{1}{2} \left\langle \left(\mathcal{A}_{\text{tot,FP}}^{\text{int}}[q_0, \eta] \right)^2 \right\rangle_c^{q_0} &= -\frac{\beta}{24} g^{\sigma\tau} \left(g^{\mu\nu} \bar{\Gamma}_{\tau\mu\kappa} \bar{\Gamma}_{\sigma\nu}{}^\kappa + 2\bar{\Gamma}_{\tau\nu}{}^\mu \bar{\Gamma}_{\sigma\mu}{}^\nu \right) \\
&\quad + \frac{\beta^2}{24} \delta(0) \, g^{\sigma\tau} \left(g^{\mu\nu} \bar{\Gamma}_{\tau\mu\kappa} \bar{\Gamma}_{\sigma\nu}{}^\kappa + \bar{\Gamma}_{\tau\mu}{}^\nu \bar{\Gamma}_{\sigma\nu}{}^\mu \right).
\end{aligned} \tag{10.626}$$

The sum of Eqs. (10.623) and (10.626) is finite and yields the same covariant result $\beta \bar{R}/24$ as in Eq. (10.552), so that we re-obtain the same covariant perturbation expansion of the effective classical Boltzmann factor as before in Eq. (10.606). Note the importance of the contribution (10.623) from the Faddeev-Popov determinant in producing the curvature scalar. Neglecting this, as done by other authors in Ref. [34], will produce in the effective classical Boltzmann factor (10.619) an additional noncovariant term $g^{\sigma\tau} T_{\sigma\tau}(q_0)/24$. This may be rewritten as a covariant divergence of a nonvectorial quantity

$$g^{\sigma\tau} T_{\sigma\tau} = \nabla_\mu V^\mu, \quad V^\mu(q_0) = g^{\sigma\tau}(q_0) \bar{\Gamma}_{\sigma\tau}^\mu(q_0). \tag{10.627}$$

As such it does not contribute to the integral over q_0^μ in Eq. (10.607), but it is nevertheless a wrong noncovariant result for the Boltzmann factor (10.606).

The appearance of a noncovariant term in a treatment where q_0^μ is the path average of $q^\mu(\tau)$ is not surprising. If the time dependence of a path shows an acceleration, the average of a path is not an invariant concept even for an infinitesimal time. One may covariantly impose the condition of a vanishing temporal average only upon fluctuation coordinates which have no acceleration. This is the case of geodesic coordinates $\xi^a(\tau)$ since their equation of motion at q_0^μ is $\ddot{\xi}^a(\tau) = 0$.

10.12.6 Particle on Unit Sphere

A special treatment exists for particle in homogeneous spaces. As an example, consider a quantum particle moving on a unit sphere in $D + 1$ dimensions. The partition function is defined by Eq. (10.554) with the Euclidean action (10.408) and the invariant measure (10.406), where the metric and its determinant are

$$g_{\mu\nu}(q) = \delta_{\mu\nu} + \frac{q_\mu q_\nu}{1 - q^2}, \qquad g(q) = \frac{1}{1 - q^2}. \tag{10.628}$$

It is, of course, possible to calculate the Boltzmann factor $B(q_0)$ with the procedure of Section 10.12.3. Instead of doing this we shall, however, exploit the homogeneity of the sphere. The invariance under reparametrizations of general Riemannian space becomes here an isometry of the metric (10.628). Consequently, the Boltzmann factor $B(q_0)$ in Eq. (10.607) becomes *independent* of the choice of q_0^μ, and the integral over q_0^μ in (10.591) yields simply the total surface of the sphere times the Boltzmann factor $B(q_0)$. The homogeneity of the space allows us to treat paths $q^\mu(\tau)$ themselves as small quantum fluctuations around the origin $q_0^\mu = 0$, which extremizes the path integral (10.554). The possibility of this expansion is due to the fact that $\Gamma_{\sigma\tau}^\mu = q^\mu g_{\sigma\tau}$ vanishes at $q^\mu(\tau) = 0$, so that the movement is at this point free of acceleration, this being similar to the situation in geodesic coordinates. As before we now take account of the fact that there are other equivalent saddle-points due to isometries of the metric (10.628) on the sphere (see, e.g., [54]). The infinitesimal transformations of a small vector q^μ:

$$q_\varepsilon^\mu = q^\mu + \varepsilon^\mu \sqrt{1 - q^2}, \quad \varepsilon^\mu = \text{const}, \quad \mu = 1, \dots, D \tag{10.629}$$

move the origin $q_0^\mu = 0$ by a small amount on the surface of the sphere. Due to the rotational symmetry of the system in the D-dimensional space, these fluctuations have a vanishing action. There are also $D(D-1)/2$ more isometries consisting of the rotations around the origin $q^\mu(\tau) = 0$ on the surface of the sphere. These are, however, irrelevant in the present context since they leave the origin unchanged.

The transformations (10.629) of the origin may be eliminated from the path integral (10.554) by including a factor $\delta^{(D)}(\bar{q})$ to enforce the vanishing of the temporal path average $\bar{q} = \beta^{-1} \int_0^\beta d\tau q(\tau)$. The associated Faddeev-Popov determinant $\Delta^{\mathrm{FP}}[q]$ is determined by the integral

$$\Delta^{\mathrm{FP}}[q] \int d^D\varepsilon \, \delta^{(D)}\left(\bar{q}_\varepsilon\right) = \Delta^{\mathrm{FP}}[q] \int d^D\varepsilon \, \delta^{(D)}\left(\varepsilon^\mu \frac{1}{\beta} \int_0^\beta d\tau \sqrt{1 - q^2}\right) = 1. \tag{10.630}$$

The result has the exponential form

$$\Delta^{\mathrm{FP}}[q] = \left(\frac{1}{\beta} \int_0^\beta d\tau \sqrt{1 - q^2}\right)^D = e^{-\mathcal{A}^{\mathrm{FP}}[q]}, \tag{10.631}$$

where $\mathcal{A}^{\mathrm{FP}}[q]$ must be added to the action (10.408):

$$\mathcal{A}^{\mathrm{FP}}[q] = -D \log\left(\frac{1}{\beta} \int_0^\beta d\tau \sqrt{1 - q^2}\right). \tag{10.632}$$

The Boltzmann factor $B(q_0) \equiv B$ is then given by the path integral without zero modes

$$
\begin{aligned}
B &= \oint \prod_{\mu,\tau} \left[dq^\mu(\tau) \sqrt{g(q(\tau))} \right] \delta^{(D)}(\bar{q}) \Delta^{\mathrm{FP}}[q] e^{-\mathcal{A}[q]} \\
&= \oint \mathcal{D}'^D q \sqrt{g(q(\tau))} \Delta^{\mathrm{FP}}[q] e^{-\mathcal{A}[q]} ,
\end{aligned}
\tag{10.633}
$$

where the measure $\mathcal{D}'^D q$ is defined as in Eq. (2.445). This can also be written as

$$
B = \oint \mathcal{D}'^D q \, e^{-\mathcal{A}[q] - \mathcal{A}^J[q] - \mathcal{A}^{\mathrm{FP}}[q]} ,
\tag{10.634}
$$

where $\mathcal{A}^J[q]$ is a contribution to the action (10.408) coming from the product

$$
\prod_\tau \sqrt{g(q(\tau))} \equiv e^{-\mathcal{A}^J[q]}.
\tag{10.635}
$$

By inserting (10.628), this becomes

$$
\mathcal{A}^J[q] = -\int_0^\beta d\tau \, \frac{1}{2}\delta(0) \log g(q) = \int_0^\beta d\tau \, \frac{1}{2}\delta(0) \log(1 - q^2) .
\tag{10.636}
$$

The total partition function is, of course, obtained from B by multiplication with the surface of the unit sphere in $D + 1$ dimensions $2\pi^{(D+1)/2}/\Gamma(D+1)/2)$. To calculate B from (10.634), we now expand $\mathcal{A}[q]$, $\mathcal{A}^J[q]$ and $\mathcal{A}^{\mathrm{FP}}[q]$ in powers of $q^\mu(\tau)$. The metric $g_{\mu\nu}(q)$ and its determinant $g(q)$ in Eq. (10.628) have the expansions

$$
g_{\mu\nu}(q) = \delta_{\mu\nu} + q_\mu q_\nu + \dots , \quad g(q) = 1 + q^2 + \dots ,
\tag{10.637}
$$

and the unperturbed action reads

$$
\mathcal{A}^{(0)}[q] = \int_0^\beta d\tau \, \frac{1}{2}\dot{q}^2(\tau).
\tag{10.638}
$$

In the absence of the zero eigenmodes due to the δ-function over $\bar{q}$ in Eq. (10.633), we find as in Eq. (10.601) the free Boltzmann factor

$$
B_0 = 1.
\tag{10.639}
$$

The free correlation function looks similar to (10.603):

$$
\langle q^\mu(\tau) q^\nu(\tau') \rangle = \delta^{\mu\nu} \bar{\Delta}(\tau, \tau').
\tag{10.640}
$$

The interactions coming from the higher expansions terms in Eq. (10.637) begin with

$$
\mathcal{A}^{\mathrm{int}}_{\mathrm{tot}}[q] = \mathcal{A}^{\mathrm{int}}[q] + \mathcal{A}^J[q] = \int_0^\beta d\tau \, \frac{1}{2}\left[(q\dot{q})^2 - \delta(0)q^2 \right].
\tag{10.641}
$$

To the same order, the Faddeev-Popov interaction (10.632) contributes

$$\mathcal{A}^{\mathrm{FP}}[q] = \frac{D}{2\beta} \int_0^\beta d\tau\, q^2. \tag{10.642}$$

This has an important effect upon the two-loop perturbation expansion of the Boltzmann factor

$$B(q_0) = 1 - \left\langle \mathcal{A}_{\mathrm{tot}}^{\mathrm{int}}[q] \right\rangle^{q_0} - \left\langle \mathcal{A}^{\mathrm{FP}}[q] \right\rangle^{q_0} + \ldots = B(0) \equiv B. \tag{10.643}$$

Performing the Wick contractions with the correlation function (10.640) with the properties (10.565)–(10.567), we find from Eqs. (10.641), (10.642)

$$
\begin{aligned}
\left\langle \mathcal{A}_{\mathrm{tot}}^{\mathrm{int}}[q] \right\rangle^{q_0} &= \frac{1}{2} \int_0^\beta d\tau \left\{ D\,\dot{\bar{\Delta}}(\tau,\tau)\bar{\Delta}(\tau,\tau) + D(D+1)\,\dot{\bar{\Delta}}^2(\tau,\tau) - \delta(0)D\bar{\Delta}(\tau,\tau) \right\} \\
&= \frac{1}{2} \int_0^\beta d\tau \left\{ -\frac{D}{\beta}\bar{\Delta}(\tau,\tau) + D(D+1)\,\dot{\bar{\Delta}}^2(\tau,\tau) \right\} = -\frac{D}{24}\beta, \tag{10.644}
\end{aligned}
$$

and

$$\left\langle \mathcal{A}^{\mathrm{FP}}[q] \right\rangle^{q_0} = \frac{D}{2\beta} \int_0^\beta d\tau\, D\,\bar{\Delta}(\tau,\tau) = \frac{D^2}{24}\beta. \tag{10.645}$$

Their combination in Eq. (10.643) yields the high-temperature expansion

$$B = 1 - \frac{D(D-1)}{24}\beta + \ldots\,. \tag{10.646}$$

This is in perfect agreement with Eqs. (10.494) and (10.606), since the scalar curvature for a unit sphere in $D+1$ dimensions is $\bar{R} = D(D-1)$. It is remarkable how the contribution (10.645) of the Faddeev-Popov determinant has made the noncovariant result (10.644) covariant.

10.13 Covariant Effective Action for Quantum Particle with Coordinate-Dependent Mass

The classical behavior of a system is completely determined by the extrema of the classical action. The quantum-mechanical properties can be found from the extrema of the effective action (see Subsection 3.22.5). This important quantity can in general only be calculated perturbatively. This will be done here for a particle with a coordinate-dependent mass. The calculation [44] will make use of the background method of Subsection 3.23.6 combined with the techniques developed earlier in this chapter. From the one-particle-irreducible (1PI) Feynman diagrams with no external lines we obtain an expansion in powers of the Planck constant $\hbar$. The result will be applicable to a large variety of interesting physical systems, for instance compound nuclei, where the collective Hamiltonian, commonly derived from a microscopic description via time-dependent Hartree-Fock theory [45], contains coordinate-dependent mass parameters.

10.13.1   Formulating the Problem

Consider a particle with coordinate-dependent mass $m(q)$ moving as in the one-dimensional potential $V(q)$. We shall study the Euclidean version of the system where the paths $q(t)$ are continued to an imaginary times $\tau = -it$ and the Lagrangian for $q(\tau)$ has the form

$$L(q, \dot{q}) = \frac{1}{2} m(q) \dot{q}^2 + V(q). \tag{10.647}$$

The dot stands for the derivative with respect to the imaginary time. The q-dependent mass may be written as $mg(q)$ where $g(q)$ plays the role of a one-dimensional dynamical metric. It is the trivial 1×1 Hessian metric of the system [recall the definition (1.12) and Eq. (1.384)]. In D-dimensional configuration space, the kinetic term would read $mg_{\mu\nu}(q)\dot{q}^\mu\dot{q}^\nu/2$, having the same form as in the curved-space action (10.186).

Under an arbitrary single-valued coordinate transformation $q = q(\tilde{q})$, the potential $V(q)$ is assumed to transform like a scalar whereas the metric $m(q)$ is a one-dimensional tensor of rank two:

$$V(q) = V(q(\tilde{q})) \equiv \tilde{V}(\tilde{q}), \qquad m(q) = \widetilde{m}(\tilde{q}) \left[d\tilde{q}(q)/dq \right]^2. \tag{10.648}$$

This coordinate transformation leaves the Lagrangian (10.647) and thus also the classical action

$$\mathcal{A}[q] = \int_{-\infty}^{\infty} d\tau \, L(q, \dot{q}) \tag{10.649}$$

invariant. Quantum theory has to possess the same invariance, exhibited automatically by Schrödinger theory. It must be manifest in the effective action. This will be achieved by combining the background technique in Subsection 3.23.6 with the techniques of Sections 10.6–10.10. In the background field method [46] we split all paths into $q(\tau) = Q(\tau) + \delta q(\tau)$, where $Q(\tau)$ is the final extremal orbit and δq describes the quantum fluctuations around it. At the one-loop level, the covariant effective action $\Gamma[Q]$ becomes a sum of the classical Lagrangian $L(Q, \dot{Q})$ and a correction term ΔL. It is defined by the path integral [recall (3.773)]

$$e^{-\Gamma[Q]/\hbar} = \int \mathcal{D}\mu(\delta q) e^{-(1/\hbar)\left\{ \mathcal{A}[Q+\delta q] - \int d\tau \, \delta q \, \delta\Gamma[Q]/\delta Q \right\}}, \tag{10.650}$$

where the measure of functional integration $\mathcal{D}\mu(\delta q)$ is obtained from the initial invariant measure $\mathcal{D}\mu(q) = Z^{-1} \prod_\tau dq(\tau) \sqrt{m(q)}$ and reads

$$\mathcal{D}\mu(\delta q) = Z^{-1} \prod_\tau d\delta q(\tau) \sqrt{m(Q)} \, e^{(1/2)\delta(0) \int d\tau \log[m(Q+\delta q)/m(Q)]}, \tag{10.651}$$

with Z being some normalization factor. The generating functional (10.650) possesses the same symmetry under reparametrizations of the configuration space as the classical action (10.649).

We now calculate $\Gamma[Q]$ in Eq. (10.650) perturbatively as a power series in $\hbar$:

$$\Gamma[Q] = \mathcal{A}[Q] + \hbar\Gamma_1[Q] + \hbar^2\Gamma_2[Q] + \dots . \tag{10.652}$$

The quantum corrections to the classical action (10.649) are obtained by expanding $\mathcal{A}[Q + \delta q]$ and the measure (10.651) *covariantly* in powers of δq:

$$\mathcal{A}[Q + \delta q] = \mathcal{A}[Q] + \int d\tau \frac{D\mathcal{A}}{\delta Q(\tau)} \delta x(\tau) + \frac{1}{2} \int d\tau \int d\tau' \frac{D^2\mathcal{A}}{\delta Q(\tau)\delta Q(\tau')} \delta x(\tau)\,\delta x(\tau')$$

$$+ \frac{1}{6} \int d\tau \int d\tau' \int d\tau'' \frac{D^3\mathcal{A}}{\delta Q(\tau)\delta Q(\tau')\delta Q(\tau'')} \delta x(\tau)\,\delta x(\tau')\,\delta x(\tau'') + \dots . \tag{10.653}$$

The expansion is of the type (10.101), i.e., the expressions δx are *covariant fluctuations* related to the ordinary variations δq in the same way as the normal coordinates Δx^μ are related to the differences Δq^μ in the expansion (10.98). The symbol $D/\delta Q$ denotes the *covariant* functional derivative in one dimension. To first order, this is the ordinary functional derivative

$$\frac{D\mathcal{A}[Q]}{\delta Q(\tau)} = \frac{\delta\mathcal{A}[Q]}{\delta Q(\tau)} = V'(Q) - \frac{1}{2} m'(Q)\,\dot{Q}^2(\tau) - m(Q)\,\ddot{Q}(\tau). \tag{10.654}$$

This vanishes for the classical orbit $Q(\tau)$. The second covariant derivative is [compare (10.100)]

$$\frac{D^2\mathcal{A}[Q]}{\delta Q(\tau)\delta Q(\tau')} = \frac{\delta^2\mathcal{A}[Q]}{\delta Q(\tau)\delta Q(\tau')} - \Gamma(Q(\tau)) \frac{\delta\mathcal{A}[Q]}{\delta Q(\tau')}, \tag{10.655}$$

where $\Gamma(Q) = m'(Q)/2\,m(Q)$ is the one-dimensional version of the Christoffel symbol for the metric $g_{\mu\nu} = \delta_{\mu\nu}m(Q)$. More explicitly, the result is

$$\frac{\delta^2\mathcal{A}[Q]}{\delta Q(\tau)\delta Q(\tau')} = -\left[m(Q)\partial_\tau^2 + m'(Q)\dot{Q}\partial_\tau + m'(Q)\ddot{Q} + \frac{1}{2}m''(Q)\dot{Q}^2 - V''(Q) \right]\delta(\tau - \tau'). \tag{10.656}$$

The validity of the expansion (10.653) follows from the fact that it is equivalent by a coordinates transformation to an ordinary functional expansion in Riemannian coordinates where the Christoffel symbol vanishes for the particular background coordinates.

The inverse of the functional matrix (10.655) supplies us with the free correlation function $G(\tau, \tau')$ of the fluctuations $\delta x(\tau)$. The higher derivatives define the interactions. The expansion terms $\Gamma_n[Q]$ in (10.652) are found from all one-particle-irreducible vacuum diagrams (3.781) formed with the propagator $G(\tau, \tau')$ and the interaction vertices. The one-loop correction to the effective action is given by the simple harmonic path integral

$$e^{-\Gamma_1[Q]} = \int \mathcal{D}\delta x \sqrt{m(Q)}\, e^{-\mathcal{A}^{(2)}[Q,\delta x]}, \tag{10.657}$$

with the quadratic part of the expansion (10.653):

$$\mathcal{A}^{(2)}[Q,\delta x] = \frac{1}{2}\int_{-\infty}^{\infty} d\tau\, d\tau'\,\delta x(\tau)\,\frac{D^2\mathcal{A}[Q]}{\delta Q(\tau)\delta Q(\tau')}\,\delta x(\tau')\,. \tag{10.658}$$

The presence of $m(Q)$ in the free part of the covariant kinetic term (10.658) and in the measure in Eq. (10.657) suggests exchanging the fluctuation δx by the new coordinates $\delta\tilde{x} = h(Q)\delta x$, where $h(Q) \equiv \sqrt{m(Q)}$ is the one-dimensional version of the triad (10.12) $e(Q) = \sqrt{m(Q)}$ associated with the metric $m(Q)$. The fluctuations $\delta\tilde{x}$ correspond to the differences Δx^i in (10.97). The covariant derivative of $e(Q)$ vanishes $D_Q e(Q) = \partial_Q e(Q) - \Gamma(Q)\,e(Q) \equiv 0$ [recall (10.40)]. Then (10.658) becomes

$$\mathcal{A}^{(2)}[Q,\delta x] = \frac{1}{2}\int_{-\infty}^{\infty} d\tau\, d\tau'\,\delta\tilde{x}(\tau)\,\frac{D^2\mathcal{A}[Q]}{\delta\tilde{Q}(\tau)\delta\tilde{Q}(\tau')}\,\delta\tilde{x}(\tau')\,, \tag{10.659}$$

where

$$\frac{D^2\mathcal{A}[Q]}{\delta\tilde{Q}(\tau)\delta\tilde{Q}(\tau')} = e^{-1}(Q)\,\frac{D^2\mathcal{A}[Q]}{\delta Q(\tau)\delta Q(\tau')}\,e^{-1}(Q) = \left[-\frac{d^2}{d\tau^2} + \omega^2(Q(\tau))\right]\delta(\tau-\tau')\,, \tag{10.660}$$

and

$$\begin{aligned}
\omega^2(Q) &= e^{-1}(Q)\,D^2V(Q)\,e^{-1}(Q) = e^{-1}(Q)\,DV'(Q)\,e^{-1}(Q) \\
&= \frac{1}{m(Q)}\,[V''(Q) - \Gamma(Q)\,V'(Q)]\,.
\end{aligned} \tag{10.661}$$

Note that this is the one-dimensional version of the Laplace-Beltrami operator (1.377) applied to $V(Q)$:

$$\omega^2(Q) = \Delta V(Q) = \frac{1}{\sqrt{m(Q)}}\frac{d}{dQ}\left[\sqrt{m(Q)}\left(\frac{V'(Q)}{m(Q)}\right)\right]\,. \tag{10.662}$$

Indeed, $e^{-2}D^2$ is the one-dimensional version of $g^{\mu\nu}D_\mu D_\nu$ [recall (10.38)] Since $V(Q)$ is a scalar, so is $\Delta V(Q)$.

Equation (10.660) shows that the fluctuations $\delta\tilde{x}$ behave like those of a harmonic oscillator with the time-dependent frequency $\omega^2(Q)$. The functional measure of integration in Eq. (10.657) simplifies in terms of $\delta\tilde{x}$:

$$\prod_\tau d\delta x(\tau)\sqrt{m(Q)} = \prod_\tau d\delta\tilde{x}(\tau)\,. \tag{10.663}$$

This allows us to integrate the Gaussian path integral (10.657) trivially to obtain the one-loop quantum correction to the effective action

$$\Gamma_1[Q] = \frac{1}{2}\mathrm{Tr}\log\left[-\partial_\tau^2 + \omega^2(Q(\tau))\right]\,. \tag{10.664}$$

Due to the τ-dependence of ω^2, this cannot be evaluated explicitly. For sufficiently slow motion of $Q(\tau)$, however, we can resort to a gradient expansion which yields asymptotically a local expression for the effective action.

10.13.2   Gradient Expansion

The gradient expansion of the one-loop effective action (10.664) has the general form

$$\Gamma_1[Q] = \int_{-\infty}^{\infty} d\tau \left[V_1(Q) + \frac{1}{2} Z_1(Q)\, \dot{Q}^2 + \cdots \right]. \tag{10.665}$$

It is found explicitly by recalling the gradient expansion of the trace of the logarithm derived in Eq. (4.307):

$$\mathrm{Tr} \log\left[-\partial_\tau^2 + \omega^2(\tau) \right] \equiv \int_{-\infty}^{\infty} d\tau \left\{ \omega(\tau) + \frac{[\partial_\tau \omega^2(\tau)]^2}{32\omega^5(\tau)} + \cdots \right\}. \tag{10.666}$$

Inserting $\Omega(\tau) = \omega(Q(\tau))$, we identify

$$V_1(Q) = \hbar\omega(Q)/2, \qquad Z_1(Q) = \frac{(D\omega^2)^2(Q)}{32\,\omega^5(Q)}, \tag{10.667}$$

and obtain the effective action to order $\hbar$ for slow motion

$$\Gamma^{\mathrm{eff}}[Q] = \int_{-\infty}^{\infty} d\tau \left[\frac{1}{2} m^{\mathrm{eff}}(Q)\, \dot{Q}^2 + V^{\mathrm{eff}}(Q) \right], \tag{10.668}$$

where the bare metric $m^{\mathrm{eff}}(Q)$ and the potential $V^{\mathrm{eff}}(Q)$ are related to the initial classical expressions by

$$m^{\mathrm{eff}}(Q) = m(Q) + \hbar \frac{(D\omega^2)^2(Q)}{32\,\omega^5(Q)}, \tag{10.669}$$

$$V^{\mathrm{eff}}(Q) = V(Q) + \hbar \frac{\omega(Q)}{2}. \tag{10.670}$$

For systems in which only the mass is Q-independent, the result has also been obtained by Ref. [48].

The range of validity of the expansion is determined by the characteristic time scale $1/\omega$. Within this time, the particle has to move only little.

Appendix 10A   Nonholonomic Gauge Transformations in Electromagnetism

To introduce the subject, let us first recall the standard treatment of magnetism. Since there are no magnetic monopoles, a magnetic field $\mathbf{B}(\mathbf{x})$ satisfies the identity $\nabla \cdot \mathbf{B}(\mathbf{x}) = 0$, implying that only two of the three field components of $\mathbf{B}(\mathbf{x})$ are independent. To account for this, one usually expresses a magnetic field $\mathbf{B}(\mathbf{x})$ in terms of a vector potential $\mathbf{A}(\mathbf{x})$, setting $\mathbf{B}(\mathbf{x}) = \nabla \times \mathbf{A}(\mathbf{x})$. Then Ampère's law, which relates the magnetic field to the electric current density $\mathbf{j}(\mathbf{x})$ by $\nabla \times \mathbf{B} = 4\pi \mathbf{j}(\mathbf{x})$, becomes a second-order differential equation for the vector potential $\mathbf{A}(\mathbf{x})$ in terms of an electric current

$$\nabla \times [\nabla \times \mathbf{A}(\mathbf{x})] = \mathbf{j}(\mathbf{x}). \tag{10A.1}$$

The vector potential $\mathbf{A}(\mathbf{x})$ is a *gauge field*. Given $\mathbf{A}(\mathbf{x})$, any locally gauge-transformed field

$$\mathbf{A}(\mathbf{x}) \rightarrow \mathbf{A}'(\mathbf{x}) = \mathbf{A}(\mathbf{x}) + \nabla \Lambda(\mathbf{x}) \tag{10A.2}$$

yields the same magnetic field $\mathbf{B}(\mathbf{x})$. This reduces the number of physical degrees of freedom in the gauge field $\mathbf{A}(\mathbf{x})$ to two, just as those in $\mathbf{B}(\mathbf{x})$. In order for this to hold, the transformation function must be single-valued, i.e., it must have commuting derivatives

$$(\partial_i\partial_j - \partial_j\partial_i)\Lambda(\mathbf{x}) = 0. \tag{10A.3}$$

The equation for absence of magnetic monopoles $\nabla \cdot \mathbf{B} = 0$ is ensured if the vector potential has commuting derivatives

$$(\partial_i\partial_j - \partial_j\partial_i)\mathbf{A}(\mathbf{x}) = 0. \tag{10A.4}$$

This integrability property makes $\nabla \cdot \mathbf{B} = 0$ the *Bianchi identity* in this gauge field representation of the magnetic field.

In order to solve (10A.1), we remove the gauge ambiguity by choosing a particular gauge, for instance the *transverse gauge* $\nabla \cdot \mathbf{A}(\mathbf{x}) = 0$ in which $\nabla \times [\nabla \times \mathbf{A}(\mathbf{x})] = -\nabla^2\mathbf{A}(\mathbf{x})$, and obtain

$$\mathbf{A}(\mathbf{x}) = \int d^3x' \frac{\mathbf{j}(\mathbf{x}')}{|\mathbf{x} - \mathbf{x}'|}. \tag{10A.5}$$

The associated magnetic field is

$$\mathbf{B}(\mathbf{x}) = \int d^3x' \frac{\mathbf{j}(\mathbf{x}') \times \mathbf{R}'}{R'^3}, \qquad \mathbf{R}' \equiv \mathbf{x}' - \mathbf{x}. \tag{10A.6}$$

This standard representation of magnetic fields is not the only possible one. There exists another one in terms of a scalar potential $\Lambda(\mathbf{x})$, which must, however, be multivalued to account for the two physical degrees of freedom in the magnetic field.

10A.1   Gradient Representation of Magnetic Field of Current Loops

Consider an infinitesimally thin closed wire carrying an electric current I along the line L. It corresponds to a current density

$$\mathbf{j}(\mathbf{x}) = I\,\boldsymbol{\delta}(\mathbf{x}; L), \tag{10A.7}$$

where $\boldsymbol{\delta}(\mathbf{x}; L)$ is the δ-function on the line L:

$$\boldsymbol{\delta}(\mathbf{x}; L) = \int_L d\mathbf{x}' \delta^{(3)}(\mathbf{x} - \mathbf{x}'). \tag{10A.8}$$

For a closed line L, this function has zero divergence:

$$\nabla \cdot \boldsymbol{\delta}(\mathbf{x}; L) = 0. \tag{10A.9}$$

This follows from the property of the δ-function on an arbitrary line L connecting the points $\mathbf{x}_1$ and $\mathbf{x}_2$:

$$\nabla \cdot \boldsymbol{\delta}(\mathbf{x}; L) = \delta(\mathbf{x}_2) - \delta(\mathbf{x}_1). \tag{10A.10}$$

For closed loops, the right-hand side vanishes.

From Eq. (10A.5) we obtain the associated vector potential

$$\mathbf{A}(\mathbf{x}) = I \int_L d\mathbf{x}' \frac{1}{|\mathbf{x} - \mathbf{x}'|}, \tag{10A.11}$$

yielding the magnetic field

$$\mathbf{B}(\mathbf{x}) = -I \int_L \frac{d\mathbf{x}' \times \mathbf{R}'}{R'^3}, \qquad \mathbf{R}' \equiv \mathbf{x}' - \mathbf{x}. \tag{10A.12}$$

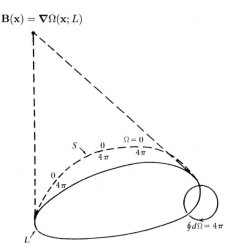

$$\mathbf{B}(\mathbf{x}) = \boldsymbol{\nabla}\Omega(\mathbf{x}; L)$$

Figure 10.5 Infinitesimally thin closed current loop L. The magnetic field $\mathbf{B}(\mathbf{x})$ at the point $\mathbf{x}$ is proportional to the solid angle $\Omega(\mathbf{x})$ under which the loop is seen from $\mathbf{x}$. In any single-valued definition of $\Omega(\mathbf{x})$, there is some surface S across which $\Omega(\mathbf{x})$ jumps by 4π. In the multivalued definition, this surface is absent.

Let us now derive the same result from a scalar field. Let $\Omega(\mathbf{x}; S)$ be the solid angle under which the current loop L is seen from the point $\mathbf{x}$ (see Fig. 10.5). If S denotes an arbitrary smooth surface enclosed by the loop L, and $d\mathbf{S}'$ a surface element, then $\Omega(\mathbf{x}; S)$ can be calculated from the surface integral

$$\Omega(\mathbf{x}; S) = \int_S \frac{d\mathbf{S}' \cdot \mathbf{R}'}{R'^3}. \tag{10A.13}$$

The argument S in $\Omega(\mathbf{x}; S)$ emphasizes that the definition depends on the choice of the surface S. The range of $\Omega(\mathbf{x}; S)$ is from -2π to 2π, as can be most easily be seen if L lies in the xy-plane and S is chosen to lie in the same place. Then we find for $\Omega(\mathbf{x}; S)$ the value 2π for $\mathbf{x}$ just below S, and -2π just above. Let us calculate from (10A.13) the vector field

$$\mathbf{B}(\mathbf{x}; S) = I\boldsymbol{\nabla}\Omega(\mathbf{x}; S). \tag{10A.14}$$

For this we rewrite

$$\boldsymbol{\nabla}\Omega(\mathbf{x}; S) = \int_S dS'_k \boldsymbol{\nabla}\frac{R'_k}{R'^3} = -\int_S dS'_k \boldsymbol{\nabla}'\frac{R'_k}{R'^3}, \tag{10A.15}$$

which can be rearranged to

$$\boldsymbol{\nabla}\Omega(\mathbf{x}; S) = -\left[\int_S \left(dS'_k \partial'_i \frac{R'_k}{R'^3} - dS'_i \partial'_k \frac{R'_k}{R'^3}\right) + \int_S dS'_i \partial'_k \frac{R'_k}{R'^3}\right]. \tag{10A.16}$$

With the help of Stokes' theorem

$$\int_S (dS_k \partial_i - dS_i \partial_k) f(\mathbf{x}) = \epsilon_{kil} \int_L dx_l f(\mathbf{x}), \tag{10A.17}$$

and the relation $\partial'_k (R'_k/R'^3) = 4\pi\delta^{(3)}(\mathbf{x} - \mathbf{x}')$, we obtain

$$\boldsymbol{\nabla}\Omega(\mathbf{x}; S) = -\left[\int_L \frac{d\mathbf{x}' \times \mathbf{R}'}{R'^3} + 4\pi \int_S d\mathbf{S}'\delta^{(3)}(\mathbf{x} - \mathbf{x}')\right]. \tag{10A.18}$$

Multiplying the first term by I, we reobtain the magnetic field (10A.12) of the current I. The second term yields the singular magnetic field of an infinitely thin magnetic dipole layer lying on the arbitrarily chosen surface S enclosed by L.

The second term is a consequence of the fact that the solid angle $\Omega(\mathbf{x}; S)$ was defined by the surface integral (10A.13). If $\mathbf{x}$ crosses the surface S, the solid angle jumps by 4π.

It is useful to re-express Eq. (10A.15) in a slightly different way. By analogy with (10A.8) we define a δ-function on a surface as

$$\boldsymbol{\delta}(\mathbf{x}; S) = \int_S d\mathbf{S}'\, \delta^{(3)}(\mathbf{x} - \mathbf{x}'), \tag{10A.19}$$

and observe that Stokes' theorem (10A.17) can be written as an identity for δ-functions:

$$\boldsymbol{\nabla} \times \boldsymbol{\delta}(\mathbf{x}; S) = \boldsymbol{\delta}(\mathbf{x}; L), \tag{10A.20}$$

where L is the boundary of the surface S. This equation proves once more the zero divergence (10A.9).

Using the δ-function (10A.19) on the surface S, Eq. (10A.15) can be rewritten as

$$\boldsymbol{\nabla}\Omega(\mathbf{x}; S) = -\int d^3x'\, \delta_k(\mathbf{x}'; S)\boldsymbol{\nabla}' \frac{R'_k}{R'^3}, \tag{10A.21}$$

and if we used also Eq. (10A.20), we find from (10A.18) a magnetic field

$$\mathbf{B}_i(\mathbf{x}; S) = -I\left[\int d^3x'[\boldsymbol{\nabla} \times \boldsymbol{\delta}(\mathbf{x}; S)] \times \frac{\mathbf{R}'}{R'^3} + 4\pi\boldsymbol{\delta}(\mathbf{x}'; S)\right]. \tag{10A.22}$$

Stokes theorem written in the form (10A.20) displays an important property. If we move the surface S to S' with the same boundary, the δ-function $\boldsymbol{\delta}(\mathbf{x}; S)$ changes by

$$\boldsymbol{\delta}(\mathbf{x}; S) \to \boldsymbol{\delta}(\mathbf{x}; S') = \boldsymbol{\delta}(\mathbf{x}; S) + \boldsymbol{\nabla}\delta(\mathbf{x}; V), \tag{10A.23}$$

where

$$\delta(\mathbf{x}; V) \equiv \int d^3x'\, \delta^{(3)}(\mathbf{x} - \mathbf{x}'), \tag{10A.24}$$

and V is the volume over which the surface has swept. Under this transformation, the curl on the left-hand side of (10A.20) is invariant. Comparing (10A.23) with (10A.2) we identify (10A.23) as a novel type of gauge transformation.[9] The magnetic field in the first term of (10A.22) is invariant under this, the second is not. It is then obvious how to find a gauge-invariant magnetic field: we simply subtract the singular S-dependent term and form

$$\mathbf{B}(\mathbf{x}) = I\left[\boldsymbol{\nabla}\Omega(\mathbf{x}; S) + 4\pi\boldsymbol{\delta}(\mathbf{x}; S)\right]. \tag{10A.25}$$

This field is independent of the choice of S and coincides with the magnetic field (10A.12) derived in the usual gauge theory. Hence the description of the magnetic field as a gradient of field $\Omega(\mathbf{x}; S)$ is completely equivalent to the usual gauge field description in terms of the vector potential $\mathbf{A}(\mathbf{x})$. Both are gauge theories, but of a completely different type.

The gauge freedom (10A.23) can be used to move the surface S into a standard configuration. One possibility is the make *gauge fixing*. choose S so that the third component of $\boldsymbol{\delta}(\mathbf{x}; S)$ vanishes. This is called the *axial gauge*. If $\boldsymbol{\delta}(\mathbf{x}; S)$ does not have this property, we can always shift S by a volume V determined by the equation

$$\delta(V) = -\int_{-\infty}^z \delta_z(\mathbf{x}; S), \tag{10A.26}$$

---

[9]For a discussion of this gauge freedom, which is independent of the electromagnetic one, see Ref. [56].

and the transformation (10A.23) will produce a $\delta(\mathbf{x}; S)$ in the axial gauge, to be denoted by $\delta^{\mathrm{ax}}(\mathbf{x}; S)$

Equation (10A.25) suggests defining a solid angle $\Omega(\mathbf{x})$ which is independent of S and depends only on the boundary L of S:

$$\boldsymbol{\nabla}\Omega(\mathbf{x}; L) \equiv \boldsymbol{\nabla}\Omega(\mathbf{x}; S) + 4\pi\boldsymbol{\delta}(\mathbf{x}; S). \tag{10A.27}$$

This is is the analytic continuation of $\Omega(\mathbf{x}; S)$ through the surface S which removes the jump and produces a *multivalued function* $\Omega(\mathbf{x}; L)$ ranging from $-\infty$ to ∞. At each point in space, there are infinitely many Riemann sheets starting from a singularity at L. The values of $\Omega(\mathbf{x}; L)$ on the sheets differ by integer multiples of 4π. From this multivalued function, the magnetic field (10A.12) can be obtained as a simple gradient:

$$\mathbf{B}(\mathbf{x}) = I\,\boldsymbol{\nabla}\Omega(\mathbf{x}; L). \tag{10A.28}$$

Ampère's law (10A.1) implies that the multivalued solid angle $\Omega(\mathbf{x}; L)$ satisfies the equation

$$(\partial_i\partial_j - \partial_j\partial_i)\Omega(\mathbf{x}; L) = 4\pi\epsilon_{ijk}\delta_k(\mathbf{x}; L). \tag{10A.29}$$

Thus, as a consequence of its multivaluedness, $\Omega(\mathbf{x}; L)$ violates the Schwarz integrability condition as the coordinate transformations do in Eq. (10.19). This makes it an unusual mathematical object to deal with. It is, however, perfectly suited to describe the physics.

To see explicitly how Eq. (10A.29) is fulfilled by $\Omega(\mathbf{x}; L)$, let us go to two dimensions where the loop corresponds to two points (in which the loop intersects a plane). For simplicity, we move one of them to infinity, and place the other at the coordinate origin. The role of the solid angle $\Omega(\mathbf{x}; L)$ is now played by the azimuthal angle $\varphi(\mathbf{x})$ of the point $\mathbf{x}$:

$$\varphi(\mathbf{x}) = \arctan\frac{x^2}{x^1}. \tag{10A.30}$$

The function $\arctan(x^2/x^1)$ is usually made unique by cutting the $\mathbf{x}$-plane from the origin along some line C to infinity, preferably along a straight line to $\mathbf{x} = (-\infty, 0)$, and assuming $\varphi(\mathbf{x})$ to jump from π to $-\pi$ when crossing the cut. The cut corresponds to the magnetic dipole surface S in the integral (10A.13). In contrast to this, we shall take $\varphi(\mathbf{x})$ to be the *multivalued* analytic continuation of this function. Then the derivative ∂_i yields

$$\partial_i\varphi(\mathbf{x}) = -\epsilon_{ij}\frac{x_j}{(x^1)^2 + (x^2)^2}. \tag{10A.31}$$

With the single-valued definition of $\partial_i\varphi(\mathbf{x})$, there would have been a δ-function $\epsilon_{ij}\delta_j(C; \mathbf{x})$ across the cut C, corresponding to the second term in (10A.18). When integrating the curl of (10A.31) across the surface s of a small circle c around the origin, we obtain by Stokes' theorem

$$\int_s d^2x(\partial_i\partial_j - \partial_j\partial_i)\varphi(\mathbf{x}) = \int_c dx_i\partial_i\varphi(\mathbf{x}), \tag{10A.32}$$

which is equal to 2π in the multivalued definition of $\varphi(\mathbf{x})$. This result implies the violation of the integrability condition as in (10A.41):

$$(\partial_1\partial_2 - \partial_2\partial_1)\varphi(\mathbf{x}) = 2\pi\delta^{(2)}(\mathbf{x}), \tag{10A.33}$$

whose three-dimensional generalization is (10A.29). In the single-valued definition with the jump by 2π across the cut, the right-hand side of (10A.32) would vanish, making $\varphi(\mathbf{x})$ satisfy the integrability condition (10A.29).

On the basis of Eq. (10A.33) we may construct a Green function for solving the corresponding differential equation with an arbitrary source, which is a superposition of infinitesimally thin line-like currents piercing the two-dimensional space at the points $\mathbf{x}_n$:

$$j(\mathbf{x}) = \sum_n I_n\delta^{(2)}(\mathbf{x} - \mathbf{x}_n), \tag{10A.34}$$

where I_n are currents. We may then easily solve the differential equation

$$(\partial_1\partial_2 - \partial_2\partial_1)f(\mathbf{x}) = j(\mathbf{x}), \tag{10A.35}$$

with the help of the Green function

$$G(\mathbf{x}, \mathbf{x}') = \frac{1}{2\pi}\varphi(\mathbf{x} - \mathbf{x}') \tag{10A.36}$$

which satisfies

$$(\partial_1\partial_2 - \partial_2\partial_1)G(\mathbf{x} - \mathbf{x}') = \delta^{(2)}(\mathbf{x} - \mathbf{x}'). \tag{10A.37}$$

The solution of (10A.35) is obviously

$$f(\mathbf{x}) = \int d^2\mathbf{x}'\, G(\mathbf{x}, \mathbf{x}')j(\mathbf{x}). \tag{10A.38}$$

The gradient of $f(\mathbf{x})$ yields the magnetic field of an arbitrary set of line-like currents vertical to the plane under consideration.

It is interesting to realize that the Green function (10A.36) is the imaginary part of the complex function $(1/2\pi)\log(z - z')$ with $z = x^1 + ix^2$, whose real part $(1/2\pi)\log|z - z'|$ is the Green function $G_\Delta(\mathbf{x} - \mathbf{x}')$ of the two dimensional Poisson equation:

$$(\partial_1^2 + \partial_2^2)G_\Delta(\mathbf{x} - \mathbf{x}') = \delta^{(2)}(\mathbf{x} - \mathbf{x}'). \tag{10A.39}$$

It is important to point out that the superposition of line-like currents cannot be smeared out into a continuous distribution. The integral (10A.38) yields the superposition of multivalued functions

$$f(\mathbf{x}) = \frac{1}{2\pi}\sum_n I_n \arctan\frac{x^2 - x_n^2}{x^1 - x_n^1}, \tag{10A.40}$$

which is properly defined only if one can clearly continue it analytically into the all parts of the Riemann sheets defined by the endpoints of the cut at the origin. If we were to replace the sum by an integral, this possibility would be lost. Thus it is, strictly speaking, impossible to represent arbitrary continuous magnetic fields as gradients of superpositions of scalar potentials $\Omega(\mathbf{x}; L)$. This, however, is not a severe disadvantage of this representation since any current can be approximated by a superposition of line-like currents with any desired accuracy, and the same will be true for the associated magnetic fields.

The arbitrariness of the shape of the jumping surface is the origin of a further interesting gauge structure which has interesting physical consequences discussed in Subsection 10A.5.

10A.2 Generating Magnetic Fields by Multivalued Gauge Transformations

After this first exercise in multivalued functions, we now turn to another example in magnetism which will lead directly to our intended geometric application. We observed before that the local gauge transformation (10A.2) produces the same magnetic field $\mathbf{B}(\mathbf{x}) = \boldsymbol{\nabla} \times \mathbf{A}(\mathbf{x})$ only, as long as the function $\Lambda(\mathbf{x})$ satisfies the Schwarz integrability criterion (10A.29)

$$(\partial_i\partial_j - \partial_j\partial_i)\Lambda(\mathbf{x}) = 0. \tag{10A.41}$$

Any function $\Lambda(\mathbf{x})$ violating this condition would change the magnetic field by

$$\Delta B_k(\mathbf{x}) = \epsilon_{kij}(\partial_i\partial_j - \partial_j\partial_i)\Lambda(\mathbf{x}), \tag{10A.42}$$

thus being no proper gauge function. The gradient of $\Lambda(\mathbf{x})$

$$\mathbf{A}(\mathbf{x}) = \boldsymbol{\nabla}\Lambda(\mathbf{x}) \tag{10A.43}$$

would be a *nontrivial* vector potential.

By analogy with the multivalued coordinate transformations violating the integrability conditions of Schwarz as in (10A.29), the function $\Lambda(\mathbf{x})$ will be called *nonholonomic gauge function*.

Having just learned how to deal with multivalued functions we may change our attitude towards gauge transformations and decide to generate *all* magnetic fields approximately in a field-free space by such improper gauge transformations $\Lambda(\mathbf{x})$. By choosing for instance

$$\Lambda(\mathbf{x}) = \Phi\Omega(\mathbf{x}; L), \tag{10A.44}$$

we find from (10A.29) that this generates a field

$$B_k(\mathbf{x}) = \epsilon_{kij}(\partial_i\partial_j - \partial_j\partial_i)\Lambda(\mathbf{x}) = \Phi\delta_k(\mathbf{x}; L). \tag{10A.45}$$

This is a magnetic field of total flux Φ inside an infinitesimal tube. By a superposition of such infinitesimally thin flux tubes analogous to (10A.38) we can obviously generate a discrete approximation to any desired magnetic field in a field-free space.

10A.3   Magnetic Monopoles

Multivalued fields have also been used to describe magnetic monopoles [10, 13, 14]. A monopole charge density $\rho_{\rm m}(\mathbf{x})$ is the source of a magnetic field $\mathbf{B}(\mathbf{x})$ as defined by the equation

$$\boldsymbol{\nabla}\cdot\mathbf{B}(\mathbf{x}) = 4\pi\rho_{\rm m}(\mathbf{x}). \tag{10A.46}$$

If $\mathbf{B}(\mathbf{x})$ is expressed in terms of a vector potential $\mathbf{A}(\mathbf{x})$ as $\mathbf{B}(\mathbf{x}) = \boldsymbol{\nabla}\times\mathbf{A}(\mathbf{x})$, equation (10A.46) implies the noncommutativity of derivatives in front of the vector potential $\mathbf{A}(\mathbf{x})$:

$$\frac{1}{2}\epsilon_{ijk}(\partial_i\partial_j - \partial_j\partial_i)A_k(\mathbf{x}) = 4\pi\rho_{\rm m}(\mathbf{x}). \tag{10A.47}$$

Thus $\mathbf{A}(\mathbf{x})$ must be multivalued. Dirac in his famous theory of monopoles [16] made the field single-valued by attaching to the world line of the particle a jumping world surface, whose intersection with a coordinate plane at a fixed time forms the *Dirac string*, along which the magnetic field of the monopole is imported from infinity. This world surface can be made physically irrelevant by quantizing it appropriately with respect to the charge. Its shape in space is just as irrelevant as that of the jumping surface S in Fig. 10.5. The invariance under shape deformations constitute once more a second gauge structure of the type mentioned earlier and discussed in Refs. [2, 6, 7, 10, 12].

Once we allow ourselves to work with multivalued fields, we may easily go one step further and express also $\mathbf{A}(\mathbf{x})$ as a gradient of a scalar field as in (10A.43). Then the condition becomes

$$\epsilon_{ijk}\partial_i\partial_j\partial_k\Lambda(\mathbf{x}) = 4\pi\rho_{\rm m}(\mathbf{x}). \tag{10A.48}$$

Let us construct the field of a magnetic monopole of charge g at a point $\mathbf{x}_0$, which satisfies (10A.46) with $\rho_m(\mathbf{x}) = g\,\delta^{(3)}(\mathbf{x}-\mathbf{x}_0)$. Physically, this can be done only by setting up an infinitely thin solenoid along an arbitrary line $L_0^\uparrow$ whose initial point lies at $\mathbf{x}_0$ and the final anywhere at infinity. The superscript $\uparrow$ indicates that the line has a strating point $\mathbf{x}_0$. Inside this solenoid, the magnetic field is infinite, equal to

$$\mathbf{B}_{\rm inside}(\mathbf{x}; L) = 4\pi g\,\boldsymbol{\delta}(\mathbf{x}; L_0^\uparrow), \tag{10A.49}$$

where $\boldsymbol{\delta}(\mathbf{x}; L_0^\uparrow)$ is the open-ended version of (10A.8)

$$\boldsymbol{\delta}(\mathbf{x}; L_0^\uparrow) = \int_{L_0^\uparrow, \mathbf{x}_0} d^3x'\,\delta^{(3)}(\mathbf{x}-\mathbf{x}'). \tag{10A.50}$$

The divergence of this function is concentrated at the starting point:

$$\boldsymbol{\nabla} \cdot \boldsymbol{\delta}(\mathbf{x}; L_0^{\uparrow}) = \delta^{(3)}(\mathbf{x} - \mathbf{x}_0). \tag{10A.51}$$

This follows from (10A.10) by moving the end point to infinity. By analogy with the curl relation (10A.20) we observe a further gauge invariance. If we deform the line L, at fixed initial point $\mathbf{x}$, the δ-function (10A.50) changes as follows:

$$\boldsymbol{\delta}(\mathbf{x}; L_0^{\uparrow}) \rightarrow \boldsymbol{\delta}(\mathbf{x}; L_0^{\prime\uparrow}) = \boldsymbol{\delta}(\mathbf{x}; L_0^{\uparrow}) + \boldsymbol{\nabla} \times \boldsymbol{\delta}(\mathbf{x}; S), \tag{10A.52}$$

where S is the surface over which $L^{\uparrow}$ has swept on its way to $L'^{\uparrow}$. Under this gauge transformation, the relation (10A.51) is obviously invariant. We shall call this *monopole gauge invariance*. The flux (10A.49) inside the solenoid is therefore a *monopole gauge field*. It is straightforward to construct from it the ordinary gauge field $\mathbf{A}(\mathbf{x})$ of the monopole. First we define the $L_0^{\uparrow}$-dependent field

$$\mathbf{A}(\mathbf{x}; L_0^{\uparrow}) = -g \int d^3x' \frac{\boldsymbol{\nabla}' \times \boldsymbol{\delta}(\mathbf{x}'; L_0^{\uparrow})}{R'} = g \int d^3x' \, \boldsymbol{\delta}(\mathbf{x}'; L_0^{\uparrow}) \times \frac{\mathbf{R}'}{R'^3}. \tag{10A.53}$$

The curl of the first expression is

$$\boldsymbol{\nabla} \times \mathbf{A}(\mathbf{x}; L_0^{\uparrow}) = -g \int d^3x' \frac{\boldsymbol{\nabla}' \times [\boldsymbol{\nabla}' \times \boldsymbol{\delta}(\mathbf{x}'; L_0^{\uparrow})]}{R'}, \tag{10A.54}$$

and consists of two terms

$$-g \int d^3x' \frac{\boldsymbol{\nabla}'[\boldsymbol{\nabla}' \cdot \boldsymbol{\delta}(\mathbf{x}'; L_0^{\uparrow})]}{R'} + g \int d^3x' \frac{\boldsymbol{\nabla}'^2 \boldsymbol{\delta}(\mathbf{x}'; L_0^{\uparrow})}{R'}. \tag{10A.55}$$

After an integration by parts, and using (10A.51), the first term is $L_0^{\uparrow}$-independent and reads

$$g \int d^3x' \delta^{(3)}(\mathbf{x} - \mathbf{x}_0) \boldsymbol{\nabla}' \frac{1}{R'} = g \frac{\mathbf{x} - \mathbf{x}_0}{|\mathbf{x} - \mathbf{x}_0|^3}. \tag{10A.56}$$

The second term becomes, after two integration by parts,

$$-4\pi g \, \boldsymbol{\delta}(\mathbf{x}'; L_0^{\uparrow}). \tag{10A.57}$$

The first term is the desired magnetic field of the monopole. Its divergence is $\delta(\mathbf{x} - \mathbf{x}_0)$, which we wanted to archive. The second term is the monopole gauge field, the magnetic field inside the solenoid. The total divergence of this field is, of course, zero.

By analogy with (10A.25) we now subtract the latter term and find the magnetic field of the monopole

$$\mathbf{B}(\mathbf{x}) = \boldsymbol{\nabla} \times \mathbf{A}(\mathbf{x}; L_0^{\uparrow}) + 4\pi g \, \boldsymbol{\delta}(\mathbf{x}; L_0^{\uparrow}). \tag{10A.58}$$

This field is independent of the string $L_0^{\uparrow}$. It depends only on the source point $\mathbf{x}_0$ and satisfies $\boldsymbol{\nabla} \cdot \mathbf{B}(\mathbf{x}) = 4\pi g \, \delta^{(3)}(\mathbf{x} - \mathbf{x}_0)$.

Let us calculate the vector potential for some simple choices of $\mathbf{x}_0$ and $L_0^{\uparrow}$, for instance $\mathbf{x}_0 = \mathbf{0}$ and $L_0^{\uparrow}$ along the positive z-axis, so that $\boldsymbol{\delta}(\mathbf{x}; L_0^{\uparrow})$ becomes $\hat{\mathbf{z}} \Theta(z)\delta(x)\delta(y)$, where $\hat{\mathbf{z}}$ is the unit vector in z-direction. Inserting this into the second expression in (10A.53) yields

$$\begin{aligned}
\mathbf{A}^{(g)}(\mathbf{x}; L_0^{\uparrow}) &= -g \int_0^{\infty} dz' \frac{\hat{\mathbf{z}} \times \mathbf{x}}{\sqrt{x^2 + y^2 + (z' - z)^2}^{3/2}} \\
&= -g \frac{\hat{\mathbf{z}} \times \mathbf{x}}{r(r - z)} = g \frac{(y, -x, 0)}{r(r - z)}.
\end{aligned} \tag{10A.59}$$

If $L_0^\uparrow$ runs to $-\infty$, so that $\boldsymbol{\delta}(\mathbf{x}; L_0^\uparrow)$ is equal to $-\hat{\mathbf{z}}\,\Theta(-z)\delta(x)\delta(y)$, we obtain

$$\mathbf{A}^{(g)}(\mathbf{x}; L_0^\uparrow) = g \int_{-\infty}^{0} dz' \frac{\hat{\mathbf{z}} \times \mathbf{x}}{\sqrt{x^2 + y^2 + (z' - z)^2}^{\,3/2}}$$

$$= g \frac{\hat{\mathbf{z}} \times \mathbf{x}}{r(r + z)} = -g \frac{(y, -x, 0)}{r(r + z)}. \qquad (10A.60)$$

The vector potential has only azimuthal components. If we parametrize (x, y, z) in terms of spherical angles θ, φ as $r(\sin\theta\cos\varphi, \sin\theta\sin\varphi, \cos\theta)$, these are

$$A_\varphi^{(g)}(\mathbf{x}; L_0^\uparrow) = \frac{g}{r\sin\theta}(1 - \cos\theta) \quad \text{or} \quad A_\varphi^{(g)}(\mathbf{x}; L_0^\uparrow) = -\frac{g}{r\sin\theta}(1 + \cos\theta), \qquad (10A.61)$$

respectively.

The shape of the line $L_0^\uparrow$ can be brought to a standard form, which corresponds to fixing a gauge of the field $\boldsymbol{\delta}(\mathbf{x}; L_0^\uparrow)$. For example, we may always choose $L_0^\uparrow$ to run from $\mathbf{x}_0$ into the z-direction.

Also here, there exists an equivalent formulation in terms of a multivalued $\mathbf{A}$-field with infinitely many Riemann sheets around the line L. For a detailed discussion of the physics of multivalued fields see Refs. [2, 6, 7, 10, 12].

An interesting observation is the following: If the gauge function $\Lambda(\mathbf{x})$ is considered as a nonholonomic displacement in some fictitious crystal dimension, then the magnetic field of a current loop which gives rise to noncommuting derivatives $(\partial_i\partial_j - \partial_j\partial_i)\Lambda(\mathbf{x}) \neq 0$ is the analog of a dislocaton [compare (10.23)], and thus implies torsion in the crystal. A magnetic monopole, on the other hand, arises from noncommuting derivatives $(\partial_i\partial_j - \partial_j\partial_i)\partial_k\Lambda(\mathbf{x}) \neq 0$ in Eq. (10A.48). It corresponds to a disclination [see (10.57)] and implies curvature. The defects in the multivalued description of magnetism are therefore similar to those in a crystal where dislocations much more abundantly observed than disclinations. They are opposite to those in general relativity which is governed by curvature alone, with no evidence for torsion so far [11].

10A.4 Minimal Magnetic Coupling of Particles from Multivalued Gauge Transformations

Multivalued gauge transformations are the perfect tool to minimally couple electromagnetism to any type of matter. Consider for instance a free nonrelativistic point particle with a Lagrangian

$$L = \frac{M}{2}\dot{\mathbf{x}}^2. \qquad (10A.62)$$

The equations of motion are invariant under a gauge transformation

$$L \to L' = L + \boldsymbol{\nabla}\Lambda(\mathbf{x})\,\dot{\mathbf{x}}, \qquad (10A.63)$$

since this changes the action $\mathcal{A} = \int_{t_a}^{t_b} dt\, L$ merely by a surface term:

$$\mathcal{A}' \to \mathcal{A} = \mathcal{A} + \Lambda(\mathbf{x}_b) - \Lambda(\mathbf{x}_a). \qquad (10A.64)$$

The invariance is absent if we take $\Lambda(\mathbf{x})$ to be a multivalued gauge function. In this case, a nontrivial vector potential $\mathbf{A}(\mathbf{x}) = \boldsymbol{\nabla}\Lambda(\mathbf{x})$ (working in natural units with $e = 1$) is created in the field-free space, and the nonholonomically gauge-transformed Lagrangian corresponding to (10A.63),

$$L' = \frac{M}{2}\dot{\mathbf{x}}^2 + \mathbf{A}(\mathbf{x})\,\dot{\mathbf{x}}, \qquad (10A.65)$$

describes correctly the dynamics of a free particle in an external magnetic field.

The coupling derived by multivalued gauge transformations is automatically invariant under additional ordinary single-valued gauge transformations of the vector potential

$$\mathbf{A}(\mathbf{x}) \rightarrow \mathbf{A}'(\mathbf{x}) = \mathbf{A}(\mathbf{x}) + \boldsymbol{\nabla}\Lambda(\mathbf{x}), \tag{10A.66}$$

since these add to the Lagrangian (10A.65) once more the same pure derivative term which changes the action by an irrelevant surface term as in (10A.64).

The same procedure leads in quantum mechanics to the minimal coupling of the Schrödinger field $\psi(\mathbf{x})$. The action is $\mathcal{A} = \int dt d^3x\, L$ with a Lagrange density (in natural units with $\hbar = 1$)

$$L = \psi^*(\mathbf{x}) \left(i\partial_t + \frac{1}{2M}\boldsymbol{\nabla}^2 \right) \psi(\mathbf{x}). \tag{10A.67}$$

The physics described by a Schrödinger wave function $\psi(\mathbf{x})$ is invariant under arbitrary local phase changes

$$\psi(\mathbf{x}, t) \rightarrow \psi'(\mathbf{x}) = e^{i\Lambda(\mathbf{x})}\psi(\mathbf{x}, t), \tag{10A.68}$$

called local U(1) transformations. This implies that the Lagrange density (10A.67) may equally well be replaced by the gauge-transformed one

$$L = \psi^*(\mathbf{x}, t) \left(i\partial_t + \frac{1}{2M}\mathbf{D}^2 \right) \psi(\mathbf{x}, t), \tag{10A.69}$$

where $-i\mathbf{D} \equiv -i\boldsymbol{\nabla} - \boldsymbol{\nabla}\Lambda(\mathbf{x})$ is the operator of physical momentum.

We may now go over to nonzero magnetic fields by admitting gauge transformations with multivalued $\Lambda(\mathbf{x})$ whose gradient is a nontrivial vector potential $\mathbf{A}(\mathbf{x})$ as in (10A.43). Then $-i\mathbf{D}$ turns into the covariant momentum operator

$$\hat{\mathbf{P}} = -i\mathbf{D} = -i\boldsymbol{\nabla} - \mathbf{A}(\mathbf{x}), \tag{10A.70}$$

and the Lagrange density (10A.69) describes correctly the magnetic coupling in quantum mechanics.

As in the classical case, the coupling derived by multivalued gauge transformations is automatically invariant under ordinary single-valued gauge transformations under which the vector potential $\mathbf{A}(\mathbf{x})$ changes as in (10A.66), whereas the Schrödinger wave function undergoes a local U(1)-transformation (10A.68). This invariance is a direct consequence of the simple transformation behavior of $\mathbf{D}\psi(\mathbf{x}, t)$ under gauge transformations (10A.66) and (10A.68) which is

$$\mathbf{D}\psi(\mathbf{x}, t) \rightarrow \mathbf{D}\psi'(\mathbf{x}, t) = e^{i\Lambda(\mathbf{x})}\mathbf{D}\psi(\mathbf{x}, t). \tag{10A.71}$$

Thus $\mathbf{D}\psi(\mathbf{x}, t)$ transforms just like $\psi(\mathbf{x}, t)$ itself, and for this reason, $\mathbf{D}$ is called gauge-covariant derivative. The generation of magnetic fields by a multivalued gauge transformation is the simplest example for the power of the nonholonomic mapping principle.

After this discussion it is quite suggestive to introduce the same mathematics into differential geometry, where the role of gauge transformations is played by reparametrizations of the space coordinates. This is precisely what is done in Subsection (10.2.2).

10A.5   Gauge Field Representation of Current Loops and Monopoles

In the previous subsections we have given examples for the use of multivalued fields in describing magnetic phenomena. The nonholonomic gauge transformations by which we created line-like nonzero field configurations were shown to be the natural origin of the minimal couplings to the classical actions as well as to the Schrödinger equation. It is interesting to observe that there exists a fully fledged theory of magnetism (which is easily generalized to electromagnetism) with these multivalued fields, if we properly handle the freedom in choosing the jumping surfaces S whose boundary represents the physical current loop in Eq. (10A.12).

To understand this we pose ourselves the problem of setting up an action formalism for calculating the magnetic energy of a current loop in the gradient representation of the magnetic field. In this Euclidean field theory, the action is the field energy:

$$\mathcal{E} = \frac{1}{8\pi} \int d^3x \, \mathbf{B}^2(\mathbf{x}). \tag{10A.72}$$

Inserting the gradient representation (10A.28) of the magnetic field, we can write

$$\mathcal{E} = \frac{I^2}{8\pi} \int d^3x \, [\boldsymbol{\nabla}\Omega(\mathbf{x})]^2. \tag{10A.73}$$

This holds for the multivalued solid angle $\Omega(\mathbf{x})$ which is independent of S. In order to perform field theoretic calculations, we must go over to the single-valued representation (10A.25), so that

$$\mathcal{E} = \frac{I^2}{8\pi} \int d^3x \, [\boldsymbol{\nabla}\Omega(\mathbf{x}; S) + 4\pi\boldsymbol{\delta}(\mathbf{x}; S)]^2. \tag{10A.74}$$

The δ-function removes the unphysical field energy on the artificial magnetic dipole layer on S. Let us calculate the magnetic field energy of the current loop from the action (10A.74). For this we rewrite the action (10A.74) in terms of an *auxiliary* vector field $\mathbf{B}(\mathbf{x})$ as

$$\mathcal{E} = \int d^3x \left\{ -\frac{1}{8\pi}\mathbf{B}^2(\mathbf{x}) - \frac{I}{4\pi}\mathbf{B}(\mathbf{x}) \cdot [\boldsymbol{\nabla}\Omega(\mathbf{x}; S) + 4\pi\boldsymbol{\delta}(\mathbf{x}; S)] \right\}. \tag{10A.75}$$

A partial integration brings the second term to

$$\int d^3x \, \frac{I}{4\pi} \boldsymbol{\nabla} \cdot \mathbf{B}(\mathbf{x}) \, \Omega(\mathbf{x}; S).$$

Extremizing this in $\Omega(\mathbf{x})$ yields the equation

$$\boldsymbol{\nabla} \cdot \mathbf{B}(\mathbf{x}) = 0, \tag{10A.76}$$

implying that the field lines of $\mathbf{B}(\mathbf{x})$ form closed loops. This equation may be enforced identically (as a Bianchi identity) by expressing $\mathbf{B}(\mathbf{x})$ as a curl of an auxiliary vector potential $\mathbf{A}(\mathbf{x})$, setting

$$\mathbf{B}(\mathbf{x}) \equiv \boldsymbol{\nabla} \times \mathbf{A}(\mathbf{x}). \tag{10A.77}$$

With this ansatz, the equation which brings the action (10A.75) to the form

$$\mathcal{E} = \int d^3x \left\{ -\frac{1}{8\pi}[\boldsymbol{\nabla} \times \mathbf{A}(\mathbf{x})]^2 - I\,[\boldsymbol{\nabla} \times \mathbf{A}(\mathbf{x})] \cdot \boldsymbol{\delta}(\mathbf{x}; S) \right\}. \tag{10A.78}$$

A further partial integration leads to

$$\mathcal{E} = \int d^3x \left\{ -\frac{1}{8\pi}[\boldsymbol{\nabla} \times \mathbf{A}(\mathbf{x})]^2 - I\,\mathbf{A}(\mathbf{x}) \cdot [\boldsymbol{\nabla} \times \boldsymbol{\delta}(\mathbf{x}; S)] \right\}, \tag{10A.79}$$

and we identify in the linear term in $\mathbf{A}(\mathbf{x})$ the *auxiliary current*

$$\mathbf{j}(\mathbf{x}) \equiv I\,\boldsymbol{\nabla} \times \boldsymbol{\delta}(\mathbf{x}; S) = I\,\boldsymbol{\delta}(\mathbf{x}; L), \tag{10A.80}$$

due to Stoke's law (10A.20). According to Eq. (10A.9), this current is conserved for closed loops L.

By extremizing the action (10A.78), we obtain Ampère's law (10A.1). Thus the auxiliary quantities $\mathbf{B}(\mathbf{x})$, $\mathbf{A}(\mathbf{x})$, and $\mathbf{j}(\mathbf{x})$ coincide with the usual magnetic quantities with the same name. If we insert the explicit solution (10A.5) of Ampère's law into the energy, we obtain the *Biot-Savart* energy for an arbitrary current distribution

$$\mathcal{E} = \frac{1}{2} \int d^3x\, d^3x'\, \mathbf{j}(\mathbf{x})\, \frac{1}{|\mathbf{x} - \mathbf{x}'|}\, \mathbf{j}(\mathbf{x}'). \tag{10A.81}$$

Note that the action (10A.78) is invariant under two mutually dual gauge transformations, the usual magnetic one in (10A.2), by which the vector potential receives a gradient of an arbitrary scalar field, and the gauge transformation (10A.23), by which the irrelevant surface S is moved to another configuration S'.

Thus we have proved the complete equivalence of the gradient representation of the magnetic field to the usual gauge field representation. In the gradient representation, there exists a new type of gauge invariance which expresses the physical irrelevance of the jumping surface appearing when using single-valued solid angles.

The action (10A.79) describes magnetism in terms of a *double gauge theory* [8], in which both the gauge of $\mathbf{A}(\mathbf{x})$ and the shape of S can be changed arbitrarily. By setting up a grand-canonical partition function of many fluctuating surfaces it is possible to describe a large family of phase transitions mediated by the proliferation of line-like defects. Examples are vortex lines in the superfluid-normal transition in helium and dislocation and disclination lines in the melting transition of crystals [2, 6, 7, 10, 12].

Let us now go through the analogous calculation for a gas of monopoles at $\mathbf{x}_n$ from the magnetic energy formed with the field (10A.58):

$$\mathcal{E} = \frac{1}{8\pi} \int d^3x \left[\boldsymbol{\nabla} \times \mathbf{A} + 4\pi g \sum_n \boldsymbol{\delta}(\mathbf{x}; L_n^\uparrow) \right]^2. \tag{10A.82}$$

As in (10A.75) we introduce an auxiliary magnetic field and rewrite (10A.82) as

$$\mathcal{E} = \int d^3x \left\{ -\frac{1}{8\pi} \mathbf{B}^2(\mathbf{x}) - \frac{1}{4\pi} \mathbf{B}(\mathbf{x}) \cdot \left[\boldsymbol{\nabla} \times \mathbf{A} + g \sum_n \boldsymbol{\delta}(\mathbf{x}; L_n^\uparrow) \right] \right\}. \tag{10A.83}$$

Extremizing this in $\mathbf{A}$ yields $\boldsymbol{\nabla} \times \mathbf{B} = 0$, so that we may set $\mathbf{B} = \boldsymbol{\nabla}\Lambda$, and obtain

$$\mathcal{E} = \int d^3x \left\{ -\frac{1}{8\pi} \left[\boldsymbol{\nabla}\Lambda(\mathbf{x}) \right]^2 + g\Lambda(\mathbf{x}) \sum_n \boldsymbol{\nabla} \cdot \boldsymbol{\delta}(\mathbf{x}; L_n^\uparrow) \right\}. \tag{10A.84}$$

Recalling (10A.51), the extremal Λ field is

$$\Lambda(\mathbf{x}) = -\frac{4\pi g}{\boldsymbol{\nabla}^2} \sum_n \delta(\mathbf{x} - \mathbf{x}_n) = g \sum_n \frac{1}{|\mathbf{x} - \mathbf{x}_n|}, \tag{10A.85}$$

which leads, after reinsertion into (10A.84), to the Coulomb interaction energy

$$\mathcal{E} = \frac{g^2}{2} \sum_{n,n'} \frac{1}{|\mathbf{x}_n - \mathbf{x}_{n'}|}. \tag{10A.86}$$

Appendix 10B   Comparison of Multivalued Basis Tetrads and Vierbein Fields

The standard *tetrads* or vierbein fields were introduced a long time ago in gravitational theories of spinning particles both in purely Riemann [16] as well as in Riemann-Cartan spacetimes [17, 18,

19, 20, 2]. Their mathematics is described in detail in the literature [21]. Their purpose was to define at every point a local Lorentz frame by means of another set of coordinate differentials

$$dx^\alpha = h^\alpha_{\ \lambda}(q)dq^\lambda, \quad \alpha = 0,\ 1,\ 2,\ 3, \tag{10B.1}$$

which can be contracted with Dirac matrices γ^α to form locally Lorentz invariant quantities. Local Lorentz frames are reached by requiring the induced metric in these coordinates to be Minkowskian:

$$g_{\alpha\beta} = h_\alpha^{\ \mu}(q)h_\beta^{\ \nu}(q)g_{\mu\nu}(q) = \eta_{\alpha\beta}, \tag{10B.2}$$

where $\eta_{\alpha\beta}$ is the flat Minkowski metric (10.30). Just like $e^i_{\ \mu}(q)$ in (10.12), these vierbeins possess reciprocals

$$h_\alpha^{\ \mu}(q) \equiv \eta_{\alpha\beta}g^{\mu\nu}(q)h^\beta_{\ \nu}(q), \tag{10B.3}$$

and satisfy orthonormality and completeness relations as in (10.13):

$$h_\alpha^{\ \mu}h^\beta_{\ \mu} = \delta_\alpha^{\ \beta}, \qquad h^\alpha_{\ \mu}h_\alpha^{\ \nu} = \delta_\mu^{\ \nu}. \tag{10B.4}$$

They also can be multiplied with each other as in (10.14) to yield the metric

$$g_{\mu\nu}(q) = h^\alpha_{\ \mu}(q)h^\beta_{\ \nu}(q)\eta_{\alpha\beta}. \tag{10B.5}$$

Thus they constitute another "square root" of the metric. The relation between these square roots is some linear transformation

$$e^a_{\ \mu}(q) = e^a_{\ \alpha}(q)h^\alpha_{\ \mu}(q), \tag{10B.6}$$

which must necessarily be a local Lorentz transformation

$$\Lambda^a_{\ \alpha}(q) = e^a_{\ \alpha}(q), \tag{10B.7}$$

since this matrix connects the two Minkowski metrics (10.30) and (10B.2) with each other:

$$\eta_{ab}\Lambda^a_{\ \alpha}(q)\Lambda^b_{\ \beta}(q) = \eta_{\alpha\beta}. \tag{10B.8}$$

The different local Lorentz transformations allow us to choose different local Lorentz frames which distinguish fields with definite spin by the irreducible representations of these transformations. The physical consequences of the theory must be independent of this local choice, and this is the reason why the presence of spinning fields requires the existence of an additional gauge freedom under local Lorentz transformations, in addition to Einstein's invariance under general coordinate transformations. Since the latter may be viewed as local translations, the theory with spinning particles are locally Poincaré invariant.

The vierbein fields $h^\alpha_{\ \mu}(q)$ have in common with ours that both violate the integrability condition as in (10.22), thus describing nonholonomic coordinates dx^α for which there exists only a differential relation (10B.1) to the physical coordinates q^μ. However, they differ from our multivalued tetrads $e^a_{\ \mu}(q)$ by being *single-valued* fields satisfying the integrability condition

$$(\partial_\mu\partial_\nu - \partial_\nu\partial_\mu)h^\alpha_{\ \lambda}(q) = 0, \tag{10B.9}$$

in contrast to our multivalued tetrads $e^i_{\ \lambda}(q)$ in Eq. (10.23).

In the local coordinate system dx^α, curvature arises from a violation of the integrability condition of the local Lorentz transformations (10B.7).

The simple equation (10.24) for the torsion tensor in terms of the multivalued tetrads $e^i_{\ \lambda}(q)$ must be contrasted with a similar-looking, but geometrically quite different, quantity formed from the vierbein fields $h^\alpha_{\ \lambda}(q)$ and their reciprocals, the *objects of anholonomy* [21]:

$$\Omega_{\alpha\beta}^{\ \ \gamma}(q) = \frac{1}{2}h_\alpha^{\ \mu}(q)h_\beta^{\ \nu}(q)\left[\partial_\mu h^\gamma_{\ \nu}(q) - \partial_\nu h^\gamma_{\ \mu}(q)\right]. \tag{10B.10}$$

$$dx^a = e^a{}_\mu dq^\mu \qquad \Gamma_{ab}{}^c = 0$$

$$dx^\alpha = h^\alpha{}_\mu dq^\mu \quad \Gamma_{\alpha\beta}{}^\gamma = h^\gamma{}_\lambda h_\alpha{}^\mu h_\beta{}^\nu K_{\mu\nu}{}^\lambda - \overset{h}{K}{}_{\alpha\beta}{}^\gamma$$

$$dq^\mu \qquad \Gamma_{\mu\nu}{}^\lambda = e_a{}^\lambda \partial_\mu e^a{}_\nu$$

Figure 10.6 Coordinate system q^μ and the two sets of local nonholonomic coordinates dx^α and dx^a. The intermediate coordinates dx^α have a Minkowski metric only at the point q, the coordinates dx^a in an entire small neighborhood (at the cost of a closure failure).

A combination of these similar to (10.28),

$$\overset{h}{K}{}_{\alpha\beta}{}^\gamma(q) = \Omega_{\alpha\beta}{}^\gamma(q) - \Omega_\beta{}^\gamma{}_\alpha(q) + \Omega^\gamma{}_{\alpha\beta}(q), \tag{10B.11}$$

appears in the so-called *spin connection*

$$\Gamma_{\alpha\beta}{}^\gamma = h^\gamma{}_\lambda h_\alpha{}^\mu h_\beta{}^\nu (K_{\mu\nu}{}^\lambda - \overset{h}{K}{}_{\mu\nu}{}^\lambda), \tag{10B.12}$$

which is needed to form a covariant derivative of local vectors

$$v_\alpha(q) = v_\mu(q) h_\alpha{}^\mu(q), \qquad v^\alpha(q) = v^\mu(q) h^\alpha{}_\mu(q). \tag{10B.13}$$

These have the form

$$D_\alpha v_\beta(q) = \partial_\alpha v_\beta(q) - \Gamma_{\alpha\beta}{}^\gamma(q) v_\gamma(q), \qquad D_\alpha v^\beta(q) = \partial_\alpha v^\beta(q) + \Gamma_{\alpha\gamma}{}^\beta(q) v^\gamma(q). \tag{10B.14}$$

For details see Ref. [2, 6]. In spite of the similarity between the defining equations (10.24) and (10B.10), the tensor $\Omega_{\alpha\beta}{}^\gamma(q)$ bears no relation to torsion, and $\overset{h}{K}{}_{\alpha\beta}{}^\gamma(q)$ is independent of the contortion $K_{\alpha\beta}{}^\gamma$. In fact, the objects of anholonomy $\Omega_{\alpha\beta}{}^\gamma(q)$ are in general nonzero in the absence of torsion [22], and may even be nonzero in flat spacetime, where the matrices $h^\alpha{}_\mu(q)$ degenerate to local Lorentz transformations. The quantities $\overset{h}{K}{}_{\alpha\beta}{}^\gamma(q)$, and thus the spin connection (10B.12), characterize the orientation of the local Lorentz frames.

The nonholonomic coordinates dx^α transform the metric $g_{\mu\nu}(q)$ to a Minkowskian form η_{ab} at any given point q^μ. They correspond to a small "falling elevator" of Einstein in which the gravitational forces vanish precisely at the center of mass, the neighborhood still being subject to tidal forces. In contrast, the nonholonomic coordinates dx^a flatten the spacetime *in an entire neighborhood* of the point. This is at the expense of producing defects in spacetime (like those produced when flattening an orange peel by stepping on it), as will be explained in Section IV. The affine connection $\Gamma_{ab}{}^c(q)$ in the latter coordinates dx^a vanishes identically.

The difference between our multivalued tetrads and the usual vierbeins is illustrated in the diagram of Fig. 10.6.

Appendix 10C   Cancellation of Powers of $\delta(0)$

There is a simple way of proving the cancellation of all UV-divergences $\delta(0)$. Consider a free particle whose mass depends on the time with an action

$$\mathcal{A}_{\text{tot}}[q] = \int_0^\beta d\tau \left[\frac{1}{2} Z(\tau) \dot{q}^2(\tau) - \frac{1}{2} \delta(0) \log Z(\tau) \right], \tag{10C.1}$$

where $Z(\tau)$ is some function of τ but independent now of the path $q(\tau)$. The last term is the simplest nontrivial form of the Jacobian action in (10.408). Since it is independent of q, it is conveniently taken out of the path integral as a factor

$$J = e^{(1/2)\delta(0)\int_0^\beta d\tau \log Z(\tau)}. \tag{10C.2}$$

We split the action into a sum of a free and an interacting part

$$\mathcal{A}^{(0)} = \int_0^\beta d\tau \, \frac{1}{2}\dot{q}^2(\tau), \quad \mathcal{A}^{\text{int}} = \int_0^\beta d\tau \, \frac{1}{2}[Z(\tau) - 1]\,\dot{q}^2(\tau), \tag{10C.3}$$

and calculate the transition amplitude (10.411) as a sum of all connected diagrams in the cumulant expansion

$$
\begin{aligned}
(0\,\beta|0\,0) &= J \int \mathcal{D}q(\tau) e^{-\mathcal{A}^{(0)}[q] - \mathcal{A}_{\text{int}}[q]} = J \int \mathcal{D}q(\tau) e^{-\mathcal{A}^{(0)}[q]} \left(1 - \mathcal{A}_{\text{int}} + \frac{1}{2}\mathcal{A}_{\text{int}}^2 - \cdots \right) \\
&= (2\pi\beta)^{-1/2} J \left[1 - \langle \mathcal{A}_{\text{int}} \rangle + \frac{1}{2}\langle \mathcal{A}_{\text{int}}^2 \rangle - \cdots \right] \\
&= (2\pi\beta)^{-1/2} J \, e^{-\langle \mathcal{A}_{\text{int}} \rangle_c + \frac{1}{2}\langle \mathcal{A}_{\text{int}}^2 \rangle_c - \cdots}.
\end{aligned} \tag{10C.4}
$$

We now show that the infinite series of $\delta(0)$-powers appearing in a Taylor expansion of the exponential (10C.2) is precisely compensated by the sum of all terms in the perturbation expansion (10C.4). Being interested only in these singular terms, we may extend the τ-interval to the entire time axis. Then Eq. (10.394) yields the propagator $\dot{\Delta}(\tau, \tau') = \delta(\tau - \tau')$, and we find the first-order expansion term

$$\langle \mathcal{A}_{\text{int}} \rangle_c = \int d\tau \, \frac{1}{2}[Z(\tau) - 1]\,\dot{\Delta}(\tau, \tau) = -\frac{1}{2}\delta(0) \int d\tau \, [1 - Z(\tau)]. \tag{10C.5}$$

To second order, divergent integrals appear involving products of distributions, thus requiring an intermediate extension to d dimensions as follows

$$
\begin{aligned}
\langle \mathcal{A}_{\text{int}}^2 \rangle_c &= \iint d\tau_1 \, d\tau_2 \, \frac{1}{2}(Z-1)_1 \frac{1}{2}(Z-1)_2 \, 2 \, \dot{\Delta}(\tau_1, \tau_2) \, \dot{\Delta}(\tau_2, \tau_1) \\
&\to \iint d^d x_1 \, d^d x_2 \, \frac{1}{2}(Z-1)_1 \frac{1}{2}(Z-1)_2 \, 2 \, {}_\mu\Delta_\nu(x_1, x_2) \, {}_\nu\Delta_\mu(x_2, x_1) \\
&= \iint d^d x_1 \, d^d x_2 \, \frac{1}{2}(Z-1)_1 \frac{1}{2}(Z-1)_2 \, 2 \, \Delta_{\mu\mu}(x_2, x_1) \, \Delta_{\nu\nu}(x_1, x_2),
\end{aligned} \tag{10C.6}
$$

the last line following from partial integrations. For brevity, we have abbreviated $[Z(\tau_i) - 1]$ by $(Z - 1)_i$. Using the field equation (10.418) and going back to one dimension yields, with the further abbreviation $(Z - 1)_i \to z_i$:

$$\langle \mathcal{A}_{\text{int}}^2 \rangle_c = \frac{1}{2} \iint d\tau_1 \, d\tau_2 \, z_1 \, z_2 \, \delta^2(\tau_1, \tau_2). \tag{10C.7}$$

To third order we calculate

$$
\begin{aligned}
\langle \mathcal{A}_{\text{int}}^3 \rangle_c &= \iiint d\tau_1 \, d\tau_2 \, d\tau_3 \, \frac{1}{2}z_1 \frac{1}{2}z_2 \frac{1}{2}z_3 \, 8 \, \dot{\Delta}(\tau_1, \tau_2) \, \dot{\Delta}(\tau_2, \tau_3) \, \dot{\Delta}(\tau_3, \tau_1) \\
&\to \iiint d^d x_1 \, d^d x_2 \, d^d x_3 \, \frac{1}{2}z_1 \frac{1}{2}z_2 \frac{1}{2}z_3 \, 8 \, {}_\mu\Delta_\nu(x_1, x_2) \, {}_\nu\Delta_\sigma(x_2, x_3) \, {}_\sigma\Delta_\mu(x_3, x_1) \\
&= -\iiint d^d x_1 \, d^d x_2 \, d^d x_3 \, \frac{1}{2}z_1 \frac{1}{2}z_2 \frac{1}{2}z_3 \, 8 \, \Delta_{\mu\mu}(x_3, x_1) \, \Delta_{\nu\nu}(x_1, x_2) \, \Delta_{\sigma\sigma}(x_2, x_3).
\end{aligned} \tag{10C.8}
$$

Applying again the field equation (10.418) and going back to one dimension, this reduces to

$$\langle \mathcal{A}_{\text{int}}^3 \rangle_c = \int \int \int d\tau_1 \, d\tau_2 \, d\tau_3 \, z_1 \, z_2 \, z_3 \delta(\tau_1, \tau_2) \, \delta(\tau_2, \tau_3) \, \delta(\tau_3, \tau_1). \tag{10C.9}$$

Continuing to n-order and substituting Eqs. (10C.5), (10C.7), (10C.9), etc. into (10C.4), we obtain in the exponent of Eq. (10C.4) a sum

$$-\langle \mathcal{A}_{\text{int}} \rangle_c + \frac{1}{2} \langle \mathcal{A}_{\text{int}}^2 \rangle_c - \frac{1}{3!} \langle \mathcal{A}_{\text{int}}^3 \rangle_c + \ldots = \frac{1}{2} \sum_1^\infty (-1)^n \frac{c_n}{n}, \tag{10C.10}$$

with

$$c_n = \int d\tau_1 \ldots d\tau_n \, C(\tau_1, \tau_2) \, C(\tau_2, \tau_3) \, \ldots \, C(\tau_n, \tau_1) \tag{10C.11}$$

where

$$C(\tau, \tau') = [Z(\tau) - 1] \, \delta(\tau, \tau'). \tag{10C.12}$$

Substituting this into Eq. (10C.11) and using the rule (10.369) yields

$$c_n = \int \int d\tau_1 d\tau_n \, [Z(\tau_1) - 1]^n \, \delta^2(\tau_1 - \tau_n) = \delta(0) \int d\tau \, [Z(\tau) - 1]^n. \tag{10C.13}$$

Inserting these numbers into the expansion (10C.10), we obtain

$$\begin{aligned}
-\langle \mathcal{A}_{\text{int}} \rangle_c + \frac{1}{2} \langle \mathcal{A}_{\text{int}}^2 \rangle_c - \frac{1}{3!} \langle \mathcal{A}_{\text{int}}^3 \rangle_c + \ldots &= \frac{1}{2} \delta(0) \int d\tau \sum_1^\infty (-1)^n \frac{[Z(\tau) - 1]^n}{n} \\
&= -\frac{1}{2} \delta(0) \int d\tau \, \log Z(\tau), \tag{10C.14}
\end{aligned}$$

which compensates precisely the Jacobian factor J in (10C.4).

Notes and References

Path integrals in spaces with curvature but no torsion have been discussed by
B.S. DeWitt, Rev. Mod. Phys. **29**, 377 (1957).
We do not agree with the measure of path integration in this standard work since it produces a physically incorrect $\hbar^2 \bar{R}/12M$-term in the energy. This has to be subtracted from the Lagrangian before summing over all paths to obtain the correct energy spectrum on surfaces of spheres and on group spaces. Thus, DeWitt's action in the short-time amplitude of the path integral is non-classical, in contrast to the very idea of path integration. A similar criticism holds for
K.S. Cheng, J. Math. Phys. **13**, 1723 (1972),
who has an extra $\hbar^2 \bar{R}/6M$-term.
See also related problems in
H. Kamo and T. Kawai, Prog. Theor. Phys. **50**, 680 (1973); M.B. Menskii, Theor. Math. Phys **18**, 190 (1974);
T. Kawai, Found. Phys. **5**, 143 (1975);
J.S. Dowker, *Functional Integration and Its Applications*, Clarendon, 1975;
M.M. Mizrahi, J. Math. Phys. **16**, 2201 (1975);
C. Hsue, J. Math. Phys. **16**, 2326 (1975);
J. Hartle and S. Hawking, Phys. Rev. D **13**, 2188 (1976);
M. Omote, Nucl. Phys. B **120**, 325 (1977);
H. Dekker, Physica A **103**, 586 (1980);

G.M. Gavazzi, Nuovo Cimento A **101**, 241 (1981);
T. Miura, Prog. Theor. Phys. **66**, 672 (1981);
C. Grosche and F. Steiner, J. Math. Phys. **36**, 2354 (1995).
A review on the earlier variety of ambiguous attempts at quantizing such systems is given in the article by
M.S. Marinov, Physics Reports **60**, 1 (1980).
A measure in the phase space formulation of path integrals which avoids an $\bar{R}$-term was found by
K. Kuchar, J. Math. Phys. **24**, 2122 (1983).

The nonholonomic mapping principle is discussed in detail in Ref. [6].

The classical variational principle which yields autoparallels rather than geodesic particle trajectories was found by
P. Fiziev and H. Kleinert, Europhys. Lett. **35**, 241 (1996) (hep-th/9503074). The new principle made it possible to derive the Euler equation for the spinning top from *within* the body-fixed reference frame. See
P. Fiziev and H. Kleinert, Berlin preprint 1995 (hep-th/9503075).

The development of perturbatively defined path integrals after Section 10.6 is due to Refs. [23, 24, 25].

The individual citations refer to

[1] H. Kleinert, Mod. Phys. Lett. A **4**, 2329 (1989) (http://www.physik.fu-berlin.de/~kleinert/199).

[2] H. Kleinert, *Gauge Fields in Condensed Matter, Vol. II, Stresses and Defects*, World Scientific, Singapore, 1989 (*ibid*.http/b2).

[3] P. Fiziev and H. Kleinert, Berlin preprint 1995 (hep-th/9503075).

[4] H. Kleinert und A. Pelster, Gen. Rel. Grav. **31**, 1439 (1999) (gr-qc/9605028).

[5] H. Kleinert, Phys. Lett. B **236**, 315 (1990) (*ibid*.http/202).

[6] H. Kleinert, *Nonholonomic Mapping Principle for Classical and Quantum Mechanics in Spaces with Curvature and Torsion*, Gen. Rel. Grav. **32**, 769 (2000) (*ibid*.http/258); Act. Phys. Pol. B **29**, 1033 (1998) (gr-qc/9801003).

[7] H. Kleinert, *Theory of Fluctuating Nonholonomic Fields and Applications: Statistical Mechanics of Vortices and Defects and New Physical Laws in Spaces with Curvature and Torsion*, in: Proceedings of NATO Advanced Study Institute on Formation and Interaction of Topological Defects, Plenum Press, New York, 1995, S. 201–232.

[8] H. Kleinert, *Double Gauge Theory of Stresses and Defects,* Phys. Lett. A **97**, 51 (1983) (*ibid*.http/107).

[9] The theory of multivalued functions developed in detail in the textbook [12] was in 1991 so unfamiliar to field theorists that the prestigious Physical Review Letters accepted, amazingly, a Comment paper on Eq. (10A.33) by
C.R. Hagen, Phys. Rev. Lett. **66**, 2681 (1991),
and a reply by
R. Jackiw and S.-Y. Pi, Phys. Rev. Lett. **66**, 2682 (1991).

[10] H. Kleinert, Int. J. Mod. Phys. A **7**, 4693 (1992) (*ibid*.http/203).

[11] H. Kleinert, *Gravity and Defects*, World Scientific, Singapore, 2007 (in preparation).

[12] H. Kleinert, *Gauge Fields in Condensed Matter, Vol. I, Superflow and Vortex Lines*, World Scientific, Singapore, 1989 (*ibid*.http/b1).

[13] H. Kleinert, Phys. Lett. B **246**, 127 (1990) (*ibid*.http/205).

[14] H. Kleinert, Phys. Lett. **B 293**, 168 (1992) (*ibid*.http/211).

[15] P.A.M. Dirac, Proc. Roy. Soc. A **133**, 60 (1931); Phys. Rev. **74**, 817 (1948), Phys. Rev. **74**, 817 (1948). See also J. Schwinger, *Particles, Sources and Fields*, Vols. 1 and 2, Addison Wesley, Reading, Mass., 1970 and 1973.

[16] S. Weinberg, *Gravitation and Cosmology – Principles and Applications of the General Theory of Relativity*, John Wiley & Sons, New York, 1972.

[17] R. Utiyama, Phys. Rev. **101**, 1597 (1956).

[18] T.W.B. Kibble, J. Math. Phys. **2**, 212 (1961).

[19] F.W. Hehl, P. von der Heyde, G.D. Kerlick, and J.M. Nester, Rev. Mod. Phys. **48**, 393 (1976).

[20] F.W. Hehl, J.D. McCrea, E.W. Mielke, and Y. Ne'eman, Phys. Rep. **258**, 1 (1995).

[21] J.A. Schouten, *Ricci-Calculus*, Springer, Berlin, Second Edition, 1954.

[22] These differences are explained in detail in pp. 1400–1401 of [2].

[23] H. Kleinert and A. Chervyakov, Phys. Lett. B **464**, 257 (1999) (hep-th/9906156).

[24] H. Kleinert and A. Chervyakov, Phys. Lett. B **477**, 373 (2000) (quant-ph/9912056).

[25] H. Kleinert and A. Chervyakov, Eur. Phys. J. C **19**, 743 (2001) (quant-ph/0002067).

[26] J.A. Gracey, Nucl. Phys. B **341**, 403 (1990); Nucl. Phys. B **367**, 657 (1991).

[27] N.D. Tracas and N.D. Vlachos, Phys. Lett. B **257**, 140 (1991).

[28] W. Magnus, F. Oberhettinger, and R. P. Soni, *Formulas and Theorems for the Special Functions of Mathematical Physics*, Springer-Verlag, Berlin, Heidelberg, New York, 1966, p. 85, or H. Bateman and A. Erdelyi, *Higher transcendental functions*, v.2, McGraw-Hill Book Company, Inc., 1953, p. 83, Formula 27.

[29] G. 't Hooft and M. Veltman, Nucl. Phys. B **44**, 189 (1972).

[30] H. Kleinert and V. Schulte-Frohlinde, *Critical Properties of ϕ^4-Theories*, World Scientific, Singapore, 2001 (*ibid*.http/b8).

[31] J.L. Gervais and A. Jevicki, Nucl. Phys. B **110**, 93 (1976).

[32] P. Salomonson, Nucl. Phys. B **121**, 433 (1977).

[33] J. de Boer, B. Peeters, K. Skenderis and P. van Nieuwenhuizen, Nucl. Phys. B **446**, 211 (1995) (hep-th/9504097); Nucl. Phys. B **459**, 631 (1996) (hep-th/9509158). See in particular the extra terms in Appendix A of the first paper required by the awkward regularization of these authors.

[34] K. Schalm, and P. van Nieuwenhuizen, Phys. Lett. B **446**, 247 (1998) (hep-th/9810115); F. Bastianelli and A. Zirotti, Nucl. Phys. B **642**, 372 (2002) (hep-th/0205182).

[35] F. Bastianelli and O. Corradini, Phys. Rev. D **60**, 044014 (1999) (hep-th/9810119).

[36] A. Hatzinikitas, K. Schalm, and P. van Nieuwenhuizen, *Trace and chiral anomalies in string and ordinary field theory from Feynman diagrams for nonlinear sigma models*, Nucl. Phys. B **518**, 424 (1998) (hep-th/9711088).

[37] H. Kleinert and A. Chervyakov, *Integrals over Products of Distributions and Coordinate Independence of Zero-Temperature Path Integrals*, Phys. Lett. A **308**, 85 (2003) (quant-ph/0204067).

[38] H. Kleinert and A. Chervyakov, Int. J. Mod. Phys. A **17**, 2019 (2002) (*ibid*.http/330).

[39] H. Kleinert, Gen. Rel. Grav. **32**, 769 (2000) (*ibid*.http/258).

[40] B.S. DeWitt, *Dynamical Theory of Groups and Fields*, Gordon and Breach, New-York, London, Paris, 1965.

[41] R.T. Seeley, Proc. Symp. Pure Math. **10**, 589 (1967);
H.P. McKean and I.M. Singer, J. Diff. Geom. **1**, 43 (1967).

[42] This is a slight modification of the discussion in
H. Kleinert, Phys. Lett. A **116**, 57 (1986) (*ibid.*http/129). See Eq. (27).

[43] N.N. Bogoliubov and D.V. Shirkov, *Introduction to the Theory of Quantized Fields*, Interscience, New York, 1959.

[44] H. Kleinert and A. Chervyakov, Phys. Lett. A **299**, 319 (2002) (quant-ph/0206022).

[45] K. Goeke and P.-G. Reinhart, Ann. Phys. **112**, 328 (1978).

[46] L. Alvarez-Gaumé, D.Z. Freedman, and S. Mukhi, Ann. of Phys. **134**, 85 (1981);
J. Honerkamp, Nucl. Phys. B **36**, 130 (1972);
G. Ecker and J. Honerkamp, Nucl. Phys. B **35** 481 (1971);
L. Tataru, Phys. Rev. D **12**, 3351 (1975);
G.A. Vilkoviski, Nucl. Phys. B **234**, 125 (1984);
E.S. Fradkin and A.A. Tseytlin, Nucl. Phys. B **234**, 509 (1984);
A.A. Tseytlin, Phys. Lett. B **223**, 165 (1989);
E. Braaten, T.L. Curtright, and C.K. Zachos, Nucl. Phys. B **260**, 630 (1985);
P.S. Howe, G. Papadopoulos, and K.S. Stelle, Nucl. Phys. B **296**, 26 (1988);
P.S. Howe and K.S. Stelle, Int. J. Mod. Phys. A **4**, 1871 (1989);
V.V. Belokurov and D.I. Kazakov, Particles & Nuclei **23**, 1322 (1992).

[47] C.M. Fraser, Z. Phys. C **28**, 101 (1985).
See also:
J. Iliopoulos, C. Itzykson, and A. Martin, Rev. Mod. Phys. **47**, 165 (1975);
K. Kikkawa, Prog. Theor. Phys. **56**, 947 (1976);
H. Kleinert, Fortschr. Phys. **26**, 565 (1978);
R. MacKenzie, F. Wilczek, and A. Zee, Phys. Rev. Lett. **53**, 2203 (1984);
I.J.R. Aitchison and C.M. Fraser, Phys. Lett. B **146**, 63 (1984).

[48] F. Cametti, G. Jona-Lasinio, C. Presilla, and F. Toninelli, *Comparison between quantum and classical dynamics in the effective action formalism,* Proc. of the Int. School of Physics "Enrico Fermi", CXLIII Ed. by G. Casati, I. Guarneri, and U. Smilansky, Amsterdam, IOS Press, 2000, pp. 431-448 (quant-ph/9910065).
See also
B.R. Frieden and A. Plastino, Phys. Lett. A **287**, 325 (2001) (quant-ph/0006012).

[49] L.D. Faddeev and V.N. Popov, Phys. Lett. B **25**, 29 (1967).

[50] C. Schubert, Phys. Rep. **355**, 73 (2001) (hep-th/0101036).

[51] H. Kleinert and A. Chervyakov, *Perturbatively Defined Effective Classical Potential in Curved Space*, (quant-ph/0301081) (*ibid.*http/339).

[52] L. Alvarez-Gaumé and E. Witten, Nucl. Phys. B **234**, 269 (1984).

[53] R.P. Feynman and H. Kleinert, Phys. Rev. A **34**, 5080 (1986) (*ibid.*http/159).

[54] S. Weinberg, *Gravitation and cosmology: principles and applications of the general theory of relativity*, (John Wiley & Sons, New York, 1972). There are $D(D+1)/2$ independent Killing vectors $l_\varepsilon^\mu(q)$ describing the isometries.

[55] See Part IV in the textbook [2] dealing with the differential geometry of defects and gravity with torsion, pp. 1427–1431.

[56] H. Kleinert, Phys. Lett. B *246*, 127 (1990) (*ibid.*http/205); Int. J. Mod. Phys. A **7**, 4693 (1992) (*ibid.*http/203); and Phys. Lett. B *293*, 168 (1992) (*ibid.*http/211).
See also the textbook [2].

11

Schrödinger Equation in General Metric-Affine Spaces

We now use the path integral representation of the last chapter to find out which Schrödinger equation is obeyed by the time evolution amplitude in a space with curvature and torsion. If there is only curvature, the result establishes the connection with the operator quantum mechanics described in Chapter 1. In particular, it will properly reproduce the energy spectra of the systems in Sections 1.14 and 1.15 — a particle on the surface of a sphere and a spinning top — which were quantized there via group commutation rules. If the space carries torsion also, the Schrödinger operator emerging from our formulation will be a *prediction*. Its correctness will be verified in Chapter 13 by an application to the path integral of the Coulomb system which can be transformed into a harmonic oscillator by a nonholonomic mapping involving curvature and torsion.

11.1   Integral Equation for Time Evolution Amplitude

Consider the time-sliced path integral Eq. (10.146)

$$\langle q|e^{-i(t-t')\hat{H}/\hbar}|q'\rangle = \frac{1}{\sqrt{2\pi i \hbar \epsilon/M}^D} \prod_{n=2}^{N+1}\left[\int d^D \Delta q_n \frac{\sqrt{g(q_n)}}{\sqrt{2\pi i \epsilon \hbar/M}^D}\right]e^{i\sum_{n=1}^{N+1}(\mathcal{A}^\epsilon+\mathcal{A}_J^\epsilon)/\hbar},(11.1)$$

with the integrals over Δq_n to be performed successively from $n = N$ down to $n = 1$.

Let us study the effect of the last Δq_n-integration upon the remaining product of integrals. We denote the entire product briefly by $\psi(q_{N+1}, t_{N+1}) \equiv \psi(q, t)$ and the product without the last factor by

$$\psi(q_N, t_N) = \psi(q_{N+1} - \Delta q_{N+1}, t_{N+1} - \epsilon) \equiv \psi(q - \Delta q, t - \epsilon).$$

Since the initial coordinate q_0 and time t_0 of the amplitude are kept fixed in the sequel, they are not shown in the arguments. We assume N to be so large that the amplitude has had time to develop from the initial state localized at q' to a smooth function of $\psi(q - \Delta q, t - \epsilon)$, smooth compared to the width of the last short-time amplitude, which is of the order $\sqrt{\hbar\epsilon\,\mathrm{tr}\,(g_{\mu\nu})/M}$.

From Eq. (11.1) we deduce the recursion relation

$$\psi(q,t) = \sqrt{g(q)} \int \frac{d^D\Delta q}{\sqrt{2\pi i\epsilon\hbar/M}^D} \exp\left[\frac{i}{\hbar}(\mathcal{A}^\epsilon + \mathcal{A}_J^\epsilon)\right] \psi(q - \Delta q, t - \epsilon). \qquad (11.2)$$

This is an integral equation

$$\psi(q,t) = \int d^D\Delta q\, K^\epsilon(\Delta q)\, \psi(q - \Delta q, t - \epsilon), \qquad (11.3)$$

with an integral kernel

$$K^\epsilon(\Delta q) = \frac{\sqrt{g(q)}}{\sqrt{2\pi i\epsilon\hbar/M}^D} \exp\left[\frac{i}{\hbar}(\mathcal{A}^\epsilon + \mathcal{A}_J^\epsilon)\right]. \qquad (11.4)$$

The integral equation (11.3) will now be turned into a Schrödinger equation. This will be done in two ways, a short way which gives direct insight into the relevance of the different terms in the mapping (10.96), and a historic more tedious way, which is useful for comparing our path integral with previous alternative proposals in the literature (cited at the end).

11.1.1   From Recursion Relation to Schrödinger Equation

The evaluation of (11.3) is much easier if we take advantage of the simplicity of the integral kernel $K^\epsilon(\Delta q)$ and the measure when expressed in terms of the variables Δx^i. Thus we introduce into (11.3) the integration variables $\Delta\xi^\mu \equiv \Delta x^i e_i{}^\mu$, with $e_i{}^\mu$ evaluated at the postpoint q. The explicit relation between $\Delta\xi^\mu$ and Δq^μ follows directly from (10.96). In terms of $\Delta\xi^\mu$, we rewrite (11.3) as

$$\psi(q,t) \;=\; \int d^D\Delta\xi\, K_0^\epsilon(\Delta\xi)\psi(q - \Delta q(\Delta\xi), t - \epsilon),$$

with the zeroth-order kernel

$$K_0^\epsilon(\Delta\xi) = \frac{\sqrt{g(q)}}{\sqrt{2\pi i\epsilon\hbar/M}^D} \exp\left[\frac{i}{\hbar}\frac{M}{2\epsilon}g_{\mu\nu}(q)\Delta\xi^\mu\Delta\xi^\nu\right] \qquad (11.5)$$

of unit normalization

$$\int d^D\Delta\xi\, K_0^\epsilon(\Delta\xi) = 1. \qquad (11.6)$$

To perform the integrals in (11.2), we expand the wave function as

$$\psi(q - \Delta q, t - \epsilon) = \left(1 - \Delta q^\mu \partial_\mu + \frac{1}{2}\Delta q^\mu \Delta q^\nu \partial_\mu \partial_\nu + \dots\right)\psi(q, t - \epsilon), \qquad (11.7)$$

and the coordinate differences Δq^μ in powers of $\Delta \xi$ by inverting Eq. (10.96):

$$\Delta q^\lambda = \left[\Delta \xi^\lambda + \frac{1}{2!}\Gamma_{\mu\nu}{}^\lambda \Delta \xi^\mu \Delta \xi^\nu - \frac{1}{3!}(\partial_\sigma \Gamma_{\mu\nu}{}^\lambda - \Gamma_{\mu\nu}{}^\tau \Gamma_{\{\sigma\tau\}}{}^\lambda)\Delta \xi^\mu \Delta \xi^\nu \Delta \xi^\sigma + \ldots\right]. (11.8)$$

All affine connections are evaluated at the postpoint q. Including in (11.2) only the relevant expansion terms, we find the integral equation

$$\psi(q,t) = \int d^D \Delta \xi \, K_0^\epsilon(\Delta \xi) \tag{11.9}$$
$$\times \left[1 - \left(\Delta \xi^\mu + \frac{1}{2!}\Gamma_{\nu\lambda}{}^\mu \Delta \xi^\nu \Delta \xi^\lambda\right)\partial_\mu + \frac{1}{2}\Delta \xi^\mu \Delta \xi^\nu \partial_\mu \partial_\nu + \ldots\right]\psi(q,t-\epsilon).$$

The evaluation requires only the normalization integral (11.6) and the two-point correlation function

$$\langle \Delta \xi^\mu \Delta \xi^\nu \rangle = \int d^D \Delta \xi \, K_0^\epsilon(\Delta \xi)\Delta \xi^\mu \Delta \xi^\nu = \frac{i\hbar\epsilon}{M}g^{\mu\nu}(q). \tag{11.10}$$

The result is

$$\psi(q,t) = \left[1 + i\epsilon \frac{\hbar^2}{2M}(g^{\mu\nu}\partial_\mu \partial_\nu - \Gamma_\nu{}^{\nu\mu}\partial_\mu) + \ldots\right]\psi(q,t-\epsilon). \tag{11.11}$$

The differential operator in parentheses is proportional to the covariant Laplacian of the field $\psi(q,t-\epsilon)$:

$$D_\mu D^\mu \psi \equiv g^{\mu\nu}D_\mu D_\nu \psi = g^{\mu\nu}D_\mu \partial_\nu \psi = (g^{\mu\nu}\partial_\mu \partial_\nu - \Gamma_\nu{}^{\nu\mu}\partial_\mu)\psi. \tag{11.12}$$

In a space with no torsion, this is equal to the Laplace-Beltrami operator applied to the field ψ:

$$\Delta \psi = \frac{1}{\sqrt{g}}\partial_\mu \sqrt{g} g^{\mu\nu}\partial_\nu \psi. \tag{11.13}$$

In a more general space, the relation between the two operators is obtained by working out the derivatives

$$\Delta = g^{\mu\nu}\partial_\mu \partial_\nu + \left(\frac{1}{\sqrt{g}}\partial_\mu \sqrt{g}\right)g^{\mu\nu}\partial_\nu + (\partial_\mu g^{\mu\nu})\partial_\nu. \tag{11.14}$$

Using

$$\left(\frac{1}{\sqrt{g}}\partial_\mu \sqrt{g}\right) = \frac{1}{2}g^{\sigma\tau}\partial_\mu g_{\sigma\tau} = \bar{\Gamma}_{\mu\nu}{}^\nu,$$
$$\partial_\mu g^{\sigma\nu} = -g^{\sigma\lambda}g^{\nu\kappa}\partial_\mu g_{\lambda\kappa}, \tag{11.15}$$
$$\partial_\mu g^{\mu\nu} = -\bar{\Gamma}_\mu{}^{\mu\nu} - \bar{\Gamma}_\mu{}^{\nu\mu},$$

we see that

$$\frac{1}{\sqrt{g}}(\partial_\mu g^{\mu\nu}\sqrt{g}) = -\bar{\Gamma}_\mu{}^{\mu\nu}, \tag{11.16}$$

and hence

$$\Delta\psi = (g^{\mu\nu}\partial_\mu\partial_\nu - \bar{\Gamma}_\mu{}^{\mu\nu}\partial_\nu)\psi = \bar{D}_\mu\bar{D}^\mu\psi. \tag{11.17}$$

Thus, the relation between the Laplacian and the Laplace-Beltrami operator is given by

$$D_\mu D^\mu\psi = (\bar{D}_\mu\bar{D}^\mu - K_\mu{}^{\mu\nu}\partial_\nu)\psi = (\bar{D}_\mu\bar{D}^\mu - 2S^\nu\partial_\nu)\psi, \tag{11.18}$$

where $\bar{D}_\mu$ denotes the covariant derivative formed with the Riemannian affine connection, the Christoffel symbol $\bar{\Gamma}_{\mu\nu}{}^\lambda$, and S_μ is the contracted torsion

$$S_\mu \equiv S_{\mu\nu}{}^\nu. \tag{11.19}$$

As a result, the amplitude $\psi(q,t)$ in (11.2) satisfies the equation

$$\psi(q,t) = \left(1 + \frac{i\epsilon\hbar}{2M}D_\mu D^\mu\right)\psi(q,t-\epsilon) + \mathcal{O}(\epsilon^2). \tag{11.20}$$

In the limit $\epsilon \to 0$, this leads to the Schrödinger equation

$$i\hbar\partial_t\psi(q,t) = \hat{H}_0\psi(q,t), \tag{11.21}$$

where $\hat{H}_0$ is the free-particle Schrödinger operator

$$\hat{H}_0 = -\frac{\hbar^2}{2M}D_\mu D^\mu. \tag{11.22}$$

It is the naively expected generalization of the flat-space operator

$$\hat{H}_0 = -\frac{\hbar^2}{2M}\partial_i^2, \tag{11.23}$$

from which (11.22) arises by transforming the derivatives with respect to Cartesian coordinates ∂_i to the general coordinate derivatives ∂_μ via the nonholonomic transformation

$$\partial_i = e_i{}^\mu\partial_\mu. \tag{11.24}$$

The result is

$$\partial_i^2 = e_i{}^\mu\partial_\mu e^{i\nu}\partial_\nu = g^{\mu\nu}\partial_\mu\partial_\nu - \Gamma_\mu{}^{\mu\nu}\partial_\nu, \tag{11.25}$$

which coincides with the Laplacian $D_\mu D^\mu$ when applied to a scalar field. Note that the operator (11.22) contains no extra term proportional to the scalar curvature R allowed by other theories.

11.1.2 Alternative Evaluation

For completeness, we also present an alternative evaluation of the q-integrals in Eq. (11.3) which is more tedious but facilitates comparison with previous work. First, the postpoint action $\mathcal{A}^\epsilon$ is conveniently split into the leading term

$$\mathcal{A}_0^\epsilon = \frac{M}{2\epsilon} g_{\mu\nu}(q) \Delta q^\mu \Delta q^\nu \tag{11.26}$$

and a remainder

$$\Delta \mathcal{A}^\epsilon \equiv \mathcal{A}^\epsilon - \mathcal{A}_0^\epsilon. \tag{11.27}$$

Correspondingly, we introduce as in (11.5) the zeroth-order kernel

$$K_0^\epsilon(\Delta q) = \frac{\sqrt{g(q)}}{\sqrt{2\pi i \epsilon \hbar / M}^D} \exp\left(\frac{i}{\hbar} \mathcal{A}_0^\epsilon\right), \tag{11.28}$$

with the unit normalization

$$\int d^D \Delta q \; K_0^\epsilon(\Delta q) = 1, \tag{11.29}$$

and expand $K^\epsilon(\Delta q)$ around $K_0^\epsilon(\Delta q)$ with a series of correction terms of higher order in Δq:

$$K^\epsilon(\Delta q) = K_0^\epsilon(\Delta q)[1 + C(\Delta q)] \equiv K_0^\epsilon(\Delta q) \left[1 + \sum_{n=1}^\infty c_n (\Delta q)^n \right]. \tag{11.30}$$

Under the smoothness assumptions above, the wave function $\psi(q - \Delta q, t - \epsilon)$ can be expanded into a Taylor series around the endpoint q, so that the integral equation (11.2) reads

$$\psi(q,t) = \int d^D \Delta q \, K_0^\epsilon(\Delta q) \left[1 + \sum_{n=1}^\infty c_n (\Delta q)^n \right]$$

$$\times \left(1 - \Delta q^\mu \partial_\mu + \frac{1}{2} \Delta q^\mu \Delta q^\nu \partial_\mu \partial_\nu + \ldots \right) \psi(q, t - \epsilon). \tag{11.31}$$

Due to the normalization property (11.6), the leading term simply reproduces $\psi(q, t - \epsilon)$. To calculate the correction terms $c_n(\Delta q)$, we expand

$$C(\Delta q) = \exp\left[\frac{i}{\hbar} (\Delta \mathcal{A}^\epsilon + \mathcal{A}_J^\epsilon) \right] - 1 \tag{11.32}$$

in powers of Δq^μ. After inserting here $\Delta \mathcal{A}^\epsilon$ from (11.27) with $\mathcal{A}_0^\epsilon$ from (11.26), we expand $\mathcal{A}^\epsilon$ as in (10.107) [recalling (10.121)]. By separating the expansion for C into even and odd powers of Δq,

$$C = C^e + C^o, \tag{11.33}$$

we find for the odd terms

$$C^{\varnothing} = -\Gamma_{\{\mu\nu\}}{}^{\nu}\Delta q^{\mu} - \frac{i}{\hbar}\frac{M}{2\epsilon}\Gamma_{\mu\nu\lambda}\Delta q^{\mu}\Delta q^{\nu}\Delta q^{\lambda} + \ldots , \tag{11.34}$$

and the even terms

$$C^{\mathrm{e}} = \sum_{a=1}^{4} C_{a}^{\mathrm{e}} + \ldots , \tag{11.35}$$

with

$$C_{1}^{\mathrm{e}} = \frac{1}{2}[\partial_{\{\mu}\Gamma_{\nu\lambda\}}{}^{\lambda} + \Gamma_{\{\nu\kappa}{}^{\sigma}\Gamma_{\{\sigma|\mu\}\}}{}^{\kappa} + \Gamma_{\{\mu\sigma}{}^{\sigma}\Gamma_{\{\nu\lambda\}}{}^{\lambda} - \Gamma_{\{\nu\kappa\}}{}^{\sigma}\Gamma_{\{\mu\sigma\}}{}^{\kappa}]\Delta q^{\mu}\Delta q^{\nu},$$

$$C_{2}^{\mathrm{e}} = \frac{iM}{2\hbar\epsilon}\Gamma_{\{\mu\nu\}}{}^{\nu}\Gamma_{\sigma\lambda\kappa}\Delta q^{\mu}\Delta q^{\sigma}\Delta q^{\lambda}\Delta q^{\kappa},$$

$$C_{3}^{\mathrm{e}} = \frac{iM}{2\hbar\epsilon}\left[\frac{1}{3}g_{\kappa\tau}(\partial_{\lambda}\Gamma_{\mu\nu}{}^{\tau} + \Gamma_{\mu\nu}{}^{\sigma}\Gamma_{\{\lambda\sigma\}}{}^{\tau}) + \frac{1}{4}\Gamma_{\mu\nu}{}^{\sigma}\Gamma_{\lambda\kappa\sigma}\right]\Delta q^{\mu}\Delta q^{\nu}\Delta q^{\lambda}\Delta q^{\kappa},$$

$$C_{4}^{\mathrm{e}} = -\frac{1}{2}\frac{M^{2}}{4\hbar^{2}\mathrm{e}^{2}}\Gamma_{\mu\nu\lambda}\Gamma_{\sigma\tau\kappa}\Delta q^{\mu}\Delta q^{\nu}\Delta q^{\lambda}\Delta q^{\sigma}\Delta q^{\tau}\Delta q^{\kappa}. \tag{11.36}$$

The dots denote terms of higher order in Δq^{μ} which do not contribute to the limit $\epsilon \to 0$.

The evaluation now proceeds perturbatively and requires the harmonic expectation values

$$\langle\mathcal{O}(\Delta q)\rangle_{0} \equiv \int d^{D}\Delta q\, K_{0}^{\epsilon}(\Delta q)\,\mathcal{O}(\Delta q). \tag{11.37}$$

The relevant correlation functions are

$$\langle\Delta q^{\mu}\Delta q^{\nu}\rangle_{0} = \frac{i\hbar\epsilon}{M}g^{\mu\nu}, \tag{11.38}$$

$$\langle\Delta q^{\mu}\Delta q^{\nu}\Delta q^{\lambda}\Delta q^{\kappa}\rangle_{0} = \left(\frac{i\hbar\epsilon}{M}\right)^{2}g^{\mu\nu\lambda\kappa}, \tag{11.39}$$

$$\langle\Delta q^{\mu}\Delta q^{\nu}\Delta q^{\lambda}\Delta q^{\kappa}\Delta q^{\sigma}\Delta q^{\tau}\rangle_{0} = \left(\frac{i\hbar\epsilon}{M}\right)^{3}g^{\mu\nu\lambda\kappa\sigma\tau}. \tag{11.40}$$

The tensor $g^{\mu\nu\lambda\kappa}$ in the second expectation (11.39) collects three Wick contractions [recall (3.302)] and reads

$$g^{\mu\nu\lambda\kappa} \equiv g^{\mu\nu}g^{\lambda\kappa} + g^{\mu\lambda}g^{\nu\kappa} + g^{\mu\kappa}g^{\nu\lambda}. \tag{11.41}$$

The tensor $g^{\mu\nu\lambda\kappa\sigma\tau}$ in the third expectation (11.40) collecting 15 Wick contractions is obtained recursively following the rule (3.303) by expanding

$$g^{\mu\nu\lambda\kappa\sigma\tau} = g^{\mu\nu}g^{\lambda\kappa\sigma\tau} + g^{\mu\lambda}g^{\nu\kappa\sigma\tau} + g^{\mu\kappa}g^{\nu\lambda\sigma\tau} + g^{\mu\sigma}g^{\nu\lambda\kappa\tau} + g^{\mu\tau}g^{\nu\lambda\kappa\sigma}. \tag{11.42}$$

A product of $2n$ factors Δq results in $(2n-1)!!$ pair contractions.

Let us collect all contributions in (11.31) relevant to order ϵ. Obviously, the highest derivative term of $\psi(q, t - \epsilon)$ is $\frac{1}{2}\Delta q^\mu \Delta q^\nu \partial_\mu \partial_\nu \psi(q, t - \epsilon)$. It receives only a leading contribution from $K_0^\epsilon(\Delta q)$,

$$i\epsilon \frac{\hbar^2}{2M} g^{\mu\nu}(q) \partial_\mu \partial_\nu \psi(q, t - \epsilon), \tag{11.43}$$

with no more corrections from $C(\Delta q)$. The term with one derivative ∂_μ on $\psi(q, t - \epsilon)$ in (11.31) becomes

$$A^\mu \partial_\mu \psi(q, t - \epsilon), \tag{11.44}$$

where A^μ is the expectation involving the odd correction terms

$$A^\mu = -\langle C^\varnothing \Delta q^\mu \rangle_0. \tag{11.45}$$

Using the rules (11.38) and (11.39), we find

$$
\begin{aligned}
A^\mu &= \left\langle \left(\Gamma_{\{\lambda\nu\}}{}^\nu \Delta q^\lambda + \frac{iM}{2\hbar\epsilon} \Gamma_{\sigma\tau\lambda} \Delta q^\sigma \Delta q^\tau \Delta q^\lambda \right) \Delta q^\mu + \ldots \right\rangle_0 \\
&= i\epsilon \frac{\hbar}{M} \left[\Gamma^{\{\mu\nu\}}{}_\nu - \frac{1}{2}(\Gamma^{\mu\nu}{}_\nu + \Gamma^{\nu\mu}{}_\nu + \Gamma_\nu{}^{\nu\mu}) \right] + \ldots \\
&= -i\epsilon \frac{\hbar}{2M} \Gamma_\nu{}^{\nu\mu} + \ldots .
\end{aligned}
\tag{11.46}
$$

In combination with (11.43), this produces the Laplacian $D_\mu D^\mu$ of the field $\psi(q, t - \epsilon)$ as in Eq. (11.11).

We now turn to the remaining contributions in (11.31) which contain no more derivatives of $\psi(q, t - \epsilon)$. They are all due to the expectation value $\langle C^e \rangle_0$ of the even correction terms. Let us define

$$V_{\text{eff}} \equiv \frac{i\hbar}{\epsilon} \langle C \rangle_0 = \frac{i\hbar}{\epsilon} \langle C^e \rangle_0, \tag{11.47}$$

to be called the effective potential caused by the correction terms $\langle C \rangle_0$ of (11.32). Using the expectation values (11.38)–(11.40) we find

$$V_{\text{eff}} = \frac{\hbar^2}{M} v \equiv \frac{\hbar^2}{M} \sum_{A,B} v_A{}^B, \tag{11.48}$$

where the sum runs over the six terms

$$
\begin{aligned}
v_2{}^1 &= -\frac{1}{2}(\Gamma_{\{\mu\sigma\}}{}^\sigma \Gamma_{\{\nu\lambda\}}{}^\lambda - \Gamma_{\{\nu\kappa\}}{}^\sigma \Gamma_{\{\mu\sigma\}}{}^\kappa)g^{\mu\nu}, \\
v_2{}^2 &= \frac{1}{8}\Gamma_{\{\mu\nu\}}{}^\tau \Gamma_{\lambda\sigma\tau} g^{\mu\nu\sigma\lambda}, \\
v_2{}^3 &= \frac{1}{2}\Gamma_{\{\mu\kappa\}}{}^\kappa \Gamma_{\nu\tau\lambda} g^{\mu\nu\tau\lambda}, \\
v_2{}^4 &= -\frac{1}{8}\Gamma_{\mu\nu\lambda} \Gamma_{\sigma\tau\kappa} g^{\mu\nu\lambda\sigma\tau\kappa}, \\
v_3{}^1 &= -\frac{1}{2}(\partial_{\{\mu} \Gamma_{\nu\lambda\}}{}^\lambda + \Gamma_{\{\nu\kappa\}}{}^\sigma \Gamma_{\{\sigma|\mu\}\}}{}^\kappa)g^{\mu\nu}, \\
v_3{}^2 &= \frac{1}{6}g_{\mu\tau}(\partial_\kappa \Gamma_{\lambda\nu}{}^\tau + \Gamma_{\lambda\nu}{}^\sigma \Gamma_{\{\kappa\sigma\}}{}^\tau)g^{\mu\nu\lambda\kappa}.
\end{aligned}
\tag{11.49}
$$

The subscripts 2 and 3 distinguish contributions coming from the quadratic and the cubic terms in the expansion (10.96) of Δx^i. By inserting on the right-hand sides the explicit expansions (11.42) and (11.41), we find after some algebra that the sum of all $v_A{}^B$-terms is zero. In fact, the $v_2{}^B$- and $v_3{}^B$-terms disappear separately. A simple structural reason for this is given in Appendix 11A.

Explicitly, the cancellation is rather obvious for $v_3{}^B$ after inserting (11.41). For $v_2{}^B$, the proof requires more work which is relegated to Appendix 11A.

Note that in a space without curvature and torsion, the above manipulations are equivalent to a direct transformation of the flat-space integral equation

$$
\begin{aligned}
\psi(\mathbf{x}, t) &= \int \frac{d^D \Delta x}{\sqrt{2\pi i \epsilon \hbar / M}^D} \exp\left[i\epsilon \frac{M}{2} (\Delta x^i)^2 \right] \\
&\times \ (1 - \Delta x^i \partial_{x^i} + \tfrac{1}{2} \Delta x^i \Delta x^j \partial_{x^i} \partial_{x^j} + \ldots) \psi(\mathbf{x}, t - \epsilon) \\
&= \left[1 + \frac{i\epsilon\hbar}{2M} \partial_i^2 + \mathcal{O}(\epsilon^2) \right] \psi(\mathbf{x}, t - \epsilon),
\end{aligned}
\tag{11.50}
$$

to the variable Δq by a coordinate transformation. In a general metric-affine space, the wave function $\psi(q, t)$ has no counter image in x-space so that (11.50) cannot be used as a starting point for a nonholonomic transformation.

11.2 Equivalent Path Integral Representations

From the derivation of the Schrödinger equation in Subsection 11.1.1 we learn an important lesson. When deriving the transformation law (10.96) between the finite coordinate differences Δx^i and Δq^μ by evaluating the integral equation (10.60) along the autoparallel, the cubic terms in Δq, which make the action and measure lengthy, can be dropped altogether. A completely equivalent path integral representation of the time evolution amplitude is obtained by transforming the flat-space path integral (10.89) into the general metric-affine one (10.146) with the help of the shortened transformation

$$
\Delta x^i = e^i{}_\lambda \left(\Delta q^\lambda - \frac{1}{2!} \Gamma_{\mu\nu}{}^\lambda \Delta q^\mu \Delta q^\nu \right).
\tag{11.51}
$$

This has the simple Jacobian

$$
J = \frac{\partial(\Delta x)}{\partial(\Delta q)} = \det\left(e^i_\kappa \right) \det\left(\delta^\kappa{}_\mu - e_i{}^\kappa e^i_{\{\mu,\nu\}} \Delta q^\nu \right),
\tag{11.52}
$$

whose effective action reads

$$
\frac{i}{\hbar} \mathcal{A}_J^\epsilon = -e_i{}^\kappa e^i{}_{\kappa,\nu} \Delta q^\nu - \frac{1}{2} e_i{}^\mu e^i_{\{\kappa,\nu\}} e_j{}^\kappa e^j_{\{\mu,\lambda\}} \Delta q^\nu \Delta q^\lambda + \ldots \ .
\tag{11.53}
$$

With the help of (10.16), this is expressed in terms of the connection yielding

$$
\frac{i}{\hbar} \mathcal{A}_J^\epsilon = -\Gamma_{\{\nu\mu\}}{}^\mu \Delta q^\nu - \frac{1}{2} \Gamma_{\{\nu\kappa\}}{}^\sigma \Gamma_{\{\mu,\sigma\}}{}^\kappa \Delta q^\nu \Delta q^\mu + \ldots \ .
\tag{11.54}
$$

The mapping (11.51) has, however, an unattractive feature: The short-time action following from (11.51)

$$\mathcal{A}^\epsilon = \frac{M}{2\epsilon}(\Delta x^i)^2 = \frac{M}{2\epsilon}\Big(g_{\mu\nu}\Delta q^\mu \Delta q^\nu - \Gamma_{\mu\nu\lambda}\Delta q^\mu \Delta q^\nu \Delta q^\lambda$$

$$+ \frac{1}{4}\Gamma_{\lambda\kappa}{}^\sigma \Gamma_{\mu\nu\sigma}\Delta q^\mu \Delta q^\nu \Delta q^\lambda \Delta q^\kappa + \ldots \Big) \tag{11.55}$$

is no longer equal to the classical action (10.107) [recall the convention (10.121)] evaluated along the autoparallel.

This was also a feature of another mapping which is the most convenient for calculations. Instead of deriving the relation between Δx^i and Δq^μ by evaluating (10.60) along the autoparallel, one may assume, for the moment, the absence of curvature and torsion and expand $\Delta x^i = x^i(q) - x^i(q - \Delta q)$ in powers of Δq:

$$\Delta x^i = e^i{}_\mu \Delta q^\mu - \frac{1}{2}e^i{}_{\mu,\nu}\Delta q^\mu \Delta q^\nu + \frac{1}{3!}e^i{}_{\mu,\nu\lambda}\Delta q^\mu \Delta q^\nu \Delta q^\lambda + \ldots . \tag{11.56}$$

After this, curvature and torsion are introduced by allowing the functions $x(q)$ and $\partial_\mu x(q)$ to be nonintegrable in the sense of the Schwartz criterion, i.e., the second derivatives of $x(q)$ and $e^i{}_\mu(q)$ need not commute with each other [implying that the right-hand side of (11.56) can no longer be written as $x^i(q) - x^i(q - \Delta q)$]. The expansion (11.56) is then a *definition* of the transformation from Δx^i to Δq^μ. Using the identities (10.16), (10.131), and (10.132), the transformation (11.56) turns into

$$\Delta x^i = e^i{}_\lambda \Big[\Delta q^\lambda - \frac{1}{2!}\Gamma_{\mu\nu}{}^\lambda \Delta q^\mu \Delta q^\nu \tag{11.57}$$

$$+ \frac{1}{3!}(\partial_\sigma \Gamma_{\mu\nu}{}^\lambda + \Gamma_{\mu\nu}{}^\tau \Gamma_{\sigma\tau}{}^\lambda)\Delta q^\mu \Delta q^\nu \Delta q^\sigma + \ldots \Big].$$

This differs from the correct one (10.96) by the third-order term

$$\Delta' x^i = \frac{1}{3!}e^i{}_{[\tau,\sigma]}e_k{}^\tau e^k{}_{\nu,\mu}\Delta q^\mu \Delta q^\nu \Delta q^\sigma = \frac{1}{3!}e^i{}_\lambda S_{\sigma\tau}{}^\lambda \Gamma_{\mu\nu}{}^\tau \Delta q^\mu \Delta q^\nu \Delta q^\sigma, \tag{11.58}$$

which vanishes if the q-space has no torsion. The Jacobian associated with (11.56) is

$$J = \frac{\partial(\Delta x)}{\partial(\Delta q)} = \det(e^i_\kappa)\det\Big(\delta^\kappa{}_\mu - e_i{}^\kappa e^i{}_{\{\mu,\nu\}}\Delta q^\nu + \frac{1}{2}e_i{}^\kappa e^i{}_{\{\mu,\nu\lambda\}}\Delta q^\nu \Delta q^\lambda + \ldots\Big), (11.59)$$

and corresponds to the effective action

$$\frac{i}{\hbar}\mathcal{A}^\epsilon_J = -e_i{}^\kappa e^i{}_{\kappa,\nu}\Delta q^\nu + \frac{1}{2}[e_i{}^\mu e^i{}_{\{\mu,\nu\lambda\}} - e_i{}^\mu e^i{}_{\{\kappa,\nu\}}e_j{}^\kappa e^j{}_{\{\mu,\lambda\}}]\Delta q^\nu \Delta q^\lambda + \ldots . \tag{11.60}$$

With (10.16), (10.131), and (10.132), this becomes

$$\frac{i}{\hbar}\mathcal{A}^\epsilon_J = -\Gamma_{\{\nu\mu\}}{}^\mu \Delta q^\nu + \frac{1}{2}(\partial_{\{\mu}\Gamma_{\nu,\kappa\}}{}^\kappa + \Gamma_{\{\nu,\kappa}{}^\sigma \Gamma_{\mu\},\sigma}{}^\kappa - \Gamma_{\{\nu\kappa}{}^\sigma \Gamma_{\{\mu,\sigma\}}{}^\kappa)\Delta q^\nu \Delta q^\mu + \ldots ,$$

$$\tag{11.61}$$

which differs from the proper Jacobian action (10.145) only by one index symmetrization.

To find the short-time action following from the mapping (11.56), we form

$$\mathcal{A}^\epsilon = \frac{M}{2\epsilon}(\Delta x^i)^2 = \frac{M}{2\epsilon}\Big[g_{\mu\nu}\Delta q^\mu \Delta q^\nu - e^i{}_\mu e^i{}_{\nu,\lambda}\Delta q^\mu \Delta q^\nu \Delta q^\lambda \tag{11.62}$$
$$+ \Big(\frac{1}{3}e^i{}_\mu e^i{}_{\nu,\lambda\kappa} + \frac{1}{4}e^i{}_{\mu,\nu}e^i{}_{\lambda,\kappa}\Big)\Delta q^\mu \Delta q^\nu \Delta q^\lambda \Delta q^\kappa + \dots\Big],$$

and use the identities (10.16), (10.131), and (10.132) to obtain

$$\mathcal{A}^\epsilon = \frac{M}{2\epsilon}\Big\{g_{\mu\nu}\Delta q^\mu \Delta q^\nu - \Gamma_{\mu\nu\lambda}\Delta q^\mu \Delta q^\nu \Delta q^\lambda \tag{11.63}$$
$$+ \Big[\frac{1}{3}g_{\mu\tau}(\partial_\kappa\Gamma_{\lambda\nu}{}^\tau + \Gamma_{\lambda\nu}{}^\delta\Gamma_{\kappa\delta}{}^\tau) + \frac{1}{4}\Gamma_{\lambda\kappa}{}^\sigma\Gamma_{\mu\nu\sigma}\Big]\Delta q^\mu \Delta q^\nu \Delta q^\lambda \Delta q^\kappa + \dots\Big\}.$$

This differs from the proper short-time action (10.107) [recall the convention (10.107)] only by the absence of the symmetrization in the indices κ and δ in the fourth term, the difference vanishing if the q-space has no torsion.

The equivalent path integral representation in which (11.2) contains the short-time action (11.63) and the Jacobian action (11.61) will be useful in Chapter 13 when solving the path integral of the Coulomb system.

The equivalence of different time-sliced path integral representations manifests itself in certain moment properties of the integral kernel (11.4). The derivations of the Schrödinger equation in Subsections. 11.1.1 and 11.1.2 have made use only of the following three moment properties of the kernel:

$$\int d^D\Delta q\, K^\epsilon(\Delta q) = 1 + \dots, \tag{11.64}$$

$$\int d^D\Delta q\, K^\epsilon(\Delta q)\Delta q^\nu = -i\epsilon\frac{\hbar}{2M}\Gamma_\mu{}^{\mu\nu} + \dots, \tag{11.65}$$

$$\int d^D\Delta q\, K^\epsilon(\Delta q)\Delta q^\mu \Delta q^\nu = i\epsilon\frac{\hbar}{M}g^{\mu\nu} + \dots, \tag{11.66}$$

evaluated at fixed postpoint q. The omitted terms indicated by the dots and all higher moments contribute to higher orders in ϵ which are irrelevant for the derivation of the differential equation obeyed by the amplitude. Any kernel $K^\epsilon(\Delta q)$ with these properties leads to the same Schrödinger equation. If a kernel is written as

$$K^\epsilon(\Delta q) = K_0^\epsilon(\Delta q)[1 + C(\Delta q)], \tag{11.67}$$

where $K_0^\epsilon(\Delta q)$ is the free-particle postpoint kernel

$$K_0^\epsilon(\Delta q) = \frac{\sqrt{g(q)}}{\sqrt{2\pi i\epsilon\hbar/M}^D}\exp\Big[\frac{i}{\hbar}g_{\mu\nu}(q)\,\Delta q^\mu\,\Delta q^\nu\Big], \tag{11.68}$$

the moment properties (11.64)–(11.66) are equivalent to

$$\langle C \rangle_0 = 0 + \dots , \tag{11.69}$$

$$\langle C\,\Delta q^\mu \rangle_0 = i\epsilon \frac{\hbar}{2M} \Gamma_\mu{}^{\mu\nu} + \dots , \tag{11.70}$$

where the expectation values are taken with respect to the kernel $K_0^\epsilon(\Delta q)$, the dots on the right-hand side indicating terms of the order ϵ^2. Note that the third of the three moment properties is trivially true since it receives only a contribution from the leading part of the kernel $K^\epsilon(\Delta q)$, i.e., from $K_0^\epsilon(\Delta q)$. The verification of the other two requires some work, in particular the first, which is the normalization condition.

Two kernels $K_1^\epsilon, K_2^\epsilon$ are equivalent if their correction terms C_1, C_2 have expectations which are small of the order ϵ^2:

$$\langle C_1 \rangle_0 = \langle C_2 \rangle_0 = \mathcal{O}(\epsilon^2), \tag{11.71}$$

$$\langle (C_1 - C_2)\Delta q^\mu \rangle_0 = \mathcal{O}(\epsilon^2). \tag{11.72}$$

These are necessary and sufficient conditions for the equivalence. Many possible correction terms C lead to the same moment integrals. All of them are physically equivalent, being associated with the same Schrödinger equation. The simplest possibilities are

$$K^\epsilon(\Delta q) = K_0^\epsilon(\Delta q) \left[1 + \frac{1}{2}\Gamma_\mu{}^\mu{}_\nu \Delta q^\nu \right], \tag{11.73}$$

or

$$K^\epsilon(\Delta q) = K_0^\epsilon(\Delta q) \left[1 - \frac{i}{D+2}\frac{M}{2\hbar\epsilon}\Gamma_\mu{}^\mu{}_\nu\, \Delta q^\nu g_{\lambda\kappa}\Delta q^\lambda \Delta q^\kappa \right], \tag{11.74}$$

where D is the space dimension. The zero-order kernel satisfies automatically the first and third moment condition, (11.64) and (11.66), while the additional term enforces precisely the second condition, (11.65), without changing the others. The equivalent kernels can also be considered as the result of working with Jacobian actions

$$\frac{i}{\hbar}\mathcal{A}_J^\epsilon = \frac{1}{2}\Gamma_\mu{}^\mu{}_\nu \Delta q^\nu - \frac{1}{8}(\Gamma_\mu{}^\mu{}_\nu \Delta q^\nu)^2, \tag{11.75}$$

$$\frac{i}{\hbar}\mathcal{A}_J^\epsilon = -\frac{i}{D+2}\frac{M}{2\hbar\epsilon}\Gamma_\mu{}^\mu{}_\nu \Delta q^\nu\, g_{\lambda\kappa}\, \Delta q^\lambda \Delta q^\kappa, \tag{11.76}$$

instead of the original one (10.145). Indeed, the second term in (11.75) can further be reduced by perturbation theory to

$$-\frac{1}{8}(\Gamma_\mu{}^\mu{}_\nu \Delta q^\nu)^2 \;\rightarrow\; -\frac{1}{8}\Gamma_\mu{}^\mu{}_\nu \Gamma_\lambda{}^\lambda{}_\kappa \langle \Delta q^\nu \Delta q^\kappa \rangle_0$$

$$= -i\epsilon\frac{\hbar}{8M}(\Gamma_\mu{}^\mu{}_\nu)^2, \tag{11.77}$$

yielding an alternative and most useful form for the Jacobian action

$$\frac{i}{\hbar}\mathcal{A}_J^\epsilon = \frac{1}{2}\Gamma_\mu{}^\mu{}_\nu\Delta q^\nu - i\epsilon\frac{\hbar}{8M}(\Gamma_\mu{}^\mu{}_\nu)^2. \tag{11.78}$$

Remarkably, this expression involves only the connection contracted in the first two indices:

$$\Gamma_\mu{}^{\mu\nu} = g^{\mu\lambda}\Gamma_{\mu\lambda}{}^\nu. \tag{11.79}$$

11.3 Potentials and Vector Potentials

It is straightforward to find the effect of external potentials and vector potentials upon the Schrödinger equation. For this, we merely observe that the time-sliced potential term

$$\mathcal{A}_{\text{pot}}^\epsilon = \frac{e}{c}A_\mu\Delta q^\mu - \frac{e}{2c}\partial_\nu A_\mu\,\Delta q^\mu\Delta q^\nu - \epsilon V(q) + \dots \tag{11.80}$$

derived in Eq. (10.182) appears in the kernel $K^\epsilon(\Delta q)$ via a factor $e^{i\mathcal{A}_{\text{pot}}^\epsilon/\hbar}$. This factor can be combined with the postpoint expansion of the wave function in the integral equation (11.31), which becomes

$$\begin{aligned}
\psi(q,t) &= \int d^D\Delta q\, K_0^\epsilon(\Delta q)\left[1 + C(\Delta q)\right] \\
&\quad \times e^{i\mathcal{A}_{\text{pot}}^\epsilon/\hbar}\left(1 - \Delta q^\mu\partial_\mu + \frac{1}{2}\Delta q^\mu\Delta q^\nu\partial_\mu\partial_\nu\right)\psi(q,t-\epsilon) + \dots \\
&= \int d^D\Delta q\, K_0^\epsilon(\Delta q)\left[1 + C(\Delta q)\right]\left[1 - \Delta q^\mu\left(\partial_\mu - i\frac{e}{\hbar c}A_\mu\right)\right. \\
&\quad \left.+\frac{1}{2}\Delta q^\mu\Delta q^\nu\left(\partial_\mu - i\frac{e}{\hbar c}A_\mu\right)\left(\partial_\nu - i\frac{e}{\hbar c}A_\nu\right) - i\epsilon V(q)\right]\psi(q,t-\epsilon) + \dots\,.
\end{aligned} \tag{11.81}$$

By going through the steps of Subsection 11.1.1 or 11.1.2, we obtain the same Schrödinger equation as in (11.21),

$$i\hbar\partial_t\psi(q,t) = \hat{H}\psi(q,t). \tag{11.82}$$

The Hamiltonian operator $\hat{H}$ differs from the free operator $\hat{H}_0$ of (11.22),

$$\hat{H}_0 = -\frac{\hbar^2}{2M}D_\mu D^\mu, \tag{11.83}$$

in two ways. First, a potential energy $V(q)$ is added. Second, the covariant derivatives D_μ are replaced by

$$D_\mu^A \equiv D_\mu - i\frac{e}{\hbar c}A_\mu. \tag{11.84}$$

This is the Schrödinger version of the minimal substitution in Eq. (2.642).

The minimal substitution extends the covariance of D_μ with respect to coordinate changes to a covariance with respect to gauge transformations of the vector potential A_μ. The subtraction of $iA_\mu/\hbar$ on the right-hand side of (11.84) reflects the fact that $P_\mu = p_\mu - A_\mu$ rather than p^μ is the gauge-invariant physical momentum of a particle in the presence of an electromagnetic field. Only the use of P_μ guarantees the gauge invariance of the electromagnetic interaction, just as in the flat-space action (2.641).

Let us briefly verify that the Schrödinger equation (11.82) with the covariant derivative (11.84) is invariant under gauge transformations. If the amplitude is multiplied by a space-dependent phase

$$\psi(q) \to e^{-i(e/\hbar c)\Lambda(q)}\psi(q), \tag{11.85}$$

the covariant derivative (11.84) is multiplied by precisely the same phase:

$$D_\mu\psi(q) \to e^{-i(e/\hbar c)\Lambda(q)}D_\mu\psi(q), \tag{11.86}$$

if the vector potential is gauge tranformed as follows:

$$A_\mu \to A_\mu + \partial_\mu\Lambda(q). \tag{11.87}$$

Under these joint transformations, the Schrödinger equation (11.82) is multiplied by an overall phase factor $e^{-i\Lambda(q)}$, and thus gauge invariant.

Adding a potential $V(q)$, the Hamilton operator in the Schrödinger equation (11.82) is therefore

$$\hat{H} = -\frac{\hbar^2}{2M}D_\mu^A D^{A\mu} + V(q). \tag{11.88}$$

Observe that the mixed terms containing derivative and vector potential appears in the symmetrized form

$$-\frac{\hbar}{2Mc}\left(\hat{p}_\mu A^\mu + A^\mu \hat{p}_\mu\right). \tag{11.89}$$

This corresponds to a symmetric time slicing of the interaction term $\dot{q}^\mu A_\mu$ which was derived in Section 10.5 by using the equation of motion in calculation of the short-time action. Here we see that this time slicing guarantees the gauge invariance of the Schrödinger equation.

Note further that there is no extra R-term in the Schrödinger equation (11.88).

11.4 Unitarity Problem

The appearance of the Laplace operator $D_\mu D^\mu$ in the free-particle Schrödinger equation (11.82) is in conflict with the traditional physical scalar product between two wave functions $\psi_1(q)$ and $\psi_2(q)$:

$$\langle\psi_2|\psi_1\rangle \equiv \int d^D q\sqrt{g(q)}\psi_2^*(q)\psi_1(q). \tag{11.90}$$

In such a scalar product, only the Laplace-Beltrami operator (11.13),

$$\Delta = \frac{1}{\sqrt{g}}\partial_\mu \sqrt{g}\, g^{\mu\nu}\partial_\nu, \tag{11.91}$$

is a Hermitian operator, not the Laplacian $D_\mu D^\mu$. The bothersome term is the contracted torsion term

$$(D_\mu D^\mu - \Delta)\psi = -2S^\mu \partial_\mu \psi. \tag{11.92}$$

This term ruins the Hermiticity and thus also the unitarity of the time evolution operator of a particle in a space with curvature and torsion.

For presently known physical systems in spaces with curvature and torsion, the unitarity problem is fortunately absent. Consider first field theories of gravity with torsion. There, the torsion field $S_{\mu\nu}{}^\lambda$ is generated by the spin current density of the fundamental matter fields. The requirement of renormalizability restricts these fields to carry spin $1/2$. However, the spin current density of spin-$1/2$ particles happens to be a completely antisymmetric tensor.[1] This property carries over to the torsion tensor. Hence, the torsion field in the universe satisfies $S^\mu = 0$. This implies that for a particle in a universe with curvature and torsion, the Laplacian always degenerates into the Laplace-Beltrami operator, assuring unitarity after all.

In Chapter 13 we shall witness another way of escaping the unitarity problem. The path integral of the three-dimensional Coulomb system is solved by a multi-valued transformation to a four-dimensional space with torsion where the physical scalar product is

$$\langle \psi_2 | \psi_1 \rangle_{\text{phys}} \equiv \int d^D q \sqrt{g}\, w(q)\psi_2^*(q)\psi_1(q), \tag{11.93}$$

with some scalar weight function $w(q)$. This scalar product is different from the naive scalar product (11.90). It is, however, reparametrization-invariant, and $w(q)$ makes the Laplacian $D_\mu D^\mu$ a Hermitian operator.

The characteristic property of torsion in the transformed Coulomb system is that $S_\mu(q) = S_{\mu\nu}{}^\nu$ can be written as a gradient of a scalar function: $S_\mu(q) = \partial_\mu \sigma(q)$ [see Eq. (13.138)]. Such torsion fields admit a Hermitian Laplace operator of a scalar field in a scalar product (11.93) with the weight

$$w(q) = e^{-2\sigma(q)}. \tag{11.94}$$

Thus, the physical scalar product can be expressed in terms of the naive one as follows:

$$\langle \psi_2 | \psi_1 \rangle_{\text{phys}} \equiv \int d^D q \sqrt{g(q)}\, e^{-2\sigma(q)}\psi_2^*(q)\psi_1(q). \tag{11.95}$$

---

[1]See, for example, H. Kleinert, *Gauge Fields in Condensed Matter*, op. cit., Vol. II, Part IV, *Differential Geometry of Defects and Gravity with Torsion*, p. 1432 (http://www.physik.fu-berlin.de/~kleinert/b8).

To prove the Hermiticity, we observe that within the naive scalar product (11.93), a partial integration changes the covariant derivative $-D_\mu$ into

$$D_\mu^* \equiv (D_\mu + 2S_\mu). \tag{11.96}$$

Consider, for example, the scalar product

$$\int d^D q \sqrt{g}\, U^{\mu\nu_1...\nu_n} D_\mu V_{\nu_1...\nu_n}. \tag{11.97}$$

A partial integration of the derivative term ∂_μ in D_μ gives

$$\text{surface term} \; - \int d^D dq [(\partial_\mu \sqrt{g}\, U^{\mu\nu_1...\nu_n}) V_{\nu_1...\nu_n}$$
$$- \sum_i \sqrt{g} U^{\mu\nu_1...\nu_i...\nu_n} \Gamma_{\mu\nu_i}{}^{\lambda_i} V_{\nu_1...\lambda_i...\nu_n}]. \tag{11.98}$$

Now we use

$$\partial_\mu \sqrt{g} = \sqrt{g}\, \bar{\Gamma}_{\mu\nu}{}^\nu = \sqrt{g}(2S_\mu + \Gamma_{\mu\nu}{}^\nu), \tag{11.99}$$

to rewrite (11.98) as

$$\text{surface term} \; - \int d^D q \sqrt{g}\, [(\partial_\mu U^{\mu\nu_1...\nu_n}) V_{\nu_1...\nu_n}$$
$$- \sum_i \Gamma_{\mu\nu_i}{}^{\lambda_i} U^{\mu\nu_1...\nu_i...\nu_n} V_{\nu_1...\lambda_i...\nu_n} \; -2S_\mu U^{\mu\nu_1...\nu_n} V_{\nu_1...\nu_n}], \tag{11.100}$$

which is equal to

$$\text{surface term} \; - \int d^D q \sqrt{g} (D_\mu^* U^{\mu\nu_1...\nu_n}) V_{\nu_1...\nu_n}. \tag{11.101}$$

In the physical scalar product (11.95), the corresponding operation is

$$\int d^D q \sqrt{g} e^{-2\sigma(q)} U^{\mu\nu_1...\nu_n} D_\mu V_{\nu_1...\nu_n} =$$
$$= \text{surface term} - \int d^D q \sqrt{g} (D_\mu^* e^{-2\sigma(q)} U^{\mu\nu_1...\nu_n}) V_{\nu_1...\nu_n}$$
$$= \text{surface term} - \int d^D q \sqrt{g} e^{-2\sigma(q)} (D_\mu \sqrt{g} U^{\mu\nu_1...\nu_n}) V_{\nu_1...\nu_n}. \tag{11.102}$$

Hence, iD_μ is a Hermitian operator, and so is the Laplacian $D_\mu D^\mu$.

For spaces with an arbitrary torsion, the correct scalar product has yet to be found. Thus the quantum equivalence principle is so far only applicable to spaces with arbitrary curvature and gradient torsion.

11.5 Alternative Attempts

Our procedure has to be contrasted with earlier proposals for constructing path integrals in spaces with curvature, in which torsion was always assumed to be absent. In the notable work of DeWitt,[2] the measure is taken to be proportional to

$$\prod_{n=1}^{N} \int dq_n \sqrt{g(q_{n-1})} = \prod_{n=1}^{N} \int dq_n \sqrt{g(q_n - \Delta q)} \tag{11.103}$$

so that the expansion in powers Δq gives a Jacobian action $\mathcal{A}^{\epsilon}_{J_0}$ of Eq. (10.133).

If one uses this action instead of the correct expression $\mathcal{A}^{\epsilon}_{J}$ of Eq. (10.145), the amplitude obeys a Schrödinger equation

$$i\hbar \partial_t \psi(q, t) = (\hat{H}_0 + V_{\text{eff}})\psi(q, t), \tag{11.104}$$

where

$$\hat{H}_0 = -\frac{\hbar^2}{2M}\Delta \tag{11.105}$$

contains the Laplace-Beltrami operator Δ, and V_{eff} is an additional effective potential

$$V_{\text{eff}} = \frac{\hbar^2}{6M}\bar{R}. \tag{11.106}$$

The Schrödinger equation (11.104) differs from ours in Eq. (11.21) derived by the nonholonomic mapping procedure by the extra R-term. The derivation is reviewed in Appendix 11A. Note that our sign convention for $\bar{R}$ is such that the surface of a sphere of radius R has $\bar{R} = 2/R^2$.

There is definite experimental evidence for the absence which will of such a term

In the amplitude proposed by DeWitt, the short-time amplitude carries an extra semiclassical prefactor, a curved-space version of the Van Vleck-Pauli-Morette determinant in Eq. (4.125). It contributes another term proportional to $\bar{R}$ which reduces (11.106) to $(\hbar^2/12M)\bar{R}$ (see Appendix 11B). Other path integral prescriptions lead even to additional noncovariant terms.[3] All such nonclassical terms proportional to $\hbar^2$ have to be subtracted from the classical action to arrive at the correct amplitude which satisfies the Schrödinger equation (11.104) without an extra V_{eff}.

There are two compelling arguments in favor of our construction principle: On the one hand, if the space has only curvature and no torsion, our path integral gives, as we have seen in Sections 8.7–8.9 and 10.4, the correct time evolution amplitude of a particle on the surface of a sphere in D dimensions and on group spaces. In contrast to other proposals, only the classical action appears in the short-time amplitude. In spaces with constant curvature, just as in flat space, our amplitude agrees with

---

[2]See Section 11.5.
[3]See the review article by M.S. Marinov quoted at the end of the previous chapter.

that obtained in operator quantum mechanics by quantizing the theory via the commutation rules of the generators of the group of motion.

In the presence of torsion the result is new. It will be tested by the integration of the path integral of the Coulomb system in Chapter 13. This requires a coordinate transformation to an auxiliary space with curvature and torsion, which reduces the system to a harmonic oscillator. The new quantum equivalence principle leads to the correct result.

The solution is so far the only indirect evidence for the question first raised by Bryce DeWitt in his fundamental 1957 paper [14], whether the Hamiltonian operator for a particle in curved space contains merely the Laplace-Beltrami operator Δ in the kinetic energy, or whether there exists an additional term proportional to $\hbar^2 R$. Recall the various older path integral literature on the subject cited in Chapter 10. From the measure generated by the nonholonomic mapping principle in Subsection 10.3.2 it follows that there is no extra $\hbar^2 R$-term. See the discussion in Section 11.5. It would, of course, be more satisfactory to have a direct experimental evidence, but so far all experimentally accessible systems in curved space have either a very small R caused by gravitation, whose detection is presently impossible, or a constant R which does not change level spacings, an example for the latter being the spinning symmetric and asymmetric top discussed in the context of Eq. (1.468). We show now that if we assume the presence of an extra $\hbar^2 R$ in the momentum space formulation of the path integral of the Coulomb system, such an extra term would cause experimentally wrong *level spacings* in the hydrogen atom [15].

The solution is so far the only indirect evidence for the question first raised by Bryce DeWitt in his fundamental 1957 paper [14], whether the Hamiltonian operator for a particle in curved space contains merely the Laplace-Beltrami operator Δ in the kinetic energy, or whether there exists an additional term proportional to $\hbar^2 R$. Recall the various older path integral literature on the subject cited in Chapter 10. From the measure generated by the nonholonomic mapping principle in Subsection 10.3.2 it follows that there is no extra $\hbar^2 R$-term. See the discussion in Section 11.5. It would, of course, be more satisfactory to have a direct experimental evidence, but so far all experimentally accessible systems in curved space have either a very small R caused by gravitation, whose detection is presently impossible, or a constant R which does not change level spacings, an example for the latter being the spinning symmetric and asymmetric top discussed in the context of Eq. (1.468). We show now that if we assume the presence of an extra $\hbar^2 R$ in the momentum space formulation of the path integral of the Coulomb system, such an extra term would cause experimentally wrong *level spacings* in the hydrogen atom [15].

11.6   DeWitt-Seeley Expansion of Time Evolution Amplitude

An important tool for comparing the results of path integrals in curved space with operator results of Schrödinger theory is the short-time expansion of the imaginary-time evolution amplitude $(q, \beta \mid q', 0)$. In Schrödinger theory, the amplitude is given by the matrix elements of the evolution operator $e^{-\beta \hat{H}} = e^{\beta \Delta \beta / 2}$, with the Laplace-Beltrami operator (11.13). This expansion has first been given by DeWitt and by Seeley[4] and reads

$$(q\,\beta \mid q'\,0) = (q \mid e^{\beta \Delta / 2} \mid q') = \frac{1}{\sqrt{2\pi \beta}^D} \, e^{-g_{\mu\nu} \Delta q^\mu \Delta q^\nu / 2\beta} \sum_{k=0}^{\infty} \beta^k a_k(q, q'). \quad (11.107)$$

The expansion coefficients are, up to fourth order in Δq^μ,

$$a_0(q, q') \equiv 1 + \frac{1}{12} \bar{R}_{\mu\nu} \Delta q^\mu \Delta q^\nu + \left(\frac{1}{360} \bar{R}^\mu{}_\kappa{}^\nu{}_\lambda \bar{R}_{\mu\sigma\nu\tau} + \frac{1}{288} \bar{R}_{\kappa\lambda} \bar{R}_{\sigma\tau} \right) \Delta q^\kappa \Delta q^\lambda \Delta q^\sigma \Delta q^\tau,$$

$$a_1(q, q') \equiv \frac{1}{12} \bar{R} + \left(\frac{1}{144} \bar{R} \bar{R}_{\mu\nu} + \frac{1}{360} \bar{R}^{\kappa\lambda} \bar{R}_{\kappa\mu\lambda\nu} + \frac{1}{360} \bar{R}^{\kappa\lambda\sigma}{}_\mu \bar{R}_{\kappa\lambda\sigma\nu} - \frac{1}{180} \bar{R}^\kappa{}_\mu \bar{R}_{\kappa\nu} \right) \Delta q^\mu \Delta q^\nu,$$

$$a_2(q, q') \equiv \frac{1}{288} \bar{R}^2 + \frac{1}{720} \bar{R}^{\mu\nu\kappa\lambda} \bar{R}_{\mu\nu\kappa\lambda} - \frac{1}{720} \bar{R}^{\mu\nu} \bar{R}_{\mu\nu}, \quad (11.108)$$

where $\Delta q^\mu \equiv (q - q')^\mu$ and all curvature tensors are evaluated at q. For $\Delta q^\mu = 0$ this simplifies to

$$(q\,\beta \mid q\,0) = \frac{1}{\sqrt{2\pi \beta}^D} \left\{ 1 + \frac{\beta}{12} \bar{R} + \frac{\beta^2}{2} \left[\frac{1}{144} \bar{R}^2 + \frac{1}{360} \left(\bar{R}^{\mu\nu\kappa\lambda} \bar{R}_{\mu\nu\kappa\lambda} - \bar{R}^{\mu\nu} \bar{R}_{\mu\nu} \right) \right] + \dots \right\}. \quad (11.109)$$

This can also be written in the cumulant form as

$$(q\,\beta \mid q\,0) = \frac{1}{\sqrt{2\pi \beta}^D} \exp \left[\frac{\beta}{12} \bar{R} + \frac{\beta^2}{720} \left(\bar{R}^{\mu\nu\kappa\lambda} \bar{R}_{\mu\nu\kappa\lambda} - \bar{R}^{\mu\nu} \bar{R}_{\mu\nu} \right) + \dots \right]. \quad (11.110)$$

The derivation goes as follows. In a neighborhood of some arbitrary point q_0^μ we expand the Laplace-Beltrami operator in normal coordinates where the metric and its determinant have the expansions (10.477) and (10.478) as

$$\Delta = \partial^2 - \frac{1}{3} \bar{R}_{ik_1 jk_2}(q_0)(q - q_0)^{k_1}(q - q_0)^{k_2} \partial_\mu \partial_\nu - \frac{2}{3} \bar{R}_{\mu\nu}(q_0)(q - q_0)^\mu \partial_\nu. \quad (11.111)$$

The time evolution operator $\hat{H} = -\Delta / 2$ in the exponent of Eq. (11.107) is now separated into a free part $\hat{H}_0$ and an interaction part $\hat{H}_{\text{int}}$ as follows

$$\hat{H}_0 = -\frac{1}{2} \partial^2, \quad (11.112)$$

$$\hat{H}_{\text{int}} = \frac{1}{6} \bar{R}_{ik_1 jk_2}(q - q_0)^{k_1}(q - q_0)^{k_2} \partial_\mu \partial_\nu + \frac{1}{3} \bar{R}_{\mu\nu}(q - q_0)^\mu \partial_\nu. \quad (11.113)$$

We now recall Eq. (1.294) and see that the transition amplitude (11.107) satisfies the integral equation

$$(q\,\beta \mid q'\,0) = \langle q \mid e^{-\beta(\hat{H}_0 + \hat{H}_{\text{int}})} \mid q' \rangle = \langle q \mid e^{-\beta \hat{H}_0} \left[1 - \int_0^\beta d\sigma e^{\sigma \hat{H}_0} \hat{H}_{\text{int}} e^{-\sigma \hat{H}} \right] \mid q' \rangle$$

$$= (q\,\beta \mid q'\,0)_0 - \int_0^\beta d\sigma \int d^D \bar{q} \, (q\,\beta - \sigma \mid \bar{q}\,0)_0 \, \hat{H}_{\text{int}}(\bar{q}) \, (\bar{q}\,\sigma \mid q\,0), \quad (11.114)$$

[4] B.S. DeWitt, *Dynamical Theory of Groups and Fields*, Gordon and Breach, New-York, 1965. R.T. Seeley, Proc. Symp. Pure Math. 10 1967 589. See also H.P. McKean and I.M. Singer. J. Diff. Geom. *1*, 43 (1967).

where

$$(q\,\beta \mid q'\,0)_0 = \langle q \mid e^{-\beta \hat{H}_0} \mid q' \rangle = \frac{1}{\sqrt{2\pi\beta}^n}\, e^{-(\Delta q)^2/2\beta}. \tag{11.115}$$

To first order in $\hat{H}_{\text{int}}$ we can replace $\hat{H}$ in the last exponential of Eq. (11.114) by $\hat{H}_0$ and obtain

$$(q\,\beta \mid q'\,0) \approx (q\,\beta \mid q'\,0)_0 - \int_0^\beta d\sigma \int d^D \bar{q}\, (q\,\beta - \sigma \mid \bar{q}\,0)_0\, \hat{H}_{\text{int}}(\bar{q})\, (\bar{q}\,\sigma \mid q\,0)_0. \tag{11.116}$$

Inserting (11.113) and choosing $q_0 = q'$, we find

$$
\begin{aligned}
(q\,\beta \mid q'\,0) =&\ (q\,\beta \mid q'\,0)_0 \left\{ 1 + \int_0^\beta d\sigma \int \frac{d^D(\Delta\bar{q})}{\sqrt{2\pi a}^D}\, e^{-[\Delta\bar{q} - (\sigma/\beta)\Delta q]^2/2a} \right. \\
&\ \times \left. \left[-\frac{1}{6}\bar{R}_{\mu\kappa\nu\lambda}\Delta\bar{q}^\kappa \Delta\bar{q}^\lambda \left(-\frac{\delta^{\mu\nu}}{\sigma} + \frac{\Delta\bar{q}^\mu \Delta\bar{q}^\nu}{\sigma^2} \right) + \frac{1}{3}\bar{R}_{\mu\nu}\frac{\Delta\bar{q}^\mu \Delta\bar{q}^\nu}{\sigma} \right] \right\},
\end{aligned} \tag{11.117}
$$

where we have replaced the integrating variable $\bar{q}$ by $\Delta\bar{q} = \bar{q} - q'$ and introduced the variable $a \equiv (\beta - \sigma)\sigma/\beta$. There is initially also a term of fourth order in $\Delta\bar{q}$ which vanishes, however, because of the antisymmetry of $\bar{R}_{\mu\kappa\nu\lambda}$ in $\mu\kappa$ and $\nu\lambda$. The remaining Gaussian integrals are performed after shifting $\Delta\bar{q} \to \Delta\bar{q} + \sigma\,\Delta q/\beta$, and we obtain

$$
\begin{aligned}
(q\,\beta \mid q'\,0) =&\ (q\,\beta \mid q'\,0)_0 \left\{ 1 + \frac{1}{6}\int_0^\beta d\sigma \left[\frac{\sigma}{\beta^2}\bar{R}_{\mu\nu}(q')\Delta q^\mu \Delta q^\nu + \frac{a}{\sigma}\bar{R}(q') \right] \right\} \\
=&\ (q\,\beta \mid q'\,0)_0 \left[1 + \frac{1}{12}\bar{R}_{\mu\nu}(q')\Delta q^\mu \Delta q^\nu + \frac{\beta}{12}\bar{R}(q') \right]. \tag{11.118}
\end{aligned}
$$

Note that all geometrical quantities are evaluated at the initial point q'. They can be re-expressed in power series around the final position q using the fact that in normal coordinates

$$g_{\mu\nu}(q') = g_{\mu\nu}(q) + \frac{1}{3}\bar{R}_{ik_1 jk_2}(q)\Delta q^{k_1} \Delta q^{k_2} + \dots\ , \tag{11.119}$$

$$g_{\mu\nu}(q')\Delta q^\mu \Delta q^\nu = g_{\mu\nu}(q)\Delta q^\mu \Delta q^\nu, \tag{11.120}$$

the latter equation being true to all orders in Δq due to the antisymmetry of the tensors $\bar{R}_{\mu\nu\kappa\lambda}$ in all terms of the expansion (11.119), which is just another form of writing the expansion (10.477) up to the second order in Δq^μ.

Going back to the general coordinates, we obtain all coefficients of the expansion (11.107) linear in the curvature tensor

$$(q\,\beta \mid q'\,0) \simeq \frac{1}{\sqrt{2\pi\beta}^n}\, e^{-g_{\mu\nu}(q)\Delta q^\mu \Delta q^\nu/2\beta} \left[1 + \frac{1}{12}\bar{R}_{\mu\nu}(q)\Delta q^\mu \Delta q^\mu + \frac{\beta}{12}\bar{R}(q) \right]. \tag{11.121}$$

The higher terms in (11.107) can be derived similarly, although with much more effort.

A simple cross check of the expansion (11.107) to high orders is possible if we restrict the space to the surface of a sphere of radius r in D dimensions which has $D-1$ dimensions. Then

$$\bar{R}_{\mu\nu\kappa\lambda} = \frac{1}{r^2} \left(g_{\mu\lambda} g_{\nu\kappa} - g_{\mu\kappa} g_{\nu\lambda} \right), \quad \mu, \nu = 1, 2, \ldots, D-1. \qquad (11.122)$$

Contractions yield Ricci tensor and scalar curvature

$$\bar{R}_{\mu\nu} = \bar{R}_{\kappa\mu\nu}{}^{\kappa} = \frac{D-2}{r^2} g_{\mu\nu}, \qquad \bar{R} = \bar{R}_{\mu}{}^{\mu} = \frac{(D-1)(D-2)}{r^2} \qquad (11.123)$$

and further:

$$\bar{R}^2_{\mu\nu\kappa\lambda} = \frac{2(D-1)(D-2)}{r^4}, \qquad \bar{R}^2_{\mu\nu} = \frac{(D-1)(D-2)^2}{r^4}. \qquad (11.124)$$

Inserting these into (11.108), we obtain the DeWitt-Seeley short-time expansion of the diagonal amplitude for any q, up to order β^2:

$$(q\,\beta \mid q\,0) = \frac{1}{\sqrt{2\pi\beta}^{D-1}} \left[1 + (D-1)(D-2)\frac{\beta}{12r^2} \right.$$
$$\left. + (D-1)(D-2)(5D^2 - 17D + 18)\frac{\beta^2}{1440r^4} + \ldots \right]. \qquad (11.125)$$

This expansion may easily be reproduced by a simple direct calculation[5] of the partition function for a particle on the surface of a sphere

$$Z(\beta) = \sum_{l=0}^{\infty} d_l \exp[-l(l+D-2)\beta/2r^2], \qquad (11.126)$$

where $-l(l+D-2)$ are the eigenvalues of the Laplace-Beltrami operator on a sphere [recall (10.165)] and $d_l = (2l+D-2)(l+D-3)!/l!(D-2)!$ their degeneracies [recall (8.113)]. Since the space is homogeneous, the amplitude $(q\,\beta \mid q\,0)$ is obtained from this by dividing out the constant surface of a sphere:

$$(q\,\beta \mid q\,0) = \frac{\Gamma(D/2)}{2\pi^{D/2}r^{D-1}} Z(\beta). \qquad (11.127)$$

For any given D, the sum in (11.126) easily be expanded in powers of β. As an example, take $D = 3$ where

$$Z(\beta) = \sum_{l=0}^{\infty} (2l+1) \exp[-l(l+1)\beta/2r^2]. \qquad (11.128)$$

---

[5]H. Kleinert, Phys. Lett. A *116*, 57 (1986) (http://www.physik.fu-berlin.de/~kleinert/129). See Eq. (27).

In the small-β limit, the sum (11.128) is evaluated as follows

$$Z(\beta) = \int_0^\infty d\left[l(l+1)\right] \exp[-l(l+1)\beta/2r^2] + \sum_{l=0}^\infty (2l+1)\left[1 - l(l+1)\beta/2r^2 + \ldots\right].$$

(11.129)

The integral is immediately done and yields

$$\int_0^\infty dz \, \exp(-z\beta/2r^2) = \frac{2r^2}{\beta}.$$

(11.130)

The sums are divergent but can be evaluated by analytic continuation from negative powers of l to positive ones with the help of Riemann zeta functions $\zeta(z) = \sum_{n=1}^\infty n^{-z}$, which vanishes for all even negative arguments. Thus we find

$$\sum_{l=0}^\infty (2l+1) = 1 + \sum_{l=1}^\infty (2l+1) = 1 + 2\zeta(-1) - \frac{1}{2} = \frac{1}{3},$$

(11.131)

$$-\frac{\beta}{2r^2}\sum_{l=0}^\infty (2l+1)l(l+1) = -\frac{\beta}{2r^2}\sum_{l=1}^\infty (2l^3 + l) = -\frac{\beta}{2r^2}\left[2\zeta(-3) + \zeta(-1)\right] = \frac{\beta}{30r^2}.$$

(11.132)

Substituting these into (11.129) yields

$$Z(\beta) = \frac{2r^2}{\beta}\left(1 + \frac{\beta}{6r^2} + \frac{\beta^2}{60r^4} + \ldots\right).$$

(11.133)

Dividing out the constant surface of a sphere $4\pi r^2$ as required by Eq. (11.127), we obtain indeed the expansion (11.125) for the surface of a sphere in three dimensions.

Appendix 11A   Cancellations in Effective Potential

Here we demonstrate the cancellation of the terms $v_2{}^B$ and $v_3{}^B$ in formula (11.48) for the effective potential. First we give a simple reason why the cancellation occurs *separately* for the contributions stemming from the second and third terms in the expansion (10.96) of Δx^i. Consider the model integral

$$\int \frac{d\Delta x}{\sqrt{2\pi\epsilon}} \exp\left[-\frac{(\Delta x)^2}{2\epsilon}\right],$$

(11A.1)

and assume that Δx has an expansion of the type (10.96):

$$\Delta x = \Delta q[1 + a_2\Delta q + a_3(\Delta q)^2 + \ldots].$$

(11A.2)

The integral transforms into

$$\int \frac{d\Delta q}{\sqrt{2\pi\epsilon}}[1 + 2a_2\Delta q + 3a_3(\Delta q)^2 + \ldots]\exp\left\{-\frac{(\Delta q)^2}{2\epsilon}[1 + 2a_2\Delta q + 2a_3(\Delta q)^2 + a_2^2(\Delta q)^2 + \ldots]\right\},$$

(11A.3)

and is evaluated perturbatively via the expansion

$$\int \frac{d\Delta q}{\sqrt{2\pi\epsilon}} \exp\left[-\frac{(\Delta q)^2}{2\epsilon}\right] \left[1 - a_2 \frac{(\Delta q)^3}{\epsilon} - a_3 \frac{(\Delta q)^4}{\epsilon} - a_2^2 \frac{(\Delta q)^4}{2\epsilon}\right.$$
$$\left. + a_2^2 \frac{(\Delta q)^6}{2\epsilon} - 2a_2^2 \frac{(\Delta q)^4}{\epsilon} + 3a_3(\Delta q)^2 + \dots\right]. \tag{11A.4}$$

If $\langle \mathcal{O} \rangle_0$ denotes the harmonic expectation value

$$\langle \mathcal{O} \rangle_0 \equiv \int \frac{d\Delta q}{\sqrt{2\pi i\epsilon}} \mathcal{O} \exp[-(\Delta q)^2/2\epsilon], \tag{11A.5}$$

one has

$$\langle (\Delta q)^2 \rangle_0 = \epsilon, \quad \langle (\Delta q)^4 \rangle_0 = 3!\epsilon^2, \quad \langle (\Delta q)^6 \rangle_0 = 5!\epsilon^3, \dots . \tag{11A.6}$$

Using these values we find that the a_2- and a_3-terms cancel separately. Precisely this cancellation mechanism is active in the separate cancellation of the more complicated expressions $v_1{}^B, v_2{}^B$ in Eq. (11.49).

To demonstrate the cancellations explicitly, consider first the derivative terms in $v_3{}^B$. They are

$$v_3{}^{\partial\Gamma} = -\frac{1}{2}g^{\mu\nu}\partial_{\{\mu}\Gamma_{\nu\lambda\}}{}^{\lambda} + \frac{1}{6}g_{\mu\tau}\partial_{\kappa}\Gamma_{\lambda\nu}{}^{\tau}\left(g^{\mu\nu}g^{\lambda\kappa} + g^{\mu\lambda}g^{\nu\kappa} + g^{\mu\kappa}g^{\nu\lambda}\right). \tag{11A.7}$$

Due to the symmetrization of the first term in $\mu\nu\lambda$, this gives zero. The cancellation of the remaining terms in $v_3{}^B$ which are quadratic in Γ is most easily shown by writing all indices as subscripts, inserting $g_{\mu\nu\lambda\kappa}$ from (11.41), and working out the contractions.

To calculate the $v_2{}^B$-terms, it is useful to introduce the notation $\Gamma_{1\mu} \equiv \Gamma_{\mu\nu}{}^{\nu}$, $\Gamma_{2\mu} \equiv \Gamma_{\nu\mu}{}^{\nu}$, $\Gamma_{3\mu} \equiv \Gamma_{\nu}{}^{\nu}{}_{\mu}$ and similarly the matrices $\tilde{\Gamma}_{1\mu}\hat{=}(\Gamma_{\mu})_{\lambda\kappa}$, $\tilde{\Gamma}_{1\mu}^T\hat{=}(\Gamma_{\mu})_{\kappa\lambda}$, $\tilde{\Gamma}_{2\mu}\hat{=}\Gamma_{\lambda\mu\kappa}$, $\tilde{\Gamma}_{2\mu}^T\hat{=}\Gamma_{\kappa\mu\lambda}$, $\tilde{\Gamma}_{3\mu}\hat{=}\Gamma_{\lambda\kappa\mu}$, $\tilde{\Gamma}_{3\mu}^T\hat{=}\Gamma_{\kappa\lambda\mu}$. For contractions such as $\Gamma_{1\mu}\Gamma_{1\mu}$ we write $\Gamma_1\Gamma_1$, and for $\Gamma_{\mu\nu\lambda}\Gamma_{\mu\nu\lambda}$ we write $\tilde{\Gamma}_1\tilde{\Gamma}_1 = \tilde{\Gamma}_2\tilde{\Gamma}_2 = \tilde{\Gamma}_3\tilde{\Gamma}_3$, whichever is most convenient. Similarly, $\Gamma_{\mu\nu\lambda}\Gamma_{\lambda\mu\nu} = \tilde{\Gamma}_1\tilde{\Gamma}_2^T = \tilde{\Gamma}_2\tilde{\Gamma}_3^T = \tilde{\Gamma}_3\tilde{\Gamma}_1$. Then we work out

$$v_2{}^1 = -\frac{1}{8}[(\Gamma_1 + \Gamma_2)^2 - \tilde{\Gamma}_3(\tilde{\Gamma}_1 + \tilde{\Gamma}_1^T + \tilde{\Gamma}_2 + \tilde{\Gamma}_2^T)], \tag{11A.8}$$

$$v_2{}^2 = \frac{1}{8}[\Gamma_3\Gamma_3 + \tilde{\Gamma}_3(\tilde{\Gamma}_3 + \tilde{\Gamma}_3^T)],$$

$$v_2{}^3 = \frac{1}{4}[(\Gamma_1 + \Gamma_2)^2 + \Gamma_3(\Gamma_1 + \Gamma_2)],$$

$$v_2{}^4 = -\frac{1}{8}[\Gamma_1{}^2 + \Gamma_2{}^2 + \Gamma_3{}^2 + 2(\Gamma_1\Gamma_2 + \Gamma_2\Gamma_3 + \Gamma_3\Gamma_1)$$
$$+ \tilde{\Gamma}_3(\tilde{\Gamma}_1 + \tilde{\Gamma}_1^T + \tilde{\Gamma}_2 + \tilde{\Gamma}_2^T + \Gamma_3 + \tilde{\Gamma}_3^T)].$$

It is easy to check that the sum of these $v_2{}^B$-terms vanishes.

Incidentally, if the symmetrizations in (11.49) following from our Jacobian action had been absent, we would find the additional terms

$$\Delta v_3{}^{\partial\Gamma} = \frac{1}{6}\bar{R} - \frac{2}{3}\partial_\mu S^\mu + \frac{1}{6}(\tilde{\Gamma}_3\tilde{\Gamma}_2^T - \Gamma_3\Gamma_2), \tag{11A.9}$$

$$\Delta v_3{}^{\Gamma^2} = -\frac{1}{2}\tilde{\Gamma}_3\tilde{\Gamma}_2 + \frac{1}{6}(\tilde{\Gamma}_3\tilde{\Gamma}_2 + \tilde{\Gamma}_3\tilde{\Gamma}_2^T + \Gamma_3\Gamma_2), \tag{11A.10}$$

whose sum yields the additional contribution to the $v_3{}^B$-terms

$$\Delta v_3 = \frac{1}{6}\bar{R} - \frac{2}{3}\partial_\mu S^\mu + \frac{2}{3}\tilde{\Gamma}_3\tilde{S}_1, \tag{11A.11}$$

after having used the identity

$$\tilde{S}_3 \tilde{\Gamma}_2 = -\tilde{\Gamma}_3 \tilde{S}_1. \tag{11A.12}$$

The first term in (11A.11) is the R-term derived by K.S. Cheng[6] as an effective potential in the Schrödinger equation.

For $v_2{}^B$, we would find the extra terms

$$\Delta v_2{}^1 = -\frac{1}{2}(\Gamma_1 \Gamma_1 - \tilde{\Gamma}_3 \tilde{\Gamma}_2) + \frac{1}{8}[(\Gamma_1 + \Gamma_2)^2 - \tilde{\Gamma}_3(\tilde{\Gamma}_1 + \tilde{\Gamma}_1^T + \tilde{\Gamma}_2 + \tilde{\Gamma}_2^T)], \tag{11A.13}$$

$$\Delta v_2{}^3 = \frac{1}{4}(\Gamma_1 - \Gamma_2)(\Gamma_1 + \Gamma_2 + \Gamma_3), \tag{11A.14}$$

which add up to

$$\Delta v_2 = -\frac{1}{2} S_1 S_1 + \frac{1}{2}\Gamma_3 S_1 - \frac{1}{3}\tilde{\Gamma}_3 \tilde{S}_1 + \frac{1}{2}\tilde{S}_1 \tilde{S}_3, \tag{11A.15}$$

where we have written S_1 for S_μ and used some trivial identities such as

$$\tilde{\Gamma}_3 \tilde{\Gamma}_2^T = \tilde{\Gamma}_3 \tilde{\Gamma}_1. \tag{11A.16}$$

Thus we would obtain an additional effective potential $V_{\text{eff}} = (\hbar^2/M)v$ with

$$v = \frac{1}{6} R - \frac{2}{3} g^{\mu\nu} \partial_\mu S_\nu - \frac{1}{2}(S_1^2 - \Gamma_3 S_1) + \frac{1}{2}\tilde{S}_1 \tilde{S}_3 - \frac{1}{3}\tilde{\Gamma}_3 \tilde{S}_1. \tag{11A.17}$$

The second and the fourth term can be combined to

$$-\frac{2}{3} D_\mu S^\mu - \frac{1}{6}\Gamma_3 S_1. \tag{11A.18}$$

Due to the presence of Γ's in v, this is a noncovariant expression which cannot possibly be physically correct. In the absence of torsion, however, v happens to be reparametrization-invariant, and this is the reason why the resulting effective potential $V_{\text{eff}} = \hbar^2 \bar{R}/6M$ appeared acceptable in earlier works. A procedure which has no reparametrization-invariant extension to spaces with torsion cannot be correct.

Appendix 11B   DeWitt's Amplitude

Bryce DeWitt, in his frequently quoted paper,[7] attempted to quantize the motion of a particle in a curved space using the naive measure of path integration as in Eq. (10.153), but with the short-time amplitude (with $q \equiv q_n$, $q' \equiv q_{n-1}$)

$$\frac{\sqrt{g(q)g(q')}^{-1/2}(\epsilon/M)^{D/2}\mathcal{D}^{1/2}}{\sqrt{2\pi i\epsilon\hbar/M}^D} \exp\left(\frac{i}{\hbar}\mathcal{A}^\epsilon\right), \tag{11B.1}$$

where $\mathcal{D}$ is the curved-space analog of the Van Vleck-Pauli-Morette determinant [defined in flat space after Eq. (4.124)]:

$$\mathcal{D} = \det_D\left[-\partial_{q^\mu}\partial_{q'^\nu}\mathcal{A}^\epsilon(q, q')\right]. \tag{11B.2}$$

After taking the Jacobian action $\mathcal{A}^\epsilon_{J_0}$ into account, this leads to an integral kernel $K^\epsilon(\Delta q)$ which differs from our correct one in Eq. (11.4) by an extra factor $\sqrt{g(q)g(q')}^{-1/2}(\epsilon/M)^{D/2}\mathcal{D}^{1/2}$. This has

---

[6]K.S. Cheng, J. Math. Phys. *13*, 1723 (1972).
[7]B.S. DeWitt, Rev. Mod. Phys. *29*, 377 (1957).

the postpoint expansion $1 + \frac{1}{12}\bar{R}_{\mu\nu}\Delta q^{\mu}\Delta q^{\nu} + \ldots$. When treated perturbatively, the extra term is equivalent to $\epsilon\hbar\bar{R}/12M$, reducing the effective potential (11.106) by a factor $1/2$. In order to obtain the correct amplitude, DeWitt had to add a nonclassical term to the action in the path integral. This term is proportional to $\hbar^2$ and removes the unwanted term V_{eff}, i.e., $\Delta_{\text{DW}}\mathcal{A}^{\epsilon} = \epsilon V_{\text{eff}}$. Such a correction procedure must be rejected on the grounds that it runs contrary to the very essence of the entire path integral approach, in which the contribution of each path is controlled entirely by the phase $e^{i\mathcal{A}/\hbar}$ with the classical action in the exponent.

The short-time kernel proposed by Cheng is the same as DeWitt's, except that it does not include the extra Van Vleck-Pauli-Morette determinant in (11B.1). In this case, the result is the full effective potential (11.106), which must be artificially subtracted from the classical action to obtain the correct amplitude.

Notes and References

The first path integral in a curved space was written down by
B.S. DeWitt, Rev. Mod. Phys. **29**, 377 (1957),
making use of previous work by
C. DeWitt-Morette, Phys. Rev. **81**, 848 (1951).
A modified ansatz is due to
K.S. Cheng, J. Math. Phys. **13**, 1723 (1972).
For recent discussions with results different from ours see
H. Kamo and T. Kawai, Prog. Theor. Phys. **50**, 680, (1973);
T. Kawai, Found. Phys. **5**, 143 (1975);
H. Dekker, Physica A **103**, 586 (1980);
G.M. Gavazzi, Nuovo Cimento **101**A, 241 (1981).
A good survey of most attempts is given by
M.S. Marinov, Phys. Rep. **60**, 1 (1980).

In a space without torsion, C. DeWitt-Morette, working with stochastic differential equations, postulates a Hamiltonian operator without an extra $\bar{R}$-term. See her lectures presented at the 1989 Erice Summer School on *Quantum Mechanics in Curved Spacetime*, ed. by V. de Sabbata, Plenum Press, 1990, and references quoted therein, in particular
C. DeWitt-Morette, K.D. Elworthy, B.L. Nelson, and G.S. Sammelman, Ann. Inst. H. Poincaré A **32**, 327 (1980).

A path measure in a phase space path integral which does not produce any $\bar{R}$-term was proposed by
K. Kuchar, J. Math. Phys. **24**, 2122 (1983).

The short derivation of the Schrödinger equation in Subsection 11.1.1 is due to
P. Fiziev and H. Kleinert, J. Phys. A **29**, 7619 (1996) (hep-th/9604172).

To reach a depth profounder still, and still
Profounder, in the fathomless abyss.
WILLIAM COWPER (1731-1800), The Winter Morning Walk

12

New Path Integral Formula for Singular Potentials

In Chapter 8 we have seen that for systems with a centrifugal barrier, the Euclidean form of Feynman's original time-sliced path integral formula diverges for certain attractive barriers. This happens even if the quantum statistics of the systems is well defined. The same problem arises for a particle in an attractive Coulomb potential, and thus in any atomic system.

In this chapter we set up a new and more flexible path integral formula which is free of this problem for any singular potential. This has recently turned out to be the key for a simple solution of many other path integrals which were earlier considered intractable.

12.1 Path Collapse in Feynman's formula for the Coulomb System

The attractive Coulomb potential $V(r) = -e^2/r$ has a singularity at the coordinate origin $r = 0$. This singularity is weaker than that of the centrifugal barrier, but strong enough to cause a catastrophe in the Euclidean path integral. Recall that an attractive centrifugal barrier does not even possess a classical partition function. The same thing is true for the attractive Coulomb potential where formula (2.350) reads

$$Z_{\rm cl} = \int \frac{d^3x}{\sqrt{2\pi\hbar^2\beta/M}^3} \exp\left(\beta\frac{e^2}{r}\right).$$

The integral diverges near the origin. In addition, there is a divergence at large r. The leading part of the latter can be removed by subtracting the free-particle partition function and forming

$$Z'_{\rm cl} \equiv Z_{\rm cl} - Z_{\rm cl}|_{e=0} = \int \frac{d^3x}{\sqrt{2\pi\hbar^2\beta/M}^3} \left[\exp\left(\beta\frac{e^2}{r}\right) - 1\right], \tag{12.1}$$

leaving only a quadratic divergence. In a realistic many-body system with an equal number of oppositely charged particles, this disappears by screening effects. Thus we shall not worry about it any further and concentrate only on the remaining small-r divergence. In a real atom, this singularity is not present since the nucleus is not a point particle but occupies a finite volume. However, this "physical regularization" of the singularity is not required for quantum-mechanical stability. The Schrödinger equation is perfectly solvable for the singular pure $-e^2/r$ potential. We should therefore be able to recover the existing Schrödinger results from the path integral formalism without any short-distance regularization.

On the basis of Feynman's original time-sliced formula, this is impossible. If a path consists of a *finite* number of straight pieces, its Euclidean action

$$\mathcal{A} = \int_{\tau_a}^{\tau_b} d\tau \left[\frac{M}{2} \mathbf{x}'^2(\tau) - \frac{e^2}{r(\tau)} \right] \tag{12.2}$$

can be lowered indefinitely by a path with an almost stretched configuration which corresponds to a slowly moving particle sliding down into the $-e^2/r$ abyss. We call this phenomenon a *path collapse*. In nature, this catastrophe is prevented by quantum fluctuations. In order to understand how this happens, it is useful to reinterpret the paths in the Euclidean path integral as random lines parametrized by $\tau \in (\tau_a, \tau_b)$. Their distribution is governed by the "Boltzmann factor" $e^{-\mathcal{A}/\hbar}$, whose effective "quantum" temperature is $T_{\mathrm{eff}} \equiv \hbar/k_B(\tau_b - \tau_a)$.[1] The logarithm of the Euclidean amplitude $(\mathbf{x}_b \tau_b | \mathbf{x}_a \tau_a)$ multiplied by $-T_{\mathrm{eff}}$ defines a free energy

$$F = E - k_B T S$$

of the random line with fixed endpoints. The quantum fluctuations equip the path with a configurational entropy S. This must be sufficiently singular to produce a regular free energy bounded from below. Obviously, such a mechanism can only work if the exact path integral contains an *infinite* number of infinitesimally small sections. Only these can contain enough configurational entropy near the singularity to halt the collapse.

The variational approach in Section 5.10 has shown an important effect of the configurational entropy of quantum fluctuations. It smoothes the singular Coulomb potential producing an effective classical potential that is finite at the origin. A path collapse was avoided by defining the path integral as an infinite product of integrals over all Fourier coefficients. The infinitely high-frequency components were integrated out and this produced the desired stability. These high-frequency components are absent in a finitely time-sliced path with a finite number of pieces, where frequencies Ω_m are bounded by twice the inverse slice thickness $1/\epsilon$ [recall Eqs. (2.106), (2.107)].

Unfortunately, the path measure used in the variational approach is unsuitable for exact calculations of nontrivial path integrals. Except for the free particle and

---

[1]This amounts to viewing the path as a polymer with configurational fluctuations in space, a possibility which is a major topic in Chapters 15 and 16.

the harmonic oscillator, these are all based on solving a finite number of ordinary integrals in a time-sliced formula. We therefore need a more powerful time-sliced path integral formula which avoids a collapse in singular potentials.

For the Coulomb system, such a formula has been found in 1979 by Duru and Kleinert.[2] It has become the basis for solving the path integral of many other nontrivial systems. Here we describe the most general extension of this formula which will later be applied to a number of systems. For the attractive Coulomb potential and other singular potentials, such as attractive centrifugal barriers, it will not only halt the collapse, but also be the key to an analytic solution.

The derived stabilization is achieved by introducing a path-independent width of the time slices. If the path approaches an abyss, the widths decrease and the number of slices increases. This enables the configurational entropy of Eq. (12.3) to grow large enough to cancel the singularity in the energy. To see the cancellation mechanism, consider a random line with n links which has, on a simple cubic lattice in D dimensions, $(2D)^n$ configurations with an entropy

$$S = n \log(2D). \tag{12.3}$$

If the number of time slices n increases near the $-e^2/r$ singularity like const/r, then the entropy is proportional to $1/r$. A path section which slides down into the abyss must stretch itself to make the kinetic energy small. But then it gives up a certain entropy S, and this raises the free energy by $k_B T_{\mathrm{eff}} S$ according to (12.3). This compensates for the singularity in the potential and halts the collapse. The purpose of this chapter is to set up a path integral formula in which this stabilizing mechanism is at work.

It should be pointed out that no instability problem would certainly arise if we were to *define* the imaginary-time path integral for the time evolution amplitude in the continuum

$$(\mathbf{x}_b \tau_b | \mathbf{x}_a \tau_a) \equiv \int \mathcal{D}^D x(\tau) \int \frac{\mathcal{D}^D p(\tau)}{(2\pi\hbar)^D} \exp\left\{ \frac{1}{\hbar} \int_{\tau_a}^{\tau_b} d\tau [i\mathbf{p}\dot{\mathbf{x}} - H(\mathbf{p}, \mathbf{x})] \right\} \tag{12.4}$$

without any time slicing as the solution of the Schrödinger differential equation

$$(\hbar\partial_\tau + \hat{H})(\mathbf{x}\,\tau | \mathbf{x}_a \tau_a) = \hbar\delta(\tau - \tau_a)\delta^{(D)}(\mathbf{x} - \mathbf{x}_a) \tag{12.5}$$

[compare Eq. (1.304)]. After solving the Schrödinger equation $\hat{H}\psi_n(\mathbf{x}) = E_n \psi_n(\mathbf{x})$, the spectral representation (1.319) renders directly the amplitude (12.4).

All subtleties described above are due to the finite number of time slices in the path integral. As explained at the end of Section 2.1, the explicit sum over all paths is an essential ingredient of Feynman's global approach to the phenomena of

---

[2]I.H. Duru and H. Kleinert, Phys. Lett. B *84*, 30 (1979) (http://www.physik.fu-berlin.de/~kleinert/65); Fortschr. Phys. *30*, 401 (1982) (*ibid.*http/83). See also the historical remarks in the preface.

quantum fluctuations. Within this approach, the finite time slicing is essential for being able to perform this sum in any nontrivial system.

We now present a general solution to the stability problem of time-sliced quantum-statistical path integrals.

12.2 Stable Path Integral with Singular Potentials

Consider the fixed-energy amplitude (9.1) which is the local matrix element

$$(\mathbf{x}_b|\mathbf{x}_a)_E = \langle \mathbf{x}_b|\hat{R}|\mathbf{x}_a\rangle \tag{12.6}$$

of the resolvent operator (1.315):

$$\hat{R} = \frac{i\hbar}{E - \hat{H} + i\eta}. \tag{12.7}$$

Recall that the $i\eta$-prescription ensures the causality of the Fourier transform of (12.6), making it vanish for $t_b < t_a$ [see the discussion after Eq. (1.323)].

The fixed-energy amplitude has poles of the form

$$(\mathbf{x}_b|\mathbf{x}_a)_E = \sum_n \frac{i\hbar}{E - E_n + i\eta}\psi_n(\mathbf{x}_b)\psi_n^*(\mathbf{x}_a) + \dots$$

at the bound-state energies, and a cut along the continuum part of the energy spectrum. The energy integral over the discontinuity across the singularities yields the completeness relation (1.326).

The new path integral formula is based on the following observation. If the system possesses a Feynman path integral for the time evolution amplitude, it does so also for the fixed-energy amplitude. This is seen after rewriting the latter as an integral

$$(\mathbf{x}_b|\mathbf{x}_a)_E = \int_{t_a}^{\infty} dt_b \langle \mathbf{x}_b|\hat{U}_E(t_b - t_a)|\mathbf{x}_a\rangle \tag{12.8}$$

involving the modified time evolution operator

$$\hat{U}_E(t) \equiv e^{-it(\hat{H}-E)/\hbar}, \tag{12.9}$$

that is associated with the modified Hamiltonian

$$\hat{H}_E \equiv \hat{H} - E. \tag{12.10}$$

Obviously, as long as the matrix elements of the ordinary time evolution operator $\hat{U}(t) = e^{-it\hat{H}/\hbar}$ can be represented by a time-sliced Feynman path integral, the same is true for the matrix elements of the modified operator $\hat{U}_E(t) = e^{-it\hat{H}_E/\hbar}$. Its explicit form is obtained, as in Section 2.1, by slicing the t-variable into $N+1$ pieces, factorizing $\exp(-it\hat{H}_E/\hbar)$ into the product of $N+1$ factors,

$$e^{-it\hat{H}_E/\hbar} = e^{-i\epsilon\hat{H}_E/\hbar} \cdots e^{-i\epsilon\hat{H}_E/\hbar}, \tag{12.11}$$

and inserting a sequence of N completeness relations

$$\prod_{n=1}^{N} \int d^D x_n |\mathbf{x}_n\rangle\langle\mathbf{x}_n| = 1 \qquad (12.12)$$

(omitting the continuum part of the spectrum). In this way, we have arrived at the path integral for the time-sliced amplitude with $t_b - t_a = \epsilon(N+1)$

$$\langle\mathbf{x}_b|\hat{U}_E^N(t_b - t_a)|\mathbf{x}_a\rangle = \prod_{n=1}^{N} \left[\int d^D x_n\right] \prod_{n=1}^{N+1} \left[\int \frac{d^D p_n}{(2\pi\hbar)^D}\right] \exp\left(\frac{i}{\hbar} A_E^N\right), \qquad (12.13)$$

where A_E^N is the sliced action

$$A_E^N = \sum_{n=1}^{N+1} \{\mathbf{p}_n(\mathbf{x}_n - \mathbf{x}_{n-1}) - \epsilon[H(\mathbf{p}_n, \mathbf{x}_n) - E]\}. \qquad (12.14)$$

In the limit of large N at fixed $t_b - t_a = \epsilon(N+1)$, this defines the path integral

$$\langle\mathbf{x}_b|\hat{U}_E(t)|\mathbf{x}_a\rangle = \int \mathcal{D}^D x(t') \int \frac{\mathcal{D}^D p(t')}{(2\pi\hbar)^D} \exp\left\{\frac{i}{\hbar}\int_0^t dt'[\mathbf{p}\dot{\mathbf{x}}(t') - H_E(\mathbf{p}(t'), \mathbf{x}(t'))]\right\}. \qquad (12.15)$$

It is easy to derive a finite-N approximation also for the fixed-energy amplitude $(\mathbf{x}_b|\mathbf{x}_a)_E$ of Eq. (12.8). The additional integral over $t_b > t_a$ can be approximated at the level of a finite N by an integral over the slice thickness ϵ:

$$\int_{t_a}^{\infty} dt_b = (N+1)\int_0^{\infty} d\epsilon. \qquad (12.16)$$

The resulting finite-N approximation to the fixed-energy amplitude,

$$(\mathbf{x}_b|\mathbf{x}_a)_E^N \equiv (N+1)\int_0^{\infty} d\epsilon\langle\mathbf{x}_b|\hat{U}_E^N(\epsilon(N+1))|\mathbf{x}_a\rangle = \int_{t_a}^{\infty} dt_b\langle\mathbf{x}_b|\hat{U}_E^N(t_b - t_a)|\mathbf{x}_a\rangle, \qquad (12.17)$$

converges against the correct limit $(\mathbf{x}_b|\mathbf{x}_a)_E$. As an example, take the free-particle case where

$$(\mathbf{x}_b|\mathbf{x}_a)_E^N = (N+1)\int_0^{\infty} d\epsilon\frac{1}{\sqrt{2\pi i(N+1)\epsilon\hbar/M}^D}$$

$$\times \exp\left[i\frac{M}{2(N+1)\epsilon}(\mathbf{x}_b - \mathbf{x}_a)^2 + iE(N+1)\epsilon\right]. \qquad (12.18)$$

After a trivial change of the integration variable, this is the same integral as in (1.339) whose result was given in (1.344) and (1.351), depending on the sign of the energy E. The N-dependence happens to disappear completely as observed in Section 2.2.4. In the general case of an arbitrary smooth potential, the convergence is still assured by the dominance of the kinetic term in the integral measure.

The time-sliced path integral formula for the fixed-energy amplitude $(\mathbf{x}_b|\mathbf{x}_a)_E$ given by (12.17), (12.13), (12.14) has apparently the same range of validity as the original Feynman path integral for the time evolution amplitude $(\mathbf{x}_b t_b | \mathbf{x}_a t_a)$. Thus, so far nothing has been gained. However, the new formula has an important advantage over Feynman's. Due to the additional time integration it possesses a new *functional* degree of freedom. This can be exploited to find a path-integral formula without collapse at imaginary times. The starting point is the observation that the resolvent operator $\hat{R}$ in Eq. (12.7) may be rewritten in the following three ways:

$$\hat{R} = \frac{i\hbar}{\hat{f}_l(E - \hat{H} + i\eta)}\hat{f}_l,$$

(12.19)

or

$$\hat{R} = \hat{f}_r\frac{i\hbar}{(E - \hat{H} + i\eta)\hat{f}_r},$$

(12.20)

or, more generally,

$$\hat{R} = \hat{f}_r\frac{i\hbar}{\hat{f}_l(E - \hat{H} + i\eta)\hat{f}_r}\hat{f}_l,$$

(12.21)

where $\hat{f}_l, \hat{f}_r$ are arbitrary operators which may depend on $\hat{\mathbf{p}}$ and $\hat{\mathbf{x}}$. They are called *regulating functions*. In the subsequent discussion, we shall avoid operator-ordering subtleties by assuming $\hat{f}_l, \hat{f}_r$ to depend only on $\hat{\mathbf{x}}$, although the general case can also be treated along similar lines. Moreover, in the specific application to follow in Chapters 13 and 14, the operators $\hat{f}_l, \hat{f}_r$ to be assumed consist of two different powers of one and the same operator $\hat{f}$, i.e.,

$$\hat{f}_l = \hat{f}^{1-\lambda}, \quad \hat{f}_r = \hat{f}^{\lambda},$$

(12.22)

whose product is

$$\hat{f}_l\hat{f}_r = \hat{f}.$$

(12.23)

Taking the local matrix elements of (12.21) renders the alternative representations for the fixed-energy amplitude

$$\langle \mathbf{x}_b | \hat{R} | \mathbf{x}_a \rangle = (\mathbf{x}_b | \mathbf{x}_a)_E = \int_{s_a}^{\infty} ds_b \langle \mathbf{x}_b | \hat{\mathcal{U}}_E(s_b - s_a) | \mathbf{x}_a \rangle,$$

(12.24)

where $\hat{\mathcal{U}}_E(s)$ is the generalization of the modified time evolution operator (12.9), to be called the *pseudotime evolution operator*,

$$\hat{\mathcal{U}}_E(s) \equiv f_r(\mathbf{x})e^{-isf_l(\mathbf{x})(\hat{H}-E)f_r(\mathbf{x})}f_l(\mathbf{x}).$$

(12.25)

The operator in the exponent,

$$\hat{\mathcal{H}}_E \equiv f_l(\mathbf{x})(\hat{H} - E)f_r(\mathbf{x}),$$

(12.26)

may be considered as an auxiliary Hamiltonian which drives the state vectors $|\mathbf{x}\rangle$ of the system along a pseudotime s-axis, with the operator $e^{-is\hat{\mathcal{H}}_E(\mathbf{p},\mathbf{x})/\hbar}$. Note that $\hat{\mathcal{H}}_E$ is in general not Hermitian, in which case $\hat{\mathcal{U}}_E(s)$ is not unitary.

As usual, we convert the expression (12.24) into a path integral by slicing the pseudotime interval $(0, s)$ into $N+1$ pieces, factorizing $\exp(-is\hat{\mathcal{H}}_E/\hbar)$ into a product of $N + 1$ factors, and inserting a sequence of N completeness relations. The result is the approximate integral representation for the fixed-energy amplitude,

$$(\mathbf{x}_b|\mathbf{x}_a)_E \ \approx \ (N+1)\int_0^\infty d\epsilon_s\langle\mathbf{x}_b|\hat{\mathcal{U}}_E^N\left(\epsilon_s(N+1)\right)|\mathbf{x}_a\rangle, \tag{12.27}$$

with the path integral for the pseudotime-sliced amplitude

$$\langle\mathbf{x}_b|\hat{\mathcal{U}}_E^N\left(\epsilon_s(N+1)\right)|\mathbf{x}_a\rangle = f_r(\mathbf{x}_b)f_l(\mathbf{x}_a)\prod_{n=1}^{N}\left[\int d^D x_n\right]\prod_{n=1}^{N+1}\left[\int\frac{d^D p_n}{(2\pi\hbar)^D}\right]e^{i\mathcal{A}_E^N/\hbar}, \tag{12.28}$$

whose time-sliced action reads

$$\mathcal{A}_E^N = \sum_{n=1}^{N+1}\left\{\mathbf{p}_n(\mathbf{x}_n - \mathbf{x}_{n-1}) - \epsilon_s f_l(\mathbf{x}_n)[H(\mathbf{p}_n, \mathbf{x}_n) - E]f_r(\mathbf{x}_n)\right\}. \tag{12.29}$$

These equations constitute the desired generalization of the formulas (12.13)–(12.17). In the limit of large N, we can write the fixed-energy amplitude as an integral

$$(\mathbf{x}_b|\mathbf{x}_a)_E = \int_0^\infty dS\langle\mathbf{x}_b|\hat{\mathcal{U}}_E(S)|\mathbf{x}_a\rangle \tag{12.30}$$

over the amplitude

$$\langle\mathbf{x}_b|\hat{\mathcal{U}}_E(S)|\mathbf{x}_a\rangle = f_r(\mathbf{x}_b)f_l(\mathbf{x}_a)\int\mathcal{D}x(s)\int\frac{\mathcal{D}p(s)}{2\pi\hbar}\exp\left\{\frac{i}{\hbar}\int_0^S ds[\mathbf{p}\mathbf{x}' - \mathcal{H}_E(\mathbf{p}, \mathbf{x})]\right\}. \tag{12.31}$$

The prime on $\mathbf{x}(s)$ denotes the derivative with respect to the pseudotime s.

For a standard Hamiltonian of the form

$$H = T(\mathbf{p}) + V(\mathbf{x}), \tag{12.32}$$

with the kinetic energy

$$T(\mathbf{p}) = \frac{\mathbf{p}^2}{2M}, \tag{12.33}$$

the momenta $\mathbf{p}_n$ in (12.28) can be integrated out and we obtain the configuration space path integral

$$\langle\mathbf{x}_b|\hat{\mathcal{U}}_E^N\left(\epsilon_s(N+1)\right)|\mathbf{x}_a\rangle = \frac{f_r(\mathbf{x}_b)f_l(\mathbf{x}_a)}{\sqrt{2\pi i\epsilon_s f_l(\mathbf{x}_b)f_r(\mathbf{x}_a)\hbar/M}^D} \tag{12.34}$$

$$\times\prod_{n=1}^{N}\left[\int\frac{d^D x_n}{\sqrt{2\pi i\epsilon_s f(\mathbf{x}_n)\hbar/M}^D}\right]e^{i\mathcal{A}_E^N/\hbar},$$

with the sliced action

$$\mathcal{A}_E^N = \sum_{n=1}^{N+1} \left\{ \frac{M}{2\epsilon_s f_l(\mathbf{x}_n) f_r(\mathbf{x}_{n-1})} (\mathbf{x}_n - \mathbf{x}_{n-1})^2 - \epsilon_s f_l(\mathbf{x}_n)[V(\mathbf{x}_n) - E] f_r(\mathbf{x}_{n-1}) \right\}. \quad (12.35)$$

In the limit of large N, this may be written as a path integral

$$\langle \mathbf{x}_b | \hat{\mathcal{U}}_E(S) | \mathbf{x}_a \rangle = f_r(\mathbf{x}_b) f_l(\mathbf{x}_a) \int \mathcal{D}x(s) \exp\left\{ \frac{i}{\hbar} \int_0^S ds \left[\frac{M}{2 f_l f_r} \mathbf{x}'^2 - f_l(V - E) f_r \right] \right\}, \quad (12.36)$$

with the slicing specification (12.35).

The path integral formula for the fixed-energy amplitude based on Eqs. (12.30) and (12.36) is independent of the particular choice of the functions $f_l(\mathbf{x}), f_r(\mathbf{x})$, just like the most general operator expression for the resolvent (12.21). Feynman's original time-sliced formula is, of course, recovered with the special choice $f_l(\mathbf{x}) \equiv f_r(\mathbf{x}) \equiv 1$.

When comparing (12.25) with (12.9), we see that for each infinitesimal pseudo-time slice, the thickness of the true time slices has the space-dependent value

$$dt = ds\, f_l(\mathbf{x}_n) f_r(\mathbf{x}_{n-1}). \quad (12.37)$$

The freedom in choosing $f(\mathbf{x})$ amounts to an *invariance under path-dependent time reparametrizations* of the fixed-energy amplitude (12.30). Note that the invariance is exact in the general operator formula (12.21) for the resolvent and in the continuum path integral formula based on (12.30) and (12.36). However, the finite pseudotime slicing in (12.34), (12.35) used to define the path integral, destroys this invariance. At a finite value of N, different choices of $f(\mathbf{x})$ produce different approximations to the matrix element of the operator $\hat{\mathcal{U}}_E(s) = f_r(\hat{\mathbf{x}}) e^{-is\hat{\mathcal{H}}_E(\hat{\mathbf{p}}, \hat{\mathbf{x}})/\hbar} f_l(\hat{\mathbf{x}})$. Their quality can vary greatly. In fact, if the potential is singular and the regulating functions $f_r(\mathbf{x}), f_l(\mathbf{x})$ are not suitably chosen, the Euclidean pseudotime-sliced expression may not exist at all. This is what happens in the Coulomb system if the functions $f_l(\mathbf{x})$ and $f_r(\mathbf{x})$ are both chosen to be unity as in Feynman's path integral formula.

The new reparametrization freedom gained by the functions $f_l(\mathbf{x}), f_r(\mathbf{x})$ is therefore not just a luxury. It is *essential* for stabilizing the Euclidean time-sliced orbital fluctuations in singular potentials.

In the case of the Coulomb system, any choice of the regulating functions $f_l(\mathbf{x}), f_r(\mathbf{x})$ with $f(\mathbf{x}) = r$ leads to a regular auxiliary Hamiltonian $\mathcal{H}_E$, and the path integral expressions (12.27)–(12.36) are all well defined. This was the important discovery of Duru and Kleinert in 1979, to be described in detail in Chapter 13, which has made a large class of previously non-existing Feynman path integrals solvable. By a similar Duru-Kleinert transformation with $f_l(\mathbf{x}), f_r(\mathbf{x})$ with $f(\mathbf{x}) = \sqrt{f_l(\mathbf{x}) f_r(\mathbf{x})} = r^2$, the earlier difficulties with the centrifugal barrier are resolved, as will be seen in Chapter 14.

12.3 Time-Dependent Regularization

Before treating specific cases, let us note that there exists a further generalization of the above path integral formula which is useful in systems with a time-dependent Hamiltonian $H(\mathbf{p}, \mathbf{x}, t)$. There we introduce an auxiliary Hamiltonian

$$\hat{\mathcal{H}} = f_l(\mathbf{x}, t)[H(\hat{\mathbf{p}}, \mathbf{x}, t) - \hat{E}]f_r(\mathbf{x}, t), \tag{12.38}$$

where $\hat{E}$ is the differential operator for the energy which is canonically conjugate to the time t:

$$\hat{E} \equiv i\hbar\partial_t. \tag{12.39}$$

The auxiliary Hamiltonian acts on an extended Hilbert space, in which the states are localized in space *and* time. These states will be denoted by $|\mathbf{x}, t\}$. They satisfy the orthogonality and completeness relations

$$\{\mathbf{x}\, t | \mathbf{x}'\, t'\} = \delta^{(D)}(\mathbf{x} - \mathbf{x}')\delta(t - t'), \tag{12.40}$$

and

$$\int d^D x \int dt |\mathbf{x}\, t\}\{\mathbf{x}\, t| = 1, \tag{12.41}$$

respectively. By construction, the Hamiltonian $\mathcal{H}$ does not depend explicitly on the pseudotime s. The pseudotime evolution operator is therefore obtained by a simple exponentiation, as in (12.25),

$$\hat{\mathcal{U}}(s) \equiv f_r(\mathbf{x}, t)e^{-isf_l(\mathbf{x}, t)(\hat{H} - \hat{E})f_r(\mathbf{x}, t)}f_l(\mathbf{x}, t). \tag{12.42}$$

The derivation of the path integral is then completely analogous to the time-independent case. The operator (12.42) is sliced into $N + 1$ pieces, and N completeness relations (12.41) are inserted to obtain the path integral

$$\{\mathbf{x}_b t_b | \hat{\mathcal{U}}^N(s) | \mathbf{x}_a t_a\} = f_r(\mathbf{x}_b, t_b)f_l(\mathbf{x}_a, t_a)$$
$$\times \prod_{n=1}^{N} \left[\int d^D x_n dt_n\right] \prod_{n=1}^{N+1} \left[\int \frac{d^D p_n}{(2\pi\hbar)^D} \frac{dE_n}{2\pi\hbar}\right] e^{i\mathcal{A}^N/\hbar}, \tag{12.43}$$

with the pseudotime-sliced action

$$\mathcal{A}^N = \sum_{n=1}^{N+1} \{\mathbf{p}_n(\mathbf{x}_n - \mathbf{x}_{n-1}) - E_n(t_n - t_{n-1})$$
$$- f_l(\mathbf{x}_n, t_n)\left[H(\mathbf{p}_n, \mathbf{x}_n, t_n) - E_n\right]f_r(\mathbf{x}_{n-1}, t_{n-1})\}, \tag{12.44}$$

where $\mathbf{x}_b = \mathbf{x}_{N+1}$, $t_b = t_{N+1}$; $\mathbf{x}_a = \mathbf{x}_0$, $t_a = t_0$. This describes orbital fluctuations in the phase space of spacetime which contains fluctuating worldlines $\mathbf{x}(s), t(s)$ and their canonically conjugate spacetime $\mathbf{p}(s), E(s)$. In the limit $N \to \infty$ we write this as

$$
\begin{aligned}
\{\mathbf{x}_b t_b | \hat{\mathcal{U}}(S) | \mathbf{x}_a t_a\} &= f_r(\mathbf{p}_b, \mathbf{x}_b, t_b) f_l(\mathbf{p}_a, \mathbf{x}_a, t_a) \\
&\times \int \mathcal{D}^D x(s) \mathcal{D}t(s) \int \frac{\mathcal{D}^D p(s)}{(2\pi\hbar)^D} \frac{\mathcal{D}E(s)}{2\pi\hbar} e^{i\mathcal{A}/\hbar},
\end{aligned} \quad (12.45)
$$

with the continuous action

$$
\begin{aligned}
\mathcal{A}[\mathbf{p}, \mathbf{x}, E, t] &= \int_0^S ds \{ \mathbf{p}(s) \mathbf{x}'(s) - E(s) t'(s) \\
&- f_l(\mathbf{p}(s), \mathbf{x}(s), t(s)) \left[H(\mathbf{p}(s), \mathbf{x}(s), t(s)) - E(s) \right] f_r(\mathbf{p}(s), \mathbf{x}(s), t(s)) \}.
\end{aligned} \quad (12.46)
$$

In the pseudotime-sliced formula (12.43), we can integrate out all intermediate energy variables E_n and obtain

$$
\begin{aligned}
\{\mathbf{x}_b t_b | \hat{\mathcal{U}}(S) | \mathbf{x}_a t_a\} &= \prod_{n=1}^{N} \left[\int d^D x_n \right] \prod_{n=1}^{N+1} \left[\int \frac{d^D p_n}{(2\pi\hbar)^D} \right] \\
&\times \delta \left(t_b - t_a - \epsilon_s \sum_{n=1}^{N+1} f_l(\mathbf{p}_n, \mathbf{x}_n, t_n) f_r(\mathbf{p}_{n-1}, \mathbf{x}_{n-1}, t_{n-1}) \right) e^{i\tilde{\mathcal{A}}^N / \hbar}
\end{aligned} \quad (12.47)
$$

with the action

$$
\tilde{\mathcal{A}}^N = \sum_{n=1}^{N+1} [\mathbf{p}_n (\mathbf{x}_n - \mathbf{x}_{n-1}) - \epsilon_s f_l(\mathbf{p}_n \mathbf{x}_n, t_n) H(\mathbf{p}_n, \mathbf{x}_n, t_n) f_r(\mathbf{p}_{n-1}, \mathbf{x}_{n-1}, t_{n-1})]. \quad (12.48)
$$

This looks just like an ordinary time-sliced action with a time-dependent Hamiltonian. The constant width of the time slices $\epsilon = (t_b - t_a)/(N+1)$, however, has now become variable and depends on phase space and time:

$$
\epsilon \to \epsilon_s f_l(\mathbf{p}_n \mathbf{x}_n, t_n) f_r(\mathbf{p}_{n-1}, \mathbf{x}_{n-1}, t_{n-1}). \quad (12.49)
$$

The δ-function in (12.47) ensures the correct relation between the pseudotime s and the physical time t. In the continuum limit we may write (12.47) as

$$
\{\mathbf{x}_b t_b | \hat{\mathcal{U}}(S) | \mathbf{x}_a t_a\} = \int \mathcal{D}^D x(s) \int \frac{\mathcal{D}^D p(s)}{(2\pi\hbar)^D} \delta(t_b - t_a - \int_0^S ds \, f(\mathbf{x}, t)) e^{i\tilde{\mathcal{A}}/\hbar}, \quad (12.50)
$$

with the pseudotime action

$$
\tilde{\mathcal{A}}[\mathbf{p}, \mathbf{x}, t] = \int_0^S ds \left[\mathbf{p} \mathbf{x}' - f_l(\mathbf{x}, t) H(\mathbf{p}, \mathbf{x}, t) f_r(\mathbf{x}, t) \right], \quad (12.51)
$$

which is a functional of the s-dependent paths $\mathbf{x}(s), \mathbf{p}(s), t(s)$. Note that in the continuum formula, the splitting of the regulating function $f(\mathbf{x}, t)$ into $f_l(\mathbf{x}, t)$ and $f_r(\mathbf{x}, t)$ according to the parameter λ in Eq. (12.22) cannot be expressed properly since f_l, H, and f_r are commuting c-number functions. We have written them in a way indicating their order in the time-sliced expressions (12.47), (12.48).

The integral over S yields the original time evolution amplitude

$$(\mathbf{x}_b t_a | \mathbf{x}_a t_a) = \int_0^\infty dS \{\mathbf{x}_b t_b | \hat{\mathcal{U}}(S) | \mathbf{x}_a t_a\} = \left\{ \mathbf{x}_b t_b \left| \frac{i\hbar}{\hat{H} - \hat{E}} \right| \mathbf{x}_a t_a \right\}. \tag{12.52}$$

Indeed, by Fourier decomposing the scalar products $\{\mathbf{x}_b t_b | \mathbf{x}_a t_a\}$,

$$\{\mathbf{x}_b t_b | \mathbf{x}_a t_a\} = \int \frac{d^D p}{(2\pi\hbar)^D} \int \frac{dE}{2\pi\hbar} e^{i\mathbf{p}(\mathbf{x}_b - \mathbf{x}_a)/\hbar - iE(t_b - t_a)/\hbar}, \tag{12.53}$$

we see that the right-hand side satisfies the same Schrödinger equation as the left-hand side:

$$[H(-i\hbar\partial_\mathbf{x}, \mathbf{x}, t) - i\hbar\partial_t] (\mathbf{x} t | \mathbf{x}_a t_a) = -i\hbar\delta^{(D)}(\mathbf{x} - \mathbf{x}_a)\delta(t - t_a) \tag{12.54}$$

[recall (1.304) and (12.5)]. If the δ-function in (12.47) is written as a Fourier integral, we obtain a kind of spectral decomposition of the amplitude (12.52),

$$(\mathbf{x}_b t_a | \mathbf{x}_a t_a) = \int_{-\infty}^\infty dE e^{-iE(t_b - t_a)/\hbar} \int_0^\infty dS \{\mathbf{x}_b t_b | \hat{\mathcal{U}}_E(S) | \mathbf{x}_a t_a\}, \tag{12.55}$$

with the pseudotime evolution amplitude:

$$\hat{\mathcal{U}}_E(s) \equiv f_r(\hat{\mathbf{p}}, \mathbf{x}, t) e^{-is f_l(\hat{\mathbf{p}}, \mathbf{x}, t)(\hat{H} - E) f_r(\hat{\mathbf{p}}, \mathbf{x}, t)} f_l(\hat{\mathbf{p}}, \mathbf{x}, t). \tag{12.56}$$

12.4   Relation to Schrödinger Theory. Wave Functions

For completeness, consider also the ordinary Schrödinger quantum mechanics described by the pseudo-Hamiltonian $\hat{\mathcal{H}}$. This operator is the generator of translations of the system along the pseudotime axis s. Let $\phi(\mathbf{x}, t, s)$ be a solution of the pseudotime Schrödinger equation

$$\mathcal{H}(\hat{\mathbf{p}}, \mathbf{x}, \hat{E}, t)\phi(\mathbf{x}, t, s) = i\hbar\partial_s\phi(\mathbf{x}, t, s), \tag{12.57}$$

written more explicitly as

$$f_l(\mathbf{x}, t) [H(\hat{\mathbf{p}}, \mathbf{x}, t) - i\hbar\partial_t] f_r(\mathbf{x}, t)\phi(\mathbf{x}, t, s) = i\hbar\partial_s\phi(\mathbf{x}, t, s). \tag{12.58}$$

Since the left-hand side is independent of s, the s-dependence of $\phi(\mathbf{x}, t, s)$ can be factored out:

$$\phi(\mathbf{x}, t, s) = \phi_\mathcal{E}(\mathbf{x}, t)e^{-i\mathcal{E}s/\hbar}. \tag{12.59}$$

If H is independent of the time t, it is always possible to stabilize the path integral by a time-independent reparametrization function $f(\mathbf{x})$. Then we remove an oscillating factor $e^{-iEt/\hbar}$ from $\phi_\mathcal{E}(\mathbf{x}, t)$ and factorize

$$\phi_\mathcal{E}(\mathbf{x}, t) = \phi_{\mathcal{E}, E}(\mathbf{x})e^{-iEt/\hbar}. \tag{12.60}$$

This leaves us with the time- and pseudotime-independent equation

$$\mathcal{H}(\hat{\mathbf{p}}, \mathbf{x}, E)\phi_{\mathcal{E},E}(\mathbf{x}) = f_l(\mathbf{x})\left[H(\hat{\mathbf{p}}, \mathbf{x}) - E\right]f_r(\mathbf{x})\phi_{\mathcal{E},E}(\mathbf{x})$$
$$= \mathcal{E}\phi_{\mathcal{E},E}(\mathbf{x}). \tag{12.61}$$

For each value of E, there will be a different spectrum of eigenvalues $\mathcal{E}_n$. This is indicated by writing the eigenvalues $\mathcal{E}_n$ as $\mathcal{E}_n(E)$ and the associated eigenstates $\phi_{\mathcal{E}_n,E}(\mathbf{x})$ as $\phi_{\mathcal{E}_n(E)}$.

Suppose that we possess a complete set of such eigenstates at a fixed energy E labeled by a quantum number n (which is here assumed to take discrete values although it may include continuous values, as usual). We can then write down a spectral representation for the local matrix elements of the pseudotime evolution amplitude (12.25):

$$\langle \mathbf{x}_b|\hat{\mathcal{U}}_E(S)|\mathbf{x}_a\rangle = f_r(\mathbf{x}_b)f_l(\mathbf{x}_a)\sum_n \phi_{\mathcal{E}_n(E)}(\mathbf{x}_b)\phi^*_{\mathcal{E}_n(E)}(\mathbf{x}_a)e^{-iS\mathcal{E}_n(E)/\hbar}. \tag{12.62}$$

From this we find the expansion for the fixed-energy amplitude (12.24):

$$(\mathbf{x}_b|\mathbf{x}_a)_E = f_r(\mathbf{x}_b)f_l(\mathbf{x}_a)\sum_n \phi_{\mathcal{E}_n(E)}(\mathbf{x}_b)\phi^*_{\mathcal{E}_n(E)}(\mathbf{x}_a)\frac{i\hbar}{\mathcal{E}_n(E)}.$$

The time evolution amplitude is given by the Fourier transform

$$(\mathbf{x}_b t_b|\mathbf{x}_a t_a) = f_r(\mathbf{x}_b)f_l(\mathbf{x}_a)\int_{-\infty}^{\infty}\frac{dE}{2\pi\hbar}e^{iE(t_b-t_a)/\hbar}\sum_n \phi_{\mathcal{E}_n(E)}(\mathbf{x}_b)\phi^*_{\mathcal{E}_n(E)}(\mathbf{x}_a)\frac{i\hbar}{\mathcal{E}_n(E)}. \tag{12.63}$$

This is to be compared with the usual spectral representation of this amplitude for the time-independent Hamiltonian $\hat{H}$

$$(\mathbf{x}_b t_b|\mathbf{x}_a t_a) = \sum_n \psi_n(\mathbf{x}_b)\psi^*_n(\mathbf{x}_a)e^{-iE_n(t_b-t_a)/\hbar}, \tag{12.64}$$

where $\psi_n(\mathbf{x})$ are the solutions of the ordinary time-independent Schrödinger equation:

$$H(\hat{\mathbf{p}}, \mathbf{x})\psi_n(\mathbf{x}) = E_n\psi_n(\mathbf{x}). \tag{12.65}$$

The relation between the two representations (12.63) and (12.64) is found by observing that for the energy E coinciding with the energy E_n, the eigenvalue $\mathcal{E}_n(E)$ vanishes, i.e., $i\hbar/\mathcal{E}_n(E)$ has poles at $E = E_n$ of the form

$$\frac{i\hbar}{\mathcal{E}_n(E)} \approx \frac{1}{\mathcal{E}'_n(E_n)}\frac{i\hbar}{E - E_n + i\eta}. \tag{12.66}$$

These contribute to the energy integral in (12.63) with a sum

$$(\mathbf{x}_b t_b|\mathbf{x}_a t_a) \sim f_r(\mathbf{x}_b)f_l(\mathbf{x}_a)\sum_n \phi_{\mathcal{E}_n(E_n)}(\mathbf{x}_b)\phi^*_{\mathcal{E}_n(E_n)}(\mathbf{x}_a)e^{-iE_n(t_b-t_a)/\hbar}. \tag{12.67}$$

A comparison with (12.64) shows the relation between the bound-state wave functions of the ordinary and the pseudotime Schrödinger equation. In general, the function $i\hbar/\mathcal{E}_n(E)$ also has cuts whose discontinuities contain the continuum wave functions of the Schrödinger equation (12.65).

These observations will become more transparent in Section 13.8 when treating in detail the bound and continuum wave functions of the Coulomb system.

Notes and References

The general path integral formula with time reparametrization was introduced by
H. Kleinert, Phys. Lett. A **120**, 361 (1987) (http://www.physik.fu-berlin.de/~kleinert/163).
The stability aspects are discussed in
H. Kleinert, Phys. Lett. B **224**, 313 (1989) (*ibid.*http/195).

Et modo quae fuerat semita, facta via est.
What was only a path is now made a high road.
MARTIAL, Epig., Book 7, 60

13

Path Integral of Coulomb System

One of the most important successes of Schrödinger quantum mechanics is the explanation of the energy levels and transition amplitudes of the hydrogen atom. Within the path integral formulation of quantum mechanics, this fundamental system has resisted for many years all attempts at a solution. An essential advance was made in 1979 when Duru and Kleinert [1] recognized the need to work with a generalized pseudotime-sliced path integral of the type described in Chapter 12. After an appropriate coordinates transformation the path integral became harmonic and solvable. A generalization of this two-step transformation has meanwhile led to the solution of many other path integrals to be presented in Chapter 14. The final solution of the problem turned out to be quite subtle due to the nonholonomic nature of the subsequent coordinate transformation which required the development of a correct path integral in spaces with curvature and torsion [2], as done in Chapter 10. Only this made it possible to avoid unwanted fluctuation corrections in the Duru-Kleinert transformation of the Coulomb system, a problem in all earlier attempts.

The first consistent solution was presented in the first edition of this book in 1990.

13.1 Pseudotime Evolution Amplitude

Consider the path integral for the time evolution amplitude of an electron-proton system with a Coulomb interaction. If m_e and m_p denote the masses of the two particles whose reduced mass is $M = m_e m_p/(m_e + m_p)$, and if e is the electron charge, the system is governed by the Hamiltonian

$$H = \frac{p^2}{2M} - \frac{e^2}{r}. \tag{13.1}$$

The formal continuum path integral for the time evolution amplitude reads

$$(\mathbf{x}_b t_b | \mathbf{x}_a t_a) = \int \mathcal{D}^3 x(t) \exp\left[\frac{i}{\hbar} \int_{t_b}^{t_a} dt (\mathbf{p}\dot{\mathbf{x}} - H)\right]. \tag{13.2}$$

As observed in the last chapter, its Euclidean version cannot be time-sliced into a finite number of integrals since the paths would collapse. The paths would stretch

931

out into a straight line with $\dot{x} \approx 0$ and slide down into the $1/r$-abyss. A path integral whose Euclidean version is stable can be written down using the pseudotime evolution amplitude (12.28). A convenient family of regulating functions is

$$f_l(\mathbf{x}) = f(\mathbf{x})^{1-\lambda}, \quad f_r(\mathbf{x}) = f(\mathbf{x})^\lambda, \tag{13.3}$$

whose product satisfies $f_l(\mathbf{x})f_r(\mathbf{x}) = f(\mathbf{x}) = r$. Since the path integral represents the general resolvent operator (12.21), all results must be independent of the splitting parameter λ after going to the continuum limit. This independence is useful in checking the calculations.

Thus we consider the fixed-energy amplitude

$$(\mathbf{x}_b|\mathbf{x}_a)_E = \int_0^\infty dS \langle \mathbf{x}_b|\hat{\mathcal{U}}_E(S)|\mathbf{x}_a \rangle, \tag{13.4}$$

with the pseudotime-evolution amplitude

$$\langle \mathbf{x}_b|\hat{\mathcal{U}}_E(S)|\mathbf{x}_a \rangle = r_b^\lambda \, r_a^{1-\lambda} \int \mathcal{D}^D x(s) \int \frac{\mathcal{D}^D p(s)}{(2\pi\hbar)^D} \exp\left\{ \frac{i}{\hbar} \int_0^S ds[\mathbf{p}\mathbf{x}' - r^{1-\lambda}(H - E)r^\lambda] \right\}, \tag{13.5}$$

where the prime denotes the derivative with respect to the pseudotime argument s. For the sake of generality, we have allowed for a general dimension D of orbital motion. After time slicing and with the notation $\Delta\mathbf{x}_n \equiv \mathbf{x}_n - \mathbf{x}_{n-1}$, $\epsilon_s \equiv S/(N+1)$, the amplitude (13.5) reads

$$\langle \mathbf{x}_b|\hat{\mathcal{U}}_E(S)|\mathbf{x}_a \rangle \approx r_b^\lambda r_a^{1-\lambda} \prod_{n=2}^{N+1} \left[\int_{-\infty}^\infty d^D\Delta x_n \right] \prod_{n=1}^{N+1} \left[\int_{-\infty}^\infty \frac{d^D p_n}{(2\pi\hbar)^D} \right] e^{i\mathcal{A}_E^N/\hbar}, \tag{13.6}$$

where the action is

$$\mathcal{A}_E^N[\mathbf{p}, \mathbf{x}] = \sum_{n=1}^{N+1} \left[\mathbf{p}_n \Delta\mathbf{x}_n - \epsilon_s \, r_n^{1-\lambda} r_{n-1}^\lambda \left(\frac{\mathbf{p}_n^2}{2M} - E \right) + \epsilon_s \, e^2 \right]. \tag{13.7}$$

The term $\epsilon_s \, e^2$ carries initially a factor $(r_{n-1}/r_n)^\lambda$ which is dropped, since it is equal to unity in the continuum limit. When integrating out the momentum variables, $N + 1$ factors $1/(r_n^{1-\lambda} r_{n-1}^\lambda)^{D/2}$ appear. After rearranging these, the configuration space path integral becomes

$$\langle \mathbf{x}_b|\hat{\mathcal{U}}_E(S)|\mathbf{x}_a \rangle \approx \frac{r_b^\lambda r_a^{1-\lambda}}{\sqrt{2\pi \, i\epsilon_s\hbar \, r_b^{1-\lambda} r_a^\lambda/M}^D} \prod_{n=2}^{N+1} \left[\int \frac{d^D\Delta x_n}{\sqrt{2\pi i\epsilon_s\hbar r_{n-1}/M}^D} \right] e^{i\mathcal{A}_E^N[\mathbf{x}, \mathbf{x}']/\hbar}, \tag{13.8}$$

with the pseudotime-sliced action

$$\mathcal{A}_E^N[\mathbf{x}, \mathbf{x}'] = (N + 1)\epsilon_s e^2 + \sum_{n=1}^{N+1} \left[\frac{M}{2} \frac{(\Delta\mathbf{x}_n)^2}{\epsilon_s r_n^{1-\lambda} r_{n-1}^\lambda} + \epsilon_s \, r_n E \right]. \tag{13.9}$$

In the last term, we have replaced $r_n^{1-\lambda} r_{n-1}^{\lambda}$ by r_n without changing the continuum limit. The limiting action can formally be written as

$$\mathcal{A}_E[\mathbf{x}, \mathbf{x}'] = e^2 S + \int_0^S ds \left(\frac{M}{2r} \mathbf{x}'^2 + Er \right). \tag{13.10}$$

We now solve the Coulomb path integral first in two dimensions, assuming that the movement of the electron is restricted to a plane while the electric field extends into the third dimension. Afterwards we proceed to the physical three-dimensional system. The case of an arbitrary dimension D will be solved in Chapter 14. The one-dimensional case will not be treated here. The exact energy levels were found before at the and of Section 4.1 from a semiclassical expansion. For a long time, the one-dimensional Coulomb system was only of mathematical interest. Recently, however, it has received increased attention due to the possibility of forming hydrogen ätoms" in a so-called *quantum wire* [3].

13.2 Solution for the Two-Dimensional Coulomb System

First we observe that the kinetic pseudoenergy has a scale dimension $[rp^2] \sim [r^{-1}]$ which is precisely opposite to that of the potential term $[r^{+1}]$. The dimensional situation is similar to that of the harmonic oscillator, where the dimensions are $[p^2] = [r^{-2}]$ and $[r^{+2}]$, respectively. It is possible make the correspondence perfect by describing the Coulomb system in terms of "square root coordinates", i.e., by transforming $r \to u^2$. In two dimensions, the appropriate square root is given by the *Levi-Civita transformation*

$$
\begin{aligned}
x^1 &= (u^1)^2 - (u^2)^2, \\
x^2 &= 2u^1 u^2.
\end{aligned} \tag{13.11}
$$

If we imagine the vectors $\mathbf{x}$ and $\mathbf{u}$ to move in the complex planes parametrized by $x = x^1 + ix^2$ and $u = u^1 + iu^2$, the transformed variable u corresponds to the complex square root:

$$u = \sqrt{x}. \tag{13.12}$$

Let us also introduce the matrix

$$A(\mathbf{u}) = \begin{pmatrix} u^1 & -u^2 \\ u^2 & u^1 \end{pmatrix}, \tag{13.13}$$

and write (13.11) as a matrix equation:

$$\mathbf{x} = A(\mathbf{u})\mathbf{u}. \tag{13.14}$$

The Levi-Civita transformation is an integrable coordinate transformation which carries the flat x^i-space into a flat u^μ-space. We mention this fact since in the later treatment of the three-dimensional hydrogen atom, the transition to the "square root coordinates" will require a nonintegrable (nonholonomic) coordinate transformation defined only differentially. As explained in Chapter 10, such mappings change, in general, a flat Euclidean space into a space with curvature and torsion. The generation of torsion is precisely the reason why the three-dimensional system remained unsolved until 1990. In two dimensions, this phenomenon happens to be absent.

If we write the transformation (13.11) in terms of a basis dyad $e^i{}_\mu(\mathbf{u})$ as $dx^i = e^i{}_\mu(\mathbf{u})\,du^\mu$, this is given by

$$e^i{}_\mu(\mathbf{u}) = \frac{\partial x^i}{\partial u^\mu}(\mathbf{u}) = 2A^i{}_\mu(\mathbf{u}), \tag{13.15}$$

with the reciprocal dyad

$$e_i{}^\mu(\mathbf{u}) = \frac{1}{2}(A^{-1})^T{}_i{}^\mu(\mathbf{u}) = \frac{1}{2\mathbf{u}^2}A^i{}_\mu(\mathbf{u}). \tag{13.16}$$

The associated affine connection

$$\Gamma_{\mu\nu}{}^\lambda = e_i{}^\lambda\,\partial_\mu\,e^i{}_\nu = \frac{1}{\mathbf{u}^2}[(\partial_\mu A)^T A]_{\nu\lambda} \tag{13.17}$$

has the matrix elements $(\Gamma_\mu)_\nu{}^\lambda = \Gamma_{\mu\nu}{}^\lambda$:

$$(\Gamma_1)_\mu{}^\nu = \frac{1}{\mathbf{u}^2}\left(\begin{array}{cc} u^1 & -u^2 \\ u^2 & u^1 \end{array}\right)_\mu{}^\nu = \frac{1}{2\mathbf{u}^2}A(\mathbf{u})^\mu{}_\nu,$$

$$(\Gamma_2)_\mu{}^\nu = \frac{1}{\mathbf{u}^2}\left(\begin{array}{cc} u^2 & u^1 \\ -u^1 & u^2 \end{array}\right)_\mu{}^\nu. \tag{13.18}$$

The affine connection satisfies the important identity

$$\Gamma_\mu{}^{\mu\nu} \equiv 0, \tag{13.19}$$

which follows from the defining relation

$$\Gamma_\mu{}^{\mu\nu} \equiv g^{\mu\nu}e_i{}^\lambda\,\partial_\mu\,e^i{}_\nu, \tag{13.20}$$

by inserting the obvious special property of $e^i{}_\mu$

$$\partial_\mu e^i{}_\mu = \partial_\mathbf{u}^2 x^i(\mathbf{u}) = 0, \tag{13.21}$$

using the diagonality of $g^{\mu\nu} = \delta^{\mu\nu}/4r$.

The identity (13.19) will be shown in Section 13.6 to be the essential geometric reason for the absence of the time slicing corrections.

The torsion and the Riemann-Cartan curvature tensor vanish identically, the former because of the specific form of the matrix elements (13.18), the latter due to the linearity of the basis dyads $e^i{}_\mu(\mathbf{u})$ in $\mathbf{u}$ which guarantees trivially the integrability conditions, i.e.,

$$e_i{}^\lambda\left(\partial_\mu e^i{}_\nu - \partial_\nu e^i{}_\mu\right) \equiv 0, \tag{13.22}$$

$$e_i{}^\kappa(\partial_\mu\partial_\nu - \partial_\nu\partial_\mu)e^i{}_\lambda \equiv 0, \tag{13.23}$$

and thus $S_{\mu\nu}{}^{\lambda} \equiv 0$, $R_{\mu\nu\lambda}{}^{\kappa} \equiv 0$.

In the continuum limit, the Levi-Civita transformation converts the action (13.10) into that of a harmonic oscillator. With

$$\mathbf{x}'^2 = 4\mathbf{u}^2 \mathbf{u}'^2 = 4r\,\mathbf{u}'^2 \tag{13.24}$$

we find

$$\mathcal{A}[\mathbf{x}] = e^2 S + \int_0^S ds \left(\frac{4M}{2}\mathbf{u}'^2 + E\mathbf{u}^2 \right). \tag{13.25}$$

Apart from the trivial term $e^2 S$, this is the action of a harmonic oscillator

$$\mathcal{A}_{\mathrm{os}}[\mathbf{u}] = \int_0^S ds \frac{\mu}{2}(\mathbf{u}'^2 - \omega^2 \mathbf{u}^2), \tag{13.26}$$

which oscillates in the pseudotime s with an effective mass

$$\mu = 4M, \tag{13.27}$$

and a pseudofrequency

$$\omega = \sqrt{-E/2M}. \tag{13.28}$$

Note that ω has the dimension $1/s$ corresponding to $[\omega] = [r/t]$ (in contrast to a usual frequency whose dimension is $[1/t]$).

The path integral is well defined only as long as the energy E of the Coulomb system is negative, i.e., in the bound-state regime. The amplitude in the continuum regime with positive E will be obtained by analytic continuation.

In the regularized form, the pseudotime-sliced amplitude is calculated as follows. Choosing a splitting parameter $\lambda = 1/2$ and ignoring for the moment all complications due to the finite time slicing, we deduce from (13.14) that

$$d\mathbf{x} = 2A(\mathbf{u})d\mathbf{u}, \tag{13.29}$$

and hence

$$d^2 x_n = 4\mathbf{u}_n{}^2 d^2 u_n. \tag{13.30}$$

Since the $\mathbf{x}$- and the $\mathbf{u}$-space are both Euclidean, the integrals over $\Delta \mathbf{x}_n$ in (13.8) can be rewritten as integrals over $\mathbf{x}_n$, and transformed directly to $\mathbf{u}_n$ variables. The result is

$$\langle \mathbf{x}_b | \hat{\mathcal{U}}_E(S) | \mathbf{x}_a \rangle = \frac{1}{4} e^{ie^2 S/\hbar} [(\mathbf{u}_b S | \mathbf{u}_a 0) + (-\mathbf{u}_b S | \mathbf{u}_a 0)], \tag{13.31}$$

where $(\mathbf{u}_b S | \mathbf{u}_a 0)$ denotes the time-sliced oscillator amplitude

$$(\mathbf{u}_b S | \mathbf{u}_a 0) \approx \frac{1}{2\pi i \hbar \epsilon_s / \mu} \prod_{n=1}^{N} \left[\int \frac{d^2 u_n}{2\pi i \hbar \epsilon_s / \mu} \right] \tag{13.32}$$

$$\times \exp \left\{ \frac{i}{\hbar} \sum_{n=1}^{N} \frac{\mu}{2} \left(\frac{1}{\epsilon_s} \Delta \mathbf{u}_n{}^2 - \epsilon_s \omega^2 \mathbf{u}_n{}^2 \right) \right\}.$$

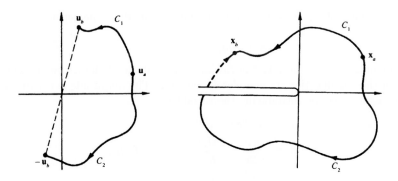

Figure 13.1 Illustration of associated final points in **u**-space, to be summed in the oscillator amplitude. In **x**-space, the paths run from $\mathbf{x}_a$ to $\mathbf{x}_b$ once directly and once after crossing the cut into the second sheet of the complex function $u = \sqrt{x}$.

The evaluation of the Gaussian integrals yields, in the continuum limit,

$$(\mathbf{u}_b S | \mathbf{u}_a 0) = \frac{\mu\omega}{2\pi i\hbar \sin \omega S} \exp\left\{ \frac{i}{\hbar} \frac{\mu\omega}{\sin \omega S} [(\mathbf{u}_b{}^2 + \mathbf{u}_a{}^2)\cos \omega S - 2\mathbf{u}_b \mathbf{u}_a] \right\}. \quad (13.33)$$

The symmetrization in $\mathbf{u}_b$ in Eq. (13.31) is necessary since for each path from $\mathbf{x}_a$ to $\mathbf{x}_b$, there are two paths in the square root space, one from $\mathbf{u}_a$ to $\mathbf{u}_b$ and one from $\mathbf{u}_a$ to $-\mathbf{u}_b$ (see Fig. 13.1).

The fixed-energy amplitude is obtained by the integral (13.4) over the pseudotime evolution amplitude (12.18):

$$(\mathbf{x}_b | \mathbf{x}_a)_E = \int_0^\infty dS\, e^{ie^2 S/\hbar} \frac{1}{4}[(\mathbf{u}_b S | \mathbf{u}_a 0) + (-\mathbf{u}_b S | \mathbf{u}_a 0)]. \quad (13.34)$$

By inserting (13.33), this becomes

$$
\begin{aligned}
(\mathbf{x}_b | \mathbf{x}_a)_E &= \frac{1}{2} \int_0^\infty dS \exp(ie^2 S/\hbar) F^2(S) \\
&\quad \times \exp\left[-\pi F^2(S)(\mathbf{u}_b^2 + \mathbf{u}_a^2)\cos \omega S\right] \cosh\left[2\pi F^2(S)\mathbf{u}_b \mathbf{u}_a\right],
\end{aligned} \quad (13.35)
$$

with the abbreviation

$$F(S) = \sqrt{\mu\omega/2\pi i\hbar \sin \omega S}, \quad (13.36)$$

for the one-dimensional fluctuation factor [recall (2.169)]. The coordinates $\mathbf{u}_b$ and $\mathbf{u}_a$ on the right-hand side are related to $\mathbf{x}_b \mathbf{x}_a$ on the left-hand side by

$$\mathbf{u}_{a,b}^2 = r_{a,b}, \quad \mathbf{u}_b \mathbf{u}_a = \sqrt{(r_b r_a + \mathbf{x}_b \mathbf{x}_a)/2}. \quad (13.37)$$

When performing the integral over S, we have to pass around the singularities in $F(S)$ in accordance with the $i\eta$-prescription, replacing $\omega \to \omega - i\eta$. Equivalently, we can rotate the contour of S-integration to make it run along the negative imaginary semi-axis,

$$S = -i\sigma, \qquad \sigma \in (0, \infty).$$

This amounts to going over to the Euclidean amplitude of the harmonic oscillator in which the singularities are completely avoided. The amplitude is rewritten in a more compact form by introducing the variables

$$\varrho \equiv e^{-2i\omega S} = e^{-2\omega\sigma}, \tag{13.38}$$

$$\kappa \equiv \frac{\mu\omega}{2\hbar} = \frac{2M\omega}{\hbar} = \sqrt{-2ME/\hbar^2}, \tag{13.39}$$

$$\nu \equiv \frac{e^2}{2\omega\hbar} = \sqrt{\frac{e^4 M}{-2\hbar^2 E}}. \tag{13.40}$$

Then

$$\pi F^2(S) = \kappa \frac{2\sqrt{\varrho}}{1 - \varrho}, \tag{13.41}$$

$$e^{ie^2 s/\hbar} F^2(S) = \frac{2}{\pi} \kappa \frac{\varrho^{1/2-\nu}}{1 - \varrho}, \tag{13.42}$$

and the fixed-energy amplitude of the two-dimensional Coulomb system takes the form

$$(\mathbf{x}_b|\mathbf{x}_a)_E = -i\frac{M}{\pi\hbar} \int_0^1 d\varrho \, \frac{\varrho^{-1/2-\nu}}{1 - \varrho} \cos\left[2\kappa \frac{2\sqrt{\varrho}}{1 - \varrho} \sqrt{(r_b r_a + \mathbf{x}_b \mathbf{x}_a)/2}\right]$$
$$\times \exp\left[-\kappa \frac{1 + \varrho}{1 - \varrho}(r_b + r_a)\right]. \tag{13.43}$$

This can be used to find the energy spectrum and the wave functions as shown in Section 13.8.

Note that the integral converges only for $\nu < 1/2$. It is possible to write down another integral representation converging for all $\nu \neq 1/2, 3/2, 5/2, \ldots$. To do this we change the variable of integrations to

$$\zeta \equiv \frac{1 + \varrho}{1 - \varrho}, \tag{13.44}$$

so that

$$\frac{d\varrho}{(1 - \varrho)^2} = \frac{1}{2} d\zeta, \qquad \varrho = \frac{\zeta - 1}{\zeta + 1}. \tag{13.45}$$

This leads to

$$(\mathbf{x}_b|\mathbf{x}_a)_E = -i\frac{M}{\pi\hbar} \frac{1}{2} \int_1^\infty d\zeta (\zeta - 1)^{-\nu-1/2} (\zeta + 1)^{\nu-1/2}$$
$$\times \cos\left\{2\kappa \sqrt{\zeta^2 - 1} \sqrt{(r_b r_a + \mathbf{x}_b \mathbf{x}_a)/2}\right\} e^{-\kappa\zeta(r_b+r_a)}. \tag{13.46}$$

The integrand has a cut in the complex ζ-plane extending from $z = -1$ to $-\infty$ and from $\zeta = 1$ to ∞. The integral runs along the right-hand cut. The integral is transformed into an integral along a contour C which encircles the right-hand cut in the clockwise sense. Since the cut is of the type $(\zeta - 1)^{-\nu-1/2}$, we may replace

$$\int_1^\infty d\zeta(\zeta - 1)^{-\nu-1/2}\ldots \to \frac{\pi e^{i\pi(\nu+1/2)}}{\sin[\pi(\nu + 1/2)]}\frac{1}{2\pi i}\int_C d\zeta(\zeta - 1)^{-\nu-1/2}\ldots , \quad (13.47)$$

and the fixed-energy amplitude reads

$$(\mathbf{x}_b|\mathbf{x}_a)_E = -i\frac{M}{\pi\hbar}\frac{1}{2}\frac{\pi e^{i\pi(\nu+1/2)}}{\sin[\pi(\nu + 1/2)]}\int_C \frac{d\zeta}{2\pi i}(\zeta - 1)^{-\nu-1/2}(\zeta + 1)^{\nu-1/2}$$

$$\times \cos\left[2\kappa\sqrt{\zeta^2 - 1}\sqrt{(r_b r_a + \mathbf{x}_b\mathbf{x}_a)/2}\right]e^{-\kappa\zeta(r_b+r_a)}. \quad (13.48)$$

13.3   Absence of Time Slicing Corrections for $D = 2$

We now convince ourselves that the finite thickness of the pseudotime slices in the intermediate formulas does not change the time evolution amplitude obtained in the last section [5]. The reader who is unaware of the historic difficulties which had to be overcome may not be interested in the upcoming technical discussion. He may skip this section and be satisfied with a brief argument given in Section 13.6.

The potential term in the action (13.9) can be ignored since it is of order ϵ_s and the time slicing can only produce higher than linear correction terms in ϵ_s which do not contribute in the continuum limit $\epsilon_s \to 0$. The crucial point where corrections might enter is in the transformation of the measure and the pseudotime-sliced kinetic terms in (13.8), (13.9). In vector notation, the coordinate transformation reads, at every time slice n,

$$\mathbf{x}_n = A(\mathbf{u}_n)\mathbf{u}_n. \quad (13.49)$$

Among the equivalent possibilities offered in Section 11.2 to transform a time-sliced path integral we choose the Taylor expansion (11.56) to map $\Delta\mathbf{x}$ into $\Delta\mathbf{u}$. After inserting (13.15) into (11.56) we find

$$\Delta x^i = 2A^i{}_\mu(\mathbf{u})\Delta u^\mu - \partial_\nu A^i{}_\mu(\mathbf{u})\Delta u^\mu\Delta u^\nu. \quad (13.50)$$

Since the mapping $\mathbf{x}(\mathbf{u})$ is quadratic in $\mathbf{u}$, there are no higher-order expansion terms. Note that due to the absence of curvature and torsion in the $\mathbf{u}$-space, the coordinate transformation is holonomic and $\Delta\mathbf{x}$ can also be calculated directly from $\mathbf{x}(\mathbf{u}) - \mathbf{x}(\mathbf{u} - \Delta\mathbf{u})$. Indeed, using the linearity of $A(\mathbf{u})$ in $\mathbf{u}$, we find from (13.49)

$$\Delta\mathbf{x}_n = A(\mathbf{u}_n)\mathbf{u}_n - A(\mathbf{u}_{n-1})\mathbf{u}_{n-1} = 2A(\mathbf{u}_n - \tfrac{1}{2}\Delta\mathbf{u}_n)\Delta\mathbf{u}_n = 2A(\bar{\mathbf{u}}_n)\Delta\mathbf{u}_n, \quad (13.51)$$

where $\bar{\mathbf{u}}_n$ is average across the slice. The Taylor expansion of $A(\mathbf{u}_n - \tfrac{1}{2}\Delta\mathbf{u}_n)$ has only two terms and leads to (13.50).

Using (13.51) we can write

$$(\Delta\mathbf{x}_n)^2 = 4\bar{\mathbf{u}}_n^2(\Delta\mathbf{u}_n)^2, \quad (13.52)$$

$$\bar{\mathbf{u}}_n \equiv (\mathbf{u}_n + \mathbf{u}_{n-1})/2. \quad (13.53)$$

The kinetic term of the short-time action in the nth time slice of Eq. (13.9) therefore becomes

$$\mathcal{A}^\epsilon = \frac{M}{2\epsilon_s} \frac{(\Delta \mathbf{x}_n)^2}{r_n^{1-\lambda} r_{n-1}^\lambda} = \frac{M}{2\epsilon_s} \frac{4\bar{\mathbf{u}}_n^2}{(\mathbf{u}_n^2)^{1-\lambda}(\mathbf{u}_{n-1}^2)^\lambda}(\Delta \mathbf{u}_n)^2. \tag{13.54}$$

This is expanded around the postpoint and yields

$$\bar{\mathbf{u}}_n = \mathbf{u}_n - \frac{1}{2}\Delta \mathbf{u}_n, \tag{13.55}$$

$$\mathbf{u}_{n-1} = \mathbf{u}_n - \Delta \mathbf{u}_n, \tag{13.56}$$

$$\frac{\bar{\mathbf{u}}_n^2}{(\mathbf{u}_n^2)^{1-\lambda}(\mathbf{u}_{n-1}^2)^\lambda} = 1 + (2\lambda - 1)\frac{\mathbf{u}_n \Delta \mathbf{u}_n}{\mathbf{u}_n^2} + \left(\frac{1}{4} - \lambda\right)\frac{\Delta \mathbf{u}_n^2}{\mathbf{u}_n^2}$$

$$+ 2\lambda^2 \left(\frac{\mathbf{u}_n \Delta \mathbf{u}_n}{\mathbf{u}_n^2}\right)^2. \tag{13.57}$$

It is useful to separate the short-time action into a leading term

$$\mathcal{A}_0^\epsilon(\Delta \mathbf{u}_n) = 4M \frac{(\Delta \mathbf{u}_n)^2}{2\epsilon_s}, \tag{13.58}$$

plus correction terms

$$\Delta \mathcal{A}^\epsilon = 4M \frac{(\Delta \mathbf{u}_n)^2}{2\epsilon_s} \tag{13.59}$$

$$\times \left[(2\lambda - 1)\frac{\mathbf{u}_n \Delta \mathbf{u}_n}{\mathbf{u}_n^2} + \left(\frac{1}{4} - \lambda\right)\frac{\Delta \mathbf{u}_n^2}{\mathbf{u}_n^2} + 2\lambda^2 \left(\frac{\mathbf{u}_n \Delta \mathbf{u}_n}{\mathbf{u}_n^2}\right)^2\right],$$

which will be treated perturbatively.

In order to perform the transformation of the measure of integration in (13.8), we expand $\Delta \mathbf{x}$ accordingly:

$$\Delta x^i = 2A^i_{\ \mu}(\mathbf{u} - \frac{1}{2}\Delta \mathbf{u})\Delta u^\mu$$

$$= 2A^i_{\ \mu}(\mathbf{u})\Delta u^\mu - \partial_\nu A^i_{\ \mu}(\mathbf{u})\Delta u^\mu \Delta u^\nu. \tag{13.60}$$

The indices n have been suppressed, for brevity. This expression has, of course, the general form (10.96), after inserting there

$$e^i_{\ \mu}(\mathbf{u}) = 2A^i_{\ \mu}(\mathbf{u}). \tag{13.61}$$

Since the transformation matrix $A^i_{\ \mu}(\mathbf{u})$ in (13.15) is linear in $\mathbf{u}$, the matrix $e^i_{\ \mu}(\mathbf{u})$ has no second derivatives, and the Jacobian action (11.60) and (10.145) reduce to

$$\frac{i}{\hbar}\mathcal{A}_J^\epsilon = -e_i^{\ \mu}e^i_{\ \{\mu,\nu\}}\Delta u^\nu - \frac{1}{2}e_i^{\ \mu}e^i_{\ \{\kappa,\nu\}}e_j^{\ \kappa}e^j_{\ \mu,\lambda}\Delta u^\nu \Delta u^\lambda$$

$$= -\Gamma_{\{\nu\mu\}}^{\ \ \mu}\Delta u^\nu - \frac{1}{2}\Gamma_{\{\nu\kappa\}}^{\ \ \mu}\Gamma_{\{\mu\lambda\}}^{\ \ \kappa}\Delta u^\nu \Delta u^\lambda. \tag{13.62}$$

The expansion coefficients are easily calculated using the reciprocal basis dyad

$$e_i^{\ \kappa} = \frac{1}{2\mathbf{u}^2}e^i_{\ \kappa}, \tag{13.63}$$

as

$$\Gamma_{\nu\mu}^{\ \ \mu} = e_i^{\ \mu}\partial_\nu e^i_{\ \mu} = \frac{2u^\nu}{\mathbf{u}^2},$$

$$\Gamma_{\mu\nu}^{\ \ \mu} = -e^i_{\ \nu}\partial_\mu e_i^{\ \mu} = \frac{2u^\nu}{\mathbf{u}^2}, \tag{13.64}$$

$$\Gamma_{\nu\kappa}^{\ \ \mu}\Gamma_{\lambda\mu}^{\ \ \kappa} = -\partial_\lambda e_i^{\ \kappa}\partial_\nu e^i_{\ \kappa} = -\frac{2}{\mathbf{u}^4}(\delta^{\nu\lambda}\mathbf{u}^2 - 2u^\nu u^\lambda).$$

The second equation is found directly from

$$- \partial_\mu e_i{}^\mu = -\partial_\mu (2\mathbf{u}^2)^{-1} e^i{}_\mu = e_i{}^\mu \, 2u^\mu \mathbf{u}^{-2}, \qquad (13.65)$$

which, in turn, follows from the obvious identity $\partial_\mu e^i{}_\mu = 0$. Note that the third expression in (13.64) is automatically equal to $\Gamma_{\{\nu\kappa\}}{}^\sigma \Gamma_{\{\lambda\sigma\}}{}^\kappa$, i.e., of the form required in (13.62), since the u^μ-space has no torsion and $\Gamma_{\nu\kappa}{}^\sigma = \Gamma_{\kappa\nu}{}^\sigma$. After inserting Eqs. (13.64) into the right-hand side of (13.62), we find the postpoint expansion

$$\frac{i}{\hbar} \mathcal{A}_J^\epsilon = - \left[2 \frac{\mathbf{u}_n \Delta \mathbf{u}_n}{\mathbf{u}_n^2} - \frac{\Delta \mathbf{u}_n{}^2}{\mathbf{u}_n^2} + 2 \left(\frac{\mathbf{u} \Delta \mathbf{u}_n}{\mathbf{u}_n^2} \right)^2 + \ldots \right]. \qquad (13.66)$$

The measure of integration in (13.8) contains additional factors r_b, r_n, r_a which require a further treatment. First we rewrite it as

$$\frac{(r_b/r_a)^{2\lambda-1}}{2\pi i \epsilon_s \hbar} \prod_{n=1}^{N} \left[\int \frac{d^2 \Delta x_n}{2\pi i \epsilon_s r_{n-1}/M} \right] \approx \frac{1}{2\pi i \epsilon_s \hbar} \prod_{n=1}^{N} \left[\int \frac{d^2 \Delta x_n}{2\pi i \epsilon_s \hbar r_n / M} \right] \prod_{n=1}^{N+1} \left(\frac{r_n}{r_{n-1}} \right)^{2\lambda}$$

$$= \frac{1}{2\pi i \epsilon_s \hbar} \prod_{n=2}^{N+1} \left[\int \frac{d^2 \Delta x_n}{2\pi i \epsilon_s \hbar r_n / M} \right] e^{i \mathcal{A}_f^N / \hbar}. \qquad (13.67)$$

On the left-hand side, we have shifted the labels n by one unit making use of the fact that with $\Delta x_n = x_n - x_{n-1}$ we can certainly write $\prod_{n=2}^{N+1} \int d^2 \Delta x_n = \prod_{n=1}^{N} \int d^2 \Delta x_n$. In the first expression on the right-hand side we have further shifted the subscripts of the factors $1/r_{n-1}$ in the integral measure from $n-1$ to n and compensated for this by an overall factor $\prod_{n=1}^{N+1} (r_n/r_{n-1})$. Together with the prefactor $(r_b/r_a)^{2\lambda-1}$, this can be expressed as a product $\prod_{n=1}^{N+1} (r_n/r_{n-1})^{2\lambda}$. There is only a negligible error of order ϵ_s^2 at the upper end [this being the reason for writing the symbol $\approx$ rather than $=$ in (13.67)]. In the last part of the equation we have introduced an effective action

$$\mathcal{A}_f^N \equiv \sum_{n=1}^{N+1} \mathcal{A}_f^\epsilon \qquad (13.68)$$

due to the $(r_n/r_{n-1})^{2\lambda}$ factors, with

$$\frac{i}{\hbar} \mathcal{A}_f^\epsilon = 2\lambda \log \frac{r_n^2}{r_{n-1}^2} = 2\lambda \log \frac{\mathbf{u}_n^2}{\mathbf{u}_{n-1}^2}. \qquad (13.69)$$

The subscript f indicates that the general origin of this term lies in the rescaling factors $f_l(\mathbf{x}_b), f_r(\mathbf{x}_a)$.

We now go over from $\Delta \mathbf{x}_n$- to $\Delta \mathbf{u}_n$-integrations using the relation

$$d^2 \Delta x = 4\mathbf{u}^2 d^2 \Delta u \, \exp \left(\frac{i}{\hbar} \mathcal{A}_J^\epsilon \right). \qquad (13.70)$$

The measure becomes

$$\frac{1}{2} \times \frac{4}{2 \cdot 2\pi i \epsilon_s \hbar} \prod_{n=1}^{N} \left[\int \frac{4 d^2 \Delta u_n}{2 \cdot 2\pi i \epsilon_s \hbar / M} \right] \exp \left[\frac{i}{\hbar} (\mathcal{A}_J^N + \mathcal{A}_f^N) \right], \qquad (13.71)$$

where $\mathcal{A}_J^N$ is the sum over all time-sliced Jacobian action terms $\mathcal{A}_J^\epsilon$ of (13.66):

$$\mathcal{A}_J^N \equiv \sum_{n=1}^{N+1} \mathcal{A}_J^\epsilon. \qquad (13.72)$$

The extra factors 2 in the measure denominators of (13.71) are introduced to let the $\mathbf{u}_n$-integrations run over the entire $\mathbf{u}$-space, in which case the $\mathbf{x}$-space is traversed twice.

The time-sliced expression (13.71) has an important feature which was absent in the continuous formulation. It receives dominant contributions not only from the region neighborhood $\mathbf{u}_n \sim \mathbf{u}_{n-1}$, in which case $(\Delta \mathbf{u}_n)^2$ is of order ϵ_s, but also from $\mathbf{u}_n \sim -\mathbf{u}_{n-1}$ where $(\bar{\mathbf{u}}_n)^2$ is of order ϵ_s. This is understandable since both configurations correspond to $\mathbf{x}_n$ being close to $\mathbf{x}_{n-1}$ and must be included. Fortunately, for symmetry reasons, they give identical contributions so that we need to discuss only the case $\mathbf{u}_n \sim \mathbf{u}_{n-1}$, the contribution from the second case being simply included by dropping the factors 2 in the measure denominators.

To process the measure further, we expand the action $\mathcal{A}_f^\epsilon$ (13.71) around the postpoint and find

$$\frac{i}{\hbar} \mathcal{A}_f^\epsilon = 2\lambda \log \left(\frac{\mathbf{u}_n^2}{\mathbf{u}_{n-1}^2} \right) = 2\lambda \left[2 \frac{\mathbf{u}_n \Delta \mathbf{u}_n}{\mathbf{u}_n^2} - \frac{\Delta \mathbf{u}_n^2}{\mathbf{u}_n^2} + 2 \left(\frac{\mathbf{u} \Delta \mathbf{u}_n}{\mathbf{u}_n^2} \right)^2 + \dots \right]. \tag{13.73}$$

A comparison with (13.66) shows that adding $(i/\hbar)\mathcal{A}_f^\epsilon$ and $(i/\hbar)\mathcal{A}_J^\epsilon$ merely changes 2λ in $\mathcal{A}_f^\epsilon$ into $2\lambda - 1$.

Thus, altogether, the time-slicing produces the short-time action

$$\mathcal{A}^\epsilon = \mathcal{A}_0^\epsilon + \Delta_{\text{corr}} \mathcal{A}^\epsilon, \tag{13.74}$$

with the leading free-particle action

$$\mathcal{A}_0^\epsilon(\Delta \mathbf{u}_n) = 4M \frac{(\Delta \mathbf{u}_n)^2}{2\epsilon_s}, \tag{13.75}$$

and the total correction term

$$\begin{aligned}
\frac{i}{\hbar} \Delta_{\text{corr}} \mathcal{A}^\epsilon &\equiv \frac{i}{\hbar} (\Delta \mathcal{A}^\epsilon + \mathcal{A}_J^\epsilon + \mathcal{A}_f^\epsilon) \\
&= \frac{i}{\hbar} 4M \frac{\Delta \mathbf{u}_n^2}{2\epsilon_s} \left[(2\lambda - 1) \frac{\mathbf{u}_n \Delta \mathbf{u}_n}{\mathbf{u}_n^2} + \left(\frac{1}{4} - \lambda \right) \frac{\Delta \mathbf{u}_n^2}{\mathbf{u}_n^2} + 2\lambda^2 \left(\frac{\mathbf{u}_n \Delta \mathbf{u}_n}{\mathbf{u}_n^2} \right)^2 \right] \\
&\quad + (2\lambda - 1) \left[2 \frac{\mathbf{u}_n \Delta \mathbf{u}_n}{\mathbf{u}_n^2} - \frac{\Delta \mathbf{u}_n^2}{\mathbf{u}_n^2} + 2 \left(\frac{\mathbf{u}_n \Delta \mathbf{u}_n}{\mathbf{u}_n^2} \right)^2 \right] + \dots \, .
\end{aligned} \tag{13.76}$$

We now show that the action $\Delta_{\text{corr}} \mathcal{A}^\epsilon$ is equivalent to zero by proving that the kernel associated with the short-time action

$$K^\epsilon(\Delta \mathbf{u}) = \frac{4}{2 \cdot 2\pi i \epsilon_s \hbar / M} \exp \left[\frac{i}{\hbar} (\mathcal{A}_0^\epsilon + \Delta_{\text{corr}} \mathcal{A}^\epsilon) \right] \tag{13.77}$$

is equivalent to the zeroth-order free-particle kernel

$$K_0^\epsilon(\Delta \mathbf{u}) = \frac{4}{2 \cdot 2\pi i \epsilon_s \hbar / M} \exp \left[\frac{i}{\hbar} \mathcal{A}_0^\epsilon \right]. \tag{13.78}$$

The equivalence is established by checking the equivalence relations (11.71) and (11.72). For the kernel (13.77), the correction (11.71) is

$$C_1 = C \equiv \exp \left(\frac{i}{\hbar} \Delta_{\text{corr}} \mathcal{A}^\epsilon \right) - 1. \tag{13.79}$$

It has to be compared with the trivial factor of the kernel (13.78):

$$C_2 = 0. \tag{13.80}$$

Thus, the equivalence requires showing that

$$\langle C \rangle_0 = 0,$$
$$\langle C \, (\mathbf{p}\Delta\mathbf{u}) \rangle_0 = 0. \tag{13.81}$$

The basic correlation functions due to $K_0^\epsilon(\Delta\mathbf{u})$ are

$$\langle \Delta u^\mu \Delta u^\nu \rangle_0 \equiv \frac{i\hbar\epsilon_s}{4M}\delta^{\mu\nu}, \tag{13.82}$$

$$\langle \Delta u^{\mu_1} \cdots \Delta u^{\mu_{2n}} \rangle_0 = \left(\frac{i\hbar\epsilon_s}{4M}\right)^n \delta^{\mu_1 \cdots \mu_{2n}}, \quad n > 1, \tag{13.83}$$

where the contraction tensors $\delta^{\mu_1 \cdots \mu_{2n}}$ of Eq. (8.64), determined recursively from

$$\delta^{\mu_1 \cdots \mu_{2n}} \equiv \delta^{\mu_1\mu_2}\delta^{\mu_3\mu_4 \cdots \mu_{2n}} + \delta^{\mu_1\mu_3}\delta^{\mu_2\mu_4 \cdots \mu_{2n}} + \ldots + \delta^{\mu_1\mu_{2n}}\delta^{\mu_2\mu_3 \cdots \mu_{2n-1}}. \tag{13.84}$$

They consist of $(2n-1)!!$ products of pair contractions $\delta^{\mu_i\mu_j}$. More specifically, we encounter, in calculating (13.81), expectations of the type

$$\langle (\Delta\mathbf{u})^{2k}(\mathbf{u}\Delta\mathbf{u})^{2l} \rangle_0 = \left(\frac{i\hbar\epsilon_s}{4M}\right)^{k+l} \frac{[D + 2(k + l - 1)]!!}{(D + 2l - 2)!!}(2l-1)!!(\mathbf{u}^2)^l, \tag{13.85}$$

and

$$\langle (\Delta\mathbf{u})^{2k}(\mathbf{u}\Delta\mathbf{u})^{2l}(\mathbf{u}\Delta\mathbf{u})(\mathbf{p}\Delta\mathbf{u}) \rangle_0 = \left(\frac{i\hbar\epsilon_s}{4M}\right)^{k+l+1} \frac{[D + 2(k + l)]!!}{(D + 2l)!!}(2l-1)!!(\mathbf{up}), \tag{13.86}$$

where we have allowed for a general $\mathbf{u}$-space dimension D. Expanding (13.81) we now check that, up to first order in ϵ_s, the expectations $\langle C \rangle_0$ and $\langle C \, (\mathbf{p}\Delta\mathbf{u}) \rangle_0$ vanish:

$$\langle C \rangle_0 = \frac{i}{\hbar}\langle \Delta_{\text{corr}}\mathcal{A}^\epsilon \rangle_0 + \frac{1}{2!}\left(\frac{i}{\hbar}\right)^2 \langle (\Delta_{\text{corr}}\mathcal{A}^\epsilon)^2 \rangle_0 = 0, \tag{13.87}$$

$$\langle C \, (\mathbf{p}\Delta\mathbf{u}) \rangle_0 = \frac{i}{\hbar}\langle \Delta_{\text{corr}}\mathcal{A}^\epsilon \, (\mathbf{p}\Delta\mathbf{u}) \rangle_0 = 0. \tag{13.88}$$

Indeed, the first term in (13.87) becomes

$$\frac{i}{\hbar}\langle \Delta_{\text{corr}}\mathcal{A}^\epsilon \rangle_0 = 2i\frac{\hbar\epsilon_s}{M}\left[-\left(\frac{1}{4} - \lambda\right)\frac{(D+2)D}{16} - 2\lambda^2\frac{D+2}{16}\right], \tag{13.89}$$

and reduces for $D = 2$ to

$$\frac{i}{\hbar}\langle \Delta_{\text{corr}}\mathcal{A}^\epsilon \rangle_0 = -i\frac{\hbar\epsilon_s}{M}\left(\lambda - \frac{1}{2}\right)^2. \tag{13.90}$$

This is canceled identically in λ by the second term in (13.87), which is equal to

$$\frac{1}{2!}\left(\frac{i}{\hbar}\right)^2 \langle (\Delta_{\text{corr}}\mathcal{A}^\epsilon)^2 \rangle_0 = \frac{i}{2}\frac{\hbar\epsilon_s}{M}\left[4(2\lambda-1)^2\frac{(D+4)(D+2)}{64} + 4(2\lambda-1)^2\frac{1}{4} - 8(2\lambda-1)^2\frac{D+2}{16}\right], \tag{13.91}$$

and reduces for $D = 2$ to

$$\frac{1}{2!}\left(\frac{i}{\hbar}\right)^2 \langle (\Delta_{\text{corr}}\mathcal{A}^\epsilon)^2 \rangle_0 = i\frac{\hbar\epsilon_s}{M}\left(\lambda - \frac{1}{2}\right)^2. \tag{13.92}$$

Similarly, the expectation (13.88),

$$\langle \Delta_{\text{corr}}\mathcal{A}^\epsilon \, (\mathbf{p}\Delta\mathbf{u}) \rangle_0 = -\frac{\hbar^2\epsilon_s}{4M}[(2\lambda-1)(D+2)/4 - (2\lambda-1)], \tag{13.93}$$

is seen to vanish identically in λ for $D = 2$.

Thus there is no finite time slicing correction to the naive transformation formula (13.34) for the Coulomb path integral in two dimensions.

13.4 Solution for the Three-Dimensional Coulomb System

We now turn to the physically relevant Coulomb system in three dimensions. The first problem is to find again some kind of "square root" coordinates to convert the potential $-Er$ in the pseudotime Hamiltonian in the exponent of Eq. (13.5) into a harmonic potential. In two dimensions, the answer was a complex square root. Here, it is a "quaternionic square root" known as the *Kustaanheimo-Stiefel transformation*, which was used extensively in celestial mechanics [6]. To apply this transformation, the three-vectors $\mathbf{x}$ must first be mapped into a four-dimensional u^μ-space ($\mu = 1, 2, 3, 4$) via the equations

$$x^i = \bar{z}\sigma^i z, \qquad r = \bar{z}z. \tag{13.94}$$

Here σ^i are the Pauli matrices (1.445), and $z, \bar{z}$ the two-component objects

$$z = \begin{pmatrix} z_1 \\ z_2 \end{pmatrix}, \quad \bar{z} = (z_1^*, z_2^*), \tag{13.95}$$

called "spinors". Their components are related to the four-vectors u^μ by

$$z_1 = (u^1 + iu^2), \quad z_2 = (u^3 + iu^4). \tag{13.96}$$

The coordinates u^μ can be parametrized in terms of the spherical angles of the three-vector $\mathbf{x}$ and an additional arbitrary angle γ as follows:

$$
\begin{aligned}
z_1 &= \sqrt{r}\cos(\theta/2)e^{-i[(\varphi+\gamma)/2]}, \\
z_2 &= \sqrt{r}\sin(\theta/2)e^{i[(\varphi-\gamma)/2]}.
\end{aligned}
$$

In Eqs. (13.94), the angle γ obviously cancels. Each point in $\mathbf{x}$-space corresponds to an entire curve in u^μ-space along which the angle γ runs through the interval $[0, 4\pi]$.

We can write (13.94) also in a matrix form

$$\begin{pmatrix} x^1 \\ x^2 \\ x^3 \end{pmatrix} = A(\vec{u}) \begin{pmatrix} u^1 \\ u^2 \\ u^3 \\ u^4 \end{pmatrix}, \tag{13.97}$$

with the 3×4 matrix

$$A(\vec{u}) = \begin{pmatrix} u^3 & u^4 & u^1 & u^2 \\ u^4 & -u^3 & -u^2 & u^1 \\ u^1 & u^2 & -u^3 & -u^4 \end{pmatrix}. \tag{13.98}$$

Since

$$r = (u^1)^2 + (u^2)^2 + (u^3)^2 + (u^4)^2 \equiv (\vec{u})^2, \tag{13.99}$$

this transformation certainly makes the potential $-Er$ in the pseudotime Hamiltonian harmonic in $\vec{u}$. The arrow on top indicates the four-vector nature of u^μ.

Consider now the kinetic term $\int ds(M/2r)(d\mathbf{x}/ds)^2$ in the action (13.10). Each path $\mathbf{x}(s)$ is associated with an infinite set of paths $\vec{u}(s)$ in $\vec{u}$-space, depending on the choice of a dummy path $\gamma(s)$ in parameter space. The mapping of the tangent vectors du^μ into dx^i is given by

$$
\begin{pmatrix} dx^1 \\ dx^2 \\ dx^3 \end{pmatrix} = 2 \begin{pmatrix} u^3 & u^4 & u^1 & u^2 \\ u^4 & -u^3 & -u^2 & u^1 \\ u^1 & u^2 & -u^3 & -u^4 \end{pmatrix} \begin{pmatrix} du^1 \\ du^2 \\ du^3 \\ du^4 \end{pmatrix}. \tag{13.100}
$$

To make the mapping unique we must prescribe at least some differential equation for the dummy angle $d\gamma$. This is done most simply by replacing $d\gamma$ by a parameter which is more naturally related to the components dx^i on the left-hand side. We embed the tangent vector (dx^1, dx^2, dx^3) into a fictitious four-dimensional space and define a new, fourth component dx^4 by an additional fourth row in the matrix $A(\vec{u})$, thereby extending (13.29) to the four-vector equation

$$
d\vec{x} = 2A(\vec{u})d\vec{u}. \tag{13.101}
$$

The arrow on top of x indicates that x has become a four-vector. For symmetry reasons, we choose the 4×4 matrix $A(\vec{u})$ as

$$
A(\vec{u}) = \begin{pmatrix} u^3 & u^4 & u^1 & u^2 \\ u^4 & -u^3 & -u^2 & u^1 \\ u^1 & u^2 & -u^3 & -u^4 \\ u^2 & -u^1 & u^4 & -u^3 \end{pmatrix}. \tag{13.102}
$$

The fourth row implies the following relation between dx^4 and $d\gamma$:

$$
\begin{aligned}
dx^4 &= 2(u^2 du^1 - u^1 du^2 + u^4 du^3 - u^3 du^4) \\
&= r(\cos\theta \, d\varphi + d\gamma).
\end{aligned} \tag{13.103}
$$

Now we make the important observation that this relation is not integrable since $\partial x^4/\partial u^1 = 2u^2$, $\partial x^4/\partial u^2 = -2u^1$, and hence

$$
(\partial_{u^1}\partial_{u^2} - \partial_{u^2}\partial_{u^1})x^4(u^\mu) = -4, \qquad (\partial_{u^3}\partial_{u^4} - \partial_{u^4}\partial_{u^3})x^4(u^\mu) = -4, \tag{13.104}
$$

implying that $x^4(u^\mu)$ does not satisfy the integrability criterion of Schwarz [recall (10.19)]. The mapping is nonholonomic and changes the Euclidean geometry of the four-dimensional $\vec{x}$-space into a non-Euclidean $\vec{u}$-space with curvature and torsion. This will be discussed in detail in the next section. The impossibility of finding a unique mapping between the *points* of $\vec{x}$- and $\vec{u}$-space has the consequence that the mapping between *paths* is multivalued with respect to the initial point. After having chosen a specific image for the initial point, the mapping (13.102) determines the image path uniquely.

We now incorporate the dummy fourth dimension into the action by replacing $\mathbf{x}$ in the kinetic term by the four-vector $\vec{x}$ and extending the kinetic action to

$$\mathcal{A}_{\text{kin}}^N \equiv \sum_{n=1}^{N+1} \frac{M}{2} \frac{(\vec{x}_n - \vec{x}_{n-1})^2}{\epsilon_s r_n^{1-\lambda} r_{n-1}^\lambda}. \tag{13.105}$$

The additional contribution of the fourth components $x_n^4 - x_{n-1}^4$ can be eliminated trivially from the final pseudotime evolution amplitude by integrating each time slice over dx_{n-1}^4 with the measure

$$\prod_{n=1}^{N+1} \int_{-\infty}^\infty \frac{d(\Delta x^4)_n}{\sqrt{2\pi i \epsilon_s \hbar r_n^{1-\lambda} r_{n-1}^\lambda / M}}. \tag{13.106}$$

Note that in these integrals, the radial coordinates r_n are fixed numbers. In contrast to the spatial integrals $d^3 x_{n-1}$, the fourth coordinate must be integrated also over the initial auxiliary coordinate $x_0^4 = x_a^4$. Thus we use the trivial identity

$$\prod_{n=1}^{N+1} \left[\int_{-\infty}^\infty \frac{d(\Delta x^4)_n}{\sqrt{2\pi i \epsilon_s \hbar r_n^{1-\lambda} r_{n-1}^\lambda / M}} \right] \exp\left[\frac{i}{\hbar} \sum_{n=1}^{N+1} \frac{M}{2} \frac{(\Delta x_n^4)^2}{\epsilon_s r_n^{1-\lambda} r_{n-1}^\lambda} \right] = 1. \tag{13.107}$$

Hence the pseudotime evolution amplitude of the Coulomb system in three dimensions can be rewritten as the four-dimensional path integral

$$\langle \mathbf{x}_b | \hat{\mathcal{U}}_E(S) | \mathbf{x}_a \rangle = \int dx_a^4 \frac{r_b^\lambda r_a^{1-\lambda}}{(2\pi i \epsilon_s \hbar r_b^{1-\lambda} r_a^\lambda / M)^2}$$
$$\times \prod_{n=2}^{N+1} \left[\int_{-\infty}^\infty \frac{d^4 \Delta x_n}{(2\pi i \epsilon_s \hbar r_{n-1}/M)^2} \right] \exp\left(\frac{i}{\hbar} \mathcal{A}_E^N \right), \tag{13.108}$$

where $\mathcal{A}_E^N$ is the action (13.9) in which the three-vectors $\mathbf{x}_n$ are replaced by the four-vectors $\vec{x}_n$, although r is still the length of the spatial part of $\vec{x}$. By distributing the factors r_b, r_n, r_a evenly over the intervals, shifting the subscripts n of the factors $1/r_n$ in the measure to $n+1$, and using the same procedure as in Eq. (13.67), we arrive at the pseudotime evolution amplitude

$$\langle \mathbf{x}_b | \hat{\mathcal{U}}_E(S) | \mathbf{x}_a \rangle = \frac{1}{(2\pi i \epsilon_s \hbar / M)^2} \int_{-\infty}^\infty \frac{dx_a^4}{r_a}$$
$$\times \prod_{n=2}^{N+1} \left[\int \frac{d^4 \Delta \vec{x}_n}{(2\pi i \epsilon_s \hbar r_n^2 / M)^2} \right] \exp\left[\frac{i}{\hbar} (\mathcal{A}_E^N + \mathcal{A}_f^N) \right], \tag{13.109}$$

with the sliced action

$$\mathcal{A}_E^N[\vec{x}, \vec{x}'] = (N+1)\epsilon_s e^2 + \sum_{n=1}^{N+1} \left[\frac{M}{2} \frac{(\Delta \vec{x}_n)^2}{\epsilon_s r_n^{1-\lambda} r_{n-1}^\lambda} + \epsilon_s \, r_n^{1-\lambda} r_{n-1}^\lambda \, E \right]. \tag{13.110}$$

The action $\mathcal{A}_f^N$ accounts for all remaining factors in the integral measure. The prefactor is now $(r_b/r_a)^{3\lambda-2}$ and can be written as a product $\prod_1^{N+1}(r_n/r_{n-1})^{3\lambda-2}$. The index shift in the factor $1/r$ changes the power $3\lambda - 2$ to 3λ

$$\frac{i}{\hbar}\mathcal{A}_f^N = 3\lambda \sum_{n=1}^{N+1} \log\left(\frac{\vec{u}_n^2}{\vec{u}_{n-1}^2}\right) \tag{13.111}$$

[compare (13.68)]. As in the two-dimensional case we shall at first ignore the subtleties due to the time slicing. Thus we set $\lambda = 0$ and apply the transformation formally to the continuum limit of the action $\mathcal{A}_E^N$, which has the form (13.10), except that $\mathbf{x}$ is replaced by $\vec{x}$. Using the properties of the matrix (13.102)

$$\begin{aligned} A^T &= 4\vec{u}^2 A^{-1}, \\ \det A &= \sqrt{\det(AA^T)} = 16r^2, \end{aligned} \tag{13.112}$$

we see that

$$\begin{aligned} \vec{x}'^2 &= 4\vec{u}^2\vec{u}'^2 = 4r\vec{u}'^2, \tag{13.113}\\ d^4x &= 16r^2 d^4u. \tag{13.114} \end{aligned}$$

In this way, we find the formal relation

$$\langle \mathbf{x}_b|\hat{\mathcal{U}}_E(S)|\mathbf{x}_a\rangle = e^{ie^2S/\hbar}\frac{1}{16}\int\frac{dx_a^4}{r_a}(\vec{u}_bS|\vec{u}_a0) \tag{13.115}$$

to the time evolution amplitude of the four-dimensional harmonic oscillator

$$(\vec{u}_bS|\vec{u}_a0) = \int\mathcal{D}^4u(s)\exp\left(\frac{i}{\hbar}\mathcal{A}_{\text{os}}\right), \tag{13.116}$$

with the action

$$\mathcal{A}_{\text{os}} = \int_0^S ds\frac{\mu}{2}(\vec{u}'^2 - \omega^2\vec{u}^2). \tag{13.117}$$

The parameters are, as in (13.27) and (13.28),

$$\mu = 4M, \quad \omega = \sqrt{-E/2M}. \tag{13.118}$$

The relation (13.115) is the analog of (13.31). Instead of a sum over the two images of each point $\mathbf{x}$ in $\mathbf{u}$-space, there is now an integral $\int dx_a^4/r_a$ over the infinitely many images in the four-dimensional $\vec{u}$-space. This integral can be rewritten as an integral over the third Euler angle γ using the relation (13.103). Since $\mathbf{x}$ and thus the polar angles θ, φ remain fixed during the integration, we have directly $\int dx_a^4/r_a = \int d\gamma_a$. As far as the range of integration is concerned, we observe that it may be restricted to a single period $\gamma_a \in [0, 4\pi]$. The other periods can be included in the oscillator amplitude. By specifying a four-vector $\vec{u}_b$, all paths are summed which run either to the final Euler angle γ_b or to all its periodic repetitions [which by (13.97) have the same $\vec{u}_b$]. This was the lesson learned in Section 6.1. Equation (13.115) contains, instead, a sum over all initial periods which is completely equivalent to this. Thus the relation (13.115) reads, more specifically,

$$\langle \mathbf{x}_b | \hat{U}_E(S) | \mathbf{x}_a \rangle = e^{ie^2 S/\hbar} \frac{1}{16} \int_0^{4\pi} d\gamma_a (\vec{u}_b S | \vec{u}_a 0). \tag{13.119}$$

The reason why the other periods in (13.115) must be omitted can best be understood by comparison with the two-dimensional case. There we observed a two-fold degeneracy of contributions to the time-sliced path integral which cancel all factors 2 in the measure (13.71). Here the same thing happens except with an infinite degeneracy: When integrating over all images $d^4 u_n$ of $d^4 x_n$ in the oscillator path integral we cover the original $\mathbf{x}$-space once for $\gamma_n \in [0, 4\pi]$ and repeat doing so for all periods $\gamma_n \in [4\pi l, 4\pi(l+1)]$. This suggests that each volume element $d^4 u_n$ must be divided by an infinite factor to remove this degeneracy. However, this is not necessary since the gradient term produces precisely the same infinite factor. Indeed,

$$(\vec{u}_n + \vec{u}_{n-1})^2 (\vec{u}_n - \vec{u}_{n-1})^2 \tag{13.120}$$

is small for $\vec{x}_n \approx \vec{x}_{n-1}$ at infinitely many places of $\gamma_n - \gamma_{n-1}$, once for each periodic repetition of the interval $[0, 4\pi]$. The infinite degeneracy cancels the infinite factor in the denominator of the measure. The only place where this cancellation does not occur is in the integral $\int dx_a^4 / r_a$. Here the infinite factor in the denominator is still present, but it can be removed by restricting the integration over γ_a in (13.119) to a single period [7].

Note that a shift of γ_a by a half-period 2π changes $\vec{u}$ to $-\vec{u}$ and thus corresponds to the two-fold degeneracy in the previous two-dimensional system.

The time-sliced path integral for the harmonic oscillator can, of course, be done immediately, the amplitude being the four-dimensional version of (13.33):

$$
\begin{aligned}
(\vec{u}_b S | \vec{u}_a 0) &= \frac{1}{(2\pi i \hbar \epsilon_s / \mu)^2} \prod_{n=1}^N \left[\int \frac{d^4 \Delta u_n}{2\pi i \hbar \epsilon_s / \mu} \right] \exp \left[\frac{i}{\hbar} \sum_{n=1}^N \frac{\mu}{2} \left(\frac{1}{\epsilon_s} \Delta \vec{u}_n^{\ 2} - \epsilon_s \omega^2 \vec{u}_n^2 \right) \right] \\
&= \frac{\omega^2}{(2\pi i \hbar \sin \omega S / \mu)^2} \exp \left\{ \frac{i}{\hbar} \frac{\mu \omega}{\sin \omega S} [(\vec{u}_b^{\ 2} + \vec{u}_a^2) \cos \omega S - 2\vec{u}_b \vec{u}_a] \right\}. \tag{13.121}
\end{aligned}
$$

To find the fixed-energy amplitude we have to integrate this over S:

$$(\mathbf{x}_b | \mathbf{x}_a)_E = \int_0^\infty dS e^{ie^2 S/\hbar} \frac{1}{16} \int_0^{4\pi} d\gamma_a (\vec{u}_b S | \vec{u}_a 0). \tag{13.122}$$

Just like (13.35), the integral is written most conveniently in terms of the variables (13.38), (13.40), so that we obtain the fixed-energy amplitude of the three-dimensional Coulomb system

$$
\begin{aligned}
(\mathbf{x}_b | \mathbf{x}_a)_E &= \frac{1}{16} \int_0^\infty dS e^{ie^2 S/\hbar} \int_0^{4\pi} d\gamma_a (\vec{u}_b s | \vec{u}_a \ 0) \\
&= -i \frac{\omega M^2}{2\pi^2 \hbar^2} \int_{-\infty}^\infty \frac{dx_a^4}{r_a} \int_0^1 d\varrho \frac{\varrho^{-\nu}}{(1-\varrho)^2} \exp \left(2\kappa \frac{2\sqrt{\varrho}}{1-\varrho} \vec{u}_b \vec{u}_a \right) \exp \left[-\kappa \frac{1+\varrho}{1-\varrho} (r_b + r_a) \right].
\end{aligned}
$$

In order to perform the integral over dx_b^4, we now express $\vec{u}_b\vec{u}_a$ in terms of the polar angles

$$\vec{u}_b\vec{u}_a = \sqrt{r_b r_a}\,\{\cos(\theta_b/2)\cos(\theta_a/2)\cos[(\varphi_b - \varphi_a + \gamma_b - \gamma_a)/2] \\ + \sin(\theta_b/2)\sin(\theta_a/2)\cos[(\varphi_b - \varphi_a - \gamma_b + \gamma_a)/2]\}. \qquad (13.123)$$

A trigonometric rearrangement brings this to the form

$$\vec{u}_b\vec{u}_a = \sqrt{r_b r_a}\,\{\cos[(\theta_b - \theta_a)/2]\cos[(\varphi_b - \varphi_a)/2]\cos[(\gamma_b - \gamma_a)/2] \\ - \cos[(\theta_b + \theta_a)/2]\sin[(\varphi_b - \varphi_a)/2]\sin[(\gamma_b - \gamma_a)/2]\}, \qquad (13.124)$$

and further to

$$\vec{u}_b\vec{u}_a = \sqrt{(r_b r_a + \mathbf{x}_b\mathbf{x}_a)/2}\,\cos[(\gamma_b - \gamma_a + \beta)/2], \qquad (13.125)$$

where β is defined by

$$\tan\frac{\beta}{2} = \frac{\cos[(\theta_b + \theta_a)/2]\sin[(\varphi_b - \varphi_a)/2]}{\cos[(\theta_b - \theta_a)/2]\cos[(\varphi_b - \varphi_a)/2]}, \qquad (13.126)$$

or

$$\cos\frac{\beta}{2} = \cos\frac{\theta_b - \theta_a}{2}\cos\frac{\varphi_b - \varphi_a}{2}\frac{r_b r_a}{\sqrt{(r_b r_a + \mathbf{x}_b\mathbf{x}_a)/2}}. \qquad (13.127)$$

The integral $\int_0^{4\pi} d\gamma_a$ can now be done at each fixed $\mathbf{x}$. This gives the fixed-energy amplitude of the Coulomb system [1, 8, 10, 9].

$$(\mathbf{x}_b|\mathbf{x}_a)_E = -i\frac{M\kappa}{\pi\hbar}\int_0^1 d\varrho\,\frac{\varrho^{-\nu}}{(1-\varrho)^2}I_0\left(2\kappa\frac{2\sqrt{\varrho}}{1-\varrho}\sqrt{(r_b r_a + \mathbf{x}_b\mathbf{x}_a)/2}\right) \\ \times\,\exp\left[-\kappa\frac{1+\varrho}{1-\varrho}(r_b + r_a)\right], \qquad (13.128)$$

where κ and ν are the same parameters as in Eqs. (13.40).

The integral converges only for $\nu < 1$, as in the two-dimensional case. It is again possible to write down another integral representation which converges for all $\nu \neq 1, 2, 3, \ldots$ by changing the variables of integration to $\zeta \equiv (1 + \varrho)/(1 - \varrho)$ and transforming the integral over ζ into a contour integral encircling the cut from $\zeta = 1$ to ∞ in the clockwise sense. Since the cut is now of the type $(\zeta - 1)^{-\nu}$, the replacement rule is

$$\int_1^\infty d\zeta\,(\zeta - 1)^{-\nu}\ldots \rightarrow \frac{\pi e^{i\pi\nu}}{\sin\pi\nu}\int_C \frac{d\zeta}{2\pi i}(\zeta - 1)^{-\nu}\ldots. \qquad (13.129)$$

This leads to the representation

$$(\mathbf{x}_b|\mathbf{x}_a)_E = -i\frac{M}{\pi\hbar}\frac{\kappa}{2}\frac{\pi e^{i\pi\nu}}{\sin\pi\nu}\int_C \frac{d\zeta}{2\pi i}(\zeta - 1)^{-\nu}(\zeta + 1)^{\nu} \\ \times I_0(2\kappa\sqrt{\zeta^2 - 1}\sqrt{(r_b r_a + \mathbf{x}_b\mathbf{x}_a)/2})e^{-\kappa\zeta(r_b + r_a)}. \qquad (13.130)$$

13.5 Absence of Time Slicing Corrections for $D = 3$

Let us now prove that for the three-dimensional Coulomb system also, the finite time-slicing procedure does not change the formal result of the last section. The reader not interested in the details is again referred to the brief argument in Section 13.6. The action $\mathcal{A}_E^N$ in the time-sliced path integral has to be supplemented, in each slice, by the Jacobian action [as in (13.62)]

$$
\begin{aligned}
\frac{i}{\hbar}\mathcal{A}_J^\epsilon &= -e^\mu e^i{}_{\{\mu,\nu\}}\Delta u^\nu - e_i{}^\mu e^i{}_{\{\kappa,\nu\}}e_j{}^\kappa e^i{}_{\{\mu,\lambda\}}\Delta u^\nu \Delta u^\lambda \\
&= -\Gamma_{\{\nu\mu\}}{}^\mu \Delta u^\nu - \frac{1}{2}\Gamma_{\{\nu\kappa\}}{}^\sigma \Gamma_{\{\lambda\sigma\}}{}^\kappa \Delta u^\nu \Delta u^\lambda .
\end{aligned}
\tag{13.131}
$$

The basis tetrad

$$
e^i{}_\mu = \partial x^i / \partial u^\mu = 2A^i{}_\mu(\vec{u}), \quad i = 1, 2, 3, 4,
\tag{13.132}
$$

is now given by the 4×4 matrix (13.102), with the reciprocal tetrad

$$
e_i{}^\mu = \frac{1}{2\vec{u}^2}e^i{}_\mu .
\tag{13.133}
$$

From this we find the matrix components of the connection [compare (13.18)]

$$
\begin{aligned}
(\Gamma_1)_\mu{}^\nu &= \frac{1}{\vec{u}^2}\begin{pmatrix} u^1 & u^2 & -u^3 & -u^4 \\ -u^2 & u^1 & -u^4 & u^3 \\ u^3 & u^4 & u^1 & u^2 \\ u^4 & -u^3 & -u^2 & u^1 \end{pmatrix}_\mu , \\[4pt]
(\Gamma_2)_\mu{}^\nu &= \frac{1}{\vec{u}^2}\begin{pmatrix} u^2 & -u^1 & u^4 & -u^3 \\ u^1 & u^2 & -u^3 & -u^4 \\ -u^4 & u^3 & u^2 & -u^1 \\ u^3 & u^4 & u^1 & u^2 \end{pmatrix}_\mu , \\[4pt]
(\Gamma_3)_\mu{}^\nu &= \frac{1}{\vec{u}^2}\begin{pmatrix} u^3 & u^4 & u^1 & u^2 \\ -u^4 & u^3 & u^2 & -u^1 \\ -u^1 & -u^2 & u^3 & u^4 \\ -u^2 & u^1 & -u^4 & u^3 \end{pmatrix}_\mu , \\[4pt]
(\Gamma_4)_\mu{}^\nu &= \frac{1}{\vec{u}^2}\begin{pmatrix} u^4 & -u^3 & -u^2 & u^1 \\ u^3 & u^4 & u^1 & u^2 \\ u^2 & -u^1 & u^4 & -u^3 \\ -u^1 & -u^2 & u^3 & u^4 \end{pmatrix}_\mu .
\end{aligned}
\tag{13.134}
$$

As in the two-dimensional case [see Eq. (13.19)], the connection satisfies the important identity

$$
\Gamma_\mu{}^{\mu\nu} \equiv 0,
\tag{13.135}
$$

which is again a consequence of the relation [compare (13.21)]

$$
\partial_\mu e^i{}_\mu = 0.
\tag{13.136}
$$

In Section 13.6, this will be shown to be the essential reason for the absence of the time slicing corrections being proved in this section.

However, there is now an important difference with respect to the two-dimensional case. The present mapping $dx^i = e^i{}_\mu(u)du^\mu$ is not integrable. Taking the antisymmetric part of $\Gamma_{\mu\nu}{}^\lambda$ we find the u^μ-space to carry a torsion $S_{\mu\nu}{}^\lambda$ whose only nonzero components are

$$
S_{12}{}^\lambda = S_{34}{}^\lambda = \frac{1}{\vec{u}^2}(-u^2, u^1, -u^4, u^3)^\lambda .
\tag{13.137}
$$

The once-contracted torsion is

$$S_\mu = S_{\mu\nu}{}^\nu = \frac{u^\mu}{\vec{u}^2}. \tag{13.138}$$

For this reason, the contracted connections

$$\begin{aligned}
\Gamma_{\nu\mu}{}^\mu &= e_i{}^\mu \partial_\nu e^i{}_\mu = \frac{4u^\nu}{\vec{u}^2}, \\
\Gamma_{\mu\nu}{}^\mu &= -e^i{}_\nu \partial_\mu e_i{}^\mu = \frac{2u^\nu}{\vec{u}^2}
\end{aligned} \tag{13.139}$$

are no longer equal, as they were in (13.64). Symmetrization in the lower indices gives

$$\Gamma_{\{\nu\mu\}}{}^\mu = \frac{3u^\nu}{\vec{u}^2}. \tag{13.140}$$

Due to this, the $\Delta u^\nu \Delta u^\lambda$-terms in (13.131) are, in contrast to the two-dimensional case, not given directly by

$$\Gamma_{\nu\kappa}{}^\sigma \Gamma_{\lambda\sigma}{}^\kappa = -\frac{4}{\vec{u}^4}(\delta^{\nu\lambda}\vec{u}^2 - 2u^\nu u^\lambda). \tag{13.141}$$

The symmetrization in the lower indices is necessary and yields

$$\begin{aligned}
\Gamma_{\{\nu\kappa\}}{}^\sigma \Gamma_{\{\lambda\sigma\}}{}^\kappa &= \Gamma_{\nu\kappa}{}^\sigma \Gamma_{\lambda\sigma}{}^\kappa - 2\Gamma_{\nu\kappa}{}^\sigma S_{\lambda\sigma}{}^\kappa + S_{\nu\kappa}{}^\sigma S_{\lambda\sigma}{}^\kappa \\
&= \Gamma_{\nu\kappa}{}^\sigma \Gamma_{\lambda\sigma}{}^\kappa - 2(-\delta_{\nu\lambda}\vec{u}^2 + 2u_\nu u_\lambda)/\vec{u}^4 + u_\nu u_\lambda/\vec{u}^4.
\end{aligned} \tag{13.142}$$

Collecting all terms, the Jacobian action (13.131) becomes

$$\frac{i}{\hbar}\mathcal{A}_J^\epsilon = -\left[3\frac{\vec{u}_n \Delta \vec{u}_n}{\vec{u}_n^2} - \frac{\Delta \vec{u}_n^2}{\vec{u}_n^2} + \frac{5}{2}\left(\frac{\vec{u}_n \Delta \vec{u}_n}{\vec{u}_n^2}\right)^2 + \dots\right]. \tag{13.143}$$

In contrast to the two-dimensional equation (13.66), this cannot be incorporated into $\mathcal{A}_f^\epsilon$. Although the two expressions contain the same terms, their coefficients are different [see (13.111)]:

$$\begin{aligned}
\frac{i}{\hbar}\mathcal{A}_f^\epsilon &= 3\lambda \log\left(\frac{\vec{u}_n^2}{\vec{u}_{n-1}^2}\right) \\
&= 3\lambda \log\left[2\frac{\vec{u}_n \Delta \vec{u}_n}{\vec{u}_n^2} - \frac{\Delta \vec{u}_n^2}{\vec{u}_n^2} + 2\left(\frac{\vec{u}\Delta \vec{u}_n}{\vec{u}_n^2}\right)^2 + \dots\right].
\end{aligned} \tag{13.144}$$

It is then convenient to rewrite (omitting the subscripts n)

$$\begin{aligned}
\frac{i}{\hbar}\mathcal{A}_J^\epsilon &= -2\log\left[\frac{\vec{u}^2}{(\vec{u} - \Delta \vec{u})^2}\right] + \frac{\vec{u}\Delta \vec{u}}{\vec{u}^2} \\
&\quad -\frac{\Delta \vec{u}^2}{\vec{u}^2} + \frac{3}{2}\left(\frac{\vec{u}\Delta \vec{u}}{\vec{u}^2}\right)^2 + \dots,
\end{aligned} \tag{13.145}$$

and absorb the first term into $\mathcal{A}_f^\epsilon$, which changes 3λ to $(3\lambda - 2)$. Thus, we obtain altogether the additional action [to be compared with (13.76)]

$$\begin{aligned}
\frac{i}{\hbar}\Delta_{\text{corr}}\mathcal{A}^\epsilon &= \frac{i}{\hbar}4M\frac{\Delta \vec{u}^2}{2\epsilon}\left[(2\lambda - 1)\frac{\vec{u}\Delta \vec{u}}{\vec{u}^2} + (\frac{1}{4} - \lambda)\frac{\Delta \vec{u}^2}{\vec{u}^2} + 2\lambda^2\left(\frac{\vec{u}\Delta \vec{u}}{\vec{u}^2}\right)^2\right] \\
&\quad + (3\lambda - 2)\left[2\frac{\vec{u}\Delta \vec{u}}{\vec{u}^2} - \frac{\Delta \vec{u}^2}{\vec{u}^2} + 2\left(\frac{\vec{u}\Delta \vec{u}}{\vec{u}^2}\right)^2\right] \\
&\quad + \frac{\vec{u}\Delta \vec{u}}{\vec{u}^2} - \frac{(\Delta \vec{u})^2}{\vec{u}^2} + \frac{3}{2}\left(\frac{\vec{u}\Delta \vec{u}}{\vec{u}^2}\right)^2 + \dots.
\end{aligned} \tag{13.146}$$

Using this we now show that the expansion of the correction term

$$C = \exp\left(\frac{i}{\hbar}\Delta_{\text{corr}}\mathcal{A}^\epsilon\right) - 1 \tag{13.147}$$

has the vanishing expectations

$$\langle C \rangle_0 = 0,$$
$$\langle C \, (\vec{p}\Delta\vec{u}) \rangle_0 = 0, \tag{13.148}$$

i.e.,

$$\frac{i}{\hbar}\langle\Delta_{\text{corr}}\mathcal{A}^\epsilon\rangle_0 + \frac{1}{2}\left(\frac{i}{\hbar}\right)^2 \langle(\Delta_{\text{corr}}\mathcal{A}^\epsilon)^2\rangle_0 = 0, \tag{13.149}$$

$$\frac{i}{\hbar}\langle\Delta_{\text{corr}}\mathcal{A} \, (\vec{p}\Delta\vec{u})\rangle_0 = 0, \tag{13.150}$$

as in (13.87), (13.88). In fact, using formula (13.86) the expectation (13.150) is immediately found to be proportional to

$$i\left[-2(2\lambda-1)\frac{D+2}{16} + 2(3\lambda-2)\frac{1}{4} + \frac{1}{4}\right], \tag{13.151}$$

which vanishes identically in λ for $D = 4$. Similarly, using formula (13.85), the first term in (13.149) has an expectation proportional to

$$i\left[-2\left(\frac{1}{4}-\lambda\right)\frac{(D+2)D}{16} - 4\lambda^2\frac{D+2}{16} - (3\lambda-2)\left(\frac{D}{4}-\frac{2}{4}\right) - \left(\frac{D}{4}-\frac{3}{8}\right)\right], \tag{13.152}$$

i.e., for $D = 4$,

$$-i\frac{3}{8}(2\lambda-1)^2, \tag{13.153}$$

to which the second term adds

$$i\frac{1}{2}\left[4(2\lambda-1)^2\frac{(D+4)(D+2)}{64} + 9(2\lambda-1)^2\frac{1}{4} - 12(2\lambda-1)^2\frac{D+2}{16}\right], \tag{13.154}$$

which cancels (13.154) for $D = 4$. Thus the sum of all time slicing corrections vanishes also in the three-dimensional case.

13.6 Geometric Argument for Absence of Time Slicing Corrections

As mentioned before, the basic reason for the absence of the time slicing corrections can be shown to be the property of the connection

$$\Gamma_\mu{}^{\mu\lambda} = g^{\mu\nu}e_i{}^\lambda\partial_\mu e^i{}_\nu = 0, \tag{13.155}$$

which, in turn, follows from the basic identity $\partial_\mu e^i_\mu = 0$ satisfied by the basis tetrad, and from the diagonality of the metric $g^{\mu\nu} \propto \delta^{\mu\nu}$. Indeed, it is possible to apply the techniques of Sections 10.1, 10.2 to the general pseudotime evolution amplitude (12.28) with the regulating functions

$$f_l = f(\mathbf{x}), \quad f_r \equiv 1. \tag{13.156}$$

Since this regularization affects only the postpoints at each time slice, it is straightforward to repeat the derivation of an equivalent short-time amplitude given in Section 11.3. The result can be expressed in the form (dropping subscripts n)

$$K^\epsilon(\Delta q) = \frac{\sqrt{g(q)}}{\sqrt{2\pi i\epsilon\hbar f/M}^D} \exp\left[\frac{i}{\hbar}(\mathcal{A}^\epsilon + \mathcal{A}^\epsilon_J)\right], \tag{13.157}$$

where f abbreviates the postpoint value $f(\mathbf{x}_n)$ and $\mathcal{A}^\epsilon$ is the short-time action

$$\mathcal{A}^\epsilon = \frac{M}{2\epsilon f}g_{\mu\nu}(q)\Delta q^\mu \Delta q^\nu. \tag{13.158}$$

There exists now a simple expression for the Jacobian action. Using formula (11.75), it becomes simply

$$\frac{i}{\hbar}\mathcal{A}^\epsilon_J = \frac{1}{2}\Gamma_\mu{}^\mu{}_\nu\Delta q^\nu - i\epsilon\frac{\hbar f}{8M}(\Gamma_\mu{}^\mu{}_\nu)^2. \tag{13.159}$$

In the postpoint formulation, the measure needs no further transformation. This can be seen directly from the time-sliced expression (13.8) for $\lambda = 0$ or, more explicitly, from the vanishing of the extra action $\mathcal{A}^\epsilon_f$ in (13.69) for $D = 2$ and in (13.144) for $D = 3$. As a result, the vanishing contracted connection appearing in (13.159) makes *all* time-slicing corrections vanish. Only the basic short-time action (13.158) survives:

$$\Delta\mathcal{A}^\epsilon = 4M\frac{(\Delta\mathbf{u})^2}{2\epsilon_s}. \tag{13.160}$$

Thanks to this fortunate circumstance, the formal solution found in 1979 by Duru and Kleinert happens to be correct.

13.7 Comparison with Schrödinger Theory

For completeness, let us also show the significance of the geometric property $\Gamma_\mu{}^{\mu\lambda} = 0$ within the Schrödinger theory. Consider the Schrödinger equation of the Coulomb system

$$\left(-\frac{1}{2M}\hbar^2\boldsymbol{\nabla}^2 - E\right)\psi(\mathbf{x}) = \frac{e^2}{r}\psi(\mathbf{x}), \tag{13.161}$$

to be transformed to that of a harmonic oscillator. The postpoint regularization of the path integral with the functions (13.156) corresponds to multiplying the Schrödinger equation with $f_l = r$ from the left. This gives

$$\left(-\frac{1}{2M}\hbar^2 r\boldsymbol{\nabla}^2 - Er\right)\psi(\mathbf{x}) = e^2\psi(\mathbf{x}). \tag{13.162}$$

We now go over to the square root coordinates u^μ transforming $-Er$ into the harmonic potential $-E(u^\mu)^2$ and the Laplacian $\boldsymbol{\nabla}^2$ into $g^{\mu\nu}\partial_\mu\partial_\nu - \Gamma_\mu{}^{\mu\lambda}\partial_\lambda$. The geometric property $\Gamma_\mu{}^{\mu\lambda} = 0$ ensures now the absence of the second term and the result is simply $g^{\mu\nu}\partial_\mu\partial_\nu$. Since $g^{\mu\nu} = \delta^{\mu\nu}/4r$, the Schrödinger equation (13.162) takes the simple form

$$\left[-\frac{1}{8M}\hbar^2\partial_\mu^2 - E(u^\mu)^2\right]\psi(u^\mu) = e^2\psi(u^\mu). \tag{13.163}$$

Due to the factor $(u^\mu)^2$ accompanying the energy E, the physical scalar product in which the states of different energies are orthogonal to each other is given by

$$\langle \psi' | \psi \rangle = \int d^4 u \, \psi'(u^\mu)(u^\mu)^2 \psi(u). \tag{13.164}$$

This corresponds precisely to the scalar product given in Eq. (11.95) with the purpose of making the Laplace operator (here $\Delta = (1/4\mathbf{u}^2)\partial_\mathbf{u}^2$) hermitian in the u^μ-space with torsion. Indeed, the once-contracted torsion tensor $S_\mu = S_{\mu\nu}{}^\nu$ can be written as a gradient of a scalar function:

$$S_\mu = \partial_\mu \sigma(\vec{u}), \qquad \sigma(\vec{u}) = \frac{1}{2} \log \vec{u}^2. \tag{13.165}$$

Quite generally, we have shown in Eq. (11.104) that if $S_\mu(q)$ is a partial derivative of a scalar field $\sigma(q)$, the physical scalar product is given by (11.95):

$$\langle \psi_2 | \psi_1 \rangle_{\text{phys}} \equiv \int d^D q \sqrt{g(q)} e^{-2\sigma(q)} \psi_2^*(q) \psi_1(q). \tag{13.166}$$

From (13.132), we have

$$\sqrt{g} = 16\vec{u}^4, \tag{13.167}$$

so that the physical scalar product is

$$\langle \psi_2 | \psi_1 \rangle_{\text{phys}} = \int d^4 u \sqrt{g} e^{-2\sigma} \psi_2^*(\vec{u}) \psi_1(\vec{u}) = \int d^4 u \, 16\vec{u}^2 \, \psi_2^*(\vec{u}) \psi_1(\vec{u}). \tag{13.168}$$

The Laplace operator obtained from $\partial_{\vec{x}}^2$ by the nonholonomic Kustaanheimo-Stiefel transformation is $\Delta = (1/4\vec{u}^2)\partial_\mu^2$. This is Hermitian in the physical scalar product (13.168), but not in the naive one (11.90) with the integral measure $\int d^4 u \, 16\vec{u}^4$.

In two dimensions, the torsion vanishes and the physical scalar product reduces to the naive one:

$$\langle \psi_2 | \psi_1 \rangle_{\text{phys}} = \int d^2 u \sqrt{g} \psi_2^*(\mathbf{u}) \psi_1(\mathbf{u}) = \int d^2 u \, 4\mathbf{u}^2 \psi_2^*(\mathbf{u}) \psi_1(\mathbf{u}). \tag{13.169}$$

With $\mu = 4M$ and $-E = \mu\omega^2/2$, Eq. (13.163) is the Schrödinger equation of a harmonic oscillator:

$$\left[-\frac{1}{2\mu} \hbar^2 \partial_\mu^2 + \frac{\mu}{2} \omega^2 (u^\mu)^2 \right] \psi(u^\mu) = \mathcal{E}\psi(u^\mu). \tag{13.170}$$

The eigenvalues of the pseudoenergy $\mathcal{E}$ are

$$\mathcal{E}_N = \hbar\omega(N + D_u/2), \tag{13.171}$$

where $D_u = 4$ is the dimension of u^μ-space,

$$N = \sum_{i=1}^{D_u} n_i \tag{13.172}$$

sums up the integer principal quantum numbers of the factorized wave functions in each direction of the u^μ-space. The multivaluedness of the mapping from $\mathbf{x}$ to u^μ allows only symmetric wave functions to be associated with Coulomb states. Hence N must be even and can be written as $N = 2(n-1)$. The pseudoenergy spectrum is therefore

$$\mathcal{E}_n = \hbar\omega\, 2(n + D_u/4 - 1), \quad n = 1, 2, 3, \ldots \ . \tag{13.173}$$

According to (13.170), the Coulomb wave functions must all have a pseudoenergy

$$\mathcal{E}_n = e^2. \tag{13.174}$$

The two equations are fulfilled if the oscillator frequency has the discrete values

$$\omega = \omega_n \equiv \frac{e^2}{2(n + D_u/4 - 1)}, \quad n = 1, 2, 3 \ldots \ . \tag{13.175}$$

With $\omega^2 = -E/2M$ and $D_u = 4$, this yields the Coulomb,energies

$$E_n = -2M\omega_n^2 = -\frac{Me^4}{\hbar^2}\frac{1}{2n^2}, \tag{13.176}$$

showing that the number $N/2 = n - 1$ corresponds to the usual principal quantum number of the Coulomb wave functions.

Let us now focus our attention upon the three-dimensional Coulomb system where $D_u = 4$. In this case, not all even oscillator wave functions correspond to Coulomb bound-state wave functions. This follows from the fact that the Coulomb wave functions do not depend on the dummy fourth coordinate x^4 (or the dummy angle γ). Thus they satisfy the constraint $\partial_{x^4}\psi = 0$, implying in u^μ-space [recall (13.132)]

$$
\begin{aligned}
-ir\partial_{x^4}\psi(\mathbf{x}) &= -ire_4{}^\mu \partial_\mu \psi(u^\mu) = -i\frac{1}{2}\left[(u^2\partial_1 - u^1\partial_2) + (u^4\partial_3 - u^3\partial_4)\right]\psi(u^\mu) \\
&= -i\partial_\gamma \psi(u^\mu) = 0.
\end{aligned} \tag{13.177}
$$

The explicit construction of the oscillator and Coulomb bound-state wave functions is most conveniently done in terms of the complex coordinates (13.96). In terms of these, the constraint (13.177) reads

$$\frac{1}{2}\left[\bar{z}\partial_{\bar{z}} - z\partial_z\right]\psi(z, z^*) = 0. \tag{13.178}$$

This will be used below to select the Coulomb states.

To solve the Schrödinger equation (13.170) we simplify the notation by going over to *atomic natural units*, where $\hbar = 1, M = 1, e^2 = 1, \mu = 4M = 4$. All lengths are measured in units of the Bohr radius (4.344), whose numerical value is $a_H = \hbar^2/Me^2 = 5.2917 \times 10^{-9}$ cm, all energies in units of $E_H \equiv e^2/a_H =$

$Me^4/\hbar^2 = 4.359 \times 10^{-11}$ erg $= 27.210$ eV, and all frequencies ω in units of $\omega_H \equiv Me^4/\hbar^3 = 4.133 \times 10^{16}/\text{sec}(= 4\pi \times$ Rydberg frequency ν_R). Then (13.170) reads (after multiplication by $4M/\hbar^2$)

$$\hat{h}\psi(u^\mu) \equiv \frac{1}{2}\left[-\partial_\mu^2 + 16\omega^2(u^\mu)^2\right]\psi(u^\mu) = 4\psi(u^\mu). \qquad (13.179)$$

The spectrum of the operator $\hat{h}$ is obviously $4\omega(N+2) = 8\omega n$. To satisfy the equation, the frequency ω has to be equal to $\omega_n = 1/2n$.

We now observe that the operator $\hat{h}$ can be brought to the standard form

$$\hat{h}^s = \frac{1}{2}\left[-\partial_\mu^2 + 4(u^\mu)^2\right], \qquad (13.180)$$

with the help of the ω-dependent transformation

$$\hat{h} = 4\omega e^{i\vartheta \hat{D}}\hat{h}^s e^{-i\vartheta \hat{D}}, \qquad (13.181)$$

in which the operator $\hat{D}$ is an infinitesimal *dilation operator* which in this context is called *tilt operator* [11, 12]:

$$\hat{D} \equiv -\frac{1}{2}iu^\mu\partial_\mu, \qquad (13.182)$$

and ϑ is the *tilt angle*

$$\vartheta = \log(2\omega). \qquad (13.183)$$

The Coulomb wave functions are therefore given by the rescaled solutions of the standardized Schrödinger equation (13.180):

$$\psi(u^\mu) = e^{i\vartheta \hat{D}}\psi^s(u^\mu) = \psi^s(\sqrt{2\omega}u^\mu). \qquad (13.184)$$

Note that for a solution with a principal quantum number n the scale parameter $\sqrt{2\omega}$ depends on n:

$$\psi_n(u^\mu) = \psi_n^s(u^\mu/\sqrt{n}). \qquad (13.185)$$

The standardized wave functions $\psi_n^s(u^\mu)$ are constructed most conveniently by means of four sets of creation and annihilation operators $\hat{a}_1^\dagger, \hat{a}_2^\dagger, \hat{b}_1^\dagger, \hat{b}_2^\dagger$, and $\hat{a}_1, \hat{a}_2, \hat{b}_1, \hat{b}_2$. They are combinations of z_1, z_2, their complex-conjugates, and the associated differential operators $\partial_{z_1}, \partial_{z_2}, \partial_{z_1^*}, \partial_{z_2^*}$. The combinations are the same as in (9.127), (9.128), written down once for z_1 and once for z_2. In addition, we choose the indices so that a_i and b_i transform by the same spinor representation of the rotation group. If c_{ij} is the 2×2 matrix

$$c = i\sigma^2 = \begin{pmatrix} 0 & 1 \\ -1 & 0 \end{pmatrix}, \qquad (13.186)$$

then $c_{ij}z_j$ transforms like z_i^*. We therefore define the creation operators

$$\hat{a}_1^\dagger \equiv -\frac{1}{\sqrt{2}}(-\partial_{z_2^*} + z_2), \quad \hat{b}_1^\dagger \equiv \frac{1}{\sqrt{2}}(-\partial_{z_1} + z_1^*),$$

$$\hat{a}_2^\dagger \equiv \frac{1}{\sqrt{2}}(-\partial_{z_1^*} + z_1), \quad \hat{b}_2^\dagger \equiv \frac{1}{\sqrt{2}}(-\partial_{z_2} + z_2^*), \qquad (13.187)$$

and the annihilation operators

$$\hat{a}_1 \equiv -\frac{1}{\sqrt{2}}(\partial_{z_2} + z_2^*), \quad \hat{b}_1 \equiv \frac{1}{\sqrt{2}}(\partial_{z_1^*} + z_1),$$

$$\hat{a}_2 \equiv \frac{1}{\sqrt{2}}(\partial_{z_1} + z_1^*), \quad \hat{b}_2 \equiv \frac{1}{\sqrt{2}}(\partial_{z_2^*} + z_2). \qquad (13.188)$$

Note that $\partial_z^\dagger = -\partial_{z^*}$. The standardized oscillator Hamiltonian is then

$$\hat{h}^s = 2(\hat{a}^\dagger \hat{a} + \hat{b}^\dagger \hat{b} + 2), \qquad (13.189)$$

where we have used the same spinor notation as in (13.94). The ground state of the four-dimensional oscillator is annihilated by $\hat{a}_1, \hat{a}_2$ and $\hat{b}_1, \hat{b}_2$. It has therefore the wave function

$$\langle z, z^* | 0 \rangle = \psi_{s,0000}(z, z^*) = \frac{1}{\sqrt{\pi}} e^{-z_1 z_1^* - z_2 z_2^*} = \frac{1}{\sqrt{\pi}} e^{-(u^\mu)^2}. \qquad (13.190)$$

The complete set of oscillator wave functions is obtained, as usual, by applying the creation operators to the ground state,

$$|n_1^a, n_2^a, n_1^b, n_2^b\rangle = N_{n_1^a, n_2^a, n_1^b, n_2^b} \hat{a}_1^{\dagger n_1^a} \hat{a}_2^{\dagger n_2^a} \hat{b}_1^{\dagger n_1^b} \hat{b}_2^{\dagger n_2^b} |0\rangle, \qquad (13.191)$$

with the normalization factor

$$N_{n_1^a, n_2^a, n_1^b, n_2^b} = \frac{1}{\sqrt{n_1^a! n_2^a! n_1^b! n_2^b!}}. \qquad (13.192)$$

The eigenvalues of $\hat{h}^s$ are obtained from the sum of the number of a- and b-quanta as

$$2(n_1^a + n_2^a + n_1^b + n_2^b + 2) = 2(N + 2) = 4n. \qquad (13.193)$$

The Coulomb bound-state wave functions are in one-to-one correspondence with those oscillator wave functions which satisfy the constraint (13.178), which may now be written as

$$\hat{L}_{05} = -\frac{1}{2}(\hat{a}^\dagger \hat{a} - \hat{b}^\dagger \hat{b}) \psi^s = 0. \qquad (13.194)$$

These states carry an equal number of a- and b-quanta. They diagonalize the (mutually commuting) a- and b-spins

$$\hat{L}_i^a \equiv \frac{1}{2} \hat{a}^\dagger \sigma_i \hat{a}, \quad \hat{L}_i^a \equiv \frac{1}{2} \hat{b}^\dagger \sigma_i \hat{b}, \qquad (13.195)$$

with the quantum numbers

$$l^a = (n_1^a + n_2^a)/2, \quad m^a = (n_1^a - n_2^a)/2,$$
$$l^b = (n_1^b + n_2^b)/2, \quad m^b = (n_1^b - n_2^b)/2, \tag{13.196}$$

where l, m are the eigenvalues of $\hat{L}^2, \hat{L}_3$. By defining

$$n_1^a \equiv n_1 + m, \quad n_2^a \equiv n_2, \quad n_1^b = n_2 + m, \quad n_2^b = n_1, \qquad \text{for } m \geq 0,$$
$$n_1^a \equiv n_1, \quad n_2^a \equiv n_2 - m, \quad n_1^b = n_2, \quad n_2^b = n_1 - m, \qquad \text{for } m \leq 0, \tag{13.197}$$

we establish contact with the eigenstates $|n_1, n_2, m\rangle$ which arise naturally when diagonalizing the Coulomb Hamiltonian in parabolic coordinates. The relation between these states and the usual Coulomb wave function of a given angular momentum $|nlm\rangle$ is obvious since the angular momentum operator $\hat{L}_i$ is equal to the sum of a- and b-spins. The rediagonalization is achieved by the usual vector coupling coefficients (see the last equation in Appendix 13A).

Note that after the tilt transformation (13.184), the exponential behavior of the oscillator wave functions $\psi_n^s(u^\mu) \propto \text{polynomial}(u^\mu) \times e^{-(u^\mu)^2}$ goes correctly over into the exponential r-dependence of the Coulomb wave functions $\psi(\mathbf{x}) \propto \text{polynomial}(\mathbf{x}) \times e^{-r/n}$.

It is important to realize that although the dilation operator $\hat{D}$ is Hermitian and the operator $e^{i\vartheta\hat{D}}$ at a *fixed* angle ϑ is unitary, the Coulomb bound states ψ_n, arising from the complete set of oscillator states ψ_n^s by applying $e^{i\vartheta\hat{D}}$, do *not* span the Hilbert space. Due to the n-dependence of the tilt angle $\vartheta_n = \log(1/n)$, a section of the Hilbert space is not reached. The continuum states of the Coulomb system, which are obtained by tilting another complete set of states, precisely fill this section. Intuitively we can understand this incompleteness simply as follows. The wave functions $\psi_n^s(u^\mu)$ have for increasing n spatial oscillations with shorter and shorter wavelength. These allow the completeness sum $\sum_n \psi_n^s(u^\mu)\psi_n^{s*}(u^\mu)$ to build up a δ-function which is necessary to span the Hilbert space. In contrast, when forming the sum of the dilated wave functions

$$\sum_n \psi_n^s(u^\mu/\sqrt{n})\psi_n^{s*}(u^\mu/\sqrt{n}),$$

the terms of larger n have increasingly stretched spatial oscillations which are *not* sufficient to build up an infinitely narrow distribution.

A few more algebraic properties of the creation and annihilation operator representation of the Coulomb wave functions are collected in Appendix 13A and 13.10.

13.8   Angular Decomposition of Amplitude, and Radial Wave Functions

Let us also give an angular decomposition of the fixed-energy amplitude. This serves as a convenient starting point for extracting the radial wave functions of the

Coulomb system which will, in Chapter 14, enable us to find the Coulomb amplitude to D dimensions. We begin with the expression (13.128),

$$(\mathbf{x}_b|\mathbf{x}_a)_E = -i\frac{M\kappa}{\pi\hbar}\int_0^1 d\varrho\frac{\varrho^{-\nu}}{(1-\varrho)^2}I_0\left(2\kappa\frac{2\sqrt{\varrho}}{1-\varrho}\sqrt{(r_b r_a + \mathbf{x}_b\mathbf{x}_a)/2}\right)$$
$$\times \exp\left\{-\kappa\frac{1+\varrho}{1-\varrho}(r_b + r_a)\right\}, \tag{13.198}$$

and rewrite the Bessel function as $I_0(z\cos(\theta/2))$, where θ is the relative angle between $\mathbf{x}_a$ and $\mathbf{x}_b$, and

$$z \equiv 2\kappa\sqrt{r_b r_a}\frac{2\sqrt{\varrho}}{1-\varrho}. \tag{13.199}$$

Now we make use of the expansion[1]

$$\left(\frac{1}{2}kz\right)^{\mu-\nu}I_\nu(kz) = k^\mu\sum_{l=0}^\infty\frac{1}{l!}\frac{\Gamma(l+\mu)}{\Gamma(1+\nu)}(2l+\mu)F(-l,l+\mu;1+\nu;k^2)(-)^l I_{2l+\mu}(z). \tag{13.200}$$

Setting $k = \cos(\theta/2)$, $\nu = 2q > 0$, $\mu = 1 + 2q$, and using formulas (1.442), (1.443) for the rotation functions, this becomes

$$I_{2q}(z\cos(\theta/2)) = \frac{2}{z}\sum_{l=|q|}^\infty(2l+1)d_{qq}^l(\theta)I_{2l+1}(z), \tag{13.201}$$

reducing for $q = 0$ to

$$I_0(z\cos(\theta/2)) = \frac{2}{z}\sum_{l=0}^\infty(2l+1)P_l(\cos\theta)I_{2l+1}(z). \tag{13.202}$$

After inserting this into (13.128) and substituting $y = -\frac{1}{2}\log\varrho$, so that

$$\varrho = e^{-2y}, \quad z = 2\kappa\sqrt{r_b r_a}\frac{1}{\sinh y}, \tag{13.203}$$

we expand the fixed-energy amplitude into spherical harmonics

$$(\mathbf{x}_b|\mathbf{x}_a)_E = \frac{1}{r_b r_a}\sum_{l=0}^\infty(r_b|r_a)_{E,l}\frac{2l+1}{4\pi}P_l(\cos\theta)$$
$$= \frac{1}{r_b r_a}\sum_{l=0}^\infty(r_b|r_a)_{E,l}\sum_{m=-l}^l Y_{lm}(\hat{\mathbf{x}}_b)Y_{lm}^*(\hat{\mathbf{x}}_a), \tag{13.204}$$

---

[1] G.N. Watson, *Theory of Bessel Functions*, Cambridge University Press, London, 1966, 2nd ed., p.140, formula (3).

with the radial amplitude

$$(r_b|r_a)_{E,l} = -i\sqrt{r_b r_a}\frac{2M}{\hbar}\int_0^\infty dy\frac{1}{\sinh y}e^{2\nu y} \tag{13.205}$$

$$\times \exp\left[-\kappa \coth y(r_b + r_a)\right] I_{2l+1}\left(2\kappa\sqrt{r_b r_a}\frac{1}{\sinh y}\right).$$

Now we apply the integral formula (9.29) and find

$$(r_b|r_a)_{E,l} = -i\frac{M}{\hbar\kappa}\frac{\Gamma(-\nu+l+1)}{(2l+1)!}W_{\nu,l+1/2}(2\kappa r_b) M_{\nu,l+1/2}(2\kappa r_a). \tag{13.206}$$

On the right-hand side the energy E is contained in the parameters $\kappa = \sqrt{-2ME/\hbar^2}$ and $\nu = e^2/2\omega\hbar = \sqrt{-e^4 M/2\hbar^2 E}$. The Gamma function has poles at $\nu = n$ with $n = l+1, l+2, l+3, \dots$. These correspond to the bound states of the Coulomb system. Writing

$$\kappa = \frac{1}{a_H}\frac{1}{\nu}, \tag{13.207}$$

with the Bohr radius

$$a_H \equiv \frac{\hbar^2}{Me^2} \tag{13.208}$$

(for the electron, $a_H \approx 0.529 \times 10^{-8}$cm), we have the approximations near the poles at $\nu \approx n$,

$$\Gamma(-\nu+l+1) \approx -\frac{(-)^{n_r}}{n_r!}\frac{1}{\nu-n},$$

$$\frac{1}{\nu-n} \approx \frac{2}{n}\frac{\hbar^2\kappa^2}{2M}\frac{1}{E-E_n},$$

$$\kappa \approx \frac{1}{a_H}\frac{1}{n}, \tag{13.209}$$

where $n_r = n - l - 1$. Hence

$$-i\Gamma(-\nu+l+1)\frac{M}{\hbar\kappa} \approx \frac{(-)^{n_r}}{n^2 n_r!}\frac{1}{a_H}\frac{i\hbar}{E-E_n}. \tag{13.210}$$

Let us expand the pole parts of the spectral representation of the radial fixed-energy amplitude in the form

$$(r_b|r_a)_{E,l} = \sum_{n=l+1}^\infty \frac{i\hbar}{E-E_n}R_{nl}(r_b)R_{nl}(r_a) + \dots. \tag{13.211}$$

The radial wave functions defined by this expansion correspond to the normalized bound-state wave functions

$$\psi_{nlm}(\mathbf{x}) = \frac{1}{r}R_{n,l}(r)Y_{lm}(\hat{\mathbf{x}}). \tag{13.212}$$

By comparing the pole terms of (13.206) and (13.211) [using (13.210) and formula (9.48) for the Whittaker functions, together with (9.50)], we identify the radial wave functions as

$$
\begin{aligned}
R_{nl}(r) &= \frac{1}{a_H^{1/2} n} \frac{1}{(2l+1)!} \sqrt{\frac{(n+l)!}{(n-l-1)!}} \\
&\times (2r/na_H)^{l+1} e^{-r/na_H} M(-n+l+1, 2l+2, 2r/na_H) \\
&= \frac{1}{a_H^{1/2} n} \sqrt{\frac{(n-l-1)!}{(n+l)!}} e^{-r/na_H} (2r/na_H)^{l+1} L_{n-l-1}^{2l+1}(2r/na_H). \quad (13.213)
\end{aligned}
$$

To obtain the last expression we have used formula (9.53).[2] It must be noted that the normalization integrals of the wave functions $R_{nl}(r)$ differ by a factor $z/2n = (2r/na_H)/2n$ from those of the harmonic oscillator (9.54), which are contained in integral tables. However, due to the recursion relation for the Laguerre polynomials

$$
z L_n^\mu(z) = (2n + \mu + 1) L_n^\mu(z) - (n + \mu) L_{n-1}^\mu(z) - (n+1) L_{n+1}^\mu(z), \quad (13.214)
$$

the factor $z/2n$ leaves the values of the normalization integrals unchanged. The orthogonality of the wave functions with different n is much harder to verify since the two Laguerre polynomials in the integrals have different arguments. Here the group-theoretic treatment of Appendix 13A provides the simplest solution. The orthogonality is shown in Eq. (13A.28).

We now turn to the continuous wave functions. The fixed-energy amplitude has a cut in the energy plane for positive energy where $\kappa = -ik$ and $\nu = i/a_H k$ are imaginary. In this case we write $\nu = i\nu'$. From the discontinuity we can extract the scattering wave functions. The discontinuity is given by

$$
\begin{aligned}
&\text{disc} \, (r_b|r_a)_{E,l} = (r_b|r_a)_{E+i\eta,l} - (r_b|r_a)_{E-i\eta,l} \\
&= \frac{M}{\hbar k} \left[\frac{\Gamma(-i\nu'+l+1)}{(2l+1)!} W_{i\nu',l+1/2}(-2ikr_b) M_{i\nu',l+1/2}(-2ikr_a) + (\nu' \to -\nu') \right]. (13.215)
\end{aligned}
$$

In the second term, we replace

$$
M_{i\nu',l+1/2}(-2ikr) = e^{-i\pi(l+1)} M_{-i\nu',l+1/2}(2ikr), \quad (13.216)
$$

and use the relation, valid for $\arg z \in (-\pi/2, 3\pi/2)$, $2\mu \neq -1, -2, -3, \ldots$,

$$
M_{\lambda,\mu}(z) = \frac{\Gamma(2\mu+1)}{\Gamma(\mu+\lambda+1/2)} e^{i\pi\lambda} e^{-i\pi(\mu+1/2)} W_{\lambda,\mu}(z) + \frac{\Gamma(2\mu+1)}{\Gamma(\mu-\lambda+1/2)} e^{i\pi\lambda} W_{-\lambda,\mu}(e^{i\pi} z),
$$
$$
(13.217)
$$

---

[2]Compare L.D. Landau and E.M. Lifshitz, *Quantum Mechanics*, Pergamon, London, 1965, p. 119. Note the different definition of our Laguerre polynomials $L_n^\mu = [(-)^\mu/(n+\mu)!] L_{n+\mu}^\mu|_{\text{L.L.}}$.

to find

$$\text{disc}\,(r_b|r_a)_{E,l} = \frac{M}{\hbar k}\frac{|\Gamma(-i\nu' + l + 1)|^2}{(2l+1)!^2}e^{\pi\nu'}M_{i\nu',l+1/2}\left(-2ikr_b\right)M_{-i\nu',l+1/2}\left(2ikr_a\right).$$
$$(13.218)$$

The continuum states enter the completeness relation as

$$\int_0^\infty \frac{dE}{2\pi\hbar}\text{disc}\,(r_b|r_a)_{E,l} + \sum_{n=l+1}^\infty R_{nl}(r_b)R_{nl}^*(r_a) = \delta(r_b - r_a) \qquad (13.219)$$

[compare (1.326)]. Inserting (13.215) and replacing the continuum integral $\int_0^\infty dE/2\pi\hbar$ by the momentum integral $\int_{-\infty}^\infty dk k\hbar/2\pi M$, the continuum part of the completeness relation becomes

$$\int_{-\infty}^\infty dk R_{kl}(r_b)R_{kl}^*(r_a), \qquad (13.220)$$

with the radial wave functions

$$R_{kl}(r) = \sqrt{\frac{1}{2\pi}}\frac{|\Gamma(-i\nu' + l + 1)|}{(2l+1)!}e^{\pi\nu'/2}M_{i\nu',l+1/2}(-2ikr). \qquad (13.221)$$

By expressing the Whittaker function $M_{\lambda,\mu}(z)$ in terms of the confluent hypergeometric functions, the Kummer functions $M(a,b,z)$, as

$$M_{\lambda,\mu}(z) = z^{\mu+1/2}e^{-z/2}M(\mu - \lambda + 1/2, 2\mu + 1, z), \qquad (13.222)$$

we recover the well-known result of Schrödinger quantum mechanics:[3]

$$R_{kl}(r) = \sqrt{\frac{1}{2\pi}}\frac{|\Gamma(-i\nu' + l + 1)|}{(2l+1)!}e^{\pi\nu'/2}e^{ikr}(-2ikr)^{l+1}M(-i\nu' + l + 1, 2l + 2, -2ikr).$$
$$(13.223)$$

13.9 Remarks on Geometry of Four-Dimensional u^μ-Space

A few remarks are in order on the Riemann geometry of the $\vec{u}$-space in four dimensions with the metric $g_{\mu\nu} = 4\vec{u}^2\delta_{\mu\nu}$. As in two dimensions, the Cartan curvature tensor $R_{\mu\nu\lambda}{}^\kappa$ vanishes trivially since $e^i{}_\mu(\vec{u})$ is linear in $\vec{u}$:

$$(\partial_\mu\partial_\nu - \partial_\nu\partial_\mu)e^i{}_\lambda(\vec{u}) = 0. \qquad (13.224)$$

In contrast to two dimensions, however, the Riemann curvature tensor $\bar{R}_{\mu\nu\lambda}{}^\kappa$ is nonzero. The associated Ricci tensor [see (10.41)], has the matrix elements

$$\bar{R}_{\nu\lambda} = \bar{R}_{\mu\nu\lambda}{}^\mu$$
$$= -\frac{3}{2\vec{u}^6}(\delta_{\nu\lambda}\vec{u}^2 - \vec{u}_\nu\vec{u}_\lambda), \qquad (13.225)$$

---

[3]L.D. Landau and E.M. Lifshitz, op. cit., p. 120.

yielding the scalar curvature

$$\bar{R} = g^{\nu\lambda}\bar{R}_{\nu\lambda} = -\frac{9}{2\vec{u}^4}. \tag{13.226}$$

In general, a diagonal metric of the form

$$g_{\mu\nu}(q) = \Omega^2(q)\delta_{\mu\nu} \tag{13.227}$$

is called *conformally flat* since it can be obtained from a flat space with a unit metric $g_{\mu\nu} = \delta_{\mu\nu}$ by a *conformal transformation* a la Weyl

$$g_{\mu\nu}(q) \rightarrow \Omega^2(q)g_{\mu\nu}(q). \tag{13.228}$$

Under such a transformation, the Christoffel symbol changes as follows:

$$\bar{\Gamma}_{\mu\nu}{}^{\lambda} \rightarrow \bar{\Gamma}_{\mu\nu}{}^{\lambda} + \Omega_{,\mu}\delta_{\nu}{}^{\lambda} + \Omega_{,\nu}\delta_{\mu}{}^{\lambda} - g_{\mu\nu}g^{\lambda\kappa}\Omega_{,\kappa}, \tag{13.229}$$

the subscript separated by a comma indicating a differentiation, i.e., $\Omega_{,\mu} \equiv \partial_{\mu}\Omega$.

In D dimensions, the Ricci tensor changes according to

$$\begin{aligned}
\bar{R}_{\mu\nu} \rightarrow\ & \Omega^{-2}\bar{R}_{\mu\nu} - (D-2)(\Omega^{-3}\Omega_{;\mu\nu} - 2\Omega^{-4}\Omega_{,\mu}\Omega_{,\nu}) \\
& -g_{\mu\nu}g^{\lambda\kappa}\left[(D-3)\Omega^{-4}\Omega_{,\lambda}\Omega_{,\kappa} + \Omega^{-3}\Omega_{;\lambda\kappa}\right] \\
=\ & \Omega^{-2}\bar{R}_{\mu\nu} + (D-2)\Omega^{-1}(\Omega^{-1})_{;\mu\nu} - g_{\mu\nu}(D-2)^{-1}\Omega^{-D}(\Omega^{D-2})_{;\lambda\kappa}g^{\lambda\kappa}. \tag{13.230}
\end{aligned}$$

A subscript separated by a semicolon denotes the covariant derivative formed with Riemann connection, i.e.,

$$\Omega_{;\mu\nu} = D_{\nu}\Omega_{,\mu} = \Omega_{\mu\nu} - \bar{\Gamma}_{\mu\nu}{}^{\lambda}\Omega_{,\lambda}. \tag{13.231}$$

The curvature scalar goes over into

$$\bar{R} \rightarrow \bar{R}^{\Omega} = \Omega^{-2}\left[\bar{R} - 2(D-1)\Omega^{-1}\Omega_{;\mu\nu}g^{\mu\nu} - (D-1)(D-4)\Omega^{-2}\Omega_{,\mu}\Omega_{,\nu}g^{\mu\nu}\right]. \tag{13.232}$$

The metric $g_{\mu\nu} = 4\vec{u}^2\delta_{\mu\nu}$ in the $\vec{u}$-space description of the hydrogen atom is conformally flat, so that we can use the above relations to obtain all geometric quantities from the initially trivial metric $g_{\mu\nu} = \delta_{\mu\nu}$ with $\bar{R}_{\mu\nu} = 0$ by inserting $\Omega = 2|\vec{u}|$, so that

$$\Omega_{,\mu} = 2\frac{u^{\mu}}{|\vec{u}|}, \qquad \Omega_{,\mu\nu} = \frac{2}{|\vec{u}|^3}(\delta^{\mu\nu}\vec{u}^2 - u^{\mu}u^{\nu}). \tag{13.233}$$

From the right-hand sides of (13.230) and (13.232) we obtain

$$\begin{aligned}
\bar{R}_{\mu\nu} &= -3(D-2)\frac{1}{4\vec{u}^6}(\delta_{\mu\nu}\vec{u}^2 - \vec{u}_{\mu}\vec{u}_{\nu}), \\
\bar{R} &= -3(D-1)(D-2)\frac{1}{4\vec{u}^4}. \tag{13.234}
\end{aligned}$$

For $D = 4$, these agree with (13.225) and (13.226). For $D = 2$, they vanish.

In the Coulomb system, the conformally flat metric (13.228) arose from the nonholonomic coordinate transformation (13.101) with the basis tetrads (13.132) and their inverses (13.133), which produced the torsion tensor (13.137). In the notation (13.228), the torsion tensor has a contraction

$$S_\mu(q) \equiv S_{\mu\nu}{}^\lambda(q) = \frac{1}{2\Omega^2(q)}\partial_\mu\Omega^2(q). \tag{13.235}$$

Note that although $S_\mu(q)$ is the gradient of a scalar field, the torsion tensor (13.137) is not a so-called *gradient torsion*, which is defined by the general form

$$S_{\mu\nu}{}^\lambda(q) = \frac{1}{2}\left[\delta_\mu{}^\lambda\partial_\nu s(q) - \delta_\nu{}^\lambda\partial_\mu s(q)\right]. \tag{13.236}$$

For a gradient torsion, $S_\mu(q)$ is also a gradient: $S_\mu(q) = \partial_\mu\sigma(q)$ where $\sigma(q) = (D-1)s(q)/2$. But it is, of course, not the only tensor, for which $S_\mu(q)$ is a gradient.

Note that under a conformal transformation a la Weyl, a massless scalar field $\phi(q)$ is transformed as

$$\phi(q) \to \Omega^{1-D/2}(q)\phi(q). \tag{13.237}$$

The Laplace-Beltrami differential operator $\Delta = \bar{D}^2$ applied to $\phi(q)$ goes over into $\Omega^{1-D/2}(q)\Delta^\Omega\phi(q)$ where

$$\Delta^\Omega = \Omega^{-2}\left[\Delta - \frac{1}{2}(D-2)\Omega^{-1}\Omega_{;\mu\nu}g^{\mu\nu} - \frac{1}{4}(D-2)(D-4)\Omega^{-2}\Omega_{,\mu}\Omega_{,\nu}g^{\mu\nu}\right]. \tag{13.238}$$

Comparison with (13.234) shows that there exists a combination of $\Delta = \bar{D}^2$ and the Riemann curvature scalar $\bar{R}$ which may be called *Weyl-covariant Laplacian*. This combination is

$$\Delta - \frac{1}{4}\frac{D-2}{D-1}\bar{R}. \tag{13.239}$$

When applied to the scalar field it transforms as

$$\left(\Delta - \frac{1}{4}\frac{D-2}{D-1}\bar{R}\right)\phi(q) \longrightarrow \Omega^{-1-D/2}\left(\Delta - \frac{1}{4}\frac{D-2}{D-1}\bar{R}\right)\phi(q). \tag{13.240}$$

Thus we can define a massless scalar field in a conformally invariant way by requiring the vanishing of (13.240) as a wave equation. This symmetry property has made the combination (13.239) a favorite Laplacian operator in curved spaces [13].

13.10 Runge-Lenz-Pauli Group of Degeneracy

A well-known symmetry of the Kepler problem was used by Pauli to find the spectrum of the Coulomb problem by purely algebraic manipulations. The vector operator constructed from the Hamilton operator $\hat{H}$ [see (13.1)], the momentum operator $\mathbf{p}$, the angular momentum operator $\hat{\mathbf{L}} \equiv \mathbf{x} \times \hat{\mathbf{p}}$, and the operator $\hat{p}_E \equiv \sqrt{-2M\hat{H}}$:

$$\hat{\mathbf{M}} = \frac{M}{\hat{p}_E}\left[\frac{1}{2M}(\hat{\mathbf{p}} \times \hat{\mathbf{L}} - \hat{\mathbf{L}} \times \hat{\mathbf{p}}) - e^2\frac{\mathbf{r}}{r}\right], \tag{13.241}$$

commutes with $\hat{H}$ and is therefore a conserved quantity. Together with $\hat{\mathbf{L}}$ it forms the algebra

$$[\hat{L}_i, \hat{L}_j] = i\epsilon_{ijk}\hat{L}_k, \quad [\hat{L}_i, \hat{M}_j] = i\epsilon_{ijk}\hat{M}_k, \quad [\hat{M}_i, \hat{M}_j] = i\epsilon_{ijk}\hat{M}_k. \tag{13.242}$$

It is orthogonal to $\hat{\mathbf{L}}$,

$$\hat{\mathbf{M}} \cdot \hat{\mathbf{L}} = \hat{\mathbf{L}} \cdot \hat{\mathbf{M}} = 0, \tag{13.243}$$

and satisfies

$$\hat{\mathbf{L}}^2 + \hat{\mathbf{M}}^2 + \hbar^2 = -e^4 \frac{M}{2\hat{H}} = e^4 \frac{M^2}{\hat{p}_E^2}. \tag{13.244}$$

The combinations

$$\hat{\mathbf{J}}^{(1,2)} \equiv \frac{1}{2}(\hat{\mathbf{L}} \pm \hat{\mathbf{M}}) \tag{13.245}$$

generate commuting angular momenta so that the squares $\left(\hat{\mathbf{J}}^{(1,2)}\right)^2$ have the eigenvalues $j^{(1,2)}(j^{(1,2)} + 1)$. The condition (13.243) implies that $j^{(1)} = j^{(2)} = j$. Hence (13.244) becomes

$$\hat{\mathbf{L}}^2 + \hat{\mathbf{M}}^2 + \hbar^2 = 4(\hat{\mathbf{L}} \pm \hat{\mathbf{M}})^2 + \hbar^2 = [4j(j+1)+1]\hbar^2 = e^4\frac{M^2}{\hat{p}_E^2} = -\hbar^2\alpha^2\frac{Mc^2}{2\hat{H}}. \tag{13.246}$$

From this follows that $\hat{p}_E$ has the eigenvalues $(2j + 1)\alpha Mc$ and $\hat{H}$ the eigenvalues $E_j = -Mc^2\alpha^2/(2j + 1)^2$. Thus we identify the principal quantum number as $n = 2j + 1 = 0, 1, 2, \ldots$. For each n, the magnetic quantum numbers m_1 and m_2 of $\hat{\mathbf{J}}^{(1)}$ and $\hat{\mathbf{J}}^{(2)}$ can run from $-j$ to j, so that each level is $(2j + 1)^2 = n^2$-times degenerate.

The wave functions of the hydrogen atom of principal quantum number n are direct products of eigenstates of $\hat{\mathbf{J}}^{(1)}$ and $\hat{\mathbf{J}}^{(2)}$:

$$|n\, m_1 m_2\rangle = |jm_1\rangle^{(1)} \otimes |jm_2\rangle^{(2)}. \tag{13.247}$$

In atomic physics, one prefers combinations of these which diagonalize the orbital angular momentum $\hat{\mathbf{L}} = \hat{\mathbf{J}}^{(1)} + \hat{\mathbf{J}}^{(2)}$. This done with the help of the Clebsch-Gordan coefficients $(j, m_1; j, m_2|l, m)$ [17] which couple spin j with spin j to spin $l = 0, 1, \ldots 2j$:

$$|n\, lm\rangle = \sum_{m_1, m_2 = -j, \ldots, j} |jm_1\rangle^{(1)} \otimes |jm_2\rangle^{(2)} (j, m_1; j, m_2|l, m). \tag{13.248}$$

13.11 Solution in Momentum Space

The path integral for a point particle in a Coulomb potential can also be solved in momentum space. As in the coordinate space treatment in Section 13.1, we shall calculate the matrix elements of the pseudotime displacement amplitude associated with the the resolvent operator $\hat{R} \equiv i/(E - \hat{H})$. As in Eq. (12.19), we shall rewrite it in the form

$$\hat{R} = \frac{i}{\hat{f}(E - \hat{H})} \hat{f} \tag{13.249}$$

where f is an arbitrary function of space, momentum. Rewriting the the path integral (12.31) in the momentum space representation (2.34), we shall evaluate the canonical Euclidean path integral for the pseudotime displacement amplitude (in atomic units specified on p. 954):

$$\langle \mathbf{p}_b | \hat{\mathcal{U}}_E(S) | \mathbf{p}_a \rangle = \int \mathcal{D}^3 x(s) \int \frac{\mathcal{D}^3 p(s)}{2\pi\hbar}$$
$$\times \exp\left\{ i \int_0^S ds \left[-\mathbf{p}' \cdot \mathbf{x} - f\left(\frac{\mathbf{p}^2}{2} - E\right) + f\frac{\alpha}{r} \right] \right\} f_a. \tag{13.250}$$

From this we find the fixed-energy amplitude via the integral [compare (13.4)]

$$(\mathbf{p}_b | \mathbf{p}_a)_E^f = \int_0^\infty dS \, \langle \mathbf{p}_b | \hat{\mathcal{U}}_E(S) | \mathbf{p}_a \rangle. \tag{13.251}$$

The left-hand side carries a superscript f to remind us of the presence of f on the right-hand side, although the amplitude does not really depend on f. This freedom of choice may be viewed as a gauge invariance [16] of (13.251) under $f \to f'$. Such an invariance permits us to subject (13.251) to an additional path integration over f, as long as a gauge-fixing functional $\Phi[f]$ ensures that only a specific "gauge" contributes. Thus we shall calculate the amplitude (13.251) as a path integral

$$(\mathbf{p}_b | \mathbf{p}_a)_E = \int \mathcal{D}f \, \Phi[f] \, (\mathbf{p}_b | \mathbf{p}_a)_E^f. \tag{13.252}$$

The only condition on $\Phi[f]$ is that it must be normalized to have a unit integral: $\int \mathcal{D}f \, \Phi[f] = 1$. The choice which leads to the desired solution of the path integral is

$$\Phi[f] = \prod_s \frac{1}{r} \exp\left\{ -\frac{i}{2r^2} \left[f - r^2 \left(\frac{\mathbf{p}^2}{2} - E \right) \right]^2 \right\}. \tag{13.253}$$

With this, the total action in the path integral (13.252) becomes

$$\mathcal{A}[\mathbf{p}, \mathbf{x}, f] = \int_0^S ds \left[-\mathbf{p}' \cdot \mathbf{x} - \frac{r^2}{2} \left(\frac{\mathbf{p}^2}{2} - E \right)^2 - \frac{1}{2r^2} f^2 + \frac{f}{r} \alpha \right]. \tag{13.254}$$

The path integrals over f and $\mathbf{x}$ in (13.252) are Gaussian and can be done, in this order, yielding a new action

$$\mathcal{A}[\mathbf{p}] = \frac{1}{2} \int_0^S ds \left[\frac{4\mathbf{p}'^2}{(\mathbf{p}^2 + p_E^2)^2} + \alpha^2 \right], \tag{13.255}$$

where we have introduced $p_E \equiv \sqrt{-2E}$, assuming E to be negative. The positive regime can later be obtained by analytic continuation.

At this point we go to a more symmetric coordinate system in momentum space by projecting the three-vectors $\mathbf{p}$ stereographically to the four-dimensional unit vectors $\vec{\pi} \equiv (\boldsymbol{\pi}, \pi_4)$:

$$\boldsymbol{\pi} \equiv \frac{2p_E \mathbf{p}}{\mathbf{p}^2 + p_E^2}, \qquad \pi_4 \equiv \frac{\mathbf{p}^2 - p_E^2}{\mathbf{p}^2 + p_E^2}. \tag{13.256}$$

This brings (13.255) to the form

$$\mathcal{A}[\vec{\pi}] = \frac{1}{2} \int_0^S ds \left(\frac{1}{p_E^2} \vec{\pi}'^2 + \alpha^2 \right). \tag{13.257}$$

The vector $\vec{\pi}$ describes a point particle with pseudomass $\mu = 1/p_E^2$ moving on a four-dimensional unit sphere. The pseudotime evolution amplitude of this system is

$$(\vec{\pi}_b S | \vec{\pi}_a 0) = \int \frac{\mathcal{D}\vec{\pi}}{(2\pi)^{3/2} p_E^3} e^{i\mathcal{A}[\vec{\pi}]}. \tag{13.258}$$

Let us see how the measure arises. When integrating out the spatial fluctuations in going from (13.254) to (13.255), the canonical measure in each time slice $[d^3 p_n/(2\pi)^3] d^3 x_n$ becomes $[d^3 p_n/(2\pi)^3][(2\pi)^{1/2}/(\mathbf{p}_n^2 + p_E^2)]^3$. From the stereographic projection (13.256) we see that this is equal to $d\vec{\pi}_n/(2\pi)^{3/2} p_E^3$, where $d\vec{\pi}_n$ denotes the product of integrals over the solid angle on the surface of the unit sphere in four dimensions. The integral $\int d\vec{\pi}$ yields the total surface $2\pi^2$. Alternatively we may rewrite as in Eq. (1.557), $\int d\vec{\pi}_n = \int d^4 \pi_n \delta(|\vec{\pi}_n| - 1) = 2 \int d^4 \pi_n \delta(\vec{\pi}_n^2 - 1)$, or use an explicit angular form of the type (8.119) or (8.123).

The expression (13.258) was obtained by purely formal manipulations on the continuum pseudotime axis and needs, therefore, pseudotime-slicing corrections similar to Section 13.5. For the motion on a spherical surface these have to be evaluated by the methods of Chapter 10. There we learned that the proper sliced path integral we know that in a curved space, the time-sliced measure of path integration is given by the product of invariant integrals $\int dq_n \sqrt{g(q_n)}$ at each time slice, multiplied by an effective action contribution $\exp(i\mathcal{A}_{\text{eff}}^\epsilon(q_n)) = \exp(i\epsilon \bar{R}(q_n)/6\mu)$, where $\bar{R}$ is the scalar curvature. For a sphere of radius r in D dimensions, $\bar{R} = (D-1)(D-2)/r^2$, implying here for $D = 4$ that $\exp(i\mathcal{A}_{\text{eff}}^\epsilon) = \exp(i\epsilon/\mu) = \exp(i\epsilon p_E^2)$. Thus, when transforming the time-sliced measure in the path integral (13.250) to the time-sliced measure on the sphere in (13.258), a factor $e^{iS p_E^2}$ is by definition contained in the measure of the path integral (13.258) [compare (10.153), (10.154)]. This has to be compensated by a prefactor $e^{-iS p_E^2}$. A careful calculation of the time slicing corrections gives an additional factor $e^{iS p_E^2/2}$. The correct version of (13.258) is therefore

$$(\vec{\pi}_b S | \vec{\pi}_a 0) = e^{-iS p_E^2/2} \int \frac{\mathcal{D}\vec{\pi}}{(2\pi)^{3/2} p_E^3} e^{i\mathcal{A}[\vec{\pi}]}, \tag{13.259}$$

with the integral measure defined as in Eqs. (10.153) and (10.154).

The path integral for the motion near the surface of a sphere in four dimensions was solved Subsection 8.7. The energies were brought on the sphere in Section 8.9. The spectral representation was given explicitly in Eq. (8.161) in terms of rotation matrices. Here we shall use the ultra-pherical harmonics $Y_{lm_1m_2}(\vec{\pi})$ defined in Eq. (8.124). From these we construct combinations appropriate for the hydrogen atom to be denoted by $Y_{n,l,m}(\vec{\pi})$, where n, l, m are the quantum numbers of the hydrogen atom with the well-known ranges ($n = 1, 2, 3, \ldots$, $l = 0, \ldots, n-1$, $m = -l, \ldots, l$). Explictly, these combinations are formed with the help of Clebsch-Gordan coefficients $(j, m_1; j, m_2 | l, m)$ [17] which couple spin j with spin j to spin $l = 0, 1, \ldots 2j$, where j is related to the principal quantum number by $n = 2j + 1$ [compare (13.248)]:

$$Y_{nlm}(\vec{\pi}) = \sum_{m_1, m_2 = -j, \ldots, j} Y_{2j, m_1, m_2}(\vec{\pi})\, (j, m_1; j, m_2 | l, m). \tag{13.260}$$

The orthonormality and completeness relations are

$$\int d\vec{\pi}\, Y^*_{n'l'm'}(\vec{\pi}) Y_{nlm}(\vec{\pi}) = \delta_{nn'}\delta_{ll'}\delta_{mm'}, \quad \sum_{n,l,m} Y_{nlm}(\vec{\pi}') Y_{nlm}(\vec{\pi}) = \delta^{(4)}(\vec{\pi}' - \vec{\pi}), \tag{13.261}$$

where the δ-function satisfies $\int d\vec{\pi}\, \delta^{(4)}(\vec{\pi}' - \vec{\pi}) = 1$. When restricting the complete sum to l and m only we obtain the four-dimensional analog of the Legendre polynomial:

$$\sum_{l,m} Y_{nlm}(\vec{\pi}') Y_{nlm}(\vec{\pi}) = \frac{n^2}{2\pi^2} P_n(\cos\vartheta), \qquad P_n(\cos\vartheta) = \frac{\sin n\vartheta}{n \sin\vartheta}, \tag{13.262}$$

where ϑ is the angle between the four-vectors $\vec{\pi}_b$ and $\vec{\pi}_a$:

$$\cos\vartheta = \vec{\pi}_b \vec{\pi}_a = \frac{(\mathbf{p}_b^2 - p_E^2)(\mathbf{p}_a^2 - p_E^2) + 4p_E^2 \mathbf{p}_b \cdot \mathbf{p}_a}{(\mathbf{p}_b^2 + p_E^2)(\mathbf{p}_a^2 + p_E^2)}. \tag{13.263}$$

Adapting the solution of the path integral for a particle on the surface of a sphere in Eqs. (8.161) with the energy correction of Section 10.4 to the present case we obtain for the path integral in Eq. (13.259) the spectral representaion

$$(\vec{\pi}_b S | \vec{\pi}_a 0) = (2\pi)^{3/2} p_E^3 \sum_{n=1}^{\infty} \frac{n^2}{2\pi^2} P_n(\cos\vartheta) \exp\left\{\left[-i(p_E^2 n^2 - \alpha^2)\right]\frac{S}{2}\right\}. \tag{13.264}$$

For the path integral in (13.259) itself, the exponential contains the eigenvalues of the squared angular-momentum operator $\hat{L}^2/2\mu$ which in D dimensions are $l(l + D - 2)/2\mu$, $l = 0, 1, 2, \ldots$. For the particle on a sphere in four dimensions and with the identification $l = 2j = n - 1$, the eigenvalues of $\hat{L}^2$ are $n^2 - 1$, leading to an exponential $e^{-i[p_E^2(n^2-1)-\alpha^2]S/2}$. Together with the exponential prefactor in (13.259), we obtain the exponential in (13.264).

In serting the spectral representation (13.264) into (13.251), we can immediately perform the integral over S, and arrive at the amplitude at zero fixed pseudoenergy

$$(\vec{\pi}_b|\vec{\pi}_a)_0 \;=\; (2\pi)^{3/2}p_E^3 \sum_{n=1}^{\infty} \frac{n^2}{2\pi^2} P_n(\cos\vartheta) \frac{2i}{2En^2 + \alpha^2}. \qquad (13.265)$$

This has poles displaying the hydrogen spectrum at energies:

$$E_n = -\frac{1}{2n^2}, \quad n = 1,2,3,\dots \ . \qquad (13.266)$$

13.11.1   Another Form of Action

Consider the following generalization of the action (13.255) containing an arbitrary function h depending on $\mathbf{p}$ and s:

$$A_e[\mathbf{p}] = \frac{1}{2}\int_0^S ds \left[\frac{1}{h}\frac{4\mathbf{p}'^2}{(\mathbf{p}^2 + p_E^2)^2} - \alpha^2 h\right]. \qquad (13.267)$$

This action is invariant under reparametrizations $s \to s'$ if one transforms simultaneously $h \to h\, ds/ds'$. The path integral with the action (13.255) in the exponent may thus be viewed as a path integral with the gauge-invariant action (13.267) and an additional path integral $\int df\, \Phi[f]$ with an arbitrary gauge-fixing functional $\Phi[h]$. Going back to a real-pseudotime parameter $s = i\tau$, the action corresponding to the Euclidean expression (13.267) describes the dynamics of the point particle in the Coulomb potential reads

$$A[\mathbf{p}] = \frac{1}{2}\int_{\tau_a}^{\tau_b} d\tau \left[\frac{1}{h}\frac{4\dot{\mathbf{p}}^2}{(\mathbf{p}^2 + p_E^2)^2} + \alpha^2 h\right]. \qquad (13.268)$$

At the extremum in h, this action reduces to

$$A[\mathbf{p}] = 2\alpha \int_{\tau_a}^{\tau_b} d\tau \sqrt{\frac{\dot{\mathbf{p}}^2}{(\mathbf{p}^2 + p_E^2)^2}}. \qquad (13.269)$$

This is the manifestly reparametrization invariant form of an action in a curved space with a metric $g^{\mu\nu} = \delta^{\mu\nu}/(\mathbf{p}^2 + p_E^2)^2$. In fact, this action coincides with the classical eikonal in momentum space:

$$S(\mathbf{p}_b, \mathbf{p}_a; E) = -\int_{\mathbf{p}_a}^{\mathbf{p}_b} d\tau\, \dot{\mathbf{p}} \cdot \mathbf{x}. \qquad (13.270)$$

Observing that the central attractive force makes $\dot{\mathbf{p}}$ point in the direction $-\mathbf{x}$, and inserting $r = \alpha(\mathbf{p}^2 + p_E^2)/2$, we find precisely the action (13.269). In fact, the canonical quantization of a system with the action (13.269) à la Dirac leads directly to a path integral with action (13.268) (see also the discussion in Chapter 19). The eikonal (13.270), and thus the action (13.269), determines the classical orbits via the first extremal principle of theoretical mechanics found in 1744 by Maupertius (see p. 380).

Appendix 13A Dynamical Group of Coulomb States

The subspace of oscillator wave functions $\psi^s(u^\mu/\sqrt{n})$ in the standardized form (13.184), which do not depend on x^4 (i.e., on γ), is obtained by applying an equal number of creation operators $a^\dagger$ and $b^\dagger$ to the ground state wave function (13.190). They are equal to the scalar products between the localized bra states $\langle z, z^*|$ and the ket states $|n_1^a, n_2^a, n_1^b, n_2^b\rangle$ of (13.191).

These ket states form an irreducible representation of the dynamical group O(4,2), the orthogonal group of six-dimensional flat space whose metric g_{AB} has four positive and two negative entries $(1, 1, 1, 1, -1, -1)$.

The 15 generators $\hat{L}_{AB} \equiv -\hat{L}_{BA}$, $A, B = 1, \ldots 6$, of this group are constructed from the spinors

$$\hat{a} \equiv \begin{pmatrix} \hat{a}_1 \\ \hat{a}_2 \end{pmatrix}, \quad \hat{b} \equiv \begin{pmatrix} \hat{b}_1 \\ \hat{b}_2 \end{pmatrix}, \tag{13A.1}$$

and their Hermitian-adjoints, using the Pauli σ-matrices and $c \equiv i\sigma^2$, as follows (since L_{ij} carry subscripts, we define $\sigma_i \equiv \sigma^i$):

$$\hat{L}_{ij} = \tfrac{1}{2}\left(\hat{a}^\dagger \sigma_k \hat{a} + \hat{b}^\dagger \sigma_k \hat{b}\right) \quad i, j, k = 1, 2, 3 \text{ cyclic,}$$

$$\hat{L}_{i4} = \tfrac{1}{2}\left(\hat{a}^\dagger \sigma_i \hat{a} - \hat{b}^\dagger \sigma_i \hat{b}\right),$$

$$\hat{L}_{i5} = \tfrac{1}{2}\left(\hat{a}^\dagger \sigma_i c \hat{b}^\dagger - \hat{a} c \sigma_i \hat{b}\right),$$

$$\hat{L}_{i6} = \tfrac{i}{2}\left(\hat{a}^\dagger \sigma_i c \hat{b}^\dagger + \hat{a} c \sigma_i \hat{b}\right),$$

$$\hat{L}_{45} = \tfrac{1}{2i}\left(\hat{a}^\dagger c \hat{b}^\dagger - \hat{a} c \hat{b}\right),$$

$$\hat{L}_{46} = \tfrac{1}{2}\left(\hat{a}^\dagger c \hat{b}^\dagger + \hat{a} c \hat{b}\right),$$

$$\hat{L}_{56} = \tfrac{1}{2}\left(\hat{a}^\dagger \hat{a} + \hat{b}^\dagger \hat{b} + 2\right). \tag{13A.2}$$

The eigenvalues of $\hat{L}_{56}$ on the states with an equal number of a- and b-quanta are obviously

$$\frac{1}{2}\left(n_1^a + n_2^a + n_1^b + n_2^b + 2\right) = n. \tag{13A.3}$$

The commutation rules between these operators are

$$[\hat{L}_{AB}, \hat{L}_{AC}] = ig_{AA}\hat{L}_{BC}, \tag{13A.4}$$

where n is the principal quantum number [see (13.193)]. It can be verified that the following combinations of position and momentum operators in a three-dimensional Euclidean space are elements of the Lie algebra of O(4,2):

$$\begin{aligned} r &= \hat{L}_{56} - \hat{L}_{46}, \\ x^i &= \hat{L}_{i5} - \hat{L}_{i4}, \\ -i(\mathbf{x}\partial_\mathbf{x} + 1) &= \hat{L}_{45}, \\ -ir\partial_{x^i} &= \hat{L}_{i6}. \end{aligned} \tag{13A.5}$$

The last equation follows from the transformation formula [recall (13.133)]

$$\partial_{x^i} = \frac{1}{2\vec{u}^2} e^i{}_\mu \partial_\mu \tag{13A.6}$$

together with

$$\begin{aligned} u^1 &= \tfrac{1}{2}(z_1 + z_1^*), \quad u^2 = \tfrac{1}{2i}(z_1 - z_1^*), \\ u^3 &= \tfrac{1}{2}(z_2 + z_2^*), \quad u^4 = \tfrac{1}{2i}(z_2 - z_2^*), \end{aligned} \tag{13A.7}$$

and

$$\partial_1 = (\partial_{z_1} + \partial_{z_1^*}), \quad \partial_2 = i(\partial_{z_1} - \partial_{z_1^*}),$$
$$\partial_3 = (\partial_{z_2} + \partial_{z_2^*}), \quad \partial_4 = i(\partial_{z_2} - \partial_{z_2^*}). \tag{13A.8}$$

Hence

$$- ir\partial_{x^i} = -\tfrac{i}{2}(\bar{z}\sigma_i\partial_{\bar{z}} + \partial_z\sigma_i z). \tag{13A.9}$$

By analogy with (13A.2), the generators $\hat{L}_{AB}$ can be expressed in terms of the z, z^*-variables as follows:

$$
\begin{aligned}
\hat{L}_{ij} &= \tfrac{1}{2}(\bar{z}\sigma_k\partial_{\bar{z}} - \partial_z\sigma_k z), \\
\hat{L}_{i4} &= -\tfrac{1}{2}(\bar{z}\sigma_i z - \partial_z\sigma_i\partial_{\bar{z}}), \\
\hat{L}_{i5} &= \tfrac{1}{2}(\bar{z}\sigma_i z + \partial_z\sigma_i\partial_{\bar{z}}), \\
\hat{L}_{i6} &= -\tfrac{i}{2}(\bar{z}\sigma_i\partial_{\bar{z}} + \partial_z\sigma_i z), \\
\hat{L}_{45} &= -\tfrac{i}{2}(\bar{z}\partial_{\bar{z}} + \partial_z z), \\
\hat{L}_{46} &= -\tfrac{1}{2}(\bar{z}z + \partial_z\partial_{\bar{z}}), \\
\hat{L}_{56} &= \tfrac{1}{2}(\bar{z}z - \partial_z\partial_{\bar{z}}).
\end{aligned} \tag{13A.10}
$$

Going over to the operators x^i, ∂_{x^i}, they become

$$
\begin{aligned}
\hat{L}_{ij} &= -i(x_i\partial_{x^j} - x_j\partial_{x^i}), \\
\hat{L}_{i4} &= \tfrac{1}{2}\left(-x^i\partial_{\mathbf{x}}^2 - x^i + 2\partial_{x^i}\mathbf{x}\partial_{\mathbf{x}}\right), \\
\hat{L}_{i5} &= \tfrac{1}{2}\left(-x^i\partial_{\mathbf{x}}^2 + x^i + 2\partial_{x^i}\mathbf{x}\partial_{\mathbf{x}}\right), \\
\hat{L}_{i6} &= -ir\partial_{x^i}, \\
\hat{L}_{45} &= -i(x^i\partial_{x^i} + 1), \\
\hat{L}_{46} &= \tfrac{1}{2}(-r\partial_{\mathbf{x}}^2 - r), \\
\hat{L}_{56} &= \tfrac{1}{2}(-r\partial_{\mathbf{x}}^2 + r),
\end{aligned} \tag{13A.11}
$$

where the purely spatial operators $\partial_{\mathbf{x}}^2$ and $\mathbf{x}\partial_{\mathbf{x}}$ are equal to $\partial_{x^\mu}^2$ and $x^\mu\partial_{x^\mu}$ because of the constraint (13.177).

The Lie algebra of the differential operators (13A.11) is isomorphic to the Lie algebra of the conformal group in four spacetime dimensions, which is an extension of the *inhomogeneous Lorentz group* or *Poincaré group*, defined by the commutators in Minkowski space ($\mu, \nu = 0, 1, 2, 3$), whose metric has the diagonal elements $(+1, -1, -1, -1)$,

$$[P_\mu, P_\nu] = 0, \tag{13A.12}$$
$$[L_{\mu\nu}, P_\lambda] = -i(g_{\mu\lambda}P_\nu - g_{\nu\lambda}P_\mu), \tag{13A.13}$$
$$[L_{\mu\nu}, L_{\lambda\kappa}] = -i(g_{\mu\lambda}L_{\nu\kappa} - g_{\nu\lambda}L_{\mu\kappa} - g_{\mu\kappa}L_{\nu\lambda} - g_{\nu\kappa}L_{\mu\lambda}). \tag{13A.14}$$

The extension involves the generators D of *dilatations* $x^\mu \to \rho x^\mu$ and K_μ of *special conformal transformations*[4]

$$x^\mu \to \frac{x^\mu - c^\mu x^2}{1 - 2cx + c^2x^2}, \tag{13A.15}$$

---

[4]Note the difference with respect to the conformal transformations à la Weyl in Eq. (13.228). They correspond to local dilatations.

with the additional commutation rules

$$[D, P_\mu] = -iP_\mu, \quad [D, K_\mu] = iK_\mu, \quad [D, L_{\mu\nu}] = 0, \tag{13A.16}$$

$$[K_\mu, K_\nu] = 0, \quad [K_\mu, P_\nu] = -2i(g_{\mu\nu}D + L_{\mu\nu}), \quad [K_\mu, L_{\nu\lambda}] = i(g_{\mu\nu}K_\lambda - g_{\mu\lambda}K_\nu). \tag{13A.17}$$

The commutation rules can be represented by the differential operators

$$\hat{P}_\mu = i\partial_\mu, \quad \hat{M}_{\mu\nu} = i(x_\mu\partial_\nu - x_\nu\partial_\mu), \quad \hat{D} = ix^\mu\partial_\mu, \tag{13A.18}$$

$$\hat{K}_\mu = i(2x_\mu x^\nu\partial_\nu - x^2\partial_\mu). \tag{13A.19}$$

Their combinations

$$J_{\mu\nu} \equiv L_{\mu\nu}, \quad J_{\mu 5} \equiv \tfrac{1}{2}(P_\mu - K_\mu), \quad J_{\mu 6} \equiv \tfrac{1}{2}(P_\mu + K_\mu), \quad J_{56} \equiv D, \tag{13A.20}$$

satisfy the commutation relation of O(4,2):

$$[J_{AB}, J_{CD}] = -i(\bar{g}_{AC}J_{BD} - \bar{g}_{BC}J_{AD} + \bar{g}_{BD}J_{AC} - \bar{g}_{BC}J_{AD}), \tag{13A.21}$$

where the metric $\bar{g}_{AB}$ has the diagonal values $(+1, -1, -1, -1, -1, +1)$.

When working with oscillator wave functions which are factorized in the four u^μ-coordinates, the most convenient form of the generators is

$$
\begin{aligned}
\hat{L}_{12} &= i(u^1\partial_2 - u^2\partial_1 - u^3\partial_4 + u^4\partial_3)/2, \\
\hat{L}_{13} &= i(u^1\partial_3 + u^2\partial_4 - u^3\partial_1 - u^4\partial_2)/2, \\
\hat{L}_{14} &= -(u^1u^3 + u^2u^4) + (\partial_1\partial_3 + \partial_2\partial_4)/4, \\
\hat{L}_{15} &= (u^1u^3 + u^2u^4) + (\partial_1\partial_3 + \partial_2\partial_4)/4, \\
\hat{L}_{16} &= -i(u^1\partial_3 + u^2\partial_4 + u^3\partial_1 + u^4\partial_2)/2, \\
\hat{L}_{23} &= i(u^1\partial_4 - u^2\partial_3 + u^3\partial_2 - u^4\partial_1)/2, \\
\hat{L}_{24} &= -(u^1u^4 - u^2u^3) + (\partial_1\partial_4 - \partial_2\partial_3)/4, \\
\hat{L}_{25} &= (u^1u^4 + u^2u^3) + (\partial_1\partial_4 - \partial_2\partial_3)/4, \\
\hat{L}_{26} &= -i(u^1\partial_4 - u^2\partial_3 - u^3\partial_2 + u^4\partial_1)/2, \\
\hat{L}_{34} &= [(u^1)^2 + (u^2)^2 - (u^3)^2 - (u^4)^2]/2 + (\partial_1^2 + \partial_2^2 - \partial_3^2 - \partial_4^2)/8, \\
\hat{L}_{35} &= -[(u^1)^2 + (u^2)^2 - (u^3)^2 - (u^4)^2]/2 + (\partial_1^2 + \partial_2^2 - \partial_3^2 - \partial_4^2)/8, \\
\hat{L}_{36} &= -i(u^1\partial_1 + u^2\partial_2 - u^3\partial_3 - u^4\partial_4)/2, \\
\hat{L}_{45} &= -i(u^1\partial_1 + u^2\partial_2 + u^3\partial_3 + u^4\partial_4 + 2)/2, \\
\hat{L}_{46} &= -(u^\mu)^2/2 - \partial_\mu^2/8, \\
\hat{L}_{56} &= (u^\mu)^2/2 - \partial_\mu^2/8.
\end{aligned}
\tag{13A.22}
$$

The commutation rules (13A.4) between these generators make the solution of the Schrödinger equation very simple. Rewriting (13.162) as

$$\left(-\frac{a_H}{2}r\nabla^2 - \frac{E}{E_H}\frac{r}{a_H} - 1\right)\psi(\mathbf{x}) = 0, \tag{13A.23}$$

and going to atomic natural units with $a_H = 1, E_H = 1$, we express $r\partial_\mathbf{x}^2$ and r in terms of $\hat{L}_{46}, \hat{L}_{56}$ via (13A.11). This gives

$$\left[\frac{1}{2}(\hat{L}_{56} + \hat{L}_{46}) - E(\hat{L}_{56} - \hat{L}_{46}) - 1\right]\psi = 0. \tag{13A.24}$$

With the help of Lie's expansion formula

$$e^{i\hat{A}}\hat{B}e^{-i\hat{A}} = 1 + i[\hat{A}, \hat{B}] + \frac{i^2}{2!}[\hat{A}, [\hat{A}, \hat{B}]] + \ldots$$

for $\hat{A} = \hat{L}_{45}$ and $\hat{B} = \hat{L}_{56}$ and the commutators $[\hat{L}_{45}, \hat{L}_{56}] = i L_{45}$ and $[\hat{L}_{45}, \hat{L}_{46}] = i\hat{L}_{56}$, this can be rewritten as

$$\left[e^{i\vartheta \hat{L}_{45}} \hat{L}_{56} e^{-i\vartheta \hat{L}_{45}} - 1 \right] \psi = 0, \tag{13A.25}$$

with

$$\vartheta = \frac{1}{2} \log(-2E). \tag{13A.26}$$

If ψ_n denotes the eigenstates of $\hat{L}_{56}$ with an eigenvalue n, the solutions of (13A.25) are obviously given by the tilted eigenstates $e^{i\vartheta L_{45}} \psi_n$ of the generator $\hat{L}_{56}$ whose eigenvalues are $n = 1, 2, 3, \ldots$ [as follows directly from the representation (13A.2)]. For these states, the parameter ϑ takes the values

$$\vartheta = \vartheta_n = -\log n, \tag{13A.27}$$

with the energies $E_n = -1/2n^2$.

Since the energy E in the Schrödinger equation (13A.24) is accompanied by a factor $\hat{L}_{46} - \hat{L}_{56}$, the physical scalar product between Coulomb states is

$$\langle \psi'^{H}_{n'} | \psi^{H}_{n} \rangle_{\text{phys}} \equiv \langle \psi'^{s}_{n'} | (\hat{L}_{56} - \hat{L}_{46}) | \psi^{s}_{n} \rangle = \delta_{n'n}. \tag{13A.28}$$

Within this scalar product, the Coulomb wave functions

$$\psi^{H}_{n}(\mathbf{x}) = \frac{1}{\sqrt{n}} e^{i\vartheta_n \hat{D}} \psi^{s}_{n}(u^\mu) = \frac{1}{\sqrt{n}} \psi^{s}_{n}(u^\mu/\sqrt{n}) \tag{13A.29}$$

are orthonormal.

The physical scalar product (13A.28) agrees of course with the scalar product (13.164) and with the scalar product (11.95) derived for a space with torsion in Section 11.4, apart from a trivial constant factor.

It is now easy to calculate the physical matrix elements of the dipole operator x^i and the momentum operator $-i\partial_{x^i}$ using the representations (13A.5). Only operations within the Lie algebra of the group O(4,2) have to be performed. This is why O(4,2) is called the *dynamical group* of the Coulomb system [12].

For completeness, let us state the relation between the states in the oscillator basis $|n_1 n_2 m\rangle$ and the eigenstates of a fixed angular momentum $|nlm\rangle$ of (13.213), which is analogous to the combination (13.248)

$$|nlm\rangle = \sum_{n_1 + n_2 + m = (n-1)/2}$$
$$\times \ |n_1 n_2 m\rangle \langle \tfrac{1}{2}(n-1), \tfrac{1}{2}(n_2 - n_1 + m); \tfrac{1}{2}(n-1), \tfrac{1}{2}(n_1 - n_2 + m)|l, m\rangle. \tag{13A.30}$$

Notes and References

For remarks on the history of the solution of the path integral of the Coulomb system see Preface. The advantages of four-dimensional u^μ-space in describing the three-dimensional Coulomb system was first exploited by
P. Kustaanheimo and E. Stiefel, J. Reine Angew. Math. **218**, 204 (1965).
See also the textbook by
E. Stiefel and G. Scheifele, *Linear and Regular Celestial Mechanics*, Springer, Berlin, 1971.
Within Schrödinger's quantum mechanics, an analog transformation was introduced in
E. Schrödinger, Proc. R. Irish Acad. **46**, 183 (1941).
See also

L. Infeld and T.E. Hull, Rev. Mod. Phys. **23**, 21 (1951).
Among the numerous applications of the transformation to the Schrödinger equation, see
M. Boiteux, Physica **65**, 381 (1973);
A.O. Barut, C.K.E. Schneider, and R. Wilson, J. Math. Phys. **20**, 2244 (1979);
J. Kennedy, Proc. R. Irish Acad. A **82**, 1 (1982).

The individual citations refer to

[1] I.H. Duru and H. Kleinert, Phys. Lett. B **84**, 30 (1979) (`http://www.physik.fu-ber-lin.de/~kleinert/65`); Fortschr. Phys. **30**, 401 (1982) (*ibid.*`http/83`). See also the historical remarks in the preface.

[2] H. Kleinert, Mod. Phys. Lett. A **4**, 2329 (1989) (*ibid.*`http/199`).

[3] See the paper by
S. Sakoda, *Exactness in the Path Integral of the Coulomb Potential in One Space Dimension*, (arXiv:0808.1600),
and references therein.

[4] H. Kleinert, Gen. Rel. Grav. **32**, 769 (2000) (*ibid.*`http/258`); Act. Phys. Pol. B **29**, 1033 (1998) (gr-qc/9801003).

[5] H. Kleinert, Phys. Lett. B **189**, 187 (1987) (*ibid.*`http/162`).

[6] For the interpretation of this transformation as a quaternionic square root, see
F.H.J. Cornish, J. Phys. A **17**, 323, 2191 (1984).

[7] This restriction was missed in a paper by
R. Ho and A. Inomata, Phys. Rev. Lett. **48**, 231 (1982).

[8] Within Schrödinger's quantum mechanics, this expression had earlier been obtained by
L.C. Hostler, J. Math. Phys. **5**, 591 (1964).

[9] Note also a calculation of the time evolution amplitude of the Coulomb system by
S.M. Blinder, Phys. Rev. A **43**, 13 (1993).
His result is given in terms of an infinite series which is, unfortunately, as complicated as the well-known spectral representation $\sum_n \psi_n(\mathbf{x}_b)\psi_n^*(\mathbf{x}_b)e^{-iE_n(t_b-t_a)/\hbar}$.

[10] There exists an interesting perturbative solution of the path integral of the *integrated* Coulomb amplitude $\int d^3x\,(\mathbf{x}_b t_b|\mathbf{x}_a t_a)$ by
M.J. Goovaerts and J.T. Devreese, J. Math. Phys. **13**, 1070 (1972).
There exists a related perturbative solution for the potential $\delta(x)$:
M.J. Goovaerts, A. Babcenco, and J.T. Devreese, J. Math. Phys. **14**, 554 (1973).

[11] For the introduction and extensive use of the tilt operator in calculating transition amplitudes, see
H. Kleinert, *Group Dynamics of the Hydrogen Atom*, Lectures presented at the 1967 Boulder Summer School, in *Lectures in Theoretical Physics*, Vol. X B, pp. 427–482, ed. by W.E. Brittin and A.O. Barut, Gordon and Breach New York, 1968 (*ibid.*`http/4`).

[12] H. Kleinert, Fortschr. Phys. **6**, 1 (1968) (*ibid.*`http/1`).

[13] N.D. Birell and P.C.W. Davies, *Quantum Fields in Curved Space*, Cambridge University Press, Cambridge, 1982.

[14] B.S. DeWitt, Rev. Mod. Phys. **29**, 377 (1957).

[15] H. Kleinert, Phys. Lett A **252**, 277 (1999) (quant-ph/9807073).

[16] K. Fujikawa, Prog. Theor. Phys. **96** 863 (1996) (hep-th/9609029); (hep-th/9608052).

[17] A.R. Edmonds, *Angular Momentum in Quantum Mechanics*, Princeton University Press, 1960.

Acribus, ut ferme talia, initiis, incurioso fine.
As is usual in such matters, keen in commencing, negligent at the end.
TACITUS, Annales, Book 6, 17

14

Solution of Further Path Integrals by Duru-Kleinert Method

The combination of a path-dependent time reparametrization and a compensating coordinate transformation, used by Duru and Kleinert to transform the Coulomb path integral into a harmonic-oscillator path integral, can be generalized to relate a variety of path integrals to each other. In this way, many unknown path integrals can be solved by their relation to known path integrals. In this chapter, the method is explained for a typical sample of one-dimensional path integrals as well as for a more involved three-dimensional system. The latter describes a generalization of the Coulomb system consisting of two particles which carry both electric and magnetic charges. It is commonly referred to as the *dionium* atom (by analogy with the positronium atom, the bound state between electron and positron). We also discuss further possible generalizations of the solution method.

14.1 One-Dimensional Systems

In one space dimension, the general relation to be established is the following: Let $\hat{H}$ be a Hamiltonian operator

$$\hat{H} = \hat{T} + \hat{V}, \tag{14.1}$$

with the kinetic term $\hat{T} = \hat{p}^2/2M$, and define the auxiliary Hamiltonian operator

$$\hat{H}_E = \hat{H} - E, \tag{14.2}$$

with the associated time evolution amplitude

$$\langle x_b | \hat{U}_E(t) | x_a \rangle \equiv \langle x_b | e^{-it\hat{H}_E/\hbar} | x_a \rangle. \tag{14.3}$$

An integration of this amplitude over all $t > 0$ yields the fixed-energy amplitude

$$(x_b | x_a)_E = \int_{t_a}^{\infty} dt_b \langle x_b | \hat{U}_E(t_b - t_a) | x_a \rangle \tag{14.4}$$

974

[recall (12.8)]. This can formally be written as a path integral

$$(x_b|x_a)_E = \int_{t_a}^{\infty} dt_b \int \mathcal{D}x(t) e^{i\mathcal{A}_E[x]/\hbar}, \qquad (14.5)$$

with an action

$$\mathcal{A}_E[x] = \int_{t_a}^{t_b} dt \left[\frac{M}{2} \dot{x}^2(t) - V(x(t)) + E \right]. \qquad (14.6)$$

As in the Coulomb system, another path integral representation is found for the amplitude (14.5) by making use of the more general representation (12.21) of the resolvent operator. By choosing two arbitrary regulating functions $f_l(x)$, $f_r(x)$ whose product is $f(x)$, we introduce the modified auxiliary Hamiltonian operator

$$\hat{\mathcal{H}}_E = f_l(x)(\hat{H} - E) f_r(x). \qquad (14.7)$$

The associated pseudotime evolution amplitude

$$\langle x_b | \hat{\mathcal{U}}_E(S) | x_a \rangle \equiv f_r(x_b) f_l(x_a) \langle x_b | e^{-iS\hat{\mathcal{H}}_E/\hbar} | x_a \rangle \qquad (14.8)$$

yields upon integration over all $S > 0$ the same fixed-energy amplitude as (14.4):

$$(x_b|x_a)_E = \int_0^{\infty} dS \langle x_b | \hat{\mathcal{U}}_E(S) | x_a \rangle \qquad (14.9)$$

[recall (12.30)]. The amplitude can therefore be calculated from the path integral

$$(x_b|x_a)_E = \int_0^{\infty} dS \left[f_r(x_b) f_l(x_a) \int \mathcal{D}x(s) e^{i\mathcal{A}_E^f[x]/\hbar} \right], \qquad (14.10)$$

with the modified action

$$\mathcal{A}_E^f[x] = \int_0^S ds \left\{ \frac{M}{2f(x(s))} x'^2(s) - f(x(s))[V(x(s)) - E] \right\}. \qquad (14.11)$$

As observed in (12.37), this action is obtained from (14.6) by a path-dependent time reparametrization satisfying

$$dt = ds \, f(x(s)). \qquad (14.12)$$

The introduction of $f(x)$ has brought the kinetic term to an inconvenient form containing a space-dependent mass $M/f(x)$. This space dependence is removed by a coordinate transformation

$$x = h(q). \qquad (14.13)$$

Since the coordinate differentials are related by

$$dx = h'(q) dq, \qquad (14.14)$$

we require the function $h(q)$ to satisfy

$$h'^2(q) = f(h(q)). \tag{14.15}$$

Then the action (14.11) reads, in terms of the new coordinate q,

$$\mathcal{A}_E^{f,q} = \int_0^S ds \left\{ \frac{M}{2} q'^2(s) - f(q(s))[V(q(s)) - E] \right\}, \tag{14.16}$$

with the obvious notation

$$f(q) \equiv f(h(q)), \quad V(q) \equiv V(h(q)). \tag{14.17}$$

In the transformed action (14.16), the kinetic term has the usual form.

The important fact to be proved and exploited in the sequel is the following: The initial fixed-energy amplitude (14.5) can be related to the fixed-pseudoenergy amplitude associated with the transformed action (14.16), if this action is extended by an effective potential

$$V_{\text{eff}}(q) = -\frac{\hbar^2}{4M} \left[\frac{h'''}{h'} - \frac{3}{2} \left(\frac{h''}{h'} \right)^2 \right]. \tag{14.18}$$

The quantity in backets is known is the Schwartz derivative $\{h, q\}$ of Eq. (14.18) encountered in the semiclassical expansion coefficient $q_2(x)$. The effective potential is caused by time slicing effects and will be derived in the next section. Thus, instead of the naively transformed action (14.16), the fixed-pseudoenergy amplitude $(q_b|q_a)_\mathcal{E}$ is obtained from the extended action

$$\mathcal{A}_{E,\mathcal{E}}^{\text{DK}}[q] = \int_0^S ds \left\{ \frac{M}{2} q'^2(s) - f(q(s))[V(q(s)) - E] - V_{\text{eff}}(q(s)) + \mathcal{E} \right\}, \tag{14.19}$$

by calculating the path integral

$$(q_b|q_a)_\mathcal{E} = \int_0^\infty dS \int \mathcal{D}q(s) e^{i\mathcal{A}_{E,\mathcal{E}}^{\text{DK}}[q]}. \tag{14.20}$$

The relation to be derived which leads to a solution of many nontrivial path integrals is

$$(x_b|x_a)_E = [f(x_b)f(x_a)]^{1/4}(q_b|q_a)_{\mathcal{E}=0}. \tag{14.21}$$

The procedure is an obvious generalization of the Duru-Kleinert transformation of the Coulomb path integral in Section 13.1, as indicated by the superscript DK on the transformed actions. Correspondingly, the actions $\mathcal{A}_E[x]$ and $\mathcal{A}_{E,\mathcal{E}}^{\text{DK}}[q]$, whose path integrals (14.5) and (14.20) producing the same fixed-energy amplitude $(x_b|x_a)_E$ via the relation (14.21), are called *DK-equivalent*.

The prefactor on the right-hand side has its origin in the normalization properties of the states. With $dx = dq\, h'(q) = dq\, f(h(q))^{1/2}$, the completeness relation

$$\int dx |x\rangle\langle x| = 1 \tag{14.22}$$

goes over into

$$\int dq \sqrt{f(q)} |h(q)\rangle\langle h(q)| = 1. \tag{14.23}$$

We want the transformed states $|q\rangle$ to satisfy the completeness relation

$$\int dq |q\rangle\langle q| \equiv 1. \tag{14.24}$$

This implies the relation between new and old states:

$$|x\rangle = f(q)^{-1/4}|q\rangle. \tag{14.25}$$

At first sight it appears as though the normalization factor in (14.21) should have the opposite power $-1/4$, but the sign is correct as it is. The reason lies, roughly speaking, in a factor $[f(x_b)f(x_a)]^{1/2}$ by which the pseudotimes dt and ds in the integrals (14.4) and (14.20) differ from each other. This causes the fixed-energy amplitude to be no longer proportional to the dimensions of the states, in which case Eq. (14.21) would have indeed carried a factor $[f(x_b)f(x_a)]^{-1/4}$. The extra factor $[f(x_b)f(x_a)]^{1/2}$ arising from the pseudotime integration *inverts* the naively expected prefactor.

In applications, the situation is usually as follows: There exists a solved path integral for a system with a singular potential. The time-sliced action is *not* the naively sliced classical action, but a more complicated regularized one which is free of path collapse problems. The most important example is the radial path integral (8.36) which involves a logarithm of a Bessel function rather than a centrifugal barrier. Further examples are the path integrals (8.173) and (8.206) of a particle near the surface of a sphere in $D = 3$ and $D = 4$ dimensions, where angular barriers are regulated by Bessel functions. In these examples, the explicit form of the time-sliced path integral without collapse as well as its solution are obtained from an angular momentum projection of a simple Euclidean path integral. In the first step of the solution procedure, the introduction of a path-dependent new time s via $dt = ds\, f(x(s))$ removes the dangerous singularities by an appropriate choice of the regulating function $f(x)$. The transformed system has a regular potential and possesses a time-sliced path integral, but it has an unconventional kinetic term. In the second step, the coordinate transformation brings the kinetic term to the conventional form. The final fixed-pseudoenergy amplitude $(q_b|q_a)_{\mathcal{E}}$ evaluated at $\mathcal{E} = 0$ coincides with the known amplitude of the initial system, apart from the above-discussed factor which is inversely related to the normalization of the states. Note that with (14.15), the relation (14.21) can also be written as

$$(x_b|x_a)_E = [h'(q_b)h'(q_a)]^{1/2}(q_b|q_a)_{\mathcal{E}=0}. \tag{14.26}$$

This transformation formula will be used to find a number of path integrals. First, however, we shall derive the effective potential (14.18) as promised.

14.2   Derivation of the Effective Potential

In order to derive the effective potential (14.18), we consider the pseudotime-sliced path integral associated with the regularized pseudotime evolution operator (14.8):

$$\langle x_b | \hat{\mathcal{U}}_E(S) | x_a \rangle$$
$$\approx \frac{f_r(x_b) f_l(x_a)}{\sqrt{2\pi i \epsilon_s \hbar f_l(x_b) f_r(x_a)/M}} \prod_{n=1}^{N} \left[\int \frac{dx_n}{\sqrt{2\pi i \epsilon_s \hbar f_n/M}} \right] \exp\left(\frac{i}{\hbar} \mathcal{A}^N \right), \qquad (14.27)$$

where

$$\mathcal{A}^N = \sum_{n=1}^{N+1} \left\{ \frac{M}{2\epsilon_s} \frac{(\Delta x_n)^2}{f_l(x_n) f_r(x_{n-1})} + \epsilon_s [E - V(x_n)] f_l(x_n) f_r(x_{n-1}) \right\}. \qquad (14.28)$$

In the measure, we have used the abbreviation $f_n \equiv f(x_n) = f_l(x_n) f_r(x_n)$. From now on, the potential $V(x)$ is omitted as being inessential to the discussion. By shifting the product index and the subscripts of f_n by one unit, and by compensating for this with a prefactor, the integration measure in (14.27) acquires the postpoint form

$$\frac{[f(x_b) f(x_a)]^{1/4}}{\sqrt{2\pi i \epsilon_s \hbar/M}} \left[\frac{f_r(x_a)}{f_r(x_b)} \right]^{-5/4} \left[\frac{f_l(x_a)}{f_l(x_b)} \right]^{1/4} \prod_{n=2}^{N+1} \int \frac{d\Delta x_n}{\sqrt{2\pi i \epsilon_s \hbar f_n/M}}, \qquad (14.29)$$

where the integrals over $\Delta x_n = x_n - x_{n-1}$ are done successively from high to low n, each at a fixed postpoint position x_n.

We now go over to the new coordinate q with a transformation function $x = h(q)$ satisfying (14.15) which makes the leading kinetic term simple:

$$\mathcal{A}_0^N = \sum_{n=1}^{N+1} \frac{M}{2\epsilon_s} (\Delta q_n)^2. \qquad (14.30)$$

The postpoint expansion of Δx_n reads at each n (omitting the subscripts)

$$\Delta x = x(q) - x(q - \Delta q) = e_1 \, \Delta q - \frac{1}{2} e_2 \, (\Delta q)^2 + \frac{1}{6} e_3 \, (\Delta q)^3 + \dots, \qquad (14.31)$$

with the expansion coefficients

$$e_1 \equiv h' = f^{1/2}, \ e_2 \equiv h'', \ e_3 \equiv h''', \dots \qquad (14.32)$$

evaluated at the postpoint q_n. The expansion (14.31) is the one-dimensional analog of the expansion (11.56), the coefficients corresponding to the basis triads $e^i{}_\mu$ in Eq. (10.12) and their derivatives $(e_1 \hat{=} e^i{}_\mu, e_2 \hat{=} e^i{}_{\mu,\nu}, \dots)$. Let us also introduce the analog of the reciprocal triad $e_i{}^\mu$ defined in (10.12):

$$\bar{e} \equiv 1/e_1 = 1/h' = 1/f^{1/2}. \qquad (14.33)$$

With it, we expand the kinetic term in (14.28) as

$$\frac{(\Delta x_n)^2}{2\epsilon_s f_l(x_n) f_r(x_{n-1})} = \frac{(\Delta q)^2}{2\epsilon_s} \left\{ 1 - \bar{e}e_2\,\Delta q + \left[\frac{1}{3}\bar{e}e_3 + \frac{1}{4}(\bar{e}e_2)^2 \right](\Delta q)^2 + \dots \right\}$$
$$\times \left\{ 1 + \frac{f'_r}{f_r}\Delta q + \left[\left(\frac{f'_r}{f_r}\right)^2 - \frac{1}{2}\frac{f''_r}{f_r} \right](\Delta q)^2 + \dots \right\}, \qquad (14.34)$$

where $f'_r \equiv df_r/dq$. From Eq. (14.31) we see that the transformation of the measure has the Jacobian

$$J = \frac{\partial \Delta x}{\partial \Delta q} = f^{1/2} \left[1 - \bar{e}e_2\Delta q + \frac{1}{2}\bar{e}e_3(\Delta q)^2 + \dots \right] \qquad (14.35)$$

[this being a special case of (11.59)]. Since the subsequent algebra is tedious, we restrict the regulating functions $f_l(x)$ and $f_r(x)$ somewhat as in Eq. (13.3) by assuming them to be different powers $f_l(x) = f(x)^{1-\lambda}$ and $f_r(x) = f(x)^\lambda$ of a single function $f(x)$, where λ is an arbitrary splitting parameter. Then the measure (14.29) becomes

$$\frac{f_b^{3\lambda/2} f_a^{(1-3\lambda)/2}}{\sqrt{2\pi i\epsilon_s\hbar/M}} \prod_{n=2}^{N+1} \int \frac{d\Delta x_n}{\sqrt{2\pi i\epsilon_s\hbar f_n/M}}, \qquad (14.36)$$

with the obvious notation $f_b \equiv f(x_b)$, $f_a \equiv f(x_a)$. We now distribute the prefactor $f_b^{3\lambda/2} f_a^{(1-3\lambda)/2}$ evenly over the time interval by writing

$$f_b^{3\lambda/2} f_a^{(1-3\lambda)/2} = f_b^{1/4} f_a^{1/4} \prod_{n=1}^{N+1} \left(\frac{f_{n-1}}{f_n} \right)^{1/4-3\lambda/2}. \qquad (14.37)$$

Then the path integral (14.27) becomes

$$\langle x_b | \hat{\mathcal{U}}_E(s) | x_a \rangle \approx \frac{f_b^{1/4} f_a^{1/4}}{\sqrt{2\pi i\epsilon_s\hbar/M}} \prod_{n=1}^{N} \left[\int \frac{d\Delta q_n}{\sqrt{2\pi i\epsilon_s\hbar/M}} \right] \qquad (14.38)$$
$$\times \exp \left\{ \frac{i}{\hbar} \left[\sum_{n=1}^{N+1} \frac{M}{2\epsilon_s}(\Delta q_n)^2 + \epsilon_s f(q_n)[E - V(q_n)] + \dots \right] \right\} [1 + C(q_n, \Delta q_n)],$$

where $1 + C$ is a correction factor arising from the three-step transformation

$$1 + C \equiv (1 + C_{\text{meas}})(1 + C_f)(1 + C_{\text{act}}). \qquad (14.39)$$

Dropping irrelevant higher orders in Δq, the three contributions on the right-hand side have the following origins:

The transformation of the measure (14.35) gives rise to the time slicing correction

$$C_{\text{meas}} = -\bar{e}e_2\,\Delta q + \frac{1}{2}\bar{e}e_3\,(\Delta q)^2 + \dots. \qquad (14.40)$$

The rearrangement of the f-factors in (14.37) produces

$$C_f = \left(\frac{1}{4} - \frac{3\lambda}{2}\right)\left[-\frac{f'}{f}\Delta q + \frac{1}{2}\frac{f''}{f}(\Delta q)^2\right]$$
$$-\frac{1}{2}\left(\frac{3}{4} + \frac{3\lambda}{2}\right)\left(\frac{1}{4} - \frac{3\lambda}{2}\right)\left(\frac{f'}{f}\right)^2 (\Delta q)^2 + \ldots . \qquad (14.41)$$

The transformation of the pseudotime-sliced kinetic term (14.34) yields

$$C_{\mathrm{act}} = \frac{i}{\hbar}M\frac{(\Delta q)^2}{2\epsilon_s}\left\{-\left(\bar{e}e_2 - \lambda\frac{f'}{f}\right)\Delta q\right.$$
$$+\left[\frac{1}{3}\bar{e}e_3 + \frac{1}{4}(\bar{e}e_2)^2 + \frac{1}{2}\left(-\lambda\frac{f''}{f} + \lambda(\lambda+1)\left(\frac{f'}{f}\right)^2\right) - \lambda\bar{e}e_2\frac{f'}{f}\right](\Delta q)^2\right\}$$
$$-\frac{M^2}{2\hbar^2}\frac{(\Delta q)^4}{4\epsilon_s^2}\left(\bar{e}e_2 - \lambda\frac{f'}{f}\right)^2(\Delta q)^2 + \ldots . \qquad (14.42)$$

We now calculate an equivalent kernel according to Section 11.2. The correction terms are evaluated perturbatively using the expectation values

$$\langle(\Delta q)^{2n}\rangle_0 = \left(\frac{i\hbar}{M}\right)^n (2n-1)!!. \qquad (14.43)$$

First we find the expectation value (11.70). Listing only the relevant terms of order ϵ_s, we obtain

$$\langle C\Delta q\rangle_0 = i\hbar\epsilon_s\left[-\bar{e}e_2 - \left(\frac{1}{4} - \frac{3\lambda}{2}\right)\frac{f'}{f} + \frac{3}{2}\left(\bar{e}e_2 - \lambda\frac{f'}{f}\right)\right]. \qquad (14.44)$$

The λ-terms cancel each other identically. The remainder vanishes upon using the relation (14.15), which reads in the present notation

$$e_1^2 = f, \qquad (14.45)$$

implying that

$$2e_1e_2 = f', \qquad 2\bar{e}e_2 = f'/f, \qquad (14.46)$$

and yielding indeed $\langle C\Delta q\rangle_0 \equiv 0$.

We now turn to the expectation $\langle C\rangle_0$ which determines the effective potential via Eq. (11.47). By differentiating the second equation in (14.46), we see that

$$f''/f = 2\left[(\bar{e}e_2)^2 + \bar{e}e_3\right]. \qquad (14.47)$$

By expressing f and f'' in Eqs. (14.41) and (14.42) in terms of the e-functions, we obtain

$$C_f = \left(\frac{1}{4} - \frac{3\lambda}{2}\right)\left\{-2\bar{e}e_2\,\Delta q + [(\bar{e}e_2)^2 + \bar{e}e_3](\Delta q)^2\right\}$$
$$-2\left(\frac{3}{4} + \frac{3\lambda}{2}\right)\left(\frac{1}{4} - \frac{3\lambda}{2}\right)(\bar{e}e_2)^2(\Delta q)^2 + \ldots , \qquad (14.48)$$

$$C_{\text{act}} = i\frac{M}{\hbar}\frac{(\Delta q)^2}{2\epsilon_s}\Bigg\{ -(1-2\lambda)\bar{e}e_2\Delta q$$

$$+\left[\frac{1}{3}\bar{e}e_3 + \frac{1}{4}(\bar{e}e_2)^2 - \lambda(\bar{e}e_2 + \bar{e}e_3) + 2\lambda(\lambda+1)(\bar{e}e_2)^2 - 2\lambda(\bar{e}e_2)^2\right](\Delta q)^2\Bigg\}$$

$$-\frac{M^2}{2\hbar^2}\frac{(\Delta q)^4}{4\epsilon_s^2}(1-2\lambda)^2(\bar{e}e_2)^2(\Delta q)^2 + \dots \quad (14.49)$$

After forming the product (14.39), the total correction reads

$$C = \bar{e}e_2\left(\frac{1}{2}-\lambda\right)\Delta q\left[-\frac{iM}{\hbar\epsilon_s}(\Delta q)^2 - 3\right]$$

$$+(\bar{e}e_2)^2\left[\frac{9}{2}\left(\lambda-\frac{1}{6}\right)\left(\lambda-\frac{1}{2}\right)(\Delta q)^2 + i\left(4\lambda^2 - \frac{7}{2}\lambda + \frac{7}{8}\right)\frac{M}{\hbar\epsilon_s}(\Delta q)^4\right.$$

$$\left.-\frac{1}{2}\left(\lambda-\frac{1}{2}\right)^2\frac{M^2}{\hbar^2\epsilon_s^2}(\Delta q)^6\right]$$

$$+\bar{e}e_3\left[-\frac{3}{2}\left(\lambda-\frac{1}{2}\right)(\Delta q)^2 - \frac{1}{2}\left(\lambda-\frac{1}{3}\right)i\frac{M}{\hbar\epsilon_s}(\Delta q)^4\right] + \dots \quad (14.50)$$

Using (14.43), we find the expectation to the relevant order ϵ_s:

$$\langle C\rangle_0 = -\frac{\epsilon_s\hbar}{M}\left[\frac{1}{4}\bar{e}e_3 - \frac{3}{8}(\bar{e}e_2)^2\right]. \quad (14.51)$$

It amounts to an effective potential

$$V_{\text{eff}} = -\frac{i\hbar^2}{M}\left[\frac{1}{4}\bar{e}e_3 - \frac{3}{8}(\bar{e}e_2)^2\right]. \quad (14.52)$$

By inserting (14.32) and (14.33), this turns into the expression (14.18) which we wanted to derive.

In summary we have shown that the kernel in (14.38)

$$K^{\epsilon_s}(\Delta q) = \frac{1}{\sqrt{2\pi i\epsilon_s\hbar/M}}\exp\left\{\frac{i}{\hbar}\sum_{n=1}^{N+1}\left[\frac{M}{2\epsilon_s}(\Delta q_n)^2 + \epsilon_s E f(q_n)\right]\right\}[1+C] \quad (14.53)$$

can be replaced by the simpler equivalent kernel

$$K^{\epsilon_s}(\Delta q) = \frac{1}{\sqrt{2\pi i\epsilon_s\hbar/M}}\exp\left\{\frac{i}{\hbar}\sum_{n=1}^{N+1}\left[\frac{M}{2\epsilon_s}(\Delta q_n)^2 + \epsilon_s E f(q_n) - \epsilon_s V_{\text{eff}}\right]\right\}, \quad (14.54)$$

in which the correction factor $1+C$ is accounted for by the effective potential V_{eff} of Eq. (14.18). This result is independent of the splitting parameter λ [1]. The same result emerges, after a lengthier algebra, for a completely general splitting of the regulating function $f(x)$ into a product $f_l(x)f_r(x)$.

14.3   Comparison with Schrödinger Quantum Mechanics

The DK transformation of the action (14.11) into the action (14.19) has of course a correspondence in Schrödinger quantum mechanics. In analogy with the introduction of the pseudotime evolution amplitude (14.8), we multiply the Schrödinger equation

$$\left[-\frac{\hbar^2}{2M}\partial_x^2 - E \right] \psi(x,t) = i\hbar\partial_t\psi(x,t) \tag{14.55}$$

from the left by an arbitrary regulating function $f_l(x)$, and obtain

$$\left[-\frac{\hbar^2}{2M}f_l(x)\partial_x^2 f_r(x) - Ef(x) \right] \psi_f(x,t) = f(x)i\hbar\partial_t\psi_f(x,t), \tag{14.56}$$

with the transformed wave function $\psi_f(x,t) \equiv f_r(x)^{-1}\psi(x,t)$. After the coordinate transformation (14.14), we arrive at

$$\left[-\frac{\hbar}{2M}f_l(q)\left(\frac{1}{h'(q)}\partial_q\right)^2 f_r(q) - Ef(q) \right] \psi_f(q,t) = f(q)i\hbar\partial_t\psi_f(x,t), \tag{14.57}$$

having used the notation $f(q) \equiv f(h(q))$ as in (14.17). Inserting $h'^2(q) = f(h(q)) = f_l(h(q))f_r(h(q))$ from (14.15), the Schrödinger equation becomes

$$\left[-\frac{\hbar^2}{2M}f_r^{-1}(q)\left(\partial_q^2 - \frac{h''}{h'}\partial_q\right) f_r(q) - Ef(q) \right] \psi_f(q,t) = f(q)i\hbar\partial_t\psi_f(q,t). \tag{14.58}$$

After going from $\psi_f(q,t)$ to a new wave function

$$\phi(q,t) = f_r^{3/4}(q)f_l^{-1/4}(q)\psi_f(q,t)$$

related to the initial one by $\psi(x,t) \equiv f_r(q)\psi_f(q,t) = f^{1/4}(q)\phi(q,t)$, the Schrödinger equation takes the form

$$\left[-\frac{\hbar^2}{2M}h'(q)^{-1/2}\left(\partial_q^2 - \frac{h''}{h'}\partial_q\right)h'(q)^{1/2} - Ef(q) \right]\phi(q,t)$$

$$= \left[-\frac{1}{2M}\partial_q^2 + V_{\text{eff}} - Ef(q) \right]\phi(q,t) = f(q)i\hbar\partial_t\phi(q,t), \tag{14.59}$$

where V_{eff} is precisely the effective potential (14.18).

For the special coordinate transformation $r = h(q) = e^q$, we obtain

$$V_{\text{eff}} = -\frac{\hbar^2}{4M}\left[\frac{h'''}{h'} - \frac{3}{2}\left(\frac{h''}{h'}\right)^2 \right] = \frac{\hbar^2}{2M}\frac{1}{4}, \tag{14.60}$$

as first pointed out by Langer [2] when improving the WKB approximation of Schrödinger equations with a centrifugal barrier $\hbar^2 l(l+1)/2Mr^2$. This is too singular to allow for a semiclassical treatment. The transformation $r = h(q) = e^q$ leads to a smooth potential problem on the entire q-axis, in which the centrfugal barrier is replaced by $\hbar^2[l(l+1) + \frac{1}{4}]/2M$. Langer conluded from this that the original Schrödinger equation in r can be treated semiclassically if $l(l+1)$ is replaced by $l(l+1) + \frac{1}{4}$. The additional $\frac{1}{4}$ is knonw as the *Langer correction*.

The operator $f(q)\partial_t$ on the right-hand side of (14.59) plays the role of pseudotime derivative ∂_s.

14.4 Applications

We now present some typical solutions of path integrals via the DK method. The initial fixed-energy amplitudes will all have the generic action

$$\mathcal{A}_E = \int dt \left[\frac{M}{2} \dot{x}^2(t) - V(x) + E \right], \tag{14.61}$$

with different potentials $V(x)$ which usually do not allow for a naive time slicing. The associated path integrals are known from certain projections of Euclidean path integrals. In the sequel, we omit the subscript E for brevity (since we want to use its place for another subscript referring to the potential under consideration). The solution follows the general two-step procedure described in Section 14.4.

14.4.1 Radial Harmonic Oscillator and Morse System

Consider the action of a harmonic oscillator in D dimensions with an angular momentum $l_\mathcal{O}$ at a fixed energy $E_\mathcal{O}$:

$$\mathcal{A}_\mathcal{O} = \int dt \left[\frac{M}{2} \dot{r}^2 - \hbar^2 \frac{\mu_\mathcal{O}^2 - 1/4}{2Mr^2} - \frac{M}{2} \omega^2 r^2 + E_\mathcal{O} \right]. \tag{14.62}$$

Here $\mu_\mathcal{O}$ is an abbreviation for

$$\mu_\mathcal{O} = D_\mathcal{O}/2 - 1 + l_\mathcal{O} \tag{14.63}$$

[recall (8.137)], $D_\mathcal{O}$ denotes the dimension, and $l_\mathcal{O}$ the orbital angular momentum of the system. The subscript $\mathcal{O}$ indicates that we are dealing with the harmonic oscillator. A free particle is described by the $\omega \to 0$ -limit of this action.

Due to the centrifugal barrier, the time evolution amplitude possesses only a complicated time-sliced path integral involving Bessel functions. According to the rule (8.139), the centrifugal barrier requires the regularization

$$\epsilon \hbar^2 \frac{\mu_\mathcal{O}^2 - 1/4}{2Mr_n^2} \longrightarrow i\hbar \log \tilde{I}_{\mu_\mathcal{O}} \left(\frac{M}{i\hbar\epsilon} r_n r_{n-1} \right). \tag{14.64}$$

This smoothens the small-r fluctuations and prevents a path collapse in the Euclidean path integral with $\mu_\mathcal{O} = 0$. The time-sliced path integral can then be solved using the formula (8.14). The final amplitude is obtained most simply, however, by solving the harmonic oscillator in D_O Cartesian coordinates, and by projecting the result into a state of fixed angular momentum l_O. The result was given in Eq. (9.32), and reads for $r_b > r_a$

$$(r_b|r_a)_{E_O,l_O} = -i\frac{1}{\omega}\frac{1}{\sqrt{r_b r_a}}\frac{\Gamma((1+\mu)/2-\nu)}{\Gamma(\mu+1)}W_{\nu,\mu/2}\left(\frac{M\omega}{\hbar}r_b^2\right)M_{\nu,\mu/2}\left(\frac{M\omega}{\hbar}r_a^2\right), \quad (14.65)$$

where the parameters on the right-hand side are

$$\nu = \nu_O \equiv \frac{E_O}{2\omega\hbar}, \quad \mu = \mu_O. \quad (14.66)$$

A stable pseudotime evolution amplitude exists after a path-dependent time transformation with the regulating function

$$f(r) = r^2. \quad (14.67)$$

The time-transformed Hamiltonian

$$\mathcal{H}_O = r^2\frac{p^2}{M} + \hbar^2\frac{\mu_O^2 - 1/4}{2M} + \frac{M}{2}\omega^2 r^4 - E_O r^2 \quad (14.68)$$

is free of the barrier singularity. Thus, when time-slicing the action

$$\mathcal{A}_O^{f=r^2} = \int_0^S ds\left(\frac{M}{2}\frac{r'^2}{r^2} - \frac{\mu_O^2 - 1/4}{2M} - \frac{M}{2}\omega^2 r^4 + E_O r^2\right) \quad (14.69)$$

associated with $\mathcal{H}_O$, no Bessel functions are needed.

Note that the factor $1/r^2$ accompanying $r'^2 = [dr(s)/ds]^2$ does not produce additional problems. It merely diminishes the fluctuations at small r. However, the r-dependence of the kinetic term is undesirable for an evaluation of the time-sliced path integral. We therefore go over to a new coordinate x via the transformation

$$r = h(x) \equiv e^x, \quad (14.70)$$

the transformation function $h(x)$ being related to the regulating function $f(x)$ by (14.15):

$$h'^2 = e^{2x} = f(r) = r^2. \quad (14.71)$$

The resulting effective potential (14.18) happens to be a constant:

$$V_{\text{eff}} = -\frac{\hbar^2}{M}\left[\frac{1}{4}\frac{h'''}{h'} - \frac{3}{8}\left(\frac{h''}{h'}\right)^2\right] = \frac{\hbar^2}{8M}. \quad (14.72)$$

Together with this constant, the DK-transformed radial oscillator action becomes

$$\mathcal{A}_O^{\text{DK}} = \int_0^S ds\left[\frac{M}{2}x'^2 - \frac{\mu_O^2}{2M} - \frac{M\omega^2}{2}e^{4x} + E_O e^{2x}\right]. \quad (14.73)$$

The effective potential (14.72) has changed the initial centrifugal barrier term from $(\mu_\mathcal{O}^2 - 1/4)/2M$ to $\mu_\mathcal{O}^2/2M$. We have omitted the pseudoenergy $\mathcal{E}$ since it is set equal to zero in the final DK relation (14.26). With the identifications

$$
\begin{aligned}
A &= \frac{M}{2}\omega^2, & (14.74) \\
B &= -E_\mathcal{O}, \\
C &= \frac{\hbar^2 \mu_\mathcal{O}^2}{2M} + E_\mathcal{M},
\end{aligned}
$$

the action (14.73) goes over into

$$
\mathcal{A}_\mathcal{M} = \int_0^S ds \left[\frac{M}{2} x'^2 - (V_\mathcal{M} - E_\mathcal{M}) \right]. \tag{14.75}
$$

This is the action for the so-called *Morse potential*

$$
V_\mathcal{M}(x) = A e^{4x} + B e^{2x} + C. \tag{14.76}
$$

Its fixed-energy amplitude

$$
(x_b | x_a)_{E_\mathcal{M}} = \int_0^\infty dS \int \mathcal{D}x(s) e^{i\mathcal{A}_\mathcal{M}/\hbar} \tag{14.77}
$$

is therefore equivalent to the radial amplitude of the oscillator (14.65) via the DK relation (14.26), which now reads

$$
(r_b | r_a)_{E_\mathcal{O}, l} = e^{(x_b + x_a)/2} (x_b | x_a)_{E_\mathcal{M}}, \tag{14.78}
$$

where $r = e^x$.

14.4.2 Radial Coulomb System and Morse System

By a similar argument, the completely different path integral of the radial Coulomb system can be shown to be DK-equivalent to the path integral of the Morse potential. The action is

$$
\mathcal{A}_C = \int dt \left[\frac{M}{2} \dot{r}^2 - \hbar^2 \frac{\mu_C^2 - 1/4}{2Mr^2} + \frac{e^2}{r} + E_C \right], \tag{14.79}
$$

where

$$
\mu_C = D_C/2 - 1 + l_C. \tag{14.80}
$$

For $e^2 = 0$, the action describes a free particle moving in a centrifugal barrier potential. As in the previous example, the action (14.79) does not lead to a time-sliced amplitude of the Feynman type, but involves Bessel functions. We must again remove the barrier via a path-dependent time transformation with

$$
f(r) = r^2 \tag{14.81}
$$

by introducing the pseudotime s satisfying $dt = ds\, r^2(s)$. This leads to the time-transformed action

$$\mathcal{A}_C^{f=r^2} = \int_0^S ds \left[\frac{M}{2} \frac{r'^2}{r^2} - \hbar^2 \frac{\mu_C^2 - 1/4}{2M} + e^2 r + E_C r^2 \right]. \tag{14.82}$$

To bring the kinetic term to the standard form, we change the variable r to x via

$$r = e^x. \tag{14.83}$$

This introduces the same effective potential as in (14.72),

$$V_{\text{eff}} = \frac{\hbar^2}{2M} \frac{1}{4}, \tag{14.84}$$

canceling the 1/4-term in the former centrifugal barrier [2]. Thus we arrive at the DK transform of the radial Coulomb action

$$\mathcal{A}_C^{\text{DK}} = \int_0^S ds \left[\frac{M}{2} x'^2 - \hbar^2 \frac{\mu_C^2}{2M} + e^2 e^x + E_C e^{2x} \right]. \tag{14.85}$$

A trivial change of variables

$$\begin{aligned} x &= 2\bar{x}, \\ M &= \bar{M}/4, \\ \mu_C &= 2\bar{\mu}, \end{aligned} \tag{14.86}$$

brings this to the form

$$\mathcal{A}_C^{\text{DK}} = \int_0^S ds \left[\frac{\bar{M}}{2} \bar{x}'^2 - \hbar^2 \frac{\bar{\mu}^2}{2\bar{M}} + e^2 e^{2\bar{x}} + E_C e^{4\bar{x}} \right], \tag{14.87}$$

and establishes contact with the Morse action (14.75). Upon replacing $\bar{x}$ by x we see that

$$(r_b|r_a)_{E_C,l_C} = \frac{1}{2} e^{(x_b + x_a)} (x_b|x_a)_{E_M}, \tag{14.88}$$

with $r = e^{2x}$. The factor 1/2 accounts for the fact that the normalized states are related by $|x\rangle = |\bar{x}\rangle/2$. The identification of the parameters is now

$$\begin{aligned} A &= -E_C, \\ B &= -e^2, \\ C &= \hbar^2 \frac{\bar{\mu}^2}{2\bar{M}} + E_M. \end{aligned} \tag{14.89}$$

14.4.3 Equivalence of Radial Coulomb System and Radial Oscillator

Since the radial oscillator and the radial Coulomb system are both DK-equivalent to a Morse system, they are DK-equivalent to each other. The relation between the parameters is

$$
\begin{aligned}
M_{\mathcal{O}} &= 4M_{\mathcal{C}}, \\
\mu_{\mathcal{O}} &= 2\mu_{\mathcal{C}}, \\
E_{\mathcal{O}} &= e^2, \\
-\frac{M_{\mathcal{O}}}{2}\omega^2 &= E_{\mathcal{C}}, \\
r_{\mathcal{O}} &= \sqrt{r_{\mathcal{C}}}.
\end{aligned}
\tag{14.90}
$$

We have added subscripts $\mathcal{O}, \mathcal{C}$ also to the masses M to emphasize the systems to which they belong. The relation $\mu_{\mathcal{O}} = 2\mu_{\mathcal{C}}$ implies

$$
D_{\mathcal{O}}/2 - 1 + l_{\mathcal{O}} = 2(D_{\mathcal{C}}/2 - 1 + l_{\mathcal{C}})
\tag{14.91}
$$

for all dimensions and angular momenta of the two systems. Due to the square root relation $r_{\mathcal{O}} = \sqrt{r_{\mathcal{C}}}$, the orbital angular momenta satisfy

$$
l_{\mathcal{O}} = 2l_{\mathcal{C}}.
\tag{14.92}
$$

For the dimensions, this implies

$$
D_{\mathcal{O}} = 2D_{\mathcal{C}} - 2.
\tag{14.93}
$$

In the cases $D_{\mathcal{C}} = 2$ and 3, there is complete agreement with Chapter 13 where the dimensions of the DK-equivalent oscillators were 2 and 4, respectively.

To relate the amplitudes with each other we find it useful to keep the notation as close as possible to that of Chapter 13 and denote the radial coordinate of the radial oscillator by u. Then the DK relation for the pseudotime evolution amplitudes states that

$$
(r_b|r_a)_{E_{\mathcal{C}},\mu_{\mathcal{C}}} = \frac{1}{2}\sqrt{u_b u_a}\,(u_b|u_a)_{E_{\mathcal{O}},\mu_{\mathcal{O}}} ,
\tag{14.94}
$$

with the right-hand side given by (14.65) (after replacing r by u, $M_{\mathcal{O}}$ by $4M_{\mathcal{C}}$, and $M_{\mathcal{O}}\omega u^2/\hbar$ by $2\kappa r$).

Note once more that the prefactor on the right-hand side has a dimension opposite to what one might have expected from the quantum-mechanical completeness relation

$$
\int_0^\infty dr\,|r\rangle\langle r| = 1,
\tag{14.95}
$$

whose u-space version reads

$$
\int_0^\infty du\,2u\,|r\rangle\langle r| = \int du\,|u\rangle\langle u| = 1.
\tag{14.96}
$$

As explained in Section 14.1, the reason lies in the different dimensions (by a factor r) of the pseudotimes over which the evolution amplitudes are integrated when going to the fixed-energy amplitudes. A further factor $1/4$ contained in (14.94) is due to the mass relation $M_O = 4M_C$.

Let us check the relation (14.94) for $D_C = 3$. The fixed-energy amplitude of the Coulomb system has the partial-wave expansion

$$(\mathbf{x}_b|\mathbf{x}_a)_{E_C} = \sum_{l_C=0}^{\infty} \sum_{m=-l_C}^{l_C} \frac{1}{r_b r_a} (r_b|r_a)_{E_C,l_C} Y_{l_C m}(\theta_b, \varphi_b) Y_{l_C m}^*(\theta_a, \varphi_a). \quad (14.97)$$

The four-dimensional oscillator, on the other hand, has

$$(\vec{u}_b|\vec{u}_a)_{E_O} = \sum_{l_O=0}^{\infty} (u_b|u_a)_{E_O,l_O} \quad (14.98)$$

$$\times \frac{l_O+1}{2\pi^2} \sum_{m_1,m_2=-l_O/2}^{l_O/2} \mathcal{D}_{m_1 m_2}^{l_O/2}(\varphi_n, \theta_n, \gamma_n) \mathcal{D}_{m_1 m_2}^{l_O/2 *}(\varphi_{n-1}, \theta_{n-1}, \gamma_{n-1}).$$

We now take Eq. (13.122),

$$(\mathbf{x}_b|\mathbf{x}_a)_{E_C} = \int_0^{\infty} dS e^{ie^2 S/\hbar} \frac{1}{16} \int_0^{4\pi} d\gamma_a (\vec{u}_b S|\vec{u}_a 0), \quad (14.99)$$

and observe that the integral $\int_0^{4\pi} d\gamma_a$ over the sum of angular wave functions

$$\frac{l_O+1}{2\pi^2} \sum_{m_1,m_2=-l_O/2}^{l_O/2} d_{m_1 m_2}^{l_O/2}(\theta_b) d_{m_1 m_2}^{l_O/2}(\theta_a) e^{im_1(\varphi_b-\varphi_a)+im_2(\gamma_b-\gamma_a)} \quad (14.100)$$

produces a sum

$$8 \sum_m^{l_O/2} Y_{l_O/2,m,0}(\theta_b, \phi_b) Y_{l_O/2,m,0}^*(\theta_a, \phi_a), \quad (14.101)$$

with the spherical harmonics

$$Y_{l_O/2,m}(\theta, \phi) = \sqrt{\frac{l_O+1}{4\pi}} e^{im\phi} d_{m,0}^{l_O/2}(\theta). \quad (14.102)$$

Only even l_O-values survive the integration, and we identify $l_C = l_O/2$

Recalling the radial amplitude of the harmonic oscillator (9.32), we find from (14.94) the radial amplitude of the Coulomb system in any dimension D_C for $r_b > r_a$:

$$(r_b|r_a)_{E_C,l_C} = -i \frac{M_C}{\hbar\kappa} \frac{\Gamma(-\nu + l_C + (D_C-1)/2)}{(2l_C + D_C - 2)!} W_{\nu,l_C+D_C/2-1}(2\kappa r_b) M_{\nu,l_C+D_C/2-1}(2\kappa r_a),$$

$$(14.103)$$

where $\kappa = \sqrt{-2ME/\hbar^2}$ and $\nu = \sqrt{-e^4 M_{\mathcal{C}}/2\hbar^2 E_{\mathcal{C}}}$ as in (13.39) and (13.40). For $D_{\mathcal{C}} = 3$, this agrees with (13.206).

The full $D_{\mathcal{C}}$-dimensional amplitude is given by the sum over partial waves

$$(\mathbf{x}_b|\mathbf{x}_a)_{E_{\mathcal{C}},l_{\mathcal{C}}} = \frac{1}{(r_b r_a)^{(D_{\mathcal{C}}-1)/2}} \sum_{l=0}^{\infty} (r_b|r_a)_{E_{\mathcal{C}},l_{\mathcal{C}}} \sum_{\mathbf{m}} Y_{l\mathbf{m}}(\hat{\mathbf{x}}_b) Y_{l\mathbf{m}}^*(\hat{\mathbf{x}}_a),$$

which becomes with (8.125)

$$(\mathbf{x}_b|\mathbf{x}_a)_{E_{\mathcal{C}},l_{\mathcal{C}}} = \frac{1}{(r_b r_a)^{(D_{\mathcal{C}}-1)/2}} \sum_{l_{\mathcal{C}}=0}^{\infty} (r_b|r_a)_{E_{\mathcal{C}},l_{\mathcal{C}}} \frac{2l_{\mathcal{C}} + D_{\mathcal{C}} - 2}{D_{\mathcal{C}} - 2} \frac{1}{S_{D_{\mathcal{C}}}} C_{l_{\mathcal{C}}}^{(D_{\mathcal{C}}/2-1)}(\cos \Delta\vartheta_n).$$

$$(14.104)$$

It is easy to perform the sum if we make use of an integral representation of the radial amplitude obtained by DK-transforming the integral representation (9.25) of the radial oscillator amplitude. Replacing the imaginary time by the new variable of integration $\varrho = e^{-2\omega(\tau_b - \tau_a)}$, the radial variables r by u, and the oscillator mass M by $M_{\mathcal{O}}$ to match the notation of Chapter 13, the amplitude (9.25) can be rewritten as

$$(u_b|u_a)_{E_{\mathcal{O}},l_{\mathcal{O}}} = -i\frac{M_{\mathcal{O}}}{\hbar}\sqrt{u_b u_a}\int_0^1 \frac{d\varrho}{2\varrho}\varrho^{-\nu} e^{-\kappa(u_b^2+u_a^2)\frac{1+\varrho}{1-\varrho}} I_{l_{\mathcal{O}}+D_{\mathcal{O}}/2-1}\left(2\kappa u_b u_a \frac{2\sqrt{\varrho}}{1-\varrho}\right), (14.105)$$

with

$$\kappa \equiv \frac{M_{\mathcal{O}}\omega}{2\hbar}, \qquad \nu \equiv E_{\mathcal{O}}/2\hbar\omega. \qquad (14.106)$$

From the DK relation (14.94) we obtain $(r_b|r_a)_{l_{\mathcal{C}},E_{\mathcal{C}}}$ and insert it into (14.104). Then we recall the summation formula (suppressing all subscripts $\mathcal{C}$)

$$\left(\frac{1}{2}kz\right)^{D/2-1/2} I_{D/2-3/2}(kz) = k^{D-2}\sum_{l=0}^{\infty}\frac{1}{l!}\frac{\Gamma(l+D-2)}{\Gamma(D/2-1/2)}(2l+D-2)$$

$$\times F(-l, l+D-2; D/2-1/2; (1+k^2)/2)(-)^l I_{2l+D-2}(z), \qquad (14.107)$$

which follows from Eq. (13.200) for $\nu = D/2 - 3/2$ and $\mu = D - 2$. After expressing the right-hand side in terms of the Gegenbauer polynomial $C_l^{D/2-1}$ with the help of (8.105), the summation formula becomes

$$\frac{1}{2}\frac{1}{(2\pi)^{D/2-1/2}}\left(\frac{z}{2}\right)^{D-2} I_{D/2-3/2}(kz)/(kz)^{D/2-3/2}$$

$$= \sum_{l_{\mathcal{C}}=0}^{\infty}\frac{2l_{\mathcal{C}} + D - 2}{D - 2}\frac{1}{S_D}C_{l_{\mathcal{C}}}^{(D/2-1)}((1+k^2)/2)I_{2l_{\mathcal{C}}+D-2}(z). \qquad (14.108)$$

Setting

$$z \equiv 2\kappa u_b u_a \frac{2\sqrt{\varrho}}{1-\varrho}, \qquad k \equiv \cos(\vartheta/2), \qquad (14.109)$$

the sum over the partial waves in (14.104) is easily performed, and we obtain for the fixed-energy amplitude of the Coulomb system in D dimensions the generalization of the integral representations (13.43) and (13.128) in two and three dimensions:

$$
\begin{aligned}
(\mathbf{x}_b|\mathbf{x}_a)_E &= -i\frac{M}{\hbar}\frac{\kappa^{D-2}}{(2\pi)^{(D-1)/2}}\int_0^1\frac{d\varrho}{(1-\varrho)^2}\varrho^{-\nu} \\
&\times \left(\frac{2\sqrt{\varrho}}{1-\varrho}\right)^{(D-3)/2}e^{-\kappa\frac{1+\varrho}{1-\varrho}(r_b+r_a)}I_{D/2-3/2}\left(kz\right)/(kz)^{D/2-3/2},
\end{aligned} \tag{14.110}
$$

where

$$
kz = 2\kappa\frac{2\sqrt{\varrho}}{1-\varrho}\sqrt{(r_b r_a + \mathbf{x}_b\mathbf{x}_a)/2}, \tag{14.111}
$$

and κ, ν are the Coulomb parameters (13.40).

By changing the integration variable to $\zeta = (1+\varrho)/(1-\rho)$ as in (13.44), the integral in (14.110) is transformed into a contour integral encircling the cut from $\zeta = 1$ to ∞ in the clockwise sense. Then the amplitude reads [3]

$$
\begin{aligned}
(\mathbf{x}_b|\mathbf{x}_a)_E &= -i\frac{M}{2\hbar}\frac{\kappa^{D-2}}{(2\pi)^{(D-1)/2}}\frac{\pi e^{i\pi(\nu-D/2+3/2)}}{\sin[\pi(\nu-D/2+3/2)]} \\
&\times \int_C\frac{d\zeta}{2\pi i}(\zeta-1)^{-\nu+D/2-3/2}(\zeta+1)^{\nu+D/2-3/2}e^{-\kappa\zeta(r_b+r_a)}I_{D/2-3/2}\left(z\right)/z^{D/2-3/2}.
\end{aligned} \tag{14.112}
$$

This expression generalizes the integral representations (13.48) and (13.130) for $D_C = 2$ and $D_C = 3$, respectively.

It is worth emphasizing that due to the catastrophic centrifugal barriers, there is no way of establishing this relation for the time-sliced radial amplitudes without the intermediate Morse potential. This has been attempted in the literature [4] by using the DK transformation with the regulating function $f(r) = r$ and a pseudotime s satisfying $dt = ds\,r(s)$ (which were successful in two and three dimensions). Although this transformation removes the Coulomb singularity, it weakens the centrifugal barrier insufficiently to a still catastrophic $1/r$-singularity. Let us exhibit the place where such an attempt fails. The starting point is the pseudotime-sliced amplitude (13.8),

$$
\langle\mathbf{x}_b|\hat{\mathcal{U}}_E(s)|\mathbf{x}_a\rangle \approx \frac{r_b^\lambda r_a^{1-\lambda}}{\sqrt{2\pi\epsilon_s\hbar r^{1-\lambda}r^\lambda/M}^{D/2}}\prod_{n=1}^N\left[\int\frac{d^D\Delta x_n}{\sqrt{2\pi\epsilon_s\hbar r_n/M}^D}\right]e^{-\mathcal{A}_E^N/\hbar}, \tag{14.113}
$$

with the action

$$
\mathcal{A}_E^N = -(N+1)\epsilon_s e^2 + \sum_{n=1}^{N+1}\left[\frac{M}{2\epsilon_s}\frac{(x_n-x_{n-1})^2}{r_n^{1-\lambda}r_{n-1}^\lambda}+\epsilon_s E r_n^{1-\lambda}r_{n-1}^\lambda\right]. \tag{14.114}
$$

In contrast to Chapter 13, we work here conveniently with an imaginary-time. In any dimension D, the amplitude has the angular decomposition

$$
\langle\mathbf{x}_b|\hat{\mathcal{U}}_E(s)|\mathbf{x}_a\rangle = \frac{1}{(r_b r_a)^{D-1/2}}\sum\langle r_b|\hat{\mathcal{U}}_E(s)|r_a\rangle_l Y_{lm}(\hat{\mathbf{x}}_b)Y_{lm}^*(\hat{\mathbf{x}}_a). \tag{14.115}
$$

The action for the radial amplitude is obtained by decomposing

$$\mathcal{A}_E^N = -(N+1)\epsilon_s e^2 + \sum_{n=1}^{N+1}\left[\frac{M}{2\epsilon_s r_n^{1-\lambda}r_{n-1}^{\lambda}}(r_n^2+r_{n-1}^2-2r_nr_{n-1}\cos\vartheta_n)+\epsilon_s Er_n\right], \tag{14.116}$$

where ϑ_n is the angle between $\mathbf{x}_n$ and $\mathbf{x}_{n-1}$. We have replaced $Er_n^{1-\lambda}r_{n-1}^{\lambda}$ by Er_n since the difference is of order ϵ_s^2 and thus negligible.

We now go through the same steps as in Section 8.5. For an individual time slice, the ϑ_n-part of the exponential is expanded as

$$\exp\left(\frac{M}{\epsilon_s}r_n^{\lambda}r_{n-1}^{1-\lambda}\cos\vartheta_n\right) = e^h\sum_{l=0}^{\infty}\tilde{a}_l(h)\sum_{\mathbf{m}}Y_{l\mathbf{m}}(\hat{\mathbf{x}}_b)Y_{l\mathbf{m}}^*(\hat{\mathbf{x}}_a), \tag{14.117}$$

with

$$a_l(h) = \left(\frac{2\pi}{h}\right)^{(D-1)/2}\tilde{I}_{D/2-1+l}(h), \quad h = \frac{M}{\hbar\epsilon_s}r_n^{\lambda}r_{n-1}^{1-\lambda} \tag{14.118}$$

[recall (8.129) and (8.101)]: The radial part of the propagator is then

$$\langle r_b|\hat{\mathcal{U}}_E(s)|r_a\rangle_l \approx \frac{r_b^{\lambda}r_a^{1-\lambda}}{\sqrt{2\pi\hbar\epsilon_s r_b^{1-\lambda}r_a^{\lambda}/M}}\prod_{n=2}^{N+1}\left[\int\frac{d\Delta r_n\ r_{n-1}^{-1/2}}{\sqrt{2\pi\hbar\epsilon_s/M}}\right]e^{-\mathcal{A}_E^N/\hbar}, \tag{14.119}$$

with the radial action

$$\mathcal{A}_E^N = -(N+1)\epsilon_s e^2 + \sum_{n=1}^{N+1}\left[\frac{M}{2\epsilon_s}\frac{(r_n-r_{n-1})^2}{r_n^{1-\lambda}r_{n-1}^{\lambda}}-\hbar\log\tilde{I}_{D/2-1+l}\left(\frac{M}{\hbar\epsilon_s}r_n^{\lambda}r_{n-1}^{1-\lambda}\right)-\epsilon_s Er_n\right]. \tag{14.120}$$

At this place we simplify the calculation by choosing the symmetric splitting parameter $\lambda = 1/2$. Going over to square root coordinates

$$u_n = \sqrt{r_n}, \tag{14.121}$$

we calculate

$$\begin{aligned}\Delta r_n &= (u_n+u_{n-1})\Delta u_n, \\ &= 2u_n(1-\Delta u_n/2u_n)\Delta u_n, \\ \frac{\partial\Delta r_n}{\partial\Delta u_n} &= 2u_n(1-\Delta u_n/u_n), \\ r_{n-1}^{-1/2} &= u_n^{-1}(1-\Delta u_n/u_n)^{-1},\end{aligned} \tag{14.122}$$

transforming the measure of integration into

$$\frac{\sqrt{u_bu_a}}{\sqrt{2\pi\hbar\epsilon_s/M}}\prod_{n=1}^{N}\int\frac{d\Delta u_n}{\sqrt{2\pi\hbar\epsilon_s/M}}. \tag{14.123}$$

Note that there are no higher Δu_n correction terms. The kinetic energy is

$$\frac{4\bar{u}_n{}^2(\Delta u_n)^2}{2\epsilon_s u_n u_{n-1}} = \frac{4}{2\epsilon_s}\left[(\Delta u_n)^2 + \frac{1}{4}\frac{(\Delta u_n)^4}{u_n{}^2} + \cdots\right]. \tag{14.124}$$

The $(\Delta u_n)^4$-term can be replaced right away by its expectation value and renders an effective potential

$$V_{\text{eff}}(u_n{}^2) = \epsilon_s \hbar^2 \frac{1}{2 \cdot 4M}\frac{3}{4u_n{}^2}. \tag{14.125}$$

The radial amplitude becomes simply

$$\langle r_b|\hat{\mathcal{U}}_E(s)|r_a\rangle_l \approx \frac{\sqrt{u_b u_a}}{\sqrt{2\pi\hbar\epsilon_s/M}} \prod_{n=2}^{N+1}\left[\int_0^\infty \frac{du_n}{\sqrt{2\pi\hbar\epsilon_s/M}}\frac{2}{}\right] e^{-\mathcal{A}_E^N/\hbar}, \tag{14.126}$$

with

$$\mathcal{A}_E^N = -(N+1)\epsilon_s + \sum_{n=1}^{N+1}\left[\frac{4M}{2}\frac{(\Delta u_n)^2}{2\epsilon_s} + V_{\text{eff}}(u_n{}^2) - \hbar\log\tilde{I}_{D/2-1+l}\left(\frac{M}{\hbar\epsilon_s}u_n u_{n-1}\right)\right]. \tag{14.127}$$

Due to the $1/u_n{}^2$-singularity in $V_{\text{eff}}(u_n{}^2)$, the time-sliced path integral does not exist. Apart from the $\epsilon_s/u_n{}^2$-term, there should be infinitely many terms of increasing order of the type $(\epsilon_s/u_n{}^2)^2, \ldots$, whose resummation is needed to obtain the correct threshold small-u_n behavior of the amplitude as discussed in Section 8.2. To have the usual kinetic term of the harmonic oscillator $M_{\mathcal{O}}(\Delta u_n)^2/2\epsilon_s$, we must identify $4M$ with the oscillator mass $M_{\mathcal{O}}$ [called μ in (13.27); see also (14.90) with $M_{\mathcal{C}} \equiv M$],

$$M_{\mathcal{O}} = 4M. \tag{14.128}$$

The centrifugal barrier in (14.127) resides in

$$-\hbar\log\tilde{I}_{D/2-1+l}\left(\frac{M_{\mathcal{O}}/4}{\hbar\epsilon_s}u_n u_{n-1}\right) + \epsilon_s\hbar^2\frac{3}{8M_{\mathcal{O}}u_n^2} + \cdots, \tag{14.129}$$

and is given by

$$\epsilon_s\frac{\hbar^2}{2M_{\mathcal{O}}}\frac{4}{u_n u_{n-1}}\left[\left(\frac{D}{2}-1+l\right)^2 - \frac{1}{4}\right] + \epsilon_s\hbar^2\frac{3}{8M_{\mathcal{O}}u_n^2} + \cdots. \tag{14.130}$$

This can be rewritten more explicitly as

$$\epsilon_s\frac{\hbar}{2M_{\mathcal{O}}}\frac{1}{u_n u_{n-1}}\left[(D_{\mathcal{C}} - 2 + 2l_{\mathcal{C}})^2 - \frac{1}{4}\right], \tag{14.131}$$

where we have added the subscript $\mathcal{C}$ to D to record its being the dimension of the Coulomb system. The expression in parentheses is identified with the parameter $\mu_{\mathcal{O}}$

of the harmonic oscillator, which appears in the subscript of the Bessel function in (8.139). This implies

$$\mu_{\mathcal{O}} = 2\mu_{\mathcal{C}}, \tag{14.132}$$

in agreement with the relation (14.91). Indeed, the higher terms in the expansion (14.130) must all conspire to sum up to the Bessel-regulated centrifugal barrier in the time-sliced radial amplitude of the harmonic oscillator

$$- \hbar \log \tilde{I}_{D-2+2l} \left(\frac{M_{\mathcal{O}}}{\hbar \epsilon_s} u_n u_{n-1} \right). \tag{14.133}$$

This is quite hard to verify term by term, although it must happen.

Using the stronger regulating function $f = r^2$, these difficulties are avoided. Instead of the pseudotime evolution amplitude (14.113), we have

$$\langle \mathbf{x}_E | \hat{\mathcal{U}}_e(s) | \mathbf{x}_a \rangle \approx \frac{r_b^2 r_a^{2-2\lambda}}{\sqrt{2\pi\epsilon_s \hbar r_b^{2-2\lambda} r_a^{2\lambda}/M}^{D/2}} \prod_{n=1}^{N} \left[\int \frac{d^D \Delta x_n}{\sqrt{2\pi\epsilon_s \hbar r_n^2/M}^D} \right] e^{-\mathcal{A}_E^{fN}/\hbar}, \tag{14.134}$$

with the time-sliced transformed action

$$\mathcal{A}_E^{fN} = -(N+1)\epsilon_s e^2 + \sum_{n=1}^{N+1} \left[\frac{M}{2\epsilon_s r_n^{2-2\lambda} r_{n-1}^{2\lambda}} (r_n^2 + r_{n-1}^2 - 2r_n r_{n-1} \cos \vartheta_n) - \epsilon_s E r_n^2 \right]. \tag{14.135}$$

For $\lambda = 1/2$, the $\cos \Delta \vartheta_n$-term is now free of the radial variables $r_n, r_{n-1}, r_n, r_{n-1}$, and the angular decomposition of the amplitude as in (14.115)–(14.120) gives the radial amplitude with a time-sliced action

$$\mathcal{A}_E^{fN} = -(N+1)\epsilon_s e^2 + \sum_{n=1}^{N+1} \left[\frac{M}{2\epsilon_s} \frac{(r_n - r_{n-1})^2}{r_n r_{n-1}} - \hbar \log \tilde{I}_{D/2-1+l} \left(\frac{M}{\hbar \epsilon_s} \right) - \epsilon_s E r_n^2 \right]. \tag{14.136}$$

Since r_n, r_{n-1} are absent in the Bessel function, the limit of small ϵ_s is now uniform in the integration variables r_n and the logarithmic term in the energy can directly be replaced by

$$\epsilon_s \frac{\hbar}{2M_{\mathcal{C}}} \left[(D_{\mathcal{C}}/2 - 1 + l_{\mathcal{C}})^2 - 1/4 \right], \tag{14.137}$$

where we have added the subscripts $\mathcal{C}$, for clarity. To perform the integration over the r_n variables, one goes over to new coordinates x with

$$r = h(x) = e^x. \tag{14.138}$$

The measure of integration is

$$\frac{\sqrt{r_b r_a}}{\sqrt{2\pi\hbar\epsilon_s/M}^{N+1}} \prod_{n=2}^{N+1} \int \frac{d\Delta r_n}{r_{n-1}}. \tag{14.139}$$

Expanding $1/r_{n-1}$ around the postpoint r_n gives

$$\frac{1}{r_{n-1}} = \frac{1}{r_n}\left(\frac{r_n}{r_{n-1}}\right) = \frac{e^{\Delta x_n}}{e^{x_n}}. \tag{14.140}$$

We now write (dropping subscripts n)

$$\Delta r = e^x - e^{x-\Delta x} = e^x(1 - e^{-\Delta x}), \tag{14.141}$$

and find the Jacobian

$$\frac{\partial \Delta r}{\partial \Delta x} = e^x e^{-\Delta x}. \tag{14.142}$$

In the x-coordinates, the measure becomes simply

$$\frac{e^{(x_b+x_a)/2}}{\sqrt{2\pi\hbar\epsilon_s/M}^{N+1}} \prod_{n=2}^{N+1} \int d\Delta x_n. \tag{14.143}$$

The kinetic term in the action turns into

$$\mathcal{A}_E^N = \sum_{n=1}^{N+1} \frac{M}{\epsilon_s}(1 - \cos\Delta x_n), \tag{14.144}$$

and has the expansions

$$\mathcal{A}_E^N = \sum_{n=1}^{N+1} \frac{M}{2\epsilon_s}\left[(\Delta x)^2 - \frac{1}{12}(\Delta x_n)^4 + \ldots\right]. \tag{14.145}$$

The higher-order terms contribute with higher powers of ϵ_s *uniformly* in x. They can be treated as usual. This is why the path-dependent time transformation of the radial Coulomb system to a radial oscillator with the regulating function $f = r^2$ is free of problems.

14.4.4   Angular Barrier near Sphere, and Rosen-Morse Potential

For another application of the solution method, consider the path integral for a mass point near the surface of a sphere in three dimensions, projected into a state of fixed azimuthal angular momentum $m = 0, \pm1, \pm2, \ldots$. The projection generates an angular barrier $\propto (m^2 - 1/4)/\sin^2\theta$ which is a potential of the Pöschl-Teller type. With $\mu = Mr^2$, the real-time action is

$$\mathcal{A}_{PT} = \int dt\left[\frac{\mu}{2}\dot{\theta}^2 + \frac{\hbar^2}{8\mu} - \frac{\hbar^2}{2\mu}\frac{\text{"}m^2 - 1/4\text{"}}{\sin^2\theta} + E_{PT}\right]. \tag{14.146}$$

The quotation marks are defined in analogy with those of the centrifugal barrier in Eq. (8.139). The precise meaning is given by the proper time-sliced expression in Eq. (8.174) whose limiting form for narrow time slices is (8.176). After an analytic continuation of the parameter m to arbitrary real numbers μ, the resulting amplitude was given in (8.186). In the sequel we refrain from using the symbol μ for the noninteger m-values to avoid confusion with the mass parameter μ.

The spectral representation of the associated fixed-energy amplitude is easily written down; it arises by simply integrating (8.186) over $-id\tau_b$ and reads

$$(\theta_b|\theta_a)_{m,E_{\mathcal{PT}}} = \sqrt{\sin\theta_b\sin\theta_a}\sum_{n=0}^{\infty}\frac{i\hbar}{E_{\mathcal{PT}}-\hbar^2 L_2/2\mu} \tag{14.147}$$
$$\times\frac{2n+2m+1}{2}\frac{(n+2m)!}{n!}P_{n+m}^{-m}(\cos\theta_b)P_{n+m}^{-m}(\cos\theta_a),$$

where $L_2 = l(l+1)$ with $l = n+m$ [recall (8.224) for $D = 3$]. The sum over n can be done using the so-called Sommerfeld-Watson transformation [5]. The sum is re-expressed as a contour integral in the complex n-plane and deformed in such a way that only the Regge poles at

$$n+m = l = l(E_{\mathcal{PT}}) \equiv -\frac{1}{2}+\sqrt{\frac{1}{4}+\frac{2\mu E_{\mathcal{PT}}}{\hbar^2}} \tag{14.148}$$

contribute, with both signs of the square root. The result for $\theta_b > \theta_a$ is [6]

$$(\theta_b|\theta_a)_{m,E_{\mathcal{PT}}} = \sqrt{\sin\theta_b\sin\theta_a}\frac{-i\mu}{\hbar}\Gamma(m-l(E_{\mathcal{PT}}))\Gamma(l(E_{\mathcal{PT}})+m+1)$$
$$\times P_{l(E_{\mathcal{PT}})}^{-m}(-\cos\theta_b)P_{l(E_{\mathcal{PT}})}^{-m}(\cos\theta_a). \tag{14.149}$$

Here we shall consider m as a free parameter characterizing the interaction strength of the Pöschl-Teller potential [7]

$$V_{\mathcal{PT}}(\theta) = \frac{\hbar^2}{2\mu}\frac{m^2}{\sin^2\theta}. \tag{14.150}$$

The regulating function removing the angular barrier is

$$f(\theta) = \sin^2\theta, \tag{14.151}$$

and the time-transformed action reads with $dt = ds\,\sin^2\theta(s)$

$$\mathcal{A}_{\mathcal{PT}}^{f=\sin^2\theta} = \int_0^S ds\left[\frac{\mu}{2\sin^2\theta}\theta'^2 + \frac{\hbar^2}{8\mu}\sin^2\theta - \frac{\hbar^2}{2\mu}(m^2-1/4) + E_{\mathcal{PT}}\sin^2\theta\right]. \tag{14.152}$$

We now bring the kinetic term to the conventional form by the variable change

$$\sin\theta = \frac{1}{\cosh x}, \qquad \cos\theta = -\tanh x, \tag{14.153}$$

which maps the interval $\theta \in (0,\pi)$ into $x \in (-\infty,\infty)$. Then we have

$$h'(x) = \sin\theta = \frac{1}{\cosh x}. \tag{14.154}$$

Forming the higher derivatives

$$h''(x) = -\frac{\tanh x}{\cosh x}, \quad h'''(x) = -\frac{1}{\cosh x}\left(1 - 2\tanh^2 x\right), \tag{14.155}$$

the effective potential is found to be

$$V_{\text{eff}} = \frac{\hbar^2}{8\mu}\left(1 + \frac{1}{\cosh^2 x}\right). \tag{14.156}$$

The DK-transformed action is therefore

$$\mathcal{A}_{\mathcal{PT}}^{\text{DK}} = \int_0^S ds \left[\frac{\mu}{2}x'^2 - \frac{\hbar^2 m^2}{2\mu} + E_{\mathcal{PT}}\frac{1}{\cosh^2 x}\right]. \tag{14.157}$$

It describes the motion of a mass point in a smooth potential well known as the *Rosen-Morse potential* (also called the *modified Pöschl-Teller potential*) [8]. The standard parametrization is

$$V_{\mathcal{RM}}(x) = -\frac{\hbar^2}{2\mu}\frac{s(s+1)}{\cosh^2 x}. \tag{14.158}$$

This corresponds to $l(E_{\mathcal{PT}})$ in (14.148) having the value s. The energy of the Rosen-Morse potential determines the parameter m in the action (14.157), and we identify

$$m = m(E_{\mathcal{RM}}) = \sqrt{-2\mu E_{\mathcal{RM}}/\hbar^2}. \tag{14.159}$$

It is obvious that the time-sliced amplitude of the Rosen-Morse potential has no path collapse problems. Its fixed-energy amplitude is thus DK-equivalent to the Pöschl-Teller amplitude (14.149), with the precise relation being

$$(\theta_b|\theta_a)_{m,E_{\mathcal{PT}}} = \sqrt{\sin\theta_b \sin\theta_a}(x_b|x_a)_{m,E_{\mathcal{RM}}}, \tag{14.160}$$

where $\tanh x = -\cos\theta$, $\theta \in (0,\pi)$, $x \in (-\infty,\infty)$. Inserting (14.149), the amplitude of the Rosen-Morse system reads explicitly

$$\begin{aligned}(x_b|x_a)_{m(E_{\mathcal{RM}})} &= \frac{-i\mu}{\hbar}\Gamma(m(E_{\mathcal{RM}}) - s)\Gamma(s + m(E_{\mathcal{RM}}) + 1) \\ &\times P_s^{-m(E_{\mathcal{RM}})}(\tanh x_b)P_s^{-m(E_{\mathcal{RM}})}(-\tanh x_a).\end{aligned} \tag{14.161}$$

The bound states lie at the poles of the first Gamma function where

$$m(E_{\mathcal{RM}}) = s - n, \quad n = 0, 1, 2, \ldots, [s], \tag{14.162}$$

with $[s]$ denoting the largest integer number $\leq s$. From the residues we extract the normalized wave functions [6]

$$\psi_n(x) = \sqrt{\Gamma(2s - n + 1)(s - n)/n}P_s^{n-s}(\tanh x). \tag{14.163}$$

For noninteger values of s, these are not polynomials. However, the identity between hypergeometric functions (1.450)

$$F(a, b; c; z) = (1 - z)^{c-a-b} F(c - a, c - b; c; z) \qquad (14.164)$$

permits relating them to polynomials:

$$P_s^{n-s}(\tanh x) = \frac{2^{n-s}}{\Gamma(s - n + 1)} \frac{1}{\cosh^{s-n} x} F(-n, 1 + 2s - n; s - n + 1; (1 - \tanh x)/2).$$
$$(14.165)$$

The continuum wave functions are obtained from (14.163) by an appropriate analytic continuation of m to $-ik$. This amounts to replacing n by $s + ik$.

14.4.5 Angular Barrier near Four-Dimensional Sphere, and General Rosen-Morse Potential

Let us extend the previous path integral of a mass point moving near the surface of a sphere from $D = 3$ to $D = 4$ dimensions. By projecting the amplitude into a state of fixed azimuthal angular momenta m_1 and m_2, an angular barrier is generated in the Euler angle θ proportional to $(m_1^2 + m_2^2 - 1/4 - 2m_1 m_2 \cos\theta)/\sin^2\theta$. This is again a potential of the Pöschl-Teller type, although of a more general form to be denoted by a subscript $\mathcal{PT}'$. The action (8.211) is, with $\mu = Mr^2/4$,

$$\mathcal{A}_{\mathcal{PT}'} = \int dt \left[\frac{\mu}{2} \dot\theta^2 + \frac{\hbar^2}{32\mu} - \frac{\hbar^2}{2\mu} \frac{\text{``} m_1^2 + m_2^2 - 2m_1 m_2 \cos\theta - 1/4 \text{''}}{\sin^2\theta} + E_{\mathcal{PT}'} \right], \qquad (14.166)$$

where the quotation marks indicate the need to regularize the angular barrier via Bessel functions as specified in (8.207). The projected amplitude was given in Eq. (8.202) and continued to arbitrary real values of $m_1 = \mu_1$, $m_2 = \mu_2$ with $\mu_1 \geq \mu_2 \geq 0$ in (8.212). As in subsection 14.4.4, we shall also use the parameters m_1, m_2 when they have noninteger values.

The most general Pöschl-Teller potential

$$V_{\mathcal{PT}'}(\theta) = \frac{\hbar^2}{2\mu} \left[\frac{s_1(s_1 + 1)}{\sin^2(\theta/2)} + \frac{s_2(s_2 + 1)}{\cos^2(\theta/2)} \right] \qquad (14.167)$$

can easily be mapped onto the above angular barrier, up to a trivial additive constant. The fixed-energy amplitude is obtained directly from Eq. (8.212) by an integration over $-id\tau_b$. It reads for $m_1 \geq m_2$

$$(\theta_b | \theta_a)_{m_1, m_2, E_{\mathcal{PT}'}} = \sqrt{\sin\theta_b \sin\theta_a}$$
$$\times \sum_{n=0}^{\infty} \frac{i\hbar}{E_{\mathcal{PT}'} - \hbar^2 L_2/8\mu} \frac{2n + 2m_1 + 1}{2} d_{m_1, m_2}^{n+m_1}(\theta_b) d_{m_1, m_2}^{n+m_1}(\theta_a), \qquad (14.168)$$

where L_2 is given by $L_2 = (l + 1)^2 - 1/4$ with $l = 2n + 2m_1$ [recall (8.219) with (8.224)].

As in Eq. (14.147), the sum over n can be performed with the help of a Sommerfeld-Watson transformation by rewriting the sum as a contour integral in the complex n-plane. After deforming the contour in such a way that only the Regge poles at

$$2n + 2m_1 = l = l(E_{\mathcal{PT}'}) \equiv -1 + 2\sqrt{\frac{1}{16} + \frac{2\mu E_{\mathcal{PT}'}}{\hbar^2}} \tag{14.169}$$

contribute, with both signs of the square root, we find for $\theta_b > \theta_a$:

$$(\theta_b | \theta_a)_{m_1, m_2, E_{\mathcal{PT}'}} = \sqrt{\sin \theta_b \sin \theta_a} \frac{-2i\mu}{\hbar} \tag{14.170}$$

$$\times \Gamma(m_1 - l(E_{\mathcal{PT}'})/2) \Gamma(l(E_{\mathcal{PT}'})/2 - m_1 + 1) \frac{1}{2} d_{m_1, -m_2}^{l(E_{\mathcal{PT}'})/2}(\theta_b - \pi) d_{m_1, m_2}^{l(E_{\mathcal{PT}'})/2}(\theta_a),$$

with arbitrary real parameters m_1, m_2 characterizing the interaction strength.

The regulating function which removes the angular barrier is

$$f(\theta) = \sin^2 \theta, \tag{14.171}$$

and the time-transformed action reads, with $dt = ds \sin^2 \theta(s)$,

$$\mathcal{A}_{\mathcal{PT}'}^{f = \sin^2 \theta} = \int_0^S ds \left[\frac{\mu}{2 \sin^2 \theta} \theta'^2 + \frac{\hbar^2}{32\mu} \sin^2 \theta \right.$$
$$\left. - \frac{\hbar^2}{2\mu} (m_1^2 + m_2^2 - 1/4 - 2m_1 m_2 \cos \theta) + E_{\mathcal{PT}'} \sin^2 \theta \right]. \tag{14.172}$$

We now bring the kinetic term to the conventional form by the variable change

$$\sin \theta = \pm \frac{1}{\cosh x}, \quad \cos \theta = -\tanh x. \tag{14.173}$$

As in the previous case, this leads to the effective potential

$$V_{\text{eff}} = \frac{\hbar^2}{8\mu} \left(1 + \frac{1}{\cosh^2 x} \right). \tag{14.174}$$

The DK-transformed action is then

$$\mathcal{A}_{\mathcal{PT}'}^{\text{DK}} = \int_0^S ds \left[\frac{\mu}{2} x'^2 - \frac{\hbar^2}{2\mu} (m_1^2 + m_2^2 + 2m_1 m_2 \tanh x) + \left(E_{\mathcal{PT}'} - \frac{3\hbar^2}{32\mu} \right) \frac{1}{\cosh^2 x} \right]. \tag{14.175}$$

It contains a smooth potential well near the origin known as the general Rosen-Morse potential [8]. A convenient general parametrization is

$$V_{\mathcal{RM}'}(x) = \frac{\hbar^2}{2\mu} \left[-\frac{s(s+1)}{\cosh^2 x} + 2c \tanh x \right], \tag{14.176}$$

which amounts to choosing

$$E_{\mathcal{PT}'} = \frac{\hbar^2}{2\mu}[s(s+1) + 3/32], \qquad m_1 m_2 = c,$$

in (14.175). Inserting this into (14.169) makes $l(E_{\mathcal{PT}'})/2$ equal to s. The energy of the general Rosen-Morse potential fixes the third parameter to

$$E_{\mathcal{RM}'} = \frac{\hbar^2}{2\mu}(m_1^2 + c^2/m_1^2). \tag{14.177}$$

The solution of this equation will be a function $m_1(E_{\mathcal{PM}'})$. Correspondingly, we define $m_2(E_{\mathcal{PM}'}) \equiv c/m_1(E_{\mathcal{PM}'})$.

Feynman's time-sliced amplitude certainly exists for this potential, and the fixed-energy amplitude is determined in terms of the angular-projected amplitude (14.170) of a mass point near the surface of a sphere which describes the motion in a general Pöschl-Teller potential. The relation is [9]

$$(\theta_b|\theta_a)_{m_1,m_2,E_{\mathcal{PT}'}} = \sqrt{\sin\theta_b \sin\theta_a}(x_b|x_a)_{m_1,m_2,E_{\mathcal{RM}'}}, \tag{14.178}$$

with $\tanh x = -\cos\theta$, $\theta \in (0, \pi)$, $x \in (-\infty, \infty)$. Explicitly we have

$$(x_b|x_a)_{E_{\mathcal{PT}'}} = \frac{-2i\mu}{\hbar}\Gamma(m_1(E_{\mathcal{RM}'}) - s)\Gamma(s - m_1(E_{\mathcal{RM}'}) + 1) \tag{14.179}$$

$$\times \frac{1}{2}d^s_{m_1(E_{\mathcal{RM}'}),-m_2(E_{\mathcal{RM}'})}(\theta_b(x_b) - \pi)d^s_{m_1(E_{\mathcal{RM}'}),m_2(E_{\mathcal{RM}'})}(\theta_a(x_a)).$$

The bound states lie at the poles of the first Gamma function. With the energy-dependent function $m_1(E_{\mathcal{RM}'})$ defined by (14.177), they are given by the solutions of the equation

$$m_1(E_{\mathcal{RM}'}) = s - n, \qquad n = 0, 1, \ldots, [s]. \tag{14.180}$$

The residues in (14.179) render the normalized wave functions

$$\Psi_n(x) = \sqrt{\frac{m_1^2 - m_2^2}{m_1}\frac{\Gamma(s+1-m_1)n!}{\Gamma(s+1-m_2)\Gamma(s+1+m_2)}} \tag{14.181}$$

$$\times [\tfrac{1}{2}(1 + \tanh x)]^{(m_1-m_2)/2}[\tfrac{1}{2}(1 - \tanh x)]^{(m_1+m_2)/2} P_n^{(m_1-m_2,m_1+m_2)}(-\tanh x),$$

or, expressed in terms of hypergeometric functions,

$$\Psi_n(x) = \sqrt{\frac{m_1^2 - m_2^2}{m_1}\frac{\Gamma(s+1+m_1)\Gamma(s+1-m_2)}{n!\Gamma(1+m_1-m_2)^2\Gamma(s+1+m_2)}}$$

$$\times [\tfrac{1}{2}(1 + \tanh x)]^{(m_1-m_2)/2}[\tfrac{1}{2}(1 - \tanh x)]^{(m_1+m_2)/2}$$

$$\times F\left(2s - n + 1, -n; 1 + m_1 - m_2; \tfrac{1}{2}(1 + \tanh x)\right), \tag{14.182}$$

with $m_1 = s - n$ and $m_2 = c/m_1$ [10]. The continuum wave functions are obtained from these by an appropriate analytic continuation of m_1 to complex values $-ik$ satisfying the relation $k^2 = -(m_1^2 + c^2/m_1^2)$ [compare (14.177)].

14.4.6 Hulthén Potential and General Rosen-Morse Potential

For a further application of the solution method, consider the path integral of a particle moving along the positive r-axis with the singular Hulthén potential

$$V_{\mathcal{H}}(r) = g\frac{1}{e^{r/a} - 1}, \tag{14.183}$$

where g and a are energy and length parameters. Note that this potential contains the Coulomb system in the limit $a \to \infty$ at $ag = e^2 =$ fixed.

The fixed-energy amplitude is controlled by the action

$$\mathcal{A}_{\mathcal{H}} = \int dt \left[\frac{M}{2}\dot{r}^2 - V_{\mathcal{H}}(r) + E_{\mathcal{H}}\right]. \tag{14.184}$$

The potential is singular at $r = 0$, and for $g < 0$, the Euclidean time-sliced amplitude does not exist due to path collapse. A regulating function which stabilizes the fluctuations is

$$f(r) = 4(1 - e^{-r/a})^2. \tag{14.185}$$

The time-transformed action is therefore

$$\mathcal{A}_{\mathcal{H}}^f = \int_0^\infty ds \left[\frac{M}{2}\frac{r'^2}{4(1-e^{-r/a})^2} - g\,4e^{-r/a}(1 - e^{-r/a}) + E_{\mathcal{H}}\,4(1 - e^{-r/a})^2\right]. \tag{14.186}$$

The coordinate transformation leading to a conventional kinetic energy in terms of the new variable x is found by solving the differential equation

$$\frac{dr}{dx} = h'(x), \tag{14.187}$$

with

$$h' = \sqrt{f} = 2(1 - e^{-r/a}). \tag{14.188}$$

The solution is

$$\frac{r}{a} = x + a\log[2\cosh(x/a)] = \log(e^{2x/a} + 1), \tag{14.189}$$

so that

$$h'(x) = 2\frac{e^{2x/a}}{e^{2x/a} + 1} = \frac{e^{x/a}}{\cosh(x/a)}. \tag{14.190}$$

The semi-axis $r \in (0, \infty)$ is mapped into the entire x-axis.

To find the effective potential we calculate the derivatives

$$h''(x) = \frac{1}{a}\frac{1}{\cosh^2(x/a)} = \frac{1}{a}\frac{e^{x/a}}{\cosh(x/a)}[1 - \tanh(x/a)],$$

$$h'''(x) = -\frac{2}{a^2}\frac{\sinh x}{\cosh^3 x} = -\frac{2}{a^2}\frac{e^{x/a}}{\cosh(x/a)}[\tanh(x/a) - \tan^2(x/a)], \tag{14.191}$$

and obtain

$$\frac{h''}{h'} = \frac{1}{a}\frac{e^{-x/a}}{\cosh(x/a)} = \frac{1}{a}[1 - \tanh(x/a)], \tag{14.192}$$

$$\frac{h'''}{h'} = -\frac{2}{a^2}\frac{e^{-x/a}\sinh(x/a)}{\cosh^2(x/a)} = -\frac{2}{a^2}\tanh(x/a)[1 - \tanh(x/a)],$$

so that the effective potential becomes

$$V_{\text{eff}} = \frac{\hbar^2}{8Ma^2}\left[2 - 2\tanh(x/a) - \frac{4}{\cosh^2(x/a)}\right]. \tag{14.193}$$

After adding this to the time-transformed potential, the DK-transformed action is found to be

$$\mathcal{A}_\mathcal{H}^{\text{DK}} = \int_0^S ds \left\{ \frac{M}{2}x'^2 - \left(g + E_\mathcal{H} - \frac{\hbar^2}{2Ma^2}\right)\frac{1}{\cosh^2(x/a)} \right.$$
$$\left. + \left(2E_\mathcal{H} + \frac{\hbar^2}{4Ma^2}\right)\tanh(x/a) + \left(2E_\mathcal{H} - \frac{\hbar^2}{4Ma^2}\right) \right\}. \tag{14.194}$$

This is the action governing the fixed-energy amplitude of the general Rosen-Morse potential (14.176)

$$V_{\mathcal{RM}'}(x/a) = \frac{\hbar^2}{2\mu}\left[-\frac{s(s+1)}{\cosh^2(x/a)} + c\tanh(x/a)\right]. \tag{14.195}$$

Since this potential is smooth, there exists a time-sliced path integral of the Feynman type. The relation between the fixed-energy amplitudes is

$$(r_b|r_a)_{E_\mathcal{H}} = e^{(x_b+x_a)/2a}\left[\cosh(x_b/a)\cosh(x_a/a)\right]^{-1/2}(x_b|x_a)_{E_{\mathcal{RM}'}}, \tag{14.196}$$

with $r/a = \log(e^{2x/a}+1) \in (0,\infty)$, $x \in (-\infty,\infty)$. The amplitude on the right-hand side is known from the last section; it is related to the amplitude for the motion of a mass point on the surface of a sphere in four dimensions, projected into a state of fixed azimuthal angular momenta m_1 and m_2. Only a simple rescaling of x/a to x is necessary to make the relation explicit.

In the literature, a solution of the time-sliced path integral with the action (14.184) has been attempted using a regulating function [11]

$$f = a^2(e^{r/a} - 1). \tag{14.197}$$

This implies going to the new variables

$$\frac{r}{a} = -2\log\cos(\theta/2), \tag{14.198}$$

so that

$$f = a^2\tan^2(\theta/2) = a^2\left[\frac{1}{\cos^2(\theta/2)} - 1\right]. \tag{14.199}$$

Note that this does not lead to a solution of the time-sliced path integral, since the transformed potential is still singular. Indeed, with $h' = a \tan(\theta/2)$, $h'' = a/[2\cos^2(\theta/2)]$, $h''' = a\sin(\theta/2)/[2\cos^3(\theta/2)]$, we would find the effective potential

$$V_{\text{eff}}(\theta) = \frac{\hbar^2}{8Ma^2}\frac{1}{\sin^2\theta}(1 + 2\cos\theta) = \frac{\hbar^2}{32Ma^2}\left[\frac{3}{\sin^2(\theta/2)} - \frac{1}{\cos^2(\theta/2)}\right], \quad (14.200)$$

and a transformed action

$$\tilde{\mathcal{A}}_{\mathcal{H}}^{\text{DK}} = \int_0^S (ds/a^2)\left\{\frac{Ma^4}{2}\theta'^2 - g + E_{\mathcal{H}}\left[\frac{1}{\cos^2(\theta/2)} - 1\right] + V_{\text{eff}}(\theta)\right\}, \quad (14.201)$$

which is of the general Pöschl-Teller type (14.167). Due to the presence of the $1/\cos^2(\theta/2)$-term, the Euclidean time evolution amplitude cannot be time-sliced. Only by starting from the particle near the surface of a sphere with the particular Bessel function regularization of (8.207), can a well-defined time-sliced amplitude be written down whose action looks like (14.201) in the continuum limit. It would be impossible, however, to invent this regularization when starting from the continuum action (14.201).

14.4.7 Extended Hulthén Potential and General Rosen-Morse Potential

The alert reader will have noticed that the regulating function (14.183) overkills the ga/r singularity of the Hulthén potential (14.183). In fact, we may add to the potential a term

$$\Delta V_{\mathcal{H}} = \frac{g'}{(e^{r/a} - 1)^2} \quad (14.202)$$

without loosing the stability of the path integral. In the limit $a \to \infty$, the extended potential contains the radial Coulomb system plus a centrifugal barrier, if we set $ga = -e^2 = \text{const}$ and $g'a^2 = \hbar^2 l(l + 1)/2M$. The potential (14.202) adds to the time-transformed action (14.186) a term

$$\Delta \mathcal{A}_{\mathcal{H}}^f = -\int_0^S ds\, g'4e^{-2r/a}, \quad (14.203)$$

which winds up in the final DK-transformed action as

$$\Delta \mathcal{A}_{\mathcal{H}}^{\text{DK}} = -\int_0^S ds\, g'\left[2 - 2\tanh^2(x/a) - \frac{1}{\cosh^2(x/a)}\right]. \quad (14.204)$$

Therefore, the extended Hulthén potential is again DK-equivalent to the general Rosen-Morse potential with the same relation (14.196) between the amplitudes, but with different relations between the constants.

14.5   *D*-Dimensional Systems

Let us now perform the path-dependent time transformation in D dimensions. The fixed-energy amplitude is given by the integral

$$(\mathbf{x}_b|\mathbf{x}_a)_E = \int_0^\infty dS \langle \mathbf{x}_b|\hat{\mathcal{U}}_E(S)|\mathbf{x}_a\rangle, \qquad (14.205)$$

with the pseudotime evolution amplitude

$$\langle \mathbf{x}_b|\hat{\mathcal{U}}_E(S)|\mathbf{x}_a\rangle = f_r(\mathbf{x}_b)f_l(\mathbf{x}_a)\langle \mathbf{x}_b| \exp\left[-\frac{i}{\hbar}Sf_l(\mathbf{x})(\hat{H} - E)f_r(\mathbf{x})\right]|\mathbf{x}_a\rangle. \quad (14.206)$$

It has the time-sliced path integral

$$\langle \mathbf{x}_b|\hat{\mathcal{U}}_E(S)|\mathbf{x}_a\rangle \approx \qquad (14.207)$$

$$\frac{f_r(\mathbf{x}_b)f_l(\mathbf{x}_a)}{\sqrt{2\pi i\epsilon_s\hbar f_l(\mathbf{x}_b)f_r(\mathbf{x}_a)/M}^D} \prod_{n=1}^{N}\left[\int \frac{dx_n}{\sqrt{2\pi i\epsilon_s\hbar f_n/M}}\right]\exp\left\{\frac{i}{\hbar}\mathcal{A}^N\right\},$$

with the action

$$\mathcal{A}^N = \sum_{n=1}^{N+1}\left\{\frac{M}{2\epsilon_s}\frac{(\Delta\mathbf{x}_n)^2}{f_l(\mathbf{x}_n)f_r(\mathbf{x}_{n-1})} + \epsilon_s[E - V(\mathbf{x}_n)]f_l(\mathbf{x}_n)f_r(\mathbf{x}_{n-1})\right\}, \quad (14.208)$$

where the integration measure contains the abbreviation $f_n \equiv f(r_n) = f_l(\mathbf{x}_n)f_r(\mathbf{x}_n)$. The time-transformed measure of path integration reads

$$\frac{f_r(\mathbf{x}_b)f_l(\mathbf{x}_a)}{\sqrt{2\pi i\epsilon_s\hbar f_l(\mathbf{x}_b)f_r(\mathbf{x}_a)/M}^D} \prod_{n=1}^{N}\int \frac{d^D x_n}{\sqrt{2\pi i\epsilon_s\hbar f_n/M}^D}. \qquad (14.209)$$

By shifting the product index and the subscripts of f_n by one unit, and by compensating for this with a prefactor, the integration measure in (14.27) acquires the postpoint form

$$\frac{f_r(\mathbf{x}_b)f_l(\mathbf{x}_a)}{\sqrt{2\pi i\epsilon_s\hbar f_l(\mathbf{x}_b)f_r(\mathbf{x}_a)/M}^D}\sqrt{\frac{f(\mathbf{x}_b)}{f(\mathbf{x}_a)}}^D \prod_{n=2}^{N+1}\int \frac{d^D\Delta x_n}{\sqrt{2\pi i\epsilon_s\hbar f_n/M}^D}. \qquad (14.210)$$

The integrals over each coordinate difference $\Delta\mathbf{x}_n = \mathbf{x}_n - \mathbf{x}_{n-1}$ are done at fixed postpoint positions $\mathbf{x}_n$.

To simplify the subsequent discussion, it is preferable to work only with the postpoint regularization in which $f_l(\mathbf{x}) = f(\mathbf{x})$ and $f_r(\mathbf{x}) \equiv 1$. Then the measure becomes simply

$$\frac{f(\mathbf{x}_a)}{\sqrt{2\pi i\epsilon_s f(\mathbf{x}_a)\hbar/M}^D} \prod_{n=2}^{N+1}\int \frac{d^D\Delta x_n}{\sqrt{2\pi i\epsilon_s\hbar f_n/M}^D}. \qquad (14.211)$$

We now introduce the coordinate transformation. In D dimensions it is given by

$$x^i = h^i(q). \tag{14.212}$$

The differential mapping may be written as in Chapter 10 as

$$dx^i = \partial_\mu h^i(q) = e^i{}_\mu(q) dq^\mu. \tag{14.213}$$

The transformation of a single time slice in the path integral can be done following the discussion in Sections 10.3 and 10.4. This leads to the path integral

$$(\mathbf{x}_b|\mathbf{x}_a)_E \approx \frac{f(q_a)}{\sqrt{2\pi i\epsilon_s f(q_a)\hbar/M}^D} \int_0^\infty dS \prod_{n=2}^{N+1} \left[\int \frac{d^D \Delta q_n g^{1/2}(q_n)}{\sqrt{2\pi i\epsilon_s \hbar f_n/M}^D} \right] e^{i\mathcal{A}_{\text{tot}}/\hbar}, \tag{14.214}$$

with the total time-sliced action

$$\mathcal{A}_{\text{tot}} = \sum_{n=1}^{N+1} \mathcal{A}_{\text{tot}}^\epsilon. \tag{14.215}$$

Each slice contains three terms

$$\mathcal{A}_{\text{tot}}^\epsilon = \mathcal{A}^\epsilon + \mathcal{A}_J^\epsilon + \mathcal{A}_{\text{pot}}^\epsilon. \tag{14.216}$$

In the postpoint form, the first two terms were given in (13.158) and (13.159). They are equal to

$$\mathcal{A}^\epsilon + \mathcal{A}_J^\epsilon = \frac{M}{2\epsilon f} g_{\mu\nu}(q) \Delta q^\mu \Delta q^\nu - i\frac{\hbar}{2}\Gamma_\mu{}^\mu{}_\nu \Delta q^\nu - \epsilon_s f \frac{\hbar^2}{8M}(\Gamma_\mu{}^\mu{}_\nu)^2. \tag{14.217}$$

The third term contains the effect of a potential and a vector potential as derived in (10.183). After the DK transformation, it reads

$$\mathcal{A}_{\text{pot}}^\epsilon = A_\mu \Delta q^\mu - i\epsilon_s f \frac{\hbar}{2M}(A_\nu \Gamma_\mu{}^{\mu\nu} + D_\mu A^\mu) - \epsilon_s f V(q). \tag{14.218}$$

14.6 Path Integral of the Dionium Atom

We now apply the generalized D-dimensional Duru-Kleinert transformation to the path integral of a dionium atom in three dimensions. This is a system of two particles with both electric and magnetic charges (e_1, g_1) and (e_2, g_2) [12]. Its Lagrangian for the relative motion reads

$$L = \frac{M}{2}\dot{\mathbf{x}}^2 + \mathbf{A}(\mathbf{x})\dot{\mathbf{x}} - V(\mathbf{x}), \tag{14.219}$$

where $\mathbf{x}$ is the distance vector pointing from the first to the second article, M the reduced mass, $V(\mathbf{x})$ a Coulomb potential

$$V(\mathbf{x}) = -\frac{e^2}{r}, \tag{14.220}$$

and $\mathbf{A}(\mathbf{x})$ the vector potential

$$\mathbf{A}(\mathbf{x}) = \hbar q \frac{\hat{\mathbf{z}} \times \mathbf{x}}{r} \left(\frac{1}{r-z} - \frac{1}{r+z} \right) = \hbar q \frac{(x\hat{\mathbf{y}} - y\hat{\mathbf{x}})z}{r(x^2 + y^2)}. \tag{14.221}$$

The coupling constants are $q \equiv -(e_1 g_2 - e_2 g_1)/\hbar c$ and $e^2 \equiv -(e_1 e_2 + g_1 g_2)$. The vector potential (14.221) implies an obvious generalization of the magnetic monopole interaction (8.299) with an electric charge [recall Appendix 10A.3] The potenial If we take the coupling as and $e^2 \equiv -e_1 e_2 - g_1 g_2$ in (14.221) we allow for the two particles to carry both electric and magnetic charges of the two particles, if we take for $V(\mathbf{x})$ the potential

$$V(\mathbf{x}) = -\frac{e^2}{r}. \tag{14.222}$$

The hydrogen atom is a special case of the dionium atom with $e_1 = -e_2 = e$ and $q = 0, l_0 = 0$. An electron around a pure magnetic monopole has $e_1 = e$, $g_2 = g$, $e_2 = g_1 = 0$.

In the vector potential (14.221) we have made use of the gauge freedom $\mathbf{A} \to \mathbf{A}(\mathbf{x}) + \nabla\Lambda(\mathbf{x})$ to enforce the transverse gauge $\nabla\mathbf{A}(\mathbf{x}) = 0$. In addition, we have taken advantage of the extra *monopole gauge invariance* which allows us to choose the shape of the Dirac string that imports the magnetic flux to the monopoles. The field $\mathbf{A}(\mathbf{x})$ in (14.221) has two strings of equal strength importing the flux, one along the positive x^3-axis from minus infinity to the origin, the other along the negative x^3-axis from plus infinity to the origin. It is the average of the vector potentials (10A.59) and (10A.60).

For the sake of generality, we shall assume the potential $V(\mathbf{x})$ to contain an extra $1/r^2$-potential:

$$V(\mathbf{x}) = -\frac{e^2}{r} + \frac{\hbar^2 l_0^2}{2Mr^2}. \tag{14.223}$$

The extra potential is parametrized as a centrifugal barrier with an effective angular momentum $\hbar l_0$.

At the formal level, i.e., without worrying about path collapse and time slicing corrections, the amplitude has been derived in Ref. [13]. Here we reproduce the derivation and demonstrate, in addition, that the time slicing produces no corrections.

14.6.1   Formal Solution

We extend the action of the type (14.11) by a dummy fourth coordinate as in the Coulomb system and go over to $\vec{u}$-coordinates depending on the radial coordinate $u = \sqrt{r}$ and the Euler angles θ, φ, γ as given in Eq. (13.97). Then the action reads

$$\mathcal{A} = \int dt \left\{ \frac{M}{2} 4u^2 \dot{u}^2 + \frac{M}{2} u^4 \left[\dot{\theta}^2 + \dot{\varphi}^2 + \dot{\gamma}^2 + 2 \left(\dot{\gamma} + \frac{\hbar q}{Mu^4} \right) \dot{\varphi} \cos\theta \right] - \frac{e^2}{u^2} - \frac{\hbar^2 l_0^2}{2Mu^4} + E \right\}. \tag{14.224}$$

By performing the Duru-Kleinert time reparametrization $dt = ds\, r(s)$ and changing the mass to $\mu = 4M$, the action takes the form

$$\mathcal{A}^{\mathrm{DK}} = \int_0^S ds\, \frac{\mu}{2}\left\{ u'^2 + \frac{u^2}{4}\left[\theta'^2 + \varphi'^2 + \gamma'^2 + 2\left(\gamma' + \frac{4\hbar q}{\mu u^2}\right)\varphi' \cos\theta\right] - \frac{4\hbar^2 l_0^2}{2\mu u^2} + Eu^2\right\}.$$

(14.225)

This can be rewritten in a canonical form

$$\mathcal{A} = \int_0^S ds(p_u u' + p_\theta \theta + p_\varphi \varphi' + p_\gamma \gamma' - H),$$

(14.226)

with the Hamiltonian

$$H = \frac{1}{2\mu}\left\{p_u^2 + \frac{4}{u^2}\left[p_\theta^2 + \frac{1}{\sin^2\theta}\left(p_\varphi^2 + (p_\gamma + \hbar q)^2 - 2(p_\gamma + \hbar q)p_\varphi \cos\theta\right)\right]\right\}$$
$$+ \frac{4}{2\mu u^2}\left[-2\hbar q p_\gamma + \hbar^2(l_0^2 - q^2)\right].$$

(14.227)

In the canonical path integral, the momenta are dummy integration variables so that we can replace $p_\gamma + \hbar q$ by p_γ. Then the action becomes

$$\mathcal{A} = \int_0^S ds[p_u u' + p_\theta \theta + p_\varphi \varphi' + (p_\gamma - \hbar q)\gamma' - \bar{H}],$$

(14.228)

with the Hamiltonian

$$\bar{H} = \frac{1}{2\mu}\left\{p_u^2 + \frac{4}{u^2}\left[p_\theta^2 + \frac{1}{\sin^2\theta}\left(p_\varphi^2 + p_\gamma^2 - 2p_\gamma p_\varphi \cos\theta\right)\right]\right\}$$
$$+ \frac{4}{2\mu u^2}\left[-2\hbar q(p_\gamma - \hbar q) + \hbar^2(l_0^2 - q^2)\right].$$

(14.229)

This differs from the pure Coulomb case in three ways:

First, the Hamiltonian has an extra centrifugal barrier proportional to the charge parameter $4q$:

$$V(r) = \frac{-8\hbar q(p_\gamma - \hbar q)}{2\mu u^2}.$$

(14.230)

Second, there is an extra centrifugal barrier

$$V(r) = \frac{\hbar^2 l_{\mathrm{extra}}^2}{2\mu u^2},$$

(14.231)

whose effective quantum number of angular momentum is given by

$$l_{\mathrm{extra}}^2 \equiv 4(l_0^2 - q^2).$$

(14.232)

Third, the action (14.228) contains an additional term

$$\Delta \mathcal{A} = -\hbar q \int_0^S ds \gamma'. \tag{14.233}$$

Fortunately, this is a pure surface term

$$\Delta \mathcal{A} = -\hbar q (\gamma_b - \gamma_a). \tag{14.234}$$

In the case $q^2 = l_0^2$, the extra centrifugal barrier vanishes, making it straightforward to write down the fixed-energy amplitude $(\mathbf{x}_b|\mathbf{x}_a)_E$ of the system. It is given by a simple modification of the relation (13.122) that expresses the fixed-energy amplitude of the Coulomb system $(\mathbf{x}_b|\mathbf{x}_a)_E$ in terms of the four-dimensional harmonic oscillator amplitude $(\vec{u}_b S|\vec{u}_a 0)$. Due to (14.233), the modification consists of a simple extra phase factor $e^{-iq(\gamma_b - \gamma_a)}$ in the integral over γ_a so that

$$(\mathbf{x}_b|\mathbf{x}_a)_E = \int_0^\infty dS e^{ie^2 S/\hbar} \frac{1}{16} \int_0^{4\pi} d\gamma_a e^{-iq(\gamma_a - \gamma_b)} (\vec{u}_b S|\vec{u}_a 0). \tag{14.235}$$

The integral over γ_a forces the momentum p_γ in the canonical action (14.228) to take the value $\hbar q$. This eliminates the term proportional to $p_\gamma - \hbar$ in (14.229).

In the general case $l_0 \neq q$, the amplitude becomes

$$(\mathbf{x}_b|\mathbf{x}_a)_E = \int_0^\infty dS e^{ie^2 S/\hbar} \frac{1}{16} \int_0^{4\pi} d\gamma_a e^{-iq(\gamma_a - \gamma_b)} (\vec{u}_b S|\vec{u}_a 0)_{l_{\text{extra}}}, \tag{14.236}$$

where the subscript l_{extra} indicates the presence of the extra centrifugal barrier potential in the harmonic oscillator amplitude. This amplitude was given for any dimension D in Eqs. (8.132) with (8.143). In the present case of $D = 4$, it has the partial-wave expansion [compare (8.161)]

$$(\vec{u}_b S|\vec{u}_a 0)_{l_{\text{extra}}} = \frac{1}{(u_b u_a)^{3/2}} \sum_{l_O=0}^\infty (u_b S|u_a 0)_{\tilde{l}_O} \frac{l_O + 1}{2\pi^2} \tag{14.237}$$

$$\times \sum_{m_1, m_2 = -l_O/2}^{l_O/2} d_{m_1 m_2}^{l_O/2}(\theta_b) d_{m_1 m_2}^{l_O/2}(\theta_a) e^{im_1(\varphi_b - \varphi_a) + im_2(\gamma_b - \gamma_a)},$$

with the radial amplitude

$$(u_b S|u_a 0)_{\tilde{l}_O} = \frac{M_O}{i\hbar} \frac{\omega \sqrt{u_b u_a}}{\sin \omega S} e^{i(M_O \omega / 2\hbar)(u_b^2 + u_a^2) \cot \omega S} I_{\tilde{l}_O + 1}\left(\frac{M_O \omega u_b u_a}{i\hbar \sin \omega S}\right). \tag{14.238}$$

This differs from the pure oscillator amplitude [compare (8.141) for $D = 4$] by having the index $l_O + 1$ of the Bessel function replaced by the square root of the "shifted square" as in (8.145):

$$\tilde{l}_O + 1 \equiv \sqrt{(l_O + 1)^2 + l_{\text{extra}}^2} = \sqrt{(l_O + 1)^2 + 4(l_0^2 - q^2)} = 2\sqrt{(j_D + 1/2)^2 + l_0^2 - q^2}. \tag{14.239}$$

The expansion (14.237) is inserted into (14.236) with the variables u_b, u_a replaced by $\sqrt{r_b}, \sqrt{r_a}$. Just as in the Coulomb case in (14.101) and (14.102), the integral $\int_0^{4\pi} d\gamma_a e^{-iq(\gamma_b - \gamma_a)}$ over the sum of angular wave functions

$$\frac{l_O + 1}{2\pi^2} \sum_{m_1, m_2 = -l/2}^{l_O/2} d_{m_1 m_2}^{l_O/2}(\theta_b) d_{m_1 m_2}^{l_O/2}(\theta_a) e^{im_1(\varphi_b - \varphi_a) + im_2(\gamma_b - \gamma_a)} \qquad (14.240)$$

can immediately be done, resulting in

$$8 \sum_m^{l_O/2} Y_{m,q}^{l_O/2}(\theta_b, \varphi_b) Y_{m,q}^{l_O/2*}(\theta_a, \varphi_a), \qquad (14.241)$$

where $Y_{m,q}^{l_O/2}(\theta, \varphi)$ are the monopole spherical harmonics (8.277). They coincide with the wave functions of a spinning symmetric top which possesses a spin q along the body axis. Physically, this spin is caused by the field's momentum density $\boldsymbol{\pi} = (\mathbf{E} \times \mathbf{B})/4\pi c$ encircling the radial distance vector $\mathbf{x}$. The *Poynting* vector yielding the energy density is $\mathbf{S} = \mathbf{E} \times \mathbf{B}/4\pi$. If a magnetically charged particle lies at the origin and electrically charged particle orbits around it at $\mathbf{x}$, the total angular momentum carried by the fields is [14]

$$\mathbf{J} = \int d^3x' \, \mathbf{x}' \times \boldsymbol{\pi}(\mathbf{x}') = \frac{1}{4\pi c} \int d^3x' \, \mathbf{x}' \times \left[\frac{g\,\mathbf{x}'}{|\mathbf{x}'|^3} \times \frac{e(\mathbf{x}' - \mathbf{x})}{|\mathbf{x}' - \mathbf{x}|^3} \right] = \frac{eg}{c} \hat{\mathbf{x}}. \qquad (14.242)$$

The quantization of the angular momentum

$$\frac{eg}{c} = n\frac{\hbar}{2}, \qquad n = \text{integer} \qquad (14.243)$$

is Dirac's famous charge quantization condition [16] [see also Eq. (8.303)].

Thus we arrive at the fixed-energy amplitude of the dionium atom, labeled by the subscript $\mathcal{D}$,

$$(\mathbf{x_b}|\mathbf{x_a})_{E_{\mathcal{D}}} = \frac{1}{r_b r_a} \sum_{j_{\mathcal{D}}} (r_b|r_a)_{E_{\mathcal{D}}, j_{\mathcal{D}}} \sum_{m=-j_{\mathcal{D}}}^{j_{\mathcal{D}}} Y_{m,q}^{j_{\mathcal{D}}}(\theta_b, \varphi_b) Y_{m,q}^{j_{\mathcal{D}}*}(\theta_a, \varphi_a), \qquad (14.244)$$

where the sum over $j_{\mathcal{D}} = l_O/2$ runs over integer or half-integer values depending on q, and with the radial amplitude given by the pseudotime integral over the radial oscillator amplitude of mass $M_O = 4M_C$ and frequency $\omega = \sqrt{-E/2M_C}$ [recall (13.118)]:

$$(r_b|r_a)_{E_{\mathcal{D}}, j_{\mathcal{D}}} = \frac{1}{2} \int_0^\infty dS e^{ie^2 S/\hbar} \frac{M_O \omega \sqrt{r_b r_a}}{i\hbar \sin \omega S} I_{\tilde{l}_O + 1}\left(\frac{M_O \omega \sqrt{r_b r_a}}{i\hbar \sin \omega S} \right)$$
$$\times \exp\left[\frac{iM_O \omega}{2\hbar}(r_b + r_a) \cot \omega S \right], \qquad (14.245)$$

where $\kappa = \sqrt{-2ME_{\mathcal{D}}/\hbar^2}$ and $\nu = \sqrt{-e^4 M_{\mathcal{D}}/2\hbar^2 E_{\mathcal{D}}}$ as in (13.39) and (13.40). Note that the dionium atom can be a fermion, even if the constituent particles are both bosons (or both fermions).

After the variable changes $e^2/\hbar = 2\omega\nu$, $\omega S = -iy$, we do the S-integral as in (9.29) and find for $r_b > r_a$ the radial amplitude of the dionium atom

$$(r_b|r_a)_{E_D, \tilde{j}_D} = -i\frac{M_D}{\hbar\kappa}\frac{\Gamma(-\nu + \tilde{j}_D + 1)}{(2\tilde{j}_D + 1)!}W_{\nu, \tilde{j}_D + 1/2}(2\kappa r_b)M_{\nu, \tilde{j}_D + 1/2}(2\kappa r_a), \quad (14.246)$$

where $\tilde{j}_D = \tilde{l}_O/2 = \sqrt{(j_D + \frac{1}{2})^2 + l_0^2 - q^2} - \frac{1}{2}$. For $q = 0$ and $l_0 = 0$, thus reduces properly to the three-dimensional Coulomb amplitude (14.103).

The energy eigenvalues are obtained from the poles of the Gamma function at

$$\nu = \nu_n \equiv \tilde{j}_D + n, \quad n = 1, 2, 3, \dots, \quad (14.247)$$

which yield

$$E_n = -Mc^2\alpha^2\frac{1}{2\left[n - \frac{1}{2} + \sqrt{(j_D + \frac{1}{2})^2 + l_0^2 - q^2}\right]^2}. \quad (14.248)$$

From the residues at the poles and the discontinuity across the cut at $E > 0$ in (14.246), we can extract the bound and continuum radial wave functions by the same method as in Section 13.8 from Eqs. (13.211)–(13.223).

14.6.2 Absence of Time Slicing Corrections

Let us now show that the above formal manipulations receive no correction in a proper time-sliced treatment [15]. Due to the presence of centrifugal and angular barriers, a path collapse can be avoided only after an appropriate regularization of both singularities. This is achieved by the path-dependent time transformation $dt = ds\, f(\mathbf{x}(s))$ with the postpoint regulating functions

$$f_l(\mathbf{x}) = f(\mathbf{x}) = r^2\sin^2\theta, \quad f_r(\mathbf{x}) \equiv 1. \quad (14.249)$$

After the extension of the path integral by an extra dummy dimension x^4, the time-sliced time-transformed fixed-energy amplitude to be studied is [see (14.205)–(14.211)]

$$(\mathbf{x}_b|\mathbf{x}_a)_E \approx \int dx_a^4\frac{1}{r_a^2\sin^2\theta_a}\int_0^\infty dS\frac{1}{(2\pi i\hbar\epsilon_s/M)^2}$$
$$\times \prod_{n=2}^{N+1}\left[\int\frac{d^4\Delta x_n}{(2\pi i\hbar\epsilon_s/M)^2 r_n^4\sin^4\theta_n}\right]e^{i\mathcal{A}^N/\hbar}, \quad (14.250)$$

with the sliced postpoint action

$$\mathcal{A}^N = \sum_{n=1}^{N+1}\left\{\frac{M}{2\epsilon_s}\frac{(\Delta x_n^i)^2}{r_n^2\sin^2\theta_n} - \epsilon_s r_n^2\sin^2\theta_n\left[V(\mathbf{x}_n) - E\right] + A_i(\mathbf{x}_n)\Delta x^i - \epsilon_s\frac{\hbar}{2M}A_{i,i}(\mathbf{x}_n)\right\}. \quad (14.251)$$

On this action, we now perform the coordinate transformation in two steps. First we go through the nonholonomic Kustaanheimo-Stiefel transformation as in Section 13.4 and express the four-dimensional $\vec{u}$-space in terms of $r = |\mathbf{x}|$ and the Euler angles θ, φ, γ. Explicitly,

$$\begin{aligned}x^1 &= r\sin\theta\cos\varphi,\\ x^2 &= r\sin\theta\sin\varphi,\\ x^3 &= r\cos\theta,\\ dx^4 &= r\cos\theta\, d\varphi + r\, d\gamma. \quad (14.252)\end{aligned}$$

Only the last equation is nonholonomic. If $q^\mu = 1, 2, 3, 4$ denotes the components $r, \beta, \varphi, \gamma$, the transformation matrix reads

$$
e^i{}_\mu = \frac{\partial x^i}{\partial q^\mu} = \begin{pmatrix} \sin\theta\cos\varphi & r\cos\theta\cos\varphi & -r\sin\theta\sin\varphi & 0 \\ \sin\theta\sin\varphi & r\cos\theta\sin\varphi & r\sin\theta\cos\varphi & 0 \\ \cos\theta & -r\sin\theta & 0 & 0 \\ 0 & 0 & r\cos\theta & r \end{pmatrix}.
$$

It has the metric

$$
g_{\mu\nu} = e^i{}_\mu e_{i\nu} = \begin{pmatrix} 1 & 0 & 0 & 0 \\ 0 & r^2 & 0 & 0 \\ 0 & 0 & r^2 & r^2\cos\theta \\ 0 & 0 & r^2\cos\theta & r^2 \end{pmatrix}, \tag{14.253}
$$

with an inverse

$$
g^{\mu\nu} = \begin{pmatrix} 1 & 0 & 0 & 0 \\ 0 & 1/r^2 & 0 & 0 \\ 0 & 0 & 1/r^2\sin^2\theta & -\cos\theta/r^2\sin^2\theta \\ 0 & 0 & -\cos\theta/r^2\sin^2\theta & 1/r^2\sin^2\theta \end{pmatrix}. \tag{14.254}
$$

The regulating function $f(\mathbf{x}) = r^2\sin^2\theta$ removes the singularities in $g^{\mu\nu}$ and thus in the free part of the Hamiltonian $(1/2M)g^{\mu\nu}p_\mu p_\nu$; there is no more danger of path collapse in the Euclidean amplitude.

In a second step we go to new coordinates

$$
r = e^\xi, \quad \sin\theta = 1/\cosh\beta, \quad \cos\theta = -\tanh\beta, \tag{14.255}
$$

as in the treatment of the angular barriers for $D = 3$ and $D = 4$ in Section 14.4. With $q^\mu = 1, 2, 3, 4$ denoting the coordinates $\xi, \beta, \varphi, \gamma$, respectively, the combined transformation matrix reads

$$
e^i{}_\mu = \begin{pmatrix} e^\xi\cosh^{-1}\beta\cos\varphi & -e^\xi\frac{\sinh\beta}{\cosh^2\beta}\cos\varphi & -e^\xi\cosh^{-1}\beta\sin\varphi & 0 \\ e^\xi\cosh^{-1}\beta\sin\varphi & -e^\xi\frac{\sinh\beta}{\cosh^2\beta}\sin\varphi & e^\xi\cosh^{-1}\beta\cos\varphi & 0 \\ -e^\xi\tanh\beta & -e^\xi\cosh^{-2}\beta & 0 & 0 \\ 0 & 0 & -e^\xi\tanh\beta & e^\xi \end{pmatrix}, \tag{14.256}
$$

with the metric

$$
g_{\mu\nu} = \begin{pmatrix} e^{2\xi} & 0 & 0 & 0 \\ 0 & e^{2\xi}\cosh^{-2}\beta & 0 & 0 \\ 0 & 0 & e^{2\xi} & -e^{2\xi}\tanh\beta \\ 0 & 0 & -e^{2\xi}\tanh\beta & e^{2\xi} \end{pmatrix}, \tag{14.257}
$$

and the determinant

$$
g = e^{8\xi}/\cosh^4\beta. \tag{14.258}
$$

The inverse metric is completely regular:

$$
g^{\mu\nu} = \begin{pmatrix} e^{-2\xi} & 0 & 0 & 0 \\ 0 & e^{-2\xi}\cosh^2\beta & 0 & 0 \\ 0 & 0 & e^{-2\xi}\cosh^2\beta & e^{-2\xi}\sinh\beta\cosh\beta \\ 0 & 0 & e^{-2\xi}\sinh\beta\cosh\beta & e^{-2\xi}\cosh^2\beta \end{pmatrix}. \tag{14.259}
$$

We now calculate the transformed actions (14.217) and (14.218). The relevant quantities which could contain time slicing corrections are $D_\mu A^\mu$ and $\Gamma_\mu{}^{\mu\nu}$. The former quantity, being equal to

$\partial_i A_i$, vanishes in the transverse gauge under consideration. The calculation of $\Gamma_\mu{}^{\mu\nu}$ is somewhat tedious (see Appendix 14A) but yields a surprisingly simple result:

$$\Gamma_\mu{}^\mu{}_\nu = (-1, 0, 0, 0). \tag{14.260}$$

Because of this simplicity, the transformed sliced action is easily written down. It is split into two parts,

$$\mathcal{A}^\epsilon_{\text{tot}} = \mathcal{A}^\epsilon_{\varphi\beta} + \mathcal{A}^\epsilon_{\varphi\gamma}, \tag{14.261}$$

one containing only the coordinates ξ, β,

$$\mathcal{A}^\epsilon_{\xi\beta} = \frac{M}{2\epsilon_s}[(\Delta\xi)_n^2 \cosh^2\beta_n + (\Delta\beta_n)^2] + \frac{i\hbar}{2}\Delta\xi_n \tag{14.262}$$
$$-\epsilon_s\left[-\frac{e^2 e^{\xi_n}}{\cosh^2\beta_n} + \frac{\hbar^2(l_{\text{extra}}^2 + 1/4)}{2M\cosh^2\beta_n} - \frac{E e^{2\xi_n}}{\cosh^2\beta_n}\right],$$

the other dealing predominantly with φ, γ,

$$\mathcal{A}^\epsilon_{\varphi\gamma} = \frac{M\cosh^2\beta_n}{2\epsilon_s}[(\Delta\varphi_n)^2 + (\Delta\gamma_n)^2 - 2\Delta\gamma_n\Delta\varphi_n\tanh\beta_n]$$
$$+\hbar q\tanh\beta_n\Delta\varphi_n - \epsilon_s\frac{\hbar^2 q^2}{2M\cosh^2\beta_n}. \tag{14.263}$$

Hence the fixed-energy amplitude becomes

$$(\mathbf{x}_b|\mathbf{x}_a)_E \approx \int_0^\infty dS \int_0^{4\pi} d\gamma_a \frac{1}{2\pi i\hbar\epsilon_s r_a^2/M\cosh\beta_a} \prod_{n=1}^N\left[\int \frac{dr_n d\beta_n}{2\pi i\hbar\epsilon_s/M\cosh^2\beta_{n+1}}\right]$$
$$\times \exp\left(\frac{i}{\hbar}\sum_{n=1}^{N+1}\mathcal{A}^\epsilon_{\xi\beta}\right)(\varphi_b\,\gamma_b\,S|\varphi_a\,\gamma_a\,0)_{[\beta]}. \tag{14.264}$$

The last factor is a pseudotime evolution amplitude in the angles φ, γ which is still a functional of $\beta(t)$, as indicated by the subscript $[\beta]$,

$$(\varphi_b\,\gamma_b\,S|\varphi_a\,\gamma_a\,0)_{[\beta]} \approx \sum_{\varphi_{N+1}=\varphi_b+2\pi l_b^\varphi}\sum_{\gamma_{N+1}=\gamma_b+4\pi l_b^\gamma} \tag{14.265}$$
$$\times \frac{1}{2\pi i\hbar\epsilon_s/M\cosh\beta_b}\prod_{n=1}^N\left[\int \frac{d\varphi_n d\gamma_n}{2\pi i\hbar\epsilon_s/M\cosh\beta_n}\right]\exp\left(\frac{i}{\hbar}\sum_{n=1}^{N+1}\mathcal{A}^\epsilon_{\varphi\gamma}\right).$$

The sums over l_b^φ, l_b^γ account for the cyclic properties of the angles φ and γ with the periods 2π and 4π, respectively, at a fixed coordinate $\mathbf{x}_b$ (as in the examples in Section 6.1).

We now introduce auxiliary momentum variables p_n^φ, p_n^γ and go over to the canonical form of the amplitude (14.265):

$$(\varphi_b\gamma_b S|\varphi_a\gamma_a 0)_{[\beta]} \approx \prod_{n=1}^N\left[\int d\varphi_n\right]\prod_{n=1}^{N+1}\left[\int \frac{dp_n^\varphi}{2\pi\hbar}\right]\prod_{n=1}^N\left[\int d\gamma_n\right]\prod_{n=1}^{N+1}\left[\int \frac{dp_n^\gamma}{2\pi\hbar}\right]$$
$$\times \exp\left[\frac{i}{\hbar}\sum_{n=1}^{N+1}\left(p_n^\varphi\Delta\varphi_n + p_n^\gamma\Delta\gamma_n\right.\right. \tag{14.266}$$
$$\left.\left.-\frac{\epsilon_s}{2M}[(p_n^\varphi)^2 + (p_n^\gamma + \hbar q)^2 + 2p_n^\varphi(p_n^\gamma + \hbar q)\tanh\beta_n] + \epsilon_s\frac{p_n^\gamma\hbar q}{M\cosh^2\beta_n}\right)\right].$$

The momenta p_n^γ are dummy integration variables and can be replaced by $p_n^\gamma - \hbar q$. The $d\varphi_n, d\gamma_n$-integrals run over the full extended zone schemes $\varphi_n, \gamma_n \in (-\infty, \infty)$ and enforce the equality of all p_n^γ. At the end, only the integrals over a common single momentum p^φ, p^γ remain and we arrive at

$$(\varphi_b \gamma_b \, S | \varphi_a \gamma_a \, 0)_{[\beta]} \tag{14.267}$$

$$\approx e^{-iq(\gamma_b - \gamma_a)} \sum_{l_b^\varphi = -\infty}^{\infty} \sum_{l_b^\gamma = -\infty}^{\infty} \int \frac{dp_\varphi}{2\pi\hbar} \int \frac{dp_\gamma}{2\pi\hbar} e^{ip_\varphi (\varphi_b + 2\pi l_b^\varphi - \varphi_a)/\hbar} e^{ip_\gamma (\gamma_b + 4\pi l_b^\gamma - \gamma_a)/\hbar}$$

$$\times \exp \left\{ -\frac{i}{\hbar} \sum_{n=1}^{N+1} \left[\frac{1}{2M\epsilon_s} (p_\varphi^2 + p_\gamma^2 + 2p_\varphi p_\gamma \tanh \beta_n) - \epsilon_s \frac{(p_\gamma - \hbar q)\hbar q}{M \cosh^2 \beta_n} \right] \right\}.$$

We can now do the sums over l_b^φ, l_b^γ which force the momenta p_φ to integer values and p_γ to half-integer values by Poisson's formula, so that

$$(\varphi_b \gamma_b \, S | \varphi_a \gamma_a \, 0)_{[\beta]} = e^{-iq(\gamma_b - \gamma_a)} \sum_{m_1, m_2} \frac{1}{2\pi} e^{im_1 (\varphi_b - \varphi_a)} \frac{1}{4\pi} e^{im_2 (\gamma_b - \gamma_a)} \tag{14.268}$$

$$\times \exp \left\{ -\frac{i}{\hbar} \sum_{n=1}^{N+1} \left[\frac{\hbar^2}{2M\epsilon_s} (m_1^2 + m_2^2 + 2m_1 m_2 \tanh \beta_n) - \frac{\epsilon_s \hbar^2 (m_2 - q)q}{M \cosh^2 \beta_n} \right] \right\}.$$

With this, the expression for the fixed-energy amplitude (14.264) of the dionium atom contains the magnetic charge only at three places: the extra centrifugal barrier in $\mathcal{A}_{\xi\beta}^\epsilon$, the phase factor of the remaining integral over γ_a, and the last term in (14.268). This last term, however, can be dropped since the integral over γ_a forces the half-integer number m_2 to become equal to $\hbar q$. The γ_b-integral over the remaining functional of $\beta(t)$ gives, incidentally,

$$\int_0^{4\pi} d\gamma_a (\varphi_b \gamma_b \, S | \varphi_a \gamma_a \, 0)_{[\beta]} = \sum_{m_1} \frac{1}{2\pi} e^{im_1 (\varphi_b - \varphi_a)}$$

$$\times \exp \left\{ -\frac{i}{\hbar} \sum_{n=1}^{N+1} \left[\frac{\hbar^2}{2M\epsilon_s} [m_1^2 + q^2 + 2m_1 q \tanh \beta_n] \right] \right\}. \tag{14.269}$$

The time-sliced expression has the parameter q at precisely the same places as the previous formal one. This proves that formula (14.236) with (14.237) is unchanged by time slicing corrections, thus completing the solution of the path integral of the dionium atom.

Note that after inserting (14.269), the time-sliced path integral (14.264) is a combination of a general Rosen-Morse system in β and a Morse system in ξ.

Let us end this discussion by the remark that like the Coulomb system, the dionium atom can be treated in a purely group-theoretic way, using only operations within the dynamical group O(4,2). This is explained in Appendix 14B.

14.7 Time-Dependent Duru-Kleinert Transformation

By generalizing the above transformation method to time-dependent regulating functions, we can derive further relations between amplitudes of different physical systems. In the path-dependent time transformation $dt = ds \, f_l(\mathbf{x}) f_r(\mathbf{x})$ regularizing the path integrals, we may allow for functions $f_l(\mathbf{p}, \mathbf{x}, t)$ and $f_r(\mathbf{p}, \mathbf{x}, t)$ depending on positions, momenta, and time. Such functions complicate the subsequent transformation to new coordinates $\mathbf{q}$, in which the kinetic term of the amplitude (12.47) with respect to the pseudotime s has the standard form $(M/2)\mathbf{q}'^2(s)$. In particular, a momentum dependence of f_l and f_r leads to involved formulas, which is the reason why this case has not yet been investigated (just like the even more general case where the right-hand side of the transformation $dt = ds\, f$ contains terms proportional to dx). If one restricts the transformation to depend only on time and uses the special splitting with the regulating functions $f_l = f$ and $f_r = 1$, or $f_l = 1$ and $f_r = f$, the result in one spatial dimension is relatively simple. On the basis of Section 12.3, the following relation is found [instead of (14.26)] between the time evolution amplitude of an initial system and a fixed-energy amplitude of the transformed systems at $\mathcal{E} = 0$:

$$(x_b t_b | x_a t_a) = g(q_b, t_b) g(q_a, t_a) \{q_b t_b | q_a t_a\}_{\mathcal{E}=0}, \qquad (14.270)$$

where $\{q_b t_b | q_a t_a\}_{\mathcal{E}=0}$ denotes the spacetime extension of the fixed-pseudoenergy amplitude. It is calculated by time-slicing the expression

$$\int_0^\infty dS \{x_b t_b | \hat{\mathcal{U}}_E(S) | x_a t_a\} \qquad (14.271)$$

on the right-hand side of (12.55), transforming the coordinates x to q, and adapting the normalization to the completeness relation of the states

$$\int dx \int dt |q\, t\} \{q\, t| = 1. \qquad (14.272)$$

This leads to the path integral representation

$$\{q_b t_b | q_a t_a\}_{\mathcal{E}} = \int_0^\infty dS \int dE\, e^{-iE(t_b - t_a)/\hbar} \int \mathcal{D}q(s) e^{i\mathcal{A}_{E,\mathcal{E}}^{\mathrm{DK}}/\hbar}, \qquad (14.273)$$

with the DK-transformed action

$$\mathcal{A}_{E,\mathcal{E}}^{\mathrm{DK}} = \int_0^S ds \Big\{ \frac{M}{2} q'^2(s) - f(q(s), t(s)) [V(q(s), t) - E] + \mathcal{E}$$
$$- V_{\mathrm{eff}}(q(s), t(s)) - \Delta V_{\mathrm{eff}}(q(s), t(s)) \Big\}. \qquad (14.274)$$

Note that the initial potential may depend explicitly on time. The function $t(s)$ is now given by the time-dependent differential equation

$$\frac{dt}{ds} = f(x, t). \qquad (14.275)$$

The coordinate transformation also depends on time,

$$x = h(q, t), \qquad (14.276)$$

and satisfies the equation

$$h'^2(q, t) = f(h(q, t), t), \qquad (14.277)$$

where $h'(q, t) \equiv \partial_q h(q, t)$ [compare (14.15)]. The function $f(q(s), t(s))$ used in (14.274) is an abbreviation for $f(h(q, t), t)$ evaluated at the time $t = t(s)$.

In addition to the effective potential V_{eff} determined by Eq. (14.18), there is now a further contribution which is due to the time dependence of $h(q,t)$ [17]:

$$\Delta V_{\text{eff}} = Mh'^2 \int dq\, h'\dot{h} \mp i\hbar\dot{h}'h'. \tag{14.278}$$

The upper sign must be used if the relation between t and s is calculated from the time-sliced postpoint recursion relation

$$t_{n+1} - t_n = \epsilon_s f(q_{n+1}, t_{n+1}). \tag{14.279}$$

The lower sign holds when solving the prepoint relation

$$t_{n+1} - t_n = \epsilon_s f(q_n, t_n). \tag{14.280}$$

Note that the first term of ΔV_{eff} contributes even at the classical level. If a function $h(q,t)$ is found satisfying Newton's equation of motion

$$M\ddot{h} = -\frac{\partial V(h,t)}{\partial h}, \tag{14.281}$$

with $V(h) \equiv V(x)|_{x=h}$, then the first term eliminates the potential in the action (14.274), and the transformed system is classically free. This happens if the new coordinate $q(t)$ associated with x,t is identified with the initial value, at some time t_0, of the classical orbit running through x,t. These are trivially time-independent and therefore behave like the coordinates of a free particle (see the subsequent example).

The normalization factor $g(q,t)$ is determined by the differential equation

$$\frac{g'}{g} = \frac{1}{2}\frac{h''}{h'} + i\frac{M}{\hbar}h'\dot{h}. \tag{14.282}$$

The solution reads

$$g(q,t) = e^{i\Lambda(q,t)}\sqrt{h'(q,t)}, \tag{14.283}$$

with

$$\Lambda(q,t) = \pm\frac{M}{\hbar}\int^q dq\, h'\dot{h}. \tag{14.284}$$

Thus, in addition to the normalization factor in (14.26), the time-dependent DK relation (14.270) also contains a phase factor.

As an example [18], we transform the amplitude of a harmonic oscillator to that of a free particle. The classical orbits are given by $x(t) = x_0\cos\omega t$, so that the transformation $x(t) = h(q,t) = q\cos\omega t$ leads to a coordinate $q(t)$ which moves without acceleration. For brevity, we write $\cos\omega t$ as $c(t)$. Obviously, $f(q,t) = c^2(t)$ is a pure function of the time [19], and the differential relation between the time t and the pseudotime s is integrated to

$$S = \frac{1}{\omega}\frac{\sin\omega(t_b - t_a)}{c(t_b)c(t_a)}. \tag{14.285}$$

This equation can be solved for $t_a(S)$ at fixed t_b, or for $t_b(S)$ at fixed t_a. The solution $t_a(S)$ is obtained from the time-sliced postpoint recursion relation (14.279), while $t_b(S)$ arises from the prepoint recursion (14.280). The DK action (14.274) is simplified in these two cases to

$$\mathcal{A}_{E,\mathcal{E}}^{\mathrm{DK}} = \frac{M}{2}\frac{(q_b - q_a)^2}{S} \pm i\hbar \log \frac{c(t_b)}{c(t_a)} + E \left\{ \begin{array}{c} [t_b - t_a(S)] \\ {[t_b(S) - t_a]} \end{array} \right\} + \mathcal{E}S. \qquad (14.286)$$

The E-integration in (14.273) yields in the first case

$$\{q_b t_b | q_a t_a\}_{\mathcal{E}} = \int_0^{\infty} dS\, \delta(t_b - t_a(S)) \frac{c(t_a)}{c(t_b)} \frac{e^{(i/\hbar)M(q_b - q_a)^2/2S}}{\sqrt{2\pi\hbar iS/M}}, \qquad (14.287)$$

and the integration over S using $-dt_a(S)/dS = c^2(t_a)$ results in

$$\{q_b t_b | q_a t_a\}_{\mathcal{E}} = \frac{1}{c(t_b)c(t_a)} \frac{e^{(i/\hbar)M(q_b - q_a)^2/2S}}{\sqrt{2\pi\hbar iS/M}}. \qquad (14.288)$$

The same amplitude is obtained for the lower sign in (14.286) with $dt_b(S)/dS = c^2(t_b)$. After inserting this together with (18.671) and (14.284) into (14.270) [the integration there gives $\Lambda(q,t) = (M/\hbar)q^2 c(t)\dot{c}(t)/2$], we obtain

$$(x_b t_b | x_a t_a) = e^{(i/\hbar)M[q_b^2 c(t_b)\dot{c}(t_b) - q_a^2 c(t_a)\dot{c}(t_a)]/2} \frac{1}{\sqrt{c(t_b)c(t_a)}} \frac{e^{(i/\hbar)M(q_b - q_a)^2/2S}}{\sqrt{2\pi\hbar iS/M}}. \qquad (14.289)$$

Since q_b and q_a are equal to $x_b/c(t_b)$ and $x_a/c(t_a)$, respectively, a few trigonometric identities lead to the well-known expression (2.173) for the amplitude of the harmonic oscillator.

It is obvious that a combination of this transformation with a time-independent Duru-Kleinert transformation makes it possible to reduce also the path integral of the Coulomb system to that of a free particle.

It will be interesting to find out which hitherto unsolved path integrals can be integrated by means of such generalized DK transformations.

Appendix 14A Affine Connection of Dionium Atom

From the transformation matrices (14.256), we calculate the derivatives [setting $q^\mu = (\xi, \beta, \varphi, \gamma)$ and using the abbreviation $f_{,\xi} \equiv \partial_\xi f$]

$$e^i{}_{\mu,\xi} = e^i{}_\mu, \qquad (14A.1)$$

$$e^i{}_{\mu,\beta} = \begin{pmatrix} -e^\xi \frac{\sinh\beta}{\cosh^2\beta}\cos\phi & -e^\xi \frac{1-\sinh^2\beta}{\cosh^3\beta}\cos\phi & e^\xi \frac{\sinh\beta}{\cosh^2\beta}\sin\phi & 0 \\ -e^\xi \frac{\sinh\beta}{\cosh^2\beta}\sin\phi & -e^\xi \frac{1-\sinh^2\beta}{\cosh^3\beta}\sin\phi & -e^\xi \frac{\sinh\beta}{\cosh^2\beta}\cos\phi & 0 \\ -e^\xi \cosh^{-2}\beta & 2e^\xi \frac{\sinh\beta}{\cosh^3\beta} & 0 & 0 \\ 0 & 0 & -e^\xi \cosh^{-2}\beta & 0 \end{pmatrix}, \qquad (14A.2)$$

$$e^i{}_{\mu,\phi} = \begin{pmatrix} -e^\xi \cosh^{-1}\beta\sin\phi & e^\xi \frac{\sinh\beta}{\cosh^2\beta}\sin\phi & -e^\xi \cosh^{-1}\beta\cos\phi & 0 \\ e^\xi \cosh^{-1}\beta\cos\phi & -e^\xi \frac{\sinh\beta}{\cosh^2\beta}\cos\phi & -e^\xi \cosh^{-1}\beta\sin\phi & 0 \\ 0 & 0 & 0 & 0 \\ 0 & 0 & 0 & 0 \end{pmatrix}, \qquad (14A.3)$$

$$e^i{}_{\mu,\alpha} = 0. \tag{14A.4}$$

From these we find $\Gamma_{\mu\nu\lambda} = e^i{}_\lambda e^i{}_{\nu,\mu}$ by contraction with $e^i{}_\lambda$:

$$\Gamma_{\xi\mu\nu} = g_{\mu\nu}, \tag{14A.5}$$

$$\Gamma_{\beta\mu\nu} = \begin{pmatrix} 0 & e^{2\xi}\cosh^{-2}\beta & 0 & 0 \\ -e^{2\xi}\cosh^{-2} & -e^{2\xi}\frac{\sinh\beta}{\cosh^3\beta} & 0 & 0 \\ 0 & 0 & 0 & -e^{2\xi}\cosh^{-2}\beta \\ 0 & 0 & 0 & 0 \end{pmatrix}, \tag{14A.6}$$

$$\Gamma_{\phi\mu\nu} = \begin{pmatrix} 0 & 0 & e^{2\xi}\cosh^{-2}\beta & 0 \\ 0 & 0 & -e^{2\xi}\frac{\sinh\beta}{\cosh^3\beta} & 0 \\ -e^{2\xi}\cosh^{-2} & e^{2\xi}\frac{\sinh\beta}{\cosh^3\beta} & 0 & 0 \\ 0 & 0 & 0 & 0 \end{pmatrix}, \tag{14A.7}$$

$$\Gamma_{\alpha\mu\nu} = 0. \tag{14A.8}$$

A contraction with the inverse metric $g^{\mu\nu}$ yields $\Gamma_\mu{}^\mu{}_\nu = (-1,0,0,0)$, as stated in Eq. (14.260).

Appendix 14B   Algebraic Aspects of Dionium States

In Appendix 13A we have shown that certain combinations of x^μ and ∂_μ operators satisfy the commutation rules of the Lie algebra of the group O(4,2) [see Eqs. (13A.11)]. This permits solving all dynamical problems via group operations. In the case $l_0 = q$ (i.e., $l_{\text{extra}} = 0$), the group-theoretic approach can be extended to include the dionium atom. In fact, it is easy to see [20] that the Lie algebra of O(4,2) remains the same if the generators L_{AB} $(A, B = 1, \ldots, 6)$ of Eq. (13A.11) are extended to $(x^i \equiv x_i)$

$$\begin{aligned}
\hat{L}_{ij} &= -\frac{i}{2}(x_i\partial_{x^j} - x_j\partial_{x^i}) + q\frac{r}{\mathbf{x}_\perp^2}x^k, \\
\hat{L}_{i4} &= \frac{1}{2}\left[-x^i\partial_{\mathbf{x}}^2 - x^i + 2\partial_{x^i}\mathbf{x}\partial_{\mathbf{x}} + 2iq\frac{r}{\mathbf{x}_\perp^2}(\mathbf{x}_\perp \times \boldsymbol{\nabla})_i - (-)^{\delta_{i3}}q^2\frac{x_i}{\mathbf{x}_\perp^2}\right], \\
\hat{L}_{i5} &= \frac{1}{2}\left[-x^i\partial_{\mathbf{x}}^2 + x^i + 2\partial_{x^i}\mathbf{x}\partial_{\mathbf{x}} + 2iq\frac{r}{\mathbf{x}_\perp^2}(\mathbf{x}_\perp \times \boldsymbol{\nabla})_i - (-)^{\delta_{i3}}q^2\frac{x_i}{\mathbf{x}_\perp^2}\right], \\
\hat{L}_{i6} &= -ir\partial_{x^i} - \frac{q}{\mathbf{x}_\perp^2}(\mathbf{x} \times \mathbf{x}_\perp)_i, \\
\hat{L}_{45} &= -i(\mathbf{x}\partial_{\mathbf{x}} + 1), \\
\hat{L}_{46} &= \frac{1}{2}\left[-r\partial_{\mathbf{x}}^2 - r + 2iq\frac{z}{\mathbf{x}_\perp^2}(\mathbf{x} \times \boldsymbol{\nabla})_3 + q^2\frac{r}{\mathbf{x}_\perp^2}\right], \\
\hat{L}_{56} &= \frac{1}{2}\left[-r\partial_{\mathbf{x}}^2 + r + 2iq\frac{z}{\mathbf{x}_\perp^2}(\mathbf{x} \times \boldsymbol{\nabla})_3 + q^2\frac{r}{\mathbf{x}_\perp^2}\right].
\end{aligned} \tag{14B.1}$$

The representation space is now characterized by the eigenvalue of the operator $\hat{L}_{05} = -ir\partial_{x^4} = -i\partial_\gamma = -\frac{1}{2}(\hat{a}^\dagger\hat{a} - \hat{b}^\dagger\hat{b})$ being equal to q. The wave functions are generalizations of the $q = 0$ -wave functions of the Coulomb system.

Notes and References

[1] The special case $\lambda = 1/2$ has also been treated by
   N.K. Pak and I. Sokmen, Phys. Rev. A **30**, 1692 (1984).

[2] R.E. Langer, Phys. Rev. **51**, 669 (1937);
W.H. Furry, Phys. Rev. **71**, 360 (1947);
P.M. Morse and H. Feshbach, *Methods of Theoretical Physics*, McGraw-Hill, 1953, pp. 1092ff.

[3] L.C. Hostler, J. Math. Phys. **11**, 2966 (1970).

[4] F. Steiner, Phys. Lett. A **106**, 256, 363 (1984).

[5] A. Sommerfeld, *Partial Differential Equations in Physics, Lectures in Theoretical Physics*,
Vol. 6, Academic, New York, 1949;
T. Regge, Nuovo Cimento **14**, 951 (1959);
F. Calogero, Nuovo Cimento **28**, 761 (1963);
A.O. Barut, *The Theory of the Scattering Matrix*, MacMillan, New York, 1967, p. 140;
P.D.B. Collins and E.J. Squires, *Regge Poles in Particle Physics*, Springer Tracts in Modern
Physics, Vol. 49, Springer, Berlin 1968.

[6] H. Kleinert and I. Mustapic, J. Math. Phys. **33**, 643 (1992) (http://www.physik.fu-
berlin.de/ kleinert/207).

[7] G. Pöschl and E. Teller, Z. Phys. **83**, 1439 (1933). See also S. Flügge, *Practical Quantum
Mechanics*, Springer, Berlin, 1974, p. 89.

[8] N. Rosen, P.M. Morse, Phys. Rev. **42**, 210 (1932); S. Flügge, *op. cit.*, p. 94. See also
L.D. Landau and E.M. Lifshitz, *Quantum Mechanics*, Pergamon, London, 1965, §23.

[9] This DK relation between these amplitudes was first given by
G. Junker and A. Inomata, in *Path Integrals From meV to MeV*, edited by M.C. Gutzwiller,
A. Inomata, J.R. Klauder, and L. Streit, World Scientific, Singapore, 1986, p. 315, and by
M. Böhm and G. Junker, J. Math. Phys. **28**, 1978 (1987).
Watch out for mistakes. For instance, in Eq. (3.28) of the first paper, the authors claim to
have calculated the fixed-energy amplitude, but give only its imaginary part restricted to
the bound-state poles.Their result (3.33) lacks the continuum states. Further errors in their
Section V have been pointed out in Footnote 20 of Chapter 8.

[10] For the explicit extraction of the wave functions see Ref. [6], Section IV B.

[11] J.M. Cai, P.Y. Cai, and A. Inomata, Phys. Rev. A **34**, 4621 (1986).

[12] J. Schwinger, *A Magnetic Model of Matter*, Science, **165**, 717 (1969).
See also the review article
K.A. Milton, *Theoretical and Experimental Status of Monopoles*, (hep-ex/0602040).

[13] H. Kleinert, Phys. Lett. A **116**, 201 (1989).

[14] J.J. Thomson, *On Momentum in the Electric field*, Philos. Mag. **8**, 331 (1904).

[15] This proof was done in collaboration with my undergraduate student J. Zaun.

[16] P.A.M. Dirac, Proc. Roy. Soc. A **133**, 60 (1931); Phys. Rev. **74**, 817 (1948), Phys. Rev. **74**,
817 (1948).

[17] A. Pelster and A. Wunderlin, Zeitschr. Phys. B **89**, 373 (1992).
See also the similar generalization of the DK transformation in stochastic differential equa-
tions by
S.N. Storchak, Phys. Lett. A **135**, 77 (1989).

[18] For other examples see
C. Grosche, Phys. Lett. A **182**, 28 (1952).

[19] This special case was treated by
P.Y. Cai, A. Inomata, and P. Wang, Phys. Lett. **91**, 331 (1982).
Note that the transformation to free particles is based on a general observation in
H. Kleinert, Phys. Lett. B **94**, 373 (1980) (http://www.physik.fu-berlin.de/~kleinert/71).

[20] A.O. Barut, C.K.E. Schneider, and R. Wilson, J. Math. Phys. **20**, 2244 (1979).

Thou com'st in such a questionable shape,
That I will speak to thee.
WILLIAM SHAKESPEARE, Hamlet

15

Path Integrals in Polymer Physics

The use of path integrals is not confined to the quantum-mechanical description point particles in spacetime. An important field of applications lies in polymer physics where they are an ideal tool for studying the statistical fluctuations of line-like physical objects.

15.1 Polymers and Ideal Random Chains

A polymer is a long chain of many identical molecules connected with each other at joints which allow for spatial rotations. A large class of polymers behaves approximately like an idealized *random chain*. This is defined as a chain of N links of a fixed length a, whose rotational angles occur all with equal probability (see Fig. 15.1). The probability distribution of the end-to-end distance vector $\mathbf{x}_b - \mathbf{x}_a$ of such an object is given by

$$P_N(\mathbf{x}_b - \mathbf{x}_a) = \prod_{n=1}^{N} \left[\int d^3 \Delta x_n \frac{1}{4\pi a^2} \delta(|\Delta \mathbf{x}_n| - a) \right] \delta^{(3)}(\mathbf{x}_b - \mathbf{x}_a - \sum_{n=1}^{N} \Delta \mathbf{x}_n). \quad (15.1)$$

The last δ-function makes sure that the vectors $\Delta \mathbf{x}_n$ of the chain elements add up correctly to the distance vector $\mathbf{x}_b - \mathbf{x}_a$. The δ-functions under the product enforce the fixed length of the chain elements. The length a is also called the *bond length* of the random chain.

The angular probabilities of the links are spherically symmetric. The factors $1/4\pi a^2$ ensure the proper normalization of the individual one-link probabilities

$$P_1(\Delta \mathbf{x}) = \frac{1}{4\pi a^2} \delta(|\Delta \mathbf{x}| - a) \quad (15.2)$$

in the integral

$$\int d^3 x_b \, P_1(\mathbf{x}_b - \mathbf{x}_a) = 1. \quad (15.3)$$

The same normalization holds for each N:

$$\int d^3 x_b \, P_N(\mathbf{x}_b - \mathbf{x}_a) = 1. \quad (15.4)$$

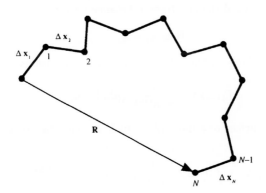

Figure 15.1 Random chain consisting of N links $\Delta\mathbf{x}_n$ of length a connecting $\mathbf{x}_a = \mathbf{x}_0$ and $\mathbf{x}_b = \mathbf{x}_N$.

If the second δ-function in (15.1) is Fourier-decomposed as

$$\delta^{(3)}\left(\mathbf{x}_b - \mathbf{x}_a - \sum_{n=1}^{N}\Delta\mathbf{x}_n\right) = \int \frac{d^3k}{(2\pi)^3} e^{i\mathbf{k}(\mathbf{x}_b-\mathbf{x}_a)-i\mathbf{k}\sum_{n=1}^{N}\Delta\mathbf{x}_n}, \qquad (15.5)$$

we see that $P_N(\mathbf{x}_b - \mathbf{x}_a)$ has the Fourier representation

$$P_N(\mathbf{x}_b - \mathbf{x}_a) = \int \frac{d^3k}{(2\pi)^3} e^{i\mathbf{k}(\mathbf{x}_b-\mathbf{x}_a)} \tilde{P}_N(\mathbf{k}), \qquad (15.6)$$

with

$$\tilde{P}_N(\mathbf{k}) = \prod_{n=1}^{N} \left[\int d^3\Delta x_n \frac{1}{4\pi a^2} \delta(|\Delta\mathbf{x}_n| - a) e^{-i\mathbf{k}\Delta\mathbf{x}_n} \right]. \qquad (15.7)$$

Thus, the Fourier transform $\tilde{P}_N(\mathbf{k})$ factorizes into a product of N Fourier-transformed one-link probabilities:

$$\tilde{P}_N(\mathbf{k}) = \left[\tilde{P}_1(\mathbf{k})\right]^N. \qquad (15.8)$$

These are easily calculated:

$$\tilde{P}_1(\mathbf{k}) = \int d^3\Delta x \frac{1}{4\pi a^2} \delta(|\Delta\mathbf{x}| - a) e^{-i\mathbf{k}\Delta\mathbf{x}} = \frac{\sin ka}{ka}. \qquad (15.9)$$

The desired end-to-end probability distribution is then given by the integral

$$
\begin{aligned}
P_N(\mathbf{R}) &= \int \frac{d^3k}{(2\pi)^3} [\tilde{P}_1(\mathbf{k})]^N e^{i\mathbf{k}\mathbf{R}} \\
&= \frac{1}{2\pi^2 R} \int_0^\infty dk\, k\, \sin kR \left[\frac{\sin ka}{ka}\right]^N,
\end{aligned}
\qquad (15.10)
$$

where we have introduced the end-to-end distance vector

$$\mathbf{R} \equiv \mathbf{x}_b - \mathbf{x}_a. \tag{15.11}$$

The generalization of the one-link distribution to D dimensions is

$$P_1(\Delta\mathbf{x}) = \frac{1}{S_D a^{D-1}} \delta(|\Delta\mathbf{x}| - a), \tag{15.12}$$

with S_D being the surface of a unit sphere in D dimensions [see Eq. (1.555)].

To calculate

$$\tilde{P}_1(\mathbf{k}) = \int d^D \Delta\mathbf{x} \, \frac{1}{S_D a^{D-1}} \delta(|\Delta\mathbf{x}| - a) e^{-i\mathbf{k}\Delta\mathbf{x}}, \tag{15.13}$$

we insert for $e^{-i\mathbf{k}\Delta\mathbf{x}} = e^{-ik|\Delta\mathbf{x}|\cos\Delta\vartheta}$ the expansion (8.129) with (8.101), and use $Y_{0,0}(\hat{\mathbf{x}}) = 1/\sqrt{S_D}$ to perform the integral as in (8.249). Using the relation between modified and ordinary Bessel functions $J_\nu(z)$[1]

$$I_\nu(e^{-i\pi/2} z) = e^{-i\pi/2} J_\nu(z), \tag{15.14}$$

this gives

$$\tilde{P}_1(\mathbf{k}) = \frac{\Gamma(D/2)}{(ka/2)^{D/2-1}} J_{D/2-1}(ka), \tag{15.15}$$

where $J_\mu(z)$ is the Bessel function.

15.2 Moments of End-to-End Distribution

The end-to-end distribution of a random chain is, of course, invariant under rotations so that the Fourier-transformed probability can only have even Taylor expansion coefficients:

$$\tilde{P}_N(\mathbf{k}) = [\tilde{P}_1(\mathbf{k})]^N = \sum_{l=0}^{\infty} P_{N,2l} \frac{(ka)^{2l}}{(2l)!}. \tag{15.16}$$

The expansion coefficients provide us with a direct measure for the even moments of the end-to-end distribution. These are defined by

$$\langle R^{2l} \rangle \equiv \int d^D R \, R^{2l} P_N(\mathbf{R}). \tag{15.17}$$

The relation between $\langle R^{2l} \rangle$ and $P_{N,2l}$ is found by expanding the exponential under the inverse of the Fourier integral (15.6):

$$\tilde{P}_N(\mathbf{k}) = \int d^D R \, e^{-i\mathbf{k}\mathbf{R}} P_N(\mathbf{R}) = \sum_{n=0}^{\infty} \int d^D R \frac{(-i\mathbf{k}\mathbf{R})^n}{n!} P_N(\mathbf{R}), \tag{15.18}$$

---

[1] I.S. Gradshteyn and I.M. Ryzhik, *op. cit.*, Formula 8.406.1.

and observing that the angular average of $(\mathbf{kR})^n$ is related to the average of R^n in D dimensions by

$$\langle(\mathbf{kR})^n\rangle = k^n\langle R^n\rangle \begin{cases} 0 & , \quad n = \text{odd}, \\ \dfrac{(n-1)!!(D-2)!!}{(D+n-2)!!} & , \quad n = \text{even}. \end{cases} \tag{15.19}$$

The three-dimensional result $1/(n+1)$ follows immediately from the angular average $(1/2)\int_{-1}^{1} d\cos\theta\,\cos^n\theta$ being $1/(n+1)$ for even n. In D dimensions, it is most easily derived by assuming, for a moment, that the vectors $\mathbf{R}$ have a Gaussian distribution $P_N^{(0)}(\mathbf{R}) = (D/2\pi Na^2)^{3/2}e^{-\mathbf{R}^2 D/2Na^2}$. Then the expectation values of all products of R_i can be expressed in terms of the pair expectation value

$$\langle R_i R_j\rangle^{(0)} = \frac{1}{D}\delta_{ij}\,a^2 N \tag{15.20}$$

via Wick's rule (3.302). The result is

$$\langle R_{i_1} R_{i_2}\cdots R_{i_n}\rangle^{(0)} = \frac{1}{D^{n/2}}\delta_{i_1 i_2 i_3\ldots i_n}\,a^n N^{n/2}, \tag{15.21}$$

with the contraction tensor $\delta_{i_1 i_2 i_3\ldots i_n}$ of Eqs. (8.64) and (13.84), which has the recursive definition

$$\delta_{i_1 i_2 i_3\ldots i_n} = \delta_{i_1 i_2}\delta_{i_3 i_4\ldots i_n} + \delta_{i_1 i_3}\delta_{i_2 i_4\ldots i_n} + \ldots \delta_{i_1 i_n}\delta_{i_2 i_3\ldots i_{n-1}}. \tag{15.22}$$

A full contraction of the indices gives, for even n, the Gaussian expectation values:

$$\langle R^n\rangle^{(0)} = \frac{(D+n-2)!!}{(D-2)!!D^{n/2}}\,a^n N^{n/2} = \frac{\Gamma(D/2+n/2)}{\Gamma(D/2)}\frac{2^{n/2}}{D^{n/2}}a^n N^{n/2}, \tag{15.23}$$

for instance

$$\left\langle R^4\right\rangle^{(0)} = \frac{(D+2)}{D}a^4 N^2, \qquad \left\langle R^6\right\rangle^{(0)} = \frac{(D+2)(D+4)}{D^2}a^6 N^3. \tag{15.24}$$

By contracting (15.21) with $k_{i_1} k_{i_2}\cdots k_{i_n}$ we find

$$\langle(\mathbf{kR})^n\rangle^{(0)} = (n-1)!!\,\frac{1}{D^{n/2}}(ka)^n N^{n/2} = k^n\left\langle(R)^n\right\rangle^{(0)} d_n, \tag{15.25}$$

with

$$d_n = \frac{(n-1)!!(D-2)!!}{(D+n-2)!!}. \tag{15.26}$$

Relation (15.25) holds for *any* rotation-invariant size distribution of $\mathbf{R}$, in particular for $P_N(\mathbf{R})$, thus proving Eq. (15.17) for random chains. Hence, the expansion coefficients $P_{N,2l}$ are related to the moments $\langle R^{2l}\rangle_N$ by

$$P_{N,2l} = (-1)^l d_{2l}\langle R^{2l}\rangle, \tag{15.27}$$

and the moment expansion (15.16) becomes

$$\tilde{P}_N(\mathbf{k}) = \sum_{l=0}^{\infty} \frac{(-1)^l (k)^{2l}}{(2l)!} d_{2l} \langle R^{2l} \rangle. \tag{15.28}$$

Let us calculate the even moments $\langle R^{2l} \rangle$ of the polymer distribution $P_N(\mathbf{R})$ explicitly for $D = 3$. We expand the logarithm of the Fourier transform $\tilde{P}_N(\mathbf{k})$ as follows:

$$\log \tilde{P}_N(\mathbf{k}) = N \log \tilde{P}_1(\mathbf{k}) = N \log \left(\frac{\sin ka}{ka} \right) = N \sum_{l=1}^{\infty} \frac{2^{2l}(-1)^l B_{2l}}{(2l)!2l} (ka)^{2l}, \tag{15.29}$$

where B_l are the Bernoulli numbers $B_2 = 1/6, B_4 = -1/30, \ldots$. Then we note that for a Taylor series of an arbitrary function $y(x)$

$$y(x) = \sum_{n=1}^{\infty} \frac{a_n}{n!} x^n, \tag{15.30}$$

the exponential function $e^{y(x)}$ has the expansion

$$e^{y(x)} = \sum_{n=1}^{\infty} \frac{b_n}{n!} x^n, \tag{15.31}$$

with the coefficients

$$\frac{b_n}{n!} = \sum_{\{m_i\}} \prod_{i=0}^{n} \frac{1}{m_i!} \left(\frac{a_i}{i!} \right)^{m_i}. \tag{15.32}$$

The sum over the powers $m_i = 0, 1, 2, \ldots$ obeys the constraint

$$n = \sum_{i=1}^{n} i \cdot m_i. \tag{15.33}$$

Note that the expansion coefficients a_n of $y(x)$ are the cumulants of the expansion coefficients b_n of $e^{y(x)}$ as defined in Section 3.17. For the coefficients a_n of the expansion (15.29),

$$a_n = \begin{cases} -N 2^{2l}(-1)^l B_{2l}/2l & \text{for } n = 2l, \\ 0 & \text{for } n = 2l + 1, \end{cases} \tag{15.34}$$

we find, via the relation (15.27), the moments

$$\langle R^{2l} \rangle = a^{2l}(-1)^l (2l+1)! \sum_{\{m_i\}} \prod_{i=1}^{l} \frac{1}{m_i!} \left[\frac{N 2^{2i}(-1)^i B_{2i}}{(2i)!2i} \right]^{m_i}, \tag{15.35}$$

with the sum over $m_i = 0, 1, 2, \ldots$ constrained by

$$l = \sum_{i=1}^{l} i \cdot m_i. \tag{15.36}$$

For $l = 1$ and 2 we obtain the moments

$$\langle R^2 \rangle = a^2 N, \quad \langle R^4 \rangle = \frac{5}{3} a^4 N^2 \left(1 - \frac{2}{5N}\right). \tag{15.37}$$

In the limit of large N, the leading behavior of the moments is the same as in (15.20) and (15.24). The linear growth of $\langle R^2 \rangle$ with the number of links N is characteristic for a random chain. In the presence of interactions, there will be a different power behavior expressed as a so-called *scaling law*

$$\langle R^2 \rangle \propto a^2 N^{2\nu}. \tag{15.38}$$

The number ν is called the *critical exponent* of this scaling law. It is intuitively obvious that ν must be a number between $\nu = 1/2$ for a random chain as in (15.37), and $\nu = 1$ for a completely stiff chain.

Note that the knowledge of all moments of the end-to-end distribution determines completely the shape of the distribution by an expansion

$$P_L(\mathbf{R}) = \frac{1}{S_D R^{D-1}} \sum_{n=0}^{\infty} \langle R^n \rangle \frac{(-1)^n}{n!} \partial_R^n \delta(R). \tag{15.39}$$

This can easily be verified by calculating the integrals (15.17) using the integrals formula

$$\int dz \, z^n \partial_z^n \delta(z) = (-1)^n n!. \tag{15.40}$$

15.3   Exact End-to-End Distribution in Three Dimensions

Consider the Fourier representation (15.10), rewritten as

$$P_N(\mathbf{R}) = \frac{i}{4\pi^2 a^2 R} \int_{-\infty}^{\infty} d\eta \, \eta \, e^{-i\eta R/a} \left(\frac{\sin \eta}{\eta}\right)^N, \tag{15.41}$$

with the dimensionless integration variable $\eta \equiv ka$. By expanding

$$\sin^N \eta = \frac{1}{(2i)^N} \sum_{n=0}^{N} (-1)^n \binom{N}{n} \exp\left[i(N - 2n)\eta\right], \tag{15.42}$$

we find the finite series

$$P_N(\mathbf{R}) = \frac{1}{2^{N+2} i^{N-1} \pi^2 a^2 R} \sum_{n=0}^{N} (-1)^n \binom{N}{n} I_N(N - 2n - R/a), \tag{15.43}$$

where $I_N(x)$ are the integrals

$$I_N(x) \equiv \int_{-\infty}^{\infty} d\eta \frac{e^{i\eta x}}{\eta^{N-1}}. \tag{15.44}$$

For $N \geq 2$, these integrals are all singular. The singularity can be avoided by noting that the initial integral (15.41) is perfectly regular at $\eta = 0$. We therefore replace the expression $(\sin \eta / \eta)^N$ in the integrand by $[\sin(\eta - i\epsilon)/(\eta - i\epsilon)]^N$. This regularizes each term in the expansion (15.43) and leads to well-defined integrals:

$$I_N(x) = \int_{-\infty}^{\infty} d\eta \frac{e^{ix(\eta - i\epsilon)}}{(\eta - i\epsilon)^{N-1}}. \tag{15.45}$$

For $x < 0$, the contour of integration can be closed by a large semicircle in the lower half-plane. Since the lower half-plane contains no singularity, the residue theorem shows that

$$I_N(x) = 0, \quad x < 0. \tag{15.46}$$

For $x > 0$, on the other hand, an expansion of the exponential function in powers of $\eta^{-i\epsilon}$ produces a pole, and the residue theorem yields

$$I_N(x) = \frac{2\pi i^{N-1}}{(N-2)!} x^{N-2}, \quad x > 0. \tag{15.47}$$

Hence we arrive at the finite series

$$P_N(\mathbf{R}) = \frac{1}{2^{N+1}(N-2)!\pi a^2 R} \sum_{0 \leq n \leq (N-R/a)/2} (-1)^n \binom{N}{n} (N - 2n - R/a)^{N-2}. \tag{15.48}$$

The distribution is displayed for various values of N in Fig. 15.2, where we have plotted $2\pi R^2 \sqrt{N} P_N(\mathbf{R})$ against the rescaled distance variable $\rho = R/\sqrt{N}a$. With this N-dependent rescaling all curves have the same unit area. Note that they converge rapidly towards a universal zero-order distribution

$$P_N^{(0)}(\mathbf{R}) = \sqrt{\frac{3}{2\pi Na^2}}^3 \exp\left\{-\frac{3R^2}{2Na^2}\right\} \to P_L^{(0)}(\mathbf{R}) = \sqrt{\frac{D}{2\pi aL}}^D e^{-DR^2/2aL}. \tag{15.49}$$

In the limit of large N, the length L will be used as a subscript rather than the diverging N. The proof of the limit is most easily given in Fourier space. For large N at finite $k^2 a^2 N$, the Nth power of the quantity $\tilde{P}_1(\mathbf{k})$ in Eq. (15.15) can be approximated by

$$[\tilde{P}_1(\mathbf{k})]^N \sim e^{-Nk^2a^2/2D}. \tag{15.50}$$

Then the Fourier transform (15.10) is performed with the zero-order result (15.49). In Fig. 15.2 we see that this large-N limit is approached uniformly in $\rho = R/\sqrt{N}a$. The approach to this limit is studied analytically in the following two sections.

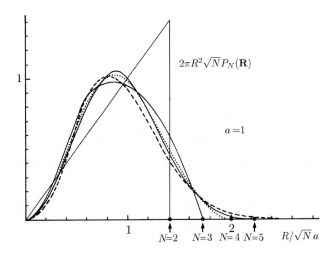

Figure 15.2 End-to-end distribution $P_N(\mathbf{R})$ of random chain with N links. The functions $2\pi R^2\sqrt{N}P_N(\mathbf{R})$ are plotted against $R/\sqrt{N}\,a$ which gives all curves the same unit area. Note the fast convergence for growing N. The dashed curve is the continuum distribution $P_L^{(0)}(\mathbf{R})$ of Eq. (15.49). The circles on the abscissa mark the maximal end-to-end distance.

15.4   Short-Distance Expansion for Long Polymer

At finite N, we expect corrections which are expandable in the form

$$P_N(\mathbf{R}) = P_N^{(0)}(\mathbf{k})\left[1 + \sum_{n=1}^{\infty}\frac{1}{N^n}C_n(R^2/Na^2)\right], \tag{15.51}$$

where the functions $C_n(x)$ are power series in x starting with x^0.

Let us derive this expansion. In three dimensions, we start from (15.29) and separate the right-hand side into the leading k^2-term and a remainder

$$\tilde{C}(\mathbf{k}) \equiv \exp\left[N\sum_{l=2}^{\infty}\frac{2^{2l}(-1)^l B_{2l}}{(2l)!\,2l}(k^2a^2)^l\right]. \tag{15.52}$$

Exponentiating both sides of (15.29), the end-to-end probability factorizes as

$$\tilde{P}_N(\mathbf{k}) = e^{-Na^2k^2/6}\tilde{C}(\mathbf{k}). \tag{15.53}$$

The function $\tilde{C}(\mathbf{k})$ is now expanded in a power series

$$\tilde{C}(\mathbf{k}) = 1 + \sum_{\substack{n=1,2,\dots \\ l=2n,2n+1,\dots}}\tilde{C}_{n,l}N^n(a^2k^2)^l, \tag{15.54}$$

with the lowest coefficients

$$
\begin{aligned}
\tilde{C}_{1,2} &= -\frac{1}{180}, & \tilde{C}_{1,3} &= -\frac{1}{2835}, \\
\tilde{C}_{1,4} &= -\frac{1}{37800}, & \tilde{C}_{2,4} &= \frac{1}{64800}, & \dots \; .
\end{aligned}
\tag{15.55}
$$

For any dimension D, we factorize

$$
\tilde{P}_N(\mathbf{k}) = e^{-Na^2 k^2/2D}\tilde{C}(\mathbf{k})
\tag{15.56}
$$

and find $\tilde{C}(\mathbf{k})$ by expanding (15.15) in powers of k and proceeding as before. This gives the coefficients

$$
\begin{aligned}
\tilde{C}_{1,2} &= -1/4D^2(D+2), \\
\tilde{C}_{1,3} &= -1/3D^3(D+2)(D+4), \\
\tilde{C}_{1,4} &= -(5D+12)/8D^4(D+2)^2(D+4)(D+6), \\
\tilde{C}_{2,4} &= 1/32D^4(D+2)^2.
\end{aligned}
\tag{15.57}
$$

We now Fourier-transform (15.56). The leading term in $\tilde{C}(\mathbf{k})$ yields the zero-order distribution (15.49) in D dimensions,

$$
P_N^{(0)}(\mathbf{R}) = \sqrt{\frac{D}{2\pi Na^2}}^D e^{-DR^2/2Na^2},
\tag{15.58}
$$

or, written in terms of the reduced distance variable $\rho = R/Na$,

$$
P_N^{(0)}(\mathbf{R}) = \sqrt{\frac{D}{2\pi Na^2}}^D e^{-DN\rho^2/2}.
\tag{15.59}
$$

To account for the corrections in $\tilde{C}(\mathbf{k})$ we take the expansion (15.54), emphasize the dependence on $k^2 a^2$ by writing

$$
\tilde{C}(\mathbf{k}) = \bar{C}(k^2 a^2),
$$

and observe that in the Fourier transform

$$
P_N(\mathbf{R}) = \int \frac{d^3 k}{(2\pi)^3} e^{i\mathbf{k}\mathbf{R}} P_N(\mathbf{k}) = \int \frac{d^3 k}{(2\pi)^3} e^{i\mathbf{k}\mathbf{R}} e^{-Nk^2 a^2/2D}\bar{C}(k^2 a^2),
\tag{15.60}
$$

the series can be pulled out of the integral by replacing each power $(k^2 a^2)^p$ by $(-2D\partial_N)^p$. The result has the form

$$
P_N(\mathbf{R}) = \bar{C}(-2D\partial_N)\int \frac{d^3 k}{(2\pi)^3} e^{i\mathbf{k}\mathbf{R}} e^{-Nk^2 a^2/2D} = \bar{C}(-2D\partial_N)P_N^{(0)}(\mathbf{R}).
\tag{15.61}
$$

Going back to coordinate space, we obtain an expansion

$$
P_N(\mathbf{R}) = \left(\frac{D}{2\pi Na^2}\right)^{D/2} e^{-DN\rho^2/2}C(R).
\tag{15.62}
$$

For $D = 3$, the function $C(R)$ is given by the series

$$C(R) = 1 + \sum_{\substack{n=1 \\ l=0,\dots,2n}}^{\infty} C_{n,l} N^{-n} (N\rho^2)^l, \tag{15.63}$$

with the coefficients

$$C_{1,1} = \left(-\frac{3}{4}, \frac{3}{2}, -\frac{9}{20}\right), \quad C_{2,1} = \left(\frac{29}{160}, -\frac{69}{40}, \frac{981}{400}, -\frac{1341}{1400}, \frac{81}{800}\right). \tag{15.64}$$

For any D, we find the coefficients

$$
\begin{aligned}
C_{1,l} &= \left(-\frac{D}{4}, \frac{D}{2}, -\frac{D^2}{4(D+2)}\right), \\
C_{2,l} &= \left(\frac{(3D^2 - 2D + 8)D}{96(D+2)}, -\frac{(D^2 + 2D + 8)D}{8(D+2)}, \frac{(3D^2 + 14D + 40)D^2}{16(D+2)^2}, \right. \\
&\quad \left. -\frac{(3D^2 + 22D + 56)D^3}{24(D+2)^2(D+4)}, \frac{D^4}{32(D+2)^2}\right).
\end{aligned}
\tag{15.65}
$$

15.5 Saddle Point Approximation to Three-Dimensional End-to-End Distribution

Another study of the approach to the limiting distribution (15.49) proceeds via the saddle point approximation. For this, the integral in (15.41) is rewritten as

$$\int_{-\infty}^{\infty} d\eta \, \eta \, e^{-Nf(\eta)}, \tag{15.66}$$

with

$$f(\eta) = i\frac{R}{Na}\eta - \log\left(\frac{\sin\eta}{\eta}\right). \tag{15.67}$$

The extremum of f lies at $\eta = \bar{\eta}$, where $\bar{\eta}$ solves the equation

$$\coth(i\bar{\eta}) - \frac{1}{i\bar{\eta}} = \frac{R}{Na}. \tag{15.68}$$

The function on the left-hand side is known as the *Langevin function*:

$$L(x) \equiv \coth x - \frac{1}{x}. \tag{15.69}$$

The extremum lies at the imaginary position $\bar{\eta} \equiv -i\bar{x}$ with $\bar{x}$ being determined by the equation

$$L(\bar{x}) = \frac{R}{Na}. \tag{15.70}$$

The extremum is a minimum of $f(\eta)$ since

$$f''(\bar{\eta}) = L'(\bar{x}) = -\frac{1}{\sinh^2 \bar{x}} + \frac{1}{\bar{x}^2} > 0. \tag{15.71}$$

By shifting the integration contour vertically into the complex η plane to make it run through the minimum at $-i\bar{x}$, we obtain

$$
\begin{aligned}
P_N(\mathbf{R}) &\approx -\frac{1}{4i\pi^2 a^2 R} e^{-Nf(\bar{\eta})} \int_{-\infty}^{\infty} d\eta \, (-i\bar{x} + \eta) \exp\left\{ -\frac{N}{2} f''(\bar{\eta})\eta^2 \right\} \\
&= \frac{\bar{\eta}}{4\pi^2 a^2 R} \sqrt{\frac{2\pi}{Nf''(\bar{x})}} e^{-Nf(\bar{\eta})}.
\end{aligned}
\tag{15.72}
$$

When expressed in terms of the reduced distance $\rho \equiv R/Na \in [0,1]$, this reads

$$P_N(\mathbf{R}) \approx \frac{1}{(2\pi Na^2)^{3/2}} \frac{L^{\mathrm{i}}(\rho)^2}{\rho \left\{ 1 - [L^{\mathrm{i}}(\rho)/\sinh L^{\mathrm{i}}(\rho)]^2 \right\}^{1/2}} \left\{ \frac{\sinh L^{\mathrm{i}}(\rho)}{L^{\mathrm{i}}(\rho) \exp[\rho L^{\mathrm{i}}(\rho)]} \right\}^N. \tag{15.73}$$

Here we have introduced the inverse Langevin function $L^{\mathrm{i}}(\rho)$ since it allows us to express $\bar{x}$ as

$$\bar{x} = L^{i}(\rho) \tag{15.74}$$

[inverting Eq. (15.70)]. The result (15.73) is valid in the entire interval $\rho \in [0,1]$ corresponding to $R \in [0, Na]$; it ignores corrections of the order $1/N$. By expanding the right-hand side in a power series in ρ, we find

$$P_N(\mathbf{R}) = \mathcal{N} \left(\frac{3}{2\pi Na^2} \right)^{3/2} \exp\left(-\frac{3R^2}{2Na^2} \right) \left(1 + \frac{3R^2}{2N^2a^2} - \frac{9R^4}{20N^3a^4} + \dots \right), \tag{15.75}$$

with some normalization constant $\mathcal{N}$. At each order of truncation, $\mathcal{N}$ is determined in such a way that $\int d^3R \, P_N(\mathbf{R}) = 1$. As a check we take the limit $\rho^2 \ll 1/N$ and find powers in ρ which agree with those in (15.62), for $D = 3$, with the expansion (15.63) of the correction factor.

15.6 Path Integral for Continuous Gaussian Distribution

The limiting end-to-end distribution (15.49) is equal to the imaginary-time amplitude of a free particle in natural units with $\hbar = 1$:

$$(\mathbf{x}_b \tau_b | \mathbf{x}_a \tau_a) = \frac{1}{\sqrt{2\pi(\tau_b - \tau_a)/M}^D} \exp\left[-\frac{M}{2} \frac{(\mathbf{x}_b - \mathbf{x}_a)^2}{\tau_b - \tau_a} \right]. \tag{15.76}$$

We merely have to identify

$$\mathbf{x}_b - \mathbf{x}_a \equiv \mathbf{R}, \tag{15.77}$$

and replace

$$\tau_b - \tau_a \quad \rightarrow \quad Na, \tag{15.78}$$

$$M \quad \rightarrow \quad D/a. \tag{15.79}$$

Thus we can describe a polymer with $R^2 \ll Na^2$ by the path integral

$$P_L(\mathbf{R}) = \int \mathcal{D}^D x \, \exp\left\{-\frac{D}{2a} \int_0^L ds \, [\mathbf{x}'(s)]^2\right\} = \sqrt{\frac{D}{2\pi a}} \, e^{-DR^2/2La}. \tag{15.80}$$

The number of time slices is here N [in contrast to (2.64) where it was $N+1$), and the total length of the polymer is $L = Na$.

Let us calculate the Fourier transformation of the distribution (15.80):

$$\tilde{P}_L(\mathbf{q}) = \int d^D R \, e^{-i\mathbf{q}\cdot\mathbf{R}} P_L(\mathbf{R}). \tag{15.81}$$

After a quadratic completion the integral yields

$$\tilde{P}_L(\mathbf{q}) = e^{-Laq^2/2D}, \tag{15.82}$$

with the power series expansion

$$\tilde{P}_L(\mathbf{q}) = \sum_{l=0}^{\infty} (-1)^l \frac{q^{2l}}{l!} \left(\frac{La}{2D}\right)^l. \tag{15.83}$$

Comparison with the moments (15.23) shows that we can rewrite this as

$$\tilde{P}_L(\mathbf{q}) = \sum_{l=0}^{\infty} (-1)^l \frac{q^{2l}}{l!} \frac{\Gamma(D/2)}{2^{2l}\Gamma(D/2+l)} \left\langle R^{2l} \right\rangle. \tag{15.84}$$

This is a completely general relation: the expansion coefficients of the Fourier transform yield directly the moments of a function, up to trivial numerical factors specified by (15.84).

The end-to-end distribution determines rather directly the structure factor of a dilute solution of polymers which is observable in static neutron and light scattering experiments:

$$S(\mathbf{q}) = \frac{1}{L^2} \int_0^L ds \int_0^L ds' \left\langle e^{i\mathbf{q}\cdot[\mathbf{x}(s)-\mathbf{x}(s')]} \right\rangle. \tag{15.85}$$

The average over all polymers running from $\mathbf{x}(0)$ to $\mathbf{x}(L)$ can be written, more explicitly, as

$$\left\langle e^{i\mathbf{q}\cdot[\mathbf{x}(s)-\mathbf{x}(s')]} \right\rangle = \int d^D x(L) \int d^D(x(s')-x(s)) \int d^D x(0) \tag{15.86}$$

$$\times P_{L-s'}(\mathbf{x}(L)-\mathbf{x}(s')) e^{-i\mathbf{q}\cdot\mathbf{x}(s')} P_{s'-s}(\mathbf{x}(s')-\mathbf{x}(s)) e^{i\mathbf{q}\cdot\mathbf{x}(s)} P_{s-0}(\mathbf{x}(s)-\mathbf{x}(0)).$$

The integrals over initial and final positions give unity due to the normalization (15.4), so that we remain with

$$\left\langle e^{i\mathbf{q}\cdot[\mathbf{x}(s)-\mathbf{x}(s')]}\right\rangle = \int d^D R\, e^{-i\mathbf{q}\cdot\mathbf{R}} P_{s'-s}(\mathbf{R}). \tag{15.87}$$

Since this depends only on $L' \equiv |s' - s|$ and not on $s + s'$, we decompose the double integral in over s and s' in (15.85) into $2\int_0^L dL'(L - L')$ and obtain

$$S(\mathbf{q}) = \frac{2}{L^2}\int_0^L dL'(L - L')\int d^D R\, e^{i\mathbf{q}\cdot\mathbf{R}(L')} P_{L'}(\mathbf{R}), \tag{15.88}$$

or, recalling (15.81),

$$S(\mathbf{q}) = \frac{2}{L^2}\int_0^L dL'(L - L')\tilde{P}_{L'}(\mathbf{q}). \tag{15.89}$$

Inserting (15.82) we obtain the *Debye structure factor* of Gaussian random paths:

$$S^{\text{Gauss}}(\mathbf{q}) = \frac{2}{x^2}\left(x - 1 + e^{-x}\right), \quad x \equiv \frac{q^2 aL}{2D}. \tag{15.90}$$

This function starts out like $1 - x/3 + x^2/12 + \ldots$ for small q and falls of like q^{-2} for $q^2 \gg 2D/aL$. The Taylor coefficients are determined by the moments of the end-to-end distribution. By inserting (15.84) into (15.89) we obtain:

$$S(\mathbf{q}) = \sum_{l=0}^{\infty}(-1)^l q^{2l}\frac{\Gamma(D/2)}{2^{2l}l!\Gamma(l + D/2)}\frac{2}{L^2}\int_0^L dL'(L - L')\left\langle R^{2l}\right\rangle. \tag{15.91}$$

Although the end-to-end distribution (15.80) agrees with the true polymer distribution (15.1) for $R \ll \sqrt{N}a$, it is important to realize that the nature of the fluctuations in the two expressions is quite different. In the polymer expression, the length of each link $\Delta\mathbf{x}_n$ is fixed. In the sliced action of the path integral Eq. (15.80),

$$\mathcal{A}^N = a\sum_{n=1}^{N}\frac{M}{2}\frac{(\Delta\mathbf{x}_n)^2}{a^2}, \tag{15.92}$$

on the other hand, each small section fluctuates around zero with a mean square

$$\langle(\Delta\mathbf{x}_n)^2\rangle_0 = \frac{a}{M} = \frac{a^2}{D}. \tag{15.93}$$

Yet, if the end-to-end distance of the polymer is small compared to the completely stretched configuration, the distributions are practically the same. There exists a qualitative difference only if the polymer is almost completely stretched. While the polymer distribution vanishes for $R > Na$, the path integral (15.80) gives a nonzero value for arbitrarily large R. Quantitatively, however, the difference is insignificant since it is exponentially small (see Fig. 15.2).

15.7   Stiff Polymers

The end-to-end distribution of real polymers found in nature is never the same as
that of a random chain. Usually, the joints do not allow for an equal probability
of all spherical angles. The forward angles are often preferred and the polymer is
stiff at shorter distances. Fortunately, if averaged over many links, the effects of the
stiffness becomes less and less relevant. For a very long random chain with a finite
stiffness one finds the same linear dependence of the square end-to-end distance on
the length $L = Na$ as for ideal random chains which has, according to Eq. (15.23),
the Gaussian expectations:

$$\langle R^2 \rangle = aL, \qquad \langle R^{2l} \rangle = \frac{(D + 2l - 2)!!}{(D - 2)!!D^l}(aL)^l. \tag{15.94}$$

For a stiff chain, the expectation value $\langle R^2 \rangle$ will increase aL to $a_{\text{eff}}L$, where a_{eff} is
the *effective bond length* In the limit of a very large stiffness, called the *rod limit*,
the law (15.94) turns into

$$\langle R^2 \rangle \equiv L^2, \qquad \langle R^{2l} \rangle \equiv L^{2l}, \tag{15.95}$$

i.e., the effective bond length length a_{eff} increases to L. This intuitively obvious
statement can easily be found from the normalized end-to-end distribution, which
coincides in the rod limit with the one-link expression (15.12):

$$P_L^{\text{rod}}(\mathbf{R}) = \frac{1}{S_D R^{D-1}}\delta(R - L), \tag{15.96}$$

and yields [recall (15.17)]

$$\langle R^n \rangle = \int d^D R \, R^n P_L^{\text{rod}}(\mathbf{R}) = \int_0^\infty dR \, R^n \, \delta(R - L) = L^n. \tag{15.97}$$

By expanding $P_L^{\text{rod}}(\mathbf{R})$ in powers of L, we obtain the series

$$P_L^{\text{rod}}(\mathbf{R}) = \frac{1}{S_D R^{D-1}} \sum_{n=0}^\infty L^n \langle R^n \rangle \frac{(-1)^n}{n!} \partial_R^n \delta(R). \tag{15.98}$$

An expansion of this form holds for any stiffness: the moments of the distribution
are the Taylor coefficients of the expansion of $P_L(\mathbf{R})$ into a series of derivatives of
$\delta(R)$-functions.

Let us also calculate the Fourier transformation (15.81) of this distribution. Recalling (15.15), we find

$$\tilde{P}_L^{\text{rod}}(\mathbf{q}) = \tilde{P}^{\text{rod}}(qL) \equiv \frac{\Gamma(D/2)}{(qL/2)^{D/2-1}}J_{D/2-1}(qL). \tag{15.99}$$

For an arbitrary rotationally symmetric $P_L(\mathbf{R}) = P_L(R)$, we simply have to super-
impose these distributions for all R:

$$\tilde{P}_L(\mathbf{q}) = S_D \int_0^\infty dR \, R^{D-1} \tilde{P}^{\text{rod}}(qL) P_L(R). \tag{15.100}$$

This is simply proved by decomposing and performing the Fourier transformation
(15.81) on $P_L(R) = \int_0^\infty dR'\, \delta(R-R') P_{R'}^{\text{rod}}(\mathbf{R}) = S_D \int_0^\infty dR'\, R'^{D-1} P_{R'}^{\text{rod}}(\mathbf{R}) P_L(R')$, and
performing the Fourier transformation (15.81) on $P_{R'}^{\text{rod}}(\mathbf{R})$. In $D=3$ dimensions,
(15.100) takes the simple form:

$$\tilde{P}_L(\mathbf{q}) = 4\pi \int_0^\infty dR\, R^2\, \frac{\sin qR}{qR} P_L(R). \tag{15.101}$$

Inserting the power series expansion for the Bessel function[2]

$$J_\nu(z) = \left(\frac{z}{2}\right)^\nu \sum_{l=0}^\infty \frac{(-1)^k (z/2)^{2l}}{l!\,\Gamma(\nu+l+1)} \tag{15.102}$$

into (15.99), and this into (15.100), we obtain

$$\tilde{P}_L(\mathbf{q}) = \sum_{l=0}^\infty (-1)^l \left(\frac{q}{2}\right)^{2l} \frac{\Gamma(D/2)}{l!\,\Gamma(D/2+l)} S_D \int_0^\infty dR\, R^{D-1}\, R^{2l} P_L(R), \tag{15.103}$$

in agreement with the general expansion (15.84). The same result is obtained by inserting into (15.103) the expansion (15.98) and using the integrals $\int_0^\infty dR\, R^m \partial_R^n \delta(R) = \delta_{mn}(-1)^n n!$, which are proved by n partial integrations.

The structure factor of a completely stiff polymer (rod limit) is obtained by inserting (15.99) into (15.89). The resulting $S^{\text{rod}}(\mathbf{q})$ depends only on qL:

$$S^{\text{rod}}(qL) = \frac{4-2D}{q^2 L^2} + \left(\frac{2}{qL}\right)^{D/2} \Gamma(D/2) J_{D/2-2}(qL) + 2F(1/2; 3/2, D/2; -q^2 L^2/4), \tag{15.104}$$

where $F(a;b,c;z)$ is the hypergeometric function (1.450). For $D=3$, the integral (15.89) reduces to $(2/L^2)\int_0^L dL'\,(L-L')(\sin qL')/qL'$, as in the similar equation (15.9), and the result is simply

$$S^{\text{rod}}(z) = \frac{2}{z^2}\left[\cos z - 1 + z\,\mathrm{Si}(z)\right], \quad \mathrm{Si}(z) \equiv \int_0^z \frac{dt}{t}\sin t. \tag{15.105}$$

This starts out like $1 - z^2/36 + z^4/1800 + \dots$. For large z we use the limit of the sine integral[3] $\mathrm{Si}(z) \to \pi/2$ to find $S^{\text{rod}}(\mathbf{q}) \to \pi/qL$.

For of an arbitrary rotationally symmetric end-to-end distribution $P_L(R)$, the structure factor can be expressed, by analogy with (15.100), as a superposition of rod limits:

$$S_L(\mathbf{q}) = S_D \int_0^\infty dR\, R^{D-1} S^{\text{rod}}(\mathbf{q}R) P_L(R). \tag{15.106}$$

When passing from long to short polymers at a given stiffness, there is a crossover between the moments (15.94) and (15.95) and the behaviors of the structure function. Let us study this in detail.

---

[2] I.S. Gradshteyn and I.M. Ryzhik, *op. cit.*, Formula 8.440.
[3] ibid., Formulas 8.230 and 8.232.

15.7.1   Sliced Path Integral

The stiffness of a polymer pictured in Fig. 15.1 may be parameterized by the bending energy

$$E_{\text{bend}}^{N} = \frac{\kappa}{2a} \sum_{n=1}^{N} (\mathbf{u}_n - \mathbf{u}_{n-1})^2, \tag{15.107}$$

where $\mathbf{u}_n$ are the unit vectors specifying the directions of the links. The initial and final link directions of the polymer have a distribution

$$(\mathbf{u}_b L | \mathbf{u}_a 0) = \frac{1}{A} \prod_{n=1}^{N-1} \left[\int \frac{d\mathbf{u}_n}{A} \right] \exp\left[-\frac{\kappa}{2ak_BT} \sum_{n=1}^{N} (\mathbf{u}_n - \mathbf{u}_{n-1})^2 \right], \tag{15.108}$$

where A is some normalization constant, which we shall choose such that the measure of integration coincides with that of a time-sliced path integral *near* a unit sphere in Eq. (8.150). Comparison if the bending energy (15.107) with the Euclidean action (8.151) we identify

$$\frac{Mr^2}{\hbar\epsilon} = \frac{\kappa}{ak_BT}, \tag{15.109}$$

and see that we must replace $N \to N-1$ and set

$$A = \sqrt{2\pi ak_BT/\kappa}^{-D-1}. \tag{15.110}$$

The result of the integrations in (15.108) is then known from Eq. (8.155):

$$(\mathbf{u}_b L | \mathbf{u}_a 0) = \sum_{l=0}^{\infty} \left[\tilde{I}_{l+D/2-1}(h) \right]^N \sum_{\mathbf{m}} Y_{l\mathbf{m}}(\mathbf{u}_b) Y_{l\mathbf{m}}^*(\mathbf{u}_a), \quad h \equiv \frac{\kappa}{ak_BT}, \tag{15.111}$$

with the modified Bessel function $\tilde{I}_{l+D/2-1}(z)$ of Eq. (8.11).

The partition function of the polymer is obtained by integrating over all final and averaging over all initial link directions [1]:

$$Z_N = \int \frac{d\,\mathbf{u}_a}{S_D} \prod_{n=1}^{N} \left[\int \frac{d\mathbf{u}_n}{A} \right] \exp\left[-\frac{\kappa}{2ak_BT} \sum_{n=1}^{N} (\mathbf{u}_n - \mathbf{u}_{n-1})^2 \right]$$

$$= \int d\,\mathbf{u}_b \int \frac{d\,\mathbf{u}_a}{S_D} (\mathbf{u}_b L | \mathbf{u}_a 0). \tag{15.112}$$

Inserting here the spectral representation (15.111) we find

$$Z_N = \left[\tilde{I}_{D/2-1}\left(\frac{\kappa}{ak_BT} \right) \right]^N = \left[\sqrt{\frac{2\pi\kappa}{ak_BT}} e^{-\kappa/ak_BT} I_{D/2-1}\left(\frac{\kappa}{ak_BT} \right) \right]^N. \tag{15.113}$$

Knowing this we may define the normalized distribution function

$$P_N(\mathbf{u}_b, \mathbf{u}_a) = \frac{1}{Z_N} (\mathbf{u}_b L | \mathbf{u}_a 0), \tag{15.114}$$

whose integral over $\mathbf{u}_a$ as well as over $\mathbf{u}_a$ is unity:

$$\int d\mathbf{u}_b\, P_N(\mathbf{u}_b, \mathbf{u}_a) = \int d\mathbf{u}_a\, P_N(\mathbf{u}_b, \mathbf{u}_a) = 1. \tag{15.115}$$

15.7.2 Relation to Classical Heisenberg Model

The above partition function is closely related to the partition function of the one-dimensional classical Heisenberg model of ferromagnetism which is defined by

$$
Z_N^{\text{Heis}} \equiv \int \frac{d\,\mathbf{u}_a}{S_D} \prod_{n=1}^{N} \left[\int d\mathbf{u}_n \right] \exp\left[\frac{J}{k_B T} \sum_{n=1}^{N} \mathbf{u}_n \cdot \mathbf{u}_{n-1} \right], \tag{15.116}
$$

where J are interaction energies due to exchange integrals of electrons in a ferromagnet. This differs from (15.112) by a trivial normalization factor, being equal to

$$
Z_N^{\text{Heis}} = \left[\sqrt{\frac{2\pi J}{k_B T}}^{2-D} I_{D/2-1}\left(\frac{J}{k_B T} \right) \right]^N. \tag{15.117}
$$

Identifying $J \equiv \kappa/a$ we may use the Heisenberg partition functions for all calculations of stiff polymers. As an example take the correlation function between neighboring tangent vectors $\langle \mathbf{u}_n \cdot \mathbf{u}_{n-1} \rangle$. In order to calculate this we observe that the partition function (15.116) can just as well be calculated exactly with a slight modification that the interaction strength J of the Heisenberg model depends on the link n. The result is the corresponding generalization of (15.117):

$$
Z_N^{\text{Heis}}(J_1, \ldots, J_N) = \prod_{n=1}^{N} \left[\sqrt{\frac{2\pi J_n}{k_B T}}^{2-D} I_{D/2-1}\left(\frac{J_n}{k_B T} \right) \right]. \tag{15.118}
$$

This expression may be used as a generating function for expectation values $\langle \mathbf{u}_n \cdot \mathbf{u}_{n-1} \rangle$ which measure the degree of alignment of neighboring spin directions. Indeed, we find directly

$$
\langle \mathbf{u}_n \cdot \mathbf{u}_{n-1} \rangle = (k_B T) \left. \frac{dZ_N^{\text{Heis}}(J_1, \ldots, J_N)}{dJ_n} \right|_{J_n \equiv J} = \frac{I_{D/2}(J/k_B T)}{I_{D/2-1}(J/k_B T)}. \tag{15.119}
$$

This expectation value measures directly the internal energy per link of te chain. Indeed, since the free energy is $F_N = -k_B T \log Z_N^{\text{Heis}}$, we obtain [recall (1.545)]

$$
E_N = N \langle \mathbf{u}_n \cdot \mathbf{u}_{n-1} \rangle = N \frac{I_{D/2}(J/k_B T)}{I_{D/2-1}(J/k_B T)}. \tag{15.120}
$$

Let us also calculate the expectation value of the angle between next-to-nearest neighbors $\langle \mathbf{u}_{n+1} \cdot \mathbf{u}_{n-1} \rangle$. We do this by considering the expectation value

$$
\langle (\mathbf{u}_{n+1} \cdot \mathbf{u}_n)(\mathbf{u}_n \cdot \mathbf{u}_{n-1}) \rangle = (k_B T)^2 \left. \frac{d^2 Z_N^{\text{Heis}}(J_1, \ldots, J_N)}{dJ_{n+1} dJ_n} \right|_{J_n \equiv J} = \left[\frac{I_{D/2}(J/k_B T)}{I_{D/2-1}(J/k_B T)} \right]^2. \tag{15.121}
$$

Then we prove that the left-hand side is in fact equal to the desired expectation value $\langle \mathbf{u}_{n+1} \cdot \mathbf{u}_{n-1} \rangle$. For this we decompose the last vector $\mathbf{u}_{n+1}$ into a component

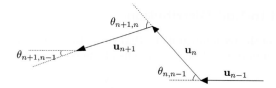

Figure 15.3 Neighboring links for the calculation of expectation values.

parallel to $\mathbf{u}_n$ and a component perpendicular to it: $\mathbf{u}_{n+1} = (\mathbf{u}_{n+1}\cdot\mathbf{u}_n)\,\mathbf{u}_n + \mathbf{u}_{n+1}^{\perp}$. The corresponding decomposition of the expectation value $\langle\mathbf{u}_{n+1}\cdot\mathbf{u}_{n-1}\rangle$ is $\langle\mathbf{u}_{n+1}\cdot\mathbf{u}_{n-1}\rangle = \langle(\mathbf{u}_{n+1}\cdot\mathbf{u}_n)(\mathbf{u}_n\cdot\mathbf{u}_{n-1})\rangle + \langle\mathbf{u}_{n+1}^{\perp}\cdot\mathbf{u}_n\rangle$. Now, the energy in the Boltzmann factor depends only on $(\mathbf{u}_{n+1}\cdot\mathbf{u}_n) + (\mathbf{u}_n\cdot\mathbf{u}_{n-1}) = \cos\theta_{n+1,n} + \cos\theta_{n,n-1}$, so that the integral over $\mathbf{u}_{n+1}^{\perp}$ runs over a surface of a sphere of radius $\sin\theta_{n+1,n}$ in $D-1$ dimensions [recall (8.116)] and the Boltzmann factor does not depend on the angles. The integral receives therefore equal contributions from $\mathbf{u}_{n+1}^{\perp}$ and $-\mathbf{u}_{n+1}^{\perp}$, and vanishes. This proves that

$$\langle\mathbf{u}_{n+1}\cdot\mathbf{u}_{n-1}\rangle = \langle\mathbf{u}_{n+1}\cdot\mathbf{u}_n\rangle\langle\mathbf{u}_n\cdot\mathbf{u}_{n-1}\rangle = \left[\frac{I_{D/2}(J/k_BT)}{I_{D/2-1}(J/k_BT)}\right]^2, \qquad (15.122)$$

and further, by induction, that

$$\langle\mathbf{u}_l\cdot\mathbf{u}_k\rangle = \left[\frac{I_{D/2}(J/k_BT)}{I_{D/2-1}(J/k_BT)}\right]^{|l-k|}. \qquad (15.123)$$

For the polymer, this implies an exponential falloff

$$\langle\mathbf{u}_l\cdot\mathbf{u}_k\rangle = e^{-|l-k|a/\xi}, \qquad (15.124)$$

where ξ is the *persistence length*

$$\xi = -a/\log\left[\frac{I_{D/2}(\kappa/ak_BT)}{I_{D/2-1}(\kappa/ak_BT)}\right]. \qquad (15.125)$$

For $D = 3$, this is equal to

$$\xi = -\frac{a}{\log\left[\coth(\kappa/ak_BT) - ak_BT/\kappa\right]}. \qquad (15.126)$$

Knowing the correlation functions it is easy to calculate the magnetic susceptibility. The total magnetic moment is

$$\mathbf{M} = a\sum_{n=0}^{N}\mathbf{u}_n, \qquad (15.127)$$

so that we find the total expectation value

$$\langle\mathbf{M}^2\rangle = a^2(N+1)\frac{1+e^{-a/\xi}}{1-e^{-a/\xi}} - 2a^2e^{-a/\xi}\frac{1-e^{-(N+1)a/\xi}}{(1-e^{-a/\xi})^2}. \qquad (15.128)$$

The susceptibility is directly proportional to this. For more details see [2].

15.7.3 End-to-End Distribution

A modification of the path integral (15.108) yields the distribution of the end-to-end distance at given initial and final directions of the polymer links:

$$\mathbf{R} = \mathbf{x}_b - \mathbf{x}_a = a \sum_{n=1}^{N} \mathbf{u}_n \tag{15.129}$$

of the stiff polymer:

$$P_N(\mathbf{u}_b, \mathbf{u}_a; \mathbf{R}) = \frac{1}{Z_N} \frac{1}{A} \prod_{n=2}^{N-1} \left[\int \frac{d\mathbf{u}_n}{A} \right] \delta^{(D)}(\mathbf{R} - a \sum_{n=1}^{N} \mathbf{u}_n)$$
$$\times \exp\left[-\frac{\kappa}{2ak_BT} \sum_{n=1}^{N-1} (\mathbf{u}_{n+1} - \mathbf{u}_n)^2 \right], \tag{15.130}$$

whose integral over $\mathbf{R}$ leads back to the distribution $P_N(\mathbf{u}_b, \mathbf{u}_a)$:

$$\int d^D R \, P_N(\mathbf{u}_b, \mathbf{u}_a; \mathbf{R}) = P_N(\mathbf{u}_b, \mathbf{u}_a). \tag{15.131}$$

If we integrate in (15.130) over all final directions and average over the initial ones, we obtain the physically more accessible end-to-end distribution

$$P_N(\mathbf{R}) = \int d\mathbf{u}_b \int \frac{d\mathbf{u}_a}{S_D} P_N(\mathbf{u}_b, \mathbf{u}_a; \mathbf{R}). \tag{15.132}$$

The sliced path integral, as it stands, does not yet give quite the desired probability. Two small corrections are necessary, the same that brought the integral *near* the surface of a sphere to the path integral *on* the sphere in Section 8.9. After including these, we obtain a proper overall normalization of $P_N(\mathbf{u}_b, \mathbf{u}_a; \mathbf{R})$.

15.7.4 Moments of End-to-End Distribution

Since the $\delta^{(D)}$-function in (15.130) contains vectors $\mathbf{u}_n$ of unit length, the calculation of the complete distribution is not straightforward. Moments of the distribution, however, which are defined by the integrals

$$\langle R^{2l} \rangle = \int d^D R \, R^{2l} \, P_N(\mathbf{R}), \tag{15.133}$$

are relatively easy to find from the multiple integrals

$$\langle R^{2l} \rangle = \frac{1}{Z_N} \int d^D R \frac{1}{A} \int d\mathbf{u}_b \prod_{n=2}^{N-1} \left[\int \frac{d\mathbf{u}_n}{A} \right] \left[\int \frac{d\mathbf{u}_a}{S_D} \right] \delta^{(D)}(\mathbf{R} - \sum_{n=1}^{N} a\mathbf{u}_n)$$
$$\times \quad R^{2l} \exp\left[-\frac{\kappa}{2ak_BT} \sum_{n=1}^{N-1} (\mathbf{u}_{n+1} - \mathbf{u}_n)^2 \right]. \tag{15.134}$$

Performing the integral over $\mathbf{R}$ gives

$$\langle R^{2l} \rangle = \frac{1}{Z_N} \frac{1}{A} \int d\mathbf{u}_b \prod_{n=2}^{N-1} \left[\int \frac{d\mathbf{u}_n}{A} \right] \left[\int \frac{d\mathbf{u}_a}{S_D} \right] \left(a \sum_{n=1}^{N} \mathbf{u}_n \right)^{2l}$$

$$\times \exp \left[-\frac{\kappa}{2ak_BT} \sum_{n=1}^{N-1} (\mathbf{u}_{n+1} - \mathbf{u}_n)^2 \right] . \tag{15.135}$$

Due to the normalization property (15.115), the trivial moment is equal to unity:

$$\langle 1 \rangle = \int d^D R \, P_N(\mathbf{R}) = \int d\mathbf{u}_b \int \frac{d\mathbf{u}_a}{S_D} P_N(\mathbf{u}_b, \mathbf{u}_a | L) = 1. \tag{15.136}$$

15.8 Continuum Formulation

Some properties of stiff polymers are conveniently studied in the continuum limit of the sliced path integral (15.108), in which the bond length a goes to zero and the link number to infinity so that $L = Na$ stays constant. In this limit, the bending energy (15.107) becomes

$$E_{\text{bend}} = \frac{\kappa}{2} \int_0^L ds \, (\partial_s \mathbf{u})^2 , \tag{15.137}$$

where

$$\mathbf{u}(s) = \frac{d}{ds} \mathbf{x}(s), \tag{15.138}$$

is the unit tangent vector of the space curve along which the polymer runs. The parameter s is the arc length of the line elements, i.e., $ds = \sqrt{d\mathbf{x}^2}$.

15.8.1 Path Integral

If the continuum limit is taken purely formally on the product of integrals (15.108), we obtain a path integral

$$(\mathbf{u}_b L | \mathbf{u}_a 0) = \int \mathcal{D}\mathbf{u} \, e^{-(\kappa/2k_BT) \int_0^L ds \, [\mathbf{u}'(s)]^2}, \tag{15.139}$$

This coincides with the Euclidean version of a path integral for a particle on the surface of a sphere. It is a nonlinear σ-model (recall p. 738).

The result of the integration has been given in Section 8.9 where we found that it does not quite agree with what we would obtain from the continuum limit of the discrete solution (15.111) using the limiting formula (8.156), which is

$$P(\mathbf{u}_b, \mathbf{u}_a | L) = \sum_{l=0}^{\infty} \exp \left(-L \frac{k_BT}{2\kappa} L_2 \right) \sum_{\mathbf{m}} Y_{l\mathbf{m}}(\mathbf{u}_b) Y_{l\mathbf{m}}^*(\mathbf{u}_a), \tag{15.140}$$

with

$$L_2 = (D/2 - 1 + l)^2 - 1/4. \tag{15.141}$$

For a particle on a sphere, the discrete expression (15.130) requires a correction since it does not contain the proper time-sliced action and measure. According the Section 8.9 the correction replaces L_2 by the eigenvalues of the square angular momentum operator in D dimensions, $\hat{L}^2$,

$$L_2 \to \hat{L}^2 = l(l + D - 2). \tag{15.142}$$

After this replacement the expectation of the trivial moment $\langle 1 \rangle$ in (15.136) is equal to unity since it gives the distribution (15.140) the proper normalization:

$$\int d\mathbf{u}_b \, P(\mathbf{u}_b, \mathbf{u}_a | L) = 1. \tag{15.143}$$

This follows from the integral

$$\int d\mathbf{u}_b \sum_{\mathbf{m}} Y_{l\mathbf{m}}^*(\mathbf{u}_b) Y_{l\mathbf{m}}(\mathbf{u}_a) = \delta_{l0}, \tag{15.144}$$

which was derived in (8.249). Thus, with $\hat{L}^2$ in (15.140) instead of L_2, no extra normalization factor is required. The sum over $Y_{l\mathbf{m}}(\mathbf{u}_2) Y_{l\mathbf{m}}^*(\mathbf{u}_1)$ may furthermore be rewritten in terms of Gegenbauer polynomials using the addition theorem (8.125), so that we obtain

$$P(\mathbf{u}_b, \mathbf{u}_a | L) = \sum_{l=0}^{\infty} \exp\left(-L\frac{k_B T}{2\kappa} \hat{L}^2\right) \frac{1}{S_D} \frac{2l + D - 2}{D - 2} C_l^{(D/2-1)}(\mathbf{u}_2 \mathbf{u}_1). \tag{15.145}$$

15.8.2 Correlation Functions and Moments

We are now ready to evaluate the expectation values of R^{2l}. In the continuum approximation we write

$$R^{2l} = \left[\int_0^L ds \, \mathbf{u}(s)\right]^{2l}. \tag{15.146}$$

The expectation value of the lowest moment $\langle R^2 \rangle$ is given by the double-integral over the correlation function $\langle \mathbf{u}(s_2)\mathbf{u}(s_1) \rangle$:

$$\langle R^2 \rangle = \int_0^L ds_2 \int_0^L ds_1 \langle \mathbf{u}(s_2)\mathbf{u}(s_1) \rangle = 2 \int_0^L ds_2 \int_0^{s_2} ds_1 \langle \mathbf{u}(s_2)\mathbf{u}(s_1) \rangle. \tag{15.147}$$

The correlation function is calculated from the path integral via the composition law as in Eq. (3.298), which yields here

$$\langle \mathbf{u}(s_2)\mathbf{u}(s_1) \rangle = \int d\mathbf{u}_b \int \frac{d\mathbf{u}_a}{S_D} \int d\mathbf{u}_2 \int d\mathbf{u}_1$$
$$\times \; P(\mathbf{u}_b, \mathbf{u}_2 | L - s_2) \, \mathbf{u}_2 \, P(\mathbf{u}_2, \mathbf{u}_1 | s_2 - s_1) \, \mathbf{u}_1 \, P(\mathbf{u}_1, \mathbf{u}_a | s_1). \tag{15.148}$$

The integrals over $\mathbf{u}_a$ and $\mathbf{u}_b$ remove the initial and final distributions via the normalization integral (15.143), leaving

$$\langle \mathbf{u}(s_2)\mathbf{u}(s_1) \rangle = \int d\mathbf{u}_2 \int \frac{d\mathbf{u}_1}{S_D} \, \mathbf{u}_2\mathbf{u}_1 P(\mathbf{u}_2, \mathbf{u}_1 | s_2 - s_1). \tag{15.149}$$

Due to the manifest rotational invariance, the normalized integral over $\mathbf{u}_1$ can be omitted. By inserting the spectral representation (15.145) with the eigenvalues (15.142) we obtain

$$\langle \mathbf{u}(s_2)\mathbf{u}(s_1) \rangle = \int d\mathbf{u}_2 \, \mathbf{u}_2\mathbf{u}_1 P(\mathbf{u}_2, \mathbf{u}_1 | s_2 - s_1)$$

$$= \sum_l e^{-(s_2 - s_1)k_B T \hat{L}^2/2\kappa} \left[\int d\mathbf{u}_2 \, \mathbf{u}_2\mathbf{u}_1 \frac{1}{S_D} \frac{2l + D - 2}{D - 2} C_l^{(D/2-1)}(\mathbf{u}_2\mathbf{u}_1) \right]. \tag{15.150}$$

We now calculate the integral in the brackets with the help of the recursion relation (15.152) for the Gegenbauer functions[4]

$$zC_l^{(\nu)}(z) = \frac{1}{2(\nu + l)} \left[(2\nu + l - 1)C_{l-1}^{(\nu)}(z) + (l + 1)C_{l+1}^{\nu}(z) \right]. \tag{15.151}$$

Obviously, the integral over $\mathbf{u}_2$ lets only the term $l = 1$ survive. This, in turn, involves the integral

$$\int d\mathbf{u}_2 \, C_0^{(D/2-1)}(\cos\theta) = S_D. \tag{15.152}$$

The $l = 1$-factor $D/(D-2)$ in (15.150) is canceled by the first $l = 1$-factor in the recursion (15.151) and we obtain the correlation function

$$\langle \mathbf{u}(s_2)\mathbf{u}(s_1) \rangle = \exp\left[-(s_2 - s_1)\frac{k_B T}{2\kappa}(D - 1) \right], \tag{15.153}$$

where $D - 1$ in the exponent is the eigenvalue of $\hat{L}^2 = l(l + D - 2)$ at $l = 1$. The correlation function (15.153) agrees with the sliced result (15.124) if we identify the continuous version of the persistence length (15.126) with

$$\xi \equiv 2\kappa/k_B T(D - 1). \tag{15.154}$$

Indeed, taking the limit $a \to 0$ in (15.125), we find with the help of the asymptotic behavior (8.12) precisely the relation (15.154).

After performing the double-integral in (15.147) we arrive at the desired result for the first moment:

$$\langle R^2 \rangle = 2 \left\{ \xi L - \xi^2 \left[1 - e^{-L/\xi} \right] \right\}. \tag{15.155}$$

---

[4]I.S. Gradshteyn and I.M. Ryzhik, op. cit., Formula 8.933.1.

This is valid for *all* D. The result may be compared with the expectation value of the squared magnetic moment of the Heisenberg chain in Eq. (15.128), which reduces to this in the limit $a \to 0$ at fixed $L = (N+1)a$.

For small L/ξ, the second moment (15.155) has the large-stiffness expansion

$$\langle R^2 \rangle = L^2 \left[1 - \frac{1}{3}\frac{L}{\xi} + \frac{1}{12}\left(\frac{L}{\xi}\right)^2 - \frac{1}{60}\left(\frac{L}{\xi}\right)^3 + \dots \right], \qquad (15.156)$$

the first term being characteristic for a completely stiff chain [see Eq. (15.95)]. For large L/ξ, on the other hand, we find the small-stiffness expansion

$$\langle R^2 \rangle \approx 2\xi L \left(1 - \frac{\xi}{L} \right) + \dots , \qquad (15.157)$$

where the dots denote exponentially small terms. The first term agrees with relation (15.94) for a random chain with an effective bond length

$$a_{\text{eff}} = 2\xi = \frac{4}{D-1}\frac{\kappa}{k_B T}. \qquad (15.158)$$

The calculation of higher expectations $\langle R^{2l} \rangle$ becomes rapidly complicated. Take, for instance, the moment $\langle R^4 \rangle$, which is given by the quadruple integral over the four-point correlation function

$$\langle R^4 \rangle = 8 \int_0^L ds_4 \int_0^{s_4} ds_3 \int_0^{s_3} ds_2 \int_0^{s_2} ds_1 \, \delta_{i_4 i_3 i_2 i_1} \langle u_{i_4}(s_4) u_{i_3}(s_3) u_{i_2}(s_2) u_{i_1}(s_1) \rangle, \qquad (15.159)$$

with the symmetric pair contraction tensor of Eq. (15.22):

$$\delta_{i_4 i_3 i_2 i_1} \equiv \left(\delta_{i_4 i_3}\delta_{i_2 i_1} + \delta_{i_4 i_2}\delta_{i_3 i_1} + \delta_{i_4 i_1}\delta_{i_3 i_2} \right). \qquad (15.160)$$

The factor 8 and the symmetrization of the indices arise when bringing the integral

$$R^4 = \int_0^L ds_4 \int_0^L ds_3 \int_0^L ds_2 \int_0^L ds_1 \, (\mathbf{u}(s_4)\mathbf{u}(s_3))(\mathbf{u}(s_2)\mathbf{u}(s_1)) \qquad (15.161)$$

to the s-ordered form in (15.159). This form is needed for the s-ordered evaluation of the $\mathbf{u}$-integrals which proceeds by a direct extension of the previous procedure for $\langle R^2 \rangle$. We write down the extension of expression (15.148) and perform the integrals over $\mathbf{u}_a$ and $\mathbf{u}_b$ which remove the initial and final distributions via the normalization integral (15.143), leaving $\delta_{i_4 i_3 i_2 i_1}$ times an integral [the extension of (15.149)]:

$$\langle \mathbf{u}_{i_4}(s_4)\mathbf{u}_{i_3}(s_3)\mathbf{u}_{i_2}(s_2)\mathbf{u}_{i_1}(s_1) \rangle = \int d\mathbf{u}_4 \int d\mathbf{u}_3 \int d\mathbf{u}_2 \int \frac{d\mathbf{u}_1}{S_D} \qquad (15.162)$$
$$\times u_{i_4} u_{i_3} u_{i_2} u_{i_1} P(\mathbf{u}_4, \mathbf{u}_3 | s_4 - s_3) P(\mathbf{u}_3, \mathbf{u}_2 | s_3 - s_2) P(\mathbf{u}_2, \mathbf{u}_1 | s_2 - s_1) .$$

The normalized integral over $\mathbf{u}_1$ can again be omitted. Still, the expression is complicated. A somewhat tedious calculation yields

$$\langle R^4 \rangle = \frac{4(D+2)}{D} L^2 \xi^2 - 8L\xi^3 \left(\frac{D^2 + 6D - 1}{D^2} - \frac{D-7}{D+1} e^{-L/\xi} \right) \tag{15.163}$$

$$+ 4\xi^4 \left[\frac{D^3 + 23D^2 - 7D + 1}{D^3} - 2\frac{(D+5)^2}{(D+1)^2} e^{-L/\xi} + \frac{(D-1)^5}{D^3(D+1)^2} e^{-2DL/(D-1)\xi} \right].$$

For small values of L/ξ, we find the large-stiffness expansion

$$\langle R^4 \rangle = L^4 \left[1 - \frac{2}{3}\frac{L}{\xi} + \frac{25D - 17}{90(D-1)} \left(\frac{L}{\xi} \right)^2 - 4\frac{7D^2 - 8D + 3}{315(D-1)^2} \left(\frac{L}{\xi} \right)^3 + \ldots \right], \tag{15.164}$$

the leading term being equal to (15.95) for a completely stiff chain.

In the opposite limit of large L/ξ, the small-stiffness expansion is

$$\langle R^4 \rangle = 4\frac{D+2}{D} L^2 \xi^2 \left[1 - 2\frac{D^2 + 6D - 1}{D(D+2)}\frac{\xi}{L} + \frac{D^3 + 23D^2 - 7D + 1}{D^2(D+2)} \left(\frac{\xi}{L} \right)^2 \right] + \ldots, \tag{15.165}$$

where the dots denote exponentially small terms. The leading term agrees again with the expectation $\langle R^4 \rangle$ of Eq. (15.94) for a random chain whose distribution is (15.49) with an effective link length $a_{\text{eff}} = 2\xi$ of Eq. (15.158). The remaining terms are corrections caused by the stiffness of the chain.

It is possible to find a correction factor to the Gaussian distribution which maintains the unit normalization and ensures that the moment $\langle R^2 \rangle$ has the small-ξ exact expansion (15.157) whereas $\langle R^2 \rangle$ is equal to (15.165) up to the first correction term in ξ/L. This has the form

$$P_L(\mathbf{R}) = \sqrt{\frac{D}{4\pi L\xi}}^D e^{-DR^2/4L\xi} \left\{ 1 - \frac{2D-1}{4}\frac{\xi}{L} + \frac{3D-1}{4}\frac{R^2}{L^2} - \frac{D(4D-1)}{16(D+2)}\frac{R^4}{\xi L^3} \right\}. \tag{15.166}$$

In three dimensions, this was first written down by Daniels [3]. It is easy to match also the moment $\langle R^4 \rangle$ by adding in the curly brackets the following terms

$$\frac{1 - 7D + 23D^2 + D^3}{D+1} \left[\frac{D+2}{8D}\frac{\xi^2}{L^2} \left(1 + \frac{R^2}{\xi L} \right) + \frac{1}{32}\frac{R^4}{L^4} \right]. \tag{15.167}$$

These terms do not, however, improve the fits to Monte Carlo data for $\xi > 1/10L$, since the expansion is strongly divergent.

From the approximation (15.166) with the additional term (15.167) we calculate the small-stiffness expansion of all *even and odd* moments as follows:

$$\langle R^n \rangle = \frac{2^n \Gamma(D/2 + n/2)}{D^{n/2}\Gamma(D/2)} L^n \xi^n \left[1 + A_1 \frac{\xi}{L} + A_2 \left(\frac{\xi}{L} \right)^2 + \ldots \right], \tag{15.168}$$

where

$$A_1 = n\frac{n - 2 - 2d^2 - 4d(n-1)}{4d(2+d)}, \qquad A_2 = n(n-2)\frac{1 - 7d + 23d^2 + d^3}{8d^2(1+d)}. \tag{15.169}$$

15.9 Schrödinger Equation and Recursive Solution for End-to-End Distribution Moments

The most efficient way of calculating the moments of the end-to-end distribution proceeds by setting up a Schrödinger equation satisfied by (15.112) and solving it recursively with similar methods as developed in 3.19 and Appendix 3C.

15.9.1 Setting up the Schrödinger Equation

In the continuum limit, we write (15.112) as a path integral [compare (15.139)]

$$P_L(\mathbf{R}) \propto \int d\mathbf{u}_b \int d\mathbf{u}_a \int \mathcal{D}^{D-1}\mathbf{u} \; \delta^{(D)}\left(\mathbf{R} - \int_0^L ds \, \mathbf{u}(s)\right) e^{-(\bar{\kappa}/2)\int_0^L ds \, [\mathbf{u}'(s)]^2}, \qquad (15.170)$$

where we have introduced the reduced stiffness

$$\bar{\kappa} = \frac{\kappa}{k_B T} = (D-1)\frac{\xi}{2}, \qquad (15.171)$$

for brevity. After a Fourier representation of the δ-function, this becomes

$$P_L(\mathbf{R}) \propto \int_{-i\infty}^{i\infty} \frac{d^D\lambda}{2\pi i} e^{\bar{\kappa}\boldsymbol{\lambda}\cdot\mathbf{R}/2} \int d\mathbf{u}_b \int d\mathbf{u}_a (\mathbf{u}_b L|\mathbf{u}_a\, 0)^{\boldsymbol{\lambda}}, \qquad (15.172)$$

where

$$(\mathbf{u}_b\, L|\mathbf{u}_a\, 0)^{\boldsymbol{\lambda}} \equiv \int_{\mathbf{u}(0)=\mathbf{u}_a}^{\mathbf{u}(L)=\mathbf{u}_b} \mathcal{D}^{D-1}\mathbf{u} \, e^{-(\bar{\kappa}/2)\int_0^L ds\left\{[\mathbf{u}'(s)]^2 + \boldsymbol{\lambda}\cdot\mathbf{u}(s)\right\}} \qquad (15.173)$$

describes a point particle of mass $M = \bar{\kappa}$ moving on a unit sphere. In contrast to the discussion in Section 8.7 there is now an additional external field $\boldsymbol{\lambda}$ which prevents us from finding an exact solution. However, all even moments $\langle R_{i_1} R_{i_2} \cdots R_{i_{2l}}\rangle$ of the end-to-end distribution (15.172) can be extracted from the expansion coefficients in powers of λ_i of the integral $\int d\mathbf{u}_b \int d\mathbf{u}_a$ over (15.173). The presence of these directional integrals permits us to assume the external electric field $\boldsymbol{\lambda}$ to point in the z-direction, or the Dth direction in D-dimensions. Then $\boldsymbol{\lambda} = \lambda\hat{z}$, and the moments $\langle R^{2l}\rangle$ are proportional to the derivatives $(2/\bar{\kappa})^{2l}\partial_\lambda^{2l} \int d\mathbf{u}_b \int d\mathbf{u}_a (\mathbf{u}_b L|\mathbf{u}_a\, 0)^{\boldsymbol{\lambda}}$. The proportionality factors have been calculated in Eq. (15.84). It is unnecessary to know these since we can always use the rod limit (15.95) to normalize the moments.

To find these derivatives, we perform a perturbation expansion of the path integral (15.173) around the solvable case $\lambda = 0$.

In natural units with $\bar{\kappa} = 1$, the path integral (15.173) solves obviously the imaginary-time Schrödinger equation

$$\left(-\frac{1}{2}\Delta_\mathbf{u} + \frac{1}{2}\boldsymbol{\lambda}\cdot\mathbf{u} + \frac{d}{d\tau}\right)(\mathbf{u}\,\tau|\mathbf{u}_a\, 0)^{\boldsymbol{\lambda}} = 0, \qquad (15.174)$$

where $\Delta_{\mathbf{u}}$ is the Laplacian on a unit sphere. In the probability distribution (15.211), only the integrated expression

$$\psi(z, \tau; \lambda) \equiv \int d\mathbf{u}_a (\mathbf{u}\,\tau | \mathbf{u}_a\, 0)^\lambda \tag{15.175}$$

appears, which is a function of $z = \cos\theta$ only, where θ is the angle between $\mathbf{u}$ and the electric field $\boldsymbol{\lambda}$. For $\psi(z, \tau; \lambda)$, the Schrödinger equation reads

$$\hat{H}\,\psi(z, \tau; \lambda) = -\frac{d}{d\tau}\psi(z, \tau; \lambda), \tag{15.176}$$

with the simpler Hamiltonian operator

$$\begin{aligned}
\hat{H} &\equiv \hat{H}_0 + \lambda\,\hat{H}_I = -\frac{1}{2}\Delta + \lambda\,z \\
&= -\frac{1}{2}\left[(1-z^2)\frac{d^2}{dz^2} - (D-1)z\frac{d}{dz}\right] + \frac{1}{2}\lambda z\,.
\end{aligned} \tag{15.177}$$

Now the desired moments (15.135) can be obtained from the coefficient of $\lambda^{2l}/(2l)!$ of the power series expansion of the z-integral over (15.175) at imaginary time $\tau = L$:

$$f(L; \lambda) \equiv \int_{-1}^{1} dz\,\psi(z, L; \lambda). \tag{15.178}$$

15.9.2 Recursive Solution of Schrödinger Equation.

The function $f(L; \lambda)$ has a spectral representation

$$f(L; \lambda) = \sum_{l=0}^{\infty} \frac{\int_{-1}^{1} dz\,\varphi^{(l)\dagger}(z)\,\exp\left(-E^{(l)}L\right)\int_{-1}^{1} dz_a\,\varphi^{(l)}(z_a)}{\int_{-1}^{1} dz\,\varphi^{(l)\dagger}(z)\,\varphi^{(l)}(z)}, \tag{15.179}$$

where $\varphi^{(l)}(z)$ are the solutions of the time-independent Schrödinger equation $\hat{H}\varphi^{(l)}(z) = E^{(l)}\varphi^{(l)}(z)$. Applying perturbation theory to this problem, we start from the eigenstates of the unperturbed Hamiltonian $\hat{H}_0 = -\Delta/2$, which are given by the Gegenbauer polynomials $C_l^{D/2-1}(z)$ with the eigenvalues $E_0^{(l)} = l(l+D-2)/2$. Following the methods explained in 3.19 and Appendix 3C we now set up a recursion scheme for the perturbation expansion of the eigenvalues and eigenfunctions [4].Starting point is the expansion of energy eigenvalues and wave-functions in powers of the coupling constant λ:

$$E^{(l)} = \sum_{j=0}^{\infty} \epsilon_j^{(l)}\,\lambda^j, \qquad |\varphi^{(l)}\rangle = \sum_{l',i=0}^{\infty} \gamma_{l',i}^{(l)}\,\lambda^i\,\alpha_{l'}\,|l'\rangle\,. \tag{15.180}$$

The wave functions $\varphi^{(l)}(z)$ are the scalar products $\langle z|\varphi^{(l)}(\lambda)\rangle$. We have inserted extra normalization constants $\alpha_{l'}$ for convenience which will be fixed soon. The unperturbed state vectors $|l\rangle$ are normalized to unity, but the state vectors $|\varphi^{(l)}\rangle$ of the interacting system will be normalized in such a way, that $\langle \varphi^{(l)}|l\rangle = 1$ holds to all orders, implying that

$$\gamma_{l,i}^{(l)} = \delta_{i,0} \qquad \gamma_{k,0}^{(l)} = \delta_{l,k} \,. \tag{15.181}$$

Inserting the above expansions into the Schrödinger equation, projecting the result onto the base vector $\langle k|\alpha_k$, and extracting the coefficient of λ^j, we obtain the equation

$$\gamma_{k,i}^{(l)}\epsilon_0^{(k)} + \sum_{j=0}^{\infty} \frac{\alpha_j}{\alpha_k} V_{k,j}\,\gamma_{j,i-1}^{(l)} = \sum_{j=0}^{i} \epsilon_j^{(l)}\gamma_{k,i-j}^{(l)} \,, \tag{15.182}$$

where $V_{k,j} = \lambda\langle k|z|j\rangle$ are the matrix elements of the interaction between unperturbed states. For $i = 0$, Eq. (15.182) is satisfied identically. For $i > 0$, it leads to the following two recursion relations, one for $k = l$:

$$\epsilon_i^{(l)} = \sum_{n=\pm 1} \gamma_{l+n,i-1}^{(l)} W_n^{(l)}, \tag{15.183}$$

the other one for $k \neq l$:

$$\gamma_{k,i}^{(l)} = \frac{\displaystyle\sum_{j=1}^{i-1} \epsilon_j^{(l)}\gamma_{k,i-j}^{(l)} - \sum_{n=\pm 1} \gamma_{k+n,i-1}^{(l)} W_n^{(l)}}{\epsilon_0^{(k)} - \epsilon_0^{(l)}}, \tag{15.184}$$

where only $n = -1$ and $n = 1$ contribute to the sums over n since

$$W_n^{(l)} \equiv \frac{\alpha_{l+n}}{\alpha_l} \langle l|\,z\,|l+n\rangle = 0, \text{ for } n \neq \pm 1. \tag{15.185}$$

The vanishing of $W_n^{(l)}$ for $n \neq \pm 1$ is due to the band-diagonal form of the matrix of the interaction z in the unperturbed basis $|n\rangle$. It is this property which makes the sums in (15.183) and (15.184) finite and leads to recursion relations with a finite number of terms for all $\epsilon_i^{(l)}$ and $\gamma_{k,i}^{(l)}$. To calculate $W_n^{(l)}$, it is convenient to express $\langle l|\,z\,|l+n\rangle$ as matrix elements between unnormalized noninteracting states $|n\}$ as

$$\langle l|\,z\,|l+n\rangle = \frac{\{l|z|l+n\}}{\sqrt{\{l|l\}\{l+n|l+n\}}}, \tag{15.186}$$

where expectation values are defined by the integrals

$$\{k|F(z)|l\} \equiv \int_{-1}^{1} C_k^{D/2-1}(z)\, F(z)\, C_l^{D/2-1}(z)(1-z^2)^{(D-3)/2}\,dz, \tag{15.187}$$

from which we find[5]

$$\{l|l\} = \frac{2^{4-D}\,\Gamma(l+D-2)\,\pi}{l!\,(2l+D-2)\,\Gamma(D/2-1)^2}. \tag{15.188}$$

---

[5]I.S. Gradshteyn and I.M. Ryzhik, *op. cit.*, Formulas 7.313.1 and 7.313.2.

Expanding the numerator of (15.186) with the help of the recursion relation (15.151) for the Gegenbauer polynomials written now in the form

$$(l + 1)|l + 1\} = (2l + D - 2) z |l\} - (l + D - 3)|l - 1\}, \qquad (15.189)$$

we find the only non-vanishing matrix elements to be

$$\{l + 1|z|l\} = \frac{l + 1}{2l + D - 2}\{l + 1|l + 1\}, \qquad (15.190)$$

$$\{l - 1|z|l\} = \frac{l + D - 3}{2l + D - 2}\{l - 1|l - 1\} . \qquad (15.191)$$

Inserting these together with (15.188) into (15.186) gives

$$\langle l|z|l - 1\rangle = \sqrt{\frac{l(l + D - 3)}{(2l + D - 2)(2l + D - 4)}}, \qquad (15.192)$$

and a corresponding result for $\langle l|z|l + 1\rangle$. We now fix the normalization constants $\alpha_{l'}$ by setting

$$W_1^{(l)} = \frac{\alpha_{l+1}}{\alpha_l} \langle l| z |l + 1\rangle = 1 \qquad (15.193)$$

for all l, which determines the ratios

$$\frac{\alpha_l}{\alpha_{l+1}} = \langle l| z |l + 1\rangle = \sqrt{\frac{(l + 1)\,(l + D - 2)}{(2l + D)\,(2l + D - 2)}}. \qquad (15.194)$$

Setting further $\alpha_1 = 1$, we obtain

$$\alpha_l = \left[\prod_{j=1}^{l} \frac{(2l + D - 2)(2l + D - 4)}{l(l + D - 3)} \right]^{1/2} . \qquad (15.195)$$

Using this we find from (15.185) the remaining nonzero $W_n^{(l)}$ for $n = -1$:

$$W_{-1}^{(l)} = \frac{l(l + D - 3)}{(2l + D - 2)(2l + D - 4)} . \qquad (15.196)$$

We are now ready to solve the recursion relations of (15.183) and (15.184) $\gamma_{k,i}^{(l)}$ and $\epsilon_i^{(l)}$ order by order in i. For the initial order $i = 0$, the values of the $\gamma_{k,i}^{(l)}$ are given by Eq. (15.181). The coefficients $\epsilon_i^{(l)}$ are equal to the unperturbed energies $\epsilon_0^{(l)} = E_0^{(l)} = l(l + D - 2)/2$. For each $i = 1, 2, 3, \ldots$, there is only a finite number of non-vanishing $\gamma_{k,j}^{(l)}$ and $\epsilon_j^{(l)}$ with $j < i$ on the right-hand sides of (15.183) and (15.184) which allows us to calculate $\gamma_{k,i}^{(l)}$ and $\epsilon_i^{(l)}$ on the left-hand sides. In this way it is easy to find the perturbation expansions for the energy and the wave functions to high orders.

Inserting the resulting expansions (15.180) into Eq. (15.179), only the totally symmetric parts in $\varphi^{(l)}(z)$ will survive the integration in the numerators, i.e., we may insert only

$$\varphi^{(l)}_{\text{symm}}(z) = \langle z|\varphi^{(l)}_{\text{symm}}\rangle = \sum_{i=0}^{\infty} \gamma^{(l)}_{0,i} \lambda^i \langle z|0\rangle . \tag{15.197}$$

The denominators of (15.179) become explicitly $\sum_{l',i} |\gamma^{(l)}_{l',i} \alpha_{l'}|^2 \lambda^{2i}$, where the sum over i is limited by power of λ^2 up to which we want to carry the perturbation series; also l' is restricted to a finite number of terms only, because of the band-diagonal structure of the $\gamma^{(l)}_{l',i}$.

Extracting the coefficients of the power expansion in λ from (15.179) we obtain all desired moments of the end-to-end distribution, in particular the second and fourth moments (15.155) and (15.164). Higher even moments are easily found with the help of a MATHEMATICA program, which is available for download in notebook form [5]. The expressions are too lengthy to be written down here. We may, however, expand the even moments $\langle R^n\rangle$ in powers of L/ξ to find a general large-stiffness expansion valid for all even *and odd* n:

$$\frac{\langle R^n\rangle}{L^n} = 1 - \frac{n}{6}\frac{L}{\xi} + \frac{n\left(-13 - n + 5D\left(1+n\right)\right)}{360\left(D-1\right)}\frac{L^2}{\xi^2} - a_3\frac{L^3}{\xi^3} + a_4\frac{L^4}{\xi^4} + \ldots , \tag{15.198}$$

where

$$a_3 = n\frac{444 - 63n + 15n^2 + 7D^2\left(4 + 15n + 5n^2\right) + 2D\left(-124 - 141n + 7n^2\right)}{45360(D-1)^2},$$

$$a_4 = \frac{n}{5443200(d-1)^3}\left(D_0 + D_1 d + D_2 d^2 + D_3 d^3\right), \tag{15.199}$$

with

$$D_0 = 3\left(-5610 + 2921n - 822n^2 + 67n^3\right), D_1 = 8490 + 12103n - 3426n^2 + 461n^3,$$

$$D_2 = 45\left(-2 - 187n - 46n^2 + 7n^3\right), \qquad D_4 = 35\left(-6 + 31n + 30n^2 + 5n^3\right). \tag{15.200}$$

The lowest odd moments are, up to order l^4,

$$\frac{\langle R\rangle}{L} = 1 - \frac{l}{6} + \frac{5D-7}{180(D-1)}l^2 - \frac{33 - 43D + 14D^2}{3780(D-1)^2}l^3 - \frac{861 - 1469D + 855D^2 - 175D^3}{453600\left(D-1\right)^3}l^4 \ldots ,$$

$$\frac{\langle R^3\rangle}{L^3} = 1 - \frac{l}{2} + \frac{5D-4}{30\left(D-1\right)}l^2 - \frac{195 - 484D + 329D^2}{7560\left(-1+d\right)^2}l^3 - \frac{609 - 2201D + 2955D^2 - 1435D^3}{151200\left(D-1\right)^3}l^4 \ldots .$$

15.9.3 From Moments to End-to-End Distribution for $D=3$

We now use the recursively calculated moments to calculate the end-to-end distribution itself. It can be parameterized by an analytic function of $r = R/L$ [4]:

$$P_L(\mathbf{R}) \propto r^k(1 - r^\beta)^m, \tag{15.201}$$

whose moments are exactly calculable:

$$\langle r^{2l} \rangle = \frac{\Gamma\left(\dfrac{3+k+2l}{\beta}\right)\Gamma\left(\dfrac{3+k}{\beta}+m+1\right)}{\Gamma\left(\dfrac{3+k}{\beta}\right)\Gamma\left(\dfrac{3+k+2l}{\beta}+m+1\right)}. \tag{15.202}$$

We now adjust the three parameters k, β, and m to fit the three most important moments of this distribution to the exact values, ignoring all others. If the distances were distributed uniformly over the interval $r \in [0,1]$, the moments would be $\langle r^{2l} \rangle^{\mathrm{unif}} = 1/(2l+2)$. Comparing our exact moments $\langle r^{2l} \rangle(\xi)$ with those of the uniform distribution we find that $\langle r^{2l} \rangle(\xi)/\langle r^{2l} \rangle^{\mathrm{unif}}$ has a maximum for n close to $n_{\max}(\xi) \equiv 4\xi/L$. We identify the most important moments as those with $n = n_{\max}(\xi)$ and $n = n_{\max}(\xi) \pm 1$. If $n_{\max}(\xi) \le 1$, we choose the lowest even moments $\langle r^2 \rangle$, $\langle r^4 \rangle$, and $\langle r^6 \rangle$. In particular, we have fitted $\langle r^2 \rangle$, $\langle r^4 \rangle$ and $\langle r^6 \rangle$ for small persistence length $\xi < L/2$. For $\xi = L/2$, we have started with $\langle r^4 \rangle$, for $\xi = L$ with $\langle r^8 \rangle$ and for $\xi = 2L$ with $\langle r^{16} \rangle$, including always the following two higher even moments. After these adjustments, whose results are shown in Fig. 15.4, we obtain the distributions shown in Fig.15.6 for various persistence lengths ξ. They are in excellent agreement with the Monte Carlo data (symbols) and better than the one-loop perturbative results (thin curves) of Ref. [6], which are good only for very stiff polymers. The MATHEMATICA program to do these fits are available from the internet address given in Footnote 5.

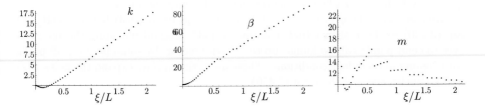

Figure 15.4 Paramters k, β, and m for a best fit of end-to-end distribution (15.201).

For small persistence lengths $\xi/L = 1/400,\ 1/100,\ 1/30$, the curves are well approximated by Gaussian random chain distributions on a lattice with lattice constant $a_{\mathrm{eff}} = 2\xi$, i.e., $P_L(\mathbf{R}) \to e^{-3R^2/4L\xi}$ [recall (15.75)]. This ensures that the lowest moment $\langle R^2 \rangle = a_{\mathrm{eff}}L$ is properly fitted. In fact, we can easily check that our fitting program yields for the parameters k,β,m in the end-to-end distribution (15.201) the $\xi \to 0$ behavior: $k \to -\xi$, $\beta \to 2 + 2\xi$, $m \to 3/4\xi$, so that (15.201) tends to the correct Gaussian behavior.

In the opposite limit of large ξ, we find that $k \to 10\xi - 7/2$, $\beta \to 40\xi + 5$, $m \to 10$, which has no obvious analytic approach to the exact limiting behavior $P_L(\mathbf{R}) \to (1 - r)^{-5/2}e^{-1/4\xi(1-r)}$, although the distribution at $\xi = 2$ is fitted numerically extremely well.

The distribution functions can be inserted into Eq. (15.89) to calculate the structure factors shown in Fig. 15.5. They interpolate smoothly between the Debye limit (15.90) and the stiff limit (15.105).

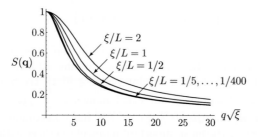

Figure 15.5 Structure functions for different persistence lengths $\xi/L = 1/400, 1/100, 1/30, 1/10, 1/5, 1/2, 1, 2$, (from bottom to top) following from the end-to-end distributions in Fig. 15.6. The curves with low ξ almost coincide in this plot over the ξ-dependent absissa. The very stiff curves fall off like $1/q$, the soft ones like $1/q^2$ [see Eqs. (15.105) and (15.90)].

15.9.4 Large-Stiffness Approximation to End-to-End Distribution

The full end-to-end distribution (15.132) cannot be calculated exactly. It is, however, quite easy to find a satisfactory approximation for large stiffness [6].

We start with the expression (15.170) for the end-to-end distribution $P_L(\mathbf{R})$. In Eq. (3.230) we have shown that a harmonic path integral including the integrals over the end points can be found, up to a trivial factor, by summing over all paths with Neumann boundary conditions. These are satisfied if we expand the fields $\mathbf{u}(s)$ into a Fourier series of the form (2.450):

$$\mathbf{u}(s) = \mathbf{u}_0 + \boldsymbol{\eta}(s) = \mathbf{u}_0 + \sum_{n=1}^{\infty} \mathbf{u}_n \cos \nu_n s, \qquad \nu_n = n\pi/L. \tag{15.203}$$

Let us parametrize the unit vectors $\mathbf{u}$ in D dimensions in terms of the first $D-1$-dimensional coordinates $u^\mu \equiv q^\mu$ with $\mu = 1, \ldots, D-1$. The Dth component is then given by a power series

$$\sigma \equiv \sqrt{1-q^2} \approx 1 - q^2/2 - (q^2)^2/8 + \ldots . \tag{15.204}$$

The we approximate the action harmonically as follows:

$$
\begin{aligned}
\mathcal{A} &= \mathcal{A}^{(0)} + \mathcal{A}^{\text{int}} = \frac{\bar{\kappa}}{2} \int_0^L ds\, [\mathbf{u}'(s)]^2 + \frac{1}{2}\delta(0)\log(1-q^2) \\
&\approx \frac{\bar{\kappa}}{2} \int_0^L ds\, [q'(s)]^2 - \frac{1}{2}\delta(0)\int_0^L ds\, q^2.
\end{aligned}
\tag{15.205}
$$

The last term comes from the invariant measure of integration $d^{D-1}q/\sqrt{1-q^2}$ [recall (10.636) and (10.641)].

Assuming, as before, that $\mathbf{R}$ points into the z, or Dth, direction we factorize

$$\delta^{(D)}\left(\mathbf{R}-\int_0^L ds\,\mathbf{u}(s)\right) = \delta\left(R-L+\int_0^L ds\,\left\{\frac{1}{2}q^2(s)+\frac{1}{8}[q^2(s)]^2+\dots\right\}\right)$$
$$\times\ \delta^{(D-1)}\left(\int_0^L ds\,q(s)\right),\qquad(15.206)$$

where $R \equiv |\mathbf{R}|$. The second δ-function on the right-hand side enforces

$$\bar{q}=L^{-1}\int_0^L ds\,q^\mu(s)=0\,,\quad \mu=1,\dots,d-1\,,\qquad(15.207)$$

and thus the vanishing of the zero-frequency parts q_0^μ in the first $D-1$ components of the Fourier decomposition (15.203).

It was shown in Eqs. (10.632) and (10.642) that the last δ-function has a distorting effect upon the measure of path integration which must be compensated by a Faddeev-Popov action

$$\mathcal{A}_e^{\mathrm{FP}}=\frac{D-1}{2L}\int_0^L ds\,q^2,\qquad(15.208)$$

where the number D of dimensions of q^μ-space (10.642) has been replaced by the present number $D-1$.

In the large-stiffness limit we have to take only the first harmonic term in the action (15.205) into account, so that the path integral (15.170) becomes simply

$$P_L(\mathbf{R})\propto\int_{\mathrm{NBC}}\mathcal{D}'^{D-1}q\,\delta\left(R-L+\int_0^L ds\,\frac{1}{2}q^2(s)\right)e^{-(\bar{\kappa}/2)\int_0^L ds\,[q'(s)]^2}.\qquad(15.209)$$

The subscript of the integral emphasizes the Neumann boundary conditions. The prime on the measure of the path integral indicates the absence of the zero-frequency component of $q^\mu(s)$ in the Fourier decomposition due to (15.207). Representing the remaining δ-function in (15.209) by a Fourier integral, we obtain

$$P_L(\mathbf{R})\propto\bar{\kappa}\int_{-i\infty}^{i\infty}\frac{d\omega^2}{2\pi i}e^{\bar{\kappa}\omega^2(L-R)}\int_{\mathrm{NBC}}\mathcal{D}'^{D-1}q\,\exp\left[-\frac{\bar{\kappa}}{2}\int_0^L ds\,\left(q'^2+\omega^2q^2\right)\right].\qquad(15.210)$$

The integral over all paths with Neumann boundary conditions is known from Eq. (2.454). At zero average path, the result is

$$\int_{\mathrm{NBC}}\mathcal{D}'^{D-1}q\,\exp\left[-\frac{\bar{\kappa}}{2}\int_0^L ds\,\left(q'^2+\omega^2q^2\right)\right]\propto\left(\frac{\omega L}{\sinh\omega L}\right)^{(D-1)/2},\qquad(15.211)$$

so that

$$P_L(\mathbf{R})\propto\bar{\kappa}\int_{-i\infty}^{i\infty}\frac{d\omega^2}{2\pi i}e^{\bar{\kappa}\omega^2(L-R)}\left(\frac{\omega L}{\sinh\omega L}\right)^{(D-1)/2}.\qquad(15.212)$$

The integral can easily be done in $D = 3$ dimensions using the original product representation (2.453) of $\omega L / \sinh \omega L$, where

$$P_L(\mathbf{R}) \propto \bar{\kappa} \int_{-i\infty}^{i\infty} \frac{d\omega^2}{2\pi i} e^{\bar{\kappa}\omega^2(L-R)} \prod_{n=1}^{\infty} \left(1 + \frac{\omega^2}{\nu_n^2}\right)^{-1}. \tag{15.213}$$

If we shift the contour of integration to the left we run through poles at $\omega^2 = -\nu_k^2$ with residues

$$\sum_{k=1}^{\infty} \nu_k^2 \prod_{n(\neq k)=1}^{\infty} \left(1 - \frac{k^2}{n^2}\right)^{-1}. \tag{15.214}$$

The product is evaluated by the limit procedure for small ϵ:

$$\left\{\prod_{n=1}^{\infty}\left[1 - \frac{(k+\epsilon)^2}{n^2}\right]^{-1}\right\}\left[1 - \frac{(k+\epsilon)^2}{k^2}\right] \rightarrow \frac{(k+\epsilon)\pi}{\sin(k+\epsilon)\pi}\frac{-2\epsilon}{k} \rightarrow \frac{-2\epsilon\pi}{\sin(k+\epsilon)\pi}$$

$$\rightarrow -\frac{2\epsilon\pi}{\cos k\pi \sin \epsilon\pi} \rightarrow 2(-1)^{k+1}. \tag{15.215}$$

Hence we obtain [6]

$$P_L(\mathbf{R}) \propto \sum_{k=1}^{\infty} 2(-1)^{k+1}\bar{\kappa}\nu_k^2\, e^{-\bar{\kappa}\nu_k^2(L-R)}. \tag{15.216}$$

It is now convenient to introduce the reduced end-to-end distance $r \equiv R/L$ and the *flexibility* of the polymer $l \equiv L/\xi$, and replace $\bar{\kappa}\nu_k^2(L-R) \rightarrow k^2\pi^2(1-r)/l$ so that (15.216) becomes

$$P_L(\mathbf{R}) = \mathcal{N}L(\mathbf{R}) = \mathcal{N}\sum_{k=1}^{\infty}(-1)^{k+1}k^2\pi^2 e^{-k^2\pi^2(1-r)/l}, \tag{15.217}$$

where $\mathcal{N}$ is a normalization factor determined to satisfy

$$\int d^3R\, P_L(\mathbf{R}) = 4\pi L^3 \int_0^{\infty} dr\, r^2\, P_L(\mathbf{R}) = 1. \tag{15.218}$$

The sum must be evaluated numerically, and leads to the distributions shown in Fig. 15.6.

The above method is inconvenient if $D \neq 3$, since the simple pole structure of (15.213) is no longer there. For general D, we expand

$$\left(\frac{\omega L}{\sinh \omega L}\right)^{(D-1)/2} = (\omega L)^{(D-1)/2}\sum_{k=0}^{\infty}(-1)^k\binom{-(D-1)/2}{k}e^{-(2k+(D-1)/2)\omega L}, \tag{15.219}$$

and obtain from (15.212):

$$P_L(\mathbf{R}) \propto \sum_{k=0}^{\infty}(-1)^k\binom{-(D-1)/2}{k}I_k(R/L), \tag{15.220}$$

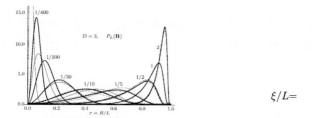

$\xi/L=$

Figure 15.6   Normalized end-to-end distribution of stiff polymer according to our analytic formula (15.201), plotted for persistence lengths $\xi/L =$ 1/400, 1/100, 1/30, 1/10, 1/5, 1/2, 1, 2 (fat curves). They are compared with the Monte Carlo calculations (symbols) and with the large-stiffness approximation (15.217) (thin curves) of Ref. [6] which fits well for $\xi/L = 2$ and 1 but becomes bad small $\xi/L < 1$. For very small values such as $\xi/L = 1/400, 1/100, 1/30$, our theoretical curves are well approximated by Gaussian random chain distributions on a lattice with lattice constant $a_{\mathrm{eff}} = 2\xi$ of Eq. (15.158) which ensures that the lowest moments $\langle R^2 \rangle = a_{\mathrm{eff}}L$ are properly fitted. The Daniels approximation (15.166) fits our theoretical curves well up to larger $\xi/L \approx 1/10$.

with the integrals

$$I_k(r) \equiv \int_{-i\infty}^{i\infty} \frac{d\bar{\omega}}{2\pi i} \bar{\omega}^{(D+1)/2} e^{-[2k+(D-1)/2]\bar{\omega}+(D-1)\bar{\omega}^2(1-r)/2l}, \qquad (15.221)$$

where $\bar{\omega}$ is the dimensionless variable ωL. The integrals are evaluated with the help of the formula[6]

$$\int_{-i\infty}^{i\infty} \frac{dx}{2\pi i} x^{\nu} e^{\beta x^2/2 - qx} = \frac{1}{\sqrt{2\pi}\beta^{(\nu+1)/2}} e^{-q^2/4\beta} D_{\nu}\left(q/\sqrt{\beta}\right), \qquad (15.222)$$

where $D_{\nu}(z)$ is the parabolic cylinder functions which for integer ν are proportional to Hermite polynomials:[7]

$$D_n(z) = \frac{1}{\sqrt{2}^n} e^{-z^2/4} H_n(z/\sqrt{2}). \qquad (15.223)$$

Thus we find

---

[6]I.S. Gradshteyn and I.M. Ryzhik, *op. cit.*, Formulas 3.462.3 and 3.462.4.
[7]*ibid.*, Formula 9.253.

$$I_k(r) = \frac{1}{\sqrt{2\pi}} \left[\frac{l}{(D-1)(1-r)} \right]^{(D+3)/4} e^{-\frac{[2k+(D-1)/2]^2}{4(D-1)(1-r)/l}} D_{(D+1)/2} \left(\frac{2k+(D-1)/2}{\sqrt{(D-1)(1-r)/l}} \right),$$

$$(15.224)$$

which becomes for $D = 3$

$$I_k(r) = \frac{1}{2\sqrt{2\pi}} \frac{1}{\sqrt{2(1-r)/l}^3} e^{-\frac{(2k+1)^2}{4(1-r)/l}} H_2 \left(\frac{2k+1}{2\sqrt{(1-r)/l}} \right).$$

$$(15.225)$$

If the sum (15.220) is performed numerically for $D = 3$, and the integral over $P_L(\mathbf{R})$ is normalized to satisfy (15.218), the resulting curves fall on top of those in Fig. 15.6 which were calculated from (15.217). In contrast to (15.217), which converges rapidly for small r, the sum (15.220) convergent rapidly for r close to unity.

Let us compare the low moments of the above distribution with the exact moments in Eqs. (15.156) and (15.164). in the large-stiffness expansion. We set $\bar{\omega} \equiv \omega L$ and expand

$$f(\bar{\omega}^2) \equiv \sqrt{\frac{\bar{\omega}}{\sinh \bar{\omega}}}^{D-1}$$

$$(15.226)$$

in a power series

$$f(\bar{\omega}^2) = 1 - \frac{D-1}{2^2 \cdot 3} \bar{\omega}^2 + \frac{(D-1)(5D-1)}{2^5 \cdot 3^2 \cdot 5} \bar{\omega}^4 - \frac{(D-1)(15+14D+35D^2)}{2^7 \cdot 3^4 \cdot 5 \cdot 7} \bar{\omega}^6 + \dots .(15.227)$$

Under the integral (15.212), each power of $\bar{\omega}^2$ may be replaced by a differential operator

$$\bar{\omega}^2 \rightarrow \hat{\bar{\omega}}^2 \equiv -\frac{L}{\kappa} \frac{d}{dr} = -\frac{2l}{D-1} \frac{d}{dr}.$$

$$(15.228)$$

The expansion $f(\hat{\bar{\omega}}^2)$ can then be pulled out of the integral, which by itself yields a δ-function, so that we obtain $P_L(\mathbf{R})$ in the form $f\left(\hat{\bar{\omega}}^2\right) \delta(r-1)$, which is a series of derivatives of δ-functions of $r-1$ staring with

$$P_L(\mathbf{R}) \propto \left[1 + \frac{l}{6} \frac{d}{dr} + \frac{(-1+5D) \, l^2}{360 \, (D-1)} \frac{d^2}{dr^2} + \frac{(15+14\,D+35\,D^2)\, l^3}{45360 \, (D-1)^2} \frac{d^3}{dr^3} + \dots \right] \delta(r-1).$$

$$(15.229)$$

From this we find easily the moments

$$\langle R^m \rangle = \int d^D R R^m \, P_L(\mathbf{R}) \propto \int_0^\infty dr \, r^{D-1} \, r^m P_L(\mathbf{R}).$$

$$(15.230)$$

Let us introduce auxiliary expectation values with respect to the simple integrals $\langle f(r) \rangle_1 \propto \int dr \, f(r) P_L$, rather than to D-dimensional volume integrals. The unnormalized moments $\langle r^m \rangle$ are then given by $\langle r^{D-1+m} \rangle_1$. Within these one-dimensional expectations, the moments of $z = r - 1$ are

$$\langle (r-1)^n \rangle_1 \propto \int_0^\infty dr \, (r-1)^n P_L(\mathbf{R}).$$

$$(15.231)$$

The moments $\langle r^m \rangle \langle [1+(r-1)]^m \rangle = \langle [1+(r-1)]^{D-1+m} \rangle_1$ are obtained by expanding the binomial in powers of $r-1$ and using the integrals $\int dz\, z^m \delta^{(n)}(z) = (-1)^m m!\, \delta_{mn}$ to find, up to the third power in $L/\xi = l$,

$$\langle R^0 \rangle = \mathcal{N} \left[1 - \frac{D-1}{6}l + \frac{(5D-1)(D-2)}{360}l^2 - \frac{(35D^3+14D+15)(D-2)(D-3)}{45360(D-1)}l^3 \right],$$

$$\langle R^2 \rangle = \mathcal{N}L^2 \left[1 - \frac{D+1}{6}l + \frac{(5D-1)D(D+1)}{360(D-1)}l^2 - \frac{(35D^2+14D+15)D(D+1)}{45360(D-1)}l^3 \right],$$

$$\langle R^4 \rangle = \mathcal{N}L^4 \left[1 - \frac{D+3}{6}l + \frac{(5D-1)(D+2)(D+3)}{360(D-1)}l^2 - \frac{(35D^2+14D+15)(D+1)(D+2)(D+3)}{45360(D-1)^2}l^3 \right].$$

The zeroth moment determines the normalization factor $\mathcal{N}$ to ensure that $\langle R^0 \rangle = 1$. Dividing this out of the other moments yields

$$\langle R^2 \rangle = L^2 \left[1 - \frac{1}{3}l + \frac{13D-9}{180(D-1)}l^2 - \frac{8}{945}l^3 + \dots \right], \tag{15.232}$$

$$\langle R^4 \rangle = L^4 \left[1 - \frac{2}{3}l + \frac{23D-11}{90(D-1)}l^2 - \frac{123D^2-98D+39}{1890(D-1)^2}l^3 + \dots \right]. \tag{15.233}$$

These agree up to the l-terms with the exact expansions (15.156) and (15.164) [or with the general formula (15.198)]. For $D = 3$, these expansions become

$$\langle R^2 \rangle = L^2 \left[1 - \frac{1}{3}l + \frac{1}{12}l^2 - \frac{8}{945}l^3 + \dots \right], \tag{15.234}$$

$$\langle R^4 \rangle = L^4 \left[1 - \frac{2}{3}l + \frac{29}{90}l^2 - \frac{71}{630}l^3 + \dots \right]. \tag{15.235}$$

Remarkably, these happen to agree in one more term with the exact expansions (15.156) and (15.164) than expected, as will be understood after Eq. (15.295).

15.9.5 Higher Loop Corrections

Let us calculate perturbative corrections to the large-stiffness limit. For this we replace the harmonic path integral (15.210) by the full expression

$$P(r;L) = S_D^{-1} \int_{\text{NBC}} \mathcal{D}'^{D-1} q(s) \delta\left(r - L^{-1} \int_0^L ds\, \sqrt{1-q^2(s)} \right) e^{-\mathcal{A}^{\text{tot}}[q] - \mathcal{A}^{\text{cor}}[q_b, q_a]}, \tag{15.236}$$

where

$$\mathcal{A}_{\text{tot}}[q] = \frac{1}{2\varepsilon} \int_0^L ds\, [g_{\mu\nu}(q)\,\dot{q}^\mu(s)\dot{q}^\nu(s) - \varepsilon\delta(s,s)\log g(q(s))] + \mathcal{A}^{\text{FP}} - \varepsilon L \frac{R}{8}. \tag{15.237}$$

For convenience, we have introduced here the parameter $\varepsilon = k_B T/\kappa = 1/\bar{\kappa}$, the reduced *inverse stiffness*, related to the flexibility $l \equiv L/\xi$ by $l = \varepsilon L(d-1)/2$. We also have added the correction term $\varepsilon L\, R/8$ to have a unit normalization of the partition function

$$Z = S_D \int_0^\infty dr\, r^{D-1}\, P(r;L) = 1. \tag{15.238}$$

The Faddeev-Popov action, whose harmonic approximation was used in (15.208), is now

$$\mathcal{A}^{\text{FP}}[q] = -(D-1)\log\left(L^{-1}\int_0^L ds\,\sqrt{1-q^2(s)}\right),\tag{15.239}$$

To perform higher-order calculations we have added an extra action which corrects for the omission of fluctuations of the velocities at the endpoints when restricting the paths to Neumann boundary conditions with zero end-point velocities. The extra action contains, of course, only at the endpoints and reads [7]

$$\mathcal{A}^{\text{cor}}[q_b, q_a] = -\log J[q_b, q_a] = -[q^2(0) + q^2(L)]/4.\tag{15.240}$$

Thus we represent the partition function (15.238) by the path integral with Neumann boundary conditions

$$Z = \int_{\text{NBC}} \mathcal{D}'^{D-1}q(s)\exp\left\{-\mathcal{A}_{\text{tot}}[q] - \mathcal{A}^{\text{cor}}[q_b, q_a] - \mathcal{A}^{\text{FP}}[q]\right\}.\tag{15.241}$$

A similar path integral can be set up for the moments of the distribution. We express the square distance $\mathbf{R}^2$ in coordinates (15.207) as

$$\mathbf{R}^2 = \int_0^L ds \int_0^L ds'\,\mathbf{u}(s)\cdot\mathbf{u}(s') = \left(\int_0^L ds\,\sqrt{1-q^2(s)}\right)^2 = R^2,\tag{15.242}$$

we find immediately the following representation for all, even and odd, moments [compare (15.147)]

$$\langle(\mathbf{R}^2)^n\rangle = \left\langle\left[\int_0^L\int_0^L ds\,ds'\,\mathbf{u}(s)\cdot\mathbf{u}(s')\right]^n\right\rangle\tag{15.243}$$

in the form

$$\langle R^n\rangle = \int_{\text{NBC}} \mathcal{D}'^{D-1}q(s)\exp\left\{-\mathcal{A}_{\text{tot}}[q] - \mathcal{A}_{\text{cor}}[q_b, q_a] - \mathcal{A}_n^{\text{FP}}[q]\right\}.\tag{15.244}$$

This differs from Eq. (15.241) only by in the Faddeev-Popov action for these moments, which is

$$\mathcal{A}_n^{\text{FP}}[q] = -(n+D-1)\log\left(L^{-1}\int_0^L ds\,\sqrt{1-q^2(s)}\right),\tag{15.245}$$

rather than (15.239).

There is no need to divide the path integral (15.244) by Z since this has unit normalization, as will be verified order by order in the perturbation expansion. The Green function of the operator d^2/ds^2 with these boundary conditions has the form

$$\Delta_{\text{N}}'(s, s') = \frac{L}{3} - \frac{|s-s'|}{2} - \frac{(s+s')}{2} + \frac{(s^2+s'^2)}{2L}.\tag{15.246}$$

The zero temporal average (15.207) manifests itself in the property

$$\int_0^L ds\,\Delta_{\text{N}}'(s, s') = 0.\tag{15.247}$$

In the following we shall simply write $\Delta(s, s')$ for $\Delta_{\text{N}}'(s, s')$, for brevity.

Partition Function and Moments Up to Four Loops

We are now prepared to perform the explicit perturbative calculation of the partition function and all even moments in powers of the inverse stiffness ε up to order $\varepsilon^2 \propto l^2$. This requires evaluating Feynman diagrams up to four loops. The associated integrals will contain products of distributions, which will be calculated unambiguously with the help of our simple formulas in Chapter 10.

For a systematic treatment of the expansion parameter ε, we rescale the coordinates $q^\mu \to \varepsilon q^\mu$, and rewrite the path integral (15.244) as

$$\langle R^n \rangle = \int_{\text{NBC}} \mathcal{D}'^{D-1} q(s) \exp\left\{-\mathcal{A}_{\text{tot},n}[q; \varepsilon].\right\}, \tag{15.248}$$

with the total action

$$\mathcal{A}_{\text{tot},n}[q; \varepsilon] = \int_0^L ds \left[\frac{1}{2}\left(\dot{q}^2 + \varepsilon \frac{(q\dot{q})^2}{1 - \varepsilon q^2}\right) + \frac{1}{2}\delta(0)\log(1 - \varepsilon q^2)\right]$$
$$- \sigma_n \log\left[\frac{1}{L}\int_0^L ds \sqrt{1 - \varepsilon q^2}\right] - \frac{\varepsilon}{4}\left[q^2(0) + q^2(L)\right] - \varepsilon L \frac{R}{8}. \tag{15.249}$$

The constant σ_n is an abbreviation for

$$\sigma_n \equiv n + (d - 1). \tag{15.250}$$

For $n = 0$, the path integral (15.248) must yield the normalized partition function $Z = 1$.

For the perturbation expansion we separate

$$\mathcal{A}_{\text{tot},n}[q; \varepsilon] = \mathcal{A}^{(0)}[q] + \mathcal{A}_n^{\text{int}}[q; \varepsilon], \tag{15.251}$$

with a free action

$$\mathcal{A}^{(0)}[q] = \frac{1}{2}\int_0^L ds\, \dot{q}^2(s), \tag{15.252}$$

and a large-stiffness expansion of the interaction

$$\mathcal{A}_n^{\text{int}}[q; \varepsilon] = \varepsilon \mathcal{A}_n^{\text{int1}}[q] + \varepsilon^2 \mathcal{A}_n^{\text{int2}}[q] + \dots. \tag{15.253}$$

The free part of the path integral (15.248) is normalized to unity:

$$Z^{(0)} \equiv \int_{\text{NBC}} \mathcal{D}'^{D-1} q(s)\, e^{-\mathcal{A}^{(0)}[q]} = \int_{\text{NBC}} \mathcal{D}'^{D-1} q(s)\, e^{-(1/2)\int_0^L ds\, \dot{q}^2(s)} = 1. \tag{15.254}$$

The first expansion term of the interaction (15.253) is

$$\mathcal{A}_n^{\text{int1}}[q] = \frac{1}{2}\int_0^L ds\left\{[q(s)\dot{q}(s)]^2 - \rho_n(s)\, q^2(s)\right\} - L\frac{R}{8}, \tag{15.255}$$

where

$$\rho_n(s) \equiv \delta_n + [\delta(s) + \delta(s - L)]/2, \qquad \delta_n \equiv \delta(0) - \sigma_n/L. \tag{15.256}$$

The second term in $\rho_n(s)$ represents the end point terms in the action (15.249) and is important for canceling singularities in the expansion.

The next expansion term in (15.253) reads

$$\mathcal{A}_n^{\text{int2}}[q] = \frac{1}{2}\int_0^L ds\left\{[q(s)\dot{q}(s)]^2 - \frac{1}{2}\left[\delta(0) - \frac{\sigma_n}{2L}\right]q^2(s)\right\}q^2(s)$$
$$+ \frac{\sigma_n}{8L^2}\int_0^L ds \int_0^L ds'\, q^2(s)q^2(s'). \tag{15.257}$$

The perturbation expansion of the partition function in powers of ε consists of expectation values of the interaction and its powers to be calculated with the free partition function (15.254). For an arbitrary functional of $q(s)$, these expectation values will be denoted by

$$\langle F[q] \rangle_0 \equiv \int_{\mathrm{NBC}} \mathcal{D}'^{D-1} q(s)\, F[q]\, e^{-(1/2)\int_0^L ds\,\dot{q}^2(s)}. \tag{15.258}$$

With this notation, the perturbative expansion of the path integral (15.248) reads

$$\langle R^n \rangle/L^n = 1 - \langle \mathcal{A}_n^{\mathrm{int}}[q;\varepsilon] \rangle_0 + \frac{1}{2}\langle \mathcal{A}_n^{\mathrm{int}}[q;\varepsilon]^2 \rangle_0 - \ldots$$

$$= 1 - \varepsilon \langle \mathcal{A}_n^{\mathrm{int1}}[q] \rangle_0 + \varepsilon^2 \left(-\langle \mathcal{A}_n^{\mathrm{int2}}[q] \rangle_0 + \frac{1}{2}\langle \mathcal{A}_n^{\mathrm{int1}}[q]^2 \rangle_0 \right) - \ldots . \tag{15.259}$$

For the evaluation of the expectation values we must perform all possible Wick contractions with the basic propagator

$$\langle q^\mu(s) q^\nu(s') \rangle_0 = \delta^{\mu\nu} \Delta(s,s'), \tag{15.260}$$

where $\Delta(s,s')$ is the Green function (15.246) of the unperturbed action (15.252). The relevant loop integrals I_i and H_i are calculated using the dimensional regularization rules of Chapter 10. They are listed in Appendix 15A and Appendix 15B.

We now state the results for various terms in the expansion (15.259):

$$\langle \mathcal{A}_n^{\mathrm{int1}}[q] \rangle_0 = \frac{(D-1)}{2}\left[\frac{\sigma_n}{L} I_1 + D I_2 - \frac{1}{2}\Delta(0,0) - \frac{1}{2}\Delta(L,L) \right] - L\frac{R}{8} = L\frac{(D-1)n}{12},$$

$$\langle \mathcal{A}_n^{\mathrm{int2}}[q] \rangle_0 = \frac{(D^2-1)}{4}\left[\left(\delta(0) + \frac{\sigma_n}{2L}\right)I_3 + 2(D+2)I_4 \right] + \frac{(D-1)\sigma_n}{8L^2}\left[(D-1)I_1^2 + 2I_5 \right]$$

$$= L^3\frac{(D^2-1)}{120}\delta(0) + L^2\frac{(D-1)}{1440}\left[(25D^2+36D+23) + n(11D+5) \right],$$

$$\frac{1}{2}\langle \mathcal{A}_n^{\mathrm{int1}}[q]^2 \rangle_0 = \frac{L^2}{2}\left\{ \frac{(D-1)L}{12}\left[\left(\delta(0) + \frac{D}{2L}\right) - \left(\delta_n + \frac{2}{L}\right) \right] - \frac{R}{8} \right\}^2$$

$$+ \frac{(D-1)}{4}\left[H_1^n - 2(H_2^n + H_3^n - H_5) + H_6 - 4D(H_4^n - H_7 - H_{10}) \right.$$

$$+ \left. H_{11} + 2D^2(H_8 + H_9) \right] + \frac{(D-1)}{4}\left[DH_{12} + 2(D+2)H_{13} + DH_{14} \right]$$

$$= \frac{L^2}{2}\left[\frac{(D-1)n}{12} \right]^2 + L^3\frac{(D-1)}{120}\delta(0)$$

$$+ L^2\frac{(D-1)}{1440}\left[(25D^2 - 22D + 25) + 4(n + 4D - 2) \right]$$

$$+ L^3\frac{(D-1)D}{120}\delta(0) + L^2\frac{(D-1)}{720}(29D - 1). \tag{15.261}$$

Inserting these results into Eq. (15.259), we find all *even and odd moments* up to order $\varepsilon^2 \propto l^2$

$$\langle R^n \rangle/L^n = 1 - \varepsilon L\frac{(D-1)n}{12} + \varepsilon^2 L^2\left[\frac{(D-1)^2 n^2}{288} + \frac{(D-1)(4n + 5D - 13)n}{1440} \right] - \mathcal{O}(\varepsilon^3)$$

$$= 1 - \frac{n}{6}l + \left[\frac{n^2}{72} + \frac{(4n + 5D - 13)n}{360(D-1)} \right] l^2 - \mathcal{O}(l^3). \tag{15.262}$$

For $n = 0$ this gives the properly normalized partition function $Z = 1$. For all n it reproduces the large-stiffness expansion (15.198) up to order l^4.

Correlation Function Up to Four Loops

As an important test of the correctness of our perturbation theory we calculating the correlation function up to four loops and verify that it yields the simple expression (15.153), which reads in the present units

$$G(s, s') = e^{-|s-s'|/\xi} = e^{-|s-s'|l/L}.$$ (15.263)

Starting point is the path integral representation with Neumann boundary conditions for the two-point correlation function

$$G(s, s') = \langle \mathbf{u}(s) \cdot \mathbf{u}(s') \rangle = \int_{\mathrm{NBC}} \mathcal{D}'^{D-1} q(s) \, f(s, s') \exp\left\{ -\mathcal{A}_{\mathrm{tot}}^{(0)}[q; \varepsilon] \right\},$$ (15.264)

with the action of Eq. (15.249) for $n = 0$. The function in the integrand $f(s, s') \equiv f(q(s), q(s'))$ is an abbreviation for the scalar product $\mathbf{u}(s) \cdot \mathbf{u}(s')$ expressed in terms of independent coordinates $q^\mu(s)$:

$$f(q(s), q(s')) \equiv \mathbf{u}(s) \cdot \mathbf{u}(s') = \sqrt{1 - q^2(s)} \sqrt{1 - q^2(s')} + q(s) \, q(s').$$ (15.265)

Rescaling the coordinates $q \to \sqrt{\varepsilon}\, q$, and expanding in powers of ε yields:

$$f(q(s), q(s')) = 1 + \varepsilon f_1(q(s), q(s')) + \varepsilon^2 f_2(q(s), q(s')) + \dots,$$ (15.266)

where

$$f_1(q(s), q(s')) = q(s) \, q(s') - \frac{1}{2} q^2(s) - \frac{1}{2} q^2(s'),$$ (15.267)

$$f_2(q(s), q(s')) = \frac{1}{4} q^2(s) \, q^2(s') - \frac{1}{8} [q^2(s)]^2 - \frac{1}{8} [q^2(s')]^2.$$ (15.268)

We shall attribute the integrand $f(q(s), q(s'))$ to an interaction $\mathcal{A}^f[q; \varepsilon]$ defined by

$$f(q(s), q(s')) \equiv e^{-\mathcal{A}^f[q; \varepsilon]},$$ (15.269)

which has the ε-expansion

$$\begin{aligned}
\mathcal{A}^f[q; \varepsilon] &= -\log f(q(s), q(s')) \\
&= -\varepsilon f_1(q(s), q(s')) + \varepsilon^2 \left[-f_2(q(s), q(s')) + \frac{1}{2} f_1^2(s, s') \right] - \dots,
\end{aligned}$$ (15.270)

to be added to the interaction (15.253) with $n = 0$. Thus we obtain the perturbation expansion of the path integral (15.264)

$$G(s, s') = 1 - \left\langle \left(\mathcal{A}_0^{\mathrm{int}}[q; \varepsilon] + \mathcal{A}^f[q; \varepsilon] \right) \right\rangle_0 + \frac{1}{2} \left\langle \left(\mathcal{A}_0^{\mathrm{int}}[q; \varepsilon] + \mathcal{A}^f[q; \varepsilon] \right)^2 \right\rangle_0 - \dots.$$ (15.271)

Inserting the interaction terms (15.253) and (15.271), we obtain

$$G(s, s') = 1 + \varepsilon \langle f_1(q(s), q(s')) \rangle_0 + \varepsilon^2 \left[\langle f_2(q(s), q(s')) \rangle_0 - \langle f_1(q(s), q(s')) \, \mathcal{A}_0^{\mathrm{int1}}[q] \rangle_0 \right] + \dots,$$ (15.272)

and the expectation values can now be calculated using the propagator (15.260) with the Green function (15.246).

In going through this calculation we observe that because of translational invariance in the pseudotime, $s \to s + s_0$, the Green function $\Delta(s, s') + C$ is just as good a Green function satisfying Neumann boundary conditions as $\Delta(s, s')$. We may demonstrate this explicitly by setting $C =$

$L(a-1)/3$ with an arbitrary constant a, and calculating the expectation values in Eq. (15.272) using the modified Green function. Details are given in Appendix 15C [see Eq. (15C.1)], where we list various expressions and integrals appearing in the Wick contractions of the expansion Eq. (15.272). Using these results we find the a-independent terms up to second order in ε:

$$\langle f_1(q(s), q(s'))\rangle_0 = -\frac{(D-1)}{2}|s-s'|,\tag{15.273}$$

$$\langle f_2(q(s), q(s'))\rangle_0 - \langle f_1(q(s), q(s'))\mathcal{A}_0^{\text{int1}}[q]\rangle_0$$

$$= (D-1)\left[\frac{1}{2}D_1 - \frac{1}{8}(D+1)D_2^2 - K_1 - DK_2 - \frac{(D-1)}{L}K_3 + \frac{1}{2}K_4 + \frac{1}{2}K_5\right]$$

$$= \frac{1}{8}(D-1)^2(s-s')^2.\tag{15.274}$$

This leads indeed to the correct large-stiffness expansion of the exact two-point correlation function (15.263):

$$G(s, s') = 1 - \varepsilon\frac{D-1}{2}|s-s'| + \varepsilon^2\frac{(D-1)^2}{8}(s-s')^2 + \ldots = 1 - \frac{|s-s'|}{\xi} + \frac{(s-s')^2}{2\xi^2} - \ldots.\tag{15.275}$$

Radial Distribution up to Four Loops

We now turn to the most important quantity characterizing a polymer, the radial distribution function. We eliminate the δ-function in Eq. (15.236) enforcing the end-to-end distance by considering the Fourier transform

$$P(k; L) = \int dr\, e^{ik(r-1)}\, P(r; L).\tag{15.276}$$

This is calculated from the path integral with Neumann boundary conditions

$$P(k; L) = \int_{\text{NBC}} \mathcal{D}'^{D-1}q(s)\exp\left\{-\mathcal{A}_k^{\text{tot}}[q; \varepsilon]\right\},\tag{15.277}$$

where the action $\mathcal{A}_k^{\text{tot}}[q; \varepsilon]$ reads, with the same rescaled coordinates as in (15.249),

$$\mathcal{A}_k^{\text{tot}}[q; \varepsilon] = \int_0^L ds\left\{\frac{1}{2}\left[\dot{q}^2 + \varepsilon\frac{(q\dot{q})^2}{1 - \varepsilon q^2}\right] + \frac{1}{2}\delta(0)\log(1 - \varepsilon q^2) - \frac{ik}{L}\left(\sqrt{1 - \varepsilon q^2} - 1\right)\right\}$$

$$- \frac{1}{4}\varepsilon\left[q^2(0) + q^2(L)\right] - \varepsilon L\frac{R}{8} \equiv \mathcal{A}^0[q] + \mathcal{A}_k^{\text{int}}[q; \varepsilon].\tag{15.278}$$

As before in Eq. (15.253), we expand the interaction in powers of the coupling constant ε. The first term coincides with Eq. (15.255), except that σ_n is replaced by

$$\rho_k(s) = \delta_k + [\delta(s) + \delta(s - L)]/2, \qquad \delta_k = \delta(0) - ik/L,\tag{15.279}$$

so that

$$\mathcal{A}_k^{\text{int1}}[q] = \int_0^L ds\frac{1}{2}\left\{[q(s)\dot{q}(s)]^2 - \rho_k(s)q^2(s)\right\} - L\frac{R}{8}.\tag{15.280}$$

The second expansion term $\mathcal{A}_k^{\text{int2}}[q]$ is simpler than the previous (15.257) by not containing the last nonlocal term:

$$\mathcal{A}_k^{\text{int2}}[q] = \int_0^L ds\frac{1}{2}\left\{[q(s)\dot{q}(s)]^2 - \frac{1}{2}\left(\delta(0) - \frac{ik}{2L}\right)q^2(s)\right\}q^2(s).\tag{15.281}$$

Apart from that, the perturbation expansion of (15.277) has the same general form as in (15.259):

$$P(k; L) = 1 - \varepsilon \langle \mathcal{A}_k^{\text{int1}}[q] \rangle_0 + \varepsilon^2 \left(-\langle \mathcal{A}_k^{\text{int2}}[q] \rangle_0 + \frac{1}{2} \langle \mathcal{A}_k^{\text{int1}}[q]^2 \rangle_0 \right) - \dots . \tag{15.282}$$

The expectation values can be expressed in terms of the same integrals listed in Appendix 15A and Appendix 15B as follows:

$$\begin{aligned}
\langle \mathcal{A}_{,k}^{\text{int1}}[q] \rangle_0 &= \frac{(D-1)}{2} \left[\frac{ik}{L} I_1 + D I_2 - \frac{1}{2} \Delta(0,0) - \frac{1}{2} \Delta(L,L) \right] - L \frac{R}{8} \\
&= -L \frac{(D-1)\left[(D-1) - ik\right]}{12},
\end{aligned} \tag{15.283}$$

$$\begin{aligned}
\langle \mathcal{A}_{,k}^{\text{int2}}[q] \rangle_0 &= \frac{(D^2-1)}{4} \left[\left(\delta(0) + \frac{ik}{2L} \right) I_3 + 2(D+2) I_4 \right] \\
&= L^3 \frac{(D^2-1)}{120} \delta(0) + L^2 \frac{(D^2-1)\left[7(D+2) + 3ik\right]}{720},
\end{aligned} \tag{15.284}$$

$$\begin{aligned}
\frac{1}{2} \langle \mathcal{A}_{,k}^{\text{int1}}[q]^2 \rangle_0 &= \frac{L^2}{2} \left\{ \frac{(D-1)L}{12} \left[\left(\delta(0) + \frac{D}{2L} \right) - \left(\delta_k + \frac{2}{L} \right) \right] - \frac{R}{8} \right\}^2 \\
&+ \frac{(D-1)}{4} \left[H_1^k - 2(H_2^k + H_3^k - H_5) + H_6 - 4D(H_4^k - H_7 - H_{10}) \right. \\
&+ \left. H_{11} + 2D^2(H_8 + H_9) \right] + \frac{(D-1)}{4} \left[D H_{12} + 2(D+2) H_{13} + D H_{14} \right] \\
&= L^2 \frac{(D-1)^2 \left[(D-1) - ik\right]^2}{2 \cdot 12^2} \\
&+ L^3 \frac{(D-1)}{120} \delta(0) + L^2 \frac{(D-1)}{1440} \left[(13D^2 - 6D + 21) + 4ik(2D + ik) \right] \\
&+ L^3 \frac{(D-1)D}{120} \delta(0) + L^2 \frac{(D-1)}{720} (29D - 1).
\end{aligned} \tag{15.285}$$

In this way we find the large-stiffness expansion up to order ε^2:

$$\begin{aligned}
P(k; L) &= 1 + \varepsilon L \frac{(D-1)}{12} \left[(D-1) - ik\right] + \varepsilon^2 L^2 \frac{(D-1)}{1440} \\
&\times \left[(ik)^2 (5D-1) - 2ik(5D^2 - 11D + 8) + (D-1)(5D^2 - 11D + 14) \right] + \mathcal{O}(\varepsilon^3).
\end{aligned} \tag{15.286}$$

This can also be rewritten as

$$P(k; L) = P_{1\,\text{loop}}(k; L) \left\{ 1 + \frac{(D-1)}{6} l + \left[\frac{(D-3)}{180(D-1)} ik + \frac{(5D^2 - 11D + 14)}{360} \right] l^2 + \mathcal{O}(l^3) \right\}, \tag{15.287}$$

where the prefactor $P_{1\,\text{loop}}(k; L)$ has the expansion

$$P_{1\,\text{loop}}(k; L) = 1 - \varepsilon L \frac{(D-1)}{2^2 \cdot 3} (ik) + \varepsilon^2 L^2 \frac{(D-1)(5D-1)}{2^5 \cdot 3^2 \cdot 5} (ik)^2 - \dots . \tag{15.288}$$

With the identification $\bar{\omega}^2 = ik\varepsilon L$, this is the expansion of *one-loop* functional determinant in (15.227). By Fourier-transforming (15.286), we obtain the radial distribution function

$$\begin{aligned}
P(r; l) &= \delta(r-1) + \frac{l}{6} \left[\delta'(r-1) + (d-1)\delta(r-1) \right] + \frac{l^2}{360(d-1)} \left[(5d-1)\delta''(r-1) \right. \\
&+ \left. 2(5d^2 - 11d + 8)\delta'(r-1) + (d-1)(5d^2 - 11d + 14)\delta(r-1) \right] + \mathcal{O}(l^3).
\end{aligned} \tag{15.289}$$

As a crosscheck, we can calculate from this expansion once more the even and odd moments

$$\langle R^n \rangle = L^n \int dr \, r^{n+(D-1)} \, P(r;l), \qquad (15.290)$$

and find that they agree with Eq. (15.262).

Using the higher-order expansion of the moments in (15.198) we can easily extend the distribution (15.289) to arbitrarily high orders in l. Keeping only the terms up to order l^4, we find that the one-loop end-to-end distribution function (15.212) receives a correction factor:

$$P(r;l) \propto \int_{-\infty}^{\infty} \frac{d\hat\omega^2}{2\pi} e^{-i\hat\omega^2(r-1)(D-1)/2l} \left(\frac{\bar\omega}{\sinh\bar\omega} \right)^{(D-1)/2} e^{-V(l,\hat\omega^2)}, \qquad (15.291)$$

with

$$V(l,\hat\omega^2) \equiv V_0(l) + \bar V(l,\hat\omega^2) = V_0(l) + V_1(l)\frac{\hat\omega^2}{l} + V_2(l)\frac{\hat\omega^4}{l^2} + V_3(l)\frac{\hat\omega^6}{l^3} + \dots \, . \qquad (15.292)$$

The first term

$$
\begin{aligned}
V_0(l) \; = \; & -\frac{d-1}{6}l + \frac{d-9}{360}l^2 + \frac{(d-1)\,\left(32 - 13\,d + 5\,d^2\right)}{6480}l^3 \\
& -\frac{34 - 272\,d + 259\,d^2 - 110\,d^3 + 25\,d^4}{259200}l^4 + \dots
\end{aligned}
\qquad (15.293)
$$

contributes only to the normalization of $P(r;l)$, and can be omitted in (15.291). The remainder has the expansion coefficients

$$
\begin{aligned}
V_1(l) \; &= \; -\frac{d-3}{360}l^2 + \frac{(-5+9\,d)}{7560\,(-1+d)}l^3 + \frac{\left(-455 + 431\,d + 91\,d^2 + 5\,d^3\right)}{907200\,(d-1)^2}l^4 + \dots \, , \\
V_2(l) \; &= \; -\frac{(5-3\,d)\,l^3}{7560} - \frac{\left(-31 + 42\,d + 25\,d^2\right)\,l^4}{907200\,(d-1)} + \dots \, , \\
V_3(l) \; &= \; -\frac{(d-1)\,l^4}{18900} + \dots \, .
\end{aligned}
\qquad (15.294)
$$

In the physical most interesting case of three dimensions, the first nonzero correction arises to order l^3. This explains the remarkable agreement of the moments in (15.234) and (15.234) up to order l^2.

The correction terms $\bar V(l,\hat\omega^2)$ may be included perturbatively into the sum over k in Eq. (15.220) by noting that the expectation value of powers of $\hat\omega^2/l$ within the $\hat\omega$-integral (15.221) are

$$\langle \hat\omega^2/l \rangle = a_k^2 \equiv \frac{2k+(D-1)/2}{(D-1)(1-r)}, \quad \langle \hat\omega^2/l \rangle^2 = 3a_k^4, \quad \langle \hat\omega^2/l \rangle^3 = 15a_k^6, \qquad (15.295)$$

so that we obtain an extra factor e^{-f_k}

$$f_k = V_1(l)a_k^2 + \left[3V_2(l) - V_1^2(l)\right] a_k^4 + \left[15V_3(l) - 12V_1(l)V_2(l) + \frac{4}{3}V_1^3(l)\right] a_k^6 + \dots \, , \qquad (15.296)$$

where up to order l^4:

$$
\begin{aligned}
3V_2(l) - V_1^2(l) \; &= \; \frac{3D-5}{2520}l^3 + \frac{156 - 231D - 26D^2 - 7D^3}{907200\,(D-1)}l^4 + \dots \, , \\
15V_3(l) - 12V_1(l)V_2(l) + \frac{4}{3}V_1^3(l) \; &= \; -\frac{D-1}{1260}l^4 + \dots \, .
\end{aligned}
\qquad (15.297)
$$

15.10   Excluded-Volume Effects

A significant modification of these properties is brought about by the interactions between the chain elements. If two of them come close to each other, the molecular forces prevent them from occupying the same place. This is called the *excluded-volume effect*. In less than four dimensions, it gives rise to a scaling law for the expectation value $\langle R^2 \rangle$ as a function of L:

$$\langle R^2 \rangle \propto L^{2\nu}, \tag{15.298}$$

as stated in (15.38). The critical exponent ν is a number between the random-chain value $\nu = 1/2$ and the stiff-chain value $\nu = 1$.

To derive this behavior we consider the polymer in the limiting path integral approximation (15.80) to a random chain which was derived for $R^2/La \ll 1$ and which is very accurate whenever the probability distribution is sizable. Thus we start with the time-sliced expression

$$P_N(\mathbf{R}) = \frac{1}{\sqrt{2\pi a/M}^D} \prod_{n=1}^{N-1} \left[\int \frac{d^D x_n}{\sqrt{2\pi a/M}^D} \right] \exp\left(-\mathcal{A}^N/\hbar \right), \tag{15.299}$$

with the action

$$\mathcal{A}^N = a \sum_{n=1}^{N} \frac{M}{2} \frac{(\Delta \mathbf{x}_n)^2}{a^2}, \tag{15.300}$$

and the mass parameter (15.79). In the sequel we use natural units in which energies are measured in units of $k_B T$, and write down all expressions in the continuum limit. The probability (15.299) is then written as

$$P_L(\mathbf{R}) = \int \mathcal{D}^D x \, e^{-\mathcal{A}^L[\mathbf{x}]}, \tag{15.301}$$

where we have used the label $L = Na$ rather than N. From the discussion in the previous section we know that although this path integral represents an ideal random chain, we can also account for a finite stiffness by interpreting the number a as an effective length parameter a_{eff} given by (15.158). The total Euclidean time in the path integral $\tau_b - \tau_a = \hbar/k_B T$ corresponds to the total length of the polymer L.

We now assume that the molecules of the polymer repel each other with a two-body potential $V(\mathbf{x}, \mathbf{x}')$. Then the action in the path integral (15.301) has to be supplemented by an interaction

$$\mathcal{A}_{\text{int}} = \frac{1}{2} \int_0^L d\tau \int_0^L d\tau' \, V(\mathbf{x}(\tau), \mathbf{x}(\tau')). \tag{15.302}$$

Note that the interaction is of a purely spatial nature and does not depend on the parameters τ, τ', i.e., it does not matter which two molecules in the chain come close to each other.

The effects of an interaction of this type are most elegantly calculated by making use of a Hubbard-Stratonovich transformation. Generalizing the procedure in Subsection 7.15.1, we introduce an auxiliary fluctuating field variable $\varphi(\mathbf{x})$ at every space point $\mathbf{x}$ and replace $\mathcal{A}_{\text{int}}$ by

$$\mathcal{A}_{\text{int}}^{\varphi} = \int_0^L d\tau \, \varphi(\mathbf{x}(\tau)) - \frac{1}{2} \int d^D x d^D x' \, \varphi(\mathbf{x}) V^{-1}(\mathbf{x}, \mathbf{x}') \varphi(\mathbf{x}'). \tag{15.303}$$

Here $V^{-1}(\mathbf{x}, \mathbf{x}')$ denotes the inverse of $V(\mathbf{x}, \mathbf{x}')$ under functional multiplication, defined by the integral equation

$$\int d^D x' \, V^{-1}(\mathbf{x}, \mathbf{x}') V(\mathbf{x}', \mathbf{x}'') = \delta^{(D)}(\mathbf{x} - \mathbf{x}''). \tag{15.304}$$

To see the equivalence of the action (15.303) with (15.302), we rewrite (15.303) as

$$\mathcal{A}_{\text{int}}^{\varphi} = \int d^D x \, \rho(\mathbf{x}) \varphi(\mathbf{x}) - \frac{1}{2} \int d^D x d^D x' \, \varphi(\mathbf{x}) V^{-1}(\mathbf{x}, \mathbf{x}') \varphi(\mathbf{x}'), \tag{15.305}$$

where $\rho(\mathbf{x})$ is the particle density

$$\rho(\mathbf{x}) \equiv \int_0^L d\tau \, \delta^{(D)}(\mathbf{x} - \mathbf{x}(\tau)). \tag{15.306}$$

Then we perform a quadratic completion to

$$\mathcal{A}_{\text{int}}^{\varphi} = -\frac{1}{2} \int d^D x d^D x' \left[\varphi'(\mathbf{x}) V^{-1}(\mathbf{x}, \mathbf{x}') \varphi'(\mathbf{x}') - \rho(\mathbf{x}) V(\mathbf{x}, \mathbf{x}') \rho(\mathbf{x}') \right], \tag{15.307}$$

with the shifted field

$$\varphi'(\mathbf{x}) \equiv \varphi(\mathbf{x}) - \int d^D x' \, V(\mathbf{x}, \mathbf{x}') \rho(\mathbf{x}'). \tag{15.308}$$

Now we perform the functional integral

$$\int \mathcal{D}\varphi(\mathbf{x}) \, e^{-\mathcal{A}_{\text{int}}^{\varphi}} \tag{15.309}$$

integrating $\varphi(\mathbf{x})$ at each point $\mathbf{x}$ from $-i\infty$ to $i\infty$ along the imaginary field axis. The result is a constant functional determinant $[\det V^{-1}(\mathbf{x}, \mathbf{x}')]^{-1/2}$. This can be ignored since we shall ultimately normalize the end-to-end distribution to unity. Inserting (15.306) into the surviving second term in (15.307), we obtain precisely the original interaction (15.302).

Thus we may study the excluded-volume problem by means of the equivalent path integral

$$P_L(\mathbf{R}) \propto \int \mathcal{D}^D x(\tau) \int \mathcal{D}\varphi(\mathbf{x}) \, e^{-\mathcal{A}}, \tag{15.310}$$

where the action $\mathcal{A}$ is given by the sum

$$\mathcal{A} = \mathcal{A}^L[\mathbf{x}, \dot{\mathbf{x}}, \varphi] + \mathcal{A}[\varphi], \tag{15.311}$$

of the line and field actions

$$A^L[\mathbf{x}, \varphi] \equiv \int_0^L d\tau \left[\frac{M}{2} \dot{\mathbf{x}}^2 + \varphi(\mathbf{x}(\tau)) \right], \tag{15.312}$$

$$A[\varphi] \equiv -\frac{1}{2} \int d^D x d^D x' \, \varphi(\mathbf{x}) V^{-1}(\mathbf{x}, \mathbf{x}') \varphi(\mathbf{x}'), \tag{15.313}$$

respectively. The path integral (15.310) over $\mathbf{x}(\tau)$ and $\varphi(\mathbf{x})$ has the following physical interpretation. The line action (15.312) describes the orbit of a particle in a
space-dependent random potential $\varphi(\mathbf{x})$. The path integral over $\mathbf{x}(\tau)$ yields the end-
to-end distribution of the fluctuating polymer in this potential. The path integral
over all potentials $\varphi(\mathbf{x})$ with the weight $e^{-A[\varphi]}$ accounts for the repulsive cloud of
the fluctuating chain elements. To be convergent, all $\varphi(\mathbf{x})$ integrations in (15.310)
have to run along the imaginary field axis.

To evaluate the path integrals (15.310), it is useful to separate $\mathbf{x}(\tau)$- and $\varphi(\mathbf{x})$-
integrations and to write end-to-end distributions as an average over φ-fluctuations

$$P_L(\mathbf{R}) \propto \int \mathcal{D}\varphi(\mathbf{x}) \, e^{-A[\varphi]} P_L^\varphi(\mathbf{R}, \mathbf{0}), \tag{15.314}$$

where

$$P_L^\varphi(\mathbf{R}, \mathbf{0}) = \int \mathcal{D}^D x(\tau) \, e^{-A^L[\mathbf{x}, \varphi]} \tag{15.315}$$

is the end-to-end distribution of a random chain moving in a *fixed* external potential
$\varphi(\mathbf{x})$. The presence of this potential destroys the translational invariance of P_L^φ. This
is why we have recorded the initial and final points $\mathbf{0}$ and $\mathbf{R}$. In the final distribution
$P_L(\mathbf{R})$ of (15.314), the invariance is of course restored by the integration over all
$\varphi(\mathbf{x})$.

It is possible to express the distribution $P_L^\varphi(\mathbf{R}, \mathbf{0})$ in terms of solutions of an
associated Schrödinger equation. With the action (15.312), this equation is obviously

$$\left[\frac{\partial}{\partial L} - \frac{1}{2M} \partial_{\mathbf{R}}^2 + \varphi(\mathbf{R}) \right] P_L^\varphi(\mathbf{R}, \mathbf{0}) = \delta^{(D)}(\mathbf{R} - \mathbf{0})\delta(L). \tag{15.316}$$

If $\psi_E^\varphi(R)$ denotes the time-independent solutions of the Hamiltonian operator

$$\hat{H}^\varphi = -\frac{1}{2M} \partial_{\mathbf{R}}^2 + \varphi(\mathbf{R}), \tag{15.317}$$

the probability $P_L^\varphi(\mathbf{R})$ has a spectral representation of the form

$$P_L^\varphi(\mathbf{R}, \mathbf{0}) = \int dE e^{-EL} \psi_E^\varphi(\mathbf{R}) \psi_E^{\varphi\,*}(\mathbf{0}), \quad L > 0. \tag{15.318}$$

From now on, we assume the interaction to be dominated by the simplest possible
repulsive potential proportional to a δ-function:

$$V(\mathbf{x}, \mathbf{x}') = va^D \delta^{(D)}(\mathbf{x} - \mathbf{x}'). \tag{15.319}$$

Then the functional inverse is

$$V^{-1}(\mathbf{x}, \mathbf{x}') = v^{-1}a^{-D}\delta^{(D)}(\mathbf{x} - \mathbf{x}'),\tag{15.320}$$

and the φ-action (15.312) reduces to

$$\mathcal{A}[\varphi] = -\frac{v^{-1}a^{-D}}{2}\int d^D x\,\varphi^2(\mathbf{x}).\tag{15.321}$$

The path integrals (15.314), (15.315) can be solved approximately by applying the semiclassical methods of Chapter 4 to both the $\mathbf{x}(\tau)$- and the $\varphi(\mathbf{x})$-path integrals. These are dominated by the extrema of the action and evaluated via the leading saddle point approximation. In the $\varphi(\mathbf{x})$-integral, the saddle point is given by the equation

$$v^{-1}a^{-D}\varphi(\mathbf{x}) = \frac{\delta}{\delta\varphi(\mathbf{x})}\log P_L^\varphi(\mathbf{R}, 0).\tag{15.322}$$

This is the semiclassical approximation to the exact equation

$$v^{-1}a^{-D}\langle\varphi(\mathbf{x})\rangle = \langle\rho(\mathbf{x})\rangle \equiv \left\langle \int_0^L d\tau\,\delta^{(D)}(\mathbf{x} - \mathbf{x}(\tau))\right\rangle_{\mathbf{x}},\tag{15.323}$$

where $\langle\ldots\rangle_{\mathbf{x}}$ is the average over all line fluctuations calculated with the help of the probability distribution (15.315).

The exact equation (15.323) follows from a functional differentiation of the path integral for P_L^φ with respect to $\varphi(\mathbf{x})$:

$$\frac{\delta}{\delta\varphi(\mathbf{x})}P_L^\varphi(\mathbf{R}) = \int \mathcal{D}\varphi\frac{\delta}{\delta\varphi(\mathbf{x})}\int \mathcal{D}^D x\, e^{-\mathcal{A}^L[\mathbf{x},\varphi]-\mathcal{A}[\varphi]} = 0.\tag{15.324}$$

By anchoring one end of the polymer at the origin and carrying the path integral from there to $\mathbf{x}(\tau)$, and further on to $\mathbf{R}$, the right-hand side of (15.323) can be expressed as a convolution integral over two end-to-end distributions:

$$\left\langle \int_0^L d\tau\,\delta^{(D)}(\mathbf{x} - \mathbf{x}(\tau))\right\rangle_{\mathbf{x}} = \int_0^L dL'\, P_{L'}^\varphi(\mathbf{x})P_{L-L'}^\varphi(\mathbf{R} - \mathbf{x}).\tag{15.325}$$

With (15.323), this becomes

$$v^{-1}a^{-D}\langle\varphi(\mathbf{x})\rangle_{\mathbf{x}} = \int_0^L dL'\, P_{L'}^\varphi(\mathbf{x})P_{L-L'}^\varphi(\mathbf{R} - \mathbf{x}),\tag{15.326}$$

which is the same as (15.323).

According to Eq. (15.322), the extremal $\varphi(\mathbf{x})$ depends really on two variables, $\mathbf{x}$ and $\mathbf{R}$. This makes the solution difficult, even at the semiclassical level. It becomes simple only for $\mathbf{R} = 0$, i.e., for a closed polymer. Then only the variable $\mathbf{x}$ remains and, by rotational symmetry, $\varphi(\mathbf{x})$ can depend only on $r = |\mathbf{x}|$. For $\mathbf{R} \neq 0$, on the other hand, the rotational symmetry is distorted to an ellipsoidal geometry, in which a closed-form solution of the problem is hard to find. As an approximation, we may use a rotationally symmetric ansatz $\varphi(\mathbf{x}) \approx \varphi(r)$ also for $\mathbf{R} \neq 0$ and calculate the end-to-end probability distribution $P_L(\mathbf{R})$ via the semiclassical approximation to the two path integrals in Eq. (15.310).

The saddle point in the path integral over $\varphi(\mathbf{x})$ gives the formula [compare (15.314)]

$$P_L(\mathbf{R}) \sim P_L^\varphi(\mathbf{R}, 0) = \int \mathcal{D}^D x \exp\left\{-\int_0^L d\tau \left[\frac{M}{2}\dot{\mathbf{x}}^2 + \varphi(r(\tau))\right]\right\}. \qquad (15.327)$$

Thereby it is hoped that for moderate $\mathbf{R}$, the error is small enough to justify this approximation. Anyhow, the analytic results supply a convenient starting point for better approximations.

Neglecting the ellipsoidal distortion, it is easy to calculate the path integral over $\mathbf{x}(\tau)$ for $P_L^\varphi(\mathbf{R}, 0)$ in the saddle point approximation. At an arbitrary given $\varphi(r)$, we must find the classical orbits. The Euler-Lagrange equation has the first integral of motion

$$\frac{M}{2}\dot{\mathbf{x}}^2 - \varphi(r) = E = \text{const.} \qquad (15.328)$$

At fixed L, we have to find the classical solutions for all energies E and all angular momenta l. The path integral reduces an ordinary double integral over E and l which, in turn, is evaluated in the saddle point approximation. In a rotationally symmetric potential $\varphi(r)$, the leading saddle point has the angular momentum $l = 0$ corresponding to a symmetric polymer distribution. Then Eq. (15.328) turns into a purely radial differential equation

$$d\tau = \frac{dr}{\sqrt{2[E + \varphi(r)]/M}}. \qquad (15.329)$$

For a polymer running from the origin to $\mathbf{R}$ we calculate

$$L = \int_0^R \frac{dr}{\sqrt{2[E + \varphi(r)]/M}}. \qquad (15.330)$$

This determines the energy E as a function of L. It is a functional of the yet unknown field $\varphi(r)$:

$$E = E_L[\varphi]. \qquad (15.331)$$

The classical action for such an orbit can be expressed in the form

$$\begin{aligned}
\mathcal{A}_{\text{cl}}[\mathbf{x}, \varphi] &= \int_0^L d\tau \left[\frac{M}{2}\dot{\mathbf{x}}^2 + \varphi(r(\tau))\right] \\
&= -\int_0^L d\tau \left[\frac{M}{2}\dot{\mathbf{x}}^2 - \varphi(r(\tau))\right] + \int_0^L d\tau M\dot{\mathbf{x}}^2 \\
&= -EL + \int_0^R dr \sqrt{2M[E + \varphi(r)]}.
\end{aligned} \qquad (15.332)$$

In this expression, we may consider E as an *independent* variational parameter. The relation (15.330) between $E, L, R, \varphi(r)$, by which E is fixed, reemerges when extremizing the classical expression $\mathcal{A}_{\mathrm{cl}}[\mathbf{x}, \varphi]$:

$$\frac{\partial}{\partial E}\mathcal{A}_{\mathrm{cl}}[\mathbf{x}, \dot{\mathbf{x}}, \varphi] = 0. \tag{15.333}$$

The classical approximation to the entire action $\mathcal{A}[\mathbf{x}, \varphi] + \mathcal{A}[\varphi]$ is then

$$\mathcal{A}_{\mathrm{cl}} = -EL + \int_0^r dr' \sqrt{2M[E + \varphi(r')]} - \frac{1}{2}v^{-1}a^{-D}\int d^D x \varphi^2(r). \tag{15.334}$$

This action is now extremized independently in $\varphi(r), E$. The extremum in $\varphi(r)$ is obviously given by the algebraic equation

$$\varphi(r') = \begin{cases} 0 \\ Mva^D S_D^{-1} r'^{1-D}/\sqrt{2M[E + \varphi(r')]} \end{cases} \text{for} \begin{array}{l} r' > r, \\ r' < r, \end{array} \tag{15.335}$$

which is easily solved. We rewrite it as

$$E + \varphi(r) = \xi^3 \varphi^{-2}(r), \tag{15.336}$$

with the abbreviation

$$\xi^3 = \alpha r^{-2\delta}, \tag{15.337}$$

where

$$\delta \equiv D - 1 > 0 \tag{15.338}$$

and

$$\alpha \equiv \frac{M}{2}v^2 a^{2D} S_D^{-2}. \tag{15.339}$$

For large $\xi \gg 1/E$, i.e., small $r \ll \alpha^{2/\delta}E^{-6/\delta}$, we expand the solution as follows

$$\varphi(r) = \xi - \frac{E}{3} + \frac{E^2}{9} + \dots. \tag{15.340}$$

This expansion is reinserted into the classical action (15.334), making it a power series in E. A further extremization in E yields $E = E(L, r)$. The extremal value of the action yields an approximate distribution function of a monomer in the closed polymer (which runs through the origin):

$$P_L(\mathbf{R}) \propto e^{-\mathcal{A}_{\mathrm{cl}}(L,R)}. \tag{15.341}$$

To see how this happens consider first the noninteracting limit where $v = 0$. Then the solution of (15.335) is $\varphi(r) \equiv 0$, and the classical action (15.334) becomes

$$\mathcal{A}_{\mathrm{cl}} = -EL + \sqrt{2MER}. \tag{15.342}$$

The extremization in E gives

$$E = \frac{M}{2}\frac{R^2}{L^2},\qquad(15.343)$$

yielding the extremal action

$$\mathcal{A}_{\text{cl}} = \frac{M}{2}\frac{R^2}{L}.\qquad(15.344)$$

The approximate distribution is therefore

$$P_N(\mathbf{R}) \propto e^{-\mathcal{A}_{\text{cl}}} = e^{-MR^2/2L}.\qquad(15.345)$$

The interacting case is now treated in the same way. Using (15.336), the classical action (15.334) can be written as

$$\mathcal{A}_{\text{cl}} = -EL + \sqrt{\frac{M}{2}}\int_0^R dr' \left[\sqrt{\varphi + E} - \frac{1}{2}\xi^{3/2}\frac{1}{\varphi + E}\right].\qquad(15.346)$$

By expanding this action in a power series in E [after having inserted (15.340) for φ], we obtain with $\epsilon(r) \equiv E/\xi = E\alpha^{-1/3}r^{2\delta/3}$

$$\mathcal{A}_{\text{cl}} = -EL + \sqrt{\frac{M}{2}}\alpha^{1/6}\int_0^R dr'\,r'^{-\delta/3}\left[\frac{3}{2} + \epsilon(r') - \frac{1}{6}\epsilon^2(r') + \dots\right].\qquad(15.347)$$

As long as $\delta < 3$, i.e., for

$$D < 4,\qquad(15.348)$$

the integral exists and yields an expansion

$$\mathcal{A}_{\text{cl}} = -EL + a_0(R) + a_1(R)E - \frac{1}{2}a_2(R)E^2 + \dots,\qquad(15.349)$$

with the coefficients

$$\begin{aligned}
a_0(R) &= -\frac{M}{2}\sqrt{\frac{9}{2}}R^{1-\delta/3}\alpha^{1/6}\frac{1}{\delta - 3},\\[4pt]
a_1(R) &= 3\sqrt{\frac{M}{2}}R^{1+\delta/3}\alpha^{-1/6}\frac{1}{\delta + 3},\\[4pt]
a_2(R) &= \frac{1}{3}\sqrt{\frac{M}{2}}R^{1+\delta}\alpha^{-1/2}\frac{1}{\delta + 1}.
\end{aligned}\qquad(15.350)$$

Extremizing $\mathcal{A}_{\text{cl}}$ in E gives the action

$$\mathcal{A}_{\text{cl}} = a_0(R) + \frac{1}{2a_2(R)}[L - a_1(R)]^2 + \dots.\qquad(15.351)$$

The approximate end-to-end distribution function is therefore

$$P_L(\mathbf{R}) \approx \mathcal{N} \exp\left\{-a_0(R) - \frac{1}{2a_2(R)}[L - a_1(R)]^2\right\}, \tag{15.352}$$

where $\mathcal{N}$ is an appropriate normalization factor. The distribution is peaked around

$$L = 3\sqrt{\frac{M}{2}} R^{1+\delta/3} \alpha^{-1/6} \frac{1}{\delta + 3}. \tag{15.353}$$

This shows the most important consequence of the excluded-volume effect: The average value of R^2 grows like

$$\langle R^2 \rangle \approx \alpha^{1/(D+2)} \left(\frac{D+2}{3}\sqrt{\frac{2}{M}}L\right)^{6/(D+2)}. \tag{15.354}$$

Thus we have found a scaling law of the form (15.298) with the critical exponent

$$\nu = \frac{3}{D+2}. \tag{15.355}$$

The repulsion between the chain elements makes the excluded-volume chain reach out further into space than a random chain [although less than a completely stiff chain, which is always reached by the solution (15.354) for $D = 1$].

The restriction $D < 4$ in (15.348) is quite important. The value

$$D^{\text{uc}} = 4 \tag{15.356}$$

is called the *upper critical dimension*. Above it, the set of all possible intersections of a random chain has the measure zero and any short-range repulsion becomes irrelevant. In fact, for $D > D^{\text{uc}}$ it is possible to show that the polymer behaves like a random chain without any excluded-volume effect satisfying $\langle R^2 \rangle \propto L$.

15.11 Flory's Argument

Once we expect a power-like scaling law of the form

$$\langle R^2 \rangle \propto L^{2\nu}, \tag{15.357}$$

the critical exponent (15.355) can be derived from a very simple dimensional argument due to Flory. We take the action

$$\mathcal{A} = \int_0^L d\tau \frac{M}{2}\dot{\mathbf{x}}^2 - \frac{va^D}{2}\int_0^L d\tau \int_0^L d\tau'\, \delta^{(D)}(\mathbf{x}(\tau) - \mathbf{x}(\tau')), \tag{15.358}$$

with $M = D/a$, and replace the two terms by their dimensional content, L for the τ-variable and R- for each x-component. Then

$$\mathcal{A} \sim \frac{M}{2}L\frac{R^2}{L^2} - \frac{va^D}{2}\frac{L^2}{R^D}. \tag{15.359}$$

Extremizing this expression in R at fixed L gives

$$\frac{R}{L} \sim R^{-D-1}L^2, \tag{15.360}$$

implying

$$R^2 \sim L^{6/(D+2)}, \tag{15.361}$$

and thus the critical exponent (15.355).

15.12 Polymer Field Theory

There exists an alternative approach to finding the power laws caused by the excluded-volume effects in polymers which is superior to the previous one. It is based on an intimate relationship of polymer fluctuations with field fluctuations in a certain somewhat artificial and unphysical limit. This limit happens to be accessible to approximate analytic methods developed in recent years in quantum field theory. According to Chapter 7, the statistical mechanics of a many-particle ensemble can be described by a single fluctuating field. Each particle in such an ensemble moves through spacetime along a fluctuating orbit in the same way as a random chain does in the approximation (15.80) to a polymer in Section 15.6. Thus we can immediately conclude that *ensembles* of polymers may also be described by a single fluctuating field. But how about a single polymer? Is it possible to project out a single polymer of the ensemble in the field-theoretic description? The answer is positive. We start with the result of the last section. The end-to-end distribution of the polymer in the excluded-volume problem is rewritten as an integral over the fluctuating field $\varphi(\mathbf{x})$:

$$P_L(\mathbf{x}_b, \mathbf{x}_a) = \int \mathcal{D}\varphi\, e^{-\mathcal{A}[\varphi]} P_L^\varphi(\mathbf{x}_b, \mathbf{x}_a), \tag{15.362}$$

with an action for the auxiliary $\varphi(\mathbf{x})$ field [see (15.312)]

$$\mathcal{A}[\varphi] = -\frac{1}{2} \int d^D x\, d^D x'\, \varphi(\mathbf{x}) V^{-1}(\mathbf{x}, \mathbf{x}') \varphi(\mathbf{x}'), \tag{15.363}$$

and an end-to-end distribution [see (15.315)]

$$P_L^\varphi(\mathbf{x}_b, \mathbf{x}_a) = \int \mathcal{D}x \exp\left\{ -\int_0^L d\tau \left[\frac{M}{2}\dot{\mathbf{x}}^2 + \varphi(\mathbf{x}(\tau)) \right] \right\}, \tag{15.364}$$

which satisfies the Schrödinger equation [see (15.316)]

$$\left[\frac{\partial}{\partial L} - \frac{1}{2M}\boldsymbol{\nabla}^2 + \varphi(\mathbf{x}) \right] P_L^\varphi(\mathbf{x}, \mathbf{x}') = \delta^{(3)}(\mathbf{x} - \mathbf{x}')\delta(L). \tag{15.365}$$

Since P_L and P_L^φ vanish for $L < 0$, it is convenient to go over to the Laplace transforms

$$P_{m^2}(\mathbf{x}, \mathbf{x}') = \frac{1}{2M} \int_0^\infty dL e^{-Lm^2/2M} P_L(\mathbf{x}, \mathbf{x}'), \qquad (15.366)$$

$$P_{m^2}^\varphi(\mathbf{x}, \mathbf{x}') = \frac{1}{2M} \int_0^\infty dL e^{-Lm^2/2M} P_L^\varphi(\mathbf{x}, \mathbf{x}'). \qquad (15.367)$$

The latter satisfies the L-independent equation

$$[-\boldsymbol{\nabla}^2 + m^2 + 2M\varphi(\mathbf{x})] P_{m^2}^\varphi(\mathbf{x}, \mathbf{x}') = \delta^{(3)}(\mathbf{x} - \mathbf{x}'). \qquad (15.368)$$

The quantity $m^2/2M$ is, of course, just the negative energy variable $-E$ in (15.332):

$$-E \equiv \frac{m^2}{2M}. \qquad (15.369)$$

The distributions $P_{m^2}^\varphi(\mathbf{x}, \mathbf{x}')$ describe the probability of a polymer of any length running from $\mathbf{x}'$ to $\mathbf{x}$, with a Boltzmann-like factor $e^{-Lm^2/2M}$ governing the distribution of lengths. Thus $m^2/2M$ plays the role of a chemical potential.

We now observe that the solution of Eq. (15.368) can be considered as the correlation function of an auxiliary fluctuating complex field $\psi(\mathbf{x})$:

$$
\begin{aligned}
P_{m^2}^\varphi(\mathbf{x}, \mathbf{x}') &= G_0^\varphi(\mathbf{x}, \mathbf{x}') = \langle \psi^*(\mathbf{x})\psi(\mathbf{x}') \rangle_\varphi \\
&\equiv \frac{\int \mathcal{D}\psi^*(\mathbf{x})\mathcal{D}\psi(\mathbf{x}) \, \psi^*(\mathbf{x})\psi(\mathbf{x}') \exp\{-\mathcal{A}[\psi^*, \psi, \varphi]\}}{\int \mathcal{D}\psi^*(\mathbf{x})\mathcal{D}\psi(\mathbf{x}) \exp\{-\mathcal{A}[\psi^*, \psi, \varphi]\}},
\end{aligned} \qquad (15.370)
$$

with a field action

$$\mathcal{A}[\psi^*, \psi, \varphi] = \int d^D x \left\{ \boldsymbol{\nabla}\psi^*(\mathbf{x})\boldsymbol{\nabla}\psi(\mathbf{x}) + m^2\psi^*(\mathbf{x})\psi(\mathbf{x}) + 2M\varphi(\mathbf{x})\psi^*(\mathbf{x})\psi(\mathbf{x}) \right\}. \quad (15.371)$$

The second part of Eq. (15.370) defines the expectations $\langle \ldots \rangle_\psi$. In this way, we express the Laplace-transformed distribution $P_{m^2}(\mathbf{x}_b, \mathbf{x}_a)$ in (15.366) in the purely field-theoretic form

$$
\begin{aligned}
P_{m^2}(\mathbf{x}, \mathbf{x}') &= \int \mathcal{D}\varphi \exp\{-\mathcal{A}[\varphi]\} \langle \psi^*(\mathbf{x})\psi(\mathbf{x}') \rangle_\varphi \\
&= \int \mathcal{D}\varphi \exp\left\{ \frac{1}{2} \int d^D y \, d^D y' \varphi(\mathbf{y}) V^{-1}(\mathbf{y}, \mathbf{y}') \varphi(\mathbf{y}') \right\} \\
&\quad \times \frac{\int \mathcal{D}\psi^* \mathcal{D}\psi \, \psi^*(\mathbf{x})\psi(\mathbf{x}') \exp\{-\mathcal{A}[\psi^*, \psi, \varphi]\}}{\int \mathcal{D}\psi^* \int \mathcal{D}\psi \exp\{-\mathcal{A}[\psi^*, \psi, \varphi]\}}.
\end{aligned} \qquad (15.372)
$$

It involves only a fluctuating field which contains all information on the path fluctuations. The field $\psi(\mathbf{x})$ is, of course, the analog of the second-quantized field in Chapter 7.

Consider now the probability distribution of a single monomer in a closed polymer chain. Inserting the polymer density function

$$\rho(\mathbf{R}) \equiv \int_0^L d\tau \, \delta^{(D)}(\mathbf{R} - \mathbf{x}(\tau)) \qquad (15.373)$$

into the original path integral for a closed polymer

$$P_L(\mathbf{R}) = \int_0^L d\tau \int \mathcal{D}^D x \int \mathcal{D}\varphi \exp\left\{-\mathcal{A}_L - \mathcal{A}[\varphi]\right\} \delta^{(D)}(\mathbf{R} - \mathbf{x}(\tau)), \qquad (15.374)$$

the δ-function splits the path integral into two parts

$$P_L(\mathbf{R}) = \int \mathcal{D}\varphi \exp\left\{-\mathcal{A}[\varphi]\right\} \int_0^L d\tau P_{L-\tau}^{\varphi}(\mathbf{0}, \mathbf{R}) P_{\tau}^{\varphi}(\mathbf{R}, \mathbf{0}). \qquad (15.375)$$

When going to the Laplace transform, the convolution integral factorizes, yielding

$$P_{m^2}(\mathbf{R}) = \int \mathcal{D}\varphi(\mathbf{x}) \exp\left\{-\mathcal{A}[\varphi]\right\} P_{m^2}^{\varphi}(\mathbf{0}, \mathbf{R}) P_{m^2}^{\varphi}(\mathbf{R}, \mathbf{0}). \qquad (15.376)$$

With the help of the field-theoretic expression for $P_{m^2}(\mathbf{R})$ in Eq. (15.370), the product of the correlation functions can be rewritten as

$$P_{m^2}^{\varphi}(\mathbf{0}, \mathbf{R}) P_{m^2}^{\varphi}(\mathbf{0}, \mathbf{R}) = \langle \psi^*(\mathbf{R}) \psi(\mathbf{0}) \rangle_{\varphi} \langle \psi^*(\mathbf{0}) \psi(\mathbf{R}) \rangle_{\varphi}. \qquad (15.377)$$

We now observe that the field ψ appears only quadratically in the action $\mathcal{A}[\psi^*, \psi, \varphi]$. The product of correlation functions in (15.377) can therefore be viewed as a term in the Wick expansion (recall Section 3.10) of the four-field correlation function

$$\langle \psi^*(\mathbf{R}) \psi(\mathbf{R}) \psi^*(\mathbf{0}) \psi(\mathbf{0}) \rangle_{\varphi}. \qquad (15.378)$$

This would be equal to the sum of pair contractions

$$\langle \psi^*(\mathbf{R}) \psi(\mathbf{R}) \rangle_{\varphi} \langle \psi^*(\mathbf{0}) \psi(\mathbf{0}) \rangle_{\varphi} + \langle \psi^*(\mathbf{R}) \psi(\mathbf{0}) \rangle_{\varphi} \langle \psi^*(\mathbf{0}) \psi(\mathbf{R}) \rangle_{\varphi}. \qquad (15.379)$$

There are no contributions containing expectations of two ψ or two ψ^* fields which could, in general, appear in this expansion. This allows the right-hand side of (15.377) to be expressed as a difference between (15.378) and the first term of (15.379):

$$P_{m^2}^{\varphi}(\mathbf{0}, \mathbf{R}) P_{m^2}^{\varphi}(\mathbf{0}, \mathbf{R}) = \langle \psi^*(\mathbf{R}) \psi(\mathbf{R}) \psi^*(\mathbf{0}) \psi(\mathbf{0}) \rangle_{\varphi} - \langle \psi^*(\mathbf{R}) \psi(\mathbf{R}) \rangle_{\varphi} \langle \psi^*(\mathbf{0}) \psi(\mathbf{0}) \rangle_{\varphi}. \qquad (15.380)$$

The right-hand side only contains correlation functions of a collective field, the *density field* [8]

$$\rho(\mathbf{R}) = \psi^*(\mathbf{R}) \psi(\mathbf{R}), \qquad (15.381)$$

in terms of which

$$P_{m^2}^{\varphi}(\mathbf{0}, \mathbf{R}) P_{m^2}^{\varphi}(\mathbf{0}, \mathbf{R}) = \langle \rho(\mathbf{R}) \rho(\mathbf{0}) \rangle_{\varphi} - \langle \rho(\mathbf{R}) \rangle_{\varphi} \langle \rho(\mathbf{0}) \rangle_{\varphi}. \qquad (15.382)$$

Now, the right-hand side is the *connected* correlation function of the density field $\rho(\mathbf{R})$:

$$\langle \rho(\mathbf{R}) \rho(\mathbf{0}) \rangle_{\varphi,c} \equiv \langle \rho(\mathbf{R}) \rho(\mathbf{0}) \rangle_{\varphi} - \langle \rho(\mathbf{R}) \rangle_{\varphi} \langle \rho(\mathbf{0}) \rangle_{\varphi}. \qquad (15.383)$$

In Section 3.10 we have shown how to generate all connected correlation functions: The action $\mathcal{A}[\psi, \psi^*, \varphi]$ is extended by a source term in the density field $\rho(\mathbf{x})$

$$\mathcal{A}_{\text{source}}[\psi^*, \psi, K] = -\int d^D x K(\mathbf{x})\rho(\mathbf{x}) = \int d^D x K(\mathbf{x})\psi^*(\mathbf{x})\psi(\mathbf{x})), \quad (15.384)$$

and one considers the partition function

$$Z[K, \varphi] \equiv \int \mathcal{D}\psi \mathcal{D}\psi^* \exp\left\{-\mathcal{A}[\psi^*, \psi, \varphi] - \mathcal{A}_{\text{source}}[\psi^*, \psi, K]\right\}. \quad (15.385)$$

This is the generating functional of all correlation functions of the density field $\rho(\mathbf{R}) = \psi^*(\mathbf{R})\psi(\mathbf{R})$ at a fixed $\varphi(\mathbf{x})$. They are obtained from the functional derivatives

$$\langle\rho(\mathbf{x}_1)\cdots\rho(\mathbf{x}_n)\rangle_\varphi = Z[K, \varphi]^{-1}\frac{\delta}{\delta K(\mathbf{x}_1)}\cdots\frac{\delta}{\delta K(\mathbf{x}_n)}Z[K, \varphi]\Big|_{K=0}. \quad (15.386)$$

Recalling Eq. (3.556), the connected correlation functions of $\rho(\mathbf{x})$ are obtained similarly from the logarithm of $Z[K, \varphi]$:

$$\langle\rho(\mathbf{x}_1)\cdots\rho(\mathbf{x}_n)\rangle_{\varphi,c} = \frac{\delta}{\delta K(\mathbf{x}_1)}\cdots\frac{\delta}{\delta K(\mathbf{x}_n)}\log Z[K, \varphi]\Big|_{K=0}. \quad (15.387)$$

For $n = 2$, the connectedness is seen directly by performing the differentiations according to the chain rule:

$$\begin{aligned}
\langle\rho(\mathbf{R})\rho(\mathbf{0})\rangle_{\varphi,c} &= \frac{\delta}{\delta K(\mathbf{R})}\frac{\delta}{\delta K(\mathbf{0})}\log Z[K, \varphi]\Big|_{K=0} \\
&= \frac{\delta}{\delta K(\mathbf{R})}Z^{-1}[K, \varphi]\frac{\delta}{\delta K(\mathbf{0})}Z[K, \varphi]\Big|_{K=0} \\
&= \langle\rho(\mathbf{R})\rho(\mathbf{0})\rangle_\varphi - \langle\rho(\mathbf{R})\rangle_\varphi\langle\rho(\mathbf{0})\rangle_\varphi.
\end{aligned} \quad (15.388)$$

This agrees indeed with (15.383). We can therefore rewrite the product of Laplace-transformed distributions (15.382) at a fixed $\varphi(\mathbf{x})$ as

$$P^\varphi_{m^2}(\mathbf{0}, \mathbf{R})P^\varphi_{m^2}(\mathbf{0}, \mathbf{R}) = \frac{\delta}{\delta K(\mathbf{R})}\frac{\delta}{\delta K(\mathbf{0})}\log Z[K, \varphi]\Big|_{K=0}. \quad (15.389)$$

The Laplace-transformed monomer distribution (15.376) is then obtained by averaging over $\varphi(\mathbf{x})$, i.e., by the path integral

$$P_{m^2}(\mathbf{R}) = \frac{\delta}{\delta K(\mathbf{R})}\frac{\delta}{\delta K(\mathbf{0})}\int \mathcal{D}\varphi(\mathbf{x})\exp\left\{-\mathcal{A}[\varphi]\right\}\log Z[K, \varphi]\Big|_{K=0}. \quad (15.390)$$

Were it not for the logarithm in front of Z, this would be a standard calculation of correlation functions within the combined ψ, φ field theory whose action is

$$
\begin{aligned}
\mathcal{A}_{\text{tot}}[\psi^*, \psi, \varphi] &= \mathcal{A}[\psi^*, \psi, \varphi] + \mathcal{A}[\varphi] \\
&= \int d^D x \left(\boldsymbol{\nabla}\psi^* \boldsymbol{\nabla}\psi + m^2 \psi^* \psi + 2M\varphi\psi^*\psi \right) \\
&\quad - \frac{1}{2} \int d^D x \, d^D x' \varphi(\mathbf{x}) V^{-1}(\mathbf{x}, \mathbf{x}') \varphi(\mathbf{x}').
\end{aligned} \tag{15.391}
$$

To account for the logarithm we introduce a simple mathematical device called the *replica trick* [9]. We consider $\log Z[K, \varphi]$ in (15.388)–(15.390) as the limit

$$
\log Z = \lim_{n \to 0} \frac{1}{n} \left(Z^n - 1 \right), \tag{15.392}
$$

and observe that the nth power of the generating functional, Z^n, can be thought of as arising from a field theory in which every field ψ occurs n times, i.e., with n identical replica. Thus we add an extra internal symmetry label $\alpha = 1, \ldots, n$ to the fields $\psi(\mathbf{x})$ and calculate Z^n formally as

$$
Z^n[K, \varphi] = \int \mathcal{D}\psi_\alpha^* \mathcal{D}\psi_\alpha \exp\left\{ -\mathcal{A}[\psi_\alpha^*, \psi_\alpha, \varphi] - \mathcal{A}[\varphi] - \mathcal{A}_{\text{source}}[\psi_\alpha^*, \psi_\alpha, K] \right\}, \tag{15.393}
$$

with the replica field action

$$
\mathcal{A}[\psi_\alpha, \psi_\alpha^*, \varphi] = \int d^D x \left(\boldsymbol{\nabla}\psi_\alpha^* \boldsymbol{\nabla}\psi_\alpha + m^2 \psi_\alpha^* \psi_\alpha + 2M\varphi \, \psi_\alpha^* \psi_\alpha \right), \tag{15.394}
$$

and the source term

$$
\mathcal{A}_{\text{source}}[\psi_\alpha^*, \psi_\alpha, K] = - \int d^D x \, \psi_\alpha^*(\mathbf{x}) \psi_\alpha(\mathbf{x}) K(\mathbf{x}). \tag{15.395}
$$

A sum is implied over repeated indices α. By construction, the action is symmetric under the group $U(n)$ of all unitary transformations of the replica fields ψ_α.

In the partition function (15.393), it is now easy to integrate out the $\varphi(\mathbf{x})$-fluctuations. This gives

$$
Z^n[K, \varphi] = \int \mathcal{D}\psi_\alpha^* \mathcal{D}\psi_\alpha \exp\left\{ -\mathcal{A}^n[\psi_\alpha^*, \psi_\alpha] - \mathcal{A}_{\text{source}}[\psi_\alpha^*, \psi_\alpha, K] \right\}, \tag{15.396}
$$

with the action

$$
\begin{aligned}
\mathcal{A}^n[\psi_\alpha^*, \psi_\alpha] &= \int d^D x \left(\boldsymbol{\nabla}\psi_\alpha^* \boldsymbol{\nabla}\psi_\alpha + m^2 \psi_\alpha^* \psi_\alpha \right) \\
&\quad + \frac{1}{2}(2M)^2 \int d^D x \, d^D x' \psi_\alpha^*(\mathbf{x}) \psi_\alpha(\mathbf{x}) V(\mathbf{x}, \mathbf{x}') \psi_\beta^*(\mathbf{x}') \psi_\beta(\mathbf{x}').
\end{aligned} \tag{15.397}
$$

It describes a self-interacting field theory with an additional $U(n)$ symmetry.

In the special case of a local repulsive potential $V(\mathbf{x}, \mathbf{x}')$ of Eq. (15.319), the second term becomes simply

$$\mathcal{A}_{\text{int}}[\psi_\alpha^*, \psi_\alpha] = \frac{1}{2}(2M)^2 v a^D \int d^D x \, [\psi_\alpha^*(\mathbf{x}) \psi_\alpha(\mathbf{x})]^2 . \tag{15.398}$$

Using this action, we can find $\log Z[K, \varphi]$ via (15.392) from the functional integral

$$\log Z[K, \varphi] \equiv \lim_{n \to 0} \frac{1}{n} \left(\int \mathcal{D}\psi_\alpha^* \mathcal{D}\psi_\alpha \exp\left\{ -\mathcal{A}^n[\psi_\alpha^*, \psi_\alpha] - \mathcal{A}_{\text{source}}[\psi_\alpha^*, \psi_\alpha, K] \right\} - 1 \right). \tag{15.399}$$

This is the generating functional of the Laplace-transformed distribution (15.390) which we wanted to calculate.

A polymer can run along the same line in two orientations. In the above description with complex replica fields it was assumed that the two orientations can be distinguished. If they are indistinguishable, the polymer fields $\Psi_\alpha(\mathbf{x})$ have to be taken as real.

Such a field-theoretic description of a fluctuating polymer has an important advantage over the initial path integral formulation based on the analogy with a particle orbit. It allows us to establish contact with the well-developed theory of critical phenomena in field theory. The end-to-end distribution of long polymers at large L is determined by the small-E regime in Eqs. (15.330)–(15.349), which corresponds to the small-m^2 limit of the system here [see (15.369)]. This is precisely the regime studied in the quantum field-theoretic approach to critical phenomena in many-body systems [10, 11]. It can be shown that for D larger than the upper critical dimension $D^{\text{uc}} = 4$, the behavior for $m^2 \to 0$ of all Green functions coincides with the free-field behavior. For $D = D^{\text{uc}}$, this behavior can be deduced from scale invariance arguments of the action, using naive dimensional counting arguments. The fluctuations turn out to cause only logarithmic corrections to the scale-invariant power laws. One of the main developments in quantum field theory in recent years was the discovery that the scaling powers for $D < D^{\text{uc}}$ can be calculated via an expansion of all quantities in powers of

$$\epsilon = D^{\text{uc}} - D, \tag{15.400}$$

the so-called ϵ-expansion. The ϵ-expansion for the critical exponent ν which rules the relation between R^2 and the length of a polymer L, $\langle R^2 \rangle \propto L^{2\nu}$, can be derived from a real ϕ^4-theory with n replica as follows [12]:

$$\begin{aligned}
\nu^{-1} = {}& 2 + \frac{(n+2)\epsilon}{n+8} \left\{ -1 - \frac{\epsilon}{2(n+8)^2}(13n+44) \right. \\
& + \frac{\epsilon^2}{8(n+8)^4} \left[3n^3 - 452n^2 - 2672n - 5312 \right. \\
& \qquad + \zeta(3)(n+8) \cdot 96(5n+22) \right] \\
& + \frac{\epsilon^3}{32(n+8)^6} \left[3n^5 + 398n^4 - 12900n^3 - 81552n^2 - 219968n - 357120 \right. \\
& \qquad + \zeta(3)(n+8) \cdot 16(3n^4 - 194n^3 + 148n^2 + 9472n + 19488) \\
& \qquad + \zeta(4)(n+8)^3 \cdot 288(5n+22)
\end{aligned}$$

$$- \zeta(5)(n+8)^2 \cdot 1280(2n^2 + 55n + 186)]$$
$$+\frac{\epsilon^4}{128(n+8)^8}[3n^7 - 1198n^6 - 27484n^5 - 1055344n^4$$
$$-5242112n^3 - 5256704n^2 + 6999040n - 626688$$
$$- \zeta(3)(n+8) \cdot 16(13n^6 - 310n^5 + 19004n^4 + 102400n^3$$
$$-381536n^2 - 2792576n - 4240640)$$
$$- \zeta^2(3)(n+8)^2 \cdot 1024(2n^4 + 18n^3 + 981n^2 + 6994n + 11688)$$
$$+ \zeta(4)(n+8)^3 \cdot 48(3n^4 - 194n^3 + 148n^2 + 9472n + 19488)$$
$$+ \zeta(5)(n+8)^2 \cdot 256(155n^4 + 3026n^3 + 989n^2 - 66018n - 130608)$$
$$- \zeta(6)(n+8)^4 \cdot 6400(2n^2 + 55n + 186)$$
$$+ \zeta(7)(n+8)^3 \cdot 56448(14n^2 + 189n + 526)]\Big\} , \qquad (15.401)$$

where $\zeta(x)$ is Riemann's zeta function (2.519). As shown above, the single-polymer properties must emerge in the limit $n \to 0$. There, ν^{-1} has the ϵ-expansion

$$\nu^{-1} = 2 - \frac{\epsilon}{4} - \frac{11}{128}\epsilon^2 + 0.114\,425\,\epsilon^3 - 0.287\,512\,\epsilon^4 + 0.956\,133\,\epsilon^5. \qquad (15.402)$$

This is to be compared with the much simpler result of the last section

$$\nu^{-1} = \frac{D+2}{3} = 2 - \frac{\epsilon}{3}. \qquad (15.403)$$

A term-by-term comparison is meaningless since the field-theoretic ϵ-expansion has a grave problem: The coefficients of the ϵ^n-terms grow, for large n, like $n!$, so that the series does not converge anywhere! Fortunately, the signs are alternating and the series can be resummed [13]. A simple first approximation used in ϵ-expansions is to re-express the series (15.402) as a ratio of two polynomials of roughly equal degree

$$\nu^{-1}|_{\text{rat}} = \frac{2. + 1.023\,606\,\epsilon - 0.225\,661\,\epsilon^2}{1. + 0.636\,803\,\epsilon - 0.011\,746\,\epsilon^2 + 0.002\,677\,\epsilon^3}, \qquad (15.404)$$

called a *Padé approximation*. Its Taylor coefficients up to ϵ^5 coincide with those of the initial series (15.402). It can be shown that this approximation would converge with increasing orders in ϵ towards the exact function represented by the divergent series. In Fig. 15.7, we plot the three functions (15.403), (15.402), and (15.404), the last one giving the most reliable approximation

$$\nu^{-1} \approx 0.585. \qquad (15.405)$$

Note that the simple Flory curve lies very close to the Padé curve whose calculation requires a great amount of work.

There exists a general scaling relation between the exponent ν and another exponent appearing in the total number of polymer configurations of length L which behaves like

$$S = L^{\alpha - 2}. \qquad (15.406)$$

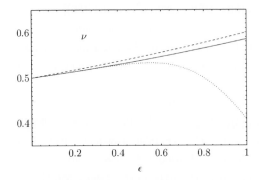

Figure 15.7 Comparison of critical exponent ν in Flory approximation (dashed line) with result of divergent ϵ-expansion obtained from quantum field theory (dotted line) and its Padé resummation (solid line). The value of the latter gives the best approximation $\nu \approx 0.585$ at $\epsilon = 1$.

The relation is

$$\alpha = 2 - D\nu. \tag{15.407}$$

Direct enumeration studies of random chains on a computer suggest a number

$$\alpha \sim \frac{1}{4}, \tag{15.408}$$

corresponding to $\nu = 7/12 \approx 0.583$, very close to (15.405).

The Flory estimate for the exponent α reads, incidentally,

$$\alpha = \frac{4 - D}{D + 2}. \tag{15.409}$$

In three dimensions, this yields $\alpha = 1/5$, not too far from (15.408).

The discrepancies arise from inaccuracies in both treatments. In the first treatment, they are due to the use of the saddle point approximation and the fact that the δ-function does not completely rule out the crossing of the lines, as required by the true self-avoidance of the polymer. The field theoretic ϵ-expansion, on the other hand, which in principle can give arbitrarily accurate results, has the problem of being divergent for any ϵ. Resummation procedures are needed and the order of the expansion must be quite large ($\approx \epsilon^5$) to extract reliable numbers.

15.13   Fermi Fields for Self-Avoiding Lines

There exists another way of enforcing the self-avoiding property of random lines [14]. It is based on the observation that for a polymer field theory with n fluctuating complex fields Ψ_α and a $U(n)$-symmetric fourth-order self-interaction as in the action

(15.397), the symmetric incorporation of a set of m anticommuting Grassmann fields removes the effect of m of the Bose fields. For free fields this observation is trivial since the functional determinant of Bose and Fermi fields are inverse to one-another. In the presence of a fourth-order self-interaction, where the replica action has the form (15.397), we can always go back, by a Hubbard-Stratonovich transformation, to the action involving the auxiliary field $\varphi(\mathbf{x})$ in the exponent of (15.393). This exponent is purely quadratic in the replica field, and each path integral over a Fermi field cancels a functional determinant coming from the Bose field.

This boson-destructive effect of fermions allows us to study theories with a negative number of replica. We simply have to use more Fermi than Bose fields. Moreover, we may conclude that a theory with $n = -2$ has necessarily trivial critical exponents. From the above arguments it is equivalent to a single complex Fermi field theory with fourth-order self-interaction. However, for anticommuting Grassmann fields, such an interaction vanishes:

$$(\theta^\dagger \theta)^2 = [(\theta_1 - i\theta_2)(\theta_1 + i\theta_2)]^2 = [2i\theta_1\theta_2]^2 = 0. \qquad (15.410)$$

Looking back at the ϵ-expansion for the critical exponent ν in Eq. (15.401) we can verify that up to the power ϵ^5 all powers in ϵ do indeed vanish and ν takes the mean-field value $1/2$.

Appendix 15A   Basic Integrals

$$\Delta(0,0) = \Delta(L,L) = L/3, \qquad (15A.1)$$

$$I_1 = \int_0^L ds\, \Delta(s,s) = L^2/6, \qquad (15A.2)$$

$$I_2 = \int_0^L ds\, \dot{\Delta}^2(s,s) = L/12, \qquad (15A.3)$$

$$I_3 = \int_0^L ds\, \Delta^2(s,s) = L^3/30, \qquad (15A.4)$$

$$I_4 = \int_0^L ds\, \Delta(s,s)\, \dot{\Delta}^2(s,s) = 7L^2/360, \qquad (15A.5)$$

$$I_5 = \int_0^L ds \int_0^L ds'\, \Delta^2(s,s') = L^4/90, \qquad (15A.6)$$

$$\Delta(0,L) = \Delta(L,0) = -L/6, \qquad (15A.7)$$

$$I_6 = \int_0^L ds\, [\Delta^2(s,0) + \Delta^2(s,L)] = 2L^3/45, \qquad (15A.8)$$

$$I_7 = \int_0^L ds \int_0^L ds'\, \Delta(s,s)\dot{\Delta}^2(s,s') = L^3/45, \qquad (15A.9)$$

$$I_8 = \int_0^L ds \int_0^L ds'\, \dot{\Delta}(s,s)\Delta(s,s')\dot{\Delta}(s,s') = L^3/180, \qquad (15A.10)$$

$$I_9 = \int_0^L ds\, \Delta(s,s)\,[\dot{\Delta}^2(s,0) + \dot{\Delta}^2(s,L)] = 11L^2/90, \qquad (15A.11)$$

$$I_{10} = \int_0^L ds\, \dot{\Delta}(s,s)\,[\Delta(s,0)\dot{\Delta}(s,0) + \Delta(s,L)\dot{\Delta}(s,L)] = 17L^2/360, \qquad (15A.12)$$

$$\dot{\Delta}(0,0) = -\dot{\Delta}(L,L) = -1/2, \tag{15A.13}$$

$$I_{11} = \int_0^L ds \int_0^L ds'\, \Delta(s,s)\, \dot{\Delta}(s,s')\dot{\Delta}(s',s') = L^3/360, \tag{15A.14}$$

$$I_{12} = \int_0^L ds \int_0^L ds'\, \Delta(s,s')\dot{\Delta}^2(s,s') = L^3/90. \tag{15A.15}$$

Appendix 15B   Loop Integrals

We list here the Feynman integrals evaluated with dimensional regularization rules whenever necessary. Depending whether they occur in the calculation of the moments from the expectations (15.261)–(15.261) of from the expectations (15.283)–(15.285) we encounter the integrals depending on $\rho_n(s) \equiv \delta_n + [\delta(s) + \delta(s-L)]/2$ with $\delta_n = \delta(0) - \sigma_n/L$ or $\rho_k(s) = \delta_k + [\delta(s) + \delta(s-L)]/2$ with $\delta_k = \delta(0) - ik/L$:

$$H_1^{n(k)} = \int_0^L ds \int_0^L ds'\, \rho_{n(k)}(s)\rho_{n(k)}(s')\Delta^2(s,s'),$$
$$= \delta_{n(k)}^2\, I_5 + \delta_{n(k)}\, I_6 + \frac{1}{2}\left[\Delta^2(0,0) + \Delta^2(0,L)\right], \tag{15B.1}$$

$$H_2^{n(k)} = \int_0^L ds \int_0^L ds'\, \rho_{n(k)}(s')\Delta(s,s)\dot{\Delta}^2(s,s') = \delta_{n(k)}\, I_7 + \frac{I_9}{2}, \tag{15B.2}$$

$$H_3^{n(k)} = \int_0^L ds \int_0^L ds'\, \rho_{n(k)}(s')\dot{\Delta}d(s,s)\Delta^2(s,s') = \left[\delta_{n(k)}\, I_5 + \frac{I_6}{2}\right]\delta(0), \tag{15B.3}$$

$$H_4^{n(k)} = \int_0^L ds \int_0^L ds'\, \rho_{n(k)}(s')\dot{\Delta}(s,s)\dot{\Delta}(s,s')\Delta(s,s') = \delta_{n(k)}\, I_8 + \frac{I_{10}}{2}. \tag{15B.4}$$

When calculating $\langle R^n \rangle$, we need to insert here $\delta_n = \delta(0) - \sigma_n/L$, thus obtaining

$$H_1^n = \frac{L^4}{90}\delta^2(0) + \frac{L^3}{45}(3-D-n)\,\delta(0) + \frac{L^2}{360}\left[(45-24D+4D^2)-4n(6-2D-n)\right], \tag{15B.5}$$

$$H_2^n = \frac{L^3}{45}\delta(0) + \frac{L^2}{180}(15-4D-4n), \tag{15B.6}$$

$$H_3^n = \frac{L^4}{90}\delta^2(0) + \frac{L^3}{90}(3-D-n)\delta(0), \tag{15B.7}$$

$$H_4^n = \frac{L^3}{180}\delta(0) + \frac{L^2}{720}(21-4D-4n), \tag{15B.8}$$

where the values for $n=0$ correspond to the partition function $Z = \langle R^0 \rangle$. The substitution $\delta_k = \delta(0) - ik/L$ required for the calculation of $P(k;L)$ yields

$$H_1^k = \frac{L^4}{90}\delta^2(0) + \frac{L^3}{45}\left[2+(-ik)\right]\delta(0) + \frac{L^2}{90}(-ik)\left[4+(-ik)\right] + \frac{5L^2}{72}, \tag{15B.9}$$

$$H_2^k = \frac{L^3}{45}\delta(0) + \frac{L^2}{180}\left[11+4(-ik)\right], \tag{15B.10}$$

$$H_3^k = \frac{L^4}{90}\delta^2(0) + \frac{L^3}{90}\left[2+(-ik)\right]\delta(0), \tag{15B.11}$$

$$H_4^k = \frac{L^3}{180}\delta(0) + \frac{L^2}{720}\left[17+4(-ik)\right]. \tag{15B.12}$$

The other loop integrals are

$$H_5 = \int_0^L ds \int_0^L ds'\, \Delta(s,s)\dot{\Delta}^2(s,s')\dot{\Delta}d(s',s') = \delta(0)I_7 = \frac{L^3}{45}\delta(0), \tag{15B.13}$$

$$H_6 = \int_0^L ds \int_0^L ds' \Delta\, d(s,s) \Delta^2(s,s') \Delta\, d(s',s') = \delta^2(0) I_5 = \frac{L^4}{90}\delta^2(0), \tag{15B.14}$$

$$H_7 = \int_0^L ds \int_0^L ds' \Delta\,{}^{.}(s,s) \Delta\,{}^{.}(s,s')\Delta(s,s') \Delta\, d(s',s') = \delta(0) I_8 = \frac{L^3}{180}\delta(0), \tag{15B.15}$$

$$H_8 = \int_0^L ds \int_0^L ds' \Delta\,{}^{.}(s,s) \Delta\,{}^{.}(s,s') \Delta\,{}^{.}(s,s') \Delta\,{}^{.}(s',s') = -\frac{L^2}{720}, \tag{15B.16}$$

$$H_9 = \int_0^L ds \int_0^L ds' \Delta\,{}^{.}(s,s) \Delta(s,s') \Delta\, d(s,s') \Delta\,{}^{.}(s',s') = \frac{I_{10}}{2} - \frac{I_8}{L} - H_8 = \frac{7L^2}{360}, \tag{15B.17}$$

$$H_{10} = \int_0^L ds \int_0^L ds' \Delta(s,s) \Delta\,{}^{.}(s,s') \Delta\, d(s,s') \Delta\,{}^{.}(s',s') = \frac{I_9}{4} - \frac{I_7}{2L} = \frac{7L^2}{360}, \tag{15B.18}$$

$$H_{11} = \int_0^L ds \int_0^L ds' \Delta(s,s) \Delta\, d^2(s,s') \Delta(s',s') = \delta(0) I_3 + \left(\frac{I_1}{L}\right)^2 - \frac{2(I_3 - I_{11})}{L} - 2I_4,$$
$$+ 2\left[\Delta^2(L,L)\, \Delta\,{}^{.}(L,L) - \Delta^2(0,0)\, \Delta\,{}^{.}(0,0)\right] - 2H_{10} = \frac{L^3}{30}\delta(0) + \frac{L^2}{9}, \tag{15B.19}$$

$$H_{12} = \int_0^L ds \int_0^L ds' \Delta\,{}^{.2}(s,s') \Delta\,{}^{.2}(s,s') = \frac{L^2}{90}, \tag{15B.20}$$

$$H_{13} = \int_0^L ds \int_0^L ds' \Delta(s,s') \Delta\,{}^{.}(s,s') \Delta\,{}^{.}(s,s') \Delta\, d(s',s) = \frac{I_4}{2} - \frac{I_{12}}{2L} - \frac{H_{12}}{2} = -\frac{L^2}{720}, \tag{15B.21}$$

$$H_{14} = \int_0^L ds \int_0^L ds' \Delta^2(s,s') \Delta\, d^2(s,s') = \delta(0) I_3 - \frac{2(I_3 - I_{12})}{L} - 2I_4 + \frac{I_5}{L^2},$$
$$+ 2\left[\Delta^2(L,L)\, \Delta\,{}^{.}(L,L) - \Delta^2(0,0)\, \Delta\,{}^{.}(0,0)\right] - 2H_{13} = \frac{L^3}{30}\delta(0) + \frac{11L^2}{72}. \tag{15B.22}$$

Appendix 15C Integrals Involving Modified Green Function

To demonstrate the translational invariance of results showed in the main text we use the slightly modified Green function

$$\Delta(s,s') = \frac{L}{3}a - \frac{|s - s'|}{2} - \frac{(s + s')}{2} + \frac{(s^2 + s'^2)}{2L}, \tag{15C.1}$$

containing an arbitrary constant a. The following combination yields the standard Feynman propagator for the infinite interval [compare (3.246)]

$$\Delta_F(s,s') = \Delta_F(s - s') = \Delta(s,s') - \frac{1}{2}\Delta(s,s) - \frac{1}{2}\Delta(s',s') = -\frac{1}{2}(s - s'). \tag{15C.2}$$

Other useful relations fulfilled by the Green function (15C.1) are, assuming $s \geq s'$,

$$D_1(s,s') = \Delta^2(s,s') - \Delta(s,s)\Delta(s',s') = (s - s')\left[\frac{s(L-s)}{L} + \frac{(s-s')(s+s')^2}{4L^2} - \frac{La}{3}\right],$$
$$\tag{15C.3}$$

$$D_2(s,s') = \Delta(s,s) - \Delta(s',s') = \frac{(s-s')(s+s'-L)}{L}. \tag{15C.4}$$

The following integrals are needed:

$$J_1(s,s') = \int_0^L dt\, \Delta(t,t)\, \Delta\,{}^{.}(t,s)\, \Delta\,{}^{.}(t,s') = \frac{(s^4 + s'^4)}{4L^2} - \frac{(2s^3 + s'^3)}{3L},$$
$$+ \frac{((a+3)s^2 + as'^2)}{6} - \frac{sa}{3}L + \frac{(20a - 9)}{180}L^2, \tag{15C.5}$$

$$J_2(s,s') = \int_0^L dt \, \dot\Delta(t,t)\Delta(t,s)\dot\Delta(t,s') = \frac{(-2s^4+6s^2s'^2+3s'^4)}{24L^2},$$

$$+ \frac{(3s^3-3s^2s'-6ss'^2-s'^3)}{12L} - \frac{(5s^2-12ss'-4as'^2)}{24} - \frac{s'a}{6}L + \frac{(20a-3)}{720}L^2,$$

(15C.6)

$$J_3(s,s') = \int_0^L dt \, \Delta(t,s)\Delta(t,s') = -\frac{(s^4+6s^2s'^2+s'^4)}{60L} + \frac{(s^2+3s'^2)s}{6},$$

$$- \frac{(s^2+s'^2)}{6}L + \frac{(5a^2-10a+6)}{45}L^3.$$

(15C.7)

These are the building blocks for other relations:

$$\langle f_2(q(s),q(s'))\rangle = \frac{(d-1)}{2}\left[D_1(s,s') - \frac{(d+1)}{4}D_2^2(s,s')\right] = \frac{(d-1)(s-s')}{4}$$

$$\times \left[\frac{2La}{3} + \frac{(d-3)s-(d+1)s'}{2} - \frac{(d-1)s^2-(d+1)s'^2}{L} + \frac{d(s-s')(s+s')^2}{2L^2}\right],$$

(15C.8)

and

$$K_1(s,s') = J_1(s,s') - \frac{1}{2}J_1(s,s) - \frac{1}{2}J_1(s',s'),$$

$$= -(s-s')\left[\frac{La}{6} - \frac{(s+s')}{4} + \frac{(s^2+ss'+s'^2)}{6L}\right],$$

(15C.9)

$$K_2(s,s') = J_2(s,s') + J_2(s',s) - J_2(s,s) - J_2(s',s'),$$

$$= -(s-s')^2\left[\frac{1}{4} - \frac{(5s+7s')}{12L} + \frac{(s+s')^2}{4L^2}\right],$$

(15C.10)

$$K_3(s,s') = J_3(s,s') - \frac{1}{2}J_3(s,s) - \frac{1}{2}J_3(s',s') = -(s-s')^2\left[\frac{(s+2s')}{6} - \frac{(s+s')^2}{8L}\right],$$

(15C.11)

$$K_4(s,s') = \Delta(0,s)\Delta(0,s') - \frac{1}{2}\Delta^2(0,s) - \frac{1}{2}\Delta^2(0,s') = -\frac{(s-s')^2(s+s'-2L)^2}{8L^2},$$

(15C.12)

$$K_5(s,s') = \Delta(L,s)\Delta(L,s') - \frac{1}{2}\Delta^2(L,s) - \frac{1}{2}\Delta^2(L,s') = -\frac{(s^2-s'^2)^2}{8L^2}.$$

(15C.13)

Notes and References

Random chains were first considered by
K. Pearson, Nature **77**, 294 (1905)
who studied the related problem of a random walk of a drunken sailor.
A.A. Markoff, *Wahrscheinlichkeitsrechnung*, Leipzig 1912;
L. Rayleigh, Phil. Mag. **37**, 3221 (1919);
S. Chandrasekhar, Rev. Mod. Phys. **15**, 1 (1943).
The exact solution of $P_N(\mathbf{R})$ was found by
L.R.G. Treloar, Trans. Faraday Soc. **42**, 77 (1946);
K. Nagai, J. Phys. Soc. Japan, **5**, 201 (1950);
M.C. Wang and E. Guth, J. Chem. Phys. **20**, 1144 (1952);
C. Hsiung, A. Hsiung, and A. Gordus, J. Chem. Phys. **34**, 535 (1961).
The path integral approach to polymer physics has been advanced by
S.F. Edwards, Proc. Phys. Soc. Lond. **85**, 613 (1965); **88**, 265 (1966); **91**, 513 (1967); **92**, 9 (1967).
See also
H.P. Gilles, K.F. Freed, J. Chem. Phys. **63**, 852 (1975)
and the comprehensive studies by
K.F. Freed, Adv. Chem. Phys. **22**,1 (1972);

K. Itō and H.P. McKean, *Diffusion Processes and Their Simple Paths*, Springer, Berlin, 1965.

An alternative model to stimulate stiffness was formulated by
A. Kholodenko, Ann. Phys. **202**, 186 (1990); J. Chem. Phys. **96**, 700 (1991)
exploiting the statistical properties of a Fermi field.
Although the physical properties of a stiff polymer are slightly misrepresented by this model, the
distribution functions agree quite well with those of the Kratky-Porod chain. The advantage of
this model is that many properties can be calculated analytically.

The important role of the upper critical dimension $D^{uc} = 4$ for polymers was first pointed out by
R.J. Rubin, J. Chem. Phys. **21**, 2073 (1953).
The simple scaling law $\langle R^2 \rangle \propto L^{6/(D+2)}$, $D < 4$ was first found by
P.J. Flory, J. Chem. Phys. **17**, 303 (1949) on dimensional grounds.
The critical exponent ν has been deduced from computer enumerations of all self-avoiding polymer
configurations by
C. Domb, Adv. Chem. Phys. **15**, 229 (1969);
D.S. Kenzie, Phys. Rep. **27**, 35 (1976).
For a general discussion of the entire subject, in particular the field-theoretic aspects, see
P.G. DeGennes, *Scaling Concepts in Polymer Physics*, Cornell University Press, Ithaca, N.Y., 1979.
The relation between ensembles of random chains and field theory is derived in detail in
H. Kleinert, *Gauge Fields in Condensed Matter*, World Scientific, Singapore, 1989,
Vol. I, Part I; *Fluctuating Fields and Random Chains*, World Scientific, Singapore, 1989
(http://www.physik.fu-berlin.de/~kleinert/b1).

The particular citations in this chapter refer to the publications

[1] The path integral (15.112) for stiff polymers was proposed by
    O. Kratky and G. Porod, Rec. Trav. Chim. **68**, 1106 (1949).
    The lowest moments were calculated by
    J.J. Hermans and R. Ullman, Physica **18**, 951 (1952);
    N. Saito, K. Takahashi, and Y. Yunoki, J. Phys. Soc. Japan **22**, 219 (1967),
    and up to order 6 by
    R. Koyama, J. Phys. Soc. Japan, **34**, 1029 (1973).
    Still higher moments are found numerically by
    H. Yamakawa and M. Fujii, J. Chem. Phys. **50**, 6641 (1974),
    and in a large-stiffness expansion in
    T. Norisuye, H. Murakama, and H. Fujita, Macromolecules **11**, 966 (1978).
    The last two papers also give the end-to-end distribution for large stiffness.

[2] H.E. Stanley, Phys. Rev. **179**, 570 (1969).

[3] H.E. Daniels, Proc. Roy. Soc. Edinburgh **63A**, 29 (1952);
    W. Gobush, H. Yamakawa, W.H. Stockmayer, and W.S. Magee, J. Chem. Phys. **57**, 2839
    (1972).

[4] B. Hamprecht and H. Kleinert, J. Phys. A: Math. Gen. **37**, 8561 (2004) (*ibid*.http/345).
    MATHEMATICA program can be downloaded from *ibid*.http/b5/prm15.

[5] The MATHEMATICA notebook can be obtained from *ibid*.http/b5/pgm15. The program
    needs less than 2 minutes to calculate $\langle R^{32} \rangle$.

[6] J. Wilhelm and E. Frey, Phys. Rev. Lett. **77**, 2581 (1996).
    See also:
    A. Dhar and D. Chaudhuri, Phys. Rev. Lett **89**, 065502 (2002);
    S. Stepanow and M. Schuetz, Europhys. Lett. **60**, 546 (2002);
    J. Samuel and S. Sinha, Phys. Rev. E **66** 050801(R) (2002).

[7] H. Kleinert and A. Chervyakov, *Perturbation Theory for Path Integrals of Stiff Polymers*, Berlin Preprint 2005 (*ibid*.http/358).

[8] For a detailed theory of such fields with applications to superconductors and superfluids see H. Kleinert, *Collective Quantum Fields*, Fortschr. Phys. **26**, 565 (1978) (*ibid*.http/55).

[9] S.F. Edwards and P.W. Anderson, J. Phys. F: Metal Phys. 965 (1975).
Applications of replica field theory to the large-L behavior of $\langle R^2 \rangle$ and other critical exponents were given by
P.G. DeGennes, Phys. Lett. A **38**, 339 (1972);
J. des Cloizeaux, Phys. Rev. **10**, 1665 (1974); J. Phys. (Paris) Lett. **39** L151 (1978).

[10] See D.J. Amit, *Renormalization Group and Critical Phenomena*, World Scientific, Singapore, 1984; G. Parisi, *Statistical Field Theory*, Addison-Wesley, Reading, Mass., 1988.

[11] H. Kleinert and V. Schulte-Frohlinde, *Critical Properties of ϕ^4-Theories*, World Scientific, Singapore, 2000 (*ibid*.http/b8).

[12] For the calculation of such quantities within the ϕ^4-theory in $4 - \epsilon$ dimensions up to order ϵ^5, see
H. Kleinert, J. Neu, V. Schulte-Frohlinde, K.G. Chetyrkin, and S.A. Larin, Phys. Lett. B **272**, 39 (1991) (hep-th/9503230).

[13] For a comprehensive list of the many references on the resummation of divergent ϵ-expansions obtained in the field-theoretic approach to critical phenomena, see the textbook [11].

[14] A.J. McKane, Phys. Lett. A **76**, 22 (1980).

Take any shape but that, and my firm nerves
Shall never tremble.
WILLIAM SHAKESPEARE, Macbeth

16

Polymers and Particle Orbits in Multiply Connected Spaces

In the previous chapter, the binary interaction potential between the polymer elements was approximated by a δ-function. Quantum-mechanically, this potential is not completely impenetrable. Correspondingly, a polymer with such an interaction has a finite probability of self-intersections. This is only a rough approximation to the situation in nature where the atomic potential is of the hard-core type and self-intersections are extremely rare. In a grand-canonical ensemble, a polymer is often entangled with itself and with others. It can be disentangled only if it has open ends. Macroscopic fluctuations are required to achieve this in the form of worm-like creeping processes. Compared with local fluctuations, these take a very long time. For closed polymers, disentangling is impossible without breaking bonds at the cost of large activation energies. In order to study such entanglement phenomena in their purest form, it is useful to idealize the strongly repulsive interaction potential, as in Section 15.10, to a topological constraint of the type discussed in Chapter 6.

Entanglement phenomena play an important role also in quantum mechanics. Fluctuating particle orbits may get entangled with magnetic flux tubes or with other particle orbits. In fact, the statistical properties of Bose and Fermi particles may be viewed as entanglement phenomena, as will be shown in this chapter.

16.1 Simple Model for Entangled Polymers

Consider the simplest model system with a topological constraint producing entanglement phenomena: a fixed polymer stretched out along the z-axis and a fluctuating second polymer. Arbitrary entanglements with the straight polymer may occur. The possible entanglements of the fluctuating polymer with itself are ignored, for simplicity. Let us study the end-to-end distribution of the fluctuating second polymer. At first we neglect the third dimension, which can be trivially included at a later stage, imagining the movement to be confined entirely to the xy-plane. If the polymer along the z-axis is infinitely thin, the total end-to-end distribution of the fluctuating polymer in the plane is certainly independent of the presence of the central polymer. In the random-chain approximation it reads for not too large R

$$P_N(\mathbf{x}_b - \mathbf{x}_a) = \sqrt{\frac{2}{2\pi La}}^2 e^{-(\mathbf{x}_b - \mathbf{x}_a)^2/2La}, \qquad (16.1)$$

where $\mathbf{x}$ is a planar vector.

In the presence of the central polymer, an interesting new problem arises: How does the end-to-end distribution decompose with respect to the number of times by which the fluctuating polymer is wrapped around the central polymer? To define this number, we choose an arbitrary reference line from the origin to infinity, say the x-axis. For each path from $\mathbf{x}_a$ to $\mathbf{x}_b$, we count how often it crosses this line, including a minus sign for opposite directions of the crossings. In this way, each path receives an integer-valued label n which depends on the position of the reference line.

A property independent of the choice of the reference line exists for the pairs of paths with fixed ends. The difference path is closed. The number n of times by which a closed path encircles the origin is a topological invariant called the *winding number*. Let us find the decomposition of the probability distribution of a closed polymer $P_N(\mathbf{x}_b - \mathbf{x}_a)$ with respect to n:

$$P_N(\mathbf{x}_b - \mathbf{x}_a) = \sum_{n=-\infty}^{\infty} P_N^n(\mathbf{x}_b, \mathbf{x}_a). \qquad (16.2)$$

The topological constraint destroys the translational invariance of the total distribution on the left-hand side, so that the different fixed-n distributions on the right-hand side depend separately on both $\mathbf{x}_b$ and $\mathbf{x}_a$.

In a path integral it is easy to keep track of the number of crossings n. The angular difference between initial and final points $\mathbf{x}_b$ and $\mathbf{x}_a$ is given by the integral

$$\varphi_b - \varphi_a = \int_{t_a}^{t_b} dt\, \dot{\varphi}(t) = \int_{t_a}^{t_b} dt\, \frac{x_1 \dot{x}_2 - x_2 \dot{x}_1}{x_1^2 + x_2^2} = \int_{\mathbf{x}_a}^{\mathbf{x}_b} \frac{\mathbf{x} \times d\mathbf{x}}{\mathbf{x}^2}. \qquad (16.3)$$

Given two paths C_1 and C_2 connecting $\mathbf{x}_a$ and $\mathbf{x}_b$, this integral differs by an integer multiple of 2π. The winding number is therefore given by the contour integral over the closed difference path C:

$$n = \frac{1}{2\pi} \oint_C \frac{\mathbf{x} \times d\mathbf{x}}{\mathbf{x}^2}. \qquad (16.4)$$

In order to decompose the end-to-end distribution (16.2) with respect to the winding number, we recall the angular decomposition of the imaginary-time evolution amplitude in Eqs. (8.9) and (8.17) of a free particle in two dimensions

$$P_L(\mathbf{x}_b - \mathbf{x}_a) = \sum_m \frac{1}{\sqrt{r_b r_a}} (r_b \tau_b | r_a \tau_a)_m \frac{1}{2\pi} e^{im(\varphi_b - \varphi_a)}, \qquad (16.5)$$

with the radial amplitude

$$(r_b\tau_b|r_a\tau_a)_m = 2\frac{\sqrt{r_b r_a}}{La}e^{-(r_b^2+r_a^2)/La}I_m\left(2\frac{r_b r_a}{La}\right).\tag{16.6}$$

We have inserted the polymer parameters following the rules of Section 15.6, replacing $M/2\hbar(\tau_b - \tau_a)$ by $1/La$, and using the label $L = Na$ in P_L rather than N, as in Eq. (15.301).

We now recall that according to Section 6.1, an angular path integral consisting of a product of integrals

$$\prod_{n=1}^{N}\int_{-\pi}^{\pi}\frac{d\varphi_n}{2\pi},\tag{16.7}$$

whose conjugate momenta are integer-valued, can be converted into a product of ordinary integrals

$$\prod_{n=1}^{N}\int_{-\infty}^{\infty}\frac{d\varphi_n}{2\pi},\tag{16.8}$$

whose conjugate momenta are continuous. These become independent of the time slice n by momentum conservation, and the common momentum is eventually restricted to its proper integer values by a final sum over an integer number n occurring in the Poisson formula [see (6.9)]

$$\sum_{n}e^{ik(\varphi_b+2\pi n-\varphi_a)} = \sum_{m=-\infty}^{\infty}\delta(k-m)e^{im(\varphi_b-\varphi_a)}.\tag{16.9}$$

Obviously, the number n on the left-hand side is precisely the winding number by which we want to sort the end-to-end distribution. The desired restricted probability $P_L^n(\mathbf{x}_b, \mathbf{x}_a)$ for a given winding number n is therefore obtained by converting the sum over m in Eq. (16.5) into an integral over μ and another sum over n as in Eq. (1.205), and by omitting the sum over n. The result is:

$$P_L^n(\mathbf{x}_b, \mathbf{x}_a) = \frac{2}{La}\int_{-\infty}^{\infty}d\mu e^{-(r_b^2+r_a^2)/La}I_{|\mu|}\left(2\frac{r_b r_a}{La}\right)\frac{1}{2\pi}e^{i\mu(\varphi_b-\varphi_a+2\pi n)}.\tag{16.10}$$

From this we find the desired probability of a closed polymer running through a point $\mathbf{x}$ with various winding numbers n around the central polymer:

$$P_L^n(\mathbf{x}, \mathbf{x}) = \frac{2}{La}\int_{-\infty}^{\infty}d\mu e^{-2r^2/La}I_{|\mu|}\left(2\frac{r^2}{La}\right)\frac{1}{2\pi}e^{i2\pi\mu n}.\tag{16.11}$$

Let us also calculate the partition function of a closed polymer with a given winding number n. To make the partition function finite, we change the system by adding a harmonic oscillator potential centered at the origin.[1] If ω is measured in units 1/length, the above probability becomes

$$P_L^n(\mathbf{x}, \mathbf{x}) = \frac{2}{a} \frac{\omega}{\sinh \omega L} \int_{-\infty}^{\infty} d\mu e^{-2(r^2/a)\omega \coth \omega L} I_{|\mu|}\left(\frac{2}{a} \frac{r^2\omega}{\sinh \omega L}\right) \frac{1}{2\pi} e^{i2\pi\mu n}. \quad (16.12)$$

This can be integrated over the entire space using the formula (2.473). The result is

$$P_L^n \equiv \int d^2x P_L^n(\mathbf{x}, \mathbf{x}) = \frac{1}{2\sinh \omega L} \int_{-\infty}^{\infty} d\mu e^{-|\mu|\omega L} e^{2\pi i \mu n}. \quad (16.13)$$

To check this formula, we sum both sides over all n. Then the integral over μ is reduced, via Poisson's formula, to a sum over integers $\mu = m = 0, \pm 1, \pm 2, \dots$, and we find

$$\begin{aligned}
P_L &\equiv \int d^2x P_L(\mathbf{x}, \mathbf{x}) = \sum_{n=-\infty}^{\infty} P_L^n \\
&= \frac{1}{2\sinh \omega L}\left(\frac{2}{1 - e^{-\omega L}} - 1\right) = \frac{1}{[2\sinh(\omega L/2)]^2}.
\end{aligned} \quad (16.14)$$

As we should have expected, this is the partition function of the two-dimensional harmonic oscillator.

To find the contribution of the various winding numbers, we perform the integral over μ and obtain

$$P_L^n = \frac{\omega L}{\sinh \omega L} \frac{1}{4\pi^2 n^2 + \omega^2 L^2}. \quad (16.15)$$

The right-hand factor is recognized as a term arising in the expansion

$$\begin{aligned}
\frac{1}{2\omega L} \coth(\omega L/2) &= \sum_{n=-\infty}^{\infty} \frac{1}{4\pi^2 n^2 + (\omega L)^2} \\
&= \frac{1}{L^2} \sum_{n=-\infty}^{\infty} \frac{1}{\omega_n^2 + \omega^2}.
\end{aligned} \quad (16.16)$$

The quantities $\omega_n \equiv 2\pi n/L$ are the polymer analogs of the Matsubara frequencies. Thus we may write

$$P_L^n = P_L \cdot \alpha_n, \quad (16.17)$$

where α_n is the relative probability of finding the winding number n (with the normalization $\Sigma_n \alpha_n = 1$),

$$\begin{aligned}
\alpha_n &= \frac{1}{\omega_n^2 + \omega^2}\left[\sum_{n=-\infty}^{\infty} \frac{1}{\omega_n^2 + \omega^2}\right]^{-1} \\
&= \frac{1}{L^2} \frac{1}{\omega_n^2 + \omega^2}\left[\frac{1}{2\omega L} \coth \frac{\omega L}{2}\right]^{-1}.
\end{aligned} \quad (16.18)$$

[1] Alternatively, we may add a magnetic field with the Landau frequency $\omega = -eB/Mc$, as done in (16.33). Then the amplitude contains $\omega/2$ instead of ω, and an extra factor $e^{m\omega L/2}$.

16.2 Entangled Fluctuating Particle Orbit: Aharonov-Bohm Effect

The entanglement of a fluctuating polymer around a straight central polymer has an interesting quantum-mechanical counterpart known as the *Aharonov-Bohm effect*. Consider a free nonrelativistic charged particle moving through a space containing an infinitely thin tube of finite magnetic flux along the z-direction:

$$B_3 = \frac{g}{2\pi} \, \epsilon_{3jk}\partial_j\partial_k\varphi = g \, \delta^{(2)}(\mathbf{x}_\perp), \qquad (16.19)$$

where $\mathbf{x}_\perp$ is the transverse vector $\mathbf{x}_\perp \equiv (x_1, x_2)$. Let us study the associated path integral. The magnetic interaction is given by [recall Eq. (2.633)]

$$\mathcal{A}_{\mathrm{mag}} = \frac{e}{c} \int_{t_a}^{t_b} dt \, \dot{\mathbf{x}} \cdot \mathbf{A}, \qquad (16.20)$$

where e is the charge and $\mathbf{A}$ the vector potential. The flux tube (16.19) is obtained from the components in the xy-plane.

$$A_i = \frac{g}{2\pi} \partial_i\varphi, \quad (i = 1, 2), \qquad (16.21)$$

where φ is the azimuthal angle around the tube:

$$\varphi(\mathbf{x}) \equiv \arctan(x_2/x_1). \qquad (16.22)$$

Note that the derivatives in front of φ in (16.19) commute everywhere, except at the origin where Stokes' theorem yields

$$\int d^2x \, (\partial_1\partial_2 - \partial_2\partial_1)\varphi = \oint d\varphi = 2\pi. \qquad (16.23)$$

The total magnetic flux through the tube is defined by the integral

$$\Phi = \int d^2x \, B_3. \qquad (16.24)$$

Inserting (16.19) we see that the total flux is equal to g:

$$\Phi = g. \qquad (16.25)$$

With the vector potentoal (16.21), the interaction (16.20) takes the form

$$\mathcal{A}_{\mathrm{mag}} = -\hbar\mu_0 \int_{t_a}^{t_b} dt \, \dot{\varphi}, \qquad (16.26)$$

where μ_0 is the dimensionless number

$$\mu_0 \equiv -\frac{eg}{2\pi\hbar c}. \qquad (16.27)$$

The minus sign is a matter of convention.

Since the particle orbits are present at all times, their worldlines in spacetime can be considered as being closed at infinity, and the integral

$$n = \frac{1}{2\pi} \int_{t_a}^{t_b} dt \, \dot{\varphi} \tag{16.28}$$

is the topological invariant (16.4) with integer values of the winding number n. The magnetic interaction (16.26) is therefore a purely topological one, its value being

$$\mathcal{A}_{\mathrm{mag}} = -\hbar\mu_0 \, 2\pi n. \tag{16.29}$$

After adding this to the action of a free particle in the radial decomposition (8.9) of the quantum-mechanical path integral, we rewrite the sum over the azimuthal quantum numbers m via Poisson's summation formula as in (16.10), and obtain

$$(\mathbf{x}_b T_b | \mathbf{x}_a T_a) = \int_{-\infty}^{\infty} d\mu \frac{1}{\sqrt{r_b r_a}} (r_b T_b | r_a T_a)_\mu \tag{16.30}$$

$$\times \sum_{n=-\infty}^{\infty} \frac{1}{2\pi} e^{i(\mu-\mu_0)(\varphi_b+2\pi n-\varphi_a)}.$$

Since the winding number n is often not easy to measure experimentally, let us extract observable consequences which are independent of n. The sum over all n forces μ to be equal to μ_0 modulo an arbitrary integer number $m = 0, \pm 1, \pm 2, \dots$. The result is

$$(\mathbf{x}_b T_b | \mathbf{x}_a T_a) = \sum_{m=-\infty}^{\infty} \frac{1}{\sqrt{r_b r_a}} (r_b T_b | r_a T_a)_{m+\mu_0} \frac{1}{2\pi} e^{im(\varphi_b-\varphi_a)}, \tag{16.31}$$

with the radial amplitude

$$(r_b T_b | r_a T_a)_{m+\mu_0} = \sqrt{r_b r_a} \frac{M}{\hbar} \frac{1}{(T_b - T_a)} \exp\left\{ -\frac{M}{2\hbar} \frac{r_b^2 + r_a^2}{T_b - T_a} \right\} I_{|m+\mu_0|}\left(\frac{M}{\hbar} \frac{r_b r_a}{T_b - T_a} \right). \tag{16.32}$$

For the sake of generality, we allow for the presence of a homogeneous magnetic field B whose Landau frequency is $\omega = -eB/Mc$. In analogy with the parameter μ_0 in (16.27), it is defined with a minus sign. Using the radial amplitude (9.105), we see that (16.32) is simply generalized to

$$(r_b T_b | r_a T_a)_{m+\mu_0} = \sqrt{r_b r_a} \frac{M\omega}{2\hbar\eta} \frac{\eta}{\sinh\eta} \exp\left[-\frac{M}{2\hbar} \frac{\omega}{2} \coth\eta (r_b^2 + r_a^2) \right]$$

$$\times I_{|m+\mu_0|}\left(\frac{M\omega r_b r_a}{2\hbar \sinh\eta} \right) e^{(m+\mu_0)\eta}, \tag{16.33}$$

where $\eta \equiv \omega(T_b - T_a)/2$.

At this point we can make an interesting observation: If μ_0 is an integer number, i.e., if

$$\frac{eg}{2\pi\hbar c} = \text{integer}, \tag{16.34}$$

the quantum-mechanical particle distribution function $(\mathbf{x}\,t_b|\mathbf{x}\,t_a)$ in (16.31) becomes *independent* of the magnetic flux tube along the z-axis. The condition implies that the magnetic flux is an integer multiple of the fundamental flux quantum

$$\Phi_0 \equiv g_0 \equiv \frac{2\pi\hbar c}{e} = \frac{hc}{e}. \tag{16.35}$$

We recognize this infinitely thin tube as a *Dirac string*. Such undetectable strings were used in Sections 8.12, Appendix 10A.3, and Section 14.6 to import the magnetic flux of a magnetic monopole from infinity to a certain point where the magnetic field lines emerge radially. In Appendix 10A.3 we have made the string invisible mathematically imposing monopole gauge invariance. The present discussion shows explicitly that the flux quantization makes Dirac strings indeed undetectable by any charged particle. This observation inspired Dirac his speculation on the existence of magnetic monopoles.

In low-temperature physics, a quantization of magnetic flux is observable in type-II superconductors. Superconductors are perfect diamagnets which expel magnetic fields. Those of type II have the property that above a certain critical external field called H_{c_1}, the expulsion is not perfect but they admit quantized magnetic tubes of flux Φ_0 (*Shubnikov phase*). For increasing fields, there are more and more such flux tubes. They are squeezed together and can form a periodically arranged bundle. When cut across in the xy-plane, the bundle looks like a hexagonal planar flux lattice [4]. If the central magnetic flux tube in (16.19) carries an amount of flux that is not an integer multiple of Φ_0, the amplitude of particles passing the tube displays an interesting interference pattern. This was initially a surprise since the space is free of magnetic fields. To calculate this pattern we consider the fixed-energy amplitude of a free particle in two dimensions (9.12), decomposed into partial waves via the addition theorem (9.14) for Bessel functions as in (9.15):

$$(\mathbf{x}_b|\mathbf{x}_a)_E = -\frac{2iM}{\hbar} \sum_{m=-\infty}^{\infty} I_m(\kappa r_<) K_m(\kappa r_>) \frac{1}{2\pi} e^{im(\varphi_b - \varphi_a)}. \tag{16.36}$$

Comparing this with (16.30) and repeating the arguments leading to (16.31), (16.32), we can immediately write down the fixed-energy amplitude in the presence of a flux Φ_0:

$$(\mathbf{x}_b|\mathbf{x}_a)_E = -\frac{2iM}{\hbar} \sum_{m=-\infty}^{\infty} I_{|m+\mu_0|}(\kappa r_<) K_{|m+\mu_0|}(\kappa r_>) \frac{1}{2\pi} e^{im(\varphi_b - \varphi_a)}. \tag{16.37}$$

The wave functions are now easily extracted. In the complex E-plane, the right-hand side has a discontinuity across the positive real axis. By going through the same steps as in (9.15)–(9.22), we derive for the discontinuity

$$\int_{-\infty}^{\infty} \frac{dE}{2\pi\hbar} \operatorname{disc}(\mathbf{x}_b|\mathbf{x}_a)_E = \sum_{m=-\infty}^{\infty} \int_0^{\infty} dk\,k\,J_{|m+\mu_0|}(kr_b) J_{|m+\mu_0|}(kr_a) \frac{1}{2\pi} e^{im(\varphi_b - \varphi_b)}. \tag{16.38}$$

The integration measure $\int (dE/2\pi\hbar)(2\pi M/\hbar)$ has been replaced by $\int_0^\infty dk\,k$ according to Eq. (9.23).

In the absence of the flux tube, the amplitude (16.38) reduces to that of a free particle, which has the decomposition

$$
\int_{-\infty}^{\infty} \frac{dE}{2\pi\hbar} \mathrm{disc}(\mathbf{x}_b|\mathbf{x}_a)_E = \int \frac{d^2k}{(2\pi)^2} e^{i\mathbf{k}(\mathbf{x}_b - \mathbf{x}_a)} = \frac{1}{2\pi} \int_0^\infty dk\,k\, J_0(k|\mathbf{x}_b - \mathbf{x}_a|)
$$

$$
= \sum_{m=-\infty}^{\infty} \int_0^\infty dk\,k\, J_m(kr_b) J_m(kr_a) \frac{1}{2\pi} e^{im(\varphi_b - \varphi_a)}. \qquad (16.39)
$$

If a flux tube is present, a beam of incoming charged particles is deflected even though the space around the z-axis contains no magnetic field. Let us calculate the scattering amplitude and the ensuing cross section from the fixed-energy amplitude (16.37). Recall the results of the quantum-mechanical scattering theory due to Lippmann and Schwinger. In this theory one studies the effect of an interaction upon an incoming free-particle state $\varphi_{\mathbf{k}}$ of wave vector $\mathbf{k}$. The result is the scattering state $\psi_{\mathbf{k}}$ obtained from the *Lippmann-Schwinger* integral equation

$$
\psi_{\mathbf{k}} = \varphi_{\mathbf{k}} + \frac{1}{E - \hat{H}_0 + i\eta} \hat{V} \psi_{\mathbf{k}}
$$

$$
= \varphi_{\mathbf{k}} - \frac{i}{\hbar} \hat{R}(E)\, \hat{V} \varphi_{\mathbf{k}}, \qquad (16.40)
$$

where E is the energy of the incoming particle, $\hat{V}$ the potential, and $\hat{R}(E)$ the resolvent operator (1.315). The scattering states $\psi_{\mathbf{k}}$ are solutions of the Schrödinger equation

$$
\hat{H}\psi_{\mathbf{k}} = (\hat{H}_0 + \hat{V})\psi_{\mathbf{k}} = E\psi_{\mathbf{k}}. \qquad (16.41)
$$

In $\mathbf{x}$-space, the Lippmann-Schwinger equation reads

$$
\psi_{\mathbf{k}}(\mathbf{x}) = \varphi_{\mathbf{k}}(\mathbf{x}) - \frac{i}{\hbar} \int d^D x'\, (\mathbf{x}|\mathbf{x}')_E V(\mathbf{x}')\varphi_{\mathbf{k}}(\mathbf{x}'). \qquad (16.42)
$$

The first term describes the impinging particles, the second the scattered ones. For the scattering amplitude, only the large-$\mathbf{x}$ behavior of the second term matters. One usually normalizes $\varphi_{\mathbf{k}}(\mathbf{x})$ to $e^{i\mathbf{k}\mathbf{x}}$ and factorizes the second term asymptotically into a product of an outgoing spherical wave times a scattering amplitude. In three dimensions, the asymptotic behavior far away from the scattering center is

$$
\psi_{\mathbf{k}}(\mathbf{x}) \xrightarrow{|\mathbf{x}|\to\infty} e^{i\mathbf{k}x} + \frac{e^{i|\mathbf{k}||\mathbf{x}|}}{|\mathbf{x}|} f(\theta, \varphi) + \ldots, \qquad (16.43)
$$

where θ and φ are the scattering angles of the outgoing beam and f is the scattering amplitude. Its square gives directly the differential cross section

$$
\frac{d\sigma}{d\Omega} = |f(\theta, \varphi)|^2. \qquad (16.44)
$$

In two dimensions, the corresponding splitting is

$$\psi_{\mathbf{k}}(\mathbf{x}) \xrightarrow{|\mathbf{x}|\to\infty} e^{ikx} + \frac{e^{i|\mathbf{k}||\mathbf{x}|}}{\sqrt{|\mathbf{x}|}} f(\varphi) + \ldots \, . \tag{16.45}$$

The scattering amplitude $f(\varphi)$ which depends only on the azimuthal angle $\varphi = \arctan(x_2/x_1)$ yields the differential cross section

$$\frac{d\sigma}{d\varphi} = |f(\varphi)|^2. \tag{16.46}$$

To calculate $f(\varphi)$, we observe that the most general solution $\Psi(\mathbf{x})$ of the Schrödinger equation (16.41) is obtained by forming the convolution integral of the discontinuity of the resolvent with an arbitrary wave function $\phi(\mathbf{x})$:

$$\psi(\mathbf{x}) = \int d^D x' \, \mathrm{disc}(\mathbf{x}|\mathbf{x}')_E \, \phi(\mathbf{x}'). \tag{16.47}$$

Using (16.38), this becomes some linear combination of wave functions $J_{|m+\mu_0|}(kr)$

$$\psi(\mathbf{x}) = \sum_{m=-\infty}^{\infty} a_m J_{|m+\mu_0|}(kr) e^{im\varphi}, \tag{16.48}$$

which certainly satisfies the Schrödinger equation (16.41). The coefficients a_m have to be chosen to satisfy the scattering boundary condition at spatial infinity. Suppose that the incident particles carry a wave vector $\mathbf{k} = (-k, 0)$. In the incoming region $x \to \infty$, they are described by a wave function

$$\lim_{x\to\infty} \psi(\mathbf{x}) = e^{-ikx} e^{-i\mu_0\varphi}. \tag{16.49}$$

The extra phase factor is necessary for the correct wave vector since in the presence of the gauge field

$$eA_i = -\hbar c \mu_0 \partial_i \varphi, \tag{16.50}$$

the physical momentum $\mathbf{p} = \hbar \mathbf{k}$ is not given by the usual derivative operator $-i\hbar\boldsymbol{\nabla}$ but by the gauge-invariant momentum operator

$$\hat{\mathbf{P}} = -i\hbar\boldsymbol{\nabla} - \frac{e}{c}\mathbf{A} = -i\hbar(\boldsymbol{\nabla} + i\mu_0\boldsymbol{\nabla}\varphi). \tag{16.51}$$

The corresponding incident gauge-invariant particle current is

$$\mathbf{j}(\mathbf{x}) = -i\frac{\hbar}{2M}\psi^\dagger \overset{\leftrightarrow}{\boldsymbol{\nabla}} \psi(\mathbf{x}) - \frac{e}{Mc}\mathbf{A}(\mathbf{x})\psi^\dagger\psi(\mathbf{x}). \tag{16.52}$$

We demonstrate below that the correct choice for the coefficients a_m is

$$a_m = (-i)^{|m+\mu_0|}, \tag{16.53}$$

leading to the scattering amplitude

$$f(\varphi) = \frac{1}{\sqrt{2\pi}} e^{-i\pi/4} \sin \pi\mu_0 \frac{e^{-i\varphi/2}}{\cos(\varphi/2)}, \tag{16.54}$$

i.e., to the cross section

$$\frac{d\sigma}{d\varphi} = \frac{1}{2\pi} \sin^2 \pi\mu_0 \frac{1}{\cos^2(\varphi/2)}. \tag{16.55}$$

It has a strong peak near the forward direction $\varphi \approx \pi$. For $\mu_0 =$ integer, there is no scattering at all and the flux tube becomes an invisible Dirac string.

To derive (16.53) and (16.54) we may assume that $\mu_0 \in (0,1)$. Otherwise, we could simply shift the sum over m in (16.37) by an integer Δm, and this would merely produce an overall factor $e^{i\Delta m(\varphi_b - \varphi_a)}$ in $\mathrm{disc}(\mathbf{x}_b|\mathbf{x}_a)_E$. This would wind up as a factor $e^{i\Delta m\varphi}$ in $\psi(\mathbf{x})$. For $\mu_0 \in (0,1)$, we split the wave function (16.48) into three parts:

$$\psi_\mathbf{k} = \psi^{(1)} + \psi^{(2)} + \psi^{(3)}. \tag{16.56}$$

The first collects the terms with positive m,

$$\psi^{(1)} = \sum_{m=1}^{\infty} (-i)^{m+\mu_0} J_{m+\mu_0} e^{im\varphi}, \tag{16.57}$$

the second those with negative m,

$$\begin{aligned}
\psi^{(2)} &= \sum_{m=-\infty}^{-1} (-i)^{m+\mu_0} J_{|m+\mu_0|} e^{im\varphi} \\
&= \sum_{m=1}^{\infty} (-i)^{m-\mu_0} J_{m-\mu_0} e^{-im\varphi},
\end{aligned} \tag{16.58}$$

and the third contains only the term $m = 0$,

$$\psi^{(3)} = (-i)^{|\mu_0|} J_{|\mu_0|}. \tag{16.59}$$

Obviously, $\psi^{(2)}$ may be obtained from $\psi^{(1)}$ via the identity

$$\psi^{(2)}(r, \varphi, \mu_0) = \psi^{(1)}(r, -\varphi, -\mu_0). \tag{16.60}$$

Thus, the wave function (16.56) requires only a calculation of $\psi^{(1)}$.

As a first step we observe that the sum (16.57) has an integral representation

$$\psi^{(1)} = \frac{1}{2}(-i)^{\mu_0} e^{-i\rho\cos\varphi} I(\rho), \tag{16.61}$$

with $I(\rho)$ being the integral

$$I(\rho) \equiv \int_0^\rho d\rho' e^{i\rho'\cos\varphi} \left(J_{1+\mu_0} - i J_{\mu_0} e^{i\varphi} \right). \tag{16.62}$$

We have set $kr \equiv \rho$ such that $\mathbf{kx} \equiv \rho \cos \varphi$. To prove the integral representation, we differentiate (16.61) and find the differential equation

$$\partial_\rho \psi^{(1)} = -i \cos \varphi \; \psi^{(1)} + \frac{1}{2}(-i)^{\mu_0} \left(J_{1+\mu_0} - i J_{\mu_0} e^{i\varphi} \right), \tag{16.63}$$

with all functions depending only on $kr \equiv \rho$. Precisely the same equation is obeyed by the sum (16.57):

$$
\begin{aligned}
\partial_\rho \psi^{(1)} &= \sum_{m=1}^{\infty} (-i)^{m+\mu_0} \partial_\rho J_{m+\mu_0} e^{im\varphi} \\
&= \sum_{m=1}^{\infty} (-i)^{m+\mu_0} \frac{1}{2} \left(J_{m+\mu_0-1} - J_{m+\mu_0+1} \right) e^{im\varphi} \\
&= -\frac{i}{2} \sum_{m=1}^{\infty} (-i)^{m+\mu_0} J_{m+\mu_0} e^{im\varphi} \left(e^{i\varphi} + e^{-i\varphi} \right) + \frac{1}{2}(-i)^{\mu_0} \left(J_{1+\mu_0} - i J_{\mu_0} e^{ie} \right).
\end{aligned}
\tag{16.64}
$$

Thus, the two expressions (16.57) and (16.61) for $\psi^{(1)}$ can differ at most by an integration constant. However, the constant must be zero since both expressions vanish at $\rho = 0$. This proves the integral representation (16.61).

In order to derive the scattering amplitude for the magnetic flux tube, we have to find the asymptotic of the wave function. This is done by splitting $\psi_\infty^{(1)}$ into two terms, a contribution $\psi_\infty^{(1)}$ in which the integral I is carried all the way to infinity, to be denoted by I_∞, and a remainder $\Delta \psi^{(1)}$ which vanishes for $r \to \infty$. The integral I_∞ can be calculated analytically using the formula

$$\int_0^\infty d\rho \, e^{i\beta\rho} J_\alpha(k\rho) = \frac{1}{(k^2 - \beta^2)^{1/2}} e^{i\alpha \arcsin(\beta/k)}, \quad 0 < \beta < k, \; \alpha > -2. \tag{16.65}$$

This gives

$$
\begin{aligned}
I_\infty &\equiv \int_0^\infty d\rho' e^{i\rho' \cos\varphi} \left(J_{1+\mu_0} - i J_{\mu_0} e^{i\varphi} \right) = \frac{1}{|\sin\varphi|} \left[e^{i\mu_0(\pi/2-|\varphi|)} - i e^{i\varphi} e^{i(1+\mu_0)(\pi/2-|\varphi|)} \right] \\
&= \frac{i}{|\sin\varphi|} e^{i\mu_0(\pi/2-|\varphi|)} \left(e^{-i|\varphi|} - e^{i\varphi} \right) = \begin{cases} 0, & \varphi < 0, \\ e^{-i\mu_0\varphi} 2i^{\mu_0}, & \varphi > 0, \end{cases}
\end{aligned}
\tag{16.66}
$$

with $\varphi \in (-\pi, \pi)$. Hence we have

$$\psi_\infty^{(1)} = \begin{cases} 0, & \varphi < 0, \\ e^{-ikx} e^{-i\mu_0\varphi}, & \varphi > 0. \end{cases} \tag{16.67}$$

Using (16.60), we find

$$\psi_\infty^{(2)} = \begin{cases} e^{-ikx} e^{-i\mu_0\varphi}, & \varphi < 0, \\ 0, & \varphi > 0. \end{cases} \tag{16.68}$$

The sum $\psi_\infty^{(1)} + \psi_\infty^{(2)}$ represents the incoming wave (16.49). The scattered wave must therefore reside in the remainder

$$\psi_{\mathrm{sc}} = \Delta\psi^{(1)} + \Delta\psi^{(2)} + \psi^{(3)}. \tag{16.69}$$

For the scattering amplitude, only the leading $1/\sqrt{r}$-behavior of the three terms is relevant. To find it for $\Delta\psi^{(1)}$, we take (16.61) and write the remainder of the integral (16.62) as

$$\Delta I(\rho) \equiv I(\rho) - I_\infty = \int_\rho^\infty d\rho' e^{i\rho' \cos\varphi} \left(J_{1+\mu_0} - ie^{i\varphi} J_{\mu_0}\right). \tag{16.70}$$

At large ρ, the asymptotic expansion

$$J_\alpha(\rho) \sim \sqrt{2/\pi\rho} \, \cos(\rho - \alpha/2 - \pi/4) \tag{16.71}$$

renders

$$\Delta I(\rho) = \sqrt{\frac{2}{\pi}} [A(\rho) + B(\rho)], \tag{16.72}$$

with the integrals

$$A(\rho) = \int_\rho^\infty \frac{d\rho'}{\sqrt{\rho'}} e^{i\rho' \cos\varphi} \cos\left[\rho' - (1+\mu_0)/2 - \pi/4\right],$$

$$B(\rho) = -ie^{i\varphi} \int_{\rho'}^\infty \frac{d\rho'}{\sqrt{\rho'}} e^{i\rho' \cos\varphi} \cos\left[\rho' - \mu_0/2 - \pi/4\right]. \tag{16.73}$$

Separating the cosine into exponentials and changing the variable ρ' in the two terms to $\rho' = t^2/(1 \pm \cos\varphi)$, we find

$$A(\rho) = \left[\frac{(-i)^{1/2+\mu_0}}{\sqrt{1+\cos\theta}} \int_{\sqrt{\rho(1+\cos\varphi)}}^\infty dt e^{it^2} + \frac{i^{3/2+\mu_0}}{\sqrt{1-\cos\theta}} \int_{\sqrt{\rho(1-\cos\varphi)}}^\infty dt e^{-it^2}\right], \tag{16.74}$$

and a corresponding expression for $B(\rho)$. The asymptotic expansion of the error function

$$\int_x^\infty dt e^{\pm it^2} = \pm \frac{i}{2} \frac{\exp(\pm ix^2)}{x} + \ldots \tag{16.75}$$

leads to

$$A(\rho) = \frac{1}{2}\left[(-i)^{\frac{1}{2}+\mu_0} \frac{e^{i\rho}}{\sqrt{\rho(1+\cos\varphi)^2}} + i^{\frac{1}{2}+\mu_0} \frac{e^{-i\rho}}{\sqrt{\rho(1-\cos\varphi)^2}}\right] e^{i\rho\cos\varphi},$$

$$B(\rho) = (-i)\frac{e^{i\varphi}}{2} \tag{16.76}$$

$$\times \left[(-i)^{-\frac{1}{2}+\mu_0} \frac{e^{i\rho}}{\sqrt{\rho(1+\cos\varphi)^2}} + i^{-\frac{1}{2}+\mu_0} \frac{e^{-i\rho}}{\sqrt{\rho(1-\cos\varphi)^2}}\right] e^{i\rho\cos\varphi}.$$

Adding the two terms together in (16.72) and inserting everything into (16.61) gives the asymptotic behavior

$$\Delta\psi^{(1)} = \frac{\sqrt{-i}}{2\sqrt{2\pi\rho}}\left[(-1)^{\mu_0} e^{i\rho} \frac{1+e^{i\varphi}}{1+\cos\varphi} + ie^{-i\rho} \frac{1-e^{i\varphi}}{1-\cos\varphi}\right]. \tag{16.77}$$

Together with $\Delta\psi^{(2)}$ found via (16.60), we obtain

$$\Delta\psi^{(1)} + \Delta\psi^{(2)} = \frac{\sqrt{-i}}{\sqrt{2\pi\rho}} \left[e^{i\rho} \frac{\cos(\pi\mu_0 - \varphi/2)}{\cos(\varphi/2)} + ie^{-i\rho} \right] + e^{-i(\rho\cos\varphi + \mu_0\varphi)}.$$

$$(16.78)$$

Adding further $\psi^{(3)}$ from (16.59) with the asymptotic limit given by (16.71), the total wave function is seen to behave like

$$\psi(\mathbf{x}) \to e^{-i(\rho\cos\varphi + \mu_0\varphi)} + \psi_{\rm sc}(\mathbf{x}), \qquad (16.79)$$

with the scattered wave

$$\psi_{\rm sc} = \frac{1}{\sqrt{2\pi i \rho}} e^{i\rho} \frac{\sin\pi\mu_0}{\cos(\varphi/2)} e^{-i\varphi/2}. \qquad (16.80)$$

This corresponds precisely to the scattering amplitude (16.54) with the cross section (16.55).

Let us mention that for half-integer values of μ_0, the solution of the Schrödinger equation has the simple integral representation

$$\psi(\mathbf{x}) = \sqrt{\frac{i}{2}} e^{-i(\varphi/2 + \rho\cos\varphi)} \int_0^{\sqrt{\rho(1+\cos\varphi)}} dt\, e^{it^2}. \qquad (16.81)$$

It vanishes on the line $\varphi = \pi$, i.e., directly behind the flux tube and is manifestly single-valued.

16.3   Aharonov-Bohm Effect and Fractional Statistics

It was noted in Section 7.5 and it is worth mentioning once more in this context that the amplitude for the relative motion of two fermion orbits can be obtained from the amplitude of the Aharonov-Bohm effect.

For this, we take the amplitude with $\mu_0 = 1$ and sum it over the final states with φ_b, $\varphi_b + \pi$, to account for particle identity. The result is

$$(\mathbf{x}_b|\mathbf{x}_a)_E + (-\mathbf{x}_b|\mathbf{x}_a)_E = -\frac{2iM}{\hbar} \sum_m I_{|m+1|}(\kappa r_<) K_{|m+1|}(\kappa r_>)$$

$$\times \frac{1}{2\pi} \left[e^{im(\varphi_b - \varphi_a)} + (-)^m e^{im(\varphi_b - \varphi_a)} \right]$$

$$= -\frac{4iM}{\hbar} e^{-i(\varphi_b - \varphi_a)} \sum_{m={\rm odd}} I_{|m|}(\kappa r_<) K_{|m|}(\kappa r_>) \frac{1}{2\pi} e^{im(\varphi_b - \varphi_a)}. \qquad (16.82)$$

The sum over the two identical final states selects only the odd wave functions, as in (7.267)–(7.268). When calculating observable quantities such as particle densities or partition functions which involve only the trace of the amplitude, the phase factor $e^{-i(\varphi_b - \varphi_a)}$ has no observable consequences and can be omitted.

For $\mu_0 \neq 1$, the resulting amplitudes may be interpreted as describing particles in two dimensions obeying an unusual fractional statistics. This interpretation has recently come to enjoy great popularity.[2] since it has led to an understanding of the experimental data of the fractional quantum Hall effect. The data can be explained by the following assumption: The excitations of a gas of electrons with Coulomb interactions in a quasi-two-dimensional material traversed by a strong magnetic field can be viewed, to lowest approximation, as a gas of quasi-particles which has *no* Coulomb interactions, but a new effective pair interaction. Each pair behaves as if one partner were accompanied by a thin magnetic flux tube of a certain value of μ_0. While the quasi-particles of the ground state carry an integer-valued μ_0 and act statistically like ordinary electrons, the elementary excitations carry a fractional value of μ_0 and display fractional statistics (more in Section 16.11).

To study the fundamental thermodynamic properties of an ensemble of such particles, we calculate the partition function of a particle running around a thin flux tube along the z-axis. For finiteness, we assume the presence of an additional homogeneous magnetic field in the z-direction. Ignoring the third dimension, we take the amplitude (16.33), integrate it over all space, and find

$$
\begin{aligned}
Z &= \int d^2x (\mathbf{x}_b \tau_b | \mathbf{x}_a \tau_a) \\
&= \frac{1}{2} e^{\mu_0 \eta} \sum_{m=-\infty}^{\infty} e^{m\eta} \int_0^{\infty} d\xi e^{-\xi \cosh \eta} I_{|m+\mu_0|}(\xi),
\end{aligned}
\tag{16.83}
$$

where

$$
\xi \equiv M\omega r^2 / 2\hbar \sinh \eta
\tag{16.84}
$$

[recall that $\omega = -eB/Mc$ and $\eta = \omega(\tau_b - \tau_a)/2$]. To calculate the partition function, the difference between the Euclidean times τ_b, τ_a is set equal to $\tau_b - \tau_a = \hbar/k_B T = \hbar\beta$, so that $\eta = \omega \hbar \beta / 2$.

To deal with two identical particles, we also need the integral in which $\mathbf{x}_b$ is exchanged by $-\mathbf{x}_b$:

$$
Z_{\mathrm{ex}} = \int d^2x (-\mathbf{x}_b \tau_b | \mathbf{x}_a \tau_a) = \frac{1}{2} e^{\mu_0 \eta} \sum_{m=-\infty}^{\infty} (-)^m e^{m\eta} \int_0^{\infty} d\xi e^{-\xi \cosh \eta} I_{|m+\mu_0|}(\xi).
\tag{16.85}
$$

In order to facilitate writing joint equations for both expansions, let us denote Z and Z_{ex} by Z_1 and $Z_{1_{\mathrm{ex}}}$, respectively. The integrals are performed with the help of formula (2.472), yielding the sums

$$
Z_{1,1_{\mathrm{ex}}} = \frac{1}{2} \sum_{m=-\infty}^{\infty} (\pm)^m e^{\eta(m+\mu_0)} \frac{1}{\sinh \eta} e^{-\eta|m+\mu_0|}.
\tag{16.86}
$$

These sums are obviously periodic under $\mu_0 \rightarrow \mu_0 + 2$. Because of translational invariance, the partition function $Z \equiv Z_1$ diverges with the total area $V = \int d^2x$ as

---

[2]See Notes and References at the end of Chapter 7.

an overall factor. To enforce convergence, we multiply the volume elements d^2x with an exponential regulating factor $e^{-\epsilon\xi}$. Then the area integrals can be extended over all space. In terms of the variable ξ of (16.84), the measure in the above rotationally symmetric integrals can be written as

$$d^2x = l_e^2(T)\frac{\sinh\eta}{\eta}d\xi, \tag{16.87}$$

with the thermal length $l_e(T) \equiv \sqrt{2\pi\hbar^2/k_BTM}$ introduced in Eq. (2.352). The role of the total area $V = \int d^2x$ is now played by the finite quantity

$$V \equiv \int d^2x\, e^{-\epsilon\xi} = \frac{l_e^2(T)}{\epsilon}\frac{\sinh\eta}{\eta}. \tag{16.88}$$

Inserting the factor $e^{-\epsilon\xi}$ into the integrals in Eqs. (16.83) and (16.85), and defining a variable η' slightly different from η by

$$\cosh\eta' \equiv \epsilon + \cosh\eta, \tag{16.89}$$

which has the expansion

$$\begin{aligned}
e^{\eta'} &= \cosh\eta' + \sqrt{\cosh^2\eta' - 1}\\
&= e^{\eta}\left(1 + \frac{\epsilon}{\sinh\eta} - \frac{1}{2}e^{-\eta}\frac{\epsilon^2}{\sinh^3\eta} + \ldots\right), \quad \eta > 0, \tag{16.90}
\end{aligned}$$

the regulated sums (16.86) for Z_1 and $Z_{1\mathrm{ex}}$ look almost the same as before:

$$Z_{1,1\mathrm{ex}} = \frac{1}{2}\sum_{m=-\infty}^{\infty}(\pm)^m e^{\eta(m+\mu_0)}\frac{1}{\sinh\eta'}e^{-\eta'|m+\mu_0|}. \tag{16.91}$$

Separating positive and negative values of $m + \mu_0$, the two sums can be done for $\mu_0 \in (0,1)$ and for $\mu_0 \in (-1,0)$. In the combined interval $\mu_0 \in (-1,1)$, we find

$$Z_{1,1\mathrm{ex}} = \frac{1}{2}e^{\eta\mu_0}\frac{1}{\sinh\eta'}\left\{\frac{e^{-\eta'\mu_0}}{1\mp a} + \frac{e^{\eta'\mu_0}}{1\mp b} - e^{\eta'|\mu_0|}\right\}, \tag{16.92}$$

where

$$a \equiv e^{\eta-\eta'}, \quad b \equiv e^{-\eta-\eta'}.$$

Two identical particles have the partition function

$$Z = \frac{1}{2}(Z_1 + Z_{1\mathrm{ex}}) = \frac{1}{2}e^{\eta\mu_0}\frac{1}{\sinh\eta'}\left\{\frac{e^{-\eta'\mu_0}}{1-a^2} + \frac{e^{\eta'\mu_0}}{1-b^2} - e^{\eta'|\mu_0|}\right\}. \tag{16.93}$$

It is symmetric under the simultaneous exchange $\mu_0 \to -\mu_0$, $\eta \to -\eta$. Outside the interval $\mu_0 \in (-1,1)$, it is defined by periodic extension.

In the absence of the thin flux tube we may take $Z_{1,1_{\mathrm{ex}}}$ directly from the amplitude (2.666). For $\mathbf{x}_b = \mathbf{x}_a$ and $\mathbf{x}_b = -\mathbf{x}_a$, this yields with the present variables

$$(\mathbf{x}_a T_b | \mathbf{x}_a T_a) = l_e^{-2}(T)\frac{\eta}{\sinh \eta}, \quad (-\mathbf{x}_a T_b | \mathbf{x}_a T_a) = l_e^{-2}(T)\frac{\eta}{\sinh \eta}e^{-2\cosh \eta\, \xi}. \quad (16.94)$$

Their regulated spatial integrals are

$$
\begin{aligned}
Z_{1,0} &= \frac{1}{2}\int_0^\infty d\xi\, e^{-\epsilon\xi} = \frac{1}{2\epsilon}, \\
Z_{1_{\mathrm{ex}},0} &= \frac{1}{2}\int_0^\infty d\xi\, e^{-(\epsilon+2\cosh \eta)\xi} = \frac{1}{2(\epsilon + 2\cosh \eta)}.
\end{aligned} \quad (16.95)
$$

The subtracted partition functions $\Delta Z_{1,1_{\mathrm{ex}}} \equiv Z_{1,1_{\mathrm{ex}}} - \frac{1}{2}Z_{1,0}$ have a finite $\epsilon \to 0$-limit, and $\Delta Z = Z - \frac{1}{2}Z_{1,0}$ becomes for $\mu_0 \in (0,2)$

$$\Delta Z = -\frac{1}{8\sinh \eta}\left[\coth\eta + 2(\mu_0 - 1) - 2e^{2(\mu_0+1)\eta}\frac{1}{\sinh 2\eta} + 4e^{2\eta\mu_0}\right]. \quad (16.96)$$

These results can be used to calculate the second coefficient appearing in a *virial expansion* of the equation of state. For a dilute gas of particles with the above magnetic interactions it reads

$$\frac{pV}{Nk_BT} = 1 + \sum_{r=2}^\infty B_r n^{r-1}. \quad (16.97)$$

Here n is the number density of the particles N/V. In many-body theory it is shown that the coefficient B_2 depends on the two-body partition function Z_2 as follows:

$$B_2 = V\left(\frac{1}{2} - \frac{Z_2}{Z_1^2}\right), \quad (16.98)$$

where Z_1 is the two-dimensional single particle partition function of mass M. In the presence of the homogeneous magnetic field, Z_1 is given by $Z_{1,0}$ of Eq. (16.95). Without the regulating factor, the spatial integral over the imaginary-time amplitude in (16.94) gives directly

$$Z_1 = V\frac{\eta}{l_e^2(T)\sinh \eta}. \quad (16.99)$$

Separating the center of mass from the relative motion, we see that $Z_2 = 2Z_1 Z_{\mathrm{rel}}$ and obtain

$$B_2 = \frac{V}{Z_1}(Z_1/2 - 2Z_{\mathrm{rel}}) = \frac{l_e^2(T)\sinh \eta}{2\eta}(Z_1 - 4Z_{\mathrm{rel}}). \quad (16.100)$$

The difference on the right-hand side is convergent for $V \to \infty$. It can be evaluated using *any* regulator for the area integration, in particular the exponential regulator of Eq. (16.88).

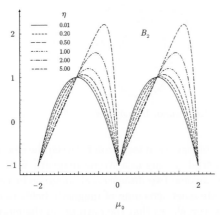

Figure 16.1 Second virial coefficient B_2 as function of flux μ_0 for various external magnetic field strengths parametrized by $\eta = -(eB/2Mc)\hbar\beta$. For a better comparison, each curve has been normalized to unity at $\mu_0 = 0$.

The partition function for the relative motion of two identical particles is obtained from Z by replacing M by the reduced mass, i.e., $M \to M/2$. This renders a factor $1/2$ via $l_e^{-2}(T)$. Hence $Z_1/2 - 2Z_{\mathrm{rel}}$ becomes equal to $-\Delta Z$ and (16.100) yields [5]

$$B_2 = \frac{l_e^2(T)}{4\eta}\left[\coth\eta + 2(\mu_0 - 1) - 2e^{2(\mu_0+1)\eta}\frac{1}{\sinh(2\eta)} + 4e^{2\eta\mu_0}\right]. \qquad (16.101)$$

The behavior of B_2 as a function of μ_0 is shown in Fig. 16.1. In the absence of a magnetic field, it reduces to

$$B_2 = \frac{l_e^2(T)}{4}\left[1 - 2(1 - |\mu_0|^2)^2\right], \quad \mu_0 \in (-1, 1). \qquad (16.102)$$

As μ_0 grows from zero to infinity, B_2 oscillates with a period 2 between $B_2 = -l_e^2(T)/4$ for even values of μ_0, and $B_2 = l_e^2(T)/4$ for odd values. These are the well-known second *virial coefficients* of free bosons and fermions. They can, of course, be obtained in a simpler and more direct way from the symmetric and antisymmetric combinations of (16.95), $Z_0 = (1/2)(Z_{1,0} \pm Z_{1_{\mathrm{ex}},0})$. Subtracting $Z_{1,0}$ leaves $\pm Z_{1_{\mathrm{ex}},0}/2$ which reduces for $\eta = 0$ to $\pm 1/8$. Accounting for the factor 2 in the reduced mass, this yields $B_2 = \mp l_e^2(T)/4$.

The expression (16.102) can be interpreted as the *virial coefficient* of particles which are neither bosons nor fermions, but obey the laws of fractional statistics. These particles are the *anyons* introduced in Section 7.5. Unfortunately, there are at present no experimental data for the virial coefficients to which the theoretical expressions (16.102) [or (16.101)] could be compared.

At this point we should mention older, meanwhile discarded speculations that the phenomenon of high-temperature superconductivity might be explained by fractional statistics of some elementary excitations. Indeed, the change in statistics can

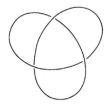

Figure 16.2 Lefthanded trefoil knot in polymer.

be derived from the electromagnetic interaction between the electrons in a quasi-two-dimensional layer of material moving in a strong magnetic field. Also, it is possible to construct a model of anyonic two-dimensional superconductivity in which topological effects lead to a Meissner screening of magnetic fields (see Section 16.13). A closer investigation, however, shows that the currents in the model show dissipation after all.

It must be emphasized that the equality between electromagnetic and statistical interaction used in the above calculations is restricted to two-particle systems and cannot be extended to arbitrarily many particles as in Section 7.5. Although it is possible to distribute the magnetic flux in a many-particle system equally between the constituents producing the desired behavior under particle exchange, an equal distribution of the charges would create an unwanted additional Coulomb potential, and the purely topological character of the interaction would be destroyed. The problem does not arise for charged particles such as electrons. Nevertheless, there is a definite need for a better and universally applicable theoretical description of anyons. This will be presented in Section 16.7.

16.4 Self-Entanglement of Polymer

An interesting consequence of the excluded-volume properties of polymers is the possibility of a self-entanglement of a closed polymer. Since its line elements are forbidden to cross each other, the fluctuations are unable to explore all possible configurations. An initially circular polymer, for example, can never turn into a *trefoil knot* of the form shown in Fig. 16.2 without breaking a molecular bond.
In the chemical formation process of a large number of polymers, many

entangled configurations arise. It is an interesting problem to find the distribution of the various independent topology classes. Until recently, the lack of theoretical methods has made analytic work almost impossible, restricting it to classification questions. Only Monte Carlo methods have yielded quantitative insights. Since 1989, however, interesting new quantum field-theoretic methods have been developed promising significant progress in the near future. Here we survey these methods and indicate how to derive analytic results. First, we introduce the relevant topological concepts.

A closed polymer will in general form a knot. A circular polymer represents a trivial knot. Two knots are called *equivalent* if they can be deformed into each

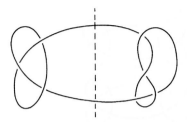

Figure 16.3 Nonprime (compound) knot. The dashed line separates two pieces. After closing the open ends, the pieces form two prime knots.

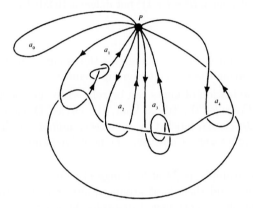

Figure 16.4 Illustration of multiplication law $a_1 a_2 \approx a_3$ in knot group. The loops a_1 and a_2 are equivalent, $a_1 \approx a_2$, while $a_4 \approx a_1^{-1}$.

other without breaking any line. Such deformations are called *isotopic*. A first step towards the classification consists in separating the equivalence classes into irreducible and reducible ones, defining *prime* or *simple* knots and *nonprime* or *compound* knots, respectively. A *compound* knot is characterized by the existence of a plane which is intersected twice (after some isotopic deformation) (see Fig. 16.3). By closing the open ends on each side of the plane one obtains two new knots. These may or may not be reduced further in the same way until one arrives at simple knots. One important step towards distinguishing different equivalence classes of simple and compound knots is the *knot group* defined as follows. In the multiply connected space created by a certain knot, choose an arbitrary point P (see Fig. 16.4). Then consider all possible closed loops starting from P and ending again at P. Two such loops are said to be *equivalent* if they can be deformed into each other without crossing the lines of the knot under consideration. The loops are, however, allowed to have arbitrary self-intersections. The classes of equivalent knots form the knot group. Group multiplication is defined by running through any two loops of two equivalence classes successively. The class whose loops can be contracted into the

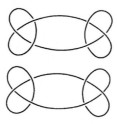

Figure 16.5 Inequivalent (compound) knots possessing isomorphic knot groups. The upper is the *granny knot*, the lower the *square knot*. They are stereoisomers characterized by the same Alexander polynomial $(t^2 - t + 1)^2$ but different HOMFLY polynomials [see (16.126)].

point P is defined as the unit element e. Changing the orientation of the loops in a class corresponds to inverting the associated group element.

In this way, the classification of knots can be related to the classification of all possible knot groups. Consider the trivial knot, the circle. Obviously, the closed loops through P fall into classes labeled by an integer number n. The associated knot group is the group of integers. Conversely, no nontrivial knot is associated with this group.

Although this trivial example might at first suggest a one-to-one correspondence between the simple knots and their knot groups, there is none. Many examples are known where inequivalent knots have isomorphic knot groups. In particular, all mirror-reflected knots which usually are inequivalent to the original ones (such as the right- and left-handed trefoil), have the same knot group. Thus, the knot groups necessarily yield an *incomplete* classification of knots. An example is shown in Fig. 16.5.

Fortunately, the degeneracies are quite rare. Only a small fraction of inequivalent knots cannot be distinguished by their knot groups.

The easiest way of picturing a knot in 3 dimensions is by drawing its projection onto the paper plane. The lines in the projection show a number of *crossings*, and the drawing must distinguish the top from the bottom line. The knot is then deformed isotopically until the projection has the minimal number of crossings. In the projection, all isotopic deformations can be decomposed into a succession of three elementary types, the so-called *Reidemeister moves* shown in Fig. 16.6.

A picture of all simple knots up to $n = 8$ is shown in Fig. 16.7. The numbers of inequivalent simple and compound knots with a given number n of minimal crossings are listed in Table 16.1.

The projected pictures can be used to construct an important algebraic quantity characterizing the knot group, called the *Alexander polynomial* discovered in 1928. It reduces the classification of knot groups to that of polynomials. This type of work is typical of the field of algebraic topology.

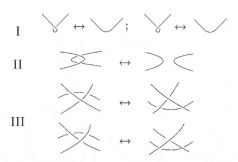

Figure 16.6 *Reidemeister moves* in projection image of knot which do not change its class of isotopy. They define the movements of *ambient isotopy*. For ribbons, only the second and third movements are allowed, defining the *regular isotopy*. The first movement is forbidden since it changes its writhe [for the definition see Eq. (16.110)].

Table 16.1 Numbers of simple and compound knots with n minimal crossings in a projected plane.

n	simple knots	compound knots
0	1	0
1	0	0
2	0	0
3	1	0
4	1	0
5	2	0
6	3	1
7	7	1
8	21	3
9	49	5
10	166	10
11	548	37
12	–	154
13	–	484
14	–	1115

We explain the construction for the trefoil knot. Attaching a directional arrow to the polymer and selecting an arbitrary starting point, we follow the arrow until we run into a first underpass. This point is denoted by 1. Now we continue to the next underpass denoted by 2, etc., up to n (see Fig. 16.8). The polymer sections between two successive underpasses i and $i+1$ are named x_{i+1}. At each underpass from x_i to x_{i+1}, we record (see Table 16.2) whether the overpassing section x_k runs from right to left (type r) or from left to right (type l). We now set up a matrix A_{ij}. Each underpass with label i defines a row A_{ij}, $j = 1, 2, 3, \ldots$ according to the

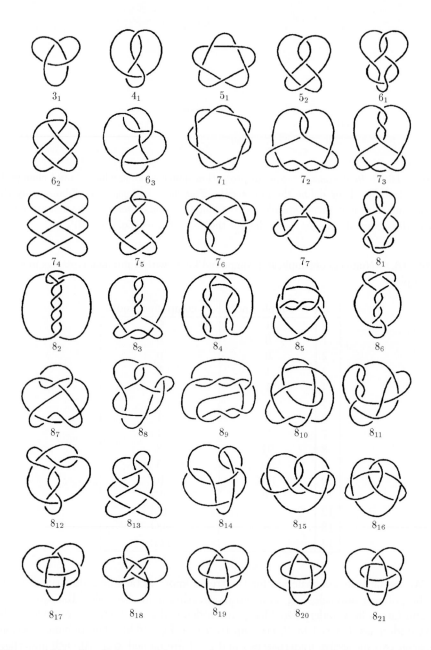

Figure 16.7 Simple knots with up to 8 minimal crossings. The number of crossings under each picture carries a subscript enumerating the equivalence classes.

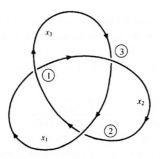

Figure 16.8 Labeling of underpasses for construction of Alexander polynomial $t^2 - t + 1$ of the left-handed trefoil.

Table 16.2 Tables of underpasses (under) and directions (dir) r or l ofoverpassing lines (over), for trefoil knot 3_1 and knot 4_1 of Fig. 16.7.

3_1 :

under	over	dir
x_1	x_3	r
x_2	x_1	r
x_3	x_2	r

4_1 :

under	over	dir
x_1	x_4	r
x_2	x_1	l
x_3	x_2	r
x_4	x_3	l

following rules: Let x_k be the overpassing section. If x_k coincides with x_i or x_{i+1} the underpass is called *trivial*. In this, case the ith row of the matrix A_{ij} has the elements

$$A_{ii} = -1, \quad A_{i,i+1} = 1. \tag{16.103}$$

All other row elements A_{ij} vanish. If the underpass is nontrivial, the nonvanishing row elements are

$$A_{ii} = 1, \ A_{ii+1} = -t, \ A_{ik} = t - 1, \quad \text{type r (right to left)}, \tag{16.104}$$
$$A_{ii} = -t, \ A_{ii+1} = 1, \ A_{ik} = t - 1, \quad \text{type l (left to right)}. \tag{16.105}$$

In this way, we find the matrix of the trefoil knot:

$$A_{ij} = \begin{pmatrix} 1 & -t & t-1 \\ t-1 & 1 & -t \\ -t & t-1 & 1 \end{pmatrix}. \tag{16.106}$$

As another example, the knot 4_1 in Fig. 16.7 has the matrix

$$A_{ij} = \begin{pmatrix} 1 & -t & 0 & t-1 \\ t-1 & -t & 1 & 0 \\ 0 & t-1 & 1 & -t \\ 1 & 0 & t-1 & -t \end{pmatrix}. \tag{16.107}$$

Table 16.3 Alexander, Jones, and HOMFLY polynomials for smallest simple knots up to 8 crossings. The numbers specify the coefficients; for instance, the knot 7_1 has the Alexander polynomial $A(t) = 1 - t + t^2 - t^3 + t^4 - t^5 + t^6$ and the knot 8_8 has the Jones polynomial $J(t) = t^{-3}(1 - t + 2t^2 - t^3 + t^4)$. For the HOMFLY polynomial $H(t, \alpha)$, see the explanation on p. 1113.

	$A(t)$	$J(t)$	$H(t,\alpha)$	
3_1	1−11	(0)1	([0]2−1)([0]1)	
4_1	1−31	(−2)−1	(1[−1]1)([−1])	
5_1	1−11−11	(0)1101	([0]03−2)([0]041)([0]01)	
5_2	2−32	(0)101	([0]11−1)([0]11)	
6_1	2−52	(−2)−10−1	(1[0]−11)([−1]−1)	
6_2	1−33−31	(−1)−11−1	([2]−21)(1[−3]1)([0]1)	
6_3	1−35−31	(−3)1−11	(−1[3]−1)(−1[3]−1)(1)	s
7_1	1−11−11−11	(0)1111101	([0]004−3)([0]0010−4)([0]006−1)([0]001)	
7_2	3−53	(0)10101	([0]10−11)([0]111)	
7_3	2−33−32	(0)110201	(−22−10[0])(−1330[0])(1100[0])	
7_4	4−74	(0)10201	(−1020[0])(121[0])	
7_5	2−45−42	(0)1102−11	([0]020−1)([0]032−1)([0]011)	
7_6	1−57−51	(−1)−12−11	([1]−12−1)([1]−22)([0]−1)	
7_7	1−59−52	(−3)1−21−1	(1−2[2])(−2[2]−1)([1])	
8_1	3−73	(−2)−10−10−1	(1−10[0]1)(−1−1[−1])	
8_2	1−33−33−31	(1)1−11−1	(1−33[0])(3−74[0])(1−51[0])(−10[0])	
8_3	4−94	(−4)−10−20−1	(10[−1]01)(−1[−2]−1)	s
8_4	2−55−52	(−3)−10−21−1	(2[−2]01)(1[−3]−21)([−1]−1)	
8_5	1−34−55−31	(1)1−21−1	(2−54[0])(3−84[0])(1−51[0])(−10[0])	
8_6	2−67−62	(−1)−11−21−1	(1−1−1[2])(1−2−2[1])(−1−1[0])	
8_7	1−35−55−31	(−2)1−12−11	([−1]4−2)([−3]8−3)([−1]5−1)([0]1)	
8_8	2−69−62	(−3)1−12−11	(−1[2]1−1)(−1[2]2−1)([1]1)	
8_9	1−35−75−31	(−4)−11−21−1	(2[−3]2)(3[−8]3)(1[−5]1)([−1])	s
8_{10}	1−36−76−31	(−2)1−13−11	([−2]6−3)([−3]9−3)([−1]5−1)([0]1)	
8_{11}	2−79−72	(−1)−12−21−1	(1−21[1])(1−2−1[1])(−1−1[0])	
8_{12}	1−7(13)−71	(−4)−11−31−1	(1−1[1]−11)(−2[1]−2)([1])	s
8_{13}	2−7(11)−72	(−3)1−22−11	([0]2−1)(−1[1]2−1)([1]1)	
8_{14}	2−8(11)−82	(−1)−12−22−1	([1])(1−1−1[1])(1−1[0])	
8_{15}	3−8(11)−83	(0)1103−22−1	(1−4310[0])(−3520[0])(210[0])	
8_{16}	1−48−98−41	(−2)1−23−21	([0]2−1)([−2]5−2)([−1]4−1)([0]1)	
8_{17}	1−48−(11)8−41	(−4)−12−32−1	(1[−1]1)(2[−5]2)(1[−4]1)([−1])	s
8_{18}	1−5(10)−(13)(10)−51	(−4)−13−33−1	(−1[3]−1)(1[−1]1)(1[−3]1)([−1])	s
8_{19}	1−1010−11	(0)11111	(1−5500[0])(−5(10)00[0])(−1600[0])(100[0])	
8_{20}	1−23−21	(−1)101	([−1]4−2)([−1]4−1)([0]1)	
8_{21}	1−45−41	(0)1−11−1	(1−33[0])(1−32[0])(−10[0])	

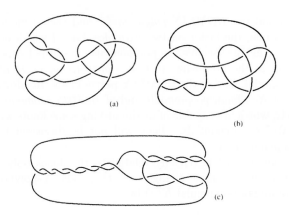

Figure 16.9 Exceptional knots found by Kinoshita and Terasaka (a), Conway (b), and Seifert (c), all with same Alexander polynomial as trivial knot $[|A(t)| \equiv 1]$.

Having set up the $n \times n$-matrix A_{ij}, we choose an arbitrary subdeterminant (minor) of order $n - 1$. It is a polynomial in t with integer coefficients. This polynomial is divided by a suitable power of t to make it start out with a constant. The result is the Alexander polynomial $A(t)$. It is independent of the choice of the subdeterminant.

For the left-handed trefoil, the matrix (16.106) yields

$$A(t) = t^2 - t + 1. \tag{16.108}$$

For the knot 4_1, we find from (16.107)

$$A(t) = t^2 - 3t + 1. \tag{16.109}$$

The Alexander polynomials of the simple knots in Fig. 16.7 are shown in Table 16.3. Note that the replacement $t \to 1/t$ leaves the Alexander polynomial invariant (after renormalizing it back by some power of t to start out with a constant).

The Alexander polynomial of a composite knot factorizes into those of the simple knots it is composed of. If two knots are mirror images of each other, they have the same knot group and the same Alexander polynomials. Due to the factorization property, two composite knots whose simple parts differ by mirror reflection (*stereoisomers*; see Fig. 16.5) have the same polynomial. Thus the Alexander polynomial cannot render a complete classification of inequivalent knots. This is true even after removing degeneracies of the above type. In Fig. 16.9, we give the simplest examples of knots with an Alexander polynomial $A(t) \equiv \pm 1$ of the trivial knot. Up to 11 crossings, these are the only examples. Since the total number of simple knots up to $n = 11$ is 795, the exceptions are indeed very few.

Recent years have witnessed the development of simpler construction procedures and more efficient polynomials for the classification of knots and links of several knots, the Jones and the *HOMFLY polynomials*[3] and their generalizations. The former depend on one, the latter on two variables, one of which occurs also with inverse powers, i.e., in this variable the polynomials are of the Laurent type. Other polynomials found in the literature, such as *Conway, X-*, or *Kauffman's bracket polynomials*, are special cases of the HOMFLY polynomials. In addition, there exist a different type of *Kauffman polynomials* and of *BLMHo polynomials* $F(a, z)$ and $Q(x)$, respectively, which are capable of distinguishing some knots with accidentally degenerate HOMFLY polynomials. They will not be discussed here. For their definition see Appendix 16B.

The X-polynomial $X(a)$ is trivially related to the Jones polynomial $J(t)$, to which it reduces after a variable change $a = t^{1/4}$. The X-polynomial is closely related to the Kauffman polynomial $K(a)$ by

$$X(a) = (-a)^{-3w} K(a). \tag{16.110}$$

The number w is the *cotorsion*, also called *twist number, Tait number*, or *writhe* [6]. It is defined by giving the loop or link an orientation and attributing to each crossing a number 1 or -1 according to the following rule. At each crossing follows the overpass along the direction of orientation. If the underpass runs from right to left, the crossing carries the number 1, otherwise -1. The sum of these numbers is the cotorsion w. In the trefoil knot in Fig. 16.2 each crossing carries a -1 so that $w = -3$.

The Kauffman polynomial is found by a very simple construction procedure. A set of n trivial loops is defined to have the Laurent polynomial

$$K_n(a) = -(a^2 + a^{-2})^{n-1}. \tag{16.111}$$

Every knot or link can be reduced to such loops by changing the crossings recursively into two new configurations according to the graphical rule shown in Fig. 16.10.

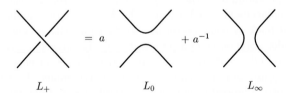

$$L_+ \qquad\qquad L_0 \qquad\qquad L_\infty$$

Figure 16.10 Graphical rule for removing crossing in generating Kauffman polynomial.

The first configuration is associated with a factor a, the second with a factor a^{-1}. The configuration receiving the factor a is most easily identified by approaching the crossing on the underpassing curve and taking a right turn. The two new

---

[3]The word "HOMFLY" collects the initials of the authors (Hoste, Ocneanu, Millet, Freyd, Lickorish, Yetter). The papers are quoted in Notes and References.

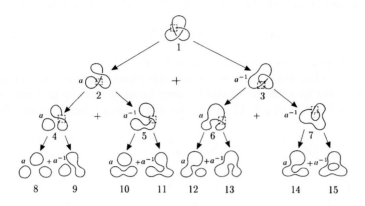

Figure 16.11 Kauffman decomposition of trefoil knot. The configuration 3 is the Hopf link. The calculation of the associated polynomials is shown in Table 16.4.

Table 16.4 Kauffman polynomials in decomposition of trefoil knot.

link	bracket polynomial	rule
15	$-a^2 - a^{-2}$	Eq. 16.111
14	1	Eq. 16.111
13	1	Eq. 16.111
12	$-a^2 - a^{-2}$	Eq. 16.111
11	1	Eq. 16.111
10	$-a^2 - a^{-2}$	Eq. 16.111
9	$-a^2 - a^{-2}$	Eq. 16.111
8	$-a^4 - 2 - a^{-4}$	Eq. 16.111
7	$-a^{-3}$	$a\langle 14 \rangle + a^{-1}\langle 15 \rangle$, Fig. 16.10
6	$-a^3$	$a\langle 12 \rangle + a^{-1}\langle 13 \rangle$, Fig. 16.10
5	$-a^3$	$a\langle 10 \rangle + a^{-1}\langle 11 \rangle$, Fig. 16.10
4	$a^5 + a$	$a\langle 8 \rangle + a^{-1}\langle 9 \rangle$, Fig. 16.10
3	$-a^4 - a^{-4}$	$a\langle 6 \rangle + a^{-1}\langle 7 \rangle$, Fig. 16.10
2	a^6	$a\langle 4 \rangle + a^{-1}\langle 5 \rangle$, Fig. 16.10
1	$a^7 - a^3 - a^{-5}$	$a\langle 2 \rangle + a^{-1}\langle 3 \rangle$, Fig. 16.10

configurations are processed further in the same way and so on until one arrives only at trivial loops. By applying these rules to a trefoil knot, we obtain a knot and a link known as the *Hopf link*. These are decomposed further as shown in Fig. 16.11. The Kauffman polynomials of each part are listed in Table 16.4. The polynomial of the trefoil knot is

$$K(a) = a^7 - a^3 - a^{-5}. \tag{16.112}$$

Since $w = -3$, we obtain with the factor $(-a)^{-3w} = -a^9$ the X-polynomial

$$X(a) = -a^{16} + a^{12} + a^4. \tag{16.113}$$

This corresponds to a Jones polynomial $J(t) = t + t^3 - t^4$.

For the Jones polynomials, there exists a simple direct construction. According to J.H. Conway, any knot can be related to lower knots or links by performing the *skein operations* shown in Fig. 16.12 on any crossing in the projection plane. Either

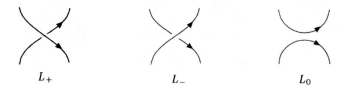

Figure 16.12   Skein operations relating higher knots to lower ones.

a crossing L_+ is transformed into L_- and L_+, or L_+ is transformed into L_- and L_0. For knots related in this way one defines the Jones polynomial $J(t)$ recursively by the *skein relation*

$$\frac{1}{t} J_{L_+}(t) - t J_{L_-}(t) = \left(\sqrt{t} - \frac{1}{\sqrt{t}}\right) J_{L_0}(t). \tag{16.114}$$

The circular loop is defined to have the trivial polynomial $J(t) \equiv 1$. By applying the skein operations to two disjoint unknotted loops in Fig. 16.13, one finds the Jones polynomial

$$J_2(t) = -(\sqrt{t} + 1/\sqrt{t}). \tag{16.115}$$

Upon carrying this procedure to n such loops, we find

$$J_n(t) = [-(\sqrt{t} + 1/\sqrt{t})]^{n-1}, \tag{16.116}$$

in agreement with (16.111). For the lowest knots, the Jones polynomials are listed in Table 16.3. Up to nine crossings, the Jones polynomials distinguish mirror-symmetric knots.

Conway discovered the first skein relation in 1970 when trying to develop a computer program for calculating Alexander polynomials. He found the Alexander polynomials to obey modulo the normalization convention, the skein relation

$$A_{L_+}(t) - A_{L_-}(t) = (\sqrt{t} - 1/\sqrt{t})A_{L_0}(t), \tag{16.117}$$

which eventually reduces the polynomials of all knots to the trivial one $A_1(t) = 1$. The skein relation simplifies the procedure so much that Conway was able to work out by hand all polynomials known at that time. Because of the simplicity of

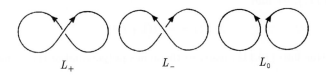

Figure 16.13 Skein operations for calculating Jones polynomial of two disjoint unknotted loops.

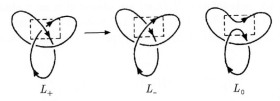

Figure 16.14 Skein operation for calculating Jones polynomial of trefoil knot.

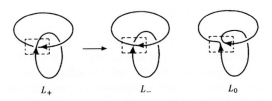

Figure 16.15 Skein operation for calculating Jones polynomial of Hopf link.

this procedure, the Alexander polynomials are now often referred to as *Alexander-Conway polynomials*.

Let us now calculate the Jones polynomial for the trefoil knot 3_1 of Fig. 16.7. First we apply the skein operation shown in Fig. 16.14. The loop L_- is unknotted and has a unit polynomial. Thus we obtain the polynomial relation

$$J_{\text{trefoil}}(t) J_{L_+}(t) = t^2 \cdot 1 + t(\sqrt{t} - 1/\sqrt{t}) J_{L_0}(t). \tag{16.118}$$

The configuration L_0 is known as a *Hopf link*. It needs one more reduction[4] via the operation shown in Fig. 16.15, resulting in the relation

$$\frac{1}{t} J_{L_0}(t) = t J_2(t) + (\sqrt{t} - 1/\sqrt{t}) J_1(t). \tag{16.119}$$

---

[4]The Kauffman bracket polynomial of the Hopf link is $K_{L_0}(a) = -a^4 - a^{-4}$. Together with the cotorsion $w = -2$, this amounts to an X-polynomial $X(a) = -a^{10} - a^2$ and the Jones polynomial $J_{L_0}(t) = -\sqrt{t}(1 + t^2)$ as in (16.120).

Using (16.111), we find

$$J_{L_0}(t) = -\sqrt{t}(1 + t^2). \tag{16.120}$$

Inserting this into (16.118) leads to the Jones polynomial of the trefoil knot

$$J_{\text{trefoil}} = t + t^3 - t^4. \tag{16.121}$$

It differs from the result found above for the left-handed trefoil by the substitution $t \to t^{-1}$.

The HOMFLY polynomials $H_L(t, \alpha)$ are obtained from a slight generalizations of the skein relation (16.114) of the Jones polynomials: The factor $(\sqrt{t} - 1/\sqrt{t})$ on the right-hand side is replaced by an arbitrary parameter α, leading to the skein relation

$$\frac{1}{t}H_{L_+}(t, \alpha) - tH_{L_-}(t, \alpha) = \alpha H_{L_0}(t, \alpha). \tag{16.122}$$

The trivial knot is defined to have the trivial polynomial $H_1(t, \alpha) = 1$. For two independent loops, the relation yields

$$H_2(t, \alpha) = (t^{-1} - t)\alpha^{-1}. \tag{16.123}$$

The HOMFLY polynomials $H(t, \alpha)$ transform under a mirror reflection of the knot into $H(-t^{-1}, \alpha)$. Note that $H_2(t, \alpha)$ is mirror-symmetric [$H_1(t, \alpha)$ is trivially so].[5] In general, the HOMFLY polynomials give reliable information on a possible mirror symmetry. There are, however, a few exceptions, i.e., mirror-related pairs of knots possessing the same HOMFLY polynomial.[6]

Examples for HOMFLY polynomials are

$$\begin{aligned}
H_{\text{trefoil(rh)}}(t, \alpha) &= -t^4 + 2t^2 + t^2\alpha^2, \\
H_{\text{trefoil(lh)}}(t, \alpha) &= -t^{-4} + 2t^{-2} + t^{-2}\alpha^2, \\
H_{\text{Hopf(rh)}}(t, \alpha) &= (t - t^3)\alpha^{-1} + t\alpha, \\
H_{\text{knot } 4_1}(t, \alpha) &= t^{-2} - 1 + t^2 - \alpha^2.
\end{aligned} \tag{16.124}$$

Setting $\alpha = t^{1/2} - t^{-1/2}$ produces the Jones polynomials, while $t \to 1, \alpha \to t^{1/2} - t^{-1/2}$ leads, with appropriate powers of t as normalization factors, back to the Alexander-Conway polynomials.

In Table 16.3, the HOMFLY polynomials are listed for knots up to 8 crossings. For mirror-unsymmetric knots, only one partner is recorded. The reflected polynomial is obtained by the substitution $t \to -t^{-1}$. The meaning of the entries is best explained with an example: The knot 7_1 has an entry $([0]004 - 3)([0]00(10) - 4)([0]006 - 1)([0]001)$, which stands for the polynomial

---

[5]For the Kauffman polynomials $F(a, x)$ defined in Appendix 16B, mirror reflection implies $F(a, x) \to F(a^{-1}, x)$.

[6]The first degeneracy of this type occurs for a link of 3 loops with 8 crossings.

$H(t, \alpha) = (4t^6 - 3t^8) + (10t^6 - 4t^8)\alpha^2 + (6t^6 - t^8)\alpha^4 + t^6\alpha^6$. A bracket marks the position and coefficient of the zeroth power in t^2; the numbers to the right and the left of it specify the coefficients of the adjacent higher and lower powers of t^2, respectively. Numbers with more than one digit are put in parentheses. The polynomial of the reflected knot is obtained by reflecting the numbers in parentheses on the associated bracket. The knots marked by an s are mirror-symmetric.

The Alexander-Conway polynomials are special cases of the HOMFLY polynomials. A comparison with the skein relation (16.117) shows that they are obtained from them by setting $t = 1$ and replacing α by $t^{1/2} - t^{-1/2}$:

$$A_L(t) = H_L(1, t^{1/2} - t^{-1/2}). \tag{16.125}$$

The reducible granny and square knots in Fig. 16.5 are distinguished by the Jones and the HOMFLY polynomials; the latter are

$$
\begin{aligned}
H_{\text{granny}}(t, \alpha) &= (2t^2 - t^4 + t^2\alpha^2)^2, \\
H_{\text{square}}(t, \alpha) &= (2t^2 - t^4 + t^2\alpha^2)(2t^{-2} - t^{-4} + t^2\alpha^2);
\end{aligned}
\tag{16.126}
$$

the former are obtained by inserting $\alpha = t^{1/2} - t^{-1/2}$.

Up to now, there exists no complete algebraic classification scheme. For example, the Jones polynomials of the knots with 10 and 13 crossings shown in Fig. 16.16 are

Figure 16.16   Knots with 10 and 13 crossings, not distinguished by Jones polynomials.

the same.[7] For further details, see the mathematical literature quoted at the end of the chapter.

Even with the incomplete classification of knots, it has until now been impossible to calculate the probability distribution of the various equivalence classes of knots. Modern computers allow us to enumerate the different topological configurations for not too long polymers and to simulate their distributions by Monte Carlo methods. In Fig. 16.17 we show the result of a simulation by Michels and Wiegel, where they measure the fraction f_N of unknotted polymers of N links. They fit their curve by a power law

$$f_N = C\mu^N N^\alpha, \tag{16.127}$$

---

[7]The HOMFLY polynomials have their first degeneracy for prime knots with 9 crossings. It was checked that up to 13 crossings (amounting to 12 965 knots) no polynomial of a nontrivial knot is accidentally degenerate with the trivial polynomial of a circle.

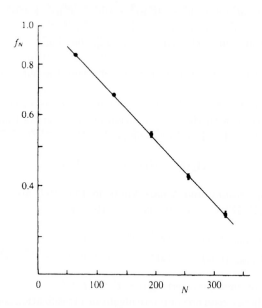

Figure 16.17   Fraction f_N of unknotted closed polymers in ensemble of fixed length $L = Na$.

with the parameters $C \approx 1.026$, $\mu \approx 0.99640$, $\alpha \approx 0.0088$. Thus f_N falls off exponentially in N like μ^N with $\mu < 1$. The exponent α is extremely small. A more recent simulation by S. Windwer takes account of the fact that the line elements are self-avoiding. It yields the parameters[8]

$$C \approx 1.2325, \quad \mu \approx 0.9949, \quad \alpha \approx 0. \tag{16.128}$$

For a polymer enclosed in a sphere of radius R, the distribution has the finite-size dependence

$$f_N(R) = e^{-A(N^\beta l/R)^\gamma}, \tag{16.129}$$

where l is the length of a link. The "critical exponents" are $\beta \approx 0.76$ and $\gamma \approx 3$.

16.5   The Gauss Invariant of Two Curves

For any analytic calculation of topological properties one needs a functional of the polymer shape which is capable of distinguishing the different knot classes. Initially,

---

[8]A first theoretical determination of these parameters has recently been given by mapping the problem onto a four-state Potts model. A presentation of this method which does not involve path integrals would go beyond the scope of this book. See the papers by A. Kholodenko quoted at the end of the chapter.

a hopeful candidate was the one-loop version of the contour integral introduced almost two centuries ago by Gauss for a pair of closed curves C and C':

$$G(C, C') = \frac{1}{4\pi} \oint_C \oint_{C'} [d\mathbf{x} \times d\mathbf{x}'] \cdot \frac{\mathbf{x} - \mathbf{x}'}{|\mathbf{x} - \mathbf{x}'|^3}. \tag{16.130}$$

Gauss proved this to be a topological invariant. In fact, we may rewrite (16.134) with the help of the δ-function (10A.8) as

$$\oint_C d\mathbf{x} \oint_{C'} \frac{d\mathbf{x}' \times (\mathbf{x} - \mathbf{x}')}{|\mathbf{x} - \mathbf{x}'|^3} = -\int d^3x\, \boldsymbol{\delta}(\mathbf{x}; C) \cdot \left[\int d^3x'\, \boldsymbol{\delta}(\mathbf{x}'; C') \times \frac{\mathbf{R}'}{R'^3} \right] \tag{16.131}$$

The second integral is recognized as the gradient of the multivalued field $\Omega(\mathbf{x}; C')$ defined by Eqs. (10A.18), which is the solid angle under which the contour C' is seen from the point $\mathbf{x}$, so that

$$G(C, C') = -\frac{1}{4\pi} \int d^3x\, \boldsymbol{\delta}(\mathbf{x}; C) \cdot \boldsymbol{\nabla}\Omega(\mathbf{x}; C'). \tag{16.132}$$

Inserting here Eq. (10A.27), where S' is any surface enclosed by the contour L', and using the fact that

$$\int d^3x\, \boldsymbol{\delta}(\mathbf{x}; C) \cdot \boldsymbol{\nabla}\Omega(\mathbf{x}; S') = -\int d^3x\, \boldsymbol{\nabla} \cdot \boldsymbol{\delta}(\mathbf{x}; C)\Omega(\mathbf{x}; S') = 0, \tag{16.133}$$

due to (10A.9) and the fact that $\Omega(\mathbf{x}; S) = 0$ is single-valued, we obtain

$$G(C, C') = -\int d^3x\, \boldsymbol{\delta}(\mathbf{x}; C) \cdot \boldsymbol{\delta}(\mathbf{x}; S'). \tag{16.134}$$

This is a purely topological integral. By rewriting it as

$$G(C, C') = -\oint_C dx_i\, \delta_i(\mathbf{x}; S'), \tag{16.135}$$

we see that $G(C, C')$ gives the linking number of C and C'. It is defined as the number of times by which one of the curves, say C', perforates the surface S spanned by the other.

Alternative expressions for the Gauss integral (16.134) are

$$G(C, C') = -\frac{1}{4\pi} \oint_{C'} d\Omega'(\mathbf{x}'; C) = -\frac{1}{4\pi} \oint_C d\Omega(\mathbf{x}, C'), \tag{16.136}$$

where where $\Omega'(\mathbf{x}'; S)$ is the solid angle under which the curve C is seen from the point $\mathbf{x}'$.

The values of the Gauss integral for various pairs of linked curves up to 8 crossings are given in the third column of Table 16.5. All the intertwined pairs of curves labeled by $2_1, 7_1, 7_2, 8_7$, for instance, have a Gauss integral $G(C, C') = \pm 1$.

Let us end this section by another interpretation of the Gauss integral. According to Section 10A.1, the solid angle Ω is equal to the magnetic potential of a current 4π running through the curve C'. Its gradient is the magnetic field

$$B_i = \partial_i \Omega. \tag{16.137}$$

Hence we can write

$$G(C, C') = -\oint_C dx_i B_i = -\oint_{C'} dx'_i B'_i. \tag{16.138}$$

According to this expression, $G(C, C')$ gives the total work required to move a unit magnetic charge around the closed orbit C in the presence of the magnetic field due to a unit electric current along C'.

Unfortunately, there exists no such topologically invariant integral for a single closed polymer. If we identify the curves C and C', the Gauss integral ceases to be a topological invariant. It can, however, be used to classify self-entangled *ribbons*. These possess two separate edges identified with C and C'. Such ribbons play an important role in biophysics. The molecules of DNA, the carriers of genetic information on the structure of living organisms, can be considered as ribbons. They consist of two chains of molecules connected by weak hydrogen bridges. These can break up thermally or by means of enzymes decomposing the ribbon into two single chains.

16.6 Bound States of Polymers and Ribbons

Two or more polymers may line up parallel to each other and form a bound state. The most famous example is the molecule of DNA. It is a bound state of two long chains of molecules which may contain a few thousand up to several billion links. The distance d between the two chains is about 20Å. In equilibrium, the two chains are twisted up in the form of a double helix, rising by about 20Å (i.e., about 10 monomers) per turn (see Fig. 16.18). The DNA molecule may be idealized as an infinitesimally thin ribbon. The ribbon is always *two-sided* since the edges of the ribbon are made up of different phosphate groups whose chemical structure makes the binding unique. One-sided structures formed by a Möbius strip are excluded.

Circular DNA molecules have interesting topological properties. In the double helix, one edge passes through the other an integer number of times. This is the *linking number* L_k of the double helix. Being a topological invariant, it does not change if the two closed edges become unbound and distorted into an arbitrary shape.

If the total number of windings N_w in the DNA helix is different from the linking number L_k, a circular helix is always under mechanical stress. It can relax by forming a supercoil (see Fig. 16.19). The number of excess turns

$$\tau = L_k - N_w, \tag{16.139}$$

provides a measure for the *supercoiling density* which is defined by the ratio

$$\sigma \equiv \frac{\tau}{N_w}. \tag{16.140}$$

In natural DNA, the supercoiling density is usually $-\sigma \sim 0.03--0.1$. The negative sign implies that the natural twist of the double helix is slightly *decreased* by the supercoiling. The negative sign seems to be essential in the main biological process, the *replication*. It may be varied by an enzyme, called DNA gyrase. A cell has a large arsenal of enzymes which can break one of the chains in the helix and unwind the linking number L_k by one or more units, changing the topology. Such enzymes run under the name of *topoisomerases* of type I. There is also one of a type II which breaks *both* chains and can tie or untie knots in the double helix of DNA as a whole.

The biophysical importance of the supercoil derives from the fact that the stress carried by such a configuration can be relaxed by breaking a number of bonds between the two chains. In fact, a number $\theta = -\sigma$ of broken bonds leads to a complete relaxation. During a cell division, all bonds are broken. Note that this process would be energetically unfavorable if the supercoiling density were positive.

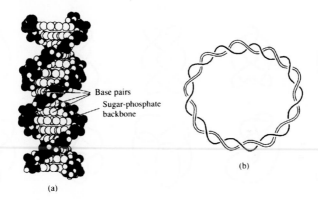

Base pairs
Sugar-phosphate
backbone

(a)

(b)

Figure 16.18   Small section and idealized view of circular DNA molecule. The link number L_k (defined as number of times one chain passes through arbitrary surface spanned by the other) is $L_k \approx 9$.

Figure 16.19   A supercoiled DNA molecule. This is the natural shape when carefully extracted from a cell. The supercoiling is negative.

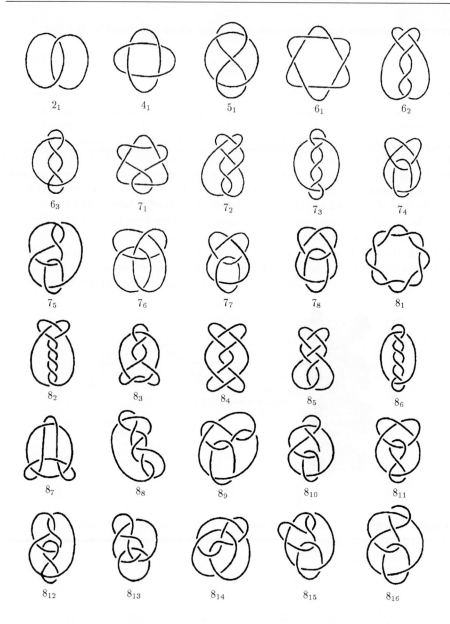

Figure 16.20 Simple links of two polymers up to 8 crossings.

Just as for knots, no complete topological invariants are known for such types of links of two (or more) closed polymers. Historically, generalized Alexander polynomials were used to achieve an approximate classification of links. They are polynomials of two variables. To construct them, we take one of the two polymers and label all underpasses in the same way as for knots. The same thing is done for the other polymer. For each underpass, a row of the Alexander matrix A_{ij} is written down with two variables s and t. The Alexander polynomial $A(s,t)$ is again given by any $(n-1) \times (n-1)$-subdeterminant of the $n \times n$-matrix A_{ij}. The links up to 8 crossings are shown in Fig. 16.20. The Alexander polynomials associated with these are listed in Table 16.5. Note that the replacements $s \to 1/s, t \to 1/t$, or both, leave the Alexander polynomial invariant (due to the prescription of renormalizing the lowest coefficients to an integer). For unlinked polymers one has $A(s,t) = 0$.

There is no need to go through the details of the procedure since the more recent and powerful Jones and HOMFLY polynomials can be constructed for arbitrary links without additional prescriptions. The latter are tabulated in the fourth column of Table 16.5. In many cases, a change in the orientation of the second loop gives rise to an inequivalent link. Then the table shows two entries underneath each other. For the knot 7_2, the upper entry $\{-1\}(1*-1)(1*0-11)(*-1-1)$ indicates the polynomial $H(t,\alpha) = \alpha^{-1}(t^{-1}-t) + \alpha(t^{-1}-t^2+t^3) + \alpha^3(-t-t^3)$. The curly bracket shows the lowest power of α and the star marks the position of the zeroth power of t. The coefficients of $t, t^3, \ldots$ stand to the right of it, those of $\ldots, t^{-3}, t^{-1}$ to the left of it.

For the special case of a circular ribbon such as a DNA molecule, the Gauss integral over the two edges, being a topological invariant, renders also a classification of the ribbon as a whole. As shown in Eq. (16.135), the Gauss integral yields precisely the linking number L_{k}.

It is useful to calculate $G(C, C')$ for a ribbon in the limit of a very small edge-to-edge distance d. This will also clarify why the Gauss integral $G(C, C')$ in which both C and C' run over the same single loop is not a topological invariant. We start with the Gauss integral for the two edges C, C' of the ribbon

$$G(C,C') = \frac{1}{4\pi} \oint_C \oint_{C'} [d\mathbf{x} \times d\mathbf{x}'] \cdot \frac{\mathbf{x} - \mathbf{x}'}{|\mathbf{x} - \mathbf{x}'|^3} \tag{16.141}$$

and shift the two neighboring integration contours C, C' both towards the ribbon axis called $\bar{C}$. Let ϵ measure the distance between the two edges, and let $\mathbf{n}(\tau)$ be the unit vector orthogonal to the axis pointing from C to C'. Then we write

$$G(C,C') = \frac{1}{4\pi} \oint_{\bar{C}} d\tau \oint_{\bar{C}} d\tau' \, [\dot{\mathbf{x}}(\tau) \times (\dot{\mathbf{x}}(\tau') + \epsilon\dot{\mathbf{n}}(\tau'))] \cdot \frac{\mathbf{x}(\tau) - \mathbf{x}(\tau') - \epsilon\mathbf{n}(\tau')}{|\mathbf{x}(\tau) - \mathbf{x}(\tau') - \epsilon\mathbf{n}(\tau')|^3}. \tag{16.142}$$

In the limit $\epsilon \to 0$, $G(C, C')$ does not just become equal to $G(\bar{C}, \bar{C})$ [which then would be the same as $G(C, C)$ or $G(C', C')$]. A careful limiting procedure performed below shows that there is a remainder T_{w},

$$L_{\mathrm{k}} = G(\bar{C}, \bar{C}) + T_{\mathrm{w}}. \tag{16.143}$$

Table 16.5 Alexander polynomials $A(s,t)$ and the coefficients of HOMFLY polynomials $H(t,\alpha)$ for simple links of two closed curves up to 8 minimal crossings, labeled as in Fig. 16.20. The value $|A(1,1)|$ is equal to the absolute value of the Gauss integral $|G(C,C')|$ for the two curves. The entries in the last column are explained in the text.

	$A(s,t)$	$|A|$	$H(t,\alpha)$
2_1	1	1	$\{-1\}(*1-1)(*1)$
4_1	$s+t$	2	$\{-1\}(*01-1)(*11)$ $\{-1\}(*01-1)(*03-1)(*01)$
5_1	$(s-1)(t-1)$	0	$\{-1\}(1*-1)(-12*-1)(1*)$
6_1	s^2+t^2+st	3	$\{-1\}(*001-1)(*006-3)(*005-1)(*001)$ $\{-1\}(*001-1)(*111)$
6_2	$st(s+t)-st+s+t$	3	$\{-1\}(*001-1)(*022-1)(*011)$
6_3	$2st-(s+t)+2$	2	$\{-1\}(*01-1)(*021-1)(*011)$ $\{-1\}(1-10*)(-21*-1)(1*)$
7_1	$s^2t^2-st(s+t)+st-(s+t)+1$	1	$\{-1\}(1-1*)(-24-3*)(10*)$ $\{-1\}(*1-1)(-11*2-1)(1*1)$
7_2	$st(s+t)-t^2-s^2-3st+s+t$	1	$\{-1\}(*1-1)(1*0-11)(*-1-1)$ $\{-1\}(*1-1)(-2*5-2)(-1*4-1)(*1)$
7_3	$2(s-t)(t-1)$	0	$\{-1\}(1*-1)(1*-1-1)(*-1-1)$
7_4	$(s-1)(t-1)(s^2+1)$	0	$\{-1\}(-13-2*)(-25-3*)(-14-1*)(10*)$
7_5	$2s^3t-t^2-s+2$	2	$\{-1\}(*002-3)(*014-3)(*012)$ $\{-1\}(-1*3-2)(-2*6-2)(-1*4-1)(*1)$
7_6	$(s+1)^2(s-1)(t-1)$	0	$\{-1\}(1*-1)(1-2*1)(1-3*1)(-1*)$
7_7	s^3+t	2	$\{-1\}(-1*3-2)(-1*4-1)(*1)$ $\{-1\}(*002-31)(*006-4)(*005-1)(*001)$
7_8	$(s-t)(t-1)$	0	$\{-1\}(-13-2*)(-13-2*)(10*)$
8_1	$(s+t)(s^2+t^2)$	4	$\{-1\}(*0001-1)(*00010-6)(*00015-5)(*0007-1)(*0001)$ $\{-1\}(*0001-1)(*1111)$
8_2	$st(s+t-1)(st+1)+s+t$	4	$\{-1\}(*0001-1)(*0034-3)(*0044-1)(*0011)$ $\{-1\}(*0001-1)(*0212-1)(*0111)$
8_3	$2s^2t^2-st(s+t)+3st-(s+t)+2$	3	$\{-1\}(*001-1)(*0043-1)(*0011)$ $\{-1\}(1-100*)(-200*-1)(11*)$
8_4	$s^2t^2(s+t)-2s^2t^2+2st(s+t)-2st+s+t$	4	$\{-1\}(*0001-1)(*0131-1)(*0121)$ $\{-1\}(*0001-1)(*0042-2)(*0043-1)(*0011)$
8_5	$s^2t^2-2st(s+t)+3st-2(s+t)+1$	3	$\{-1\}(*001-1)(*0130-1)(*0121)$ $\{-1\}(1-100*)(1-42-2*)(-23-1*)(10*)$
8_6	$2st-3(s+t)+2$	2	$\{-1\}(*01-1)(*0201-1)(*0111)$ $\{-1\}(1-10*)(-20*1-1)(1*1)$
8_7	$s^2t^2-2st(s+t)+s^2+3st+t^2-2(s+t)+1$	1	$\{-1\}(1-1*)(1-4*3-1)(-2*3-1)(*1)$ $\{-1\}(*1-1)(*2-33-1)(*1-32)(*0-1)$
8_8	$s^2t^2-2st(s+t)+s^2+st+t^2-2(s+t)+1$	3	$\{-1\}(*1-1)(-1*4-31)(-1*3-2)(*1)$ $\{-1\}(*1-22-1)(1*1-33)(*-1-2)$
8_9	$s^3+2s^2t-4s^2-4st+s+2t$	2	$\{-1\}(*1-22-1)(*-1-2)$ $\{-1\}(*1-22-1)(*2-34-1)(*1-32)(*0-1)$
8_{10}	$(s^2-1)(t-1)$	0	$\{-1\}(1-2*2-1)(1-4*4-1)(1-3*2)(-1*)$
8_{11}	$s^3t-2s^2(s+t)+2s(s+t)-2(s+t)+1$	2	$\{-1\}(*002-31)(*005-2-1)(*0042-1)(*0011)$ $\{-1\}(2-3*1)(1-5*3-1)(-2*3-1)(*1)$
8_{12}	$(s^2-1)(t-1)$	0	$\{-1\}(-13-2*)(-14-4*1)(2-3*1)(-1*)$
8_{13}	$(s^2+1)(s-1)(t-1)$	0	$\{-1\}(1*-1)(*1-21)(-1*2-2)(*1)$
8_{14}	$s^3t-4s^2t+4s^2+4st-4s+1$	2	$\{-1\}(*002-31)(*005-2-1)(*0131)$ $\{-1\}(-1*3-2)(1-3*5-1)(-2*3-1)(*1)$
8_{15}	$(s-1)(t-1)$	0	$\{-1\}(1-2*2-1)(-2*3-1)(*1)$
8_{16}	$s^3-2s(s+1)+1$	2	$\{-1\}(*1-22-1)(*2-22)(*0-1)$ $\{-1\}(*1-22-1)(*3-43)(*1-41)(*0-1)$

The remainder is called the *twist* of the ribbon, defined by

$$T_\mathrm{w} \equiv \frac{1}{2\pi} \oint_{\bar{C}} d\tau \dot{\mathbf{x}}(\tau) \cdot [\mathbf{n}(\tau) \times \dot{\mathbf{n}}(\tau)]/|\dot{\mathbf{x}}(\tau)|. \tag{16.144}$$

Incidentally, this integral makes sense also for a single curve if $\mathbf{n}(\tau)$ is taken to be the principal normal vector of the curve. Then T_w gives what is called the *total integrated torsion* of the single curve. The first term in (16.143), the Gauss integral for a single closed curve $\bar{C}$, is called in this context the *writhing number* of the curve

$$W_\mathrm{r} \equiv G(\bar{C}, \bar{C}) = \frac{1}{4\pi} \oint_{\bar{C}} d\tau \oint_{\bar{C}} d\tau' \, [\dot{\mathbf{x}}(\tau) \times \dot{\mathbf{x}}(\tau')] \cdot \frac{\mathbf{x}(\tau) - \mathbf{x}(\tau')}{|\mathbf{x}(\tau) - \mathbf{x}(\tau')|^3}. \tag{16.145}$$

Thus one writes the relation (16.143) commonly in the form

$$L_\mathrm{k} = W_\mathrm{r} + T_\mathrm{w}. \tag{16.146}$$

Only the sum $W_\mathrm{r} + T_\mathrm{w}$ is a topological invariant, with T_w containing the information on the ribbon structure of the closed loop $\bar{C}$. This formula was found by Calagareau in 1959 and generalized by White in 1969.

From what we have seen above in Eq. (16.136), the writhing number may also be written as an integral

$$W_\mathrm{r} = -\frac{1}{4\pi} \oint_{\bar{C}} d\Omega(\mathbf{x}), \tag{16.147}$$

where $\Omega(\mathbf{x})$ is the solid angle under which the axis of the ribbon is seen from another point on the axis. When rewritten in the form (16.138), it has the magnetic interpretation stated there.

This interpretation is relevant for understanding the spacetime properties of the dionium atom which in turn may be viewed as a world ribbon whose two edges describe an electric and a magnetic charge. We have pointed out in Section 14.6 that for a half-integer charge parameter q, a dionium atom consisting of two bosons is a fermion. For this reason, a path integral over a fluctuating ribbon can be used to describe the quantum mechanics of a fermion in three spacetime dimensions [7].

Let us derive the relation (16.146). We split the integral (16.142) over τ into two parts: a small neighborhood of the point τ', i.e.,

$$\tau \in (\tau' - \delta, \tau' + \delta) \tag{16.148}$$

and the remainder, for which the integrand is regular. In the regular part, we can set the distance ϵ between the curves C and C' equal to zero, and obtain the Gauss integral $G(\bar{C}, \bar{C})$, i.e., the writhing number W_r. In the singular part, we approximate $x(\tau)$ within the small neighborhood (16.148) by the straight line

$$\begin{aligned}
\mathbf{x}(\tau) &\approx \mathbf{x}(\tau') + \dot{\mathbf{x}}(\tau')(\tau - \tau'), \\
\dot{\mathbf{x}}(\tau) &\approx \dot{\mathbf{x}}(\tau').
\end{aligned} \tag{16.149}$$

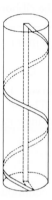

Figure 16.21 Illustration of Calagareau-White relation (16.146). The number of windings around the cylinder is $L_k = 2$, while $T_w = L_k - p/\sqrt{p^2 + R^2}$, where p is the pitch of the helix and R the radius of the cylinder.

Then the τ-integral can be performed. For $\epsilon \ll \delta \ll 1$, we find

$$
\begin{aligned}
T_w &= -\frac{1}{4\pi} \oint_{C'} d\tau' \left[\mathbf{n}(\tau') \times \dot{\mathbf{x}}(\tau')\right] \cdot \dot{\mathbf{n}}(\tau') \epsilon^2 \int_{\tau'-\delta}^{\tau'+\delta} d\tau \frac{1}{\sqrt{|\dot{\mathbf{x}}(\tau')|^2(\tau - \tau')^2 + \epsilon^2}} \\
&= \frac{1}{2\pi} \oint_{C'} d\tau' \left[\dot{\mathbf{x}}(\tau') \times \mathbf{n}(\tau')\right] \cdot \dot{\mathbf{n}}(\tau')/|\dot{\mathbf{x}}(\tau')|.
\end{aligned}
\tag{16.150}
$$

It is worth emphasizing that in contrast to the Gauss integral for two curves C, C', the value of the Gauss integral for $\bar{C}$ is *not* an integer but a continuous number. It depends on the shape of the ribbon, changing continuously when the ribbon is deformed isotopically. If one section of the ribbon passes through the other, however, it changes by 2 units. The defining equation shows that W_r vanishes if the ribbon axis has a center or a plane of symmetry. To give an example for a circular closed ribbon with an integer number L_k and an *arbitrary writhing number* W_r we follow Brook-Fuller and Crick. A cylinder with a closed ribbon is wound flat around the surface of a cylinder, returning along the cylinder axis (see Fig. 16.21). While the two edges of the ribbon perforate each other an integer number of times such that $L_k = 2$, the ribbon axis has a noninteger Gauss integral W_r depending on the ratio between pitch and radius, $W_r = 2 - p/\sqrt{p^2 + R^2}$.

16.7 Chern-Simons Theory of Entanglements

The Gauss integral has a form very similar to the Biot-Savart law of magnetostatics found in 1820. That law supplies an action-at-a-distance formula for the interaction energy of two currents I, I' running along the curves C and C':

$$E = -\frac{II'}{c^2} \oint_C \oint_{C'} dx \cdot dx' \frac{1}{|x - x'|}, \tag{16.151}$$

where c is the light velocity. It was a decisive conceptual advance of Maxwell's theory to explain formula (16.151) by means of a *local* field energy arising from a vector potential $A(x)$. In view of the importance of the Gauss integral for the topological classification of entanglements, it is useful to derive a local field theory producing the Gauss integral as a topological action-at-a-distance. Imagine the two contours C and C' carrying stationary currents of some pseudo-charge which we normalize at first to unity. These currents are coupled to a vector potential $A(x)$, which is now unrelated to magnetism and which will be called *statisto-magnetic vector potential*:

$$\mathcal{A}_{\text{e,curr}} = -i \oint_C dx A(x) - i \oint_{C'} dx A(x). \tag{16.152}$$

The action is of the Euclidean type as indicated by the subscript e (recall the relation with the ordinary action $\mathcal{A} = i\mathcal{A}_e$). We now construct a field action for $A(x)$ so that the field equations render an interaction between the two currents which is precisely of the form of the Gauss integral. This field action reads

$$\mathcal{A}_{\text{e,CS}} = \frac{i}{2} \int d^3x \, A \cdot (\nabla \times A) \tag{16.153}$$

and is called the *Chern-Simons action*. It shares with the ordinary Euclidean magnetic field action the quadratic dependence on the vector potential $A(x)$, as well as the invariance under local gauge transformations

$$A(x) \rightarrow A(x) + \nabla \Lambda(x), \tag{16.154}$$

which is obvious when transforming the second vector potential in (16.153). The gauge transformation of the first vector potential produces no change after a partial integration. Also the coupling in (16.152) to the contours C and C' is gauge-invariant after a partial integration, since the contours are closed, satisfying

$$\nabla \cdot \oint dx = 0. \tag{16.155}$$

In contrast to the magnetic field energy, however, the action (16.153) is purely imaginary. The factor i is important for the applications in which the Chern-Simons action will give rise to phase factors of the form $e^{i2\theta G(C,C')}$.

By extremizing the combined action, we obtain the field equation

$$\nabla \times A(x) = \left(\oint_C + \oint_{C'} \right) dx. \tag{16.156}$$

Its solution is

$$A_i(\mathbf{x}) = \left(\oint_C + \oint_{C'} \right) G_{ij}(\mathbf{x}, \mathbf{x}') dx'_j, \tag{16.157}$$

where $G_{ij}(\mathbf{x}, \mathbf{x}')$ is a suitable Green function solving the inhomogeneous field equation (16.157) with a δ-function source instead of the current. Due to the gauge invariance of the left-hand side, however, there is no unique solution. Given a solution $G_{ij}(\mathbf{x}, \mathbf{x}')$, any gauge-transformed Green function

$$G_{ij}(\mathbf{x}, \mathbf{x}') \rightarrow G_{ij}(\mathbf{x}, \mathbf{x}') + \nabla_i \Lambda_j(\mathbf{x}, \mathbf{x}') + \nabla_j \Lambda_i(\mathbf{x}, \mathbf{x}')$$

will give the same curl $\nabla \times \mathbf{A}$. Only the transverse part of the vector potential (16.157) is physical, and the Green function has to satisfy

$$\epsilon_{ijk} \nabla_j G_{kl}(\mathbf{x}, \mathbf{x}') = \delta_{ij}^{(3)}(\mathbf{x} - \mathbf{x}')^T, \tag{16.158}$$

where

$$\delta_{ij}^{(3)}(\mathbf{x} - \mathbf{x}')^T = \left(\delta_{ij} - \frac{\nabla_i \nabla_j}{\nabla^2} \right) \delta^{(3)}(\mathbf{x} - \mathbf{x}') \tag{16.159}$$

is the transverse δ-function. The vector potential is then obtained from (16.157) in the transverse gauge with

$$\nabla \cdot \mathbf{A}(\mathbf{x}) = 0. \tag{16.160}$$

The solution of the differential equation (16.158) is easily found in the two-dimensional transverse subspace. The Fourier transform of Eq. (16.158) reads

$$i \epsilon_{ijk} p_j G_{kl}(\mathbf{p}) = \delta_{il} - \frac{p_j p_l}{\mathbf{p}^2}, \tag{16.161}$$

and this is obviously solved by

$$G_{ij}(\mathbf{p}) = i \epsilon_{ikj} p_k \frac{1}{\mathbf{p}^2}. \tag{16.162}$$

The transverse gauge (16.160) may be enforced in an action formalism by adding to the action (16.153) a *gauge-fixing term*

$$\mathcal{A}_{\mathrm{GF}} = \frac{1}{2\alpha} (\nabla \cdot \mathbf{A})^2, \tag{16.163}$$

with an intermediate gauge parameter α which is taken to zero at the end. The field equation (16.158) is then changed to

$$\left(\epsilon_{ijk} \nabla_j + \frac{i}{\alpha} \nabla_i \nabla_k \right) G_{kl}(\mathbf{x}, \mathbf{x}') = \delta_{ik} \delta^{(3)}(\mathbf{x} - \mathbf{x}'), \tag{16.164}$$

which reads in momentum space

$$\left(i\epsilon_{ijk}p_j - \frac{i}{\alpha}p_ip_k \right) G_{kl}(\mathbf{p}) = \delta_{ik}. \tag{16.165}$$

This has the unique solution

$$G_{ik}(\mathbf{p}) = \left(i\epsilon_{ijk}p_j + i\alpha \frac{p_ip_k}{\mathbf{p}^2} \right) \frac{1}{\mathbf{p}^2}, \tag{16.166}$$

whose $\alpha \to 0$ -limit is (16.162).

Going back to configuration space, the Green function becomes

$$G_{ij}(\mathbf{x}, \mathbf{x}') = \int \frac{d^3p}{(2\pi)^3} e^{i\mathbf{p}(\mathbf{x}-\mathbf{x}')} \frac{i\epsilon_{ikj}p_k}{\mathbf{p}^2} = \frac{1}{4\pi}\epsilon_{ikj}\nabla_k \frac{1}{|\mathbf{x}-\mathbf{x}'|} = \frac{1}{4\pi}\epsilon_{ijk}\frac{(x-x')_k}{|\mathbf{x}-\mathbf{x}'|^3}. \tag{16.167}$$

Inserting this into $\mathcal{A}_{e,\text{curr}} + \mathcal{A}_{e,\text{CS}}$ yields the interaction between the currents[9]

$$\mathcal{A}_{e,\text{int}} = -i \oint_C \oint_{C'} dx_i dx_j' G_{ij}(\mathbf{x}, \mathbf{x}'). \tag{16.168}$$

Up to the prefactor $-i$, this is precisely the Gauss integral $G(C, C')$ of the two curves C, C'. In addition there are the self-interactions of the two curves

$$\mathcal{A}_{e,\text{int}} = -\frac{i}{2} \left(\oint_C \oint_C + \oint_{C'} \oint_{C'} \right) dx_i dx_j' G_{ij}(\mathbf{x}, \mathbf{x}'), \tag{16.169}$$

which are equal to $-(i/2)[G(C,C) + G(C',C')]$. Due to their nontopological nature discussed earlier, these have no quantized values and must be avoided.

Such unquantized self-interactions can be avoided by considering systems whose orbits are subject to appropriate restrictions. They may, for instance, contain only lines which are not entangled with themselves and run in a preferred direction from $-\infty$ to ∞. Ensembles of nonrelativistic particles in two space dimensions have precisely this type of worldlines in three-dimensional spacetime. They are therefore an ideal field for the Chern-Simons theory, as we shall see below in more detail.

Another way of avoiding unquantized self-interactions is based on a suitable limiting procedure. If the lines C and C' coincide, the Gaussian integral $G(C, C')$ over $C = C'$ may be spread over a large number N of parallel running lines C_i $(i = 1, \ldots, N)$, each of which carries a topological charge $1/N$. The sum $(1/N^2) \sum_{ij} G(C_i, C_j)$ contains N-times the same self-interaction and $N(N-1)$-times the same integer-valued linking number L_k of pairs of lines. In the limit $N \to \infty$, only the number L_k survives. The result coincides with the Gauss integral for the two frame lines C_1 and C_N of the ribbon. The number L_K may therefore be called the *frame linking number*. This number depends obviously on the choice of the framing. There is a preferred choice for which L_k vanishes. This eliminates the self-interaction trivially.

----

[9]Compare this with the derivation of Eq. (3.244).

Although the limiting procedure makes the self-interaction topological, the arbitrariness of the framing destroys all information on the knot classes. This information can be salvaged by means of a generalization of the above topological action leading to a nonabelian version of the Chern-Simons theory. That theory has the same arbitrariness in the choice of the framing. However, by choosing the framing to be the same as in the abelian case and calculating only ratios of observable quantities, it is possible to eliminate the framing freedom. This will enable us to distinguish the different knot classes after all.

16.8 Entangled Pair of Polymers

For a pair of polymers, the above problems with self-entanglement can be avoided by a slight modification of the Chern-Simons theory. This will allow us to study the entanglement of the pair. In particular, we shall be able to calculate the *second topological moment* of the entanglement, which is defined as the expectation value $\langle m^2 \rangle$ of the square of the linking number m [8]. The self-entanglements will of the individual polymers will be ignored.

The result will apply approximately to a polymer in an ensemble of many others, since these may be considered roughly as a single very long effective polymer.

Consider two polymers running along the contours C_1 and C_2 which statistically can be linked with each other any number of times $m = 0, 1, 2, \ldots$. The situation is illustrated in Fig. 16.22 for $m = 2$. The linking number (16.134) for these two polymers can be calculated with the help of

Figure 16.22 Closed polymers along the contours C_1, C_2 respectively.

a slight modification of the Chern-Simons action (16.153) and the couplings (16.152). We simply introduce two vector potentials and the Euclidean action

$$\mathcal{A}_e = \mathcal{A}_{e,\text{CS}12} + \mathcal{A}_{e,\text{curr}}, \tag{16.170}$$

where

$$\mathcal{A}_{e,\text{CS}12} = i \int d^3x\, \mathbf{A}_1 \cdot (\boldsymbol{\nabla} \times \mathbf{A}_2) \tag{16.171}$$

and

$$\mathcal{A}_{e,\text{curr}} = -i \oint_{C_1} d\mathbf{x}\mathbf{A}_1(\mathbf{x}) - i \oint_{C_2} d\mathbf{x}\mathbf{A}_2(\mathbf{x}). \tag{16.172}$$

If we choose the gauge fields to be transverse, as in (16.160), we obtain with the same technique as before the correlation functions of the gauge fields

$$D_{ab}^{\mu\nu}(\mathbf{x}, \mathbf{x}') \equiv \langle A_a^\mu(\mathbf{x}) A_b^\nu(\mathbf{x}') \rangle, \quad a, b = 1, 2 \tag{16.173}$$

are

$$D_{11}^{ij}(\mathbf{x}, \mathbf{x}') = 0, \qquad\qquad D_{22}^{ij}(\mathbf{x}, \mathbf{x}') = 0, \tag{16.174}$$

$$D_{12}^{ij}(\mathbf{x}, \mathbf{x}') = D_{21}^{ij}(\mathbf{x}, \mathbf{x}') = \int \frac{d^3 p}{(2\pi)^3} e^{i\mathbf{p}(\mathbf{x}-\mathbf{x}')} \frac{i\epsilon_{ikj}k^k}{\mathbf{p}^2} = \frac{1}{4\pi}\epsilon_{ikj}\nabla_k \frac{1}{|\mathbf{x}-\mathbf{x}'|}$$

$$= \frac{1}{4\pi}\epsilon_{ij\kappa}\frac{(x-x')^k}{|\mathbf{x}-\mathbf{x}'|^3}. \tag{16.175}$$

The transverse gauge is enforced by adding to the action (16.171) a gauge-fixing term

$$\mathcal{A}_{\mathrm{GF}} = \frac{1}{2\alpha}[(\boldsymbol{\nabla}\cdot\mathbf{A}_1)^2 + (\boldsymbol{\nabla}\cdot\mathbf{A}_2)^2], \tag{16.176}$$

and taking the limit $\alpha \to 0$ at the end. Extremizing the extended action produces the Gaussian linking number $-iG(C_1, C_2)$ of Eq. (16.134). The calculation is completely analogous to that leading to Eq. (16.168).

The partition function of the two polymers and the gauge fields is given by the path integral

$$Z = \int_{C_1} \mathcal{D}\mathbf{x}_1 \int_{C_2} \mathcal{D}\mathbf{x}_2 \int \mathcal{D}\mathbf{A}_1 \mathcal{D}\mathbf{A}_2 e^{-\mathcal{A}_{\mathrm{e}} - \mathcal{A}_{\mathrm{GF}}}. \tag{16.177}$$

Performing the functional integral over the gauge fields, we obtain from the extremum

$$Z = \mathrm{const} \times \int_{C_1} \mathcal{D}\mathbf{x}_1 \int_{C_2} \mathcal{D}\mathbf{x}_2 \, e^{iG(C_1, C_2)}, \tag{16.178}$$

where the constant is the trivial fluctuation factor of the vector potentials. The Gaussian integral $G(C_1, C_2)$ has the values $m = 0, \pm 1, \pm 2, \dots$ of the linking numbers.

In order to analyze the distribution of linking number in the two-polymer system we must be able to fix a certain linking number. This is possible by replacing the phase factor e^{im} in the path integral (16.178) by $e^{im\lambda}$, and calculating $Z(\kappa)$. An integral $\int d\kappa e^{-i\bar{m}\lambda} Z(\lambda)$ will then select any specific linking number $\bar{m}$. But a phase factor $e^{im\lambda}$ is simply produced in the partition function (16.178) by attaching to one of the current couplings in the interaction (16.172), say to that of C_2, a factor λ, thus changing (16.172) to

$$\mathcal{A}_{\mathrm{e,curr},\lambda} = -i\oint_{C_1} d\mathbf{x}\,\mathbf{A}_1(\mathbf{x}) - i\lambda\oint_{C_2} d\mathbf{x}\,\mathbf{A}_2(\mathbf{x}). \tag{16.179}$$

The λ-dependent partition function is then

$$Z(\lambda) = \int_{C_1} \mathcal{D}\mathbf{x}_1 \int_{C_2} \mathcal{D}\mathbf{x}_2 \int \mathcal{D}\mathbf{A}_1 \mathcal{D}\mathbf{A}_2 e^{-\mathcal{A}_{\mathrm{e,CS12}} - \mathcal{A}_{\mathrm{e,curr},\lambda} - \mathcal{A}_{\mathrm{GF}}}$$

$$= \mathrm{const} \times \int_{C_1} \mathcal{D}\mathbf{x}_1 \int_{C_2} \mathcal{D}\mathbf{x}_2 \, e^{im\lambda}. \tag{16.180}$$

Ultimately, we want to find the probability distribution of the linking numbers m as a function of the lengths of C_1 and C_2. The solution of this two-polymer problem may be considered as an approximation to a more interesting physical problem in which a particular polymer is linked to any number N of polymers, which are effectively replaced by a single long "effective" polymer [9]. Unfortunately, the full distribution of m is very hard to calculate. Only a calculation of the second topological moment is possible with limited effort. This quantity is given by the expectation value $\langle m^2 \rangle$ of the square of the linking number m.

Let $P_{L_1, L_2}(\mathbf{x}_1, \mathbf{x}_2; m)$ be the configurational probability to find the polymer C_1 of length L_1 with fixed coinciding endpoints at $\mathbf{x}_1$ and the polymer C_2 of length L_2 with fixed coinciding endpoints at $\mathbf{x}_2$, entangled with a Gaussian linking number m. The second moment $\langle m^2 \rangle$ is given by the ratio of integrals

$$\langle m^2 \rangle = \frac{\int d^3x_1 d^3x_2 \int_{-\infty}^{+\infty} dm \; m^2 P_{L_1,L_2}(\mathbf{x}_1,\mathbf{x}_2;m)}{\int d^3x_1 d^3x_2 \int_{-\infty}^{+\infty} dm \; P_{L_1,L_2}(\mathbf{x}_1,\mathbf{x}_2;m)}, \tag{16.181}$$

performed for either of the two probabilities. The integrations in $d^3\mathbf{x}_1 d^3\mathbf{x}_2$ covers all positions of the endpoints. The denominator plays the role of a partition function of the system:

$$Z \equiv \int d^3x_1 d^3x_2 \int_{-\infty}^{+\infty} dm \; P_{L_1,L_2}(\mathbf{x}_1,\mathbf{x}_2;m). \tag{16.182}$$

Due to translational invariance of the system, the probabilities depend only on the differences between the endpoint coordinates:

$$P_{L_1,L_2}(\mathbf{x}_1,\mathbf{x}_2;m) = P_{L_1,L_2}(\mathbf{x}_1 - \mathbf{x}_2;m). \tag{16.183}$$

Thus, after the shift of variables, the spatial double integrals in (16.181) can be rewritten as

$$\int d^3x_1 d^3x_2 P_{L_1,L_2}(\mathbf{x}_1,\mathbf{x}_2;m) = V \int d^3x \; P_{L_1,L_2}(\mathbf{x};m), \tag{16.184}$$

where V denotes the total volume of the system.

16.8.1 Polymer Field Theory for Probabilities

The calculation of the path integral over all line configurations is conveniently done within the polymer field theory developed in Section 15.12. It permits us to rewrite the partition function (16.180) as a functional integral over two $\psi_1^{\alpha_1}(\mathbf{x}_1)$ and $\psi_2^{\alpha_2}(\mathbf{x}_2)$ with n_1 and n_2 replica ($\alpha_1 = 1,\dots,n_1$, $\alpha_2 = 1,\dots,n_2$). At the end we shall take $n_1, n_2 \to 0$ to ensure that these fields describe only one polymer each, es explained in Section 15.12. For these fields we define an auxiliary probability $P_{\vec{z}}(\vec{\mathbf{x}}_1,\vec{\mathbf{x}}_2;\lambda)$ to find the polymer C_1 with open ends at $\mathbf{x}_1,\mathbf{x}_1'$ and the polymer C_2 with open ends at $\mathbf{x}_2,\mathbf{x}_2'$. The double vectors $\vec{\mathbf{x}}_1 \equiv (\mathbf{x}_1,\mathbf{x}_1')$ and $\vec{\mathbf{x}}_2 \equiv (\mathbf{x}_2,\mathbf{x}_2')$ collect initial and final endpoints of the two polymers C_1 and C_2. The auxiliary probability $P_{\vec{z}}(\vec{\mathbf{x}}_1,\vec{\mathbf{x}}_2;\lambda)$ is given by a functional integral

$$P_{\vec{z}}(\vec{\mathbf{x}}_1,\vec{\mathbf{x}}_2;\lambda) = \lim_{n_1,n_2 \to 0} \int \mathcal{D}(\text{fields}) \; \psi_1^{\alpha_1}(\mathbf{x}_1)\psi_1^{*\alpha_1}(\mathbf{x}_1')\psi_2^{\alpha_2}(\mathbf{x}_2)\psi_2^{*\alpha_2}(\mathbf{x}_2')e^{-\mathcal{A}}, \tag{16.185}$$

where $\mathcal{D}(\text{fields})$ indicates the measure of functional integration, and $\mathcal{A}$ the total action (16.180) governing the fluctuations. The expectation value is calculated for any fixed pair (α_1, α_2) of replica labels, i.e., replica labels are not subject to Einstein's summation convention of repeated indices. The action $\mathcal{A}$ consists of kinetic terms for the fields, a quartic interaction of the fields to account for the fact that two monomers of the polymers cannot occupy the same point, the so-called *excluded-volume effect*, and a Chern-Simons field to describe the linking number m. Neglecting at first the excluded-volume effect and focusing attention on the linking problem only, the action reads

$$\mathcal{A} = \mathcal{A}_{\text{CS12}} + \mathcal{A}_{\text{e,curr}} + \mathcal{A}_{\text{pol}} + \mathcal{A}_{\text{GF}}, \tag{16.186}$$

with a polymer field action

$$\mathcal{A}_{\text{pol}} = \sum_{i=1}^{2} \int d^3\mathbf{x} \left[|\bar{\mathbf{D}}^i \psi_i|^2 + m_i^2 |\Psi_i|^2 \right]. \tag{16.187}$$

They are coupled to the polymer fields by the covariant derivatives

$$\mathbf{D}^i = \mathbf{\nabla} + i\gamma_i \mathbf{A}^i, \tag{16.188}$$

with the coupling constants $\gamma_{1,2}$ given by

$$\gamma_1 = 1, \qquad \gamma_2 = \lambda. \tag{16.189}$$

The square masses of the polymer fields are given by

$$m_i^2 = 2M z_i. \tag{16.190}$$

where $M = 3/a$, with a being the length of the polymer links [recall (15.79)], and z_i the chemical potentials of the polymers, measured in units of the temperature. The chemical potentials are conjugate variables to the length parameters L_1 and L_2, respectively. The symbols Ψ_i collect the replica of the two polymer fields

$$\Psi_i = \left(\psi_i^1, \ldots, \psi_i^{n_i} \right), \tag{16.191}$$

and their absolute squares contain the sums over the replica

$$|\mathbf{D}^i \bar{\Psi}_i|^2 = \sum_{\alpha_i=1}^{n_i} |\mathbf{D}^i \psi_i^{\alpha_i}|^2, \qquad |\Psi_i|^2 = \sum_{\alpha_i=1}^{n_i} |\psi_i^{\alpha_i}|^2. \tag{16.192}$$

Having specified the fields, we can now write down the measure of functional integration in Eq. (16.185):

$$\mathcal{D}(\text{fields}) = \int \mathcal{D}A_1^i \mathcal{D}A_2^j \mathcal{D}\Psi_1 \mathcal{D}\Psi_1^* \mathcal{D}\Psi_2 \mathcal{D}\Psi_2^*. \tag{16.193}$$

By Eq. (16.180), the parameter λ is conjugate to the linking number m. We can therefore calculate the probability $P_{L_1,L_2}(\vec{x}_1, \vec{x}_2; m)$ in which the two polymers are open with different endpoints from the auxiliary one $P_{\vec{z}}(\vec{x}_1, \vec{x}_2; \lambda)$ by the following Laplace integral over $\vec{z} = (z_1, z_1)$:

$$P_{L_1,L_2}(\vec{x}_1, \vec{x}_2; m) = \lim_{\substack{\mathbf{x}_1' \to \mathbf{x}_1 \\ \mathbf{x}_2' \to \mathbf{x}_2}} \int_{c-i\infty}^{c+i\infty} \frac{M dz_1}{2\pi i} \frac{M dz_2}{2\pi i} e^{z_1 L_1 + z_2 L_2} \int_{-\infty}^{\infty} dk e^{-im\lambda} P_{\vec{z}}(\vec{x}_1, \vec{x}_2; \lambda).$$

$$\tag{16.194}$$

16.8.2 Calculation of Partition Function

Let us use the polymer field theory to calculate the partition function (16.182). By Eq. (16.194), it is given by the integral over the auxiliary probabilities

$$Z = \int d^3x_1 d^3x_2 \lim_{\substack{\mathbf{x}_1' \to \mathbf{x}_1 \\ \mathbf{x}_2' \to \mathbf{x}_2}} \int_{c-i\infty}^{c+i\infty} \frac{M dz_1}{2\pi i} \frac{M dz_2}{2\pi i} e^{z_1 L_1 + z_2 L_2} \int_{-\infty}^{\infty} dm \int_{-\infty}^{+\infty} d\lambda e^{-im\lambda} P_{\vec{z}}(\vec{x}_1, \vec{x}_2; \lambda).$$

$$\tag{16.195}$$

The integration over m is trivial and gives $2\pi\delta(\lambda)$, enforcing $\lambda = 0$, so that

$$Z = \int d^3x_1 d^3x_2 \lim_{\substack{\mathbf{x}_1' \to \mathbf{x}_1 \\ \mathbf{x}_2' \to \mathbf{x}_2}} \int_{c-i\infty}^{c+i\infty} \frac{M dz_1 M dz_2}{2\pi i} e^{z_1 L_1 + z_2 L_2} P_{\vec{z}}(\vec{x}_1, \vec{x}_2; 0). \tag{16.196}$$

To calculate $P_{\vec{z}}(\vec{x}_1, \vec{x}_2; 0)$, we observe that the action $\mathcal{A}$ in Eq. (16.186) depends on λ only via the polymer part (16.187), and is quadratic in λ. Let us expand $\mathcal{A}$ as

$$\mathcal{A} = \mathcal{A}_0 + \lambda \mathcal{A}_1 + \lambda^2 \mathcal{A}_2, \tag{16.197}$$

with the λ-independent part

$$\mathcal{A}_0 \equiv \mathcal{A}_{\mathrm{CS12}} + \mathcal{A}_{\mathrm{GF}} + \int d^3x \left[|\mathbf{D}_1 \Psi_1|^2 + |\boldsymbol{\nabla}\Psi_2|^2 + \sum_{i=1}^{2} |\Psi_i|^2 \right], \tag{16.198}$$

a linear coefficient

$$\mathcal{A}_1 \equiv \int d^3x \; \mathbf{j}_2(\mathbf{x}) \cdot \mathbf{A}_2(\mathbf{x}) \tag{16.199}$$

containing a pseudo-current of the second polymer field

$$\mathbf{j}_2(\mathbf{x}) = i\Psi_2^*(\mathbf{x})\boldsymbol{\nabla}\Psi_2(\mathbf{x}), \tag{16.200}$$

and a quadratic coefficient

$$\mathcal{A}_2 \equiv \frac{1}{4}\int d^3\mathbf{x} \; \mathbf{A}_2^2 |\Psi_2(\mathbf{x})|^2. \tag{16.201}$$

With these definitions we write with the help of (16.198):

$$P_{\vec{z}}(\vec{\mathbf{x}}_1, \vec{\mathbf{x}}_2; 0) = \int \mathcal{D}(\text{fields})e^{-\mathcal{A}_0}\psi_1^{\alpha_1}(\mathbf{x}_1)\psi_1^{*\alpha_1}(\mathbf{x}_1')\psi_2^{\alpha_2}(\mathbf{x}_2)\psi_2^{\alpha_2}(\mathbf{x}'). \tag{16.202}$$

In the action (16.198), the fields Ψ_2, Ψ_2^* are obviously free, whereas the fields Ψ_1, Ψ_1^* are apparently not because of the couplings with the Chern-Simons fields in the covariant derivative $\mathbf{D}^1$. This coupling is, however, without physical consequences. Indeed, by integrating out A_2^i in (16.202), we find from $\mathcal{A}_{\mathrm{CS12}}$ the flatness condition:

$$\boldsymbol{\nabla} \times \mathbf{A}_1 = 0. \tag{16.203}$$

On a flat space with vanishing boundary conditions at infinity this implies $\mathbf{A}_1 = 0$. As a consequence, the functional integral (16.202) factorizes as follows [compare (15.370)]

$$P_{\vec{z}}(\vec{\mathbf{x}}_1, \vec{\mathbf{x}}_2; 0) = G_0(\mathbf{x}_1 - \mathbf{x}_1'; z_1)G_0(\mathbf{x}_2 - \mathbf{x}_2'; z_2), \tag{16.204}$$

where $G_0(\mathbf{x}_i - \mathbf{x}_i'; z_i)$ are the free correlation functions of the polymer fields:

$$G_0(\mathbf{x}_i - \mathbf{x}_i'; z_i) = \langle \psi_i^{\alpha_i}(\mathbf{x}_i)\psi_i^{*\alpha_i}(\mathbf{x}_i')\rangle. \tag{16.205}$$

In momentum space, the correlation functions are

$$\langle \tilde{\psi}_i^{\alpha_i}(\mathbf{k}_i)\tilde{\psi}_i^{*\alpha_i}(\mathbf{k}_i')\rangle = \delta^{(3)}(\mathbf{k}_i - \mathbf{k}_i')\frac{1}{\mathbf{k}_i^2 + m_i^2}, \tag{16.206}$$

such that

$$G_0(\mathbf{x}_i - \mathbf{x}_i'; z_i) = \int \frac{d^3k}{(2\pi)^3}e^{i\mathbf{k}\cdot\mathbf{x}}\frac{1}{\mathbf{k}_i^2 + m_i^2}, \tag{16.207}$$

and

$$\begin{aligned} G_0(\mathbf{x}_i - \mathbf{x}_i'; L_i) &= \int_{c-i\infty}^{c+i\infty} \frac{M dz_i}{2\pi i}e^{z_i L_i}G_0(\mathbf{x}_i - \mathbf{x}_i'; z_i) \\ &= \frac{1}{2}\left(\frac{M}{4\pi L_i}\right)^{3/2}e^{-M(\mathbf{x}_i - \mathbf{x}_i')/2L_i}. \end{aligned} \tag{16.208}$$

The partition function (16.196) is then given by the integral

$$Z = 2\pi \int d^3x_1 d^3x_2 \lim_{\substack{\mathbf{x}_1' \to \mathbf{x}_1 \\ \mathbf{x}_2' \to \mathbf{x}_2}} G_0(\mathbf{x}_1 - \mathbf{x}_1'; L_1) G_0(\mathbf{x}_2 - \mathbf{x}_2'; L_2). \tag{16.209}$$

The integrals at coinciding endpoints can easily be performed, yielding

$$Z = \frac{2\pi M^3 V^2}{(8\pi)^3} (L_1 L_2)^{-3/2}. \tag{16.210}$$

It is important to realize that in Eq. (16.195) the limits of coinciding endpoints $\mathbf{x}_i' \to \mathbf{x}_i$ and the inverse Laplace transformations do *not* commute unless a proper renormalization scheme is chosen to eliminate the divergences caused by the insertion of the composite operators $|\psi^\alpha(\mathbf{x})|^2$. This can be seen for a single polymer. If we were to commuting the limit of coinciding endpoints with the Laplace transform, we would obtain

$$\int_{c-i\infty}^{c+i\infty} \frac{dz}{2\pi} e^{zL} \lim_{\mathbf{x}' \to \mathbf{x}} G_0(\mathbf{x} - \mathbf{x}'; z) = \int_{c-i\infty}^{c+i\infty} \frac{dz}{2\pi i} e^{zL} G_0(0, z), \tag{16.211}$$

where

$$G_0(0; z) = \langle |\psi(\mathbf{x})|^2 \rangle. \tag{16.212}$$

This expectation value, however, is linearly divergent:

$$\langle |\psi(\mathbf{x})|^2 \rangle = \int \frac{d^3k}{k^2 + m^2} \to \infty. \tag{16.213}$$

16.8.3   Calculation of Numerator in Second Moment

Let us now turn to the numerator in Eq. (16.181):

$$N \equiv \int d^3x_1 d^3x_2 \int_{-\infty}^{\infty} dm\, m^2\, P_{L_1, L_2}(\mathbf{x}_1, \mathbf{x}_2; m). \tag{16.214}$$

We shall set up a functional integral for N in terms of the auxiliary probability $P_{\vec{z}}(\vec{\mathbf{x}}_1, \vec{\mathbf{x}}_2; 0)$ analogous to Eq. (16.195). First we observe that

$$\begin{aligned} N = {}& \int d^3x_1 d^3r_2 \int_{-\infty}^{\infty} dm\, m^2 \lim_{\substack{\mathbf{x}_1' \to \mathbf{x}_1 \\ \mathbf{x}_2' \to \mathbf{x}_2}} \int_{c-i\infty}^{c_\tau i\infty} \frac{M dz_i}{2\pi i} \frac{M dz_2}{2\pi i} \\ &\times e^{z_1 L_1 + z_2 L_2} \int_{-\infty}^{\infty} d\lambda e^{-im\lambda} P_{\vec{z}}(\vec{\mathbf{x}}_1, \vec{\mathbf{x}}_2; \lambda). \end{aligned} \tag{16.215}$$

The integration in m is easily performed after noting that

$$\int_{-\infty}^{\infty} dm\, m^2 e^{-im\lambda} P_{\vec{z}}(\vec{\mathbf{x}}_1, \vec{\mathbf{x}}_2; \lambda) = -\int_{-\infty}^{\infty} dm \left(\frac{\partial^2}{\partial\lambda^2} e^{-im\lambda} \right) P_{\vec{z}}(\vec{\mathbf{x}}_1, \vec{\mathbf{x}}_2; \lambda). \tag{16.216}$$

After two integrations by parts in λ, and an integration in m, we obtain

$$\begin{aligned} N = {}& \int d^3x_1 d^3x_2 \lim_{\substack{\mathbf{x}_1' \to \mathbf{x}_1 \\ \mathbf{x}_2' \to \mathbf{x}_2}} (-1) \int_{c-i\infty}^{c+i\infty} \frac{M dz_1}{2\pi i} \frac{M dz_2}{2\pi i} e^{z_1 L_1 + z_2 L_2} \\ &\times \int_{-\infty}^{\infty} d\lambda\, \delta(\lambda) \left[\frac{\partial^2}{\partial\lambda^2} P_{\vec{z}}(\vec{\mathbf{x}}_1, \vec{\mathbf{x}}_2; \lambda) \right]. \end{aligned} \tag{16.217}$$

Performing the now the trivial integration over λ yields

$$N = \int d^3x_1 d^3x_2 \lim_{\substack{\mathbf{x}_1' \to \mathbf{x}_1 \\ \mathbf{x}_2' \to \mathbf{x}_2}} (-1) \int_{c-i\infty}^{c+i\infty} \frac{M dz_1}{2\pi i} \frac{M dz_2}{2\pi i} e^{z_1 L_1 + z_2 L_2} \left[\frac{\partial^2}{\partial\lambda^2} P_{\bar{z}}(\bar{\mathbf{x}}_1, \bar{\mathbf{x}}_2; 0) \right]. \qquad (16.218)$$

To compute the term in brackets, we use again (16.197) and Eqs. (16.198)–(16.223), to find

$$\begin{aligned}
N =\ & \int d^3x_1 d^3x_2 \lim_{\substack{n_1 \to 0 \\ n_2 \to 0}} \int_{c-i\infty}^{c+i\infty} \frac{M dz_1}{2\pi i} \frac{M dz_2}{2\pi i} e^{z_1 L_1 + z_2 L_2} \\
& \times \int \mathcal{D}(\text{fields})\ \exp(-\mathcal{A}_0) |\psi_1^{\alpha_1}(\mathbf{x}_1)|^2 |\psi_2^{\alpha_2}(\mathbf{x}_2)|^2 \\
& \times \left[\left(\int d^3x\, \mathbf{A}_2 \cdot \mathbf{\Psi}_2^* \mathbf{\nabla} \Psi_2 \right)^2 + \frac{1}{2} \int d^3x\, \mathbf{A}_2^2 |\Psi_2|^2 \right].
\end{aligned} \qquad (16.219)$$

In this equation we have taken the limits of coinciding endpoint inside the Laplace integral over z_1, z_2. This will be justified later on the grounds that the potentially dangerous Feynman diagrams containing the insertions of operations like $|\Psi_i|^2$ vanish in the limit $n_1, n_2 \to 0$.

In order to calculate (16.219), we decompose the action into a free part

$$\mathcal{A}_0^0 \equiv \mathcal{A}_{\text{CS}} + \int d^3x \left[|\mathbf{D}^1 \Psi_1|^2 + |\mathbf{\nabla} \Psi_2|^2 + \sum_{i=1}^{2} 2|\Psi_i|^2 \right], \qquad (16.220)$$

and interacting parts

$$\mathcal{A}_1^0 \equiv \int d^3x\, \mathbf{j}_1(\mathbf{x}) \cdot \mathbf{A}_1(\mathbf{x}), \qquad (16.221)$$

with a "current" of the first polymer field

$$\mathbf{j}_1(\mathbf{x}) \equiv i\Psi_1^*(\mathbf{x}) \mathbf{\nabla} \Psi_1(\mathbf{x}), \qquad (16.222)$$

and

$$\mathcal{A}_0^2 \equiv \frac{1}{4} \int d^3\mathbf{x}\, \mathbf{A}_1^2 |\Psi_1(\mathbf{x})|^2. \qquad (16.223)$$

Expanding the exponential

$$e^{\mathcal{A}_0} = e^{\mathcal{A}_0^0 + \mathcal{A}_0^1 + \mathcal{A}_0^2} = e^{\mathcal{A}_0} \left[1 - \mathcal{A}_0^1 + \frac{(\mathcal{A}_0^1)^2}{2} - \mathcal{A}_0^2 + \cdots \right], \qquad (16.224)$$

and keeping only the relevant terms, the functional integral (16.219) can be rewritten as a purely Gaussian expectation value

$$\begin{aligned}
N =\ & \kappa^2 \int d^3x_1 d^3x_2 \lim_{\substack{n_1 \to 0 \\ n_2 \to 0}} \int_{c-i\infty}^{c+i\infty} \frac{M dz_1}{2\pi i} \frac{M dz_2}{2\pi i} e^{z_1 L_1 + z_2 L_2} \\
& \times \int \mathcal{D}(\text{fields})\ \exp(-\mathcal{A}_0^0) |\psi_1^{\alpha_1}(\mathbf{x}_1)|^2 |\psi_2^{\alpha_2}(\mathbf{x}_2)|^2 \\
& \times \left[\left(\int d^3x\, \mathbf{A}_1 \cdot \mathbf{\Psi}_1^* \mathbf{\nabla} \Psi_1 \right)^2 + \frac{1}{2} \int d^3x\, \mathbf{A}_1^2 |\Psi_1|^2 \right] \\
& \times \left[\left(\int d^3x\, \mathbf{A}_2 \cdot \mathbf{\Psi}_2^* \mathbf{\nabla} \Psi_2 \right)^2 + \frac{1}{2} \int d^3x\, \mathbf{A}_2^2 |\Psi_2|^2 \right].
\end{aligned} \qquad (16.225)$$

Figure 16.23 Four diagrams contributing in Eq. (16.225). The lines indicate correlation functions of Ψ_i-fields. The crossed circles with label i denote the insertion of $|\Psi_i(\mathbf{x}_i)|^2$.

Note that the initially asymmetric treatment of polymers C_1 and C_2 in the action (16.187) has led to a completely symmetric expression for the second moment.

Only four diagrams shown in Fig. 16.23 contribute in Eq. (16.225). The first diagram is divergent due to the divergence of the loop formed by two vector correlation functions. This infinity may be absorbed in the four-Ψ interaction accounting for the excluded volume effect which we do not consider at the moment. We now calculate the four diagrams separately.

16.8.4 First Diagram in Fig. 16.23

From Eq. (16.225) one has to evaluate the following integral

$$N_1 = \frac{\kappa^2}{4} \lim_{\substack{n_1 \to 0 \\ n_2 \to 0}} \int_{c-i\infty}^{c+i\infty} \frac{M dz_1}{2\pi i} \frac{M dz_2}{2\pi i} e^{z_1 L_1 + z_2 L_2} \int d^3 x_1 d^3 x_2 \int d^3 x_1' d^3 x_2' \tag{16.226}$$

$$\times \left\langle |\psi_1^{\alpha_1}(\mathbf{x}_1)|^2 |\psi_2^{\alpha_2}(\mathbf{x}_2)|^2 \left(|\Psi_1|^2 \mathbf{A}_1^2 \right)_{\mathbf{x}_1'} \left(|\Psi_2|^2 \mathbf{A}_2^2 \right)_{\mathbf{x}_2'} \right\rangle .$$

As mentioned before, there is an ultraviolet-divergent contribution which must be regularized. The system has, of course, a microscopic scale, which is the size of the monomers. This, however, is not the appropriate short-distance scale to be uses here. The model treats the polymers as random chains. However, the monomers of a polymer in the laboratory are usually not freely movable, so that polymers have a certain stiffness. This gives rise to a certain persistence length ξ_0 over which a polymer is stiff. This length scale is increased to $\xi > \xi_0$ by the excluded-volume effects. This is the length scale which should be used as a proper physical short-distance cutoff. We may impose this cutoff by imagining the model as being defined on a simple cubic lattice of spacing ξ. This would, of course, make analytical calculations quite difficult. Still, as we shall see, it is possible to estimate the dependence of the integral N_1 and the others in the physically relevant limit in which the lengths of the polymers are much larger than the persistence length ξ.

An alternative and simpler regularization is based on cutting off all ultraviolet-divergent continuum integrals at distances smaller than ξ.

After such a regularization, the calculation of N_1 is rather straightforward. Replacing the expectation values by the Wick contractions corresponding to the first diagram in Fig. 16.23, and performing the integrals as shown in Appendix 16A, we obtain

$$N_1 = \frac{V}{4\pi} \frac{M^4}{(8\pi)^6} (L_1 L_2)^{-\frac{1}{2}} \int_0^1 ds \, [(1-s)s]^{-\frac{3}{2}} \int d^3 x e^{-M\mathbf{x}^2/2s(1-s)} \tag{16.227}$$

$$\times \int_0^1 dt \, [(1-t)t]^{-\frac{3}{2}} \int d^3 y e^{-M\mathbf{y}^2/2t(1-t)} \int d^3 x_1'' \frac{1}{|\mathbf{x}_1''|^4} .$$

The variables $\mathbf{x}$ and $\mathbf{y}$ have been rescaled with respect to the original ones in order to extract the behavior of N_1 in L_1 and L_2. As a consequence, the lattices where $\mathbf{x}$ and $\mathbf{y}$ are defined have now spacings $\xi/\sqrt{L_1}$ and $\xi/\sqrt{L_2}$ respectively.

The $\mathbf{x}, \mathbf{y}$ integrals may be explicitly computed in the physical limit $L_1, L_2 \gg \xi$, in which the above spacings become small. Moreover, it is possible to approximate the integral in $\mathbf{x}_1''$ with an integral over a continuous variable l and a cutoff in the ultraviolet region:

$$\int d^3 x_1'' \frac{1}{|\mathbf{x}_1''|^4} \quad \sim \quad 4\pi^2 \int_\xi^\infty \frac{dl}{l^2}. \tag{16.228}$$

After these approximations, we finally obtain

$$N_1 = V\pi^{1/2} \frac{M}{(4\pi)^3} (L_1 L_2)^{-1/2} \xi^{-1}. \tag{16.229}$$

16.8.5 Second and Third Diagrams in Fig. 16.23

Here we have to calculate

$$N_2 = \kappa^2 \lim_{\substack{n_1 \to 0 \\ n_2 \to 0}} \int_{c-i\infty}^{c+i\infty} \frac{M dz_1}{2\pi i} \frac{M dz_2}{2\pi i} e^{z_1 L_1 + z_2 L_2} \int d^3 x_1 d^3 x_2 \int d^3 x_1' d^3 x_1'' d^3 x_2'$$
$$\times \left\langle |\psi_1^{\alpha_1}(\mathbf{x}_1)|^2 |\psi_2^{\alpha_2}(\mathbf{x}_2)|^2 \left(\mathbf{A}_1 \cdot \mathbf{\Psi}_1^* \boldsymbol{\nabla} \mathbf{\Psi}_1 \right)_{\mathbf{x}_1'} \left(\mathbf{A}_1 \cdot \mathbf{\Psi}_1^* \boldsymbol{\nabla} \mathbf{\Psi}_1 \right)_{\mathbf{x}_1''} \left(\mathbf{A}_2^2 |\mathbf{\Psi}_2|^2 \right)_{\mathbf{x}_2'} \right\rangle. \tag{16.230}$$

The above amplitude has no ultraviolet divergence, so that no regularization is required. The Wick contractions pictured in the second Feynman diagrams of Fig. 16.23 lead to the integral

$$N_2 = -4\sqrt{2} V L_2^{-1/2} L_1^{-1} \frac{M^3}{\pi^6} \int_0^1 dt \int_0^t dt' C(t, t'), \tag{16.231}$$

where $C(t, t')$ is a function independent of L_1 and L_2:

$$C(t, t') = [(1-t)t'(t-t')]^{-3/2} \int d^3 x d^3 y d^3 z \, e^{-M(\mathbf{y}-\mathbf{x})^2/2(1-t)}$$
$$\times \left(\boldsymbol{\nabla}_\mathbf{y}^j e^{-M\mathbf{y}^2/2t'} \right) \left(\boldsymbol{\nabla}_\mathbf{x}^i e^{-M\mathbf{x}^2/2(t-t')} \right) \frac{[\delta_{ij}\mathbf{z} \cdot (\mathbf{z}+\mathbf{x}) - (z+x)_i \, z_j]}{|\mathbf{z}|^3 |\mathbf{z}+\mathbf{x}|^3}. \tag{16.232}$$

As in the previous section, the variables $\mathbf{x}, \mathbf{y}, \mathbf{z}$ have been rescaled with respect to the original ones in order to extract the behavior in L_1.

If the polymer lengths are much larger than the persistence length one can ignore the fact that the monomers have a finite size and it is possible to compute $C(t, t')$ analytically, leading to

$$N_2 = -\frac{V L_2^{-1/2} L_1^{-1}}{(2\pi)^6} M^{3/2} 4K, \tag{16.233}$$

where K is the constant

$$K \equiv \frac{1}{6} B\left(\frac{3}{2}, \frac{1}{2}\right) + \frac{1}{2} B\left(\frac{5}{2}, \frac{1}{2}\right) - B\left(\frac{7}{2}, \frac{1}{2}\right) + \frac{1}{3} B\left(\frac{9}{2}, \frac{1}{2}\right) = \frac{19\pi}{384} \approx 0.154, \tag{16.234}$$

and $B(a, b) = \Gamma(a)\Gamma(b)/\Gamma(a+b)$ is the Beta function. For large $L_1 \to \infty$, this diagram gives a negligible contribution with respect to N_1.

The third diagram in Fig. 16.23 give the same as the second, except that L_1 and L_2 are interchanged.

$$N_3 = N_2|_{L_1 \leftrightarrow L_2}. \tag{16.235}$$

16.8.6   Fourth Diagram in Fig. 16.23

Here we have the integral

$$
N_4 = -4\kappa^2 \frac{1}{2} \lim_{\substack{n_1 \to 0 \\ n_2 \to 0}} \int_{c-i\infty}^{c+i\infty} \frac{M dz_1}{2\pi i} \frac{M dz_2}{2\pi i} e^{z_1 L_1 + z_2 L_2} \int d^3 x_1 d^3 x_2 \int d^3 x_1' d^3 x_2' d^3 x_1'' d^3 x_2''
$$

$$
\times \left\langle |\psi_1^{\alpha_1}(\mathbf{x}_1)|^2 |\psi_2(\mathbf{x}_2^{\alpha_2})|^2 (\mathbf{A}_1 \cdot \Psi_1^* \boldsymbol{\nabla} \Psi_1)_{\mathbf{x}_1'} (\mathbf{A}_1 \cdot \Psi_1^* \boldsymbol{\nabla} \Psi_1)_{\mathbf{x}_1''} \right.
$$

$$
\left. \times (\mathbf{A}_2 \cdot \Psi_2^* \boldsymbol{\nabla} \Psi_2)_{\mathbf{x}_2'} (\mathbf{A}_2 \cdot \Psi_2^* \boldsymbol{\nabla} \Psi_2)_{\mathbf{x}_2''} \right\rangle, \tag{16.236}
$$

which has no ultraviolet divergence. After some analytic effort we find

$$
N_4 = -\frac{1}{16} \frac{M^5 V}{(2\pi)^{11}} (L_1 L_2)^{-1/2} \int_0^1 ds \int_0^s ds' \int_0^1 dt \int_0^t dt' C(s, s', t, t'), \tag{16.237}
$$

where

$$
C(s, s'; t, t') = \left[(1-s)s'(s-s') \right]^{-3/2} \left[(1-t)t'(t-t') \right]^{-3/2}
$$

$$
\times \int \frac{d^3 p}{(2\pi)^3} \left[\epsilon_{ik\alpha} \frac{p^\alpha}{\mathbf{p}^2} \epsilon_{jl\beta} \frac{p^\beta}{\mathbf{p}^2} + \epsilon_{il\alpha} \frac{p^\alpha}{\mathbf{p}^2} \epsilon_{jk\beta} \frac{p^\beta}{\mathbf{p}^2} \right] \tag{16.238}
$$

$$
\times \left[\int d^3 x' d^3 y' e^{-i\sqrt{L_1}\mathbf{p}(\mathbf{x'}-\mathbf{y'})} e^{-M\mathbf{x'}^2/2(1-s)} \left(\boldsymbol{\nabla}_{\mathbf{y'}}^j e^{-M\mathbf{y'}^2/2t'} \right) \left(\boldsymbol{\nabla}_{\mathbf{x'}}^i e^{-M(\mathbf{x}-\mathbf{y})^2/2(s-s')} \right) \right]
$$

$$
\times \left[\int d^3 u' d^3 v' e^{-i\sqrt{L_2}\mathbf{p}(\mathbf{u'}-\mathbf{v'})} e^{-M\mathbf{v'}^2/2(1-t)} \left(\boldsymbol{\nabla}_{\mathbf{u'}}^l e^{-M\mathbf{u'}^2/2t'} \right) \left(\boldsymbol{\nabla}_{\mathbf{v'}}^k e^{-M(\mathbf{u'}-\mathbf{v'})^2/2(t-t')} \right) \right],
$$

and $\mathbf{x'}, \mathbf{y'}$ are scaled variables. To take into account the finite persistence length, they should be defined on a lattice with spacing $\xi/\sqrt{L_1}$. Similarly, $\mathbf{u'}, \mathbf{v'}$ should be considered on a lattice with spacing $\xi/\sqrt{L_2}$. Without performing the space integrations $d^3 x' d^3 y' d^3 u' d^3 v'$, the behavior of N_4 as a function of the polymer lengths can be easily estimated in the following limits:

1. $L_1 \gg 1; L_1 \gg L_2$

$$
N_4 \propto L_1^{-1} \tag{16.239}
$$

2. $L_2 \gg 1; L_2 \gg L_1$

$$
N_4 \propto L_2^{-1} \tag{16.240}
$$

3. $L_1, L_2 \gg 1, \quad L_2/L_1 = \alpha = \text{finite}$

$$
N_4 \propto L_1^{-3/2}. \tag{16.241}
$$

Moreover, if the lengths of the polymers are considerably larger than the persistence length, the function $C(s, s', t, t')$ can be computed in a closed form:

$$
N_4 \approx -\frac{128V}{\pi^5} \frac{M}{\pi^{3/2}} (L_1 L_2)^{-1/2} \int_0^1 ds \int_0^1 dt (1-s)(1-t)(st)^{1/2}
$$

$$
\times [L_1 t(1-s) + L_2(1-t)s]^{-1/2}. \tag{16.242}
$$

It is simple to check that this expression has exactly the above behaviors.

16.8.7 Second Topological Moment

Collecting all contributions we obtain the result for the second topological moment:

$$\langle m^2 \rangle = \frac{N_1 + N_2 + N_3 + N_4}{Z}, \tag{16.243}$$

with N_1, N_2, N_3, N_4, Z given by Eqs. (16.210), (16.229), (16.233), (16.235), and (16.231).

In all formulas, we have assumed that the volume V of the system is much larger than the size of the volume occupied by a single polymer, i.e., $V \gg L_1^3$

To discuss the physical content of the result (16.232), we assume C_2 to be a long effective polymer representing all polymers in a uniform solution. We introduce the polymer concentration l as the average mass density of the polymers per unit volume:

$$l = \frac{M}{V}, \tag{16.244}$$

where M is the total mass of the polymers

$$M = \sum_{i=1}^{N_p} m_a \frac{L_k}{a}. \tag{16.245}$$

Here m_a is the mass of a single monomer of length a, L_k is the length of polymer C_k, and N_p is the total number of polymers. Thus L_k/a is the number of monomers in the polymer C_k. The polymer C_1 is singled out as any of the polymers C_k, say $C_{\bar{k}}$, of length $L_1 = L_{\bar{k}}$. The remaining ones are replaced by a long effective polymer C_2 of length $L_2 = \sigma_{k \neq \bar{k}} L_k$. From the above relations we may also write

$$L_2 \approx \frac{aVl}{m_a}. \tag{16.246}$$

In this way, the length of the effective molecule C_2 is expressed in terms of physical parameters, the concentration of polymers, the monomer length, and the mass and volume of the system. Inserting (16.246) into (16.232), with N_1, N_2, N_3, N_4, Z given by Eqs. (16.210), (16.229), (16.233), (16.235), and (16.231). and keeping only the leading terms for $V \gg 1$, we find for the second topological moment $\langle m^2 \rangle$ the approximation

$$\langle m^2 \rangle \approx \frac{N_1 + N_2}{Z}, \tag{16.247}$$

and this has the approximate form

$$\langle m^2 \rangle = \frac{al}{m_a} \left[\frac{\xi^{-1} L_i}{2\pi^{1/2} M^2} - \frac{2K L_1^{1/2}}{\pi^4 M^{3/2}} \right], \tag{16.248}$$

with K of (16.234).

Thus we have succeeded in setting up a topological field theory to describe two fluctuating polymers C_1 and C_2, and calculated the second topological moment for the linking number m between C_1 and C_2. The result is used as an approximation for the second moment for a single polymer with respect to all others in a solution of many polymers.

An interesting remaining problem is to calculate the effect of the excluded volume.

16.9 Chern-Simons Theory of Statistical Interaction

The Chern-Simons theory (16.153) together with the coupling (16.152) generates the desired topological interaction corresponding to the Gaussian integral between pairs of curves C and C'. We now demonstrate that this topological interaction is the same as the statistical interaction introduced in Eq. (7.279) and encountered again in (16.26), where it governed the physics of a charged particle running around a magnetic flux tube. This observation will make the Gaussian integral and thus the Chern-Simons action relevant for the description of the statistical properties of nonrelativistic particle orbits. In contrast to the electromagnetic generation of fractional statistics for particle orbits in Section 16.2 via the Aharonov-Bohm effect, the field theory involving the statisto-magnetic vector potential has the advantage of removing the asymmetry between the particles, of which one had to carry a charge, the other one a magnetic flux. An arbitrary number of identical particle orbits can now be endowed with a fractional statistics, the same that was produced by topological interaction (7.279).

To prove the equality between the two topological interactions in two space and one time dimension, consider an electron in a plane encircling an infinitely thin magnetic "flux tube" at the origin (the word flux tube stands between quotation marks since the "tube" is only a point in the two-dimensional space). In a Euclidean spacetime, the worldline of the electron C winds itself around the straight "flux tube" C' along the τ-axis. For this geometry, the integral over C' in (16.134) can easily be done using the formula $\int_{-\infty}^{\infty} dt/\sqrt{t^2 + d^2}^3 = 2$. The result is

$$G(C, C') = \frac{1}{2\pi} \int d\tau \, \dot{\mathbf{x}}(\tau) \boldsymbol{\nabla} \varphi(\mathbf{x}(\tau)) = \frac{1}{2\pi} \int d\tau \, \dot{\varphi}(\mathbf{x}(\tau)), \qquad (16.249)$$

where $\varphi(\mathbf{x}(\tau))$ denotes the azimuthal angle of the electron with respect to the "flux tube" at the time τ

Up to a factor 2π, the expression (16.249) agrees with the statistical interaction in (16.26) and (7.279). In two space and one time dimension, the behavior under particle exchange can be assigned to an amplitude at will by adding to the Euclidean action a Gaussian integral with a suitable prefactor. A phase factor $e^{i\theta}$ is produced by the exchange when choosing the following Euclidean action-at-a-distance:

$$\mathcal{A}_{\mathrm{e,int}} = i2\hbar\theta G(C, C'). \qquad (16.250)$$

This topological interaction is generated by the Chern-Simons action

$$\mathcal{A}_{\mathrm{e,CS}} = \frac{1}{4\theta\hbar i} \int d^3x \, \mathbf{A} \cdot (\boldsymbol{\nabla} \times \mathbf{A}). \qquad (16.251)$$

The phase angle θ is related to the former parameter μ_0 of the statistical interactions (16.26) and (7.279) by

$$\theta = \pi\mu_0. \qquad (16.252)$$

For $\mu_0 = \pm 1, \pm 3, \pm 5, \ldots$, the particle orbits have Fermi statistics; for $\mu_0 = 0, \pm 2, \pm 4, \ldots$, they have Bose statistics. Fractional values of μ_0 lead to fractional statistics. In contrast to the magnetic generation of fractional statistics, the Chern-Simons mechanism applies to any number of particle orbits. By one of the methods discussed after Eq. (16.169) it must, however, be assured that the Gaussian "self-energy" does not render any undesirable nontopological contributions.

To maintain the analogy with the magnetic interactions as far as possible, we write the Chern-Simons action for a gas of electrons in the form

$$\mathcal{A}_{\text{e,CS}} = \frac{1}{4\pi i} \frac{e^2}{c^2 \hbar \mu_0} \int d^3x \, \mathbf{A} \times (\boldsymbol{\nabla} \times \mathbf{A}), \tag{16.253}$$

and the coupling of the statisto-magnetic vector potential to the particle orbits as

$$\mathcal{A}_{\text{e,curr}} = -i\frac{e}{c} \oint_C d\mathbf{x} \, \mathbf{A}(\mathbf{x}) - i\frac{e}{c} \oint_{C'} d\mathbf{x} \, \mathbf{A}(\mathbf{x}). \tag{16.254}$$

This looks precisely like the Euclidean coupling of an ordinary vector potential to electrons. For an arbitrary number of orbits we define the Euclidean two-dimensional current density

$$\mathbf{j}(\mathbf{x}) \equiv ec \sum_\alpha \oint_{C_\alpha} d\mathbf{x}_\alpha \, \delta^{(3)}(\mathbf{x} - \mathbf{x}_\alpha) \tag{16.255}$$

and write the interaction (16.254) as

$$\mathcal{A}_{\text{e,curr}} = -i\frac{1}{c^2} \int d^3x \, \mathbf{j}(\mathbf{x})\mathbf{A}(\mathbf{x}). \tag{16.256}$$

The curl of the vector potential

$$\mathbf{B} \equiv \boldsymbol{\nabla} \times \mathbf{A} \tag{16.257}$$

is referred to as a *statisto-magnetic field*. By varying (16.253) plus (16.256) with respect to $\mathbf{A}(\mathbf{x})$, we obtain the field equation

$$\mathbf{B}(\mathbf{x}) = \mu_0 \frac{2\pi\hbar c}{e^2} \, \mathbf{j}(\mathbf{x}). \tag{16.258}$$

With the help of the elementary flux quantum Φ_0, this can also be written as

$$\mathbf{B}(\mathbf{x}) = \mu_0 \Phi_0 \frac{1}{e} \mathbf{j}(\mathbf{x}). \tag{16.259}$$

To apply the above formulas, we must transform them from the three-dimensional Euclidean spacetime to the Minkowski space, where the curves C_α become particle orbits in two space dimensions whose coordinates $\mathbf{x}_\perp = (x, y)$ are functions of the time t. Specifically, we substitute the three coordinates (x_1, x_2, x_3) as follows:

$$(x_1, x_2) \;\rightarrow\; \mathbf{x}_\perp \equiv (x, y),$$
$$x_3 \;\rightarrow\; ix_0 \equiv ict.$$

The Euclidean field components $A_{1,2,3}$ go over into the Minkowskian *statisto-electric potential* ϕ and two spatial components $A_{x,y}$. The three fields $B_{1,2,3}$ turn into the Minkowskian *statisto-electric fields* E_y, E_x and a statisto-magnetic field B_z:

$$A_3 = i\phi = iA_0, \quad A_1 = A_x, \quad A_2 = A_y,$$
$$B_3 = iB_z, \quad B_1 = -iE_y, \quad B_2 = iE_x. \tag{16.260}$$

The Euclidean currents become, up to a factor i, the two-dimensional charge and current densities:

$$j_3 = ij_0 = ic\rho(\mathbf{x}_\perp), \quad \rho(\mathbf{x}) \equiv e\sum_\alpha \delta^{(2)}(\mathbf{x}_\perp - \mathbf{x}_{\perp\alpha}),$$

$$j_1 = ij_x(\mathbf{x}_\perp) = e\sum_\alpha \dot{x}_\alpha \delta^{(2)}(\mathbf{x}_\perp - \mathbf{x}_{\perp\alpha}),$$

$$j_2 = ij_y(\mathbf{x}_\perp) = e\sum_\alpha \dot{y}_\alpha \delta^{(2)}(\mathbf{x}_\perp - \mathbf{x}_{\perp\alpha}), \tag{16.261}$$

where ρ is the particle density per unit area. The motion of a particle in an external field ϕ, A_x, A_y is then governed by the interaction

$$\mathcal{A}_{\text{int}} = \int dt d^2x \left[\rho\phi - \frac{1}{c}(j_x A_x + j_x A_y) \right]. \tag{16.262}$$

Conversely, particles with fractional statistics in a 2+1-dimensional spacetime generate Minkowskian *statisto-electromagnetic fields* following the equations

$$B_z = \mu_0 \Phi_0 \rho, \quad E_x = \mu_0 \Phi_0 \frac{1}{c} j_y, \quad E_y = \mu_0 \Phi_0 \frac{1}{c} j_x. \tag{16.263}$$

The electromagnetic normalization in Eq. (16.262) has the advantage that a charged particle cannot distinguish a statisto-magnetic field from a true magnetic field. This property forms the basis for a simple interpretation of the fractional quantum Hall effect as will be seen in Section 18.9.

16.10   Second-Quantized Anyon Fields

After the developments in Chapter 7, we should expect that the phase factor $e^{i\mu_0\pi}$, appearing in the path integral upon exchanging the endpoints of two anyonic orbits, can also be found in a second-quantized operator formulation. To verify this, we consider a free Bose field with the action (7.286) and couple it with a statisto-electromagnetic field subject to a Chern-Simons action. The resulting free anyon action reads

$$\mathcal{A}_{\text{anyon}} = \mathcal{A}_{\text{CS}} + \mathcal{A}_{\text{boson}}, \tag{16.264}$$

where

$$\mathcal{A}_{\text{boson}} = \int d^2x \int_{t_a}^{t_b} dt \left\{ \psi^*(\mathbf{x}, t) \left[i\hbar \left(\partial_t + i\frac{e}{\hbar}\phi(\mathbf{x}, t) \right) + \mu \right] \psi(\mathbf{x}, t) \right.$$
$$\left. - \frac{\hbar^2}{2M} \left| \left[\boldsymbol{\nabla} - i\frac{e}{\hbar c}\mathbf{A}(\mathbf{x}, t) \right] \psi(\mathbf{x}, t) \right|^2 \right\}. \tag{16.265}$$

The latter corresponds to the action (7.286) in the continuum limit with the derivatives replaced by covariant derivatives according to the usual minimal substitution rule (2.642):

$$p_0 \to p_0 - \frac{e}{c}\phi, \quad \mathbf{p} \to \mathbf{p} - \frac{e}{c}\mathbf{A}. \tag{16.266}$$

The first field equation in (16.263) now reads

$$B_z(\mathbf{x}, t) = \mu_0 \Phi_0 \psi^\dagger(\mathbf{x}, t)\psi(\mathbf{x}, t), \tag{16.267}$$

so that the vector potential satisfies the differential equation

$$\partial_x A_y(\mathbf{x}, t) - \partial_y A_x(\mathbf{x}, t) = \mu_0 \Phi_0 \psi^\dagger(\mathbf{x}, t)\psi(\mathbf{x}, t). \tag{16.268}$$

It determines (A_x, A_y) up to the gauge freedom $(\partial_x \Lambda, \partial_y \Lambda)$, where $\Lambda(\mathbf{x}, t)$ is an arbitrary single-valued function satisfying Schwarz' integrability condition

$$(\partial_x \partial_y - \partial_y \partial_x)\Lambda(\mathbf{x}, t) = 0. \tag{16.269}$$

In the present case, it is useful to allow for a violation of this condition by searching for a multivalued function $\alpha(\mathbf{x}, t)$ whose gradient is equal to a given vector potential:

$$(A_x, A_y) = (\partial_x \alpha, \partial_y \alpha). \tag{16.270}$$

This function must obey the differential equation (recall the discussion in Appendix 10A)

$$(\partial_x \partial_y - \partial_y \partial_x)\alpha(\mathbf{x}, t) = \mu_0 \frac{2\pi\hbar c}{e}\psi^\dagger(\mathbf{x}, t)\psi(\mathbf{x}, t). \tag{16.271}$$

The Green function of this differential equation, which is the elementary building block for the construction of all multivalued functions in two dimensions, is the function used before in Eq. (16.249) [see also (10A.30)]:

$$\varphi(\mathbf{x} - \mathbf{x}') \equiv \arctan[(y - y')/(x - x')]. \tag{16.272}$$

It gives the angle between the vectors $\mathbf{x}$ and $\mathbf{x}'$ and violates the Schwarz integrability condition at the points where the vectors coincide [recall (10A.33)]:

$$(\partial_x \partial_y - \partial_y \partial_x)\varphi(\mathbf{x} - \mathbf{x}') = 2\pi\delta^{(2)}(\mathbf{x} - \mathbf{x}'). \tag{16.273}$$

To satisfy this equation, the cut of the arctan in the complex plane must be avoided, which is always possible by deforming it appropriately. The function $\varphi(\mathbf{x} - \mathbf{x}')$ has the important property

$$\varphi(\mathbf{x} - \mathbf{x}') - \varphi(\mathbf{x}' - \mathbf{x}) = \pi. \tag{16.274}$$

In the two terms on the left-hand side, the point $\mathbf{x}'$ is moved around the point $\mathbf{x}$ in the anticlockwise sense.

With the help of the multivalued Green function (16.272) we find immediately the solution of Eq. (16.271):

$$\alpha(\mathbf{x}, t) = \mu_0 \frac{\hbar c}{e} \int d^2 x \varphi(\mathbf{x} - \mathbf{x}') \psi^\dagger(\mathbf{x}, t) \psi(\mathbf{x}, t). \tag{16.275}$$

The relation (16.270) permits the elimination of the statisto-magnetic field from the action. Actually, this statement is true in general. One can always multiply the fields $\psi(\mathbf{x}, t)$ by a phase factor

$$\exp\left[-i\frac{e}{\hbar c} \int^{\mathbf{x}} d\mathbf{x}' \mathbf{A}(\mathbf{x}', t)\right],$$

where the contour integral is taken to $\mathbf{x}$ from any fixed point along some fixed path. In front of the transformed field

$$\Psi(\mathbf{x}, t) = e^{-i(e/\hbar c) \int^{\mathbf{x}} d\mathbf{x}' \mathbf{A}(\mathbf{x}', t)} \psi(\mathbf{x}, t), \tag{16.276}$$

the covariant derivatives

$$D_i \psi(\mathbf{x}, t) = (\partial_i - i\frac{e}{\hbar c} A_i) \psi(\mathbf{x}, t) \tag{16.277}$$

become ordinary derivatives ∂_i. Unfortunately, the new field $\Psi(\mathbf{x}, t)$ depends on the vector potential in a complicated nonlocal way so that this transformation is in general not worthwhile. In the present case, however, the equation of motion for the vector potential is so simple that the transformation *can* be done explicitly. In fact, the nonlocality has precisely the desired property of changing the statistics of the fields from Bose statistics to any statistics.

We show this by considering the field operators $\hat{\psi}(\mathbf{x}, t)$ which are canonically quantized according to Eq. (7.294). In the continuum limit they satisfy the commutation rules

$$
\begin{aligned}
[\hat{\psi}(\mathbf{x}, t), \hat{\psi}^\dagger(\mathbf{x}', t)] &= \delta^{(2)}(\mathbf{x} - \mathbf{x}'), \\
[\hat{\psi}^\dagger(\mathbf{x}, t), \hat{\psi}^\dagger(\mathbf{x}', t)] &= 0, \\
[\hat{\psi}(\mathbf{x}, t), \hat{\psi}(\mathbf{x}', t)] &= 0.
\end{aligned}
\tag{16.278}
$$

The transformed field operators satisfy the corresponding commutation rules modified by a phase factor $e^{i\mu_0 \pi}$:

$$
\begin{aligned}
\hat{\psi}(\mathbf{x}, t)\hat{\psi}^\dagger(\mathbf{x}', t) - e^{i\pi\mu_0}\hat{\psi}^\dagger(\mathbf{x}', t)\hat{\psi}(\mathbf{x}, t) &= \delta^{(2)}(\mathbf{x} - \mathbf{x}'), \\
\hat{\psi}^\dagger(\mathbf{x}, t)\hat{\psi}^\dagger(\mathbf{x}', t) - e^{i\pi\mu_0}\hat{\psi}^\dagger(\mathbf{x}', t)\hat{\psi}^\dagger(\mathbf{x}, t) &= 0, \\
\hat{\psi}(\mathbf{x}, t)\hat{\psi}(\mathbf{x}', t) - e^{i\pi\mu_0}\hat{\psi}(\mathbf{x}', t)\hat{\psi}(\mathbf{x}, t), &= 0.
\end{aligned}
\tag{16.279}
$$

As in (16.274), the vector $\mathbf{x}'$ on the left-hand side has to be carried around $\mathbf{x}$ in the anticlockwise sense. Using the relation (16.270), the integral in the prefactor of

(16.276) can immediately be performed and the transformed fields are simply given by

$$\tilde{\psi}(\mathbf{x}, t) = e^{-i\frac{e}{\hbar c}\alpha(\mathbf{x},t)}\psi(\mathbf{x}, t), \quad \tilde{\psi}^\dagger(\mathbf{x}, t) = \psi(\mathbf{x}, t)e^{i\frac{e}{\hbar c}\alpha(\mathbf{x},t)}. \tag{16.280}$$

The same relations hold for the second-quantized field operators. This makes it quite simple to prove the commutation rules (16.279). We do this here only for the second rule which controls the behavior of the many-body wave functions under the exchange of any two-particle coordinates:

$$\hat{\psi}^\dagger(\mathbf{x}, t)\hat{\psi}^\dagger(\mathbf{x}', t) - e^{i\pi\mu_0}\hat{\psi}^\dagger(\mathbf{x}', t)\hat{\psi}^\dagger(\mathbf{x}, t) = 0.$$

This amounts to the relation

$$\hat{\psi}^\dagger(\mathbf{x}, t)e^{i\frac{e}{\hbar c}\hat{\alpha}(\mathbf{x},t)}\hat{\psi}^\dagger(\mathbf{x}', t)e^{i\frac{e}{\hbar c}\hat{\alpha}(\mathbf{x}',t)} = e^{i\pi\mu_0}\hat{\psi}^\dagger(\mathbf{x}', t)e^{i\frac{e}{\hbar c}\hat{\alpha}(\mathbf{x}',t)}\hat{\psi}^\dagger(\mathbf{x}, t)e^{i\frac{e}{\hbar c}\hat{\alpha}(\mathbf{x},t)}. \tag{16.281}$$

The phase factors in the middle can be taken to the right-hand side by using the transformation formula

$$e^{i\int d^2x' f(\mathbf{x}',t)\hat{\psi}^\dagger(\mathbf{x},t)\hat{\psi}(\mathbf{x},t)}\hat{\psi}^\dagger(\mathbf{x}, t)e^{-i\int d^2x' f(\mathbf{x}',t)\hat{\psi}^\dagger(\mathbf{x},t)\hat{\psi}(\mathbf{x},t)} = e^{if(\mathbf{x},t)}\hat{\psi}^\dagger(\mathbf{x}, t), \tag{16.282}$$

which follows from the Lie expansion [recall (1.432)]

$$e^{i\hat{A}}\hat{B}e^{-i\hat{A}} = 1 + i[\hat{A}, \hat{B}] + \frac{i^2}{2!}[\hat{A}, [\hat{A}, \hat{B}]] + \dots . \tag{16.283}$$

Setting $f(\mathbf{x})$ equal to $\mu_0\varphi(\mathbf{x} - \mathbf{x}')$, Eq. (16.281) goes over into

$$\hat{\psi}^\dagger(\mathbf{x}, t)\hat{\psi}^\dagger(\mathbf{x}', t)e^{i\mu_0\varphi(\mathbf{x}-\mathbf{x}')}e^{i\frac{e}{\hbar c}[\hat{\alpha}(\mathbf{x},t)+\hat{\alpha}(\mathbf{x}',t)]}$$
$$= e^{i\pi\mu_0}\hat{\psi}^\dagger(\mathbf{x}', t)\hat{\psi}^\dagger(\mathbf{x}, t)e^{i\mu_0\varphi(\mathbf{x}'-\mathbf{x})}e^{i\frac{e}{\hbar c}[\hat{\alpha}(\mathbf{x}',t)+\hat{\alpha}(\mathbf{x},t)]}. \tag{16.284}$$

The correctness of this equation follows directly from the property (16.274) of the $\varphi(\mathbf{x})$ field and from the commutativity of the Bose fields $\psi^\dagger(\mathbf{x}, t)$ with each other. This proves the second of the anyon commutation rules (16.279). The others are obtained similarly.

Note that we could just as well have constructed the anyon fields from Fermi fields by shifting the exchange phase by an angle π.

16.11 Fractional Quantum Hall Effect

If particles obeying fractional statistics move in an ordinary magnetic field, they are also subject to a *statisto-magnetic field*. As observed earlier, this acts upon each particle in the same way as an additional true magnetic field. This observation provides a key for the understanding of the *fractional quantum Hall effect*. The arguments will now be sketched.

To measure the effect experimentally, a thin slab of conducting material (the original experiment used the compound $Al_xGa_{1-x}As$) is placed at low temperatures

(≈ 0.5 K) in the xy-plane traversed by a strong magnetic field B_z (between 10 and 200 kG) along the z-axis. An electric field E_x is applied in the x-direction and an electric *Hall current* j_y per length unit is measured in the y-direction. Such a current is expected in a dissipative electron gas with a number density (per unit area) ρ, where the fields satisfy the relation

$$E_x = -\frac{1}{\rho ec} j_y B_z \qquad (16.285)$$

(see Appendix 16E). The transverse resistance defined by

$$R_{xy} \equiv \frac{B_z}{\rho ec} \qquad (16.286)$$

rises linearly in B_z. Its dimension is sec/cm. In contrast with this naive expectation, the experimental data for R_{xy} rise stepwise with a number of plateaus whose resistance take the values $h/e^2\nu$, where ν is a rational number with odd denominators:

$$\nu = \tfrac{1}{5}, \tfrac{2}{7}, \tfrac{1}{3}, \tfrac{2}{5}, \tfrac{2}{3}, \ldots \ . \qquad (16.287)$$

We have omitted the observed number $\nu = \tfrac{5}{2}$ since its theoretical explanation requires additional physical considerations (see the references at the end of the chapter).

Similar plateaus had been observed at integer values of ν. Those are explained as follows.

In an ideal Fermi liquid at zero temperature, the electron orbits have energies $\mathbf{p}^2/2M$. Their momenta fill a Fermi sphere of radius p_F. The size of p_F is determined from the particle number per unit area ρ via the phase space integral

$$\rho L_x L_y = 2 \times \int \frac{dp_x dp_y L_x L_y}{(2\pi\hbar)^2}, \qquad (16.288)$$

where L_x and L_y are the lengths of the rectangular layered material in the x- and y-directions. The factor 2 accounts for the two spin orientations. The rotationally invariant integration up to p_F yields

$$p_F = \sqrt{2\pi\rho}\,\hbar. \qquad (16.289)$$

By switching on a magnetic field B_z, the rotational invariance is destroyed and the electrons circle with a velocity $v = \omega r$ on Landau orbits around the z-direction with the cyclotron frequency $\omega = eB/Mc$. In quantum mechanics, the system corresponds to an ensemble of harmonic oscillators which in the gauge $\mathbf{A} = (0, Bx, 0)$ [see Eq. (9.93)] move back and forth in the x-direction and have a spectrum $(n + 1/2)\hbar\omega$ [see Eq. (9.100)]. The phase space integral in the x-direction $\int dp_x L_x/(2\pi\hbar)$ becomes therefore a sum over n. The center of oscillations is $x_0 = p_y/M\omega$ [see

Eq. (9.95)], so that the remaining phase space integral $\int dp_y L_y/(2\pi\hbar)$ can be integrated to $M\omega L_x/(2\pi\hbar)$. Thus (16.288) gives, for each spin orientation,

$$\rho = M\omega \frac{1}{2\pi\hbar} \sum_{n=0}^{n_F} . \tag{16.290}$$

The number of filled levels is $\nu = n_F + 1$. In the vacuum, the levels of one orientation are degenerate with those of the opposite orientation at a neighboring n (up to radiative corrections of the order $\alpha \approx 1/137$). This is due to the anomalous magnetic moment of the spin-1/2 electron being equal to one Bohr magneton $\mu_B = e\hbar/2Mc \approx 0.927 \times 10^{-20}$ erg/gauss (i.e., twice as large as classically expected, the factor 2 being caused by the relativistic Thomas precession). Due to the factor 2, the energy levels for the two orientations are split by $\omega\hbar$, which is precisely equal to the energy difference between levels of neighboring n. In a solid material, however, the anomalous magnetic moment is strongly renormalized and the degeneracy is removed. There, every level has a definite spin orientation.

According to Eq. (16.290), the highest level is occupied completely if each level has taken up a particle number corresponding to its maximal filling density

$$\rho_{\max} = \frac{M\omega}{2\pi\hbar}. \tag{16.291}$$

At smaller magnetic fields, this density is small and the electrons are spread over many levels whose number ν is given by

$$\rho = \frac{M\omega}{2\pi\hbar}\nu. \tag{16.292}$$

Expressing ω in terms of B_z leads to

$$\frac{B_z}{\rho} = \frac{hc}{e}\frac{1}{\nu}. \tag{16.293}$$

Using the flux quantum $\Phi_0 = hc/e$, this equation states that the magnetic flux per electron

$$\Phi \equiv \frac{B_z}{\rho} \tag{16.294}$$

has the value

$$\frac{\Phi}{\Phi_0} = \frac{1}{\nu}. \tag{16.295}$$

If the magnetic field is increased, the Landau levels can accommodate more electrons which then reside in a decreasing number ν of levels. By inserting into Eq. (16.286) the values of B_z at which the highest level becomes depleted, one obtains precisely the experimentally observed quantized Hall resistances

$$R_{xy} = \frac{h}{e^2}\frac{1}{\nu} \tag{16.296}$$

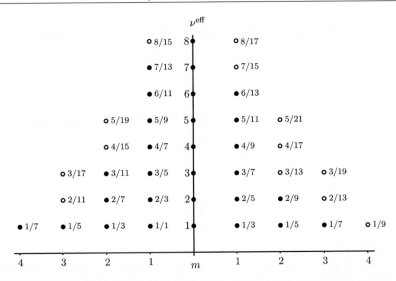

Figure 16.24 Values of parameter ν, at which plateaus in fractional quantum Hall resistance $h/e^2\nu$ are expected theoretically. The right-hand side shows the values $\nu^{\mathrm{eff}}/(2m\nu^{\mathrm{eff}}+1)$, the left-hand side $\nu^{\mathrm{eff}}/(2m\nu^{\mathrm{eff}}-1)$. The full circles indicate the values found experimentally.

with integer values of ν. The assumption of a statisto-magnetic interaction makes it possible to explain the fractional quantum Hall effect by reducing it to the ordinary quantum Hall effect.

In the fractional quantum Hall effect, the magnetic field is so strong that even the lowest Landau level is only partially filled. This is why one did not expect any plateaus at all. According to a simple idea due to Jain, however, it is possible to relate the fractional plateaus to the integer plateaus. For this one assumes that the electrons in the ground state of the fractional quantum Hall effect carry an even statisto-magnetic flux $-2m\Phi_0$ due to the presence of a Chern-Simons action. For the wave function, this amounts to a statistical phase factor $e^{i2\pi m}$ under the exchange of two particle coordinates; it leaves the Fermi statistics of the electrons unchanged. Now one takes advantage of the observation made in the last section that the electrons cannot distinguish a statisto-magnetic field from an external magnetic field. They move in Landau orbits enforced by the combined field

$$B_z^{\mathrm{eff}} = B_z - B_z^{\mathrm{stat}}, \qquad B_z^{\mathrm{stat}} = 2m\Phi_0\rho. \tag{16.297}$$

The cyclotron frequency of the electrons in their Landau orbits is

$$\omega^{\mathrm{eff}} = eB_z^{\mathrm{eff}}/Mc. \tag{16.298}$$

Since the effective field is now much smaller than the external field, the Landau levels possess a greatly reduced capacity. Thus the electrons must be distributed over several levels in spite of the large magnetic field. The number decreases as the field grows further. The steps appear at those places where the effective magnetic field has its integer quantum Hall plateaus, i.e., at the effective magnetic fields

$$B_z^{\text{eff}} = \pm\rho\Phi_0/\nu^{\text{eff}}, \quad \nu^{\text{eff}} = 1, 2, 3, \dots . \tag{16.299}$$

The values of ν_{eff} are related to the ν-values of the external magnetic field as follows:

$$\pm\frac{1}{\nu^{\text{eff}}} = \frac{1}{\nu} - 2m. \tag{16.300}$$

From this one has

$$\nu = \frac{\nu^{\text{eff}}}{2m\nu^{\text{eff}} \pm 1}. \tag{16.301}$$

The resulting values of ν on the integer-valued plane spanned by the numbers m and ν^{eff} are shown in Fig. 16.24. Only odd denominators are allowed. The values of ν found by this simple hypothesis agree well with those of the lower experimental levels (16.287).

16.12 Anyonic Superconductivity

At the end of Section 16.3 we have mentioned that an ensemble of particles with fractional statistics in 2+1 spacetime dimensions exhibits Meissner screening. This has given rise to speculations that the presently poorly understood phenomenon of *high-temperature superconductivity* may be explained by anyons physics. The new kind of superconductivity is observed in materials which contain pronounced layer structures, and it is conceivable that the currents move in these two-dimensional subspaces without dissipation. With some effort it can indeed be shown that in 2+1 dimensions a Chern-Simons action may be generated in principle[10] by integrating out Fermi fields. Accepting this, we can easily derive that an addition of this action to the usual electromagnetic field action gives the magnetic field a finite range, i.e., a finite penetration depth. The usual electromagnetic action reads

$$\mathcal{A} = \frac{1}{8\pi} \int dt d^3x [\mathbf{E}^2 - (\boldsymbol{\nabla} \times \mathbf{A})^2], \tag{16.302}$$

where $\mathbf{E}$ is the electric field

$$\mathbf{E} = -\frac{1}{c}\frac{\partial \mathbf{A}}{\partial t} - \boldsymbol{\nabla} A_0. \tag{16.303}$$

In the Euclidean formulation with $x_4 = ict$, the action becomes

$$\mathcal{A}_{\text{e}} = \frac{1}{8\pi c} \int d^4x [\mathbf{E}^2 + (\boldsymbol{\nabla} \times \mathbf{A})^2]. \tag{16.304}$$

---

[10]See Notes and References at the end of the chapter.

To add the Chern-Simons action, we restrict the spacetime dimensionality to 3. The restriction is imposed by considering a system in 4 spacetime dimensions and assuming it to be translationally invariant along the fourth coordinate direction x_4. Then there are no electric fields, and the Euclidean action becomes

$$\mathcal{A}_{\mathrm{e}} = \frac{L}{8\pi c} \int d^3x (\boldsymbol{\nabla} \times \mathbf{A})^2, \tag{16.305}$$

where L denotes the length of the system in the x_4-direction. To this, we now add the Chern-Simons action (16.253) and the current coupling (16.256). By extremizing the total action, we obtain the field equation:

$$\frac{L}{4\pi c} \boldsymbol{\nabla} \times (\boldsymbol{\nabla} \times \mathbf{A}) + i \frac{e^2}{2\pi c^2 \hbar \mu_0} \boldsymbol{\nabla} \times \mathbf{A} = i \frac{1}{c^2} \mathbf{j}. \tag{16.306}$$

For the magnetic field $\mathbf{B} = \boldsymbol{\nabla} \times \mathbf{A}$, the equation reads

$$\frac{L}{4\pi c} (\boldsymbol{\nabla} \times \mathbf{B} + i\lambda^{-1}\mathbf{B}) = i \frac{1}{c^2} \mathbf{j}, \tag{16.307}$$

where the parameter λ denotes the following length ($\alpha = e^2/\hbar c = $ fine-structure constant $\approx 1/137$):

$$\lambda \equiv \frac{c\hbar\mu_0}{2e^2} L = \frac{\mu_0}{2\alpha} L. \tag{16.308}$$

By multiplying (16.307) vectorially with $\boldsymbol{\nabla}$ and using the equation once more, we obtain

$$\frac{L}{4\pi c} (-\boldsymbol{\nabla}^2 + \lambda^{-2})\mathbf{B} = i\frac{1}{c^2} \boldsymbol{\nabla} \times \mathbf{j} + \frac{1}{c^2}\lambda^{-1}\mathbf{j} . \tag{16.309}$$

In the current-free case, the magnetic field is seen to have only a finite penetration depth λ into the material. In an ordinary superconductor, this phenomenon is known as the *Meissner effect*. There it can be understood as a consequence of the induction of supercurrents in an ideal (i.e., incompressible and frictionless) liquid of charged particles, which lowers the invading magnetic field according to Lenz' rule. In the absence of friction, there is a complete extinction.

Recall that a superconductor with time-dependent currents and fields is governed by the characteristic London equation (see Appendix 16D)

$$\boldsymbol{\nabla} \times \mathbf{j} \propto \mathbf{B}. \tag{16.310}$$

For a two-dimensional superconductor, this amounts to

$$(\boldsymbol{\nabla} \times \mathbf{j}_\perp)_z \propto B_z. \tag{16.311}$$

The above anyonic system shows a similar induction phenomenon. In the absence of currents, Eq. (16.309) determines the magnetic field B_z from the particle density by

$$B_z = \mu_0 \Phi_0 \rho. \tag{16.312}$$

If there are currents in the xy-plane $\mathbf{j}_\perp = (j_x, j_y)$, the magnetic field is increased by ΔB_z in accordance with the equation

$$\frac{L}{4\pi c}(-\boldsymbol{\nabla}^2 + \lambda^{-2})\Delta B_z = \frac{1}{c^2}(\boldsymbol{\nabla} \times \mathbf{j}_\perp)_z. \tag{16.313}$$

This is the desired relation between the magnetic field and the curl of the current which indicates the superconducting character of the system expressed before in the London equation (16.311). The contact with the London equation is established by a restriction to smooth field configurations in which the first term in (16.313) can be ignored.

Thus we conclude that the currents and magnetic fields in a two-dimensional system of anyons show Meissner screening. This is not sufficient to make the system superconductive since it does not automatically imply the absence of dissipation. In a usual superconductor, the existence of an energy gap makes the dissipative part of the current-current correlation function vanish for wave vectors smaller than some value k_c. This value determines the *critical current strength* above which super-conductivity returns to normal. In the anyonic system, the absence of dissipation was proved in an approximation. Recent studies of higher corrections, however, have shown the presence of dissipation after all, destroying the hope for an anyonic superconductor.

16.13 Non-Abelian Chern-Simons Theory

The topological field interaction (16.153) can be generalized to nonabelian gauge groups. For the local symmetry group $SU(N)$ it reads

$$\mathcal{A}_{\mathrm{e,CS}} = \frac{k}{4\pi i} \int d^3 x \epsilon_{ijk} \mathrm{tr}_N \left(A_i \nabla_j A_k + \frac{2}{3} A_i A_j A_k \right), \tag{16.314}$$

where A_i are Hermitian traceless $N \times N$-matrices and tr_N denotes the associated trace. In the nonabelian theory, the gauge transformations are

$$A_i \leftrightarrow U A_i U^{-1} + i(\partial_i U) U^{-1}. \tag{16.315}$$

It can be shown that they transform $\mathcal{A}_{\mathrm{e,CS}}$ as follows [10]:

$$\mathcal{A}_{\mathrm{e,CS}} \to \mathcal{A}_{\mathrm{e,CS}} + 2\pi i n k\hbar, \quad n = \text{integer}. \tag{16.316}$$

Thus, the action is not completely gauge-invariant. For integer values of k, however, the additional $2\pi i n k\hbar$ does not have any effect upon the phase factor $e^{-\mathcal{A}_{\mathrm{e,CS}}/\hbar}$ in the path integral associated with the orbital fluctuations. Thus there is gauge invariance for integer values of k (in contrast to the abelian case where k is arbitrary).

In the nonabelian theory, gauge fixing is a nontrivial issue. It is no longer possible to simply add a gauge fixing functional of the type (16.163). The reason is that the volume in the field space of gauge transformations depends on the gauge field. For a consistent gauge fixing, this volume has to be divided out of the gauge-fixing functional as was first shown by Fadeev and Popov [11]. For an adequate discussion of this interesting topic which lies beyond the scope of this quantum-mechanical text the reader is referred to books on quantum field theory.

As in the abelian case, the functional derivative of the Chern-Simons action with respect to the vector potential gives the field strength

$$B_i \equiv \epsilon_{ijk} F_{jk}, \tag{16.317}$$

where F_{ij} is the nonabelian version of the curl:

$$F_{ij} \equiv \partial_i A_j - \partial_j A_i - i[A_i, A_j]. \tag{16.318}$$

In 1989, Witten found an important result: The expectation value of a gauge-invariant integral defined for any loop L,

$$W_L[\mathbf{A}] \equiv \mathrm{tr}\,_N \hat{W}[\mathbf{A}] \equiv \mathrm{tr}\,_N \hat{P} e^{i \oint_L d\mathbf{x} \mathbf{A}}, \tag{16.319}$$

the so-called *Wilson loop integral*, possesses a close relationship with the Jones polynomials of knot theory. The loop L can consist of several components linked in an arbitrary way, in which case the integral in $W_L[\mathbf{A}]$ runs successively over all components. The operator $\hat{P}$ in front of the exponential function denotes the *path-ordering operator*. It is defined in analogy with the time-ordering operator $\hat{T}$ in (1.241): If the exponential function in (16.319) is expanded into a Taylor series, it specifies the order in which the $N \times N$-matrices $A_i(\mathbf{x})$, which do not commute for different $\mathbf{x}$ and i, appear in the products. If the path is labeled by a "time parameter", the earlier matrices stand to the right of the later ones. The fluctuations of the vector potential are controlled by the Chern-Simons action (16.314). Expectation values of the loop integrals are defined by the functional integral

$$\langle W_L[\mathbf{A}] \rangle \equiv \frac{\int \mathcal{D} A_i e^{-\mathcal{A}_{\mathrm{e,CS}}/\hbar} W_L[\mathbf{A}]}{\int \mathcal{D} A_i e^{-\mathcal{A}_{\mathrm{e,CS}}/\hbar}}. \tag{16.320}$$

To calculate the self-interaction of a loop, we proceed as described in the abelian case after Eq. (16.169) by spreading the line out into an infinitely thin ribbon of parallel lines. The borders of the ribbon are positioned in such a way that their linking number L_{k} vanishes. If 0 denotes a circle, i.e., a trivial knot, one can show that

$$\langle W_0[\mathbf{A}] \rangle = \frac{q^{N/2} - q^{-N/2}}{q^{1/2} - q^{-1/2}}, \tag{16.321}$$

with

$$q \equiv e^{-2\pi i/(N+k)}. \tag{16.322}$$

For an arbitrary link L one finds the skein relation (see Appendix 16C)

$$q^{N/2} \langle W_{L_+}[\mathbf{A}] \rangle - q^{-N/2} \langle W_{L_-}[\mathbf{A}] \rangle = (q^{1/2} - q^{-1/2}) \langle W_{L_0}[\mathbf{A}] \rangle. \tag{16.323}$$

If $N = 2$, this agrees up to the sign of the right-hand side with the relation (16.114) for the Jones polynomials $J_L(t)$. For general values of $N \neq 2$, we obviously obtain with $t = q^{-N/2}$ and $\alpha = q^{1/2} - q^{-1/2} = -(t^{1/N} - t^{-1/N})$ the skein relations (16.122) of the HOMFLY polynomials. The important relation is

$$\frac{\langle W_L[\mathbf{A}]\rangle}{\langle W_0[\mathbf{A}]\rangle} = H_L(t, -(t^{1/N} - t^{-1/N})), \qquad t = e^{\pi i/k}. \tag{16.324}$$

Since the second variable in $H_L(t, \alpha)$ appears only in even or odd powers, $H_L(t, -(t^{1/2} - t^{-1/2}))$ is a Jones polynomial up to a sign $(-1)^{s+1}$, where s is the number of loops in L.

A favored choice of framing is one in which the self-linking number L_{k} of each component is equal to the twist number or writhe w introduced in Eq. (16.110). Then the ribbon lies flat on the projection plane of the knot. This framing can easily be drawn on the blackboard by splitting the line L into two parallel running lines; it is therefore called the *blackboard framing*. Incidentally, each choice of framing can be drawn as a blackboard framing if one adds to the loop L an appropriate number of windings via a Reidemeister move of type I. These are trivial for lines and nontrivial only for ribbons (see Fig. 16.6). In the blackboard framing, each such winding changes the values L_{k} and w simultaneously by one unit. Thus $L_{\mathrm{k}} = w$ can be brought to any desired value. Take, for example, the trefoil knot in Fig. 16.2. In the blackboard framing it has the self-linking number $L_{\mathrm{k}} = w = -3$. This can be brought to zero by adding three windings via a Reidemeister move of type I.[11]

In the framing $L_{\mathrm{k}} = w$, the right-hand side of (16.324) carries an extra phase factor c^w, where

$$c = e^{-i2\pi(N^2-1)/2Nk}. \tag{16.325}$$

For comparison: In the abelian Chern-Simons theory, the phase factor is $c = e^{-2\pi i/k}$, and the expectation $\langle W_L[\mathbf{A}]\rangle$ has the value $c^{\Sigma_{i\neq j} L_{\mathrm{k}ij}}$ for a link of several loops labeled by i with vanishing individual self-linking numbers $L_{\mathrm{k}i}$. In the framing $L_{\mathrm{k}i} = w_i$, the value is $c^{\Sigma_{i\neq j} L_{\mathrm{k}ij} + \Sigma_i w_i}$.

The investigation of the properties of loops with nonabelian topological interactions is an interesting task of present-day research.

Appendix 16A   Calculation of Feynman Diagrams in Polymer Entanglement

For the calculation of the amplitudes $N_1, \ldots, N_4$ in Eqs. (16.226), (16.230), (16.235), and (16.236), we need the following simple tensor formulas involving two completely antisymmetric tensors ε^{ijl}:

$$\varepsilon_{ijk}\varepsilon^{imn} = \delta_j^m \delta_k^n - \delta_j^n \delta_k^m, \qquad \varepsilon_{ijk}\varepsilon^{ijl} = 2\delta_k^l. \tag{16A.1}$$

The Feynman diagrams shown in Fig. 16.23 corresponds to integrals over products of the polymer correlation functions G_0 defined in Eq. (16.213), which have to be integrated over space and

---

[11]Mathematicians usually prefer another framing in which the ribbons lie flat on the so-called *Seifert surfaces*.

Laplace transformed. For the latter we make use of the convolution property of the integral over two Laplace transforms $\tilde{f}(z)$ and $\tilde{g}(z)$ of the functions f, g:

$$\int_{c-i\infty}^{c+i\infty} \frac{M dz}{2\pi i} e^{zL} \tilde{f}(z)\tilde{g}(z) = \int_0^L ds f(s)g(L-S). \tag{16A.2}$$

All spatial integrals are Gaussian of the form

$$\int d^3x e^{-a\mathbf{x}^2 + 2b\mathbf{x}\cdot\mathbf{y}} = (2\pi)^{3/2} a^{-3/2} e^{b^2\mathbf{y}^2/a}, \quad a > 0. \tag{16A.3}$$

Contracting the fields in Eq. (16.226), and keeping only the contributions which do not vanish in the limit of zero replica indices, we find with the help of Eqs. (16A.1) and (16A.2):

$$\begin{aligned} N_1 &= \int d^3x_1, d^3x_2 \int_0^{L_1} ds \int_0^{L_2} dt \int d^3x_1' d^3x_2' \, G_0(\mathbf{x}_1 - \mathbf{x}_1'; s)G_0(\mathbf{x}_1' - \mathbf{x}_1; L_1 - s) \\ &\times \ G_0(\mathbf{x}_2 - \mathbf{x}_2'; t)G_0(\mathbf{x}_2' - \mathbf{x}_2; L_2 - t)\frac{l}{|\mathbf{x}_1' - \mathbf{x}_2'|^4}. \end{aligned} \tag{16A.4}$$

Performing the changes of variables

$$s' = \frac{s}{L_1}, \quad t' = \frac{t}{L_2}, \quad \mathbf{x} = \frac{\mathbf{x}_1 - \mathbf{x}_1'}{\sqrt{L_1}}, \quad \mathbf{y} = \frac{\mathbf{x}_2 - \mathbf{x}_2'}{\sqrt{L_2}}, \tag{16A.5}$$

and setting $\mathbf{x}_1'' \equiv \mathbf{x}_1' - \mathbf{x}_2'$, we easily derive (16.227).

For small $\xi/\sqrt{L_1}$ and $\xi/\sqrt{L_2}$, we use the approximation (16.228). The space integrals can be done using the formula (16A.3). After some work we obtain the result (16.240).

For the amplitude N_2 in Eq. (16.230) we obtain likewise the integral

$$\begin{aligned} N_2 &= \int d^3x_1 d^3x_2 \int d^3x_1' d^3x_1'' d^3x_2' \\ &\times \left[\int_0^{L_1} ds \int_0^S ds' G_0(\mathbf{x}_1' - \mathbf{x}_1; L_1 - s) \, \boldsymbol{\nabla}_{x_1''}^j G_0(\mathbf{x}_1 - \mathbf{x}_1''; s') \boldsymbol{\nabla}_{x_1'}^i G_0(\mathbf{x}_1'' - \mathbf{x}_1'; s - s') \right] \\ &\times D_{ik}(\mathbf{x}_1' - \mathbf{x}_2) D_{jk}(\mathbf{x}_1'' - \mathbf{x}_2') \left[\int_0^{L_2} dt \, G_0(\mathbf{x}_2 - \mathbf{x}_2'; L_2 - t)G_0(\mathbf{x}_2' - \mathbf{x}_2; t) \right], \end{aligned} \tag{16A.6}$$

where $D_{ij}(\mathbf{x}, \mathbf{x}')$ are the correlation functions (16.174) and (16.175) of the vector potentials. Setting $\mathbf{x}_2 \equiv \sqrt{L_2}\mathbf{u} + \mathbf{x}_2'$ and supposing that $\xi/\sqrt{L_2}$ is small, the integral over $\mathbf{u}$ can be easily evaluated with the help of the Gaussian integral (16A.3). After the substitutions $\mathbf{x}_1'' = \sqrt{L_1}\mathbf{y} + \mathbf{x}_1$, $\mathbf{x}_1' = \sqrt{L_1}(\mathbf{y} - \mathbf{x}) + \mathbf{x}_1$, $\mathbf{x}_2' = \sqrt{L_1}(\mathbf{y} - \mathbf{x} - \mathbf{z}) + \mathbf{x}_1$ and a rescaling of the variables s, s' by a factor L_1^{-1}, we derive Eq. (16.231) with (16.232).

For small $\xi/\sqrt{L_1}, \xi/\sqrt{L_2}$, the spatial integrals are easily evaluated leading to:

$$N_2 = \frac{-\sqrt{2}VL_2^{-1/2}L_1^{-1}M^{-1/2}}{(4n)^6} \int_0^1 dt \int_0^t dt' t'(1-t)\sqrt{\frac{t-t'}{1-t+t'}}. \tag{16A.7}$$

After the change of variable $t' \to t'' = t - t'$, the double integral is reduced to a sum of integrals the type

$$c(n, m) = \int_0^1 dt\, t^m \int_0^t dt'\, t'^n \sqrt{\frac{t'}{1-t'}}, \quad m, n = \text{integers}.$$

These can be simplified by replacing t^m by $dt^{m+1}/dt(m+1)$, and doing the integrals by parts. In this way, we end up with a linear combination of integrals of the form:

$$\int_0^1 dt \frac{t^{\kappa+\frac{1}{2}}}{\sqrt{1-t}} = B\left(\kappa + \frac{3}{2}, \frac{1}{2}\right). \tag{16A.8}$$

The calculations of N_3 and N_4 are very similar, and are therefore omitted.

Appendix 16B   Kauffman and BLM/Ho Polynomials

The Kauffman polynomials are given by $F(a, x) = a^{-w}\Lambda(a, x)$, where w is the writhe and $\Lambda(a, x)$ satisfies the skein relation

$$\Lambda_{L_+}(a, x) + \Lambda_{L_-}(a, x) = x[\Lambda_{L_0}(a, x) + \Lambda_{L_\infty}(a, x)]. \tag{16B.1}$$

The subscripts refer to the same loop configurations as in Figs. 16.10 and 16.12. The trivial loop has

$$\Lambda(a, x) = \frac{a + a^{-1}}{z} - 1. \tag{16B.2}$$

While the Kauffman polynomial is a knot invariant, the Λ-polynomial is only a ribbon invariant.[12] If a winding L_{T+} or L_{T-} is removed from a loop with the help of a Reidemeister move of type I in Fig. 16.6 (which for infinitely thin lines would be trivial while changing the writhe of a ribbon by one unit) then $\Lambda(a, x)$ receives a factor a or a^{-1}, respectively (see Fig. 16.25).

$$L_{T+} \qquad L_{T0} \qquad\qquad L_{T-} \qquad L_{T0}$$

Figure 16.25   Trivial windings L_{T+} and L_{T-}. Their removal by means of Reidemeister move of type I decreases or increases writhe w by one unit.

The Kauffman polynomials arise from Wilson loop integrals of a nonabelian Chern-Simons theory, if the action (16.314) is SO(N)- rather than SU(N)-symmetric. A list of these polynomials can be found in papers by Lickorish and Millet and by Doll and Hoste quoted at the end of the chapter.

The BLMHo polynomials are special cases of the Kauffman polynomials. The relation between them is $Q(x) \equiv F(1, x)$.

Appendix 16C   Skein Relation between Wilson Loop Integrals

Here we sketch the derivation of the skein relation (16.323) for the expectation values of Wilson's loop integrals (16.319). Let us decompose A_i in terms of the $N^2 - 1$ generators T_a of the group SO(N):

$$A_i = \sum_a A_i^a T_a. \tag{16C.1}$$

They satisfy the commutation rules

$$[T_a, T_b] = i f_{abc} T_c. \tag{16C.2}$$

---

[12]More precisely, $F(a, x)$ is invariant under the three Reidemeister moves which, in the projected picture of the knot in Fig. 16.6, define the ambient isotopy, whereas Λ changes under the first Reidemeister move, associated only with regular isotopy. whereas

For simplicity, we assume k to be very large so that we can restrict the treatment to the lowest order in $1/k$. To avoid inessential factors of the constants $e, c, \hbar$, we set these equal to 1. Under a small variation of the fields one has

$$\frac{\delta \hat{W}_L[\mathbf{A}]}{\delta A_i^a(\mathbf{x})} = i\hat{P} \int_L dx_i' \delta^{(3)}(\mathbf{x} - \mathbf{x}') T_a(\mathbf{x}') \hat{W}_L[\mathbf{A}], \tag{16C.3}$$

where the path-ordering operator $\hat{P}$ arranges the expression to its right in such a way that T_a is situated in $\hat{W}_L$ at the correct path-ordered place. To emphasize this, we have recorded the position of T_a by means of an $\mathbf{x}$-argument. More precisely, if we discretize the loop integral and write

$$\hat{W}_L[\mathbf{A}] = e^{iA_i(\bar{\mathbf{x}}^1)\Delta x_i^1} e^{iA_i(\bar{\mathbf{x}}^2)\Delta x_i^2} \cdots e^{iA_i(\bar{\mathbf{x}}^n)\Delta x_i^n} \cdots, \tag{16C.4}$$

where $\bar{\mathbf{x}}^n$ are the midpoints of the intervals $\Delta \mathbf{x}^n$, a differentiation with respect to one of the $A_i(\bar{\mathbf{x}}^n)$-fields replaces the associated factor $e^{iA_i(\bar{\mathbf{x}}_n)\Delta x_i^n}$ by $iT_a e^{iA_i(\bar{\mathbf{x}}_n)\Delta x_i^n}$. With the δ-function on a line L defined in Eq. (10A.8), we write (16C.3) as

$$\frac{\delta \hat{W}_L[\mathbf{A}]}{\delta A_i^a(\mathbf{x})} = i\hat{P} \delta_i(\mathbf{x}, L) T_a(\mathbf{x}) \hat{W}_L[\mathbf{A}]. \tag{16C.5}$$

For simplicity, we assume $\mathbf{x}$ to be only once traversed by the loop L.

If the shape of the loop is deformed infinitesimally by $dS_i = \epsilon_{ijk} dx_i d'x_j$, then $\hat{W}_L$ changes by

$$\delta \hat{W}_L = i dx_i d'x_j \hat{P} F_{ij}^a(\mathbf{x}) T_a(\mathbf{x}) \hat{W}_L, \tag{16C.6}$$

where F_{ij}^a are the $N^2 - 1$ components of the nonabelian field strengths

$$F_{ij} = \partial_i A_j - \partial_j A_i - i[A_i, A_j] \tag{16C.7}$$

and $\mathbf{x}$ the midpoints of the parallelograms spanned by $d\mathbf{x}$ and $d'\mathbf{x}$. The derivation of Eq. (16C.6) is based on the observation that a change of the path by a small parallelogram adds to the line integral $\hat{W}_L$ a factor $\hat{W}_\square$, which is a Wilson loop integral around the small parallelogram. The latter is evaluated as follows:

$$
\begin{aligned}
\hat{W}_\square &= e^{iA_i(\mathbf{x}-d'\mathbf{x}/2)dx_i} e^{iA_j(\mathbf{x}+d\mathbf{x}/2)d'x_j} e^{-iA_i(\mathbf{x}+d'\mathbf{x}/2)dx_i} e^{-iA_j(\mathbf{x}-d\mathbf{x}/2)d'x_j} \\
&= e^{i[A_i(\mathbf{x})dx_i - \partial_j A_i(\mathbf{x})dx_i d'x_j + \ldots]} e^{i[A_j(\mathbf{x})d'x_j + \partial_i A_j(\mathbf{x})dx_i d'x_j + \ldots]} \\
&\quad \times e^{-i[A_i(\mathbf{x})dx_i + \partial_j A_i(\mathbf{x})dx_i d'x_j + \ldots]} e^{-i[A_j(\mathbf{x})d'x_j - \partial_i A_j(\mathbf{x})dx_i d'x_j + \ldots]} \\
&= e^{iF_{ij}(\mathbf{x})dx_i d'x_j}.
\end{aligned} \tag{16C.8}
$$

The last line is found with the help of the Baker-Hausdorff formula $e^A e^B = e^{A+B+[A,B]/2+\cdots}$ (recall Appendix 2A).

Let us denote the Chern-Simons functional integral over $\hat{W}_L[\mathbf{A}]$ by $\overline{\hat{W}}_L$. Their $N \times N$-traces are $W_L[A]$ and $\overline{W}_L$. The latter differs from the expectation $\langle W_L[\mathbf{A}] \rangle$ in (16.320) by not containing the normalizing denominator, i.e.,

$$\overline{W}_L \equiv \int \mathcal{D}\mathbf{A} e^{-\mathcal{A}_{e,\text{CS}}} W_L[\mathbf{A}]. \tag{16C.9}$$

This changes under the loop deformation by

$$
\begin{aligned}
\delta \overline{W}_L &= \int \mathcal{D}\mathbf{A} \delta W_L[\mathbf{A}] e^{-\mathcal{A}_{e,\text{CS}}} \\
&= i dx_i d'x_j \int \mathcal{D}\mathbf{A} F_{ij}^a(\mathbf{x}) T_a(\mathbf{x}) W_L[\mathbf{A}] e^{-\mathcal{A}_{e,\text{CS}}},
\end{aligned} \tag{16C.10}
$$

with the tacit agreement that a generator $T_a(\mathbf{x})$ written in front of the trace has to be evaluated within the trace at the correct path-ordered position. Now we observe that F_{ij}^a can also be obtained by applying a functional derivative to the Chern-Simons action (16.314):

$$i\frac{4\pi}{k}\epsilon_{ijk}\frac{\delta\mathcal{A}_{e,\mathrm{CS}}}{\delta A_k^a(\mathbf{x})} = F_{ij}^a(\mathbf{x}). \tag{16C.11}$$

This allows us to rewrite (16C.10) as

$$-\frac{4\pi}{k}\int\mathcal{D}\mathbf{A}\int dS_i\frac{\delta\mathcal{A}_{e,\mathrm{CS}}}{\delta A_i^a(\mathbf{x})}T_a(\mathbf{x})W_L[\mathbf{A}]e^{-\mathcal{A}_{e,\mathrm{CS}}}$$

and further as

$$\frac{4\pi}{k}\int\mathcal{D}\mathbf{A}dS_iT_a(\mathbf{x})W_L\frac{\delta}{\delta A_i^a(\mathbf{x})}e^{-\mathcal{A}_{e,\mathrm{CS}}}.$$

A partial functional integration produces

$$-\frac{4\pi}{k}\int\mathcal{D}\mathbf{A}\int dS_iT_a(\mathbf{x})\frac{\delta W_L[\mathbf{A}]}{\delta A_i^a(\mathbf{x})}e^{-\mathcal{A}_{e,\mathrm{CS}}},$$

which brings the variation to the form

$$\delta\overline{W}_L = -\frac{4\pi i}{k}\int\mathcal{D}\mathbf{A}dS_i\delta_i(\mathbf{x},L)T_a(\mathbf{x})T_a(\mathbf{x})W_L[\mathbf{A}]e^{-\mathcal{A}_{e,\mathrm{CS}}}. \tag{16C.12}$$

The expectation of Wilson's loop integral $\overline{W}_L$ changes under a deformation only if the loop crosses another line element. This property makes $\overline{W}_L$ a ribbon invariant, i.e., an invariant of regular isotopy.

For a finite deformation, the right-hand side has to be integrated over the area S across which the line has swept. Using the integral formula

$$\int_S dS_i\delta_i(\mathbf{x},L) = \left\{\begin{matrix}1\\0\end{matrix}\right\} \text{ if the line } L \left\{\begin{matrix}\text{pierces } S\\\text{misses } S\end{matrix}\right\}, \tag{16C.13}$$

we obtain for each crossing

$$\overline{W}_{L_+} - \overline{W}_{L_-} \equiv \Delta\overline{W}_L = -\frac{4\pi i}{k}\int\mathcal{D}\mathbf{A}T_a(\mathbf{x})T_a(\mathbf{x})W_L[\mathbf{A}]e^{-\mathcal{A}_{e,\mathrm{CS}}}. \tag{16C.14}$$

The two generators $T_a(\mathbf{x})$ lie path-ordered on the different line pieces of the crossing. To establish contact with the knot polynomials, the left-hand sides have been labeled by the loop subscripts L_+ and L_- appearing in the skein relations of Fig. 16.114.

The product of the generators on the right-hand side is the Casimir operator of the $N \times N$ -representation of $SO(N)$:

$$(T_a)_{\alpha\beta}(T_a)_{\gamma\delta} = \frac{1}{2}\delta_{\alpha\delta}\delta_{\beta\gamma} - \frac{1}{2N}\delta_{\alpha\beta}\delta_{\gamma\delta}. \tag{16C.15}$$

When inserted into Eq. (16C.14), we obtain the graphical relation:

The second graph on the right-hand side can be decomposed into

Taking these two terms to the left-hand side of (16C.14), we obtain the skein relation

$$\left(1 - \frac{\pi i}{Nk}\right)\overline{W}_{L_+} - \left(1 + \frac{\pi i}{Nk}\right)\overline{W}_{L_-} = -\frac{2\pi i}{k}\overline{W}_{L_0}. \tag{16C.16}$$

We now apply this relation to the windings displayed in Fig. 16.25. They decompose into a line and a circle. Due to the trace operation in $\overline{W}_{L_0}$, the circle contributes a factor N. Thus we obtain the relation

$$\left(1 - \frac{\pi i}{Nk}\right)\overline{W}_{L_{T+}} - \left(1 + \frac{\pi i}{Nk}\right)\overline{W}_{L_{T-}} = -\frac{2\pi i}{k}N\,\overline{W}_{L_{T0}}. \tag{16C.17}$$

Now we remove on the left-hand side the windings according to the graphical rules of Fig. 16.25. Under this operation, the Wilson loop integral is not invariant. Like BLMHo polynomials, it acquires a factor a or a^{-1}:

$$\overline{W}_{L_{T+}} = a\overline{W}_{L_{T0}}, \qquad \overline{W}_{L_{T-}} = a^{-1}\overline{W}_{L_{T0}}. \tag{16C.18}$$

To be compatible with (16C.17), the parameter a must satisfy

$$a = 1 - \frac{\pi i}{Nk}(N^2 - 1), \qquad a^{-1} = 1 + \frac{\pi i}{Nk}(N^2 - 1). \tag{16C.19}$$

Due to (16C.18), the Wilson loop integral is only a ribbon invariant exhibiting regular isotopy. A proper knot invariant which distinguishes ambient isotopy classes arises when multiplying $\overline{W}_L$ by a^{-w}. The polynomials $H_L \equiv e^{-w}\overline{W}_L$ satisfy the skein relation

$$\left[\left(1 - \frac{\pi i}{Nk}\right)a\right]H_{L_+} - \left[\left(1 + \frac{\pi i}{Nk}\right)a^{-1}\right]H_{L_-} = -\frac{2\pi i}{k}H_{L_0}. \tag{16C.20}$$

The prefactors on the left-hand side can be written for large k as $1 - 2\pi i N/k \approx q^{N/2}$ and $1 + 2\pi i N/k \approx q^{-N/2}$ with $q = 1 - 2\pi i \pi/k$. The prefactor on the right-hand side is equal to $q^{1/2} - q^{-1/2}$. To leading order in $1/k$, we have thus derived the skein relation (16.323) for the HOMFLY polynomials H_L.

Appendix 16D　London Equations

Consider an ideal fluid of charged particles. By definition, it is non-viscous and incompressible, satisfying $\nabla \cdot \mathbf{v} = 0$. If the charge of the particles is e (which we take to be negative for electrons), the electric current density is

$$\mathbf{j} = \rho e\mathbf{v}, \tag{16D.1}$$

where ρ is the particle density. The current is obviously conserved.

The equation of motion of the particles in an electric and magnetic field is governed by the Lorentz force and reads

$$M\dot{\mathbf{v}} = e\left(\mathbf{E} + \frac{1}{c}\mathbf{v} \times \mathbf{B}\right). \tag{16D.2}$$

Using the kinematic identity

$$\frac{d\mathbf{v}}{dt} = \frac{\partial \mathbf{v}}{\partial t} + (\mathbf{v} \cdot \nabla)\mathbf{v} = \frac{\partial \mathbf{v}}{\partial t} + \nabla\left(\frac{1}{2}\mathbf{v}^2\right) - \mathbf{v} \times (\nabla \times \mathbf{v}), \tag{16D.3}$$

this leads to the partial differential equation for the velocity field $\mathbf{v}(\mathbf{x}, t)$

$$M\frac{\partial \mathbf{v}}{\partial t} + \boldsymbol{\nabla}\left(\frac{M}{2}\mathbf{v}^2\right) = e\mathbf{E} + M\mathbf{v}\cdot\left(\boldsymbol{\nabla}\times\mathbf{v} + \frac{e}{Mc}\mathbf{B}\right). \tag{16D.4}$$

Consider the time dependence of the vector field on the right-hand side

$$\mathbf{X} = \boldsymbol{\nabla}\times\mathbf{v} + \frac{e}{Mc}\mathbf{B}. \tag{16D.5}$$

Using Maxwell's equation

$$\frac{\partial}{\partial t}\mathbf{B} = -c\boldsymbol{\nabla}\times\mathbf{E}, \tag{16D.6}$$

we derive

$$\frac{\partial}{\partial t}\mathbf{X} = \boldsymbol{\nabla}\times(\boldsymbol{\nabla}\times\mathbf{X}). \tag{16D.7}$$

Suppose now that there is initially no $\mathbf{B}$-field in the ideal fluid at rest which therefore has $\mathbf{X} \equiv 0$ everywhere. If a magnetic field is turned on, Eq. (16D.7) guarantees that $\mathbf{X}$ remains zero at all times. This implies that

$$\boldsymbol{\nabla}\times\mathbf{j} = -\frac{\rho e^2}{Mc}\mathbf{B}, \tag{16D.8}$$

which is the first London equation. By inserting the first London equation into Eq. (16D.4), we find the second London equation

$$\frac{\partial}{\partial t}\mathbf{v} + \boldsymbol{\nabla}\left(\frac{M}{2}\mathbf{v}^2\right) = e\mathbf{E}. \tag{16D.9}$$

If the vector potential is taken in the transverse gauge $\boldsymbol{\nabla}\cdot\mathbf{A} = 0$ (which in this context is also called *London gauge*), then the first London equation can be solved and yields

$$\mathbf{j} = -\frac{\rho e^2}{Mc}\mathbf{A}. \tag{16D.10}$$

By inserting this equation into the Maxwell equation with no electric field $\mathbf{E}$, we obtain

$$\boldsymbol{\nabla}\times\mathbf{B} = \frac{4\pi}{c}\mathbf{j} = -\frac{\rho 4\pi e^2}{Mc^2}\mathbf{A}. \tag{16D.11}$$

When rewritten in the form

$$\boldsymbol{\nabla}\times(\boldsymbol{\nabla}\times\mathbf{A}) + \lambda^{-2}\mathbf{A} = 0, \tag{16D.12}$$

with

$$\lambda^{-2} = \frac{\rho 4\pi e^2}{Mc^2}, \tag{16D.13}$$

the equation exhibits directly the finite penetration depth λ of a magnetic field into the fluid, the celebrated Meissner effect. It is the ideal manifestation of the Lenz rule, according to which an incoming magnetic field induces currents reducing the magnetic field — in the present case to extinction.

Appendix 16E   Hall Effect in Electron Gas

A gas of electrons with a density ρ carries an electric current

$$\mathbf{j} = \rho e \mathbf{v}. \tag{16E.1}$$

In a magnetic field, the particle velocities change due to the Lorentz force by

$$M\dot{\mathbf{v}} = e\left(\mathbf{E} + \frac{1}{c}\mathbf{v} \times \mathbf{B}\right). \tag{16E.2}$$

If σ_0 denotes the conductivity of the system without a magnetic field, the electric current is obviously given by

$$\begin{aligned}
\mathbf{j} &= \sigma_0\left(\mathbf{E} + \frac{1}{c}\mathbf{v} \times \mathbf{B}\right) \\
&= \sigma_0\left(\mathbf{E} + \frac{1}{\rho e c}\mathbf{j} \times \mathbf{B}\right).
\end{aligned} \tag{16E.3}$$

The second term shows the classical Hall resistance (16.286).

Notes and References

For the Aharonov-Bohm effect, see the original work by
Y. Aharonov and D. Bohm, Phys. Rev. **115**, 485 (1959).
For a review see
S. Ruijsenaars, Ann. Phys. (N.Y.) **146**, 1 (1983).
See also the papers
A. Inomata and V.A. Singh, J. Math. Phys. **19**, 2318 (1978);
E. Corinaldesi and F. Rafeli, Am. J. Phys. **46**, 1185 (1978);
M.V. Berry, Eur. J. Phys. **1**, 240 (1980);
S. Ruijsenaars, Ann. Phys. **146**, 1 (1983);
G. Morandi and E. Menossi, Eur. J. Phys. **5**, 49 (1984);
R. Jackiw, Ann. Phys. **201**, 83 (1990); and in "M. A. B. Bég Memorial Volume" (A. Ali and P. Hoodbhoy, Eds.), World Scientific, Singapore, 1991;
G. Amelino-Camelia, Phys. Lett. B **326**, 282 (1994); Phys. Rev. D **51**, 2000 (1995);
C. Manuel and R. Tarrach, Phys. Lett. B **328**, 113 (1994);
S. Ouvry, Phys. Rev. D **50**, 5296 (1994);
C.R. Hagen, Phys. Rev. D **31**, 848 (1985); D **52** 2466 (1995);
P. Giacconi, F. Maltoni, and R. Soldati, Phys. Rev. D **53**, 952 (1996);
R. Jackiw and S.-Y. Pi, Phys. Rev. D **42**, 3500 (1990);
O. Bergman and G. Lozano, Ann. Phys. (N.Y.) **229**, 416 (1994);
M. Boz, V. Fainberg, and N.K. Pak, Phys. Lett. A **207**,1 (1995); Ann. Phys (N.Y.) **246**, 347 (1996);
M. Gomes, J.M.C. Malbouisson, and A.J. da Silva, Phys. Lett. A **236**, 373 (1997); Int. J. Mod. Phys. A **13**, 3157 (1998); (hep-th/0007080).

Path integrals in multiply connected spaces and their history are discussed in the textbook
L.S. Schulman, *Techniques and Applications of Path Integration*, Wiley, New York, 1981.

Details on Lippmann-Schwinger equation is found in most standard textbooks, say
S.S. Schweber, *Relativistic Quantum Field Theory*, Harper and Row, New York, 1961, Section 11b.

In chemistry, the properties of self-entangled polymer rings, called *catenanes*, were first investigated by

H.L. Frisch and E. Wasserman, J. Am. Chem. Soc. **83**, 3789 (1961).
Their existence was proved mass-spectroscopically by
R. Wolovsky, J. Am. Chem. Soc. **92**, 2132 (1961);
D.A. Ben-Efraim, C. Batich, and E. Wasserman, J. Am. Chem. Soc. **92**, 2133 (1970).

In optics, the Kirchhoff diffraction formula can be rewritten as a path integral formula with linking terms:
J.H. Hannay, Proc. Roy. Soc. Lond. A **450**, 51 (1995),

In biophysics,
J.C. Wang, Accounts Chem. Res. D **10**, 2455 (1974)
showed that DNA molecules can get entangled and must be disentangled during replication.

The path integral approach to the entanglement problem in polymer systems was pioneered by
S.F. Edwards, Proc. Phys. Soc. **91**, 513 (1967);
S.F. Edwards, J. Phys. A **1**, 15 (1968).
See also
M.G. Brereton and S. Shaw, J. Phys. A **13**, 2751 (1980)
and later works of these authors.

Investigations via Monte Carlo simulations were made by
A.V. Vologodskii, A.V. Lukashin, M.D. Frank-Kamenetskii, and V.V. Anshelevin, Sov. Phys. JETP **39**, 1095 (1974);
A.V. Vologodskii, A.V. Lukashin, and M.D. Frank-Kamenetskii, Sov. Phys.-JETP **40**, 932 (1975).
See also the review article by
M.D. Frank-Kamenetskii and A.V. Vologodskii, Sov. Phys. Usp. **24**, 679 (1982).
This article also discussed ribbons.

For further computer work on knot distributions see
J.P.J. Michels and F.W. Wiegel, Phys. Lett. A **9**, 381 (1982); Proc. Roy. Soc. A **403**, 269 (1986),
and references therein. The work is summarized in the textbook by
F.W. Wiegel, *Introduction to Path-Integral Methods in Physics and Polymer Science*, World Scientific, Singapore, 1986.
See also
S. Windwer, J. Chem. Phys. **93**, 765 (1990).
The parameter C at the end of Section 6.4 was found by
A. Kholodenko, Phys. Lett. A **159**, 437 (1991),
who mapped the problem onto a q-state Potts model with $q = 4$. This mapping gives $\alpha = 0$ and $C = 2e^{-\pi/6} \approx 1.18477$.

For the Gauss integral as a topological invariant of links see the original paper by
G.F. Gauss, Koenig. Ges. Wiss. Goettingen **5**, 602 (1877).
The *writhing number* W_r was introduced by
F.B. Fuller, Proc. Nat. Acad. Sci. USA **68**, 815 (1971),
who applied the mathematical relation to DNA. See also
F.H.C. Crick, Proc. Nat. Acad. Sci. USA **68**, 2639 (1976).
The relation $L_k = T_w + W_r$ was first written down by
G. Calagareau, Rev. Math. Pur. et Appl. **4**, 58 (1959); Czech. Math. J. **4**, 588 (1961),
and extended by
J.H. White, Am. J. Math. **90**, 1321 (1968).

In particle physics, ribbons are used to construct path integrals over fluctuating fermion orbits:
A.M. Polyakov, Mod. Phys. Lett. A **3**, 325 (1988).
For more details see

C.H. Tze, Int. J. Mod. Phys. A **3**, 1959 (1988).

The construction of the Alexander polynomial of links is described in
A.V. Vologodskii, A.V. Lukashin, and M.D. Frank-Kamenetskii, JETP **40**, 932 (1974).
Their derivation from the skein relations is shown in
J.H. Conway, *An Enumeration of Knots and Links*, Pergamon, London, 1970, pp. 329–358;
L.H. Kauffman, Topology **20**, 101 (1981).
In the mathematical literature, the various knot polynomials are discussed by
L.H. Kauffman, Topology **26**, 395 (1987); Contemporary Mathematics AMS **78**, 283 (1988); Trans.
Amer. Math. Soc. 318, 417 (1990); *On Knots*, Princeton University Press, Princeton, 1987; *Knots
and Physics*, World Scientific, Singapore, 1991; J. Math. Phys. **36**, 2402 (1995).
V. Jones, Bull. Am. Math. Soc. **12**, 103 (1985); Ann. Math. **126**, 335 (1987);
P. Freyd, D. Yetter, J. Hoste, W. B. R. Lickorish, K.C. Millet, and A. Ocneanu, Bull. Am. Math.
Soc. **12**, 239 (1985);
W. B. R. Lickorish and K.C. Millet, Math. Magazine **61**, 3 (1987).
The lower HOMFLY polynomials are tabulated in the text. For the higher ones see the microfilm
accompanying the article
H. Doll and J. Hoste, Math. of Computation **57**, 747 (1991)
and the unpublished tables by
M.B. Thistlethwaite, University of Knoxville, Tennessee.
The author is grateful for a copy of them.
A collection of many relevant articles is found in
T. Kuhno (ed.), *New Developments in the Theory of Knots*, World Scientific, Singapore, 1990.
A short introduction to the classification problem of knots is given in the popular articles
W.F.R. Jones, Scientific American, November 1990, p. 52,
I. Stewart, Spektrum der Wissenschaft, August 1990, p. 12.

The Chern-Simons actions have in recent years received increasing attention due to their relevance
for explaining the fractional quantum Hall effect and a possible presence in high-temperature superconductivity . Actions of this type were first observed in four-dimensional quantum field theories
in the form of so-called anomalies by
J. Wess and B. Zumino, Phys. Lett. B **36**, 95 (1971).
The action (16.253) in three spacetime dimensions was first analyzed by
S. Deser, R. Jackiw, and S. Templeton, Ann. Phys. **140**, 372 (1982),
who pointed out the connection with the Chern classes of differential geometry described by
S. Chern, *Complex Manifolds without Potential Theory*, Springer, Berlin, 1979.
In particular they found the mass of the electromagnetic field which was conjectured to be the
origin of the Meissner effect in high-temperature superconductors. See
A.L. Fetter, C. Hanna, and R.B. Laughlin, Phys. Rev. B **39**, 9679 (1989);
Y.-H. Chen and F. Wilczek, Int. J. Mod. Phys. B **3**, 117 (1989);
Y.-H. Chen, F. Wilczek, E. Witten, and B.I. Halperin, Int. J. Mod. Phys. B **3**, 1001 (1989);
A. Schakel, Phys. Rev. D **44**, 1198 (1992).
However, the recent finding of dissipation in anyonic systems by
D.V. Khveshchenko and I.I. Kogan, Int. J. Mod. Phys. B **5**, 2355 (1991) speaks against an anyon
mechanism of this phenomenon.

A Chern-Simons type of action appeared when integrating out fermions in
H. Kleinert, Fortschr. Phys. **26**, 565 (1978) (http://www.physik.fu-berlin.de/~kleinert/55).

The relation with Chern classes was recognized by
M.V. Berry, Proc. Roy. Soc. A **392**, 45 (1984);
B. Simon, Phys. Rev. Lett. **51**, 2167 (1983).
The Chern-Simons action in the text was derived for a degenerate electron liquid in two dimensions
by

T. Banks and J.D. Lykken, Nucl. Phys. B **336**, 500 (1990);
S. Randjbar-Daemi, A. Salam, and J. Strathdee, Nucl. Phys. B **340**, 403 (1990),
P.K. Panigrahi, R. Ray, and B. Sakita, Phys. Rev. B **42**, 4036 (1990).
There is related work by
M. Stone, Phys. Rev. D **33**, 1191 (1986);
I.J.R. Aitchison, Acta Physica Polonica B **18**, 207 (1987).
See also the reprints of many papers on this subject in
A. Shapere and F. Wilczek, *Geometric Phases in Physics*, World Scientific, Singapore, 1989.
F. Wilczek, *Fractional Statistics and Anyon Superconductivity*, World Scientific, Singapore, 1990,
which itself provides a clear introduction to the subject and contains many important reprints. A
good review is also contained in the lectures
J.J. Leinaas, *Topological Charges in Gauge Theories*, Nordita Preprint, 79/43, ISSN 0106-2646.

Textbooks on this subject are
A. Lerda, *Anyons-Quantum Mechanics of Particles with Fractional Statistics*, Lecture Notes in
Physics, m14, Springer, Berlin 1992;
A. Khare, *Fractional Statistics and Quantum Theory*, World Scientific, Singapore, 1997.
The Lerda book contains many useful examples and explains the origin of difficulties in treating
interacting anyons. The Khare book provides a well-motivated treatment and includes a brief
introduction to the Braid group. Both include discussions of the Quantum Hall Effect and Anyon
Superconductivity.

For the relation between the Chern-Simons theory and knot polynomials see
E. Witten, Comm. Math. Phys. **121**, 351 (1989), Nucl. Phys. B **330**, 225 (1990).
See also
P. Cotta-Ramusino, E. Guadagnini, M. Martellini, and M. Mintchev, Nucl. Phys. B **330**, 557
(1990);
G.V. Dunne, R. Jackiw, and C. Trugenburger, Ann. Phys. **194**, 197 (1989);
A. Polychronakos, Ann. Phys. **203**, 231 (1990);
E. Guadagnini, I. J. Mod. Phys. A **7**, 877 (1992).

The integer quantum Hall effect was found by
K. vonKlitzing, G. Dorda, and M. Pepper, Phys. Rev. Lett. **45**, 494 (1980);
the fractional one by
D.C. Tsui, H.L. Stormer, and A.C. Gossard, Phys. Rev. Lett. **48**, 1559 (1980).
Theoretical explanations are given in
R.B. Laughlin, Phys. Rev. Lett. **50**, 1395 (1983), Phys. Rev. B **23**, 3383 (1983);
F.D.M. Haldane, Phys. Rev. Lett. **51**, 605 (1983);
B.I. Halperin, Phys. Rev. Lett. **52**, 1583 (1984);
D.P. Arovas, J.R. Schrieffer, F. Wilczek, Phys. Rev. Lett. **53**, 722 (1984);
D.P. Arovas, J.R. Schrieffer, F. Wilczek, and A. Zee, Nucl. Phys. B **251**, 117 (1985);
J.K. Jain, Phys. Rev. Lett. **63**, 199 (1989).
The exceptional filling factor $\nu = \frac{5}{2}$ is discussed in
R. Willet et al., Phys. Rev. Lett. **59**, 17765 (1987);
S. Kivelson, C. Kallin, D.P. Arovas, and J.R. Schrieffer, Phys. Rev. Lett. **56**, 873 (1986).

The Chern-Simons path integral is treated semiclassically in
D.H. Adams, Phys. Lett. B **417**, 53 (1998) (hep-th/9709147).

For a simple discussion of the change from Bose to Fermi statistics at the level of creation and
annihilation operators via a topological interaction see
E. Fradkin, Phys. Rev. Lett. **63**, 322 (1989); *Field Theories of Condensed Matter Physics*,
Addison-Wesley, 1991.

The lattice form of the action $\sum_{\mathbf{x}} \epsilon_{\mu\nu\lambda} A^{\mu}(\mathbf{x}) \nabla_{\nu} A_{\lambda}(\mathbf{x})$ used by that author is not correct since

The lattice form of the action $\sum_{\mathbf{x}} \epsilon_{\mu\nu\lambda} A^\mu(\mathbf{x}) \nabla_\nu A_\lambda(\mathbf{x})$ used by that author is not correct since it violates gauge invariance. This can, however, easily be restored without destroying the results by replacing the first $A_\mu(\mathbf{x})$-field by $A_\mu(\mathbf{x} - \mathbf{e}_\mu)$, where $\mathbf{e}_\mu$ is the unit vector in the μ-direction. See the general discussion of lattice gauge transformations in
H. Kleinert, *Gauge Fields in Condensed Matter*, Vol. I, World Scientific, Singapore, 1989, Chapter 8 (http://www.physik.fu-berlin.de/~kleinert/b1).

See furthermore
D. Eliezer and G.W. Semenoff, *Anyonization of Lattice Chern-Simons Theory*, Ann. Phys. **217**, 66 (1992).

For the London equations see the original paper by
F. London and H. London, Proc. Roy. Soc. A **147**, 71 (1935)
and the extension thereof:
A.B. Pippard, *ibid.*, A **216**, 547 (1953).

The individual citations refer to

[1] P.A.M. Dirac, Proc. Roy. Soc. A **133**, 60 (1931); Phys. Rev. **74**, 817 (1948), Phys. Rev. **74**, 817 (1948).
    See also
    J. Schwinger, *Particles, Sources and Fields*, Vols. 1 and 2, Addison Wesley, Reading, Mass., 1970 and 1973.

[2] For a review, see
    G. Giacomelli, in *Monopoles in Quantum Field Theory*, World Scientific, Singapore, 1982, edited by N.S. Craigie, P. Goddard, and W. Nahm, p. 377.

[3] B. Cabrera, Phys. Rev. Lett. **48**, 1378 (1982).

[4] For a detailed discussion of the physics of vortex lines in superconductors, see
    H. Kleinert, *Gauge Fields in Condensed Matter*, World Scientific, Singapore, 1989, Vol. I, p. 331 (http://www.physik.fu-berlin.de/~kleinert/b1).

[5] See the reprint collection
    A. Shapere and F. Wilczek, *Geometric Phases in Physics*, World Scientific, Singapore, 1989. In particular the paper by
    D.P. Arovas, *Topics in Fractional Statistics*, p.284.

[6] The Tait number or writhe must not be confused with the writhing number W_r introduced in Section 16.6 which is in general noninteger. See
    P.G. Tait, *On Knots I, II, and III*, Scientific Papers, Vol. 1. Cambridge, England: University Press, pp. 273-347, 1898.

[7] See Ref. [7] of Chapter 19. There exists, unfortunately, no obvious extension to four space-time dimensions.

[8] The development in this Section follows
    F. Ferrari, H. Kleinert, and I. Lazzizzera, Phys. Lett. A **276**, 1 (2000) (cond-mat/0002049); Eur. Phys. J. B **18**, 645 (2000) (cond-mat/0003355); nt. Jour. Mod. Phys. B **14**, 3881 (2000) (cond-mat/0005300); *Topological Polymers: An Application of Chern-Simons Field Theories*, in K. Lederer and N. Aust (eds.), *Chemical and Physical Aspects of Polymer Science and Engineering*, 5-th Oesterreichische Polymertage, Leoben 2001, Macromolecular Symposia, 1st Edition, ISBN 3-527-30471-1, Wiley-VCH, Weinheim, 2002.

[9] M.G. Brereton and S. Shah, J. Phys. A: Math. Gen. *15*, 989 (1982).

[10] R. Jackiw, in *Current Algebra and Anomalies*, ed. by S.B. Treiman, R. Jackiw, B. Zumino, and E. Witten, World Scientific, Singapore, 1986, p. 211.

[11] L.D. Faddeev and V.N. Popov, Phys. Lett. B *25*, 29 (1967).

So many paths, that wind and wind,
While just the art of being kind
Is all the sad world needs.

ELLA WILCOX (1855-1919), The World's Needs

17

Tunneling

Tunneling processes govern the decay of metastable atomic and nuclear states, as well as the transition of overheated or undercooled thermodynamic phases to a stable equilibrium phase. Path integrals are an important tool for describing these processes theoretically. For high tunneling barriers, the decay proceeds slowly and its properties can usually be explained by a semiclassical expansion of a simple model path integral. By combining this expansion with the variational methods of Chapter 5, it is possible to extend the range of applications far into the regime of low barriers.

In this chapter we present a novel theory of tunneling through high and low barriers and discuss several typical examples in detail. A useful fundamental application arises in the context of perturbation theory since the large-order behavior of perturbation expansions is governed by semiclassical tunneling processes. Here the new theory is used to calculate perturbation coefficients to any order with a high degree of accuracy.

17.1 Double-Well Potential

A simple model system for tunneling processes is the symmetric double-well potential of Eq. (5.78). It may be rewritten in the form

$$V(x) = \frac{\omega^2}{8a^2}(x-a)^2(x+a)^2, \tag{17.1}$$

which exhibits the two degenerate symmetric minima at $x = \pm a$ (see Fig. 17.1). The coupling strength is

$$g = \omega^2/2a^2. \tag{17.2}$$

Near the minima, the potential looks approximately like a harmonic oscillator potential $V_\pm(x) = \omega^2(x \mp a)^2/2$:

$$V(x) = \frac{\omega^2}{2}(x \mp a)^2 \left(1 \pm \frac{x \mp a}{a} + \dots\right) \equiv V_\pm(x) + \Delta V_\pm(x) + \dots. \tag{17.3}$$

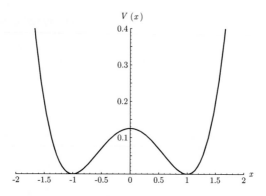

Figure 17.1 Plot of symmetric double-well potential $V(x) = (x - a)^2(x + a)^2\omega^2/8a^2$ for $\omega = 1$ and $a = 1$.

The height of the potential barrier at the center is

$$V_{\mathrm{max}} = \frac{(\omega a)^2}{8}. \tag{17.4}$$

In the limit $a \to \infty$ at a fixed frequency ω, the barrier height becomes infinite and the system decomposes into a sum of two independent harmonic-oscillator potentials widely separated from each other. Correspondingly, the wave functions of the system should tend to two separate sets of oscillator wave functions

$$\psi_n(\Delta x_\pm) \to \left(\frac{\omega}{\pi\hbar}\right)^{1/4} \frac{1}{2^{n/2}\sqrt{n!}} e^{-\omega(\Delta x_\pm)^2/2\hbar} H_n(\Delta x_\pm\sqrt{\omega/\hbar}), \tag{17.5}$$

where the quantities

$$\Delta x_\pm \equiv x \pm a \tag{17.6}$$

measure the distances of the point x from the respective minima.

A similar separation occurs in the time evolution amplitude which decomposes into the sum of the amplitudes of the individual oscillators

$$(x_b t_b | x_a t_a) \xrightarrow{a \to \infty} (x_b t_b | x_a t_a)_- + (x_b t_b | x_a t_a)_+ \tag{17.7}$$

$$\equiv \int \mathcal{D}x(t) \exp\left\{\frac{i}{\hbar}\int_{t_a}^{t_b} dt\frac{1}{2}[\dot{x}^2 - \omega^2(x + a)^2]\right\} + (a \to -a).$$

For convenience, we have assumed a unit particle mass $M = 1$ in the Lagrangian of the system:

$$L = \frac{\dot{x}^2}{2} - V(x). \tag{17.8}$$

If a is no longer infinite, a particle in either of the two oscillator wells has a nonvanishing amplitude for tunneling through the barrier to the other well, and the wave functions of the right- and left-hand oscillators are mixed with each other. Since the action is symmetric under the mirror reflection $x \to -x$, the solutions of the Schrödinger equation

$$\hat{H}\psi(x,t) = H(-i\partial_x, x)\psi(x,t) = i\hbar\partial_t\psi(x,t), \tag{17.9}$$

with the Hamiltonian

$$H(p,x) = \frac{p^2}{2} + V(x), \tag{17.10}$$

can be separated into symmetric and antisymmetric wave functions. As usual, the symmetric states have a lower energy than the antisymmetric ones since a smaller number of nodes implies less kinetic energy for the particles.

If the distance parameter a is very large, then, to leading order in $a \to \infty$, the lowest two wave functions coincide approximately with the symmetric and antisymmetric combinations of the harmonic-oscillator wave functions

$$\psi_{\mathrm{s,a}} \approx \frac{1}{\sqrt{2}}[\psi_0(x-a) \pm \psi_0(x+a)]. \tag{17.11}$$

Due to tunneling, the lowest two energies show some deviation from the harmonic ground state value

$$E_{\mathrm{s,a}}^{(0)} = \frac{1}{2}\hbar\omega + \Delta E_{\mathrm{s,a}}^{(0)}. \tag{17.12}$$

At a large distance parameter a, this deviation is very small. In quantum mechanics, the level shifts $\Delta E_{\mathrm{s,a}}$ can be calculated in lowest-order perturbation theory by inserting the approximate wave functions (17.11) into the formula

$$\Delta E_{\mathrm{s,a}} = \int dx \psi_{\mathrm{s,a}} \hat{H} \psi_{\mathrm{s,a}}. \tag{17.13}$$

Since the wave functions $\psi_0(x \pm a)$ of the individual potential wells fall off exponentially like $e^{-x^2/2}$ at large x, the level shifts $\Delta E_{\mathrm{s,a}}$ are exponentially small in the square distance a^2.

In this chapter we derive the level shifts $\Delta E_{\mathrm{s}}, \Delta E_{\mathrm{a}}$ and the related tunneling amplitudes from the path integral of the system. For large a, this will be relatively simple since we can have recourse to the semiclassical approximation developed in Chapter 4 which becomes exact in the limit $a \to \infty$. As long as we are interested only in the lowest two states, the problem can immediately be simplified. We take the spectral representation of the amplitude

$$
\begin{aligned}
(x_b t_b | x_a t_a) &= \int \mathcal{D}x(t) e^{(i/\hbar)\int_{t_a}^{t_b} dt[\dot{x}^2/2 - V(x)]} \\
&= \sum_n \psi_n(x_b)\psi_n(x_a) e^{-iE_n(t_b - t_a)/\hbar}
\end{aligned} \tag{17.14}
$$

to imaginary times $t_{a,b} \to \tau_{a,b} = \mp iL/2$, where it becomes

$$
\begin{aligned}
(x_b \ L/2|x_a - L/2) &= \int \mathcal{D}x(\tau) e^{-(1/\hbar) \int_{-L/2}^{L/2} d\tau [x'^2/2 + V(x)]} \\
&= \sum_n \psi_n(x_b) \psi_n(x_a) e^{-E_n L/\hbar},
\end{aligned}
\tag{17.15}
$$

with the notation $x'(\tau) \equiv dx(\tau)/d\tau$. In the limit of large L, the spectral sum (17.15) is obviously most sensitive to the lowest energies, the contributions of the higher energies E_n being suppressed exponentially. Thus, to calculate the small level shifts of the two lowest states, $\Delta E_{s,a}$, we have only to find the leading and subleading exponential behaviors. Since the wave functions are largest close to the bottoms of the double well at $x \sim \pm a$, we may consider the amplitudes with the initial and final positions x_a and x_b lying precisely at the bottoms, once on the same side of the potential barrier,

$$
(a \ L/2|a \ -L/2) = (-a \ L/2|-a \ -L/2),
\tag{17.16}
$$

and once on the opposite sides

$$
(a \ L/2|-a \ -L/2) = (-a \ L/2|a \ -L/2).
\tag{17.17}
$$

For these amplitudes we now calculate the semiclassical approximation in the limit $L \to \infty$. The results will lead to level shift formula in Section 17.7.

17.2   Classical Solutions — Kinks and Antikinks

According to Chapter 4, the leading exponential behavior of the semiclassical approximation is obtained from the classical solutions to the path integral. The fluctuation factor requires the calculation of the quadratic fluctuation correction. The result has the form

$$
\sum_{\text{class. solutions}} \exp\{-\mathcal{A}_{\text{cl}}/\hbar\} \times F,
\tag{17.18}
$$

where $\mathcal{A}_{\text{cl}}$ denotes the action of each classical solution and F the fluctuation factor.

The amplitude (17.16), which contains the bottom of the same well on either side, is dominated by a trivial classical solution which remains all the time at the same bottom:

$$
x(\tau) \equiv \pm a.
\tag{17.19}
$$

Classical solutions exist also for the other amplitudes (17.17) which connect the different bottoms at $-a$ and a. These solutions cross the barrier and read, in the limit $L \to \infty$,

$$
x(\tau) = x_{\text{cl}}^{\pm}(\tau) \equiv \pm a \tanh[\omega(\tau - \tau_0)/2],
\tag{17.20}
$$

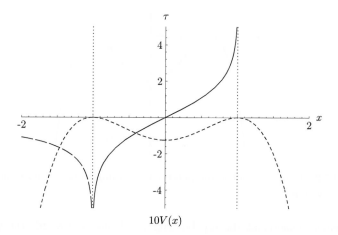

Figure 17.2 Classical kink solution (solid curve) in double-well potential (short-dashed curve with units marked on the lower half of the vertical axis). The solution connects the two degenerate maxima in the reversed potential. The long-dashed curve shows a solution which starts out at a maximum and slides down into the adjacent abyss.

with an arbitrary parameter τ_0 specifying the point on the imaginary time axis where the crossing takes place. The crossing takes place within a time of the order of $2/\omega$. For large positive and negative τ, the solution approaches $\pm a$ exponentially (see Fig. 17.2). Alluding to their shape, the solutions $x_{\mathrm{cl}}^{\pm}(\tau)$ are called *kink* and *antikink* solutions, respectively.[1]

To derive these solutions, consider the equation of motion in real time,

$$\ddot{x}(t) = -V'(x(t)), \qquad (17.21)$$

where $V'(x) \equiv dV(x)/dx$. In the Euclidean version with $\tau = -it$, this reads

$$x''(\tau) = V'(x(\tau)). \qquad (17.22)$$

Since the differential equation is of second order, there is merely a sign change in front of the potential with respect to the real-time differential equation (17.21). The Euclidean equation of motion corresponds therefore to a usual equation of motion of a point particle in real time, whose potential is turned upside down with respect to Fig. 17.1. This is illustrated in Fig. 17.3. The reversed potential allows obviously for a classical solution which starts out at $x = -a$ for $\tau \to -\infty$ and arrives at $x = a$ for $\tau \to +\infty$. The particle needs an infinite time to leave the initial potential mountain and to climb up to the top of the final one. The movement through the central valley proceeds within the finite time $\approx 2/\omega$. If the particle does not start

---

[1]In field-theoretic literature, such solutions are also referred to as *instanton* or *anti-instanton* solutions, since the valley is crossed within a short time interval. See the references quoted at the end of the chapter.

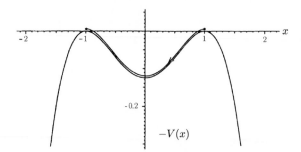

Figure 17.3 Reversed double-well potential governing motion of position x as function of imaginary time τ.

its movement exactly at the top but slightly displaced towards the valley, say at $x = -a + \epsilon$, it will reach $x = a - \epsilon$ after a finite time, then return to $x = -a + \epsilon$, and oscillate back and forth forever. In the limit $\epsilon \to 0$, the period of oscillation goes to infinity and only a single crossing of the valley remains.

To calculate this movement, the differential equation (17.22) is integrated once after multiplying it by $x' = dx/d\tau$ and rewriting it as

$$\frac{1}{2}\frac{d}{d\tau}x'^2 = \frac{d}{d\tau}V(x(\tau)). \tag{17.23}$$

The integration gives

$$\frac{x'^2}{2} + [-V(x(\tau))] = \text{const}. \tag{17.24}$$

If τ is reinterpreted as the physical time, this is the law of *energy conservation* for the motion in the reversed potential $-V(x)$. Thus we identify the integration constant in (17.24) as the total energy E in the reversed potential:

$$\text{const} \equiv E. \tag{17.25}$$

Integrating (17.24) further gives

$$\tau - \tau_0 = \pm\frac{1}{\sqrt{2}}\int_{x(\tau_0)}^{x(\tau)}\frac{dx}{\sqrt{E + V(x)}}. \tag{17.26}$$

A look at the potential in Fig. 17.3 shows that an orbit starting out with the particle at rest for $\tau \to -\infty$ must have $E = 0$. Inserting the explicit potential (17.1) into (17.26), we obtain for $|x| < a$

$$\begin{aligned}
\tau - \tau_0 &= \pm\frac{2a}{\omega}\int_0^x\frac{dx'}{(a-x')(x'+a)} = \pm\frac{1}{\omega}\log\frac{a+x}{a-x} \\
&= \pm\frac{2}{\omega}\text{arctanh}\frac{x}{a}.
\end{aligned} \tag{17.27}$$

Thus we find the kink and antikink solutions crossing the barrier:

$$x_{cl}(\tau) = \pm a \tanh[(\tau - \tau_0)\omega/2]. \tag{17.28}$$

The Euclidean action of such a classical object can be calculated as follows [using (17.24) and (17.25)]:

$$
\begin{aligned}
\mathcal{A}_{cl} &= \int_{-\infty}^{\infty} d\tau \left[\frac{x_{cl}'^2}{2} + V(x_{cl}(\tau)) \right] = \int_{-\infty}^{\infty} d\tau (x_{cl}'^2 - E) \\
&= -EL + \int_{-a}^{a} dx \sqrt{2[E + V(x)]}.
\end{aligned} \tag{17.29}
$$

The kink has $E = 0$, so that

$$\sqrt{2[E + V(x)]} = \frac{\omega}{2a}(a^2 - x^2), \tag{17.30}$$

and the classical action becomes

$$\mathcal{A}_{cl} = \frac{\omega}{2a} \int_{-a}^{a} dx (a^2 - x^2) = \frac{2}{3}a^2\omega = \frac{\omega^3}{3g}. \tag{17.31}$$

Note that for $E = 0$, the classical action is also given by the integral

$$\mathcal{A}_{cl} = \int_{-\infty}^{\infty} d\tau\, x_{cl}'^2. \tag{17.32}$$

There are also solutions starting out at the top of either mountain and sliding down into the adjacent exterior abyss, for instance (see again Fig. 17.3)

$$
\begin{aligned}
\tau - \tau_0 &= \mp \frac{2a}{\omega} \int_{x}^{\infty} \frac{dx'}{(x' - a)(x' + a)} = \pm \frac{1}{\omega} \log \frac{x + a}{x - a} \\
&= \pm \frac{2}{\omega} \operatorname{arccoth} \frac{x}{a}.
\end{aligned} \tag{17.33}
$$

However, these solutions cannot connect the bottoms of the double well with each other and will not be considered further.

Being in the possession of the classical solutions (17.19) and (17.28) with a finite action, we are now ready to write down the classical contributions to the amplitudes (17.16) and (17.17). According to the semiclassical formula (17.18), they are

$$(a\, L/2|a\, -L/2) = 1 \times F_\omega(L) \tag{17.34}$$

and

$$(a\, L/2|-a\, -L/2) = e^{-\mathcal{A}_{cl}/\hbar} \times F_{cl}(L). \tag{17.35}$$

The factor 1 in (17.34) emphasizes the vanishing action of the trivial classical solution (17.19). The exponential $e^{-\mathcal{A}_{cl}/\hbar}$ contains the action of the kink solutions (17.28).

The degeneracy of the solutions in τ_0 is accounted for by the fluctuation factor $F_{\rm cl}(L)$, as will be shown below.

Actually, the classical kink and antikink solutions (17.28) do not occur exactly in (17.35) since they reach the well bottoms at $x = \pm a$ only at infinite Euclidean times $\tau \to \pm\infty$. For the amplitude to be calculated we need solutions for which x is equal to $\pm a$ at large but finite values $\tau = \pm L/2$. Fortunately, the error can be ignored since for large L the kink and antikink solutions approach $\pm a$ exponentially fast. As a consequence the action of a proper solution which would reach $\pm a$ at a finite L differs from the action $\mathcal{A}_{\rm cl}$ only by terms which tend to zero like $e^{-\omega L}$. Since we shall ultimately be interested only in the large-L limit we can neglect such exponentially small deviations.

In the following section we determine the fluctuation factors $F_{\rm cl}$.

17.3 Quadratic Fluctuations

The semiclassical limit includes the effects of the quadratic fluctuations. These are obtained after approximating the potential around each minimum by a harmonic potential and keeping only the lowest term in the expansion (17.3). The fluctuation factor of a pure harmonic oscillator of frequency ω and unit mass has been calculated in Section 2.3 with the result

$$F_\omega(L) = \sqrt{\frac{\omega}{2\pi\hbar \sinh \omega L}} \sim \sqrt{\frac{\omega}{\pi\hbar}} e^{-\omega L/2} + \mathcal{O}(e^{-3\omega L/2}). \qquad (17.36)$$

The leading exponential at large L displays the ground state energy $\omega/2$, while the corrections contain all information on the exited states whose energy is $(n + 1/2)\omega$ with $n = 1, 2, 3, \ldots$.

Note that according to the spectral representation of the amplitude (17.15), the factor $\sqrt{\omega/\pi\hbar}$ in (17.36) must be equal to the square of the ground state wave function $\Psi_0(\Delta x_\pm)$ at the potential minimum. This agrees with (17.5).

Consider now the fluctuation factor of a single kink contribution. It is given by the path integral over the fluctuations $y(\tau) \equiv \delta x(\tau)$

$$F_{\rm cl}(L) = \int \mathcal{D}y(\tau) e^{-(1/\hbar)\int_{-L/2}^{L/2} d\tau (1/2)[y'^2 + V''(x_{\rm cl}(\tau))y^2]}, \qquad (17.37)$$

where $x_{\rm cl}(\tau)$ is the kink solution, and $y(\tau)$ vanishes at the endpoints:

$$y(L/2) = y(-L/2) = 0. \qquad (17.38)$$

Suppose for the moment that $L = \infty$. Then the kink solution is given by (17.28) and we obtain the fluctuation potential

$$\begin{aligned}
V''(x_{\rm cl}(\tau)) &= \frac{3}{2}\frac{\omega^2}{a^2}x_{\rm cl}^2(\tau) - \frac{1}{2}\omega^2 = \omega^2\left(\frac{3}{2}\tanh^2[\omega(\tau - \tau_0)/2] - \frac{1}{2}\right) \\
&= \omega^2\left(1 - \frac{3}{2}\frac{1}{\cosh^2[\omega(\tau - \tau_0)/2]}\right).
\end{aligned} \qquad (17.39)$$

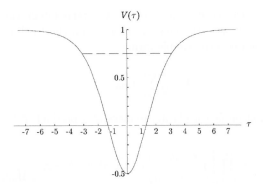

Figure 17.4 Potential (17.41) for quadratic fluctuations around kink solution (17.28) in Schrödinger equation. The dashed lines indicate the bound states at energies 0 and $3\omega^2/4$.

Thus, the quadratic fluctuations are governed by the Euclidean action

$$\mathcal{A}_{\mathrm{fl}}^0 = \int_{-L/2}^{L/2} d\tau \frac{1}{2} \left[y'^2 + \omega^2 \left(1 - \frac{3}{2} \frac{1}{\cosh^2[\omega(\tau - \tau_0)/2]} \right) y^2 \right]. \tag{17.40}$$

The rules for doing a functional integral with a quadratic exponent were explained in Chapter 2. The paths $y(\tau)$ are expanded in terms of eigenfunctions of the differential equation

$$\left[-\frac{d^2}{d\tau^2} + \omega^2 \left(1 - \frac{3}{2} \frac{1}{\cosh^2[\omega(\tau - \tau_0)/2]} \right) \right] y_n(\tau) = \lambda_n y_n(\tau), \tag{17.41}$$

with λ_n being the eigenvalues. This is a Schrödinger equation for a particle moving along the τ-axis in an attractive potential well of the Rosen-Morse type [compare (14.157) and see Fig. 17.4]:

$$V(\tau) = \omega^2 \left(1 - \frac{3}{2} \frac{1}{\cosh^2[\omega(\tau - \tau_0)/2]} \right). \tag{17.42}$$

The eigenfunctions $y_n(\tau)$ satisfy the usual orthonormality condition

$$\int_{-\infty}^{\infty} d\tau y_n(\tau) y_{n'}(\tau) = \delta_{nn'}. \tag{17.43}$$

Given a complete set of these solutions $y_n(\tau)$ with $n = 0, 1, 2, \ldots$, we now perform the *normal-mode expansion*

$$y^{\xi_0, \xi_1, \ldots}(\tau) = \sum_{n=0}^{\infty} \xi_n y_n(\tau). \tag{17.44}$$

After inserting this into (17.40), we perform a partial integration in the kinetic term and the τ-integrals with the help of (17.43), and the Euclidean action of the quadratic fluctuations takes the simple form

$$\mathcal{A} = \frac{1}{2} \sum_{n=0}^{\infty} \lambda_n \xi_n^2. \tag{17.45}$$

With this, the fluctuation factor (17.37) reduces to a product of Gaussian integrals over the normal modes

$$F_{\text{cl}}(L) = \mathcal{N} \prod_{n=0}^{\infty} \left[\int_{-\infty}^{\infty} \frac{d\xi_n}{\sqrt{2\pi\hbar}} \right] e^{-\sum_{n=0}^{\infty} \xi_n^2 \lambda_n / 2\hbar} = \mathcal{N} \frac{1}{\sqrt{\Pi_n \lambda_n}}. \tag{17.46}$$

The normalization constant $\mathcal{N}$, to be calculated below, accounts for the Jacobian which relates the time-sliced measure to the normal mode measure.

First we shall calculate the eigenvalues λ_n. For this we use the amplitude of the Rosen-Morse potential obtained via the Duru-Kleinert transformation (14.160) in Eq. (14.161). If the potential is written in the form

$$V(\tau) = \omega^2 - \frac{V_0}{\cosh^2[m(\tau - \tau_0)]}, \tag{17.47}$$

there are bound states for $n = 0, 1, 2, \ldots, n_{\text{max}} < s$, where s is defined by [1]

$$s \equiv \frac{1}{2} \left[-1 + \sqrt{1 + 4\frac{V_0}{m^2}} \right]. \tag{17.48}$$

Their wave functions are, according to (14.163), and (14.165),

$$y_n(\tau) = \sqrt{\frac{m}{n!}} \sqrt{\Gamma(s-n)\Gamma(1+2s-n)} \frac{2^{n-s}}{\Gamma(1+s-n)} \cosh^{n-s}[m(\tau - \tau_0)]$$
$$\times F(-n, 1+2s-n; s-n+1; \tfrac{1}{2}(1 - \tanh[m(\tau - \tau_0)])), \tag{17.49}$$

where $F(a, b; c; z)$ are the hypergeometric functions (1.450). In terms of s, the parameter V_0 becomes

$$V_0 = m^2 s(s+1). \tag{17.50}$$

The bound-state energies are

$$\lambda_n^2 = \omega^2 - m^2 (s-n)^2. \tag{17.51}$$

In the Schrödinger equation (17.41), we have $m = \omega/2$, $V_0 = 3\omega^2/2$, so that $s = 2$ and there are exactly two bound-state solutions. These are

$$y_0(\tau) = -\sqrt{\frac{3\omega}{8}} \frac{1}{\cosh^2[\omega(\tau - \tau_0)/2]} \tag{17.52}$$

and

$$y_1(\tau) = \sqrt{\frac{3\omega}{4}} \frac{1}{\cosh[\omega(\tau-\tau_0)/2]} F(-1,4;2;\tfrac{1}{2}(1-\tanh[\omega(\tau-\tau_0)/2]))$$

$$= \sqrt{\frac{3\omega}{4}} \frac{\sinh[\omega(\tau-\tau_0)/2]}{\cosh^2[\omega(\tau-\tau_0)/2]}. \tag{17.53}$$

The negative sign in (17.52) is a matter of convention, to give $y_1(\tau)$ the same sign as $x_{\rm cl}(\tau-\tau_0)$ in Eq. (17.89) below. The normalization factors can be checked using the formula

$$\int_0^\infty \frac{\sinh^\mu x}{\cosh^\nu x} = \frac{1}{2} B\left(\frac{\mu+1}{2}, \frac{\nu-\mu}{2}\right), \tag{17.54}$$

with $B(x,y) \equiv \Gamma(x)\Gamma(y)/\Gamma(x+y)$ being the Beta function. The corresponding eigenvalues are

$$\lambda_0 = 0, \quad \lambda_1 = 3\omega^2/4. \tag{17.55}$$

The existence of a zero-eigenvalue mode is a general property of fluctuations around localized classical solutions in a system which is translationally invariant along the τ-axis. It prevents an immediate application of the quadratic approximation, since a zero-eigenvalue mode is not controlled by a Gaussian integral as the others are in Eq. (17.46). This difficulty and its solution will be discussed in Subsection 17.3.1.

In addition to the two bound states, there are continuum wave functions with $\lambda_n \geq \omega^2$. For an energy

$$\lambda_k = \omega^2 + k^2, \tag{17.56}$$

they are given by a linear combination of

$$y_k(\tau) \propto Ae^{ik\tau} F(s+1,-s;1-ik/m;\tfrac{1}{2}(1-\tanh[m(\tau-\tau_0)])) \tag{17.57}$$

and its complex-conjugate (see the end of Subsection 14.4.4). Using the identity for the hypergeometric function[2]

$$F(a,b,c;z) = \frac{\Gamma(c)\Gamma(c-a-b)}{\Gamma(c-a)\Gamma(c-b)} F(a,b;a+b-c+1;1-z)$$

$$+ (1-z)^{c-a-b}\frac{\Gamma(c)\Gamma(-c+a+b)}{\Gamma(a)\Gamma(b)} F(c-a,c-b;c-a-b+1;1-z), \tag{17.58}$$

and $F(a,b;c;0) = 1$, we find the asymptotic behavior

$$F \xrightarrow{\tau\to\infty} 1, \tag{17.59}$$

$$F \xrightarrow{\tau\to-\infty} \frac{\Gamma(-ik/m)\Gamma(1-ik/m)}{\Gamma(-s-ik/m)\Gamma(s+1-ik/m)} + e^{-2ik\tau}\frac{\Gamma(ik/m)\Gamma(1-ik/m)}{\Gamma(-s)\Gamma(1+s)}. \tag{17.60}$$

---

[2]M. Abramowitz and I. Stegun, *op. cit.*, formula 15.3.6.

These limits determine the asymptotic behavior of the wave functions (17.57). With an appropriate choice of the normalization factor in (17.57) we fulfill the standard scattering boundary conditions

$$\psi(\tau) \to \begin{cases} e^{ik\tau} + R_k e^{-ik\tau}, & \tau \to -\infty, \\ T_k e^{ik\tau}, & \tau \to \infty. \end{cases} \tag{17.61}$$

These define the transmission and reflection amplitudes. From (17.59) and (17.60) we calculate directly

$$T_k = \frac{\Gamma(-s - ik/m)\Gamma(s + 1 - ik/m)}{\Gamma(-ik/m)\Gamma(1 - ik/m)}, \tag{17.62}$$

$$R_k = \frac{\Gamma(-s - ik/m)\Gamma(s + 1 - ik/m)}{\Gamma(-s)\Gamma(1 + s)} \frac{\Gamma(ik/m)}{\Gamma(-ik/m)} = T_k \frac{\Gamma(ik/m)\Gamma(1 - ik/m)}{\Gamma(-s)\Gamma(1 + s)}. \tag{17.63}$$

Using the relation $\Gamma(z) = \pi/\sin(\pi z)\Gamma(1 - z)$, this can be written as

$$T_k = \frac{\Gamma(s + 1 - ik/m)}{\Gamma(s + 1 + ik/m)} \frac{\Gamma(1 + ik/m)}{\Gamma(1 - ik/m)} \frac{\sin(ik/m)}{\sin(s + ik/m)}, \tag{17.64}$$

$$R_k = T_k \frac{\sin(\pi s)}{\sin(ik/m)}. \tag{17.65}$$

The scattering matrix

$$S_k = \begin{pmatrix} T_k & R_k \\ R_k & T_k \end{pmatrix} \tag{17.66}$$

is unitary since

$$R_k T_k^* + R_k^* T_k = 0, \qquad |T_k|^2 + |R_k|^2 = 1. \tag{17.67}$$

It is diagonal on the state vectors

$$\psi^e = \frac{1}{\sqrt{2}} \begin{pmatrix} 1 \\ 1 \end{pmatrix}, \quad \psi^\o = \frac{1}{\sqrt{2}} \begin{pmatrix} 1 \\ -1 \end{pmatrix}, \tag{17.68}$$

which, as we shall prove below, correspond to odd and even partial waves. The respective eigenvalues $\lambda_k^{e,\o} = e^{2i\delta_k^{e,\o}}$ define the phase shifts $\delta_k^{e,\o}$, in terms of which

$$T_k = \frac{1}{2}(e^{2i\delta_k^e} + e^{2i\delta_k^\o}), \tag{17.69}$$

$$R_k = \frac{1}{2}(e^{2i\delta_k^e} - e^{2i\delta_k^\o}). \tag{17.70}$$

Let us verify the association of the eigenvectors (17.68) with the even and odd partial waves. For this we add to the wave function (17.61) the mirror-reflected solution

$$\psi^r(\tau) \to \begin{cases} T_k e^{-ik\tau}, & \tau \to -\infty, \\ e^{-ik\tau} + R_k e^{ik\tau}, & \tau \to \infty, \end{cases} \tag{17.71}$$

and obtain

$$\psi^{\mathrm{e}}(\tau) = \psi(\tau) + \psi^{\mathrm{r}}(\tau) \rightarrow \begin{cases} e^{ik\tau} + (R_k + T_k)e^{-ik\tau}, & \tau \rightarrow -\infty, \\ e^{-ik\tau} + (R_k + T_k)e^{ik\tau}, & \tau \rightarrow \infty. \end{cases} \quad (17.72)$$

Inserting (17.69), this can be rewritten as

$$\psi^{\mathrm{e}}(\tau) \rightarrow \begin{cases} e^{i\delta_k^{\mathrm{e}}}[e^{i(k\tau - \delta_k^{\mathrm{e}})} + e^{-i(k\tau - \delta_k^{\mathrm{e}})}] = 2e^{i\delta_k^{\mathrm{e}}}\cos(k|\tau| + \delta_k^{\mathrm{e}}), & \tau \rightarrow -\infty, \\ e^{i\delta_k^{\mathrm{e}}}[e^{-i(k\tau + \delta_k^{\mathrm{e}})} + e^{i(k\tau + \delta_k^{\mathrm{e}})}] = 2e^{i\delta_k^{\mathrm{e}}}\cos(k|\tau| + \delta_k^{\mathrm{e}}), & \tau \rightarrow \infty. \end{cases} \quad (17.73)$$

The odd combination, on the other hand, gives

$$\psi^{\mathrm{o}}(\tau) = \psi(\tau) - \psi^{\mathrm{r}}(\tau) \rightarrow \begin{cases} e^{ik\tau} + (R_k - T_k)e^{-ik\tau}, & \tau \rightarrow -\infty, \\ -e^{-ik\tau} - (R_k - T_k)e^{ik\tau}, & \tau \rightarrow \infty, \end{cases} \quad (17.74)$$

and becomes with (17.69):

$$\psi^{\mathrm{o}}(\tau) \rightarrow \begin{cases} e^{i\delta_k^{\phi}}[e^{i(k\tau - \delta_k^{\phi})} - e^{-i(k\tau - \delta_k^{\phi})}] = 2ie^{i\delta_k^{\mathrm{e}}}\sin(k|\tau| + \delta_k^{\phi}), & \tau \rightarrow -\infty, \\ -e^{i\delta_k^{\mathrm{e}}}[e^{-i(k\tau + \delta_k^{\phi})} - e^{i(k\tau + \delta_k^{\phi})}] = -2ie^{i\delta_k^{\mathrm{e}}}\sin(k|\tau| + \delta_k^{\phi}), & \tau \rightarrow \infty. \end{cases} \quad (17.75)$$

From Eqs. (17.69), (17.70) we see that

$$|T_k|^2 = \cos^2(\delta_k^{\mathrm{e}} - \delta_k^{\phi}), \qquad |R_k|^2 = \sin^2(\delta_k^{\mathrm{e}} - \delta_k^{\phi}), \quad (17.76)$$

and further

$$e^{2i(\delta_k^{\mathrm{e}} + \delta_k^{\phi})} = (T_k + R_k)(T_k - R_k) = T_k^2 + \frac{R_k R_k^* T_k}{T_k^*} = T_k^2 + (1 - T_k T_k^*)\frac{T_k}{T_k^*} = \frac{T_k}{T_k^*}. \quad (17.77)$$

From this we find the explicit equation for the sum of even and odd phase shifts

$$\delta_k^{\mathrm{e}} + \delta_k^{\phi} = \frac{1}{2}\arg\frac{T_k}{T_k^*} = \arg T_k. \quad (17.78)$$

Similarly we derive the equation for the difference of the phase shifts:

$$-i\sin[2(\delta_k^{\mathrm{e}} - \delta_k^{\phi})] = T_k R_k^* - T_k^* R_k = 2T_k R_k^* = -2\frac{R_k^*}{T_k^*}|T_k|^2. \quad (17.79)$$

Dividing this by the first equation in (17.76), we obtain

$$-i\tan(\delta_k^{\mathrm{e}} - \delta_k^{\phi}) = -\frac{R_k^*}{T_k^*} = -\frac{\sinh(ik/m)}{\sin(\pi s)}, \quad (17.80)$$

and thus

$$\delta_k^{\mathrm{e}} - \delta_k^{\phi} = \arctan\frac{\sin(\pi s)}{\sinh(k/m)}\sin(\pi s). \quad (17.81)$$

For $s =$ integer, the even and odd phase shifts become equal,

$$\delta_k^e = \delta_k^o \equiv \delta_k, \tag{17.82}$$

so that the reflection amplitude vanishes, and the transmission amplitude reduces to a pure phase factor $T_k = e^{2i\delta_k}$, with the phase shift determined from (17.78) to

$$2\delta_k = -i \log T_k. \tag{17.83}$$

Now the wave functions have the simple asymptotic behavior

$$y_k(\tau) \to e^{i(k\tau \pm \delta_k)}, \quad \tau \to \pm\infty, \tag{17.84}$$

and (17.64) reduces for both even and odd phase shifts to

$$e^{2i\delta_k} = (-1)^s \frac{\Gamma(s+1-ik/m)}{\Gamma(1-ik/m)} \frac{\Gamma(1+ik/m)}{\Gamma(s+1+ik/m)}. \tag{17.85}$$

For the Schrödinger equation (17.41) with $s = 2$, this becomes simply

$$e^{2i\delta_k} = \frac{2-ik/m}{2+ik/m} \frac{1-ik/m}{1+ik/m}, \tag{17.86}$$

and hence

$$\delta_k = \arctan[k/m] + \arctan[k/2m]. \tag{17.87}$$

If we now try to evaluate the product of eigenvalues in (17.46), two difficulties arise:

1) The zero eigenvalue causes the result to be infinite.
2) The continuum states make the evaluation of the product of eigenvalues $\prod_n \sqrt{\lambda_n}$ nontrivial.

These two difficulties will be removed in the following two subsections.

17.3.1 Zero-Eigenvalue Mode

The physical origin of the infinity caused by the zero-eigenvalue solution in the fluctuation integral over ξ_0 in (17.46) lies in the time-translational invariance of the system. This fact supplies the key to removing the infinity. A kink at an imaginary time τ_0 contributes as much to the path integral as a kink at any other time τ_0'. The difference between two adjacent solutions at an infinitesimal temporal distance can be viewed as a small kink fluctuation which does not change the Euclidean action. There is a zero eigenvalue associated with this difference. If there is only a single zero-eigenvalue solution as in the Schrödinger equation (17.41) its wave function must be proportional to the derivative of the kink solution:

$$
\begin{aligned}
y_0 &= \alpha \left[\frac{x_{\mathrm{cl}}(\tau - \tau_0) - x_{\mathrm{cl}}(\tau - \tau_1)}{\tau_0 - \tau_1} \right]_{\tau_0 \to \tau_1} = \alpha x_{\mathrm{cl}}'(\tau - \tau_0) \\
&= -\alpha \frac{a\omega/2}{\cosh^2[\omega(\tau - \tau_0)/2]}.
\end{aligned} \tag{17.88}
$$

This coincides indeed with the zero-eigenvalue solution (17.52). The normalization factor α is fixed by the integral

$$\alpha = \left[\int_{-\infty}^{\infty} d\tau x_{cl}'^2 \right]^{-1/2}. \tag{17.89}$$

With the help of (17.32) the right-hand side can also be expressed in terms of the kink action:

$$\alpha = \frac{1}{\sqrt{\mathcal{A}_{cl}}}. \tag{17.90}$$

Inserting (17.2) and (17.31), this becomes

$$\alpha = \sqrt{3/2a^2\omega} = \sqrt{3g/\omega^3} \tag{17.91}$$

[the positive square root corresponding to the negative sign in (17.52)]. The zero-eigenvalue mode is associated with a shift of the position of the kink solution from τ_0 to any other place τ_0'. For an infinite length L of the τ-axis, this is the source of the infinity of the integral over ξ_0 in the product (17.46). At a finite L, on the other hand, only a τ-interval of length L is available for displacements. Therefore, the infinite $1/\sqrt{\lambda_0}$ should become proportional to L for large L:

$$\frac{1}{\sqrt{\lambda_0}} = \text{const} \cdot L. \tag{17.92}$$

What is the proportionality constant? We find it by transforming the integration measure (17.46),

$$\mathcal{N} \int_{-\infty}^{\infty} \frac{d\xi_0}{\sqrt{2\pi\hbar}} \prod_{n=1}^{\infty} \left[\int_{-\infty}^{\infty} \frac{d\xi_n}{\sqrt{2\pi\hbar}} \right], \tag{17.93}$$

to a form in which the translational degree of freedom appears explicitly

$$\mathcal{N} \frac{1}{\sqrt{2\pi\hbar}} \int_{-\infty}^{\infty} d\tau_0 \prod_{n\neq0} \int_{-\infty}^{\infty} \left[\frac{d\xi_n}{\sqrt{2\pi\hbar}} \right] \left| \frac{\partial\xi_0}{\partial\tau_0}(\xi_1, \xi_2, \ldots) \right|. \tag{17.94}$$

The Jacobian appearing in the integrand satisfies the identity

$$\int d\tau_0 \, \delta(\xi_0) \left| \frac{\partial\xi_0}{\partial\tau_0} \right| = 1. \tag{17.95}$$

Let us calculate this Jacobian using the method developed by Faddeev and Popov [2]. Given an arbitrary fluctuation $y^{\xi_0,\xi_1,\xi_2,\cdots}$, the parameter ξ_0 may be recovered by forming the scalar product with the zero-eigenvalue wave function $y_0(\tau)$:

$$\xi_0 = \int_{-\infty}^{\infty} d\tau y^{\xi_0,\xi_1,\xi_2,\cdots}(\tau) y_0(\tau). \tag{17.96}$$

Moreover, it is easy to see that the fluctuation $y^{\xi_0,\xi_1,\xi_2,\cdots}$ can be replaced by the full path

$$x^{\xi_0,\xi_1,\xi_2,\cdots}(\tau) = x_{\rm cl}(\tau) + y^{\xi_0,\xi_1,\xi_2,\cdots}(\tau), \tag{17.97}$$

so that we can also write

$$\xi_0 = \int_{-\infty}^{\infty} d\tau \, x^{\xi_0,\xi_1,\xi_2,\cdots}(\tau)y_0(\tau). \tag{17.98}$$

This follows from the fact that the additional kink solution $x_{\rm cl}(\tau)$ does not have any overlap with its derivative. Indeed,

$$\int_{-\infty}^{\infty} d\tau \, x_{\rm cl}(\tau - \tau_0)y_0(\tau - \tau_0) \;\propto\; \int_{-\infty}^{\infty} d\tau \, x_{\rm cl}(\tau - \tau_0)x'_{\rm cl}(\tau - \tau_0)$$

$$\propto \; \frac{1}{2}x_{\rm cl}^2 \Big|_{-\infty}^{\infty} = 0. \tag{17.99}$$

With (17.98), the delta function $\delta(\xi_0)$ appearing in (17.95) can be rewritten in the form

$$\delta(\xi_0) = \delta\left(\int_{-\infty}^{\infty} d\tau \, x^{\xi_0,\xi_1,\xi_2,\cdots}(\tau)y_0(\tau)\right). \tag{17.100}$$

To establish the relation between the normal coordinate ξ_0 and the kink position τ_0, we now replace $x^{\xi_0,\xi_1,\xi_2,\cdots}$ by an alternative parametrization of the paths in which the role of the variable ξ_0 is traded against the position of the kink solution, i.e., we rewrite the fluctuating path

$$x^{\xi_0,\xi_1,\xi_2,\cdots}(\tau) = x_{\rm cl}(\tau) + \sum_{n=0}^{\infty} \xi_n y_n(\tau) \tag{17.101}$$

in the form

$$x^{\tau_0,\xi_1,\xi_2,\cdots}(\tau) \equiv x_{\rm cl}(\tau - \tau_0) + \sum_{n=1}^{\infty} \xi_n y_n(\tau - \tau_0). \tag{17.102}$$

By definition the point $\tau_0 = 0$ coincides with $\xi_0 = 0$. Thus, if we insert (17.102) into (17.100) and use this δ-function in the identity (17.95), we find the condition for the Jacobian $\partial\xi_0/\partial\tau_0$:

$$\int_{-\infty}^{\infty} d\tau_0 \, \delta\left(\int_{-\infty}^{\infty} d\tau \, x^{\tau_0,\xi_1,\xi_2,\cdots}(\tau)y_0(\tau)\right)\left|\frac{\partial\xi_0}{\partial\tau_0}\right| = 1. \tag{17.103}$$

Since the δ-function has a vanishing argument at $\tau_0 = 0$, we expand the argument in powers of τ_0, keeping only the lowest order. Writing $y_0(\tau) = \alpha x'_{\rm cl}(\tau)$, we obtain

$$\int_{-\infty}^{\infty} d\tau \, x^{\tau_0,\xi_1,\xi_2,\cdots}(\tau)y_0(\tau) = -\alpha\tau_0\left[\int_{-\infty}^{\infty} d\tau \, x_{\rm cl}'^2 + \sum_{n=1}^{\infty} \xi_n \int_{-\infty}^{\infty} d\tau \, x'_{\rm cl}\, y'_n\right] + \mathcal{O}(\tau_0^2). \tag{17.104}$$

Using (17.32) and abbreviating the scalar products in the brackets as

$$r_n \equiv \int_{-\infty}^{\infty} d\tau \, x'_{\text{cl}} \, y'_n, \tag{17.105}$$

we may express the right-hand side of (17.104) more succinctly as

$$- \alpha \tau_0 \left[\mathcal{A}_{\text{cl}} + \sum_{n=1}^{\infty} \xi_n r_n \right] + \mathcal{O}(\tau_0^2). \tag{17.106}$$

Inserting this into (17.103), and using (17.90), we arrive at the Jacobian

$$\left| \frac{\partial \xi_0}{\partial \tau_0} \right| = \mathcal{A}_{\text{cl}}^{1/2} \left(1 + \mathcal{A}_{\text{cl}}^{-1} \sum_{n=1}^{\infty} \xi_n r_n \right). \tag{17.107}$$

As a consequence, the zero-eigenvalue modes contribute to the fluctuation factor (17.46) as follows:

$$
\begin{aligned}
F_{\text{cl}}(L) &= \mathcal{N} \prod_{n=1}^{\infty} \left[\int_{-\infty}^{\infty} \frac{d\xi_n}{\sqrt{2\pi\hbar}} \right] e^{-(1/2\hbar) \sum_{n=1}^{\infty} \xi_n^2 \lambda_n} \int_{-\infty}^{\infty} \frac{d\xi_0}{\sqrt{2\pi\hbar}} e^{-(1/2\hbar)\xi_0^2 \lambda_0} \\
&= \mathcal{N} \prod_{n=1}^{\infty} \left[\int_{-\infty}^{\infty} \frac{d\xi_n}{\sqrt{2\pi\hbar}} e^{-(1/2\hbar) \sum_{n=1}^{\infty} \xi_n^2 \lambda_n} \right] \int_{-\infty}^{\infty} \frac{d\tau_0}{\sqrt{2\pi\hbar}} \mathcal{A}_{\text{cl}}^{1/2} \left(1 + \mathcal{A}_{\text{cl}}^{-1} \sum_{n=1}^{\infty} \xi_n r_n \right) \\
&= \mathcal{N} \frac{1}{\sqrt{\prod_{n=1}^{\infty} \lambda_n}} \sqrt{\frac{\mathcal{A}_{\text{cl}}}{2\pi\hbar}} \int_{-\infty}^{\infty} d\tau_0.
\end{aligned} \tag{17.108}
$$

The linear terms in the large parentheses disappear at the level of quadratic fluctuations since they are odd in ξ. Thus we may use for *quadratic fluctuations* the simple *mnemonic rule*

$$- \frac{\partial \xi_0}{\partial \tau_0} = \frac{\partial x(\tau)/\partial \tau_0}{\partial x(\tau)/\partial \xi_0} = \frac{\dot{x}_{\text{cl}}(\tau) + \ldots}{\alpha \dot{x}_{\text{cl}}(\tau)} = \frac{1}{\alpha} + \ldots = \sqrt{\mathcal{A}_{\text{cl}}} + \ldots, \tag{17.109}$$

where ... stands for the irrelevant linear terms in ξ_n inside the integral (17.108).

For higher-order fluctuations, the linear terms in the large parentheses cannot be ignored. They contribute an effective Euclidean action

$$\mathcal{A}_{\text{e}}^{\text{eff}} = -\hbar \log \left[1 + \mathcal{A}_{\text{cl}}^{-1} \sum_{n=1}^{\infty} \xi_n r_n \right] = -\hbar \log \left[1 + \mathcal{A}_{\text{cl}}^{-1} \int d\tau \, x'_{\text{cl}}(\tau) y'(\tau) \right], \tag{17.110}$$

which will be needed for calculations in Section 17.8.

The integral over τ_0 is now an appropriate place to impose the finiteness of the time interval $\tau \in (-L/2, L/2)$, namely

$$\int_{-L/2}^{L/2} d\tau_0 = L. \tag{17.111}$$

A comparison of (17.108) with (17.46) shows that the correct evaluation of the formally diverging zero-eigenvalue contribution $1/\sqrt{\lambda_0}$ is equivalent to the following replacement:

$$\frac{1}{\sqrt{\lambda_0}} \equiv \int \frac{d\xi_0}{\sqrt{2\pi\hbar}} e^{-(1/2\hbar)\lambda_0 \xi_0^2} \longrightarrow \sqrt{\frac{\mathcal{A}_{\text{cl}}}{2\pi\hbar}} \int_{-L/2}^{L/2} d\tau_0 = \sqrt{\frac{\mathcal{A}_{\text{cl}}}{2\pi\hbar}} \, L. \tag{17.112}$$

17.3.2 Continuum Part of Fluctuation Factor

We now turn to the second problem, the calculation of the product of the continuum eigenvalues in (17.46). To avoid carrying around the overall normalization factor $\mathcal{N}$ it is convenient to factor out the quadratic fluctuations (17.36) in the absence of a kink solution. Let λ_n^0 be the associated eigenvalues. Their fluctuation potential $V''(x_{\mathrm{cl}}(\tau)) = \omega^2 x^2$ is harmonic [compare (17.39)] with a fluctuation factor known from (2.169). Comparing this with the expression (17.46) without a kink, we obtain for the normalization factor the equation

$$\mathcal{N}\frac{1}{\sqrt{\Pi_n \lambda_n^0}} = F_\omega(L) = \sqrt{\frac{\omega}{2\pi\hbar \sinh \omega L}} \sim \sqrt{\frac{\omega}{\pi\hbar}} e^{-\omega L/2}. \tag{17.113}$$

Pulling this factor out of the product on the right-hand side of (17.46), we are left with a ratio of eigenvalue products:

$$F_{\mathrm{cl}}(L) = \mathcal{N}\frac{1}{\sqrt{\Pi_n \lambda_n}} = \mathcal{N}\frac{1}{\sqrt{\Pi_n \lambda_n^0}}\sqrt{\frac{\Pi_n \lambda_n^0}{\Pi_n \lambda_n}} = F_\omega(L)\sqrt{\frac{\Pi_n \lambda_n^0}{\Pi_n \lambda_n}}. \tag{17.114}$$

As long as L is finite, the continuum wave functions are all discrete. Let $\partial n/\partial k$ denote their density of states per momentum interval. Then the ratio of the continuum eigenvalues can be written for large L as

$$\sqrt{\frac{\Pi_n \lambda_n^0}{\Pi_n \lambda_n}}\bigg|_{\mathrm{cont}} = \exp\left[-\frac{1}{2}\int_0^\infty dk \left(\frac{\partial n}{\partial k} - \frac{\partial n}{\partial k}\bigg|_0\right)\log \lambda_n\right]. \tag{17.115}$$

The density of states, in turn, may be extracted from the phase shifts (17.85). For this we observe that for a very large L where boundary conditions are a matter of choice, we may impose the periodic boundary condition

$$y(\tau + L) = y(\tau). \tag{17.116}$$

Together with the asymptotic forms (17.84), this implies

$$e^{i(kL/2+\delta_k)} = e^{-i(kL/2+\delta_k)}, \tag{17.117}$$

which quantizes the wave vectors k to discrete values satisfying

$$kL + 2\delta_k = 2\pi n. \tag{17.118}$$

The derivative with respect to k yields the density of states

$$\frac{\partial n}{\partial k} = \frac{L}{2\pi} + \frac{1}{\pi}\frac{d\delta_k}{dk}. \tag{17.119}$$

Since the phase shifts vanish in the absence of a kink solution, this implies

$$\frac{\partial n}{\partial k}\bigg|_0 = \frac{L}{2\pi}, \tag{17.120}$$

and the general formula (17.115) becomes simply

$$\sqrt{\frac{\prod_n \lambda_n^0}{\prod_n \lambda_n}}\Bigg|_{\text{cont}} = \exp\left[-\frac{1}{2\pi}\int_0^\infty dk \frac{d\delta_k}{dk}\log(\omega^2 + k^2)\right]. \tag{17.121}$$

To calculate the integral for our specific fluctuation problem (17.41), we use the expression (17.85) to find the derivative of the phase shift for any integer value of s:

$$\frac{d\delta_k}{dk} = -\frac{1}{m}\left[\frac{2}{4 + (k/m)^2} + \ldots + \frac{s}{s^2 + (k/m)^2}\right]. \tag{17.122}$$

For $s = 2$, the exponent in (17.121) becomes

$$\frac{1}{2\pi}\int_{-\infty}^\infty dx \left(\frac{1}{1 + x^2} + \frac{2}{4 + x^2}\right)\log\left[\omega^2(1 + x^2 m^2/\omega^2)\right]. \tag{17.123}$$

The ω^2-term in the logarithm can be separated from this using Levinson's theorem. It states that the integral $\int_0^\infty dk(\partial n/\partial k - \partial n/\partial k|_0)$ is equal to the number of bound states:

$$\int_0^\infty dk \left(\frac{\partial n}{\partial k} - \frac{\partial n}{\partial k}\bigg|_0\right) = \frac{1}{\pi}\int_0^\infty dk \frac{d\delta_k}{dk} = s. \tag{17.124}$$

This relation is obviously fulfilled by (17.122). It is a consequence of the fact that a potential with s bound states has s states less in the continuum than a free system. Using this property, the integral (17.123) can be rewritten as

$$\log \omega^2 + \int_{-\infty}^\infty \frac{dx}{2\pi}\left(\frac{1}{1 + x^2} + \frac{2}{4 + x^2}\right)\log(1 + x^2 m^2/\omega^2). \tag{17.125}$$

The rescaled integral is calculated using the formula

$$\int_{-\infty}^\infty \frac{dx}{2\pi}\frac{\log(1 + p^2 x^2)}{r^2 + s^2 x^2} = \frac{1}{rs}\log\left(1 + p\frac{r}{s}\right). \tag{17.126}$$

The result is

$$\log \omega^2 + \log\left(1 + \frac{m}{\omega}1\right) + \log\left(1 + \frac{m}{\omega}2\right). \tag{17.127}$$

When inserted into the exponent of (17.121), it yields

$$\sqrt{\frac{\prod_n \lambda_n^0}{\prod_n \lambda_n}}\Bigg|_{\text{cont}} = \omega^2\left(1 + \frac{m}{\omega}\right)\left(1 + \frac{m}{\omega}2\right). \tag{17.128}$$

In our case with $m = \omega/2$, this reduces to $3\omega^2$. Including the bound-state eigenvalue λ_1 of Eq. (17.55) in the denominator, this amounts to

$$\sqrt{\frac{\prod_n \lambda_n^0}{\prod_n' \lambda_n}} = \frac{1}{\sqrt{3\omega^2/4}}3\omega^2 = \sqrt{12}\omega \equiv K'. \tag{17.129}$$

Multiplying this with the zero-eigenvalue contribution as evaluated in Eq. (17.108), we arrive at the final result for the fluctuation factor in the presence of a kink or an antikink solution:

$$F_{\rm cl}(L) = F_\omega(L)KL, \tag{17.130}$$

with

$$K = \frac{1}{\sqrt{\lambda_0}L}K' = \sqrt{\frac{\mathcal{A}_{\rm cl}}{2\pi\hbar}}\sqrt{12}\omega. \tag{17.131}$$

17.4 General Formula for Eigenvalue Ratios

The above-calculated ratio of eigenvalue products

$$\sqrt{\frac{\prod_n \lambda_n^0}{\prod_n \lambda_n}}\bigg|_{\rm cont} \tag{17.132}$$

of the Rosen-Morse Schrödinger equation appears in many applications with different potential strength parameters s. It is therefore useful to derive a formula for this ratio which is valid for any s. The eigenvalue equation reads

$$\left[-\frac{d^2}{d\tau^2} + \omega^2 - \frac{m^2 s(s+1)}{\cosh^2 m(\tau - \tau_0)}\right] y_n(\tau) = \lambda_n y_n(\tau). \tag{17.133}$$

First we consider the case of an arbitrary integer value of s. Following the previous discussion, the ratio of eigenvalue products is found to be

$$\sqrt{\frac{\prod_n \lambda_n^0}{\prod_n \lambda_n}}\bigg|_{\rm cont} = \exp\left[-\frac{1}{2\pi}\int_0^\infty dk \frac{d\delta_k}{dk}\log(\omega^2 + k^2)\right] \tag{17.134}$$

$$= \exp\left\{-\frac{1}{2\pi}\int_0^\infty d(k/m)\sum_{n=1}^s \frac{n}{n^2 + (k/m)^2}\log[\omega^2 + (k/m)^2 m^2]\right\}.$$

The ω^2-term in the logarithm is eliminated by the generalization of (17.124) to any integer s:

$$\frac{1}{\pi}\int_{-\infty}^\infty d(k/m)\sum_{n=1}^s \frac{n}{n^2 + (k/m)^2} = s. \tag{17.135}$$

Hence

$$\sqrt{\frac{\prod_n \lambda_n^0}{\prod_n \lambda_n}}\bigg|_{\rm cont} = \omega^s \exp\left[\frac{1}{2\pi}\int_{-\infty}^\infty dx \sum_{n=1}^s \frac{n}{n^2 + x^2}\log(1 + x^2 m^2/\omega^2)\right]. \tag{17.136}$$

The integrals can be done using formula (17.126), and we obtain

$$\sqrt{\frac{\prod_n \lambda_n^0}{\prod_n \lambda_n}}\bigg|_{\rm cont} = \omega^s \prod_{n=1}^s \left(1 + \frac{m}{\omega}n\right). \tag{17.137}$$

For $s = 2$, and $m = \omega/2$, this reduces to the previous result (17.128).

Only a little more work is required to find the ratio of all discrete and continuous eigenvalue products for a noninteger value of s. Introduce a new parameter z, let s be a parameter smaller than 1 so that there are no bound states, and consider the fluctuation equation

$$\left[-\frac{d^2}{d\tau^2} + m^2\left(z - \frac{s(s+1)}{\cosh^2 m(\tau - \tau_0)}\right)\right] y_n(\tau) = \lambda_n y_n(\tau). \tag{17.138}$$

The general Schrödinger operator under consideration (17.133) corresponds to $z = \omega^2/m^2$. Since there are no bound states by assumption, the first line in formula (17.134) now gives the ratio of *all* eigenvalues:

$$\sqrt{\frac{\prod_n \lambda_n^0}{\prod_n \lambda_n}} = \exp\left[-\frac{1}{2\pi}\int_{-\infty}^{\infty} dk \frac{d\delta_k}{dk}\log(m^2 z + k^2)\right]. \tag{17.139}$$

Here δ_k is equal to the average of even and odd phase shifts $(\delta_k^e + \delta_k^o)/2$. For the same reason, we can replace $\log(m^2 z + k^2)$ by $\log(z + k^2/m^2)$ without error [using the generalization of (17.124)]. After substituting $k^2 \to \omega^2 \epsilon$, we find

$$\sqrt{\frac{\prod_n \lambda_n^0}{\prod_n \lambda_n}} = \exp\left[-\frac{1}{2\pi}\int_C d\epsilon \frac{d\delta_{\omega\sqrt{\epsilon}}}{d\epsilon}\log(z + \epsilon)\right], \tag{17.140}$$

where the contour of integration C encircles the right-hand cut clockwise in the ϵ-plane. A partial integration brings this to

$$\exp\left(\frac{1}{2\pi}\int_C d\epsilon\, \delta_{m\sqrt{\epsilon}}\frac{1}{z + \epsilon}\right). \tag{17.141}$$

For $z < 0$, the contour of integration can be deformed to encircle the only pole at $\epsilon = -z$ counterclockwise, yielding

$$\sqrt{\frac{\prod_n \lambda_n^0}{\prod_n \lambda_n}} = \exp[i\delta_{m\sqrt{-z}}]. \tag{17.142}$$

Inserting for δ_k the average of even and odd phase shifts from (17.78), we obtain

$$\sqrt{\frac{\prod_n \lambda_n^0}{\prod_n \lambda_n}} = \left[\frac{\Gamma(\sqrt{z} - s)\Gamma(\sqrt{z} + s + 1)}{\Gamma(\sqrt{z})\Gamma(\sqrt{z} + 1)}\right]^{1/2}. \tag{17.143}$$

In the fluctuation equation (17.41), the parameters m^2 and $\omega^2 = zm^2$ are such as to create a zero eigenvalue at $n = 0$ according to formula (17.51). Then $z = s^2$. In the neighborhood of this z-value, the eigenvalue

$$\lambda_0 = m^2[z - s^2] \tag{17.144}$$

is a would-be zero eigenvalue. Dividing it out of the product (17.143), we obtain an equation which remains valid in the limit $z \to s^2$:

$$\sqrt{\frac{\Pi_n \lambda_n^0}{\Pi_n' \lambda_n}} = m \left[(\sqrt{z} + s) \frac{\Gamma(\sqrt{z} - s + 1)\Gamma(\sqrt{z} + s + 1)}{\Gamma(\sqrt{z})\Gamma(\sqrt{z} + 1)} \right]^{1/2}. \qquad (17.145)$$

This can be continued analytically to arbitrary z and s, as long as z remains sufficiently close to $s = 2$. For $s = 2$ and $z = 4$ (corresponding to $m = \omega/2$) we recover the earlier result (17.129):

$$\sqrt{\frac{\Pi_n \lambda_n^0}{\Pi_n' \lambda_n}} = \sqrt{12}\omega. \qquad (17.146)$$

17.5 Fluctuation Determinant from Classical Solution

The above evaluations of the fluctuation determinant require the complete knowledge of the bound and continuum spectrum of the fluctuation equation. Fortunately, there exists a way to find the determinant which needs much less information, requiring only the knowledge of the large-τ behavior of the classical solution and the value of its action. The basis for this derivation is the Gelfand-Yaglom formula derived in Section 2.4. According to it, the fluctuation determinant of a differential operator

$$\hat{O} = -\frac{d^2}{d\tau^2} + \omega^2 - \frac{m^2 s(s+1)}{\cosh^2[m(\tau - \tau_0)]} \qquad (17.147)$$

is given by the value of the zero-eigenvalue solution $D(\tau)$ at the final τ value $\tau = L/2$

$$\det \hat{O} = \mathcal{N} D(L/2), \qquad (17.148)$$

provided that it was chosen to satisfy the initial conditions at $\tau = -L/2$:

$$D(-L/2) = 0, \quad \dot{D}(-L/2) = 1. \qquad (17.149)$$

The normalization factor $\mathcal{N}$ is irrelevant when considering ratios of fluctuation determinants, as we do in the problem at hand.[3] To satisfy the boundary conditions (17.149), we need two linearly independent solutions of zero eigenvalue. One is known from the invariance under time translations. It is proportional to the time derivative of the classical solution [see (17.88)]:

$$y_0(\tau) = \alpha x'_{\text{cl}}(\tau). \qquad (17.150)$$

---

[3]If the determinant is calculated for the time-sliced operator $\hat{O}$ with $d/d\tau$ replaced by the difference operator ∇_τ, the normalization is $\mathcal{N} = 1/\epsilon$ where ϵ is the thickness of the time slices. See Chapter 2.

In the above fluctuation problem (17.41) with $s = 2$ and $m = \omega/2$, the classical solution is $x_{\text{cl}}(\tau) = \text{arctanh}[\omega(\tau - \tau_0)/2]$. It has the asymptotic behavior

$$y_0(\tau) \rightarrow \frac{\omega}{2} e^{-\omega|\tau|} \quad \text{for} \quad \tau \rightarrow \pm\infty, \tag{17.151}$$

with a symmetric exponential falloff in both directions of the τ-axis. In the sequel it will be convenient to work with zero-eigenvalue solutions without the prefactor $\omega/2$, which behave asymptotically like a pure exponential. These will be denoted by $\xi(\tau)$ and $\eta(\tau)$, the solution $\xi(\tau)$ being proportional to y_0, i.e.,

$$\xi(\tau) \rightarrow e^{-\omega|\tau|} \quad \text{for} \quad \tau \rightarrow \pm\infty. \tag{17.152}$$

The second independent solution can be found from d'Alembert's formula (2.234). Its explicit form is not required; only its asymptotic behavior is relevant. Assuming the Lagrangian to be invariant under time reversal, which is usually the case, this asymptotic behavior is found via the following argument: Since $\phi_0^{(2)}(\tau)$ is linearly independent of $\phi_0^{(1)}(\tau)$, we can be sure that it has asymptotically the opposite exponential behavior (i.e., it grows with τ) and the opposite symmetry under time reversal (i.e., it is antisymmetric). Thus η must behave as follows:

$$\eta(\tau) \rightarrow \pm e^{\omega|\tau|} \quad \text{for} \quad \tau \rightarrow \pm\infty. \tag{17.153}$$

We now form the linear combination which satisfies the boundary conditions (17.149) for large negative $\tau = -L/2$:

$$D(\tau) = \frac{1}{W} \left[\xi(-L/2)\eta(\tau) - \eta(-L/2)\xi(\tau) \right], \tag{17.154}$$

where

$$W \equiv W[\xi(\tau)\eta(\tau)] = \xi(\tau)\dot{\eta}(\tau) - \eta(\tau)\dot{\xi}(\tau) \tag{17.155}$$

is the *Wronskian* of the two solutions. It is independent of τ and can be evaluated from the asymptotic behavior as

$$W = 2\omega. \tag{17.156}$$

Inserting (17.152) and (17.153) into (17.154), we find the solution

$$D(\tau) = \frac{1}{W} \left[e^{-\omega L/2}\eta(\tau) + e^{\omega L/2}\xi(\tau) \right]. \tag{17.157}$$

Even without knowing the solutions $\phi_0^{(1)}(\tau)$, $\phi_0^{(2)}(\tau)$ at a finite τ, the fluctuation determinant at large $\tau = L/2$ can be written down:

$$D(L/2) = \frac{2}{W} = \frac{1}{\omega}. \tag{17.158}$$

For fluctuations around the constant classical solution, the zero-eigenvalue solution with the boundary conditions (17.149) is

$$D^{(0)}(\tau) = \frac{1}{\omega} \sinh[\omega(\tau + L/2)]. \tag{17.159}$$

It behaves for large $\tau = L$ like

$$D^{(0)}(L/2) \rightarrow \frac{1}{2\omega} e^{\omega L} \qquad \text{for large } L. \tag{17.160}$$

The ratio is therefore

$$\frac{D^{(0)}(L/2)}{D(L/2)} \rightarrow \frac{1}{2} e^{\omega L} \qquad \text{for large } L. \tag{17.161}$$

This exponentially large number is a signal for the presence of a would-be zero eigenvalue in $D(L/2)$. Since the τ-interval $(-L/2, L/2)$ is finite, there exists no exactly vanishing eigenvalue. In the finite interval, the derivative (17.150) of the kink solution does not quite satisfy the Dirichlet boundary condition. If the vanishing at the endpoints was properly enforced, the particle distribution would have to be compressed somewhat, and this would shift the energy slightly upwards. The shift is exponentially small for large L, so that the would-be zero eigenvalue has an exponentially small eigenvalue $\lambda_0 \propto e^{-\omega L}$. A finite result for $L \rightarrow \infty$ is obtained by removing this mode from the ratio (17.161) and considering the limit

$$\frac{\prod_n \lambda_n{}^0}{\prod_n' \lambda_n} = \lim_{L \rightarrow \infty} \frac{D^{(0)}(L/2)}{D(L/2)} \lambda_0. \tag{17.162}$$

The leading $e^{-\omega L}$-behavior of the would-be zero eigenvalue can be found perturbatively using as before only the asymptotic behavior of the two independent solutions. To lowest order in perturbation theory, an eigenfunction satisfying the Dirichlet boundary condition at finite L is obtained from an eigenfunction ϕ_0 which vanishes at $\tau = -L/2$ by the formula

$$\phi_0^L(\tau) = \phi_0(\tau) + \frac{\lambda_0}{W} \int_{-L/2}^{\tau} d\tau' \, [\xi(\tau)\eta(\tau') - \eta(\tau)\xi(\tau')] \, \phi_0(\tau'). \tag{17.163}$$

The limits of integration ensure that $\phi_0^L(\tau)$ vanishes at $\tau = -L/2$. The eigenvalue λ_0 is determined by enforcing the vanishing also at $\tau = L/2$. Taking for $\phi_0(\tau)$ the zero-eigenvalue solution $D(\tau)$

$$\lambda_0 = -D(L/2)W \left[\xi(L/2) \int_{-L/2}^{L/2} d\tau \, \eta(\tau)D(\tau) - \eta(L/2) \int_{-L/2}^{L/2} d\tau \, \xi(\tau)D(\tau) \right]^{-1}. \tag{17.164}$$

Inserting (17.154) and using the orthogonality of $\xi(\tau)$ and $\eta(\tau)$ (following from the fact that the first is symmetric and the second antisymmetric), this becomes

$$\lambda_0 = -D(L/2)W^2 \left[\xi(-L/2)\xi(L/2) \int_{-L/2}^{L/2} d\tau \, \eta^2(\tau) + \eta(-L/2)\eta(L/2) \int_{-L/2}^{L/2} d\tau \, \xi^2(\tau) \right]^{-1}. \tag{17.165}$$

Invoking once more the symmetry of $\xi(\tau)$ and $\eta(\tau)$ and the asymptotic behavior (17.152) and (17.153), we obtain

$$\lambda_0 = -D(L/2)W^2 \left[e^{-\omega L} \int_{-L/2}^{L/2} d\tau\, \eta^2(\tau) - e^{\omega L} \int_{-L/2}^{L/2} d\tau\, \xi^2(\tau) \right]^{-1}. \qquad (17.166)$$

The first integral diverges like $e^{\omega L}$; the second is finite. The prefactor makes the second integral much larger than the first, so that we find for large L the would-be zero eigenvalue

$$\lambda_0 = D(L/2)e^{-\omega L} \frac{W^2}{\int_{-\infty}^{\infty} d\tau\, \xi^2(\tau)}. \qquad (17.167)$$

This eigenvalue is exponentially small and positive, as expected. Inserting it into (17.162) and using (17.156) and (17.160), we find the eigenvalue ratio

$$\frac{\prod_n \lambda_n^0}{\prod_n' \lambda_n} = \lim_{L \to \infty} 2\omega \frac{1}{\int_{-\infty}^{\infty} d\tau\, \xi^2(\tau)}. \qquad (17.168)$$

The determinant $D(L/2)$ has disappeared and the only nontrivial quantity to be evaluated is the normalization integral over the translational eigenfunction $\xi(\tau)$.

The normalization integral requires the knowledge of the full τ-behavior of the zero-eigenvalue solution $\phi_0^{(1)}(\tau)$; the asymptotic behavior used up to this point is insufficient. Fortunately, the classical solution $x_{\rm cl}(\tau)$ also supplies this information. The normalized solution is $y_0 = \alpha x'_{\rm cl}(\tau)$ behaving asymptotically like $2a\alpha\omega e^{-\omega\tau}$. Imposing the normalization convention (17.152) for $\phi_0^{(1)}(\tau)$, we identify

$$\xi(\tau) = \frac{1}{2a\omega\alpha} \alpha x'_{\rm cl}(\tau). \qquad (17.169)$$

Using the relation (17.32), the normalization integral is simply

$$\int_{-\infty}^{\infty} d\tau\, \xi(\tau)^2 = \frac{\mathcal{A}_{\rm cl}}{4a^2\omega^2}. \qquad (17.170)$$

With it the eigenvalue ratio (17.168) becomes

$$\frac{\prod_n \lambda_n^0}{\prod_n' \lambda_n} = 2\omega \frac{4a^2\omega}{\mathcal{A}_{\rm cl}}. \qquad (17.171)$$

By inserting the value of the classical action $\mathcal{A}_{\rm cl} = 2a^2\omega/3$ from (17.31), we obtain

$$\frac{\prod_n \lambda_n^0}{\prod_n' \lambda_n} = 12\omega^2, \qquad (17.172)$$

just as in (17.146) and (17.129).

It is remarkable that the calculation of the ratio of the fluctuation determinants with this method requires only the knowledge of the classical solution $x_{\rm cl}(\tau)$.

17.6   Wave Functions of Double-Well

The semiclassical result for the amplitudes

$$(a \ L/2|a \ -L/2), \quad (a \ L/2| -a \ -L/2), \tag{17.173}$$

with the endpoints situated at the bottoms of the potential wells can easily be extended to variable endpoints $x_b \neq a, x_a \neq \pm a$, as long as these are situated near the bottoms. The extended amplitudes lead to approximate particle wave functions for the lowest two states. The extension is trivial for the formula (17.34) without a kink solution. We simply multiply the fluctuation factor by the exponential $\exp(-\mathcal{A}_{cl}/\hbar)$ containing the classical action of the path from x_a to x_b. If x_a and x_b are both near one of the bottoms of the well, the entire classical orbit remains near this bottom. If the distance of the orbit from the bottom is less than $1/a\sqrt{\omega}$, the potential can be approximated by the harmonic potential $\omega^2 x^2$. Thus near the bottom at $x = a$, we have the simple approximation to the action

$$\mathcal{A}_{\rm cl} \approx \frac{\omega}{2\hbar \sinh \omega L} \left\{ [(x_a - a)^2 + (x_b - a)^2] \cosh \omega L - 2(x_b - a)(x_a - a) \right\}. \tag{17.174}$$

For a very long Euclidean time L, this tends to

$$\mathcal{A}_{\rm cl} \approx \frac{\omega}{2\hbar} [(x_b - a)^2 + (x_a - a)^2]. \tag{17.175}$$

The amplitude (17.34) can therefore be generalized to

$$(x_b \ L/2|x_a \ L/2) \approx \sqrt{\frac{\omega}{\pi\hbar}} e^{-(\omega/2\hbar)[(x_b-a)^2+(x_a-a)^2]} e^{-\omega L/2\hbar}.$$

This can also be written in terms of the bound-state wave functions (17.5) for $n = 0$ as

$$(x_b \ L/2|x_a \ L/2) \approx \psi_0(x_b - a)\psi_0(x_a - a)e^{-\omega L/2\hbar}. \tag{17.176}$$

For the amplitude (17.35) with the path running from one potential valley to the other, the construction is more subtle. The approximate solution is obtained by combining a harmonic classical path running from $(x_a, -L/2)$ to $(a, -L/4)$, a kink solution running from $(a, -L/4)$ to $(a, L/4)$, and a third harmonic classical path running from $(a, L/4)$ to $(x_b, L/2)$. This yields the amplitude

$$\sqrt{\frac{\omega}{\pi\hbar}} e^{-(\omega/2\hbar)(x_b-a)^2} KL e^{-\mathcal{A}_{cl}/\hbar} e^{-(\omega/2\hbar)(x_a-a)^2}. \tag{17.177}$$

Note that by patching the three pieces together, it is impossible to obtain a true classical solution. For this we would have to solve the equations of motion containing a kink with the modified boundary conditions $x(-L/2) = x_a$, $x(L/2) = x_b$. From the exponential convergence of $x(\tau) \to \pm a$ (like $e^{-\omega L}$) it is, however, obvious that the true classical action differs from the action of the patched path only by exponentially small terms.

As before, the prefactor in (17.177) can be attributed to the ground state wave functions $\psi_0(x)$, and we find the amplitude for x_b close to $-a$ and x_a close to a:

$$(x_b \, L/2|x_a \, - L/2) \approx \psi_0(x_b + a)\psi_0(x_a - a)KLe^{-\mathcal{A}_{\rm cl}/\hbar}e^{-\omega L/2\hbar}. \qquad (17.178)$$

17.7 Gas of Kinks and Antikinks and Level Splitting Formula

The above semiclassical treatment is correct to leading order in $e^{-\omega L}$. This accuracy is not sufficient to calculate the degree of level splitting between the two lowest states of the double well caused by tunneling. Further semiclassical contributions to the path integral must be included. These can be found without further effort. For very large L, it is quite easy to accommodate many kinks and antikinks along the τ-axis without a significant deviation of the path from the equation of motion. Due to the fast approach to the potential bottoms $x = \pm a$ near each kink or antikink solution, an approximate solution can be constructed by smoothly combining a number of individual solutions as long as they are widely separated from each other. The deviations from a true classical solution are all exponentially small if the separation distance $\Delta\tau$ on the τ-axis is much larger than the size of an individual kink (i.e., $\Delta\tau \gg 1/\omega$). The combined solution may be thought of as a very dilute gas of kinks and antikinks on the τ-axis. This situation is referred to as the *dilute-gas limit*. Consider such an "almost-classical solution" consisting of N kink-antikink solutions $x_{\rm cl}(\tau) = \pm a\tanh[\omega(\tau - \tau_i)/2]$ in alternating order positioned at, say, $\tau_1 \gg \tau_2 \gg \tau_3 \gg \ldots \gg \tau_N$ and smoothly connected at some intermediate points $\bar\tau_1, \ldots, \bar\tau_{N-1}$. In the dilute-gas approximation, the combined action is given by the sum of the individual actions. For the amplitude (17.34) in which the paths connect the same potential valleys, the number of kinks must be equal to the number of antikinks. The action combined is then an even multiple of the single kink action:

$$\mathcal{A}_{2n} \approx 2n\mathcal{A}_{\rm cl}. \qquad (17.179)$$

For the amplitude (17.35), where the total number is odd, the combined action is

$$\mathcal{A}_{2n+1} \approx (2n + 1)\mathcal{A}_{\rm cl}. \qquad (17.180)$$

As the kinks and antikinks are localized objects of size $2/\omega$, it does not matter how they are distributed on the large-τ interval $[-L/2, L/2]$, as long as their distances are large compared with their size. In the dilute-gas limit, we can neglect the sizes. In the path integral, the translational degree of freedom of widely spaced N kinks and antikinks leads, via the zero-eigenvalue modes, to the multiple integral

$$\int_{-L/2}^{L/2} d\tau_N \int_{-L/2}^{\tau_N} d\tau_{N-1} \cdots \int_{-L/2}^{\tau_1} d\tau_1 = \frac{L^N}{N!}. \qquad (17.181)$$

The Jacobian associated with these N integrals is [see (17.112)]

$$\sqrt{\frac{\mathcal{A}_{\mathrm{cl}}}{2\pi\hbar}}^{-N}. \tag{17.182}$$

The fluctuations around the combined solution yield a product of the individual fluctuation factors. For a given set of connection points we have

$$\frac{1}{\sqrt{\Pi'_n \lambda_n}\big|_{\bar{L}_N}} \frac{1}{\sqrt{\Pi'_n \lambda_n}\big|_{\bar{L}_{N-1}}} \times \ldots \times \frac{1}{\sqrt{\Pi'_n \lambda_n}\big|_{\bar{L}_1}}, \tag{17.183}$$

where $\bar{L}_i \equiv \bar{\tau}_i - \bar{\tau}_{i-1}$ are the patches on the τ-axis in which the individual solutions are exact. Their total sum is

$$L = \sum_{i=1}^{N} \bar{L}_i. \tag{17.184}$$

We now include the effect of the fluctuations at the intermediate times $\bar{\tau}_i$ where the individual solutions are connected. Remembering the amplitudes (17.176), we see that the fluctuation factor for arbitrary endpoints x_i, x_{i-1} near the bottom of the potential valley must be multiplied at each end with a wave function ratio $\psi_0(x \pm a)/\psi_0(0)$. Thus we have to replace

$$\frac{1}{\sqrt{\Pi_n \lambda_n}} \rightarrow \frac{\psi_0(x_i \pm a)}{\psi_0(0)} \frac{1}{\sqrt{\Pi_n \lambda_n}} \frac{\psi_0^\dagger(x_{i-1} \pm a)}{\psi_0^\dagger(0)}. \tag{17.185}$$

The adjacent x_i-values of all fluctuation factors are set equal and integrated out, giving

$$\frac{1}{\sqrt{\Pi_n \lambda_n}}\bigg|_{L} = \int dx_N \cdots dx_1 \frac{1}{\sqrt{\Pi_n \lambda_n}}\bigg|_{L_N} \frac{\psi_0(x_{N-1} - a)\psi_0^\dagger(x_{N-1} - a)}{|\psi_0(0)|^2} \frac{1}{\sqrt{\Pi_n \lambda_n}}\bigg|_{L_{N-1}}$$
$$\times \ldots \times \frac{\psi_0(x_1 - a)\psi_0^\dagger(x_1 - a)}{|\psi_0(0)|^2} \frac{1}{\sqrt{\Pi_n \lambda_n}}\bigg|_{L_1}. \tag{17.186}$$

Due to the unit normalization of the ground state wave functions, the integrals are trivial. Only the $|\psi_0(0)|^2$-denominators survive. They yield a factor

$$\frac{1}{|\psi_0(0)|^{2(N-1)}} = \sqrt{\frac{\omega}{\pi\hbar}}^{-(N-1)}. \tag{17.187}$$

It is convenient to multiply and divide the result by the square root of the product of eigenvalues of the harmonic kink-free fluctuations, whose total fluctuation factor is known to be

$$\frac{1}{\sqrt{\Pi_n \lambda_n^0}\big|_{L}} = \sqrt{\frac{\omega}{\pi\hbar}} e^{-\omega L/2\hbar}. \tag{17.188}$$

Then we obtain the total corrected fluctuation factor

$$\sqrt{\frac{\omega}{\pi\hbar}}^{-(N-2)} e^{-\omega L/2\hbar} \sqrt{\prod_n \lambda_n^0}\Big|_L \frac{1}{\sqrt{\prod_n' \lambda_n}\big|_{L_1}} \frac{1}{\sqrt{\prod_n' \lambda_n}\big|_{L_2}} \times \ldots \times \frac{1}{\sqrt{\prod_n' \lambda_n}\big|_{L_N}}. \quad (17.189)$$

We now observe that the harmonic fluctuation factor (17.188) for the entire interval $\sqrt{\prod_n \lambda_n^0}\big|_L = \sqrt{\omega/\pi\hbar}\exp(-\omega L/2\hbar)$ can be factorized into a product of such factors for each interval $\bar\tau_i, \bar\tau_{i-1}$ as follows:

$$\sqrt{\prod_n \lambda_n^0}\Big|_L = \sqrt{\frac{\omega}{\pi\hbar}}^{-(N-1)} \sqrt{\prod_n \lambda_n^0}\Big|_{L_1} \cdots \sqrt{\prod_n \lambda_n^0}\Big|_{L_N}. \quad (17.190)$$

The total corrected fluctuation factor can therefore be rewritten as

$$\sqrt{\frac{\omega}{\pi\hbar}} e^{-\omega L/2\hbar} \sqrt{\frac{\prod_n \lambda_n^0}{\prod_n' \lambda_n}}\Big|_{L_1} \times \ldots \times \sqrt{\frac{\prod_n \lambda_n^0}{\prod_n' \lambda_n}}\Big|_{L_N}. \quad (17.191)$$

Each eigenvalue ratio gives the L_i-independent result

$$K' = \sqrt{\frac{\prod_n \lambda_n^0}{\prod_n' \lambda_n}}\Big|_{L_i}, \quad (17.192)$$

with K' of Eq. (17.131). Expressing K' in terms of K via

$$K' = \sqrt{\frac{\mathcal{A}_{\rm cl}}{2\pi\hbar}}^{-1} K, \quad (17.193)$$

the factors $\sqrt{\mathcal{A}_{\rm cl}/2\pi\hbar}^{-1}$ remove the Jacobian factors (17.182) arising from the positional integrals (17.181). Altogether, the total fluctuation factor of N kink-antikink solutions with all possible distributions on the τ-axis is

$$\sqrt{\frac{\omega}{\pi\hbar}} e^{-\omega L/2\hbar} \frac{L^N}{N!} K^N e^{-N\mathcal{A}_{\rm cl}/\hbar}. \quad (17.194)$$

Summing over all even and odd kink-antikink configurations, we thus obtain

$$(a\,L/2|\pm a\,-L/2) = \sqrt{\frac{\omega}{\pi\hbar}} e^{-\omega L/2\hbar} \sum_{\substack{\rm even \\ \rm odd}} \frac{1}{N!}(KLe^{-\mathcal{A}_{\rm cl}/\hbar})^N. \quad (17.195)$$

This can be summed up to

$$(a\,L/2|\pm a\,-L/2) = \sqrt{\frac{\omega}{\pi\hbar}} e^{-\omega L/2\hbar}$$
$$\times \frac{1}{2}\left[\exp\left(KLe^{-\mathcal{A}_{\rm cl}/\hbar}\right) \pm \exp\left(-KLe^{-\mathcal{A}_{\rm cl}/\hbar}\right)\right]. \quad (17.196)$$

As in the previous section, we generalize this result to positions x_b, x_a near the potential minima (with a maximal distance of the order of $\sqrt{\hbar/\omega}$). Using the classical action (17.175) and expressing it in terms of ground state wave functions, we can now add the contribution of the amplitudes for all possible configurations, arriving at

$$(x_b \ L/2|x_a \ -L/2) = e^{-\omega L/2\hbar} \tag{17.197}$$
$$\times \left\{ \psi_0(x_b - a)\psi_0(x_a - a)\frac{1}{2} \left[\exp(Ke^{-\mathcal{A}_{cl}/\hbar}L) + \exp(-Ke^{-\mathcal{A}_{cl}/\hbar}L) \right] \right.$$
$$+\psi_0(x_b - a)\psi_0(x_a - a)\frac{1}{2} \left[\exp(Ke^{-\mathcal{A}_{cl}/\hbar}L) - \exp(-Ke^{-\mathcal{A}_{cl}/\hbar}L) \right]$$
$$\left. +(x_b \to -x_b) + (x_a \to -x_a) + (x_b \to -x_b, x_a \to -x_a) \right\}.$$

The right-hand side is recombined to

$$\frac{1}{\sqrt{2}}[\psi_0(x_b - a) + \psi_0(x_b + a)] \times \frac{1}{\sqrt{2}}[\psi_0(x_a - a) + \psi_0(x_a + a)]$$
$$\times \exp\left[-\left(\frac{\omega}{2} - Ke^{-\mathcal{A}_{cl}/\hbar} \right) L \right]$$
$$+ \frac{1}{\sqrt{2}}[\psi_0(x_b - a) - \psi_0(x_b + a)] \times \frac{1}{\sqrt{2}}[\psi_0(x_a - a) - \psi_0(x_a + a)]$$
$$\times \exp\left[-\left(\frac{\omega}{2} + Ke^{-\mathcal{A}_{cl}/\hbar} \right) L \right]. \tag{17.198}$$

Here we identify the ground state wave function as the symmetric combination of the ground state wave functions of the individual wells

$$\Psi_0(x) = \frac{1}{\sqrt{2}}[\psi_0(x - a) + \psi_0(x + a)]. \tag{17.199}$$

Its energy is

$$\mathcal{E}^{(0)} = E^{(0)} - \frac{\Delta E^{(0)}}{2} = \left(\omega/2 - Ke^{-\mathcal{A}_{cl}/\hbar} \right)\hbar. \tag{17.200}$$

The first excited state has the antisymmetric wave function

$$\Psi_1(x) = \frac{1}{\sqrt{2}}[\psi_0(x - a) - \psi_0(x + a)] \tag{17.201}$$

and the slightly higher energy

$$\mathcal{E}^{(1)} = E^{(0)} + \frac{\Delta E^{(0)}}{2} = \left(\omega/2 + Ke^{-\mathcal{A}_{cl}/\hbar} \right)\hbar. \tag{17.202}$$

The level splitting is therefore

$$\Delta E = 2K\hbar e^{-\mathcal{A}_{cl}/\hbar}. \tag{17.203}$$

Inserting K from (17.131), we obtain the formula

$$\Delta E = 4\sqrt{3}\sqrt{\frac{\mathcal{A}_{\mathrm{cl}}}{2\pi\hbar}}\hbar\omega e^{-\mathcal{A}_{\mathrm{cl}}/\hbar}, \tag{17.204}$$

with $\mathcal{A}_{\mathrm{cl}} = (2/3)a^2\omega$. When expressing the action in terms of the height of the potential barrier $V_{\max} = a^2\omega^2/8 = 3\omega\mathcal{A}_{\mathrm{cl}}/16$, the formula reads

$$\Delta E = 4\sqrt{3}\sqrt{\frac{8V_{\max}}{3\pi\omega\hbar}}\hbar\omega e^{-16V_{\max}/3\hbar\omega}. \tag{17.205}$$

The level splitting decreases exponentially with increasing barrier height. Note that $V_{\max}$ is related to the coupling constant of the x^4-interaction by $V_{\max} = \omega^4/16g$.

To ensure the consistency of the approximation we have to check that the assumption of a low density gas of kinks and antikinks is self-consistent. When looking at the series (17.195) for the exponential (17.196), we see that the average number of contributing terms is given by

$$\bar{N} \approx KLe^{-\mathcal{A}_{\mathrm{cl}}/\hbar} = \frac{\Delta E}{2\hbar}L. \tag{17.206}$$

The associated average separation between kinks and antikinks is

$$\Delta L \equiv 2\hbar/\Delta E. \tag{17.207}$$

If we compare this with their size $2/\omega$, we find the ratio

$$\frac{\mathrm{distance}}{\mathrm{size}} \approx \frac{\hbar\omega}{\Delta E}. \tag{17.208}$$

For increasing barrier height, the level splitting decreases and the dilution increases exponentially. Thus the dilute-gas approximation becomes exact in the limit of infinite barrier height.

17.8   Fluctuation Correction to Level Splitting

Let us calculate the first fluctuation correction to the level splitting formula (17.204). For this we write the potential (17.1) as in (5.78):

$$V(x) = -\frac{\omega^2}{4}x^2 + \frac{g}{4}x^4 + \frac{1}{4g}, \tag{17.209}$$

with the interaction strength

$$g \equiv \frac{\omega^2}{2a^2}. \tag{17.210}$$

Expanding the action around the classical solution, we obtain the action of the fluctuations $y(\tau) = x(\tau) - x_{\mathrm{cl}}(\tau)$. Its quadratic part was given in Eq. (17.40) which we write as

$$\mathcal{A}_{\mathrm{fl}}^0 = \frac{1}{2}\int d\tau d\tau'\, y(\tau)\mathcal{O}_\omega(\tau, \tau')y(\tau'), \tag{17.211}$$

with the functional matrix

$$\mathcal{O}_\omega(\tau, \tau') \equiv \left[-\frac{d^2}{d\tau^2} + \omega^2 \left(1 - \frac{3}{2} \frac{1}{\cosh^2[\omega(\tau - \tau_0)/2]} \right) \right]' \delta(\tau - \tau') \qquad (17.212)$$

associated with the Schrödinger operator for a particle in a Rosen-Morse potential (14.158). The prime indicates the absence of the zero eigenvalue in the spectral decomposition of $\mathcal{O}_\omega(\tau, \tau')$. Since the associated mode does not perform Gaussian fluctuations, it must be removed from $y(\tau)$ and treated separately. At the semiclassical level, this was done in Subsection 17.3.1, and the zero eigenvalue appeared in the level splitting formula (17.204) as a factor (17.112). The removal gave rise to an additional effective interaction (17.110):

$$\mathcal{A}_e^{\text{eff}} = -\hbar \log \left[1 + \mathcal{A}_{\text{cl}}^{-1} \int d\tau\, x_{\text{cl}}'(\tau) y'(\tau) \right]. \qquad (17.213)$$

With (17.88)–(17.91), this can be rewritten after a partial integration as

$$\mathcal{A}_e^{\text{eff}} = -\hbar \log \left[1 - \sqrt{\frac{3g}{\omega^3}} \int d\tau\, y_0'(\tau) y(\tau) \right]. \qquad (17.214)$$

The interaction between the fluctuations is

$$\mathcal{A}_{\text{fl}}^{\text{int}} = \frac{g}{4} \int d\tau \left[y^4(\tau) + 4x_{\text{cl}}(\tau) y^3(\tau) \right]. \qquad (17.215)$$

In the path integral, we now perform a Taylor series expansion of the exponential $e^{-(\mathcal{A}_{\text{fl}}^{\text{int}} + \mathcal{A}_e^{\text{eff}})/\hbar}$ in powers of the coupling strength g. A perturbative evaluation of the correlation functions of the fluctuations $y(\tau)$ according to the rules of Section 3.20 produces a correction factor to the path integral

$$C = \left[1 - (I_1 + I_2 + I_3) \frac{g\hbar}{\omega^3} + \mathcal{O}(g^2) \right], \qquad (17.216)$$

where I_1, I_2, and I_3 are the dimensionless integrals running over the entire τ-axis:

$$\begin{aligned}
I_1 &= \frac{\omega^3}{4\hbar^2} \int d\tau\, \langle y^4(\tau) \rangle_{\mathcal{O}_\omega}, \\
I_2 &= -\frac{\omega^3 g}{2\hbar^3} \int d\tau d\tau'\, x_{\text{cl}}(\tau) \langle y^3(\tau) y^3(\tau') \rangle_{\mathcal{O}_\omega} x_{\text{cl}}(\tau'), \\
I_3 &= -\frac{\omega^3}{\hbar^2} \sqrt{\frac{3g}{\omega^3}} \int d\tau d\tau'\, y_0'(\tau) \langle y(\tau) y^3(\tau') \rangle_{\mathcal{O}_\omega} x_{\text{cl}}(\tau').
\end{aligned} \qquad (17.217)$$

In order to check the dimensions we observe that the classical solution (17.28) can be written with (17.210) as $x_{\text{cl}}(\tau) = \sqrt{\omega^2/2g}\, \tanh[\omega(\tau - \tau_0)/2]$, while $y(\tau)$ and τ have the dimensions $\sqrt{\hbar/\omega}$ and $1/\omega$, respectively. The Dirac brackets $\langle \ldots \rangle_{\mathcal{O}_\omega}$ denote the expectation with respect to the quadratic fluctuations controlled by the action (17.211). Due to the absence of a zero eigenvalue, the fluctuations are harmonic. The expectation values of the various powers of $y(\tau)$ can therefore be expanded according to the Wick rule of Section 3.17 into a sum of pair contractions involving products of Green functions

$$G'_{\mathcal{O}_\omega}(\tau, \tau') = \langle y(\tau) y(\tau') \rangle_{\mathcal{O}_\omega} = \hbar \mathcal{O}_\omega^{-1}(\tau, \tau'), \qquad (17.218)$$

where $\mathcal{O}_\omega^{-1}(\tau, \tau')$ denotes the inverse of the functional matrix (17.212).

The first term in (17.217) gives rise to three Wick contractions and becomes

$$I_1 = \frac{3\omega^3}{4\hbar^2} \int d\tau\, G'^2_{\mathcal{O}_\omega}(\tau, \tau). \qquad (17.219)$$

The integrand contains an asymptotically constant term which produces a linear divergence for large L. This divergence is subtracted out as follows:

$$I_1 = L\frac{3\omega}{16} + \frac{3\omega^3}{4\hbar^2} \int d\tau \left[G_{\mathcal{O}_\omega}^{\prime 2}(\tau, \tau) - \frac{\hbar^2}{4\omega^2} \right]. \tag{17.220}$$

The first term is part of the first-order fluctuation correction *without* the classical solution, i.e., it contributes to the constant background energy of the classical solution. It is obtained by replacing

$$G_{\mathcal{O}_\omega}^{\prime 2}(\tau, \tau') \rightarrow \hbar G_\omega(\tau - \tau') = \frac{\hbar}{2\omega} e^{-\omega|\tau - \tau'|} \tag{17.221}$$

[recall (3.301) and (3.246)]. In the amplitudes (17.195), the background energy changes only the exponential prefactor $e^{-\omega L/2\hbar}$ to $e^{-(1+3g\hbar/16\omega^3)\omega L/2\hbar}$ and does not contribute to the level splitting. The level splitting formula receives a correction factor

$$C' = \left[1 - c_1 \frac{g\hbar}{\omega^3} + \dots \right] = \left[1 - (I_1' + I_2' + I_3') \frac{g\hbar}{\omega^3} + \mathcal{O}(g^2) \right], \tag{17.222}$$

in which all contributions proportional to L are removed. Thus I_1 is replaced by its subtracted part $I_1' \equiv I_1 - L3\omega/16$.

The integral I_2 has 15 Wick contractions which decompose into two classes:

$$\begin{aligned} I_2 &\equiv I_{21} + I_{22} = -\frac{g\omega^3}{2\hbar^3} \int d\tau d\tau' \\ &\times x_{\text{cl}}(\tau) \left[6G_{\mathcal{O}_\omega}^{\prime 3}(\tau, \tau') + 9G_{\mathcal{O}_\omega}'(\tau, \tau)G_{\mathcal{O}_\omega}'(\tau, \tau')G_{\mathcal{O}_\omega}'(\tau', \tau') \right] x_{\text{cl}}(\tau'). \end{aligned} \tag{17.223}$$

Each of the two subintegrals I_{21} and I_{22} contains a divergence with L which can again be found via the replacement (17.221). The subtracted integrals in (17.222) are $I_{21}' = I_{21} + \omega L/8$ and $I_{22}' = I_{22} + 3\omega L/16$. Thus, altogether, the exponential prefactor $e^{-\omega L/2\hbar}$ in the amplitudes (17.195) is changed to $e^{-[1/2+(3/16-1/8-9/16)g\hbar/\omega^3]\omega L/2\hbar} = e^{-(1/2-g\hbar/2\omega^3)\omega L/2\hbar}$, in agreement with (5.258). To compare the two expressions, we have to set $\omega = \sqrt{2}$ since the present ω is the frequency at the bottom of the potential wells whereas the ω in Chapter 5 [which is set equal to 1 in (5.258)] parametrized the negative curvature at $x = 0$.

The Wick contractions of the third term lead to the finite integral

$$I_3 = I_3' = -3\frac{\omega^3}{\hbar^2}\sqrt{\frac{3g}{\omega^3}} \int d\tau d\tau' \, y_0'(\tau)G_{\mathcal{O}_\omega}'(\tau, \tau')G_{\mathcal{O}_\omega}'(\tau', \tau')x_{\text{cl}}(\tau'). \tag{17.224}$$

The correction factor (17.216) can be pictured by means of Feynman diagrams as

$$C = 1 - 3 \,\, \bigcirc\bigcirc \,\, + \frac{1}{2!}\left(6 \,\,\rightsquigarrow\!\!\bigcirc\!\!\rightsquigarrow\,\, + 9 \,\, \rightsquigarrow\!\!\bigcirc\bigcirc\!\!\rightsquigarrow \,\, \right)$$

$$+ \, 3 \,\, \rightsquigarrow\!\!\bigcirc\!\!\rightsquigarrow \,\, + \mathcal{O}(g^2), \tag{17.225}$$

where the vertices and lines represent the analytic expressions shown in Fig. 17.5.

For the evaluation of the integrals we need an explicit expression for $G_{\mathcal{O}_\omega}'(\tau, \tau')$. This is easily found from the results of Section 14.4.4. In Eq. (14.161), we gave the fixed-energy amplitude $(x_b|x_a)_{E_{\mathcal{RM}}, E_{\mathcal{PT}}}$ solving the Schrödinger equation

$$\left(-\frac{\hbar^2}{2\mu}\frac{d^2}{dx^2} - E_{\mathcal{RM}} + \frac{\hbar^2}{2\mu}\frac{E_{\mathcal{PT}}}{\cosh^2 x} \right)(x_b|x_a)_{E_{\mathcal{RM}}, E_{\mathcal{PT}}} = -i\hbar\delta(x_b - x_a). \tag{17.226}$$

Inserting $E_{\mathcal{PT}} = (\hbar^2/2\mu)s(s+1)$, the amplitude reads for $x_b > x_a$

$$
(x_b|x_a)_{E_{\mathcal{RM}},E_{\mathcal{PT}}} = \frac{-i\mu}{\hbar}\Gamma(m(E_{\mathcal{RM}}) - s)\Gamma(s + m(E_{\mathcal{RM}}) + 1)
$$
$$
\times P_s^{-m(E_{\mathcal{RM}})}(\tanh x_b)P_s^{-m(E_{\mathcal{RM}})}(-\tanh x_a), \qquad (17.227)
$$

with

$$
m(E_{\mathcal{RM}}) = \sqrt{1 - 2\mu E_{\mathcal{RM}}/\hbar^2}. \qquad (17.228)
$$

After a variable change $x = \omega\tau/2$ and $\hbar^2/\mu = \omega^2/2$, we set $s = 2$ and insert the energy $E_{\mathcal{RM}} = -3\omega^2/4$. Then the operator in Eq. (17.226) coincides with $\mathcal{O}_\omega(\tau,\tau')$ of Eq. (17.212), and we obtain the desired Green function for $\tau > \tau'$

$$
G_{\mathcal{O}_\omega}(\tau,\tau') = \frac{\hbar}{\omega}\Gamma(m-2)\Gamma(m+3)\,P_2^{-m}(\tanh\frac{\omega\tau}{2})\,P_2^{-m}(-\tanh\frac{\omega\tau'}{2}), \qquad (17.229)
$$

with $m = 2$. Due to translational invariance along the τ-axis, this Green function has a pole at $E_{\mathcal{RM}} = -3\omega^2/4$ which must be removed before going to this energy. The result is the subtracted Green function $G'_{\mathcal{O}_\omega}(\tau,\tau')$ which we need for the perturbation expansion. The subtraction procedure is most easily performed using the formula $G'_{\mathcal{O}_\omega} = (d/dE_{\mathcal{RM}})E_{\mathcal{RM}}G_{\mathcal{O}_\omega}|_{E_{\mathcal{RM}}=-3\omega^2/4}$. In terms of the parameter m, this amounts to

$$
G'_{\mathcal{O}_\omega}(\tau,\tau') = \frac{1}{2m}\frac{d}{dm}(m^2 - 4)G_{\mathcal{O}_\omega}(\tau,\tau')\bigg|_{m=2}. \qquad (17.230)
$$

Inserting into (17.227) the Legendre polynomials from (14.165),

$$
P_2^{-m}(z) = \frac{1}{\Gamma(1+m)}\left(\frac{1+z}{1-z}\right)^{-m/2}\left[1 - \frac{3}{1+m}(1-z) + \frac{3}{(1+m)(2+m)}(1-z)^2\right], \qquad (17.231)
$$

the Green function (17.230) can be written as

$$
G'_{\mathcal{O}_\omega}(\tau,\tau') = \hbar[Y_0(\tau_>)y_0(\tau_<) + y_0(-\tau_>)Y_0(-\tau_<)], \qquad (17.232)
$$

where $\tau_>$ and $\tau_<$ are the greater and the smaller of the two times τ and τ', respectively, and $y_0(\tau), Y_0(\tau)$ are the wave functions

$$
y_0(\tau) = -2\sqrt{6\omega}P_2^{-2}\left(-\tanh\frac{\omega\tau}{2}\right) = -\sqrt{\frac{3\omega}{8}}\frac{1}{\cosh^2\frac{\omega\tau}{2}}, \qquad (17.233)
$$

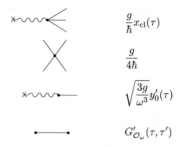

Figure 17.5   Vertices and lines of Feynman diagrams for correction factor C in Eq. (17.225).

$$Y_0(\tau) = \frac{1}{2\sqrt{6}\omega} \frac{1}{2\omega m} \left\{ \frac{1}{2} \left[\frac{d}{dm}(m^2-4)\Gamma(m-2)\Gamma(m+3) \right] P_2^{-m}\left(\tanh\frac{\omega\tau}{2} \right) \right.$$
$$\left. + \left[(m^2-4)\Gamma(m-2)\Gamma(m+3) \right] \frac{d}{dm} P_2^{-m}\left(\tanh\frac{\omega\tau}{2} \right) \right\} \Bigg|_{m=2}. \qquad (17.234)$$

From (17.231) we see that

$$\frac{d}{dm} P_2^{-m}\left(\tanh\frac{\omega\tau}{2} \right) \Bigg|_{m=2} = \frac{\sqrt{6}}{144} y_0(\tau)[6(3-2\gamma+\omega\tau) - e^{-\omega\tau}(8+e^{-\omega\tau})], \qquad (17.235)$$

where $\gamma \approx 0.5773156649$ is the Euler-Mascheroni constant (2.467). Hence

$$Y_0(\tau) = \frac{1}{12\omega^2} y_0(\tau)[e^{-\omega\tau}(e^{-\omega\tau}+8) - 2(2+3\omega\tau)]. \qquad (17.236)$$

For $\tau = \tau'$, the Green function is

$$G'_{\mathcal{O}_\omega}(\tau,\tau) = \frac{\hbar}{2\omega} \frac{1}{\cosh^4\frac{\omega\tau}{2}} \left(\cosh^4\frac{\omega\tau}{2} + \cosh^2\frac{\omega\tau}{2} - \frac{11}{8} \right). \qquad (17.237)$$

Note that an application of the Schrödinger operator (17.212) to the wave functions $Y_0(\tau)$ and $y_0(\tau)$ produces $-y_0(\tau)$ and 0, respectively. These properties can be used to construct the Green function $G'_{\mathcal{O}_\omega}(\tau,\tau')$ by a slight modification of the Wronski method of Chapter 3. Instead of the differential equation $\mathcal{O}_\omega G(\tau,\tau') = \hbar\delta(\tau-\tau')$, we must solve the projected equation

$$\mathcal{O}'_\omega G'_{\mathcal{O}_\omega}(\tau,\tau') = \hbar[\delta(\tau-\tau') - y_0(\tau)y_0(\tau')], \qquad (17.238)$$

where the right-hand side is the completeness relation without the zero-eigenvalue solution:

$$\sum_{n\neq 0} y_n(\tau)y_n(\tau') = \delta(\tau-\tau') - y_0(\tau)y_0(\tau'). \qquad (17.239)$$

The solution of the projected equation (17.238) is precisely given by the combination (17.232) of the solutions $Y_0(\tau)$ and $y_0(\tau)$ with the above-stated properties.

The evaluation of the Feynman integrals I_1, I_{21}, I_{22}, I_3 is somewhat tedious and is therefore described in Appendix 17A. The result is

$$I'_1 = \frac{97}{560}, \quad I'_{21} = \frac{53}{420}, \quad I'_{22} = \frac{117}{560}, \quad I_3 = \frac{49}{20}. \qquad (17.240)$$

These constants yield for the correction factor (17.222)

$$C' = \left[1 - \frac{71}{24}\frac{g\hbar}{\omega^3} + \mathcal{O}(g^2) \right], \qquad (17.241)$$

modifying the level splitting formula (17.204) for the ground state energy to

$$\Delta E^{(0)} = 4\sqrt{3}\sqrt{\frac{\omega^3/3g}{2\pi\hbar}}\hbar\omega e^{-\omega^3/3g\hbar - 71g\hbar/24\omega^3 + \cdots}. \qquad (17.242)$$

This expression can be compared with the known energy eigenvalues of the lowest two double-well states. In Section 5.15, we have calculated the variational approximation $W_3(x_0)$ to the effective classical potential of the double well and obtained for small g an energy (see Fig. 5.24) which did not yet incorporate the effects of tunneling. We now add to this the level shifts $\pm\Delta E^{(0)}/2$ from Eq. (17.242) and obtain the curves also shown in Fig. 5.24. They agree reasonably well with the Schrödinger energies.

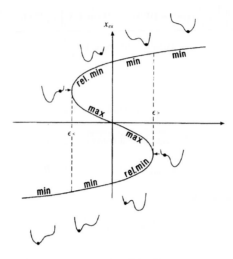

Figure 17.6 Positions of extrema x_{ex} in asymmetric double-well potential, plotted as function of asymmetry parameter ϵ. If rotated by 90^0, the plot shows the typical cubic shape. Between $\epsilon_>$ and $\epsilon_<$, there are two minima and one central maximum. The branches denoted by "min" are absolute minima; those denoted by "rel min" are relative minima.

17.9 Tunneling and Decay

The previous discussion of level splitting leads us naturally to another important tunneling phenomenon of quantum theory: the decay of metastable states. Suppose that the potential is not completely symmetric. For definiteness, let us add to $V(x)$ of Eq. (17.1) a linear term which breaks the symmetry $x \to -x$:

$$\Delta V = -\epsilon \frac{x - a}{2a}. \tag{17.243}$$

For small $\epsilon > 0$, this slightly depresses the left minimum at $x = -a$. The positions of the extrema are found from the cubic equation

$$V'(x_{\text{ex}}) = \frac{\omega^2 a}{2}\left[\left(\frac{x_{\text{ex}}}{a}\right)^3 - \frac{x_{\text{ex}}}{a} - \frac{\epsilon}{\omega^2 a^2}\right] = 0. \tag{17.244}$$

They are shown in Fig. 17.6. For large ϵ, there is only one extremum, and this is always a minimum. In the region where x_{ex} has three solutions, say x_-, x_0, x_+, the branches denoted by "rel min." in Fig. 17.6 correspond to relative minima which lie higher than the absolute minimum. The central branch corresponds to a maximum. As ϵ decreases from large positive to large negative values, a classical particle at rest at the minimum follows the upper branch of the curve and drops to the lower branch as ϵ becomes smaller than $\epsilon_<$. Quantum-mechanically, however, there is tunneling to the lower state before $\epsilon_<$. Tunneling sets in as soon as ϵ becomes negative, i.e., as soon as the initial minimum at x_+ comes to lie higher than the other minimum at x_-. The state whose wave packet is localized initially around x_+ decays into the lower minimum around x_-. After some finite time, the wave packet is concentrated around x_-.

A state with a finite lifetime is described analytically by an energy which lies in the lower half of the complex energy plane, i.e., which carries a negative imaginary part E^{im}. The imaginary part gives half the decay rate $\Gamma/2\hbar$. This follows directly from the temporal behavior of a wave function with an energy $E = E^{\mathrm{re}} + iE^{\mathrm{im}}$ which is given by

$$\psi(\mathbf{x})e^{-iEt/\hbar} = \psi(\mathbf{x})e^{-iE^{\mathrm{re}}t/\hbar}e^{E^{\mathrm{im}}t/\hbar} = \psi(\mathbf{x})e^{-iE^{\mathrm{re}}t/\hbar}e^{-\Gamma t/2\hbar}. \tag{17.245}$$

The last factor leads to an exponential decay of the norm of the state

$$\int d^3x|\psi(\mathbf{x})|^2 = e^{-\Gamma t/\hbar}, \tag{17.246}$$

which shows that $\hbar/\Gamma$ is the lifetime of the state. A positive sign of the imaginary part of the energy is ruled out since it would imply the state to have an exponentially growing norm.

We are now going to calculate Γ for the lowest state.[4] If ϵ has a small negative value, the initial probability is concentrated in the potential valley around the right-hand minimum $x = x_+ \approx a$. We assume the potential barrier to be high compared to the ground state energy. Then a semiclassical treatment is adequate. In this approximation we evaluate the amplitude

$$(x_+t_b|x_+t_a).$$

It contains the desired information on the lifetime of the lowest state by behaving, for large $t_b - t_a$, as

$$(x_+t_b|x_+t_a) \sim \psi_0(0)\psi_0(0)e^{-iE^{\mathrm{re}}(t_b-t_a)/\hbar}e^{-\Gamma(t_b-t_a)/2\hbar}.$$

As before, it is convenient to work with the Euclidean amplitude with $\tau_a = -L/2$ and $\tau_b = L/2$,

$$(x_+ \ L/2|x_+ - L/2), \tag{17.247}$$

which behaves for large L as

$$(x_+ \ L/2|x_+ - L/2) \sim \psi_0(0)\psi_0(0)e^{-E^{\mathrm{re}}L/\hbar}e^{i\Gamma L/2\hbar}. \tag{17.248}$$

The classical approximation to this amplitude is dominated by the path solving the imaginary-time equation of motion which corresponds to a real-time motion in the reversed potential $-V(x)$ (see Fig. 17.7). The particle starts out at $x = x_+$ for

---

[4]Due to the finite lifetime this state is not stationary. For sufficiently long lifetimes, however, it is approximately stationary for a finite time.

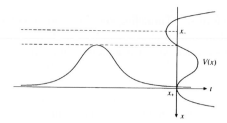

Figure 17.7 Classical bubble solution in reversed asymmetric quartic potential for $\epsilon < 0$, starting out at the potential maximum at x_+, crossing the valley, and returning to the maximum.

$\tau = -L/2$, traverses the minimum of $-V(x)$ at some finite value $\tau = \tau_0$, and comes back to x_+ at $\tau = L/2$. This solution is sometimes called a *bounce* solution, because of its returning to the initial point.

There exists an important application of the tunneling theory to the vaporization process of overheated water, to be discussed in Section 17.11. There the same type of solutions plays the role of *critical bubbles* triggering the phase transition. Since bounce solutions were first discussed in this context [3], we shall call them *bubble solutions* or critical bubbles.

We now proceed as in the previous section, i.e., we calculate
a) the classical action of a bubble solution,
b) the quadratic fluctuations around a bubble solution,
c) the sum over infinitely many bubble solutions.

By following these three steps naively, we obtain the amplitude

$$(x_+ \ L/2|x_+ - L/2) = \sqrt{\frac{\omega}{\pi\hbar}} e^{-\omega L/2\hbar} \exp\left[\sqrt{\mathcal{A}_{\mathrm{cl}}/2\pi\hbar} K' L e^{-\mathcal{A}_{\mathrm{cl}}/\hbar}\right]. \qquad (17.249)$$

Here $\mathcal{A}_{\mathrm{cl}}$ is the action of the bubble solution and K' collects the fluctuations of all nonzero-eigenvalue modes in the presence of the bubble solution as in (17.192):

$$K' = \sqrt{\left.\frac{\Pi_n^0 \lambda_n}{\Pi'_n \lambda_n}\right|_L}. \qquad (17.250)$$

The translational invariance makes the imaginary part in the exponent proportional to the total length L of the τ-axis.

From the large-L behavior of the amplitude (17.249) we obtain the ground state energy

$$E^{(0)} = \left(\frac{\omega}{2} - \sqrt{\frac{\mathcal{A}_{\mathrm{cl}}}{2\pi\hbar}} K' e^{-\mathcal{A}_{\mathrm{cl}}/\hbar}\right). \qquad (17.251)$$

In order to deduce the finite lifetime of the state from this formula we note that, just like the kink solution, the bubble solution has a zero-eigenvalue fluctuation associated with the time translation invariance of the system. As before, its wave function is given by the time derivative of the bubble solution

$$y_0(\tau) = \frac{1}{\sqrt{\mathcal{A}_{\rm cl}}} x'_{\rm cl}(\tau). \tag{17.252}$$

In contrast to the kink solution, however, the bubble solution returns to the initial position, implying that $x_{\rm cl}(\tau)$ has a maximum. Thus, the zero-eigenvalue mode $\propto x'_{\rm cl}(\tau)$ contains a sign change (see Fig. 17.7). In wave mechanics, such a place is called a *node* of the wave function. A wave function with a node cannot be the ground state of the Schrödinger equation governing the fluctuations

$$\left[-\frac{d^2}{d\tau^2} + V''(x_{\rm cl}(\tau)) \right] y_n(\tau) = \lambda_n y_n(\tau). \tag{17.253}$$

A symmetric wave function without a node must exist, which will have a lower energy than the zero-eigenvalue mode, i.e., it will have a negative eigenvalue $\lambda_{-1} < 0$. The associated wave function is denoted by $y_{-1}(\tau)$. It corresponds to a size fluctuation of the bubble solution. The nodeless wave function $y_{-1}(\tau)$ is the ground state. There can be no further negative-eigenvalue solution [4].

It is instructive to trace the origin of the negative sign within the efficient calculation method of the fluctuation determinant in Section 17.5. In contrast to the instanton treated there, the bubble solution has opposite symmetry, with an antisymmetric translational mode $x'_{\rm cl}(\tau)$. From this we may construct again two linearly independent solutions to find the determinant $D(\tau)$ to be used in Eq. (17.154).

The negative eigenvalue λ_{-1} enters in the calculation of the functional integral (17.46) via a fluctuation integral

$$\int \frac{d\xi_1}{\sqrt{2\pi\hbar}} e^{-(1/2\hbar)\xi_1^2 \lambda_{-1}}. \tag{17.254}$$

This integral diverges. The harmonic fluctuations of the integration variable take place around a maximum; they are unstable. At first sight one might hope to obtain a correct result by a naive analytic continuation doing first the integral for $\lambda_{-1} > 0$, where it gives

$$\int \frac{d\xi_{-1}}{\sqrt{2\pi\hbar}} e^{-(1/2\hbar)\xi_{-1}^2 \lambda_{-1}} = \frac{1}{\sqrt{\lambda_{-1}}}, \tag{17.255}$$

and then continuing the right-hand side analytically to negative λ_{-1}. The result would be

$$\int \frac{d\xi_{-1}}{\sqrt{2\pi\hbar}} e^{-(1/2\hbar)\xi_{-1}^2 \lambda_{-1}} = \pm \frac{i}{\sqrt{|\lambda_{-1}|}}. \tag{17.256}$$

From (17.250) and (17.251) we then might expect the formula for the decay rate to be

$$\frac{1}{\hbar}\Gamma = -2i\sqrt{\frac{\mathcal{A}_{\rm cl}}{2\pi\hbar}} K' e^{-\mathcal{A}_{\rm cl}/\hbar} \quad \text{(wrong)}, \tag{17.257}$$

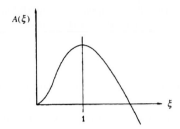

Figure 17.8 Action of deformed bubble solution as function of deformation parameter ξ. The maximum at $\xi = 1$ represents the critical bubble.

with

$$K' = i|K'| = \sqrt{\left|\frac{\prod_n^0 \lambda_n}{\prod_{n \neq 0, -1} \lambda_n}\right|_L} \frac{i}{\sqrt{|\lambda_{-1}|}}. \qquad (17.258)$$

However, this naive manipulation does not quite give the correct result. As we shall see immediately, the error consists in a missing factor $1/2$ which has a simple physical explanation. A more careful analytic continuation is necessary to find this factor [3]. As a function of ξ, it behaves as shown in Fig. 17.8.

For a proper analytic continuation, consider a continuous sequence of paths in the functional space and parametrize it by some variable ξ. Let the trivial path

$$x(\tau) \equiv x_+ \qquad (17.259)$$

correspond to $\xi = 0$, and the bubble solution

$$x(\tau) = x_{\mathrm{cl}}(\tau) \qquad (17.260)$$

to $\xi = 1$. The action of the trivial path is zero, that of the bubble solution is $\mathcal{A} = \mathcal{A}_{\mathrm{cl}}$. As the parameter ξ increases to values > 1, the bubble solution is deformed with a growing portion of the curve moving down towards the bottom of the lower potential valley (see Fig. 17.9). This lowers the action more and more.

There is a maximum at the bubble solution $\xi = 1$. The negative eigenvalue $\lambda_{-1} < 0$ of the fluctuation equation (17.253) is proportional to the negative curvature at the maximum. Since there exists only a single negative eigenvalue, the fluctuation determinant of the remaining modes is positive. It does not influence the process of analytic continuation. Thus we may study the analytic continuation within a simple model integral designed to have the qualitative behavior described above:

$$Z = \int_0^\infty \frac{d\xi}{\sqrt{2\pi}} e^{\lambda(\xi^2 + \alpha \xi^3)}. \qquad (17.261)$$

The parameter λ stands for the negative eigenvalue λ_{-1}, whereas α is an auxiliary parameter to help perform the analytic continuation. For $\alpha > 0$, the integral is

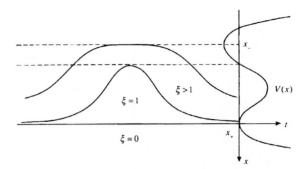

Figure 17.9 Sequence of paths as function of parameter ξ, starting out at $\xi = 0$, with a constant solution in the metastable valley $x(\tau) \equiv x \equiv x_+$, reaching the extremal bubble solution $x(\tau) \equiv x_{\rm cl}(\tau)$ for $\xi = 1$, and sliding more and more down towards the stable minimum for $1 < \xi \to \infty$.

stable and well defined. For $\alpha < 0$, the "Euclidean action" in the exponent $\mathcal{A} = -\lambda(\xi^2 + \alpha\xi^3)$ has a maximum at

$$\xi_m = -\frac{2}{3\alpha}. \tag{17.262}$$

Near the maximum, it has the expansion

$$\mathcal{A} = -\lambda \left[\frac{4}{27\alpha^2} - (\xi - \xi_m)^2 + \ldots \right]. \tag{17.263}$$

The second term possesses a negative curvature λ which represents the negative eigenvalue λ_{-1}. The parameter α^2 plays the role of $\hbar/(-\lambda \mathcal{A}_{\rm cl})$ in the bubble discussion, and the semiclassical expansion of the path integral corresponds to an expansion of the model integral in powers of α^2. We want to show that the lowest two orders of this expansion yield an imaginary part

$$\operatorname{Im} Z \sim e^{\lambda 4/27\alpha^2} \frac{1}{2} \frac{1}{\sqrt{|\lambda|}}, \tag{17.264}$$

where the exponential is the classical contribution and the factor contains the fluctuation correction. To derive (17.264), we continue the integral (17.261) analytically from $\alpha > 0$, where it is well defined, into the complex α-plane. It is convenient to introduce a new variable $t = \alpha\xi$. Then Z becomes

$$Z = \frac{1}{\alpha} \int_0^\infty \frac{dt}{\sqrt{2\pi}} \exp\left[\frac{\lambda}{\alpha^2}(t^2 + t^3) \right]. \tag{17.265}$$

Since $\lambda < 0$ this integral converges for $\alpha > 0$. To continue it to negative real values of α, we set $\alpha \equiv |\alpha|e^{i\varphi}$ and increase the angle φ from zero to $\pi/2$. While doing so,

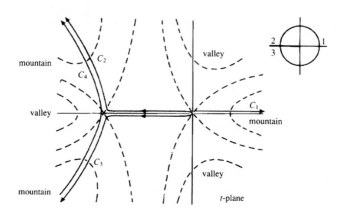

Figure 17.10 Lines of constant $\mathrm{Re}\,(t^2 + t^3)$ in complex t-plane and integration contours C_i for various phase angles of α (shown in the insert) which maintain convergence of the integral (17.265).

we deform the contour in the t-plane in order to maintain convergence. Thus we introduce an auxiliary real variable t' and set

$$t = e^{i2\varphi/3}t', \quad t' \in (0, \infty). \tag{17.266}$$

The continued integral is then performed as $\int dt = e^{i2\varphi/3} \int_0^\infty dt'$. From the geometric viewpoint, the convergence is maintained for the following reason: For $\alpha > 0$, the real part of the "action" $-(\lambda/\alpha^2)(t^2 + t^3)$ has asymptotically three mountains at azimuthal angles $\varphi = 0, 2\pi/3, 4\pi/3$, and three valleys at $\varphi = \pi/3, \pi, 5\pi/3$ (see Fig. 17.10). As α is rotated by the phase $e^{i\varphi}$, these mountains rotate with 2/3 of the angle φ anticlockwise in the t-plane. Since the contour keeps running up the same mountain, the integral continues to converge, rendering an analytic function of α. After α has been rotated to $e^{i\pi}\alpha = -\alpha$, the exponent in (17.265) takes back the original form, but the contour C runs up the mountain at $\varphi = 2\pi/3$. It does not matter which particular shape is chosen for the contour in the finite regime. We may deform the contour to the shape C_2 shown in Fig. 17.10. Next we observe that the point $-\alpha$ can also be reached by rotating α in the clockwise sense with $-\varphi$ increasing to π. In this case the final contour will run like C_3 in Fig. 17.10. The difference between the two analytic continuations is

$$\Delta Z \equiv Z(|\alpha|e^{-i\pi}) - Z(|\alpha|e^{i\pi}) = \frac{1}{|\alpha|}\int_{C_4} \frac{dt}{\sqrt{2\pi}}\exp\left\{\frac{\lambda}{\alpha^2}(t^2 + t^3)\right\}, \tag{17.267}$$

where the contour $C_4 = C_2 - C_3$ connects the mountain at $\varphi = 4\pi/3$ with that at $\varphi = 2\pi/3$. The convergence of the combined integral is most rapid if the contour

C_4 is chosen to run along the line of steepest slope. This traverses the minimum at $t = -1$ vertically in the complex t-plane.

The fact that ΔZ is nonzero implies that the partition function has a cut in the complex α-plane along the negative real axis. Since Z is real for $\alpha > 0$, it is a real analytic function in the complex α-plane and the difference ΔZ gives a purely imaginary discontinuity across the cut:

$$\Delta Z \equiv \operatorname{disc} Z = Z(-|\alpha| - i\eta) - Z(-|\alpha| + i\eta). \tag{17.268}$$

Let us calculate the discontinuity in the limit of small α where the dominant contribution comes from the neighborhood of the point $t = -1$. While the action at this point has a local maximum along the real t-axis, it has a local minimum along the vertical contour in the complex t-plane. For small α^2, the integral can be found via the saddle point approximation calculating the local minimum in the quadratic approximation:

$$\operatorname{disc} Z \approx e^{\lambda 4/27\alpha^2} \int_{-i\infty}^{i\infty} \frac{d\xi}{\sqrt{2\pi}} e^{\lambda(\xi - \xi_0)^2} \tag{17.269}$$

$$= e^{\lambda 4/27\alpha^2} \frac{i}{\sqrt{-\lambda}}.$$

Due to the real analyticity of Z, the imaginary part of Z is equal to one half of this:

$$\operatorname{Im} Z(-|\alpha| \mp i\eta) = \pm e^{\lambda 4/27\alpha^2} \frac{1}{2\sqrt{-\lambda}}. \tag{17.270}$$

The contour leading up to the extremal point adds only a real part to Z. The result (17.270) is therefore the exact leading contributing to the imaginary part in the limit $\alpha^2 \to 0$, corresponding to the semiclassical limit $\hbar \to 0$.

The exponent in (17.270) is the action of the model integral at the saddle point. The second factor produces the desired imaginary part. For a sequence of paths in functional space whose action depends on ξ as in Fig. 17.9, the result can be phrased as follows:

$$\int_0^\infty \frac{d\xi}{\sqrt{2\pi\hbar}} e^{-\mathcal{A}(\xi)/\hbar} = \int_0^1 \frac{d\xi}{\sqrt{2\pi\hbar}} e^{-\mathcal{A}(\xi)/\hbar} + e^{-\mathcal{A}(1)/\hbar} \int_1^{1-i\infty} \frac{d\xi}{\sqrt{2\pi\hbar}} e^{-\mathcal{A}''(1)(\xi-1)^2/2\hbar}$$

$$\approx \int_0^1 \frac{d\xi}{\sqrt{2\pi\hbar}} e^{-\mathcal{A}(\xi)/\hbar} + \frac{i}{2} e^{-\mathcal{A}(1)/\hbar} \frac{1}{\sqrt{-\mathcal{A}''(1)}}. \tag{17.271}$$

After translating this result to the form (17.254), we conclude that the integration over the negative-eigenvalue mode

$$\int \frac{d\xi_{-1}}{\sqrt{2\pi\hbar}} e^{-\xi_{-1}^2 \lambda_{-1}/2\hbar} \tag{17.272}$$

becomes, for $\lambda_{-1} < 0$ and after a proper analytic continuation,

$$\int \frac{d\xi_{-1}}{\sqrt{2\pi\hbar}} e^{-\xi_{-1}^2 \lambda_{-1}/2\hbar} = \frac{i}{2} \frac{1}{\sqrt{|\lambda_{-1}|}}. \tag{17.273}$$

It is easy to give a physical interpretation to the factor $1/2$ appearing in this formula, in contrast to the naively continued formula (17.287). At the extremum, the classical solution, which plays the role of a critical bubble, can equally well contract or expand in size. In the first case, the path $x(\tau)$ returns towards to the original valley and the bubble disappears. In the second case, the path moves more and more towards the lower valley at $x = -x_-$, thereby transforming the system into the stable ground state. The factor $1/2$ accounts for the fact that only the *expansion* of the bubble solution produces a stable ground state, not the *contraction*.

The factor $1/2$ multiplies the naively calculated imaginary part of the partition function which becomes

$$\operatorname{Im} Z(-|g| - i\eta) \approx \frac{1}{2}\sqrt{\mathcal{A}_{\mathrm{cl}}/2\pi\hbar}|K'|Le^{-\mathcal{A}_{\mathrm{cl}}/\hbar}. \tag{17.274}$$

The summation over an infinite number of bubble solutions moves the imaginary contribution to Z into the exponent as follows:

$$\operatorname{Re} Z + \operatorname{Im} Z = \operatorname{Re} Z(1 + \operatorname{Im} Z/\operatorname{Re} Z) \xrightarrow[\text{infinite sum}]{} \operatorname{Re} Z e^{\operatorname{Im} Z/\operatorname{Re} Z} \tag{17.275}$$

as in Eqs. (17.195), (17.196). By comparison with (17.248), we obtain the correct semiclassical *tunneling rate formula* [rather than (17.257)]:

$$\frac{1}{\hbar}\Gamma = \sqrt{\frac{\mathcal{A}_{\mathrm{cl}}}{2\pi\hbar}}|K'|e^{-\mathcal{A}_{\mathrm{cl}}/\hbar}, \tag{17.276}$$

where K' is the square root of the eigenvalue ratios, with the zero-eigenvalue mode removed. The prefactor has the dimension of a frequency. It defines the *bubble decay frequency*

$$\omega_{\mathrm{att}} = \sqrt{\frac{\mathcal{A}_{\mathrm{cl}}}{2\pi\hbar}}\,|K'|. \tag{17.277}$$

The exponential in (17.276) is a "quantum Boltzmann factor" which suppresses the formation of a bubble triggering the tunneling process via its expansion. The subscript indicates that the frequency plays the role of an *attempt frequency* by which the metastable state attempts to tunnel through the barrier into the stable ground state.

17.10 Large-Order Behavior of Perturbation Expansions

The above semiclassical approach of the decay rate of a metastable state has an important fundamental application. At the end of Chapter 3 we have remarked that the perturbation expansion of the anharmonic oscillator has a zero radius of convergence. This property is typical for many quantum systems. The precise form of the divergence is controlled by the tunneling rate formula (17.276), as we shall see now.

17.10.1   Growth Properties of Expansion Coefficients

As a specific, but typical, example we consider the anharmonic oscillator with the action

$$\mathcal{A} = \int_{-L/2}^{L/2} d\tau \left[\frac{x'^2}{2} - \frac{\omega^2}{2} x^2 - \frac{g}{4} x^4 \right], \tag{17.278}$$

and study the partition function as an analytic function of g. It is given by the path integral at large L (which now represents the imaginary time $\beta = 1/k_B T$, setting $\hbar = 1$)

$$Z(g) = \int \mathcal{D}x(\tau) e^{\mathcal{A}}. \tag{17.279}$$

The L-dependence of the partition function follows from the spectral representation

$$Z(g) = \sum_n e^{-E^{(n)}(g)L}, \tag{17.280}$$

where $E^{(n)}(g)$ are the energy eigenvalues of the system. In the limit $L \to \infty$, this becomes an expansion for the ground state energy $E^{(0)}(g)$. In the limit $L \to \infty$, $Z(g)$ behaves like

$$Z(g) \to e^{-E^{(0)}(g)L}, \tag{17.281}$$

exhibiting directly the ground state energy.

Since the path integral can be done exactly at the point $g = 0$, it is suggestive to expand the exponential in powers of g and to calculate the perturbation series

$$Z(g) = \sum_{k=0} Z_k \left(\frac{g}{\omega^3} \right)^k. \tag{17.282}$$

As shown in Section 3.20, the expansion coefficients are given by the path integrals

$$
\begin{aligned}
Z_k &= \frac{(-g)^k}{k!} \int \mathcal{D}x(\tau) \left[\int_{-L/2}^{L/2} d\tau \, x^4(\tau) \right]^k \exp\left[-\int_{-L/2}^{L/2} d\tau \left(\frac{1}{2}\dot{x}^2 + \frac{\omega^2}{2} x^2 \right) \right] \\
&= Z^{-1} \frac{(-g)^k}{k!} \left\langle \left(\int_{-L/2}^{L/2} d\tau \, x^4(\tau) \right)^k \right\rangle_\omega. \tag{17.283}
\end{aligned}
$$

By selecting the connected Feynman diagrams in Fig. 3.7 contributing to this path integral, we obtain the perturbation expansion in powers of g for the free energy F. In the limit $L \to \infty$, this becomes an expansion for the ground state energy $E^{(0)}(g)$, in accordance with (17.281). By following the method in Section 3.18, we find similar expansions for all excited energies $E^{(n)}(g)$ in powers of g. For $g = 0$, the energies are, of course, those of a harmonic oscillator, $E_0^{(n)} = \omega(n + 1/2)$. In general, we find the series

$$E^{(n)}(g) = \sum_{k=0}^{\infty} E_k^{(n)} \left(\frac{g}{4} \right)^k. \tag{17.284}$$

Most perturbation expansions have the grave deficiency observed in Eq. (3C.27). Their coefficients grow for large order k like a factorial $k!$ causing a vanishing radius of convergence. They can yield approximate results only for very small values of g. Then the expansion terms $E_k^{(n)} (g/4)^k$ decrease at least for an initial sequence of k-values, say for $k = 0, \ldots, N$. For large k-values, the factorial growth prevails. Such series are called *asymptotic*. Their optimal evaluation requires a truncation after the smallest correction term. In general, the large-order behavior of perturbation expansions may be parametrized as

$$E_k = \gamma p^{\beta+1} k^\beta (-4a)^k (pk)! \left[1 + \frac{\gamma_1}{k} + \frac{\gamma_2}{k^2} + \ldots \right], \tag{17.285}$$

where the leading term $(pk)!$ grows like

$$(pk)! = (k!)^p (p^p)^k k^{(1-p)/2} \frac{\sqrt{p}}{(2\pi)^{(p-1)/2}} \left[1 + \mathcal{O}(1/k) \right]. \tag{17.286}$$

This behavior is found by approximating $n!$ via Stirling's formula (5.204). It is easy to see that the kth term of the series (17.284) is minimal at

$$k \approx k_{\min} \equiv \frac{1}{p(a|g|)^{1/p}}. \tag{17.287}$$

This is found by applying Stirling's formula once more to $(k!)^p$ and by minimizing $\gamma(k!)^p k^{\beta'} (p^p a|g|)^k$ with $\beta' = \beta + (1 - p)/2$, which yields the equation

$$p \log k + \log(p^p a|g|) + (\beta + p/2)/k + \ldots = 0. \tag{17.288}$$

An equivalent way of writing (17.285) is

$$E_k = \gamma p(-4a)^k \Gamma(pk + \beta + 1) \left[1 + \frac{c_1}{pk + \beta} + \frac{c_2}{(pk + \beta)(pk + \beta - 1)} + \ldots \right]. \tag{17.289}$$

The simplest example for a function with such strongly growing expansion coefficients can be constructed with the help of the exponential integral

$$E_1(g) = \int_g^\infty \frac{dt}{t} e^{-t}. \tag{17.290}$$

Defining

$$E(g) \equiv \frac{1}{g} e^{1/g} E_1(1/g) = \int_0^\infty dt \frac{1}{1 + gt} e^{-t}, \tag{17.291}$$

this has the diverging expansion

$$E(g) = 1 - g + 2! g^2 - 3! g^3 + \ldots + (-1)^N N! g^N + \ldots. \tag{17.292}$$

At a small value of g, such as $g = 0.05$, the series can nevertheless be evaluated quite accurately if truncated at an appropriate value of N. The minimal correction is reached at $N = 1/g = 20$ where the relative error with respect to the true value $E \approx 0.9543709099$ is equal to $\Delta E/E \approx 1.14 \cdot 10^{-8}$. At a somewhat larger value $g = 0.2$, on the other hand, the optimal evaluation up to $N = 5$ yields the much larger relative error $\approx 1.8\%$, the true value being $E \approx 0.852110880$.

The integrand on the right-hand side of (17.291), the function

$$B(t) = \frac{1}{1+t}, \tag{17.293}$$

is the so-called *Borel transform* of the function $E(g)$. It has a power series expansion which can be obtained from the divergent series (17.292) for $E(g)$ by removing in each term the catastrophically growing factor $k!$. This produces the convergent series

$$B(t) = 1 - t + t^2 - t^3 + \dots, \tag{17.294}$$

which sums up to (17.293). The integral

$$F(g) = \int_0^\infty \frac{dt}{g} e^{-t/g} B(t) \tag{17.295}$$

restores the original function by reinstalling, in each term t^k, the removed $k!$-factor.

Functions $F(g)$ of this type are called *Borel-resummable*. They possess a convergent Borel transform $B(t)$ from which $F(g)$ can be recovered with the help of the integral (17.295). The resummability is ensured by the fact that $B(t)$ has no singularities on the integration path $t \in [0, \infty)$, including a wedge-like neighborhood around it. In the above example, $B(t)$ contains only a pole at $t = -1$, and the function $E(g)$ is Borel-resummable. Alternating signs of the expansion coefficients of $F(g)$ are a typical signal for the resummability.

The best-known quantum field theory, *quantum electrodynamics*, has divergent perturbation expansions, as was first pointed out by Dyson [5]. The expansion parameter g in that theory is the *fine-structure constant*

$$\alpha = 1/137.035963(15) \approx 0.0073. \tag{17.296}$$

Fortunately, this is so small that an evaluation of observable quantities, such as the anomalous magnetic moment of the electron

$$a_{\mathrm{e}} = \frac{\Delta\mu}{\mu} = \frac{1}{2}\frac{\alpha}{\pi} - 0.328\,478\,965\,7\left(\frac{\alpha}{\pi}\right)^2 + 1.1765(13)\left(\frac{\alpha}{\pi}\right)^3 + \dots, \tag{17.297}$$

gives an extremely accurate result:

$$a_e^{\mathrm{theor}} = (1\,159\,652\,478 \pm 140) \cdot 10^{-12}. \tag{17.298}$$

The experimental value differs from this only in the last three digits, which are 200 ± 40. The divergence of the series sets in only after the 137th order.

A function $E(g)$ with factorially growing expansion coefficients cannot be analytic at the origin. We shall demonstrate below that it has a left-hand cut in the complex g-plane. Thus it satisfies a dispersion relation

$$E(g) = \frac{1}{2\pi i} \int_0^\infty dg' \frac{\operatorname{disc} E(-g')}{g' + g}, \tag{17.299}$$

where $\operatorname{disc} E(g')$ denotes the discontinuity across the left-hand cut

$$\operatorname{disc} E(g) \equiv E(g - i\eta) - E(g + i\eta). \tag{17.300}$$

It is then easy to see that the above large-order behavior (17.289) is in one-to-one correspondence with a discontinuity which has an expansion, around the tip of the cut,

$$\operatorname{disc} E(-|g|) = 2\pi i \gamma (a|g|)^{-(\beta+1)/p} e^{-1/(a|g|)^p} [1 + c_1 (a|g|)^{1/p} + c_2 (a|g|)^{2/p} + \ldots]. \tag{17.301}$$

The parameters are the same as in (17.289). The one-to-one correspondence is proved by expanding the dispersion relation (17.299) in powers of $g/4$, giving

$$E_k = (-4)^k \int_0^\infty \frac{dg'}{2\pi i} \frac{1}{g'^{k+1}} \operatorname{disc} E(-g'). \tag{17.302}$$

The expansion coefficients are given by moment integrals of the discontinuity with respect to the inverse coupling constant $1/g$. Inserting (17.301) and using the integral formula[5]

$$\int_0^\infty dg \frac{1}{|g|^{\alpha+1}} e^{-1/(a|g|)^{(1/p)}} = a^\alpha p \Gamma(p\alpha), \tag{17.303}$$

we indeed recover (17.289).

From the strong-coupling limit of the ground state energy of the anharmonic oscillator Eq. (5.168) we see that the discontinuity grow for large g like $g^{1/3}$. In this case, the dispersion relation (17.304) needs a subtraction and reads

$$E(g) = E(0) + \frac{g}{2\pi i} \int_0^\infty \frac{dg'}{g'} \frac{\operatorname{disc} E(-g')}{g' + g}. \tag{17.304}$$

This does not influence the moment formula (17.302) for the expansion coefficients, except that the lowest coefficient is no longer calculable from the discontinuity. Since the lowest coefficient is known, there is no essential restriction.

17.10.2 Semiclassical Large-Order Behavior

The large-order behavior of many divergent perturbation expansions can be determined with the help of the tunneling theory developed above. Consider the potential

---

[5]I.S. Gradshteyn and I.M. Ryzhik, *op. cit.*, Formula 3.478.

of the anharmonic oscillator at a small negative coupling constant g (see Fig. 17.11). The minimum at the origin is obviously metastable so that the ground state has only a finite lifetime. There are barriers to the right and left of the metastable minimum, which are very high for very small negative coupling constants. In this limit, the lifetime can be calculated accurately with the semiclassical methods of the last section. The fluctuation determinant yields an imaginary part of $Z(g)$ of the form (17.270), which determines the imaginary part of the ground state energy via (17.276), which is accurate near the tip of the left-hand cut in the complex g-plane. From this imaginary part, the dispersion relation (17.302) determines the large-order behavior of the perturbation coefficients.

The classical equation of motion as a function of τ is

$$x''(\tau) - V'(x(\tau)) = 0. \tag{17.305}$$

The differential equation is integrated as in (17.26), using the first integral of motion

$$\frac{1}{2}x'^2 - \frac{1}{2}\omega^2 x^2 - \frac{g}{4}x^4 = E = \text{const}, \tag{17.306}$$

from which we find the solutions for $E = 0$

$$\tau - \tau_0 = \pm \frac{1}{\omega} \int dx \frac{1}{x\sqrt{1 - (|g|/2\omega^2)x^2}} = \mp \frac{1}{\omega} \text{arcosh}\left(\sqrt{\frac{2\omega^2}{|g|}}\frac{1}{x}\right), \tag{17.307}$$

or

$$x(\tau) = x_{\text{cl}}(\tau) \equiv \pm \sqrt{\frac{2\omega^2}{|g|}}\frac{1}{\cosh[\omega(\tau - \tau_0)]}. \tag{17.308}$$

They represent excursions towards the abysses outside the barriers and correspond precisely to the bubble solutions of the tunneling discussion in the last section. The excursion towards the abyss on the right-hand side is illustrated in Fig. 17.11. The associated action is calculated as in (17.29):

$$\begin{aligned} \mathcal{A}_{\text{cl}} &= \int_{-L/2}^{L/2} d\tau \left[\frac{1}{2}x_{\text{cl}}'^2(\tau) + V(x_{\text{cl}}(\tau))\right] = 2\int_0^{L/2} d\tau \left[x_{\text{cl}}'^2(\tau) - E\right] \\ &= 2\int_0^{x_m} dx\sqrt{2(E+V)} - EL, \end{aligned} \tag{17.309}$$

where x_m is the maximum of the solution. The bubble solution has $E = 0$, so that

$$\mathcal{A}_{\text{cl}} = 2\int_0^{x_m} dx\sqrt{2V} = \frac{4\omega^3}{3|g|}. \tag{17.310}$$

Inserting the fluctuating path $x(\tau) = x_{\text{cl}}(\tau) + y(\tau)$ into the action (17.278) and expanding it in powers of $y(\tau)$, we find an action for the quadratic fluctuations of the same form as in Eq. (17.211), but with a functional matrix

$$\begin{aligned} O_\omega(\tau, \tau') &= \left[-\frac{d^2}{d\tau^2} + \omega^2 + 3gx_{\text{cl}}^2(\tau)\right]' \delta(\tau - \tau') \\ &= \left[-\frac{d^2}{d\tau^2} + \omega^2\left(1 - \frac{6}{\cosh^2[\omega(\tau - \tau_0)]}\right)\right]' \delta(\tau - \tau'). \end{aligned} \tag{17.311}$$

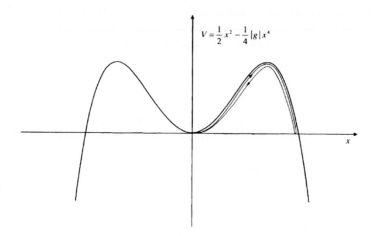

Figure 17.11 Potential of anharmonic oscillator (17.278) for small negative coupling g. The ground state centered at the origin is metastable. It decays via a classical solution which makes an excursion towards the abyss as indicated by the oriented curve.

This is once more the operator of the Rosen-Morse type encountered in Eq. (17.138) with $m = \omega, z = 1$, and $s = 2$. The subscript ω on the operator symbol indicates the asymptotic harmonic form of the potential. The potential accommodates again two bound states with the normalized wave functions[6] and energies (see Fig. 17.12)

$$y_0(\tau) = -\sqrt{\frac{3\omega}{2}} \frac{\sinh[\omega(\tau - \tau_0)]}{\cosh^2[\omega(\tau - \tau_0)]} \quad \text{with } \lambda_0 = 0, \tag{17.312}$$

$$y_{-1}(\tau) = \sqrt{\frac{3\omega}{4}} \frac{1}{\cosh^2[\omega(\tau - \tau_0)]} \quad \text{with } \lambda_{-1} = -3\omega^2. \tag{17.313}$$

These are the same functions as in (17.52), (17.53), apart from the fact that m is now ω rather than $\omega/2$. However, the energies are shifted with respect to the earlier case. Now the first excited state has a zero eigenvalue so that the ground state has a negative eigenvalue. This is responsible for the finite lifetime of the ground state.

The fluctuation determinant is obtained by any of the above procedures, for instance from the general formula (17.143),

$$\frac{\prod_n \lambda_n^0}{\prod_n \lambda_n} = \frac{\Gamma(\sqrt{z} - s)\Gamma(\sqrt{z} + s + 1)}{\Gamma(\sqrt{z})\Gamma(\sqrt{z} + 1)}, \tag{17.314}$$

---

[6]The sign of y_0 is chosen to agree with that of $x'_{\text{cl}}(\tau)$ in accordance with (17.88).

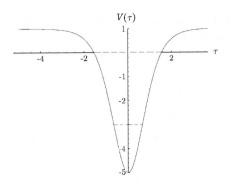

Figure 17.12 Rosen-Morse Potential for fluctuations around the classical bubble solution.

by inserting the parameters $z = 1$ and $s = 2$. The zero eigenvalue is removed by multiplying this with $(z - 1)\omega^2$, resulting in the eigenvalue ratio

$$\frac{\prod_n \lambda_n^0}{\prod_n' \lambda_n} = \lim_{z \to 1}(\sqrt{z} - 1)(\sqrt{z} + 1)\omega^2 \left[\frac{\Gamma(\sqrt{z} - 2)\Gamma(\sqrt{z} + 3)}{\Gamma(\sqrt{z})\Gamma(\sqrt{z} + 1)}\right] = -12\omega^2. \quad (17.315)$$

The negative sign due to the negative-eigenvalue solution in (17.313) accounts for the instability of the fluctuations.

Using formula (17.274), we find the imaginary part of the partition function

$$\text{Im}\, Z(-|g| - i\eta) \approx \sqrt{\frac{6}{\pi}}\sqrt{\frac{4\omega^3}{3|g|}}\omega L e^{-4\omega^3/3|g|}e^{-\omega L/2}. \quad (17.316)$$

After summing over all bubble solutions, as in (17.275), we obtain the imaginary part of the ground state energy

$$\text{Im}\, E^{(0)}(-|g| - i\eta) = -\omega\sqrt{\frac{6}{\pi}}\sqrt{\frac{4\omega^3}{3|g|}}e^{-4\omega^3/3|g|}. \quad (17.317)$$

A comparison of this with (17.301) fixes the growth parameters of the large-order perturbation coefficients to

$$a = 3/4\omega^3, \quad \beta = -\frac{1}{2}, \quad \gamma = -\frac{\omega}{\pi}\sqrt{\frac{6}{\pi}}, \quad p = 1. \quad (17.318)$$

Recalling the one-to-one correspondence between (17.301) and (17.289), we see that the large-order behavior of the perturbation coefficients of the ground state energy $E^{(0)}(g)$ is

$$E_k^{(0)} = -\frac{\omega}{\pi}\sqrt{\frac{6}{\pi}}(-3/\omega^3)^k\Gamma(k + 1/2). \quad (17.319)$$

It is just as easy to find the large-k behavior of the excited states. Their decay is triggered by a periodic classical solution with a very long but finite Euclidean period L, which moves back and forth between positions $x_< \neq 0$ and $x_> < \sqrt{2\omega^2/|g|}$. Its action is approximately given by

$$\mathcal{A}'_{\rm cl} \approx \frac{4\omega^3}{3|g|}(1 - 12e^{-\omega L}). \tag{17.320}$$

In comparison with the limit $L \to \infty$, the Boltzmann-like factor $e^{-\mathcal{A}_{\rm cl}}$ of this solution is replaced by

$$e^{-\mathcal{A}'_{\rm cl}} = e^{-\mathcal{A}_{\rm cl}} \sum_{n=0}^{\infty} \mathcal{A}_{\rm cl}^n \frac{12^n}{n!} e^{-n\omega L}. \tag{17.321}$$

The exponentials in the sum raise the reference energies in the imaginary part of $Z(-|g|-i\eta)$ in (17.316) from $\omega/2$ to $\omega(n+1/2)$. The imaginary parts for the energies to the nth excited states become

$$\operatorname{Im} E^{(n)}(-|g| - i\eta) = -\frac{12^n}{n!}\omega\sqrt{\frac{6}{\pi}}\sqrt{\frac{4\omega^3}{3|g|}}^{-1+2n} e^{-4\omega^3/3|g|}, \tag{17.322}$$

implying an asymptotic behavior of the perturbation coefficients:

$$E_k^{(n)} = -\frac{\omega}{\pi}\sqrt{\frac{6}{\pi}}\frac{12^n}{n!}(-3/\omega^3)^k\Gamma(k + n + 1/2). \tag{17.323}$$

It is worth mentioning that within the semiclassical approximation, the dispersion integrals for the energies can be derived directly from the path integral (17.279). This can obviously be rewritten as

$$\begin{aligned} Z(g) &= \int_{-i\infty}^{i\infty}\frac{d\lambda}{2\pi i}\int_0^\infty \frac{da}{4}e^{-(ga+\lambda a)/4} \\ &\quad \times \int \mathcal{D}x(\tau)\exp\left\{-\int_{L/2}^{L/2}d\tau\left[\frac{1}{2}x'^2 + \frac{\omega^2}{2}x^2 - \frac{\lambda}{4}x^4\right]\right\}. \end{aligned} \tag{17.324}$$

The integration over λ generates a δ-function $4\delta(\int d\tau\, x^4(\tau) - a)$ which eliminates the additionally introduced a-integration. The integral over a is easily performed. It yields a factor $1/(\lambda + g)$, so that we obtain the integral formula

$$Z(g) = \int_{-i\infty}^{i\infty}\frac{d\lambda}{2\pi i}\frac{1}{\lambda + g}Z(-\lambda). \tag{17.325}$$

The integrand has a pole at $\lambda = -g$ and a cut on the positive real λ-axis. We now deform the contour of integration in λ until it encloses the cut tightly in the clockwise sense. In the semiclassical approximation, the discontinuity across the cut is given by Eq. (17.316), i.e., with the present variable λ:

$$\operatorname{Im} Z(-|\lambda| - i\eta) = \sqrt{\frac{6}{\pi}}\sqrt{\frac{4\omega^3}{3\lambda}}e^{-4\omega^3/3\lambda}e^{-\omega L/2}. \tag{17.326}$$

On the upper branch of the cut, $\sqrt{\lambda}$ is positive, on the lower negative. Thus we arrive at a simple dispersion integral from $\lambda = 0$ to $\lambda = \infty$:

$$Z(g) = 2\omega \int_0^\infty \frac{d\lambda}{2\pi} \frac{1}{\lambda + g} \sqrt{\frac{6}{\pi}} \sqrt{\frac{4\omega^3}{3\lambda}} e^{-4\omega^3/3\lambda} e^{-\omega L/2}. \tag{17.327}$$

For the ground state energy, this implies

$$E^{(0)}(g) = -2\omega \int_0^\infty \frac{d\lambda}{2\pi} \frac{1}{\lambda + g} \sqrt{\frac{6}{\pi}} \sqrt{\frac{4\omega^3}{3\lambda}} e^{-4\omega^3/3\lambda}. \tag{17.328}$$

Of course, this expression is just an approximation, since the integrand is valid only at small λ. In fact, the integral converges only in this approximation. If the full imaginary part is inserted, the integral diverges. We shall see below that for large λ, the imaginary part grows like $\lambda^{1/3}$. Thus a subtraction is necessary. A convergent integral representation exists for $E^{(0)}(g) - E^{(0)}(0)$. With $E^{(0)}(0)$ being equal to $\omega/2$, we find the convergent dispersion integral

$$E^{(0)}(g) = \frac{\omega}{2} + 2\omega g \int_0^\infty \frac{d\lambda}{2\pi} \frac{1}{\lambda(\lambda + g)} \sqrt{\frac{6}{\pi}} \sqrt{\frac{4\omega^3}{3\lambda}} e^{-4\omega^3/3\lambda}. \tag{17.329}$$

The subtraction is advantageous also if the initial integral converges since it suppresses the influence of the large-λ regime on which the semiclassical tunneling calculation contains no information.

After substituting $\lambda \to 4g/3t\omega^3$, the integral (17.329) is seen to become a Borel integral of the form (17.295).

By expanding $1/(\lambda + g)$ in a power series in g

$$\frac{1}{\lambda + g} = \sum_{k=0}^\infty (-1)^k g^k \lambda^{-k-1}, \tag{17.330}$$

we obtain the expansion coefficients as the moment integrals of the imaginary part as a function of $1/g$:

$$E_k^{(0)} = -2\omega \, (-4)^k \int_0^\infty \frac{d\lambda}{2\pi} \frac{1}{\lambda^{k+1}} \sqrt{\frac{6}{\pi}} \sqrt{\frac{4\omega^3}{3\lambda}} e^{-4\omega^3/3\lambda}. \tag{17.331}$$

This leads again to the large-k behavior (17.319).

The direct treatment of the path integral has the virtue that it can be generalized also to systems which do not possess a Borel-resummable perturbation series. As an example, one may derive and study the integral representation for the level splitting formula in Section 17.7.

17.10.3 Fluctuation Correction to the Imaginary Part and Large-Order Behavior

It is instructive to calculate the first nonleading term $c_1 a|g|$ in the imaginary part (17.301), which gives rise to a correction factor $1 + c_1/k$ in the large-order behavior

(17.289). As in Section 17.8, we expand the action around the classical solution. The interaction between the fluctuations $y(\tau)$ is the same as before in (17.215). The quadratic fluctuations are now governed by the differential operator

$$\mathcal{O}_\omega(\tau, \tau') = \left[-\frac{d^2}{d\tau^2} + \omega^2 \left(1 - \frac{6}{\cosh^2 \omega(\tau - \tau_0)} \right) \right]' \delta(\tau - \tau'), \tag{17.332}$$

the prime indicating the absence of the zero eigenvalue. Its removal gives rise to the factor

$$\mathcal{A}_e^{\text{eff}} = -\hbar \log \left[1 - \sqrt{\frac{3|g|}{4\omega^3}} \int d\tau \, y_0'(\tau) y(\tau) \right]. \tag{17.333}$$

After expanding, in the path integral, the exponential $e^{-(\mathcal{A}_\hbar^{\text{int}} + \mathcal{A}_e^{\text{eff}})/\hbar}$ in powers of the interaction up to the second order, a perturbative evaluation of the correlation functions of the fluctuations $y(\tau)$ according to the rules of Section 3.20 yields a correction factor

$$C = \left[1 + (I_1 + I_2 + I_3) \frac{|g|\hbar}{\omega^3} + \mathcal{O}(g^2) \right], \tag{17.334}$$

with the same τ-integrals as in Eqs. (17.217), (17.219), (17.223), and (17.224), after replacing g by $|g|$. The correction parameter C has again a diagrammatic expansion (17.225), where the vertices stand for the same analytic expressions as in Fig. 17.5, except for the third vertex, which is now

$$\times\!\!\!\sim\!\!\!\sim\!\!\!\bullet\!\!\!-\!\!\!- \qquad \sqrt{\frac{3|g|}{4\omega^3}} y_0'(\tau). \tag{17.335}$$

The lines represent the subtracted Green function

$$G'_{\mathcal{O}_\omega}(\tau, \tau') = \langle y(\tau) y(\tau') \rangle_{\mathcal{O}_\omega} = \hbar \mathcal{O}_\omega^{-1}(\tau, \tau'), \tag{17.336}$$

where $\mathcal{O}_\omega^{-1}(\tau, \tau')$ is the inverse of the functional matrix (17.332).

In contrast to the level splitting calculation in Section 17.8, only the integral I_1 requires a subtraction,

$$I_1 = \frac{3\omega^3}{4\hbar^2} \int d\tau \, G'^2_{\mathcal{O}_\omega}(\tau, \tau) = L \frac{3\omega}{16} + \frac{3\omega^3}{4\hbar^2} \int d\tau \left[G'^2_{\mathcal{O}_\omega}(\tau, \tau) - \frac{\hbar^2}{4\omega^2} \right], \tag{17.337}$$

and Eq. (17.334) assumes that I_1 is subtracted, i.e., I_1 should be replaced by $I_1' \equiv I_1 - L3\omega/16$. The correction factor for the tunneling rate reads, therefore,

$$C' = \left[1 + (I_1' + I_2 + I_3) \frac{|g|\hbar}{\omega^3} + \mathcal{O}(g^2) \right]. \tag{17.338}$$

The subtracted integral contributes only to the real part of the ground state energy which we know to be $(1/2 + 3g\hbar/16\omega^3)\hbar\omega$.

As in Section 17.8, the explicit Green function $G'_{\mathcal{O}_\omega}(\tau, \tau')$ is found from the amplitude (17.227). By a change of variables $x = \omega\tau$ and $\hbar^2/2\mu = \omega^2$, setting $s = 2$, the Schrödinger operator in (17.226) coincides with that in (17.212), provided we set $E_{\mathcal{RM}} = 0$. The amplitude (17.227) then yields the Green functions for $\tau > \tau'$

$$G_{\mathcal{O}_\omega}(\tau, \tau') = \frac{\hbar}{2\omega}\Gamma(m - 2)\Gamma(m + 3) \times P_2^{-m}(\tanh\omega\tau)P_2^{-m}(-\tanh\omega\tau'), \quad (17.339)$$

with $m = 1$. Due to translational invariance along the τ-axis, this Green function has a pole at $E_{\mathcal{RM}} = 0$ [just like the Green function (17.229)]. The pole must be removed before going to this energy, and the result is the subtracted Green function $G'_{\mathcal{O}_\omega}(\tau, \tau')$, given by

$$G'_{\mathcal{O}_\omega}(\tau, \tau') = \frac{1}{2m}\frac{d}{dm}(m^2 - 1)G_{\mathcal{O}_\omega}(\tau, \tau')\bigg|_{m=1}. \quad (17.340)$$

Using (17.231), we find the subtracted Green function

$$G'_{\mathcal{O}_\omega}(\tau, \tau') = \hbar[Y_0(\tau_>)y_0(\tau_<) + y_0(-\tau_>)Y_0(-\tau_<)], \quad (17.341)$$

with

$$y_0(\tau) = 2\sqrt{\frac{3\omega}{2}}P_2^{-1}(-\tanh\omega\tau) = -\sqrt{\frac{3\omega}{2}}\frac{\sinh\omega\tau}{\cosh^2\omega\tau}, \quad (17.342)$$

$$\begin{aligned}
Y_0(\tau) &= \sqrt{\frac{2}{3\omega}}\frac{1}{8\omega m}\left\{\frac{1}{2}\left[\frac{d}{dm}(m^2 - 1)\Gamma(m - 2)\Gamma(m + 3)\right]P_2^{-m}(\tanh\omega\tau)\right. \\
&\quad \left. + \left[(m^2 - 1)\Gamma(m - 2)\Gamma(m + 3)\right]\frac{d}{dm}P_2^{-m}(\tanh\omega\tau)\right\}\bigg|_{m=1} \\
&= -\sqrt{\frac{2}{3\omega^3}}\left[\frac{3}{4}\frac{1}{\cosh\omega\tau} + \left(-\frac{3}{4}\omega\tau - \frac{1}{8}\right)\frac{\sinh\omega\tau}{\cosh^2\omega\tau} - \frac{1}{4}e^{-\omega\tau}\right]. \quad (17.343)
\end{aligned}$$

For $\tau = \tau'$

$$G'_{\mathcal{O}_\omega}(\tau, \tau) = \frac{\hbar}{2\omega}\frac{1}{\cosh^2\omega\tau}(\cosh^2\omega\tau - 1)(\cosh^2\omega\tau - 1/2). \quad (17.344)$$

The evaluation of the integrals $I'_1, I_{21}, I_{22}, I_3$ proceeds as in Section 17.8 (performed in Appendix 17A), yielding [6]

$$I'_1 = -\frac{11 \cdot 29}{2^4 \cdot 5 \cdot 7}, \quad I_{21} = -\frac{71}{2^5 \cdot 3 \cdot 7}, \quad I_{22} = \frac{3 \cdot 13}{2^4 \cdot 7}, \quad I_3 = -\frac{53}{2^4 \cdot 5}. \quad (17.345)$$

The correction factor (17.338) is therefore

$$C' = \left[1 - \frac{95}{72}\frac{3|g|\hbar}{4\omega^3} + \mathcal{O}(g^2)\right]. \quad (17.346)$$

Using the one-to-one correspondence between (17.289) and (17.301), this yields the large-k behavior of the expansion coefficients of the ground state energy:

$$E_k^{(0)} = -\frac{\omega}{\pi}\sqrt{\frac{6}{\pi}}(-3/\omega^3)^k\Gamma(k + 1/2)[1 - 95/72k + \ldots]. \quad (17.347)$$

17.10.4 Variational Approach to Tunneling. Perturbation Coefficients to All Orders

The semiclassical calculations of tunneling amplitudes are valid only for very high barriers. It is possible to remove this limitation with the help of a variational approach [7] similar to the one described in Chapter 5. For simplicity, we discuss here only the case of an anharmonic oscillator at zero temperature. For the lowest energy levels we shall derive highly accurate imaginary parts over the entire left-hand cut in the coupling constant plane. The accuracy can be tested by inserting these imaginary parts into the dispersion relation (17.329) to recover the perturbation coefficients of the energies. These turn out to be in good agreement with the exact ones to all orders.

For the path integral of the anharmonic oscillator

$$Z(g) = \int \mathcal{D}x(\tau) \exp\left\{-\int_{-L/2}^{L/2} d\tau \left[\frac{1}{2}x'^2 + \frac{\omega^2}{2}x^2 + \frac{g}{4}x^4\right]\right\}, \qquad (17.348)$$

the variational energy (5.32) at zero temperature is given by

$$W_1 = \frac{\Omega}{2} + \frac{\omega^2 - \Omega^2}{2}a^2 + \frac{3g}{4}a^4, \qquad (17.349)$$

with $a^2 = 1/2\Omega$. We have omitted the path average argument x_0 since, by symmetry of the potential, the minimum lies at $x_0 = 0$. The energy has to be extremized in Ω^2. This yields the cubic equation $\Omega^3 - \omega^2\Omega - 3g/2 = 0$. The physically relevant solution starts out with ω at $g = 0$ and has two branches: For $g \in (-g^{(0)}, 0)$ with $g^{(0)} = 4\omega^3/9\sqrt{3}$ [compare (5.163)], it is given by

$$\Omega = \frac{2\omega}{\sqrt{3}} \cos\left[\frac{\pi}{3} - \frac{1}{3}\arccos\left(-g/g^{(0)}\right)\right]. \qquad (17.350)$$

For large negative coupling constants $g < -g^{(0)}$, the solution is

$$\Omega^{\mathrm{re}} = \frac{\omega}{\sqrt{3}} \cosh(\gamma/3), \quad \Omega^{\mathrm{im}} = \omega \sinh(\gamma/3); \quad \gamma = \mathrm{arcosh}\left(-g/g^{(0)}\right). \qquad (17.351)$$

In this regime, the ground state energy acquires an imaginary part

$$\mathrm{Im}\, W_1 = \frac{1}{4}\Omega^{\mathrm{i}}(1 - 1/|\Omega|^2) - \frac{3g}{4}\Omega^{\mathrm{re}}\Omega^{\mathrm{im}}/2|\Omega|^4. \qquad (17.352)$$

This imaginary part describes the instability of the system to *slide* down into the two abysses situated at large positive and negative x. In this regime of coupling constants, the barriers to the right and left of the origin are no obstacle to the decay since they are smaller than the zero-point energy.

In the first regime of small negative coupling constants $g \in (-g^{(0)}, 0)$, the barriers are high enough to prevent at least one long-lived ground state from sliding down. Its energy is approximately given by the minimum of (17.349). It can decay towards the abysses via an extremal excursion across the trial potential $\Omega^2 x^2/2 + gx^4/4$. The associated bubble solution reads, according to (17.308),

$$x(\tau) = x_{\mathrm{cl}}(\tau) \equiv \pm\sqrt{2\Omega^2/|g|}\frac{1}{\cosh[\Omega(\tau - \tau_0)]}. \qquad (17.353)$$

It has the action $\mathcal{A}_{\mathrm{cl}} = 4\Omega^3/3|g|$. Its fluctuation determinant is given by (17.315), if ω is replaced by the trial frequency Ω. Translations contribute a factor $\beta\Omega\sqrt{\mathcal{A}_{\mathrm{cl}}/2\pi}$. Thus, the partition function has an imaginary part

$$\mathrm{Im}\, Z(-|g| - i\eta) = \beta\Omega\sqrt{\frac{6}{\pi}}\sqrt{\frac{4\Omega^3}{3|g|}}e^{-\beta\Omega/2 - 4\Omega^3/3|g|}. \qquad (17.354)$$

In the variational approach, this replaces the semiclassical expression (17.316), which will henceforth be denoted by $Z_{\mathrm{sc}}^{\mathrm{im}}(g)$.

The expression (17.354) receives fluctuation corrections. To lowest order, they produce a factor $\exp(-\langle\mathcal{A}_{\mathrm{fl,tot}}^{\mathrm{int}}\rangle_{\mathcal{O}_\Omega})$, where the action $\mathcal{A}_{\mathrm{fl,tot}}^{\mathrm{int}}$ contains the interaction terms (17.215) and (17.213) of the fluctuations, with ω replaced by Ω, plus additional terms arising from the variational ansatz. They compensate for the fact that we are using the trial potential $\Omega^2 x^2/2$ rather than the proper $\omega^2 x^2/2$ as the zeroth-order potential for the perturbation expansion. These compensation terms have the action

$$\begin{aligned}\mathcal{A}_{\mathrm{fl,var}}^{\mathrm{int}} &= \int_{-\infty}^{\infty} d\tau\, \frac{\omega^2 - \Omega^2}{2}x^2(\tau) \\ &= \int_{-\infty}^{\infty} d\tau\, \frac{\omega^2 - \Omega^2}{2}[x_{\mathrm{cl}}^2(\tau) + 2x_{\mathrm{cl}}(\tau)y(\tau) + y^2(\tau)]. \qquad (17.355)\end{aligned}$$

The expectations $\langle\ldots\rangle_{\mathcal{O}_\Omega}$ in the perturbation correction are calculated with respect to fluctuations governed by the operator (17.332), in which ω is replaced by Ω. As before, all correlation functions are expanded by Wick's rule into sums of products of the simple correlation functions $G'_{\mathcal{O}_\Omega}(\tau, \tau')$ of (17.336). Using the integral formula (17.54), we have

$$\int_{-\infty}^{\infty} d\tau\, x_{\mathrm{cl}}^2(\tau) = 4\Omega/|g|. \qquad (17.356)$$

The expectation of $\int_{-\infty}^{\infty} d\tau\, y(\tau)^2$ is found with the help of (17.344) as

$$\begin{aligned}\int_{-\infty}^{\infty} d\tau\, \langle y^2(\tau)\rangle_{\mathcal{O}_\Omega} &= L\frac{1}{2\Omega} + \frac{1}{\Omega}\int_{-\infty}^{\infty} d\tau\, \left[G'_{\mathcal{O}_\Omega}(\tau, \tau) - 1/2\right] \\ &= L\frac{1}{2\Omega} - \frac{7}{6\Omega^2}. \qquad (17.357)\end{aligned}$$

The second term can be obtained quite simply by differentiating the logarithm of (17.314) with respect to $\Omega^2 z$.

The linearly divergent term $L/2\Omega$ contributes to the earlier-calculated term proportional to L in the integral (17.337) (with ω replaced by Ω); together they yield

L-times W_1 of (17.349). Thus we can remove a factor e^{-LW_1} from $\text{Im}\,Z$, write Z as $\approx \text{Re}\,Z e^{\text{Im}\,Z/\text{Re}\,Z} = e^{-LW_1 + \text{Im}\,Z/\text{Re}\,Z}$ [as in (17.275)], and deduce the imaginary part of the energy from the exponent.

We now go over to the cumulants in accordance with the rules of perturbation theory in Eqs. (3.480)–(3.484) involving the integrals (17.217) (with g and ω replaced by $|g|$ and Ω, respectively). Using (17.345) we find the correction factor $e^{-A_0 - A_1}$ with

$$A_0 = \frac{95\,|g|}{96\,\Omega^3}, \qquad A_1 = \frac{1}{2}(\omega^2 - \Omega^2)\left(\frac{4\Omega}{|g|} - \frac{7}{6\Omega^2}\right). \qquad (17.358)$$

If we want to find all terms contributing to the imaginary part up to the order g, we must continue the perturbation expansion to the next order. This yields a further factor

$$\exp\left\{\frac{1}{2}[\langle\mathcal{A}^{\text{int}\,2}_{\text{fl,tot}}\rangle_{\mathcal{O}_\Omega} - \langle\mathcal{A}^{\text{int}}_{\text{fl,tot}}\rangle^2_{\mathcal{O}_\Omega}]\right\} = \exp(-A_2 - A_3 - A_4), \qquad (17.359)$$

with the integrals

$$A_2 = -\frac{1}{2}(\omega^2 - \Omega^2)^2 \int d\tau d\tau'\, x_{\text{cl}}(\tau)\langle y(\tau)y(\tau')\rangle_{\mathcal{O}_\Omega} x_{\text{cl}}(\tau'),$$

$$A_3 = -(\omega^2 - \Omega^2)\sqrt{\frac{3|g|}{4\Omega^3}} \int d\tau d\tau'\, y_0'(\tau)\langle y(\tau)y(\tau')\rangle_{\mathcal{O}_\Omega} x_{\text{cl}}(\tau'),$$

$$A_4 = (\omega^2 - \Omega^2)|g| \int d\tau d\tau'\, x_{\text{cl}}(\tau)\langle y(\tau)y^3(\tau')\rangle_{\mathcal{O}_\Omega} x_{\text{cl}}(\tau'). \qquad (17.360)$$

Performing the Wick contractions in the correlation functions, the integrals are conveniently rewritten as

$$A_2 = -\frac{1}{2}(\omega^2 - \Omega^2)^2 \frac{1}{\Omega|g|} a_2,$$

$$A_3 = -(\omega^2 - \Omega^2)\frac{1}{\Omega^2} a_3,$$

$$A_4 = (\omega^2 - \Omega^2)\frac{1}{\Omega^2} a_4, \qquad (17.361)$$

where a_2, a_3, a_4 are given by

$$a_2 = |g|\Omega \int d\tau d\tau'\, x_{\text{cl}}(\tau) G'_{\mathcal{O}_\Omega}(\tau, \tau') x_{\text{cl}}(\tau'),$$

$$a_3 = \Omega^2 \sqrt{\frac{3|g|}{4\Omega^3}} \int d\tau d\tau'\, y_0'(\tau) G'_{\mathcal{O}_\Omega}(\tau, \tau') x_{\text{cl}}(\tau'),$$

$$a_4 = 3|g|\Omega^2 \int d\tau d\tau'\, x_{\text{cl}}(\tau) G'_{\mathcal{O}_\Omega}(\tau, \tau') G'_{\mathcal{O}_\Omega}(\tau', \tau') x_{\text{cl}}(\tau'). \qquad (17.362)$$

In terms of these, the imaginary part of the energy reads

$$\text{Im}\,E(-|g| - i\eta) = -\Omega\sqrt{\frac{6}{\pi}}\sqrt{\frac{4\Omega^3}{3|g|}} e^{-4\Omega^3/3|g| - c_1 3|g|/4\Omega^3}$$

$$\times \exp\left[-\frac{\omega^2 - \Omega^2}{2}\left(\frac{4\Omega}{|g|} - \frac{7}{6\Omega^2} - 2\frac{a_3 - a_4}{\Omega^2}\right) + \frac{(\omega^2 - \Omega^2)^2}{2\Omega|g|} a_2\right], \qquad (17.363)$$

evaluated at the Ω-value (17.350).

To best visualize the higher-order effect of fluctuations, we factorize (17.363) into the semiclassical part (17.317) and a correction factor $\varepsilon_i(g)$,

$$\mathrm{Im}\, E(-|g| - i\eta) = -\omega \sqrt{\frac{6}{\pi}} \sqrt{\frac{4\omega^3}{3|g|}}\, e^{-4\omega^3/3|g|} \varepsilon_i(g), \qquad (17.364)$$

where

$$\begin{aligned}
\varepsilon_i(g) &= \left(\frac{\Omega}{\omega}\right)^{5/2} \exp\left[-4\frac{\Omega^3 - \omega^3}{3|g|} - c_1\frac{3|g|}{4\Omega^3}\right. \\
&\left. - \frac{\omega^2 - \Omega^2}{2}\left(\frac{4\Omega}{|g|} - \frac{7}{6\Omega^2} - 2\frac{a_3 - a_4}{\Omega^2}\right) + \frac{(\omega^2 - \Omega^2)^2}{2\Omega|g|}a_2\right]. \qquad (17.365)
\end{aligned}$$

The calculation of the integrals (17.362) proceeds as in Appendix 17A, yielding

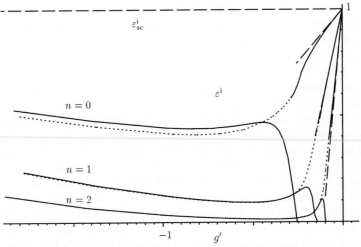

Figure 17.13 Reduced imaginary part of lowest three energy levels of anharmonic oscillator for negative couplings plotted against $g' \equiv g/\omega^3$. The semiclassical limit corresponds to $\varepsilon_i \equiv 1$. The small-$|g'|$ branch is due to tunneling, the large-$|g'|$ branch to direct decay (sliding). Solid and dotted curves show the imaginary parts of the variational approximations W_1 and W_3, respectively; dashed straight lines indicate the exactly known slopes.

$a_2 = -1, a_3 = 3/4, a_4 = 1/12$.

The result is shown in Fig. 17.13. The slope of $\varepsilon_i(g)$ at $g = 0$ maintains the first-order value $c_1 3/4 = 95/96$, i.e., the additional terms in the exponent of (17.365) cancel each other to first order in g.

There exists a short derivation of this result using the same method as in the derivation of Eq. (5.190). We take the fluctuation-corrected semiclassical approximation at the frequency ω

$$\text{Im}\, E(-|g| - i\eta) = -\omega\sqrt{\frac{6}{\pi}}\sqrt{\frac{4\omega^3}{3|g|}}e^{-4\omega^3/3|g|-3c_1|g|/4\omega^3}, \qquad (17.366)$$

move the ω-dependent prefactor into the exponent with the help of the logarithm, replace everywhere ω by $\sqrt{\Omega^2 - (\Omega^2 - \omega^2)} = \sqrt{\Omega^2 - gr^2/2}$ with $r^2 = 2(\Omega^2 - \omega^2)/g$, and expand the exponent in powers of g including all orders of g to which the exponent of (17.366) is known (treating r as a quantity of order unity). This leads again to (17.364) with (17.365).

The imaginary part is inserted into the dispersion relation (17.329) and yields for positive g the energy

$$E^{(0)}(g) = \frac{\omega}{2} + 2\omega g \int_0^\infty \frac{d\lambda}{2\pi} \frac{1}{\lambda(\lambda+g)}\sqrt{\frac{6}{\pi}}\sqrt{\frac{4\omega^3}{3\lambda}}e^{-4\omega^3/3\lambda}\varepsilon_i(\lambda). \qquad (17.367)$$

Expanding the integrand in powers of g gives an integral formula for the perturbation coefficients analogous to (17.331). Its evaluation yields the numbers shown in Table 17.1. They are compared with exact previous larger-order values (17.319) which follow from $\varepsilon_i \equiv 1$. The improvement of our knowledge on the imaginary part of the energy makes it possible to extend the previous large-order results to low orders. Even the lowest coefficient with $k = 1$ is reproduced very well [8].

The high degree of accuracy of the low-order coefficients is improved further by going to the higher variational approximation W_3 of Eq. (5.192) and extracting from it the imaginary part $\text{Im}\, W_3(0)$ at zero temperature [9]. When continuing the coupling constant g to the sliding regime, we obtain the dotted curve in Fig. 17.13. It merges rather smoothly into the tunneling branch at $g \approx -0.24$. Plotting the merging regime with more resolution, we find two closely lying intersections at $g' = -0.229$ and $g' = -0.254$. We choose the first of these to cross over from one branch to the other. After inserting the imaginary part into the integral (17.331), we obtain the fifth column in Table 17.1. For $k = 1$, the accuracy is now better than 0.05%. To make the approximation completely consistent, the tunneling amplitude should also be calculated to the corresponding order. This would yield a further improvement in the low-order coefficients.

It is instructive to test the accuracy of our low-order results by evaluating the dispersion relation (17.367) for the g-dependent ground state energy $E^{(0)}(g)$. The results shown in Fig. 17.14 compare well with the exact curves. They are only slightly worse than the original Feynman-Kleinert approximation W_1 evaluated at positive values of g. We do not show the approximation W_3 since it is indistinguishable from the exact energy on this plot.

The approximation obtained from the dispersion relation has the advantage of possessing the properly diverging power series expansion and a reliable information on the analytic cut structure in the complex g-plane. Also here, the third-order

Table 17.1 Comparison between exact perturbation coefficients, semiclassical ones, and those obtained from moment integrals over the imaginary parts consisting of (17.363) in the tunneling regime and the analytic continuation of the variational approximations W_1 and W_3 in the sliding regime. An alternating sign $(-1)^{k-1}$ is omitted and ω is set equal to 1.

k	E_k	E_k^{sc}	$E_k^{\mathrm{var1+disp}}$	$E_k^{\mathrm{var3+disp}}$
1	0.75	1.16954520	0.76306206	0.74932168
2	2.625	5.26295341	2.49885978	2.61462012
3	20.8125	39.4721506	18.3870038	20.7186128
4	241.289063	414.457581	205.886443	240.857317
5	3580.98047	5595.17734	3093.38043	3590.69587
6	63982.8135	92320.4261	57436.2852	64432.5387
7	1329733.73	1800248.43	1244339.99	1342857.03
8	31448214.7	40505587.0	30397396.0	31791078.0
9	833541603	1032892468	822446267	842273537
10	24478940700	29437435332	24420208763	24703889150

result $E^{(0)}_{\mathrm{var3+disp}}$ based on the imaginary part of W_3 for $g < 0$ is so accurate that it cannot be distinguished from the exact ones on the plot.

The strong-coupling behavior is well reproduced by our curves. Recall the limiting expression for the middle curve given in Eq. (5.77) and the exact one (5.226) with the coefficients of Table 5.9.

The calculation of the imaginary part in the sliding regime can be accelerated by removing from the perturbation coefficients the portion which is due to the imaginary part of the tunneling amplitude. By adding the energy associated with this portion in the form of a dispersion relation it is possible to find variational approximations which for positive coupling constants are not only numerically accurate but which also have power series expansions with the correct large-order behavior [which was not the case for the earlier approximations $W_N(g)$].

The entire treatment can be generalized to excited states. The variational energies are then replaced by the minima of the expressions derived in Section 5.19,

$$W_1^{(n)} = \Omega n_2 + \frac{\omega^2 - \Omega^2}{2}\frac{n_2}{\Omega} + \frac{g}{4}\frac{n_4}{\Omega^2}, \tag{17.368}$$

with $n_2 = n + 1/2$ and $n_4 = (3/2)(n^2 + n + 1/2)$. The optimal Ω-values are given by the solutions (17.350), (17.351), with $g^{(0)}$ replaced by $g^{(n)} = 2n_2/3\sqrt{3}n_4$. For $g \in (-g^{(n)}, 0)$, the energies are real; for $g < -g^{(n)}$ they possess the imaginary part

$$\operatorname{Im} W_1^{(n)} = \frac{1}{2}\Omega^{\mathrm{i}}\left(1 - \frac{\omega^2}{|\Omega|^2}\right)n_2 - \frac{g}{2}\Omega^{\mathrm{re}}\Omega^{\mathrm{im}}\frac{n_4}{|\Omega|^4}. \tag{17.369}$$

For $g \in (-g^{(n)}, 0)$, the imaginary part arises from the bubble solution. In the semiclassical limit it produces a factor $12^n \mathcal{A}_{\mathrm{cl}}^n/n!$ for $n > 0$ as in Eq. (17.322) (with

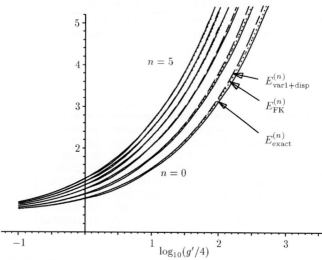

Figure 17.14 Energies of anharmonic oscillator as function of $g' \equiv g/\omega^3$, obtained from variational imaginary part, and the dispersion relation (17.328) as a function of the coupling constant g. Comparison is made with the exact curve and the Feynman-Kleinert variational energy for $g > 0$.

ω replaced by Ω). Also here, the variational approach can easily be continued higher order approximations W_2, W_3, ...

To first order in g, the imaginary part is known from a WKB calculation [10]. It reads

$$\operatorname{Im} E^{(n)}(-|g| - i\eta) = -\frac{12^n}{n!}\omega\sqrt{\frac{6}{\pi}}\left[\frac{4\omega^3}{3|g|}\right]^{-1+2n} e^{-4\omega^3/3|g|-c_1^{(n)}3|g|/4\omega^3}, \qquad (17.370)$$

with the slope parameter

$$c_1^{(n)} = \frac{d\varepsilon_i^{(n)}}{d(g/\omega^3)} = \left(\frac{95}{96} + \frac{29}{16}n + \frac{17}{16}n^2\right)\omega^{-3}. \qquad (17.371)$$

Following the procedure described after Eq. (17.366), we obtain from this a variational expression for the imaginary part which generalizes Eq. (17.364) to any n:

$$\operatorname{Im} E^{(n)}(-|g| - i\eta) = -\frac{12^n}{n!}\omega\sqrt{\frac{6}{\pi}}\left[\frac{4\omega^3}{3|g|}\right]^{-1+2n} e^{-4\omega^3/3|g|}\varepsilon_i^{(n)}(g), \qquad (17.372)$$

with a correction factor

$$\varepsilon_i^{(n)}(g) = \left(\frac{\Omega}{\omega}\right)^{3n+5/2} \exp\left[-4\frac{\Omega^3 - \omega^3}{3|g|} - c_1^{(n)}\frac{3|g|}{4\Omega^3}\right.$$
$$\left. - \frac{\omega^2 - \Omega^2}{2}\left(\frac{4\Omega}{|g|} - \frac{3n+5/2}{\Omega^2}\right) - \frac{(\omega^2 - \Omega^2)^2}{2\Omega|g|}\right]. \qquad (17.373)$$

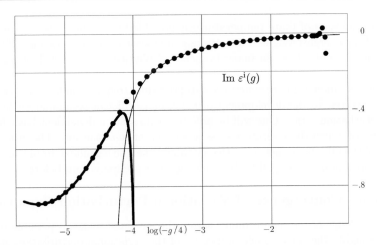

Figure 17.15 Reduced imaginary part of ground state energy of anharmonic oscillator from variational perturbation theory plotted for small negative g against $\log(-g/4)$. The fat curve is the analytic continuation of the strong-coupling expansion (5.226) with the expansion coefficients up to the 22nd order listed in Table 5.9. The thin curve is the divergent semiclassical expansion of the contribution of the classical solution in Eq. (17.374).

Inserting for Ω the optimal value $\Omega^{(n)}$, we obtain the solid curves shown in Fig. 17.13. Their slopes have the exact values (17.371).

The sliding regime for the excited states can be obtained from an analytic continuation of the variational energies. For $n = 1, 2$ the resulting imaginary parts are shown as dotted curves in Fig. 17.13. They merge smoothly with the corresponding tunneling branches obtained from W_1.

If we extend the variational evaluation of the perturbation expansions to high orders in g, we find the imaginary part over the left-hand cut extending deeper and deeper into the regime dominated by the classical solution and the fluctuations around it [9]. This is shown in the double-logarithmic plot of Fig. 17.15.

The result may be compared with the the divergent semiclassical expansion around the classical solution [12]

$$\log \varepsilon^{\mathrm{i}}(g) = -\frac{4}{3g} + k_1 \frac{g}{4} + k_2 \left(\frac{g}{4}\right)^2 + \dots . \tag{17.374}$$

also plotted in Fig. 17.15. The coefficients are listed in Table 17.2.

Table 17.2 Coefficients k_n of semiclassical expansion (17.374) around classical solution.

1	2	3	4	5
3.95833	19.3437500	174.2092014	2177.286133	34045.58329

6	7	8	9	10
632817.0536	1.357206×10^7	3.2924×10^8	8.92×10^9	2.65×10^{11}

The inclusion of finite temperatures is possible by summing over the imaginary parts of the energies weighted by a Boltzmann factor with these energies. This opens the road to applications in many branches of physics where tunneling phenomena are relevant.

It will be interesting to generalize this procedure to quantum field theories, where it can give rise to the development of much more efficient resummation techniques for perturbation series. One will be able to set up system-dependent basis functions in terms of which these series possess a convergent re-expansion. The critical exponents of the $O(N)$-symmetric φ^4-theory should then be calculable from the presently known five-loop results [13] with a much greater accuracy than before.

17.10.5　Convergence of Variational Perturbation Expansion

The knowledge of the discontinuity across the left-hand makes it possible to understand roughly the convergence properties of the variational perturbation expansion developed in Section 5.14. The ground state energy satisfies the subtracted dispersion relation [compare (17.299)]

$$E^{(0)}(g) = \frac{\omega}{2} - \frac{g}{2\pi i} \int_0^{-\infty} \frac{dg'}{g'} \frac{\operatorname{disc} E^{(0)}(g')}{g' - g}, \tag{17.375}$$

where $\operatorname{disc} E^{(0)}(g')$ denotes the discontinuity across the left-hand cut in the complex g-plane. An expansion of the integrand in powers of g yields the perturbation series

$$E^{(0)}(g) = \omega \sum_{k=0}^{N} E_k^{(0)} \left(\frac{g}{4\omega^3} \right)^k. \tag{17.376}$$

The associated variational energy has the form [compare (5.206)]

$$W_N^{\Omega}(g) = \Omega \sum_{k=0}^{N} \varepsilon_k^{(0)} \left(\frac{g}{4\Omega^3} \right)^k. \tag{17.377}$$

It is obtained from (17.376) by the replacement (5.188) and a re-expansion in powers of g. In the present context, we write this replacement as

$$\omega \longrightarrow \Omega(1 - \sigma \hat{g})^{1/2}, \tag{17.378}$$

where $\hat{g}$ is the dimensionless coupling constant g/Ω^3, and

$$\sigma = \Omega(\Omega^2 - 1)/g \tag{17.379}$$

[recall Eqs. (5.213) and (5.208)].

There is a simple way of obtaining the same re-expansion from the dispersion relation (17.375). Introducing the dimensionless coupling constant $\bar{g} \equiv g/\omega^3$, the replacement (17.378) amounts to

$$\bar{g} \longrightarrow \tilde{g}(\hat{g}) \equiv \frac{\hat{g}}{(1 - \sigma \hat{g})^{3/2}}. \tag{17.380}$$

Since Eq. (17.375) represents an energy, it can be written as ω times a dimensionless function of $\bar{g}$. Apart from the replacement (17.380) in the argument, it receives an overall factor $\Omega/\omega = (1 - \sigma\hat{g})^{1/2}$. We introduce the reduced energies

$$\hat{E}(\hat{g}) \equiv E(g)/\Omega, \qquad (17.381)$$

which depends only on the reduced coupling constant $\hat{g}$, the dispersion relation (17.375) for $E^{(0)}(g)$ implies a dispersion relation for $\hat{E}^{(0)}(\hat{g})$:

$$\hat{E}^{(0)}(g) = (1 - \sigma\hat{g})^{1/2}\left[\frac{1}{2} + \frac{\tilde{g}(\hat{g})}{2\pi i}\int_0^{-\infty}\frac{d\bar{g}'}{\bar{g}'}\frac{\text{disc }\bar{E}^{(0)}(\bar{g}')}{\bar{g}' - \tilde{g}(\hat{g})}\right]. \qquad (17.382)$$

The resummed perturbation series is obtained from this by an expansion in powers of $\hat{g}/4$ up to order N.

It should be emphasized that only the truncation of the expansion causes a difference between the two expressions (17.375) and (17.382), since $\bar{g}$ and $\tilde{g}$ are the same numbers, as can be verified by inserting (17.379) into the right-hand side of (17.380).

To find the re-expansion coefficients we observe that the expression (17.382) satisfies a dispersion relation in the complex $\hat{g}$-plane. If C denotes the cuts in this plane and $\text{disc}_C E(\hat{g})$ is the discontinuity across these cuts, the dispersion relation reads

$$\hat{E}^{(0)}(\hat{g}) = \frac{1}{2} + \frac{\hat{g}}{2\pi i}\int_C \frac{d\hat{g}'}{\hat{g}'}\frac{\text{disc}_C\hat{E}^{(0)}(\hat{g}')}{\hat{g}' - \hat{g}}. \qquad (17.383)$$

We have changed the argument of the energy from $\bar{g}$ to $\hat{g}$ since this will be the relevant variable in the sequel.

When expanding the denominator in the integrand in powers of $\hat{g}/4$, the expansion coefficients $\varepsilon_l^{(0)}$ are found to be moment integrals with respect to the inverse coupling constant $1/\hat{g}$ [compare (17.302)]:

$$\varepsilon_k^{(0)} = -\frac{4^k}{2\pi i}\int_C \frac{d\hat{g}}{\hat{g}^{k+1}}\text{disc}_C\hat{E}^{(0)}(\hat{g}). \qquad (17.384)$$

In the complex $\hat{g}$-plane, the integral (17.382) has in principle cuts along the contours $C_1, C_{\bar{1}}, C_2, C_{\bar{2}}$, and C_3, as shown in Fig. 17.16. The first four cuts are the images of the left-hand cut in the complex g-plane; the curve C_3 is due to the square root of $1 - \sigma\hat{g}$ in the mapping (17.380) and the prefactor of (17.382).

Let $\bar{D}(\bar{g})$ abbreviate the reduced discontinuity in the original dispersion relation (17.375):

$$\bar{D}(\bar{g}) \equiv \text{disc }\bar{E}^{(0)}(\bar{g}) = 2i\text{Im }\bar{E}^{(0)}(\bar{g} - i\eta), \quad \bar{g} \leq 0. \qquad (17.385)$$

Then the discontinuities across the various cuts are

$$\underset{C_{1,\bar{1},2,\bar{2}}}{\text{disc}} \hat{E}^{(0)}(\hat{g}) = (1 - \sigma\hat{g})^{1/2}\bar{D}(\hat{g}(1 - \sigma\hat{g})^{-3/2}), \qquad (17.386)$$

$$\underset{C_3}{\text{disc}} \hat{E}^{(0)}(\hat{g}) = -2i(\sigma\hat{g} - 1)^{1/2}$$
$$\times\left[\frac{1}{2} - \int_0^\infty \frac{d\bar{g}'}{2\pi}\frac{\hat{g}(\sigma\hat{g} - 1)^{-3/2}}{\bar{g}'^2 + \hat{g}^2(\sigma\hat{g} - 1)^{-3}}\bar{D}(-\bar{g}')\right]. \qquad (17.387)$$

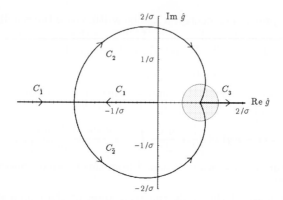

Figure 17.16 Cuts in complex $\hat{g}$-plane whose moments with respect to inverse coupling constant determine re-expansion coefficients. The cuts inside the shaded circle happen to be absent due to the convergence of the strong-coupling expansion for $g > g_s$.

For small negative $\bar{g}$, the discontinuity is given by the semiclassical limit (17.317):

$$\bar{D}\bar{g}) \approx -2i\sqrt{\frac{6}{\pi}}\sqrt{\frac{4}{-3\bar{g}}}e^{4/3\bar{g}}. \tag{17.388}$$

We denote by $\varepsilon_k^{(0)}(C_i)$ the contributions of the different cuts to the integral (17.384) for the coefficients. After inserting (17.388) into Eq. (17.386), we obtain from the cut along C_1 the semiclassical approximation

$$\varepsilon_k^{(0)}(C_1) \approx -2\,4^k \int_{C_1} \frac{d\hat{g}}{2\pi} \frac{1}{\hat{g}^{k+1}} \sqrt{\frac{6}{\pi}} \sqrt{-\frac{4(1-\sigma\hat{g})^{5/2}}{3\hat{g}}} e^{4(1-\sigma\hat{g})^{3/2}/3\hat{g}}. \tag{17.389}$$

For the kth term S_k of the series this yields an estimate

$$S_k \propto \left[\int_{C_\gamma} \frac{d\gamma}{2\pi} e^{f_k(\gamma)}\right](\sigma\hat{g})^k, \tag{17.390}$$

where $f_k(\gamma)$ is the function of $\gamma \equiv \sigma\hat{g}$

$$f_k(\gamma) = -\left(k+\frac{3}{2}\right)\log(-\gamma) + \frac{4\sigma}{3\gamma}(1-\gamma)^{3/2}. \tag{17.391}$$

For large k, the integral may be evaluated via the saddle point approximation of Subsection 4.2.1. The extremum of $f_k(\gamma)$ satisfies the equation

$$-k+\frac{3}{2} = \frac{4\sigma}{3\gamma}(1-\gamma)^{1/2}\left(1+\tfrac{1}{2}\gamma\right), \tag{17.392}$$

which is solved by

$$\gamma \xrightarrow[k\to\infty]{} \gamma_k = -4\sigma/3k. \tag{17.393}$$

At the extremum, $f_k(\gamma)$ has the value

$$f_k \xrightarrow[k\to\infty]{} k\log(3k/4e\sigma) - 2\sigma. \tag{17.394}$$

The constant -2σ in this limiting expression arises when expanding the second term of Eq. (17.391) into a Taylor series, $(4\sigma/3\gamma)(1-\gamma)^{3/2} = 4\sigma/3\gamma_k - 2\sigma + \ldots$. Only the first two terms survive the large-k limit.

Thus, to leading order in k, the kth term of the re-expanded series becomes

$$S_k \propto e^{-2\sigma}\left(\frac{-3k}{e}\right)^k\left(\frac{\hat{g}}{4}\right)^k. \tag{17.395}$$

The corresponding re-expansion coefficients are

$$\varepsilon_k^{(0)} \propto e^{-2\sigma}E_k^{(0)}. \tag{17.396}$$

They have the remarkable property of growing in precisely the same manner with k as the initial expansion coefficients $E_k^{(0)}$, except for an overall suppression factor $e^{-2\sigma}$. This property was found empirically in Fig. 5.20b.

In order to estimate the convergence of the variational perturbation expansion, we note that with

$$\sigma = \frac{\Omega(\Omega^2-1)}{g} \tag{17.397}$$

and $\hat{g}$ from (5.213), we have

$$\sigma\hat{g} = 1 - \frac{1}{\Omega^2}. \tag{17.398}$$

For large Ω, this expression is smaller than unity. Hence the powers $(\sigma\hat{g})^k$ alone yield a convergent series. An optimal re-expansion of the energy can be achieved by choosing, for a given large maximal order N of the expansion, a parameter σ proportional to N:

$$\sigma \approx \sigma_N \equiv cN. \tag{17.399}$$

Inserting this into (17.391), we obtain for large $k = N$

$$f_N(\gamma) \approx N\left[-\log(-\gamma) + \frac{4c}{3\gamma}(1-\gamma)^{3/2}\right]. \tag{17.400}$$

The extremum of this function lies at

$$1 + \frac{4c}{3\gamma}(1-\gamma)^{1/2}(1+\tfrac{1}{2}\gamma) = 0. \tag{17.401}$$

The constant c is now chosen in such a way that the large exponent proportional to N in the exponential function $e^{f_N(\gamma)}$ due to the first term in (17.400) is canceled by an equally large contribution from the second term, i.e., we require at the extremum

$$f_N(\gamma) = 0. \qquad (17.402)$$

The two equations (17.401) and (17.402) are solved by

$$\gamma = -0.242\,964\,029\,973\,520\,\ldots, \quad c = 0.186\,047\,272\,987\,975\,\ldots\,. \qquad (17.403)$$

In contrast to the extremal γ in Eq. (17.393) which dominates the large-k limit, the extremal γ of the present limit, in which k is also large but of the order of N, remains finite (the previous estimate holds for $k \gg N$). Accordingly, the second term $(4c/3\gamma)(1-\gamma)^{3/2}$ in $f_N(\gamma)$ contributes in full, not merely via the first two Taylor expansion terms of $(1-\gamma)^{3/2}$, as it did in (17.394).

Since $f_N(\gamma)$ vanishes at the extremum, the Nth term in the re-expansion has the order of magnitude

$$S_N \propto (\sigma_N \hat{g}_N)^N = \left(1 - \frac{1}{\Omega_N^2}\right)^N. \qquad (17.404)$$

According to (17.397) and (17.399), the frequency Ω_N grows for large N like

$$\Omega_N \sim \sigma_N^{1/3} g^{1/3} \sim (cNg)^{1/3}. \qquad (17.405)$$

As a consequence, the last term of the series decreases for large N like

$$S_N(C_1) \propto \left[1 - \frac{1}{(\sigma_N g)^{2/3}}\right]^N \approx e^{-N/(\sigma g)^{2/3}} \approx e^{-N^{1/3}/(cg)^{2/3}}. \qquad (17.406)$$

This estimate does not yet explain the convergence of the variational perturbation expansion in the strong-coupling limit observed in Figs. 5.21 and 5.22. For the contribution of the cut C_1 to S_N, the derivation of such a behavior requires including a little more information into the estimate. This information is supplied by the empirically observed property, that the best Ω_N-values lie for finite N on a curve [recall Eq. (5.211)]:

$$\sigma_N \sim cN\left(1 + \frac{6.85}{N^{2/3}}\right). \qquad (17.407)$$

Thus the asymptotic behavior (17.399) receives, at a finite N, a rather large correction. By inserting this σ_N into $f_N(\gamma)$ of (17.400), we find an extra exponential factor

$$\begin{aligned}
e^{\Delta f_N} &\approx \exp\left[N\frac{4c}{3}\frac{(1-\gamma)^{3/2}}{\gamma}\frac{6.85}{N^{2/3}}\right] \\
&= \exp\left[-N\log(-\gamma)\frac{6.85}{N^{2/3}}\right] \approx e^{-9.7N^{1/3}}. \qquad (17.408)
\end{aligned}$$

This reduces the size of the last term due to the cut C_1 in (17.406) to

$$S_N(C_1) \propto e^{-[9.7+(cg)^{-2/3}]N^{1/3}}, \tag{17.409}$$

which agrees with the convergence seen in Figs. 5.21 and 5.22.

There is no need to evaluate the effect of the shift in the extremal value of γ caused by the correction term in (17.407), since this would be of second order in $1/N^{2/3}$.

How about the contributions of the other cuts? For $C_{\bar{1}}$, the integrals in (17.384) run from $\hat{g} = -2/\sigma$ to $-\infty$ and decrease like $(-2/\sigma)^{-k}$. The associated last term $S_N(C_{\bar{1}})$ is of the negligible order $e^{-N \log N}$. For the cuts $C_{2,\bar{2},3}$, the integrals (17.384) start at $\hat{g} = 1/\sigma$ and have therefore the leading behavior

$$\varepsilon_k^{(0)}(C_{2,\bar{2},3}) \sim \sigma^k. \tag{17.410}$$

This implies a contribution to the Nth term in the re-expansion of the order of

$$S_N(C_{2,\bar{2},3}) \sim (\sigma \hat{g})^N, \tag{17.411}$$

which decreases merely like (17.406) and does not explain the empirically observed convergence in the strong-coupling limit. As before, an additional information produces a better estimate. The cuts in Fig. 17.16 do not really reach the point $\sigma \hat{g} = 1$. There exists a small circle of radius $\Delta \hat{g} > 0$ in which $\hat{E}^{(0)}(\hat{g})$ has no singularities at all. This is a consequence of the fact unused up to this point that the strong-coupling expansion (5.231) converges for $g > g_s$. For the reduced energy, this expansion reads:

$$\hat{E}^{(0)}(\hat{g}) = \left(\frac{\hat{g}}{4}\right)^{1/3} \left\{ \alpha_0 + \alpha_1 \left[\frac{\hat{g}}{4\omega^3} \frac{1}{(1-\sigma\hat{g})^{3/2}} \right]^{-2/3} + \alpha_2 \left[\frac{\hat{g}}{4\omega^3} \frac{1}{(1-\sigma\hat{g})^{3/2}} \right]^{-4/3} + \dots \right\}. \tag{17.412}$$

The convergence of (5.231) for $g > g_s$ implies that (17.412) converges for all $\sigma \hat{g}$ in a neighborhood of the point $\sigma \hat{g} = 1$ with a radius

$$\Delta(\sigma \hat{g}) \sim \left(\frac{\hat{g}}{-\bar{g}_s} \right)^{2/3} = \left\{ \frac{1}{-\sigma \bar{g}_s} [1 + \Delta(\sigma \hat{g})] \right\}^{2/3}, \tag{17.413}$$

where $\bar{g}_s \equiv g_s/\omega^3$. For large N, $\Delta(\sigma \hat{g})$ goes to zero like $1/(N|\bar{g}_s|c)^{2/3}$. Thus the integration contours of the moment integrals (17.384) for the contributions $\varepsilon_k^{(0)}(C_i)$ of the other cuts do not begin at the point $\sigma \hat{g} = 1$, but a little distance $\Delta(\sigma \hat{g})$ away from it. This generates an additional suppression factor

$$(\sigma \hat{g})^{-N} \sim [1 + \Delta(\sigma \hat{g})]^{-N}. \tag{17.414}$$

Let us set $-\bar{g}_s = |\bar{g}_s| \exp(i\varphi_s)$ and $x_s \equiv (-\hat{g}/\bar{g}_s)^{2/3} = -|x_s| \exp(i\theta)$, and introduce the parameter $a \equiv 1/[|\bar{g}_s|c]^{2/3}$. Since there are two complex conjugate contributions we obtain, for large N a last term of the re-expanded series the order of

$$S_N(C_{2,\bar{2},3}) \approx e^{-N^{1/3}a \cos\theta} \cos(N^{1/3}a \sin\theta). \tag{17.415}$$

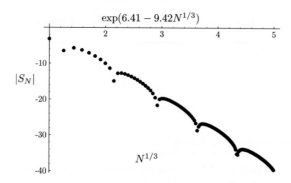

Figure 17.17 Theoretically obtained convergence behavior of Nth approximants for α_0, to be compared with the empirically found behavior in Fig. 5.21.

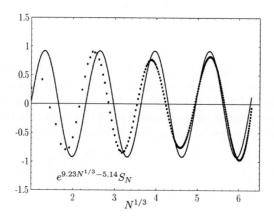

Figure 17.18 Theoretically obtained oscillatory behavior around exponentially fast asymptotic approach of α_0 to its exact value as a function of the order N of the approximant, to be compared with the empirically found behavior in Fig. 5.22, averaged between even and odd orders.

By choosing

$$|\bar{g}_{\rm s}| \sim 0.160, \qquad \theta \sim -0.467, \tag{17.416}$$

we obtain the curves shown in Figs. 17.17 and 17.18 which agree very well with the observed Figs. 5.21 and 5.22. Their envelope has the asymptotic falloff $e^{-9.23N^{1/3}}$.

Let us see how the positions of the leading Bender-Wu singularities determined by (17.416) compare with what we can extract directly from the strong-coupling series (5.231) up to order 22. For a pair of square root singularities at $x_{\rm s} = -|x_{\rm s}| \exp(\pm i\theta)$, the coefficients of a power series $\sum \alpha_n x^n$ have the asymptotic ratios $R_n \equiv \alpha_{n+1}/\alpha_n \sim$

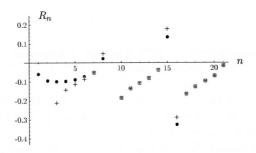

Figure 17.19 Comparison of ratios R_n between successive expansion coefficients of strong-coupling expansion (dots) with ratios R_n^{as} of expansion of superposition of two singularities at $g = 0.156 \times \exp(\pm 0.69)$ (crosses).

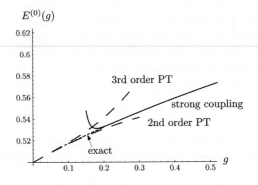

Figure 17.20 Strong-coupling expansion of ground state energy in comparison with exact values and perturbative results of 2nd and 3rd order. The convergence radius in $1/g$ is larger than $1/0.2$.

$R_n^{\mathrm{as}} \equiv -\cos[(n+1)\theta + \delta]/|x_{\mathrm{s}}|\cos(n\theta + \delta)$. In Fig. 17.19 we have plotted these ratios against the ratios R_n obtained from the coefficients α_n of Table 5.9 . For large n, the agreement is good if we choose

$$|x_{\mathrm{s}}| = 1/0.117, \quad \theta = -0.467, \tag{17.417}$$

with an irrelevant phase angle $\delta = -0.15$. The angle θ is in excellent agreement with the value found in (17.416). From $|x_{\mathrm{s}}|$ we obtain $|\bar{g}_{\mathrm{s}}| = 4|1/x_{\mathrm{s}}|^{3/2} = 0.160$, again in excellent agreement with (17.416).

This convergence radius is compatible with the heuristic convergence of the strong-coupling series up to order 22, as can be seen in Fig. 17.19 by comparing the curves resulting from the series with the exact curve.

It is possible to extend the convergence proof to the more general divergent power series discussed in Section 5.17, whose strong-coupling expansions have the more general growth parameters p and q [14]. The convergence is assured for $1/2 < 2/q < 1$ [15]. If the interaction of the anharmonic oscillator is $\int d\tau\, x^n(\tau)$ with $n \neq 4$, the dimensionless expansion parameter for the energies is $g/\omega^{n/2+1}$ rather than g/ω^3. Then $q = n/2 + 1$, such that for $n \geq 6$ the convergence is lost. This can be verified by trying to resum the expansions for the ground state energies of $n = 6$ and $n = 8$, for example. For $n = 6$, the cut in Fig. 17.16 becomes circular such that there is no more shaded circle C_3 in which the strong-coupling series converges.

17.11   Decay of Supercurrent in Thin Closed Wire

An important physical application of the above tunneling theory explains the temperature behavior of the resistance of a thin[7] superconducting wire. The superconducting state is described by a complex order parameter $\psi(z)$ depending on the spatial variable z along the wire. We then speak of an *order field*. The variable z plays the role of the Euclidean time τ in the previous sections. We shall consider a closed wire where $\psi(z)$ satisfies the periodic boundary condition

$$\psi(z) = \psi(z + L). \tag{17.418}$$

The energy density of the system is described approximately by a *Ginzburg-Landau expansion* in powers of ψ and its gradients containing only the terms

$$\varepsilon(z) = |\partial_z \psi(z)|^2 + m^2 |\psi(z)|^2 + \frac{g}{4}|\psi(z)|^4. \tag{17.419}$$

The total fluctuating energy is given by the functional

$$E[\psi^*, \psi] = \int_{-L/2}^{L/2} dz\, \varepsilon(z), \tag{17.420}$$

---

[7]A superconducting wire is called *thin* if it is much smaller than the coherence length to be defined in Eq. (17.425).

and the probability of each fluctuation is determined by the Boltzmann factor $\exp\{-E[\psi^*, \psi]/k_B T\}$. The parameter m^2 in front of $|\psi(z)|^2$ is called the *mass term* of the field. It vanishes at the critical temperature T_c and behaves near T_c like

$$m^2 \approx m_0^2 \left(\frac{T}{T_c} - 1\right). \tag{17.421}$$

Below T_c, the square mass is negative and the wire becomes superconducting. One can easily estimate, that each term in the Landau expansion is of the order of $|1 - T/T_c|^2$ and any higher expansion term in (17.419) would be smaller than that by at least a power $|1 - T/T_c|^{1/2}$.

The partition function of the system is given by the path integral

$$Z = \int \mathcal{D}\psi^*(z)\mathcal{D}\psi(z)e^{-E[\psi^*, \psi]/k_B T}. \tag{17.422}$$

If T does not lie too close to T_c [although close enough to justify the Landau expansion, i.e., the neglect of higher expansion terms in (17.419) suppressed by a factor $|1 - T/T_c|^{1/2}$], this path integral can be treated semiclassically in the way described earlier in this chapter [16].

The basic microscopic mechanism responsible for the phenomenon of superconductivity will be irrelevant for the subsequent discussion. Let us only recall the following facts: A superconductor is a metal at low temperatures whose electrons near the surface of the Fermi sea overcome their Coulomb repulsion due to *phonon exchange*. This enables them to form bound states between two electrons of opposite spin orientations in a relative *s*-wave, the celebrated *Cooper pairs*.[8] The attraction which binds the Cooper pairs is extremely weak. This is why the temperature has to be very small to keep the pairs from being destroyed by thermal fluctuations. The critical temperature T_c, where the pairs break up, is related to the binding energy of the Cooper pairs by $E_{\text{pair}} = k_B T_c$. The field-theoretic process called phonon exchange is a way of describing the accumulation of positive ions along the path of an electron which acts as an attractive potential wake upon another electron while screening the Coulomb repulsion. The attraction is very weak and leads to a bound state only in the *s*-wave (the centrifugal barrier $\propto l(l+1)/r^2$ preventing the formation of a bound state in higher partial waves). The potential between the electrons may well be approximated by a δ-function potential $V(x) \approx -g\delta(r)$. The critical temperature T_c, usually a few degrees Kelvin, is found to satisfy the characteristic exponential relation

$$T_c k_B = \mu e^{-1/g}. \tag{17.423}$$

The parameter μ denotes the upper energy cutoff of the phonon spectrum $T_D k_B$, where T_D is the Debye temperature of the lattice vibration.

---

[8]We consider here only with old-fashioned superconductivity which sets in below a very small critical temperature of a few-degree Kelvin. The physics of the recently discovered high-temperature superconductors is at present not sufficiently understood to be discussed along the same lines.

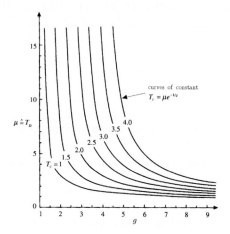

Figure 17.21   Renormalization group trajectories in the g, μ plane of superconducting electrons (g=attractive coupling constant, μ=Debye temperature, $k_B = 1$). Curves with same T_c imply identical superconducting properties. The renormalization group determines the reparametrizations of a fixed superconductive system along any of these curves.

An important result of the theory, confirmed by experiment, is that all T-dependent characteristic equilibrium properties of the superconductor near T_c depend only on the single parameter T_c. Thus, many quite different systems with different microscopic parameters $\mu \equiv T_D$ and g will have the *same* superconducting properties (see Fig. 17.21). The critical temperature is an important prototype for the understanding of the so-called *dimensionally transmuted coupling constant* in quantum field theories, which plays a completely analogous role in specifying the system. In quantum field theory, an arbitrary mass parameter μ is needed to define the coupling strength of a renormalized theory and physical quantities depend only on the combination[9]

$$M_c = \mu e^{-1/g(\mu)}. \tag{17.424}$$

The set of all changes μ which are accompanied by a simultaneous change of $g(\mu)$ such as to stay on a fixed curve with M_c from the *renormalization group*. The curve $\mu, g(\mu)$ is called the *renormalization group trajectory* [15].

If one works in natural units with $\hbar = k_B = M = 1$, the critical temperature corresponds to a length of $\approx 1000\text{Å}$. This length sets the scale for the spatial correlations of the Cooper pairs near the critical point via the relation

$$\xi(T) = \frac{\text{const.}}{T_c} \left(1 - \frac{T}{T_c}\right)^{-1/2} \approx 1000\text{Å} \left(1 - \frac{T}{T_c}\right)^{-1/2}. \tag{17.425}$$

---

[9]In quantum chromodynamics, this dimensionally transmuted coupling constant is of the order of the pion mass and usually denoted by Λ.

The Cooper pairs are much larger than the lattice spacing, which is of the order of 1Å. Their size is determined by the ratio $\hbar^2 k_F / m_e \pi k_B T_c$, where k_F is the wave number of electrons of mass m_e on the surface of the Fermi sphere. The temperature T_c in conventional superconductors of the order of 1 K corresponds to $1/11604.447$ eV. Thus the thermal energy $k_B T_c$ is smaller than the atomic energy $E_H = 27.210$ eV (recall the atomic units defined on p. 13.7) by a factor 2.345×10^{-3}, and we find that $\hbar^2 / m_e \pi k_B T_c$ is of the order of $10^2 a_H^2$. Since a_H is of the order of $1/k_F$, we estimate the size of the Cooper pairs as being roughly 100 times larger than the lattice spacing.

This justifies a posteriori the δ-function approximation for the attractive potential, whose range is just a few lattice spacings, i.e., much smaller than $\xi(T)$. The presence of such large bound states causes the superconductor to be coherent over the large distance $\xi(T)$. For this reason, $\xi(T)$ is called the *coherence length*.

Similar Cooper pairs exist in other low-temperature fermion systems such as ^{3}He, where they give rise to the phenomenon of *superfluidity*. There, the interatomic potential contains a hard repulsive core for $r < 2.7$Å. This prevents the formation of an s-wave bound state. In addition it produces a strong spin-spin correlation in the almost fully degenerate Fermi liquid, with a preference of parallel spin configurations. Because of the necessary antisymmetry of the pair wave function of the electrons, this amounts to a repulsion in any even partial wave. For this reason, Cooper pairs can only exist in the p-wave spin triplet state. The binding energy is much weaker than in a superconductor, suppressing the critical temperature by roughly a factor thousand. Experimentally, one finds $T_c = 27$mK at a pressure of $p = 35$ bar. Since the masses of the ^{3}He atoms are larger than those of the electrons by about the same factor thousand, the coherence length ξ has the same order of magnitude in both systems, i.e., $1/T_c$ has the same length when measured in units of Å.

The theoretical description of the behavior of the condensate is greatly simplified by re-expressing the fundamental Euclidean action in terms of a Cooper *pair field* which is the composite field

$$\psi_{\text{pair}}(\mathbf{x}) = \psi_{\text{e}}(\mathbf{x})\psi_{\text{e}}(\mathbf{x}). \tag{17.426}$$

Such a change of field variables can easily be performed in a path integral formulation of the field theory. The method is very similar to the introduction of the auxiliary field $\varphi(\mathbf{x})$ in the polymer field theory of Section 15.12. Since this subject has been treated extensively elsewhere[10] we shall not go into details. The partition function of the system reads

$$Z = \int \mathcal{D}\psi_{\text{e}}^*(\mathbf{x}) \mathcal{D}\psi_{\text{e}}(\mathbf{x}) e^{-\mathcal{A}[\psi_{\text{e}}^*, \psi_{\text{e}}]}. \tag{17.427}$$

---

[10]The way to describe the pair formation by means of path integrals is explained in H. Kleinert, *Collective Quantum Fields*, Fortschr. Phys. *26*, 565 (1978) (http://www.physik.fu-berlin.de/~kleinert/55).

By going from integration variables ψ_e to ψ_{pair}, we can derive the alternative pair partition function

$$Z = \int \mathcal{D}\psi^*_{pair}(\mathbf{x})\mathcal{D}\psi_{pair}(\mathbf{x})e^{-\mathcal{A}[\psi^*_{pair},\psi_{pair}]}, \tag{17.428}$$

where ψ_{pair} is the Cooper pair field (17.426).

In general, the new action is very complicated. For temperatures close to T_c, however, it can be expanded in powers of the field ψ_{pair} and its derivatives, leading to a Landau expansion of the type (17.419). For static fields the Euclidean field action is

$$\mathcal{A}[\psi^*_{pair}, \psi_{pair}] = E/k_B T = \frac{1}{k_B T}\int d^3x\, \varepsilon(\mathbf{x}) \tag{17.429}$$

$$= \frac{1}{k_B T}\int d^3x \left[\left(-\log\frac{\mu}{T} + \frac{1}{g^2}\right)|\psi_{pair}|^2 + \frac{1}{2T_c^2}|\psi_{pair}|^4 + \frac{1}{T_c^2}|\boldsymbol{\nabla}\psi_{pair}|^2 + \dots\right],$$

where the dots denote the omitted higher powers of ψ_{pair} and of their derivatives, each accompanied by an additional factor $1/T_c$.

Let us discuss the path integral (17.429) first in the classical limit. We observe that with the critical temperature (17.423), the mass term in the energy can be written as

$$-\log\frac{T_c}{T}\,|\psi_{pair}|^2 \sim -\left(1 - \frac{T}{T_c}\right)|\psi_{pair}|^2. \tag{17.430}$$

It has the "wrong sign" for $T < T_c$, so that the field has no stable minimum at $\psi_{pair} = 0$. It fluctuates around one of the infinitely many nonzero values with the fixed absolute value

$$|\psi_{pair,0}| = T_c\sqrt{1 - \frac{T}{T_c}}. \tag{17.431}$$

It is then useful to take a factor $T_c\left(1 - T/T_c\right)^{1/2}$ out of the field ψ_{pair}, define

$$\psi(\mathbf{x}) \equiv \psi_{pair}(\mathbf{x})\frac{1}{T_c(1 - T/T_c)^{1/2}}, \tag{17.432}$$

and write the renormalized energy density as

$$\varepsilon(\mathbf{x}) = |\boldsymbol{\nabla}\psi|^2 - |\psi|^2 + \frac{1}{2}|\psi|^4. \tag{17.433}$$

Here we have made use of the coherence length (17.425) to introduce a dimensionless space variable $\mathbf{x}$, replacing $\mathbf{x} \to \mathbf{x}\,\xi$. We also have dropped an overall energy density factor proportional to $(1 - T/T_c)^2 T_c^2$.

In the rescaled form (17.433), the minimum of the energy lies at $|\psi_0| = 1$, where it has the density

$$\varepsilon = \varepsilon_c = -1/2. \tag{17.434}$$

The negative energy accounts for the binding of the Cooper pairs in the condensate (in the present natural units) and is therefore called the *condensation energy*. In terms of (17.433), the partition function in equilibrium can be written as

$$Z = \int \mathcal{D}\psi^*(\mathbf{x})\mathcal{D}\psi(\mathbf{x})e^{-(1/T)\int d^3x\,\varepsilon(\mathbf{x})}. \tag{17.435}$$

We are now prepared to discuss the flow properties of an electric current of the system carried by the Cooper pairs. It is carried by the divergenceless pair current [compare (1.102)]

$$\mathbf{j}(\mathbf{x}) = \frac{1}{2i}\psi^*(\mathbf{x})\overset{\leftrightarrow}{\boldsymbol{\nabla}}\psi(\mathbf{x}) \tag{17.436}$$

associated with the transport of the number of pairs, apart from a charge factor of the pairs, which is equal to twice the electron charge.

The important question to be understood by the theory is: How can this current become "super", and stay alive for a very long time (in practice ranging from hours to years, as far as the patience of the experimentalist may last) [17]. To see this let us set up a current in a long circular wire and assume that the wire thickness is much smaller than the coherence length $\xi(T)$. Then transverse variations of the pair field $\psi(\mathbf{x})$ are strongly suppressed with respect to longitudinal ones (by the gradient terms $|\boldsymbol{\nabla}\psi(\mathbf{x})|^2$ in the Boltzmann factor) and the system depends mainly on the coordinate z *along* the wire, so that the above formalism can be applied. If the cross section of the wire is absorbed into the inverse temperature prefactor in the Boltzmann factor in (17.435), we may simply study the partition function (17.435) for a one-dimensional problem along the z-axis. The energy density (17.433) is precisely of the form announced in the beginning in Eq. (17.419). It is convenient to decompose the complex field $\psi(z)$ into polar coordinates

$$\psi(z) = \rho(z)e^{i\gamma(z)}, \tag{17.437}$$

in terms of which the energy density reads

$$\varepsilon(z) = -\rho^2 + \frac{1}{2}\rho^4 + \rho_z^2 + \rho^2\gamma_z^2, \tag{17.438}$$

where the subscript z indicates a derivative with respect to z. The field equations are

$$j(z) = \rho^2(z)\gamma_z(z) = \text{const} \tag{17.439}$$

and

$$\rho_{zz} = -\rho + \rho^3 + \frac{j^2}{\rho^3}. \tag{17.440}$$

If z is reinterpreted as an imaginary "time", the latter equation can be interpreted as describing the mechanical motion of a mass point at the position $\rho(z)$ moving as a function of the "time" z in the potential

$$-V(\rho) \equiv \rho^2 - \frac{1}{2}\rho^4 + \frac{j^2}{\rho^2}, \tag{17.441}$$

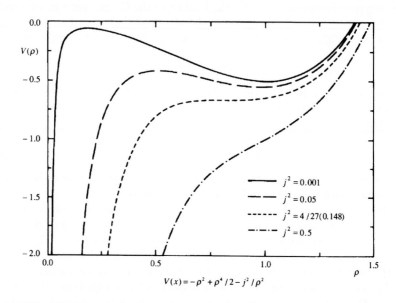

Figure 17.22 Potential $V(\rho) = -\rho^2 + \rho^4/2 - j^2/\rho^2$ showing barrier in superconducting wire to the left of ρ_0 to be penetrated if the supercurrent is to relax.

which is the potential shown in Fig. 17.22 turned upside down.

Certainly, the time-sliced path integral in ρ would suffer from the phenomenon of path collapse described in Chapter 8. At the level of the semiclassical approximation to be performed here, however, this does not happen. There are two types of extremal solutions. The trivial solutions are

$$\gamma(z) = kz, \qquad \rho(z) \equiv \rho_0 = \sqrt{1 - k^2}. \qquad (17.442)$$

Since the wire is closed, the phase $\gamma(z)$ has to be periodic over the total length L of the wire. This implies the quantization of the wave number k,

$$k_n = \frac{2\pi}{L}n, \quad n = 0, \pm 1, \pm 2, \dots . \qquad (17.443)$$

The current associated with these solutions is

$$j = \rho_0^2 k = (1 - k^2)k. \qquad (17.444)$$

As a function of k, this has an absolute maximum at the so-called *critical current* j_c, i.e.,

$$|j| < j_c \equiv \frac{2}{3\sqrt{3}}. \qquad (17.445)$$

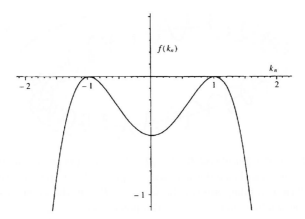

Figure 17.23 The condensation energy as function of velocity parameter $k_n = 2\pi n/L$.

No solution of the field equations can carry a larger current than this. The critical wave number is

$$k_c \equiv \frac{1}{\sqrt{3}}, \tag{17.446}$$

and the energy density:

$$e_c(k) = V(\rho_0) = -\frac{1}{2}(1 - k^2)^2. \tag{17.447}$$

It is plotted in Fig. 17.23. Note that the k-values (17.446) for which a supercurrent can exist between the turning points. The energy $e_c(k)$ represents the negative condensation energy of the state in the presence of the current. For $k \to 0$ it goes against the current-free value (17.434).

We can now understand why all states of current j_n smaller than j_c are, in fact, "super" in the sense of having an extremely long lifetime. At each value of k_n, the wire carries a metastable current which can only decay by a slow tunneling. To see this, we picture the field configuration as a spiral of radius ρ wound around the wire with the azimuthal angle representing the phase $\gamma(z) = k_n z$ (see Fig. 17.24). At zero temperature, the size ρ of the order parameter is frozen at ρ_0 and the winding number is *absolutely stable* on topological grounds. Then, each metastable state with wave number k_n has an infinite lifetime. If the current is to relax by one unit of n it is necessary that at some place z, thermal fluctuations carry $\rho(z)$ to zero. There the phase becomes undefined and may slip by 2π. At the typical low temperatures of these systems, such *phase slips* are extremely rare. To have a local excursion of $\rho(z)$ to $\rho \approx 0$ at one place z, with an appreciable measure in the functional integral (17.435), it must start from a nontrivial solution of the equations of motion which carries $\rho(z)$ as closely as possible to zero. From our experience with the mechanical

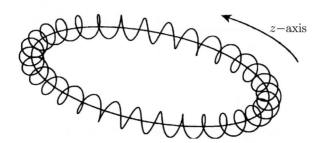

Figure 17.24 Order parameter $\Delta(z) = \rho(z)e^{i\gamma(z)}$ of superconducting thin circular wire neglecting fluctuations. The order parameter is pictured as a spiral of radius ρ_0 and pitch $\partial\gamma(z)/\partial z = 2\pi n/L$ winding around the wire. At $T = 0$, the supercurrent is absolutely stable since the winding number n is fixed topologically.

motion of a mass point in a potential such as $-V(\rho)$ of Eq. (17.441), it is easily realized that there exists such a solution. It carries $\rho(z)$ from $\rho_0 = \sqrt{1 - k^2}$ at $z = -\infty$ across the potential barrier to the small value $\rho_1 = \sqrt{2}k$ and back once more across the barrier to ρ_0 at $z = \infty$ (see Fig. 17.25). Using the first integral of motion of the differential equation (17.440), the law of energy conservation

$$\frac{1}{2}\rho_z^2 - \frac{1}{2}V(\rho) = E = -\frac{1}{2}V(\rho_0) = \frac{1}{4}\rho_0(\rho_0 + 2\rho_1) \qquad (17.448)$$

leads to the equation

$$\rho_z = \sqrt{2E + V(\rho)}. \qquad (17.449)$$

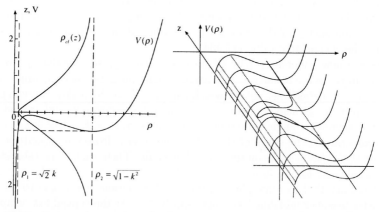

Figure 17.25 Extremal excursion of order parameter in superconducting wire. It corresponds to a mass point starting out at ρ_0, rolling under the influence of "negative gravity" up the mountain unto the point $\rho_1 = \sqrt{2}k$, and returning back to ρ_0, with the variable z playing the role of a time variable.

This is solved by the integral

$$
\begin{aligned}
z - z_1 &= \sqrt{2} \int_{\rho_1}^{\rho} \frac{\rho d\rho}{\sqrt{\rho^6 - 2\rho^4 + 4E\rho^2 - 2j^2}} \\
&= \frac{1}{\sqrt{2}} \int_{\rho_1^2}^{\rho^2} \frac{d\rho^2}{\sqrt{(\rho^2 - \rho_1^2)(\rho^2 - \rho_0^2)}},
\end{aligned}
\tag{17.450}
$$

yielding

$$
z - z_1 = -\frac{2}{\sqrt{2(\rho_0^2 - \rho_1^2)}} \operatorname{arctanh} \sqrt{\frac{\rho^2 - \rho_1^2}{\rho_0^2 - \rho_1^2}}.
\tag{17.451}
$$

Inverting this, we find the bubble solution

$$
\rho_{\text{cl}}^2(z) = 1 - k^2 - \frac{\omega^2/2}{\cosh^2[\omega(z - z_1)/2]},
\tag{17.452}
$$

where

$$
\omega = \sqrt{2(\rho_0^2 - \rho_1^2)}
\tag{17.453}
$$

is the curvature of $V(\rho)$ close to ρ_0, i.e.,

$$
V(\rho) \approx \omega^2 (\rho - \rho_0)^2 + \dots .
\tag{17.454}
$$

The extra energy of the bubble solution is

$$
E_{\text{cl}} = \int_0^L dz \left[e(\rho_{\text{cl}}) - e_c(k) \right] = \frac{4}{3}\omega = \frac{4}{3}\sqrt{2(1 - 3k^2)}.
\tag{17.455}
$$

The explicit solution (17.452) reaches the point of smallest ρ at z_1, where its value is

$$
\rho_1 \equiv \rho(z_1) = \sqrt{2}k.
\tag{17.456}
$$

This value is still nonzero and does not yet permit a phase slip. However, we shall now demonstrate that quadratic fluctuations around the solution (17.456) do, in fact, to reduce the current. For this, we insert the fluctuating order field

$$
\rho(z) = \rho_{\text{cl}}(z) + \delta\rho(z)
\tag{17.457}
$$

into the free energy. With ρ_{cl} being extremal, the lowest variation of E is of second order in $\delta\rho(z)$

$$
\delta^2 E = \int_0^L dz \, \delta\rho(z)[-\partial_z^2 + V''(\rho)]\delta\rho(z).
\tag{17.458}
$$

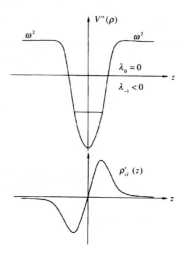

Figure 17.26 Infinitesimal translation of critical bubble yields antisymmetric wave function of zero energy ρ'_{cl} solving differential equation (17.509). Since this wave function has a node, there must be a negative-energy bound state.

This expression is not positive definite as can be verified by studying the eigenvalue problem

$$[-\partial_z^2 + V''(\rho_{cl})]\psi_n(z) = \left[-\partial_z^2 - 1 + 3\rho_{cl}^2 - 3\frac{j^4}{\rho_{cl}^4}\right]\psi_n(z) = \lambda_n\psi(z). \quad (17.459)$$

The potential $V''(\rho_{cl}(z))$ has asymptotically the value ω^2. When approaching $z = z_1$ from the right, it develops a minimum at a negative value (see Fig. 17.26). After that it goes again against ω^2. The energy eigenvalues λ_0 and λ_{-1} lie as indicated in the figure. The fact that there is precisely one negative eigenvalue λ_{-1} can be proved without an explicit solution by the same physical argument that was used to show the instability of the fluctuation problem (17.253): A small temporal translation of the classical solution corresponds to a wave function which has no energy and a zero implying the existence of precisely one lower wave function with $\lambda_{-1} < 0$ and no zero.

The negative eigenvalue makes the critical bubble solution unstable against contraction or expansion. The former makes the fluctuation return to the spiral classical solution (17.452) of Fig. 17.24, the second removes one unit from the winding number of the spiral and reduces the supercurrent. For the precise calculation of the decay rate, the reader is referred to the references quoted at the end of the chapter. Here we only give the final result which is [18]

$$\text{rate} = \text{const} \times L\,\omega(k)e^{-E_{cl}/k_BT}, \quad (17.460)$$

with the k-dependent prefactor

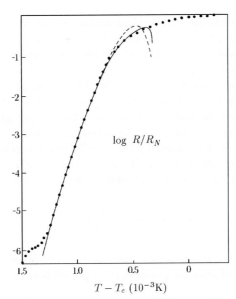

Figure 17.27 Logarithmic plot of resistance of thin superconducting wire as function of temperature at current $0.2\mu A$ in comparison with experimental data (vertical axis is normalized by the Ohmic resistance $R_n = 0.5\Omega$ measured at $T > T_c$, see papers quoted at end of chapter).

$$\omega(k) = 2|\lambda'_{-1}|\frac{(1-3k^2)^{7/4}}{(1-k^2)^{1/2}}\exp\left[-\frac{3\sqrt{2}k}{\sqrt{1-3k^2}}\arctan\left(\frac{\sqrt{1-3k^2}}{\sqrt{2}k}\right)\right], \qquad (17.461)$$

where

$$\lambda'_{-1} \equiv -\frac{1}{2}\left\{\left[(1+k^2)^2 + 3(1-3k^2)^2\right]^{1/2} - (1+k^2)\right\} < 0 \qquad (17.462)$$

is the negative eigenvalue of the fluctuations in the complex field $\psi(z)$ [which is not directly related to λ_{-1} of Eq. (17.459) and requires a separate discussion of the initial path integral (17.435)]. This complicated-looking expression has a simple quite accurate approximation which had previously been deduced from a numerical evaluation of the fluctuation determinant [19]:

$$\omega(k) \approx (1 - \sqrt{3}k)^{15/4}(1 + k^2/4). \qquad (17.463)$$

Both expressions vanish at the critical value $k = k_c = 1/\sqrt{3}$.

The resistance of a thin superconducting wire following from this calculation is compared with experimental data in Fig. 17.27.

17.12   Decay of Metastable Thermodynamic Phases

A generalization of this   decay mechanism can be found in the first-order phase transitions of many-particle systems. These possess some order parameter with an effective potential which has two minima corresponding to two different thermodynamic phases. Take, for instance, water near the boiling point. At the boiling temperature, the liquid and gas phases have the same energy. This situation corresponds to the symmetric potential. At a slightly higher temperature, the liquid phase is overheated and becomes metastable. The potential is now slightly asymmetric. The decay of the overheated phase proceeds by the formation of critical bubbles [3]. Their outside consists of the metastable water phase, their inside is filled with vapor lying close to the stable minimum of the potential. The radius of the critical bubble is determined by the equilibrium between the gain in volume energy and the cost in surface energy. If σ is the surface tension and ϵ the difference in energy density, the energy of bubble solution depends on the radius as follows:

$$E \propto \sigma\, 4\pi R^2 - \epsilon \frac{4\pi}{3} R^3. \tag{17.464}$$

A plot of this energy in Fig. 17.28 looks just like that of the action $\mathcal{A}(\xi)$ in Fig. 17.8. Thus the role of the deformation parameter ξ is played here by the bubble radius R.

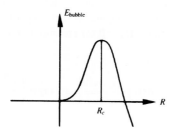

Figure 17.28 Bubble energy as function of its radius R.

At the critical bubble, the energy has a maximum. The fluctuations of the critical bubble must therefore have a negative eigenvalue. This negative eigenvalue mode accounts for the fact that the critical bubble is unstable against expansion and contraction. When expanding, the bubble transforms the entire liquid into the stable gas phase. When contracting, the bubble disappears and the liquid remains in the overheated phase. Only the first half of the fluctuations have to be counted when calculating the lifetime of the overheated phase.

It is instructive to take a comparative look at the instability of a critical bubble to see how the different spatial dimensions modify the properties of the solution. We shall discuss first the case of three space dimensions. As in the case of superconductivity, the description of the liquid-vapor phase transition makes use of a space-dependent order parameter, the real order field $\varphi(\mathbf{x})$. The two minima of the potential $V(\varphi)$ describe the two phases of the system. The kinetic term $\mathbf{x}'^2$ in the path integral is now a field gradient term $[(\partial_{\mathbf{x}}\varphi(\mathbf{x}))^2]$ which ensures finite correlations between neighboring field configurations. The Euclidean action controlling the fluctuations is therefore of the form

$$A[\varphi] = \int d^3x \left\{ \frac{1}{2}[\nabla \varphi(\mathbf{x})]^2 + V(\varphi) \right\}, \tag{17.465}$$

where $V(\varphi)$ is the same potential as in Eq. (17.1), but it is extended by the asymmetric energy (17.243). Within classical statistics, the thermal fluctuations are controlled by the path integral for the partition function

$$Z = \int \mathcal{D}\varphi(\mathbf{x}) e^{-A[\varphi]/T}. \tag{17.466}$$

Here T is the temperature measured in multiples of the Boltzmann constant k_B. The path integral $\int \mathcal{D}\varphi(\mathbf{x})$ is defined by cutting the three-dimensional space into small cubes of size ϵ and performing one field integration at each point.

The critical bubble extremizes the action. Assuming spherical symmetry, the bubble satisfies in D dimensions the classical Euler-Lagrange field equation

$$\left(-\frac{d^2}{dr^2} - \frac{D-1}{r}\frac{d}{dr} \right) \varphi_{\rm cl} + V'(\varphi_{\rm cl}(r)) = 0. \tag{17.467}$$

This differs from the equation (17.305) for the one-dimensional bubble solution by the extra gradient term $-[(D-1)/r]\partial_r\varphi_{\rm cl}(r)$. Such a term is an obstacle to an exact solution of the equation via the energy conservation law (17.306). The relevant qualitative properties of the solution can nevertheless be seen in a similar way as for the bubble solution. As in Fig. 17.11 we plot the reversed potential and imagine the solution $\varphi(r)$ to describe the motion of a mass point in this potential with $-r$ playing the role of a "time". Setting $\varphi_{\rm cl}(r) = x(-t)$, the field equation (17.467) takes the form

$$\ddot{x}(t) - \frac{D-1}{t}\dot{x}(t) - V'(x(t)) = 0. \tag{17.468}$$

In this notation, the second term, i.e., the term $-[(D-1)/r]\partial_r\varphi_{\rm cl}(r)$ in (17.467) plays the role of a negative "friction" accelerating motion of the particle along $x(t)$. This effect decreases with time like $1/t$. With our everyday experience of mechanical systems, the qualitative behavior of the solution can immediately be plotted qualitatively as shown in Fig. 17.29. For $D = 1$, the energy conservation makes the particle reach the right-hand zero of the potential. For $D > 1$, the "antifriction" makes the trajectory overshoot. At $r = 0$, the solution is closest to the stable minimum (the maximum of the reversed potential) on the left-hand side. In the superheated water system, this corresponds to the inside of the bubble being filled with vapor. As r moves outward in the bubble, the state moves closer to the metastable state, i.e., it becomes more and more liquid. The antifriction term has the effect that the point of departure on the left-hand side lies energetically below the final value of the metastable state.

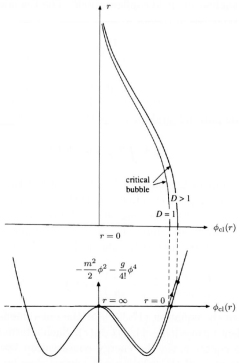

Figure 17.29 Qualitative behavior of critical bubble solution as function of its radius.

Consider now the fluctuations of such a critical bubble in $D = 3$ dimensions. Suppose that the field deviates from the solution of the field equation (17.467) by $\delta\varphi(\mathbf{x})$. The deviations satisfy the differential equation

$$\left[-\frac{d^2}{dr^2} - \frac{2}{r}\frac{d}{dr} + \frac{\hat{L}^2}{r^2} + V''(\varphi_{\text{cl}}(r))\right]\delta\varphi(\mathbf{x}) = \lambda\,\delta\varphi(\mathbf{x}), \qquad (17.469)$$

where $\hat{L}^2$ is the differential operator of orbital angular momentum (in units $\hbar = 1$). Taking advantage of rotational invariance, we expand $\delta\varphi(\mathbf{x})$ into eigenfunctions of angular momentum φ_{nlm}, the spherical harmonics $Y_{lm}(\hat{\mathbf{x}})$:

$$\phi(\mathbf{x}) = \sum_{nlm}\varphi_{nlm}(r)Y_{lm}(\hat{\mathbf{x}}). \qquad (17.470)$$

The coefficients φ_{nlm} satisfy the radial differential equation

$$\left[-\frac{d^2}{dr^2} - \frac{2}{r}\frac{d}{dr} + \frac{l(l+1)}{r^2} + V''(\varphi_{\text{cl}}(r))\right]\varphi_{nlm}(r) = \lambda_{nl}\varphi_{nlm}(r), \qquad (17.471)$$

with

$$V''(\varphi_{\text{cl}}(r)) = -\frac{\omega^2}{2} + \frac{3}{2}\frac{\omega^2}{a^2}\varphi_{\text{cl}}^2(r). \qquad (17.472)$$

One set of solutions is easily found, namely those associated with the translational motion of the classical solution. Indeed, if we take the bubble at the origin,

$$\varphi_{cl}(\mathbf{x}) = \varphi_{cl}(r), \tag{17.473}$$

to another place $\mathbf{x} + \mathbf{a}$, we find, to lowest order in $\mathbf{a}$,

$$\begin{aligned} \varphi_{cl}(\mathbf{x} + \mathbf{a}) &= \varphi_{cl}(\mathbf{x}) + \mathbf{a}\partial_{\mathbf{x}}\varphi_{cl}(\mathbf{x}) \\ &= \varphi_{cl}(r) + \mathbf{a}\hat{\mathbf{x}}\partial_r\varphi_{cl}(r). \end{aligned} \tag{17.474}$$

But $\hat{\mathbf{x}}$ is just the Cartesian way of writing the three components of the spherical harmonics $Y_{1m}(\hat{x})$. If we introduce the *spherical components* of a vector as follows

$$\begin{pmatrix} x_0 \\ x_1 \\ x_{-1} \end{pmatrix} \equiv \begin{pmatrix} x_3 \\ (x_1 + ix_2)/\sqrt{2} \\ -(x_1 - ix_2)/\sqrt{2} \end{pmatrix}, \tag{17.475}$$

we see that

$$\hat{x}_m = \sqrt{\frac{4\pi}{3}} Y_{1m}(\hat{\mathbf{x}}). \tag{17.476}$$

Thus, $\delta\varphi(\mathbf{x}) = \mathbf{a}\hat{\mathbf{x}}\partial_r\varphi(\mathbf{x})$ must be a solution of Eq. (17.469) with zero eigenvalue λ. This can easily be verified directly: The factor $\hat{\mathbf{x}}$ causes $\hat{L}^2$ to have the eigenvalue 2, and the accompanying radial derivative $\delta\varphi(\mathbf{x}) = \partial_r\varphi_{cl}(r)$ is a solution of Eq. (17.471) for $l = 1$ and $\lambda_{nl} = 0$, as is seen by differentiating the Euler-Lagrange equation (17.467) with respect to r. Choosing the principal quantum number of these translational modes to be $n = 1$, we assign the three components of $\hat{\mathbf{x}}\partial_r\varphi_{cl}(r)$ to represent the eigenmodes $\varphi_{1,1,m}$.

As long as the bubble radius is large compared to the thickness of the wall, which is of the order $1/\omega$, the $1/r^2$ -terms will be very small. There exists then an entire family of solutions $\varphi_{1lm}(\mathbf{x})$ with all possible values of l which all have approximately the same radial wave function $\partial_r\varphi_{cl}(r)$. Their eigenvalues are found by a perturbation expansion. The perturbation consists in the centrifugal barrier but with the $l = 1$ barrier subtracted since it is already contained in the derivative $\partial_r\varphi_{cl}(r)$, i.e.,

$$V_{\text{pert}} = [l(l+1) - 2]/2r^2. \tag{17.477}$$

The bound-state wave functions φ_{1lm} are normalizable and differ appreciably from zero only in the neighborhood of the bubble wall. To lowest approximation, the perturbation expansion produces therefore an energy

$$\lambda_{nl} \approx \frac{l(l+1) - 2}{r_c^2}, \tag{17.478}$$

where r_c is the radius of the critical bubble. As a consequence, the lowest $l = 0$ eigenstate has a negative energy

$$\lambda_{00} \approx -\frac{1}{r_c^2}. \tag{17.479}$$

Physically, this single $l = 0$ -mode corresponds to an infinitesimal radial vibration of the bubble. As already explained above it is not astonishing that a radial vibration has a negative eigenvalue. The critical bubble lies at a maximum of the action. Expansion or contraction is energetically favorable. Since $Y_{00}(\mathbf{x})$ is a constant, the wave function is proportional to $(d/dr)\varphi_{cl}(r)$ itself without an angular factor. This is seen directly by performing an infinitesimal radial contraction

$$\varphi_{cl}((1 - \epsilon)r) = \varphi_{cl}(r) - \epsilon r \partial_r \varphi_{cl}(r). \tag{17.480}$$

The variation $r\partial_r\varphi_{cl}(r)$ is almost zero except in the vicinity of the critical radius r_c, so that $r\partial_r\varphi_{cl}(r) \approx r_c\partial_r\varphi_{cl}(r)$ which is the above wave function. Being the ground state of the Schrödinger equation (17.469), it should be denoted by $\varphi_{000}(r)$. Since it solves approximately the Schrödinger equation (17.471) with $l = 1$, it also solves this equation approximately with $l = 0$ and the energy (17.479).

Finally let us point out that in $D > 1$ dimensions, the value of the negative eigenvalue can be calculated very simply from a phenomenological consideration of the bubble action. Since the inside of the bubble is very close to the true ground state of the system whose energy density lies lower than that of the metastable one by ϵ, the volume energy of a bubble of an arbitrary radius R is

$$E_V = -S_D \frac{R^D}{D} \epsilon, \tag{17.481}$$

where $S_D R^{D-1}$ is the surface of the bubble and $S_D R^D/D$ its volume. The surface energy can be parametrized as

$$E_S = S_D R^{D-1} \sigma, \tag{17.482}$$

where σ is a constant proportional to the surface tension. Adding the two terms and differentiating with respect to R, we obtain a critical bubble radius at

$$R = r_c = (D - 1)\sigma/\epsilon, \tag{17.483}$$

with a critical bubble energy

$$E_c = \frac{S_D}{D} R_c^{D-1} \sigma = \frac{S_D}{D(D - 1)} R_c^D \epsilon = \frac{S_D}{D}(D - 1)^{D-1} \frac{\sigma^D}{\epsilon^{D-1}}. \tag{17.484}$$

The second derivative with respect to the radius R is, at the critical radius,

$$\left.\frac{d^2 E}{dR^2}\right|_{R=r_c} = -DE_c \frac{D - 1}{r_c^2}. \tag{17.485}$$

Identifying the critical bubble energy E_c with the classical Euclidean action $\mathcal{A}_{\mathrm{cl}}$ we find the variation of the bubble action as

$$\delta^2 \mathcal{A}_{\mathrm{cl}} \approx -\frac{1}{2}(\delta R)^2 D \, \mathcal{A}_{\mathrm{cl}} \frac{D-1}{r_c^2}. \tag{17.486}$$

We now express the dilational variation of the bubble radius in terms of the normal coordinate of (17.470). The normalized wave function is obviously

$$\varphi_{000}(r) = \frac{\partial_r \varphi_{\mathrm{cl}}(r)}{\sqrt{\int d^D x (\partial_r \varphi_{\mathrm{cl}})^2}}. \tag{17.487}$$

But the expression under the square root is exactly D times the action of the critical bubble

$$\int d^D x (\partial_r \varphi_{\mathrm{cl}})^2 = D \mathcal{A}_{\mathrm{cl}}. \tag{17.488}$$

To prove this we introduce a scale factor s into the solution of the bubble and evaluate the action

$$
\begin{aligned}
\tilde{\mathcal{A}}_{\mathrm{cl}} &= \int d^D x \left\{ \frac{1}{2}([\partial_r \varphi_{\mathrm{cl}}(sr)]^2 + V(\varphi_{\mathrm{cl}}(sr)) \right\} \\
&= \frac{1}{s^D} \int d^D x \left\{ \frac{s^2}{2}[\partial_r \varphi_{\mathrm{cl}}(r)]^2 + V(\varphi_{\mathrm{cl}}(r)) \right\}.
\end{aligned} \tag{17.489}
$$

Since $\tilde{\mathcal{A}}_{\mathrm{cl}}$ is extremal at $s=1$, it has to satisfy

$$\left. \frac{\partial \tilde{\mathcal{A}}_{\mathrm{cl}}}{\partial s} \right|_{s=1} = 0, \tag{17.490}$$

or

$$\int d^D x \left\{ (D-2)\frac{1}{2}[\partial_r \varphi_{\mathrm{cl}}]^2 + DV(\varphi_{\mathrm{cl}}(r)) \right\} = 0. \tag{17.491}$$

Hence

$$\int d^D x V(\varphi_{\mathrm{cl}}(r)) = -\frac{D-2}{D} \int d^D x \frac{1}{2}[\partial_r \varphi_{\mathrm{cl}}(r)]^2, \tag{17.492}$$

implying that

$$
\begin{aligned}
\mathcal{A}_{\mathrm{cl}} &= \left(\frac{1}{2} - \frac{D-2}{2D} \right) \int d^D x [\partial_r \varphi_{\mathrm{cl}}(r)]^2 \\
&= \frac{1}{D} \int d^D x [\partial_r \varphi_{\mathrm{cl}}(r)]^2.
\end{aligned} \tag{17.493}
$$

With (17.487), the φ_{000} contribution to $\delta\varphi(\mathbf{x})$ reads

$$\delta\varphi(\mathbf{x}) = \xi_{000}\varphi_{000}(r) = \xi_{000}\frac{\partial_r \varphi}{\sqrt{D\mathcal{A}_{\mathrm{cl}}}}, \tag{17.494}$$

and we arrive at

$$\delta R = \frac{\xi_{000}}{\sqrt{D\mathcal{A}_{\rm cl}}}. \tag{17.495}$$

Inserting this into (17.486) shows that the second variation of the Euclidean action $\delta^2 \mathcal{A}_{\rm cl}$ can be written in terms of the normal coordinates associated with the normalized fluctuation wave function φ_{000} as

$$\delta^2 \mathcal{A}_{\rm cl} = -\xi_{000}^2 \frac{D-1}{2r_c^2}. \tag{17.496}$$

From this relation, we read off the negative eigenvalue

$$\lambda_{00} = -\frac{D-1}{2r_c^2}. \tag{17.497}$$

For $D = 3$, this is in agreement with the $D = 3$ value (17.479). For general D, the eigenvalue corresponding to (17.479) would have been derived with the arguments employed there from the derivative term $-[(D-1)/r]d/dr$ in the Lagrangian and would also have resulted in (17.497).

All other multipole modes φ_{nlm} have a positive energy. Close to the bubble wall (as compared with the radius), the classical solutions $(1/r)\varphi_{nlm}(r)$ can be taken approximately from the solvable one-dimensional equation

$$\left[-\frac{1}{2}\frac{d^2}{dr^2} + \frac{\omega^2}{2}\left(1 - \frac{3}{2}\frac{1}{\cosh^2[\omega(r-r_c)/2]} \right) \right] \left(\frac{1}{r}\varphi_{nlm} \right) \approx \tilde{\lambda}_n \left(\frac{1}{r}\varphi_{nlm} \right). \tag{17.498}$$

The wave functions with $n = 0$ are

$$\varphi_{0lm} \approx \sqrt{\frac{3\omega}{8}} \frac{1}{\cosh^2[\omega(r-r_c)/2]}, \tag{17.499}$$

and have the eigenvalues

$$\lambda_{0l} \approx \frac{l(l+1)-2}{2r_c^2}. \tag{17.500}$$

The $n = 1$ -bound states are

$$\varphi_{1lm} \approx \sqrt{\frac{3\omega}{4}} \frac{\sinh[\omega(r-r_c)/2]}{\cosh^2[\omega(r-r_c)/2]}, \tag{17.501}$$

with eigenvalues

$$\lambda_{1l} \approx \frac{3}{8}\omega^2 + \frac{l(l+2)-2}{2r_c^2}. \tag{17.502}$$

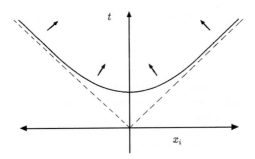

Figure 17.30 Decay of metastable false vacuum in Minkowski space. It proceeds as a shock wave which after some time traverses the world almost with light velocity, converting the false into the true vacuum.

17.13 Decay of Metastable Vacuum State in Quantum Field Theory

The theory of decay presented in the last section has an interesting quantum field-theoretic application. Consider a metastable scalar field system in a D-dimensional Euclidean spacetime at temperature zero. At a fixed time, there will be a certain average number of bubbles, regulated by the "quantum Boltzmann factor" $\exp(-\mathcal{A}_{\mathrm{cl}}/\hbar)$. If the bubble gas is sufficiently dilute (i.e., if the distances between bubbles is much larger than the radii), each bubble is described quite accurately by the classical solution. In Minkowski space, a Euclidean radius $r = \sqrt{\mathbf{x}^2 + c^2\tau^2}$ corresponds to $r = \sqrt{\mathbf{x}^2 - c^2t^2}$, where c is the light velocity. The critical bubble has therefore the spacetime behavior

$$\varphi_{\mathrm{cl}}(\mathbf{x}, t) = \varphi_{\mathrm{cl}}(r = \sqrt{\mathbf{x}^2 - c^2t^2}). \tag{17.503}$$

From the above discussion in Euclidean space we know that φ will be equal to the metastable false vacuum in the outer region $r > r_c$, i.e., for

$$\mathbf{x}^2 - c^2t^2 > r_c^2. \tag{17.504}$$

The inside region

$$\mathbf{x}^2 - c^2t^2 < r_c^2 \tag{17.505}$$

contains the true vacuum state with the lower energy. Thus a critical bubble in spacetime has the hyperbolic structure drawn in Fig. 17.30. Therefore, the Euclidean critical bubble describes in Minkowski space the growth of a bubble as a function of time. The bubble starts life at some time $t = r_c/c$ and expands almost instantly to a radius of order r_c. The position of the shock wave is described by

$$\mathbf{x}^2 - c^2t^2 = r_c^2. \tag{17.506}$$

This implies that a shock wave that runs through space with a velocity

$$v = \frac{|\mathbf{x}|}{t} = \frac{c}{\sqrt{1 - r_c^2/c^2 t^2}} \tag{17.507}$$

and converts the metastable into the stable vacuum — a global catastrophe. A Euclidean bubble centered at another place $\mathbf{x}_b, \tau_b$ would correspond to the same process starting at $\mathbf{x}_b$ and a time

$$t_b = r_c/c + \tau_b. \tag{17.508}$$

A finite time after the creation of a bubble, of the order r_c/c, the velocity of the shock wave approaches the speed of light (in many-body systems the speed of sound). Thus, we would hardly be able to see precursors of such a catastrophe warning us ahead of time. We would be annihilated with the present universe before we could even notice.

17.14 Crossover from Quantum Tunneling to Thermally Driven Decay

For completeness, we discuss here the difference between a decay caused by a quantum-mechanical tunneling process at $T = 0$, and a pure thermally driven decay at large temperatures. Consider a one-dimensional system possessing, at some place x_*, a high potential barrier, much higher than the thermal energy $k_B T$, with a shape similar to Fig. 17.10. Let the well to the left of the barrier be filled with a grand-canonical ensemble of noninteracting particles of mass M in a nearly perfect equilibrium. Their distribution of momenta and positions in phase space is governed by the Boltzmann factor $e^{-\beta[p^2/2M+V(x)]}$. The rate, at which the particles escape across the barrier, is given by the classical statistical integral

$$\Gamma_{\rm cl} = Z_{\rm cl}^{-1} \int dx \int \frac{dp}{2\pi\hbar} e^{-\beta[p^2/2M+V(x)]} \delta(x - x_*) \frac{p}{M} \Theta(p), \tag{17.509}$$

where $Z_{\rm cl}$ is the classical partition function

$$Z_{\rm cl} = \int dx \int \frac{dp}{2\pi\hbar} e^{-\beta[p^2/2M+V(x)]}. \tag{17.510}$$

The step function $\Theta(p)$ selects the particles running to the right across the top of the potential barrier. Performing the phase space path integral in (17.509) yields

$$\Gamma_{\rm cl} = \frac{Z_{\rm cl}^{-1}}{2\pi\hbar\beta} e^{-V(x_*)}. \tag{17.511}$$

If the metastable minimum of the potential is smooth, $V(x)$ can be replaced approximately in the neighborhood of x_0 by the harmonic expression

$$V(x) \approx \frac{M}{2} \omega_0^2 (x - x_0)^2. \tag{17.512}$$

The classical partition function is then given approximately by

$$Z_{cl} \approx \frac{1}{\hbar \beta \omega_0},$$
(17.513)

and the decay rate follows the simple formula

$$\Gamma_{cl} \approx \frac{\omega_0}{2\pi} e^{-\beta V(x_*)}.$$
(17.514)

Let us compare this result with the decay rate due to pure quantum tunneling. In the limit of small temperatures, the decay proceeds from the ground state, and the partition function is approximately equal to

$$Z \approx e^{-\beta(E^{(0)} - i\hbar\Gamma/2)}.$$
(17.515)

The decay rate is given by the small imaginary part of the partition function:

$$\Gamma \xrightarrow[T \to 0]{} \frac{2}{\hbar\beta} \frac{\operatorname{Im} Z}{\operatorname{Re} Z}.$$
(17.516)

In contrast to this, the thermal rate formula (17.511) implies for the high-temperature regime, where Γ becomes equal to Γ_{cl}, the relation:

$$\Gamma \xrightarrow[T \to \infty]{} \frac{\omega_*}{\pi} \frac{\operatorname{Im} Z_{cl}}{\operatorname{Re} Z_{cl}}.$$
(17.517)

The frequency ω_* is determined by the curvature of the potential at the top of the barrier, where it behaves like

$$V(x) \approx -\frac{M}{2}\omega_*^2(x - x_*)^2.$$
(17.518)

The relation (17.517) follows immediately by calculating in the integral (17.510) the contribution of the neighborhood of the top of the barrier in the saddle point approximation. As in the integral, this is done (17.261) by rotating the contour of integration which starts at $x = x_*$ into the upper complex half-plane. Writing $x = x_* + iy$ this leads to the following integral:

$$\operatorname{Im} Z_{cl} \approx \int_0^\infty \frac{dy}{\sqrt{2\pi\hbar^2\beta/M}} e^{-\beta[V(x_*) + \frac{M}{2}\omega_*^2 y^2]} \approx \frac{1}{2\hbar\beta\omega_*} e^{-\beta V(x_*)}.$$
(17.519)

Since the real part is given by (17.513) we find the ratio

$$\frac{\operatorname{Im} Z_{cl}}{\operatorname{Re} Z_{cl}} \approx \frac{\omega_0}{2\omega_*} e^{-\beta V(x_*)},$$
(17.520)

so that (17.511) is equivalent to (17.517).

The two formulas (17.516) and (17.517) are derived for the two extreme regimes $T \gg T_0$ and $T \ll T_0$, respectively, where T_0 denotes the characteristic temperature associated with the curvature of the potential at the metastable minimum $T_0 = \hbar\omega_0/k_B$. Numerical studies have shown that the applicability extends into the close neighborhood of T_0 on each side of the temperature axis. The crossover regime is quite small, of the order $\mathcal{O}(\hbar^{3/2})$.

Note that given the knowledge of the imaginary parts of all excited states, which can be obtained as in Section 17.9, it will be possible to calculate the average lifetime of a metastable state at all temperatures without the restrictions of the semiclassical approximation. This remains to be done.

Appendix 17A   Feynman Integrals for Fluctuation Correction

For the integral (17.219) we obtain with (17.237) immediately the result stated in Eq. (17.240)

$$I_1' = \frac{97}{560}. \tag{17A.1}$$

To calculate the remaining three double integrals I_{21}, I_{22}, I_3 in (17.223) and (17.224) we observe that because of the symmetry of the Green function $G_{\mathcal{O}_\omega}(\tau, \tau')$ in τ, τ' the measure of integration can be rewritten as $2\int_{-\infty}^{\infty} d\tau \int_{-\infty}^{t} d\tau'$. We further introduce the dimensionless classical functions

$$\tilde{x}_{\rm cl}(\tau) \equiv \sqrt{\frac{g}{2\omega^2}}x_{\rm cl}(\tau) = \tanh(\tau/2),$$

$$\tilde{y}_0(\tau) \equiv \sqrt{\frac{8}{3\omega}}y_0(\tau) = \frac{1}{\cosh^2(\omega\tau/2)}, \tag{17A.2}$$

use natural units with $\omega = 1, g = 1$, since these quantities cancel in all integrals, and define

$$x_{\rm G}(\tau) \equiv \tilde{x}_{\rm cl}(\tau)G_{\mathcal{O}_\omega}(\tau, \tau)$$

$$x_{\rm KG}(\tau) \equiv \left[\int_{-\infty}^{\tau} G_{\mathcal{O}_\omega}(\tau, \tau')x_{\rm G}(\tau')d\tau'\right]_A$$

$$x_{\rm K3G}(\tau) \equiv \left[\int_{-\infty}^{\tau} G_{\mathcal{O}_\omega}^3(\tau, \tau')x_{\rm G}(\tau')dt'\right]_A, \tag{17A.3}$$

where the subscript A denotes the antisymmetric part in τ. Because of the antisymmetry of $x_{\rm cl}$, the symmetric part gives no contribution to the integrals which can be written as

$$I_{21} = 6\int_0^\infty d\tau\, \tilde{x}_{\rm cl}(\tau)x_{\rm K3G}(\tau), \tag{17A.4}$$

$$I_{22} = 9\int_0^\infty d\tau\, x_{\rm G}(\tau)x_{\rm KG}(\tau), \tag{17A.5}$$

$$I_3 = 24\int_0^\infty d\tau\, y_0'(\tau)x_{\rm KG}(\tau). \tag{17A.6}$$

When evaluating the integrals in (17A.3) the antisymmetry of $y_0(\tau)$ is useful. We easily find

$$x_{\rm KG}(\tau) = \frac{1}{4}\frac{12\tau + \tanh(\tau/2)}{12\cosh^2(\tau/2)}. \tag{17A.7}$$

Inserting this into I_{22} and I_3 we encounter integrals of two types

$$\int_0^\infty d\tau\, \sinh^m(\tau/2)/\cosh^n(\tau/2) \quad \text{and} \quad \int_0^\infty \tau \sinh^m(\tau/2)/\cosh^n(\tau/2).$$

The former can be performed with the help of formula (17.54), the latter require integrations by parts of the type

$$\int_0^\infty f(\tau/2)\tanh\frac{\tau}{2}\,d\tau = -\int_0^\infty f'(\tau/2)\ln(2\,\cosh\frac{\tau}{2})\,d\tau, \tag{17A.8}$$

which lead to a finite sum of integrals of the first type plus integrals of the type[11]

$$\int_0^\infty \log\cosh(\tau/2)\sinh^m(\tau/2)/\cosh^n(\tau/2).$$

These, in turn, are equal to $-\partial_\nu \int_0^\infty d\tau\,\sinh^m(\tau/2)/\cosh^{\nu+1}(\tau/2)$ so that it is evaluated again via (17.54). After performing the subtraction in I_{22} we obtain the values given in Eq. (17.240).

The evaluation of the integral I_{21} is more tedious since we must integrate over the third power of the Green function, which is itself a lengthy function of τ. It is once more useful to exploit the symmetry properties of the integrand. We introduce the abbreviations

$$f_n^{S/A}(\tau) \equiv \frac{1}{2}[H_R^n(\tau) \pm H_R^n(-\tau)]y_0^{3-n}(\tau),$$

$$F_n^{S/A}(\tau) \equiv \int_0^\tau f_n^{S/A}(\tau')\tilde{x}_{cl}(\tau')d\tau', \tag{17A.9}$$

$$N_n \equiv \int_0^\infty H_R^n(\tau)y_0^{3-n}(\tau')\tilde{x}_{cl}(\tau')d\tau',$$

and find x_{K3G} in the form

$$x_{K3G}(\tau) = f_3^A(\tau)[N_0 - F_0^S(\tau)] + 3f_2^A(\tau)[N_1 - F_1^S(\tau)] + 3f_1^A(\tau)[N_2 - F_2^S(\tau)]$$
$$+ 3f_2^S(\tau)F_1^A(\tau) + 3f_1^S(\tau)F_2^A(\tau) + f_0^S(\tau)F_3^A(\tau), \tag{17A.10}$$

with

$$N_0 = \frac{3}{32}, \quad N_1 = -\frac{7}{128}, \quad N_2 = \frac{203}{512} - \frac{\log 2}{2}. \tag{17A.11}$$

Explicitly:

$$x_{K3G}(\tau) = \frac{\mathrm{sech}^7\frac{\tau}{2}}{3\cdot 2^9}\left[3\tau\left(58\cosh\frac{\tau}{2} - 27\cosh\frac{3\tau}{2} - 3\cosh\frac{5\tau}{2}\right)\right.$$
$$-\left(753\sinh\frac{\tau}{2} + 48\sinh\frac{3\tau}{2} + 22\sinh\frac{5\tau}{2} + \sinh\frac{7\tau}{2}\right)$$
$$+ 36\cosh\frac{\tau}{2}\ln(2\cosh\frac{\tau}{2})\,(6\tau + 8\sinh\tau + \sinh 2\tau)$$
$$\left. - 108\cosh\frac{\tau}{2}\left(\int_0^\tau d\tau'\,\tau'\tanh\frac{\tau'}{2}\right)\right]. \tag{17A.12}$$

The integrals to be performed are of the same types as before, except for the one involving the last term which requires one further partial integration:

$$-\int_0^\infty d\tau\,\tilde{x}_{cl}(\tau)\mathrm{sech}^6\frac{\tau}{2}\int_0^\tau d\tau'\,\tau'\tanh\frac{\tau'}{2} = \frac{1}{3}\int_0^\infty d\tau\,\frac{d}{d\tau}\left[\mathrm{sech}^6\frac{\tau}{2}\right]\int_0^\tau d\tau'\,\tau'\tanh\frac{\tau'}{2}$$
$$= -\frac{1}{3}\int_0^\infty d\tau\,\mathrm{sech}^6\frac{\tau}{2}\,\tau\tanh\frac{\tau}{2} = -\frac{16}{135}. \tag{17A.13}$$

---

[11]I.S. Gradshteyn and I.M. Ryzhik, *op. cit.*, Formulas 2.417.

After the necessary subtraction of the divergent term we find the value of I_{21} given in Eq. (17.240).

The final result for the first coefficient of the Taylor expansion of the subtracted fluctuation factor C' in Eq. (17.222) is therefore

$$c_1 = \frac{71}{24} \approx 2.958. \tag{17A.14}$$

This number was calculated in Ref. [12] by solving the Schrödinger equation [20].

With the help of the WKB approximation he derived a recursion relation for the higher coefficients c_k of the expansion of C':

$$C' = \left[1 - c_1 \frac{g\hbar}{\omega^3} - c_2 \left(\frac{g\hbar}{\omega^3} \right)^2 - c_3 \left(\frac{g\hbar}{\omega^3} \right)^3 + \dots \right]. \tag{17A.15}$$

From this he calculated the next nine coefficients

$$c_2 = \frac{315}{32} \approx 9.84376, \quad c_3 = \frac{65953}{1152} \approx 57.2509. \tag{17A.16}$$

Their large behavior is

$$c_k \sim \frac{9}{\pi} \left(\frac{3}{2} \right)^k k! [\ln(6k) + \gamma]. \tag{17A.17}$$

Notes and References

The path integral theory of tunneling was first discussed by
A.I. Vainshtein, *Decaying Systems and the Divergence of Perturbation Series*, Novosibirsk Report (1964), in Russian (unpublished),
and in the context of the nucleation of first-order phase transitions by
J.S. Langer, Ann. Phys. **41**, 108 (1967).
The subject was studied further in the field-theoretic literature:

M.B. Voloshin, I.Y. Kobzarev, L.B. Okun, Yad. Fiz. **20**. 1229 (1974); Sov. J. Nucl. Phys. **20**, 644 (1975);
R. Rajaraman, Phys. Rep. **21**, 227 (1975),
R. Rajaraman, Phys. Rep. **21**, 227 (1975),
P. Frampton, Phys. Rev. Lett. **37**, 1378 (1976) and Phys. Rev. D **15**, 2922 (1977),
S. Coleman, Phys. Rev. D **15**, 2929 (1977); also in *The Whys of Subnuclear Physics*, Erice Lectures 1977, Plenum Press, 1979, ed. by A. Zichichi,
I. Affleck, Phys. Rev. Lett. **46**, 388 (1981).

For a finite-temperature discussion see
L. Dolan and J. Kiskies, Phys. Rev. D **20**, 505-513 (1979).

For multidimensional tunneling processes see
H. Kleinert and R. Kaul, J. Low Temp. Phys. **38**, 539 (1979) (http://www.physik.fu-berlin.de/~kleinert/66),
A. Auerbach and S. Kivelson, Nucl. Phys. B **257**, 799 (1985).

The fact that tunneling calculations can be used to derive the growth behavior of large-order perturbation coefficients was first noticed by
A.I. Vainshtein, Novosibirsk Preprint 1964, unpublished.
A different but closely related way of deriving this behavior was proposed by
L.N. Lipatov, JETP Lett. **24**, 157 (1976); **25**, 104 (1977); **44**, 216 (1977); **45**, 216 (1977).

A review on this subject is given in
J. Zinn-Justin, Phys. Rep. **49**, 205 (1979), and in *Recent Advances in Field Theory and Statistical Mechanics*, Les Houches Lectures 1982, Elsevier Science 1984, ed. by J.-B. Zuber and R. Stora.
J. Zinn-Justin, *Quantum Field Theory and Critical Phenomena*, Clarendon, Oxford, 1990.

The important applications to the ϵ-expansion of critical exponents in O(N)-symmetric φ^4 field theories were made by
E. Brezin, J.C. Le Guillou, and J. Zinn-Justin, Phys. Rev. D **15**, 1544, 1558 (1977).
E. Brezin and G. Parisi, J. Stat. Phys. **19**, 269 (1978).
The perturbation expansion of the φ^4-theory up to fifth-order was calculated in Ref. [13].
For two quartic interactions of different symmetries (one O(N)-symmetric, the other of cubic symmetry) see:
H. Kleinert and V. Schulte-Frohlinde, Phys. Lett. B **342**, 284 (1995) (cond-mat/9503038)
A detailed discussion and a comprehensive list of the references to the original papers are contained in the textbook [15].

The calculation of the anomalous magnetic moments in quantum electrodynamics is described in
T. Kinoshita and W.B. Lindquist, Phys. Rev. D **27**, 853 (1983).
M.J. Levine and R. Roskies, in *Proceedings of the Second International Conference on Precision Measurements and Fundamental Constants*, ed. by B.N. Taylor and W.D. Phillips, Natl. Bur. Std. US, Spec. Publ. **617** (1981).

The analytic result for the semiclassical decay rate of a supercurrent in a thin wire was found by
I.H. Duru, H. Kleinert, and N. Ünal, J. Low Temp. Phys. **42**, 137 (1981) (*ibid.*http/74),
H. Kleinert and T. Sauer, J. Low Temp. Physics **81**, 123 (1990) (*ibid.*http/204).
The experimental situation is explained in
M. Tinkham, *Introduction to Superconductivity*, McGraw-Hill, New York, 1975.
See, in particular, Chapter 7, Sections 7.1–7.3.
For quantum corrections to the decay rate see
N. Giordano, Phys. Rev. Lett. **61**, 2137 (1988);
N. Giordano and E.R. Schuler, Phys. Rev. Lett. **63**, 2417 (1989).

For thermally driven tunneling processes see the review article
P. Hänggi, P. Talkner, and M. Borkovec, Rev. Mod. Phys. **62**, 251 (1990).
Tunneling processes with dissipation were first discussed at $T = 0$ by
A.O. Caldeira and A.J. Leggett, Ann. Phys. **149**, 374 (1983), **153**, 445 (1973) (Erratum)
and for $T \neq 0$ by
H. Grabert, U. Weiss and P. Hänggi, Phys. Rev. Lett. **52**, 2193 (1984),
A.I. Larkin and Y.N. Ovchinnikov, Sov. Phys. JETP **59**, 420 (1984).
Papers on coherent tunneling:
A.J. Leggett, S. Chakravarty, A.T. Dorsey, M.P.A. Fisher, A. Garg, and W. Zwerger, Rev. Mod. Phys. **59**, 1 (1987),
U. Weiss, H. Grabert, P. Hänggi, and P. Riseborough, Phys. Rev. B **35**, 9535 (1987),
H. Grabert, P. Olschowski, and U. Weiss, Phys. Rev. B **36**, 1931 (1987),
On the use of periodic orbits see:
P. Hänggi and W. Hontscha, Ber. Bunsen-Ges. Phys. Chemie **95**, 379 (1991).

The individual citations refer to

[1] See also the solutions in textbooks such as
L.D. Landau and E.M. Lifshitz, *Quantum Mechanics*, Pergamon, London, 1965, §25, PROBLEM 4.

[2] L.D. Faddeev and V.N. Popov, Phys. Lett. B **25**, 29 (1967).

[3] J.S. Langer, Ann. Phys. **41**, 108 (1967).

[4] For a general proof see
S. Coleman, Nucl. Phys. B **298**, 178 (1988).

[5] F.J. Dyson, Phys. Rev. **85**, 631 (1952).

[6] Compare the result in Eq. (17.345) with
J.C. Collins and D.E. Soper, Ann. Phys. **112**, 209 (1978).

[7] The variational approach to tunneling was initiated in
H. Kleinert, Phys. Lett. B **300**, 261 (1993).
It has led to very precise tunneling rates in subsequent work by
R. Karrlein and H. Kleinert, Phys. Lett. A **187**, 133 (1994);
H. Kleinert and I. Mustapic, Int. J. Mod. Phys. A **11**, 4383 (1995), most precisely in Ref. [11].

[8] H. Kleinert, Phys. Lett. B **300**, 261 (1993) (*ibid*.http/214).

[9] R. Karrlein and H. Kleinert, Phys. Lett. A **187**, 133 (1994) (hep-th/9504048).

[10] C.M. Bender and T.T. Wu, Phys. Rev. D **7**, 1620 (1973), Eq. (5.22).

[11] B. Hamprecht and H. Kleinert, *Tunneling Amplitudes by Perturbation Theory*, Phys. Lett. B **564**, 111 (2003) (hep-th/0302124).

[12] J. Zinn-Justin, J. Math. Phys. **22**, 511 (1981); Table III.

[13] H. Kleinert, J. Neu, V. Schulte-Frohlinde, K.G. Chetyrkin, and S.A. Larin, Phys. Lett. B **272**, 39 (1991) (hep-th/9503230).

[14] H. Kleinert, Phys. Rev. D **57**, 2264 (1998); Addendum: Phys. Rev. D **58**, 107702 (1998) (cond-mat/9803268).

[15] For details and applications to see the textbook
H. Kleinert and V. Schulte-Frohlinde, *Critical Phenomena in ϕ^4-Theory*, World Scientific, Singapore, 2001 (*ibid*.http/b8).

[16] The path integral treatment of the decay rate of a supercurrent in a thin wire was initiated by
J.S. Langer and V. Ambegaokar, Phys. Rev. **164**, 498 (1967),
D.E. McCumber and B.I. Halperin, Phys. Rev. B **1**, 1054 (1970).

[17] M. Tinkham, *Introduction to Superconductivity*, McGraw-Hill, New York, 1975.

[18] H. Kleinert and T. Sauer, J. Low Temp. Physics **81**, 123 (1990) (*ibid*.http/204).

[19] D.E. McCumber and B.I. Halperin Phys. Rev. B **1**, 1054 (1970).

[20] Divide the numbers in the Table IV of Ref. [12] by $6n4^n$ to get our c_n.

18

Nonequilibrium Quantum Statistics

Quantum statistics described by the theoretical tools of the previous chapters is quite limited. The physical system under consideration must be in thermodynamic equilibrium, with a constant temperature enforced by a thermal reservoir. In this situation, partition function and the density matrix can be calculated from an analytic continuation of quantum-mechanical time evolution amplitudes to an imaginary time $t_b - t_a = -i\hbar/k_B T$. In this chapter we want to go beyond such equilibrium physics and extend the path integral formalism to nonequilibrium time-dependent phenomena. The tunneling processes discussed in Chapter 17 belong really to this class of phenomena, and their full understanding requires the theoretical framework of this chapter. In the earlier treatment this was circumvented by addressing only certain quasi-equilibrium questions. These were answered by applying the equilibrium formalism to the quantum system at positive coupling constant, which guaranteed perfect equilibrium, and extending the results to the quasi-equilibrium situation analytic continuation to small negative coupling constants.

Before we can set up a path integral formulation capable of dealing with true nonequilibrium phenomena, some preparatory work is useful based on the traditional tools of operator quantum mechanics.

18.1 Linear Response and Time-Dependent Green Functions for $T \neq 0$

If the deviations of a quantum system from thermal equilibrium are small, the easiest description of nonequilibrium phenomena proceeds via the *theory of linear response*. In operator quantum mechanics, this theory is introduced as follows. First, the system is assumed to have a time-independent Hamiltonian operator $\hat{H}$. The ground state is determined by the Schrödinger equation, evolving as a function of time according to the equation

$$|\Psi_S(t)\rangle = e^{-i\hat{H}t}|\Psi_S(0)\rangle \tag{18.1}$$

(in natural units with $\hbar = 1$, $k_B = 1$). The subscript S denotes the Schrödinger picture.

Next, the system is slightly disturbed by adding to $\hat{H}$ a time-dependent external interaction,

$$\hat{H} \to \hat{H} + \hat{H}^{\text{ext}}(t), \tag{18.2}$$

where $\hat{H}^{\text{ext}}(t)$ is assumed to set in at some time t_0, i.e., $\hat{H}^{\text{ext}}(t)$ vanishes identically for $t < t_0$. The disturbed Schrödinger ground state has the time dependence

$$|\Psi_S^{\text{dist}}(t)\rangle = e^{-i\hat{H}t}\hat{U}_H(t)|\Psi_S(0)\rangle, \tag{18.3}$$

where $\hat{U}_H(t)$ is the time translation operator in the Heisenberg picture. It satisfies the equation of motion

$$i\dot{\hat{U}}_H(t) = \hat{H}_H^{\text{ext}}(t)\hat{U}_H(t), \tag{18.4}$$

with[1]

$$\hat{H}_H^{\text{ext}}(t) \equiv e^{i\hat{H}t}\hat{H}^{\text{ext}}(t)e^{-i\hat{H}t}. \tag{18.5}$$

To lowest-order perturbation theory, the operator $\hat{U}_H(t)$ is given by

$$\hat{U}_H(t) = 1 - i \int_{t_0}^{t} dt' \, \hat{H}_H^{\text{ext}}(t') + \cdots . \tag{18.6}$$

In the sequel, we shall assume the onset of the disturbance to lie at $t_0 = -\infty$. Consider an arbitrary time-independent Schrödinger observable $\hat{O}$ whose Heisenberg representation has the time dependence

$$\hat{O}_H(t) = e^{i\hat{H}t}\hat{O}e^{-i\hat{H}t}. \tag{18.7}$$

Its time-dependent expectation value in the disturbed state $|\Psi_S^{\text{dist}}(t)\rangle$ is given by

$$\begin{aligned}
\langle\Psi_S^{\text{dist}}(t)|\hat{O}|\Psi_S^{\text{dist}}(t)\rangle &= \langle\Psi_S(0)|\hat{U}_H^\dagger(t)e^{i\hat{H}t}\hat{O}e^{-i\hat{H}t}\hat{U}_H(t)|\Psi_S(0)\rangle \\
&\approx \langle\Psi_S(0)|\left(1 + i\int_{-\infty}^{t} dt' \, \hat{H}_H^{\text{ext}}(t') + \ldots\right)\hat{O}_H(t) \\
&\quad \times \left(1 - i\int_{-\infty}^{t} dt' \, \hat{H}_H^{\text{ext}}(t') + \ldots\right)|\Psi_S(0)\rangle \\
&= \langle\Psi_H|\hat{O}_H(t)|\Psi_H\rangle - i\langle\Psi_H|\int_{-\infty}^{t} dt' \left[\hat{O}_H(t), \hat{H}_H^{\text{ext}}(t')\right]|\Psi_H\rangle + \ldots .
\end{aligned} \tag{18.8}$$

We have identified the time-independent Heisenberg state with the time-dependent Schrödinger state at zero time in the usual manner, i.e., $|\Psi_H\rangle \equiv |\Psi_S(0)\rangle$. Thus the expectation value of $\hat{O}$ deviates from equilibrium by

$$\begin{aligned}
\delta\langle\Psi_S(t)|\hat{O}|\Psi_S(t)\rangle &\equiv \langle\Psi_S^{\text{dist}}(t)|\hat{O}(t)|\Psi_S^{\text{dist}}(t)\rangle - \langle\Psi_S(t)|\hat{O}(t)|\Psi_S(t)\rangle \\
&= -i\int_{-\infty}^{t} dt' \, \langle\Psi_H|\left[\hat{O}_H(t), \hat{H}_H^{\text{ext}}(t')\right]|\Psi_H\rangle.
\end{aligned} \tag{18.9}$$

---

[1]Note that after the replacements $H \to H_0$, $H_H^{\text{ext}} \to H_I^{\text{int}}$, Eq. (18.4) coincides with the equation for the time evolution operator in the interaction picture to appear in Section 18.7. In contrast to that section, however, the present interaction is a nonpermanent artifact to be set equal to zero at the end, and H is the complicated total Hamiltonian, not a simple free one. This is why we do not speak of an interaction picture here.

If the left-hand side is transformed into the Heisenberg picture, it becomes

$$\delta\langle\Psi_S(t)|\hat{O}|\Psi_S(t)\rangle = \delta\langle\Psi_H|\hat{O}_H(t)|\Psi_H\rangle = \langle\Psi_H|\delta\hat{O}_H(t)|\Psi_H\rangle,$$

so that Eq. (18.9) takes the form

$$\langle\Psi_H|\delta\hat{O}_H(t)|\Psi_H\rangle = -i\int_{-\infty}^{t}dt'\,\langle\Psi_H|\left[\hat{O}_H(t),\hat{H}_H^{\text{ext}}(t')\right]|\Psi_H\rangle. \qquad (18.10)$$

It is useful to use the retarded Green function of the operators $\hat{O}_H(t)$ and $\hat{H}_H(t')$ in the state $|\Psi_H\rangle$ [compare (3.40)]:

$$G_{OH}^R(t,t') \equiv \Theta(t-t')\langle\Psi_H|\left[\hat{O}_H(t),\hat{H}_H(t')\right]|\Psi_H\rangle. \qquad (18.11)$$

Then the deviation from equilibrium is given by the integral

$$\langle\Psi_H|\delta\hat{O}_H(t)|\Psi_H\rangle = -i\int_{-\infty}^{\infty}dt'\,G_{OH}^R(t,t'). \qquad (18.12)$$

Suppose now that the observable $\hat{O}_H(t)$ is capable of undergoing oscillations. Then an external disturbance coupled to $\hat{O}_H(t)$ will in general excite these oscillations. The simplest coupling is a linear one, with an interaction energy

$$\hat{H}^{\text{ext}}(t) = -\hat{O}_H(t)\delta j(t), \qquad (18.13)$$

where $j(t)$ is some external source. Inserting (18.13) into (18.12) yields the linear-response formula

$$\langle\Psi_H|\delta\hat{O}_H(t)|\Psi_H\rangle = i\int_{-\infty}^{\infty}dt'\,G_{OO}^R(t,t')\delta j(t'), \qquad (18.14)$$

where G_{OO}^R is the retarded Green function of two operators $\hat{O}$:

$$G_{OO}^R(t,t') = \Theta(t-t')\langle\Psi_H|\left[\hat{O}_H(t),\hat{O}_H(t')\right]|\Psi_H\rangle. \qquad (18.15)$$

At frequencies where the Fourier transform of $G_{OO}(t,t')$ is singular, the slightest disturbance causes a large response. This is the well-known resonance phenomenon found in any oscillating system. Whenever the external frequency ω hits an eigenfrequency, the Fourier transform of the Green function diverges. Usually, the eigenfrequencies of a complicated N-body system are determined by calculating (18.15) and by finding the singularities in ω.

It is easy to generalize this description to a thermal ensemble at a nonzero temperature. The principal modification consists in the replacement of the ground state expectation by the thermal average

$$\langle\hat{O}\rangle_T \equiv \frac{\text{Tr}\left(e^{-\hat{H}/T}\hat{O}\right)}{\text{Tr}\left(e^{-\hat{H}/T}\right)}.$$

Using the free energy

$$F = -T \log \mathrm{Tr}\,(e^{-\hat{H}/T}),$$

this can also be written as

$$\langle \hat{O} \rangle_T = e^{F/T} \mathrm{Tr}\,(e^{-\hat{H}/T}\hat{O}). \tag{18.16}$$

In a grand-canonical ensemble, $\hat{H}$ must be replaced by $\hat{H} - \mu\hat{N}$ and F by its grand-canonical version F_G (see Section 1.17). At finite temperatures, the linear-response formula (18.14) becomes

$$\delta\langle \hat{O}(t) \rangle_T = i \int_{-\infty}^{\infty} dt'\, G_{OO}^R(t, t')\delta j(t'), \tag{18.17}$$

where $G_{OO}^R(t, t')$ is the *retarded Green function at nonzero temperature* defined by [recall (1.302)]

$$G_{OO}^R(t, t') \equiv G_{OO}^R(t - t') \equiv \Theta(t - t')\, e^{F/T} \mathrm{Tr}\,\left\{ e^{-\hat{H}/T}\left[\hat{O}_H(t), \hat{O}_H(t')\right] \right\}. \tag{18.18}$$

In a realistic physical system, there are usually many observables, say $\hat{O}_H^i(t)$ for $i = 1, 2, \ldots, l$, which perform coupled oscillations. Then the relevant retarded Green function is some $l \times l$ matrix

$$G_{ij}^R(t, t') \equiv G_{ij}^R(t - t') \equiv \Theta(t - t')\, e^{F/T} \mathrm{Tr}\,\left\{ e^{-\hat{H}/T}\left[\hat{O}_H^i(t), \hat{O}_H^j(t')\right] \right\}. \tag{18.19}$$

After a Fourier transformation and diagonalization, the singularities of this matrix render the important physical information on the resonance properties of the system.

The retarded Green function at $T \neq 0$ occupies an intermediate place between the real-time Green function of field theories at $T = 0$, and the imaginary-time Green function used before to describe thermal equilibria at $T \neq 0$ (see Subsection 3.8.2). The Green function (18.19) depends both on the real time and on the temperature via an imaginary time.

18.2   Spectral Representations of Green Functions for $T \neq 0$

The retarded Green functions are related to the imaginary-time Green functions of equilibrium physics by an analytic continuation. For two arbitrary operators $\hat{O}_H^1$, $\hat{O}_H^2$, the latter is defined by the thermal average

$$G_{12}(\tau, 0) \equiv G_{12}(\tau) \equiv e^{F/T} \mathrm{Tr}\,\left[e^{-\hat{H}/T}\hat{T}_\tau \hat{O}_H^1(\tau)\hat{O}_H^2(0) \right], \tag{18.20}$$

where $\hat{O}_H(\tau)$ is the *imaginary-time Heisenberg operator*

$$\hat{O}_H(\tau) \equiv e^{\hat{H}\tau}\hat{O}e^{-\hat{H}\tau}. \tag{18.21}$$

To see the relation between $G_{12}(\tau)$ and the retarded Green function $G_{12}^R(t)$, we take a complete set of states $|n\rangle$, insert them between the operators $\hat{O}^1$, $\hat{O}^2$, and expand $G_{12}(\tau)$ for $\tau \geq 0$ into the spectral representation

$$G_{12}(\tau) = e^{F/T} \sum_{n,n'} e^{-E_n/T} e^{(E_n - E_{n'})\tau} \langle n|\hat{O}^1|n'\rangle\langle n'|\hat{O}^2|n\rangle. \tag{18.22}$$

Since $G_{12}(\tau)$ is periodic under $\tau \to \tau + 1/T$, its Fourier representation contains only the discrete Matsubara frequencies $\omega_m = 2\pi m T$:

$$\begin{aligned}
G_{12}(\omega_m) &= \int_0^{1/T} d\tau\, e^{i\omega_m \tau} G_{12}(\tau) \\
&= e^{F/T} \sum_{n,n'} e^{-E_n/T} \left(1 - e^{(E_n - E_{n'})/T}\right) \langle n|\hat{O}^1|n'\rangle\langle n'|\hat{O}^2|n\rangle \\
&\qquad\qquad \times \frac{-1}{i\omega_m - E_{n'} + E_n}.
\end{aligned} \tag{18.23}$$

The retarded Green function satisfies no periodic (or antiperiodic) boundary condition. It possesses Fourier components with *all* real frequencies ω:

$$\begin{aligned}
G_{12}^R(\omega) &= \int_{-\infty}^{\infty} dt\, e^{i\omega t}\, \Theta(t) e^{F/T} \mathrm{Tr}\left\{ e^{-\hat{H}/T} \left[\hat{O}_H^1(t), \hat{O}_H^2(0)\right]_{\mp} \right\} \\
&= e^{F/T} \int_0^{\infty} dt\, e^{i\omega t} \sum_{n,n'} \Big[\; e^{-E_n/T} e^{i(E_n - E_{n'})t} \langle n|\hat{O}^1|n'\rangle\langle n'|\hat{O}^2|n\rangle \\
&\qquad\qquad \mp e^{-E_n/T} e^{-i(E_n - E_{n'})t} \langle n|\hat{O}^2|n'\rangle\langle n'|\hat{O}^1|n\rangle \Big].
\end{aligned} \tag{18.24}$$

In the second sum we exchange n and n' and perform the integral, after having attached to ω an infinitesimal positive-imaginary part $i\eta$ to ensure convergence [recall the discussion after Eq. (3.84)]. The result is

$$\begin{aligned}
G_{12}^R(\omega) &= e^{F/T} \sum_{n,n'} e^{-E_n/T} \left[1 - e^{(E_n - E_{n'})/T}\right] \langle n|\hat{O}^1|n'\rangle\langle n'|\hat{O}^2|n\rangle \\
&\qquad\qquad \times \frac{i}{\omega - E_{n'} + E_n + i\eta}.
\end{aligned} \tag{18.25}$$

By comparing this with (18.23) we see that the thermal Green functions are obtained from the retarded ones by replacing [1]

$$\frac{i}{\omega - E_{n'} + E_n + i\eta} \;\to\; \frac{-1}{i\omega_m - E_{n'} + E_n}. \tag{18.26}$$

A similar procedure holds for fermion operators $\hat{O}^i$ (which are not observable). There are only two changes with respect to the boson case. First, in the Fourier expansion of the imaginary-time Green functions, the bosonic Matsubara frequencies ω_m in (18.23) become fermionic. Second, in the definition of the retarded Green functions (18.19), the commutator is replaced by an anticommutator, i.e., the retarded Green function of fermion operators $\hat{O}_H^i$ is defined by

$$G_{ij}^R(t, t') \equiv G_{ij}^R(t - t') \equiv \Theta(t - t') e^{F/T} \text{Tr} \left\{ e^{-\hat{H}/T} \left[\hat{O}_H^i(t), \hat{O}_H^j(t') \right]_+ \right\}. \quad (18.27)$$

These changes produce an opposite sign in front of the $e^{(E_n - E_{n'})/T}$-term in both of the formulas (18.23) and (18.25). Apart from that, the relation between the two Green functions is again given by the replacement rule (18.26).

At this point it is customary to introduce the *spectral function*

$$
\begin{aligned}
\rho_{12}(\omega') &= \left(1 \mp e^{-\omega'/T} \right) e^{F/T} \\
&\quad \times \sum_{n,n'} e^{-E_n/T} 2\pi \delta(\omega - E_{n'} + E_n) \langle n | \hat{O}^1 | n' \rangle \langle n' | \hat{O}^2 | n \rangle,
\end{aligned}
\quad (18.28)
$$

where the upper and the lower sign hold for bosons and fermions, respectively. Under an interchange of the two operators it behaves like

$$\rho_{12}(\omega') = \mp \rho_{12}(-\omega'). \quad (18.29)$$

Using this spectral function, we may rewrite the Fourier-transformed retarded and thermal Green functions as the following spectral integrals:

$$G_{12}^R(\omega) = \int_{-\infty}^{\infty} \frac{d\omega'}{2\pi} \rho_{12}(\omega') \frac{i}{\omega - \omega' + i\eta}, \quad (18.30)$$

$$G_{12}(\omega_m) = \int_{-\infty}^{\infty} \frac{d\omega'}{2\pi} \rho_{12}(\omega') \frac{-1}{i\omega_m - \omega'}. \quad (18.31)$$

These equations show how the imaginary-time Green functions arise from the retarded Green functions by a simple analytic continuation in the complex frequency plane to the discrete Matsubara frequencies, $\omega \to i\omega_m$. The inverse problem of reconstructing the retarded Green functions in the entire upper half-plane of ω from the imaginary-time Green functions defined only at the Matsubara frequencies ω_m is not solvable in general but only if other information is available [2]. For instance, the sum rules for canonical fields to be derived later in Eq. (18.66) with the ensuing asymptotic condition (18.67) are sufficient to make the continuation unique [3].

Going back to the time variables t and τ, the Green functions are

$$G_{12}^R(t) = \Theta(t) \int_{-\infty}^{\infty} \frac{d\omega'}{2\pi} \rho_{12}(\omega') e^{-i\omega' t}, \quad (18.32)$$

$$G_{12}(\tau) = \int_{-\infty}^{\infty} \frac{d\omega'}{2\pi} \rho_{12}(\omega') T \sum_{\omega_m} e^{-i\omega_m \tau} \frac{-1}{i\omega_m - \omega'}. \quad (18.33)$$

The sum over even or odd Matsubara frequencies on the right-hand side of $G_{12}(\tau)$ was evaluated in Section 3.3 for bosons and fermions as

$$
\begin{aligned}
T \sum_n e^{-i\omega_m \tau} \frac{-1}{i\omega_m - \omega} &= G_{\omega,e}^p(\tau) = e^{-\omega(\tau - 1/2T)} \frac{1}{2 \sin(\omega/2T)} \\
&= e^{-\omega \tau} (1 + n_\omega) \quad (18.34)
\end{aligned}
$$

and

$$T \sum_n e^{-i\omega_m \tau} \frac{-1}{i\omega_m - \omega} = G^a_{\omega,e}(\tau) = e^{-\omega(\tau - 1/2T)} \frac{1}{2\cos(\omega/2T)}$$
$$= e^{-\omega\tau}(1 - n_\omega), \qquad (18.35)$$

with the Bose and Fermi distribution functions [see (3.93), (7.529), (7.531)]

$$n_\omega = \frac{1}{e^{\omega/T} \mp 1}, \qquad (18.36)$$

respectively.

18.3   Other Important Green Functions

In studying the dynamics of systems at finite temperature, several other Green functions are useful whose spectral functions we shall now derive.

In complete analogy with the retarded Green functions for bosonic and fermionic operators, we may introduce their counterparts, the so-called *advanced Green functions* (compare page 38)

$$G^A_{12}(t,t') \equiv G^A_{12}(t - t') = -\Theta(t' - t)e^{F/T}\mathrm{Tr}\left\{e^{-\hat{H}/T}\left[\hat{O}^1_H(t), \hat{O}^2_H(t')\right]_\mp\right\}. \qquad (18.37)$$

Their Fourier transforms have the spectral representation

$$G^A_{12}(\omega) = \int_{-\infty}^{\infty} \frac{d\omega'}{2\pi} \rho_{12}(\omega') \frac{i}{\omega - \omega' - i\eta}, \qquad (18.38)$$

differing from the retarded case (18.30) only by the sign of the $i\eta$-term. This makes the Fourier transforms vanish for $t > 0$, so that the time-dependent Green function has the spectral representation [compare (18.32)]

$$G^A_{12}(t) = -\Theta(-t) \int_{-\infty}^{\infty} \frac{d\omega}{2\pi} \rho_{12}(\omega)e^{-i\omega t}. \qquad (18.39)$$

By subtracting retarded and advanced Green functions, we obtain the thermal expectation value of commutator or anticommutator:

$$C_{12}(t,t') = e^{F/T}\mathrm{Tr}\left\{e^{-\hat{H}/T}\left[\hat{O}^1_H(t), \hat{O}^2_H(t')\right]_\mp\right\} = G^R_{12}(t,t') - G^A_{12}(t,t'). \qquad (18.40)$$

Note the simple relations:

$$G^R_{12}(t,t') = \Theta(t - t')C_{12}(t,t'), \qquad (18.41)$$
$$G^A_{12}(t,t') = -\Theta(t' - t)C_{12}(t,t'). \qquad (18.42)$$

When inserting the spectral representations (18.30) and (18.39) of $G_{12}^R(t)$ and $G_{12}^A(t)$ into (18.40), and using the identity (1.324),

$$\frac{i}{\omega - \omega' + i\eta} - \frac{i}{\omega - \omega' - i\eta} = 2\frac{\eta}{(\omega - \omega')^2 + \eta^2} = 2\pi\delta(\omega - \omega'), \qquad (18.43)$$

we obtain the spectral integral representation for the commutator function:[2]

$$C_{12}(t) = \int_{-\infty}^{\infty} \frac{d\omega}{2\pi} \rho_{12}(\omega)e^{-i\omega t}. \qquad (18.44)$$

Thus a knowledge of the commutator function $C_{12}(t)$ determines directly the spectral function $\rho_{12}(\omega)$ by its Fourier components

$$C_{12}(\omega) = \rho_{12}(\omega). \qquad (18.45)$$

An important role in studying the dynamics of a system in a thermal environment is played by the time-ordered Green functions. They are defined by

$$G_{12}(t, t') \equiv G_{12}(t - t') = e^{F/T} \text{Tr} \left[e^{-\hat{H}/T} \, \hat{T} \hat{O}_H^1(t) \hat{O}_H^2(t') \right]. \qquad (18.46)$$

Inserting intermediate states as in (18.23) we find the spectral representation

$$
\begin{aligned}
G_{12}(\omega) &= \int_{-\infty}^{\infty} dt\, e^{i\omega t}\, \Theta(t)\ e^{F/T} \text{Tr}\left\{ e^{-\hat{H}/T} \hat{O}_H^1(t) \hat{O}_H^2(0) \right\} \\
&+ \int_{-\infty}^{\infty} dt\, e^{i\omega t}\ \Theta(-t) e^{F/T} \text{Tr}\left\{ e^{-\hat{H}/T} \hat{O}_H^2(t) \hat{O}_H^1(0) \right\} \\
&= e^{F/T} \int_0^{\infty} dt\, e^{i\omega t} \sum_{n,n'} e^{-E_n/T} e^{i(E_n - E_{n'})t}\ \langle n|\hat{O}^1|n'\rangle\langle n'|\hat{O}^2|n\rangle \\
&\pm\ e^{F/T} \int_{-\infty}^0 dt\, e^{i\omega t} \sum_{n,n'} e^{-E_n/T} e^{-i(E_n - E_{n'})t} \langle n|\hat{O}^2|n'\rangle\langle n'|\hat{O}^1|n\rangle. \qquad (18.47)
\end{aligned}
$$

Interchanging again n and n', this can be written in terms of the spectral function (18.28) as

$$G_{12}(\omega) = \int_{-\infty}^{\infty} \frac{d\omega'}{2\pi} \rho_{12}(\omega') \left[\frac{1}{1 \mp e^{-\omega'/T}} \frac{i}{\omega - \omega' + i\eta} + \frac{1}{1 \mp e^{\omega'/T}} \frac{i}{\omega - \omega' - i\eta} \right]. \quad (18.48)$$

Let us also write down the spectral decomposition of a further operator expression complementary to $C_{12}(t)$ of (18.40), in which boson or fermion fields appear with the "wrong" commutator:

$$A_{12}(t - t') \equiv e^{F/T} \text{Tr} \left\{ e^{-\hat{H}/T} \left[\hat{O}_H^1(t), \hat{O}_H^2(t') \right]_\pm \right\}. \qquad (18.49)$$

---

[2]Due to the relation (18.41), the same representation is found by dropping the factor $\Theta(t)$ in (18.32).

This function characterizes the size of fluctuations of the operators O_H^1 and O_H^2. Inserting intermediate states, we find

$$
\begin{aligned}
A_{12}(\omega) &= \int_{-\infty}^{\infty} dt\, e^{i\omega t} e^{F/T} \mathrm{Tr}\left\{ e^{-\hat{H}/T} \left[\hat{O}_H^1(t), \hat{O}_H^2(0) \right]_{\pm} \right\} \\
&= e^{F/T} \int_{-\infty}^{\infty} dt\, e^{i\omega t} \sum_{n,n'} \Big[\, e^{-E_n/T} e^{i(E_n - E_{n'})t} \langle n|\hat{O}^1|n'\rangle \langle n'|\hat{O}^2|n\rangle \\
&\qquad\qquad \pm e^{-E_n/T} e^{-i(E_n - E_{n'})t} \langle n|\hat{O}^2|n'\rangle \langle n'|\hat{O}^1|n\rangle \Big]. \quad (18.50)
\end{aligned}
$$

In the second sum we exchange n and n' and perform the integral, which runs now over the entire time interval and gives therefore a δ-function:

$$
\begin{aligned}
A_{12}(\omega) &= e^{F/T} \sum_{n,n'} e^{-E_n/T} \left[1 \pm e^{(E_n - E_{n'})/T} \right] \langle n|\hat{O}^1|n'\rangle \langle n'|\hat{O}^2|n\rangle \\
&\qquad\qquad \times 2\pi \delta(\omega - E_{n'} + E_n). \quad (18.51)
\end{aligned}
$$

In terms of the spectral function (18.28), this has the simple form

$$
A_{12}(\omega) = \int_{-\infty}^{\infty} \frac{d\omega'}{2\pi} \tanh^{\mp 1} \frac{\omega'}{2T} \rho_{12}(\omega')\, 2\pi \delta(\omega - \omega') = \tanh^{\mp 1} \frac{\omega}{2T} \rho_{12}(\omega). \quad (18.52)
$$

Thus the expectation value (18.49) of the "wrong" commutator has the time dependence

$$
A_{12}(t,t') \equiv A_{12}(t - t') = \int_{-\infty}^{\infty} \frac{d\omega}{2\pi} \rho_{12}(\omega) \tanh^{\mp 1} \frac{\omega}{2T} e^{-i\omega(t-t')}. \quad (18.53)
$$

There exists another way of writing the spectral representation of the various Green functions. For retarded and advanced Green functions G_{12}^R, G_{12}^A, we decompose in the spectral representations (18.30) and (18.38) according to the rule (1.325):

$$
\frac{i}{\omega - \omega' \pm i\eta} = i \left[\frac{\mathcal{P}}{\omega - \omega'} \mp i\pi \delta(\omega - \omega') \right], \quad (18.54)
$$

where $\mathcal{P}$ indicates principal value integration across the singularity, and write

$$
G_{12}^{R,A}(\omega) = i \int_{-\infty}^{\infty} \frac{d\omega'}{2\pi} \rho_{12}(\omega') \left[\frac{\mathcal{P}}{\omega - \omega'} \mp i\pi \delta(\omega - \omega') \right]. \quad (18.55)
$$

Inserting (18.54) into (18.48) we find the alternative representation of the time-ordered Green function

$$
G_{12}(\omega) = i \int_{-\infty}^{\infty} \frac{d\omega'}{2\pi} \rho_{12}(\omega') \left[\frac{\mathcal{P}}{\omega - \omega'} - i\pi \tanh^{\mp 1} \frac{\omega}{2T} \delta(\omega - \omega') \right]. \quad (18.56)
$$

The term proportional to $\delta(\omega - \omega')$ in the spectral representation is commonly referred to as the *absorptive* or *dissipative part* of the Green function. The first term proportional to the principal value is called the *dispersive* or *fluctuation part*.

The relevance of the spectral function $\rho_{12}(\omega')$ in determining both the *fluctuation part* as well as the *dissipative part* of the time-ordered Green function is the content of the important *fluctuation-dissipation theorem*. In more detail, this may be restated as follows: The common spectral function $\rho_{12}(\omega')$ of the commutator function in (18.44), the retarded Green function in (18.30), and the *fluctuation part* of the time-ordered Green function in (18.56) determines, after being multiplied by a factor $\tanh^{\mp 1}(\omega'/2T)$, the *dissipative part* of the time-ordered Green function in Eq. (18.56).

The three Green functions $-iG_{12}(\omega)$, $-iG_{12}^R(\omega)$, and $-iG_{12}^A(\omega)$ have the same real parts. By comparing Eqs. (18.30) and (18.31) we found that retarded and advanced Green functions are simply related to the imaginary-time Green function via an analytic continuation. The spectral decomposition (18.56) shows this is not true for the time-ordered Green function, due to the extra factor $\tanh^{\mp 1}(\omega/2T)$ in the absorptive term.

Another representation of the time-ordered Green is useful. It is obtained by expressing $\tan^{\mp 1}$ in terms of the Bose and Fermi distribution functions (18.36) as $\tan^{\mp 1} = 1 \pm 2n_\omega$. Then we can decompose

$$G_{12}(\omega) = \int_{-\infty}^{\infty} \frac{d\omega'}{2\pi} \rho_{12}(\omega') \left[\frac{i}{\omega - \omega' + i\eta} \pm 2\pi n_\omega \, \delta(\omega - \omega') \right]. \qquad (18.57)$$

18.4 Hermitian Adjoint Operators

If the two operators $\hat{O}_H^1(t)$, $\hat{O}_H^2(t)$ are Hermitian adjoint to each other,

$$\hat{O}_H^2(t) = [\hat{O}_H^1(t)]^\dagger, \qquad (18.58)$$

the spectral function (18.28) can be rewritten as

$$\begin{aligned} \rho_{12}(\omega') = & (1 \mp e^{-\omega'/T})e^{F/T} \\ & \times \sum_{n,n'} e^{-E_n/T} 2\pi\delta(\omega' - E_{n'} + E_n)|\langle n|\hat{O}_H^1(t)|n'\rangle\|^2. \end{aligned} \qquad (18.59)$$

This shows that

$$\begin{aligned} \rho_{12}(\omega')\omega' &\geq 0 \qquad \text{for bosons,} \\ \rho_{12}(\omega') &\geq 0 \qquad \text{for fermions.} \end{aligned} \qquad (18.60)$$

This property permits us to derive several useful inequalities between various diagonal Green functions in Appendix 18A.

Under the condition (18A.7), the expectation values of anticommutators and commutators satisfy the time-reversal relations

$$\begin{aligned} G_{12}^A(t,t') &= \mp G_{21}^R(t',t)^*, & (18.61) \\ A_{12}(t,t') &= \pm A_{21}(t',t)^*, & (18.62) \\ C_{12}(t,t') &= \mp C_{21}(t',t)^*. & (18.63) \\ G_{12}(t,t') &= \pm G_{21}(t',t)^*. & (18.64) \end{aligned}$$

Examples are the corresponding functions for creation and annihilation operators which will be treated in detail below. More generally, this properties hold for any interacting nonrelativistic particle fields $\hat{O}_H^1(t) = \hat{\psi}_{\mathbf{p}}(t)$, $\hat{O}_H^2(t) = \hat{\psi}_{\mathbf{p}}^\dagger(t)$ of a specific momentum $\mathbf{p}$.

Such operators satisfy, in addition, the canonical equal-time commutation rules at each momentum

$$\left[\hat{\psi}_{\mathbf{p}}(t), \hat{\psi}_{\mathbf{p}}^\dagger(t)\right] = 1 \tag{18.65}$$

(see Sections 7.6, 7.9). Using (18.40), (18.44) we derive from this *spectral function sum rule*:

$$\int_{-\infty}^{\infty} \frac{d\omega'}{2\pi} \rho_{12}(\omega') = 1. \tag{18.66}$$

For a canonical free field with $\rho_{12}(\omega') = 2\pi\delta(\omega'-\omega)$, this sum rule is of course trivially fulfilled. In general, the sum rule ensures the large-ω behavior of imaginary-time, retarded, and advanced Green functions of canonically conjugate field operators to be the same as for a free particle, i.e.,

$$G_{12}(\omega_m) \xrightarrow[\omega_m \to \infty]{} \frac{i}{\omega_m}, \quad G_{12}^{A,R}(\omega) \xrightarrow[\omega \to \infty]{} \frac{1}{\omega}. \tag{18.67}$$

18.5 Harmonic Oscillator Green Functions for $T \neq 0$

As an example, consider a single harmonic oscillator of frequency Ω or, equivalently, a free particle at a point in the second-quantized field formalism (see Chapter 7). We shall start with the second representation.

18.5.1 Creation Annihilation Operators

The operators $\hat{O}_H^1(t)$ and $\hat{O}_H^2(t)$ are the creation and annihilation operators in the Heisenberg picture

$$\hat{a}_H^\dagger(t) = \hat{a}^\dagger e^{i\Omega t}, \quad \hat{a}_H(t) = \hat{a} e^{-i\Omega t}. \tag{18.68}$$

The eigenstates of the Hamiltonian operator

$$\hat{H} = \frac{1}{2}\left(\hat{p}^2 + \Omega^2 \hat{x}^2\right) = \frac{\omega}{2}\left(\hat{a}^\dagger \hat{a} + \hat{a}\hat{a}^\dagger\right) = \omega\left(\hat{a}^\dagger \hat{a} \pm \frac{1}{2}\right) \tag{18.69}$$

are

$$|n\rangle = \frac{1}{\sqrt{n!}}(\hat{a}^\dagger)^n|0\rangle, \tag{18.70}$$

with the eigenvalues $E_n = (n \pm 1/2)\Omega$ for $n = 0, 1, 2, 3, \ldots$ or $n = 0, 1$, if $\hat{a}^\dagger$ and $\hat{a}$ commute or anticommute, respectively [compare Eq. (7.551)]. In the second-quantized field interpretation the energies are $E_n = n\Omega$ and the final Green functions are the same. The spectral function $\rho_{12}(\omega')$ is trivial to calculate. The Schrödinger operator $\hat{O}^2 = \hat{a}^\dagger$ can connect the state $|n\rangle$ only to $\langle n+1|$, with the matrix element $\sqrt{n+1}$. The operator $\hat{O}^1 = \hat{a}$ does the opposite. Hence we have

$$\rho_{12}(\omega') = 2\pi\delta(\omega' - \Omega)(1 \mp e^{-\Omega/T})e^{F/T} \sum_{n=0}^{\infty,0} e^{-(n\pm 1/2)\Omega/T}(n+1). \tag{18.71}$$

Now we make use of the explicit partition functions of the oscillator whose paths satisfy periodic and antiperiodic boundary conditions:

$$Z_\Omega \equiv e^{-F/T} = \sum_{n=0}^{\infty,1} e^{-(n\pm 1/2)\Omega/T} = \left\{ \begin{array}{l} [2\sinh(\Omega/2T)]^{-1} \\ 2\cosh(\Omega/2T) \end{array} \right. \quad \text{for} \quad \left. \begin{array}{l} \text{bosons} \\ \text{fermions} \end{array} \right\}. \tag{18.72}$$

These allow us to calculate the sums in (18.71) as follows

$$\sum_{n=0}^{\infty} e^{-(n+1/2)\Omega/T}(n+1) = \left(-T\frac{\partial}{\partial\Omega} + \frac{1}{2}\right)e^{-F/T} = \left(1 \mp e^{-\Omega/T}\right)^{-1} e^{-F/T},$$

$$\sum_{n=0}^{0} e^{-(n-1/2)\Omega/T}(n+1) = e^{\Omega/2T} = \left(1 + e^{-\Omega/T}\right)^{-1} e^{-F/T}. \tag{18.73}$$

The spectral function $\rho_{12}(\omega')$ of the a single oscillator quantum of frequency Ω is therefore given by

$$\rho_{12}(\omega') = 2\pi\delta(\omega' - \Omega). \tag{18.74}$$

With it, the retarded and imaginary-time Green functions become

$$G_\Omega^R(t, t') = \Theta(t - t')e^{-\Omega(t-t')}, \tag{18.75}$$

$$G_\Omega(\tau, \tau') = -T \sum_{m=-\infty}^{\infty} e^{-i\omega_m(\tau-\tau')}\frac{1}{i\Omega_m - \Omega} \tag{18.76}$$

$$= e^{-\Omega(\tau-\tau')} \left\{ \begin{array}{l} 1 \pm n_\Omega \\ \pm n_\Omega \end{array} \right. \quad \text{for} \quad \tau \begin{array}{l} \geq \\ < \end{array} \tau', \tag{18.77}$$

with the average particle number n_Ω of (18.36). The commutation function, for instance, is by (18.44) and (18.74):

$$C_{12}(t, t') = e^{-i\Omega(t-t')}, \tag{18.78}$$

and the correlation function of the "wrong commutator" is from (18.53) and (18.74):

$$A_\Omega(t, t') = \tanh^{\mp 1}\frac{\Omega}{2T}e^{-i\Omega(t-t')}. \tag{18.79}$$

Of course, these harmonic-oscillator expressions could have been obtained directly by starting from the defining operator equations. For example, the commutator function

$$C_\Omega(t, t') = e^{F/T}\text{Tr}\left\{ e^{-\hat{H}/T}[\hat{a}_H(t), \hat{a}_H^\dagger(t')]_{\mp} \right\} \tag{18.80}$$

turn into (18.78) by using the commutation rule at different times

$$[\hat{a}_H(t), \hat{a}_H^\dagger(t')] = e^{-i\Omega(t-t')}, \tag{18.81}$$

which follows from (18.68). Since the right-hand side is a c-number, the thermodynamic average is trivial:

$$e^{F/T}\mathrm{Tr}\,(e^{-\hat{H}/T}) = 1. \tag{18.82}$$

After this, the relations (18.41), (18.42) determine the retarded and advanced Green functions

$$G_\Omega^R(t - t') = \Theta(t - t')e^{-i\Omega(t-t')}, \quad G_\Omega^A(t - t') = -\Theta(t' - t)e^{-i\Omega(t-t')}. \tag{18.83}$$

For the Green function at imaginary times

$$G_\Omega(\tau, \tau') \equiv e^{F/T}\mathrm{Tr}\left[e^{-\hat{H}/T}\hat{T}_\tau \hat{a}_H(\tau)\hat{a}_H^\dagger(\tau')\right], \tag{18.84}$$

the expression (18.77) is found using [see (18.85)]

$$\begin{aligned}
\hat{a}_H^\dagger(\tau) &\equiv e^{\hat{H}\tau}\hat{a}^\dagger e^{-\hat{H}\tau} = \hat{a}^\dagger e^{\Omega\tau}, \\
\hat{a}_H(\tau) &\equiv e^{\hat{H}\tau}\hat{a}e^{-\hat{H}\tau} = \hat{a}e^{-\Omega\tau},
\end{aligned} \tag{18.85}$$

and the summation formula (18.73).

The "wrong" commutator function (18.79) can, of course, be immediately derived from the definition

$$A_{12}(t - t') \equiv e^{F/T}\mathrm{Tr}\left\{e^{-\hat{H}/T}\left[\hat{a}_H(t), \hat{a}_H^\dagger(t')\right]_\pm\right\} \tag{18.86}$$

and (18.68), by inserting intermediate states.

For the temporal behavior of the time-ordered Green function we find from (18.48)

$$G_\Omega(\omega) = \left(1 \mp e^{-\Omega/T}\right)^{-1} G_\Omega^R(\omega) + \left(1 \mp e^{\Omega/T}\right)^{-1} G_\Omega^A(\omega), \tag{18.87}$$

and from this by a Fourier transformation

$$\begin{aligned}
G_\Omega(t, t') &= \left(1 \mp e^{-\Omega/T}\right)^{-1}\Theta(t - t')e^{-i\Omega(t-t')} - \left(1 \mp e^{\Omega/T}\right)^{-1}\Theta(t' - t)e^{-i\Omega(t-t')} \\
&= \left[\Theta(t - t') \pm (e^{\Omega/T} \mp 1)^{-1}\right]e^{-i\Omega(t-t')} = \left[\Theta(t - t') \pm n_\Omega\right]e^{-i\Omega(t-t')}.
\end{aligned} \tag{18.88}$$

The same result is easily obtained by directly evaluating the defining equation using (18.68) and inserting intermediate states:

$$\begin{aligned}
G_\Omega(t, t') &\equiv G_\Omega(t - t') = e^{F/T}\mathrm{Tr}\left[e^{-\hat{H}/T}\,\hat{T}\hat{a}_H(t)\hat{a}_H^\dagger(t')\right] \\
&= \Theta(t - t')\langle\hat{a}\,\hat{a}^\dagger\rangle e^{-i\Omega(t-t')} \pm \Theta(t' - t)\langle\hat{a}^\dagger\,\hat{a}\rangle e^{-i\Omega(t-t')} \\
&= \Theta(t - t')(1 \pm n_\Omega)e^{-i\Omega(t-t')} \pm \Theta(t' - t)n_\Omega e^{-i\Omega(t-t')}, \tag{18.89}
\end{aligned}$$

which is the same as (18.88). For the correlation function with a and $a^\dagger$ interchanged,

$$\bar{G}_\Omega(t,t') \equiv G_\Omega(t-t') = e^{F/T} \mathrm{Tr}\left[e^{-\hat{H}/T}\,\hat{T}\hat{a}_H^\dagger(t)\hat{a}_H(t')\right], \tag{18.90}$$

we find in this way

$$\begin{aligned}
\bar{G}_\Omega(t,t') &= \Theta(t-t')\langle\hat{a}^\dagger\hat{a}\rangle e^{-i\Omega(t-t')} \pm \Theta(t'-t)\langle\hat{a}\,\hat{a}^\dagger\rangle e^{-i\Omega(t-t')} \\
&= \Theta(t-t')n_\Omega e^{-i\Omega(t-t')} \pm \Theta(t'-t)(1\pm n_\Omega)e^{-i\Omega(t-t')}, \tag{18.91}
\end{aligned}$$

in agreement with (18.64).

18.5.2 Real Field Operators

From the above expressions it is easy to construct the corresponding Green functions for the position operators of the harmonic oscillator $\hat{x}(t)$. It will be useful to keep the discussion more general by admitting oscillators which are not necessarily mass points in space but can be field variables. Thus we shall use, instead of $\hat{x}(t)$, the symbol $\varphi(t)$, and call this a field variable. As in Eq. (7.295) we decompose the field as

$$\hat{x}(t) = \sqrt{\frac{\hbar}{2M\Omega}}\left[\hat{a}e^{-i\Omega t} + \hat{a}^\dagger e^{i\Omega t}\right]. \tag{18.92}$$

In this section we use physical units. The commutator function (18.40) is directly

$$C(t,t') \equiv \langle[\hat{\varphi}(t),\hat{\varphi}(t')]_\mp\rangle_\rho = -\frac{\hbar}{2M\Omega}2i\sin\Omega(t-t'), \tag{18.93}$$

implying a spectral function [recall (18.44)]

$$\rho(\omega') = \frac{1}{2M\Omega}2\pi\left[\delta(\omega'-\Omega) - \delta(\Omega'+\Omega)\right]. \tag{18.94}$$

The real operator $\hat{\varphi}(t)$ behaves like the difference of a particle of frequency Ω and $-\Omega$, with an overall factor $1/2M\Omega$. It is then easy to find the retarded and advanced Green functions of the operators $\hat{\varphi}(t)$ and $\hat{\varphi}(t')$:

$$G^R(t,t') = \frac{\hbar}{2M\Omega}\left[G_\Omega^R(t,t') - G_{-\Omega}^R(t,t')\right] = -\frac{\hbar}{2M\Omega}\Theta(t-t')2i\sin\Omega(t-t'), \tag{18.95}$$

$$G^A(t,t') = \frac{\hbar}{2M\Omega}\left[G_\Omega^A(t,t') - G_{-\Omega}^A(t,t')\right] = \frac{\hbar}{2M\Omega}\Theta(t-t')2i\sin\Omega(t'-t). \tag{18.96}$$

From the spectral representation (18.53), we obtain for the "wrong commutator"

$$A(t,t') = \langle[\hat{\varphi}(t),\hat{\varphi}(t')]_\mp\rangle = \frac{\hbar}{2M\Omega}\coth^{\pm1}\frac{\Omega}{2k_BT}2\cos\Omega(t-t'). \tag{18.97}$$

The relation with (18.93) is again a manifestation of the fluctuation-dissipation theorem (18.53).

The average of these two functions yields the time-dependent correlation function at finite temperature containing only the product of the operators

$$G^P(t,t') \equiv \langle \hat{\varphi}(t)\hat{\varphi}(t')\rangle = \frac{\hbar}{2M\Omega}\left[(1 \pm 2n_\Omega)\cos\Omega(t-t') - i\sin\Omega(t-t')\right], \quad (18.98)$$

with the average particle number n_Ω of (18.36). In the limit of zero temperature where $n_\Omega \equiv 0$, this reduces to

$$G^P(t,t') = \langle \hat{\varphi}(t)\hat{\varphi}(t')\rangle = \frac{\hbar}{2M\Omega}e^{-i\Omega(t-t')}. \quad (18.99)$$

The time-ordered Green function is obtained from this by the obvious relation

$$G(t,t') = \Theta(t-t')G^P(t,t') \pm \Theta(t'-t)G^P(t',t) = \frac{1}{2}\left[A(t,t') + \epsilon(t-t')C(t,t')\right], \quad (18.100)$$

where $\epsilon(t-t')$ is the step function of Eq. (1.312). Explicitly, the time-ordered Green function is

$$G(t,t') \equiv \langle \hat{T}\hat{\varphi}(t)\hat{\varphi}(t')\rangle = \frac{\hbar}{2M\Omega}\left[(1 \pm 2n_\Omega)\cos\Omega|t-t'| - i\sin\Omega|t-t'|\right], \quad (18.101)$$

which reduces for $T \to 0$ to

$$G(t,t') = \langle \hat{T}\hat{\varphi}(t)\hat{\varphi}(t')\rangle = \frac{\hbar}{2M\Omega}e^{-i\Omega|t-t'|}. \quad (18.102)$$

Thus, as a mnemonic rule, a finite temperature is introduced into a zero-temperature Green function by simply multiplying the real part of the exponential function by a factor $1 \pm 2n_\Omega$. This is another way of stating the *fluctuation-dissipation theorem*.

There is another way of writing the time-ordered Green function (18.101) in the bosonic case:

$$G(t,t') \equiv \langle \hat{T}\hat{\varphi}(t)\hat{\varphi}(t')\rangle = \frac{\hbar}{2M\Omega}\frac{\cosh\left[\frac{\Omega}{2}(\hbar\beta - i|t-t'|)\right]}{\sinh\frac{\hbar\Omega\beta}{2}}. \quad (18.103)$$

For $t - t' > 0$, this coincides precisely with the periodic Green function $G_e^p(\tau,\tau') = G_e^p(\tau - \tau')$ at imaginary-times $\tau > \tau'$ [see (3.248)], if τ and τ' are continued analytically to it and it', respectively. Decomposing (18.101) into real and imaginary parts we see by comparison with (18.100) that anticommutator and commutator functions are the doubled real and imaginary parts of the time-ordered Green function:

$$A(t,t') = 2\,\mathrm{Re}\,G(t,t'), \qquad C(t,t') = 2i\,\mathrm{Im}\,G(t,t'). \quad (18.104)$$

In the fermionic case, the hyperbolic functions cosh and sinh in numerator and denominator are simply interchanged, and the result coincides with the analytically continued antiperiodic imaginary-time Green function (3.263).

The time-reversal properties (18.61)–(18.64) of the Green functions become for real fields $\hat{\varphi}(t)$:

$$G^A(t,t') = \mp G^R(t',t), \tag{18.105}$$
$$A(t,t') = \pm A(t',t), \tag{18.106}$$
$$C(t,t') = \mp C(t',t), \tag{18.107}$$
$$G(t,t') = \pm G(t',t). \tag{18.108}$$

18.6   Nonequilibrium Green Functions

Up to this point we have assumed the system to be in intimate contact with a heat reservoir which ensures a constant temperature throughout the volume. The disturbance in (18.3) was taken to be small, so that only a small fraction of the particles could be excited. If the disturbance grows larger, large clouds of excitations can be formed in a local region. Such a system leaves thermal equilibrium, and the response is necessarily nonlinear. The system must be studied in its full quantum-mechanical time evolution. In order to describe such a process theoretically, we shall assume an initial equilibrium characterized by some density operator [compare (2.365)]

$$\hat{\rho} = \sum_n \rho_n |n\rangle\langle n|, \tag{18.109}$$

with eigenvalues

$$\rho_n = e^{-E_n/T}. \tag{18.110}$$

The disturbance sets in at some time t_0. If the initial state is out of equilibrium, the formalism to be described remains applicable, with only a few adaptations, if the initial state at t_0 is still characterized by a density operator of type (18.109), but has probabilities ρ_n different from (18.110). Of course, in the limit of very small deviations from thermal equilibrium, the formalism to be described reduces to the previously treated linear-response theory.

We first develop a perturbation theory for the time evolution of operators in a nonequilibrium situation. This serves to set up a path integral formalism for the description of the dynamical behavior of a single particle in contact with a thermal reservoir. This description can, in principle, be extended to ensembles of many particles by considering a similar path integral for a fluctuating field. After the discussion in Chapter 7, the necessary second quantization is straightforward and requires no detailed presentation.

The perturbation theory for nonequilibrium quantum-statistical mechanics to be developed now is known under the name of *closed-time path Green function formalism* (CTPGF). This formalism was developed by Schwinger [4] and Keldysh [5], and has been applied successfully to many nonequilibrium problems in statistical physics, in particular to superconductivity and plasma physics.

The fundamental problem of nonequilibrium statistical mechanics is finding the time evolution of thermodynamic averages of products of Heisenberg operators $\hat{\varphi}_H(t)$. For interesting applications it is useful to keep the formulation general and deal with relativistic *fields* of operators $\hat{\varphi}_H(\mathbf{x}, t)$. As in Section 7.6, an extra spatial argument $\mathbf{x}$ allows for a different time-dependent operator $\hat{\varphi}(t)$ at each point $\mathbf{x}$ in space. In order to prepare ourselves for the most interesting study of electromagnetic fields, we consider the simplest relativistically invariant classical action describing an observable field in D dimensions which has the form

$$\mathcal{A}_0 = \int dt d^D x \, \mathcal{L}_0(\mathbf{x}, t) \equiv \frac{1}{2} \int dt d^D x \left\{ [\dot{\varphi}(\mathbf{x}, t)]^2 - [\boldsymbol{\nabla}\varphi(\mathbf{x}, t)]^2 - m^2 \varphi^2(\mathbf{x}, t) \right\}. \quad (18.111)$$

As in Section 7.6, we go over to a countable set of infinite points $\mathbf{x}$ assuming that space is a fine lattice of spacing ϵ, with the continuum limit $\epsilon \to 0$ taken at the end. The associated Euler-Lagrange equation extremizing the action is the *Klein-Gordon equation*

$$\ddot{\varphi}(\mathbf{x}, t) + (-\partial_{\mathbf{x}}^2 + m^2)\varphi(\mathbf{x}, t) = 0. \quad (18.112)$$

This is solved by plane waves

$$f_{\mathbf{p}}(\mathbf{x}, t) = \frac{1}{\sqrt{2\omega_{\mathbf{p}} V}} e^{-i\omega_{\mathbf{p}}t + i\mathbf{p}\mathbf{x}}, \qquad \bar{f}_{\mathbf{p}}(\mathbf{x}, t) = \frac{1}{\sqrt{\omega_{\mathbf{p}} V}} e^{i\omega_{\mathbf{p}}t + i\mathbf{p}\mathbf{x}} \quad (18.113)$$

of positive and negative energy. As in Section 7.6, we imagine the system to be confined to a finite cubic volume V. Then the momenta $\mathbf{p}$ are discrete. The solutions (18.113) behave like an infinite set of harmonic oscillator solution, one for each momentum vector $\mathbf{p}$, with the $\mathbf{p}$-dependent frequencies

$$\omega_{\mathbf{p}} \equiv \sqrt{\mathbf{p}^2 + m^2}. \quad (18.114)$$

The general solution of (18.112) may be expanded as

$$\varphi(\mathbf{x}, t) = \sum_{\mathbf{p}} \frac{1}{2\omega_{\mathbf{p}} V} \left(a_{\mathbf{p}} e^{-i\omega_{\mathbf{p}} + i\mathbf{p}\mathbf{x}} + a_{\mathbf{p}}^* e^{i\omega_{\mathbf{p}}t + i\mathbf{p}\mathbf{x}} \right). \quad (18.115)$$

The canonical momenta of the field variables $\varphi(\mathbf{x}, t)$ are the field velocities

$$\pi(\mathbf{x}, t) \equiv p_{\mathbf{x}}(t) \equiv \dot{\varphi}(\mathbf{x}, t). \quad (18.116)$$

The fields are quantized by the canonical commutation rules

$$[\hat{\pi}(\mathbf{x}, t), \varphi(\mathbf{x}, t)] = -i\delta_{\mathbf{x}\mathbf{x}'}. \quad (18.117)$$

The quantum field is now expanded as in (18.115), but in terms of operators $\hat{a}_{\mathbf{p}}$ and their Hermitian adjoint operators $\hat{a}_{\mathbf{p}}^{\dagger}$. These satisfy the usual canonical commutation rules of creation and annihilation operators of Eq. (7.294):

$$[\hat{a}_{\mathbf{p}}(t), \hat{a}_{\mathbf{p}'}^{\dagger}(t)] = \delta_{\mathbf{p}\mathbf{p}'}, \qquad [\hat{a}_{\mathbf{p}}^{\dagger}(t), \hat{a}_{\mathbf{p}'}^{\dagger}(t)] = 0, \qquad [\hat{a}_{\mathbf{p}}(t), \hat{a}_{\mathbf{p}'}(t)] = 0. \quad (18.118)$$

The simplest nonequilibrium quantities to be studied are the thermal averages of one or two such field operators. More generally, we may investigate the averages of one or two fields with respect to an arbitrary initial density operator $\hat{\rho}$, the so-called ρ-*averages*:

$$\begin{aligned}
\langle \hat{\varphi}_H(x) \rangle_\rho &= \mathrm{Tr}\,[(\hat{\rho}\;\hat{\varphi}_H(x)]\,, \\
\langle \hat{\varphi}_H(x)\hat{\varphi}_H(y) \rangle_\rho &= \mathrm{Tr}\,[\hat{\rho}\;\hat{\varphi}_H(x)\hat{\varphi}_H(y)]\,.
\end{aligned} \qquad (18.119)$$

For brevity, we have gone over to a four-vector notation and use spacetime coordinates $x \equiv (\mathbf{x}, t)$ to write $\hat{\varphi}_H(\mathbf{x}, t)$ as $\hat{\varphi}_H(x)$.

In general, the fields $\varphi(\mathbf{x}, t)$ will interact with each other and with further fields, adding to (18.111) some interaction $\mathcal{A}^{\mathrm{int}}$. The behavior of an interacting field system can then be studied in perturbation theory. This is done by techniques related to those in Section 1.7. First we identify a time-independent part of the Hamiltonian for which we can solve the Schrödinger equation exactly. This is called the free part of the Hamiltonian $\hat{H}_0$. For the field $\varphi(\mathbf{x}, t)$ at hand this follows from the action (18.111) via the usual Legendre transformation (1.13). Its operator version is

$$\hat{H}_0 = \int d^D x\, \hat{\mathcal{H}}_0(\mathbf{x}, t) \equiv \frac{1}{2} \int d^D x \left\{ [\dot{\hat{\varphi}}(\mathbf{x}, t)]^2 + [\boldsymbol{\nabla}\hat{\varphi}(\mathbf{x}, t)]^2 + m^2 \hat{\varphi}^2(\mathbf{x}, t) \right\}. \quad (18.120)$$

The interaction $\mathcal{A}^{\mathrm{int}}$ gives rise to an interaction Hamiltonian $\hat{H}^{\mathrm{int}}(t)$. Then we introduce the field operators in Dirac's *interaction picture* $\hat{\varphi}(x)$. These are related to the Heisenberg operators via the free Hamiltonian $\hat{H}_0$, by

$$\hat{\varphi}(x) \equiv e^{i\hat{H}_0(t-t_0)} \hat{\varphi}_H(x, t_0) e^{-i\hat{H}_0(t-t_0)}. \qquad (18.121)$$

The operators in the two pictures are equal to each other at a time t_0 at which the density operator $\hat{\rho}$ is known. We also introduce the interaction picture for the interaction Hamiltonian[3]

$$\hat{H}_I^{\mathrm{int}}(t) \equiv e^{i\hat{H}t} \hat{H}^{\mathrm{int}}(t) e^{-i\hat{H}t}. \qquad (18.122)$$

This operator is used to set up the time evolution operator in the interaction picture

$$\hat{U}(t, t_0) \equiv \hat{T} \exp\left[i \int_{t_0}^t dt'\, \hat{H}_I^{\mathrm{int}}(t') \right]. \qquad (18.123)$$

It allows us to express the time dependence of the field operators $\hat{\varphi}(x)$ as follows:

$$\hat{\varphi}_H(x) = \hat{U}(t_0, t)\hat{\varphi}(x)\hat{U}(t, t_0). \qquad (18.124)$$

The ρ-averages of the Heisenberg fields in the interaction representation are therefore

$$\langle \hat{\varphi}_H(x) \rangle_\rho = \mathrm{Tr}\left[\hat{\rho}\,\hat{U}(t_0, t)\hat{\varphi}(x)\hat{U}(t, t_0) \right], \qquad (18.125)$$

$$\langle \hat{\varphi}_H(x)\hat{\varphi}_H(x') \rangle_\rho = \begin{cases} \mathrm{Tr}\left[\hat{\rho}\,\hat{U}(t_0, t)\hat{\varphi}(x)\hat{U}(t, t')\hat{\varphi}(x')\hat{U}(t', t_0) \right], & t > t', \\ \mathrm{Tr}\left[\hat{\rho}\,\hat{U}(t_0, t')\hat{\varphi}(x')\hat{U}(t', t)\hat{\varphi}(x)\hat{U}(t, t_0) \right], & t' > t. \end{cases} \qquad (18.126)$$

---

[3]For consistency, the field operator $\hat{\varphi}(x)$ should carry the same subscript I which is, however, omitted to shorten the notation.

Now, suppose that the interaction has been active for a very long time, i.e., we let $t_0 \to -\infty$. In this limit, (18.125) can be rewritten in terms of the scattering operator $\hat{S} \equiv \hat{U}(\infty, -\infty)$ of the system.[4] Using the time-ordering operator $\hat{T}$ of Eq. (1.241), we may write

$$\langle \hat{\varphi}_H(x) \rangle_\rho = \mathrm{Tr} \left[\hat{\rho} \, \hat{S}^\dagger \hat{T} \hat{S} \hat{\varphi}(x) \right], \tag{18.127}$$

$$\langle \hat{\varphi}_H(x) \hat{\varphi}_H(y) \rangle_\rho = \mathrm{Tr} \left[\hat{\rho} \, \hat{S}^\dagger \hat{T} \hat{S} \hat{\varphi}(x) \hat{\varphi}(y) \right]. \tag{18.128}$$

These expressions are indeed the same as those in (18.125) and (18.125); for instance

$$\hat{S}^\dagger \hat{T} \left(\hat{S} \hat{\varphi}(x) \right) = \hat{U}(-\infty, t) \hat{U}(t, \infty) \hat{T} \left(\hat{U}(\infty, t) \hat{\varphi}(x) \hat{U}(t, -\infty) \right)$$
$$= \hat{U}(-\infty, t) \hat{\varphi}(x) \hat{U}(t, -\infty). \tag{18.129}$$

For further development it is useful to realize that the operators in the expectations (18.127) and (18.128) can be reinterpreted time-ordered products of a new type, ordered along a *closed-time contour* which extends from $t = -\infty$ to $t = \infty$ *and back*. This contour is imagined to encircle the time axis in the complex t-plane as shown in Figure 18.1. The contour runs from $t = -\infty$ to $t = \infty$ above the real

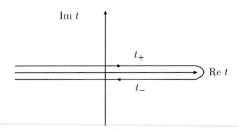

Figure 18.1   Closed-time contour in forward–backward path integrals.

time axis and returns below it. Accordingly, we distinguish t values from the upper branch and the lower branch by writing them as t_+ and t_-, respectively. Similarly we define $x(t_+) \equiv x_+$ and $x(t_-) \equiv x_-$. When viewed as a function of the closed-time contour, the operator

$$\hat{S}^\dagger \hat{T} \left(\hat{S} \hat{\varphi}(x) \right) \tag{18.130}$$

can be rewritten as

$$\hat{T}_P \left(\hat{S}^\dagger \hat{S} \hat{\varphi}(x_+) \right), \tag{18.131}$$

where $\hat{T}_P$ performs a time ordering along the closed-time contour. The coordinate x lies on the positive branch of the contour, where it is denoted by x_+. The operator $\hat{T}_P$ is called *path-ordering operator*.

---

[4]The matrix elements of $\hat{S}$ between momentum eigenstates form the so-called S-matrix.

We can then write down immediately a generating functional for an arbitrary product of field operators ordered along the closed-time path:

$$\hat{T}_P \left\{ \hat{S}^\dagger \hat{S} \exp \left[i \int dx \, j(x_+) \hat{\varphi}(x_+) \right] \right\}, \qquad (18.132)$$

where dx is short for $d^3x\,dt$. Functional differentiation with respect to $j(x_+)$ produces $\hat{\varphi}(x_+) \equiv \hat{\varphi}(x)$. For symmetry reasons it is also useful to introduce the source $j(x_-)$ coupled to the field on the lower time branch $\hat{\varphi}(x_-)$. Thus we shall work with the symmetric generating functional

$$Z[j_P] = \mathrm{Tr} \left(\hat{\rho} \, \hat{T}_P \hat{S}^\dagger \hat{S} \exp \left\{ i \left[\int dx \, j(x_+) \hat{\varphi}(x_+) + \int dx \, j(x_-) \hat{\varphi}(x_-) \right] \right\} \right).$$

It can be written as

$$Z[j_P] = \mathrm{Tr} \left\{ \hat{\rho} \, \hat{T}_P \hat{S}^\dagger \hat{S} \exp \left[i \int_P dx \, j_P(x) \hat{\varphi}_P(x) \right] \right\}, \qquad (18.133)$$

with the subscript p distinguishing the time branches.

The path-ordering symbol serves to write down a useful formal expression for the interaction representation of the operator $\hat{S}^\dagger \hat{S}$:[5]

$$\hat{S}^\dagger \hat{S} = \hat{T}_P \exp \left[-i \int_P dt \, \hat{H}_I^{\mathrm{int}}(t) \right]. \qquad (18.134)$$

In terms of this, $Z[j_P]$ takes the suggestive form

$$Z[j_P] = \mathrm{Tr} \left\{ \hat{\rho} \, \hat{T}_P \exp \left[-i \int_P dt \hat{H}_I^{\mathrm{int}}(t) + i \int_P dx \, j_P(x) \hat{\varphi}_P(x) \right] \right\}. \qquad (18.135)$$

To calculate the integrals along the closed-time contour p, it is advantageous to traverse the lower time branch in the same direction as the upper from $t = -\infty$ to ∞ (since we are used to integrating in this direction), and rewrite the closed-contour integral in the source term,

$$\int_P dx \, j_P(x) \hat{\varphi}_P(x) = \int d^3x \left[\int_{-\infty}^{\infty} dt \, j(\mathbf{x}, t_+) \hat{\varphi}(x_+) + \int_{\infty}^{-\infty} dt \, j(\mathbf{x}, t_-) \hat{\varphi}(x_-) \right], \ (18.136)$$

as

$$\int_P dx \, j_P(x) \hat{\varphi}_P(x) = \int d^3x \int_{-\infty}^{\infty} dt \, [j(\mathbf{x}, t_+) \hat{\varphi}(x_+) - j(\mathbf{x}, t_-) \hat{\varphi}(x_-)]. \qquad (18.137)$$

Obviously, the functional derivative with respect to $-j(\mathbf{x}, t_-)$ produces a factor $\hat{\varphi}(x_-)$. Correspondingly, we shall imagine the two fields $\hat{\varphi}(x_+)$, $\hat{\varphi}(x_-)$ as two components of a vector

$$\vec{\hat{\varphi}}(x) = \begin{pmatrix} \hat{\varphi}(x_+) \\ \hat{\varphi}(x_-) \end{pmatrix}, \qquad (18.138)$$

---

[5]Note that the left-hand side is equal to 1 due to $\hat{S}$ being unitary. However, this identity cannot be inserted into the path-ordered expressions (18.131)–(18.133), since the current terms require a factorization of $\hat{S}$ or $\hat{S}^\dagger$ at specific times and an insertion of field operators between the factors.

with the associated current

$$\vec{j}(x) = \begin{pmatrix} j(x_+) \\ -j(x_-) \end{pmatrix}. \tag{18.139}$$

In this vector notation, the source term reads

$$\int dx\, \vec{j}(x)\hat{\vec{\varphi}}(x), \tag{18.140}$$

and all closed-time path formulas go directly over into vector or matrix formulas whose integrals run only once along the positive time axis, for example

$$\int_{\mathrm{P}} dx\, j_{\mathrm{P}}(x) G_{\mathrm{P}}(x,x') j_{\mathrm{P}}(x') = \int dx\, \vec{j}(x) G(x,x') \vec{j}(x'), \tag{18.141}$$

where $G(x,x')$ on the right-hand side denotes the 2×2 matrix

$$G(x,y) = \begin{pmatrix} G_{++}(x,y) & G_{+-}(x,y) \\ G_{-+}(x,y) & G_{--}(x,y) \end{pmatrix} \equiv \begin{pmatrix} G(x_+,y_+) & G(x_+,y_-) \\ G(x_-,y_+) & G(x_-,y_-) \end{pmatrix}. \tag{18.142}$$

Since all formulas for j_{P} and $\hat{\varphi}_{\mathrm{P}}$ hold also for $\vec{j}$ and $\hat{\vec{\varphi}}$, we shall identify the closed-time path objects with the corresponding vectors and matrices.

Differentiating the generating functional with respect to j_{P} produces all Green functions of the theory. Forming two derivatives gives the two-point Green function

$$G_{\mathrm{P}}(x,y) = \frac{\delta}{i\delta j_{\mathrm{P}}(x)} \frac{\delta}{i\delta j_{\mathrm{P}}(y)} Z[j_{\mathrm{P}}]\Big|_{j_{\mathrm{P}}=0} = \mathrm{Tr}\left[\hat{\rho}\, \hat{T}_{\mathrm{P}} \hat{S}^\dagger \hat{S}\, \hat{\varphi}_{\mathrm{P}}(x) \hat{\varphi}_{\mathrm{P}}(y) \right], \tag{18.143}$$

which we decompose according to the branches of the closed-time contour in the same way as the matrix (18.142):

$$G_{\mathrm{P}}(x,y) = \begin{pmatrix} G_{++}(x,y) & G_{+-}(x,y) \\ G_{-+}(x,y) & G_{--}(x,y) \end{pmatrix}. \tag{18.144}$$

The four matrix elements collect precisely the four physically relevant time-dependent Green functions discussed in the last section for the case of $\hat{\rho}$ being an equilibrium density operator. Here they may be out of thermal equilibrium, formed with an arbitrary ρ-average rather than the thermal average at a given temperature. Going back from the interaction picture to the Heisenberg picture, the matrix $G_{\mathrm{P}}(x,y)$ is the expectation

$$G_{\mathrm{P}}(x,y) = \langle \hat{T}_{\mathrm{P}} \hat{\varphi}_H(x_{\mathrm{P}}) \hat{\varphi}_H(y_{\mathrm{P}}) \rangle_\rho, \tag{18.145}$$

where x_{P} can be x_+ or x_-. Considering the different components we observe that the path order is trivial as soon as x and y lie on different branches of the time axis. Since y_+ lies always before x_-, the path-ordering operator can be omitted so that

$$G_{-+}(x,y) = \langle \hat{\varphi}_H(x) \hat{\varphi}_H(y) \rangle_\rho. \tag{18.146}$$

In the opposite configuration, the path order is opposite. When reestablishing the original order, a negative sign arises for fermion fields. Hence,

$$G_{+-}(x,y) = \langle \hat{\varphi}_H(y)\hat{\varphi}_H(x)\rangle_\rho = \pm\langle \hat{\varphi}_H(x)\hat{\varphi}_H(y)\rangle_\rho. \tag{18.147}$$

In either case, a distinction of the upper and lower time branches is superfluous after an explicit path ordering.

If both x and y lie on the upper branch, the path order coincides with the usual time order so that $G_{++}(x,y)$ is equal to the expectation

$$G_{++}(x,y) = \langle \hat{T}\hat{\varphi}_H(x)\hat{\varphi}_H(y)\rangle_\rho \equiv G(x,y), \tag{18.148}$$

i.e., the ρ-average of the usual time-ordered Green function. Similarly, if x and y both lie on the lower branch, the path order coincides with the usual anti-time order and

$$G_{--}(x,y) = \langle \hat{\bar{T}}\hat{\varphi}_H(x)\hat{\varphi}_H(y)\rangle_\rho \equiv \bar{G}(x,y). \tag{18.149}$$

From these relations it is easy to see that only three of the four matrix elements of $G_P(x,y)$ are linearly independent, since there exists the relation

$$G_{++} + G_{--} = G_{+-} + G_{-+}. \tag{18.150}$$

This can be verified by writing out explicitly the time order and antiorder on the left-hand side. In the linear-response theory of Sections 18.1 and 18.2, the most convenient independent Green functions are the retarded and the advanced ones, together with the expectation of the anticommutator (the commutator for fermions). By analogy, we also define here, in the nonequilibrium case,

$$G^R(x,y) = \Theta(x-y)\langle[\hat{\varphi}_H(x),\hat{\varphi}_H(y)]_\mp\rangle_\rho, \tag{18.151}$$

$$G^A(x,y) = -\Theta(y-x)\langle[\hat{\varphi}_H(x),\hat{\varphi}_H(y)]_\mp\rangle_\rho, \tag{18.152}$$

$$A(x,y) = \langle[\hat{\varphi}_H(x),\hat{\varphi}_H(y)]_\pm\rangle_\rho. \tag{18.153}$$

As in (18.53), the last expression coincides with the absorptive or dissipative part of the Green function. The expectation of the commutator (the anticommutator for fermions),

$$C(x,y) = \langle[\hat{\varphi}_H(x),\hat{\varphi}_H(y)]_\mp\rangle_\rho, \tag{18.154}$$

is not an independent quantity. It is related to the others by

$$C(x,y) = G^R(x,y) - G^A(x,y). \tag{18.155}$$

A comparison of the Fourier decomposition of the field (18.115) with (18.92) shows that the Green functions are simple plane-wave superpositions of harmonic oscillator of all momenta $\mathbf{p}$ and frequency $\Omega = \omega_{\mathbf{p}}$. The normalization factor $\hbar/M$ becomes $1/V$. For instance

$$G^R(x, x') = \sum_{\mathbf{p}} \frac{M}{\hbar V} e^{i\mathbf{p}(\mathbf{x}-\mathbf{x}')} G^R(t, t')|_{\Omega=\omega_{\mathbf{p}}}. \tag{18.156}$$

In the continuum limit, where the sum over momenta goes over into an integral with the rule (7.558), this becomes, from (18.95),

$$G^R(x, x') = -\Theta(x - x') \int \frac{d^D p}{2\omega_{\mathbf{p}}(2\pi)^D} e^{i\mathbf{k}(\mathbf{x}-\mathbf{x}')} 2i \sin \omega_{\mathbf{p}}(t - t'). \tag{18.157}$$

Similarly we find from (18.102)

$$A(x, x') = \int \frac{d^D p}{2\omega_{\mathbf{p}}(2\pi)^D} e^{i\mathbf{k}(\mathbf{x}-\mathbf{x}')} 2 \cos \omega_{\mathbf{p}}(t - t'). \tag{18.158}$$

These and the other Green functions satisfy identities analogous to those formed from the position operator $\hat{\varphi}(t)$ of a simple harmonic oscillator in (18.105)–(18.108):

$$\begin{align}
G^A(x, x') &= \mp G^R(x', x), \tag{18.159}\\
A(x, x') &= \pm A(x', x), \tag{18.160}\\
C(x, x') &= \mp C(x', x). \tag{18.161}\\
G(x, x') &= \pm G(x', x)^*. \tag{18.162}
\end{align}$$

It is now easy to express the matrix elements of the 2×2 Green function $G_P(x, y)$ in (18.144) in terms of the three independent quantities (18.153). Since

$$\begin{align}
G^R &= G_{-+} - G_{--} = G_{++} - G_{+-},\\
G^A &= G_{+-} - G_{--} = G_{++} - G_{-+},\\
A &= G_{-+} + G_{+-} = G_{++} + G_{--},\\
C &= G_{-+} - G_{+-} = G^R - G^A,
\end{align} \tag{18.163}$$

we find

$$\begin{align}
G_{-+} &= \frac{1}{2}(A + C) = \tfrac{1}{2}(A + G^R - G^A),\\
G_{+-} &= \frac{1}{2}(A - C) = \tfrac{1}{2}(A - G^R + G^A),
\end{align} \tag{18.164}$$

and

$$\begin{align}
G_{++} &= G^R + G_{+-} = \tfrac{1}{2}(A + G^R + G^A),\\
G_{--} &= G_{+-} + G_{-+} - G_{++}\\
&= A - G_{++} = \tfrac{1}{2}(A - G^R - G^A).
\end{align} \tag{18.165}$$

The matrix $G_P(x, y)$ can therefore be written as follows:

$$G_P = \frac{1}{2} \begin{pmatrix} A + G^R + G^A & A - G^R + G^A \\ A + G^R - G^A & A - G^R - G^A \end{pmatrix}. \tag{18.166}$$

For actual calculations it is somewhat more convenient to use a transformation introduced by Keldysh [5]. It arises from the similarity transformation

$$\tilde{G} = QG_PQ^{-1}, \quad \text{with} \quad Q = \frac{1}{\sqrt{2}} \begin{pmatrix} 1 & -1 \\ 1 & 1 \end{pmatrix} = (Q^T)^{-1}, \tag{18.167}$$

producing the simpler triangular Green function matrix

$$\tilde{G}(x, y) = \frac{1}{\sqrt{2}} \begin{pmatrix} 1 & -1 \\ 1 & 1 \end{pmatrix} \frac{1}{2} \begin{pmatrix} A + G^R + G^A & A - G^R + G^A \\ A + G^R - G^A & A - G^R - G^A \end{pmatrix}$$

$$\times \frac{1}{\sqrt{2}} \begin{pmatrix} 1 & 1 \\ -1 & 1 \end{pmatrix} = \begin{pmatrix} 0 & G^A \\ G^R & A \end{pmatrix}. \tag{18.168}$$

Due to the calculational advantages it is worth re-expressing all quantities in the new basis. The linear source term, for example, becomes

$$\int_P dx\, j_P(x)\hat{\varphi}_P(x) = \int dx\, (j(x_+), -j(x_-)) \begin{pmatrix} \hat{\varphi}(x_+) \\ \hat{\varphi}(x_-) \end{pmatrix}$$

$$= \int dx\, \tilde{j}(x)\tilde{\hat{\varphi}}(x), \tag{18.169}$$

with the source vectors

$$\tilde{\hat{\varphi}}(x) \equiv Q \begin{pmatrix} \hat{\varphi}(x_+) \\ \hat{\varphi}(x_-) \end{pmatrix} = \frac{1}{\sqrt{2}} \begin{pmatrix} \hat{\varphi}(x_+) - \hat{\varphi}(x_-) \\ \hat{\varphi}(x_+) + \hat{\varphi}(x_-) \end{pmatrix}, \tag{18.170}$$

and the field vectors

$$\tilde{j}(x) \equiv \begin{pmatrix} j_1(x) \\ j_2(x) \end{pmatrix} = Q \begin{pmatrix} j(x_+) \\ -j(x_-) \end{pmatrix} = \frac{1}{\sqrt{2}} \begin{pmatrix} j(x_+) + j(x_-) \\ j(x_+) - j(x_-) \end{pmatrix}. \tag{18.171}$$

The quadratic source term

$$\int dx\, dx'\, j_P(x)G_P(x, x')j_P(x') \tag{18.172}$$

$$= \int dx\, dx'\, (j(x_+), -j(x_-)) \begin{pmatrix} G_{++} & G_{+-} \\ G_{-+} & G_{--} \end{pmatrix}(x, x') \begin{pmatrix} j(x'_+) \\ -j(x'_-) \end{pmatrix}$$

becomes

$$\int dx\, dx'\, \tilde{j}^T(x)\tilde{G}(x, x')\tilde{j}(x'). \tag{18.173}$$

The product[6] of two Green functions $\tilde{G}^{(1)}$ and $\tilde{G}^{(2)}$ has the same triangular form as each factor. The three nonzero entries are composed as follows:

$$
\begin{aligned}
\tilde{G}^{12} &= \tilde{G}^{(1)}\tilde{G}^{(2)} = QG_P^{(1)}Q^{-1}QG_P^{(2)}Q^{-1} \\
&= \begin{pmatrix} 0 & G_1^A G_2^A \\ G_1^R G_2^R & G_1^R A_2 + A_1 G_2^A \end{pmatrix}.
\end{aligned}
\tag{18.174}
$$

More details on these Green functions can be found in the literature [6].

18.7 Perturbation Theory for Nonequilibrium Green Functions

The interaction picture can be used to develop a perturbation expansion for nonequilibrium Green functions. For this we go back to the generating functional (18.135) and assume that the interaction depends only on the field operators. Usually it will be a local interaction, i.e., a spacetime integral over an interaction density:

$$
\exp\left[-i\int_P dt\,\hat{H}_I^{\text{int}}(t)\right] = \exp\left[i\int_P dt\int d^3x\,L^{\text{int}}(\hat{\varphi}_P(\mathbf{x},t))\right].
\tag{18.175}
$$

The subsequent formal development applies also to the case of a more general non-local interaction

$$
\exp\left\{i\mathcal{A}_P^{\text{int}}[\hat{\varphi}_P]\right\}.
\tag{18.176}
$$

To account for the interaction, we use the fact used in Section 3.18, that within the expectation (18.135) the field $\hat{\varphi}_P$ can be written as a differential operator $\delta/i\delta j_P(x)$ applied to the source term. In this form, the interaction term can be moved outside the thermal expectation. The result is the generating functional in the interaction picture

$$
Z[j_P] = \exp\left\{i\mathcal{A}_P^{\text{int}}\left[\delta/i\delta j_P\right]\right\} Z_0[j_P],
\tag{18.177}
$$

where

$$
Z_0[j_P] = \text{Tr}\left\{\hat{\rho}\,\hat{T}_P\exp\left[i\int_P dx\,\hat{\varphi}_P(x)\,j_P(x)\right]\right\}
\tag{18.178}
$$

is the free partition function.

To apply this formula, we have to find $Z_0[j_P]$ explicitly. By expanding the exponential in powers of $i\mathcal{A}_P^{\text{int}}[\delta/i\delta j_P]$ and performing the functional derivatives $\delta/i\delta j_P$, we obtain the desired perturbation expansion for $Z[j_P]$.

For a general density operator $\hat{\rho}$, the free partition function $Z_0[j_P]$ cannot be written down in closed form. Here we give $Z_0[j_P]$ explicitly only for a harmonic

---

[6]The product is meant in the functional sense, i.e.,
$(G^{(1)}G^{(2)})(x,y) = \int dz\,G^{(1)}(x,z)G^{(2)}(z,y)$.

system in thermal equilibrium, where the ρ-averages $\langle \ldots \rangle_\rho$ are the thermal averages $\langle \ldots \rangle_T$ calculated in Sections 18.1 and 18.2. Since the fluctuation terms in the field $\varphi(t)$ are quadratic, $Z_0[j_P]$ must have an exponent quadratic in the sources j_P. To satisfy (18.143), the functional is necessarily given by

$$Z_0[j_P] = \exp\left[-\frac{1}{2} \int dx dy \, j_P(x) G_P(x, y) j_P(y)\right]. \tag{18.179}$$

Inserting the 4×4-matrix (18.166), this becomes

$$Z_0[j_+, j_-] = \exp\left[-\frac{1}{2} \int dx \int dx' (j_+, -j_-) Q^{-1} \begin{pmatrix} 0 & G^A \\ G^R & A \end{pmatrix} Q \begin{pmatrix} j_+ \\ -j_- \end{pmatrix}\right]$$

$$= \exp\left\{-\frac{1}{4} \int dx \int dx' \left[(j_+ + j_-)(x) G^A(x, x')(j_+ - j_-)(x')\right.\right.$$
$$+ (j_+ - j_-)(x) G^R(x, x')(j_+ + j_-)(x')$$
$$\left.\left. + (j_+ - j_-)(x) A(x, x')(j_+ - j_-)(x')\right]\right\}, \tag{18.180}$$

where

$$j_+(x) \equiv j(x_+), \quad j_-(x) \equiv j(x_-). \tag{18.181}$$

The advanced Green functions are different from zero only for $t < t'$. Using relation (18.159), the second term is seen to be the same as the first. For the real field at hand, these terms are purely imaginary [see (18.156)]. The anticommutation function $A(x, x')$ is symmetric by (18.160). We therefore rewrite (18.180) as

$$Z_0[j_+, j_-] = \exp\left\{-\frac{1}{2} \int dx \int dx' \, \Theta(x' - x)\right. \tag{18.182}$$
$$\left. \times \left[(j_+ - j_-)(x) G^R(x, x')(j_+ + j_-)(x') + (j_+ - j_-)(x) A(x, x')(j_+ - j_-)(x')\right]\right\}.$$

For any given spectral function, the exponent can easily be written down explicitly using the spectral representations (18.44) and (18.53).

As an important example consider the simple case of a single harmonic oscillator of frequency Ω. Then the field $\hat\varphi(x)$ depends only on the time t, and the commutator and "wrong" commutator functions are given by (18.93) and (18.102). Reintroducing all factors $\hbar$ and k_B, we have

$$Z_0[j_+, j_-] = \exp\left\{-\frac{1}{2\hbar^2} \int dt \int dt' \, \Theta(t - t')\right. \tag{18.183}$$
$$\left. \times \left[(j_+ - j_-)(t) C(t, t')(j_+ + j_-)(t') + (j_+ - j_-)(t) A(t, t')(j_+ - j_-)(t')\right]\right\}$$

or, more explicitly,

$$Z_0[j_+, j_-] = \exp\left\{-\frac{1}{2M\Omega\hbar} \int dt \int dt' \, \Theta(t' - t)\right.$$
$$\times \left[-(j_+ - j_-)(t) \, i\sin\Omega(t - t') \, (j_+ + j_-)(t')\right. \tag{18.184}$$
$$\left.\left. + (j_+ - j_-)(t) \, \coth\frac{\hbar\Omega}{2k_BT} \cos\Omega(t - t') \, (j_+ - j_-)(t')\right]\right\}.$$

We have taken advantage of the presence of the Heaviside function to express the retarded Green function for $t > t'$ as a commutator function $C(t, t')$ [recall (18.151), (18.154)]. Together with the anticommutator function $A(t, t')$, we obtain for $t > t'$

$$G(t, t') = \frac{1}{2}[A(t, t') + C(t, t')] = \frac{\hbar}{2M\Omega} \frac{\cosh\frac{\Omega}{2}[\hbar\beta - i(t - t')]}{\sinh\frac{\hbar\Omega\beta}{2}}, \qquad (18.185)$$

which coincides with the time-ordered Green function (18.101) for $t > t'$, and thus with the analytically continued periodic imaginary-time Green function (3.248). The exponent in this generating functional is thus quite similar to the equilibrium source term (3.218).

The generating functional (18.180) can, of course, be derived without the previous operator discussion completely in terms of path integrals for the harmonic oscillator in thermal equilibrium. Using the notation $\hat{X}(t)$ for a purely time-dependent oscillator field $\hat{\varphi}(x)$, we can take the generating functional directly from Eq. (3.168):

$$
\begin{aligned}
(X_b t_b | X_a t_a)_\Omega^j &= \int \mathcal{D}X(t) \exp\left\{\frac{i}{\hbar}\int_{t_a}^{t_b} dt\left[\frac{M}{2}(\dot{X}^2 - \Omega^2 X^2) + jX\right]\right\} \\
&= e^{(i/\hbar)\mathcal{A}_{\mathrm{cl},j}} F_{\Omega,j}(t_b, t_a).
\end{aligned}
\qquad (18.186)
$$

with a total classical action

$$
\begin{aligned}
\mathcal{A}_{\mathrm{cl},j} &= \frac{1}{2}\frac{M\Omega}{\sin\Omega(t_b - t_a)}\left[(X_b^2 + X_a^2)\cos\Omega(t_b - t_a) - 2X_b X_a\right] \\
&\quad + \frac{1}{\sin\Omega(t_b - t_a)}\int_{t_a}^{t_b} dt[X_a \sin\Omega(t_b - t) + X_b \sin\Omega(t - t_a)]j(t),
\end{aligned}
\qquad (18.187)
$$

and the fluctuation factor (3.170), and express (18.187) as in (3.171) in terms of the two independent solutions $D_a(t)$ and $D_b(t)$ of the homogenous differential equations (3.48) introduced in Eqs. (2.226) and (2.227):

$$
\begin{aligned}
\mathcal{A}_{\mathrm{cl},j} &= \frac{M}{2D_a(t_b)}\left[X_b^2 \dot{D}_a(t_b) - X_a^2 \dot{D}_b(t_a) - 2X_b X_a\right] \\
&\quad + \frac{1}{D_a(t_b)}\int_{t_a}^{t_b} dt\,[X_b D_a(t) + X_a D_b(t)]j(t).
\end{aligned}
\qquad (18.188)
$$

The fluctuation factor is taken as in (3.172). Then we calculate the thermal average of the forward–backward path integral of the oscillator $X(t)$ via the Gaussian integral

$$Z_0[j_+, j_-] = \int dX_b\, dX_a\, (X_b\,\hbar\beta | X_a 0)_\Omega\, (X_b t_b | X_a t_a)_\Omega^{j+}\, (X_b t_b | X_a t_a)_\Omega^{j-*}. \qquad (18.189)$$

Here $(X_b\,\hbar\beta | X_a 0)_\Omega$ is the imaginary-time amplitude (2.409):

$$
\begin{aligned}
(X_b\hbar\beta | X_a 0) &= \frac{1}{\sqrt{2\pi\hbar/M}}\sqrt{\frac{\Omega}{\sinh\hbar\beta}} \\
&\quad \times \exp\left\{-\frac{1}{2\hbar}\frac{M\Omega}{\sinh\hbar\beta\Omega}[(X_b^2 + X_a^2)\cosh\hbar\beta\Omega - 2X_b X_a]\right\}.
\end{aligned}
\qquad (18.190)
$$

We have preferred deriving $Z_0[j_+, j_-]$ in the operator language since this illuminates better the physical meaning of the different terms in the result (18.185).

18.8 Path Integral Coupled to Thermal Reservoir

After these preparations, we can embark on a study of a simple but typical problem of nonequilibrium thermodynamics. We would like to understand the quantum-mechanical behavior of a particle coupled to a thermal reservoir of temperature T and moving in an arbitrary potential $V(x)$ [7]. Without the reservoir, the probability of going from x_a, t_a to x_b, t_b would be given by[7]

$$|(x_b t_b | x_a t_a)|^2 = \left| \int \mathcal{D}x(t) \exp\left\{ \frac{i}{\hbar} \int dt \left[\frac{M}{2} \dot{x}^2 - V(x) \right] \right\} \right|^2. \quad (18.191)$$

This may be written as a path integral over two independent orbits, to be called $x_+(t)$ and $x_-(t)$:

$$(x_b t_b | x_a t_a)(x_b t_b | x_a t_a)^* = \int \mathcal{D}x_+(t) \, \mathcal{D}x_-(t) \quad (18.192)$$

$$\times \exp\left\{ \frac{i}{\hbar} \int_{t_a}^{t_b} dt \left[\frac{M}{2}(\dot{x}_+^2 - \dot{x}_-^2) - (V(x_+) - V(x_-)) \right] \right\}.$$

In accordance with the development in Section 18.7, the two orbits are reinterpreted as two branches of a single closed-time orbit $x_P(t)$. The time coordinate t_P of the path runs from t_a to t_b slightly above the real time axis and returns slightly below it, just as in Fig. 18.1. The probability distribution (18.191) can then be written as a path integral over the closed-time contour encircling the interval (t_a, t_b):

$$|(x_b t_b | x_a t_a)|^2 = \int \mathcal{D}x_P \exp\left\{ \frac{i}{\hbar} \int_P dt \left[\frac{M}{2} \dot{x}_P^2 - V(x_P) \right] \right\}. \quad (18.193)$$

We now introduce a coupling to a thermal reservoir for which we use, as in the equilibrium discussion in Section 3.13, a bath of independent harmonic oscillators $\hat{\varphi}_i(t)$ of masses M_i and frequencies Ω_i in thermal equilibrium at temperature T. For simplicity, the coupling is assumed to be linear in $\hat{\varphi}_i(t)$ and the position of the particle $x(t)$. The bath contributes to (18.193) a factor involving the thermal expectation of the linear interaction

$$|(x_b t_b | x_a t_a)|^2 = \int \mathcal{D}x_P \exp\left\{ \frac{i}{\hbar} \int_P dt \left[\frac{M}{2} \dot{x}_P^2 - V(x_P) \right] \right\}$$

$$\times \text{Tr} \left\{ \hat{\rho} \, \hat{T}_P \exp\left[\frac{i}{\hbar} \sum_i c_i \int_P dt \, \hat{\varphi}_P^i(t) \, x_P(t) \right] \right\}. \quad (18.194)$$

---

[7]In the sequel, we display the constants $\hbar$ and k_B explicitly.

Here, $\hat{\varphi}_P^i(t)$ for $i = 1, 2, 3, \ldots$ are the position operators of the auxiliary harmonic oscillators. Since the oscillators are independent, the trace of the exponentials factorizes into a product of single-oscillator expressions

$$\text{Tr}\left\{\hat{\rho}\,\hat{T}_P \exp\left[\frac{i}{\hbar}\sum_i c_i \int_P dt\,\hat{\varphi}_P^i(t)x_P(t)\right]\right\} = \prod_i \text{Tr}\left\{\hat{\rho}\,\hat{T}_P \exp\left[\frac{i}{\hbar}c_i \int_P dt\,\hat{\varphi}_P^i(t)x_P(t)\right]\right\}. \tag{18.195}$$

The density operator $\hat{\rho}$ has the eigenvalues (18.110).

Each factor on the right-hand side is of the form (18.178) with $\hat{\varphi}(t) = c_i\hat{\varphi}_P^i(t)/\hbar$ and $j_{+,-} = x_{+,-}(t)$, so that (18.195) leads to the partition function (18.183), which reads here

$$Z_0^b[x_+, x_-] = \exp\left\{-\frac{1}{2\hbar^2}\int dt \int dt'\,\Theta(t - t')\right. \tag{18.196}$$
$$\left. \times\left[(x_+ - x_-)(t)C_b(t, t')(x_+ + x_-)(t') + (x_+ - x_-)(t)A_b(t, t')(x_+ - x_-)(t')\right]\right\},$$

where $C_b(t, t')$ and $A_b(t, t')$ collect the commutator and anticommutator functions of the bath. They are sums of correlation functions (18.93) and (18.102) of the individual oscillators of mass M_i and frequency Ω_i, each contributing with a weight c_i^2. Thus we may write

$$C_b(t, t') = \sum_i c_i^2 \langle[\hat{\varphi}_i(t), \hat{\varphi}_i(t')]\rangle_T = -\hbar \int_{-\infty}^{\infty} \frac{d\omega'}{2\pi} \rho_b(\omega') i \sin\omega'(t - t'), \tag{18.197}$$

$$A_b(t, t') = \sum_i c_i^2 \langle\{\hat{\varphi}_i(t), \hat{\varphi}_i(t')\}\rangle_T = \hbar \int_{-\infty}^{\infty} \frac{d\omega'}{2\pi} \rho_b(\omega') \coth\frac{\hbar\omega'}{2k_BT} \cos\omega'(t - t'), \tag{18.198}$$

where the ensemble averages at a fixed temperature T are now denoted by a subscript T, and

$$\rho_b(\omega') \equiv 2\pi \sum_i \frac{c_i^2}{2M_i\Omega_i}[\delta(\omega' - \Omega_i) - \delta(\omega' + \Omega_i)] \tag{18.199}$$

is the spectral function of the bath. It is the antisymmetric continuation of the spectral function (3.405) to negative ω'. Since the spectral function of the bath $\rho_b(\omega')$ of (18.199) is odd in ω', we can replace both trigonometric functions $-i\sin\omega'(t - t')$ and $\cos\omega'(t - t')$ in (18.199) by the exponentials $e^{-i\omega'(t - t')}$.

The expression in the exponent of (18.196) may be considered as an effective action in the path integral, caused by the thermal bath. We shall therefore write

$$Z_0[x_+, x_-] = \exp\left\{\frac{i}{\hbar}\mathcal{A}^{\text{FV}}[x_+, x_-]\right\} = \exp\left\{\frac{i}{\hbar}\left(\mathcal{A}_D^{\text{FV}}[x_+, x_-] + \mathcal{A}_F^{\text{FV}}[x_+, x_-]\right)\right\}, \tag{18.200}$$

where the effective action $\mathcal{A}^{\text{FV}}[x_+, x_-]$ consists of a dissipative part $\mathcal{A}_D^{\text{FV}}[x_+, x_-]$ and a fluctuation part $\mathcal{A}_F^{\text{FV}}[x_+, x_-]$. The expression $Z_0[x_+, x_-]$ is the famous *influence functional* first introduced by Feynman and Vernon.

Inserting (18.200) into (18.194) and displaying explicitly the two branches of the path $x_P(t)$ with the proper limits of time integrations, we obtain from (18.194) the probability for the particle to move from $x_a t_a$ to $x_b t_b$ as the path integral

$$|(x_b t_b | x_a t_a)|^2 = \int \mathcal{D}x_+(t) \int \mathcal{D}x_-(t) \times$$

$$\times \exp\left\{ \frac{i}{\hbar} \int_{t_a}^{t_b} dt \left[\frac{M}{2}(\dot{x}_+^2 - \dot{x}_-^2) - (V(x_+) - V(x_-)) \right] + \frac{i}{\hbar}\mathcal{A}^{\mathrm{FV}}[x_+, x_-] \right\}. \quad (18.201)$$

For a better understanding of the influence functional, we introduce an auxiliary retarded function

$$\gamma(t - t') \equiv \Theta(t - t') \frac{1}{M} \int_{-\infty}^{\infty} \frac{d\omega}{2\pi} \frac{\sigma_{\mathrm{b}}(\omega)}{\omega} e^{-i\omega(t-t')}. \quad (18.202)$$

Then we can write

$$\Theta(t - t') C_{\mathrm{b}}(t, t') = i\hbar M\dot{\gamma}(t - t') + i\hbar M \Delta\omega^2 \delta(t - t'), \quad (18.203)$$

where the quantity

$$\Delta\omega^2 \equiv -\frac{1}{M} \int_{-\infty}^{\infty} \frac{d\omega'}{2\pi} \frac{\sigma_{\mathrm{b}}(\omega')}{\omega'} = -\frac{1}{M} \sum_i \frac{c_i^2}{M_i \Omega_i^2} \quad (18.204)$$

was introduced before in Eq. (3.417). Inserting the first term of the decomposition (18.203) into (18.196), the dissipative part of the influence functional can be integrated by parts in t' and becomes

$$\mathcal{A}_D^{\mathrm{FV}}[x_+, x_-] = -\frac{M}{2} \int_{t_a}^{t_b} dt \int_{t_a}^{t_b} dt' \, (x_+ - x_-)(t)\gamma(t - t')(\dot{x}_+ + \dot{x}_-)(t')$$

$$+ \frac{M}{2} \int_{t_a}^{t_b} dt (x_+ - x_-)(t)\gamma(t - t_b)(x_+ + x_-)(t_a). \quad (18.205)$$

The δ-function in (18.203) contributes to $\mathcal{A}_D^{\mathrm{FV}}[x_+, x_-]$ a term analogous to (3.418)

$$\Delta\mathcal{A}_{\mathrm{loc}}[x_+, x_-] = \frac{M}{2} \int_{t_a}^{t_b} dt \Delta\omega^2 (x_+^2 - x_-^2)(t), \quad (18.206)$$

which may simply be absorbed into the potential terms of the path integral (18.201), renormalizing them to

$$-\frac{i}{\hbar} \int_{t_a}^{t_b} dt \, [V_{\mathrm{ren}}(x_+) - V_{\mathrm{ren}}(x_-)]. \quad (18.207)$$

This renormalization is completely analogous to that in the imaginary-time formula (3.420).

The odd bath function $\rho_{\mathrm{b}}(\omega')$ can be expanded in a power series with only odd powers of ω'. The lowest approximation

$$\rho_{\mathrm{b}}(\omega') \approx 2M\gamma\omega', \quad (18.208)$$

describes Ohmic dissipation with some friction constant γ [recall (3.424)]. For frequencies much larger than the atomic relaxation rates, the friction goes to zero. This behavior is modeled by the Drude form (3.425) of the spectral function

$$\rho_b(\omega') \approx 2M\gamma\omega'\frac{\omega_D^2}{\omega_D^2 + \omega'^2}. \tag{18.209}$$

Inserting this into Eq. (18.202), we obtain the Drude form of the function $\gamma(t)$:

$$\gamma_D^R(t) \equiv \Theta(t)\,\gamma\omega_D e^{-\omega_D t}. \tag{18.210}$$

The superscript emphasizes the retarded nature. This can also be written as a Fourier integral

$$\gamma_D^R(t) = \int_{-\infty}^{\infty} \frac{d\omega'}{2\pi} \gamma_D^R(\omega') e^{-i\omega' t}, \tag{18.211}$$

with the Fourier components

$$\gamma_D^R(\omega') = \gamma\frac{i\omega_D}{\omega' + i\omega_D}. \tag{18.212}$$

The position of the pole in the lower half-plane ensures the retarded nature of the friction term by producing the Heaviside function in (18.210) [recall (1.308)].

The imaginary-time expansion coefficients γ_m of Eq. (3.428) are related to these by

$$\gamma_m = \gamma(\omega')|_{\omega'=i|\omega_m|}, \tag{18.213}$$

by close analogy with the relation between the retarded and imaginary-time Green functions (18.30) and (18.31).

In the Ohmic limit (18.208), the dissipative part of the influence functional simplifies. Then $\gamma_D^R(t)$ becomes narrowly peaked at positive t, and may be approximated by a right-sided *retarded* δ-function as

$$\gamma_D^R(t) \to \gamma\,\delta^R(t), \tag{18.214}$$

whose superscript R indicates the retarded asymmetry of the δ-function. With this, (18.205) becomes a local action

$$A_D^{\mathrm{FV}}[x_+, x_-] = -\frac{M}{2}\gamma \int_{t_a}^{t_b} dt (x_+ - x_-)(\dot{x}_+ + \dot{x}_-)^R - \frac{M}{2}\gamma(x_+^2 - x_-^2)(t_a). \tag{18.215}$$

The right-sided nature of the function $\delta^R(t)$ causes an infinitesimal *negative* shift in the time argument of the velocities $(\dot{x}_+ + \dot{x}_-)(t)$ with respect to the factor $(x_+ - x_-)(t)$, indicated by the superscript R. It expresses the *causality* of the friction forces and will be seen to be crucial in producing a probability conserving time evolution of the probability distribution.

The second term changes only the curvature of the effective potential at the initial time, and can be ignored.

It is useful to incorporate the slope information (18.208) also into the bath correlation function $A_b(t, t')$ in (18.198), and factorize it as

$$A_b(t, t') = 2M\gamma k_B T K(t, t'), \tag{18.216}$$

where

$$K(t, t') = K(t - t') \equiv \frac{1}{2M\gamma k_B T} \sum_i c_i^2 \langle \{\hat{\varphi}_i(t), \hat{\varphi}_i(t')\} \rangle_T. \tag{18.217}$$

The prefactor in (18.216) is conveniently abbreviated by the constant

$$w \equiv 2M\gamma k_B T, \tag{18.218}$$

which is related to the so-called *diffusion constant*

$$D \equiv k_B T / M\gamma \tag{18.219}$$

by

$$w = 2\gamma^2 M^2 D. \tag{18.220}$$

The Fourier decomposition of (18.217) is

$$K(t, t') = \int_{-\infty}^{\infty} \frac{d\omega'}{2\pi} K(\omega') e^{-i\omega'(t-t')}, \tag{18.221}$$

with

$$K(\omega') \equiv \frac{1}{2M\gamma} \frac{\rho_b(\omega')}{\omega'} \frac{\hbar\omega'}{2k_B T} \coth \frac{\hbar\omega'}{2k_B T}. \tag{18.222}$$

In the limit of a purely Ohmic dissipation this simplifies to

$$K(\omega') \to K^{\mathrm{Ohm}}(\omega') \equiv \frac{\hbar\omega'}{2k_B T} \coth \frac{\hbar\omega'}{2k_B T}. \tag{18.223}$$

The function $K(\omega')$ has the normalization $K(0) = 1$, giving $K(t-t')$ a unit temporal area:

$$\int_{-\infty}^{\infty} dt\, K(t - t') = 1. \tag{18.224}$$

In the classical limit $\hbar \to 0$, the Drude spectral function (18.209) leads to

$$K_D^{\mathrm{cl}}(\omega') = \frac{\omega_D^2}{\omega'^2 + \omega_D^2}, \tag{18.225}$$

with the Fourier transform

$$K_D^{cl}(t - t') = \frac{1}{2\omega_D} e^{-\omega_D(t-t')}. \qquad (18.226)$$

In the limit of Ohmic dissipation, this becomes a δ-function. Thus $K(t - t')$ may be viewed as a δ-function broadened by quantum fluctuations and relaxation effects.

With the function $K(t, t')$, the fluctuation part of the influence functional in (18.196), (18.200), (18.201) becomes

$$\mathcal{A}_F^{FV}[x_+, x_-] = i\frac{w}{2\hbar} \int_{t_a}^{t_b} dt \int_{t_a}^{t_b} dt' \, (x_+ - x_-)(t) \, K(t, t') \, (x_+ - x_-)(t'). \qquad (18.227)$$

Here we have used the symmetry of the function $K(t, t')$ to remove the Heaviside function $\Theta(t - t')$ from the integrand, and to extend the range of t'-integration to the entire interval (t_a, t_b).

In the Ohmic limit, the probability of the particle to move from $x_a t_a$ to $x_b t_b$ is given by the path integral

$$|(x_b t_b | x_a t_a)|^2 = \int \mathcal{D}x_+(t) \int \mathcal{D}x_-(t)$$

$$\times \exp\left\{ \frac{i}{\hbar} \int_{t_a}^{t_b} dt \left[\frac{M}{2} (\dot{x}_+^2 - \dot{x}_-^2) - (V(x_+) - V(x_-)) \right] \right\}$$

$$\times \exp\left\{ -i \int_{t_a}^{t_b} dt \frac{M\gamma}{2\hbar} (x_+ - x_-)(t)(\dot{x}_+ + \dot{x}_-)^R(t) \right.$$

$$\left. - \frac{w}{2\hbar^2} \int_{t_a}^{t_b} dt \int_{t_a}^{t_b} dt' \, (x_+ - x_-)(t) \, K^{Ohm}(t, t') \, (x_+ - x_-)(t') \right\}. \qquad (18.228)$$

This is the *closed-time path integral* of a particle in contact with a thermal reservoir.

The paths $x_+(t), x_-(t)$ may also be associated with a forward and a backward movement of the particle in time. For this reason, (18.228) is also called a *forward–backward path integral*. The hyphen is pronounced as *minus*, to emphasize the opposite signs in the partial actions.

It is now convenient to change integration variables and go over to average and relative coordinates of the two paths x_+, x_-:

$$x \equiv (x_+ + x_-)/2,$$
$$y \equiv x_+ - x_-. \qquad (18.229)$$

Then (18.228) becomes

$$|(x_b t_b | x_a t_a)|^2 = \int \mathcal{D}x(t) \int \mathcal{D}y(t)$$

$$\times \exp\left\{ -\frac{i}{\hbar} \int_{t_a}^{t_b} dt \left[M \left(-\dot{y}\dot{x} + \gamma y \dot{x}^R \right) + V \left(x + \frac{y}{2} \right) - V \left(x - \frac{y}{2} \right) \right] \right.$$

$$\left. - \frac{w}{2\hbar^2} \int_{t_a}^{t_b} dt \int_{t_a}^{t_b} dt' \, y(t) K^{Ohm}(t, t') y(t') \right\}. \qquad (18.230)$$

18.9 Fokker-Planck Equation

At high temperatures, the Fourier transform of the Kernel $K(t, t')$ in Eq. (18.223) tends to unity such that $K(t, t')$ becomes a δ-function, so that the path integral (18.230) for the probability distribution of a particle coupled to a thermal bath simplifies to

$$P(x_b t_b | x_a t_a) \equiv |(x_b t_b | x_a t_a)|^2 = \int \mathcal{D}x(t) \int \mathcal{D}y(t)$$
$$\times \exp\left\{-\frac{i}{\hbar} \int_{t_a}^{t_b} dt\, y[M\ddot{x} + M\gamma\dot{x}^R + V'(x)] - \frac{w}{2\hbar^2} \int_{t_a}^{t_b} dt\, y^2\right\}. \quad (18.231)$$

The superscript R records the infinitesimal backward shift of the time argument as in Eq. (18.215). The y-variable can be integrated out, and we obtain

$$P(x_b t_b | x_a t_a) = \mathcal{N} \int \mathcal{D}x(t)\, \exp\left\{-\frac{1}{2w} \int_{t_a}^{t_b} dt\, [M\ddot{x} + M\gamma\dot{x}^R + V'(x)]^2\right\}. \quad (18.232)$$

The proportionality constant $\mathcal{N}$ can be fixed by the normalization integral

$$\int dx_b\, P(x_b t_b | x_a t_a) = 1. \quad (18.233)$$

Since the particle is initially concentrated around x_a, the normalization may also be fixed by the initial condition

$$\lim_{t_b \to t_a} P(x_b t_b | x_a t_a) = \delta(x_b - x_a). \quad (18.234)$$

The right-hand side of (18.232) looks like a Euclidean path integral associated with the Lagrangian [8]

$$L_e = \frac{1}{2w}[M\ddot{x} + M\gamma\dot{x} + V'(x)]^2. \quad (18.235)$$

The result will, however, be different, due to time-ordering of the $\dot{x}^R$-term.

Apart from this, the Lagrangian is not of the conventional type since it involves a second time derivative. The action principle $\delta\mathcal{A} = 0$ now yields the Euler-Lagrange equation

$$\frac{\partial L}{\partial x} - \frac{d}{dt}\frac{\partial L}{\partial \dot{x}} + \frac{d^2}{dt^2}\frac{\partial L}{\partial \ddot{x}} = 0. \quad (18.236)$$

This equation can also be derived via the usual Lagrange formalism by considering x and $\dot{x}$ as independent generalized coordinates x, v.

18.9.1 Canonical Path Integral for Probability Distribution

In Section 2.1 we have constructed path integrals for time evolution amplitudes to solve the Schrödinger equation. By analogy, we expect the path integral (18.232) for the probability distribution to satisfy a differential equation of the Schrödinger type. This equation is known as a *Fokker-Planck equation*. As in Section 2.1, the relation is established by rewriting the path integral in canonical form. Treating $v = \dot{x}$ as an independent dynamical variable, the canonical momenta of x and v are [9]

$$
\begin{aligned}
p &= i\frac{\partial L}{\partial \dot{x}} = i\frac{M\gamma}{w}[M\ddot{x} + M\gamma\dot{x} + V'(x)] = i\frac{M\gamma}{w}[M\dot{v} + M\gamma v + V'(x)], \\
p_v &= i\frac{\partial L}{\partial \ddot{x}} = \frac{1}{\gamma}p.
\end{aligned}
\tag{18.237}
$$

The Hamiltonian is given by the Legendre transform

$$
H(p, p_v, x, v) = L_{\mathrm{e}}(\dot{x}, \ddot{x}) - \sum_{i=1}^{2}\frac{\partial L_{\mathrm{e}}}{\partial \dot{x}_i}\dot{x}_i = L_{\mathrm{e}}(v, \dot{v}) + ipv + ip_v\dot{v},
\tag{18.238}
$$

where $\dot{v}$ has to be eliminated in favor of p_v using (18.237). This leads to

$$
H(p, p_v, x, v) = \frac{w}{2M^2}p_v^2 - ip_v\left[\gamma v + \frac{1}{M}V'(x)\right] + ipv.
\tag{18.239}
$$

The canonical path integral representation for the probability distribution reads therefore

$$
\begin{aligned}
P(x_b t_b | x_a t_a) = \int \mathcal{D}x \int \frac{\mathcal{D}p}{2\pi} \int \mathcal{D}v \int \frac{\mathcal{D}p_v}{2\pi} \\
\times \exp\left\{ \int_{t_a}^{t_b} dt\, [i(p\dot{x} + p_v\dot{v}) - H(p, p_v, x, v)] \right\}.
\end{aligned}
\tag{18.240}
$$

It is easy to verify that the path integral over p enforces $v \equiv \dot{x}$, after which the path integral over p_v leads back to the initial expression (18.232). We may keep the auxiliary variable $v(t)$ as an independent fluctuating quantity in all formulas and decompose the probability distribution $P(x_b t_b | x_a t_a)$ with respect to the content of v as an integral

$$
P(x_b t_b | x_a t_a) = \int_{-\infty}^{\infty} dv_b \int_{-\infty}^{\infty} dv_a\, P(x_b v_b t_b | x_a v_a t_a).
\tag{18.241}
$$

The more detailed probability distribution on the right-hand side has the path integral representation

$$
\begin{aligned}
P(x_b v_b t_b | x_a v_a t_a) = |(x_b v_b t_b | x_a v_a t_a)|^2 = \int \mathcal{D}x \int \frac{\mathcal{D}p}{2\pi} \int \mathcal{D}v \int \frac{\mathcal{D}p_v}{2\pi} \\
\times \exp\left\{ \int_{t_a}^{t_b} dt\, [i(p\dot{x} + p_v\dot{v}) - H(p, p_v, x, v)] \right\},
\end{aligned}
\tag{18.242}
$$

where the endpoints of v are now kept fixed at $v_b = v(t_b)$, $v_a = v(t_a)$.

We now use the relation between a canonical path integral and the Schrödinger equation discussed in Section 2.1 to conclude that the probability distribution (18.242) satisfies the Schrödinger-like differential equation [10]:

$$H(\hat{p}, \hat{p}_v, x, v)P(x\,v\,t_b|x_a v_a t_a) = -\partial_t P(x\,v\,t|x_a v_a t_a). \tag{18.243}$$

It is called the *Klein-Kramers equation* for the motion of an inert point particle with dissipation. It is a special case of a two-variable *Fokker-Planck equation* whose general version deals with N variables $x x_1, \ldots, x_N$ collected in a vector $\mathbf{x}$, and has the form

$$\partial_t P(\mathbf{x}\,t|\mathbf{x}_a v_a t_a) = \left[-\partial_i D_i(\mathbf{x}) + \partial_i \partial_j D_{ij}^{(2)}(\mathbf{x})\right] P(\mathbf{x}\,t|\mathbf{x}_a v_a t_a). \tag{18.244}$$

The above equation (18.243) is a special case of this for $N = 2$;

$$\partial_t P(\mathbf{x}\,t|t_a \mathbf{x}_a) = \left(-\kappa_{ij}\partial_i x_j + D_{ij}\partial_i\partial_j\right) P(\mathbf{x}\,t|t_a \mathbf{x}_a). \tag{18.245}$$

with the *diffusion matrix*

$$\mathbf{D} = \begin{pmatrix} 0 & 0 \\ 0 & w/2M^2 \end{pmatrix} = \begin{pmatrix} 0 & 0 \\ 0 & \gamma k_B T/M \end{pmatrix} = \begin{pmatrix} 0 & 0 \\ 0 & \gamma^2 D \end{pmatrix}, \tag{18.246}$$

and

$$\boldsymbol{\kappa} = \begin{pmatrix} 0 & -1 \\ V'(x)/M & \gamma \end{pmatrix}. \tag{18.247}$$

It must be realized that when going over from the classical Hamiltonian (18.239) to the Hamiltonian operator in the differential equation (18.243), there is an operator ordering problem. Such a problem was encountered in Section 10.5 and discussed further at the end of Section 11.3. In this respect the analogy with the simple path integrals in Section 2.1 is not perfect. When writing down Eq. (18.243) we do not know in which order the momentum operator $\hat{p}_v$ must stand with respect to v. If we were dealing with an ordinary functional integral in (18.232) we would know the order. It would be found as in the case of the electromagnetic interaction in Eq. (11.89) to have the symmetric order $-(\hat{p}_v v + v\hat{p}_v)/2$.

On physical grounds, it is easy to guess the correct order. The differential equation (18.243) has to conserve the total probability

$$\int dx\, dv\, P(x\,v\,t_b|x_a v_a t_a) = 1 \tag{18.248}$$

for all times t. This is guaranteed if all momentum operators stand to the left of all coordinates in the Hamiltonian operator. Indeed, integrating the Fokker-Planck equation (18.243) over x and v, only a left-hand position of the momentum operators leads to a vanishing integral, and thus to a time independent total probability. We suspect that this order must be derivable from the retarded nature of the velocity in the term $y\dot{x}^R$ in (18.231). This will be shown in the next section.

18.9.2 Solving the Operator Ordering Problem

The ordering problem in the Hamiltonian operator associated with (18.239) does not involve the potential $V(x)$. We may therefore study this problem most simply by considering the classical free Hamiltonian

$$\tilde{H}_0(p, p_v, x, v) = \frac{w}{2M^2} p_v^2 - i\gamma p_v v + ipv, \qquad (18.249)$$

associated with the Lagrangian path integral

$$P_0(x_b t_b | x_a t_a) = \mathcal{N} \int \mathcal{D}x(t) \exp\left\{-\frac{1}{2w} \int_{t_a}^{t_b} dt \, [M\ddot{x} + M\gamma \dot{x}^R]^2\right\}, \qquad (18.250)$$

if we ignore the operator ordering problem. We may further concentrate our attention upon the probability distribution with $x_b = x_a = 0$, and assume $t_b - t_a$ to be very large. Then the frequencies of all Fourier decompositions are continuous.

In spite of the restrictions to large $t_b - t_a$, the result to be derived will be valid for any time interval. The reason is that operator order is a property of extremely short time intervals, so that it does not matter, how long the time interval is on which we solve the problem.

Forgetting for a moment the retarded nature of the velocity $\dot{x}$, the Gaussian path integral can immediately be done and yields

$$P_0(0\, t_b | 0\, t_a) \propto \mathrm{Det}^{-1}(-\partial_t^2 - \gamma\partial_t)$$
$$\propto \exp\left[-(t_b - t_a) \int_{-\infty}^{\infty} \frac{d\omega'}{2\pi} \log(\omega'^2 - i\gamma\omega')\right], \qquad (18.251)$$

where γ is *positive*. The integral on the right-hand side diverges. This is a consequence of the fact that we have not used Feynman's time slicing procedure for defining the path integral. As for an ordinary harmonic oscillator discussed in detail in Sections 2.3 and 2.14), this would lead to a finite integral in which ω' is replaced by $\tilde{\omega}' \equiv (2 - 2\cos a\omega')/a^2$:

$$\frac{1}{2}\int_{-\infty}^{\infty} \frac{d\omega'}{2\pi} \log[\tilde{\omega}'^4 + \gamma^2\tilde{\omega}'^2] = \frac{1}{2}\int_{-\infty}^{\infty} \frac{d\omega'}{2\pi} \log\tilde{\omega}'^2 + \frac{1}{2}\int_{-\infty}^{\infty} \frac{d\omega'}{2\pi} \log[\tilde{\omega}'^2 + \gamma^2] = 0 + \frac{\gamma}{2}. \qquad (18.252)$$

For a derivation see Section 2.14, in particular the first term in Eq. (2.483). The same result can equally well be obtained without time slicing by regularizing the divergent integral in (18.251) analytically, as shown in (2.502). Recall the discussion in Section 10.6 where analytic regularization was seen to be the only method that allows to define path integrals without time slicing in such a way that they are invariant under coordinate transformations [11]. It is therefore suggestive to apply the same procedure also to the present path integrals with dissipation and to use the dimensionally regularized formula (2.539):

$$\int_{-\infty}^{\infty} \frac{d\omega'}{2\pi} \log(\omega' \pm i\gamma) = \frac{\gamma}{2}, \qquad \gamma > 0. \qquad (18.253)$$

Applying this to the functional determinant in (18.251) yields

$$\mathrm{Det}(-\partial_t^2 - \gamma\partial_t) = \mathrm{Det}(i\partial_t)\mathrm{Det}(i\partial_t + i\gamma) = \exp\left[\mathrm{Tr}\,\log(i\partial_t) + \mathrm{Tr}\,\log(i\partial_t + i\gamma)\right]$$
$$= \exp\left[(t_b - t_a)\frac{\gamma}{2}\right], \tag{18.254}$$

and thus

$$P_0(0\,t_b|0\,t_a) \propto \exp\left[-(t_b - t_a)\frac{\gamma}{2}\right]. \tag{18.255}$$

This corresponds to an energy $\gamma/2$, and an ordering $-i\gamma(\hat{p}_v v + v\hat{p}_v)/2$ in the Hamiltonian operator.

We now take the retardation of the time argument of $\dot{x}^R$ into account. Specifically, we replace the term $\gamma y\dot{x}^R$ in (18.250) by the Drude form on the left-hand side (18.214) before going to the limit $\omega_D \to \infty$:

$$\gamma y\dot{x}^R(t) \to \int dt'\, y(t)\,\gamma_D^R(t - t')\,x(t'), \tag{18.256}$$

containing now explicitly the retarded Drude function (18.210) of the friction. Then the frequency integral in (18.251) becomes

$$\int_{-\infty}^{\infty} \frac{d\omega'}{2\pi} \log\left(\omega'^2 - \gamma\frac{\omega'\omega_D}{\omega'+i\omega_D}\right) = \int_{-\infty}^{\infty} \frac{d\omega}{2\pi}\left[-\log(\omega'+i\omega_D) + \log\left(\omega'^2 + i\omega'\omega_D - \gamma\omega_D\right)\right], \tag{18.257}$$

where we have omitted a vanishing integral over $\log\omega'$ on account of (18.253). We now decompose

$$\log\left(\omega'^2 + i\omega'\omega_D - \gamma\omega_D\right) = \log(\omega'+i\omega_1) + \log(\omega'+i\omega_2), \tag{18.258}$$

with

$$\omega_{1,2} = \frac{\omega_D}{2}\left(1 \pm \sqrt{1 - \frac{4\gamma}{\omega_D}}\right), \tag{18.259}$$

and use formula (2.539) to find

$$\int_{-\infty}^{\infty} \frac{d\omega}{2\pi}\left[-\log(\omega'+i\omega_D) + \log\left(\omega'^2 + i\omega'\omega_D - \gamma\omega_D\right)\right] = -\frac{\omega_D}{2} + \frac{\omega_1}{2} + \frac{\omega_2}{2} = 0. \tag{18.260}$$

The vanishing frequency integral implies that the retarded functional determinant is trivial:

$$\mathrm{Det}(-\partial_t^2 - \gamma\partial_t^R) = \exp\left[\mathrm{Tr}\,\log(-\partial_t^2 - \gamma\partial_t^R)\right] = 1, \tag{18.261}$$

instead of (18.254) obtained from the frequency integral without the Drude modification. With the determinant (18.261), the probability becomes a constant

$$P_0(0\,t_b|0\,t_a) = \text{const.} \tag{18.262}$$

This shows that the retarded nature of the friction force has *subtracted* an energy $\gamma/2$ from the energy in (18.255). Since the ordinary path integral corresponds to a Hamiltonian operator with a symmetrized term $-i(\hat{p}_v v + v\hat{p}_v)/2$, the subtraction of $\gamma/2$ changes this term to $-i\gamma\hat{p}_v v$.

Note that the opposite case of an *advanced* velocity term $\dot{x}^A$ in (18.250) would be approximated by a Drude function $\gamma_D^A(t)$ which looks just like $\gamma_D^R(t)$ in (18.212), but with *negative* ω_D. The right-hand side of (18.260) would then become 2γ rather than zero. The corresponding formula for the functional determinant is

$$\mathrm{Det}(-\partial_t^2 - \gamma\partial_t^A) = \exp\left[\mathrm{Tr}\,\log(-\partial_t^2 - \gamma\partial_t^A)\right] = \exp\left[(t_b - t_a)\gamma\right], \qquad (18.263)$$

where $\gamma\partial_t^A$ stands for the advanced version of the functional matrix (18.256) in which ω_D is replaced by $-\omega_D$. Thus we would find

$$P_0(0\,t_b|0\,t_a) \;\propto\; \exp\left[-(t_b - t_a)\gamma\right], \qquad (18.264)$$

with an *additional* energy $\gamma/2$ with respect to the ordinary formula (18.255). This corresponds to the opposite (unphysical) operator order $-i\gamma v\hat{p}_v$ in $\hat{H}_0$, which would violate the probability conservation of time evolution twice as much as the symmetric order.

The above formulas for the functional determinants can easily be extended to the slightly more general case where $V(x)$ is the potential of a harmonic oscillator $V(x) = M\omega_0^2 x^2/2$. Then the path integral (18.232) for the probability distribution becomes

$$P_0(x_b t_b|x_a t_a) = \mathcal{N}\int \mathcal{D}x(t)\, \exp\left\{-\frac{1}{2w}\int_{t_a}^{t_b} dt\,[M\ddot{x} + M\gamma\dot{x}^R + \omega_0^2 x]^2\right\}, \qquad (18.265)$$

which we evaluate at $x_b = x_a = 0$, where it is given by the properly retarded expression

$$\begin{aligned}
P_0(0\,t_b|0\,t_a) \;&\propto\; \mathrm{Det}^{-1}(-\partial_t^2 - \gamma\partial_t + \omega_0^2)\\
&\propto\; \exp\left[-(t_b - t_a)\int_{-\infty}^{\infty}\frac{d\omega'}{2\pi}\log(\omega'^2 - i\gamma\omega' - \omega_0^2)\right]. \qquad (18.266)
\end{aligned}$$

The logarithm can be decomposed into a sum $\log(\omega' + i\omega_1) + \log(\omega' + i\omega_2)$ with

$$\omega_{1,2} = \frac{\gamma}{2}\left(1 \pm \sqrt{1 - \frac{4\omega_0^2}{\gamma^2}}\right). \qquad (18.267)$$

We now apply the analytically regularized formula (2.539) to obtain

$$\int_{-\infty}^{\infty}\frac{d\omega'}{2\pi}\left[\log(\omega' + i\omega_1) + \log(\omega' + i\omega_2)\right] = \frac{\omega_1}{2} + \frac{\omega_2}{2} = \gamma. \qquad (18.268)$$

Both under- and overdamped motion yield the same result. This is one of the situations where our remarks after Eqs. (2.542) and (2.541) concerning the cancellation of oscillatory parts apply. For the functional determinant (18.266), the result is

$$\mathrm{Det}(-\partial_t^2 - \gamma\partial_t - \omega_0^2) = \exp\left[\mathrm{Tr}\,\log(-\partial_t^2 - \gamma\partial_t - \omega_0^2)\right] = \exp\left[(t_b - t_a)\frac{\gamma}{2}\right]. \qquad (18.269)$$

Note that the result is independent of ω_0. This can simply be understood by forming the derivative of the logarithm of the functional determinant in (18.254) with respect to ω_0^2. Since $\log \operatorname{Det} M = \operatorname{Tr} \log M$, this yields the trace of the associated Green function:

$$\frac{\partial}{\partial \omega_0^2} \operatorname{Tr} \log(-\partial_t^2 - \gamma \partial_t - \omega_0^2) = -\int dt \, (-\partial_t^2 - \gamma \partial_t - \omega_0^2)^{-1}(t, t). \qquad (18.270)$$

In Fourier space, the right-hand side turns into the frequency integral

$$-\int \frac{d\omega'}{2\pi} \frac{1}{(\omega' + i\omega_1)(\omega' + i\omega_1)}. \qquad (18.271)$$

Since the two poles lie below the contour of integration, we may close it in the upper half-plane and obtain zero. Closing it in the lower half plane would initially lead to two nonzero contributions from the residues of the two poles which, however, cancel each other.

The Green function (18.270) is causal, in contrast to the oscillator Green function in Section 3.3 whose left-hand pole lies in the upper half-plane (recall Fig. 3.3). Thus it carries a Heaviside function as a prefactor [recall Eq. (1.301) and the discussion of causality there]. The vanishing of the integral (18.270) may be interpreted as being caused by the Heaviside function (1.300).

The γ-dependence of (18.269) can be calculated likewise:

$$\frac{\partial}{\partial \gamma} \log \operatorname{Det} \partial_t (-\partial_t^2 - \gamma \partial_t - \omega_0^2) = -\int dt \, [\partial_t (-\partial_t^2 - \gamma \partial_t - \omega_0^2)^{-1}](t, t). \qquad (18.272)$$

We perform the trace in frequency space:

$$i \int \frac{d\omega'}{2\pi} \frac{\omega'}{(\omega' + i\omega_1)(\omega' + i\omega_1)}. \qquad (18.273)$$

If we now close the contour of integration with an infinite semicircle in the upper half plane to obtain a vanishing integral from the residue theorem, we must subtract the integral over the semicircle $i \int d\omega'/2\pi\omega'$ and obtain $1/2$, in agreement with (18.269).

Formula (18.269) can be generalized further to time-dependent coefficients

$$\operatorname{Det}\left[-\partial_t^2 - \gamma(t)\partial_t - \Omega^2(t)\right] = \exp\left\{\operatorname{Tr} \log\left[-\partial_t^2 - \gamma(t)\partial_t - \Omega^2(t)\right]\right\} = \exp\left[\int_{t_a}^{t_b} dt \frac{\gamma(t)}{2}\right]. \qquad (18.274)$$

This follows from the factorization

$$\operatorname{Det}\left[-\partial_t^2 - \gamma(t)\partial_t - \Omega^2(t)\right] = \operatorname{Det}[\partial_t + \Omega_1(t)] \operatorname{Det}[\partial_t + \Omega_2(t)], \qquad (18.275)$$

with

$$\Omega_1(t) + \Omega_2(t) = \gamma(t), \qquad \partial_t \Omega_2(t) + \Omega_1(t)\Omega_2(t) = \Omega^2(t), \qquad (18.276)$$

and applying formula (3.134).

The probability obtained from the general path integral (18.232) without retardation of the velocity term is therefore

$$P_0(0\, t_b | 0\, t_a) \propto \exp\left[-(t_b - t_a)\frac{\gamma}{2}\right],$$

(18.277)

as in (18.255).

Let us now introduce retardation of the velocity term by using the ω'-dependent Drude expression (18.212) for the friction coefficient. First we consider again the harmonic path integral (18.265), for which (18.266) becomes

$$P_0(0\, t_b | 0\, t_a) \propto \exp\left\{-(t_b - t_a)\int_{-\infty}^{\infty}\frac{d\omega'}{2\pi}\log\left[\omega'^2 - i\gamma_D^R(\omega')\omega' - \omega_0^2\right]\right\}.$$

(18.278)

Rewriting the logarithm as $-\log(\omega' + i\omega_D) + \Sigma_{k=1}^3 \log(\omega' + i\omega_k)$ with

$$\omega_{1,2} = \frac{\gamma}{2}\left(1 \pm \sqrt{1 - \frac{4\omega_0^2}{\gamma^2}}\right), \qquad \omega_3 = \omega_D - \gamma$$

(18.279)

[recall Eq. (3.451) in the equilibrium discussion of Section 3.15], we use again formula (2.539) to find

$$\int_{-\infty}^{\infty}\frac{d\omega'}{2\pi}\left[-\log(\omega' + i\omega_D) + \sum_{k=1}^{3}\log(\omega' + i\omega_k)\right] = -\omega_D + \sum_{k=1}^{3}\frac{\omega_k}{2} = 0.$$

(18.280)

Thus γ and ω_0 disappear from the functional determinant, and we remain with

$$P_0(0\, t_b | 0\, t_a) = \text{const}.$$

(18.281)

This implies a unit functional determinant [12]

$$\text{Det}\,(\partial_t^2 + i\gamma\partial_t^R + \omega_0^2) = 1,$$

(18.282)

in contrast to the unretarded determinant (18.269). The γ-independence of this can also be seen heuristically as in (18.270) by forming the derivative with respect to γ:

$$\frac{\partial}{\partial\gamma}\text{Det}\,(-\partial_t^2 - \gamma\partial_t^R - \omega_0^2) = -\int dt\,[\partial_t^R(\partial_t^2 - \gamma\partial_t - \omega_0^2)^{-1}(t,t).$$

(18.283)

Since the retarded derivative carries a Heaviside factor $\Theta(t - t')$ of (1.300), we find zero for $t = t'$.

The result $1/2$ of the unretarded derivative in (18.272) can similarly be understood as a consequence of the average Heaviside function (1.309) at $t = t'$.

An advanced time derivative in the determinant (18.282) would, of course, have produced the result

$$\text{Det}\,(\partial_t^2 + i\gamma\partial_t^A + \omega_0^2) = \gamma.$$

(18.284)

By analogy with (18.275), the general retarded determinant is also independent of $\gamma(t)$ and $\Omega(t)$.

$$\mathrm{Det}\left[-\partial_t^2 - \gamma(t)\partial_t^R - \Omega^2(t)\right] = 1. \tag{18.285}$$

In the advanced case, we would find similarly

$$\mathrm{Det}\left[-\partial_t^2 - \gamma(t)\partial_t^A - \Omega^2(t)\right] = \exp\left[\int dt\,\gamma(t)\right]. \tag{18.286}$$

By comparing the functional determinants (18.269) and (18.282) we see that the retardation prescription can be avoided by a trivial additive change of the Lagrangian (18.235) to

$$L_e(x,\dot{x}) = \frac{1}{2w}\left[\ddot{x} + M\gamma\dot{x} + V'(x)\right]^2 - \frac{\gamma}{2}. \tag{18.287}$$

From this, the path integral can be calculated with the usual time slicing, and the result can be deduced directly from Ref. [8].

The Hamiltonian associated with this Lagangian is only slightly modified with respect to the naive form (18.249):

$$H_0(p, p_v, x, v) = \frac{w}{2M^2}p_v^2 - i\gamma p_v v + ipv - \frac{\gamma}{2}. \tag{18.288}$$

The extra $\gamma/2$ ensures that the operator version of the Hamiltonian (18.239) has $\hat{p}_v$ to the left of v.

18.9.3 Strong Damping

For $\gamma \gg V''(x)/M$, the dynamics is dominated by dissipation, and the Lagrangian (18.235) takes a more conventional form in which only x and $\dot{x}$ appear:

$$L_e(x,\dot{x}) = \frac{1}{2w}\left[M\gamma\dot{x}^R + V'(x)\right]^2 = \frac{1}{4D}\left[\dot{x}^R + \frac{1}{M\gamma}V'(x)\right]^2, \tag{18.289}$$

where $\dot{x}^R$ lies slightly *earlier* $V'(x(t))$. The probability distribution

$$P(x_b t_b | x_a t_a) = \mathcal{N}\int \mathcal{D}x\,\exp\left[-\int_{t_a}^{t_b} dt\,L_e(x,\dot{x}^R)\right] \tag{18.290}$$

looks like an ordinary Euclidean path integral for the density matrix of a particle of mass $M = 1/2D$. As such it obeys a differential equation of the Schrödinger type. Forgetting for a moment the subtleties of the retardation, we introduce an auxiliary momentum integration and go over to the canonical representation of (18.290):

$$P(x_b t_b | x_a t_a) = \int \mathcal{D}x\int \frac{\mathcal{D}p}{2\pi}\,\exp\left\{\int_{t_a}^{t_b} dt\left[ip\dot{x} - 2D\frac{p^2}{2} + ip\frac{1}{M\gamma}V'(x)\right]\right\}. \tag{18.291}$$

This probability distribution satisfies therefore the Schrödinger type of equation

$$H(\hat{p}_b, x_b)P(x_b t_b | x_a t_a) = -\partial_{t_b} P(x_b t_b | x_a t_a), \qquad (18.292)$$

with the Hamiltonian operator

$$H(\hat{p}, x) \equiv 2D\frac{\hat{p}^2}{2} - i\hat{p}\frac{1}{M\gamma}V'(x) = -D\,\partial_x \left[\partial_x + \frac{1}{DM\gamma}V'(x)\right]. \qquad (18.293)$$

In order to conserve probability, the momentum operator has to stand to the left of the potential term. Only then does the integral over x_b of Eq. (18.292) vanish. Equation (18.292) is the overdamped Klein-Kramers equation, also called the *Smoluchowski equation*. It is a special case of ordinary Fokker-Planck equation.

Without the retardation on $\dot{x}$ in (18.290), the path integral would give a symmetrized operator $-i[\hat{p}V'(x) + V'(x)\hat{p}]/2$ in $\hat{H}$. This follows from the fact that the coupling $(1/2DM\gamma)\dot{x}V'(x)$ looks precisely like the coupling of a particle to a magnetic field with a "vector potential" $A(x) = (1/2DM\gamma)V'(x)$ [see (10.171)].

Realizing this it is not difficult to account quite explicitly for the effect of retardation of the velocity in the path integral (18.289) Let us assume, for a moment, that w is very small. Then the path integral (18.290) without the retardation,

$$P_0(x_b t_b | x_a t_a) = \mathcal{N} \int \mathcal{D}x \, \exp\left\{-\frac{1}{2w}\int_{t_a}^{t_b} dt \, [M\gamma\dot{x} + V'(x)]^2\right\}, \qquad (18.294)$$

can be performed in the Gaussian approximation resulting for $x_b = x_a = 0$ in the inverse functional determinant

$$P_0(0\,t_b | 0\,t_a) = \text{Det}^{-1}\left[\partial_t + V''(x)/M\gamma\right], \qquad (18.295)$$

whose value is according to formula (3.134)

$$\text{Det}\left[\partial_t + V''(x)/M\gamma\right] = \exp\left[\int dt \, V''(x)/2M\gamma\right]. \qquad (18.296)$$

The retarded version of this determinant is trivial:

$$\text{Det}\left[\partial_t^R + V''(x)/M\gamma\right] = 1, \qquad (18.297)$$

as we learned from Eq. (18.285. The advanced version would be [compare (18.286)]

$$\text{Det}\left[\partial_t^A + V''(x)/M\gamma\right] = \exp\left[\int dt \, V''(x)/M\gamma\right]. \qquad (18.298)$$

Although the determinants (18.296), (18.297), and (18.298) were discussed here only for a large time interval $t_b - t_a$, the formulas remain true for all time intervals, due to the trivial first-order nature of the differential operator. In a short time interval, however, the second derivative is approximately time-independent. For this reason the difference between ordinary and retarded path integrals (18.290) is given by the difference between the functional determinants (18.296) and (18.297) not only if w is very small but for all w. Thus we can avoid the retardation of the velocity as in Eq. (18.287) by adding to the Lagrangian (18.289) a term containing the second derivative of the potential:

$$L_e(x, \dot{x}) = \frac{1}{4D} \left[\dot{x} + \frac{1}{M\gamma} V'(x) \right]^2 - \frac{1}{2M\gamma} V''(x). \tag{18.299}$$

From this, the path integral can be calculated with the same slicing as for the gauge-invariant coupling in Section 10.5:

$$P_0(x_b t_b | x_a t_a) = \mathcal{N} \int \mathcal{D}x(t) \exp \left[-\int_{t_a}^{t_b} dt \left\{ \frac{1}{4D} \left[\dot{x} + \frac{V'(x)}{M\gamma} \right]^2 - \frac{V''(x)}{2M\gamma} \right\} \right]. \tag{18.300}$$

As an example consider a harmonic potential $V(x) = M\omega_0^2 x^2/2$, where the Lagrangian (18.299) becomes

$$L_e(x, \dot{x}) = \frac{1}{4D} \left(\dot{x} + \kappa x \right)^2 - \frac{\kappa}{2}, \tag{18.301}$$

with the abbreviation $\kappa \equiv \omega_0^2/\gamma$. The equation of motion reads

$$-\ddot{x} + \kappa^2 x = 0, \tag{18.302}$$

and its solution connecting x_a, t_a with x_b, t_b is

$$x(t) = \frac{1}{e^{2\kappa t_a} - e^{2\kappa t_b}} \left[e^{\kappa(t+t_a)} x_a - e^{\kappa(-t+t_a+2\kappa t_b)} x_a - e^{\kappa(t+t_b)} x_b + e^{\kappa(-t+2 t_a+t_b)} x_b \right]. \tag{18.303}$$

This has the total Euclidean action

$$\mathcal{A}_e = \frac{\kappa \left(e^{\kappa t_b} x_b - e^{\kappa t_a} x_a \right)^2}{2 D \left(e^{2\kappa t_b} - e^{2\kappa t_a} \right)} - \frac{\kappa}{2}. \tag{18.304}$$

The fluctuation determinant is from Eq. (2.169), after an appropriate substitution of variables,

$$F_\kappa(t_b - t_a) = \frac{1}{\sqrt{2\pi \sinh \kappa(t_b - t_a)}}. \tag{18.305}$$

The probability distribution is then given by

$$P(x_b t_b | x_a t_a) = F_\kappa(t_b - t_a) e^{-\mathcal{A}_e} = \frac{1}{\sqrt{2\pi \sigma^2(t_b - t_a)}} \exp \left\{ -\frac{[x_b - \bar{x}(t_b - t_a)]^2}{2\sigma^2(t_b - t_a)} \right\}, \tag{18.306}$$

where $\bar{x}(t)$, $\sigma^2(t)$ are the averages

$$\bar{x}(t) \equiv \langle x(t) \rangle = x_a e^{-\kappa t}, \quad \sigma^2(t) \equiv \langle [x(t) - \bar{x}(t)]^2 \rangle = \frac{D}{\kappa} \left(1 - e^{-2\kappa t} \right), \tag{18.307}$$

obtained from the integrals[8]

$$\bar{x}(t_b - t_a) \equiv \langle x(t_b - t_a) \rangle \equiv \int_{-\infty}^{\infty} x_b P(x_b t_b | x_a t_a), \tag{18.308}$$

$$\langle [x(t_b - t_a) - \bar{x}(t_b - t_a)]^2 \rangle \equiv \int_{-\infty}^{\infty} [x_b - \bar{x}(t_b - t_a)]^2 P(x_b t_b | x_a t_a). \tag{18.309}$$

The canonical momentum associated with the Lagrangian (18.301) is $p = (\dot{x} + \kappa x)/2D$ so that the Hamiltonian operator becomes, via the Euclidean Legendre transformation (2.338), and with the operator ordering fixed as discussed after Eq. (18.247):

$$\hat{H}(p, x) = D\hat{p}^2 + i\kappa\hat{p}x, \qquad p \equiv -i\partial_x. \tag{18.310}$$

This is the same operator as in Eq. (18.293), and the Fokker-Planck equation (18.292) reads, for the harmonic potential:

$$\left(-D\partial_{x_b}^2 + \kappa\partial_{x_b}x_b \right) P(x_b t_b | x_a t_a) = -\partial_{t_b} P(x_b t_b | x_a t_a). \tag{18.311}$$

For $t_b \to t_a$, the probability distribution (18.306) starts out as a δ-function around the initial position x_a. In the limit of large $t_b - t_a$, it converges against the limiting distribution

$$\lim_{t_b \to \infty} P(x_b t_b | x_a t_a) = \sqrt{\frac{\kappa}{2\pi D}} \exp\left\{ -\kappa \frac{x_b^2}{2D} \right\}. \tag{18.312}$$

Replacing κ again by $\omega_0^2/\gamma = V''(0)/M\gamma$, and D from (18.219), this becomes

$$\lim_{t_b \to \infty} P(x_b t_b | x_a t_a) = \sqrt{\frac{V''(0)}{2\pi k_B T}} \exp\left\{ -\frac{1}{k_B T} V(x_b) \right\}. \tag{18.313}$$

Thus, the limiting distribution of (18.290) depends only on x_b. It is given by the Boltzmann factor associated with the potential $V(x)$, in which the particle moves. This result can be generalized to a large class of potentials.

An interesting related result can be derived by introducing an external source term $j_b x(t_b)$ into the Lagrangian (18.299). By repeated functional differentiation with respect to j_b we find that the expectation values

$$\langle x^n \rangle = \lim_{t_b \to \infty} \langle x^n(t_b) \rangle = \lim_{t_b \to \infty} \frac{\int \mathcal{D}x \, x^n(t_b) e^{-\int_{t_a}^{t_b} dt \, L_e(x, \dot{x}^R)}}{\int \mathcal{D}x \, e^{-\int_{t_a}^{t_b} dt \, L_e(x, \dot{x}^R)}} \tag{18.314}$$

have the large-time limit

$$\lim_{t_b \to \infty} \langle x^n(t_b) \rangle = \langle x^n \rangle = \frac{\int dx \, x^n e^{-V(x)/k_B T}}{\int dx \, e^{-V(x)/k_B T}}. \tag{18.315}$$

The generalization of this relation to quantum field theory forms the basis of *stochastic quantization* in Section 18.12.

---

[8]An alternative method for calculating such expectation values will be presented in Section 18.15.

18.10 Langevin Equations

Consider the forward–backward path integral (18.230) for high γT. Then the second exponent limits the fluctuations of y to satisfy $|y| \ll |x|$, and $K(t, t')$ will be assumed to take the Drude form (18.226), which becomes a δ-function for $\omega_D \to \infty$. Then we can expand

$$V\left(x + \frac{y}{2}\right) - V\left(x - \frac{y}{2}\right) \sim yV'(x) + \frac{y^3}{24}V'''(x) + \dots , \qquad (18.316)$$

keeping only the first term. We further introduce an auxiliary quantity $\eta(t)$ by

$$\eta(t) \equiv M\ddot{x}(t) + M\gamma\dot{x}^R(t) + V'(x(t)). \qquad (18.317)$$

With this, the exponential function in (18.230) becomes after a partial integration of the first term using the endpoint properties $y(t_b) = y(t_a) = 0$:

$$\exp\left\{-\frac{i}{\hbar}\int_{t_a}^{t_b} dt\, y\eta - \frac{w}{2\hbar^2}\int_{t_a}^{t_b} dt \int_{t_a}^{t_b} dt'\, y(t)K(t, t')y(t')\right\}. \qquad (18.318)$$

The variable y can obviously be integrated out and we find a probability distribution

$$P[\eta] \propto \exp\left\{-\frac{1}{2w}\int_{t_a}^{t_b} dt \int_{t_a}^{t_b} dt'\, \eta(t)K^{-1}(t, t')\eta(t')\right\}, \qquad (18.319)$$

where the fluctuation width w was given in (18.218), and $K^{-1}(t, t')$ denotes the inverse functional matrix of $K(t, t')$.

The defining equation (18.317) for $\eta(t)$ may be viewed as a *stochastic differential equation* to be solved for arbitrary initial positions $x(t_a) = x_a$ and velocities $\dot{x}(t_a) = v_a$. The differential equation is driven by a Gaussian random *noise* variable $\eta(t)$ with a correlation function

$$\langle\eta(t)\rangle_\eta = 0, \qquad \langle\eta(t)\eta(t')\rangle_\eta = w\,K(t - t'), \qquad (18.320)$$

where the expectation value of an arbitrary functional of $F[x]$ is defined by the path integral

$$\langle F[x]\rangle_\eta \equiv \mathcal{N}\int_{x(t_a)=x_a} \mathcal{D}x\, P[\eta]F[x]. \qquad (18.321)$$

The normalization factor $\mathcal{N}$ is fixed by the condition $\mathcal{N}\int \mathcal{D}\eta\, P[\eta] = 1$, so that $\langle 1\rangle_\eta = 1$. In the sequel, this factor will always be absorbed in the measure $\mathcal{D}\eta$.

For each noise function $\eta(t)$, the solution of the differential equation yields a path $x_\eta(x_a, x_b, t_a)$ with a final position $x_b = x_\eta(x_a, x_b, t_b)$ and velocity $v_b = \dot{x}_\eta(x_a, x_b, t_b)$, all being functionals of $\eta(t)$. From this we can calculate the distribution $P(x_b v_b t_b | x_a v_a t_a)$ of the final x_b and v_b by summing over all paths resulting from the noise functions $\eta(t)$ with the probability distribution (18.319). The result is of course the same as the distribution (18.242) obtained previously from the canonical path integral.

It is useful to exhibit clearly the dependence on initial and final velocities by separating the stochastic differential equation (18.317) into two first-order equations

$$M\dot{v}(t) + M\gamma v^R(t) + V'(x(t)) = \eta(t), \qquad (18.322)$$
$$\dot{x}(t) = v(t), \qquad (18.323)$$

to be solved for initial values $x(t_a) = x_a$ and $\dot{x}(t_a) = v_a$. For a given noise function $\eta(t)$, the final positions and velocities have the probability distribution

$$P_\eta(x_b v_b t_b | x_a v_a t_a) = \delta(x_\eta(t) - x_b)\delta(\dot{x}_\eta(t) - v_b). \qquad (18.324)$$

Given these distributions for all possible noise functions $\eta(t)$, we find the final probability distribution $P(x_b v_b t_b | x_a v_a t_a)$ from the path integral over all $\eta(t)$ calculated with the noise distribution (18.319). We shall write this in the form

$$P(x_b v_b t_b | x_a v_a t_a) = \langle P_\eta(x_b v_b t_b | x_a v_a t_a) \rangle_\eta. \qquad (18.325)$$

Let us change of integration variable from $x(t)$ to $\eta(t)$. This produces a Jacobian

$$J[x] \equiv \mathrm{Det}[\delta\eta(t)/\delta x(t')] = \det[M\partial_t^2 + M\gamma\partial_t^R + V''(x(t))]. \qquad (18.326)$$

In Eq. (18.285) we have seen that due to the retardation of ∂_t^R, this Jacobian is unity. Hence we can rewrite the expectation value (18.321) as a functional integral

$$\langle F[x] \rangle_\eta \equiv \int \mathcal{D}\eta \, P[\eta] \, F[x] \Big|_{x(t_a)=x_a}. \qquad (18.327)$$

From the probability distribution $P(x_b v_b t_b | x_a v_a t_a)$ we find the pure position probability $P(x_b t_b | x_a t_a)$ by integrating over all initial and final velocities as in Eq. (18.241). Thus we have shown that a solution of the forward–backward path integral at high temperature (18.232) can be obtained from a solution of the stochastic differential equations (18.317), or more specifically, from the pair of stochastic differential equations (18.322) and (18.323).

The stochastic differential equation (18.317) together with the correlation function (18.320) is called *semiclassical Langevin equation*. The fluctuation width w in (18.320) was given in (18.218). The attribute *semiclassical* emphasizes the truncation of the expansion (18.316) after the first term, which can be justified only for nearly harmonic potentials. For a discussion of the range of applicability of the truncation see the literature [14]. The untruncated path integral is equivalent to an operator form of the Langevin equation, the so-called *quantum Langevin equation* [15]. This equivalence will be discussed further in Subsection 18.19.

The physical interpretation of Eq. (18.320) goes as follows. For $T \to 0$ and $\hbar \to 0$ at $\hbar/T = \mathrm{const}$, the random variable $\eta(t)$ does not fluctuate at all and (18.317) reduces to the classical equation of motion of a particle in a potential $V(x)$, with an additional friction term proportional to γ. For T and $\hbar$ both finite, the particle is shaken around its classical path by thermal and quantum fluctuations. At high temperatures (at fixed $\hbar$), $K(\omega')$ reduces to

$$\lim_{T\to\infty} K(\omega') \equiv 1. \qquad (18.328)$$

Then $\eta(t)$ is an instantaneous random variable with zero average and a nonzero pair correlation function:

$$\langle \eta(t) \rangle_\eta = 0, \qquad \langle \eta(t)\eta(t') \rangle_\eta = w\,\delta(t-t'), \tag{18.329}$$

All higher correlation functions vanish. A random variable with these characteristics is referred to as *white noise*. The stochastic differential equation (18.317) with the white noise (18.329) reduces to the *classical Langevin equation with inertia* [16].

In the opposite limit of small temperatures, $K(\omega')$ diverges like

$$K(\omega') \xrightarrow[T \to 0]{} \frac{\hbar|\omega'|}{2k_B T}, \tag{18.330}$$

so that $w\,K(\omega')$ has the finite limit

$$\lim_{T \to 0} w K(\omega') = M\gamma\hbar|\omega'|. \tag{18.331}$$

To find the Fourier transformation of this, we use the Fourier decomposition of the Heaviside function (1.302)

$$\Theta(\omega') = \frac{1}{2\pi} \int_{-\infty}^{\infty} dt\, e^{-i\omega' t} \frac{i}{t+i\eta} \tag{18.332}$$

to form the antisymmetric combination

$$\begin{aligned}
\Theta(\omega') - \Theta(-\omega') &= \frac{1}{2\pi} \int_{-\infty}^{\infty} dt\, e^{-i\omega' t} \left(\frac{i}{t+i\eta} + \frac{i}{t-i\eta} \right) \\
&\equiv \frac{i}{\pi} \int_{-\infty}^{\infty} dt\, e^{-i\omega' t} \frac{\mathcal{P}}{t}.
\end{aligned} \tag{18.333}$$

A multiplication by ω' yields

$$\begin{aligned}
|\omega'| = \omega'[\Theta(\omega') - \Theta(-\omega')] &= -\frac{1}{\pi} \int_{-\infty}^{\infty} dt\, \partial_t e^{-i\omega' t} \frac{\mathcal{P}}{t} = \frac{1}{\pi} \int_{-\infty}^{\infty} dt\, e^{-i\omega' t} \partial_t \frac{\mathcal{P}}{t} \\
&= -\frac{1}{\pi} \int_{-\infty}^{\infty} dt\, e^{-i\omega' t} \frac{\mathcal{P}}{t^2}.
\end{aligned} \tag{18.334}$$

By comparison with (18.331) we see that the pure quantum limit of $K(t-t')$ can be written as

$$w K(t-t') \underset{T=0}{=} -\frac{M\gamma\hbar}{\pi} \frac{\mathcal{P}}{(t-t')^2}. \tag{18.335}$$

Hence the quantum-mechanical motion in contact with a thermal reservoir looks just like a classical motion, but disturbed by a random source with temporally long-range correlations

$$\langle \eta(t)\eta(t') \rangle_\eta = -\frac{M\gamma\hbar}{\pi} \frac{\mathcal{P}}{(t-t')^2}. \tag{18.336}$$

The temporal range is found from the temporal average

$$\langle(\Delta t)^2\rangle_t \equiv \int_{-\infty}^{\infty} d\Delta t\,(\Delta t)^2 K(\Delta t) = -\frac{\partial^2}{\partial\omega'^2}K(\omega')\Big|_{\omega'=0} = -\frac{1}{6}\left(\frac{\hbar}{k_B T}\right)^2. \quad (18.337)$$

Apart from the negative sign (which would be positive for Euclidean times), the random variable acquires more and more memory as the temperature decreases and the system moves deeper into the quantum regime. Note that no extra normalization factor is required to form the temporal average (18.337), due to the unit normalization of $K(t-t')$ in (18.224).

In the overdamped limit, the classical Langevin equation with inertia (18.317) reduced to the *overdamped Langevin equation*:

$$\dot{x}(t) = -V'(x(t))/M\gamma + \eta(t)/M\gamma. \quad (18.338)$$

At high temperature, the noise variable $\eta(t)$ has the correlation functions (18.329). Then the stochastic differential equation (18.338) is said to describe a *Wiener process*. The first term on the right-hand side $r_x(x(t)) \equiv -V'(x(t))/M\gamma$ is called the *drift* of the process.

The probability distribution of $x(t)$ resulting from this process is calculated as in Eqs. (18.325), (18.321) from the path integral

$$P(x_b t_a|x_a t_a) = \int \mathcal{D}\eta\, P[\eta]\,\delta(x_\eta(t) - x_b), \quad (18.339)$$

and $\mathcal{D}\eta$ is normalized so that $\int \mathcal{D}\eta\, P[\eta] = 1$.

A path integral representation closely related to this is obtained by using the identity

$$\int_{x(t_a)=x_a}^{x(t_b)=x_b} \mathcal{D}x\,\delta[\dot{x} - \eta] = \delta(x_\eta(t_b) - x_b), \quad (18.340)$$

which can easily be proved by time-slicing the Fourier representation of the δ-functional

$$\delta[\dot{x} - \eta] = \int \mathcal{D}p\, e^{i\int dt\, p(\dot{x}-\eta)} \quad (18.341)$$

and performing all the momentum integrals. This brings the path integral (18.339) to the form

$$P(x_b t_b|x_a t_a) = \int_{x(t_a)=x_a}^{x(t_b)=x_b} \mathcal{D}x \int \mathcal{D}\eta\, P[\eta]\,\delta[\dot{x} - \eta]. \quad (18.342)$$

For a harmonic potential $V(x) = M\omega_0^2 x^2/2$, where the overdamped Langevin equation reads

$$\dot{x}(t) = -\omega_0^2\, x(t)/M\gamma + \eta(t)/M\gamma = -\kappa\, x(t)/M\gamma + \bar{\eta}(t), \quad (18.343)$$

where the noise variable $\bar{\eta}(t) \equiv \eta(t)/M\gamma$ has the correlation functions

$$\langle\bar{\eta}(t)\rangle = 0, \quad \langle\bar{\eta}(t)\bar{\eta}(t')\rangle_\eta = \frac{w}{M^2\gamma^2}\delta(t - t') = 2D\delta(t - t'), \quad (18.344)$$

the calculation of the stochastic path integral yields, of course, once more the probability (18.306).

18.11   Path Integral Solution of Klein-Kramers Equation

For a free particle at finite temperature, there exists another way of representing the solution (18.440). Consider the original path integral for the probability in Eq. (18.240) with the Hamiltonian (18.288). Let us introduce the thermal velocity $v_T \equiv \sqrt{k_B T/M}$ and write the action as [compare (2.338)]

$$\mathcal{A}_e = \int_{t_a}^{t_b} dt \left[-i(p\dot{x} + p_v \dot{v}) + H(p, p_v, v, x) \right], \tag{18.345}$$

with

$$H(p, p_v, v, x) = \gamma v_T^2 \left(p_v - i\frac{v}{2v_T^2} \right)^2 + \frac{\gamma}{4v_T^2} \left(v + 2i\frac{v_T^2}{\gamma} p \right)^2 + \frac{v_T^2}{\gamma} p^2 + \frac{\gamma}{2}. \tag{18.346}$$

In the path integral (18.240), we may integrate out $x(t)$, which converts the path integral over $p(t)$ into an ordinary integral, so that we obtain the integral representation

$$P(x_b v_b t_b | x_a v_a t_a) = \int \frac{dp}{2\pi} P(v_b t_b | v_a p_a)_p e^{ip(x_b - x_a) - v_T^2 p^2 (t_b - t_a)/\gamma}, \tag{18.347}$$

where

$$P_p(v_b t_b | v_a t_a) = \int \mathcal{D}v \int \frac{\mathcal{D}p_v}{2\pi} \exp \left\{ \int_{t_a}^{t_b} dt \left[ip_v \dot{v} - H_p(p_v, v) \right] \right\}, \tag{18.348}$$

with the p-dependent Hamiltonian governing p_v and v:

$$H_p(p_v, v) \equiv \gamma v_T^2 \left(p_v - i\frac{v}{2v_T^2} \right)^2 + \frac{\gamma}{4v_T^2} \left(v + 2i\frac{v_T^2}{\gamma} p \right)^2 - \frac{\gamma}{2}. \tag{18.349}$$

In the associated Hamiltonian operator, the shift of p_v by $-iv/2v_T^2$ can be removed by a similarity transformation to

$$\tilde{H}_p(p_v, v) \equiv e^{v^2/4v_T^2} \hat{H}_p e^{-v^2/4v_T^2} = \gamma v_T^2 p_v^2 + \frac{\gamma}{4v_T^2} \left(v + 2i\frac{v_T^2}{\gamma} p \right)^2 - \frac{\gamma}{2}. \tag{18.350}$$

Thus we can rewrite $P_p(v_b t_b | v_a t_a)$ as

$$P_p(v_b t_b | v_a t_a) = e^{-v_b^2/4v_T^2} \tilde{P}_p(v_b t_b | v_a t_a) e^{v_a^2/4v_T^2} \tag{18.351}$$

where $\tilde{P}_p(v_b t_b | v_a t_a)$ is the probability associated with the Hamiltonian $\tilde{H}_p(p_v, v)$. This describes a harmonic oscillator with frequency γ around the p-dependent center at $v_p = -2iv_T^2 p/\gamma$. If we denote the mass of this oscillator by $m = 1/2\gamma v_T^2$, we can immediately write down the wave eigenfunctions as $\psi_n(v - v_p)$ with $\psi(x)$ of Eq. (2.300). The energies are $n\gamma$. Thus we may express the probability distribution of x and v as a spectral representation

$$P(x_b v_b t_b | x_a v_a t_a) = e^{-(v_b^2 - v_a^2)/4v_T^2} \int \frac{dp}{2\pi} \sum_{n=0}^{\infty} \psi_n(v_b - v_p) \psi_n(v_a - v_p) e^{-n\gamma(t_b - t_a)}$$

$$\times e^{ip(x_b - x_a) - v_T^2 p^2 (t_b - t_a)/\gamma}, \tag{18.352}$$

where

$$\psi_n(v) = \frac{1}{(2^n n! \sqrt{\pi})^{1/2} (\sqrt{2} v_T)^{1/2}} e^{-v^2/4v_T^2} H_n(v/\sqrt{2} v_T).$$ (18.353)

In the limit of strong damping, only $n = 0$ contributes to the sum, and we find

$$P(x_b v_b t_b | x_a v_a t_a) = \frac{e^{-(v_b^2 - v_a^2)/4v_T^2}}{\sqrt{2\pi} v_T} \int \frac{dp}{2\pi} e^{-[(v_b - v_p)^2 + (v_a - v_p)^2]/4v_T^2} e^{ip(x_b - x_a) - v_T^2 p^2 (t_b - t_a)/\gamma}.$$

Integrating this over v_b leads to

$$P(x_b t_b | x_a v_a t_a) = \int \frac{dp}{2\pi} e^{ip(x_b - x_a - v_a/\gamma) - v_T^2 p^2 (t_b - t_a)/\gamma} = \frac{1}{\sqrt{4\pi v_T^2/\gamma}} e^{-\frac{\gamma(x_b - x_a - v_a/\gamma)^2}{4v_T^2 (t_b - t_a)}}, \quad (18.354)$$

where we have neglected terms of order γ^{-2} in the exponent. A compact expression for the general solution will be derived in (18.440) by stochastic calculus.

18.12   Stochastic Quantization

In Eq. (18.314) we observed that the expectation value of powers of a classical variable x in a potential $V(x)$ can be recovered as a result of a path integral associated with the Lagrangian (18.299). From Eq. (18.339) we know that the path integral (18.314) can be replaced by the stochastic path integral:

$$\langle x^n \rangle = \lim_{s \to \infty} \langle x^n(s) \rangle = \lim_{s \to \infty} \int \mathcal{D}\eta \, x_\eta^n(s) P[\eta], \quad (18.355)$$

where

$$P[\eta] \equiv \int \mathcal{D}\eta \, e^{-(1/4k_B T) \int_{s_a}^{s} ds' \eta^2(s')}, \quad (18.356)$$

To simplify the equations, we have replaced the physical time by a rescaled parameter $s = t/M\gamma$.

Equivalently we may say that we obtain the expectation values (18.355) by solving the stochastic differential equation of the Wiener process

$$x'(s) = -V'(x) + \eta(s), \quad (18.357)$$

where $\eta(s)$ is a white noise with the pair correlation functions

$$\langle \eta(s) \rangle_T = 0, \qquad \langle \eta(s)\eta(s') \rangle_T = 2k_B T \, \delta(s - s'), \quad (18.358)$$

and going to the large-s limit of the expectation values $\langle x^n(s) \rangle$.

This can easily be generalized to Euclidean quantum mechanics. Suppose we want to calculate the correlation functions (3.295)

$$\langle x(\tau_1)x(\tau_2) \cdots x(\tau_n) \rangle \equiv Z^{-1} \int \mathcal{D}x \, x(\tau_1)x(\tau_2) \cdots x(\tau_n) \exp\left(-\frac{1}{\hbar} \mathcal{A}_e\right). \quad (18.359)$$

We introduce an additional auxiliary time variable s and set up a stochastic differential equation

$$\partial_s x(\tau; s) = -\frac{\delta \mathcal{A}_e}{\delta x(\tau; s)} + \eta(\tau; s), \tag{18.360}$$

where $\eta(\tau; s)$ has correlation functions

$$\langle \eta(\tau; s) \rangle = 0, \quad \langle \eta(\tau; s) \eta(\tau'; s') \rangle = 2\hbar \delta(\tau - \tau') \delta(s - s'). \tag{18.361}$$

The role of the thermal fluctuation width $2k_B T$ in (18.358) is now played by $2\hbar$. The correlation functions (18.359) can now be calculated from the auxiliary correlation functions of $x(\tau, s)$ in the large-s limit:

$$\langle x(\tau_1) x(\tau_2) \cdots x(\tau_n) \rangle = \lim_{s \to \infty} \langle x(\tau_1, s) x(\tau_2, s) \cdots x(\tau_n, s) \rangle. \tag{18.362}$$

Due to the extra time variable of stochastic variable $x(\tau; s)$ with respect to (18.357), the probability distribution associated with the stochastic differential equation (18.379) is a functional $P[x_b(\tau), s_b; x_a(\tau), s_a]$ given by the functional generalization of the path integral (18.300):

$$\begin{aligned} P[x_b(\tau_b), s; x_a(\tau), s_a] &= \mathcal{N} \int \mathcal{D}x(\tau; s) \\ &\times e^{-\int_{s_a}^{s_b} ds \left\{ \frac{1}{4\hbar} \int_{-\infty}^{\infty} d\tau \left[\partial_s x(\tau;s) + \frac{\delta}{\delta x(\tau;s)} \mathcal{A}_e \right] - \frac{1}{2\hbar} \frac{\delta^2}{\delta x(\tau;s)^2} \mathcal{A}_e \right\}}. \end{aligned} \tag{18.363}$$

This satisfies the functional generalization of the Fokker-Planck equation (18.292):

$$H[\hat{p}(\tau), x(\tau)] P(x(\tau) s | x_a(\tau); s_a) = -\partial_s P(x(\tau) s | x_a(\tau); s_a), \tag{18.364}$$

with the Hamiltonian

$$H[\hat{p}(\tau), x(\tau)] = \int_{-\infty}^{\infty} d\tau \left[\hbar \hat{p}^2(\tau) - i\hat{p}(\tau) \frac{\delta}{\delta x(\tau)} \mathcal{A}_e \right], \tag{18.365}$$

where $\hat{p}(\tau) \equiv \delta/\delta x(\tau)$. We have dropped the subscript b of the final state, for brevity. Explicitly, the Fokker-Planck equation (18.364) reads

$$-\int_{-\infty}^{\infty} d\tau \frac{\hbar \delta}{\delta x(\tau)} \left[\frac{\hbar \delta}{\delta x(\tau)} + \frac{\delta \mathcal{A}_e}{\delta x(\tau)} \right] P[x(\tau), s; x_a(\tau), s_a] = -\hbar \partial_s P[x(\tau), s; x_a(\tau), s_a]. \tag{18.366}$$

For $s \to \infty$, the distribution becomes independent of the initial path $x_a(\tau)$, and has the limit [compare (18.314)]

$$\lim_{s \to \infty} P[x(\tau), s; x_a(\tau), s_a] = \frac{e^{-\mathcal{A}_e[x]/\hbar}}{\int \mathcal{D}x(\tau) \, e^{-\mathcal{A}_e[x]/\hbar}}, \tag{18.367}$$

and the correlation functions (18.378) are given by the usual path integral, apart from the normalization which is here such that $\langle 1 \rangle = 1$.

As an example, consider a harmonic oscillator where Eq. (18.360) reads

$$\partial_s x(\tau; s) = -M(-\partial_\tau^2 + \omega^2)x(\tau; s) + \eta(\tau; s). \tag{18.368}$$

This is solved by

$$x(\tau; s) = \int_0^s ds'\, e^{-M(-\partial_\tau^2 + \omega^2)(s' - s)}\eta(\tau; s'). \tag{18.369}$$

The correlation function reads, therefore,

$$\langle x(\tau_1; s_1)x(\tau_2; s_2)\rangle = \int_0^{s_1} ds_1' \int_0^{s_2} ds_2'\, e^{M(-\partial_\tau^2 + \omega^2)(s_1' + s_2' - s_1 - s_2)}\langle \eta(\tau_1; s_1')\eta(\tau_2; s_2')\rangle. \tag{18.370}$$

Inserting (18.361), this becomes

$$\langle x(\tau_1; s_1)x(\tau_2; s_2)\rangle = \hbar \int_0^\infty ds\, \left[e^{-M(-\partial_\tau^2 + \omega^2)(s + |s_1 - s_2|)} - e^{-M(-\partial_\tau^2 + \omega^2)(s + s_1 + s_2)} \right], \tag{18.371}$$

or

$$\langle x(\tau_1; s_1)x(\tau_2; s_2)\rangle = \frac{\hbar}{M} \frac{1}{-\partial_\tau^2 + \omega^2} \left[e^{-M(-\partial_\tau^2 + \omega^2)|s_1 - s_2|} - e^{-M(-\partial_\tau^2 + \omega^2)(s_1 + s_2)} \right]. \tag{18.372}$$

For Dirichlet boundary conditions ($x_b = x_a = 0$) where operator $(-\partial_\tau^2 + \omega^2)$ has the sinusoidal eigenfunctions of the form (3.63) with eigenfrequencies (3.64), this has the spectral representation

$$
\begin{aligned}
\langle x(\tau_1; s_1)x(\tau_2; s_2)\rangle &= \frac{\hbar}{M} \frac{2}{t_b - t_a} \sum_{n=1}^\infty \frac{1}{\nu_n^2 + \omega^2} \sin \nu_n(\tau_1 - \tau_a) \sin \nu_n(\tau_2 - \tau_a) \\
&\quad \times \left[e^{-M(\nu_n^2 + \omega^2)|s_1 - s_2|} - e^{-M(-\nu_n^2 + \omega^2)(s_1 + s_2)} \right].
\end{aligned}
\tag{18.373}
$$

For large s_1, s_2, the second term can be omitted. If, in addition, $s_1 = s_2$, we obtain the imaginary-time correlation function [compare (3.69), (3.301), and (3.36)]:

$$
\begin{aligned}
\lim_{s_1 = s_2 \to \infty} \langle x(\tau_1; s)x(\tau_2; s)\rangle &= \langle x(\tau_1)x(\tau_2)\rangle = \frac{\hbar}{M} \frac{1}{-\partial_\tau^2 + \omega^2}(\tau_1, \tau_2) \\
&= \frac{\hbar}{M} \frac{\sinh \omega(\tau_b - \tau_>) \sinh \omega(\tau_< - \tau_a)}{\omega \sinh \omega(\tau_b - \tau_a)}.
\end{aligned}
\tag{18.374}
$$

We can use these results to calculate the time evolution amplitude according to an imaginary-time version of Eq. (3.315):

$$(x_b\tau_b|x_a\tau_a) = C(x_b, x_a)e^{-A_e(x_b, x_a; \tau_b - \tau_a)/\hbar}e^{-\int_{\tau_a}^{\tau_b} \frac{M}{2} d\tau_b' \langle L_{e,\mathrm{fl}}(x_b, \dot{x}_b)\rangle/\hbar}, \tag{18.375}$$

where $A_e(x_b, x_a; \tau_b - \tau_a)$ is the Euclidean version of the classical action (4.87). If the Lagrangian has the standard form, then

$$\langle L_{e,\mathrm{fl}}(x_b, \dot{x}_b)\rangle = \frac{M}{2}\langle \delta\dot{x}_b^2\rangle, \tag{18.376}$$

and we obtain the imaginary-time evolution amplitude in an expression like (3.315). The constant of integration is determined by solving the differential equation (3.316), and a similar equation for $\mathbf{x}_a$. From this we find as before that $C(\mathbf{x}_b, \mathbf{x}_a)$ is independent of $\mathbf{x}_b$ and $\mathbf{x}_a$.

For the harmonic oscillator with Dirichlet boundary conditions we calculate from this

$$\frac{M}{2}\langle \delta \dot{\mathbf{x}}_b^2 \rangle = \frac{\hbar \omega}{2} D \coth \omega (\tau_b - \tau_a). \tag{18.377}$$

Integrating this over τ_b yields $\hbar(D/2)\log[2\sinh\omega(\tau_b - \tau_a)]$, so that the second exponential in (18.375) reduces to the correct fluctuation factor in the D-dimensional imaginary-time amplitude [compare (2.409)].

The formalism can easily be carried over to real-time quantum mechanics. We replace $t \to -i\tau$ and $\mathcal{A}_e \to -i\mathcal{A}$, and find that the real-time correlation functions are obtained from the large-s limit

$$\langle x(t_1)x(t_2)\cdots x(\tau_n)\rangle = \lim_{s\to\infty}\langle x(t_1, s)x(t_2, s)\cdots x(t_n, s)\rangle, \tag{18.378}$$

where $x(t; s)$ satisfies the stochastic differential equation

$$\hbar\partial_s x(t; s) = i\frac{\delta\mathcal{A}}{\partial x(t; s)} + \eta(t; s), \tag{18.379}$$

where the noise $\eta(t; s)$ has the same correlation functions as in (18.361), if we replace τ by t. This procedure of calculating quantum-mechanical correlation functions is called *stochastic quantization* [17].

18.13 Stochastic Calculus

The relation between Langevin and Fokker-Planck equations is a major subject of the so-called *stochastic calculus*. Given a Langevin equation, the time order of the potential $V(x)$ with respect to $\dot{x}$ and $\ddot{x}$ is a matter of choice. Different choices form the basis of the *Itō* or the *Stratonovich calculus*. The retarded position which appears naturally in the derivation from the forward–backward path integral favors the use of the Itō calculus. A midpoint ordering as in the gauge-invariant path integrals in Section 10.5 corresponds to the Stratonovich calculus.

18.13.1 Kubo's stochastic Liouville equation

It is worthwhile to trace how the retarded operator order of the friction term enters the framework of stochastic calculus. Thus we assume that the stochastic differential equations (18.322) and (18.323) have been solved for a specific noise function $\eta(t)$ such that we know the probability distribution $P_\eta(x\,v\,t|x_a v_a t_a)$ in (18.324). Now we observe that the time dependence of this distribution is governed by a simple

differential equation known as *Kubo's stochastic Liouville equation* [18], which is
derived as follows [19]. A time derivative of (18.324) yields

$$\partial_t P_\eta(x\,v\,t|x_a v_a t_a) = \dot{x}_\eta(t)\delta'(x_\eta(t) - x)\delta(\dot{x}_\eta(t) - v) + \ddot{x}_\eta(t)\delta(x_\eta(t) - x)\delta'(\dot{x}_\eta(t) - v).$$
$$(18.380)$$

The derivatives of the δ-functions are initially with respect to the arguments $x_\eta(t)$
and $\dot{x}_\eta(t)$. These can, however, be expressed in terms of derivatives with respect to
$-x$ and $-v$. However, since $\ddot{x}_\eta(t)$ depends on $\dot{x}_\eta(t)$ we have to be careful where to put
the derivative $-\partial_v$. The general formula for such an operation may be expressed as
follows: Given an arbitrary dynamical variable $z(t)$ which may be any local function
(local in the temporal sense) of $x(t)$ and $\dot{x}(t)$, and whose derivative is some function
of $z(t)$, i.e., $\dot{z}(t) = F(z(t))$, then

$$\frac{d}{dt}\delta(z(t) - z) = \dot{z}(t)\frac{\partial}{\partial z(t)}\delta(z(t) - z) = -\frac{\partial}{\partial z}[\dot{z}(t)\delta(z(t) - z)] = -\frac{\partial}{\partial z}[F(z)\delta(z(t) - z)].$$
$$(18.381)$$

To prove this formula, we multiply each expression by an arbitrary smooth test
function $g(z)$ and integrate over z. Each integral yields indeed the same result
$\dot{g}(z(t)) = \dot{z}(t)g'(z(t)) = F(z)g'(z(t))$. Applying the identity (18.381) to (18.380), we
obtain an equation for $P_\eta(x\,v\,t|x_a v_a t_a)$:

$$\partial_t P_\eta(x\,v\,t|x_a v_a t_a) = -[\partial_x \dot{x}_\eta(t) + \partial_v \ddot{x}_\eta(t)]P_\eta(x\,v\,t|x_a v_a t_a).$$
$$(18.382)$$

We now express $\ddot{x}_\eta(t)$ with the help of the Langevin equation (18.317) in terms of the
friction force $-M\gamma\dot{x}_\eta(t)$, the force $-V'(x_\eta(t))$, and the noise $\eta(t)$. In the presence
of the δ-function $\delta(\dot{x}_\eta(t) - v)$, the velocity $\dot{x}_\eta(t)$ can everywhere be replaced by v,
and Eq. (18.382) becomes

$$\partial_t P_\eta(x\,v\,t|x_a v_a t_a) = -\left\{v\partial_x + \frac{1}{M}\left[\eta(t) + f(x, v)\right]\right\}P_\eta(x\,v\,t|x_a v_a t_a),$$
$$(18.383)$$

where

$$f(x, v) \equiv -M\gamma v - V'(x)$$
$$(18.384)$$

is the sum of potential and friction forces. This is Kubo's stochastic Liouville equation which, together with the correlation function (18.218) of the noise variable and
the prescription (18.325) of forming expectation values, determines the temporal
behavior of the probability distribution $P(x\,v\,t|x_a v_a t_a)$.

18.13.2   From Kubo's to Fokker-Planck Equations

Let us calculate the expectation value of $P_\eta(x\,v\,t|x_a v_a t_a)$ with respect to noise fluc-
tuations and show that $P(x\,v\,t|x_a v_a t_a)$ of Eq. (18.325) satisfies the Fokker-Planck
equation with inertia (18.243). First we observe that in a Gaussian expectation

value (18.321), the multiplication of a functional $F[\eta]$ by η produces the same result as the functional differentiation with respect to η with a subsequent functional multiplication by the correlation function $\langle \eta(t)\eta(t') \rangle$:

$$\langle \eta(t)F[\eta] \rangle_\eta = \int dt' \langle \eta(t)\eta(t') \rangle_\eta \left\langle \frac{\delta\eta(t)}{\delta\eta(t')} F[\eta] \right\rangle_\eta. \tag{18.385}$$

This follows from the fact that $\eta(t)$ can be obtained from a functional derivative of the Gaussian distribution in (18.321) as:

$$\eta(t)e^{-\frac{1}{2w}\int dtdt'\eta(t)K^{-1}(t,t')\eta(t')} = -w\int dt' K(t,t')\frac{\delta}{\delta\eta(t')}e^{-\frac{1}{2w}\int dtdt'\eta(t)K^{-1}(t,t')\eta(t')}. \tag{18.386}$$

Inside the functional integral (18.321) over $\eta(t)$, an integration by parts moves the functional derivative $-\delta/\delta\eta(t')$ in front of $F[\eta]$ with a sign change. The surface terms can be discarded since the integrand decrease exponentially fast for large noises $\eta(t)$. Thus we obtain indeed the useful formula (18.385).

With the goal of a Gaussian average (18.321) in mind, we can therefore replace Eq. (18.383) by

$$\partial_t P_\eta(x\,v\,t|x_a v_a t_a) = -\left\{ v\partial_x + \frac{1}{M}\partial_v \left[w\int dt' K(t,t')\frac{\delta}{\delta\eta(t')} + f(x,v) \right] \right\} P_\eta(x\,v\,t|x_a v_a t_a). \tag{18.387}$$

After this, we observe that

$$\frac{\delta}{\delta\eta(t')}\delta(x_\eta(t)-x)\delta(\dot{x}_\eta(t)-v) = -\left[\frac{\delta x_\eta(t)}{\delta\eta(t')}\partial_x + \frac{\delta \dot{x}_\eta(t)}{\delta\eta(t')}\partial_v \right]\delta(x_\eta(t)-x)\delta(\dot{x}_\eta(t)-v). \tag{18.388}$$

From the stochastic differential equation (18.317) we deduce the following behavior of the functional derivatives:

$$\frac{\delta\ddot{x}_\eta(t)}{\delta\eta(t')} = \frac{1}{M}\delta(t-t') - \gamma\Theta(t-t') + \text{ smooth function of } t-t', \tag{18.389}$$

$$\frac{\delta\dot{x}_\eta(t)}{\delta\eta(t')} = \frac{1}{M}\Theta(t-t') + \mathcal{O}(t-t'), \tag{18.390}$$

$$\frac{\delta x_\eta(t)}{\delta\eta(t')} = \mathcal{O}((t-t')^2). \tag{18.391}$$

Inserting (18.380) with (18.390) and (18.391) into (18.387), the functional derivatives (18.390) and (18.391) are multiplied by $K(t,t')$ and integrated over t'.

Consider now the regime of large temperatures. There the function $K(t,t')$ is narrowly peaked around $t = t'$, forming almost a δ-function [recall the unit normalization (18.328)]. We shall emphasize this by writing $K(t,t') \equiv \delta_\epsilon(t-t')$, with the subscript indicating the width ϵ of $K(t,t')$ which goes to zero like $\hbar/k_B T$ for large T

[recall (18.337)]. In this limit, the contribution of the derivative (18.391) vanishes, whereas (18.390) contributes to (18.387) a term

$$\int dt' K(t,t') \frac{\delta}{\delta \eta(t')} \delta(x_\eta(t)-x)\delta(\dot{x}_\eta(t)-v) \tag{18.392}$$

$$= -\int dt' \delta_\epsilon(t-t') \frac{\delta \dot{x}_\eta(t)}{\delta \eta(t')} \partial_v \delta(x_\eta(t)-x)\delta(\dot{x}_\eta(t)-v) = -\frac{1}{2M} \partial_v \delta(x_\eta(t)-x)\delta(\dot{x}_\eta(t)-v).$$

The factor $1/2$ on the right-hand side arises from the fact that the would-be δ-function $\delta_\epsilon(t-t')$ is symmetric in $t-t'$, so that its convolution with the Heaviside function $\Theta(t-t')$ is nonzero only over half the peak. Taking the noise average (18.325), we obtain from (18.387) the Fokker-Planck equation with inertia (18.243):

$$\partial_t P(x\,v\,t|x_a v_a t_a) = \left\{ -v\partial_x + \frac{1}{M}\partial_v \left[\frac{w}{2M}\partial_v - f(x,v) \right] \right\} P(x\,v\,t|x_a v_a t_a). \tag{18.393}$$

Note that the differential operators have precisely the same order as in Eq. (18.239) as a consequence of formula, here (18.381).

In the overdamped limit, the derivation of the Fokker-Planck equation becomes simple. Then we have to consider only the pure x-space distribution

$$P_\eta(x\,t|x_a v_a t_a) = \int dv\, P_\eta(x\,v\,t|x_a v_a t_a) = \delta(x_\eta(t)-x), \tag{18.394}$$

whose time derivative is given by

$$\begin{aligned} \partial_t P_\eta(x\,t|x_a v_a t_a) &= -\partial_x \dot{x}_\eta(t) P_\eta(x\,t|x_a v_a t_a) \\ &= -\frac{1}{M\gamma}\partial_x [\eta(t) - V'(x)] P_\eta(x\,t|x_a v_a t_a). \end{aligned} \tag{18.395}$$

After treating the noise term $\eta(t)$ according to the rule (18.385),

$$\eta(t) \rightarrow w \int dt' \delta_\epsilon(t-t') \frac{\delta}{\delta \eta(t')}, \tag{18.396}$$

we use

$$\frac{\delta}{\delta \eta(t')} \delta(x_\eta(t) - x) = -\frac{\delta x_\eta(t)}{\delta \eta(t')} \delta(x_\eta(t) - x) \tag{18.397}$$

and

$$\begin{aligned} \frac{\delta \dot{x}_\eta(t)}{\delta \eta(t')} &= \frac{1}{M\gamma}\delta(t-t') + \text{ smooth function of } t-t', \\ \frac{\delta x_\eta(t)}{\delta \eta(t')} &= \frac{1}{M\gamma}\Theta(t-t') + \mathcal{O}(t-t'), \end{aligned} \tag{18.398}$$

to find the overdamped Fokker-Planck equation (18.292):

$$\partial_t P(x\,t|x_a t_a) = \left[D\partial^2 + \frac{1}{M\gamma}V'(x) \right] P(x\,t|x_a t_a). \tag{18.399}$$

The distributions $P(x\,t|x_a t_a)$ and $P(x\,v\,t|x_a v_a t_a)$ develop from initial δ-function distributions $P(x\,t_a|x_a t_a) = \delta(x - x_a)$ and $P(x\,v\,t_a|x_a t_a) = \delta(x - x_a)\delta(v - x_a)$.

Let us multiply these δ-functions with arbitrary initial probabilities $P(x, t_a)$ and $P(x\,v, t_a)$ and integrate over x and v. Then we obtain the stochastic path integrals

$$P(x\,,t) = \int \mathcal{D}\eta\, e^{-(1/2w)\int dt dt'\,\eta(t)K^{-1}(t,t')\eta(t')} P(x_{a\eta}(t), t_a), \qquad (18.400)$$

$$P(x\,v,t) = \int \mathcal{D}\eta\, e^{-(1/2w)\int dt dt'\,\eta(t)K^{-1}(t,t')\eta(t')} P(x_{a\eta}(t), v_{a\eta}, t), \qquad (18.401)$$

where $x_{a\eta}$ and $v_{a\eta}$ are initial positions and velocities of paths which arrive at the final x and v following the equation of motion with a fixed noise $\eta(t)$:

$$x_{a\eta}(t) = x - \int_{t_a}^t dt'\,\dot{x}(t'), \qquad v_{a\eta}(t) = x - \int_{t_a}^t dt'\,\dot{v}(t'). \qquad (18.402)$$

At high temperatures, the overdamped equation can be written with (18.338) as

$$P(x\,,t) = \int \mathcal{D}\eta\, e^{-(1/2w)\int dt\,\eta^2(t)} P\left(x - \frac{1}{M\gamma}\int_{t_a}^t dt'\,[\eta(t') - V'(x(t'))], t\right). \qquad (18.403)$$

The time evolution equation (18.399) follows from this by calculating for a short time increment ϵ:

$$P(x\,,t + \epsilon) = \int \mathcal{D}\eta\, e^{-(1/2w)\int dt\,\eta^2(t)} \left\{ -\frac{\epsilon}{M\gamma}\int_t^{t+\epsilon} dt'\,[\eta(t') - V'(x(t'))]\,\partial_x \right.$$

$$+ \frac{1}{2M^2\gamma^2}\int_t^{t+\epsilon} dt'\int_t^{t+\epsilon} dt''\,[\eta(t') - V'(x(t'))][\eta(t'') - V'(x(t''))]\,\partial_x^2 + \dots \Big\}$$

$$\times P\left(x - \frac{1}{M\gamma}\int_{t_a}^t dt'\,[\eta(t') - V'(x(t'))], t\right). \qquad (18.404)$$

We now use the correlation functions (18.329), ignore all powers higher than linear in ϵ, and find in the limit $\epsilon \to 0$ directly the equation (18.399).

18.13.3   Itō's Lemma

An important tool for dealing with stochastic variables is suplied by *Itō's lemma*. Let $x(t)$ be a stochastic variable following a Wiener process with a drift $r_x(x(t), t)$ which is supposed to be a smooth function of $x(t)$ and t [compare (18.338)], i.e., $\dot{x}(t)$ fluctuates harmonically with a white noise around its average $\langle\dot{x}(t)\rangle = r_x(x(t), t)$ according to a stochastic differential equation

$$\dot{x}(t) = \langle\dot{x}(t)\rangle + \eta(t) = r_x + \eta(t). \qquad (18.405)$$

We shall omit the smooth dependence of r_x on its arguments since this will be irrelevant for the subsequent arguments. The white noise has zero average $\langle\eta(t)\rangle = 0$, and its only nonzero correlation function is

$$\langle \eta(t)\eta(t') \rangle = \sigma^2 \delta(t - t'). \tag{18.406}$$

The value of $x(t)$ at a slightly later time $t + \epsilon$ is $x(t + \epsilon) = x(t) + \Delta x(t)$, where

$$\Delta x(t) \equiv \int_t^{t+\epsilon} dt' \, \dot{x}(t') = \epsilon r_x + \int_t^{t+\epsilon} dt' \, \eta(t'). \tag{18.407}$$

Consider now an arbitrary function $f(x(t))$. Its value at the time $t + \epsilon$ has the Taylor expansion

$$\begin{aligned} f(x(t + \epsilon)) &= f(x(t)) + f'(x(t))\Delta x(t) \\ &+ \frac{1}{2}f''(x(t))[\Delta x(t)]^2 + \frac{1}{3!}f^{(3)}[\Delta x(t)]^3 + \dots . \end{aligned} \tag{18.408}$$

The linear term in $\Delta x(t)$ on the right-hand side of (18.408) has the average

$$\langle \Delta x(t) \rangle = \int_t^{t+\epsilon} dt' \, \langle \dot{x}(t') + \eta(t') \rangle = \int_t^{t+\epsilon} dt' \, \langle \dot{x}(t') \rangle \approx \epsilon r_x, \tag{18.409}$$

where we have omitted the arguments $x(t)$ and t of $r_x(x(t), t)$, since the variation of $r_x(x(t), t)$ in the small interval $(t, t + \epsilon)$ can be neglected to lowest order in ϵ.

The average of the quadratic term $\langle [\Delta x(t)]^2 \rangle$ is

$$\begin{aligned} \langle [\Delta x(t)]^2 \rangle &= \int_t^{t+\epsilon} dt_1 \int_t^{t+\epsilon} dt_2 \, \langle [\langle \dot{x}(t_1) \rangle + \eta(t_1)] [\langle \dot{x}(t_2) \rangle + \eta(t_2)] \rangle \\ &\approx \epsilon^2 r_x^2 + \langle \eta(t_1)\eta(t_2) \rangle. \end{aligned}$$

The second term is of the order ϵ due to the δ-function in the correlation function (18.406). Thus we find

$$\langle [\Delta x(t)]^2 \rangle = \epsilon \sigma^2 + \mathcal{O}(\epsilon^2). \tag{18.410}$$

The average of the cubic term $\langle [\Delta x(t)]^3 \rangle$ is given by the integral

$$\begin{aligned} & \int_t^{t+\epsilon} dt_1 \int_t^{t+\epsilon} dt_2 \int_t^{t+\epsilon} dt_3 \, \langle [\langle \dot{x}(t_1) \rangle + \eta(t_1)] [\langle \dot{x}(t_2) \rangle + \eta(t_2)] [\langle \dot{x}(t_3) \rangle + \eta(t_2)] \rangle \\ &= \int_t^{t+\epsilon} dt_1 \int_t^{t+\epsilon} dt_2 \int_t^{t+\epsilon} dt_3 \, \Big[\langle \dot{x}(t_1) \rangle \langle \dot{x}(t_2) \rangle \langle \dot{x}(t_3) \rangle + \langle \dot{x}(t_1) \rangle \langle \eta(t_2)\eta(t_3) \rangle \\ &\qquad\qquad + \langle \dot{x}(t_2) \rangle \langle \eta(t_1)\eta(t_3) \rangle + \langle \dot{x}(t_3) \rangle \langle \eta(t_1)\eta(t_2) \rangle \Big] \\ &= \epsilon^3 r_x^3 + 3\epsilon^2 r_x \sigma^2 = \mathcal{O}(\epsilon^2). \end{aligned} \tag{18.411}$$

The averages of the higher powers $[\Delta x(t)]^n$ are obviously at least of order $\epsilon^{n/2}$. Thus we find in the limit $\epsilon \to 0$ the simple formula

$$\langle \dot{f}(x(t)) \rangle = \langle f'(x(t)) \rangle \langle \dot{x}(t) \rangle + \frac{\sigma^2}{2} \langle f''(x(t)) \rangle. \tag{18.412}$$

Note that in a time-sliced formulation, $f(x(t))\dot{x}(t)$ has the form $f(x_n)(x_{n+1} - x_n)/\epsilon$, with independently fluctuating x_n and x_{n+1}, so that we may treat x_n and $(x_{n+1} - x_n)/\epsilon$ as independent fluctuating variables. In the continuum limit $x(t)$ and $\dot{x}(t)$ become independent.

The important point noted by Itō is now that this result is not only true for the averages but also for the derivative $\dot{f}(x(t))$ itself, i.e., $f(x(t))$ obeys the stochastic differential equation

$$\dot{f}(x(t)) = f'(x(t))\,\dot{x}(t) + \frac{\sigma^2}{2}f''(x(t)), \tag{18.413}$$

which is known as Itō's lemma.

In order to prove this we must show that the omitted fluctuations in the higher powers $[\Delta x(t)]^n$ for $n \geq 2$ are of higher order in ϵ than the leading fluctuation of $\Delta x(t)$ which is of order ϵ. Indeed, let us denote the fluctuating part of $[\Delta x(t)]^n$ by $z_n(t)$. For $n = 1, 2$, these are

$$z_1(t) = \int_t^{t+\epsilon} dt\,\eta(t), \qquad z_2(t) \equiv [z_{2,1}(t) + z_{2,2}(t)], \tag{18.414}$$

where the two parts of $z_2(t)$ are

$$z_{2,1}(t) = 2\int_t^{t+\epsilon} dt_1\,\langle\dot{x}(t_1)\rangle\,z_1(t) \approx 2\epsilon r_x\,z_1(t), \quad z_{2,2}(t) = [z_1(t)]^2. \tag{18.415}$$

The fluctuations of $z_{2,1}(t)$ are smaller than the leading ones of $z_1(t)$ by a factor ϵ. They can therefore be ignored in the limit $\epsilon \to 0$.

The size of the fluctuations $z_{2,2}(t)$ are estimated by calculating its variance $\langle[z_{2,2}(t)]^2\rangle - \langle z_{2,2}(t)\rangle^2$. The first of the two expectation values is

$$\left\langle[z_{2,2}(t)]^2\right\rangle = \int_t^{t+\epsilon} dt_1 \int_t^{t+\epsilon} dt_2 \int_t^{t+\epsilon} dt_3 \int_t^{t+\epsilon} dt_4\,\langle\eta(t_1)\eta(t_2)\eta(t_3)\eta(t_4)\rangle. \tag{18.416}$$

According to Wick's rule (3.302) for harmonic fluctuations, the expectation value on the right-hand sid is equal to the sum of three pair contractions

$$\langle\eta(t_1)\eta(t_2)\rangle\langle\eta(t_3)\eta(t_4)\rangle + \langle\eta(t_1)\eta(t_3)\rangle\langle\eta(t_2)\eta(t_4)\rangle\langle\eta(t_1)\eta(t_4)\rangle\langle\eta(t_2)\eta(t_3)\rangle. \tag{18.417}$$

Inserting (18.406) and performing the integrals yields

$$\left\langle[z_{2,2}(t)]^2\right\rangle = 3\epsilon^2\sigma^4. \tag{18.418}$$

The second term in the variance of $z_{2,2}(t)$ is

$$\langle z_{2,2}(t)\rangle^2 = \langle z_1^2(t)\rangle^2 = \left[\int_t^{t+\epsilon} dt_1 \int_t^{t+\epsilon} dt_2\,\langle\eta(t_1)\eta(t_2)\rangle\right]^2 = \epsilon^2\sigma^4. \tag{18.419}$$

Hence we obtain for the variance of $z_{2,2}(t)$:

$$\langle[z_{2,2}(t)]^2\rangle - \langle z_{2,2}(t)\rangle^2 = 2\sigma^4\epsilon^2. \tag{18.420}$$

This must be compared with the variance of the leading fluctuations $z_1(t)$ in (18.413):

$$\langle [z_1(t)]^2 \rangle - \langle z_1(t) \rangle^2 = \int_t^{t+\epsilon} dt_1 \int_t^{t+\epsilon} dt_2 \, \langle \eta(t_1)\eta(t_2) \rangle = \epsilon \sigma^2, \qquad (18.421)$$

which implies that $z_1(t)$ is of the order of $\sigma\sqrt{\epsilon}$. Thus the fluctuating part of $[\Delta x(t)]^2$ is by a factor $\sqrt{\epsilon}$ smaller than that of $\Delta x(t)$, so that it can be ignored in the continuum limit $\epsilon \to 0$.

Thus we have proved that not only the expectation value $\langle [\Delta x(t)]^2 \rangle$ becomes equal to $\epsilon\sigma^2$ as stated in Eq. (18.410), but also the fluctuating quantity $[\Delta x(t)]^2$ itself:

$$[\Delta x(t)]^2 = \epsilon\sigma^2 + \mathcal{O}(\epsilon^2). \qquad (18.422)$$

In a similar way we can derive the estimates $[\Delta x(t)]^n = \mathcal{O}((\sigma\sqrt{\epsilon})^n)$ for all higher fluctuations $z_n(t)$ in the Taylor expansion (18.408). These can all be neglected compared to $z_1(t)$, thus proving Itō's lemma (18.413).

For an exponential function, Itō's lemma yields

$$\frac{d}{dt}e^{Px} = \left(P\dot{x} + \frac{\sigma^2 P^2}{2} \right) e^{Px}. \qquad (18.423)$$

This can be integrated to

$$e^{Px} = e^{\int_0^t dt' \, P\dot{x}} \, e^{P^2\sigma^2 t/2}. \qquad (18.424)$$

The expectation value of this can also be formulated as a rule for calculating the expectation value of an exponential of an integral over a Gaussian noise variable with zero average:

$$\left\langle e^{P \int_0^t dt' \, \eta(t')} \right\rangle = e^{P^2 \int_0^t dt' \int_0^t dt'' \langle \eta(t')\eta(t'') \rangle} = e^{P^2\sigma^2 t/2}. \qquad (18.425)$$

This rule can also be derived directly from Wick's rule (3.307). The right-hand side corresponds to the *Debye-Waller factor* introduced in solid-state physics to describe the reduction of the intensities of Bragg peaks by thermal fluctuations of the atomic positions [see Eq. (3.308)].

There is a simple mnemonic way of formalizing this derivation of Eq. (18.413) in a sloppy differential notation. We expand

$$f(x(t+dt)) = f(x(t) + \dot{x}dt) = f(x(t)) + f'(x(t))\dot{x}(t)dt + \frac{1}{2}f''(x(t))\dot{x}^2(t)dt^2 + \dots, \qquad (18.426)$$

and insert in the higher-order expansion terms $\dot{x} = \langle \dot{x} \rangle + \eta(t)$ where $\langle \eta(t) \rangle = 0$ and the expectation

$$\langle \eta^2(t) \rangle \, dt = \sigma^2, \qquad (18.427)$$

which expresses infinitesimally the correct equation

$$\int_t^{t+\epsilon} dt' \, \langle \eta(t')\eta(t) \rangle = \int_t^{t+\epsilon} dt' \, \sigma^2 \, \delta(t'-t) = \sigma^2. \qquad (18.428)$$

The variable $\dot{x}^2(t)dt^2$ has an expectation value $\sigma^2 dt$ and a variance $\langle [\dot{x}^2(t)dt^2]^2 - \langle \dot{x}^2(t)dt \rangle^2 \rangle = 2\sigma^2 dt^2$, so that $\dot{x}^2(t)dt^2$ in (18.426) can be replaced as follows:

$$\dot{x}^2(t)dt^2 \to \sigma^2 dt/2. \qquad (18.429)$$

A corresponding estimate holds for all higher powers:

$$z_n \approx \mathcal{O}((\sigma\sqrt{\epsilon})^n). \qquad (18.430)$$

or

$$\dot{x}^n(t)dt^n \approx \mathcal{O}((\sigma\sqrt{dt})^n). \qquad (18.431)$$

These can all be omitted in the expansion (18.426), thus leading back to Itō's rule (18.412).

It must be realized that Itō's lemma is valid only in the limit $\epsilon \to 0$. For a discrete time axis with small but finite time intervals $\Delta t = \epsilon$, the fluctuations of $z_n(t)$ cannot strictly be ignored but are only suppressed by a small factor $\sigma\sqrt{\Delta T}$. The discrete version of Itō's lemma expands the fluctuating difference $\Delta f(x(t_n)) \equiv f(x(t_{n+1})) - f(x(t_n))$ as follows:

$$\frac{\Delta f(x(t_n))}{\Delta t} = f'(x(t_n))\frac{\Delta x(t_n)}{\Delta t} + \frac{\sigma^2}{2}f''(x(t_n)) + \mathcal{O}(\sigma\sqrt{\Delta t}). \qquad (18.432)$$

18.14 Solving the Langevin Equation

In Eq. (18.306) we have found the probability distribution for the motion of a particle with large dissipation by solving the path integral (18.300) for the harmonic oscillator potential $V(x) = \omega_0^2 x^2/2$. For completeness, let us calculated the same result within stochastic calculus. The stochasti differential equation associated with the Lagrangian (18.301) is

$$\dot{x}(t) = -\kappa x(t) + \bar{\eta}(t), \qquad (18.433)$$

where

$$\langle \bar{\eta}(t) \rangle_\eta, \quad \langle \bar{\eta}(t)\bar{\eta}(t') \rangle_\eta = 2D\delta(t-t'). \qquad (18.434)$$

This equation is solved by

$$x(t) = x_0 e^{-\kappa t} + \int_0^t dt_1 \, e^{-\kappa(t-t_1)}\bar{\eta}(t_1), \qquad (18.435)$$

so that we obtain $\langle x(t)\rangle_\eta = x_0 e^{-\gamma t}$ and

$$\langle x(t)x(t')\rangle_\eta = x_0^2 e^{-\kappa(t+t')} + 2D \int_0^t dt_1 \, e^{-\kappa(t-t_1)} \int_0^{t'} dt_2 \, e^{-(t'-t_2)} \delta(t_1 - t_2)$$
$$= x_0^2 e^{-\kappa(t+t')} + \kappa^{-1} D \left(e^{-\kappa|t-t'|} - e^{-\kappa(t+t')} \right), \tag{18.436}$$

and the mean-square deviation

$$\langle [x(t) - \langle x(t)\rangle]^2 \rangle_\eta = \kappa^{-1} D \left(1 - e^{-2\kappa t} \right). \tag{18.437}$$

From these expectation values we recover immediately the previous distribution function (18.306).

This result can easily be generalized to a D-component Langevin equation

$$\dot{\mathbf{x}}(t) = -\boldsymbol{\kappa}\,\mathbf{x}(t) + \bar{\boldsymbol{\eta}}(t), \tag{18.438}$$

where $\boldsymbol{\kappa}$ is a matrix, and $\bar{\boldsymbol{\eta}}(t)$ a noise vector. Its correlation functions may be expressed in terms of a diffusion matrix $\mathbf{D}$ as

$$\langle \bar{\boldsymbol{\eta}}(t)\rangle = 0, \quad \langle \bar{\boldsymbol{\eta}}(t)\bar{\boldsymbol{\eta}}^T(t')\rangle_\eta = 2\mathbf{D}\delta(t - t'), \tag{18.439}$$

to be compared with the one-dimensional expressions (18.329).

Then the probability (18.306) becomes

$$P(\mathbf{x}_b t_b | \mathbf{x}_a t_a) = \frac{1}{\sqrt{2\pi}^D} \frac{1}{\sqrt{\det\left[\sigma^2(t_b - t_a)\right]}^D}$$
$$\times \exp\left\{ -\frac{1}{2}[x_b - \bar{x}(t_b-t_a)]^i [\sigma_{ij}^2(t_b-t_a)]^{-1} [x_b - \bar{x}(t_b-t_a)]^j \right\}, \tag{18.440}$$

where

$$\bar{\mathbf{x}}(t) = e^{-\kappa t}\,\mathbf{x}_a, \tag{18.441}$$

and

$$\sigma_{ij}^2(t) \equiv \langle [x(t) - \bar{x}(t)]^i [x(t) - \bar{x}(t)]^j \rangle_\eta. \tag{18.442}$$

The probability (18.440) solves the Fokker-Planck equation (18.245)

$$\partial_t P(\mathbf{x}\,t | t_a \mathbf{x}_a) = \left(-\kappa_{ij}\partial_i x_j + D_{ij}\partial_i\partial_j \right) P(\mathbf{x}\,t | t_a \mathbf{x}_a). \tag{18.443}$$

The D-component result (18.440) allows us to solve the Langevin equation with inertia in Eq. (18.317). We simply rewrite the equivalent pair of equations (18.322) and (18.323) in the matrix form (18.438) with $x_1 = x$ and $x_2 = v$, and identify

$$\boldsymbol{\kappa} = \begin{pmatrix} 0 & -1 \\ \omega_0^2 & \gamma \end{pmatrix}, \quad \bar{\boldsymbol{\eta}}(t) = \frac{1}{M}\begin{pmatrix} 0 \\ \eta(t) \end{pmatrix}, \tag{18.444}$$

so that the diffusion matrix takes the form (18.246).

The eigenvalues of the nonhermitian matrix $\boldsymbol{\kappa}$ in (18.441) are $\kappa_{1,2} = \frac{1}{2}(\gamma \pm \sqrt{\gamma^2 - 4\omega_0^2})$. The associated eigenvectors $\mathbf{u}^{(1,2)}$ satisfying $\boldsymbol{\kappa}\mathbf{u}^{(1,2)} = \kappa_{1,2}\mathbf{u}^{(1,2)}$ are $(-1, \kappa_1)$ and $(1, -\kappa_2)$, respectively, while those to the left satisfying $\mathbf{v}^{(1,2)}\boldsymbol{\kappa} = \kappa_{(1,2)}\mathbf{v}^{(1,2)}$ are $(\kappa_2, 1)/(\kappa_1 - \kappa_2)$ and $(\kappa_1, 1)/(\kappa_1 - \kappa_2)$, respectively. The two sets of eigenvectors are mutually orthonormal and complete: $\mathbf{u}^{(i)} \cdot \mathbf{v}^{(j)} = \delta^{ij}$, $\sum_k v_i^{(k)} u_j^{(k)} = \delta_{ij}$. The matrix $\boldsymbol{\kappa}$ has then the spectral representation $\kappa_{ij} = \sum_k \kappa_k u_i^{(k)} v_j^{(k)}$, and an exponential $(e^{-\boldsymbol{\kappa}t})_{ij} = \sum_k e^{-\kappa_k t} u_i^{(k)} v_j^{(k)}$. which reads explicitly

$$e^{-\boldsymbol{\kappa}t} = \frac{1}{\kappa_1 - \kappa_2} \begin{pmatrix} \kappa_1 e^{-\kappa_2 t} - \kappa_2 e^{-\kappa_1 t} & e^{-\kappa_2 t} - e^{-\kappa_1 t} \\ \omega_0^2 (e^{-\kappa_1 t} - e^{-\kappa_2 t}) & \kappa_1 e^{-\kappa_1 t} - \kappa_2 e^{-\kappa_2 t} \end{pmatrix}. \tag{18.445}$$

The inverse matrix $[\sigma_{ij}^2(t_b - t_a)]^{-1}$ is given by

$$[\sigma_{ij}^2(t)]^{-1} = \left[\det \sigma_{ij}^2(t)\right]^{-1} \begin{pmatrix} \sigma_{vv}^2(t) & -\sigma_{xv}^2(t) \\ -\sigma_{xv}^2(t) & \sigma_{xx}^2(t) \end{pmatrix} \tag{18.446}$$

where the matrix elements $\sigma_{ij}^2(t_b - t_a)$ are calculated from the expectation values (18.442). This is done by expressing the solution of (18.438) as in (18.435) in the form

$$\mathbf{x}(t) = e^{-\boldsymbol{\kappa}t} \mathbf{x}_a + \int_{t_a}^{t} dt\, \bar{\boldsymbol{\eta}}(t), \tag{18.447}$$

and using the correlation functions (18.439) to find

$$\sigma_{xx}^2(t) = \frac{\gamma^2 D}{(\kappa_1 - \kappa_2)^2} \left[\frac{1}{\kappa_1} \left(1 - e^{-2\kappa_1 t}\right) + \frac{1}{\kappa_2} \left(1 - e^{-2\kappa_2 t}\right) - \frac{4}{\kappa_1 + \kappa_2} \left(1 - e^{-\kappa_1 + \kappa_2)t}\right) \right],$$

$$\sigma_{xv}^2(t) = \frac{\gamma^2 D}{(\kappa_1 - \kappa_2)^2} \left(e^{-\kappa_1 t} - e^{-\kappa_2 t}\right)^2, \tag{18.448}$$

$$\sigma_{vv}^2(t) = \frac{\gamma^2 D}{(\kappa_1 - \kappa_2)^2} \left[\kappa_1 \left(1 - e^{-2\kappa_1 t}\right) + \kappa_2 \left(1 - e^{-2\kappa_2 t}\right) - \frac{4}{\kappa_1^{-1} + \kappa_2^{-1}} \left(1 - e^{-\kappa_1 + \kappa_2)t}\right) \right].$$

After a long time, these converge to

$$\sigma_{xx}^2(t) \rightarrow \frac{\gamma D}{\kappa_1 \kappa_2} = \frac{\gamma D}{\omega_0^2}, \qquad \sigma_{xv}^2(t) \rightarrow 0, \qquad \sigma_{vv}^2(t) \rightarrow \gamma D, \tag{18.449}$$

so that the determinant $\det \sigma_{ij}^2(t)$ becomes $\gamma^2 D^2/\omega_0^2$, and the distribution (18.440) turns into the Boltzmann distribution

$$\lim_{t_b \to \infty} P(x_b v_b t_b | x_a v_a t_a) = \frac{\omega_0}{2\pi\gamma D} e^{-(v_b^2 + \omega_0^2 x_b^2)/2\gamma D} = \frac{M\omega_0}{2\pi k_B T} e^{-M(v_b^2 + \omega_0^2 x_b^2)/2k_B T}. \tag{18.450}$$

The velocity shows the well-known *Maxwell distribution*:

$$P(v_b) = \frac{1}{\sqrt{2\pi\gamma D}} e^{-v_b^2/2\gamma D} = \frac{1}{\sqrt{2\pi k_B T/M}} e^{-M v_b^2/2k_B T} = \frac{1}{\sqrt{2\pi v_T}} e^{-v_b^2/2v_T^2}. \tag{18.451}$$

which exhibits an average thermal velocity

$$v_T \equiv \sqrt{k_B T/M}. \tag{18.452}$$

If we integrate the two-dimensional result $P(x_b v_b t_b | x_a v_a t_a)$ over all final velocities, we obtain

$$P(x_b t_b | x_a v_a t_a) = \int dv_b \, P(x_b v_b t_b | x_a v_a t_a) = \frac{1}{\sqrt{2\pi\sigma_{xx}^2(t_b - t_a)}} \exp\left\{ -\frac{1}{2} \frac{[x_b - \bar{x}(t_b - t_a)]^2}{\sigma_{xx}^2(t_b - t_a)} \right\}. \tag{18.453}$$

Note that this depends on v_a via $\bar{x}(t_b - t_a) = x_a + \gamma^{-1}(1 - e^{-\gamma(t_b - t_a)})v_a$.

In the absence of an external potential, i.e. for $\omega_0 = 0$, the eigenvalues $\kappa_{1,2}$ are γ and 0, respectively, and the matrix $e^{-\kappa t}$ reduces to

$$e^{-\kappa t} = \begin{pmatrix} 1 & \gamma^{-1}(1 - e^{-\gamma t}) \\ 0 & e^{-\gamma t} \end{pmatrix}. \tag{18.454}$$

The matrix elements $\sigma_{ij}^2(t)$ are simply[9]

$$\sigma_{xx}^2(t) = \gamma^{-1}D(2\gamma t - 3 + 4e^{-\gamma t} - e^{-2\gamma t}), \quad \sigma_{xv}^2(t) = D(1 - e^{-\gamma t})^2, \quad \sigma_{vv}^2(t) = \gamma D(1 - e^{-2\gamma t}), \tag{18.455}$$

whose determinant is

$$\det \sigma_{ij}^2(t) = D^2 \left[2\gamma t(1 - e^{-2\gamma t}) + (1 - e^{-\gamma t})^2(-4 - 2e^{-\gamma t} + e^{-3\gamma t}) \right]. \tag{18.456}$$

In the large-time limit, these become

$$\sigma_{xx}^2(t) \to 2Dt, \quad \sigma_{xv}^2(t) \to D, \quad \sigma_{vv}^2(t) \to \gamma D, \quad \det \sigma_{ij}^2(t) \to 2\gamma t D^2. \tag{18.457}$$

The last result can, of course, be derived by integrating the pair of Langevin equations with inertia (18.322) and (18.323) for zero potential $V(x)$ successively. First the equation for $v(t)$ which reads $v(t) = v_0 e^{-\gamma t} + \int_0^t dt_1 \, e^{-\gamma(t-t_1)}\eta(t_1)/M$, and yields $\langle v(t)v(t')\rangle_\eta = v_0^2 e^{-\gamma(t+t')} + \gamma D\left(e^{-\gamma|t-t'|} - e^{-\gamma(t+t')}\right)$, using the white noise correlation functions (18.329). The equations for $x(t)$ are obtained from these by integration over t.

18.15 Heisenberg Picture for Probability Evolution

It is possible to develop a Heisenberg operator description of the time dependence of thermal expectations. This goes by complete analogy with the development in Section 2.23 for the quantum-mechanical time evolution amplitude. Consider the thermal expectations of x and x^2 for a particle which sits at the initial time $t = t_a$ at x_a. They are given by the integrals

$$\langle x \rangle \equiv \int_{-\infty}^{\infty} dx_b \, x_b P(x_b t_b | x_a t_a), \tag{18.458}$$

$$\langle x^2 \rangle \equiv \int_{-\infty}^{\infty} dx_b \, x_b^2 P(x_b t_b | x_a t_a). \tag{18.459}$$

---

[9]See Section 18.15 for an alternative method of calculating the expectation values (18.442).

For simplicity, let us first look at the case of a dominant friction term. As in quantum mechanics, it is useful to introduce a bra-ket notation, but for the *probabilities* rather than the amplitudes,

$$\langle x_b t_b | x_a t_a \rangle \equiv |(x_b t_b | x_a t_a)|^2. \tag{18.460}$$

The fact that this probability satisfies the Fokker-Planck equation implies that we can write it as

$$\langle x_b t_b | x_a t_a \rangle = e^{-(t_b - t_a) H(\hat{p}_b, x_b)} \delta(x_b - x_a). \tag{18.461}$$

Thus we may introduce time-independent basis vectors $|x_a\rangle$ satisfying

$$\langle x_b | x_a \rangle = \delta(x_b - x_a). \tag{18.462}$$

On this basis, the operators $\hat{p}$, $\hat{x}$ are defined in the usual way. They satisfy

$$\langle x_b | \hat{x} = x_b \langle x_b |, \qquad \langle x_b | \hat{p} = -i \frac{\partial}{\partial x_b} \langle x_b |. \tag{18.463}$$

Then we may rewrite (18.461) in bra-ket notation as

$$\langle x_b t_b | x_a t_a \rangle = \langle x_b | e^{-H(\hat{p}, \hat{x})(t_b - t_a)} | x_a \rangle. \tag{18.464}$$

The expectation value of any function $f(x)$ is calculated as follows

$$
\begin{aligned}
\langle f(x) \rangle &= \int_{-\infty}^{\infty} dx_b \, f(x_b) \langle x_b | e^{-(t_b - t_a) H(\hat{p}, \hat{x})} | x_a \rangle \\
&= \int_{-\infty}^{\infty} dx_b \, \langle x_b | f(\hat{x}) e^{-(t_b - t_a) H(\hat{p}, \hat{x})} | x_a \rangle \\
&= \int_{-\infty}^{\infty} dx_b \int_{-\infty}^{\infty} dx \, \langle x_b | e^{-(t_b - t_a) H(\hat{p}, \hat{x})} | x \rangle \langle x | f(\hat{x}(t_b - t_a)) | x_a \rangle.
\end{aligned}
\tag{18.465}
$$

In the last term we have introduced the time-dependent Heisenberg type of operator

$$\hat{x}(t) \equiv e^{tH(\hat{p}, \hat{x})} \hat{x} e^{-tH(\hat{p}, \hat{x})}. \tag{18.466}$$

The probability $P(x_b t_b | x_a t_a)$ satisfies the normalization condition

$$
\begin{aligned}
\int_{-\infty}^{\infty} dx_b \, P(x_b t_b | x_a t_a) &= \int_{-\infty}^{\infty} dx_b \, \langle x_b t_b | x_a t_a \rangle \\
&= \int_{-\infty}^{\infty} dx_b \, \langle x_b | e^{-(t_b - t_a) H(\hat{p}, \hat{x})} | x_a \rangle = 1.
\end{aligned}
\tag{18.467}
$$

Appliying this to the last line of (18.465), we arrive at the simple formula

$$\langle f(x) \rangle = \int_{-\infty}^{\infty} dx_b \, \langle x_b | f(\hat{x}(t_b - t_a)) | x_a \rangle. \tag{18.468}$$

For the Brownian motion of a point particle where

$$L_e = \frac{\dot{x}^2}{4D}, \qquad H = Dp^2, \tag{18.469}$$

the Heisenberg operators are

$$\hat{p}(t) = \hat{p}, \qquad \hat{x}(t) = e^{\hat{H}t} \hat{x} e^{-\hat{H}t} = \hat{x} - i2D\hat{p}t, \tag{18.470}$$

and

$$\hat{x}^2(t) = \hat{x}^2 - i2D \cdot (\hat{p}\hat{x} + \hat{x}\hat{p})t - 4D^2\hat{p}^2t^2,$$
$$= \hat{x}^2 + 2Dt - i2D \cdot 2\hat{x}\hat{p} - 4D^2\hat{p}^2t. \tag{18.471}$$

It is easy to calculate the following matrix elements:

$$\int_{-\infty}^{\infty} dx_b \, \langle x_b|\hat{x}|x_a\rangle = x_a,$$

$$\int_{-\infty}^{\infty} dx_b \, \langle x_b|\hat{p}|x_a\rangle = -i\int_{-\infty}^{\infty} dx_b \, \frac{\partial}{\partial x_b}\delta(x_b - x_a) = 0,$$

$$\int_{-\infty}^{\infty} dx_b \, \langle x_b|\hat{x}^2|x_a\rangle = \int_{-\infty}^{\infty} dx_b \, x_b{}^2\delta(x_b - x_a) = x_a{}^2, \tag{18.472}$$

$$\int_{-\infty}^{\infty} dx_b \, \langle x_b|\hat{p}^2|x_a\rangle = -\int_{-\infty}^{\infty} dx_b \, \frac{\partial^2}{\partial x_b{}^2}\delta(x_b - x_a) = 0,$$

$$\int_{-\infty}^{\infty} dx_b \, \langle x_b|\hat{p}\hat{x}|x_a\rangle = -i\int_{-\infty}^{\infty} dx_b \, \frac{\partial}{\partial x_b}\delta(x_b - x_a)x_a = 0.$$

The vanishing integrals reflect the translational invariance of the integrated bra state $\int_{-\infty}^{\infty} dx_b \, \langle x_b|$, which is therefore annihilated by a translational operator $\hat{p}_b$ on its right:

$$\int_{-\infty}^{\infty} dx_b \, \langle x_b|\hat{p} = 0. \tag{18.473}$$

With the help of Eqs. (18.472), we obtain

$$\langle x\rangle = x_a, \qquad \langle x^2\rangle = x_a{}^2 + 2D(t_b - t_a), \tag{18.474}$$

and

$$\langle(x - x_a)^2\rangle = 2D(t_b - t_a). \tag{18.475}$$

Clearly, a similar formalism can be developed for the general case with the Lagrangian containing $\ddot{x}$-terms. All we have to do is define time-dependent Heisenberg operators for both sets of canonical coordinates x, p, v, p_v. For instance, consider the case of a free particle, where $V(x) = 0$ and the Hamiltonian (18.239) reduces to

$$H = \frac{w}{2M^2}p_v^2 - i\gamma p_v v + ipv. \tag{18.476}$$

If we want to calculate expectations $\langle f(x, v)\rangle$ for a particle initially at x_a with an initial velocity $\dot{x}_a = v_a$, we now have to evaluate integrals of the form

$$\langle f(x, v)\rangle = \int_{-\infty}^{\infty} dx_b \int_{-\infty}^{\infty} dv_b \, f(x_b, v_b)P(x_bv_bt_b|x_av_at_a)$$

$$= \int_{-\infty}^{\infty} dx_b \int_{-\infty}^{\infty} dv_b \, \langle x_bv_b|f(\hat{x}(t_b - t_a), \hat{v}(t_b - t_a))|x_av_a\rangle. \tag{18.477}$$

Here we introduce basis vectors $|xv\rangle$ which diagonalize the operators $\hat{x}$, $\hat{v}$. The momentum operators satisfy

$$\langle xv|\hat{p} = -\frac{\partial}{\partial x}\langle xv|, \qquad \langle xv|\hat{p}_v = -\frac{\partial}{\partial v}\langle xv|. \tag{18.478}$$

Then we can write

$$\langle x^2\rangle = \int_{-\infty}^{\infty} dx_b \int_{-\infty}^{\infty} dv_b \, \langle x_bv_b|\hat{x}^2(t_b - t_a)|x_av_a\rangle, \tag{18.479}$$

where $\hat{x}(t)$ is the Heisenberg operator defined by

$$\hat{x}(t) = e^{tH(\hat{p},\hat{p}_v,\hat{x},\hat{v})}\hat{x}e^{-tH(\hat{p},\hat{p}_v,\hat{x},\hat{v})}. \tag{18.480}$$

The Heisenberg equations of motions are

$$\begin{aligned}
\dot{\hat{p}}(t) &= [\hat{H},\hat{p}(t)] = 0, \\
\dot{\hat{p}}_v(t) &= [\hat{H},\hat{p}_v(t)] = \gamma\hat{p}_v(t) - \hat{p}(t), \\
\dot{\hat{x}}(t) &= [\hat{H},\hat{x}(t)] = \hat{v}(t), \\
\dot{\hat{v}}(t) &= [\hat{H},\hat{v}(t)] = -i\frac{w}{M^2}\hat{p}_v(t) - \gamma\hat{v}(t).
\end{aligned} \tag{18.481}$$

According to the first equation, $\hat{p}(t)$ is a constant operator:

$$\hat{p}(t) \equiv \hat{p} = \text{const.}$$

The second equation is solved by

$$\hat{p}_v(t) = \hat{p}_v e^{\gamma t} - \frac{1}{\gamma}\hat{p}(e^{\gamma t} - 1), \tag{18.482}$$

where $\hat{p}_v$ is the initial value of $\hat{p}_v(t)$ at $t = 0$. With this, the fourth equation in (18.481) can be integrated to give

$$\begin{aligned}
\hat{v}(t) &= \hat{v}e^{-\gamma t} - i\frac{w}{M^2}\int_0^t dt'\, e^{-\gamma(t-t')}\hat{p}_v(t') \\
&= \hat{v}e^{-\gamma t} - i\frac{w}{\gamma M^2}\left[\hat{p}_v\sinh\gamma t - \frac{1}{\gamma}\hat{p}(\cosh\gamma t - 1)\right].
\end{aligned} \tag{18.483}$$

Inserting this into the third equation in (18.481) we obtain immediately

$$\hat{x}(t) = \hat{x} + \hat{v}\frac{1}{\gamma}(1 - e^{-\gamma t}) - i\frac{w}{\gamma M^2}\left[\hat{p}_v\cosh\gamma t - \frac{1}{\gamma}p(\sinh\gamma t - \gamma t)\right]. \tag{18.484}$$

Using now the relations extending (18.473):

$$\int_{-\infty}^{\infty} dx_b \int_{-\infty}^{\infty} dx_{2b}\, \langle x_b x_{2b}| \left\{ \begin{array}{c} \hat{p} \\ \hat{p}_v \end{array} \right\} = 0 \tag{18.485}$$

to express the translational invariance of the integrated bra state, we find directly

$$\langle x \rangle = x_a + \dot{x}_a\frac{1}{\gamma}\left(1 - e^{-\gamma(t_b - t_a)}\right), \qquad \langle v \rangle = v_a e^{-\gamma(t_b - t_a)}, \tag{18.486}$$

in agreement with (18.441) and (18.454). The expectations values of the quadratic cumulants $\langle(x - \langle x\rangle)^2\rangle$, $\langle(x - \langle x\rangle)(v - \langle v\rangle)\rangle$, $\langle(v - \langle v\rangle)^2\rangle$ are found to be the same as in (18.457).

18.16 Supersymmetry

An interesting new symmetry can be derived from the functional determinant (18.296) which causes the extra last term in the exponent of the path integral (18.300). Let us rewriting this implicitly as

$$P_0(x_b t_b | x_a t_a) \propto \int \mathcal{D}x(t)\, \text{Det}\left[\partial_t + \frac{V''(x)}{M\gamma}\right]\exp\left\{-\int_{t_a}^{t_b} dt\, \frac{1}{4D}\left[\dot{x} + \frac{V'(x)}{M\gamma}\right]^2\right\}. \tag{18.487}$$

In this expression, the time ordering of the velocity $\dot{x}$ with respect to $V'(x)/M\gamma$ is arbitrary. It may be quantum-mechanical (Stratonovich-like), but equally well retarded (Itō-like), or advanced, as long as the same ordering is used in both the Lagrangian and the determinant.

The new symmetry arises if one generates the determinant with the help of an auxiliary fermion field $c(t)$ from a path integral over $c(t)$:

$$\det\left[\partial_t + V''(x(t))/M\gamma\right] \propto \int \mathcal{D}c\mathcal{D}\bar{c}\, e^{-\int dt\bar{c}(t)\left[M\gamma\partial_t + V''(x(t))\right]c(t)}. \tag{18.488}$$

In quantum field theory, such auxiliary fermionic fields are referred to as *ghost fields*. With these we can rewrite the path integral (18.290) for the probability distribution as an ordinary path integral

$$P(x_b t_b | x_a t_a) = \int \mathcal{D}x \int \mathcal{D}c\mathcal{D}\bar{c}\, \exp\left\{-\mathcal{A}_{\mathrm{PS}}[x, c, \bar{c}]\right\}, \tag{18.489}$$

where $\mathcal{A}_{\mathrm{PS}}$ is the Euclidean action

$$\mathcal{A}_{\mathrm{PS}} = \frac{1}{2DM^2\gamma^2} \int_{t_a}^{t_b} dt \left\{ \frac{1}{2}\left[M\gamma\dot{x} + V'(x)\right]^2 + \bar{c}(t)\left[M\gamma\partial_t + V''(x(t))\right]c(t) \right\}, \tag{18.490}$$

first written down by Parisi and Sourlas [20] and by McKane [21]. This action has a particular property. If we denote the expression in the first brackets by

$$U_x \equiv M\gamma\partial_t x + V'(x), \tag{18.491}$$

the operator between the Grassmann variables in (18.490) is simply the functional derivative of U_x:

$$U_{xy} \equiv \frac{\delta U_x}{\delta y} = M\gamma\partial_t + V''(x). \tag{18.492}$$

Thus we may write

$$\mathcal{A}_{\mathrm{PS}} = \frac{1}{2D} \int_{t_a}^{t_b} dt \left[\frac{1}{2}U_x^2 + \bar{c}(t)\, U_{xy}\, c(t) \right], \tag{18.493}$$

where $U_{xy}c(t)$ is the usual short notation for the functional matrix multiplication $\int dt' U_{xy}(t, t')c(t')$. The relation between the two terms makes this action *supersymmetric*. It is invariant under transformations which mix the Fermi and Bose degrees of freedom. Denoting by ε and $\bar{\varepsilon}$ a small anticommuting Grassmann variable and its conjugate (see Section 7.10), the action is invariant under the field transformations

$$\delta x(t) = \bar{\varepsilon}c(t) + \bar{c}(t)\varepsilon, \tag{18.494}$$

$$\delta\bar{c}(t) = -\bar{\varepsilon}U_x, \tag{18.495}$$

$$\delta c(t) = U_x\varepsilon. \tag{18.496}$$

The invariance follows immediately after observing that

$$\delta U_x = \bar{\varepsilon}U_{xy}c(t) + \bar{c}(t)U_{xy}\varepsilon. \tag{18.497}$$

Formally, a similar construction is also possible for a particle with inertia in the path integral (18.232), which is an ordinary path integral involving the Lagrangian (18.287). Here we can write

$$P(x_b t_b | x_a t_a) = \mathcal{N} \int \mathcal{D}x\, J[x] \exp\left\{ -\frac{1}{2w} \int_{t_a}^{t_b} dt \left[M\ddot{x} + M\gamma\dot{x} + V'(x)\right]^2 \right\}, \tag{18.498}$$

where $J[x]$ abbreviates the determinant

$$J[x] = \det\left[M\partial_t^2 + M\gamma\partial_t + V''(x(t))\right], \tag{18.499}$$

which is known from formula (18.274). The path integral (18.498) is valid for *any* ordering of the velocity term, as long as it is the same in the exponent and the functional determinant.

We may now express the functional determinant as a path integral over fermionic ghost fields

$$J[x] = \det\left[M\partial_t^2 + M\gamma\partial_t + V''(x(t))\right] \propto \int \mathcal{D}c\mathcal{D}\bar{c}\, e^{-\int dt\, \bar{c}(t)[M\partial_t^2 + M\gamma\partial_t + V''(x(t))]c(t)},$$

(18.500)

and rewrite the probability distribution $P(x_b t_b | x_a t_a)$ as an ordinary path integral

$$P(x_b t_b | x_a t_a) \propto \int \mathcal{D}x \int \mathcal{D}c\mathcal{D}\bar{c}\, \exp\{-\mathcal{A}^{KS}[x, {}_,\bar{c}]\},$$

(18.501)

where $\mathcal{A}[x, {}_,\bar{c}]$ is the Euclidean action

$$\mathcal{A}^{KS}[x, {}_,\bar{c}] \equiv \int_{t_a}^{t_b} dt \left\{ \frac{1}{2w}[M\ddot{x} + M\gamma\dot{x} + V'(x)]^2 + \bar{c}(t)\left[M\partial_t^2 + M\gamma\partial_t + V''(x(t))\right]c(t) \right\}.$$

(18.502)

This formal expression contains subtleties arising from the boundary conditions when calculating the Jacobian (18.500) from the functional integral on the right-hand side. It is necessary to factorize the second-order operator in the functional determinant and express the determinant of each first-order factor as a functional integral over Grassmann variables as in (18.488). At the end, the action is again supersymmetric, but there are twice as many auxiliary Fermi fields [22].

As a check of this formula, we may let the coupling to the thermal reservoir go to zero, $\gamma \to 0$. Then the first factor in (18.501),

$$\exp\left(-\int_{t_a}^{t_b} dt \left\{ \frac{1}{2w}[M\ddot{x} + M\gamma\dot{x} + V'(x)]^2 \right\}\right)$$

becomes proportional to a δ-functional $\delta[M\ddot{x} + V'(x)]$. The argument is simply the functional derivative of the original action of the quantum system in (18.191), so that we obtain in the limit $\delta[\delta\mathcal{A}/\delta x]$. The functional matrix between the Grassmann fields in (18.501), on the other hand, reduces to $\delta^2\mathcal{A}/\delta x(t)\delta x(t')$, and we arrive at the path integral

$$P(x_b t_b | x_a t_a) \underset{\gamma \to 0}{\propto} \int \mathcal{D}x\, \delta[\delta\mathcal{A}/\delta x]$$

$$\times \int \mathcal{D}c\mathcal{D}\bar{c}\, \exp\left\{-\int_{t_a}^{t_b} dt \int_{t_a}^{t_b} dt'\, \bar{c}(t)\delta^2\mathcal{A}/\partial x(t)\partial x(t')c(t')\right\}.$$

(18.503)

Performing the integral over the Grassmann variables yields

$$P(x_b t_b | x_a t_a) \underset{\gamma \to 0}{\propto} \int \mathcal{D}x\, \delta[\delta\mathcal{A}/\delta x]\, \text{Det}\left[\delta^2\mathcal{A}/\partial x(t)\partial x(t')\right].$$

(18.504)

The δ-functional selects from all paths only those which obey the Euler-Lagrange equations of motion. With the help of the functional identity

$$\delta[M\ddot{x} + V'(x)] = \delta[x - x_{\text{cl}}] \times \text{Det}^{-1}[M\ddot{x} + V''(x)],$$

(18.505)

which generalizes identity $\delta(f(x)) = \delta(x)/f'(x)$ if $f(0) = 0$, the above path integral becomes simply

$$P(x_b t_b | x_a t_a) \underset{\gamma \to 0}{\propto} \int \mathcal{D}x\, \delta[x - x_{\text{cl}}],$$

(18.506)

which is the correct probability distribution of classical physics. Note the important difference with respect to the classical amplitude in Eq. (4.96), where the concentration of the path integral on the classical path is enforced by a strongly oscillating complex expression requiring the semiclassical fluctuation factor in Eq. (4.97) for proper normalization. In the probability (18.506) this is achieved by a real δ-functional.

Note that by a Fourier decomposition of the δ-functional (18.503) we obtain the alternative path integral representation of classical physics

$$P(x_b t_b | x_a t_a) \underset{\gamma \to 0}{\propto} \int \mathcal{D}x \mathcal{D}\lambda \mathcal{D}c \mathcal{D}\bar{c} \, e^{-\int_{t_a}^{t_b} dt \delta \mathcal{A}/\delta x(t) \lambda(t) - \int_{t_a}^{t_b} dt \int_{t_a}^{t_b} dt' \bar{c}(t) \delta^2 \mathcal{A}/\delta x(t) \delta x(t') c(t')}. \qquad (18.507)$$

This is supersymmetric under the transformations

$$\delta x = \bar{\varepsilon} c, \quad \delta c = 0, \quad \delta \bar{c} = -\bar{\varepsilon} \lambda, \quad \delta \lambda = 0, \qquad (18.508)$$

as observed by Gozzi [23].

There exists a compact way of rewriting the action using *superfields*. We define a three-dimensional *superspace* consisting of time and two auxiliary Grassmann variables θ and $\bar{\theta}$. Then we define a superfield

$$X(t) \equiv x(t) + i\bar{\theta} c(t) - i\theta \bar{c}(t) - \bar{\theta} \theta \lambda(t). \qquad (18.509)$$

We now consider the superaction

$$\mathcal{A}^{\text{super}} \equiv \int d\bar{\theta} d\theta \mathcal{A}[X] \equiv \int d\bar{\theta} d\theta \mathcal{A}[x + i\bar{\theta} c - i\theta \bar{c} - \bar{\theta} \theta \lambda] \qquad (18.510)$$

and expand the action into a functional Taylor series:

$$\int d\bar{\theta} d\theta \left\{ \mathcal{A}[x] + \frac{\delta \mathcal{A}}{\delta x} (i\bar{\theta} c - i\theta \bar{c} - \bar{\theta} \theta \lambda) + \frac{1}{2} (i\bar{\theta} c - i\theta \bar{c} - \bar{\theta} \theta \lambda) \frac{\partial^2 \mathcal{A}}{\delta x \delta x} (i\bar{\theta} c - i\theta \bar{c} c - \bar{\theta} \theta \lambda) \right\}.$$

Due to the nilpotency (7.375) of the Grassmann variables, the expansion stops after the second term. Recalling now the integration rules (7.378) and (7.379), this becomes

$$\frac{\delta \mathcal{A}}{\delta x} \lambda + \frac{1}{2} \bar{c} \frac{\partial^2 \mathcal{A}}{\delta x \delta x} c,$$

which is precisely the short-hand functional notation for the negative exponent in the path integral (18.507).

18.17   Stochastic Quantum Liouville Equation

At lower temperatures, where quantum fluctuations become important, the forward–backward path integral (18.230) does not allow us to derive a Schrödinger-like differential equation for the probability distribution $P(x \, v \, t | x_a v_a t_a)$. To see the obstacle, we go over to the canonical representation of (18.230):

$$|(x_b t_b | x_a t_a)|^2 = \int \mathcal{D}x \mathcal{D}y \int \frac{\mathcal{D}p}{2\pi} \frac{\mathcal{D}p_y}{2\pi} \exp \left\{ \frac{i}{\hbar} \int_{t_a}^{t_b} dt \, [p\dot{x} + p_y \dot{y} - H_T] \right\}, \qquad (18.511)$$

where

$$H_T = \frac{1}{M} p_y p_x + \gamma p_y y + V(x + y/2) - V(x - y/2) - i\frac{w}{2\hbar} y \hat{K}^{\text{Ohm}} y \qquad (18.512)$$

plays the role of a temperature-dependent quasi-Hamiltonian for an Ohmic system associated with the Lagrangian of the forward–backward path integral (18.230). The notation $\hat{K}^{\text{Ohm}} y(t)$ abbreviates the product of the functional matrix $K^{\text{Ohm}}(t, t')$ with the functional vector $y(t')$ defined by $\hat{K}^{\text{Ohm}}y(t) \equiv \int dt' K^{\text{Ohm}}(t,t')y(t')$. Hence is H_T a nonlocal object (in the temporal sense), and this is the reason for calling it quasi-Hamiltonian.

It is useful to omit y-integrations at the endpoints in the path integral (18.511), and set up a path integral representation for the product of amplitudes

$$U(x_b y_b t_b | x_a y_a t_a) \equiv (x_b + y_b/2\ t_b | x_a + y_a/2\ t_a)(x_b - y_b/2\ t_b | x_a - y_a/2\ t_a)^*. \tag{18.513}$$

Given some initial density matrix $\rho(x_+, x_-; t) = \rho(x + y/2, x - y/2; t)$ at time $t = t_a$, which may actually be in equilibrium and time-independent, as in Eq. (2.365), the functional matrix $U(x_b y_b t_b | x_a y_a t_a)$ allows us to calculate $\rho(x_+, x_-; t)$ at any time by the time evolution equation

$$\rho(x + y/2, x - y/2; t) = \int dx_a\, dy_a\, U(x\,y\,t | x_a y_a t_a)\, \rho(x_a + y_a/2, x_a - y_a/2; t_a). \tag{18.514}$$

Recall that the Fourier transform of $\rho(x + y/2, x - y/2; t)$ with respect to y is the Wigner function (1.224).

When considering the change of $U(x\,y\,t | x_a y_a t_a)$ over a small time interval ϵ, the momentum variables p and p_y have the same effect as differential operators $-i\partial_{x_b}$ and $-i\partial_{y_b}$, respectively. The last term in H_T, however, is nonlocal in time, thus preventing a derivation of a Schrödinger-like differential equation.

The locality problem can be removed by introducing a noise variable $\eta(t)$ with the correlation function determined by (18.321):

$$\langle \eta(t)\eta(t')\rangle_T = \frac{w}{2}[K^{\text{Ohm}}]^{-1}(t,t'). \tag{18.515}$$

Then we can define a temporally local η-dependent Hamiltonian operator

$$\hat{H}_\eta \equiv \frac{1}{M}\left(\hat{p}_x + \gamma y\right)\hat{p}_y + V(x + y/2) - V(x - y/2) - y\eta, \tag{18.516}$$

which governs the evolution of η-dependent versions of the amplitude products (18.513) via the *stochastic Schrödinger equation*

$$i\hbar\partial_t U_\eta(x\,y\,t | x_a y_a t_a) = \hat{H}_\eta\, U_\eta(x\,y\,t | x_a y_a t_a). \tag{18.517}$$

The same equation is obeyed by the noise-dependent density matrix $\rho_\eta(x, y; t)$.

Averaging these equation over η with the distribution (18.321) yields for $y_a = y_b = 0$ the same probability distribution as the forward–backward path integral (18.230):

$$|(x_b t_b | x_a t_a)|^2 = U(x_b\,0\,t_b | x_a\,0\,t_a) \equiv \langle U(x_b\,0\,t_b | x_a\,y_a\,t_a)\rangle_\eta. \tag{18.518}$$

At high temperatures, the noise averaged stochastic Schrödinger equation (18.517) takes the form

$$i\hbar\partial_t U(x\,y\,t | x_a\,y_a\,t_a) = \hat{\bar{H}}_T U(x\,y\,t | x_a\,y_a\,t_a), \tag{18.519}$$

where $\hat{\bar{H}}$ is now a local (in the temporal sense)

$$\hat{\bar{H}}_T \equiv \frac{1}{M}\hat{p}_y\hat{p}_x + \gamma y\hat{p}_y + V(x + y/2) - V(x - y/2) - i\frac{w}{2\hbar}y^2, \tag{18.520}$$

arising from the Hamiltonian (18.512) in the high-temperature limit $K^{\text{Ohm}} \to 1$ [recall (18.223)]. In terms of the separate path positions $x_\pm = x \pm y/2$ where $p_x = \partial_+ + \partial_-$ and $p_y = (\partial_+ - \partial_-)/2$, this takes the more familiar form [24]

$$\hat{\bar{H}}_T \equiv \frac{1}{2M}\left(\hat{p}_+^2 - \hat{p}_-^2\right) + V(x_+) - V(x_-) + \frac{\gamma}{2}(x_+ - x_-)(\hat{p}_+ - \hat{p}_-) - i\frac{w}{2\hbar}(x_+ - x_-)^2. \tag{18.521}$$

The last term is often written as $-i\hbar\Lambda(x_+ - x_-)^2$, where Λ is the so-called *decoherence rate per square distance*

$$\Lambda \equiv \frac{w}{2\hbar^2} = \frac{M\gamma k_B T}{\hbar^2}. \tag{18.522}$$

It is composed of the damping rate γ and the squared thermal length (2.351):

$$\Lambda = \frac{2\pi\gamma}{l_e^2(\hbar\beta)}, \tag{18.523}$$

and controls the decay of interference peaks [25].

Note that the order of the operators in the mixed term of the form $y\hat{p}_y$ in Eq. (18.520) is opposite to the mixed term $-i\hat{p}_v v$ in the differential operator (18.239) of the Fokker-Planck equation. This order is necessary to guarantee the conservation of probability. Indeed, multiplying the time evolution equation (18.519) by $\delta(y)$, and integrating both sides over x and y, the left-hand side vanishes.

The correctness of this order can be verified by calculating the fluctuation determinant of the path integral for the product of amplitudes (18.513) in the Lagrangian form, which looks just like (18.230), except that the difference between forward and backward trajectories $y(t) = x_+(t) - x_-(t)$ is nonzero at the endpoints. For the fluctuation which vanish at the endpoints, this is irrelevant. As explained before, the order is a short-time issue, and we can take $t_b - t_a \to \infty$. Moreover, since the order is independent of the potential, we may consider only the free case $V(x \pm y/2) \equiv 0$. The relevant fluctuation determinant was calculated in formula (18.254). In the Hamiltonian operator (18.520), this implies an additional energy $-i\gamma/2$ with respect to the symmetrically ordered term $\gamma\{y, \hat{p}_y\}/2$, which brings it to $\gamma y\hat{p}_y$, and thus the order in (18.521).

18.18 Master Equation for Time Evolution

In the high-temperature limit, the Hamiltonian (18.521) becomes local. Then the evolution equation (18.514) for the density matrix $\rho(x_{+a}, x_{-a}; t_a)$ can be converted into an operator equation

$$i\hbar\partial_t \rho(x_+, x_-; t_a) = \hat{\tilde{H}}_T \rho(x_+, x_-; t_a), \tag{18.524}$$

where $\hat{\tilde{H}}_T$ is the operator version of the temperature-dependent Hamiltonian (18.521). Such an equation does not exist at low temperatures, due to the nonlocality of the last term in (18.512). Then one cannot avoid solving the stochastic Schrödinger equation (18.517) with the subsequent averaging (18.518). For moderately high temperatures, however, a Hamiltonian formalism can still be set up, although it requires solving a recursion relation. For this purpose we write down the quasi-Hamiltonian in D dimensions

$$\hat{\tilde{H}}_T \equiv \frac{1}{2M}\left(\hat{\mathbf{p}}_+^2 - \hat{\mathbf{p}}_-^2\right) + V(\mathbf{x}_+) - V(\mathbf{x}_-) + \frac{M\gamma}{2}(\hat{\mathbf{x}}_+ - \hat{\mathbf{x}}_-)(\hat{\dot{\mathbf{x}}}_+ + \hat{\dot{\mathbf{x}}}_-)^R$$
$$-i\frac{w}{2\hbar}(\hat{\mathbf{x}}_+ - \hat{\mathbf{x}}_-)\hat{K}^{\mathrm{Ohm}}(\hat{\mathbf{x}}_+ - \hat{\mathbf{x}}_-), \tag{18.525}$$

where the Fourier transform of $K^{\mathrm{Ohm}}(t, t')$ is expanded in powers of ω' [recall (18.223)]

$$K^{\mathrm{Ohm}}(\omega') = 1 + \frac{1}{3}\left(\frac{\hbar\omega'}{2k_B T}\right)^2 + \dots . \tag{18.526}$$

In this way we find for the last term the locally looking high-temperature expansion

$$-i(\hat{\mathbf{x}}_+ - \hat{\mathbf{x}}_-)\hat{K}^{\mathrm{Ohm}}(\hat{\mathbf{x}}_+ - \hat{\mathbf{x}}_-) = -i(\hat{\mathbf{x}}_+ - \hat{\mathbf{x}}_-)^2 + i\frac{w\hbar}{24(k_B T)^2}(\hat{\dot{\mathbf{x}}}_+ - \hat{\dot{\mathbf{x}}}_-)^2 + \dots . \tag{18.527}$$

The expression is not really local, since the operator $\hat{\dot{\mathbf{x}}}$ is defined implicitly as an abbreviations for the commutator

$$\hat{\dot{\mathbf{x}}} \equiv \frac{i}{\hbar}[\hat{\bar{H}}_T, \hat{\mathbf{x}}]. \tag{18.528}$$

If the expansion (18.527) is carried further, higher derivatives of $\mathbf{x}$ arise, which are all defined recursively:

$$\hat{\ddot{\mathbf{x}}} \equiv \frac{i}{\hbar}[\hat{\bar{H}}_T, \hat{\dot{\mathbf{x}}}], \quad \hat{\dddot{\mathbf{x}}} \equiv \frac{i}{\hbar}[\hat{\bar{H}}_T, \hat{\ddot{\mathbf{x}}}], \ \dots \ . \tag{18.529}$$

Thus Eq. (18.525) with the expansion (18.527) is a recursive equation for the Hamiltonian operator $\hat{\bar{H}}_T$. For small γ (and thus $w = 2M\gamma k_B T$), the recursion can be solved iteratively, in the first step by inserting $\hat{\dot{\mathbf{x}}} \approx \hat{\mathbf{p}}/M$ into Eq. (18.530).

It is useful to re-express (18.524) in the Dirac operator form where the density matrix has a bra–ket representation $\hat{\rho}(t) = \sum_{mn} \rho_{mn}(t)|m\rangle\langle n|$. Denoting $\hat{\mathbf{p}}^2/2M + \hat{V}$ in (18.525) by $\hat{H}$, we obtain with the expansion (18.527) the local *master equation*:

$$i\hbar\partial_t\hat{\rho} = \hat{\bar{H}}_T\,\hat{\rho} \quad \equiv \quad [\hat{H}, \hat{\rho}] + \frac{M\gamma}{2}\left(\hat{\mathbf{x}}\hat{\dot{\mathbf{x}}}\hat{\rho} - \hat{\rho}\hat{\dot{\mathbf{x}}}\hat{\mathbf{x}} + \hat{\mathbf{x}}\,\hat{\rho}\,\hat{\dot{\mathbf{x}}} - \hat{\dot{\mathbf{x}}}\,\hat{\rho}\,\hat{\mathbf{x}}\right)$$

$$- \frac{iw}{2\hbar}[\hat{\mathbf{x}}, [\hat{\mathbf{x}}, \hat{\rho}]] - \frac{iw\hbar^2}{24(k_BT)^2}[\hat{\dot{\mathbf{x}}}, [\hat{\dot{\mathbf{x}}}, \hat{\rho}]] + \dots \ . \tag{18.530}$$

The validity of the above iterative procedure is most easily proved in the time-sliced path integral. The final slice of infinitesimal width ϵ reads

$$U(\mathbf{x}_{+b}, \mathbf{x}_{-b}, t_b | \mathbf{x}_{+a}, \mathbf{x}_{-a}, t_b - \epsilon)$$
$$= \int \frac{d\mathbf{p}_+(t_b)}{(2\pi)^3} \int \frac{d\mathbf{p}_-(t_b)}{(2\pi)^3} e^{\frac{i}{\hbar}\{\mathbf{p}_+(t_b)[\mathbf{x}_+(t_b) - \mathbf{x}_+(t_b - \epsilon)] - \mathbf{p}_- \cdot \mathbf{x}_- - \bar{H}_T(t_b)\}}. \tag{18.531}$$

Consider now a term of the generic form $\dot{F}_+(\mathbf{x}_+(t))F_-(\mathbf{x}_-(t))$ in $\bar{H}_T(t)$. When differentiating $U(\mathbf{x}_{+b}, \mathbf{x}_{-b}, t_b | \mathbf{x}_{+a}, \mathbf{x}_{-a}, t_b - \epsilon)$ with respect to the final time t_b, the integrand receives a factor $-\bar{H}_T(t_b)$. At t_b, the term $\dot{F}_+(\mathbf{x}_+(t))F_-(\mathbf{x}_-(t))$ in $\bar{H}_T(t)$ has the explicit form $\epsilon^{-1}[F_+(\mathbf{x}_+(t_b)) - F_+(\mathbf{x}_+(t_b - \epsilon))]F_-(\mathbf{x}_-(t_b))$. It can be taken out of the integral, yielding

$$\epsilon^{-1}[F_+(\mathbf{x}_+(t_b))U - UF_+(\mathbf{x}_+(t_b - \epsilon))]F_-(\mathbf{x}_-(t_b)). \tag{18.532}$$

In operator language, the amplitude U is associated with $\hat{U} \approx 1 - i\epsilon\hat{\bar{H}}_T/\hbar$, such the term $\dot{F}_+(\mathbf{x}_+(t))F_-(\mathbf{x}_-(t))$ in $\hat{\bar{H}}_T$ yields a Schrödinger operator

$$\frac{i}{\hbar}\left[\hat{\bar{H}}_T, \hat{F}_+(\mathbf{x}_+)\right]F_-(\mathbf{x}_-) \tag{18.533}$$

in the time evolution equation (18.530).

For functions of the second derivative $\ddot{\mathbf{x}}$ we have to split off the last two time slices in (18.531) and convert the two intermediate integrals over $\mathbf{x}$ into operator expressions, which obviously leads to the repeated commutator of $\hat{\bar{H}}_T$ with $\hat{\mathbf{x}}$, and so on.

The operator order in the terms in the parentheses of Eq. (18.530) is fixed by the retardation of $\dot{\mathbf{x}}_\pm$ with respect to $\mathbf{x}_\pm$ in (18.521). This implies that the associated operator $\hat{\dot{\mathbf{x}}}(t)$ has a time argument which lies slightly *before* that of $\hat{\mathbf{x}}_\pm$, thus acting upon $\hat{\rho}$ before $\hat{\mathbf{x}}$. This puts $\hat{\dot{\mathbf{x}}}(t)$ to the right of $\hat{\mathbf{x}}$, i.e., next to $\hat{\rho}$. On the right-hand side of $\hat{\rho}$, the time runs in the opposite direction such that $\hat{\dot{\mathbf{x}}}$ must lie to the left of $\hat{\mathbf{x}}$, again next to $\hat{\rho}$. In this way we obtain an operator order which ensures that Eq. (18.530) conserves the total probability.

This property and the positivity of $\hat{\rho}$ are actually guaranteed by the observation, that the master equation (18.530) can be written in the *Lindblad form* [26]

$$\partial_t \hat{\rho} = -\frac{i}{\hbar}[\hat{H},\hat{\rho}] - \sum_{n=1}^{2}\left(\frac{1}{2}\hat{L}_n\hat{L}_n^\dagger\hat{\rho} + \frac{1}{2}\hat{\rho}\hat{L}_n\hat{L}_n^\dagger - \hat{L}_n^\dagger\hat{\rho}\hat{L}_n\right), \tag{18.534}$$

with the two Lindblad operators [27]

$$\hat{L}_1 \equiv \frac{\sqrt{w}}{2\hbar}\hat{\mathbf{x}}, \qquad \hat{L}_2 \equiv \frac{\sqrt{3w}}{2\hbar}\left(\hat{\mathbf{x}} - i\frac{\hbar}{3k_BT}\dot{\hat{\mathbf{x}}}\right). \tag{18.535}$$

Note that the operator order in Eq. (18.530) prevents the term $\hat{\mathbf{x}}\dot{\hat{\mathbf{x}}}\hat{\rho}$ from being a pure divergence. If we rewrite it as a sum of a commutator and an anticommutator, $[\hat{\mathbf{x}},\dot{\hat{\mathbf{x}}}]/2 + \{\hat{\mathbf{x}},\dot{\hat{\mathbf{x}}}\}/2$, then the latter term is a pure divergence, and we can think of the first two γ-terms in (18.530) as being due to an additional anti-Hermitian term in the Hamiltonian operator $\hat{H}$, the *dissipation operator*

$$\hat{H}_\gamma = \gamma M \frac{1}{4}[\hat{\mathbf{x}},\dot{\hat{\mathbf{x}}}]. \tag{18.536}$$

18.19   Relation to Quantum Langevin Equation

The stochastic Liouville equation (18.517) can also be derived from an operator version of the Langevin equation (18.317), the so-called *Quantum Langevin equation*

$$M\ddot{\hat{x}}(t) + M\gamma\dot{\hat{x}}(t) + V'(\hat{x}(t)) = \hat{\eta}(t), \tag{18.537}$$

where $\hat{\eta}(t)$ is an operator noise variable with the commutation rule

$$[\hat{\eta}_t,\hat{\eta}_{t'}] = w\frac{i\hbar}{k_BT}\partial_t\delta(t-t'), \tag{18.538}$$

and the correlation function [28]

$$\frac{1}{2}\langle[\hat{\eta}_t,\hat{\eta}_{t'}]_+\rangle_{\hat{\eta}} = w\,K(t,t'). \tag{18.539}$$

The commutator (18.538) and the correlation function (18.539) are related to each other as required by the fluctuation-dissipation theorem: By omitting the factor $\coth(\hbar\omega/2k_BT)$ in Eq. (18.223), the Fourier integral (18.221) for $K(t,t')$ reduces to $(\hbar/2k_BT)\partial_t\delta(t-t')$. A comparison with the general spectral representation (18.53) shows that the expectation value (18.539) has the spectral function

$$\rho_b(\omega') = 2M\gamma\hbar\omega'. \tag{18.540}$$

By inserting this into the spectral representation (18.53) we obtain the right-hand side of the commutator equation (18.538).

A noise variable with the properties (18.538) and (18.539) can be constructed explicitly by superimposing quantized oscillator velocities of frequencies ω as follows:

$$\hat{\eta}(t) = -i\sqrt{\frac{M\hbar\gamma}{\pi}}\int_0^\infty d\Omega'\sqrt{\Omega'}[a_{\Omega'}e^{-i\omega't} - a_{\omega'}^\dagger e^{i\omega't}]. \tag{18.541}$$

It is worth pointing out that there exists a direct derivation of the quantum Langevin equation (18.537), whose noise operator $\hat{\eta}(t)$ satisfies the commutator and fluctuation properties (18.538) and (18.539), from Kubo's stochastic Liouville equation, and thus from the forward–backward path integral (18.230) [29].

18.20 Electromagnetic Dissipation and Decoherence

There exists a thermal bath of particular importance: atoms are usually observed at a finite temperature where they interact with a grand-canonical ensemble of photons in thermal equilibrium. This interaction will broaden the natural line width of atomic levels even if all major mechanisms for the broadening are removed. To study this situation, let us set up a forward–backward path integral description for a bath of photons, and derive from it a master equation for the density matrix which describes electromagnetic dissipation and decoherence. As an application, we shall calculate the Wigner-Weisskopf formula for the natural line width of an atomic state at zero temperature, find the finite-temperature effects, and calculate the Lamb shift between atomic s- and p-wave states of principal quantum number $n = 2$ with the term notation $2S_{1/2}$ and $2P_{1/2}$. The master equation may eventually have applications to dilute interstellar gases or to few-particle systems in cavities.

18.20.1 Forward–Backward Path Integral

With the application to atomic physics in mind, we shall consider a three-dimensional quantum system described by a time-dependent quantum-mechanical density matrix $\rho(\mathbf{x}_+, \mathbf{x}_-; t)$. In contrast to Eq. (18.514), we use here the forward and backward variables as arguments, and write the time evolution equation as

$$\rho(\mathbf{x}_{+b}, \mathbf{x}_{-a}; t_b) = \int d\mathbf{x}_{+a}\, d\mathbf{x}_{-a}\, U(\mathbf{x}_{+b}, \mathbf{x}_{-b}, t_b | \mathbf{x}_{+a}, \mathbf{x}_{-a}, t_a) \rho(\mathbf{x}_{+a}, \mathbf{x}_{-a}; t_a). \tag{18.542}$$

In an external electromagnetic vector potential $\mathbf{A}(\mathbf{x}, t)$, the time-evolution kernel is determined by a forward–backward path integral of the type (18.192), in which the forward and backward paths start at different initial and final points $\mathbf{x}_{+a}, \mathbf{x}_{-a}$ and $\mathbf{x}_{+b}, \mathbf{x}_{-b}$, respectively:

$$U(\mathbf{x}_{+b}, \mathbf{x}_{-b}, t_b | \mathbf{x}_{+a}, \mathbf{x}_{-a}, t_a) \equiv (\mathbf{x}_{+b}, t_b | \mathbf{x}_{+a}, t_a)(\mathbf{x}_{-b}, t_b | \mathbf{x}_{-a}, t_a)^* = \int \mathcal{D}\mathbf{x}_+ \mathcal{D}\mathbf{x}_-$$

$$\times \exp\left\{ \frac{i}{\hbar} \int_{t_a}^{t_b} \left[\frac{M}{2}(\dot{\mathbf{x}}_+^2 - \dot{\mathbf{x}}_-^2) - V(\mathbf{x}_+) + V(\mathbf{x}_-) - \frac{e}{c}\dot{\mathbf{x}}_+ \mathbf{A}(\mathbf{x}_+, t) + \frac{e}{c}\dot{\mathbf{x}}_- \mathbf{A}(\mathbf{x}_-, t) \right] \right\}. \tag{18.543}$$

The vector potential $\mathbf{A}(\mathbf{x}, t)$ is a superposition of oscillators $\mathbf{X}_{\mathbf{k}}(t)$ of frequency $\Omega_{\mathbf{k}} = c|\mathbf{k}|$ in a volume V:

$$\mathbf{A}(\mathbf{x}, t) = \sum_{\mathbf{k}} c_{\mathbf{k}}(\mathbf{x}) \mathbf{X}_{\mathbf{k}}(t), \qquad c_{\mathbf{k}} = \frac{e^{i\mathbf{k}\mathbf{x}}}{\sqrt{2\Omega_{\mathbf{k}} V}}, \qquad \sum_{\mathbf{k}} = \int \frac{d^3 k V}{(2\pi)^3}. \tag{18.544}$$

At a finite temperature T, these oscillators are assumed to be in equilibrium, where we shall write their time-ordered correlation functions as

$$G_{\mathbf{k}\mathbf{k}'}^{ij}(t, t') = \langle \hat{T} \hat{X}_{\mathbf{k}}^i(t), \hat{X}_{-\mathbf{k}'}^j(t') \rangle = \delta_{\mathbf{k}\mathbf{k}'}^{ij\,\mathrm{tr}} G_{\Omega_{\mathbf{k}}}(t, t') \equiv \delta_{\mathbf{k}\mathbf{k}'} P_{\mathbf{k}}^{\perp ij} G_{\Omega_{\mathbf{k}}}(t, t'). \tag{18.545}$$

The transverse projection matrix is the result of the sum over the transverse polarization vectors of the photons:

$$P_{\mathbf{k}}^{\perp ij} = \sum_{h=\pm} \epsilon^i(\mathbf{k}, h) \epsilon^{j*}(\mathbf{k}, h) = (\delta^{ij} - k^i k^j / \mathbf{k}^2). \tag{18.546}$$

The function $G_{\Omega_{\mathbf{k}}}(t, t')$ on the right-hand side of (18.545) is the Green function (18.185) of a single oscillator of frequency $\Omega_{\mathbf{k}}$. It is decomposed into real and imaginary parts, defining $A_{\Omega_{\mathbf{k}}}(t, t')$ and $C_{\Omega_{\mathbf{k}}}(t, t')$ as in (18.185), which are commutator and anticommutator functions of the oscillator at temperature T: $C_{\Omega_{\mathbf{k}}}(t, t') \equiv \langle [\hat{X}(t), \hat{X}(t')] \rangle_T$ and $A_{\Omega_{\mathbf{k}}}(t, t') \equiv \langle [\hat{X}(t), \hat{X}(t')] \rangle_T$, respectively.

The thermal average of the evolution kernel (18.543) is then given by the forward–backward path integral

$$
U(\mathbf{x}_{+b}, \mathbf{x}_{-b}, t_b | \mathbf{x}_{+a}, \mathbf{x}_{-a}, t_a) = \int \mathcal{D}\mathbf{x}_+(t) \int \mathcal{D}\mathbf{x}_-(t)
$$

$$
\times \exp\left\{ \frac{i}{\hbar} \int_{t_a}^{t_b} dt \left[\frac{M}{2}(\dot{\mathbf{x}}_+^2 - \dot{\mathbf{x}}_-^2) - (V(\mathbf{x}_+) - V(\mathbf{x}_-)) \right] + \frac{i}{\hbar} \mathcal{A}^{\mathrm{FV}}[\mathbf{x}_+, \mathbf{x}_-] \right\}, \tag{18.547}
$$

where $\exp\{i\mathcal{A}^{\mathrm{FV}}[\mathbf{x}_+, \mathbf{x}_-]/\hbar\}$ is the Feynman-Vernon influence functional defined in Eq. (18.200). The influence action $\mathcal{A}^{\mathrm{FV}}[\mathbf{x}_+, \mathbf{x}_-]$ is the sum of a dissipative and a fluctuating part $\mathcal{A}_D^{\mathrm{FV}}[\mathbf{x}_+, \mathbf{x}_-]$ and $\mathcal{A}_F^{\mathrm{FV}}[\mathbf{x}_+, \mathbf{x}_-]$, whose explicit forms are now

$$
\mathcal{A}_D^{\mathrm{FV}}[\mathbf{x}_+, \mathbf{x}_-] = \frac{ie^2}{2\hbar c^2} \int dt \int dt' \, \Theta(t - t')
$$

$$
\times \left[\dot{\mathbf{x}}_+(t) \mathbf{C}_\mathrm{b}(\mathbf{x}_+ t, \mathbf{x}_+' t') \dot{\mathbf{x}}_+(t') - \dot{\mathbf{x}}_+(t) \mathbf{C}_\mathrm{b}(\mathbf{x}_+ t, \mathbf{x}_-' t') \dot{\mathbf{x}}_-(t') \right.
$$

$$
\left. - \dot{\mathbf{x}}_-(t) \mathbf{C}_\mathrm{b}(\mathbf{x}_- t, \mathbf{x}_+' t') \dot{\mathbf{x}}_+(t') + \dot{\mathbf{x}}_-(t) \mathbf{C}_\mathrm{b}(\mathbf{x}_- t, \mathbf{x}_-' t') \dot{\mathbf{x}}_-(t') \right], \tag{18.548}
$$

and

$$
\mathbf{A}_F^{\mathrm{FV}}[\mathbf{x}_+, \mathbf{x}_-] = \frac{ie^2}{2\hbar c^2} \int dt \int dt' \, \Theta(t - t')
$$

$$
\times \left[\dot{\mathbf{x}}_+(t) \mathbf{A}_\mathrm{b}(\mathbf{x}_+ t, \mathbf{x}_+' t') \dot{\mathbf{x}}_+(t') + \dot{\mathbf{x}}_+(t) \mathbf{A}_\mathrm{b}(\mathbf{x}_+ t, \mathbf{x}_-' t') \dot{\mathbf{x}}_-(t') \right.
$$

$$
\left. + \dot{\mathbf{x}}_-(t) \mathbf{A}_\mathrm{b}(\mathbf{x}_- t, \mathbf{x}_+' t') \dot{\mathbf{x}}_+(t') + \dot{\mathbf{x}}_-(t) \mathbf{A}_\mathrm{b}(\mathbf{x}_- t, \mathbf{x}_-' t') \dot{\mathbf{x}}_-(t') \right], \tag{18.549}
$$

with $\mathbf{C}_\mathrm{b}(\mathbf{x}_- t, \mathbf{x}_-' t')$ and $\mathbf{A}_\mathrm{b}(\mathbf{x}_- t, \mathbf{x}_-' t')$ collecting the 3×3 commutator and anticommutator functions of the bath of photons. They are sums of correlation functions over the bath of the oscillators of frequency $\Omega_\mathbf{k}$, each contributing with a weight $c_\mathbf{k}(\mathbf{x}) c_{-\mathbf{k}}(\mathbf{x}') = e^{i\mathbf{k}(\mathbf{x}-\mathbf{x}')}/2\Omega_\mathbf{k} V$. Thus we may write, generalizing (18.197) and (18.198),

$$
C_\mathrm{b}^{ij}(\mathbf{x} t, \mathbf{x}' t') = \sum_\mathbf{k} c_{-\mathbf{k}}(\mathbf{x}) c_\mathbf{k}(\mathbf{x}') \left\langle [\hat{X}_{-\mathbf{k}}^i(t), \hat{X}_\mathbf{k}^j(t')] \right\rangle_T
$$

$$
= -i\hbar \int \frac{d\omega' d^3 k}{(2\pi)^4} \rho_\mathbf{k}(\omega') P_\mathbf{k}^{\perp\, ij} e^{i\mathbf{k}(\mathbf{x}-\mathbf{x}')} \sin \omega'(t - t'), \tag{18.550}
$$

$$
A_\mathrm{b}^{ij}(\mathbf{x} t, \mathbf{x}' t') = \sum_\mathbf{k} c_{-\mathbf{k}}(\mathbf{x}) c_\mathbf{k}(\mathbf{x}') \left\langle \left\{ \hat{X}_{-\mathbf{k}}^i(t), \hat{X}_\mathbf{k}^j(t') \right\} \right\rangle_T
$$

$$
= \hbar \int \frac{d\omega' d^3 k}{(2\pi)^4} \rho_\mathbf{k}(\omega') P_\mathbf{k}^{\perp\, ij} \coth \frac{\hbar\omega'}{2k_B T} e^{i\mathbf{k}(\mathbf{x}-\mathbf{x}')} \cos \omega'(t - t'), \tag{18.551}
$$

where $\rho_\mathbf{k}(\omega')$ is the spectral density contributed by the oscillator of momentum $\mathbf{k}$:

$$
\rho_\mathbf{k}(\omega') \equiv \frac{2\pi}{2\Omega_\mathbf{k}} [\delta(\omega' - \Omega_\mathbf{k}) - \delta(\omega' + \Omega_\mathbf{k})]. \tag{18.552}
$$

At zero temperature, we recognize in (18.550) and (18.551) twice the imaginary and real parts of the Feynman propagator of a massless particle for $t > t'$, which in four-vector notation with $k = (\omega/c, \mathbf{k})$ and $x = (ct, \mathbf{x})$ reads

$$
G(x, x') = \frac{1}{2}[A(x, x') + C(x, x')] = \int \frac{d^4 k}{(2\pi)^4} e^{ik(x-x')} \frac{i\hbar}{k^2 + i\eta}
$$

$$
= \int \frac{d\omega \, d^3 k}{(2\pi)^4} \frac{ic\hbar}{\omega^2 - \Omega_\mathbf{k}^2 + i\eta} e^{-i[\omega(t-t') - \mathbf{k}(\mathbf{x}-\mathbf{x}')]}, \tag{18.553}
$$

where η is an infinitesimally small number > 0.

We shall now focus attention upon systems which are so small that the effects of retardation can be neglected. Then we can ignore the $\mathbf{x}$-dependence in (18.551) and (18.552) and find

$$C_{\rm b}^{ij}(\mathbf{x}\,t, \mathbf{x}'\,t') \approx C_{\rm b}^{ij}(t, t') = i\frac{\hbar}{2\pi c}\frac{2}{3}\delta^{ij}\partial_t\delta(t - t'). \tag{18.554}$$

Inserting this into (18.548) and integrating by parts, we obtain two contributions. The first is a diverging term

$$\Delta\mathcal{A}_{\rm loc}[\mathbf{x}_+, \mathbf{x}_-] = \frac{\Delta M}{2}\int_{t_a}^{t_b} dt\,(\dot{\mathbf{x}}_+^2 - \dot{\mathbf{x}}_-^2)(t), \tag{18.555}$$

where

$$\Delta M \equiv -\frac{e^2}{c^2}\int\frac{d\omega' d^3k}{(2\pi)^4}\frac{\sigma_{\mathbf{k}}(\omega')}{\omega'}\delta_{\mathbf{kk}}^{ij\,{\rm tr}} = -\frac{e^2}{3\pi^2 c^3}\int_0^\infty dk \tag{18.556}$$

diverges linearly. This simply renormalizes the kinetic terms in the path integral (18.547), renormalizing them to

$$\frac{i}{\hbar}\int_{t_a}^{t_b} dt\,\frac{M_{\rm ren}}{2}\left(\dot{\mathbf{x}}_+^2 - \dot{\mathbf{x}}_-^2\right). \tag{18.557}$$

By identifying M with $M_{\rm ren}$ this renormalization may be ignored.

The second term has the form [compare (18.205)]

$$\mathcal{A}_D^{\rm FV}[\mathbf{x}_+, \mathbf{x}_-] = -\gamma\frac{M}{2}\int_{t_a}^{t_b} dt\,(\dot{\mathbf{x}}_+ - \dot{\mathbf{x}}_-)(t)(\ddot{\mathbf{x}}_+ + \ddot{\mathbf{x}}_-)^R(t), \tag{18.558}$$

with the friction constant of the photon bath encountered before in Eq. (3.441):

$$\gamma \equiv \frac{e^2}{6\pi c^3 M} = \frac{2}{3}\frac{\alpha}{\omega_M}, \tag{18.559}$$

where $\alpha \equiv e^2/\hbar c \approx 1/137$ is the fine-structure constant (1.502) and $\omega_M \equiv Mc^2/\hbar$ the Compton frequency associated with the mass M. Note once more that in contrast to the usual friction constant γ in Section 3.13, this has the dimension 1/frequency.

As discussed in Section 18.8, the retardation enforced by the Heaviside function in the exponent of (18.548) removes the left-hand half of the δ-function [see (18.214)]. It ensures the *causality* of the dissipation forces, which has been shown in Section 18.9.2 to be crucial for producing a probability conserving time evolution of the probability distribution [13]. The superscript R in (18.558) shifts the acceleration $(\ddot{\mathbf{x}}_+ + \ddot{\mathbf{x}}_-)(t)$ slightly towards an earlier time with respect to the velocity factor $(\dot{\mathbf{x}}_+ - \dot{\mathbf{x}}_-)(t)$.

We now turn to the anticommutator function. Inserting (18.552) and the friction constant γ from (18.559), it becomes

$$\frac{e^2}{c^2}A_{\rm b}(\mathbf{x}\,t, \mathbf{x}'\,t') \approx 2\gamma k_B T K^{\rm Ohm}(t, t'), \tag{18.560}$$

as in Eq. (18.216), with the same function $K^{\rm Ohm}(t, t')$ as in Eq. (18.223), whose high-temperature expansion starts out as in Eq. (18.526).

In terms of the function $K^{\rm Ohm}(t, t')$, the fluctuation part of the influence functional in (18.549), (18.548), (18.547) becomes [compare (18.227)]

$$\mathcal{A}_F^{\rm FV}[\mathbf{x}_+, \mathbf{x}_-] = i\frac{w}{2\hbar}\int_{t_a}^{t_b} dt\int_{t_a}^{t_b} dt'\,(\dot{\mathbf{x}}_+ - \dot{\mathbf{x}}_-)(t)\,K^{\rm Ohm}(t, t')\,(\dot{\mathbf{x}}_+ - \dot{\mathbf{x}}_-)(t'). \tag{18.561}$$

Here we have used the symmetry of the function $K^{\text{Ohm}}(t, t')$ to remove the Heaviside function $\Theta(t - t')$ from the integrand, extending the range of t'-integration to the entire interval (t_a, t_b). We also have introduced the constant

$$w \equiv 2Mk_BT\gamma, \tag{18.562}$$

for brevity.

In the high-temperature limit, the time evolution amplitude for the density matrix is given by the path integral

$$U(\mathbf{x}_{+b}, \mathbf{x}_{-b}, t_b | \mathbf{x}_{+a}, \mathbf{x}_{-a}, t_a) = \int \mathcal{D}\mathbf{x}_+(t) \int \mathcal{D}\mathbf{x}_-(t)$$

$$\times \exp\left\{\frac{i}{\hbar} \int_{t_a}^{t_b} dt \left[\frac{M}{2}(\dot{\mathbf{x}}_+^2 - \dot{\mathbf{x}}_-^2) - (V(\mathbf{x}_+) - V(\mathbf{x}_-))\right]\right\} \tag{18.563}$$

$$\times \exp\left\{-\frac{i}{2\hbar} M\gamma \int_{t_a}^{t_b} dt \, (\dot{\mathbf{x}}_+ - \dot{\mathbf{x}}_-)(\ddot{\mathbf{x}}_+ + \ddot{\mathbf{x}}_-)^R - \frac{w}{2\hbar^2} \int_{t_a}^{t_b} dt \, (\dot{\mathbf{x}}_+ - \dot{\mathbf{x}}_-)^2\right\},$$

where the last term is now local since $K^{\text{Ohm}}(t, t') \to \delta(t - t')$. In this limit (as in the classical limit $\hbar \to 0$), this term squeezes the forward and backward paths together. The density matrix (18.563) becomes diagonal. The γ-term, however, remains and describes classical radiation damping.

At moderately high temperature, we should include also the first correction term in (18.526) which adds to the exponent an additional term

$$-\frac{w}{24(k_BT)^2} \int_{t_a}^{t_b} dt \, (\ddot{\mathbf{x}}_+ - \ddot{\mathbf{x}}_-)^2. \tag{18.564}$$

The extended expression is the desired *closed-time path integral* of a particle in contact with a thermal reservoir.

18.20.2 Master Equation for Time Evolution in Photon Bath

It is possible to derive a master equation for the evolution of the density matrix $\rho(\mathbf{x}_{+a}, \mathbf{x}_{-a}; t_a)$ analogous to Eq. (18.530) for a quantum particle in a photon bath. Since the dissipative and fluctuating parts of the influence functional in Eq. (18.555) and (18.561) coincide with the corresponding terms in (18.230), except for an extra dot on top of the coordinates, the associated temperature-dependent Hamiltonian operator is directly obtained from (18.525) with the expansion (18.527) by adding the extra dots: In the high-temperature limit we obtain

$$\hat{\mathcal{H}} \equiv \frac{1}{2M}(\hat{\mathbf{p}}_+^2 - \hat{\mathbf{p}}_-^2) + V(\mathbf{x}_+) - V(\mathbf{x}_-) + \frac{M\gamma}{2}(\hat{\dot{\mathbf{x}}}_+ - \hat{\dot{\mathbf{x}}}_-)(\hat{\dot{\mathbf{x}}}_+ + \hat{\dot{\mathbf{x}}}_-)^R - i\frac{w}{2\hbar}(\hat{\dot{\mathbf{x}}}_+ - \hat{\dot{\mathbf{x}}}_-)^2,$$

$$\tag{18.565}$$

extended at moderately high temperatures by the Hamiltonian corresponding to (18.564):

$$\Delta\hat{\tilde{H}}_T \equiv i\frac{w\hbar}{24(k_BT)^2}(\hat{\ddot{\mathbf{x}}}_+ - \hat{\ddot{\mathbf{x}}}_-)^2. \tag{18.566}$$

The master equation corresponding to the Ohmic equation (18.530) reads now

$$i\hbar\partial_t\hat{\rho} = \hat{\tilde{H}}_T\,\hat{\rho} \equiv [\hat{H}, \hat{\rho}] + \frac{M\gamma}{2}\left(\hat{\dot{\mathbf{x}}}\hat{\dot{\mathbf{x}}}\hat{\rho} - \hat{\rho}\hat{\dot{\mathbf{x}}}\hat{\dot{\mathbf{x}}} + \hat{\dot{\mathbf{x}}}\,\hat{\rho}\,\hat{\dot{\mathbf{x}}} - \hat{\dot{\mathbf{x}}}\,\hat{\rho}\,\hat{\dot{\mathbf{x}}}\right)$$

$$- \frac{iw}{2\hbar}[\hat{\dot{\mathbf{x}}}, [\hat{\dot{\mathbf{x}}}, \hat{\rho}]] - \frac{iw\hbar^2}{24(k_BT)^2}[\hat{\ddot{\mathbf{x}}}, [\hat{\ddot{\mathbf{x}}}, \hat{\rho}]]. \tag{18.567}$$

The conservation of total probability and the positivity of $\hat{\rho}$ are ensured by the observation, that Eq. (18.567) can be written in the *Lindblad form*

$$\partial_t\hat{\rho} = -\frac{i}{\hbar}[\hat{H}, \hat{\rho}] - \sum_{n=1}^{2}\left(\frac{1}{2}\hat{L}_n\hat{L}_n^\dagger\hat{\rho} + \frac{1}{2}\hat{\rho}\hat{L}_n\hat{L}_n^\dagger - \hat{L}_n^\dagger\hat{\rho}\hat{L}_n\right), \tag{18.568}$$

with the two Lindblad operators

$$\hat{L}_1 \equiv \frac{\sqrt{w}}{2\hbar}\hat{\mathbf{x}}, \qquad \hat{L}_2 \equiv \frac{\sqrt{3w}}{2\hbar}\left(\hat{\mathbf{x}} - i\frac{\hbar}{3k_BT}\dot{\hat{\mathbf{x}}}\right). \tag{18.569}$$

As noted in the discussion of Eq. (18.530), the operator order in (18.567) prevents the term $\hat{\mathbf{x}}\dot{\hat{\mathbf{x}}}\hat{\rho}$ from being a pure divergence. By rewriting it as a sum of a commutator and an anticommutator, $[\hat{\mathbf{x}}, \dot{\hat{\mathbf{x}}}]/2 + \{\hat{\mathbf{x}}, \dot{\hat{\mathbf{x}}}\}/2$, the latter term is a pure divergence, and we can think of the first two γ-terms in (18.567) as being due to an additional anti-Hermitian *dissipation operator*

$$\hat{H}_\gamma = \gamma M \frac{1}{4}[\hat{\mathbf{x}}, \dot{\hat{\mathbf{x}}}]. \tag{18.570}$$

For a free particle with $V(\mathbf{x}) \equiv 0$ and $[\hat{H}, \hat{\mathbf{p}}] = 0$, one has $\dot{\hat{\mathbf{x}}}_\pm = \hat{\mathbf{p}}_\pm/M$ to all orders in γ, such that the time evolution equation (18.567) becomes

$$i\hbar\partial_t\hat{\rho} = [\hat{H}, \hat{\rho}] - \frac{iw}{2M^2\hbar}[\hat{\mathbf{p}}, [\hat{\mathbf{p}}, \hat{\rho}]]. \tag{18.571}$$

In the momentum representation of the density matrix $\hat{\rho} = \sum_{\mathbf{p}\mathbf{p}'} \rho_{\mathbf{p}\mathbf{p}'}|\mathbf{p}\rangle\langle\mathbf{p}'|$, the last term simplifies to $-i\Gamma \equiv -iw(\mathbf{p} - \mathbf{p}')^2/2M^2\hbar^2$ multiplying $\hat{\rho}$, which shows that a free particle does not dissipate energy by radiation, and that the off-diagonal matrix elements decay with the rate Γ.

For small e^2, the implicit equation Eq. (18.565) with the expansion term (18.566) can be solved approximately in a single iteration step, inserting $\dot{\hat{\mathbf{x}}} \approx \hat{\mathbf{p}}/M$ and $\ddot{\hat{\mathbf{x}}} \approx -\boldsymbol{\nabla}V/M$.

18.20.3 Line Width

Let us apply the master equation (18.567) to atoms, where $V(\mathbf{x})$ is the Coulomb potential, assuming it to be initially in an eigenstate $|i\rangle$ of H, with a density matrix $\hat{\rho}(0) = |i\rangle\langle i|$. Since atoms decay rather slowly, we may treat the γ-term in (18.567) perturbatively. It leads to a time derivative of the density matrix

$$\partial_t\langle i|\hat{\rho}(t)|i\rangle = -\frac{\gamma}{\hbar M}\langle i|[\hat{H}, \hat{\mathbf{p}}]\,\hat{\mathbf{p}}\,\hat{\rho}(0)|i\rangle = \frac{\gamma}{M}\sum_{f\neq i}\omega_{if}\langle i|\mathbf{p}|f\rangle\langle f|\mathbf{p}|i\rangle$$

$$= -M\gamma\sum_f\omega_{if}^3\,|\mathbf{x}_{fi}|^2, \tag{18.572}$$

where $\hbar\omega_{if} \equiv E_i - E_f$, and $\mathbf{x}_{fi} \equiv \langle f|\mathbf{x}|i\rangle$ are the matrix elements of the dipole operator.

An extra width comes from the last two terms in (18.567):

$$\partial_t\langle i|\hat{\rho}(t)|i\rangle = -\frac{w}{M^2\hbar^2}\langle i|\mathbf{p}^2|i\rangle - \frac{w}{12M^2(k_BT)^2}\langle i|\dot{\mathbf{p}}^2|i\rangle$$

$$= -w\sum_n\omega_{if}^2\left[1 + \frac{\hbar^2\omega_{if}^2}{12(k_BT)^2}\right]|\mathbf{x}_{fi}|^2. \tag{18.573}$$

This time dependence is caused by spontaneous emission and induced emission and absorption. To identify the different contributions, we rewrite the spectral decompositions (18.550) and (18.551) in the $\mathbf{x}$-independent approximation as

$$C_b(t,t') + A_b(t,t') \tag{18.574}$$
$$= \frac{4\pi}{3}\hbar\int\frac{d\omega'd^3k}{(2\pi)^4}\frac{\pi}{2M\Omega_{\mathbf{k}}}\left\{1 + \coth\frac{\hbar\omega'}{2k_BT}\right\}[\delta(\omega' - \Omega_{\mathbf{k}}) - \delta(\omega' + \Omega_{\mathbf{k}})]\,e^{-i\omega'(t-t')},$$

or

$$C_b(t,t') + A_b(t,t') \tag{18.575}$$
$$= \frac{4\pi}{3}\hbar\int\frac{d\omega'd^3k}{(2\pi)^4}\frac{\pi}{2M\Omega_{\mathbf{k}}}\left\{2\delta(\omega' - \Omega_{\mathbf{k}}) + \frac{2}{e^{\hbar\Omega_{\mathbf{k}}/k_BT} - 1}[\delta(\omega' - \Omega_{\mathbf{k}}) + \delta(\omega' + \Omega_{\mathbf{k}})]\right\}e^{-i\omega'(t-t')}.$$

Following Einstein's intuitive interpretation, the first term in curly brackets is due to spontaneous emission, the other two terms accompanied by the Bose occupation function account for induced emission and absorption. For high and intermediate temperatures, (18.575) has the expansion

$$
\frac{4\pi}{3}\hbar \int \frac{d\omega' d^3k}{(2\pi)^4} \frac{\pi}{2M\Omega_{\mathbf{k}}} \left\{ 2\delta(\omega' - \Omega_{\mathbf{k}}) \right.
$$
$$
\left. + \left(\frac{2k_B T}{\hbar \Omega_{\mathbf{k}}} - 1 + \frac{1}{6}\frac{\hbar \Omega_{\mathbf{k}}}{k_B T} \right) [\delta(\omega' - \Omega_{\mathbf{k}}) + \delta(\omega' + \Omega_{\mathbf{k}})] \right\} e^{-i\omega'(t-t')}. \qquad (18.576)
$$

The first term in curly brackets corresponds to the spontaneous emission. It contributes to the rate of change $\partial_t \langle i | \hat{\rho}(t) | i \rangle$ a term $-2M\gamma \sum_{f<i} \omega_{if}^3 |\mathbf{x}_{fi}|^2$. This differs from the right-hand side of Eq. (18.572) in two important respects. First, the sum is restricted to the lower states $f < i$ with $\omega_{if} > 0$, since the δ-function allows only for decays. Second, there is an extra factor 2. Indeed, by comparing (18.574) with (18.576) we see that the spontaneous emission receives equal contributions from the 1 and the $\coth(\hbar\omega'/2k_B T)$ in the curly brackets of (18.574), i.e., from dissipation and fluctuation terms $C_{\mathrm{b}}(t, t')$ and $A_{\mathrm{b}}(t, t')$.

Thus our master equation yields for the natural line width of atomic levels the equation

$$
\Gamma = 2M\gamma \sum_{f<i} \omega_{if}^3 |\mathbf{x}_{fi}|^2, \qquad (18.577)
$$

in agreement with the historic *Wigner-Weisskopf formula*.

In terms of Γ, the rate (18.572) can therefore be written as

$$
\partial_t \langle i | \hat{\rho}(t) | i \rangle = -\Gamma + M\gamma \sum_{f<i} \omega_{if}^3 |\mathbf{x}_{fi}|^2 + M\gamma \sum_{f>i} |\omega_{if}|^3 |\mathbf{x}_{fi}|^2. \qquad (18.578)
$$

The second and third terms do not contribute to the total rate of change of $\langle i | \hat{\rho}(t) | i \rangle$ since they are canceled by the induced emission and absorption terms associated with the -1 in the big parentheses of the fluctuation part of (18.576). The finite lifetime changes the time dependence of the state $|i, t\rangle$ from $|i, t\rangle = |i, 0\rangle e^{-iEt}$ to $|i, 0\rangle e^{-iEt - \Gamma t/2}$.

Note that due to the restriction to $f < i$ in (18.577), there is no operator local in time whose expectation value is Γ. Only the combination of spontaneous and induced emissions and absorptions in (18.578) can be obtained from a local operator, which is in fact the dissipation operator (18.570).

For all temperatures, the spontaneous and induced transitions together lead to the rate of change of $\langle i | \hat{\rho}(t) | i \rangle$:

$$
\partial_t \langle i | \hat{\rho}(t) | i \rangle = -2M\gamma \left(\sum_{f<i} \omega_{if}^3 + \sum_f \omega_{if}^3 \frac{1}{e^{\hbar\omega_{if}/k_B T} - 1} \right) |\mathbf{x}_{fi}|^2. \qquad (18.579)
$$

For a state with principal quantum number n the temperature effects become detectable only if T becomes larger than $-1/(n+1)^2 + 1/n^2 \approx 2/n^3$ times the Rydberg temperature $T_{\mathrm{Ry}} = 157886.601 K$. Thus we have to go to $n \gtrsim 20$ to have observable effects at room temperature.

18.20.4 Lamb shift

For atoms, the Feynman influence functional (18.547) allows us to calculate the celebrated Lamb shift. Being interested in the time behavior of the pure-state density matrix $\rho = |i\rangle\langle i|$, we may calculate the effect of the actions (18.548) and (18.549) perturbatively. For this, consider the dissipative part of the influence action (18.548), and in it the first term involving $\mathbf{x}_+(t)$ and $\mathbf{x}_+(t')$, and integrate the external positions in the path integral (18.547) over the initial wave functions, forming

$$U_{ii,t_b;ii,t_a} = \int d\mathbf{x}_{+b}\, d\mathbf{x}_{-b} \int d\mathbf{x}_{+a}\, d\mathbf{x}_{-a} \langle i|\mathbf{x}_{+b}\rangle\langle i|\mathbf{x}_{-b}\rangle$$
$$\times\, U(\mathbf{x}_{+b}, \mathbf{x}_{-b}, t_b|\mathbf{x}_{+a}, \mathbf{x}_{-a}, t_a)\langle \mathbf{x}_{+b}|i\rangle\langle \mathbf{x}_{-b}|i\rangle. \tag{18.580}$$

To lowest order in γ, the effect of the $\mathbf{C}_b$-term in (18.548) can be evaluated in the local approximation (18.554) as follows. We take the linear approximation to the exponential $\exp[\int dtdt'\mathcal{O}(t,t')] \approx 1 + \int dtdt'\mathcal{O}(t,t')$ and propagate the initial state with the help of the amplitude $U_{ii,t';ii,t_a}$ to the first time t', then with $U_{fi,t;fi,t'}$ to the later time t, and finally with $U_{ii,t_a;ii,t}$ to the final time t_b. The intermediate state between the times t and t' are arbitrary and must be summed. Details how to do such a perturbation expansion are given in Section 3.17. Thus we find

$$\Delta_C U_{ii,t_b;ii,t_a} = i\frac{e^2}{2\hbar^2 c^2}\int_{t_a}^{t_b} dtdt' \sum_f \int d\mathbf{x}_+ \int d\mathbf{x}'_+\, U_{ii,t_a;ii,t}\langle i|\mathbf{x}_+\rangle\mathbf{x}_+\langle \mathbf{x}_+|f\rangle$$
$$\times [\partial_t \partial_{t'} \mathbf{C}_b(t,t')]U_{fi,t;fi,t'}\langle f|\mathbf{x}'_+\rangle\mathbf{x}'_+\langle \mathbf{x}'_+|i\rangle U_{ii,t';ii,t_a}. \tag{18.581}$$

Inserting $U_{ii,t_a;ii,t} = e^{-iE_i(t_a-t)/\hbar}$ etc., this becomes

$$\Delta_C U_{ii,t_b;ii,t_a} = -\frac{e^2}{2\hbar^2 c^2}\int_{t_a}^{t_b} dtdt' \langle i|\hat{\mathbf{x}}(t)\, [\partial_t \partial_{t'}\mathbf{C}_b(t,t')]\, \hat{\mathbf{x}}(t')|i\rangle$$
$$= -\frac{e^2}{2\hbar^2 c^2}\sum_f \int_{t_a}^{t_b} dtdt'\, e^{i\omega_{if}(t-t')}\langle i|\hat{\mathbf{x}}|f\rangle\mathbf{C}_b(t,t')\langle f|\hat{\mathbf{x}}|i\rangle. \tag{18.582}$$

Expressing $C_b^{ij}(t,t')$ of Eq. (18.554) in the form

$$C_b^{ij}(t,t') = \frac{\hbar}{2\pi c}\frac{2}{3}\delta^{ij}\int \frac{d\omega}{2\pi}\, \omega\, e^{-i\omega(t-t')}, \tag{18.583}$$

the integration over t and t' yields

$$\Delta_C U_{ii,t_b;ii,t_a} = -i\frac{e^2}{4\pi\hbar c^3}\frac{2}{3}\int_{t_a}^{t_b} dt \int \frac{d\omega}{2\pi}\sum_f \frac{\omega}{\omega-\omega_{if}-i\eta}|\hat{\mathbf{x}}_{fi}|^2. \tag{18.584}$$

The same treatment is applied to the A_b in the action (18.549), where the first term involving $\mathbf{x}_+(t)$ and $\mathbf{x}_+(t')$ changes (18.585) to

$$\Delta U_{ii,t_b;ii,t_a} = -i\frac{e^2}{4\pi\hbar c^3}\frac{2}{3}\int_{t_a}^{t_b} dt \int \frac{d\omega}{2\pi}\sum_f \frac{\omega}{\omega-\omega_{if}+i\eta}\left(1+\coth\frac{\hbar\omega}{2k_BT}\right)|\hat{\mathbf{x}}_{fi}|^2. \tag{18.585}$$

The ω-integral is conveniently split into a zero-temperature part

$$I(\omega_{if},0) \equiv \int_0^\infty \frac{d\omega}{\pi}\sum_f \frac{\omega}{\omega-\omega_{if}+i\eta}, \tag{18.586}$$

and a finite-temperature correction

$$\Delta I_T(\omega_{if},T) \equiv 2\int_0^\infty \frac{d\omega}{\pi}\sum_f \frac{\omega}{\omega-\omega_{if}+i\eta}\frac{1}{e^{\hbar\omega/k_BT}-1}. \tag{18.587}$$

Decomposing $1/(\omega_{if}-\omega+i\eta) = \mathcal{P}/(\omega-\omega_{if}) - i\pi\delta(\omega_{if}-\omega)$, the imaginary part of the ω-integral yields half of the natural line width in (18.572). The other half comes from the part of the integral

(18.548) involving $\mathbf{x}_-(t)$ and $\mathbf{x}_-(t')$. The principal-value part of the zero-temperature integral diverges linearly, the divergence yielding again the mass renormalization (18.556). Subtracting this divergence from $I(\omega_{if}, 0)$, the remaining integral has the same form as $I(\omega_{if}, 0)$, but with ω in the numerator replaced by $\omega_{if} = 0$. This integral diverges logarithmically like $(\omega_{if}/\pi) \log[(\Lambda - \omega_{if})/|\omega_{if}|]$, where Λ is Bethe's cutoff [30]. For $\Lambda \gg |\omega_{if}|$, the result (18.585) implies an energy shift of the atomic level $|i\rangle$:

$$\Delta E_i = \frac{e^2}{4\pi c^3} \frac{2}{3\pi} \sum_f \omega_{if}^3 |\hat{\mathbf{x}}_{fi}|^2 \log \frac{\Lambda}{|\omega_{if}|}, \tag{18.588}$$

which is the *Lamb shift*.

Usually, the weakly varying logarithm is approximated by a weighted average $L = \log[\Lambda/\langle|\omega_{if}|\rangle]$ over energy levels and taken out of the integral. Then contribution of the term (18.585) can be attributed to an extra term

$$\hat{H}_{\mathrm{LS}} \approx -i\frac{L}{\pi} \gamma M \frac{1}{4}[\hat{\mathbf{x}}, \dot{\hat{\mathbf{x}}}] \tag{18.589}$$

in the Hamiltonian (18.567). In this form, the Lamb shift appears as a Hermitian logarithmically divergent correction to the operator (18.570) governing the spontaneous emission of photons.

To lowest order in γ, the commutator is for a Coulomb potential $V(\mathbf{x}) = -e^2/r$ equal to

$$-\frac{i}{M^2}[\hat{\mathbf{p}}, \dot{\hat{\mathbf{p}}}] = \frac{\hbar}{M^2} \nabla^2 V(\mathbf{x}) = \frac{\hbar^2 c\alpha}{M^2} 4\pi\delta^{(3)}(\mathbf{x}), \tag{18.590}$$

leading to

$$\Delta E_i = \frac{4\alpha^2 \hbar^3}{3M^2 c} \langle i|\delta^{(3)}(\mathbf{x})|i\rangle. \tag{18.591}$$

For an atomic state of principal quantum number n with a wave function $\psi_n(\mathbf{x})$, this becomes

$$\Delta E_n = \frac{4\alpha \hbar^3}{3M^2 c^2} \alpha L |\psi_n(\mathbf{0})|^2. \tag{18.592}$$

Only atomic s-states can contribute, since the wave functions of all other angular momenta vanish at the origin. Explicitly, the s-states of the hydrogen atom (13.213) have the value at the origin

$$\psi_n(\mathbf{0}) = \frac{1}{\sqrt{n^3\pi}} \left(\frac{1}{a_H}\right)^{3/2}, \tag{18.593}$$

where $a_H = \hbar/Mc\alpha$ is the Bohr radius (4.344). If the nuclear charge is Z, then a_{B}, is diminished by this factor. Thus we obtain the energy shift

$$\Delta E_n = \frac{4\alpha^2 \hbar^3}{3M^2 c} \left(\frac{Mc\alpha}{\hbar}\right)^3 \frac{L}{n^3\pi}. \tag{18.594}$$

For a hydrogen atom with $n = 2$, this becomes

$$\Delta E_2 = \frac{\alpha^3}{6\pi} \alpha^2 M c^2 L. \tag{18.595}$$

The quantity $Mc^2\alpha^2$ is the unit energy of atomic physics determining the hydrogen spectrum to be $E_n = -Mc^2\alpha^2/2n^2$. Thus

$$M\alpha^2 = 4.36 \times 10^{-11}\mathrm{erg} = 27.21\mathrm{eV} = 2\,\mathrm{Ry} = 2 \cdot 3.288 \times 10^{15}\mathrm{Hz}. \tag{18.596}$$

Inserting this together with $\alpha \approx 1/137.036$ into (18.595) yields[10]

$$\Delta E_2 \approx 135.6\text{MHz} \times L. \tag{18.597}$$

The constant L can be calculated approximately as

$$L \approx 9.3, \tag{18.598}$$

leading to the estimate

$$\Delta E_2 \approx 1261\text{MHz}. \tag{18.599}$$

The experimental Lamb shift

$$\Delta E_{\text{Lamb shift}} \approx 1057 \text{ MHz} \tag{18.600}$$

is indeed contained in this range. In this calculation, two effects have been ignored: the vacuum polarization of the photon and the form factor of the electron caused by radiative corrections. They reduce the frequency (18.599) by $(27.3 + 51)$MHz bringing the theoretical number closer to experiment. The vacuum polarization will be discussed in detail in Section 19.4.

At finite temperature, (18.588) changes to

$$\Delta E_i = \frac{e^2}{4\pi c^3} \frac{2}{3\pi} \sum_f \omega_{if}^3 |\hat{\mathbf{x}}_{fi}|^2 \left[\log \frac{\Lambda}{|\omega_{if}|} + \left(\frac{k_B T}{\hbar \omega_{if}}\right)^2 J\left(\frac{\hbar \omega_{if}}{k_B T}\right) \right], \tag{18.601}$$

where $J(z)$ denotes the integral

$$J(z) \equiv z \int_0^\infty dz \, \frac{\mathcal{P}}{z' - z} \frac{z'}{e^{z'} - 1}, \tag{18.602}$$

which has the low-temperature (large-z) expansion $J(z) = -\pi^2/6 - 2\zeta(3)/z + \ldots$, and goes to zero for high temperature (small z) like $-z \log z$, as shown in Fig. 18.2.

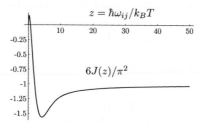

Figure 18.2 Behavior of function $6J(z)/\pi^2$ in finite-temperature Lamb shift.

The above equations may have applications to dilute interstellar gases or, after a reformulation in a finite volume, to few-particle systems contained in cavities. So far, a master equation has been set up only for a finite number of modes [31].

---

[10]The precise value of the Lamb constant $\alpha^4 M/6\pi$ is 135.641 ± 0.004 MHz.

18.20.5   Langevin Equations

For high γT, the last term in the forward–backward path integral (18.563) makes the size of the fluctuations in the difference between the paths $\mathbf{y}(t) \equiv \mathbf{x}_+(t) - \mathbf{x}_-(t)$ very small. It is then convenient to introduce the average of the two paths as $\mathbf{x}(t) \equiv [\mathbf{x}_+(t) + \mathbf{x}_-(t)]/2$, and expand

$$V\left(\mathbf{x} + \frac{\mathbf{y}}{2}\right) - V\left(\mathbf{x} - \frac{\mathbf{y}}{2}\right) \sim \mathbf{y} \cdot \nabla V(\mathbf{x}) + \mathcal{O}(y^3)\dots , \qquad (18.603)$$

keeping only the first term. We further introduce an auxiliary quantity $\boldsymbol{\eta}(t)$ by

$$\dot{\boldsymbol{\eta}}(t) \equiv M\dddot{\mathbf{x}}(t) - M\gamma\ddot{\mathbf{x}}(t) + \nabla V(\mathbf{x}(t)). \qquad (18.604)$$

With this, the exponential function in (18.563) becomes

$$\exp\left[-\frac{i}{\hbar}\int_{t_a}^{t_b} dt\,\dot{\mathbf{y}}\boldsymbol{\eta} - \frac{w}{2\hbar^2}\int_{t_a}^{t_b} dt\,\dot{\mathbf{y}}^2(t)\right], \qquad (18.605)$$

where w is the constant (18.562).

Consider now the diagonal part of the amplitude (18.603) with $\mathbf{x}_{+b} = \mathbf{x}_{-b} \equiv \mathbf{x}_b$ and $\mathbf{x}_{+a} = \mathbf{x}_{-a} \equiv \mathbf{x}_a$, implying that $\mathbf{y}_b = \mathbf{y}_a = 0$. It represents a probability distribution

$$P(\mathbf{x}_b\,t_b|\mathbf{x}_a\,t_a) \equiv |(\mathbf{x}_b, t_b|\mathbf{x}_a, t_a)|^2 \equiv U(\mathbf{x}_b, \mathbf{x}_b, t_b|\mathbf{x}_a, \mathbf{x}_a, t_a). \qquad (18.606)$$

Now the variable $\mathbf{y}$ can simply be integrated out in (18.605), and we find the probability distribution

$$P[\boldsymbol{\eta}] \propto \exp\left[-\frac{1}{2w}\int_{t_a}^{t_b} dt\,\boldsymbol{\eta}^2(t)\right]. \qquad (18.607)$$

The expectation value of an arbitrary functional of $F[x]$ can be calculated from the path integral

$$\langle F[\mathbf{x}]\rangle_\eta \equiv \mathcal{N}\int \mathcal{D}\mathbf{x}\, P[\boldsymbol{\eta}]F[\mathbf{x}], \qquad (18.608)$$

where the normalization factor $\mathcal{N}$ is fixed by the condition $\langle 1\rangle = 1$. By a change of integration variables from $x(t)$ to $\eta(t)$, the expectation value (18.608) can be rewritten as a functional integral

$$\langle F[\mathbf{x}]\rangle_\eta \equiv \mathcal{N}\int \mathcal{D}\boldsymbol{\eta}\, P[\boldsymbol{\eta}]\, F[\mathbf{x}]. \qquad (18.609)$$

Note that the probability distribution (18.607) is $\hbar$-independent. Hence in the approximation (18.603) we obtain the classical Langevin equation. In principle, the integrand contains a factor $J^{-1}[x]$, where $J[x]$ is the functional Jacobian

$$J[\mathbf{x}] \equiv \mathrm{Det}\,[\delta\eta^i(t)/\delta x^j(t')] = \det\left[\left(M\partial_t^2 - M\gamma\partial_t^{3R}\right)\delta_{ij} + \nabla_i\nabla_j V(\mathbf{x}(t))\right]. \qquad (18.610)$$

By the same procedure as in Section 18.9.2 it can be shown that the determinant is unity, due to the retardation of the friction term, thus justifying its omission in (18.609).

The path integral (18.609) may be interpreted as an expectation value with respect to the solutions of a *stochastic differential equation* (18.604) driven by a Gaussian random *noise* variable $\eta(t)$ with a correlation function

$$\langle \eta^i(t)\eta^j(t')\rangle_T = \delta^{ij} w\,\delta(t - t'). \qquad (18.611)$$

Since the dissipation carries a third time derivative, the treatment of the initial conditions is nontrivial and will be discussed elsewhere. In most physical applications γ leads to slow decay rates. In this case the simplest procedure to solve (18.604) is to write the stochastic equation as

$$M\dddot{\mathbf{x}}(t) + \nabla V(\mathbf{x}(t)) = \dot{\boldsymbol{\eta}}(t) + M\gamma\ddot{\mathbf{x}}(t), \qquad (18.612)$$

and solve it iteratively, first without the γ-term, inserting the solution on the right-hand side, and such a procedure is equivalent to a perturbative expansion in γ in Eq. (18.563).

Note that the lowest iteration of Eq. (18.612) with $\eta \equiv 0$ can be multiplied by $\dot{x}$ and leads to the equation for the energy change of the particle

$$\frac{d}{dt}\left[\frac{M}{2}\dot{x}^2 + V(\mathbf{x}) - M\gamma\dot{x}\ddot{x}\right] = -M\gamma\ddot{x}^2. \tag{18.613}$$

The right-hand side is the classical electromagnetic power radiated by an accelerated particle. The extra term in the brackets is known as *Schott term* [32].

18.21 Fokker-Planck Equation in Spaces with Curvature and Torsion

According to the new equivalence principle found in Chapter 10, equations of motion can be transformed by a nonholonomic transformation $dx^i = e^i{}_\mu(q)dq^\mu$ into spaces with curvature and torsion, where they are applicable to the diffusion of atoms in crystals with defects [33]. If we denote $g_{\mu\nu}\dot{q}^\nu$ by v_μ, the Langevin equation (18.317) goes over into

$$\dot{v}_\mu = F_\mu(q, v) + e^i{}_\mu(q)\eta_i \tag{18.614}$$

where $F_\mu(q, v)$ is the sum of all forces after the nonholonomic transformation:

$$F_\mu(q, v) \equiv M\left[\Gamma_{\nu\lambda\mu}(q)v^\nu v^\lambda - \gamma v_\mu\right] - \partial_\mu V(q). \tag{18.615}$$

In addition to the transformed force (18.384), $F_\mu(q, v)$ contains the apparent forces resulting from the coordinate transformation. For a distribution

$$P_\eta(qvt|q_a v_a t_a) = \delta(q_\eta(t) - q)\delta(\dot{q}_\eta - v_\mu) \tag{18.616}$$

one obtains, instead of (18.383), the Kubo equation

$$\partial_t P_\eta(q\,v\,t|q_a v_a t_a) = \left\{-\partial_\mu g^{\mu\nu}(q)v_\nu - \frac{1}{M}\partial^\nu_\mu\left[e^i{}_\mu(q)\eta_i + F_\mu(q, v)\right]\right\}P_\eta(q\,v\,t|q_a v_a t_a), \tag{18.617}$$

and from this the generalization of the Fokker-Planck equation (18.393) to spaces with curvature and torsion:

$$\partial_t P(x\,v\,t|x_a v_a t_a) = \left\{-\partial_\mu g^{\mu\nu}v_\nu + \frac{1}{M}\partial^\nu_\mu\left[\frac{w}{2M}\partial^\nu_\mu - F_\mu(q, v)\right]\right\}P(x\,v\,t|x_a v_a t_a). \tag{18.618}$$

In the overdamped limit, the integrated probability distributions

$$P(q\,t|q_a t_a) \equiv \int d^D v P(q\,v\,t|q_a v_a t_a) \tag{18.619}$$

satisfies the equation [generalizing (18.399)]

$$\partial_t P(q\,t|q_a t) = \left[D\partial_\mu e_i{}^\mu \partial_\nu e_i{}^\nu + \frac{1}{M\gamma}\partial_\nu g^{\mu\nu}V_\nu(q)\right]P(q\,t|q_a t), \tag{18.620}$$

where $V_\nu(q) \equiv \partial_\nu V(q)$.

In the Fokker-Planck equations (18.617) and (18.620), the probability distributions $P(q\,v\,t|q_a v_a t_a)$ and $P(qt|q_a t)$ have the unit normalizations

$$\int d^D q d^D v\, P(q\,v\,t|q_a v_a t_a) = 1, \quad \int d^D q\, P(q\,t|q_a t_a) = 1, \tag{18.621}$$

as can be seen from the definitions (18.616) and (18.619). For distributions normalized with the invariant volume integral $\int d^D q \sqrt{g}$, to be denoted by $P^{\text{inv}}(q t | q_a t_a) \equiv \sqrt{g}^{-1} P(qt|q_a t)$, we obtain from (18.620) the following invariant Fokker-Planck equation:

$$\partial_t P^{\text{inv}}(q\,t|q_a t_a) = \left\{ \frac{D}{\sqrt{g}} \partial_\mu g^{\mu\nu} \sqrt{g} \left[\partial_\nu + 2S_\nu + \frac{1}{k_B T} V_\nu(x) \right] \right\} P^{\text{inv}}(q\,t|q_a t_a). \qquad (18.622)$$

The first term on the right-hand side contains the Laplace-Beltrami operator (11.13).

With the help of the covariant derivative D_μ^* defined in Eq. (11.96), which arises from D_μ by a partial integration, this equation can also be written as

$$\partial_t P^{\text{inv}}(q\,t|q_a t_a) = \left[D\, g^{\mu\nu} D_\mu^* D_\nu^* + \frac{1}{M\gamma} \bar{D}_\mu V^\mu(x) \right] P^{\text{inv}}(q\,t|q_a t_a), \qquad (18.623)$$

where $\bar{D}_\mu$ is the covariant derivative (10.37) associated with the Christoffel symbol.

18.22 Stochastic Interpretation of Quantum-Mechanical Amplitudes

In the last section we have seen that the probability distribution $|(x_b, t_b | x_a, t_a)|^2$ is the result of a stochastic differential equation describing a classical path disturbed by a noise term $\eta(t)$ with the correlation function (18.320). It is interesting to observe that the quantum-mechanical amplitude $(x_b, t_b | x_a, t_a)$ possesses quite a similar stochastic interpretation, albeit with some imaginary factors i and an unsatisfactory aspect as we shall see. Recall the path integral representation of the time evolution amplitude in Eq. (2.710). It involves the action $A(x, t; x_a, t_a)$ from the initial point x_a to the actual particle position x. Recalling the definition of the fluctuation factor $F(x_b, x_a; t_b - t_a)$ in Eq. (4.97), we see that this factor is given by the path integral

$$F(x_b, x_a; t_b - t_a) = \int_{(x_a, t_a) \rightsquigarrow (x_b, t_b)} \mathcal{D}x \exp\left[\frac{i}{\hbar} \int_{t_a}^{t_b} dt \frac{M}{2} (\dot{x} - v)^2 \right], \qquad (18.624)$$

where $v(x, t) = (1/M) \partial_x A(x, x_a; t_b - t_a)$ is the classical particle velocity. Up to a factor i and the absence of the retardation symbol, this path integral has the same form as the one for the probability (18.290) at large damping. As in (2.711) we introduce the momentum variable $p(t)$ and obtain the canonical path integral

$$F(x_b, x_a; t_b - t_a) = \int_{(x_a, t_a) \rightsquigarrow (x_b, t_b)} \mathcal{D}'x \frac{\mathcal{D}p}{2\pi\hbar} e^{(i/\hbar) \int_{t_a}^{t_b} dt \{ p(t)[\dot{x}(t) - v(x(t), t)] - p^2(t)/2M \}}, \qquad (18.625)$$

which looks similar to the stochastic path integral (18.290). The role of the diffusion constant $D = k_B T / M \gamma$ is now played by $\hbar/2$. By analogy with the path integral of a particle in a magnetic field in (2.652), the fluctuation factor satisfies a Schrödinger-like equation

$$\left(\frac{\hat{p}_b^2}{2M} + \frac{1}{2} \{ \hat{p}_b, v_b \} \right) F(x_b, x_a; t_b - t_a) = i\hbar F(x_b, x_a; t_b - t_a). \qquad (18.626)$$

This can easily be verified for the free particle, where $v_b = (x_b - x_a)/(t_b - t_a)$, and the fluctuation factor is from (2.125) $F(x_b, x_a; t_b - t_a) = \sqrt{2\pi i \hbar (t_b - t_a)}$. Note the symmetric operator order of the product $\hat{p}v$ in accordance with the time slicing in Section 10.5, and the ensuing operator order observed in Eq. (11.88). By reordering the Hamiltonian operator on the left-hand side of (18.626) to position $\hat{p}$ to the left of the velocity,

$$\hat{H} \to \frac{\hat{p}^2}{2M} + \hat{p}v + \frac{i}{2} \nabla v. \qquad (18.627)$$

Without the last term, the path integral (18.625) would describe fluctuating paths obeying the stochastic differential equation analogous to the classical Langevin equation (18.338) [34]:

$$\dot{x}(t) - v(x(t), t) = p(t)/M. \tag{18.628}$$

The momentum variable $p(t)$ plays the role of the noise variable $\eta(t)$. Up to a factor i, this *quantum noise* has the same correlation functions as a white-noise variable:

$$\langle p(t)p(t') \rangle = -iM\hbar\delta(t - t'). \tag{18.629}$$

The "Fokker-Planck equation" associated with this "process" would be the ordinary Schrödinger equation for the amplitude $(x_b t_b | x_a t_a)$.

For a free particle, the ordering problem can be solved by noting that in the path integral (18.624), the constant v can be removed from the path integral leaving

$$F(x_b, x_a; t_b - t_a) = e^{-iA(x_b, x_a; t_b - t_a)/\hbar} \int_{(x_a, t_a) \rightsquigarrow (x_b, t_b)} \mathcal{D}x \, \exp\left(\frac{i}{\hbar} \int_{t_a}^{t_b} dt \frac{M}{2} \dot{x}^2 \right), \tag{18.630}$$

the right-hand factor being the path integral for the amplitude $(x_b t_b | x_a t_a)$ itself. This is identical with the path integral for the probability distribution of Brownian motion, and quantum-mechanical fluctuations are determined by the process

$$\dot{x}(t) = p(t)/M, \tag{18.631}$$

with the quantum noise (18.629).

But also in the presence of a potential, it is possible to specify a process which properly represents quantum-mechanical fluctuations, although the situation is more involved [35]. To find it we rewrite the action in (18.624) as

$$\mathcal{A} = \int_{t_a}^{t_b} dt \frac{M}{2} (\dot{x} - v)^2 = \int_{t_a}^{t_b} dt \frac{M}{2} \left[(\dot{x} - s)^2 - \frac{i\hbar}{2} s'^2 \right], \tag{18.632}$$

with some as yet unknown function $s(x)$. The associated Hamiltonian is now

$$\hat{H} \rightarrow \frac{\hat{p}^2}{2M} + \frac{1}{2}\{\hat{p}, s\} + i\hbar s'^2 = \frac{\hat{p}^2}{2M} + \hat{p}s, \tag{18.633}$$

with the proper operator order. In order for (18.632) to hold, the function $s(x)$ must satisfy the equations

$$v = s', \quad -i\hbar s'^2 + s^2 = v^2. \tag{18.634}$$

Recalling Eqs. (4.12) and (4.5) we see that the equations in (18.634) can be satisfied with the help of the full eikonal $S(x)$:

$$s(x) = S(x)/M. \tag{18.635}$$

The process which describes the quantum-mechanical fluctuations is therefore

$$\dot{x}(t) - S(x)/M = p(t)/M. \tag{18.636}$$

The analogy is, however, not really satisfactory, since the full eikonal contains information on all fluctuations. Indeed, by the definition (4.4), it is given by the logarithm of the amplitude $(x\,t | x_a t_a)$, or any superposition $\psi(x, t) = \int dx_a \, (x\,t | x_a t_a)\psi(x_a t_a)$ of it:

$$S(x) = -i\hbar \log(x\,t | x_a t_a). \tag{18.637}$$

For the fluctuation factor this implies

$$s(x) - v(x) = \delta v(x) \equiv -\frac{i}{M}\hbar F(x\,t|x_a t_a).$$

(18.638)

The path integral representation (18.624) for the fluctuation factor which has maximal analogy
with the stochastic path integral (18.300) is therefore

$$F(x_b t_b, x_a t_a) = \int_{(x_a,t_a)\rightsquigarrow(x_b,t_b)} \mathcal{D}x \exp\left\{\frac{i}{\hbar}\int_{t_a}^{t_b} dt\,\frac{M}{2}\left[\dot{x}^R - v + \frac{i}{M}\hbar\log F(x_b t_b, x_a t_a)\right]^2\right\}.$$

(18.639)

Since we have to know $F(x_b t_b, x_a t_a)$ to describe quantum-mechanical fluctuations as a process,
this representation is of little practical use. The initial representation (18.624) which does not
correspond to a proper process can, however, be used to solve quantum-mechanical problems.

18.23 Stochastic Equation for Schrödinger Wave Function

It is possible to write a stochastic type of path integral for the Schrödinger wave function $\psi(\mathbf{x}, t)$
in D dimensions. By close analogy with Eq. (18.403), it reads

$$\psi(\mathbf{x}_b, t_b) = \int \mathcal{D}\mathbf{v}\, e^{(i/\hbar)\int_{t_a}^{t_b} dt\,[(M/2)\mathbf{v}^2(t) - V(\mathbf{x}_\mathbf{v}(t_b,t))]}\psi\left(\mathbf{x}_\mathbf{v}(t_b, t_a), t_a\right),$$

(18.640)

where

$$\mathbf{x}_\mathbf{v}(t_b, t) \equiv \mathbf{x}(t_b) - \int_t^{t_b} dt'\, \mathbf{v}(t')$$

(18.641)

is a functional of $\mathbf{v}(t')$ parameterizing all possible fluctuating paths arriving at the fixed final point
$\mathbf{x}(t_b)$ after having started from an arbitrary initial point $\mathbf{x}(t)$. They are *Brownian bridges* between
the two points. The variables $\mathbf{v}(t)$ are the independently fluctuating velocities of the particle. The
natural appearance of the velocities in the measure of the stochastic path integral (18.640) is in
agreement with our observation in Eq. (10.141) that the time-sliced measure should contain the
coordinate *differences* $\Delta\mathbf{x}_n$ as the integration variables rather than the coordinates themselves,
which was the starting point for the nonholonomic coordinate transformations to spaces with
curvature and torsion.

We easily verify that (18.640) satisfies the Schrödinger equation by calculating the wave function at a slightly later time $t_b + \epsilon$, and expanding the right-hand side in powers of ϵ. Using the
correlation functions

$$\langle v^i(t)\rangle = 0, \qquad \langle v^i(t)v^j(t')\rangle = i\hbar\delta^{ij}\,\delta(t - t'),$$

(18.642)

we find, via a similar intermediate step as in (18.404), the desired result:

$$i\partial_t\psi(\mathbf{x}, t) = \left[-\frac{\hbar^2}{2M}\partial_\mathbf{x}^2 + V(\mathbf{x})\right]\psi(\mathbf{x}, t).$$

(18.643)

One may also write down a corresponding path integral for the time evolution amplitude

$$(\mathbf{x}_b t_b|\mathbf{x}_a t_a) = \int \mathcal{D}\mathbf{v}\, e^{(i/\hbar)\int_{t_a}^{t_b} dt\,[(M/2)\mathbf{v}^2(t) - V(\mathbf{x}_\mathbf{v}(t_b,t))]}\delta^{(D)}\left(\mathbf{x}_a - \mathbf{x}_b + \int_{t_a}^{t_b} dt\,\mathbf{v}(t)\right),$$

(18.644)

which returns the Schrödinger amplitude (18.640) after convolution with $\psi(\mathbf{x}_a, t_a)$ (and reduces to
$\delta^{(D)}(\mathbf{x}_a - \mathbf{x}_b)$ for $t_b \to t_a$, as it should).

The addition of an interaction with a vector potential is nontrivial. The electromagnetic interaction in (10.168)

$$\mathcal{A}_{\text{em}} = \int_{t_a}^{t} dt \, \mathbf{A}(\mathbf{x}(t)) \cdot \dot{\mathbf{x}} \tag{18.645}$$

cannot be simply inserted into the exponent of the path integral (18.640) since in the evaluation via the correlation functions (18.642) assumes the independence of the noise variables $\mathbf{v}(t)$. This, however, is not true in the interaction (18.645). Recall the discussion in Section 10.6 which showed that the time-sliced version of the interaction (18.645) must contain the midpoint ordering of the vector potential with respect to the intervals $\Delta\mathbf{x}$ to be compatible with the classical field equation. In Section 11.3 we have furthermore seen that this guaranteed gauge invariance. For the time-sliced short-time action, this implies that (18.645) has the form [see (10.179)]

$$\mathcal{A}_{\text{em}}^{\epsilon} = \mathbf{A}(\bar{\mathbf{x}}) \cdot \Delta\mathbf{x}. \tag{18.646}$$

In this expression, a variation of $\Delta\mathbf{x}$ changes also $\bar{\mathbf{x}}$, implying that in the sum over all sliced actions, the $\Delta\mathbf{x}$ are not independent. This is only achieved by the re-expanded postpoint interaction [see 10.178]. In the continuum, we shall indicate the postpoint product as before in (18.231) by a retardation symbol R, and rewrite (18.645) as

$$\mathcal{A}_{\text{em}} = \int_{t_a}^{t} dt \left[\mathbf{A}(\mathbf{x}(t))\dot{\mathbf{x}}^R(t) - i\epsilon\frac{\hbar}{2M}\boldsymbol{\nabla}\cdot\mathbf{A}(\mathbf{x}(t)) \right]. \tag{18.647}$$

In the theory of stochastic differential equations, this postpoint expression is called an *Itō integral*. The ordinary midpoint integral (18.645) is referred to as *Stratonovich integral*.

The Itō integral can now be added to the action in (18.644) with $\dot{\mathbf{x}}(t)$ replaced by $\mathbf{v}(t)$, and we obtain

$$(\mathbf{x}_b t_b | \mathbf{x}_a t_a) = \int \mathcal{D}\mathbf{v} \exp\left\{ \frac{i}{\hbar}\int_{t_a}^{t_b} dt \left[\frac{M}{2}\mathbf{v}^2(t) + \mathbf{A}(\mathbf{x}_\mathbf{v}(t_b,t))\mathbf{v}^R(t) - V(\mathbf{x}_\mathbf{v}(t_b,t)) \right] \right\}$$
$$\times \exp\left\{ \frac{i}{\hbar}\int_{t_a}^{t_b} dt \left[-i\epsilon\frac{\hbar}{2M}\boldsymbol{\nabla}\cdot\mathbf{A}(\mathbf{x}_\mathbf{v}(t_b,t)) \right] \right\} \delta^{(D)}\left(\mathbf{x}_a - \mathbf{x}_b + \int_{t_a}^{t_b} dt \, \mathbf{v}(t)\right). \tag{18.648}$$

Expanding the functional integrand in powers of $\mathbf{v}(t)$ as in (18.403) and using the correlation functions (18.642) we obtain the Schrödinger equation

$$i\partial_t(\mathbf{x}\,t | \mathbf{x}_a t_a) = \left[-\frac{\hbar^2}{2M}[\boldsymbol{\nabla} - i\mathbf{A}(\mathbf{x})]^2 + V(\mathbf{x}) \right] (\mathbf{x}\,t | \mathbf{x}_a t_a). \tag{18.649}$$

The advantage of the Itō integral is that such a calculation becomes quite simple using the correlation functions (18.642). The integral itself, however, is awkward to handle since it cannot be modified by partial integration. This is only possible for the ordinary, Stratonovich integral.

18.24 Real Stochastic and Deterministic Equation for Schrödinger Wave Function

The noise variable in the previous stochastic differential equation had an imaginary correlation function (18.629). It is possible to set up a completely real stochastic differential equation and modify this into a simple deterministic model which possesses the quantum properties of a particle in an arbitrary potential. In particular, the model has a discrete energy spectrum with a definite ground state energy, in this respect going beyond an earlier model by 't Hooft [36], whose spectrum was unbounded from below.

Let $\mathbf{u}(\mathbf{x}) = \left(u^1(\mathbf{x}), u^2(\mathbf{x})\right)$ be a time-independent field in two dimensions to be called *mother field*. The reparametrization freedom of the spatial coordinates is fixed by choosing harmonic coordinates in which

$$\boldsymbol{\nabla}^2 u(\mathbf{x}) = 0, \tag{18.650}$$

where $\boldsymbol{\nabla}^2$ is the Laplace operator. Equivalently, the components $u^1(\mathbf{x})$ and $u^2(\mathbf{x})$ may be assumed to satisfy the Cauchy-Riemann equations

$$\partial_\mu u^\nu = \epsilon_\mu{}^\rho \epsilon^\nu{}_\sigma \partial_\rho u^\sigma, \quad (\mu, \nu, \ldots = 1, 2), \tag{18.651}$$

where $\epsilon_{\mu\nu}$ is the antisymmetric Levi-Civita pseudotensor. The metric is $\delta_{\mu\nu}$, so that indices can be sub- or superscripts.

18.24.1 Stochastic Differential Equation

Consider now a point particle in contact with a heat bath of "temperature" $\hbar$. Its classical orbit $\mathbf{x}(t)$ is assumed to follow a stochastic differential equation consisting of a fixed rotation and a random translation in the diagonal direction $\mathbf{n} \equiv (1, 1)$:

$$\dot{\mathbf{x}}(t) = \boldsymbol{\omega} \times \mathbf{x}(t) + \mathbf{n}\,\eta(t), \tag{18.652}$$

where $\boldsymbol{\omega}$ is the rotation vector of length ω pointing orthogonal to the plane, and $\eta(t)$ a white-noise variable with zero expectation and the correlation function

$$\langle \eta(t)\eta(t') \rangle = \hbar\,\delta(t - t'). \tag{18.653}$$

For a particle starting at $\mathbf{x}(0) = \mathbf{x}$, the position $\mathbf{x}(t)$ at a later time t is a function of $\mathbf{x}$ and a *functional* of the noise variable $\eta(t')$ for $0 < t' < t$:

$$\mathbf{x}(t) = \mathbf{X}_\eta(\mathbf{x}, t). \tag{18.654}$$

As earlier in Section 18.13, a subscript η is used to indicate the functional dependence on the noise variable ν.

We now use the orbits ending at all possible final points $\mathbf{x} = \mathbf{x}(t)$ to define a time-dependent field $\mathbf{u}(\mathbf{x}; t)$ which is equal to $\mathbf{u}(\mathbf{x})$ at $t = 0$, and evolves with time as follows:

$$\mathbf{u}(\mathbf{x}; t) = \mathbf{u}_t[\mathbf{x}; \eta] \equiv \mathbf{u}\left(\mathbf{X}_0[t, \mathbf{x}; \eta]\right), \tag{18.655}$$

where the notation $\mathbf{u}_t[\mathbf{x}; \eta]$ indicates the variables as in (18.654).

As a consequence of the dynamic equation (20A.1), the change of the field $\mathbf{u}(\mathbf{x}, t)$ in a small time interval from $t = 0$ to $t = \Delta t$ has the expansion

$$\Delta \mathbf{u}_\eta(\mathbf{x}, 0) = \Delta t\,[\boldsymbol{\omega} \times \mathbf{x}] \cdot \boldsymbol{\nabla}\,\mathbf{u}_\eta(\mathbf{x}, 0) + \int_0^{\Delta t} dt'\,\eta(t')\,(\mathbf{n} \cdot \boldsymbol{\nabla})\,\mathbf{u}_\eta(\mathbf{x}, 0)$$

$$+ \frac{1}{2} \int_0^{\Delta t} dt' \int_0^{\Delta t} dt''\,\eta(t')\eta(t'')\,(\mathbf{n} \cdot \boldsymbol{\nabla})^2\,\mathbf{u}_\eta(\mathbf{x}, 0) + \ldots\,. \tag{18.656}$$

The omitted terms are of order $\Delta t^{3/2}$.

18.24.2 Equation for Noise Average

We now perform the noise average of Eq. (20A.2), defining the average field

$$\mathbf{u}(\mathbf{x}, t) \equiv \langle \mathbf{u}_\eta(\mathbf{x}, t) \rangle. \tag{18.657}$$

Using the vanishing average of $\eta(t)$ and the correlation function (18.653), we obtain in the limit $\Delta t \to 0$ the time derivative

$$\partial_t \mathbf{u}(\mathbf{x}, t) = \hat{\mathcal{H}} \mathbf{u}(\mathbf{x}, t), \quad \text{at} \ \ t = 0, \tag{18.658}$$

with the time evolution operator

$$\hat{\mathcal{H}} \equiv \{[\boldsymbol{\omega} \times \mathbf{x}] \cdot \boldsymbol{\nabla}\} + \frac{\hbar}{2}(\mathbf{n} \cdot \boldsymbol{\nabla})^2. \tag{18.659}$$

The average over η has made the operator $\hat{\mathcal{H}}$ time-independent. For this reason, the average field $\mathbf{u}(\mathbf{x}, t)$ at an arbitrary time t is obtained by the operation

$$\mathbf{u}(\mathbf{x}, t) = \hat{\mathcal{U}}(t) \mathbf{u}(\mathbf{x}, 0), \tag{18.660}$$

where $\hat{\mathcal{U}}(t)$ is a simple exponential

$$\hat{\mathcal{U}}(t) \equiv e^{\hat{\mathcal{H}}t}, \tag{18.661}$$

as follows immediately from (20A.11) and the trivial property $\hat{\mathcal{H}}\hat{\mathcal{U}}(t) = \hat{\mathcal{U}}(t)\hat{\mathcal{H}}$.

Note that the operator $\hat{\mathcal{H}}$ commutes with the Laplace operator $\boldsymbol{\nabla}^2$, thus ensuring that the harmonic property (18.650) of $\mathbf{u}(\mathbf{x})$ remains true for all times, i.e.,

$$\boldsymbol{\nabla}^2 \mathbf{u}(\mathbf{x}, t) \equiv 0. \tag{18.662}$$

18.24.3 Harmonic Oscillator

We now show that Eq. (20A.11) describes the quantum mechanics of a harmonic oscillator. Let us restrict our attention to the line with arbitrary $x_1 \equiv x$ and $x_2 = 0$. Applying the Cauchy-Riemann equations (18.651), we can rewrite Eq. (20A.11) in the pure x-form

$$\partial_t u^1(x, t) = \omega x \, \partial_x u^2(x, t) - \frac{\hbar}{2}\partial_x^2 u^2(x, t), \tag{18.663}$$

$$\partial_t u^2(x, t) = -\omega x \, \partial_x u^1(x, t) + \frac{\hbar}{2}\partial_x^2 u^1(x, t), \tag{18.664}$$

where we have omitted the second spatial coordinates $x_2 = 0$. Now we introduce a complex field

$$\psi(x, t) \equiv e^{-\omega x^2/2\hbar} \left[u^1(x, t) + i u^2(x, t) \right]. \tag{18.665}$$

This satisfies the differential equation

$$i\hbar \partial_t \psi(x, t) = \left(-\frac{\hbar^2}{2}\partial_x^2 + \frac{\omega^2}{2}x^2 - \frac{\hbar\omega}{2} \right) \psi(x, t), \tag{18.666}$$

which is the Schrödinger equation of a harmonic oscillator with the discrete energy spectrum $E_n = (n + 1/2)\hbar\omega$, $n = 0, 1, 2, \ldots$.

18.24.4 General Potential

The method can easily be generalized to an arbitrary potential. We simply replace (20A.1) by

$$\dot{x}^1(t) = -\partial_2 S^1(\mathbf{x}(t)) + n^1 \eta(t),$$
$$\dot{x}^2(t) = -\partial_1 S^1(\mathbf{x}(t)) + n^2 \eta(t), \tag{18.667}$$

where $\mathbf{S}(\mathbf{x})$ shares with $\mathbf{u}(\mathbf{x})$ the harmonic property (18.650):

$$\boldsymbol{\nabla}^2 \mathbf{S}(\mathbf{x}) = 0, \tag{18.668}$$

i.e., the functions $S^\mu(\mathbf{x})$ with $\mu = 1, 2$ fulfill Cauchy-Riemann equations like $u^\mu(\mathbf{x})$ in (18.651). Repeating the above steps we find, instead of the operator (18.659),

$$\hat{\mathcal{H}} \equiv -(\partial_2 S^1)\partial_1 - (\partial_1 S^1)\partial_2 + \frac{\hbar}{2}(\mathbf{n} \cdot \boldsymbol{\nabla})^2, \tag{18.669}$$

and Eqs. (18.663) and (20A.14) become:

$$\partial_t u^1(x,t) = (\partial_x S^1)\partial_x u^2(x,t) - \frac{\hbar}{2}\partial_x^2 u^2(x,t), \tag{18.670}$$

$$\partial_t u^2(x,t) = -(\partial_x S^1)\partial_x u^1(x,t) + \frac{\hbar}{2}\partial_x^2 u^1(x,t). \tag{18.671}$$

This time evolution preserves the harmonic nature of $\mathbf{u}(\mathbf{x})$. Indeed, using the harmonic property $\boldsymbol{\nabla}^2 \mathbf{S}(\mathbf{x}) = 0$ we can easily derive the following time dependence of the Cauchy-Riemann combinations in Eq. (18.651):

$$\partial_t(\partial_1 u^1 - \partial_2 u^2) = \hat{\mathcal{H}}(\partial_1 u^1 - \partial_2 u^2) - \partial_2\partial_1 S^1(\partial_1 u^1 - \partial_2 u^2) + \partial_2^2 S^1(\partial_2 u^1 + \partial_1 u^2),$$

$$\partial_t(\partial_2 u^1 + \partial_1 u^2) = \hat{\mathcal{H}}(\partial_2 u^1 + \partial_1 u^2) - \partial_2\partial_1 S^1(\partial_2 u^1 + \partial_1 u^2) - \partial_2^2 S^1(\partial_1 u^1 - \partial_2 u^2).$$

Thus $\partial_1 u^1 - \partial_2 u^2$ and $\partial_2 u^1 + \partial_1 u^2$ which are zero at any time remain zero at all times.

On account of Eqs. (18.671), the combination

$$\psi(x,t) \equiv e^{-S^1(x)/\hbar}\left[u^1(x,t) + iu^2(x,t)\right] \tag{18.672}$$

satisfies the Schrödinger equation

$$i\hbar\partial_t\psi(x,t) = \left[-\frac{\hbar^2}{2}\partial_x^2 + V(x)\right]\psi(x,t), \tag{18.673}$$

where the potential is related to $S^1(x)$ by the Riccati differential equation

$$V(x) = \frac{1}{2}[\partial_x S^1(x)]^2 - \frac{\hbar}{2}\partial_x^2 S^1(x). \tag{18.674}$$

The harmonic oscillator is recovered for the pair of functions

$$S^1(\mathbf{x}) + iS^2(\mathbf{x}) = \omega(x^1 + ix^2)^2/2. \tag{18.675}$$

18.24.5 Deterministic Equation

The noise $\eta(t)$ in the stochastic differential equation Eq. (18.667) can also be replaced by a source composed of deterministic classical oscillators $q_k(t)$, $k = 1, 2, \ldots$ with the equations of motion

$$\dot{q}_k = p_k, \qquad \dot{p}_k = -\omega_k^2 q_k, \tag{18.676}$$

as

$$\eta(t) \equiv \sum_k \dot{q}_k(t). \tag{18.677}$$

The initial positions $q_k(0)$ and momenta $p_k(0)$ are assumed to be randomly distributed with a Boltzmann factor $e^{-\beta H_{\text{osc}}/\hbar}$, such that

$$\langle q_k(0)q_k(0)\rangle = \hbar/\omega_k^2, \qquad \langle p_k(0)p_k(0)\rangle = \hbar. \tag{18.678}$$

Using the equation of motion

$$\dot{q}_k(t) = \omega_k q_k(0)\sin\omega_k t + p_k(0)\sin\omega_k t, \tag{18.679}$$

we find the correlation function

$$
\begin{aligned}
\langle \dot{q}_k(t)\dot{q}_k(t')\rangle &= \omega_k^2 \cos\omega_k t \cos\omega_k t' \langle q_k(0)q_k(0)\rangle + \sin\omega_k t \sin\omega_k t\langle p_k(0)p_k(0)\rangle \\
&= \cos\omega_k(t - t'). \tag{18.680}
\end{aligned}
$$

We may now assume that the oscillators $q_k(t)$ are the Fourier components of a massless field, for instance the gravitational field whose frequencies are $\omega_k = k$, and whose random initial conditions are caused by the big bang. If the sum over k is simply a momentum integral $\int_{-\infty}^{\infty} dk$, then (18.680) yields a white-noise correlation function (18.653) for $\eta(t)$.

Thus it is indeed possible to simulate the quantum-mechanical wave functions $\psi(x, t)$ and the energy spectrum of an arbitrary potential problem by deterministic equations with random initial conditions at the beginning of the universe.

It remains to solve the open problem of finding a classical origin of the second important ingredient of quantum theory: the theory of quantum measurement to be extracted from the wave function $\psi(x, t)$. Only then shall we understand how God throws dice [37].

Appendix 18A   Inequalities for Diagonal Green Functions

Let us introduce several diagonal Green functions consisting of thermal averages of equal-time commutators and anticommutators of bosonic and fermionic field operators, elementary or composite. For brevity, we write

$$
\langle \ldots \rangle_T = \mathrm{Tr}\left[\exp(-\hat{H}/T)\ldots\right]\Big/ \mathrm{Tr}\left[\exp(-\hat{H}/T)\right] = \langle \ldots \rangle_T \tag{18A.1}
$$

and define the averages with obvious spectral representations

$$
\begin{aligned}
c &\equiv \langle [\hat{\psi}, \hat{\psi}^\dagger]_\mp \rangle_T = \int_{-\infty}^{\infty} \frac{d\omega}{2\pi} \rho_{12}(\omega), \\
a &\equiv \langle [\hat{\psi}, \hat{\psi}^\dagger]_\pm \rangle_T = \int_{-\infty}^{\infty} \frac{d\omega}{2\pi} \rho_{12}(\omega) \tanh^{\mp 1}\frac{\omega}{2T}.
\end{aligned} \tag{18A.2}
$$

We shall also introduce a quantity obtained by integrating the imaginary-time Green function over a period $\tau \in [0, 1/T)$. This gives for boson and fermion fields the nonnegative expression [see (18.23)]

$$
\begin{aligned}
g &\equiv G(\omega_m = 0) = \int_0^{1/T} d\tau \, \langle \hat{\psi}(\tau)\hat{\psi}^\dagger(0)\rangle_T \\
&= \int_{-\infty}^{\infty} \frac{d\omega}{2\pi} \rho_{12}(\omega) \frac{1}{\omega} \left\{ \begin{matrix} 1 \\ \tanh(\omega/2T) \end{matrix} \right\} \geq 0. \tag{18A.3}
\end{aligned}
$$

Note that for fermion fields, the spectral weight in this integral is accompanied by an extra factor $\tanh(\omega/2T)$. This is due to the fact that $\omega_m = 0$ is *no* fermionic Matsubara frequency, but a "wrong" bosonic one. In fact, the sum (18.23) for $G(\omega_m)$ contains a factor $1 - e^{(E_n - E_{n'})/T}$ for both bosons and fermions, while $\rho_{12}(\omega)$ in the spectral representation (18A.3) introduces, via (18.59), a factor $1 - e^{-\omega/T}$ for bosons and a factor $1 + e^{\omega/T}$ for fermions, thus explaining the relative factor $\tanh(\omega/2T)$ in (18A.3).

The integration over τ leads to the factor $1/\omega$ in (18A.3). This factor is also found by integrating the retarded Green function $G_{12}(t)$ and the commutator function $C_{12}(t)$ over all real times, resulting in the spectral representations

$$
\begin{aligned}
i\int_{-\infty}^{\infty} dt\, \Theta(t)\langle [\hat{\psi}(t), \hat{\psi}^\dagger(0)]_\mp \rangle_T &= \int \frac{d\omega}{2\pi} \rho_{12}(\omega)\frac{1}{\omega}, \\
i\int_{-\infty}^{\infty} dt\, \Theta(t)\langle [\hat{\psi}(t), \hat{\psi}^\dagger(0)]_\pm \rangle_T &= \int \frac{d\omega}{2\pi} \rho_{12}(\omega)\frac{1}{\omega} \tanh^{\mp 1}\frac{\omega}{2T}.
\end{aligned} \tag{18A.4}
$$

Another set of thermal expectation values involves products of field operators with time derivatives rather than integrals. Their spectral representations contain an extra factor ω. For example, the τ-derivative of the expectation value in (18A.3) leads to

$$-\langle \dot{\hat{\psi}}(0), \hat{\psi}^\dagger(0)\rangle_T = \int_{-\infty}^\infty \frac{d\omega}{2\pi} \rho_{12}(\omega)\omega(1 \pm n_\omega). \tag{18A.5}$$

The real-time derivatives of the expectation values in (18A.4) have the spectral integrals

$$
\begin{aligned}
d &\equiv i\langle [\dot{\hat{\psi}}(0), \hat{\psi}^\dagger(0)]_\mp\rangle_T = \int_{-\infty}^\infty \frac{d\omega}{2\pi} \rho_{12}(\omega)\omega, \\
e &\equiv i\langle [\dot{\hat{\psi}}(0), \hat{\psi}^\dagger(0)]_\pm\rangle_T = \int_{-\infty}^\infty \frac{d\omega}{2\pi} \rho_{12}(\omega)\omega \tanh^{\mp 1} \frac{\omega}{2T}.
\end{aligned}
\tag{18A.6}
$$

The expectation values c, a, g, d, e satisfy several rigorous inequalities. To derive these, we observe that

$$\mu(\omega) = \frac{1}{g}\frac{1}{2\pi}\rho_{12}(\omega)\frac{1}{\omega}\left\{ \begin{array}{c} 1 \\ \tanh(\omega/2T) \end{array} \right\} \tag{18A.7}$$

is a positive function. This follows directly from (18.59), according to which $\rho_{12}(\omega)$ at negative ω is negative for bosons and positive for fermions. Having divided out the total integral g defined in (18A.3), the integral over $\mu(\omega)$ is normalized to unity,

$$\int_{-\infty}^\infty d\omega\, \mu(\omega) = 1, \tag{18A.8}$$

for both bosons and fermions. Using $\mu(\omega)$, we form the following ratios:

$$\frac{c}{g} = \int_{-\infty}^\infty d\omega\, \mu(\omega)\omega \left\{ \begin{array}{c} 1 \\ \coth(\omega/2T) \end{array} \right\}, \tag{18A.9}$$

$$\frac{a}{g} = \int_{-\infty}^\infty d\omega\, \mu(\omega)\omega \left\{ \begin{array}{c} \coth(\omega/2T) \\ 1 \end{array} \right\}, \tag{18A.10}$$

$$\frac{d}{g} = \int_{-\infty}^\infty d\omega\, \mu(\omega)\omega^2 \left\{ \begin{array}{c} 1 \\ \coth(\omega/2T) \end{array} \right\}, \tag{18A.11}$$

$$\frac{e}{g} = \int_{-\infty}^\infty d\omega\, \mu(\omega)\omega^2 \left\{ \begin{array}{c} \coth(\omega/2T) \\ 1 \end{array} \right\}. \tag{18A.12}$$

The inequalities to be derived are based on the Jensen-Peierls inequality for convex functions derived in Chapter [5]. Recall that a convex function $f(\omega)$ satisfies

$$f\left(\frac{\omega_1 + \omega_2}{2}\right) \le \frac{f(\omega_1) + f(\omega_2)}{2}, \tag{18A.13}$$

which is generalized to

$$f\left(\sum_i \mu_i \omega_i\right) \le \sum_i \mu_i f(\omega_i), \tag{18A.14}$$

where μ_i is an arbitrary set of positive numbers with $\sum_i \mu_i = 1$. In the continuum limit, this becomes

$$f\left(\int_{-\infty}^\infty d\omega\, \mu(\omega)\omega\right) \le \int_{-\infty}^\infty d\omega\, \mu(\omega)f(\omega). \tag{18A.15}$$

It is obvious that a similar Jensen-Peierls inequality holds also for concave functions with inequality sign in the opposite direction.

The Jensen-Peierls inequality (18A.15) is now applied to the function

$$f(\omega) = \omega \coth \frac{\omega}{2T}, \tag{18A.16}$$

which looks like a slightly distorted hyperbola coming from infinity along the diagonal lines $|\omega|$ and crossing the f-axis at $\omega = 0$, $f(0) = 2T$. The second derivative of $f(\omega)$ is positive everywhere, ensuring the convexity. The function (18A.16) appears in the integrand of the boson part of Eq. (18A.10). The right-hand side of (18A.15) can therefore be written as a/g. The left-hand side is obviously equal to $(c/g) \coth(c/2Tg)$. Hence we arrive at the inequality

$$c \coth \frac{c}{2Tg} \leq a. \tag{18A.17}$$

In terms of the original field operators, this amounts to

$$\langle [\hat{\psi}, \hat{\psi}^\dagger]_+ \rangle_T \geq \langle [\hat{\psi}, \hat{\psi}^\dagger]_- \rangle_T \coth \frac{\langle [\hat{\psi}, \hat{\psi}^\dagger]_- \rangle_T}{2T \int_0^{1/T} d\tau \, \langle \hat{\psi}(\tau) \hat{\psi}^\dagger(0) \rangle_T}. \tag{18A.18}$$

In the special case that $\hat{\psi}$ is a *canonical* interacting boson field of momentum $\mathbf{p}$, the commutator is simply $[\hat{\psi}, \hat{\psi}^\dagger]_- = 1$, and the inequality becomes

$$1 + 2\langle \hat{\psi}_{\mathbf{p}}^\dagger \hat{\psi}_{\mathbf{p}} \rangle_T \geq \coth(1/2Tg) = 1 + \frac{2}{e^{1/gT} - 1} = 1 + 2n_{g^{-1}}, \tag{18A.19}$$

i.e.,

$$\langle \hat{\psi}_{\mathbf{p}}^\dagger \hat{\psi}_{\mathbf{p}} \rangle_T \geq \frac{1}{e^{1/gT} - 1} \equiv n_{g^{-1}}, \tag{18A.20}$$

where $n_{g^{-1}}$ is the free-boson distribution function (18.36) for an energy g^{-1}.

This is quite an interesting relation. The quantity g is the Euclidean equilibrium Green function $G(\omega_m, \mathbf{p})$ at $\omega_m = 0$. For free particles in contact with a reservoir, it is given by

$$g^{-1} = G(0, \mathbf{p})^{-1} = \frac{\mathbf{p}^2}{2M} - \mu \equiv \xi(\mathbf{p}), \tag{18A.21}$$

i.e., it is equal to the particle energy measured with respect to the chemical potential μ. Moreover, we know that for free particles

$$\langle \hat{\psi}_{\mathbf{p}}^\dagger \hat{\psi}_{\mathbf{p}} \rangle_T = n_{\xi(\mathbf{p})}, \tag{18A.22}$$

so that the inequality (18A.20) becomes an equality. The content of the inequality (18A.20) may therefore be phrased as follows: For any interaction, the occupation of a state with momentum $\mathbf{p}$ is never *smaller* than for a free boson level of energy $g^{-1} = G(0, \mathbf{p})$.

Another inequality can be derived from the concave function

$$\bar{f}(y) = \sqrt{y} \coth \frac{\sqrt{y}}{2T}, \tag{18A.23}$$

using $y = \omega^2$ and the measure

$$\int_{-\infty}^{\infty} d\omega \, \mu(\omega) = \int_0^{\infty} \frac{dy}{\sqrt{y}} \mu(\sqrt{y}) = 1. \tag{18A.24}$$

As argued before, concave functions satisfy the inequality opposite to (18A.15), from which we derive the inequality

$$\bar{f}\left(\int_0^\infty \frac{dy}{\sqrt{y}}\,\mu(\sqrt{y})y\right) \geq \int_0^\infty \frac{dy}{\sqrt{y}}\,\mu(\sqrt{y})\bar{f}(y), \tag{18A.25}$$

which can be rewritten as

$$\bar{f}\left(\int_{-\infty}^\infty d\omega\,\mu(\omega)\omega^2\right) \geq \int_{-\infty}^\infty d\omega\,\mu(\omega)\bar{f}(\omega^2). \tag{18A.26}$$

Again, the right-hand side is a/g, but now it is bounded from above by

$$\frac{a}{g} \leq \sqrt{\frac{d}{g}}\,\coth\left(\frac{1}{2T}\sqrt{\frac{d}{g}}\right). \tag{18A.27}$$

The combined inequality

$$c\,\coth\frac{c}{2Tg} \leq a \leq \sqrt{dg}\,\coth\left(\frac{1}{2T}\sqrt{\frac{d}{g}}\right) \tag{18A.28}$$

may be used to derive further inequalities:

$$\begin{aligned}
c^2 &\leq dg, \\
c\coth(d/2Tc) &\leq a, \\
g &\leq \coth(c/2Ta), \\
c &\leq d\tanh(c/2Ta), \\
c &\leq a\tanh(d/2Tc).
\end{aligned} \tag{18A.29}$$

For fermion fields we see that an inequality like (18A.17) holds with c and a interchanged, i.e.,

$$a\coth\frac{a}{2Tg} \leq c, \tag{18A.30}$$

which leads to

$$\langle[\hat{\psi},\hat{\psi}^\dagger]_-\rangle_T \leq \langle[\hat{\psi},\hat{\psi}^\dagger]_+\rangle_T \tanh\frac{\langle[\hat{\psi},\hat{\psi}^\dagger]_+\rangle_T}{2T\int_0^{1/T} d\tau\,\langle\hat{\psi}(\tau)\hat{\psi}^\dagger(0)\rangle_T}. \tag{18A.31}$$

For canonical fermion fields with $[\hat{\psi},\hat{\psi}^\dagger]_+ = 1$, this becomes

$$1 - 2\langle\hat{\psi}^\dagger\hat{\psi}\rangle_T \leq \tanh(1/2gT) = 1 - \frac{2}{e^{1/gT}+1}, \tag{18A.32}$$

i.e., the fermionic counterpart of (18A.20):

$$\langle\hat{\psi}_{\mathbf{p}}^\dagger\hat{\psi}_{\mathbf{p}}\rangle_T \leq \frac{1}{e^{1/gT}+1} = n_{g^{-1}}, \tag{18A.33}$$

where $n_{g^{-1}}$ is the free-fermion distribution function (18.36) at an energy g^{-1}. As in the Bose case, free particles fulfill

$$\langle\hat{\psi}_{\mathbf{p}}^\dagger\hat{\psi}_{\mathbf{p}}\rangle_T = n_{\xi(\mathbf{p})}, \tag{18A.34}$$

with $g^{-1} = \xi(\mathbf{p})$, so that the inequality (18A.33) becomes an equality. The inequality implies that an interacting Fermi level is never occupied *more* than a free fermion level of energy $g^{-1} = G(0,\mathbf{p})^{-1}$.

Also the second inequality in (18A.29) can be taken over to fermions which amounts to (18A.30), but with a and d replaced by c and e.

Appendix 18B   General Generating Functional

For a field operator $a(t)$ of frequency Ω and its Hermitian conjugate $a^\dagger(t)$, the retarded and advanced Green functions and the expectation values of commutators and anticommutators were derived in Eqs. (18.68)–(18.77):

$$
\begin{aligned}
G_\Omega^R(t,t') &= \Theta(t-t')e^{-i\Omega(t-t')}, \\
G_\Omega^A(t,t') &= -\Theta(t'-t)e^{-i\Omega(t-t')}, \\
C_\Omega(t,t') &= e^{-i\Omega(t-t')}, \\
A_\Omega(t,t') &= \left(\tanh\frac{\Omega}{2T}\right)^{\mp 1} e^{-i\Omega(t-t')}.
\end{aligned}
\tag{18B.1}
$$

Introducing complex sources $\eta(t)$ and $\eta^\dagger(t)$ associated with these operators, the generating functional for these functions is

$$
Z_0[\eta_{\rm P}, \eta_{\rm P}^\dagger] = {\rm Tr}\left\{ \hat\rho\, \hat T_{\rm P} \exp\left[-i\int_{t_a}^{t_b} dt(\hat a^\dagger \eta + \eta^\dagger \hat a)\right] \right\}.
\tag{18B.2}
$$

The complex sources are distinguished according to the closed-time contour branches by a subscript P. The generating functional can then be written down immediately as

$$
Z_0[\eta_{\rm P}, \eta_{\rm P}^\dagger] = \exp\left\{ -\int dt \int dt'\, \eta_{\rm P}^\dagger(t) G_{\rm P}(t,t') \eta_{\rm P}(t') \right\},
\tag{18B.3}
$$

generalizing (18.179), where the matrix

$$
G_{\rm P} = \frac{1}{2}\left(
\begin{array}{cc}
A_\Omega + G_\Omega^R + G_\Omega^A & A_\Omega - G_\Omega^R + G_\Omega^A \\
A_\Omega + G_\Omega^R - G_\Omega^A & A_\Omega - G_\Omega^R - G_\Omega^A
\end{array}
\right)
\tag{18B.4}
$$

contains the following operator expectations on the two time branches:

$$
\begin{aligned}
G_{\rm P}(t,t') &= \left(
\begin{array}{cc}
\langle \hat T_{\rm P} \hat a_H(t_+)\hat a_H^\dagger(t'_+)\rangle_T & \langle \hat T_{\rm P} \hat a_H(t_+)\hat a_H^\dagger(t'_-)\rangle_T \\
\langle \hat T_{\rm P} \hat a_H(t_-)\hat a_H^\dagger(t'_+)\rangle_T & \langle \hat T_{\rm P} \hat a_H(t_-)\hat a_H^\dagger(t'_-)\rangle_T
\end{array}
\right) \\
&= \left(
\begin{array}{cc}
\langle \hat T \hat a_H(t_+)\hat a_H^\dagger(t'_+)\rangle_T & \pm\langle \hat a_H^\dagger(t'_-)\hat a_H(t_+)\rangle_T \\
\langle \hat a_H(t_-)\hat a_H^\dagger(t'_+)\rangle_T & \langle \hat{\bar T} \hat a_H(t_-)\hat a_H^\dagger(t'_-)\rangle_T
\end{array}
\right).
\end{aligned}
\tag{18B.5}
$$

Note that $\hat a_H(t_+), \hat a_H^\dagger(t_+)$ and $\hat a_H(t_-), \hat a_H^\dagger(t_-)$ obey the Heisenberg equations of motion with the Hamiltonians

$$
\begin{aligned}
\hat H_+ &\equiv \frac{\Omega}{2}\left[\hat a_H^\dagger(t_+)\hat a_H(t_+) \pm \hat a_H(t_+)\hat a_H^\dagger(t_+)\right], \\
\hat H_- &\equiv -\frac{\Omega}{2}\left[\hat a_H^\dagger(t_-)\hat a_H(t_-) \pm \hat a_H(t_-)\hat a_H^\dagger(t_-)\right].
\end{aligned}
\tag{18B.6}
$$

In the second-quantized field interpretation, they read

$$
\begin{aligned}
\hat H_+ &\equiv \frac{\Omega}{2}\hat a_H^\dagger(t_+)\hat a_H(t_+), \\
\hat H_- &\equiv -\frac{\Omega}{2}\hat a_H^\dagger(t_-)\hat a_H(t_-).
\end{aligned}
$$

The explicit time dependence of the matrix elements of $G_{\rm P}(t,t')$ in Eq. (18B.5) is

$$
G_{\rm P}(t,t') = e^{-i\Omega(t-t')}\left(
\begin{array}{cc}
\Theta(t-t') \pm n_\Omega & \pm n_\Omega \\
1 \pm n_\Omega & \Theta(t'-t) \pm n_\Omega
\end{array}
\right),
\tag{18B.7}
$$

with $n_\Omega = (e^{\Omega/T} \pm 1)^{-1}$.

This Green function can, incidentally, be decomposed as

$$G_P^0(t,t') + G_P^N(t,t'),\tag{18B.8}$$

where $G_P^0(t,t')$ is the Green function at zero temperature, i.e., the expression (18B.7) for $n_\Omega \equiv 0$. The matrix $G^N(t,t')$ contains the expectations of the *normal products*:

$$\begin{aligned}
G_P^N(t,t') &\equiv \begin{pmatrix} \langle \hat{N}\hat{a}_H(t_+)\hat{a}_H^\dagger(t'_+)\rangle_T & \langle \hat{N}\hat{a}_H(t_+)\hat{a}_H^\dagger(t'_-)\rangle_T \\ \langle \hat{N}\hat{a}_H(t_-)\hat{a}_H^\dagger(t'_+)\rangle_T & \langle \hat{N}\hat{a}_H(t_-)\hat{a}_H^\dagger(t'_-)\rangle_T \end{pmatrix} \\
&\equiv \pm \begin{pmatrix} \langle \hat{a}_H^\dagger(t'_+)\hat{a}_H(t_+)\rangle_T & \langle \hat{a}_H^\dagger(t'_-)\hat{a}_H(t_+)\rangle_T \\ \langle \hat{a}_H^\dagger(t'_+)\hat{a}_H(t_-)\rangle_T & \langle \hat{a}_H^\dagger(t'_-)\hat{a}_H(t_-)\rangle_T \end{pmatrix}.
\end{aligned}\tag{18B.9}$$

For an arbitrary product of operators, the normal product $\hat{N}(\ldots)$ is defined by reordering the operators so that all annihilation operators come to act first upon the state on the right-hand side. At the end, the product receives the phase factor $(-)^F$, where F is the number of fermion permutations to arrive at the normal order. A similar decomposition exists at the operator level *before* taking expectation values. For any pair of operators $\hat{A}(t)$, $\hat{B}(t')$ which are linear combinations of creation and annihilation operators, the time-ordered product can be decomposed as

$$\hat{T}\hat{A}(t)\hat{B}(t') = \langle \hat{T}\hat{A}(t)\hat{B}(t')\rangle_0 + \hat{N}\hat{A}(t)\hat{B}(t'),\tag{18B.10}$$

where $\langle \ldots \rangle_0 \equiv \mathrm{Tr}\left(|0\rangle\langle 0|\ldots\right)$ denotes the zero-temperature expectation. This decomposition is proved in Appendix 18C, where it is also generalized to products of more than two operators.

Let us also go to the Keldysh basis here:

$$\tilde{\eta}_P = Q\eta_P = \frac{1}{\sqrt{2}}\begin{pmatrix} 1 & -1 \\ 1 & 1 \end{pmatrix}\hat{\eta}_P.\tag{18B.11}$$

Then the generating functional becomes [instead of (18.180)]:

$$\begin{aligned}
Z_0[\eta_P^*, \eta_P] &= \exp\left[-\int dt \int dt'\, (\eta_+^*, -\eta_-^*)Q^{-1}\begin{pmatrix} 0 & G_\Omega^A \\ G_\Omega^R & A \end{pmatrix}Q\begin{pmatrix} \eta_+ \\ -\eta_- \end{pmatrix}\right] \\
&= \exp\left\{-\frac{1}{2}\int dt \int dt'\,\left[(\eta_+^* - \eta_-^*)(t)G_\Omega^R(t-t')(\eta_+ + \eta_-)(t')\right.\right. \\
&\qquad\qquad\qquad + (\eta_+^* + \eta_-^*)(t)G_\Omega^A(t-t')(\eta_+ - \eta_-)(t') \\
&\qquad\qquad\qquad \left.\left. + (\eta_-^* - \eta_-^*)(t)A_\Omega(t-t')(\eta_+ - \eta_-)(t')\right]\right\},
\end{aligned}\tag{18B.12}$$

where we have used the notation

$$\eta_+(t) \equiv \eta(t_+), \quad \eta_-(t) \equiv \eta(t_-).\tag{18B.13}$$

Expression (18B.12) can be simplified [as before (18.180)] using the time reversal relation (18.62) in the form

$$A_\Omega(t,t') = \Theta(t,t')A_\Omega(t,t') \pm \Theta(t'-t)A_\Omega(t',t),\tag{18B.14}$$

leading to

$$\begin{aligned}
Z_0[\eta_P^*, \eta_P] = \exp\Big\{ &-\frac{1}{2}\int_{-\infty}^\infty dt \int_{-\infty}^t dt' \\
&\times \Big[(\eta_+ - \eta_-)^*(t)G_\Omega^R(t,t')(\eta_+ + \eta_-)(t') \\
&\quad - (\eta_+ - \eta_-)(t)G_\Omega^R(t,t')^*(\eta_+ + \eta_-)^*(t') \\
&\quad + (\eta_+ - \eta_-)^*(t)A_\Omega(t,t')(\eta_+ - \eta_-)(t') \\
&\quad + (\eta_+ - \eta_-)(t)A_\Omega(t,t')^*(\eta_+ - \eta_-)^*(t')\Big]\Big\}.
\end{aligned}\tag{18B.15}$$

For the case of a second-quantized field, this is the most useful generating functional.

The expression (18B.15) can be used to derive the generating functional for correlation functions between one or more $\varphi(t)$ and the associated canonically-conjugate momenta. As an example, consider immediately a harmonic oscillator with $\varphi(t) = x(t)$ and the momentum $p(t)$. We would like to find the generating functional

$$Z[j_P, k_P] = \mathrm{Tr} \left(\hat{\rho}\, \hat{T}_P \exp \left\{ i \int_P dx \, [j_P(t) x_P(t) + k_P(t) p_P(t)] \right\} \right). \tag{18B.16}$$

The position variable $x(t)$ is decomposed as in (18.92) into a sum of creation and an annihilation operators:

$$\hat{x}(t) = \sqrt{\frac{\hbar}{2M\Omega}} \left[\hat{a} e^{-i\Omega t} + \hat{a}^\dagger e^{i\Omega t} \right]. \tag{18B.17}$$

The inverse of this decomposition is

$$\left\{ \begin{array}{c} \hat{a} \\ \hat{a}^\dagger \end{array} \right\} = (M\Omega\, \hat{\varphi} \pm i\hat{p})/\sqrt{2M\Omega\hbar}, \tag{18B.18}$$

and there is an analogous relation of the complex sources:

$$\left\{ \begin{array}{c} \eta \\ \eta^\dagger \end{array} \right\} = (j \pm iM\Omega\, k)/\sqrt{2M\Omega\hbar}. \tag{18B.19}$$

Inserting these sources into (18B.15), we obtain the generating functional

$$Z_0[j_P, k_P] = \exp \left\{ -\frac{1}{2M\Omega} \int_{-\infty}^{\infty} dt \int_{-\infty}^{t} dt'\, (j_+ - j_-)(t) \right.$$
$$\times \left\{ [\mathrm{Re}\, A_\Omega(t,t') + i\mathrm{Im}\, G_\Omega^R(t,t')] j_+(t') \right.$$
$$\left. - [\mathrm{Re}\, A_\Omega(t,t') - i\mathrm{Im}\, G_\Omega^R(t,t')] j_-(t') \right\} \tag{18B.20}$$
$$-\frac{1}{2} \int_{-\infty}^{\infty} dt \int_{-\infty}^{t} dt'\, (k_+ - k_-)(t) \left\{ [\mathrm{Im}\, A_\Omega(t,t') - i\mathrm{Re}\, G_\Omega^R(t,t')] j_+(t') \right.$$
$$\left. \left. - [\mathrm{Im}\, A_\Omega(t,t') + i\mathrm{Re}\, G_\Omega^R(t,t')] j_-(t') \right\} + (j \leftrightarrow kM\Omega) \right\}.$$

Here it is useful to introduce the quantities

$$\alpha(t,t') = \frac{1}{2M\Omega} \left[\mathrm{Re}\, A_\Omega(t,t') + i\mathrm{Im}\, G_\Omega^R(t,t') \right],$$
$$\beta(t,t') = \frac{1}{2M\Omega} \left[\mathrm{Im}\, A_\Omega(t,t') - i\mathrm{Re}\, G_\Omega^R(t,t') \right]. \tag{18B.21}$$

Then the generating functional reads

$$Z_0[j_+, j_-, k_+, k_-] = \exp \left\{ -\int_{-\infty}^{\infty} dt \int_{-\infty}^{t} dt'\, (j_+ - j_-)(t)\, [\alpha(t,t') j_+(t') - \alpha^*(t,t') j_-(t')] + (j \leftrightarrow kM\Omega) \right.$$
$$\left. -M\Omega \int_{-\infty}^{\infty} dt \int_{-\infty}^{t} dt'\, (k_+ - k_-)(t)\, [\beta(t,t') j_+(t') - \beta^*(t,t') j_-(t')] + (j \leftrightarrow kM\Omega) \right\}. \tag{18B.22}$$

If the oscillator is coupled only to the real source j, i.e., if its generating functional reads

$$Z[j_P] = \mathrm{Tr} \left\{ \hat{\rho}\, \hat{T}_P \exp \left[i \int_P dx\, j_P(t) \hat{x}_P(t) \right] \right\}, \tag{18B.23}$$

we can drop all but the first line in the exponent of (18B.22) and have

$$Z_0[j_+, j_-] = \exp\left\{ -\int_{-\infty}^{\infty} dt \int_{-\infty}^{t} dt' \, (j_+ - j_-)(t) \left[\alpha(t,t')j_+(t') - \alpha^*(t,t')j_-(t')\right] \right\}.$$
(18B.24)

Since (18B.22) and (18B.24) contain only the causal temporal order $t > t'$, the retarded Green function $G_\Omega^R(t,t')$ in (18B.21) can be replaced by the expectation value of the commutator [see (18.40), (18.41), and (18.42)]. Thus, for $t > t'$, the functions $\alpha(t,t')$ and $\beta(t,t')$ are equal to[11]

$$\alpha(t,t') = \frac{1}{2M\Omega} \left[\mathrm{Re}\, A_\Omega(t,t') + i\mathrm{Im}\, C_\Omega(t,t')\right], \qquad t > t',$$

$$\beta(t,t') = \frac{1}{2M\Omega} \left[\mathrm{Im}\, A_\Omega(t,t') - i\mathrm{Re}\, C_\Omega(t,t')\right], \qquad t > t'.$$
(18B.25)

For a single oscillator of frequency Ω, we use the spectral function (18.74) properties (18.44) and (18.53) of $A_\Omega(t,t')$ and $C_\Omega(t,t')$, and find the simple expressions:

$$\alpha(t,t') = \frac{1}{2M\Omega} \left[\mathrm{Re}\, e^{-i\Omega(t-t')} \left\{ \begin{array}{c} \coth\frac{\Omega}{2T} \\ \tanh\frac{\Omega}{2T} \end{array} \right\} + i\mathrm{Im}\, e^{-i\Omega(t-t')} \right]$$

$$= \frac{1}{2M\Omega} \left[\cos\Omega(t-t') \left\{ \begin{array}{c} \coth\frac{\Omega}{2T} \\ \tanh\frac{\Omega}{2T} \end{array} \right\} - i\sin\Omega(t-t') \right],$$
(18B.26)

$$\beta(t,t') = \frac{1}{2M\Omega} \left[\mathrm{Im}\, e^{-i\Omega(t-t')} \left\{ \begin{array}{c} \coth\frac{\Omega}{2T} \\ \tanh\frac{\Omega}{2T} \end{array} \right\} - i\mathrm{Re}\, e^{-i\Omega(t-t')} \right]$$

$$= -\frac{1}{2M\Omega} \left[\sin\Omega(t-t') \left\{ \begin{array}{c} \coth\frac{\Omega}{2T} \\ \tanh\frac{\Omega}{2T} \end{array} \right\} + i\cos\Omega(t-t') \right].$$
(18B.27)

Note that real and imaginary parts of the functions $\bar{\alpha}(t-t')$ can be combined into a single expression $(\beta = 1/T)$

$$\bar{\alpha}(t-t') = \frac{1}{2M\Omega} \left\{ \begin{array}{ll} \dfrac{\cosh[\Omega(\beta/2 - i(t-t')]}{\sinh(\Omega\beta/2)} & \text{for \quad bosons,} \\[3mm] \dfrac{\sinh[\Omega(\beta/2 - i(t-t')]}{\cosh(\Omega\beta/2)} & \text{for \quad fermions.} \end{array} \right.$$
(18B.28)

The bosonic function agrees with the time-ordered Green function (18.101) for $t > t'$ and continues it analytically to $t < t'$.

In Fourier space, the functions (18B.26) and (18B.27) correspond to

$$\alpha(\omega') = \frac{\pi}{2M\Omega} \left(\coth^{\pm 1}\frac{\omega'}{2T} + 1\right) [\delta(\omega' - \Omega) - \delta(\omega' + \Omega)],$$

$$\beta(\omega') = -\frac{i\pi}{2M\Omega} \left(\coth^{\pm 1}\frac{\omega'}{2T} + 1\right) [\delta(\omega' - \Omega) + \delta(\omega' + \Omega)].$$

Let us split these functions into a zero-temperature contribution plus a remainder

$$\alpha(\omega') = \frac{\pi}{M\Omega} \left\{\delta(\omega' - \Omega) \pm \frac{1}{e^{\Omega/T} \mp 1}[\delta(\omega' - \Omega) + \delta(\omega' + \Omega)]\right\},$$

$$\beta(\omega') = \frac{\pi}{M\Omega} \left\{\delta(\omega' - \Omega) \pm \frac{1}{e^{\Omega/T} \mp 1}[\delta(\omega' - \Omega) - \delta(\omega' + \Omega)]\right\}.$$

---

[11]Note that $\alpha(t,t') = \langle x(t)x(t')\rangle_T$.

On the basis of this formula, Einstein first explained the induced emission and absorption of light by atoms which he considered as harmonically oscillating dipoles in contact with a thermal reservoir. He imagined them to be harmonically oscillating dipole moments coupled to a thermal bath consisting of the Fourier components of the electromagnetic field in thermal equilibrium. Such a thermal bath is called a *black body*. The first purely dissipative and temperature-independent term in $\alpha(\omega')$ was attributed by Einstein to the *spontaneous emission* of photons. The second term is caused by the bath fluctuations, making energy go in and out via *induced emission and absorption* of photons. It is proportional to the occupation number of the oscillator state $n_\Omega = (e^{-\Omega/T} \mp 1)^{-1}$. The equality of the prefactors in front of the two terms is the important manifestation of the fluctuation-dissipation theorem found earlier [see (18.53)].

Appendix 18C Wick Decomposition of Operator Products

Consider two operators $\hat{A}(t)$ and $\hat{B}(t)$ which are linear combinations of creation and annihilation operators

$$
\begin{aligned}
\hat{A}(t) &= \alpha_1 \hat{a}(t) + \alpha_2 \hat{a}^\dagger(t), \\
\hat{B}(t) &= \beta_1 \hat{a}(t) + \beta_2 \hat{a}^\dagger(t).
\end{aligned}
\tag{18C.1}
$$

We want to show that the time-ordered product of two operators has the decomposition quoted in Eq. (18B.10):

$$
\hat{T}\hat{A}(t)\hat{B}(t) = \langle \hat{T}\hat{A}(t)\hat{B}(t)\rangle_0 + \hat{N}\hat{A}(t)\hat{B}(t).
\tag{18C.2}
$$

The first term on the right-hand side is the thermal expectation of the time-ordered product at zero temperature; the second term is the normal product of the two operators.

If $\hat{A}$ and $\hat{B}$ are both creation or annihilation operators, the statement is trivial with $\langle \hat{T}\hat{A}\hat{B}\rangle_0 = 0$. If one of the two, say $\hat{A}(t)$, is a creation operator and the other, $\hat{B}(t)$, an annihilation operator, then

$$
\begin{aligned}
\hat{T}\hat{a}(t)\hat{a}^\dagger(t') &= \Theta(t-t')\hat{a}(t)\hat{a}^\dagger(t') \pm \Theta(t'-t)\hat{a}^\dagger(t')\hat{a}(t) \\
&= \Theta(t-t')[\hat{a}(t)\hat{a}^\dagger(t')]_\mp \pm \hat{a}^\dagger(t')\hat{a}(t).
\end{aligned}
\tag{18C.3}
$$

Due to the commutator (anticommutator) the first term is a c-number. As such it is equal to the expectation value of the time-ordered product at zero temperature. The second term is a normal product, so that we can write

$$
\hat{T}\hat{a}(t)\hat{a}^\dagger(t') = \langle \hat{T}\hat{a}(t)\hat{a}^\dagger(t')\rangle_0 + \hat{N}\hat{a}(t)\hat{a}^\dagger(t').
\tag{18C.4}
$$

The same thing is true if a and $a^\dagger$ are interchanged (such an interchange produces merely a sign change on both sides of the equation). The general statement for $\hat{A}(t)\hat{B}(t')$ follows from the bilinearity of the product.

The decomposition (18C.2) of the time-ordered product of two operators can be extended to a product of n operators, where it reads

$$
\hat{T}\hat{A}(t_1)\dots\hat{A}(t_n) = \sum_{i=2}^{n} \hat{N}\overset{\cdot\cdot}{\hat{A}}(t_1)\dots\overset{\cdot\cdot}{\hat{A}}(t_i)\dots\hat{A}(t_n).
\tag{18C.5}
$$

A common pair of dots on top of a pair of operators denotes a *Wick contraction* of Section 3.10. It indicates that the pair of operators has been replaced by the expectation $\langle \hat{T}\hat{A}(t_1)\hat{A}(t_i)\rangle_0$, multiplied by a factor $(-)^F$, if F = fermion permutations were necessary to bring the contracted operator to the adjacent positions. The remaining factors are contracted further in the same way. In this way, any time-ordered product

$$
\hat{T}\hat{A}(t_1)\cdots\hat{A}(t_n)
\tag{18C.6}
$$

can be expanded into a sum of normal products of these operators containing successively one, two, three, etc. pairs of contracted operators.

The expansion rule can be phrased most compactly by means of a generating functional

$$\hat{T} e^{i \int_{-\infty}^{\infty} dt \hat{A}(t) j(t)} = e^{-\frac{1}{2} \int_{-\infty}^{\infty} dt dt' j(t) \langle \hat{T} \hat{A}(t) \hat{A}(t') \rangle_0 j(t')} \, \hat{N} \left(e^{i \int_{-\infty}^{\infty} dt \hat{A}(t) j(t)} \right). \qquad (18C.7)$$

Differentiations with respect to the source $j(t)$ on both sides produce precisely the above decompositions.

By going to thermal expectation values of (18C.7) at a temperature T, we find

$$\left\langle \hat{T} e^{i \int_{-\infty}^{\infty} dt \hat{A}(t) j(t)} \right\rangle_T = e^{-\frac{1}{2} \int_{-\infty}^{\infty} dt dt' j(t) G(t,t') j(t')}, \qquad (18C.8)$$

with

$$G(t, t') = \langle \hat{T} \hat{A}(t) \hat{A}(t') \rangle_0 + \langle \hat{N} \hat{A}(t) \hat{A}(t') \rangle_T. \qquad (18C.9)$$

The first term on the right-hand side is calculated at zero temperature. All finite temperature effects reside in the second term.

Notes and References

The fluctuation-dissipation theorem was first formulated by
H.B. Callen and T.A. Welton, Phys. Rev. **83**, 34 (1951).
It generalizes the relation between the diffusion constant and the viscosity discovered by
A. Einstein, Ann. Phys. (Leipzig) **17**, 549 (1905),
and an analogous relation for induced light emission in
A. Einstein, "Strahlungs-Emission und -Absorption nach der Quantentheorie", Verhandlungen der Deutschen Physikalischen Gesellschaft **18**, 318 (1916),
where he derived Planck's black-body formula. See also the functioning of this theorem in the thermal noise in a resistor:
H. Nyquist, Phys. Rev. **32**, 110 (1928).

K.V. Keldysh, Z. Eksp. Teor. Fiz. **47**, 1515 (1964); Sov. Phys. JETP **20**, 1018 (1965).
See also
V. Korenman, Ann. Phys. (N. Y.) **39**, 72 (1966);
D. Dubois, in *Lectures in Theoretical Physics*, Vol. IX C, ed. by W.E. Brittin, Gordon and Breach, New York, 1967;
D. Langreth, in *Linear and Nonlinear Electronic Transport in Solids*, ed. by J.T. Devreese and V. Van Doren, Plenum, New York, 1976;
A.M. Tremblay, B. Patton, P.C. Martin, and P. Maldague, Phys. Rev. A **19**, 1721 (1979).
For the derivation of the Langevin equation from the forward–backward path integral see
S.A. Adelman, Chem. Phys. Lett. **40**, 495 (1976);
and especially
A. Schmid, J. Low Temp. Phys. **49**, 609 (1982).
To solve the operator ordering problem, Schmid assumes that a time-sliced derivation of the forward–backward path integral would yield a sliced version of the stochastic differential equation (18.317) $\eta_n \equiv (M/\epsilon)(x_n - 2x_{n-1} + x_{n-2}) + (M\gamma/2)(x_n - x_{n-2}) + \epsilon V'(x_{n-1})$. The matrix $\partial \eta / \partial x$ has a constant determinant $(M/\epsilon)^N (1 + \epsilon \gamma/2)^N$. His argument [cited also in the textbook by
U. Weiss, *Quantum Dissipative Systems*, World Scientific, 1993,
in the discussion following Eq. (5.93)] is unacceptable for two reasons: First, his slicing is not derived. Second, the resulting determinant has the wrong continuum limit proportional to

$\exp\left[\int dt\,\gamma/2\right]$ for $\epsilon \to 0$, $N = (t_b - t_a)/\epsilon \to \infty$, corresponding to the unretarded functional determinant (18.274), whereas the correct limit should be γ-independent, by Eq. (18.282). The above textbook by U. Weiss contains many applications of nonequilibrium path integrals.

More on Langevin and Fokker-Planck equations can be found in
S. Chandrasekhar, Rev. Mod. Phys. **15**, 1 (1943);
N.G. van Kampen, *Stochastic Processes in Physics and Chemistry*, North-Holland, Amsterdam, 1981;
P. Hänggi and H. Thomas, Phys. Rep. **88**, 207 (1982);
C.W. Gardiner, *Handbook of Stochastic Methods*, Springer Series in Synergetics, 1983, Vol. 13;
H. Risken, *The Fokker-Planck Equation, ibid.*, 1983, Vol. 18;
R. Kubo, M. Toda, and N. Hashitsume, *Statistical Physics II*, Springer, Berlin, 1985;
H. Grabert, P. Schramm, and G.-L. Ingold, Phys. Rep. **168**, 116 (1988).

The stochastic Schrödinger equation with the Hamiltonian operator (18.521) was derived by
A.O. Caldeira and A.J. Leggett, Physica A **121**, 587 (1983); A **130** 374(E) (1985).
See also
A.O. Caldeira and A.J. Leggett, Phys. Rev. A **31**, 1059 (1985).
A recent discussion of the relation between time slicing and Itō versus Stratonovich calculus can be found in
H. Nakazato, K. Okano, L. Schülke, and Y. Yamanaka, Nucl. Phys. B **346**, 611 (1990).
For the Heisenberg operator approach to stochastic calculus see
N. Saito and M. Namiki, Progr. Theor. Phys. **16**, 71 (1956).
Recent applications of the Langevin equation to decay problems and quantum fluctuations are discussed in
U. Eckern, W. Lehr, A. Menzel-Dorwarth, F. Pelzer.
See also their references and those quoted at the end of Chapter 3.

The quantum Langevin equation is discussed in
G.W. Ford, J.T. Lewis und R.F. O'Connell, Phys. Rev. Lett. **55**, 42273 (1985); Phys. Rev. A **37**, 4419 (1988); Ann. of Phys. **185**, 270 (1988).

Deterministic models for Schrödinger wave functions are discussed in
G. 't Hooft, Class. Quant. Grav. **16**, 3263 (1999) (gr-gc/9903084); hep-th/0003005; Int. J. Theor. Phys. **42**, 355 (2003) (hep-th/0104080); hep-th/0105105; Found. Phys. Lett. **10**, 105 (1997) (quant-ph/9612018).
See also the Lecture Ref. [37].
The representation in Section 18.24 is due to
Z. Haba and H. Kleinert, Phys. Lett. A **294**, 139 (2002) (quant-ph/0106095).
See also
F. Haas, *Stochastic Quantization of the Time-Dependent Harmonic Oscillator*, Int. J. Theor. Phys. **44**, 1 (2005) (quant/ph-0406062).
M. Blasone, P. Jizba, and H. Kleinert, Phys. Rev. A **71**, 2005; Braz. J. Phys. **35**, 479 (2005) (quant/ph-0504047); Annals Phys. **320**, 468 (2005) (quant/ph-0504200).
Another improvement is due to
M. Blasone, P. Jizba, G. Vitiello, Phys. Lett. A **287**, 205 (2001) (hep-th/0007138);
M. Blasone, E. Celeghini, P. Jizba, G. Vitiello, *Quantization, Group Contraction and Zero-Point Energy*, Phys. Lett. A **310**, 393 (2003) (quant-ph/0208012).

The individual citations refer to

[1] Some authors define $G_{12}(\tau)$ as having an extra minus sign and the retarded Green function with a factor $-i$, so that the relation is more direct: $G_{12}^R(\omega) = G_{12}(\omega_m = -i\omega + \eta)$. See

A.A. Abrikosov, L.P. Gorkov, and I.E. Dzyaloshinski, Sov. Phys. JETP **9**, 636 (1959); or *Methods of Quantum Field Theory in Statistical Physics*, Dover, New York, 1975; also A.L. Fetter and J.D. Walecka, *Quantum Theory of Many-Particle Systems*, McGraw-Hill, New York, 1971.
Our definition (18.20) without a minus sign conforms with the definition of the fixed-energy amplitude in Chapter 9, Eq. (1.321), which is also a retarded Green function.

[2] E.S. Fradkin, *The Green's Function Method in Quantum Statistics*, Sov. Phys. JETP **9**, 912 (1959).

[3] G. Baym and D. Mermin, J. Math. Phys. **2**, 232 (1961).
One extrapolation uses Padé approximations:
H.J. Vidberg and J.W. Serene, J. Low Temp. Phys. **29**, 179 (1977); W.H. Press, S.A. Teukolsky, W.T. Vetterling, and B.P. Flannery, *Numerical Recipes in Fortran*, Cambridge Univ. Press (1992), Chapter 12.5.
Since the thermal Green function are usually known only approximately, the continuation is not unique. A maximal-entropy method by
R.N. Silver, D.S. Sivia, and J.E. Gubernatis, Phys. Rev. B **41**, 2380 (1990)
selects the most reliable result.

[4] J. Schwinger, J. Math. Phys. **2**, 407 (1961).

[5] K.V. Keldysh, Z. Eksp. Teor. Fiz. **47**, 1515 (1964); Sov. Phys. JETP **20**, 1018 (1965).

[6] K.-C. Chou, Z.-B. Su, B.-L. Hao, and L. Yu, Phys. Rep. **118**, 1 (1985);
also K.-C. Chou et al., Phys. Rev. B **22**, 3385 (1980).

[7] R.P. Feynman and F.L. Vernon, Ann. Phys. **24**, 118 (1963);
R.P. Feynman and A.R. Hibbs, *Quantum Mechanics and Path Integrals*, McGraw-Hill, New York, 1965, Sections 12.8 and 12.9.

[8] The solution of path integrals with second time derivatives in the Lagrangian is given in
H. Kleinert, J. Math. Phys. **27**, 3003 (1986) (http://www.physik.fu-berlin.de/~klei-nert/144).

[9] H. Kleinert, *Gauge Fields in Condensed Matter*, op. cit., Vol. II, Section 17.3 (*ibid.*http/b2), and references therein.

[10] See the review paper by
S. Chandrasekhar, Rev. Mod. Phys. **15**, 1 (1943), or the original papers:
A.A. Fokker, Ann. Phys. (Leipzig) **43**, 810 (1914);
M. Planck, Sitzber. Preuss. Akad. Wiss. p. 324 (1917);
O. Klein, Arkiv Mat. Astr. Fysik **16**, No. 5 (1922);
H.A. Kramers, Physica **7**, 284 (1940);
M. Smoluchowski, Ann. Phys. (Leipzig) **48** , 1103 (1915).

[11] H. Kleinert, A. Chervyakov, Phys. Lett. B **464**, 257 (1999) (hep-th/9906156); Phys. Lett. B **477**, 373 (2000) (quant-ph/9912056); Eur. Phys. J. C **19**, 743-747 (2001) (quant-ph/0002067); Phys. Lett. A **273**, 1 (2000) (quant-ph/0003095); Int. J. Mod. Phys. A **17**, 2019 (2002) (quant-ph/0208067); Phys. Lett. A **308**, 85 (2003) (quant-ph/0204067); Int. J. Mod. Phys. A **18**, 5521 (2003) (quant-ph/0301081) .

[12] In a first attempt to show that this functional the determinant is unity,
A. Schmid, J. Low Temp. Phys. **49**, 609 (1982). contrived a suitable time slicing of the differential operator in (18.326) to achieve this goal. However, since this was not derived from a time slicing of the initial forward–backwards path integral (18.230), this procedure cannot be considered as a proof.

[13] The operator-ordering problem was first solved by
H. Kleinert, Ann. of Phys. **291**, 14 (2001) (quant-ph/0008109).

[14] R. Benguria and M. Kac, Phys. Rev. Lett. **46**, 1 (1981);
Y.C. Chen, M.P.A. Fisher and A.J. Leggett, J. Appl. Phys. **64**, 3119 (1988).

[15] G.W. Ford, M. Kac, and P. Mazur, J. Math. Phys. **6**, 504 (1965).
G.W. Ford and M. Kac, J. Stat. Phys. **46**, 803 (1987).

[16] P. Langevin, Comptes Rendues **146**, 530 (1908).

[17] G. Parisi and Y.S. Wu, Scientia Sinica **24**, 483 (1981).

[18] R. Kubo, J.Math.Phys. **4**, 174 (1963);
R. Kubo, M. Toda, and N. Hashitsume, *Statistical Physics II (Nonequilibrium Statistical Mechanics)*, Springer-Verlag, Berlin, 1985 (Chap. 2).

[19] J. Zinn-Justin, *Critical Phenomena*, Clarendon, Oxford, 1989.

[20] G. Parisi and N. Sourlas, Phys. Rev. Lett. **43**, 744 (1979); J. de Phys. **41**, L403 (1981);
Nucl. Phys. B *206*, 321 (1982);

[21] A.J. McKane, Phys. Lett. A *76*, 22 (1980).

[22] H. Kleinert and S. Shabanov, Phys. Lett. A **235**, 105 (1997) (quant-ph/9705042).

[23] E. Gozzi, Phys. Lett. **B** 201, 525 (1988).

[24] A.O. Caldeira and A.J. Leggett, Physica A **121**, 587 (1983); A *130* 374(E) (1985).

[25] More on this subject is found in the collection of articles
D. Giulini, E. Joos, C. Kiefer, J. Kupsch, I.O. Stamatescu, H.D. Zeh, *Decoherence and the Appearance of a Classical World in Quantum Theory*, Springer, Berlin, 1996.

[26] G. Lindblad, Comm. Math. Phys. **48**, 119 (1976). This paper shows that the form (18.534) of the master equation (18.524) guarantees the positivity of the probabilities derived from the solutions. The right-hand side can be more generally $\sum_{mn} h_{mn} \left(\frac{1}{2} \hat{L}_m \hat{L}_n \hat{\rho} + \frac{1}{2} \hat{\rho} \hat{L}_m \hat{L}_n - \hat{L}_n \hat{\rho} \hat{L}_m \right) + \text{h.c.}$.

[27] L. Diosi, Europhys. Lett. **22**, 1 (1993).

[28] C.W. Gardiner, IBM J. Res. Develop. **32**, 127 (1988).

[29] H. Kleinert and S. Shabanov, Phys. Lett. A **200**, 224 (1995) (quant-ph/9503004);
K. Tsusaka, Phys. Rev. E **59**, 4931 (1999).

[30] H.A. Bethe, Phys. Rev. **72**, 339 (1947).

[31] C. Cohen-Tannoudji, J. Dupont-Roc, G. Grynberg, *Photons and Atoms: Introduction to Quantum Electrodynamics*, Wiley, New York, 1992.

[32] R. Rohrlich, Am. J. Phys. **68**, 1109 (2000).

[33] H. Kleinert and S. Shabanov, J. Phys. A: Math. Gen. **31**, 7005 (1998) (cond-mat/9504121);
R. Bausch, R. Schmitz, and L.A. Turski, Phys. Rev. Lett. **73**, 2382 (1994); Z. Phys. B **97**, 171 (1995).

[34] M. Roncadelli, Europhys. Lett. **16**, 609 (1991); J. Phys. A **25**, L997 (1992);
A. Defendi and M. Roncadelli, Europhys. Lett. **21**, 127 (1993).

[35] Z. Haba, Lett. Math. Phys. **37**, 223 (1996).

[36] G. 't Hooft, Found. Phys. Lett. **10**, 105 (1997) (quant-ph/9612018).

[37] G. 't Hooft, *How Does God Throw Dice?* in *Fluctuating Paths and Fields - Dedicated to Hagen Kleinert on the Occasion of his 60th Birthday*, Eds. W. Janke, A. Pelster, H.-J. Schmidt, and M. Bachmann, World Scientific, Singapore, 2001 (http://www.physik.fu-berlin.de/~kleinert/fest.html).

Agri non omnes frugiferi sunt.
Not all fields are fruitful.
CICERO, Tusc. Quaest., 2, 5, 13

19

Relativistic Particle Orbits

Particles moving at large velocities near the speed of light are called *relativistic* particles. If such particles interact with each other or with an external potential, they exhibit quantum effects which cannot be described by fluctuations of a single particle orbit. Within short time intervals, additional particles or pairs of particles and antiparticles are created or annihilated, and the total number of particle orbits is no longer invariant. Ordinary quantum mechanics which always assumes a fixed number of particles cannot describe such processes. The associated path integral has the same problem since it is a sum over a given set of particle orbits. Thus, even if relativistic kinematics is properly incorporated, a path integral cannot yield an accurate description of relativistic particles. An extension becomes necessary which includes an arbitrary number of mutually linked and *branching* fluctuating orbits.

Fortunately, there exists a more efficient way of dealing with relativistic particles. It is provided by *quantum field theory*. We have demonstrated in Section 7.14 that a grand-canonical ensemble of particle orbits can be described by a functional integral of a single fluctuating field. Branch points of newly created particle lines are accounted for by anharmonic terms in the field action. The calculation of their effects proceeds by perturbation theory which is systematically performed in terms of Feynman diagrams with calculation rules very similar to those in Section 3.18. There are again lines and interaction vertices, and the main difference lies in the lines which are correlation functions of fields rather than position variables $x(t)$. The lines and vertices represent direct pictures of the topology of the worldlines of the particles and their possible collisions and creations.

Quantum field theory has been so successful that it is generally advantageous to describe the statistical mechanics of many completely different types of line-like objects in terms of fluctuating fields. One important example is the polymer field theory in Section 15.12. Another important domain where field theory has been extremely successful is in the theory of line-like defects in crystals, superfluids, and superconductors. In the latter two systems, the defects occur in the form of quantized vortex lines or quantized magnetic flux lines, respectively. The entropy of their classical shape fluctuations determines the temperature where the phase transitions take place. Instead of the usual way of describing these systems as ensembles of particles with their interactions, a field theory has been developed whose Feynman diagrams are the direct pictures of the line-like defects, called *disorder field theory* [1].

The most important advantage of field theory is that it can describe most easily phase transitions, in which particles form a condensate. The disorder theory is therefore particularly suited to understand phase transitions in which defect-, vortex-, or flux-lines proliferate, which happens in the processes of crystal melting, superfluid to normal, or superconductor to normal transitions, respectively. In fact, the disorder theory is so far the only theory in which the critical behavior of the superconductor near the transition is properly understood [2].

A particular quantum field theory, called *quantum electrodynamics* describes with great success the electromagnetic interactions of electrons, muons, quarks, and photons. It has been extended successfully to include the weak interactions among these particles and, in addition, neutrinos, using only a few quantized Dirac fields and a quantized electromagnetic vector potential. The inclusion of a nonabelian gauge field, the gluon field, is a good candidate for explaining all known features of strong interactions.

It is certainly unnecessary to reproduce in an orbital formulation the great amount of results obtained in the past from the existing field theory of weak, electromagnetic, and strong interactions. The orbital formulation was, in fact, proposed by Feynman back in 1950 [3], but never pursued very far due to the success of quantum field theory. Recently, however, this program was revived in a number of publications [4, 5]. The main motivation for this lies in another field of fundamental research: the *string theory* of fundamental particles. In this theory, all elementary particles are supposed to be excitations of a single line-like object with tension, and various difficulties in obtaining a consistent theory in the physical spacetime have led to an extension by fermionic degrees of freedom, the result being the so-called *superstring*. Strings moving in spacetime form worldsurfaces rather than worldlines. They do not possess a second-quantized field theoretic formulation. Elaborate rules have been developed for the functional integrals describing the splitting and merging of strings. If one cancels one degree of freedom in such a superstring, one has a theory of splitting and merging particle worldlines. As an application of the calculation rules for strings, processes which have been known from calculations within the quantum field theory have been recalculated using these reduced superstring rules. In this textbook, we shall give a small taste of such calculations by evaluating the change in the vacuum energy of electromagnetic fields caused by fluctuating relativistic spinless and spin-1/2 particles.

It should be noted that since up to now, no physical result has emerged from superstring theory,[1] there is at present no urgency to dwell deeper into the subject.

By giving a short introduction into this subject we shall be able to pay tribute to some historic developments in quantum mechanics, where the relativistic generalization of the Schrödinger equation was an important step towards the development of

---

[1]This theory really deserves a price for having the highest popularity-per-physicality ratio in the history of science, enjoying a great amount of financial support. The situation is very similar to the geocentric medieval picture of the world.

quantum field theory [6]. For this reason, many textbooks on quantum field theory begin with a discussion of relativistic quantum mechanics. By analogy, we shall incorporate relativistic kinematics into path integrals.

It should be noted that an esthetic possibility to give a path Fermi statistics is based on the Chern-Simons theory of entanglement of Chapter 16. However, this approach is still restricted to $2 + 1$ spacetime dimensions [7], and an extension to the physical $3 + 1$ spacetime dimensions is not yet in sight.

19.1 Special Features of Relativistic Path Integrals

Consider a free point particle of mass M moving through the $3+1$ spacetime dimensions of Minkowski space at relativistic velocity. Its path integral description is conveniently formulated in four-dimensional Euclidean spacetime where the fluctuating worldlines look very similar to the fluctuating polymers discussed in Chapter 15.

Thus, time is taken to be imaginary, i.e.,

$$t = -i\tau = -ix^4/c, \tag{19.1}$$

and the length of a four-vector $x = (\mathbf{x}, x^4)$ is given by

$$x^2 = \mathbf{x}^2 + (x^4)^2 = \mathbf{x}^2 + c^2\tau^2. \tag{19.2}$$

If $x^\mu(\lambda)$ is an arbitrarily parametrized orbit, the well-known classical Euclidean action is proportional to the invariant length of the orbit in spacetime:

$$S = \int_{\lambda_a}^{\lambda_b} d\lambda \sqrt{x'^2(\lambda)}, \tag{19.3}$$

and reads

$$\mathcal{A}_{\text{cl,e}} = McS, \tag{19.4}$$

or, explicitly,

$$\mathcal{A}_{\text{cl,e}} = Mc \int_{\lambda_a}^{\lambda_b} ds(\lambda), \tag{19.5}$$

with

$$ds(\lambda) \equiv d\lambda \sqrt{x'^2(\lambda)} = d\lambda \sqrt{\mathbf{x}'^2(\lambda) + c^2\tau'^2(\lambda)}. \tag{19.6}$$

The prime denotes the derivative with respect to the parameter λ. The action is independent of the choice of the parametrization. If λ is replaced by a new parameter

$$\lambda \to \bar{\lambda} = f(\lambda), \tag{19.7}$$

then

$$x'^2 \quad \to \quad \frac{1}{f'^2} x'^2, \tag{19.8}$$

$$d\lambda \quad \to \quad d\lambda \, f', \tag{19.9}$$

so that ds and the action remain invariant.

We now calculate the Euclidean amplitude for the worldline of the particle to run from the spacetime point $x_a = (\mathbf{x}_a, c\tau_a)$ to $x_b = (\mathbf{x}_b, c\tau_b)$. For the sake of generality, we treat the case of D Euclidean spacetime dimensions. Before starting we observe that the action (19.5) does not lend itself easily to a calculation of the path integral over $e^{-\mathcal{A}_{\rm cl,e}/\hbar}$. There exists an alternative form for the classical action that is more suitable for this purpose. It involves an auxiliary field $h(\lambda)$ and reads:

$$\bar{\mathcal{A}}_{\rm e} = \int_{\lambda_a}^{\lambda_b} d\lambda \left[\frac{Mc}{2h(\lambda)} x'^2(\lambda) + h(\lambda)\frac{Mc}{2} \right]. \tag{19.10}$$

This has the advantage of containing the particle orbit quadratically as in a free nonrelativistic action. The auxiliary field $h(\lambda)$ has been inserted to make sure that the classical orbits of the action (19.10) coincide with those of the original action (19.5). Indeed, extremizing $\bar{\mathcal{A}}_{\rm e}$ in $h(\lambda)$ gives the relation

$$h(\lambda) = \sqrt{x'^2(\lambda)}. \tag{19.11}$$

Inserting this back into $\bar{\mathcal{A}}_{\rm e}$ renders the classical action

$$\mathcal{A}_{\rm cl,e} = Mc \int_{\lambda_a}^{\lambda_b} d\lambda \sqrt{x'^2(\lambda)}, \tag{19.12}$$

which is the same as (19.5).

At this point the reader will worry that although the new action (19.10) describes the same classical physics as the original action (19.12), it may lead to a completely different quantum physics of a relativistic partile. With a little effort, however, it can be shown that this is not so. Since the proof is quite technical, it will be given in Appendix 19A.

The new action (19.10) shares with the old action (19.12) the reparametrization invariance (19.7) for arbitrary fluctuating path configurations. We only have to assign an appropriate transformation behavior to the extra field $h(\lambda)$. If λ is replaced by a new parameter $\bar{\lambda} = f(\lambda)$, then x'^2 and $d\lambda$ transform as in (19.8) and (19.9), and the action remains invariant, if $h(\lambda)$ is simultaneously changed as

$$h \to h/f'. \tag{19.13}$$

We now set up the path integral of a relativistic particle associated with the action (19.10). First we sum over the orbital fluctuations at a *fixed* $h(\lambda)$. To find the correct measure of integration, we use the canonical formulation in which the Euclidean action reads

$$\bar{\mathcal{A}}_{\rm e}[p,x] = \int_{\lambda_a}^{\lambda_b} d\lambda \left[-ipx' + \frac{h(\lambda)}{2Mc} \left(p^2 + M^2c^2 \right) \right]. \tag{19.14}$$

This must be sliced in the length parameter λ. We form $N+1$ slices as usual, choosing arbitrary small parameter differences $\epsilon_n = \lambda_n - \lambda_{n-1}$ depending on n, and write the sliced action as

$$\bar{\mathcal{A}}_{\rm e}^N[p,x] = \sum_{n=1}^{N+1} \left[-ip_n(x_n - x_{n-1}) + h_n\epsilon_n \frac{p_n^2}{2Mc} + \epsilon_n h_n \frac{Mc}{2} \right]. \tag{19.15}$$

This path integral has a universal phase space measure [recall (2.28)]

$$\int \mathcal{D}^D x \int \frac{\mathcal{D}^D p}{(2\pi\hbar)^D} e^{-\bar{\mathcal{A}}_e[p,x]/\hbar} \approx \prod_{n=1}^{N} \left[\int d^D x_n \right] \prod_{n=1}^{N+1} \left[\int \frac{d^D p_n}{(2\pi\hbar)^D} e^{-\bar{\mathcal{A}}_e^N[p,x]/\hbar} \right]. \quad (19.16)$$

The momentum variables p_n are integrated out to give the configuration space integrals (setting $\lambda_{N+1} \equiv \lambda_b$, $h_{N+1} \equiv h_b$) [compare (2.74)]

$$\frac{1}{\sqrt{2\pi\hbar\epsilon_b h_b/Mc}^D} \prod_{n=1}^{N} \left[\int \frac{d^D x_n}{\sqrt{2\pi\hbar\epsilon_n h_n/Mc}^D} \right] \exp\left(-\frac{1}{\hbar} \bar{\mathcal{A}}_e^N[x] \right), \quad (19.17)$$

with the time-sliced action in configuration space

$$\bar{\mathcal{A}}_e^N[x] = \sum_{n=1}^{N+1} \left[\frac{Mc}{2h_n\epsilon_n} (\Delta x_n)^2 + \epsilon_n h_n \frac{Mc}{2} \right]. \quad (19.18)$$

The Gaussian integrals over x_n in (19.17) can now be done successively using Formula (2.69), and we find [as in (2.75)]

$$\frac{1}{\sqrt{2\pi\hbar L/Mc}^D} \exp\left[-\frac{Mc}{2\hbar} \frac{(x_b - x_a)^2}{L} - \frac{Mc}{2\hbar} L \right], \quad (19.19)$$

where L is the total sliced length of the orbit

$$L \equiv \sum_{n=1}^{N+1} \epsilon_n h_n, \quad (19.20)$$

whose continuum limit is

$$L = \int_{\lambda_a}^{\lambda_b} d\lambda \, h(\lambda). \quad (19.21)$$

The result (19.19) does not depend on the function $h(\lambda)$ but only on L, this being a consequence of the reparametrization invariance of the path integral. While the total λ-interval changes under the transformation, the total length L of (19.21) is invariant under the joint transformations (19.7) and (19.13). This invariance permits only the invariant length L to appear in the integrated expression (19.19), and the path integral over $h(\lambda)$ can be reduced to a simple integral over L. The appropriate path integral for the time evolution amplitude reads

$$(x_b|x_a) = \mathcal{N} \int_0^\infty dL \int \mathcal{D}h \, \Phi[h] \int \mathcal{D}^D x \, e^{-\bar{\mathcal{A}}_e/\hbar}, \quad (19.22)$$

where $\mathcal{N}$ is some normalization factor and $\Phi[h]$ an appropriate *gauge-fixing functional*.

19.1.1 Simplest Gauge Fixing

The simplest choice for the latter is a δ-functional,

$$\Phi[h] = \delta[h - c], \tag{19.23}$$

which fixes $h(\lambda)$ to be equal to the light velocity everywhere, and relates

$$L = c(\lambda_b - \lambda_a). \tag{19.24}$$

This relation makes the dimension of the parameter λ timelike. This Lorentz-invariant time parameter is the so-called proper time of special relativity, and should not be confused with the parameter time τ contained in the Dth component $x^D = c\tau$ [see Eq. (19.1) for $D = 4$]. By analogy with the discussion of thermodynamics in Chapter 2 we shall then denote $\lambda_b - \lambda_a$ by $\hbar\beta$ and write (19.24) as

$$L = c(\lambda_b - \lambda_a) \equiv c\hbar\beta. \tag{19.25}$$

If we further use translational invariance to set $\lambda_a = 0$, we arrive at the gauge-fixed path integral

$$(x_b|x_a) = \mathcal{N}c\hbar \int_0^\infty d\beta\, e^{-\beta M c^2/2} \int \mathcal{D}^D x\, e^{-\mathcal{A}_{0,e}/\hbar}, \tag{19.26}$$

with

$$\mathcal{A}_{0,e} = \int_0^{\hbar\beta} d\lambda\, \frac{M}{2}\dot{x}^2. \tag{19.27}$$

Since λ is now timelike, we use a dot to denote the derivative: $\dot{x}(\lambda) \equiv dx(\lambda)/d\lambda$. Remarkably, the gauge-fixed action coincides with the action of a free nonrelativistic particle in D Euclidean spacetime dimensions. Having taken the trivial term $\int_0^{\hbar\beta} d\lambda\, Mc/2\hbar$ out of the action (19.14), the expression (19.26) contains a Boltzmann weight $e^{-\beta M c^2/2}$ multiplying each particle orbit of mass M.

The solution of the path integral is then given by

$$(x_b|x_a) = \mathcal{N}c\hbar \int_0^\infty d\beta\, \frac{1}{\sqrt{2\pi\hbar^2\beta/M}^D} \exp\left[-\frac{M}{2\hbar}\frac{(x_b - x_a)^2}{\hbar\beta} - \beta\frac{Mc^2}{2}\right]. \tag{19.28}$$

By Fourier-transforming the x-dependence, this amplitude can also be written as

$$(x_b|x_a) = \mathcal{N}c\hbar \int_0^\infty d\beta\, e^{-\beta M c^2/2} \int \frac{d^D k}{(2\pi)^D} \exp\left[ik(x_b - x_a) - \beta\frac{\hbar^2 k^2}{2M}\right], \tag{19.29}$$

and evaluated to

$$(x_b|x_a) = \mathcal{N}\frac{2Mc}{\hbar} \int \frac{d^D k}{(2\pi)^D} \frac{1}{k^2 + M^2 c^2/\hbar^2} e^{ik(x_b - x_a)}. \tag{19.30}$$

Upon setting $\mathcal{N} = \lambda_M^C/2$, where

$$\lambda_M^C \equiv \hbar/Mc, \qquad (19.31)$$

is the well-known Compton wavelength of a particle of mass M [recall Eq. (4.345)], this becomes the Green function of the Klein-Gordon field equation in Euclidean time:

$$(-\partial_b^2 + M^2 c^2/\hbar^2)(x_b|x_a) = \delta^{(D)}(x_b - x_a). \qquad (19.32)$$

In the Fourier representation (19.30), the integral over k [or the integral over β in (19.28)] can be performed with the explicit result for the Green function

$$(x_b|x_a) = \frac{1}{(2\pi)^{D/2}} \left(\frac{Mc}{\hbar\sqrt{x^2}} \right)^{D/2-1} K_{D/2-1}\left(Mc\sqrt{x^2}/\hbar \right), \qquad (19.33)$$

where $K_\nu(z)$ denotes the modified Bessel function and $x \equiv x_b - x_a$. In the nonrelativistic limit $c \to \infty$, the asymptotic behavior $K_\nu(z) \to \sqrt{\pi/2z}\,e^{-z}$ [see Eq. (1.353)] leads to

$$(x_b|x_a) = (\mathbf{x}_b \tau_a | \mathbf{x}_a \tau_a) \xrightarrow{c\to\infty} \frac{\hbar}{2Mc} e^{-Mc^2(\tau_b - \tau_a)/\hbar} (\mathbf{x}_b \tau_b | \mathbf{x}_a \tau_a)_{\text{Schr}}, \qquad (19.34)$$

with the usual Euclidean time evolution amplitude of the free Schrödinger equation

$$(\mathbf{x}_b \tau_b | \mathbf{x}_a \tau_a)_{\text{Schr}} = \frac{1}{\sqrt{2\pi\hbar(\tau_b - \tau_a)/M}^{D-1}} \exp\left\{ -\frac{M}{2\hbar} \frac{(\mathbf{x}_b - \mathbf{x}_a)^2}{\tau_b - \tau_a} \right\}. \qquad (19.35)$$

The exponential prefactor in (19.34) contains the effect of the rest energy Mc^2 which is ignored in the nonrelativistic Schrödinger theory.

Note that the same limit may be calculated conveniently in the saddle point approximation to the β-integral (19.28). For $c \to \infty$, the exponent has a sharp extremum at

$$\beta = \frac{\sqrt{(x_b - x_a)^2}}{c\hbar} = \frac{\sqrt{(\mathbf{x}_b - \mathbf{x}_a)^2 + c^2(\tau_b - \tau_a)^2}}{c\hbar} \xrightarrow{c\to\infty} \frac{\tau_b - \tau_a}{\hbar} + \frac{(\mathbf{x}_b - \mathbf{x}_a)^2}{2c^2\hbar(\tau_b - \tau_a)} + \ldots, (19.36)$$

and the β-integral can be evaluated in a quadratic approximation around this value. This yields once again (19.34).

19.1.2 Partition Function of Ensemble of Closed Particle Loops

The diagonal amplitude (19.26) with $x_b = x_a$ contains the sum over all lengths and shapes of a *closed* particle loop in spacetime. This sum can be made a partition function of a closed loop if we remove a degeneracy factor proportional to $1/L$ from the integral over L. Then all cyclic permutations of the points of the loop are counted only once. Apart from an arbitrary normalization factor to be fixed later, the partition function of a single closed loop reads

$$Z_1 = \int_0^\infty \frac{d\beta}{\beta} e^{-\beta Mc^2/2} \int \mathcal{D}^D x \, e^{-\mathcal{A}_{0,e}/\hbar}. \tag{19.37}$$

Inserting the right-hand integral in (19.29) for the path integral (with $x_b = x_a$), this becomes

$$Z_1 = V_D \int_0^\infty \frac{d\beta}{\beta} e^{-\beta Mc^2/2} \int \frac{d^D k}{(2\pi)^D} \exp\left(-\beta \frac{\hbar^2 k^2}{2M}\right), \tag{19.38}$$

where V_D is the total volume of spacetime. This can be evaluated immediately. The Gaussian integral gives for each of the D dimensions a factor $1/\sqrt{2\pi\hbar^2\beta/M}$, after which formula (2.496) leads to

$$Z_1 = V_D \int_0^\infty \frac{d\beta}{\beta} e^{-\beta Mc^2/2} \frac{1}{\sqrt{2\pi\hbar^2\beta/M}^D} = \frac{V_D}{\lambda_M^C \, ^D} \frac{1}{(4\pi)^{D/2}} \Gamma(1 - D/2), \tag{19.39}$$

where λ_M^C is the Compton wavelength (19.31). With the help of the sloppy formula (2.504) of analytic regularization which implies the minimal subtraction explained in Subsection 2.15.1, the right-hand side of (19.38) can also be written as

$$Z_1 = -V_D \int \frac{d^D k}{(2\pi)^D} \log\left(k^2 + M^2 c^2/\hbar^2\right). \tag{19.40}$$

The right hand side can be expressed in functional form as

$$Z_1 = -\text{Tr} \log\left(-\partial^2 + M^2 c^2/\hbar^2\right) = -\text{Tr} \log\left(-\hbar^2 \partial^2 + M^2 c^2\right), \tag{19.41}$$

the two expressions being equal in the analytic regularization of Section 2.15, since a constant inside the logarithm gives no contribution by Veltman's rule (2.506).

The partition function of a grand-canonical ensemble is obtained by exponentiating this:

$$Z = e^{Z_1} = e^{-\text{Tr} \log\left(-\hbar^2 \partial^2 + M^2 c^2\right)}. \tag{19.42}$$

In order to interpret this expression physically, we separate the integral $\int d^D k/(2\pi)^D$ into an integral over the temporal component k^D and a spatial remainder, and write

$$k^2 + M^2 c^2/\hbar^2 = \left(k^D\right)^2 + \omega_{\mathbf{k}}^2/c^2, \tag{19.43}$$

with the frequencies

$$\omega_{\mathbf{k}} \equiv c\sqrt{\mathbf{k}^2 + M^2 c^2/\hbar^2}. \tag{19.44}$$

Recalling the result (2.503) of the integral (2.489) we obtain

$$Z_1 = -2V_D \int \frac{d^{D-1}k}{(2\pi)^{D-1}} \frac{\hbar\omega_{\mathbf{k}}}{2c}. \tag{19.45}$$

The exponent is the sum of two ground state energies of oscillators of energy $\hbar\omega_{\mathbf{k}}/2$, which are the vacuum energies associated with two relativistic particles. In quantum field theory one learns that these are particles and antiparticles. Many neutral particles are identical to their antiparticles, for example photons, gravitons, and the pion with zero charge. For these, the factor 2 is absent. Then the integral (19.38) contains a factor $1/2$ accounting for the fact that paths running along the same curve in spacetime but in the opposite sense are identified.

Comparing (19.42) with (3.556) and (3.619) for $j = 0$ we identify $-Z_1\hbar$ with $-W[0]$ and the Euclidean effective action Γ of the ensemble of loops:

$$-Z_1 = -W[0]/\hbar = \Gamma_{\mathrm{e}}/\hbar. \tag{19.46}$$

19.1.3   Fixed-Energy Amplitude

The fixed-energy amplitude is related to (19.22) by a Laplace transformation:

$$(\mathbf{x}_b|\mathbf{x}_a)_E \equiv -i \int_{\tau_a}^{\infty} d\tau_b \, e^{E(\tau_b-\tau_a)/\hbar} \, (x_b|x_a), \tag{19.47}$$

where τ_b, τ_a are once more the time components in x_b, x_a. As explained in Chapter 9, the poles and the cut along the energy axis in this amplitude contain all information on the bound and continuous eigenstates of the system. The fixed-energy amplitude has the reparametrization-invariant path integral representation [here with the conventions of Eq. (19.10)]

$$(\mathbf{x}_b|\mathbf{x}_a)_E = \frac{\hbar}{2Mc} \int_0^{\infty} dL \int \mathcal{D}h \, \Phi[h] \int \mathcal{D}^D x \, e^{-\bar{\mathcal{A}}_{e,E}/\hbar}, \tag{19.48}$$

with the Euclidean action

$$\bar{\mathcal{A}}_{e,E} = \int_{\lambda_a}^{\lambda_b} d\lambda \left[\frac{Mc}{2h(\lambda)} \mathbf{x}'^2(\lambda) - h(\lambda) \frac{E^2}{2Mc^3} + h(\lambda) \frac{Mc}{2} \right]. \tag{19.49}$$

To prove this, we write the temporal x^D-part of the sliced D-dimensional action (19.18) in the canonical form (19.15). In the associated path integral (19.16), we integrate out all x_n^D-variables, producing N δ-functions. These remove the integrals over N momentum variables p_n^D, leaving only a single integral over a common p^D. The Laplace transform (19.47), finally, eliminates also this integral making p^D equal to $-iE/c$. In the continuum limit, we thus obtain the action (19.49).

The path integral (19.48) forms the basis for studying relativistic potential problems. Only the physically most relevant example will be treated here.

19.2 Tunneling in Relativistic Physics

Relativistic harbors several new tunneling phenomena, of which we want to discuss two especially interesting ones.

19.2.1 Decay Rate of Vacuum in Electric Field

In relativistic physics, an empty Minkowski space with a constant electric field $\mathbf{E}$ is unstable. There is a finite probability that a particle-antiparticle pair can be created. For particles of mass M, this requires the energy

$$E_{\text{pair}} = 2Mc^2. \tag{19.50}$$

This energy can be supplied by the external electric field. If the pair of charge $\pm e$ is separated by a distance which is roughly twice the Compton wavelength $\lambda_M^C = \hbar/Mc$ of Eq. (19.31), it gains an energy $2|\mathbf{E}|\lambda_M^C e$. The decay will therefore become significant when

$$|\mathbf{E}| > E_c = \frac{M^2 c^3}{e\hbar}. \tag{19.51}$$

Euclidean Action

In Chapter 17 we have shown that in the semiclassical limit where the decay-rate is small, it is proportional to a Boltzmann-like factor $e^{-\mathcal{A}_{\text{cl,e}}/\hbar}$, where $\mathcal{A}_{\text{cl,e}}$ is the action of a Euclidean classical solution mediating the decay. Such a solution is easily found. We use the classical action in the form (19.12) and choose the parameter λ to measure the imaginary time $\lambda = \tau = it = x^4/c$. Then the action takes the form

$$\mathcal{A}_{\text{cl,e}} = \int_{\tau_a}^{\tau_b} d\tau \left[Mc^2\sqrt{1 + \dot{\mathbf{x}}^2(\tau)/c^2} - e\,\mathbf{E} \cdot \mathbf{x}(\tau) \right]. \tag{19.52}$$

This is extremized by the classical equation of motion

$$M\frac{d}{d\tau}\frac{\dot{\mathbf{x}}(\tau)}{\sqrt{1 + \dot{\mathbf{x}}^2(\tau)/c^2}} = -e\mathbf{E}, \tag{19.53}$$

whose solutions are circles in spacetime of a fixed E-dependent radius l_E:

$$(\mathbf{x} - \mathbf{x}_0)^2 + c^2(\tau - \tau_0)^2 = l_E^2 \equiv \left(\frac{Mc^2}{eE}\right)^2, \quad E \equiv |\mathbf{E}|. \tag{19.54}$$

To calculate their action we parametrize the circles in the $\hat{\mathbf{E}} - \tau$ -plane, where $\hat{\mathbf{E}}$ is the unit vector in the direction of $\mathbf{E}$, by an angle θ as

$$\mathbf{x}(\theta) = l_E\hat{\mathbf{E}}\cos\theta + \mathbf{x}_0, \quad \tau(\theta) = \frac{l_E}{c}\sin\theta + \tau_0. \tag{19.55}$$

A closed circle has an action

$$\mathcal{A}_{\mathrm{cl,e}} = Mc^2 \frac{l_E}{c} \int_0^{2\pi} d\theta \cos\theta \left[\frac{1}{\cos\theta} - \cos\theta \right] = Mc\,l_E\pi = \hbar \frac{E_c}{E}\pi. \tag{19.56}$$

The decay rate of the vacuum is therefore proportional to

$$\Gamma \propto e^{-\pi E_c/E}. \tag{19.57}$$

The circles (19.54) are, of course, the space-time pictures of the creation and the annihilation of particle-antiparticle pairs at times $\tau_0 - l_E/c$ and $\tau_0 + l_E/c$ and positions $\mathbf{x}_0$, respectively (see Fig. 19.1). A particle can also run around the circle

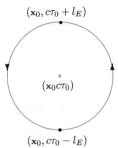

$(\mathbf{x}_0, c\tau_0 + l_E)$

$(\mathbf{x}_0 c\tau_0)$

$(\mathbf{x}_0, c\tau_0 - l_E)$

Figure 19.1 Spacetime picture of pair creation at the point $\mathbf{x}_0$ and time $\tau_0 - l_E/c$ and annihilation at the later time $\tau_0 + l_E/c$.

repeatedly. This leads to the formula

$$\Gamma \propto \sum_{n=1}^{\infty} F_n\, e^{-n\pi E_c/E}, \tag{19.58}$$

with fluctuation factors F_n which we are now going to determine.

Fluctuations

As explained above, fluctuations must be calculated with the relativistic Euclidean action (19.10), in which we have to include the electric field by minimal coupling:

$$\bar{\mathcal{A}}_{\mathrm{e}} = \int_{\lambda_a}^{\lambda_b} d\lambda \left[\frac{M}{2h(\lambda)} x'^2(\lambda) + h(\lambda)\frac{Mc}{2} - i\frac{e}{c}A(x(\lambda))\,x'(\lambda) \right]. \tag{19.59}$$

The vacuum will decay by the creation of an ensemble of pairs which in Euclidean spacetime corresponds to an ensemble of particle loops. For free particles, the partition function Z was given in (19.42) as an exponential of the one-loop partition function Z_1 in (19.37). The corresponding Z_1 in the presence of an electromagnetic field has the one-loop partition function

$$Z_1 = \int_0^{\infty} \frac{d\beta}{\beta}\, e^{-\beta Mc^2/2} \int \mathcal{D}^4 x\, e^{-\bar{\mathcal{A}}_{\mathrm{e}}/\hbar}, \tag{19.60}$$

with the Euclidean action

$$\bar{\mathcal{A}}_e = \int_0^{\hbar\beta} d\lambda \left[\frac{M}{2} x'^2(\lambda) - i\frac{e}{c} A(x(\lambda)) x'(\lambda) \right]. \tag{19.61}$$

The equations of motion are now

$$\mathbf{x}'' = \frac{e}{Mc} \left(x_4' \, \mathbf{E} - i\mathbf{x}' \times \mathbf{B} \right), \qquad x_4'' = -\frac{e}{Mc} \mathbf{x}' \cdot \mathbf{E}, \tag{19.62}$$

If both a constant electric and a constant magnetic field are present, the vector potential is $A_\mu = -F_{\mu\nu} x^\nu/2$ and the action (19.61) takes the simple quadratic form

$$\bar{\mathcal{A}}_e = \int_0^{\hbar\beta} d\lambda \left[\frac{M}{2} x'^2 - i\frac{e}{2c} F_{\mu\nu} x^\mu x'^\nu \right], \tag{19.63}$$

and the equations of motion (19.62) are

$$x''^\mu = -i\frac{e}{c} F^{\mu\nu} x'_\nu, \quad \text{with} \quad F_{ij} = -\epsilon_{ijk} B^k, \quad F^{i4} = iF^{i0} = iE^i. \tag{19.64}$$

In a purely electric field, the solutions are circular orbits:

$$\mathbf{x}(\lambda) = \hat{\mathbf{E}} A \cos \omega_L^E(\lambda - \lambda_0) + c_2, \quad x_4(\lambda) = A \sin \omega_L^E(\lambda - \lambda_0) + c_4, \tag{19.65}$$

with the electric version of the Landau or cyclotron frequency (2.646)

$$\omega_L^E \equiv \frac{eE}{Mc}. \tag{19.66}$$

The circular orbits are the same as those in the previous formulation (19.55). If $\mathbf{E}$ points in the z-direction, the action (19.63) decomposes into two decoupled quadratic actions $\bar{\mathcal{A}}_e^{(12)} + \bar{\mathcal{A}}_e^{(34)}$ for the motion in the $x^1 - x^2$ and $x^3 - x^4$ -planes, respectively, and the one-loop partition function (19.60) factorizes as follows:

$$\begin{aligned} Z_1 &= \int_0^\infty \frac{d\beta}{\beta} e^{-\beta Mc^2/2} \int \mathcal{D}^2 x^{(12)} \, e^{-\bar{\mathcal{A}}_e^{(12)}/\hbar} \int \mathcal{D}^2 x^{(34)} \, e^{-\bar{\mathcal{A}}_e^{(34)}/\hbar} \\ &\equiv \int_0^\infty \frac{d\beta}{\beta} e^{-\beta Mc^2/2} Z^{(12)}(0) Z^{(34)}(E). \end{aligned} \tag{19.67}$$

The path integral for $Z^{(12)}(0)$ collecting the fluctuations in the $x^1 - x^2$ -plane have the trivial action

$$\bar{\mathcal{A}}_e^{(12)} = \int_0^{\hbar\beta} d\lambda \frac{M}{2} (x_1'^2 + x_2'^2), \tag{19.68}$$

with the trivial fluctuation determinant $\text{Det}\,(-\partial_\lambda^2) = 1$, so that we obtain for $Z^{(12)}(0)$ the free-particle partition function in two dimensions

$$Z^{(12)}(0) = \Delta x_1 \Delta x_2 \sqrt{\frac{M}{2\pi\hbar^2\beta}}^{-2}. \tag{19.69}$$

The factor $\Delta x_1 \Delta x_2$ is the total area of the system in the $x^1 - x^2$ -plane. Note that upper and lower indices are the same in the present Euclidean metric.

For the motion in the $x^3 - x^4$ -plane, the quadratic fluctuations with periodic boundary conditions have a functional determinant

$$\mathrm{Det} \begin{pmatrix} -\partial_\lambda^2 & -i\omega_L^E \partial_\lambda \\ i\omega_L^E \partial_\lambda & -\partial_\lambda^2 \end{pmatrix} = \mathrm{Det}\left(-\partial_\lambda^2\right) \times \mathrm{Det}\left(-\partial_\lambda^2 + \omega_L^{E2}\right) = 1 \times \frac{\sinh \omega_L^E \hbar\beta}{\omega_L^E \hbar\beta}, \quad (19.70)$$

leading to the partition function becomes

$$Z^{(34)}(E) = \Delta x_3 \Delta x_4 \sqrt{\frac{M}{2\pi\hbar^2\beta}}^{-2} \frac{\sinh \omega_L^E \hbar\beta}{\omega_L^E \hbar\beta}. \quad (19.71)$$

This result can, of course, also be obtained without calculation from the observation that the Euclidean electric path integral is completely analogous to the real-time magnetic path integral solved in Section 2.18. Indeed, with the **E**-field pointing in the z-direction, the action (19.63) becomes for the motion in the $x^3 - x^4$ -plane

$$\bar{\mathcal{A}}_e^{(34)} = \int_0^{\hbar\beta} d\lambda \frac{M}{2} \left[x_3'^2 + x_4'^2 + \frac{e}{c}E(x_3 x_4' - x_4 x_3')\right]. \quad (19.72)$$

This coincides with the magnetic action (2.633) in real time, if we insert there the magnetic vector potential (2.634) and replace B by E. The equations of motion (19.62) reduces to

$$x_3'' = \omega_L^E x_4', \qquad x_4'' = -\omega_L^E x_3', \quad (19.73)$$

in agreement with the magnetic equations of motion (2.670) in real time. Thus the motion in the $x_3 - x_4$ -plane as a function of the pseudotime λ is the same as the real-time motion in the $x - y$ -plane, if the magnetic field B in the z-direction is exchanged by an E-field of the same size pointing in the x_3-axis. The amplitude can therefore be taken directly from (2.666) — we must merely replace the real time difference $t_b - t_a$ by $\hbar\beta$.

Inserting (19.69) and (19.71) into (19.67), we obtain the partition function of a single closed particle orbit in four Euclidean spacetime dimensions:

$$Z_1 = \Delta x_4 V \int_0^\infty \frac{d\beta}{\beta} \sqrt{\frac{M}{2\pi\hbar^2\beta}}^{-4} \frac{\omega_L^E \hbar\beta/2}{\sin \omega_L^E \hbar\beta/2} e^{-\beta Mc^2/2}, \quad (19.74)$$

where $V \equiv \Delta x_1 \Delta x_2 \Delta x_3$ is the total spatial volume. We now go over to real times by setting $\Delta x_4 = ic\Delta t$. By exponentiating the subtracted expression (19.74) as in Eq. (19.42), we go to a grand-canonical ensemble, and may identify Z_1 with i times the effective electromagnetic action caused by fluctuating ensemble of particle loops

$$Z_1 = i\Delta\mathcal{A}^{\mathrm{eff}}/\hbar = i\Delta t V \Delta\mathcal{L}^{\mathrm{eff}}/\hbar. \quad (19.75)$$

The integral over β in (19.74) is divergent. In order to make it convergent, we perform two subtractions. In the subtracted expression we change the integration variable to the dimensionless $\zeta = \beta Mc^2/2$, and obtain the effective Lagrangian density

$$\Delta \mathcal{L}^{\text{eff}} = \hbar c \left(\frac{Mc}{\hbar}\right)^4 \frac{1}{4(2\pi)^2} \int_0^\infty \frac{d\zeta}{\zeta^3} \left[\frac{E\zeta/E_c}{\sin E\zeta/E_c} - 1 - \frac{1}{6}\left(\frac{E}{E_c}\right)^2\right] e^{-\zeta}. \quad (19.76)$$

The first subtraction has removed the divergence coming from the $1/\zeta^3$-singularity in the integrand. This produces a real infinite field-independent contribution to the effective action which can be omitted since it is unobservable by electromagnetic experiments.[2] After subtracting this divergence, the integral still contains a logarithmic divergence. It can be interpreted as a contribution proportional to E^2 to the Lagrangian density

$$\Delta \mathcal{L}^{\text{eff}}_{\text{div}} = \hbar c \left(\frac{Mc}{\hbar}\right)^4 \frac{1}{24(2\pi)^2} \left(\frac{E}{E_c}\right)^2 \int_0^\infty \frac{d\zeta}{\zeta} e^{-\zeta} = \frac{\alpha}{24\pi} \int_0^\infty \frac{d\zeta}{\zeta} e^{-\zeta}, \quad (19.77)$$

which changes the original Maxwell term $E^2/2$ to $Z_A E^2/2$ with

$$Z_A = 1 + \frac{\alpha}{12\pi} \int_0^\infty \frac{d\zeta}{\zeta} e^{-\zeta}. \quad (19.78)$$

According to the rules of renormalization theory, the prefactor is removed by renormalizing the field strength, replacing $E \to E/Z_A^{1/2}$, and identifying the replaced field with the physical, *renormalized* field $E/Z_A^{1/2} \equiv E_R$.

Due to the presence of the affective action, the vacuum is no longer time independent, put depends on time like $e^{-i(H-i\Gamma\hbar/2)\Delta t/\hbar}$. Thus the decay rate of the vacuum per unit volume is given by the imaginary part of the effective Lagrangian density

$$\frac{\Gamma}{V} = \frac{2}{\hbar} \text{Im}\, \Delta \mathcal{L}^{\text{eff}}. \quad (19.79)$$

In order to calculate this we replace $E\zeta/E_c$ by z in (19.76), and expand in the integrand

$$\frac{z}{\sin z} = 1 + 2\sum_{n=1}^\infty (-1)^n \frac{z^2}{z^2 - n^2\pi^2} = 2\sum_{n=1}^\infty (-1)^n \frac{\zeta^2}{\zeta^2 - \zeta_n^2}, \quad \zeta_n \equiv n\pi \frac{E_c}{E}. \quad (19.80)$$

Adding to the poles the usual infinitesimal $i\eta$-shifts in the complex plane (see p. 115), we replace with the help of the decomposition (1.325)

$$\frac{\zeta}{\zeta^2 - \zeta_n^2} \to \frac{\zeta}{\zeta^2 - \zeta_n^2 + i\eta} = i\frac{\pi}{2}\delta(\zeta + \zeta_n) - i\frac{\pi}{2}\delta(\zeta - \zeta_n) + \zeta\frac{\mathcal{P}}{\zeta^2 - \zeta_n^2}. \quad (19.81)$$

---

[2]This energy would, however, be observable in the cosmological evolution to be discussed in Subsection 19.2.3.

The δ-functions yield imaginary parts and thus directly the decay rate

$$
\begin{aligned}
\frac{\Gamma}{V} &= \frac{2}{\hbar} \operatorname{Im} \mathcal{L}^{\text{eff}} = c \left(\frac{Mc}{\hbar}\right)^4 \left(\frac{E}{E_c}\right)^2 \frac{1}{4\pi^3} \frac{1}{2} \sum_{n=1}^{\infty} (-1)^{n-1} \frac{(-1)^{n-1}}{n^2} e^{-n\pi E_c/E} \\
&= \frac{1}{8\pi^3} \frac{e^2}{\hbar c} E^2 \sum_{n=1}^{\infty} \frac{(-1)^{n-1}}{n^2} e^{-n\pi E_c/E}.
\end{aligned} \tag{19.82}
$$

The principal values produce a real effective Lagrangian density

$$
\Delta\mathcal{L}_{\mathcal{P}}^{\text{eff}} = \hbar c \left(\frac{Mc}{\hbar}\right)^4 \frac{1}{4(2\pi)^2} \mathcal{P} \int_0^{\infty} \frac{d\zeta}{\zeta^3} \left(\frac{E\zeta/E_c}{\sin E\zeta/E_c} - 1 - \frac{E^2\zeta^2}{6E_c^2}\right) e^{-\zeta}. \tag{19.83}
$$

If we expand

$$
\frac{z}{\sin z} = 1 + \frac{z^2}{6} + \frac{7}{360}z^4 + \frac{31}{15120}z^6 + \dots \tag{19.84}
$$

and perform the integrals over ζ, we find

$$
\Delta\mathcal{L}^{\text{eff}} = \hbar c \left(\frac{Mc}{\hbar}\right)^4 \frac{1}{4(2\pi)^2} \left[\frac{7}{360}\left(\frac{E}{E_c}\right)^4 + \frac{31}{2520}\left(\frac{E}{E_c}\right)^6 + \dots\right], \tag{19.85}
$$

or

$$
\mathcal{L}^{\text{eff}} = \frac{1}{2}\left\{E^2 + \frac{7\alpha^2}{180}\frac{(\hbar c)^3}{(Mc^2)^4}E^4 + \frac{31\pi\alpha^3}{315}\left[\frac{(\hbar c)^3}{(Mc^2)^4}\right]^2 E^6 + \dots\right\}, \tag{19.86}
$$

The subscript $\mathcal{P}$ of $\mathcal{L}$ has been omitted since the imaginary part of the effective Lagrangian density (19.82) has a vanishing Taylor expansion. Each coefficient in (19.99) is exact to leading order in α.

The extra expansion terms in (19.99) imply that the physical vacuum has a nontrivial E-dependent dielectric constant $\epsilon(E)$. This is caused by the virtual creation and annihilation of particle-antiparticle pairs. Since the dielectric displacement $D(E)$ is obtained from the first derivative of $\mathcal{L}^{\text{eff}}$, the dielectric constant is given by $\epsilon(E) = D(E)/E = \partial\mathcal{L}^{\text{eff}}/E\partial E$. From (19.99) we find the lowest expansion terms

$$
\epsilon(E) = 1 + \frac{7\alpha^2}{90}\frac{(\hbar c)^3}{(Mc^2)^4}E^2 + \frac{31\pi\alpha^3}{105}\left[\frac{(\hbar c)^3}{(Mc^2)^4}\right]^2 E^4 + \dots. \tag{19.87}
$$

Such a term gives rise to a small amplitude for photon-photon scattering in the vacuum, a process which has been observed in the laboratory.

Including Constant Magnetic Field parallel to Electric Field

Let us see how the decay rate (19.82) and the effective Lagrangian (19.76) are influenced by the presence of an additional constant **B**-field. This will at first be assumed to be parallel to the **E**-field, with both fields pointing in the z-direction. Then the action (19.68) for the motion in the $x^1 - x^2$ -plane becomes

$$\bar{\mathcal{A}}_e^{(12)} = \int_0^{\hbar\beta} d\lambda \frac{M}{2} \left[x_1'^2 + x_2'^2 + \frac{e}{c} iB(x_1 x_2' - x_2 x_1') \right]. \tag{19.88}$$

The partition function in the $x^1 - x^2$ -plane will therefore have the same form as (19.71), except with ω_L^E replaced by $i\omega_L^B$:

$$Z^{(12)}(B) = Z^{(34)}(iB). \tag{19.89}$$

Thus the B-field changes the partition function (19.74) of a single closed orbit to

$$Z_1 = \Delta x_4 \, V \int_0^\infty \frac{d\beta}{\beta} \sqrt{\frac{M}{2\pi\hbar^2\beta}}^4 \frac{\omega_L^E \hbar\beta/2}{\sin \omega_L^E \hbar\beta/2} \frac{\omega_L^B \hbar\beta/2}{\sinh \omega_L^B \hbar\beta/2} e^{-\beta Mc^2/2}, \tag{19.90}$$

and the effective Lagrangian (19.76) becomes

$$\Delta\mathcal{L}^{\text{eff}} = \hbar c \left(\frac{Mc}{\hbar}\right)^4 \frac{1}{4(2\pi)^2} \int_0^\infty \frac{d\zeta}{\zeta^3} \left[\frac{E\zeta/E_c}{\sin E\zeta/E_c} \frac{B\zeta/E_c}{\sinh B\zeta/E_c} - 1 - \frac{(E^2-B^2)\zeta^2}{6E_c^3} \right] e^{-\zeta}. \tag{19.91}$$

In the subtracted free-field term (19.77), the term E^2 is changed to the Lorentz-invariant combination $E^2 - B^2$. The decay rate (19.82) is modified to

$$\frac{\Gamma}{V} = \frac{2}{\hbar} \text{Im} \, \Delta\mathcal{L}^{\text{eff}} = c \left(\frac{Mc}{\hbar}\right)^4 \left(\frac{E}{E_c}\right)^2 \frac{1}{4\pi^3} \frac{1}{2} \sum_{n=1}^\infty (-1)^{n-1} \frac{1}{n^2} \frac{n\pi B/E}{\sinh n\pi B/E} e^{-n\pi E_c/E}. \tag{19.92}$$

Including Constant Magnetic Field in any Direction

The case of a constant magnetic field pointing in any direction can be reduced to the parallel case by a simple Lorentz transformation. It is always possible to find a Lorentz frame in which the fields become parallel. This special frame will be called *center-of-fields* frame, and the transformed fields in this frame will be denoted by $\mathbf{B}_{\text{CF}}$ and $\mathbf{E}_{\text{CF}}$. The transformation has the form

$$\mathbf{E}_{\text{CF}} = \gamma \left(\mathbf{E} + \frac{\mathbf{v}}{c} \times \mathbf{B} \right) - \frac{\gamma^2}{\gamma+1} \frac{\mathbf{v}}{c} \left(\frac{\mathbf{v}}{c} \cdot \mathbf{E} \right), \tag{19.93}$$

$$\mathbf{B}_{\text{CF}} = \gamma \left(\mathbf{B} - \frac{\mathbf{v}}{c} \times \mathbf{E} \right) - \frac{\gamma^2}{\gamma+1} \frac{\mathbf{v}}{c} \left(\frac{\mathbf{v}}{c} \cdot \mathbf{B} \right), \tag{19.94}$$

with a velocity of the transformation determined by

$$\frac{\mathbf{v}/c}{1 + (|\mathbf{v}|/c)^2} = \frac{\mathbf{E} \times \mathbf{B}}{|\mathbf{E}|^2 + |\mathbf{B}|^2}, \tag{19.95}$$

and $\gamma \equiv [1 - (|\mathbf{v}|/c)^2]^{-1/2}$. The fields $|\mathbf{E}_{\text{CF}}| \equiv \varepsilon$ and $|\mathbf{B}_{\text{CF}}| \equiv \beta$ are, of course, Lorentz-invariant quantities which can be expressed in terms of the two quadratic

Lorentz invariants of the electromagnetic field: the scalar S and the pseudoscalar P defined by

$$S \equiv -\frac{1}{4} F_{\mu\nu} F^{\mu\nu} = \frac{1}{2} \left(\mathbf{E}^2 - \mathbf{B}^2 \right) = \frac{1}{2} \left(\varepsilon^2 - \beta^2 \right), \quad P \equiv -\frac{1}{4} F_{\mu\nu} \tilde{F}^{\mu\nu} = \mathbf{E}\,\mathbf{B} = \varepsilon\beta.$$
(19.96)

Solving these equation yields

$$\left\{ \begin{matrix} \varepsilon \\ \beta \end{matrix} \right\} \equiv \sqrt{\sqrt{S^2 + P^2} \pm S} = \frac{1}{\sqrt{2}} \sqrt{\sqrt{(\mathbf{E}^2 - \mathbf{B}^2)^2 + 4(\mathbf{E}\,\mathbf{B})^2} \pm (\mathbf{E}^2 - \mathbf{B}^2)}. \quad (19.97)$$

As a result we find that the formulas (19.91) and (19.92) are valid for arbitrary constant fields $\mathbf{E}$ and $\mathbf{B}$ if we replace E and B by the Lorentz invariants ε and β. After this we may expand the integrand in (19.91) in powers of ε and β using Eq. (19.84),

$$\frac{1}{\tau^3} \frac{e\beta\tau}{\sin e\beta\tau} \frac{e\varepsilon\tau}{\sinh e\varepsilon\tau} = \frac{1}{\tau^3} - \frac{e^2}{6\tau} \left(\varepsilon^2 - \beta^2 \right) + e^4 \frac{\tau}{360} \left(7\varepsilon^4 - 10\varepsilon^2\beta^2 + 7\beta^4 \right)$$
$$- e^6 \frac{\tau^3}{1520} \left(31\varepsilon^6 - 49\varepsilon^4\beta^2 + 49\varepsilon^2\beta^4 - 31\beta^6 \right) + \dots . \quad (19.98)$$

and obtain the effective Lagrangian the fields generalizing (19.99):

$$\mathcal{L}^{\text{eff}} = \frac{1}{2} \left\{ (\mathbf{E}^2 - \mathbf{B}^2) + \frac{7\alpha^2}{180} \frac{(\hbar c)^3}{(Mc^2)^4} (\mathbf{E}^2 - \mathbf{B}^2)^2 \right.$$
$$\left. + \frac{31\pi\alpha^3}{315} \left[\frac{(\hbar c)^3}{(Mc^2)^4} \right]^2 (\mathbf{E}^2 - \mathbf{B}^2)[2(\mathbf{E}^2 - \mathbf{B}^2)^2 - 4(\mathbf{EB})^2] + \dots \right\}, \quad (19.99)$$

For strong fields, a saddle point approximation to the generalized integral (19.91) yields the asymptotic form

$$\mathcal{L}^{\text{eff}} \equiv -\frac{e^2}{192\pi^2} (\mathbf{E}^2 - \mathbf{B}^2) \log \left[-4e^2 \frac{(\hbar c)^3}{(Mc^2)^4} (\mathbf{E}^2 - \mathbf{B}^2) \right] + \dots . \quad (19.100)$$

Spin-1/2 Particles

We anticipate the small modification which is necessary to obtain the analogous result for spin-1/2 fermions: The tools for this will be developed in Subsections 19.5.1–19.5.3. The relevant formula has actually been derived before in Eq. (7.512). The bosonic result receives for fermions a factor -1 and a fluctuation factor which is the square root of the functional determinant

$$4 \operatorname{Det} \left(-\delta_{\mu\nu} \, \partial_\lambda - i \frac{e}{Mc} F_{\mu\nu} \right) = 4 \det \left(\cosh \frac{e}{Mc} F_{\mu\nu} \hbar\beta \right). \quad (19.101)$$

The normalization factor follows from Eqs. (7.415) and (7.418), where we found that the path integral of single complex fermion field carries a normalization factor 2. For a purely electric field, the square-root of the right-hand side of (19.101) is $2\det^{1/2}\left[\cos(e/Mc)F_{\mu\nu}\right] = 2\cos(eE/Mc)$. Multiplying the cos-factor into the expansion (19.80), this is modified to

$$z\frac{\cos z}{\sin z} = 1 + 2\sum_{n=1}^{\infty}\frac{z^2}{z^2 - n^2\pi^2} = 2\sum_{n=1}^{\infty}\frac{\zeta^2}{\zeta^2 - \zeta_n^2}, \quad \zeta_n \equiv n\pi\frac{E_c}{E}. \quad (19.102)$$

Performing now the singular integral over ζ in (19.76), we obtain for the decay rate the same formula as in (19.82), except that the alternating signs are absent. The factor 4 of the fermionic determinant reduces to a factor 2 for the effective action. The resulting effective Lagrangian density for spin-1/2 fermions is therefore

$$\Delta\mathcal{L}_{\text{spin}\frac{1}{2}}^{\text{eff}} = -\hbar c\left(\frac{Mc}{\hbar}\right)^4\frac{1}{2(2\pi)^2}\int_0^{\infty}\frac{d\zeta}{\zeta^3}\left(\frac{E\zeta/E_c}{\tan E\zeta/E_c} - 1 + \frac{E^2\zeta^2}{3E_c^3}\right)e^{-\zeta}, \quad (19.103)$$

a result first derived by Heisenberg and Euler in 1935 [8]. Its imaginary part yields the decay rate of the vacuum due to pair creation

$$\begin{aligned}
\frac{\Gamma_{\text{spin}\frac{1}{2}}}{V} &= \frac{2}{\hbar}\,\text{Im}\,\Delta\mathcal{L}_{\text{spin}\frac{1}{2}}^{\text{eff}} = c\left(\frac{Mc}{\hbar}\right)^4\left(\frac{E}{E_c}\right)^2\frac{1}{4\pi^3}\sum_{n=1}^{\infty}\frac{1}{n^2}e^{-n\pi E_c/E} \\
&= \frac{1}{4\pi^3}\frac{e^2}{\hbar c}E^2\sum_{n=1}^{\infty}\frac{1}{n^2}e^{-n\pi E_c/E}.
\end{aligned} \quad (19.104)$$

The reason for the reduction from 4 to 2 is that the sum over bosonic paths has to be first divided by a factor 2 to remove their orientation, before the fermionic factor 4 is applied. This procedure is not so obvious at this point but will be understood later in Subsection 19.5.2. The remaining factor 2 accounts for the two spin orientations of the charged particles.

The Taylor series

$$\frac{z}{\tan z} = 1 - \frac{z^2}{3} - \frac{1}{45}z^4 - \frac{2}{945}z^6 - \ldots \quad (19.105)$$

in the integrand of (19.103) leads to the expansion

$$\Delta\mathcal{L}^{\text{eff}} = \hbar c\left(\frac{Mc}{\hbar}\right)^4\frac{2}{16\pi^2}\left[\frac{1}{45}\left(\frac{E}{E_c}\right)^4 + \frac{4}{315}\left(\frac{E}{E_c}\right)^6 + \ldots\right], \quad (19.106)$$

so that

$$\mathcal{L}^{\text{eff}} = \frac{1}{2}\left\{E^2 + \frac{4\alpha^2}{45}\frac{(\hbar c)^3}{(Mc^2)^4}E^4 + \frac{64\pi\alpha^3}{315}\left[\frac{(\hbar c)^3}{(Mc^2)^4}\right]^2 E^6 + \ldots\right\}. \quad (19.107)$$

The term proportional to α^2-term implies a small amplitude for photon-photon scattering which can be observed in the laboratory [9].

As in the boson result (19.99), each coefficient is exact to leading order in α, and the virtual creation and annihilation of fermion-antifermion pairs gives the physical vacuum a nontrivial E-dependent dielectric constant

$$\epsilon(E) = \frac{1}{E}\frac{\partial \mathcal{L}^{\text{eff}}}{\partial E} = 1 + \frac{8\alpha^2}{45}\frac{(\hbar c)^3}{(Mc^2)^4}E^2 + \frac{64\pi\alpha^3}{105}\left[\frac{(\hbar c)^3}{(Mc^2)^4}\right]^2 E^4 + \ldots \; . \quad (19.108)$$

If we admit also a constant **B**-field parallel to **E**, the formulas (19.104) and (19.103) for the spin-$\frac{1}{2}$-particles are modified in the same way as the bosonic formulas (19.82) and (19.83), except that the determinant (19.101) in spinor space introduces a further factor $\cosh(eB/Mc)$. Thus we obtain

$$\Delta\mathcal{L}^{\text{eff}}_{\text{spin}\frac{1}{2}} = -\hbar c\left(\frac{Mc}{\hbar}\right)^4\frac{1}{2(2\pi)^2}\int_0^\infty \frac{d\zeta}{\zeta^3}\left[\frac{E\zeta/E_c}{\tan E\zeta/E_c}\frac{B\zeta/E_c}{\tanh B\zeta/E_c} - 1 + \frac{(E^2-B^2)\zeta^2}{3E_c^3}\right]e^{-\zeta}.$$
$$(19.109)$$

For a general combination of constant electric and magnetic fields, we simply exchange E and B by the Lorentz invariants ε and β. From the imaginary part we obtain the decay rate

$$\frac{\Gamma_{\text{spin}\frac{1}{2}}}{V} = \frac{2}{\hbar}\,\text{Im}\,\Delta\mathcal{L}^{\text{eff}}_{\text{spin}\frac{1}{2}} = c\left(\frac{Mc}{\hbar}\right)^4\left(\frac{\varepsilon}{E_c}\right)^2\frac{1}{4\pi^3}\sum_{n=1}^\infty \frac{1}{n^2}\frac{n\pi\beta/\varepsilon}{\tanh n\pi\beta/\varepsilon}e^{-n\pi E_c/\varepsilon}. \quad (19.110)$$

For strong fields, a saddle point approximation to the generalized integral (19.103) yields the asymptotic form

$$\mathcal{L}^{\text{eff}} \equiv -\frac{e^2}{48\pi^2}(\mathbf{E}^2 - \mathbf{B}^2)\log\left[-4e^2\frac{(\hbar c)^3}{(Mc^2)^4}(\mathbf{E}^2 - \mathbf{B}^2)\right] + \ldots \; . \quad (19.111)$$

19.2.2 Birth of Universe

A similar tunneling phenomenon could explain the birth of the expanding universe [10].

As an idealization of the observed density of matter, the universe is usually assumed to be isotropic and homogeneous. Then it is convenient to describe it by a coordinate frame in which the metric is rotationally invariant. To account for the expansion, we have to allow for an explicit time dependence of the spatial part of the metric. In the spatial part, we choose coordinates which participate in the expansion. They can be imagined as being attached to the gas particles in a homogenized universe. Then the time passing at each coordinate point is the proper time. In this context it is the so-called *cosmic standard time* to be denoted by t. We imagine being an observer at a coordinate point with $dx^i/dt = 0$, and measure t by counting the number of orbits of an electron around a proton in a hydrogen atom, starting from the moment of the big bang (forgetting the fact that in the early time of the universe, the atom does not yet exist).

Geometry

With this time calibration, the component g_{00} of the metric tensor is identically equal to unity

$$g_{00}(x) \equiv 1, \tag{19.112}$$

such that at a fixed coordinate point, the proper time coincides with the coordinate time, $d\tau = dt$. Moreover, since all clocks in space follow the same prescription, there is no mixing between time and space coordinates, a property called *time orthogonality*, so that

$$g_{0i}(x) \equiv 0. \tag{19.113}$$

As a consequence, the Christoffel symbol $\bar{\Gamma}_{00}{}^{\mu}$ [recall (10.7)] vanishes identically:

$$\bar{\Gamma}_{00}{}^{\mu} \equiv \frac{1}{2} g^{\mu\nu} (\partial_0 g_{0\nu} + \partial_0 g_{0\nu} - g_{00}) \equiv 0. \tag{19.114}$$

This is the mathematical way of expressing the fact that a particle sitting at a coordinate point, which has $dx^i/dt = 0$, and thus $dx^\mu/dt = u^\mu = (c, 0, 0, 0)$, experiences no acceleration

$$\frac{du^\mu}{d\tau} = -\bar{\Gamma}_{00}{}^{\mu} c^2 = 0. \tag{19.115}$$

The coordinates themselves are trivially comoving.

Under the above condition, the invariant distance has the general form

$$ds^2 = c^2 dt^2 - {}^{(3)}g_{ij}(x) dx^i dx^j. \tag{19.116}$$

We now impose the spatial isotropy upon the spatial metric g_{ij}. Denote the spatial length element by dl, so that

$$dl^2 = {}^{(3)}g_{ij}(x) dx^i dx^j. \tag{19.117}$$

The isotropy and homogeneity of space is most easily expressed by considering the spatial curvature ${}^{(3)}R_{ijk}{}^{l}$ calculated from the spatial metric ${}^{(3)}g_{ij}(x)$. The space corresponds to a spherical surface. If its radius is a, the curvature tensor is, according to Eq. (10.161),

$$^{(3)}R_{ijkl}(x) = \frac{1}{a^2} \left[{}^{(3)}g_{il}(x) {}^{(3)}g_{jk}(x) - {}^{(3)}g_{ik}(x) {}^{(3)}g_{jl}(x) \right]. \tag{19.118}$$

The derivation of this expression in Section 10.4 was based on the assumption of a spherical space whose curvature $K \equiv 1/a^2$ is positive. If we allow also for hyperbolic and parabolic spaces with negative and vanishing curvature, and characterize these by a constant

$$k = \left\{ \begin{array}{ll} 1 & \text{spherical} \\ 0 & \text{parabolic} \\ -1 & \text{hyperbolic} \end{array} \right\} \text{ universe,} \tag{19.119}$$

then the prefactor $1/a^2$ in (19.118) is replaced by $K \equiv k/a^2$. For $k = -1$ and 0, the space has an open topology and an infinite total volume.

The Ricci tensor and curvature scalar are in these three cases [compare (10.163) and (10.156)]

$$^{(3)}R_{il} = k\frac{2}{a^2} g_{il}(x), \qquad ^{(3)}R = k\frac{6}{a^2}. \tag{19.120}$$

By construction it is obvious that for $k = 1$, the three-dimensional space has a closed topology and a finite spatial volume which is equal to the surface of a sphere of radius a in four dimensions

$$^4S^a = 2\pi^2 a^3. \tag{19.121}$$

A circle in this space has maximal radius a and a maximal circumference $2\pi a$. A sphere with radius $r_0 < a$ has a volume

$$\begin{aligned}
^{(3)}V_{r_0}^a &= \int_0^{2\pi} d\varphi \int_0^\pi d\theta \sin\theta \int_0^r dr \frac{r^2}{\sqrt{1 - r^2/a^2}} \tag{19.122} \\
&= 4\pi \left(\frac{a^3}{2} \arcsin\frac{r_0}{a} - \frac{a^2 r_0}{2}\sqrt{1 - \frac{r_0^2}{a^2}} \right).
\end{aligned}$$

For small r_0, the curvature is irrelevant and the volume depends on r_0 like the usual volume of a sphere in three dimensions:

$$^{(3)}V_{r_0}^a \approx V_{r_0} = \frac{4\pi}{3}r_0^2. \tag{19.123}$$

For $r_0 \to a$, however, $^{(3)}V_{r_0}^a$ approaches a saturation volume $2\pi a^3$.

The analogous expressions for negative and zero curvature are obvious.

Robertson-Walker Metric

In spherical coordinates, the four-dimensional invariant distance (19.116) defines the *Robertson-Walker metric*.

$$ds^2 = c^2 dt^2 - dl^2 \tag{19.124}$$

$$dl^2 = \frac{dr^2}{1 - kr^2/a^2} + r^2(d\theta^2 + \sin^2\theta d\varphi^2). \tag{19.125}$$

It will be convenient to introduce, instead of r, the angle α on the surface of the four-sphere, such that

$$r = a\sin\alpha. \tag{19.126}$$

Then the metric has the four-dimensional angular form

$$ds^2 = c^2 dt^2 - a^2(t)[d\alpha^2 + f^2(\alpha)(d\theta^2 + \sin^2\theta d\varphi^2)], \tag{19.127}$$

where for spherical, parabolic, and hyperbolic spaces, $f(\alpha)$ is equal to

$$f(\alpha) = \begin{cases} \sin \alpha & k = 1, \\ \alpha & k = 0, \\ \sinh \alpha & k = -1. \end{cases} \tag{19.128}$$

In order to have maximal symmetry, it is useful to absorb $a(t)$ into the time and define a new timelike variable η by

$$c \, dt = a(\eta) \, d\eta, \tag{19.129}$$

so that the invariant distance is measured by

$$ds^2 = a^2(\eta)[d\eta^2 - d\alpha^2 - f^2(\alpha)(d\theta^2 + \sin^2 \theta d\varphi^2)]. \tag{19.130}$$

Then the metric is simply

$$g_{\mu\nu} = a^2(\eta) \begin{pmatrix} 1 & & & \\ & -1 & & \\ & & -f^2(\alpha) & \\ & & & -f^2(\alpha)\sin^2\theta, \end{pmatrix} \tag{19.131}$$

and the Christoffel symbols become

$$\Gamma_{00}{}^0 = \frac{a_\eta}{a}, \quad \Gamma_{00}{}^i = 0, \quad \Gamma_{0i}{}^0 = 0, \quad \Gamma_{0i}{}^j = \frac{a_\eta}{a}\delta_i{}^j, \quad \Gamma_{ij}{}^0 = -\frac{a_\eta}{a^3}g_{ij}, \quad \Gamma_{ij}{}^k = 0, \tag{19.132}$$

where the subscripts denote derivatives with respect to the corresponding variables:

$$a_\eta \equiv \frac{da}{d\eta} = \frac{a}{c}\frac{da}{dt} \equiv \frac{a}{c}a_t. \tag{19.133}$$

We now calculate the 00-component of the Ricci tensor:

$$R_{00} = \partial_\mu \Gamma_{00}{}^\mu - \partial_0 \Gamma_{\mu 0}{}^\mu - \Gamma_{\mu 0}{}^\nu \Gamma_{0\nu}{}^\mu + \Gamma_{00}{}^\mu \Gamma_{\nu\mu}{}^\nu. \tag{19.134}$$

Inserting the Christoffel symbols (19.132) we find

$$\partial_\mu \Gamma_{00}{}^\mu - \partial_0 \Gamma_{\mu 0}{}^\mu = -\partial_0 \Gamma_{i0}{}^i = -3\frac{d}{d\eta}\frac{a_\eta}{a} = -3\frac{1}{a^2}\left(a_{\eta\eta}a - a_\eta^2\right), \tag{19.135}$$

$$\Gamma_{\mu 0}{}^\nu \Gamma_{0\nu}{}^\mu = \Gamma_{00}{}^0\Gamma_{00}{}^0 + \Gamma_{00}{}^i\Gamma_{0i}{}^0 + \Gamma_{i0}{}^0\Gamma_{00}{}^i + \Gamma_{i0}{}^j\Gamma_{0j}{}^i = \left(\frac{a_\eta}{a}\right)^2 + 3\left(\frac{a_\eta}{a}\right)^2, \tag{19.136}$$

$$\Gamma_{00}{}^\mu \Gamma_{\nu\mu}{}^\nu = \Gamma_{00}{}^0\Gamma_{00}{}^0 + \Gamma_{00}{}^0\Gamma_{i0}{}^i + \Gamma_{00}{}^i\Gamma_{0i}{}^0 + \Gamma_{00}{}^i\Gamma_{ki}{}^k = \left(\frac{a_\eta}{a}\right)^2 + 3\left(\frac{a_\eta}{a}\right)^2, \tag{19.137}$$

so that

$$R_{00} = -\frac{3}{a^2}(aa_{\eta\eta} - a_\eta^2), \quad R_0{}^0 = g^{00}R_{00} = -\frac{3}{a^4}(aa_{\eta\eta} - a_\eta^2). \tag{19.138}$$

The other components can be determined by relating them to the three-dimensional curvature tensor $^{(3)}R_{ij}$ which has the simple form (19.118). So we calculate

$$
\begin{aligned}
R_{ij} = R_{\mu ij}{}^{\mu} &= R_{kij}{}^{k} + R_{0ij}{}^{0} \\
&= {}^{(3)}R_{ij} - \Gamma_{kj}{}^{0}\Gamma_{i0}{}^{k} + \Gamma_{ij}{}^{0}\Gamma_{k0}{}^{k} + R_{0ij}{}^{0}.
\end{aligned}
\tag{19.139}
$$

Inserting

$$
R_{0ij}{}^{0} = \partial_{0}\Gamma_{ij}{}^{0} - \partial_{i}\Gamma_{0j}{}^{0} - \Gamma_{0j}{}^{l}\Gamma_{il}{}^{0} - \Gamma_{0j}{}^{0}\Gamma_{i0}{}^{0} + \Gamma_{ij}{}^{l}\Gamma_{0l}{}^{0} + \Gamma_{ij}{}^{0}\Gamma_{00}{}^{0},
\tag{19.140}
$$

$$
{}^{(3)}R_{ij} = k\,\frac{2}{a^{2}}g_{ij}
\tag{19.141}
$$

and the above Christoffel symbols (19.132) gives

$$
R_{ij} = -\frac{1}{a^{4}}(2ka^{2} + a_{\eta}^{2} + aa_{\eta\eta})g_{ij}
\tag{19.142}
$$

and thus a curvature scalar

$$
\begin{aligned}
R &= g^{00}R_{00} + g^{ij}R_{ij} = -\frac{1}{a^{2}}\left[\frac{3}{a^{2}}(aa_{\eta\eta} - a_{\eta}^{2})\right] - \frac{3}{a^{4}}(2ka^{2} + a_{\eta}^{2} + aa_{\eta\eta}) \\
&= -\frac{6}{a^{3}}(a_{\eta\eta} + ka).
\end{aligned}
\tag{19.143}
$$

Action and Field Equation

In the absence of matter, the *Einstein-Hilbert action* of the gravitational field is

$$
\overset{f}{\mathcal{A}} = \int d^{4}x\sqrt{-g}\,\overset{f}{\mathcal{L}} = -\frac{1}{2\kappa}\int d^{4}x\sqrt{-g}(R + 2\lambda),
\tag{19.144}
$$

were κ is related to Newton's gravitational constant

$$
G_{N} \approx 6.673 \cdot 10^{-8}\ \text{cm}^{3}\text{g}^{-1}\text{s}^{-2}
\tag{19.145}
$$

by

$$
\frac{1}{\kappa} = \frac{c^{3}}{8\pi G_{N}}.
\tag{19.146}
$$

A natural length scale of gravitational physics is the *Planck length*, which can be formed from a combination of Newton's gravitational constant (19.145), the light velocity $c \approx 3 \times 10^{10}$ cm/s, and Planck's constant $\hbar \approx 1.05459 \times 10^{-27}$:

$$
l_{\mathrm{P}} = \left(\frac{c^{3}}{G_{N}\hbar}\right)^{-1/2} \approx 1.615 \times 10^{-33}\ \text{cm}.
\tag{19.147}
$$

It is the Compton wavelength $l_{\rm P} \equiv \hbar/m_{\rm P}c$ associated with the Planck mass

$$m_{\rm P} = \left(\frac{c\hbar}{G_N}\right)^{1/2} \approx 2.177 \times 10^{-5}\,{\rm g} = 1.22 \times 10^{22}\,{\rm MeV}/c^2. \tag{19.148}$$

The constant $1/\kappa$ in the action (19.144) can be expressed in terms of the Planck length as

$$\frac{1}{\kappa} = \frac{\hbar}{8\pi l_P^2}. \tag{19.149}$$

If we add to (19.144) a matter action and vary to combined action with respect to the metric $g_{\mu\nu}$, we obtain the *Einstein equation*

$$\frac{1}{\kappa}\left(R_{\mu\nu} - \frac{1}{2}g_{\mu\nu}R - \lambda g_{\mu\nu}\right) = T_{\mu\nu}, \tag{19.150}$$

where $T_{\mu\nu}$ is the energy-momentum tensor of matter. The constant λ is called the *cosmological constant*. It is believed to arise from the zero-point oscillations of all quantum fields in the universe.

A single field contributes to the Lagrangian density $\overset{f}{\mathcal{L}}$ in (19.144) a term $-\Lambda \equiv -\lambda/\kappa$ which is typically of the order of $\hbar/l_{\rm P}^4$. For bosons, the sign is positive, for fermions negative, reflecting the filling of all negative-energy in the vacuum. A constant of this size is much larger than the present experimental estimate. In the literature one usually finds estimates for the dimensionless quantity

$$\Omega_{\lambda 0} \equiv \frac{\lambda c^2}{3H_0^2}. \tag{19.151}$$

where H_0 is the *Hubble constant*, whose inverse is roughly the lifetime of the universe

$$H_0^{-1} \approx 14 \times 10^9\,{\rm years}. \tag{19.152}$$

Present fits to distant supernovae and other cosmological data yield the estimate [11]

$$\Omega_{\lambda 0} \approx 0.68 \pm 0.10. \tag{19.153}$$

The associated cosmological constant λ has the value

$$\lambda = \Omega_{\lambda 0}\frac{3H_0^2}{c^2} \approx \frac{\Omega_{\lambda 0}}{(6.55 \times 10^{27}\,{\rm cm})^2} \approx \frac{\Omega_{\lambda 0}}{(6.93 \times 10^9\,{\rm ly})^2} \approx \frac{\Omega_{\lambda 0}}{(2.14\,R_{\rm universe})^2}. \tag{19.154}$$

Note that in the presence of λ, the Schwarzschild solution around a mass M has the metric

$$ds^2 = B(r)c^2dt^2 - B^{-1}dr^2 - r^2d\theta^2 - r^2\sin^2\theta d\phi^2, \tag{19.155}$$

with

$$B(r) = 1 - \frac{2MG_N}{c^2 r} - \frac{2}{3}\lambda r^2 = 1 - \frac{M}{m_P}\frac{l_P}{r} - \frac{2}{3}\Omega_{\lambda 0}\frac{r^2}{(2.14\,R_{\text{universe}})^2}. \quad (19.156)$$

The λ-term adds a small repulsion to Newton's force between mass points. if the distances are the order of the radius of the universe.

The value of the constant Λ associated with (19.153) is

$$\Lambda = \frac{\lambda}{\kappa} = \Omega_{\lambda 0}\frac{3H_0^2}{c^2}\frac{l_P^2}{8\pi} \approx 10^{-122}\frac{\hbar}{l_P^4}. \quad (19.157)$$

Such a small prefactor can only arise from an almost perfect cancellation of the contributions of boson and fermion fields. This cancellation is the main reason for postulating a broken supersymmetry in the universe, in which every boson has a fermionic counterpart. So far, the known particle spectra show no trace of such a symmetry. Thus there is need to explain it by some other not yet understood mechanism.

The simplest model of the universe governed by the action (19.144) is called *Friedmann model* or *Friedmann universe*.

19.2.3 Friedmann Model

Inserting (19.138) and (19.143) into the 00-component of the Einstein equation (19.150), we obtain the equation for the energy

$$\frac{3}{a^4}\left(a_\eta^2 + ka^2\right) - \lambda = \kappa T_0^{\,0}. \quad (19.158)$$

In terms of the cosmic standard time t, the general equation reads

$$3\left[\left(\frac{a_t}{a}\right)^2 + k\frac{c^2}{a^2}\right] - \lambda c^2 = c^2\kappa T_0^{\,0}. \quad (19.159)$$

The simplest Friedmann model is based on an energy-momentum tensor $T_0^{\,0}$ of an ideal pressure-less gas of mass density ρ:

$$T_\mu^{\,\nu} = c\rho u_\mu u^\nu, \quad (19.160)$$

where u^μ are the velocity four-vectors $u^\mu = (\gamma, \gamma\mathbf{v}/c)$ of the particles whose components are $u^\mu = (\gamma, \gamma\mathbf{v}/c)$ with $\gamma \equiv 1/\sqrt{1 - v^2/c^2}$ if the four components transform like $(dx^0 = cdt, d\mathbf{x})$. The gas is assumed to be at rest in our comoving coordinates. so that only the $T_0^{\,0}$-component is nonzero:

$$T_0^{\,0} = c\rho. \quad (19.161)$$

This component is invariant under the time transformation (19.129).

As a fortunate accident, this component of the Einstein equation has no $a_{\eta\eta}a$ term. Thus we may simply study the first-order differential equation

$$\frac{3}{a^4}\left(a_\eta^2 + ka^2\right) - \lambda = c\kappa\rho. \tag{19.162}$$

Since the total volume of the universe is $2\pi a^3$, we can express ρ in terms of the total mass M as follows

$$\rho = \frac{M}{2\pi^2 a^3}. \tag{19.163}$$

In this way we arrive at the differential equation

$$\frac{3}{a^4}(a_\eta^2 + ka^2) - \lambda = \frac{\kappa Mc}{2\pi^2 a^3} = \frac{4G_{\mathrm{N}}M}{\pi c^2 a^3}. \tag{19.164}$$

This equation of motion can also be obtained in another way. We express the action (19.144) in terms of $a(\eta)$ using the equation (19.178) for R. We use the volume (19.121) and the relation (19.129) to rewrite the integration measure as

$$\int d^4x \sqrt{-g} = \int dt\, {}^{(4)}S^a = 2\pi^2 \int d\eta\, a^4(\eta), \tag{19.165}$$

so that

$$\overset{f}{\mathcal{A}} = \frac{2\pi^2}{2\kappa} \int d\eta \left[6a(a_{\eta\eta} + ka) - 2\lambda a^4\right] \doteq \frac{2\pi^2}{\kappa} \int d\eta \left[-3a_\eta^2 + 3ka^2 - \lambda a^4\right]. \tag{19.166}$$

The second expression arises from the first by a partial integration and ignoring the boundary terms which do not influence the equation of motion. The above matter is described by the action

$$\overset{m}{\mathcal{A}} = -\int d^4x \sqrt{-g}c\rho = -2\pi^2 \int d\eta\, \frac{Mc}{2\pi^2} a(\eta). \tag{19.167}$$

Variation with respect to a yields the Euler-Lagrange equation

$$6(a_{\eta\eta} + ka) - 4\lambda a^3 - \frac{\kappa Mc}{2\pi^2} = 0. \tag{19.168}$$

Note that in terms of the Robertson-Walker time t [recall (19.129)], the equation of motion reads

$$\ddot{a} = \frac{\lambda}{3}a - \frac{1}{6}\frac{\kappa Mc}{2\pi^2 a^2}. \tag{19.169}$$

As one should expect, the cosmological expansion is slowed down by matter, due to the gravitational attraction. A positive cosmological constant, on the other hand, accelerates the expansion. At the special value

$$\lambda = \lambda_{\mathrm{Einstein}} \equiv \frac{\kappa Mc}{2\pi^2 a^3} = \frac{4G_{\mathrm{N}}M}{\pi c^2 a^3} = \frac{4\pi G_{\mathrm{N}}\rho}{c}, \tag{19.170}$$

the two effects cancel each other and there exist a time-independent solution at a radius a and a density ρ. This is the cosmological constant which Einstein chose before Hubble's discovery of the expanding universe to agree with Hoyle's steady-state model of the universe (a choice which he later called the biggest blunder of his life).

Multiplying (19.168) by a_η and integration over η yields the pseudo-energy conservation law

$$3(a_\eta^2 + ka^2) - \lambda a^4 - \frac{\kappa Mc}{2\pi^2}a = \text{const}, \tag{19.171}$$

in agreement with (19.164) for a vanishing pseudo-energy.

This equation may also be written as

$$a_\eta^2 + ka^2 - a_{\max}a - \frac{\lambda}{3}a^4 = 0, \tag{19.172}$$

where

$$a_{\max} \equiv \frac{\kappa Mc}{6\pi^2} = \frac{4G_N M}{3\pi c^2}. \tag{19.173}$$

This looks like the energy conservation law for a point particle of mass 2 in an effective potential of the universe

$$V^{\text{univ}}(a) = ka^2 - a_{\max}a - \frac{\lambda}{3}a^4, \tag{19.174}$$

at zero total energy. The potential is plotted for the spherical case $k = 1$ in Fig. 19.2.

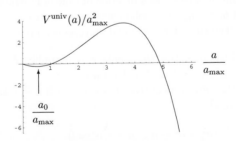

Figure 19.2 Potential of closed Friedman universe as a function of the reduced radius $a/a_{\max}$ for $\lambda a_{\max}^2 = 0.1$. Note the metastable minimum which leads to a possible solution $a \equiv a_0$. A tunneling process to the right leads to an expanding universe.

Friedmann neglects the cosmological constant and considers the equation

$$a_\eta^2 + ka^2 - a_{\max}a = 0. \tag{19.175}$$

The solution of the differential equation for this trajectory is found by direct integration. Assuming $k = 1$ we obtain

$$\eta = \int \frac{da}{\sqrt{-V^{\text{univ}}(a)}} = \int \frac{da}{\sqrt{-(a - a_{\text{max}}/2)^2 + a_{\text{max}}^2/4}} = -\arccos \frac{2a}{a_{\text{max}}}. \tag{19.176}$$

With the initial condition $a(0) = 0$, this implies

$$a(\eta) = \frac{a_{\text{max}}}{2} (1 - \cos \eta). \tag{19.177}$$

Integrating Eq. (19.129), we find the relation between η and the physical (=proper) time

$$t = \frac{1}{c} \int d\eta \, a(\eta) = \frac{a_{\text{max}}}{2c} \int d\eta \, (1 - \cos \eta) = \frac{a_{\text{max}}}{2c} (\eta - \sin \eta). \tag{19.178}$$

The solution $a(t)$ is the cycloid pictured in Fig. 19.3. The radius of the universe bounces periodically with period $t_0 = \pi a_{\text{max}}/c$ from zero to a_{max}. Thus it emerges from a big bang, expands with a decreasing expansion velocity due to the gravitational attraction, and recontracts to a point.

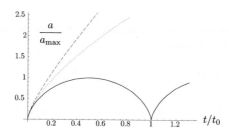

Figure 19.3   Radius of universe as a function of time in Friedman universe, measured in terms of the period $t_0 \equiv \pi a_{\text{max}}/c$. (solid curve=closed, dashed curve=hyperbolic, dotted curve =parabolic). The curve for the closed universe is a cycloid.

Certainly, for high densities the solution is inapplicable since the pressure-less ideal gas approximation (19.161) breaks down.

Consider now the case of negative curvature with $k = -1$. Then the differential equation (19.175) reads

$$a_\eta^2 - a^2 - a_{\text{max}} a - \frac{\lambda}{3} a^4 = 0. \tag{19.179}$$

In order to compare the curves we shall again introduce a mass parameter M and rewrite the density as in (19.163), although M has no longer the meaning of the total mass of the universe (which is now infinite). The solution for $\lambda = 0$ is now

$$a(\eta) = \frac{a_{\text{max}}}{2} (\cosh \eta - 1), \tag{19.180}$$

$$t = \frac{a_{\text{max}}}{2c} (\sinh \eta - \eta). \tag{19.181}$$

The solution is again depicted in Fig. 19.3. After a big bang, the universe expands forever, although with decreasing speed, due to the gravitational pull. The quantity $a_{\max}$ is no longer the maximal radius nor is t_0 the period.

Consider finally the parabolic case $k = 0$, where the equation of motion (19.175) reads

$$a_\eta^2 - a_{\max}a - \frac{\lambda}{3}a^4 = 0, \tag{19.182}$$

where M is a mass parameter defined as before in the negative curvature case. Now the solution for $\lambda = 0$ is simply

$$\eta = 2\sqrt{\frac{a}{a_{\max}}}, \tag{19.183}$$

which is inverted to

$$a(\eta) = a_{\max}\frac{\eta^2}{4}. \tag{19.184}$$

Now we solve (19.129) with

$$t = \frac{a_{\max}}{12c}\eta^3, \tag{19.185}$$

so that

$$a(t) = \left(\frac{9}{4}a_{\max}\right)^{1/3}(ct)^{2/3}. \tag{19.186}$$

This solution is simply the continuation of leading term in the previous two solutions to large t.

19.2.4 Tunneling of Expanding Universe

It is now interesting to observe that the potential for the spherical universe in Fig. 19.2 allows for a time-independent solution in which the radius lies at the metastable minimum which we may call a_0. The solution $a \equiv a_0$ is a time-independent universe. We may now imagine that the expanding universe arises from this by a tunneling process towards the abyss on the right of the potential [10]. Its rate can be calculated from the Euclidean action of the associated classical solution in imaginary time corresponding to the motion from a_0 towards the right in the inverted potential $-V^{\mathrm{univ}}(a)$.

Observe that this birth can only lead to universe of positive curvature. For a negative curvature, where the a^2-term in (19.174) has the opposite sign, the metastable minimum is absent.

19.3 Relativistic Coulomb System

An external time-independent potential $V(\mathbf{x})$ is introduced into the path integral (19.48) by substituting the energy E by $E - V(\mathbf{x})$. In the case of an attractive Coulomb potential, the second term in the action (19.49) becomes

$$\mathcal{A}^{\text{int}} = -\int_{\lambda_a}^{\lambda_b} d\lambda \, h(\lambda) \frac{(E + e^2/r)^2}{2Mc^2}, \qquad (19.187)$$

where $r = |\mathbf{x}|$. The associated path integral is calculated via a Duru-Kleinert transformation as follows [12].

Consider the three-dimensional Coulomb system where the spacetime dimension is $D = 4$. Then we increase the three-dimensional space in a trivial way by a dummy fourth component x^4, just as in the nonrelativistic treatment in Section 13.4. The additional variable x^4 is eliminated at the end by an integral $\int dx_a^4/r_a = \int d\gamma_a$, as in (13.115) and (13.122). Then we perform a Kustaanheimo-Stiefel transformation (13.101) $dx^\mu = 2A(u)^\mu{}_\nu du^\nu$. This changes $x'^{\mu 2}$ into $4\vec{u}^2\vec{u}'^2$, with the vector symbol indicating the four-vector nature. The transformed action reads:

$$\tilde{\mathcal{A}}_{\text{e},E} = \int_{\lambda_a}^{\lambda_b} d\lambda \left\{ \frac{4M\vec{u}^2}{2h(\lambda)} \vec{u}'^2(\lambda) + \frac{h(\lambda)}{2Mc^2\vec{u}^2} \left[(M^2c^4 - E^2)\vec{u}^2 - 2Ee^2 - \frac{e^4}{\vec{u}^2} \right] \right\}. \qquad (19.188)$$

We now choose the gauge $h(\lambda) = 1$, and go from λ to a new parameter s via the path-dependent time transformation $d\lambda = f ds$ with $f = \vec{u}^2$. Result is the DK-transformed action

$$\bar{\mathcal{A}}_{\text{e},E}^{\text{DK}} = \int_{s_a}^{s_b} ds \left\{ \frac{4M}{2}\vec{u}'^2(s) + \frac{1}{2Mc^2} \left[\left(M^2c^4 - E^2 \right) \vec{u}^2 - 2Ee^2 - \frac{e^4}{\vec{u}^2} \right] \right\}. \qquad (19.189)$$

It describes a particle of mass $\mu = 4M$ moving as a function of the "pseudotime" s in a harmonic oscillator potential of frequency

$$\omega = \frac{1}{2Mc}\sqrt{M^2c^4 - E^2}. \qquad (19.190)$$

The oscillator possesses an additional attractive potential $-e^4/2Mc^2\vec{u}^2$, which is conveniently parametrized in the form of a centrifugal barrier

$$V_{\text{extra}} = \hbar^2 \frac{l_{\text{extra}}^2}{2\mu\vec{u}^2}, \qquad (19.191)$$

whose squared angular momentum has the negative value

$$l_{\text{extra}}^2 \equiv -4\alpha^2. \qquad (19.192)$$

Here α denotes the fine-structure constant $\alpha \equiv e^2/\hbar c \approx 1/137$. In addition, there is also a trivial constant potential

$$V_{\text{const}} = -\frac{E}{Mc^2}e^2. \qquad (19.193)$$

If we ignore, for the moment, the centrifugal barrier V_{extra}, the solution of the path integral can immediately be written down [see (13.122)]:

$$(\mathbf{x}_b|\mathbf{x}_a)_E = -i\frac{\hbar}{2Mc}\frac{1}{16}\int_0^\infty dL\ e^{e^2 EL/Mc^2\hbar}\int_0^{4\pi} d\gamma_a\,(\vec{u}_b L|\vec{u}_a 0)\,, \tag{19.194}$$

where $(\vec{u}_b L|\vec{u}_a 0)$ is the time evolution amplitude of the four-dimensional harmonic oscillator.

There are no time slicing corrections for the same reason as in the three-dimensional case. This is ensured by the affine connection of the Kustaanheimo-Stiefel transformation satisfying

$$\Gamma_\mu{}^{\mu\lambda} = g^{\mu\nu}e_i{}^\lambda\partial_\mu e^i{}_\nu = 0 \tag{19.195}$$

(see the discussion in Section 13.6).

Performing the integral over γ_a in (19.194), we obtain

$$\begin{aligned}
(\mathbf{x}_b|\mathbf{x}_a)_E = {}& -i\frac{\hbar}{2Mc}\frac{M\kappa}{\pi\hbar}\int_0^1 d\varrho\frac{\varrho^{-\nu}}{(1-\varrho)^2}I_0\left(2\kappa\frac{2\sqrt{\varrho}}{1-\varrho}\sqrt{(r_b r_a + \mathbf{x}_b\mathbf{x}_a)/2}\right) \\
& \times \exp\left[-\kappa\frac{1+\varrho}{1-\varrho}(r_b + r_a)\right],
\end{aligned} \tag{19.196}$$

with the variable

$$h \equiv e^{-2\omega L}, \tag{19.197}$$

and the parameters

$$\nu = \frac{e^2}{2\omega\hbar}\frac{E}{Mc^2} = \frac{\alpha}{\sqrt{M^2c^4/E^2 - 1}}, \tag{19.198}$$

$$\kappa = \frac{\mu\omega}{2\hbar} = \frac{1}{\hbar c}\sqrt{M^2c^4 - E^2} = \frac{E}{\hbar c}\frac{\alpha}{\nu}. \tag{19.199}$$

As in the further treatment of (13.198), the use of formula (13.202)

$$I_0(z\cos(\theta/2)) = \frac{2}{z}\sum_{l=0}^\infty (2l+1)P_l(\cos\theta)I_{2l+1}(z) \tag{19.200}$$

provides us with a partial wave decomposition

$$\begin{aligned}
(\mathbf{x}_b|\mathbf{x}_a)_E = {}& \frac{1}{r_b r_a}\sum_{l=0}^\infty (r_b|r_a)_{E,l}\frac{2l+1}{4\pi}P_l(\cos\theta) \\
= {}& \frac{1}{r_b r_a}\sum_{l=0}^\infty (r_b|r_a)_{E,l}\sum_{m=-l}^l Y_{lm}(\hat{\mathbf{x}}_b)Y_{lm}^*(\hat{\mathbf{x}}_a).
\end{aligned} \tag{19.201}$$

The radial amplitude is normalized slightly differently from (13.205):

$$(r_b|r_a)_{E,l} = -i\frac{\hbar}{2Mc}\sqrt{r_b r_a}\frac{2M}{\hbar}\int_0^\infty dy\frac{1}{\sinh y}e^{2\nu y}$$

$$\times \exp\left[-\kappa\coth y(r_b + r_a)\right]I_{2l+1}\left(2\kappa\sqrt{r_b r_a}\frac{1}{\sinh y}\right). \tag{19.202}$$

At this place, we incorporate the additional centrifugal barrier via the replacement

$$2l + 1 \rightarrow 2\tilde{l} + 1 \equiv \sqrt{(2l+1)^2 + l_{\text{extra}}^2}, \tag{19.203}$$

as in Eqs. (8.145) and (14.239). The integration over y according to (9.29) yields

$$(r_b|r_a)_{E,l} = -i\frac{\hbar}{2Mc}\frac{M}{\hbar\kappa}\frac{\Gamma(-\nu+\tilde{l}+1)}{(2\tilde{l}+1)!}W_{\nu,\tilde{l}+1/2}(2\kappa r_b)M_{\nu,\tilde{l}+1/2}(2\kappa r_a). \tag{19.204}$$

This expression possesses poles in the Gamma function whose positions satisfy the equations $\nu - \tilde{l} - 1 = 0, 1, 2, \ldots$. These determine the bound states of the Coulomb system. To simplify subsequent expressions, we introduce the small positive l-dependent parameter

$$\delta_l \equiv l - \tilde{l} = l + 1/2 - \sqrt{(l+1/2)^2 - \alpha^2} \approx \frac{\alpha^2}{2l+1} + \mathcal{O}(\alpha^4). \tag{19.205}$$

Then the pole positions satisfy $\nu = \tilde{n}_l \equiv n - \delta_l$, with $n = l+1, l+2, l+3, \ldots$. Using the relation (19.198), we obtain the bound-state energies:

$$\begin{aligned}
E_{nl} &= \pm Mc^2\left[1 + \frac{\alpha^2}{(n-\delta_l)^2}\right]^{-1/2} \\
&\approx \pm Mc^2\left[1 - \frac{\alpha^2}{2n^2} - \frac{\alpha^4}{n^3}\left(\frac{1}{2l+1} - \frac{3}{8n}\right) + \mathcal{O}(\alpha^6)\right].
\end{aligned} \tag{19.206}$$

Note the appearance of the plus-minus sign as a characteristic property of energies in relativistic quantum mechanics. A correct interpretation of the negative energies as positive energies of antiparticles is straightforward only within quantum field theory, and will not be discussed here. Even if we ignore the negative energies, there is poor agreement with the experimental spectrum of the hydrogen atom. The spin of the electron must be included to get more satisfactory results.

To find the wave functions, we approximate near the poles $\nu \approx \tilde{n}_l$:

$$\begin{aligned}
\Gamma(-\nu+\tilde{l}+1) &\approx -\frac{(-)^{n_r}}{n_r!}\frac{1}{\nu-\tilde{n}_l}, \\
\frac{1}{\nu-\tilde{n}_l} &\approx \frac{2}{\tilde{n}_l}\frac{\hbar^2\kappa^2}{2M}\left(\frac{E}{Mc^2}\right)^2\frac{2Mc^2}{E^2 - E_{nl}^2}, \\
\kappa &\approx \frac{E}{Mc^2}\frac{1}{a_H}\frac{1}{\tilde{n}_l},
\end{aligned} \tag{19.207}$$

with the radial quantum number $n_r = n - l - 1$. By analogy with the nonrelativistic equation (13.207), the last equation can be rewritten as

$$\kappa = \frac{1}{\tilde{a}_H} \frac{1}{\nu}, \tag{19.208}$$

where

$$\tilde{a}_H \equiv a_H \frac{Mc^2}{E} \tag{19.209}$$

denotes a modified energy-dependent Bohr radius [compare (4.344)]. It sets the length scale of relativistic bound states in terms of the energy E. Instead of being $1/\alpha \approx 137$ times the Compton wavelength of the electron $\hbar/Mc$, the modified Bohr radius is equal to $1/\alpha$ times $\hbar c/E$.

Near the positive-energy poles, we now approximate

$$- i\Gamma(-\nu + \tilde{l} + 1) \frac{M}{\hbar \kappa} \approx \frac{(-)^{n_r}}{\tilde{n}_l^2 n_r!} \frac{1}{\tilde{a}_H} \left(\frac{E}{Mc^2} \right)^2 \frac{2Mc^2 i\hbar}{E^2 - E_{nl}^2}. \tag{19.210}$$

Using this behavior and formula (9.48) for the Whittaker functions [together with (9.50)] we write the contribution of the bound states to the spectral representation of the fixed-energy amplitude as

$$(r_b|r_a)_{E,l} = \frac{\hbar}{Mc} \sum_{n=l+1}^{\infty} \left(\frac{E}{Mc^2} \right)^2 \frac{2Mc^2 i\hbar}{E^2 - E_{nl}^2} R_{nl}(r_b) R_{nl}(r_a) + \dots . \tag{19.211}$$

A comparison between the pole terms in (19.204) and (19.211) renders the radial wave functions

$$\begin{aligned}
R_{nl}(r) &= \frac{1}{\tilde{a}_H^{1/2} \tilde{n}_l} \frac{1}{(2\tilde{l}+1)!} \sqrt{\frac{(\tilde{n}_l + \tilde{l})!}{(n-l-1)!}} \\
&\quad \times (2r/\tilde{n}_l \tilde{a}_H)^{\tilde{l}+1} e^{-r/\tilde{n}_l \tilde{a}_H} M(-n+l+1, 2\tilde{l}+2, 2r/\tilde{n}_l \tilde{a}_H) \tag{19.212} \\
&= \frac{1}{\tilde{a}_H^{1/2} \tilde{n}_l} \sqrt{\frac{(n-l-1)!}{(\tilde{n}+\tilde{l})!}} e^{-r/\tilde{n}\tilde{a}_H} (2r/\tilde{n}_l \tilde{a}_H)^{\tilde{l}+1} L_{\tilde{n}_l - l - 1}^{2\tilde{l}+1}(2r/\tilde{n}_l \tilde{a}_H).
\end{aligned}$$

The properly normalized total wave functions are

$$\psi_{nlm}(\mathbf{x}) = \frac{1}{r} R_{nl}(r) Y_{lm}(\hat{\mathbf{x}}). \tag{19.213}$$

The continuous wave functions are obtained in the same way as from the nonrelativistic amplitude in formulas (13.215)–(13.223).

19.4 Relativistic Particle in Electromagnetic Field

Consider now the relativistic particle in a general spacetime-dependent electromagnetic vector field $A^\mu(x)$.

19.4.1 Action and Partition Function

An electromagnetic field $A^\mu(x)$ is included into the canonical action (19.14) in the usual way by the minimal substitution (2.642):

$$\bar{\mathcal{A}}_e[p,x] = \int_{\lambda_a}^{\lambda_b} d\lambda \left\{ -ipx' + \frac{h(\lambda)}{2Mc} \left[\left(p - \frac{e}{c}A\right)^2 + M^2c^2 \right] \right\}, \qquad (19.214)$$

and the amplitude (19.22):

$$(x_b|x_a) = \frac{\hbar}{2Mc} \int_0^\infty dL \int \mathcal{D}h\, \Phi[h] \int \mathcal{D}^D x\, e^{-\bar{\mathcal{A}}_e/\hbar}, \qquad (19.215)$$

with the minimally coupled action [compare (2.704)]

$$\bar{\mathcal{A}}_e = \int_{\lambda_a}^{\lambda_b} d\lambda \left[\frac{Mc}{2h(\lambda)} x'^2(\lambda) + i\frac{e}{c}x'(\lambda)A(x(\lambda)) + h(\lambda)\frac{Mc}{2} \right], \qquad (19.216)$$

which reduces in the simplest gauge (19.23) to the obvious extension of (19.26):

$$(x_b|x_a) = \frac{\hbar^2}{2M} \int_0^\infty d\beta\, e^{-\beta Mc^2/2} \int \mathcal{D}^4 x\, e^{-\mathcal{A}_e}, \qquad (19.217)$$

with the action

$$\mathcal{A}_e = \mathcal{A}_{e,0} + \mathcal{A}_{e,\text{int}} \equiv \int_0^{\hbar\beta} d\tau \left[\frac{M}{2}\dot{x}^2(\tau) + i\frac{e}{c}\dot{x}(\tau)A(x(\tau)) \right]. \qquad (19.218)$$

The partition function of a single closed particle loop of all shapes and lengths in an external electromagnetic field is from (19.37)

$$Z_1 = \int_0^\infty \frac{d\beta}{\beta}\, e^{-\beta Mc^2/2} \int \mathcal{D}^D x\, e^{-\mathcal{A}_e/\hbar}. \qquad (19.219)$$

As in (19.45) and (19.46) this yields, up to a factor $1/\hbar$ the effective action of an ensemble of closed particle loops in an external electromagnetic field.

19.4.2 Perturbation Expansion

Since the electromagnetic coupling is rather small, we can split the exponent $e^{-\mathcal{A}/\hbar}$ into $e^{-\mathcal{A}_0/\hbar}e^{-\mathcal{A}_{\text{int}}/\hbar}$ and expand the second factor in powers of $\mathcal{A}_{\text{int}}$:

$$e^{-\mathcal{A}_{\text{int}}/\hbar} = \sum_{n=0}^\infty \frac{(-ie/\hbar c)^n}{n!} \prod_{i=1}^n \left[\int_0^{\hbar\beta} d\tau_i\, \dot{x}(\tau_i)A(x(\tau_i)) \right]. \qquad (19.220)$$

If the noninteracting effective action implied by Eqs. (19.39), (19.37), and (19.46) is denoted by

$$\frac{\Gamma_{e,0}}{\hbar} \equiv -Z_1 = -\int_0^\infty \frac{d\beta}{\beta}\, e^{-\beta Mc^2/2} \frac{V_D}{\sqrt{2\pi\hbar^2\beta/M}^{\,D}} = -\frac{V_D}{\lambda_M^{C\,D}} \frac{1}{(4\pi)^{D/2}} \Gamma(1-D/2), \qquad (19.221)$$

with the Compton wavelength λ_M^C of Eq. (19.31), we obtain the perturbation expansion

$$\frac{\Gamma_e}{\hbar} = \frac{\Gamma_{e,0}}{\hbar} - \int_0^\infty \frac{d\beta}{\beta} e^{-\beta Mc^2/2} \frac{V_D}{\sqrt{2\pi\hbar^2\beta/M}^D} \sum_{n=1}^\infty \frac{(-ie/c)^n}{n!} \left\langle \prod_{i=1}^n \left[\int_0^{\hbar\beta} d\tau_i \, \dot{x}(\tau_i) A(x(\tau_i)) \right] \right\rangle_0,$$

$$(19.222)$$

where $\langle \ldots \rangle_0$ denotes the free-particle expectation values [compare (3.480)–(3.483)] taken in the free path integral with periodic paths with a fixed β [compare (19.38) and (19.41)]:

$$\langle \mathcal{O}[x] \rangle_0 \equiv \frac{\int \mathcal{D}^D x \, \mathcal{O}[x] \, e^{-\mathcal{A}_{e,0}/\hbar}}{\int \mathcal{D}^D x \, e^{-\mathcal{A}_{e,0}/\hbar}}. \qquad (19.223)$$

The denominator is equal to $V_D/\sqrt{2\pi\hbar^2\beta/M}^D$.

The free effective action in the expansion (19.222) can be omitted by letting the sum start with $n = 0$.

The evaluation of the cumulants proceeds by Fourier decomposing the vector fields as

$$A(x) = \int \frac{d^D k}{(2\pi)^D} e^{ikx} A(k), \qquad (19.224)$$

and rewriting (19.222) as

$$\frac{\Gamma_e}{\hbar} = \frac{\Gamma_{e,0}}{\hbar} - \int_0^\infty \frac{d\beta}{\beta} e^{-\beta Mc^2/2} \frac{V_D}{\sqrt{2\pi\hbar^2\beta/M}^D}$$

$$\times \sum_{n=1}^\infty \frac{(-ie/\hbar c)^n}{n!} \prod_{i=1}^n \left[\int \frac{d^D k_i}{(2\pi)^D} A^{\mu_i}(k_i) \right] \left\langle \prod_{i=1}^n \left[\int_0^{\hbar\beta} d\tau_i \, \dot{x}^{\mu_i}(\tau_i) e^{ik_i x(\tau_i)} \right] \right\rangle_0. (19.225)$$

First we evaluate the expectation values

$$\left\langle \dot{x}(\tau_1) e^{ik_1 x(\tau_1)} \cdots \dot{x}(\tau_n) e^{ik_n x(\tau_n)} \right\rangle_0. \qquad (19.226)$$

Due to the periodic boundary conditions, we separate, as in Section 3.25, the path average $x_0 = \bar{x}(\tau)$ [recall (3.804)], writing

$$x(\tau) = x_0 + \delta x(\tau), \qquad (19.227)$$

and factorize (19.226) as

$$\left\langle e^{i(k_1+\ldots+k_n)x_0} \right\rangle_0 \left\langle \delta\dot{x}(\tau_1) e^{ik_1\delta x(\tau_1)} \cdots \delta\dot{x}(\tau_n) e^{ik_n\delta x(\tau_n)} \right\rangle_0. \qquad (19.228)$$

The first average can be found as an average with respect to the x_0-part of the path integral whose measure was given in Eq. (3.808). It yields a δ function ensuring the conservation of the total energy and momenta of the n photons involved:

$$\left\langle e^{i(k_1+\ldots+k_n)x_0} \right\rangle_0 = \frac{1}{V_D} (2\pi)^D \delta^{(D)}(k_1 + \ldots + k_n). \qquad (19.229)$$

The denominator comes from the normalization of the expectation value which has an integral $\int d^D x_0$ in the denominator.

The second average is obtained using Wick's theorem. The correlation function $\langle \delta x^\mu(\tau_1) \delta x^\nu(\tau_2) \rangle_0$, is obtained from Eq. (3.839) in the limit $\Omega \to 0$ for $\tau_1, \tau_2 \in (0, \beta)$. It is the periodic propagator with subtracted zero mode:

$$\langle \delta x^\mu(\tau_1) \delta x^\nu(\tau_2) \rangle_0 = \delta^{\mu\nu} G(\tau_1, \tau_2) = \delta^{\mu\nu} \frac{\hbar}{M} \bar\Delta(\tau_1, \tau_2), \tag{19.230}$$

where $\bar\Delta(\tau_1 - \tau_2)$ is the subtracted periodic Green function $\quad G^{\mathrm{a}}_{\omega,e}(\tau - \tau')$ of the differential operator $-\partial_\tau^2$ calculated in Eq. (3.251) in the short notation of Subsection 10.12.1 [see Eq. (10.565)]. In the presently used physical units it reads:

$$\bar\Delta(\tau, \tau') \equiv \bar\Delta(\tau - \tau') = \frac{(\tau - \tau')^2}{2\hbar\beta} - \frac{\tau - \tau'}{2} + \frac{\hbar\beta}{12}, \quad \tau \in [0, \hbar\beta]. \tag{19.231}$$

The time derivatives of (19.231) are from (10.566):

$$\dot{\bar\Delta}(\tau, \tau') = -\bar\Delta^{\cdot}(\tau, \tau') \equiv \frac{\tau - \tau'}{\hbar\beta} - \frac{\epsilon(\tau - \tau')}{2}, \quad \tau, \tau' \in [0, \hbar\beta]. \tag{19.232}$$

With these functions it is straightforward to calculate the expectation value using the Wick rule (3.307) for $j(\tau) = \sum_i^n k_i \delta(\tau - \tau_i)$:

$$\left\langle e^{ik_1 \delta x(\tau_1)} \cdots e^{ik_n \delta x(\tau_n)} \right\rangle_0 = e^{-\frac{1}{2} \sum_{i,j=1}^n k_i k_j G(\tau_i, \tau_j)}. \tag{19.233}$$

By rewriting the right-hand side as

$$e^{-\frac{1}{2} \sum_{i,j=1}^n k_i k_j G(\tau_i, \tau_j)} = e^{-\frac{1}{2} \sum_{i,j=1}^n k_i k_j [G(\tau_i, \tau_j) - G(\tau_i, \tau_i)] - \frac{1}{2} \left(\sum_{i=1}^n k_i\right)^2 G(\tau_i, \tau_i)}, \tag{19.234}$$

we see that if the momenta k_i add up to zero, $\sum_{i=1}^n k_i = 0$, we can replace (19.233) by

$$\left\langle e^{ik_1 \delta x(\tau_1)} \cdots e^{ik_n \delta x(\tau_n)} \right\rangle_0 = \exp\left\{ -\sum_{i<j}^n k_i k_j [G(\tau_i, \tau_j) - G(\tau_i, \tau_i)] \right\}. \tag{19.235}$$

It is therefore useful to introduce the subtracted Green function

$$G'(\tau_i, \tau_j) \equiv G(\tau_i, \tau_j) - G(\tau_i, \tau_i). \tag{19.236}$$

Recall the similar situation in the evaluation (5.377).

An obvious extension of (19.233) is

$$\left\langle e^{i[k_1 \delta x(\tau_1) + q_1 \dot x(\tau_1)]} \cdots e^{i[k_n \delta x(\tau_n) + q_n \dot x(\tau_n)]} \right\rangle_0 \tag{19.237}$$
$$= e^{-\frac{1}{2} \sum_{i,j=1}^n k_i k_j G(\tau_i, \tau_j) - \frac{1}{2} \sum_{i,j=1}^n q_i k_j \, {}^{\cdot}G(\tau_i, \tau_j) - \frac{1}{2} \sum_{i,j=1}^n k_i q_j \, G^{\cdot}(\tau_i, \tau_j) - \frac{1}{2} \sum_{i,j=1}^n q_i q_j \, {}^{\cdot}G^{\cdot}(\tau_i, \tau_j)},$$

where the dots have the same meaning as in (10.395).

19.4.3   Lowest-Order Vacuum Polarization

Consider the lowest nontrivial case $n = 2$. By differentiating (19.235) with respect to iq_1 and iq_2, and setting $q_i = 0$, we obtain for $k_2 = -k_1 = -k$:

$$\left\langle \dot{x}(\tau_1)e^{ik\delta x(\tau_1)}\dot{x}(\tau_2)e^{-ik\delta x(\tau_2)}\right\rangle_0 = \left[G^{\cdot\cdot}(\tau_1,\tau_2) + k^2 G^{\cdot}(\tau_1,\tau_2)G^{\cdot}(\tau_1,\tau_2)\right] e^{k^2[G(\tau_1,\tau_2)-G(\tau_1,\tau_1)]}.$$
(19.238)

Inserting this into (19.225) after factorization according to (19.228), we obtain the lowest correction to the effective action

$$\Delta\Gamma_e = -\frac{e^2}{2\hbar c^2}\int_0^\infty \frac{d\beta}{\beta}e^{-\beta Mc^2/2}\frac{1}{\sqrt{2\pi\hbar^2\beta/M}^D}\prod_{i=1}^2\left[\int\frac{d^D k_i}{(2\pi)^D}\right](2\pi)^D\delta^{(D)}(k_1+k_2)A^\mu(k_1)A^\nu(k_2)$$

$$\times \int_0^{\hbar\beta}d\tau_1\int_0^{\hbar\beta}d\tau_2\left[G^{\cdot\cdot}(\tau_1,\tau_2)\delta^{\mu\nu} + k_1^\mu k_1^\nu\, G^{\cdot}(\tau_1,\tau_2)G^{\cdot}(\tau_1,\tau_2)\right]e^{k_1^2 G'(\tau_1,\tau_2)},\qquad (19.239)$$

where we have displayed the proper vector indices. A partial integration over τ_1 brings the second line to the form

$$-\int_0^{\hbar\beta}d\tau_1\int_0^{\hbar\beta}d\tau_2\left(k_1^2\delta^{\mu\nu} - k_1^\mu k_1^\nu\right)\, G^{\cdot}(\tau_1,\tau_2)\,G^{\cdot}(\tau_1,\tau_2)\,e^{k_1^2 G'(\tau_1,\tau_2)}.\qquad (19.240)$$

Expressing $G'(\tau_1,\tau_2)$ as $(\hbar/M)\left[\bar{\Delta}(\tau_1 - \tau_2) - \bar{\Delta}(0)\right]$ and using the periodicity in τ_2, we calculate the integral

$$\frac{\hbar^2}{M^2}\left(k_1^2\delta^{\mu\nu} - k_1^\mu k_1^\nu\right)\hbar\beta\int_0^{\hbar\beta}d\tau\,\dot{\Delta}'^{\,2}_p(\tau)e^{\hbar k_1^2 M[\Delta'_p(\tau)-\Delta'_p(0)]}.\qquad (19.241)$$

We now introduce reduced times $u \equiv \tau/\hbar\beta$ and rewrite the Green functions for $\tau \in (0,\hbar\beta)$, $u \in (0,1)$ as

$$\bar{\Delta}(\tau_1 - \tau_2) = -\frac{\hbar\beta}{2}\left[u(1-u) - \frac{1}{6}\right],\qquad (19.242)$$

$$\dot{\bar{\Delta}}(\tau_1 - \tau_2) = u - \frac{1}{2},\qquad (19.243)$$

such that the integral in (19.241) becomes

$$\frac{1}{4}\int_0^1 du\,(2u-1)^2 e^{-\beta\frac{\hbar^2 k_1^2}{2M}u(1-u)}.\qquad (19.244)$$

Inserting this into (19.239) and dropping the irrelevant subscript of k_1 we arrive at

$$\Delta\Gamma_e = \frac{e^2}{2\hbar c^2}\int_0^\infty\frac{d\beta}{\beta}e^{-\beta Mc^2/2}\frac{1}{\sqrt{2\pi\hbar^2\beta/M}^D}\int\frac{d^D k}{(2\pi)^D}A^\mu(k)A^\nu(-k)$$

$$\times\left(k^2\delta^{\mu\nu} - k^\mu k^\nu\right)\frac{\hbar^4\beta^2}{4M^2}\int_0^1 du\,(2u-1)^2 e^{-\beta\frac{\hbar^2 k^2}{2M}u(1-u)}.\qquad (19.245)$$

After replacing $\hbar^2\beta/2M \to \beta$, the integral over β can easily be performed using the formula (2.496) with the result

$$\Delta\Gamma_{\rm e} = \frac{e^2\hbar}{c^2}\frac{1}{(4\pi)^{D/2}}\frac{1}{2c}\int\frac{d^Dk}{(2\pi)^D}A^\mu(k)A^\nu(-k)\left(k^2\delta^{\mu\nu} - k^\mu k^\nu\right)$$
$$\times\,\Gamma\left(2 - D/2\right)\int_0^1 du\,(2u-1)^2\left[u(1-u)k^2 + M^2c^2/\hbar^2\right]^{D/2-2}. \quad (19.246)$$

In the prefactor we recognize the fine structure constant $\alpha = e^2/\hbar c$ [recall (1.502)]. The momentum integral can be rewritten as

$$\frac{1}{2}\int\frac{d^Dk}{(2\pi)^D}A^\mu(k)A^\nu(-k)\left(k^2\delta^{\mu\nu} - k^\mu k^\nu\right) = \frac{1}{4}\int\frac{d^Dk}{(2\pi)^D}F_{\mu\nu}(-k)F_{\mu\nu}(k), \quad (19.247)$$

where

$$F_{\mu\nu}(x) = \partial_\mu A_\nu(x) - \partial_\nu A_\mu(x) \quad (19.248)$$

is the tensor of electromagnetic field strengths. We now abbreviate the integral over u as follows:

$$\Pi(k^2) \equiv \alpha\frac{4\pi}{(4\pi)^{D/2}}\Gamma\left(2-D/2\right)\int_0^1 du\,(2u-1)^2\left[u(1-u)k^2 + M^2c^2/\hbar^2\right]^{D/2-2}.(19.249)$$

This allows us to re-express (19.246) in configuration space:

$$\Delta\Gamma_{\rm e} = \frac{1}{16\pi c}\int d^4x\,F_{\mu\nu}(x)\Pi(-\partial^2)F_{\mu\nu}(x), \quad (19.250)$$

where $F_{\mu\nu}(x)$ is the Euclidean version of the gauge-invariant 4-dimensional curl of the vector potential:

$$F_{\mu\nu}(x) = \partial_\mu A_\nu(x) - \partial_\nu A_\mu(x). \quad (19.251)$$

In Minkowski space, the components of $F_{\mu\nu}$ are the electric and magnetic fields:

$$F_{0i} = -F^{0i} = -\partial^0 A^i + \partial^i A^0 = -\partial_0 A^i - \partial_i A^0 = -E^i, \quad (19.252)$$
$$F_{ij} = F^{ij} = \partial^i A^j + \partial^j A^i = -\partial_i A^j + \partial_j A^i = -\epsilon_{ijk}B^k. \quad (19.253)$$

This is in accordance with the electrodynamic definitions

$$\mathbf{E} \equiv -\frac{1}{c}\dot{\mathbf{A}} - \boldsymbol{\nabla}\phi, \qquad \mathbf{B} \equiv \boldsymbol{\nabla}\times\mathbf{A}, \quad (19.254)$$

where $A^0(x)$ is identified with the electric potential $\phi(x)$.

In terms of $F_{\mu\nu}(x)$, the Maxwell action in the presence of a charge density $\rho(x)$ and a electric current density $\mathbf{j}(x)$

$$\mathcal{A}^{\rm em} = \int dt\,d^3x\left\{\frac{1}{4\pi}\left[\mathbf{E}^2(x) - \mathbf{B}^2(x)\right] - \left[\rho(x)\phi(x) - \frac{1}{c}\mathbf{j}(x)\cdot\mathbf{A}(x)\right]\right\}, \quad (19.255)$$

can be written covariantly as

$$A^{\text{em}} = -\int d^4x \left[\frac{1}{8\pi c} F^2_{\mu\nu}(x) + \frac{1}{c^2} j^\mu(x) A_\mu(x) \right], \qquad (19.256)$$

where

$$j_\mu(x) = (c\rho(x), \mathbf{j}(x)) \qquad (19.257)$$

is the four-vector formed by charge density and electric current.

By extremizing the action (19.256) in the vector field $A^\mu(x)$ we find the Maxwell equations in the covariant form

$$\partial_\nu F^{\nu\mu}(x) = \frac{1}{c} j^\mu(x), \qquad (19.258)$$

whose zeroth and spatial components reduce to the time-honored laws of Gauss and Ampère:

$$\mathbf{\nabla} \cdot \mathbf{E} = 4\pi\rho \quad \text{(Gauss's law)}, \qquad (19.259)$$

$$\mathbf{\nabla} \times \mathbf{B} = \frac{4\pi}{c} \mathbf{j} \quad \text{(Ampère's law)}. \qquad (19.260)$$

Expressing $\mathbf{E}(x)$ in terms of the potential using Eq. (19.254) and inserting this into Gauss's law, we obtain for a static point charge e at the origin the Poisson equation

$$-\mathbf{\nabla}^2 \phi(\mathbf{x}) = 4\pi e \delta^{(3)}(\mathbf{x}). \qquad (19.261)$$

An electron of charge $-e$ experiences an attractive mechanical potential $V(\mathbf{x}) = -e\phi(\mathbf{x})$. In momentum space this satisfies the equation reads

$$\mathbf{k}^2 V(\mathbf{k}) = -4\pi e^2. \qquad (19.262)$$

From this we find directly the Coulomb potential of a hydrogen atom

$$V(\mathbf{x}) = (\mathbf{\nabla}^2)^{-1} 4\pi e \delta^{(3)}(\mathbf{x}) = -\int \frac{d^3k}{(2\pi)^3} \frac{4\pi e^2}{\mathbf{k}^2} = -\frac{e^2}{r}, \quad r \equiv |\mathbf{x}|, \qquad (19.263)$$

where e^2 can be expressed in terms of the fine-structure constant α as $e^2 = \hbar c \alpha$. (1.502).

The Euclidean result (19.250) implies that a fluctuating closed particle orbit changes the first term in the Maxwell action (19.256) to

$$A^{\text{eff}}_{\text{em}} = -\int d^D x \frac{1}{16\pi c} F_{\mu\nu}(x) \left[1 + \Pi(-\partial^2) \right] F_{\mu\nu}(x). \qquad (19.264)$$

The quantity $\Pi(-\partial^2)$ is the *self-energy* of the electromagnetic field caused by the fluctuating closed particle orbit.

The self-energy changes the Maxwell equations (19.259) and (19.260) into

$$\left[1 + \Pi(-\partial^2)\right] \nabla \cdot \mathbf{E} = 4\pi\rho,$$
$$\left[1 + \Pi(-\partial^2)\right] \nabla \times \mathbf{B} = \frac{4\pi}{c} \mathbf{j}. \tag{19.265}$$

The static equation (19.262) for the atomic potential changes therefore into

$$\left[1 + \Pi(\mathbf{k}^2)\right] \mathbf{k}^2 V(\mathbf{k}) = -4\pi e^2. \tag{19.266}$$

Since $\Pi(\mathbf{k}^2)$ is of order $\alpha \approx 1/137$, this can be solved approximately by

$$V(\mathbf{k}) \equiv -4\pi e^2 \left[1 - \Pi(\mathbf{k}^2)\right] \frac{1}{\mathbf{k}^2}. \tag{19.267}$$

In real space, the attractive atomic potential is changed to lowest order in α as

$$-\frac{\alpha}{r} \to -\left[1 - \Pi(\nabla^2)\right] \frac{\alpha}{r}. \tag{19.268}$$

Let us calculate this change explicitly. For small ϵ and k^2, we expand the self-energy (19.249) in $D = 4 - \epsilon$ dimensions as

$$\Pi(k^2) = \frac{\alpha}{24\pi} \left[-\frac{2}{\epsilon} + \log \frac{M^2 c^2 e^\gamma}{4\pi\hbar^2}\right] - \frac{\alpha\hbar^2 k^2}{160\pi M^2 c^2} + \mathcal{O}\left(\epsilon, \frac{k^2}{M^2 c^2/\hbar^2}\right). \tag{19.269}$$

Inserting this into (19.267), or into (19.267) and using the Poisson equation $-\nabla^2 \times 1/r = 4\pi\delta^{(3)}(\mathbf{x})$, we see that the self energy changes the Coulomb potential as follows:

$$-\frac{\alpha}{r} \to -\left[1 - \Pi(\nabla^2)\right] \frac{\alpha}{r} \approx -\left\{1 - \frac{\alpha}{24\pi^2} \left[-\frac{2}{\epsilon} + \log \frac{M^2 c^2 e^\gamma}{4\pi\hbar^2}\right]\right\} \frac{\alpha}{r} - \frac{\alpha^2 \hbar^2}{40 M^2 c^2} \delta^{(3)}(\mathbf{x}). \tag{19.270}$$

The first term amounts to a small renormalization of the electromagnetic coupling by the factor in curly brackets, which is close to unity for finite ϵ since α is small. We are, however, interested in the result in $D = 4$ spacetime dimensions where $\epsilon \to 0$ and (19.270) diverges. The physical resolution of this divergence problem is to assume the initial point charge e_0 in the electromagnetic interaction to be different from the experimentally observed e to precisely compensate the renormalization factor, i.e.,

$$e_0^2 = e^2 \left\{1 + \frac{\alpha}{24\pi^2} \left[-\frac{2}{\epsilon} + \log \frac{M^2 c^2 e^\gamma}{4\pi\hbar^2}\right]\right\}. \tag{19.271}$$

Thus, the result Eq. (19.270) is really obtained in terms of e_0, i.e., with α replaced by α_0. Then, using (19.271), we find that up to order α^2, the atomic potential is

$$V^{\text{eff}}(\mathbf{x}) = -\frac{\alpha}{r} - \frac{\alpha^2 \hbar^2}{40 M^2 c^2} \delta^{(3)}(\mathbf{x}). \tag{19.272}$$

The second is an additional attractive contact interaction. It shifts the energies of the s-wave bound states in Eq. (19.206) slightly downwards.

19.5   Path Integral for Spin-1/2 Particle

For particles of spin 1/2 the path integral formulation becomes algebraically more involved. Let us first recall a few facts from Dirac's theory of the electron.

19.5.1   Dirac Theory

In the Dirac theory, electrons are described by a four-component field $\psi_\alpha(x)$ in spacetime parametrized by $x^\mu = (ct, \mathbf{x})$ with $\mu = 0, 1, 2, 3$. The field satisfies the wave equation

$$(i\hbar\slashed{\partial} - Mc)\,\psi(x) = 0, \tag{19.273}$$

where $\slashed{\partial}$ is a short notation for $\gamma^\mu \partial_\mu$ and γ^μ are 4×4 Dirac matrices satisfying the anticommutation rules

$$\{\gamma^\mu, \gamma^\nu\} = 2g^{\mu\nu}, \tag{19.274}$$

where $g_{\mu\nu}$ is now the Minkowski metric

$$g_{\mu\nu} = \begin{pmatrix} 1 & 0 & 0 & 0 \\ 0 & -1 & 0 & 0 \\ 0 & 0 & -1 & 0 \\ 0 & 0 & 0 & -1 \end{pmatrix}. \tag{19.275}$$

An explicit representation of these rules is most easily written in terms of the Pauli matrices (1.445):

$$\gamma^0 = \begin{pmatrix} 1 & 0 \\ 0 & -1 \end{pmatrix}, \quad \gamma^i = \begin{pmatrix} 0 & \sigma^i \\ -\sigma^i & 0 \end{pmatrix}, \tag{19.276}$$

where 1 is a 2×2 unit matrix. The anticommutation rules (19.274) follow directly from the multiplication rules for the Pauli matrices:

$$\sigma^i \sigma^j = \delta^{ij} + i\epsilon^{ijk}\sigma^k. \tag{19.277}$$

The action of the Dirac field is

$$\mathcal{A} = \int d^4x\, \bar{\psi}(x)\,(i\hbar\slashed{\partial} - Mc)\,\psi(x), \tag{19.278}$$

where the conjugate field $\bar{\psi}(x)$ is defined as

$$\bar{\psi}(x) \equiv \psi^\dagger(x)\gamma^0. \tag{19.279}$$

It can be shown that this makes $\bar{\psi}(x)\psi(x)$ a scalar field under Lorentz transformations, $\bar{\psi}(x)\gamma^\mu\psi(x)$ a vector field, and $\mathcal{A}$ an invariant. If we decompose $\psi(x)$ into its Fourier components

$$\psi(x) = \sum_{\mathbf{k}} \frac{1}{\sqrt{V}} e^{i\mathbf{k}\mathbf{x}} \psi_{\mathbf{k}}(t), \tag{19.280}$$

where V is the spatial volume, the action reads

$$\mathcal{A} = \int_{t_a}^{t_b} dt \sum_{\mathbf{k}} \psi_{\mathbf{k}}^\dagger(t) \left[i\hbar\partial_t - H(\hbar\mathbf{k}) \right] \psi_{\mathbf{k}}(t), \tag{19.281}$$

with the 4×4 Hamiltonian matrix

$$H(\mathbf{p}) \equiv \gamma^0\boldsymbol{\gamma}\,\mathbf{p}\,c + \gamma^0 Mc^2. \tag{19.282}$$

This can be rewritten in terms of 2×2-submatrices as

$$H(\mathbf{p}) = \begin{pmatrix} Mc & \mathbf{p}\boldsymbol{\sigma} \\ -\mathbf{p}\boldsymbol{\sigma} & -Mc \end{pmatrix} c. \tag{19.283}$$

Since the matrix is Hermitian, it can be diagonalized by a unitary transformation to

$$H^{\mathrm{d}}(\mathbf{p}) = \begin{pmatrix} \varepsilon_{\mathbf{k}} & 0 \\ 0 & -\varepsilon_{\mathbf{k}} \end{pmatrix}, \tag{19.284}$$

where

$$\varepsilon_{\mathbf{k}} \equiv c\sqrt{\mathbf{p}^2 + M^2 c^2} \tag{19.285}$$

are energies of the relativistic particles of mass M and momentum $\mathbf{p}$. Each entry in (19.284) is a 2×2-submatrix.

This is achieved by the *Foldy-Wouthuysen transformation*

$$H^{\mathrm{d}} = e^{iS} H e^{-iS}, \tag{19.286}$$

where

$$S = -i\boldsymbol{\gamma} \cdot \boldsymbol{\zeta}/2, \quad \boldsymbol{\zeta} \equiv \arctan\left(\mathbf{v}/c\right), \quad \mathbf{v} \equiv \mathbf{p}/M = \text{velocity}. \tag{19.287}$$

The vector $\boldsymbol{\zeta}$ points in the direction of the velocity $\mathbf{v}$ and has the length $\zeta = \arctan(v/c)$, such that

$$\cos\zeta = \frac{Mc}{\sqrt{\mathbf{p}^2 + M^2 c^2}}, \quad \sin\zeta = \frac{|\mathbf{p}|}{\sqrt{\mathbf{p}^2 + M^2 c^2}}. \tag{19.288}$$

A function of a vector $\mathbf{v}$ is defined by its Taylor series where even powers of $\mathbf{v}$ are scalars $\mathbf{v}^{2n} = v^{2n}$ and odd powers are vectors $\mathbf{v}^{2n+1} = v^{2n}\mathbf{v}$. If $\hat{\mathbf{v}}$ denotes as usual the direction vector $\hat{\mathbf{v}} \equiv \mathbf{v}/|\mathbf{v}|$, the matrix $\boldsymbol{\gamma} \cdot \hat{\mathbf{v}}$ has the property that all even powers of it are equal to a 4×4 unity matrix up to an alternating sign: $(\boldsymbol{\gamma} \cdot \hat{\mathbf{v}})^{2n} = (-1)^n$. Thus $S = -i\boldsymbol{\gamma} \cdot \hat{\mathbf{v}}\zeta$ and the Taylor series of e^{iS} reads explictly

$$e^{iS} = \sum_{n=0,2,4,\dots} \frac{(-1)^n}{n!} \left(\frac{\zeta}{2}\right)^n + (\boldsymbol{\gamma} \cdot \hat{\mathbf{v}}) \sum_{n=1,3,5,\dots} \frac{(-1)^{n-1}}{n!} \left(\frac{\zeta}{2}\right)^n = \cos\frac{\zeta}{2} + \boldsymbol{\gamma} \cdot \hat{\mathbf{v}} \sin\frac{\zeta}{2}. \tag{19.289}$$

Now, S commutes trivially with $\boldsymbol{\gamma} \cdot \mathbf{p} = \boldsymbol{\gamma} \cdot \hat{\boldsymbol{\zeta}} |\mathbf{p}|$, while anticommuting with γ^0 due to the anticommutation rules (19.274). Hence we can move the right-hand transformation in (19.286) simply to the left-hand side with a sign change of S, and obtain

$$H^{\mathrm{d}} = e^{2iS} H. \tag{19.290}$$

It is easy to calculate e^{2iS}: we merely have to double the rapidity in (19.289) and obtain

$$e^{2iS} = \cos \zeta + \boldsymbol{\gamma} \cdot \hat{\mathbf{v}} \sin \zeta = \frac{Mc}{\sqrt{\mathbf{p}^2 + M^2 c^2}} \left(1 + \boldsymbol{\gamma} \cdot \mathbf{p}/Mc \right). \tag{19.291}$$

Hence we obtain

$$H^{\mathrm{d}} = e^{2iS} H = \frac{Mc}{\sqrt{\mathbf{p}^2 + M^2 c^2}} \left(1 + \boldsymbol{\gamma} \cdot \mathbf{p}/Mc \right) Mc^2 \gamma^0 \left(1 + \boldsymbol{\gamma} \cdot \mathbf{p}/Mc \right). \tag{19.292}$$

Taking the right-hand parentheses to the left of γ^0 changes the sign of $\boldsymbol{\gamma}$. The product $\left(1 + \boldsymbol{\gamma} \cdot \mathbf{p}/Mc \right) \left(1 - \boldsymbol{\gamma} \cdot \mathbf{p}/Mc \right)$ is simply $1 + \mathbf{p}^2/M^2 c^2$, such that

$$H^{\mathrm{d}} = c\sqrt{\mathbf{p}^2 + M^2 c^2}\, \gamma^0 = \varepsilon_{\mathbf{k}}\, \gamma^0 = \hbar \omega_{\mathbf{k}}\, \gamma^0. \tag{19.293}$$

Remembering γ^0 from Eq. (19.276) shows that H^{d} has indeed the diagonal form (19.284).

Going to the diagonal fields $\psi_{\mathbf{k}}^{\mathrm{d}}(t) = e^{iS} \psi_{\mathbf{k}}(t)$, the action becomes

$$\mathcal{A} = \int_{t_a}^{t_b} dt \sum_{\mathbf{k}} \psi_{\mathbf{k}}^{\mathrm{d}\dagger}(t) \left[i\hbar \partial_t - H^d(\hbar \mathbf{k}) \right] \psi_{\mathbf{k}}^{\mathrm{d}}(t). \tag{19.294}$$

Thus a Dirac field is equivalent to a sum of infinitely many momentum states, each being associated with four harmonic oscillators of the Fermi type. The path integral is a product of independent harmonic path integrals of frequencies $\pm \omega_{\mathbf{k}}$.

It is then easy to calculate the quantum-mechanical partition function using the result (7.419) for each oscillator, continued to real times:

$$Z_{\mathrm{QM}} = \prod_{\mathbf{k}} \left\{ 2 \cosh^4[\omega_{\mathbf{k}}(t_b - t_a)/2] \right\}. \tag{19.295}$$

This can also be written as

$$Z_{\mathrm{QM}} = \exp \left(4 \sum_{\mathbf{k}} \log \left\{ 2 \cosh[\omega_{\mathbf{k}}(t_b - t_a)/2] \right\} \right), \tag{19.296}$$

or as

$$Z_{\mathrm{QM}} = \exp \left[4 \sum_{\mathbf{k}} \operatorname{Tr} \log \left(i\hbar \partial_t - \hbar \omega_{\mathbf{k}} \right) \right] = \exp \left\{ \sum_{\mathbf{k}} \operatorname{Tr} \log \left[i\hbar \partial_t - H^d(\hbar \mathbf{k}) \right] \right\}. \tag{19.297}$$

Since the trace is invariant under unitary transformations, we can rewrite this as

$$Z_{\rm QM} = \exp\left\{\sum_{\bf k} {\rm Tr}\, \log\left[i\hbar\partial_t - H(\hbar{\bf k})\right]\right\}, \tag{19.298}$$

or, since the determinant of γ^0 is unity, as

$$Z_{\rm QM} = \exp\left\{\sum_{\bf k} {\rm Tr}\, \log\left[i\hbar\gamma^0\partial_t - \gamma^0 H^{\rm d}(\hbar{\bf k})\right]\right\} = \exp\left\{\sum_{\bf k} {\rm Tr}\, \log\left[i\hbar\gamma^0\partial_t - \hbar c\boldsymbol{\gamma}\,{\bf k} - Mc^2\right]\right\}.$$

If we include the spatial coordinates into the functional trace, this can also be written as

$$Z_{\rm QM} = \exp\left[{\rm Tr}\, \log\left(i\hbar\gamma^0\partial_t - i\hbar c\boldsymbol{\gamma}\,\boldsymbol{\nabla} - Mc^2\right)\right] = \exp\left\{{\rm Tr}\, \log\left[c\left(i\hbar\slashed{\partial} - Mc\right)\right]\right\}.$$

In analytic regularization of Section 2.15, the factor c in the tracelog can be dropped. Moreover, there exists a simple algebraic identity

$$(i\hbar\slashed{\partial} + Mc)(i\hbar\slashed{\partial} - Mc) = -\hbar^2\partial^2 - M^2c^2. \tag{19.299}$$

The factors on the left-hand side have the same functional determinant since [compare (7.338) and (7.420)]

$${\rm Det}\,(i\hbar\slashed{\partial} - Mc) = e^{{\rm Tr}\, \log(i\hbar\slashed{\partial} - Mc)} = e^{V_4 \int \frac{d^4p}{(2\pi)^4} {\rm Tr}\, \log(\hbar\slashed{p} - Mc)} = e^{V_4 \int \frac{d^4p}{(2\pi)^4} {\rm Tr}\, \log(-\hbar\slashed{p} - Mc)}$$

$$= {\rm Det}\,(-i\hbar\slashed{\partial} - Mc). \tag{19.300}$$

This allows us, as a generalization of (7.420), to write

$${\rm Det}\,(i\hbar\slashed{\partial} - Mc) = {\rm Det}\,(i\hbar\slashed{\partial} + Mc) = \sqrt{{\rm Det}\,(-\hbar^2\partial^2 - M^2c^2)1_{4\times4}}, \tag{19.301}$$

where $1_{4\times4}$ is a 4×4 unit matrix. In this way we arrive at the quantum-mechanical partition function

$$Z_{\rm QM} = \exp\left[4 \times \frac{1}{2} {\rm Tr}\, \log(-\hbar^2\partial^2 - M^2c^2)\right] \equiv e^{i\Gamma_0^{\rm f}/\hbar}. \tag{19.302}$$

The factor 4 comes from the trace in the 4×4 matrix space, whose indices have disappeared in the formula. The exponent determines the effective action Γ of the quantum system by analogy with the Euclidean relation Eq. (19.46).

The Green function of the Dirac equation (19.273) is a 4×4 -matrix defined by

$$(i\hbar\slashed{\partial} - Mc)_{\alpha\beta}\,(x|x_a)_{\beta\gamma} = i\hbar\delta^{(D)}(x - x_a)\delta_{\alpha\gamma}. \tag{19.303}$$

Suppressing the Dirac indices, it has the spectral representation

$$(x_b|x_a) = \int \frac{d^4p}{(2\pi\hbar)^4} \frac{i\hbar}{\slashed{p} - Mc + i\eta} e^{-ip(x-x')/\hbar}. \tag{19.304}$$

This can be written more formally as a functional matrix

$$(x_b|x_a) = \langle x_b| \frac{i\hbar}{i\hbar\slashed{\partial} - Mc} |x_a\rangle, \tag{19.305}$$

which obviously satisfies the differential equation (19.303).

19.5.2 Path Integral

It is straightforward to write down a path integral representation for the amplitude
(19.305):

$$(x_b|x_a) = \int_0^\infty dS \int_{x_a=x(\tau_a)}^{x_b=x(\tau_b)} \mathcal{D}^4 x \int \frac{\mathcal{D}^4 p}{(2\pi\hbar)^4} e^{i\mathcal{A}/\hbar}, \qquad (19.306)$$

with the action

$$\mathcal{A}[x,p] = \int_0^S d\tau \left[-p\dot{x} + (\not{p} - Mc) \right], \qquad (19.307)$$

where $\dot{x} \equiv dx(\tau)/d\tau$. We are using the proper time τ to parametrize the orbits.[3]
Thus S is the total time, in contrast to the length parameter L in the Euclidean
discussion in the previous section [see Eq. (19.3)]. The minus sign in front of $p\dot{x}$ is
necessary to have the positive sign for the spatial part $\mathbf{p}\dot{\mathbf{x}}$ in the Minkowski metric
(19.275).

The action (19.307) can immediately be generalized to

$$\bar{\mathcal{A}}[x,p] = \int_0^S d\tau \left[-p\dot{x} + h(\tau)(\not{p} - Mc) \right], \qquad (19.308)$$

with any function $h(\tau) > 0$. This makes it invariant under the reparametrization

$$\tau \to f(\tau), \qquad h(\tau) \to h(\tau)/f(\tau). \qquad (19.309)$$

The path integral (19.306) contains then an extra functional integration over $h(\tau)$
with some gauge-fixing functional $\Phi[h]$, as in (19.22), which has been chosen in
(19.307) as $\Phi[h] = \delta[h - 1]$.

The path integral alone yields an amplitude

$$\langle x|e^{iS(i\hbar\not{\partial} - Mc)/\hbar}|x_a\rangle, \qquad (19.310)$$

and the integral over S in (19.306) produces indeed the propagator (19.305). In
evaluating this we must assume, as usual, that the mass carries an infinitesimal
negative imaginary part $i\eta$. This is also necessary to guarantee the convergence of
the path integral (19.306).

Electromagnetism is introduced as usual by the minimal substitution (2.642). In
the operator version, we have to substitute

$$\partial_\mu \longrightarrow \partial_\mu + i\frac{e}{\hbar c}A_\mu. \qquad (19.311)$$

Thus we obtain the gauge-invariant action

$$\bar{\mathcal{A}}[x,p] = \int_0^S d\tau \left[-p\dot{x} + h(\tau)\left(\not{p} - \frac{e\hbar}{c}\not{A} - Mc \right) \right]. \qquad (19.312)$$

---

[3]It corresponds to the parameter λ in Subsection 19.2.1. From now on we prefer the letter τ,
since there will be no danger of confusing the proper time with the coordinate time in Subsection
19.2.1.

Another path integral representation which is closer to the spinless case is obtained by rewriting (19.305) as

$$(x|x_a) = (i\hbar\slashed{\partial} + Mc)\langle x|\frac{i\hbar}{-\hbar^2\partial^2 - M^2c^2}|x_a\rangle, \qquad (19.313)$$

where we have omitted the negative infinitesimal imaginary part $-i\hbar$ of the mass, for brevity, and used the fact that

$$(i\hbar\slashed{\partial} + Mc)(i\hbar\slashed{\partial} - Mc) = -\hbar^2\partial^2 - M^2c^2, \qquad (19.314)$$

on account of the anticommutation relation (19.274). By rewriting (19.313) as a proper-time integral

$$(x|x_a) = \frac{1}{2M}(i\hbar\slashed{\partial} + Mc)\int_0^\infty dS\langle x|e^{iS(-\hbar^2\partial^2 - M^2c^2)/2M\hbar}|x_a\rangle, \qquad (19.315)$$

we find immediately the canonical path integral

$$(x|x_a) = \frac{1}{2Mc}(i\hbar\slashed{\partial} + Mc)\int_0^\infty dS \int_{x_a=x(\tau_a)}^{x=x(\tau_b)} \mathcal{D}^4 x \int \frac{\mathcal{D}^4 p}{(2\pi\hbar)^4} e^{i\mathcal{A}/\hbar}, \qquad (19.316)$$

with the action

$$\mathcal{A}[x, p] = \int_0^S d\tau \left[-p\dot{x} + \frac{1}{2M}\left(p^2 - M^2c^2\right)\right]. \qquad (19.317)$$

The suppressed Dirac indices of the 4×4 -amplitude on the left-hand side, $(x|x_a)_{\alpha\beta}$, are entirely due to the prefactor $(i\hbar\slashed{\partial} + Mc)_{\alpha\beta}$ on the right-hand side.

As in the generalization of (19.307) to (19.307), this action can be generalized to

$$\bar{\mathcal{A}}[x, p] = \int_0^S d\tau \left[-p\dot{x} + \frac{h(\tau)}{2Mc}\left(p^2 - M^2c^2\right)\right], \qquad (19.318)$$

with any function $h(\tau) > 0$, thus becoming invariant under the reparametrization (19.309), and the path integral (19.306) contains then an extra functional integration $\int \mathcal{D}h(\tau)\,\Phi[h]$. The action (19.318), is precisely the Minkowski version of the path integral of a spinless particle of the previous section [see Eq. (19.14)].

Introducing here electromagnetism by the minimal substitution (19.311) in the prefactor of (19.313) and on the left-hand side of (19.314), the latter becomes then

$$\left(i\hbar\slashed{\partial} - \frac{e}{c}\slashed{A} + Mc\right)\left(i\hbar\slashed{\partial} - \frac{e}{c}\slashed{A} - Mc\right) = \hbar^2\left[\left(i\partial - \frac{e}{\hbar c}A\right)^2 - \frac{e}{\hbar c}\Sigma^{\mu\nu}F_{\mu\nu}\right] - M^2c^2, \qquad (19.319)$$

where

$$\Sigma^{\mu\nu} \equiv \frac{i}{4}[\gamma^\mu, \gamma^\nu] = -\Sigma^{\nu\mu} \qquad (19.320)$$

are the generators of Lorentz transformations in the space of Dirac spinors. For any fixed index μ, they satisfy the commutation rules:

$$[\Sigma^{\mu\nu}, \Sigma^{\mu\kappa}] = ig^{\mu\mu}\Sigma^{\nu\kappa}. \tag{19.321}$$

Due to the antisymmetry in the two indices, this determines all nonzero commutators of the Lorentz group.

Using Eqs. (19.252), we can write the last interaction term in (19.319) as

$$\Sigma^{\mu\nu} F_{\mu\nu} = -2\Sigma^i B^i + 2\Sigma^{0i} E^i, \tag{19.322}$$

where Σ^i are the generators of rotation

$$\Sigma^i \equiv \frac{1}{2}\epsilon_{ijk}\Sigma^{jk} = \frac{1}{2}\begin{pmatrix} \sigma^i & 0 \\ 0 & \sigma^i \end{pmatrix}, \tag{19.323}$$

and

$$\Sigma^{0i} \equiv i\alpha^i \equiv i\gamma^0\gamma^i = i\begin{pmatrix} -\sigma^i & 0 \\ 0 & \sigma^i \end{pmatrix} \tag{19.324}$$

are the generators of rotation-free Lorentz transformations. Thus

$$\Sigma^{\mu\nu} F_{\mu\nu} = -\begin{pmatrix} \boldsymbol{\sigma}(\mathbf{B} + i\mathbf{E}) & 0 \\ 0 & \boldsymbol{\sigma}(\mathbf{B} - i\mathbf{E}) \end{pmatrix}. \tag{19.325}$$

19.5.3 Amplitude with Electromagnetic Interaction

The obvious generalization of the path integral (19.316) which includes minimal electromagnetic interactions is then

$$(x|x_a) = \frac{1}{2M}\left[\left(i\hbar\slashed{\partial} - \frac{e}{c}\slashed{A}\right) + Mc\right]\int_0^\infty dS \int \mathcal{D}h(\tau)\,\Phi[h]\int_{x_a=x(\tau_a)}^{x=x(\tau_b)} \mathcal{D}^4 x \int \frac{\mathcal{D}^4 p}{(2\pi\hbar)^4}\hat{T}e^{i\mathcal{A}/\hbar}, \tag{19.326}$$

with the action

$$\bar{\mathcal{A}}[x, p] = \int_0^S d\tau \left\{-p\dot{x} + \frac{h(\tau)}{2Mc}\left[\left(p - \frac{e}{c}A\right)^2 - \frac{\hbar e}{c}\Sigma^{\mu\nu} F_{\mu\nu} - M^2 c^2\right]\right\}. \tag{19.327}$$

The symbol $\hat{T}$ is the time-ordering operator defined in (1.241), now with respect to the proper time τ, which has to be present to account for the possible noncommutativity of $F_{\mu\nu}\Sigma^{\mu\nu}/2$ at different τ. Integrating out the momentum variables yields the configuration-space path integral

$$(x|x_a) = \frac{1}{2M}\left[\left(i\hbar\slashed{\partial} - \frac{e}{c}\slashed{A}\right) + Mc\right]\int_0^\infty dS \int \mathcal{D}h(\tau)\,\Phi[h]\int_{x_a=x(\tau_a)}^{x=x(\tau_b)} \mathcal{D}^4 x\,\hat{T}e^{i\mathcal{A}/\hbar}, \tag{19.328}$$

with the action

$$\bar{A}[x] = \int_0^S d\tau \left[-\frac{Mc}{2h(\tau)} \dot{x}^2 - \frac{e}{c} \dot{x} A - h(\tau) \frac{\hbar e}{2Mc^2} \Sigma^{\mu\nu} F_{\mu\nu} - h(\tau) \frac{Mc}{2} \right]. \quad (19.329)$$

The coupling to the magnetic field adds to the rest energy Mc^2 an interaction energy

$$H_{\text{int}} = -\frac{\hbar e}{Mc} \boldsymbol{\sigma} \cdot \mathbf{B}. \quad (19.330)$$

From this we extract the magnetic moment of the electron. We compare (19.330) with the general interaction energy (8.315), and identify the magnetic moment as

$$\boldsymbol{\mu} = \frac{\hbar e}{Mc} \boldsymbol{\sigma}. \quad (19.331)$$

Recall that in 1926, Uhlenbeck and Goudsmit explained the observed Zeeman splitting of atomic levels by attributing to an electron a half-integer spin. However, the magnetic moment of the electron turned out to be roughly twice as large as what one would expect from a charged rotating sphere of angular momentum $\mathbf{L}$, whose magnetic moment is

$$\boldsymbol{\mu} = \mu_B \frac{\mathbf{L}}{\hbar}, \quad (19.332)$$

where $\mu_B \equiv \hbar e/Mc$ is the Bohr magneton (2.647). On account of this relation, it is customary to parametrize the magnetic moment of an elementary particle of spin $\mathbf{S}$ as follows:

$$\boldsymbol{\mu} = g\mu_B \frac{\mathbf{S}}{\hbar}. \quad (19.333)$$

The dimensionless ratio g with respect to (19.332) is called the *gyromagnetic ratio* or *Landé factor*. For a spin-1/2 particle, $\mathbf{S}$ is equal to $\boldsymbol{\sigma}/2$, and comparison with (19.331) yields the gyromagnetic ratio

$$g = 2, \quad (19.334)$$

the famous result found first by Dirac, predicting the intrinsic magnetic moment μ of an electron to be equal to the Bohr magneton μ_B, thus being twice as large as expected from the relation (19.332), if we insert there the spin 1/2 for the orbital angular momentum.

In quantum electrodynamics one can calculate further corrections to this Dirac result as a perturbation expansion in powers of the fine-structure constant α [recall (1.502)]. The first correction to g due to one-loop Feynman diagrams was found by Schwinger:

$$g = 2 \times \left(1 + \frac{\alpha}{2\pi}\right) \approx 2 \times 1.001161, \quad (19.335)$$

where α is the fine-structure constant (1.502). Experimentally, the gyromagnetic ratio has been measured to an incredible accuracy:

$$g = 2 \times 1.001\,159\,652\,193(10), \tag{19.336}$$

in excellent agreement with (19.335). If the perturbation expansion is carried to higher orders, one is able to reach agreement up to the last experimentally known digits [14].

In the literature, there exist other representations of path integrals for Dirac particles involving Grassmann variables. For this we recall the discussion in Subsection 7.11.3 that a path integral over four real Grassmann fields θ^μ, $\mu = 0, 1, 2, 3$

$$\int \mathcal{D}^4\theta \, \exp\left[\frac{i}{\hbar}\int d\tau \, \frac{i}{2}\theta_\mu \dot\theta^\mu\right], \tag{19.337}$$

generates a matrix space corresponding to operators $\hat\theta^\mu$ with the anticommutation rules

$$\left\{\hat\theta^\mu, \hat\theta^\nu\right\} = g^{\mu\nu}, \tag{19.338}$$

and the matrix elements

$$\langle\beta|\hat\theta^\mu|\alpha\rangle = (\gamma_5\gamma^\mu)_{\beta\alpha}, \qquad \beta, \alpha = 1, 2, 3, 4. \tag{19.339}$$

It is then possible to replace path integral (19.328) by

$$\begin{aligned}
(x|x_a) &= \frac{1}{2M}\int_0^\infty dS \int \mathcal{D}h \, \Phi[h] \\
&\times \int \mathcal{D}\chi\Phi[\chi]\int \mathcal{D}^4\theta \int \mathcal{D}x \int \frac{\mathcal{D}p}{(2\pi\hbar)^4} \, e^{i\left(\bar{\mathcal{A}}[x]+\mathcal{A}_{\mathrm{G}}[\theta^\mu, A]\right)/\hbar},
\end{aligned} \tag{19.340}$$

with the action of a relativistic spinless particle [the action (19.329) without the spin coupling]

$$\bar{\mathcal{A}}[x, p] = \int_0^S d\tau \left\{-p\dot{x} + \frac{h(\tau)}{2Mc}\left[\left(p - \frac{e}{c}A\right)^2 - M^2c^2\right]\right\}, \tag{19.341}$$

and an action involving the Grassmann fields θ^μ:

$$\mathcal{A}_{\mathrm{G}}[\theta^\mu, A] = \int_0^S d\tau \left\{\frac{i\hbar}{4}\theta_\mu(\tau)\dot\theta^\mu(\tau) - h(\tau)\frac{i\hbar e}{4Mc^2}F_{\mu\nu}(x(\tau))\theta^\mu(\tau)\theta^\nu(\tau)\right\}. \tag{19.342}$$

This follows directly from Eq. (7.508). The function $h(\tau)$ is the same as in the bosonic actions (19.14) and the path integral (19.22) guaranteeing the reparametrization invariance (19.13).

After integrating out the momentum variables in the path integral (19.340), the canonical action is of course replaced by the configuration space action (19.216). In the simplest gauge (19.23), the total action reads

$$\bar{\mathcal{A}}[x, \theta^\mu] = \int_0^S d\tau \left[-\frac{M}{2}\dot{x}^2 - \frac{e}{c}\left(\dot{x}A + i\frac{\hbar}{4M}F_{\mu\nu}\theta^\mu\theta^\nu\right) - \frac{Mc^2}{2} + \frac{i\hbar}{4}\theta_\mu(\tau)\dot{\theta}^\mu(\tau)\right].$$
(19.343)

Note that the Grassmann variables can always be integrated out, yielding the functional determinant [compare with the real-time formula (7.512)]

$$\int \mathcal{D}x \, e^{-\frac{i}{4\hbar}\int d\tau\left[\theta_\mu(\tau)\dot{\theta}^\mu(\tau)+(e/4Mc)F_{\mu\nu}\theta^\mu\theta^\nu\right]} = 4\mathrm{Det}^{1/2}\left[\delta_{\mu\nu}\,\partial_\tau - i\frac{e}{Mc}F_{\mu\nu}(x(\tau))\right].$$
(19.344)

For a constant field tensor, and with the usual antiperiodic boundary conditions, the result has been given before in Eq. (19.101).

19.5.4 Effective Action in Electromagnetic Field

In the absence of electromagnetism, the effective action of the fermion orbits is given by (19.302). Its Euclidean version differs from the Klein-Gordon expression in (19.38) only by a factor -2:

$$\frac{\Gamma_{e,0}^f}{\hbar} = -2\,\mathrm{Tr}\,\log\left[-\hbar^2\partial^2 + M^2c^2\right].$$
(19.345)

Explicitly we have from (19.39), (19.41), and (19.46):

$$\frac{\Gamma_{e,0}^f}{\hbar} = 2V_D \int_0^\infty \frac{d\beta}{\beta} e^{-\beta Mc^2/2}\frac{1}{\sqrt{2\pi\hbar^2\beta/M}^D} = 2\frac{V_D}{\lambda_M^C{}^D}\frac{1}{(4\pi)^{D/2}}\Gamma(1 - D/2).$$
(19.346)

The factor 2 may be thought of as $4 \times 1/2$ where the factor 4 comes from the free path integral over the Grassmann field,

$$\int \mathcal{D}^D\theta \, e^{-\mathcal{A}_{e,0}[\theta]/\hbar} = 4.$$
(19.347)

counts the four components of the Dirac field. Recall that by (19.284), the Dirac field carries four modes, one of energy $\hbar\omega_\mathbf{k}$, with two spin degrees of freedom, the other of energy $-\hbar\omega_\mathbf{k}$ with two spin degrees. The latter are shown in quantum field theory to correspond to an antiparticle with spin 1/2. The path integral over $x(\tau)$ which counts paths in opposite directions with the ground state energy (19.39) describes particles and antiparticles [recall the remarks after Eq. 19.45]. This explains why only the spin factor 2 remains in (19.346).

By including the vector potential via the minimal substitution $\hat{p} \rightarrow \hat{p} - (e/c)A$, we obtain the Euclidean effective action from Eq. (19.219), and thus obtain immediately the path integral representation

$$\frac{\tilde{\Gamma}_e^f}{\hbar} = 2\int_0^\infty \frac{d\beta}{\beta} e^{-\beta Mc^2/2}\int \mathcal{D}^D x \, e^{-\mathcal{A}_e/\hbar},$$
(19.348)

with the Euclidean action (19.218).

This is not yet the true partition function Γ_e of the spin-$1/2$ particle, since the proper path integral contains the additional Grassmann terms of the action (19.343). In the Euclidean version, the full interaction is

$$\mathcal{A}_{\mathrm{e,int}}[x,\theta] = \int_0^{\hbar\beta} d\tau \, \frac{e}{c}\left[i\,\dot{x}_\mu(\tau)A_\mu(x(\tau)) - \frac{i}{4M}F_{\mu\nu}(x(\tau))\theta^\mu(\tau)\theta^\nu(\tau)\right]. \quad (19.349)$$

Thus we obtain the path integral representation

$$\frac{\Gamma_e^{\mathrm{f}}}{\hbar} = 2\int_0^\infty \frac{d\beta}{\beta}\, e^{-\beta Mc^2/2}\int \mathcal{D}^D x \int \mathcal{D}^D \theta\, e^{-\{\mathcal{A}_{\mathrm{e,0}}[x,\theta]+\mathcal{A}_{\mathrm{e,int}}[x,\theta]\}/\hbar}, \quad (19.350)$$

where the free part of the Euclidean action is

$$\mathcal{A}_{\mathrm{e,0}}[x,\theta] = \mathcal{A}_{\mathrm{e,0}}[x] + \mathcal{A}_{\mathrm{e,0}}[\theta] \equiv \int_0^{\hbar\beta} d\tau \, \frac{M}{2}\dot{x}^2(\tau) + \int_0^{\hbar\beta} d\tau \, \frac{\hbar}{4}\theta^\mu(\tau)\dot{\theta}^\mu(\tau). \quad (19.351)$$

19.5.5 Perturbation Expansion

The perturbation expansion is a straightforward generalization of the expansion (19.222):

$$\frac{\Gamma_e^{\mathrm{f}}}{\hbar} = \frac{\Gamma_{\mathrm{e,0}}^{\mathrm{f}}}{\hbar} + \int_0^\infty \frac{d\beta}{\beta}\, e^{-\beta Mc^2/2}\frac{2V_D}{\sqrt{2\pi\hbar^2\beta/M}^D}\sum_{n=1}^\infty \frac{(-ie/c)^n}{n!} \quad (19.352)$$

$$\times \left\langle \prod_{i=1}^n \left\{\int_0^{\hbar\beta} d\tau_i \left[\dot{x}_\mu(\tau_i)A_\mu(x(\tau_i)) - \frac{\hbar}{4M}F_{\mu\nu}(x(\tau_i))\theta^\mu(\tau_i)\theta^\nu(\tau_i)\right]\right\}\right\rangle_0.$$

The leading free effective action coincides, of course, with the $n=0$-term of the sum [compare (19.346)].

The expectation values are now defined by the Grassmann extension of the Gaussian path integral (19.223):

$$\langle \mathcal{O}[x,\theta]\rangle_0 \equiv \frac{\int \mathcal{D}^D x \int \mathcal{D}^D \theta\, \mathcal{O}[x,\theta]\, e^{-\mathcal{A}_{\mathrm{e,0}}[x,\theta]/\hbar}}{\int \mathcal{D}^D x\, e^{-\mathcal{A}_{\mathrm{e,0}}[x]/\hbar}\int \mathcal{D}^D \theta\, e^{-\mathcal{A}_{\mathrm{e,0}}[\theta]/\hbar}}. \quad (19.353)$$

where the denominator is equal to $(1/2)V_D/\sqrt{2\pi\hbar^2\beta/M}^D \times 4$.

There exists also an expansion analogous to (19.225), where the vector potentials have been Fourier decomposed according to (19.224). Then we obtain an expansion just like (19.225), except for a factor -2 and with the expectation values replaced as follows:

$$\left\langle \prod_{i=1}^n \left[\int_0^{\hbar\beta} d\tau_i\, \dot{x}^{\mu_i}(\tau_i)e^{ik_i x(\tau_i)}\right]\right\rangle_0$$

$$\rightarrow \left\langle \prod_{i=1}^n \left\{\int_0^{\hbar\beta} d\tau_i \left[\dot{x}^{\mu_i}(\tau_i) + \frac{i\hbar}{2M}k_i^{\nu_i}\theta^{\nu_i}(\tau_i)\theta^{\mu_i}(\tau_i)\right]e^{ik_i x(\tau_i)}\right\}\right\rangle_0. \quad (19.354)$$

The evaluation of these expectation values proceeds as in Eqs. (19.226)–(19.237), except that we also have to form Wick contractions of Grassmann variables which have the free correlation functions

$$\langle \theta^\mu(\tau)\theta^\nu(\tau')\rangle = 2\delta^{\mu\nu}G^a_{\omega,e}(\tau - \tau'),\tag{19.355}$$

where

$$G^a_{\omega,e}(\tau - \tau') = \frac{1}{2}\epsilon(\tau), \qquad \tau \in [-\hbar\beta, \hbar\beta]\tag{19.356}$$

is the Euclidean version of the antiperiodic Green function (3.109) solving the inhomogeneous equation

$$\partial_\tau G^a_{\omega,e}(\tau) = \delta(\tau).\tag{19.357}$$

Outside the basic interval $[-\hbar\beta, \hbar\beta]$ the function is to be continued antiperiodically. in accordance with the fermionic nature of the Grassmann variables.

In operator language, the correlation function (19.355) is the time-ordered expectation value $\langle \hat{T}\hat{\theta}(\tau)\hat{\theta}(\tau')\rangle_0$ [recall (3.296)]. By letting $\tau \to \tau'$ once from above and once from below, the correlation function shows agreement with the anticommutation rule (19.338). In verifying this we must use the fact that the time ordered product of fermion operators is defined by the following modification of the bosonic definition in Eq. (1.241):

$$\hat{T}(\hat{O}_n(t_n)\cdots\hat{O}_1(t_1)) \equiv \epsilon_P \hat{O}_{i_n}(t_{i_n})\cdots\hat{O}_{i_1}(t_{i_1}),\tag{19.358}$$

where $t_{i_n},\ldots,t_{i_1}$ are the times $t_n,\ldots,t_1$ relabeled in the causal order, so that

$$t_{i_n} > t_{i_{n-1}} > \ldots > t_{i_1}.\tag{19.359}$$

The difference lies in the sign factor ϵ_P which is equal to 1 for an even and -1 for an odd number of permutations of fermion variables.

19.5.6 Vacuum Polarization

Let us see how the fluctuations of an electron loop change the electromagnetic field action. To lowest order, we must form the expectation value (19.354) for $n = 0$ and $k_1 = -k_2 \equiv k$:

$$\left\langle \left[\dot{x}^{\mu_1}(\tau_1) + \frac{i\hbar}{2M}k^{\nu_1}\theta^{\nu_1}(\tau_1)\theta^{\mu_1}(\tau_1)\right]e^{ik\delta x(\tau_1)}\left[\dot{x}^{\mu_2}(\tau_2) - \frac{i\hbar}{2M}k^{\nu_2}\theta^{\nu_2}(\tau_2)\theta^{\mu_2}(\tau_2)\right]e^{-ik\delta x(\tau_2)}\right\rangle_0.\tag{19.360}$$

From the contraction of the velocities $\dot{x}^{\mu_1}(\tau_1)$ and $\dot{x}^{\mu_2}(\tau_2)$ we obtain again the spinless result (19.238) leading in (19.240) to the integrand

$$\left(k_1^2\delta^{\mu_1\nu_2} - k_1^{\mu_1}k_1^{\mu_2}\right)\dot{G}^2(\tau_1,\tau_2) = \left(k_1^2\delta^{\mu_1\nu_2} - k_1^{\mu_1}k_1^{\mu_2}\right)\frac{\hbar^2}{M^2}(u - 1/2)^2.\tag{19.361}$$

In addition, there are the Wick contractions of the Grassmann variables:

$$\left\langle \left[\frac{\hbar}{2M} k^{\nu_1} \theta^{\nu_1}(\tau_1) \theta^{\mu_1}(\tau_1) \right] e^{ik\delta x(\tau_1)} \left[\frac{i\hbar}{2M} k^{\nu_2} \theta^{\nu_2}(\tau_2) \theta^{\mu_2}(\tau_2) \right] e^{-ik\delta x(\tau_2)} \right\rangle_0$$
$$= -\left(k_1^2 \delta^{\mu_1 \nu_2} - k_1^{\mu_1} k_1^{\mu_2} \right) \frac{\hbar^2}{M^2} \frac{1}{4} \epsilon^2(\tau_1 - \tau_2). \qquad (19.362)$$

Since $\epsilon^2(\tau_1 - \tau_2) = 1$, this changes the spinless result (19.361) to

$$\left(k_1^2 \delta^{\mu_1 \nu_2} - k_1^{\mu_1} k_1^{\mu_2} \right) {}^\cdot G^2(\tau_1, \tau_2) = \left(k_1^2 \delta^{\mu_1 \nu_2} - k_1^{\mu_1} k_1^{\mu_2} \right) \frac{\hbar^2}{M^2} [(u - 1/2)^2 - 1/4]. \quad (19.363)$$

Remembering the factor -2 in the expansion (19.353) with respect to the spinless one, we find that the vacuum polarization due to fluctuating spin-1/2 orbits is obtained from the spinless result (19.249) by changing the factor $4(u - 1/2)^2 = (2u - 1)^2$ in the integrand to $-2 \times 4u(u - 1) = 8u(1 - u)$. The resulting function $\Pi(k^2)$ has the expansion

$$\Pi(k^2) = \frac{1}{3\pi} \left[\frac{2}{\epsilon} - \log \frac{M^2 c^2 e^\gamma}{4\pi \hbar^2} \right] - \frac{\hbar^2 k^2}{15\pi M^2 c^2} + \mathcal{O}\left(\epsilon, \frac{k^2}{M^2 c^2/\hbar^2} \right). \qquad (19.364)$$

The first term produces a renormalization of the charge which is treated as in the bosonic case [recall (19.269)–(19.272)], which causes an additional contact interaction

$$-\frac{\alpha}{r} \rightarrow -\frac{\alpha}{r} - \frac{4\alpha^2 \hbar^2}{15 M^2 c^2} \delta^{(3)}(\mathbf{x}). \qquad (19.365)$$

There, the vacuum polarization has the effect of lowering the state $2S_{1/2}$, which is the s-state of principal quantum number $n = 2$, against the p-state $2P_{1/2}$ by 27.3 MHz. The experimental frequency shift is positive $\approx 1057\,\mathrm{MHz}$ [recall Eq. (18.600)], and is mainly due to the effect of the electron moving through a bath of photons as calculated in Eq. (18.599).

The effect of vacuum polarization was first calculated by Uehling [13], who assumed it to be the main cause for the Lamb shift. He was disappointed to find only 3% of the experimental result, and a wrong sign.

The situation in muonic atoms is different. There the vacuum polarization *does* produce the dominant contribution to the Lamb shift for a simple reason: The other effects contain in a factor M/M_μ^2, where M_μ^2 is the mass of the muon, whereas the vacuum polarization still involves an electron loop containing only the electron mass M, thus being enhanced by a factor $(M_\mu/M)^2 \approx 210^2$ over the others.

The calculations for the electron in an atom have been performed to quite high orders [14] within quantum electrodynamics. We have gone through the above calculation only to show that it is possible to re-obtain quantum field-theoretic result within the path integral formalism. More details are given in the review article [5],

As mentioned in the beginning, the above calculations are greatly simplified version of analogous calculations within superstring theory, which so far have not produced any physical results. If this ever happens, one should expect that also in this field a second-quantized field theory would be extremely useful to extract efficiently observable consequences. Such a theory still need development [15].

19.6 Supersymmetry

It is noteworthy that the various actions for a spin-1/2 particle is invariant under certain supersymmetry transformations.

19.6.1 Global Invariance

Consider first the fixed-gauge action (19.343). Its appearance can be made somewhat more symmetric by absorbing a factor $\sqrt{\hbar/2M}$ into the Grassmann variables $\theta^\mu(\tau)$, so that it reads

$$\bar{A}[x, \theta^\mu] = \int_0^S d\tau \left[-\frac{M}{2}\dot{x}^2 - \frac{e}{c}\left(\dot{x}A + \frac{i}{2}F_{\mu\nu}\theta^\mu\theta^\nu\right) - \frac{Mc^2}{2} + \frac{M}{2}i\theta_\mu(\tau)\dot{\theta}^\mu(\tau) \right]. \quad (19.366)$$

The correlation functions (19.355) of the θ-variables are now

$$\langle \theta^\mu(\tau)\theta^\nu(\tau')\rangle = \delta^{\mu\nu}G^{\mathrm{f}}(\tau, \tau') \equiv \delta^{\mu\nu}\frac{\hbar}{M}\Delta_0^{\mathrm{f}}(\tau - \tau'), \quad (19.367)$$

with $\Delta_0^{\mathrm{f}}(\tau - \tau') = \epsilon(\tau - \tau')/2$. In this normalization, $G^{\mathrm{f}}(\tau, \tau')$ coincides, up to a sign, with the first term in the derivative $\dot{G}(\tau, \tau')$ of the bosonic correlation function [recall (19.230) and the first term in (19.232)].

Let us apply to the variables the infinitesimal transformations

$$\delta x^\mu(\tau) = i\alpha\theta^\mu(\tau), \qquad \delta\theta^\mu(\tau) = \alpha\dot{x}^\mu(\tau). \quad (19.368)$$

where α is an arbitrary Grassmann parameter. For the free terms this is obvious. The interacting terms change by

$$-i\alpha \int d^4x \frac{e}{c}\left(\dot{\theta}^\mu A_\mu + F_{\mu\nu}\dot{x}^\mu\theta^\nu\right). \quad (19.369)$$

Inserting $F_{\mu\nu}\dot{x}^\mu(\tau) = dA_\nu(x(\tau))/d\tau - \partial_\nu[A^\mu(x(\tau))\dot{x}^\mu(\tau)]$, the first term cancels and the second is a pure surface term, such that the action is indeed invariant.

Supersymmetric theories have a compact representation in an extended space called *superspace*. This space is formed by pairs (τ, ζ), where ζ is a Grassmann variable playing the role of a supersymmetric partner of the time parameter τ. The coordinates $x^\mu(\tau)$ are extended likewise by defining

$$X^\mu(\tau) \equiv x^\mu(\tau) + i\zeta\theta^\mu(\tau). \quad (19.370)$$

A supersymmetric derivative is defined by

$$DX^\mu(\tau) \equiv \left(\frac{\partial}{\partial\zeta} + i\zeta\frac{\partial}{\partial\tau}\right) X^\mu(\tau) = i\theta^\mu(\tau) + i\zeta\dot{x}^\mu(\tau). \tag{19.371}$$

If we now form the integral, using the Grassmann formula (7.379),

$$\int d\tau \frac{d\zeta}{2\pi} \, i\dot{X}_\mu(\tau)DX^\mu(\tau) = \int d\tau \frac{d\zeta}{2\pi} \, i\left[\dot{x}(\tau) + i\zeta\dot{\theta}^\mu(\tau)\right]\left[i\theta^\mu(\tau) + i\zeta\dot{x}^\mu(\tau)\right], \tag{19.372}$$

we find

$$\int d\tau \left(-\dot{x}^2 + i\theta_\mu\dot{\theta}^\mu\right), \tag{19.373}$$

which proportional to the free part of the action (19.343). As a curious property of differentiations in superspace we note that

$$D^2 X^\mu(\tau) = i\dot{x}^\mu(\tau) - \zeta\dot{\theta}^\mu(\tau), \qquad D^3 X^\mu(\tau) = -\dot{\theta}^\mu(\tau) - \zeta\ddot{x}(\tau), \tag{19.374}$$

such that the kinetic term (19.372) can also be written as

$$-\int d\tau \frac{d\zeta}{2\pi} X_\mu(\tau)D^3 X^\mu(\tau). \tag{19.375}$$

The interaction is found from the integral in superspace

$$i\int d\tau \frac{d\zeta}{2\pi} A^\mu(X(\tau))DX(\tau)$$

$$= i\int d\tau \frac{d\zeta}{2\pi} \left[A^\mu(x(\tau)) + i\partial_\nu A^\mu(x(\tau))\theta^\nu(\tau)\right]\left[i\theta^\mu(\tau) + i\zeta\dot{x}^\mu(\tau)\right], \tag{19.376}$$

which is equal to

$$-\int d\tau \left[A^\mu(\tau)\dot{x}(\tau) + \frac{i}{2}F_{\mu\nu}\theta^\mu(\tau)\theta^\nu(\tau)\right],$$

thus reproducing the interaction in (19.343). The action in superspace can therefore be written in the simple form

$$\mathcal{A}[X] = i\int d\tau \frac{d\zeta}{2\pi} \left[-\frac{M}{2}X_\mu(\tau)D^3 X^\mu(\tau) + \frac{e}{c}A^\mu(X(\tau))DX(\tau)\right]. \tag{19.377}$$

19.6.2   Local Invariance

A larger class of supersymmetry transformations exists for the action without gauge fixing which is the sum of the free part (19.341) and the interacting part (19.342). Absorbing again the factor $\sqrt{\hbar/2M}$ into the Grassmann variable $\theta^\mu(\tau)$, and rescaling in addition $h(\tau)$ by a factor $1/c$, the reparametrization-invariant action reads

$$\bar{\mathcal{A}}[x,p,\theta,h] = \int_0^S d\tau \left\{-p\dot{x} + \frac{h(\tau)}{2M}\left[\left(p - \frac{e}{c}A\right)^2 - M^2 c^2\right]\right.$$

$$\left. + \frac{M}{2}i\theta_\mu(\tau)\dot{\theta}^\mu(\tau) - ih(\tau)\frac{e}{c}F_{\mu\nu}(x(\tau))\theta^\mu(\tau)\theta^\nu(\tau)\right\}. \tag{19.378}$$

Let us now compose the action from invariant building blocks. For simplicity, we ignore the electromagnetic interaction. In a first step we also omit the mass term. The extra variable $h(\tau)$ requires an extra Grassmann partner $\chi(\tau)$ for symmetry, and we form the action

$$\bar{\mathcal{A}}_1[x,p,\theta,h,\chi] = \int_0^S d\tau \left\{ -p\dot{x} + \frac{h(\tau)}{2M}p^2 + \frac{M}{2}i\theta_\mu(\tau)\dot{\theta}^\mu(\tau) + \frac{i}{2}\chi(\tau)\theta^\mu(\tau)p_\mu(\tau) \right\}. \quad (19.379)$$

This action possesses a *local supersymmetry*. If we now perform τ-dependent versions of the supersymmetry transformations (19.368)

$$\delta x^\mu = i\alpha(\tau)\theta^\mu, \qquad \delta\theta^\mu = \alpha(\tau)p, \qquad \delta p = 0,$$
$$\delta h = i\alpha(\tau)\chi, \qquad \delta\chi = 2\dot{\alpha}(\tau). \qquad (19.380)$$

If we integrate out the momenta in the path integral, the action (19.379) goes over into

$$\bar{\mathcal{A}}_1[x,\theta,h,\chi] = \int_0^S d\tau \left\{ -\frac{\dot{x}^2}{2h(\tau)} + \frac{M}{2}i\theta_\mu(\tau)\dot{\theta}^\mu(\tau) + \frac{i}{2h(\tau)}\chi(\tau)\theta^\mu(\tau)\dot{x}_\mu(\tau) \right\}, \quad (19.381)$$

where a term proportional to $\chi^2(\tau)$ has been omitted since it vanishes due to the nilpotency (7.375). This action is locally supersymmetric under the transformations

$$\delta x^\mu = i\alpha(\tau)\theta^\mu, \qquad \delta\theta^\mu = \frac{\alpha(\tau)}{h(\tau)}\left[\dot{x} - \frac{i}{2}\chi\,\theta^\mu \right],$$
$$\delta h = i\alpha(\tau)\chi, \qquad \delta\chi = 2\dot{\alpha}(\tau). \qquad (19.382)$$

We now add the mass term

$$\mathcal{A}_M = -\frac{1}{2}\int_0^S d\tau\, h(\tau)Mc^2. \qquad (19.383)$$

This needs a supersymmetric partner to compensate the variation of $\mathcal{A}_m$ under (19.382).

$$\mathcal{A}_5 = \frac{i}{2}\int_0^S d\tau\, \left[\theta_5(\tau)\dot{\theta}_5(\tau) + Mc\chi(\tau)\theta_5(\tau) \right]. \qquad (19.384)$$

Indeed, add to (19.382) the transformation

$$\delta\theta_5 = Mc\,\alpha(\tau), \qquad (19.385)$$

we see that the sum $\mathcal{A}_M + \mathcal{A}_5$ is invariant. Adding this to (19.379), we obtain the locally invariant canonical action

$$\bar{\mathcal{A}}[x,p,\theta,\theta_5,h,\chi] = \int_0^S d\tau \left\{ -p\dot{x} + \frac{h(\tau)}{2M}p^2 - \frac{h(\tau)}{2}Mc + \frac{M}{2}i\left[\theta_\mu(\tau)\dot{\theta}^\mu(\tau) + \theta_5(\tau)\dot{\theta}_5(\tau) \right] \right.$$
$$\left. + \frac{i}{2}\chi(\tau)\left[\theta^\mu(\tau)p_\mu(\tau) + Mc\theta_5(\tau) \right] \right\}. \qquad (19.386)$$

Appendix 19A Proof of Same Quantum Physics of Modified Action

Consider the sliced path integral for a relativistic point particle associated with the original action (19.12). If we set the initial and final paramters λ_a and λ_b equal to λ_0 and λ_{N+1}, and slice the λ-axis at the places λ_n $(n = 1, 2, \ldots, N)$, the action becomes

$$\mathcal{A}_{\mathrm{cl,e}} = Mc \sum_{n=1}^{N+1} |x_n - x_{n-1}|, \tag{19A.1}$$

where $|x_n - x_{n-1}| = \sqrt{(x_n - x_{n-1})^2}$ are the Euclidean distances [recall (19.2)]. The Euclidean amplitude for the particle to run from $x_a = x_0$ to $x_b = x_{N+1}$ is therefore given by the product of integrals in D spacetime dimensions

$$(x_b|x_a) = \mathcal{N} \prod_{n=1}^{N} \left[\int d^D x_n \right] e^{-Mc \sum_{n=1}^{N+1} |x_n - x_{n-1}|/\hbar}. \tag{19A.2}$$

As a consequence of the reparamterization invariance of (19.12), this expression is independent of the thickness $\lambda_n - \lambda_{n-1}$ of the slices, which we shall denote by rename as $\equiv h_n \epsilon$, where ϵ is some fixed small number.

We now factorize the exponential of the sum into a product of $N+1$ exponentials and represent each factor as an integral [using Formulas (1.343) and (1.345)]

$$e^{-Mc|x_n - x_{n-1}|/\hbar} = \sqrt{\frac{\epsilon M c}{2\pi \hbar}} \int_0^\infty dh_n h_n^{-1/2} e^{-\epsilon h_n Mc/2\hbar - Mc(x_n - x_{n-1})^2/2h_n \epsilon \hbar}. \tag{19A.3}$$

Absorbing constants in the normalization factor $\mathcal{N}$, we arrive at

$$
\begin{aligned}
(x_b|x_a) = \ &\mathcal{N} \prod_{n=1}^{N+1} \left[\int_0^\infty dh_n h_n^{(D-1)/2} \right] e^{-Mc\epsilon \sum_{n=1}^{N+1} h_n/2} \\
&\times \frac{1}{\sqrt{2\pi \epsilon h_{N+1}/Mc}^D} \prod_{n=1}^{N} \left[\int \frac{d^D x_n}{\sqrt{2\pi \epsilon h_n/Mc}^D} \right] e^{-Mc\epsilon \sum_{n=1}^{N+1} (x_n - x_{n-1})^2/2h_n \epsilon},
\end{aligned} \tag{19A.4}
$$

The second line contains only harmonic integrals over x_n, which can all be done with the help of the formulas of Appendix 2B, with the result

$$(x_b L|x_a 0) = \frac{1}{\sqrt{2\pi L \hbar/Mc}^D} e^{-Mc(x_b - x_a)^2/2L\hbar}, \tag{19A.5}$$

where L is the total parameter length

$$L \equiv \sum_{n=1}^{N+1} \epsilon h_n, \tag{19A.6}$$

Replacing (19A.5) by its Fourier representation [compare with (1.329) and (1.337)], we can rewrite (19A.4) as

$$(x_b|x_a) = \mathcal{N} \prod_{n=1}^{N+1} \left[\int_0^\infty dh_n h_n^{(D-1)/2} \right] e^{-McL/2\hbar} \int \frac{d^D p}{(2\pi \hbar)^D} e^{-Lp^2/2Mc + ip(x_b - x_a)/\hbar}. \tag{19A.7}$$

Thus we are left with the product of integrals over h_n. Before we can perform these, we must make sure to respect the sum (19A.6). This is done by inserting an auxiliary unit integral that separates out an integral over the total length L:

$$1 = \int_0^\infty dL \, \delta(\Sigma_{n=1}^{N+1} \epsilon h_n - L) = \int_0^\infty \frac{dL}{2\lambda_C} \int_{-i\infty}^{\infty} \frac{d\sigma}{2\pi i} e^{-\sigma(\Sigma_{n=1}^{N+1} \epsilon h_n - L)/2\lambda_C}, \tag{19A.8}$$

where λ_C is the Compton wavelength (19.31). Then the product of integrals over h_n in the brackets of (19A.7) can be rewritten as follows:

$$\int_0^\infty \frac{dL}{2\lambda_C} \left\{ \int_{-i\infty}^{i\infty} \frac{d\sigma}{2\pi i} e^{\sigma L/2\lambda_C} \prod_{n=1}^{N+1} \left[\int_0^\infty dh_n h_n^{(D-1)/2} e^{-\sigma \epsilon h_n/2\lambda_C} \right] \right\} e^{-L/2\lambda_C} \tag{19A.9}$$

Setting $h_n = r_n^2$, we treat the curly backets as

$$\int_{-i\infty}^{i\infty} \frac{d\sigma}{2\pi i} e^{\sigma L/2\lambda_C} \prod_{n=1}^{N+1} \left[\int_{-\infty}^\infty dr_n r_n^D e^{-\sigma \epsilon r_n^2/2\lambda_C} \right]$$

$$= \left[\frac{2\pi \lambda_C}{\Gamma(D+1)\epsilon} \right]^{N+1} \int_{-i\infty}^{i\infty} \frac{d\sigma}{2\pi i} e^{\sigma L/2\lambda_C} \sigma^{-(N+1)(D+1)/2}. \tag{19A.10}$$

For large N, the integral over σ can be approximated by the Gaussian integral around the neighborhood of the saddle point at $\sigma = (N+1)(D+1)\lambda_C/L$:

$$\int_{-i\infty}^{i\infty} \frac{d\sigma}{2\pi i} e^{\sigma L/2\lambda_C} \sigma^{-(N+1)(D+1)/2} \underset{\text{large } N}{\approx} \frac{1}{\sqrt{2\pi}} \left[\frac{L}{(N+1)(D+1)\lambda_C} \right]^{(N+1)(D+1)/2}. \tag{19A.11}$$

While letting N tend to infinity, we keep $L/(N+1) \equiv \bar{\epsilon}$ fixed. Then we may write the right-hand side as an exponential

$$\frac{1}{\sqrt{2\pi}} \left[\frac{(D+1)\lambda_C}{\bar{\epsilon}} \right]^{-L(D+1)/2\bar{\epsilon}} = \frac{1}{\sqrt{2\pi}} e^{-zL/2\lambda_C}, \tag{19A.12}$$

where

$$z \equiv \nu \log \nu \quad \text{with} \quad \nu \equiv (D+1)\lambda_C/\bar{\epsilon} \tag{19A.13}$$

is a large number. Inserting (19A.12) back into the curly brackets of (19A.9), the constant z can be absorbed into the mass of the particle by replacing M by the renormalized quantity $M_1 = M(1+z)$. With this, (19A.9) becomes

$$\int_0^\infty \frac{dL}{2\lambda_C} e^{-LM_1 c/2\hbar}, \tag{19A.14}$$

and the path integral (19A.4) reduces to

$$(x_b|x_a) = \mathcal{N}'' \int_0^\infty \frac{dL}{2\lambda_C} e^{-M_1 cL/2\hbar} \int \frac{d^D p}{(2\pi\hbar)^D} e^{-Lp^2/2Mc + ip(x_b - x_a)/\hbar}, \tag{19A.15}$$

where all irrelevant factors are contained in the normalization factor $\mathcal{N}''$. The remaining integral over L can now be done and yields

$$(x_b|x_a) = \mathcal{N}'' \int \frac{d^D k}{(2\pi)^D} \frac{1}{k^2 + M_R^2 c^2/\hbar^2} e^{ik(x_b - x_a)},$$

where M_R is the renomalized mass

$$M_R = M(1+z)^{1/2}. \tag{19A.16}$$

If we choose the normalization factor $\mathcal{N}'' = 1$, this is exactly the same result as that obtained before in Eq. (19.30) from the modified action (19.10), if we use in the original action (19.12) the mass $M/(1+z)^{1/2}$ rather than M for the calculation.

Notes and References

Relativistic quantum mechanics is described in detail in
J.D. Bjorken and S.D. Drell, *Relativistic Quantum Mechanics*, McGraw-Hill, New York, 1964,
relativistic quantum field theory in
S.S. Schweber, *Introduction to Relativistic Quantum Field Theory*, Harper and Row, New York, 1962;
J.D. Bjorken and S.D. Drell, *Relativistic Quantum Fields*, McGraw-Hill, New York, 1965;
C. Itzykson and J.-B. Zuber, *Quantum Field Theory*, McGraw-Hill, New York, 1985.

The individual citations refer to

[1] For the development and many applications see the textbook
H. Kleinert, *Gauge Fields in Condensed Matter*, World Scientific, Singapore, 1989;
Vol. I, *Superflow and Vortex Lines (Disorder Fields, Phase Transitions)*; Vol. II, *Stresses and Defects (Differential Geometry, Crystal Melting)* (wwwK/b1, where wwwK is short for (http://www.physik.fu-berlin.de/~kleinert).

[2] See Vol. I of the textbook [1] and the original paper
H. Kleinert, Lett. Nuovo Cimento **35**, 405 (1982) (*ibid*.http/97).
The theoretical prediction of this paper was confirmed only 20 years later in
S. Mo, J. Hove, A. Sudbø, Phys. Rev. B **65**, 104501 (2002) (cond-mat/0109260); Phys. Rev. B **66**, 064524 (2002) (cond-mat/0202215).

[3] R.P. Feynman, Phys. Rev. **80**, 440 (1950).

[4] There are basically two types of approach towards a worldline formulation of spinning particles: one employs auxiliary Bose variables:
R.P. Feynman, Phys. Rev. **84**, 108 (1989);
A.O. Barut and I.H. Duru, Phys. Rep. **172**, 1 (1989);
the other anticommuting Grassmann variables:
E.S. Fradkin, Nucl. Phys. **76**, 588 (1966);
R. Casalbuoni, Nuov. Cim. A **33**, 389 (1976); Phys. Lett. B **62**, 49 (1976);
F.A. Berezin and M.S. Marinov, Ann. Phys. **104**, 336 (1977);
L. Brink, S. Deser, B. Zumino, P. DiVecchia, and P.S. Howe, Phys. Lett. B **64**, 435 (1976);
L. Brink, P. DiVecchia, and P.S. Howe, Nucl. Phys. B **118**, 76 (1976).
These worldline formulations were used to recalculate processes of electromagnetic and strong interactions by
M.B. Halpern, A. Jevicki, and P. Senjanovic, Phys. Rev. D **16**, 2476 (1977);
M.B. Halpern and W. Siegel, Phys. Rev. D **16**, 2486 (1977);
Z. Bern and D.A. Kosower, Nucl. Phys. B **362**, 389 (1991); **379**, 451 (1992);
M. Strassler, Nucl. Phys. B **385**, 145 (1992);
M.G. Schmidt and C. Schubert, Phys. Lett. B **331**, 69 (1994); Nucl. Phys. Proc. Suppl. B,C **39**, 306 (1995); Phys. Rev. D **53**, 2150 (1996) (hep-th/9410100).
For many more references see the review article in Ref. [5].

[5] C. Schubert, Phys. Rep. **355**, 73 (2001);
G.V. Dunne, Phys. Rep. **355**, 73 (2002);

[6] As a curiosity of history, Schrödinger invented first the relativistic Klein-Gordon equation and extracted the Schrödinger equation from this by taking its nonrelativistic limit similar to Eqs. (19.34) and (19.35).

[7] In particle physics, the Chern-Simons theory of ribbons explained in Section 16.7 was used to construct path integrals over fluctuating fermion orbits:
A.M. Polyakov, Mod. Phys. Lett. A **3**, 325 (1988).

For more details see
C.H. Tze, Int. J. Mod. Phys. A **3**, 1959 (1988).

[8] W. Heisenberg and H. Euler, Z. Phys. **98**, 714 (1936).
English translation available at `wwwK/files/heisenberg-euler.pdf`.
J. Schwinger, Phys. Rev. **84**, 664 (1936); **93**, 615; **94**, 1362 (1954).

[9] E. Lundström, G. Brodin, J. Lundin, M. Marklund, R. Bingham, J. Collier, J.T. Mendonca, and P. Norreys, Phys. Rev. Lett. **96**, 083602 (2006).

[10] A. Vilenkin, Phys. Rev. D **27**, 2848 (1983).

[11] See the internet page `http://super.colorado.edu/~michaele/Lambda/links.html`.

[12] The path integral of the relativistic Coulomb system was solved by
H. Kleinert, Phys. Lett. A **212**, 15 (1996) (hep-th/9504024).
The solution method possesses an inherent supersymmetry as shown by
K. Fujikawa, Nucl. Phys. B **468**, 355 (1996).

[13] E.A. Uehling, Phys. Rev. **49**, 55 (1935).

[14] T. Kinoshita (ed.), *Quantum Electrodynamics*, World Scientific, Singapore, 1990.

[15] For a first attempt see
H. Kleinert, Lettere Nuovo Cimento **4**, 285 (1970) (`wwwK/24`).
New developments can be traced back from the recent papers
I.I. Kogan and D. Polyakov, Int. J. Mod. Phys. A **18**, 1827 (2003) (hep-th/0208036);
D. Juriev, Alg. Groups Geom. **11**, 145 (1994);
R. Dijkgraaf, G. Moore, E. Verlinde, and H. Verlinde, Comm. Math. Phys. **185**, 197 (1997).

Quandoquidem inter nos sanctissima divitiarum maiestas
Since the majesty of wealth is most sacred with us
JUVENAL, Sat. 1, 113

20

Path Integrals and Financial Markets

An important field of applications for path integrals are financial markets. The prices of assets fluctuate as a function of time and, if the number of participants in the market is large, the fluctuations are pretty much random. Then the time dependence of prices can be modeled by fluctuating paths.

20.1   Fluctuation Properties of Financial Assets

Let $S(t)$ denote the price of a stock or another financial asset. Over long time spans, i.e., if data recording frequency is low, the average over many stock prices has a time behavior that can be approximated by pieces of exponentials. This is why they are usually plotted on a logarithmic scale. This is best illustrated by a plot of the Dow-Jones industrial index over 60 years in Fig. 20.1. The fluctuations of the index have a certain average width called the *volatility* of the market. Over

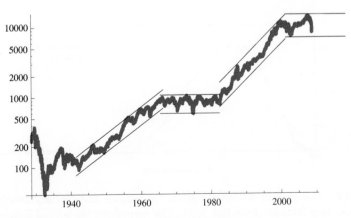

Figure 20.1   Logarithmic plot of Dow Jones industrial index over 80 years. There are four roughly linear regimes, two of exponential growth, two of stagnation [1].

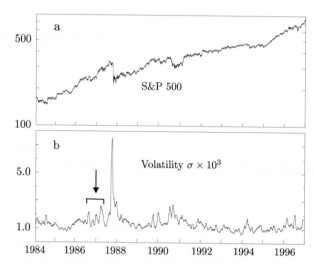

Figure 20.2 (a) Index S&P 500 for 13-year period Jan. 1, 1984 — Dec. 14, 1996, recorded every minute, and (b) volatility in time intervals 30 min (from Ref. [2]).

long times, the volatility is not constant but changes stochastically, as illustrated by the data of the S&P 500 index over the years 1984-1997, as shown in Fig. 20.2 [3]. In particular, there are strong increases shortly before a market crash.

The theory to be developed will at first ignore these fluctuations and assume a constant volatility. Attempts to include them have been made in the literature [3]–[79] and a promising version will be described in Section 20.4.

The volatilities follow approximately a Gamma dstribution, as illustrated in Fig. 20.3.

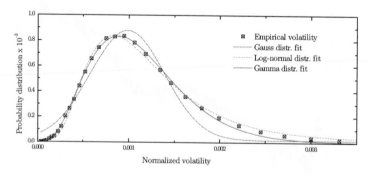

Figure 20.3 Comparison of best Gaussian, log-normal, and Gamma distribution fits to volatilities over 300 min (from Ref. [80]). The normalized log-normal distribution has the form $D^{\log-normal}(z) = (2\pi\sigma^2 z^2)^{-1/2} e^{-(\log z - \mu)^2/2\sigma^2}$. The Gamma distribution will be discussed further in Subsection 20.1.5.

An individual stock will in general be more volatile than an averge market index, especially when the associated company is small and only few shares are traded per day.

20.1.1   Harmonic Approximation to Fluctuations

To lowest approximation, the stock price $S(t)$ satisfies a stochastic differential equation for exponential growth

$$\frac{\dot{S}(t)}{S(t)} = r_S + \eta(t), \tag{20.1}$$

where r_S is the growth rate, and $\eta(t)$ is a white noise variable defined by the correlation functions

$$\langle \eta(t) \rangle = 0, \qquad \langle \eta(t)\eta(t') \rangle = \sigma^2 \delta(t - t'). \tag{20.2}$$

The standard deviation σ is a precise measure for the volatility of the stock price. The squared volatility $v \equiv \sigma^2$ is called the *variance*.

The quantity $dS(t)/S(t)$ is called the *return* of the asset. From financial data, the return is usually extracted for finite time intervals Δt rather than the infinitesimal dt since prices $S(t)$ are listed for certain discrete times $t_n = t_0 + n\Delta t$. There are, for instance, abundant tables of daily closing prices of the market $S(t_n)$, from which one obtains the *daily returns* $\Delta S(t_n)/S(t_n) = [S(t_{n+1}) - S(t_n)]/S(t_n)$. The set of available $S(t_n)$ is called the *time series* of prices.

For a suitable choice of the time scales to be studied, the assumption of a white noise is fulfilled quite well by actual fluctuations of asset prices, as illustrated in Fig. 20.4.

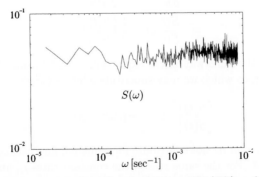

Figure 20.4 Fluctuation spectrum of exchange rate DM/US\$ as function of frequency in units 1/sec, showing that the noise driving the stochastic differential equation (20.1) is approximately white (from [13]).

For the logarithm of the stock or asset price[1]

$$x(t) \equiv \log S(t) \tag{20.3}$$

this implies a stochastic differential equation for linear growth [14, 15, 16, 17]

$$\dot{x}(t) = \frac{\dot{S}}{S} - \frac{1}{2}\sigma^2 = r_x + \eta(t), \tag{20.4}$$

where

$$r_x \equiv r_S - \frac{1}{2}\sigma^2 \tag{20.5}$$

is the drift of the process [compare (18.405)]. A typical set of solutions of (20.4) is shown in Fig. 20.5.

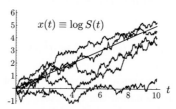

Figure 20.5 Behavior of logarithmic stock price following the stochastic differential equation (20.3).

The finite differences $\Delta x(t_n) = x(t_{n+1}) - x(t_n)$ and the corresponding differentials dx are called *log-returns*.

The extra term $\sigma^2/2$ in (20.5) is due to Itō's Lemma (18.413) for functions of a stochastic variable $x(t)$. Recall that the formal expansion in powers of dt:

$$\begin{aligned} dx(t) &= \frac{dx}{dS}dS(t) + \frac{1}{2}\frac{d^2x}{dS^2}dS^2(t) + \dots \\ &= \frac{\dot{S}(t)}{S(t)}dt - \frac{1}{2}\left[\frac{\dot{S}(t)}{S(t)}\right]^2 dt^2 + \dots \end{aligned} \tag{20.6}$$

may be treated in the same way as the expansion (18.426) using the mnemonic rule (18.429), according to which we may substitute $\dot{x}^2 dt \rightarrow \langle \dot{x}^2 \rangle dt = \sigma^2$, and thus

$$\left[\frac{\dot{S}(t)}{S(t)}\right]^2 dt \rightarrow \dot{x}^2(t)dt = \sigma^2. \tag{20.7}$$

The higher powers in dt do not contribute for Gaussian fluctuations since they carry higher powers of dt. For the same reason the constant rates r_S and r_x in $\dot{S}(t)/S(t)$ and $\dot{x}(t)$ do not show up in $[\dot{S}(t)/S(t)]^2 dt = \dot{x}^2(t)dt$.

---

[1]To form the logarithm, the stock or asset price $S(t)$ is assumed to be dimensionless, i.e. the numeric value of the price in the relevant currency.

In charts of stock prices, relation (20.5) implies that if we fit a straight line through a plot of the logarithms of the prices with slope r_x, the stock price itself grows on the average like

$$\langle S(t) \rangle = S(0)\, e^{rst} = S(0)\langle e^{r_x t + \int_0^t dt'\, \eta(t')} \rangle = S(0)\, e^{(r_x + \sigma^2/2)t}. \tag{20.8}$$

This result is, of course, a direct consequence of Eq. (18.425).

The description of the logarithms of the stock prices by Gaussian fluctuations around a linear trend is only a rough approximation to the real stock prices. The volatilities depend on time. If observed at small time intervals, for instance every minute or hour, they have distributions in which frequent events have an exponential distribution [see Subsection 20.1.6]. Rare events, on the other hand, have a much higher probability than in Gaussian distributions. The observed probability distributions possess *heavy tails* in comparison with the extremely light tails of Gaussian distributions. This was first noted by Pareto in the 19th century [18], reemphasized by Mandelbrot in the 1960s [19], and investigated recently by several authors [20, 22]. The theory needs therefore considerable refinement. As an intermediate generalization we shall introduce, beside the heavy power-like tails, also the so-called *semi-heavy tails*, which drop off faster than any power, such as $e^{-x^a} x^b$ with arbitrarily small $a > 0$ and any $b > 0$. We shall see later in Section 20.4 that semi-heavy tails of financial distribution may be viewed as a consequence of Gaussian fluctuations with fluctuating volatilities. Before we come to these we may fit the data phenomenologically with various non-Gaussian distributions and explore the consequences.

20.1.2   Lévy Distributions

Following Pareto and Mandelbrot we may attempt to fit the distributions of the price changes $\Delta S_n = S(t_{n+1}) - S(t_n)$, the returns $\Delta S_n / S(t_n)$, and the log-returns $\Delta x_n = x(t_{n+1}) - x(t_n)$ for a certain time difference $\Delta t = t_{n+1} - t_n$ approximately with the help of Lévy distributions [19, 22, 24, 23]. For brevity we shall, from now on, use the generic variable z to denote any of the above differences. The Lévy distributions are defined by the Fourier transform

$$\tilde{L}^\lambda_{\sigma^2}(z) \equiv \int_{-\infty}^{\infty} \frac{dp}{2\pi}\, e^{ipz}\, L^\lambda_{\sigma^2}(p), \tag{20.9}$$

with

$$L^\lambda_{\sigma^2}(p) \equiv \exp\left[-(\sigma^2 p^2)^{\lambda/2}/2 \right]. \tag{20.10}$$

For an arbitrary distribution $\tilde{D}(z)$, we shall write the Fourier decomposition as

$$\tilde{D}(z) = \int \frac{dp}{2\pi}\, e^{ipz} D(p) \tag{20.11}$$

and the Fourier components $D(p)$ as an exponential

$$D(p) \equiv e^{-H(p)}, \tag{20.12}$$

where $H(p)$ plays a similar role as the Hamiltonian in quantum statistical path integrals. By analogy with this we shall also define $\tilde{H}(z)$ so that

$$\tilde{D}(z) = e^{-\tilde{H}(z)}. \tag{20.13}$$

An equivalent definition of the Hamiltonian is

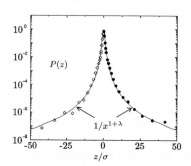
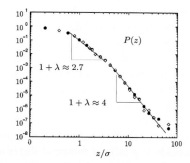

Figure 20.6 Left: Lévy tails of the S&P 500 index (1 minute log-returns) plotted against z/δ. Right: Double-logarithmic plot exhibiting power-like tail regions of the S&P 500 index (1 minute log-returns) (after Ref. [23])

$$e^{-H(p)} \equiv \langle e^{-ipz} \rangle. \tag{20.14}$$

For the Lévy distributions (20.9), the Hamiltonian is

$$H(p) = \frac{1}{2}(\sigma^2 p^2)^{\lambda/2}. \tag{20.15}$$

The Gaussian distribution is recovered in the limit $\lambda \to 2$ where the Hamiltonian simply becomes $\sigma^2 p^2/2$.

For large z, the Lévy distribution (20.9) falls off with the characteristic power-law

$$\tilde{L}_{\sigma^2}^{\lambda}(z) \to A_{\sigma^2}^{\lambda} \frac{\lambda}{|z|^{1+\lambda}}. \tag{20.16}$$

This power falloff is the heavy tail of the distribution discussed above. It is also called *power tail*, *Paretian tail*, or *Lévy tail*). The size of the tails is found by approximating the integral (20.9) for large z, where only small momenta contribute, as follows:

$$\tilde{L}_{\sigma^2}^{\lambda}(z) \approx \int_{-\infty}^{\infty} \frac{dp}{2\pi} e^{ipz} \left[1 - \frac{1}{2}(\sigma^2 p^2)^{\lambda/2} \right] \xrightarrow[z \to \infty]{} A_{\sigma^2}^{\lambda} \frac{\lambda}{|z|^{1+\lambda}}, \tag{20.17}$$

with

$$A_{\sigma^2}^\lambda = -\frac{\sigma^\lambda}{2\lambda} \int_0^\infty \frac{dp'}{\pi} p'^\lambda \cos p' = \frac{\sigma^\lambda}{2\pi\lambda} \sin(\pi\lambda/2)\, \Gamma(1+\lambda). \tag{20.18}$$

The stock market data are fitted best with λ between 1.2 and 1.5 [13], and we shall use $\lambda = 3/2$ most of the time, for simplicity, where one has

$$A_{\sigma^2}^{3/2} = \frac{1}{4} \frac{\sigma^{3/2}}{\sqrt{2\pi}}. \tag{20.19}$$

The full Taylor expansion of the Fourier transform (20.10) yields the asymptotic series

$$\tilde{L}_{\sigma^2}^\lambda(z) = \sum_{n=0}^\infty \frac{(-1)^n}{n!} \int_0^\infty \frac{dp}{\pi} \frac{\sigma^{\lambda n} p^{\lambda n}}{2^n} \cos pz = \sum_{n=0}^\infty \frac{(-1)^{n+1}}{n!} \frac{\sigma^{\lambda n}}{2^n \pi} \Gamma(1+n\lambda) \frac{\sin \frac{\pi\lambda}{2}}{|z|^{1+\lambda}}. \tag{20.20}$$

This series is not useful for practical calculations since it diverges. In particular, it is unable to reproduce the pure Gaussian distribution in the limit $\lambda \to 2$.

There also exists an asymmetric Lévy distribution whose Hamiltonian is

$$H_{\lambda,\sigma,\beta}(p) = \frac{1}{2} |\sigma p|^\lambda \left[1 - i\beta\epsilon(p)F_{\lambda,\sigma}(p)\right], \tag{20.21}$$

where $\epsilon(p)$ is the step function (1.312), and

$$F_{\lambda,\sigma}(p) = \begin{cases} \tan(\pi\lambda/2) & \text{for } \lambda \neq 1, \\ -(1/\pi)\log p^2 & \text{for } \lambda = 1. \end{cases} \tag{20.22}$$

The large-$|z|$ behavior of this distribution is again given by (20.17), except that the prefactor (20.18) is multiplied by a factor $(1+\beta)$.

20.1.3 Truncated Lévy Distributions

Mathematically, an undesirable property of the Lévy distributions is that their fluctuation width diverges for $\lambda < 2$, since the second moment

$$\sigma^2 = \langle z^2 \rangle \equiv \int_{-\infty}^\infty dz\, z^2\, \tilde{L}_{\sigma^2}^\lambda(z) = -\frac{d^2}{dp^2} L_{\sigma^2}^\lambda(p) \bigg|_{p=0} \tag{20.23}$$

is infinite. If one wants to describe data which show heavy tails for large log-returns but have finite widths one must make them fall off at least with semi-heavy tails at very large returns. Examples are the so-called *truncated Lévy distributions* [22]. They are defined by

$$\tilde{L}_{\sigma^2}^{(\lambda,\alpha)}(z) \equiv \int_{-\infty}^\infty \frac{dp}{2\pi} e^{ipz} L_{\sigma^2}^{(\lambda,\alpha)}(p) = \int_{-\infty}^\infty \frac{dp}{2\pi} e^{ipz - H(p)}, \tag{20.24}$$

with a Hamiltonian which generalizes the Lévy Hamiltonian (20.15) to

$$
\begin{aligned}
H(p) &\equiv \frac{\sigma^2}{2} \frac{\alpha^{2-\lambda}}{\lambda(1-\lambda)} \left[(\alpha+ip)^\lambda + (\alpha-ip)^\lambda - 2\alpha^\lambda \right] \\
&= \sigma^2 \frac{(\alpha^2+p^2)^{\lambda/2} \cos[\lambda \arctan(p/\alpha)] - \alpha^\lambda}{\alpha^{\lambda-2}\lambda(1-\lambda)}.
\end{aligned}
\tag{20.25}
$$

The asymptotic behavior of the truncated Lévy distributions differs from the power behavior of the Lévy distribution in Eq. (20.17) by an exponential factor $e^{-\alpha z}$ which guarantees the finiteness of the width σ and of all higher moments. A rough estimate of the leading term is again obtained from the Fourier transform of the lowest expansion term of the exponential function $e^{-H(p)}$:

$$
\begin{aligned}
\tilde{L}^{(\lambda,\alpha)}_{\sigma^2}(z) &\approx e^{2s\alpha^\lambda} \int_{-\infty}^\infty \frac{dp}{2\pi} e^{ipz} \left\{ 1 - s \left[(\alpha+ip)^\lambda + (\alpha-ip)^\lambda \right] \right\} \\
&\xrightarrow{z\to\infty} e^{2s\alpha^\lambda} \Gamma(1+\lambda) \frac{\sin(\pi\lambda)}{\pi} s \frac{e^{-\alpha|z|}}{|z|^{1+\lambda}},
\end{aligned}
\tag{20.26}
$$

where

$$
s \equiv \frac{\sigma^2}{2} \frac{\alpha^{2-\lambda}}{\lambda(1-\lambda)}.
\tag{20.27}
$$

The integral follows directly from the formulas [25]

$$
\int_{-\infty}^\infty \frac{dp}{2\pi} e^{ipz} (\alpha+ip)^\lambda = \frac{\Theta(z)}{\Gamma(-\lambda)} \frac{e^{-\alpha z}}{z^{1+\lambda}}, \quad \int_{-\infty}^\infty \frac{dp}{2\pi} e^{ipz} (\alpha-ip)^\lambda = \frac{\Theta(-z)}{\Gamma(-\lambda)} \frac{e^{-\alpha|z|}}{|z|^{1+\lambda}}, \tag{20.28}
$$

and the identity for Gamma functions[2]

$$
\frac{1}{\Gamma(-z)} = -\Gamma(1+z)\sin(\pi z)/\pi.
\tag{20.29}
$$

The full expansion is integrated with the help of the formula [26]

$$
\begin{aligned}
&\int_{-\infty}^\infty \frac{dp}{2\pi} e^{ipz} (\alpha+ip)^\lambda (\alpha-ip)^\nu \\
&= (2\alpha)^{\lambda/2+\nu/2} \frac{1}{|z|^{1+\lambda/2+\nu/2}}
\begin{cases}
\dfrac{1}{\Gamma(-\lambda)} W_{(\nu-\lambda)/2,(1+\lambda+\nu)/2}(2\alpha z) & z > 0, \\[2mm]
\dfrac{1}{\Gamma(-\nu)} W_{(\lambda-\nu)/2,(1+\lambda+\nu)/2}(2\alpha z) & z < 0,
\end{cases}
\quad \text{for}
\end{aligned}
\tag{20.30}
$$

where the Whittaker functions $W_{(\nu-\lambda)/2,(1+\lambda+\nu)/2}(2\alpha z)$ can be expressed in terms of Kummer's confluent hypergeometric function $_1F_1(a;b;x)$ of Eq. (9.45) as

$$
\begin{aligned}
W_{\delta,\kappa}(x) &= \frac{\Gamma(-2\kappa)}{\Gamma(1/2-\kappa-\delta)} x^{\kappa+1/2} e^{-x/2} {}_1F_1(1/2+\kappa-\delta; 2\kappa+1; x) \\
&+ \frac{\Gamma(2\kappa)}{\Gamma(1/2+\kappa-\delta)} x^{-\kappa+1/2} e^{-x/2} {}_1F_1(1/2-\kappa-\delta; -2\kappa+1; x),
\end{aligned}
\tag{20.31}
$$

---

[2]M. Abramowitz and I. Stegun, op. cit., Formula 6.1.17.

as can be seen from (9.39), (9.46) and Ref. [27]. For $\nu = 0$, only $z > 0$ gives a nonzero integral (20.30), which reduces, with $W_{-\lambda/2,1/2+\lambda/2}(z) = z^{-\lambda/2}e^{-z/2}$, to the left equation in (20.28). Setting $\lambda = \nu$ we find

$$\int_{-\infty}^{\infty} \frac{dp}{2\pi} e^{ipz} (\alpha^2 + p^2)^{\nu} = (2\alpha)^{\nu/2} \frac{1}{|z|^{1+\nu}} \frac{1}{\Gamma(-\nu)} W_{0,1/2+\nu}(2\alpha|z|). \tag{20.32}$$

Inserting

$$W_{0,1/2+\nu}(x) = \sqrt{\frac{2z}{\pi}} K_{1/2+\nu}(x/2), \tag{20.33}$$

we may write

$$\int_{-\infty}^{\infty} \frac{dp}{2\pi} e^{ipz} (\alpha^2 + p^2)^{\nu} = \left(\frac{2\alpha}{|z|}\right)^{1/2+\nu} \frac{1}{\sqrt{\pi}\Gamma(-\nu)} K_{1/2+\nu}(\alpha|z|). \tag{20.34}$$

For $\nu = -1$ where $K_{-1/2}(x) = K_{1/2}(x) = \sqrt{\pi/2x}\,e^{-x}$, this reduces to

$$\int_{-\infty}^{\infty} \frac{dp}{2\pi} e^{ipz} \frac{1}{\alpha^2 + p^2} = \frac{1}{2\alpha} e^{-\alpha|z|}. \tag{20.35}$$

Summing up all terms in the expansion of the exponential function $e^{-H(p)}$:

$$\tilde{L}_{\sigma^2}^{(\lambda,\alpha)}(z) \approx e^{2s\alpha^\lambda} \int_{-\infty}^{\infty} \frac{dp}{2\pi} \left\{ 1 + \sum_{n=1}^{\infty} \frac{(-s)^n}{n!} \left[(\alpha + ip)^\lambda + (\alpha - ip)^\lambda \right]^n \right\} e^{ipz} \tag{20.36}$$

yields the true asymptotic behavior

$$\tilde{L}_{\sigma^2}^{(\lambda,\alpha)}(z) \xrightarrow[z\to\infty]{} e^{(2-2^\lambda)s\alpha^\lambda} \Gamma(1+\lambda) \frac{\sin(\pi\lambda)}{\pi} s \frac{e^{-\alpha|z|}}{|z|^{1+\lambda}}, \tag{20.37}$$

which differs from the estimate (20.26) by a constant factor (see Appendix 20A for details) [28]. Hence the tails are semi-heavy.

In contrast to Gaussian distributions which are characterized completely by their width σ, the truncated Lévy distributions contain three parameters σ, λ, and α. Best fits to two sets of fluctuating market prices are shown in Fig. 20.7. For the S&P 500 index we plot the cumulative distributions

$$P_<(z) = \int_{-\infty}^{z} dz' \, \tilde{L}_{\sigma^2}^{(\lambda,\alpha)}(z'), \quad P_>(z) = \int_{z}^{\infty} dz' \, \tilde{L}_{\sigma^2}^{(\lambda,\alpha)}(z') = 1 - P_<(z), \tag{20.38}$$

for the price differences $z = \Delta S$ over $\Delta t = 15$ minutes. For the ratios of the changes of the currency rates DM/\$ we plot the returns $z = \Delta S/S$ with the same Δt. The plot shows the negative and positive branches $P_<(-z)$, and $P_>(z)$ both plotted on the positive z axis. By definition:

$$P_<(-\infty) = 0, \ \ P_<(0) = 1/2, \ \ P_<(\infty) = 1,$$
$$P_>(-\infty) = 1, \ \ P_>(0) = 1/2, \ \ P_>(\infty) = 0. \tag{20.39}$$

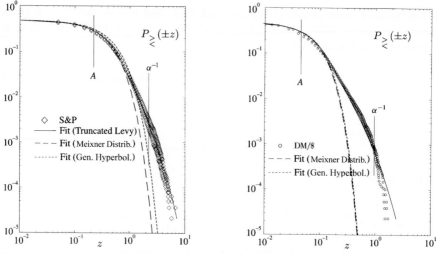

Figure 20.7 Best fit of cumulative versions (20.38) of truncated Lévy distribution to financial data. For the S&P 500 index, the fluctuating variable z is directly the index change ΔS every Δt=15 minutes (fit with $\sigma^2 = 0.280$ and $\kappa = 12.7$). For the DM/US\$ exchange ratio, the variable z is equal to $100\Delta S/S$ every fifteen minutes (fit with $\sigma^2 = 0.0163$ and $\kappa = 20.5$). The negative fluctuations lie on a slightly higher curve than the positive ones. The difference is often neglected. The parameters A and α are the size and truncation parameters of the distribution. The best value of λ is $3/2$ (from [13]). The dashed curves show the best fits of generalized hyperbolic functions (20.117) (l.h.s. $\lambda = 1.46$, $\alpha = 4.93$, $\beta = 0$, $\delta = 0.52$; r.h.s. $\lambda = 1.59$, $\alpha = 32.1$, $\beta = 0$, $\delta = 0.221$).

The fits are also compared with those by other distributions explained in the figure captions. A fit to most data sequences is possible with a rather universal parameter λ close to $\lambda = 3/2$. The remaining two parameters fix all expansion coefficients of Hamiltonian (20.25):

$$H(p) = \frac{1}{2}c_2\,p^2 - \frac{1}{4!}c_4\,p^4 + \frac{1}{6!}c_6\,p^6 - \frac{1}{8!}c_8\,p^8 + \ldots . \tag{20.40}$$

The numbers $c_{2n} = -(-1)^n H^{(2n)}(0)$ are the *cumulants* of the truncated Lévy distribution [compare (3.584)], also denoted by $\langle z^n \rangle_c$. Here they are equal to

$$
\begin{aligned}
\langle z^2 \rangle_c &= c_2 = \sigma^2, \\
\langle z^4 \rangle_c &= c_4 = \sigma^2(2 - \lambda)(3 - \lambda)\alpha^{-2}, \\
\langle z^6 \rangle_c &= c_6 = \sigma^2(2 - \lambda)(3 - \lambda)(4 - \lambda)(5 - \lambda)\alpha^{-4}, \\
&\ \ \vdots \\
\langle z^{2n} \rangle_c &= c_{2n} = \sigma^2\frac{\Gamma(2n - \lambda)}{\Gamma(2 - \lambda)}\alpha^{2-2n}.
\end{aligned}
\tag{20.41}
$$

The first cumulant c_2 determines the quadratic fluctuation width

$$\langle z^2 \rangle \equiv \int_{-\infty}^{\infty} dz\, z^2\, \tilde{L}_{\sigma^2}^{(\lambda,\alpha)}(z) = \left. -\frac{d^2}{dp^2} e^{-H(p)} \right|_{p=0} = c_2 = \sigma^2, \qquad (20.42)$$

the second the expectation of the fourth power of z

$$\langle z^4 \rangle \equiv \int_{-\infty}^{\infty} dz\, z^4\, \tilde{L}_{\sigma^2}^{(\lambda,\alpha)}(z) = \left. \frac{d^4}{dp^4} e^{-H(p)} \right|_{p=0} = c_4 + 3c_2^2, \qquad (20.43)$$

and so on:

$$\langle z^6 \rangle = c_6 + 15c_4 c_2 + 15c_2^3, \quad \langle z^8 \rangle = c_8 + 28c_6 c_2 + 35c_4^2 + 210c_4 c_2^2 + 105c_2^4, \dots . (20.44)$$

In a first analysis of the data, one usually determines the so-called *kurtosis*, which is the normalized fourth-order cumulant

$$\kappa \equiv \bar{c}_4 \equiv \frac{c_4}{c_2^2} = \frac{\langle z^4 \rangle_c}{\langle z^2 \rangle_c^2} = \frac{\langle z^4 \rangle_c}{\sigma^4}. \qquad (20.45)$$

It depends on the parameters σ, λ, α as follows

$$\kappa = \frac{(2-\lambda)(3-\lambda)}{\sigma^2 \alpha^2}. \qquad (20.46)$$

Given the volatility σ and the kurtosis κ, we extract the Lévy parameter α from the equation

$$\alpha = \frac{1}{\sigma} \sqrt{\frac{(2-\lambda)(3-\lambda)}{\kappa}}. \qquad (20.47)$$

In terms of κ and λ, the normalized expansion coefficients are

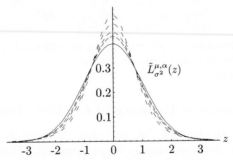

Figure 20.8   Change in shape of truncated Lévy distributions of width $\sigma = 1$ with increasing kurtoses $\kappa = 0$ (Gaussian, solid curve), $1, 2, 5, 10$.

$$\bar{c}_4 = \kappa, \quad \bar{c}_6 = \kappa^2 \frac{(5-\lambda)(4-\lambda)}{(3-\lambda)(2-\lambda)}, \quad \bar{c}_8 = \kappa^2 \frac{(7-\lambda)(6-\lambda)(5-\lambda)(4-\lambda)}{(3-\lambda)^2(2-\lambda)^2},$$

$$\vdots$$

$$\bar{c}_n = \kappa^{n/2-1} \frac{\Gamma(n-\lambda)/\Gamma(4-\lambda)}{(3-\lambda)^{n/2-2}(2-\lambda)^{n/2-2}}. \qquad (20.48)$$

For $\lambda = 3/2$, the second equation in (20.47) becomes simply

$$\alpha = \frac{1}{2}\sqrt{\frac{3}{\sigma^2 \kappa}}, \tag{20.49}$$

and the coefficients (20.50):

$$\bar{c}_4 = \kappa, \quad \bar{c}_6 = \frac{5 \cdot 7}{3}\kappa^2, \quad \bar{c}_8 = 5 \cdot 7 \cdot 11\,\kappa^2,$$

$$\vdots$$

$$\bar{c}_n = \frac{\Gamma(n-3/2)/\Gamma(5/2)}{3^{n/2-2}/2^{n-4}}\kappa^{n/2-1}. \tag{20.50}$$

At zero kurtosis, the truncated Lévy distribution reduces to a Gaussian distribution of width σ. The change in shape for a fixed width and increasing kurtosis is shown in Fig. 20.8.

From the S&P and DM/US$ data with time intervals $\Delta t = 15$ min one extracts $\sigma^2 = 0.280$ and 0.0163, and the kurtoses $\kappa = 12.7$ and 20.5, respectively. This implies $\alpha \approx 0.46$ and $\alpha \approx 1.50$, respectively. The other normalized cumulants $(\bar{c}_6, \bar{c}_8, \ldots)$ are then all determined to be $(1881.72, 788627.46, \ldots)$ and $(-4902.92, 3.3168 \times 10^6, \ldots)$, respectively. The cumulants increase rapidly showing that the expansion needs resummation.

The higher normalized cumulants are given by the following ratios of expectation values

$$\bar{c}_6 = \frac{\langle z^6 \rangle}{\langle z^2 \rangle^3} - 15\frac{\langle z^4 \rangle}{\langle z^2 \rangle^2} + 30,$$

$$\bar{c}_8 = \frac{\langle z^8 \rangle}{\langle z^2 \rangle^4} - 28\frac{\langle z^6 \rangle}{\langle z^2 \rangle^3} - 35\frac{\langle z^4 \rangle^2}{\langle z^2 \rangle^4} + 420\frac{\langle z^4 \rangle}{\langle z^2 \rangle^2} - 630, \ldots . \tag{20.51}$$

In praxis, the high-order cumulants cannot be extracted from the data since they are sensitive to the extremely rare events for which the statistics is too low to fit a distribution function.

20.1.4 Asymmetric Truncated Lévy Distributions

We have seen in the data of Fig. 20.7 that the price fluctuations have a slight asymmetry: Price drops are slightly larger than rises. This is accounted for by an asymmetric truncated Lévy distribution. It has the general form [24]

$$L_{\sigma^2}^{(\lambda,\alpha,\beta)}(p) \equiv e^{-H(p)}, \tag{20.52}$$

with a Hamiltonian function

$$H(p) \equiv \frac{\sigma^2}{2}\frac{\alpha^{2-\lambda}}{\lambda(1-\lambda)}\left[(\alpha+ip)^\lambda(1+\beta) + (\alpha-ip)^\lambda(1-\beta) - 2\alpha^\lambda\right]$$

$$= \sigma^2\frac{(\alpha^2+p^2)^{\lambda/2}\left\{\cos[\lambda\arctan(p/\alpha)] + i\beta\sin[\lambda\arctan(p/\alpha)]\right\} - \alpha^\lambda}{\alpha^{\lambda-2}\lambda(1-\lambda)}. \tag{20.53}$$

This has a power series expansion

$$H(p) = ic_1 p + \frac{1}{2} c_2 p^2 - i \frac{1}{3!} c_3 p^3 - \frac{1}{4!} c_4 p^4 + i \frac{1}{5!} c_5 p^5 + \dots . \tag{20.54}$$

There are now even and odd cumulants $c_n = -i^n H^{(n)}(0)$ with the values

$$c_n = \sigma^2 \frac{\Gamma(n-\lambda)}{\Gamma(2-\lambda)} \alpha^{2-n} \begin{cases} 1 \\ \beta \end{cases} \text{ for } \begin{array}{l} n = \text{even}, \\ n = \text{odd}. \end{array} \tag{20.55}$$

The even cumulants are the same as before in (20.41). Similarly, the even expectation values (20.42)–(20.44) are extended by the odd expectation values:

$$\langle z \rangle \equiv \int_{-\infty}^{\infty} dz\, z\, \tilde{L}_{\sigma^2}^{(\lambda,\alpha,\beta)}(z) = i \frac{d}{dp} e^{-H(p)} \Big|_{p=0} = c_1,$$

$$\langle z^2 \rangle \equiv \int_{-\infty}^{\infty} dz\, z^2\, \tilde{L}_{\sigma^2}^{(\lambda,\alpha,\beta)}(z) = -\frac{d^2}{d^2 p} e^{-H(p)} \Big|_{p=0} = c_2 + c_1^2,$$

$$\langle z^3 \rangle \equiv \int_{-\infty}^{\infty} dz\, z^3\, \tilde{L}_{\sigma^2}^{(\lambda,\alpha,\beta)}(z) = -i \frac{d^3}{d^3 p} e^{-H(p)} \Big|_{p=0} = c_3 + 3c_2 c_1 + c_1^3,$$

$$\langle z^4 \rangle \equiv \int_{-\infty}^{\infty} dz\, z^4\, \tilde{L}_{\sigma^2}^{(\lambda,\alpha,\beta)}(z) = \frac{d^4}{d^4 p} e^{-H(p)} \Big|_{p=0} = c_4 + 4c_3 c_1 + 3c_2^2 + 6c_2 c_1^2 + c_1^4,$$

$$\vdots \tag{20.56}$$

The inverse relations are

$$\begin{aligned}
c_1 &= \langle z \rangle_c = \langle z \rangle, \\
c_2 &= \langle z^2 \rangle_c = \langle z^2 \rangle - \langle z \rangle^2 = \langle (z - \langle z \rangle_c) \rangle^2, \\
c_3 &= \langle z^3 \rangle_c = \langle z^3 \rangle - 3\langle z \rangle \langle z^2 \rangle + 2\langle z \rangle^3 = \langle (z - \langle z \rangle_c) \rangle^3, \\
c_4 &= \langle z^4 \rangle_c = \langle z^4 \rangle - 3\langle z^2 \rangle^2 - 4\langle z \rangle \langle z^3 \rangle + 12\langle z \rangle^2 \langle z^2 \rangle - 6\langle z \rangle^4 \\
&= \langle (z - \langle z \rangle_c) \rangle^4 - 3\langle z^2 - \langle z \rangle_c^2 \rangle^2 = \langle (z - \langle z \rangle_c) \rangle^4 - 3c_2^2.
\end{aligned} \tag{20.57}$$

These are, of course, just simple versions of the cumulant expansions (3.582) and (3.584).

The distribution is now centered around a nonzero average value:

$$\mu \equiv \langle z \rangle = c_1. \tag{20.58}$$

The fluctuation width is given by

$$\sigma^2 \equiv \langle z^2 \rangle - \langle z \rangle^2 = \langle (z - \langle z \rangle)^2 \rangle = c_2. \tag{20.59}$$

For large z, the asymmetric truncated Lévy distributions exhibit semi-heavy tails, obtained by a straightforward modification of (20.28):

$$\tilde{L}_{\sigma^2}^{(\lambda,\alpha)}(z) \approx \int_{-\infty}^{\infty} \frac{dp}{2\pi} e^{ipz} \left\{ 1 - \frac{\sigma^2}{2} \frac{\alpha^{2-\lambda}}{\lambda(1-\lambda)} \left[(\alpha + ip)^\lambda (1+\beta) + (\alpha - ip)^\lambda (1-\beta) - 2\alpha^\lambda \right] \right\}$$

$$\xrightarrow[z \to \infty]{} \sigma^2 e^{2s\alpha^\lambda} \Gamma(1+\lambda) \frac{\sin(\pi\lambda)}{\pi} s \frac{e^{-\alpha|z|}}{|z|^{1+\lambda}} [1 + \beta \operatorname{sgn}(z)]. \tag{20.60}$$

In analyzing the data, one uses the *skewness*

$$s \equiv \frac{\langle (z - \langle z \rangle)^3 \rangle}{\sigma^3} = \bar{c}_3 = \frac{c_3}{c_2^{3/2}}. \tag{20.61}$$

It depends on the parameters σ, λ, β, and α or κ as follows

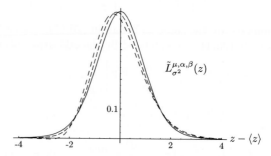

Figure 20.9 Change in shape of truncated Lévy distributions of width $\sigma = 1$ and kurtosis $\kappa = 1$ with increasing skewness $s = 0$ (solid curve), $0.4, 0.8$. The curves are centered around $\langle z \rangle$.

$$s = \frac{(2 - \lambda)\beta}{\sigma\alpha}. \tag{20.62}$$

The kurtosis can also be defined by [compare (20.45)][3]

$$\kappa \equiv \bar{c}_4 \equiv \frac{c_4}{c_2^2} = \frac{\langle z^4 \rangle_c}{\langle z^2 \rangle_c^2} = \frac{\langle z^4 \rangle_c}{\sigma^4} = \frac{\langle (z - \langle z \rangle)^4 \rangle}{\sigma^4} - 3. \tag{20.63}$$

From the data one extracts the three parameters volatility σ, skewness s, and kurtosis κ, which completely determine the asymmetric truncated Lévy distribution. The data are then plotted against $z - \langle z \rangle = z - \mu$, so that they are centered at the average position. This centered distribution will be denoted by $\bar{L}_{\sigma^2}^{(\lambda,\alpha,\beta)}(z)$, i.e.

$$\bar{L}_{\sigma^2}^{(\lambda,\alpha,\beta)}(z) \equiv \tilde{L}_{\sigma^2}^{(\lambda,\alpha,\beta)}(z - \mu). \tag{20.64}$$

The Hamiltonian associated with this zero-average distribution is

$$\bar{H}(p) \equiv H(p) - H'(0)p, \tag{20.65}$$

and its expansion in power of the momenta starts out with p^2, i.e. the first term in (20.54) is subtracted.

---

[3]Some authors call the ratio $\langle (z - \langle z \rangle) \rangle^4 / \sigma^4$ in (20.63) kurtosis and the quantity κ the *excess kurtosis*. Their kurtosis is equal to 3 for a Gaussian distribution, ours vanishes.

In terms of σ, s, and κ, the normalized expansion coefficients are

$$\bar{c}_n = \kappa^{n/2-1} \frac{\Gamma(n-\lambda)/\Gamma(4-\lambda)}{(3-\lambda)^{n/2-2}(2-\lambda)^{n/2-2}} \begin{cases} 1 & n = \text{even}, \\ \sqrt{(3-\lambda)/(2-\lambda)\kappa}\, s & n = \text{odd}. \end{cases} \quad \text{for} \quad (20.66)$$

The change in shape of the distributions of a fixed width and kurtosis with increasing skewness is shown in Fig. 20.9. We have plotted the distributions centered around the average position $z = \langle c_1 \rangle$ which means that we have removed the linear term $ic_1 p$ from $H(p)$ in (20.52), (20.53), and (20.54). This subtracted Hamiltonian whose power series expansion begins with the term $c_2 p^2/2$ will be denoted by

$$\bar{H}(p) \equiv H(p) - H'(0)p = \frac{1}{2}c_2\, p^2 - i\frac{1}{3!}c_3 p^3 - \frac{1}{4!}c_4\, p^4 + i\frac{1}{5!}c_5 p^5 + \dots . \quad (20.67)$$

20.1.5 Gamma Distribution

For a Hamiltonian

$$H_+(p) = \nu \log\left(1 - ip/\mu\right) \quad (20.68)$$

one obtains the normalized *Gamma distribution* of mathematical statistics:

$$\tilde{D}_{\mu,\nu}^{\mathrm{Gamma}}(z) = \frac{1}{\Gamma(\nu)}\mu^\nu z^{\nu-1} e^{-\mu z}, \qquad \int_0^\infty dz\, \tilde{D}_{\mu,\nu}^{\mathrm{Gamma}}(z) = 1, \quad (20.69)$$

which is restricted to positive variables z.

Expanding $H_+(p)$ in a power series $-\sum_{n=1}^\infty i^n \nu \mu^{-n} p^n/n = -\sum_{n=1}^\infty i^n c_n p^n/n!$, we identify the cumulants

$$c_n = (n-1)!\, \nu/\mu^n, \quad (20.70)$$

so that the lowest moments are

$$\bar{z} = \nu/\mu, \quad \sigma^2 = \nu/\mu^2, \quad s = 2/\sqrt{\nu}, \quad \kappa = 6/\nu. \quad (20.71)$$

The maximum of the distribution lies slightly below the average value $\bar{z}$ at

$$z_{\mathrm{max}} = (\nu-1)/\mu = \bar{z} - 1/\mu. \quad (20.72)$$

The Gamma distribution was seen in Fig. 20.3 to yield an optimal fit to the fluctuating volatilities when is it is plotted as a distribution of the volatility $\sigma = \sqrt{z}$:

$$\tilde{D}_{\mu,\nu}^{\mathrm{Chi}}(\sigma) \equiv 2\sigma \tilde{D}_{\mu,\nu}^{\mathrm{Gamma}}(\sigma^2), \qquad \int_0^\infty d\sigma\, \tilde{D}_{\mu,\nu}^{\mathrm{Chi}}(\sigma) = 1. \quad (20.73)$$

As a normalized distribution of the volatility σ rather than the variance $v = \sigma^2$, one calls this function a *Chi distribution*.

In the limit $\nu \to \infty$ at fixed $\bar{z} = \nu/\mu$, the Gamma distribution becomes a Dirac δ-function:

$$\tilde{D}^{\text{Gamma}}_{\mu,\nu}(z) \to \delta\left(z - \frac{\nu}{\mu}\right). \tag{20.74}$$

If we add to $H_+(p)$ the Hamiltonian

$$H_-(p) = \nu \log(1 + ip/\mu) \tag{20.75}$$

we obtain the distribution of negative z-values:

$$\tilde{D}^{\text{Gamma}}_{\mu,\nu}(z) = \frac{1}{\Gamma(\nu)} \mu^{-\nu} |z|^{\nu-1} e^{-\mu|z|}, \quad z \le 0. \tag{20.76}$$

By combining the two Hamiltonians with different parameters μ one can set up a skewed two-sided distribution.

20.1.6 Boltzmann Distribution

The highest-frequency returns ΔS of NASDAQ 100 and S&P 500 indices have a special property: they display a purely exponential behavior for positive as well as negative z, as long as the probability is rather large [32]. The data are fitted by the Boltzmann distribution

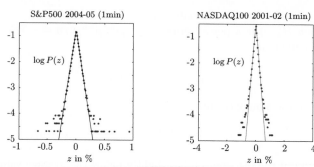

Figure 20.10 Boltzmann distribution of S&P 500 and NASDAQ 100 high-frequency log-returns recorded by the minute.

$$\tilde{B}(z) = \frac{1}{2T} e^{-|z|/T}. \tag{20.77}$$

We can see in Fig. 20.10, that only a very small set of rare events of large $|z|$ does not follow the Boltzmann law, but displays heavy tails. This allows us to assign a temperature to the stock markets [33]. The temperature depends on the volatility of the selected stocks and changes only very slowly with the general economic and political environment. Near a crash it reaches maximal values, as shown in Fig. 20.11.

It is interesting to observe the historic development of Dow Jones temperature over the last 78 years (1929-2006) in Fig. 20.12. Although the world went through a lot of turmoil and economic development in the 20th century, the temperature remained almost a constant except for short heat bursts. The hottest temperatures occurred in the 1930's, the time of the great depression. These temperatures were never reached again. An especially hot burst occurred during the crash year 1987. Thus, in the long run, the market temperature is not related to the value of the index but indicates the riskiness of the market. Only in bubbles, high temperatures go along with high stock prices. An extraordinary increase in market temperature before a crash may be useful to investors as a signal to go short.

The Fourier transform of the Boltzmann distribution is

$$B(p) = \int_{-\infty}^{\infty} dz \, e^{ipz} \frac{1}{2T} e^{-|z|/T} = \frac{1}{1 + (Tp)^2} = e^{-H(p)}, \qquad (20.78)$$

so that we identify the Hamiltonian as

$$H(p) = \log[1 + (Tp)^2]. \qquad (20.79)$$

This has only even cumulants:

$$c_{2n} = -(-1)^n H^{(2n)}(0) = 2(2n-1)! \, T^{2n}, \quad n = 1, 2, \ldots . \qquad (20.80)$$

The Boltzmann distribution is a special case of a two-sided Gamma distribution discussed in the previous subsection. It has semi-heavy tails.

We now observe that the Fourier transform can be rewritten as an integral [recall (2.497)]

$$B(p) = e^{-H(p)} = \frac{1}{1 + (Tp)^2} = \int_0^{\infty} d\tau \, e^{-\tau(1 + T^2 p^2)}. \qquad (20.81)$$

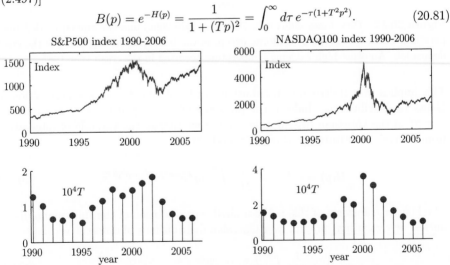

Figure 20.11 Market temperatures of S&P 500 and NASDAQ 100 indices from 1990 to 2006. The crash in the year 2000 occurred at the maximal temperatures $T_{\text{S\&P500}} \approx 2 \times 10^{-4}$ and $T_{\text{NASDAQ}} \approx 4 \times 10^{-4}$.

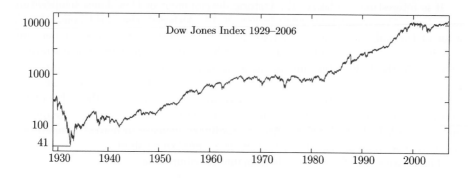

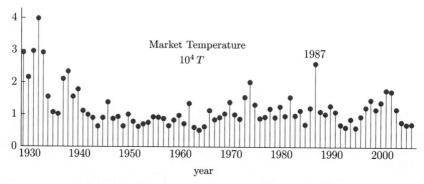

Figure 20.12 Dow Jones index over 78 years (1929-2006) and the annual market temperature, which is remarkably uniform, except in the 1930's, in the beginning of the great depression. An other high extreme temperature occurred in the crash year 1987.

This implies that Boltzmann distribution can be obtained from a superposition of Gaussian distributions. Indeed, the prefactor of p^2 is equal to twice the variance $v = \sigma^2$ of the Gaussian distribution. Hence we change the variable of integration from τ to σ^2, substituting $\tau = \sigma^2/2T^2$, and obtain

$$B(p) = e^{-H(p)} = \frac{1}{2T^2} \int_0^\infty d\sigma^2 \, e^{-\sigma^2/2T^2} e^{-\sigma^2 p^2/2}. \tag{20.82}$$

The Fourier transform of this is the desired representation of the Boltzmann distribution as a volatility integral over Gaussian distributions of different widths:

$$\tilde{B}(z) = \frac{1}{2T^2} \int_0^\infty d\sigma^2 \, e^{-\sigma^2/2T^2} \frac{1}{\sqrt{2\pi\sigma^2}} e^{-z^2/2\sigma^2}. \tag{20.83}$$

The volatility distribution function is recognized to be a special case of the Gamma distribution (20.69) for $\mu = 1/2T^2$, $\nu = 1$.

20.1.7   Student or Tsallis Distribution

Recently, another non-Gaussian distribution has become quite popular through work of Tsallis [58, 59, 60]. It has been proposed as a good candidate for describing heavy tails a long time ago under the name Student distribution by Praetz [61] and by Blattberg and Gonedes [62]. This distribution can be written as

$$\tilde{D}_\delta(z) = N_\delta \frac{1}{\sqrt{2\pi\sigma_\delta^2}} e_\delta^{-z^2/2\sigma_\delta^2}, \tag{20.84}$$

where N_δ is a normalization factor

$$N_\delta = \frac{\sqrt{\delta}\,\Gamma(1/\delta)}{\Gamma(1/\delta - 1/2)}, \tag{20.85}$$

and $\sigma_K \equiv \sigma\sqrt{1 - 3\delta/2}$, where σ is the volatility of the distribution. The function $e_\delta(z)$ is an approximation of the exponential function called δ-exponential:

$$e_\delta^z \equiv [1 - \delta z]^{-1/\delta}. \tag{20.86}$$

In the limit $\delta \to 0$, this reduces to the ordinary exponential function e^z.

   Remarkably, the distribution of a fixed total amount of money W between N persons of equal earning talents follows such a distribution. The partition functions is given by

$$Z_N(W) = \left[\prod_{n=1}^N \int_0^W dw_n\right] \delta(w_1 + \ldots + w_N - W). \tag{20.87}$$

After rewriting this as

$$Z_N(W) = \int_{-\infty}^\infty \frac{d\lambda}{2\pi} \left[\prod_{n=1}^N \int_0^W dw_n\right] e^{-i\lambda(w_1 + \ldots + w_N)} e^{i\lambda W} = \int_{-\infty}^\infty \frac{d\lambda}{2\pi} \frac{(e^{-i\lambda W} - 1)^N e^{i\lambda W}}{(-i\lambda + \epsilon)^N}, \tag{20.88}$$

where ϵ is an infinitesimal positive number, and a binomial expansion of $(e^{-i\lambda W} - 1)^N e^{i\lambda W}$, we can perform the integral over λ, using the formula[4]

$$\int_{-\infty}^\infty \frac{d\lambda}{2\pi} \frac{1}{(-i\lambda + \epsilon)^\nu} e^{-ipx} = \frac{p^{\nu-1}}{\Gamma(\nu)} e^{-\epsilon p}\, \Theta(p), \tag{20.89}$$

where $\Theta(z)$ is the Heaviside function (1.300), we obtain

$$Z_N(W) = \frac{W^{N-1}}{\Gamma(N)} \sum_{k=0}^{N-1} (-1)^k \binom{N}{N-k} (N - k - 1)^{N-1} \tag{20.90}$$

---

[4]See I.S. Gradshteyn and I.M. Ryzhik, *op. cit.*, Formula 3.382.7.

The sum over binomial coefficients adds up to unity, due to a well-known identity for binomial coefficients[5], so that

$$Z_N(W) = \frac{W^{N-1}}{\Gamma(N)}. \qquad (20.91)$$

If we replace W by $W - w_n$, this partition function gives us the unnormalized probability that an individual owns the part w_n of total wealth. The normalized probability is then

$$P_N(w_n) = Z_N^{-1} \frac{(W - w_n)^{N-2}}{\Gamma(N-1)} = \frac{N-1}{W}\left(1 - \frac{w_n}{E}\right)^{N-2}. \qquad (20.92)$$

Defining $w \equiv W/(N-2)$, which for large N is the average wealth per person, $P_N(w_n)$ can be expressed in terms of the δ-exponential (20.86) as

$$P_N(w_n) = \frac{N-1}{W}\left[1 - \frac{w_n}{(N-2)w}\right]^{N-2} = \frac{N-1}{W}e_{-1/(N-2)}^{-w_n/w}. \qquad (20.93)$$

In the limit of infinitely many individuals, this turns into the normalized Boltzmann distribution $w^{-1}e^{-w_n/w}$.

The *Student-Tsallis distribution* (20.84) is simply a normalized δ-exponential. Instead of the parameter δ one often uses the Tsallis parameter $q = \delta + 1$. A plot of these functions for different δ-values is shown on the left hand of Fig. 20.13. From Eq. (20.86) we see that the Student-Tsallis distribution with $\delta > 0$ has heavy tails with a power behavior $1/z^{1/\delta} = 1/z^{1/(q-1)}$. A fit of the log-returns of 10 NYSE

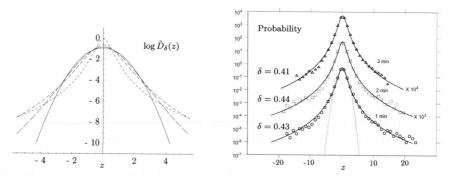

Figure 20.13 Left: Logarithmic plot of the normalized δ-exponential (20.84) (Student-Tsallis distribution) for $\delta = 0$ (Gaussian), 0.2, 0.4, 0.6, all for $\sigma = 1$. Right: Fit of the log-returns of 20 NYSE top-volume stocks over short time scales from 1 to 3 minutes by this distribution (after Ref. [60]). Dotted line is Gaussian distribution.

top-volume stocks is shown in Fig. 20.13.

---

[5]See I.S. Gradshteyn and I.M. Ryzhik, *op. cit.*, Formula 0.154.6.

Remarkably, the δ-exponential in (20.84) can be written as a superposition of Gaussian distributions. With the help of the integral formula (2.497) we find

$$e_\delta^{-z^2/2\sigma_\delta^2} = \frac{1}{\Gamma(1/\delta)} \int_0^\infty \frac{ds}{s} \, s^{1/\delta} e^{-s} e^{-s\delta \, z^2/2\sigma_\delta^2} . \tag{20.94}$$

After a change of variables this can be rewritten as

$$e_\delta^{-z^2/2\sigma_\delta^2} = \frac{\mu^{1/\delta}}{\Gamma(1/\delta)} \int_0^\infty \frac{dv}{v} \, v^{1/\delta} e^{-\mu v} e^{-vz^2/2} , \qquad \mu = \sigma_\delta^2/\delta. \tag{20.95}$$

This is a superposition of Gaussian functions $e^{-vz^2/2}$ whose *inverse variances* $v = 1/\sigma^2$ occur with a weight that is precisely the Gamma distribution of the variable v with $\nu = 1/\delta$:

$$\tilde{D}_{\mu,1/\delta}^{\mathrm{Gamma}}(v) = \frac{1}{\Gamma(1/\delta)} \mu^{1/\delta} v^\nu e^{-\mu v}. \tag{20.96}$$

20.1.8 Tsallis Distribution in Momentum Space

When fitting the volatility distributions of the S&P 500 index in Fig. 20.3, we observed that the best fit was obtained by a Gamma distribution. Inspired by the discussion in the last section we observe that the δ-exponential (20.86) in momentum space

$$e^{-H_{\delta,\beta}(p)} \equiv e_\delta^{-\beta p^2/2} = \left[1 + \beta\delta \, p^2/2\right]^{-1/\delta} , \tag{20.97}$$

has precisely such a decomposition. We simply rewrite it by analogy with (20.94) as

$$e_\delta^{-\beta p^2/2} = \frac{1}{\Gamma(1/\delta)} \int_0^\infty \frac{ds}{s} \, s^{1/\delta} e^{-s} e^{-s\beta\delta p^2/2} , \qquad \beta \equiv \nu/\mu = 1/\mu\delta, \tag{20.98}$$

and change the variable of integration to obtain

$$e_\delta^{-\beta p^2/2} = \frac{\mu^\nu}{\Gamma(\nu)} \int_0^\infty \frac{dv}{v} \, v^\nu e^{-\mu v} e^{-vp^2/2}, \qquad \nu = 1/\delta, \ \mu = \nu/\beta = 1/\beta\delta. \tag{20.99}$$

This is a superposition of Gaussian distributions whose variances $v = \sigma^2$ are weighted by the Gamma distribution (20.69):

$$e_\delta^{-\beta p^2/2} = \int_0^\infty dv \, D_{\mu,\nu}^{\mathrm{Gamma}}(v) e^{-vp^2/2}, \qquad \nu = 1/\delta, \ \mu = \nu/\beta = 1/\beta\delta. \tag{20.100}$$

The average of the Gamma distribution is $\bar{v} = \nu/\mu = \beta$ [recall (20.71)], so that the left-hand side can also be written as $e_\delta^{-\bar{v}p^2/2}$.

The lowest cumulants in the Hamiltonian $H_{\delta,\beta}(p) = \frac{1}{2!}c_2 \, p^2 - \frac{1}{4!}c_4 \, p^4 + \dots$ are

$$c_2 = \frac{\nu}{\mu}, \ c_4 = 3\frac{\nu}{\mu^2}, \ c_6 = \frac{5!!}{\mu^3} \left(2 + 3\nu - 2\nu^2 - \nu^3\right), \ c_8 = \frac{7!!}{\mu^3} \left(2 + \nu - 3\nu^2 + \nu^3 + n u^4\right). \tag{20.101}$$

For $\mu = 1/2T^2$ and $\nu = 1$, these reduce to those of the Boltzmann distribution in Eq. (20.80).

The superposition (20.99) can immediately be Fourier transformed to

$$\tilde{D}_{\delta,\beta}(z) = \frac{\mu^{1/\delta}}{\Gamma(1/\delta)} \int_0^\infty \frac{dv}{v} \, v^{1/\delta} e^{-\mu v} \frac{1}{\sqrt{2\pi v}} e^{-z^2/2v}, \qquad \mu = 1/\beta\delta. \qquad (20.102)$$

which is a superposition of Gaussians whose variances $v = \sigma^2$ follow a Gamma distribution (20.69) centered around $\bar{v} = 1/\delta\mu$ of width $\overline{(v - \bar{v})^2} = 1/\delta\mu^2$. Hence δ is given by the ratio $\delta = \overline{(v - \bar{v})^2}/\bar{v}^2$. Recalling Eq. (20.71) we see that δ determines the kurtosis of the distribution to be $\kappa = 6\delta$.

The integral over v in (20.102) can be done using Formula (2.557), yielding

$$\tilde{D}_{\delta,\beta}(z) = \frac{\sqrt{\mu}}{\Gamma(1/\delta)} \frac{1}{\sqrt{2\pi}} \left(\frac{\mu z^2}{2}\right)^{1/2\delta - 1/4} 2K_{1/\delta - 1/2}(\sqrt{2\mu}z), \qquad \mu = 1/\beta\delta. (20.103)$$

For $\delta = 1$ and $\mu = 1/2T^2$, we use (1.345) and recover the Boltzmann distribution (20.83)

The small-z behavior of $K_\nu(z)$ is $(1/2)\Gamma(\nu)(z/2)^{-\nu}$ for $\operatorname{Re}\nu > 0$ [recall Eq. (1.347)]. If we assume $\delta < 2$, we obtain from (20.103) the value of the distribution function at the origin:

$$\tilde{D}_{\delta,\beta}(0) = \frac{\sqrt{\mu}\,\Gamma(1/\delta - 1/2)}{\Gamma(1/\delta)}. \qquad (20.104)$$

The same result can, of course, be found directly from Eq. (20.102). This value diverges at $\delta = 2/(1 - 2n)$ $(n = 0, 1, 2, \ldots)$.

20.1.9 Relativistic Particle Boltzmann Distribution

A physically important distribution in momentum space is the Boltzmann distribution $e^{-\beta E(p)}$ of relativistic particle energies $E(p) = \sqrt{p^2 + M^2}$, where β is the inverse temperature $1/k_B T$. The exponential can be expressed as a superposition of Gaussian distributions

$$e^{-\beta\sqrt{p^2+M^2}} = \int_0^\infty dv \, \omega_\beta(v) e^{-\beta v(p^2+M^2)/2} \qquad (20.105)$$

with a weight function

$$\omega_\beta(v) \equiv \sqrt{\frac{\beta}{2\pi v^3}} e^{-\beta/2v}. \qquad (20.106)$$

This is a special case of a *Weibull distribution*

$$\tilde{D}_W(x) = \frac{b}{\Gamma(a/b)} x^{a-1} e^{-x^b}, \qquad (20.107)$$

which has the simple moments

$$\langle x^n \rangle = \Gamma((a+n)/b)/\Gamma(a/b). \qquad (20.108)$$

20.1.10   Meixner Distributions

Quite reasonable fits to financial data are provided by the *Meixner distributions* [34, 35] which read in configuration and momentum space:

$$\tilde{M}(z) = \frac{[2\cos(b/2)]^{2d}}{2a\pi\Gamma(2d)} |\Gamma(d+iz/a)|^2 \exp[bz/a], \tag{20.109}$$

$$M(p) = \left\{ \frac{\cos(b/2)}{\cosh[(ap-ib)/2]} \right\}^{2d}. \tag{20.110}$$

They have the same semi-heavy tail behavior as the truncated Lévy distributions

$$\tilde{M}(z) \to C_{\pm}|z|^{\rho} e^{-\sigma_{\pm}|z|} \quad \text{for} \quad z \to \pm\infty, \tag{20.111}$$

with

$$C_{\pm} = \frac{[2\cos(b/2)]^{2d}}{2a\pi\Gamma(2d)} \frac{2\pi}{a^{2d-1}} e^{\pm 2\pi d \tan(b/2)}, \ \rho = 2d - 1, \ \sigma_{\pm} \equiv (\pi \pm b)/a. \tag{20.112}$$

The moments are

$$\mu = ad\tan(b/2), \quad \sigma^2 = a^2d/2\cos^2(b/2), \quad s = \sqrt{2}\sin(b/2)/\sqrt{d}, \quad \kappa = [2 - \cos b]/d, \tag{20.113}$$

such that we can calculate the parameters from the moments as follows:

$$a^2 = \sigma^2\left(2\kappa - 3s^2\right), \quad d = \frac{1}{\kappa - s^2}, \quad b = 2\arcsin\left(s\sqrt{d/2}\right). \tag{20.114}$$

As an example for the parameters of the distribution, a good fit of the daily Nikkei-225 index is possible with

$$a = 0.029828, \ b = 0.12716, \ d = 0.57295, \ \langle z \rangle = -0.0011243. \tag{20.115}$$

The curve has to be shifted in z by Δz to make $\Delta z + \mu$ equal to $\langle z \rangle$. Such a Meixner distribution has been used for option pricing in Ref. [35].

The Meixner distributions can be fitted quite well to the truncated Lévy distribution in the regime of large probability. In doing so we observe that the variance σ^2 and the kurtosis κ are not the best parameters to match the two distributions. The large-probability regime of the distributions can be matched perfectly by choosing, in the symmetric case, the value and the curvature at the origin to be the same in both curves. This is seen in Fig. 20.14.

In the asymmetric case we have to match also the first and third derivatives. The derivatives of the Meixner distribution are:

$$\tilde{M}(0) = \frac{2^{2d-1}\Gamma^2(d)}{\pi a\Gamma(2d)},$$

$$\tilde{M}'(0) = b\frac{2^{2d-1}\Gamma^2(d)[1 - d\psi(d)]}{\pi a^2\Gamma(2d)},$$

$$\tilde{M}''(0) = -\frac{2^{2d}\Gamma^2(d)\psi(d)}{\pi a^3\Gamma(2d)},$$

$$\tilde{M}^{(3)}(0) = -\frac{b}{2}\frac{2^{2d}\Gamma^2(d)\left[6\psi(d) - 6d\psi^2(d) - d\psi^{(3)}(d)\right]}{\pi a^4\Gamma(2d)},$$

$$\tilde{M}^{(4)}(0) = \frac{2^{2d}\Gamma^2(d)\left[6\psi^2(d) + \psi^{(3)}(d)\right]}{\pi a^5\Gamma(2d)}, \tag{20.116}$$

where $\psi^{(n)}(z) \equiv d^{n+1}\log\Gamma(z)/dz^{n+1}$ are the Polygamma functions.

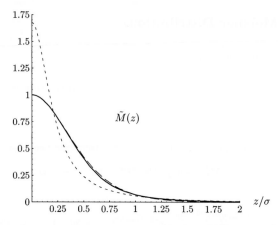

Figure 20.14 Comparison of best fit of Meixner distribution to truncated Lévy distributions. One of them (short dashed) has the same volatility σ and kurtosis κ. The other (long-dashed) has the same size and curvature at the origin. The parameters are $\sigma^2 = 0.280$ and $\kappa = 12.7$ as in the left-hand cumulative distribution in Fig. 20.7. The Meixner distribution with the same σ^2 and κ has parameters $a = 2.666$, $d = 0.079$, $b = 0$, the distribution with the same value and curvature at the origin has $a = 0.6145$, $d = 1.059$, $b = 0$. The very large σ-regime, however, is not fitted well as can be seen in the cumulative distributions which reach out to z of the order of 10σ in Fig. 20.7.

20.1.11 Generalized Hyperbolic Distributions

Another non-Gaussian distributions proposed in the literature is the so-called *generalized hyperbolic distributions*..[6] As the truncated Lévy and Meixner distributions, these have simple analytic forms both in z- and p-space:

$$\tilde{H}_G(z) = \frac{\left(\gamma^2 - \beta^2\right)^{\lambda/2} e^{\beta z}}{\gamma^{\lambda-1/2}\delta^\lambda \sqrt{2\pi}} \left[\delta^2 + z^2\right]^{\lambda/2 - 1/4} \frac{K_{\lambda-1/2}\left(\gamma\sqrt{\delta^2 + z^2}\right)}{K_\lambda\left(\delta\sqrt{\gamma^2 - \beta^2}\right)} \tag{20.117}$$

and

$$G(p) = \frac{\left(\delta\sqrt{\gamma^2 - \beta^2}\right)^\lambda}{K_\lambda\left(\delta\sqrt{\gamma^2 - \beta^2}\right)} \frac{K_\lambda\left(\delta\sqrt{\gamma^2 - (\beta + ip)^2}\right)}{\left[\delta\sqrt{\gamma^2 - (\beta + ip)^2}\right]^\lambda}, \tag{20.118}$$

the latter defining another Hamiltonian

$$H_G(p) \equiv -\log G(p). \tag{20.119}$$

But in contrast to the former, this set of functions is not closed under time evolution. The distributions at a later time t are obtained from the Fourier transforms $e^{-H(p)t}$. For truncated Lévy distributions and the Meixner distributions the factor t can simply be absorbed in the parameters of the functions ($\sigma^2 \to t\sigma^2$ in the first and $d \to td$ in the second case). For the generalized hyperbolic distributions, this is no longer true since $e^{-H_G(p)t} = [G(p)]^t$ involves higher powers of

---

[6]See the publications [36]–[74].

Bessel functions whose analytic Fourier transform cannot be found. In order to describe a complete temporal evolution we must therefore close the set of functions by adding all Fourier transforms of $e^{-H_G(p)t}$. In praxis, this is not a serious problem — it merely leads to a slowdown of numeric calculations which always involve a numeric Fourier transformation.

The asymptotic behavior of the generalized hyperbolic distributions has semi-heavy tails. From the large-z behavior of the Bessel function $K_\nu(z) \to \sqrt{\pi/2z}e^{-z}$ we find

$$\tilde{H}_G(z) \to \sqrt{\frac{\pi}{2\gamma}} \frac{(\gamma^2 - \beta^2)^{\lambda/2} e^{\beta z}}{\gamma^{\lambda-1/2}\delta^\lambda \sqrt{2\pi}} \frac{1}{K_\lambda\left(\delta\sqrt{\gamma^2 - \beta^2}\right)} z^{\lambda-1} e^{-\gamma z}. \tag{20.120}$$

Introducing the variable $\zeta \equiv \delta\sqrt{\gamma^2 - \beta^2}$, this can again be expanded in powers of p as in Eq. (20.54), yielding the first two cumulants:

$$c_1 = \beta \frac{\delta^2}{\zeta} \frac{K_{1+\lambda}(\zeta)}{K_\lambda(\zeta)}; \tag{20.121}$$

$$c_2 = \frac{\delta^2}{\zeta} \frac{K_{1+\lambda}(\zeta)}{K_\lambda(\zeta)} + \frac{\beta^2\delta^4}{\zeta^2} \left\{ \frac{K_{2+\lambda}(\zeta)}{K_\lambda(\zeta)} - \left[\frac{K_{1+\lambda}(\zeta)}{K_\lambda(\zeta)}\right]^2 \right\}. \tag{20.122}$$

Using the identity [50]

$$K_{\nu+1}(z) - K_{\nu-1}(z) = \frac{2\nu}{z} K_\nu(z), \tag{20.123}$$

the latter equation can be expressed entirely in terms of

$$\rho = \rho(\zeta) = \frac{K_{1+\lambda}(\zeta)}{K_\lambda(\zeta)} \tag{20.124}$$

as

$$c_2 = \frac{\delta^2}{\zeta}\rho + \frac{\beta^2\delta^4}{\zeta^3} \left[\zeta + 2(1+\lambda)\rho - \zeta\rho^2\right]. \tag{20.125}$$

Usually, the asymmetry of the distribution is small implying that c_1 is small, implying a small β. It is then useful to introduce the symmetric variance

$$\sigma_s^2 \equiv \delta^2\rho/\zeta, \tag{20.126}$$

and write

$$c_1 = \beta\sigma_s, \quad c_2 = \sigma^2 = \sigma_s^2 + \beta^2 \left[\frac{\delta^4}{\zeta^2} + 2(1+\lambda)\frac{\delta^2}{\zeta^2}\sigma_s - \sigma_s^2\right]. \tag{20.127}$$

The cumulants c_3 and c_4 are most compactly written as

$$c_3 = \beta \left[\frac{3\delta^4}{\zeta^2} + 6(1+\lambda)\frac{\delta^2}{\zeta^2}\sigma_s^2 - 3\sigma_s^4\right]$$
$$+ \beta^3 \left\{ 2(2+\lambda)\frac{\delta^6}{\zeta^4} + \left[4(1+\lambda)(2+\lambda) - 2\zeta^2\right]\frac{\delta^4}{\zeta^4}\sigma_s^2 - 6(1+\lambda)\frac{\delta^2}{\zeta^2}\sigma_s^4 + 2\sigma_s^6 \right\} \tag{20.128}$$

and

$$c_4 = \kappa\sigma^4 = \frac{3\delta^4}{\zeta^2} + \frac{6\delta^2}{\zeta^2}(1+\lambda)\sigma_s^2 - 3\sigma_s^4$$
$$+ 6\beta^2 \left\{ 2(2+\lambda)\frac{\delta^6}{\zeta^4} + \left[4(1+\lambda)(2+\lambda) - 2\zeta^2\right]\frac{\delta^4}{\zeta^4}\sigma_s^2 - 6(1+\lambda)\frac{\delta^2}{\zeta^2}\sigma_s^4 + 2\sigma_s^6 \right\}$$
$$+ \beta^4 \left\{ \left[4(2+\lambda)(3+\lambda) - \zeta^2\right]\frac{\delta^8}{\zeta^6} + \left[4(1+\lambda)(2+\lambda)(3+\lambda) - 2(5+4\lambda)\zeta^2\right]\frac{\delta^6}{\zeta^6}\sigma_s^2 \right.$$
$$\left. -2\left[(1+\lambda)(11+7\lambda) - 2\zeta^2\right]\frac{\delta^4}{\zeta^4}\sigma_s^4 + 12(1+\lambda)\frac{\delta^2}{\zeta^2}\sigma_s^6 - 3\sigma_s^8 \right\}. \tag{20.129}$$

The first term in c_4 is equal to σ_s^4 times the kurtosis of the symmetric distribution

$$\kappa_s \equiv \frac{3\delta^4}{\zeta^2\sigma_s^4} + \frac{6\delta^2}{\zeta^2\sigma_s^2}(1+\lambda) - 3. \tag{20.130}$$

Inserting here σ_s^2 from (20.126), we find

$$\kappa_s \equiv \frac{3}{r^2(\zeta)} + (1+\lambda)\frac{6}{\zeta\,r(\zeta)} - 3. \tag{20.131}$$

Since all Bessel functions $K_\nu(z)$ have the same large-z behavior $K_\nu(z) \to \sqrt{\pi/2z}e^{-z}$ and the small-z behavior $K_\nu(z) \to \Gamma(\nu)/2(z/2)^\nu$, the kurtosis starts out at $3/\lambda$ for $\zeta = 0$ and decreases monotonously to 0 for $\zeta \to \infty$. Thus a high kurtosis can be reached only with a small parameter λ.

The first term in c_3 is $\beta\kappa_s\sigma_s^4$, and the first two terms in c_4 are $\kappa_s\sigma_s^4 + 6\left(c_3/\beta - \kappa_s\sigma_s^4\right)$. For a symmetric distribution with certain variance σ_s^2 and kurtosis κ_s we select some parameter $\lambda < 3/\kappa_s$, and solve the Eq. (20.131) to find ζ. This is inserted into Eq. (20.126) to determine

$$\delta^2 = \frac{\sigma_s^2\zeta}{\rho(\zeta)}. \tag{20.132}$$

If the kurtosis is larger, it is not an optimal parameter to determine generalized hyperbolic distributions. A better fit to the data is reached by reproducing correctly the size and shape of the distribution near the peak and allow for some deviations in the tails of the distribution, on which the kurtosis depends quite sensitively.

For distributions which are only slightly asymmetric, which is usually the case, it is sufficient to solve the above symmetric equations and determine the small parameter β approximately by the skew $s = c_3/\sigma^3$ from the first line in (20.128) as

$$\beta \approx \frac{s}{\kappa_s\sigma_s}. \tag{20.133}$$

This approximation can be improved iteratively by reinserting β into the second equation in (20.127) to determine, from the variance σ^2 of the data, an improved value of σ_s^2. Then Eq. (20.129) is used to determine from the kurtosis κ of the data an improved value of κ_s, and so on.

For the best fit near the origin where probabilities are large we use the derivatives

$$G(0) = \left(\frac{\delta}{\gamma}\right)^{\lambda-1/2}\zeta^{-\lambda}k_-, \tag{20.134}$$

$$G'(0) = \beta G(0), \tag{20.135}$$

$$G''(0) = -\left(\frac{\delta}{\gamma}\right)^{\lambda-3/2}\zeta^{-\lambda}k_+ + \left(\frac{\delta}{\gamma}\right)^{\lambda-1/2}\delta^{-2}\zeta^{-\lambda}\left(1 - 2\lambda - \beta^2\delta^2\right)k_-,$$

$$G^{(3)}(0) = -\frac{\beta}{2}\left[3\left(\frac{\delta}{\gamma}\right)^{\lambda-3/2}\zeta^{-\lambda}k_+ + \left(\frac{\delta}{\gamma}\right)^{\lambda-1/2}\delta^{-2}\zeta^{-\lambda}\left(3 - 6\lambda - \beta^2\delta^2\right)\right]k_-, \tag{20.136}$$

where we have abbreviated

$$k_\pm \equiv \frac{1}{\sqrt{2\pi}}\frac{K_{\lambda\pm1/2}(\delta\gamma)}{K_\lambda(\delta\gamma)}. \tag{20.137}$$

The generalized hyperbolic distributions with $\lambda = 1$ are called *hyperbolic distributions*. The option prices following from these can be calculated by inserting the appropriate parameters into an interactive internet page (see Ref. [51]). Another special case used frequently in the literature is $\lambda = -1/2$ in which case one speaks of a *normal inverse Gaussian distributions* also abbreviated as NIGs.

20.1.12   Debye-Waller Factor for Non-Gaussian Fluctuations

At the end of Section 3.10 we calculated the expectation value of an exponential function $\langle e^{Pz} \rangle$ of a Gaussian variable which led to the Debye-Waller factor of Bragg scattering (3.308). This factor was introduced in solid state physics to describe the reduction of intensity of Bragg peaks due to the thermal fluctuations of the atomic positions. It is derivable from the Fourier representation

$$\langle e^{Pz} \rangle \equiv \int dz \, \frac{1}{\sqrt{2\pi\sigma^2}} e^{-z^2/2\sigma^2} e^{Pz} = \int dz \int \frac{dp}{2\pi} e^{-\sigma^2 p^2/2} e^{ipz+Pz} = e^{\sigma^2 P^2/2}. \quad (20.138)$$

There exists a simple generalization of this relation to non-Gaussian distributions, which is

$$\langle e^{Pz} \rangle \equiv \int dz \int \frac{dp}{2\pi} e^{-H(p)} e^{ipz+Pz} = e^{-H(iP)}. \quad (20.139)$$

20.1.13   Path Integral for Non-Gaussian Distribution

Let us calculate the properties of the simplest process whose fluctuations are distributed according to any of the general non-Gaussian distributions. We consider the stochastic differential equation for the logarithms of the asset prices

$$\dot{x}(t) = r_x + \eta(t), \quad (20.140)$$

where the noise variable $\eta(t)$ is distributed according to an arbitrary distribution. The constant drift r_x in (20.140) is uniquely defined only if the average of the noise variable vanishes: $\langle \eta(t) \rangle = 0$. The general distributions discussed above can have a nonzero average $\langle x \rangle = c_1$ which has to be subtracted from $\eta(t)$ to identify r_x. The subsequent discussion will be simplest if we imagine r_x to have replaced c_1 in the above distributions, i.e. if the power series expansion of the Hamiltonian (20.54) is replaced as follows:

$$
\begin{aligned}
H(p) \to H_{r_x}(p) &\equiv H(p) - H'(0)p + ir_x p \equiv \bar{H}(p) + ir_x p \\
&\equiv ir_x p + \frac{1}{2} c_2 \, p^2 - i\frac{1}{3!} c_3 p^3 - \frac{1}{4!} c_4 \, p^4 + i\frac{1}{5!} c_5 p^5 + \dots \quad (20.141)
\end{aligned}
$$

Thus we may simply work with the original expansion (20.54) and replace at the end

$$c_1 \to r_x. \quad (20.142)$$

The stochastic differential equation (20.140) can be assumed to read simply

$$\dot{x}(t) = \eta(t). \quad (20.143)$$

With the ultimate replacement (20.142) in mind, the probability distribution of the endpoints $x_b = x(t_b)$ for the paths starting at a certain initial point $x_a = x(t_a)$ is given by a path integral of the form (18.342):

$$P(x_b t_b | x_a t_a) = \int \mathcal{D}\eta \int_{x(t_a)=x_a}^{x(t_b)=x_b} \mathcal{D}x \exp\left[-\int_{t_a}^{t_b} dt \, \tilde{H}(\eta(t)) \right] \delta[\dot{x} - \eta]. \quad (20.144)$$

The function $\tilde{H}(\eta)$ is the negative logarithm of the chosen distribution of log-returns [recall (20.13)]

$$\tilde{H}(\eta) = -\log \tilde{D}(\eta). \tag{20.145}$$

For example, $\tilde{H}(\eta)$ is given by $-\log \tilde{L}_{\sigma^2}^{(\lambda,\alpha)}(\eta)$ of Eq. (20.24) for the truncated Lévy distribution or by $-\log \tilde{B}(\eta)$ of Eq. (20.77) or the Boltzmann distributions.

In the mathematical literature, the measure in the path integral over the noise

$$\mathcal{D}\mu \equiv \mathcal{D}\eta\, P[\eta] = \mathcal{D}\eta\, e^{-\int_{t_a}^{t_b} dt\, \tilde{H}(\eta(t))} \tag{20.146}$$

of the probability distribution (20.144) is called the *measure of the process* $\dot{x}(t) = \eta(t)$. The path integral

$$\int_{x(t_a)=x_a}^{x(t_b)=x_b} \mathcal{D}x\, \delta[\dot{x} - \eta] \tag{20.147}$$

is called the *filter* which determines the distribution of x_b at time t_b for all paths $x(t)$ starting out at x_a at time t_a.

A measure differing from (20.146) only by the drift

$$\mathcal{D}\mu' = \mathcal{D}\eta\, P[\eta - r] = \mathcal{D}\eta\, e^{-\int_{t_a}^{t_b} dt\, \tilde{H}(\eta(t)-r)} \tag{20.148}$$

is called *equivalent measure*. The ratio

$$\mathcal{D}\mu'/\mathcal{D}\mu = e^{-\int_{t_a}^{t_b}\left[\tilde{H}(\eta-r)-\tilde{H}(\eta)\right]} \tag{20.149}$$

is called the *Radon-Nikodym derivative*. For a Gaussian noise, this is simply

$$\mathcal{D}\mu'/\mathcal{D}\mu = e^{\int_{t_a}^{t_b}\left[r\eta(t)-r^2t/2\right]}. \tag{20.150}$$

If one wants to calculate the expectation value of any function $f(x(t))$, one has to split the filter into a product

$$\left[\int_{x(t)=x}^{x(t_b)=x_b} \mathcal{D}x\, \delta[\dot{x} - \eta]\right] \times \left[\int_{x(t_a)=x_a}^{x(t)=x} \mathcal{D}x\, \delta[\dot{x} - \eta]\right], \tag{20.151}$$

and perform an integral over $f(x)$ with this filter in the path integral (20.146). Going over to the probabilities $P(x_b t_b | x_a t_a)$, we obtain the integral

$$\langle f(x(t))\rangle = \int dx\, P(x_b t_b | x\, t)\, f(x)\, P(x\, t | x_a t_a). \tag{20.152}$$

The correlation functions of the noise variable $\eta(t)$ in the path integral (20.144) are given by a straightforward functional generalization of formulas (20.56). For this purpose, we express the noise distribution $P[\eta] \equiv \exp\left[-\int_{t_a}^{t_b} dt\, \tilde{H}(\eta(t))\right]$ in (20.144) as a Fourier path integral

$$P[\eta] = \int \frac{\mathcal{D}p}{2\pi} \exp\left\{\int_{t_a}^{t_b} dt\, [ip(t)\eta(t) - H(p(t))]\right\}, \tag{20.153}$$

and note that the correlation functions can be obtained from the functional derivatives

$$\langle \eta(t_1) \cdots \eta(t_n) \rangle = (-i)^n \int \mathcal{D}\eta \int \frac{\mathcal{D}p}{2\pi} \left[\frac{\delta}{\delta p(t_1)} \cdots \frac{\delta}{\delta p(t_n)} e^{i \int_{t_a}^{t_b} dt\, p(t)\eta(t)} \right] e^{-\int_{t_a}^{t_b} dt\, H(p(t))}.$$

After n partial integrations, this becomes

$$\langle \eta(t_1) \cdots \eta(t_n) \rangle = i^n \int \mathcal{D}\eta \int \frac{\mathcal{D}p}{2\pi} e^{i \int_{t_a}^{t_b} dt\, p(t)\eta(t)} \frac{\delta}{\delta p(t_1)} \cdots \frac{\delta}{\delta p(t_n)} e^{-\int_{t_a}^{t_b} dt\, H(p(t))}$$

$$= i^n \left[\frac{\delta}{\delta p(t_1)} \cdots \frac{\delta}{\delta p(t_n)} e^{-\int_{t_a}^{t_b} dt\, H(p(t))} \right]_{p(t)\equiv 0}. \tag{20.154}$$

By expanding the exponential $e^{-\int_{t_a}^{t_b} dt\, H(p(t))}$ in a power series (20.54) we find immediately the lowest correlation functions

$$\langle \eta(t_1) \rangle \equiv Z^{-1} \int \mathcal{D}\eta\, \eta(t_1) \exp\left[-\int_{t_a}^{t_b} dt\, \tilde{H}(\eta(t)) \right] = 0, \tag{20.155}$$

$$\langle \eta(t_1)\eta(t_2) \rangle \equiv Z^{-1} \int \mathcal{D}\eta\, \eta(t_1)\eta(t_2) \exp\left[-\int_{t_a}^{t_b} dt\, \tilde{H}(\eta(t)) \right]$$
$$= c_2 \delta(t_1 - t_2) + c_1^2, \tag{20.156}$$

$$\langle \eta(t_1)\eta(t_2)\eta(t_3) \rangle \equiv Z^{-1} \int \mathcal{D}\eta\, \eta(t_1)\eta(t_2)\eta(t_3) \exp\left[-\int_{t_a}^{t_b} dt\, \tilde{H}(\eta(t)) \right]$$
$$= c_3 \delta(t_1 - t_2)\delta(t_1 - t_3)$$
$$+ c_2 c_1 [\delta(t_1 - t_2) + \delta(t_2 - t_3) + \delta(t_1 - t_3)] + c_1^3, \tag{20.157}$$

$$\langle \eta(t_1)\eta(t_2)\eta(t_3)\eta(t_4) \rangle \equiv Z^{-1} \int \mathcal{D}\eta\, \eta(t_1)\eta(t_2)\eta(t_3)\eta(t_4) \exp\left[-\int_{t_a}^{t_b} dt\, \tilde{H}(\eta(t)) \right]$$
$$= c_4 \delta(t_1 - t_2)\delta(t_1 - t_3)\delta(t_1 - t_4)$$
$$+ c_3 c_1 [\delta(t_1 - t_2)\delta(t_1 - t_3) + 3\,\text{cyclic perms}]$$
$$+ c_2^2 [\delta(t_1 - t_2)\delta(t_3 - t_4) + \delta(t_1 - t_3)\delta(t_2 - t_4) + \delta(t_1 - t_4)\delta(t_2 - t_3)]$$
$$+ c_2 c_1^2 [\delta(t_1 - t_2) + 5\,\text{pair terms}] + c_1^4, \tag{20.158}$$

where

$$Z \equiv \int \mathcal{D}\eta \exp\left[-\int_{t_a}^{t_b} dt\, \tilde{H}(\eta(t)) \right]. \tag{20.159}$$

The higher correlation functions are obvious generalizations of (20.44). The different contributions on the right-hand side of (20.156)–(20.262) are distinguishable by their connectedness structure.

Note that the term proportional to c_3 in the three-point correlation function (20.158) and the terms proportional to c_3 and to c_4 in the four-point correlation function (20.262) do not obey Wick's rule (3.302) since they contain contributions are caused by the non-Gaussian terms $-ic_3 p^3/3!$ $-c_4 p^4/4!$ in the Hamiltonian (20.141).

20.1.14 Time Evolution of Distribution

The δ-functional in Eq. (20.144) may be represented by a Fourier integral leading to the path integral

$$P(x_b t_b | x_a t_a) = \int \mathcal{D}\eta \int \mathcal{D}x \int \frac{\mathcal{D}p}{2\pi} \exp\left\{\int_{t_a}^{t_b} dt \left[ip(t)\dot{x}(t) - ip(t)\eta(t) - \tilde{H}(\eta(t))\right]\right\}. \quad (20.160)$$

Integrating out the noise variable $\eta(t)$ amounts to performing the inverse Fourier transform of the type (20.24) at each instant of time. Then we obtain

$$P(x_b t_b | x_a t_a) = \int \mathcal{D}x \int \frac{\mathcal{D}p}{2\pi} \exp\left\{\int_{t_a}^{t_b} dt \left[ip(t)\dot{x}(t) - H(p(t))\right]\right\}. \quad (20.161)$$

Integrating over all $x(t)$ with fixed endpoints enforces a constant momentum along the path, and we remain with the single momentum integral

$$P(x_b t_b | x_a t_a) = \int_{-\infty}^{\infty} \frac{dp}{2\pi} \exp\left[ip(x_b - x_a) - (t_b - t_a)H(p)\right]. \quad (20.162)$$

Given an arbitrary noise distribution $\tilde{D}(z)$ extracted from the data of a given frequency $1/\Delta t$, we simply identify the Hamiltonian $H(p)$ from the Fourier representation

$$\tilde{D}(z) = \int_{-\infty}^{\infty} \frac{dp}{2\pi} e^{ikz - H(p)}, \quad (20.163)$$

and insert $H(p)$ into (20.170) to find the time dependence of the distribution. The time is then measured in units of the time interval Δt.

For a truncated Lévy distribution, the result is

$$P(x_b t_b | x_a t_a) = \tilde{L}^{(\lambda, \alpha)}_{\sigma^2(t_b - t_a)}(x_b - x_a). \quad (20.164)$$

This is a truncated Lévy distribution of increasing width. The result for other distributions is analogous.

Since the distribution (20.170) depends only on $t = t_b - t_a$ and $x = x_b - x_a$, we shall write it shorter as

$$P(x, t) = \int_{-\infty}^{\infty} \frac{dp}{2\pi} \exp\left[ipx - tH(p)\right]. \quad (20.165)$$

20.1.15 Central Limit Theorem

For large t, the distribution $P(x, t)$ becomes more and more Gaussian (see Fig. 20.18 for the Boltzmann distribution). This is a manifestation of the *central limit theorem* of statistical mechanics which states that the convolution of infinitely many arbitrary distribution functions of finite width always approaches a Gaussian distribution. This is easily proved. We simply note that after an integer number t of convolutions, a probability distribution $\tilde{D}(z)$ with Hamiltonian $H(p)$ has the Fourier components $[D(p)]^t = e^{-tH(p)}$, so that it is given by the integral

$$\tilde{D}(z,t) = \int \frac{dp}{2\pi} e^{ipz} e^{-tH(p)}. \tag{20.166}$$

For large t, the integral can be evaluated in the saddle point approximation (4.51). Denoting the momentum at extremum of the exponent by p_z, which is determined implicitly by $tH'(p_z) = iz$, and setting $\sigma^2 = H''(p_z)$, we obtain the Gaussian distribution

$$\tilde{D}(z,t) \underset{t \text{ large}}{\longrightarrow} e^{ip_z z - tH(p_z)} \int \frac{dp}{2\pi} e^{i(p-p_z)z - t\sigma^2(p-p_z)^2/2} = \frac{e^{ip_z z - tH(p_z)}}{\sqrt{2\pi\sigma^2}} e^{-z^2/2t\sigma^2}$$

$$= \frac{e^{\sigma^2 p_z^2/2 - tH(p_z)}}{\sqrt{2\pi\sigma^2}} e^{-(z - t\sigma^2 p_z)^2/2t\sigma^2}. \tag{20.167}$$

The same procedure can be applied to the integral (20.165).

The transition from Boltzmann to Gaussian distributions is shown for the S&P 500 index in Fig. 20.15.

If a distribution has a heavy power tail $\propto |z|^{-\lambda-1}$ with $\lambda < 2$, so that the volatility σ is infinite, the Hamiltonian starts out for small p like $|p|^\lambda$, and the saddle point approximation is governed by this term rather than the quadratic term p^2. For an asymmetric distribution, the leading terms in the Hamiltonian are those of the asymmetric Lévy distribution (20.21) plus a possible linear term irp accounting for a drift, and the asymptotic z-behavior will be given by Eq. (20.17) with an extra factor $(1 + \beta)$ [21].

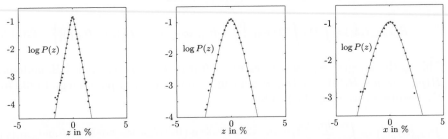

Figure 20.15 Fits of Gaussian distribution to S&P 500 log-returns recorded in intervals of 60 min, 240 min, and 1 day.

If a distribution has no second moments, the central limit theorem does not apply. Such distributions tend to another limit, the so-called *Pareto-Lévy-stable distribution*. Their Hamiltonian has precisely the generic form (20.21), except for a possible additional linear drift term [21]:

$$H(p) = -irp + H_{\lambda,\sigma,\beta}(p). \tag{20.168}$$

In this context, the Lévy parameter λ is also called the *stability parameter*. The parameter α has to be in the interval $(0, 2]$. The parameter β is called *skewness parameter*. For plots of $\tilde{D}(z, 1)$ see Ref. [75].

The convergence of variable with heavy-tail distributions is the content of the *generalized central limit theorem*.

20.1.16 Additivity Property of Noises and Hamiltonians

At this point we make the useful observation that each term in the Hamiltonian (20.141) can be attributed to an independent noise term in the stochastic differential equation (20.140). Indeed, if this differential equation has two noise terms

$$\dot{x}(t) = r_x + \eta_1(t) + \eta_2(t) \tag{20.169}$$

whose fluctuations are governed by two different Hamiltonians, the probability distribution (20.144) is replaced by

$$P(x_b t_b | x_a t_a) = \int \mathcal{D}\eta_1 \int \mathcal{D}\eta_2 \int \mathcal{D}x \exp\left[-\int_{t_a}^{t_b} dt [\tilde{H}_1(\eta_1(t)) + \tilde{H}_2(\eta_2(t))]\right] \delta[\dot{x} - r_x - \eta_1 - \eta_2].$$

After a Fourier decomposition of the δ-functional, this becomes

$$P(x_b t_b | x_a t_a) = \int \mathcal{D}p_1 \int \mathcal{D}p_2 \int \mathcal{D}\eta_1 \int \mathcal{D}\eta_2 \int \mathcal{D}x \int \mathcal{D}p e^{-\int_{t_a}^{t_b} dt\, [ip(\dot{x} - r_x - \eta_1 - \eta_2)]}$$
$$\times \exp\left\{-\int_{t_a}^{t_b} dt\, [H_1(p_1) - ip_1\eta_1(t) + H_2(p_2) - ip_2\eta_2(t)]\right\},$$

after which the path integrals over $\eta_1(t)$, $\eta_2(t)$ lead to

$$P(x_b t_b | x_a t_a) = \int \mathcal{D}p \int \mathcal{D}x \exp\left\{\int_{t_a}^{t_b} dt\, [ip\dot{x} - ir_x p - H_1(p) - H_2(p)]\right\}. \tag{20.170}$$

This is the integral representation Eq. (20.161) with the combined Hamiltonian $H(p) = ir_x p + H_1(p) + H_2(p)$, which can be rewritten as a pure integral (20.170). By rewriting this integral trivially as

$$P(x_b t_b | x_a t_a) = \int_{-\infty}^{\infty} \frac{dp_1}{2\pi} \frac{dp_2}{2\pi} e^{ip_1(x_b - x_c) - (t_b - t_a)[ir_x p + H_1(p_1)]} e^{ip_2(x_c - x_a) - (t_b - t_a)H_2(p_2)}, \tag{20.171}$$

we see that the probability distribution associated with a sum of two Hamiltonians is the convolution of the individual probability distributions:

$$P(x_b t_b | x_a t_a) = \int_{-\infty}^{\infty} dx dx_c P_1(x_b t_b | x_c t_a) P_2(x_c t_b | x_a t_a). \tag{20.172}$$

Thus we may calculate the probability distributions associated with the noises $\eta_1(p)$ and $\eta_2(p)$ separately, and combine the results at the end by a convolution.

20.1.17   Lévy-Khintchine Formula

It is sometimes useful to represent the Hamiltonian in the form of a Fourier integral

$$H(p) = \int dz\, e^{ipz} F(z). \tag{20.173}$$

Due to the special significance of the linear term in $H(p)$ governing the drift, this is usually subtracted out of the integral by rewriting (20.173) as

$$H_r(p) = irp + \int dz \left(e^{ipz} - 1 - ipz \right) F(z). \tag{20.174}$$

The first subtraction ensures the property $H_r(0) = 0$ which guarantees the unit normalization of the distribution. This subtracted representation is known as the *Lévy-Khintchine formula*, and the function $F(z)$ is the so-called *Lévy weight* of the distribution. Some people also subtract out the quadratic term and write

$$H_r(p) = irp + \frac{\sigma^2}{2} p^2 + \int dz \left(e^{ipz} - 1 \right) \bar{F}(z). \tag{20.175}$$

They employ a weight $\bar{F}(z)$ which has no first and second moment, i.e., $\int dz\, \bar{F}(z) z = 0$, $\int dz\, \bar{F}(z) z^2 = 0$, to avoid redundancy in the representation.

Note that according to the central limit theorem (20.167), the large-time probability distribution of x will become Gaussian for large t. Thus, in the Lévy-Khintchine decomposition (20.175) of the Hamiltonian $H(p)$, only the first two terms contribute for large t leading to the distribution:

$$P(x,t) \underset{t \text{ large}}{\to} \frac{e^{-(x-tr)^2/2t\sigma^2}}{\sqrt{2\pi\sigma^2}}. \tag{20.176}$$

The Lévy measure has been calculated explicitly for many non-Gaussian distributions. As an example, the generalized hyperbolic case has for $\lambda \geq 0$ a Lévy measure [76]

$$F(z) = \frac{e^{\beta z}}{|z|} \left\{ \frac{1}{\pi^2} \int_0^\infty \frac{dy}{y} \frac{e^{-\sqrt{y+\gamma^2}|z|}}{J_\lambda^2(\delta\sqrt{y}) + Y_\lambda^2(\delta\sqrt{y})} + \lambda e^{-\gamma|z|} \right\}, \tag{20.177}$$

where $J_\lambda(z)$ and $Y_\lambda(z)$ are standard Bessel functions.

The decomposition of the Hamiltonian according the Lévy-Khintchine formula and the additivity of the associated noises form the basis of the Lévy-Itō theorem which states that an arbitrary stochastic differential equation with Hamiltonian (20.175) can be decomposed in the form

$$\dot{x} = r_x t + \eta_G + \eta_{\leq 1} + \eta_{>1}, \tag{20.178}$$

where η_G is a Gaussian noise

$$\eta_{\leq 1} = \int_{|x|\leq 1} dz \left(e^{ipz} - 1 \right) \bar{F}(z) \tag{20.179}$$

a superposition of discrete noises called Poisson point process with jumps smaller than or equal to unity, and

$$\eta_{>1} = \int_{|x|\leq 1} dz \left(e^{ipz} - 1 \right) \bar{F}(z) \tag{20.180}$$

a noise with jumps larger than unity.

Consider the simplest noise of the type $\eta_{\leq 1}$ arising from a Lévy weight $\bar{F}_Z(z) = \tau_Z \delta(z + Z)$ with $0 < Z \leq 1$ in (20.179) so that the Hamiltonian is $H_Z(p) = \tau_Z(e^{-iZp} - 1)$. The associated distribution function $\tilde{D}_Z(z)$ is, according to (20.11),

$$\tilde{D}_Z(z) = \int \frac{dp}{2\pi} e^{ipz} e^{\tau_Z(e^{-ipZ}-1)}. \tag{20.181}$$

Expanding the last exponential in powers of e^{ipZ}, we obtain

$$\tilde{D}_Z(z) = \int \frac{dp}{2\pi} e^{ipz} \sum_{n=0}^{\infty} e^{-\tau z} \frac{\tau_Z^n}{n!} \tau_Z^n e^{-ipnZ} = \sum_{n=0}^{\infty} e^{-\tau z} \frac{\tau_Z^n}{n!} \delta(z - nZ). \tag{20.182}$$

This function describes jumps by nZ ($n = 0, 1, 2, 3, \ldots$) whose probability follows a *Poisson distribution*

$$P(n, \tau_Z) = e^{-\tau z} \frac{\tau_Z^n}{n!}, \tag{20.183}$$

which is properly normalized to unity: $\sum_{n=0}^{\infty} P(n, \tau_Z) = 1$. The expectation values of powers of the jump number are

$$\langle n^k \rangle = \sum_{n=0}^{\infty} n^k P(n, \tau_Z) = e^{-\tau z} (\tau_Z \partial_{\tau_z})^k e^{\tau z} \sum_{n=0}^{\infty} P(n, \tau_Z) = e^{-\tau z} (\tau_Z \partial_{\tau_z})^k e^{\tau z} = \frac{\Gamma(\lambda_Z + k)}{\Gamma(\lambda_Z)}. \tag{20.184}$$

Thus $\langle n \rangle = \lambda_Z$, $\langle n^2 \rangle = \lambda_Z(\lambda_Z + 1)$, $\langle n^3 \rangle = \lambda_Z(\lambda_Z + 1)(\lambda_Z + 2)$, $\langle n^4 \rangle = \lambda_Z(\lambda_Z + 1)(\lambda_Z + 2)(\lambda_Z + 3)$, so that $\sigma^2 = \lambda_Z$, $s = 2/\sqrt{\lambda_Z}$, and $\kappa = 6/\lambda_Z$.

As typical noise curve is displayed in Fig. 20.16. An arbitrary Lévy weight $\bar{F}(z)$ in $\eta_{\leq 1}$ may

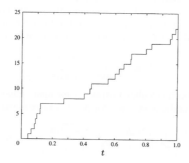

Figure 20.16 Typical Noise of Poisson Process.

always be viewed as a superposition $\bar{F}(z) = \int_{-1}^{1} dZ\, F(Z)\delta(z - Z)$ so that the distribution function becomes a superposition of Poisson noises:

$$\tilde{D}(z) = \int_{-1}^{1} dZ\, F(Z) \sum_{n=0}^{\infty} e^{-\tau z} \frac{\tau_Z^n}{n!} \delta(z - nZ). \tag{20.185}$$

20.1.18 Semigroup Property of Asset Distributions

An important property of the probability (20.144) is that it satisfies the semigroup equation

$$P(x_c t_c | x_a t_a) = \int_{-\infty}^{\infty} dx_b\, P(x_c t_c | x_b t_b) P(x_b t_b | x_a t_a). \tag{20.186}$$

In the stochastic context this property is referred to as *Chapman-Kolmogorov equation* or *Smoluchowski equation*. It is a general property of processes which a without memory. Such processes are referred to a *Markovian* [12]. Note that the semigroup property (20.221) implies the initial condition

$$P(x_c t_a | x_a t_a) = \delta(x_b - x_a). \tag{20.187}$$

In Fig. 20.17 we show that the property (20.221) is satisfied reasonably well by experimental asset distributions, except for small deviations in the low-probability tails.

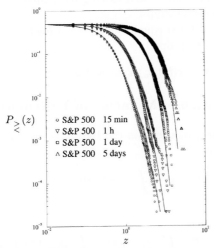

Figure 20.17   Cumulative distributions obtained from repeated convolution integrals of the 15-min distribution (from Ref. [13]). Apart from the far ends of the tails, the semigroup property (20.221) is reasonably well satisfied.

One may verify the semigroup property also using the high-frequency distributions of S&P 500 and NASDAQ 100 indices in Fig. 20.10 recorded by the minute, which display the Boltzmann distribution (20.77). If we convolute the minute distribution an integer number t times, the result aggress very well with the distribution of t-minute data. Only at the heavy tails of rare events do not follow this pattern

Note that due to the semigroup property (20.221) of the probabilities one has the trivial identity

$$
\begin{aligned}
\langle f(x(t_b), t_b) \rangle &= \int dx_b \, f(x_b, t_b) P(x_b t_b | x_a t_a) \\
&= \int dx \left[\int dx_b \, f(x_b, t_b) P(x_b t_b | x \, t) \right] P(x \, t | x_a t_a). \tag{20.188}
\end{aligned}
$$

In the mathematical literature, the expectation value on the left-hand side of Eq. (20.188) is written as

$$\mathbb{E}[f(x(t_b), t_b) | x_a t_a] \equiv \int dx_b \, f(x_b, t_b) P(x_b t_b | x_a t_a). \tag{20.189}$$

With this notation, the second line can be re-expressed in the form

$$\int dx\, \mathbb{E}[f(x_b,t_b)|x\,t]P(x\,t|x_at_a) = \mathbb{E}[\mathbb{E}[f(x_b,t_b)|x\,t]|x_at_a], \qquad (20.190)$$

so that we obtain the property of expectation values

$$\mathbb{E}[f(x(t_b,t_b))|x_at_a] = \mathbb{E}[\mathbb{E}[f(x_b,t_b)|x\,t]|x_at_a]. \qquad (20.191)$$

In mathematical finance, this complicated-looking but simple property is called the *towering property* of expectation values.

Note that since $P(x_bt_b|x_at_a)$ is equal to $\delta(x_b - x_a)$ for $t_a = t_b$, the expectation value (20.189) has the obvious property

$$\mathbb{E}[f(x(t_a),t_a)|x_at_a] = f(x_a,t_a) \qquad (20.192)$$

20.1.19 Time Evolution of Moments of Distribution

From the time-dependent distribution (20.165) it is easy to calculate the time dependence of the moments:

$$\langle x^n \rangle(t) \equiv \int_{-\infty}^{\infty} dx\, x^n\, P(x,t) \qquad (20.193)$$

Inserting (20.165), we obtain

$$\langle x^n \rangle(t) = \int_{\infty}^{\infty} \frac{dp}{2\pi} e^{-tH(p)} \int_{-\infty}^{\infty} dx\, x^n e^{ipx} = \int_{\infty}^{\infty} dp\, e^{-tH(p)} (-i\partial_p)^n \delta(p). \quad (20.194)$$

After n partial integrations, this becomes

$$\langle x^n \rangle(t) = (i\partial_p)^n e^{-tH(p)} \Big|_{p=0}. \qquad (20.195)$$

All expansion coefficients c_n of $H(p)$ in Eq. (20.141) receive the same factor t, so that the cumulants of the moments all grow linearly in time:

$$\langle x^n \rangle_c(t) = -tH^{(n)}(0) = t\langle x^n \rangle_c(1) = tc_n. \qquad (20.196)$$

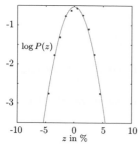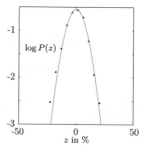

Figure 20.18 Gaussian distributions of S&P 500 and NASDAQ 100 weekly log-returns.

20.1.20   Boltzmann Distribution

As a an example, consider the Boltzmann distribution (20.77) of the minute data. The subsequent time-dependent variance and kurtosis are found by inserting the cumulants Eq. (20.80) into (20.196), yielding

$$\langle x^2 \rangle_c(t) = t\, 2T^2, \quad \kappa(t) = \frac{c_4}{t c_2^2} = \frac{3}{t}. \tag{20.197}$$

The first increases linearly in time, the second decreases like one over time, which makes the distribution more and more Gaussian, as required by the central limit theorem (20.167). The two quantities are plotted in Figs. 20.19 and 20.20. The agreement with the data is seen to be excellent.

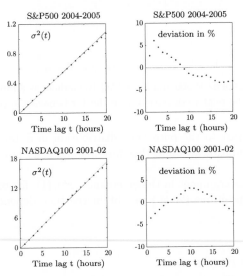

Figure 20.19   Variance of S&P 500 and NASDAQ 100 indices as a function of time. The slopes $\langle x^2 \rangle_c(t)/t$ are roughly $1.1/20 \times 1/60\,\mathrm{min}$ and $18.1/20 \times 1/60\,\mathrm{min}$, respectively, so that Eq. (20.198) yields the temperatures $T_{\mathrm{SP500}} \approx 0.075$ and $T_{\mathrm{NASDAQ100}} \approx 0.15$. The right-hand side shows the relative deviation from the linear shape in percent.

Note that if the minute data are not available, but only data taken with a frequency $1/t_0$ (in units $1/\mathrm{min}$) and an expectation $\langle x^2 \rangle_c(t_0)$, then we can find the temperature of the minute distribution from the formula

$$T = \sqrt{\langle x^2 \rangle_c(t_0)/2t_0}. \tag{20.198}$$

Since κ goes to zero like $1/t$, the distribution becomes increasingly Gaussian as the time grows, this being a manifestation of the central limit theorem (20.167) of statistical mechanics according to which the convolution of infinitely many arbitrary distribution functions of finite width always approaches a Gaussian distribution.

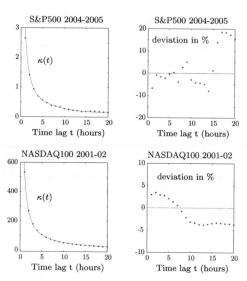

Figure 20.20 Kurtosis of S&P 500 and NASDAQ 100 indices as a function of time. The right-hand side shows the relative deviation from the $1/t$ behavior in percent.

This result is in contrast to the pure Lévy distribution in Subsection 20.1.2 with $\lambda < 2$, which has no finite width and therefore maintains its power falloff at large distances.

If we omit the time argument in the cumulants $\langle x^n \rangle_c(1)$, for brevity, and insert these into the decompositions (20.56) we obtain the time dependence of the moments:

$$\langle x \rangle(t) = t\langle x \rangle_c,$$

$$\langle x^2 \rangle(t) = t\langle x^2 \rangle_c + t^2\langle x \rangle_c^2,$$

$$\langle x^3 \rangle(t) = t\langle x^3 \rangle_c + 3t^2\langle x \rangle_c\langle x^2 \rangle_c + t^3\langle x \rangle_c^3,$$

$$\langle x^4 \rangle(t) = t\langle x^4 \rangle_c + 3t^2\langle x^2 \rangle_c^2 - 4t^2\langle x \rangle_c\langle x^3 \rangle_c + 6t^3\langle x \rangle_c^2\langle x^2 \rangle_c + t^4\langle x \rangle_c^4, \qquad (20.199)$$

$$\vdots$$

Let us now calculate the time evolution of the Boltzmann distribution (20.77) of the minute data. The Hamiltonian $H(p)$ was identified in Eq. (20.79), so that

$$e^{-tH(p)} = \frac{1}{[1 + (Tp)^2]^t}. \qquad (20.200)$$

The time-dependent distribution is therefore given by the Fourier integral

$$P(x,t) = \int_{-\infty}^{\infty} \frac{dp}{2\pi} e^{ipx - tH(p)} = \int_{-\infty}^{\infty} \frac{dp}{2\pi} \frac{1}{(1 + T^2 p^2)^t} e^{ipx}. \qquad (20.201)$$

The calculation proceeds most easily rewriting (20.200), by analogy with (20.81) and using again formula (2.497), as an integral

$$e^{-tH(p)} = \frac{1}{[1 + (Tp)^2]^t} = \frac{1}{\Gamma(t)} \int_0^\infty \frac{d\tau}{\tau} \tau^t e^{-\tau(1+T^2 p^2)}. \tag{20.202}$$

The prefactor of p^2 is equal to $t\sigma^2 = tv$ a Gaussian distribution in x-space. We therefore change the variable of integration from τ to v, substituting $\tau = tv/2T^2$, and obtain [compare (20.82)]

$$e^{-tH(p)} = \left(\frac{t}{2T^2}\right)^t \frac{1}{\Gamma(t)} \int_0^\infty \frac{dv}{v} v^t e^{-tv/2T^2} e^{-tvp^2/2}. \tag{20.203}$$

The Fourier transform of this yields the time evolution of the Boltzmann distribution as a time-dependent superposition of Gaussian distributions of different widths [compare (20.83)]:

$$P(x,t) = \left(\frac{t}{2T^2}\right)^t \frac{1}{\Gamma(t)} \int_0^\infty \frac{dv}{v} v^t e^{-tv/2T^2} \frac{1}{\sqrt{2\pi tv}} e^{-x^2/2tv}. \tag{20.204}$$

The integral can be performed using the integral formula (2.557), and we find the time-dependent distribution

$$P(x,t) = \int_{-\infty}^\infty \frac{dp}{2\pi} e^{ipx - tH(p)} = \frac{1}{T\sqrt{\pi}\Gamma(t)} \left(\frac{|x|}{2T}\right)^{t-1/2} K_{t-1/2}(|x|/T), \tag{20.205}$$

where t is measured in minutes.

For $t = 1$, we reobtain the fundamental distribution (20.77), recalling the explicit form of $K_{1/2}(z)$ in Eq. (1.345).

In the limit of large t, the integral over v in (20.204) can be evaluated in the saddle-point approximation of Section 4.2. We expand the weight function in the integrand around the maximum at $v_m = 2T^2(1 - 1/t)$ as

$$v^{t-1} e^{-tv/2T^2} = \left[2T^2(1 - 1/t)\right]^{t-1} e^{-(t-1)} e^{-t\delta v^2/2[2T^2(1-1/t)]^2} \left[1 + \mathcal{O}(\delta v^3)\right], \tag{20.206}$$

where $\delta v = v - v_m$, and $\mathcal{O}(\delta v^3)$ is equal to $-t\delta v^3/3[2T^2(1-1/t)]^3$. Then we observe that for large t, the Gaussian can be expanded into derivatives of δ-functions as follows:

$$e^{-tz^2/2} \left[1 + \frac{a}{3} z^3 + \dots\right] = \sqrt{2\pi/t} \left[\delta(z) + \frac{1}{2t}\delta''(z) + \frac{a}{t^2}\delta'(z)\dots\right]. \tag{20.207}$$

This can easily be verified by multiplying both sides with $f(z) = f(0) + z f'(0) + z^2 f''(0)/2 + \dots$ and integrating over z. Thus we find to leading order

$$e^{-t\delta v^2/2[2T^2(1-1/t)]^2} \to \sqrt{\frac{2\pi}{t}} 2T^2(1 - 1/t)\, \delta(\delta v). \tag{20.208}$$

Using the large-t limit $(1 - 1/t)^t \to e^{-1}$, and including the first correction from (20.207), we obtain

$$v^{t-1}e^{-tv/2T^2} \to \sqrt{\frac{2\pi}{t}}(2T^2)^t e^{-t}\left\{\delta(\delta v) + \frac{[2T^2(1-1/t)]^2}{2t}\delta''(\delta v) + \dots\right\}. \quad (20.209)$$

If we now employ Stirling's formula (17.286) to approximate $t^t/\Gamma(t) \to \sqrt{t/2\pi}e^t$, we see that the integral (20.204) converges for large t against the Gaussian distribution $e^{-x^2/2tv_m}/\sqrt{2\pi tv_m}$ with the saddle point variance $v_m \to 2T^2$.

The behavior near the peak of the distribution can be calculated from the small-z expansion[7]

$$\left(\frac{z}{2}\right)^\nu K_\nu(z) = \frac{\pi}{2\sin\pi\nu\,\Gamma(1-\nu)}\left[1 + \frac{\Gamma(1-\nu)}{1!\Gamma(2-\nu)}\frac{z^2}{4} + \mathcal{O}(z^4, z^{2\nu})\right]. \quad (20.210)$$

For large ν and small z, this can be approximated by [recall (20.29)]

$$\left(\frac{z}{2}\right)^\nu K_\nu(z) \approx \frac{\Gamma(\nu)}{2}e^{-z^2/4(t-3/2)}. \quad (20.211)$$

We now use Stirling's formula (17.286) to find the large-t limit $\Gamma(t-1/2)/\Gamma(t) \to 1$, leading to the Gaussian behavior near the peak:

$$P(x,t) \underset{\text{small }|x|}{\approx} \frac{1}{\sqrt{2\pi\,2T^2t}}e^{-x^2/2\,2T^2(t-3/2)}. \quad (20.212)$$

This corresponds to the approximation of the Fourier transform (20.200) by $e^{-tT^2p^2}$. The Gaussian shape reaches out to $x \approx T\sqrt{t}$. For very large t, it holds everywhere, as it should by virtue of the central limit theorem (20.167).

20.1.21 Fourier-Transformed Tsallis Distribution

For the Hamiltonian defined in Eq. (20.97), the time evolution is given by the Fourier transform

$$e^{-tH_{\delta,\beta}(p)} = \left[1 - \beta\delta\,p^2/2\right]^{-t/\delta} = \frac{1}{\Gamma(t/\delta)}\int_0^\infty \frac{ds}{s}\,s^{t/\delta}e^{-s}e^{-s\beta\delta\,p^2/2}, \quad \beta \equiv \frac{\nu}{\mu} = \frac{1}{\mu\delta}, (20.213)$$

which can be rewritten, by analogy with (20.99), as

$$e^{-tH_{\delta,\beta}(p)} = \frac{\mu^{t/\delta}}{\Gamma(t/\delta)}\int_0^\infty \frac{dv}{v}\,v^{t/\delta}e^{-\mu v}e^{-vp^2/2}, \quad \mu = 1/\beta\delta. \quad (20.214)$$

Taking the Fourier transform of this yields the time-dependent distribution function

$$P_{\delta,\beta}(x,t) = \frac{\mu^{t/\delta}}{\Gamma(t/\delta)}\int_0^\infty \frac{dv}{v}\,v^{t/\delta}e^{-\mu v}\frac{1}{\sqrt{2\pi v}}e^{-x^2/2v}, \quad \mu = 1/\beta\delta, \quad (20.215)$$

---

[7]M. Abramowitz and I. Stegun, op. cit., Formulas 9.6.2 and 9.6.10.

which becomes with the help of Formula (2.497):

$$P_{\delta,\beta}(x,t) = \frac{\mu^{t/\delta}}{\Gamma(t/\delta)} \frac{1}{\sqrt{2\pi}} \left(\frac{x^2}{2\mu}\right)^{t/2\delta - 1/4} 2K_{t/\delta - 1/2}(\sqrt{2\mu}x), \quad \mu = \frac{1}{\beta\delta}. \quad (20.216)$$

For $\delta = 1$ and $\mu = 1/2T^2$, this reduces to the time-dependent Boltzmann distribution (20.205). At the origin, its value is [compare (20.104)]

$$P_{\delta,\beta}(0,t) = \sqrt{\mu}\,\Gamma(t/\delta - 1/2)/\Gamma(t/\delta). \quad (20.217)$$

20.1.22　Superposition of Gaussian Distributions

All time-dependent distributions whose Fourier transforms have the form $e^{-tH(p)}$ possess a path integral representation (20.161) and which obeys the semigroup equation (20.221). In several examples we have seen that the Fourier transforms $e^{-H(p)}$ can be obtained from a superposition of Gaussian distributions of different variances $e^{-vp^2/2}$ with some weight function $w(v)$:

$$e^{-H(p)} = \int_0^\infty dv\, w(v) e^{-vp^2/2}. \quad (20.218)$$

This was true for the Boltzmann distribution in Eq. (20.81), for the Fourier-transformed Tsallis distribution in Eq. (20.100), and for the Boltzmann Distribution of a relativistic particle in Eq. (20.105). In all these cases it was possible to write a similar superposition for the time-dependent distributions:

$$e^{-tH(p)} = \int_0^\infty dv'\, w_t(v') e^{-v'p^2/2} = \int_0^\infty dv\, \omega_t(v) e^{-tvp^2/2}, \quad (20.219)$$

where we have introduced the time-dependent weight function

$$\omega_t(v) \equiv t\, w_t(vt). \quad (20.220)$$

See Eqs. (20.202), (20.214), and (20.105) for the above three cases.

The question arises as to general property of the time dependence of the weight function $w_t(v)$ which ensures that that the Fourier transform of the superposition (20.219) obtained as in Eq. (20.164) satisfies the semigroup condition (20.221) [52]. For this, the superposition has to depend on time as

$$e^{-(t_2+t_1)H(p)} = e^{-t_2 H(p)} e^{-t_1 H(p)},$$

or $e^{-tH(p)} = [e^{-H(p)}]^t$. This implies that the weight factor $w_t(v)$ has the property

$$\int dv_{12} w_{t_1+t_2}(v_{12}) e^{-v_{12}p^2/2} = \int dv_2 w_{t_2}(v_2) e^{-v_2 p^2/2} \int dv_1 w_{t_1}(v_1) e^{-v_1 p^2/2}. \quad (20.221)$$

For the Laplace transforms $\tilde{w}_t(v)$ of $w_t(v)$

$$\tilde{w}_t(p_v) \equiv \int dv\, e^{-p_v v} w_t(v) \quad (20.222)$$

this amounts to the factorization property

$$\tilde{w}_{t_1+t_2}(p_v) = \tilde{w}_{t_2}(p_v)\tilde{w}_{t_1}(p_v), \tag{20.223}$$

which is fulfilled by the exponential

$$\tilde{w}_t(p_v) = e^{-tH_v(p_v)}, \tag{20.224}$$

with some Hamiltonian $H_v(p_v)$. Since the integral over v in (20.222) runs only over the positive v-axis, the Hamiltonian $H_v(p_v)$ is analytic in the upper half-plane of p_v.

It is easy to relate $H_v(p_v)$ to $H(p)$. For this we form the inverse Laplace transformation, the so-called *Bromwich integral* [53]

$$w_t(v) = \int_{\gamma-i\infty}^{\gamma+i\infty} \frac{dp_v}{2\pi i} e^{p_v v - t H_v(p_v)}, \tag{20.225}$$

in which γ is some real number larger than the real parts of all singularities in $e^{-p_v v - t H_v(p_v)}$. Inserting this into (20.219) and performing the v-integral yields

$$e^{-tH(p)} = \int_{\gamma-i\infty}^{\gamma+i\infty} \frac{dp_v}{2\pi i} \frac{1}{p^2/2 - p_v} e^{-tH_v(p_v)}. \tag{20.226}$$

Closing the integral over p_v in the complex p_v-plane and deforming the contour to an anticlockwise circle around $p_v = p^2/2$ we find the desired relation:

$$H_v(p^2/2) = H(p). \tag{20.227}$$

The time-dependent distribution associated with any Hamiltonian $H(p)$ via the Fourier integral (20.170) can be represented as a superposition of Gaussian distributions with the help of Eq. (20.219) if we merely choose $H_v(p_v) = H(\sqrt{2p_v})$.

Recalling the derivation of the Fourier representation (20.165) from the path integral (20.153) and the descendence of that path integral from the stochastic differential equation (20.140) with the noise Hamiltonian $H(p)$ we conclude that the Hamiltonian $H_v(ip_v)$ governs the noise in a stochastic differential equation for the volatility fluctuations.

Take, for example, the superposition of Gaussians (20.105). The Laplace transform of the weight function $w_\beta(v)$ is

$$\tilde{w}_\beta(p_v) = \int dv \, e^{-vp_v} w_\beta(v) = \int dv \, e^{-vp_v} \sqrt{\frac{\beta}{2\pi v^3}} e^{-\beta/2v} = e^{-\sqrt{2\beta p_v}}. \tag{20.228}$$

The Laplace transform of $w_\beta(v)$ is related to this by

$$\int dv' e^{-v'p_v} w_\beta(v') = \int dv \, e^{-\beta v p_v} w_\beta(v) = \tilde{w}_\beta(\beta p_v). \tag{20.229}$$

Hence it is given by $\tilde{w}_\beta(p_v) = e^{-\beta\sqrt{2p_v}} = e^{-\beta H_v(p_v)}$, which satisfies the relation (20.223). Moreover, $H_v(p^2/2)$ is equal to $H(p) = \sqrt{p^2}$, as required by Eq. (20.227) and in accordance with Eq. (20.105) for $M = 0$.

Note that instead of the *Bromwich integral* (20.226) it is sometimes more convenient to use *Post's Laplace inversion formula* [54]

$$w_t(v) = \lim_{k\to\infty} \frac{(-1)^k}{k!} x^{k+1} \frac{\partial^k \tilde{w}_t(x)}{\partial x^k}\bigg|_{x=k/v}. \tag{20.230}$$

20.1.23　Fokker-Planck-Type Equation

From the Fourier representation (20.170) it is easy to prove that the probability satisfies a Fokker-Planck-type equation

$$\partial_t P(x_b t_b | x_a t_a) = -H(-i\partial_x) P(x_b t_b | x_a t_a). \tag{20.231}$$

Indeed, the general solution $\psi(x, t)$ of this differential equation with the initial condition $\psi(x, 0)$ is given by the path integral generalizing (20.144)

$$\psi(x, t) = \int \mathcal{D}\eta \exp\left[-\int_{t_a}^{t_b} dt\, \tilde{H}(\eta(t))\right] \psi\left(x - \int_{t_a}^{t} dt'\eta(t')\right). \tag{20.232}$$

This satisfies the Fokker-Planck-type equation (20.231). To show this, we take $\psi(x, t)$ at a slightly later time $t + \epsilon$ and expand

$$
\begin{aligned}
\psi(x, t + \epsilon) &= \int \mathcal{D}\eta \exp\left[-\int_{t_a}^{t_b} dt\, \tilde{H}(\eta(t))\right] \psi\left(x - \int_{t_a}^{t} dt'\eta(t') - \int_{t}^{t+\epsilon} dt'\eta(t')\right). \\
&= \int \mathcal{D}\eta \exp\left[-\int_{t_a}^{t_b} dt\, \tilde{H}(\eta(t))\right] \left\{\psi\left(x - \int_{t_a}^{t} dt'\eta(t')\right)\right. \\
&\quad - \psi'\left(x - \int_{t_a}^{t} dt'\eta(t')\right) \int_{t}^{t+\epsilon} dt'\eta(t') \\
&\quad + \frac{1}{2}\psi''\left(x - \int_{t_a}^{t} dt'\eta(t')\right) \int_{t}^{t+\epsilon} dt_1 dt_2\, \eta(t_1)\eta(t_2) \\
&\quad - \frac{1}{3!}\psi'''\left(x - \int_{t_a}^{t} dt'\eta(t')\right) \int_{t}^{t+\epsilon} dt_1 dt_2 dt_3\, \eta(t_1)\eta(t_2)\eta(t_3) \\
&\quad \left. + \frac{1}{4!}\psi^{(4)}\left(x - \int_{t_a}^{t} dt'\eta(t')\right) \int_{t}^{t+\epsilon} dt_1 dt_2 dt_3 dt_4\, \eta(t_1)\eta(t_2)\eta(t_3)\eta(t_4) + \dots \right\}.
\end{aligned}
\tag{20.233}
$$

Using the correlation functions (20.155)–(20.262) we obtain

$$
\begin{aligned}
\psi(x, t + \epsilon) &= \int \mathcal{D}\eta \exp\left[-\int_{t_a}^{t_b} dt\, \tilde{H}(\eta(t))\right] \\
&\quad \times \left[-\epsilon c_1 \partial_x + \left(\epsilon c_2 + \epsilon^2 c_1\right)\frac{1}{2}\partial_x^2 - \left(\epsilon c_3 + 3\epsilon^2 c_2 c_1\right)\frac{1}{3!}\partial_x^3 \right. \\
&\quad \left. + \left(\epsilon c_4 + \epsilon^2 4 c_3 c_1 + \epsilon^2 3 c_2^2 + \epsilon^3 c_2 c_1^2 + \epsilon^4 c_1^2\right)\frac{1}{4!}\partial_x^4 + \dots\right] \psi\left(x - \int_{t_a}^{t} dt'\eta(t')\right).
\end{aligned}
\tag{20.234}
$$

In the limit $\epsilon \to 0$, only the linear terms in ϵ contribute, which are all due to the connected parts of the correlation functions of $\eta(t)$. The differential operators in the brackets can now be pulled out of the integral and we find the differential equation

$$\partial_t \psi(x, t) = \left[-c_1 \partial_x + \frac{c_2}{2!}\partial_x^2 - \frac{c_3}{3!}\partial_x^3 + \frac{c_4}{4!}\partial_x^4 + \dots\right] \psi(x, t). \tag{20.235}$$

We now replace $c_1 \to r_x$ and express using (20.141) the differential operators in brackets as Hamiltonian operator $-H_{r_x}(-i\partial_x)$. This leads to the Schrödinger-like equation [55]

$$\partial_t \psi(x, t) = -H_{r_x}(-i\partial_x) \psi(x, t). \tag{20.236}$$

Due to the many derivatives in $H(i\partial_x)$, this equation is in general non-local. This can be made explicit with the help of the Lévy-Khintchine weight function $F(x)$ in the Fourier representation (20.173). In this case, the right-hand side

$$-H(-i\partial_x)\psi(x,t) = \int dx'\, e^{-x'\partial_x}\, F(x')\psi(x,t) = \int dx'\, F(x')\psi(x-x',t), \quad (20.237)$$

and the Fokker-Planck-like equation (20.236) takes the form of an integro-differential equation. Some people like to use the subtracted form (20.175) of the Lévy-Khintchine formula and arrive at the integro-differential equation

$$\partial_t\psi(x,t) = \left[-c_1\partial_x - \frac{c_2}{2}\partial_x^2\right]\psi(x,t) + \int dx'\,\bar{F}(x')\psi(x-x',t). \quad (20.238)$$

The integral term can then be treated as a perturbation to an ordinary Fokker-Planck equation.

The initial condition of the probability is, of course, always given by (20.187), as a consequence of the semigroup property (20.221).

20.1.24 Kramers-Moyal Equation

The Fokker-Planck-type equation (20.231) is a special case of the most general time evolution equation satisfied by a probability distribution $P(x_b, t_b, |x_a t_a)$ which fulfills the Chapman-Kolmogorov or Smoluchowski equation (20.221). For a short time interval $t_c - t_b = \epsilon$, that equation can be written as

$$P(x_c\, t_b+\epsilon|x_a t_a) = \int_{-\infty}^{\infty} dx_b\, P(x_c t_b + \epsilon|x_b t_b)P(x_b t_b|x_a t_a). \quad (20.239)$$

We now re-express $P(x_c t_b+\epsilon|x_b t_b)$ as a trivial integral

$$P(x_c t_b+\epsilon|x_b t_b) = \int dx\, \delta(x-x_b+x_b-x_c)P(x\, t_b+\epsilon|x_b t_b), \quad (20.240)$$

and expand the δ-function in powers of $x - x_b$ to obtain

$$P(x_c t_b+\epsilon|x_b t_b) = \int dx\, \sum_{n=0}^{\infty} \frac{(x-x_b)^n}{n!}P(x\, t_b+\epsilon|x_b t_b)\, \partial_{x_b}^n\delta(x_b - x_c). \quad (20.241)$$

Introducing the moments

$$C_n(x_b t_b) \equiv \int dx\, (x-x_b)^n\partial_{t_b}P(x\, t_b|x_b t_b), \quad (20.242)$$

and inserting (20.241) into Eq. (20.239) the right-hand side reads

$$P(x_c t_b|x_b t_b) + \epsilon\sum_{n=1}^{\infty}\int dx_b\, \frac{C_n(x_b t_b)}{n!}[\partial_{x_b}^n\delta(x_b - x_c)]\partial_{t_b}P(x_b t_b|x_a t_a) + \mathcal{O}(\epsilon^2). \quad (20.243)$$

Taking the $n = 0$ -term to the left-hand side of (20.239) and going to the limit $\epsilon \to 0$, we obtain the *Kramers-Moyal equation*:

$$\partial_{t_b} P(x_b t_b | x_a t_a) = -H(-i\partial_{x_b}, x_b, t_b) P(x_b t_b | x_a t_a). \qquad (20.244)$$

with the Hamiltonian operator

$$H(-i\partial_{x_b}, x_b, t_b) \equiv - \sum_{n=1}^{\infty} (-\partial_{x_b})^n \frac{C_n(x_b t_b)}{n!}, \qquad (20.245)$$

which is the generalization of the Hamiltonian operator in Eqs. (20.235) and (20.231).

There is a simple formal solution of the time evolution equation (20.244) with the initial condition (20.187). It can immediately be written down by recalling the similar time evolution equation (1.233):

$$P(x_b t_b | x_a t_a) = \hat{T} e^{- \int_{t_a}^{t_b} dt\, H(-i\partial_{x_b}, x_b, t)} \delta(x_b - x_a), \qquad (20.246)$$

where $\hat{T}$ is the time-ordering operator (1.241). Using Dirac bra-ket states $\langle x_b |$ and $|x_a\rangle$ with the scalar product $\langle x_b | x_a \rangle = \delta(x_b - x_a)$ as in Eq. (18.462), and the rules (18.463), we can rewrite (20.246) as

$$P(x_b t_b | x_a t_a) = \langle x_b | \hat{T} e^{- \int_{t_a}^{t_b} dt\, H(-i\partial_{x_b}, x_b, t)} | x_a \rangle. \qquad (20.247)$$

Due to the initial condition (20.187), the moments (20.242) can also be calculated from the short-time limit

$$C_n(x\,t) \underset{\epsilon \to 0}{=} \frac{1}{\epsilon} \int dx\, (x' - x)^n P(x' t + \epsilon | x t) \underset{\epsilon \to 0}{=} \frac{1}{\epsilon} \langle [x(t+\epsilon) - x(t)]^n \rangle, \quad n \geq 1. \ (20.248)$$

Another way of writing the expectation value on the right-hand side is

$$C_n(x\,t) \underset{\epsilon \to 0}{=} \frac{1}{\epsilon} \int_t^{t+\epsilon} dt_1 \cdots \int_t^{t+\epsilon} dt_n\, \langle \dot{x}(t_1) \cdots \dot{x}(t_n) \rangle, \quad n \geq 1. \qquad (20.249)$$

If $x(t)$ follows the Langevin equation (20.169) with the correlation functions (20.155)–(20.262), we obtain

$$C_n(x\,t) = c_n, \qquad (20.250)$$

and the Kramers-Moyal equation (20.248) reduces to out previous Fokker-Planck-type equation (20.231).

For a Langevin equation

$$\dot{x}(t) = r(x, t) + \sigma(x, t)\eta(t), \qquad (20.251)$$

with a white noise variable $\eta(t)$, we find instead

$$C_1(x\,t) = a(x,t) + b'(x,t)b(x,t) = a(x,t) + \frac{\partial_x C_2(x\,t)}{2}, \quad C_2(x\,t) = b^2(x,t). \quad (20.252)$$

According to a theorem due to Pawula [56], the positivity of the probability in the Kramers-Moyal equation (20.244) requires that the expansion (20.245) can only be truncated after the first or second terms, or must go on to infinity.

20.2 Itō-like Formula for Non-Gaussian Distributions

By a procedure similar to that in Subsection 18.13.3 it is possible to derive a generalization of the Itō-like expansions (18.412) and (18.432) for the fluctuations of functions containing non-Gaussian noise.

20.2.1 Continuous Time

As in (18.408) we expand $f(x(t + \epsilon))$:

$$
\begin{aligned}
f(x(t + \epsilon)) = {} & f(x(t)) + f'(x(t))\Delta x(t) \\
& + \frac{1}{2}f''(x(t))[\Delta x(t)]^2 + \frac{1}{3!}f^{(3)}[\Delta x(t)]^3 + \dots \quad (20.253)
\end{aligned}
$$

where $\dot{x}(t) = \eta(t)$ is the stochastic differential equation with a nonzero expectation value $\langle \eta(t) \rangle = c_1$. In contrast to the expansion (18.426), which for $\epsilon \to 0$ had to be carried only up to the second order in $\Delta x(t) \equiv \int_t^{t+\epsilon} dt' \, \dot{x}(t')$, we must now keep *all* orders. Evaluating the noise averages of the multiple integrals on the right-hand side with the help of the correlation functions (20.155)–(20.262), we find the time dependence of the expectation value of an arbitrary function of the fluctuating variable $x(t)$

$$
\begin{aligned}
\langle f(x(t+\epsilon)) \rangle = {} & \langle f(x(t)) \rangle + \langle f'(x(t)) \rangle \epsilon c_1 + \frac{1}{2}\langle f''(x(t)) \rangle (\epsilon c_2 + \epsilon^2 c_1^2) \\
& + \frac{1}{3!}\langle f^{(3)}(x(t)) \rangle (\epsilon c_3 + \epsilon^2 c_2 c_1 + \epsilon^3 c_1^3) + \dots \quad (20.254) \\
= {} & \langle f(x(t)) \rangle + \epsilon \left[c_1 \partial_x + c_2 \frac{1}{2}\partial_x^2 + c_3 \frac{1}{3!}\partial_x^3 + \dots \right] \langle f(x(t)) \rangle + \mathcal{O}(\epsilon^2).
\end{aligned}
$$

After the replacement $c_1 \to r_x$, the function $f(x(t))$ obeys therefore the following equation:

$$
\langle \dot{f}(x(t)) \rangle = -H_{r_x}(i\partial_x)\langle f(x(t)) \rangle. \quad (20.255)
$$

Separating the lowest-derivative term, this takes a form generalizing Eq. (18.412):

$$
\langle \dot{f}(x(t)) \rangle = \langle f'(x(t)) \rangle \langle \dot{x}(t) \rangle - \bar{H}_{r_x}(i\partial_x)\langle f(x(t)) \rangle. \quad (20.256)
$$

This might be viewed as the expectation value of the stochastic differential equation

$$
\dot{f}(x(t)) = f'(x(t))\dot{x}(t) - \bar{H}_{r_x}(i\partial_x)f(x(t)), \quad (20.257)
$$

which, if valid, would be a simple direct generalization of Itō's Lemma (18.413). However, this conclusion is not allowed. The reason lies in the increased size of the fluctuations of the higher expansion terms $[\Delta x(t)]^n$ for $n \geq 2$. In the harmonic case of Subsection 18.13.3, these were negligible in comparison with the leading term $z_1(t)$ in Eq. (18.413). Let us see what happens in the present case, where all higher cumulants c_n in the Hamiltonian (20.141) may be nonzero. For the argument we proceed as in Eq. (18.405) by working with the stochastic differential equation

$$\dot{x}(t) = \langle \dot{x}(t) \rangle + \eta(t) = c_1 + \eta(t). \tag{20.258}$$

rather that (20.169). Then the correlation functions of $\eta(t)$ consist only of the connected parts of (20.155)–(20.262):

$$\langle \eta(t_1) \rangle = c_1, \tag{20.259}$$
$$\langle \eta(t_1)\eta(t_2) \rangle = c_2\delta(t_1 - t_2) \tag{20.260}$$
$$\langle \eta(t_1)\eta(t_2)\eta(t_3) \rangle = c_3\delta(t_1 - t_2)\delta(t_1 - t_3), \tag{20.261}$$
$$\langle \eta(t_1)\eta(t_2)\eta(t_3)\eta(t_4) \rangle = c_4\delta(t_1 - t_2)\delta(t_1 - t_3)\delta(t_1 - t_4) \tag{20.262}$$
$$+ c_2^2[\delta(t_1 - t_2)\delta(t_3 - t_4) + \delta(t_1 - t_3)\delta(t_2 - t_4) + \delta(t_1 - t_4)\delta(t_2 - t_3)].$$
$$\vdots$$

We now estimate the size of the fluctuations z_n of $[\Delta x(t)]^n$. The first contribution $z_{2,1}$ to z_2 defined in Eq. (18.415) is still negligible since it is smaller than the leading one $z_1(t) \equiv \int_t^{t+\epsilon} dt'\, \eta(t')$ by a factor ϵ. The second contribution $z_{2,3}$ to z_2 defined in Eq. (18.415), however, has now a larger variance, due to the c_4-term in (20.262), which contains one more δ-function than the c_2^2-term, so that

$$\left\langle [z_{2,2}(t)]^2 \right\rangle = \int_t^{t+\epsilon} dt_1 \int_t^{t+\epsilon} dt_2 \int_t^{t+\epsilon} dt_3 \int_t^{t+\epsilon} dt_4 \, \langle \eta(t_1)\eta(t_2)\eta(t_3)\eta(t_4) \rangle = \epsilon c_4 + \epsilon^2 c_2^2. \tag{20.263}$$

This is larger than the harmonic estimate (18.420) by a factor $1/\epsilon$, which makes $z_{2,2}(t)$, in general, as large as the leading fluctuation $z_1(t)$ of $x_1(t)$. It can therefore *not* be ignored. The subtraction of the second term $\langle z_{2,2}(t) \rangle^2 = \epsilon^2 c_2^2$ in the variance does not help since that is ignorable.

A similar estimate holds for the higher powers. Take for instance $[\Delta x(t)]^3$:

$$[\Delta x(t)]^3 = \epsilon^3 c_1^3 + 3\epsilon^2 c_1^2 z_1(t) + \epsilon c_1 [z_1(t)]^2 + [z_1(t)]^3, \tag{20.264}$$

and calculate the variance of the strongest fluctuating last term in this: $\langle \{[z_1(t)]^3\}^2 \rangle - \langle [z_1(t)]^3 \rangle^2$. The first contribution is equal to

$$\int_t^{t+\epsilon} dt_1 \int_t^{t+\epsilon} dt_2 \int_t^{t+\epsilon} dt_3 \int_t^{t+\epsilon} dt_4 \int_t^{t+\epsilon} dt_5 \int_t^{t+\epsilon} dt_6 \, \langle \eta(t_1)\eta(t_2)\eta(t_3)\eta(t_4)\eta(t_5)\eta(t_6) \rangle, \tag{20.265}$$

and becomes, after an obvious extension of the expectation values (20.155)–(20.262) to the six-point function:

$$\langle \{[z_1(t)]^3\}^2 \rangle = \epsilon c_6 + \mathcal{O}(\epsilon^2). \tag{20.266}$$

The second contribution $\langle [z_1(t)]^3 \rangle^2$ in (20.264) is equal to $\epsilon^2 c_3^2$ and can be ignored for $\epsilon \to 0$. Thus, due to (20.266), the size of the fluctuations $[z_1(t)]^3$ is of the same order as of the leading one $x_1(t)$ [compare again (18.421)].

This is an important result which is the reason why we cannot derive a direct generalization (20.257) of the Itō formula (18.413) to non-Gaussian fluctuations, but only the weaker formula (20.256) for the expectation value.

For an exponential function $f(x) = e^{Px}$ this implies the relation

$$\frac{d}{dt}\langle e^{Px(t)}\rangle = \left[P\langle\dot{x}(t)\rangle - \bar{H}_{r_x}(iP)\right]\langle e^{Px(t)}\rangle, \tag{20.267}$$

and it is not allowed to drop the expectation values.

A consequence of the weaker Eq. (20.267) for $P = 1$ is that the rate r_S with which the average of the stock price $S(t) = e^{x(t)}$ grows is given by formula (20.1), where the rate r_x is related to r_S by

$$r_S = r_x - \bar{H}(i) = r_x - [H(i) - iH'(0)] = -H_{r_x}(i), \tag{20.268}$$

which replaces the simple Itō relation $r_S = r_x + \sigma^2/2$ in Eq. (20.5). Recall the definition of $\bar{H}(p) \equiv H(p) - H'(0)p$ in Eq. (20.141). The corresponding version of the left-hand part of Eq. (20.4) reads

$$\left\langle\frac{\dot{S}}{S}\right\rangle = \langle\dot{x}(t)\rangle - \bar{H}(i) = \langle\dot{x}(t)\rangle - [H(i) - iH'(0)] = \langle\dot{x}(t)\rangle - r_x - H_{r_x}(i). \tag{20.269}$$

The forward price of a stock must therefore be calculated with the generalization of formula (20.8), in which we assume again $\eta(t)$ to fluctuate around zero rather than r_x:

$$\langle S(t)\rangle = S(0)e^{r_S t} = S(0)\langle e^{r_x t + \int_0^t dt'\,\eta(t')}\rangle = S(0)e^{-H_{r_x}(i)t} = S(0)e^{\{r_x t - [H(i) - iH'(0)]t\}}. \tag{20.270}$$

If $\eta(t)$ fluctuates around r_x this takes the simpler form

$$\langle S(t)\rangle = S(0)e^{r_S t} = S(0)\langle e^{\int_0^t dt'\,\eta(t')}\rangle = S(0)e^{-H_{r_x}(i)t}. \tag{20.271}$$

This result may be viewed as a consequence of the following generalization of the Gaussian Debye-Waller factor (18.425):

$$\left\langle e^{P\int_0^t dt'\,\eta(t')}\right\rangle = e^{-H_{r_x}(iP)t}. \tag{20.272}$$

Note that we may derive the differential equation (20.255) of an arbitrary function $f(x(t))$ from a simple mnemonic rule, expanding sloppily [similar to Eq. (18.426) but restricted to the expectation values]

$$\langle f(x(t+dt))\rangle = \langle f(x(t))\rangle + \langle f'(x(t))\rangle\langle\dot{x}(t)\rangle\,dt + \frac{1}{2}\langle f''(x(t))\rangle\langle\dot{x}^2(t)\rangle\,dt^2$$
$$+ \frac{1}{3!}\langle f^{(3)}(x(t))\rangle\langle\dot{x}^3(t)\rangle dt^3 + \dots, \tag{20.273}$$

and replacing

$$\langle\dot{x}(t)\rangle dt \to c_1 dt, \quad \langle\dot{x}^2(t)\rangle dt^2 \to c_2 dt, \quad \langle\dot{x}^3(t)\rangle dt^3 \to c_3 dt,\dots. \tag{20.274}$$

In contrast to Eq. (18.429), this replacement holds now only on the average.

For the same reason, portfolios containing assets with non-Gaussian fluctuations cannot be made risk-free in the continuum limit $\epsilon \to 0$, as will be seen in Section 20.7.

20.2.2 Discrete Times

The prices of financial assets are recorded in the form of a discrete time series $x(t_n)$ in intervals $\Delta t = t_{n+1} - t_n$, rather than the continuous function $x(t)$. For stocks with a large turnover, or for market indices, the smallest time interval is typically $\Delta t = 1$ minute. We have seen at the end of Section 18.13.3 that this makes Itō's Lemma (18.413) an approximate statement. Without the limit $\Delta t \to 0$, the fluctuations of the higher expansion terms in (20.253) no longer disappear but are merely suppressed by a factor $\sigma\sqrt{\Delta t_n}$. In quiet economic periods this is usually quite small for $\Delta t = 1$ minute, so that the higher-order fluctuations can usually be neglected after all.

While this approximate validity is a disadvantage of the discrete time series over the continuous one, it is an advantage with respect to processes with non-Gaussian noise, at least as long as the financial markets are not in turmoil. Then the non-Gaussian higher-order fluctuations of the expansion terms in (20.253) are suppressed by the same order $\sigma\sqrt{\Delta t}$ as the corrections to Itō's rule in the Gaussian case [recall (18.432)]. As a typical example, take the Boltzmann distribution (20.77). Its cumulants c_n carry a factor T^n [see Eq. (20.80)] where the market temperature T is a small number of the order of a few percent in the natural time units of one minute used in this context [see Fig. (20.11)]. The smallness of T is, of course, a consequence that the minute data do not usually possess large volatility, except near a crash. In the natural units of minutes, the time interval ϵ in the calculations after Eq. (20.257) is unity, so that powers of ϵ can no longer be used for size estimates. This role is now taken over by T. In fact, since c_n is of order T^n, the fluctuations of $z_n(t)$ are of the order T^n. Thus, in non-Gaussian fluctuations of a discrete time series, the smallness of T leads to a suppression of the higher fluctuations after all. Moreover, the suppression is just as good as for time series with Gaussian fluctuations, where the corrections are of the order $(\sigma\sqrt{\Delta t})^{n-1}$ for $z_n(t)$ with $n \geq 2$. These have the size of T^{n-1} for the Boltzmann distribution [recall (20.198)]. A similar estimate holds for all non-Gaussian noise distributions with semi-heavy tails as defined on at the end of Subsection 20.1.1.

For all semi-heavy tails for which the cumulants c_n decrease with power T^n of a small parameter such as the temperature T, we may drop the expectations values of the Itō-like expansion (20.256), and remain with an approximate Itō formula for the fluctuating difference $\Delta f(x(t_n)) \equiv f(x(t_{n+1}))$:

$$\Delta f(x(t_n)) = f'(x(t_n))\Delta x(t_n) - \bar{H}_{r_x}(i\partial_x)f(x(t_n)) + \mathcal{O}(\sigma\sqrt{\Delta t}), \qquad (20.275)$$

which is a direct generalization of the discrete version (18.432) of Itō's Lemma (18.413).

A characteristic property of the Boltzmann distribution (20.77) and others with semi-heavy tails is that their Hamiltonian possesses a power series in p^2 of the type (20.141) with finite cumulants c_n of the order of σ^n, where σ is the volatility of the minute data. This property is violated only by power tails of price distributions which do exist in the data (see Fig. 20.10). These do not go away upon multiple convolution which turns the Boltzmann distribution more and more into a Gaussian as required by the central limit theorem (20.167). These heavy tails are caused by drastic price changes in short times observed in nervous markets near a crash. If these are taken into account, the discrete version (18.432) of Itō's Lemma (18.413) is no longer valid. This will be an obstacle to setting up a risk-free portfolio in Section 20.7.

20.3 Martingales

In financial mathematics, an often-encountered concept is that of a *martingale*. The name stems from a casino strategy in which a gambler doubles his stake each time a bet is lost. A stochastic variable $m(t)$ is called a martingale, if its expectation value is time-independent.

A trivial martingale is provided by any noise variable $m(t) = \eta(t)$.

20.3.1 Gaussian Martingales

For a harmonic noise variable, the exponential $m(t) = e^{\int_0^t dt' \eta(t') - \sigma^2 t/2}$ is a nontrivial martingale, due to Eq. (20.8). For the same reason, a stock price $S(t) = e^{x(t)}$ with $x(t)$ obeying the stochastic differential equation (20.140) can be made a martingale by a time-dependent multiplicative factor associated with the average growth rate r_S, i.e.,

$$e^{-r_S t} S(t) = e^{-r_S t} e^{x(t)} = e^{-r_S t} e^{r_x t + \int_0^t dt' \, \eta(t')} \tag{20.276}$$

is a martingale. The prefactor $e^{-r_S t}$ is referred to as a *discount factor* with the rate r_S. If we calculate the probability distribution (20.170) associated with the stochastic differential equation (20.140), which for harmonic fluctuations of standard deviation σ corresponds to a Hamiltonian

$$H(p) = i r_x p + \frac{\sigma^2}{2} p^2, \tag{20.277}$$

The integral representation (20.170) for the probability distribution,

$$P^{r_x}(x_b t_b | x_a t_a) = \int_{-\infty}^{\infty} \frac{dp}{2\pi} \exp\left[ip(x_b - x_a) - \Delta t \left(i r_x p + \sigma^2 p^2 / 2 \right) \right] \tag{20.278}$$

with $\Delta t = t_b - t_a$ has obviously a time-independent expectation value of $e^{-r_S t} S(t) = e^{-r_S t} e^{x(t)}$. Indeed, the expectation value at the time t_b is given by the integral over x_b

$$e^{-r_S t_b} \int dx_b \, e^{x_b} P^{r_x}(x_b t_b | x_a t_a) = e^{-r_S t_b} \int dx_b \, e^{x_b} \int_{-\infty}^{\infty} \frac{dp}{2\pi} \exp\left[ip(x_b - x_a) - \Delta t \, \sigma^2 p^2 / 2 \right],$$

which yields

$$e^{-r_S t_b} \int_{-\infty}^{\infty} \frac{dp}{2\pi} \delta(p - i) e^{-ipx_a} e^{\Delta t \, \sigma^2 p^2 / 2} = e^{-r_S t} e^{x_a} e^{(r_x - \sigma^2/2)\Delta t} = e^{-r_S t_a} e^{x_a}, \tag{20.279}$$

where we have used the Itō relation $r_S = r_x + \sigma^2/2$ of Eq. (20.5). The result implies that

$$\langle e^{-r_S t_b} S(t_b) \rangle^{r_x} = \langle e^{-r_S t_b} e^{x(t_b)} \rangle^{r_x} = e^{-r_S t_a} S(t_a). \qquad (20.280)$$

Since this holds for all t_b we may drop the subscripts b, thus proving the time independence and thus the martingale nature of $e^{-r_S t} S(t)$.

In mathematical finance, where the expectation value in Eq. (20.280) is written according to the definition (20.189) as $\mathbb{E}[e^{-r_S(t_b - t_a)} e^{x_b}|x_a t_a]$, the martingale property of $e^{-r_S(t_b - t_a)} S(t)$ is expressed as

$$\mathbb{E}[e^{-r_S(t_b - t_a)} S(t_b)|x_a t_a] = S(t_a) \qquad (20.281)$$

or as $\mathbb{E}[e^{-r_S(t_b - t_a)} e^{x_b}|x_a t_a] = e^{x_a}$.

For the martingale $f(x_b, t_b) = e^{-r_S(t_b - t_a)} e^{x_b}$, the expectation values on the right-hand side of the towering formula (20.191) reduces with the help of (20.281) and (20.192) to

$$\mathbb{E}[\mathbb{E}[f(x_b, t_b)|x\,t]|x_a t_a] = \mathbb{E}[\mathbb{E}[f(x, t)|x\,t]|x_a t_a] = \mathbb{E}[f(x, t)|x_a t_a]. \qquad (20.282)$$

Using once more the relations (20.281) and (20.192) to leads to

$$\mathbb{E}[\mathbb{E}[f(x_b, t_b)|x\,t]|x_a t_a] = f(x_a, t_a). \qquad (20.283)$$

Performing the momentum integral in (20.278) gives the explicit distribution

$$P^{r_x}(x_b t_b|x_a t_a) \equiv \frac{1}{\sqrt{2\pi\sigma^2(t_b - t_a)}} \exp\left\{ -\frac{[x_b - x_a - r_x(t_b - t_a)]^2}{2\sigma^2(t_b - t_a)} \right\}. \qquad (20.284)$$

We may incorporate the discount factor $e^{-r_S(t_b - t_a)}$ into the probability distribution (20.284) and define a martingale distribution for the stock price

$$P^{M, r_x}(x_b t_b|x_a t_a) = e^{-r_S(t_b - t_a)} P^{r_x}(x_b t_b|x_a t_a), \qquad (20.285)$$

whose normalization falls off like $e^{-r_S(t_b - t_a)}$. If we define expectation values with respect to $P^{M, r_x}(x_b t_b|x_a t_a)$ by the integral (without normalization)

$$\langle f(x_b) \rangle^{M, r_x} \equiv \int dx_b\, f(x_b) P^{M, r_x}(x_b t_b|x_a t_a), \qquad (20.286)$$

then the stock price itself is a martingale:

$$\langle e^{x_b} \rangle^{M, r_x} = e^{x_a}. \qquad (20.287)$$

Note that there exists an entire family of equivalent martingale distributions

$$P^{(M, r)}(x_b t_b|x_a t_a) = e^{-r\Delta t} \int_{-\infty}^{\infty} \frac{dp}{2\pi} \exp\left[ip(x_b - x_a) - \Delta t\, H_r(p)\right], \qquad (20.288)$$

with an *arbitrary* rate r and $r_x \equiv r + \bar{H}(i)$. Indeed, multiplying this with e^{x_b} and integrating over x_b gives rise to a δ-function $\delta(p - i)$ and produces the same result e^{x_a} for any time difference $\Delta t = t_b - t_a$. Performing the integral yields

$$P^{(M, r)}(x_b t_b|x_a t_a) \equiv \frac{e^{-r(t_b - t_a)}}{\sqrt{2\pi\sigma^2(t_b - t_a)}} \exp\left\{ -\frac{[x_b - x_a - r_x(t_b - t_a)]^2}{2\sigma^2(t_b - t_a)} \right\}. \qquad (20.289)$$

20.3.2 Non-Gaussian Martingale Distributions

For $S(t) = e^{x(t)}$ with an arbitrary non-Gaussian noise $\eta(t)$, there are many ways of constructing distributions which make the stock price a martingale.

Natural Martingale Distribution

The relation (20.268) allows us to write down immediately the simplest martingale distribution. It is given by an obvious generalization of the Gaussian expression (20.276):

$$e^{-r_S t} S(t) = e^{-r_S t} e^{r_x t + \int_0^t dt' \eta(t')} \tag{20.290}$$

where r_S and r_x are related by $r_S = r_x - \bar{H}(i) = -H_{r_x}(i)$. It is easy to prove this. The distribution function associated with the stochastic differential equation (20.140) with non-Gaussian fluctuations is given by

$$P^{r_x}(x_b t_b | x_a t_a) = \int \mathcal{D}\eta \int \mathcal{D}x \exp\left\{ -\int_{t_a}^{t_b} dt\, \tilde{H}_{r_x}(\eta(t)) \right\} \delta[\dot{x} - r_x(t_b - t_a)\eta], \tag{20.291}$$

where $\Delta t \equiv (t_b - t_a)$ and $r_x = r_S + \bar{H}(i)$. As explained in Subsection 20.1.23, the path integral is solved by the Fourier integral [compare (20.170)]

$$P^{r_x}(x_b t_b | x_a t_a) = \int_{-\infty}^{\infty} \frac{dp}{2\pi} \exp\left[ip(x_b - x_a) - \Delta t\, H_{r_x}(p) \right]. \tag{20.292}$$

Using this distribution we can again calculate the time independence of the expectation value (20.280), so that $P^{M, r_x}(x_b t_b | x_a t_a) = e^{-r_S(t_b - t_a)} P^{r_x}(x_b t_b | x_a t_a)$ is a martingale distribution which makes the stock price time-independent, as in Eq. (20.287).

Note that there exists an entire family of equivalent martingale distributions

$$P^{(M, r)}(x_b t_b | x_a t_a) = e^{-r\Delta t} \int_{-\infty}^{\infty} \frac{dp}{2\pi} \exp\left[ip(x_b - x_a) - \Delta t\, H_{r_x}(p) \right], \tag{20.293}$$

with an *arbitrary* rate r and $r_x \equiv r + \bar{H}(i)$. Indeed, multiplying this with e^{x_b} and integrating over x_b gives rise to a δ-function $\delta(p - i)$ and produces the same result e^{x_a} for any time difference $\Delta t = t_b - t_a$.

For non-Gaussian fluctuations there exists an infinite set martingale distributions for the stock price of which we are now going to dicuss the one proposed by Esscher.

Esscher Martingales

In the literature on mathematical finance, much attention is paid to another family of equivalent martingale distributions. It has been used a long time ago to estimate risks of actuaries [67] and introduced more recently into the theory of option prices [68, 69] where it is now of wide use [70]–[76]. This family is constructed as follows. Let $\tilde{D}(z)$ be an arbitrary distribution function with a Fourier transform

$$\tilde{D}(z) = \int_{-\infty}^{\infty} \frac{dp}{2\pi} e^{-H(p)} e^{ipz}, \tag{20.294}$$

and $H(0) = 0$, to guarantee a unit normalization $\int dz \tilde{D}(z) = 1$. We now introduce an *Esscher-transformed* distribution function. It is obtained by slightly tilting the initial distribution $\tilde{D}(z)$, multiplying it with an asymmetric exponential factor $e^{\theta z}$:

$$D^\theta(z) \equiv e^{H(i\theta)} \, e^{\theta z} \tilde{D}(z). \tag{20.295}$$

The constant prefactor $e^{H(i\theta)}$ is necessary to conserve the total probability. This distribution can be written as a Fourier transform

$$D^\theta(z) = \int_{-\infty}^{\infty} \frac{dp}{2\pi} e^{-H^\theta(p)} e^{ipz}, \tag{20.296}$$

with the Esscher-transformed Hamiltonian

$$H^\theta(p) \equiv H(p + i\theta) - H(i\theta). \tag{20.297}$$

Since $H^\theta(0) = 0$, the transformed distribution is properly normalized: $\int dz D^\theta(z) = 1$. We now define the Esscher-transformed expectation value

$$\langle F(z) \rangle^\theta \equiv \int dz D^\theta(z) F(z). \tag{20.298}$$

It is related to the original expectation value by

$$\langle F(z) \rangle^\theta \equiv e^{H(i\theta)} \langle e^{\theta z} F(z) \rangle. \tag{20.299}$$

For the specific function $F(z) = e^z$, Eq. (20.299) becomes

$$\langle e^z \rangle^\theta = e^{-H^\theta(i)} \equiv e^{H^\theta(i\theta)} \langle e^{(\theta+1)z} \rangle = e^{H^\theta(i\theta) - H^\theta(i\theta+i)}. \tag{20.300}$$

Applying the transformation (20.296) to each time slice in the general path integral (20.144), we obtain the *Esscher-transformed* path integral

$$P^\theta(x_b t_b | x_a t_a) = e^{-r_S \Delta t} e^{H_{r_x}(i\theta)\Delta t} \int \mathcal{D}\eta \int \mathcal{D}x \exp \left\{ \int_{t_a}^{t_b} dt \left[\theta\eta(t) - \tilde{H}_{r_x}(\eta(t)) \right] \right\} \delta[\dot{x} - \eta], \tag{20.301}$$

where $\Delta t \equiv t_b - t_a$ is the time interval. This leads to the Fourier integral [compare (20.293)]

$$P^\theta(x_b t_b | x_a t_a) = e^{-r_S \Delta t} e^{H_{r_x}(i\theta)\Delta t} \int_{-\infty}^{\infty} \frac{dp}{2\pi} \exp \left[ip(x_b - x_a) - \Delta t \, H_{r_x}(p + i\theta) \right]. \tag{20.302}$$

Let us denote the expectation values calculated with this probability by $\langle \ldots \rangle^\theta$. Then we find for $S(t) = e^{x(t)}$ the time dependence

$$\langle S(t) \rangle^\theta = e^{-H_{r_x}^\theta(i)t}. \tag{20.303}$$

This equation shows that the exponential of a stochastic variable $x(t)$ can be made a martingale with respect to any Esscher-transformed distribution if we remove the exponential growth factor $\exp(r^\theta \Delta t)$ with

$$r^\theta \equiv -H_{r_x}^\theta(i) = -H_{r_x}(i + i\theta) + H_{r_x}(i\theta). \tag{20.304}$$

Thus, a family of equivalent martingale distributions for the stock price $S(t) = e^{x(t)}$ is

$$P^{M\theta}(x_b t_b | x_a t_a) \equiv e^{-r^\theta t} P^\theta(x_b t_b | x_a t_a), \tag{20.305}$$

for any choice of the parameter θ.

For a harmonic distribution function (20.284), the Esscher martingales and the previous ones are equivalent. Indeed, starting from (20.284) in which $r_S = r_x + \sigma^2/2$, the Esscher transform leads, after a quadratic completion, to the family of natural martingales (20.284) with the rate parameter $r = r_x + \theta\sigma^2$.

Other Non-Gaussian Martingales

Many other non-Gaussian martingales have been discussed in the literature. Mathematicians have invented various sophisticated criteria under which one would be preferable over the others for calculating financial risks. Davis, for instance, has introduced a so-called *utility function* [77] which is supposed to select optimal martingales for different purposes. There exists also a so-called minimal martingale [83], but the mathematical setup in these discussions is hard to understand.

For the upcoming development of a theory of option pricing, only the initial *natural martingale* will be relevant.

20.4   Origin of Semi-Heavy Tails

We have indicated at the end of Subsection 20.1.1 that semi-heavy tails of the type used in the previous discussions may be viewed as a phenomenological description of a nontrivial Gaussian process with fluctuating volatility. This has been confirmed in Subsection 20.1.22 where we have expressed a number of non-Gaussian distributions as a superposition of Gaussian distributions. We have further seen that the weight functions $w_t(v)$ for the superpositions can be Laplace-transformed to find a Hamiltonian $H_v(p_v)$ from which we can derive which governs the noise distribution a stochastic differential equation for the volatility whose noise distribution is which governed by $H_v(ip_v)$.

There exist a simple solvable model due to Heston [78] which shows this explicitly.[8]

---

[8]The discussion in this section follows a paper by Drăgulescu and Yakovenko [79].

20.4.1   Pair of Stochastic Differential Equations

Starting point is a coupled pair of ordinary stochastic differential equations with Gaussian noise, one for the fluctuating logarithm $x(t)$ of the stock price, and one for the fluctuating time-dependent variance $\sigma^2(t)$. Since this will appear frequently in the upcoming discussion, we shall name it $v(t)$. For simplicity, we remove the growth rate of the share price. Replacing σ^2 in the stochastic differential equation (20.4) by the time-dependent variance $v(t)$ we obtain

$$\dot{x}(t) = -\frac{v(t)}{2} + \sqrt{v(t)}\eta(t), \qquad (20.306)$$

where the noise variable $\eta(t)$ has zero average and *unit* volatility

$$\langle \eta(t) \rangle = 0, \qquad \langle \eta(t)\eta(t') \rangle = \delta(t - t'). \qquad (20.307)$$

The variance is assumed to satisfy an equation with another noise variable $\eta_v(t)$ [7]

$$\dot{v}(t) = -\gamma[v(t) - \bar{v}] + \varepsilon\sqrt{v(t)}\eta_v(t). \qquad (20.308)$$

The time in $\sqrt{v(t)}$ is supposed to lie slightly *before* the time in the noise $\eta_v(t)$. The parameter ε determines the volatility of the variance. The parameter $\bar{v}$ will become the average of the variance at large times. It is proportional to the variance σ^2 of the Gaussian distribution to which the distribution converges in the long-time limit by the central limit theorem. The precise relation will be given in Eq. (20.363).

In general, the noise variables $\eta(t)$ and $\eta_v(t)$ are expected to exhibit correlations which are accounted for by introducing an independent noise variable $\eta'(t)$ and setting

$$\eta_v(t) \equiv \rho\,\eta(t) + \sqrt{1 - \rho^2}\eta'(t). \qquad (20.309)$$

The pair correlation functions are then

$$\langle \eta(t)\eta(t') \rangle = \delta(t - t'), \quad \langle \eta(t)\eta_v(t') \rangle = \rho\delta(t - t'), \quad \langle \eta_v(t)\eta_v(t') \rangle = \delta(t - t'). \quad (20.310)$$

20.4.2   Fokker-Planck Equation

If $\boldsymbol{\eta}(t)$ denotes the pair of noise variables $(\eta(t), \eta_v(t))$, the probability distribution analogous to (18.324) for paths $x(t)$, $v(t)$ starting out at x_a, v_a and arriving at x, v is

$$P_{\boldsymbol{\eta}}(x\,v\,t\,|x_a v_a t_a) = \delta(x_\eta(t) - x)\delta(v_{\eta_v}(t) - v). \qquad (20.311)$$

The time evolution of this is calculated using the differentiation rule (18.381) as

$$\partial_t P_{\boldsymbol{\eta}}(x\,v\,t\,|x_a v_a t_a) = -\left[\partial_x \dot{x}_\eta(t) + \partial_v \dot{v}_{\eta_v}(t)\right] P_{\boldsymbol{\eta}}(x\,v\,t\,|x_a v_a t_a), \qquad (20.312)$$

or more explicitly

$$\partial_t P_\eta(x\,v\,t\,|x_a v_a t_a) = \left\{ \partial_x \left[\frac{v(t)}{2} - \sqrt{v(t)}\eta(t) \right] \right.$$
$$\left. + \partial_v \left[\gamma\left(v(t) - \bar{v}\right) - \varepsilon\sqrt{v(t)}\eta_v(t) \right] \right\} P_\eta(x\,v\,t\,|x_a v_a t_a). \quad (20.313)$$

We now take the noise average (18.325) with the distribution

$$P[\eta] = \exp\left\{ -\frac{1}{2}\int dt \left[\eta^2(t) + \eta'^2(t) \right] \right\} = \exp\left\{ -\frac{1}{1-\rho^2}\int dt \left[\eta^2(t) + \eta_v^2(t) - 2\rho\,\eta\eta_v \right] \right\}. \quad (20.314)$$

Applying the rules (18.385) and (18.386), we may replace $\eta(t)$ and $\eta_v(t)$ on the right-hand side of (20.313):

$$\eta(t) \quad \to \quad -\delta/\delta\eta(t) - \rho\,\delta/\delta\eta_v(t) \quad (20.315)$$
$$\eta_v(t) \quad \to \quad -\rho\,\delta/\delta\eta(t) - \delta/\delta\eta_v(t). \quad (20.316)$$

The functional derivatives are evaluated by analogy with (18.388) as

$$\frac{\delta}{\delta\eta(t')}\delta(x_\eta(t) - x)\delta(v_{\eta_v}(t) - v) = -\left[\frac{\delta x_\eta(t)}{\delta\eta(t')}\partial_x + \frac{\delta v_{\eta_v}(t)}{\delta\eta(t')}\partial_v \right]\delta(x_\eta(t) - x)\delta(v_{\eta_v}(t) - v), \quad (20.317)$$

with a similar equation for $\delta/\delta\eta_v(t)$. Under the above-made assumption that $\sqrt{v(t)}$ lies before $\eta_v(t)$, we find the evolution equation

$$\partial_t P(x\,v\,t\,|x_a v_a t_a) = -\hat{H}\,P(x\,v\,t\,|x_a v_a t_a), \quad (20.318)$$

where $\hat{H}$ is the Hamiltonian operator

$$\hat{H} = -\frac{1}{2}\partial_x^2\,v - \frac{1}{2}\partial_x\,v - \frac{\varepsilon^2}{2}\partial_v^2\,v - \gamma\partial_v(v - \bar{v}) - \rho\varepsilon\partial_x\partial_v\,v. \quad (20.319)$$

The probability distribution can thus be written as a path integral

$$P(x_b v_b t_b|x_a v_a t_a) = \int \mathcal{D}x \int \frac{\mathcal{D}p}{2\pi} \int \mathcal{D}v \int \frac{\mathcal{D}p_v}{2\pi}$$
$$\times \exp\left\{ \int_{t_a}^{t_b} dt\,[i(p\dot{x} + p_v\dot{v}) - H(p, p_v, v)] \right\}, \quad (20.320)$$

with the Hamiltonian

$$H(p, p_v, v) = \frac{p^2}{2}v - i\frac{1}{2}pv + \varepsilon^2\frac{p_v^2}{2}v - i\gamma p_v(v - \bar{v}) + \rho\varepsilon\,pp_v\,v - \frac{3\varepsilon^2}{4}ip_v - \frac{\gamma}{2} - \frac{i}{2}\rho\varepsilon p. \quad (20.321)$$

The last two terms arise from the fact that the order of the operators in $\hat{H}$ obtained from the path integral is always symmetric in v and p_v, such that the order in Eq. (20.319) is reached only after the replacement of $p_v v \to (1/2)\{\hat{p}_v, v\} = \hat{p}_v v + i/2$

in the fourth and fifth terms of (20.321). Recall the discussions in Sections 10.5, 11.3, and Subsection 18.9.2.

The third-last term in (20.321) is present to ensure the appearance of the operator $\partial_v^2 v$ in (20.319) with no extra ∂_v due to operator ordering. The term $p_v^2 v$ in (20.321) can be understood as a kinetic term $g_{\mu\nu} p^\mu p^\nu$ in a one-dimensional Riemannian space with a metric $g^{\mu\nu} = \delta^{\mu\nu}/v$. According to Subsection 11.1.1, this turns into the Laplace-Beltrami operator $\Delta = \sqrt{v}\partial_v^2 \sqrt{v}\partial_v = v\partial_v^2 + \frac{1}{2}\partial_v = \partial_v^2 v - \frac{3}{2}\partial_v$. The extra ∂_v is canceled by the third-last term in (20.319).

The stochastic differential equation (20.308) can, in principle, lead to negative variances. Since this would be unphysical, we must guarantee that this does not happen. The condition for this is that the fluctuation width of the variance is sufficiently small, satisfying

$$\frac{\varepsilon^2}{2\gamma\bar{v}} \leq 1. \tag{20.322}$$

Then an initially positive $v(t)$ will always stay positive. Indeed, consider the partial differential equation (20.318) for $v = 0$:

$$(\partial_t + \alpha\partial_v) P(x\,v\,t\,|x_a v_a t_a) = (\gamma + \rho\varepsilon\partial_x) P(x\,v\,t\,|x_a v_a t_a), \tag{20.323}$$

where $\alpha \equiv \gamma\bar{v} - \varepsilon^2/2$. It is solved by some function

$$P(x\,v\,t\,|x_a v_a t_a) = f(v - \alpha t, x), \tag{20.324}$$

which shows that a nonvanishing f for positive v can never propagate to negative-v.[9]

The time dependence of the variance is given by the integral

$$P(v\,t\,|v_a t_a) \equiv \int dx\, P(x\,v\,t\,|x_a v_a t_a). \tag{20.325}$$

It satisfies the equation

$$\frac{\partial}{\partial t} P(v\,t\,|v_a t_a) = \left[\frac{\varepsilon^2}{2}\partial_v^2 v + \gamma\partial_v(v - \bar{v})\right] P(v\,t\,|v_a t_a), \tag{20.326}$$

which follows from (20.318) by integration over x. After a long time, the solution becomes stationary and reads

$$P^*(v) = \frac{\mu^\nu}{\Gamma(\nu)} v^{\nu-1} e^{-\mu v}, \quad \text{where} \quad \mu \equiv \frac{2\gamma}{\varepsilon^2}, \quad \nu \equiv \mu\bar{v}. \tag{20.327}$$

This is the Gamma distribution found in Fig. 20.3. Hence the pair of stochastic differential equations (20.306) and (20.308) produces precisely the type of volatility fluctuations observed in the previous financial data.

The maximum of $P^*(v)$ lies slightly below $\bar{v}$ at $v_{\max} = (\nu - 1)/\mu = \bar{v} - \varepsilon^2/2\gamma$ [recall (20.72)]. If we define the w of $P^*(v)$ by the curvature at the maximum, we

---

[9]See also p. 67 in the textbook [8].

find $w = \sqrt{v_{\max}\varepsilon^2/2\gamma}$. The shape of $P^*(v)$ is characterized by the dimensionless ratio

$$\frac{v_{\max}}{w} = \sqrt{\frac{2\gamma\bar{v}}{\varepsilon^2} - 1}. \qquad (20.328)$$

The size of the fluctuations of the variance are restricted by the condition (20.322) which guarantees positive $v(t)$ for all times. The distribution (20.327) is plotted in Fig. 20.21. As a cross check, we go to the limit $\varepsilon^2/2\gamma\bar{v} \to 0$ where the fluctuations of the variance are frozen out. The ratio $v_{\max}/w$ diverges and the distribution $P^*(v)$ tends to $\delta(v - \bar{v})$, as expected.

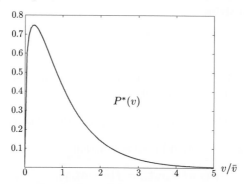

Figure 20.21 Stationary distribution (20.327) of variances for parameters of fits to the Dow-Jones data in Fig. 20.22 listed in Table 20.1. Compare with Fig. 20.3.

20.4.3 Solution of Fokker-Planck Equation

Since the Hamiltonian operator (20.319) does not depend explicitly on x, we can take advantage of translational invariance and write $P(x\,v\,t\,|x_av_at_a)$ as a Fourier integral

$$P(x\,v\,t\,|x_av_at_a) = \int \frac{dp}{2\pi} e^{ip(x-x_a)}\, \bar{P}_p(v\,t\,|v_at_a). \qquad (20.329)$$

The fixed-momentum probability satisfies the Fokker-Planck equation

$$\partial_t\bar{P}_p(v\,t\,|v_at_a) = \left[\gamma\frac{\partial}{\partial v}(v-\bar{v}) - \frac{p^2-ip}{2}v - i\rho\varepsilon p\frac{\partial}{\partial v}v - \frac{\varepsilon^2}{2}\frac{\partial^2}{\partial v^2}v\right]\bar{P}_p(v\,t\,|v_at_a). \qquad (20.330)$$

This partial differential equation is of second-order. The variable v occurs only linearly. It is therefore convenient to go to Fourier space also in v and write

$$\bar{P}_p(v\,t\,|v_at_a) = \int \frac{dp_v}{2\pi} e^{ip_vv}\, \tilde{P}_p(p_v\,t\,|v_at_a), \qquad (20.331)$$

which solves the first-order partial differential equation

$$\left[\frac{\partial}{\partial t} + \left(\Gamma p_v + \frac{i\varepsilon^2}{2}p_v^2 + \frac{ip^2+p}{2}\right)\frac{\partial}{\partial p_v}\right]\tilde{P}_p(p_v\,t\,|v_a t_a) = -i\gamma\bar{v}\,p_v\,\tilde{P}_p(p_v\,t\,|v_a t_a), \quad (20.332)$$

where we have used the abbreviation

$$\Gamma(p) \equiv \gamma + i\rho\varepsilon p. \quad\quad (20.333)$$

Since the initial condition for $\bar{P}_p(v\,t\,|v_a t_a)$ is $\delta(v-v_a)$, the Fourier transform has the initial condition

$$\tilde{P}_p(p_v\,t_a|v_a t_a) = e^{-ip_v v_a}. \quad\quad (20.334)$$

The solution of the first-order partial differential equation (20.332) is found by the method of characteristics [9]:

$$\tilde{P}_p(p_v\,t|v_a t_a) = \exp\left[-i\tilde{p}_v(t_a)v_a - i\gamma\bar{v}\int_{t_a}^t dt'\,\tilde{p}_v(t')\right], \quad\quad (20.335)$$

where the function $\tilde{p}_v(t)$ is the solution of the characteristic differential equation

$$\frac{d\tilde{p}_v(t)}{dt} = \Gamma(p)\tilde{p}_v(t) + \frac{i\varepsilon^2}{2}\tilde{p}_v^2(t) + \frac{i}{2}(p^2 - ip), \quad\quad (20.336)$$

with the boundary condition $\tilde{p}_v(t_b) = p_v$. The differential equation is of the Riccati type with constant coefficients [10], and its solution is

$$\tilde{p}_v(t) = -i\frac{2\Omega(p)}{\varepsilon^2}\frac{1}{\zeta(p,p_v)e^{\Omega(p)(t_b-t)} - 1} + i\frac{\Gamma(p) - \Omega(p)}{\varepsilon^2}, \quad\quad (20.337)$$

where we introduced the complex frequency

$$\Omega(p) = \sqrt{\Gamma^2(p) + \varepsilon^2(p^2 - ip)}, \quad\quad (20.338)$$

and the coefficient are

$$\zeta(p,p_v) = 1 - i\frac{2\Omega(p)}{\varepsilon^2 p_v - i[\Gamma(p) - \Omega(p)]}. \quad\quad (20.339)$$

Taking the Fourier transform we obtain the solution of the original Fokker-Planck equation (20.318):

$$P(x\,v\,t\,|x_a v_a t_a) = \iint_{-\infty}^{+\infty}\frac{dp\,dp_v}{2\pi\,2\pi}e^{ipx+ip_v v}$$
$$\times \exp\left\{-i\tilde{p}_v(t_a)v_a + \frac{\gamma\bar{v}[\Gamma(p)-\Omega(p)]}{\varepsilon^2}\Delta t - \frac{2\gamma\bar{v}}{\varepsilon^2}\ln\frac{\zeta(p,p_v)-e^{-\Omega(p)\Delta t}}{\zeta(p,p_v)-1}\right\}, \quad (20.340)$$

where $\Delta t \equiv t - t_a$ is the time interval.

20.4.4   Pure x-Distribution

We now show that upon averaging over the variance we obtain a non-Gaussian distribution of x with semi-heavy tails of the type observed in actual financial data. Thus we go over to the probability distribution

$$P(x\,t\,|x_a v_a t_a) \equiv \int_{-\infty}^{+\infty} dv\, P(x\,v\,t\,|x_a v_a t_a), \qquad (20.341)$$

where the final variable v is integrated out. The integration of (20.340) over v generates the delta-function $\delta(p_v)$ which enforces $p_v = 0$. Thus we obtain

$$P(x\,t\,|x_a v_a t_a) = \int_{-\infty}^{+\infty} \frac{dp}{2\pi}\, e^{ipx - \frac{p^2 - ip}{\Gamma + \Omega \coth (\Omega \Delta t/2)} v_a + \frac{\gamma \Gamma \bar{v}}{\varepsilon^2} \Delta t - \frac{2\gamma \bar{v}}{\varepsilon^2} \ln \left(\cosh \frac{\Omega \Delta t}{2} + \frac{\Gamma}{\Omega} \sinh \frac{\Omega \Delta t}{2} \right)}, \qquad (20.342)$$

where we have omitted the arguments of $\Gamma(p)$ and $\Omega(p)$, for brevity. As a cross check of this result we consider the special case $\varepsilon = 0$ where the time evolution of the variance is deterministic:

$$v(t) = \bar{v} + (v_a - \bar{v})e^{-\gamma \Delta t}. \qquad (20.343)$$

Performing the integral over p in Eq. (20.342) we obtain

$$P(x\,t\,|x_a t_a v_a) = \frac{1}{\sqrt{2\pi(t - t_a)\bar{v}(t)}} \exp \left\{ -\frac{[x + \bar{v}(t)(t - t_a)/2]^2}{2\bar{v}(t)} \right\}, \qquad (20.344)$$

where $\bar{v}(t)$ denotes the time-averaged variance

$$\bar{v}(t) \equiv \frac{1}{\Delta t} \int_{t_a}^{t} dt'\, v(t'). \qquad (20.345)$$

The distribution becomes Gaussian in this limit. The same result could, of course, have been obtained directly from the stochastic differential equation (20.306).

The result (20.342) cannot be compared directly with financial time series data, because it depends on the unknown initial variance v_a. Let us assume that v_a has initially a stationary probability distribution $P^*(v_a)$ of Eq. (20.327).[10] Then we evaluate the probability distribution $P(x\,t\,|x_a t_a)$ by averaging (20.342) over v_a with the weight $P^*(v_a)$:

$$P(x\,t\,|x_a t_a) = \int_0^{\infty} dv_a\, P^*(v_a)\, P(x\,t\,|x_a t_a v_a). \qquad (20.346)$$

The final result has the Fourier representation

$$P(x\,t\,|x_a t_a) = \int_{-\infty}^{+\infty} \frac{dp}{2\pi}\, e^{ip\Delta x - \Delta t H(p, \Delta t)}, \qquad (20.347)$$

---

[10]For a determination of the empirical probability distribution of volatilities for the S&P 500 index see Ref. [2].

where $\Delta x \equiv x - x_a$ and $H(p, \Delta t)$ is the Hamiltonian

$$H(p,\Delta t) = -\frac{\gamma\Gamma(p)\bar{v}}{\varepsilon^2} + \frac{2\gamma\bar{v}}{\varepsilon^2\Delta t}\ln\left[\cosh\frac{\Omega(p)\Delta t}{2} + \frac{\Omega^2(p)-\Gamma^2(p)+2\gamma\Gamma(p)}{2\gamma\Omega(p)}\sinh\frac{\Omega(p)\Delta t}{2}\right].$$
(20.348)

In the absence of correlations between the fluctuations of stock price and variance, i.e. for $\rho = 0$, the second term in the right-hand side of Eq. (20.348) vanishes. The simplified result is discussed in Appendix 20B.

The general Hamiltonian (20.348) vanishes for $p = 0$. This guarantees the unit normalization of the distribution (20.347) at all times: $\int dx\, P(x\,t\,|x_a t_a) = 1$. The first expansion coefficient of $H(p, \Delta t)$ in powers of p is [recall the definition in Eq. (20.54)]

$$c_1(\Delta t) = -\frac{\bar{v}}{2} - \frac{\rho}{\epsilon\Delta t}\left(1 - e^{-\gamma\Delta t}\right).$$
(20.349)

We have added the argument Δt to emphasize the time dependence. By adding $ic_1(\Delta t)p$ to $H(p, \Delta t)$, we obtain the Hamiltonian $\bar{H}(p, \Delta t)$ which starts out quadratically in p in accordance with our general definition in Eq. (20.65):

$$\bar{H}(p,\Delta t) \equiv H(p,\Delta t) - ic_1(\Delta t)p = H(p,\Delta t) + i\left[\frac{\bar{v}}{2} + \frac{\rho}{\epsilon\Delta t}\left(1 - e^{-\gamma\Delta t}\right)\right]p.$$
(20.350)

The distribution $P(x\,t\,|x_a t_a)$ is real since $\mathrm{Re}\,H$ is an even function and $\mathrm{Im}\,H$ is an odd function of p.

In general, the integral in (20.347) must be calculated numerically. The interesting limit regimes $\Delta t \gg 1$ and $x \gg 1$, however, can be understood analytically. These will be treated in Subsections 20.4.5 and 20.4.6. In Fig. 20.22, the calculated distributions $P(x\,t\,|x_a t_a)$ with increasing time intervals are displayed as solid curves and compared with the corresponding Dow-Jones data indicated by dots. The technical details of the data analysis are explained in Appendix 20C. The figure demonstrates that with a fixed set of the five parameters γ, $\bar{v}$, ε, μ, and ρ, the distribution (20.347) with (20.348) reproduce extremely well the data for *all* time scales Δt. In the logarithmic plot the far tails of the distributions fall off linearly.

20.4.5   Long-Time Behavior

According to Eq. (20.308), the variance approaches the equilibrium value $\bar{v}$ within the characteristic relaxation time $1/\gamma$. Let us study the limit in which the time interval Δt is much longer than the relaxation time: $\gamma\Delta t \gg 1$. According to Eqs. (20.333) and (20.338), this condition also implies that $\Omega(p)\Delta t \gg 1$. Then Eq. (20.348) reduces to

$$H(p,\Delta t) \approx \frac{\gamma\bar{v}}{\varepsilon^2}[\Omega(p) - \Gamma(p)].$$
(20.351)

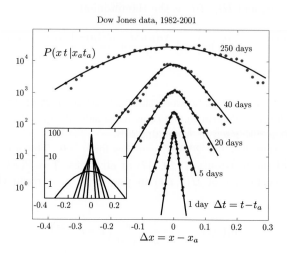

Figure 20.22 Probability distribution of logarithm of stock price for different time scales (from Ref. [79]). For a better comparison of the shapes, the data points for increasing Δt are shifted upwards by a factor 10 each. The unshifted positions are shown in the insert.

The Fourier integral (20.347) can easily be performed after changing the variable of integration to

$$p = C \tilde{p} + i p_0, \tag{20.352}$$

where

$$C \equiv \frac{\omega_0}{\varepsilon \sqrt{1 - \rho^2}}, \quad \omega_0 \equiv \sqrt{\gamma^2 + \varepsilon^2 (1 - \rho^2) p_0^2}, \quad p_0 \equiv \frac{\varepsilon - 2\rho\gamma}{2\varepsilon(1 - \rho^2)}. \tag{20.353}$$

Then the integral (20.347) takes the form

$$P(x\,t\,|x_a t_a) = \frac{C}{\pi} e^{-p_0 \Delta x + \Lambda \Delta t} \int_0^\infty d\tilde{p} \, \cos(A\tilde{p}) e^{-B\sqrt{1+\tilde{p}^2}}, \tag{20.354}$$

where

$$A = C \left(\Delta x + \rho \frac{\gamma \bar{v}}{\varepsilon} \Delta t \right), \quad B = \frac{\gamma \bar{v} \omega_0}{\varepsilon^2} \Delta t, \tag{20.355}$$

and

$$\Lambda = \frac{\gamma \bar{v}}{2\varepsilon^2} \frac{2\gamma - \rho\varepsilon}{1 - \rho^2}. \tag{20.356}$$

The integral in (20.354) is equal to[11] $BK_1(\sqrt{A^2 + B^2})/\sqrt{A^2 + B^2}$, where $K_1(y)$ is a modified Bessel function, such that the probability distribution (20.347) for $\gamma \Delta t \gg 1$ can be represented in the scaling form

$$P(x\,t\,|x_a t_a) = N(\Delta t)\, e^{-p_0 \Delta x} F^*(y), \quad F^*(y) = K_1(y)/y, \tag{20.357}$$

---

[11]See I.S. Gradshteyn and I.M. Ryzhik, *op. cit.*, Formula 3.914.

with the argument

$$y \equiv \sqrt{A^2 + B^2} = \frac{\omega_0}{\varepsilon} \sqrt{\frac{(\Delta x + \rho\gamma\bar{v}\Delta t/\varepsilon)^2}{1 - \rho^2} + \left(\frac{\gamma\bar{v}\Delta t}{\varepsilon}\right)^2}, \quad (20.358)$$

and the time-dependent normalization factor

$$N(\Delta t) = \frac{\omega_0^2 \gamma\bar{v}}{\pi\varepsilon^3\sqrt{1 - \rho^2}} \Delta t\, e^{\Lambda\Delta t}. \quad (20.359)$$

Thus, up to the factors $N(\Delta t)$ and $e^{-p_0\Delta x}$, the dependence of $P(x\,t\,|x_a t_a)$ on the arguments Δx and Δt is given by the function $F^*(y)$ of the single scaling argument y. When plotted as a function of y, the data for different Δx and Δt should collapse on the single universal curve $F^*(y)$. This does indeed happen as illustrated in Fig. 20.23, where the Dow-Jones data for different time differences Δt follows the curve $F^*(y)$ for seven decades.

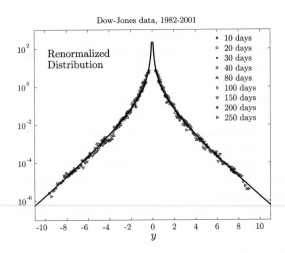

Dow-Jones data, 1982-2001

Figure 20.23 Renormalized distribution function $P(x\,t\,|x_a t_a)e^{p_0\Delta x}/N(\Delta t)$ plotted as a function of the scaling argument y defined in Eq. (20.358). The solid curve shows the universal function $F^*(y) = K_1(y)/y$ (from Ref. [79]).

In the limit $y \gg 1$, we can use the asymptotic expression $K_1(y) \approx e^{-y}\sqrt{\pi/2y}$ and find the asymptotic behavior

$$\ln \frac{P(x\,t\,|x_a t_a)}{N(\Delta t)} \approx -p_0\Delta x - y. \quad (20.360)$$

Let us examine this expression for large and small $|\Delta x|$. In the first case $|\Delta x| \gg \gamma\bar{v}\Delta t/\varepsilon$, and Eq. (20.358) shows that $y \approx \omega_0|\Delta x|/\varepsilon\sqrt{1 - \rho^2}$, so that (20.360) becomes

$$\ln \frac{P(x\,t\,|x_a t_a)}{N(\Delta t)} \approx -p_0\Delta x - c|\Delta x|. \quad (20.361)$$

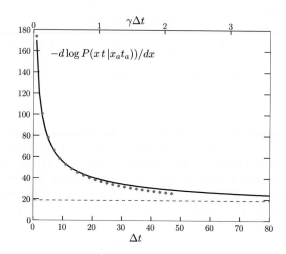

Figure 20.24 Solid curve showing slope $-d\log P(x\,t\,|x_a t_a)/dx$ of exponential tail of distribution as a function of time. The dots indicate the analytic short-time approximation (20.369) to the curve (from [79]).

This shows that the probability distribution $P(x\,t\,|x_a t_a)$ has exponential tails (20.361) for large $|\Delta x|$. Note that in the present limit $\gamma\Delta t \gg 1$ the slopes of the exponential tails are independent of Δt. The presence of p_0 causes the slopes for positive and negative Δx to be different, so that the distribution $P(x\,t\,|x_a t_a)$ is not up-down symmetric with respect to price changes. From the definition of p_0 in Eq. (20.353) we see that this asymmetry increases for a negative correlation $\rho < 0$ between stock price and variance.

In the second case $|\Delta x| \ll \gamma\bar{v}\Delta t/\varepsilon$, by Taylor-expanding y in (20.358) near its minimum and substituting the result into (20.360), we obtain

$$\ln\frac{P(x\,t\,|x_a t_a)}{N'(\Delta t)} \approx -p_0\Delta x - \frac{\omega_0(\Delta x + \rho\gamma\bar{v}\Delta t/\varepsilon)^2}{2(1-\rho^2)\gamma\bar{v}\Delta t}, \tag{20.362}$$

where $N'(\Delta t) = N(\Delta t)\exp(-\omega_0\gamma\bar{v}\Delta t/\varepsilon^2)$. Thus, for small $|\Delta x|$, the probability distribution $P(x\,t\,|x_a t_a)$ is a Gaussian, whose width

$$\sigma^2 = \frac{(1-\rho^2)\gamma\bar{v}}{\omega_0}\Delta t \tag{20.363}$$

grows linearly with Δt. The maximum of $P(x\,t\,|x_a t_a)$ lies at

$$\Delta x_m(t) = \Delta r_S\Delta t, \quad \text{with} \quad \Delta r_S \equiv -\frac{\gamma\bar{v}}{2\omega_0}\left[1 + 2\frac{\rho(\omega_0 - \gamma)}{\varepsilon}\right]. \tag{20.364}$$

The position moves with a constant rate Δr_S which adds to the average growth rate r_S of $S(t)$ removed at the beginning of the discussion. The true final growth rate of $S(t)$ is $\bar{r}_S = r_S + \Delta r_S$.

The above discussion explains the property of the data in Fig. 20.22 that the logarithmic plots of $P(x\,t\,|x_a t_a)$ are linear in the tails and quadratic near the peaks with the parameters specified in Eqs. (20.361) and (20.362).

As time progresses, the distribution broadens in accordance with the scaling form (20.357) and (20.358). In the limit $\Delta t \to \infty$, the asymptotic expression (20.362) is valid for all Δx and the distribution becomes a pure Gaussian, as required by the central limit theorem [22].

It is interesting to quantify the fraction $f(\Delta t)$ of the total probability contained in the Gaussian portion of the curve. This fraction is plotted in Fig. 20.25. The precise way of defining and calculating the fraction $f(\Delta t)$ is explained in Appendix 20B. The inset in Fig. 20.25 illustrates that the time dependence of the probability density at the maximum x_m approaches $\Delta t^{-1/2}$ for large time, a characteristic property of evolution of Gaussian distributions.

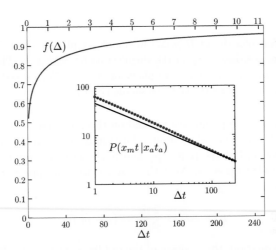

Figure 20.25   Fraction $f(\Delta t)$ of total probability contained in Gaussian part of $P(x\,t\,|x_a t_a)$ as function of time interval Δt. The inset shows the time dependence of the probability density at maximum $P(x_m t\,|x_a t_a)$ (points), compared with the falloff $\propto \Delta t^{-1/2}$ of a Gaussian distribution (solid curve).

20.4.6   Tail Behavior for all Times

For large $|\Delta x|$, the integrand in (20.347) oscillates rapidly as a function of p, so that the integral can be evaluated in the saddle point approximation of Section 4.2. As in the evaluation of the integral (17.9) we shift the contour of integration in the complex p-plane until it passes through the leading saddle point of the exponent $ip\Delta x - \Delta t H(p, \Delta t)$. To determine its position we note that the function $H(p, \Delta t)$ in Eq. (20.348) has singularities in the complex p-plane, where the argument of the

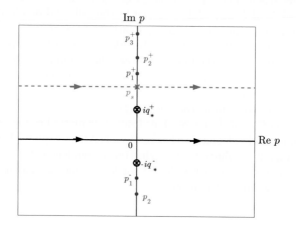

Figure 20.26 Singularities of $H(p, \Delta t)$ in complex p-plane (dots). Circled crosses indicate the limiting positions $\pm iq_*^{\pm}$ of the singularities for $\gamma \Delta t \gg 1$. The cross shows the saddle point p_s located in the upper half-plane for $\Delta x > 0$. The dashed line is the shifted contour of integration to pass through the saddle point p_s.

logarithm vanish. These points are located on the imaginary p-axis and are shown by dots in Fig. 20.26. The relevant singularities are those lying closest to the real axis. They are located at the points p_1^+ and p_1^-, where the argument of the second logarithm in (20.348) vanishes. Near these zeros, we can approximate $H(p, \Delta t)$ by the singular term:

$$\bar{H}(p, \Delta t) \approx \frac{2\gamma\bar{v}}{\varepsilon^2 \Delta t} \log(p - p_1^{\pm}). \tag{20.365}$$

With this approximation, the position of the saddle point $p_s = p_s(\Delta x)$ is determined by the equation

$$i\Delta x = \Delta t \frac{dH(p, \Delta t)}{dp}\bigg|_{p=p_s} \approx \frac{2\gamma\bar{v}}{\varepsilon^2} \times \begin{cases} \dfrac{1}{p_s - p_1^+}, & \Delta x > 0, \\[2mm] \dfrac{1}{p_s - p_1^-}, & \Delta x < 0. \end{cases} \tag{20.366}$$

The solutions are indicated in Fig. 20.26 by the cross. The approximation (20.366) is obviously applicable since for a large $|\Delta x|$ satisfying the condition $|\Delta x p_1^{\pm}| \gg \gamma\bar{v}/\varepsilon^2$, the saddle point p_s is very close to one of the singularities. Inserting the approximation (20.366) into the Fourier integral (20.347), we obtain the asymptotic expression for the probability distribution

$$P(x\,t\,|x_a t_a) \sim \begin{cases} e^{-\Delta x\, q_t^+}, & \Delta x > 0, \\ e^{-|\Delta x|\, q_t^-}, & \Delta x < 0, \end{cases} \tag{20.367}$$

where $q_t^\pm \equiv \mp i p_1^\pm(t)$ are real and positive. Thus the tails of the probability distribution $P(x\,t\,|x_a t_a)$ for large $|\Delta x|$ are exponential for all times t. The slopes of the logarithmic plots in the tails $q^\pm \equiv \mp d\log(x\,t\,|x_a t_a)/dx$ are determined by the positions $p_1^\pm$ of the singularities closest to the real axis.

These positions $p_1^\pm$ depend on the time interval Δt. For times much shorter than the relaxation time $(\gamma t \ll 1)$, the singularities lie far away from the real axis. As time increases, the singularities approach the real axis. For times much longer than the relaxation time $(\gamma \Delta t \gg 1)$, the singularities approach the limiting points $p_1^\pm \to \pm i q_*^\pm$, marked in Fig. 20.26 by circled crosses. The limiting values are

$$q_*^{\pm} = \pm p_0 + \frac{\omega_0}{\varepsilon\sqrt{1-\rho^2}} \quad \text{for} \quad \gamma\Delta t \gg 1. \tag{20.368}$$

The slopes $q^\pm(\Delta t)$ approach these limiting slopes monotonously from above. The behavior is shown in Fig. 20.24. The slopes (20.368) are of course in agreement with Eq. (20.361) in the limit $\gamma\Delta t \gg 1$. In the opposite limit of short time $(\gamma\Delta t \ll 1)$, we find the analytic time behavior

$$q^{\pm}(\Delta t) \approx \pm p_0 + \sqrt{p_0^2 + \frac{4\gamma}{\varepsilon^2(1-\rho^2)\Delta t}} \quad \text{for} \quad \gamma\Delta t \ll 1. \tag{20.369}$$

This approximation is shown in Fig. 20.24 as dots.

20.4.7 Path Integral Calculation

Instead of solving the Fokker-Planck equation (20.330) we may also study directly the path integral for the probability distribution using the Hamiltonian (20.321). Note the extra two terms at the end in comparison with the operator expression (20.319). They account for the symmetric operator order implied by the path integral. The distribution $\bar{P}_p(v_b, t_b | v_a t_a)$ at fixed momentum introduced in Eq. (20.329) has the path integral representation

$$\bar{P}_p(v_b, t_b | v_a t_a) = \int \mathcal{D}v\, \frac{\mathcal{D}p_v}{2\pi}\, e^{\mathcal{A}_p[p_v, v]}, \tag{20.370}$$

with the action

$$\mathcal{A}_p[p_v, v] = \int_{t_a}^{t_b} dt\, [i p_v \dot{v} - H(p, p_v, v)]. \tag{20.371}$$

The path integral (20.370) sums over all paths $p_v(t)$ and $v(t)$ with the boundary conditions $v(t_a) = v_a$ and $v(t_b) = v_b$.

It is convenient to integrate the first term in the (20.371) by parts, and to separate $H(p, p_v, v)$ into a v-independent part $i\gamma\bar{v}p_v - i\gamma/2 - i\rho\epsilon p/2$ and a linear term $[\partial H(p, p_v, v)/\partial v]v$. Thus we write

$$\begin{aligned}
\mathcal{A}_p[p_v, v] &= i[p_v(t_b)v_b - p_v(t_a)v_a] - i\gamma\bar{v}\int_{t_a}^{t_b} dt\, p_v(t)(t_b - t_a) \\
&\quad - \int_{t_a}^{t_b} dt\, \left[i\dot{p}_v(t) + \frac{\partial H}{\partial v(t)} \right] v(t).
\end{aligned} \tag{20.372}$$

Since the path $v(t)$ appears linearly in this expression, we can integrate it out to obtain a delta-functional $\delta\left[p_v(t) - \tilde{p}_v(t)\right]$, where $\tilde{p}_v(t)$ is the solution of the Hamilton equation $\dot{p}_v(t) = i\partial H/\partial v(t)$. This, however, coincides exactly with the characteristic differential equation (20.336) which was solved by Eq. (20.337) with a boundary condition $\tilde{p}_v(t_b) = p_v$. Taking the path integral over $\mathcal{D}p_v$ removes the delta-functional, and we find

$$\bar{P}_p(v_b, t_b | v_a t_a) = \int_{-\infty}^{+\infty} \frac{dp_v}{2\pi} J\, e^{i[p_v v - \tilde{p}_v(t_a)v_a] - i\gamma\theta \int_0^t dt\, \tilde{p}_v(t) + (\gamma - i\rho\epsilon p)(t_b - t_a)/2}, \quad (20.373)$$

where J denotes the Jacobian

$$J = \text{Det}^{-1}\left(i\partial_t + \frac{\partial^2 H(p, p_v, v)}{\partial p_v \partial v}\right). \quad (20.374)$$

From (20.321) we see that

$$\frac{\partial^2 H(p, p_v, v)}{\partial p_v \partial v} = -i\gamma + \rho\epsilon p. \quad (20.375)$$

According to Formula (18.254), this is equal to

$$J = e^{-(\gamma - i\rho\epsilon p)(t_b - t_a)/2}, \quad (20.376)$$

thus canceling the last term in the exponent of the integrand in (20.373). The result for $\bar{P}_p(v_b, t_b | v_a t_a)$ is therefore the same as the one obtained from the Fokker-Planck equation in Eq. (20.331) with the Fourier transform (20.335).

20.4.8   Natural Martingale Distribution

Let us calculate the natural martingales associated with the Hamiltonian (20.348). Reinserting the initially removed drift r_S, the total Hamiltonian reads

$$H^{\text{tot}}(p, \Delta t) = H(p, \Delta t) + i r_S p. \quad (20.377)$$

To construct the martingale according to the rule of Subsection 20.1.11, we need to know the value of $\bar{H}(p, \Delta t)$ at the momentum $p = i$. The expression is somewhat complicated, but the analysis of the Dow-Jones data in Appendix 20C shows that the parameter ρ determining the strength of correlations between the noise functions $\eta(t)$ and $\eta_v(t)$ [see (20.310)] is small. It is therefore sufficient to find only the first two terms of the Taylor series of $H(i, \Delta t)$ in powers of ρ:

$$H(i, \Delta t) \approx \frac{\bar{v}}{\epsilon\Delta t}\left(1 - e^{-\gamma\Delta t}\right)\rho + \frac{\bar{v}}{4\gamma\Delta t}\left[3 - 2e^{-\gamma\Delta t}\left(1 + 2\gamma\Delta t\right) - e^{-2\gamma\Delta t}\right]\rho^2. \quad (20.378)$$

From this we obtain

$$\bar{H}(i, \Delta t) = H(i, \Delta t) + c_1(\Delta t). \quad (20.379)$$

The analog of the martingale (20.291) is given by a Fourier integral (20.292) with $H_{r_x}(p)$ replaced by $H^{\text{tot}}_{r_x(\Delta t)}(p, \Delta t)$ of Eq. (20.350):

$$P^{(M,r_S)}(x_b t_b | x_a t_a) = e^{-r(\Delta t)\Delta t} \int_{-\infty}^{\infty} \frac{dp}{2\pi} \exp\left[ip(x_b - x_a) - \Delta t\, H^{\text{tot}}_{r_x(\Delta t)}(p, \Delta t)\right]. \quad (20.380)$$

The linear term $i r_x(\Delta t)p$ in $H^{\text{tot}}_{r_x(\Delta t)}(p, \Delta t)$ is now time-dependent:

$$r_x(\Delta t) = r_S + c_1(\Delta t) + \bar{H}(i, \Delta t). \quad (20.381)$$

This makes the rate in the exponential prefactor, by which the distribution becomes a martingale distribution, a time-dependent quantity:

$$r(\Delta t) = r_S + c_1(\Delta t). \quad (20.382)$$

By analogy with (20.293), there exists again an entire family of natural martingales, which is obtained by replacing r_S by an *arbitrary* growth rate r in which case $r(\Delta t)$ becomes equal to $r + c_1(\Delta t)$.

20.5 Time Series

True market prices are not continuous functions of time but recorded in discrete time intervals. For stocks which have a low turnover, only the daily prices are recorded. For others which are traded in large amounts, the prices change by the minute. They are listed as time series $S(t_n)$. In the year 2003, Robert Engle received the Nobel prize for his description of time series with the help of the so called *ARCH models* (autoregressive conditional heteroskedasticity model) and its generalization proposed by Tim Bollerslev [85], the *GARCH models*. The first contains an integer parameter q and is defined by a discrete modification of the pair of stochastic differential equations (20.306) and (20.308) for log-return and variance

$$x(t_n) = \sqrt{v(t_{n-1})}\eta(t_n), \quad v(t_n) = v_0 + \sum_{k=1}^{q} \alpha_k(t_{n-k})x^2(t_{n-k}), \quad (20.383)$$

with a white noise variable $\eta(t_n)$. The drift is omitted so that $\langle x(t_n) \rangle = 0$. This is called the ARCH(q) model. The second is a generalization of this in which the equation for the variance contains extra terms involving the past p values of σ^2:

$$v(t_n) = v_0 + \sum_{k=1}^{q} \alpha_k(t_{n-k})x^2(t_{n-k}) + \sum_{k=1}^{p} \beta_k(t_{n-k})v(t_{n-k}). \quad (20.384)$$

The ARCH(1) process has the time-independent expectations of variance and kurtosis

$$\sigma^2 = \langle v(t_n) \rangle = \frac{v_0}{1 - \alpha_1}, \quad \kappa = \frac{\langle x^4(t_n) \rangle_c}{\langle x^2(t_n) \rangle_c^2} = \frac{6\alpha_1^2}{1 - 3\alpha_1^2}. \quad (20.385)$$

For the GARCH(1,1) process, these quantities are

$$\sigma^2 = \langle v(t_n) \rangle = \frac{v_0}{1 - \alpha_1 - \beta_1}, \quad \kappa = \frac{\langle x^4(t_n) \rangle_c}{\langle x^2(t_n) \rangle_c^2} = \frac{6\alpha_1^2}{1 - 3\alpha_1^2 - 2\alpha_1\beta_1 - \beta_1^2}. \quad (20.386)$$

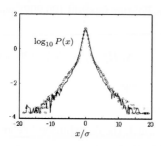 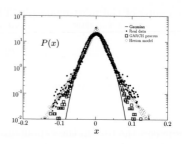

Figure 20.27 Left: Comparison of GARCH(1,1) process with $v_0 = 2.3 \times 10^{-5}$, $\alpha_1 = 0.091$, $\beta = 0.9$ with S&P 500 index (minute data). Right: Comparison of GARCH(1,1), Heston model, and Gaussian of the same σ with daily market data. Parameters of GARCH process are $v_0 = 7.7 \times 10^{-6}$, $\alpha_1 = 0.07906$, $\beta_1 = 0.90501$ and of Heston model $\gamma = 4.5 \times 10^{-2}$, $\bar{v} = 8.62 \times 10^{-8}$, $r_S = 5.67 \times 10^{-4}$, $\varepsilon = 10.3 \times 10^{-3}$ [79, 81, 82].

20.6 Spectral Decomposition of Power Behaviors

Plots of the log return curves taken over short time intervals reveal a richer structure than the simple model functions discussed so far. In fact, it seems reasonable to distinguish different types of stocks according to the characters of the investors. For instance, the bubble of the computer stocks was partly caused by quite inexperienced people who were not investing mainly to develop an industry but who wanted to earn fast money. There was, on the other hand also a substantial portion of institutional investors who poured in money more steadily. It appears that one should distinguish the sources of noise coming from the different groups of investors.

We therefore improve the stochastic differential equation (20.4) by extending it to

$$\dot{x}(t) = r_x + \sum_{\lambda} \eta_\lambda(t), \qquad (20.387)$$

where the sum runs over different groups of investors, each producing a different a noise $\eta_\lambda(t)$ with Lévy distributions falling off with different powers $|x|^{-1-\lambda_i}$. Their probability distributions are

$$P_\lambda[\eta_\lambda] = \exp\left[-\int_{t_a}^{t_b} dt\, \tilde{H}_\lambda(\eta_\lambda(t))\right] = \int \frac{\mathcal{D}p}{2\pi} \exp\left\{\int_{t_a}^{t_b} dt\, [ip(t)\eta_\lambda(t) - H_\lambda(p(t))]\right\}, \qquad (20.388)$$

with a Hamiltonian [compare (20.10)]

$$H_\lambda(p) \equiv \frac{\sigma_\lambda^2 |p|^\lambda}{2}. \qquad (20.389)$$

The probability (20.144) of returns becomes now

$$P(x_b t_b | x_a t_a) = \prod_{\lambda} \left\{\int \mathcal{D}\eta_\lambda \int \mathcal{D}x \exp\left[-\int_{t_a}^{t_b} dt\, \tilde{H}_\lambda(\eta_\lambda(t))\right]\right\} \delta[\dot{x} - \sum_{\lambda} \eta_\lambda]. \qquad (20.390)$$

If we insert here a Fourier representation of the δ-functional, we obtain

$$P(x_b t_b | x_a t_a) = \int_{-\infty}^{\infty} \frac{\mathcal{D}p}{2\pi} \prod_\lambda \left[\int \mathcal{D}\eta_\lambda \right] \int \mathcal{D}x \, e^{\int_{t_a}^{t_b} dt \left[ip(t)\dot{x}(t) - \tilde{H}_\lambda(\eta_\lambda(t)) \right]} e^{-i \sum_\lambda \int_{-\infty}^{\infty} dt \, p(t)\eta_\lambda(t)}.$$

$$(20.391)$$

Performing the path integrals over $\eta_\lambda(t)$, this becomes

$$P(x_b t_b | x_a t_a) = \int_{-\infty}^{\infty} \frac{\mathcal{D}p}{2\pi} \int \mathcal{D}x \, e^{\int_{t_a}^{t_b} dt \left[ip(t)\dot{x}(t) - H_\lambda(p) \right]}, \qquad (20.392)$$

where the Hamiltonian is the sum of the group Hamiltonians

$$H(p) \equiv \sum_\lambda H_\lambda(p). \qquad (20.393)$$

The continuous generalization of this is

$$H(p) = H(|p|) \equiv \int_0^{\infty} d\lambda \, \sigma_\lambda^2 |p|^\lambda. \qquad (20.394)$$

The spectral function σ_λ^2 must be extracted from the data by forming the integral

$$\sigma_\lambda^2 = \int_{-i\infty}^{i\infty} \frac{d\log|p|}{2\pi i} \, p^{-\lambda} H(p). \qquad (20.395)$$

20.7 Option Pricing

Historically, the most important use of path integrals in financial markets was made in the context of determining a fair price of financial derivatives, in particular options.[12] Options are an ancient financial tool. They are used for speculative purposes or for hedging major market transactions against unexpected changes in the market environment. These can sometimes produce dramatic price explosions or erosions, and options are supposed to prevent the destruction of large amounts of capital. Ancient Romans, Grecians, and Phoenicians traded options against outgoing cargos from their local seaports. In financial markets, options are contracted between two parties in which one party has the right but not the obligation to do something, usually to buy or sell some underlying asset. Having rights without obligations has a value, so option holders must pay a price for acquiring them. The price depends on the value of the associated asset, which is why they are also called *derivative assets* or briefly *derivatives*. *Call options* are contracts giving the option holder the right to buy something, while *put options* entitle the holder to sell something. The price of an option is called *premium*. Usually, options are associated with stock, bonds, or commodities like oil, metals or other raw materials. In the sequel we shall consider call options on stocks, to be specific.

Modern option pricing techniques have their roots in early work by Charles Castelli who published in 1877 a book entitled *The Theory of Options in Stocks*

---

[12] An introduction is found on the site http://bradley.bradley.edu/~arr/bsm/model.html.

and Shares. This book presented an introduction to the hedging and speculation aspects of options. Twenty three years later, Louis Bachelier offered the earliest known analytical valuation for options in his dissertation at the Sorbonne [86]. Remarkably, Bachelier discovered the treatment of stochastic phenomena five years before Einstein's related but much more famous work on Brownian motion [87], and twenty-three years before Wiener's mathematical development [88]. The stochastic differential equations considered by him still had an important defect of allowing for negative security prices, and for option prices exceeding the price of the underlying asset. Bachelier's work was continued by Paul Samuelson, who wrote in 1955 an unpublished paper entitled *Brownian Motion in the Stock Market.* During that same year, Richard Kruizenga, one of Samuelson's students, cited Bachelier's work in his dissertation *Put and Call Options: A Theoretical and Market Analysis.* In 1962, a dissertation by A. James Boness entitled *A Theory and Measurement of Stock Option Value* developed a more satisfactory pricing model which was further improved by Fischer Black and Myron Scholes. In 1973 they published their famous *Black and Scholes Model* [89] which, together with the improvements introduced by Robert Merton, earned them the Nobel prize in 1997.[13]

As discussed before, the Gaussian distribution severely underestimates the probability of large jumps in asset prices and this was the main reason for the catastrophic failure in the early fall of 1998 of the hedge fund *Long Term Capital Management*, which had Scholes and Merton on the advisory board (and as shareholders). The fund contained derivatives with a notional value of 1,250 Billion US$. The fund collected 2% for administrative expenses and 25% of the profits, and was initially extremely profitable. It offered its shareholders returns of 42.8% in 1995, 40.8% in 1996, and 17.1% even in the disastrous year of the Asian crisis 1997. But in September 1998, after mistakenly gambling on a convergence in interest rates, it almost went bankrupt. A number of renowned international banks and Wall Street institutions had to bail it out with 3.5 Billion US$ to avoid a chain reaction of credit failures.

In spite of this failure, the simple model is still being used today for a rough but fast orientation on the fairness of an option price.

20.7.1 Black-Scholes Option Pricing Model

In the early seventies, Fischer Black was working on a valuation model for stock warrants and observed that his formulas resembled very much the well-known equations for heat transfer. Soon after this, Myron Scholes joined Black and together they discovered an approximate option pricing model which is still of wide use.

The Black and Scholes Model is based on the following assumptions:

1. The returns are normally distributed.

   The shortcomings of this assumption have been discussed above in Section 20.1. The appropriate improvement of the model will be developed below.

---

[13]For F. Black the prize came too late — he had died two years earlier.

2. Markets are efficient.

This assumption implies that the market operates continuously with share prices following a continuous stochastic process without memory. It also implies that different markets have the same asset prices.

This is not quite true. Different markets do in general have slightly different prices. Their differences are kept small by the existence of arbitrage dealers. There also exist correlations over a short time scale which make it possible, in principle, to profit without risk from the so-called *statistical arbitrage*. This possibility is, however, strongly limited by transaction fees.

3. No commissions are charged.

This assumption is not satisfied. Usually, market participants have to pay a commission to buy or sell assets. Even floor traders pay some kind of fee, although this is usually very small. The fees payed by individual investors is more substantial and can distort the output of the model.

4. Interest rates remain constant and known.

The Black and Scholes model assumes the existence of a *riskfree interest rate* to represent this constant and known rate. In reality there is no such thing as the riskfree rate. As an approximation, one uses the discount rate on U.S. Government Treasury Bills with 30 days left until maturity. During periods of rapidly changing interest rates, these 30 day rates are often subject to change, thereby violating one of the assumptions of the model.

5. The stock pays no dividends during the option's life.

Most companies pay dividends to their share holders, so this is a limitation to the model since higher dividends lead to lower call premiums. There is, however, a simple possibility of adjusting the model to the real situation by subtracting the discounted value of a future dividend from the stock price.

6. European exercise terms are used.

European exercise terms imply the exercise of an option only on the expiration date. This is in contrast to the American exercise terms which allow for this at any time during the life of the option. This greater flexibility makes an American option more valuable than the European one.

The difference is, however, not dramatic in praxis because very few calls are ever exercised before the last few days of their life, since an early exercise means giving away the remaining time value on the call. Different exercise times towards the end of the life of a call are irrelevant since the remaining time value is very small and the intrinsic value has a small time dependence, barring a dramatic event right before expiration date.

Since 1973, the original Black and Scholes Option Pricing Model has been improved and extended considerably. In the same year, Robert Merton [90] included the effect of dividends. Three years later, Jonathan Ingerson relaxed the assumption of no taxes or transaction costs, and Merton removed the restriction of constant interest rates. In recent years, the model has been generalized to determine the prices of options with many different properties.

The relevance of path integrals to this field was recognized in 1988 by the theoretical physicist J.W. Dash, who wrote two unpublished papers on the subject entitled *Path Integrals and Options I* and *II* [91]. Since then many theoretical physicists have entered the field, and papers on this subject have begun appearing on the Los Alamos server [13, 92, 93].

20.7.2 Evolution Equations of Portfolios with Options

The option price $O(t)$ has larger fluctuations than the associated stock price. It usually varies with a slope $\partial O(S(t),t)/\partial S(t)$ which is commonly denoted by $\Delta(S(t),t)$ and called the *Delta* of the option. If $\Delta(S(t),t)$ depends only weakly on $S(t)$ and t it is possible, in the ideal case of Gaussian price fluctuations, to guarantee a steady growth of a portfolio. One merely has to mix a suitable number $N_S(t)$ stocks with $N_O(t)$ options and short-term bonds whose number is denoted by $N_B(t)$. As mentioned before, these are typically U.S. Government Treasury Bills with 30 days left to maturity which have only small price fluctuations. The composition $[N_S(t), N_O(t), N_B(t)]$ is referred to as the *strategy* of the portfolio manager. The total wealth has the value

$$W(t) = N_S(t)S(t) + N_O(t)O(S,t) + N_B(t)B(t). \qquad (20.396)$$

The goal is to make $W(t)$ grow with a smooth exponential curve *without fluctuations*

$$\dot{W}(t) \approx r_W W(t). \qquad (20.397)$$

As an idealization, the short-term bonds are assumed to grow deterministically without any fluctuations:

$$\dot{B}(t) \approx r_B B(t). \qquad (20.398)$$

The rate r_B is the earlier-introduced riskfree interest rate encountered in true markets only if there are no events changing excessively the value of short-term bonds.

The existence of arbitrage dealers will ensure that the growth rate r_W is equal to that of the short-term bonds

$$r_W \approx r_B. \qquad (20.399)$$

Otherwise the dealers would change from one investment to the other.

In the decomposition (20.396), the desired growth (20.397) reads

$$N_S(t)\dot{S}(t) + N_O(t)\dot{O}(S,t) + N_B(t)\dot{B}(t) + \dot{N}_S(t)S(t) + \dot{N}_O(t)O(S,t) + \dot{N}_B(t)B(t)$$
$$= r_W \left[N_S(t)S(t) + N_O(t)O(S,t) + N_B(t)B(t) \right]. \qquad (20.400)$$

Due to (20.398) and (20.399), the terms containing $N_B(t)$ without a dot drop out. Moreover, if no extra money is inserted into or taken from the system, i.e., if stocks, options, and bonds are only traded against each other, this does not change the total wealth, assuming the absence of commissions. This so-called *self-financing strategy* is expressed in the equation

$$\dot{N}_S(t)S(t) + \dot{N}_O(t)O(S,t) + \dot{N}_B(t)B(t) = 0. \qquad (20.401)$$

Thus the growth equation (20.397) translates into

$$\dot{W}(t) = N_S\dot{S} + N_O\dot{O} + N_B\dot{B} = r_W\left(N_SS + N_OO + N_BB\right). \qquad (20.402)$$

Due to the equality of the rates $r_W = r_B$ and Eq. (20.398), the entire contribution of $B(t)$ cancels, and we obtain

$$N_S\dot{S} + N_O\dot{O} = r_W\left(N_SS + N_OO\right). \qquad (20.403)$$

The important observation is now that there exists an optimal ratio between the number of stocks N_S and the number of options N_O, which is equal to the negative slope $\Delta(S(t),t)$:

$$\frac{N_S(t)}{N_O(t)} = -\Delta(S(t),t) = -\frac{\partial O(S(t),t)}{\partial S(t)}. \qquad (20.404)$$

Then Eq. (20.403) becomes

$$N_S\dot{S} + N_O\dot{O} = N_O r_W\left(-\frac{\partial O}{\partial x} + O\right). \qquad (20.405)$$

The two terms on the left-hand side are treated as follows: First we use the relation (20.404) to rewrite

$$N_S\dot{S} = -N_O\frac{\partial O(S,t)}{\partial S}\dot{S} = -N_O\frac{\partial O(S,t)}{\partial x}\frac{\dot{S}}{S}. \qquad (20.406)$$

In the second term on the left-hand side of (20.405), we expand the total time dependence of the option price in a Taylor series

$$\begin{aligned}
\frac{dO}{dt} &= \frac{1}{dt}\left[O(x(t) + \dot{x}(t)\,dt, t + dt) - O(x(t),t)\right] \\
&= \frac{\partial O}{\partial t} + \frac{\partial O}{\partial x}\dot{x} + \frac{1}{2}\frac{\partial^2 O}{\partial x^2}\dot{x}^2\,dt + \frac{1}{3!}\frac{\partial^3 O}{\partial x^3}\dot{x}^3\,dt^2 + \dots . \qquad (20.407)
\end{aligned}$$

We have gone over to the logarithmic stock price variable $x(t)$ rather than $S(t)$ itself. In financial mathematics, the lowest derivatives on the right-hand side are all denoted by special symbols. We have already introduced the name Delta for the slope $\partial O/\partial S$. The curvature $\Gamma \equiv \partial^2 O/\partial S^2 = (\partial^2 O/\partial x^2 - \partial O/\partial x)/S^2$ is called the *Gamma* of an option. Another derivative with a standard name is the *Vega*

$V \equiv \partial^2 O / \partial \sigma^2$. The partial time derivative $\partial O / \partial t$ is denoted by Θ. The set of these quantities is commonly called the *Greeks*.

In general, the expansion (20.407) is carried to arbitrary powers of $\dot{x}$ as in (20.273). It is, of course, only an abbreviated notation for the proper expansion in powers of a stochastic variable to be performed as in Eq. (20.273). After inserting (20.407) and (20.406) on the left-hand side of Eq. (20.405), this becomes

$$
N_S \dot{S} + N_O \dot{O} = - N_O \frac{\partial O}{\partial x} \frac{\dot{S}}{S}
$$

$$
+ N_O \left(\frac{\partial O}{\partial t} + \frac{\partial O}{\partial x} \dot{x} + \frac{1}{2} \frac{\partial^2 O}{\partial x^2} \dot{x}^2 dt + \frac{1}{3!} \frac{\partial O}{\partial x} \dot{x}^3 dt + \dots \right) \quad (20.408)
$$

$$
= N_O \left[\frac{\partial O}{\partial t} + \left(\dot{x} - \frac{\dot{S}}{S} \right) \frac{\partial O}{\partial x} + \frac{1}{2} \frac{\partial^2 O}{\partial x^2} \dot{x}^2 dt + \frac{1}{3!} \frac{\partial^3 O}{\partial x^3} \dot{x}^3 dt + \dots \right].
$$

Replacing further the left-hand side by the right-hand side of (20.405) we obtain

$$
\frac{\partial O}{\partial t} = -r_W O - \left(\dot{x} - \frac{\dot{S}}{S} + r_W \right) \frac{\partial O}{\partial x} - \frac{1}{2} \frac{\partial^2 O}{\partial x^2} \dot{x}^2 dt - \frac{1}{3!} \frac{\partial^3 O}{\partial x^3} \dot{x}^3 dt + \dots = 0. \quad (20.409)
$$

The crucial observation which earned Black and Scholes the Nobel prize is that for Gaussian fluctuations with a Hamiltonian $H(p) = \sigma^2 p^2 / 2$, the equation (20.409) becomes very simple. First, due to the Itō relation (20.4), the prefactor of $-\partial O / \partial x$ becomes a constant

$$
r_W - \frac{\sigma^2}{2} \equiv r_{xW}, \quad (20.410)
$$

where the notation r_{xW} is chosen by analogy with r_x in the Itō relation (20.5). Thus there are no more fluctuations in the prefactor of $\partial O / \partial x$.

Moreover, also the fluctuations of all remaining terms

$$
-\frac{1}{2} \frac{\partial^2 O}{\partial x^2} \dot{x}^2 \, dt - \frac{1}{6} \frac{\partial^3 O}{\partial x^3} \dot{x}^3 \, dt^2 + \dots \quad (20.411)
$$

can be neglected due to the result (18.429) and the estimates (18.431), which make truncate the expansion (20.411) and make it equal to $-(\sigma^2/2) \partial^2 O / \partial x^2$. Thus Eq. (20.409) looses its stochastic character, and the fair option price $O(x, t)$ is found to obey the Fokker-Planck-like differential equation

$$
\frac{\partial O}{\partial t} = r_W O - r_{xW} \frac{\partial O}{\partial x} - \frac{\sigma^2}{2} \frac{\partial^2 O}{\partial x^2}, \quad (20.412)
$$

At the same time, the total wealth (20.402) looses its stochastic character and grows with the riskfree rate r_W following the deterministic Eq. (20.398). The cancellation of the fluctuations is a consequence of choosing the ratio between options and stocks according to Eq. (20.404). The portfolio is now hedged against fluctuation.

The Fokker-Planck equation (20.412) can be expressed completely in terms of the Greeks of the option, using the relation (20.410):

$$\Theta = r_W O - r_{xw} S\Delta - \frac{\sigma^2}{2}\left(S\frac{\partial O}{\partial S} + S^2 \frac{\partial^2 O}{\partial S^2}\right) = r_W O - r_W S\Delta - \frac{\sigma^2}{2}S^2\Gamma. \qquad (20.413)$$

The strategy to make a portfolio riskfree by balancing the fluctuations of N_S stocks by N_O options to satisfy Eq. (20.404) is called Delta-hedging. Hedging is of course imperfect. Since $\Delta(S(t),t)$ depends on the stock price, a Delta hedge requires frequent re-balancing of the portfolio, even several times per day. After a time span δt, the Delta has changed by

$$\delta\Delta = \frac{d}{dt}\frac{\partial^2 O(S(t),t)}{\partial S(t)}\delta t = \left[\frac{\partial}{\partial t}\frac{\partial O(S(t),t)}{\partial S(t)} + \frac{\partial^2 O(S(t),t)}{\partial S(t)^2}\right]\delta t = \left[\frac{\partial\Theta}{\partial S} + \Gamma\right]\delta t. \tag{20.414}$$

So one has to readjust the ratio of stocks and options in Eq. (20.404) by

$$\delta\frac{N_S}{N_O} = -\delta\Delta = -\left(\frac{\partial\Theta}{\partial S} + \Gamma\right)\delta t. \qquad (20.415)$$

This procedure is called *dynamical Δ-hedging*. Since buying and selling costs money, δt cannot be made too small, otherwise dynamical hedging becomes too expensive.

In principle it is possible to buy a third asset to hedge also the curvature Gamma. This procedure is not so popular, again because of the transaction costs.

For non-Gaussian noise, the differential equation (20.409) is still stochastic due to remaining fluctuations in the expansion terms (20.411). This is an obstacle to building a riskfree portfolio with deterministically growing total wealth $W(t)$. As in Eq. (20.255) we can only derive the equation of motion for the average value

$$\left\langle\frac{1}{2}\frac{\partial^2 O}{\partial x^2}\dot{x}^2\,dt + \frac{1}{6}\frac{\partial^3 O}{\partial x^3}\dot{x}^3\,dt^2 + \ldots\right\rangle = -\bar{H}(i\partial_x)O. \qquad (20.416)$$

Hence a fair option price can only be calculated on the average from the Fokker-Planck-like differential equation

$$\frac{\partial}{\partial t}\langle O\rangle = \left[r_W - r_{xw}\frac{\partial}{\partial x} + \bar{H}(i\partial_x)\right]\langle O\rangle, \qquad (20.417)$$

where we have defined, by analogy with (20.268) and (20.410), an auxiliary rate parameter

$$r_{xw} \equiv r_W + \bar{H}(i). \qquad (20.418)$$

However, as pointed out in Subsection 20.2.2, this limitation is not really stringent for discrete short-time data if the tails of the noise distribution are non-Gaussian with semi-heavy tails. Then the higher expansion terms in Eq. (20.411) are suppressed by a small factor $\sigma\sqrt{\Delta t}$, and the expectation values in Eqs. (20.416) and (20.417) can again be dropped approximately.

For Gaussian fluctuations, the Fokker-Planck equation (20.412) can easily be solved. If we rename $t \to t_a$, and $x \to x_a$, for symmetry reasons, the solution which starts out, at some time t_b, like $\delta(x_a - x_b)$, has the Fourier representation

$$P(x_b t_b | x_a t_a) = e^{-r_W(t_b - t_a)} \int_{-\infty}^{\infty} \frac{dp}{2\pi} e^{ip(x_b - x_a)} e^{-\left(\sigma^2 p^2/2 + ir_{xW}p\right)(t_b - t_a)}. \tag{20.419}$$

A convergent integral exists for $t_b > t_a$.

For non-Gaussian fluctuations with semi-heavy tails, there exists an approximate solution whose Fourier representation is

$$P(x_b t_b | x_a t_a) = e^{-r_W(t_b - t_a)} \int_{-\infty}^{\infty} \frac{dp}{2\pi} e^{ip(x_b - x_a)} e^{-\left[\bar{H}(p) + ir_{xW}p\right](t_b - t_a)}, \tag{20.420}$$

with $\bar{H}(p)$ of Eq. (20.141).

Recalling the discussion in Section 20.3, this distribution function is recognized as a member of the equivalent family of martingale distributions (20.293) for the stock price $S(t) = e^{x(t)}$. It is the particular distribution in which the discount factor r coincides with the riskfree interest rate r_W.

20.7.3 Option Pricing for Gaussian Fluctuations

For Gaussian fluctuations where $H(p) = \sigma^2 p^2/2$, the integral in (20.419) can easily be performed and yields

$$P(x_b t_b | x_a t_a) = \Theta(t_b - t_a) \frac{e^{-r_W(t_b - t_a)}}{\sqrt{2\pi\sigma^2(t_b - t_a)}} \exp\left\{ -\frac{[x_b - x_a - r_{xW}(t_b - t_a)]^2}{2\sigma^2(t_b - t_a)} \right\}. \tag{20.421}$$

This probability distribution is obviously the solution of the path integral

$$P(x_b t_b | x_a t_a) = \Theta(t_b - t_a) e^{-r_W(t_b - t_a)} \int \mathcal{D}x \, \exp\left\{ -\frac{1}{2\sigma^2} \int_{t_a}^{t_b} [\dot{x} - r_{xW}]^2 \right\}. \tag{20.422}$$

The distribution function (20.421) is recognized as the riskfree member of the family of Gaussian martingale distributions (20.285) for the stock price $S(t) = e^{x(t)}$. It is the particular distribution in which the discount factor r equals the riskfree interest rate r_W, i.e., the expression (20.421) coincides with the martingale distribution $P^{(M, r_W)}(x_b t_b | x_a t_a)$ of Eq. (20.288), whose explicit form is the expression (20.289) fro $r = r_W$. This distribution is referred to as the *risk-neutral* martingale distribution.

An option is written for a certain *strike price* E of the stock. The value of the option at its expiration date t_b is given by the difference between the stock price on expiration date and the strike price:

$$O(x_b, t_b) = \Theta(S_b - E)(S_b - E) = \Theta(x_b - x_E)(e^{x_b} - e^{x_E}), \tag{20.423}$$

where

$$x_E \equiv \log E. \tag{20.424}$$

The Heaviside function in (20.423) accounts for the fact that only for $S_b > E$ it is worthwhile to execute the option.

From (20.423) we calculate the option price at an arbitrary earlier time using the time evolution probability (20.421)

$$O(x_a, t_a) = \int_{-\infty}^{\infty} dx_b \, O(x_b, t_b) \, P^{(M, r_W)}(x_b t_b | x_a t_a). \tag{20.425}$$

Inserting (20.423) we obtain the sum of two terms

$$O(x_a, t_a) = O_S(x_a, t_a) - O_E(x_a, t_a), \tag{20.426}$$

where

$$O_S(x_a, t_a) = \frac{e^{-r_W(t_b - t_a)}}{\sqrt{2\pi\sigma^2(t_b - t_a)}} \int_{x_E}^{\infty} dx_b \, e^{x_b} \exp\left\{-\frac{[x_b - x_a - r_{xW}(t_b - t_a)]^2}{2\sigma^2(t_b - t_a)}\right\}, \tag{20.427}$$

and

$$O_E(x_a, t_a) = E e^{-r_W(t_b - t_a)} \frac{1}{\sqrt{2\pi\sigma^2(t_b - t_a)}} \int_{x_E}^{\infty} dx_b \, \exp\left\{-\frac{[x_b - x_a - r_{xW}(t_b - t_a)]^2}{2\sigma^2(t_b - t_a)}\right\}. \tag{20.428}$$

In the second integral we set

$$x_- \equiv x_a + r_{xW}(t_b - t_a) = x_a + \left(r_W - \frac{1}{2}\sigma^2\right)(t_b - t_a), \tag{20.429}$$

and obtain

$$O_E(x_a, t_a) = E \frac{e^{-r_W(t_b - t_a)}}{\sqrt{2\pi\sigma^2(t_b - t_a)}} \int_{x_E - x_-}^{\infty} dx_b \, \exp\left\{-\frac{x_b^2}{2\sigma^2(t_b - t_a)}\right\}. \tag{20.430}$$

After rescaling the integration variable $x_b \to -\xi\sigma\sqrt{t_b - t_a}$, this can be rewritten as

$$O_E(x_a, t_a) = e^{-r_W(t_b - t_a)} E \, N(y_-), \tag{20.431}$$

where $N(y)$ is the cumulative Gaussian distribution function

$$N(y) \equiv \int_{-\infty}^{y} \frac{d\xi}{\sqrt{2\pi}} e^{-\xi^2/2}, \tag{20.432}$$

evaluated at

$$
\begin{aligned}
y_- &\equiv \frac{x_- - x_E}{\sqrt{\sigma^2(t_b - t_a)}} = \frac{\log[S(t_a)/E] + r_{xW}(t_b - t_a)}{\sqrt{\sigma^2(t_b - t_a)}} \\
&= \frac{\log[S(t_a)/E] + \left(r_W - \frac{1}{2}\sigma^2\right)(t_b - t_a)}{\sqrt{\sigma^2(t_a - t_b)}}.
\end{aligned} \tag{20.433}
$$

The integral in the first contribution (20.427) to the option price is found after completing the exponent in the integrand quadratically as follows:

$$
x_b - \frac{[x_b - x_a - r_{xW}(t_b - t_a)]^2}{2\sigma^2(t_b - t_a)}
$$
$$
= -\frac{[x_b - x_a - (r_{xW} + \sigma^2)(t_b - t_a)]^2 - 2r_W\sigma^2(t_b - t_a) - 2x_a\sigma^2(t_b - t_a)}{2\sigma^2(t_b - t_a)}. \quad (20.434)
$$

Introducing now

$$
x_+ \equiv x_a + \left(r_{xW} + \sigma^2\right)(t_b - t_a) = x_a + \left(r_W + \frac{1}{2}\sigma^2\right)(t_b - t_a), \quad (20.435)
$$

and rescaling x_b as before, we obtain

$$
O_S(x_a, t_a) = S(t_a)N(y_+), \quad (20.436)
$$

with

$$
y_+ \equiv \frac{x_+ - x_E}{\sqrt{\sigma^2(t_a - t_b)}} = \frac{\log[S(t_a)/E] + (r_{xW} + \sigma^2)(t_b - t_a)}{\sqrt{\sigma^2(t_a - t_b)}}
$$
$$
= \frac{\log[S(t_a)/E] + \left(r_W + \frac{1}{2}\sigma^2\right)(t_b - t_a)}{\sqrt{\sigma^2(t_a - t_b)}}. \quad (20.437)
$$

The combined result

$$
O(x_a, t_a) = S(t_a)N(y_+) - e^{-r_W(t_b - t_a)}E\,N(y_-) \quad (20.438)
$$

is the celebrated *Black-Scholes formula* of option pricing.

In Fig. 20.28 we illustrate how the dependence of the call price on the stock price varies with different times to expiration $t_b - t_a$ and with different volatilities σ.

Floor dealers of stock markets use the Black-Scholes formula to judge how expensive options are, so they can decide whether to buy or to sell them. For a given riskfree interest rate r_W and time to expiration $t_b - t_a$, and a set of option, stock, and strike prices O, S, E, they calculate the volatility from (20.438). The result is called the *implied volatility*, and denoted by $\Sigma(x - x_E)$. As typical plot as a function of $x - x_E$ is shown in Fig. 20.29. If the Black-Scholes formula were exactly valid, the data should lie on a horizontal line $\Sigma(x - x_E) = \sigma$. Instead, they scatter around a parabola which is called the *smile* of the option. The smile indicates the presence of a nonzero kurtosis in the distribution of the returns, as we shall see in Subsection 20.7.7.

It goes without saying that if the integral (20.427) is carried out over the entire x_b axis, it becomes independent of time, due to the martingale character of the risk-neutral distribution $P^{(M,r_W)}(x_b t_b | x_a t_a)$.

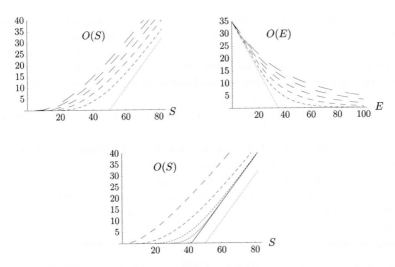

Figure 20.28   Left: Dependence of call price O on stock price S for different times before expiration date (increasing dash length: $1, 2, 3, 4, 5$ months). The parameters are $E = 50$ US\$, $\sigma = 40\%$, $r_W = 6\%$ per month. Right: Dependence on the strike price E for fixed stock price 35 US\$ and the same times to expiration (increasing with dash length). Bottom: Dependence on the volatilities (from left to right: $80\%, 60\%, 20\%, 10\%, 1\%$) at a fixed time $t_b - t_a = 3$ months before expiration.

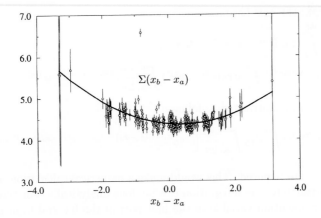

Figure 20.29   Smile deduced from options (see [13]).

20.7.4 Option Pricing for Boltzmann Distribution

The above result can easily be extended to price the options for assets whose returns obey the Boltzmann distribution (20.204). Since this is a superposition of Gaussian distributions, we merely have to perform the same superposition over the Black-Scholes formula (20.438). Thus we insert (20.438) into the integral (20.204) and obtain the option price for the Boltzmann-distributed assets of variance σ^2:

$$O^B(x_a, t_a) = \left(\frac{t}{\sigma^2}\right)^t \frac{1}{\Gamma(t)} \int_0^\infty \frac{dv}{v} \, v^t e^{-tv/\sigma^2} O^v(x_a, t_a), \qquad (20.439)$$

where $O^v(x_a, t_a)$ is the Black-Scholes option price (20.438) of variance v. The superscript indicates that the variance σ^2 in the variables of y_+ and y_- in (20.433) and (20.437) is now exchanged by the integration variable v.

Since the Boltzmann distribution for the minute data turns rapidly into a Gaussian (recall Fig. 20.15), the price changes with respect to the Black-Scholes formula are relevant only for short-term options running less than a week. The changes can most easily be estimated by using the expansion (20.209) to write

$$O^B(x_a, t_a) = O^v(x_a, t_a) + \frac{[2T^2(1-1/t)]^2}{2t} \partial_v^2 O^v(x_a, t_a) + \dots . \qquad (20.440)$$

20.7.5 Option Pricing for General Non-Gaussian Fluctuations

For general non-Gaussian fluctuations with semi-heavy tails, the option price must be calculated numerically from Eqs. (20.425) and (20.423). Inserting the Fourier representation (20.419) and using the Hamiltonian

$$H_{r_{x_W}}(p) \equiv \bar{H}(p) + ir_{x_W} p \qquad (20.441)$$

defined as in (20.141), this becomes

$$
\begin{aligned}
O(x_a, t_a) &= \int_{x_E}^\infty dx_b \, (e^{x_b} - e^{x_E}) P(x_b t_b | x_a t_a) \\
&= e^{-r_W(t_b - t_a)} \int_{x_E}^\infty dx_b \, (e^{x_b} - e^{x_E}) \int_{-\infty}^\infty \frac{dp}{2\pi} e^{ip(x_b - x_a) - H_{r_{x_W}}(p)(t_b - t_a)}. \quad (20.442)
\end{aligned}
$$

The integrand can be rearranged as follows:

$$O(x_a, t_a) = e^{-r_W(t_b - t_a)} \int_{x_E}^\infty dx_b \int_{-\infty}^\infty \frac{dp}{2\pi} \left[e^{x_a} e^{i(p-i)(x_b - x_a)} - e^{x_E} e^{ip(x_b - x_a)} \right] e^{-H_{r_{x_W}}(p)(t_b - t_a)}.$$

$$(20.443)$$

Two integrations are required. This would make a numerical calculation quite time consuming. Fortunately, one integration can be done analytically. For this purpose we change the momentum variable in the first part of the integral from p to $p + i$ and rewrite the integral in the form

$$O(x_a, t_a) = e^{-r_W(t_b - t_a)} \int_{x_E}^\infty dx_b \int_{-\infty}^\infty \frac{dp}{2\pi} e^{ip(x_b - x_a)} f(p), \qquad (20.444)$$

with

$$f(p) \equiv e^{x_a} e^{-H_{r_x W}(p+i)(t_b-t_a)} - e^{x_E} e^{-H_{r_x W}(p)(t_b-t_a)}. \tag{20.445}$$

We have suppressed the arguments $x_a, x_E, t_b - t_a$ in $f(p)$, for brevity. The integral over x_b in (20.444) runs over the Fourier transform

$$\tilde{f}(x_b - x_a) = \int_{-\infty}^{\infty} \frac{dp}{2\pi} e^{ip(x_b-x_a)} f(p) \tag{20.446}$$

of the function $f(p)$. It is then convenient to express the integral $\int_{x_E}^{\infty} dx_b$ in terms of the Heaviside function $\Theta(x_b - x_E)$ as $\int_{-\infty}^{\infty} dx_b \Theta(x_b - x_E)$ and use the Fourier representation (1.308) of the Heaviside function to write

$$\int_{x_E}^{\infty} dx_b \, \tilde{f}(x_b - x_a) = \int_{-\infty}^{\infty} dx_b \int_{-\infty}^{\infty} \frac{dq}{2\pi} \frac{i}{q+i\eta} e^{-iq(x_b-x_E)} \tilde{f}(x_b - x_a). \tag{20.447}$$

Inserting here the Fourier representation (20.446), we can perform the integral over x_b and obtain the momentum space representation of the option price

$$O(x_a, t_a) = e^{-r_W(t_b-t_a)} \int_{-\infty}^{\infty} \frac{dp}{2\pi} e^{ip(x_E-x_a)} \frac{i}{p+i\eta} f(p). \tag{20.448}$$

For numerical integrations, the singularity at $p = 0$ is inconvenient. We therefore employ the decomposition (18.54)

$$\frac{1}{p+i\eta} = \frac{\mathcal{P}}{p} - i\pi\delta(p), \tag{20.449}$$

to write

$$O(x_a, t_a) = e^{-r_W(t_b-t_a)} \left[\frac{1}{2} f(0) + i \int_{-\infty}^{\infty} \frac{dp}{2\pi} \frac{e^{ip(x_E-x_a)} f(p) - f(0)}{p} \right]. \tag{20.450}$$

We have used the fact that the principal value of the integral over $1/p$ vanishes to subtract the constant $f(0)$ from $e^{ip(x_E-x_a)} f(p)$. After this, the integrand is regular and does not need any more the principal-value specification, and allows for a numerical integration.

For x_a very much different from x_E, we may approximate

$$\int_{-\infty}^{\infty} \frac{dp}{2\pi} \frac{e^{ip(x_E-x_a)} f(p) - f(0)}{p} \approx \frac{1}{2} \epsilon(x_a - x_E) f(0), \tag{20.451}$$

where $\epsilon(x) \equiv 2\Theta(x) - 1$ is the step function (1.311), and obtain

$$O(x_a, t_a) \approx \frac{e^{-r_W(t_b-t_a)}}{2} \left[1 + \Theta(x_a - x_E) \right] f(0). \tag{20.452}$$

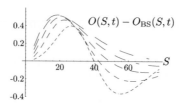

Figure 20.30 Difference between call price $O(S,t)$ obtained from truncated Lévy distribution with kurtosis $\kappa = 4$ and Black-Scholes price $O_{\mathrm{BS}}(S,t)$ with $\sigma^2 = \bar{v}$ as function of stock price S for different times before expiration date (increasing dash length: $1,2,3,4,5$ months). The parameters are $E = 50$ US\$, $\sigma = 40\%$, $r_W = 6\%$ per month.

Using (20.418) we have $e^{-H_{r_{xW}}(i)} = e^{r_W}$, and since $e^{-H_{r_{xW}}(0)} = 1$ we see that $O(x_a, t_a)$ goes to zero for $x_a \to -\infty$ and has the large-x_a behavior

$$O(x_a, t_a) \approx e^{x_a} - e^{x_E} e^{-r_W (t_b - t_a)} = S(t_a) - e^{-r_W(t_b - t_a)} E. \qquad (20.453)$$

This is the same behavior as in the Black-Scholes formula (20.438).

In Fig. 20.30 we display the difference between the option prices emerging from our formula (20.450) with a truncated Lévy distribution of kurtosis $\kappa = 4$, and the Black-Scholes formula (20.438) for the same data as in the upper left of Fig. 20.28.

For truncated Lévy distributions, the Fourier integral in Eq. (20.442) can be expressed directly in terms of the original distribution function which is the Fourier transform of (20.52):

$$\tilde{L}_{\sigma^2}^{(\lambda, \alpha, \beta)}(x) = \int_{-\infty}^{\infty} \frac{dp}{2\pi} e^{ipx - H(p)}. \qquad (20.454)$$

By inspecting Eq. (20.53) we see that the factor $t_b - t_a$ multiplying $H_{r_{xW}}(p)$ in (20.442) can be absorbed into the parameters $\sigma, \lambda, \alpha, \beta$ of the truncated Lévy distributions by replacing

$$\sigma^2 \to \sigma^2(t_b - t_a), \qquad r_{xW} \to r_{xW}(t_b - t_a). \qquad (20.455)$$

Let us denote the truncated Lévy distribution with zero average by $\bar{L}_{\sigma^2}^{(\lambda, \alpha, \beta)}(x)$. It is the Fourier transform of $e^{-\bar{H}(p)}$:

$$\bar{L}_{\sigma^2}^{(\lambda, \alpha, \beta)}(x) = \int_{-\infty}^{\infty} \frac{dp}{2\pi} e^{ipx - \bar{H}(p)}. \qquad (20.456)$$

The Fourier transform of $e^{-\bar{H}(p)(t_b - t_a)}$ is then simply given by $\bar{L}_{\sigma^2(t_b - t_a)}^{(\lambda, \alpha, \beta)}(x)$. The additional term r_{xW} in the exponent of the integral (20.442) via (20.441) leads to a drift r_{xW} in the distribution, and we obtain

$$\int_{-\infty}^{\infty} \frac{dp}{2\pi} e^{ipx - [\bar{H}(p) + i r_{xW} p](t_b - t_a)} = \tilde{L}_{\sigma^2(t_b - t_a)}^{(\lambda, \alpha, \beta)}(x - r_{xW}(t_b - t_a)). \qquad (20.457)$$

Inserting this into (20.419), we find the riskfree martingale distribution to be inserted into (20.442):

$$P(x_b t_b | x_a t_a) = e^{-r_W(t_b - t_a)} \bar{L}^{(\lambda,\alpha,\beta)}_{\sigma^2(t_b - t_a)}(x_b - x_a - r_{xW}(t_b - t_a)). \tag{20.458}$$

The result is therefore a truncated Lévy distribution of increasing width and uniformly moving average position. Since all expansion coefficients c_n of $H(p)$ in Eq. (20.40) receive the same factor $t_b - t_a$, the kurtosis $\kappa = c_4/c_2^2$ decreases inversely proportional to $t_b - t_a$. As time proceeds, the distribution becomes increasingly Gaussian, this being a manifestation of the central limit theorem of statistical mechanics. This is in contrast to the pure Lévy distribution which has no finite width and therefore maintains its power falloff at large distances.

Explicitly, the formula (20.442) for the option price becomes

$$O(x_a, t_a) = e^{-r_W(t_b - t_a)} \int_{x_E}^{\infty} dx_b \left(e^{x_b} - e^{x_E} \right) \bar{L}^{(\lambda,\alpha,\beta)}_{\sigma^2(t_b - t_a)}(x_b - x_a - r_{xW}(t_b - t_a)). \tag{20.459}$$

This and similar equations derived from any of the other non-Gaussian models lead to fairer formulas for option prices.

20.7.6   Option Pricing for Fluctuating Variance

If the fluctuations of the variance are taken into account, the dependence of the price of an option on $v(t)$ needs to be considered in the derivation of a time evolution equation for the option price. Instead of Eq. (20.407) we write the time evolution as

$$\frac{dO}{dt} = \frac{1}{dt}\Big[O(x(t) + \dot{x}(t)\,dt, v(t) + \dot{v}(t)\,dt, t + dt) - O(x(t), v(t), t) \Big]$$

$$= \frac{\partial O}{\partial t} + \frac{\partial O}{\partial x}\dot{x} + \frac{\partial O}{\partial v}\dot{v} + \frac{1}{2}\frac{\partial^2 O}{\partial x^2}\dot{x}^2\,dt + \frac{\partial^2 O}{\partial x \partial v}\dot{x}\dot{v}\,dt + \frac{1}{2}\frac{\partial^2 O}{\partial v^2}\dot{v}^2 dt + \ldots . \tag{20.460}$$

The expansion can be truncated after the second derivative due to the Gaussian nature of the fluctuations. We use Itō's rule to replace

$$\dot{x}^2 \longrightarrow v(t), \qquad \dot{v}^2 \longrightarrow \epsilon^2 v(t), \qquad \dot{x}\dot{v} \longrightarrow \rho\epsilon v(t). \tag{20.461}$$

These replacements follow directly from Eqs. (20.306) and (20.308) and the correlation functions (20.310). Thus we obtain

$$\frac{dO}{dt} = \frac{1}{dt}\Big[O(x(t) + \dot{x}(t)\,dt, v(t) + \dot{v}(t)\,dt, t + dt) - O(x(t), v(t), t) \Big]$$

$$= \frac{\partial O}{\partial t} + \frac{\partial O}{\partial x}\dot{x} + \frac{\partial O}{\partial v}\dot{v} + \frac{1}{2}\frac{\partial^2 O}{\partial x^2}v + \frac{\partial^2 O}{\partial x \partial v}\rho\epsilon v + \frac{1}{2}\frac{\partial^2 O}{\partial v^2}\epsilon^2 v. \tag{20.462}$$

This is inserted into Eq. (20.403). If we adjust the portfolio according to the rule (20.404), and use the Itō relation $\dot{S}/S = \dot{x} + v/2$, we obtain the equation [compare (20.409)]

$$N_O r_W \left(-\frac{\partial O}{\partial x} + O \right) = -N_O \frac{\partial O}{\partial x}\left(\dot{x} + \frac{v^2}{2} \right) + N_O \dot{O}$$

$$= N_O \left[-\frac{v^2}{2} \frac{\partial O}{\partial x} + \frac{\partial O}{\partial v} \dot{v} + \frac{\partial O}{\partial t} + \frac{1}{2} \frac{\partial^2 O}{\partial x^2} v + \frac{\partial^2 O}{\partial x \partial v} \rho \epsilon v + \frac{1}{2} \frac{\partial^2 O}{\partial v^2} \epsilon^2 v + \dots \right]. \quad (20.463)$$

As before in Eq. (20.409), the noise in $\dot{x}$ has disappeared. In contrast to the single-variable treatment, however, the noise variable η_v remains in the equation. It can only be removed if we trade a financial asset V whose price is equal to the variance directly on the markets. Then we can build a *riskfree portfolio* containing four assets

$$W(t) = N_S(t)S(t) + N_O(t)O(S,t) + N_V(t)V(t) + N_B(t)B(t), \quad (20.464)$$

instead of (20.396). Indeed, by adjusting

$$\frac{N_V(t)}{N_O(t)} = -\frac{\partial O(S(t), v(t), t)}{\partial v(t)}, \quad (20.465)$$

we could cancel the term $\dot{v}$ in Eq. (20.463). There is definitely need to establish trading in such an asset. Without this, we can only reach an approximate freedom of risk by ignoring the noise $\eta_v(t)$ in the term $\dot{v}$ and replacing $\dot{v}$ by the deterministic first term in the stochastic differential equation (20.308):

$$\dot{v}(t) \longrightarrow -\gamma[v(t) - \bar{v}]. \quad (20.466)$$

In addition, we can account for the fact that the option price rises with the variance as in the Black-Scholes formula by adding on the right hand side of (20.466) a phenomenological correction term $-\lambda v$ called *price of volatility risk* [78, 11]. Such a term has simply the effect of renormalizing the parameters γ and $\bar{v}$ to

$$\gamma^* = \gamma + \lambda, \quad \text{and} \quad \bar{v}^* = \gamma \bar{v}/\gamma^*. \quad (20.467)$$

Thus we find the Fokker-Planck-like differential equation [compare (20.417)]

$$\frac{\partial O}{\partial t} = r_W O - \left(r_W - \frac{v}{2} \right) \frac{\partial O}{\partial x} + \gamma^*[v(t) - \bar{v}^*] \frac{\partial O}{\partial v} - \frac{v}{2} \frac{\partial^2 O}{\partial x^2} - \rho \epsilon v \frac{\partial^2 O}{\partial x \partial v} - \frac{\epsilon^2 v}{2} \frac{\partial^2 O}{\partial v^2}. \quad (20.468)$$

On the right-hand side we recognize the Hamiltonian operator (20.319), with γ and $\bar{v}$ replaced by γ^* and $\bar{v}^*$, in terms of which we can write

$$\frac{\partial O}{\partial t} = r_W (O - \partial_x O) + \left(\hat{H}^* + \gamma^* + \rho \epsilon \partial_x + \epsilon^2 \partial_v \right) O. \quad (20.469)$$

The solution of this equation can easily be expressed as a slight modification of the solution $P(x_b, v_b, t_b | x_a v_a t_a)$ in Eq. (20.340) of the differential equation (20.318). Since it contains only additional first-order derivatives with respect to (20.318), we simply find, with $x \equiv x_a$ and $t \equiv t_a$, the solution $P^v(x_b, v_b, t_b | x_a v_a t_a)$ satisfying the initial condition

$$P^v(x_b, v_b, t_b | x_a v_a t_a) = \delta(x_b - x_a)\delta(v_b - v_a) \quad (20.470)$$

as follows:

$$P^v(x_b, v_b, t_b | x_a v_a t_a) = e^{-(r_W + \gamma^*)\Delta t} P_{\text{sh}}(x_b, v_b, t_b | x_a v_a t_a), \qquad (20.471)$$

where the subscript sh indicates that the arguments $x_b - x_a$ and $v_b - v_a$ are shifted:

$$x_b - x_a \rightarrow x_b - x_a - (r_W - \rho \epsilon), \qquad v_b - v_a \rightarrow v_b - v_a - \epsilon^2. \qquad (20.472)$$

The distribution (20.470) may be inserted into an equation of the type (20.425) to find the option price at the time t_a from the price (20.423) at the expiration date t_b. If we assume the variance v_a to be equal to $\bar{v}$, and the remaining parameters to be

$$\gamma^* = 2, \quad \bar{v} = 0.01, \quad \epsilon = 0.1, r_W = 0, \qquad (20.473)$$

the price of an option with strike price $E = 100$ one half year before expiration with the stock price $S = E$ (this is called an option *at-the-money*) is 2.83 US\$ for $\rho = -0.5$ and 2.81 US\$ for $\rho = 0.5$. The difference with respect to the Black-Scholes price is shown in Fig. 20.31.

20.7.7 Perturbation Expansion and Smile

A perturbative treatment of any non-Gaussian distributions $\tilde{D}(x)$, which we assume to be symmetric, for simplicity, starts from the expansion

$$D(p) = \left[1 + \frac{a_4}{4!}p^4 - \frac{a_6}{6!}p^6 + \frac{c_8}{8!}p^8 - \dots - \dots\right] e^{-\sigma^2 p^2/2}, \qquad (20.474)$$

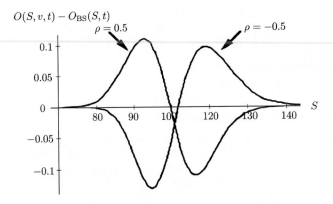

Figure 20.31 Difference between option price $O(S, v, t)$ with fluctuating volatility and Black-Scholes price $O_{\text{BS}}(S, t)$ with $\sigma^2 = \bar{v}$ for option of strike price 100 US\$. The parameters are given in Eq. (20.473). The noise correlation parameter is once $\rho = -0.5$ and once $\rho = 0.5$. For an at-the-money option the absolute value is 2.83 US\$ for $\rho = -0.5$ and 2.81 US\$ for $\rho = 0.5$ (after Ref. [78]).

where

$$a_4 = c_4, \quad a_6 = c_6, \quad a_8 = c_8 + 35c_4^2, \quad a_{10} = c_{10} + 210c_4c_6, \dots \,, \tag{20.475}$$

which can also be expressed as a series

$$D(p) = \left[1 + \frac{a_4}{4!} 2^2 \left(\frac{\partial}{\partial \sigma^2} \right)^2 + \frac{a_6}{6!} 2^3 \left(\frac{\partial}{\partial \sigma^2} \right)^3 + \frac{a_8}{8!} 2^4 \left(\frac{\partial}{\partial \sigma^2} \right)^4 + \dots \right] e^{-\sigma^2 p^2/2}. \tag{20.476}$$

By taking the Fourier transform we obtain the expansion of the distributions in x-space

$$
\begin{aligned}
\tilde{D}(x) &= \left[1 + \frac{a_4}{4!} 2^2 \left(\frac{\partial}{\partial \sigma^2} \right)^2 + \frac{a_6}{6!} 2^3 \left(\frac{\partial}{\partial \sigma^2} \right)^3 + \frac{a_8}{8!} 2^4 \left(\frac{\partial}{\partial \sigma^2} \right)^4 + \dots \right] \frac{e^{-x^2/2\sigma^2}}{\sqrt{2\pi\sigma^2}} \\
&= \left[1 + \frac{\bar{c}_4}{8} - \frac{\bar{c}_6}{48} + \frac{35\bar{c}_4^2}{384} + \frac{\bar{c}_8}{384} + \frac{x^2}{\sigma^2} \left(-\frac{\bar{c}_4}{4} + \frac{\bar{c}_6}{16} - \frac{35\bar{c}_4^2}{96} - \frac{\bar{c}_8}{96} \right) \right. \\
&\quad + \frac{x^4}{\sigma^4} \left(\frac{\bar{c}_4}{24} - \frac{\bar{c}_6}{48} + \frac{35\bar{c}_4^2}{192} + \frac{\bar{c}_8}{192} \right) + \frac{x^6}{\sigma^6} \left(\frac{\bar{c}_6}{720} - \frac{35\bar{c}_4^2}{1440} - \frac{\bar{c}_8}{1440} \right) \\
&\quad \left. + \frac{x^8}{\sigma^8} \left(\frac{35\bar{c}_4^2}{40320} + \frac{\bar{c}_8}{40320} \right) + \dots \right] \frac{e^{-x^2/2\sigma^2}}{\sqrt{2\pi\sigma^2}}.
\end{aligned}
\tag{20.477}
$$

The quantities $\bar{c}_n$ contain the kurtosis κ, in the case of a truncated Lévy distribution with the powers $\varepsilon^{n/2-1}$. If the distribution is close to a Gaussian, we may re-expand all expressions in powers of the higher cumulants. In the case of a truncated Lévy distribution, we may keep systematically all terms up to a certain maximal power of ε and find

$$
\begin{aligned}
\tilde{D}(x) &= \left\{ 1 - \frac{x^2}{2\sigma^2} \left(\frac{\bar{c}_4}{2} - \frac{\bar{c}_6}{8} + \frac{2\bar{c}_4^2}{3} + \frac{\bar{c}_8}{48} + \frac{5\bar{c}_4\bar{c}_6}{192} - \frac{33\bar{c}_4^3}{256} \right) \right. \\
&\quad + \frac{x^4}{\sigma^4} \left(\frac{\bar{c}_4}{24} - \frac{\bar{c}_6}{48} + \frac{17\bar{c}_4^2}{96} + \frac{\bar{c}_8}{192} + \frac{\bar{c}_4\bar{c}_6}{288} - \frac{239\bar{c}_4^3}{9216} \right) \\
&\quad + \frac{x^6}{\sigma^6} \left(\frac{\bar{c}_6}{720} - \frac{7\bar{c}_4^2}{288} - \frac{\bar{c}_8}{1440} - \frac{\bar{c}_4\bar{c}_6}{5760} + \frac{7\bar{c}_4^3}{2304} \right) \\
&\quad \left. + \frac{x^8}{\sigma^8} \left(\frac{\bar{c}_4^2}{1152} + \frac{\bar{c}_8}{40320} - \frac{\bar{c}_4^3}{9216} \right) + \dots \right\} \frac{e^{-x^2/2\sigma^2}}{\sqrt{2\pi\sigma_1^2}},
\end{aligned}
\tag{20.478}
$$

where we have introduced the modified width

$$\tilde{\sigma}_1^2 \equiv \sigma^2 \left(1 - \frac{\bar{c}_4}{4} + \frac{\bar{c}_6}{24} - \frac{13\bar{c}_4^2}{96} - \frac{\bar{c}_8}{192} - \frac{\bar{c}_4\bar{c}_6}{64} - \frac{31\bar{c}_4^3}{512} + \dots \right). \tag{20.479}$$

The prefactor can be taken into the exponent yielding

$$
\begin{aligned}
\tilde{D}(x) &= \exp \left\{ -\frac{x^2}{2\sigma^2} \left(1 + \frac{\bar{c}_4}{2} + \frac{2\bar{c}_4^2}{3} - \frac{33\bar{c}_4^3}{256} - \frac{\bar{c}_6}{8} + \frac{5\bar{c}_4\bar{c}_6}{192} + \frac{\bar{c}_8}{48} \right) \right. \\
&\quad + \frac{x^4}{\sigma^4} \left(\frac{\bar{c}_4}{24} + \frac{7\bar{c}_4^2}{48} - \frac{1007\bar{c}_4^3}{9216} - \frac{\bar{c}_6}{48} + \frac{11\bar{c}_4\bar{c}_6}{576} + \frac{\bar{c}_8}{192} \right) \\
&\quad + \frac{x^6}{\sigma^6} \left(-\frac{\bar{c}_4^2}{72} + \frac{43\bar{c}_4^3}{768} + \frac{\bar{c}_6}{720} - \frac{23\bar{c}_4\bar{c}_6}{2880} - \frac{\bar{c}_8}{1440} \right) \\
&\quad \left. + \frac{x^8}{\sigma^8} \left(-\frac{101\bar{c}_4^3}{9216} + \frac{7\bar{c}_4\bar{c}_6}{5760} + \frac{\bar{c}_8}{40320} \right) + \dots \right\} \frac{1}{\sqrt{2\pi\sigma_1^2}}.
\end{aligned}
\tag{20.480}
$$

Introducing a second modified width

$$\tilde{\sigma}_2^2 \equiv \sigma^2 \left(1 - \frac{\bar{c}_4}{2} + \frac{\bar{c}_6}{8} - \frac{5\bar{c}_4^2}{12} - \frac{\bar{c}_8}{48} - \frac{29\bar{c}_4\bar{c}_6}{192} + \frac{515\bar{c}_4^3}{768} + \dots \right), \tag{20.481}$$

the exponential can be brought to the form

$$
\tilde{D}(x) = \exp\left\{ -\frac{x^2}{\tilde{\sigma}_2^2} + \frac{x^4}{\tilde{\sigma}_2^4}\left(\frac{\bar{c}_4}{24} - \frac{\bar{c}_6}{48} + \frac{5\bar{c}_4^2}{48} + \frac{\bar{c}_8}{192} + \frac{29\bar{c}_4\bar{c}_6}{576} - \frac{2576\bar{c}_4^3}{9216}\right) \right.
$$
$$
+ \frac{x^6}{\tilde{\sigma}_2^6}\left(\frac{\bar{c}_6}{720} - \frac{\bar{c}_4^2}{72} - \frac{8\bar{c}_8}{1440} - \frac{29\bar{c}_4\bar{c}_6}{2880} + \frac{59\bar{c}_4^3}{768}\right)
$$
$$
\left. + \frac{x^8}{\tilde{\sigma}_2^8}\left(\frac{\bar{c}_8}{40320} + \frac{7\bar{c}_4\bar{c}_6}{5760} - \frac{101\bar{c}_4^3}{9216}\right) + \dots \right\} \frac{1}{\sqrt{2\pi\tilde{\sigma}_1^2}}. \tag{20.482}
$$

The exponential can be re-expressed compactly in the quasi-Gaussian form

$$
\tilde{D}(x) = \frac{1}{\sqrt{2\pi\tilde{\sigma}_1^2}} \exp\left[-\frac{x^2}{2\Sigma^2(x)} \right], \tag{20.483}
$$

where we have defined an x-dependent width

$$
\tilde{\Sigma}(x) \equiv \sigma_2^2\left[1 + \frac{x^2}{\sigma_2^2}\left(\frac{\bar{c}_4}{12} + \frac{5\bar{c}_4^2}{24} - \frac{2575\bar{c}_4^3}{4608} - \frac{\bar{c}_6}{24} + \frac{29\bar{c}_4\bar{c}_6}{288} + \frac{\bar{c}_8}{96}\right) \right.
$$
$$
+ \frac{x^4}{\sigma_2^4}\left(\frac{-\bar{c}_4^2}{48} + \frac{217\bar{c}_4^3}{1152} - \frac{1375\bar{c}_4^4}{27648} + \frac{\bar{c}_6}{360} - \frac{13\bar{c}_4\bar{c}_6}{480} - \frac{\bar{c}_4^2\bar{c}_6}{1728} + \frac{\bar{c}_6^2}{576} - \frac{\bar{c}_8}{720} + \frac{\bar{c}_4\bar{c}_8}{576}\right)
$$
$$
\left. + \frac{x^6}{\sigma_2^6}\left(-\frac{359\bar{c}_4^3}{13824} + \frac{127\bar{c}_4^4}{6912} + \frac{5\bar{c}_4\bar{c}_6}{1728} - \frac{13\bar{c}_4^2\bar{c}_6}{17280} - \frac{\bar{c}_6^2}{4320} + \frac{\bar{c}_8}{20160} - \frac{\bar{c}_4\bar{c}_8}{4320}\right) + \dots \right]. \tag{20.484}
$$

For small x the deviation of $\Sigma^2(x)$ from a constant σ^2 is dominated by the quadratic term. If one plots the fluctuation width $\Sigma(x)$ for the logarithms of the observed option prices one finds the smile parabola shown before in Fig. 20.29. On the basis of the expansion (20.476) it is possible to derive an approximate option price formula for assets fluctuating according to the truncated Lévy distribution. We simply apply the differential operator in front of the Gaussian distribution

$$
\mathcal{O} \equiv \left[1 + \frac{a_4}{4!}2^2\left(\frac{\partial}{\partial\sigma^2}\right)^2 + \frac{a_6}{6!}2^3\left(\frac{\partial}{\partial\sigma^2}\right)^3 + \frac{a_8}{8!}2^4\left(\frac{\partial}{\partial\sigma^2}\right)^4 + \dots \right] \tag{20.485}
$$

to the Gaussian distribution (20.421). For $\lambda = 3/2$, the coefficients are

$$
a_4 = \frac{\varepsilon}{M^2}, \quad a_6 = \frac{5 \cdot 7\,\varepsilon^2}{3M^3}, \quad a_8 = \frac{5 \cdot 7 \cdot 11\,\varepsilon^2}{M^4} + \frac{5 \cdot 7\,\varepsilon^2}{M^4},
$$
$$
a_{10} = \frac{5^2 \cdot 7 \cdot 11 \cdot 13\,\varepsilon^4}{3M^5} + \frac{2^2 \cdot 5 \cdot 7 \cdot 17\,\varepsilon^4}{3M^5}, \dots. \tag{20.486}
$$

The operator $\mathcal{O}$ winds up in front of the Black-Scholes expression (20.438), such that we obtain the formal result

$$
O^L(x_a, t_a) = \mathcal{O}O(x_a, t_a) = \mathcal{O}\left[S(t_a)N(y_+) - e^{-(r_{xw} + \sigma^2/2)(t_b - t_a)}E\,N(y_-) \right], \tag{20.487}
$$

where we have used (20.418) to exhibit the full σ-dependence on the right-hand side. This expression may now be expanded in powers of the kurtosis κ. The term proportional to a_4 yields a first correction to the Black-Scholes formula, linear in ε,

$$
O_1(x_a, t_a) = -\frac{\varepsilon}{12M^2}\left\{ \left(Se^{-y_+^2/2} - e^{-r_W(t_b - t_a)}Ee^{-y_-^2/2}\right)\frac{y_m}{\sqrt{2\pi}\sigma^2} \right.
$$
$$
\left. -e^{-r_W(t_b - t_a)}(t_b - t_a)EN(y_-) \right\}. \tag{20.488}
$$

The term proportional to a_6 adds a correction proportional to ε^2:

$$O_2(x_a, t_a) = \frac{35\varepsilon^2}{3M^3} \frac{1}{2^3 \cdot 3^2 \cdot 5} \left\{ \left[S e^{-y_+^2/2} \left(y_+ y_-^2 - 3y_+ + 4\sqrt{\sigma^2(t_b - t_a)} \right) \right. \right.$$
$$\left. - e^{-rw(t_b - t_a)} E e^{-y_-^2/2} \left(y_-^3 - 3y_- - 2\sigma^2(t_b - t_a)y_- \right) \right] \frac{1}{\sqrt{2\pi}\sigma^4}$$
$$\left. + e^{-rw(t_b - t_a)}(t_b - t_a)^2 E N(y_-) \right\}. \tag{20.489}$$

The next term proportional to a_8 adds corrections proportional to ε^2 and ε^3:

$$O_3(x_a, t_a) = \frac{385\varepsilon^3 + 35\varepsilon^2}{M^4} \frac{1}{2^6 \cdot 3^2 \cdot 5 \cdot 7}$$
$$\times \left\{ \left[S e^{-y_+^2/2} \left(-y_-^3 y_+^2 + 9y_- y_+^2 + y_-^3 - 15y_+ + 18 + \sqrt{\sigma^2(t_b - t_a)}(12y_- y_+ + 18) \right) \right. \right.$$
$$- e^{-rw(t_b - t_a)} E e^{-y_-^2/2} \left(y_-^5 - 10y_-^3 + 15y_- + \sigma^2(t_b - t_a)(3y_-^3 - 9y_-) \right.$$
$$\left. \left. + \sigma^4(t_b - t_a)^2 \, 3y_- \right) \right] \frac{1}{\sqrt{2\pi}\sigma^6} + e^{-rw(t_b - t_a)}(t_b - t_a)^3 E N(y_-) \right\}. \tag{20.490}$$

Since the expansion is asymptotic, an efficient resummation scheme will be needed for practical applications.

Appendix 20A Large-x Behavior of Truncated Lévy Distribution

Here we derive the divergent asymptotic expansion in the large-x regime. Using the variable $y \equiv p/\alpha$ and the constant $a \equiv -s\alpha^\lambda$, we may write the Fourier integral (20.24) as

$$\tilde{L}_{\sigma^2}^{(\lambda,\alpha)}(x) = \alpha e^{-2a} \int_{-\infty}^{\infty} \frac{dy}{2\pi} e^{i\alpha yx} e^{a[(1-iy)^\lambda + (1+iy)^\lambda]}. \tag{20A.1}$$

Expanding the last exponential in a Taylor series, we obtain

$$\tilde{L}_{\sigma^2}^{(\lambda,\alpha)}(x) = \alpha e^{-2a} \int_{-\infty}^{\infty} \frac{dy}{2\pi} e^{iy\alpha x} \sum_{n=0}^{\infty} \frac{a^n}{n!} [(1-iy)^\lambda + (1+iy)^\lambda]^n =$$
$$= \alpha e^{-2a} \int_{-\infty}^{\infty} \frac{dy}{2\pi} e^{i\alpha xy} \sum_{n=0}^{\infty} \frac{a^n}{n!} \sum_{m=0}^{n} \binom{n}{m} (1-iy)^{\lambda(n-m)}(1+iy)^{\lambda m} \tag{20A.2}$$

containing binominal coefficients. Changing the order of summations yields

$$\tilde{L}_{\sigma^2}^{(\lambda,\alpha)}(x) = \alpha e^{-2a} \sum_{n=0}^{\infty} \frac{a^n}{n!} \sum_{m=0}^{n} \binom{n}{m} \int_{-\infty}^{\infty} \frac{dy}{2\pi} e^{i\alpha xy}(1-iy)^{\lambda(n-m)}(1+iy)^{\lambda m}. \tag{20A.3}$$

This can be written with the help of the Whittaker functions (20.30) as

$$\tilde{L}_{\sigma^2}^{(\lambda,\alpha)}(x) = \alpha e^{-2a} \sum_{n=0}^{\infty} \frac{a^n}{n!} \sum_{m=0}^{n} \binom{n}{m} \frac{(\alpha x)^{-1-\lambda n/2} 2^{\lambda n/2}}{\Gamma(-\lambda m)} W_{\lambda n/2 - \lambda m, (\lambda n + 1)/2}(2\alpha x). \tag{20A.4}$$

After converting the Gamma functions of negative arguments into those of positive arguments, this becomes

$$\tilde{L}_{\sigma^2}^{(\lambda,\alpha)}(x) = -\frac{\alpha}{\pi}e^{-2a}\sum_{n=1}^{\infty}a^n\sum_{m=0}^{n}\frac{2^{\lambda n/2}\Gamma(1+\lambda m)\sin(\pi\lambda m)}{(\alpha x)^{1+\lambda n/2}m!(n-m)!}W_{\lambda n/2-\lambda m,(\lambda n+1)/2}(2\alpha x).$$

(20A.5)

The Whittaker functions $W_{\lambda,\gamma}(x)$ have the following asymptotic expansion

$$W_{\lambda,\gamma}(x) = e^{-x/2}x^{\lambda}\left\{1+\sum_{k=1}^{\infty}\frac{1}{k!x^k}\prod_{j=1}^{k}\left[\gamma^2-(\lambda-j+1/2)^2\right]\right\}.$$

(20A.6)

For $\gamma \equiv (\lambda n+1)/2$ and $\lambda \equiv \lambda(n-2m)/2$, the product takes the form

$$\prod_{j=1}^{k}[\gamma^2-(\lambda-j+1/2)^2] = \prod_{j=1}^{k}\left\{\left(\frac{\lambda n+1}{2}\right)^2-\left[\frac{\lambda(n-2m)}{2}-j+\frac{1}{2}\right]^2\right\}$$

$$= \prod_{j=1}^{k}(\lambda m+j)(\lambda n+1-\lambda m-j).$$

(20A.7)

Inserting this into (20A.6) and the result into (20A.5), we obtain the asymptotic expansion for large x:

$$\tilde{L}_{\sigma^2}^{(\lambda,\alpha)}(x) = -\frac{1}{\pi}e^{-2a}\frac{e^{-\alpha x}}{x}\sum_{n=1}^{\infty}a^n\,2^{\lambda n}\sum_{m=1}^{n}\frac{\Gamma(1+\lambda m)\sin(\pi\lambda m)}{m!(n-m)!}(2\alpha x)^{-\lambda m}$$

$$\times\left[1+\sum_{k=1}^{\infty}\frac{\prod_{j=1}^{k}(\lambda m+j)(\lambda n+1-\lambda m-j)}{k!(2\alpha x)^k}\right].$$

(20A.8)

We have raised the initial value of the index of summation m by one unit since $\sin(\pi\lambda m)$ vanishes for $m=0$. If we define a product of the form $\prod_{j=1}^{0}\ldots$ to be equal to unity, we can write the term in the last bracket as

$$\sum_{k=0}^{\infty}\frac{1}{k!(2\alpha x)^k}\prod_{j=1}^{k}(j+\lambda m)(1-\lambda m-j+\lambda n).$$

(20A.9)

Rearranging the double sum in (20A.8), we write $\tilde{L}_{\sigma^2}^{(\lambda,\alpha)}(x)$ as

$$\tilde{L}_{\sigma^2}^{(\lambda,\alpha)}(x) = -\frac{1}{\pi}e^{-2a}\frac{e^{-\alpha x}}{x}\sum_{m=1}^{\infty}\frac{\Gamma(1+\lambda m)\sin(\pi\lambda m)}{m!(2\alpha x)^{\lambda m}}\sum_{n=m}^{\infty}\frac{a^n\,2^{\lambda n}}{(n-m)!}$$

$$\times\sum_{k=0}^{\infty}\frac{1}{k!(2\alpha x)^k}\prod_{j=1}^{k}(j+\lambda m)(1-\lambda m-j+\lambda n),$$

(20A.10)

and further as

$$\tilde{L}_{\sigma^2}^{(\lambda,\alpha)}(x) = -\frac{1}{\pi}e^{-2a}\frac{e^{-\alpha x}}{x}\sum_{m=1}^{\infty}\frac{\Gamma(1+\lambda m)\sin(\pi\lambda m)}{m!}(2\alpha x)^{-\lambda m}$$

$$\times\sum_{n=0}^{\infty}(2^{\lambda}a)^{n+m}\frac{1}{n!}\sum_{k=0}^{\infty}\frac{1}{k!(2\alpha x)^k}\prod_{j=1}^{k}(j+\lambda m)(1-j+\lambda n) =$$

$$= -\frac{1}{\pi}e^{-2a}\frac{e^{-\alpha x}}{x}\sum_{k=0}^{\infty}\frac{1}{k!(2\alpha x)^k}\sum_{m=1}^{\infty}\frac{\Gamma(1+\lambda m)\sin(\pi\lambda m)(2^\lambda a)^m}{(2\alpha x)^{\lambda m}m!}$$

$$\times \prod_{j'=1}^{k}(j'+\lambda m)\sum_{n=0}^{\infty}\frac{(2^\lambda a)^n}{n!}\prod_{j=1}^{k}(\lambda n+1-j). \qquad (20A.11)$$

The last sum over n in this expression can be re-expressed more efficiently with the help of a generating function

$$f^{(k)}(y^\lambda)\equiv\frac{d^k}{dy^k}e^{y^\lambda}, \qquad (20A.12)$$

whose Taylor series is

$$f^{(k)}(y^\lambda)=\frac{d^k}{dy^k}\sum_{n=0}^{\infty}\frac{y^{\lambda n}}{n!}=\sum_{n=0}^{\infty}\frac{1}{n!}\prod_{j=1}^{k}(\lambda n-j+1)y^{\lambda n-k}, \qquad (20A.13)$$

leading to

$$\tilde{L}_{\sigma^2}^{(\lambda,\alpha)}(x) = \frac{1}{\pi}e^{-2a}\frac{e^{-\alpha x}}{x}\sum_{k=0}^{\infty}\frac{a^{k/\lambda}2^k f^{(k)}(2^\lambda a)}{k!(2\alpha x)^k}$$

$$\times \sum_{m=1}^{\infty}\frac{\Gamma(1+\lambda m)\sin(\pi\lambda m)(2^\lambda a)^m}{m!}(2\alpha x)^{-\lambda m}\prod_{j=1}^{k}(\lambda m+j).$$

$$(20A.14)$$

This can be rearranged to

$$\tilde{L}_{\sigma^2}^{(\lambda,\alpha)}(x) = -\frac{e^{-2a}}{\pi}e^{-\alpha x}\sum_{k=0}^{\infty}\frac{(-s)^{k/\lambda}f^{(k)}(-s(2\alpha)^\lambda)}{k!}$$

$$\times \sum_{m=1}^{\infty}\frac{\Gamma(1+\lambda m+k)\sin(\pi\lambda m)(-s)^m}{m!x^{1+\lambda m+k}}.$$

$$(20A.15)$$

Thus we obtain the asymptotic expansion

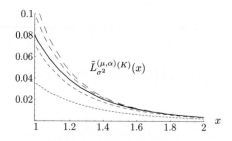

Figure 20.32 Comparison of large-x expansions containing different numbers of terms (with $K=0$, 1, 2, 3, 4, 5, with increasing dash length) with the tails of the truncated Lévy distribution for $\lambda=\sqrt{2}$, $\sigma=0.5$, $\alpha=1$.

$$\tilde{L}_{\sigma^2}^{(\lambda,\alpha)}(x) = -\frac{e^{-2a}}{\pi} \frac{e^{-\alpha x}}{x} \sum_{k=0}^{\infty} A_k \sum_{m=1}^{\infty} B_{km} \frac{(-s)^m}{x^{\lambda m+k}}, \tag{20A.16}$$

where

$$A_k = \frac{(-s)^{k/\lambda} f^{(k)}(-s(2\alpha)^\lambda)}{k!}, \qquad B_{km} = \frac{\Gamma(1+\lambda m+k)\sin(\pi\lambda m)}{m!}. \tag{20A.17}$$

We shall denote by $\tilde{L}_{\sigma^2}^{(\lambda,\alpha)(K)}(x)$ the approximants in which the sums over K are truncated after the Kth term. To have a definite smallest power in $|x|$, the sum over m is truncated after the smallest integer larger than $(K-k+\lambda)/\lambda$. The leading term is

$$\begin{aligned}
\tilde{L}_{\sigma^2}^{(\lambda,\alpha)(0)}(x) &= -\frac{e^{-2a}}{\pi} e^{-\alpha x} f^{(0)}(-s(2\alpha)^\lambda) \frac{\Gamma(1+\lambda)\sin(\pi\lambda)(-s)}{x^{1+\lambda}} \\
&= s\frac{e^{\alpha^\lambda s(2-2^\lambda)}}{\pi} \frac{\Gamma(1+\lambda)\sin(\pi\lambda)}{x^{1+\lambda}} e^{-\alpha x}. \tag{20A.18}
\end{aligned}$$

The large-x approximations $\tilde{L}_{\sigma^2}^{(\lambda,\alpha)(0)}(x)$ are compared with the numerically calculated truncated Lévy distribution in Fig. 20.32.

Appendix 20B   Gaussian Weight

For simplicity, let us study the Gaussian content in the final distribution (20.347) only for the simpler case $\rho = 0$. Then we shift the contour of integration in (20.347) to run along $p + i/2$, and we study the Fourier integral

$$P(x\,t\,|x_a t_a) = e^{-\Delta x/2} \int_{-\infty}^{+\infty} \frac{dp}{2\pi} e^{ip\Delta x - \bar{H}(p,\Delta t)}, \tag{20B.1}$$

where

$$\bar{H}(p,\Delta t) = -\frac{\gamma^2 \bar{v} t}{\kappa^2} + \frac{2\gamma\bar{v}}{\kappa^2} \ln\left[\cosh\frac{\Omega t}{2} + \frac{\Omega^2 + \gamma^2}{2\gamma\Omega} \sinh\frac{\Omega t}{2}\right], \tag{20B.2}$$

and

$$\Omega = \sqrt{\gamma^2 + \kappa^2(p^2 + 1/4)}. \tag{20B.3}$$

The function $\bar{H}(p,\Delta t)$ is real and symmetric in p. The integral (20B.1) is therefore a symmetric function of Δx. The only source of asymmetry of $P(x\,t\,|x_a t_a)$ in Δx is the exponential prefactor in (20B.1).

Let us expand the integral in (20B.1) for small Δx:

$$P(x\,t\,|x_a t_a) \approx e^{-\Delta x/2}\left[\mu_0 - \frac{1}{2}\mu_2(\Delta x)^2\right] \approx \mu_0 e^{-\Delta x/2} e^{-\mu_2(\Delta x)^2/2\mu_0}, \tag{20B.4}$$

where the coefficients are the first and the second moments of $\exp[\bar{H}(p,\Delta t)]$

$$\mu_0(\Delta t) = \int_{-\infty}^{+\infty} \frac{dp}{2\pi} e^{-\bar{H}(p,\Delta t)}, \qquad \mu_2(\Delta t) = \int_{-\infty}^{+\infty} \frac{dp}{2\pi} p^2 e^{-\bar{H}(p,\Delta t)}. \tag{20B.5}$$

If we ignore the existence of semi-heavy tails and extrapolate the Gaussian expression on the right-hand side to $\Delta x \in (-\infty, \infty)$, the total probability contained in such a Gaussian extrapolation will be the fraction

$$f(\Delta t) = \int_{-\infty}^{+\infty} d\Delta x\, \mu_0 e^{-\Delta x/2 - \mu_2 \Delta x^2/2\mu_0} = \sqrt{\frac{2\pi\mu_0^3}{\mu_2}} e^{\mu_0/8\mu_2}. \tag{20B.6}$$

This is always less than 1 since the integral (20B.6) ignores the probability contained in the semi-heavy tails. The difference $1 - f(\Delta t)$ measures the relative contribution of the semi-heavy tails. The parameters $\mu_0(\Delta t)$ and $\mu_2(\Delta t)$ are calculated numerically and the resulting fraction $f(\Delta t)$ is plotted in Fig. 20.25 as a function of Δt. For $\Delta t \to \infty$, the distribution becomes Gaussian, whereas for small Δt, it becomes a broad function of p.

Appendix 20C   Comparison with Dow-Jones Data

For the comparison of the theory in Section 20.4 with actual financial data shown in Fig. 20.22, the authors of Ref. [79] downloaded the daily closing values of the Dow-Jones industrial index for the period of 20 years from 1 January 1982 to 31 December 2001 from the Web site of Yahoo [108]. The data set contained 5049 points $S(t_n)$, where the discrete time variable t_n parametrizes the days. Short days before holidays were ignored. For each t_n, they compiled the log-returns $\Delta x(t_n) = \ln S(t_{n+1})/S(t_n)$. Then they partitioned the x-axis into equally spaced intervals of width Δx and counted the number of log-returns $\Delta x(t_n)$ falling into each interval. They omitted all intervals with occupation numbers less than five, which they considered as too few to rely on. Only less than 1% of the entire data set was omitted in this way. Dividing the occupation number of each bin by Δr and by the total occupation number of all bins, they obtained the probability density for a given time interval $\Delta t = 1$ day. From this they found $P^{(DJ)}(x\,t\,|x_a t_a)$ by replacing $\Delta x \to \Delta x - r_S \Delta t$.

Assuming that the system is ergodic, so that ensemble averaging is equivalent to time averaging, they compared $P^{(DJ)}(x\,t\,|x_a t_a)$ with the calculated $P(x\,t\,|x_a t_a)$ in Eq. (20.347). The parameters of the model were determined by minimizing the mean-square deviation $\sum_{\Delta x, \Delta t} |\log P^{(DJ)}(x\,t\,|x_a t_a) - \log P(x\,t\,|x_a t_a)|^2$, with the sum taken over all available Δx and over $\Delta t = 1, 5, 20, 40$, and 250 days. These values of Δt were selected because they represent different regimes: $\gamma \Delta t \ll 1$ for $t = 1$ and 5 days, $\gamma \Delta t \approx 1$ for $t = 20$ days, and $\gamma \Delta t \gg 1$ for $t = 40$ and 250 days. As Figs. 20.22 and 20.23 illustrate, the probability density $P(x\,t\,|x_a t_a)$ calculate from the Fourier integral (20.347) with components (20.348) agrees with the data very well, not only for the selected five values of time t, but for the whole time interval from 1 to 250 trading days. The comparison cannot be extended to Δt longer than 250 days, which is approximately 1/20 of the entire range of the data set, because it is impossible to reliably extract $P^{(DJ)}(x\,t\,|x_a t_a)$ from the data when Δt is too long.

The best fits for the four parameters γ, $\bar{v}$, ε, μ are given in Table 20.1. Within the scattering of the data, there are no discernible differences between the fits with the correlation coefficient ρ being zero or slightly different from zero. Thus the correlation parameter ρ between the noise terms for stock price and variance in Eq. (20.310) is practically zero. This conclusion is in contrast with the value $\rho = -0.58$ found in [109] by fitting the leverage correlation function introduced in [110]. Further study is necessary to understand this discrepancy. All theoretical curves shown in the above figures are calculated for $\rho = 0$, and fit the data very well.

The parameters γ, $\bar{v}$, ε, μ have the dimensionality of 1/time. One row in Table 20.1 gives their values in units of 1/day, as originally determined in our fit. The other row shows the annualized values of the parameters in units of 1/year, where one year is here equal to the average number of 252.5 trading days per calendar year. The relaxation time of variance is equal to $1/\gamma = 22.2$ trading days = 4.4 weeks $\approx$ 1 month, where 1 week = 5 trading days. Thus one finds that the variance has a rather long relaxation time, of the order of one month, which is in agreement with an earlier conclusion in Ref. [109].

Using the numbers given in Table 20.1, the value of the parameter $\varepsilon^2/2\gamma\bar{v}$ is ≈ 0.772 thus satisfying the smallness condition Eq. (20.322) ensuring that v never reaches negative values.

The stock prices have an apparent growth rate determined by the position $x_m(t)$ where the probability density is maximal. Adding this to the initially subtracted growth rate r_S we find that the apparent growth rate is $\bar{r}_S = r_S - \gamma\bar{v}/2\omega_0 = 13\%$ per year. This number coincides with the apparent average growth rate of the Dow-Jones index obtained by a simple fit of the data points S_{t_n} with an exponential function of t_n. The apparent growth rate $\bar{r}_S$ is comparable to the

Table 20.1 Parameters of equations with fluctuating variance obtained from fits to Dow-Jones data. The fit yields $\rho \approx 0$ for the correlation coefficient and $1/\gamma = 22.2$ trading days for the relaxation time of variance.

Units	γ	$\bar{v}$	ε	μ
1/day	4.50×10^{-2}	8.62×10^{-5}	2.45×10^{-3}	5.67×10^{-4}
1/year	11.35	0.022	0.618	0.143

average stock volatility after one year $\sigma = \sqrt{\bar{v}} = 14.7\%$. Moreover, the parameter (20.328) which characterizes the width of the stationary distribution of variance is equal to $v_{\max}/w = 0.54$. This means that the distribution of variance is broad, and variance can easily fluctuate to a value twice greater than the average value $\bar{v}$. As a consequence, even though the average growth rate of the stock index is positive, there is a substantial probability of $\int_{-\infty}^{0} d\Delta x P(x\,t\,|x_a t_a) \approx 17.7\%$ to have negative growth for $\Delta t = 1$ year.

According to (20.368), the asymmetry between the slopes of exponential tails for positive and negative Δx is given by the parameter p_0, which is equal to $1/2$ when $\rho = 0$ [see also the discussion of Eq. (20B.1) in Appendix 20B]. The origin of this asymmetry can be traced back to the transformation from $\dot{S}(t)/S(t)$ to $\dot{x}(t)$ using Itô's formula. This produces a term $v(t)/2$ in Eq. (20.306), which leads to the first term in the Hamiltonian operator (20.319). For $\rho = 0$ this is the only source of asymmetry in Δx of $P(x\,t\,|x_a t_a)$. In practice, the asymmetry of the slopes $p_0 = 1/2$ is quite small (about 2.7%) compared to the average slope $q_*^{\pm} \approx \omega_0/\varepsilon = 18.4$.

Notes and References

Option pricing beyond Black and Scholes via path integrals was discussed in Ref. [111], where strategies are devised to minimize risks in the presence of extreme fluctuations, as occur on real markets. See also Ref. [112].
Recently, a generalization of path integrals to functional integrals over surfaces has been proposed in Ref. [113] as an alternative to the Heath-Jarrow-Morton approach of modeling yield curves (see http://risk.ifci.ch/00011661.htm).

The individual citations refer to

[1] The development of the Dow Jones industrial index up to date can be found on internet sites such as http://stockcharts.com/charts/historical/djia1900.html. Alternatively they can be plotted directly using Stephen Wolfram's program *Mathematica* (v.6) using the small program
```
t=FinancialData["^DJI",All];l = Length[t]
tt = Table[t[[k]][[1]][[1]]+(t[[k]][[1]][[2]]-1)/12,t[[k]][[2]], {k,1,1}]
ListLogPlot[tt]
```
Note that the last stagnation period was *predicted* in the 3rd edition of this book in 2004.

[2] P. Fizeau, Y. Liu, M. Meyer, C.-K. Peng, and H.E. Stanley, *Volatility Distribution in the S&P500 Stock Index*, Physica A **245**, 441 (1997) (cond-mat/9708143).

[3] B.E. Baaquie, *A Path Integral Approach to Option Pricing with Stochastic Volatility: Some Exact Results*, J. de Physique I **7**, 1733 (1997) (cond-mat/9708178).

[4] J.P. Fouqué, G. Papanicolaou, and K.R. Sircar, *Derivatives in Financial Markets with Stochastic Volatility* (Cambridge University Press, Cambridge, 2000); International Journal of Theoretical and Applied Finance, **3**, 101 (2000).

[5] J. Hull and A. White, Journal of Finance **42**, 281 (1987); C.A. Ball and A. Roma, Journal of Financial and Quantitative Analysis **29**, 589 (1994); R. Schöbel and J. Zhu, European Finance Review **3**, 23 (1999).

[6] E.M. Stein and J.C. Stein, Review of Financial Studies **4**, 727 (1991).

[7] J.C. Cox, J.E. Ingersoll, and S.A. Ross, Econometrica **53**, 385 (1985).
     In the mathematical literature such an equation runs under the name *Feller process*. See
     Ref. [4] and
     W. Feller, *Probability Theory and its Applications*, sec. ed., Vol. II., John Wiley & Sons,
     1971.
     See also related work in turbulence by
     B. Holdom, Physica A **254**, 569 (1998) (cond-mat/9709141).

[8] P. Wilmott, *Derivatives*, John Wiley & Sons, New York, 1998.

[9] R. Courant and D. Hilbert, *Methods of Mathematical Physics, vol. 2* (John Wiley & Sons,
     New York, 1962).

[10] C.M. Bender and S.A. Orszag, *Advanced Mathematical Methods for Scientists and Engineers*,
     Springer, New York, 1999.

[11] C.G. Lamoureux and W.D. Lastrapes, *Forecasting Stock-Return Variance: Toward an Un-
     derstanding of Stochastic Implied Volatilities*, Rev. of Financial Studies **6**, 293 (1993). See
     also
     D.T. Breeden, *An Intertemporal Asset Pricing Model with Stochastic Consumption and In-
     vertment Opportunities*, Jour. of Financial Economics **7**, 265 (1979).

[12] This is named after the Russian mathematician Andrei Andrejewitsch Markov who published
     in 1912 the German book on Probalility Theory cited on p. 1081 with the German spelling
     Markoff.

[13] J.-P. Bouchaud and M. Potters, *Theory of Financial Risks, From Statistical Physics to Risk
     Management*, Cambridge University Press, 2000.

[14] J. Hull, *Options, Futures and Other Derivatives*, Prentice-Hall Int., 1997.

[15] R. Rebonato, *Interest-Rate Option Models*, John Wiley & Sons, Chichester, 1996.

[16] M.W. Baxter and A.J.O. Rennie, *Financial Calculus*, Cambridge Univ. Press, Cambridge,
     1996.

[17] J. Voigt, *The Statistical Mechanics of Financial Markets*, Springer, Berlin, 2001.

[18] V. Pareto, Giornale degli Economisti, Roma, January 1895; and *Cours d'économie politique*,
     F. Rouge Editeur, Lausanne and Paris, 1896; reprinted in an edition of his complete works
     (Vol III) under the title *Écrits sur la courbe de la répartition de la richesse*, Librairie Droz,
     Geneva, 1965 (http://213.39.120.146:8200/droz/FMPro).

[19] B.B. Mandelbrot, *Fractals and Scaling in Finance*, Springer, Berlin, 1997; J. of Business
     36, 393 (1963).

[20] T. Lux, Appl. Financial Economics **6**, 463 (1996);
     M. Loretan and P.C.B. Phillips, J. Empirical Finance **1**, 211 (1994).

[21] J.P. Nolan, *Stable Distributions*, American University (Wahington D.C.) lecture notes 2004
     http://academic2.american.edu/~jpnolan/stable/chap1.pdf.

[22] R.N. Mantegna and H.E. Stanley, *Stochastic Process with Ultraslow Convergence to a Gaus-
     sian: The Truncated Lévy Flight*, Phys. Rev. Lett. **73**, 2946 (1994).

[23] P. Gopikrishnan, M. Meyer, L.A.N. Amaral, and H.E. Stanley, Europ. Phys. Journ. B **3**,
     139 (1998);
     P. Gopikrishnan, V. Plerou, L.A.N. Amaral, M. Meyer, and H.E. Stanley, *Scaling of the
     distribution of fluctuations of financial market indices*, Phys. Rev. E **60**, 5305 (1999);
     V. Plerou, P. Gopikrishnan, L.A.N. Amaral, M. Meyer, and H.E. Stanley, *Scaling of the*

distribution of price fluctuations of individual companies, Phys. Rev. E **60**, 6519 (1999);
V. Plerou, P. Gopikrishnan, X. Gabaix, H.E. Stanley, *On the Origin of Power-Law Fluctuations in Stock Prices*, Quantitative Finance **4**, C11 (2004).

[24] I. Koponen, *Analytic Approach to the Problem of Convergence of Truncated Lévy Flights Towards the Gaussian Stochastic Process*, Phys. Rev. E **52**, 1197-1199 (1995).

[25] I.S. Gradshteyn and I.M. Ryzhik, *op. cit.*, Formulas 3.382.6 and 3.382.7.

[26] ibid., *op. cit.*, Formula 3.384.9.

[27] ibid., *op. cit.*, Formula 9.220.3 and 9.220.4.

[28] This result was calculated by A. Lyashin when he visited my group in Berlin. Note that other authors have missed the prefactor in the asymptotic behavior (20.37), for example A. Matacz, *Financial Modeling and Option Theory with the Truncated Lévy Process*, cond-mat/9710197. The prefactor can be dropped only for $\alpha = 0$.

[29] ibid., *op. cit.*, Formulas 3.462.3.

[30] ibid., *op. cit.*, Formula 9.246.

[31] ibid., *op. cit.*, Formula 9.247.2.

[32] Exponential short-time behavior has been observed in Ref. [13] and numerous other authors:
L.C. Miranda and R. Riera, Physica A **297**, 509 (2001);
J.L. McCauley and G.H. Gunaratne, Physica A **329**, 178 (2003);
T. Kaizoji, Physica A **343**, 662 (2004);
R. Remer and R. Mahnke, Physica A **344**, 236 (2004);
D. Makowiec, Physica A **344**, 36 (2004);
K. Matia, M. Pal, H. Salunkay, and H.E. Stanley, Europhys. Lett. **66**, 909 (2004);
A.C. Silva, R.E. Prange, and V.M. Yakovenko, Physica A **344**, 227 (2004);
R. Vicente, C.M. de Toledo, V.B.P. Leite, and N. Caticha, Physica A **361**, 272 (2006);
A.C. Silva and V.M. Yakovenko, (physics/0608299).

[33] H. Kleinert and T. Xiao-Jiang to be published.

[34] B. Grigelionis, *Processes of Meixner Type*, Lith. Math. J. **39**, 33 (1999).

[35] W. Schoutens, *Meixner Processes in Finance*, Report 2001-002, EURANDOM, Eindhoven (www.eurandom.tue.nl/reports/2001/002wsreport.ps).

[36] O. Barndorff-Nielsen, T. Mikosch, S. Resnick, eds., *Lévy Processes — Theory and Applications* Birkhäuser, 2001.

[37] O. Barndorff-Nielsen, *Infinite Divisibility of the Hyperbolic and Generalized Inverse Gaussian Distributions*, Zeitschrift für Wahrscheinlichkeitstheorie und verwandte Gebiete **38**, 309-312 (1977).

[38] O. Barndorff-Nielsen, *Processes of Normal Inverse Gaussian Type*, Finance & Stochastics, 2, No. 1, 41-68 (1998).

[39] O. Barndorff-Nielsen, *Normal Inverse Gaussian Distributions and Stochastic Volatility Modeling*, Scandinavian Journal of Statistics **24**, 1-13 (1977).

[40] O. Barndorff-Nielsen, N. Shephard, *Modeling with Lévy Processes for Financial Econometrics*, MaPhySto Research Report **No. 16**, University of Aarhus, (2000).

[41] O. Barndorff-Nielsen, N. Shephard, *Incorporation of a Leverage Effect in a Stochastic Volatility Model*, MaPhySto Research Report **No. 18**, University of Aarhus, (1998).

[42] O. Barndorff-Nielsen, N. Shephard, *Integrated Ornstein Uhlenbeck Processes*, Research Report, Oxford University, (2000).

[43] J. Bertoin, (1996) Lévy Processes, Cambridge University Press.

[44] J. Bretagnolle, *Processus à incréments indépendants*, Ecole d'Eté de Probabilités, Lecture Notes in Mathematics, Vol. **237**, pp 1-26, Berlin, Springer, (1973).

[45] P.P. Carr, D. Madan, *Option Valuation using the Fast Fourier Transform*, Journal of Computational Finance **2**, 61-73 (1998).

[46] P.P. Carr, H. Geman, D. Madan, M. Yor, *The Fine Structure of Asset Returns: an Empirical Investigation*, Working Paper, (2000).

[47] T. Chan, *Pricing Contingent Claims on Stocks Driven by Lévy Processes*, Annals of Applied Probability **9**, 504-528, (1999).

[48] R. Cont, *Empirical Properties of Asset Returns: Stylized Facts and Statistical Issues*, Quantitative Finance **1**, No. 2, (2001).

[49] R. Cont, J.-P. Bouchaud, M. Potters, *Scaling in Financial Data: Stable Laws and Beyond*, in B. Dubrulle, F. Graner & D. Sornette (eds.): Scale invariance and beyond, Berlin, Springer, (1997).

[50] ibid., *op. cit.*, Formula 8.486.10.

[51] See http://www.fdm.uni-freiburg.de/UK.

[52] P. Jizba, H. Kleinert, *Superposition of Probability Distributions*, Phys. Rev. E **78**, 031122 (arXiv:0802.069).

[53] G. Arfken, *Mathematical Methods for Physicists*, 3rd ed., Academic Press, Orlando, FL. See ¸15.12 *Inverse Laplace Transformation*", pp. 853-861, 1985.

[54] E. Post, Trans. Amer. Math. Soc. **32** (1930) 723.

[55] H. Kleinert, *Stochastic Calculus for Assets with Non-Gaussian Price Fluctuations*, Physica A **311**, 538 (2002) (cond-mat/0203157).

[56] R.F. Pawula, Phys. Rev. **162**, 186 (1967).

[57] See http://www.physik.fu-berlin.de/~kleinert/b5/files .

[58] L. Borland, *A Theory of Non-Gaussian Option Pricing*, Quantitative Finance **2**, 415 (2002) (cond-mat/0205078).

[59] C. Tsallis, J. Stat. Phys. **52**, 479 (1988); E.M.F. Curado and C. Tsallis, J. Phys. A **24**, L69 (1991); 3187 (1991); A **25**, 1019 (1992).

[60] C. Tsallis, C. Anteneodo, L. Borland, R. Osorio, *Nonextensive Statistical Mechanics and Economics*, Physica A **324**, 89 (2003) (cond-mat/030130).

[61] P. Praetz *The Distribution of Share Price Changes*, Journal of Business **45**, 49 (1972).

[62] R. Blattberg and N. Gonedes, *A Comparison of the Stable and Student Distributions as Statistical Models of Stock Prices*, Journal of Business **47**, 244 (1972).

[63] R.S. Liptser and A.N. Shiryaev, *Theory of Martingales*, Kluwer, 1989.

[64] D. Duffie, *Dynamic asset pricing theory* , Princeton University Press, 2001, p. 22.

[65] J.M. Steele, *Stochastic Calculus and Financial Applications*, Springer, New York, 2001, p. 50.

[66] H. Kleinert, *Option Pricing from Path Integral for Non-Gaussian Fluctuations. Natural Martingale and Application to Truncated Lévy Distributions*, Physica A **312**, 217 (2002) (cond-mat/0202311).

[67] F. Esscher, *On the Probability Function in the Collective Theory of Risk*, Skandinavisk Aktuarietidskrift **15**, 175 (1932).

[68] H.U. Gerber and E.S.W. Shiu, *Option Pricing by Esscher Transforms*, Trans. Soc. Acturaries **46**, 99 (1994).

[69] J.M. Harrison and S.R. Pliska, *Martingales and Stochastic Integrals in the Theory of Continuous Trading*, Stoch. Proc. Appl. **11**, 215 (1981); *A Stochastic Calculus Model of Continuous Trading Complete Markets*, **13**, 313 (1981);

[70] E. Eberlein, J. Jacod, *On the Range of Options Prices*, Finance and Stochastics **1**, 131, (1997).

[71] E. Eberlein, U. Keller, *Hyperbolic Distributions in Finance*, Bernoulli **1**, 281-299, (1995).

[72] K. Prause, *The Generalized Hyperbolic Model: Estimation, Financial Derivatives, and Risk Measures*, Universität Freiburg Dissertation, 1999 (http://www.freidok.uni-freiburg.de/volltexte/15/pdf/15_1.pdf).

[73] E. Eberlein, U. Keller, K. Prause, *New Insights into Smile, Mispricing and Value at Risk: the Hyperbolic Model*, Journal of Business **71**, No. 3, 371-405, (1998).

[74] E. Eberlein, S. Raible, *Term Structure Models Driven by General Lévy Processes*, Mathematical Finance **9**, 31-53, (1999).

[75] See the Wikipedia page
en.wikipedia.org/wiki/Levy_skew_alpha-stable_distribution.

[76] S. Raible, *Lévy Processes in Finance*, Ph.D. Thesis, Univ. Freiburg, 2000 (http://www.freidok.uni-freiburg.de/volltexte/15/pdf/51_1.pdf).

[77] M.H.A. Davis, *A General Option Pricing Formula*, Preprint, Imperial College, London (1994). See also Ref. [70] and
T. Chan, *Pricing Contingent Claims on Stocks Driven by Lévy Processes*, Ann. Appl. Probab. **9**, 504 (1999).

[78] S.L. Heston, *A Closed-Form Solution for Options with Stochastic Volatility with Applications to Bond and Currency Options*, Review of Financial Studies **6**, 327 (1993).

[79] A.A. Drăgulescu and V.M. Yakovenko, Quantitative Finance **2**, 443 (2002) (cond-mat/0203046).

[80] P. Jizba, H. Kleinert, and P. Haener, *Perturbation Expansion for Option Pricing with Stochastic Volatility*, Berlin preprint 2007 (arXiv:0708.3012).

[81] S. Miccichè, G. Bonanno, F. Lillo, R.N. Mantegna, Physica A **314**, 756 (2002).

[82] D. Valenti, B. Spagnolo, and G. Bonanno, Physica A **314**, 756 (2002).

[83] H. Föllmer and M. Schweizer, *Hedging and Contingent Claims under Incomplete Information*, in *Applied Stochastic Analysis*, edited by M.H.A. Davis and R.J. Elliot, 389 Gordon and Breach 1991.

[84] R.F. Engle, *Autoregressive conditional heteroskedasticity with estimates of the variance of UK inflation*, Econometrica **50**, 987 (1982); *Dynamic conditional correlation–A simple class of multivariate GARCH models*, J. of Business and Econ. Stat., **20**, 339 (3002) (http://www.physik.fu-berlin.de/~kleinert/finance/engle1.pdf);
R.F. Engle and K.F. Kroner, *Multivariate simultaneous generalized ARCH*, Econometric Theory **11**, 122 (1995);
R.F. Engle and K. Sheppard, *Theoretical and empirical properties of dynamic conditional correlation multivariate GARCH*, Nat. Bur. of Standards working paper No. 8554 (2001)

(http://www.nber.org/papers/w8554.pdf).
See also the GARCH toolbox for calculations at http://www.kevinsheppard.com/research/ucsd_garch/ucsd_garch.aspx.

[85] T. Bollerslev *Generalized autoregressive conditional heteroskedasticity*, J. of Econometrics **31**, 307 (1986); *Modelling the coherence in short-run nominal exchange rates: A multivariate generalized ARCH model*, Rev. Economics and Statistics **72**, 498 (1990).

[86] L. Bachelier, *Théorie de la Spéculation*, L. Gabay, Sceaux, 1995 (reprinted in P. Cootner (ed.), *The Random Character of Stock Market Prices*, MIT Press, Cambridge, Ma, 1964, pp. 17–78.

[87] A. Einstein, *Über die von der molekularkinetischen Theorie der Wärme geforderte Bewegung von in ruhenden Flüssigkeiten suspendierten Teilchen*, Annalen der Physik **17**, 549 (1905).

[88] N. Wiener, *Differential-Space*, J. of Math. and Phys. **2**, 131 (1923).

[89] F. Black and M. Scholes, J. Pol. Economy **81**, 637 (1973).

[90] R.C. Merton, *Theory of Rational Option Pricing*, Bell J. Econ. Management Sci. **4**, 141 (1973).

[91] These papers are available as CNRS preprints CPT88/PE2206 (1988) and CPT89/PE2333 (1989). Since it takes some effort to obtain them I have placed them on the internet where they can be downloaded as files dash1.pdf and dash2.pdf from http://www.physik.fu-berlin.de/~kleinert/b3/papers.

[92] S. Fedotov and S. Mikhailov, preprint (cond-mat/9806101).

[93] M. Otto, preprints (cond-mat/9812318) and (cond-mat/9906196).
B.E. Baaquie, L.C. Kwek, and M. Srikant, *Simulation of Stochastic Volatility using Path Integration: Smiles and Frowns*, cond-mat/0008327

[94] R. Cond, *Scaling and Correlation in Financial Data*, (cond-mat/9705075).

[95] A.J. McKane, H.C. Luckock, and A.J. Bray, *Path Integrals and Non-Markov Processes. I. General Formalism*, Phys. Rev. A **41**, 644 (1990);
A.N. Drozdov and J.J. Brey, *Accurate Path Integral Representations of the Fokker-Planck Equation with a Linear Reference System: Comparative Study of Current Theories*, Phys. Rev. E **57**, 146 (1998);
V. Linetsky, *The Path Integral Approach to Financial Modeling and Options Pricing*, Computational Economics **11** 129 (1997);
E.F. Fama, *Efficient Capital Markets*, Journal of Finance **25**, 383 (1970).
A. Pagan, *The Econometrics of Financial Markets*, Journal of Empirical Finance **3**, 15 (1996).
C.W.J. Granger, Z.X. Ding, *Stylized Facts on the Temporal Distribution Properties of Daily Data from Speculative Markets*, University of San Diego Preprint, 1994.

[96] H. Geman, D. Madan, M. Yor, *Time Changes in Subordinated Brownian Motion*, Preprint, (2000).

[97] H. Geman, D. Madan, M. Yor, *Time Changes for Lévy Processes*, Preprint (1999).

[98] J. Jacod, A.N. Shiryaev, *Limit Theorems for Stochastic Processes*, Berlin, Springer, (1987).

[99] P. Lévy, *Théorie de l'addition des variables aliatoires*, Paris, Gauthier Villars, (1937).

[100] D. Madan, F. Milne, *Option Pricing with Variance Gamma Martingale Components*, Mathematical finance, 1, 39-55, (1991).

[101] D. Madan, P.P. Carr, E.C. Chang, *The Variance Gamma Process and Option Pricing*, European Finance Review 2, 79-105, (1998).

[102] D. Madan, E. Seneta, *The Variance Gamma Model for Share Market Returns*, Journal of Business **63**, 511-524, (1990).

[103] B.B. Mandelbrot, *Fractals and Scaling in Finance*, Berlin, Springer, (1997).

[104] P. Protter, *Stochastic Integration and Differential Equations: a new approach*, Berlin, Springer, (1990).

[105] T.H. Rydberg, *The Normal Inverse Gaussian Lévy Process: Simulation and Approximation*, Commun. Stat., Stochastic Models **13(4)**, 887-910 (1997).

[106] G. Samorodnitsky, M. Taqqu, *Stable Non-Gaussian Random Processes*, New York, Chapman and Hall (1994).

[107] K. Sato, *Lévy Processes and Infinitely Divisible Distributions*, Cambridge University Press, (1999).

[108] Yahoo Finance `http://finance.yahoo.com`. To download data, enter in the symbol box: ^DJI, and then click on the link: *Download Spreadsheet*.

[109] J. Masoliver and J. Perelló, Physica A **308**, 420 (2002) (cond-mat/0111334); Phys. Rev. E **67**, 037102 (2003) (cond-mat/0202203); (physics/0609136).

[110] J.-P. Bouchaud, A. Matacz, and M. Potters, Phys. Rev. Letters **87**, 228701 (2001).

[111] J.-P. Bouchaud and D. Sornette, *The Black-Scholes Option Pricing Problem in mathematical finance: Generalization and extensions for a large class of stochastic processes*, J. de Phys. **4**, 863 (1994); *Reply to Mikheev's Comment on the Black-Scholes Pricing Problem*, J. de Phys. **5**, 219 (1995);

[112] J.-P. Bouchaud, G. Iori, and D. Sornette, *Real-World Options*, Risk **9**, 61 (1996) (`http://xxx.lanl.gov/abs/cond-mat/9509095`); J.-P. Bouchaud, D. Sornette, and M. Potters, *Option Pricing in the Presence of Extreme Fluctuations*, in *Mathematics of Derivative Securities*, ed. by M.A.H. Dempster and S.R. Pliska, Cambridge University Press, 1997, pp. 112-125; *The Black-Scholes Option Pricing Problem in Mathematical Finance: Generalization and Extensions for a Large Class of Stochastic Processes*, J. de Phys. **4**, 863 (1994).

[113] P. Santa-Clara and D. Sornette, *The Dynamics of the Forward Interest Rate Curve with Stochastic String Shocks*, Rev. of Financial Studies **14**, 149 (2001) (cond-mat/9801321).

Index